Molecular Cell Biology

Molecular Cell Biology

SECOND EDITION

JAMES DARNELL

Vincent Astor Professor
Rockefeller University

▲

HARVEY LODISH

Member of the Whitehead Institute for
Biomedical Research
Professor of Biology, Massachusetts
Institute of Technology

▲

DAVID BALTIMORE

President (designate)
Rockefeller University

SCIENTIFIC
AMERICAN
BOOKS

Distributed by W. H. Freeman and Company, New York

Cover illustration by Tomo Narashima

Library of Congress Cataloging-in-Publication Data

Darnell, James E.
  Molecular cell biology / James Darnell, Harvey Lodish, David
  Baltimore.—2d ed.
    p.   cm.
  Includes bibliographical references.
  ISBN 0-7167-1981-9:—ISBN 0-7167-2078-7
(international student ed.):
    1. Cytology.   2. Molecular biology.   I. Lodish, Harvey F.
II. Baltimore, David.   III. Title.
  [DNLM: 1. Cells.   2. Molecular Biology.   QH 581.2 D223m]
QH581.2.D37   1990
574.87′6042—dc20
DNLM/DLC   89-70096
for Library of Congress   CIP

Printed in the United States of America

Scientific American Books is a subsidiary of Scientific American, Inc.
Distributed by W. H. Freeman and Company, 41 Madison Avenue,
New York, New York 10010 and 20 Beaumont Street,
Oxford OX1 2NQ England

234567890   KP   99876543210

To our wives and children
and to our mentors
and teachers, particularly

Sydney Brenner
Harry Eagle
Richard Franklin
Salvador Luria
Norton Zinder

Preface

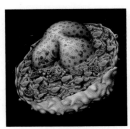

We asserted in the preface to the first edition of this book that the reductionist approach and the new techniques of molecular biology would soon unify all experimental biology. Now, four years later, perhaps the only surprise is the speed and completeness with which biologists from fields formerly considered distant have embraced the new experimental approaches.

In retrospect, the traditional separation among physiology, cell biology, biochemistry, developmental biology, genetics, and even much of neurobiology was more apparent than real. The quest in all these fields was the same: to discover proteins that could carry out specific biologically important tasks—that could, for example, regulate ion flow, cause cell motility or contractility, catalyze a degradative or synthetic reaction, induce an embryonic structure, or send or receive a chemical signal. Ultimately, the quest led to an understanding of the genes that encode the proteins. Now, with the development of molecular genetics, genes can be purified, sequenced, mutated at will, and reintroduced into cells, laying open golden opportunities for every biologist who learns to use these new tools.

Our contention four years ago, reaffirmed now, is that the teaching of biology must reflect a unified experimental approach. The education of biology students should be shaped by it from the beginning. Thus we have maintained our emphasis on the fundamental experimental approaches that we regard as the necessary foundation of this new integrated science of *molecular cell biology* and at the same time have been able to bring our coverage of the material as up-to-the-moment as possible.

We were grateful to be asked by our publisher to prepare a second edition: it signified to us as least some measure of acceptance of our first edition by teachers and students. And, in the years since publication of the first edition, we have been highly appreciative of the comments we have received from readers. Along with praise came suggestions for improvement. Perhaps the criticism of the first edition that we took most to heart was the advice of several dedicated teachers that the book was too detailed, too difficult. On reflection, we came to see that many important ideas could be expressed better by being expressed more simply. Thus we have tried our best to clarify unclear explanations and to remove overlong discussions that seemed, on balance, too arcane for most students. When detail was essential, we tried to make the main points and conclusions stand out more clearly.

But we stated an aim in our first preface from which we have not retreated: we believe biology professors and their students should have available comprehensive books in biology that are neither more nor less demanding than the standard comprehensive texts in chemistry and physics.

In schools where an introduction to biology or chemistry or both is followed by a series of one-semester courses in modern biology, this book serves well for one-semester courses in cell biology, molecular biology, and cellular biochemistry and physiology. In the growing number of schools where biology is taught as the experimentally most active science, this book can also form the basis for a year-long introduction to modern biology.

But now to the details of the new edition. We believe that we have succeeded in incorporating a number of key improvements into the book. The most obvious is the use of full color to make the illustrations clearer and easier to follow. We have also integrated more plant biology into the text, recognizing the increased depth of understanding that modern biology has brought to the study of the plant kingdom. A third improvement is that, rather than isolating developmental biology and cellular differentiation into a separate chapter, we have integrated it with the other material in this edition. We did this because we recognized that this is not a text for teaching development but that gene control and cellular alteration must be seen in the context of developmental strategies. Although quite similar in general form to the first edition, the book is thoroughly updated in content and organization, reflecting the order in which we ourselves have taught the material. Special care has been taken to organize the material for the greatest flexibility.

The introductory chapters contain a brief (but, we hope, sufficient) description of molecules that make up all cells and of the principles of chemical interactions that allow these molecules to function. Part I also outlines modern techniques for studying cells. In Part II the student is taught how genes work. We stress RNA (Chapters 7 and 8) before we discuss gene structure and chromosomal organization (Chapters 9 and 10) because the individuality of organisms, while surely encoded in DNA, is brought to reality by the programmed expression of genes during the differentiation of the cells of each organism. Thus we concentrate on genes as transcription units and deal extensively with how they are controlled (Chapter 11) as well as how they are replicated and repaired (Chapter 12). Moreover, Chapter 11, now completely rewritten, contains much new information on the regulation of gene expression during development.

In Part III we turn to the ways in which proteins—the ultimate gene products—work together to make a living cell. Instructing the new biology student in the names and shapes of the parts of a cell can by itself be a satisfying experience because of the wealth of pictorial detail available. But it is the integration of the structural detail with molecular function that gives meaning to cell activity, and many aspects of this integration have been brought to an advanced level within the last few years. Because the plasma—or cell-surface—membrane is so important to the structure and function of a cell, it is the subject of two chapters: Chapter 13 describes the structure of cell membranes and membrane proteins; Chapter 14 examines the role of membrane proteins in the transport of small molecules into and out of the cell, and also in endocytosis, the process by which proteins and particles are brought into the cell. Cellular events require energy, which is usually supplied by the generation of ATP in the cytosol and mitochondria (Chapter 15). A separate chapter on photosynthesis (Chapter 16) discusses this key plant process. Chapters 17 and 18 describe the assembly of the membranes and organelles that make up a cell, focusing on the targeting of proteins to their correct destinations. The complex network of cell-to-cell communication required to coordinate differentiation, growth, metabolism, and behavior—systems for chemical and electrical signal transmission—is covered in Chapters 19 and 20. Chapters 21 and 22 turn to the complex system of fibers—the cytoskeleton—that is responsible for the cell's shape and motility. Finally, Chapter 23 focuses on the many proteins, polysaccharides, proteoglycans, and other polymers that form the extracellular matrix surrounding animal cells and the plant cell wall and on how these molecules influence growth and differentiation of cells.

Topics of particular interest and concern—cancer biology, immunology, and cellular evolution—conclude the book (Part IV).

A *Student's Companion* by five outstanding teachers—David Rintoul, Ruth Welti, Robert Van Buskirk, Brian Storrie, and Muriel Lederman—is available for use along with the main text; we trust both students and instructors will find it useful. The *Companion* will allow students to apply what they have learned and develop mastery through five different types of problems.

A project of this scale, involving an entire rewriting of over a thousand pages, is not possible without the invaluable assistance of many different people. We thank both our colleagues who made major contributions to specific chapters in their special areas of interest and those who read chapters or generously provided illustrations: Barry Alpert, Atlantex and Zieler Instrument Corporation; David Asai, Purdue University; Karl Aufderheide, Texas A & M University; Amy Bakken, University of Washington, Seattle; Thomas Bibring, Vanderbilt University; Mina Bissell, Lawrence Berkeley Laboratory; Jef Boeke, Johns Hopkins University; Jonathan Braun, UCLA; Thomas R. Cech, University of Colorado; Carolyn Cohen, Brandeis University; The Cold Spring Harbor Press; Susan E. Conrad, Michigan State University; Francis Crick, Salk Institute; Robert Darnell, Sloan Kettering Institute; Anthony De Franco, University of California, San Francisco; Gary E. Dean, University of Cincinnati College of Medicine; Steven Di Nardo, Rockefeller University; Herman Eisen, MIT; Ronald M. Evans, Salk Institute; Nina Federoff, Carnegie Institution of Washington; Gerald R. Fink, MIT and Whitehead Institute; Jeffrey Flier, Beth Israel Hospital, Boston; Steven Harrison, Harvard University; Yoav Henis, Tel Aviv University, Israel; Merrill B. Hille, University of Washington, Seattle; Susan J. Hunter, University of Maine; Richard Hynes, MIT; Andre Jagendorf, Cornell University; Flora Katz, Southwestern Medical Center, Dallas; Lon S. Kaufman, University of Illinois, Chicago Circle; Hans Kende, Michigan State University; Michael Klymkowsky, University of Colorado, Boulder; Monty Krieger, MIT; Ueli Laemmli, University of Geneva; Jeanne Lawrence, University of Massachusetts, Worcester; Michael Lerner, Yale University; Michael Levine, Columbia University; Thomas Linsenmayer, Tufts University School of Medicine; Thomas Maniatis, Harvard University; Manuela Martins-Green, Lawrence-Berkeley Laboratory; Paul Matsudaira, MIT and Whitehead Institute; Nancy S. Milburn, Tufts University; Peter Model, Rockefeller University; Richard M. Myers, University of California, San Francisco, School of Medicine; Bernardo Nadal-Ginard, Harvard Medical School; Leslie Orgel, Salk Institute; Joanne Otto, Purdue University; David Parry, Massey University, New Zealand; Brian Poole, Tufts University School of Medicine; Newtol Press, The University of Wisconsin, Milwaukee; Lola Reid, Albert Einstein College of Medicine, Yeshiva University; J. F. Richards, University of British Columbia; Thomas M. Roberts, Florida State University; Robert Roeder, Rockefeller University; Rodney Rothstein, Columbia College of Physicians and Surgeons; David W. Russell, Southwestern Medical Center, Dallas; Klaus-Dieter Scharff, Institute of Plant Biology, Halle, German Democratic Republic; Barbara B. Sears, Michigan State University; Phillip A. Sharp, MIT; Joel Sheffield, Temple University; James Sidow, Duke University; Robert H. Singer, University of Massachusetts, Worcester; Frank Solomon, MIT; Peter Stambrook, University of Cincinnati College of Medicine; Thomas Steitz, Yale University; Bruce Stillman, Cold Spring Harbor Laboratory; Thomas Stossel, Massachusetts General Hospital; Bruce Tidor, Whitehead Institute; Bernard Trumpower, Dartmouth Medical School; Robert Van Buskirk, SUNY Binghamton; Richard Van Etten, Whitehead Institute; Patricia Wadsworth, University of Massachusetts at Amherst; D. B. Walden, The University of Western Ontario; James Wang, Harvard University; Robert Weinberg, MIT and Whitehead Institute; Ruth Welti, Kansas State University; Gene Williams, Indiana University, Bloomington; Lewis T. Williams, Kansas State University; D. L. Worcester, University of Missouri, Columbia; Michael W. Young, Rockefeller University.

No group of authors could have had a more encouraging and helpful publisher and publishing colleagues. Linda Chaput, the President of W. H. Freeman and Company not only has put enormous resources at our disposal but has also played a very personally supportive role. We owe her an enormous debt and hope we used her confidence well.

The staff at Freeman during the three years of preparation that this edition required have been—to a person—devoted and highly competent. We note particularly the help of Lloyd Black and Linda Davis in the general planning of the book. Andrew Kudlacik, James Funston, and Ruth Steyn gave us expert editorial assistance; Sonia DiVittorio was absolutely indispensable not only as the final arbiter on editorial questions but also as the mainspring in the coordination of the art program. Diane Maass proofread all stages of the art program, and Philip McCaffrey directed the efforts of an expert team of copy editors and proofreaders. Travis Amos was extremely energetic and valuable in locating and developing sources for photographs. Mike Suh designed the interior as well as the cover and, together with John Hatzakis and Howard Johnson, achieved a skillful blend of text and illustration on each page. Bill Page ably coordinated the exceptional work of the illustrators—Tomo Narashima, George Kelvin, Tom Cardamone, Louis Pappas, Vantage Art, Inc., and York Graphic Services. Julia De Rosa heroically organized all parts of the production process into a coherent whole.

Our own office staff, including Miriam Boucher, Lois Cousseau, Audrey English, Nancy Kong, and Ginger Pierce have become extremely adept at organizing reprints and helping us locate specific information and were unfailingly patient as we assembled our manuscript.

All these people and many others made the enormous

task of writing this new edition, if not actually pleasant, at least tolerable.

Again to our families we say simply, thank you.

We believe that a comprehension of modern biology is needed both by those who use biological concepts professionally and by the general public, who increasingly will be faced with decisions about integrating new biological understanding into the fabric of their lives. We hope that this edition of *Molecular Cell Biology* will help both groups to better comprehend the revolution in understanding of living systems that is being generated by research laboratories around the world.

Jim Darnell
Harvey Lodish
David Baltimore

February 1990

From the Preface to the First Edition

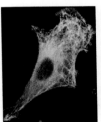

Biology today is scarcely recognizable as the subject that biologists knew and taught 10 years ago. A decade ago, gene structure and function in the simple cells of bacteria were known in considerable detail. But now we also know that a different set of molecular rules governs gene organization and expression in all eukaryotic cells, including those of humans. We are learning about the genes and regulatory proteins that control not only single metabolic steps but complicated developmental events such as the formation of a limb, a wing, or an eye. In addition to these advances in understanding the genetic machinery and its regulation, great progress has been made in the study of the structure and function of cell organelles and of specialized cell proteins. To comprehend fully what has been learned requires a reformulation of a body of related information formerly classified under the separate headings of genetics, biochemistry, and cell biology. *Molecular Cell Biology* aims to present the essential elements of this new biology.

Traditionally, the sciences of genetics, biochemistry, and cell biology—the three areas in which the greatest progress has been made in the last 25 to 30 years—used different experimental approaches and often different experimental material. Classical geneticists sought mutations in specific genes to begin identifying the gene products and characterizing their physiological functions. Biochemists tried to understand the actions of proteins, especially enzymes, from their sequences and three-dimensional structures. Cell biologists attempted to discover how specific proteins took part in the construction and operation of specialized cell structures. These subjects were taught as three courses, albeit with varying degrees of overlap.

A group of techniques collectively referred to as molecular genetics is mainly responsible for unifying the three disciplines. Not only do these techniques provide a powerful analytic force, but they also serve to unify all experimental biology in its language and concerns. With the tools of molecular genetics, genes for all types of proteins—enzymes, structural proteins, regulatory proteins—can be purified, sequenced, changed at will, reintroduced into individual cells of all kinds (even into the germ lines of organisms) and expressed there as proteins. Most of experimental biology now relies heavily on molecular genetics.

In addition to the outstanding advances in molecular genetics, comparable advances have been made in culturing the cells of vertebrates, invertebrates, and plants, including the cells that produce various individual monoclonal antibodies. The use of cell cultures has greatly unified and simplified experimental designs. Finally, very sophisticated instrumentation has become available. Powerful electron microscopes and advanced techniques in electron microscopy have greatly improved our understanding of cell substructure. Modern computers have arrived in time to store—and then to compare—rapidly accumulating information (such as protein and nucleic acid sequences), as well as to present graphic displays of molecular structures. Of equal importance, computers rapidly complete elaborate calculations so that x-ray crystallographic analysis (or other kinds of image analysis) can be performed in days instead of months or years.

Those who teach biology at the undergraduate or graduate level and in medical schools can convey this comprehensive and integrated experimental approach in the classroom only when they have access to appropriate teaching materials. Our book is intended to fill the need for such materials: We wrote it to solve our own problems as teachers. It was our purpose to teach a one-year course that integrates molecular biology with biochemistry, cell biology, and genetics and that applies this coherent insight to such fascinating problems as development, immunology, and cancer. We hope that the availability of this material in a unified form will stimulate the teaching of molecular cell biology as an integral subject and that such integrated courses will be offered to students as early as possible in their undergraduate education. Only then will students be truly able to grasp the findings of the new biology and its relation to the specialized areas of cell biology, genetics, and biochemistry.

We have aimed to provide a college textbook that is no more difficult than the basic textbooks encountered by undergraduate physics and chemistry students in their respective programs of study. That there will be complaints about the scope and depth of a textbook this large seems inevitable. But in addition to dividing the book into parts, we have clearly identified the parts of the chapters themselves, by means of descriptive subheadings. This organization will enable students and teachers to be selective in their reading. We recognize that some teachers of one-semester courses may wish to continue teaching molecular biology and cell biology as separate courses. The book is organized so that, in such situations, emphasis can easily be given to either the gene or the cell.

Whichever path students and teachers choose to follow, we believe that the focus of teaching an experimental science such as biology should be the experiments themselves. We have devoted much space to presentations of experiments. Scientists make advances by phrasing the unknown as an experimental question, designing an experiment to answer the question, and assembling the experimental results to produce a coherent answer. Students who see how biological progress is intimately connected to experimentation will be more likely to keep pace with the progress of biology in the future and perhaps even to contribute to it.

Over the past seven years, from the earliest planning stages to completion, this book has occupied a significant portion of our energies. During this period we have often exchanged views and read each others' work, so that our book is truly a joint responsibility and a joint product.

We have called upon our friends and colleagues very liberally and cannot possibly mention all of those who gave us useful advice at various stages of development of the manuscript. We specifically thank: Wayne M. Becker, University of Wisconsin, Madison; Stephen Benson, California State University, Hayward; Sherman Beychok, Columbia University; David Bloch, University of Texas, Austin; David Clayton, Rockefeller University; Francis Crick, Salk Institute; Robert Davenport, MIT; David De Rosier, Brandeis University; Ford Doolittle, Dalhousie University; Ernest DuBrul, University of Toledo; Charles Emerson, University of Virginia; Richard Firtel, University of California, San Diego; Ursula Goodenough, Washington University; Jeffrey Flier, Harvard Medical School; Barry Gumbiner, University of California Medical Center at San Francisco; James Hageman, New Mexico State University; Maija Hinkle, Cornell University; Johns Hopkins, Washington University; Richard Hynes, MIT; Warren Jelinek, New York University Medical School; David Kabat, Oregon Health Sciences University; Flora Katz, University of California Medical Center at San Francisco; Thomas Kreis, European Molecular Biology Laboratory, Heidelberg; Scott Landfear, Harvard School of Public Health; John Lis, Cornell University; Maurice Liss, Boston College; Mary Nijhout, Duke University; Daniel O'Kane, University of Pennsylvania; Nan Orme-Johnson, Tufts University School of Medicine; Leslie Orgel, Salk

Institute; Larry Puckett, Immuno Nuclear Corporation; Charles Richardson, Harvard Medical School; Karin Rodland, Reed College; Alan Schwartz, Dana Farber Cancer Center; Richard D. Simoni, Stanford University; Roger Sloboda, Dartmouth College; Frank Solomon, MIT; David R. Soll, University of Iowa; Pamela Sperry, California State Polytechnic Institute at Pomona; Rocky Tuan, University of Pennsylvania; Joseph Viles, Iowa State University; Michael Young, Rockefeller University; and Norton Zinder, Rockefeller University for reading one or more chapters. In addition, many scientists have generously supplied drawings and photo-acknowledgement in the appropriate places in the book.

Our editorial colleagues at Scientific American Books have provided the very substantial resources necessary to complete such a large project and given us their most thoughtful, careful advice and professional skills in unlimited amounts. When our spirits or energies sagged, Patty Mittelstadt, Andrew Kudlacik, Janet Wagner, Donna McIvor, Sally Immerman, James Funston, Betsy Galbraith, Faye Webern, and in particular, Neil Patterson and Linda Chaput could always be counted on to get us over a rough spot. In many, many places in the book not only have outright mistakes been removed but a clearer sentence, a more orderly line of thinking, a better picture or drawing exists because of their tireless efforts. The errors and infelicities that remain we must acknowledge are our own. We are also thankful for the talented work of the illustrators, Shirley Baty and George Kelvin, and the lively cooperation of the production team: Mike Suh, Margo Dittmer, Melanie Neilson, and Ellen Cash. Finally we wish to give a very special vote of thanks to our secretaries—Miriam Boucher, Lois Cousseau, Audrey English, Ginger Pierce, and Marilyn Smith—for their endless patience in dealing with the many necessary drafts of the book.

To our friends and associates in our laboratories and, most of all, to our families we apologize for the long absences and the vacant stares that frequently came in the wake of long hours of working on the book. If our efforts are successful in helping to unify the teaching of molecular cell biology we will be deeply grateful.

Jim Darnell
Harvey Lodish
David Baltimore

April 1986

Contents in Brief

Contents

▼ ▼ ▼

PART *I*

Molecules, Cells, Proteins, and Experimental Techniques: A Primer

▼   ▼   ▼

Basal Bodies and Centrioles: Structure and Properties — 840

Centrioles and Basal Bodies Are Built of Microtubules — 841

Centrioles and Basal Bodies Contain a Unique Small DNA — 842

Centrioles Can Convert into Basal Bodies and Vice Versa — 842

Function of Microtubules in Mitosis — 844

Light-Microscope Techniques Reveal the Mitotic Spindle in Living Cells — 844

Bundles of Microtubules Form the Mitotic Spindle — 847

Kinetochore Microtubules Connect the Chromosomes to the Poles — 848

Dynamic Instability Explains the Morphogenesis of the Mitotic Spindle — 849

Many Events in Mitosis Do Not Depend on the Mitotic Spindle — 850

Balanced Forces Align Metaphase Chromosomes at the Equator of the Spindle — 850

Anaphase Consists of Two Distinct Motile Events — 851

Poleward Chromosome Movement (Anaphase A) Is Powered by Microtubule Disassembly at the Kinetochore and Requires No External Energy Source — 851

Separation of the Poles (Anaphase B) Involves Sliding of Adjacent Microtubules Powered by ATP Hydrolysis — 852

Cytokinesis Is the Final Separation of the Daughter Cells — 853

Summary — 855

References — 855

CHAPTER *22 Actin, Myosin, and Intermediate Filaments: Cell Movements and Cell Shape* — 859

Actin and Myosin Filaments — 860

Actin Monomers Polymerize into Long Helical Filaments — 860

Actin Filaments Are Intrinsically More Stable Than Microtubules — 861

Myosin Is a Bipolar, Fibrous Molecule That Binds Actin — 862

Driven by ATP Hydrolysis, Myosin Heads Move along Actin Filaments — 863

Muscle Structure and Function — 865

Striated Muscle Consists of a Regular Array of Actin and Myosin Filaments — 865

Thick and Thin Filaments Move Relative to Each Other during Contraction — 867

ATP Hydrolysis Powers the Contraction of Muscle — 868

Release of Calcium from the Sarcoplasmic Reticulum Triggers Contraction — 870

Calcium Activation of Actin, Mediated by Tropomyosin and Troponin, Regulates Contraction in Striated Muscle — 872

Calcium Activation of Myosin Light Chains Regulates Contraction in Smooth Muscle and Invertebrate Muscle — 873

cAMP, 1,2-Diacylglycerol, and Caldesmon Also Affect Contractability of Smooth Muscle — 873

Smooth and Striated Muscles Contain Functionally Different Myosin Light Chains and Tropomyosins — 875

Proteins Anchor Actin Filaments to the Plasma Membrane or the Z Disk — 875

Long Proteins Organize the Sarcomere — 878

Dystrophin Is a Muscle Protein Identified by Study of a Genetic Disease — 878

Phosphorylated Compounds in Muscle Act as a Reservoir for ATP Needed for Contraction — 879

Actin and Myosin in Nonmuscle Cells — 879

All Vertebrates Have Multiple Actin Genes and Actin Proteins — 880

Many Actin-binding Proteins Are Present in Nonmuscle Cells — 880

Noncontractile Bundles of Actin Filaments Maintain Microvilli Structure — 881

Actin and Myosin Are Essential for Cytokinesis in Nonmuscle Cells — 884

Movements of the Endoplasmic Reticulum along Actin Filaments Power Cytoplasmic Streaming — 885

Polymerization of Actin Monomers Is Controlled by Specific Actin-binding Proteins in Nonmuscle Cells — 885

Movement of Amebas and Macrophages Involves Reversible Gel-Sol Transitions of an Actin Network — 888

Movements of Fibroblasts and Nerve Growth Cones Involve Controlled Polymerization and Rearrangements of Actin Filaments — 890

Actin Stress Fibers Permit Cultured Cells to Attach to Surfaces — 892

Intermediate Filaments — 894

Chapter-opening Illustrations

Introduction A helical segment of the dimeric bacteriophage λ repressor (green) is bound in the major groove of two successive turns of the DNA helix (blue); computer representation based on x-ray crystallographic data. *Courtesy of Jane and David Richardson.*

Chapter 1 A computer-generated representation of an ATP molecule. The dots show the van der Waals radii of the atoms, and the sticks show the bonds connecting the atoms. The fused rings of the purine are easily visible at the left, and the phosphates are at the right. *Courtesy of Computer Graphics Laboratory, University of California, San Francisco. © Regents, University of California.*

Chapter 2 Crystals of the digestive enzyme chymotrypsin. These crystals form when highly concentrated protein solutions are prepared under special conditions. Such crystals were used to determine the structure of chymotrypsin using x-ray crystallographic methods. *Courtesy of Alexander McPherson.*

Chapter 3 The crystal structure of *Escherichia coli* glutaminyl-tRNA synthetase (blue) complexed with its cognate glutaminyl transfer RNA (red and yellow) and ATP (green). The structure reveals that the enzyme recognizes this specific tRNA, discriminating against all others, through extensive interactions with the tRNA extending from the anticodon to the acceptor stem. *From M. A. Rould, J. J. Perona, D. Söll, and T. A. Steitz, 1989, Structure of E. coli glutaminyl-tRNA synthetase complexed with tRNA^{Gln} and ATP at 2.8 Å resolution, Science, 246:1135–1142. Courtesy of M. A. Rould, J. J. Perona, P. Vogt, and T. A. Steitz.*

Chapter 4 A starfish oocyte showing an enlarged nucleus, also referred to as a germinal vesicle. From *Cell*, 21 October 1988. *Courtesy of Laurent Meijer.*

Chapter 5 HeLa (human cervical carcinoma) cells in culture. This is the most commonly used of all human (or vertebrate) cell lines in culture. *Courtesy of M. G. Gabridge, cytoGraphics Inc./Biological Photo Service.*

Chapter 6 A front view of a commercially available oligo-deoxynucleotide (DNA) synthesizer. The bottles contain each of the four nucleotide derivatives as well as other reagents required for DNA synthesis. These are automatically pumped into the reaction chamber as required. *Courtesy of Applied Biosystems.*

Chapter 7 *Escherichia coli* with pili. Two newly divided cells show projecting pili that can serve as receptors for filamentous viruses. *Courtesy of the Pasteur Institute.*

Chapter 8 A thin section of a cell showing prominent nucleoli (dark-staining granular areas) within the nucleus. The nucleoli are the sites of rRNA precursor formation and ribosomal maturation, while tRNA and mRNA are formed in the remainder of the nucleus. *Courtesy of David M. Phillips/Visuals Unlimited.*

Chapter 9 Harlequin chromosomes testify to the reciprocal exchange of genetic material between sister chromatids. Metaphase chromatids were chemically altered so that one fluoresces more brightly, when it is stained with bromodeoxyuridine (BudR; yellow), than its sister chromatid. An agent that damages DNA fostered the multiple exchanges that can be seen in the photograph. *Courtesy of Sheldon Wolff, Laboratory of Radiobiology and Environmental Health, University of California, San Francisco.*

Chapter 10 Metaphase human chromosomes hybridized with a biotin-labeled fluorescent oliogonucleotide probe (yellow) that specifically detects telomeric DNA. A cell in interphase (*bottom*) shows that the telomeric structures, which can be as long as 10^6 base pairs, have compact intranuclear localization. *Courtesy of Robert Moyzis.*

Chapter 11 Three-dimensional computer model of a single "zinc finger" domain of a DNA-binding protein. The central zinc atom (light blue) is bound ("coordinated") by cysteine and histidine residues at the base of the loop of amino acids (light purple tube). *Courtesy of Michael Pique, Research Institute of Scripps Clinic.*

Chapter 12 The genome of *Escherichia coli* emptied from a single bacterium after partial cell wall digestion and detergent lysis of the lipid membrane. Whether the anchored loops seen in the released DNA exist inside the cell is not known, but evidence for such anchoring sites exists. *Courtesy of Dr. Gopal Murti, Science Photo Library/Photo Researchers, Inc.*

Chapter 13 The bacterial photosynthetic reaction center, the only integral membrane protein whose three-dimensional structure is known to atomic detail. For the determination of this structure, Hartmut Michel, Robert Huber, and Johann Deisenhofer were awarded the Nobel Prize in Chemistry in 1988. *Courtesy of Johann Deisenhofer.*

Chapter 14 Acidic endosomes and lysosomes. CHO (Chinese hamster ovary) cells were incubated with several fluorescein-labeled proteins, which were taken up by endocytosis and accumulated in endosomes and lysosomes. The ability of light at 450 nm and 490 nm to excite the fluorescence of fluorescein varies with pH; thus the living cells were viewed by image intensification fluorescence microscopy, and the pH of individual vesicles was determined and converted to colors corresponding to the scale at the bottom of the picture. *Courtesy of Fred Maxfield.*

Chapter 15 Two populations of mitochondria in a living human fibroblast, treated with the lipophilic cation JC-1, fluoresce red and green. Formation of an aggregate of the cation within the mitochondria is caused by a high mitochondrial membrane potential and generates the red fluorescence. *Courtesy of Lan-Bo Chen.*

Chapter 16 Abundant chloroplasts, each ~ 10 μm in diameter, are seen in the cells in a section of a leaf from the snake plant *Sansevieria trifasciata. Courtesy of Runk/Schoenberger/Grant Heilman Photography.*

Chapter 17 Secretion vesicles in mast cells. The cell on the left contains abundant vesicles filled with histamine. Secretion of histamine-filled vesicles by the cell on the right was induced by the binding to mast cell surface receptors of IgE antibodies specific for an antigen to birch pollen. *From* The Body Victorious, *Dell Publishing Co. Courtesy of Lennart Nilsson.*

Chapter 18 Leber's hereditary optic neuropathy—a maternally inherited genetic disease that results in retinal swelling, "tortuous" blood vessels (so called because of their distorted shape), and blindness—is due to a point mutation in the mitochondrial DNA gene encoding NADH-CoQ reductase subunit 4. *From J. L. Smith et al., 1973,* Archives of Ophthalmology *90:349–354. Courtesy of Douglas C. Wallace.*

Chapter 19 The passage of a wave of elevated calcium (Ca) through a field of astrocytes. When confluent cultures of hippocampal astrocytes are stimulated by the neurotransmitter glutamate, oscillatory increases in cytoplasmic free Ca are one result. These can take the form of waves of Ca increase that propagate between cells at ~ 20 μm/s. Ca changes measured using the fluorescent indicator fluo-3 were sampled at 4-s intervals. Areas exceeding a threshold level of increase at a given time were added as a single overlay color to an image showing baseline cytoplasmic fluorescence. The color overlays reflect successive 4-s sample periods, proceeding through a spectrally coded time sequence. A wave begins in the violet patch at the left of center, proceeds through a light blue area to the center of the field, and then radiates upward through green, yellow, orange, and red arcs. *Courtesy of Stephen Smith.*

Chapter 20 A scanning electron micrograph showing three skeletal muscle fibers and the axon of a motor neuron. *From* Behold Man, *Little, Brown and Co. Courtesy of Lennart Nilsson.*

Chapter 21 Confocal fluorescence micrograph of a fibroblast at the anaphase stage of mitosis. Tubulin is detected by a red-fluorescing antibody and DNA by the yellow-green dye quinacrine. *Courtesy of John M. Murray.*

Chapter 22 The actin cytoskeleton in a fibroblast cell cultured on a dish coated with fragments of the extracellular matrix protein fibronectin. The long actin fibers generate the elongated shape of the cell and attach to the plasma membrane at points that contact the substratum. [See M. Obara et al., 1988, Cell 53:649.] *Courtesy of Kenneth Yamada.*

Chapter 23 Section of the edge of a rat skin wound, showing granulation tissue (*bottom*) covered by migrating epithelium (*top*). Fluorescent labeling of the matrix protein tenascin (green) and nuclei (blue). [See E. Mackie et al., 1988, J. Cell Biol. 107:2757. *Courtesy of Eleanor Mackie.*

Chapter 24 A ribbon diagram representing the backbone chain of the *ras* protein p21 with the location of a bound GTP molecule indicated. The *ras* gene, when mutated, is an oncogene and is able to transform normal cells into cancer cells. The oncogene form of the *ras* protein is only slightly different in structure from the normal *ras* protein. *Courtesy of Emil F. Pai and Fred Wittinghofer, Max-Planck-Institut für medizinische Forschung.*

Chapter 25 Computer-generated model of an antigen binding to an antibody molecule. The antibody has two identical arms, each of which could bind an antigen. One antigen, a small protein, is shown (*upper left*) ready to be bound to the antibody. *Courtesy of Petiteformat/IPRP Photo Researchers.*

Chapter 26 Electron micrograph of "fossil" bacteria. Thin slices of sedimentary rock of up to 3×10^9 years of age, which were probably formed at ocean edges and therefore would be expected to have embedded bacteria, show images that are thought to be ancient bacteria. *Courtesy of S. M. Awramik, University of California/Biological Photo Service.*

Molecular Cell Biology

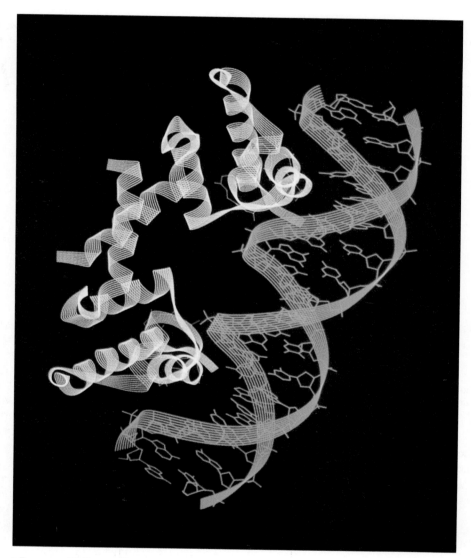

λ-Repressor-DNA complex*

The History of Molecular Cell Biology

The aim of modern biology is to interpret the properties of the organism by the structure of its constituent molecules.

FRANÇOIS JACOB
The Logic of Life
Translated by Betty E. Spillman
Copyright 1973 by Pantheon Books,
a Division of Random House, Inc.

[W]e have complete confidence that further research, of the intensity recently given to genetics, will eventually provide man with the ability to describe with completeness the essential features that constitute life.

JAMES D. WATSON
The Molecular Biology of the Gene,
3d ed., The Benjamin/Cummings
Publishing Company, 1976

M*odern biology aims at no less than a full understanding of cell function in molecular terms. The statements (left) by two architects of modern biology presage the determination of today's biologists to carry the spectacular successes of the 1950s and 1960s—the discoveries of the structure of DNA, the roles of RNA in protein synthesis, the genetic code, and the nature of gene regulation in bacteria—into studies of the cells and organs of higher organisms, including human beings. The molecular approach to biology affects every traditional biological discipline: histology, cytology, anatomy, embryology, genetics, physiology, and evolution. Most of these areas of biology deal with* eukaryotic *cells (cells with a nucleus). However, in early molecular studies, biologists relied on* prokaryotic *cells (cells without a nucleus)—particularly a single bacterial species,* Escherichia coli *(E. coli). It became customary in the 1960s to introduce the biology student to molecular events in cell function mainly through descriptions of these bacterial experiments. E. coli and the viruses that infect it became the core subject matter of molecular biology.*

*For a more complete description of chapter opening images, see the annotated listing that follows the table of contents.

The use of bacterial models to illustrate the most elementary and general principles by which all cells function remains logical and sound. But now many of the molecular characteristics that make the cells from higher organisms different from bacterial cells are also known. A molecular cell biology that embraces all types of cells is upon us. Many general principles of eukaryotic cell structure and function—and the proteins responsible for this structure and function—have no counterparts in bacterial cells. Not only do the cells of higher organisms contain proteins and cellular structures not found in bacteria, but there are many differences in the organization and expression of their genes. To serve as a foundation for the study of cytology, embryology, physiology, and evolution, subjects concerned mainly with higher organisms built of eukaryotic cells, a molecular description of cell function should feature the current knowledge about eukaryotic cells, as this book aims to do.

This introductory section briefly traces the historical background for the discoveries that illuminated the chemical nature of the gene—the elementary unit of heredity—and the general principles of eukaryotic cell structure. The union of genetics with cell biology and biochemistry is the cause of much excitement in modern biology. No longer does the microscopist simply display the wonderful internal structures of cells. Biochemists have purified and identified the protein components of these structures and now know much about their functions. Molecular biologists have used the exciting new techniques of molecular genetics to isolate hundreds of the genes responsible for many important elements of cell structure and function. The merging of these related studies, ranging from genetics to protein chemistry, has created a new discipline. The more comprehensive approach constitutes molecular cell biology. ▲

Evolution and the Cell Theory

Three central questions about life and living things have recurred since ancient times: How did life originate? Why does "like beget like"? How does an individual animal or plant develop from a fertilized egg or seed? Until the nineteenth century, these questions were addressed primarily through religion or philosophy, or both. In the last half of the nineteenth century, the *theory of evolution* and the *cell theory* catalyzed the conversion of biology from an observational pastime to an active experimental science.

The Theory of Evolution Arises from Naturalistic Studies

In their theories of evolution, Charles Darwin (Figure I-1) and Alfred Wallace made the first modern scientific responses to the three central questions just noted. By charting the geographic distributions of animals and

▲ **Figure I-1** Charles Darwin (1809–1882) four years after his epical voyage on *H.M.S. Beagle*. He had already begun private notebooks formulating his concept of evolution which would be published in *Origin of Species* (1859). *Courtesy of John Moss/Black Star.*

plants and by observing the anatomic similarities and differences among closely related groups of organisms, Darwin and Wallace recognized the inconstancy of the biological world: when examined over long periods of time, the extant species changed. They hypothesized that certain forces of the environment—changing land masses, fluctuations in local temperature and rainfall, and long-term climatic trends—act as agents of "natural selection" on succeeding generations of organisms in which heritable changes occur. In such a selective environment, new species continuously arise and old species no longer suited to the environment die out. To quote from Darwin's *Origin of Species* (1859):

As many more individuals of each species are born than can possibly survive, and as consequently there is a frequently recurring struggle for existence, it follows that any being, if it vary in any manner profitable to itself, under the complex and sometimes varying conditions of life, will have a better chance of survival and thus be *naturally selected* [italics ours]. From the strong principle of inheritance, any selected variety will tend to propagate its new and modified form.

We will see in many places throughout this book that the molecular comparison of genes between species strongly affirms Darwinian predictions and is beginning to reveal what types of genes are modified when evolutionary changes occur.

The Cell Theory Comes to Prominence through Improved Microscopic Techniques and Recognition That Single Cells Can Grow and Divide

In the last half of the seventeenth century, the Dutch naturalist Antonie van Leeuwenhoek first glimpsed the world at the microscopic level using a simple magnifying lens. In ordinary pond water, he saw a bewildering array of particles too small to be visible to the naked eye. These structures, which Leeuwenhoek called "animalcules," were later recognized as single-celled organisms. (Some of Leeuwenhoek's original samples, which still contain algal and other cells, have been recovered: see Figure I-2.) At about the same time, Robert Hooke, in his famous treatise *Micrographia* (1665), described the microscopic units that make up cork (a dead tissue). Hooke termed these units "cells" (Figure I-3). However, not until nearly two centuries later was the living cell recognized as the basic unit of life. When improved microscopes, techniques for preserving tissues, and tools for slicing thin tissue sections became available in the nineteenth century, investigators

(a)

(b)   20μm

▲ **Figure I-2** (a) Examples of the dried specimens prepared by Antonie van Leeuwenhoek and sent to the Royal Society of London between 1674 and 1687. They were discovered by Brian J. Ford in 1981. (b) A scanning electron micrograph of one of the specimens reveals some of the items—algal cells (Al) and diatom chains (D)—that Leeuwenhoek must have seen in his samples of brackish water. [See *Nature*, 1981, **292**:407.] *Courtesy of Brian J. Ford.*

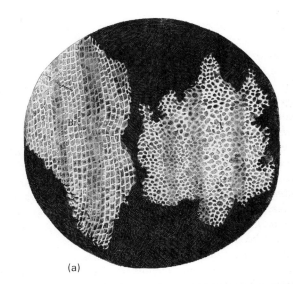

(a)

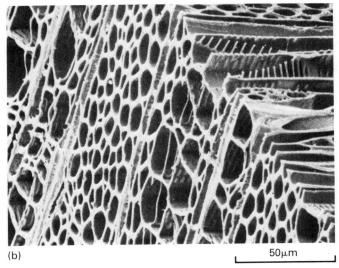

(b)   50μm

▲ **Figure I-3** (a) In 1665, Robert Hooke drew the cut surface of cork that he observed through a light microscope; he called the spaces in the pattern "cells." (b) A scanning electron micrograph of a cut surface of wood. *Part (a) courtesy of Chapin Library, Williams College; part (b) courtesy of B. A. Meylan.*

observed that tissues were composed of units or cells and that these cells could *divide.* It began to appear that each individual cell possessed life. In 1839 Theodor Schwann described the construction of tissues:

The elementary parts of all tissues are formed of cells in an analogous, though very diversified, manner, so that it may be asserted that there is one universal principle of development for the elementary parts of organisms, however different, and that this principle is the formation of cells.

A similar formulation had been elaborated in 1838 by Matthias Jakob Schleiden, who studied plants:

Each cell leads a double life, one independent pertaining to its development, the other an intermediary, since it has become an integrated part of a plant.

A few years later, in 1858, the German pathologist Rudolph Virchow articulated what soon became the accepted form of the cell theory:

Every animal appears as a sum of vital units, each of which bears in itself the complete characteristics of life.

Up to the time of Schleiden and Schwann, many philosophers and scientists believed in *vitalism.* According to the vitalists, no single part of an organism was alive; instead, properties of living matter were somehow shared by the whole organism. Further, the germinal material of most organisms was thought to be a primitive protoplasm, a "primitive blastema." The cell theory greatly weakened these beliefs with the proposition that organisms could arise from the growth of constituent cells. One implication of the cell theory was most important: if individual cells could grow and divide, they were proper subjects for the study of living organisms.

Before Louis Pasteur, single-celled organisms of the sort Leeuwenhoek had seen were thought to arise from spontaneous generation and were not considered to have an inheritance of their own. But Pasteur showed that microbes would grow in a culture medium only if a few microbes were seeded into the medium. His classic experiments weighed heavily against the concept of the spontaneous generation of single-cell organisms. By the end of the nineteenth century, the cell theory was widely accepted and the foundations of modern biology were in place.

Classical Biochemistry and Genetics

The theory of evolution combined with the cell theory provided the intellectual framework on which biology developed as an experimental science. Two major branches of the new experimental science grew independently for many years: (1) the isolation and chemical characterization of cell substances, or *biochemistry,* and (2) the study of the inheritance of characteristics by whole animals and plants, or *genetics.*

Biochemistry Begins with the Demonstration That Chemical Transformations Can Take Place in Cell Extracts

Early in the nineteenth century, it was discovered that extracts of both plant and animal cells, when heated or mixed with acid, yield a fibrous precipitate that contains carbon, hydrogen, oxygen, and nitrogen in approximately equal amounts. G. J. Mulder concluded in 1838 that this fibrous material was

without doubt the most important of the known components of living matter, and it would appear that without it life would not be possible. This substance has been named protein.

The question of how protein and other cellular compounds were formed became controversial. Did the laws of chemistry apply to cells? One important approach to these problems was to try to synthesize organic compounds in the laboratory. Friedrich Wöhler, a German chemist, was able to form urea (a substance present in the blood and excreted in urine) and oxalic acid (a prominent constituent of spinach and many other vegetables) from simpler organic substances. Marcelin Berthelot succeeded in making acetylene, a simple hydrocarbon, from carbon and hydrogen, proving that the most widely occurring linkage in organic matter could be made in the laboratory. Finally, in 1897 Eduard Buchner demonstrated that organic chemical transformation could be performed by cell extracts. He ground up yeast cells, carefully filtered out any remaining living cells, and showed that glucose could be converted into ethyl alcohol by the cell extract, just as occurs in the production of wine by yeast cells.

At its start, the science of biochemistry had the dual goals of the chemical analysis of cell constituents and the analysis of the chemical reactions carried out by these components. An important aspect of early biochemistry was the research on the nature of the chemical units that make up protein. By 1900, 16 of the 20 standard amino acids that serve as the building blocks of proteins were known. Threonine, the last to be discovered, was not isolated until 1935. By 1900 Emil Fischer had proposed the correct mechanism for the formation of the chemical links in protein: the peptide bonds between adjacent amino acids. Fischer deduced that these bonds are created by the elimination of water when the α-amino group of one amino acid is joined to the carboxyl group of its neighbor.

The other major cell constituents—lipids, carbohydrates, and nucleic acids—were recognized and partially purified from various sources in the last half of the nineteenth century. By 1871 Friedrich Miescher, in his studies of the constituents of cells, had isolated what must have been primarily deoxyribonucleic acid (DNA) from the nuclei of dead white blood cells. But almost 50 years were to pass before biochemists resumed the investigations that would link a specific cellular component to heredity.

Classical Genetics Begins with the Controlled Breeding Studies of Gregor Mendel

It is hard to overemphasize the methodological and conceptual leaps that the Austrian monk Gregor Mendel (Figure I-4) made in discovering the basic rules of heredity. First, Mendel chose a simple organism, the garden pea, which could be easily grown and fertilized. The flowers normally self-pollinate, but they can be opened up and artificially pollinated without damaging them or their ability to form seed. Mendel was able to make experimental genetic crosses by fertilizing the ovules of one strain of pea with the pollen from another. Second, he was careful to select a particular variety of garden pea (*Pisum sativum*) that had been purified by repeated self-

pollination, so that each pure strain always "bred true"; that is, all plants produced had identical characteristics. Third, the "characters," or traits, that Mendel chose to study—such as the color and texture of seeds and seed coats, the position of flowers on the stems, and the length of the stems—could be determined unambiguously, so that no confusion arose in counting the number of offspring of each type after each cross (Figure I-5). Finally,

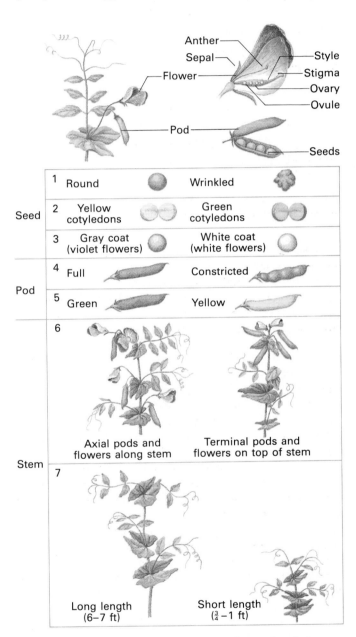

▲ **Figure I-5** The seven "characters" observed and described by Mendel. To cross two strains, the flower of one strain is opened before self-pollination occurs, the anthers are removed, and the ovules are dusted with pollen from the other strain. The offspring of such a cross are called the F_1 generation. *From M. W. Strickberger, 1976, Genetics, 2d ed., Macmillan, p. 115.*

▲ **Figure I-4** Gregor Mendel (1822–1884), the father of genetics. *Courtesy of the Moravian Museum, Brno.*

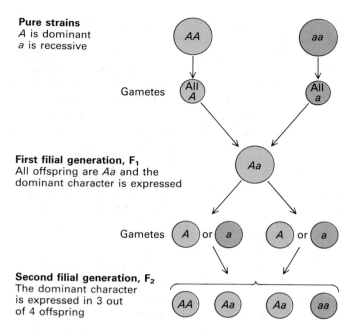

Pure strains
A is dominant
a is recessive

Gametes

First filial generation, F₁
All offspring are Aa and the
dominant character is expressed

Gametes

Second filial generation, F₂
The dominant character
is expressed in 3 out
of 4 offspring

▲ **Figure I-6** Mendel's experiments revealed that an individual organism has two alternative heredity units for a given trait: the dominant unit (*A*) and the recessive unit (*a*). In crossing pure strains of peas, he discovered that although all F₁ plants expressed the dominant unit, the recessive unit was apparently still present because it emerged in the F₂ generation. Mendel concluded that the two heredity units do not "blend" in the parent plants but remain discrete and segregate in the formation of pollen and ovules (gametes). Each ovule or grain of pollen receives only one of the two parental units. This conclusion has come to be known as Mendel's first law.

and most importantly, Mendel reached his conclusions with the aid of analyses of large numbers of descendants of his experimental crosses.

Mendel used his pure strains of peas to demonstrate that if certain traits are expressed in either parent, all offspring will express those traits. Such traits are *dominant*. Other traits are expressed in all offspring only if they are expressed in both parents (each a pure strain). These *recessive* traits are alternative forms of the dominant traits. Of crucial importance was Mendel's discovery that if two pure strains—one having the dominant form of a trait and the other the recessive form—are crossed, the recessive form disappears in the first filial (F₁) generation; however, if F₁ plants are crossed, the recessive form of the trait reappears in one-fourth of the plants of the second filial (F₂) generation (Figure I-6). Because of this masking and reappearance of recessive traits, Mendel concluded that each parental plant has two hereditary units for each trait but that these units segregate during the formation of gametes (pollen or ovules), so that each gamete has only one unit.

Mendel went on to follow several dominant/recessive trait pairs simultaneously through two generations of crossbreeding. He showed that different trait pairs could behave independently of one another, which led him to formulate a law of independent assortment of hereditary units for different traits (Figure I-7).

These remarkably clear experiments were described fully by Mendel in 1865, but his conclusions were intellectually so far ahead of his contemporaries that they were ignored. Not until 1900 were they rediscovered and widely accepted. The cell theory had to become more firmly rooted before early cell biologists could make the connection between Mendel's genetics and cell division.

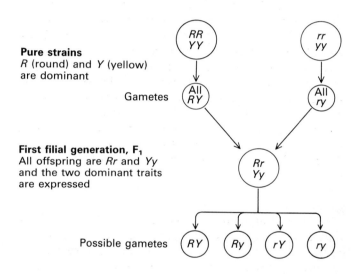

Pure strains
R (round) and Y (yellow)
are dominant

Gametes

First filial generation, F₁
All offspring are Rr and Yy
and the two dominant traits
are expressed

Possible gametes

	RY	Ry	rY	ry
Possible combinations in second filial generation, F₂ RY	RR YY	RR Yy	Rr YY	Rr Yy
Each gametic type crossed with each other type Ry	RR Yy	RR yy	Rr Yy	Rr yy
Plants are obtained in the following 9:3:3:1 ratio: 9 round and yellow, 3 round and green, 3 wrinkled and yellow, and 1 wrinkled and green rY	Rr YY	Rr Yy	rr YY	rr Yy
ry	Rr Yy	Rr yy	rr Yy	rr yy

▲ **Figure I-7** When Mendel crossed strains of peas that were pure for each of two traits—say, round (*RR*) and yellow (*YY*) peas with wrinkled (*rr*) and green (*yy*) peas—the plants of the F₂ generation exhibited four different combinations of the two traits, which always appeared in the ratio 9:3:3:1. Mendel accounted for this ratio by proposing that the two traits (form and color) were assorted among the offspring independently of one another. This conclusion has come to be known as Mendel's second law.

Chromosomes Are Identified as the Carriers of the Mendelian Units of Heredity

Once it was widely accepted that each cell within an organism is the product of the division of another cell, attention turned to the sperm and the egg cells, the union of which represents the first step in all cell division in higher organisms. In animals, the egg cell is comparatively large and consists mostly of cytoplasm and stored food. The much smaller sperm cell is composed mainly of a nucleus. It was reasoned that if the egg and the sperm both make equal hereditary contributions to the offspring, the nucleus—not the cytoplasm—must hold the key to genetic transfer.

Soon after the nucleus became an important subject of study, filamentous structures were observed microscopically in the nuclei of both plant and animal cells. These ubiquitous structures, the *chromosomes,* are constant in size and number within the somatic (general body) cells of a given species, but their sizes and shapes vary from species to species. Each somatic cell contains two copies of each morphologic type of chromosome; that is, the chromosomes of somatic cells exist in homologous pairs. During every somatic-cell division, the chromosomes double in number and separate into two groups, so that each daughter cell receives the same number and kinds of chromosomes as the mother cell: two of each morphologic type.

In 1903 Walter Sutton observed that in the cell division that produces germ cells (sperm and egg cells, or gametes), the threadlike chromosomes divide in such a way that each gamete receives only *one* chromosome of each morphologic type. Since Mendelian crossbreeding had shown that each parent contributes one hereditary unit for each trait in the offspring, Sutton recognized the chromosomes as the carriers of Mendel's units of heredity and reasoned that the parental sperm and egg each contributes one set of chromosomes to every new individual.

The Reduction in Chromosome Numbers in Meiosis That Forms Germ Cells Is Crucial to the Development of the Chromosome Theory of Heredity

The meaning of the difference in chromosomal behavior between somatic cells and germ cells was then given close attention. Both the sperm cells and the egg cells of a given species have one chromosome of each type. Each gamete thus contributes equally to the fertilized egg, which has twice as many chromosomes as either gamete alone. (In calling this contribution of chromosomes "equal," we ignore for the moment the different sex chromosomes contributed by sperm and egg; this topic will be discussed in Chapters 5 and 9.)

Before the first cell division of a fertilized egg and before all divisions thereafter, each chromosome duplicates to form a pair of threadlike structures (*chromatids*) attached at one point, the *centromere*. These "sister" chromatids split longitudinally at division, so that each daughter cell receives the same number of chromosomes contained in the fertilized egg. The number of chromosomes in a gamete is called the *haploid* number n for that particular species; the number in a somatic cell is called the *diploid* number $2n$. The equal partitioning of sister chromatids at cell division, called *mitosis*, was first described around 1880 by Walter Flemming, an important early cytologist.

As Sutton noted, the formation of sperm or egg cells entails a qualitatively different kind of chromosomal division. As in mitosis, this process, called *meiosis*, begins with diploid cells whose chromosomes have duplicated to form pairs of sister chromatids attached at their centromeres. In the first meiotic division, the sister chromatids do not separate; each of the two daughter cells receives both copies of either the maternal or paternal member of a homologous pair. In the second meiotic division, the chromatids do separate longitudinally. Each of the resulting four nuclei contains only one copy of each chromosome type, producing the haploid number of chromosomes. Although four germ cells can arise from each cell that enters the meiotic cycle, often only one mature egg cell results; the other three meiotic products degenerate.

Sutton realized that meiotic cell division is the basis for the distribution of traits in Mendelian heredity. He described the assortment of chromosomes in the course of gamete formation for a hypothetical organism with somatic cells containing four homologous pairs of chromosomes: the paternal contributions A, B, C, and D and the maternal ones a, b, c, and d:

Each of the ripe germ cells arising from the reduction divisions must receive one member from each of the "synaptic pairs" [the sister chromatids, which are still coupled after the first meiotic division]. And there are 16 possible combinations of originally maternal and paternal chromosomes that will form a complete series: to wit, a, B, C, D; A, b, C, D; A, B, c, D; A, B, C, d; a, b, C, D; a, B, C, d; a, b, c, d; and their conjugates, A, b, c, d; etc.

Further, Sutton noted that

the number of possible combinations in the germ-products of a single individual of any species is represented by the simple formula 2^n, in which n represents the number of chromosomes in the reduced [gametic] series.

Chromosomes Are Shown to Contain Linear Arrays of Genes That Can Undergo Reordering

The study of the assortment of Mendelian units of hered-ity, or *genes* (as they were named by the Danish geneticist Wilhelm Johannsen in 1909), and of the relation of this assortment to chromosomal structure was greatly ex-panded by Thomas Hunt Morgan (Figure I-8) and his students in the early 1900s. Morgan chose the fruit fly, *Drosophila melanogaster,* for his studies because it ma-tures through an entire developmental cycle from fertil-ized egg to adult in about 2 weeks. Early studies showed that a particular chromosome is responsible for sex deter-mination. Morgan's group also discovered, by crossing (interbreeding) genetically marked flies, that the genes of *Drosophila* segregate as four sets, termed *linkage groups.* The presence of four linkage groups in the genetic crosses was correlated with the presence of four visible chromo-somes in *Drosophila* cells. In studies of corn, the number of genetic linkage groups (10) corresponded to the num-ber of chromosomes; similar results were obtained for peas (seven linkage groups and seven chromosomes). By

▲ **Figure I-9** Barbara McClintock (1902–) lecturing on her genetic work with corn, which provided the first micro-scopic evidence of chromosome breakage and reunion during recombination. Later McClintock showed that some chromo-some segments are "mobile" in the genome, for which she received the Nobel Prize in 1983. *Courtesy of Nik Klein-berg/Picture Group.*

▲ **Figure I-8** Thomas Hunt Morgan (1866–1945), the founder of modern experimental genetics, in his laboratory at Columbia University. *Courtesy of Mrs. Curt Stern.*

1920 the chromosome theory of heredity had become the common currency of the new field of genetics.

Another important accomplishment of the *Drosophila* geneticists was the discovery of how genes are arranged in chromosomes. Both by genetic techniques and by visible observation, the *Drosophila* chromosomes were deter-mined to be *regular* linear arrays in which particular genes are positioned consistently. Subsequently, it was found that gene order in the chromosomes could change. In 1931 Barbara McClintock (Figure I-9) correlated visi-ble rearrangements of chromosome segments with the redistribution of specific genetic traits in strains of corn (*Zea mays*). These classical genetic studies gave the genes a physical reality; by the mid-1930s genes were no longer remote or "theoretical" entities but widely accepted de-terminants of biological specificity. The discovery of the chemical nature of the gene was a consequence of the merging of genetics and biochemistry between 1930 and 1950.

The Merging of Genetics and Biochemistry

The first relation between a genetic defect and a biochemical abnormality was discovered in the study of a human disease. By 1909 Archibald Garrod realized that the disease alkaptonuria was caused by a rare recessive mutation inherited according to Mendelian rules. Patients having this disease eventually suffer severe arthritis, but their condition can be diagnosed prior to the onset of arthritic pain because they excrete a large amount of homogentisic acid in their urine. This phenolic compound, which turns the urine black, was recognized during the 1890s as a breakdown product of the amino acids tyrosine and phenylalanine, each of which contains a six-membered aromatic ring. Garrod surmised the nature of the genetic defect in his treatise on "Inborn Errors of Metabolism":

> We may further conceive that the splitting of the benzene ring in normal metabolism is the work of a special enzyme, [and] that in congenital alkaptonuria this enzyme is wanting.

Garrod was correct: the missing enzyme is homogentisic acid oxidase—predominantly a liver enzyme that catalyzes breakage of the benzene ring during the degradation of phenolic compounds. But it was too early in the history of genetics or biochemistry for the realization that genes do in fact control the structures of enzymes.

Appropriately enough, an important early step in the unification of genetics and biochemistry was the result of an experiment with fruit flies by George W. Beadle and Boris Euphrussi.

Drosophila Studies Establish the Connection between Gene Activity and Biochemical Action; Neurospora Experiments Confirm That One Gene Controls One Enzyme

Experiments with fruit flies yielded much of the early information about genes. Eye color was one of the most intensively studied genetic characteristics of *Drosophila*. The normal color is a dark red due, it is now known, to the production of two chemically distinct pigments. Morgan and his colleagues collected many flies bearing changes, or mutations, in the gene(s) controlling eye color. Flies with one of two unlinked mutant genes, cinnabar or vermillion, have eyes that are a much lighter red. Beadle and Euphrussi grafted eye tissue from larvae of one or the other of these mutant stocks of flies onto normal, or wild-type, larvae. The extra eyes that developed in the grafts were normal in color. Beadle and Euphrussi concluded that some metabolic product from the sur-

rounding normal tissue diffuses into the mutant graft and allows the normal pigments to form in the grafted tissue. Thus a metabolic product could correct a genetic defect.

Because a graft of cinnabar larval tissue onto a vermillion larva (or vice versa) did not correct the eye color in the grafted tissue, the experimenters reasoned that cinnabar and vermillion represent mutations in different enzymes that are both necessary for pigment formation. One of the components of the dark red pigment was then shown to be a metabolic product of the amino acid tryptophan. It finally proved true that the mutations cinnabar and vermillion produce defects in two different enzymes that catalyze two different chemical reactions required to transform tryptophan into the dark red pigment. *The connection between gene activity and biochemical action was established.*

Seeking a genetic system even simpler than *Drosophila*—one in which biochemistry could be brought still closer to genetics—Beadle and Edward L. Tatum turned to the bread mold *Neurospora crassa*, a fungus that derives genetic material from two parents, as higher organisms do. Like *Pisum sativum* and *Drosophila*, *Neurospora* could be used in genetic crosses. But *Neurospora* offered the enormous advantage that it can be cultured in the laboratory from a single spore on a few simple substances: water, a sugar (a carbon source), salts containing ammonium ions (a nitrogen source), and biotin. Each wild-type *Neurospora* spore gives rise to a single colony that can grow equally well on a simple or an enriched medium. However, a rare spore (one in approximately 1000) requires an enriched medium for growth. Beadle and Tatum found that a number of such mutants resumed normal growth on a simple medium to which a single substance, such as an amino acid or a vitamin, was added. They concluded that each mutant cell that could be restored to growth by the addition of a single compound carried a defect in a single gene that impaired the production of an enzyme necessary for a single metabolic step; in other words, *one gene was responsible for one enzyme.*

Beadle and Tatum's powerful conclusion that genes somehow control protein (enzyme) structure led to studies of the chemical structure of a gene—the first step in elucidating the molecular basis for the genetic control of protein synthesis. Successful studies of gene structure and the genetic control of protein synthesis (to be fully discussed in later chapters) form the recent history of molecular cell biology. Once Beadle and Tatum unveiled the power of microbial systems in genetics, the use of bacterial cells soon followed. In the mid-1940s Joshua Lederberg (Figure I-10) found that the genes of bacteria also could be defined by nutritional mutations. In addition, he observed that genes could be transferred from one bacterium to another when they were allowed to contact or "mate." The use of bacteria as the simplest genetic system quickly became extremely popular. A biochemical experi-

▲ **Figure I-10** Joshua Lederberg (1925–) around 1958, when he shared a Nobel Prize with Edward L. Tatum (1909–1975) and George W. Beadle (1903–). Lederberg's work on genetic exchange between strains of *E. coli* helped to establish it as the most widely used organism in molecular biology. *Courtesy of J. Lederberg.*

ment with bacteria that just predated Lederberg's work led to a discovery essential to our understanding of the chemical nature of the gene.

DNA Is Identified as the Genetic Material, Paving the Way for the Study of the Molecular Basis of Gene Structure and Function

In 1944 Oswald Avery, Colin MacLeod, and MacLyn McCarty were studying *Streptococcus pneumoniae*, a bacterium frequently isolated from people with pneumonia. The more active disease-causing strains are called *smooth* (S) because a gelatinous outer capsule on each cell makes the colonies glisten. The less virulent strains, called *rough* (R), lack this capsule and form more ragged-looking colonies. The two strains can also be distinguished by antibacterial antibodies that react with the S strains in the serum of convalescing patients. Avery and his colleagues— following the lead of British physician Fred Griffiths, who showed that the R strain of *S. pneumoniae* could be converted into the S strain during infection of mice— extracted the deoxyribonucleic acid (DNA) from a culture of S bacteria and mixed it with cells of the R type in a test tube. When the DNA-treated pneumococci were grown into colonies on bacterial plates, some of the bacteria had smooth (S) characteristics. These cells continued

thereafter to exhibit the S type and were said to be "transformed." The substance that caused the transformation— the *transforming principle*—was shown to be DNA (Figure I-11).

The proof was obtained not only by biochemical testing (for example, for the presence of deoxyribose and the absence of ribose) but also by developing another experimental plan that would be repeated often in later years: Avery, MacLeod, and McCarty used an enzyme to prove a biological point. They found that adding a variety of proteases (enzymes that digest proteins) did not alter the transforming principle. However, tiny amounts of purified deoxyribonuclease (DNase), a then newly recognized enzyme that destroys DNA, immediately inactivated the transforming principle. This proved that DNA was the transforming principle that could alter bacteria genetically. Like Mendel's work 80 years earlier and Garrod's work 40 years earlier, this experiment was not widely appreciated by the scientific community, although it did help Lederberg decide to use bacteria for genetic studies. After several more years of experience in genetic studies with bacteria, biologists and geneticists were ready to accept another experiment that also used bacterial systems and showed DNA to be the genetic material.

Bacterial cells, like almost all other cells in nature, are susceptible to viral infections. Many viruses contain DNA that has the ability, once inside a host cell, to dictate the reproduction of more virus; the only nucleic acid in most bacterial viruses, or bacteriophages, is DNA. These viruses exhibit heritable traits (as examples, fast or slow growth, growth only on certain cells, or growth in a restricted temperature range) that can identify specific viral strains. Moreover, different strains of bacteriophages can exchange genes when two strains infect the same cell, indicating that the infecting viruses participate in a genetic process within the host cell. In 1952 Alfred Hershey and

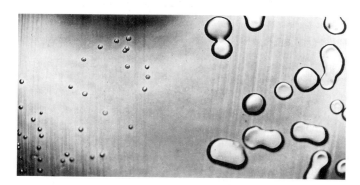

▲ **Figure I-11** Colonies of *Streptococcus pneumoniae*, showing small populations that appear rather dull or "rough" (*left*) and large, glistening, "smooth" colonies (*right*). The DNA of the organisms in smooth colonies can transform rough organisms into the smooth form. *From O. T. Avery, C. M. MacLeod, and M. McCarty, 1944,* J. Exp. Med. *78:137.*

Martha Chase showed that only the DNA of the bacterial virus—and not its protein portion—must enter the host bacterium to initiate infection. Thus the genetic information that causes new virus production resides in the DNA, not in the protein of the bacterial virus. Hershey and Chase's result, plus the earlier findings of Avery and his colleagues on bacterial transformation, established DNA as the genetic material. [Ribonucleic acid (RNA) is the genetic material in some viruses. The nature of viruses as cell-dependent genetic systems will be discussed in Chapter 5.]

This conclusion spawned some tentative answers to age-old questions about the origins of diversity in life, about heredity, and about animal and plant development. The knowledge that DNA was the universal genetic material of cells suggested that (1) early evolution must have depended on the development of a cell carrying sufficient instructions in its DNA to grow; (2) the random variation and selection that, according to the ideas of Darwin and Wallace, led to changes in species must have resulted from random changes in the DNA; (3) the faithful reproduction of DNA from generation to generation causes "like to beget like"; and (4) the programmed unfurling of the genetic endowment in the DNA underlies the development of every new plant or animal.

The conceptual advances that occurred between 1859, when Charles Darwin's *Origin of Species* was published, and 1952, when a small group of informed scientists knew that DNA was the controlling molecule of life, were of enormous consequence. No sooner had the biological importance of DNA been recognized than its physical structure was discovered.

The Birth of Molecular Biology

Watson and Crick Deduce the Double-Helical Structure of DNA

The modern era of molecular cell biology, which has been mainly concerned with how genes govern cell activity and how proteins carry out specialized tasks including DNA and RNA formation, began in 1953 when James D. Watson and Francis H. C. Crick deduced the double-helical structure of DNA (Figure I-12). Two major clues led Watson and Crick to build a correct model of the DNA molecule. First, Erwin Chargaff had separated and measured the four nucleic acid bases—thymine, cytosine, adenine, and guanine—in DNA from various sources. He found that the amount of adenine always equals that of thymine and that the amount of guanine always equals that of cytosine; this *base pairing* is conventionally written A = T and G = C. [Note that total A + T does not always equal total G + C; the (A + T)/(G + C) ratio can vary in DNA from different sources.] Second, Maurice

▲ **Figure I-12** (*left*) James D. Watson (1928–) and (*right*) Francis H. C. Crick (1916–) with the double-helical model of DNA they constructed in 1952–1953. *From J. D. Watson,* The Double Helix, *Atheneum, p. 215, copyright 1968 by J. D. Watson. Courtesy of A. C. Barrington Brown.*

Wilkins and his collaborators, particularly Rosalind Franklin, had obtained x-ray diffraction patterns of DNA fibers. These images, formed by the refractions of an x-ray beam by the regularly spaced atoms of the DNA molecule, revealed extensive details about the structure: its helical nature; the diameter of the linear molecule (2 nanometers, or nm); the distance between adjacent bases (0.34 nm); and because one turn equals 3.4 nm, the approximate number of bases per turn of the helix (10). By fitting together the base-pairing data and the structural data obtained from x-rays, Watson and Crick built an accurate model of DNA.

The simple, self-complementary organization of DNA suggested a means by which it could be copied in each generation. (DNA replication is discussed in Chapter 12.) Using the principles of base complementarity, scientists also discovered how information in DNA is accurately transferred first into RNA and then into protein (see Chapters 3 and 7–11).

X-Ray Crystallography Facilitates the Construction of Three-Dimensional Models of Complex Biological Molecules

Before its use in determining the structure of DNA, x-ray crystallography had been applied to crystals of simple substances and to pure proteins. Linus Pauling had sug-

◄ **Figure I-13** (*left*) Max Perutz (1914–) holding a three-dimensional model of hemoglobin and (*right*) John Kendrew (1917–) holding one of myoglobin, both developed using x-ray crystallographic techniques. *Courtesy of Keystone Press Agency.*

gested a regular helical arrangement for certain parts of protein chains in 1951. At Cambridge, England, Max Perutz and John Kendrew (Figure I-13) had been analyzing x-ray patterns produced by crystals of small proteins for more than a decade when, in 1959, Kendrew completed the determination of the spatial position of each atom in sperm whale myoglobin; the structure indeed contained helixes. In molecular biology, x-ray crystallography contributes to three-dimensional models of complex biological molecules and shows how the protein structure facilitates specific biochemical capacity. For example, from the three-dimensional study of hemoglobin, the oxygen-carrying protein of blood, Perutz and his colleagues determined the exact nature of the oxygen-binding site: a "pocket" in which the iron-containing heme molecule is bound. When oxygen is bound by the heme in the pocket, the protein chains assume a different physical arrangement, or conformation, from that when no oxygen is bound. By now well over 100 three-dimensional protein structures, as well as the first RNA crystal structures for transfer RNA (tRNA), have been determined at a resolution of 0.2–0.5 nm. How proteins aid in speeding chemical reactions is understood now in considerable detail.

Biochemical Experiments Have Elucidated the Relationship between Enzymes and Metabolic Pathways

The development of molecular cell biology also depended on traditional biochemistry accomplished from 1910 to 1960. Experiments in which cells are broken apart and

assays are devised for specific proteins, which are then purified, are the domain of biochemists. Our understanding of the nature of chemical reactions and the discovery of individual enzymes have depended greatly on research in which critical biochemical steps are reconstructed from purified proteins in the test tube. Since Eduard Buchner's early experiments showing that yeast-cell extracts could ferment glucose to ethanol, virtually all of the enzymes responsible for the most important metabolic pathways have been identified through the efforts of hundreds of biochemists. Vitamins have been recognized as enzyme cofactors. In addition, the importance of high-energy phosphates has been proved, particularly of adenosine triphosphate (ATP), recognized by the great biochemist Fritz Lipmann as the "energy currency of the cell." The mechanisms of ATP generation from light energy and carbohydrate metabolism have been elucidated. We examine some of these crucial topics of biochemistry and how cell structures carry out these biochemical functions in Chapters 1, 2, 15, and 16.

A Modern View of Cell Structure

Advances in Electron Microscopy Reveal the Commonality of Structures within Eukaryotic Cells

Throughout the 1930s and early 1940s, while geneticists were seeking the identity of the gene and biochemists were finding enzymes and interpreting the cell's chemistry, relatively little attention was paid to the organization of the higher structures within the cell. The light microscope imposed a great physical limitation on inquiry into cell structure. Resolution of the best light microscopes is dictated by the average wavelength of visible light; objects less than 500 nm in diameter could not be resolved from one another. In that era only the larger intracellular structures—the nuclei, nucleoli, and chloroplasts in plant cells and mitochondria in animal and plant cells—had been identified with ordinary microscopes. The electron microscope lifted a veil from the eyes of biologists. Even in its early forms, its practical resolving power was 1–5 nm; its theoretical resolving power is 10 times higher (0.1–0.5 nm). A world of structure never before observed was suddenly displayed, and a striking fact emerged: all eukaryotic cells have certain cell structures in common.

This important conclusion, which has been elaborated over the past several years, is discussed in Chapters 4 and 13–23.

Biochemical Activities Can Be Assigned to Specific Subcellular Structures

By the end of the 1950s, the fundamental observations made independently by geneticists, biochemists, and structural cell biologists were ready to be stitched together into a coherent pattern that would become the design for molecular cell biology. Two different experimental approaches provided key contributions.

First, a correlation between cell structure and biochemical activity was achieved by breaking open plant and animal cells and separating, or fractionating, the broken cell extracts. Each fraction was studied under the electron microscope and through biochemical assays. Much of the early electron-microscopic identification of cell structures and biochemical characterization of cell fractions was the work of Albert Claude and his associates Christian de Duve, George Palade, and Keith Porter (Figure I-14). The current sophisticated picture of cell structure and function is very largely due to their research.

The Activity of Genes Is Highly Regulated by the Protein Products of Other Genes

The second experimental approach that had a great unifying effect was genetic analysis. The work of Beadle, Tatum, Lederberg, and others in the 1940s and early 1950s had made it clear that genes encode proteins, but the means by which gene activity was controlled was still unclear. Yet it was widely supposed that the differences between cells having the same genes were due to differences in the activities of those genes. Two French scientists, François Jacob and Jacques Monod, proposed that the protein products of certain genes regulate the activities of other genes. This principle finally unified the diverse approaches to the subject that we now call molecular cell biology. The concept of the cell had evolved a long way from its original characterization as a simple unit of living matter: the cell had become an organism in which the controlled and integrated actions of genes produce specific sets of proteins that build characteristic structures and carry out characteristic enzymatic activities.

The Molecular Approach Is Applied to Eukaryotic Cells

Soon after Jacob and Monod presented their findings on regulatory genes in 1961, three significant advances occurred: (1) proof that messenger RNA (mRNA) carries information from DNA to the protein-synthesizing machinery; (2) the discovery of the genetic code, by which information is stored in the nucleic acids; and (3) the discovery that proteins are translated by transfer RNA (tRNA) and ribosomes (small nucleoprotein granules first discovered by electron microscopy).

Much of the research that contributed to these achievements was done by biochemists and molecular geneticists working with *E. coli*. However, two developments in the 1960s led an ever-increasing number of workers into the study of mammalian cells: (1) the successful culturing of mammalian cells in vitro, particularly by such pioneers as Harry Eagle, who perfected the first defined growth me-

(a)

(b)

▲ **Figure I-14**   The architects of modern cell biology. (a: *left to right*) George Palade (1912–), Albert Claude (1899–1983), and Keith Porter (1912–). (b) Christian de Duve (1917–). Dr. Claude's group at Rockefeller University was the first to achieve success in the use of cell fractionation and biochemical analysis of fractions (mainly the work of Drs. Claude, de Duve, and Palade) and in electron microscopy (primarily done by Drs. Porter and Palade). *Part (a) courtesy of Columbia University/Manny Warman; part (b) courtesy of Rockefeller University.*

◄ **Figure I-15** (*left to right*) Harry Eagle (1905–), Renato Dulbecco (1914–), and Theodore T. Puck (1916–) in 1972, when they were honored for their work on the successful in vitro growth of animal cells and the development of quantitative techniques in animal virology. *Courtesy of Harry Eagle.*

dium, and Theodore Puck, who led the way in culturing single mammalian cells, and (2) the emergence of animal virology as a quantitative science, notably through the work of Renato Dulbecco and his students (Figure I-15).

Although many aspects of protein and nucleic acid synthesis and of cell growth and cell function were elucidated in the 1960s and early 1970s, a lack of detailed genetic information about higher organisms impeded progress. Since the late 1970s, however, new techniques have revolutionized the study of DNA and the genes that it embodies. The DNA from any organism can be cut into reproducible pieces with site-specific endonucleases called restriction enzymes; the pieces can then be linked to bacterial vectors (DNA viruses or plasmids capable of independent growth) and introduced into bacterial hosts. In this way, almost any DNA segment from any organism can be isolated and produced in any amount simply by culturing the bacteria. These procedures are collectively referred to as *gene cloning* or *recombinant DNA technology*. Rapid DNA sequencing techniques were worked out. Suddenly, biologists were ushered into a new era: instead of just counting chromosomes or identifying gene mutations and mapping genes imprecisely by means of breeding experiments, they could now obtain the exact DNA sequence of individual genes and thus their exact location and coding capacity.

Not only has the isolation of particular DNA segments and their sequencing become commonplace, but molecular biologists can now introduce specific mutations at will; altered genes can be reintroduced into cells or even into the germline of organisms, so that the function of individual genes can be studied in particular cells. These fantastic strides forward, combined with the ever-improving methods for examining cells and their components, both biochemically and structurally, have aroused great optimism about the success of future research. We stand on the brink of the formerly unthinkable achievement of knowing the DNA sequence of the entire human genome, perhaps by the end of this century. Surely few problems in molecular cell biology, even finding the cause for cancer or determining the molecular basis of differentiation and development, will remain unsolved much longer.

References

Classical Biochemistry and Genetics

CREIGHTON, H. S., and B. MCCLINTOCK. 1931. A correlation of cytological and genetical crossing-over in *Zea mays. Proc. Nat'l Acad. Sci. USA* 17:492–497. Reprinted in J. A. Peters, 1959, *Classic Papers in Genetics*, Prentice-Hall.

CHRISTENSEN, A. K. 1987. Studying cells: from light to electrons. In *The American Association of Anatomists, 1888–1987,* J. E. Pauly, ed. William & Wilkins.

GABRIEL, M. L., and S. FOGEL, eds. 1955. *Great Experiments in Biology.* Prentice-Hall.

JACOB, F. 1973. *The Logic of Life: A History of Heredity.* Pantheon.

MENDEL, G. 1865. Versuche über Pflanzen-Hybriden. *Verh. Naturforschung Ver. Brünn* 4:3–47. Translated by W. A. Bateson as "Experiments in plant hybridization," and reprinted in J. A. Peters, 1959, *Classic Papers in Genetics,* Prentice-Hall.

MORGAN, T. H. 1910. Sex linked inheritance in *Drosophila. Science* 32:120–122. Reprinted in J. A. Peters, 1959, *Classic Papers in Genetics,* Prentice-Hall.

PETERS, J. A. 1959. *Classic Papers in Genetics.* Prentice-Hall.

SUTTON, W. S. 1903. The chromosomes in heredity. *Biol. Bull.* 4:231–251. Reprinted in J. A. Peters, 1959, *Classic Papers in Genetics,* Prentice-Hall.

The Merging of Genetics and Biochemistry

GARROD, A. 1909. *Inborn Errors of Metabolism.* Oxford University Press.

MORGAN, T. H. 1926. *The Theory of the Gene.* Yale University Press.

WILSON, E. B. 1925. *The Cell in Development and Inheritance,* 3d ed. Macmillan.

Genes Encode Enzymes

BEADLE, G. W., and E. L. TATUM. 1941. Genetic control of biochemical reactions in *Neurospora. Proc. Nat'l Acad. Sci. USA* 27:499–506. Reprinted in J. A. Peters, 1959, *Classic Papers in Genetics,* Prentice-Hall.

LEDERBERG J., and E. L. TATUM. 1946. Gene recombination in *Escherichia coli. Nature* 158:588. Reprinted in J. A. Peters, 1959, *Classic Papers in Genetics,* Prentice-Hall.

ZINDER, N. D., and J. LEDERBERG. 1952. Genetic exchange in *Salmonella. J. Bacteriol.* 64:679–699. Reprinted in J. A. Peters, 1959, *Classic Papers in Genetics,* Prentice-Hall.

DNA Is Identified as the Genetic Material

AVERY, O. T., C. M. MACLEOD, and M. MCCARTY. 1944. Studies on the chemical nature of the substance inducing transformation of pneumococcal types. *J. Exp. Med.* 79:137–158.

HERSHEY, A. D., and M. CHASE. 1952. Independent functions of viral protein and nucleic acid in growth of bacteriophage. *J. Gen. Physiol.* 36:39–56.

MCCARTY, M., and O. T. AVERY. 1946. I. Studies on the chemical nature of the substance inducing transformation of pneumococcal types. II. Effect of deoxyribonuclease on the biological activity of the transforming substance. *J. Exp. Med.* 83:89–96.

MCCARTY, M. 1985. *The Transforming Principle: Discovering That Genes Are Made of DNA.* Norton.

The Birth of Molecular Biology

CAIRNS, J., G. S. STENT, and J. D. WATSON, eds. 1966. *Phage and the Origins of Molecular Biology.* Cold Spring Harbor Laboratory.

CHARGAFF, E. 1950. Chemical specificity of the nucleic acids and mechanisms of their enzymatic degradation. *Experimentia* 6:201–240.

KENDREW, J. C. 1963. Myoglobin and the structure of proteins. *Science* 139:1259–1266.

PERUTZ, M. F. 1964. The hemoglobin molecule. *Sci. Am.* 211(5):64–76.

STENT, G. S., and R. CALENDAR. 1978. *Molecular Genetics: An Introductory Narrative,* 2d ed. W. H. Freeman and Company.

WATSON, J. D. 1968. *The Double Helix.* Atheneum.

WATSON, J. D., and F. H. C. CRICK. 1953. General implications of the structure of deoxyribonucleic acid. *Nature* 171:964–967.

WATSON, J. D., and F. H. C. CRICK. 1953. A structure for deoxyribose nucleic acid. *Nature* 171:737–738.

A Modern View of Cell Structure

Biochemical Activities Can Be Assigned to Specific Subcellular Structures

CLAUDE, A. 1946. Fractionation of mammalian liver cells by differential centrifugation, I and II. *J. Exp. Med.* 84:51–171.

CLAUDE, A. 1975. The coming of age of the cell. *Science* 189:433–435.

DE DUVE, C. 1975. Exploring cells with a centrifuge. *Science* 189:186–194.

PALADE, G. 1975. Intracellular aspects of the process of protein synthesis. *Science* 189:347–358.

PORTER, K. R., A. CLAUDE, and E. FULLAM. 1945. A study of tissue culture cells by electron microscopy: methods and preliminary observations. *J. Exp. Med.* 81:233–243.

The Activity of Genes Is Highly Regulated

JACOB, F., and J. MONOD. 1961. Genetic regulatory mechanisms in the synthesis of proteins. *J. Mol. Biol.* 3:318–356.

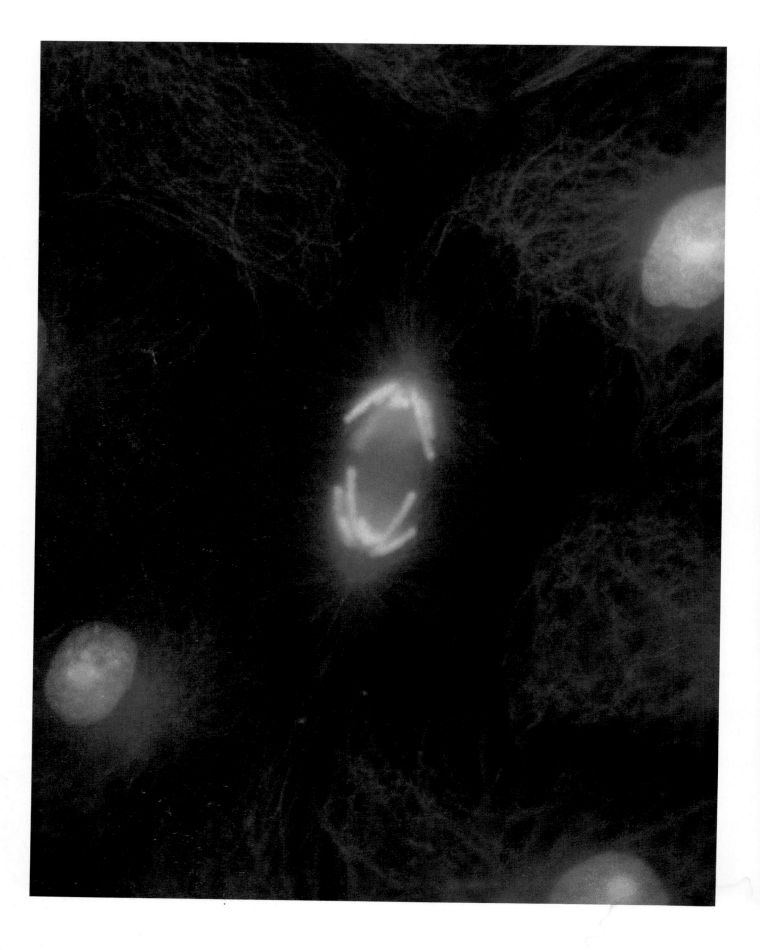

I

Molecules, Cells, Proteins, and Experimental Techniques: A Primer

◀ *Conventional fluorescence micrograph of a fibroblast at the anaphase stage of mitosis. The cell has been stained with a fluorescent antibody against microtubules (red) and with quinacrine (yellow-green) to label DNA in the chromosomes.* Courtesy of J. M. Murray.

This book is about an active experimental science that burst forth from a union of the older fields of biochemistry, molecular biology, cell biology, and genetics. From recent successes in molecular cell biology we now have a detailed picture of how DNA and RNA function; we have also gained a great depth of knowledge about all the major classes of cell proteins and about the mechanisms by which the proteins carry out cell functions. A purpose of this book is to impart all this information. Simply delivering the conclusions, however, while omitting the excitement of experimental discovery would miss the essence of modern biology. We therefore present experimental work in detail in many places in Parts II–IV. Part I gives the background material needed for comprehension of the discussions that follow in later chapters. This background is a depiction jointly composed by biochemists, molecular biologists, cell biologists, and geneticists. With it firmly in mind, the reader may share the excitement generated as discoveries add fine detail to the picture.

A cell works in accord with chemical principles. Chapter 1 reviews those principles of basic chemistry that apply to the workings of cells. Cells are constructed of and contain molecules. Chapter 2 describes the molecular building blocks of life—amino acids, nucleotides, sugars, and lipids—and the biological macromolecules—proteins, DNA, and RNA. Proteins are seen to serve as the key working molecules of biological systems and nucleic acids to encode information for protein synthesis. Cells synthesize molecules. Chapter 3 outlines how cells make nucleic acids and proteins and shows that RNA plays three separate roles in protein synthesis. Cells have an organized substructure from which scientists have learned much about the evolution of life on earth as well as about the particular functions of cells. Among the many types of cells that exist, there is great commonality in substructure. Chapter 4 reviews the major structures of cells. It also explains

the experimental techniques of fractionation and microscopy, which have unlocked many cellular secrets.

Other techniques have also proved powerful in adding to our understanding of cells. Chapters 5 and 6 describe these. How is information about cells obtained from experimental organisms? Chapter 5 tells how biologists grow cells, how they apply genetic techniques to cells, and how they use cellular research to address the study of whole organisms. How do molecular cell biologists conduct experiments? What chemical and physical techniques do they use? Chapter 6 provides answers: they use radioisotope markers, they separate proteins and nucleic acids according to length by various biochemical procedures, and they sequence, cut, manipulate, and hybridize macromolecules using techniques and equipment developed only since the late 1970s.

Students who have a thorough understanding of introductory chemistry and biology may find that a quick reading of Chapters 1–4 is all the preparation they need for understanding what follows. Students should read Chapters 5 and 6 in sequence and carefully enough that they may return easily to relevant sections for review as needed for understanding experiments discussed later in the book.

▽ ▽ ▽

Chemical Foundations

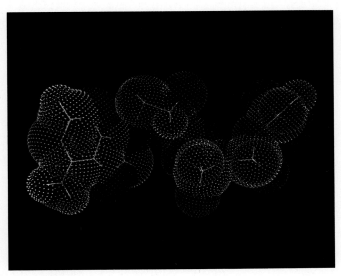

ATP molecule

Chapter 1 is a review of the concepts required to comprehend the workings of a cell. A knowledge of introductory college-level chemistry is assumed. Students who have taken more advanced chemistry courses may wish to examine this material to refresh their memories. The topics addressed here relate importantly to concepts and experiments presented later in the book.

The chapter begins with a discussion of energy and its transformations. It then reviews the different types of bonds that can be formed between individual atoms in a molecule, between groups of atoms within larger molecules, and between different molecules. Because water is the major constituent of cells and of the spaces between cells, the next topics treated are the structure of water and the reactions of chemicals dissolved in it—in particular, the reactions of acids and bases. Finally, the factors that control the direction and rate of chemical transformations are discussed, laying the groundwork for later chapters on cellular metabolism. Chapter 2 will consider the chemistry of the major classes of molecules that make up living systems. ▲

Energy

The production of energy, its storage, and its use are as central to the economy of the cell as they are to the management of the world's resources. These processes govern the strength of chemical bonds and determine the direction and rate of chemical reactions.

Energy may be defined as the ability to do work—a concept that is easy to grasp when it is applied to automobile engines and electric power plants. When we consider the energy associated with chemical bonds and chemical reactions within cells, however, the concept of work becomes less intuitive. Cells require energy to do all of their work, including the synthesis of glucose from carbon dioxide and water in photosynthesis, the contraction of muscles, and the replication of DNA.

There are two principal forms of energy: kinetic and potential. *Kinetic energy* is the energy of movement—the motion of a car, for example, or the motion of molecules. *Heat,* or thermal energy, is what we call the energy of the molecular motion of matter. For heat to do work, it must flow from a region of higher temperature to one of lower temperature. Differences in temperature often exist between the internal and external environments of cells; however, cells generally cannot harness these heat differentials to do work. Even in higher animals that have evolved a mechanism for thermoregulation, the kinetic energy of molecules is used chiefly to maintain constant organismic temperatures.

The energy that usually concerns us when we study biological or chemical systems is *potential energy,* or stored energy. The potential energy in atoms and molecules is their ability to undergo energy-releasing chemical reactions (to form and break chemical bonds). The sugar glucose, for example, is high in potential energy. Cells degrade it continually; the energy released when its chemical bonds are broken is harnessed to do many kinds of work.

Radiant energy can be considered the kinetic energy of photons, or waves of light. Electric energy can be either kinetic (the energy of moving electrons or other charged particles) or potential (the energy of charge separation, such as electrons from protons). All forms of energy—both in inanimate objects and in the living world—are interconvertible, in accordance with the first law of thermodynamics, which states that *energy is neither created nor destroyed.* In photosynthesis, for example, the radiant energy of light is transformed into the chemical potential energy of the bonds between the atoms in a glucose molecule. In muscles and nerves, chemical potential energy is transformed, respectively, into mechanical and electric energy. Most of the biochemical reactions described in this book involve the making or breaking of at least one chemical bond.

Another form of potential energy to which we shall refer often is that of concentration gradients. When a substance is present on one side of a barrier, such as a membrane, in a concentration different from that on the other side, the result is a concentration gradient. All cells form concentration gradients by selectively taking up nutrients and ions from their surroundings. The energy in chemical bonds is utilized for this process.

Because all forms of energy are interconvertible, they can be expressed in the same units of measurement—namely, the calorie (cal) or the kilocalorie (1 kcal = 1000 cal).*

Chemical Bonds

There are two kinds of chemical bonds: strong, covalent bonds and weaker, noncovalent bonds. *Covalent bonds,* the bonds of organic chemistry, hold the atoms in an individual molecule together. *Noncovalent bonds* determine the three-dimensional architecture of large biological molecules and molecular complexes through cooperativity: although no one bond is strong, the effect of many weak bonds functioning together can be very powerful. Also noncovalent bonds are more easily broken, making them the basis of many dynamic biological processes.

The Most Stable Bonds between Atoms Are Covalent

In a covalent bond, two atoms are held close together because electrons in their outermost shells move in orbitals that are shared by both atoms. Most of the molecules in living systems contain only six different atoms: hydrogen, carbon, nitrogen, phosphorus, oxygen, and sulfur. The outer electron shell of each atom has a characteristic number of electrons:

$$ \text{H} \quad \cdot\text{C}\cdot \quad \cdot\text{N}\cdot \quad \cdot\text{P}\cdot \quad \cdot\text{O}\cdot \quad \cdot\text{S}\cdot $$

These atoms readily form covalent bonds with other atoms and rarely exist as isolated entities. As a rule, each type of atom forms a characteristic number of covalent bonds with other atoms. A hydrogen atom, with one electron in its outer shell, forms only one bond. A carbon

---

*A calorie is defined as the amount of thermal energy required to heat 1 cm^3 of water at 14°C by 1°C. Many biochemistry textbooks use the joule (J) instead: 1 cal = 4.18 J, so the two systems can be interconverted quite readily. The energy changes in chemical reactions, such as the making or breaking of chemical bonds, are measured in kilocalories per mole (kcal/mol) in this book. One mole of any substance is the amount that contains 6.02×10^{23} molecules, which is known as Avogadro's number. The weight of a mole of a substance in grams (g) is the same as its molecular weight. For example, the molecular weight of water is 18, so 1 mol of water weighs 18 g.

atom, with four electrons, generally forms four bonds, as in methane (CH_4):

$$H:\overset{\displaystyle H}{\underset{\displaystyle H}{C}}:H \quad \text{or} \quad H-\overset{\displaystyle H}{\underset{\displaystyle H}{C}}-H$$

Nitrogen and phosphorus have five electrons in their outer shells. These atoms can form either three covalent bonds, as in ammonia (NH_3), or five, as in phosphoric acid (H_3PO_4):

$$\overset{\displaystyle H}{\underset{\displaystyle H\quad H}{N}} \qquad HO-\overset{\displaystyle O}{\underset{\displaystyle OH}{P}}-OH$$

Although oxygen and sulfur contain six electrons in their outermost shells, an atom of oxygen or sulfur usually forms only two covalent bonds, as in molecular oxygen (O_2) or hydrogen sulfide (H_2S):

$$O::O \quad \text{or} \quad O{=}O, \quad H-\overset{\displaystyle H}{S}$$

A sulfur atom can, however, form as many as six covalent bonds, as in sulfur trioxide (SO_3) or sulfuric acid (H_2SO_4):

$$\underset{\displaystyle O\quad O}{\overset{\displaystyle O}{S}} \qquad HO-\overset{\displaystyle O}{\underset{\displaystyle O}{S}}-OH$$

Large Energy Changes Are Associated with the Making and Breaking of Covalent Bonds Covalent bonds are very stable links between atoms; the energy required to break a covalent bond is much larger than the thermal energy available at room or body temperatures (25°C or 37°C). Thermal energy at 25°C is less than 1 kilocalorie per mole (kcal/mol); the energy required to break a C—C bond in ethane, for instance, is about 83 kcal/mol:

$$H_3C:CH_3 \longrightarrow \cdot CH_3 + \cdot CH_3 \quad \Delta E = +83 \text{ kcal/mol}$$

where ΔE represents the difference between the energy of the reactants and that of the products. Such a high energy of scission (breakage) of the ethane bond means that, at room temperature (25°C), well under 1 in 10^{12} ethane molecules exists as a pair of $\cdot CH_3$ radicals. Covalent bonds generally are this strong (Table 1-1).

Chemical changes take place because the energy needed to break bonds is supplied by the formation of other bonds. The energy differentials between reactants and products typically range from 1–20 kcal/mol (much lower than the energies listed in Table 1-1). In the course of a reaction, though, molecules usually adopt temporary states of much higher energy, and the energy requirement

Table 1-1 The strengths of some covalent bonds important in biological systems*

Type of bond	Energy (kcal/mol)	Type of bond	Energy (kcal/mol)
SINGLE BOND		DOUBLE BOND	
O—H	110	C=O	170
H—H	104	C=N	147
P—O	100	C=C	146
C—H	99	P=O	120
V—H	93	TRIPLE BOND	
C—O	84	C≡C	195
C—C	83		
S—H	81		
C—N	70		
C—S	62		
N—O	53		
S—S	51		

*The value given for each bond is the amount of energy required to break it. Note that double and triple bonds are stronger than single bonds.

for achieving these states, often in the range of 100 kcal/mol, is the true basis of chemical stability. It is the reason that the covalent structures of molecules remain intact over very long periods of time. This high energy input, called the *activation energy,* is discussed in detail at the end of this chapter. Biological systems act specifically to overcome activation energies and thus channel chemical reactions in specific directions.

Covalent Bonds Exhibit Precise Orientations
When two or more atoms form covalent bonds with another central atom, these bonds are oriented at precise angles to one another. The angles are determined by the mutual repulsion of the outer electron orbitals of the central atom. In methane, for example, the central carbon atom is bonded to four hydrogen atoms whose positions define the four points of a tetrahedron, so that the angle between any two bonds is 109.5° (Figure 1-1). However, if a carbon atom is linked to only three other atoms, two of its outer electrons participate in the formation of a double bond:

$$\overset{\diagdown}{\underset{\diagup}{C}}{=}$$

In this case, the carbon atom and all three atoms linked to it lie in the same plane, because atoms connected by a double bond cannot rotate freely about the bond axis (Figure 1-2). The rigid planarity imposed by double bonds has enormous significance for the shape of large biological molecules such as proteins and nucleic acids.

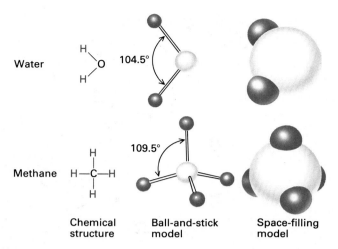

Water

Methane

| Chemical structure | Ball-and-stick model | Space-filling model |

▲ **Figure 1-1** Bond orientations in a water molecule and in a methane molecule. Each molecule is represented in three ways. The atoms in the ball-and-stick models are smaller than they actually are in relation to bond length to show the bond angles clearly. The sizes of the electron clouds in the space-filling models are more accurate.

Table 1-2 **The electronegativities of some atoms important in biological systems**

Element	Electronegativity
Fluorine (F)	4.0 (most electronegative)
Oxygen (O)	3.5
Chlorine (Cl)	3.0
Nitrogen (N)	3.0
Bromine (Br)	2.8
Sulfur (S)	2.5
Carbon (C)	2.5
Iodine (I)	2.5
Phosphorus (P)	2.1
Hydrogen (H)	2.1
Magnesium (Mg)	1.2
Calcium (Ca)	1.0
Lithium (Li)	1.0
Sodium (Na)	0.9
Potassium (K)	0.8 (least electronegative)

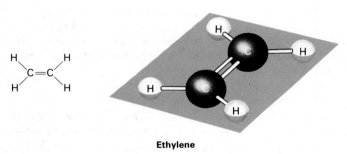

Ethylene

▲ **Figure 1-2** In an ethylene molecule, the carbon atoms are connected by a double bond, causing all the atoms to lie in the same plane. (Atoms connected by a single bond can rotate freely about the bond axis; those connected by a double bond cannot.)

The outer electron orbitals not involved in covalent bond formation also contribute to molecular configurations. The outer shell of the oxygen atom has two such pairs of electrons. The orbitals of the nonbonding electrons in the water molecule have a specific spatial orientation and a high electron density that compresses the angle between the covalent H—O—H bonds to 104.5° (Figure 1-1).

Dipoles Result from Unequal Sharing of Electrons in Covalent Bonds

In a covalent bond, one or more pairs of electrons are shared between two atoms. In certain cases, the bonded atoms exert different attractions for the electrons of the bond, resulting in unequal sharing of the electrons. The power of an atom in a molecule to attract electrons to itself, called *electronegativity*, is mea-

sured on a scale from 4.0 (for fluorine, the most electronegative atom) to a hypothetical zero (Table 1-2). In a covalent bond in which the atoms either are identical or have the same electronegativity, the bonding electrons are shared equally. Such a bond is said to be *nonpolar*. This is the case for C—C bonds. However, if two atoms differ in electronegativity, one will exert a greater attraction on the electrons of the bond. A bond of this type is said to be *dipolar;* one end will be slightly negatively charged (δ^-), and the other end will be slightly positively charged (δ^+). In a water molecule, for example, the oxygen atom, with an electronegativity of 3.5, attracts the electrons of the bonds more than do the hydrogen atoms, each of which has an electronegativity of 2.1; that is, the bonding electrons spend more time orbiting the oxygen atom than the hydrogens. Because both hydrogen atoms are on the same side of the oxygen atom, that side of the molecule has a slight positive charge, whereas the other side has a slight negative charge. A molecule that incorporates separated positive and negative charges is called a *dipole* (Figure 1-3).

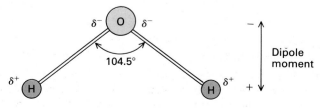

▲ **Figure 1-3** The water molecule as an electrical dipole. The symbol δ represents a partial charge (a weaker charge than the one on an electron or a proton). The size and direction of the charge separation is termed the dipole moment.

Noncovalent Bonds Stabilize the Structures of Biological Molecules

Many of the bonds that maintain the structures of large molecules, such as proteins and nucleic acids, are not covalent. The forces that stabilize the three-dimensional architecture of individual large molecules and that bind one molecule to another are often much weaker. The energy released in the formation of these noncovalent bonds is only 1–5 kcal/mol. Because the average kinetic energy of molecules at room temperature (25°C) is about 0.6 kcal/mol, many molecules will have enough energy to break these weak bonds. At physiological temperatures (25–37°C), weak bonds have a transient existence, but many can act together to produce highly stable structures (Figure 1-4).

Because noncovalent bonds play such crucial roles in biological structures, any student of the cell should be well acquainted with them. The four main types are the hydrogen bond, the ionic bond, the van der Waals interaction, and the hydrophobic bond or interaction. We shall consider each of these in turn.

The Hydrogen Bond   A hydrogen atom normally forms a covalent bond with only one other atom at a time. However, a covalently bonded hydrogen atom may form an additional bond: a *hydrogen bond* is a weak association between an electronegative atom (the *acceptor* atom) and a hydrogen atom covalently bonded to another atom (the *donor* atom). The hydrogen atom is closer to the donor (D) than to the acceptor (A):

$$D—H + A \rightleftharpoons D—H \cdots A$$
$$\text{hydrogen}$$
$$\text{bond}$$

The covalent bond between the donor and the hydrogen atom must be dipolar, and the outer shell of the acceptor atom must have nonbonding electrons that attract the δ^+

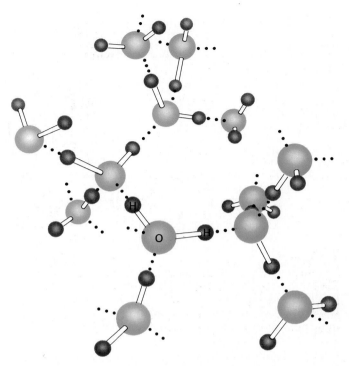

▲ **Figure 1-5**   In liquid water, each H_2O molecule apparently forms transient hydrogen bonds with several others, creating a fluid network of hydrogen-bonded molecules. The precise structure of liquid water is still not known with certainty.

charge of the hydrogen atom. The hydrogen bond in water is a classic example: a hydrogen atom in one molecule is attracted to a pair of electrons in the outer shell of an oxygen atom in an adjacent molecule (Figure 1-5). The strength of a hydrogen bond in water is about 5 kcal/mol, which is much weaker than a covalent H—O bond; the strengths of hydrogen bonds between other donor and acceptor atoms are similar.

An important feature of all hydrogen bonds is directionality. In the strongest hydrogen bonds, the donor, the hydrogen atom, and the acceptor all lie in a straight line. The distance between the nuclei of the hydrogen and oxygen atoms of adjacent hydrogen-bonded molecules in water is approximately 0.27 nm, about twice the length of covalent H—O bonds in water. Most hydrogen bonds are 0.26–0.31 nm long (Table 1-3). The length of the covalent bond between the donor and the hydrogen atom of a hydrogen bond is a bit longer than it would be if there were no acceptor for the hydrogen, because the acceptor "pulls" the hydrogen away from the donor.

Because all covalent N—H bonds are dipolar, nitrogen can act as a donor in a hydrogen bond; so can oxygen in the O—H bonds of molecules other than water. Nitrogen and oxygen can also be acceptors in the formation of hydrogen bonds with adjacent molecules. The amino (—NH₂) and hydroxyl (—OH) groups are most fre-

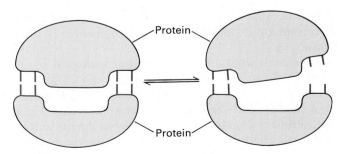

▲ **Figure 1-4**   The importance of multiple weak bonds in stabilizing an association between two large molecules. In the complex on the left, four noncovalent bonds bind two protein molecules together. Even if two of the bonds are broken, as in the complex on the right, the remaining two bonds will stabilize the structure, facilitating the re-formation of the broken bonds.

Table 1-3 Typical hydrogen bond lengths*

Bond	Length (nm)
OH $\cdots$ O$^-$	0.26
OH $\cdots$ O	0.27
OH $\cdots$ N	0.28
$^+$NH $\cdots$ O	0.29
NH $\cdots$ O	0.30
NH $\cdots$ N	0.31

*The values listed are the distances between the nuclei of the donor and the acceptor atom.

quently involved in hydrogen bond formation in biological systems. The presence of such groups makes many molecules soluble in water. For instance, these groups in methanol (CH_3OH) and methylamine (CH_3NH_2) can form many hydrogen bonds with water, enabling the molecules to dissolve in it to high concentrations (Figure 1-6).

The strength of hydrogen bonds in biological molecules in aqueous solution is much weaker than the theoretical 5 kcal/mol because, when such a hydrogen bond is broken, the released partners can reform hydrogen bonds with water. Much of the energy released by the breakage is regained, and the net change may be only 1 or 2 kcal/mol, not much more than the 0.6 kcal/mol of available thermal energy. It is only because of the aggregate strength of multiple hydrogen bonds that they play a central role in the architecture of large biological molecules.

The Structures of Liquid Water and Ice

Hydrogen bonding among water molecules is of crucial importance because all life requires an aqueous environment. The mutual attraction of its molecules causes water to have

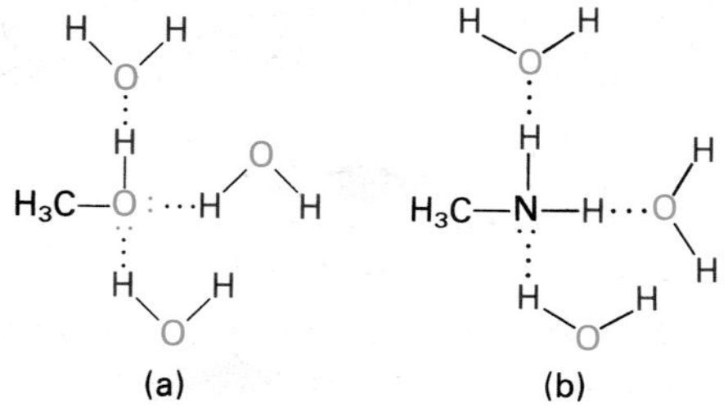

▲ **Figure 1-6** Hydrogen bonds between (a) methanol and water and (b) methylamine and water. Each of the two pairs of nonbonding electrons in the outer shell of oxygen can accept a hydrogen atom in a hydrogen bond. Similarly, the single pair of unshared electrons in the outer shell of nitrogen is capable of becoming an acceptor in a hydrogen bond. The —OH oxygen and the —NH$_2$ nitrogen can also donate hydrogen bonds to oxygen atoms in H_2O.

melting and boiling points at least 100°C higher than they would be if water were nonpolar; in the absence of these intermolecular attractions, water on earth would exist primarily as a gas.

Ordinary ice is a crystal. Each oxygen atom forms two covalent bonds with two hydrogen atoms and two hydrogen bonds with hydrogen atoms of adjacent molecules. The exact structure of liquid water is still unknown. It is believed to contain many icelike, maximally hydrogen-bonded networks that are presumably so transient and small that stable crystals do not form (see Figure 1-5). Most likely, water molecules are in rapid motion, constantly making and breaking hydrogen bonds with adjacent molecules. As the temperature of water increases toward 100°C, the kinetic energy of its molecules becomes greater than the energy of the hydrogen bonds connecting them, and the gaseous form of water appears.

Ionic Bonds

Covalent bonds between different atoms are generally dipolar, because one of the atoms is usually more electronegative than the other. When a dipolar bond breaks, the bonding electrons often stay with the more electronegative atom, which then becomes a negatively charged ion, or an *anion;* the other part of the molecule becomes a positively charged ion, or a *cation*. In some compounds, the atoms are so different in electronegativity that the bonding electrons are always found around the more electronegative atom—that is, the electrons are never shared. In sodium chloride (NaCl), for example, the bonding electron contributed by the sodium atom is completely transferred to the chlorine atom. Even in solid crystals of NaCl, the sodium and chlorine atoms are ionized, so it is more accurate to write the formula for the compound as Na$^+$Cl$^-$.

Because electrons are not shared, the bonds in such compounds cannot be considered covalent. These *ionic bonds* result from the attraction of a positive charge for a negative charge. Unlike covalent or hydrogen bonds, ionic bonds do not have fixed or specific geometric orientations because the electrostatic field around an ion—its attraction for an opposite charge—is uniform in all directions.

In aqueous solutions, simple ions of biological significance, such as Na$^+$, K$^+$, Ca^{2+}, Mg^{2+}, and Cl$^-$, do not exist as free, isolated entities. Instead, each is surrounded by a stable, tightly held shell of water molecules (Figure 1-7). Primary interaction occurs between the ion and the oppositely charged end of the water dipole; for example:

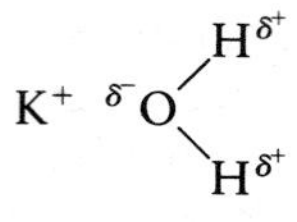

An estimation of the size of such an ion should include its shell of water molecules. Ions play an important biological role when they pass through narrow pores, or chan-

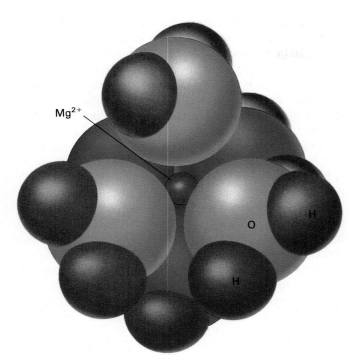

▲ **Figure 1-7** In water, six H_2O molecules cluster around a magnesium ion (Mg^{2+}). These molecules are held in place by electrostatic interactions between the positive Mg^{2+} and the partially negative oxygen of the water dipole.

nels, in membranes. Ionic movements through membranes are essential for the conduction of nerve impulses and for the stimulation of muscle contraction.

Most ionic compounds are quite soluble in water because a large amount of energy is released when ions tightly bind water molecules. Oppositely charged ions are shielded from one another by the water and tend not to recombine. Molecules with dipolar bonds also can attract water molecules, as can molecules that easily form hydrogen bonds. Such *polar molecules* can dissolve in water and are said to be *hydrophilic* (from the Greek for "water-loving"). Typical chemical groups that interact well with water are the hydroxyl —OH, the amino —NH_2, the peptide bond

$$\overset{O}{\underset{|}{-C-}}\overset{H}{\underset{|}{N-}}$$

and the ester bond

$$\overset{O}{\underset{||}{-C-}}O-$$

Van der Waals Interactions When two atoms approach one another closely, they create a nonspecific, weak attractive force that produces a *van der Waals interaction,* named for Dutch physicist Johannes Diderik van der Waals (1837–1923), who first described it. Momentary random fluctuations in the distribution of the electrons of any atom give rise to transient dipoles. If two

noncovalently bonded atoms are close enough together, the transient dipole in one atom will perturb the electron cloud of the other. This perturbation generates a transient dipole in the second atom, and the two dipoles will attract each other weakly. All types of molecules, both polar and nonpolar, exhibit van der Waals interactions. In particular, they are responsible for the cohesion among molecules of nonpolar liquids and solids, such as heptane

$$H-\overset{\overset{\displaystyle H}{|}}{\underset{\underset{\displaystyle H}{|}}{C}}-\overset{\overset{\displaystyle H}{|}}{\underset{\underset{\displaystyle H}{|}}{C}}-\overset{\overset{\displaystyle H}{|}}{\underset{\underset{\displaystyle H}{|}}{C}}-\overset{\overset{\displaystyle H}{|}}{\underset{\underset{\displaystyle H}{|}}{C}}-\overset{\overset{\displaystyle H}{|}}{\underset{\underset{\displaystyle H}{|}}{C}}-\overset{\overset{\displaystyle H}{|}}{\underset{\underset{\displaystyle H}{|}}{C}}-\overset{\overset{\displaystyle H}{|}}{\underset{\underset{\displaystyle H}{|}}{C}}-H$$

which cannot form ionic or hydrogen bonds with other molecules. Attraction decreases rapidly with increasing distance and is effective only when atoms are quite close to one another. However, if atoms get too close together, they become repelled by the negative charges in their outer electron shells.

When the van der Waals attraction between two atoms exactly balances the repulsion between their two electron clouds, the atoms are said to be in *van der Waals contact.* Each type of atom has a *van der Waals radius* at which it is in van der Waals contact with other atoms. Two covalently bonded atoms are closer together than two atoms that are merely in van der Waals contact (Figure 1-8). Table 1-4 lists the van der Waals and covalent radii of some important atoms in biological systems.

The energy of the van der Waals interaction is about 1 kcal/mol, only slightly higher than the average thermal energy of molecules at 25°C. Thus the van der Waals interaction is even weaker than the hydrogen bond, which

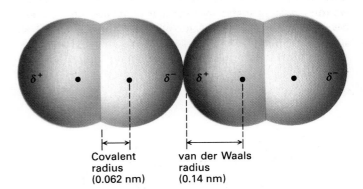

Covalent radius (0.062 nm) van der Waals radius (0.14 nm)

▲ **Figure 1-8** Two oxygen molecules in van der Waals contact. They are attracted by transient dipoles in their electron clouds. Each type of atom has a van der Waals radius, at which van der Waals interactions with other atoms are optimal. Because atoms repel one another if they are close enough together for their outer electron shells to overlap, the van der Waals radius is a measure of the size of the electron cloud surrounding an atom. (Note that the covalent radius indicated here is for the double bond of O=O; the single-bond covalent radius of oxygen is slightly longer, as shown in Table 1-4.)

Table 1-4   Van der Waals radii and covalent (single-bond) radii of some biologically important atoms*

Atom	Van der Waals radius (nm)	Covalent radius for a single bond (nm)
H	0.10	0.030
O	0.14	0.074
F	0.14	0.071
N	0.15	0.073
C	0.17	0.077
S	0.18	0.103
Cl	0.18	0.099
Br	0.20	0.114
I	0.22	0.133

*The internuclear distance for a covalent bond or a van der Waals interaction is approximately the sum of the values for the two participating atoms. Note that the van der Waals radius is about twice as long as the covalent radius.

typically has an energy of 1–2 kcal/mol in aqueous solutions. The attraction between two large molecules can be appreciable, however, if they have precisely complementary shapes, so that they make many van der Waals contacts when they come into proximity. The van der Waals contacts are among the interactions that take place between an antibody molecule and its specific antigen and between many enzymes and their specifically bound substrates.

Hydrophobic Interactions   *Nonpolar molecules* contain neither ions nor dipolar bonds and do not become hydrated. Because they are insoluble or almost insoluble in water, they are said to be *hydrophobic* (from the Greek for "water-fearing"). The covalent bonds between two carbon atoms and between carbon and hydrogen atoms are the most common nonpolar bonds in biological sys-

tems. *Hydrocarbons*—molecules made up only of carbon and hydrogen—are virtually insoluble in water. The large hydrophobic molecule tristearin (Figure 1-9), a component of animal fat, is also insoluble in water, even though its six oxygen atoms participate in some slightly dipolar bonds between carbon and oxygen. The core or central section of all biological membranes, including the surface membranes of cells, is composed almost exclusively of the hydrocarbon portions of molecules and contains few, if any, water molecules or other polar molecules.

We have all heard the axiom that oil and water don't mix. The force that causes hydrophobic molecules or nonpolar portions of molecules to aggregate rather than to dissolve in water is called the hydrophobic bond or, more precisely, the *hydrophobic interaction*. This is not a separate bonding force; rather, it is the result of the energy required to insert a nonpolar molecule into water. A nonpolar molecule cannot form hydrogen bonds with water, so it distorts the usual water structure, forcing the water to make a cage of bonds around it, but not with it. Conversely, nonpolar molecules bond together comfortably through van der Waals interactions. The result is a very powerful tendency for hydrophobic molecules to bond together and not dissolve in water. Small hydrocarbons such as butane

$$CH_3—CH_2—CH_2—CH_3$$

are slightly soluble in water [90 g/l (grams per liter) at 25°C and 1 atm (atmosphere) of pressure], because they can dissolve without disrupting the water lattice appreciably. Note, however, that 1-butanol

$$CH_3—CH_2—CH_2—CH_2OH$$

mixes completely with water in all proportions. Thus the replacement of just one hydrogen atom with the dipolar —OH group allows the molecule to form hydrogen bonds with water and greatly increases its solubility.

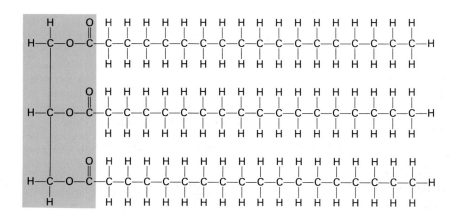

◀ **Figure 1-9**   The chemical structure of tristearin, or tristearoyl glycerol, a component of natural fats, contains three molecules of stearic acid, $CH_3(CH_2)_{16}COOH$, esterified to one molecule of glycerol, $HOCH_2CH(OH)CH_2OH$. The atoms at the polar end of the molecule (blue) are hydrophilic; the rest of the molecule is highly hydrophobic.

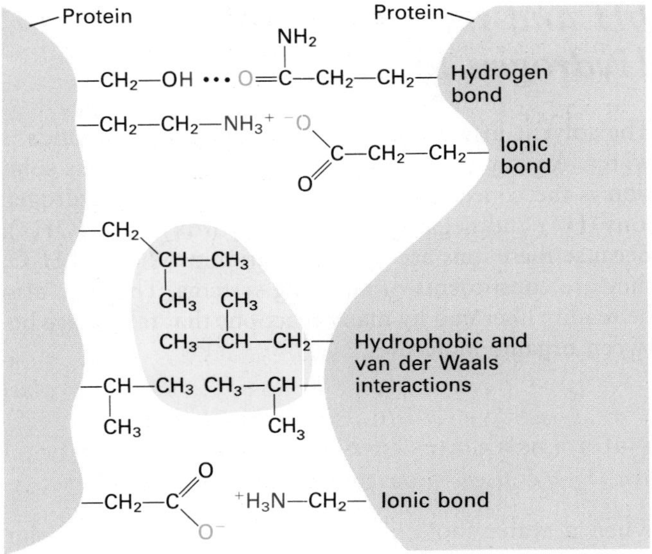

▲ **Figure 1-10** The binding of a hypothetical pair of proteins by two ionic bonds, one hydrogen bond, and one large combination of hydrophobic and van der Waals interactions. The structural complementarity of the surfaces of the two molecules gives rise to this particular combination of weak bonds and hence to the specificity of binding between the molecules.

Binding Specificity Can Be Conferred by Weak Interactions We know that multiple noncovalent bonds can contribute to the stability of large biological molecules. These interactions can also confer *specificity* by determining how large molecules will fold or which regions of different molecules will bind together. All weak bonds are effective only over a short range and require close contact between the reacting groups. For noncovalent bonds to form properly, there must be a complementarity between the sites on the two interacting surfaces. Figure 1-10 illustrates how several different weak bonds and interactions can bind two protein chains together. Almost any other arrangement of the same groups on the two surfaces would not allow the molecules to bind so tightly.

Chemical Equilibrium

Any chemical—whether or not it is in a cell—can, in principle, undergo various chemical reactions. The *rate* at which these reactions can take place under given conditions (concentration, temperature, pressure, and so on) and the *extent* to which they can proceed dictate which reactions actually occur. When reactants first come together—before any products have been formed—their rate of reaction is determined in part by their initial concentrations. As the reaction products accumulate, the concentrations of reactants decrease and so does the reaction rate. Meanwhile, some of the product molecules begin to participate in the reverse reaction, which re-forms the reactants. This reaction is slow at first but speeds up as the concentrations of products increase. Eventually, the rates of the forward and reverse reactions become equal, so that the concentrations of reactants and products stop changing. The mixture is then said to be in *chemical equilibrium.*

The ratio of products to reactants at equilibrium depends on the nature of the compounds as well as on temperature and pressure. Under standard physical conditions (25°C and 1 atm), the ratio of products to reactants at equilibrium, which is always the same for a given reaction, is termed the *equilibrium constant* (K_{eq}). For the simple reaction $A + B \rightleftharpoons X + Y$, the equilibrium constant is given by

$$K_{eq} = \frac{[X][Y]}{[A][B]}$$

where brackets indicate equilibrium concentrations. In general, for a reaction

$$aA + bB + cC + \cdots \rightleftharpoons zZ + yY + xX + \cdots$$

where capital letters represent particular molecules or atoms and lowercase letters represent the number of each in the reaction formula, the equilibrium constant is given by

$$K_{eq} = \frac{[Z]^z[Y]^y[X]^x \cdots}{[A]^a[B]^b[C]^c \cdots}$$

To illustrate several points concerning equilibrium, we shall use a fairly simple biochemical reaction: the interconversion of the compounds glyceraldehyde 3-phosphate (G3P) and dihydroxyacetone phosphate (DHAP). This reaction, which occurs in glycolysis, is catalyzed by the enzyme triosephosphate isomerase:

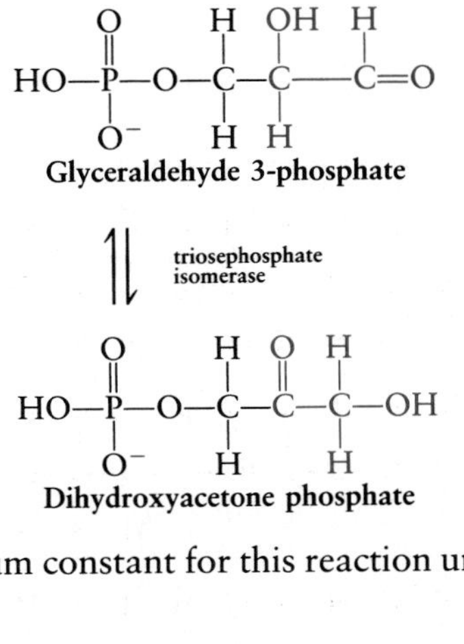

Glyceraldehyde 3-phosphate

triosephosphate isomerase

Dihydroxyacetone phosphate

The equilibrium constant for this reaction under standard conditions is

$$K_{eq} = \frac{[DHAP]}{[G3P]} = 22.2$$

Thus the ratio of the concentrations of G3P and DHAP is 1:22.2 when the reaction reaches equilibrium.

In reactions involving a single reactant and a single product, the ratio of product to reactant at equilibrium is equal to the equilibrium constant K_{eq} and is independent of the amounts of product and reactant initially present. This ratio is also independent of the reaction rate. In the presence of an enzyme or other catalyst, the reaction rate may increase, but the final ratio of product to reactant will always be the same. The magnitude of this constant has no bearing on the rate of the reaction or on whether the reaction will take place at all under normal conditions. Despite the large equilibrium constant for the conversion of G3P to DHAP, for example, in an aqueous solution in which the enzyme catalyst is absent, the high energy input necessary to rearrange the bonds slows the reaction to the point that it is undetectable.

In a reaction involving multiple reactants and/or products, the equilibrium concentration of a *particular* product or reactant depends on the initial amounts of all reactants and products as well as on the equilibrium constant. Consider, for example, the hydrolysis (cleavage by addition of water) of a quantity of the dipeptide glycylalanine (GA) to glycine (G) and alanine (A):

$$^+NH_3-CH_2-\overset{\overset{O}{\|}}{C}-NH-\underset{\underset{CH_3}{|}}{CH}-\overset{\overset{O}{\|}}{C}-O^- + H_2O$$

Glycylalanine

$$\Updownarrow$$

$$^+NH_3-CH_2-\overset{\overset{O}{\|}}{C}-O^- \quad + \quad ^+NH_3-\underset{\underset{CH_3}{|}}{CH}-\overset{\overset{O}{\|}}{C}-O^-$$

Glycine   **Alanine**

where

$$K_{eq} = \frac{[G][A]}{[GA]}$$

(The concentration of water is, by convention, not included in the calculation of such equilibrium ratios.) The equilibrium is strongly in the direction of the formation of glycine and alanine. In other words, most of the glycylalanine is hydrolyzed at equilibrium. However, suppose that the initial reaction mixture contains a small amount of glycylalanine and a large amount of alanine. As the reaction proceeds, the concentration of alanine [A] will always greatly exceed the concentration of glycine [G] produced by hydrolysis. This must reduce the equilibrium ratio of [G] to [GA], because K_{eq} remains constant. In other words, excess alanine will drive the reaction in the reverse direction; thus, at equilibrium, more glycine might be found within the dipeptide than in free form— that is, most of the glycylalanine might *not* be hydrolyzed.

pH and the Concentration of Hydrogen Ions

The solvent inside cells and in all extracellular fluids is water. An important characteristic of any aqueous solution is the concentration of positively charged hydrogen ions (H^+) and negatively charged hydroxyl ions (OH^-). Because these ions are the dissociation products of H_2O, they are constituents of all living systems. They can also be readily liberated by many reactions that take place between organic molecules.

Water Dissociates into Hydronium and Hydroxyl Ions

When a water molecule dissociates, one of its dipolar H—O bonds breaks. The resulting hydrogen ion—a proton—has a short lifetime as a free particle and quickly combines with a water molecule to form an *hydronium ion* (H_3O^+). For convenience, however, we refer to the concentration of hydrogen ions in a solution, [H^+], even though we really mean the concentration of hydronium ions, [H_3O^+].

The dissociation of water is a reversible reaction,

$$H_2O \rightleftharpoons H^+ + OH^-$$

and at 25°C,

$$[H^+][OH^-] = 10^{-14}\, M^2$$

where M symbolizes molarity, or moles per liter (mol/l). In pure water, [H^+] = [OH^-] = $10^{-7}\, M$.

The concentration of hydrogen ions in a solution is expressed conventionally as its pH:

$$pH = -\log [H^+] = \log \frac{1}{[H^+]}$$

In pure water at 25°C, [H^+] = $10^{-7}\, M$, so

$$pH = -\log 10^{-7} = 7.0$$

On the pH scale from zero to 14, 7.0 is considered neutral: pH values below 7.0 indicate acidic solutions; values above 7.0 indicate basic (alkaline) solutions (Table 1-5). In a 0.1 M solution of hydrogen chloride (HCl) in water, [H^+] = 0.1 M because all of the HCl has dissociated into H^+ and Cl^- ions. For this solution

$$pH = -\log 0.1 = 1.0$$

One of the most important properties of a biological fluid is its pH. The pH of the cytoplasm of cells is normally about 7.2. Cells of higher organisms contain organelles, such as lysosomes, in which the pH is much lower (about 5). This value of [H^+] is more than 100 times higher than its value in the cytoplasm. Lysosomes contain many degradative enzymes that function optimally in an

Table 1-5 The pH scale

	Concentration of hydrogen ions (mol/l)	pH	Example
	10^{-0}	0	
	10^{-1}	1	Gastric fluids
	10^{-2}	2	Lemon juice
Increasing acidity	10^{-3}	3	Vinegar
	10^{-4}	4	Acid soil
	10^{-5}	5	Lysosomes
	10^{-6}	6	Cytoplasm of contracting muscle
Neutral	10^{-7}	7	Pure water and cytoplasm
	10^{-8}	8	Sea water
	10^{-9}	9	Very alkaline natural soil
	10^{-10}	10	Alkaline lakes
Increasing alkalinity	10^{-11}	11	Household ammonia
	10^{-12}	12	Lime (saturated solution)
	10^{-13}	13	
	10^{-14}	14	

acidic environment, whereas their action is inhibited in the near-neutral environment of the cytoplasm. Maintenance of a specific pH is imperative for some cellular structures to function properly. On the other hand, dramatic shifts in cellular pH may play an important role in controlling cellular activity. For example, the pH of the cytoplasm of an unfertilized sea urchin egg is 6.6. Within 1 minute of fertilization, however, the pH rises to 7.2 ($[H^+]$ decreases to about one-fourth its original value). The change in pH appears to trigger the growth and division of the egg.

Acids Release Hydrogen Ions and Bases Combine with Hydrogen Ions

In general, any molecule or ion that can release a hydrogen ion is called an *acid* and any molecule or ion that can combine with a hydrogen ion is called a *base*. Thus hydrogen chloride is an acid. The hydroxyl ion is a base, as is ammonia (NH_3), which readily picks up a hydrogen ion to become an ammonium ion (NH_4^+). Many organic molecules are acidic because they have a carboxyl group ($-COOH$) that tends to dissociate to form the negatively charged carboxylate ion ($-COO^-$):

$$X-C\overset{\displaystyle O}{\underset{\displaystyle OH}{\big<}} \; \rightleftharpoons \; X-C\overset{\displaystyle O}{\underset{\displaystyle O^-}{\big<}} \; + H^+$$

where X represents the rest of the molecule. The amino group ($-NH_2$), a part of many important biological mol-

ecules, is a base because, like ammonia, it can take up a hydrogen ion:

$$X-NH_2 + H^+ \rightleftharpoons X-NH_3^+$$

When acid is added to a solution, $[H^+]$ increases (the pH goes down). Consequently, $[OH^-]$ decreases because the hydroxyl ions combine with the hydrogen ions. Conversely, when a base is added to a solution, $[H^+]$ decreases (the pH goes up). Because $[H^+][OH^-] = 10^{-14} \, M^2$, any increase in $[H^+]$ is coupled with a reduction in $[OH^-]$, and vice versa. No matter how acidic or alkaline a solution is, it always contains both ions: neither $[OH^-]$ nor $[H^+]$ is ever zero. For example, if $[H^+] = 0.1 \, M$ (pH = 1.0), then $[OH^-] = 10^{-13} \, M$.

The degree to which a dissolved acid releases hydrogen ions or to which a base takes them up depends partly on the pH of the solution. Amino acids have the general formula

$$H-\overset{\displaystyle NH_2}{\underset{\displaystyle R}{\overset{|}{\underset{|}{C}}}}-COOH$$

(where R represents the rest of the molecule), but in neutral solutions (pH = 7.0) they exist predominantly in the doubly ionized form

$$H-\overset{\displaystyle NH_3^+}{\underset{\displaystyle R}{\overset{|}{\underset{|}{C}}}}-COO^-$$

Such a dipolar ion is called a *zwitterion*. In solutions at low pH, carboxylate ions recombine with the abundant hydrogen ions, so that the predominant form of amino acid molecule is

$$H-\overset{\displaystyle NH_3^+}{\underset{\displaystyle R}{\overset{|}{\underset{|}{C}}}}-COOH$$

At high pH, the scarcity of hydrogen ions decreases the chance that an amino group or a carboxylate ion will pick up a hydrogen ion, so that the predominant form of amino acid molecule is

$$H-\overset{\displaystyle NH_2}{\underset{\displaystyle R}{\overset{|}{\underset{|}{C}}}}-COO^-$$

Many Biological Molecules Contain Multiple Acidic or Basic Groups

Many chemicals used by cells have multiple acidic or basic groups. In the laboratory, it is often essential to know the precise state of dissociation of each of these groups at various pH values. The dissociation of a simple acid HA, such as acetic acid (CH_3COOH), is described by

$$HA \rightleftharpoons H^+ + A^-$$

The equilibrium constant K_a for this reaction is

$$K_a = \frac{[H^+][A^-]}{[HA]}$$

By taking the logarithm of both sides and rearranging the result, we can derive the very important relation known as the *Henderson-Hasselbalch equation:*

$$\log K_a = \log \frac{[H^+][A^-]}{[HA]}$$

$$\log K_a = \log [H^+] + \log \frac{[A^-]}{[HA]}$$

$$-\log [H^+] = -\log K_a + \log \frac{[A^-]}{[HA]}$$

Substituting pH for $-\log [H^+]$ and pK_a for $-\log K_a$, we have

$$pH = pK_a + \log \frac{[A^-]}{[HA]}$$

The pK_a of any acid is equal to the pH at which one-half of the molecules are dissociated and one-half are neutral. This can be derived by observing that if $pK_a = pH$, then $\log ([A^-]/[HA]) = 0$, or $[A^-] = [HA]$. The Henderson-Hasselbalch equation allows us to calculate the degree of dissociation of an acid if both the pH of the solution and the pK_a of the acid are known.

A *titration curve* shows how the fraction of molecules in the undissociated form (HA) depends on pH (Figure 1-11). At one pH unit below the pK_a of an acid, 91 percent of the molecules are in the HA form; at one pH unit above the pK_a, 91 percent are in the A^- form. If we add additional acid or base to a solution of an acid (or a base)

at its pK_a value, the pH of the solution changes, but to a lesser degree than it would if the original acid (or base) were not present. Protons released by the added acid are taken up by the A^- form of the original acid, or the hydroxyl ions generated by the added base are neutralized by protons released by the original acid. This capacity of acids and bases, called *buffering*, declines rapidly at more than one pH unit from the pK_a.

Phosphoric acid is physiologically important because its multiple groups are capable of dissociating:

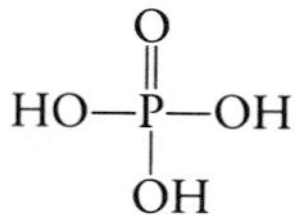

$$\begin{array}{c} O \\ \parallel \\ HO-P-OH \\ | \\ OH \end{array}$$

The three protons do not dissociate simultaneously; loss of each can be described by a discrete pK_a constant:

$$H_3PO_4 \rightleftharpoons H_2PO_4^- + H^+ \qquad pK_a = 2.1$$
$$H_2PO_4^- \rightleftharpoons HPO_4^{2-} + H^+ \qquad pK_a = 7.2$$
$$HPO_4^{2-} \rightleftharpoons PO_4^{3-} + H^+ \qquad pK_a = 12.7$$

Phosphate ions are present in considerable quantities in cells and are an important factor in maintaining, or buffering, the pH of the cytoplasm. The pK_a for the dissociation of the second proton (pH 7.2) is similar to the pH of the cytoplasm. Because $pK_a = pH = 7.2$, then, according to the second equation and the Henderson-Hasselbalch equation, 50 percent of cellular phosphate is $H_2PO_4^-$ and 50 percent is HPO_4^{2-}.

In nucleic acids, phosphate is found as a diester. It is linked to two carbon atoms of adjacent ribose sugars:

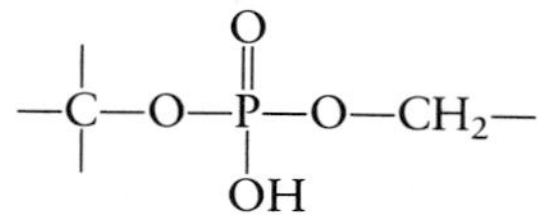

$$\begin{array}{c} O \\ | \quad \parallel \\ -C-O-P-O-CH_2- \\ | \quad | \\ \quad OH \end{array}$$

The pK_a for the dissociation of the $-OH$ proton is about 3, which is similar to the pK_a for the dissociation of the first proton from phosphoric acid. Thus, at neutral pH, each phosphate residue in deoxyribonucleic acid (DNA) or ribonucleic acid (RNA) is negatively charged:

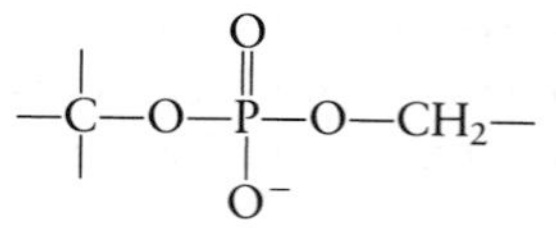

$$\begin{array}{c} O \\ | \quad \parallel \\ -C-O-P-O-CH_2- \\ | \quad | \\ \quad O^- \end{array}$$

This is why DNA and RNA are called nucleic acids.

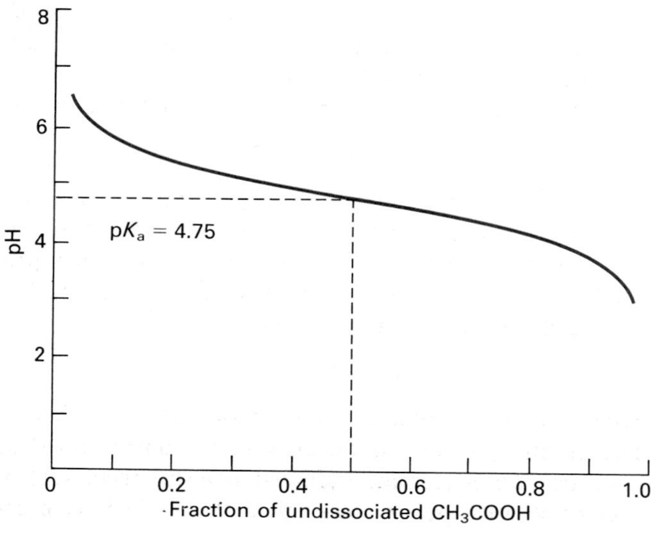

◀ **Figure 1-11** The titration curve of acetic acid (CH_3COOH). The pK_a for the dissociation of acetic acid to H^+ and CH_3COO^- is 4.75. Because pH is measured on a logarithmic scale, the solution changes from 91 percent CH_3COOH at pH 3.75 to 9 percent CH_3COOH at pH 5.75. The acid has maximum buffering capacity in this pH range.

The Direction of Chemical Reactions

Because biological systems are held at constant temperature and pressure, it is possible to predict the direction of a chemical reaction by using a measure of potential energy called *free energy, G*, after American chemist Josiah Willard Gibbs (1839–1903), a founder of the science of thermodynamics.

The Change in Free Energy ΔG Determines the Direction of a Chemical Reaction

We are interested in what happens to the free energy when one molecule or molecular configuration is changed into another. Thus our concern is with relative, rather than absolute, values of free energy—in particular, with the difference between the values before and after the change, written ΔG. Gibbs showed that at constant temperature and pressure, "all systems change in such a way that free energy is minimized."

For our purposes, a system is an entity that does not exchange mass but that can exchange heat with its surroundings. Thus a test tube containing a chemical reaction is a system if it is surrounded by a water bath at constant temperature; heat can be exchanged through the walls of the test tube, but matter cannot be. A cell can be a system if it does not exchange molecules with the extracellular solution.

In mathematical terms, Gibbs's law that systems change to minimize free energy is a statement about ΔG. If ΔG is negative for a chemical reaction or mechanical process, the reaction or process will occur spontaneously; if ΔG is positive, it will not.

The ΔG of a Reaction Depends on Changes in Heat and Entropy
The ΔG of a reaction (at any constant temperature and pressure) is a composite of two factors: the change in the heat content between reactants and products and the change in the randomness of the system. The heat or *enthalpy, H*, of reactants or products is equal to their total bond energies. Enthalpy is released or absorbed in a chemical reaction when bonds are formed or broken; thus the overall change in enthalpy ΔH is equal to the overall change in bond energies. According to the first law of thermodynamics, the total amount of energy in the system and its surroundings cannot change. In an *exothermic* reaction, heat is given off and ΔH is negative (the products contain less energy than the reactants did); in an *endothermic* reaction, heat is absorbed and ΔH is positive. Reactions tend to proceed if they liberate energy (if $\Delta H < 0$), but this is only one of two important parameters of free energy to consider.

Entropy, S, is a measure of the degree of randomness or disorder of a system. A change in entropy is denoted by ΔS. Entropy increases as a system becomes more disordered and decreases as it becomes more structured. According to the second law of thermodynamics, a reaction tends to occur spontaneously when the total entropy of the system and its surroundings increases. Consider, for example, the potential energy stored in a concentration gradient. An important biological process—the diffusion of solutes from one solution to a solution in which their concentration is lower—is driven only by an increase in entropy. Suppose that a 0.1 *M* solution of NaCl is separated from a 0.01 *M* solution by a membrane through which Na^+ and Cl^- ions can diffuse. The ions from the 0.1 *M* solution can diffuse over a larger volume than they could before, as can the ions from the more dilute solution; thus the disorder of the system increases. Maximum entropy is achieved when all ions can diffuse freely over the largest possible volume—that is, when the concentrations of all ions are the same on both sides of the membrane. If the two solutions are of equal volume and the degree of hydration of Na^+ does not change significantly on dilution, ΔH will be approximately zero and the negative free energy of the reaction

$$(0.01 \; M \text{ solution} \overset{\text{membrane}}{\mid} 0.1 \; M \text{ solution}) \rightarrow 0.055 \; M \text{ solution}$$

will be due solely to the positive value of ΔS.

The relatively high entropy of pure water promotes the hydrophobic interaction between hydrocarbon molecules in it. In a flask of water, all H_2O molecules move randomly, making and breaking hydrogen bonds. If a long hydrocarbon molecule, such as heptane or tristearin, is dissolved in the water, the presence of the hydrocarbon, which cannot participate in hydrogen bonds with H_2O, forces the water to form a cage around the foreign molecule, restricting the free motion of some H_2O molecules. This imposes a high degree of order on their arrangement and lowers the entropy of the system ($\Delta S < 0$). Because the entropy change is negative, hydrophobic molecules do not dissolve well in aqueous solutions and rather tend to stay associated with each other.

Gibbs showed that free energy can be defined as

$$G = H - TS$$

where H is the enthalpy (heat energy) of the system, T is its temperature in kelvins (K), and S is its entropy. If temperature and pressure remain constant, a reaction proceeds spontaneously only if there is a decrease in free energy (a negative ΔG) in the equation

$$\Delta G = \Delta H - T \Delta S$$

For example, an exothermic reaction ($\Delta H < 0$) that increases entropy ($\Delta S > 0$) occurs spontaneously ($\Delta G < 0$). An endothermic reaction ($\Delta H > 0$) can still occur spontaneously if ΔS is positive enough that the $T \Delta S$ term can overcome the positive ΔH. If $\Delta G = 0$ (if there is no change in free energy in the conversion of reactants into products), then the system will be at equilibrium: any conversion of reactants to products will be balanced by an equal conversion of products to reactants.

Under any conditions, the second law of thermodynamics states that the change in entropy of a system and its surroundings must be positive for a reaction to proceed; that is, a reaction can take place only if $\Delta S_{total} > 0$,

$$\Delta S_{total} = \Delta S_{system} + \Delta S_{surroundings}$$

In other words, the *overall* degree of disorder in the system and its surroundings must increase.

Many biological reactions lead to an increase in order rather than to an increase in entropy. An obvious example is the reaction that links amino acids together to form a protein. A solution of protein molecules has a lower entropy than does a solution of the same amino acids unlinked, because the free movement of any amino acid in a protein is restricted when it is bound in a long chain. For the linking reaction to proceed, a compensatory increase in entropy must occur elsewhere in the system or its surroundings.

We now have two equivalent criteria for predicting whether a reaction will proceed: $\Delta S_{total} > 0$, and $\Delta G_{system} < 0$. Because S_{total} refers to the system plus its surroundings, it is not as easy to measure or calculate, but the two parameters are intimately related. A loss of heat from the system ($\Delta H_{system} < 0$) represents a gain of heat, which is manifested in the increased molecular motion, or entropy, of the surroundings. In fact

$$\Delta H_{system} = -T\,\Delta S_{surroundings}$$

where T is again temperature in kelvins. Combining this equation with Gibbs's formula for ΔG gives us

$$\Delta G_{system} = \Delta H_{system} - T\,\Delta S_{system}$$
$$= -T\,\Delta S_{surroundings} - T\,\Delta S_{system} = -T\,\Delta S_{total}$$

Thus ΔG is always negative if ΔS_{total} is positive, indicating that any reaction that increases the entropy of a system plus its surroundings will tend to proceed spontaneously.

Temperature, Concentrations of Reactants, and Other Parameters Affect the ΔG of a Reaction   The

change in free energy for a reaction is influenced by temperature, pressure, and the initial concentrations of reactants and products. Most biological reactions—like others that take place in aqueous solutions—also are affected by the pH of the solution. References on biochemistry give values for $\Delta G^{\circ\prime}$, the *standard free-energy change* of a reaction under the conditions of 298 K (25°C), 1 atm, pH 7.0 (as in pure water), and initial concentrations of 1 *M* for all reactants and products except protons, which are kept at pH 7.0 (Table 1-6). The sign of $G^{\circ\prime}$ depends on the direction in which the reaction is writ-

Table 1-6   Values of $\Delta G^{\circ\prime}$, the standard free-energy change, for some important biochemical reactions

Reaction	$\Delta G^{\circ\prime}$ (kcal/mol)
HYDROLYSIS	
Acid anhydrides:	
Acetic anhydride + $H_2O \longrightarrow$ 2 acetate	−21.8
$PP_i + H_2O \longrightarrow 2P_i$*	−8.0
ATP + $H_2O \longrightarrow$ ADP + P_i	−7.3
Esters:	
Ethylacetate + $H_2O \longrightarrow$ ethanol + acetate	−4.7
Glucose 6-phosphate + $H_2O \longrightarrow$ glucose + P_i	−3.3
Amides:	
Glutamine + $H_2O \longrightarrow$ glutamate + NH_4^+	−3.4
Glycylglycine + $H_2O \longrightarrow$ 2 glycine (a peptide bond)	−2.2
Glycosides:	
Sucrose + $H_2O \longrightarrow$ glucose + fructose	−7.0
Maltose + $H_2O \longrightarrow$ 2 glucose	−4.0
ESTERIFICATION	
Glucose + $P_i \longrightarrow$ glucose 6-phosphate + H_2O	+3.3
REARRANGEMENT	
Glucose 1-phosphate $\longrightarrow$ glucose 6-phosphate	−1.7
Fructose 6-phosphate $\longrightarrow$ glucose 6-phosphate	−0.4
Glyceraldehyde 3-phosphate $\longrightarrow$ dihydroxyacetone phosphate	−1.8
ELIMINATION	
Malate $\longrightarrow$ fumarate + H_2O	+0.75
OXIDATION	
Glucose + $6O_2 \longrightarrow 6CO_2 + 6H_2O$	−686
Palmitic acid + $23O_2 \longrightarrow 16CO_2 + 16H_2O$	−2338
PHOTOSYNTHESIS	
$6CO_2 + 6H_2O \longrightarrow$ glucose + $6O_2$	+686

*PP_i = pyrophosphate; P_i = phosphate.

SOURCE: A. L. Lehninger, 1975, *Biochemistry*, 2d ed., Worth, p. 397.

ten. If the reaction A → B has a $\Delta G^{\circ\prime}$ of $-x$ kcal/mol, then the reverse reaction B → A will have a $\Delta G^{\circ\prime}$ value of $+x$ kcal/mol.

The conditions of most biological reactions, particularly the concentrations of reactants, differ from standard conditions. However, we can estimate free-energy changes for different temperatures and initial concentrations using

$$\Delta G = \Delta G^{\circ\prime} + RT \ln Q = \Delta G^{\circ\prime} + RT \ln \frac{[\text{products}]}{[\text{reactants}]}$$

where R is the gas constant of 1.987 cal/(degree · mol), T is the temperature (in kelvins), and Q is the initial ratio of products to reactants (constructed like the equilibrium constant, page 27). For example, in the interconversion

$$\text{G3P} \rightleftharpoons \text{DHAP}$$

we have

$$Q = \frac{[\text{DHAP}]}{[\text{G3P}]} \quad \text{and} \quad \Delta G^{\circ\prime} = -1840 \text{ cal/mol}$$

The equation for ΔG then becomes

$$\Delta G = -1840 + 1.987T \ln \frac{[\text{DHAP}]}{[\text{G3P}]}$$

If the initial concentrations of both [DHAP] and [G3P] are 1 M, then $\Delta G = \Delta G^{\circ\prime} = -1840$ cal/mol, because $RT \ln 1 = 0$. The reaction will tend to proceed from left to right, in the direction of formation of DHAP. If, however, [DHAP] begins as 0.1 M and [G3P] as 0.001 M, with other conditions being standard, then $Q = 0.1/0.001 = 100$, and

$$\begin{aligned} \Delta G &= -1840 + 1.987(298) \ln 100 \\ &= -1840 + 1.987(298)(4.605) \\ &= -1840 + 2727 = 887 \text{ cal/mol} \end{aligned}$$

Clearly, the reaction will now proceed in the direction of formation of G3P.

In a reaction $A + B \rightleftharpoons C$, in which two molecules combine to form a third, the equation for ΔG becomes

$$\Delta G = \Delta G^{\circ\prime} + RT \ln \frac{[\text{C}]}{[\text{A}][\text{B}]}$$

The direction of the reaction will shift more toward the right (toward formation of C) if *either* [A] or [B] is increased.

The Standard Free-Energy Change $\Delta G^{\circ\prime}$ and the Equilibrium Constant K_{eq} Are Related

A chemical mixture at equilibrium is already in a state of minimal free energy: no free energy is being generated or released. Thus, for a system at equilibrium, we can write

$$0 = \Delta G = \Delta G^{\circ\prime} + RT \ln Q$$

At equilibrium, however, the value of Q is the equilibrium constant K_{eq}, so that

$$\Delta G^{\circ\prime} = -RT \ln K_{eq}$$

Expressed in terms of base 10 logarithms, this equation becomes

$$\Delta G^{\circ\prime} = -2.3RT \log K_{eq}$$

or

$$-\frac{\Delta G^{\circ\prime}}{2.3RT} = \log K_{eq}$$

or, by the definition of logarithms

$$K_{eq} = 10^{-\Delta G^{\circ\prime}/2.3RT}$$

This simple but important relation between the standard free energy of a reaction and the equilibrium constant makes it possible to determine values of $\Delta G^{\circ\prime}$ by

Table 1-7   Values of $\Delta G^{\circ\prime}$ for some values of K_{eq}

K_{eq}	$\Delta G^{\circ\prime}$ (cal/mol)*
0.001	4086
0.01	2724
0.1	1362
1.0	0
10.0	−1362
100.0	−2724
1000.0	−4086

*Calculated from the formula $\Delta G^{\circ\prime} = -2.3RT \log K_{eq}$.

simply measuring the concentrations of chemicals at equilibrium, rather than by trying to measure changes in entropy or free energy directly. Note that if $\Delta G^{\circ\prime}$ is negative, then $K_{eq} > 1$; that is, the formation of products from reactants is favored (Table 1-7).

Although a chemical equilibrium appears to be unchanging and static, it is actually a dynamic state. Two opposing reactions continue to proceed at the same rate, thereby canceling each other out. The point of equilibrium does not depend on the speed of the reactions. When an enzyme or some other catalyst speeds up a reaction, it also speeds up the reverse reaction; thus equilibrium is reached sooner than it is if the reaction is not catalyzed. However, *the equilibrium constant and $\Delta G^{\circ\prime}$ are the same with or without catalysis.*

An equilibrium constant or a $\Delta G^{\circ\prime}$ value is unaffected by any other reactions going on in the same mixture, even if they change the concentrations of reactants or products. Values of ΔG, however, are affected by the concentrations of reactants and products.

The Generation of a Concentration Gradient Requires an Expenditure of Energy

Cells must often accumulate chemicals, such as glucose and K^+ ions, in greater concentrations than exist in their environment. Consequently, they must transport these chemicals against a concentration gradient. To find the amount of energy required to transfer 1 mol of a substance from outside the cell, where its concentration is C_1, to inside the cell, where its concentration is C_2, we employ the equation

$$\Delta G = RT \ln \frac{C_2}{C_1}$$

The "uphill" transport of molecules against a concentration gradient ($C_2 > C_1$) clearly does not take place spontaneously (so $\Delta G > 0$); it requires the input of cellular chemical energy. If the concentrations differ by a factor of 10, then, at 25°C, $\Delta G = RT \ln 10 = 1.36$ kcal/mol of substance transported. Such calculations assume that a molecule of a given substance inside a cell is identical to a molecule of that substance outside (that the substance is not sequestered, bound, or chemically changed by the transport).

Many Cellular Processes Involve the Transfer of Electrons in Oxidation-Reduction Reactions

Many chemical reactions result in the transfer of electrons from one atom or molecule to another. This may or may not accompany the formation of new chemical bonds. The loss of electrons from an atom or a molecule is called *oxidation*, and the gain of electrons by an atom or a molecule is called *reduction*. Because electrons are neither created nor destroyed in a chemical reaction, if an atom or molecule is oxidized, another must be reduced. For example, oxygen draws electrons from Fe^{2+} (ferrous) ions to form Fe^{3+} (ferric) ions, a reaction that occurs when energy is generated in cellular particles called mitochondria. Each oxygen atom receives two electrons, one from each of two Fe^{2+} ions:

$$2Fe^{2+} + \frac{1}{2}O_2 \longrightarrow 2Fe^{3+} + O^{2-}$$

Fe^{2+} is oxidized, and O_2 is reduced. Oxygen accepts electrons in many oxidation reactions in aerobic animal cells. Reduced oxygen (O^{2-}) readily acquires two protons to yield H_2O.

In the transformation of succinate into fumarate, which also takes place in mitochondria, succinate loses two hydrogen atoms (Figure 1-12), which is equivalent to a loss of two protons and two electrons. Thus succinate is said to be oxidized in its conversion into fumarate, and another molecule—flavin adenine dinucleotide (FAD), which accepts the electrons—is reduced (to $FADH_2$). Many biologically important oxidation and reduction reactions occur because of the removal or the addition of hydrogen atoms (protons plus electrons) rather than the transfer of isolated electrons.

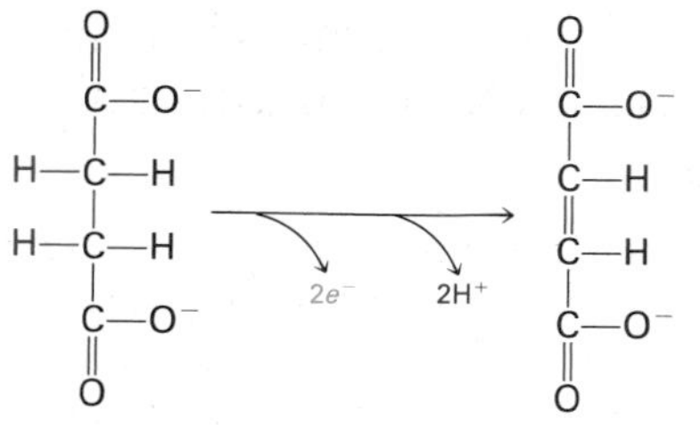

▲ **Figure 1-12** The conversion of succinate to fumarate is an oxidation reaction: two electrons and two protons are released. This reaction occurs in mitochondria as part of the citric acid cycle, which functions in the final stages of the oxidation of glucose by oxygen to form carbon dioxide.

When ferrous ion (Fe^{2+}) and oxygen (O_2) combine, the Fe^{2+} is oxidized and the O_2 is reduced by the following mechanism:

$$2Fe^{2+} \longrightarrow 2Fe^{3+} + 2e^-$$
$$2e^- + \frac{1}{2}O_2 \longrightarrow O^{2-}$$
$$O^{2-} + 2H^+ \longrightarrow H_2O$$

The readiness with which an atom or a molecule takes up an electron is its *reduction potential, E*. Reduction potentials are measured in volts (V) from an arbitrary zero point set at the reduction potential of the following reaction under standard conditions (25°C, 1 atm, and reactants at 1 M):

$$H^+ + e^- \underset{\text{oxidation}}{\overset{\text{reduction}}{\rightleftharpoons}} \frac{1}{2}H_2$$

The value of E for a molecule or an atom under standard conditions is its *standard reduction potential, E'_0* (Table 1-8). Standard reduction potentials may differ somewhat from those found under the conditions in a cell, because the concentrations of reactants in a cell are not 1 M. A positive reduction potential means that a molecule or ion (say, Fe^{3+}) has a higher affinity for electrons than the H^+ ion does in the standard reaction. A negative reduction

Table 1-8 Values of the standard reduction potential E'_0 for some important oxidation-reduction reactions (pH 7.0, 25°C)*

Oxidant	Reductant	n	E'_0(V)
Succinate + CO_2	α-Ketoglutarate	2	−0.67
Acetate	Acetaldehyde	2	−0.60
Ferredoxin (oxidized)	Ferredoxin (reduced)	1	−0.43
$2H^+$	H_2	2	−0.42
NAD^+	$NADH + H^+$	2	−0.32
$NADP^+$	$NADPH + H^+$	2	−0.32
Glutathione (oxidized)	Glutathione (reduced)	2	−0.23
Acetaldehyde	Ethanol	2	−0.20
Pyruvate	Lactate	2	−0.19
Fumarate	Succinate	2	0.03
Cytochrome b (+3)	Cytochrome b (+2)	1	0.07
Ubiquinone (oxidized)	Ubiquinone (reduced)	2	0.10
Cytochrome c (+3)	Cytochrome c (+2)	1	0.22
Fe^{3+}	Fe^{2+}	1	0.77
$\frac{1}{2}O_2 + 2H^+$	H_2O	2	0.82

*E'_0 refers to the partial reaction written as

$$\text{Oxidant} + e^- \longrightarrow \text{reductant}$$

and n is the number of electrons transferred.

SOURCE: L. Stryer, 1988, *Biochemistry*, 3d ed., W. H. Freeman and Company, p. 400.

potential means that the substance—for example, acetate (CH_3COO^-) in its reduction to acetaldehyde (CH_3CHO)—has a lower affinity for electrons. In an oxidation-reduction reaction, electrons move spontaneously toward atoms or molecules having more positive reduction potentials. In other words, a compound having a more negative reduction potential can reduce one having a more positive potential.

In an oxidation-reduction reaction, the total voltage change (change in electric potential) ΔE is the sum of the voltage changes (reduction potentials) of the individual oxidation or reduction steps. Because all forms of energy are interconvertible, we can express ΔE as a change in chemical free energy ΔG. The charge in 1 mol (6×10^{23} molecules) of electrons is 96,500 coulombs (96,500 joules per volt)—a quantity known as the faraday constant ($\mathcal{F}$), after British physicist Michael Faraday (1791–1867). Thus we can write

$$\Delta G \text{ (cal/mol)} = -n\mathcal{F}\,\Delta E = -n\left(\frac{96{,}500}{4.18}\right)\Delta E \text{ (volts)}$$

where n is the number of electrons transferred and 4.18 is the factor used to convert joules into calories.

The reduction potential is customarily used to describe the electric energy change that occurs when an atom or a molecule gains an electron. In an oxidation-reduction reaction, we also use the oxidation potential—the voltage change that takes place when an atom or molecule *loses* an electron—which is simply the negative of the reduction potential:

Reduction: $Cu^{2+} + e^- \longrightarrow Cu^+$ $\quad \Delta E_0' = +0.35$ V

Oxidation: $Cu^+ \longrightarrow Cu^{2+} + e^-$ $\quad \Delta E_0' = -0.35$ V

The voltage change in a complete oxidation-reduction reaction, in which one molecule is reduced and another is oxidized, is simply the sum of the oxidation and reduction potentials of the atoms or molecules in the partial reactions. Consider, for example, the change in electric potential (and, correspondingly, in standard free energy) when succinate is oxidized by oxygen:

$$\text{Succinate} + \tfrac{1}{2}O_2 \rightleftharpoons \text{fumarate} + H_2O$$

In this case, the partial reactions are:

Succinate $\rightleftharpoons$ fumarate + $2H^+ + 2e^-$	$\Delta E_0' = -0.03$ V $\Delta G^{\circ\prime} = +1.39$ kcal/mol ($n = 2$)
$\tfrac{1}{2}O_2 + 2e^- \rightleftharpoons O^{2-}$	$\Delta E_0' = +0.82$ V $\Delta G^{\circ\prime} = -37.88$ kcal/mol ($n = 2$)
$2H^+ + O^{2-} \rightleftharpoons H_2O$	
Sum: Succinate + $\tfrac{1}{2}O_2 \rightleftharpoons$ fumarate + H_2O	$\Delta E^{\circ\prime} = +0.79$ V $\Delta G^{\circ\prime} = -36.49$ kcal/mol

A positive $\Delta E_0'$ signifies a negative $\Delta G^{\circ\prime}$ and thus, under standard conditions, a reaction that can occur spontaneously.

An Unfavorable Chemical Reaction Can Proceed if It Is Coupled with an Energetically Favorable Reaction

Many chemical reactions in cells have a positive ΔG: they are energetically unfavorable and will not proceed spontaneously. One example is the synthesis of small peptides, such as glycylalanine, from amino acids. How can such a reaction proceed? The cell's solution is to couple a reaction that has a positive ΔG to a reaction that has a negative ΔG of larger magnitude, so that the sum of the two reactions has a negative ΔG. Suppose that the reaction

$$A \rightleftharpoons B + X$$

has a $\Delta G^{\circ\prime}$ of +5 kcal/mol and that the reaction

$$X \rightleftharpoons Y + Z$$

has a $\Delta G^{\circ\prime}$ of −10 kcal/mol. In the absence of the second reaction, there would be much more A than B at equilibrium. The occurrence of the second process, by which X becomes Y + Z, changes that outcome: because it is such a favorable reaction, it will pull the first process toward the formation of B and the consumption of A.

The $\Delta G^{\circ\prime}$ of the overall reaction $A \rightleftharpoons B + Y + Z$ will be the sum of the $\Delta G^{\circ\prime}$ values of each of the two partial reactions:

$A \rightleftharpoons B + X$	$\Delta G^{\circ\prime} = +\ 5$ kcal/mol
$X \rightleftharpoons Y + Z$	$\Delta G^{\circ\prime} = -10$ kcal/mol
Sum: $A \rightleftharpoons B + Y$	$\Delta G^{\circ\prime} = -\ 5$ kcal/mol

The overall reaction releases energy. In cells, energetically unfavorable reactions of the type $A \rightleftharpoons B + X$ are often coupled to the hydrolysis of the compound adenosine triphosphate (ATP).

Hydrolysis of the Phosphoanhydride Bonds in ATP Releases Substantial Free Energy

Cells extract energy from foods through a series of reactions that exhibit negative free-energy changes. Much of the free energy released is not allowed to dissipate as heat but is captured in chemical bonds formed by other molecules for use throughout the cell. In almost all organisms, the most important molecule for capturing and transferring free energy is *adenosine triphosphate,* or ATP (Figure 1-13).

The useful free energy in an ATP molecule is contained in high-energy *phosphoanhydride bonds,* which are

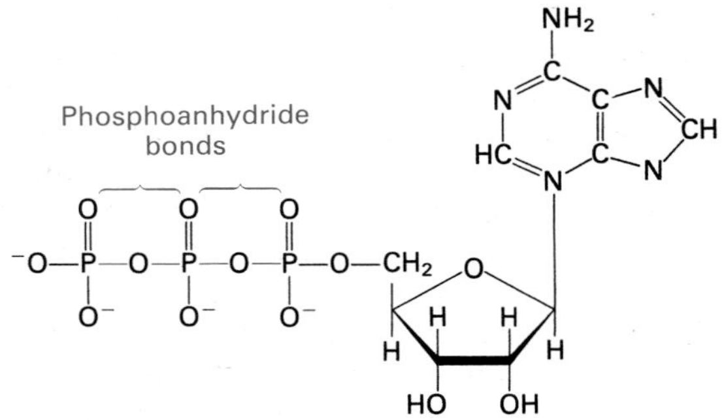

▲ **Figure 1-13** The structure of adenosine triphosphate, in which two high-energy phosphoanhydride bonds (red) link the three phosphate groups.

formed from the condensation of two molecules of phosphate by the loss of water:

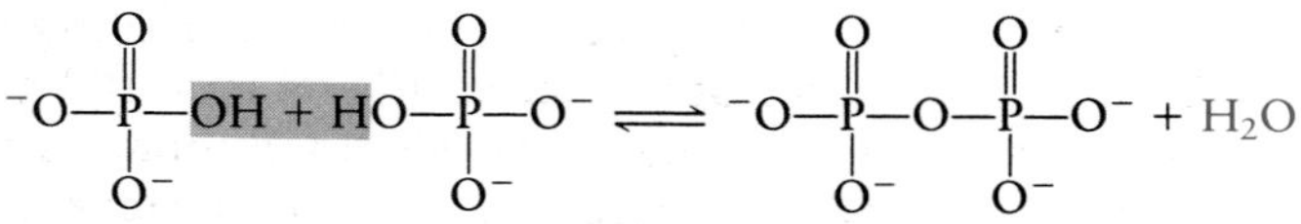

An ATP molecule has two phosphoanhydride bonds and is often written

$$\text{Adenosine—p} \sim \text{p} \sim \text{p}$$

or simply

$$\text{Ap} \sim \text{p} \sim \text{p}$$

where p stands for a phosphate group and $\sim$ denotes a high-energy bond. This is an ordinary covalent bond; we call it a high-energy bond because it releases about 7.3 kcal/mol of free energy (under standard biochemical conditions) when it is broken, as in hydrolysis:

$$\text{Ap} \sim \text{p} \sim \text{p} + H_2O \longrightarrow \text{Ap} \sim \text{p} + P_i + H^+$$

or

$$\text{Ap} \sim \text{p} \sim \text{p} + H_2O \longrightarrow \text{Ap} + PP_i + H^+$$

or

$$\text{Ap} \sim \text{p} + H_2O \longrightarrow \text{Ap} + P_i + H^+$$

where P_i stands for inorganic phosphate and PP_i for inorganic pyrophosphate, two phosphate groups linked by a phosphoanhydride bond. Removal of a phosphate or a pyrophosphate group from ATP leaves adenosine diphosphate (ADP) or adenosine monophosphate (AMP), respectively. In cells, the free energy released by the breaking of high-energy bonds is transferred, usually along with a phosphate group, to other molecules. This supplies them with enough free energy to undergo reactions that would otherwise be unfavorable. Cells contrive to keep the ratio of ATP to ADP and AMP high, often as high as 10:1. Thus reactions in which the terminal phosphate group of ATP is transferred to another molecule will be driven even further along.

Table 1-9 Values of $\Delta G^{\circ\prime}$ for the hydrolysis of various biologically important phosphate compounds*

Compound	$\Delta G^{\circ\prime}$ (kcal/mol)
PHOSPHOENOLPYRUVATE 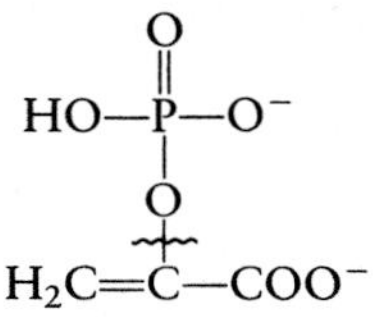	−14.8
CREATINE PHOSPHATE 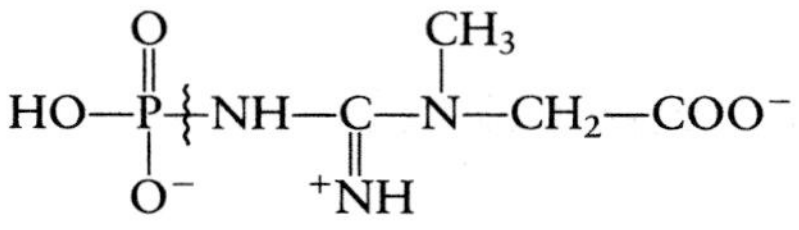	−10.3
PYROPHOSPHATE 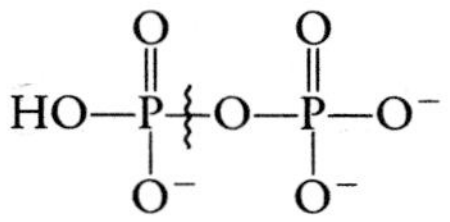	−8.0
ATP (to ADP + P_i)	−7.3
ATP (to AMP + PP_i)	−7.3
GLUCOSE 1-PHOSPHATE 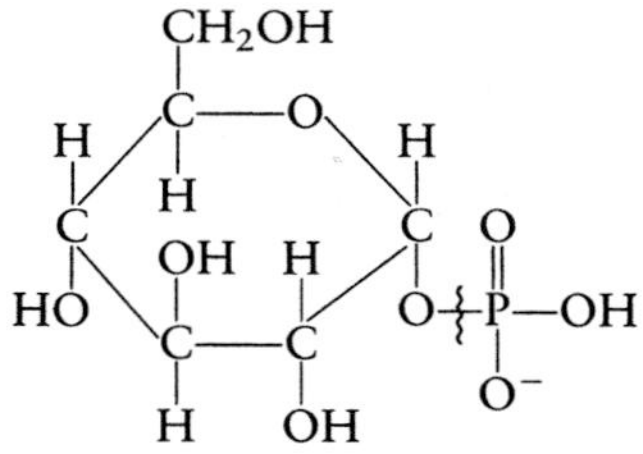	−5.0
GLUCOSE 6-PHOSPHATE 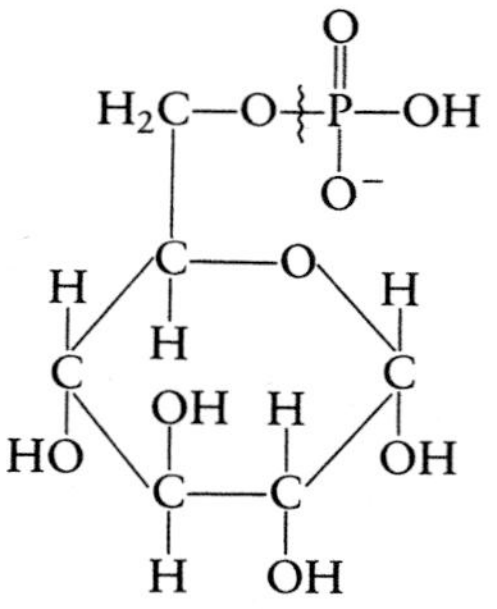	−3.3
GLYCEROL 3-PHOSPHATE 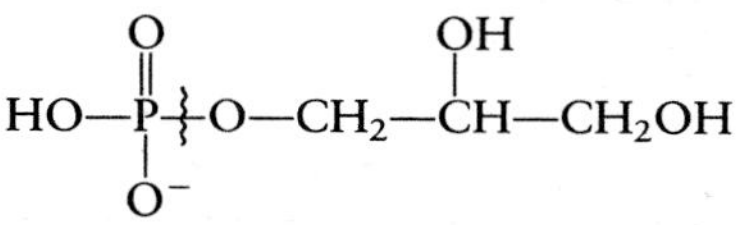	−2.2

*The bond that is cleaved is indicated by the wavy line.

Other bonds—particularly those between a phosphate group and some other substance—have the same high-energy character of phosphoanhydride bonds. The interphosphate bond of ATP is not the most or the least energetic of these bonds (Table 1-9). Why, then, is ATP the most important cellular molecule for capturing and transferring free energy? The free energy of hydrolysis of ATP is sufficiently great that reactions in which the terminal phosphate group is transferred to another molecule have a substantially negative $\Delta G^{\circ\prime}$. However, if the magnitude of the $\Delta G^{\circ\prime}$ for hydrolysis of the phosphoanhydride bond were much higher than it is, cells might require too much energy to form this bond in the first place.

ATP Is Used to Fuel Many Cellular Processes

If the terminal phosphoanhydride bond of ATP ruptures by hydrolysis to produce ADP and P_i, energy is released in the form of heat. In the presence of specific cellular enzymes, however, much of this energy is converted to more useful forms, some of which are summarized in Figure 1-14. Energy is used to synthesize large cellular molecules—for example, to link amino acids to form proteins, or to connect single sugars (monosaccharides) to form oligosaccharides. Energy is also used to synthesize many of the small molecules required by the cell. The hydrolysis of ATP supplies the energy needed to move individual cells from one location to another and to contract muscle cells. This energy also plays an important role in the transport of molecules into or out of the cell, usually against a concentration gradient. The generation of gradi-

ents of ions, such as Na^+ and K^+, across a cellular membrane is the basis for the electric activity of cells and, in particular, for the conduction of impulses by nerves.

What energy sources are required to form high-energy bonds? Plants and microorganisms trap the energy in light through *photosynthesis*. In the chloroplasts of plant cells and in photosynthetic bacteria, chlorophyll pigments absorb the energy of light, which is then used to synthesize ATP from ADP and P_i. Our current understanding of the mechanism of these photosynthetic processes is described in Chapter 16. Much of the ATP produced in photosynthesis is used to help convert carbon dioxide to glucose:

$$6CO_2 + 6H_2O \underset{\text{ATP} \quad \text{ADP} + P_i}{\xrightarrow{\hspace{1.5cm}}} C_6H_{12}O_6 + 6O_2$$

Animals obtain free energy by oxidizing food molecules through *respiration*. All synthesis of ATP in animal cells and in nonphotosynthetic microorganisms results from the chemical transformation of energy-rich dietary or storage molecules. The predominant source of energy in cells is the six-carbon sugar glucose. When 1 mol (180 g) of glucose reacts with oxygen under standard conditions according to the following reaction, 686 kcal of energy is released:

$$C_6H_{12}O_6 + 6O_2 \longrightarrow 6CO_2 + 6H_2O$$
$$\Delta G^{\circ\prime} = -686 \text{ kcal/mol}$$

If glucose is simply burned in air, all of this energy is released as heat. In the cell, through an elaborate set of enzyme-catalyzed reactions, the metabolism of one mole-

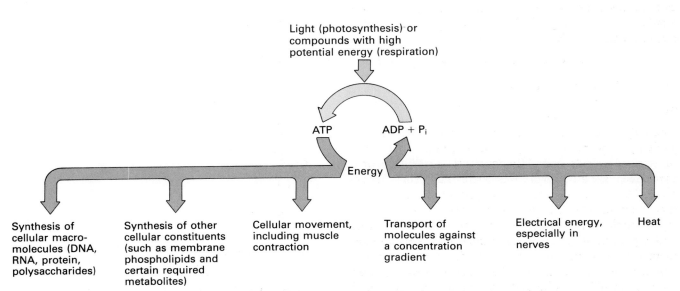

▲ **Figure 1-14** The ATP cycle. ATP is formed from ADP and P_i by photosynthesis in plants and by the metabolism of energy-rich compounds in animals. The hydrolysis of ATP to ADP and P_i is linked to many key cellular functions; the free energy released by the breaking of the phosphoanhydride bond is trapped as usable energy.

cule of glucose is coupled to the synthesis of as many as 36 molecules of ATP from 36 of ADP:

$$C_6H_{12}O_6 + 6O_2 + 36P_i + 36ADP \longrightarrow$$
$$6CO_2 + 6H_2O + 36ATP$$

Because one high-energy phosphoanhydride bond in ATP represents 7.3 kcal/mol, about 263 kcal of energy is conserved in ATP per mole of glucose metabolized (an efficiency of 263/686, or about 38 percent). This type of cellular metabolism is termed *aerobic* because it is dependent on the oxygen in the air. Aerobic *catabolism* (degradation) of glucose is found in all higher plant and animal cells and in many bacterial cells. The overall result of glucose respiration:

$$C_6H_{12}O_6 + 6O_2 \longrightarrow 6CO_2 + 6H_2O$$

is an exact reversal of the photosynthetic reaction in which glucose is formed:

$$6CO_2 + 6H_2O \longrightarrow C_6H_{12}O_6 + 6O_2$$

except that the energy of light is essential for the photosynthetic reaction. Respiration and photosynthesis are the two major processes constituting the *carbon cycle* in nature: glucose and oxygen produced by plants are the raw materials for respiration and the generation of ATP by plant and animal cells alike; the end products of respiration, CO_2 and H_2O, are the raw materials for the photosynthetic production of glucose and oxygen. The only net source of energy in this cycle is sunlight. Directly or indirectly, photosynthesis is the source of chemical energy for almost all cells.

Activation Energy and Reaction Rate

Many chemical reactions that exhibit a negative standard free-energy change do not proceed unaided at a measurable rate. For example, in pure aqueous solutions, glyceral-

dehyde 3-phosphate is a fairly stable compound that reacts very slowly or not at all; yet, in cells it can undergo several different reactions (Figure 1-15), each of which has a negative $\Delta G^{\circ\prime}$.

Similarly, a mixture of hydrogen and oxygen gases sealed in a flask at room temperature will remain quiescent indefinitely. If, however, the gases are exposed to an electric spark or a flame, the mixture explodes. The hydrogen gas burns vigorously, combining with the oxygen to form water:

$$2H_2 + O_2 \longrightarrow 2H_2O$$

This reaction releases a considerable amount of energy (57.8 kcal/mol H_2O formed).

To understand this seemingly ambiguous behavior, let's consider the mechanism of the reaction of hydrogen with oxygen. Three molecules, two H_2 and one O_2, must come together in such a way that the bonds can rearrange to form the products. At the high temperature in a hydrogen flame, many molecules are moving so fast that when two of them collide, one breaks up into single atoms:

$$2H_2 \longrightarrow H_2 + 2H$$

This reaction is *endergonic:* it absorbs energy. Some of the energy of the two colliding molecules is used to break one of their covalent bonds; thus the products of the reaction do not have as much kinetic energy as the intact molecules did. The kinetic energy used to break the bond has been transformed into chemical potential energy; the single hydrogen atoms react very easily with oxygen molecules:

$$4H + O_2 \longrightarrow 2H_2O$$

This reaction is *exergonic:* it releases energy—more energy, in fact, than was absorbed by the endergonic reaction in which the H_2 was broken apart. If we consider the two reactions as separate stages of an overall reaction, then the overall reaction is exergonic because the second stage releases more energy than the first absorbs.

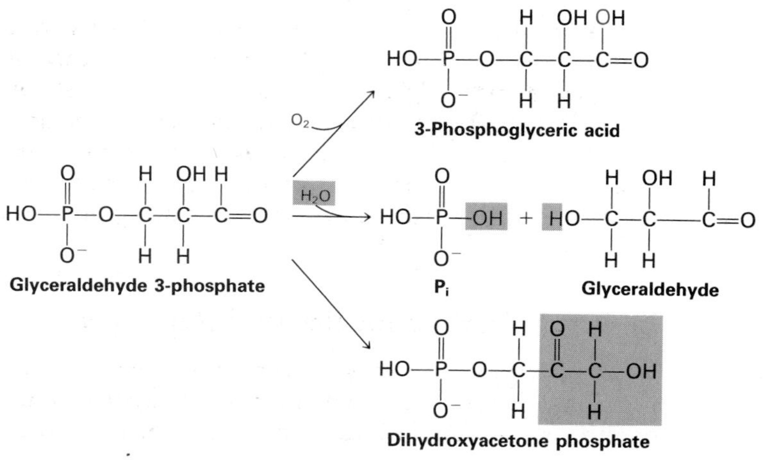

◀ **Figure 1-15** Glyceraldehyde 3-phosphate, like most cellular molecules, can undergo any of several exothermic reactions: oxidation (by O_2) to 3-phosphoglyceric acid, hydrolysis to glyceraldehyde and phosphate (P_i), and rearrangement to dihydroxyacetone phosphate. In the absence of enzymes or other catalysts, however, these reactions cannot occur in aqueous solution. Different enzymes catalyze each reaction; the presence of a specific enzyme is required for a particular reaction to proceed.

Energy Is Required to Initiate a Reaction

The input of energy required to initiate a reaction is called the *activation energy*. In a burning flame, the kinetic energy of many molecules of H_2 or O_2 is great enough to serve as the activation energy. At room or body temperature, however, the average kinetic energy of a typical gas such as H_2 is about 1.5 kcal/mol. Although many molecules will have more kinetic energy than this average, the chances that an H_2 molecule will have enough kinetic energy to equal the activation energy (~100 kcal/mol) and thereby initiate the reaction are virtually zero. Thus a mixture of hydrogen and oxygen will not react until it receives enough energy—say, from a spark or a flame—to achieve the activation energy. The energetic relation between the initial reactants and the products of an exergonic reaction can usually be described by a diagram (Figure 1-16).

Exergonic biochemical reactions (similar to the formation of H_2O) require activation energy. In some, bonds must be broken before new bonds can form; in others, electrons must be excited before they will pair up in a covalent bond. In still others, molecules need only enough energy to overcome the mutual repulsion of their electron clouds to get close enough to react. For example, the conversion of glyceraldehyde 3-phosphate (G3P) to dihydroxyacetone phosphate (DHAP) involves at least one

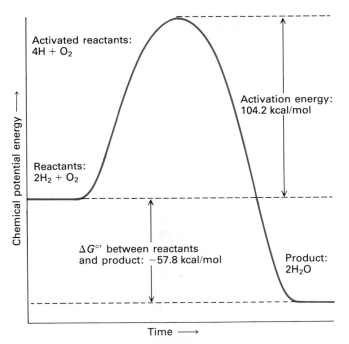

▲ **Figure 1-16** The reaction of hydrogen with oxygen requires an initial input of 104.2 kcal/mol (the activation energy) even though the products have a much lower free energy than the reactants do. The energy content of the reactants are depicted at each stage.

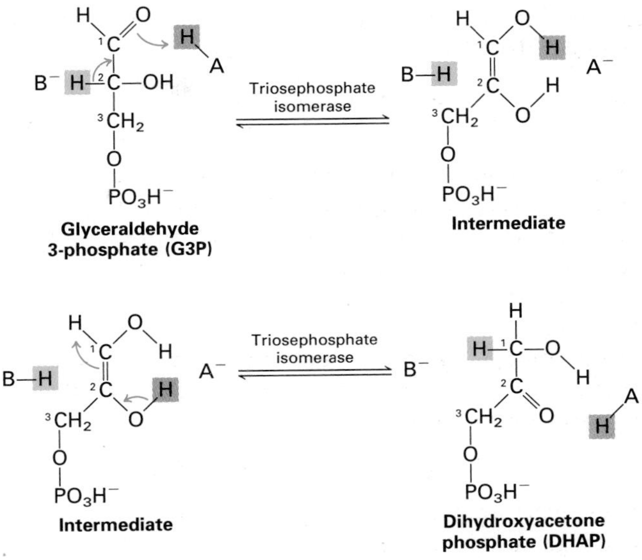

▲ **Figure 1-17** The conversion of glyceraldehyde 3-phosphate (G3P) to dihydroxyacetone phosphate (DHAP) involves an intermediate. Two groups, a base B^- and an acid HA, are parts of triosephosphate isomerase, the enzyme that catalyzes this reaction. To form the intermediate, B^- abstracts a proton (blue) from the 2 carbon of G3P; HA adds a proton (red) to the keto-oxygen on the 1 carbon. To convert the intermediate to DHAP, BH donates its proton to the 1 carbon (regenerating the original B^-) and A^- abstracts a proton from the —OH on the 2 carbon (regenerating HA). The curved arrows denote the movements of pairs of electrons that accompany the making and breaking of these bonds. [See D. Straus et al., 1985, *Proc. Nat'l Acad. Sci.* **82:**2272.]

intermediate (Figure 1-17). As the intermediate forms, the following changes take place simultaneously: a proton is removed from one carbon, another proton is donated to an oxygen, and pairs of electrons move from one bond to another. The activation energy required by each of these partial reactions contributes to the activation energy necessary to form this reaction intermediate.

Thus each stage in a multistep reaction has its own activation energy (Figure 1-18). For the overall reaction to proceed, the highest activation energy must be achieved. Because biochemical reactions occur at moderate temperatures, the kinetic energy of colliding molecules is generally insufficient to provide the necessary activation energy; in most cases, reactants must meet this requirement in some other way.

Enzymes Catalyze Biochemical Reactions

The rate at which a chemical reaction proceeds depends on several factors. The temperature of the system is important in nonbiochemical reactions, which proceed

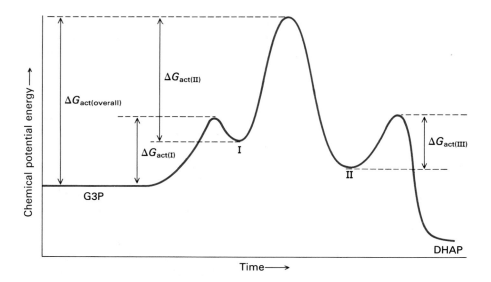

◂ **Figure 1-18** Hypothetical energy changes in the conversion of glyceraldehyde 3-phosphate (G3P) to dihydroxyacetone phosphate (DHAP). The troughs in the curve represent stable intermediates in the reaction. The vertical distance from a trough to the succeeding crest represents the activation energy (ΔG_{act}) required for one intermediate to be converted to the next or the final product (for example, $\Delta G_{act(II)}$ represents the activation energy required for the conversion of intermediate I to intermediate II). The total activation energy $\Delta G_{act(overall)}$ is the difference between the free energy of the reactants and that represented by the highest crest along the pathway.

faster as the temperature increases and an increasing number of molecules in the reaction mixture attain the activation energy. In reactions with two or more reactants, an increase in temperature will also increase the frequency of encounters between molecules or atoms. However, most cells experience such small variations in temperature that they are not important in regulating the rate of biochemical reactions.

Two significant rate-regulating factors for biological systems are the concentrations and the pH of the reactants. A reaction involving two or more different molecules proceeds faster at high concentrations because the molecules are more likely to encounter one another.

However, the most important determinants of biochemical reaction rates are *enzymes,* which are proteins that act as catalysts. A *catalyst* is a substance that brings reactants together but that does not end up among the products of the reaction—a function aptly reflected in the Chinese term for catalyst, *tsoo mei,* which literally means "marriage broker." A catalyst increases the rate of a reaction but is not itself permanently changed. It does not alter the change in free energy or the equilibrium constant; it accelerates the rates of forward and reverse reactions by the same factor. In fact, whether or not a catalyst is present, the equilibrium constant is equal to

$$K_{eq} = \frac{\text{rate of forward reaction}}{\text{rate of reverse reaction}}$$

Enzymes and all other catalysts act by reducing the activation energy required to make a reaction proceed (Figure 1-19). Some enzymes bind a *substrate,* the substance on which they act, in a way that strains its bonds and

makes it easy for the substrate to react. Other enzymes bind multiple substrates in a way that brings them together so they can react readily with one another. In each case, the overall effect is to reduce the activation energy needed for the reaction to take place. Triosephosphate isomerase, the enzyme that catalyzes the conversion of

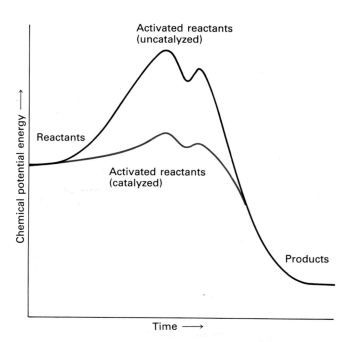

▴ **Figure 1-19** A catalyst accelerates the rate of a reaction by reducing the activation energy. Catalysts do not alter the free energy of reactants or products or affect their equilibrium concentrations.

G3P to DHAP, binds G3P as its substrate. Once G3P is bound, the requisite movements of its hydrogen atoms, protons, and electrons are all facilitated by specific chemical groups on the parts of the enzyme adjacent to the bound substrate (see Figure 1-17).

An enzyme binds either a single substrate or a set of similar substrates. Each enzyme catalyzes a single reaction on the bound substrate. Thus the presence or absence of particular enzymes in a cell or in extracellular fluids determines which of many possible chemical reactions will occur.

Summary

Atoms in a molecule are held together in a fixed orientation chiefly by covalent chemical bonds. Such bonds have relatively high energies of formation (50–200 kcal/mol) and consist of pairs of electrons shared by two atoms. Many covalent bonds between unlike atoms are dipolar, meaning that the electrons spend more time around the more electronegative atom.

A number of weaker noncovalent chemical bonds and interactions help to determine the shape of many large biological molecules and to stabilize complexes composed of different molecules. In a hydrogen bond, a hydrogen atom covalently bonded to a donor atom associates with an acceptor atom (whose nonbonding electrons attract the hydrogen). Hydrogen bonds among water molecules are largely responsible for the properties of both the liquid and the crystalline solid (ice) forms. Ionic bonds result from the electrostatic attraction between the positive and negative charges of ions. In aqueous solutions, all cations and anions are surrounded by a tightly bound shell of water molecules. Weak and relatively nonspecific van der Waals interactions are created whenever any two atoms approach each other closely. Hydrophobic interactions occur among nonpolar molecules, such as hydrocarbons, in an aqueous environment. These "bonds" result from the lack of interaction between water and the nonpolar molecules and from the van der Waals forces that occur among the nonpolar molecules. Although any single noncovalent bond may be very weak (most release little energy when they form), several such bonds between molecules or between the parts of one molecule can result in very stable structures. The attraction between two large molecules can be quite strong if they have complementary sites on their surfaces.

The most useful measure of chemical reactions in biological systems is the change in free energy ΔG. This value depends on the change in heat or enthalpy ΔH (chiefly the overall change in bond energies) and the change in entropy ΔS (the randomness of molecular motion): $\Delta G = \Delta H - T\,\Delta S$. Chemical reactions tend to proceed in the direction for which ΔG is negative.

The direction of a chemical reaction can be predicted if both its standard $\Delta G^{\circ\prime}$ value and the concentrations of the reactants and the products are known. The equilibrium constant K_{eq} and $\Delta G^{\circ\prime}$ are mathematically related: $K_{eq} = 10^{-\Delta G^{\circ\prime}/2.3RT}$. A chemical reaction having a positive ΔG can proceed spontaneously if it is coupled with a reaction having a negative ΔG of larger magnitude.

In cells, the two phosphoanhydride bonds in adenosine triphosphate (ATP) are a principal source of chemical potential energy. The hydrolysis of one or both of these bonds is often coupled with another energetically unfavorable reaction. Such reactions are involved in the synthesis of proteins from amino acids, the synthesis of other molecules, cellular movement, and the transport of compounds into or out of a cell against a concentration gradient. All of these reactions are fueled by the hydrolysis of ATP. In plant cells, ATP is generated from adenosine diphosphate (ADP) and inorganic phosphate (P_i) through the use of energy absorbed from light; much of the ATP is consumed in the synthesis of glucose from CO_2 and H_2O. In animal cells and in nonphotosynthetic microorganisms, most of the ATP is generated during the oxidation (by O_2) of glucose to CO_2 and H_2O. Directly or indirectly, photosynthesis is the source of chemical energy for almost all cells.

Whether a reaction will actually proceed depends on its activation energy. Sometimes the kinetic energy of the reactant molecules is sufficient to overcome the activation-energy barrier. If the activation energy is too high to permit the reaction to occur at a measurable rate, catalysts can speed up the reaction by decreasing the activation energy. A catalyst accelerates both the forward and the reverse reactions to the same extent; it does not change ΔG, $\Delta G^{\circ\prime}$, or K_{eq}. Enzymes are biological catalysts that generally facilitate only one of the many possible transformations that a molecule can undergo.

References

ATKINS, P. W. 1989. *General Chemistry*. Scientific American Books.

BUTLER, J. 1964. *Solubility and pH Calculations*. Addison-Wesley.

CANTOR, C. R., and P. R. SCHIMMEL. 1980. *Biophysical Chemistry*, part 1. W. H. Freeman and Company. Chapter 5.

DAVENPORT, H. W. 1974. *ABC of Acid-Base Chemistry*, 6th ed. University of Chicago Press.

EDSALL, J. T., and J. WYMAN. 1958. *Biophysical Chemistry*, vol. 1. Academic Press.

EISENBERG, D., and W. KAUZMANN. 1969. *The Structure and Properties of Water*. Oxford University Press.

HILL, T. L. 1977. *Free Energy Transduction in Biology*. Academic Press.

KLOTZ, I. M. 1978. *Energy Changes in Biochemical Reactions*. Academic Press.

LEHNINGER, A. L. 1982. *Principles of Biochemistry.* Worth. Chapters 1, 3, 4, 13, and 14.

PAULING, L. 1960. *The Nature of the Chemical Bond,* 3d ed. Cornell University Press.

STRYER, L. 1988. *Biochemistry,* 3d ed. W. H. Freeman and Company. Chapters 1, 8, 13, and 17.

TANFORD, C. 1980. *The Hydrophobic Effect: Formation of Micelles and Biological Membranes,* 2d ed. Wiley.

WATSON, J. D., N. M. HOPKINS, J. W. ROBERTS, J. A. STEITZ, and A. M. WEINER. 1988. *Molecular Biology of the Gene,* 4th ed. Benjamin-Cummings. Chapters 2 and 5.

WOOD, W. B., J. H. WILSON, R. M. BENBOW, and L. E. HOOD. 1981. *Biochemistry: A Problems Approach,* 2d ed. Benjamin-Cummings. Chapters 1, 5, and 9.

ZUBAY, C. 1988. *Biochemistry,* 2d ed. Macmillan.

Molecules in Cells

Protein crystal, chymotrypsin

Cells are constructed of two different types of chemical substances—small molecules and polymers—that are distinguished by size and organization. Small molecules generally consist of fewer than 50 atoms; each small molecule has its characteristic structure. Polymers are composed of many copies of a few small molecules linked in chains by covalent bonds; these polymer subunits are called monomers, or residues. The principal cellular polymers—the primary focus of this chapter—are nucleic acids, the substances that preserve and transmit genetic information, and proteins, the products generated from the transmitted information. Proteins are constructed from amino acids; nucleic acids are built of monomers called nucleotides. The arrangement of monomers in these biopolymers is linear (Figure 2-1).

Nucleic acids are made from four different nucleotides, linked together in chains that may be millions of units long. Because these subunits can be linked in any order, the number of possible nucleic acids n units long is 4^n (Figure 2-1b). A 10-unit nucleic acid has 4^{10} (more than 1 million) possible structures; a 100-unit nucleic acid has 4^{100} (more than 10^{60}).

The chemical reactions that constitute life processes are directed and controlled by proteins. There are 20 different amino acids in proteins. Thus a 100-unit protein has 20^{100} (more than 10^{130}) possible structures. This enormous variability means that cells and organisms can differ greatly in structure and function even though they are constructed of the same types of biopolymers produced by similar chemical reactions.

Starch (a storage form of glucose in plant cells), cellulose (a constituent of plant cell walls), and glycogen (a storage form of glucose in liver and muscle cells) are examples of another important type of biopolymer: the poly-saccharide, *which is built of sugar monomers (Figure 2-1). At least 15 different monomeric sugars can be bonded in multiple ways to form various polysaccharides; thus many polysaccharides are nonlinear, branched molecules.*

Monomers are not the only small molecules important to cell structure. The lipids, for example, form the basic structure of cell membranes. Lipids cohere noncovalently in very large sheetlike complexes; the membranes thus formed are as crucial to living systems as are the biopolymers.

This chapter deals with the structures and some functions of biopolymers and small molecules; later chapters describe how the polymers are made and consider many of their other functions and interactions. ▲

Proteins

Proteins are the working molecules of the cell. They catalyze an extraordinary range of chemical reactions, provide structural rigidity, control membrane permeability, regulate the concentrations of metabolites, recognize and noncovalently bind other biomolecules, cause motion, and control gene function. These incredibly diverse tasks are performed by molecules constructed from only 20 different amino acids.

Amino Acids—the Building Blocks of Proteins—Differ Only in Their Side Chains

The monomers that make up proteins are called amino acids because, with one exception, each contains an *amino group* (—NH$_2$) and an acidic carboxyl group (—COOH). The exception, proline, has an *imino group* (—NH—) instead of an amino group. At typical pH values in cells, the amino and carboxyl groups are ionized as —NH$_3^+$ and —COO$^-$. All amino acids are constructed according to a basic design: a central carbon atom, called the α carbon C$_\alpha$ (because it is adjacent to the acidic carboxyl group), is bonded to an amino (or imino) group, to the carboxyl group, to a hydrogen atom, and to one variable group, called a *side chain* or R *group* (Figure 2-2). The side chains give the amino acids their individuality.

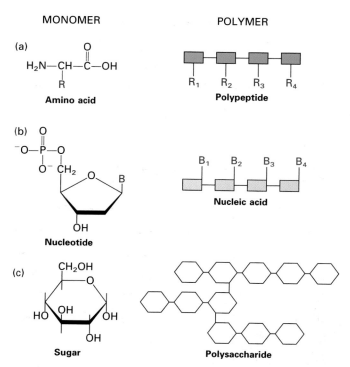

▲ **Figure 2-1** (a) Proteins, linear biopolymers called polypeptides, are formed from monomeric subunits termed amino acids. Each of the 20 different amino acids has a different R group, or side chain. Thus the polypeptide shown here, which is constructed of four amino acids, has 20^4, or 160,000, possible structures. (b) Nucleic acids, also linear biopolymers, are formed from four monomers termed nucleotides, each of which has a different nitrogen-containing base structure (B). The nucleic acid shown here has 4^4, or 256, possible structures. (c) Polysaccharides are built of monomeric saccharide (sugar) subunits. Because sugar residues can bind to one another at different positions, nonlinear branching polymers are often formed. The rings in (b) and (c) are depicted as Haworth projections (planar structures with a hint of perspective).

The amino acids represent the alphabet in which linear proteins are "written"; any student of biology must be familiar with the special properties of each letter of this alphabet. These letters can be classified into a few distinct categories.

The side chains of four of the amino acids are highly ionized and therefore charged at neutral pH. *Arginine* and *lysine* are positively charged; *aspartic acid* and *glutamic acid* are negatively charged and exist as asparate and glutamate. The side chain of a fifth amino acid, *histidine,* is positively charged, but only weakly at neutral pH. In many cases, arginine may substitute for lysine, or aspartate for glutamate, with little effect on the structure or function of the protein.

Serine and *threonine,* whose side chains have an —OH group, can interact strongly with water by forming hydrogen bonds. The side chains of *asparagine* and *gluta-*

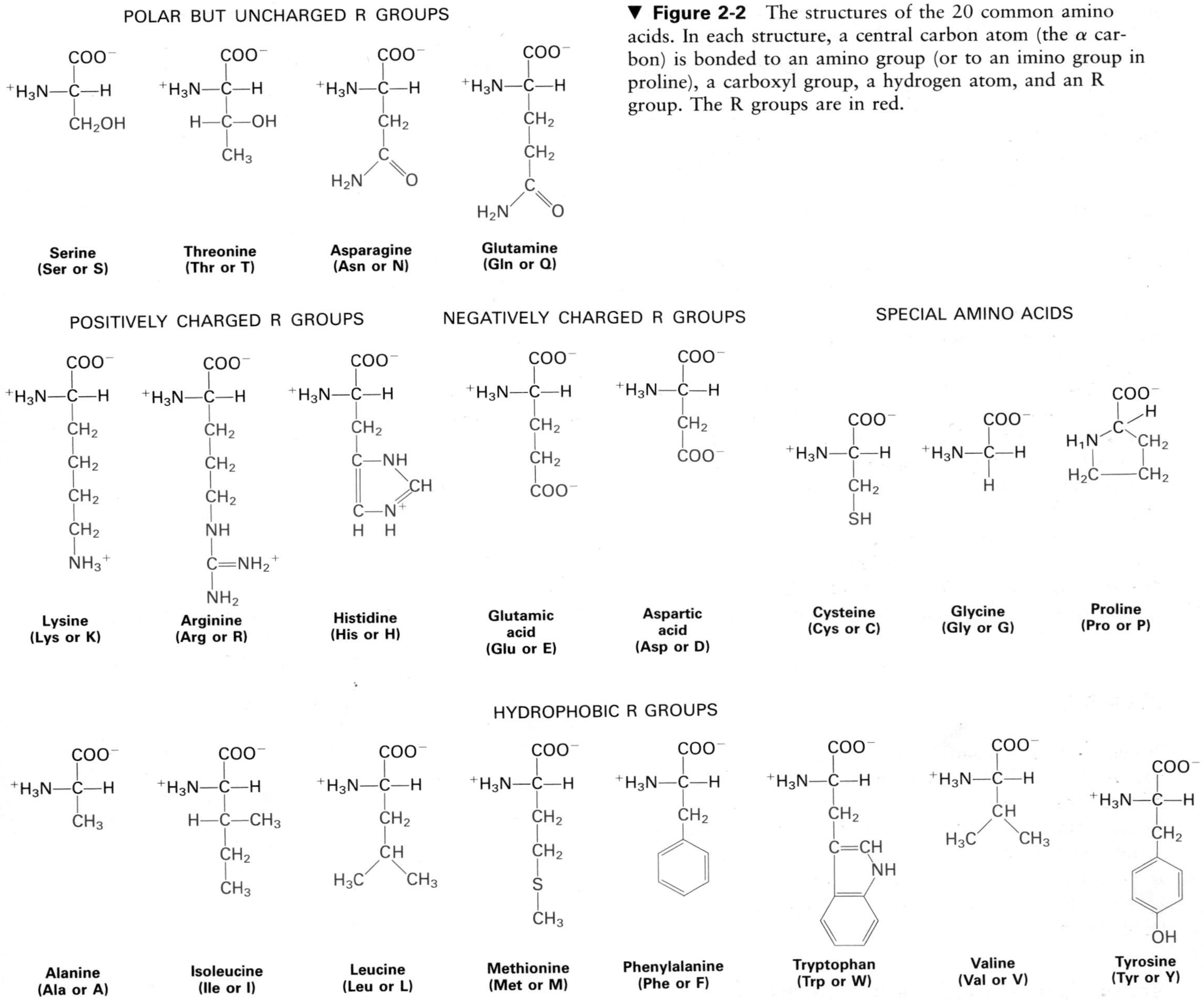

POLAR BUT UNCHARGED R GROUPS

**Serine
(Ser or S)**

**Threonine
(Thr or T)**

**Asparagine
(Asn or N)**

**Glutamine
(Gln or Q)**

▼ **Figure 2-2** The structures of the 20 common amino acids. In each structure, a central carbon atom (the α carbon) is bonded to an amino group (or to an imino group in proline), a carboxyl group, a hydrogen atom, and an R group. The R groups are in red.

POSITIVELY CHARGED R GROUPS

**Lysine
(Lys or K)**

**Arginine
(Arg or R)**

**Histidine
(His or H)**

NEGATIVELY CHARGED R GROUPS

**Glutamic
acid
(Glu or E)**

**Aspartic
acid
(Asp or D)**

SPECIAL AMINO ACIDS

**Cysteine
(Cys or C)**

**Glycine
(Gly or G)**

**Proline
(Pro or P)**

HYDROPHOBIC R GROUPS

**Alanine
(Ala or A)**

**Isoleucine
(Ile or I)**

**Leucine
(Leu or L)**

**Methionine
(Met or M)**

**Phenylalanine
(Phe or F)**

**Tryptophan
(Trp or W)**

**Valine
(Val or V)**

**Tyrosine
(Tyr or Y)**

mine have polar amide groups with even more extensive hydrogen-bonding capacities. Together with the charged amino acids, these amino acids constitute the nine hydrophilic or polar amino acids.

The side chains of several other amino acids—*alanine, isoleucine, leucine, methionine, phenylalanine, tryptophan,* and *valine*—consist only of hydrocarbons, except for the sulfur atom in methionine and the nitrogen atom in tryptophan. These nonpolar amino acids are hydrophobic; their side chains are only slightly soluble in water. *Tyrosine* is also strongly hydrophobic because of its benzene ring, but its hydroxyl group allows it to interact with water, making its properties somewhat ambiguous.

Cysteine plays a special role in proteins because its —SH group allows it to dimerize through an —S—S— bond to a second cysteine, thus covalently linking regions of polypeptide to one another. When the —SH remains free, cysteine is quite hydrophobic.

Two other special amino acids are *glycine* and *proline.* Glycine has a hydrogen atom as its R group; thus it is the smallest amino acid and has no special hydrophobic or hydrophilic character. Proline, as an imino acid, is very rigid and creates a fixed kink in a polypeptide chain. It is quite hydrophobic.

The structure of all amino acids except glycine are asymmetrically arranged around the α carbon, because it is bonded to four different atoms or groups of atoms

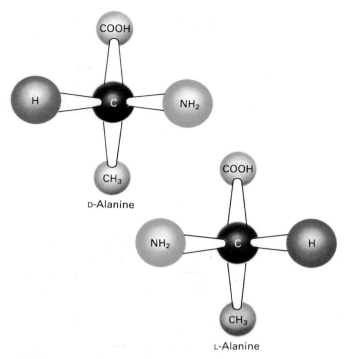

▲ **Figure 2-3** Stereoisomers of the amino acid alanine. The α carbon is black.

(—NH$_2$, —COOH, —H, and —R). Thus all amino acids except glycine can have one of two stereoisomeric forms. By convention, these mirror-image structures are called the D and the L forms of the amino acid (Figure 2-3). They cannot be interconverted without breaking a chemical bond. With rare exceptions, only the L forms of amino acids are found in proteins.

Polypeptides Are Polymers Composed of Amino Acids Connected by Peptide Bonds

The *peptide bond,* the chemical bond that connects two amino acids in a polymer, is formed between the amino group of one amino acid and the carboxyl group of another. This reaction, called *condensation,* liberates a water molecule:

$$^+H_3N-\overset{\overset{\displaystyle H}{|}}{\underset{\underset{\displaystyle R_1}{|}}{C_\alpha}}-\overset{\overset{\displaystyle O}{\|}}{C}-O^- + {}^+H_3N-\overset{\overset{\displaystyle H}{|}}{\underset{\underset{\displaystyle R_2}{|}}{C_\alpha}}-\overset{\overset{\displaystyle O}{\|}}{C}-O^- \xrightarrow{H_2O}$$

$$^+H_3N-\overset{\overset{\displaystyle H}{|}}{\underset{\underset{\displaystyle R_1}{|}}{C_\alpha}}-\overset{\overset{\displaystyle O}{\|}}{C}-N-\overset{\overset{\displaystyle H}{|}}{\underset{\underset{\displaystyle R_2}{|}}{C_\alpha}}-\overset{\overset{\displaystyle O}{\|}}{C}-O^-$$

Peptide bond

Because the carboxyl carbon and oxygen atoms are connected by a double bond, the peptide bond between car-

bon and nitrogen exhibits a partial double-bond character, as shown by the resonance structures

$$-\overset{|}{\underset{|}{C_\alpha}}-\overset{\overset{\displaystyle O}{\|}}{C}-NH-\overset{|}{\underset{|}{C_\alpha}}- \longleftrightarrow -\overset{|}{\underset{|}{C_\alpha}}-\overset{\overset{\displaystyle O^-}{|}}{C}=\overset{+}{N}H-\overset{|}{\underset{|}{C_\alpha}}-$$

making it shorter than the typical C—N single bond. The six atoms of the peptide group (the two carbons of the adjacent amino acids and the carbon, oxygen, nitrogen, and hydrogen atoms of the bond) lie in the same plane (Figure 2-4a). However, adjacent peptide groups are not necessarily coplanar, due to rotation about the C—C$_\alpha$ and N—C$_\alpha$ bonds (Figure 2-4b).

A single linear array of amino acids connected by peptide bonds is called a *polypeptide*. If the polypeptide is short (fewer than 30 amino acids long), it may be called an *oligopeptide* or just a *peptide*. Polypeptides in living cells differ greatly in length; they generally contain between 40 and 1000 amino acids. Each polypeptide has a free amino group at one end (the N-terminus) and a free carboxyl group at the other (the C-terminus):

$$^+H_3N-\overset{\overset{\displaystyle H}{|}}{\underset{\underset{\displaystyle R_1}{|}}{C_\alpha}}-\overset{\overset{\displaystyle O}{\|}}{C}-\overset{\overset{\displaystyle H}{|}}{N}-\overset{\overset{\displaystyle H}{|}}{\underset{\underset{\displaystyle H}{|}}{C_\alpha}}-\overset{\overset{\displaystyle H}{|}}{\underset{\underset{\displaystyle R_2}{|}}{C}}-\overset{\overset{\displaystyle H}{|}}{N}-\overset{\overset{\displaystyle H}{|}}{\underset{\underset{\displaystyle R_3}{|}}{C_\alpha}}-\overset{\overset{\displaystyle O}{\|}}{C}-\cdots$$

Amino end
(N-terminus)   **Carboxyl end**
                                    (C-terminus)

A protein is not merely a linear string of amino acids. The polypeptide folds up to form a specific three-dimensional structure that can be a long rod, as in the *fibrous proteins* that give tissues their rigidity, or a compact ball called a *globular protein,* as in many proteins that catalyze chemical reactions (enzymes), or a combination of balls and rods. The polypeptide can be modified further by the covalent or noncovalent attachment of additional small molecules.

A protein adopts a stable, folded conformation mainly through noncovalent (ionic, hydrogen, van der Waals, and hydrophobic) interactions. Its stability is also enhanced by the formation of covalent disulfide bonds between cysteines in different parts of the chain. Proteins may also consist of multiple polypeptide chains held together by noncovalent forces and, in some cases, by disulfide bonds. A well-characterized example is the hemoglobin molecule, which consists of four chains: two identical α chains and two identical β chains (Figure 2-5).

Three-dimensional Protein Structure Is Determined through X-ray Crystallography

The detailed three-dimensional structures of numerous proteins have been established by the painstaking efforts of many workers—notably, Max Perutz and John Kendrew, who perfected the x-ray crystallography of

(a)

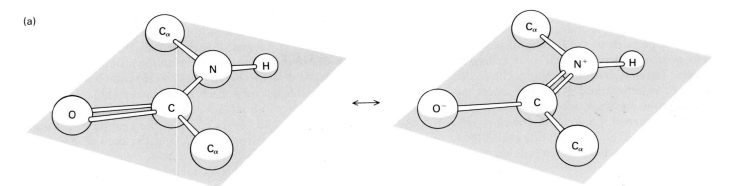

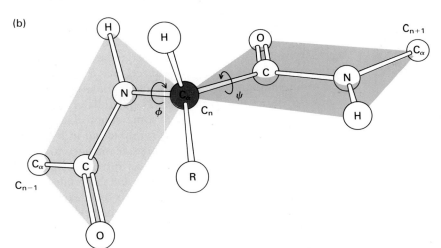

(b)

◀ **Figure 2-4** (a) Because the carbon-nitrogen peptide bond has a partial double-bond character, the peptide group is planar. (b) However, there is considerable flexibility in the geometry of polypeptides: rotation is possible about the two covalent single bonds that connect each α carbon to the two adjacent planar peptide units. But some restrictions do apply to the values of ψ and ϕ. For example, if the pictured adjacent peptide groups were coplanar, then certain oxygen and hydrogen atoms would be separated by less than their van der Waals radii and would repel one another.

proteins, in which beams of x-rays are passed through a crystal of protein. The wavelengths of x-rays are about 0.1–0.2 nanometers (nm)—short enough to resolve the atoms in the protein crystal. The three-dimensional structure of the protein can be deduced from the *diffraction pattern* of discrete spots that is produced when the scattered radiation is intercepted by photographic film. Such patterns are extremely complex; as many as 25,000 diffraction spots can be obtained from a small protein. Elaborate calculations and modifications of the protein (such as binding of heavy metal) must be made to interpret the diffraction pattern and to solve the structure of the protein.

Recently, three-dimensional structures of some small proteins have been determined by nuclear magnetic resonance (nmr) methods. An advantage of this approach is that it avoids the need to crystallize the protein. A disadvantage is that it is limited to relatively small proteins (up to about 20,000 molecular weight).

The Structure of a Polypeptide Can Be Described at Four Levels

The structures adopted by polypeptides can be divided into four levels of organization. *Primary structure* refers to the linear arrangement of amino acid residues along a

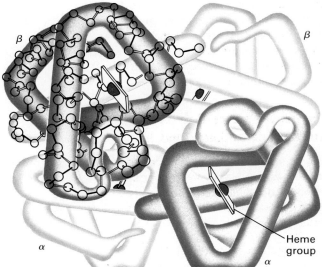

▲ **Figure 2-5** The conformations assumed by the two α and two β chains in a molecule of hemoglobin. Each chain forms several α helices (see Figure 2-6). Only the backbones formed by the carbon and nitrogen atoms of the chains are shown here. A multitude of noncovalent interactions stabilize the conformations of the individual chains and the contacts between them. A heme group is bound to each chain. *After R. E. Dickerson and I. Geis, 1969, The Structure and Action of Proteins, Benjamin-Cummings, p. 56. Copyright 1969 by Irving Geis.*

polypeptide chain and to the locations of covalent bonds (mainly —S—S— bonds) between chains. *Secondary structure* pertains to the folding of parts of these chains into regular structures, such as α helices and β pleated sheets. *Tertiary structure* includes the folding of regions between α helices and β pleated sheets, as well as the combination of these secondary features into compact shapes (domains). *Quaternary structure* refers to the organization of several polypeptide chains into a single protein molecule, such as in hemoglobin.

Two Regular Secondary Structures Are Particularly Important

The α Helix Although some regions of proteins are held in unique and irregular conformations, much protein structure involves repeated use of a limited number of regular configurations. One common structure, the *α helix,* was first described by Linus Pauling and Robert B. Corey in 1951. Through careful model building, these scientists came to realize that polypeptide seg-

ments composed of certain amino acids tend to arrange themselves in regular helical conformations. In an α helix, the carboxyl oxygen of each peptide bond is hydrogen-bonded to the hydrogen on the amino group of the fourth amino acid away (Figure 2-6), so that the helix has 3.6 amino acids per turn. Each amino acid residue represents an advance of about 1.5 Å along the axis of the helix. Every C=O and N—H group in the peptide bonds participates in a hydrogen bond, and the rigid planarity of the peptide bonds contributes to the rigid shape of the helix. In this inflexible, stable arrangement of amino acids, the side chains are positioned along the outside of a cylinder. The hydrogen-bonding potential of the peptide bonds is entirely satisfied internally, so that the polar or nonpolar quality of the cylindrical surface is determined entirely by the side chains. At least some of the amino acids in most proteins are organized into α helices.

Certain amino acid sequences adopt the α-helical conformation more readily than others. What determines this propensity is complicated, but some simple factors are evident. For instance, proline is rarely found in α-helical regions because it cannot use its peptide nitrogen to make a hydrogen bond. Glycine also is an infrequent participant. Another inhibiting factor can be the tendency of multiple identically charged residues to repel each other.

The α helix is a rodlike element of protein structure that serves many functions. A globular protein can be made up of short α-helical rods connected by bends that allow the rods to interact with each other; hemoglobin, for instance, is 70 percent α helical (see Figure 2-5). Alternatively, a single rod can span a long distance, as in the protein on the surface of the influenza virus (Figure 2-7a). Even in extended molecules, a,b,c the α helix is usually found packed against other elements of protein, not as an isolated structure. Long fibers, such as the skin protein keratin or the muscle protein myosin (Figure 2-7b), can be formed by two or three α helices that wrap gently around each other to form *coiled coils.* Small rods of α helix interact with DNA in some DNA-binding proteins (Figure 2-7c). A helical rod bearing only hydrophobic side chains can span lipid membranes well because the hydrophilic peptide bonds are buried inside the helix.

Many α helices are *amphipathic:* they expose hydrophilic side chains on one face and hydrophobic side chains on another face. Looking down the central axis of an α helix (Figure 2-8a), the amino acid residues are arranged in a wheel; if the helix is amphipathic, most or all

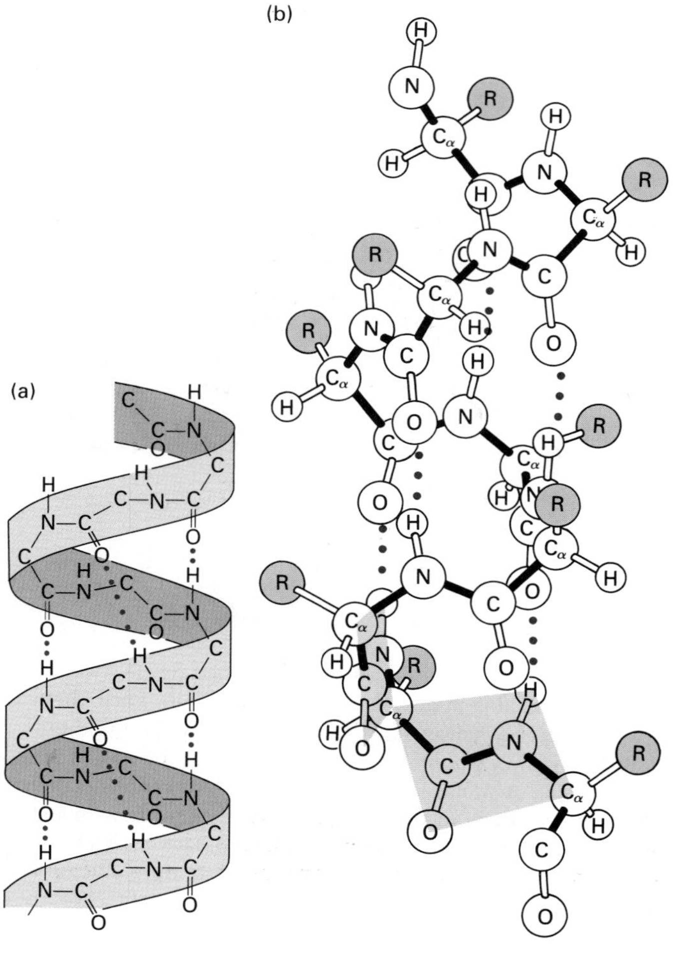

◀ **Figure 2-6** Models of the α helix. (a) This ribbonlike representation without R groups emphasizes the helical form. (b) This ball-and-stick representation emphasizes the role of the individual atoms and shows the R groups (green) that protrude from the helix body at regular intervals. Some of the planes of the C$_\alpha$—CO—NH groups are shaded orange. *Part (b) after L. Stryer, 1988,* Biochemistry, *3d ed., W. H. Freeman and Company, p. 26.*

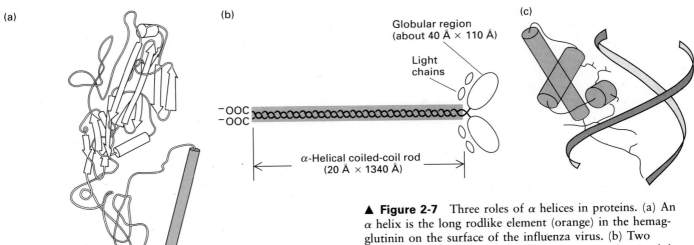

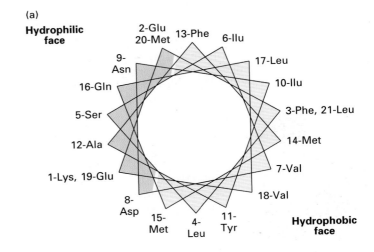

▲ **Figure 2-7** Three roles of α helices in proteins. (a) An α helix is the long rodlike element (orange) in the hemagglutinin on the surface of the influenza virus. (b) Two α helices make up the long coiled-coil structural unit of the major muscle protein, myosin. (c) Three α helices form the DNA-binding domain of a protein involved in controlling gene expression (the λ phage repressor). Here, the complex of the protein is bound to a specific stretch of DNA. *Part (a) from I. A. Wilson, J. J. Skehel, and D. C. Wiley, 1981, Nature **289**:366; courtesy of D. C. Wiley. Part (b) from L. Stryer, 1988, Biochemistry, 3d ed., W. H. Freeman and Company, p. 924. Part (c) from S. R. Jordon and C. O. Pabo, 1988, Science 242: 893; courtesy of C. O. Pabo.*

of the charged side chains will be on one side of the wheel and the hydrophobic side chains will be on the other. Amphipathic helices function as important structural elements in water-soluble proteins: their hydrophilic surfaces face the water and their hydrophobic surfaces face the interior of the protein, which is generally a hydrophobic environment.

In α-helical coiled coils, the individual right-handed α helices wrap around each other in a left-handed fashion, and hydrophobic side chains are usually found along the contact surface. The amino acid sequence of each α helix contains a seven-residue repeat unit ("heptad"); hydrophobic residues predominate at positions 1 and 4 in each heptad. Thus the hydrophilic/hydrophobic characteristics of coiled coils are periodic.

▲ **Figure 2-8** A method of representing an α helix that easily identifies amphipathic segments. (a) In a view down the axis of the helix, the amino acid residues are arranged in the form of a wheel; if the helix makes one turn every 3.6 residues, the pattern is repeated every 18 residues. (b) In a

linear representation of the polypeptide, the hydrophilic residues (blue) seem to be randomly arranged; if the wheel shows most of the hydrophilic residues on one side, however, the helix is amphipathic.

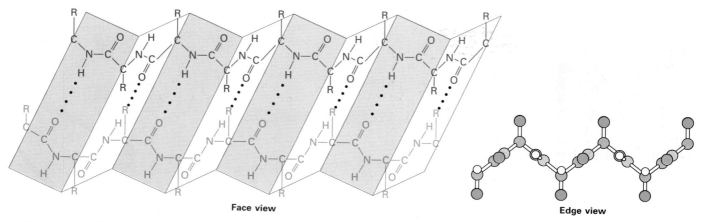

Face view

Edge view

▲ **Figure 2-9** Hydrogen bonding between polypeptide strands allows a β pleated sheet to form. The planarity of the peptide bond forces the structure to be pleated. A side view of a β pleated sheet shows how the R groups (green) protrude from the faces.

The β Pleated Sheet   A second regular structural element of many proteins is the *β pleated sheet,* which is created by a series of hydrogen bonds between the backbone atoms of the peptide bonds in different polypeptide chains or between the peptide bonds in different sections of a folded polypeptide. Adjacent polypeptide chains in β pleated sheets can be *antiparallel* or *parallel,* depending on their relative direction (Figure 2-9). β pleated sheets are always curved; if many polypeptides participate, the sheet can close up to form a barrel. Multiple pleated sheets provide toughness and rigidity in many structural proteins. Silk fibroin, for example, consists almost entirely of stacks of antiparallel β pleated sheets.

Like the α helix, the β pleated sheet pairs up all of the potential hydrogen bonding atoms of the peptide bond. The side chains, which project alternately from the two faces of the sheet, determine the hydrophilic or hydrophobic character of the particular face.

Many Proteins Are Organized into Domains

Most proteins have a compact, globular structure. The α helices, β pleated sheets, and other common secondary structural elements are packed together, along with more idiosyncratic structural elements, into a relatively rigid, ball-like form. Because the polypeptide is a linear string, packing it into a globular form requires that the chain change direction frequently at the surfaces. A complete reversal of direction may be accomplished by the *β turn*—a rigid, hairpin-like structural element that often connects strands of a β pleated sheet.

Longer polypeptides are often folded into several globular units rather than into one huge unit (Figure 2-10). Each independently folded unit is called a *domain.* Most domains contain between 50 and 300 amino acids. Poly-peptide segments shorter than 50 residues ordinarily cannot form a folded structure with a distinct hydrophobic interior and hydrophilic surface. The residues forming a single domain generally are found in one continuous segment of the primary structure, although sequentially noncontinuous segments do in some cases contribute to a single domain. Such noncontinuous segments may be linked by a flexible hinge, or they may be held in precise relative orientation by extensive noncovalent interactions. A regular substructure that occurs in otherwise different domains is termed a *motif.* The helix-turn-helix motif of DNA-binding proteins is an example. The polypeptide segment composing a motif would not in general fold stably into the motif structure in the absence of the rest of the domain. Domains of similar structure are found in different proteins, and structurally related domains often serve similar functions in different proteins. These observations suggest that domains of a particular structural class may have evolved from a common precursor.

These features of domain organization can be illustrated by various proteins whose structures are known. For example, each subunit of hemoglobin (see Figure 2-5) is a small polypeptide folded into one domain; four separate subunits fit tightly together to form the active protein. In each subunit, many short α-helical segments are connected by turns that fold them together compactly. Examples of more complex domain organization are shown in Figure 2-10. The enzyme triosephosphate isomerase (TIM) is a single-domain protein having an eight-stranded β barrel as its central organizing structure. The enzyme pyruvate kinase (PK) has three domains, one of which is a "TIM barrel." Note that the TIM barrel in PK is an example of a domain formed by two noncontinuous segments of the polypeptide chain. The intervening segment forms a second domain, and the C-terminal part of the protein forms the third domain. The DNA-binding protein CAP (catabolite activator protein) is dimeric, and

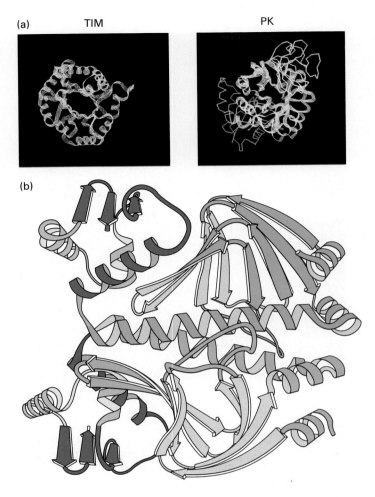

(a) TIM　　PK

(b)

◀ **Figure 2-10** (a) Comparison of triosephosphate isomerase (TIM) and pyruvate kinase (PK). TIM, shown in a ribbon diagram, has eight α helices on the outside and eight β sheet strands in the center; this structural motif is called an αβ barrel. The core domain of PK is also an αβ barrel (green), clearly evolved from TIM or a common ancestral protein. It is evident that PK gained two protein domains (purple) that are hung from the αβ barrel. This sequence shows evolution at work at the molecular level (the genes for TIM and PK also reflect this evolution). (b) Catabolite activator protein (CAP). Each subunit of this dimeric protein folds into two domains—an N-terminal domain (blue and green) that binds cAMP and a C-terminal domain (red) that binds to DNA. The helix-turn-helix motifs within the DNA-binding domain are shown in orange. *Part (a) courtesy of Bruce Tidor; part (b) courtesy of Jane Richardson.*

oil drops, but much less fluid. Exterior residues, by contrast, include many hydrophilic amino acids, which facilitate solubility of the protein in water. The sites of multiple interactions between polypeptides or domains are often hydrophobic patches on an exterior face of the folded polypeptide.

Regions of Similar Architecture Often Have Similar Sequences

Because x-ray crystallography is a lengthy procedure and many proteins are difficult to crystallize, it is much easier to determine the primary structure of a protein than it is to determine its architecture. Thus any aid to predicting the form of a protein from its linear organization is welcome. We might imagine that a computer program could contain rules for folding the linear structure into its three-dimensional form, but the number of possible interactions both within the protein and between it and its water solvent is so great that no accurate computational prediction is yet possible.

The greatest aid to predicting structure is sequence similarity. Two proteins with common architectural features often have related amino acid sequences. A molecule of hemoglobin, for example, consists of two copies each of two types of chains: α and β. The α and β chains fold in similar ways, primarily using α-helical segments and turns. Comparison of the linear structures of the human α and β chains shows their similarities (Figure 2-11). Identical amino acids are found at 64 of the 139 comparable sites. At many other sites, the differences are *conservative;* different amino acids are chemically similar (aspartic acid and glutamic acid, arginine and lysine, serine and threonine, leucine and isoleucine or valine). Although conservative amino acid replacements may be benign in some locations, only one amino acid is allowed at a given position in some parts of proteins. This can be true even in the interior of domains: if the "oil drop" requires close packing, the loss or gain of even one methyl group can destabilize the structure.

each subunit contains two domains. The activity of CAP is regulated by a small molecule, cyclic AMP (cAMP). One domain of CAP binds cAMP: it has a folded structure composed of eight β strands in two sheets, forming a barrel-like arrangement, and a long α helix, which makes the principal contacts with the other subunit in the dimer. The other domain, which binds to DNA, contains the helix-turn-helix motif. Each bound cAMP contacts both subunits, and binding is believed to influence their relative orientation. This change in quaternary structure is probably responsible for determining whether the protein can bind specifically to DNA. Antibodies provide another excellent example of the way distinct functions are performed by distinct domains in a molecule; antibodies are introduced later in this chapter and discussed in detail in Chapter 25.

Because a protein is a three-dimensional object with a filled center, it is particularly difficult to represent completely in illustrations. Figure 2-10 is a useful compromise between space-filling models, which can show only the protein surface, and stick models, which are difficult to decipher. But it should be emphasized that the center of a protein consists of closely packed atoms, mainly from hydrophobic amino acids; thus protein interiors are like

(a)

N-termini											
α	Val (1)		Leu	Ser	Pro	Ala (5)	Asp	Lys			
β	Val (1)	His	Leu	Thr	Pro	Glu (6)	Glu	Lys			
α Thr	Asn	Val (10)	Lys	Ala	Ala	Trp	Gly	Lys	Val	Gly	
β Ser	Ala	Val (10)	Thr	Ala	Leu	Trp	Gly	Lys	Val	Asn	
α Ala	His (20)	Ala	Gly	Glu	Tyr	Gly	Ala	Glu	Ala	Leu	
β	Val (20)	Asp	Glu	Val	Gly	Gly	Glu	Ala	Leu		
α Glu	Arg	Met	Phe	Leu	Ser	Phe	Pro	Thr	Thr	Lys (40)	
β Gly	Arg (30)	Leu	Leu	Val	Val	Tyr	Pro	Trp	Thr	Gln	
α Thr	Tyr	Phe	Pro	His	Phe		Asp	Leu	Ser	His (50)	
β Arg (40)	Phe	Phe	Glu	Ser	Phe	Gly	Asp	Leu	Ser	Thr (50)	
α Gly	Ser	Ala					Gln	Val	Lys		
β Pro	Asp	Ala	Val	Met	Gly	Asn	Pro	Lys	Val (60)	Lys	
α Gly	His	Gly	Lys (60)	Lys	Val	Ala	Asp	Ala	Leu	Thr	
β Ala	His	Gly	Lys	Lys	Val	Leu	Gly	Ala (70)	Phe	Ser	
α Asn	Ala	Val (70)	Ala	His	Val	Asp	Asp	Met	Pro	Asn	
β Asp	Gly	Leu	Ala	His	Leu	Asp	Asn (80)	Leu	Lys	Gly	
α Ala	Leu (80)	Ser	Ala	Leu	Ser	Asp	Leu	His	Ala	His	
β Thr	Phe	Ala	Thr	Leu	Ser	Glu (90)	Leu	His	Cys	Asp	
α Lys (90)	Leu	Arg	Val	Asp	Pro	Val	Asn	Phe	Lys	Leu (100)	
β Lys	Leu	His	Val	Asp	Pro (100)	Glu	Asn	Phe	Arg	Leu	
α Leu	Ser	His	Cys	Leu	Leu	Val	Thr	Leu	Ala (110)	Ala	
β Leu	Gly	Asn	Val	Leu (110)	Val	Cys	Val	Leu	Ala	His	
α His	Leu	Pro	Ala	Glu	Phe	Thr	Pro	Ala (120)	Val	His	
β His	Phe	Gly	Lys (120)	Glu	Phe	Thr	Pro	Pro	Val	Gln	
α Ala	Ser	Leu	Asp	Lys	Phe	Leu	Ala (130)	Ser	Val	Ser	
β Ala	Ala	Tyr (130)	Gln	Lys	Val	Val	Ala	Gly	Val	Ala	
α Thr	Val	Leu	Thr	Ser	Lys	Tyr (140)	Arg				
β Asp	Ala (140)	Leu	Ala	His	Lys	Tyr	His				

C-termini

(b)

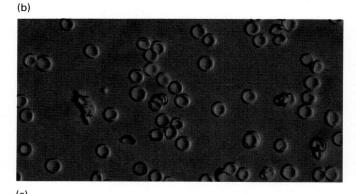

(c)

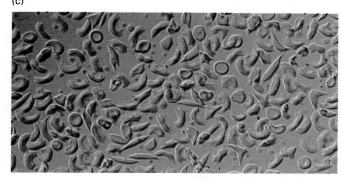

▲ **Figure 2-11** (a) Sequences of amino acids in the α and β chains of normal adult human hemoglobin. The α chain contains 141 amino acids; the β chain, 146. The sequences are aligned to show the regions of identity between the chains. The gaps at a few points presumably represent insertions or deletions of DNA sequences during the evolution of the α and β globin genes from a common ancestor. Sickle-cell hemoglobin differs from normal hemoglobin only by having a valine instead of a glutamate at position 6 of one or both of the β chains. Comparison of normal red blood cells (b) to those from a patient with sickle-cell disease (c) shows the effect of the mutation in hemoglobin on the structure of the cells. *Part (a) courtesy of M. Murayama, Murayama Research Labs/Biological Photo Service. Part (b) courtesy of A. Marmont and E. Damasio, Division of Hematology, S. Martino's Hospital, Genoa, Italy.*

If regions of common form have common sequence, do regions of similar sequence have similar structure? This relation also has been borne out; it is the most powerful way to predict protein structure. Again, antibody molecules are a potent example: similar sequences appear in a vast number of independently produced molecules, indicating that all antibodies have the same general structural organization; small differences give individual character to each type of antibody. In fact, the antibody form is adopted by many other proteins. In general, once a single type of protein structure has evolved, proteins evolve from it as slight variations of the original. Nature, like man, finds that reinventing the wheel is not an economic form of progress and chooses to adapt previous forms to new uses rather than to evolve them de novo. Those interested in protein structure reap the benefit of nature's conservatism, which in favorable cases allows the prediction of three-dimensional form from linear structure.

Many Proteins Contain Tightly Bound Prosthetic Groups

One influence on the shape of a folded polypeptide is the presence or absence of a *prosthetic group*: a small molecule that is not a peptide but that tightly binds to the protein and plays a crucial role in its function. For example, each of the four chains of hemoglobin binds and enfolds a prosthetic group called *heme*, which actually carries the oxygen (see Figure 2-5). Prosthetic groups can be linked to proteins noncovalently, as in hemoglobin, or covalently. Not all proteins have prosthetic groups.

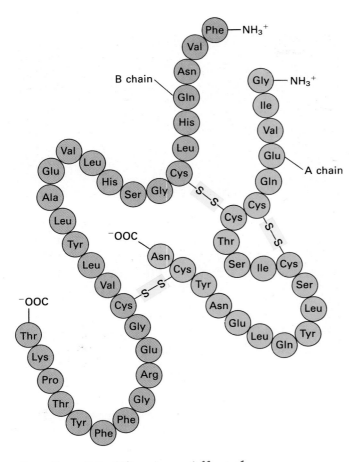

Disulfide bridges often play an important role in the structures of *extracellular proteins* (proteins that function outside cells). The hormone insulin, for example, is composed of two polypeptides: A and B chains containing 21 and 30 residues, respectively. One disulfide bridge connects two cysteine residues of the A chain; two others link the A and B chains together (Figure 2-12).

Intracellular proteins generally have few disulfide bridges, because small molecules containing —SH groups are highly concentrated inside cells. The tripeptide glutathione—the most important such molecule—reduces cysteine-cysteine disulfide bridges in intracellular proteins. If we let G stand for the remainder of the glutathioine molecule, the reaction is

$$2 \text{ G—SH} + \text{—Cys—S—S—} \longrightarrow \text{G—S—S—G} + 2 \text{ —Cys—SH}$$

The oxidized glutathione molecules linked in pairs by disulfide bonds (G—S—S—G) are reduced in turn to free molecules (G—SH) in a reaction catalyzed by the enzyme glutathione reductase.

Proteins undergo other forms of covalent modification, the most common being the removal of the N-terminal methionine that initiates the synthesis of all proteins (Chapter 3). Also, a protein made as a precursor polypeptide in many cases is then cleaved to form the functional protein. The single polypeptide chain preproinsulin, for example, undergoes several successive cleavages to produce insulin (Figure 2-13).

Due to the nature of protein synthesis, only 20 protein side chains are initially available, but many proteins can make use of more than that number. A vast number of stable protein modifications can be made, including hydroxylation of proline and lysine, mainly in collagen; acetylation of the amino group of lysine; addition of various groups at the N-terminus of proteins and methylation of histidine, mainly in actin (Figure 2-14). Carbohydrate groups are attached to asparagine, serine, or threonine side chains in many secreted or cell-surface proteins. Some proteins have attached lipids.

Many protein functions are controlled by reversible side-chain modifications. A key one is the substitution of phosphate groups for hydroxyl groups on serine, threonine, or tyrosine:

Phosphoserine

The activity of many enzymes is regulated by their state of *phosphorylation*.

Covalent Modifications Affect the Structures and Functions of Proteins

An important determinant of the shapes of proteins is the covalent *disulfide bridge* between two cysteine residues in the same polypeptide chain or in different chains. Such —S—S— bonds stabilize folded conformations of proteins. Chemically, the disulfide bridge is formed by an oxidation reaction in which two hydrogen atoms (two protons and two electrons) are lost:

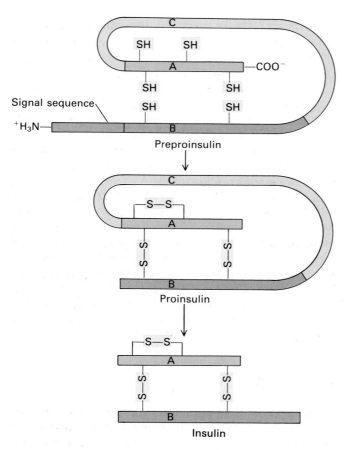

▲ **Figure 2-13** The processing of preproinsulin into proinsulin and of proinsulin into insulin. The cleavage of preproinsulin, which occurs immediately after the synthesis of its chain of 108 amino acids is complete, removes 24 amino acids (collectively termed the *signal sequence*) from the amino end of the molecule. The remaining 84 amino acids constitute proinsulin—a molecule in which all of the correct disulfide bridges have been formed. While the hormone is being packaged for secretion, the 33 residues (the C chain) between the A and the B chains are removed; what remains is insulin. *After L. Stryer, 1988, Biochemistry, 3d ed., W. H. Freeman and Company, p. 995.*

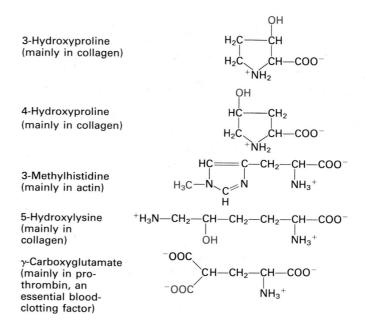

▲ **Figure 2-14** The structures of some modified amino acids found in specific proteins. In each case, the substituted group (red) has replaced a hydrogen atom.

A number of treatments can cause polypeptides to unfold by disrupting the weak bonds that hold a protein together. When this happens, the protein is said to be *denatured*. If the native protein is heated, for example, thermal energy can break the weak bonds. Extremes of pH can alter the charges on amino acid side chains, disrupting ionic and hydrogen bonds. Reagents such as 8 *M* urea

$$\underset{H_2N-\overset{\displaystyle O}{\overset{\displaystyle \|}{C}}-NH_2}{}$$

disrupt both hydrogen and hydrophobic bonds. Most denatured proteins precipitate: attractions between hydrophobic groups that would normally be buried inside the molecules clump them together.

The Native Conformation of a Protein Can Be Denatured by Heat or Chemicals

Any polypeptide chain could, in principle, be folded into a virtually infinite number of conformations. In general, however, all molecules of any protein species adopt the same conformation, called the *native state* of the molecule. The most stable conformation of a folded protein molecule is that of lowest free energy achieved through noncovalent bonds between the amino acids and any prosthetic groups and between the protein and its environment. This condition is usually the native state.

Many Denatured Proteins Can Renature into Their Native State

Many proteins that have been denatured by other techniques can be completely unfolded and dissolved in 8 *M* urea. If the urea is then slowly removed by dialysis, or diluted away, the denatured proteins often *renature* (refold) into their native state, reforming the same weak bonds. In this way, the proteins can be carried through a denaturation-renaturation cycle that first destroys and then reestablishes their original structure and function (Figure 2-15).

In general, renaturation is most complete if a protein is made up of a single polypeptide chain and lacks a noncovalently bound prosthetic group. For example, staphylococcal nuclease—a bacterial enzyme of 149 residues that degrades DNA and RNA—is totally denatured in acid but renatures to its native conformation within 0.1 s after the solution is neutralized. The three-dimensional architecture of this protein is solely a consequence of interactions among its amino acids and with its aqueous environment. In such cases, the genetic program of the cell must only define the primary structure of the protein— the amino acid sequence—and the tertiary structure is assured. With care, most proteins can be carried through a denaturation-renaturation cycle. Thus it is generally true that linear structure determines three-dimensional architecture.

The native form of some proteins is not the conformation with the lowest free energy and consequently cannot be completely restored on renaturation. This is particu

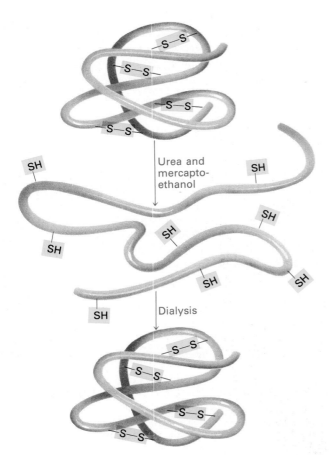

▲ **Figure 2-15** Denaturation and renaturation of a protein. Most polypeptides can be completely unfolded by treatment with an 8 *M* urea solution containing mercaptoethanol ($HSCH_2CH_2OH$). The urea breaks intramolecular hydrogen and hydrophobic bonds, and the mercaptoethanol reduces each disulfide bridge to two —SH groups. When these chemicals are removed by dialysis, the —SH groups on the unfolded chain oxidize spontaneously to re-form disulfide bridges, and the polypeptide chain simultaneously refolds into its native configuration.

larly true of multichain proteins. The two chains of insulin, for example, can be separated by a combination of reducing agents (to break the disulfide bridges) and concentrated solutions of such chemicals as urea (to disrupt hydrogen and hydrophobic bonds). When the insulin renatures in the presence of oxidizing agents that promote the formation of disulfide bridges, a number of stable multichain aggregates do form, but *native* insulin molecules make up only a minor proportion of them. In the others, the re-formed disulfide bridges connect inappropriate parts of the chain.

Insulin is formed by the partial proteolysis (breaking down) of proinsulin, its larger precursor (see Figure 2-13). Denatured proinsulin, as opposed to the denatured two-chain form of insulin, can renature to form the native structure of proinsulin with a high efficiency. Presumably, within the cell, either proinsulin or preproinsulin folds in such a way that the correct disulfide bridges form at the lowest free energy. The cell utilizes these intermediate stages to form insulin, whose stable conformation is not the one of lowest free energy.

Enzymes

Protein catalysts called *enzymes* are mediators of the dynamic events of life; almost every chemical reaction in a cell is catalyzed by an enzyme. Like other catalysts, enzymes increase the rates of reactions that are already energetically favorable; more precisely, enzymes increase the rates of forward and reverse reactions by the same factor. The name of an enzyme usually indicates its function: the suffix *-ase* is commonly appended to the name of the type of molecule on which the enzyme acts. Thus proteases degrade proteins, phosphatases remove phosphate residues, and ribonuclease cleaves RNA molecules.

The chemicals that undergo a change in a reaction catalyzed by an enzyme are the *substrates* of that enzyme. Because little free energy may be liberated in either direction in reversible reactions, the distinction between chemicals that are substrates and those that are products is often arbitrary.

Most enzymes are found inside cells, but a number are secreted by cells and function in the blood, the digestive tract, or other extracellular spaces. In microbial species, some enzymes function outside the organism. The number of different types of chemical reactions in any one cell is very large: an animal cell, for example, normally contains 1000–4000 different types of enzymes, each of which catalyzes a single chemical reaction or set of closely related reactions. Certain enzymes are found in the majority of cells because they catalyze common cellular reactions—the synthesis of proteins, nucleic acids, and phospholipids and the conversion of glucose and oxygen into carbon dioxide and water, which produces most of the chemical energy used in animal cells. Other enzymes are found only in a particular type of cell within an organism, such as a liver cell or a nerve cell, because they carry out some chemical reaction unique to that cell. Also, many mature cells, including erythrocytes (red blood cells) and epidermal (skin) cells, may no longer be capable of making proteins or nucleic acids yet these cells still contain specific sets of enzymes that they synthesized at an earlier stage of differentiation.

Certain Amino Acids in Enzymes Bind Substrates: Others Catalyze Reactions on the Bound Substrates

Two striking properties characterize all enzymes: their enormous *catalytic power* and their *specificity*. Quite often, the rate of an enzymatically catalyzed reaction is 10^6-10^{12} times that of an uncatalyzed reaction under otherwise similar conditions. The specificity of an enzyme is determined by the different rates at which it catalyzes closely similar chemical reactions or by its ability to distinguish between closely similar substrates.

Certain amino acid side chains of an enzyme are important in determining its specificity and its ability to accelerate the reaction rate. The properties of an enzyme are thus functions of its linear arrangement of amino acids and of the appropriate foldings of the peptide chain. Enzyme molecules have two important regions, or sites: one that recognizes and binds the substrate(s), and one that catalyzes the reaction once the substrate(s) have been bound. The amino acids in each of these key regions do not need to be adjacent in the linear polypeptide; they are brought into proximity in the folded molecule. In some enzymes, the catalytic site is part of the substrate-binding site. These two regions are called, collectively, the *active site*.

The binding of a substrate to an enzyme usually involves the formation of multiple noncovalent ionic, hydrogen, and hydrophobic bonds and van der Waals interactions (Figure 2-16). The array of chemical groups in the active site of the enzyme is precisely arranged so that the specific substrate can be more tightly bound than any other molecule (with the exception of some enzyme inhibitors) and the reaction can occur readily. In catalysis, covalent bonds between the enzyme and the substrate may be formed (and then broken) to reduce the activation energy for the reaction.

Trypsin and Chymotrypsin Are Well-characterized Proteolytic Enzymes

The proteolytic (protein-digesting) enzymes trypsin and chymotrypsin are synthesized in the pancreas and secreted into the small intestine as inactive precursors, or *zymogens*, called trypsinogen and chymotrypsinogen, respectively. These zymogens are not activated until they reach the small intestine where they hydrolyze peptide

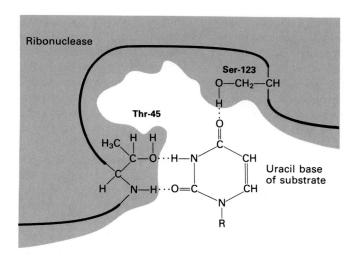

▲ **Figure 2-16** The specific binding of a substrate to an enzyme involves the formation of multiple noncovalent bonds. Here, two amino acid residues of the enzyme ribonuclease bind uracil, part of its substrate, by three hydrogen bonds. Substrates without the two C=O groups and one N—H group in the appropriate positions would be unable to bind or would bind less tightly. Other regions of the enzyme, not depicted here, bind other parts of the RNA substrate by hydrogen bonds and van der Waals interactions.

bonds of ingested proteins—a step in their digestion to single amino acids (Figure 2-17). The delay in activation serves an important regulatory purpose: it prevents the enzyme from digesting the pancreatic tissue in which it was made. Two irreversible proteolytic cleavages activate chymotrypsin. One cleavage removes serine 14 (the serine at position 14) and arginine 15 from chymotrypsinogen; the other removes threonine 147 and asparagine 148

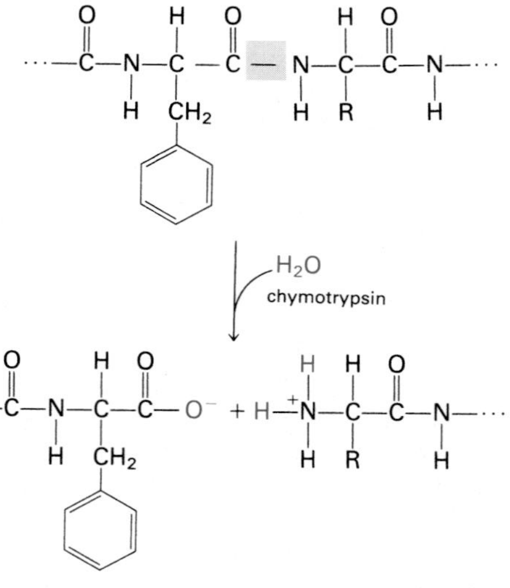

▲ **Figure 2-17** The hydrolysis of a peptide bond by chymotrypsin.

The hydrolysis reaction proceeds in two main steps. First, the peptide bond is broken and the carboxyl group is transferred to the hydroxyl residue of serine 195:

$$\text{Enz—(Ser-195)—OH} + \text{R}_1\text{—NH—}\overset{\displaystyle O}{\overset{\|}{C}}\text{—R}_2 \longrightarrow$$
Enzyme   **Substrate**

$$\text{Enz—(Ser-195)—O—}\overset{\displaystyle O}{\overset{\|}{C}}\text{—R}_2 + \text{R}_1\text{—NH}_2$$
Acylenzyme

Second, this *acylenzyme* intermediate is hydrolyzed:

$$\text{Enz—(Ser-195)—O—}\overset{\displaystyle O}{\overset{\|}{C}}\text{—R}_2 + \text{H}_2\text{O} \longrightarrow$$

$$\text{Enz—(Ser-195)—OH} + {}^-\text{O—}\overset{\displaystyle O}{\overset{\|}{C}}\text{—R}_2 + \text{H}^+$$

Note that the second step restores the enzyme to its original state.

Aspartate 102 and histidine 57 facilitate the acylation reaction by removing the proton from serine 195 and adding it to the nitrogen of the departing amino group (Figure 2-20). In a similar manner, aspartate 102 and histidine 57 facilitate the hydrolysis of the acylenzyme. These enzymatically catalyzed steps—transfer of a proton from the enzyme to the substrate, formation of a covalent acylserine intermediate, and hydrolysis of the acylenzyme—all drastically reduce the overall activation energy of the proteolysis reaction.

The hydroxyl group on serine 195 is unusually reactive. The concept of an "active" serine residue at the active site predated the determination of the crystal structure of chymotrypsin. It was already known, for example, that the compound diisopropylfluorophosphate is a potent inhibitor of chymotrypsin; it reacts only with the hydroxyl on serine 195 to form a stable covalent compound that irreversibly inactivates the enzyme:

$$\text{Enz—(Ser-195)—OH} + \text{F—}\overset{\displaystyle O}{\underset{\displaystyle \underset{\text{HC(CH}_3)_2}{|}}{\overset{\|}{P}}}\text{OCH(CH}_3)_2 \longrightarrow$$
Diisopropylfluorophosphate

$$\text{Enz—(Ser-195)—O—}\overset{\displaystyle O}{\underset{\displaystyle \underset{\text{HC(CH}_3)_2}{|}}{\overset{\|}{P}}}\text{OCH(CH}_3)_2 + \text{HF}$$

Trypsin and Chymotrypsin Have Different Substrate-binding Sites A comparison of trypsin and chymotrypsin will emphasize the nature of the specificity of enzymatically catalyzed reactions. About 40 percent of the amino acids in these two molecules are the same; in

particular, the amino acid sequences in the vicinity of the key serine residue are identical:

$$\overset{\displaystyle 195}{\text{—Gly—Asp—Ser—Gly—Gly—Pro}}$$

The three-dimensional structures and catalytic mechanisms of these two enzymes are also quite similar, indicating that they evolved from a common polypeptide. The major difference between trypsin and chymotrypsin is found in the side chains of the amino acids that line the substrate-binding site. The negatively charged amino acids in this area of the trypsin molecule facilitate the binding of only positively charged (lysine or arginine) residues, instead of hydrophobic ones.

Other Hydrolytic Enzymes Contain Active Serine Other, mostly unrelated, hydrolytic enzymes also contain an active serine residue that is essential for catalysis. For example, acetylcholinesterase catalyzes the hydrolysis of the neurotransmitter acetylcholine to acetate and choline:

$$\text{H}_3\text{C—}\overset{\displaystyle O}{\overset{\|}{C}}\text{—O—CH}_2\text{—CH}_2\text{—}\overset{+}{\text{N}}\text{(CH}_3)_3 + \text{H}_2\text{O} \longrightarrow$$

$$\text{H}_3\text{C—}\overset{\displaystyle O}{\overset{\|}{C}}\text{—O}^- + \text{HO—CH}_2\text{—CH}_2\text{—}\overset{+}{\text{N}}\text{(CH}_3)_3 + \text{H}^+$$

Diisopropylfluorophosphate is a potent, irreversible inhibitor of acetylcholinesterase as well as of chymotrypsin. The compound is lethal to animals because it blocks nerve transmission by causing a buildup of the transmitter substance. (The action of this transmitter is discussed in Chapter 20.)

Coenzymes Are Essential for Certain Enzymatically Catalyzed Reactions

Many enzymes contain a *coenzyme*—a tightly bound small molecule or prosthetic group essential to enzymatic activity. Vitamins required in trace amounts in the diet are often converted to coenzymes. Coenzyme A, for instance, is derived from the vitamin pantothenic acid; the coenzyme pyridoxal phosphate is derived from vitamin B_6. To cite just one example of how coenzymes function, we consider pyridoxal phosphate. The aldehyde group

$$\overset{\displaystyle O}{\overset{\|}{\text{—C—H}}}$$

can form a covalent complex called a *Schiff base* with an —NH$_2$ group of an amino acid, which facilitates or lowers the activation energy for the breaking of bonds to the carbon of the amino acid. Figure 2-21 shows how pyridoxal phosphate catalyzes the decarboxylation of histidine to form histamine—a potent dilator of small blood vessels. Histamine is released by certain cells in the course of allergenic hypersensitivity.

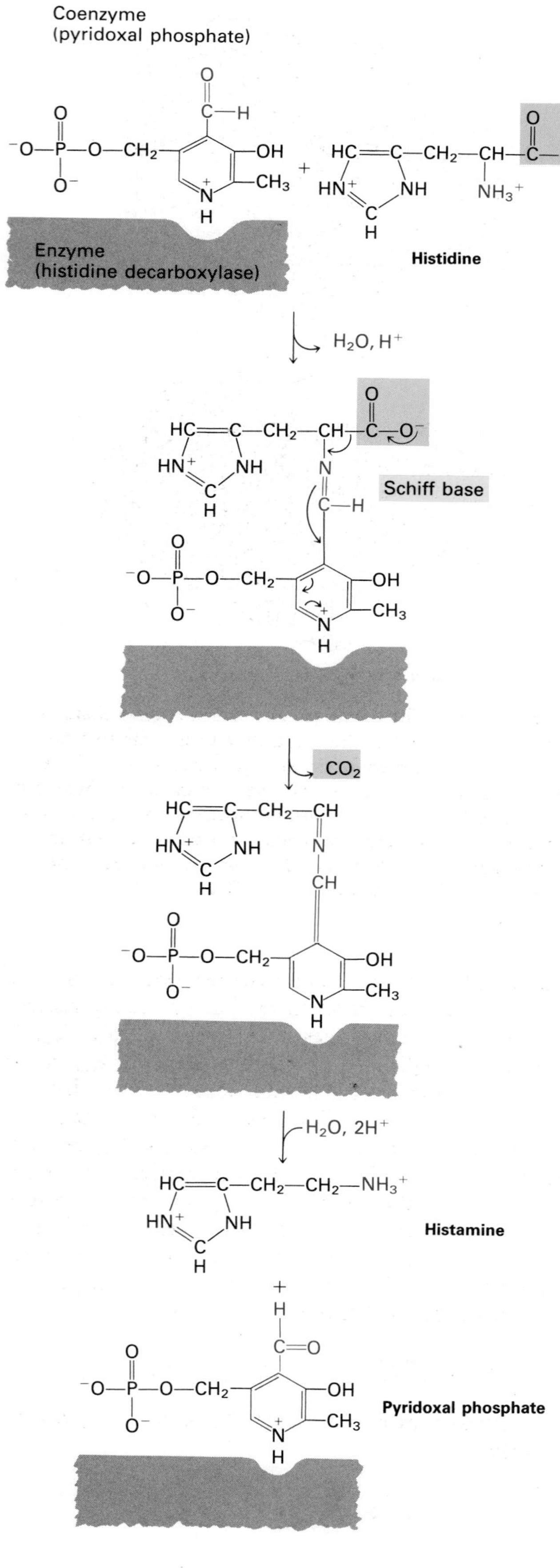

Coenzyme
(pyridoxal phosphate)

Enzyme
(histidine decarboxylase)

Histidine

H₂O, H⁺

Schiff base

CO₂

H₂O, 2H⁺

Histamine

Pyridoxal phosphate

◀ **Figure 2-21** Pyridoxal phosphate, a coenzyme, participates in many reactions involving amino acids. When it is bound to histidine decarboxylase, as in this example, it forms a Schiff base with the α amino group of histidine. The positive charge on the nitrogen of pyridoxal phosphate then attracts the electrons from the carboxylate group of the histidine, via a charge relay system. This weakens the bond between the α carbon of the histidine and the carboxylate group, causing the release of CO_2. Finally, histamine, the reaction product, is hydrolyzed from the pyridoxal complex.

Substrate Binding May Induce a Conformational Change in the Enzyme

When a substrate binds to an enzyme, molecules of complementary charge or shape, or both, may simply fit together into a complex stabilized by a variety of noncovalent bonds. Such an interaction resembles the fitting of a key into a lock and is said to occur by a *lock-and-key* mechanism (Figure 2-22a).

In some enzymes, the binding of the substrate induces a conformational change in the enzyme that causes the catalytic residues to become positioned correctly. Molecules that attach to the substrate-binding site, or *recognition site,* of the enzyme but that do not induce a conformational change are not substrates of that enzyme. Thus an enzyme differentiates between a substrate and a nonsubstrate in two ways: Does the potential substrate bind to the enzyme? If so, does it induce the correct conformational change? When both criteria are met, the enzyme-substrate complex is said to demonstrate *induced fit* (Figure 2-22b).

An important example of induced fit is provided by the enzyme hexokinase, which catalyzes the transfer of a phosphate residue from ATP to a specific carbon atom of glucose:

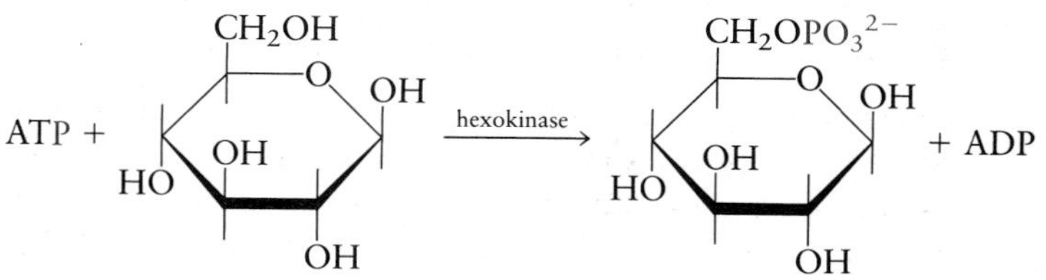

Glucose   Glucose-6-phosphate

This is the first step in the degradation of glucose by cells. X-ray crystallography has shown that hexokinase consists of two domains. The binding of glucose induces a major conformational change that brings these domains closer together and creates a functional catalytic site (Figure 2-23). Only glucose and closely related molecules can induce this conformational change, ensuring that the enzyme is used to phosphorylate only the correct substrates. Molecules such as glycerol, ribose, and even water may bind to the enzyme at the recognition site but cannot induce the requisite conformational change, so they are not substrates for the enzyme.

(a) LOCK-AND-KEY

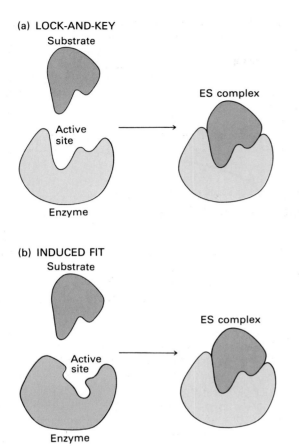

▲ **Figure 2-22** Two mechanisms for the interaction of an enzyme and a substrate. (a) In the lock-and-key mechanism, the substrate fits directly into the binding site of the enzyme. (b) If binding occurs by induced fit, the substrate induces a conformational change in the enzyme that appropriately positions the substrate for catalysis.

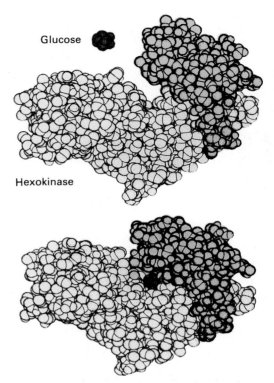

Glucose–hexokinase complex

▲ **Figure 2-23** The conformation of hexokinase changes markedly when it binds the substrate glucose: the two domains of the enzyme come closer together to surround the substrate. Molecules such as the five-carbon sugar ribose can also bind to hexokinase by forming specific hydrogen bonds with groups in the substrate-binding pocket of the enzyme, but only glucose can form all of the bonds that cause the enzyme to change its conformation. *Courtesy of Dr. Thomas A. Steitz.*

The Catalytic Activity of an Enzyme Can Be Characterized by a Few Numbers

Enzymatic specificity is usually quantified by discrimination ratios: a good substrate may be cleaved 10,000 times as fast as a poor substrate. The catalytic power of an enzyme on a given substrate involves two numbers: K_m, which measures the affinity of the enzyme for its substrate, and V_{max}, which measures the maximal velocity of enzymatic catalysis. Equations for K_m and V_{max} are most easily derived by considering the simple reaction

$$S \longrightarrow P$$
$$(\text{substrate} \longrightarrow \text{product})$$

in which the rate of product formation depends on [S], the concentration of the substrate, and on [E], the concentration of the catalytic enzyme. For an enzyme with a single catalytic site, Figure 2-24(a) shows how $d[P]/dt$, the rate of product production, depends on [S] when [E] is kept constant.

At low concentrations of S, the reaction rate is proportional to [S]; as [S] is increased the rate does not increase indefinitely in proportion to [S] but eventually reaches V_{max}, at which it becomes independent of [S]. V_{max} is proportional to [E] and to a catalytic constant k_{cat} that is an intrinsic property of the individual enzyme; halving [E] reduces the rate at all values of [S] by one-half.

When interpreting curves such as those in Figure 2-24, bear in mind that all enzymatically catalyzed reactions include at least three steps: (1) the binding of the substrate (S) to the enzyme (E) to form an enzyme-substrate complex (ES); (2) the conversion of ES to the enzyme-product complex (EP); and (3) the release of the product (P) from EP, to yield free P:

$$E + S \xrightleftharpoons{\text{binding}} ES \xrightarrow{\text{catalysis}} EP \xrightarrow{\text{release}} P + E$$

In the simplest case, the release of P is so rapid that we can write

$$E + S \underset{k_2}{\overset{k_1}{\rightleftharpoons}} ES \xrightarrow{k_{cat}} E + P$$

The reaction rate $d[P]/dt$ is proportional to the concentration of ES and to the catalytic constant k_{cat} for the given enzyme:

$$\frac{d[P]}{dt} = k_{cat}[ES] \qquad (1)$$

To calculate [ES], we assume the reaction is in a steady state, so that $k_1[E][S]$, the formation rate of [ES], is equal to the rate of its consumption, either by dissociation of uncatalyzed substrate at a rate of $k_2[ES]$ or by catalysis at a rate of $k_{cat}[ES]$:

$$k_1[E][S] = (k_2 + k_{cat})[ES] \qquad (2)$$

If

$$[E]_{tot} = [E] + [ES] \qquad (3)$$

(where $[E]_{tot}$ is the sum of the free and the complexed enzyme, or the total amount of enzyme), then we can combine equations (2) and (3) to obtain

$$[E]_{tot} = [E] + [ES] = \frac{(k_2 + k_{cat})}{k_1[S]}[ES] + [ES]$$

$$= [ES]\left[1 + \left(\frac{k_2 + k_{cat}}{k_1}\right)\left(\frac{1}{[S]}\right)\right]$$

If we define K_m, called the *Michaelis* constant, as

$$\frac{k_2 + k_{cat}}{k_1} \qquad (4)$$

then

$$[ES] = \frac{[E]_{tot}}{1 + K_m/[S]}$$

Thus

$$\frac{d[P]}{dt} = k_{cat}[ES] = k_{cat}[E]_{tot}\frac{1}{1 + K_m/[S]}$$

$$= k_{cat}[E]_{tot}\frac{[S]}{[S] + K_m} \qquad (5)$$

This equation fits the curves shown in Figure 2-24a. V_{max}, which is equal to $k_{cat}[E]_{tot}$, is the maximal rate of product formation if all recognition sites on the enzyme are filled with substrate. K_m is equivalent to the substrate concentration at which the reaction rate is half-maximal. (If $[S] = K_m$, then from equation (5) we calculate the rate of product formation to be $\frac{1}{2}k_{cat}[E]_{tot} = \frac{1}{2}V_{max}$.) For most enzymes, the slowest step is the catalysis of [ES] to [E] + [P]. In these cases, k_{cat} is much less than k_2, so that $K_m = (k_2 + k_{cat})/k_1 \cong k_2/k_1$ is equal to the equilibrium constant for binding S to E. Thus the parameter K_m describes the affinity of an enzyme for its substrate. The smaller the value of K_m, the more avidly the enzyme can bind the substrate from a dilute solution (Figure 2-24b) and the lower the value of [S] needed to reach half-maximal velocity. The concentrations of the various

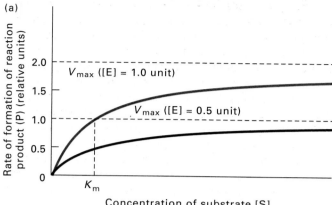

(a)

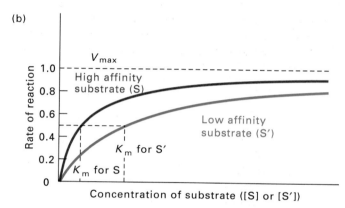

(b)

▲ **Figure 2-24** (a) The rate of a hypothetical enzymatically catalyzed reaction S → P for two different concentrations of enzyme [E] as a function of the concentration of substrate [S]. The substrate concentration that yields a half-maximal reaction rate is denoted by K_m. Doubling the amount of enzyme causes a proportional increase in the rate of the reaction, so that the maximal velocity V_{max} is doubled. The K_m, however, is unaltered. (b) The rates of reactions catalyzed by an enzyme with a substrate S, for which the enzyme has a high affinity, and with a substrate S′, for which the enzyme has a low affinity. The V_{max} value is the same for S and S′, but K_m is higher for S′.

small molecules in a cell vary widely, as do the K_m values for the different enzymes that act on them. Generally, the intracellular concentration of a substrate is approximately the same as or greater than the K_m value of the enzyme to which it binds.

The Actions of Most Enzymes Are Regulated

Many reactions in cells do not occur at a constant rate. Instead, the catalytic activity of the enzymes is *regulated* so that the amount of reaction product is just sufficient to meet the needs of the cell.

An Enzyme Can Be Feedback Inhibited in a Reaction Pathway   Consider a series of reactions leading to the synthesis of the amino acid isoleucine, which is primarily used by cells as a monomer in the synthesis of proteins. The amount of isoleucine needed depends on the rate of protein synthesis in the cell. The first step in the synthesis of isoleucine is the elimination of an amino group, which converts the amino acid threonine to the compound α-ketobutyrate. Threonine deaminase—the enzyme that catalyzes this reaction—plays a key role in regulating the level of isoleucine. In addition to its substrate-binding sites for threonine, threonine deaminase contains a binding site for isoleucine. When isoleucine is bound there, the enzyme molecule undergoes a conformational change, so that it cannot function as efficiently. Thus isoleucine acts as an *inhibitor* of the reaction for the conversion of threonine. If the isoleucine concentration in the cell is high, the binding of isoleucine to the enzyme temporarily reduces the rate of isoleucine synthesis:

α-Ketobutyrate   Isoleucine

This is an example of *feedback inhibition*, whereby an enzyme that catalyzes one of a series of reactions is inhibited by the ultimate product of the pathway.

In isoleucine synthesis, as in most cases of feedback inhibition, the final product in the reaction pathway inhibits the enzyme that catalyzes the first step that does not also lead to other products. The suppression of enzyme function is not permanent. If the concentration of free isoleucine is lowered, bound isoleucine dissociates from the enzyme, which then reverts to its active conformation. The binding of the inhibitor isoleucine to the enzyme and its subsequent release can be described by the equilibrium-binding constant K_i, which is similar to the constant K_m used for substrate binding:

$$[E \cdot Ile]_{inactive} \xrightleftharpoons{K_i} [Ile] + [E]_{active}$$

$$K_i = \frac{[Ile][E]_{active}}{[E \cdot Ile]_{inactive}}$$

Many Enzymes Have Multiple Binding Sites for Regulatory Molecules   Some enzymes have binding sites for small molecules that affect their catalytic activity; a stimulator molecule is called an *activator*. Enzymes may even have multiple sites for recognizing more than one activator or inhibitor. In a sense, enzymes are like microcomputers; they can detect concentrations of a variety of molecules and use that information to vary their own activities. Molecules that bind to enzymes and increase or decrease their activities are called *effectors*. Effectors can modify enzymatic activity because enzymes can assume both active and inactive conformations: activators are positive effectors; inhibitors are negative effectors. Effectors bind at *regulatory sites,* or *allosteric sites* (from the Greek for "another shape"), a term used to emphasize that the regulatory site is an element of the enzyme distinct from the catalytic site and to differentiate this form of regulation from competition between substrates and inhibitors at the catalytic site.

Multimeric Organization Permits Cooperative Interactions among Subunits   Many enzymes and some other proteins are multimeric—that is, they contain several copies, or subunits, of one or more distinct polypeptide chains. Some multimeric enzymes contain identical subunits, each of which has a catalytic site and possibly an effector site. In other enzymes, regulatory sites and catalytic sites are located on different subunits, each with a particular structure. On binding an activator, inhibitor, or substrate, a subunit undergoes a conformational change, usually small, that triggers a change in quaternary structure. This quaternary rearrangement favors a similar conformational change in the other subunits, thereby increasing their affinity for the type of ligand initially bound (Figure 2-25). When several subunits interact cooperatively, a given increase or decrease in substrate or effector concentration causes a larger change in the rate of an enzymatic reaction than would occur if the subunits acted independently. Because of such *cooperative interactions*, a small change in the concentration of an effector or substrate can lead to large changes in catalytic activity.

Cooperative interactions among the four subunits in hemoglobin demonstrate clearly the advantages of multimeric organization. The binding of an O_2 molecule to any one of the four chains (each hemoglobin chain binds one O_2) induces a local conformational change in that subunit. This change can in turn induce a large change in quaternary structure. The quaternary change involves a rearrangement of the positions of the two α and two β chains in the tetramer. The local conformational changes that accompany O_2 binding can then occur more readily in the remaining subunits, increasing their affinity for oxygen. The binding of a second O_2 makes the quaternary structural change even more likely. The cooperative

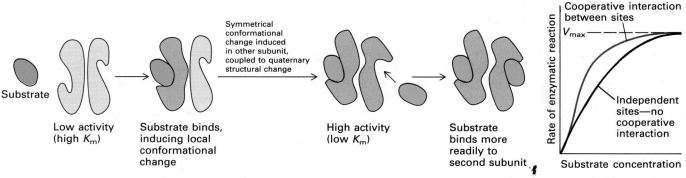

▲ **Figure 2-25** A cooperative interaction between active sites (two identical subunits of a hypothetical enzyme). The binding of a substrate to one subunit of a multimeric enzyme induces a conformational change in the adjacent subunit, which lowers the K_m for the binding of the substrate there. Thus a small change in the substrate concentration can cause a much larger increase in the reaction rate than would occur if there were no cooperative interactions between active sites.

interaction between the chains causes the molecule to take up or lose four O_2 molecules over a much narrower range of oxygen pressures than it would otherwise. As a result, hemoglobin is almost completely oxygenated at the oxygen pressure in the lungs and largely deoxygenated at the oxygen pressure in the tissue capillaries (Figure 2-26).

The contrast between hemoglobin and myoglobin is revealing. Myoglobin is a single-chain oxygen-binding protein found in muscle. The oxygen-binding curve of myoglobin has the characteristics of a simple equilibrium reaction:

$$E + O_2 \underset{K_{O_2}}{\rightleftharpoons} E—K_{O_2}$$

Myoglobin has a greater binding affinity for O_2 (a lower K_{O_2}) than hemoglobin at all oxygen pressures. Thus, at

the oxygen pressure in capillaries, O_2 moves from hemoglobin into the muscle cells, where it binds to myoglobin, ensuring the efficient transfer of O_2 from blood to tissues.

The quaternary-structure rearrangements associated with multimeric organization also provide a way for the effects of activator or inhibitor binding at an allosteric site to be transmitted to a distant catalytic site without large changes in the secondary or tertiary structure of an enzyme, which would be incompatible with the principle that a particular primary structure must adopt a unique folded conformation. Thus, for example, small conformational changes in a domain in response to binding of an effector molecule would produce a quaternary-structure change, which amplifies the conformational signal and allows it to be transmitted robustly to other parts of the enzyme, where it would induce a small conformational change affecting enzymatic activity. Membrane-embedded receptor proteins that must transmit a conformational signal from one side of a membrane to the other are also likely to be multimeric; they transmit the signal by quaternary-structure rearrangement or by an effector-induced shift in the monomer-multimer equilibrium.

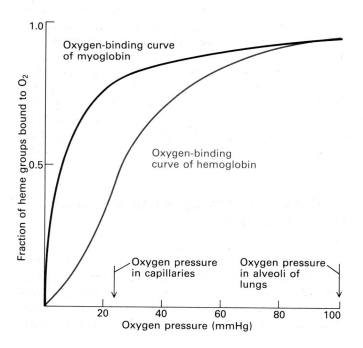

◀ **Figure 2-26** The binding of oxygen to hemoglobin depends on cooperative interactions between the four chains. The graph shows the fraction of heme groups in hemoglobin and in myoglobin bound to O_2 as a function of the oxygen pressure. Note that the binding activity of hemoglobin increases sharply over a narrow range of oxygen pressures (20–40 mmHg). Hemoglobin is saturated with O_2 in the lungs, but it releases much of its bound O_2 at the low oxygen pressure in the tissue capillaries. At any oxygen pressure, myoglobin has a higher affinity for O_2 than hemoglobin does. As myoglobin is a principal muscle protein, this property allows oxygen to be transferred from blood to muscle.

Enzymes Are Regulated in Many Ways   The activities of enzymes are extensively regulated so that the numerous enzymes in a cell work together harmoniously. All metabolic pathways are closely controlled at all times. Synthetic reactions occur when the products of these reactions are needed; degradative reactions occur when molecules must be broken down. Kinetic controls affecting the activities of key enzymes determine which pathways are going to be used and the rates at which they will function.

Regulation of cellular processes involves more than simply turning enzymes on and off, however. Some regulation is accomplished through *compartmentation*. Many enzymes are localized in specific compartments of the cell, such as the mitochondria or lysosomes, thereby restricting the substrates, effectors, and other enzymes with which an enzyme can interact. In addition, compartmentation permits reactions that might otherwise compete with one another in the same solution to occur simultaneously in different parts of a cell. Cellular processes are also regulated through the control of the rates of enzyme synthesis and destruction.

Antibodies

Enzymes are not the only proteins that bind tightly and specifically to smaller compounds. The insulin receptor on the surface of a liver cell, for example, can bind to insulin so tightly that the receptors on a cell are half-saturated when the insulin concentration is only 10^{-9} *M*. This protein does not bind to most other compounds present in blood; it mediates the specific actions of insulin on liver cells. A molecule other than an enzyme substrate that can bind specifically to a macromolecule is often called a *ligand* of that macromolecule.

The capacity of proteins to distinguish among different molecules is developed even more highly in blood proteins called *antibodies,* or *immunoglobulins,* than in enzymes. Animals produce antibodies in response to the invasion of an infectious agent, such as a bacterium or a virus. Antibodies will be discussed at length in Chapter 25. We introduce them here because they will appear as critical reagents in the discussions of many intervening chapters.

The recognition site of an antibody can bind tightly to very specific sites—generally on proteins or carbohydrates—on the surface of the infectious agent. Experimentally, animals produce antibodies in response to the injection of almost any foreign polymer; such antibodies bind specifically and tightly to the invading substance but, like enzymes, do not bind to dissimilar molecules. The antibody acts as a signal for the elimination of infectious agents. When it binds to a bacterium, virus, or virus-infected cell, certain white blood cells (leucocytes) recognize the invading body as foreign and respond by

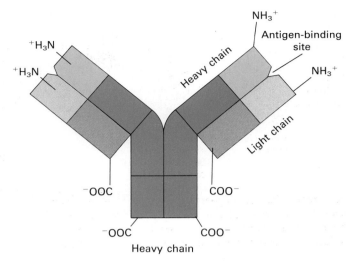

▲ **Figure 2-27**   The structure of an antibody molecule illustrated in an immunoglobulin (IgG) made of four polypeptide chains: two identical heavy chains (blue) and two identical light chains (orange). Each antigen-binding site is formed by the N-terminal segments of a heavy and a light chain. The N-termini are highly variable in sequence, giving rise to the wide range of antibody specificity.

destroying it. The specificity of antibodies is exquisite: they can distinguish between proteins that differ by only a single amino acid and between the cells of different individual members of the same species.

All vertebrates can produce a large variety of antibodies, including ones that bind to chemically synthesized molecules. Exposure to an antibody-producing agent, called an *antigen,* causes an organism to make a large quantity of different antibody proteins, each of which may bind to a slightly different region of the antigen. For a given antigen, these constellations of antibodies may differ from one member of a species to another.

Antibodies are formed from two types of polypeptides: heavy chains, each of which is folded into four domains, and light chains, each of which is folded into two domains (Figure 2-27). The N-terminal domains of both heavy and light chains are highly variable in sequence, giving rise to the specific binding characteristics of antibodies.

Antibodies Can Distinguish among Closely Similar Molecules

The sequence of bovine insulin is identical to that of human insulin, except at three amino acids. Yet when bovine insulin is injected into people, some individuals respond by synthesizing antibodies that specifically recognize the specific amino acids in the bovine molecule, even though human beings generally do not produce antibodies that recognize their own insulin. Injecting mouse albumin—the major serum protein—into mice does not elicit the production of antialbumin antibodies. However, if a small molecule, such as 2,4-dinitrophenol (DNP)

$$O_2N\text{—}\bigcirc\text{—OH}$$
$$NO_2$$

is coupled to the albumin, mice do produce antibodies that bind specifically to the modified region of the protein—in this case, to the dinitrophenyl group. A small group capable of eliciting antibody production is called a *hapten*. The anti-DNP/albumin antibody will not bind to albumin that is not complexed with DNP or that is modified by other haptens (even phenyl groups with different substituents).

Antibodies Are Valuable Tools for Identifying and Purifying Proteins

Because they bind so selectively to proteins, antibodies can be used experimentally to isolate one protein from a complex mixture. In one technique, *affinity chromatography*, a pure antibody is chemically coupled to tiny plastic beads, which are then placed in a small column. When a protein solution is applied, only the protein to which the antibody is directed adheres to the column; all nonadherent proteins pass through the column unimpeded (Figure 2-28a). The adherent protein can then be eluted by adding a solution that disrupts the binding between the protein and the antibody (Figure 2-28b). Similarly, antibodies can be used to detect specific proteins in cells or other biological materials.

Nucleic Acids

Cells receive instructions about which proteins to synthesize and in what quantities from *nucleic acids*—the molecules that store and transmit information in cells. As in many systems of communication, this information is processed in the form of a code. The translation of this code is described in Chapter 3. Here, we examine the chemical structures of the molecules that store the encoded information.

Nucleic Acids Are Linear Polymers of Nucleotides Connected by Phosphodiester Bonds

Cells have two closely related information-carrying molecules: deoxyribonucleic acid (DNA) and ribonucleic acid (RNA). Like proteins, DNA and RNA are linear polymers. However, the number of monomers in a nucleic acid is generally much greater than the number of amino acids in a protein. Cellular RNAs range in length from tens to thousands of units. The number of units in a DNA molecule can be in the millions.

DNA and RNA each consist of only four different monomers, called *nucleotides*. A nucleotide has three parts: a phosphate group, a *pentose* (a five-carbon sugar molecule), and an organic *base* (Figure 2-29). In RNA,

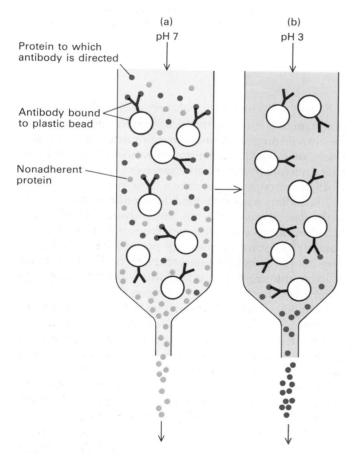

Figure 2-28 The purification of a protein from a mixture by affinity chromatography. (a) In step 1, the mixture is filtered through a column containing antibody molecules that are specific for the desired protein. Only that protein binds to the antibody matrix; any other proteins in the mixture are eluted. (b) In step 2, a solution such as acetic acid is added to disrupt the antigen-antibody complex, so that a pure protein is eluted from the column.

Protein to which antibody is directed

Antibody bound to plastic bead

Nonadherent protein

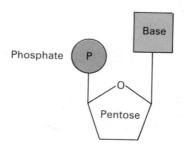

◀ **Figure 2-29** A schematic diagram of the structure of a nucleotide.

Base

Phosphate   P

Pentose

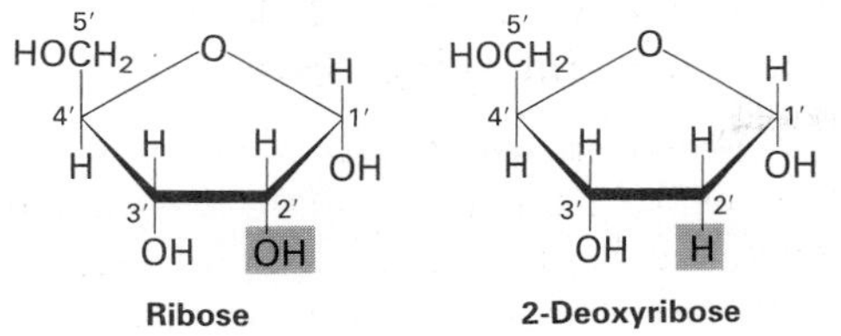

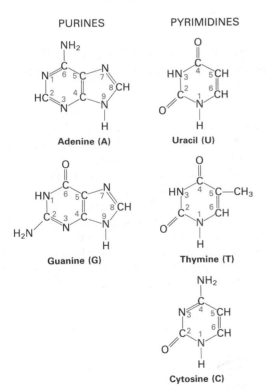

▲ **Figure 2-30** Haworth projections of the structures of ribose and deoxyribose. By convention, the carbon atoms of the pentoses are numbered with primes. In nucleotides and nucleic acids, the 5′ carbon is linked in an ester bond to the phosphate and the 1′ carbon is linked to the base.

the pentose is always *ribose;* in DNA, it is *deoxyribose* (Figure 2-30). The only other difference between the DNA and RNA monomers is found in one of their bases.

The base components of nucleic acids are either *purines* or *pyrimidines* (Figure 2-31). A purine has a pair of fused rings; a pyrimidine has only one ring. Both purines and pyrimidines are heterocyclic: the rings are built of more than one kind of atom—in this case, nitrogen in addition to carbon. The presence of nitrogen atoms gives these molecules their basic character, but none is protonated (ionic) at neutral pH. The acidic character of nucleotides

PURINES

Adenine (A)

Guanine (G)

PYRIMIDINES

Uracil (U)

Thymine (T)

Cytosine (C)

▲ **Figure 2-31** The chemical structures of the principal bases in nucleic acids. In nucleic acids and nucleotides, the 9 nitrogen atom of purines and the 1 nitrogen atom of pyrimidines (red) are bonded to the 1′ carbon of ribose or deoxyribose.

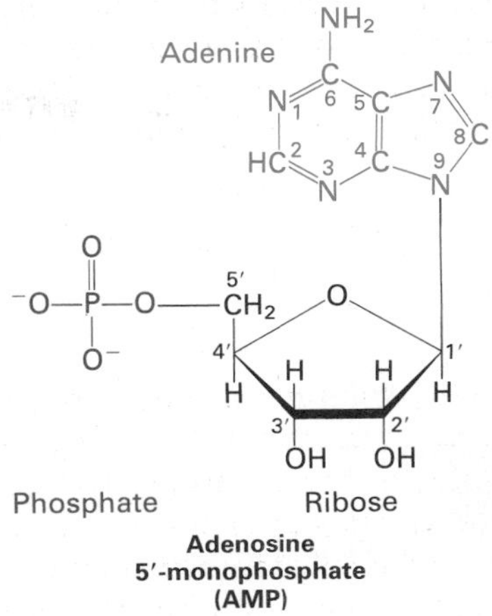

Adenine

Phosphate   Ribose

**Adenosine
5′-monophosphate
(AMP)**

▲ **Figure 2-32** The chemical structures of a typical nucleotide, adenosine 5′-monophosphate. The C—N linkage between the sugar and the base shown here is a β linkage, meaning that the base lies above the plane of the ribose, as conventionally drawn. (In the α linkage, the base would lie below the plane.) Only the β linkage is found in cellular nucleotides.

is due to the presence of phosphate, which dissociates under physiological conditions, freeing hydrogen ions. The bases adenine, guanine, and cytosine (found in both DNA and RNA); thymine (found only in DNA); and uracil (found only in RNA) are often abbreviated A, G, C, T, and U.

The sugar component of a nucleotide is the link between the base and the phosphate group. Whether the sugar is ribose or deoxyribose, its 1′ carbon atom is attached to the 9 nitrogen of a purine or to the 1 nitrogen of a pyrimidine. The hydroxyl group on the 5′ carbon atom is replaced by an ester bond to the phosphate group (Figure 2-32).

Cells contain several types of molecules that differ from nucleotides by lack of an attached phosphate group. A combination of a base and a sugar without a phosphate group is called a *nucleoside.* The four *ribonucleosides* and *deoxyribonucleosides* are named in Table 2-1.

Nucleotides, which have one, two, or three attached phosphate groups, are referred to as *nucleoside phosphates.* The phosphates are generally esterified at the 5′ hydroxyl (see Figure 2-32). *Nucleoside monophosphates* have a single esterified phosphate, *diphosphates* contain a pyrophosphate group

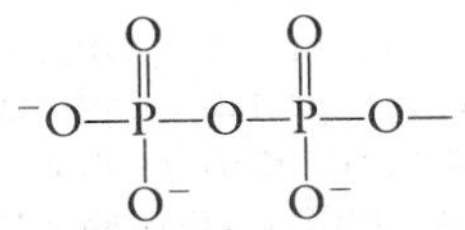

and *triphosphates* have a third phosphate (see Figure 1-13). Table 2-1 names all of the various forms of nucleo-

Table 2-1   Naming nucleosides and nucleotides

| | | Bases | | |
| | | Purines | | Pyrimidines |
		Adenine (A)	Guanine (G)	Cytosine (C)	Uracil (U) (Thymine [T])
Nucleosides	in RNA	Adenosine	Guanosine	Cytidine	Uridine
	in DNA	Deoxyadenosine	Deoxyguanosine	Deoxycytidine	Deoxythymidine
Nucleotides	in RNA	Adenylate	Guanylate	Cytidylate	Uridylate
	in DNA	Deoxyadenylate	Deoxyguanylate	Deoxycytidylate	Thymidylate
Nucleoside monophosphates		AMP	GMP	CMP	UMP
Nucleoside diphosphates		ADP	GDP	CDP	UDP
Nucleoside triphosphates		ATP	GTP	CTP	UTP
Deoxynucleoside mono-, di-, and triphosphates		dAMP, etc.			

side phosphates. A supply of nucleoside triphosphates is necessary for the synthesis of nucleic acids; the triphosphate ATP is the most widely used energy carrier in the cell.

When nucleotides polymerize to form nucleic acids, the hydroxyl group attached to the 3' carbon of a sugar of one nucleotide forms an ester bond to the phosphate of another nucleotide, eliminating a molecule of water:

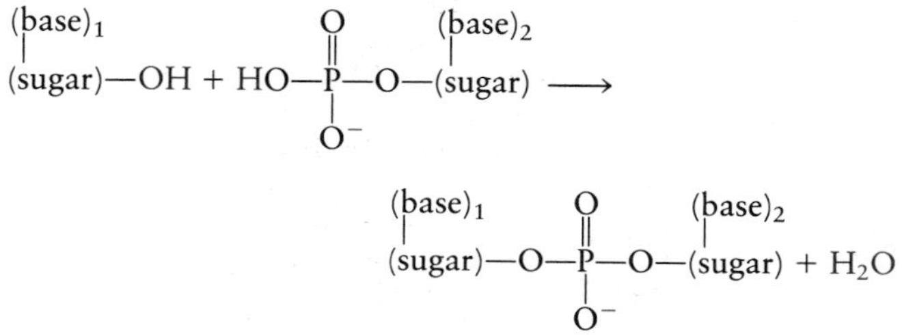

$$\text{(base)}_1 \qquad\qquad O \qquad \text{(base)}_2$$
$$\text{(sugar)}\!-\!OH + HO\!-\!\overset{\|}{\underset{O^-}{P}}\!-\!O\!-\!\text{(sugar)} \longrightarrow$$

$$\text{(base)}_1 \qquad\qquad O \qquad \text{(base)}_2$$
$$\text{(sugar)}\!-\!O\!-\!\overset{\|}{\underset{O^-}{P}}\!-\!O\!-\!\text{(sugar)} + H_2O$$

This condensation is similar to the reaction that forms a peptide bond. Thus a single nucleic acid strand is a phosphate-pentose polymer (a polyester) with purine and pyrimidine bases as side groups. The links between the nucleotides are called *phosphodiester bonds*. Like a polypeptide, a nucleic acid strand has a chemical orientation: the *3' end* has a free hydroxyl group attached to the 3' carbon of a sugar; the *5' end* has a free hydroxyl or phosphate group attached to the 5' carbon of a sugar (Figure 2-33). This directionality has given rise to the convention that polynucleotide sequences are written and read in the 5' → 3' direction (from left to right); for example, the sequence AUG is assumed to be (5')AUG(3'). (Although, strictly speaking, the letters A, G, C, T, and U stand for bases, they are also often used in diagrams to represent the whole nucleotides containing these bases.) The orientation of a nucleic acid strand is an extremely important property of the molecule, as we shall see.

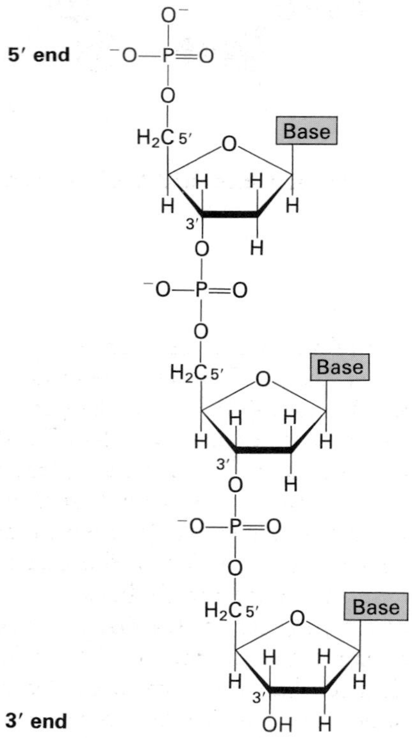

DNA

The modern era of molecular biology began in 1953, when James D. Watson and Francis H. C. Crick elucidated the double-helical structure of DNA by the detailed analysis of x-ray diffraction patterns and careful model building. Both this event and the structure of the double

▲ **Figure 2-33**   The chemical structure of a trinucleotide, a single strand of DNA containing only three nucleotides. The nucleotide at the 3' end has a free 3' deoxyribose hydroxyl group (a hydroxyl that is not bonded to another nucleotide). Similarly, the 5' end has a free 5' hydroxyl or phosphate.

helix itself are quite familiar by now to anyone who reads newspapers and magazines. A closer look at the structure of the "thread of life," as the DNA molecule is sometimes called, will reveal why its discovery was so important.

The Native State of DNA Is a Double Helix of Two Antiparallel Chains with Complementary Nucleotide Sequences

Double-helical DNA consists of two sugar-phosphate strands that wind around each other; the base pairs are stacked in between the strands (Figure 2-34). The orientation of the two strands is antiparallel (their $5' \rightarrow 3'$ directions are opposite); they are held together by hydrogen bonds and hydrophobic interactions. The bases on oppo-

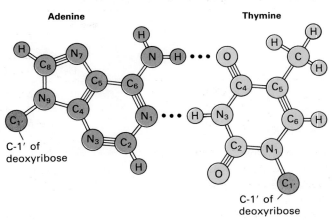

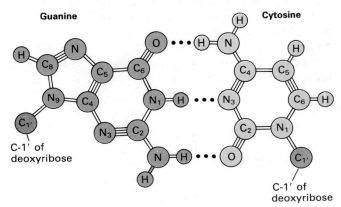

▲ **Figure 2-35**   Ball-and-stick models of A-T and G-C base pairs.

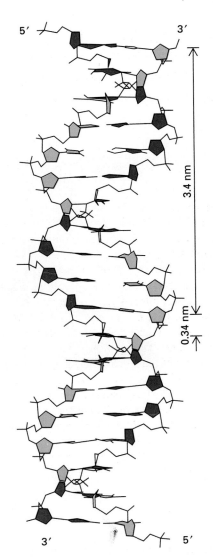

▲ **Figure 2-34**   A skeletal model of double-helical DNA. The opposite strands are different colors, and the bases are seen on edge. *From L. Stryer, 1988,* Biochemistry, *3d ed., W. H. Freeman and Company, p. 77.*

site strands are held in precise register: A is paired with T by two hydrogen bonds; G is paired with C by three hydrogen bonds (Figure 2-35). This *base-pair complementarity* is a consequence of the size, shape, and chemical composition of the bases. The bases are planar and form planar pairs, so that the pairs stack on top of one another. Hydrophobic and van der Waals interactions between adjacent base pairs in the stack contribute significantly to the overall stability of the double helix.

Because of the geometry of the double helix, a purine must always pair with a pyrimidine. The rules are a bit more complicated than this, however. For example, a guanine residue (a purine) could theoretically form hydrogen bonds with a thymine (a pyrimidine), but this would cause a minor distortion in the helix. A similar distortion would allow the formation of a base pair between two pyrimidines, cytosine and thymine. Neither G-T nor C-T base pairs are normally found in DNA.

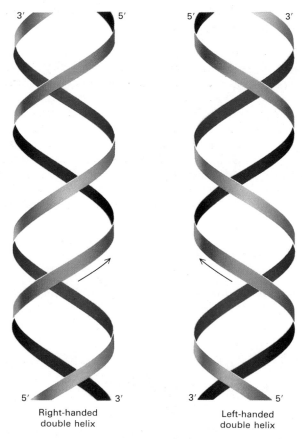

3′   **5′**

5′   **3′**

5′   **3′**

3′   **5′**

Right-handed
double helix

Left-handed
double helix

▲ **Figure 2-36** Two possible helical forms of DNA are mirror images of each other. The geometry of the sugar-phosphate backbone of DNA favors the right-handed double helix. (Right-handed and left-handed are defined by convention.)

Two polynucleotide strands can, in principle, form either a right-handed or a left-handed helix (Figure 2-36), but the geometry of the sugar-phosphate backbone is more compatible with the former and therefore natural DNA is right handed. The x-ray diffraction pattern of DNA indicates that the stacked bases are regularly spaced 0.34 nm apart along the helix axis (see Figure 2-34). The helix makes a complete turn every 3.4 nm; thus there are about 10 pairs per turn. On the outside of the molecule, the spaces between intertwined strands form two helical grooves of different widths (Figure 2-37a). Consequently, part of each base is accessible to molecules from outside the helix. Although the multitude of hydrogen and hydrophobic bonds between the strands provide stability and rigidity, the double helix is somewhat flexible because, unlike the α helix in proteins, it has no hydrogen bonds between successive residues in a strand (Figure 2-37b).

Recent studies have shown that certain short DNA polymers in crystals can adopt an alternative left-handed configuration: the Z-DNA. Whether such left-handed helices occur in nature is a matter of intense current interest;

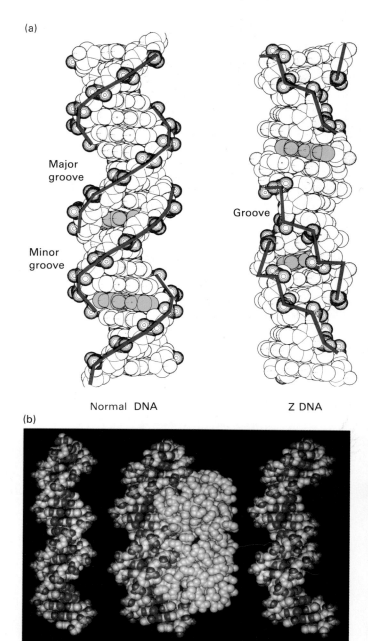

(a)

Major
groove

Minor
groove

Groove

Normal DNA

Z DNA

(b)

▲ **Figure 2-37** (a) Space-filling models of the normal right-handed Watson Crick DNA and the rarer left-handed Z-DNA. The red lines connect the phosphate residues along the chain. In both forms of DNA, the same base pairs are used; they are nearly perpendicular to the axis of the helix (two base pairs in each helix are blue). Normal DNA has two grooves (one major and one minor). Note the irregularity of the Z-DNA backbone; the single groove of Z-DNA is quite deep, extending to the axis of the helix. (b) The bending of DNA by a bound protein. The linear DNA *(left)* can be bound by a protein *(middle)*, which can put a bend in the DNA *(right)*. The inherent flexibility of DNA allows it to be curved by the force of protein binding. *Part (a) courtesy of A. Rich;* part (b) from A. K. Aggarwal, D. W. Rogers, M. Drottar, M. Ptashne, and S. C. Harrison, 1988, *Science* **242**:899; *courtesy of S. C. Harrison.*

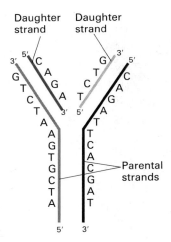

▲ **Figure 2-38** A schematic diagram of DNA replication. The bases in the two strands of the parental double helix complement one another. When the strands separate in replication, bases complementary to those in the parental strands are added to form the new daughter strands; this ensures that the daughter DNA molecules are the same as the parents.

there is growing evidence that certain DNA regions rich in guanine and cytosine within long molecules do adopt a Z-like conformation. Such an alternative structure could be a recognition signal for some important function of DNA.

In DNA replication, the strands of the helix separate; each strand then seeks to replace the hydrogen-bonded complementary nucleotides it has lost. The newly hydrogen-bonded nucleotides become linked together, resulting in two double helices identical to the original one (Figure 2-38). This process is described in detail in Chapter 3.

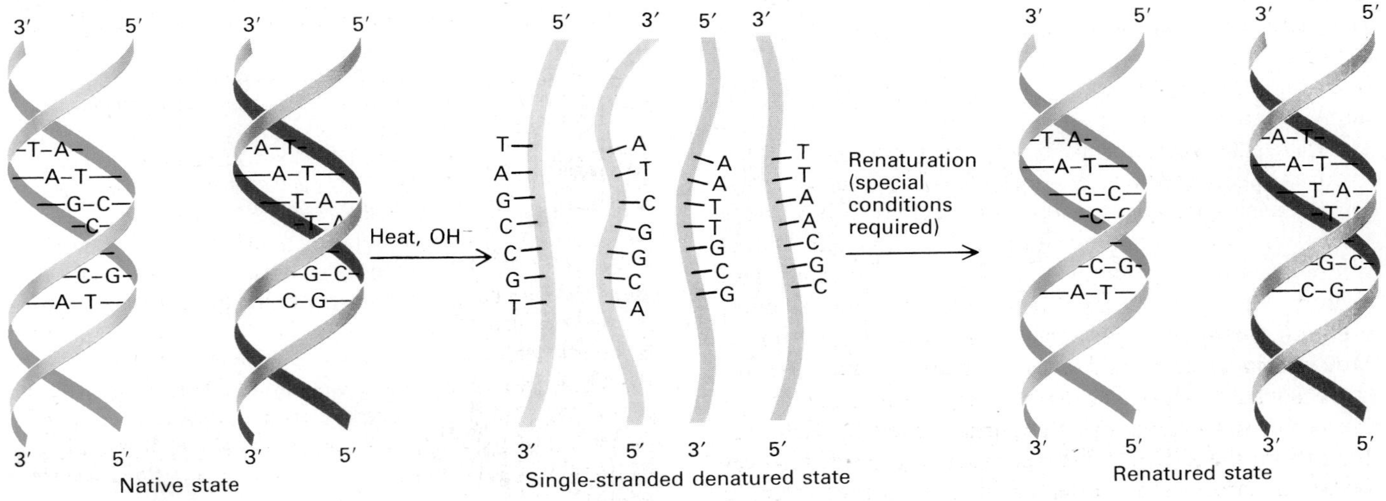

▼ **Figure 2-39** The denaturation and renaturation of two double-stranded DNA molecules.

DNA Is Denatured When the Two Strands Are Made to Separate

The unwinding and separation of DNA strands can be observed experimentally: if a solution of DNA is heated, the thermal energy of molecular motion eventually breaks the hydrogen bonds and other forces that stabilize the double helix, and the strands of each pair separate, or *denature* (Figure 2-39). One way to detect this "melting" of DNA is to take advantage of the fact that native double-stranded DNA absorbs much less light of 260-nm wavelength (a wavelength of ultraviolet) than does the equivalent amount of single-stranded DNA (Figure 2-40). Thus, as DNA denatures, its absorption of ultravi-

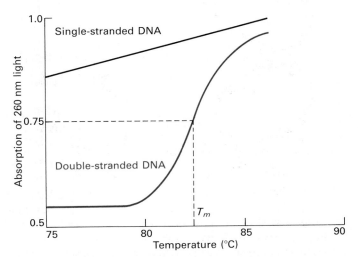

▲ **Figure 2-40** The absorption of ultraviolet light of 260 nm wavelength by solutions of single-stranded and double-stranded DNA. As regions of double-stranded DNA unpair, the absorption of light by those regions increases proportionally. The temperature at which one-half of the bases of double-stranded DNA denature is denoted by T_m. Light absorption by single-stranded DNA changes little as the temperature is increased.

olet light increases. Near the denaturation temperature, a small increase in temperature completely denatures DNA because the double helix is held together by multiple weak cooperative interactions so that a little more thermal energy destabilizes the entire structure.

The temperature at which the strands of DNA will separate depends on several factors. Molecules that contain a greater number of G-C pairs require higher temperatures to denature (Figure 2-41), because G-C pairs are more stable with three bonds than A-T pairs are with two. Solutions with a low salt concentration tend to destabilize the double helix and cause it to melt at lower temperatures. DNA is also denatured by exposure to other agents that destabilize hydrogen bonds, such as alkaline solutions and concentrated solutions of formamide or urea:

$$\underset{\text{Formamide}}{\overset{\overset{\displaystyle O}{\|}}{HC-NH_2}} \qquad \underset{\text{Urea}}{\overset{\overset{\displaystyle O}{\|}}{H_2N-C-NH_2}}$$

Using precise, tiny increments of temperature, A-T–rich regions of DNA can be melted while the entire strand of DNA is still held together by residual G-C–rich regions.

The single-stranded molecules that result from denaturation are stable; even if the temperature is lowered,

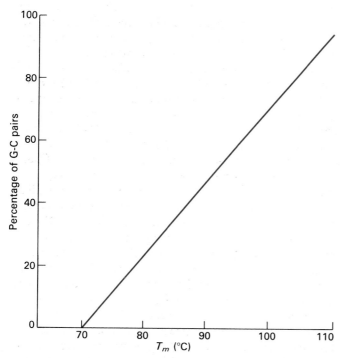

▲ **Figure 2-41** A plot of the G-C content of 38 different DNA samples against T_m. An increased percentage of G-C base pairs is correlated with an increased resistance to thermal denaturation of the double helix. For an unknown DNA sample, T_m can be used to determine the percentage of G + C.

they generally do not renature into the native double-stranded molecule but form tangled balls. However, by adjusting the temperature and the salt concentration, the two complementary strands can be made to renature in time (see Figure 2-39). This property is the basis of the powerful technique of *nucleic acid hybridization*.

Many DNA Molecules Are Circular

For DNA to denature, the two strands must be free to unwind from each other. Linear DNA molecules, which have free ends, denature quite readily. Some DNA molecules, however, are circular; the two strands form a closed, endless structure. All bacterial and many viral DNAs are circular molecules. Mitochondrial DNA, a small DNA molecule that encodes several specialized proteins, is also a closed circle.

When a circular DNA molecule is subjected to elevated temperatures or pH, its hydrogen bonds are broken and other stabilizing interactions are decreased. Because the two intertwined circular strands cannot unwind freely, they denature into a tangled mass (Figure 2-42a). When the solution temperature or pH is returned to normal, the molecule rapidly renatures to form the original base-paired, double-stranded structure.

Renaturation is easy for circular DNA molecules because the two strands remain partially aligned even under denaturing conditions, and thus complementary regions can readily locate each other. Once a short region of complementary base-paired sequence is formed, the rest of the sequence "zips" together. (Sometimes, under extreme conditions, denaturation can be so extensive that the DNA will become too entangled to renature.) When a linear DNA is denatured, in contrast, the two strands separate in solution and their rate of renaturation is slowed by the amount of time required for the regions of complementary sequence to collide in the correct register and nucleate "zipping" of the entire strand.

If the native circular DNA is *nicked* (if one of the strands is cut), the two strands unwind and separate when the molecule is denatured: one strand remains circular while the other becomes linear (Figure 2-42b). The nicking of circular DNA can occur naturally during the process of DNA replication or experimentally by cleaving a single phosphodiester bond in a circle with deoxyribonuclease (a DNA-degrading enzyme).

Many Closed Circular DNA Molecules Are Supercoiled

Closed circular DNA molecules isolated from mitochondria, viruses, and bacteria are often supercoiled (see Figure 12-23). If one strand of a supercoiled DNA is nicked, the two strands become free to wind, or unwind, around

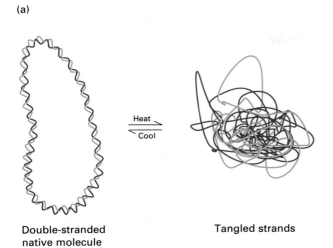

Double-stranded native molecule   **Tangled strands**

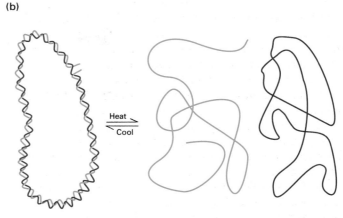

Nick in one strand   **Strands completely separated**

▲ **Figure 2-42** Denaturation of circular DNA. (a) If both strands are closed circles, denaturation disrupts the double helix, but the two single strands become tangled about each

other and cannot separate. (b) If one or both strands are nicked, however, the two strands will separate on thermal denaturation.

each other. In this event, the supercoil "relaxes" into a circular DNA with no supercoils and can be resealed into the relaxed state by ligating the nicked strand.

To understand the origin of these supercoils and their release when a strand is nicked, recall that free energy is minimal in the double-helical conformation of DNA in solution (see Figure 2-34). This means that DNA spontaneously winds to achieve the most stable conformation—one in which all bases are in a right-handed, double-helical arrangement. Suppose, however, that a small segment of a linear DNA is unwound and the free ends of the two strands are linked covalently to form a closed circle. The joining of the free ends fixes the DNA molecule as a whole in an underwound state. When the unwound segments pair up again to assume the double-helical conformation, the entire molecule must adjust its spatial shape by twisting, for example, into a right-handed superhelical form (Figure 2-43a). An alternate deformation is for the molecule to assume a left-handed spiral form. A supercoiled DNA that is underwound is termed a *negatively supercoiled* DNA; the shape it assumes is determined by the relative free energies of the various contorted forms.

If, instead, a linear DNA helix is wound more tightly than normal before the free ends are linked together, an overwound or *positively supercoiled* DNA is formed. A positively supercoiled DNA can assume the form of a left-handed superhelix or a right-handed spiral (Figure 2-43b).

Possibly because of their interaction with proteins such as histone, DNA helices in cells are unwound about 5–8 percent with respect to the standard right-handed, double-helical conformation of DNA in solution. When closed circular DNA is removed from cells and stripped of its protein, it becomes negatively coiled.

RNA Is Usually Single-stranded and Serves Many Different Functions

RNA is quite similar to DNA in chemical make-up: the sugar component of RNA has one more hydroxyl group (see Figure 2-30), and thymine in DNA is replaced by uracil in RNA (see Figure 2-31). These differences are relatively minor, so RNA can take on the same configurations as DNA; it can be double-stranded or single-stranded, linear or circular. It can also form a hybrid helix composed of one RNA strand and one DNA strand.

DNA stores genetic information in a monotonous double-stranded form. RNA is more versatile, having many functions and structures. Most RNA is single stranded. Small RNA molecules, such as tRNA adopt a defined three-dimensional architecture in solution (Chapter 3). Larger RNA molecules appear to have locally well-defined three-dimensional structures, with more flexible links. In these respects, RNA conformation presents some analogy to protein structure. The structure of an RNA is determined mainly by local base pairing that follows the same rules as DNA; parts of the RNA strand fold back on themselves to form short double helices. Other interactions between bases, unlike those in DNA, also occur. The bulk of RNA in cells is bound to proteins in large complexes of very specific structure.

The analogy to protein structure goes one step further: RNA can act catalytically, taking on the properties of an enzyme. Early in evolution, RNA may have served both as a genetic repository and as the first enzymes. Life could have evolved first through RNA; proteins and DNA may have occurred later. Even today, in special situations, RNA can act as genetic material—a property that may be a remnant of its function in primitive life. Single-stranded

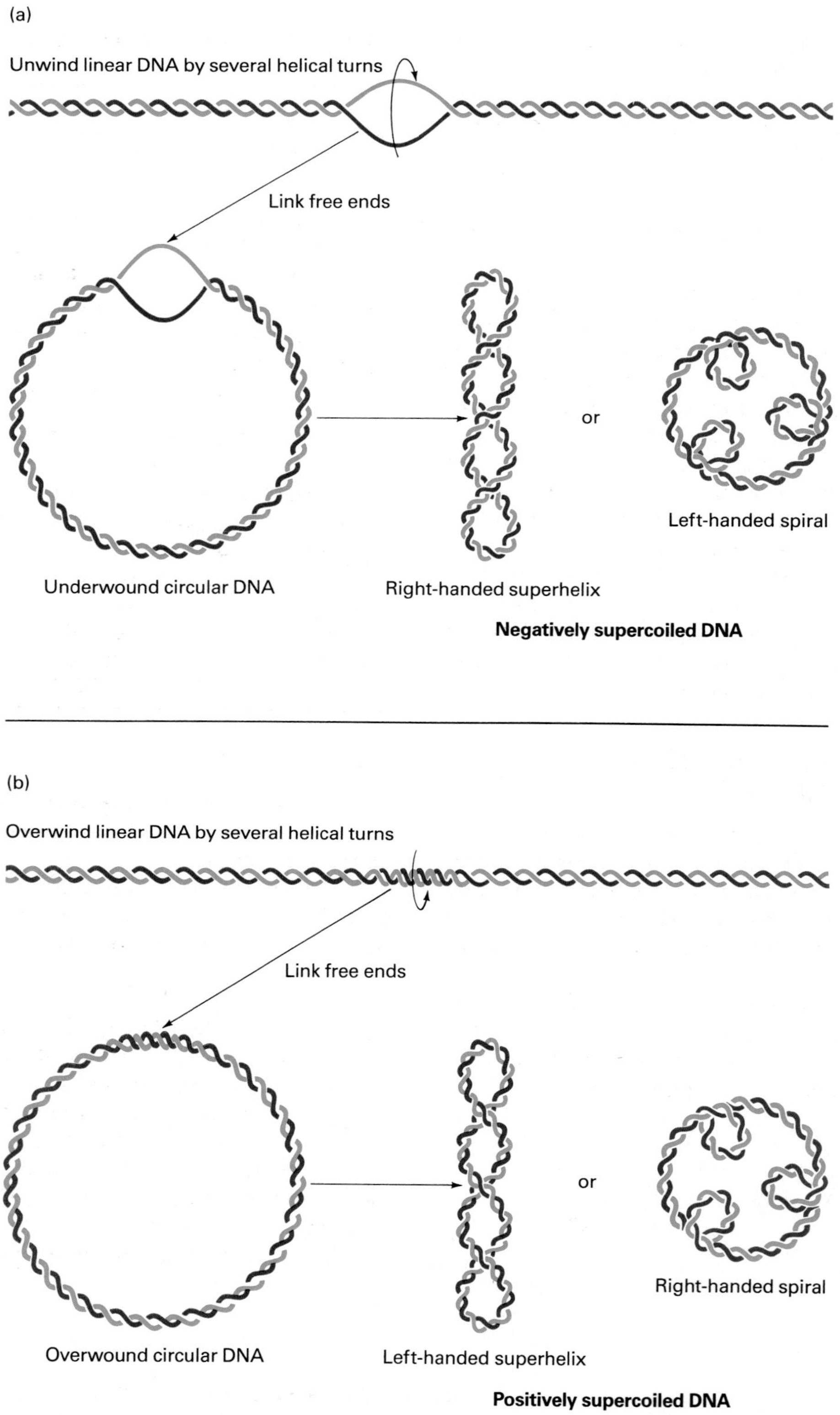

(a)

Unwind linear DNA by several helical turns

Link free ends

Underwound circular DNA

Right-handed superhelix

or

Left-handed spiral

Negatively supercoiled DNA

(b)

Overwind linear DNA by several helical turns

Link free ends

Overwound circular DNA

Left-handed superhelix

or

Right-handed spiral

Positively supercoiled DNA

▲ **Figure 2-43** Supercoiling in closed circular DNA. (a) A right-handed superhelix or a left-handed spiral is formed if the linear double-stranded molecule is slightly unwound before the ends are linked together. One supercoil results for each helical turn that is unwound. (b) Converse deformations result if the linear DNA is overwound before the ends are linked together; an overwound circular DNA will form either a left-handed superhelix or a right-handed spiral.

RNA is the genetic material of many viruses, including poliovirus, influenza virus, and tobacco mosaic virus; double-stranded RNA is the genetic material of rotaviruses, reoviruses, and the plant wound tumor virus (Chapter 5).

Lipids and Biomembranes

Proteins catalyze most of the chemical reactions in a cell, but they would be unable to coordinate their activities without *biomembranes,* which separate a cell from its surroundings, provide anchoring points for some proteins, and define the boundaries of the intracellular compartments characteristic of eukaryotic cells. The major structural elements of biomembranes—organic molecules called *lipids*—are insoluble or only slightly soluble in water. A large part of each lipid is hydrophobic, typically containing only hydrogen and carbon atoms; if it were a separate molecule, it would simply be a hydrocarbon. A pure hydrocarbon, such as heptane, is so hydrophobic that it is almost completely insoluble in water (Chapter 1). In an aqueous environment, hydrocarbons segregate into a separate nonaqueous phase.

Fatty Acids Are the Principal Components of Membranes and Lipids

A *fatty acid* molecule contains a long hydrocarbon chain attached to a carboxyl group, which is acidic in character. Fatty acids differ in length and in the extent and position of their double bonds. Most fatty acids in cells have 16 or 18 carbon atoms and zero to three double bonds (Table 2-2).

Fatty acids with no double bonds are said to be *saturated;* those with at least one double bond are *unsaturated.* A double bond in an unsaturated fatty acid has two possible configurations, cis or trans:

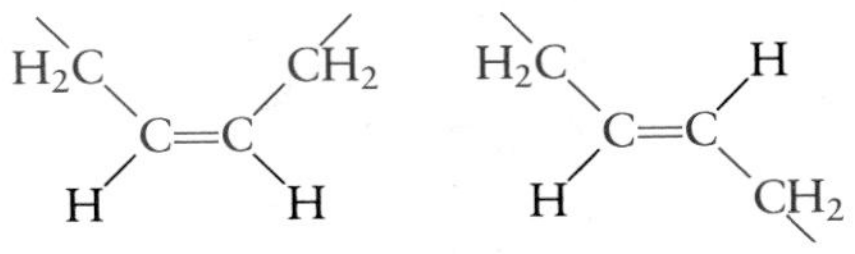

The double bonds in most unsaturated fatty acids have the cis orientation, which introduces a bend in the hydrocarbon side chain.

The *triacylglycerols,* typical storage forms of fatty acids in cells, are strongly hydrophobic molecules. A triacylglycerol is formed from one molecule of *glycerol*

$$HOCH_2—CH(OH)—CH_2OH$$

by esterification of fatty acids to each of the three hydroxyl groups. In the process, three molecules of water are eliminated (Figure 2-44). The fatty acid part of a triacylglycerol or other ester is termed an *acyl* group.

Triacylglycerols, like pure hydrocarbons, are insoluble in water and in salt solutions; thus they form lipid droplets in cells. Fat cells, or *adipose cells,* store these droplets of triacylglycerols as a source of energy for the body (Figure 2-45). When triggered by hormones such as adrena-

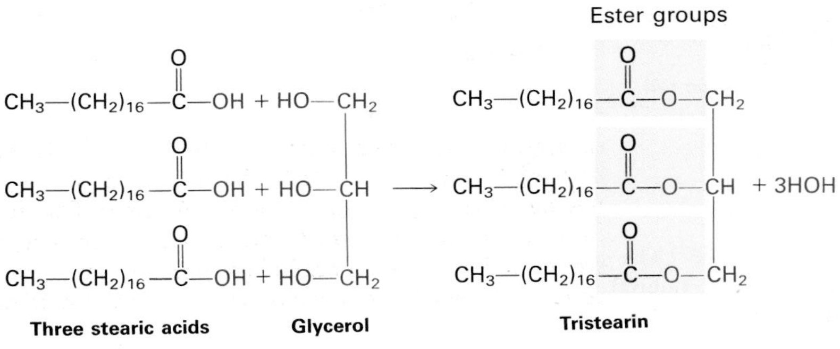

Table 2–2 Structures of some typical fatty acids found in cells

Structure	Systematic name	Common name
SATURATED FATTY ACIDS		
$CH_3(CH_2)_{10}COOH$	*n*-Dodecanoic	Lauric
$CH_3(CH_2)_{12}COOH$	*n*-Tetradecanoic	Myristic
$CH_3(CH_2)_{14}COOH$	*n*-Hexadecanoic	Palmitic
$CH_3(CH_2)_{16}COOH$	*n*-Octadecanoic	Stearic
$CH_3(CH_2)_{18}COOH$	*n*-Eicosanoic	Arachidic
$CH_3(CH_2)_{22}COOH$	*n*-Tetracosanoic	Lignoceric

Structure	Common name
UNSATURATED FATTY ACIDS	
$CH_3(CH_2)_5CH{=}CH(CH_2)_7COOH$	Palmitoleic
$CH_3(CH_2)_7CH{=}CH(CH_2)_7COOH$	Oleic
$CH_3(CH_2)_4CH{=}CHCH_2CH{=}CH(CH_2)_7COOH$	Linoleic
$CH_3CH_2CH{=}CHCH_2CH{=}CHCH_2CH{=}CH(CH_2)_7COOH$	Linolenic
$CH_3(CH_2)_4(CH{=}CHCH_2)_3CH{=}CH(CH_2)_3COOH$	Arachidonic

▶ **Figure 2-44** Tristearin, a triacylglycerol, is the product of the esterification of each of the three hydroxyl groups of glycerol with stearic acid. Triacylglycerols are a storage form of fatty acids.

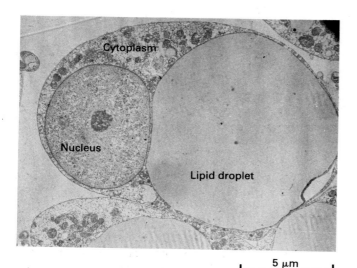

▲ **Figure 2-45** An electron micrograph of an adipose (fat) cell showing a single large lipid storage droplet, a small nucleus, and a small cytoplasm. This cell resulted from the differentiation of a line of preadipocyte cells in tissue culture. [See H. Green and M. Meuth, 1974, *Cell* 3:127–133.] *Courtesy of H. Green.*

lin, adipose cells hydrolyze the triacylglycerols into free fatty acids that are released into the blood; cells then remove them and degrade them to derive energy.

The lipids in membranes also contain long-chained fatty acyl groups, but these are linked (usually by an ester bond) to small, highly hydrophilic groups. Consequently, membrane lipids do not clump together in droplets but orient themselves in sheets to expose their hydrophilic ends to the aqueous environment. Molecules in which one end (the head) interacts with water and the other end (the tail) avoids it are said to be *amphipathic* (from the Greek for "tolerant of both").

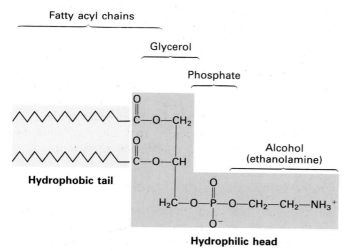

▲ **Figure 2-46** The structure of phosphatidylethanolamine, a typical phospholipid. The hydrophobic and hydrophilic parts of the molecule are yellow and blue, respectively.

Phospholipids Are Key Components of Biomembranes

The tendency of amphipathic molecules to form organized structures spontaneously in water is the key to the structure of cell membranes, which typically contain a large proportion of amphipathic lipids. The *phospholipids* are the most abundant amphipathic lipids. *Phosphoglycerides,* a principal class of phospholipids, contain fatty acyl side chains esterified to two of the three hydroxyl groups of a glycerol molecule; the third hydroxyl group is esterified to phosphate. The simplest phospholipid, *phosphatidic acid,* contains only these components. In most phospholipids, however, the phosphate group is also esterified to a hydroxyl group on a hydrophilic compound such as ethanolamine

$$HO-CH_2-CH_2-NH_3^+$$

or serine, choline, or glycerol, so that the head group is somewhat larger (Figure 2-46). Both of the fatty acyl side chains may be saturated or unsaturated, or one chain may be saturated and the other unsaturated. The negative charge on the phosphate as well as the charged groups and hydroxyl groups on the alcohol esterified to it interact strongly with water.

Certain Steroids Are Components of Biomembranes

Cholesterol and its derivatives constitute another important class of membrane lipids, the *steroids.* The basic structure of steroids is the four-ring hydrocarbon shown in Figure 2-47a. Cholesterol (Figure 2-47b) is the major steroidal constituent of animal tissues; other steroids play

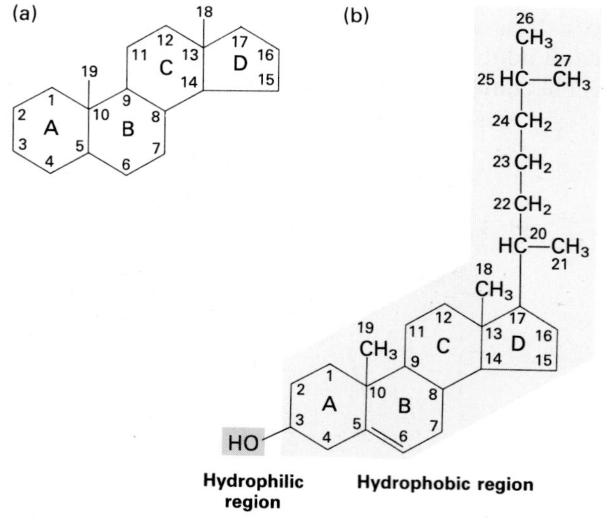

▲ **Figure 2-47** (a) The general structure of a steroid. All steroids contain the same four hydrocarbon rings, conventionally labeled A, B, C, and D, with the carbons numbered as shown. (b) The structure of cholesterol.

more important roles in plant cell membranes. Although cholesterol is almost entirely hydrocarbon in composition, it is amphipathic because it contains a hydroxyl group that interacts with water.

Phospholipids Spontaneously Form Micelles or Bilayers in Aqueous Solutions

Phospholipids assume three different forms in aqueous solutions: micelles, bilayer sheets, and liposomes (Figure 2-48). The type of structure formed by a pure phospholipid or a mixture of phospholipids depends on the length and degree of saturation of the fatty acyl chains, on the temperature, on the ionic composition of the aqueous medium, and on the mode of dispersal of the phospholipids in the solution.

When a solution of phospholipids is violently dispersed in water, some of them may cluster together in *micelles* (spherical structures about 20 nm or less in diameter). The hydrocarbon side chains are sequestered inside the micelle; the polar head groups lie on its surface, in contact with the water.

Phospholipids also spontaneously form symmetrical sheetlike structures called *phospholipid bilayers*. Bilayers are two molecules thick. The hydrocarbon side chains of each layer minimize contact with water by aligning themselves together in the center of the bilayer. Their close packing is stabilized by van der Waals attractions between the hydrocarbon side chains of adjacent phospholipids. Electrostatic and hydrogen bonds stabilize the interaction of the polar head groups with water. A phospholipid bilayer can be of almost unlimited size—from micrometers to millimeters in length or width. *Liposomes* are spherical bilayer structures, larger than micelles, that contain aqueous interiors.

Biolayers are the basis for all biological membranes that separate cells from their environment and intracellular compartments from the cytoplasm. Whether micelles play any role in the formation of membranes remains an open question.

Proteins, another essential constituent of all biomembranes, allow the various cellular membranes to function in specific ways. In many cases, all or part of a protein molecule interacts directly with the hydrophobic core of a phospholipid membrane bilayer. Such membrane-embedded proteins must adopt quite different conformations from those of water-soluble proteins. The structure and function of phospholipids and proteins in biological membranes is detailed in Chapter 13.

Carbohydrates

A carbohydrate is constructed of carbon (*carbo-*) and hydrogen and oxygen (*-hydrate*, or water). The formula for the simplest carbohydrates—the *monosaccharides*, or *simple sugars*—is $(CH_2O)_n$, where n = 3, 4, 5, 6, or 7. All monosaccharides contain hydroxyl groups and either an aldehyde or a keto group:

Aldehyde   **Keto**

Of primary importance are the sugars with five carbon atoms (n = 5), called *pentoses*, and those with six, *hexoses*. Glucose ($C_6H_{12}O_6$) is the key source of energy in all cells of higher organisms. Ribose and deoxyribose, two pentose sugars, are essential constituents of nucleic acids. Certain carbohydrates (often termed *complex carbohydrates*) consist of polymerized sugar units. *Polysaccharides* play important roles as energy-storage molecules within the cell and as structural components on the outer cell surface of or between cells. *Oligosaccharide* chains are often found attached to proteins or lipids; these modified molecules, called *glycoproteins*, or *glycolipids*, are important constituents of cell-surface membranes.

Many Important Sugars Are Hexoses

Because many biologically important sugars are hexoses, which are structurally related to D-glucose, it is important to know some of the structural features of D-glucose. (D and L sugars are mirrors of one another; except for L-fucose, sugars found in biological systems are generally of the D type.)

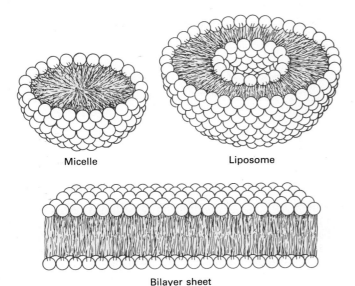

Micelle   Liposome

Bilayer sheet

▲ **Figure 2-48** Cross-sectional views of the three structures that can be formed by phospholipids in aqueous solutions: spherical micelles with hydrophobic interiors, sheets of phospholipids in a bilayer, and spherical liposomes comprising one phospholipid bilayer.

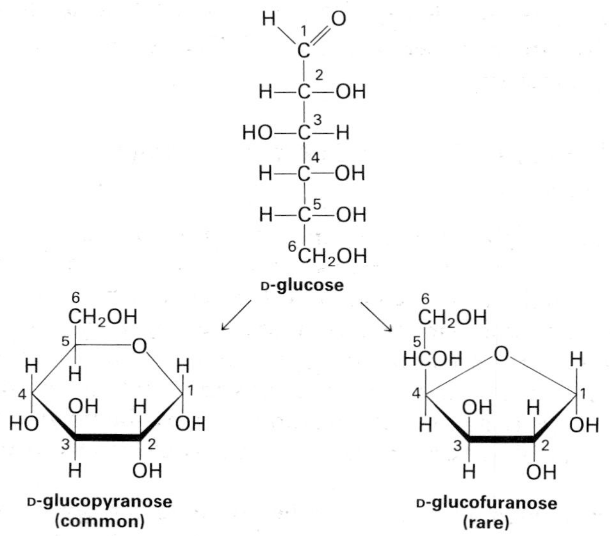

D-glucose

**D-glucopyranose
(common)**

**D-glucofuranose
(rare)**

▲ **Figure 2-49** Three alternative configurations of D-glucose. The ring forms are generated from the linear molecule by the reaction of the aldehyde at the 1 carbon with the hydroxyl on the 5 carbon or on the 4 carbon.

Glucose can exist in three different forms: a linear structure or one of two different ring structures (Figure 2-49). The aldehyde group on the 1 carbon can react with the hydroxyl group on the 5 carbon to form a hemiacetal

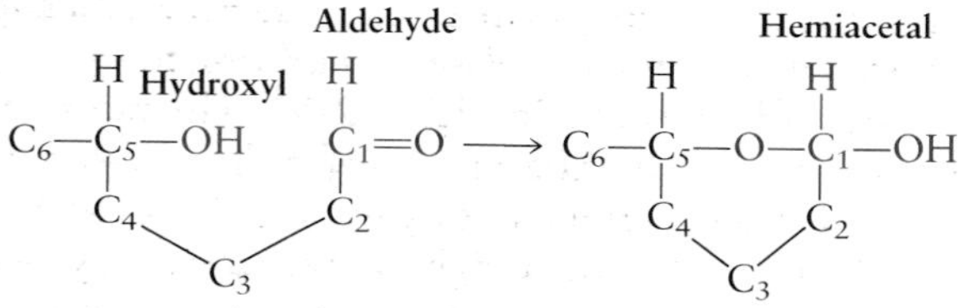

Aldehyde   Hemiacetal

Hydroxyl

that contains a six-membered ring, *glucopyranose*. Similarly, condensation of the hydroxyl group on the 4 carbon with the aldehyde results in the formation of *glucofuranose*, a hemiacetal containing a five-membered ring. Although all three forms of glucose exist in biological systems, the pyranose form is by far the most abundant.

Carbon atoms 2, 3, 4, and 5 in the linear form of glucose are *asymmetric*. Each carbon is bonded to four different atoms or groups, and the exact spatial orientation of these bonds is of great importance. For example, if the hydrogen atom and the hydroxyl group attached to the 2 carbon were interchanged, the resulting molecule could not be converted to glucose without breaking and making covalent bonds. *Mannose* is identical to glucose except for the orientation of the substituents on the 2 carbon. If the pyranose forms of glucose and mannose are shown as Haworth projections (Figure 2-50), the hydroxyl group on the 2 carbon of glucose points downward whereas that on mannose points upward. *Galactose*, another hexose, differs from glucose only in the orientation of the

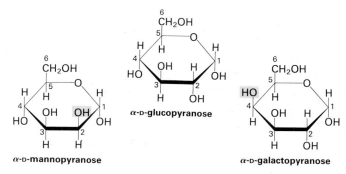

α-D-mannopyranose   **α-D-glucopyranose**   **α-D-galactopyranose**

▲ **Figure 2-50** Haworth projections of the structures of glucose, mannose, and galactose in their pyranose forms. The hydroxyl groups with different orientations from those of glucose are indicated in orange.

hydroxyl group on the 4 carbon (see Figure 2-50). Enzymes and binding proteins can distinguish between these single points of difference and assign a specific biological role to each molecule.

The Haworth projection is an oversimplification because the actual pyranose ring is not planar. Rather, the molecule adopts a conformation in which each of the ring carbons is at the center of a tetrahedron, just like the carbon in methane (see Figure 1-1). The preferred conformation of pyranose structures is the *chair*, in which the bonds from a ring carbon to other atoms may take two directions: *axial* and *equatorial* (Figure 2-51).

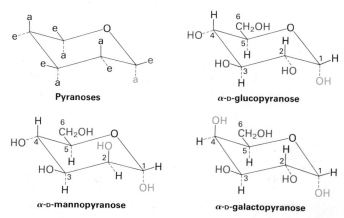

Pyranoses   **α-D-glucopyranose**

α-D-mannopyranose   **α-D-galactopyranose**

▲ **Figure 2-51** Chair conformations of glucose, mannose, and galactose in their pyranose forms. The chair is the most stable conformation of a six-membered ring. (In an alternative form, called the *boat*, both carbon 1 and carbon 4 lie above the plane of the ring.) In the generalized pyranose ring at the top, a = axial atoms and e = equatorial atoms; the four bonds at each of the ring carbon atoms are tetrahedral. In α-D-glucopyranose, all of the hydroxyl groups except the one bonded to the 1 carbon are equatorial. In α-D-mannopyranose, the hydroxyl groups bonded to the 2 and 1 carbons are axial. In α-D-galactopyranose, the hydroxyl groups bonded to the 1 and 4 carbons are axial.

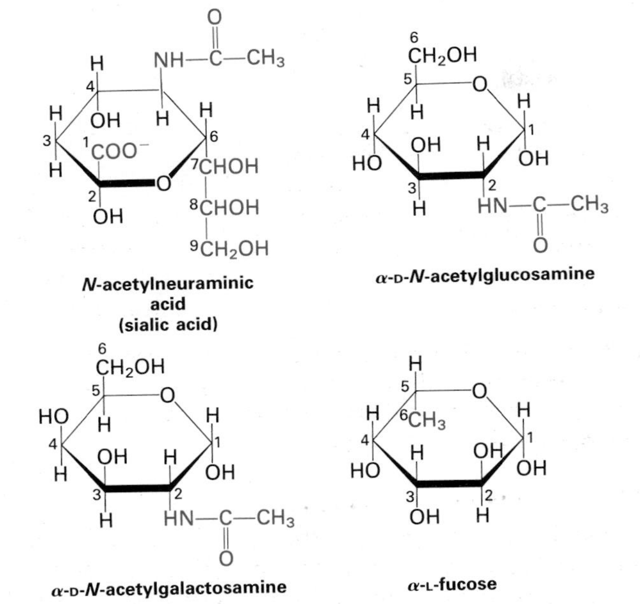

N-acetylneuraminic acid (sialic acid)

α-D-N-acetylglucosamine

α-D-N-acetylgalactosamine

α-L-fucose

▲ **Figure 2-52** The structures of four sugars commonly found in larger molecules in cells. Red highlights the modifications mentioned in the text.

Glucose is often present in cells as an independent molecule, whereas mannose, galactose, and the four modified hexoses shown in Figure 2-52 are more common components of polysaccharides, glycoproteins, or glycolipids. Note that both N-acetylglucosamine and N-Acetylgalactosamine contain an acetamide group

$$-NH-\overset{\overset{\displaystyle O}{\|}}{C}-CH_3$$

in place of the usual hydroxyl on the 2 carbon. N-Acetylneuraminic acid contains an acetamide group as well as three extra carbon atoms; because of its carboxyl group,

it is negatively charged. The 6 carbon of fucose is a methyl group rather than the CH_2OH group found in most other sugars.

Polymers of Glucose Serve as Storage Reservoirs

Because D-glucose is the principal source of cellular energy, it is important for cells to maintain a reservoir of it. Some sugars are stored as *disaccharides,* which consist of two monosaccharides linked together by a *glycosidic bond.* In the formation of a glycosidic bond, the 1 carbon atom of one sugar molecule reacts with a hydroxyl group of another; as in the formation of most biopolymers, the linkage is accompanied by the loss of water (Figure 2-53).

In principle, a large number of different glycosidic bonds can be formed between two sugar residues. Galactose can be bonded to glucose, for example, by any of the following linkages: $\alpha(1 \to 1)$, $\alpha(1 \to 2)$, $\alpha(1 \to 3)$, $\alpha(1 \to 4)$, $\alpha(1 \to 6)$, $\beta(1 \to 1)$, $\beta(1 \to 2)$, $\beta(1 \to 3)$, $\beta(1 \to 4)$, or $\beta(1 \to 6)$, where α or β specifies the conformation at the 1 carbon in galactose and the number following the arrow indicates the glucose carbon to which the galactose is bound. The theoretical number of linkages in *trisaccharides* is much larger, but only a few are actually found in nature due to the specificities of the enzymes that synthesize oligosaccharides from single sugar molecules.

Lactose is the major sugar in milk; *sucrose* is a sugar found abundantly in most plants. The formation of both disaccharides is depicted in Figure 2-53. The most common storage carbohydrate in animal cells is *glycogen,* a very long, branched polymer of glucose (Figure 2-54). As much as 10 percent of the weight of liver can be glycogen. (The synthesis and utilization of this polymer is described in Chapter 19.) *Cellulose*—a different long-chained polymer of glucose that is the major constituent of plant cell

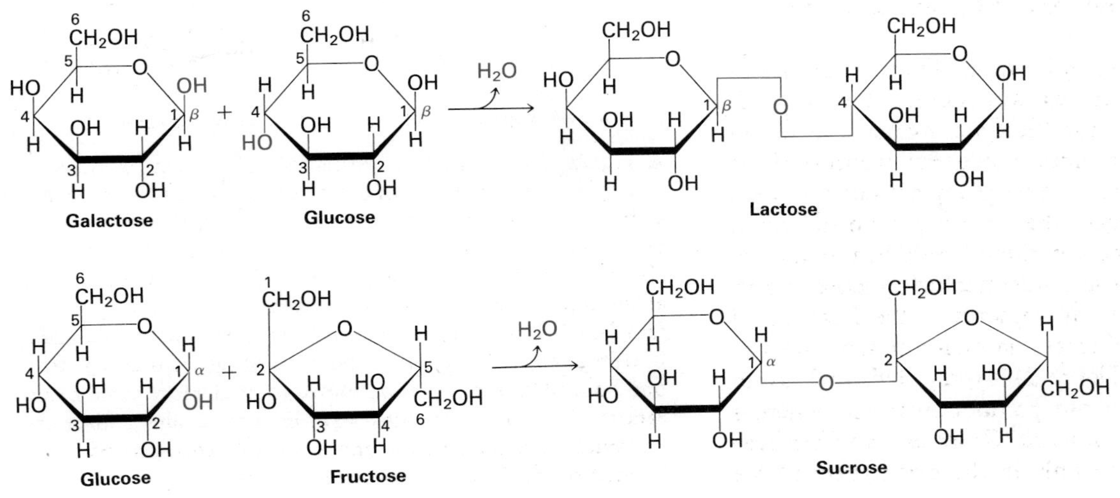

Galactose   Glucose   Lactose

Glucose   Fructose   Sucrose

◀ **Figure 2-53** The formation of glycosidic linkages to generate the disaccharides lactose and sucrose. The lactose linkage is $\beta (1 \to 4)$; the sucrose linkage is $\alpha (1 \to 2)$.

▼ **Figure 2-54** The structure of glycogen, a storage polymer of glucose. (a) Haworth projections showing that glycogen consists principally of chains of glucose residues connected by $\alpha (1 \rightarrow 4)$ linkages. The end of one chain is connected to another chain by an $\alpha (1 \rightarrow 6)$ linkage. (b) A schematic drawing of a glycogen molecule with its network of chains. *After L. Stryer, 1988, Biochemistry, 3d ed., W. H. Freeman and Company, pp. 449, 450.*

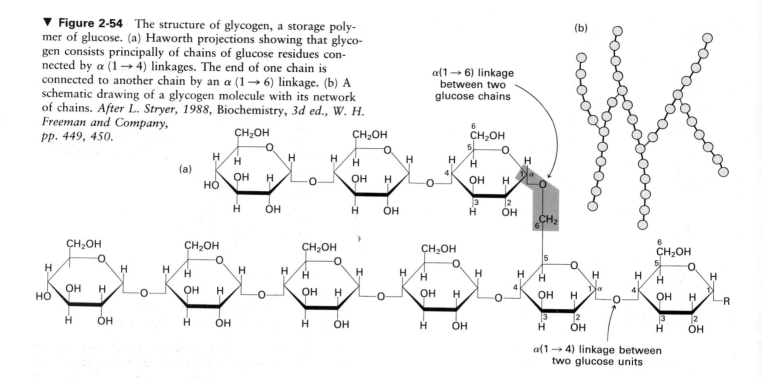

walls—is the most abundant organic chemical on earth. (The important structural and functional features of cellulose are described in Chapter 4).

Glycoproteins Are Composed of Proteins Covalently Bound to Sugars

Many membrane proteins and serum proteins contain carbohydrate chains called *glycoproteins*, which often contribute importantly to the folding and stability of the proteins as well as to their synthesis and positioning within a cell. Carbohydrates generally play no role in the catalytic functions of proteins; in fact, many synthetic glycoprotein derivatives that lack the carbohydrate chain still function normally. However, carbohydrates may increase the solubility of glycoproteins in water. The outermost sugar in a carbohydrate chain attached to a protein is often *N*-acetylneuraminic acid, which creates a negative charge on the surface of the glycoprotein and, due to charge repulsion, helps to prevent dissolved proteins from clumping together.

Sugar residues in glycoproteins are commonly linked to two different classes of amino acid residues. The sugars are classified as *O*-linked if they are bonded to the hydroxyl oxygen of serine, threonine, and (in collagen) hydroxylysine; they are classified as *N*-linked if they are bonded to the amide nitrogen of asparagine. The structures of *O*- and *N*-linked oligosaccharides are very different and usually do not contain the same sugars. *O*-linked oligosaccharides are generally shorter and more variable

than *N*-linked structures and often contain only one, two, or three sugar residues. Figure 2-55 shows the structures of some common *O*-linked sugars.

N-linked oligosaccharides, particularly in complex structures that contain galactose and *N*-acetylneuraminic acid, are found on many membrane and secreted proteins. An example of the oligosaccharide of serum antibody proteins is shown in Figure 2-56. These structures contain *N*-acetylglucosamine, mannose, fucose, galactose, and *N*-acetylneuraminic acid. The proteins differ

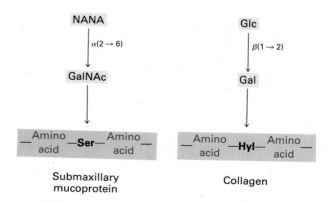

▲ **Figure 2-55** The structures of some common *O*-linked oligosaccharides (NANA = *N*-acetylneuraminic or sialic acid, GalNAc = *N*-acetylgalactosamine, Glc = glucose, Gal = galactose, Ser = serine, and Hyl = hydroxyl-lysine). The submaxillary mucoprotein, a component of saliva, is secreted by the salivary gland.

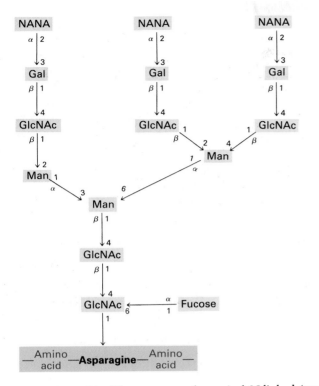

▲ **Figure 2-56** The structure of a typical *N*-linked (asparagine-linked) oligosaccharide attached to many serum proteins, such as antibodies. (NANA = *N*-acetylneuraminic acid, Gal = galactose, GlcNAc = *N*-acetylglucosamine, and Man = mannose).

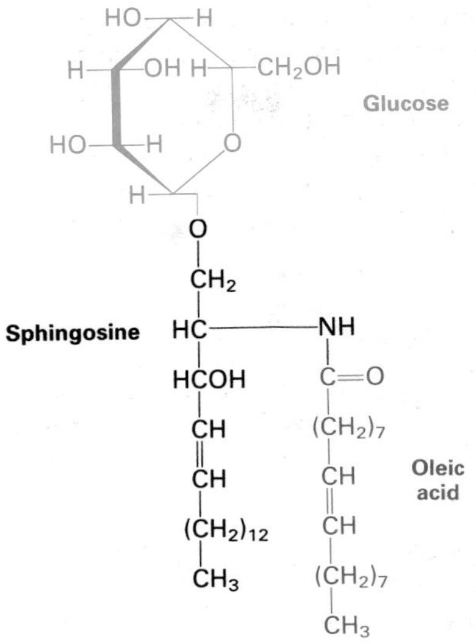

◀ **Figure 2-57** The structure of glucosylcerebroside, one of the simplest glycolipids, consists of sphingosine (an amino alcohol with a long, unsaturated hydrocarbon chain), an oleic acid side chain (red), and a single glucose residue (blue).

primarily in the number of branches (two, three, or four), in the number of *N*-acetylneuraminic acid residues (zero, one, two, or three), and in the chemical nature of the linkage of *N*-acetylneuraminic acid to galactose. Even oligosaccharides found at the same site in a single protein species are often heterogeneous.

Glycolipids of Various Structures Are Found in the Cell Surface Membrane

Other abundant constituents of cell membranes are the *glycolipids*—compounds in which carbohydrate chains of variable length are covalently linked to lipids. The structure of the simplest glycolipid, glucosylcerebroside (Figure 2-57), contains a single glucose sugar unit.

One or more of the sugars in glycolipids is often *N*-acetylneuraminic acid (see Figure 2-52); glycolipids containing this sugar are called *gangliosides*. Glycolipids are situated mainly, but not exclusively, on the surface membranes of cells, where their polar carbohydrate chains face outward (toward the environment and away from the cells). *N*-acetylneuraminic acid residues on glycolipids (and glycoproteins) give most animal cells a net negative surface charge.

Important human glycolipids and glycoproteins are the *blood group antigens,* which can trigger harmful immune reactions. In human beings and other animals, the exact structures of certain carbohydrates linked to membrane lipids and proteins are genetically determined. Thus, when a person receives a mismatched blood transfusion, the foreign carbohydrates may stimulate the production of an immune response to the foreign cells.

The carbohydrates of the human blood groups (A, B, and O) have been studied in great detail. The A, B, and O antigens are structurally related oligosaccharides that may be linked either to lipids or to proteins. The O antigen is a chain of fucose, galactose, *N*-acetylglucosamine, and glucose, with a ceramide lipid (or a hydroxyl protein side chain) linked to the glucose (Figure 2-58). The A and

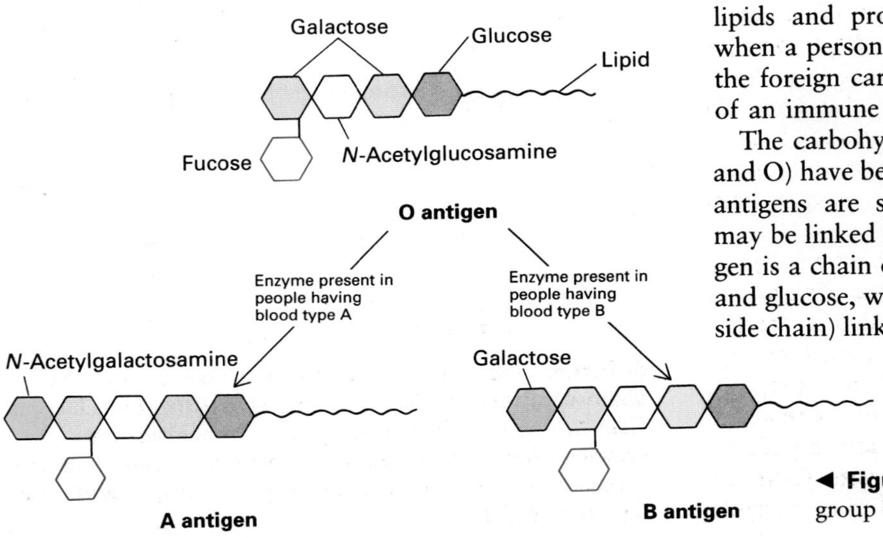

Galactose

Glucose

Lipid

Fucose

N-Acetylglucosamine

O antigen

Enzyme present in people having blood type A

Enzyme present in people having blood type B

N-Acetylgalactosamine

Galactose

A antigen

B antigen

◀ **Figure 2-58** The structures of the human blood-group antigens.

O antigens are identical, except that the A antigen contains a residue of N-acetylgalactosamine attached to the outer galactose residue; the B antigen is also similar to the O, except for an extra galactose residue attached to the outer galactose. All people have the enzymes that synthesize the O antigen. People with type A blood also have the enzyme that adds the extra N-acetylgalactosamine; those with type B blood have the enzyme that adds the extra galactose. People with type AB synthesize both A and B antigens; those with type O make only the O antigen.

People with blood types A and O normally have anti-B antibodies in their serum, which are specific for the extra galactose in the B antigen. People with blood types B and O normally make anti-A antibodies, which are specific for the N-acetylgalactosamine residue of the A antigen. Thus, when B or AB blood cells are injected into a person with blood type A or O, the anti-B antibodies bind to the injected cells and trigger their destruction by *phagocytic cells.* Similarly, A or AB cells cannot be safely injected into a person with blood type B or O. Type O blood can be safely injected into people with type O, A, B, or AB. Because type AB people lack both anti-A and anti-B antibodies, they can receive injections of A, B, AB, or O blood.

The ability of antibodies to distinguish subtle differences in macromolecular structure underscores the role that cell surface oligosaccharides play in establishing the uniqueness of cells and organisms.

The Primacy of Proteins

The major cell constituents have been considered mainly as isolated chemical species in this chapter; in fact, they function together. All cellular molecules must be able to interact with other types of molecules, especially with proteins. For example, most nucleic acids in cells have tightly adhering proteins and thus exist as *nucleoproteins;* the DNA of eukaryotes is heavily coated by basic proteins called *histones* and by a variety of nonhistone proteins essential to the proper functioning of DNA in the cell. As another example, cell membranes are never pure phospholipids; attached proteins determine many of the unique features of cellular and intracellular membranes.

Lipids, nucleic acids, and carbohydrates have fairly uniform chemical structures and physical and chemical behaviors, but regions of proteins can be either hydrophilic or hydrophobic and either negatively or positively charged. The surface of a single protein molecule can contain hydrophobic and hydrophilic regions, a nucleic acid-binding region, a catalytic region, a region that binds tightly to a sugar, and a region that binds tightly to a nucleotide. All parts of a protein fit together to make a functioning unit. It is hard to overestimate the potential variability, specificity, and functional range of proteins.

Although all of the information for synthesizing proteins is found in DNA, it is a monotonous molecule compared with protein. DNA is like a computer tape with building instructions encoded on it, whereas protein has the substantiality and complexity of the final construction.

As we shall see in later chapters, proteins have a hand in every cellular function: replication and transmission of genes, construction of the cell's physical substance, energy metabolism, cellular and muscular movement, and intracellular communication. It is amazing that a single kind of molecule capable of all these functions could evolve. When we construct a building, we use a variety of substances, each for a specific job. That one type of polymer can be responsible for so many biochemical accomplishments is one of the great wonders of life.

Summary

Proteins are linear polymers constructed from 20 common amino acids linked by peptide bonds. Amino acids differ from one another in their side chains: some are positively or negatively charged at neutral pH; some are polar; others are hydrophobic. Many proteins contain covalently modified side chains, such as the phosphate groups bound to serine, threonine, or tyrosine residues. Glycoproteins contain either shorter oligosaccharide chains linked to serine or threonine residues or longer oligosaccharides attached to asparagine side chains. Other proteins contain a prosthetic group, such as heme.

Depending on their composition, proteins fold into different shapes: some are compact, with one or more globular domains; others are fibrous. Many proteins contain one or more regions of rigid α helix or β pleated sheet. Water-soluble proteins typically fold so that hydrophobic side chains are internal (away from the aqueous medium). The shapes of many proteins are maintained by covalent interactions in the form of disulfide (—S—S—) bridges between cysteine residues in the side chains. In addition, many proteins contain multiple polypeptide chains held together by noncovalent bonds. The forces that maintain the native state of a protein can be disrupted by heat or by certain solvents so that the protein becomes denatured. The denatured species often can renature to restore the native state.

Many proteins are enzymes—catalysts that accelerate the rate of specific chemical reactions. Enzymes have at least two key regions in their active sites: one tightly binds the substrate(s), another catalyzes the chemical reaction on the bound substrate(s). A covalent enzyme-substrate complex frequently functions as an intermediate in catalysis: the binding of a substrate to an enzyme may induce a conformational change in the enzyme that positions the catalytic residues correctly. An example of catalytic specificity is provided by two proteases: trypsin and chymotrypsin; these enzymes have similar catalytic sites but recognize different amino acid side chains in the substrates that they cleave.

Most enzymes in cells do not function at a constant rate because allosteric activators or inhibitors bind at sites separate from the substrate-binding site to modulate the catalytic activities of enzymes. An enzyme that catalyzes one of a series of reactions is often inhibited by the ultimate product; this process is called feedback inhibition. A multimeric enzyme can have multiple substrate-binding sites; the binding of a substrate at one site can cause a conformational change that increases the affinity of the other site(s), making the enzyme responsive to a narrow range of substrate concentrations. Such cooperative interactions among binding sites allows a molecule of hemoglobin to bind or release four O_2 molecules over a narrow range of oxygen pressure.

Antibodies, like enzymes, bind specifically and tightly to small regions on molecules. In response to the presence of foreign molecules (antigens), vertebrates synthesize antibodies that experimental scientists can use to identify and purify the antigen.

A single strand of DNA (deoxyribonucleic acid) has a linear backbone of deoxyribose and phosphate. Attached to the 1' carbon of the sugar is one of four bases: adenine (A), guanine (G), cytosine (C), or thymine (T). Like a polypeptide, which has N- and C-termini, a nucleic acid has a chemical polarity—a 3' and a 5' end. In their native states, all cellular DNAs are double-stranded: two antiparallel strands are twisted about each other in a right-handed helix with the stacked bases on the inside, their planes perpendicular to the axis of the helix. An adenine residue in one strand is hydrogen-bonded to a thymine residue in the other; a guanine residue similarly pairs with a cytosine residue. Some DNA sequences may form an alternative, left-handed (Z) helix. Heating a linear molecule of double-stranded DNA above a critical temperature causes the two strands to unwind and separate and the molecule to denature; under appropriate conditions, the molecule can renature to form the native base-paired helix.

Natural DNAs are often closed circles, many of which are twisted further into superhelices. When a closed circular DNA is denatured, the strands cannot separate and the molecule renatures spontaneously when returned to nondenaturing conditions.

RNA (ribonucleic acid) differs from DNA in several ways. In RNA, the sugar is ribose, rather than deoxyribose, and uracil is found in place of thymine. Unlike DNA, the RNA molecule is generally single-stranded, but it can have a defined architecture determined by double-helical segments.

Long-chained fatty acids and triacylglycerols are, like pure hydrocarbons, insoluble and segregate into a separate phase in an aqueous medium. Adipose cells store droplets of triacylglycerols as a source of energy for the body. Phospholipids and cholesterol—the main structural components of cellular membranes—are amphipathic: they contain a large hydrophobic (hydrocarbon) region and a small hydrophilic end. Although micelles and liposomes can be formed from phospholipids, cellular membranes are typically sheets of phospholipid bilayers, which have a hydrophobic interior between two hydrophilic surfaces.

Carbohydrates are important as energy-storage molecules and as structural components in cells. Glucose $(C_6H_{12}O_6)$ is the most common sugar in cells. As a linear molecule, it contains four asymmetric carbon atoms. In the formation of the ring glucopyranose, another asymmetric carbon is created to yield α-D-glucopyranose or β-D-glucopyranose. Other important sugars are mannose and galactose, which differ from glucose only in the positions of substituents on one carbon atom, and modified molecules such as N-acetylglucosamine and N-acetyl-neuraminic acid. Storage forms of hexoses include the long-chained polymer glycogen and the disaccharides lactose and sucrose. Sulfated and otherwise modified polysaccharides are principal structural components of extracellular spaces in animals. Short saccharide chains are often bound to lipids or proteins to generate glycolipids or glycoproteins. The compositions and the bonding arrangements of saccharide substituents are genetically determined; as in the case of the A, B, and O blood-group antigens, this genetic programming can give rise to harmful immune reactions.

References

General References

SMITH, E. L., R. L. HILL, I. R. LEHMAN, R. J. LEFKOWITZ, P. HANDLER, and A. WHITE. 1983. *Principles of Biochemistry: General Aspects*, 7th ed. McGraw-Hill. Chapters 2–8, 10, and 11.

STRYER, L. 1988. *Biochemistry*, 3d ed. W. H. Freeman and Company. Chapters 1–4, 7–9, 12, 27.

Proteins

ANFINSEN, C. B. 1973. Principles that govern the folding of protein chains. *Science* 181:223–230.

CANTOR, C. R., and P. R. SCHIMMEL. 1980. *Biophysical Chemistry*, parts 1, 2, and 3. W. H. Freeman and Company. Includes several chapters (2, 5, 13, 17, 20, and 21) on the principles of protein folding and conformation.

CHOTHIA, C. 1984. Principles that determine the structure of proteins. *Ann. Rev. Biochem.* 53:537–572

CREIGHTON, T. E. 1984. *Proteins: Structures and Molecular Properties*. W. H. Freeman and Company.

DICKERSON, R. E., and I. GEIS. 1969. *The Structure and Action of Proteins*. Benjamin-Cummings. Contains many excellent three-dimensional structures of enzymes and other proteins.

DICKERSON, R. E., and I. GEIS. 1983. *Hemoglobin: Structure, Function, Evolution, and Pathology.* Benjamin-Cummings.

DOOLITTLE, R. 1985. Proteins. *Sci. Am.* 253(4):88–96.

KARPLUS, M., and J. A. MCCAMMON, 1986. The dynamics of proteins. *Sci. Am.* 254(4):42–51. A description of internal motions in folded proteins.

KIM, P. S., and R. L. BALDWIN. 1982. Specific intermediates in the folding reactions of small proteins and the mechanism of protein folding. *Ann. Rev. Biochem.* 51:459–489. A review of experimental results on several proteins.

LESK, A. M. 1984. Themes and contrasts in protein structures. *Trends Biochem. Sci.* 9:v–vii.

PERUTZ, M. F. 1964. The hemoglobin molecule. *Sci. Am.* 211(5):64–76 (Offprint 196). This article and the next describe the three-dimensional structure of hemoglobin and the conformational changes that occur during oxygenation and deoxygenation.

PERUTZ, M. F. 1978. Hemoglobin structure and respiratory transport. *Sci. Am.* 239(6):92–125 (Offprint 1413).

RICHARDSON, J. S. 1981. The anatomy and taxonomy of protein structure. *Adv. Protein Chem.* 34:167–339.

ROSE, G. D., A. R. GESELOWITZ, G. J. LESSER, R. H. LEE, and M. H. ZEHFUS. 1985. Hydrophobicity of amino acid residues in globular proteins. *Science* 229:834–838.

ROSSMAN, M. G., and P. ARGOS. 1981. Protein folding. *Ann. Rev. Biochem.* 50:497–532.

SCHULZ, G. E., and R. H. SCHIRMER, 1979. *Principles of Protein Structure.* Springer-Verlag. A general treatment of protein structure.

Enzymes

BENDER, M. L., R. J. BERGERON, and M. KOMIYAMA. 1984. *The Bioorganic Chemistry of Enzyme Action.* Wiley.

FERSHT, A. 1984. *Enzyme Structure and Mechanism,* 2d ed. W. H. Freeman and Company.

PHILLIPS, D. C. 1966. The three-dimensional structure of an enzyme-molecule. *Sci. Am.* 215(5):78–90 (Offprint 1055). The structure of lysozyme and the nature of the enzyme-substrate complex.

WALSH, C. 1979. *Enzymatic Reaction Mechanisms.* W. H. Freeman and Company. A detailed discussion of the chemical bases of action of many types of enzymes.

Antibodies

CAPRA, J. D., and A. B. EDMONSON. 1977. The antibody combining site. *Sci. Am.* 236(1):50–59 (Offprint 1350). A clear description of how an antibody molecule binds to an antigen.

PORTER, R. R. 1973. Structural studies on immunoglobulins. *Science* 180:713–716.

TONEGAWA, S. 1985. The molecules of the immune system. *Sci. Am.* 253(4):122–130.

Nucleic Acids

BAUER, W. R., F. H. C. CRICK, and J. H. WHITE. 1980. Supercoiled DNA. *Sci. Am.* 243(1):118–133 (Offprint 1474). A mathematical model for describing and analyzing DNA supercoiling.

CANTOR, C. R., and P. R. SCHIMMEL. 1980. *Biophysical Chemistry,* part 1. W. H. Freeman and Company. Chapters 22–24. The conformation of DNA is described using physical-chemical principles.

CECH, T. 1986. RNA as an enzyme. *Sci. Am.* 255(5):64–75.

DICKERSON, R. E. 1983. The DNA helix and how it is read. *Sci. Am.* 249(6):94–111.

FELSENFELD, G. 1985. DNA. *Sci. Am.* 253(4):58–66.

KORNBERG, A. and T. A. BAKER. In press. *DNA Replication, 2d ed.* W. H. Freeman and Company. Chapter 1. A good summary of the principles of DNA structure.

MIN JOU, W., G. HAEGEMAN, M. YSEBAERT, and W. FIERS. 1972. Nucleotide sequence of the gene coding for the bacteriophage MS-2 coat protein. *Nature* 237:82–88. A vivid demonstration of double helices within a single-stranded RNA.

RICH, A., and S.-H. KIM. 1978. The three-dimensional structure of a transfer RNA. *Sci. Am.* 238(1):52–62 (Offprint 1377).

SAENGER, W. 1988. *Principles of Nucleic Acid Structure.* Springer-Verlag. A comprehensive treatise on the structures of RNA, DNA, and their constituents.

WANG, J. C. 1980. Superhelical DNA. *Trends Biochem. Sci.* 5:219–221.

Lipids and Biomembranes

FINEAN, J. B., R. COLEMAN, and R. H. MICHELL. 1984. *Membranes and Their Cellular Functions,* 3d ed. Blackwell.

QUINN, P. J. 1976. *The Molecular Biology of Cell Membranes.* University Park Press. Chapter 2. An excellent description of the structures of membrane components.

TANFORD, C. 1980. *The Hydrophobic Effect,* 2d ed. Wiley. Includes a good discussion of the interactions of proteins and membranes.

Carbohydrates

GINSBURG, V., and P. ROBBINS, eds. 1984. *Biology of Carbohydrates.* Wiley.

KORNFELD, R., and S. KORNFELD. 1980. Structure of glycoproteins and their oligosaccharide units. In *The Biochemistry of Glycoproteins and Proteoglycans,* W. J. Lennarz, ed. Plenum.

RODEN, L. 1980. Structure and metabolism of connective tissue proteoglycans. In *The Biochemistry of Glycoproteins and Proteoglycans,* W. J. Lennarz, ed. Plenum. A clear, detailed description of sulfated oligosaccharides.

SHARON, M. 1980. Carbohydrates. *Sci. Am.* 243(5):90–116 (Offprint 1483). The structures and functions of several oligosaccharides.

3

Synthesis of Proteins and Nucleic Acids

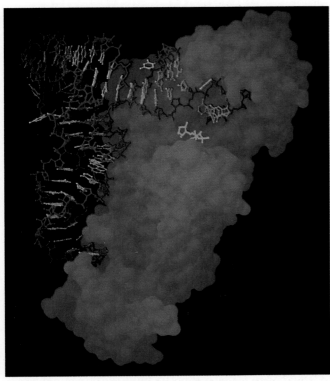

Crystal structure of *E. coli* glutaminyl-tRNA synthetase (blue) complexed with tRNA^{Gln} (red and yellow) and ATP (green)

Classical geneticists from Mendel through Morgan defined genes as the biological units of heredity and proved they are aligned linearly on the chromosomes. Microbial geneticists showed that genes control the structures of proteins. Since 1953, when Watson and Crick ushered in the modern era of molecular biology by elucidating the duplex structure of DNA, two questions have dominated research in molecular biology: how is DNA duplicated, and how does it control protein synthesis?

Faithful duplication of DNA and correct chromosomal segregation from generation to generation are the physical bases for the systematic inheritance of characteristics in all organisms. Occasional random changes, or mutations, in the DNA base sequence and recombination between chromosomes are responsible for occasional variations in inheritance—the events studied by the classical geneticists.

Many important steps in DNA synthesis have been discovered, especially through the analysis of prokaryotic (bacterial) cells. Nevertheless, it is not entirely known *how DNA synthesis starts and stops and how chromo-somal recombination is achieved, even in these simple systems. Less is known about the details of DNA synthe-sis and recombination in eukaryotic cells, although im-portant progress has been achieved, especially in under-standing DNA synthesis.*

How the genes direct protein synthesis is the story of RNA. Since the early 1950s, much successful work has been done on RNA synthesis and on how RNA directs protein synthesis in prokaryotic and eukaryotic cells. Comprehensive discussions of what is now known about RNA formation appear in later chapters. Here, we de-scribe the general reactions that occur in the synthesis of proteins and nucleic acids—the reactions by which the monomers (amino acids or nucleotides) are polymerized into long, correctly ordered chains. ▲

Rules for the Synthesis of Proteins and Nucleic Acids

The intricate relation in cells between the synthesis of DNA, RNA, and protein is circular and can be dia-grammed as follows:

$$DNA \longrightarrow RNA \longrightarrow protein$$

DNA directs the synthesis of RNA, and RNA then directs the synthesis of protein; special proteins catalyze the syn-thesis of both RNA and DNA. This cyclic flow of infor-mation occurs in all cells and has been called the "central dogma" of molecular biology. A discussion of the mecha-nisms of the synthesis of nucleic acids or proteins could begin with any one of the three kinds of polymers. In considering how the synthesis of any polymer begins, progresses, and ends, four general rules may be drawn from the experimental results:

1. *Proteins and nucleic acids are made up of a limited number of different subunits.* Although the number of amino acids is theoretically limitless and several dozen have been identified as metabolic products in various organisms, only 20 different amino acids are used in making proteins. Likewise, only four nitrogenous bases (out of a larger number) are used to construct RNA or DNA in cells. Cell-free enzyme preparations and cells in culture can be "fooled" into incorporating chemical relatives of these four bases, but this almost never happens in nature.

2. *The subunits are added one at a time.* A priori, biolog-ical polymers could be built by aligning the subunits in the correct order on a template, or mold, and simulta-neously fusing all of these units. But it does not hap-pen that way. The assembly of proteins and nucleic acids is a step-by-step procedure (Figure 3-1) in only one chemical direction: protein synthesis begins at the amino (NH_2) terminus and continues through to the

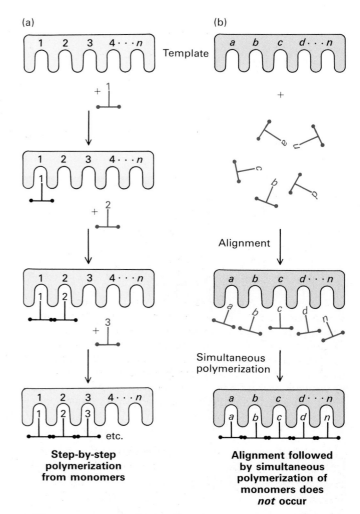

▲ **Figure 3-1**   (a) Cells produce linear polymers by adding monomers one at a time to a lengthening chain. The order of addition is directed by a template. (b) Cells do *not* produce polymers by aligning all the monomers against the template and then fusing them simultaneously.

carboxyl (COOH) terminus; nucleic acid synthesis begins at the 5′ end and proceeds to the 3′ end (Figure 3-2).

3. *Each chain has a specific starting point, and growth proceeds in one direction to a fixed terminus; this re-quires start and stop signals.* If the cellular machinery for polymer synthesis did not start and stop the pro-cess correctly, a cell would be full of partial, probably useless polymers. Elaborate cellular mechanisms en-sure correct starts and stops.

4. *The primary synthetic product is usually modified.* The functional form of a nucleic acid or a protein mol-ecule is rarely the same length as its initially synthe-sized form. The original chain is often inactive or in-complete. To make an active chain, specific enzymes act to change the length of the original chain: proteins are often shortened and RNA chains almost always

Growth of a polypeptide

$$H-N-C-C-N-C-C-N-C-C-N-C-C-OH \quad + \quad H-N-C-C-OH$$

(with side chains R_1, R_2, R_3, R_4 and R_5)

$$\longrightarrow H-N-C-C-N-C-C-N-C-C-N-C-C-N-C-C-OH + H_2O$$

(with side chains R_1, R_2, R_3, R_4, R_5)

Growth of a nucleic acid

$$N_1 \; N_2 \; N_3 \; N_4 \quad (P)(P)(P) \; 3' \; OH \; + \; N_5 \; 3' \; OH \; \longrightarrow \; N_1 \; N_2 \; N_3 \; N_4 \; N_5 \; (P)(P)(P)(P) \; 3' \; OH \; + \; pp_i$$

▲ **Figure 3-2** In the formation of proteins in the cell, one amino acid after another is added to the carboxyl end of a growing chain; thus the growth of a polypeptide chain is said to be from the NH$_2$-terminus to the COOH-terminus. In the polymerization of nucleic acids, one nucleotide at a time is added to the 3′ hydroxyl group at the end of a growing chain; with each addition, a pyrophosphate group (PP$_i$) is produced. The first nucleotide of such a chain retains its tri- phosphate (ppp) at the 5′ end. Thus nucleic acids grow from the 5′ end to the 3′ end. (In the polypeptide example, R$_1$, R$_2$, etc., denote side chains of amino acids. In the nucleic acid, N$_1$, N$_2$, etc., denote nucleic acid bases; a vertical line is the sugar ribose or deoxyribose; and (P) represents a phos- phodiester bond that connects the 3′ carbon of one sugar to the 5′ carbon of the next.)

are; DNA chains are linked together. Distant segments of long RNA chains may be spliced together, omitting the intervening sequence; protein chains may be cross- linked by covalent bonds (Figure 3-3). Primary chains can also undergo certain chemical additions, either during the formation of the chain or after its synthesis is complete. Methyl groups can be added to specific sites in DNA, RNA, and proteins; phosphate groups and a wide variety of oligosaccharides can be added to proteins. Such chemical alterations can have impor- tant functions, some of which have been elucidated. The details of this macromolecular carpentry are dis- cussed in Chapters 7–12 and Chapter 23.

Protein Synthesis: The Three Roles of RNA

Proteins are the active working components of the cellu- lar machinery. Whereas DNA stores the information for protein synthesis and RNA carries out the instructions encoded in DNA, most biological activities are carried out by proteins; their synthesis is at the heart of cellular function.

Because the linear order of amino acids in each protein determines its function, the mechanism that maintains this order during protein synthesis is critical. The design

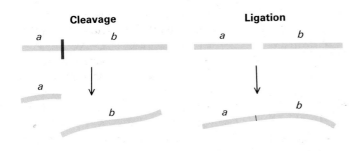
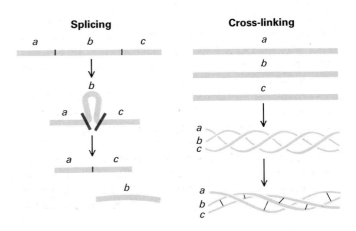

▲ **Figure 3-3** Linear biopolymers are typically modified after they are formed.

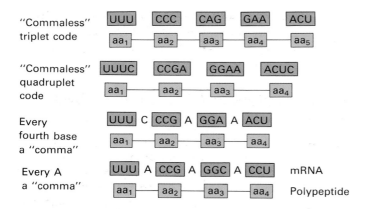

"Commaless" triplet code

"Commaless" quadruplet code

Every fourth base a "comma"

Every A a "comma"

◀ **Figure 3-4** Sequences of mRNA bases (red) interpreted by different hypothetical coding systems. The only system known to be used by organisms is the "commaless" triplet code. (The abbreviation aa denotes an amino acid. Polypeptides are shown in green.)

symbolize each amino acid. The code employed must be capable of specifying at least 20 "words."

If two nucleotides were used to code for one amino acid, then only 16 (or 4^2) different code words could be formed, which would be an insufficient number. However, if a group of three nucleotides are used for each code word, then 64 (or 4^3) code words can be formed. Any code using groups of three or more nucleotides will have more than enough units to encode 20 amino acids. Many such coding systems are mathematically possible, including systems that contain "punctuation" (Figure 3-4). For example, only three out of every four bases might be used for coding, with the fourth serving as a "comma" to separate the words. However, the actual system used by cells is a "commaless" triplet code. Each triplet is called a *codon;* 61 of the 64 possible codons specify individual amino acids (Table 3-1).

A recent surprising discovery is that some nucleotide sequences contain overlapping information, still in the

and function of the protein-synthesizing apparatus is similar in all cells: three kinds of RNA molecules perform different but cooperative roles. *Messenger RNA* (mRNA) encodes the genetic information copied from DNA in the form of a sequence of bases that specifies a sequence of amino acids. *Transfer RNA* (tRNA) is the key to the code; the amino acids specified by the base sequence of an mRNA molecule are carried to and deposited at the growing end of a polypeptide chain by specific tRNA molecules. *Ribosomal RNA* (rRNA) combines with a set of proteins to form *ribosomes,* which have binding sites for all of the interacting molecules (including mRNA, tRNA, and protein factors) necessary for protein synthesis. Ribosomes bearing tRNAs and special proteins can physically move along an mRNA chain to translate its encoded genetic information. *Translation* refers to the whole process by which the base sequence of the mRNA is used to order and to join the amino acids in a protein. These three types of RNA participate in this essential protein-synthesizing pathway in all cells; in fact, the development of the three distinct functions of RNA was probably the molecular key to the origin of life.

We begin our description of protein synthesis by explaining the genetic code and the role of mRNA in carrying coded information. Next, we describe the structure of tRNA and its elementary biochemistry to show how the language of nucleic acids is converted to the language of proteins. Finally, we summarize the present understanding of how the ribosome serves in organizing the steps of protein synthesis.

Messenger RNA Carries Information from DNA in a Three-letter Genetic Code

Nucleic acids are linear polymers composed of four mononucleotide units. RNA contains ribonucleotides of adenine, cytidine, guanine, and uridine; DNA contains deoxyribonucleotides of adenine, cytidine, guanine, and thymidine. Because four nucleotides cannot specify the linear arrangement of the 20 possible amino acids in a one-to-one manner, a *group* of nucleotides is required to

Table 3-1 The genetic code (RNA to amino acids)*

First position (5' end)	Second position				Third position (3' end)
	U	C	A	G	
U	Phe	Ser	Tyr	Cys	U
	Phe	Ser	Tyr	Cys	C
	Leu	Ser	Stop (och)	Stop	A
	Leu	Ser	Stop (amb)	Trp	G
C	Leu	Pro	His	Arg	U
	Leu	Pro	His	Arg	C
	Leu	Pro	Gln	Arg	A
	Leu	Pro	Gln	Arg	G
A	Ile	Thr	Asn	Ser	U
	Ile	Thr	Asn	Ser	C
	Ile	Thr	Lys	Arg	A
	Met (start)	Thr	Lys	Arg	G
G	Val	Ala	Asp	Gly	U
	Val	Ala	Asp	Gly	C
	Val	Ala	Glu	Gly	A
	Val (Met)	Ala	Glu	Gly	G

*Stop (och) stands for the ochre termination triplet, and Stop (amb) for the amber, named after the bacterial strains in which they were identified. AUG is the most common initiator codon; GUG usually codes for valine, but, rarely, it can also code for methionine to initiate an mRNA chain.

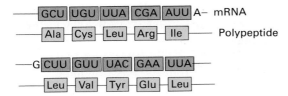

▲ **Figure 3-5** An overlapping triplet code that is read in two different frames. The mRNA is the same sequence in both lines but is read in a different "frame." Several dozen instances of such overlaps have been discovered in the viruses and cell genes of prokaryotes and eukaryotes.

form of a triplet code. The *reading frame* for any set of triplets can be shifted by moving the starting point for translation one or two bases in either direction; thus two or three different amino acid sequences can be encoded by the same region of a nucleic acid chain (Figure 3-5).

The meaning of each codon is the same in almost all known organisms—a strong argument that life on earth evolved only once. The genetic code has been found to differ for a few codons in mitochondria and ciliated protozoans. It is not known whether these rare exceptions to the general code are evolutionarily early forms or if they developed later, but the latter is more likely.

Because there are 61 codons for 20 amino acids in the general code, many amino acids have more than one codon; in fact, leucine, serine, and arginine each have six (Table 3-2). The different codons for a given amino acid are said to be *synonymous*. The code itself is termed *degenerate*, which simply means that it contains redundancies.

The "start" (*initiator*) codon AUG specifies the amino acid methionine: all protein chains in prokaryotic and eukaryotic cells begin with this amino acid. At the beginning of a few chains, methionine is encoded by GUG instead. The three codons UAA, UGA, and UAG do not specify amino acids but constitute "stop" (*terminator*) signals at the ends of protein chains. So a precise linear

array of ribonucleotides in groups of threes in an mRNA specifies a precise linear sequence of amino acids in a protein and also signals where to start and stop synthesis of the protein chain.

Synthetic mRNA and Trinucleotides Break the Code

The discovery of mRNA and how it functions led to the solution of the genetic code—one of the great triumphs of modern biochemistry. The underlying experimental work was largely carried out with cell-free extracts from bacteria. All the necessary components for protein synthesis except mRNA (tRNAs, ribosomes, amino acids, and the energy-rich nucleotides ATP and GTP) were present in these extracts. On the addition of chemically defined synthetic mRNAs, the extracts formed specific polypeptides. For example, synthetic mRNA composed only of U residues yielded polypeptides made up only of phenylalanine. Thus it was concluded that UUU codes for phenylalanine. Likewise, each of the other three homopolymers coded for a single amino acid (Figure 3-6). Next, synthetic mRNA with alternating bases was used; for example,

$$\dots A C A C A C A C A C A C A \dots$$

The polypeptides made in response to this polymer contained alternating threonine and histidine residues. A further experiment was needed to determine whether threonine was encoded by ACA and histidine by CAC or vice versa. An mRNA made of repeated sequences of AAC,

$$\dots A A C A A C A A C A A C A \dots$$

was found to stimulate the synthesis of three kinds of polypeptide chains: all asparagine, all threonine, and all glutamine. Apparently, the decoding mechanism could

Table 3-2 The degeneracy of the genetic code

Number of synonymous codons	Amino acid	Total number of codons
6	Leu, Ser, Arg	18
4	Gly, Pro, Ala, Val, Thr	20
3	Ile	3
2	Phe, Tyr, Cys, His, Gln, Glu, Asn, Asp, Lys	18
1	Met, Trp	2
Total number of codons for amino acids		61
Number of codons for termination		3
Total number of codons in genetic code		64

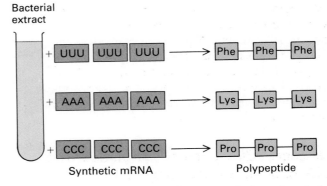

▲ **Figure 3-6** The genetic code was worked out largely by using bacterial extracts that contained all the components necessary for protein synthesis except mRNA. When synthetic mRNAs consisting entirely of a single type of nucleotide were added to the extracts, polypeptides composed of a single type of amino acid formed as indicated. [See M. W. Nirenberg and J. H. Matthei, 1961, *Proc. Nat'l Acad. Sci. USA* 47:1588.]

(a)

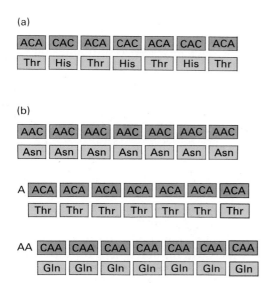

ACA	CAC	ACA	CAC	ACA	CAC	ACA
Thr	His	Thr	His	Thr	His	Thr

(b)

AAC	AAC	AAC	AAC	AAC	AAC	AAC
Asn	Asn	Asn	Asn	Asn	Asn	Asn

A
ACA	ACA	ACA	ACA	ACA	ACA	ACA
Thr	Thr	Thr	Thr	Thr	Thr	Thr

AA
CAA	CAA	CAA	CAA	CAA	CAA	CAA
Gln	Gln	Gln	Gln	Gln	Gln	Gln

▲ **Figure 3-7** (a) When synthetic mRNAs made with alternating A and C residues are added to a bacterial extract, polypeptides made of alternating threonine and histidine residues are formed. The assignment of ACA to threonine and therefore of CAC to histidine is made possible by the result shown in (b). The other two assignments (asparagine to AAC and glutamine to CAA) were derived from further experiments. [See H. G. Khorana, 1968, in *Nobel Lectures: Physiology or Medicine (1963–1970)*, Elsevier (1973), p. 341.]

start at any nucleotide and read the mRNA as three different repeated codons: all AAC, all ACA, or all CAA. The only codon in common between the two synthetic mRNAs was ACA, and the only amino acid in common in the polypeptide products was threonine. Therefore, ACA was assigned to threonine (Figure 3-7). Comparisons of the coding capacity of many such mixed polynucleotides revealed a substantial part of the genetic code.

A second type of experiment with bacterial extracts, also of great importance in the solution of the code, was based on the observation that the presence of a specific RNA trinucleotide causes tRNA bearing a specific amino acid to bind to a ribosome. Thus it was possible to match every amino acid with at least one trinucleotide (Figure 3-8). In all, 61 of the 64 possible trinucleotides were assigned as codons for the 20 amino acids. Trinucleotides UAA, UGA, and UAG did not encode amino acids.

When bacterial extracts were programmed with synthetic mRNAs, polypeptides formed much more slowly than when natural mRNAs were added to bacterial extracts and the lengths of the polypeptide chains were variable. The coding ability of the synthetic mRNA had produced reliable results in experiments designed to decipher the code, but true proteins were programmed by mRNA only when natural mRNAs were added to the bacterial extracts. The first successful synthesis of a specific protein

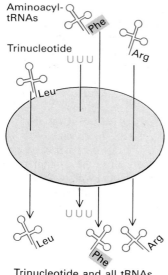

Trinucleotide and all tRNAs
pass through filter

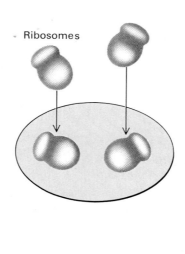

Ribosomes stick to filter

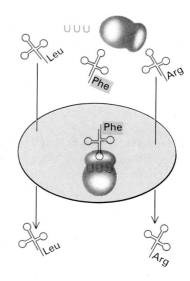

Complex of ribosome, UUU, and
Phe-tRNA sticks to filter

▲ **Figure 3-8** Marshall Nirenberg and his collaborators used extracts of *Escherichia coli* with chemically synthesized trinucleotides to decipher the entire genetic code. They prepared 20 mixtures of aminoacyl-tRNAs (tRNAs with an amino acid attached). In each mixture, a different amino acid was radioactively labeled (green); the other 19 amino acids were present on tRNAs but remained unlabeled. Aminoacyl-tRNAs and trinucleotides would pass through a nitrocellulose filter without binding (left panel); ribosomes, however, did bind to the filter (center panel). A trinucleotide was tested by adding it with ribosomes to samples from each of the 20 aminoacyl-tRNA mixtures. Each sample was then filtered; only one of them would leave a radioactively labeled complex stuck to the filter (for example, the complex of UUU, phenylalanyl-tRNA, and ribosome shown here). Because all possible trinucleotides could be synthesized and tested, each could be matched with an amino acid. [See M. W. Nirenberg and P. Leder, 1964, *Science* **145**:1399.]

(a)

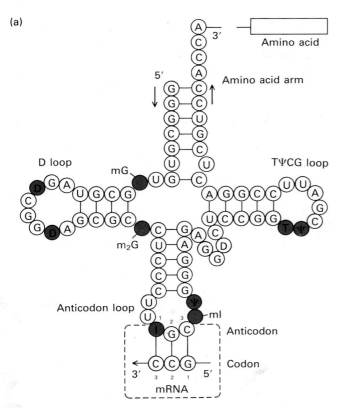

(b)

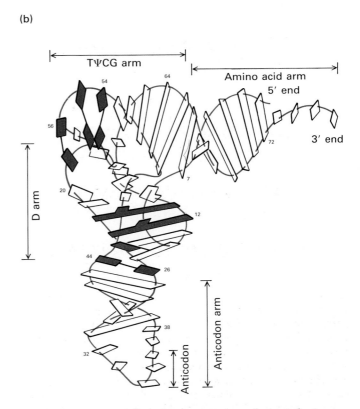

▲ **Figure 3-9** (a) The primary structure of yeast alanine tRNA (tRNA^{Ala}), the first such sequence ever determined. This molecule is synthesized from the nucleotides A, C, G, and U, but some of the nucleotides, shown in red, are modified after synthesis: D = dihydrouridine, I = inosine, T = thymine, Ψ = pseudouridine, and m = methyl group. The primary structure is a cloverleaf consisting of four base-paired stems and three loops: the D loop (for dihydrouridine, a virtually constant constituent of this loop); the anticodon loop; and the TΨCG loop (for thymidylate, pseudouridylate, cytidylate, and guanylate, which are almost always present in sequence in this loop). (b) The standard Watson-Crick base pairs shown as rectangular bars connecting segments of the folded molecule of yeast tRNA^{Phe}. A number of nonstandard molecular interactions (red) help to stabilize the L-shaped molecule. [Part (a), see R. W. Holly et al., 1965, *Science* **147**:1462. Part (b) prepared from x-ray crystallographic data; see J. L. Sussman and S. H. Kim, 1976, *Science* **192**:853.]

occurred when the mRNA of bacteriophage F2 (a virus) was added to bacterial extracts and the coat, or capsid, protein (the "packaging" protein that covers the virus particle) was formed. With the use of real mRNAs it was soon discovered that AUG encoded methionine at the start of almost all proteins and that the three trinucleotides UAA, UGA, and UAG that did not encode any amino acid but were "stop" *(terminator)* codons.

The Anticodon of Transfer RNA Decodes mRNA by Base Pairing with Its Codon

There is no evidence for direct chemical recognition between specific nucleic acid bases and specific amino acids; that is, in the synthesis of a polypeptide chain, the triplets in an mRNA molecule do not select amino acids directly. Instead, the protein-synthesizing system uses transfer RNA (tRNA) as an *adapter molecule* to translate the information in each mRNA triplet so that the appropriate amino acid is added to the chain. Each tRNA molecule must be recognized by one of 20 enzymes called *aminoacyl tRNA synthetases*. Each of these enzymes can add one of the 20 amino acids to tRNA. Once its correct amino acid is attached, the tRNA recognizes a codon in mRNA and adds its amino acid to the growing polypeptide.

Transfer RNA molecules are short (only 70 to 80 nucleotides in length) and varied: there may be as many as 30 or 40 different types in bacteria and up to 50 or so in

animal and plant cells. The sequence of a tRNA molecule always ends in CCA; the amino acid is attached to the 3' hydroxyl group of the terminal adenosine.

In solution, tRNA molecules are folded into three-dimensional structures. The backbone of the configuration is a stem-loop structure resembling a cloverleaf. The four stems are short double helixes stabilized by Watson-Crick base pairing; three of the four stems have loops of seven or eight bases at their ends (Figure 3-9a). At the center of the middle loop, three bases—the *anticodon*—

(a)

1-Methylinosine

1-Methylguanosine

N²,N²-Dimethylguanosine

N⁶,N⁶-Dimethyladenine

(b)

Uracil

Ribose

Uridine

**Dihydrouridine
(5,6-dihydrouridine)**

**Ribothymidine
(5-methyluridine)**

**Pseudouridine
(ribose on C-5)**

(c)

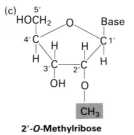

2′-O-Methylribose

◀ **Figure 3-10** The structures of some modified bases in tRNAs. (a) Methylated purine nucleosides. (b) The common nucleoside uridine and its derivatives. (c) 2′-O-Methylribose, which can occur with any RNA base.

(abbreviated Ψ), in which the ribose is attached to carbon 5 instead of to nitrogen 1. These modifications produce a characteristic TΨCG loop in an unpaired region at approximately the same position in nearly all tRNAs (see Figure 3-9). The exact role of most other tRNA modifications has not yet been elucidated. But the fact that the D, anticodon, and TΨCG loops on the tRNA structure (see Figure 3-9) are frequently modified in similar ways suggests that these sites have a common role in protein synthesis. This assumption is further supported by the similar two- and three-dimensional structures of tRNAs.

If *perfect* Watson-Crick base pairing were demanded between codon and anticodon, 61 different tRNA species (one for each codon that specifies an amino acid) would be required. However, this is not the case. A single tRNA anticodon is capable of recognizing more than one (but not necessarily every) codon that corresponds to a given amino acid. One tRNA can recognize multiple codons because of *wobble*, or nonstandard base pairing, between the third nucleotide in the codon and its partner, the first nucleotide in the anticodon. In addition to A-U and G-C, several other base-pair combinations in the wobble position form stable enough interactions to allow codon recognition (Figure 3-11). For example, the codons (5′)UUU(3′) (or UUC) in mRNA call for phenylalanyl-tRNA^Phe (a tRNA with phenylalanine attached). There are several different tRNA^Phe molecules; their anticodons are (3′)AAA(5′), (3′)AAG (5′), or (3′)AAI(5′). (I is the abbreviation for inosine, a modified nucleoside; Figure 3-12.) The bonds between U and G in the wobble positions of codon and anticodon

$$(5')UUU(3')$$
$$(3')AAG(5')$$

are strong enough to allow the (5′)GAA(3′) anticodon to recognize the codon UUU. However, although the tRNA^Phe with a GAA anticodon can pair with the codon, (5′)UUC(3′), which also specifies phenylalanine, the anticodon (3′)AAA(5′) does not recognize the same codon. In the most versatile tRNAs, the wobble position of the anticodon is occupied by inosine, a guanosine analog that lacks an amino group at the 2 carbon. In the wobble position of the tRNA anticodon, inosine can pair satisfactorily with cytosine, adenine, or uracil (see Figure 3-12). Thus one tRNA with inosine in the wobble position can

can form base pairs with the three nucleotides of the mRNA codon. It must be remembered that these three nucleotides are complementary to and in an orientation antiparallel to the codon. The stems and loops of the tRNA are folded into an L-shaped, three-dimensional form (Figure 3-9b). Different hydrogen bonds from those in the standard G-C and A-U base pairs also help to maintain the shape of the molecule.

The nucleic acid bases of tRNAs are highly modified after the tRNA has been synthesized. The most common modification is the addition of a methyl group to specific nucleic acid bases. (Examples of modified structures appear in Figure 3-10.) Most tRNAs are synthesized with a four-base sequence of UUCG near the middle of the molecule. The first uridylate is methylated to become a thymidylate; the second is rearranged into a pseudouridylate

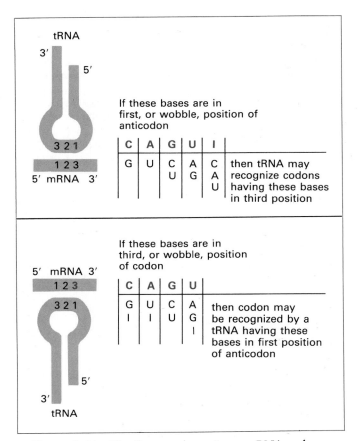

▲ Figure 3-11 The first two bases in an mRNA codon make Watson-Crick base pairs with the third and second bases of a tRNA anticodon. However, the base in the third (or wobble) position of an mRNA codon often forms a nonstandard base pair with the base in the first (or wobble) position of a tRNA anticodon. Wobble allows a tRNA to recognize more than one mRNA codon; conversely, it allows a codon to be recognized by more than one kind of tRNA, although each tRNA will bear the same amino acid. Note that a tRNA with I (inosine) in the wobble position can "read" (become paired with) three different codons, and a tRNA with G or U in the wobble position can read two codons; however, a tRNA with C or A in the wobble position can read only one codon.

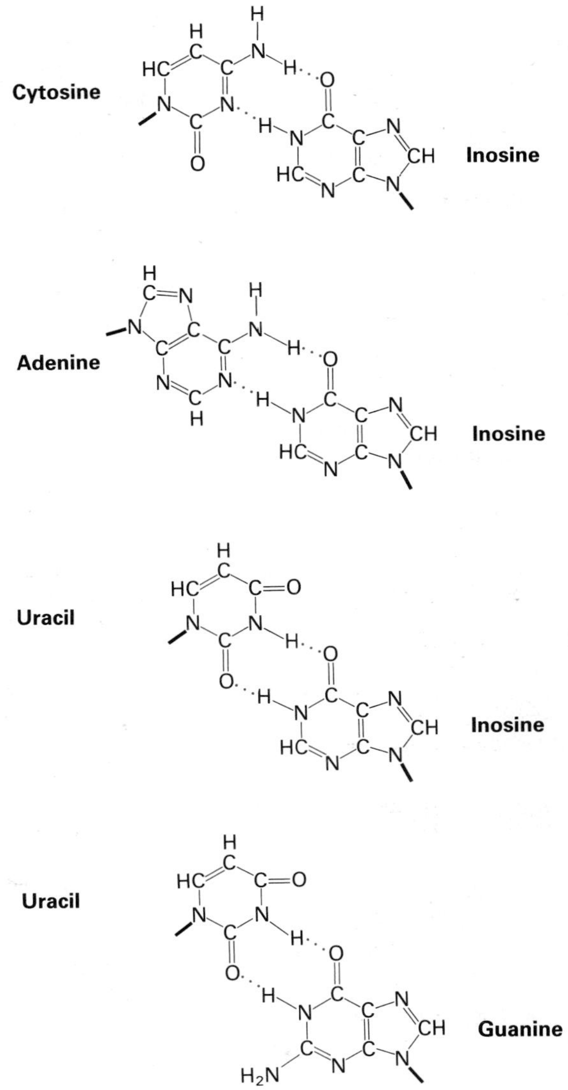

▲ Figure 3-12 The nonstandard, wobble base pairs C-I, A-I, U-I, and U-G. The heavy bonds signify the positions at which the nitrogenous bases attach to the 1′ carbons of the riboses.

decode three different codons for amino acids encoded by multiple codons that differ only as to whether C, A, or U occupies the third position (including leucine, proline, serine, valine, and arginine). The reason why wobble is allowed in the third anticodon site is still unknown, but its effect may be to speed up protein synthesis by the use of alternative tRNAs.

Aminoacyl-tRNA Synthetases Activate tRNA

How does a tRNA molecule become attached to the appropriate amino acid in the first place? Twenty different enzymes are required to specifically recognize amino

acids and their compatible, or *cognate*, tRNAs. Each enzyme can attach a particular amino acid molecule to the end of a cognate tRNA and can recognize different tRNAs for the same amino acid. These coupling enzymes, called *aminoacyl-tRNA synthetases* (AASs), link the amino acid to the free 3′ hydroxyl of the ribose of the terminal adenosine of the tRNA (Figure 3-13). The linkage occurs by a two-step reaction that requires the cleavage of an ATP molecule:

$$\text{Enzyme} + \text{amino acid} + \text{ATP} \xrightarrow{\text{Mg}^{2+}}$$
$$\text{enzyme(aminoacyl-AMP)} + \text{PP}_i \quad (1)$$

$$\text{tRNA} + \text{enzyme(aminoacyl-AMP)} \longrightarrow$$
$$\text{aminoacyl-tRNA} + \text{AMP} + \text{enzyme} \quad (2)$$

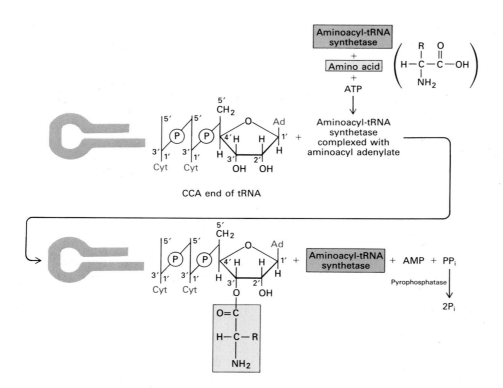

◀ **Figure 3-13** Amino acids are attached to appropriate tRNAs by aminoacyl-tRNA synthetases (enzymes that recognize one kind of amino acid and its cognate tRNAs). The aminoacylation of a tRNA requires energy from the hydrolysis of ATP. The amino acid is attached to the 3′ hydroxyl of the terminal adenylate of the tRNA or to the 2′ hydroxyl with prompt rearrangement to the 3′. The equilibrium of the reaction favors the indicated products because the pyrophosphate is converted into inorganic phosphate by a pyrophosphatase. (Ad = adenine, Cyt = cytosine)

First the enzyme, the amino acid, and ATP form a complex, releasing AMP and PP$_i$ (inorganic pyrophosphate) (1); then the aminoacyl group is transferred to the tRNA (2). The resulting aminoacyl-tRNA retains the energy of the ATP, and the amino acid residue is said to be *activated*. The equilibrium of the reaction is driven further toward activation of the amino acid because *pyrophosphatase* then splits the high-energy phosphoanhydride bond in pyrophosphate. The overall reaction is

$$\text{Amino acid} + \text{ATP} + \text{tRNA} \xrightarrow{\text{enzyme}}$$
$$\text{aminoacyl-tRNA} + \text{AMP} + 2\text{P}_i$$

Each tRNA Molecule Must Be Identifiable by a Specific tRNA Synthetase

The identification of a tRNA by its cognate AAS is just as important to the correct translation of the genetic code as codon-anticodon pairing. Once a tRNA is loaded with an amino acid, codon-anticodon pairing directs the tRNA into the peptide synthesis site; if the wrong amino acid is attached to the tRNA, an error in protein synthesis results.

A classic experiment underlies this conclusion. A cysteine residue already attached to a tRNA^{Cys} (a cysteine-specific tRNA) was chemically changed into alanine, so that the tRNA molecule became an alanyl-tRNA^{Cys} (Figure 3-14). When used in the synthesis of a polypeptide, the tRNA added its alanine residue to the growing chain in response to a cysteine codon. On complete synthesis of the polypeptide, all the usual cysteine residues had been replaced with alanine, proving that only the anticodon of an aminoacyl-tRNA—and not the amino acid—is involved in the recognition step that causes the amino acid to be incorporated into a protein.

The basis for the identification of each tRNA by its cognate AAS has not yet been completely solved, but important recent progress has been achieved. Because one

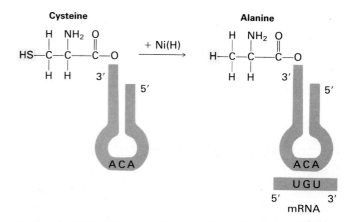

▲ **Figure 3-14** After cysteine is activated by its attachment to a tRNA^{Cys} molecule, it can be chemically altered (by treatment with a nickel compound) into alanine. If the aminoacyl-tRNA is then used in protein synthesis, the resulting polypeptide contains alanine where the mRNA codes for cysteine. [See F. Chappeville, F. Lipmann, G. von Ehrenstein, B. Weisblum, W. J. Ray, and S. Benzer, 1962, *Proc. Nat'l Acad. Sci. USA* **48**:1086.]

AAS can add the same amino acid to two (or more) different tRNAs with different anticodons encoding the same amino acid, each of these tRNAs must have a similar binding site, or *identity*, recognized by the synthetase. Machines can now accurately synthesize DNA molecules at least 100 nucleotides long, so that it is possible to make genes for tRNAs with any desired sequence. The identity of the tRNAs produced by copying these normal and mutant genes can be tested with purified AASs.

The basis for the identity of several tRNAs of *Escherichia coli* is known. No single structure or sequence governs tRNA identity. Perhaps the most logical identity site in a tRNA molecule is the anticodon itself; and for tRNA^{Met} and tRNA^{Val}, this appears to be the case. The CAU anticodon of the tRNA^{Met} that inserts methionine internally in protein (a special tRNA^{Met} carries a methionine that begins a polypeptide) was exchanged for the UAC anticodon of tRNA^{Val}. The normal forms of these two tRNAs are not recognized by the opposite AASs. After the exchange of anticodon sequences, the reverse was true: the tRNA^{Met} with the UAC substitute could now be loaded by the valine AAS but not by the methionine AAS; the converse held for the mutant tRNA^{Val}. Thus, in these two cases, the anticodons are clearly the most important part of the tRNA identity.

However, the anticodon is not the identifier in most cases. In one particularly simple situation, the G3-U70 base pair in the acceptor stem of tRNA^{Ala} is the critical identity element: this base pair must be present for this tRNA to be recognized by its cognate AAS. Furthermore, the insertion of this single base pair in the stem of a minor species of tRNA^{Phe} causes its conversion to an alanine identity. In other cases, combinations of nucleotides are found to be important (for example, in the major serine tRNA; Figure 3-15).

The existence of tRNA identity and the crystallographic solution of structures of tRNAs with their cognate AASs should soon provide a clear molecular understanding of the rules for this critical step in protein synthesis.

Ribosomes Are Protein-synthesizing Machines

The highly specific chemical interactions of translation do not take place in free solution inside a cell. This critical function, which results in the formation of more than 1 million peptide bonds each second in an average mammalian cell, requires the interaction of many chemical groups on RNAs and assisting proteins. Protein synthesis would be very inefficient if each of these participants had to react in free solution: simultaneous collisions between the necessary components of the reaction would be so rare that the rate of amino acid polymerization would be very

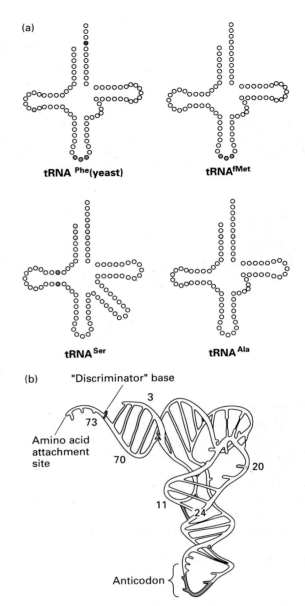

▲ **Figure 3-15** (a) Major identity elements in four tRNAs. Each base in the tRNA is represented by a circle. Colored circles indicate positions in the cloverleaf that have been shown to be identity elements for the cognate AAS. In each case, additional elements may yet be discovered. (b) Location of elements in a generalized tertiary structure of tRNA. The positions of the major identity elements shown in (a) are indicated. [See L. D. H. Schulman and J. Abelson, 1988, *Science* 240:1590.]

slow indeed. Instead, the mRNA with its encoded information and the individual tRNAs loaded with their correct amino acids are brought together by their mutual binding to the RNA-protein complex called a *ribosome*.

With the aid of the electron microscope ribosomes were first discovered as discrete, rounded structures prominent in tissues secreting large amounts of protein; initially, however, they were not known to play a role in

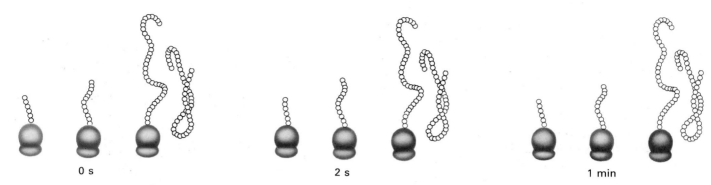

▲ **Figure 3-16**   An experiment that identified ribosomes as the sites of protein synthesis. Ribosomes with unfinished peptide chains of different lengths are shown. When cells are exposed to radioactive amino acids for very short times, the newly incorporated labeled amino acids are in growing protein chains associated with ribosomes. After long exposure times, completely labeled proteins that are free of the ribosomes contain more and more of the labeled amino acids. [See K. McQuillen, R. B. Roberts, and R. J. Britten, 1959, *Proc. Nat'l Acad. Sci. USA* **45**:1437.]

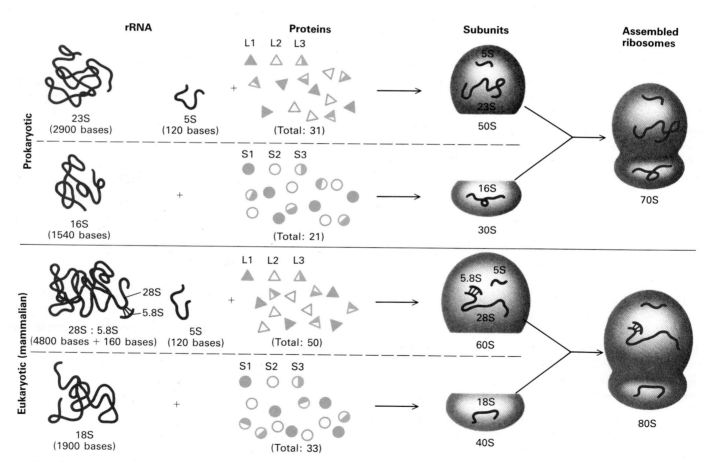

▲ **Figure 3-17**   In all cells, each ribosome consists of a large and a small subunit. The different subunits contain rRNAs of different lengths as well as a number of different proteins (indicated by different shadings). All ribosomes contain two major rRNA molecules. Prokaryotic ribosomes also have one small 5S rRNA about 120 bases long. Eukaryotic ribosomes have two small rRNAs: a 5S molecule similar to the prokaryotic 5S, and a 5.8S molecule 160 bases long. The proteins are named L1, L2, etc., and S1, S2, etc., depending on whether they are found in the large or the small subunit. Some cell organelles also have ribosomes: chloroplast ribosomes are similar to prokaryotic ribosomes; ribosomes in mitochondria have smaller RNAs and fewer proteins than prokaryotic ribosomes.

protein synthesis. When reasonably pure preparations were achieved, ribosomes were seen to be very consistent in size. Once it was demonstrated that newly synthesized polypeptides were attached to structures the size of ribosomes, experimenters concluded that ribosomes are the sites of protein synthesis (Figure 3-16).

Like the structure of tRNAs, the structure of ribosomes is similar but not identical in all species. This consistency is another reflection of the common evolutionary origin of many of the most basic constituents of living cells.

The ribosome is a complex composed of individual RNA molecules (the third major type of cellular RNA) and more than 50 proteins, organized into a large subunit and a small subunit (Figure 3-17). The proteins in the two subunits differ, as do the molecules of ribosomal RNA (rRNA). The large ribosomal subunit contains a larger rRNA molecule; the small subunit contains a small rRNA. The subunits and the rRNA molecules are commonly designated in terms of Svedberg units (for example, 23S for the large prokaryotic rRNA); the Svedberg is a measure of the sedimentation rate of suspended particles centrifuged under standard conditions. The lengths of the major rRNA molecules, the quantity of proteins in each subunit, and consequently the sizes of the subunits differ in prokaryotic and eukaryotic cells. At this point, we want to emphasize the similarities among ribosomes.

The sequences of the small ribosomal RNAs from several hundred organisms and the large ribosomal RNAs from several dozen organisms are now known. Although the primary nucleotide sequences of these rRNAs vary considerably, the overall structure of the folded rRNA is similar in most cases. By maximizing the formation of complementary base-paired stems, a two-dimensional stem-loop structure can be drawn for rRNA. Such structures have been verified by chemically cross-linking the bases in base-paired segments; the cross-linking is then detected in enzymatic digests of the RNA that leave base-paired RNA intact. Forty such stem-loops are located similarly along the sequences of prokaryotic and eukaryotic small rRNAs (Figure 3-18). When rRNA is in solu-

▶ **Figure 3-18** Secondary structural maps of small ribosomal RNAs. Several hundred primary sequences of small ribosomal RNAs are available. The rRNAs were folded into stem-loop structures to achieve maximal base-pairing. The stem-loops were proved in various ways, such as chemical cross-linking of base pairs followed by digestion of the rRNA by enzymes that destroy single-stranded RNA and isolation of linked segments. The helices (stems of the stem-loops) found in both (a) prokaryotic and (b) eukaryotic small rRNAs are numbered 1–40. In general, the length and position of the 40 helices are well conserved, although the exact sequence varies from species to species. The most highly conserved regions are represented as red lines. The numbers of stem-loops found only in prokaryotes or only in eukaryotes are preceded by P or E. *Adapted from E. Huysmans and R. DeWachter, 1987, Nucleic Acids Res.* **14:**73–118.

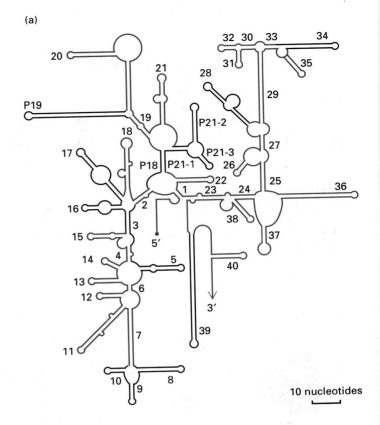

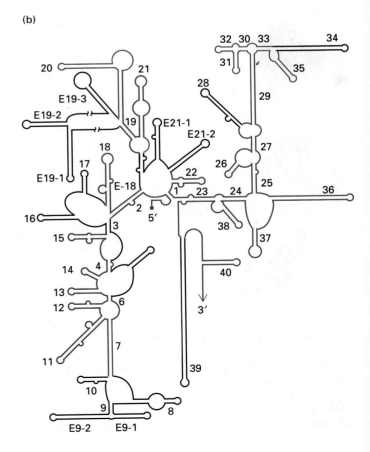

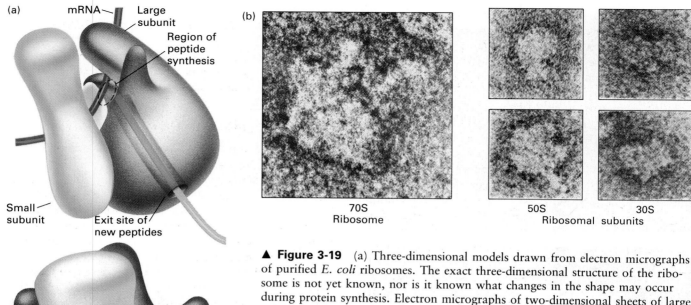

▲ **Figure 3-19** (a) Three-dimensional models drawn from electron micrographs of purified *E. coli* ribosomes. The exact three-dimensional structure of the ribosome is not yet known, nor is it known what changes in the shape may occur during protein synthesis. Electron micrographs of two-dimensional sheets of large subunits give evidence of the tunnel that is shown. The exit of new peptides is thought to be at the base of the large subunit, possibly from the tunnel. The region of peptide synthesis has been inferred from biochemical experiments that show attachment of new peptide chains to the large subunit and from proximity to positions of important proteins in the small subunit (see Figure 3-20). The path of mRNA is reasonable but is a conjecture. (b) Electron micrographs of a 70S ribosome and of 50S and 30S ribosomal subunits. The subunits are viewed from different angles. *Drawings after J. A. Lake. Photographs from J. A. Lake, 1976, J. Mol. Biol. 105:131.* [See A. Yonath, K. R. Leonard, and H. G. Wittman, 1987, *Science* **236**:813.]

tion, the stem-loop structure undoubtedly folds in three dimensions and exhibits three major domains. The stem-loop structure of the large rRNA is more complicated, but again the structures in all species examined exhibit a clear resemblance, suggesting that the fundamental machinery in all present-day cells may have arisen only once in evolution.

Although the ribosome is a very complicated assembly of proteins and RNA, great progress has been made on bacterial ribosome structure. The dimensions and overall structure, including prominent surface features, are revealed by high-resolution electron-microscope studies (Figure 3-19). Each constituent rRNA and protein has been purified and its sequence determined. Furthermore, the ribosomal particle can be reassembled when the proteins and rRNA are mixed in the correct order. This knowledge has led to experiments that reveal which ribosomal proteins are bound to which stem-loops of rRNA and which proteins are neighbors in the whole particle (Figure 3-20). For example, RNAse digests and base-specific chemical cleavage of whole ribosomes release pieces of protein plus RNA, both of which can be characterized. It is also possible to replace one of the proteins of the smaller subunit from the bacterium *E. coli* with the

same protein from another bacterium (say, a *Bacillus* strain) to produce a fully functional hybrid ribosomal subunit. The position of the *Bacillus* protein on the ribosome can be determined in the electron microscope by using tagged antibodies directed against that protein. Alternatively, by replacing some of the hydrogens of a protein with deuteriums, its location within a reassembled particle can be determined by physical techniques: a protein containing deuterium deflects a neutron beam differently than a normal protein does. When two deuterated proteins are used in the reassembly of a ribosomal subunit, the distance between them can be determined precisely. Most of the proteins in the small subunit of *E. coli* have been located by using antibodies, deuterated proteins, and the biochemical techniques of base-specific cleavages. From the estimated space required by each protein and its position in the rRNA, a detailed model of the folded RNA and protein can be made (Figure 3-20b).

All of the individual proteins of eukaryotic ribosomes have been separated, but reassembly of complete ribosomes has not yet been achieved. As with prokaryotes, proteins on the ribosomal surface and loops of rRNA at the surface must bind the various factors that assist in the various steps of protein synthesis.

(a)

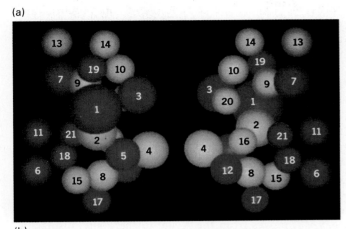

(b)

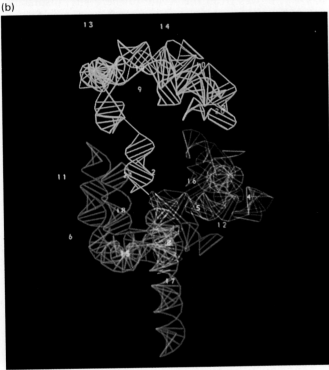

▲ **Figure 3-20** Models of the structure of the *E. coli* small ribosomal subunit. The separated rRNA and protein components can be reassociated into ribosomes and proteins located on the stem-loop structures shown in Figure 3-18 by several techniques (see text). (a) A model ("front" and "back" views) of the proteins in the small ribosomal subunit of *E. coli*. A sphere representing each protein is shown in proportion to its size. The spatial arrangement is inferred from the reassociation of isolated pairs of deuterium-labeled proteins with the small RNA, followed by neutron-scatter examination of the reconstituted particles to detect the distance between the deuterated proteins. (b) Model of folded 16S rRNA with associated proteins (green numbers) constructed from both biochemical and biophysical data. Three major domains are shown in red, yellow, and blue. [Part (a), see M. S. Capel et al., 1988, *J. Mol. Biol.* **200**:65.] *Part (b) from S. Stern, T. Powers, L.-M. Changchien, and H. Noller, 1989, Science **244**:786.* Photos courtesy of P. B. Moore and H. Noller.

The Steps in Protein Synthesis

Protein synthesis is usually considered in three stages—*initiation, elongation,* and *termination*—each of which involves distinct and important biochemical events.

AUG Is the Initiation Signal in mRNA

The first event of the initiation stage is the attachment of a free molecule of methionine to the end of a tRNA^{Met} by the specific methionyl-tRNA synthetase. There are at least two types of tRNA^{Met}. Only one, called tRNA$_i^{Met}$, can initiate protein synthesis. The same tRNA synthetase can attach methionine to other tRNAs that incorporate it within the protein chain; however, only methionyl-tRNA$_i^{Met}$ (methionine, attached to tRNA$_i^{Met}$) can bind to a small ribosomal subunit to begin the process of protein synthesis. [In bacteria, the amino group of the methionine of the methionyl-tRNA$_i^{Met}$ is modified by the addition of a formyl group and is sometimes described as *N*-formylmethionyl-tRNA$_i^{fMet}$. However, methionyl-tRNA$_i^{Met}$ (abbreviated Met-tRNA$_i^{Met}$) is a general designation for the initiator tRNA in all cells.] The Met-tRNA$_i^{Met}$ together with a molecule of GTP (a guanosine phosphate) and a small ribosomal subunit bind to the mRNA at a specific site most often located quite near the AUG initiation codon. The energy required for several different steps in protein synthesis is furnished by GTP hydrolysis; in this first case, GTP is required to position the AUG correctly on the ribosomal subunit.

In bacteria, the ribosome is guided to the initiation site by a short nucleotide sequence. This *Shine-Dalgarno sequence,* named for its discoverers, is complementary to a sequence at or very near the 3′ end of the small rRNA molecule. Thus

mRNA 5′—UAAGGAGG–(5–10 nucleotides)–AUG
$\qquad$ | | | | | | | |
$\qquad$ HO—AUUCCUCC–(~1400 nucleotides)–5′ rRNA

Every mRNA does not match this sequence exactly, but on average six out of eight nucleotides match. Bacterial ribosomes can initiate at these sites in the middle of mRNAs thousands of nucleotides long. In eukaryotic cells, the consensus sequence for ribosomal recognition signalling a nearby protein coding sequence is different:

mRNA 5′—ACCAUGG–

Whether the eukaryotic sequence functions by recognition of complementary sites in the ribosomal RNA is not yet known, but the mechanism of initial ribosomal engagement with mRNA is different; recognizing the 5′ end of the mRNA is the important first step. Over 90 percent of the AUGs used as initiation signals in eukaryotes are the first (the 5′-most) AUG in the molecule. The 5′ end of eukaryotic mRNAs has a special modified end, a methylated guanylate residue linked to the first ordinary nucleotide in a 5'-5' pyrophosphate linkage, which chemically seals off or "caps" the RNA:

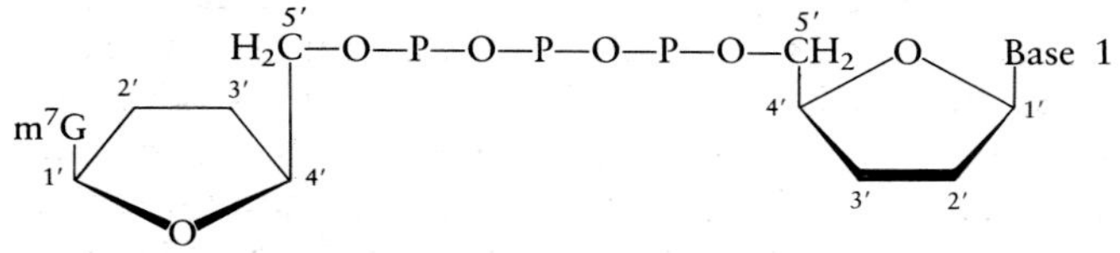

$$\text{m}^7\text{G} \quad \text{H}_2\text{C}-\text{O}-\text{P}-\text{O}-\text{P}-\text{O}-\text{P}-\text{O}-\text{CH}_2 \quad \text{O} \quad \text{Base 1}$$

Artificial circular mRNAs have been prepared, and they are not translated; mRNAs without the cap are translated very poorly. Therefore, initiation of the translation of eukaryotic mRNAs involves recognition of the cap followed by the location of a consensus sequence surrounding the AUG codon.

Initiation Factors, tRNA, mRNA, and the Small Ribosomal Subunit Form an Initiation Complex

A group of proteins called *initiation factors* (IF) (Figure 3-21) help the ribosome find an initiation site. Without these proteins, the complex of mRNA, Met-tRNA$_i^{Met}$, GTP, and the small ribosomal subunit does not form. Three initiation factors have been purified from prokaryotic (bacterial) cells, and at least five or six are known in eukaryotic cells; each has been shown to play a role in correct protein initiation. The most important step is the formation of a complex of the small subunit with met-tRNA$_i^{Met}$ and the mRNA. In prokaryotes, IF$_2$ is critical in finding the AUG. In eukaryotes, the large eIF$_4$ complex helps the small subunit bind the end of the mRNA and search for the AUG. Only after the AUG is located does the large subunit join the complex of Met-tRNA$_i^{Met}$, GTP, and the small ribosomal subunit (Figure 3-21a). In general, ribosomes and initiation factors from different bacterial species can be substituted for one another in protein synthesis in the test tube. The same is true for eukaryotic ribosomes and initiation factors, even when the mixture consists of, for example, a combined extract of human and yeast cells. But translation of prokaryotic mRNA by eukaryotic ribosomes (and vice versa) is poor.

Ribosomes Use Two tRNA-binding Sites (A and P) during Protein Elongation

For the peptide chain to begin to grow, a second aminoacyl-tRNA must be properly positioned on the ribosome. This is the job of one of the *elongation factors* (EF) (Figure 3-21b). Two ribosomal sites are occupied by tRNA molecules: the *A site* accommodates the incoming *amino*acyl-tRNA that is to contribute a new amino acid to the growing chain; the *P site* contains the *p*eptidyl-tRNA complex (the tRNA still linked to all the amino acids added to the chain so far). Alternative types of binding

between the ribosome and the tRNAs may occur as the ribosome changes shape during protein synthesis, but these could be considered intermediate stages between the two major binding conformations.

As the contributor of the first amino acid of the chain, the Met-tRNA$_i^{Met}$ enters the P position. In the hypothetical peptide shown in Figure 3-21, the incoming phenylalanyl-tRNA^{Phe} (a molecule of the second amino acid, phenylalanine, attached to a tRNA^{Phe} by phenylalanyl-tRNAsynthetase) is then bound to the A position (Figure 3-21b). A peptide bond is formed between the carboxyl group of the methionine and the amino group of the Phe-tRNA^{Phe} to make the dipeptide methionyl-phenylalanyl-tRNA^{Phe}; in this process, the peptidyl-tRNA^{Phe} moves from the A site to the P site, displacing the tRNA$_i^{Met}$. The hydrolysis of GTP furnishes the energy for this *translocation* of the peptidyl-tRNA. In our example, the third codon is (5')CUG(3'); therefore a leucinyl-tRNA^{Leu} with an anticodon of (3')GAC(5') or (3')GAI(5') is required. After the Leu-tRNA^{Leu} attaches to the A site, the leucine is incorporated into the new peptidyl chain and the translocation process repeats itself.

The reactivity of a peptidyl-tRNA and an aminoacyl-tRNA with its activated amino acid is such that if the two are brought together, a peptide bond forms spontaneously. Thus a major role of the ribosome must be to offer binding sites to aminoacyl-tRNA in such a way that the correct codon-anticodon match is made before an activated amino acid is brought close to the peptidyl-tRNA. The selection of the correct aminoacyl-tRNA for elongation of the chain is the most time-consuming part of protein synthesis. The average rate of addition is about five amino acids per second. In bacterial cells, the proteins that carry out the delivery of aminoacylated tRNAs to the correct site on the ribosome and assist in the translocation of the tRNA have been thoroughly characterized (Tu and Ts in Figure 3-21b). Recently the eukaryotic proteins equivalent to the bacterial proteins have been shown to be very similar proteins.

For polypeptides longer than the tripeptide depicted in Figure 3-21, the process of translocation continues step-by-step until all the amino acids encoded by the mRNA have been added. In each translocation step, the ribosome with its attached peptidyl-tRNA moves three nucleotides closer to the 3' end of the mRNA. The exact mechanism for this movement remains obscure; however, some protein or proteins in the ribosome probably change configuration, using the energy of GTP hydrolysis to propel the mRNA through the ribosome. Because some hydrogen bonds in rRNA (see Figure 3-18) are between distant nucleotides, translocation is believed to occur through a *contraction-relaxation* cycle in which the folding of the ribosome changes. Because of new findings on the catalytic roles of RNA molecules in other reactions, it is conjectured that the ribosomal RNA, may play an important catalytic function in peptide-bond synthesis.

UAA, UGA, and UAG Are the Termination Codons

In the mRNA for our hypothetical tripeptide in Figure 3-21, the three bases following the leucine codon are UAG. This codon, like the codons UAA and UGA, signals the release of the peptidyl-tRNA complex (Figure 3-21c) when recognized by protein *termination factors* (TF). Almost simultaneously the complex divides into an uncharged tRNA molecule lacking an attached amino acid and a newly completed protein chain that can either assume its final shape or combine with additional protein subunits. (Actually, the interaction of two peptide chains can begin while a protein chain is still growing.) After releasing its peptidyl-tRNA, the ribosome disengages from the mRNA, divides into two subunits, and is ready to start the whole cycle again.

The peptide synthesis depicted in Figure 3-21 oversimplifies the release process; a tripeptide might, in fact, not even be released. The growing peptide chain on a ribosome is "buried" within the ribosome. Experimentally, a brief proteolytic digestion of active ribosomes destroys most of each growing peptide chain, leaving an undigested piece 35 amino acids long associated with each ribosome. The logical place to accommodate and protect about 35 amino acids is the tunnel recently observed in the large subunit (see Figure 3-19). At the base of the large subunit (near the presumed end of the tunnel), growing peptides can react with added antibodies, indicating that this is the exit site for the new polypeptide.

Rare tRNAs Suppress Nonsense Mutations

Perhaps most mutations result in the substitution of one amino acid for another; these are termed *missense mutations*. However, if a mutation in a gene produces one of the termination codons UAA, UGA, or UAG, a premature termination signal results (for example, if UGG mutates to UAG, a tryptophan codon becomes a stop codon). Such a mutation is called a *nonsense mutation* because it causes the translation apparatus to stop too soon. Nonsense mutations were recognized in bacteria when some mutations were found to result in the formation of shortened forms of bacterial cell or virus proteins.

Chain-terminating mutations were themselves found to be correctable or *suppressed* if the mutant gene was experimentally transferred to special bacterial strains, termed *suppressor strains*. All the mutations that can be corrected by one particular suppressor strain are called *amber* mutations; those that can be corrected by another strain are called *ochre* mutations. Each suppressor strain carries either a mutation in the anticodon of a tRNA, which changes its reading capacity, or contains enough of a minor "suppressing" tRNA that can occasionally read a terminator codon as an amino acid codon.

For example, the amber codon (5')UAG(3') is not normally recognized by a tRNA molecule. However, this codon can be recognized by a minor tyrosine tRNA species that contains (3')AUC(5') as its anticodon. Thus in the translation of a mutant mRNA containing the chain terminator UAG in the middle of a protein-coding sequence, a tyrosine will be inserted by the suppressor tRNA and a full-length mutant protein is produced that may or may not have nearly full activity. Perhaps because of the existence of suppressor tRNAs, the 3' ends of coding regions in mRNAs often contain two or more stop codons within a short stretch. In such cases, the termination of protein synthesis has a "fail-safe" mechanism.

Suppressor tRNAs have also been discovered in yeast cells. The discovery and characterization of suppressible mutations have provided a valuable tool for the molecular geneticist. Any gene harboring a mutation that can be suppressed by a tRNA mechanism encodes a protein (rather than tRNA or rRNA). By ascertaining which suppressible site is mutant (in our example, the codon for tyrosine) the experimenter gains some information about the amino acids required for protein function.

Nucleic Acid Synthesis

The ordered assembly of the basic units in DNA (deoxyribonucleotides) or RNA (ribonucleotides) involves much simpler cellular mechanisms than the correct assembly of the amino acids in a protein chain. As Watson and Crick remarked in their proposal of the double-helical DNA structure: "It has not escaped our notice that the specific pairing we have postulated immediately suggests a possible copying mechanism for the genetic material." Their prediction that the synthesis of nucleic acids is produced by copying the DNA strands has proved to be correct. The phenomena associated with the duplication of nucleic acids have given rise to two very broad and intensive areas of research in molecular biology: DNA replication and its role in cell growth and division, and the synthesis of specific mRNAs and their role in cell adaptation and differentiation.

Nucleic Acid Synthesis Can Be Described by Five Rules

We now consider a few general principles governing the synthesis of DNA and RNA chains and briefly discuss some properties of the enzymes that carry out nucleic acid synthesis. We will need to refer to those general principles as we develop the detailed relation of these crucial events to cell growth and differentiation in Chapters 7–12, 17–18, and 24–25.

1. *Both DNA and RNA chains are produced in cells by the copying of a preexisting DNA strand according*

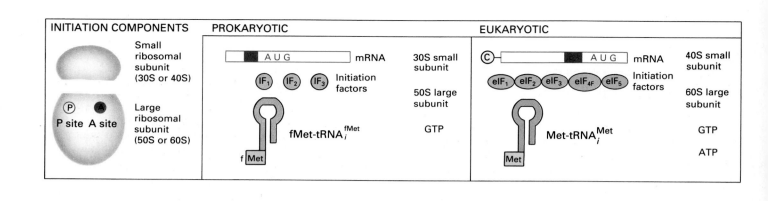

INITIATION COMPONENTS	PROKARYOTIC	EUKARYOTIC

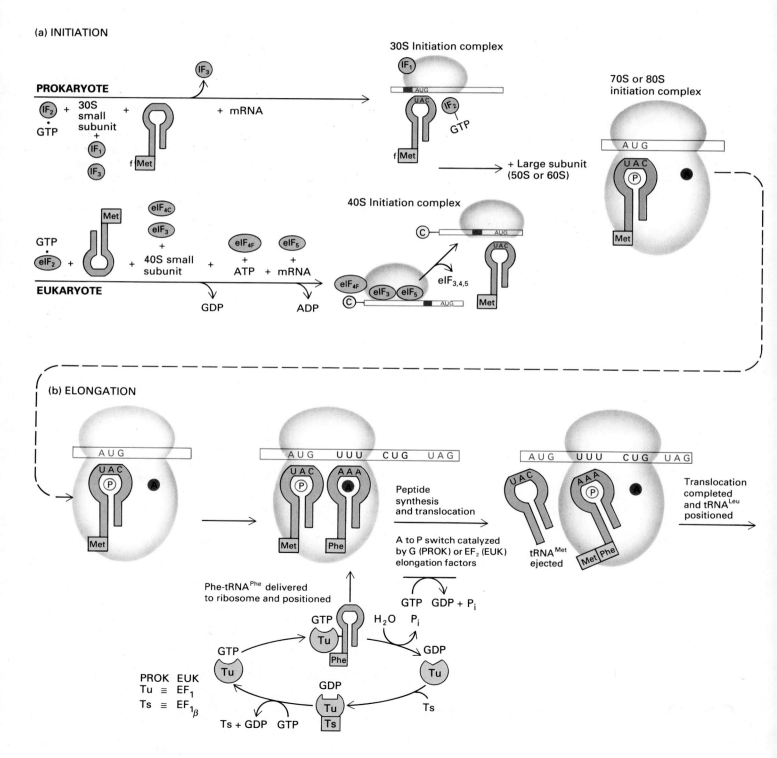

(a) INITIATION

PROKARYOTE

30S Initiation complex

70S or 80S initiation complex

IF₂ · GTP + 30S small subunit + IF₁ + IF₃ + fMet + mRNA

+ Large subunit (50S or 60S)

EUKARYOTE

40S Initiation complex

GTP · eIF₂ + Met + eIF₄C + eIF₃ + 40S small subunit + eIF₄F + ATP + eIF₅ + mRNA

GDP   ADP

eIF₃,₄,₅

(b) ELONGATION

Peptide synthesis and translocation

A to P switch catalyzed by G (PROK) or EF₂ (EUK) elongation factors

Translocation completed and tRNA^Leu positioned

tRNA^Met ejected

Phe-tRNA^Phe delivered to ribosome and positioned

GTP   GDP + Pᵢ
H₂O   Pᵢ

PROK EUK
Tu ≅ EF₁
Ts ≅ EF₁β

Ts + GDP   GTP

◄ **Figure 3-21** The three stages of the translation of the genetic message from mRNA into protein.

(a) *Initiation.* An initiation factor (IF_2 in prokaryotes and eIF_2 in eukaryotes) binds a molecule of GTP and a molecule of Met-tRNA$_i^{Met}$ to form a ternary complex. (The initiating methionine is formylated in prokaryotes.) This complex binds to other initiation factors (IF_3, eIF_3, etc.), to mRNA, and to the small ribosomal subunit to make the 30S initiation complex in prokaryotes and the 40S complex in eukaryotes. After locating the ribosome binding site (red box) near the AUG initiation codon, a large ribosomal subunit joins to complete the 70S or 80S initiation complex. (During these steps, GTP is hydrolyzed in prokaryotes.) The Met-tRNA$_i^{Met}$ bearing the first amino acid is now bound to the ribosome at the P (peptidyl-tRNA) site. The initiation complex is ready to begin synthesis of the peptide chain.

(b) *Elongation.* The growing polypeptide is always attached to the tRNA that brought in the last amino acid. A new aminoacyl-tRNA (Phe-tRNA^{Phe} here) binds to the ribosome at the A site. During elongation in prokaryotes, the protein complex Tu-Ts catalyzes the binding of each aminoacyl-tRNA to the ribosome. In eukaryotic cells, the similar proteins are called elongation factors (EF_1 and EF_{1_β}). An activated Tu-GTP complex binds to the TΨCG loop found in all tRNAs and allows the tRNA to associate correctly with the ribosome at the site. GTP is hydrolyzed, and the cycle is repeated when Ts aids in reactivating the Tu. After the incom-

ing aminoacyl-tRNA is correctly placed in the A site—that is, once the codon-anticodon pairing is correct—the peptide chain (here, the first methionine) is transferred to the amino group of the newly arrived aminoacyl-tRNA, generating a peptidyl-tRNA that has acquired an additional amino acid. (In our example, the compound is methionyl-phenyl-alanyl-tRNA^{Phe}.) At this stage, the peptidyl-tRNA is bound to the ribosome at the A site, and the ribosome moves one codon down the mRNA chain. (Here, the mRNA codons are illustrated with spaces separating them for convenience.) The translocation reaction is catalyzed (in bacteria by the elongation factor G and in eukaryotes by the elongation factor EF_2) using energy from the hydrolysis of GTP. With this movement, the empty tRNA is released from the P site and the peptidyl-tRNA is shifted to the P site. This sequence of events is repeated for every amino acid added to the growing chain. Thus two molecules of GTP are used: one to position the tRNA and one in the addition of each amino acid during translocation.

(c) *Termination.* When the ribosome arrives at the UAG codon, the translation is completed with the aid of termination factors (three proteins in bacteria, one in eukaryotes). Hydrolysis of the peptidyl-tRNA on the ribosome, with the release of the completed polypeptide and the last tRNA, is followed by dissociation of the two ribosomal subunits. This final step also requires GTP hydrolysis.

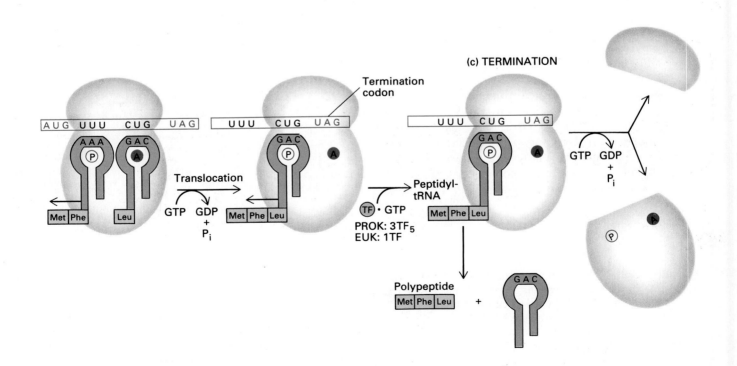

to the rules of Watson-Crick base pairing. The DNA from which the new strand is copied is called a *template.* The information in the template is preserved: although the first copy has a complementary sequence, not an identical one, a copy of the copy produces the original sequence again. In the replication of a double-helical DNA molecule, or *duplex,* both original DNA strands are copied. In some viruses, RNA molecules are produced by copying preexisting RNA molecules; in one class, the *retroviruses,* DNA is produced by copying RNA. However, the vast majority of cellular RNA and DNA synthesis in cells is from a preexisting DNA template.

2. *Nucleic acid strand growth is in one direction: $5' \rightarrow 3'$.* Because nucleic acids are phosphodiesters (polymers of nucleotides regularly linked by a phosphate group between the $5'$—OH of one sugar and the $3'$—OH of the adjacent sugar), each strand has a definite chemical orientation: a $5'$(phosphate) end and a $3'$ (hydroxyl) end. All RNA and DNA synthesis, both cellular and viral, proceeds in the same chemical direction: from the $5'$ to the $3'$ end (Figure 3-22). The nucleotides used in the construction of nucleic acid chains are $5'$-triphosphates of ribonucleosides or deoxyribonucleosides. Strand growth is energetically unfavorable but is driven by the energy available in the triphosphates. The α phosphate of the incoming nucleotide is attached to the $3'$ hydroxyl of the ribose (or deoxyribose) of the preceding residue to form a phosphodiester bond, releasing a pyrophosphate (PP_i). The equilibrium of the reaction is driven further toward chain elongation by pyrophosphatase, which catalyzes the cleavage of PP_i into two molecules of inorganic phosphate (P_i).

3. *Special enzymes called polymerases make RNA or DNA.* The enzymes that copy DNA to make more DNA are *DNA polymerases;* those that copy RNA from DNA are *RNA polymerases.* An accurate copy of a nucleic acid by a polymerase always requires a template. However, a few polymerases can catalyze the addition of bases to a chain without a template.

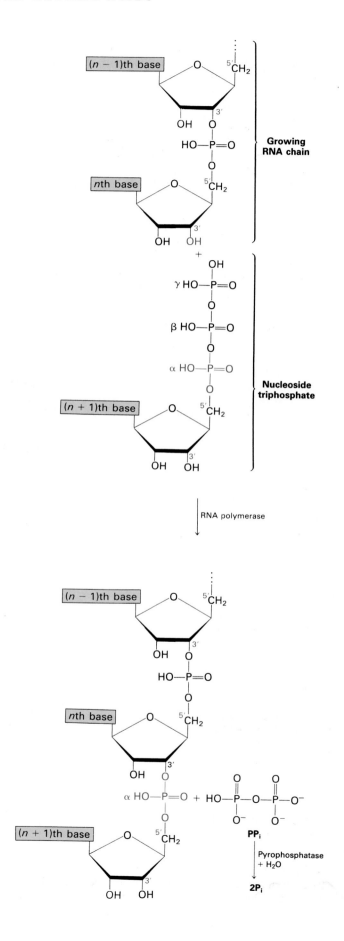

▶ **Figure 3-22** Nucleic acids grow one nucleotide at a time and always in the same direction: $5' \rightarrow 3'$. A nucleoside triphosphate bonds to the $3'$ hydroxyl end of the growing strand, releasing one pyrophosphate ion (PP_i). The remaining phosphate of the triphosphate becomes part of the backbone of the new strand. Both RNA and DNA grow by the same type of reaction: deoxyribonucleoside triphosphates are substrates for DNA growth; ribonucleoside triphosphates are substrates for RNA growth.

One important enzyme of this type is poly A polymerase, which adds a string of adenylates to the 3′ ends of eukaryotic mRNAs.

The copying of DNA to make RNA is called *transcription*. Because the two DNA strands are complementary, rather than identical, they obviously have different protein-coding potentials. It was believed for many years that only one strand of the DNA duplex gives rise to usable information when transcribed into RNA—and this is almost always true. In a few recently discovered cases, however, this rule is violated: the DNA encodes proteins on both strands over a limited section of the DNA.

Cells have several different types of DNA polymerases, but the physiological role of each of these enzymes is not fully understood, even for viral and bacterial enzymes. Apparently one type of RNA polymerase synthesizes mRNA, rRNA, and tRNA in prokaryotic (bacterial) cells. Eukaryotic cells, even single-celled organisms such as yeasts, have three distinct types of RNA polymerases, each of which is responsible for making a different kind of RNA. Eukaryotic RNA polymerase I makes ribosomal RNA, RNA polymerase II produces mainly mRNA, and RNA polymerase III makes tRNA and other small RNAs. It is not yet known why the labor of RNA synthesis in eukaryotic cells is divided among three enzymes, but the roles of these different RNAs clearly differ, as do the sites of their synthesis.

RNA polymerases can initiate a nucleic acid strand, DNA polymerases cannot. An RNA polymerase can find an appropriate initiation site on duplex DNA, bind the DNA, separate the two strands in that region, and begin generating a new RNA strand (Figure 3-23).

The nucleotide at the terminal 5′ end of an RNA strand is chemically distinct from the nucleotides

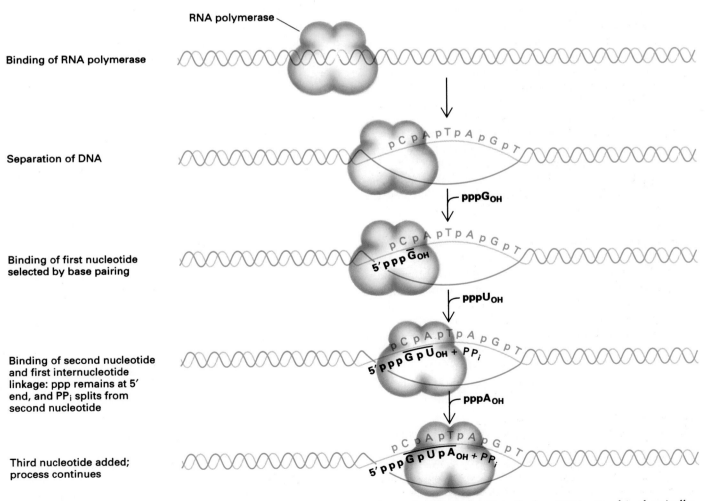

Binding of RNA polymerase

RNA polymerase

Separation of DNA

Binding of first nucleotide selected by base pairing

Binding of second nucleotide and first internucleotide linkage: ppp remains at 5′ end, and PP_i splits from second nucleotide

Third nucleotide added; process continues

▲ **Figure 3-23** The transcription of a DNA strand into an RNA strand is catalyzed by an RNA polymerase, which can initiate the synthesis of strands de novo on DNA templates. The nucleotide at the 5′ end of an RNA strand is chemically distinct from the nucleotides within the strand in that it retains all three of its phosphate groups.

within the strand in that it retains all three phosphate groups. When an additional nucleotide is added to the 3′ end of the growing strand, only the α phosphate is retained; the β and γ phosphates are lost.

Instead of initiating a nucleic acid chain de novo, DNA polymerases add nucleotides to the hydroxyl group at the 3′ end of a preexisting RNA or DNA strand, called a *primer*. If RNA is the primer, the resulting polynucleotide is RNA at the 5′ end and DNA at the 3′ end.

4. *Duplex DNA synthesis requires a special growing fork.*   Not only is a primer required to get DNA synthesis started, but DNA manufacture has other complications as well. Because duplex DNA consists of two intertwined base-paired strands (see Figure 2-52), the base-pair copying of one strand requires the unwinding of the original duplex *and* a release or absorption of the resulting torsional force—a process most likely accomplished by specific "unwinding proteins" and by topoisomerases, enzymes that nick the DNA to allow a swivel point. Synthesis of daughter strands proceeds at or near the *growing fork* between the unwound parental strands (Figure 3-24).

All nucleic acid strands grow in the 5′ → 3′ direction, but the two DNA strands of a duplex are antiparallel. On one parental strand, synthesis of a daughter strand, called the *leading strand,* proceeds continuously in the 5′ → 3′ direction, in the direction of the growing fork. On the other parental strand, discontinuous segments of DNA are also synthesized in the 5′ → 3′ direction, that is, in the direction opposite to overall movement of the growing fork. This discontinuous daughter strand is called the *lagging strand.* The short, discontinuous segments, called *Okazaki fragments* after their discoverer, are linked by *DNA ligase* to form a continuous strand. At least twelve different proteins participate in a growing fork, which makes double-stranded DNA synthesis much more complex than RNA synthesis.

5. *RNA processing is a key event in making functional RNA.*   In both prokaryotic and eukaryotic cells, the RNA strand initially produced by copying DNA is often not biologically active. In all cells, the tRNAs and rRNAs are formed by cleaving a large primary transcript and trimming the ends of the first cleavage product. In eukaryotic cells, both the 5′ and 3′ ends of all mRNA molecules must be chemically modified before the mRNA is activated. In addition, the primary RNA transcript may undergo shortening and even *splicing,* in which the RNA chain is cut on either side of an intervening sequence (the *intron*) and the flanking pieces of RNA (the *exons*) are rejoined. The removal of introns may

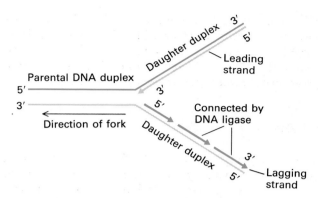

▲ **Figure 3-24**   Replication of DNA. Nucleic acid chains can grow only in the 5′ → 3′ direction. The new strand in one daughter duplex is replicated continuously. Replication in the other daughter duplex is discontinuous; DNA ligase connects these fragments to form a continuous strand.

occur in several places along the *primary transcript* to make the final functional RNA. The chemical mechanism of splicing is not the same in every case. *Self-splicing* can occur within the RNA alone, with no protein involved. In most cases of mRNA processing, however, *small ribonucleoprotein particles* assemble in a cluster, termed a *spliceosome,* that aids in the splicing event. (Splicing, its role in gene regulation, and its potential role in evolution are discussed in Chapters 8, 11, and 26, respectively.)

Chemical Differences between RNA and DNA Provide Functional Properties

Why should DNA and RNA differ at all if the addition of mononucleotides by DNA and RNA polymerases is chemically similar? One possibility is that DNA is chemically more stable and RNA molecules need to be changed during cell growth. Thus enzymes must be able to degrade RNA selectively. The 2′ hydroxyl on the ribose renders RNA more chemically unstable; it could also be a convenient recognition marker for enzymes that degrade RNA. For example, a low level of hydroxide ion (a high pH) leads to the cleavage of RNA into 2′,3′-cyclic phosphates; DNA, which lacks a 2′ hydroxyl group, is resistant to such alkaline hydrolysis.

For another example, let's consider the chemical conversion of cytosine residues in DNA or RNA to uracil by the loss of the amino group. This *deamination* process alters the sequence of the molecule. A uracil residue formed in DNA by the deamination of cytosine can be recognized by cell enzymes as a "mistake" because thymine, rather than uracil, is normally found in DNA. Specific enzymes can excise the uracil and repair this induced base change by replacing the uracil with cytosine. In other words, the enzymes recognize a G-U base pair in DNA and convert it back to the normal G-C base pair. The conversion of cytosine to uracil in RNA, however, is not recognized as an error and is not repaired. In short, the presence in DNA of thymine (rather than uracil) and of deoxyribose (rather than ribose) promotes stability.

Summary

The essential biopolymers in cells are the proteins and the nucleic acids. Certain general rules can be applied to the assembly of these polymers. Each is constructed one molecule at a time from a fixed set of subunits: proteins have 20 amino acids; nucleic acids have four nucleotides, with uracil in RNA instead of thymine in DNA. The synthesis begins and ends at particular sites. The initial product is commonly altered to make the final product.

Most biological reactions depend on the specificity in protein chains, which resides in the linear order of their amino acids. The sequence of nucleotides in DNA in the form of a triplet genetic code specifies the linear order of amino acids in proteins. The translation of genetic information requires the participation of many proteins and three kinds of RNA: messenger RNA (mRNA), which is a copy of the nucleotide sequence from DNA; transfer RNA (tRNA), which carries individual amino acids to the ribosomes, where mRNA is translated into proteins; and ribosomal RNA (rRNA), which brings mRNA and tRNA together so that the protein-synthesizing reactions can occur.

Certain protein factors aid in the three steps of protein synthesis: initiation, elongation, and termination. The highlights of these steps follow: methionine is attached to $tRNA_i^{Met}$ by methionyl-tRNA synthetase; Met-$tRNA_i^{Met}$ locates the AUG initiation codon on the mRNA; the correct amino acids (attached to their tRNAs by specific aminoacyl-tRNA synthetases) are brought to the ribosome; there, one-by-one, the correct tRNA is chosen by codon-anticodon pairing to add its amino acid to the growing peptide. Finally, a termination codon (UAA, UGA, or UAG) is reached on the mRNA, and the finished peptide chain is released.

Nucleic acid synthesis, in which the correct nucleotides are selected by Watson-Crick base pairing, is much simpler than protein synthesis: enzymes termed polymerases elongate nucleic acid chains in the $5' \rightarrow 3'$ direction by adding ribonucleotide or deoxyribonucleotide triphosphates. The synthesis of RNA is called transcription. RNA polymerases can initiate RNA chains de novo at appropriate sites (promoters) on the DNA. In the cell, transcriptional regulatory proteins assist or prevent RNA polymerase attachment to specific promoters. Prokaryotes (bacterial cells) apparently have one RNA polymerase; eukaryotes have three, each of which catalyzes the synthesis of mRNA, rRNA, or small RNAs such as tRNA. RNA primary transcripts for ribosomal RNA and tRNA are always shortened in some way before use. Eukaryotic mRNAs are always modified at both ends and may be spliced—a process by which intervening sequences are excised and the remaining sequences are rejoined.

DNA polymerases cannot initiate new chains de novo; they can only elongate from the $3'$ hydroxyl group of preexisting primer molecules. Because the strands in a molecule of DNA are antiparallel and because chain elongation proceeds in the $5' \rightarrow 3'$ direction, DNA replication requires the continuous copying of one strand and the discontinuous copying of the other, connected by DNA ligase.

References

Protein Synthesis: The Three Roles of RNA

The Genetic Code

CLAYTON, D. A. 1984. Transcription of the mammalian mitochondrial genome. *Ann. Rev. Biochem.* 53:573–594.

GAREN, A. 1968. Sense and non-sense in the genetic code. *Science* 160:149–155.

KHORANA, H. G. 1968. Nucleic acid synthesis in the study of the genetic code. In *Nobel Lectures: Physiology or Medicine (1963–1970)*, Elsevier (1973).

NIRENBERG, M. W., and P. LEDER. 1964. RNA codewords and protein synthesis. *Science* 145:1399–1407.

WOESE, C. B. 1967. *The Genetic Code.* Harper & Row.

Messenger RNA

BRENNER, S., F. JACOB, and M. MESELSON. 1961. An unstable intermediate carrying information from genes to ribosomes for protein synthesis. *Nature* 190:576–581.

NIRENBERG, M. W., and J. H. MATTHEI. 1961. The dependence of cell-free protein synthesis in *E. coli* upon naturally occurring or synthetic polyribonucleotides. *Proc. Nat'l Acad. Sci. USA* 47:1588–1602.

Transfer RNA and Aminoacyl-tRNA Synthetase

BJORK, G. R., J. V. ERICSON, C. E. D. GUSTAFSON, T. G. HAGERVALL, Y. H. JANSSON, and P. M. WIKSTRON. 1987. Transfer RNA modification. *Ann. Rev. Biochem.* 56:263–287.

CHAPPEVILLE, F., F. LIPMANN, G. VON EHRENSTEIN, B. WEISBLUM, W. J. RAY, and S. BENZER. 1962. On the role of soluble ribonucleic acid in coding for amino acids. *Proc. Nat'l Acad. Sci. USA* 48:1086–1092.

HOAGLAND, M. B., M. L. STEPHENSON, J. F. SCOTT, L. I HECHT, and P. C. ZAMECNIK. 1958. A soluble ribonucleic acid intermediate in protein synthesis. *J. Biol. Chem.* 231:241–257.

HOLLEY, R. W., J. APGAR, G. A. EVERETT, J. T. MADISON, M. MARQUISEE, S. H. MERRILL, J. R. PENSWICK, and A. ZAMIR. 1965. Structure of a ribonucleic acid. *Science* 147:1462–1465.

MCCLAIN, W. H., and K. FOSS. 1988. Changing the identity of a tRNA by introducing a G–U wobble pair near the $3'$ acceptor end. *Science* 240:793–796.

NORMALY, J., R. C. OGDEN, S. J. HORVATH, and J. ABELSON. 1986. Changing the identity of a transfer RNA. *Nature* 321:213–219.

REGAN, L., J. BOWIE, and P. SCHIMMEL. 1987. Polypeptide sequences essential for RNA recognition by an enzyme. *Science* 235:1651–1653.

RICH, A., and S.-H. KIM. 1978. The three-dimensional structure of transfer RNA. *Sci. Am.* 240(1):52–62 (Offprint 1377).

ROULD, M. A., J. J. PERONA, D. SÖLL, and T. A. STEITZ. 1989. Structure of *E. coli* glutaminyl-tRNA synthetase complexed with tRNA^{Gln} and ATP at 2.8 Å resolution. *Science* 246:1135–1142.

SCHIMMEL, P. R. 1987. Aminoacyl-tRNA synthetases: general scheme of structure-function relationships in the polypeptides and recognition of transfer RNAs. *Ann. Rev. Biochem.* 56:125–158.

SCHULMAN, L. H., and J. ABELSON. 1988. Recent excitement in understanding tRNA identity. *Science* 240:1591–1593.

SCHULMAN, L. H., and H. PELKA. 1988. Anticodon switching changes the identity of methionine and valine transfer RNAs. *Science* 242:765–768.

Ribosomes

BERNABEU, C., and J. A. LAKE. 1982. Nascent polypeptide chains emerge from the exit domain of the large ribosomal subunit. *Proc. Nat'l Acad. Sci. USA* 79:3111–3115.

CAPEL, M. S., M. KJELDGAARD, D. M. ENGELMAN, and P. A. MOORE. 1988. Positions of S2, S13, S16, S17, S19, and S21 in the 30S ribosomal subunit of *Escherichia coli*. *J. Mol. Biol.* 200:65–87.

DAHLBERG, A. E. 1989. The functional role of rRNA in protein synthesis. *Cell* 57:525–529.

GERBI, S. A. 1986. The evolution of eukaryotic ribosomal DNA. *BioSystems* 19:247–258.

HARDESTY, B., and G. KRAMER, eds. 1986. *Structure, Function, and Genetics of Ribosomes*. Springer-Verlag.

LAKE, J. A. 1981. The ribosome. *Sci. Am.* 245(2):84–97 (Offprint 1501).

NOLLER, H. F. 1984. Structure of ribosomal RNA. *Ann. Rev. Biochem.* 53:119–162.

NOMURA, M., R. GOURSE, and G. BAUGHMAN. 1984. Regulation of the synthesis of ribosomes and ribosomal components. *Ann. Rev. Biochem.* 53:75–118.

OAKES, M. I., M. W. CLARK, E. HENDERSON, and J. A. LAKE. 1986. DNA hybridization electron microscopy: ribosomal RNA nucleotides 1392–1407 are exposed in the cleft of the small subunit. *Proc. Nat'l Acad. Sci. USA* 83:275–279.

STERN, S., B. WEISER, and H. F. NOLLER. 1988. Model for the three-dimensional folding of 16S ribosomal RNA. *J. Mol. Biol.* 204:447–481.

WITTMAN, H. G. 1983. Architecture of prokaryotic ribosomes. *Ann. Rev. Biochem.* 52:35–66.

YONATH, A., K. R. LEONARD, and H. G. WITTMAN. 1987. A tunnel in the large ribosomal subunit revealed by three-dimensional image reconstitution *Science* 236:813–816.

The Steps in Protein Synthesis

CASKEY, T. H. 1980. Peptide chain termination. *Trends Biochem. Sci.* 5:234–237.

CAVENER, D. 1987. Comparison of the consensus sequences flanking translational start sites in *Drosophila* and vertebrates. *Nucl. Acids Res.* 15:1353–1361.

JACOB, W. F., M. SANTER, and A. E. DAHLBERG. 1987. A single base change in the Shine-Dalgarno region of 16S rRNA of *E. coli* affects translation of many proteins. *Proc. Nat'l Acad. Sci. USA* 84:4757–4761.

KOZAK, M. 1986. Point mutations define a sequence flanking the AUG initiator codon that modulates transcription by eukaryotic ribosomes. *Cell* 44:283–292.

KOZAK, M. 1986. Regulation of protein synthesis in virus-infected cells. *Adv. Virus Res.* 31:229–292.

MOAZED, D. AND F. NOLLER. 1989. Interaction of tRNA with 23S rRNA in the ribosomal A, P, and E sites. *Cell* 57:585–587.

MOLDAVE, K. 1985. Eukaryotic protein synthesis. *Ann. Rev. Biochem.* 54:1109–1150.

SONNENBERG, N. 1988. Cap-binding proteins of eukaryotic messenger RNA: functions in initiation and control of translation. *Prog. Nuc. Acids. Res. Mol. Biol.* 35:173–203.

Nucleic Acid Synthesis

CECH, T. R., and B. L. BASS. 1986. Biological catalysis by RNA. *Ann. Rev. Biochem.* 55:599–629.

KORNBERG, A. and T. A. BAKER. In press. *DNA Replication*, 2d ed. W. H. Freeman and Company.

KORNBERG, A. 1988. DNA replication. *J. Biol. Chem.* 223:1–4.

MCCLURE, W. R. 1985. Mechanism and control of transcription initiation in prokaryotes. *Ann. Rev. Biochem.* 54:171–204.

SWEETSER, D., M. NONET, and R. A. YOUNG. 1987. Prokaryotic and eukaryotic polymerases have homologous core subunits. *Proc. Nat'l Acad. Sci. USA* 84:1192–1196.

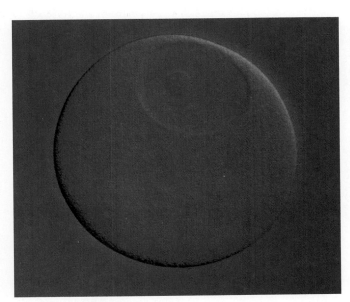

CHAPTER 4

Oocyte with germinal vesicle

The Study of Cell Organization and Subcellular Structure

*T*he discovery that all living matter is constructed of smaller units, or cells, was a notable scientific achievement of the nineteenth century. The pioneering research of Theodor Schwann and Matthias Schleiden was extended to an enormous number of microorganisms, plants, and animals; this once-surprising finding is universally accepted today. All cells in all organisms share certain common structural features, such as the architecture of their membranes. Many complicated metabolic events are also carried out in basically the same way in all organisms: the replication of DNA, the synthesis of proteins, and the production of chemical energy by the conversion of glucose to carbon dioxide. These biochemical similarities are not coincidental; they share a pedigree: all cells in the present biological universe have evolved from common ancestors.

Before the introduction of the electron microscope, all cells were generally assumed to share similar basic principles of organization as well. True, bacterial cells are much smaller than typical animal, plant, or even fungal cells (Figure 4-1); but because treatment of some bacteria with a Feulgen stain (which stains DNA) reveals central masses of nuclear material, it seemed possible that all cells possessed a defined nucleus and other internal compartments. But the limited resolution of the light microscope (only about 0.5 μm) does not provide an accurate

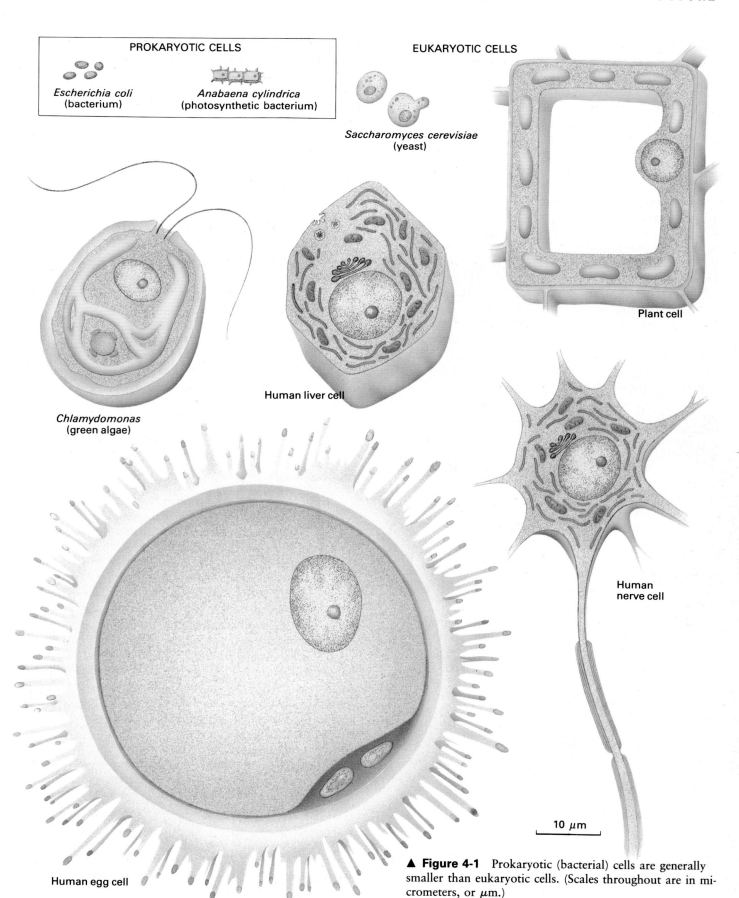

PROKARYOTIC CELLS

Escherichia coli
(bacterium)

Anabaena cylindrica
(photosynthetic bacterium)

EUKARYOTIC CELLS

Saccharomyces cerevisiae
(yeast)

Plant cell

Chlamydomonas
(green algae)

Human liver cell

Human
nerve cell

Human egg cell

10 μm

▲ **Figure 4-1** Prokaryotic (bacterial) cells are generally smaller than eukaryotic cells. (Scales throughout are in micrometers, or μm.)

image of the internal structure of bacterial cells. The question of cell structure was not answered until the 1950s, when a variety of morphological, biochemical, and genetic studies established that two types of cells— eukaryotic cells (literally, cells with a true nucleus) and prokaryotic cells (cells with no defined nucleus)—have persisted independently for perhaps a billion or more years of biological evolution. Although prokaryotes and eukaryotes differ radically in their organization, they apparently evolved from the same type of cell; the nature of this ancestral cell is unknown.

This chapter begins with an overview of cell structure that focuses on the similarities and differences between prokaryotes and eukaryotes. It then describes the methods cell biologists use to study cell organization: the classic techniques of microscopy and cell fractionation used on fixed, killed cells, which provide only a static picture, and some newer developments, which reveal the movements and internal structures of living cells. Finally, the main components of eukaryotic cells—the internal organelles and fibers—are examined; the extracellular substances that surround cells and, in some cases, give them shape and strength are also investigated. ▲

Prokaryotic and Eukaryotic Cells

Eukaryotes include all plants and animals—from the most primitive ferns to the most complex flowering plants and from simple sponges to insects and mammals. Eukaryotes also include many single-celled microorganisms, such as true algae, amebae, fungi, and molds.

Prokaryotes include all bacteria, divided into two separate lineages: *eubacteria* and *archaebacteria*. Most bacteria are eubacteria; in general, when bacterial structure or metabolism is discussed, eubacteria are the subject. The eubacteria include the *photosynthetic organisms,* formerly known as *blue-green algae,* which obtain energy from photosynthesis. Much less is known about archaebacteria, which grow in unusual environments. The *methanogens* live only in oxygen-free milieus such as swamps; these bacteria generate methane (CH_4), also known as "swamp gas," by the reduction of carbon dioxide. Other archaebacteria include the *halophiles,* which require high concentrations of salt to survive, and the *thermoacidophiles,* which grow in hot (80°C) sulfur springs, where a pH of less than 2 is common.

Prokaryotes Have a Simpler Structure Than Eukaryotes

The Plasma Membrane In general, only one type of membrane, the *plasma membrane,* forms the boundary of the cell proper in prokaryotes. Like all biological membranes, the structure of the plasma membrane is based on a phospholipid bilayer (see Figure 2-48), which is permeable to certain gases, such as oxygen and carbon dioxide, and to water (these substances can diffuse freely across it). However, it is virtually impermeable to most molecules the cell must obtain from its environment, such as sugars, amino acids, and inorganic ions (for example, K^+ or Cl^-). The plasma membrane utilizes many membrane proteins called *permeases,* or *transporters,* which form channels in the phospholipid bilayer and allow certain molecules to enter or leave the cell.

Internal Membranes In general, all internal membranes in prokaryotic cells are connected to the plasma membrane. In some bacteria, the plasma membrane has infoldings called *mesosomes* (Figure 4-2a). The extensive mesosomes of photosynthetic bacteria contain the pro-

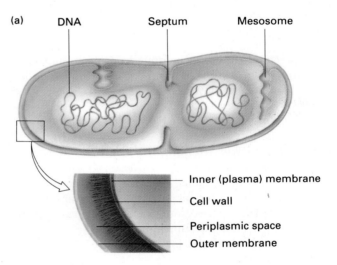

(a) DNA   Septum   Mesosome

Inner (plasma) membrane
Cell wall
Periplasmic space
Outer membrane

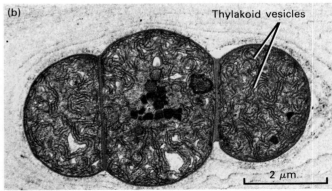

(b) Thylakoid vesicles

2 μm

▲ **Figure 4-2** (a) Diagram of the structure of a gram-negative prokaryotic cell; note the periplasmic space between the inner and outer membranes. (b) Electron micrograph of a thin section through three attached cells of the photosynthetic (blue-green) alga *Nostoc carneum.* The extensive array of *thylakoid vesicles* (internal photosynthetic membranes) is characteristic of this prokaryotic group. *Courtesy of T. E. Jensen and C. C. Bowen.*

Periplasmic space and cell wall

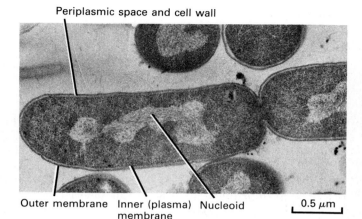

Outer membrane Inner (plasma) Nucleoid
membrane

0.5 μm

▲ **Figure 4-3** Electron micrograph of a thin section of
E. coli, a gram-negative bacterium. The micrograph shows
the inner (plasma) membrane, the outer surface membrane
that is part of the cell wall, and the nucleoid, the DNA-
containing fibrous central region of the cell. *Courtesy of
I. D. J. Burdett and R. G. E. Murray.*

teins that trap light and generate adenosine triphosphate
(ATP). Some internal photosynthetic membranes, called
thylakoid vesicles, may not connect with the plasma
membrane and may enclose small regions of the cytoplasm (Figure 4-2b).

Cell Walls Provide Strength and Rigidity Bacterial
species can be divided into two classes. *Gram-negative
bacteria* (those not stained by the Gram technique),
which include the common intestinal bacterium *Escherichia coli (E. coli),* are surrounded by two surface membranes (Figure 4-3). The inner membrane is the actual
plasma membrane—the major permeability barrier of the

Peptidoglycan   Protein array

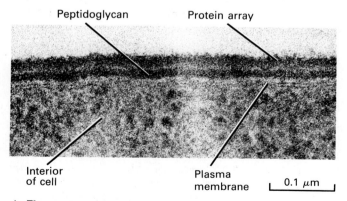

Interior
of cell   Plasma
membrane   0.1 μm

▲ **Figure 4-4** Electron micrograph of a section of the surface of *Bacillus polymyxa,* a gram-positive bacterium. Visible
are the thin inner (plasma) membrane, a layer composed of a
peptidoglycan (a linear repeating polymer of two sugar residues, crosslinked by short chains of amino acids) and tei-
choic acid (a linear polymer consisting of alternating phosphate and polyalcohol groups), and the outermost protein
array. *Courtesy of R. G. E. Murray.*

(a)

Intercellular   Plasma   Endoplasmic   Golgi
space   membrane   reticulum   vesicles

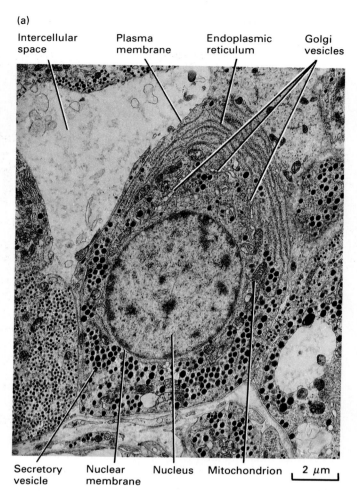

Secretory   Nuclear   Nucleus   Mitochondrion   2 μm
vesicle   membrane

▲ **Figure 4-5** (a) Electron micrograph of a thin section of
a hormone-secreting cell from the rat pituitary, showing the
subcellular features typical of many animal cells. *Facing
page:* (b) Diagram of a "typical" animal cell. Not every animal cell contains all of the organelles, granules, and fibrous
structures shown here, and other substructures can be present in some cells. Animal cells also differ considerably in
shape and in the prominence of various organelles and substructures. *Part (a) courtesy of Biophoto Associates.*

cell. The outer membrane is unusual in that it is permeable to many chemicals having a molecular weight of 1000
or more; it contains proteins called *porins,* which line
channels large enough to accommodate such molecules.
Between the two membranes lie the *cell wall,* composed
of *peptidoglycan* (a complex of proteins and oligosaccharides that gives rigidity to the cell), and the *periplasm* (a
space generally occupied by proteins secreted by the cell).
Gram-positive bacteria, such as *Bacillus polymyxa,* have
only a plasma membrane and a cell wall (Figure 4-4).
Their peptidoglycan wall is thicker than that of gram-
negative organisms.

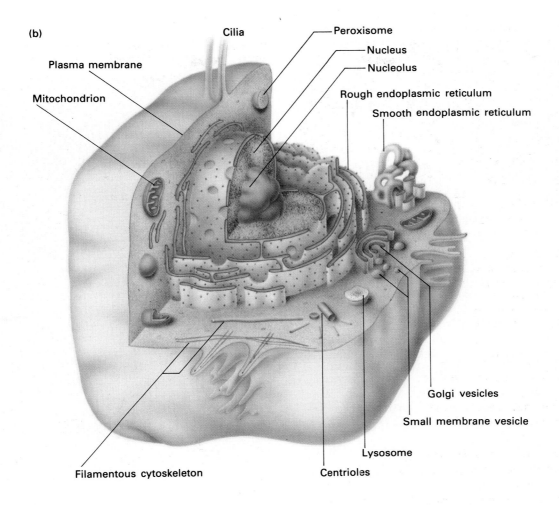

(b)

Cilia

Peroxisome

Nucleus

Nucleolus

Plasma membrane

Rough endoplasmic reticulum

Smooth endoplasmic reticulum

Mitochondrion

Golgi vesicles

Small membrane vesicle

Lysosome

Filamentous cytoskeleton

Centrioles

Eukaryotic Cells Have Complex Systems of Internal Membranes and Fibers

Both prokaryotic and eukaryotic cells are surrounded by a plasma membrane. However, unlike most prokaryotic cells, eukaryotic cells also contain extensive internal membranes, unconnected to the plasma membrane, that enclose and separate specific regions from the rest of the cytoplasm. These membranes define a collection of subcellular structures called *organelles.*

According to one definition, an organelle is any subcellular entity that can be isolated by centrifugation at a very high speed; this would include such structures as ribosomes, particles of glycogen (a polymer of glucose), and large multienzyme complexes. In this book, however, we use the term *organelle* to refer only to membrane-limited structures. Generally, the largest organelle in a cell is the *nucleus.* Many other organelles are found in most eukaryotic cells (Figures 4-5 and 4-6): the *mitochondria,* in which the oxidation of small molecules generates most cellular ATP; the *rough* and *smooth endoplasmic reticula* (ER), a network of membranes in which glycoproteins

and lipids are synthesized; *Golgi vesicles,* which direct membrane constituents to appropriate places in the cell; *peroxisomes* in all eukaryotes and *glyoxisomes* in plant seeds, which metabolize hydrogen peroxide; and assorted smaller vesicles. Animal cells contain *lysosomes,* which degrade many proteins, nucleic acids, and lipids. Plant cells contain *chloroplasts,* the site of photosynthesis. Both plant cells and certain eukaryotic microorganisms contain one or more *vacuoles* (see Figure 4-6a), large fluid-filled organelles that store many nutrient and waste molecules and also participate in the degradation of cellular proteins and other macromolecules. Each organelle plays a unique role in the growth and metabolism of the cell, and each contains a specific collection of enzymes that catalyze requisite chemical reactions. Some of this specificity resides in the organelle membranes, to which a number of the enzymes and other proteins are bound.

The *cytoplasm* of eukaryotic cells—the region lying outside the nucleus—also contains an array of fibrous proteins called, collectively, the *cytoskeleton.* Among these proteins are the *microfilaments,* built of the protein actin; the somewhat wider *microtubules,* built of tubulin;

▶ **Figure 4-6** (a) Electron micrograph of a thin section of a leaf cell from *Phleum pratense*, showing the cell wall, a large internal vacuole, and parts of five chloroplasts, structures unique to plant cells. Although a nucleus is not evident in this micrograph, plant cells do contain a nucleus and other features of eukaryotic cells, as depicted in (b)—*facing page*—the diagram of a "typical" plant cell. *Part (a) courtesy of Biophoto Associates/Myron C. Ledbetter/Brookhaven National Laboratory.*

(a)

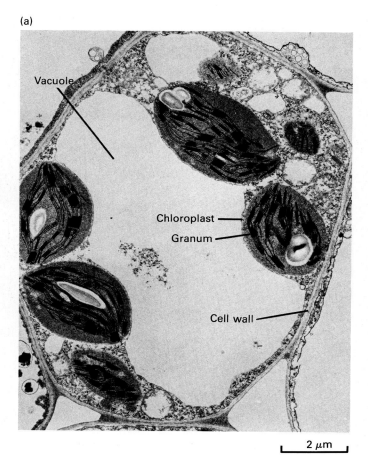

⊢ 2 μm ⊣

and the *intermediate filaments*, built of one or more rod-shaped protein subunits. Cytoskeletal fibers give the cell strength and rigidity. They also control movement within the cell; microtubules, for instance, are critical to chromosomal movement during cell division.

Plant cells are surrounded by a rigid cell wall containing cellulose and other polymers that also contributes to the strength and rigidity of the cell. Fungi are also surrounded by a cell wall of a different composition from that of bacterial or plant cells. During cell growth, the wall must expand; during cell division, a new wall must be laid down between the two daughter cells. Animal cells generally are not surrounded by walls.

Prokaryotes and Eukaryotes Contain Similar Macromolecules

The volume of a typical animal or plant cell is several hundred times that of a typical bacterial cell; yet the chemical composition of prokaryotic and eukaryotic cells is strikingly similar. By weight, about 70 percent of a typical cell is water. Other small molecules, including salts, lipids, amino acids, and nucleotides, account for another 7 percent. The remainder (approximately 23 percent) is composed of macromolecules.

HeLa cells, a line of cells derived from a human cervical carcinoma, which grow well in cell cultures, are among the most thoroughly studied mammalian cells. The concentrations of small molecules are similar in HeLa cells and in the bacterium *E. coli* (Table 4-1). A HeLa cell contains about 4×10^6 ribosomes and about 7×10^5 mRNA molecules, roughly 150 times as many of each as found in *E. coli* (Table 4-2). A HeLa cell has about 5 ×

Table 4-1 Some of the chemical components of a rapidly growing bacterial cell, *Escherichia coli*

Component	Percentage of total cell weight	Average molecular weight	Approximate number per cell	Number of different kinds
H_2O	70	18	4×10^{10}	1
Inorganic ions (Na^+, K^+, Mg^{2+}, Ca^{2+}, Fe^{2+}, Cl^-, PO_4^{3-}, SO_4^{2-}, etc.)	1	40	2.5×10^8	20
Carbohydrates and precursors	3	150	2×10^8	200
Amino acids and precursors	0.4	120	3×10^7	100
Nucleotides and precursors	0.4	300	1.2×10^7	200
Lipids and precursors	2	750	2.5×10^7	50
Other small molecules (hemes, quinones, breakdown products of food molecules, etc.)	0.2	150	1.5×10^7	200

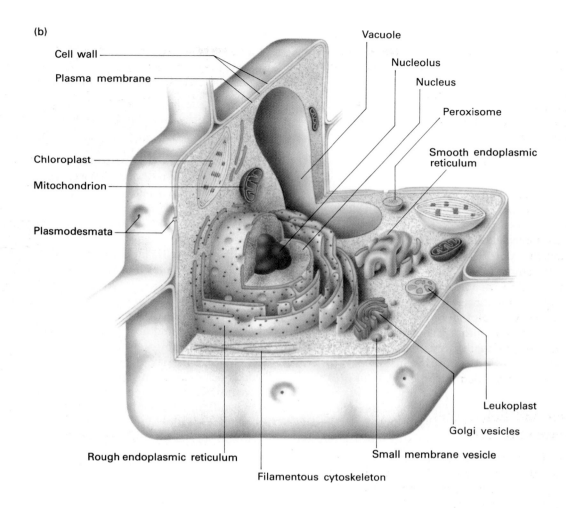

(b)

Cell wall

Plasma membrane

Vacuole

Nucleolus

Nucleus

Peroxisome

Smooth endoplasmic reticulum

Chloroplast

Mitochondrion

Plasmodesmata

Leukoplast

Golgi vesicles

Rough endoplasmic reticulum

Small membrane vesicle

Filamentous cytoskeleton

Table 4-2   The main macromolecular components of *E. coli* and HeLa cells

Component	Amount per HeLa cell	Amount per *E. coli* cell
Total DNA	15 picograms (pg)*	0.017 pg†
Total RNA	30 pg	0.10 pg
Total protein	300 pg (5×10^9 molecules, of average m.w. 40,000)	0.2 pg (3×10^6 molecules, of average m.w. 40,000)
Cytoplasmic ribosomes	4×10^6	3×10^4
Cytoplasmic tRNA molecules	6×10^7	4×10^5
Cytoplasmic mRNA molecules‡	7×10^5	4×10^3
Nuclear precursor rRNA molecules	6×10^4	
Heterogeneous nuclear RNA molecules§	1.6×10^5	
Total dry weight	400 pg	0.4 pg

*HeLa cells are hypotetraploid: they contain about four copies of each chromosome. The normal diploid human DNA complement is about 5 pg per cell.
†A rapidly growing *E. coli* cell contains, on average, four DNA genomes. Each genomic DNA weighs 0.0044 pg.
‡An average chain length of 1500 nucleotides is assumed.
§An average chain length of 6000 nucleotides is assumed; this group of molecules contains precursor mRNAs.

10^9 protein molecules, consisting of 5000–10,000 different polypeptide species. The much smaller *E. coli* cell has about 3×10^6 protein molecules, consisting of 1000–2000 different species.

Prokaryotes and Eukaryotes Differ in the Amount of DNA per Cell

The differences in genetic organization between typical prokaryotic and eukaryotic cells become obvious when we consider the amount of DNA per cell (Table 4-3). The genome of *E. coli* contains 4.4×10^{-15} g, or 0.0044 picograms (pg), of DNA, an amount equal to 4×10^6 base pairs. Because three DNA bases encode the position of each amino acid in a protein and an average protein contains about 400 amino acids, approximately 1200 DNA base pairs are used to encode each protein species. Thus *E. coli* DNA has a maximum coding capacity of about 3300 different proteins. Not all of the bacterial DNA encodes proteins, however, although a large part of it does. An *E. coli* cell may actually contain as many as 2000 different species of mRNAs and proteins.

All eukaryotic cells contain more DNA than prokaryotic cells do. Yeast cells, which have some of the smallest genomes among eukaryotes, contain about three times as much DNA as *E. coli* does. The cells of higher plants and animals typically have 40–1000 times as much DNA as *E. coli*; the genomes of some amphibians are 40,000 times as large. If all of this DNA were used to encode proteins, some animal cells would be able to encode as many as 3×10^6 different types of proteins. However, only a small fraction of the total DNA in a eukaryotic cell usually encodes proteins. It is widely believed that invertebrates can make about 5000 proteins and that humans can make about 50,000—far fewer than the theoretical capacities of their respective genomes. Yet vertebrates can make several million different types of antibody molecules. This variety is generated by rearrangements of DNA segments that encode parts of immunoglobulins and by mutations in these rearranged genes.

The Organization of DNA Differs in Prokaryotic and Eukaryotic Cells

In all prokaryotes studied to date, most or all cellular DNA is in the form of a single circular molecule. The cell is said to have a single *chromosome*, although the arrangement of DNA within this chromosome differs greatly from that within the chromosomes of eukaryotic cells.

Prokaryotic cells lack a membrane-bound nucleus. Most of the genomic DNA lies in the central region of the cell (see Figure 4-2); yet, in many prokaryotes, such as *Bacillus subtilis,* the DNA apparently is attached to the plasma membrane at many points. The majority of the ribosomes are found near the periphery of the cell, possibly because they are excluded from the central regions occupied extensively by the DNA. The DNA must be folded back on itself many times: an *E. coli* chromosomal DNA molecule stretched out to its full length would be over 1 millimeter (mm) long, or 1000 times as long as the cell itself.

Several species of prokaryotes also contain small circular DNA molecules called *plasmids,* which have 1000–30,000 base pairs. Plasmids generally encode proteins that are not essential to cell growth; many encode proteins that the organism requires to resist antibiotics or other toxic materials.

The nuclear DNA of all eukaryotic cells, in contrast, is divided between two or more chromosomes which, except during cell division, are contained in a membrane-bound nucleus. The number and size of individual chro-

Table 4-3   The DNA content of various cells

Organism	Size of DNA genome		Maximum number of proteins encoded*	Number of chromosomes (haploid)†
	Number of base pairs	Total length (mm)		
PROKARYOTIC				
Escherichia coli (bacterium)	4×10^6	1.36	3.3×10^3	1
EUKARYOTIC				
Saccharomyces cerevisiae (yeast)	1.35×10^7	4.60	1.125×10^4	17
Drosophila melanogaster (insect)	1.65×10^8	56	1.375×10^5	4
Homo sapiens (human)	2.9×10^9	990	2.42×10^6	23
Zea mays (corn)	5.0×10^9	1710	4.0×10^6	10

*Assuming 1200 base pairs per protein.
†Most insect and human cells are diploid, so they have twice the number of chromosomes shown.

mosomes vary widely among different eukaryotes (see Table 4-3). Yeasts, for example, have 12–18 chromosomes, each of which, on average, contains only 20 percent of the DNA of an *E. coli* chromosome. Human cells, at the other extreme, contain two sets of 23 chromosomes, each of which has about 30 times the amount of DNA present in an *E. coli* cell. Each chromosome is believed to contain a single, linear double-stranded DNA molecule. Some eukaryotic cells (yeasts and a few plants) also contain plasmid DNAs.

Almost all chromosomal DNA in eukaryotic cells is associated with a set of five different positively charged proteins called *histones*. Interaction between the histones and the DNA is very regular: every sequence of 150–180 base pairs of DNA is bound to one molecule of histone H1 and to two molecules each of histones H2A, H2B, H3, and H4. The amino acid sequence of some of these histones has been conserved during evolution. The H3 from peas is virtually identical to the H3 from cows. Histones are unknown in prokaryotic cells; bacterial DNA is associated with a different type of protein.

Despite their many differences, prokaryotic and eukaryotic cells have many biochemical pathways in common, and most aspects of the translation of mRNA into proteins are similar in all cells. Thus prokaryotes and eukaryotes are believed to be descended from the same primitive cell. Their divergence must have occurred before the separation of plant and animal cells. All extant prokaryotic and eukaryotic cells and organisms are the result of a total of 3.5 billion years of evolution. It is not surprising, then, that cells are so well adapted to their own environmental niches.

Light Microscopy and Cell Architecture

There is no one "correct" view of a cell; thus it is essential to understand some of the details of the key cell-viewing techniques, the types of images they produce, and their limitations. Schleiden and Schwann first identified individual cells using a primitive light microscope, and light microscopy has continued to play a major role in biological research. The development of electron microscopes has greatly extended the ability to resolve subcellular particles and yielded much new information on the organization of plant and animal tissues. The nature of the images depends on the type of light or electron microscope employed and on the way in which the cell or tissue has been prepared. Each technique is designed to emphasize particular structural features of the cell: Figure 4-7 shows how a typical cell, a human leukocyte, appears when three different techniques are used.

Although the most common application of light and electron microscopy—to visualize fixed, killed cells—

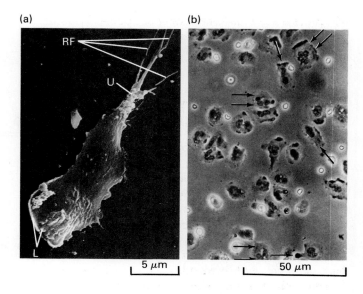

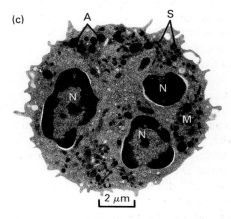

▲ **Figure 4-7** Views of human leukocytes produced by three different microscopic techniques. (This white blood cell is *polymorphonuclear*, meaning its nucleus contains many irregularly shaped lobes.) (a) A scanning electron micrograph. The three-dimensional appearance of the cell surface is characteristic of images obtained by this technique. The shape of this cell indicates that it migrates. Such a cell has a wide, flattened projection (a *lamellipodium*, L) at its leading edge and a narrow tail (the *uropod*, U) ending in *retraction fibers* (RF). (b) A light micrograph (using phase-contrast optics) of polymorphonuclear leukocytes attached to a glass slide. Some cells have spread out and become firmly attached to the glass at many points on their surfaces; these stationary cells are indicated by double arrows. Other cells are migrating along the glass; their direction of movement is indicated by a single arrow. Note that only the nuclei and a few larger cytoplasmic particles are visible with this technique. (c) A transmission electron micrograph. This thin section of a cell shows numerous granules of two types: the larger *azurophil granules* (A) and the smaller *specific granules* (S). Also evident are three lobes of the single nucleus (N) and some mitochondria (M), which are not present in large numbers in leukocytes. *Courtesy of M. J. Karnofsky; cells prepared by J. M. Robinson, Department of Pathology, Harvard Medical School.*

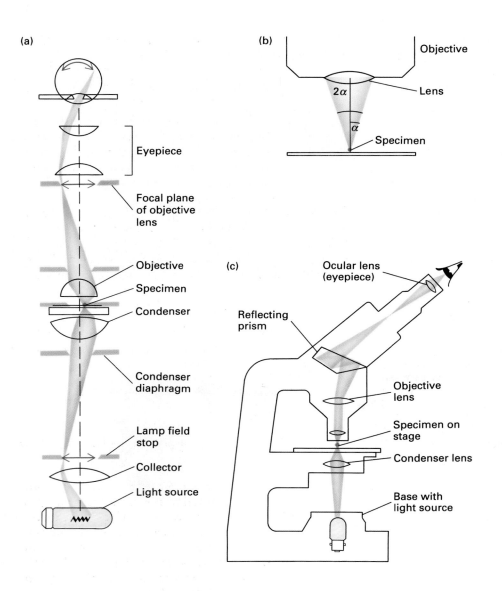

◀ **Figure 4-8** The optical pathway in a compound optical microscope. (a) Light from a bright source is focused by the collector and condenser lenses onto the stage holding the specimen. The objective lens picks up the light transmitted by the specimen and focuses it on the focal plane of the objective lens, magnifying the specimen image. The lamp field stop, the condenser diaphragm, and the other apertures restrict the amount of light entering or leaving a lens. The ocular lens (eyepiece) of the microscope is focused on this objective focal plane; it picks up the light emanating from the already magnified image of the specimen and projects it onto the plane of the human eye or a piece of photographic film, magnifying the image again. (b) An important parameter is the half-angle α of the cone of light entering the objective lens from the specimen. The larger the value of α, the finer the resolution the objective lens can provide. (c) Diagram of a modern compound microscope. *Parts (a) and (b) adapted from B. Wilson 1976,* The Science and Art of Basic Microscopy, *Figures 3-9 and 3-6.*

reveals much information, a critical question about such results is how true to life is the image of a biological specimen that has been fixed, stained, and dehydrated before examination? Thus the following two sections describe not only the classic methods used to view fixed cells but also some of the refinements that allow microscopy of unaltered or less altered specimens.

Standard Light (Bright-field) Microscopy Utilizes Fixed, Stained Specimens

The Compound Microscope The *compound microscope* (Figure 4-8), the most common microscope in use today, contains several lenses that magnify the image of a specimen under study. The specimen is usually mounted on a transparent glass slide and positioned on the movable *specimen stage* of the microscope. Light from a

lamp, normally mounted in the base of the microscope, is focused by a *condenser lens* onto the plane of the specimen. Light from the specimen is picked up by the *objective lens* and focused on its *focal plane*, creating a magnified image of the specimen. This image can be recorded directly. Usually, though, the image on the objective focal plane is magnified by the *ocular lens,* or eyepiece. The total magnification is a product of the magnification of the individual lenses: if the objective lens magnifies 100-fold (a 100× lens, the maximum usually employed) and the eyepiece magnifies 10-fold, the final magnification recorded by the human eye or on film will be 1000-fold.

Resolution The most important property of any microscopic lens is not its magnification but its power of *resolution*—its ability to distinguish between two very closely positioned objects. For any lens, the minimum distance D between two distinguishable objects depends on the *angular aperture* α, or half-angle of the cone of light entering the lens from the specimen; on the *refractive index* N of the air or fluid between the specimen and the objective lens; and on the wavelength λ of incident light:

$$D = \frac{0.61\lambda}{N \sin \alpha}$$

Improving the resolution is equivalent to decreasing the value of D, which can be done by changing λ, N, or α.

The angular aperture depends on the width of the objective lens and its distance from the specimen (see Figure 4-8c). Increasing the angle α, by moving the objective lens closer to the specimen, will increase sin α and reduce D.

The refractive index measures the degree to which a medium bends a light ray that passes into it from another medium. The refractive index of air is defined as N = 1.0. One way to improve the resolution (decrease D) is to introduce an *immersion oil* between the specimen and the objective lens. Since the refractive index of such oils is 1.5, the resolution will be improved 1.5-fold. An intuitive reason for this improvement is that media with a high refractive index between the specimen and the objective lens will "bend" more of the light emanating from the specimen into the lens.

If the wavelength of incident light is shorter, then D decreases and the resolution improves. The maximum angular aperture for the best objective lenses is 70° (sin 70° = 0.94). With the visible light of shortest wavelength (blue, λ = 450 nm) and with air (N = 1.0) above the sample, the finest resolution will be

$$D = \frac{0.61 \times 450 \text{ nm}}{1.0 \times 0.94} = 292 \text{ nm}$$

or about 0.3 micrometers (μm). If immersion oil (N = 1.5) is introduced, then

$$D = \frac{0.61 \times 450 \text{ nm}}{1.5 \times 0.94} = 194 \text{ nm}$$

or about 0.2 μm. Thus the *limit of resolution* of the light microscope, using visible light, is about 0.2 μm; no matter how many times the image is magnified, the microscope can never resolve objects less than about 0.2 μm apart or reveal details of less than that size.

Preparing Samples for Light Microscopy Specimens for light microscopy are commonly fixed with a solution containing alcohol or formaldehyde; these compounds denature most proteins and nucleic acids. Formaldehyde also cross-links amino groups on adjacent molecules:

$$R_1-NH_2 + H-\overset{\overset{\displaystyle O}{\|}}{C}-H \longrightarrow R_1-NH-\overset{\overset{\displaystyle OH}{|}}{\underset{\underset{\displaystyle H}{|}}{C}}-H$$

$$R_1-NH-\overset{\overset{\displaystyle OH}{|}}{\underset{\underset{\displaystyle H}{|}}{C}}-H + R_2-NH_2 \longrightarrow$$

$$R_1-NH-\overset{\overset{\displaystyle H}{|}}{\underset{\underset{\displaystyle H}{|}}{C}}-NH-R_2 + H_2O$$

where R_1 and R_2 represent the rest of each molecule. These covalent bonds stabilize protein-protein and protein–nucleic acid interactions and render the molecules insoluble and stable for subsequent procedures. The sample is then embedded in paraffin and cut into thin sections of one or a few micrometers. For optimum resolution, a section of the material under study is usually prepared and mounted on a glass slide (Figure 4-9).

Different cellular constituents (such as the nucleus, cytoplasm, and mitochondria) absorb about the same degree of visible light. This absence of contrast makes it difficult to distinguish them under a light microscope; often not even the nucleus of a cell can be seen. For this reason, the final step in preparing a specimen for observation is to stain it. Many chemical stains bind to molecules that have specific features: for example, *eosin* and *meth-*

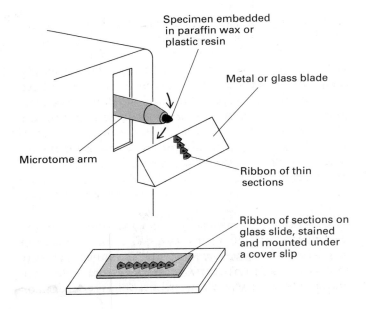

Specimen embedded in paraffin wax or plastic resin

Metal or glass blade

Microtome arm

Ribbon of thin sections

Ribbon of sections on glass slide, stained and mounted under a cover slip

◀ **Figure 4-9** Preparation of tissues for light microscopy. A piece of fixed tissue is dehydrated by soaking it in alcohol-water solutions, in pure alcohol, and finally in a solvent such as xylene. The specimen is then placed in warm, liquid paraffin, which is allowed to harden. A piece of the specimen is mounted on the arm of a microtome. Sections a few micrometers thick are cut as the arm moves up and down over the metal or glass blade.

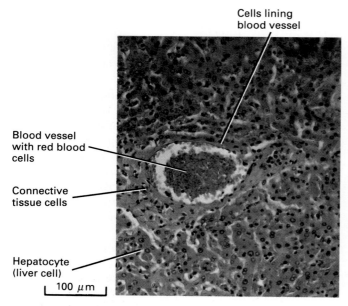

▲ **Figure 4-10**   A light-microscopic view of a section of human liver, stained with the dyes eosin and methylene blue. Note that at least four different cell types are distinguishable by this staining protocol. *Courtesy of P. G. Aitken/Biophoto Associates.*

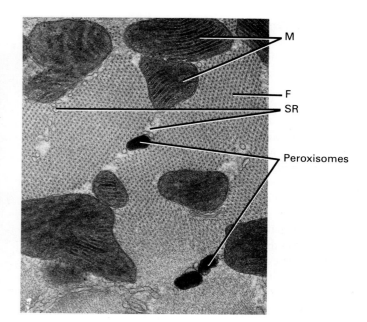

▲ **Figure 4-11**   Thin sections of rat heart muscle stained for catalase and viewed through an electron microscope. Catalase, which is found in small membrane vesicles called peroxisomes, is detected by treating cell sections with hydrogen peroxide and a dye; the catalase-catalyzed oxidation of the dye produces an insoluble dense precipitate. Also visible (unstained) are mitochondria (M); vesicles of the sarcoplasmic reticulum (SR), a set of membranes that store Ca^{2+} ions; and (in cross section) actin and myosin contractible fibers (F). *From H. D. Fahmi and S. Yokota, 1981,* International Cell Biology 1980–1981, *H. G. Schweiger, ed., Springer-Verlag, p. 640. Courtesy of H. D. Fahmi.*

ylene blue bind to many different kinds of proteins (Figure 4-10); *benzidine* binds to heme-containing proteins and nucleic acids; and the *fuchsin* used in Feulgen staining binds to DNA. If an enzyme catalyzes a reaction that produces a colored or otherwise visible precipitate from a colorless precursor, the enzyme may be detected in cell sections by staining them with the precursor. This last technique, called *cytochemical staining,* can be used with both light and electron microscopes (Figure 4-11).

Immunofluorescence Microscopy Reveals Specific Proteins and Organelles within a Cell

Perhaps the most powerful techniques for localizing proteins within a cell by light microscopy make use of the *fluorescence microscope* and antibodies specific for the desired protein. A chemical is said to be *fluorescent* if it absorbs light at one wavelength (the *exciting wavelength*) and emits light at a specific and longer wavelength within the visible spectrum. Two very useful dyes for microscopy are *rhodamine,* which emits red light, and *fluorescein,* which emits green light. In modern fluorescence microscopes (Figure 4-12), only fluorescent light emitted by the sample is used to form an image; light of the exciting wavelength induces the fluorescence but is then absorbed by filters placed between the objective lens and the eye or camera. Such dyes as fluorescein and rhodamine have a low nonspecific affinity for biological molecules but can

be chemically coupled to purified antibodies specific to almost any desired macromolecule: a fluorescent antibody added to a permeabilized cell or tissue section binds to the chosen antigens, which then light up when illuminated by the exciting wavelength (Figure 4-13).

Fluorescence microscopy can also be applied to live cells. For example, purified actin may be chemically linked to a fluorescent dye. Careful biochemical studies have established that this "tagged" molecule is indistinguishable in function from its normal counterpart. If the tagged protein is *microinjected* into a cultured cell, the cellular and tagged actin copolymerize into normal long actin fibers. This technique can also be used to study individual microtubules within a cell.

In another technique, a chemically synthesized lipid is covalently linked to a fluorescent dye and added to a cell culture, where it is taken up into all cell membranes—particularly (for unknown reasons) into the endoplasmic reticulum (ER). The fluorescent image from such labeled cells shows the lacelike ER network (Figure 4-14) that can also be seen in fixed cells by more conventional fluorescence microscopy.

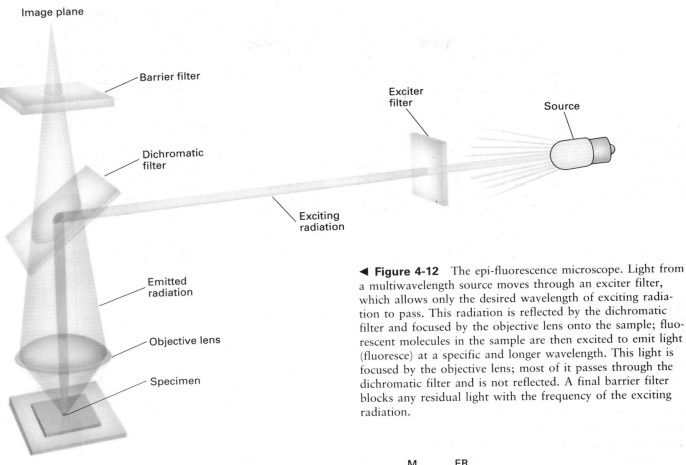

◄ Figure 4-12 The epi-fluorescence microscope. Light from a multiwavelength source moves through an exciter filter, which allows only the desired wavelength of exciting radiation to pass. This radiation is reflected by the dichromatic filter and focused by the objective lens onto the sample; fluorescent molecules in the sample are then excited to emit light (fluoresce) at a specific and longer wavelength. This light is focused by the objective lens; most of it passes through the dichromatic filter and is not reflected. A final barrier filter blocks any residual light with the frequency of the exciting radiation.

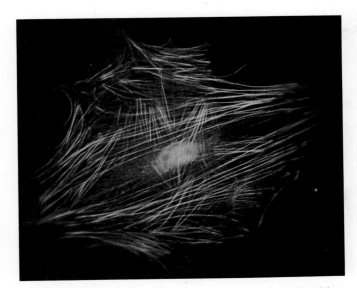

▲ Figure 4-13 Distribution of actin in a cultured fibroblast cell. A fixed human skin fibroblast was permeabilized with a detergent and stained with a fluorescent antiactin antibody. This fluorescence micrograph of a cell shows the long actin fibers within it. *Courtesy of E. Lazarides.*

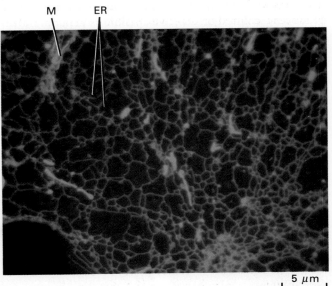

▲ Figure 4-14 Fluorescence micrograph of the endoplasmic reticulum (ER) in a flattened region of a living monkey kidney epithelial cell. The ER is a set of membrane fibers that fuse with each other to form a network, or reticulum. The larger, nonreticular fluorescent structures are mitochondria (M). [See C. Lee and L. B. Chen, 1988. *Cell* **44**:37–46.] *Courtesy of L. B. Chen.*

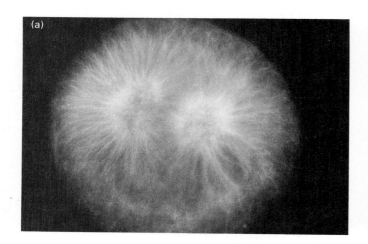

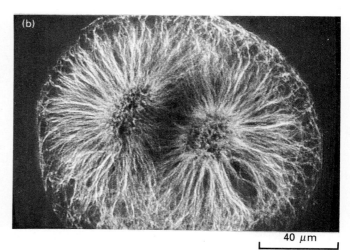

40 μm

▲ **Figure 4-15**   The advantage of confocal fluorescence microscopy. A mitotic fertilized egg from a sea urchin *(Psammechinus)* is lysed with a detergent, exposed to an anti-tubulin antibody, and then exposed to a fluorescein-tagged antibody that binds to the first antibody. (a) When viewed by conventional fluorescence microscopy, the mitotic spindle is blurred due to the background "glow" of fluorescence from tubulin above and below the plane of focus. (b) The confocal microscopic image is sharp, particularly in the center of the mitotic spindle; fluorescence is detected only from molecules in the focal plane. *From J. G. White, W. Amos, and M. Fordham, 1987,* J. Cell Biol. **104:**41–48.

The Confocal Scanning Microscope Produces Vastly Improved Fluorescent Images

Immunofluorescence microscopy has its limitations. The fixatives employed to preserve cell architecture often destroy the *antigenicity* of a protein—its ability to bind to its specific antibody. Also, the method is generally difficult to use on thin cell sections; the embedding media often fluoresce themselves, obscuring the specific signal from the antibody. In microscopy of whole cells, the fluorescent light comes from molecules above and below the plane of focus. Thus the observer sees a superposition of fluorescent images from molecules at many depths in the cell, making it difficult to determine the actual three-dimensional molecular arrangement (see Figure 4-13).

The confocal scanning microscope avoids the last problem by permitting the observer to visualize fluorescent molecules in a single plane of focus, thereby creating a vastly sharper image (Figure 4-15). At any instant in this *confocal imaging,* only a single small part of a sample is illuminated with exciting light from a focused laser beam, which rapidly moves to different spots in the sample focal

(a)   (b)   (c)   (d)

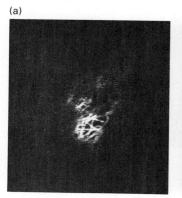
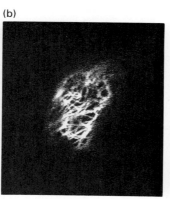
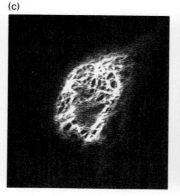
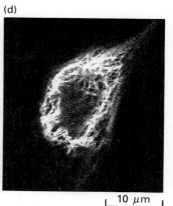

10 μm

▲ **Figure 4-16**   Optical sectioning of the tubulin cytoskeleton in a cultured kidney epithelial cell, using the confocal microscope. A permeabilized cell is treated with an antitubulin antibody; the bound antibody is then detected by a fluorescent antibody that binds to the antitubulin antibody. These fluorescent images were recorded from four depths of the sample 0.4 μm apart. *Courtesy E. Karsenti, B. Bacallo, and E. Stelzer.*

plane. Images from these spots are recorded by a computer, and the composite image is displayed on a computer screen. By a refinement known as *optical sectioning,* a computer records *serial sections*—fluorescent images of planes at different depths of the sample (Figure 4-16)—and combines them into one three-dimensional image (here, of the distribution of the protein tubulin within a cell).

Dark-field Microscopy Allows Detection of Small Refractile Objects

Standard light (bright-field) microscopy requires that the sample be colored or otherwise absorb or refract the incident light differently from the surrounding medium. Most unstained cells lack these characteristics and are poorly visible. More than one technique has been devised to solve this problem.

In *dark-field microscopy,* light is directed from the condenser to the sample at an oblique angle, so that none of the incident light enters the objective lens (creating a dark field if no specimen is present). Only light refracted (bent) or diffracted (scattered) by the specimen enters the objective lens to form the image (Figures 4-17 and 4-18). Resolution is not very good, but this technique does enable the observer to visualize small objects that refract a great deal of incident light, which appear like bright raindrops in a

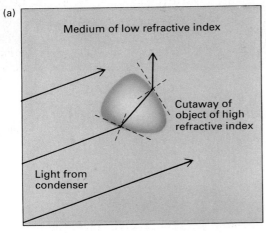

(a)

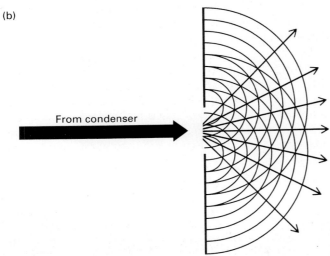

(b)

▲ **Figure 4-17** Two ways in which a specimen can redirect light from a condenser lens. (a) A beam of light is refracted (bent) once as it passes from air into a transparent object and again when it departs. (b) Light waves impinging on a pinhole in an opaque object spread out in all directions. Overlapping waves emanating from different sides of the hole reinforce each other in some directions (*straight arrows*); to an observer in one of those directions, the pinhole will seem bright. This phenomenon is called *diffraction.* Similarly, when light impinges on an opaque object, the edges diffract the light waves, and the edges will seem bright when viewed in some directions (*straight arrows*) and dark in others. Both refracted and defracted light are used to form the image in a dark-field microscope.

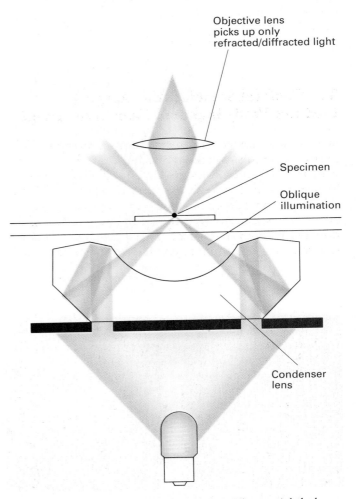

▲ **Figure 4-18** Dark-field microscopy. The special dark-field condenser lens causes light to enter the specimen at an angle, so that the direct rays of illumination do not enter the objective lens. Only diffracted or refracted light rays from the specimen are visible; the background is dark.

(a)

(b)

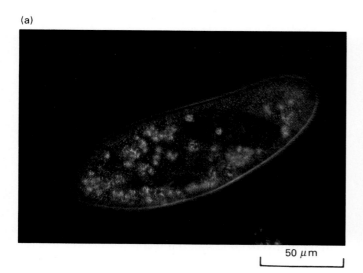
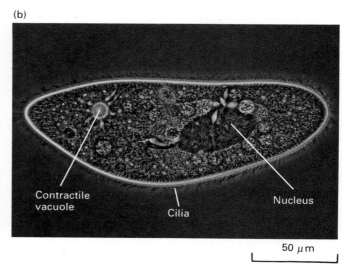

▲ **Figure 4-19** Living *Paramecium caudatum* viewed by (a) dark-field microscopy, which reveals small, refractile objects within the cell (such as mitochondria and basal bodies), and (b) phase-contrast microscopy, which reveals nonrefractile components (such as the contractile vacuole). *Courtesy of M. I. Walker/Biophoto Associates.*

(a)

(b)

(c)

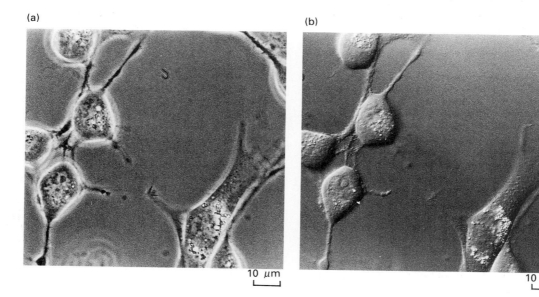

▲ **Figure 4-20** The same field of live, cultured fibroblast cells viewed by (a) phase-contrast microscopy and (b) Nomarski interference (differential interference) microscopy. (c) A newly hatched larvae of the nematode *Caenorhabditis elegans*, viewed with Nomarski optics. The individual nuclei of many of the organism's 500 cells are visible. *Parts (a) and (b) courtesy of B. Alpert and Y. Henis; part (c) from J. E. Sulston and H. R. Horvitz, 1977,* Devel. Biol. **56**:*110.*

beam of sunlight (Figure 4-19). Dark-field microscopy is widely used in microbiology to detect small cells, such as bacteria. It is also used in *autoradiography* to detect the tiny silver grains produced in a photographic emulsion by radiation.

Phase-contrast and Nomarski Interference Microscopy Visualize Living Cells

Two optical techniques are notable for giving detailed views of transparent, live, unstained cells and tissues (Figure 4-20). Both convert small differences in refractive index or thickness between parts of the specimen or between the specimen and the surrounding medium to differences of light and dark in the final image. Light moves more slowly in media of higher refractive index. Consequently, part of a light wave that passes through a specimen will be refracted or diffracted (see Figure 4-17) and will be out of phase (synchronism) with the part of the wave that does not pass through the specimen. How much their phases differ depends on the difference in refractive index along the two paths and on the thickness of the specimen. If the two parts of the light wave are recombined, the resultant light will be brighter if they are in phase and less bright if they are out of phase.

In *phase-contrast microscopy*, a glass *phase plate* between the specimen and the observer further increases the differences in contrast (Figure 4-21). Undiffracted light from the sample passes through a gray ring in the plate, where part of it is absorbed and, more importantly, the phase of the incident light is altered by one-quarter of a wavelength. This further increases the contrast, since the diffracted rays pass through the clear part of the phase plate, before the two types of light are recombined to form the image.

Nomarski interference microscopy, or *differential interference microscopy*, generates an image that looks as if the specimen is casting a shadow to one side; the "shadow" primarily represents a difference in refractive index rather than literal thickness. This technique utilizes plane-polarized light. In Nomarksi interference microscopy, a prism splits the beam of incident light so that one part of the beam passes through one region of a specimen and the other part part passes through a closely adjacent region; a second prism then reassembles the two beams. Minute differences in thickness or refractive index between adjacent parts of a sample are converted into a bright image (if the two beams are in phase when they recombine) or a dark one (if they are out of phase).

The phase-contrast microscope is especially useful in examining the structure and movement of the larger organelles, such as the nucleus and mitochondria, in live cultured cells. The greatest disadvantage of this technique is that it is suitable for observing only single cells or thin cell layers. The Nomarski interference technique, in con-

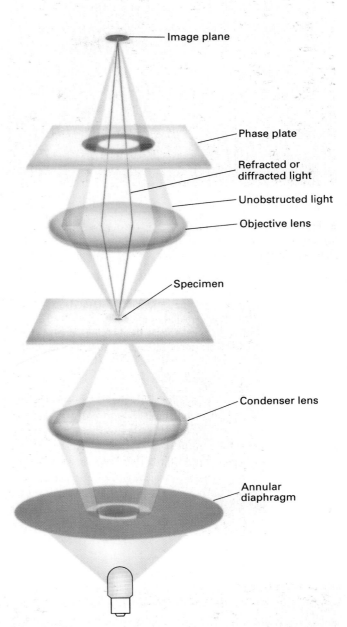

▲ **Figure 4-21** The optical pathway of the phase-contrast microscope. Incident light passes through an annular diaphragm, which focuses a circular annulus (ring) of light on the sample. Light that passes unobstructed through the specimen is focused by the objective lens onto the thick gray ring of the phase plate, which absorbs some of the direct light and alters its phase. If a specimen refracts or diffracts the light, however, the phase of some light waves is altered and they are redirected through the thin, clear region of the phase plate before recombining with the unrefracted light waves. This produces an image in which the degree of darkness or brightness of a region of the sample depends on the refractive index of that region.

Labels on figure: Image plane · Phase plate · Refracted or diffracted light · Unobstructed light · Objective lens · Specimen · Condenser lens · Annular diaphragm

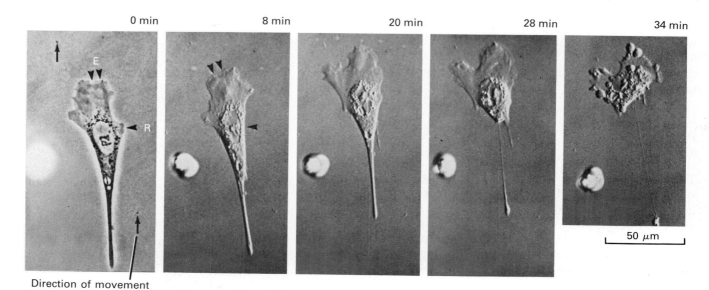

Direction of movement

▲ **Figure 4-22** Time-lapse micrographs show the movement of a cultured fibroblast cell along a glass surface. A bit of debris on the substratum serves as a reference point. The first image, at 0 min, was obtained by phase-contrast microscopy. Successive images of the same cell, obtained by Nomarski optics, show the lamella at the right of the cell retracting (R) and the lamellipodia at the leading edge of the cell extending (E). In the frame taken at 8 min, the leading edge has moved forward about 9 μm and the lamellipodia there form a thin flat sheet. By 28 min, the broad leading edge has spread and separated into two lamellae; the thin trailing edge of the cell has begun to retract into the cell body. By 34 min, retraction of the trailing edge is almost complete; only a thin thread of cytoplasm from the trail is left behind, anchored to the substratum. *From W.-T. Chen, 1981, J. Cell Sci. 49:1.*

trast, only defines the outlines of the large organelles, such as the nucleus and vacuole (see Figure 4-20b). However, thick objects can be observed with this technique by optical sectioning. *Time-lapse microscopy*, in which the same cell is photographed at regular intervals over periods of several hours, also allows the observer to study cell movement, provided the microscopic stage can control the temperature of the specimen and the gas environment (Figure 4-22).

Electron Microscopy

The fundamental principles of electron microscopy are similar to those of light microscopy: the major difference is that electromagnetic lenses, not optical lenses, focus a high-velocity electron beam instead of visible light. Because electrons are absorbed by atoms in air, the entire tube between the electron source and the viewing screen is maintained under an ultra-high vacuum.

Quantum physics teaches us that a beam of subatomic particles such as electrons can be considered either as a stream of discrete particles or as a series of waves. In typical electron microscopes, electrons have the properties of a wave with a wavelength of only 0.005 nm. Recall that the minimum distance D at which two objects can be distinguished is proportional to the wavelength of the light λ that illuminates them. Thus the maximum resolution of the electron microscope is theoretically 0.005 nm (less than the diameter of a single atom), or 40,000 times the resolution of the light microscope and 2 million times that of the unaided human eye. The effective resolution of the electron microscope in the study of biological systems is considerably less than this ideal, however; under optimal conditions, a resolution of 0.10 nm can be obtained.

Transmission Electron Microscopy Depends on the Differential Scattering of a Beam of Electrons

The *transmission electron microscope* directs a beam of electrons through a specimen. Electrons are emitted by the tungsten *cathode* when it is electrically heated. The electric potential of the cathode is kept at 50,000–100,000 volts; that of the *anode*, near the top of the tube, is zero. This drop in voltage causes the electrons to accelerate as they move toward the anode. The condenser lens, like the one in a light microscope, focuses the electron beam onto the sample. The objective lens and a *projector lens* focus the electrons that pass through the sample onto a viewing screen or a piece of photographic film (Figure 4-23).

Like the light microscope, the transmission electron microscope is used to view thin sections of a specimen,

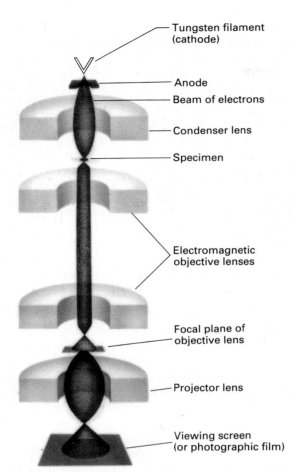

Tungsten filament (cathode)

Anode

Beam of electrons

Condenser lens

Specimen

Electromagnetic objective lenses

Focal plane of objective lens

Projector lens

Viewing screen (or photographic film)

▲ **Figure 4-23** The optical path in a transmission electron microscope. A beam of electrons emanating from a heated tungsten filament is focused onto the specimen plane by the magnetic condenser lens. The electrons passing through the specimen are focused by a series of objective and projector lenses to form a magnified image of the specimen on a fluorescent viewing screen or a piece of photographic film. The entire column, from the electron generator to the screen, is maintained at a very high vacuum.

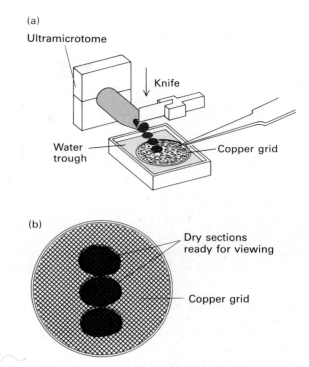

(a)

Ultramicrotome

Knife

Water trough

Copper grid

(b)

Dry sections ready for viewing

Copper grid

▲ **Figure 4-24** Preparation of a sample of tissue for transmission electron microscopy. The tissue is dissected, cut into small cubes, and plunged into a fixing solution that crosslinks and immobilizes proteins. (Glutaraldehyde is frequently used; osmium tetroxide, another fixing substance, stains intracellular membranes and certain macromolecules.) The sample is dehydrated by placing it in successively more concentrated solutions of alcohol or acetone; it is then immersed in a solution of plastic embedding medium and put in an oven. Heat causes the solution to polymerize into a hard plastic block, which is trimmed; sections less than $0.1\,\mu$m thick are then cut with an ultramicrotome, a fine slicing instrument with a diamond blade (a). The sections are floated off the blade edge onto the surface of water in a trough. A copper grid coated with carbon or some other material is used to pick up the sections, which are then dried (b).

but the fixed sections must be much thinner for electron microscopy (only 50–100 nm, about 0.2 percent of the thickness of a single cell; Figure 4-24). The image depends on variations in the scattering of the incident electrons by different molecules in the preparation. Without staining, the beam of electrons passes through a cell or tissue sample uniformly; there is little differentiation of components. Staining techniques can reveal the location and distribution of specific materials. Heavy metals, such as gold or osmium, appear dark on the micrograph; they scatter (diffract) most of the incident electrons (scattered electrons are not focused by the electromagnetic lenses and do not form the image). Osmium tetroxide preferentially stains certain cellular components, such as membranes, which appear black in the micrographs (the trans-

mission electron micrographs in Figures 4-2 to 4-7 were taken of sections stained with this substance). A recent advance is the introduction of gold particles or *ferritin*, an iron-containing protein, as electron-dense tags, which can detect antibody molecules bound to specific target proteins in thin sections (Figure 4-25).

Minute Details Can Be Visualized on Viruses and Subcellular Particles

The electron microscope is also used to obtain information about the shapes of purified viruses, fibers, enzymes, and subcellular particles. In one technique, called *metal shadowing*, a thin layer of evaporated metal, such as platinum, is laid at an angle on a biological specimen (Figure

(a)

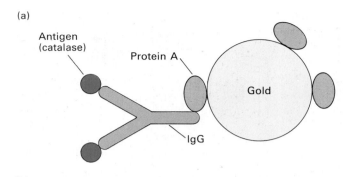

(b)

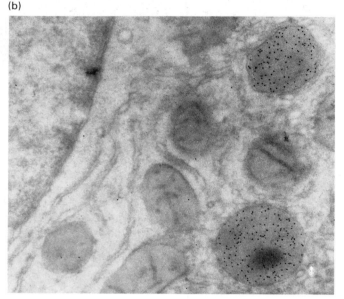

0.5 μm

◀ **Figure 4-25** The use of antibodies to detect the subcellular location of a specific protein, the enzyme catalase, by electron microscopy. A slice of liver tissue is fixed with glutaraldehyde and sectioned. An antibody (IgG) to catalase is then added to the section. When complexed with gold particles 4 nm in diameter, protein A from the bacterium *Staphylococcus aureus* binds tightly to the common Fc domain (see Figure 2-28) of the antibodies. Each gold particle makes the resulting immune complex (a) visible in the electron microscope. (b) The location of the gold particles (black dots) in the peroxisomes indicates the presence of the catalase. *From H. J. Geuze, J. W. Slot, P. A. van der Ley, and R. C. T. Scheffer, 1981, J. Cell Biol.* **89**:653. *Reproduced from the* Journal of Cell Biology *by copyright permission of The Rockefeller University Press.*

Scanning Electron Microscopy Visualizes Details on the Surface of Cells or Particles

The *scanning electron microscope* allows the investigator to view the surfaces of *unsectioned* specimens. The sample is fixed and dried and coated with a thin layer of a heavy metal, such as platinum, by evaporation in a vacuum (see Figure 4-26); in this case, the sample is rotated so that the platinum is deposited uniformly on the surface. An intense electron beam inside the microscope scans rapidly over the sample. Molecules in the specimen are excited and release secondary electrons that are focused onto a scintillation detector; the resulting signal is displayed on a cathode-ray tube. Because the number of secondary electrons produced by any one point on the sample depends on the angle of the electron beam in rela-

4-26). An acid bath dissolves the biological material, leaving a metal *replica* of its surface which can then be examined in the transmission electron microscope (Figure 4-27). If the metal is deposited on only one side of the sample, the image seems to have shadows where the metal appears dark and the shadows appear light.

Electron microscopy cannot be used to study live cells because they are generally too vulnerable to the required conditions and preparatory techniques. In particular, water is an essential component of all biological structures; in its absence, macromolecules can become denatured and nonfunctional. However, a new technique allows the observer to examine hydrated, unfixed, and unstained biological specimens directly in the electron microscope. An aqueous suspension of a sample is applied in an extremely thin film to a grid, frozen in liquid nitrogen, maintained in this state by means of a special mount, and then observed in the electron microscope. The very low temperature (−196°C) keeps the water from evaporating, even in a vacuum, and the sample can be observed in detail in its native, hydrated state without shadowing or fixing it (Figure 4-28).

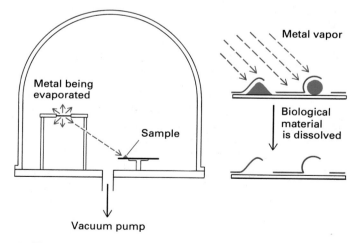

▲ **Figure 4-26** Metal shadowing. This technique makes surface details on very small particles visible in the electron microscope. The sample on a grid is placed in a vacuum container. A filament of a heavy metal, such as platinum, is heated so that the metal evaporates and some of it falls over the sample grid in a very thin film. The biological material is then dissolved by acid, so that the observer views only the metal replica of the sample.

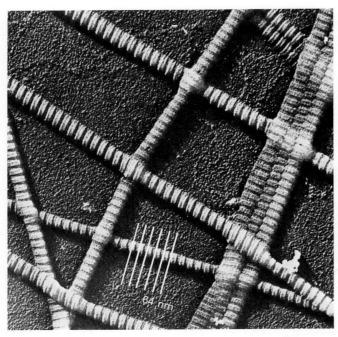

▲ Figure 4-27 A platinum-shadowed replica of the substructural fibers of calfskin collagen, the major structural element of tendons, bone, and similar tissues. The fibers are about 200 nm thick; a characteristic 64-nm repeated pattern (white parallel lines) is visible along the length of each fiber. *Courtesy of R. Bruns.*

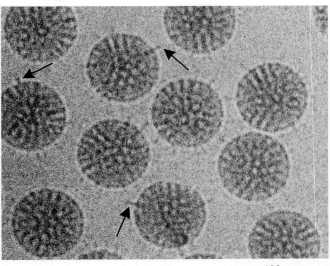

▲ Figure 4-28 Electron micrograph of unstained rotavirus particles. A thin suspension of virus particles in water is applied to an electron microscopy grid and frozen. It is then visualized in a transmission electron microscope equipped with a sample stage cooled with liquid nitrogen. The low temperature prevents the ice surrounding the particles from evaporating in the vacuum. Because many biological specimens scatter more electrons than water does, the investigator can observe a very thin specimen without fixing, staining, or dehydrating it. Note the minute spikes (*arrows*) visible on some particles. *From B. V. Venkataram Prasad, G. J. Wang, J. Clerx, and W. Chiu, 1988,* J. Mol. Biol. **199**:269–274.

tion to the surface, the scanning electron micrograph has a three-dimensional appearance (see Figure 4-7a). The resolving power of scanning electron microscopes is only about 10 nm, much less than that of transmission instruments.

Sorting Cells and Their Parts

Most animal and plant tissues contain a mixture of cell types. For experimental purposes, however, an investigator often wishes to study a pure population of one type of cell. In some cases, cells differ in some physical property that allows them to be separated. White blood cells and red cells (erythrocytes) have very different densities because erythrocytes have no nucleus. Such cells can be separated on the basis of density. Such a convenient differentiation cannot be made between most cell types, however, and other cell-separation techniques have had to be developed.

Flow Cytometry Is Used to Sort Cells

The optical technique of *flow cytometry* can identify particular cells and separate them from a mixture. Indeed, an instrument called a *fluorescence-activated cell sorter,* or *FACS,* can select a single cell from a group of many cells.

After a fluorescent dye is linked to an antibody specific to a cell-surface molecule, cells bearing this molecule will bind the antibody. In the FACS, a stream of cells flows past a laser beam and the correct wavelength of light causes the cells that contain the antibody-dye complex to fluoresce. The stream is broken into droplets that contain no more than one cell; only droplets containing a fluorescent cell are given a negative electric charge, which allows them to be separated from the others (Figure 4-29). The FACS can select one cell that bears a specific surface marker from thousands that do not; the selected cell can then be grown in culture. A machine with this optical capability can sort cells according to any absorbing or emitting characteristic.

Other uses of flow cytometry include the measurement of a cell's DNA and RNA content and the determination of its general shape and size. Most FACS instruments contain a detector that measures the amount of laser light scattered by each cell (Figure 4-29), which is proportional to the number and size of the cells. Thus the instrument can make simultaneous measurements of the size of a cell (from the amount of scattered light) and the amount of DNA it contains (from the amount of fluorescence of a DNA-binding dye). In a population of cultured yeast or

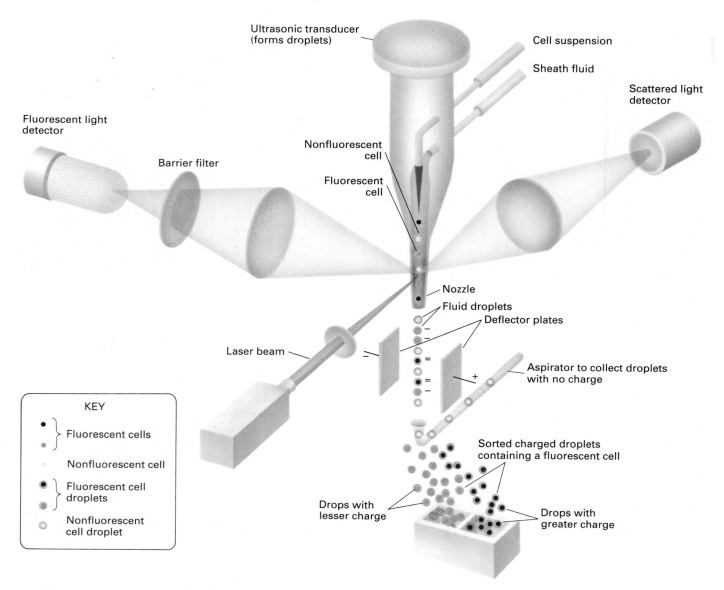

▲ **Figure 4-29** The fluorescence-activated cell sorter (FACS). A concentrated suspension of cells is allowed to react with a fluorescent antibody or dye that binds to a particle or molecule such as DNA. The suspension is then mixed with a buffer (the sheath fluid); the cells are passed single-file through a laser light beam, and the fluorescent light emitted by each cell is measured. The light scattered by each cell, from which the size and shape of the cell are determined, can be measured simultaneously. The suspension is then forced through a nozzle, which forms tiny droplets containing at most a single cell. At the time of formation, each droplet is given an electric charge proportional to the amount of fluorescence of its cell. Droplets with no charge or different electric charges (due to different amounts of bound dye) are separated by an electric field and collected. It takes only milliseconds to sort each droplet, so up to 10 million cells per hour can pass through the machine. In this way, cells that have unusual properties can be separated and then grown. *After D. R. Parks and L. A. Herzenberg, 1982, in* Methods in Cell Biology, *vol. 26, Academic Press, p. 283.*

animal cells, for example, this technique can purify cells in one of three stages of the cell cycle: G (before DNA synthesis with a defined amount x of DNA); S (during DNA replication; between x and $2x$ of DNA); and G2 (after DNA replication but before cell division; $2x$ of DNA).

Fractionation Methods Isolate Subcellular Structures

Although microscopy can localize a particular protein to a specific subcellular fiber or organelle, it is essential to isolate quantities of each of the major subcellular organelles to study their structures and metabolic functions in detail. Rat liver has been used in many classic studies because it is abundant in a single cell type (Figure 4-30). However, the same isolation principles apply to virtually all cells and tissues.

(a)

(b)

5 μm

1.0 μm

(c)

(d)

0.1 μm

0.1 μm

(e)

(f)

1.0 μm

0.5 μm

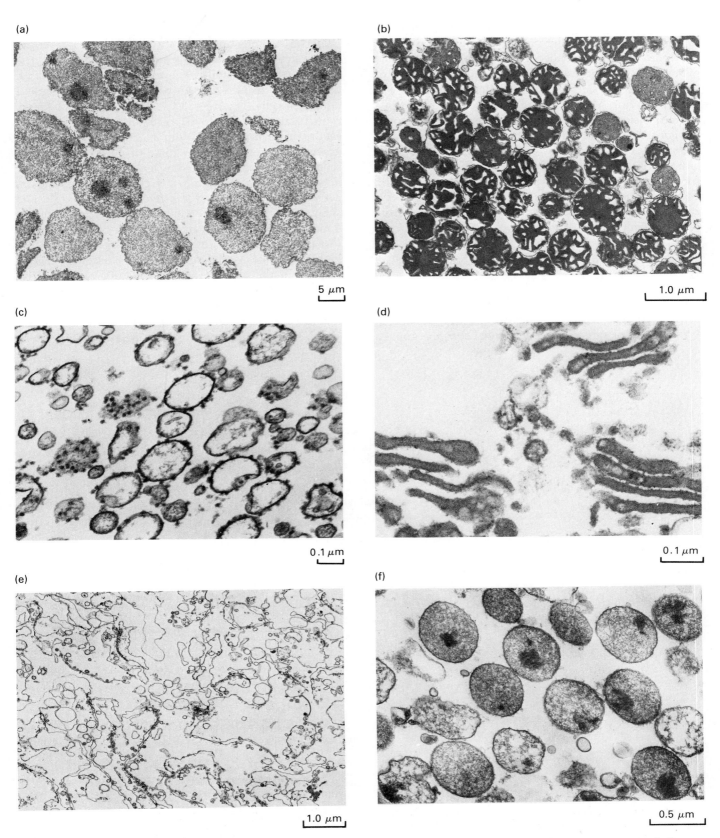

▲ **Figure 4-30** Electron micrographs of purified rat liver organelles: (a) nuclei, (b) mitochondria, (c) rough endoplasmic reticulum, sheared into smaller vesicles termed micro- somes, (d) Golgi vesicles, (e) plasma membranes, and (f) per- oxisomes, *Parts (a)–(e) courtesy of S. Fleischer and B. Fleischer; part (f) courtesy of P. Lazarow.*

The initial step for purifying subcellular structures is to rupture the cell wall, the plasma membrane, or both. First, the cells are suspended in a solution of appropriate pH and salt content, usually isotonic sucrose (0.25 M) or a combination of salts similar in composition to those in the cell interior. Many cells are then broken by stirring a cell suspension in a high-speed blender or by exposing it to high-frequency sound *(sonication).* Plasma membranes can also be sheared by special pressurized tissue homogenizers in which the cells are forced through a very narrow space between the plunger and the vessel wall. Generally, the cell solution is kept at 0°C to best preserve enzymes and other constituents after their release from the stabilizing forces of the cell.

Velocity Centrifugation Separates on the Basis of Size and Density

The various organelles differ in both size and density. Most fractionation procedures begin with *rate-zonal centrifugation,* or *differential-velocity centrifugation*—a technique in which an *ultracentrifuge* generates the most powerful sedimenting forces. Large, dense structures form a deposit in a centrifuge tube faster than small, less dense ones do. Generally, the cell homogenate is first filtered or centrifuged at relatively low speeds to remove unbroken cells (Figure 4-31). A slightly faster or longer centrifugation selectively deposits the nucleus—the largest organelle (usually 5–10 μm in diameter). The force of gravity is denoted by g. A force of 600g, necessary to sediment nuclei, is generated by a typical centrifuge rotor operating at 500 revolutions per minute (r/min). The undeposited material is then transferred to another tube and subjected to centrifugation at a higher speed (15,000g × 5 min), which results in the deposition of mitochondria (organelles 1–2 μm long), chloroplasts (also 1–2 μm long), lysosomes, and peroxisomes. A subsequent centrifugation in the ultracentrifuge (100,000g × 60 min) results in the deposition of the plasma membrane and fragments of the endoplasmic reticulum. A force of 100,000g requires about 50,000 r/min in an ultracentrifuge; at this speed, the rotor chamber is kept in a high vacuum to

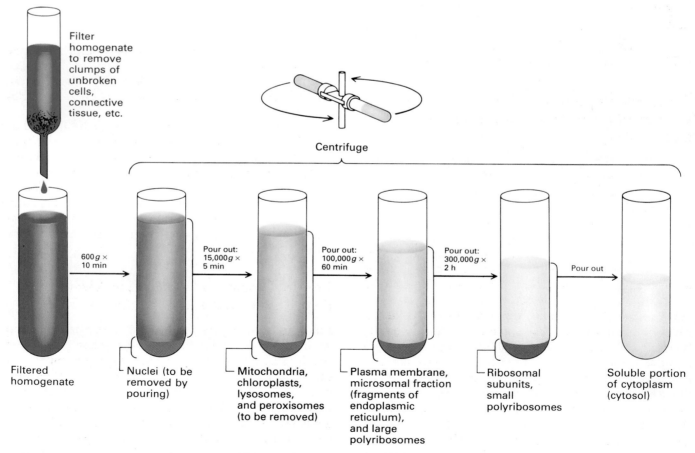

▲ **Figure 4-31** Cell fractionation by rate-zonal centrifugation. The different sedimentation rates of various cellular components make it possible to separate them partially by centrifugation. Nuclei and viral particles can sometimes be purified completely by such a procedure. Each soluble fraction can be further separated by density-gradient centrifugation.

▶ **Figure 4-32** Separation of organelles from rat liver by density-gradient centrifugation. Material in a pellet from a separation by rate-zonal centrifugation at 15,000g (see Figure 4-31) is resuspended and layered on a density gradient composed of layers of successively less-dense sucrose solutions in a centrifuge tube. Under centrifugation, each organelle migrates to and remains at its appropriate equilibrium density. Before cell disruption in this particular experiment, the liver is perfused with a solution containing a small amount of detergent, which is taken into the cells by endocytosis and transferred to the lysosomes but does not lyse them. Thus the lysosomes are less dense than they would normally be, affording a "clean" separation from the mitochondria.

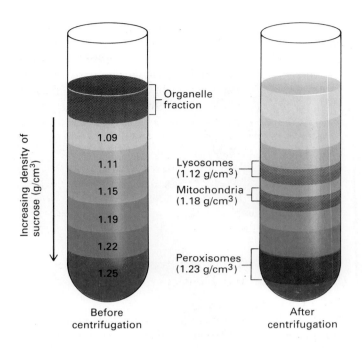

reduce friction (and subsequent heating) between air and the spinning rotor. Ribosomal subunits, small polyribosomes, and particles such as glycogen require additional centrifugation at still higher speeds to be recovered. Only the *cytosol*—the soluble aqueous portion of the cytoplasm—remains undeposited after centrifugation at 300,000g × 2 hr. (Chapter 6 contains a more detailed discussion of the principles of velocity sedimentation and illustrates how it can be used to resolve even smaller particles, such as ribosomal subunits and large proteins.)

Equilibrium Density-gradient Centrifugation Separates Materials by Density Alone

Rate-zonal centrifugation does not yield totally pure organelle fractions. One method that may be employed to process fractions further is *equilibrium density-gradient centrifugation,* in which organelles are separated by their density but not by their size. The impure organelle fraction is layered on top of a solution that contains a gradient of a dense nonionic substance, such as sucrose or glycerol. The solution is most concentrated (dense) at the bottom of the centrifuge tube (where a typical sucrose solution is about 1.25 g/cm³); the concentration decreases gradually toward the top of the tube, so that the solution is least dense at the surface. The tube is centrifuged at a high speed (about 40,000 r/min) for several hours to allow each particle to migrate to an *equilibrium position.* At this point, the density of the surrounding liquid is equal to the density of the particle and it does not move further. In typical preparations from animal cells, the rough endoplasmic reticulum has a density of 1.20 g/cm³, so that it separates well from the Golgi vesicles (with a density of 1.14 g/cm³) and the plasma membrane (with a density of 1.12 g/cm³). The higher density of the rough endoplasmic reticulum is due largely to the ribosomes bound to it. Another example is the resolution of

lysosomes, mitochondria, and peroxisomes from the same initial fraction of a rate-zonal centrifugation (Figure 4-32). Equilibrium density-gradient centrifugation can also be used to separate different kinds of nucleic acids and nucleoproteins.

The purity of organelle preparations can be assessed by morphology (through use of the electron microscope) or by their content of marker molecules. For example, the protein cytochrome *c* is found only in mitochondria; its presence in a fraction is a measure of its contamination by mitochondria. Similarly, catalase is present only in peroxisomes; acid phosphatase, only in lysosomes; and ribosomes, only in the rough endoplasmic reticulum or the cytosol.

Immunological Techniques Can Yield Pure Preparations of Certain Organelles

Many fractions are not pure even after rate-zonal and equilibrium density-gradient centrifugation. The next step is to take advantage of the fact that the membrane of each organelle contains unique enzymes and other proteins; several groups of investigators are now trying to purify organelles by immunological techniques, using antibodies specific for these organelle-membrane proteins. One example is the purification of a particular class of cellular vesicles that contain the protein *clathrin* on their outer surface (Figure 4-33). An antibody to clathrin, bound to a bacterial carrier, selectively removes these vesicles from a crude preparation of membranes.

(a)

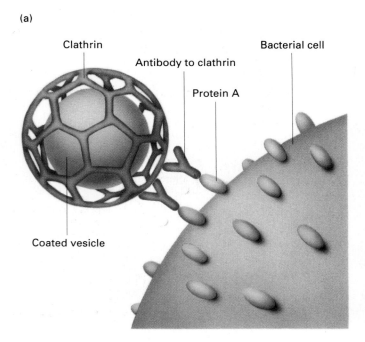

(b)

▲ **Figure 4-33** Immunological purification of clathrin-coated vesicles. (a) A suspension of membranes from rat liver is incubated with a clathrin antibody (a protein that coats the outer surface of certain cytoplasmic vesicles). The surface of the bacterium *Staphylococcus aureus* contains protein A, which binds antibodies nonspecifically (see Figure 4-25). By adding bacteria to the mixture of antibodies and membranes and recovering the bacteria by low-speed centrifugation, the clathrin-coated vesicles bound (by the antibody and protein A) to the bacterium are selectively purified. (b) A thin-section electron micrograph of clathrin-coated vesicles bound to these bacteria. [See E. Merisko, M. Farquahr, and G. Palade, 1982. *J. Cell Biol.* 93:846–858.] *Part (b) courtesy of G. Palade.*

The Organelles of the Eukaryotic Cell

Later chapters will examine the key roles played by the cytosol and the many different organelles in eukaryotic cellular metabolism. The final section of this chapter presents a brief overview of their structures and functions.

The development of the electron microscope in the early 1950s made it possible to study subcellular organization at a much finer resolution than could be achieved by using the light microscope. It immediately became clear that the internal structure of eukaryotes is much more complex than that of prokaryotes. Especially striking is the intricate structure of the eukaryotic nucleus.

The Eukaryotic Nucleus Is Bounded by a Double Membrane

The eukaryotic nucleus is surrounded by two membranes containing phospholipids. The inner nuclear membrane defines the nucleus itself. In many cells, the outer nuclear membrane is continuous with the rough endoplasmic reticulum, an extensive cytoplasmic membrane system to be discussed later. The space between the inner and outer nuclear membranes is continuous with the *lumen*, or inner cavity, of the rough endoplasmic reticulum. The membranes appear to fuse at the nuclear pores (Figure 4-34). The distribution of nuclear pores is particularly vivid when the nucleus is viewed by the *freeze-fracture technique* (Figures 4-34b and 4-35). These ringlike pores, which are constructed of a specific set of membrane proteins, function as channels, regulating the movement of material between the nucleus and the cytoplasm.

The Nucleus Contains the Nucleolus, a Fibrous Matrix, and DNA-Protein Complexes

In a nucleus that is not dividing, the chromosomes are elongated and are only about 25 nm thick; they cannot be observed in the light microscope. However, a suborganelle of the nucleus, the *nucleolus*, is easily recognized under the light microscope. The *nucleolar organizer*, a region of one or more chromosomes in the nucleolus,

(a)

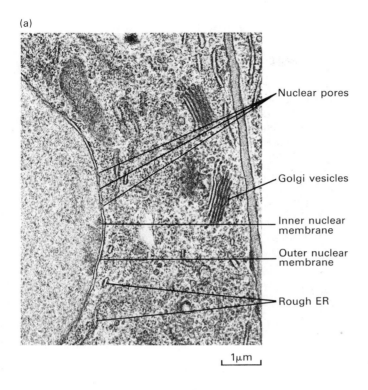

Nuclear pores

Golgi vesicles

Inner nuclear membrane

Outer nuclear membrane

Rough ER

1μm

Nucleus

(b)

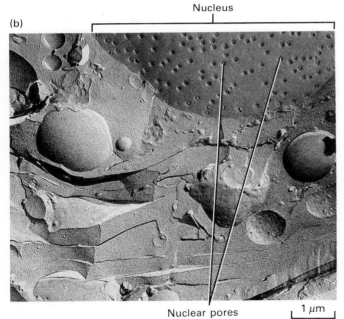

Nuclear pores

1 μm

▲ **Figure 4-34** Views of the nuclear envelope. (a) Electron micrograph of a thin section of an *Arabidopsis* (plant) cell, showing the pores between the nuclear membranes. The rough endoplasmic reticulum and stacks of Golgi vesicles are also visible. (b) A freeze-fractured preparation of an onion root-tip cell, showing the nucleus and pores in the nuclear membrane, which appear to be filled with a granular substance. *Part (a) courtesy of Biophoto Associates/Myron C. Ledbetter/Brookhaven National Laboratory; part (b) courtesy of D. Branton.*

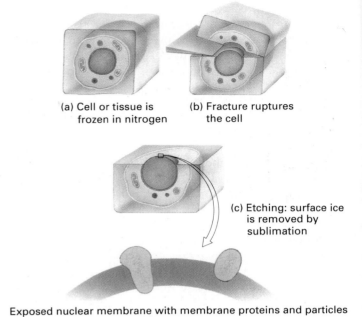

(a) Cell or tissue is frozen in nitrogen

(b) Fracture ruptures the cell

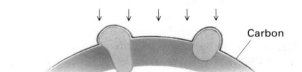

(c) Etching: surface ice is removed by sublimation

Exposed nuclear membrane with membrane proteins and particles

Carbon

(d) Carbon is added to form a continuous surface

Platinum

(e) Surface is shadowed with a thin layer of platinum

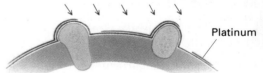

(f) Tissue is dissolved with acid; carbon-metal replica can be viewed under the electron microscope

▲ **Figure 4-35** Production of a freeze-fracture, deep-etch image of a cell. (a) A preparation of cells or tissues is quickly frozen in liquid nitrogen at −196°C, which instantly immobilizes cell components. (b) The block of frozen cells is fractured with a sharp blow from a cold knife. The fracture plane is irregular, usually occurring along the plasma membrane or surfaces of organelles. (c) The specimen is then placed in a vacuum, where the surface ice is removed by sublimation; this technique is called *deep etching* or *freeze etching*. (d) A thin layer of carbon is evaporated vertically onto the surface to produce a carbon replica. After metal shadowing (e; see Figure 4-26) with platinum, the organic material is removed by acid, leaving a carbon-metal replica of the tissue ready for examination with the electron microscope (f). In prints of the electron micrograph the image is usually reversed: carbon-coated areas appear light and platinum-shadowed areas appear dark.

contains many copies of the DNA that directs the synthesis of ribosomal RNA. Most of the cell's ribosomal RNA is synthesized in the nucleolus; some ribosomal proteins are added to ribosomal RNAs within the nucleolus as well. The finished or partly finished ribosomal subunit passes through a nuclear pore into the cytoplasm.

Investigators know little about the large-scale organization of the nucleus, although they have gained some knowledge of how DNA is packaged in the cell nucleus. In the electron microscope, the nonnucleolar regions of the nucleus, called the *nucleoplasm,* can be seen to have areas of high DNA concentration, often closely associated with the nuclear membrane. Fibrous proteins called *lamins* form a two-dimensional network along the inner surface of the inner membrane, giving it shape and apparently binding DNA to it. Beyond these rudimentary facts, the organization of materials and activities in the nucleoplasm is largely a mystery.

The major physiological function of the nucleus is to direct the synthesis of RNA. In a growing or differentiating cell, the nucleus is the site of vigorous metabolic activity. In other cells, such as resting mast cells (blood-borne cells that release histamine when triggered by allergens) and mature avian erythrocytes, the nucleus is inactive or dormant and minimal synthesis of DNA and RNA takes place.

The Cytosol Contains Many Cytoskeletal Elements and Particles

The cytosol is the nonparticulate region of the cell cytoplasm (the part not contained within any of the organelles). Initially, the cytosol was thought to be a fairly homogenous "soup" in which all of the organelles floated. However, it is now known that the cytosol of eukaryotic cells contains a cytoskeleton composed of at least three classes of fibers—the microtubules (20 nm in diameter), the microfilaments (7 nm in diameter), and the intermediate filaments (10 nm in diameter)—which helps to maintain cell shape and mobility and probably provides anchoring points for other cellular structures. At least five types of intermediate filaments, each made of a different type of protein, have been identified in various animal cells.

Despite decades of research, many structural features of the cytosol are still in question. Its protein composition is high (about 20–30 percent), so that weak protein-protein interactions are favored. Many proteins bind to each other so weakly that these interactions cannot be detected in the dilute protein solutions typically used in biochemical experiments. In particular, many multiple-enzyme aggregates may disperse at the dilution created when cells are homogenized. On the other hand, the high protein concentration in the cytosol may allow proteins

to precipitate or aggregate during fixation or preparation for microscopy, thereby creating artifactual "structures."

Because sections of cells or tissues used for standard electron microscopy must be thinner than 0.1 μm, only a small piece of a cell can be observed in any one section. Most fibrous systems are several μm in length, so they can be seen as long elements only in sections that, by chance, happen to be in the plane of the fiber bundles (Figure 4-36). Sections across this plane contain only a short bit of the fibers; single sections may contain a few or many fiber profiles but give no visual evidence for fiber length. Serial sectioning of the tissue sample by tracing a fiber from one image into the next to reconstruct its three-dimensional architecture can compensate for these shortcomings. However, serial sectioning can be tedious (100 sections are needed to examine a 20 μm-thick cell) and is used only in special cases.

A striking view of cytoplasmic fibers in an unsectioned cell (Figure 4-37) is obtained by first treating it with nonionic detergents to dissolve the plasma and organelle membranes and to remove the cytosol. The insoluble cytoskeleton is then cooled very quickly (within millisec-

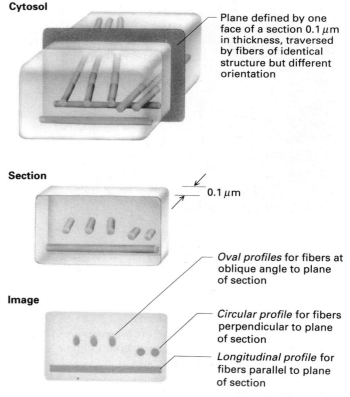

▲ **Figure 4-36** A section through a bundle of fibers can generate very different images, depending on the angle of the cut to the plane of the fibers.

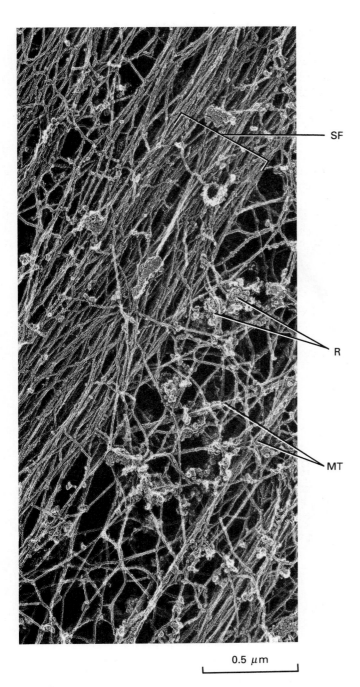

0.5 µm

▲ **Figure 4-37** Electron micrograph of a platinum replica of a cytoskeleton prepared by quick freezing and deep etching from a fibroblast immersed in the detergent Triton X-100 to remove soluble cytoplasmic proteins and membranes. Prominent are bundles of actin microfilaments termed *stress fibers* (SF), which are thought to connect segments of the plasma membrane and anchor the cell to the substratum. Also visible are two thicker microtubules (MT) and a more diffuse meshwork of filaments studded with grapelike clusters, which are probably polyribosomes (R). *From J. E. Heuser and M. Kirschner, 1980,* J. Cell Biol. *86:212. Reproduced from the* Journal of Cell Biology *by copyright permission of The Rockefeller University Press.*

onds) by the *quick-freeze technique* to the temperature of liquid helium (−269°C, or 4° above absolute zero), which prevents the formation of ice crystals and distortions in cytoskeletal architecture. While the preparation is still frozen, some of the surface water is removed in a vacuum by deep etching, resulting in exposure of the nonvolatile protein fibers. Once coated with a thin layer of platinum, these fibers are visible in the ordinary transmission electron microscope. By this technique, the cytoplasm of cultured animal cells is resolved almost exclusively into actin microfilaments, tubulin-containing microtubules, and intermediate filaments, which crisscross each other in complex patterns; at many points, different types of cytoskeletal fibers contact each other. In cultured cells, actin microfilaments often occur in bundles of long fibers that appear to be connected by small fibrous proteins.

The cytosol of many cells also contains *inclusion bodies*, granules that are not bounded by a membrane. Some cells—specifically, muscle cells and hepatocytes, the principal cell type in the liver—contain cytosolic granules of glycogen, a glucose polymer that functions as a storage form of usable cellular energy (Figure 4-38). In well-fed

Nuclear membrane

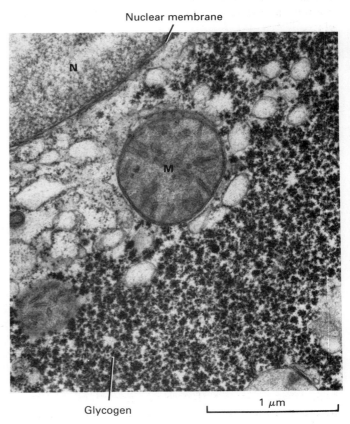

Glycogen

1 µm

▲ **Figure 4-38** Electron micrograph of a portion of the cytoplasm of a rat hepatocyte, showing glycogen rosettes, a mitochondrion (M), and part of the nucleus (N) and nuclear membrane. *Courtesy of Biophoto Associates.*

animals, glycogen can account for as much as 10 percent of the wet weight of the liver. The cytoplasm of the specialized fat cells in adipose tissue contains large droplets of almost pure triacylglycerols, a storage form of fatty acids. In all cells, the cytosol also contains a large number of different enzymes and is a major site of cellular metabolism; in many, it contains 25–50 percent of the total cell protein. Many investigators believe that the cytosol is highly organized, with most proteins bound to fibers or localized in specific regions.

The Endoplasmic Reticulum Is an Interconnected Network of Internal Membranes

Generally, the largest membrane in a eukaryotic cell is the endoplasmic reticulum (ER)—a network of interconnected closed membrane vesicles (see Figure 4-14). The endoplasmic reticulum has a number of functions in the cell but is notably important in the synthesis of many membrane lipids and proteins. Regions of the *rough endoplasmic reticulum* are studded with ribosomes; regions of the *smooth endoplasmic reticulum* lack ribosomes (Figure 4-39).

The Smooth Endoplasmic Reticulum The smooth endoplasmic reticulum is the site of synthesis and metabolism of fatty acids and phospholipids. The amount of smooth ER varies, depending on the type of cell. Many cells have very little smooth ER; in the hepatocyte, in

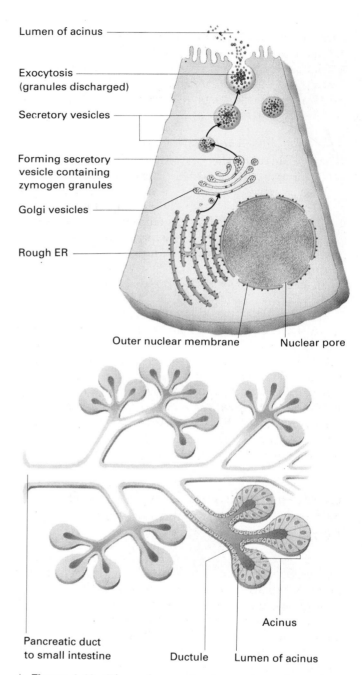

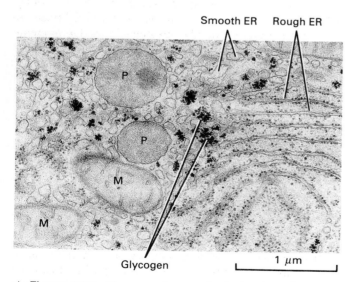

▲ **Figure 4-39** Electron micrograph of a section of a rat hepatocyte, showing two mitochondria (M), two peroxisomes (P), rough and smooth endoplasmic reticula, and glycogen rosettes. *Courtesy of P. Lazarow.*

▲ **Figure 4-40** The pathway taken by newly synthesized secretory proteins in a rat pancreatic acinar cell. Immediately after synthesis, the secretory proteins are found in the lumen of the rough ER. Transfer vesicles transport them to the Golgi vesicles (see Figure 4-41). Next they are concentrated in secretory vesicles containing granules of zymogens (pancreatic enzyme precursors, such as chymotrypsinogen). The cell is oriented so that the secretory vesicles form under its apical surface—the plasma membrane region that faces a ductule of the acinus. Exocytosis of the vesicles (fusion with the plasma membrane), triggered by hormones or nerve stimulation, releases the contents into the ductule; from there, the inactive precursors move to the intestine, where they are proteolytically activated into digestive enzymes (see Figure 2-18).

contrast, smooth ER is abundant. Enzymes in the smooth ER modify or detoxify chemicals such as pesticides or carcinogens by converting them into more water-soluble, conjugated products that can be secreted from the body. High doses of such compounds result in a large proliferation of the smooth ER.

The Rough Endoplasmic Reticulum Ribosomes bound to the rough endoplasmic reticulum synthesize certain membrane and organelle proteins and other proteins to be secreted from the cell. The ribosomes that fabricate secretory proteins are bound to the rough ER, in part by nascent polypeptide chains. As a growing secretory polypeptide emerges from the ribosome, it passes through the rough ER membrane; specific proteins in the membrane facilitate this transport. The newly made secretory proteins accumulate in the lumen (inner cavity) of the rough ER.

All eukaryotic cells contain a discernible amount of rough ER because it is needed to fabricate some plasma membrane proteins and glycoproteins. However, rough ER is abundant in specialized cells that produce secreted proteins. For example, the pancreatic acinar cells (the groups of cells surrounding the ductules) are specialized for the synthesis of digestive enzymes, such as chymotrypsin (Figure 4-40). Plasma cells, for another example, produce serum antibodies. In both types of cells, a large part of the cytosol is filled with rough ER.

Golgi Vesicles Process Secretory Proteins and Partition Cellular Proteins and Membranes

Several minutes after their synthesis, secretory proteins in the rough ER move to the lumenal cavity of another group of membrane-limited organelles, the Golgi vesicles, located near the nucleus in many cells (see Figure 4-30d). Three-dimensional reconstructions from serial sections of Golgi vesicles reveal a series of flattened membrane sacs surrounded by a number of more-or-less spherical membrane vesicles that appear to transport proteins to and from the *Golgi complex* (Figure 4-41). A secretory protein moves from the lumen of the ER to the lumen of the Golgi vesicles within small membrane vesicles (holding some of the ER contents) that bud off from regions of the rough ER not coated with ribosomes. These small vesicles then fuse with the stack of *Golgi sacs,* which have three defined regions—the *cis,* the *medial,* and the *trans.* Transfer vesicles from the rough ER apparently fuse with the cis region of the Golgi complex; the contents are then transferred to the medial and then to the trans region.

Golgi vesicles are often referred to as the "traffic police" of the cell because they play a key role in sorting many of the cell's proteins and membrane constituents

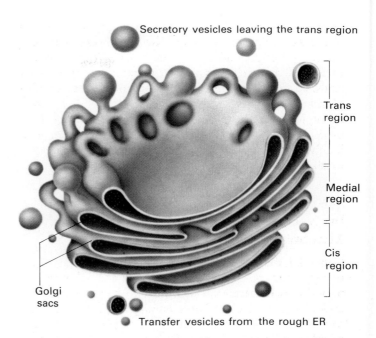

Secretory vesicles leaving the trans region

Trans region

Medial region

Cis region

Golgi sacs

Transfer vesicles from the rough ER

▲ **Figure 4-41** Three-dimensional model of the complex of Golgi vesicles, built by analyzing micrographs of serial sections through a secretory cell. The transfer vesicles, which have budded off the rough ER, fuse with the cis membranes of the Golgi complex. In acinar cells, the secretory vesicles that bud off the sacs on the trans membranes store secretory proteins, such as chymotrypsinogen, in concentrated form. *After a model by J. Kephart.*

and directing them to their proper destinations. A number of enzymes in the Golgi vesicles react with and modify secretory proteins passing through the Golgi lumen or membrane proteins and glycoproteins in the Golgi membranes *en route* to their final destinations. For example, a Golgi enzyme may add a "signal" or "tag," such as a phosphate residue, to certain proteins to direct them to their proper sites in the cell.

Proteins Are Secreted by the Fusion of an Intracellular Vesicle with the Plasma Membrane

After secretory proteins are modified in the Golgi vesicles, they are transported out of the complex by secretory vesicles, which seem to bud off the trans side of the complex (see Figure 4-40). In many cells, the membranes of these vesicles quickly fuse with the plasma membrane, releasing their contents into the extracellular space. The fusion of the membrane of an intracellular vesicle with the plasma membrane is termed *exocytosis.* In other cells, the Golgi vesicles form secretory vesicles that, in turn, fuse with similar vesicles to become intracellular membrane-limited

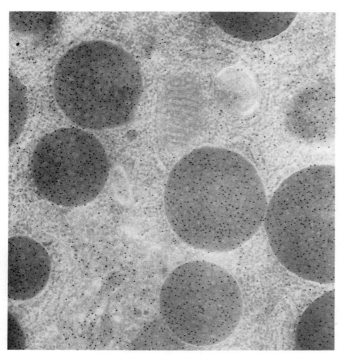

$\llcorner$ 0.5 μm $\lrcorner$

▲ **Figure 4-42** Electron micrograph showing secretory vesicles containing zymogen granules in a rat pancreatic acinar cell. To detect amylase, a protein typically found in zymogen granules, sections are treated with an antibody to amylase; 8-nm gold-particle complexes with protein A are then added to reveal the bound antibody (see Figure 4-25). *Courtesy of H. J. Geuze.*

storage reservoirs for the secretory proteins (Figure 4-42). The contents of these reservoirs are not released into the extracellular fluid until an appropriate signal (such as a hormone) stimulates the cell to fuse its reservoirs with the surface membrane. Figure 4-41 shows secretory vesicles containing zymogen granules (particles composed of pancreatic enzyme precursors) just under the apical surface of a pancreatic acinar cell.

Small Vesicles May Shuttle Membrane Constituents from One Organelle to Another

The foregoing discussion of protein secretion illustrates several important principles pertaining to the structure and function of membrane-limited organelles. Each type of organelle contains a specific set of enzymes and other proteins for the purpose of performing a specific function. Some organelles (Golgi vesicles) modify proteins; others (secretory vesicles) concentrate and store them. Organelles do not exist in isolation from other organelles;

they constantly "communicate" with one another, probably through small vesicles that shuttle membrane constituents and lumen contents between organelles. Some of these transport vesicles, termed *coated vesicles,* are surrounded by an outer protein shell composed primarily of the fibrous protein clathrin (see Figure 4-33). How intracellular transport vesicles "know" with which membranes to fuse and where to deliver their contents is still in question.

Lysosomes Contain a Battery of Degradative Enzymes That Function at pH 5

No part of a cell is immortal; even in growing cells, where macromolecules and organelles are increasing, membranes, proteins, and other constituents are constantly

Table 4-4 Acid hydrolases that have been located in lysosomes

Enzyme	Natural substrate
PHOSPHATASES	
Acid phosphatase	Most phosphomonoesters
Acid phosphodiesterase	Oligonucleotides and other phosphodiesters
NUCLEASES	
Acid ribonuclease	RNA
Acid deoxyribonuclease	DNA
PROTEASES	
Cathepsin(s)	Proteins
Collagenase	Collagen
Peptidases	Peptides
POLYSACCHARIDE- AND MUCOPOLYSACCHARIDE-HYDROLYZING ENZYMES	
β-galactosidase	Galactosides
α-glucosidase	Glycogen
α-mannosidase	Mannosides, glycoproteins
β-hexosaminidase	Glycolipids
Glucocerebrosidase	Glycolipids
β-glucuronidase	Polysaccharides and mucopolysaccharides
Lysozyme	Bacterial cell walls and mucopolysaccharides
Hyaluronidase	Hyaluronic acids; chondroitin sulfates
Arylsulfatase	Organic sulfates
LIPID-DEGRADING ENZYMES	
Esterase(s)	Fatty acyl esters
Phospholipase(s)	Phospholipids

SOURCE: D. Pitt, 1975, *Lysosomes and Cell Function,* Longman.

(a)

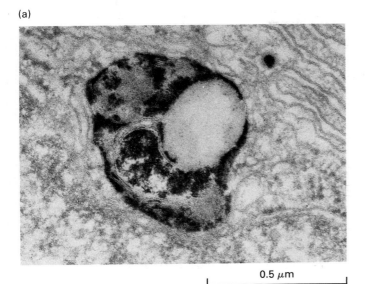

0.5 μm

(b)

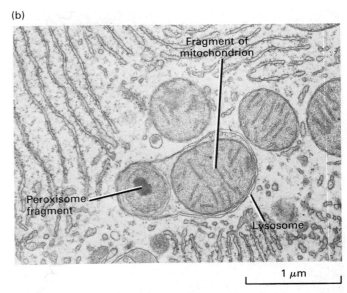

Fragment of mitochondrion

Peroxisome fragment

Lysosome

1 μm

▲ **Figure 4-43** Electron micrographs of lysosomes. (a) Part of a rat epididymis cell stained for acid phosphatase, a characteristic lysosomal enzyme. (b) The cytoplasm of a rat liver cell, illustrating an autophagic ("self-eating") vesicle (here, a secondary lysosome that probably resulted from the fusion of one or more primary lysosomes with an aged or defective mitochondrion and a similarly disabled peroxisome; note the vesicle contains bits of these organelles). *Courtesy of D. Friend.*

being degraded. The *lysosomes,* membrane-limited organelles found in animal cells, contain *acid hydrolases*—enzymes that degrade polymers into their monomeric subunits (Table 4-4). Lysosomes, which vary in size and shape (several hundred may be present in a typical cell), contain phosphatases, which remove phosphate groups from mononucleotides and other compounds, such as phospholipids; nucleases, which degrade RNA and DNA into mononucleotide building blocks; proteases, which degrade a variety of proteins and peptides; and enzymes that degrade complex polysaccharides and lipids into smaller units.

All of these enzymes work only at acid pH values and are inactive at the neutral pH values of cells and most extracellular fluids. To enable these enzymes to function, the inside of a lysosome is maintained at about pH 4.8 by a hydrogen-ion pump in the lysosomal membrane. The acid pH helps to denature proteins and make them accessible to the action of the lysosomal hydrolases, whose structures resist acid denaturation. If a lysosome releases its contents into the cytosol, where the pH is between 7.0 and 7.3, no degradation of cytosolic components takes place.

Lysosomes degrade many membranes and organelles that have outlived their usefulness to the cell. *Primary lysosomes* are roughly spherical and do not contain obvious particulate or membrane debris. *Secondary lysosomes* are larger and irregularly shaped and do contain particles or membranes that are being digested (Figure 4-43). Secondary lysosomes appear to result from the fusion of primary lysosomes with other membrane organelles. How aged or defective organelles are marked for degradation and transferred to lysosomes is not known, but occasionally a bit of mitochondrion or other membrane can be seen within a lysosome (see Figure 4-43b).

Lysosomes also play an important role in the degradation of extracellular macromolecules. Many proteins and particles are brought into the cell by *endocytosis,* the progressive infolding of a region of the plasma membrane to form a closed vesicle entirely surrounded by cytoplasm. Extracellular materials are either trapped within the lumen of the vesicle or bound to its lumenal surface. It is currently believed that these vesicles eventually fuse with lysosomes, which degrade the ingested macromolecules.

Tay-Sachs Disease Is the Result of a Defective Lysosomal Enzyme

The importance of lysosomes in the degradation of cellular membrane constituents is demonstrated by the existence of mutants defective in certain lysosomal hydrolases. Tay-Sachs disease is an inherited recessive disorder that results in mental retardation, derangement of the central nervous system, and death by the age of 5. In normal people, the ganglioside G_{M2}, a constituent of the plasma membranes of many mammalian cells (nerve cells in particular), is continually synthesized and degraded; the membranes of the brain cells in Tay-Sachs victims accumulate G_{M2} due to the absence of a specific lysosomal hydrolase, β-N-hexosaminidase A, a

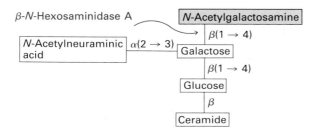

Ganglioside G$_{M2}$

▲ **Figure 4-44** The first step in the degradation of the ganglioside G$_{M2}$. The red arrow indicates the activity site of β-N-hexosaminidase A, the degradative enzyme deficient in victims of Tay-Sachs disease.

key enzyme in the normal turnover of G$_{M2}$ (Figure 4-44). The excess G$_{M2}$ is believed to cause all of the symptoms of Tay-Sachs disease. In a large number of human lysosome-storage diseases similar to Tay-Sachs, a missing lysosomal enzyme causes the lysosomes to fill up with partly degraded cellular material.

Vacuoles in Plant Cells Store Small Molecules and Enable the Cell to Elongate Rapidly

Most plant cells and many microorganisms, such as yeasts, contain at least one membrane-limited internal vacuole (see Figure 4-6a). The number and size of vacuoles depend on both the type of cell and its stage of development; a single vacuole may occupy as much as 80 percent of a mature plant cell. Within the vacuoles, plant cells store water, ions, certain waste products, and such food materials as sucrose. In particular, vacuoles store excess nitrogen-containing compounds, such as amino acids. Most organic nitrogen in plants is derived from atmospheric nitrogen by the energy-requiring process known as *nitrogen-fixation;* it is advantageous for the plant to store excess fixed nitrogen.

The concentration of dissolved materials is much higher within the vacuoles than it is in the cytosol or ex-

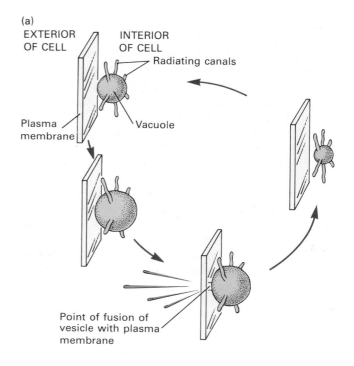

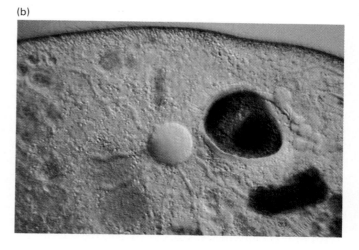

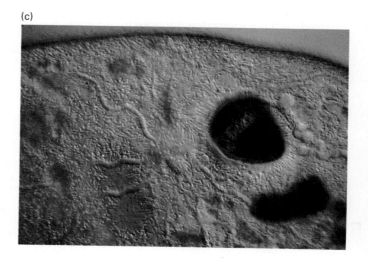

▶ **Figure 4-45** (a) The contractile vacuole in *Paramecium caudatum.* The vesicle in this typical ciliated protozoan is filled by radiating canals that collect fluid from the cytosol. When the vesicle is full, it fuses for a brief period with the plasma membrane and expels its contents. When the vesicle is nearly empty, the canals begin to refill it with fluid from the cytoplasm. (b) A full vacuole and system of radiating canals. (c) A nearly empty vacuole; the radiating canals are collecting more fluid from the cytoplasm to refill it.

tracellular fluids. The vacuolar membrane is semipermeable: it is permeable to water; it is impermeable to the small molecules stored within it. Water flows across a semipermeable membrane from a solution having a low concentration of dissolved materials to one having a high concentration. In other words, water will flow from a solution of high H_2O concentration (a low concentration of solutes) to one of low H_2O concentration (a high concentration of solutes) to attain the same concentration of water on both sides of the membrane. This movement is called *osmotic flow*.

The entry of water into the vacuole from the cytosol causes the vacuole to expand, creating hydrostatic pressure, or *turgor*, inside the cell. This pressure is balanced by the mechanical resistance of the cellulose-containing cell wall that surrounds all plant cells. Most plant cells have a turgor of 5–20 atmospheres (atm); their cell walls must be strong enough to react to this pressure in a controlled way. Unlike animal cells, plant cells can elongate extremely rapidly at rates of 20–75 μm/h. This elongation, which occurs when the somewhat elastic cell wall stretches under the pressure created by water taken into the vacuole, usually accompanies plant growth.

Vacuoles also act as receptacles for waste products and excess salts taken up by the plant and may function similarly to lysosomes in animal cells. Vacuoles have an acidic pH, maintained by a proton pump in the vacuole membrane, and contain a battery of degradative enzymes similar to those listed in Table 4-4.

Contractile Vacuoles in Certain Protozoans Function in Osmotic Regulation

Because plant, algal, fungal, and bacterial cells are surrounded by rigid cell walls, they do not swell when they are placed in a *hypotonic solution*, in which the concentration of dissolved molecules, such as salts, is lower than it is inside the cells. Most bacterial and plant cells do not absorb extra fluid even when they are exposed to pure water. Animal cells and most protozoans do not have a rigid wall. Many protozoans, however, contain a *contractile vacuole* (see Figure 4-19b) which, like the plant vacuole, takes up water from the cytosol. Unlike the plant vacuole, the contractile vacuole discharges its contents periodically through fusion with the plasma membrane (Figure 4-45). Water continuously enters the cell by osmotic flow; the contractile vacuole prevents too much water from accumulating in the cell and swelling it to the bursting point. The rupturing of the plasma membrane by a flow of water into the cytosol is termed *osmotic lysis*.

Osmotic Control in Animal Cells Animal cells do not contain vacuoles for the disposal of excess water. If animal cells are placed in a hypotonic solution, the cells will swell and possibly burst as water enters them by os-

(a) Isotonic medium

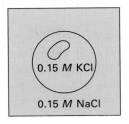

(b) Hypotonic medium

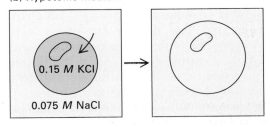

(c) Hypertonic medium

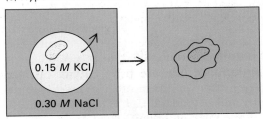

▲ **Figure 4-46** Animal cells respond to the osmotic strength of the surrounding medium. Sodium, potassium, and chloride ions do not move freely across the cell membrane, but water does (*arrows*). (a) When the medium is isotonic, there is no net flux of water into or out of the cell. (b) When the medium is hypotonic, water flows into the cell until the ion concentration inside and outside the cell is the same. Here, the initial cytosolic ion concentration is twice the extracellular ion concentration, so the cell tends to swell to twice its original volume, at which point the internal and external ion concentrations are the same. (c) When the medium is hypertonic, water flows out of the cell until the ion concentration inside and outside the cell is the same. Here, the initial cytosolic ion concentration is one-half the extracellular ion concentration, so the cell is reduced to one-half its original volume.

motic flow. Immersing animal cells in a *hypertonic solution* (one with a *higher* than normal concentration of solutes) causes them to shrink as water leaves them by osmotic flow. Consequently, it is essential that animal cells be maintained in an isotonic medium, where the concentration of solutes is close to that of the cell cytosol (Figure 4-46).

Peroxisomes Produce and Degrade Hydrogen Peroxide

The *peroxisomes* are a class of small, membrane-limited organelles found in the cytoplasm of all animal and many plant cells: *glyoxisomes* are organelles of similar structure found in plant seeds that use lipids as a source of carbon for growth. Because the morphology of these two organelles resembles that of lysosomes, for a long time they were believed to be lysosomes. However, cell-fractionation studies have established that the enzymes in these organelles differ greatly in composition and function from those in lysosomes. Peroxisomes and glyoxisomes contain enzymes that degrade fatty acids and amino acids. A byproduct of these reactions is hydrogen peroxide (H_2O_2), a corrosive substance that oxidizes many amino acid side chains, such as methionine. To counter the potentially deleterious effects of the hydrogen peroxide, peroxisomes also contain copious amounts of the enzyme catalase, which degrades hydrogen peroxide:

$$2H_2O_2 \xrightarrow{\text{catalase}} 2H_2O + O_2$$

Many peroxisomes contain a crystalline array of catalase molecules, which is seen in the electron microscope as a densely stained "core" (see Figures 4-30f and 4-39). The exact role of peroxisomes in cellular metabolism is somewhat mysterious, because many enzyme-catalyzed degradative reactions occur in other organelles without the synthesis and degradation of H_2O_2. Some investigators feel that one function of peroxisomes is the generation of heat, rather than ATP, as a product of the catabolism of energy-rich molecules, such as fatty acids.

The Mitochondrion Is the Principal Site of ATP Production in Aerobic Cells

The principal source of cellular energy in nonphotosynthetic cells is glucose. The complete aerobic degradation of glucose to CO_2 and H_2O is coupled to the synthesis of as many as 32 molecules of ATP:

$$C_6H_{12}O_6 + 6O_2 + 32P_i + 32ADP \longrightarrow$$
$$6CO_2 + 6H_2O + 32ATP + 32H_2O$$

In eukaryotic cells, the initial stages of glucose degradation occur in the cytosol, and the terminal stages, including those involving oxygen, occur in the mitochondria. Two ATPs are generated in the cytosol. As many as 30 are generated in the mitochondria, and even this value is uncertain: much of the energy released in mitochondrial oxidation can be used for other purposes, such as heat generation and transportation of molecules into or out of the mitochondrion, making less energy available for ATP synthesis. Nonetheless, the mitochondrion can still be regarded as the "power plant" of the cell.

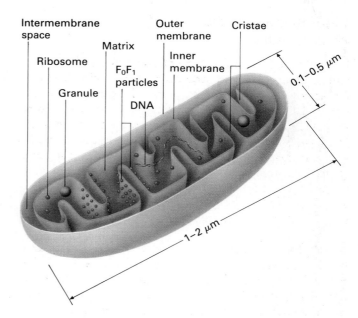

▲ **Figure 4-47** Three-dimensional diagram of a mitochondrion cut longitudinally.

The details of the process by which ATP is formed in the cytosol and mitochondria require a chapter of their own (Chapter 15); here, we describe some of the basic structural features of the organelle. Most eukaryotic cells contain many mitochondria. They are one of the largest organelles in the cell (generally only the nucleus, vacuoles, and chloroplasts are larger), occupying up to 25 percent of the volume of the cytoplasm. They are large enough to be seen under a light microscope, but the details of their structure can be viewed only with the electron microscope. Mitochondria contain two very different membranes: the outer membrane and the inner membrane (Figure 4-47; see also Figures 4-11 and 4-30b).

The outer membrane contains porins, proteins that render the membrane permeable to molecules having molecular weights as high as 10,000. In this respect, the outer membrane is uncharacteristic of biological membranes. Its composition is about half lipid and half protein.

The inner membrane is much less permeable. It is about 20 percent lipid and 80 percent protein—a higher proportion of protein than occurs in other cellular membranes. The inner membrane has a large number of infoldings, or *cristae*, that protrude into the *matrix*, or central space. The matrix and inner membrane are the sites of the enzymes that catalyze the final oxidation of sugars and lipids and the synthesis of ATP from ADP and inorganic phosphate. The mitochondrial DNA is also located in the matrix; its role in the synthesis of mitochondrial proteins is detailed in Chapter 18.

Chloroplasts Are the Sites of Photosynthesis

Except for the vacuoles, the chloroplasts are the largest and most characteristic organelles in the cells of plants and green algae (see Figure 4-6a). Like mitochondria, chloroplasts do not have fixed positions in cells and often migrate from place to place. They can be as long as 10 μm and are typically 0.5–2 μm thick, but they vary in size and shape in different cells, especially among the algae. On a weight basis, 35 percent of the chloroplast is lipid, 5 percent is protein, and 7 percent is pigment. Predominant among the pigments is *chlorophyll*, the substance that traps light and gives the chloroplast its green color.

Like the mitochondrion, the chloroplast is surrounded by an outer and an inner membrane (Figure 4-48). Chloroplasts also contain an extensive internal membrane system made up of thylakoid vesicles (interconnected vesicles flattened to form disks), often grouped in stacks called *grana* (see Figure 4-6a). The thylakoid vesicles contain chlorophyll and other pigments and enzymes that trap light and generate ATP during photosynthesis. A space called the *stroma* surrounds the thylakoids and the grana. *Carbon dioxide fixation*—the conversion of CO_2 to intermediates during the synthesis of sugars—occurs in the stromal space. Photosynthesis also requires a chapter of its own.

Chloroplasts, like mitochondria, contain DNA; this DNA is known to encode some of the key chloroplast proteins. Like mitochondrial proteins, most chloroplast proteins are made in the cytosol and incorporated in the organelle later. The synthesis of chloroplast and mitochondrial proteins is described in Chapter 18.

The Plasma Membrane Has Many Varied and Essential Roles

Some of the essential functions of the plasma membrane, the semipermeable barrier that defines the outer perimeter of the cell, are to allow nutrients to enter the cell, to filter out unwanted materials in the extracellular milieu, and to prevent metabolites and ions from leaving the cell. In all cells, the plasma membrane maintains the proper ionic composition and osmotic pressure of the cytoplasm. Like the membranes of other organelles, the plasma membrane contains specific transport or permease proteins that allow the passage of certain small molecules but not others.

Particles can enter animal cells by endocytosis of the plasma membrane. In all cells, the contents of intracellular vesicles can also be extruded through the plasma membrane by exocytosis. Clearly, regions of the plasma membrane and the lipids and proteins within them undergo a great deal of recycling.

A major function of the plasma membrane is to communicate and interact with other cells. Most cells in a multicellular animal do not exist as isolated entities; rather, groups of cells with related specializations combine to form tissues. Certain designated areas of the plasma membrane make contact with other cells to strengthen tissues and to allow the exchange of metabolites between cells. (Tissue structure is discussed in Chapters 13 and 23.) The plasma membrane also contains anchoring points for many of the fibrous structures that permeate the cytosol. Another function of the plasma membrane is to recognize extracellular signals, such as *hormones* (chemicals secreted by one cell type that act on another cell type at some distance); alternatively, molecules located on cell surfaces may act on neighboring cells. Many hormones, such as insulin, bind to *receptors* on the surface of target cells; receptor proteins recognize the specific hormone and signal the cell interior that the hormone has been received.

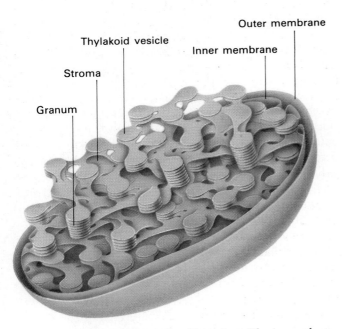

▲ **Figure 4-48** Structure of a chloroplast. The internal membrane (thylakoid) vesicles are organized into stacks, or grana, which reside in a matrix (the stroma). All the chlorophyll in the chloroplast is contained in the membranes of the thylakoid vesicles.

Cilia and Flagella Are Motile Extensions of the Eukaryotic Plasma Membrane

The surfaces of animal cells and eukaryotic microorganisms often contain a number of protuberances and extensions that serve many specific and important functions. *Cilia* and *flagella* are similar whiplike, motile structures

(a)

10 μm

(b)

Core of microtubules

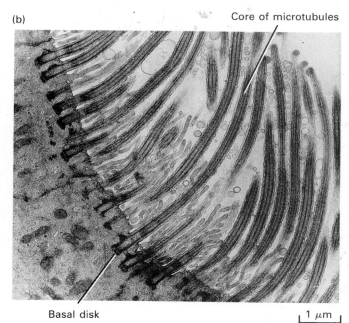

Basal disk

1 μm

▲ **Figure 4-49**   (a) Scanning electron micrograph of the surface of hamster trachea cells, showing abundant clusters of cilia, which protrude into the lumen of the trachea. (b) A micrograph of a thin section through the surface of a rabbit tracheal cell. Note that the surfaces of the cilia are continuous with the plasma membrane and that each cilium contains a core of microtubules that terminate in a basal disk structure just under this membrane. *Courtesy of E. R. Dirksen.*

that extend from the plasma membranes of unicellular organisms and many animal cells (Figure 4-49a). The bunches of microtubule filaments that run the length of the central cores of cilia and flagella (Figure 4-49b) are structurally similar to the microtubules that compose the mitotic spindle. Cilia beat backward and forward; flagella, which are longer than cilia, typically rotate in a screwlike manner. Both motions can propel a cell; the beating of the flagellum, for example, is the only means of locomotion for sperm. Cilia are also present on many fixed epithelial cells, such as trachea cells; there, the cilia propel liquids and small particles along the surface of the sheet of cells that lines the airways.

Microvilli Enhance the Absorption of Nutrients

In many animal cells specialized for the absorption of nutrients, the plasma membrane is folded into a large number of fingerlike projections called *microvilli*, which greatly increase the surface area and therefore the absorption rate of the cell. The surfaces of the epithelial cells lining the cavity of the small intestine, which is studded with microvilli, are collectively called the *brush border* (Figure 4-50). The plasma membrane of the brush border contains many enzymes that assist in the degradation of sucrose into glucose and fructose and of proteins into

amino acids. This membrane also contains specific permeases that allow the epithelial cell to absorb these nutrients from the intestine.

The Plasma Membrane Binds to the Cell Wall or the Extracellular Matrix

Many essential eukaryotic cell entities lie outside the plasma membrane. A variety of proteins and polysaccharides attached to the outer surfaces of animal cells participate in the formation of specific contacts and junctions between cells; such interactions impart strength and rigidity to multicellular tissues. Much of the volume of *connective tissue* in animals consists of the spaces between cells. Loose connective tissue forms the bedding on which most small glands or epithelial cells lie. It contains only a few cells and a number of fibers of various diameters. The blood capillaries that bring oxygen and nutrients to the cells in all parts of the body are embedded in loose connective tissue. A principal function of the amorphous spaces between cells in this tissue is to allow these nutrients to diffuse freely into the cells of the epithelia and the glands.

Collagen, the major protein in the extracellular spaces of connective tissue, is the most abundant protein in the entire animal kingdom. The type of collagen in connective tissue is distinctive in that it forms insoluble fibers that have a very high tensile strength. Collagens (see Figure 4-27) are secreted by many cells, including fibroblasts, a principal cell in connective tissue (Figure 4-51).

Connective tissue is also rich in large, complex molecules, or *proteoglycans*, which consist mostly (95 percent) of acidic polysaccharides, such as hyaluronic acid, and partly (5 percent) of proteins. Proteoglycans form the *ground substance* in which collagen and other connective-tissue fibers are embedded. The plasma membrane bears receptor proteins that bind the cells to several matrix components.

The plasma membrane of plant cells is intimately involved in the assembly of cell walls; new layers of wall are laid down next to the plasma membrane as a cell matures (Figure 4-52). The walls are built primarily of *cellulose*, a rodlike polysaccharide formed from $\beta(1 \rightarrow 4)$-linked glu-

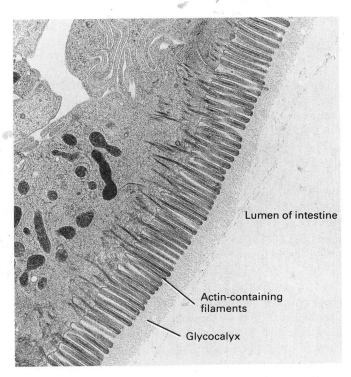

Lumen of intestine

Actin-containing filaments

Glycocalyx

▲ **Figure 4-50** Transmission electron micrograph of the brush border of an intestinal epithelial cell, showing the microvilli with their cores of actin microfilaments. Attached to the brush border is the *glycocalyx*—a fibrous network of glycoproteins containing glycosidases and peptidases that catalyze the final stages in the digestion of ingested macromolecules. *Courtesy of S. Ito.*

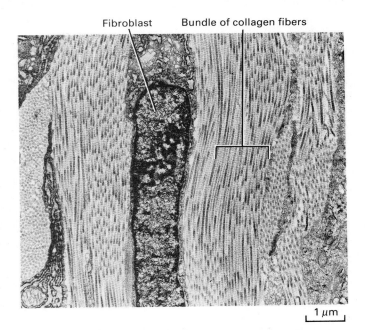

Fibroblast   Bundle of collagen fibers

1 μm

▲ **Figure 4-51** Electron micrograph of collagen fibers and the parent fibroblast cells that secreted them. *From D. Eyre, 1980, Science **207**:1314.*

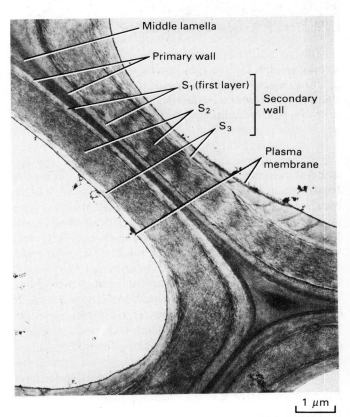

Middle lamella

Primary wall

S_1 (first layer)

S_2

S_3

Secondary wall

Plasma membrane

1 μm

▲ **Figure 4-52** Electron micrograph of a thin section showing parts of the cell walls separating three *Taxus canadensis* (plant) cells. The principal layers of each wall are evident: the middle lamella; the primary wall; and the three layers of secondary wall (S_1, S_2, and S_3). As the cell matures, the layers of cellulose fibers are laid down one by one in the sequence shown. The fibers in each layer run in a different direction from those in the preceding layer. The plasma membrane is adjacent to the S_3 layer, the youngest stratum of the cell wall. *Courtesy of Biophoto Associates/Myron C. Ledbetter/Brookhaven National Laboratory.*

cose monomers. In plant cell walls, the cellulose molecules aggregate by hydrogen bonding into bundles of fibers; other polysaccharides within the wall cross-link the cellulose fibers. In woody plants, a very different compound, *lignin* (a complex water-insoluble polymer of phenol and other aromatic monomers), imparts strength and rigidity to the cell walls. Other chemicals also are found in the walls of various plant cells; for example, waxes prevent plant tissues and proteins from drying out.

Summary

Two types of cells exist in our biological universe: prokaryotic cells with no nuclei (including bacteria of the archaebacterial and eubacterial lineages) and eukaryotic cells with nuclei (including animals, plants, and many microorganisms). Prokaryotes and eukaryotes have a similar overall chemical composition and share many metabolic pathways. Prokaryotic cells are generally smaller and differ from eukaryotic cells in terms of structural elements and genetic processes.

In most respects, the internal architecture and central metabolic pathways are similar in the cells of all plants, animals, and unicellular eukaryotic organisms. Eukaryotic cells have intracellular membranes and organelles that are never found in prokaryotes: examples are the lysosomes, which degrade cellular membranes and endocytosed materials; the peroxisomes, which synthesize and degrade hydrogen peroxide; the endoplasmic reticulum, where secretory proteins, membrane proteins, and lipids are synthesized; the Golgi vesicles, which sort and process proteins (particularly secretory proteins) and membrane constituents; the mitochondria, where the terminal stages of sugar and lipid oxidation and ATP synthesis take place; and the chloroplasts, where photosynthesis occurs in plants. The nucleus is surrounded by a nuclear membrane composed of an inner and an outer membrane; the outer membrane is continuous with the endoplasmic reticulum. Pores in the nuclear membrane allow particles to enter and leave the nucleus; lamins lining the inner surface of the inner membrane give it rigidity and bind some nuclear DNA.

Both plant and bacterial cells have walls, but they are very different in composition. Peptidoglycan, in particular, is unique to prokaryotes. Plant cell walls are built of cellulose and other polymers; the wall is the major determinant of cell shape and gives the cell rigidity. Animal cells are surrounded by an extracellular matrix consisting of collagen, proteoglycans, and other components that give strength and rigidity to tissues and organs.

Vacuoles often fill much of the plant cell and generate turgor pressure that pushes the plasma membrane against the cell wall. Small contractile vacuoles in microorganisms participate in osmotic regulation, allowing the cells to survive in water. Animal cells, in contrast, will shrink or swell, respectively, if placed in solutions of too high or too low osmotic strength.

The cytoplasm of all animal and plant cells contains a network of fibrous proteins, the cytoskeleton, which gives the cell structural stability and contributes to cell movement. Three principal types of protein filaments constitute the cytoskeleton: actin microfilaments, microtubules, and intermediate filaments.

The plasma membrane serves many functions: it selectively absorbs nutrients, expels wastes, and anchors the fibers of the intracellular cytoskeleton or the extracellular matrix or wall. The membrane contains proteins that serve as hormone receptors and that bind a cell to its neighbors. Cilia and flagella, extensions of the plasma membrane, can propel the cell or move liquids past it. Other extensions of the membrane, such as the microvilli in absorptive cells, facilitate the absorption of nutrients.

Various microscopic techniques generate different views of the cell with different resolutions. Standard (bright-field) light microscopy is best for stained or colored cells or tissue sections. Immunofluorescence microscopy allows specific proteins and organelles to be detected in fixed cells; the movements of microinjected fluorescent proteins can be followed in living cells. Confocal imaging, which allows the observer to view fluorescent molecules in a single plane of a specimen, permits optical sectioning of the sample and produces very sharp images. Dark-field microscopy allows the detection of small refractile objects, such as live bacterial cells and autoradiographic grains. Phase contrast and Nomarski optics enable scientists to view the details of live, unstained cells and to monitor cell movement.

Electron microscopy has a vastly higher resolution than light microscopy but generally requires the specimen to be fixed, sectioned, and dehydrated. After metal shadowing, the transmission electron microscope allows the investigator to visualize the structural details of such particles as viruses and collagen fibers. Unfixed, unstained specimens can be viewed in the electron microscope if they are frozen in hydrated form. The scanning electron microscope can be used to view unsectioned cells or tissues; it produces images that appear to be three dimensional. The techniques of quick freezing and deep etching generate striking micrographs of cytoskeletal fibers.

References

Many of the topics listed below are treated in more detail in later chapters of this book, where additional articles are cited.

Cells as the Basic Structural Units

MARGULIS, L., and K. V. SCHWARTZ. 1987. *Five Kingdoms: An Illustrated Guide to the Phyla of Life on Earth,* 2d ed. W. H. Freeman and Company. Electron micrographs and diagrams of representative cells from the principal groups of prokaryotes, eukaryotic microorganisms, plants, and animals.

DE DUVE, C. 1984. *A Guided Tour of the Living Cell*, vols. 1 and 2. Scientific American Library. A simple idiosyncratic introduction to cell structure and function.

Plant and Microbial Cells

BROCK, T. D., D. W. SMITH, and M. T. MADIGAN. 1984. *Biology of Microorganisms*, 4th ed. Prentice-Hall.

INGRAHAM, J. L., O. MAALOE, and F. C. NEIDHARDT. 1983. *Growth of the Bacterial Cell*. Sinauer.

LEDBETTER, M. C., and K. R. PORTER. 1970. *Introduction to the Fine Structure of Plant Cells*. Springer-Verlag.

RAVEN, P. H., R. F. EVERT, and H. CURTIS. 1986. *Biology of Plants*, 4th ed. Worth. An introduction to the structures and functions of plants and their cells.

Mammalian Cells (Histology Texts)

BLOOM, W., and D. W. FAWCETT. 1986. *A Textbook of Histology*, 11th ed. Saunders. An excellent text containing detailed descriptions of the structures of mammalian organs, tissues, and cells.

CORMACK, D. H. 1984. *Introduction to Histology*. Lippincott. Another excellent histology text.

WEISS, L., and L. LANSING, eds. 1983. *Histology—Cell and Tissue Biology*, 5th ed. Elsevier Biomedical.

Mammalian Cells (Atlases)

The following books are sources of detailed electron micrographs illustrating the structures of most mammalian cells and tissues.

CARR, K. E., and P. G. TONER, 1983. *Cell Structure—An Introduction to Biomedical Electron Microscopy*. Churchill Livingstone.

FAWCETT, D. W. 1981. *The Cell*, 2d ed. Saunders.

KESSEL, R. G., and R. H. KARDON. 1979. *Tissues and Organs: A Test Atlas of Scanning Electron Microscopy*. W. H. Freeman and Company.

PORTER, K. R., and M. A. BONNEVILLE. 1973. *Fine Structure of Cells and Tissues*, 4th ed. Lea & Febiger.

Microscopy and the Internal Architecture of Cells

The following books and reviews cover all of the standard techniques in light and electron microscopy.

Light Microscopy

POLAK, J. M., and S. VAN NOORDEN, 1984. *An Introduction to Immunocytochemistry*. Roy. Microscopic Soc. Microscopy Handbooks Ser. Oxford.

SPENCER, M. 1982. *Fundamentals of Light Microscopy*. Cambridge University Press.

WHITE, J. G., W. B. AMOS, and M. FORDHAM. 1987. An evaluation of confocal versus conventional imaging of biological structures by fluorescence light microscopy. *J. Cell Biol.* 105:41–48.

WILLINGHAM, M. C., and PASTAN, I. 1985. *An Atlas of Immunofluorescence in Cultured Cells*. Academic Press.

Electron Microscopy

CHIU, W. 1986. Electron microscopy of frozen, hydrated biological specimens. *Ann. Rev. Biophys. Biophys. Chem.* 15:237–257.

EVERHART, T. E., and T. L. HAYES. 1972. The scanning electron microscope. *Sci. Am.* 226(1):54–69.

HAYAT, M. A., ed. 1980. *Principles and Techniques of Electron Microscopy—Biological Applications*, vol. 1, 2d ed. Wiley.

HEUSER, J. 1981. Quick-freeze, deep-etch preparation of samples for 3-D electron microscopy. *Trends Biochem. Sci.* 6:64–68.

PEASE, D. C., and K. R. PORTER. 1981. Electron microscopy and ultramicrotomy. *J. Cell Biol.* 91:287s–292s.

WATT, I. M. 1985. *The Principles and Practice of Electron Microscopy*. Cambridge University Press.

WEAKLEY, B. S. 1981. *Beginner's Handbook in Biological Transmission Electron Microscopy*, 2d ed. Churchill Livingstone.

WISCHNITZER, S. 1981. *Introduction to Electron Microscopy*, 3d ed. Pergamon.

Cellular Membrane Systems and Organelles

These articles and books describe the properties of individual fibers and organelles and techniques for purifying them.

BAINTON, D. 1981. The discovery of lysosomes. *J. Cell Biol.* 91:66s–76s.

BARRANGER, J. A., and R. O. BRADY, eds. 1984. *Molecular Basis of Lysosome Storage Disorders*. Academic Press.

COURTOY, P. J., J. QUINTART, and P. BAUDHIUN. 1984. Shift of equilibrium density induced by 3,3'-diaminobenzidine cytochemistry: a new procedure for the analysis and purification of peroxidase-containing organelles. *J. Cell Biol.* 98:870–876.

DAVIS, L. I., and G. BLOBEL. 1986. Identification and characterization of a nuclear pore complex protein. *Cell* 45:699–709.

DE DUVE, C. 1975. Exploring cells with a centrifuge. *Science* 189:186–194. The Nobel Prize lecture of a pioneer in the study of cellular organelles. (See also Palade, below.)

DE DUVE, C. 1983. Microbodies in the living cell. *Sci. Am.* 248(May), 74–84.

DE DUVE, C., and H. BEAUFAY. 1981. A short history of tissue fractionation. *J. Cell Biol.* 91:293s–299s.

FARQUHAR, M., and G. PALADE. 1981. The Golgi apparatus (complex) (1954–1981) from artifact to center stage. *J. Cell Biol.* 91:77s–103s.

HOWELL, K. E., E. DEVANEY, and J. GRUENBERG. 1989. Subcellular fractionation of tissue culture cells. *Trends in Biochem. Sci.* 14:44–48.

LIPSKY, N. G., and R. E. PAGANO. 1985. A vital stain for the Golgi apparatus. *Science* 228:745–747.

LOUVARD, D., H. REGGIO, and G. WARREN. 1982. Antibodies to the Golgi complex and endoplasmic reticulum. *J. Cell Biol.* **92**:92–107.

MERISKO, E., M. E. FARQUHAR, and G. PALADE. 1982. Coated vesicle isolation by immunoadsorption on Staphylococcus aureus cells. *J. Cell Biol.* **92**:846–857.

PALADE, G. 1975. Intracellular aspects of the process of protein synthesis. *Science* **189**:347–358. The Nobel Prize lecture of a pioneer in the study of subcellular organelles. (See also de Duve, above.)

SCHEELER, P. 1981. *Centrifugation in Biology and Medical Science.* Wiley.

TOLBERT, N. E., and E. ESSNER. 1981. Microbodies: peroxisomes and glyoxysomes. *J. Cell Biol.* **91**:271s–283s.

The Organization of the Cytoplasm

The following books and articles describe the structure of the cytoplasm, focusing on the cytoskeletal fibers and on cell motility.

BERSHADSKY, A., and J. VASILEV. 1988. *Cytoskeleton.* Plenum Press.

BORISY, G. G., D. W. CLEVELAND, and D. MURPHY (eds.) 1984. *Molecular Biology of the Cytoskeleton.* Cold Spring Harbor Laboratory.

FULTON, A. B. 1980. How crowded is the cytoplasm? *Cell* **30**:345–347.

SCHLIWA, M. 1986. *The Cytoskeleton: An Introductory Survey* (Cell Biology Monographs, vol. 13). Springer-Verlag.

5

Growing and Manipulating Cells and Viruses

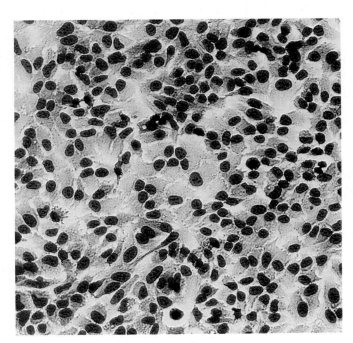

Human cervical carcinoma (HeLa) cells

Understanding current advances in molecular cell biology requires a familiarity with the most commonly used biological materials and knowledge of the latest experimental techniques. This chapter deals with how some of the most useful cells are cultured and manipulated, often with the aid of viruses or plasmids, which are also described. In Chapter 6 we describe how new techniques in molecular biology coupled with the newest cellular manipulations have created the experimental approach called molecular genetics. Particular cells or organisms are chosen for study because they provide experimentally favorable examples of an important molecular event or sequence of events—the replication, recombination, and rearrangement of DNA; the control of genes; the construction of membranes; the formation of organelles; the secretion of proteins; the structure and action of cytoskeletal elements; or the differentiation of cells and the construction of organ systems, such as the nervous system.

Two important techniques in cell biology are endowments from microbiology. First, cultured cells derived from a single cell are valuable in molecular studies because they constitute a homogeneous cell mass in which *each cell may be assumed to contain the same constituents in the same proportions. Single-celled eukaryotic organisms such as yeasts, whose entire life cycle can be studied in culture, are popular choices for basic studies; Saccharomyces cerevisiae (baker's yeast) is particularly useful because of the extensive genetic analysis that exists for this organism. Second, the infection of cells by a bacterial, animal, or plant virus affords an opportunity to study the action of a limited set of genes designed to carry out a restricted, specific molecular task. In addition to viruses, plasmids (circular DNA molecules that replicate along with the cellular DNA elements) are widely used to introduce DNA bearing genes of interest into cells through the process called* transfection *(by analogy with* infection *by viruses). Thus a homogeneous cell culture infected by a chosen virus or transfected with specific DNA molecules is often used in many molecular studies (to be described in Chapters 7, 8, and 12).*

Few cultured cells from vertebrates, however, perform the specialized tasks that differentiated cells do in the body. Therefore, the investigator must frequently turn to the whole organism to obtain cells that perform special tasks, such as the synthesis of a large amount of a specific protein or group of related proteins. Tissues such as the liver of vertebrates and the oviduct of birds that are rich in a single cell type are often used in molecular analysis. Other useful tissues (including the spleen, the thymus, the bone marrow, and the blood) are less uniform, but all of them can be separated into single cells and then sorted into individual cell types. The relative ease with which these individual cell types can be identified and isolated is advantageous, particularly to immunologists (Chapter 25).

Finally, an important focus of present-day molecular cell biology is embryologic development. Invertebrate organisms such as Drosophila melanogaster, *the fruit fly, and* Caenorhabditis elegans, *a small worm, are exceedingly popular for these studies because detailed genetic studies of them are possible. Among vertebrates, mice and rats are the most conveniently maintained and genetically well-studied mammals; frogs are frequently used to study certain aspects of early development. Important applications of studies on gene control to developmental biology are described in Chapter 11. Differentiated cell functions and the molecules that contribute to them are described in Chapters 19 (hormones), 22 (nerves), 23 (matrix), and 25.* ▲

Types of Cell Division

The most fundamental property of all cells—their growth and division—must be discussed before we describe the choice and use of specific organisms. Perhaps the most crucial aspect of cell division is the duplication of the cell in a way that every daughter cell receives an identical

copy of its genetic material. Prokaryotic and eukaryotic cells differ markedly in the coordination of DNA synthesis and in the subsequent equal partitioning of DNA during cell division. No specialized structure for division is visible in prokaryotes by light microscopy. Since the last century, however, biologists have recognized eukaryotic chromosomes in the microscope and understood their importance in cell division: *mitosis* in diploid somatic cells ensures genetic identity of equal daughter cells; *meiosis* produces haploid germ cells and ensures genetic continuity between generations in multicellular organisms.

The Cell Cycle in Prokaryotes Consists of DNA Replication Followed Immediately by Cell Division

The genome of a prokaryotic cell is a single circular molecule of DNA. If growing bacteria are briefly exposed to radioactive thymidine, almost every cell is labeled, indicating that each cell is making DNA. Thus the time a bacterium requires to divide must be quite short relative to the time it requires to make DNA. For example, *Escherichia coli* can divide in 30 min, and all but a minute or so of this time is used to complete one round of DNA synthesis. During division, a cell wall forms that divides the cell in two. It is likely that the two daughter DNA duplexes are linked at different points on the plasma membrane to ensure that one of the new DNA duplexes is delivered to each daughter (Figure 5-1). In certain bacteria, infolded plasma membrane regions termed *mesosomes* may provide anchors for DNA. Even in these organisms, however, there is no visible condensation and decondensation of the DNA, as there is in eukaryotic cells during mitosis.

Eukaryotic DNA Synthesis Occurs in a Special Phase of the Cell Cycle

Different eukaryotic cells can grow and divide at quite different rates. Yeast cells, for example, can divide every 120 min, and the first divisions of fertilized eggs in the embryonic cells of sea urchins and insects take only 15–30 min because one large preexisting cell is subdivided. However, most growing plant and animal cells take 10–20 h to double in number, and some duplicate at a much slower rate. Many cells in adults, such as nerve cells and striated muscle cells, do not divide at all; others, like the fibroblasts that assist in healing wounds, grow on demand but are otherwise quiescent.

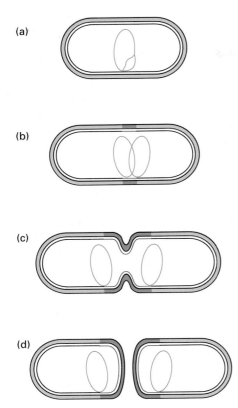

▲ **Figure 5-1** The DNA in prokaryotic cells is attached to the cell membrane and remains attached during cell division. (a) The circular chromosome (blue), which has already begun replication, is attached to the plasma membrane. (b) When DNA replication is complete, the new chromosome has an independent point of attachment to the membrane. New membrane and cell wall forms midway along the length of the cell. (c) As more sections of membrane and cell wall form, part of this growth gives rise to a *septum* dividing the cell. (d) Cell division is complete; the new chromosome is attached to the plasma membrane of each daughter cell.

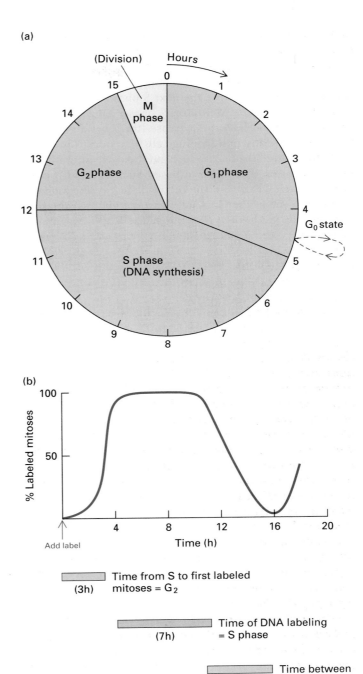

▲ **Figure 5-2** The cell cycle in a mammalian cell having a generation time of 16 h. (a) The three phases spanning the first 15 h or so—the G_1 (first gap) phase, the S (*synthetic*) phase, and the G_2 (second gap) phase—make up the interphase, during which DNA and other cellular macromolecules are synthesized. The remaining hour is the M (*mitotic*) phase, during which the cell actually divides. (b) The phases of the cell cycle were determined by exposing a culture briefly to labeled thymidine, which is incorporated into DNA, and then observing the time of appearance of labeled mitotic cells.

Still, every eukaryotic cell that divides must be ready to donate equal genetic material to two daughter cells. In contrast to bacteria, DNA synthesis in eukaryotic cells is not continual but is completed well before cell division. For example, if root tips, in which cells divide frequently, are exposed to radioactive thymidine for a few minutes, only a fraction of the cells incorporate high levels of label into DNA; the remainder are not labeled at all. Moreover, the labeled cells are not in the process of dividing. Thus DNA synthesis does not occur throughout the cell division cycle but is restricted to a part of it before mitosis.

A thorough analysis of the relationship between eukaryotic DNA synthesis and cell division has been made in mammalian cell cultures in which each cell is capable of growth and division (Figure 5-2). These experiments led to the conclusion that DNA synthesis in a eukaryotic cell occurs only during one phase of the cell cycle called the S (*synthetic*) phase. A gap of time occurs after DNA synthesis and before cell division; another gap occurs after division and before the next S phase. The cell cycle thus consists of the M (*mitotic*) phase, a G_1 phase (the first gap), the S phase, a G_2 phase (the second gap), and back to M. Many nondividing cells in tissues (for example, all resting fibroblasts) suspend the cycle after mitosis and just prior to DNA synthesis. Such "resting" cells, which have exited from the cell cycle before the S phase, are said to be in the G_0 state. Learning about regulatory events that guide the cell from phase to phase remains a major experimental challenge to modern cell biologists.

Mitosis Is the Complex Process That Apportions the New Chromosomes Equally to Daughter Cells

Mitosis has been recognized for a century as the common mechanism in eukaryotes for partitioning the genetic material equally at cell division. Beginning with the union of egg and sperm, each somatic cell of sexually reproducing organisms is *diploid,* that is, it has two copies of each morphologically distinct chromosome, one inherited from the male parent and one from the female parent. (Each somatic cell is said to be *2n,* where *n* is the number of chromosome types.) The presence of one male and one female copy of each morphologic type of chromosome at mitosis ensures each and every somatic cell of genetic equality. Between mitoses, chromosomes are not visible by light microscopy. DNA-protein complexes called *chromatin* are dispersed throughout the nucleus during this time, and a very characteristic series of microscopic events is observed during mitosis. The beginning of a mitotic cell division is signaled by the appearance of stainable chromosomes as thin threads inside the nucleus. Although the events that follow unfold continuously, they are conventionally divided into four substages: *prophase, metaphase, anaphase,* and *telophase* (Figure 5-3).

During prophase and metaphase, a chromosome can be visualized as two identical coiled filaments, the *chromatids* (often called *sister chromatids*), each of which contains one of the two new daughter DNA molecules produced in the preceding S phase, making each cell that enters metaphase *4n.* Sister chromatids are held together at their *centromeres,* or constricted regions. They become shorter and more densely packed at metaphase.

The small cylindrical particles termed *centrioles* play a key role in prophase and metaphase in animal cell division. They move apart, and microtubules are seen to radiate from them in all directions, forming *asters.* (Centrioles are themselves constructed of microtubules, and they

▸ **Figure 5-3** The stages of mitosis and cytokinesis in an animal cell.

(a) *Interphase.* During the S phase, chromosomal DNA is replicated and bound to protein but the chromosomes are not seen as distinct structures. The nucleolus is the only nuclear substructure that is visible under the light microscope. The premitotic cell is *4n* (meaning there are two copies of each chromosome) and since there are two morphologic chromosomes of each type, there are actually four copies of each chromosome. (Morphologic types are distinguished by color.)

(b) *Early prophase.* The centrioles begin moving toward opposite poles of the cell; the chromosomes can be seen as long threads. The nuclear membrane begins to disaggregate.

(c) *Middle and late prophase.* Chromosome condensation is completed; each visible chromosome structure is composed of two chromatids held together at their centromeres. Each chromatid contains one of the two newly replicated daughter DNA molecules. The microtubular spindle begins to radiate from the regions just adjacent to the centrioles, which are moving closer to their poles. Some spindle fibers reach from pole to pole; most go to chromatids and attach at kinetochores.

(d) *Metaphase.* The chromosomes move toward the equator of the cell, where they become aligned in the equatorial plane. The sister chromatids have not yet separated. This is the phase in which morphologic studies of chromosomes are usually carried out.

(e) *Anaphase.* The two sister chromatids separate into independent chromosomes. Each contains a centromere that is linked by a spindle fiber to one pole to which it moves. Thus one copy of each chromosome is donated to each daughter cell. Simultaneously, the cell elongates, as do the pole-to-pole spindles. Cytokinesis begins as the cleavage furrow starts to form.

(f) *Telophase.* New membranes form around the daughter nuclei; the chromosomes uncoil and become less distinct, the nucleolus becomes visible again, and the nuclear membrane forms around each daughter nucleus. Cytokinesis is nearly complete, and the spindle disappears as the microtubules and other fibers depolymerize. Throughout mitosis the "daughter" centriole at each pole grows until it is full-length. At telophase the duplication of each of the original centrioles is completed, and a new daughter centriole will be generated during the next interphase.

(g) *Interphase.* Upon the completion of cytokinesis, the cell enters the G_1 phase of the cell cycle and proceeds again around the cycle.

duplicate during interphase by forming daughter centrioles.) Some of these microtubules connect the centrioles with the *kinetochores,* granular regions visible in the centromeres of the chromatids. Other microtubules form a network that links the centrioles as they move apart. These microtubules, together with associated proteins, are called the *spindle.* The regions surrounding the centrioles, from which the microtubules radiate, are termed the *poles,* or polar regions, of the cell. (Most higher plant

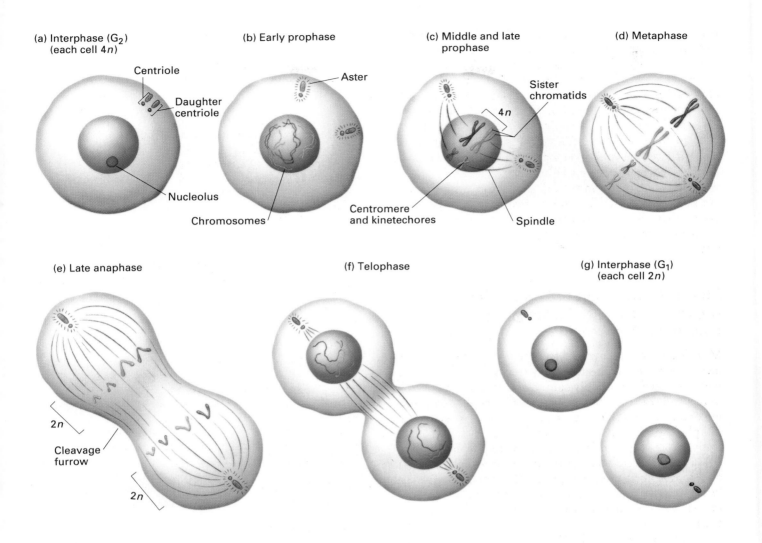

(a) Interphase (G$_2$) (each cell 4n)

Centriole

Daughter centriole

Nucleolus

Chromosomes

(b) Early prophase

Aster

(c) Middle and late prophase

Sister chromatids

4n

Centromere and kinetechores

Spindle

(d) Metaphase

(e) Late anaphase

2n

Cleavage furrow

2n

(f) Telophase

(g) Interphase (G$_1$) (each cell 2n)

cells do not contain visible centrioles. An analogous region of the cell acts as a microtubule-organizing center, from which the spindle microtubules radiate.) At the end of prophase, the nuclear membrane disappears.

During metaphase, the condensed sister chromatids, which are connected to each pole of the cell by the microtubules attached to their kinetochores, migrate to the *equatorial plane* of the cell. The microtubules attached to the chromosomes appear to play a role in orienting them.

Anaphase is marked by the separation of the two sister chromatids at their centromeres. As nearly as we know, the separation happens simultaneously to the entire set of chromosomes. The members of each chromatid pair—now independent chromosomes—migrate to opposite poles of the cell, which ensures that each new daughter cell receives an equal chromosome set. How the chromosomes are moved to opposite poles is uncertain; current ideas and experiments relevant to chromosome movement are discussed in Chapter 21.

In telophase, the chromosomes start to uncoil and become less condensed. Nuclear membrane fragments are seen joining together to become two new nuclear membranes surrounding the two sets of daughter chromosomes. Simultaneously, there is division of the cell cytoplasm and separation, or *cytokinesis,* of the two cells.

Because a plant cell is surrounded by a rigid cell wall, the shape of the cell does not change greatly in mitosis (Figure 5-4). During telophase, formation of the cell membrane and cell wall is completed in a plane separating the two nuclei.

Yeast Cells Have a Simplified Division

The yeast *S. cerevisiae* is a single-cell eukaryote that has a simplified cell-division apparatus (Figure 5-5). When division is about to begin, a bud emerges from a thickened region of the cell wall called a *plaque.* The bud gradually enlarges. The chromosomes (although too small to be easily visible under the light microscope) are attached to a structure,

(a)

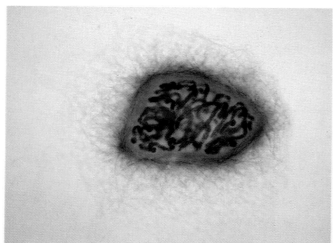

(b)

(c)

(d)

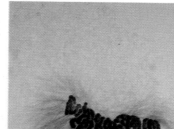

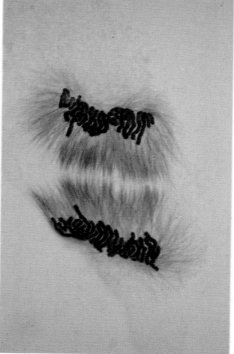

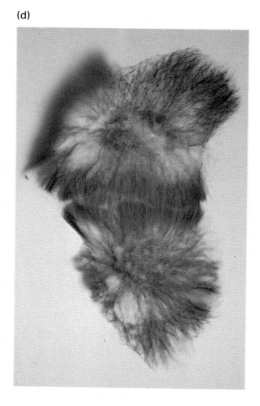

▲ **Figure 5-4** Light micrographs of mitosis in a plant. Microtubules are stained red and chromosomes are counterstained blue.

(a) *Prophase.* The chromosomes are beginning to condense.

(b) *Metaphase.* The chromosomes are aligned on the equatorial plane.

(c) *Anaphase.* The sister chromatids have separated, and the two groups of chromosomes are moving toward opposite poles of the spindle.

(d) *Telophase.* New membranes are forming around the daughter nuclei, and the chromosomes are uncoiling and becoming less distinct.
Courtesy of Andrew Bajer.

the *spindle pole body* (SPB), which duplicates like the centriole in animal cells, signaling the beginning of nuclear division. Microtubules form to guide the chromosomes: a spindle pole body enters the bud when it is large enough; the chromosomes then separate, and one set enters the bud to form the genetic complement of the new daughter cell. Finally, the bud separates from the parent cell. In yeast, the cell derived from the bud is smaller and is called the daughter and the original larger cell is the mother. This is an important distinction because the cells are not *biochemically* equal.

Meiosis Is the Form of Cell Division in Which Haploid Cells Are Produced from Diploid Cells

In all multicellular organisms and in single-celled organisms that are diploid at some phase of their life cycle, one important type of cell division departs from the plan of mitosis; this is meiosis, the division that gives rise to gametes—sperm and egg cells—in higher plants and animals. Like the body cells of most multicellular organisms, a premeiotic cell is diploid—it contains two of each

(a)

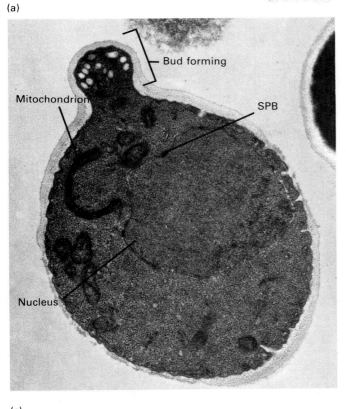

(b)

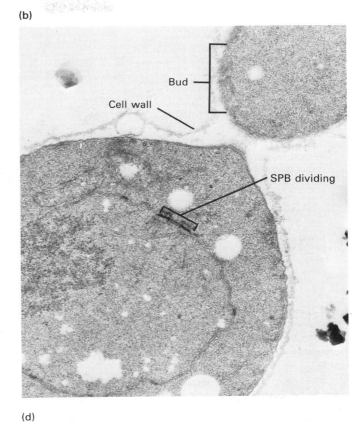

(c)

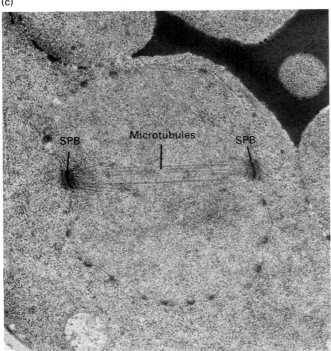

(d)

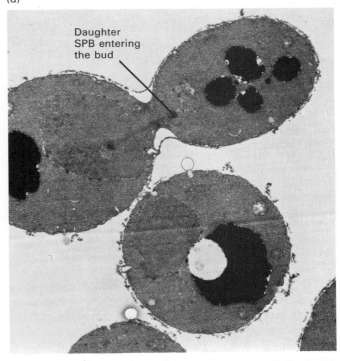

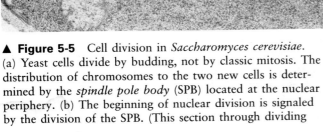

▲ **Figure 5-5** Cell division in *Saccharomyces cerevisiae.*
(a) Yeast cells divide by budding, not by classic mitosis. The distribution of chromosomes to the two new cells is determined by the *spindle pole body* (SPB) located at the nuclear periphery. (b) The beginning of nuclear division is signaled by the division of the SPB. (This section through dividing cells missed the cytoplasmic connection between the bud and the mother cell.) (c) When the two daughter SPBs separate, the microtubles connecting them can be seen. (d) A portion of the nucleus containing one SPB then enters the bud. Formation of the cell wall completes the separation of the two cells. *Photographs courtesy of B. Byers.*

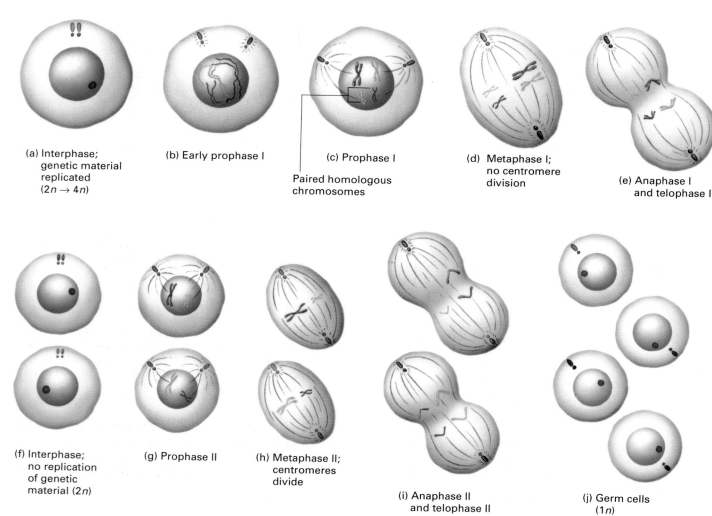

(a) Interphase; genetic material replicated ($2n \rightarrow 4n$)

(b) Early prophase I

(c) Prophase I

Paired homologous chromosomes

(d) Metaphase I; no centromere division

(e) Anaphase I and telophase I

(f) Interphase; no replication of genetic material ($2n$)

(g) Prophase II

(h) Metaphase II; centromeres divide

(i) Anaphase II and telophase II

(j) Germ cells ($1n$)

▲ **Figure 5-6** The steps in meiosis. Steps (a) through (c) are carried out as in mitosis (see Figure 5-3). Metaphase (d) differs in two important ways. Each morphologic chromosome (actually two sister chromatids) aligns at the cell equator with its homologous partner during *synapsis,* the stage of chromosome pairing that allows exchange between the two. No division of centromeres occurs, however, so *both* copies (sister chromatids) of one morphologic type go into one cell; the other homologue of each morphologic type goes into the other cell. Thus the genetic material of each of the two chromosomes of the same morphologic type separates (segregates). Because one of these chromosomes came from an egg and one originally from a sperm, parental characteristics are reassorted randomly into each new sperm and egg at the meiotic stage. In the second meiotic division, the centromeres do divide, and each cell receives one chromosome of each morphologic type.

morphologic type of chromosome. The two chromosomes of each type are descended from different parents, so they are *homologous* (genetically similar but not identical). In meiosis, *one* round of DNA replication to make the cell $4n$ is followed by *two* separate cell divisions, yielding four $1n$, or *haploid,* cells that contain only one chromosome of each homologous pair (Figure 5-6). There is an *interphase* between these two divisions that is unique in that it lacks a DNA synthesis phase.

At the first meiotic division, the centromeres do not divide. Instead, both chromatids of a chromosome travel together to one daughter cell, but each daughter cell receives a chromatid of each homologous pair (each morphologic type). The genetic contributions of the two parents, one of each homologous pair, therefore end up in different daughter cells: this failure of sister chromatid separation is responsible for the process of *genetic segregation*. In the second meiotic division, the centromeres do divide, as in mitosis, so that all four daughter cells end up with one chromosome from each of the original homologous pairs.

The Genetic Consequences of Meiosis The segregation of homologous chromosomes in meiosis is random, that is, the maternal and paternal members of each pair segregate independently. The example in Figure 5-7 shows only two morphologic types of chromosomes, a long (L) and a short (S), each of either maternal or paternal heritage (Lm, Lp, Sm, Sp). In this case, four types of gametes can be formed: LmSm, LpSp, LmSp, LpSm. The

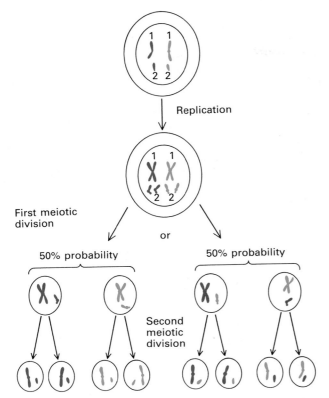

▲ **Figure 5-7** The genetic consequences of meiosis are illustrated for an organism in which the haploid number of chromosomes is 2. Each diploid cell contains pairs of homologous chromosomes, one inherited from each parent. (Here, the maternal and paternal chromosomes are distinguished by color.) In the first meiotic division, each daughter cell receives either the maternal or paternal chromosome from each homologous pair. Which chromosome is incorporated into which daughter is random, and each homologous pair of chromosomes is segregated independently of the other pairs. Thus, in this simple case, there are two possible outcomes of the first division, each having an equal probability of 50 percent. In the second meiotic division, the chromatids separate to generate a total of four haploid daughter cells, each containing one chromosome of each type.

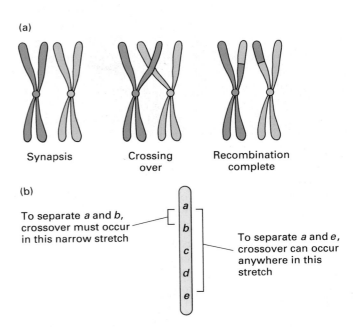

▲ **Figure 5-8** The basis of classic gene-mapping techniques. (a) Crossing over between homologous chromosomes during meiotic metaphase I. (b) The shorter the distance between two genes on a chromatid, the less likely they are to be separated by crossing over.

number of possible varieties of meiotic segregants is $2n$, where n is the number of chromosomes. For humans with 23 chromosomes the number is about 8 million.

In addition to the segregation of chromosomes, another critical genetic event occurs in most (but not all) meiotic cells. Before the first meiotic division, the chromosomes of each homologous pair align with each other, an act called *synapsis*. At this time, recombination between chromatids can occur (Figure 5-8). This swapping of material between maternally and paternally derived chromosomes is called *crossing over*. Combined with random segregation, crossing over is the source of new combinations of genes in interbreeding populations. These two phenomena are also the basis of the classic methods of genetic mapping by breeding experiments. Genetic

traits that almost always segregate together are said to be *linked* and are inferred to be controlled by genes on the same chromosome; the more frequently recombination occurs between two genes on the same chromosome, the more distant they are.

The Growth of Microorganisms and Cells in Culture

An important characteristic of many cells chosen for study is that a single cell can be readily grown into a colony, a process called *cloning*. A genetically pure strain of cells is called a *clone*. Most microorganisms that grow in nature as single cells are easy to grow in culture dishes—usually on top of *agar*, a semisolid base of plant polysaccharides. The agar is first dissolved in a heated nutrient medium; the solution then solidifies as it cools. Both prokaryotic and eukaryotic single-celled organisms are routinely grown in this manner. Furthermore, it is now possible to clone cells from many animals and plants. The techniques of *cell culture* are critically important in modern biological research.

Escherichia coli Is a Favorite Organism of Molecular Biologists

The most commonly used cell in molecular cell biology is the bacterium *Escherichia coli*. A few of the many genetic and molecular experiments with *E. coli* that help to explain bacterial growth and gene regulation are described in Chapters 7 and 12. Among the advantages of using prokaryotic cells such as *E. coli* are the simplicity of the growth medium (Table 5-1) and the rapid growth rate of the organisms (division time ranges from 20 min to 1 h). A colony of bacteria grown overnight (~12 h) from a single cell can contain 10^7–10^8 cells. However, perhaps the most important experimental advantage of bacteria is the ease with which genetically distinct populations of cells can be isolated. This advantage is further heightened by concentrating on a single bacterium, such as *E. coli*. Literally tens of thousands of genetically distinct strains of *E. coli* are available for study in laboratories throughout the world. Almost one-half of the maximum estimated number of 3000 genes in *E. coli* have been located on the circular map of the genes, which is now correlated with the physical map of the single circular DNA molecule that constitutes the *E. coli* chromosome (Figure 5-9).

▼ **Figure 5-9** Genetic and physical maps of the circular chromosome of *Escherichia coli*. The abbreviations on the genetic map (outside circle and enlarged segment) relate to the metabolic properties of the gene products; for example, *pyrE*, *asnA*, and *glnA* were each identified in a bacterial clone with a gene defect that made the addition of a pyrimidine, asparagine, or glutamine, respectively, necessary for growth. A strain with a genetic deficiency is said to be genetically *marked*, and the resulting genetic trait is termed a *marker* for that strain. A group of such markers can define a strain with great precision. Some of the regions of *E. coli* have been mapped so completely that every gene is known. An example is *ilv*, the locus for enzymes that catalyze the synthesis of isoleucine, leucine, and valine. Each capital letter is a gene name and represents a known purified enzyme. The numbers inside the genetic map indicate approximate percentages of the distance around the circular *E. coli* genome, measured by the time required for gene transfer in conjugation. The green inner circle represents a physical map of the *E. coli* DNA. The entire length of the genome is about 4×10^6 base pairs (bp), abbreviated 4000 kilobases (kb) or 4 megabases (Mb). Segment A is the largest (about 600 kb); V is the smallest (about 20 kb). The ends of each piece are sites cleaved by a restriction endonuclease (an enzyme) called Not I that recognizes a particular sequence of 8 bp. *After B. J. Bachman, K. B. Low, and A. L. Taylor, 1976, Bacteriol. Rev.* **40**:116, *and C. L. Smith, J. G. Econome, A. Schutt, S. Klco, and C. R. Cantor, 1986, Science* **236**:1448.

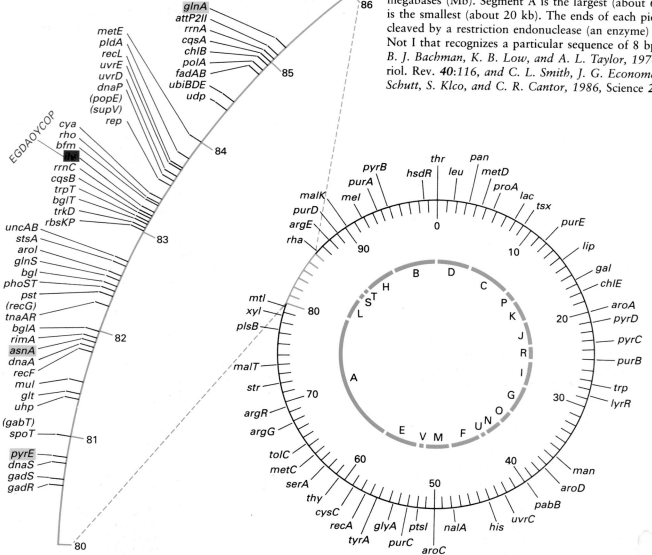

Table 5-1 Growth media for microorganisms such as *Escherichia coli* **and** *Saccharomyces cerevisiae**

MINIMAL MEDIUM[†]

Carbon source: e.g., glucose or glycerol

Nitrogen source: NH_4^+ (e.g., $NaNH_4HPO_4$) or an organic compound such as histidine

Salts: Na^+, K^+, Mg^+, Ca^{2+}, SO_4^{2-}, Cl^-, and PO_4^{3-}

Trace elements

RICH MEDIUM

Partly hydrolyzed animal or plant tissue (rich in amino acids, short peptides, and lipids)

Yeast extract (rich in vitamins and enzyme cofactors, nucleic acid precursors, and amino acids)

Carbon source, nitrogen source, and salts, as in minimal medium

---

*For more detailed information, see R. W. Davis, D. Botstein, and J. W. Roth, 1982, *A Manual for Genetic Engineering: Advanced Bacterial Genetics*, Cold Springs Harbor Laboratory.

[†]Typical for most bacteria and yeast. Some photosynthetic bacteria (e.g., *Rhodospirillum rubrum*, cyanobacteria, and blue-green algae) require CO_2 for the carbon source. Some nitrogen-fixing bacteria (e.g., *Azotobacter*) require atmospheric N_2. Other organisms have special needs: e.g., *Hemophilus* strains require factors found in whole blood.

(The physical map was constructed by the use of restriction endonucleases.) Every one of these approximately 3000 genes can be assumed to be identical in every cell in a cloned colony of *E. coli* or a liquid culture started from such a colony.

Strictly speaking, however, not all cells in a cultured clone are genetically identical. The probability of a detectable mutation is from 1 in 10^6 to 1 in 10^8 per generation for each of the 3000 genes. By the time a single cell has divided to produce a few hundred or thousand cells, some individual genes will differ from cell to cell. Because of this genetic impurity, it is a common practice either to reclone cells frequently or to subject the cultured clone to *selective* conditions to ensure that only cells with a particular desired set of characteristics will continue to survive and grow. For example, streptomycin, penicillin, or tetracycline can be added to the medium to kill all but the antibiotic-resistant cells; or a cell requiring a particular nutrient can be identified by the addition of that nutrient to a medium deficient in it (Figure 5-10).

Genes Can Be Transferred between Bacteria in Three Ways

In the 1940s and 1950s, the availability of genetically different bacterial strains led to the discovery of a number of means by which genes could be exchanged between bacteria. Each of the strains participating in these transfer experiments was marked because it possessed a different group of genetic traits (for example, dependence on one or more growth factors, plus a sensitivity or resistance to a particular chemical or virus). Although the details of this exciting period in the development of molecular genetics are beyond the scope of this book, bacterial genet-

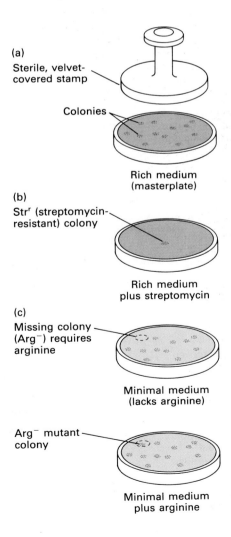

▶ **Figure 5-10** Mutations and their detection in bacteria. Random mutations in bacterial cultures may produce cells that differ in a single genetic characteristic from all other cells of the culture. Such genetic changes can be detected among colonies growing in rich medium on a *plate* (a petri dish) by *replica plating*, a technique developed by Joshua and Esther Lederberg in the 1940s. A circular velvet-covered stamp equal in size to the petri dish is pressed into it, so that the stamp picks up a sample of each bacterial colony in this dish (a). The bacteria on the stamp can then be deposited in the same arrangement on new plates. The media in the new plates are chosen to test different genetic characteristics, such as sensitivity or resistance to an antibiotic, shown in (b), or nutritional auxotrophy (the inability to grow without a specific nutrient), shown in (c). In this manner, colonies arising from single cells that have mutations in particular genes can be identified. With a doubling time of about 30 min, a colony of 10^7–10^8 cells forms overnight.

ics underlies all of molecular genetics. For this reason, the three mechanisms by which genes can be transferred between two bacterial cells are described here.

Conjugation

In *conjugation,* two *E. coli* cells, designated "male" and "female," become attached to one another and form a conjugation bridge. After a nick is made in the DNA of the male cell, one strand of the chromosome enters the female cell (Figure 5-11). The newly introduced DNA strand inserts and recombines with the DNA of the recipient cell. Subsequent DNA replication and cell division give rise to a new recombinant cell with characteristics derived from each of the parental cells.

Transduction

Viruses that invade bacterial cells are called *bacteriophages.* Most bacteriophages contain DNA. During growth and reproduction, some bacteriophages incorporate host-cell DNA into their own DNA. If a bacteriophage then infects another cell, the DNA of the first host cell may be transferred to the second. This mechanism of gene transfer by bacteriophages is called *transduction* (Figure 5-12). *Special* or *restricted* transducing bacteriophages pick up specific regions of the host bacterial chromosome; *general* transducing bacteriophages acquire DNA randomly from the bacterial chromosome. Some transducing bacteriophages incorporate so much host DNA that not enough room is left for the viral DNA sequences necessary for bacteriophage replication. These *defective* transducing phages are no longer able to kill host bacteria, but they can still invade new hosts and transfer genes to them. (A more detailed description of the different types of bacteriophages is given later in this chapter.)

Double-stranded bacterial chromosomes—with functioning *a*, *b*, and *c* genes (*left*), and with nonfunctioning *a*, *b*, and *c* genes (*right*)

A nick is made in one strand of the donor chromosome. DNA synthesis begins at nick, displacing one of the strands. The displaced strand is transferred through the conjugation bridge.

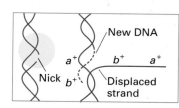

◀ **Figure 5-11**   Conjugation in *E. coli.* Two bacterial cells can join to exchange genetic information. This is accomplished when the high-frequency recombination (Hfr) strain, which is designated a "male" strain because of the presence of a fertility factor (F) in the chromosome, is mixed with the F⁻ strain, a "female" strain. The male cells possess surface projections in the form of hollow tubes, called *pili,* which form the conjugation bridge through which DNA is transferred. The male DNA is nicked at the site of the fertility factor, and only one strand of the DNA is transferred; the conjugation bridge can be broken before transfer is complete. A recombinant bacterium is formed only after the entering single strand is incorporated and replicated and cell division has occurred.

Both strands in the recipient cell begin to replicate. One strand produces a double-stranded $a^+b^+c^-$ region, whereas the other produces a double-stranded $a^-b^-c^-$ region. The cell now has two different forms of genes *a* and *b*.

Replication proceeds until complete. The two genotypes are segregated by cell division.

The bridge is broken by the experimenter before the entire Hfr DNA is transferred. In this case, conjugation is interrupted before the c^+ information is transferred. The donor strand joins the chromosome of the recipient cell at the *ab* region, replacing one old strand, which is digested.

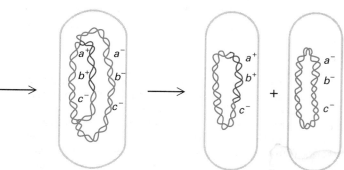

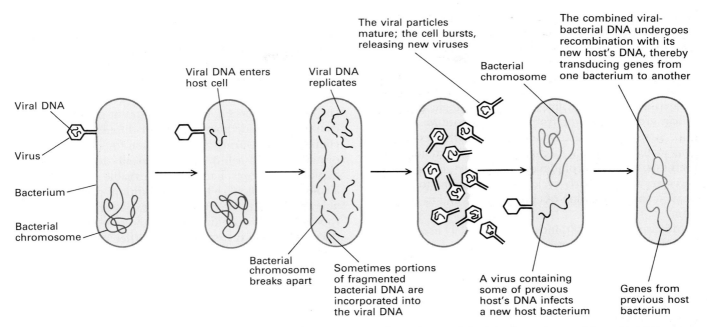

▲ **Figure 5-12** Transduction. Sometimes a bacteriophage can incorporate host-cell DNA into its own DNA during a growth cycle. Such a bacteriophage can then carry the host-cell DNA out of the host bacterium and introduce it into a new host, where it can undergo recombination with the chromosomal DNA.

Transformation *Transformation* is the genetic change of a bacterium after exposure to and recombination with isolated DNA from a genetically different bacterium (Figure 5-13). This mechanism of gene transfer, which was discovered with *Streptococcus pneumoniae,* provided a crucial piece of evidence that DNA was the genetic material. For about 10 years, however, the transformation of one bacterial cell by the DNA of another was the *only* means of experimental gene transfer, and at first only a few bacterial species could be transformed.

Recently, new treatments of bacterial cells have been developed, and now the transformation of *E. coli* and similar bacteria is possible. The cells are usually exposed to high concentrations of calcium (Ca^{2+}) ions, which somehow cause the bacterial plasma membranes to admit foreign DNA. By far the most common DNA currently used in bacterial transformation is *plasmid DNA*. *Plasmids* are small circles of DNA capable of independent replication. Plasmids that encode proteins that allow survival of the host cell (by resistance to an antibiotic, for example) are the chief *vectors* used in recombinant DNA research.

▶ **Figure 5-13** Transformation. The DNA most commonly used in transformation experiments is in the form of small, circular molecules called plasmids. These circles of DNA, which replicate autonomously in the bacterial cell, often carry genes that enable a cell to become resistant to an antibiotic. Thus a cell that has acquired a plasmid can be selected by growing cells in a medium containing the antibiotic.

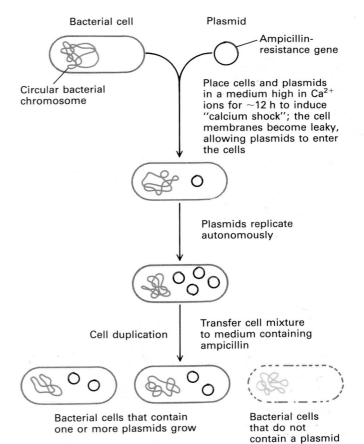

The Yeast Life Cycle Includes Haploid and Diploid Phases

Genetically homogeneous populations can be obtained with eukaryotic cells as well as with prokaryotic cells. For example, a yeast or mold culture can be grown in a simple defined medium (see Table 5-1) from a single vegetative cell (a growing cell) or from a single spore (a dormant cell). Subjects particularly well-studied with these organisms include the cell's mechanisms for controlling DNA synthesis, genetic recombination, and the gene products necessary to the orderly cycle of events leading to the replication and segregation of chromosomes and to cell division.

Most species of yeast, including the widely used *S. cerevisiae*, grow in nature as single diploid cells. Each cell has 12–18 chromosomes, depending on the species. As we noted in discussing mechanisms of cell division, growing haploid or diploid yeast cells reproduce asexually by budding. Nuclear division is not as elaborate in yeast as it is in most eukaryotic cells, yet the result is basically similar to that of mitosis. However, when a diploid yeast cell encounters adverse conditions (nutrient deprivation, for example), it undergoes a form of differentiation to become a spore. During its preparation for sporulation, the original diploid cell divides meiotically to produce four haploid *ascospores*. Each spore is capable of giving rise to a colony of haploid cells, once conditions are again favorable for growth. Alternatively, the cells arising from two spores may fuse, giving rise to a diploid cell. Cells from the same spore sac (*ascus*) or two different sacs (*asci*) can fuse. However, fusion only occurs between two cells of the opposite *mating type* (Figure 5-14a).

In nature, it is probably less common for two haploid yeast cells to fuse from different asci than from the same ascus; when it does happen, it broadens the gene pool available to the species. The fusion of opposite mating types of haploid cells occurs because each type secretes a substance, a *mating-type factor,* that stops the growth of the opposite type and leads to cell fusion. In the laboratory, the intentional fusion of two genetically different haploid cells to produce a diploid cell can be used to perform a *genetic cross*. After growth as diploids, individual cells can be caused to undergo meiosis and segregation into new haploid spores (Figure 5-15). All of the genes on a single parental chromosome tend to remain together (unless they become separated by crossing over) and thus establish *linkage groups*, of which there are 17 in *S. cere-*

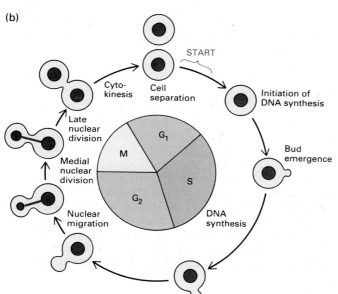

(a)

◀ Figure 5-14   (a) The life cycle of *S. cerevisiae*. The two haploid cells that unite to form a diploid cell differ in mating type. (Here blue and red circles represent the nuclei of opposite mating types called a and α.) Under favorable growth conditions, diploid cells reproduce asexually by budding. When such a cell is deprived of essential nutrients, however, it undergoes meiosis within a saclike structure called an ascus, and the yeast chromosomes assort independently in Mendelian fashion to yield four haploid ascospores. If the ascus is returned to favorable growth conditions, pairs of spores of opposite mating type within it may fuse to form diploid cells, which will proceed to grow and bud as before; this happens often in nature. On the other hand, each ascospore can remain haploid and can give rise to a haploid colony. Haploid cells from different colonies (of different mating types) can fuse or be fused into diploid cells. (b) The sequence of events in the cell division cycle of yeast based on morphologic and biochemical experiments. Mutations are available that block cells in each of the steps in the cell cycle. The four phases—G_1, S, G_2, and M—are explained in Figure 5-2. The entire cycle in rich medium takes about 2 h and about 8 h in minimal medium. *Part (b) from L. H. Hartwell et al., 1974, Science **183**:46.*

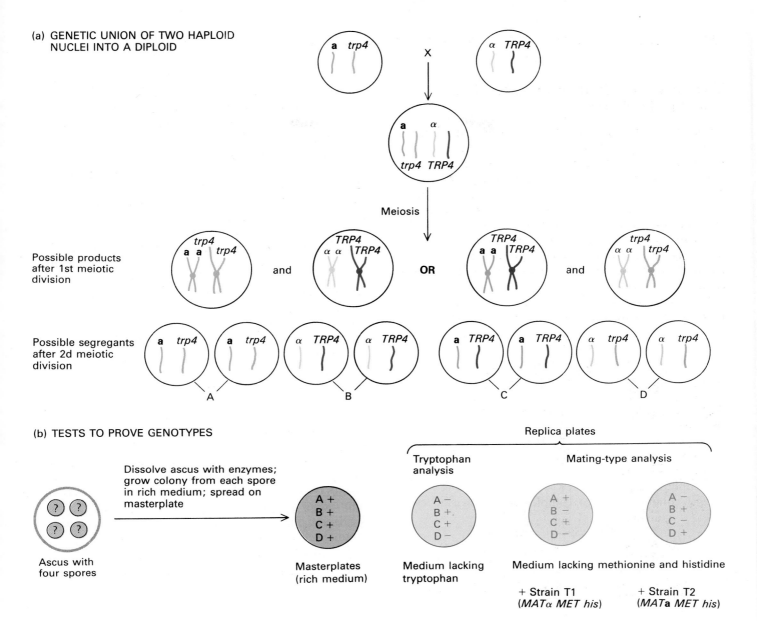

(a) GENETIC UNION OF TWO HAPLOID NUCLEI INTO A DIPLOID

Possible products after 1st meiotic division

Possible segregants after 2d meiotic division

(b) TESTS TO PROVE GENOTYPES

Ascus with four spores

Dissolve ascus with enzymes; grow colony from each spore in rich medium; spread on masterplate

Masterplates (rich medium)

Replica plates

Tryptophan analysis

Mating-type analysis

Medium lacking tryptophan

Medium lacking methionine and histidine

+ Strain T1 (*MATα MET his*)

+ Strain T2 (*MATa MET his*)

▲ **Figure 5-15** Exercises in yeast genetics.

(a) Following two markers that lie on different chromosomes. The two properties followed in the diagram are *mating type* (MATa or MATα) and tryptophan requirement (*TRP4* or *trp4*). (If the two strains were not of opposite mating types, they would not unite as a diploid and no spores would form.) A fusion between the haploid strains *MATa trp4* and *MATα TRP4* is followed by starvation to induce meiosis. The individual meiotic products can be observed in yeast (and in other fungi) because each spore can be independently grown into a colony and tested for different properties. The *TRP4* strain has normal enzymatic capacity for tryptophan synthesis; the *trp4* strain has a mutant gene that incapacitates the cell for a particular step in tryptophan biosynthesis. We can follow this property by plating on a medium lacking or containing tryptophan. The second meiotic division produces two pairs of haploid ditypes: if they are either all parental (A or B) or all nonparental (C or D) ditypes, then the two traits are carried on different chromosomes.

(b) Tests to prove genotypes. To determine the genetic makeup of the meiotic products, the experimenter first cultures each spore in a rich medium. The *TRP* spores, which can grow without tryptophan, can be identified from a single replica plate on a medium lacking tryptophan. Because only B and C spores grow in this medium, B or C are *TRP*.

Determination of mating type requires that an additional marker, such as *met* (indicating the presence of a defective gene that disrupts the cell's ability to grow without methionine), be present in both original haploid strains of yeast cells, so that all meiotic products are *met*. Cultures from each of the four meiotic products are mated with two different tester strains. The tester strains are both *MET*, and they, too, have a marker (here, *his*). Tester strain 1 (T1) is *MATα MET his*, and tester strain 2 (T2) is *MATa MET his*. When cultures of the meiotic products (which are all *met HIS*) are plated with T1 or with T2 in media lacking methionine and histidine, colonies form only where mating has occurred, because only diploid cells will be both *MET* and *HIS*. If a meiotic product mates with T1, it must be *MATa* (A or C); if it mates with T2, it must be *MATα* (B or D).

visiae. Newly discovered mutations that affect previously unrecognized genes can easily be placed on a particular chromosome by determining with which linkage group the new gene segregates during meiosis. Recombinational analysis has allowed very detailed genetic maps to be developed for this organism.

Many different types of mutant yeast-cell strains have been isolated, and the chromosomal positions of the affected genes have been located or "mapped." Hundreds of nutritional mutants (cells that require nutrients that normal cells do not) are available, as are *cdc* (*c*ell *d*ivision *c*ycle) *mutations* (strains blocked at various steps in the normal progress of the cell cycle). Independent mutations at many points in the cell cycle are available. Most of the *cdc* mutants are temperature-sensitive mutants that will not grow at 36° but will grow at 23°) Figure 5-14b shows some of the major points of interruption identified by these mutants. One of the most interesting mutations (*cdc28* by name) allows the cell to go through all the steps of the cycle, including cell division, at elevated temperature but not to start another cycle. Once a cell does pass this "start point," it traverses the entire cycle. Studies of this mutant and others favor the interpretation that the cycle does not consist of a series of events, each depending on the successful conclusion of the previous event, but that a master timer is responsible for running the show smoothly and one or a few critical events (the "start" event) commit a cell to completing the division cycle.

Cultured Animal Cells Share Certain Growth Requirements and Capacities

All animals must obtain from their environments a group of amino acids referred to as the *essential amino acids*: arginine, histidine, isoleucine, leucine, lysine, methionine,

Table 5-2 Daily requirement of essential amino acids (for college-age males)*

Amino acid	Grams
Arginine	0[†]
Histidine	unknown[‡]
Isoleucine	1.30
Leucine	2.02
Lysine	1.50
Methionine	2.02
Phenylalanine	2.02
Threonine	0.91
Tryptophan	0.46
Valine	1.50

*The amino acid requirements of men were established by feeding experiments. All of the listed amino acids are necessary to maintain body mass. However, arginine and histidine can be omitted from the diet for a short time without damaging body cells.
[†]Required by infants and growing children.
[‡]Essential, but the precise requirement is not yet established.
SOURCE: W. C. Rose, R. L. Wixom, H. B. Lockhart, and G. F. Lambert, 1955, *J. Biol. Chem.* **217**:987.

phenylalanine, threonine, tryptophan, and valine (Table 5-2). The same is true of all animal cells in culture. In addition, cells in culture require cysteine, glutamine, and tyrosine. Animals can obtain these three amino acids from their own specialized cells; as examples, the liver makes tyrosine from phenylalanine and both the liver and the kidney can make glutamine. The remaining amino acids and the imino acid proline can be synthesized by animal cells both within the organism and in culture. The other essential components of a culture medium are vitamins (which the cells cannot make, at least not in adequate amounts), salts, glucose, and serum (the noncellular part of blood, essentially a solution of various proteins) (Table 5-3).

Table 5-3 Mammalian cell media (HeLa cell)

SERUM-CONTAINING MEDIUM (EAGLE'S MEDIUM)

Essential amino acids:	(10^{-4}–10^{-5} *M*), as listed in Table 5-2, plus cysteine, glutamine, and tyrosine
Vitamins:	(1 mg/L) choline, folic acid, nicotinamide, pantothenate, pyridoxal, thiamine; (2 mg/L) inositol; (0.1 mg/L) riboflavin
Salts:	Na^+, K^+, Ca^{2+}, Mg^{2+}, Cl^-, PO_4^{3-}, HCO_3^-
Glucose	
Serum dialyzed (5–10% of total volume)*	

SERUM-FREE MEDIUM

Amino acids:	Essential as above, plus (10^{-4} *M*) L-alanine, L-asparagine, salts, vitamins, glucose
Other additions:	Linoleic acid, lipoic acid, hypoxanthine, putrescine, pyruvate, thymidine, trace elements (Mn, Mo, Ni, Se, Si, Sn, V, Cd), and hormone growth factors (insulin, transferrin, hydrocortisone, fibroblast, and epidermal growth factors)

*Serum is a mixture of hundreds of proteins with a total protein concentration of 50–70 mg/mL. Albumin is the most plentiful serum protein (30–50 mg/mL). Growth factors are present in very low concentrations; e.g., growth hormone at 34 ng (nanograms)/mL and insulin at 0.2 ng/mL.

SOURCE: H. Eagle, 1959, *Science* **130**:432; and S. E. Hutchings and G. H. Sato, 1978, *Proc. Nat'l Acad. Sci. USA* **75**:901.

(a)

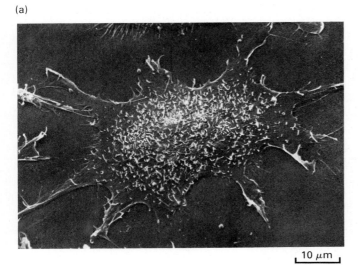

10 μm

(b)

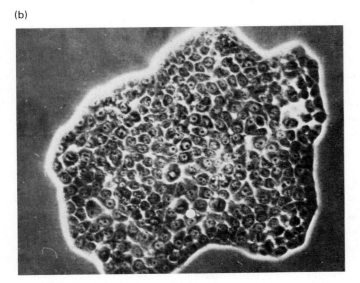

(c)

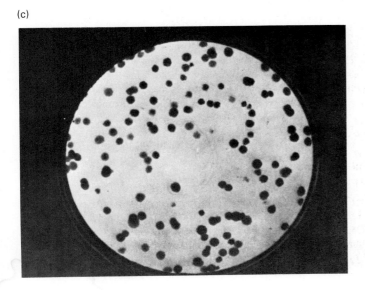

All normal animal cells in culture can synthesize the nucleic acid precursors from the simpler compounds in the medium. The role of serum, which is required by most cells in a culture medium, is not completely understood, but serum apparently supplies needed trace materials, including protein growth factors. It is now possible to grow a few types of mammalian cells in a completely defined medium supplemented with trace minerals and protein growth factors (see Table 5-3). Cells cultured from single cells on a glass or a plastic dish (Figure 5-16a) form visible colonies in 10–14 days (see Figures 5-16b and c). Some cells can be grown in suspension, which offers a considerable experimental advantage: equivalent samples are easier to obtain from suspension cultures than from colonies grown in a dish.

Primary and Transformed Cells Two types of cultured animal cells are in general use: primary and transformed. *Primary cell cultures* are derived from normal animal tissue, probably most often from skin or whole embryos. The cells in such cultures are predominantly *fibroblasts*, which are found in connective tissue in all parts of the body. One important role of fibroblasts is to heal wounds in the skin. Fibroblasts produce collagen and other material that lies between the cells of connective tissue; in this sense, they are specialized. They grow well when first placed in culture, continuing to double for 50–100 generations, at which time they reach a "crisis," grow very slowly (if at all) for a few more generations, and then cease to growth altogether. No cause has been found for this growth limitation, and no nutritional regimen has yet been discovered to cure the problem. It is interesting that fibroblasts from newborns grow for more generations in culture than do cells from older people. Epithelial cells generally produce more specialized products than fibroblasts do; some have been cultured, although they are more difficult to grow and also do not grow indefinitely.

All mammalian cells capable of *indefinite* growth in culture are derived either from tumor cells taken directly from an animal or from cultured cells that have undergone a change that causes them to behave as tumor cells. Continuously growing cells are said to be *transformed*; this term is used to describe any cells that are immortal.

◄ **Figure 5-16** Cultured mammalian cells viewed in three ways. (a) A single mouse cell attached to a plastic petri dish, viewed through a scanning electron microscope. To separate attached cells so they can be plated individually, a cell culture is treated with a protease such as trypsin. (b) A photomicrograph of a single colony of human HeLa cells, produced from a single cell after growth for 2 weeks. (c) After staining cells in a petri dish, individual colonies can easily be seen and counted. [See P. I. Marcus, S. J. Cieciura, and T. T. Puck, 1956, *J. Exp. Med.* **104**:615.] *Photograph (a) courtesy of N. K. Weller; photographs (b) and (c) courtesy of T. T. Puck.*

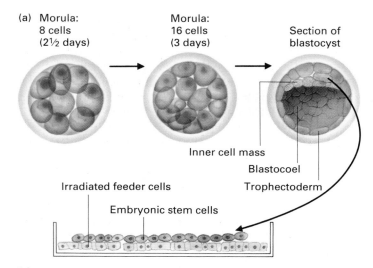

(a) Morula: 8 cells (2½ days)   Morula: 16 cells (3 days)   Section of blastocyst

Inner cell mass

Blastocoel

Trophectoderm

Irradiated feeder cells

Embryonic stem cells

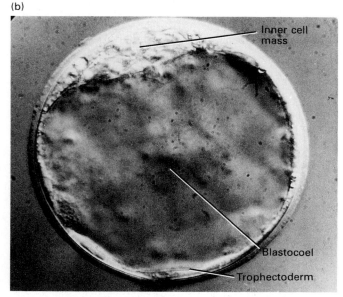

(b)

Inner cell mass

Blastocoel

Trophectoderm

(c)

Trophectoderm cell

Cell of inner cell mass

Blastocoel

Unfortunately, the term "transformation" has two different meanings in cell biology: Originally it described the process by which foreign DNA is incorporated and foreign genes are then expressed in bacteria (see Figure 5-15). Later, the term was used to refer to the process by which normal restraints on growth in vivo of animal cells are abolished, so that the cells no longer die after a limited number of divisions. This second usage is much more common at present.

Neither primary fibroblasts nor the majority of transformed cells carry out completely normal differentiated cell functions such as the production of special proteins or the organization of tissues. However, undifferentiated cultured cells have been used to advantage in general studies on the synthesis of RNA, DNA, and protein and in studies of the structural elements common to all cells. One of the most popular undifferentiated cultured cells is the HeLa cell (see Figures 5-16b and c), the first human cell to be grown continuously in the laboratory. The HeLa cell was obtained in 1952 from a malignant tumor, a carcinoma of the uterine cervix. The compositions of the HeLa cell, as a typical example of an animal cell, and of *E. coli* are compared in Table 4-2. The volume of the average animal cell is about 1000 times larger than a bacterial cell and about 10 times larger than a yeast cell.

Undifferentiated cultured cells readily serve as hosts for viral infections. In a pure cell population, every cell in the culture is equally susceptible to a given virus; thus, in biochemical terms, every cell can be expected to behave in a similar manner.

Embryonic Stem Cells   A recent advance has been made in growing a very important cell type. During embryo formation in mammals, the fertilized egg grows into a hollow ball of 20–50 cells called the *morula*. Within this ball, some cells are set aside to become the embryo proper; others will form the membranes (amnion and placenta) by which the embryo is attached to the uterine wall to be nourished. At the time of implantation, the collec-

◀ **Figure 5-17**   Preparation of embryonic stem cells. (a) Fertilized mouse eggs divide slowly at first; after about 4½ days, they form a hollow structure called a *blastocyst,* which contains about 100 cells around an inner cavity called the *blastocoel.* Only the *embryonic stem cells,* which constitute the *inner cell mass,* actually form the embryo. Other cells form the *trophectoderm,* which gives rise to auxiliary tissues such as the placenta. Embryonic stem cells released from the blastocyst will grow on top of lethally irradiated "feeder cells." Foreign genes can be introduced into these cells, which can then be reintroduced into blastocysts to participate in forming an embryo. (b) A light micrograph of a blastocyst, and (c) an electron micrograph of a section through a blastocyst. [See E. Robertson et al., 1986, *Nature* 323:445; and S. L. Mansour, K. R. Thomas, and M. R. Capeechi, 1988, *Nature* 336:348.] *Photographs courtesy of P. Calarco.*

tion of cells is called a *blastocyst* (Figure 5-17). The embryonic stem cells of a mouse blastocyst can be removed at this stage and grown in culture through many generations. These cultured cells can be treated or modified in various ways and reintroduced into a blastocyst. The experimentally manipulated cells then take part in embryo formation and ultimately can even give rise to the germ cells of the mouse.

Differentiated Cells in Culture In addition to less differentiated normal cells (such as fibroblasts) and undifferentiated cancer cells (such as HeLa cells), investigators study transformed cells that can grow in culture while continuing to perform many of the functions of specialized tissues. For example, cultured *myoblasts* (muscle precursor cells) that have been transformed to immortality will still fuse to form *myotubes*, which resemble multinucleated muscle cells (Figure 5-18) and will make many, if not all, of the specialized proteins associated with contraction. *Erythroleukemia* cells (abnormal precursors of red blood cells) can be induced to produce hemoglobin and undergo the structural changes in the cell membrane associated with red blood cell maturation. A wide variety of transformed precursors to white blood cells (lymphocytes, monocytes, macrophages, and others) can carry out at least part of the differentiated function of their normal counterparts. Cells from malignant tumors called *teratocarcinomas* can generate a variety of different tissue types during growth, but the differentiated cells are not arranged in an orderly fashion.

In addition to cases in which transformed cells have displayed or been induced to display at least partially differentiated cell function, some success has been achieved in culturing normal precursors to the cells of skin epithelium (Figure 5-19). Under appropriate culture conditions and in the presence of a hormone called the *epidermal growth factor*, the basal cells of the epidermis will divide many times. Once they form a closely packed colony, they differentiate into the *cornified cells*, or *keratinocytes*, that make up the outer surface of the skin. (The keratins

(a) Early stage of myotube   (b)

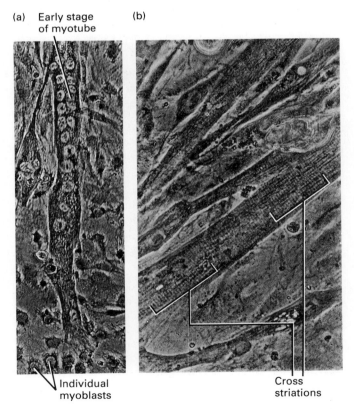

Individual myoblasts   Cross striations

▲ **Figure 5-18** (a) A transformed line of rat myoblasts will grow indefinitely as single cells in laboratory cultures. (b) When cultured cell growth is stopped (for example, by removing serum from the medium), the myoblasts fuse to produce myotubes. The characteristic cross striations of muscle cells can be seen.

Cornified cells

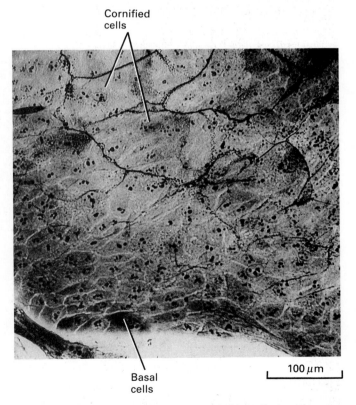

100 μm

Basal cells

▲ **Figure 5-19** The differentiation of skin cells in culture. When the epidermal layer of the skin of newborn animals is placed in culture with irradiated, nongrowing "feeder" cells, the basal cells of the skin grow and differentiate into cornified epithelial cells. The flattened and tightly packed nature of the epithelial sheet is characteristic of the cells at the body surface. [See T. T. Sun and H. Green, 1976, *Cell* 9:511.]

are a large protein family that are the major constituents of the outer layer of skin.) Fat cells have similarly been cultured from precursors (probably fibroblasts) found in subcutaneous tissue.

However, many specialized animal cells cannot be cultured at all. This is not altogether surprising, because certain fully differentiated normal cells, for example, those found in muscles, nerves, or the kidney, do not grow continually in the body. Structural changes in these cells as they differentiate during tissue formation may prevent any further growth. Muscle cells fuse in early development, so that many nuclei occupy one cell body; nerve cells develop extensions (*axons*) up to several feet long that form intricate attachments to other nerve cells. Such structural features would make it difficult for a cell to divide in a way that would yield two functioning daughter cells.

Some specialized cells, such as hepatocytes from the liver, do slowly "turn over": when they die, they are replaced by an occasional division of remaining liver cells. Moreover, if a large section of liver is removed from a mammal, the remaining cells initiate cell division and grow until the liver regains its normal size. Nevertheless, despite much effort, cultures of growing hepatocytes have not been achieved. In some cases, specialized cell functions have been successfully maintained in cultured *tissue* samples (not isolated cell samples), particularly the breast tissue of rodents and the oviduct tissue of chickens.

Cell Fusion: An Important Technique in Somatic-Cell Genetics

Because some animal cells can be cultured from single cells in a well-defined medium, it is possible to select for genetically distinct *somatic* (body) *cells,* just as is done for bacteria and yeast cells. The chromosomes in a somatic cell are large and highly visible after staining, making it easy to identify them by type and to observe their distinctive arrangements. Moreover, somatic cells can be fused so that two nuclei function in one cell, called a *heterokaryon.* The branch of cell biology called *somatic-cell genetics* uses the technique of cell fusion to produce large numbers of hybrid cells and to study gene function when nuclei are introduced into novel cellular environments.

Hybrid Cells Containing Chromosomes from Different Mammals Assist in Gene-mapping Studies

Spontaneous fusion of animal cells in culture occurs infrequently, but the rate increases greatly in the presence of certain viruses that have lipoprotein envelopes similar to the plasma membranes of animal cells. Apparently, a gly-coprotein in the viral membrane promotes cell fusion, but the mechanism is not yet fully understood. Cell fusion can also be promoted by the addition of polyethylene glycol, which causes cell plasma membranes to adhere to those of any adjacent cells (Figure 5-20). In most fused animal cells, the nuclei eventually also fuse, producing viable cells that contain chromosomes from both "parents."

Hybrids of cultured cells from different mammals—for example, human and rodent cells or cells of different rodents (rats, mice, and hamsters)—have been widely used in somatic-cell genetics. As hybrids of human and mouse

(a)

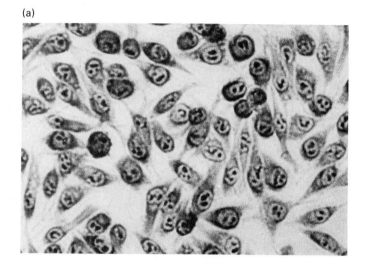

(b)

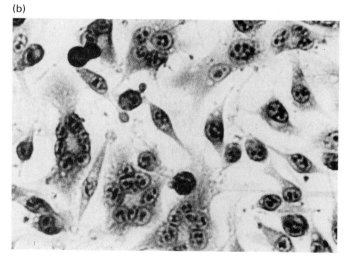

▲ **Figure 5-20** Fused cultured cells. (a) Unfused growing mouse cells. (b) Mouse cells that have fused as a result of treatment for 1 min with a 45 percent concentration of poly-ethylene glycol; there are 2–5 nuclei per cell. By varying the concentrations and the times of exposure, the investigator can maximize the number of fused cells (heterokaryons) containing only two nuclei. *From R. L. Davidson and P. S. Gerald, 1976, Som. Cell Genet. 2:165.*

cells grow and divide, they gradually lose human chromosomes in random order. In a medium that can support both mouse and human cells, the hybrids eventually lose all human chromosomes. However, in a medium in which human cells can grow but mouse cells lack one enzyme needed for growth, the human chromosome containing the gene that codes for the needed enzyme will remain.

By using various media in which mouse cells cannot grow but human cells can, *panels* of hybrid cell lines have been established. Each cell line in a panel contains a different limited number of human chromosomes (ideally, a single human chromosome) and a full set of mouse chromosomes, each of which can be identified visually under a light microscope. Thus an individual human chromosome (or a small group of them) can be probed for the presence of a particular gene—for example, by testing a cell line biochemically for a particular enzyme or immunologically (with an antibody) for a surface antigen; or DNA

hybridization techniques may be used to locate a particular DNA sequence. In this way, many human genes have been mapped to specific human chromosomes. Panels of mouse-hamster hybrid cells have also been established; in these cells, the majority of mouse chromosomes are lost, allowing the genes to be mapped in the mouse genome.

Mutants in Salvage Pathways of Purine and Pyrimidine Synthesis Are Good Selective Markers

One metabolic pathway has been particularly useful in cell-fusion experiments. Most animal cells can synthesize the purine and pyrimidine nucleotides de novo from simpler carbon and nitrogen compounds, rather than from already formed purines and pyrimidines (Figure 5-21 *top*). The folic acid antagonists *amethopterin* and *ami-*

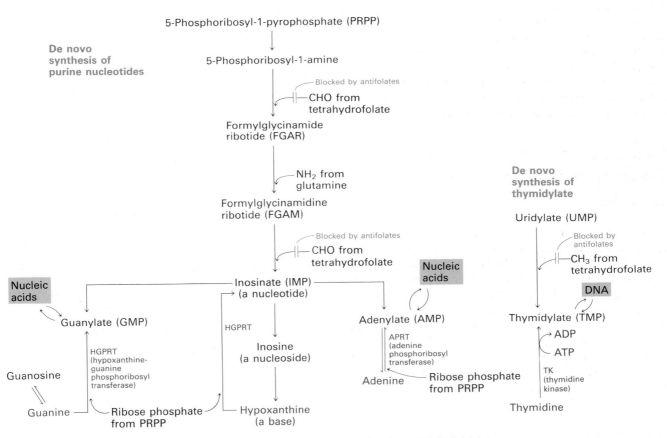

Salvage pathways to obtain purine nucleotides and thymidylate

▲ **Figure 5-21** Under normal circumstances, cultured animal cells synthesize purine nucleotides and thymidylate by *de novo pathways* that require the transfer of a methyl or formyl group from an activated form of tetrahydrofolate (for example, N^5,N^{10}-methylenetetrahydrofolate), shown in the upper portion of the diagram. Antifolates, such as aminopterin and amethopterin, block the reactivation of tetrahydro-

folate, preventing purine and thymidylate synthesis. A second mechanism for obtaining nucleotides is called a *salvage pathway* (bottom part of diagram). The enzymes of the salvage pathways are HGPRT, APRT, and TK. These pathways are not blocked by antifolates. If the medium contains purine bases or nucleosides and thymidine, most mammalian cells can use them directly to make nucleotides by these pathways.

nopterin interfere with the donation of methyl and formyl groups by tetrahydrofolic acid in the early stages of de novo synthesis of glycine, purines, nucleoside triphosphates, and thymidine triphosphate. These drugs are called *antifolates*. In addition to the biosynthetic pathways, enzymes in most cells also can use purines and thymidine directly by *salvage pathways* that bypass the metabolic blocks imposed by antifolates (Figure 5-21 *bottom*).

A number of mutant cells lines have been discovered that are unable to carry out one of the salvage steps. For example, cell cultures lacking thymidine kinase (TK) have been selected, and cultures have been established from humans who lack adenine phosphoribosyl transferase (APRT) or hypoxanthine-guanine phosphoribosyl transferase (HGPRT). These different types of salvage mutants become useful partners in cell fusions with one another or with cells that have salvage pathway enzymes but that are differentiated and cannot grow in culture by themselves. The selective medium most often used to culture such fused cells is called *HAT medium,* because it contains *h*ypoxanthine (a purine), *a*minopterin, and *t*hymidine. Normal cells can grow in HAT medium; salvage mutants cannot, but their hybrids with normal cells can.

Hybridomas Are Fused Lymphoid Cells That Make Monoclonal Antibodies

The technique of cell fusion followed by selection is widely used in the production of monoclonal antibodies. A *monoclonal antibody* is a single pure antibody produced in quantity by a cultured clone of a special cell type called a B lymphocyte. Each normal B lymphocyte from, say, a rat or mouse spleen is capable of producing a single antibody. If a mouse is injected with an antigen, B lymphocytes that make an antibody that recognizes the antigen are stimulated to grow and to produce that antibody. In the animal, each stimulated B cell forms a clone of cells in the spleen or bone marrow. However, normal lymphocytes will not grow indefinitely in culture and so cannot be used to establish an immortal productive clone.

This property of normal B lymphocytes is sidestepped by fusing normal B lymphocytes with cancerous lymphocytes called *myeloma cells* that are immortal (Figure 5-22). Many different cultured cell lines of myeloma cells from mice and rats have been established; from these, mutant myeloma cell lines that have lost the salvage pathways for purines (indicated by their inability to grow in HAT medium) have been selected. When these mutant myeloma cells are fused with normal antibody-producing cells from a rat or mouse spleen, *hybridoma cells* result. Like myeloma cells, hybridoma cells can grow indefinitely in culture; like normal spleen cells, the fused cells have purine salvage-pathway enzymes and can grow in HAT medium. If a mixture of fused and unfused cells is

placed in HAT medium, the unfused mutant myeloma cells and the unfused spleen cells die, leaving a culture of immortal hybridoma cells, each of which produces a single antibody. Clones of hybridoma cells can be tested separately for the production of a desired antibody; the clones containing that antibody then can be cultured in large amounts.

Such pure antibodies are very valuable research reagents. For example, monoclonal antibodies can recog-

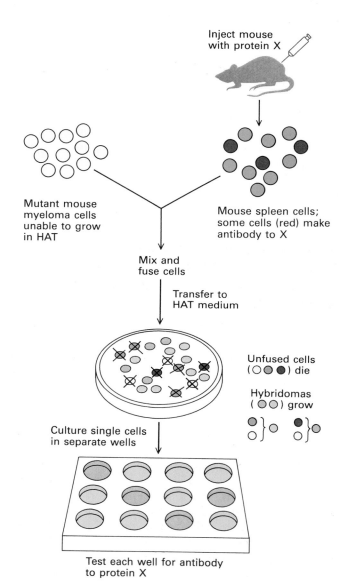

▲ **Figure 5-22** The procedure for producing a monoclonal antibody to protein X. The myeloma cells can grow indefinitely in culture (but not in HAT medium); spleen cells produce antibodies and furnish the enzymes of the salvage pathway (see Figure 5-21). The fusion products grow in HAT medium and produce antibodies. Once a hybridoma clone that produces a desired antibody is identified, the clone can be cultured to yield a large amount of that antibody.

nize a protein and so identify the location of the protein in specific cells of an organ or in specific cell fractions. Once identified, even very scarce proteins can be isolated by the technique of *antibody-affinity chromatography* in which the monoclonal antibody, bound to a solid substrate such as paper fibers, is exposed to a solution containing the desired protein. Only that protein binds to the antibody, and it can be easily purified (see Figure 2-28).

Monoclonal antibodies have become very important diagnostic tools in medicine; in the future, they may give rise to new treatments as well.

DNA Transfer into Eukaryotic Cells

The fusion of whole cells continues to be a valuable tool for bringing genes from different cells together. However, it is now much more common to introduce purified genes into cells by a variety of newer techniques. (The cloning and purification of individual genes is discussed in Chapter 6). Pure DNA is readily taken up by specially treated cells, or it can be injected directly into appropriate cells.

Yeast Cells Exhibit Homologous Recombination of Foreign DNA in Contrast to Nonspecific Integration in Mammalian Cells

Yeast cells can be treated with enzymes to remove thick outer walls; such *spheroplasts* will take up DNA added to the medium. Plant cells also can be converted to spheroplasts and will take up added DNA. Cultured mammalian cells take up DNA directly, particularly if it is first converted to a fine precipitate by treatment with calcium (Ca^{2+}) ions. In all these cells, once the foreign DNA is inside, enzymes that probably function normally in DNA repair and recombination join the ingested DNA with the cells' chromosomes. If yeast DNA is introduced into yeast cells, the recombination almost always occurs at the chromosomal site homologous to the introduced DNA segment (Figure 5-23). Because of this homologous entry, specific genes can be replaced in yeast chromosomes. Given the ability to replace a normal gene with a defective one, the role of the protein encoded by the gene can be explored. Mutant forms of a yeast gene can be prepared and integrated into the same chromosomal site previously occupied by the normal gene. Such replacements or "*gene knockouts*," are performed on diploid yeast cells, in which a normal and mutant copy can coexist. Whether the mutant gene fails to carry out an obligatory function in a haploid cell normally capable of growth is then tested by inducing meiosis and sporulation and determining whether each resulting haploid spore

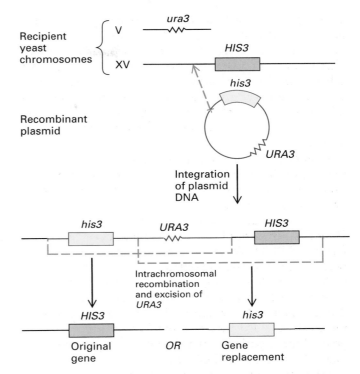

▲ **Figure 5-23** Replacement of the normal *HIS3* gene by a deletion mutant in the yeast *S. cerevisiae*. By the technique of recombinant DNA technology, a plasmid containing two yeast genes is prepared: one, *his3*, encodes only part of the sequence of an enzyme in the histidine biosynthetic pathway; the second, *URA3*, encodes an enzyme in the pathway of synthesis of the pyrimidine uracil. The *URA3* gene is included simply as a selective marker. The recipient cell (a *Ura3⁻* strain) takes up the plasmid, which can integrate either into the *ura3* gene (uninteresting for our experiment) or into the *HIS3* gene. Replica plating can now be used to determine which *URA3* cells are unable to grow without histidine because their good *HIS3* gene has been exchanged for a *his* gene. This technique can be used to replace any gene segment in a yeast cell. [See S. Scherer and R. W. Davis, 1979, *Proc. Nat'l Acad. Sci. USA* 76:4951.]

can give rise to a colony. One of the first genes tested in this way was the actin gene; haploid mutants without normal actin cannot grow (Figure 5-24).

In contrast to yeast cells, mammalian cells integrate newly introduced DNA much more often in nonspecific sites than in the correct locations, perhaps due to the much larger size of the mammalian genome. In fact, very large sections of DNA will integrate into animal cells, so selectable markers can be transferred at the same time as any desired test DNA. Different DNA samples most often become linked inside the cell and integrated together into the cell's chromosomes. A maximum of 1–10 percent of the cells in a culture integrate any DNA.

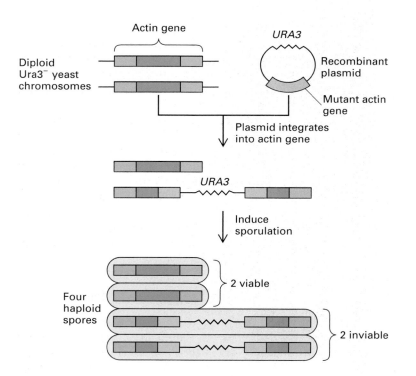

◀ **Figure 5-24** The gene that encodes actin is required by yeast. Actin is a prominent structural protein in the cytoplasm of all vertebrate and higher plant cells; it is also found in yeasts. A recombinant plasmid containing the *URA3* gene as well as a shortened yeast actin gene (light blue, corresponding to the center section within the full-length gene) is introduced into a normal diploid yeast cell that is Ura3⁻ (unable to synthesize pyrimidine). Cells that have integrated the plasmid are selected (based on growth without pyrimidine followed by restriction enzyme analysis; see Figure 5-14). Only recombinants within the actin gene on at least one chromosome are studied further. Whether the actin gene is required for viability can then be tested; meiosis and sporulation are induced by starving the cells; each cell produces four haploid spores. The wild-type spores can grow, but those with the short actin gene do not, showing their need for actin. *From D. Shortle, J. E. Haber, and D. Botstein, 1982,* Science **217**:371.

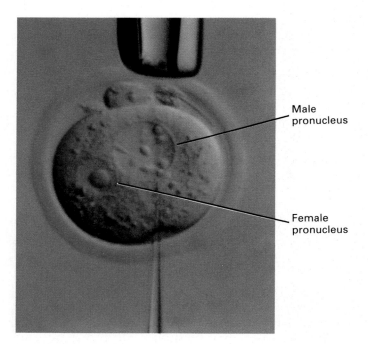

▲ **Figure 5-25** DNA can be directly injected into individual cells under the microscope through fine-tipped glass needles, or pipettes, which control pumping of small volumes (about 10^{-10} ml). In this fertilized mouse egg, the two parental nuclei have not yet fused. The egg is held by a blunt pipette, a fine-tipped pipette has been inserted into the male pronucleus, and a DNA solution injected. Such injected eggs are viable and can be transplanted into the uterus of a primed mouse; the eggs will develop into mice that contain the injected DNA sequences in every cell. *Photograph courtesy of R. L. Brinster.*

An important new method for introducing DNA is called *electroporation*. Some cells do not take up the calcium precipitate of DNA well but do take up DNA if they are exposed to a brief electric shock of several thousand volts. Apparently, the shock does not kill the cells, but it does open momentary holes in the plasma membrane; the DNA enters, and the holes reseal.

DNA can also be injected directly into the nuclei of vertebrate cells (Figure 5-25). Frog oocytes are among the most popular targets for testing immediate gene action, because they have very large nuclei. In addition, the nuclei of several lines of cultured mammalian cells, including HeLa cells, have been injected successfully with DNA.

Foreign DNA Can Be Introduced into the Germ Line of Animals to Produce Transgenic Strains

Perhaps the most dramatic use of the injection technique is the introduction of DNA into fertilized mouse or *Drosophila* eggs. For example, when foreign DNA is injected into one of the two pronuclei (the male and female haploid nuclei contributed by the parents) of a fertilized mouse egg before they fuse (Figure 5-26), the DNA is incorporated into the chromosomes of the diploid zygote. The injected eggs can be transferred to foster mothers, in which normal cell growth and differentiation occurs. About 10–30 percent of the progeny will contain the for-

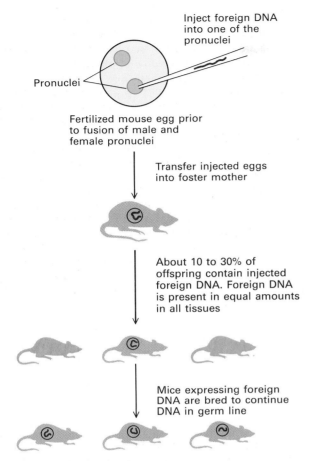

▲ **Figure 5-26**   The introduction of foreign DNA into a mouse and the establishment of a homozygous transgenic mouse strain. [See R. L. Brinster et al., 1981, *Cell* 27:223.]

success has been achieved in recombining a newly introduced mutant gene with the homologous preexisting gene (as happens with yeasts), so that animal genes can be replaced. About 1 in 100 cultured embryonic stem cells recombines injected genes at the correct site, and this will have a mutation in the gene. These cells can be selected and then reintroduced into mouse embryos. Ultimately some of the recombinant cells will enter the mouse germ line. Thus mice with replaced genes can be produced and studied. These techniques for homologous recombination are important to basic research studies in which genes of unknown function are replaced by mutant counterparts in the germ line. They are also important in the potential cure of genetic diseases where a normal gene would replace a mutant one in a somatic cell type such as a fibroblast or a blood cell precursor.

Plants Can Be Regenerated from Plant Cell Cultures

Established cell cultures exist for fewer plant than animal species, and the growth of a culture from a single plant cell is not yet as reliable as the growth of cultures from single animal cells. However, one striking property of cultured plant cells is very exciting and useful. Meristematic (growing) cells from dissected plant tissue or cells within excised parts of a plant will grow in culture to form *callus tissue*, an undifferentiated lump of cells. Under the influence of plant growth hormones, different plant parts (roots, stems, and leaves) develop from the callus and eventually grow into whole, fertile plants (Figure 5-27).

Ti Plasmids Can Introduce Genes into Plants   In nature, plant cells often live in close association with bacteria. One such bacterial species, *Agrobacterium tume-*

eign DNA in equal amounts (up to 100 copies per cell) in *all* tissues, including germ cells. Immediate breeding and backcrossing (parent-offspring mating) of the 10–20 percent of these mice that breed normally can produce *pure transgenic strains* homozygous for the *transgene,* the newly introduced gene.

This technique is of great importance in the study of embryogenesis and organ development, because the activity of injected DNA can be observed in various tissues. Similarly, foreign DNA can be incorporated into *Drosophila* by injecting it into the region of the embryo that will give rise to germ cells. The resulting fly will have some progeny that carry the injected DNA; backcrosses allow construction of strains that are homozygous for the incorporated DNA.

The production of transgenic strains of *Drosophila* and mice provide valuable information because even if the transgenes are not inserted into the correct chromosomal site, they still function to produce RNA normally during development and differentiation. In recent work with cultured embryonic stem cells and other animal cells, some

▲ **Figure 5-27**   Culture dish with beginnings of individual roots and shoots from callus tissue grown from carrot meristem cells obtained from a leaf disk. *Photograph courtesy of Runk/Schoenberger, from Grant Heilman.*

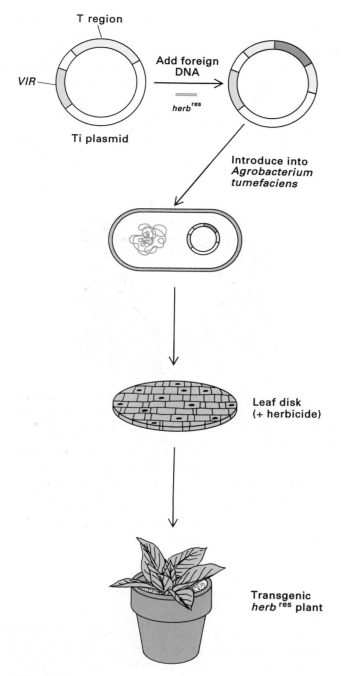

▲ **Figure 5-28** Gene transfer in plants. In nature, the Ti (tumor-inducing) plasmid in *Agrobacterium tumefaciens* gains entry to a plant and joins the plant DNA due to the action of the *VIR* (virulent) region of the plasmid. Abnormal growths (galls) result from action of the T region of the plasmid. By recombinant DNA techniques, a foreign gene is introduced into a Ti plasmid in place of the T gene. An *Agrobacterium* containing the recombinant plasmid is used to introduce the new gene into the plant DNA. When a selectable gene is used—here, one that can make a plant resistant to a herbicide (*herb*res)—recombinant plants can be selected. [See R. T. Fraley et al., 1983, *Proc. Nat'l Acad. Sci. USA* 80:4803.]

faciens, attaches to the cells of dicotyledonous plants and causes the formation of plant tumors known as *galls* (see Figure 19-42). (Plants with two leaflets from each seed are called dicotyledons, or *dicots*; plants with one leaflet are called *monocots*.) The bacterium introduces a circular DNA molecule called the *Ti* (tumor-inducing) *plasmid* into the plant cell in a manner similar to bacterial conjugation (see Figure 5-11). The plasmid DNA then combines with the plant DNA (Figure 5-28). Since the Ti plasmid has been isolated, new genes can be inserted into it using recombinant DNA techniques.

When an *Agrobacterium* containing a recombinant Ti plasmid infects a cultured plant cell, the newly incorporated foreign gene is carried into the plant genome. Thus, by using recombinant DNA technology plus the ability of plants to regenerate from adult cells in culture, it becomes possible to introduce new genes into plants. Most experiments were first done with dicots (petunia, tobacco, carrot) but now recombinant genes that function correctly in monocots are also available. In addition, direct introduction of DNA by the technique of electroporation has been successful in rice plants, and the future looks bright for the manipulation of other commercially important monocot crop plants. Also available for gene transfer experiments are cells of the tiny, rapidly growing member of the cabbage family called *Arabidopsis thaliana*. This organism is very convenient because it takes up little space, is easy to grow, and has a small genome. Many plant geneticists have begun to work with it, so many genes have been mapped on every chromosome of this organism.

Viruses: Structure and Function

A virus is a small cellular parasite that cannot reproduce by itself. Once it infects a susceptible cell, however, the virus can direct the cell machinery to produce more viral material. Each virus has either RNA or DNA as its genetic material; no known virus has both. The nucleic acid may be single- or double-stranded. The entire infectious virus particle is called a *virion* and consists of the nucleic acid and an outer shell of protein. Some viruses contain only enough RNA or DNA to encode 3–10 proteins; others can encode 100–200 proteins.

The coat, or *capsid*, that encloses the nucleic acid is composed of one or more proteins specific to each kind of virus. The capsid plus the enclosed nucleic acid is called the *nucleocapsid*. The protein subunits of capsid are arranged in one of two ways. The simpler structure is a protein helix with the RNA or DNA protected within. Tobacco mosaic virus (TMV) is a classic example of the helical nucleocapsid. In TMV the protein subunits form broken disklike structures, like lock-washers, which form

(a) Section of a helical virus

(b) A small icosahedral virus

(c) A large icosahedral virus

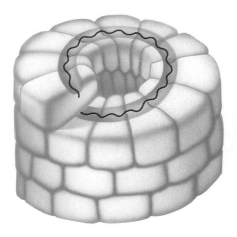

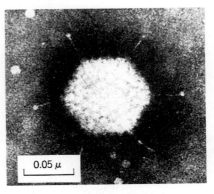

0.05 μ

▲ **Figure 5-29** The two basic geometric shapes of virions. (a) Protein subunits can take the form of helical arrays around an RNA or DNA molecule, with the nucleic acid strand (red) running in a helical groove within the enclosing protein tube. Both the drawing and the electron micrograph illustrate the helical array of the protein subunits of tobacco mosaic virus (TMV) around its RNA. (b and c) The quasi-spherical viruses are actually polyhedra assembled with icosahedral (20-sided) symmetry. The simplest and smallest have only five-fold symmetry at each vertex (b). As capsids get larger to accommodate larger genomes, the bonding between proteins not at the vertices is quasi-equivalent; the subunits on the vertices maintain five-fold symmetry, but those making up the surfaces in between exhibit six-fold symmetry (c). (The actual shape of the protein subunit does not conform to a flat triangle, but the overall effect when the subunits are assembled is of a roughly spherical structure with triangular faces.) The electron micrograph is of an adenovirus, a *naked* icosahedral virus (one having no envelope) with surface projections at its vertices. *After S. E. Luria, J. E. Darnell, Jr., D. Baltimore, and A. Campbell, 1978, General Virology, 3d ed., Wiley, pp. 39–40. Photograph (a) courtesy of R. C. Valentine; photograph (b) courtesy of J. Finch.*

the helical shell of a long rodlike virus when stacked together (Figure 5-29a). The other major structural class of viruses is quasispherical—actually, a polyhedron with icosahedral symmetry. In the simplest, each of 20 triangular faces accommodates three identical capsid subunits, making a total of 60; at each of the 12 vertices, five subunits make contact symmetrically (Figure 5-29b). Thus all protein subunits are in *equivalent* contact. Tobacco satellite necrosis virus actually has such a simple structure. However, most quasi-spherical viruses are larger and contain more proteins that form shells whose subunits are in *quasi-equivalent* contact. Here, the proteins at the icosahedral vertices remain arranged in a five-fold symmetry, but additional subunits cover the surfaces between in a pattern of six-fold symmetry (Figure 5-29c).

In recent years, the atomic structures of a number of icosahedral viruses have been determined by x-ray crystallography (Figure 5-30). The first three of these structures—tomato bushy stunt virus, poliovirus, and rhinovirus (the common cold virus)—revealed a remarkably similar common design, in terms of the rules of icosahedral symmetry as well as in the details of its surface proteins. In each virus, at the atomic resolution, clefts or "canyons" can be observed near the points at which the viral surface proteins meet. These clefts are believed to accommodate receptors when the virus attaches to a host cell. In some viruses, the symmetrically arranged nucleocapsid is covered by an external *envelope* (Figure 5-31) that consists mainly of lipids but also contains some virus-specific proteins. The lipids in the viral envelope are

(a)

Protein subunits

VP1

Picorna viruses

VP3

VP2

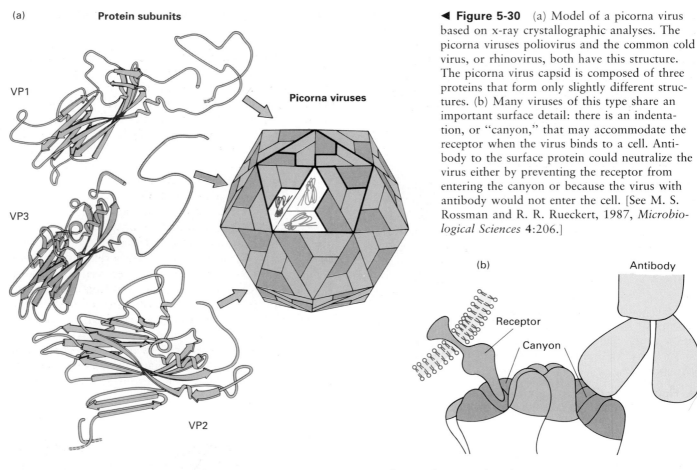

◀ **Figure 5-30** (a) Model of a picorna virus based on x-ray crystallographic analyses. The picorna viruses poliovirus and the common cold virus, or rhinovirus, both have this structure. The picorna virus capsid is composed of three proteins that form only slightly different structures. (b) Many viruses of this type share an important surface detail: there is an indentation, or "canyon," that may accommodate the receptor when the virus binds to a cell. Antibody to the surface protein could neutralize the virus either by preventing the receptor from entering the canyon or because the virus with antibody would not enter the cell. [See M. S. Rossman and R. R. Rueckert, 1987, *Microbiological Sciences* 4:206.]

(b)

Antibody

Receptor

Canyon

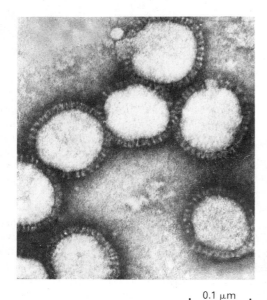

0.1 μm

▲ **Figure 5-31** The influenza virions shown here have an external envelope consisting of lipids and proteins that surrounds the nucleocapsid. *Courtesy of S. E. Luria, J. E. Darnell, Jr., D. Baltimore, and A. Campbell, 1978, General Virology, 3d ed., Wiley, p. 280.*

similar to those in the plasma membrane of an infected host cell; the viral envelope is, in fact, derived from that membrane.

The simple components of some viruses, such as the single RNA molecule and the single type of protein of TMV, undergo *self-assembly* if they are simply brought together in solution. Other viruses contain dozens of proteins and do not spontaneously assemble in the test tube. Within the cell, however, the multiple components assemble in stages, first into subviral particles and then into completed virions. Some subviral particles assemble outside cells.

Most Viral Host Ranges Are Narrow

Because the host range—the group of cell types that a virus will infect—is generally restricted, it serves as a basis for classifying the virus. A virus that infects only bacteria is called a bacteriophage, or simply a *phage*. Viruses that infect animal or plant cells are referred to generally as *animal viruses* or *plant viruses*. A few viruses can grow in both insects and plants: for example, potato yellow dwarf virus can grow in leafhoppers (insects) as well as in plants. Some other animal viruses have wide host ranges also; *vesicular stomatitis* virus grows in insects and in many different mammalian cells. Nevertheless, many animal viruses do not cross phyla, and some (for example, poliomyelitis virus) grow only in closely related species such as primates. Some viruses, such as the *h*uman *i*mmunodeficiency *v*irus (HIV)—the cause of *a*cquired *i*mmuno*d*eficiency *s*yndrome (AIDS), grow only in certain lymphocytes, monocytes, and glial cells of the central nervous system, which all have appropriate surface receptor proteins to which viral surface proteins attach.

Viruses Can Be Accurately Counted

Virology is a highly quantitative science. Very accurate methods for counting infectious particles allow the experimenter to be certain that every cell in a sample is infected or, alternatively, that only a few cells are. Each preparation grown from a single viral particle represents a viral clone in which the RNA or DNA constitutes a pure sample of nucleic acid. The number of infectious viral particles in a dilute sample can be determined accurately by culturing it on a plate of cells and then counting the local lesions, or *plaques*, on the plate. This *plaque assay technique* (Figure 5-32) is in standard use for bacterial and animal viruses. Plant viruses can be assayed similarly by counting local lesions on plant leaves innoculated with viruses.

Viral Growth Cycles Can Be Divided into Stages

A protein on the surface of a virus binds or *adsorbs* specifically to a *receptor* protein on the cell to begin an infection. This interaction determines the host range of a virus. Then, in one of various ways, the viral DNA or RNA crosses the plasma membrane into the cytoplasm. The entering genetic material may still be accompanied by inner viral proteins, but the capsid is typically left behind. The DNA of most animal viruses (with associated proteins, if any) ends up in the cell nucleus, where the cellular DNA is, of course, also found. Once inside the cell, the viral nucleic acid interacts with the host's machinery for synthesizing proteins and nucleic acids (enzymes, ribosomes, tRNA, and so on) to direct its own replication and the synthesis of viral proteins. Most viral protein products fall into three categories: special enzymes needed for viral replication, inhibitory factors that stop cell metabolism, and proteins used in the construction of new virions. The last class generally is made in much larger amounts than the other two.

After a number of new virions have been completed, most bacterial cells and some plant and animal cells rupture, or lyse, releasing all the virions at once. In many

(a)

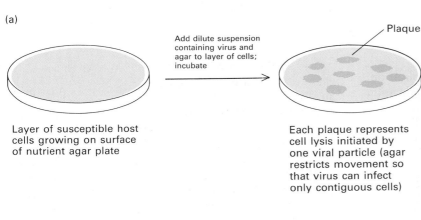

Layer of susceptible host cells growing on surface of nutrient agar plate

Add dilute suspension containing virus and agar to layer of cells; incubate

Plaque

Each plaque represents cell lysis initiated by one viral particle (agar restricts movement so that virus can infect only contiguous cells)

(b)

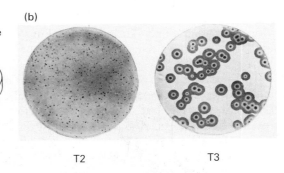

T2   T3

◀ **Figure 5-32**   (a) The plaque assay technique is used to estimate the number of infectious particles in a viral suspension. Each lesion, or *plaque*, is created by one infectious particle, which causes new viruses to form and spread to neighboring cells. The viruses in each plaque represent a pure clone of virus. (b) Plates illuminated from behind show plaques formed by T2 and T3 bacteriophages plated on *E. coli*. (c) Plaques produced by two animal viruses: (*left*) western equine encephalomyelitis virus, plated on chicken embryo fibroblasts; (*right*) poliomyelitis virus plated on HeLa cells. *Photographs in part (b) from M. Demerec and R. Fano, 1945, Genetics 30:119; part (c left) courtesy of R. Dulbecco; part (c right) from S. E. Luria, J. E. Darnell Jr., D. Baltimore, and A. Campbell, 1978, General Virology, 3d ed., Wiley, p. 26.*

(c)

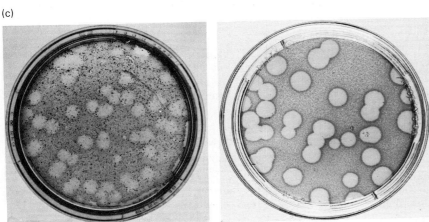

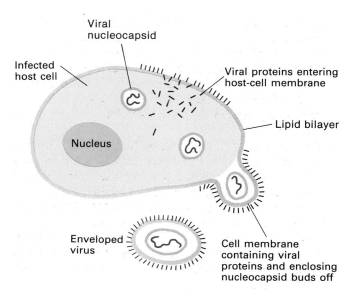

▲ **Figure 5-33**   The maturation of an enveloped virus. Inside the host cell, virus-encoded proteins are inserted into the host-cell membrane and project from its surface. The viral nucleocapsid, containing other viral proteins and nucleic acid, exits from the cell by budding through the membrane, acquiring a lipid envelope containing the inserted viral proteins.

plant and animal viral infections, however, no discrete event of lysis occurs; rather, the dead host cell releases the virions as it gradually disintegrates. Enveloped viruses bud from the host cell, thus acquiring their outermost layer (Figure 5-33).

These events—adsorption, penetration, replication, and release—describe the *lytic cycle* of viral replication (Figure 5-34). The outcome is the production of a new round of viral particles and the death of the cell. At the other extreme, a viral DNA can enter the cell and become integrated into the host chromosome, where is remains quiescent and replicates as part of the cell. Both these bacteriophages, called *temperate phages,* and a number of animal viruses can be integrated with the host-cell chromosomes and be carried along from generation to generation. Probably the most important in eukaryotes are the *retroviruses,* whose genomic RNA is copied into DNA that then integrates with the cell's DNA. (The viruses are described later in this chapter and extensively in Chapter 25.) A few phages and animal viruses infect a cell and cause new virion production without killing the cell or becoming integrated.

Bacterial Viruses Are Widely Used to Investigate Biochemical and Genetic Events

Bacterial viruses played a crucial role in the development of molecular biology. Thousands of different bacteriophages have been isolated, and almost every one is uniquely suited to the investigation of a specific biochemical or genetic event. For our present purposes, however, we will mention only four types.

1. *DNA phages of the T series in E. coli.*   These large lytic phages contain a single molecule of DNA (about 2×10^5 base pairs long in T2, T4, and T6 and about 4×10^4 base pairs long in T1, T3, T5, and T7). A T phage enters an *E. coli* cell through a "tail" (see Figure 5-34). The bacteriophage DNA then directs a program of events that produces approximately 100 new bacteriophage particles in about 20 min, at which time the infected cell lyses and releases the new phages. The formation of phage mRNA, its specification of phage proteins, and the multiplication of phage DNA are classic subjects of molecular biological study. Many principles described in later chapters were first established from the study of these viruses.

2. *Temperate phages of E. coli.*   Typical of this class of phages is the *E. coli* bacteriophage λ, which has one of the most well-studied genomes. On entering an *E. coli* cell, the λDNA can take one of two courses of action: it can, like the T phages, direct the production of new phage particles and lyse the cell (the lytic cycle), or it

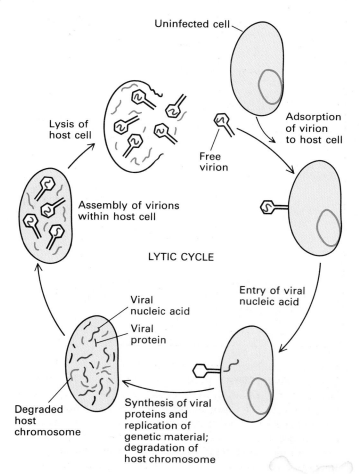

▲ **Figure 5-34**   A generalized viral lytic cycle.

can integrate with the bacterial chromosome and remain there indefinitely as the host cell grows and divides. In the latter case, termed the *lysogenic cycle*, the viral DNA forms a circle and approaches the circular host DNA at a specific site. Enzymes break both circular molecules of DNA and then rejoin the broken ends so that the viral DNA becomes inserted into the host DNA (Figure 5-35). The carefully controlled action of viral genes maintains λDNA as part of the host chromosome by suppressing the lytic functions of the phage. (This and other genetic aspects of λ-phage infection are discussed in Chapters 7 and 12.)

3. *Small DNA phages of E. coli.* These viruses, which encode only 10–12 proteins, are typified by phage φX174. The entire DNA sequences of a number of such phages have been determined, allowing all of the possible protein products to be deduced; in fact, many of these phage proteins have been purified and their precise functions are known. The viruses in this group are so simple that almost every step in the replication of their DNA requires the use of cellular machinery. For this reason, they have been useful in the identification of the cell proteins involved in DNA replication and widely used in the study of DNA synthesis in the test tube.

4. *RNA phages of E. coli.* Some *E. coli* bacteriophages contain RNA instead of DNA. Because they are easy to grow in large amounts and because their RNA genomes also serve as their mRNA, these phages are a ready source of a pure species of mRNA. Some of the earliest demonstrations that cell-free protein synthesis can be programmed by mRNA were made with RNA from these bacteriophages. Also, the first long mRNA molecule to be sequenced was the genome of an RNA bacteriophage.

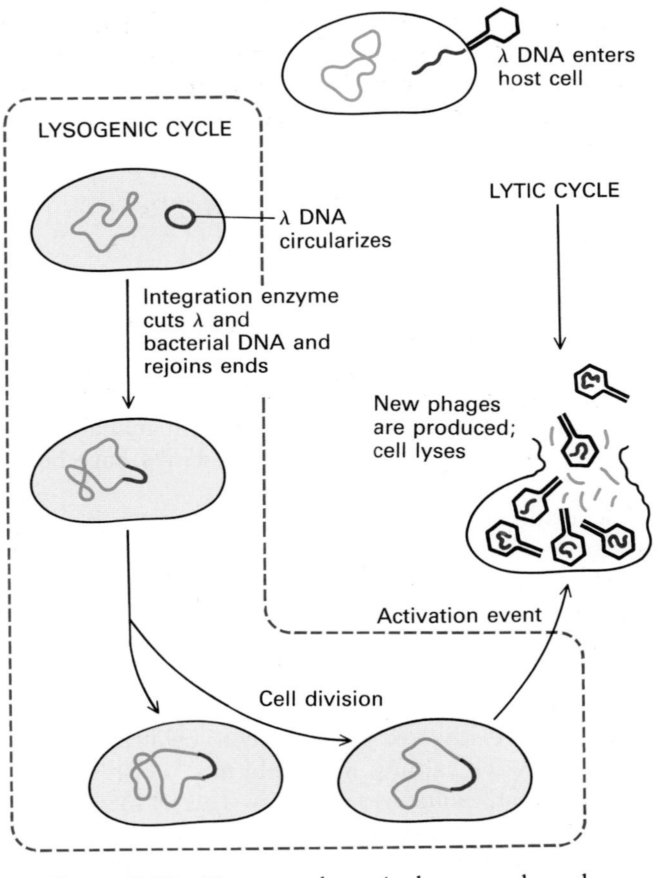

Plant Viruses Proved That RNA Can Act as a Genetic Material

The study of plant viruses inspired some of the first experiments in molecular biology. In 1935, Wendell Stanley purified and partly crystallized TMV; other plant viruses were crystallized soon after. Pure proteins had been crystallized only a short time before Stanley's work, and it was at first thought that the TMV crystals were pure protein. Later studies showed that these crystalline preparations also contained RNA. In fact, plant viruses may contain either RNA or DNA, but for many years only plant viruses containing RNA were known.

Experiments with TMV were very important in establishing nucleic acids as the informational molecules in viruses. For example, the fact that pure RNA from TMV could infect plant cells was demonstrated in 1956 by Alfred Gierer and Gerhard Schramum in Germany and by Heinz Fraenkel-Conrat and Robley Williams in the United States. Further, it was shown that TMV could be separated into protein and nucleic acid parts and then reassembled into infectious virus. If protein and RNA from different strains of TMV are reassembled, the reassembled virus is infectious, but the source of the RNA determines which of the two viral strains are produced in a host cell (Figure 5-36). These classic experiments, plus the study of self-assembly of the TMV protein and its RNA molecule, made plant viruses the most popular subjects of biophysical studies throughout the 1940s and 1950s.

▲ **Figure 5-35** Temperate bacteriophages, such as the λ phage, do not always produce new virions and lyse their host cells immediately after infection. Sometimes they enter a lysogenic cycle, in which the viral DNA is incorporated into the host-cell DNA and maintained there as the host grows and divides. Under certain conditions (for example, irradiation with ultraviolet light) λDNA is activated, separates from the host chromosome, and initiates the lytic cycle of viral replication.

Animal Viruses Are Very Diverse

Animal viruses show a bewildering array of shapes, sizes, and genetic strategies. In this book, we are concerned with viruses that exhibit at least one of two features: they

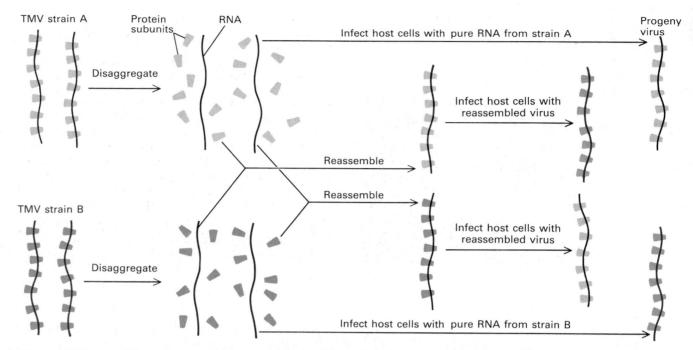

▲ **Figure 5-36**   Proof that information for TMV protein is carried in TMV RNA, not in the protein. This conclusion is based on two experimental results: (1) The RNA alone is infectious; (2) virus reassembled from RNA and protein produces progeny specified by the RNA, not the protein. The infectivity of pure RNA is quite low, whereas the infectivity of the reconstituted whole virus is quite high. [See H. Fraenkel-Conrat, 1958, *The Harvey Lectures, 1957–1958, Ser. 53,* Academic Press, p. 56.]

utilize important cellular pathways, closely mimicking a normal cellular function to form their molecules, or they can integrate their genomes into those of normal cells. The integration of viral DNA sequences into a host cell can result in the induction of cancer.

The Classification of Viruses

Viruses originally bore the names of the diseases they caused or of the animals or plants they infected. However, it was soon discovered that many different kinds of viruses can produce the same symptoms or the same apparent disease states; for example, at least a dozen different viruses can cause red eyes, a runny nose, and sneezing. Clearly, the original way of classifying viruses obscured many important differences in their structures and life cycles.

What *is* central to a viral life cycle are the types of nucleic acids formed during replication and the pathway by which mRNA is produced. The relation between the viral mRNA and the nucleic acid of the infectious particle is the basis of a simple means of classifying viruses. In this system, a viral mRNA is designated as a *plus strand* and its complementary sequence, which cannot function as an mRNA, is a *minus strand*. A strand of DNA complementary to a viral mRNA is also a *minus strand*. Production of a plus strand of mRNA requires that a minus strand of RNA or DNA be used as a template. Using this system, six classes of animal viruses are recognized. Bacteriophages and plant viruses also can be classified in this way, but the system has been used most widely in animal virology because representatives of all six classes have been identified.

The relation of the virion nucleic acid to the mRNA of the virus is summarized in Figure 5-37 for each of the six classes (see also Table 5-4). Classes I and II are DNA viruses. *Class I viruses* contain a double strand of DNA; examples to be discussed in later chapters include the *adenoviruses* (Figure 5-38), which cause upper respiratory infections in many animals, and *SV40* (simian virus 40), a monkey virus discovered accidentally to be already present in kidney-cell cultures from wild monkeys that were used to prepare poliovirus vaccines. The DNA of most animal viruses enters the cell nucleus, where the enzymes normally responsible for producing cellular mRNA are diverted to produce viral mRNA. Another group of class I viruses, the *poxviruses*, is typified by *variola* (smallpox) and *vaccinia*, an attenuated (weakened) poxvirus used in vaccinations to induce immunity to smallpox. These very large, brick-shaped viruses ($0.1 \times 0.1 \times 0.2$ μm) carry their own enzymes for making mRNA, and they replicate in the cell cytoplasm.

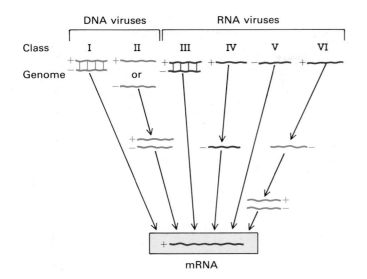

◄ **Figure 5-37** Animal viruses can be classified by the pathway of mRNA formation and by the composition of their genomes: RNA (red) or DNA (blue); single- or double-stranded. If mRNA is designated as the plus strand, the six classes of animal viruses shown here can be identified. In classes I and III, the minus strand of the double-stranded genome is copied to make mRNA. In class V, the single-stranded genome is a minus strand that is copied to make mRNA. In class II, the single initial strand of DNA is copied to make a double-stranded DNA template, the minus strand of which is copied to make mRNA. In class IV, the single plus strand of the virion is copied in the cell into a minus strand, which is then copied to make mRNA. In class VI (the retroviruses), the single plus strand of RNA is copied into a single DNA strand, which in turn is copied into duplex DNA; the mRNA is then copied from the correct strand of the double-stranded DNA molecule.

Table 5-4 Animal viruses commonly used in molecular cell biology

Virus	Known hosts	Type	Nucleic acid class*	Size (kb)†	Lipid membrane?	Research areas in which virus is used
Adenoviruses	Vertebrates	DNA	I	36	no ⎫	mRNA production and regulation, DNA replication, cell transformation
SV40	Primates	DNA	I	5.2	no ⎬	
Herpes viruses	Vertebrates	DNA	I	150	yes ⎭	
Vaccinia	Vertebrates	DNA	I	200	yes	Genome structure, mRNA synthesis by virion enzymes
Parvoviruses	Vertebrates	DNA	II	1–2	no	DNA replication
Retroviruses	Vertebrates and (?) invertebrates	RNA/ DNA	VI	5–8	yes	Cell transformation, oncogenes, acquired immunodeficiency syndrome (AIDS)
Reoviruses	Vertebrates	RNA	III	1.2–4.0‡	no	mRNA synthesis by virion enzymes, mRNA translation
Influenza	Mammals	RNA	V	1.0–3.3‡	yes ⎫	Membrane formation, glycoprotein biosynthesis, and intracellular transport
Vesicular stomatitis virus	Vertebrates	RNA	V	12	yes ⎬	
Sindbis virus	Insects and vertebrates	RNA	IV	10	yes ⎭	
Poliomyelitis virus	Primates	RNA	IV	7	no	Viral RNA replication, interruption of host translation, polyprotein cleavage

*Class refers to strategy for mRNA synthesis (see Figure 5-37).
†Size is given in kilobases (1 kb = 1000 nucleotides) for single-stranded nucleic acids or kilobase pairs for double-stranded molecules.
‡Reoviruses have 10 double-stranded RNA segments, and influenza has 8 single-stranded RNA segments; the length of each segment is in the range indicated.

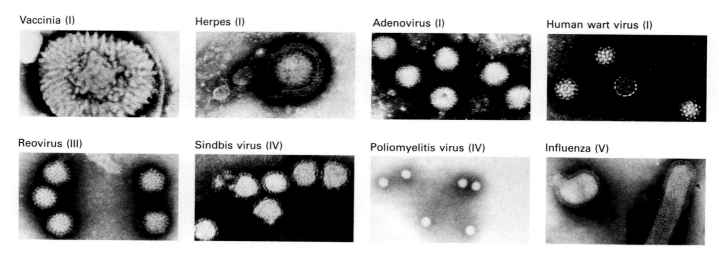

▲ **Figure 5-38** Electron micrographs of representative animal viruses, all taken at the same magnification (×50,000) and shadowed with phosphotungstic acid. The class of each virus is in parentheses. *Courtesy of P. Choppin; from P. Choppin, 1965,* Viral and Ricketsial Infections of Man, *Lippincott (frontispiece).*

Class II viruses, called *parvoviruses* (*parvo-* is Latin for "poor"), are simple viruses that contain a single strand of DNA. Some parvoviruses encapsidate (enclose) both plus and minus strands of DNA, but in separate virions. Others encapsidate only the minus strand, which is copied inside the cell into double-stranded DNA, which is then itself copied into mRNA.

Classes III–VI of animal viruses contain RNA genomes. A wide range of animals, from insects to human beings, are infected by viruses in each of these classes. *Class III viruses* contain a double-stranded RNA. The minus RNA strand acts as a template for the synthesis of plus strands of mRNA. The virions of all class III viruses known to date have segmented genomes containing 10–12 double-stranded RNA segments, each of which encodes a single polypeptide. In these viruses, the virion itself contains a complete set of enzymes that can produce mRNA in the test tube as well as in the cell cytoplasm after infection. A number of important studies have used class III viruses as a source of pure mRNA.

Class IV viruses contain a single plus strand of RNA. Because the genome RNA is identical to the mRNA, the virion RNA by itself is infectious. The mRNA is copied into a minus strand, which then produces more plus strands. Two types of class IV viruses are known. In *class IVa viruses,* typified by *poliomyelitis virus,* the RNA molecule in the virion is identical with the mRNA that encodes all viral proteins. The individual proteins are first synthesized as a single, long polypeptide strand, or *polyprotein,* which is then cleaved to yield the different functional proteins. *Class IVb viruses,* also called *togaviruses* (*toga-* is Latin for "cover") because the virions are surrounded by a lipid envelope, synthesize at least two forms of mRNA in the host cell. One of these mRNAs is the same length as the virion RNA; the other corresponds to the third of the virion RNA at the 3′ end. Both mRNAs produce polyproteins. Included in the class IVb group are a large number of rare insect-borne viruses, once called arboviruses (*arthropod-borne* viruses), that cause encephalitis in human beings.

Class V viruses contain single negative strands of RNA; that is, the virion RNA is complementary in base sequence to the mRNA. Thus the virion contains the template for making mRNA but does not itself encode proteins. Two subdivisions of class V can be distinguished. The genome of *class Va viruses* is a single molecule of RNA; a virus-specific polymerase contained in the virion synthesizes several mRNAs from different regions of this single template strand. Each class Va viral mRNA encodes one protein. *Class Vb viruses,* typified by *influenza virus,* have segmented genomes, each of which is a template for the synthesis of a different single mRNA. As with class Va viruses, the virion contains the virus-specific RNA polymerase necessary to make the mRNA; thus the minus strands of class V nucleic acids alone (in the absence of the virus-specific polymerases) are not infectious. The influenza virus RNA polymerase initiates the transcription of each mRNA by a unique mechanism. The polymerase begins its mRNA synthesis by stealing 12–15 nucleotides from the 5′ end of a cellular mRNA or mRNA precursor in the nucleus. This oligonucleotide acts as a primer for the viral RNA polymerase. The individual mRNAs made by class Vb viruses encode single proteins in most cases; however, some mRNAs can be read in two different frames to produce two distinct proteins.

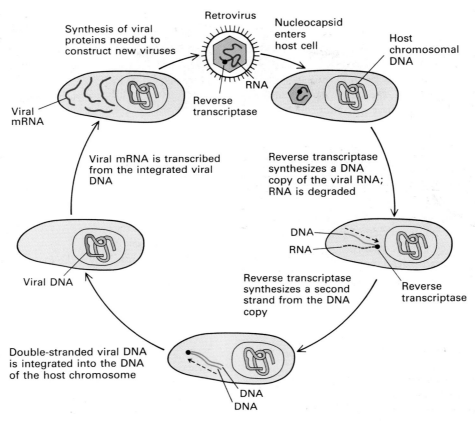

▲ **Figure 5-39** The life cycle of a retrovirus. Viral RNA is shown in red; viral DNA, in blue.

Class VI viruses are also known as *retroviruses* because the RNA of their genome (a single plus strand) directs the formation of a DNA molecule, which ultimately acts as the template for making the mRNA (Figure 5-39). First, the virion RNA is copied into a single strand of DNA, which then forms a complementary strand. This double-stranded DNA is integrated into the chromosomal DNA of the infected cell. Finally, the integrated DNA is transcribed by the cell's own machinery into RNA that either acts as a viral mRNA or becomes enclosed in a virion, thereby completing the retrovirus cycle. If the retrovirus contains cancer genes, the cell that it infects is transformed into a tumor cell. Several human retroviruses are known, including one that causes a form of leukemia and another, HIV, that attacks particular lymphocytes (T4 cells), resulting in AIDS.

The Use of Viruses in Molecular Cell Biology

Three aspects of virological studies are of particular interest in this book (see Table 5-4):

1. Cellular enzymes synthesize many of the viral molecules, particularly viral mRNAs and virus-specific proteins such as envelope proteins. Thus viruses can be used to explore the action of the cellular machinery. They have been particularly useful in studying mRNA production and the enzymes of DNA replication in the nuclei of animal cells and in the pathways of glycoprotein formation in the cell cytoplasm.

2. Viruses offer a precise means of interrupting host-cell functions; as a result, they can shed light on normal cell processes. For example, the roles of certain factors in protein synthesis have been revealed because viruses interrupt their action.

3. The fact that viral DNA incorporated into host-cell chromosomal DNA is inherited in successive generations serves as an excellent model for the recombination of foreign DNA with cellular DNA. Such viral genes can also be considered representative cellular genes, and studies of their expression can help to explain aspects of the expression of normal cell genes (such as where transcription begins and ends and what sequences are required for gene expression). Finally, when certain genes carried by cancer viruses unite with a normal cell, the normal cell becomes a cancer cell. This transformation of the growth potential of cells is one of the most intensively studied processes in molecular cell biology.

Summary

An important factor in the development of molecular cell biology has been the use of a limited number of experimental systems. Different features of organisms—their rapid growth from a single cell, the ease with which their genes can be manipulated, their production of large amounts of a specific substance, and their capacity to carry out developmental changes—govern the choice of which cell or organism is most appropriate in a given study. By far the most popular bacterium is *Escherichia coli;* because genes can be readily transferred in *E. coli* by conjugation, transduction, or transformation, this organism is the best-understood cell in the world of biology. *Saccharomyces cerevisiae,* a species of yeast, is the most frequently used single-celled eukaryote; yeast genetics also has reached a very advanced state through the use of classical and molecular techniques. Studies with bacteria and yeast established the basic principles of microbiology, including the use of various techniques designed to select one genetically different cell from among many similar cells. These principles have been adapted to cultured cells from both plants and animals.

Many different cultured animal cells are widely used. A primary cell culture is derived from animal cells that will not continue to grow indefinitely; a culture of transformed cells will grow indefinitely. Fusion of primary with transferred cells can rescue the primary cell, a technique used to select hybrid cells that form monoclonal antibodies. In addition to cultured cells, cells from specialized tissues such as bone marrow (blood cells), liver, muscle, and nerves are often isolated from animals and used directly. These tissues are often chosen because of their high content of specific cell types. Plant cell cultures are increasing in popularity, especially now that whole plants can be reconstituted from cultured cells.

Among the most important and widely used cell culture techniques are those that facilitate gene transfer between cells. This is particularly true now that gene purification has been achieved. The most common form of gene transfer is the direct uptake of DNA by the recipient cell. Genes can also be transferred by injection into some recipient cells, particularly large cells such as frog eggs, but also in certain cultured cells. After genes are transferred to growing cells, the individual cells that incorporate the genes can be selected. In addition to introducing DNA into cultured cells, it is now possible to introduce DNA into germ-line cells. Every cell in an organism produced from such germ cells can have the new DNA.

Viruses play a major role in modern cell biology both because they provide an opportunity to study small, reproducible sequences of genes in a single organism and because viral products are easily purified. An understanding of the basic elements of virology is indispensable to today's biologist. Two kinds of bacterial and animal viruses are in wide use: viruses that grow lytically (that kill cells) and viruses that are temperate (that integrate with host-cell genomes).

References

The Growth of Microorganisms and Cells in Culture

Bacterial Cells
The following titles cover the logic of selective techniques and genetic transfers, as well as the simple and elegant laboratory procedures used.

BIRGE, E. A. *Bacterial and Bacteriophage Genetics.* Springer-Verlag.

DAVIS, R. W., D. BOTSTEIN, and J. R. ROTH. 1980. *Advanced Bacterial Genetics.* Cold Spring Harbor Laboratory.

INGRAHAM, J. L., O. MAALØE, and F. C. NEIDHARDT. 1983. *Growth of the Bacterial Cell.* Sinauer Associates.

SAMBROOK, J., T. MANIATIS, and E. F. FRITSCH, eds. 1989. *Molecular Cloning* 2d ed. Cold Spring Harbor Laboratory.

Yeast Cells
These articles summarize key techniques and ideas in yeast genetics.

HARTWELL, L. H. 1978. Cell division from a genetic perspective. *J. Cell Biol.* 77:627–637.

HERSKOWITZ, I. 1987. Functional inactivation of genes by dominant negative mutations. *Nature* 329:219–222.

SHERMAN, F., G. R. FINK, and J. B. HICKS. 1987. *Methods in Yeast Genetics.* Cold Spring Harbor Laboratory.

STRATHERN, J., E. JONES, and J. BROACH, eds. 1981. *The Molecular Biology of the Yeast Saccharomyces: Life Cycle and Inheritance.* Cold Spring Harbor Laboratory.

STRUHL, K. 1983. The new yeast genetics. *Nature* 305:391–397.

Animal Cells: Growth Requirements and Capacities
The following articles explain how to grow animal cells in mass culture and as single cells and describe some specialized functions that can be studied in culture.

BARNES, D. W., D. A. SIRBASKY, and G. H. SATO, eds. 1984. *Cell Culture Methods for Molecular and Cell Biology.* Alan R. Liss.

EAGLE, H. 1955. Nutrition needs of mammalian cells in tissue culture. *Science* 122:501–504.

EAGLE, H. 1959. Amino acid metabolism in mammalian cell cultures. *Science* 130:432–437.

EVANS, M. J., and M. H. KAUFMAN. 1981. Establishment in culture of pluripotential cells from mouse embryos. *Nature* 292:154–156.

FRESHNEY, R. I. 1987. *Culture of Animal Cells: A Manual of Basic Technique.* Alan R. Liss.

FUCHS, F., and H. GREEN. 1981. Regulation of terminal differentiation of cultured human keratinocytes by vitamin A. *Cell* 25:617–625.

HAYFLICK, L., and P. S. MOORHEAD. 1961. The serial cultivation of human diploid cell strains. *Exp. Cell Res.* 25:585–621.

POLLACK, R., ed. 1981. *Readings in Mammalian Cell Culture.* 2d ed. Cold Spring Harbor Laboratory.

PUCK. T. T., and P. I. MARCUS. 1955. A rapid method for viable cell titration and clone production with HeLa cells in tissue culture: the use of x-irradiated cells to supply conditioning factors. *Proc. Nat'l Acad. Sci. USA* 41:432–437.

Cell Cycle

BEACH, D., ed. 1988. *Cell Cycle Control in Eukaryotes.* Cold Spring Harbor Laboratory.

CROSS, F., J. ROBERTS, and H. WEINTRAUB. 1989. Simple and complex cell cycles. *Ann. Rev. Cell Biol.* 5:341–395.

Somatic-Cell Genetics and Gene Mapping

These articles describe techniques for fusing cultured cells and for using mutants to select specific cells. Included are explanations of the making of hybridomas for monoclonal antibody production.

DAVIDSON, R. L., and P. S. GERALD. 1976. Improved techniques for the induction of mammalian cell hybridization of polyethylene glycol. *Som. Cell Genet.* 2:165–176.

D'EUSTACHIO, P., and F. H. RUDDLE. 1983. Somatic-cell genetics and gene families. *Science* 220:919–924.

HARLOW, E., ed. 1988. *Antibodies.* Cold Spring Habor Laboratory.

JEFFREYS, A. J., V. WILSON, and S. L. THEIN. 1985. Individual specific fingerprints of human DNA. *Nature* 316:76–78.

KOHLER, G., and C. MILSTEIN. 1975. Continuous cultures of fused cells secreting antibody of predefined specificity. *Nature* 256:495–497.

MILSTEIN, C. 1980. Monoclonal antibodies. *Sci. Am.* 243(4):66–74.

RINGERTZ, N. R., and R. E. SAVAGE. 1976. *Cell Hybrids.* Academic Press.

RUDDLE, F. H. 1982. A new era in mammalian gene mapping: somatic-cell genetics and recombinant DNA methodologies. *Nature* 294:115–119.

SORIEUL, S., and B. EPHRUSSI. 1966. Karyological demonstration of hybridization mammalian cells *in vitro. Nature* 190:653–654.

YELTON, D. E., and M. D. SCHARFF. 1981. Monoclonal antibodies: a powerful new tool in biology and medicine. *Ann. Rev. Biochem.* 50:657–680.

Mutants in Salvage Pathways

ADAMS, R. L. P., R. H. BURDON, A. M. CAMPBELL, D. P. LEADER, and R. M. S. SMELLIE. 1981. *The Biochemistry of the Nucleic Acids.* London: Chapman & Hall.

CASKEY, C. T., and G. D. KRUH. 1979. The HPRT locus. *Cell* 16:1–9.

STANBURY, J. B., J. B. WYNGAARDEN, and D. S. FREDERICKSON, eds. 1983. *The Metabolic Basis of Inherited Disease,* 5th ed., part 6: *The Disorders of Purine and Pyrimidine Metabolism.* McGraw-Hill.

DNA Transfer into Eukaryotic Cells

The following titles describe experiments in which DNA has been introduced into cultured cells and into germ-line cells in vivo.

Animal Cells

BRINSTER, R. L., H. Y. CHEN, M. TRUMBAUER, A. W. SENEAR, R. WARREN, and R. D. PALMITER. 1982. Somatic expression of herpes thymidine kinase in mice following injection of a fusion gene into eggs. *Cell* 27:223–231.

GRAHAM, F., and A. J. VAN DER EB. 1973. A new technique for the assay of infectivity of human adenovirus 5 DNA. *Virology* 52:456–467.

KLEIN, T. M., E. D. WOLF, R. WU, and J. C. SANFORD. 1987. High-velocity microprojectiles for delivering nucleic acids into living cells. *Nature* 327:70–73.

MANSOUR, S. L., K. R. THOMAS, and M. R. CAPECCHI. 1988. Disruption of the proto oncogene int-2 in mouse embryo-derived stem cells: a general strategy for targeting mutations to nonselectable genes. *Nature* 336:348–352.

PALMITER, R. D., and R. L. BRINSTER. 1986. Germ-line transformation of mice. *Ann. Rev. Genetics* 20:465–500.

PELLICER, A., D. ROBINS, B. WOLD, R. SWEET, J. JACKSON, I. LOWY, J. M. ROBERTS, G. K. SIM, S. SILVERSTEIN, and R. AXEL. 1980. Altering genotype and phenotype by DNA-mediated gene transfer. *Science* 209:1414–1422.

ROBERTSON, E., A. BRADLEY, M. KUEHN, and M. EVANS. 1986. Germ-line transmission of genes introduced into cultured pluripotential cells by retroviral vectors. *Nature* 323:445–447.

WIGLER, M., S. SILVERSTEIN, L.-S. LEE, A. PELLICER, Y. CHENG, and R. AXEL. 1977. Transfer of purified herpes virus thymidine kinase gene to cultured mouse cells. *Cell* 11:223–232.

Plant Cells

FRALEY, R. T., S. C. ROGERS, R. B. HORSCH, P. R. SANDERS, J. S. FLICK, S. P. ADAMS, M. L. BITTNER, L. A. BRAND, C. L. FINK, J. S. FRY, G. R. GALLUPPI, S. B. GOLDBERG, N. L. HOFFMAN, and S. C. WOO. 1983. Expression of bacterial genes in plant cells. *Proc. Nat'l Acad. Sci. USA* 80:4803–4807.

FRALEY, R. T., N. M. FREY, and J. S. SCHELL, eds. 1988. *Genetic Improvements of Agriculturally Important Crops: Progress and Issues.* Cold Spring Harbor Laboratory.

HELENTJAVIS, I., and B. BURR, eds. 1989. *Development and Application of Molecular Markers to Problems in Plant Genetics.* Cold Spring Harbor Laboratory.

LURQUIN, P. F. 1987. Foreign gene expression in plant cells. *Progress in Nucleic Acid Research and Molecular Biology,* vol. 34, W. E. Cohn and K. Moldave, eds. Academic Press.

Viruses: Structure and Function

The following books are useful general references that cover many aspects of virology.

FIELDS, B. N., ed. 1985. *Virology*. Raven.

FRAENKEL-CONRAT, H., and R. R. WAGNER, eds. 1974–1989. *Comprehensive Virology*. Plenum.

LAUFFER, M. A., and K. MARAMOROSCH., eds. 1953–1989. *Advances in Virus Research*, vols. 1–30. Academic Press.

LURIA, S. E., J. E. DARNELL, JR., D. BALTIMORE, and A. CAMPBELL. 1978. *General Virology*, 3d ed. Wiley.

ROSSMAN, M. G., and R. R. RUECKERT. 1987. What does the molecular structure of viruses tell us about viral functions? *Microbiol. Sci.* 4:206–214.

WEISS, R., N. M. TEICH, H. F. VARMUS, and J. M. COFFIN, eds. 1984. *RNA Tumor Viruses*. Cold Spring Harbor Laboratory.

C H A P T E R

6

Manipulating Macromolecules

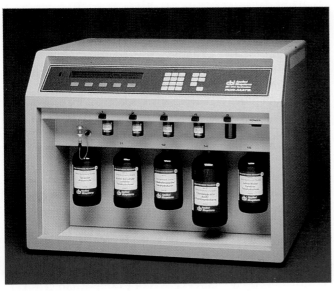

DNA synthesizer

The greatest advances in molecular cell biology in the recent past have been based on the analysis and manipulation of macromolecules, particularly DNA. For years it was clear that many deep biological secrets were locked up in the sequence of bases in DNA, but obtaining the sequences of long regions of DNA—not to mention altering these sequences at will—seemed a distant dream. An avalanche of technical advances in the 1970s drastically changed this perspective. First, enzymes were discovered that cut the DNA from any organism at specific short nucleotide sequences, generating a reproducible set of pieces. The availability of these enzymes, called restriction endonucleases, *greatly facilitated two important developments: DNA cloning and DNA sequencing.*

Two DNA molecules can be joined enzymatically and thus restriction fragments of any DNA can be inserted into a variety of vectors, often plasmid DNA, to produce recombinant DNA. The recombinant molecules can be introduced into an appropriate cell population (most often bacteria) and cells containing recombinant DNA molecules can be selected. This procedure is referred to as cloning *the chosen DNA sequence. Once a clone of cells bearing the recombinant DNA is selected, unlimited quantities of the chosen DNA can be prepared. In addition, DNA oligonucleotides (up to 100 bases long) can now be chemically synthesized entirely automatically. Not only natural oligonucleotides, but any desired mutant sequence can be produced and can be inserted in recombinant DNA.*

Rapid DNA sequencing came into being in the late 1970s. By the use of restriction endonucleases, long DNA molecules from a single organism could be broken into a reproducible array of fragments, whose order in the original molecule could then be determined. It also became possible to determine the sequence of bases in fragments containing as many as 500 nucleotides. There was no longer any obstacle to obtaining the sequence of a DNA molecule containing 10,000 or more nucleotides. Suddenly, any DNA was accessible to isolation and to sequencing. Now, computer-automated procedures for sequencing and for storing, comparing, and analyzing data will make it possible to obtain the entire sequence of the human genome in the near future.

Any desired DNA, whether natural, modified, or completely synthetic, can be reinserted into cells and tested for biological activity. Selected DNA fragments that encode proteins of particular interest have been transferred to bacteria and to other cells, where the transferred DNA directs the production of natural or mutant proteins. In addition, the direct chemical synthesis of peptides as much as 75 amino acids long is also now routine.

Almost overnight, this group of techniques, often collectively called molecular genetics, *became the dominant approach to the study of many basic biological questions, including how gene expression is regulated in eukaryotic cells and how protein or domains of proteins function. The power and success of the new technology have raised high hopes that the practical use of our ever-increasing biological knowledge will bring many benefits to mankind.*

This chapter outlines the techniques just summarized and describes some older procedures still widely used in molecular experiments today. ▲

Radioisotopes: The Indispensable Modern Means of Following Biological Activity

A major goal of classic biochemistry from the 1930s through the 1950s was to chart the metabolic pathways in cells. This work contributed enormously to our present detailed picture of cellular biochemistry. Almost all this fundamental biochemical knowledge was garnered with chemical or enzymatic assays that relied on simple tests using color indicators or measurements of light absorption at characteristic wavelengths.

Since World War II, when radioactive materials first became widely available as byproducts of work in nuclear physics, chemists and biologists have fashioned an almost limitless variety of radioactive "tracer" molecules. Today, the radioactively labeled precursors of macromolecules greatly simplify many standard biochemical assays and significantly enhance our ability to follow biochemical events in whole cells. Almost all experimental biology depends on the use of radioactive compounds.

In a labeled molecule, at least one atom is present as a *radioisotope* (in radioactive form; see Table 6-1). This radioisotope does not change the chemical properties of the molecule. For example, an enzyme—whether it is in a cell extract or a living cell—does not distinguish between a labeled and an unlabeled molecule when performing a metabolic function, such as synthesizing protein, DNA, or RNA. Because radioactive atoms can be detected when they emit a particle, they can be used to trace the activities of the labeled molecules.

Not all labeled materials can be used interchangeably in whole cells and cell-free systems. For one reason, many compounds that participate in intermediary metabolism cannot be used to study the metabolism of whole cells because they do not enter cells. For example, labeled ATP may contribute phosphorus 32 (^{32}P) in RNA synthesis in a cell-free system, but ATP does not enter cells. On the other hand, labeled orthophosphate ($^{32}PO_4^{3-}$) added to the medium does enter cells and is incorporated into nucleotides and then into cellular RNA; but $^{32}PO_4^{3-}$ is not efficiently incorporated into nucleotides in cell extracts and therefore is not used to label RNA in extracts.

Amino acids and nucleotides labeled with either carbon 14 (^{14}C) or tritium (^3H) are commercially available, as are hundreds of labeled metabolic intermediates. Methionine labeled with sulfur 35 (^{35}S) is widely used as a protein label because it is available in high specific activities. The *specific activity* is the amount of radioactivity per unit of material; for example, commercial [^{35}S]-methionine can have over 10^{15} disintegrations per minute per millimole of methionine. The magnitude of the specific activity depends on the ratio of unstable (potentially radioactive) atoms to stable (nonradioactive) atoms and on the probability of decay of the unstable atoms, indicated by their half-life.

Because phosphate in which every phosphorus atom is ^{32}P is readily obtainable, various cell-free methods of labeling nucleic acids employ nucleotides labeled with ^{32}P. Likewise, a radioactive isotope of iodine (^{125}I) is available in almost pure form; this tracer atom can be enzymatically or chemically attached to a protein or nucleic acid without drastically affecting the macromolecule.

Table 6-1   Commonly used radioisotopes

Radioisotope	Half-life	Energy of emitted particle (MeV)*	Mean path length in water (μm)	Specific activity (mCi/mA)[†]	Common specific activities for compounds (mCi/mmol)[‡]
Tritium (hydrogen 3)	12.35 yr	0.0186	0.47	2.92×10^4	$10^2 - 10^5$
Carbon 14	5730 yr	0.156	42	62.4	$1 - 10^2$
Sulfur 35	87.5 days	0.167	40	1.50×10^6	$1 - 10^6$
Phosphorus 33	25.5 days	0.248	—	5.32×10^6	$10 - 10^4$
Phosphorus 32	14.3 days	1.709	2710	9.2×10^6	$10 - 10^5$
Iodine 131	8.07 days	0.806	—	1.6×10^7	$10^2 - 10^4$
Iodine 125	60 days	0.035	—	2.2×10^6	$10^2 - 10^4$

*MeV = 10^6 electronvolts. The maximum energy for each emission is given. The particle emitted is a β particle, except in the case of ^{131}I and ^{125}I, which emit γ particles.

[†]The unit mCi (millicuries) is a measure of the number of disintegrations per time unit: 1 mCi = 2.2×10^9 disintegrations per minute. The unit mA (milliatoms) is the atomic weight of the element expressed in milligrams.

[‡]These values are for commercially available compounds that may have many carbon or hydrogen atoms.

SOURCE: New England Nuclear, Boston.

(a)

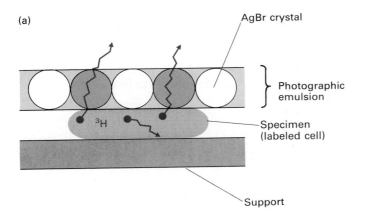

AgBr crystal

Photographic emulsion

^{3}H   Specimen (labeled cell)

Support

(b)

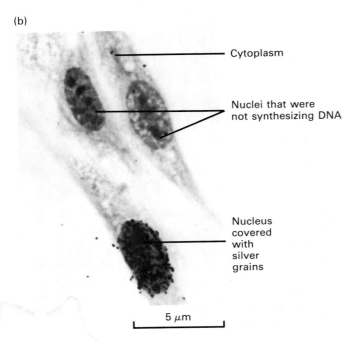

Cytoplasm

Nuclei that were not synthesizing DNA

Nucleus covered with silver grains

5 μm

Radioisotopes Are Detected by Autoradiography or by Quantitative Assays

Two detection schemes for assaying incorporated radio-activity are in general use:

1. In *autoradiography,* a cell or cell constituent is labeled and then overlaid with a photographic emulsion sensitive to radiation. Development of the emulsion reveals the distribution of labeled material. In whole cells, autoradiographic studies determine the original sites of the synthesis of macromolecules and their subsequent movements within cells. For example, incorporation of [^{3}H]thymidine identifies the nucleus as the major site of DNA synthesis and cell fractionation and histologic staining show most DNA is also in the nucleus (Figure 6-1). In contrast, the incorporation of

◀ **Figure 6-1**   The technique of autoradiography. (a) A radiation-sensitive photographic emulsion containing silver salts (AgBr) is placed over labeled cells attached to a glass slide (for the light microscope) or to a carbon-coated grid (for the electron microscope). The cell regions containing the labeled molecules emit radioactive particles, along the tracks of which silver is deposited. When the photographic emulsion is developed, the silver deposited appears as dark grains under the light microscope and as curly filaments in the electron microscope. (b) These fibroblasts from Chinese hamsters were labeled with [^{3}H]thymidine for 1 h. Two of the cells were not synthesizing DNA during this time (the larger dark areas in their nuclei are nucleoli), but one cell was. Small black grains almost entirely cover that cell nucleus, indicating the new DNA is there. *Part (a) redrawn from E. D. P. DeRobertis and E. M. F. DeRobertis, 1979,* Cell and Molecular Biology, *Saunders, p. 62; part (b) courtesy of D. M. Prescott.*

labeled uridine into RNA has shown that most RNA is first made in the nucleus but that most RNA in cell fractions is located in the cytoplasm. Incorporation of labeled amino acids has revealed that most protein is made in the cytoplasm. The transport pathway of proteins from synthesis to secretion was first documented by electron microscopic autoradiography, which allows each silver filament that results from a radioactive disintegration to be observed.

2. In *quantitative assays,* cells are labeled either in vivo or in vitro and their constituents are isolated and purified in various ways. The amount or type of radioactivity in these constituents is then measured—by a Geiger counter, which detects ions produced in a gas by the radioactive emissions, or by a scintillation counter, which counts the flashes of light generated by mixing the radioactive sample with a substance that fluoresces after absorbing the energy of a particle resulting from the decay of the nucleus of the radioactive atom.

A combination of labeling and biochemical techniques is often employed. A cell constituent may be purified before it is labeled and, after labeling, be subjected to experimental procedures. Autoradiography of the labeled products of such experiments—most often after they have been separated by gel electrophoresis (discussed later in this chapter) or by chromatography—is perhaps the most common experiment in all of modern biology.

The purpose of the experiment governs the choice of a radioisotope as well as the detection method. A labeled compound must have a high enough specific activity that the radioactivity in the cell fraction is significant enough to be studied when the compound is incorporated into cells. For example, ^3H-labeled nucleic acid precursors are available in much higher specific activities than ^{14}C-labeled samples are; the former allows RNA or DNA to be adequately labeled after a shorter time of incorporation or in a smaller cell sample.

In autoradiographic studies, the energy in the particles released by radioactive disintegrations affects the experimenter's ability to localize the site at which the radioactivity is incorporated. For example, the β particles emitted by ^{32}P are so energetic (see Table 6-1) that the streaks they make on photographic film can be as long as 1 mm, much longer than the diameters of individual cells. ^3H is highly preferred for locating radioactive substances or structures in cells: the track created on photographic film by the β particle released by ^3H decay is only about 0.47 μm long; thus ^3H-labeled structures can be located within cells to an accuracy of about 0.5–1.0 μm, or about one-fifth the diameter of the nucleus of a mammalian cell.

Pulse-Chase Experiments Must Be Designed with Knowledge of the Cell's Pool of Amino Acids and Nucleotides

In many experiments using radioactive metabolic material, a labeled compound is added to cells and the path of the labeled compound can then be traced as it moves through various compartments or molecules within cells. One type of experiment, the *pulse-chase experiment,* utilizes the brief addition (a *pulse*) of a labeled compound, followed by its removal and replacement (the *chase*) by an excess of unlabeled compound; the cells or cell constituents are examined at various times thereafter to monitor the radioactivity incorporated during the pulse.

Before an amino acid, a nucleoside, or a phosphate ion (for example) is incorporated into a protein or a nucleic acid, it enters the cell's *pool* of molecular building blocks—a collection of small molecules free to diffuse throughout the cytoplasm and nucleus of the cell but not necessarily free to diffuse into or out of membrane-bound organelles,

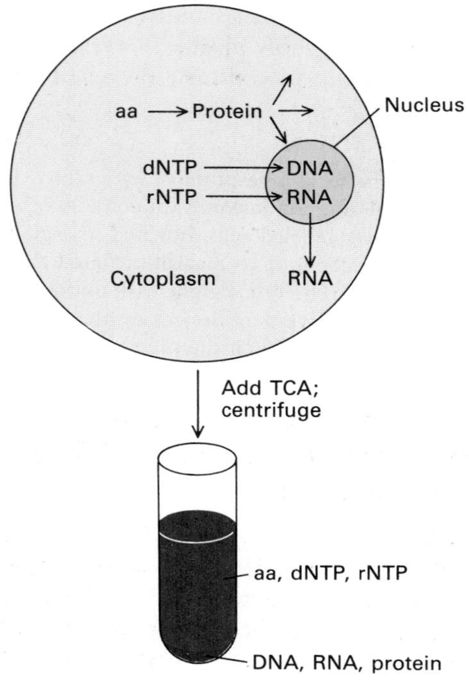

◀ **Figure 6-2** A cell's pool of small soluble molecules—amino acids (aa) and nucleotides (dNTP and rNTP)—may be separated from the macromolecules (DNA, RNA, and proteins) by adding cold acid, usually trichloroacetic acid (TCA), which destroys the cell structure and precipitates all macromolecules. Centrifugation then deposits the macromolecules in a pellet, leaving the amino acids and nucleotides in the supernatant. The rate at which cells take up labeled molecules and incorporate them into macromolecules can be determined by taking such samples at frequent intervals after the addition of labeled amino acids or nucleotide precursors to the cell-culture medium.

(a)

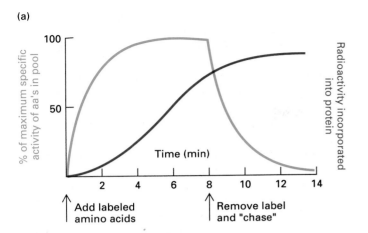

(b)

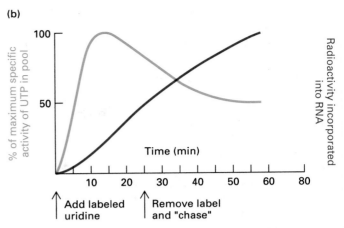

▲ **Figure 6-3** (a) If growing cells are exposed to a medium containing labeled amino acids, it takes about 5 min for the amino acids in the cell pool to reach the maximum specific activity. The accumulation of radioactivity in proteins starts more slowly, because the label must be incorporated into the amino acid pool first. However, if medium containing unlabeled amino acids is used instead, the incorporation of radioactivity into proteins stops within a few minutes due to the rapid equilibration between amino acids inside the cells and in the medium. Thus a marked pulse-chase effect is seen. (b) A pulse of labeled uridine is incorporated into UTP in the cell pool in about 10 min, and the pool can be diluted some-

what by excess unlabeled nucleosides outside the cells (drop in specific activity at 25 min). However, it takes much longer for a chase with unlabeled uridine to level off the amount of radioactivity incorporated into RNA (reduce the ongoing incorporation of radioactivity to 0). This is because the labeled uridine in the pool has been phosphorylated and is unable to escape to the medium, and there are almost 20 percent more uridine nucleotides in the pool than in cellular RNA. All of the labeled uridine is eventually incorporated into the RNA, but only after several hours. Thus no marked pulse-chase effect is seen.

such as the mitochondria or chloroplasts. Depending on the growth conditions of the cell the quantities of components of the pool can vary. Likewise, the rates at which different molecules are absorbed, utilized, and secreted by the cell can also vary (Figure 6-2).

Because of the rapid exchange of amino acids between the pool and the medium, a clear pulse-chase effect can be achieved with them. The acid-soluble pool can be made to contain radioactive amino acids in a few seconds, and they can be removed just as quickly (Figure 6-3a).

Ribonucleosides and deoxyribonucleosides, however, become phosphorylated soon after they enter the cell pool, and phosphorylated compounds do not generally leave the cell. Thus labeled nucleosides can enter the cell, but no equilibrium is established between the nucleic acid precursors in the medium and their phosphorylated counterparts in the cell. Nevertheless, a practically useful pulse-chase effect can be obtained in experiments with radioactive *deoxyribonucleosides*, because the deoxyribonucleotide content of the cell pool is sufficient for only a few minutes of DNA synthesis. Labeled thymidine, for example, can be satisfactorily chased even though it is phosphorylated, because an amount of thymidylate (TTP) equal to that in the pool is taken up every few minutes by replicating DNA.

Labeled *ribonucleosides* behave differently, because it takes several hours for enough RNA synthesis to occur to

consume the content of the ribonucleotide pool in animal cells. In most cultured animal cells, the pool does absorb a pulse of labeled ribonucleosides quickly, say within 10 min. However, a marked chase response (one that occurs within a few minutes) is not possible. Although the addition of unlabeled ribonucleosides to the exterior medium may further expand the ribonucleotide content of the cell pool, dilute the label within it, and decrease the rate of RNA labeling, the amount of incorporated label does not clearly level off until several hours after the chase begins (Figure 6-3b).

In planning and interpreting experiments that use labeled precursors of proteins, DNA, or RNA to study macromolecular synthesis, these characteristics of small molecules in the soluble pool must always be borne in mind.

Labeled Precursors Can Trace the Assembly of Macromolecules and Their Distribution in a Cell

When a radioactive building block first enters a cell, it can only label the macromolecules that are in the process of being constructed. For example, if a radioactive amino acid is added to a culture, the nascent (unfinished and still growing) protein chains are the first proteins to be la-

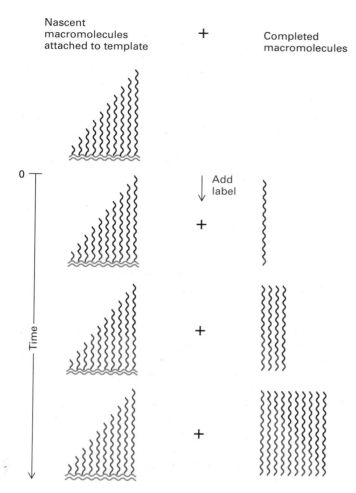

Nascent macromolecules attached to template **+** Completed macromolecules

▲ **Figure 6-4** Labeled radioactive precursors (red) first appear in nascent macromolecules. As time passes, molecules that contain more radioactive label are completed. At the end of an interval equivalent to the synthesis time of a macromolecule, the total amounts of radioactivity in all finished and all unfinished molecules are equal.

The Dintzis Experiment Proved That Proteins Are Synthesized from the Amino End to the Carboxyl End

Other very important facts—the cellular locus of synthesis of a macromolecule and the direction of its growth—can be determined by labeling growing chains. Indeed, the analysis of newly finished chains was used by Howard Dintzis in a classic experiment demonstrating the step-by-step formation of protein chains from the amino terminus to the carboxyl terminus. Over 90 percent of the protein synthesized by *reticulocytes* (the next-to-final stage in the differentiation of red blood cells in the bone marrow of mammals) consists of the α- and β-globin chains that form the protein part of hemoglobin. (Hemoglobin is composed of four globin chains: two α and two β.) Dintzis exposed reticulocytes to radioactive amino acids and then, at short intervals, collected the finished chains. He separated the α and β chains and digested each with trypsin, an enzyme that attacks on the carboxyl side of arginine and lysine residues to produce a specific set of fragments for each chain which can be separated. Dintzis knew the sequence of amino acids in both globin chains as well as the position of each fragment within the globin chains.

Dintzis reasoned that the first completed chains to contain the radioactive label would be those that were almost complete when the label was added. Thus the *first* portion of the finished chains to contain label would be near the end at which chain synthesis finished and, by extension, the *last* portion of the finished globin chains to become labeled would lie at the end where chain synthesis started. The results showed that the radioactive label always appeared in the tryptic fragments in a certain order—in the carboxyl-terminal fragment first and the amino-terminal fragment last, with intermediate fragments becoming consecutively labeled in the order in which they lay between the two termini (Figure 6-5). From this, Dintzis deduced that synthesis begins at the amino terminus of each chain and moves in a step-by-step progression to the carboxyl end of the chain.

Whereas Dintzis studied the labeling of newly finished molecules, other workers have studied nascent molecules. (Experiments on the labeling of nascent RNA and DNA are described in Chapters 8 and 12.) The logic of these studies parallels that of the Dintzis experiment: the shortest labeled molecules in a nascent set will be those whose sequence is near the start site; increasingly longer members will contain additional sequences progressively more remote from the start site.

Determining the Sizes of Nucleic Acids and Proteins

Whereas the *sequence* of the monomers in a protein or nucleic acid ultimately determines the functional capacity of the polymer, the most useful physical characteristic in the

beled. As time passes, an increasing number of completed chains contain the radioactive label. The time required to form a specific macromolecule can be estimated by sampling a labeled cell culture at very short intervals to compare the amount of radioactivity in all nascent macromolecules still attached to the templates with the amount in all free (complete) macromolecules. The first finished chains obtained after the label is added contain only a small amount of the label, because they were almost completed before it was introduced. Each nascent chain initially also contains a small amount of label; as time passes, however, more label accumulates in newly finished chains and the nascent chains become completely labeled. At this point, there is an equal amount of label in the finished and nascent chains (Figure 6-4). The time elapsed since the label was added is equal to the time required for the synthesis of one chain.

Negatively charged nucleic acids
or SDS-protein conplexes

Place mixture on an agarose
or polyacrylamide gel
Apply electric field

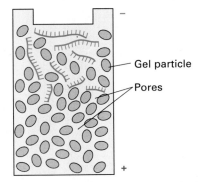

Gel particle

Pores

Molecules move through pores
in gel at a rate inversely
proportional to their chain length

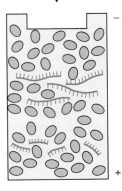

▲ **Figure 6-7** Gel electrophoresis is carried out by pouring a liquid containing either melted agarose or chemically treated polyacrylamide into a cylinder (for a round gel) or between two flat, parallel glass plates 1–2 mm apart. As the gel solidifies, it forms interconnected pores, or channels, whose size depends on the concentration of agarose or polyacrylamide. The substances to be separated are then layered on top of the gel (or at one edge of it if it lies between two plates), and an electric current is passed through the gel. In usual laboratory practice, the migration of RNA or DNA depends on the charges on the phosphates: at neutral pH, a nucleic acid bears one negative charge per phosphate. Proteins can be separated by binding sodium dodecyl sulfate (SDS) to their amino acid residues, which contributes approximately one negative charge per residue. If all the particles have about the same charge-mass ratio, they move through the gel at a rate inversely proportional to their chain length.

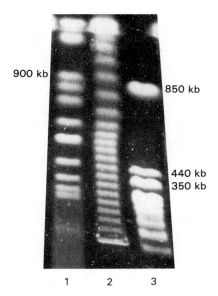

900 kb

850 kb

440 kb
350 kb

1   2   3

▲ **Figure 6-8** Pulse-field gel electrophoretic separation of large DNA molecules. In this technique, DNA molecules are moved first in one direction by application of an electric field. As they move, the molecules stretch out lengthwise in the direction of the field. The current is then stopped for a short time, and the molecules begin to "relax" into random coils; the time required for relaxation depends on the length of a molecule. The electric field can then be reapplied at 90° to the first direction or opposite to the first direction. Longer molecules relax more slowly than shorter ones, and so take longer to start moving in the new direction. Repeated alternation of field direction thus separates the molecules between the two directions and makes it possible to separate giant DNA molecules of 10^6 base pairs and more. The "ladder" in the lane 2 shows concatemers (linked units) of bacteriophage λ DNA in which each unit is 48.5 kb long (the band on the bottom is a single unit). Comparison with this ladder allows calculation of the length of other long DNA fragments. Lane 1 shows individual DNA molecules that each represent one chromosome from *Saccharomyces cerevisiae;* lane 3 shows a restriction digest with enzyme *Not*I that was used to map the *E. coli* chromosome (see Figure 5–9). [See C. L. Smith et al., 1987, *Science* **236**:1448; C. L. Smith et al., 1987, *Nuc. Acids Res.* **15**:4481.] *Photograph courtesy of C. L. Smith.*

more slowly (Figure 6-7). Even very long nucleic acids (chains containing 10,000–20,000 residues) that differ in length by only a few percentage points can be separated. In mixtures containing chains of 500 nucleotides or less, *each chain length can be resolved*, which has made DNA sequencing possible.

By employing the new technique of *pulse-field gel electrophoresis,* different-sized double-stranded DNAs in the range of 1–10 million base pairs (bp), or 1–10 megabases (Mb), can now be separated (Figure 6-8). Electrophoretic migration is begun in one direction; then the current is briefly stopped and reapplied at a 90° angle or in the opposite direction. These long molecules tend to align along the electric field when the current is on and to relax when it is off. Relaxation time is affected by pores in the gel: longer molecules take longer to relax, and so respond more slowly as the current is switched, than shorter ones do, allowing the chains to be separated. This technique is very important for purifying long DNA molecules. It is required for the analysis of cellular chromosomes, which range from the smallest yeast chromosomes (about 5×10^5 bp) to the largest animal and plant chromosomes (2 or 3×10^8 bp).

Protein chains also can be separated according to length. Before and during electrophoresis, the proteins are continuously exposed to the detergent SDS (sodium dodecylsulfate, a common commercial cleaning agent found in toothpaste). Approximately one molecule of detergent binds to each amino acid. At neutral pH, the detergent is negatively charged; the adjacent negatively charged SDS molecules repel one another, forcing the proteins with bound detergent into rodlike shapes endowed with similar charge-mass ratios. Proteins in this state are said to be *denatured*. As with nucleic acids, chain length (which reflects mass) is the determinant for the separation of proteins by electrophoresis through polyacrylamide gels (Figure 6-9). Even chains that differ in molecular weight by less than 1 percent can be separated.

Gel Electrophoresis Can Separate Most Proteins in a Cell

The traditional biochemical approach first to enzyme detection and ultimately to detailed enzyme chemistry is to detect enzymatic activity in a sample from a natural source and isolate the proteins that catalyze the activity. Biochemical methods of separating pure proteins from natural mixtures rely on differences in sedimentation rate or in charge change related to varying salt concentrations or pH. This causes the protein to bind differentially to various substances (e.g., cellulose products) and makes chromatography possible.

However, many experiments in molecular biology are designed to enumerate the polypeptides formed in a certain cell at a certain time, rather than to detect active enzymes or determine their concentrations. Sometimes just the presence of a given protein is to be detected anywhere within the cell, without purifying the protein. Or it may be important to compare the synthesis rate of a protein or a set of proteins with that of all other proteins in the cell, again without isolating any particular protein. Gel electrophoresis can often accomplish these aims.

Two-dimensional Gels Electrophoresis of all cellular proteins in one direction through a column or a thin rectangular SDS gel reveals only the major proteins. If these proteins are of interest or if a cell is producing large

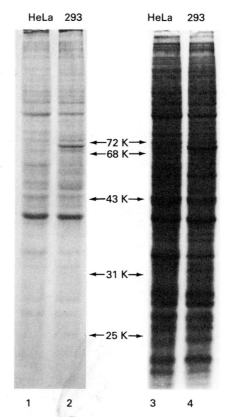

▲ **Figure 6-9** Resolution of proteins by one-dimensional gel electrophoresis. The proteins of two human cell lines—HeLa, a human cervical cancer cell, and 293, a virus-transformed embryonic fibroblast—were dissolved in SDS and subjected to electrophoresis. The newly made proteins are visible in lanes 1 and 2 by autoradiography, because the cells were labeled with [^{35}S]methionine, and in lanes 3 and 4 by the dye Coomassie blue, which stains all proteins. The designations 72 K, 68 K, etc., indicate the positions of marker proteins (proteins of known sizes) with molecular weights of 72,000, 68,000, etc. The major proteins in these two cell types are obviously quite similar. *Photographs courtesy of J. R. Nevins and C. Lawrence.*

amounts of specific proteins (as occurs during viral infection), then this one-dimensional analysis may suffice.

Resolution of virtually all proteins in the cell can be accomplished in a two-dimensional gel, which separates the proteins in a sample first by charge and then by size (Figure 6-10). Separation by charge is carried out by *isoelectric focusing* (IEF). A protein that has not been denatured with SDS has a characteristic overall charge on its surface, which varies with pH. When placed in a gradient of pH and subjected to an electric field, a protein will migrate to the pH at which its overall surface charge is neutral and remain at this *isoelectric point*. Proteins separated in a gel of this type can, while still in the gel, be layered on top of another gel soaked with SDS; thus the proteins can be separated by electrophoresis in a second

(a)

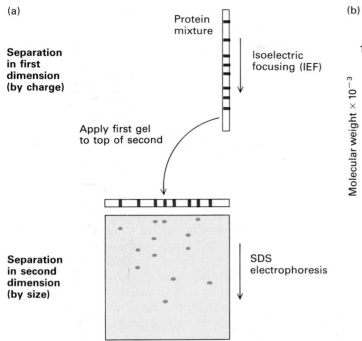

Separation in first dimension (by charge)

Protein mixture

Isoelectric focusing (IEF)

Apply first gel to top of second

Separation in second dimension (by size)

SDS electrophoresis

(b)

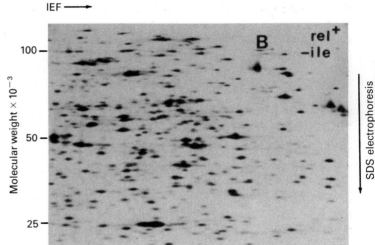

IEF ⟶

Molecular weight × 10⁻³

A   control

SDS electrophoresis

IEF ⟶

Molecular weight × 10⁻³

B   rel⁺ −ile

SDS electrophoresis

▲ **Figure 6-10**   (a) The preparation of two-dimensional protein gels by isoelectric focusing (IEF) followed by electrophoresis. (b) Labeled proteins can be detected by autoradiography. Each spot represents a single polypeptide. The spots are elongated horizontally because the average charge on a protein molecule varies somewhat during IEF. These patterns are reproducible, so that changes in individual proteins can be detected. The proteins are from cells growing on a normal medium supplemented with isoleucine *(top)* and from cells placed for a brief period in a medium that lacks isoleucine *(bottom)*. Certain spots *(circles)* are absent in the bottom photograph or are much fainter than in the top photograph; these differences represent changes in the synthesis pattern of cell proteins in response to amino acid starvation. *From P. H. O'Farrell, 1978,* Cell **14***:545. Copyright M.I.T.*

dimension on the basis of size. As many as several thousand different protein chains—virtually the total protein content of a cell—can be detected and separated by this technique. Two-dimensional gels are very useful in studying the expression of various genes in differentiated cells.

There are two widely used methods of detecting proteins in gels:

1. The total amount of each type of protein in a sample can be estimated with gel electrophoresis by staining the gels with a dye that binds approximately equally to all proteins. The intensities of the spots of dye indicate the comparative quantities of proteins of different lengths.
2. Gel electrophoresis provides a way of detecting the synthesis of any particular protein without isolating it. If whole cells are briefly labeled with radioisotopes

before they are analyzed, each newly synthesized chain can be detected in the gel by autoradiography. However, because new proteins may be secreted from the cell or may be subject to different rates of metabolic turnover, the concentration of a labeled protein in a cell may not accurately reflect its rate of synthesis.

In Vitro Protein Synthesis and Gel Electrophoresis Provide an Assay for Messenger RNA

Two general approaches are used to determine what proteins a cell can make. In one method, the contents of whole, labeled cells are examined for newly synthesized proteins (see Figures 6-9, lanes 1 and 2, and 6-10). In the other, mRNA is extracted from the cells and translated in

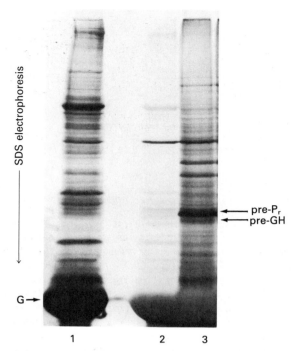

SDS electrophoresis

G→

pre-P$_r$
pre-GH

1   2   3

▲ **Figure 6-11** The translation of mRNA by mixtures of ribosomes, tRNAs, and protein synthesis factors extracted from reticulocytes. Here, the total protein produced by such reactions has been separated by electrophoresis and is visible by autoradiography because [^{35}S]methionine was added to the extract. Proteins synthesized from mRNAs in an untreated reticulocyte extract are shown in lane 1; note the large amount of globin synthesis (G). In lane 2, a bacterial nuclease (from *Micrococcus aureus*) has greatly reduced the amount of synthesis. In lane 3, the nuclease has been chemically inactivated, and mRNAs from rat pituitary cells have been added. Several prominent pituitary-specific proteins are visible, including two hormone precursors: one of prolactin (pre-P$_r$, 236 amino acids long), and one of growth hormone (pre-GH, 212 amino acids long). [See H. R. B. Pelham and R. J. Jackson, 1976, *Eur. J. Biochem.* 67:247.] *Photographs courtesy of D. Anderson.*

the presence of labeled amino acids by cell-free protein-synthesizing systems (Figure 6-11). Both approaches are actually assays for functional mRNAs. In either case, the products can be separated and identified by gel electrophoresis.

Different cell extracts can be used to label proteins in vitro (assay for active mRNAs). Bacterial cell extracts that can translate homopolymers were first widely used to break the genetic code and to examine bacterial and bacteriophage proteins; now extracts of eukaryotic cells are also commonly used. Two of the most popular cell-free systems are extracts of reticulocytes and of wheat germ, the embryo plant in a fertile wheat seed. Both are prepared by treating the cells first with a nuclease that destroys endogenous mRNA (mRNA from the source cells) and then with a chemical that blocks the nuclease so subsequently added mRNA is not destroyed. After this

treatment, very little protein synthesis by endogenous mRNA occurs (see Figure 6-11, lane 2). Thus the added mRNA is responsible for almost all protein synthesis, and the products of the added mRNA can be easily detected.

Examining the Sequences of Nucleic Acids and Proteins

The first biopolymer to be sequenced was a protein, and this discovery has great historical importance. Before Fred Sanger reported the sequence of human insulin in 1953, some biochemists were not convinced that proteins had specific sequences from end to end. The single unique sequence found in insulin implied a highly precise ordering mechanism during protein synthesis. Since that time, the coding of protein sequence by nucleic acid sequence has been made clear. Recently, it has become much easier to obtain long nucleic acid sequences than long protein sequences and thus, with the aid of the genetic code, to *deduce* the sequence of many proteins rather than actually determine them directly.

Because the functions of nucleic acids and proteins depend on the linear sequences of their monomers, research in molecular biology relies heavily on techniques that reveal and compare sequences. However, the sequence information required in experiments varies considerably in extent and type. In the simplest case, only an estimate of the degree of similarity, or *sequence relatedness,* between two samples of nucleic acid or protein is required. Often, it is necessary simply to determine whether a particular sequence is present in a given mixture of nucleic acids or proteins. Once the presence of a certain sequence in a mixture of sequences is established, a variety of other questions arise. What is the concentration or amount of the specific sequence? Where within a DNA, RNA, or protein molecule is the sequence of interest located? And finally, what is the precise nucleotide or amino acid sequence for the entire molecule? A variety of techniques are used to address these questions; each applies better to some questions than to others.

Molecular Hybridization of Two Nucleic Acid Strands Can Be Detected in Several Ways

Under the conditions of temperature and ion concentration found in cells, DNA is maintained as a duplex (two-stranded) structure by the many hydrogen bonds of the A-T and G-C base pairs. The duplexes can be *melted (denatured* into single strands) by heating them (usually in a dilute salt solution of, for example, $0.01M$ NaCl) or by raising the pH above 11. If the temperature is lowered and the ion concentration in the solution is raised, or if the pH is lowered, the single strands will *anneal,* or reassociate, to reconstitute duplexes (if their concentration in solution is great enough). In a mixture of nucleic acids, only complementary strands reassociate; the extent of their reassociation is virtually unaffected by the presence of noncomplementary strands. Such *molecular hybridization* can take place between complementary strands of either DNA or RNA or between an RNA strand and a DNA strand (Figure 6-12).

Visualization of Hybrids Electron microscopic examination of molecular hybrids conveniently reveals the sequence relatedness of two nucleic acid samples. If two melted nucleic acid samples that are complementary over only part of their length are allowed to hybridize, a *heteroduplex* results (Figure 6-13); complementary (duplex) and noncomplementary (single-stranded) regions can be distinguished in such preparations. This technique can be used not only to compare DNA strands but also to locate DNA sites complementary to RNA molecules. By this latter procedure, it is possible to distinguish and locate the regions of DNA that are transcribed into RNA. Regions of RNA-DNA hybridization create loops (called *R loops*) in the nucleic acid molecules, where the RNA sequence has base-paired with one DNA strand and displaced the other DNA strand (Figure 6-14). The fact that 1 μm of

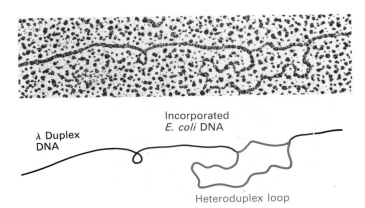

▲ **Figure 6-13** Electron micrograph *(top)* of a DNA heteroduplex. DNA molecules on a carbon grid can be distinguished as long threads when they are shadowed with heavy metals (here, platinum and palladium). This heteroduplex has formed from strands of two λ bacteriophages incorporating different but related sequences of *E. coli* DNA *(bottom)*. The λ strands form a double-stranded hybrid where the inserted *E. coli* sequences are complementary (red); the dissimilar inserted sequences remain unassociated, resulting in a heteroduplex loop of single-stranded DNA (blue). *From R. W. Davis and J. S. Parkinson, 1971,* J. Mol. Biol. *56:403.*

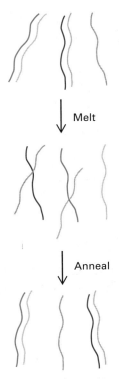

▲ **Figure 6-12** Molecular hybridization: reassociation of the complementary strands of a nucleic acid. Under conditions of high pH or temperature, the duplexes in a solution of nucleic acids melt, or separate into single strands. With an appropriate change in conditions, complementary strands reassociate. The presence of noncomplementary chains does not affect the reassociation rate of complementary chains.

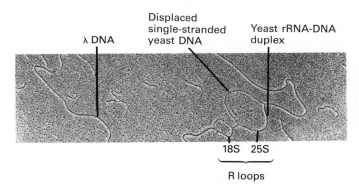

▲ **Figure 6-14** If double-stranded DNA is treated with a 50-percent solution of formamide at room temperature (or 25°C), some hydrogen bonds between the strands of the molecule break, weakening but not completely melting the duplex. If RNA that is complementary to one strand of the duplex DNA is then introduced, the RNA binds to its complementary site on one DNA strand, displacing the other DNA strand. This occurs because an RNA-DNA duplex is more stable than a DNA-DNA duplex. The hybrid duplex and the displaced stretch of single-stranded DNA are called an *R loop*. The two R loops that appear in this electron micrograph result from the hybridization of 18S and 25S ribosomal RNA from yeast with a region of bacteriophage λ DNA that contains an inserted stretch of yeast ribosomal genes. [See M. Thomas, R. L. White, and R. W. Davis, 1976, *Proc. Nat'l Acad. Sci. USA* **73**:2294.] *Photograph courtesy of R. W. Davis.*

(a)

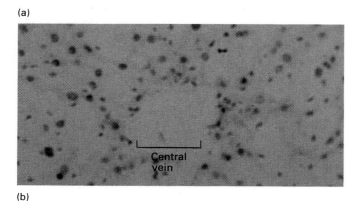

(b)

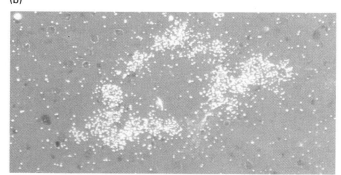

▲ **Figure 6-15** Autoradiograph showing in situ hybridization. A mouse liver section was exposed to labeled RNA complementary to glutamine synthetase mRNA. The label was allowed time to hybridize; then, unhybridized labeled RNA was washed away. (a) A light microscopic view of the autoradiograph showing cords of liver cells (hepatocytes) around a central vein. The barely visible dark grains around the central vein are in the first layer of hepatocytes. (b) The second view is a dark-field picture, which shows the grains (white dots) with much greater contrast. [See F. Kuo et al., 1988, *Mol. Cell Biol.* 8:4966.] *Photographs courtesy of F. Kuo.*

double-stranded nucleic acid contains about 3000 bases can be used to estimate the number of nucleotides in single- and double-stranded regions, and thus an accurate map of the transcribed section of DNA. The technique was crucial in proving splicing of mRNA in eukaryotes.

In Situ Hybridization Another use for molecular hybridization that has achieved great popularity is called *in situ hybridization.* Labeled RNA or DNA that is complementary to a specific mRNA is prepared. Cells or tissue slices are briefly exposed to heat or acid, which fixes the cell contents, including the mRNA, in place on a glass slide, the fixed cell or tissue is then exposed to the labeled complementary RNA for hybridization. Removal of unhybridized labeled RNA and coating the slide with a photographic emulsion is followed by autoradiography to reveal the presence and even the location of specific mRNA within individual cells (Figure 6-15).

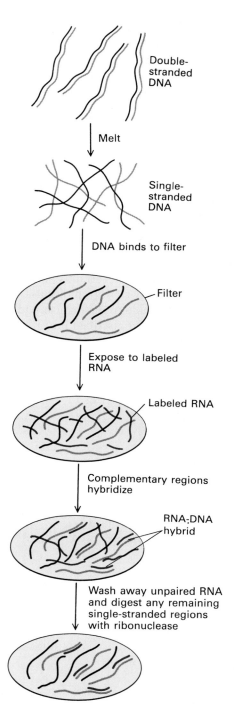

▲ **Figure 6-16** The filter-binding assay for RNA-DNA (or DNA-DNA) hybridization is an extremely popular and flexible method of detecting complementary regions. Under the proper conditions of ion strength and temperature, filter-bound single-stranded DNA is exposed to a labeled RNA (or DNA) sample. Molecules or sequences complementary to the filter-bound DNA pair with it; unpaired labeled molecules can be removed. This technique allows as little as 1 part in 10^6 of specific RNA or DNA to be detected.

Hybrids on Filter-immobilized Nucleic Acid A common method of detecting hybrids between nucleic acid samples employs a single-stranded nucleic acid attached to a solid matrix. Nitrocellulose and treated nylon membranes are the most widely used matrices; it is not known why single-stranded DNA (or RNA) binds to these substrates, but this affinity is enormously useful. The radioactive RNA or DNA that is to be tested for sequences complementary to the bound nucleic acid is allowed to hybridize with it. After sufficient time, the unhybridized single strands unassociated with the bound nucleic acid are washed away. RNA hybridized to bound DNA is resistant to ribonucleases, whereas unpaired RNA regions are digested by these enzymes; thus any remaining single-stranded RNA regions are trimmed away in such experiments. The amount of hybrid formed can then be measured by the amount of bound radioactive label present (Figure 6-16). With the appropriate choice of filter-bound nucleic acid, one specific RNA (or DNA) sequence can be detected in a mixture of many different sequence types.

In the procedure called *DNA excess hybridization,* the total RNA from cells (a very complex mixture of sequences) is labeled and exposed to unlabeled purified specific DNA. If the DNA is present in excess (if there are more than enough copies to hybridize with all complementary segments of RNA), the amount of hybrid formed is proportional to the amount of RNA input. This allows an accurate measurement of the amount of the particular RNA in a mixed sample.

It is also possible to test for the presence of a particular RNA sequence and to quantify the amount of it in different samples by *competition hybridization.* A measured sample of a specific labeled RNA is exposed to just enough complementary DNA to completely hybridize with it; a sample of unlabeled RNA is then added. If the unlabeled RNA sample contains the same sequence as the labeled RNA, they "compete" for the DNA; increasing the ratio of unlabeled to labeled samples decreases the amount of labeled RNA hybridized. The extent to which this takes place is a measure of the amount of competing RNA in the unlabeled sample.

The Rate of Nucleic Acid Hybridization Can Be a Measure of Complexity

The rate of hybridization between two complementary single-stranded nucleic acids in solution depends on the frequency with which complementary regions collide and *nucleate,* or start to form a duplex. This frequency, in turn, depends on the concentration of the two strands. If the DNA fragments of two different organisms—say, *Escherichia coli* and yeast—are incubated in amounts that yield the same *total* DNA concentration, the *complexity* of the DNA (the number of base pairs in the total genome) is about four times as great for yeast as for *E. coli.* A separated strand of *E. coli* DNA therefore encounters its correct partner

four times as often as a strand of yeast DNA does, and *E. coli* DNA reassociates at a faster rate (Figure 6-17).

From the equation for determining the quantitative relation between reassociation rate and genome complexity (given in Figure 6-17), the reassociation rate of any DNA sample can be used to calculate the relative complexity of the source genome. Experimentally, the initial concentration of DNA C_0 and the time t are varied to measure the reassociation rate, so the resulting curves are often called $C_0 t$ *(cot)* curves.

If the DNA sequences of an organism are present once per haploid genome, the reassociation curve is uniform. If some sequences are repeated, these hybridize more rapidly. Reassociation measurements have been important both in comparisons between different types of organisms

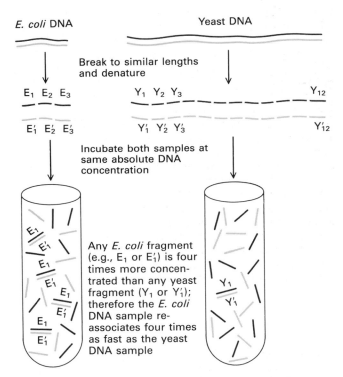

▲ **Figure 6-17** The complexity of DNA controls the rate of its reassociation. The relative hybridization rates within two samples of dissolved and melted genomic DNA depend on their relative complexity (the number of DNA base pairs in the genome of each organism), provided the samples are equal in absolute concentration (total nucleotide concentration). DNA is broken into pieces of 1000–2000 bases each, so size plays little role in the comparison. The equation for the reassociation rate is

$$C_t = \frac{1}{1 + KC_0 t}$$

where C_0 and C_t are the molar concentrations of single strands at times 0 and t, respectively, and K is the rate constant for the particular type of DNA (this constant depends on the complexity of the DNA). [See R. J. Britten and D. E. Kohne, 1968, *Science* **161**:529.]

Spot number	Sequence	Spot number	Sequence	Spot number	Sequence
1	AG	10	UAG	16	UUAG
3	CCCG	11	UCG	17	UCUG
4	AAG	12	CCUG	18	AUCUCG
5	ACCG	13	CCUACG	20	CCG
6	AAAG	14b	AAUACCG	21	CG
8	CG	14c	AUCCAG	22	pG
9	AUG	15	CCACACCACCUG	G	Gp

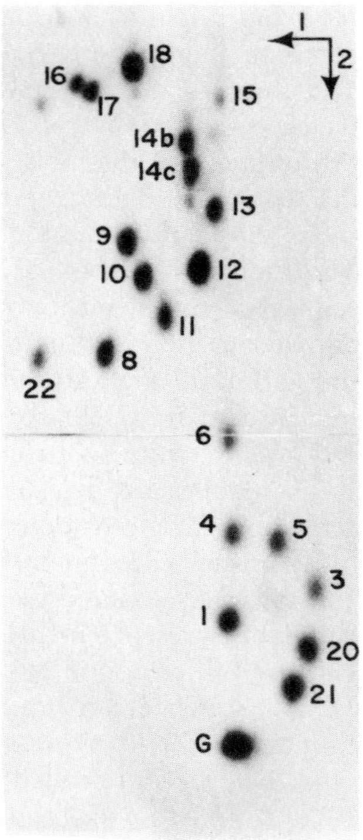

▲ **Figure 6-18** A ribonuclease T1 fingerprint of 5S ribosomal RNA from oocytes of the frog *Xenopus laevis*. This enzyme cuts RNA on the 3′ side of all guanylate residues, NpGp ↓ Np(Np)$_n$Gp ↓ Np producing fragments that all contain one Gp (guanylate) at their 3′ ends. The digest is applied to treated paper (cellulose acetate), and a two-step separation is carried out: electrophoresis in one dimension *(arrow 1)*, followed by chromatography in the other *(arrow 2)*. If the starting sample is radioactive (^{32}P-labeled RNA is often used), the oligonucleotides can be identified by autoradiography. Spots of RNA can be cut out of the paper sheet and further analyzed biochemically to determine their sequences. No spots are numbered 2, 7, 14a, or 19; these numbers were given to oligonucleotides identified in another type of 5S rRNA. *From D. D. Brown, D. Carroll, and R. D. Brown, 1977, Cell* **12**:1045. *Copyright M.I.T.*

(prokaryotic versus eukaryotic; vertebrate versus invertebrate; and so on) and in studies of the degree of repetition of certain sequences within eukaryotic genomes. (Repetitious DNA sequences are discussed in detail in Chapter 10.) In a variation of the use of reassociation curves, a trace amount of radioactive pure DNA is added to unlabeled RNA from a cell of interest. Because the DNA is present in a tiny amount, the rate of RNA-DNA hybridization depends on the concentration of complementary RNA. From that rate, the amount of complementary RNA in a sample can be estimated. Data curves from such measurements are referred to as R_0t (rot, or RNA concentration) curves.

Fingerprinting (Partial Sequence Analysis) Allows Quick Comparisons of Macromolecules

The enzymatic fragmentation of proteins and nucleic acids at specific sites provides a means of recognizing particular macromolecules quickly. As we saw in Chapter 2, the enzyme trypsin digests protein chains, cleaving them on the carboxyl side of each lysine and each arginine residue to produce a specific set of peptides from any given pure protein. Likewise, the enzyme ribonuclease T1 cuts RNA at the 3′ side of each guanylate residue to produce specific fragments ending with a guanylate. The resulting oligonucleotides are fairly short: they normally contain 2–20 nucleotides, because consecutive guanylates are usually no more than 20 bases apart. Reliable separation and detection of different peptides from a pure protein or different oligonucleotides from a pure RNA sample can

be accomplished by electrophoresis, chromatography, or both.

Because the oligonucleotides or peptides produced by a given enzyme from a pure RNA sample or protein are always the same, the pattern of separated fragments is always the same. The characteristic pattern of fragments from a primary sequence is called a *fingerprint* (Figures 6-18 and 6-19). *Fingerprinting*, or *partial sequence analysis*, allows the rapid comparison of two samples of RNA or protein when there is no need to determine the complete sequence of nucleotides or amino acids. The fingerprinting technique was first developed for proteins, which can be cleaved by both enzymes and chemical reactions. With this historic fingerprints of globin shown in Figure 6-19, Vernon Ingram demonstrated that people suffering from sickle-cell anemia, a genetic disease, have a valine substituted in one position in place of a glutamic acid in their β globin. This was the first mutant protein shown to be affected in function by a change in one amino acid residue.

Restriction Enzymes Allow the Precise Mapping of Specific Sites in DNA

The most flexible, simple, and useful technique for partial sequence analysis of DNA was made possible by the discovery of bacterial restriction endonucleases, which recognize specific short oligonucleotides from four to eight residues long in DNA and then cleave the DNA at each site (Figure 6-20a). The word "restriction" refers to the

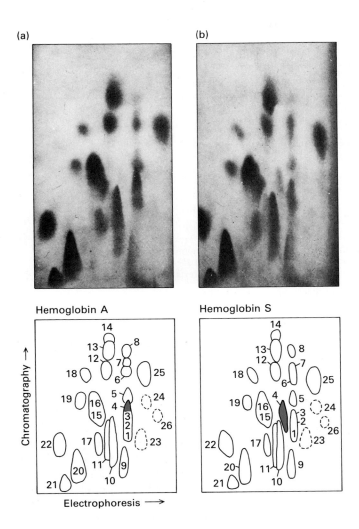

Hemoglobin A

Hemoglobin S

Chromatography →

Electrophoresis →

◀ **Figure 6-19** Fingerprints of (a) normal and (b) sickle-cell human β-globin. Proteolytic enzymes such as trypsin are used to break the peptide chain at known amino acid residues (trypsin cuts after each arginine and each lysine). The resulting set of specific fragments can then be separated by electrophoresis followed by chromatography. Individual peptide spots can be distinguished by spraying the chromatography paper with ninhydrin, a reagent that forms a purple product with free amino groups. (Spots 23, 24, and 26 show up poorly with ninhydrin.)

These fingerprints are identical, with one exception. Peptide 4 of the β chains of hemoglobin S (the hemoglobin of people with the sickle-cell disease) is found in a slightly different location than peptide 4 of normal hemoglobin A. Analysis of the two peptides has shown that hemoglobin S has a valine instead of a glutamic acid at residue 6 in the β chain; thus a single amino acid replacement is the cause of sickle-cell anemia. This represented the first demonstration that a random mutation in nature resulted in a single amino acid substitution. *From V. Ingram, 1958*, Biochim. Biophys. Acta *28:543.*

▼ **Figure 6-20** (a) *Eco*RI and many other restriction endonucleases cleave DNA so that the fragments have short complementary single-stranded segments at the ends. These "sticky ends" are important in recombinant DNA techniques because they readily pair with the ends of other cleavage fragments produced by the same restriction endonuclease. *Eco*RI recognizes the sequence shown here. (b) Most cells with restriction endonucleases also have corresponding *modification endonucleases*. *Eco*RI methylase, a modification endonuclease, catalyzes the methylation of two adenylates (shown in blue) in the recognition sequence; this prevents cleavage by *Eco*RI. Thus a cell making *Eco*RI endonuclease and methylase does not destroy its own DNA.

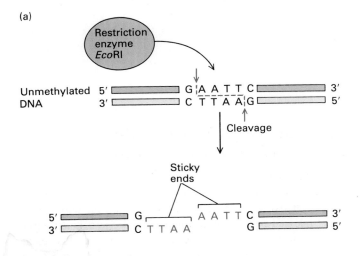

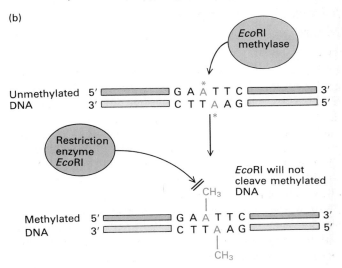

Table 6-4 Examples of the actions of restriction endonucleases

Source microorganism	Enzyme*	Recognition site (↓)†	λ (50)	Ad2 (36)	SV40 (5.2)	pBR322 (4.3)
			Number of cuts (kb)‡			
Arthrobacter luteus	*Alu*I	AG ↓ CT	143	158	34	14
Thermus aquaticus	*Taq*I	T ↓ CGA	121	50	1	13
Haemophilus parahaemolyticus	*Hph*I	GGTGA+5	168	99	4	18
Haemophilus gallinarum	*Hga*I	GACGC+8	102	87	0	12
Escherichia coli	*Eco*RI	G ↓ AATTC	5	5	1	1
Haemophilus influenzae	*Hin*dIII	A ↓ AGCTT	6	12	6	1
Nocardia otitiscaviaruns	*Not*I	GC ↓ GGCCGC	0	7	0	0
Streptomyces fimbriatus	*Sfi*I	GGCCN$_4$ ↓ NGGCC	0	3	1	0

* Enzymes are named with abbreviations of the bacterial strains from which they are isolated; the Roman numeral indicates the enzyme's priority of discovery in that strain (for example, *Alu*I was the first restriction enzyme to be isolated from *Arthrobacter luteus*).
† Recognition sequences are written 5′ → 3′ (only one strand is given). For example, G ↓ GATCC is an abbreviation for

$$
\begin{array}{c}
\downarrow \\
(5')G\overset{\downarrow}{G}ATCC(3') \\
(3')CCTAGG(5') \\
\uparrow
\end{array}
$$

The cleavage site for *Hph*I and *Hga*I occurs five or eight bases away from the recognition sequence; N indicates any base.
‡ These columns list the number of cleavage sites recognized by specific endonucleases on the DNA of bacteriophage λ (λ), adenovirus type 2 (Ad2), simian virus 40 (SV40), and an *E. coli* plasmid (pBR322). The sizes of the DNAs are given in kilobases (kb). Note that the actual number of cuts in these sequences deviates from the expected number in random sequences, which would be given by $L/4^n$, where n is the length of the site recognized by an enzyme and L is the length of the sequence.
SOURCE: R. J. Roberts, 1988, *Nuc. Acids Res.* **16**(supp):r271.

function of these enzymes in the bacteria of origin: a restriction endonuclease destroys (restricts) incoming foreign DNA (for example, bacteriophage DNA or DNA accidentally taken up during transformation) by cleaving it at these specific sites, called *restriction sites.*

Another enzyme protects a bacterium's own DNA from cleavage by modifying it at or near each potential cleavage site: a methylase adds a methyl group to one or two bases, usually within the restriction site. When a methyl group is present there, the restriction endonuclease is prevented from cutting the DNA (Figure 6-20b). Together with the restriction endonuclease, the methylating enzyme forms a *restriction-modification system* that protects the host DNA while it destroys foreign DNA.

A restriction endonuclease cuts a pure DNA sample into a consistently reproducible set of fragments that can be easily separated by gel electrophoresis (Figure 6-21). Several hundred restriction enzymes with different recognition sites are now available (see Table 6-4). If the order of nucleotides in DNA were random, the number of cuts expected would be larger for an enzyme that requires only a four-base site than for one that requires a longer site and larger for longer DNAs than for shorter ones. However, the sites for restriction endonucleases are not randomly distributed; by testing a series of enzymes that cut at different sites it is possible to cut a particular DNA many or only a few times. The most recently discovered eight-base cutters have proved to be especially useful for producing very large fragments, which then can be separated by pulse-field gel electrophoresis (see Figures 5-9

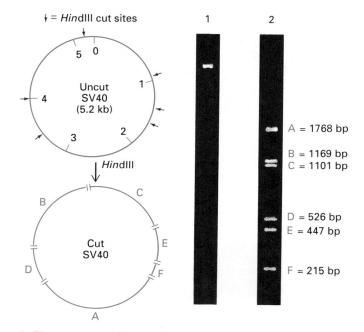

▲ **Figure 6-21** The DNA from SV40 virus can be purified and digested with the restriction endonuclease *Hin*dIII (from *Haemophilus influenzae*). The digest is then subjected to electrophoresis in a gel containing ethidium bromide, a molecule that binds to DNA and fluoresces when exposed to ultraviolet irradiation. Lane 1 represents the uncut DNA; lane 2, the digested DNA. *Hin*dIII cuts the SV40 molecule at six sites (↓), producing six fragments. By convention, the pieces of DNA released by a restriction endonuclease are labeled A–Z in order of decreasing size; the *Hin*dIII fragments of SV40 are therefore labeled A–F. *Photograph courtesy of D. Nathans.*

and 6-8). Fragments of 1–10 megabases (10^6–10^7 bp) are used to map the chromosomes of very large genomes, such as those of mouse and man.

Digestion of DNA by restriction endonucleases, followed by simple electrophoretic separation of the fragments, has revolutionized chromosome mapping. The use of two or more restriction endonucleases on a pure DNA sample can show the order of the restriction sites in a DNA sample (Figure 6-22). Also, many sites can be located by partial digestion of terminally labeled DNA with only one enzyme (Figure 6-23). In these ways, it is possible to produce a map showing the order of the restriction sites in any region of DNA. An important application of restriction endonucleases is their use to cut off one end of a DNA sample that has been end-labeled so that DNA pieces labeled at the other end are available for further study (see Figure 6-23).

Southern DNA Blots The ability to divide DNA into reproducible pieces allows the restriction sites around a particular sequence of interest to be mapped. This possibility is realized in the laboratory by determining which restriction fragments hybridize to a specific labeled *probe* sequence technique called the *Southern blot* (after its originator, Edward Southern). DNA restriction fragments from a sample are separated by gel electrophoresis; their distribution in the gel is preserved as they are denatured and transferred by blotting to a solid substrate with

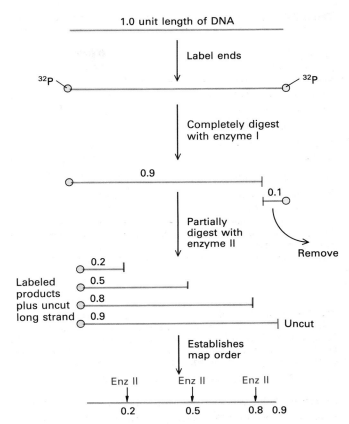

▲ **Figure 6-23** Mapping the multiple recognition sites of a restriction enzyme by partial digestion. DNA is labeled at its termini with ^{32}P, and fragments with *one* labeled terminus can be obtained by cutting off one end with an appropriate enzyme. The mapping procedure is applied to the remaining piece with a second enzyme. Complete digestion would produce only one labeled fragment (here, the 0.2-unit piece), but brief, partial digestion (in which the enzyme cuts each long piece only once, at most) produces a labeled fragment for each restriction site. From the lengths of the labeled pieces, the positions of enzyme II restriction sites can be inferred. [See H. O. Smith and M. Birnstiel, 1976, *Nuc. Acids Res.* 3:2387.]

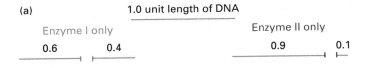

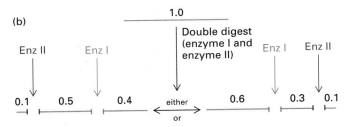

▲ **Figure 6-22** Mapping the cleavage sites of two restriction enzymes with respect to one another. (a) When a given piece of DNA is exposed separately to two restriction enzymes (I and II), each cuts the DNA once. The lengths of the fragments are determined by gel electrophoresis. (b) Digestion with *both* enzymes is used to determine the relative positions of the cuts along the DNA. The fragment lengths identify the positions of the restriction sites for enzymes I and II with respect to the ends of the DNA and therefore with respect to each other. By continuing this process with different pairs of enzymes, the investigator can construct a detailed map of restriction sites.

a charged surface (usually a nitrocellulose filter). The filter is then exposed to a specific radioactive nucleic acid sequence (the probe). The blotted DNA fragments that are complementary to the probe hybridize with them, and their location on the filter can be revealed by autoradiography (Figure 6-24). This technique is so sensitive that a DNA sequence that appears only once in the human genome (about 1 part in 10^6) can be detected in as little as 5 μg of DNA (the DNA content of about 10^6 cells).

This test is widely used in genetic studies of humans, who do not, as a rule, breed within families. Consequently, the human population shows many genetic differences, or *genetic polymorphisms*. These variations are

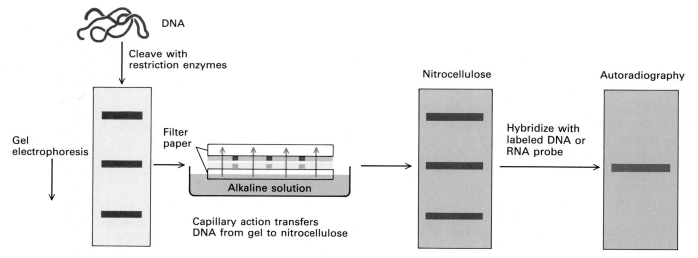

▲ **Figure 6-24**   The Southern blot technique for detecting the presence of specific DNA sequences. [See E. M. Southern, 1975, *J. Mol. Biol.* 98:508.]

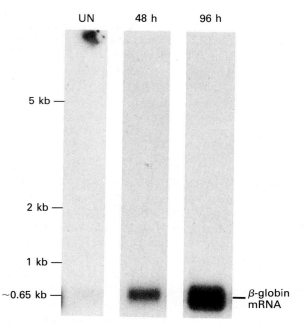

▲ **Figure 6-25**   The Northern blot technique for detecting the presence of specific mRNA molecules. Autoradiography shows the position of the complementary mRNA in the gel, and the density of the spot shows the amount of it. The photograph indicates the relative quantities in kilobases (kb) of β-globin mRNA in erythroleukemia cells at three different times: when cells are growing and have not started to make globin (lane UN, for "*un*induced"), and 48 and 96 h after they have been induced to stop growing and begin differentiating. The β-globin mRNA is barely detectable in growing cells but increases by a factor of more than 1000 in 96 h of differentiation. *Photograph courtesy of L. Kole.*

often indicated by the presence or absence of particular restriction sites in the DNA, called *restriction fragment-length polymorphisms.*

Northern (RNA) and Western (Protein) Blots

The *Northern blot,* so-named because it is patterned after the Southern blot, is used to detect the presence of specific mRNA molecules. The RNA molecules in a sample are denatured by mixing them with an agent, such as formaldehyde, to prevent hydrogen bonds between base pairs (stems) and ensure that the RNA is in unfolded, linear form. The RNA sample (often the total RNA from cells) is then separated according to size by gel electrophoresis; as in a Southern blot, it is then transferred to a nitrocellulose filter to which the extended denatured RNA will adhere. The filter is exposed to a labeled DNA probe and subjected to autoradiography. The Northern blot indicates the amount as well as the presence and size of a specific mRNA in a sample and the procedure is widely used to compare the amounts of a specific mRNA in cells under different conditions (Figure 6-25).

Another bit of laboratory jargon that has become a widely accepted name for a common technique is the *Western blot.* In this procedure, a one- or two-dimensional electrophoretic separation of proteins is carried out and the protein is then transferred, or blotted, to nitrocellulose. The nitrocellulose sheet can be exposed to radioactive antibodies against a particular protein; autoradiography reveals the presence of that protein.

Band Analysis of S1 Digests

An important method for measuring the length of complementary sequences in two nucleic acids employs the endonuclease S1, an en-

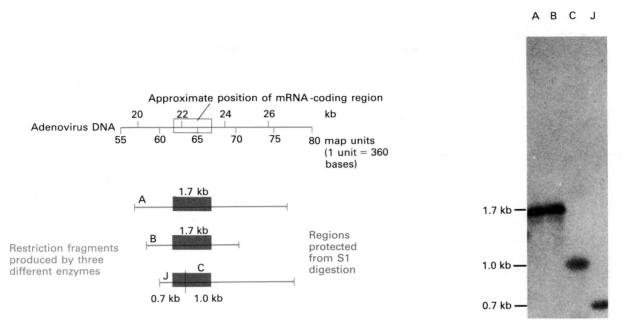

▲ **Figure 6-26** The S1 mapping technique determines the lengths of complementary sequences in two nucleic acid samples. A portion of the map of adenovirus DNA is shown. Earlier hybridization experiments established that an mRNA was complementary to a sequence in the large region spanned here by restriction fragment A. Restriction fragments A, B, C, and J were prepared from ^{32}P-labeled DNA, denatured and hybridized with a large excess of mRNA prepared from virus-infected cells. The mixture was treated with S1 to destroy any unpaired DNA that had not found an mRNA partner; the protected labeled DNA fragments (red) were then separated by gel electrophoresis. The autoradiograph of the gel shows the lengths of the segments protected by (complementary to) the mRNA. Fragments A and B were protected for 1.7 kb; C and J were protected for 1.0 and 0.7 kb. Thus the mRNA includes 1.7 kb positioned as indicated. *From A. J. Berk and P. A. Sharp, 1978,* Cell **14:**695*. Copyright M.I.T.*

zyme from the mold *Aspergillus oryzae* that destroys unpaired RNA or DNA but not double-stranded molecules (Figure 6-26). Either the RNA or the DNA in a hybrid may be labeled. For example, the total unlabeled mRNA from a cell can be exposed to a labeled DNA probe (usually consisting of one or more restriction fragments) that may include all or part of the region of DNA that is transcribed to produce one particular mRNA. The labeled RNA-DNA hybrid is then digested with S1 to remove unpaired nucleic acid strands, leaving hybrid duplexes intact. After electrophoresis, different hybrids form discrete bands, whose positions can be used to estimate the lengths of the hybrids. This technique is widely used to determine how much of a particular DNA restriction fragment is complementary to an mRNA region.

Finding the Start Site of an mRNA

It is often very important to find the point in a DNA sequence at which transcription of a particular mRNA begins. Two methods are used: the endonuclease S1 or the *primer extension technique*. First, a general region of a DNA molecule that includes the start site must have been located (DNA sequencing is described in the next section). Appropriate restriction sites can be chosen in this sequence to prepare a piece of end-labeled DNA approximately 100 nucleotides long that will hybridize with the 5′ portion of the mRNA. Figure 6-27 shows how such a piece of labeled DNA, trimmed with S1 endonuclease, can be used to locate the exact start site. The same logic applies with primer extension. A DNA oligonucleotide about 20 bases long is chosen to find a specific complementary site on the mRNA. This primer can be extended enzymatically to the beginning of the mRNA; the length of this extension product can then be determined accurately by gel electrophoresis.

Endonucleases Compared to Exonucleases

Thus far, we have been discussing endonucleases, enzymes that cut DNA or RNA (or both) *within* a chain. Restriction endonucleases have a restricted cutting specificity; S1 and the widely used pancreatic RNase and DNase digest almost all internucleotide bonds equally. The S1 endonuclease is single-stranded; the pancreatic enzymes can be single- or double-stranded.

(a)

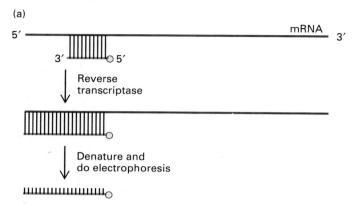

(b)

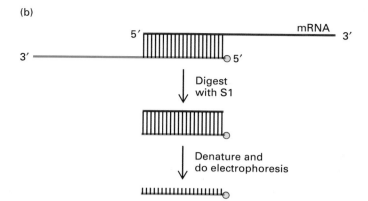

▲ **Figure 6-27** The site in DNA that encodes the first nucleotide in an mRNA molecule can be found by using primer extension or S1 endonuclease (see Figure 6-26). (a) In the primer extension technique, a short (approximately 10 nucleotides) oligodeoxynucleotide (blue) is prepared and end-labeled. After it is hybridized to the mRNA (red), it is lengthened by the enzyme reverse transcriptase until it reaches the first nucleotide of the mRNA. (b) The use of S1 to map start sites begins with the preparation of a uniquely end-labeled, short (approximately 100 nucleotides), single DNA strand, that encodes the general region of the mRNA start site and whose total sequence is known. This is hybridized with the mRNA, and unpaired nucleic acid is then digested with S1 endonuclease. Denaturation leaves a labeled DNA piece whose length accurately marks the distance of the starting nucleotide of the mRNA from the nucleotide that hybridized with the labeled DNA end.

Other enzymes called *exonucleases* remove nucleotides one at a time from the *ends* of entire RNA or DNA strands. Some act only on single strands; others remove nucleotides from either the 5′ or the 3′ ends (but not both) of duplex DNA. There are too many of these enzymes to attempt a comprehensive description of them here. We shall describe exonucleases as necessary in this and other chapters.

The Sequence of Nucleotides in Long Stretches of DNA Can Be Rapidly Determined

The discovery of restriction endonucleases was an important step that led to general methods for determining the exact nucleotide sequences in long stretches of DNA. An

▶ **Figure 6-28** DNA sequencing by the Maxam-Gilbert method. A 5′-end–labeled DNA fragment is prepared for sequencing. Four identical samples of this fragment are subjected to four different chemical reactions. Each breaks the fragment only (or mainly) at the A, G, C + T, or C residues, respectively. The reactions are controlled, so that each labeled chain is likely to be broken only once. The resulting *labeled* subfragments created by all four reactions have the label at one end and the chemical cleavage point at the other. Gel electrophoresis and autoradiography of each separate mixture yield one radioactive band for each nucleotide in the original fragment. Bands appearing in the A and G lanes can be read directly. Bands in the C + T and C lanes are read as Cs; those in the C + T lane alone, as Ts. [See A. Maxam and W. Gilbert, 1977, *Proc. Nat'l Acad. Sci. USA* 74:560.]

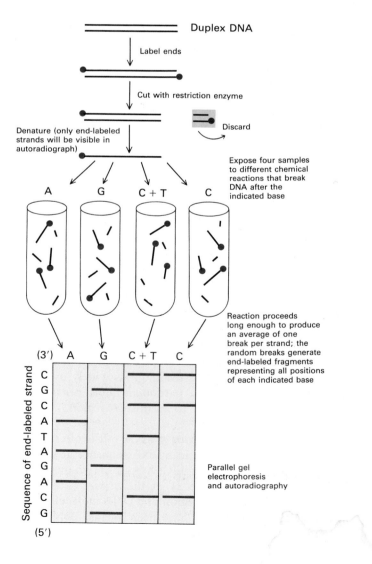

(a)

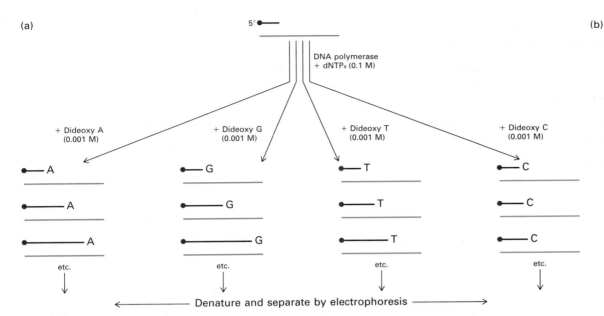

(b)

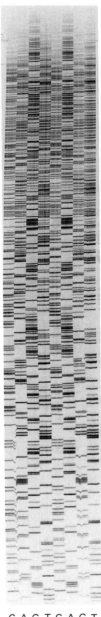

▲ **Figure 6-29** DNA sequencing by the Sanger (dideoxy) method. (a) A single strand of DNA to be sequenced is hybridized to a 5′-end–labeled deoxynucleotide primer; four separate reaction mixtures are prepared in which the primer is elongated by a DNA polymerase. Each mixture contains the four normal deoxynucleoside triphosphates plus one of the four dideoxynucleoside triphosphates in a ratio of about 1 to 100. Since a dideoxynucleotide has no 3′ hydroxyl, no further chain elongation is possible when such a residue is added to the chain. Thus, each reaction mixture will produce prematurely terminated chains ending at every occurrence of the dideoxynucleotide. Each mixture is then separated on a sequencing gel as in Figure 6-27. (b) An actual radioautogram in which over 300 bases can be read. (The enzyme used in the figure is Sequenase™, a commercial preparation of a bacteriophage-encoded polymerase.) Each reaction was carried out in duplicate, which aids in checking the sequence. *Courtesy of United States Biochemical Corporation.*

C A G T C A G T

earlier and very important advance was the recognition that careful gel electrophoretic procedures can separate every single DNA fragment in a series up to 500 bases long. Two highly successful procedures for DNA sequencing are in wide use. Both depend on the separation of a set of fragments that differ in length by only one nucleotide (Figure 6-28). The first, invented by Allan Maxam and Walter Gilbert, chemically cleaves an end-labeled DNA sample; the second, developed by Fred Sanger and his colleagues, uses enzymatic synthesis to extend a short sequence of end-labeled DNA (Figure 6-29).

Modern DNA sequencing is fairly simple and accurate over long regions; already, the total genomes of many viruses and almost all of the *E. coli* genome have been sequenced. Automation of the techniques for sequencing large pieces of DNA (see Figure 6-8) should permit sequencing of the entire human genome in 10–15 years, if not before.

Proteins Can Be Sequenced Automatically

From a DNA sequence and the genetic code, it is possible to deduce the sequences of the encoded protein. And with the aid of computers to locate "open reading frames,"

i.e., codon stretches without protein termination signals, investigators often do just that. Nevertheless, the ability to sequence protein chains directly remains a crucially important and necessary tool of molecular biology. To cite one application, the genome of a higher animal may contain a number of genes that are similar but not identical in sequence; only by knowing the protein sequence of a product of such related regions can the observer know which DNA sequence is responsible for encoding a particular protein. Even more importantly, perhaps, proteins of interest are most often isolated *before* their genes, so obtaining at least a partial amino acid sequence is a critical first step in studying many proteins. The most popular direct protein-sequencing technique in use today is the

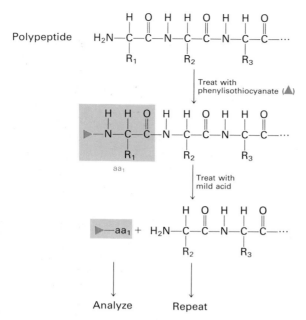

▲ **Figure 6-30**   Sequencing a protein by the Edman degradation procedure. The peptide is treated with phenylisothiocyanate, which combines with the amino-terminal residue in the peptide chain, rendering the first peptide bond in the chain labile to treatment with mild acid. The same pair of reactions is carried out repeatedly to remove the amino acids one at a time. After each step, the removed amino acid is chemically identified. In this way, the entire amino acid sequence of a short peptide can be determined.

Table 6-5   Terms used in recombinant DNA research

Genomic DNA	All DNA sequences of an organism
cDNA (complementary DNA)	DNA copied from an mRNA molecule
Plasmid	A small, circular, extrachromosomal DNA molecule capable of reproducing independently in a host cell
Vector	A plasmid or a viral DNA molecule into which either a cDNA sequence or a genomic DNA sequence is inserted
Host cell	A cell (usually a bacterium) in which a vector can be propagated
Genomic clone	A selected host cell with a vector containing a fragment of genomic DNA from a different organism
cDNA clone	A selected host cell with a vector containing a cDNA molecule from another organism
Library	A complete set of genomic clones from an organism or of cDNA clones from one cell type

Edman degradation procedure, in which amino acid residues are cleaved from a protein one by one; after each cleavage, the released amino acid is identified (Figure 6-30). Machines called sequenators can perform this reaction on tiny amounts of a pure protein; obtaining an accurate sequence of 50 amino acids is not exceptional.

Recombinant DNA: Selection and Production of Specific DNA

The essence of cell chemistry is to purify sufficient quantities of a particular substance to permit its chemical behavior to be analyzed. Segments of pure samples of identical, relatively short DNA molecules from viruses or plasmids can be isolated directly and subdivided into smaller pieces with the use of restriction endonucleases. But the human genome, for example, contains about 3×10^9 bp, so that cutting roughly at every 3000th base pair would produce a million fragments that could not be sep-

arated from each other. This obstacle to obtaining pure DNA samples from large genomes has been overcome by recombinant DNA technology.

Two widely used types of recombinant DNA preparations—genomic clones and cDNA clones—are made. A *genomic clone* contains a fragment of genomic DNA; a *cDNA clone* contains a molecule of *complementary DNA* copied from mRNA by enzymes (Table 6-5). In both, the DNA of interest is linked to a *vector*—most often a bacteriophage or a plasmid that can reproduce independently within a bacterial host. (The most widely used *host-vector systems* are E. coli as host with either a plasmid or bacteriophage λ as the vector.) Recently yeast artificial chromosomes (YACs) have been prepared that can be used as vectors in yeast cells for very large genomic fragments. A *library* consisting of a full set of genomic or cDNA clones can be prepared from the total DNA of an organism or cell type or from the set of cDNA molecules copied from all mRNAs in a cell (Figure 6-31).

The preparation and selection of cDNA and genomic clones are illustrated in the following section by a description of how recombinant DNA containing mouse globin sequences can be obtained.

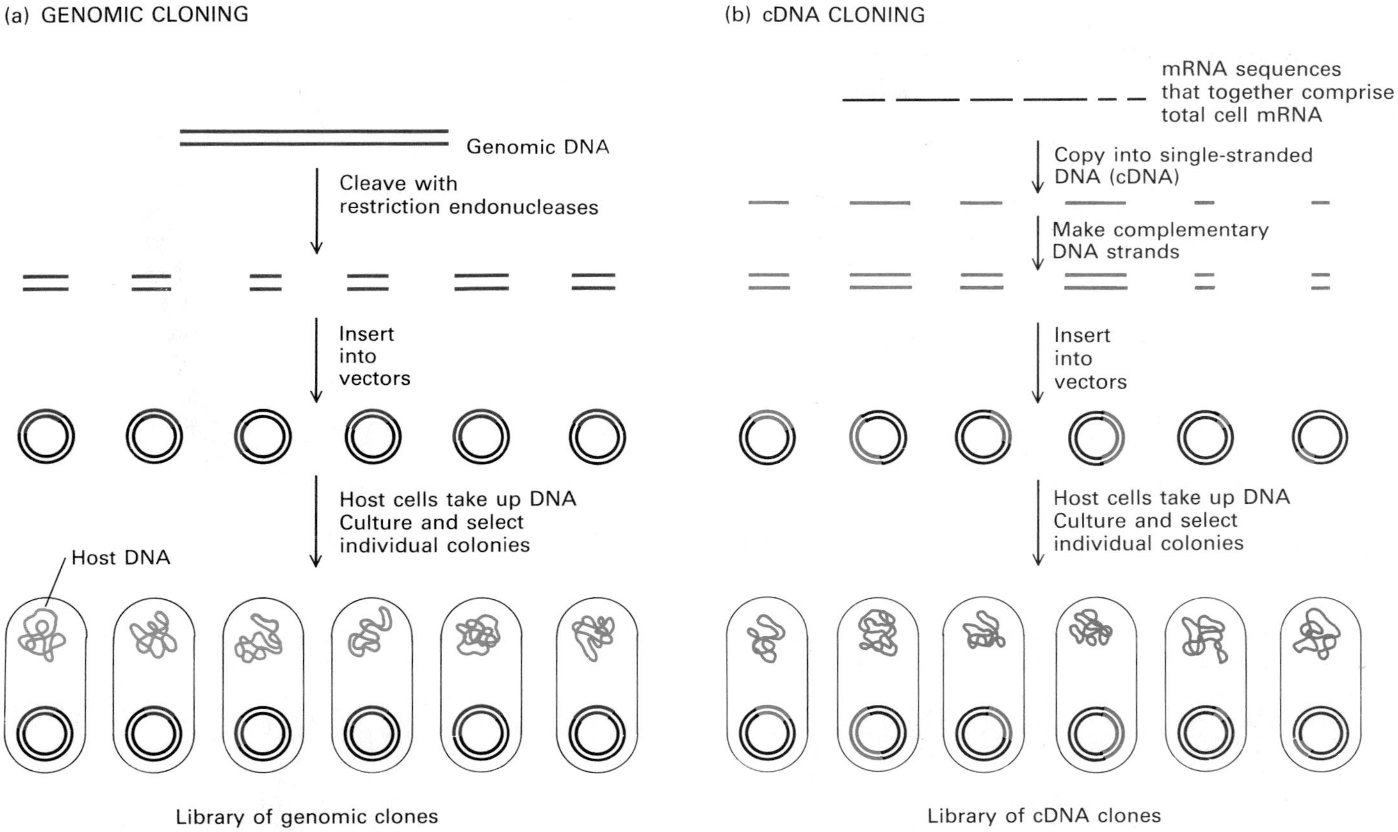

(a) GENOMIC CLONING

(b) cDNA CLONING

▲ **Figure 6-31** A comparison of genomic cloning (a) with cDNA cloning (b). In genomic cloning, the genomic DNA must be cleaved with restriction endonucleases before it can be inserted into vectors; in cDNA cloning, the mRNAs must first be copied into double-stranded DNA molecules.

cDNA Clones Are Whole or Partial Copies of mRNA

To prepare cDNA clones with globin-encoding sequences (Figure 6-32), the starting material is reticulocytes, erythrocyte (red blood cell) precursors. Over 90 percent of the proteins synthesized by these cells are α- and β-globins and therefore they are rich sources of globin mRNA.

The enzyme *reverse transcriptase* (found in retroviruses; see Figure 5-39) is used to make cDNA clones of the reticulocyte mRNAs. Like the DNA polymerases in cells, this enzyme can build a complementary nucleic acid strand on a template, but only by adding nucleotides to a primer. Thus, before the reverse transcriptase can do its work, a short primer strand must be hybridized to the nucleotides near the 3′ ends of the mRNAs. Fortunately, a single oligonucleotide primer—a string of thymidylate residues (poly T)—serves for most eukaryotic mRNAs, which end in a string of 50–250 adenylate residues (poly A).

After the cDNA copy of the mRNA has been made, the mRNA is removed by an alkali treatment that destroys RNA but does not affect DNA and a duplex DNA is made from the cDNA strand. In one technique, the 3′ end of each cDNA strand is elongated by adding several residues of a single nucleotide (say, poly C) through the action of a *terminal transferase,* an enzyme that adds bases at free 3′ ends. A poly G primer is hybridized with the terminal poly C and this G primer is then elongated by a DNA polymerase. What results is a complete double-stranded DNA copy of the original mRNA.

The next step is to insert the now double-stranded DNA into a plasmid. Plasmids, which occur naturally in almost all bacteria, were originally detected by their ability to transfer genes between bacteria (Chapter 5). It has been shown that a specific region of the plasmid circle, the *replication origin,* must be present to assure replication of the plasmid in a host bacterium.

The plasmid DNA is cleaved once with a restriction enzyme at a point that leaves the replication origin undisturbed. The double-stranded copy of the mRNA-globin is then inserted at the cut site and the circle is rejoined. The

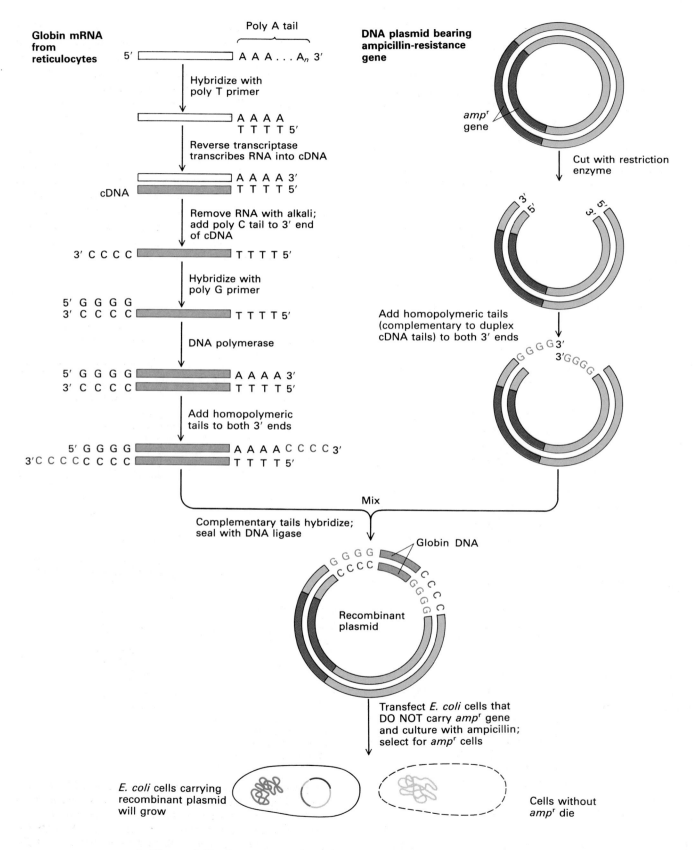

▲ **Figure 6-32** Preparation of a cDNA clone with globin encoding sequences.

first technique, still widely used, for carrying out this insertion is called *homopolymeric tailing*. A homopolymer (say, poly C) is added to the two 3′ ends of the double-stranded cDNA-globin, and a complementary homopolymer (poly G) is added to the 3′ ends of the cut plasmid. When the "tailed" plasmid and DNA-globin are mixed, their complementary single-stranded tails spontaneously hybridize; the resulting circular recombinant molecule can be resealed with the enzyme DNA ligase (Chapters 3 and 12). Specially treated *E. coli* cells take up the plasmid, and the recombinant molecule multiplies along with the cells.

If the chosen plasmid contains a gene for resistance to an antibiotic, the cells that take up the plasmid will grow and multiply in the presence of the antibiotic but the other cells will not. If, at the outset, the number of plasmids allowed to infect the *E. coli* cells is one-tenth or less of the total number of *E. coli* cells, it is very unlikely that more than one plasmid will end up in a recipient cell. As a rule, then, the recombinant DNA in all cells of a colony grown from a single cell will have descended from a single recombinant DNA molecule. In the case described here, 90 percent of the recombinant plasmids would encode α- or β-globin. Because mRNA molecules are often not completely copied, partial sequences also may be cloned. To verify exactly what the plasmid vector contains, the recombinant molecule can be sequenced.

Complementary DNA clones can be prepared from the unpurified mRNA from any cell type, but this produces a random mixture of individual recombinant clones that must be screened to isolate specific ones (see Figure 6-34). It is also possible to use an antibody that reacts with a protein to detect whether an *E. coli* colony (or a plaque if the vector is a bacteriophage) contain the protein encoded by the cloned cDNA.

Genomic Clones Are Copies of DNA from Chromosomes

The most common procedure for preparing and selecting specific clones from genomic DNA—for example, the total collection of DNA in mouse chromosomes—makes use of λ bacteriophage. The DNA of the phage is about 50 kb long, but a center section about 25 kb long can be removed and replaced with foreign DNA without impairing the ability of the phage to infect and reproduce in most *E. coli* cells. A genomic library is a collection of recombinant molecules, maintained either in phage particles or in plasmids growing in bacteria, that includes *all* DNA sequences of a given species. Once it is prepared, the library can be screened for the phage or plasmid that contains the DNA sequence of interest.

The size of a library depends on the amount of DNA in the organism's haploid genome. For example, the human and mouse genomes are between 3 and 4 × 10⁹ bp long.

If one of these genomes were divided into fragments about 20 kb long for insertion into bacteriophage λ, then 2×10^5 different recombinant bacteriophage λ particles would be required to constitute a complete library. Because the pieces of DNA are incorporated into phages by chance, about 10^6 recombinant phages are necessary to assure that each DNA piece has a 90–95 percent chance of being included.

The first step in preparing a genomic library is to extract all the organism's DNA from some cell types (Figure 6-33). Sperm cells or early embryos are often used. The DNA is then broken into fragments by a restriction endonuclease, such as *Eco*RI, which cleaves the DNA in a way that produces short, single-stranded, "sticky" ends (AATT) on every fragment (see Figure 6-20). Digestion is stopped when the average size of a fragment is approxi-

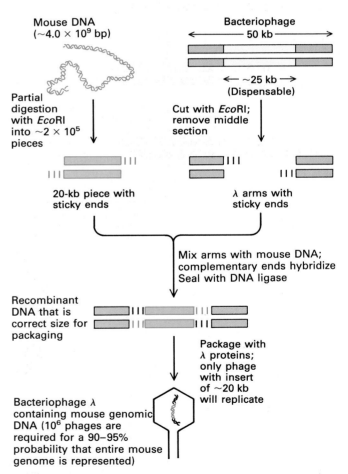

▲ **Figure 6-33** The construction of a genomic library of mouse DNA in bacteriophage λ. The total DNA from mouse cells (both sperm cells and embryonic tissue cells presumably have a complete set of sequences) is often used. A single region of the mouse genome, such as the one that encodes β-globin, would occur approximately once in 10^5 particles.

▶ **Figure 6-34** Selection of a specific genomic clone from a bacteriophage λ library. Although about 2×10^5 phages *could* contain all mouse sequences, 2×10^6 phages are plated to ensure that a phage with the desired sequence is included. This requires an area of 1000–2000 cm^2 to accommodate all the phage plaques. (In the initial plating, the plaques are not allowed to develop to a visible size. The plating can be repeated with fewer phages to obtain pure isolates.) The position of the spot on the autoradiograph identifies the desired plaque on the plate. Phage particles from that plaque can then be selected.

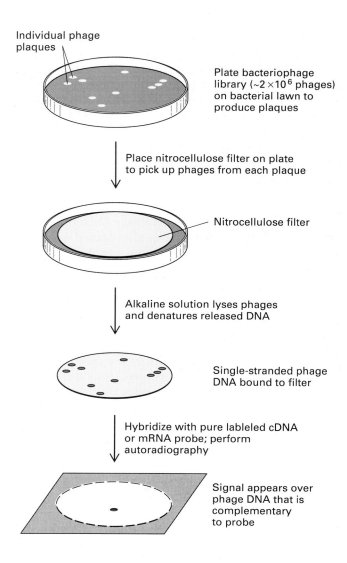

Individual phage plaques

Plate bacteriophage library (~2×10^6 phages) on bacterial lawn to produce plaques

Place nitrocellulose filter on plate to pick up phages from each plaque

Nitrocellulose filter

Alkaline solution lyses phages and denatures released DNA

Single-stranded phage DNA bound to filter

Hybridize with pure labeled cDNA or mRNA probe; perform autoradiography

Signal appears over phage DNA that is complementary to probe

mately 20 kb. The bacteriophage λ DNA also can be cut at two restriction sites by *Eco*RI to yield a center section approximately 25 kb long plus two shorter flanking ends, or *arms*. The center section of the phage DNA can then be separated from the two arms.

The λ arms and the collection of genomic DNA fragments are mixed in about equal amounts (approximately 10^6 DNA fragments and a similar number of pairs of λ arms). Because the sticky ends are complementary, molecules approximately the same length as normal phage DNA will form, but they will include a piece of mouse DNA about 20 kb in length. DNA ligase, the enzyme that normally joins DNA breaks, is used to seal the recombinant DNA molecules, which are then coated with bacteriophage proteins prepared from infected *E. coli* cells. Only DNA molecules of the correct size will be effectively coated, or *packaged,* and give rise to fully infectious λ-bacteriophages, which can be grown on a lawn of *E. coli.* Bacteriophages containing DNA sequences that code for any specific sequence (for example, globin) can be detected by hybridization of cDNA-globin sequences (prepared as described in Figure 6-32) with DNA obtained from each plaque (Figure 6-34).

Vectors for Recombining DNA Exist in Many Cell Types

Any gene that can be subjected to a hybridization assay can be purified. Once a bacteriophage or other bacterial vector containing the desired gene is prepared, an unlimited amount of the pure gene can be obtained by growing the vector and extracting the DNA. Vectors can also carry recombinant DNA molecules in yeast, higher plant cells, and human cells. The vectors most frequently used in mammalian cells are the small DNA viruses SV40 and polyoma or the slightly larger papilloma viruses that can grow as plasmids. Retroviruses are reminiscent of transducing bacteriophages in that they enter a cell, their RNA is copied into DNA, and then inserted into the host chromosomes. Defective retrovirus vectors, which can sponsor DNA copying and insertion but cannot reproduce themselves, promise to be a successful means of gene ther-

apy for individuals with single genetic defects that may be treated in somatic (body) tissue. In plant cells, the most common vector is the Ti plasmid, whose host is *Agrobacterium tumefaciens*; this bacterium fuses with and transfers the recombinant DNA to the plant cell.

For the cell biologist, the availability [through the use of recombinant DNA technology] of unlimited amounts of a pure gene offers rich opportunities for chemical and biological study. Access to vectors in yeast and in cultured mammalian cells affords the additional possibility of testing the biological functions of particular eukaryotic DNA sequences in a variety of eukaryotic cells.

Industrial microbiologists employ recombinant DNA techniques to engineer bacteria and other easily cultured organisms to make proteins for use in medicine, agriculture, and research. A number of viral proteins important in immunizations (for example, against the foot-and-mouth virus in cattle or the hepatitis virus in humans) have already been synthesized in *E. coli,* as have several hormones and enzymes (among them, insulin, growth hormone, and tissue plasminogen activator [TPA], which is used to combat heart attacks). The vectors that direct such programmed protein synthesis, called *expression vectors,* allow the experimenter to take advantage of bacterial genetic tricks that increase mRNA synthesis to produce large quantities of a desired protein.

The Polymerase Chain Reaction Amplifies Specific DNA Sequences in a Mixture

A new procedure called the *polymerase chain reaction* (PCR) can selectively and repeatedly replicate selected segments from a complex DNA mixture (Figure 6-35). This way of amplifying rare sequences from a mixture has vastly increased the sensitivity of genetic tests.

In a typical application of PCR, DNA from a small sample of blood is cut into segments with a restriction endonuclease and denatured into single strands. Oligonu-

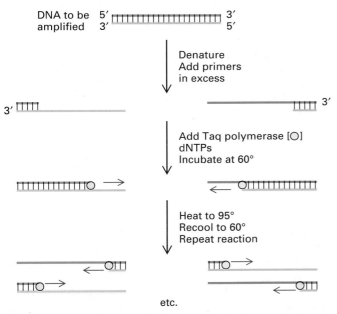

▲ **Figure 6-35** The polymerase chain reaction (PCR). *Taq polymerase,* a heat-resistant DNA polymerase from *Thermus aquaticus,* is used to extend primers between two fixed points on a DNA molecule. All the components for chain elongation (primers, deoxynucleotides, and polymerase) are heat-stable. Thus multiple heating and cooling cycles result in alternating DNA melting and synthesis. DNA between the recognition sites of the two oligonucleotide primers accumulates exponentially. Overnight, it may be amplified as much as a millionfold.

cleotide probes complementary to the 3′ ends of the DNA segment to be amplified are prepared. The probe is added in great excess to the denatured DNA at a temperature between 50° and 60°. The total genomic DNA sample, which is at a low concentration, remains denatured but the specific oligonucleotide probe hybridizes with its correct site on the DNA. The hybridized probe will then serve as a primer for DNA chain synthesis, which begins upon addition of a supply of deoxynucleotides and a temperature-resistant DNA polymerase obtained from *Thermus aquaticus* (a bacterium that lives in hot springs). This enzyme (called the *Taq polymerase*) can extend the primers at high temperatures (up to 72°). When synthesis is complete, the whole mixture is heated further (to 95°) to melt the newly formed DNA duplexes. When the temperature is lowered again, another round of synthesis can take place because excess primer is still present. This cycle of synthesizing and remelting can be repeated to amplify the sequence of interest. At each round, the number of copies of the sequences between the primer sites is doubled and therefore the desired sequence increases exponentially.

The polymerase chain reaction allows specific DNA regions from a tiny sample to be examined quickly. PCR is already in use as a diagnostic procedure in human genetics. In basic research, PCR allows recovery of entire sequences between any two ends whose sequences are known.

Controlled Deletions and Base-Specific Mutagenesis of DNA

The availability of pure DNA in unlimited amounts has permitted a variety of chemical and enzymatic techniques for altering DNA to be developed. The practice of genetics no longer depends on isolating naturally occurring mutant organisms; DNA can be changed in the test tube and reinserted into cells. Thus deletions and mutations can be introduced into genes. Determining the effects on protein structure and changing DNA sequences that may function as genetic regulatory or control elements are two of the most important uses of these techniques.

Two techniques for introducing mutations—the deletion of a short DNA sequence and the alteration of a single base—are illustrated in Figure 6-36. The function of the mutant DNA—whether it is a *deletion mutant* (Figure 6-36a) or a *point mutant* (Figure 6-36b)—can be tested by reintroducing it into a cell by injection or transformation (Chapter 5). The power of this approach is that without knowing the role of a particular sequence beforehand, the experimenter can determine its function by altering its structure and reintroducing it into the organism. Charles Weissman has termed these practices "reverse genetics."

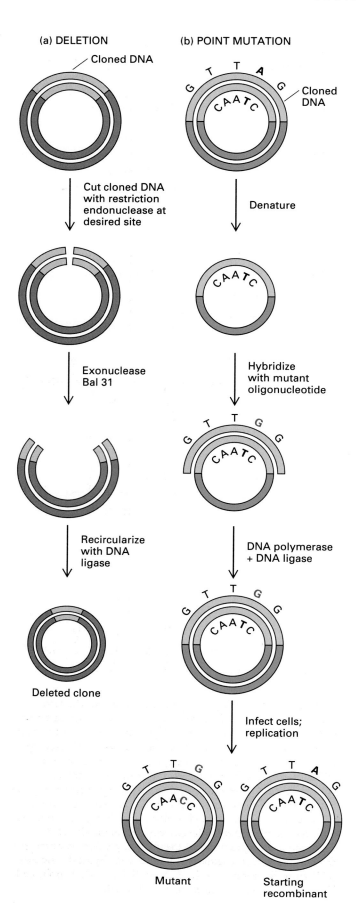

Synthetic Peptide and Nucleotide Sequences: Their Use in Isolating and Identifying Genes

As more and more primary sequences of proteins and nucleic acids become known, the special importance of certain short sequences—regulatory signals in nucleic acids and functional subsections, or "domains," in proteins—become more apparent. These sequences can be chemically synthesized. With such fragments, the function of a part of a protein, rather than the whole protein, can be tested or altered oligonucleotides can be inserted into normal cloned DNA sequences to study the effects of specific mutations (see Figure 6-36).

Another extremely valuable aspect of synthetic oligonucleotides and peptides is that they make it possible to isolate whole genes and pure proteins, respectively. Because the genetic code is universal, a nucleic acid sequence can be used to predict the exact protein sequence it encodes; with less certainty (due to degeneracy in the code), a peptide sequence can be used to predict the approximate nucleic acid sequence that encodes it. Thus it has become feasible to go back and forth between the chemical languages of nucleic acids and proteins to obtain additional information about a polymer of one type or the other (Figure 6-37).

For example, if an mRNA region for a protein that is not yet isolated is cloned and sequenced, a synthetic peptide that is part of the protein can be prepared and used to provoke an antibody that will react with a protein containing that peptide. With such an antibody, the previously unisolated protein corresponding to the already isolated RNA can be identified in cells and purified. A reciprocal selection is also possible: if a protein has been purified and a short region of peptide sequence is avail-

◀ **Figure 6-36** In vitro mutagenesis: constructing DNA deletions and point mutations through the use of recombinant DNA techniques. (a) Deletions are made in cloned DNA in a plasmid by removing entire sections of DNA between two restriction sites or by cutting at a single restriction site and using the exonuclease Bal 31, which removes nucleotides from both ends of a cut double-stranded DNA molecule. Deletions of various lengths are chosen from a collection of such truncated molecules. (b) The two strands of a cloned DNA are separated, and a chemically synthesized oligonucleotide primer (see Figure 6-39) that is mismatched at a desired site is hybridized to one of the DNA strands and then extended by a DNA polymerase. Each strand of the new double-stranded molecule is copied during replication to produce a mixed population of the original DNA and mutants, which are then separately cloned.

able, then oligonucleotides coding for that amino acid sequence can be synthesized and used to screen a genomic or cDNA library for the particular DNA sequence.

The degeneracy of the genetic code is an important consideration in choosing peptides from which to reconstruct partial mRNA sequences. For example, peptides containing arginine, leucine, or serine (six codons each) are to be avoided if possible. The best amino acids for making such probes are tryptophan and methionine (one codon apiece) and phenylalanine, tyrosine, histidine, aspartic acid, glutamic acid, asparagine, and glutamine (two codons each). The number of oligonucleotides that have to be synthesized to be certain of a perfect match with the native mRNA is multiplicative; for example, if a probe is to represent six amino acids with a total of 2, 3, 2, 1, 2, 2 codons, then 48 separate sequences are required.

Techniques for the chemical synthesis of peptides (Figure 6-38) have been available for some time; techniques for DNA oligonucleotide synthesis (Figure 6-39) are also in wide use. The basic logic of these techniques is similar, although the chemistry is different. Note that during chemical synthesis, peptide chains grow from the carboxyl terminus to the amino terminus and DNA chains grow from the 3′ to the 5′ end. Both directions are opposite to the directions of biosynthetic reactions in cells or cell extracts.

FROM GENE TO PROTEIN	FROM PROTEIN TO GENE
Take specific cells (e.g., brain or liver)	Isolate protein (an enzyme or other biologically active protein—e.g., a hormone)
Isolate mRNA; make cDNA	Obtain partial amino acid sequence
Select and sequence individual cDNA clones	Make oligonucleotides that correspond to amino acid sequence
Deduce protein sequence from DNA sequence	Use labeled oligonucleotides to select DNA clone from genomic library
Make peptides specified by sequence; inject into animals to produce antibodies	Sequence selected gene
Isolate pure protein by affinity to antibody	

▲ **Figure 6-37** It is now possible to identify an mRNA of interest (say, an mRNA present in only one part of the brain) and to use it to isolate the protein it encodes without knowing the function of that protein. On the other hand, it is possible to sequence part of a protein that has a specific function (say, an enzyme or a growth factor) and then to synthesize an oligonucleotide that can be used to identify and isolate the gene that encodes the complete protein.

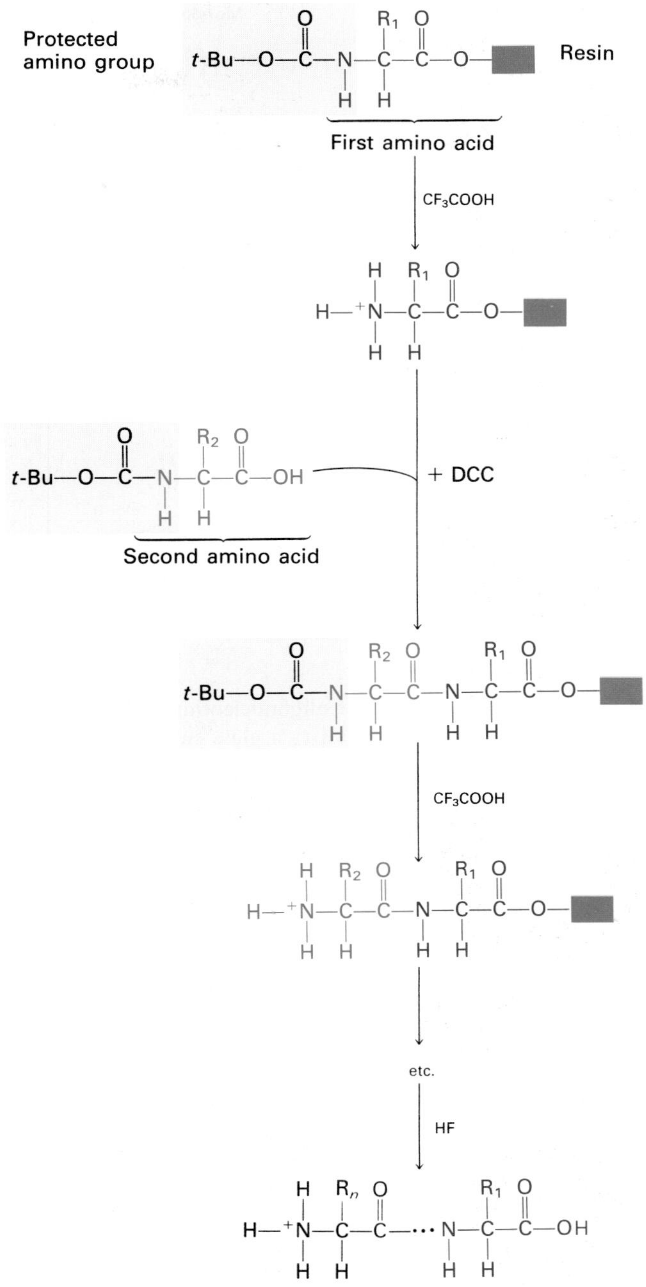

▲ **Figure 6-38** Solid-phase peptide synthesis. The first amino acid of the desired peptide is attached at its carboxyl end by esterification to a resin. The amino group of the first amino acid in the peptide under construction is blocked by the attachment of a *tert*-butyloxycarbonyl group (yellow), which is removed by treatment with trifluoroacetic acid (CF_3COOH). The resulting free amino group forms a peptide bond with a second amino acid, which is presented with a reactive carboxyl group and a blocked amino group, together with the coupling agent dicyclohexylcarbodiimide (DCC). The process is repeated until the desired product is obtained; the peptide is then chemically cleaved with hydrofluoric acid (HF) from the resin. [See R. B. Merrifield, L. D. Vizioli, and H. G. Boman, 1982, *Biochemistry* **21**:5020.]

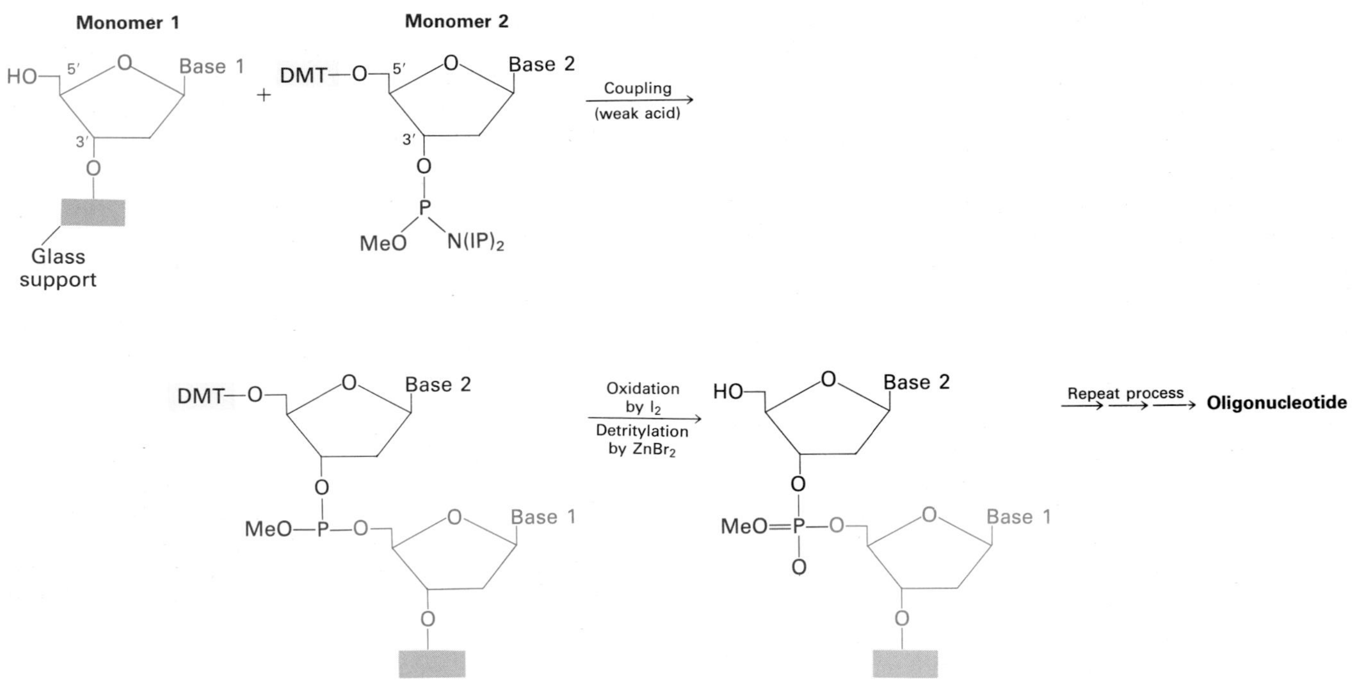

▲ **Figure 6-39** Synthesis of oligonucleotides. The first nucleotide (monomer 1) is bound to a glass support by its 3′ hydroxyl; its 5′ hydroxyl remains available. The synthesis of the first internucleotide link is carried out by mixing monomer 1 with monomer 2, which contains a reactive 3′-diisopropyl phosphoramidite [(IP)$_2$] with attached methyl group (Me), a nucleotide derivative that has the blocking group 4′,4′-dimethoxytrityl (DMT) bound to its 5′ hydroxyl. In the presence of a weak acid, the two nucleotides couple to form a phosphodiester with phorphorus in a trivalent state. Oxidation by iodine (I$_2$) yields a phosphotriester in which the P is pentavalent; detritylation with zinc bromide (ZnBr$_2$) is carried out, and the process is repeated. The methyl groups on the phosphates are all removed at alkaline pH when synthesis is finished. [See S. L. Beaucage and M. H. Caruthers, 1981. *Tetrahedron Letters* **22**:1859.]

Summary

An indispensable adjunct of modern molecular cell biology is the use of isotopes to label biologically important molecules. The isotopes may be radioactive (most commonly used are ^{3}H, ^{14}C, and ^{32}P) or density-labeled (for example, ^{15}N or ^{13}C). These tracers are widely used in cell-free biochemical experiments and in the observation of metabolic events within cells. Important considerations in the use of isotopes include the energy of the emitted particle during radioactive decay, the speed at which various labeled macromolecular precursors enter the cell, and the extent of exchange between compounds in the cell and the medium. For example, tritiated (^{3}H) compounds give the best autoradiographic images because the emitted β particle has a low energy and the image on the photographic emulsion is better defined.

Pulse-chase experiments using labeled amino acids or thymidine to study the synthesis of proteins or DNA can produce clear results because amino acids are exchanged between the cell and the medium within a minute or two; the thymidine enters a very small pool that is quickly consumed by cell growth. However, pulse-chase experiments with labeled RNA precursors are much less effective because ribonucleosides enter a large intracellular pool that is slowly consumed.

Techniques for separating purified molecules from cells have reached the level of a high art. In addition to the many varieties of chromatographic procedures, two basic methods—centrifugation and electrophoresis—are frequently applied to problems in molecular cell biology. Both techniques are most useful in separating molecules according to chain length. Separations of very large molecules that differ by less than 1 percent in size are routine. In addition, separation in two dimensions (by size and by charge) allows the total protein content of cells to be resolved into more than 5000 individual components. The use of electrophoresis to separate nucleic acids on the basis of size has become one of the most common laboratory procedures. In mixtures of chains of 500 nucleotides or less, chains of every length can be separated. These nucleic acid fragments can now be sequenced with such facility that DNA stretches thousands of nucleotides long are typically sequenced within days. Protein sequencing of shorter peptides has been entirely automated, as has the chemical synthesis of oligonucleotides and peptides of 50 units or more in length.

Two aspects of nucleic acid biochemistry—molecular hybridization and nucleic acid enzymology—used in conjunction with microbial genetics have spawned an array of revolutionary techniques for identifying, cloning, and producing natural and mutant nucleic acid sequences. Molecular hybridization (both RNA-DNA and DNA-DNA), the fundamental method of testing the identity of a nucleic acid sample, underlies many of these applications. The detection of a single gene representing perhaps as little as one part in 10^6 of the total human genome is routinely carried out by a hybridization procedure known as the Southern blot. Especially sensitive are the Northern blot, which detects specific mRNA, and the Western blot, which employs antibodies to detect individual proteins.

Among the most important discoveries that allowed gene cloning was the recognition of the restriction endonucleases that cut DNA at characteristic restriction sites of 4–8 bp, thereby generating reproducible fragments from any genome. Enzymes that synthesize DNA and RNA are widely available in highly purified forms, as are enzymes that add to or remove nucleotides from the ends of nucleic acids and enzymes that join DNA segments. The clever use of these enzymes coupled with a deep understanding of microbial genetics that provides exquisitely designed selectable *vectors* to receive tailor-made pieces of DNA has made recombinant DNA experiments commonplace. Synthetic oligonucleotides allow planned deletion and mutation of genes by substitution of sequences in recombinant DNA. Today, any gene can be purified and the functional regions of its DNA sequences can be explored by reintroducing the DNA into cells and into whole organisms. As subsequent chapters will show these fantastic techniques have completely reshaped the way biology is carried out today.

References

Radioisotopes: The Indispensable Modern Means of Following Biological Activity

FREEMAN, L. M., and M. D. BLAUFOX. 1975. *Radioimmunoassay.* Grune & Stratton.

HENDEE, W. R. 1973. *Radioactive Isotopes in Biological Research.* Wiley.

Autoradiography

HUTEN, E., A. KUROUVA, and W. J. GEHRING. 1984. Spatial distribution of transcriptase from the segmentation gene *fushi tarazu* during *Drosophila* embryonic development. *Cell* **37:**833–841.

PARDUE, M. L., and J. G. GALL. 1969. Molecular hybridization of radioactive DNA to the DNA of cytological preparations. *Proc. Nat'l Acad. Sci. USA* **64:**600–604.

ROGERS, A. W. 1979. *Techniques of Autoradiography,* 3d ed. Elsevier/North-Holland.

Labeling Cells and Pools

DINTZIS, H. 1961. Assembly of the peptide chains of hemoglobin. *Proc. Nat'l Acad. Sci. USA* **47:**247–261.

EAGLE, H., and K. A. PIEZ. 1958. The free amino acid pool of cultured human cells. *J. Biol. Chem.* **231:**533–545.

PUCKETT, L., and J. E. DARNELL JR. 1977. Essential factors in the kinetic analysis of RNA synthesis in HeLa cells. *J. Cell Phys.* **90:**521–534.

Labeling Isolated Molecules

BOLTON, A. E., and W. M. HUNTER. 1973. The labeling of proteins to high specific radioactivities by conjugation to a ^{125}I-containing acylating agent. *Biochem. J.* **133:**529–538.

RIGBY, P. W. J., M. DIECKMANN, C. RHODES, and P. BERG. 1977. Labeling deoxyribonucleic acid to high specific activity in vitro by nick translation with DNA polymerase I. *J. Mol. Biol.* **113:**237–251.

Determining Sizes of Nucleic Acids and Proteins

CANTOR, C. R., and P. R. SCHIMMEL. 1980. *Biophysical Chemistry,* part II: *Techniques for the Study of Biological Structure and Function.* W. H. Freeman and Company.

DAWKINS, H. J. S., D. J. FERRIER, and T. L. SPENCER. 1987. Field inversion gel electrophoresis (FIGE) in vertical slabs as an improved method for large DNA separation. *Nuc. Acids Res.* **15:**3634–3636.

SMITH, C. L., J. G. ECONOME, A. SCHUTT, S. KLCO, and C. R. CANTOR. 1987. A physical map of the *Escherichia coli* K12 genome. *Science* **236:**1448–1453.

Centrifugation

DE DUVE, C. 1975. Exploring cells with a centrifuge. *Science* **189:**186–194.

SCHEELER, P. 1980. *Centrifugation in Biology and Medical Science.* Wiley.

VINOGRAD, J. 1963. Sedimentation equilibrium in a buoyant density gradient. In *Methods in Enzymology,* vol. VI, S. P. Colowick and N. O. Kaplan, eds. Academic Press.

VINOGRAD, J., R. RASLOFF, and W. BAUER. 1967. A dye-buoyant density method for the detection and isolation of closed circular duplex DNA. *Proc. Nat'l Acad. Sci. USA* **57:**1514–1521.

Electrophoresis

ANDREWS, A. T. 1986. *Electrophoresis, 2d ed.* Oxford University Press.

CANTOR, C. R., C. L. SMITH, and M. K. MATTHEW. 1988. Pulsed-field gel electrophoresis of very large DNA molecules. *Ann. Rev. Biophysics & Biophysical Chem.* **17:**41–72.

CARLE, G. F., M. FRANK, and M. V. OLSON. 1986. Electrophoretic separations of large DNA molecules by periodic inversion of the electric field. *Science* **232:**65–70.

CELIS, J. E. and R. BRAVO. 1983. Two-dimensional gel electrophoresis of proteins. Academic Press.

CROTHERS, D. M. 1987. Gel electrophoresis of protein-DNA complexes. *Nature* 325:464–465.

MAIZEL, J. V. JR. 1971. Polyacrylamide gel electrophoresis of viral proteins. In *Methods in Virology*, vol. V, K. Maramorosch and H. Koprowski, eds. Academic Press.

O'FARRELL, P. H. 1975. High resolution, two-dimensional electrophoresis of proteins. *J. Biol. Chem.* 250:4007–4021.

RICKWOOD, D., and B. D. HAMES, eds. 1982. *Gel Electrophoresis of Nucleic Acids*. London: IRL Press Ltd.

Examining Sequences of Nucleic Acids and Proteins

Nucleic Acids: Hybridization

ALWINE, J. C., D. J. KEMP, and G. R. STARK. 1977. A method for detection of specific RNAs in agarose gels by transfer to diazobenzyloxymethyl-paper and hybridization with DNA probes. *Proc. Nat'l Acad. Sci. USA* 74:5350–5354.

BERK, A. J., and P. A. SHARP. 1977. Sizing and mapping of early adenovirus mRNAs by gel electrophoresis of S1 endonuclease digested hybrids. *Cell* 12:721–732.

BRITTEN, R. J., and E. D. KOHNE. 1968. Repeated sequences in DNA. *Science* 161:529–540.

DAVIS, R. W., and N. DAVIDSON. 1968. Electron microscopic visualization of deletion mutations. *Proc. Nat'l Acad. Sci. USA* 60:243–250.

HALL, B. D., and S. SPIEGELMAN. 1961. Sequence complementarity of T2-DNA and T2-specific RNA. *Proc. Nat'l Acad. Sci. USA* 47:137–146.

HU, N., and J. MESSING. 1982. The making of single-stranded probes. *Gene* 17:271–277.

THOMAS, M., R. L. WHITE, and R. W. DAVIS. 1976. Hybridization of RNA to double-stranded DNA: formation of R loops. *Proc. Nat'l Acad. Sci. USA* 73:2294–2298.

THOMAS, P. S. 1980. Hybridization of denatured RNA and small DNA fragments transferred to nitrocellulose. *Proc. Nat'l Acad. Sci. USA* 77:5201–5205.

WETMUR, J. G., and N. DAVIDSON. 1968. Kinetics of renaturation of DNA. *J. Mol. Biol.* 31:349–370.

Nucleic Acids: Restriction Enzymes

DAVIES, K. E. ed., 1988. *Genome Analysis: A Practical Approach*. Oxford: IRL Press.

NATHANS, D., and H. O. SMITH. 1975. Restriction endonucleases in the analysis and restructuring of DNA molecules. *Ann. Rev. Biochem.* 44:273–293.

NELSON, M., and M. MCCLELLAND. 1989. The effect of site-specific methylation on restriction–modification enzymes. *Nuc. Acids Res.* 17(Suppl.):r389–r415.

ROBERTS, R. J. 1989. Restriction enzymes and their isoschizomers. *Nuc. Acids Res.* 17(Suppl.):r347–r388.

SMITH, H. O. 1970. Nucleotide sequence specificity of restriction endonucleases. *Science* 205:455–462.

SMITH, H. O., and M. L. BERNSTIEL. 1976. A simple method for DNA restriction-site mapping. *Nuc. Acids Res.* 3:2387–2398.

SOUTHERN, E. M. 1975. Detection of specific sequences among DNA fragments separated by gel electrophoresis. *J. Mol. Biol.* 98:503–517.

Nucleic Acids: Sequence Analysis

BROWNLEE, G. 1972. Determination of sequences in RNA. In *Laboratory Techniques in Biochemistry and Molecular Biology*, vol. 3, T. S. Work and E. Work, eds. Elsevier/North-Holland.

CHURCH, G. M., and S. KIEFFER-HIGGINS. 1988. Multiplex DNA sequencing. *Science* 240:185–188.

DOOLITTLE, R. F. 1986. *Of Urfs and Orfs: A Primer on How to Analyze Derived Amino Acid Sequences*. University Science Books.

ERLICH, H. E., R. A. GIBBS, and H. H. KAZAZIAN JR., eds. 1989. *Polymerase Chain Reaction*. Cold Spring Harbor Laboratory.

MAXAM, A. M., and W. GILBERT. 1977. A new method for sequencing DNA. *Proc. Nat'l Acad. Sci. USA* 74:560–564.

MAXAM, A. M., and W. GILBERT. 1980. Sequencing end-labeled DNA with base-specific DNA. In *Methods in Enzymology*, vol. 65, L. Grossman and K. Moldave, eds. Academic Press.

SAIKI, R. K., D. H. GELFAND, S. STOFFEL, S. J. SCHARF, R. HIGUICHI, G. T. HORN, K. B. MULLIS, and H. A. ERLICH. 1987. Primer-directed enzymatic amplification of DNA with a thermostable DNA polymerase. *Science* 239:487–491.

SANGER, F. 1981. Determination of nucleotide sequences in DNA. *Science* 214:1205–1210.

Proteins: Sequence Analysis

CLEVELAND, D. W., S. G. FISCHER, M. W. KIRSCHNER, and U. K. LAEMMLI. 1977. Peptide mapping by limited proteolysis in sodium dodecyl sulfate and analysis by gel electrophoresis. *J. Biol. Chem.* 252:1102–1106.

HUNKAPILLER, M. W., and L. E. HOOD. 1983. Protein sequence analysis: automated microsequencing. *Science* 219:650–659.

INGRAM, V. M. 1956. A specific chemical difference between the globins of normal human and sickle-cell anaemia haemoglobin. *Nature* 178:792–794.

SANGER, F. 1952. The arrangement of amino acids in proteins. *Adv. Protein Chem.* 7:1–67. A description of original sequencing, fingerprinting, and partial-chain-analysis methods.

WALSH, K. A., L. H. ERICSSON, D. C. PARMELEE, and K. TITANI. 1981. Advances in protein sequencing. *Annu. Rev. Biochem.* 50:261–284. A review of up-to-date methods, including automated sequence analysis.

Recombinant DNA, Gene Cloning, and Mutagenesis

AUSUBEL, F. M., R. BRENT, R. E. KINGSTON, R. E. MOORE, D. D. SEIDMAN, J. G. SMITH, J. A. STRUHL, and K. STRUHL. 1987. *Current Protocols in Molecular Biology*. Wiley.

BENTON, W. D., and R. W. DAVIS. 1977. Screening λgt recombinant clones by hybridization to single plaques in situ. *Science* 196:180–183.

BERG, P. 1981. Dissections and reconstructions of genes and chromosomes. *Science* 213:296–303.

BOTSTEIN, D., and D. SHORTLE. 1985. Strategies and applications of in vitro mutagenesis. *Science* 229:1193–1201.

COHEN, S. N., A. C. Y. CHANG, H. W. BOYER, and R. B. HELLING. 1973. Construction of biologically functional bacterial plasmids in vitro. *Proc. Nat'l Acad. Sci. USA* **70**:3240–3244.

COOKE, H. 1987. Cloning in yeast: An appropriate scale for mammalian genomes. *Trends Genet.* **3**:173–175.

GRUNSTEIN, M., and D. S. HOGNESS. 1975. Colony hybridization: a method for the isolation of cloned DNAs that contain a specific gene. *Proc. Nat'l Acad. Sci. USA* **72**:3961–3965.

HOHN, B. 1979. In vitro packaging of λ and cosmid DNA. In *Methods of Enzymology: Recombinant DNA*, vol. 68, R. Wu, ed. Academic Press.

OKAYAMA, H., and P. BERG. 1982. High-efficiency cloning of full-length cDNA. *Mol. Cell. Biol.* **2**:161–170.

OLD, R. W., and S. B. PRIMROSE. 1985. *Principles of Gene Manipulation: An Introduction to Genetic Engineering.* Blackwell Scientific Publications.

SAMBROOK, J., E. F. FRITSCH, and T. MANIATIS. 1989. *Molecular Cloning.* Cold Spring Harbor Laboratory.

SINGH, H., J. H. LEBOWITZ, A. S. BALDWIN JR., and P. A. SHARP. 1988. Molecular cloning of an enhancer binding protein: isolation by screening of an expression library with a recognition-site DNA. *Cell* **52**:415–423.

SMITH, M. 1985. In vitro mutagenesis. *Ann. Rev. Genet.* **19**:423–462.

THOMAS, K. R., and M. R. CAPECCHI. 1987. Site-directed muta- genesis by gene targeting in mouse-embryo-derived stem cells. *Cell* **51**:503–512.

Synthetic Peptide and Nucleotide Sequences

Synthesis of Polypeptides

KENT, S. B. H. 1988. Chemical synthesis of peptides and proteins. *Ann. Rev. Biochem.* **57**:957–990.

MERRIFIELD, B. 1986. Solid phase synthesis. *Science* **232**:341–347.

PELHAM, H. R. B., and R. J. JACKSON. 1976. An efficient mRNA dependent translation system from reticulocyte lysates. *Eur. J. Biochem.* **67**:247–256. A description of the most frequently used in vitro technique for protein synthesis.

Synthesis of Oligonucleotides

CARUTHERS, M. H. 1985. Gene synthesis machines: DNA chemistry and its uses. *Science* **230**:281–285.

ITAKURA, K., and A. D. RIGGS. 1980. Chemical DNA synthesis and recombinant DNA studies. *Science* **209**:1401–1405.

ITAKURA, K., J. J. ROSSI, and R. B. WALLACE. 1984. Synthesis and use of synthetic oligonucleotides. *Ann. Rev. Biochem.* **53**:323–356.

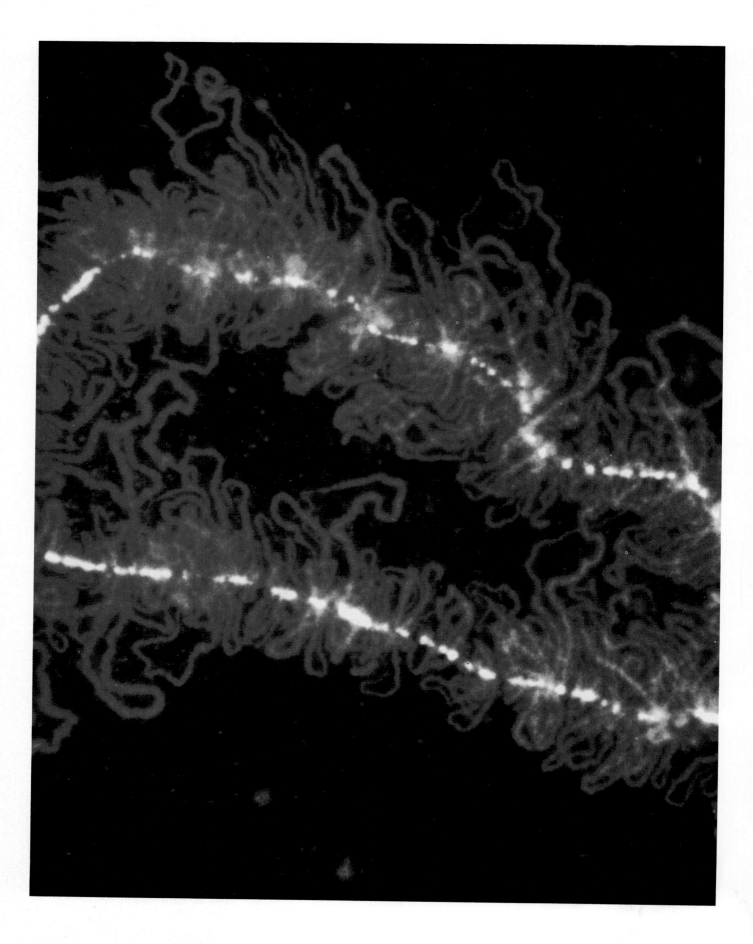

II

Gene Expression, Structure, and Replication

◀ *A portion of a lampbrush chromosome bivalent from the newt,* Notophthalmus viridescens. *The chromomere axes fluoresce white after staining with the DNA-specific dye DAPI. The lateral transcription loops fluoresce red after indirect immunofluorescent staining with a monoclonal antibody against a ribonucleoprotein.* Courtesy of M. Roth and J. Gall.

What allows cells to function? What makes some cells different from others? The idea governing all discussions of these questions is that *the actions and properties of each cell type are determined by the proteins it contains*. But what determines the amount of each protein in a cell? The concentration of the protein's corresponding messenger RNA, the frequency with which the messenger RNA is translated, and the stability of the protein itself are the determining factors. These three factors determine which genes are expressed and the extent to which they are expressed in each cell. Thus the differential activity of different genes, or *gene control*, determines the actions and properties of cells. How nucleic acid synthesis and gene control drive the activities of cells is the subject of the next six chapters.

Foundations for understanding gene control came out of studies with bacterial cells. Chapter 7 discusses these studies and the knowledge they contributed. What mechanisms of gene control characterize the relatively uncompartmentalized environment of the prokaryotic cell, where DNA, mRNA, rRNA, tRNA, and proteins all may have direct access to each other? Does the thoroughly studied and well understood subject of RNA synthesis and its control in bacteria provide useful clues for understanding gene expression in eukaryotes? With very rare exceptions, DNA in bacteria is transcribed into mature messenger RNA that requires no modification before it can be used. How similar is this to what happens in eukaryotic cells? How does the expression of genes in eukaryotes differ from that in prokaryotes? Chapter 8 describes RNA synthesis in eukaryotic cells and the processing that it must undergo to become functional.

Two chapters on chromosomes follow the chapter on eukaryotic RNA. There are two reasons for this sequence of topics. First, biologists would never have understood chromosome anatomy at the molecular level if they had not known about RNA processing. Second, molecular cell biologists study chromosomes primarily to find out how individual genes are regulated, and this, of course, demands that one first appreciate RNA synthesis. Chapter 9 discusses the general structure of eukaryotic chromosomes. What are the distinctive features of chromosomes that can be seen through a microscope? What knowledge was obtained from classic genetics about the arrangement of genes on chromosomes?

What is required for the duplication and segregation of chromosomes? How do proteins package DNA into chromosomes? How do the genes of classic genetics—the protein-coding sequences of DNA—fit into the molecular anatomy of chromosomes? Chapter 10 describes eukaryotic chromosomes and their genes at the molecular level.

From the working out of the steps of protein synthesis, the mystery of gene control in eukaryotes emerged: at what levels, or at what molecular steps, is the expression of genes in eukaryotes controlled? How is the development of a multicellular organism programmed in the turning on and off of genes? What are the signals that specify "on" and "off"? Information, described in Chapter 11, relevant to these questions has accumulated with astonishing rapidity although many tantalizing questions concerning eukaryotic gene control are still unanswered. How have molecular studies contributed understanding at a rate faster than could have been thought possible during the period of classic genetics when the questions clearly emerged? What are the molecular probes that have unlocked the secrets?

The chromosomes must, of course, duplicate in cell division. In Chapter 12 we deal with the molecular events in DNA replication and new chromosome assembly. How similar are the structures and biochemical mechanisms for replicating DNA in prokaryotic and eukaryotic cells? What complications characterize the eukaryotic system? Truly remarkable progress has been made in the understanding of bacterial replication systems, and knowledge about eukaryotic systems is rapidly increasing.

▼   ▼   ▼

CHAPTER 7

RNA Synthesis and Gene Control in Prokaryotes

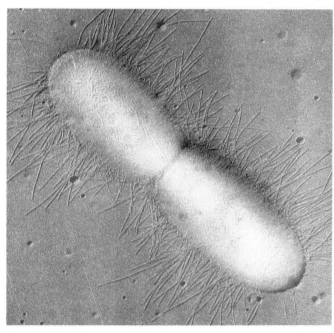

E. *coli* with pili

*O*ur discussion of gene expression and gene regulation begins with prokaryotic genes because the detailed knowledge now available about prokaryotic gene control is helpful in understanding eukaryotic gene control, which is both more complicated and not as well understood. Even in bacteria, in which gene control is carried out primarily by DNA-binding proteins that affect transcription, several different types of transcriptional control circuits exist. Eukaryotic cells use some of the most important principles in transcriptional regulation, but eukaryotic cells exhibit fundamental differences in cellular organization that affect gene expression and gene control. Because these differences must be kept firmly in mind, we review them briefly before proceeding with the discussion of bacterial gene control. ▲

In bacterial cells, which do not contain an organized cell nucleus, an mRNA molecule is completely accessible to ribosomes and other elements of the protein-synthesizing apparatus during its formation. Since both transcription of DNA into mRNA and the ribosomal translation of mRNA proceed in the 5′ → 3′ direction, bacterial protein synthesis can begin on an mRNA molecule even while it is still being formed (Figure 7-1). Furthermore, bacterial mRNA molecules are not chemically modified before they are translated. Thus, *coupled transcription-translation* is possible and indeed is the rule in bacterial cells.

In eukaryotic cells, DNA transcription occurs in the nucleus, which does not contain mature ribosomes engaged in protein synthesis, although it does contain ribosomal precursors undergoing maturation in the nucleolus. Furthermore, newly formed eukaryotic mRNA precursors are extensively and specifically modified in the nucleus before they emerge into the cytoplasm to associate with ribosomes. These modifications include the addition of chemical groups at both ends and, in most cases, the cutting of the molecule into segments and splicing together of the pieces. Thus, coupled transcription-translation cannot occur in eukaryotic cells because (1) the sites at which transcription and translation occur are structurally separated and (2) the initial RNA transcript must be modified to form an mRNA that is functional in translation.

Prokaryotes and eukaryotes also differ in the apparent purpose of gene control. In bacteria, gene control serves mainly to allow the single cell to adjust to changes in its nutritional environment so that its growth and division can be optimized. Although some genes in metazoan organisms also can respond directly to environmental changes, the most characteristic and biologically far-reaching purpose of gene control in eukaryotes is the regulation of a genetic program that underlies embryologic development and tissue differentiation.

Overall Strategy of Prokaryotic Gene Control

Because ribosomes begin translating nascent mRNA in bacteria as soon as the first ribosome-binding site on the mRNA is formed, *initiation of transcription* is a critical point of gene control in bacterial cells. However, transcription can undergo a controlled termination before the coding sequence of a gene has been reached, and sometimes transcription can terminate between two genes. Thus, control of the *termination of transcription* can be as important as control of initiation. Presumably, such controlled termination reactions spare a bacterium from producing mRNA that is not needed.

The concentration of a given mRNA in any cell depends not only on its rate of synthesis but also on its rate

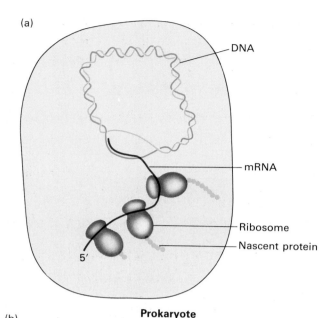

(a)

Prokaryote

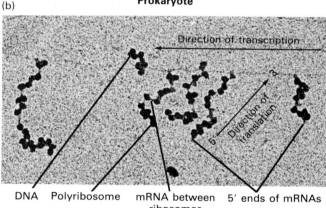

(b)

▲ **Figure 7-1** Coupled transcription-translation in prokaryotes. (a) Translation (protein synthesis) from 5′ end of mRNA molecule begins while transcription is still in process. (b) Electron micrograph of a transcribed portion of the *E. coli* chromosome with chains of ribosomes (called polyribosomes) strung along the growing mRNAs. The indicated direction of transcription is inferred because the longer mRNAs (loaded with more ribosomes) are further along in the transcription process. The direction of translation is likewise inferred because the ends of the mRNAs free of the DNA template are the 5′ ends, and mRNA is translated 5′ → 3′. *Photograph from O. L. Miller Jr., B. A. Hamkalo, and C. A. Thomas Jr., 1970, Science* **169**:*392.*

of degradation—that is, on its *metabolic stability.* Because the lifetime of mRNA in a bacterial cell usually is considerably shorter than the doubling time of the cell, once synthesis of a specific mRNA ceases, that mRNA quickly disappears and synthesis of its corresponding protein stops. A few bacterial mRNAs are known to have longer-than-average lifetimes, but even these longer-lived mRNAs last only half as long as the doubling time of bacterial cells.

In summary, then, three elements of control—transcriptional initiation, transcriptional termination, and rapid mRNA turnover—constitute the underlying strategy of bacterial gene control.

Control of Transcriptional Initiation

In this section, the main principles of the initiation of transcription in bacteria are discussed, and many important terms are introduced. The following two sections include detailed descriptions of negative control of the genes for lactose-metabolizing enzymes and positive control of the genes for arabinose-metabolizing enzymes; these two gene systems were particularly important in the historical development of knowledge about gene control. Additional aspects of the control of initiation are then discussed in sections on compound control and regulatory proteins.

Initiation of Transcription in Prokaryotes Entails Sequence Recognition by RNA Polymerase Plus Sigma Factors

Transcription of DNA in bacteria is catalyzed by a single RNA polymerase. The purified polymerase, or *core enzyme*, consisting of two identical α subunits and two similar but nonidentical subunits, β and β', can bind with low affinity to any region in double-stranded DNA. By binding loosely, releasing momentarily, and then binding again, an RNA polymerase efficiently explores the DNA. With the aid of one of a set of accessory proteins called *sigma* (σ) factors, RNA polymerase recognizes and binds to specific DNA regions called *promoters*. After the enzyme binds at a specific promoter, the bound *holoenzyme* (RNA polymerase plus sigma factor) causes the double-stranded DNA structure to *melt* or unwind, so that transcription can begin (Figure 7-2). The first or 5' RNA nucleotide is most often a purine (A or G), but pyrimidine nucleotides (C or U) are occasionally found. The sigma

(a)

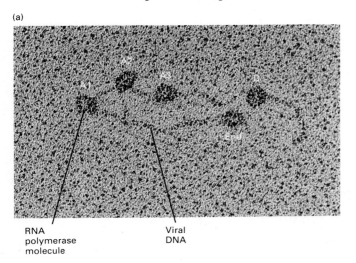

RNA polymerase molecule

Viral DNA

(b)

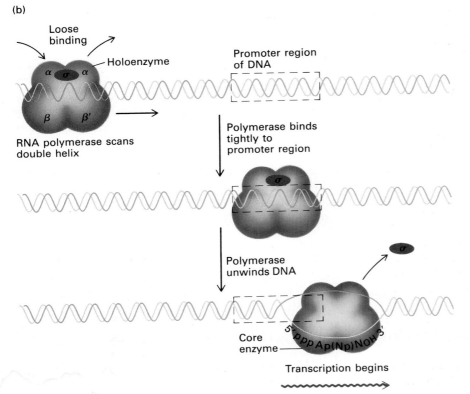

Loose binding

Holoenzyme

Promoter region of DNA

RNA polymerase scans double helix

Polymerase binds tightly to promoter region

Polymerase unwinds DNA

Core enzyme

Transcription begins

◀ **Figure 7-2** The RNA polymerase of *E. coli* is a large enzyme with a total mass of almost 500 kilodaltons (kDa). The holoenzyme, which initiates RNA synthesis at appropriate sites, has five subunits and is designated $\alpha_2\beta\beta'\sigma$; the *core enzyme* lacks the σ factor. (a) Electron micrograph of *E. coli* RNA polymerase molecules bound to DNA from bacteriophage T7. The polymerase reacts nonspecifically with the ends of all DNAs. However, the other four polymerase molecules are bound at specific promoters (A1, A2, A3, and D) on the phage DNA. (b) The holoenzyme binds weakly to and explores the DNA double helix until, with the aid of the sigma (σ) factor, it recognizes a promoter sequence to which it binds tightly. The holoenzyme melts (unwinds) a short section of DNA and begins transcription. The sigma factor is released by the time the first few nucleotides have been joined, and the core enzyme continues RNA chain elongation. [See R. C. Williams, 1977, *Proc. Nat'l Acad. Sci. USA* **74**:2311.] *Photograph courtesy of R. C. Williams.*

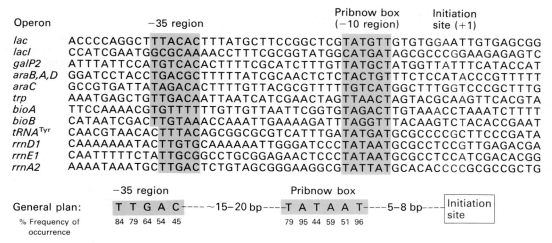

▲ **Figure 7-3** The nucleotide sequences in several promoter sites that direct transcription by *E. coli* RNA polymerase containing the major sigma factor, σ^{70}. Each sequence shown belongs to a specific operon, that is, a cluster of genes that are all controlled by the same promoter. The Pribnow box is a strongly conserved region of six bases; a sequence around the −35 position, which has a five- or six-base core, is also conserved. The bases in red type are initiation sites. Usually transcription starts at a single nucleotide; occasionally multiple, alternate sites are present (e.g., in the *lac* and *rrnA2* op-erons). The general plan of the promoters is shown at the bottom of the diagram; the percentages below each base in-dicate the frequency with which that base was present in a comparison of 112 different promoters. Note that the initia-tion site for *bioB* (a gene for the vitamin biotin) is at a thy-mine (which becomes a uracil in the mRNA) and that the site for *rrnA2* (a gene for rRNA) has three cytosines; all of the other initiation sites occur at an adenine or a guanine. [See D. Hawley and W. R. McClure, 1983, *Nucleic Acids Res.* **11**:2237.]

factor is released after a few nucleotides have been joined. About 10 to 12 nucleotides of a growing RNA chain are hydrogen-bonded to the DNA at any one time. As the chain elongates, the RNA "peels off" the DNA. Experi-ments with cross-linking reagents that attach RNA to protein have shown that newly synthesized RNA lies near the surfaces of the β and β' subunits.

The most extensively studied bacterial promoters are those in *E. coli*. They are about 40 bases long and include a common sequence of about 6 bases located upstream (i.e., in the 5′ direction) from the site at which RNA syn-thesis begins. This sequence is called the *Pribnow box* after the investigator who first recognized it. The position of the nucleotide at the initiation site is designated +1; the nucleotides that precede this site are designated with negative numbers (there is no zero position). Five to eight bases separate the Pribnow box from the initiation site, so the Pribnow box centers around the −10 position (Figure 7-3). Although not all of the common *E. coli* promoters have exactly the same sequence of bases in this region, a sequence similar to TATAAT is the rule. The first A and the final T residue of this sequence are almost always present; the correspondence at each site in the rest of the Pribnow box ranges from about 45 to 80 percent. Such a conserved region is called a *consensus sequence*. Another consensus sequence, TTGAC, centers around the −35 position; this region is also critical for the accurate and rapid initiation of transcription of most bacterial genes. In most *E. coli* genes, these consensus sequences are used

to direct transcription. In many other bacteria, including those related to *E. coli* (e.g., *Salmonella typhimurium*) and those not closely related (e.g., *Bacillus subtilis*), the same consensus sequences have been found in most genes.

As noted already, recognition of a promoter by RNA polymerase and binding of the enzyme to it are mediated by the sigma factor. Various sigma factors have been identified, and each permits RNA polymerase to recog-nize promoters that contain different consensus se-quences in the −35 region (Table 7-1). The exact mecha-nism of this differential recognition is not currently understood. However, the amino acid sequences of more than a dozen different sigma factors are now known; the presence of long stretches of similar amino acids in these proteins indicates that they are related.

In *E. coli* exposed to high temperatures, for example, normal RNA synthesis stops and production begins of a sigma factor (σ^{32}) that recognizes another set of consen-sus sequences, those associated with the "heat-shock" genes. RNA polymerase plus σ^{32} then recognizes these promoter sequences, and the heat-shock genes are tran-scribed. *Bacillus subtilis* and other bacteria that form spores employ different sigma factors to recognize pro-moters that are used during normal *vegetative* growth and during *sporulation* (the process that results in sealing of the bacterium in a dessicated heat-proof package called a *spore*). The vegetative factor (σ^{43}) is similar to the major *E. coli* factor. *E. coli* cells subjected to nitrogen starvation produce a sigma factor (σ^{54}) that recognizes

Table 7-1 Sigma (σ) factors and the promoter consensus sequences they help recognize

Nature of σ factor*	Consensus sequences	
	−35 region	−10 region
MAJOR BACTERIAL FACTORS		
Escherichia coli (σ⁷⁰)		
Salmonella typhimurium (σ⁷⁰) }	TTGACA	TATAAT
Bacillus subtilis (σ⁴³)		
MINOR BACTERIAL FACTORS		
E. coli (heat shock; σ³²)	CTTGAA	CCCGATAT
E. coli (nitrogen starvation; σ⁵⁴)		(-27)CTGGYAYRN₄TTGCA(-10)
B. subtilis (flagellar synthesis; σ²⁸)	CTAAA	CCGATAT
B. subtilis (sporulation; σ²⁹)	TT-AAA	CATATT
BACTERIOPHAGE FACTORS		
T4 (*E. coli* phage)	None	TATAAATA
SPO1-28 (*B. subtilis* phage)	TNAGGAGANNA	TTTNTTT
SPO1-33/34	CGTTTAGA	GATATT

*Superscript numbers indicate molecular weight in kilodaltons of a sigma factor.
SOURCE: R. Losick and J. Pero, 1981, *Cell* **25**:582; J. F. Briat and F. Neidhardt, 1984. *Ann. Rev. Genet.* **18**:295; and J. Helman and M. J. Chamberlain, 1988, *Ann. Rev. Biochem.* **57**:839.

promoters for genes encoding enzymes needed for recycling of organic nitrogen compounds; thus, by "turning on" synthesis of such enzymes, which normally are not needed, the organism can survive nitrogen deficit. Other sigma factors involved in the response of bacteria to various stressful conditions probably will be identified in the future. Finally, many bacteriophages take over bacterial cell metabolism by changing sigma factors, thereby causing the RNA polymerase of the cell to copy virus genes rather than bacterial genes.

RNA Polymerase "Footprints" Elegant chemical experiments with purified DNA fragments containing promoters have proved that RNA polymerase touches DNA at certain bases in the conserved regions of promoters. In these experiments, RNA polymerase is bound to DNA, which is then subjected to DNase attack or to chemical reagents that modify the chain at specific sites unprotected by the polymerase; further chemical treatment breaks the chain at the modified base. When the DNase-digested or chemically broken DNA is analyzed by gel electrophoresis, the regions protected by RNA polymerase appear as gaps ("footprints") in the array of bands (Figures 7-4 and 7-5). Mutant DNAs that have decreased promoter activity show decreased protection of

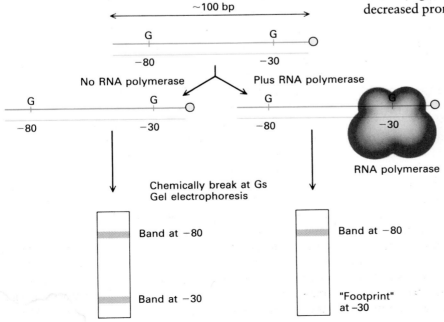

◄ **Figure 7-4** Footprint experiments are used to determine the sites of interaction between DNA and a protein. In this case, a purified double-stranded DNA fragment that is labeled at the 5' end of one strand is allowed to interact with RNA polymerase. The DNA-protein complex is then chemically treated (or subjected to DNase attack) in such a way that the chain breaks at certain bases if they are not in contact with the protein. Examination of the band pattern of the denatured DNA after gel electrophoresis indicates which of the bases on the labeled strand of DNA were protected by protein. In this diagram, only two such sites (two guanines) are shown.

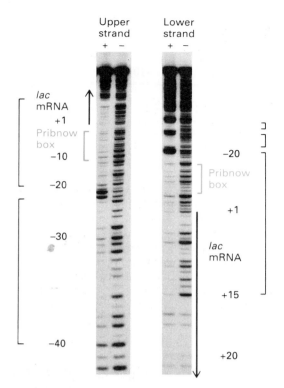

▲ **Figure 7-5** RNA polymerase association with the *lacUV5* operator demonstrated by the DNase "footprint" method. A cloned segment of the *lacUV5* promoter region was end-labeled on either the upper or lower strand. The two labeled DNA samples were then treated with DNase either with (+) or without (−) RNA polymerase being present. The DNase treatment was adjusted so that it nicked the DNA an average of once per fragment. When the unprotected DNA was examined by gel electrophoresis, a "ladder" of fragments was seen due to individual breaks introduced by the enzyme. In the samples where RNA polymerase holoenzyme was present, digestion was blocked in certain regions (*brackets*). From these experiments the regions of the fragment protected or "footprinted" by the enzyme were determined. The gel patterns show that the enzyme covers regions from about 40 nucleotides before and 15 after the start site for RNA synthesis, with regions in between where digestion is actually enhanced. [The exact assignments were made in other experiments where the digested DNA was run in parallel with a sequencing gel on the same DNA fragment; see S. Schmitz and D. J. Galas, 1979, *Nucleic Acids Res.* 6:111.] *Photograph courtesy of D. J. Galas.*

the promoter by the polymerase. Such results support the conclusion that specific contact points between promoters and RNA polymerase are important in the formation of an active transcription complex.

Footprint experiments have permitted identification of specific sites within promoters that interact with *E. coli* RNA polymerase. These are illustrated in Figure 7-6 for the *lacUV5* operon and the *A3* operon of bacteriophage T7. The polymerase seems to contact DNA asymmetri-

cally. In other words, from the hypothetical point of view of someone looking down the cylindrical double helix, most of the contact points fall into one 180° sector. When the polymerase is bound, the helix is twisted so that it unwinds from about the middle of the Pribnow box to two or three bases past the first base to be transcribed into RNA.

Operons Are Clusters of Genes Controlled at One Promoter Site

In bacteria, one promoter often serves a series of clustered genes. Such a gene cluster, which is called an *operon,* is transcribed into a single messenger RNA. Each gene of an operon is represented in the corresponding mRNA, and each section of the mRNA may be independently translated. Such mRNA molecules are *polycistronic,* meaning that they encode more than one polypeptide chain. The term *cistron* applies to the smallest genetic unit that encodes one polypeptide. The origin of the term is explained in Chapter 9.

The genes of a given operon, which often encode several enzymes active in a single metabolic pathway, are therefore controlled coordinately: either all are transcribed or none are transcribed—a remarkably economical control system. However, the coding segments of a polycistronic mRNA are translated independently; that is, each segment has its own ribosome-binding site, initiation codon, and termination codon. Consequently, the translation rate can differ among coding segments. Differences in translation rate probably are caused largely by folds (stem-loops) around initiation sites in the mRNA.

Bacterial Transcription Can Be Induced or Repressed by Specific Nutrients

Not all of the potential promoters in a bacterial cell are always available. The nutritional content of the medium in which a bacterial cell is suspended primarily determines which promoter sites (and thus which operons) are available to RNA polymerase at any given time. Thus the promoter for an operon encoding the enzymes needed to metabolize a particular nutrient becomes available only when that nutrient is present in the medium. Because of this specific response, referred to as *enzyme induction,* bacterial cells selectively manufacture those enzymes needed for the uptake and metabolism of the nutrients that are present in the medium. Enzymes synthesized in response to their substrates are said to be *induced.*

A common example of bacterial enzyme induction involves the enzymes that convert various sugars into glucose, a readily metabolized carbon source. As shown in Figure 7-7a, one set of enzymes is needed to convert galactose (a six-carbon monosaccharide) into glucose; an additional enzyme is needed to break down lactose (a disaccharide) into glucose and galactose; and a third set of

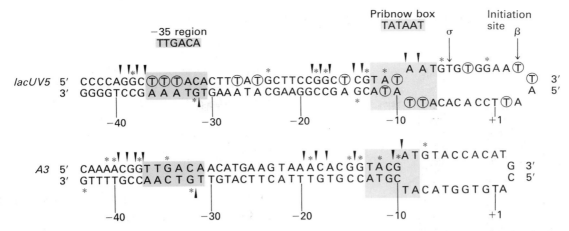

▲ **Figure 7-6** Contact points in the promoter regions of the *lacUV5* E. coli operon (*top*) and the *A3* operon of bacteriophage T7 (*bottom*) that interact with *E. coli* RNA polymerase. The DNA strands near the RNA initiation sites (+1) are separated to indicate that this DNA region is unwound when RNA polymerase is bound. Note that the two promoter sequences are similar but not identical and that the bases which interact with the polymerase are similarly spaced in the two promoters. The indicated sites were identified by various chemical reactions with end-labeled promoter DNA sequences in the presence or absence of polymerase (as described in Figures 7-4 and 7-5). The symbols used in the figure have the following meanings: (ı) a phosphate group that, when ethylated, prevents polymerase binding; (A or G) a purine that, when methylated, prevents polymerase binding; (Ⓣ) a thymine that polymerase protects; (σ or β) a site at which the σ factor or the β polymerase subunit can be chemically cross-linked to the DNA by a photochemical reaction. [See U. Siebenlist, R. B. Simpson, and W. Gilbert, 1980, *Cell* 20:269.]

(a)

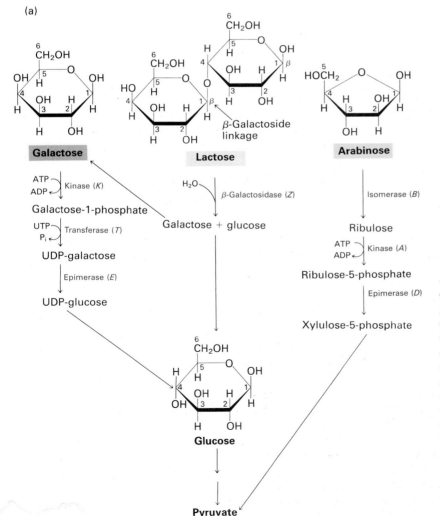

(b)

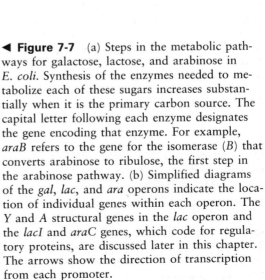

(P = Promoter)

◀ **Figure 7-7** (a) Steps in the metabolic pathways for galactose, lactose, and arabinose in *E. coli*. Synthesis of the enzymes needed to metabolize each of these sugars increases substantially when it is the primary carbon source. The capital letter following each enzyme designates the gene encoding that enzyme. For example, *araB* refers to the gene for the isomerase (B) that converts arabinose to ribulose, the first step in the arabinose pathway. (b) Simplified diagrams of the *gal*, *lac*, and *ara* operons indicate the location of individual genes within each operon. The Y and A structural genes in the *lac* operon and the *lacI* and *araC* genes, which code for regulatory proteins, are discussed later in this chapter. The arrows show the direction of transcription from each promoter.

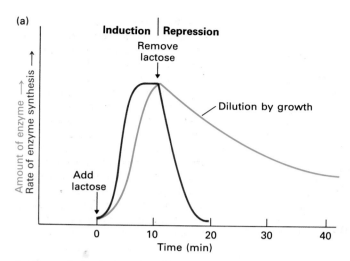

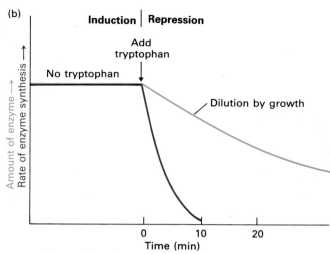

▲ **Figure 7-8** Induction and repression of bacterial enzymes. (a) Both the rate of synthesis (red) and intracellular concentration (blue) of the enzymes needed to metabolize lactose increase rapidly after lactose is added. When lactose is removed, synthesis of additional enzyme is repressed and soon stops altogether. The concentration of the remaining enzyme is then gradually diluted by cell growth. (b) Both the maximal rate of synthesis and the intracellular concentration of the enzymes needed to produce tryptophan, a required amino acid, are reached when tryptophan is absent from the medium. When tryptophan is added, enzyme synthesis is repressed immediately, and the enzyme concentration decreases as the cells grow and divide.

enzymes is required to convert arabinose (a five-carbon monosaccharide) into a glycolytic intermediate. The genes for these different groups of enzymes are clustered in different operons, called the *gal, lac,* and *ara* operons (Figure 7-7b). When galactose, lactose, or arabinose is added to the medium of a bacterial cell culture, synthesis of the corresponding enzymes increases rapidly; after removal of the sugar inducer, enzyme synthesis soon stops, so that the enzymes are diluted as the cells grow and divide (Figure 7-8a).

The lack of a nutrient also can induce the synthesis of enzymes. Many bacteria can synthesize all the amino acids necessary for growth, but the enzymes required to produce a particular amino acid are present in a bacterial cell only when they are needed, that is, when the amino acid is not present in the medium. An experiment demonstrating this phenomenon is illustrated in Figure 7-8b. When placed in a medium containing no tryptophan, the cells synthesize the five enzymes necessary for producing tryptophan at a maximal rate. When sufficient tryptophan is added to the medium, the cells take up what they need from the medium and stop making the enzymes that catalyze the synthesis of this amino acid; the preexisting enzyme proteins are then diluted by growth. In this case the *trp* operon (which encodes these enzymes) is said to be *repressed*. This repression of the *synthesis* of new enzyme molecules by a metabolic product is different from feedback inhibition, in which the final product in a metabolic pathway inhibits the *action* of existing enzyme molecules.

Regulatory Proteins Control the Access of RNA Polymerase to Promoters in Bacterial DNA

The induction and repression of bacterial enzymes occur mainly through control of transcription of the appropriate genes. To help understand how this control mechanism works, we can divide bacterial proteins and the genes that encode them into two groups. The largest group includes all the proteins that carry out reactions necessary for metabolism, cell growth, and division (e.g., enzymes, membrane proteins, and ribosomal components). The genes that encode these proteins are called *structural genes*. Although the rate of synthesis of mRNA encoded by most structural genes is carefully regulated, some structural genes are transcribed *constitutively*, that is at a rate that is more or less constant.

The second, smaller group of bacterial proteins, referred to as *regulatory proteins*, help the cell sense the environment and regulate the rate of transcription of structural genes by binding to DNA. The genes encoding regulatory proteins are referred to as *regulatory genes*. Although it was originally thought that the active product of regulatory genes might be RNA, only a few regulatory RNAs have been identified. Almost all regulatory genes studied so far encode a regulatory protein, which exerts either negative control or positive control of transcription, most often as a separate operon.

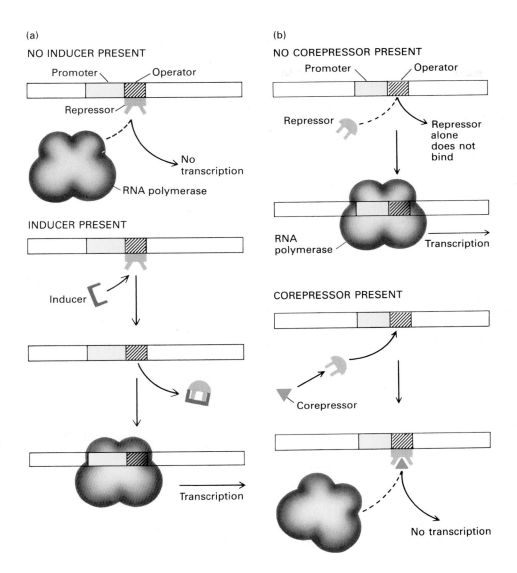

(a)

NO INDUCER PRESENT

Promoter

Operator

Repressor

No transcription

RNA polymerase

INDUCER PRESENT

Inducer

Transcription

(b)

NO COREPRESSOR PRESENT

Promoter

Operator

Repressor

Repressor alone does not bind

RNA polymerase

Transcription

COREPRESSOR PRESENT

Corepressor

No transcription

◄ **Figure 7-9** Action of two kinds of effectors—inducers and corepressors. (a) Some repressor proteins in their native states bind to DNA. Therefore, with no inducer present, such a repressor binds to its operator. This blocks effective access of RNA polymerase to the promoter region, and mRNA synthesis cannot be initiated. When an inducer is present, it combines with the repressor, which causes the repressor to change shape. As a result, the repressor disengages from the DNA molecule and the promoter region becomes accessible to RNA polymerase. (b) Another type of repressor protein is ineffective unless a corepressor is present. The repressor alone does not bind tightly to its operator, so by itself it does not prevent mRNA synthesis. When a corepressor combines with the repressor, the complex has the right shape for binding to the operator in such a way that mRNA synthesis cannot begin.

Repressors *Negative-acting regulatory proteins* bind to DNA at or near a promoter site, thus physically preventing access of RNA polymerase to the corresponding gene or operon and its subsequent transcription into mRNA. Such proteins are called *repressors;* the DNA site at which a repressor binds is called the *operator.* The operator sequence usually is close to the promoter sequence and often overlaps it.

Repressors can combine with small molecules, called *effectors,* that greatly affect the binding affinity for their operator. Two types of effectors exist. One type, *inducers,* combines with repressors to decrease binding affinity for DNA. For example, the repressor protein for the *lac* operon can bind to its operator in DNA in the absence of lactose or a metabolic product of lactose. However, in the presence of lactose or its metabolic product, a repressor-inducer complex forms. Binding of the inducer changes the shape of the repressor protein, decreasing its binding affinity for its operator. When the repressor disengages

from its operator, the genes for the corresponding enzymes are transcribed (Figure 7-9a).

Some effectors, called *corepressors,* have the opposite role. In this case, the repressor is not functional when its corepressor is absent. For example, when the amino acid tryptophan (the product of the enzymes encoded by the *trp* operon) is readily available, it combines with the *trp* repressor protein. Only then does the *trp* repressor bind tightly to the operator and inhibit transcription of the genes for the enzymes that make tryptophan. Thus, the amino acid acts as a corepressor, and the addition of tryptophan to an *E. coli* culture stops the synthesis of tryptophan-producing enzymes (Figure 7-9b).

Activators The second type of regulatory protein, *positive-acting proteins,* bind to DNA at or near a promoter site and increase the frequency with which RNA polymerase binds to the promoter. Positive-acting proteins are also called *activators,* and the DNA operator sites to

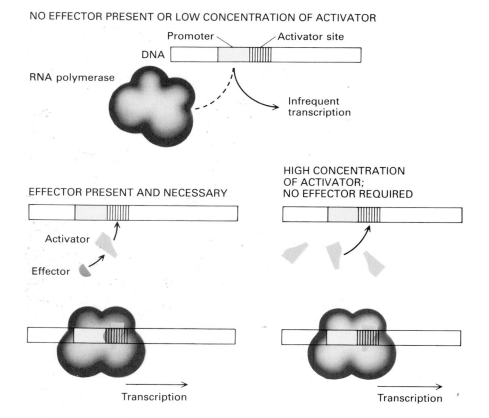

NO EFFECTOR PRESENT OR LOW CONCENTRATION OF ACTIVATOR

◄ Figure 7-10 Positive control of transcription by activators, some of which require effectors. When an effector binds to its activator or when an activator that does not need an effector is present in a high enough concentration, the activator binds to the DNA, attracting RNA polymerase to the promoter and increasing transcription of the gene.

which they bind often are called *activator sites* (Figure 7-10). Among the best-studied activators is the AraC protein, which is necessary for transcription of genes encoding the arabinose-metabolizing enzymes. Another important activator is the CAP protein; this activator regulates several operons, generally in conjunction with another specific activator or with a repressor (see section "Compound Control of Transcription").

Structure and Binding of Regulatory Proteins
Crystallographic analyses of DNA-protein crystals have revealed important details of the binding of regulatory proteins to DNA. Among the regulatory proteins that have been crystallized are Cro protein, a repressor encoded by bacteriophage λ; CAP protein, an activator for a number of *E. coli* genes; and cI protein, a bacteriophage λ protein that activates transcription of its own gene *cI* and represses transcription of other λ genes. The three-dimensional shapes of all three of these proteins have striking overall similarities. All of the proteins bind as dimers to DNA, and in each monomer of the dimers there are three characteristic α-helical regions. The structure of the Cro dimer illustrating the so-called helix-turn-helix motif is shown in Figure 7-11.

Computer-aided model building with DNA and Cro protein structures plus newer data on cocrystals of DNA and the Cro protein has shown that one of the helices (α₃)

in each Cro monomer fits very comfortably into the major groove of the DNA structure. A space-filling model of the structure of Cro bound to DNA shows how closely the dimer can bind to the double helix of DNA by fitting into two adjacent major grooves (Figure 7-11c). Mutations in the *Cro* gene that alter the amino acids in the α₃ helix disrupt binding of Cro to DNA. This finding supports the binding mechanism indicated by the models.

Some Repressors Can Recognize Several Operators: The Arginine Regulon

Since the structural genes for the enzymes needed to metabolize lactose, arabinose, and galactose in *E. coli* are clustered together into operons (single transcription units), each of the repressors and activators that regulate these operons need recognize only one operator. In eukaryotes, however, the genes encoding the enzymes in one metabolic pathway generally are not clustered into operons. A similar pattern of nonclustered genes exists for the *E. coli* genes that encode the enzymes for arginine biosynthesis. For this reason, the *E. coli* arginine genes are a valuable model for transcriptional control in eukaryotic cells in which different chromosomes may contain genes for coordinated functions.

The *arg* genes are scattered over a number of nonadjacent sites on the *E. coli* chromosome (Figure 7-12a).

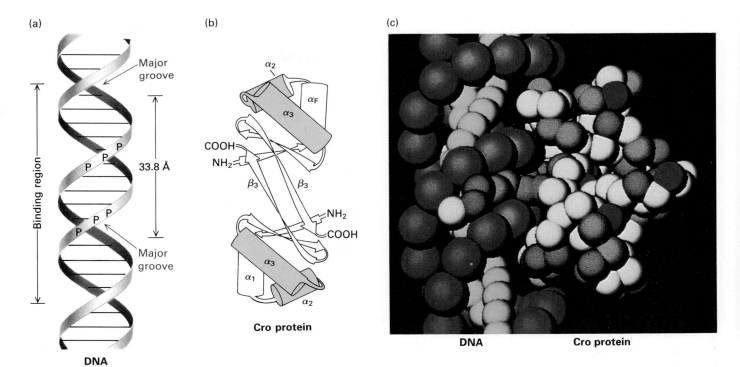

▲ **Figure 7-11** Structure of Cro protein, a negative-acting protein encoded in the genome of bacteriophage λ. (a) The double-helical structure of DNA for comparison. (b) Three-dimensional structure of the Cro protein dimer drawn to same scale as the DNA in (a); α helices are shown as cylinders and β-pleated sheets as arrows. The distance between the two α_3 helices in Cro protein is the same as the distance between the two major grooves in the protein-binding region of the DNA. (c) Three-dimensional space-filling computer model of Cro protein approaching DNA. The protein chain of the Cro (white, green, blue, and red balls) assume shapes that directly contact DNA. α Helices enter adjacent major grooves, bending the DNA (large red spheres are phosphate groups; yellow spheres represent the bottom of the major grooves). Small "arms" of the protein associate in the minor groove. [See G. Felsenfeld, *Readings from Sci. Am., The Molecules of Life,* p. 15.] *Computer image by B. W. Matthews and D. H. Ohlendorf.*

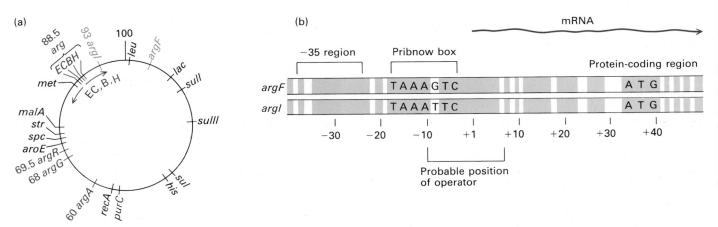

▲ **Figure 7-12** (a) The genes that encode the enzymes for arginine synthesis in *E. coli*—the *arg* F, A, G, R, E, C, B, H, and *I* genes—are widely separated from one another on the *E. coli* circular chromosome. Each of these genes was located by means of a separate mutation that rendered cells incapable of growing without arginine. The numbers represent units of the *E. coli* chromosome (actually, minutes required to transfer a gene during conjugation). The arrows indicate directions of transcription. (b) Regions of identity in the sequences of the *argF* and *argI* genes are shown in color. There is a much greater conservation of upstream sequences in these two genes than in a diverse group of unrelated genes (see Figure 7-3). The similarity continues into the first part of the coding region. The single repressor for arginine genes therefore can specifically recognize all the conserved operator regions. [See R. Cunin et al., 1986, *Microbiol. Rev.* **50**:314.]

When arginine is in plentiful supply, transcription of all of these genes is repressed by a single repressor. Such a scattered group of coordinately regulated genes is referred to as a *regulon*. Comparison of DNA sequences for the various arginine genes indicates that the presumed operator sequences around the mRNA initiation sites are very similar (Figure 7-12b). This similarity helps explain how a single repressor can recognize and bind to scattered sites in the *E. coli* chromosome.

Negative Control of Transcription: The Lactose Operon

Mutations in structural genes of bacteria alter enzyme activity; mutations in regulatory genes, on the other hand, alter the response of the bacteria to environmental changes. The identification of mutant regulatory genes in bacteria was indispensable in achieving an understanding of normal bacterial gene regulation. By use of conjugation, transduction, and transformation techniques, gene transfers can be carried out in *E. coli* so that the effects of mutations in structural and regulatory genes can be examined.

Historically, the most important experiments on bacterial gene regulation concerned *E. coli* and its metabolism of lactose, a disaccharide of glucose and galactose. If *E. coli* is grown on lactose (or on a chemically similar β-galactoside), the cellular concentrations of three proteins increase: the enzyme β-galactosidase, which splits lactose into glucose and galactose; permease, an inner-bacterial-membrane protein that increases the amount of lactose taken into the cells; and transacetylase, an enzyme whose exact role in lactose metabolism is still unclear. Together, the three genes coding for these proteins constitute the structural part of the *lactose operon*. They are arranged in the following order on the *E. coli* chromosome (see Figure 7-7b): the *Z* gene (for β-galactosidase), the *Y* gene (for permease), and the *A* gene (for transacetylase). As noted already, lactose or a metabolic product of lactose induces transcription of the *lac* operon. The classic genetic and biochemical studies on the lactose operon first defined the concept of *negative control* (i.e., control by a negative-acting regulatory protein) of the initiation of mRNA synthesis.

Early Experiments with Regulatory Mutants Suggested That Lactose Repressor Is a DNA-binding Protein

The genetic makeup of the lactose operon of a normal bacterium can be written $Z^+Y^+A^+$, which indicates that active forms of all three proteins are produced. A mutation in one of the genes can result in a bacterium that produces an altered or completely inactive form of the corresponding protein. A $Z^-Y^+A^+$ mutant, for example, would respond to the presence of lactose by producing active permease and transacetylase but no effective β-galactosidase. A most important finding was that some mutant bacteria have normal structural information for the three proteins but have lost the ability to regulate any of the three genes. For example, some regulatory mutants make the three proteins even in the absence of an inducer; these are called *constitutive* mutants, in contrast to normal, *inducible* bacteria, which make the proteins only in the presence of lactose. Genetic tests showed that some of the mutations causing constitutive production of β-galactosides were in the *I*, or inducibility, gene. The genotype of a normal, inducible bacterium is represented by I^+ and that of a constitutive mutant by I^-.

The nature of *I*-gene function was first suggested by conjugation experiments performed in the late 1950s in the laboratories of François Jacob and Jacques Monod at the Pasteur Institute in Paris (Figure 7-13). Normal donor bacteria (I^+Z^+) were crossed with mutant recipient bacteria (I^-Z^-). After conjugation, the recipient bacteria contained both sets of genes; they had become *merozygotes* (from the Greek for "joined parts"). Their genetic constitution can be represented by I^+Z^+/I^-Z^- (donor/recipient).

Before conjugation began, the donor cells did not produce β-galactosidase in the absence of inducer (lactose) but did in its presence; the recipient cells produced no active β-galactosidase in either condition because their Z gene was defective. Shortly after DNA transfer began in the absence of inducer, β-galactosidase began to be formed; however, if no inducer was added, enzyme synthesis stopped after about 2 h (Figure 7-14). Because the donor cells had been killed as soon as enough time had

▲ **Figure 7-13** Francois Jacob (1920–   , *left*) and Jacques Monod (1919–1976, *center*) at the Pasteur Institute in Paris, together with André Lwoff (1902–   , *right*), with whom they shared the Nobel Prize in 1965. *Photograph by Réné Saint-Paul, Paris; courtesy of F. Jacob.*

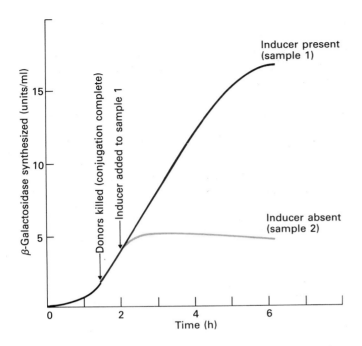

◀ **Figure 7-14** Conjugation experiments with *lac* mutants led to the hypothesis that a repressor prevents transcription of Z gene for β-galactosidase. *E. coli* cells that are I^+Z^+ produce this enzyme only in the presence of inducer; it is not produced at all by cells that are I^-Z^-. After the conjugation of I^+Z^+ (donor) cells with I^-Z^- (recipient) cells in the absence of inducer, the donors were killed. The merozygotes (I^+Z^+/I^-Z^-) produced β-galactosidase even without inducer up to about 2 h (sample 2) and then enzyme synthesis ceased. Only if inducer was added (sample 1), did enzyme synthesis continue. [See A. B. Pardee, F. Jacob, and J. Monod, 1959, *J. Mol. Biol.* 1:165.]

passed for the transfer of the I and Z genes, it was the conjugant bacteria (the merozygotes) that had produced β-galactosidase and then stopped in the absence of inducer.

To explain these results, the experimenters suggested that the donor cells, in the absence of inducer, contained a product of the I^+ gene that prevented expression of the Z^+ gene. This product would not be present in the recipient cells because their I genes were mutant and therefore inactive. Thus, when the Z^+ gene entered the recipient cells from the donors, it suddenly found itself in surroundings that lacked the I^+ product and began to direct the production of β-galactosidase. However, I^+ also had been donated to the recipient cells; in time, in the absence of inducer, the product of I^+ accumulated in the merozygotes and the activity of Z^+ was again suppressed. The concept of a gene product that acted as a repressor

(the I^+ product) originally was proposed based on the results of this conjugation experiment.

Additional genetic experiments with genes of the lactose operon greatly illuminated the function of the repressor. In these studies, different combinations of mutations in the lactose operon and in the I gene were recovered in plasmids, which were then introduced into *E. coli* cells that already had a lactose operon and an I gene in their chromosomes. (Recall that a plasmid is a small circular DNA molecule capable of independent replication within a bacterium.)

The results of these experiments are shown in Table 7-2, which lists the activities of several *E. coli* strains containing different combinations of mutant *lac* genes. First, partial diploid cells that were I^-Z^+/I^+Z^- still required inducer for β-galactosidase synthesis. This showed that the repressor encoded on one chromosome could act on the Z^+ gene of another chromosome. Then the discovery of mutations in DNA sequences called O (operator) suggested that the repressor acted on a particular DNA site; this site was shown by further genetic analysis to be located very close to the Z gene. Mutations in O sequences caused the neighboring Z^+ genes to be active even without inducer; these mutations were called O^c (operator-constitutive). When an O^c mutant with an active Z^+ gene

Table 7-2 β-Galactosidase levels in various genetically different *E. coli* strains

	Enzyme level		
Strain	**Without inducer**	**With inducer**	**Conclusion**
$O^+I^-Z^+$	Maximal	Maximal	Constitutive enzyme snythesis
$O^+I^-Z^+/O^+I^+Z^-$	Low	Maximal	Diffusible, trans-active* repressor; I^+ dominant over I^-
O^cI^+ (or I^-)Z^+	Near maximal	Maximal	Operator-constitutive and dominant over I^+; repressor unable to function
O^cI^+ (or I^-)$Z^+/O^+I^+Z^+$	Half maximal	Maximal	
O^cI^+ (or I^-)$Z^-/O^+I^+Z^+$	Low maximal	Half maximal	O^c dominant and cis-active*

*"Trans-active" means capable of activity on another chromosome. "Cis-active" means only on the same chromosome; for a mutation to be cis-active, it must be located close to the affected gene.

was used to create the partial diploid $O^c I^- Z^+/O^+ I^+ Z^+$, the cells were still partially constitutive. Thus O^c continued to cause the neighboring Z^+ to be active, even in the presence of the I-gene product.

At this point, two types of mutations that result in constitutive β-galactosidase production were recognized—mutations in the I gene and mutations in the operator. Mutations in the operator were *cis-active* (i.e., they affected only the Z gene on the same chromosome), whereas mutations in the I gene were *trans-active* (i.e., they could affect either a Z gene on the same chromosome or one on another chromosome). Thus it was concluded that a diffusible product (the I-gene product) acted at a site next to the Z gene (the operator). Discovery of temperature-sensitive mutations in the I gene suggested that its product was a protein because proteins are far more sensitive to heat than nucleic acids. In addition, the I gene was also subject to suppressible mutations (i.e., mutations that were correctable by suppressor tRNAs; see the discussion of correction of "nonsense" mutations in Chapter 3). This was strong evidence that the repressor was a protein. How did it work?

Transcription of *lac* Operon Is Regulated by Repressor: The Jacob-Monod Model

Jacob and Monod proposed in 1961 that the repressor protein encoded by the *lacI* gene specifically regulated the DNA that encoded all three enzymes of the lactose operon. They further suggested that an unstable intermediate molecule—an mRNA—conveyed the information in DNA to the protein-synthesizing machinery and that the function of the repressor was to regulate the synthesis of mRNA.

Measurements of the mRNA transcribed from the *lac* structural genes provided critical support for the Jacob-Monod theory. RNA-DNA hydridization assays with plasmid DNA containing the *lac* operon showed that normal *E. coli* cells grown in the presence of lactose contain large amounts of mRNA for β-galactosidase. In the absence of lactose, the cells contain very little mRNA complementary to the lactose operon DNA.

The lactose repressor protein proposed by Jacob and Monod has been purified, and its entire sequence of 360 amino acids is known. The repressor binds to a 25- to 30-base DNA region that includes several bases before and 20 bases after the initiation site for the synthesis of β-galactosidase mRNA (Figure 7-15). The binding site for RNA polymerase also is in this region; thus the promoter overlaps the operator. This overlap explains why binding of the repressor to the operator interferes with the initiation of mRNA synthesis. Recent evidence that RNA polymerase can attach to the promoter region even in the presence of the repressor, but cannot start RNA synthesis, indicates that the repressor prevents initiation of transcription rather than the initial binding of the polymerase.

The contact points involved in binding of the lactose repressor protein with the *lac* operator have been identified by footprint experiments similar to those used in studies of RNA polymerase binding to the promoter (see Figures 7-4 and 7-5). These studies have shown that normal *(wild-type)* repressor protein has contact points in the DNA region in which mutations are known to cause constitutive β-galactosidase synthesis. In O^c mutants, in which β-galactosidase synthesis is constituitive, binding of the repressor to DNA is impaired.

Positive Control of Transcription: The Arabinose Operon

When *E. coli* cells are grown on arabinose as an energy source, they produce the three enzymes needed to convert arabinose into xylulose-5-phosphate, which then enters the glycolytic pathway and is oxidized to supply energy

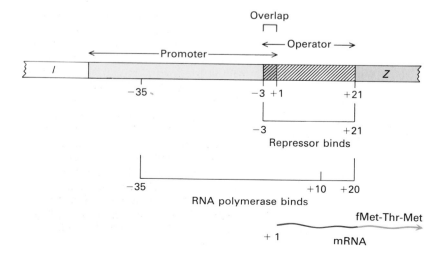

◀ **Figure 7-15** The arrangement of the promoter and operator sequences that control the *lac* operon in *E. coli*. The end of the I gene and the beginning of the Z gene are also shown. Because the promoter and operator sequences overlap, binding of the repressor prevents the initiation of transcription by RNA polymerase. [See W. S. Reznikoff and J. N. Abelson, 1978, in *The Operon*, J. H. Miller and W. S. Reznikoff, eds., Cold Spring Harbor Laboratory.]

for growth (see Figure 7-7.) These three enzymes—an isomerase, a kinase, and an epimerase—are the products of three genes (*B*, *A*, and *D*, respectively) that belong to the arabinose *(ara)* operon. Although arabinose induces the enzymes that metabolize it, just as lactose induces the enzymes needed to metabolize lactose, the regulatory "circuits" of the two operons are very different.

Early studies with *E. coli* mutants incapable of using arabinose revealed several strains with mutations in each of the structural genes (*B*, *A*, and *D*), as well as some with mutations in what proved to be a regulatory gene termed *araC*. The action of the *araC*-gene product (the protein AraC) differs from the action of the *lacI*-gene product (the lactose repressor). Mutations in *lacI* caused cells to be constitutive, whereas mutations in *araC* made cells incapable of growing on arabinose even though they still have the genes that encode enzymes for arabinose metabolism. The AraC protein was hypothesized to be a positive activator (see Figure 7-10). This was confirmed in experiments with cell extracts containing all the components of the protein-synthesizing apparatus plus RNA polymerase and DNA carrying the *ara* operon (purified

from a transducing bacteriophage). In this cell-free system, the B, A, and D enzymes were produced only when *both* the AraC protein and arabinose were present. Thus, AraC is a sugar-specific, positive-acting protein necessary for transcription of the *ara* operon into *BAD* mRNA. Other bacterial enzymes are similarly regulated by positive-acting proteins.

AraC Protein Binds at Several DNA Sites Including Sites Distant from the Initiation Site

Footprint experiments, similar to those described in Figures 7-4 and 7-5, have revealed the details of the interaction between the AraC protein and DNA. As described in the previous section, the DNA binding site for the lactose repressor is close to the start site for *lac* mRNA synthesis (see Figure 7-15). In contrast, when the AraC protein was tested for binding to DNA in the region of the *BAD* operon and in the nearby *araC* gene, a number of binding sites for the protein were observed (Figure 7-16a). The

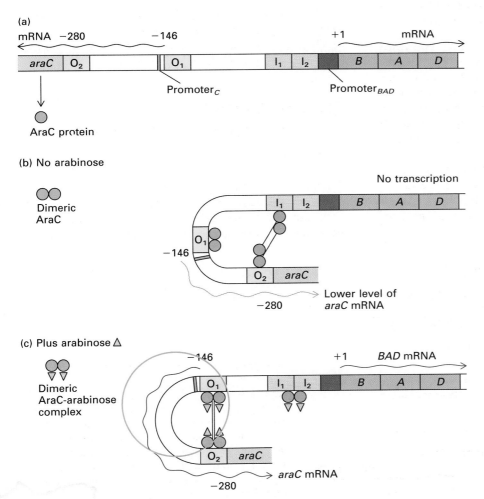

◀ **Figure 7-16** Regulation of the *ara* operon by AraC. (a) The *araC* gene of *E. coli* encodes a regulatory protein that, as a dimer, regulates transcription of the *BAD* genes, which encode the enzymes for metabolizing arabinose. Four binding sites—*ara*I$_1$, *ara*I$_2$, O$_1$, and O$_2$—are used in this regulation. (b) In the absence of arabinose, AraC dimers assume a conformation that permits them to bind to O$_1$, O$_2$, and the I sites, covering the I$_1$ but not the I$_2$ site. The dimers bound at O$_2$ and I$_1$ interact, forming a loop of about 190 nucleotides in the DNA. Because the *ara*I$_2$ site is not occupied, no transcription of the *BAD* genes occurs. A low level of *araC* mRNA is produced in this configuration. (c) In the presence of arabinose, a complex forms between the AraC dimer and the sugar. This changes the shape of the AraC dimer so that 1) the repression loop (O$_2$-I$_1$) is dispelled and 2) AraC now binds to both the *ara*I$_1$ and *ara*I$_2$ sites, which activates transcription of the *BAD* genes. In addition, *araC* mRNA is transiently formed at a higher rate until the alternate O$_1$-O$_2$ loop forms. [See K. J. Martin et al., 1986, *Proc. Nat'l Acad. Sci. USA* 83:3654; E. P. Hamilton and N. Lee, 1988, *ibid.* 85:1749; and L. Huo, K. J. Martin, and R. Schleif 1988, *ibid.* 85:5444.]

two binding sites at *ara*I (1 and 2), which lie close to the *BAD* initiation site, are both required for the AraC binding that activates transcription of the *BAD* operon in the presence of arabinose. However, two other binding sites (O_1 and O_2), located upstream from the initiation site, also exist. A mutation in the most distant binding site (O_2), which lies at -280, causes the *BAD* operon to be expressed at a fairly high level even without arabinose being present; likewise, a mutation at *ara*I$_1$ results in constitutive synthesis of the BAD enzymes. These results indicate that in the absence of arabinose the AraC protein represses the *BAD* operon by binding at two contact sites—O_2 and *ara*I$_1$—that lie more than 250 nucleotides apart. Thus it appears that the AraC protein when bound to different sites in the region of the *BAD* operon can exert either a positive or negative control over the *ara* operon. In contrast, the *lac* repressor exerts only negative control over the *lac* operon and binds to DNA only in the absence of lactose.

Regulation of *ara* Operon Involves Formation of DNA Loops

A model of *ara* regulation, now widely accepted, proposes formation of a loop in DNA. In the absence of arabinose, AraC exists in a conformation that binds to the *ara*I$_1$ and O_2 sites, and a loop forms in the DNA that is stabilized by protein-protein interactions between the AraC molecules bound at the two sites (Figure 7-16b). Because the *ara*I$_2$ site is not occupied by this form of the AraC protein, transcription of the *BAD* operon does not occur. In the presence of arabinose, AraC assumes another allosteric form that binds both *ara*I sites; hence, transcription of the *BAD* operon can proceed. The AraC-

arabinose complex also binds at O_1 and O_2, forming a different, smaller loop than is present in the absence of arabinose (Figure 7-16c). As a consequence of the formation of these DNA loops, transcription of the *araC* gene itself is affected. Slightly more AraC mRNA is synthesized in the presence of arabinose (when the DNA loop forms between O_2 and O_1) than in the absence of arabinose (when the loop forms between *ara*I$_1$ and O_2). Thus the AraC protein participates in regulating its own synthesis, in response to arabinose, in addition to regulating transcription of the *BAD* operon. Another example of such *autoregulation* is mentioned later in this chapter.

Once the importance of DNA-protein interactions at a distance from the *BAD* initiation site was recognized, a number of other distant regulatory sites were identified. One well-studied case involves a positive-acting protein that regulates the *E. coli* operon *glnALG* (which is responsible for glutamine synthesis) during nitrogen (ammonia) starvation. This DNA-binding protein called NR$_1$ (nitrogen regulator 1), is able to stimulate high transcription of the *ALG* genes even though only a few NR$_1$ molecules exist per cell. Two strong binding sites for NR$_1$, located about 110 and 140 bases upstream of the start site for transcription, are required for synthesis of high levels of *ALG* mRNA (Figure 7-17). By the techniques of recombinant DNA methodology in plasmids, the two binding sites for NR$_1$ can be moved as far away as 1800 base pairs and still retain their functional role in stimulating *ALG* mRNA formation during nitrogen starvation. Moreover, either orientation of the NR$_1$ binding sites in the DNA is functional. It seems highly likely that when NR$_1$ is bound to these sites, it helps to stabilize a loop so that assistance is given to start RNA synthesis at the now-distant promoter site.

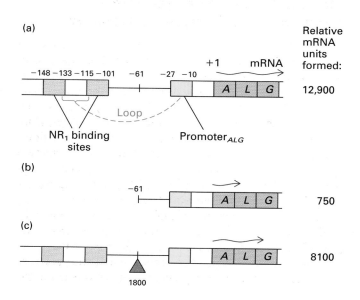

▼ **Figure 7-17** (a) Important elements in the regulation of the *glnALG* operon, which encodes proteins important in nitrogen metabolism (e.g., the *A* gene encodes glutamine synthetase). Transcription of this operon is increased during ammonia starvation. A variant sigma factor (σ^{54}) is required for RNA polymerase to start transcription of the operon (see Table 7-1). A regulatory protein NR$_1$, which has two binding sites (orange), activates transcription of the *ALG* genes over 100 bases downstream. (b) Deletion of the NR$_1$ binding sites decreases transcription (20-fold less mRNA). (c) Transcription is reduced only 30% when the NR$_1$ binding sites are moved more than 1800 base pairs from the *ALG* transcription start site by insertion of inert DNA in either orientation. This finding suggests that a long loop can form to position NR$_1$ close to the promoter for the *glnALG* operon and that in the normal gene (a) a loop also forms.

Compound Control of Transcription

The sugar normally used to grow bacteria is glucose, the simplest, most directly utilized sugar, which enters cell metabolism without requiring the induction of any new enzymes. If glucose is present in a medium to which lactose, arabinose, or any of a number of other sugars is added, the enzymes needed for metabolizing the other sugar are not induced until the glucose is used up. Apparently, glucose or a breakdown product of glucose (a *catabolite*) prevents the synthesis of mRNAs for a wide variety of sugar-metabolizing enzymes. This phenomenon is referred to as *catabolite repression*. In the original studies on the induction of the *lac* and *ara* operons, a medium lacking glucose was used to prevent catabolite repression, in order to simplify the experiments.

Interest in catabolite repression revived, however, when biochemists discovered that bacteria starved for glucose show a marked increase in the synthesis of an unusual nucleotide, cyclic adenosine-3′,5′-monophosphate (cyclic AMP, or cAMP). This molecule was originally discovered by enzymologists working with vertebrate enzymes that are controlled by the presence or absence of cAMP. The increased level of cyclic AMP in bacteria seems to be an "alert" signal for dire metabolic stress. In addition, when the bacterial culture medium contains cyclic AMP, glucose, and lactose, the induction of β-galactosidase is *not* suppressed as it is when no cyclic AMP is present. The same result is obtained for arabinose enzyme induction. It is now known that cyclic AMP plays a direct role in overcoming catabolite repression of several different enzymes and participates in initiating transcription of many genes in prokaryotes and eukaryotes.

A Single Protein, CAP, Exerts Positive Control on Several Different Operons

Once effects of cyclic AMP on transcription were discovered, it was hypothesized that one or more proteins must mediate or "interpret" the connection between high cyclic AMP levels (glucose starvation) and the need to induce other sugar-metabolizing enzymes. To test this theory, a search was made for mutant cells that could not grow in the presence of any of a wide variety of sugars even when glucose was absent from the medium. Individual mutant *E. coli* colonies that could not use lactose, galactose, maltose, or arabinose were found. By a second mutation, these cells regained the ability to use all of these sugars. In other words, a mutation in a *single* gene appeared to be the reason for the inability of the mutant cells to utilize any of the sugars.

About half of the original mutant cells did not make cyclic AMP in the absence of glucose. The other half could make cyclic AMP but did not increase their synthesis of the enzymes for metabolizing any of the sugars mentioned above in the presence of those sugars. This finding suggested that the second group of mutants contained a defective protein that could not bind to cyclic AMP. Subsequently, a protein that could bind to cyclic AMP was isolated from normal cells and was found to be missing in the second group of mutant cells. Conclusive evidence that the lactose operon is controlled by cyclic AMP and the cyclic AMP–binding protein—now called the *catabolite activator protein* (CAP)—was obtained from in vitro transcription experiments. In cell-free extracts containing RNA polymerase and purified DNA carrying the *lac* operon, the rate of synthesis of *lac* mRNA in the absence of either cyclic AMP or CAP was only about 5 percent of the rate in the presence of both. Thus the *lac* operon is under both the positive control of the CAP-cAMP complex and the negative control of the lactose repressor protein (Figure 7-18). The CAP-cAMP complex binds to the *lac* promoter, causing the DNA to bend considerably (Figure 7-18 inset). This bending somehow activates transcription, exposing a strong binding site for RNA polymerase.

The CAP protein is required for transcription of many other genes, including the arabinose *BAD* operon. That operon requires two positive activators, the CAP protein and the AraC protein, for full activation.

Although the compound control of the transcription of the *lac* and the *ara* operons is complicated, it is advantageous to the *E. coli* cell. As long as glucose is plentiful, little cyclic AMP is produced; thus CAP is not activated and the induction of enzymes that metabolize other sugars does not occur even when they are present. When glucose is absent *and* another sugar is present, transcription is initiated and the sugar can be metabolized.

The Galactose Operon Has Both a Regulated and a Constitutive Promoter

Growth of *E. coli* on the six-carbon monosaccharide galactose leads to increased synthesis of three galactose-metabolizing enzymes encoded by the *gal* operon (see Figure 7-7). The fully induced level of the three galactose enzymes is only 10 to 15 times as high as the uninduced level, whereas the maximal induced level of β-galactosidase (encoded in the *lac* operon) may be 1000 times the uninduced level. Another distinction between the two operons is that catabolite repression by glucose suppresses the *lac* operon by at least 100-fold but suppresses the *gal* operon only about 10- to 15-fold. In other words, there is a basal, noninduced level of the galactose-metabolizing enzymes regardless of the carbon source. The physiological reason why *E. coli* might benefit from this narrower range of regulation is that one of the galactose enzymes, an epimerase, produces a direct precursor for

(a) Glucose present (cAMP low); no lactose: No *lac* mRNA

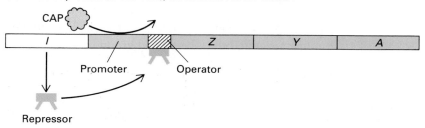

(b) Glucose present (cAMP low); lactose present

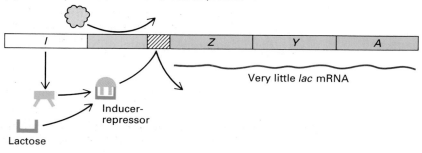

(c) No glucose (cAMP high); lactose present

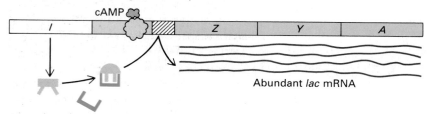

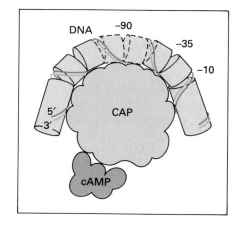

▲ **Figure 7-18** Control of the *lac* operon by two proteins—the Lac repressor and CAP, cyclic AMP–binding protein. (a) Without lactose, no *lac* mRNA is formed whether or not glucose is present, because the repressor is bound to the operator. (b) In the presence of glucose and lactose (or its metabolite, allolactose) the repressor changes shape and does not bind to the operator; however, cyclic AMP is low because glucose is present and thus CAP also does not bind. As a result, little *lac* mRNA is synthesized. (c) In the presence of lactose and the absence of glucose, maximal transcription of the *lac* operon occurs: the repressor does not bind, the con-centration of cyclic AMP increases, the resulting CAP-cAMP complex forms and binds at the −35 position (yellow) to activate transcription of the *lac* operon. *Inset:* Binding of CAP causes the DNA in the promoter region to bend more than 90°. In this diagram, the DNA is drawn as two blue lines around a bent-broken cylinder. Apparently, RNA polymerase binds most effectively when the promoter has this bent structure; hence CAP is an activator for the *lac* operon. *Redrawn from B. Gartenberg and D. M. Crothers, 1988,* Nature, *333:824. [See H-N. Lie-Johnson et al., 1986,* Cell *47:995.]*

cell wall biosynthesis; thus this enzyme is needed by the cell whether or not glucose is in the medium.

The mechanism by which the *gal* operon is regulated is a paradigm for a number of other complex control systems. In the *gal* operon, there are two overlapping promoters (P_{G1} and P_{G2}) and two separate initiation sites for RNA synthesis (Figure 7-19). The P_{G2} site binds RNA polymerase and allows the transcription of genes for the galactose-metabolizing enzymes even when glucose is present and galactose is absent; that is, this site is respon-sible for the constitutive synthesis of these enzymes. Binding of RNA polymerase to P_{G2} is inhibited by high levels of cyclic AMP and CAP. The other promoter, P_{G1}, requires high levels of both cyclic AMP and CAP in order for RNA polymerase to bind; this site is active in transcription only in the absence of glucose. Because P_{G1} has a much higher affinity for polymerase than does P_{G2}, the rate of *gal* mRNA synthesis is greater when the P_{G1} promoter is used (that is, under inducible conditions) than when the P_{G2} promoter is used.

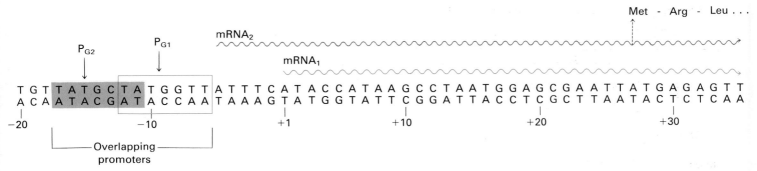

▲ **Figure 7-19** The galactose operon has two overlapping promoters and two RNA start sites. One promoter (P_{G2}) is active constitutively (i.e., without galactose and even when glucose is present). The second promoter (P_{G1}) requires ga- lactose, CAP, and high levels of cyclic AMP (brought about by the absence of glucose). [See S. Adhya, 1987, p. 1503, in *Escherichia coli and Salmonella typhimurium*, F. Neidhardt, ed., American Society of Microbiology.]

Control of Regulatory Proteins

The discovery of bacterial regulatory proteins immediately raised the question whether the regulatory proteins themselves were regulated, and if so, how. In *E. coli*, some regulatory proteins are formed constitutively; some control their own synthesis (i.e., they engage in *autogenous* regulation); and some are present in an inactive form until needed and then are converted to an active form.

LacI and Some Other Repressors Are Synthesized Continually

As we have noted, the lactose repressor is encoded by the *lacI* gene, which is situated near but not within the *lac* operon (see Figure 7-15). The *lacI* gene has its own promoter, distinct from the promoter for the *lac* structural genes. This promoter appears to be always "open" and the synthesis of mRNA for the lactose repressor occurs at a rate such that several hundred repressor molecules are present in the cell at all times. It seems likely that many negative-acting regulatory proteins are made constitutively at a low level. In other words, their rate of *synthesis* is not affected by the presence or absence of their effector, although their negative *action* is overcome when excess effector molecules enter the cell.

Synthesis of Some Regulatory Proteins Is under Autogenous Regulation

As discussed in the section on the arabinose operon, synthesis of the AraC protein is affected by the presence of arabinose (see Figure 7-16). Several other examples of autogenous regulation of regulatory proteins are known, including the repressor for the two histidine-utilization (*hut*) operons.

The enzymes encoded by the *hut* operons allow histidine to be rapidly degraded providing both a carbon and nitrogen source for the bacterium. In the absence of histidine only small amounts of these enzymes are synthesized because a repressor is present that greatly limits synthesis. This repressor is itself encoded in one of the *hut* operons, whose slightly "leaky" transcription is the basis for the formation of the repressor in the absence of histidine. When histidine is present, the *hut* operons are de-repressed and the degradative enzymes are made in a short-lived burst during which more repressor is also made (Figure 7-20). This allows the whole system to be efficiently shut down again as soon as the histidine is depleted. Thus the repressor has a *negative autogenous* effect.

Some Regulatory Proteins Are Controlled by Conversion between Active and Inactive Forms

Changing the level of synthesis of regulatory proteins is not the only way to change their activity in gene regulation. In theory, a regulatory protein also could be controlled by interconversion between an inactive and active form. One of the best-studied such examples is the regulation of the *ntr* (nitrogen-regulated) operons in *E. coli* and *Salmonella typhimurium*. These operons encode 20 to 25 proteins that are induced at low ammonia concentration and allow survival during "nitrogen starvation" (NH_3 dissolved in water forms NH_4^+, which is the preferred bacterial nitrogen source). RNA polymerase can only recognize these operons when sigma factor 54 (σ^{54}) is present; thus the first event in the nitrogen-response pathway is synthesis of σ^{54}. In addition, the positive-acting NR_1 protein binds to DNA upstream of the start sites of all the *ntr* operons and regulates their transcription (see Figure

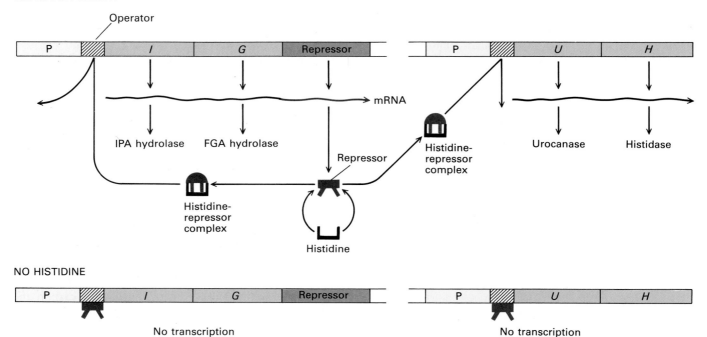

▲ **Figure 7-20** Regulation of the two *hut* operons, which encode the four enzymes needed to degrade histidine to glutamic acid. Bacteria that contain these enzymes can use histidine as their sole carbon and nitrogen source. Both *hut* operons are regulated by a single repressor, which is encoded by a gene in one of the operons. Synthesis of the four enzymes and of the repressor itself is induced by histidine. Thus the repressor exerts negative autogenous control of its own synthesis. [See B. Magasanik, 1978, *The Operon*, J. H. Miller and W. S. Reznikoff, eds., Cold Spring Harbor Laboratory.]

7-17 for an example of NR_1 effect). The NR_1 protein is present in cells at all times; that is, like the lactose repressor, it is synthesized constitutively. However, when ammonia is plentiful, NR_1 exists in an inactive form; at low ammonia concentrations, it is converted to an active form.

Although fairly complicated, the genes involved and the biochemical events in this interconversion are now understood. The conversion of the inactive to the active form of NR_1 involves the phosphorylation of NR_1 in a cyclic pathway. The decision to add or remove this phosphate, a reaction catalyzed by a second regulatory protein, NR_2, ultimately depends on the ratio of ammonia (or glutamine, which is made from ammonia) to 2-ketoglutarate, a glutamine precursor, which accumulates when ammonia is low (Figure 7-21). Phosphorylation of NR_1 by NR_2 is a phosphate transfer: a histidine residue in NR_2 accepts a γ-phosphate of ATP and transfers it to an aspartic acid residue in NR_1, giving the active form, NR_1-P_i. Thus NR_2, by itself, acts as a protein kinase. The NR_2 protein can also act as a phosphatase when it is complexed with another protein (P2). The final player in this cycle is the enzyme uridylyl transferase (UTase), which catalyzes transfer of four uridylyl residues (UMP) to and from the P2 protein. When P2 does not contain the UMP residues, it binds with NR_2; when P2 contains the UMP

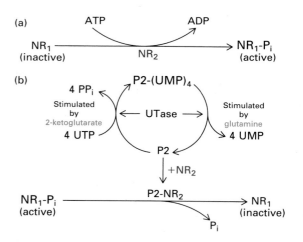

▲ **Figure 7-21** (a) Activation-inactivation cycle for NR_1, by a kinase NR_2. NR_1 is a positive-acting regulatory protein necessary for transcription of all the *ntr* operons, which encode proteins needed during nitrogen starvation (see Figure 7-17). (b) When *E. coli* is starved for ammonia, glutamine levels are low and 2-ketoglutarate levels are high. The relative amounts of these two compounds through influence on the uridylyl transferase and the cycle involving P2 phosphatase and NR_2 kinase shown here determine whether inactive NR_1 or the phosphorylated active form NR_1-P predominates. [See B. Magasanik, 1987, in *Escherichia coli and Salmonella typhimurium*, F. Neidhardt, ed., American Society of Microbiology.]

residues, it does not bind with NR_2. The addition of UMP residues to P2 by UTase is stimulated by 2-ketoglutarate; as noted previously, 2-ketoglutarate levels increase at low ammonia concentrations. Conversely, removal of the UMP residues is stimulated by glutamine, whose level is high when ammonia is present.

In summary, in the presence of ammonia, the inactive form of the gene regulator NR_1 predominates, and transcription of the *ntr* genes does not occur. At low ammonia concentrations, inactive NR_1 is converted to active NR_1 (by the kinase form of the NR_2 protein), which then activates transcription of the *ntr* genes.

This description of the NR_1 activation-inactivation cycle illustrates how complicated but how logical bacterial gene control can be. Furthermore, since the key change in the regulatory protein is a phosphorylation-dephosphorylation by a single kinase-phosphatase, this system is an instructive model for understanding events that might be similarly managed in eukaryotic cells.

Control of Transcriptional Termination

Control of transcriptional initiation is the dominant element in the flexible adaptation of bacteria to the environment. However, early in vitro transcription experiments suggested that the point at which RNA polymerase stops transcription also is important in controlling some operons. First, transcription of some operons frequently was incomplete in cell-free transcription; second, the RNA polymerase sometimes passed sites that were not passed when the same DNA was transcribed in vivo. These observations indicate that somehow intact bacterial cells can keep RNA polymerase at its task long enough to complete synthesis of polycistronic mRNAs and yet can stop transcription at appropriate *termination sites*. Several proteins that serve as termination factors (or antitermination factors) have been identified. One of the best studied is the termination factor called *rho*, which interacts with a growing mRNA chain. In some cases, however, termination can occur in the absence of the rho factor. Ribosomes also have been shown to affect chain termination by shielding the growing mRNA transcript from events that cause premature termination.

Rho-independent Chain Termination Is Associated with Certain Structural Features in the Termination Site

When RNA polymerase transcribes DNA in vitro, the polymerase pauses at regions rich in G and C residues. These pause sites can act as termination sites both in vitro and in vivo; that is, short, released mRNA chains corresponding to the DNA segment up to the pause site can be

identified. Not every such pause site observed in vitro acts as a termination site in vivo; but the reverse is true: every documented termination site is a pause site. A plausible model that explains the observation of pause sites during in vitro transcription and factor-independent termination in vivo is shown in Figure 7-22.

At many pause sites, including almost all of those that function as termination sites, the DNA has two characteristic sequences that give the transcribed RNA special features. At the termination site itself, there is a series of U residues in the mRNA transcript. Preceding the stop site, the mRNA contains a GC-rich region consisting of an inverted repeat (e.g., 5′CCCACT—AGTGGG3′). When such a self-complementary region of a growing mRNA chain is synthesized, the complementary sequences can base-pair with one another; it is thought that such self-pairing causes the polymerase to pause. The DNA behind the polymerase is now free to reanneal, displacing the polymerase and the mRNA chain from the template. This displacement may be facilitated by the relatively weak

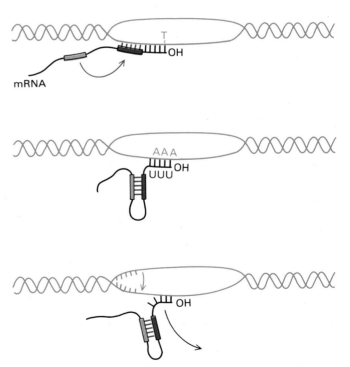

▲ **Figure 7-22** A model of rho-independent termination of transcription. The sequence of the mRNA synthesized near a termination (T) site contains a string of U residues preceded by a region of high GC composition with dyad symmetry (red boxes). The bases in the symmetrical region of the mRNA form a stem-loop by base pairing. This interaction together with the relatively weak rU-dA base pair at the termination site displaces the mRNA chain and the polymerase. [See T. Platt, 1981, *Cell* **24**:10.]

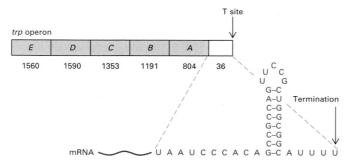

▲ **Figure 7-23** Sequence of *trp* termination (T) site, a rho-independent site. The *trp* operon is composed of five genes. The genes are followed by the base sequence shown, at the end of which termination occurs. The GC-rich stem-loop (hairpin) structure in the mRNA preceding the final four U residues is characteristic of rho-independent termination sites. [See T. Platt, 1981, *Cell* **24**:10.]

rU-dA base pairs at the end of the newly made RNA. This model is supported by the finding that mutations that weaken the dyad symmetry or that result in fewer consecutive U residues decrease termination at these sites.

An example of a rho-independent termination site occurs in the *trp* operon. Synthesis of *trp* mRNA ceases at a site 36 bases beyond the last coding region in the mRNA. At this site there are four consecutive U residues preceded by a 22-nucleotide sequence high in G and C that can fold back on itself and base-pair to form a perfect "hairpin" (Figure 7-23). These three features—high GC content, dyad symmetry (self-complementarity), and several U residues in a row—are common to almost all of the 30 or so rho-independent sites that have been analyzed. The efficiency of termination at rho-independent sites ranges from about 25 to 75 percent of that at rho-dependent sites. Thus *rho* mutants can still terminate transcription (at rho-independent sites), but they do so considerably less efficiently than normal cells.

Rho-dependent Chain Termination Requires the Presence of a Specific Protein

Sites for rho-dependent termination in *E. coli* were first recognized during in vitro studies of transcription of the DNA of bacteriophage λ. When *E. coli* RNA polymerase is mixed with λ DNA, RNA synthesis begins in a central region of the DNA at two promoter sites, P_L and P_R, which are known to function immediately after the virus infects cells. Messenger RNA is synthesized by the RNA polymerase in two directions (to the left of P_L and to the right of P_R). The λ mRNA transcripts formed in vitro are several thousand nucleotides long, much longer than those present inside cells during the early stages of λ in-

fection. However, when extracts of uninfected cells are added to the cell-free transcription system, two discrete products are formed—one containing 500 nucleotides and one containing 1000 nucleotides. These mRNAs, which ordinarily are found inside infected cells, correspond to coding regions extending from the P_L and P_R promoters to specific termination sites. In other words, in the absence of some factor in the uninfected cell extract, in vitro transcription of λDNA does not terminate at the same sites as does in vivo transcription. The protein responsible for termination at these sites in λ DNA was purified and is called *rho*. Perhaps about half the termination sites in *E. coli* require the rho factor.

Many rho-dependent chain-terminating regions have been identified in λ and *E. coli* DNA. Comparisons of the sequences of these regions have revealed no obvious similarities. Mutations in the *rho* gene itself cause RNA polymerase to read through from one operon to the next, so it is clear that the rho factor plays an important role in the termination of mRNA chains inside the cell. Although it is not known precisely how the rho factor causes termination, it is known that the protein forms a hexamer and that a 70- to 80-base segment of the growing RNA transcript wraps around the outside of the hexamer. This activates an ATPase activity that is associated with purified rho factor; the energy thus released may discharge the polymerase and the new RNA chain from the DNA template.

Attenuation Provides Secondary Control of Chain Termination

As discussed earlier, the primary regulation of the *trp* operon, which encodes the five different enzymes needed in the synthesis of tryptophan, occurs at the level of transcriptional initiation. In the presence of tryptophan, the *trp* repressor binds to the operator and prevents transcription of the tryptophan-synthesizing enzymes (see Figure 7-9b).

The discovery of *E. coli* mutants that manufactured greater-than-normal amounts of tryptophan-synthesizing enzymes ultimately led to recognition of another basic mechanism of prokaryotic gene control. One group of these mutants appeared to make a normal *trp* repressor and to have a normal *trp* operator; however, their basal level of *trp* enzymes was higher than normal, and they responded to the absence of tryptophan with an even higher rate of transcription. Comparison of the tryptophan genes from this group of mutants with those from wild-type cells showed that short stretches of DNA containing fewer than 50 base pairs were deleted in the mutants. The deletions occurred just after the promoter-operator region in the normal *trp* operon and before the coding sequences for the *trp* enzymes.

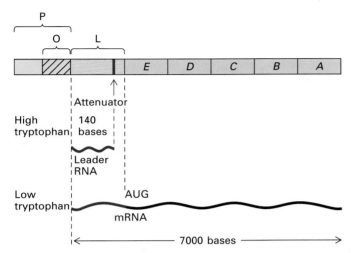

▲ **Figure 7-24** Attenuation provides a secondary control mechanism in the *trp* operon. The leader sequence (L), which lies between the operator (O) and the structural genes (*E*, *D*, *C*, *B*, *A*), contains an attenuator site at which transcription is terminated depending on the concentration of tryptophan in the medium. [See T. Platt, 1978, in *The Operon*, J. H. Miller and W. S. Reznikoff, eds., Cold Spring Harbor Laboratory.]

How could the deletion of a short region of DNA near the initiation site for mRNA^{Trp} cause unregulated synthesis of this mRNA? In the hope of answering this question, experimenters sequenced the beginning of the mRNA^{Trp} molecule and the DNA region near the *trp* promoter in both normal and mutant *E. coli* strains. This led to the discovery of a new mechanism in the regulation of bacterial transcription. This mechanism, called *attenuation*, operates in several operons that encode enzymes for amino acid synthesis.

A molecule of mRNA^{Trp} contains 162 nucleotides upstream (i.e., toward the 5' end of the molecule) from the AUG codon that constitutes the protein initiation site for the first of the five tryptophan enzymes (Figure 7-24). This stretch of mRNA is called the *leader sequence* (L). In normal *E. coli* cells, only small amounts of any mRNA^{Trp} sequences are made when tryptophan is abundant; these consist entirely of the leader RNA. When tryptophan is scarce, however, the entire *trp* operon (about 7000 nucleotides) is transcribed, including the leader sequence and all the coding sequences for the *trp* enzymes. Thus control is exercised at the level of termination, which occurs at the 140th nucleotide of mRNA^{Trp} when tryptophan is abundant.

Even when the entire mRNA^{Trp} sequence is formed in large amounts, more leader sequences than whole mRNA molecules are made. This situation has been interpreted as indicating that the leader sequence contains an *attenuator*, a site that exercises a "veto" over transcription after a certain distance. When tryptophan is scarce, about 25

to 50 percent of the RNA polymerase molecules continue transcribing past the attenuator. When tryptophan is abundant, some initiation of transcription takes place, but virtually all of the transcripts are cut short. Mutant cells that lack the attenuator produce more mRNA^{Trp} under all conditions than do normal cells. Thus the *trp* operon is controlled not only by the repressor-operator mechanism but also by attenuation of RNA synthesis at a site 140 nucleotides from the start site. The number of RNA polymerase molecules that pass the attenuator depends precisely on the concentration of tryptophan in the medium. Therefore, regulation of the *trp* operon is not an all-or-none affair; rather, secondary control by attenuation permits a cell to finely balance the amount of the tryptophan-synthesizing enzymes formed with the cell's need for tryptophan.

Whether attenuation occurs apparently depends on the formation of a particular base-paired stem-loop structure in the mRNA leader sequence. Formation of this structure depends in turn on the rate of ribosomal translation of the leader sequence, which is engaged by a ribosome soon after it is synthesized. Whether the ribosome efficiently translates the leader sequence depends on the supply of amino acids encoded by the sequence—in particular, on the supply of the amino acid that is the product of the enzymes encoded by that mRNA.

The tryptophan leader mRNA has been analyzed very thoroughly. Within the first half of this leader, there are two codons for tryptophan. When tryptophan is present in sufficient quantity, the ribosome moves quickly along the leader transcript. The portion of the transcript that has not yet been translated folds into a stem-loop structure (the 3-4 loop shown in Figure 7-25) that immediately precedes a U-rich region. In this case, rho-independent termination of the growing chain occurs by the mechanism described earlier. Attenuation is therefore maximized when tryptophan is present in abundance. On the other hand, when the supply of tryptophan is low, the ribosome pauses at each tryptophan codon in the leader sequence. While the ribosome is stalled, the "waiting" mRNA transcript folds into a different stem-loop structure (the 2-3 loop shown in Figure 7-25), preventing formation of the alternate loop which is part of the termination signal (the 3-4 loop and the U residues). When the 2-3 loop forms, termination does not occur and the RNA polymerase continues transcription. Thus attenuation is minimized when tryptophan is scarce.

Attenuation occurs in other operons in addition to the *trp* operons. For example, the mRNAs encoding the enzymes for synthesis of phenylalanine and histidine contain leaders with 7 of 14 and 7 of 16 codons for phenylalanine and histidine, respectively, preceding a protein stop codon in the first 50 bases of their sequence. Attenuation of these operons also occurs by a mechanism similar to that depicted in Figure 7-25 for the *trp* operon.

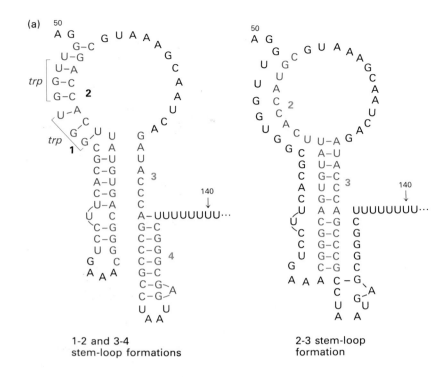

1-2 and 3-4
stem-loop formations

2-3 stem-loop
formation

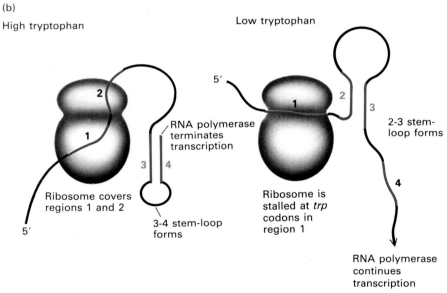

High tryptophan

Low tryptophan

RNA polymerase
terminates
transcription

Ribosome covers
regions 1 and 2

3-4 stem-loop
forms

Ribosome is
stalled at *trp*
codons in
region 1

2-3 stem-
loop forms

RNA polymerase
continues
transcription

◀ **Figure 7-25** Mechanism of attenuation in the *trp* operon. (a) Base sequences in a portion of the tryptophan leader mRNA showing the possible stem-loop structures that can form. (b) At high tryptophan concentrations, transcription is interrupted because the 3-4 stem-loop, characteristic of rho-independent termination, forms during translation of the leader region. At low tryptophan concentrations, translation stalls at the tryptophan codons and the 2-3 stem-loop forms, preventing formation of the 3-4 termination loop. [See C. Yanofsky, 1981, *Nature* **289**:751.] *Part (b) from T. Platt, 1981,* Cell *24:10.*

Antitermination Proteins Prevent Termination and Allow "Read-Through" Control

Termination at an attenuator site is spontaneous unless a ribosome prevents formation of a secondary stem-loop in the mRNA. In contrast, for rho-dependent termination to occur, the effect of specific antitermination proteins must be counteracted. Two of the best-studied cases concern early events in the infection of *E. coli* by bacteriophage λ and the transcription of the ribosomal operons of *E. coli*.

As we mentioned previously, in the early stages of phage infection, transcription begins in a central region of the λ DNA and proceeds to the left and to the right of two promoters, P_L and P_R. Because transcription is terminated at specific sites (T_L and T_R), the RNA transcripts from each promoter region are at first quite short, representing about 1000 and 500 bases from P_L and P_R, respectively (Figure 7-26a). The P_L transcript encodes a bacteriophage protein called N, which prevents termination of transcription at T_L and T_R and thus permits synthesis of long RNA transcripts corresponding to other portions of the λ genome (Figure 7-26b).

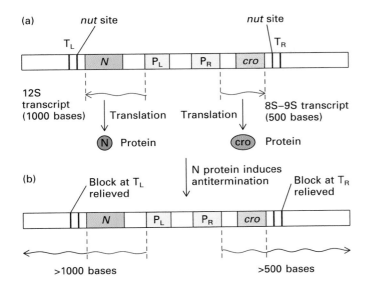

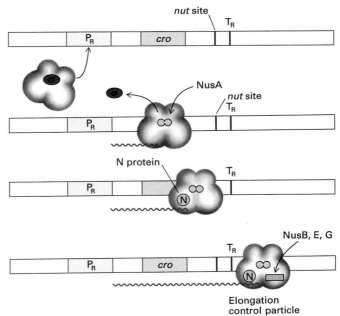

▲ **Figure 7-26** Effect of antitermination by N protein in central region of the λ genome. (a) Early in bacteriophage λ infection, transcription from the promoters P_L and P_R is interrupted at T_L and T_R by rho-dependent termination; thus, only short transcripts are produced. (b) As infection proceeds, the concentration of N protein increases. It eventually overcomes the rho-dependent termination at T_L and T_R, so longer transcripts begin to accumulate.

▲ **Figure 7-27** Mechanism of N-protein antitermination at T_R in the λ genome. Soon after *E. coli* RNA polymerase (holoenzyme) attaches at P_R and begins transcription of *cro*, the sigma factor is lost and a dimeric NusA molecule associates with the core enzyme. After this complex passes the nut_R site, an N-protein molecule becomes part of the transcription complex probably bound to the mRNA at the *nut* region. Binding of three additional Nus-protein molecules (B, E, G) produces a functional elongation control particle, which can overcome the action of the rho factor at T_R. A similar series of events occurs to the left of P_L. [See J. Horwitz et al., 1987, *Cell* **51**:631.]

Discovery of how the N protein acts as an antiterminator in the transcription of λ DNA helped illuminate the general process of antitermination in *E. coli* (Figure 7-27). When the N protein was attached to a solid substrate such as cellulose, it was able to bind to and select from bacterial extracts a specific protein called NusA. This protein can bind as a dimer to the core RNA polymerase. Detailed genetic and recombinant DNA studies led to identification of sites in λ DNA, called *nut* for *N*-protein *ut*ilization, that are required for the antiterminating action of N. Soon after an RNA polymerase molecule begins transcription, its sigma factor is lost and the dimeric NusA protein is gained. At the *nut* site in the DNA, the polymerase also picks up the N protein, which associates with NusA. At least three additional Nus proteins (B, E, and G) complete the transcription complex. This entire array of proteins, termed the *elongation control particle*, possesses the ability to cross the rho-dependent termination sites (T_L and T_R).

The *E. coli* genome contains many rho-dependent termination sites, and the Nus proteins probably are required to overcome termination at these sites. One such rho-dependent termination site has been identified in *E. coli* genes for ribosomal RNA. The elongation control particle must be assembled for complete transcription of these genes to occur. More study will probably uncover other instances of the participation of this complex during transcription.

Bacteriophage λ Infection: Alternative Physiologic States Determined by a Complex Transcriptional Control Program

In our discussion of the known transcriptional control mechanisms of bacteria—positive and negative regulation of initiation, autogenous control of regulatory proteins, and regulation of termination—bacteriophage λ has often been mentioned. Here we want to consider how all these events are integrated to produce major physiologic decisions. The entire sequence of λ DNA, which contains more than 50,000 base pairs, is known. The λ DNA encodes more than 50 proteins, and the function of most of these proteins is known. Detailed studies of the phage-specific proteins that regulate transcription of the phage's own DNA have revealed much about the integrated set of transcriptional controls responsible for critical physiologic choices in a cell containing λ DNA.

After a λ phage infects an *E. coli* cell, two alternative series of events can occur: (1) the phage DNA is replicated, additional phage particles are produced, and the cell soon lyses (*lytic cycle*); or (2) the phage DNA is integrated into the host chromosome and is replicated along with the bacterial DNA in a quiescent form (lysogenic state). If a lysogenic cell undergoes a shock that releases the dormant virus, the phage DNA leaves the host chromosome and the lytic cycle is triggered. The factors that determine whether the lytic or lysogenic pathway is dominant are fairly well understood and are instructive as a model for how transcriptional control networks might

direct eukaryotic cells into particular pathways (e.g., in developing organisms).

Let us first consider a cell just infected by a λ phage. If the lysogenic state is to be achieved, several events must occur. The replication enzymes and replication of λ DNA must be suppressed. The enzymes that catalyze integration of the λ DNA into the *E. coli* chromosome must be made. Finally, once integration is accomplished, the λ genome must be held in check, that is, repressed. This final requirement is critical if the cell is to continuously harbor phage DNA without harm. To establish and maintain a stable lysogenic state, an adequate supply of the phage-specific repressor protein cI must be produced. This protein binds to and regulates both P_R and P_L, the two major early promoters in λ DNA.

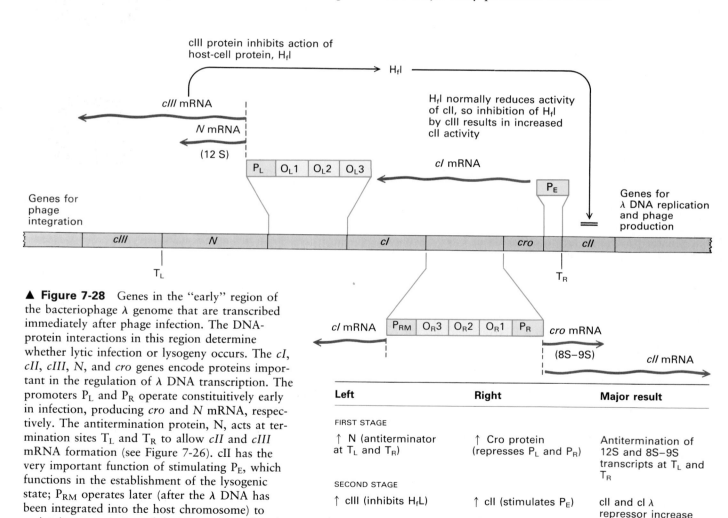

▲ **Figure 7-28** Genes in the "early" region of the bacteriophage λ genome that are transcribed immediately after phage infection. The DNA-protein interactions in this region determine whether lytic infection or lysogeny occurs. The *cI*, *cII*, *cIII*, *N*, and *cro* genes encode proteins important in the regulation of λ DNA transcription. The promoters P_L and P_R operate constituitively early in infection, producing *cro* and *N* mRNA, respectively. The antitermination protein, N, acts at termination sites T_L and T_R to allow *cII* and *cIII* mRNA formation (see Figure 7-26). cII has the very important function of stimulating P_E, which functions in the establishment of the lysogenic state; P_{RM} operates later (after the λ DNA has been integrated into the host chromosome) to maintain transcription of the *cI* gene encoding the repressor (cI). The operators O_R1, O_R2, and O_R3 bind cI (see Figure 7-29).

The events controlling the function of λ DNA are summarized in the chart. The arrows pointing up indicate increases; those pointing down, decreases. The key determinant in allowing lysogeny is the quantity of cI repressor that is formed.

Left	Right	Major result
FIRST STAGE		
↑ N (antiterminator at T_L and T_R)	↑ Cro protein (represses P_L and P_R)	Antitermination of 12S and 8S–9S transcripts at T_L and T_R
SECOND STAGE		
↑ cIII (inhibits H_fL)	↑ cII (stimulates P_E)	cII and cI λ repressor increase
↑ Integration enzymes		
THIRD STAGE		
cI dominates	cI dominates	Transcription decreases from P_R and P_L and increases from P_{RM}; lysogeny established
↓ N	↓ Cro	
↓ Integration enzymes	↓ Replication	

λ Repressor (cI Protein) Predominates in the Lysogenic State

Because no cI protein is present when a phage enters an *E. coli* cell, transcription begins in the early region of the λ genome from both P_L and P_R (Figure 7-28). The major product from P_L is the mRNA that encodes the N protein. This protein in turn produces antitermination at T_L between the N and *cIII* genes, so that the mRNA for *cIII* is produced. From P_R the earliest product is the mRNA that codes for the Cro protein; after the N protein causes antitermination at T_R, mRNA for cII also is transcribed from P_R. Two other promoters in the central region (P_E and P_{RM}) serve as promoters to make *cI* mRNA. Early in phage infection, little if any transcription occurs from P_{RM} because it is repressed by the Cro protein. P_E is a key promoter because early production of cI arises from it; the cII protein stimulates transcription from P_E.

The proteins necessary for integration of phage DNA into the host chromosome are encoded by genes downstream from *cIII*, whereas those involved in the lytic cycle are encoded downstream from *cII*. Whether lysogeny or lysis eventually occurs thus depends on the relative rates of transcription from P_L and P_R and the activity of the resulting products. What determines the balance in any individual cell is not clear, but a number of events that favor production of cI repressor, and hence the lysogenic state, have been identified. First, cIII inhibits the activity of an *E. coli* protein called H_f1, which somehow depresses the activity of cII. Inhibition of the H_f1 protein by cIII therefore leads to increased cII activity. As we noted above, cII increases transcription from P_E to give *cI* mRNA; furthermore, the mRNA initiated at P_E is translated very efficiently. Thus the concentration of cI protein increases. This cascade of events leads to a suppression of λ phage replication in favor of integration. As we discuss below in more detail, cI is a positive regulator at P_{RM} (repressor maintenance) and the integrated state depends on cI production from the P_{RM} promoter. So long as cI is available in sufficient amounts, the phage genome remains repressed.

Elegant biochemical studies on the interactions between the cI protein and three operator regions of λ DNA—O_R1, O_R2, and O_R3—have elucidated the molecular states that underlie the control of λ DNA in the lysogenic state. Each operator consists of a DNA sequence 17 base pairs long; the three O_R sites lie in front of P_R as shown in Figure 7-28. Three similar operators lie in front of P_L. The *cI* gene is located between P_L and P_R; the autogenously regulated promoter for the *cI* gene—P_{RM}—is close to the O_R3 site. The base sequences of the three O_R and three O_L sites are similar but not identical. Both the cI protein and the Cro protein bind to O_R, although normally dimeric cI binds about 10 times more strongly than does Cro.

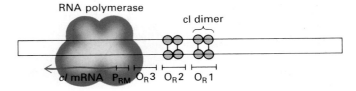

In lysogenic state, binding of cI protein activates transcription of *cI* from P_{RM} and represses transcription of *cro* from P_R

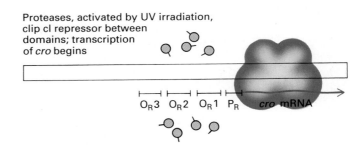

Proteases, activated by UV irradiation, clip cI repressor between domains; transcription of *cro* begins

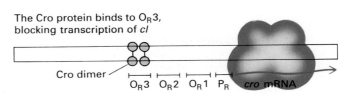

The Cro protein binds to O_R3, blocking transcription of *cI*

▲ **Figure 7-29** Maintenance and release from lysogeny by binding of cI protein (λ repressor) and Cro protein to operator regions of the λ genome. The base sequences in the three O_R sites are similar but not identical. Mutations in these sites upset the balance of the lysogenic state and can result in phage capable only of lytic growth, so-called virulent mutants. Dimeric cI binds in the order $O_R1 > 2 > 3$, whereas Cro dimers bind in the order $O_R3 > 2 > 1$. Because cI has about a 10 times higher affinity than Cro for DNA, it is preferentially bound even when present at only about 10 to 20 percent the level of Cro. [See M. Ptashne et al., 1980, *Cell* **19**:1; A. D. Johnson et al., 1981, *Nature* **294**:217.]

Both the cI and the cro protein spontaneously form dimers, which have a high affinity for DNA. The affinity of cI dimers for the binding sites increases in the order $O_R1 > O_R2 > O_R3$. Binding of cI to O_R1 cooperatively increases its binding to O_R2 (but not to O_R3) because there is protein-protein interaction between the two dimers at O_R1 and O_R2. At the concentrations of cI found in lysogenic cells, O_R1 and O_R2 are largely saturated. When cI dimers occupy O_R1 and O_R2, transcription of the cI gene from P_{RM} is significantly stimulated (Figure 7-29).

The mechanism of transcriptional stimulation by a regulatory protein bound to DNA remains unknown. However, it is known that particular acidic amino acids in the cI protein are very important for stimulation at P_{RM} (Figure 7-30). When these residues are mutated, cI can no

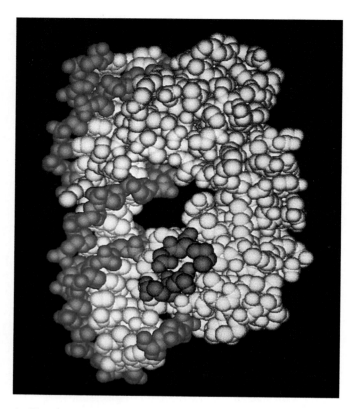

▲ **Figure 7-30** Computer model of λ repressor (cI) bound to DNA based on crystallographic analysis of the structure of cI protein cocrystallized with λ operator DNA. In this model, dimers of cI protein (brown and green) are bound to DNA (blue and white). The protein dimers reach into adjacent major grooves; the "arm" of the brown dimer wraps around the major groove facing the reader. The residues in red are acidic amino acids required for positive autoregulation of the cI gene by the protein and probably represent contact points between cI and RNA polymerase. [See S. R. Jordan and C. O. Pabo, 1988, *Science* **242**:893.] *Computer model and photograph from R. Bushman and M. Ptashne.*

longer act as a positive regulator. Mutations in *E. coli* that restore the ability of the cI in such λ mutants to act as a positive regulator have been identified and shown to involve "compensatory" mutations in the RNA polymerase itself. Therefore it is likely that these acidic residues in cI attract RNA polymerase and assist it to transcribe more often from P_{RM} when cI is bound at O_R1 and O_R2.

In addition to the positive action of cI on its own promoter (P_{RM}), the protein will eventually, at very high concentrations, fill all three operator sites, which represses synthesis of cI. Binding of a cI dimer to O_R1 prevents RNA polymerase from interacting with P_R, thereby repressing the transcription of *cro*. Thus, in the lysogenic state, the cI repressor continues to be synthesized because of autogenous positive control, producing a permanent *epigenetic* repression of the *cro* gene and activation of the cI gene.

λ Cro Protein Predominates during Lytic Cycle

As we mentioned earlier, DNA-damaging events can release the quiescent phage DNA. When a cell is subjected to ultraviolet irradiation, it undergoes a number of biochemical changes that allow it to repair DNA. One of these changes is the activation of a protease that cleaves a number of proteins, including the λ repressor. The cleaved repressor no longer dimerizes and thus loses its ability to bind to DNA (Figure 7-29b). As a result, repression of the *cro* gene is relieved and transcription of the gene, beginning at P_R, is initiated. The Cro protein, also a DNA-binding repressor protein, binds at the O_R sites in the order $O_R3 > O_R2 > O_R1$ and in so doing shuts down transcription of the cI gene at P_{RM} (Figure 7-29c). The concentration of cI decreases, relieving its repression at P_L, so that transcription of the N gene begins. Production of N protein in turn leads to synthesis of the mRNA transcripts for proteins in the lytic pathway; in the absence of an effective amount of cI repressor, the lytic cycle ensues.

"Global Controls" in E. coli

Most of the examples of gene regulation discussed so far in this chapter involve responses to specific nutritional conditions such as the presence of a disaccharide as a carbon source or the absence of an amino acid in the medium. We also have mentioned that certain drastic changes such as sudden temperature elevation (heat shock) lead to production and utilization of a new sigma factor (σ^{32}) that leads to the transcription of several (perhaps 5 to 10) previously silent operons. We also noted in describing the regulatory protein NR_1 that another sigma factor (σ^{54}) is made and used during ammonia starvation to transcribe perhaps 20 operons that normally are quiescent. Current work now centers on response to sudden O_2 deprivation and carbon source removal, and it seems likely that some type of "global control" may be required to turn sets of genes on and off during such emergency states.

An important global control discovered in *E. coli* more than 20 years ago is unlike any we have discussed so far in that it is based on a small metabolic product and not on a protein that affects individual genes. When cells are transferred from a rich medium to poor medium or when a mutant cell requiring a particular amino acid is transferred to a medium lacking it, the most immediately noticeable event is that accumulation of ribosomal RNA stops as protein synthesis slows down. This phenomenon, called the *stringent response* by Gunter Stent and Sydney Brenner, indicates that bacterial cells can regulate RNA synthesis in the face of a metabolic crisis. It is now known that synthesis of both rRNA and tRNA, the "stable

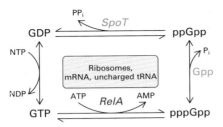

▲ **Figure 7-31** Cellular routes of (p)ppGpp synthesis and degradation. The key enzyme in the pathway, RelA, makes pppGpp from GTP when ribosomes are stalled in an "idling" reaction. The requirements include uncharged tRNA, mRNA, and ribosomes with which the RelA protein is associated. Action of pyrophosphorylase (Gpp) can interconvert (p)ppGpp and ppGpp. A 3′-phosphorylase termed *SpoT* removes 3′-phosphates, leaving the normal intermediate GDP. [See M. Cashel and K. E. Rudd, 1987, p. 1410 in *Escherichia coli and Salmonella typhimurium*, F. Neidhardt (ed.), American Society Microbiology.]

RNAs," ceases under such conditions. The stringent response is initiated by the accumulation of two guanine polyphosphates, ppGpp and pppGpp, which are synthesized by an enzyme called RelA (Figure 7-31). Mutations in the *relA* gene "relaxes" *E. coli* so that the stringent response is ablated. (However, if *relA* mutants are left for more than 30 to 60 minutes in a medium without a necessary amino acid, they die due to unbalanced growth.) RelA is not capable of making ppGpp on its own but requires association with "stalled" or "idling" ribosomes. That is, mRNA, at least one uncharged tRNA, ribosomes that are trying to translate the message, and RelA, which attaches to the ribosomes, are all required for the generation of pppGpp. GTP is used as the energy source for many steps in protein synthesis (tRNA positioning, translocation, etc.), and this use of the bound nucleotides may be related to ppGpp accumulation. The ppGpp has been shown to have a negative effect specifically on rRNA and tRNA transcription in vitro, and it obviously allows some but not all mRNAs to be synthesized and translated to meet the requirements of dwindling supplies of amino acids. Exactly how during the stringent response some mRNAs are made and others inhibited is not yet clear.

This mechanism protects the bacterium from unbalanced growth and is remarkable for its prompt and precise effects.

Stability of Biopolymers in Bacterial Cells

When bacterial cells are placed in a medium lacking an essential nutrient or containing a particular sugar, the enzymes necessary for synthesizing the missing nutrient or the available sugar are induced. Furthermore, if the nutrient is then added to the medium or the sugar is removed, the mRNAs for the induced enzymes will no longer be made. What happens to these mRNAs and to the induced enzymes after they are no longer needed?

mRNAs Are Degraded Rapidly

In general, unknown nucleases destroy bacterial mRNAs so quickly that their half-lives range from 3 to 5 min, whether or not inducers are present. In bacterial cells that grow and divide every 20 to 30 min, mRNA must be renewed five to ten times in each generation. Thus, the finely tuned mechanisms for the control of mRNA initiation plus the rapid degradation of mRNA molecules allow very efficient and flexible control of metabolism. When a bacterial cell needs certain enzymes, it rapidly begins to make them; when the need is gone, mRNA synthesis stops and the existing mRNA is destroyed quickly.

The enzymes responsible for mRNA degradation are not known. Experiments on the degradation of polycistronic mRNAs from different operons have not shown a uniform mechanism of destruction (e.g., 5′ → 3′ or endonucleolytic). It does seem certain that some polycistronic mRNAs (e.g., *lac* and *trp*) are cleaved internally to produce monocistronic molecules before their degradation is completed. One explanation of these results is that regions between coding segments of a long mRNA are not protected as well by ribosomes as are coding regions.

A few bacterial mRNAs have half-lives that are longer than the average. For example, the mRNA that encodes a particular cell membrane protein has a half-life of about 10 min. The basis for the increased longevity of this mRNA is unknown; perhaps the membrane protein translated from this mRNA enters the membrane and anchors the mRNAs there, thus protecting it from degradation.

Synthesis of Some Ribosomal Proteins Is Regulated by Control of mRNA Translation

The synthesis of ribosomal proteins and the cellular concentration of ribosomal RNAs closely parallel the rate of cell growth. In cell-free systems, specific mRNAs for ribosomal proteins can be synthesized from added DNA and translated into proteins. In the presence of excess ribosomal proteins, the genes that encode these proteins still are transcribed into mRNAs, but some of the mRNAs are not translated. This suggests that the translation of certain mRNAs into ribosomal proteins can be regulated by an excess of other ribosomal proteins.

Experiments with preparations of specific ribosomal mRNAs and specific ribosomal proteins have shown that excess ribosomal protein S8 (i.e., protein number 8 associated with the small ribosomal subunit) inhibits translation of the mRNA for protein L5 (protein number 5 associated with the large subunit) but not that for L14 or L24. A comparison of the sequences of the 16S rRNA and of the mRNA encoding the beginning of the L5 protein suggests a physical basis for the translational arrest caused by excess ribosomal proteins. In the 16S rRNA, there is a stem-loop structure whose base sequence has many similarities with the sequence at the beginning of the L5 mRNA (Figure 7-32). This region of the 16S rRNA binds tightly to the S8 protein. However, if the amount of S8 protein exceeds that of rRNA (as occurs, for example, in bacteria in growth-limiting conditions), some S8 protein will be free to bind to L5 mRNA; this binding directly inhibits translation of L5 mRNA. Because mRNA is destroyed rapidly, the inhibition of translation keeps the level of L5 mRNA low in nongrowing cells, even though it is being synthesized.

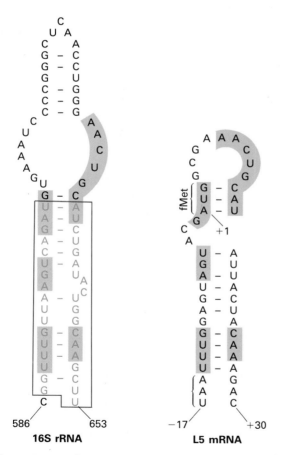

▲ **Figure 7-32** The primary and secondary structure of portions of 16S rRNA and the mRNA for ribosomal protein L5. In the 16S rRNA, the binding site for S8 ribosomal protein is indicated by the red box. Green shading highlights base sequences that are identical in the two RNAs. The striking similarities in the sequences and in the stem-loop structures in these RNAs indicate why they presumably compete for free S8 protein. [See P. O. Olins and M. Nomura, 1981, *Nucleic Acids Res.* 9:1757.]

Several cases of translational regulation of bacterial mRNAs are now known including the bacteriophage T4 protein RegA, which autogenously suppresses its own translation as well as that of some *E. coli* mRNAs.

Bacterial Proteins Are Diluted or Destroyed When Not Needed

Although induced mRNAs in bacterial cells disappear quickly when their products are no longer needed, this is not true of the corresponding proteins. Once the inducing condition for an enzyme is removed, most enzyme molecules are simply diluted by the growth and division of cells; and are not destroyed (see Figure 7-8). At one time it was thought that there was no intracellular turnover of bacterial proteins. However, if bacteria are severely deprived of nutrients, they can break down particular proteins to synthesize other proteins. For example, cells that are acutely starved for tryptophan can break down some proteins to release enough tryptophan to make the tryptophan-synthesizing enzymes. It is now known that *E. coli* has several intracellular proteases and that small quantities of proteins are broken down constantly into amino acids for the purpose of reusing these components. Although this process is not normally thought to be an important element of metabolic regulation in bacterial cells, it is necessary in time of starvation.

Summary

RNA synthesis and its control in bacteria are among the most thoroughly studied and best-understood subjects of molecular cell biology. Despite the differences between prokaryotic and eukaryotic cells, the gene-control mechanisms described in this chapter provide many useful clues for understanding gene expression in eukaryotes.

In bacterial cells, a single RNA polymerase is responsible for recognizing promoter sites in DNA. This recognition is assisted by sigma factors; different sigma factors permit the polymerase to recognize different classes of promoters. The typical *E. coli* promoter contains two consensus sequences, one centered about 10 base pairs before the start site for RNA synthesis, and one centered about 35 base pairs before this site. When an RNA polymerase molecule makes contact with the promoter and binds to it, RNA synthesis is initiated. Some bacterial genes are transcribed at a constant rate (usually infrequently) without any regulation.

Initiation of transcription of many other genes is governed by site-specific proteins that interact with DNA regions termed operators. Operator sequences most often lie near promoter sites and sometimes overlap the promoters (i.e., include some of the same base pairs). A few genes have been identified whose operators lie several hundred nucleotides away from the initiation site; the operators in these genes act by forming a loop in the DNA. The regulatory proteins that bind to operators are called repressors if they inhibit initiation and activators if they increase it. Many genes are under the complex control of both repressor and activator proteins; the repressor must be removed and the activator must be present in order for these genes to be transcribed. Such genes often have two neighboring control sites, one recognized by the positive-acting regulatory protein and one recognized by the negative-acting protein. In general, the promoter-operator sites in prokaryotic DNA control the transcription of regions encoding more than one protein. The entire DNA transcription unit is called an operon. A prokaryotic mRNA therefore often contains information for more than one protein.

Control also occurs at the level of transcriptional termination in some genes. During the process of mRNA synthesis (and concomitant translation), the RNA polymerase does not always complete transcription of all coding information in an operon. Thus the amounts of some mRNAs in bacteria are controlled by the regulated *termination of transcription*. Such regulation can occur between two genes in an operon, either by termination at that site or by antitermination—that is, prevention of termination by the binding of a specific protein (an antiterminator). Regulated termination can also occur before the coding part of a gene is reached. This premature termination, which is called attenuation, is widely used to regulate the production of mRNAs that encode enzymes for the synthesis of amino acids.

A complex sequence of events following infection of *E. coli* cells by the bacteriophage λ determines whether the lytic cycle or the lysogenic state ensues and whether the bacteriophage remains quiescent in the bacterial genome. The interactions of several transcriptional regulatory proteins illustrate how semipermanent epigenetic decisions can be made and suggest how a complex transcriptional control program might direct eukaryotic cells toward alternate physiologic states.

References

General Mechanisms of Prokaryotic Gene Control

GLASS, R. E. 1982. *Gene Function: E. coli and Its Heritable Elements*. University of California Press.

McCLURE, W. R. 1985. Mechanism and control of transcription initiation in prokaryotes. *Ann. Rev. Biochem.* 54:171–204.

NEIDHARDT, F., ed. 1987. *Escherichia coli and Salmonella typhimurium*. American Society of Microbiology.

PABO, C. T., and R. T. SAUER. 1984. Protein-DNA interactions. *Ann. Rev. Biochem.* 53:293–321.

SCHLEIF, R. 1988. DNA binding by proteins. *Science* 241:1182–1187.

RNA Polymerase

HANNA, M. M., and C. F. MEARES. 1983. Topography of transcription: path of the leading end of nascent RNA through the *E. coli* transcription complex. *Proc. Nat'l Acad. Sci. USA* 80:4238–4242.

LOSICK, R., and M. J. CHAMBERLIN, eds. 1976. *RNA Polymerase*. Cold Spring Harbor Laboratory.

SWEETSER, D., M. NONET, and R. A. YOUNG. 1987. Prokaryotic and eukaryotic RNA polymerases have homologous core subunits. *Proc. Nat'l Acad. Sci. USA* 84:1192–1197.

VON HIPPEL, P. H., D. G. BEAR, W. D. MORGAN, and J. A. McSWIGGEN. 1984. Protein–nucleic acid interaction in transcription. *Ann. Rev. Biochem.* 53:389–446.

Sigma (σ) Factors

BRIAT, J. F., M. Z. GILMAN, and M. J. CHAMBERLIN. 1985. *Bacillus subtilis* σ^{28} and *Escherichia coli* σ^{32} (htpR) are minor σ factors that display an overlapping promoter specificity. *J. Mol. Biol.* 260:2038–2041.

GRIBSKOV, M., and R. R. BURGESS. 1986. Sigma factors from *E. coli*, *B. subtilis*, phage SPO1, and phage T4 are homologous proteins. *Nuc. Acids Res.* 14:6745–6763.

GROSSMAN, A. D., D. B. STRAUSS, W. W. WALTER, and C. A. GROSS. 1987. σ^{32} synthesis can regulate the synthesis of heat shock proteins in *E. coli*. *Genes & Develop.* 1:179–184.

HELMANN, J. D., and M. J. CHAMBERLAIN. 1988. Structure and function of bacterial sigma factors. *Ann. Rev. Biochem.* 57:839–872.

Promoters

HAWLEY D. K., and W. R. McCLURE. 1983. Compilation and analysis of *Escherichia coli* promoter DNA sequences. *Nuc. Acids Res.* 11:2237–2255.

PRIBNOW, D. 1975. Bacteriophage T7 early promoters: nucleotide sequences of two RNA polymerase binding sites. *J. Mol. Biol.* 99:419–443.

SIEBENLIST, U., R. B. SIMPSON, and W. GILBERT. 1980. *E. coli* RNA polymerase interacts homologously with two different promoters. *Cell* 20:269–281.

Negative Control of Transcription

BECKWITH, L. J. 1987. Genetics at the Pasteur Institute: Substance and style. *ASM News* 53:551–555.

CUNIN, R., N. GLANSDORFF, A. PEIRARD, and V. STALON. 1986. Biosynthesis and metabolism of arginine in bacteria. *Microbiol. Rev.* 50:314–352.

GILBERT, W., and B. MÜLLER-HILL. 1967. Isolation of the *lac* repressor. *Proc. Nat'l Acad. Sci. USA* 58:2415–2421.

JACOB, F., and J. MONOD. 1961. Genetic regulatory mechanisms in the synthesis of proteins. *J. Mol. Biol.* 3:318–356.

KAISER, A. D., and F. JACOB. 1957. Recombination between related temperate bacteriophages and the genetic control of immunity and prophage localization. *Virology* 4:509–521.

PARDEE, A. B., F. JACOB, and J. MONOD. 1959. The genetic control and cytoplasmic expression of inducibility in the synthesis of β-galactosidase by *E. coli*. *J. Mol. Biol.* 1:165–178.

TAKEDA, Y., D. H. OHLENDORF, W. F. ANDERSON, and B. F. MATTHEWS. 1983. DNA-binding proteins. *Science* 221:1020–1026.

ZIPSER, D., and J. BECKWITH, eds. 1970, 1977. *The lac Operon.* Cold Spring Harbor Laboratory.

Positive Control of Transcription: Bending and Looping of DNA

ENGLESBERG, E., and G. WILCOX. 1974. Regulation: positive control. *Ann. Rev. Genet.* 8:219–242.

GARTENBERG, M. R., and D. M. CROTHERS. 1988. DNA sequence determinants of CAP-induced bending and protein binding affinity. *Nature* 333:824–829.

GRIFFITH, J., A. HOCHSCHILD, and M. PTASHNE. 1986. DNA loops induced by cooperative binding of λ repressor. *Nature* 322:750–752.

JORDAN, S. R., and C. O. PABO. 1988. Structure of the lambda complex at 2.5 A resolution: details of the repressor-operator interactions. *Science* 242:893–899.

LIU-JOHNSON, H-N., M. R. GARTENBERG, and D. M. CROTHERS. 1986. The DNA binding domain and bending angle of *E. coli* CAP protein. *Cell* 47:995–1005.

MARTIN, K., L. HUO, and R. F. SCHLIEF. 1986. The DNA loop model for *ara* repression: AraC protein occupies the proposed loop sites *in vivo* and repression negative mutations lie in these same sites. *Proc. Nat'l Acad. Sci. USA* 83:3654–3658.

RAIBAUD, O., and M. SCHWARTZ. 1984. Positive control of transcription initiation in bacteria. *Ann. Rev. Genet.,* 18:173–206.

REITZER, L. J., and B. MAGASANIK. 1986. Transcription in *glnA* in *E. coli* is stimulated by activator bound to sites far from the promoter. *Cell* 45:785–792.

SCHLEIF, R. 1988. DNA looping. *Science* 240:127–128.

STRANEY, S., and D. M. CROTHERS. 1987. Lac repressor is a transient gene-activating protein. *Cell* 51:699–707.

ZUBAY, G., D. SCHWARTZ, and J. BECKWITH. 1970. Mechanism of activation of catabolic-sensitive genes: a positive control system. *Proc. Nat'l Acad. Sci. USA* 66:104–110.

Compound Control of Transcription

CASHEL, M., and K. E. RUDD. 1987. The stringent response. In *Escherichia coli and Salmonella typhimurium,* F. Neidhardt, ed. American Society of Microbiology.

DE CROMBRUGGHE, B., and I. PASTON. 1978. Cyclic AMP, the cyclic AMP receptor protein, and their dual control of the galactose operon. In *The Operon,* J. H. Miller and W. S. Reznikoff, eds. Cold Spring Harbor Laboratory.

GOTTESMAN, S. 1984. Bacterial regulation: global regulatory networks. *Ann. Rev. Genet.* 18:415–442.

Control of Regulatory Proteins

AIBA, H. 1983. Autoregulation of the *Escherichia coli crp* [or CAP] gene. CRP is a transcriptional repressor for its own gene. *Cell* 32:141–149.

MAGASANIK, B. 1978. Regulation in the *hut* system. In *The Operon,* J. H. Miller and W. S. Reznikoff, eds. Cold Spring Harbor Laboratory.

MAGASANIK, B. 1987. The control of carbon and nitrogen utilization. In *Escherichia coli and Salmonella typhimurium,* F. Neidhardt, ed. American Society of Microbiology.

Control of Transcriptional Termination

HORWITZ, R. J., J. LI, and J. GREENBLATT. 1987. An elongation control particle containing the N gene transcriptional antitermination protein of bacteriophage lambda. *Cell* 51:631–641.

PLATT, T. 1986. Transcription termination and the regulation of gene expression. *Ann. Rev. Biochem.* 55:339–372.

ROBERTS, J. W. 1988. Phage lambda and the regulation of transcription termination. *Cell* 52:5–6.

YANOFSKY, C. 1981. Attenuation in the control of expression of bacterial operons. *Nature* 289:751–758.

The Early Events of Bacteriophage λ Infection

HERSKOWITZ, I., and D. HAGEN. 1980. The lysis-lysogeny decision of phage λ: explicit programming on responsiveness. *Ann. Rev. Genet.* 14:399–446.

HOCHSCHILD, A., N. IRWIN, and M. PTASHNE. 1983. Repressor structure and the mechanism of positive control. *Cell* 32:319–325.

PABO, C. O., and M. LEWIS. 1982. The operator-binding domain of λ repressor: structure and DNA recognition. *Nature* 298:443–447.

PTASHNE, M. 1986. A genetic switch. Gene control and phage λ. Blackwell Scientific Publications and Cell Press.

Posttranscriptional Control

BRAWERMAN, G. 1987. Determinants of mRNA stability. *Cell* 48:5–6.

CANNISTRARO, V. J., M. N. SUBBARGO, and D. KENNELL. 1986. Specific endonucleolytic cleavage for decay of *Escherichia coli* mRNA. *J. Mol. Biol.* 192:257–274.

DEAN, D., J. L. YATES, and M. NOMURA. 1981. *Escherichia coli* ribosomal protein S8 feedback regulates part of *spc* operon. *Nature* 289:89–91.

GOLD, L. 1988. Posttranscriptional regulatory mechanisms in *Escherichia coli. Ann. Rev. Biochem.* 57:199–234.

MILLER, E. S., J. KARAM, M. DAWSON, M. TROJANOWSKA, P. GAUSS, and L. GOLD. 1987. Translational repression: biological activity of plasmid-encoded bacteriophage T4 RegA protein. *J. Mol. Biol.* 194:397–410.

NOMURA, M., D. DEAN, and J. L. YATES. 1982. Feedback regulation of ribosomal protein synthesis in *Escherichia coli. Trends Biochem. Sci.* 7:92–95.

OLINS, P. O., and M. NOMURA. 1981. Translational regulation by ribosomal protein S8 in *Escherichia coli:* structural homology between rRNA binding site and feedback target on mRNA. *Nuc. Acids Res.* 9:1757–1764.

CHAPTER 8

RNA Synthesis and Processing in Eukaryotes

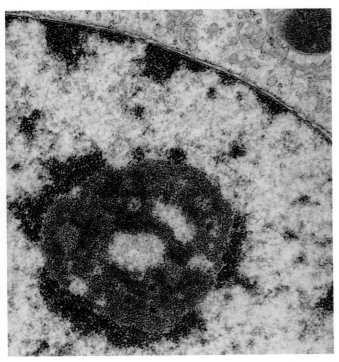

Nucleolus in nucleus of mammalian cell

*L*ike prokaryotic genes, the majority of eukaryotic genes are regulated at the level of transcription. However, a great many eukaryotic genes are also regulated by differential processing of their primary RNA transcripts, by differential rates of translation of their mRNAs in the cytoplasm, and by differential rates of degradation of these mRNAs.

Because information about the expression of eukaryotic genes came initially from biochemical studies on RNA synthesis and because knowledge of the biochemistry of nuclear RNA is necessary for understanding the structure of eukaryotic genes, RNA synthesis and processing are described in this chapter. Chapters 9 and 10 deal with the structure and organization of genes within the chromosomes in eukaryotes. These subjects pave the way for an integrated discussion of gene control in eukaryotes in Chapter 11. ▲

The expression of genes in eukaryotes differs in four significant ways from that in prokaryotes (Figure 8-1):

1. Most eukaryotic DNA is associated with proteins termed *histones* in a highly ordered *chromatin* structure. Because this association makes the DNA relatively inaccessible to transcription, changes in the packing of chromatin probably are necessary before transcription of eukaryotic genes can occur. In contrast, bacterial DNA is not bound to protein in a compact form, and the transcription of specific genes anywhere in the single bacterial chromosome can be started or stopped in seconds. A similar rapid initiation and termination of transcription can occur with many genes in single-cell eukaryotes and even with some genes in multicellular organisms; however, transcription of many genes in multicellular organisms does not respond quickly to outside stimuli but presumably depends on changes in chromatin packing that occurs sometime during the development of the organism.

2. In prokaryotes, a single polymerase is used for transcription of all genes. In contrast, eukaryotes have three different RNA polymerases: I for ribosomal RNAs, II for mRNAs, and III for small RNAs like tRNA and 5SRNA.

3. The initial mRNA transcript in eukaryotes is almost always *not* a functional mRNA that can be translated into a protein, whereas in prokaryotes it is. Rather, precursor mRNA (pre-mRNA) is produced in the nucleus and undergoes several biochemical modifications before it enters the cytoplasm as functional mRNA.

Similarly, precursor rRNAs and tRNAs must be modified to form functional molecules; such modification of precursor rRNA and tRNA also occurs in bacteria.

4. In eukaryotic cells, mRNA is synthesized in the nucleus, but it is translated into protein in the cytoplasm. Thus, coupled transcription-translation does not occur in eukaryotes as it does in prokaryotes.

Relationship of Nuclear and Cytoplasmic RNA

A central feature of gene expression in eukaryotic cells is that although proteins are synthesized in the cytoplasm and most of the RNA molecules in the cell are found in the cytoplasm, the genes (DNA molecules) that control protein synthesis are located in the nucleus, where the RNA is synthesized (Figure 8-2). In an effort to establish the relationship between nuclear RNA and cytoplasmic RNA, which takes part in protein synthesis, researchers separated newly formed nuclear RNA into classes that could be identified and followed as the RNA migrated from the nucleus to the cytoplasm. They first characterized the nuclear RNA precursors of rRNA and then the precursors of tRNA. Finally, the details of the formation of mRNA from its nuclear precursors were worked out. The fractionation schemes and experimental approaches that were developed during this early work on RNA synthesis are still widely used.

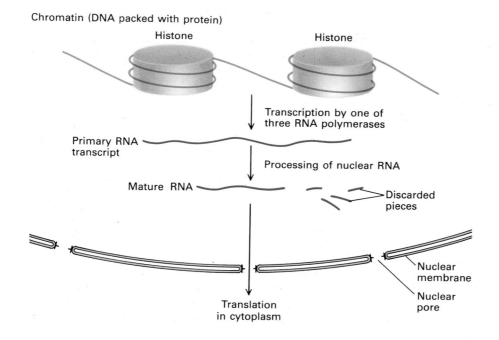

Chromatin (DNA packed with protein)

Histone   Histone

Transcription by one of three RNA polymerases

Primary RNA transcript

Processing of nuclear RNA

Mature RNA

Discarded pieces

Nuclear membrane

Nuclear pore

Translation in cytoplasm

◀ **Figure 8-1** Four features distinguish RNA production in eukaryotes from that in prokaryotes: (1) The DNA to be transcribed is wound around a histone core. (2) It is transcribed by one of three RNA polymerases. (3) The primary RNA transcript is processed within the nucleus into the finished RNA. (4) Only after transport of the RNA—presumably through a nuclear pore—does translation occur in the cytoplasm. In bacteria, all precursor rRNA and tRNA are processed but mRNA only rarely requires processing.

(a)   (b)

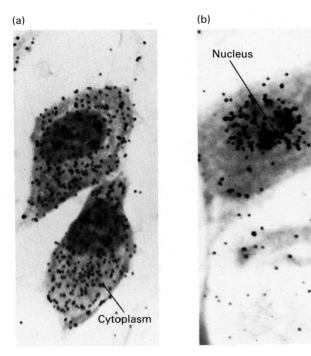

Nucleus

Cytoplasm

▲ **Figure 8-2** Demonstration that protein synthesis is cytoplasmic and RNA synthesis is nuclear. Chinese hamster cells were exposed to (a) amino acids labeled with tritium or to (b) [³H]uridine, and autoradiographs were made. Grains indicating incorporation of amino acids into proteins are located in the cytoplasm (some cytoplasm overlies the nucleus), whereas those indicating incorporation of uridine into RNA are in the nucleus. *Courtesy of D. Prescott.*

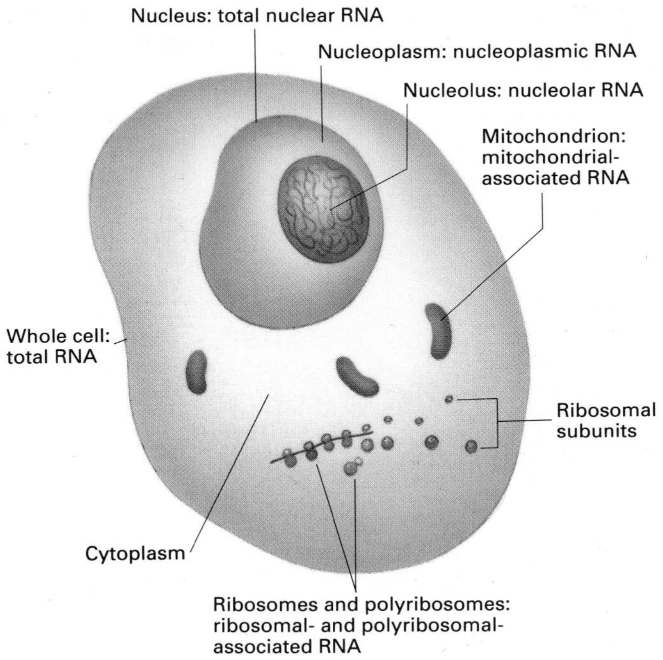

Nucleus: total nuclear RNA

Nucleoplasm: nucleoplasmic RNA

Nucleolus: nucleolar RNA

Mitochondrion: mitochondrial-associated RNA

Whole cell: total RNA

Ribosomal subunits

Cytoplasm

Ribosomes and polyribosomes: ribosomal- and polyribosomal-associated RNA

▲ **Figure 8-3** Fractionation of mammalian cells yields various components from which RNA can be extracted and purified. Commonly used RNA preparations are indicated. In plant cells, chloroplasts also can be separated from the cytoplasm, and their RNA can be extracted.

Cell Fractionation and Labeling Experiments Reveal Locations and Classes of RNA

Demonstrating the relationship of nuclear RNA to cytoplasmic RNA requires two kinds of fractionation. In one, the nucleus is physically separated from the cytoplasm so that the RNA in each fraction can be examined (Figure 8-3). The second is a fractionation in time which allows the kinetic (time) relationship between nuclear and cytoplasmic RNA to be followed. The sequential appearance of a labeled RNA precursor in different nuclear RNA molecules before its appearance in cytoplasmic products points to the initial, the intermediate, and the final products in a precursor-product pathway.

Cultured mammalian cells, particularly HeLa cells, are ideal for cell fractionation and the separation of nuclear and cytoplasmic RNAs. Centrifugation (or electrophoresis through gels) of the purified total cellular RNA separates molecules according to their size. Each size class in the total cellular RNA is detected by its absorption of

ultraviolet light; the RNA species that are present in the largest amounts are, of course, the easiest to observe. This type of analysis reveals three major species of RNA: large rRNA, small rRNA, and tRNA (Figure 8-4). The 28S rRNA, which is about 5 kilobases (kb) in length, and the 18S rRNA, which is about 2 kb, appear to make up approximately 80 percent of the total RNA, with most of the remainder being tRNA. However, RNA species that constitute only a small portion of the total RNA (e.g., mRNA, 5.8S rRNA, and 5S rRNA) are not detected as distinct components by this method.

Of the rRNA and tRNA species in fractionated cells detected by their absorption of UV light, almost all (more than 90 percent) are in the cell cytoplasm. However, if cells are exposed for 5 min or less to a labeled RNA precursor such as [³H]uridine before they are fractionated, almost all of the *labeled* RNA is in the nucleus. Furthermore, the size of the newly labeled nuclear RNA does not match that of the major species in the cytoplasm. These briefly labeled nuclear molecules are quite heterogeneous in size, but most of the label is found in molecules that

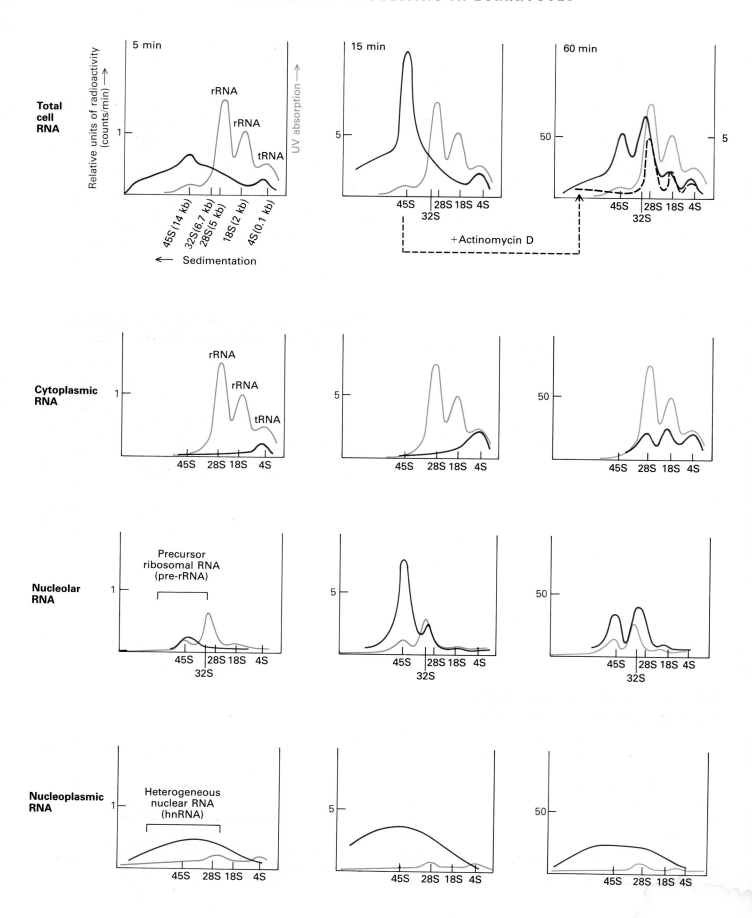

◀ **Figure 8-4** Labeling of cultured HeLa cells with [³H]uridine for 5, 15, or 60 min and subsequent analysis of whole-cell extracts and cell fractions by rate-zonal sedimentation. The preexisting (old) RNA was monitored by its absorption of ultraviolet light (blue curves); the newly synthesized RNA was monitored by its radioactivity (red curves). Equal-sized cell samples were taken at each labeling time for analysis of whole-cell extracts and cell fractions. The sequential appearance of label in various RNA species helps to identify the initial, intermediate, and final RNA products. (Note that smaller, slower sedimenting species are to the right in plots.) When cells were labeled for 15 min and then treated for 45 min with actinomycin D, a drug that blocks transcription, the 45S RNA disappeared but 28S and 18S rRNA appeared (dashed black line, top panel). This result indicates that 45S RNA is a precursor for 28S and 18S rRNA. [See J. E. Darnell, 1968, *Bacteriol. Rev.* 32:262.]

sediment as fast or faster than the 28S rRNA that is found in the cytoplasm (Figure 8-4). In addition, the majority of this briefly labeled RNA is outside of the nucleolus in the so-called nucleoplasm.

If cells are labeled for longer periods before fractionation (e.g., for 15 and 60 min), labeled nuclear RNA species of specific sizes begin to emerge in the nucleolus against the background of heterogeneous nucleoplasmic RNA. The first labeled species to be detected in any quantity is a 45S RNA that is about 14 kb long; soon another labeled species, a 32S RNA, about 6.7 kb long appears. Molecules in these two size classes constitute a large percentage of the total labeled nuclear RNA after a 60-min labeling period. A short time after the labeled 6.7-kb (32S) species appears, a radioactive label begins to appear in the cytoplasmic 28S and 18S rRNAs, which are about 5 kb and 2 kb, respectively. These results suggest that the nuclear molecules that are formed initially act as precursors for cytoplasmic RNAs. As we shall see, the 45S and 32S nucleolar RNAs are precursors to the 28S, 18S, and 5.8S rRNAs.

45S Nucleolar RNA of HeLa Cells Is a Precursor for Ribosomal RNA

Following the labeling experiments described in the previous section, chemical experiments showed that the labeled 45S RNA found in the nucleolus is related to rRNA. Its nucleotide composition is similar to that of rRNA, and it contains the same methylated oligonucleotide sequences as rRNA. Additional experiments also demonstrated that labeled 45S RNA is converted into smaller RNA molecules. This was shown by labeling cells for 15 min and then adding actinomycin D, which binds to DNA and immediately stops transcription and hence

synthesis of additional labeled 45S RNA. After addition of actinomycin D, the amount of labeled 45S RNA decreases, and the amounts of 28S and 18S rRNAs increase (Figure 8-4, top panel).

These results, which strongly suggested that 45S nucleolar RNA is a precursor for 28S and 18S rRNA, were the first to indicate that nuclear RNA molecules are refashioned, or *processed,* into cytoplasmic RNAs. The identification and characterization of pre-rRNA gave rise to two other terms commonly used in discussions of eukaryotic gene expression. A *primary RNA transcript* is a newly synthesized RNA molecule that has not yet been modified in length; that is, it is the initial unmodified RNA product. A *transcription unit* is a DNA region, bounded by an initiation site and termination site, that is transcribed to produce a primary transcript (Figure 8-5).

Heterogeneous Nuclear RNA Is a Separate Class of Rapidly Labeled Nuclear RNA

As mentioned earlier, when HeLa cells are allowed to incorporate [³H]uridine for 5 min or less (see Figure 8-4), most of the labeled nuclear RNA lies outside the nucleolus and is heterogeneous in size. Thus, it is referred to as *heterogeneous nuclear RNA* (hnRNA). These molecules range in length from a few hundred nucleotides to more than 30 kb, with an average length of about 5 kb. The average nucleotide composition of hnRNA is similar to that of total cellular DNA (44 percent G + C) and quite different from that of rRNA (60 percent G + C). As will be discussed later in this chapter, hnRNA includes the precursors to mRNAs, although not all hnRNAs are converted to mRNAs.

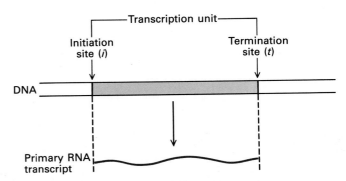

▲ **Figure 8-5** Transcription units in DNA have initiation sites and termination sites. Primary RNA transcripts are the initial unmodified RNA products.

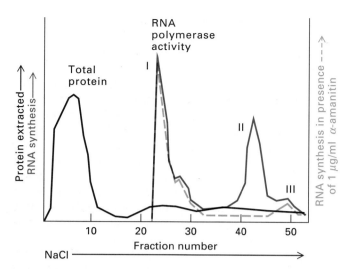

▲ **Figure 8-6** The separation and identification of the three eukaryotic RNA polymerases by chromatographic analysis. A protein extract from the nuclei of cultured frog cells was passed through a DEAE Sephadex column to which charged proteins adsorb differentially, and adsorbed proteins were eluted (black curve) with NaCl solutions of increasing concentration. Successive fractions of the eluted proteins were assayed for the ability to transcribe DNA (red curve) in the presence of the four nucleoside triphosphates (including radioactive UTP). Most of the proteins did not bind to the column, but the enzymes did. The synthesis of RNA by each fraction in the presence of 1 μg/ml of α-amanitin also was measured in (blue curve). Polymerases I and III are insensitive to the compound at that concentration, whereas polymerase II is sensitive, that is, it ceases RNA synthesis. (Polymerase III is sensitive to 10 μg/ml of α-amanitin, however, although polymerase I is unaffected even by this higher concentration.) [See R. G. Roeder, 1974, *J. Biol. Chem.* 249:241.]

Function and Structure of RNA Polymerases

All eukaryotic cells examined so far (e.g., human, rodent, frog, fruit fly, yeast, and plant cells) contain three different RNA polymerases, designated I, II, and III. Initial recognition of these separate enzymes occurred when chromatographic purification of the enzymes for RNA synthesis resulted in three fractions, eluted at different salt concentrations, each of which had RNA synthesis activity (Figure 8-6). The polymerases also can be distinguished by their sensitivity to α-amanitin, a poisonous glycoside produced by some mushrooms. Polymerase I is very insensitive to α-amanitin; polymerase II is very sensitive; and polymerase III has intermediate sensitivity.

Three Polymerases Catalyze Formation of Different RNAs

Each eukaryotic RNA polymerase catalyzes transcription of genes encoding different classes of RNA. Polymerase I is located in the nucleolus and is responsible for synthesis of ribosomal precursor RNAs. Polymerase III functions outside the nucleolus and transcribes the genes for tRNA, 5S rRNA, and for a whole array of small RNAs; some of the latter are known to function in RNA processing, but the functions of many are unknown at present. Polymerase II catalyzes transcription of all the protein-coding genes; that is, it functions in production of mRNAs. RNA polymerase II, or perhaps a specialized form of it, also produces several small RNAs that take part in RNA processing.

RNA Polymerases Have Complex Subunit Structure

All three eukaryotic RNA polymerases have a much more complicated structure than does the core *E. coli* polymerase, which contains two small identical α subunits and two large subunits, β and β'. Each of the eukaryotic polymerases contain 10 to 15 different polypeptide subunits, including two large subunits of about 140 to 150 kilodaltons (kDa) and 190 to 220 kDa (Figure 8-7).

The genes encoding the two large subunits in RNA polymerases from *E. coli,* yeasts, mice, fruit flies, and other organisms have been cloned and sequenced. Comparisons of the corresponding amino acid sequences indicate that the largest subunit (L') in all three eukaryotic polymerases shares several conserved regions in which 70 to 80 percent of the amino acids are identical; in addition, some of these similar regions are found in the β' subunit of *E. coli* polymerase (Figure 8-8). Likewise, the second largest subunit (L) is similar in all three eukaryotic polymerases and shares many similarities with the *E. coli* β subunit. The extensive homology in the amino acid sequences of the RNA polymerases from various sources indicates a common evolutionary root for all cell types. It is logical that a process so basic as copying RNA from DNA must have evolved for the earliest successful cells.

The smaller polymerase subunits also exhibit various similarities, both among the different types (I, II, and III) and among the same polymerase from different organisms. For example, the distribution of small subunits in each type of polymerase is similar in all eukaryotic organisms. Three similar-sized small subunits are present in all three polymerases; two additional subunits are present in polymerase I and III (see Figure 8-7). However, the sizes of other small subunits differ among the three polymerases. Studies with yeasts have shown that all of the half-dozen genes encoding these subunits so far tested are necessary for cell growth and survival. Thus, it seems likely

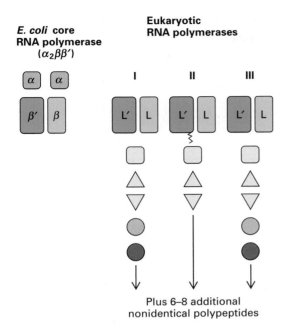

▲ **Figure 8-7** The subunits of RNA polymerases. The diagram emphasizes the relative simplicity of the *E. coli* core enzyme compared to the multisubunit composition of the three eukaryotic polymerases (I, II, and III). However, the largest subunits in *E. coli* and eukaryotic polymerases have stretches of amino acid similarity indicating a common evolutionary origin ($\beta' \cong L'$; $\beta \cong L$). The L' subunit of eukaryotic polymerase II has a special stretch of repeated amino acids *(zigzag line)* at its carboxyl terminus that is not found in polymerase I or III. Some smaller subunits are similar in each of the three polymerases, and some are specific to each polymerase. The polymerases from all eukaryotic organisms have quite similar distributions of polymerase subunits.

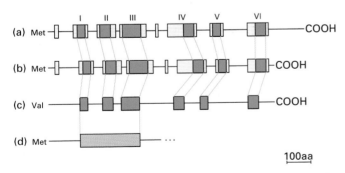

▲ **Figure 8-8** Comparisons of amino acid sequences in the largest subunit (L') of yeast RNA polymerase II (a) and III (b), the β' subunit of the *E. coli* polymerase (c), and a portion of the largest subunit of *Drosophila* polymerase II (d). The green boxes indicate regions of conserved sequence (70–80 percent amino acid identity) among all four polymerase subunits. The yellow boxes indicate regions of great similarity (80 percent amino acid identity) in the two yeast proteins. *After L. A. Allison et al., 1985, Cell **42**:599 and J. Biggs et al., 1985, Cell **42**:611.*

that every subunit must be present for eukaryotic RNA polymerase to be functional.

The carboxyl end of the largest subunit (L') of eukaryotic polymerase II contains a stretch of seven amino acids that is repeated 20 or more times in yeast and 50 times in mice. Although the function of this repeating unit is unknown, it is critical for growth. In experiments with haploid yeast cells, for example, cells with 10 or fewer repeating units did not survive; those with 10 to 12 repeats survived but grew poorly; and those with 13 or more repeats exhibited normal growth. Thus some critical function carried out by these repeats requires at least a certain number. Phosphate groups are attached to both serines and the tyrosine in the repeating unit. If the phosphate groups are lost during preparation of polymerase II, the enzyme is much less active in catalyzing mRNA synthesis. Neither polymerase I or III contains these repeating units.

Transcription Factors Assist Polymerases to Recognize Initiation Sites

Almost nothing is known yet about how any of the subunits of eukaryotic RNA polymerases interact with DNA. It is known that many different proteins help the polymerases to find the correct initiation sites and to begin transcription. Because the control of RNA synthesis is such a crucial event in cell biology, these *transcription factors* have attracted one of the largest group of investigators of any single field in modern biology.

Many transcription factors have been isolated from eukaryotes. They can be classified into three types: (1) those which are required for initiation of transcription; (2) those which participate later in elongation of the RNA transcript; and (3) those which assist in formation of an active preinitiation complex but may not be required after initiation of transcription. The first two types, called *general factors*, are required for transcription of all genes transcribed by a particular eukaryotic RNA polymerase. *Specific factors*, the third type, help regulate transcription of particular genes and are analogous to the regulatory proteins found in prokaryotes. General transcription factors for RNA polymerase I, II, and III are discussed in this chapter; specific factors are discussed in detail in Chapter 11.

Three Methods for Mapping Transcription Units

To learn which sequences in DNA may play a role in starting and stopping transcription, it was necessary to first define the transcription unit—that is, where on DNA the RNA polymerase starts and stops. In bacterial cells, the transcription units are the operons that were first defined by genetic tests. Even before detailed DNA sequences were available, purified *E. coli* RNA polymerase was found to intitiate RNA synthesis approximately at the beginning of the genetically identified operons, and these in vitro start sites were accepted as defining the correct in vivo RNA initiation sites in the operon.

In eukaryotic cells, however, similar detailed genetic experiments defining transcription units had not been performed. Furthermore, when purified RNA polymerases were incubated with nucleotides and purified DNA, only a small amount of RNA was formed, and most of the RNA synthesized simply started at the broken ends of DNA molecules that were available to copy. For these reasons, transcription units in eukaryotes were first defined by examining the labeled primary RNA transcripts extracted from whole cells or whole nuclei. Much later, recombinant DNA technology provided the means to identify the exact initiation and termination sites for transcription.

Three of the common techniques used to examine the generation of a primary transcript—nascent-chain analysis, electron microscopy, and ultraviolet-damage analysis—were first used to study pre-rRNA. Because most of our ideas about RNA synthesis, processing, and control started with mapping of the general boundaries of transcription units (especially those for ribosomal RNA), we describe each of these mapping techniques in the following three sections.

Nascent-chain Analysis Provides "Snapshots" of RNA Molecules during Synthesis

In an earlier section we described how kinetic labeling can be used to identify different classes of RNA and their location in the cell (see Figure 8-4). If cultured cells are exposed to a radioactive label for very brief times, however, only *nascent chains* become labeled. The incomplete nascent chains emerging from a single transcription unit are heterogeneous, ranging in length from one nucleotide to one less than the number of nucleotides in the complete primary transcript. Analysis of such nascent chains permits mapping of different coding regions within a transcription unit.

The first step in nascent-chain analysis is to obtain a sufficient quantity of pure DNA from which a particular RNA is transcribed in cells (e.g., viral DNA can be used, or other DNA can be cloned in a bacterial vector). By use of specific restriction endonucleases, the DNA can be cut into pieces whose successive order in the genome is known, and these DNA pieces can be used in RNA-DNA hybridization experiments to determine initiation and termination sites and the direction of RNA synthesis.

DNA fragments that lie within a transcription unit will hybridize to labeled nascent RNAs synthesized from that

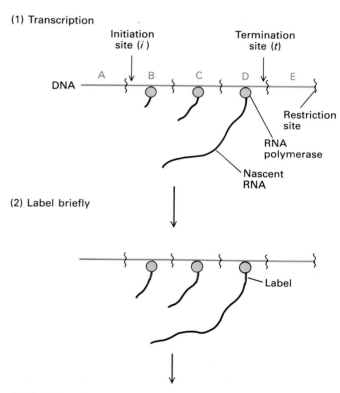

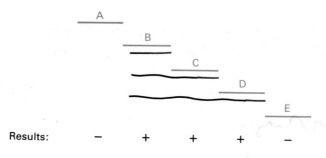

▶ **Figure 8-9** Mapping of a transcription unit by nascent-chain analysis. (1) A section of DNA with transcription start and stop sites is illustrated in the act of being transcribed. (2) The introduction of a radioactive label (red) for a brief time produces labeled nascent RNA molecules. (3) A sufficient quantity of the purified DNA (e.g., the DNA of a virus or cloned recombinant DNA) is then divided into an ordered set of fragments (A to E) by cutting at defined sites with a restriction endonuclease; the restriction sites are shown in (1). Hybridization of the labeled nascent RNA to the DNA pieces defines which fragments lie within the transcription unit and in which order: the labeled regions of the shortest RNAs will hybridize to fragment B in the diagram, and the labeled regions of the longer RNAs will hybridize to succeeding fragments, showing that RNA synthesis begins in the B fragment and continues through fragments C and D.

transcription unit; the DNA that lies on either side of the transcription unit will not hybridize to the labeled RNAs. The sizes of the labeled nascent RNAs that hybridize to various members of a set of ordered DNA fragments indicates the direction of RNA chain growth: the shortest labeled nascent RNA will hybridize to the DNA nearest the promoter; longer and longer nascent RNA chains will hybridize to successively more distant DNA fragments (Figure 8-9). The technique of nascent-chain analysis was the basis for the original mapping of transcription units for mRNAs in animal cells, as will be discussed later in this chapter.

Electron Microscopy Can Visualize Transcription Units in Action

The first transcription units to be viewed in the electron microscope were ribosomal DNA from the nucleolus of frog oocytes (cells on the pathway to becoming egg cells). Because these cells have greatly increased amounts of rDNA and increased numbers of nucleoli, all making pre-rRNA, they are excellent material with which to observe pre-rRNA synthesis in the electron microscope.

In the electron micrograph presented in Figure 8-10, for example, the transcribing RNA polymerases, trailing fibrils of pre-rRNA, can be easily seen. As many as 50 polymerases transcribing pre-rRNA from a single transcription unit have been seen in such electron micrographs. The stretches of DNA devoid of trailing fibrils are nontranscribed spacer regions between the repeated transcription units for pre-rRNA. The progression of lengths of the fibrils from short to long indicates the direction of transcription. Electron micrographs of other eukaryotic transcription units have also been obtained. So far they have been used mainly to assess the average size of these other transcription units in comparison with ribosomal transcription units.

Electron microscopy can also aid in assessing the relationship between primary transcription products and final products. Views of 45S pre-rRNA, for example, have revealed that the molecule contains reproducible and characteristic stem-loop structures (caused by intramolecular complementarity). These structures are also visible in the final products, the 28S and 18S cytoplasmic rRNAs; thus the final products can be identified within their precursors.

Effect of UV Radiation on RNA Synthesis Can Be Used to Map Transcription Units

A third technique for mapping transcription units depends on the fact that exposure of DNA to ultraviolet (UV) radiation causes formation of intrachain pyrimidine dimers, mainly between adjacent thymidylate residues. Since RNA polymerases terminate transcription at these

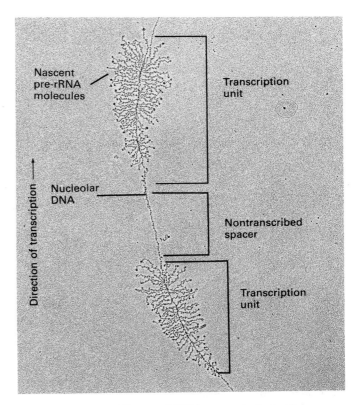

▲ **Figure 8-10** Ribosomal DNA transcription unit as viewed by electron microscopy. Frog oocytes were lysed and the nuclear contents were captured on the surface of an electron-microscope grid; this material was then dried and shadowed. Each "feather" represents a nascent pre-rRNA molecule emerging from a rDNA transcription unit. The tandemly arranged transcription units are separated by nontranscribed spacer regions. *Courtesy of Y. Osheim and O. L. Miller, Jr.*

dimers, cells exposed to UV radiation produce incomplete RNA transcripts from UV-damaged transcription units. In a long DNA molecule, the damage caused by UV radiation is distributed approximately randomly. As shown in Figure 8-11, the chances that an RNA polymerase molecule will encounter a T-T dimer increase as it transcribes regions farther and farther from the initiation site. Thus the ratio of RNA formed from a given region in a transcription unit after and before irradiation decreases exponentially as the distance from the promoter increases.

This technique has been used to map the pre-rRNA transcription unit for the position of the 18S and 28S RNA within the 45S precursor. Cells were exposed to various doses of UV radiation and then labeled with [³H]uridine; the newly formed 45S, 28S, and 18S RNA were then separated and the amount of radioactivity in each fraction was determined. As expected, the 45S pre-rRNA, which is the longest molecule and contains the 28S and 18S sequences, was the most sensitive to irradia-

(a)

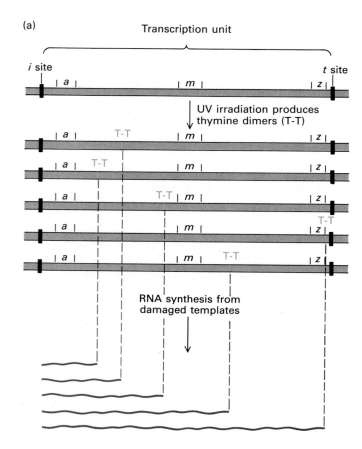

(b)

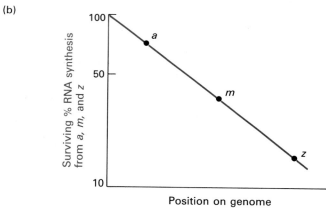

◀ **Figure 8-11** Transcription unit mapping by analysis of UV damage. Chemical changes induced in DNA by UV radiation result in the formation of thymine dimers (T-T), which are distributed approximately randomly in long DNA molecules. When a polymerase molecule reaches a thymine dimer, the synthesis of RNA stops. (a) Five UV-damaged templates are illustrated. RNA synthesis in region *a*, which is near the initiation site, is less likely to be stopped than RNA synthesis in regions *m* and *z*, which are farther from the initiation site. (b) Plot of surviving RNA synthesis after irradiation versus genome position. The equation relating the surviving RNA synthesis to the number of damage sites *(D)* in a transcription unit is

$$\frac{\text{Rate of RNA synthesis after irradiation}}{\text{Original rate of RNA synthesis}} = \frac{1 - e^{-D}}{D}$$

transcription unit. In this case, all genes following a UV-damage site would be knocked out, and both 28S and 18S synthesis would decline in parallel following irradiation. But this does not happen: after any UV dose, more 18S than 28S rRNA is synthesized. Thus initiation of transcription occurs independently in each rRNA gene.

Synthesis and Processing of Pre-rRNA

The DNA that encodes pre-rRNA has been purified from many eukaryotic species and cloned as recombinant DNA. There is a general overall similarity in this rDNA among species (see Figure 8-12). First, the preribosomal RNA transcription units are arranged in long tandem arrays. Second, three sequences within the pre-rRNA transcription unit correspond with those in ribosomal components: an 18S region, a 28S region, and a 160-nucleotide (5.8S) segment that is hydrogen-bonded to 28S rRNA within ribosomes. The 5′ to 3′ order of these three regions in a transcription unit is always the same: 18S, 5.8S, and 28S rRNA. Third, in all eukaryotic cells (and even in bacteria), the pre-RNA transcription unit is longer than the sum of the finished rRNA molecules (Table 8-1). For example, in human cells, the primary pre-rRNA transcript is about 13 to 14 kb long, whereas 28S rRNA is 5.1 kb, 5.8S RNA is 160 bases, and 18S RNA is 1.9 kb. Thus only about half of the primary transcript appears in the final ribosomal products. Because the other half of the primary transcript does not accumulate, it must be destroyed rapidly. There is no known function for the *transcribed spacer RNA* that does not become incorporated into ribosomes.

tion. The 18S survived irradiation better than the 28S, indicating that the 18S coding region is closer to the promoter than the 28S coding region.

Another more subtle but equally important point was deduced from the differential sensitivity of 18S and 28S rRNA synthesis to UV radiation. The transcription units for ribosomal RNA in vertebrate cells are arranged in long tandem repetitive arrays (Figure 8-12). Suppose that the RNA polymerase could begin transcription only at the beginning of one of these long arrays of multiple ribosomal genes and not at the beginning of each individual

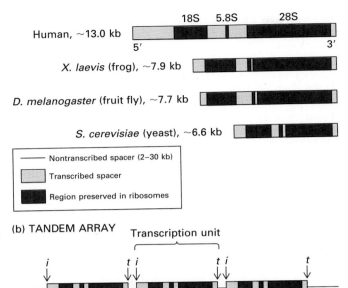

(a) RIBOSOMAL TRANSCRIPTION UNITS

▲ **Figure 8-12** (a) The ribosomal transcription units of four eukaryotes. Variations in the lengths of the transcribed spacer regions (orange bars) account for the major differences in the lengths of the transcription units. The regions that appear in ribosomes are always 18S, 5.8S, and 26 to 28S. (b) The genomes of all animals contain multiple tandem copies of the rRNA transcription units. The nontranscribed spacer regions between transcription units can vary greatly, e.g., ~2 kb in frogs to ~30 kb in humans.

The 5′ end of pre-rRNAs from different species contains a variable-sized region that is discarded; the preserved 18S region follows, and spacer regions of variable length depending on the species separate the 18S, 5.8S, and 28S regions. Within the ribosome, the small 5.8S rRNA is hydrogen-bonded to the 28S rRNA. Nucleolytic enzymes digest the spacer sequences that separate the 5.8S and 28S regions and leave the two preserved regions hydrogen-bonded to each other.

RNA Polymerase I Requires Species-specific Binding Factors to Begin Transcription

Recent work on rRNA synthesis has concentrated on how RNA polymerase I begins and ends transcription of the pre-rRNA genes. When purified rDNA is incubated with RNA polymerase I and necessary nucleoside triphosphates, no pre-rRNA is synthesized. However, if a small amount of crude nuclear extract is added to the cell-free system, transcription of the pre-rRNA genes begins at the correct sites. Purification from crude extracts led to the discovery of two DNA-binding factors that assist RNA polymerase I to correctly initiate and transcribe pre-rRNA genes. Although one of the factors (B for binding) can bind to DNA by itself, its binding affinity is greater in the presence of the second (S) factor. The DNA sites at which the factors and polymerase bind are located within about 150 nucleotides of the initiation site. The two fac-

Table 8-1 Lengths of primary rRNA transcripts and cytoplasmic rRNAs in various species

	Primary transcript*		Ribosomal RNA length*		
	S value	Length (kb)	26S–28S (kb)	16S–18S (kb)	Percentage of precursor preserved
Escherichia coli (prokaryote)	30	6.0	3.0	1.5	75
Saccharomyces cerevisiae (yeast)	37	6.6	3.8	1.7	77
Dictyostelium discoideum (slime mold)	37	7.4	4.1	1.8	80
Drosophila melanogaster (fruit fly)	34	7.7	4.1	1.8	76
Xenopus laevis (frog)	40	7.9	4.5	1.9	81
Gallus domesticus (chicken)	45	11.2	4.6	1.8	57
Mus musculus (mouse)	45	13.7	5.1	1.9	51
Homo sapiens (human)	45	13.7	5.1	1.9	51

*The lengths of the various RNA molecules are estimates based on gel electrophoresis and direct measurements of electron micrographs. The size is of the first major product with definite 5′ and 3′ ends.

SOURCE: B. Lewin, 1980, *Gene Expression*, vol. 3, Wiley, p. 867.

tors bind first, then RNA polymerase I; transcription begins once this initiation complex is formed (Figure 8-13).

The DNA sequence near the initiation site and the B and S factors are species-specific, whereas the polymerase is not. For example, in vitro transcription of mouse pre-ribosomal DNA occurs only in the presence of mouse factors, but either mouse or human RNA polymerase I can be used. In other words, in vitro transcription of pre-rRNA genes requires the presence of DNA and binding factors from the same species, although the RNA polymerase can be from another species.

rDNA Termination Site Lies Downstream and Initiation Site Lies Upstream from First Stable Transcript

As noted earlier, multiple ribosomal genes are arranged in tandem, and electron micrographs indicate that the polymerases transcribing rDNA are closely packed along each transcription unit (see Figure 8-10). These observations suggested that some mechanism keeps polymerases that have just finished transcription in the vicinity so that they might more efficiently find the next pre-rRNA gene to start transcription again. This possibility has been explored in experiments on transcription of pre-rRNA genes from the frog *Xenopus laevis*.

Nascent-chain analysis of pre-rRNA transcripts synthesized in isolated frog nuclei established that polymerase I does not stop transcription at the end of DNA encoding the 28S rRNA but continues until it is near to the next start site. In these experiments, an artificial DNA containing two tandem slightly shortened and therefore marked copies of the *Xenopus* ribosomal RNA gene was prepared and injected into frog oocytes. Both gene copies in this injected recombinant DNA were transcribed. However, if a short segment near the first stop site (and just upstream from the start site for the second gene copy) was not present in the recombinant DNA, then the second gene was no longer transcribed. This finding indicates that if a polymerase molecule cannot finish transcribing the first pre-rRNA gene in a tandem array, then it cannot correctly transcribe the next adjacent one.

Work with ribosomal genes from species as different as yeast and mouse has confirmed that RNA polymerase I transcribes past the 3′ end of the large 28S region. Cleavage of the transcript downstream of the 28S sequence occurs almost immediately after synthesis. In frogs, transcription continues to a stop site about 10 kb past the end of the 28S sequence, only about 200 nucleotides from the start site of the next transcription unit. In mice, there appear to be multiple stop sites about 2000 nucleotides (2 kb) past the end of the 28S sequence followed by a long region in which transcription cannot be detected. Given the evidence that in association with DNA-binding proteins, looping of DNA over long distances can occur, it seems entirely possible that there is some association between the termination site of one pre-rRNA transcription unit and the initiation site of its downstream neighbor, even when these sites are separated by a long untranscribed region. Although the precise mechanisms have not yet been deduced, the evidence is fairly clear that an

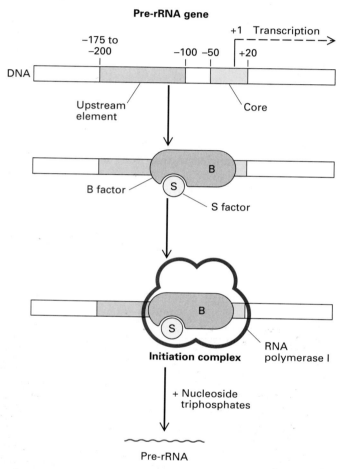

▲ **Figure 8-13** Formation of initiation complex for transcription of rDNA by RNA polymerase I. The DNA and B and S factors must be from the same species (e.g., mouse or human), although the polymerase can be from a different species. Both mouse and human pre-rRNA transcription units have a core promoter sequence of about 70 bases and a necessary upstream DNA sequence or element. Removal of the core completely abolishes transcription; removal of the upstream element decreases transcription 10- to 20-fold. These sequences are not the same in the two DNAs except for an identical sequence in the core region extending from the start site (+1) to about +20. Thus there are species-specific DNA sequences and species-specific binding factors. The S factor increases the affinity of the B factor for DNA but does not bind directly to DNA. [See S. Bell et al., 1988, *Science* **241**:1192.]

RNA polymerase I molecule can move efficiently from one pre-rRNA transcription unit to the next one, so that transcription of multiple units within a tandem array is not continually interrupted by disengagement of the polymerase.

Just as the termination site of a pre-rRNA transcription unit lies downstream from the 3′ end of the first stable transcript to accumulate in cells, the initiation site lies upstream of the 5′ end. Analyses of nucleolar pre-rRNA products in human and mouse cells have revealed that the first 5′ cleavage, which removes only a few hundred nucleotides, occurs within a minute or two of the cleavage at the 3′ end. Only about 10% of the 45S nucleolar RNA has a pppNp indicative of a newly initiated RNA. About 90% of the nucleolar 45S pre-rRNA in mammals has been cleaved at both the 3′ and 5′ ends soon after synthesis of the initial transcript. Subsequent processing events occur more slowly than the synthesis and initial cleavages of the first transcription product, 40S or 45S pre-rRNA accumulates.

In no case does the RNA encoded in rDNA between the 28S region of one transcription unit and the beginning of the next transcription unit accumulate in cells; this is also true of the other spacer regions. Cells thus must have an efficient mechanism for destroying transcribed but unused RNA.

Pre-rRNA Associates with Proteins and Is Cleaved within the Nucleolus to Form Ribosomal Subunits

Processing of pre-rRNA occurs in the nucleolus and involves several distinct steps that lead to formation of the 40S and 60S ribosomal subunits. Processing begins with the association of ribosomal proteins with pre-rRNA. The proteins in the nucleolus include proteins that are found in finished cytoplasmic particles as well as in the nucleolus. Other antigens are present only in the nucleolus (Figure 8-14). Indeed, pre-rRNA apparently binds proteins even while it is still being synthesized, and nucleolar RNA probably is never completely free of associated proteins. Despite extensive attempts, separated ribosomal proteins and rRNA from eukaryotic ribosomes have not been successfully reassociated in the test tube, suggesting that a specific order of assembly is required. Very likely, the proper rRNA-protein interactions can occur only if the proteins are added in a certain order during synthesis of pre-rRNA. The assembly of bacterial ribosomal subunits from purified, separated components has been shown to require such a stepwise addition.

Several nucleoprotein particles of different sizes have been extracted from mammalian nucleoli. The largest of these (80S) contain an intact 43S or 45S pre-rRNA mole-

(a)

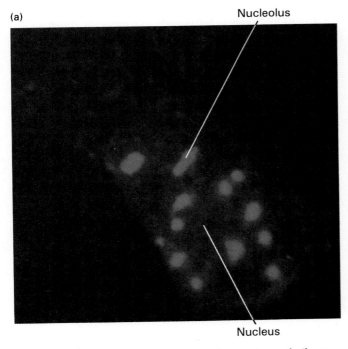

(b)

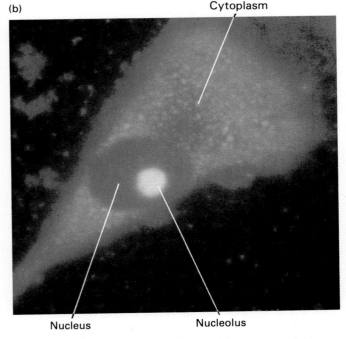

▲ **Figure 8-14** Antibody staining of nucleolar and ribosomal antigens. Serum from patients with autoimmune diseases often reacts with cellular components. (a) Antiserum against nucleolar protein, showing no cytoplasmic reaction. (b) Antiserum prepared in rabbits against human ribosomes. Note that nucleolus as well as cytoplasm is stained. *Photographs courtesy of M. R. Lerner.*

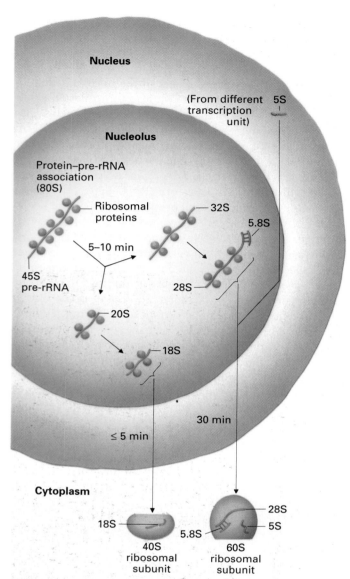

Nucleus

(From different transcription unit) 5S

Nucleolus

Protein–pre-rRNA association (80S)

Ribosomal proteins

45S pre-rRNA

5–10 min

32S

5.8S

28S

20S

18S

30 min

≤ 5 min

Cytoplasm

18S

5.8S

40S ribosomal subunit

28S

5S

60S ribosomal subunit

▲ **Figure 8-15** Cleavages and times required in processing of pre-rRNA into ribosomal subunits. Ribosomal and nucleolar proteins associate with 45S pre-rRNA soon after it is synthesized. Synthesis of 5S rRNA occurs outside of the nucleolus.

28S rRNA. In growing HeLa cells, for example, there are about five times as many 32S molecules as intact 45S primary transcripts. Although the enzymes that process pre-rRNA have not been isolated and characterized, they must be highly specific since the ends of the finished rRNA products are identical. The proteins involved in binding to and processing the accumulated ribonucleoprotein particles are visible in electron micrographs of sections through the nucleolus.

One important variation in pre-rRNA processing should be mentioned here, although it will be discussed in detail later. In a few organisms the 28S rRNA molecules are spliced together from primary transcript regions that are separated by intervening sequences. In fact, this reaction can occur in the test tube without the assistance of protein, that is, it occurs by *self-splicing;* within cells, however, self-splicing may be aided by proteins. The splicing of pre-rRNA occurs only in a few single-cell organisms (e.g., *Tetrahymena thermophila*) and in the mitochondrial rRNA of some fungi (e.g., *Neurospora crassa*). It is important to note that the spacer sequences that separate the 18S, 5.8S, and 28S regions are not in any way related to sequences removed by splicing in the cases where splicing does occur. The revolutionary finding of RNA-mediated splicing renews interest in how nucleolar processing occurs. Perhaps some other cleavages are directed by RNA.

In addition to cleavage steps that shorten pre-rRNA, processing also involves methylation of pre-rRNA with methionine acting as methyl donor. In human cells, more than 100 methyl groups are added to specific bases and to the riboses of specific ribonucleotides; most of these methyl groups can be detected even in nascent pre-rRNA molecules. In addition, a few methyl groups are added to 45S pre-rRNA molecules after synthesis is finished, and four methyl groups are added to 18S rRNA in the cytoplasm—that is, after cleavage. All of the methyl groups added to the pre-rRNA primary transcript are preserved during processing (i.e., they are found in 28S and 18S rRNA in the cytoplasm); their positions are highly conserved in vertebrate cells (frogs and humans), and many of the same sites are also present in yeast. If cells are deprived of methionine, an intermediate in the transfer of methyl groups, processing of pre-rRNA is interrupted, suggesting a role in pre-rRNA processing for methyl groups.

When assembly of ribosomal particles in the nucleolus is complete, they move—by mechanisms as yet unknown—to the cytoplasm, where they appear first as free subunits (see Figure 8-15). Since the nuclear pores (to be discussed later in this chapter) are only about 10 to 20 nm in diameter and the large ribosomal subunit is about the same size, a conformational change in one or both of these structures may occur to permit passage of the subunit to the cytoplasm.

cule. This RNA molecule is cut in a series of cleavage steps that ultimately yield the 18S, 5.8S, and 28S rRNAs found in ribosomes (Figure 8-15). The small (40S) ribosomal subunit containing 18S rRNA is finished and delivered to the cytoplasm much faster than is the large (60S) ribosomal subunit. Because of this difference in processing times, the pool of nuclear precursors for large ribosomal subunits is larger than that for small ribosomal subunits. In fact, the most prominent RNA component in the nucleolus is 32S pre-rRNA, the immediate precursor of

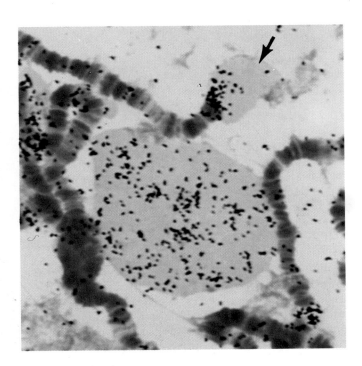

◄ **Figure 8-16** A new nucleolus in a transgenic *Drosophila* strain at the site of entry of a recombinant DNA encoding a single gene for pre-rRNA. Micrograph following autoradiography and Giemsa staining of a chromosome preparation shows the large pale normal nucleolus and the smaller new nucleolus (*arrow*) at the tip of the X chromosome. The silver grains (black dots) over each represent newly labeled pre-rRNA, which was demonstrated by in situ hybridization. [See G. H. Karpen, J. E. Schafer, and C. D. Laird, 1988, *Genes & Devel.* 2:1745.] *Photograph courtesy of G. H. Karpen.*

Ribosomal RNA Genes Act as Nucleolar Organizers

How is a nucleolus formed? In situ hybridization identified genes encoding pre-rRNA in the nucleolus some time ago, but it was not known whether any other DNA was required to form the nucleolus. Recent experiments with transgenic *Drosophila* strains have shown that introduction of recombinant DNA encoding a single complete pre-rRNA transcription unit induces formation of an active nucleolus (Figure 8-16). Thus a single pre-rRNA gene is sufficient to be a *nucleolar organizer,* and all the other components of the ribosome diffuse to the newly formed pre-rRNA. The structure of the induced nucleolus appears, at least by light microscopy, to be the same as, albeit smaller than, a normal nucleolus with 200 or so ribosomal RNA genes.

Synthesis and Processing of 5S rRNA and tRNAs

Eukaryotic cells contain many different small RNAs that are 300 nucleotides or less in length that are distinct from the much more abundant tRNAs or 5S' and 5.8S' rRNAs (Figure 8-17). Most of these molecules are confined to the nucleolus but some exist only in the cytoplasm. The genes encoding these small RNAs all lie outside of the nucleolus, and most are transcribed by RNA polymerase III. In recent years many small RNAs have been shown to be

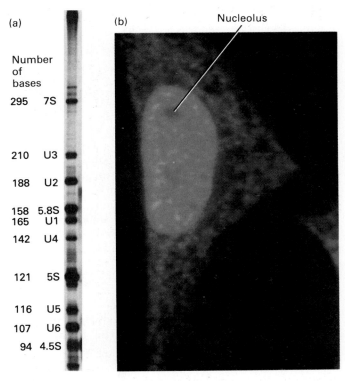

▲ **Figure 8-17** (a) The small RNA species from cultured Chinese hamster cells. RNA labeled with $^{32}PO_4$ was purified from the nuclei of cultured cells and subjected to gel electrophoresis and autoradiography. Almost all of the RNAs identified here have been sequenced. The species labeled U1 to U6 are rich in uridylates. All except U3 are involved in mRNA splicing; U3 is found mainly in the nucleolus. The 5.8S and 5S species are ribosomal RNAs; the 7S RNA belongs to a particle that assists in transporting proteins through membranes. These last three RNAs are present mainly in the cell cytoplasm. (b) Antibody stain of nuclear ribonucleoproteins that participate in splicing the antigen termed Sm is present in particles containing U1, U2, U4, U5, and U6 small RNAs. Antigen is not present in U3, which is evidenced here by an absence of staining in the nucleolus. *Part (a) courtesy of W. Jelinek and S. Haynes; part (b) courtesy of M. Lerner.*

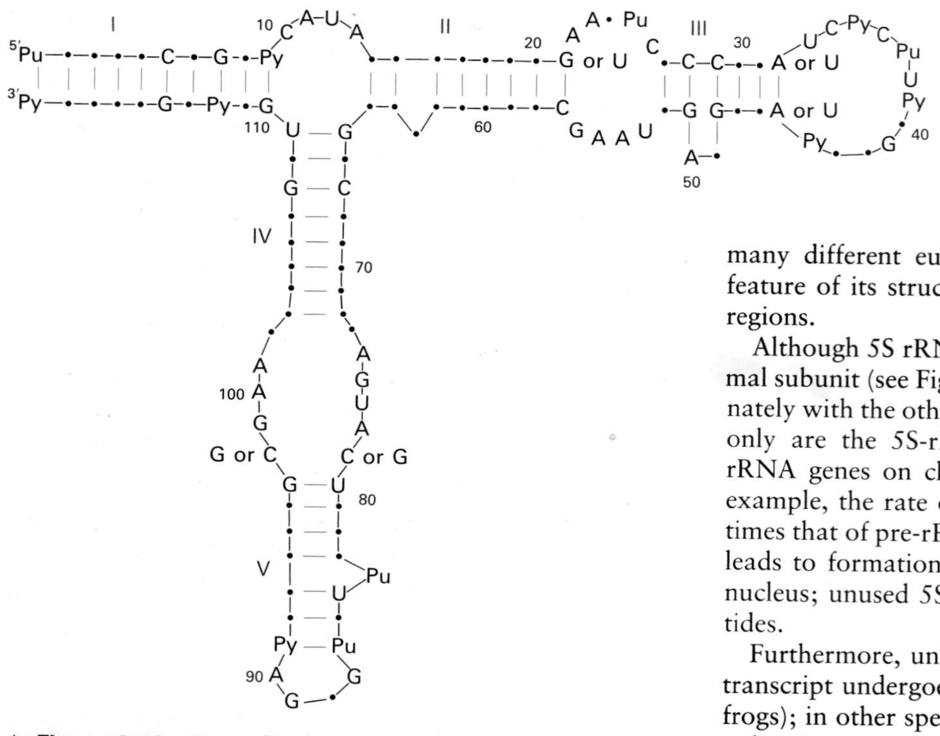

▲ **Figure 8-18** Generalized structure of 5S rRNA obtained from an analysis of over 60 different 5S rRNAs from animals, plants, and single-celled organisms. The five stem-loop regions (I to V) are a constant feature. Blue lines connect bases that are hydrogen-bonded. Pu = a purine; Py = a pyrimidine. Where dots appear for bases, the sequence may vary but the base pairing is maintained. [See N. Delihas and J. Andersen, 1982, *Nucleic Acids Res.* 10:7323.]

part of nuclear ribonucleoprotein particles (snRNPs) that function during nuclear processing. The functions of many other small RNAs, however, are as yet undiscovered. In this section, we describe the synthesis by RNA polymerase III of 5S rRNA and tRNAs; in later sections the role of other small nuclear RNAs in RNA processing is discussed.

Primary 5S-rRNA Transcript Undergoes Little or No Processing

In addition to 18S, 5.8S, and 28S rRNAs, which are derived from pre-rRNA, eukaryotic ribosomes also contain a short 120-nucleotide molecule known as 5S rRNA. This RNA is transcribed from a group of identical tandemly arranged genes that are located outside the nucleolus. Human cells contain several hundred 5S-rRNA genes; somatic frog cells also contain about 500 such genes, while frog oocytes contain 20,000 copies of a slightly different 5S-rRNA gene. The sequence and probably the secondary structure of 5S rRNA are highly conserved in

many different eukaryotes (Figure 8-18). A distinctive feature of its structure is the presence of five *stem-loop* regions.

Although 5S rRNA is a component of the large ribosomal subunit (see Figure 8-15), it is not synthesized coordinately with the other ribosomal RNAs in vertebrates. Not only are the 5S-rRNA genes separated from the pre-rRNA genes on chromosomes, but in human cells, for example, the rate of synthesis of 5S rRNA is about five times that of pre-rRNA. This differential rate of synthesis leads to formation of excess 5S-rRNA molecules in the nucleus; unused 5S molecules are degraded into nucleotides.

Furthermore, unlike pre-rRNA, the primary 5S-rRNA transcript undergoes no processing in some species (e.g., frogs); in other species, the primary transcript undergoes only minor processing involving removal of 10 to 50 nucleotides from the 3' end. The presence of a triphosphate nucleotide (pppG) at the 5' end of 5S rRNA indicates that transcription begins at this point; if the 5' end resulted from cleavage of a longer molecule, the end nucleotide would contain only one phosphate.

The First Eukaryotic Transcription Factor to Be Purified Is Required for Synthesis of 5S rRNA

Studies on synthesis of 5S rRNA led to the first detailed description of the DNA sequence and transcription factor requirements for a eukaryotic transcription unit. Accurate synthesis of 5S rRNA has been accomplished in several experimental systems. If purified 5S rDNA is injected into frog oocytes or if recombinant DNA that includes 5S rDNA is added to extracts of frog oocyte nuclei, endogenous RNA polymerase III catalyzes transcription of the 5S-rRNA genes. Also, purified RNA polymerase III added to chromatin (from lysed nuclei) DNA and associated proteins synthesizes 5S-rRNA.

Measurements of 5S-rRNA synthesis from plasmids containing either the entire 5S gene or portions of it, derived from deletion mutants in the region of the 5S gene, led to a surprising discovery: removal of any part of a sequence of 40 to 50 bases *within* the gene greatly inhibits the initiation of transcription (Figure 8-19). A transcription factor termed TF$_{III}$A—the first transcriptional-activation protein to be purified from eukaryotic cells—binds to this region, which begins about 47 bases past the initiation site for the transcription of the 5S-rRNA gene.

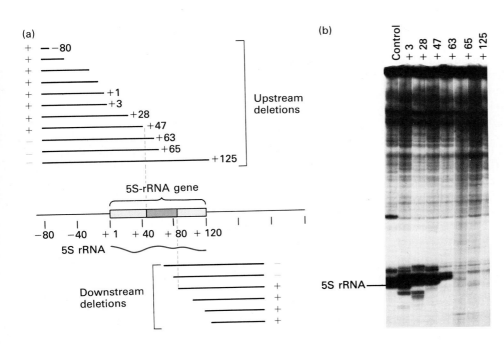

► **Figure 8-19** Mapping of activation region (blue) necessary for accurate initiation of transcription of frog 5S-rRNA gene. DNA sequences containing various upstream or downstream deletions around the 5S region were produced (a) and injected into frog oocytes along with labeled triphosphates. The resulting RNA products were analyzed by electrophoresis and autoradiography (b). The mutant genes that were transcribed are marked with a + and those that were not are marked with a − in (a). For example, in (b) it is shown that +3, +28, +47, and +63 mutants were transcribed, but +65 and +125 mutants were not. [See S. Sakonju, D. F. Bogenhagen, and D. D. Brown, 1980, *Cell* **19**:27.] *Photograph courtesy of D. D. Brown.*

Almost any recombinant DNA containing an intact binding region from within the 5S gene is transcribed by RNA polymerase III, although transcription occurs at the highest rate if the normal start sequences from the 5′ end are present.

Binding of TF$_{III}$A to the 5S gene has been conclusively demonstrated in footprint experiments. The purified transcription factor is allowed to bind to 5S DNA, and the mixture is then subjected to chemical or enzymatic attack. The presence of TF$_{III}$A protects the DNA within the 40- to 50-bp region that is required for initiation of transcription (Figure 8-20).

The gene encoding TF$_{III}$A has been cloned and sequenced. A remarkable feature of the deduced amino acid sequence is the presence of recurring cystine and/or histidine residues spaced at regular intervals. These amino acids can bind zinc ions, and the purified native protein contains an amount of zinc equal to the cystine-histidine clusters (Figure 11-13). These findings led to the proposal that TF$_{III}$A is folded so as to form "zinc fingers" that protrude from the surface of the protein and bind to the DNA. Such a folding scheme would cause the protein to have a long axis sufficient for it to stretch over the five helical turns of DNA that it contacts in its binding site.

► **Figure 8-20** Locating TF$_{III}$A-binding region on 5S-rRNA gene by footprint experiment. Purified TF$_{III}$A was mixed with 5′ end-labeled 5S rDNA, and the mixture, as well as a control without the transcription factor, was digested briefly with DNase, producing a series of fragments stretching from the 5′ end to the cleavage site. (a) Autoradiograph of DNase-treated mixture containing TF$_{III}$A *(lane 1)* and without factor *(lane 2)*. The lack of bands in the region from base 47 to base 96 on the 5S rDNA in lane 1 indicates that these bases were protected from digestion by bound protein. The four bands that appear in lane 1 *(small arrows)* but not in lane 2 represent DNA regions that become "hypersensitive" to DNase attack as a result of binding the transcription factor. (b) Diagram of 5S-rRNA gene showing regions where TF$_{III}$A and RNA polymerase III bind. [See R. G. Roeder, 1980, *Cell* **19**:717.] *Photograph courtesy of R. G. Roeder.*

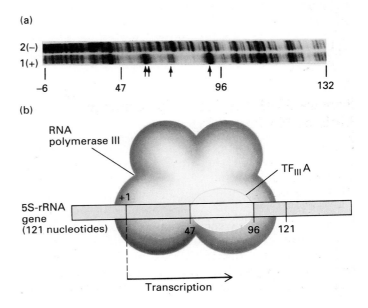

Studies with mutant TF$_{III}$ protein made by recombinant methods, have shown that the amino acid substitutions within the zinc-finger regions upset the binding of TF$_{III}$A to its binding site. The proposed zinc-finger structure is almost certainly correct in broad outline. Since its recognition in TF$_{III}$A, a similar though often less extensive zinc-finger region has been found in a number of other DNA-binding proteins.

Although TF$_{III}$A is necessary for 5S-rRNA synthesis, it is not sufficient by itself to cause purified RNA polymerase III to start transcription. Two other factors, termed TF$_{III}$B and TF$_{III}$C, also are required for transcription of 5S-rRNA genes. These factors, which also are needed for tRNA synthesis, are discussed later.

Some animals—perhaps most—have several sets of 5S-rRNA genes whose sequences differ in only a few nucleotides. For example, frogs have two major sets and one minor set of tandemly repeated 5S genes. One major set, containing 20,000 gene copies, is transcribed only during the formation of the oocyte, which produces a huge number of ribosomes from its multiple tandem arrays of ribosomal genes. The other major set of 5S genes in frogs, which contains about 500 gene copies, is transcribed in all other cells. The oocyte and somatic 5S-rRNA genes differ only in a few bases, all in the TF$_{III}$A-binding region. TF$_{III}$A binds about 5 to 10 times better to the somatic than to the oocyte 5S-rRNA genes. In addition, the somatic 5S transcription complexes are more stable to incubation in cell extracts than are oocyte 5S complexes. These differences may not be the only determinative factors in the expression of 5S genes during development, but they are at least part of the basis for the differential transcription of 5S-rRNA genes in somatic cells and oocytes.

Processing of Pre-tRNA in the Nucleus Involves Splicing and Modification of Bases

Transfer RNAs, which constitute about 15 percent of the total RNA of eukaryotic cells, average 75–80 nucleotides in length. However, all of the primary transcription products of tRNA genes studied so far are, like pre-rRNA but unlike 5S-rRNA, considerably larger than the final products. Thus, precursor tRNA molecules undergo shortening before they become mature, functional molecules.

They also undergo modifications; in fact, about 1 base in every 10 of a primary tRNA transcript is modified in some way.

Precursor tRNA (pre-tRNA) was initially recognized in pulse-labeling experiments with cultured HeLa cells. Gel electrophoresis of the RNA synthesized by cells labeled for a short time with [³H]uridine revealed the presence of products that were up to 40 bases longer than tRNA but contained some of the modified bases found in tRNA. When actinomycin D, which prevents RNA synthesis, was added to the cells, the amount of these longer products decreased, while the amount of tRNA increased. These findings indicate that the initial longer transcripts are precursors of tRNA.

In more recent work, many genes encoding tRNAs have been cloned and sequenced. Comparison of the sequences of such genes with their corresponding tRNAs has shown that some eukaryotic tRNAs are not encoded in a consecutive stretch of DNA. For example, yeast tyrosine tRNA lacks a 14-base intervening sequence that is present in the middle of the tRNA^{Tyr} gene (Figure 8-21). Clearly, a segment must be removed from the middle of the primary transcripts of some tRNAs and the ends spliced together. RNA splicing was first discovered in studies of mRNA; the biochemical events involved in the splicing of different RNAs are discussed later in this chapter.

The transcription of tDNA (the genes that encode tRNA) by RNA polymerase III has been studied in experimental systems similar to those described earlier for synthesis of 5S rRNA. For example, the tDNA sequences from a number of different eukaryotic organisms have been cloned as recombinant DNA molecules in *E. coli*. After injection into the nuclei of frog oocytes, recombi-

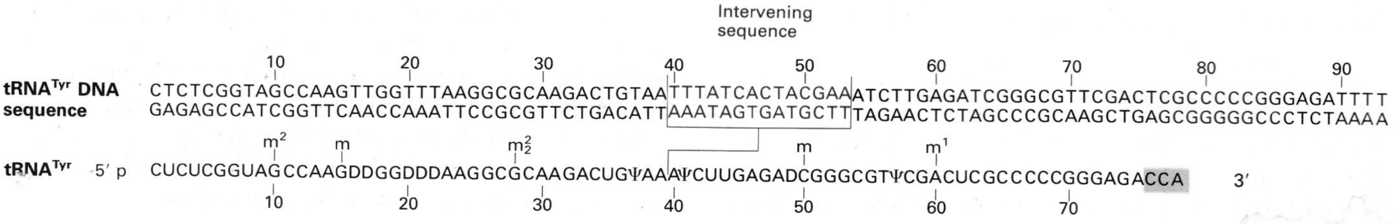

▼ **Figure 8-21** The sequences of a tyrosine tRNA gene from yeast and of corresponding tRNA^{Tyr} found in the cytoplasm of yeast cells. The DNA contains a 14-base intervening sequence that does not appear in the tRNA. (Note that the CCA 3′ terminus of the mature tRNA is not encoded in the DNA.) The modified bases in the tRNA are as follows: D = dihydrouridine, Ψ = pseudouridine, m = methyl group on ribose, m² = methyl on position 2 of base, m²₂ = dimethyl on position 2 of base, and m¹ = methyl on position 1 of base. [See H. M. Goodman, M. V. Olson, and B. D. Hall, 1977, *Proc. Nat'l Acad. Sci. USA* 74:5453.]

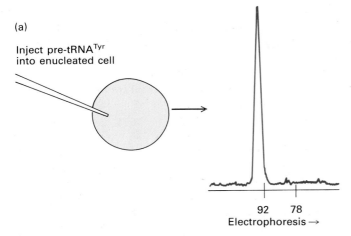

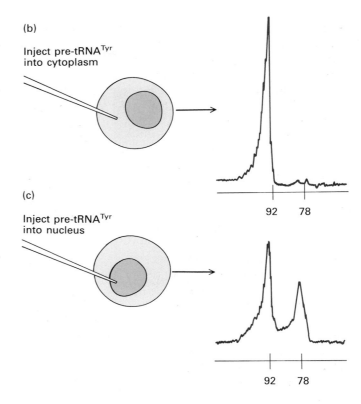

▲ **Figure 8-22** Experimental demonstration that pre-tRNA is spliced in the nuclei of frog oocytes. Labeled purified 92-base-long unspliced precursor of yeast tRNA^{Tyr} was injected into (a) frog oocytes from which the nucleus had been removed, (b) the cytoplasm of intact nucleated oocytes, or (c) the nucleus of intact oocytes. After about an hour, the RNA was extracted and analyzed by gel electrophoresis. Labeled tRNA corresponding in length to that of mature tRNA (78 nucleotides) was detected only when the precursor was injected into the nucleus.

nant DNA containing the yeast tyrosine tDNA sequence is transcribed by the endogenous frog RNA polymerase III. Moreover, if yeast pre-tRNA containing the extra intervening sequence and unprocessed 3′ end is isolated and injected into frog oocyte nuclei, it is processed into a finished, spliced molecule. In addition to splicing, tRNA processing includes removal of three residues at the 3′ end and the addition of CCA in their place; addition of methyl and isopentenyl groups; and conversion of uridine into pseudouridine residues (Figure 8-21). Recognition of yeast tDNA and pre-tRNA by the appropriate enzymes in the frog cells indicates conservation of recognition sites and enzymes over a wide evolutionary range.

Experiments with nucleated and enucleated frog oocytes have demonstrated that processing of precursor tRNA occurs only in the nucleus (Figure 8-22). Fractionation of cultured mammalian cells in nonaqueous solvents (which prevent leakage of nuclear contents) also has shown that precursor tRNA is present only in the nucleus and completed tRNA only in the cytoplasm.

RNA Polymerase III Requires Ordered Addition of Multiple Factors to Begin Transcription

Two regions within tRNA genes have been shown to be required for synthesis of tRNA in the frog oocyte assay system. The removal of either of these regions from a cloned tRNA gene greatly diminishes transcription. These two regions—which correspond to approximately +10 to +20 and +50 to +60 of the mature tRNA molecule—are highly conserved in tRNA genes from many eukaryotic species (Figure 8-23). These conserved regions encode the D and TΨCG stem-loops in the tRNA structure (see Figure 3-9), which play a key role in the transfer of amino acids by tRNA molecules. In addition, these same conserved regions in tDNA serve as binding sites for protein factors that assist RNA polymerase III in transcribing tRNA genes. Thus the same highly conserved nucleotide sequences in both tDNA and tRNA bind specific proteins.

The three transcription factors necessary for active initiation by RNA polymerase III are designated TF$_{III}$A, B, and C. As noted earlier, all three factors are required in the synthesis of 5S rRNA, whereas only two (TF$_{III}$B and TF$_{III}$C) are required for tRNA synthesis. In the case of 5S-rRNA synthesis, TF$_{III}$A (a 5S-gene-specific molecule) is added first, followed in sequence by the C and B factors; in the case of tRNA synthesis, TF$_{III}$C is added first, followed by the B factor (Figure 8-24). The B factor does not bind to DNA by itself and is thought to bind to TF$_{III}$C. Once the preinitiation complex is formed, it is stable. Added RNA polymerase III then will transcribe its 5s or tRNA template through multiple rounds of synthesis (up to 50). Templates added after a stable complex is formed are not used. In both cases, transcription ceases when the polymerase encounters a string of T residues; this is reminiscent of rho-independent termination of transcription by bacterial RNA polymerase.

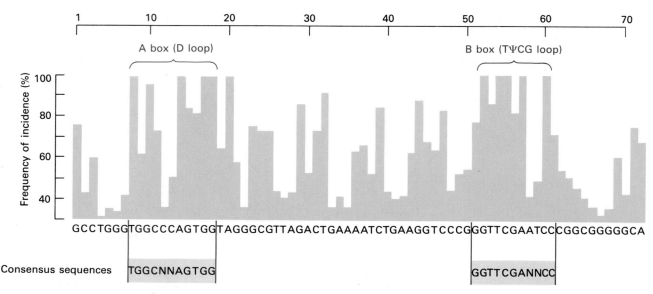

▲ **Figure 8-23** Frequency of occurrence (%) of most common nucleotides in 80 different eukaryotic tRNA genes. Two consensus sequences, called the A box and the B box, are present. The A box encodes the D loop of tRNA, and the B box encodes the TΨCG loop; both loops are constant fea- tures of tRNA. The A and B boxes also serve as parts of the promoter in tRNA genes. [See D. H. Gauss and M. Sprinzl, 1981, *Nucleic Acids Res.* 9:r1; and G. Galli, H. Hofstetter, and M. L. Birnstiel, 1981, *Nature* 294:626.]

As noted already, RNA polymerase III catalyzes the formation of many small RNAs in addition to 5S rRNA and tRNAs. Little is known about the transcription factors required for synthesis of these other small RNAs. However, the factors that aid RNA polymerase III to transcribe 5S-rRNA genes and tRNA genes are not the same as those that aid polymerase I to transcribe pre-rRNA genes. Furthermore, RNA polymerase II requires different and more varied factors to transcribe mRNA genes, as we discuss in a later section.

Synthesis and Processing of mRNAs: General Pathway

As mentioned in the first section of this chapter, heterogeneous RNA (hnRNA), like 45S pre-rRNA, is rapidly labeled in cultured HeLa cells exposed to [³H]uridine (see Figure 8-4). Although demonstrating that 45S pre-rRNA is the precursor for cytoplasmic rRNA was reasonably easy, proving that hnRNA is a precursor for mRNAs was much more difficult because the amount of any particular mRNA synthesized is much less than the amount of pre-rRNA. Some DNA-containing viruses, however, produce specific mRNAs in abundance, and the study of cells infected with such viruses was crucial in discovering how mRNAs are made. In fact, many of the special properties of eukaryotic mRNAs were first discovered by studying mRNAs made from viral genomes.

The first task in learning how mRNAs are formed was to isolate cytoplasmic mRNAs and compare them to nuclear RNA species that might be primary transcripts. By the mid-1970s, mRNAs in mammalian cells, especially after virus infection, had been found to contain specially modified 5' and 3' termini; these same modifications were found in the much longer hnRNA molecules. Adenovirus-infected cells were then used in studies mapping the first primary transcripts containing mRNA sequences. Almost immediately afterward, in 1977, splicing of pieces of the adenovirus primary mRNA transcripts in infected human cells was discovered. Within a year or two after this discovery, the widespread use of recombinant DNA techniques combined with DNA sequencing revealed the generality of splicing and led to the detailed picture we now have about eukaryotic gene structure and mRNA processing.

In this section we review the evidence that cytoplasmic mRNA is derived from nuclear hnRNA and that splicing of primary nuclear RNA transcripts occurs during formation of many mRNA molecules.

mRNAs and hnRNAs Have Similar Base Compositions but Differ in Length

The first step in studying the synthesis of eukaryotic mRNA was to isolate purified mRNA. Because active mRNA molecules are associated with ribosomes in the process of making proteins, cells pulse-labeled with radioactive amino acids can be fractionated to isolate ribo-

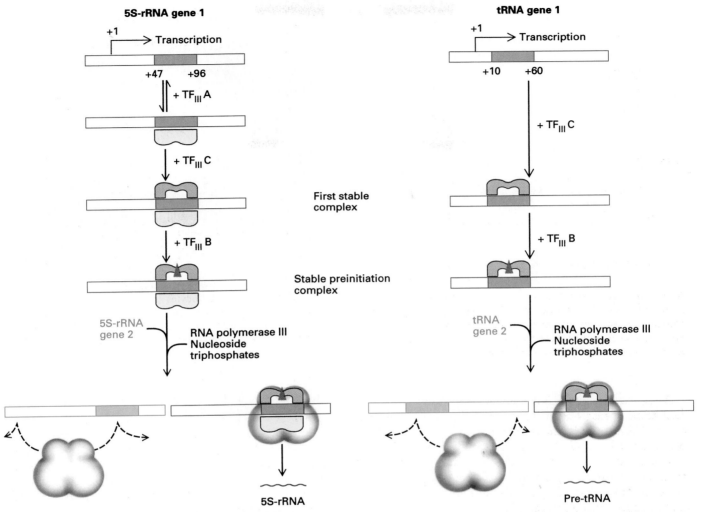

▲ **Figure 8-24** Formation of preinitiation complexes for genes transcribed by RNA polymerase III. 5S genes require factors TF$_{III}$A, B, and C for transcription by RNA polymerase III, whereas tRNA genes require only factors B and C. In both cases, after factor addition to a mixture containing excess template DNA, stable complexes are formed. At this stage, addition of any other template DNAs (e.g., 5S gene 2 or tRNA gene 2 [blue]) does not displace factors already bound to first template complexes and thus does not affect their transcription. Second templates are not transcribed. [See A. B. Lasser, P. L. Markin, and R. G. Roeder, 1983, *Science* **222**:740.]

somes containing new protein chains. Almost all of the newly labeled protein is associated with groups of ribosomes, called *polyribosomes* or *polysomes* (Figure 8-25). If protein-synthesizing cell extracts are treated briefly with ribonuclease, the mRNA is cut and polysomes are converted to single ribosomes with nascent proteins still attached.

The size of polyribosomes varies among cells making different proteins. In reticulocytes, which mainly synthesize globin chains containing about 140 amino acids, each polyribosome contains 3 to 5 ribosomes. In cultured HeLa cells, which make a broad array of proteins with an average size of 400 amino acids, the polyribosomes con-

tain 3 to about 20 ribosomes with an average of 8 to 10. HeLa cells infected with poliovirus, the mRNA of which is 7000 nucleotides long, have very large polyribosomes with 40 or more ribosomes. The sizes of these various polyribosomes determined by sedimentation analysis have been confirmed by electron microscopy (Figure 8-26). All these findings together strongly suggest that each polyribosome contains a single mRNA molecule being translated by many ribosomes and that the size of a polyribosome is dictated by the length of the attached mRNA.

Once techniques for separating labeled mRNA from polyribosomes and other labeled RNA were developed, it

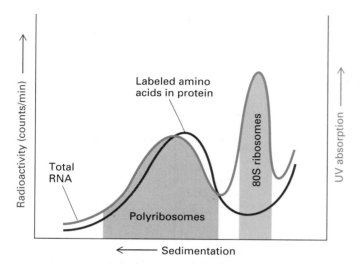

▲ **Figure 8-25** Demonstration that protein synthesis, and hence functional mRNA, is associated with polyribosomes. When cytoplasmic extract of cells is exposed to radioactive amino acids and then sedimented through a sucrose gradient, the polyribosomes, which sediment faster than single 80S ribosomes, contain most of the newly incorporated radioactivity (red curve). UV absorption at 260 nm (blue curve) detects RNA and thus indicates the position of the ribosomes and polyribosomes. [See J. Warner, P. Knopf, and A. Rich, 1963, *Proc. Nat'l Acad. Sci. USA* **49**:122; and S. Penman, K. Scherrer, and J. E. Darnell, 1963, *Proc. Nat'l Acad. Sci. USA* **49**:654.]

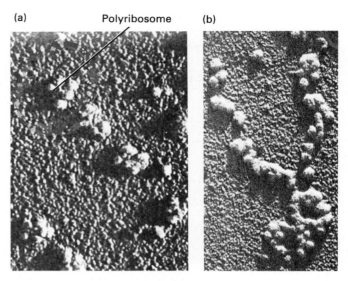

▲ **Figure 8-26** Electron micrographs of polyribosomes prepared from rabbit reticulocytes (a) and from poliovirus-infected HeLa cells (b). The polyribosomes that synthesize poliovirus proteins contain perhaps 40 ribosomes, whereas the reticulocyte polyribosomes, which synthesize α- and β-globin contain 3 to 5 ribosomes. This size difference is related to the length of the associated mRNA, which is 7000 nucleotides long in the case of the poliovirus proteins and only about 650 nucleotides long in the case of α- and β-globin. *Courtesy of A. Rich.*

became possible to compare mRNA with hnRNA. For example, if HeLa cells were exposed briefly to ^{32}P-labeled nucleotides, the labeled hnRNA outside the nucleolus and the labeled polysome-associated mRNA, uncontaminated with rRNA, could be isolated and analyzed. Although the length of mRNA (0.5–3 kb) was considerably shorter than that of hnRNA (2–30 kb), the ^{32}P distribution in the four nucleotides indicated that they have a similar base composition, which differs distinctly from the composition of pre-rRNA or rRNA. This similarity in base composition was the earliest evidence for a link between hnRNA and mRNA.

Most Eukaryotic mRNAs Are Monocistronic

Almost all mRNAs in eukaryotes are *monocistronic* (i.e., they encode a single polypeptide chain), as evidenced by their single start site for protein synthesis, whereas many (but not all) mRNAs in prokaryotes are *polycistronic* (i.e., they encode multiple polypeptide chains), as indicated by their multiple protein start and stop signals (Figure 8-27). This fundamental difference in the information content of eukaryotic and prokaryotic mRNAs is reflected in the difference in their lengths; prokaryotic mRNAs can be much longer than eukaryotic mRNAs,

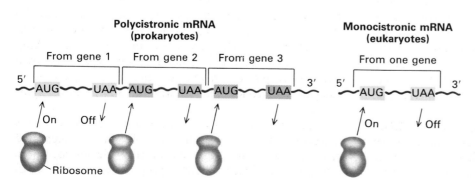

◀ **Figure 8-27** The relatively small size of mammalian mRNAs (approximately 1.5 kb on average) first suggested that eukaryotic mRNA is monocistronic, in contrast to prokaryotic mRNAs, which are polycistronic and include RNA from entire operons. A few cases of eukaryotic mRNAs with multiple protein start sites and even multiple reading frames (and therefore multiple protein products) have been discovered.

which average only 1–2 kb in length. Even in mono-cistronic mRNAs, however, the start and stop sites for protein synthesis do not correspond to the ends of the mRNA. Recent studies also have shown that some eu-karyotic mRNAs, particularly those from animal viruses, have two initiation sites and produce two different pro-teins from overlapping stretches of mRNA. It is possible that any given molecule of mRNA with the capability of two different translation frames can be translated in only one way.

Both Ends of mRNAs and hnRNAs Contain Posttranscription Modifications

Addition of Poly A 3′ End   In the early 1960s, re-searchers isolated an enzyme termed poly A polymerase that simply added adenylate residues to the ends of RNA chains. Adenylate-rich RNA also was found in cytoplas-mic extracts of mammalian cells. Subsequent studies with vaccinia, a large DNA virus that grows in the cytoplasm of mammalian cells, led to the discovery of a 3′ terminal sequence of adenylate residues, now called *poly A,* which is a distinctive feature of mRNA.

Although the mRNA from *all* eukaryotic species has a 3′ terminal poly A, formation of this sequence has been studied most in mammalian cells. About one-fourth of the hnRNA molecules in such cells have a poly A terminal segment. Moreover, the length of the poly A segment is very similar in newly labeled mRNA and hnRNA. That poly A is made in the nucleus as part of hnRNA was demonstrated by exposing cells for a very brief time (2 min or less) to [^{3}H]adenosine. At the beginning of expo-sure, the only labeled poly A in the cell is in the nucleus associated with hnRNA; later, most of the poly A is in cytoplasmic polyribosomal mRNA. This pattern of ap-pearance of label in the same-sized 3′ terminal homopoly-mer provided further evidence of a link between hnRNA and mRNA.

Poly A is not encoded in DNA, and it is not added to tRNA or rRNA. Thus, poly A is a posttranscriptional addition to hnRNA that subsequently accumulates in poly-ribosomal mRNA (Figure 8-28). The length of the poly A segment in hnRNA is quite consistent: in cultured mam-malian cells and probably in all vertebrate cells, it con-tains 200–250 nucleotides. In the cells of lower animals and plants, the initial length of the poly A segment may be shorter than in mammalian cells, but the size of the segment appears to be fairly uniform in each species. After poly A-containing mRNA reaches the cytoplasm, the poly A segment gradually is shortened as the mRNA ages; at steady state, its size varies between 30 and 250 nucleotides.

In some experimental systems, cytoplasmic mRNA molecules with a very short or no poly A segment have been separated from those containing 20 or more A resi-dues. This observation raised the possibility that some

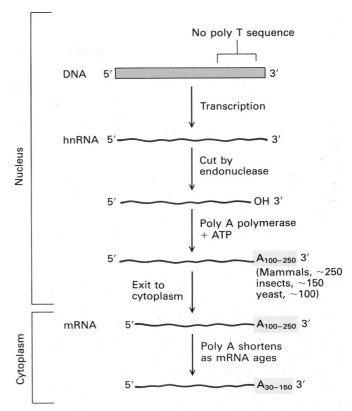

▲ **Figure 8-28** Steps in the addition of poly A during for-mation of eukaryotic mRNA. Note that the poly A sequence is not encoded in DNA. Cleavage at the 3′ end of hnRNA followed by addition of A residues one at a time occurs in the nucleus. The only known eukaryotic mRNAs whose pro-cessing does not include poly A addition are those for his-tone proteins; however, because 3′ A residues are lost as cy-toplasmic mRNAs age, mRNAs with very short poly A are observed.

poly A minus mRNA may enter the cytoplasm. However, the in vitro translation products of such purported poly A minus mRNAs are very similar to the translation prod-ucts of mRNAs that contain poly A. In fact, only one class of mRNAs in mammalian cells has been shown to lack poly A. These are the histone mRNAs, which encode DNA-packaging proteins. Thus, the presence of poly A is not an absolute requirement for an mRNA to enter the cytoplasm and direct protein synthesis. However, except for histone mRNAs, no eukaryotic mRNAs lacking poly A have been found to enter the cytoplasm; the pres-ence of cytoplasmic mRNA lacking poly A results from cleavage of A residues after entry of mRNA into the cyto-plasm.

A practical benefit of the presence of poly A in mRNA molecules is that eukaryotic mRNA can be separated from rRNA and tRNA by affinity chromatography. When homopolymers of either poly rU (polyuridylic acid)

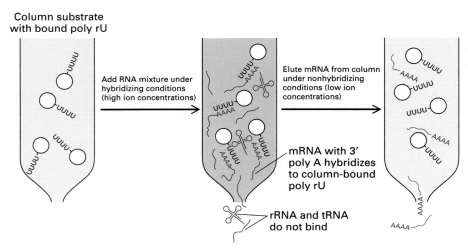

Column substrate with bound poly rU

Add RNA mixture under hybridizing conditions (high ion concentrations)

mRNA with 3′ poly A hybridizes to column-bound poly rU

rRNA and tRNA do not bind

Elute mRNA from column under nonhybridizing conditions (low ion concentrations)

◀ **Figure 8-29** The purification of eukaryotic mRNA by affinity chromatography. Messenger RNA containing poly A tails hybridizes with poly rU or dT bound to a substrate (e.g., cellulose), whereas rRNA and tRNA do not bind and can be washed away. Under nonhybridizing conditions (e.g., very low ion concentrations), the pure mRNA can be eluted from the substrate.

or poly dT (polythymidilic acid) are bound to a solid matrix, such as paper fibers, they selectively hybridize to the 3′ poly A segment of mRNA. All RNAs that do not contain poly A, such as rRNA and tRNA, do not bind to the matrix. Subsequently, the pure, bound mRNA can be released from the homopolymer by washing it with a solution of low ionic strength (Figure 8-29).

Formation of 5′ Cap Structure The purification of eukaryotic mRNA led to the discovery of a second distinctive feature—a complex methylated structure at the 5′ end of the mRNA molecule called a *cap*. This structure was first characterized in mRNAs made by enzymes in virus particles that synthesize viral mRNAs; it was later isolated from polyribosomal mRNA of cells. The cap consists of a terminal nucleotide, 7-methylguanylate (m⁷G), in a 5′ → 5′ linkage with the initial nucleotide of the mRNA chain (Figure 8-30a). Thus there is no free 5′ end of an mRNA molecule; in fact, it has 3′-hydroxyl groups on the ribose rings of the nucleotides at both ends. As we shall discuss later, the cap probably aids in translation of an mRNA molecule.

The cap structure also is present in hnRNA, and the enzymes involved in the biosynthesis of the cap in the nuclei of mammalian cells have been identified (Figure 8-30b). Both mRNA and hnRNA also contain a number of 6-methyl adenine residues in internal nucleotides.

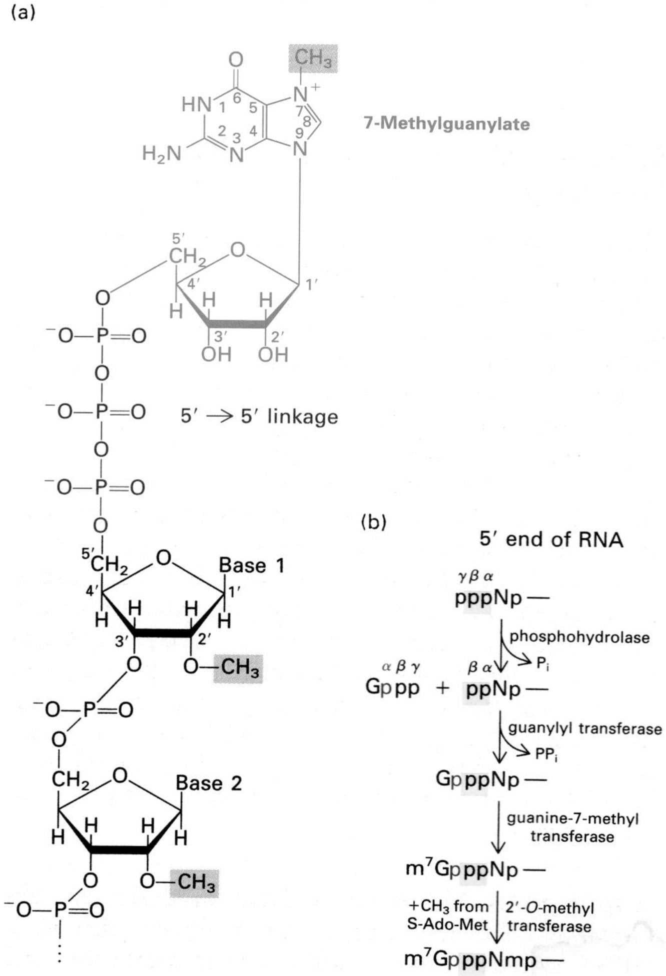

(a)

7-Methylguanylate

5′ → 5′ linkage

Base 1

Base 2

(b)

5′ end of RNA

γ β α
pppNp —

↓ phosphohydrolase
→ Pᵢ

α β γ   β α
Gppp + ppNp —

↓ guanylyl transferase
→ PPᵢ

GpppNp —

↓ guanine-7-methyl transferase

m⁷GpppNp —

+CH₃ from │ 2′-O-methyl
S-Ado-Met │ transferase

m⁷GpppNmp —

▶ **Figure 8-30** (a) The structure of the 5′ methylated cap of eukaryotic mRNA. The distinguishing chemical features are the 5′ → 5′ linkage of 7-methylguanylate to the initial nucleotide of the mRNA molecule and the methyl group at the 2′ hydroxyl of the ribose of the first nucleotide *(base 1)*. Both these features occur in all animal cells and in cells of higher plants, although yeasts lack the methyl group on base 1. The ribose of the second nucleotide *(base 2)* also is methylated in vertebrates. [See A. J. Shatkin, 1976, *Cell* 9:645.] (b) Steps in the biosynthesis of the cap structure in the nuclei of mammalian cells. Guanylyl transferase catalyzes the addition of guanylate (Gppp) only to polyphosphate (ppN or pppN) ends. The methyl (CH₃) donor is S-adenosylmethionine (S-Ado-Met). [See S. Venkatesan and B. Moss, 1982, *Proc. Nat'l Acad. Sci. USA* 79:304.]

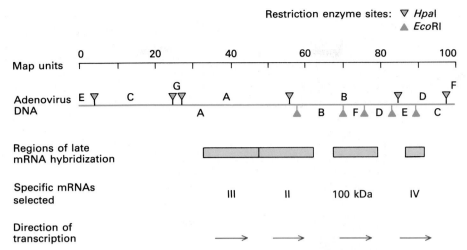

▲ **Figure 8-31** Map positions for coding regions of adenovirus proteins on the 36-kb viral DNA molecule (1 map unit = 360 base pairs). The cut sites for two restriction endonucleases (*Hpa*I and *Eco*RI) are shown on a conventional physical map that reads from left to right. The genomic positions of the DNA fragments created by these restriction enzymes were ordered from left to right, and each fragment was isolated for use in hybridization experiments (see Figures 6-23 and 6-24). The regions in the adenovirus DNA to which particular late mRNAs hybridized were used to locate coding regions for specific mRNAs (see blue bars below). III, II, and IV represent adenovirus capsid proteins; 100 kDa is a protein involved in capsid assembly. The DNA coding region must overlap these two fragments. The direction of transcription (*arrows*) was found by determining the 5′ → 3′ direction of each of the two complete DNA strands and then determining which strand hybridized to any particular mRNA. All of the four mRNAs hybridized to the same strand, thus indicating that they are synthesized in the same direction. [See J. Flint, 1977, *Cell* 10:153.]

Splicing Is the Final Step in mRNA Processing

The discussion so far has shown that hnRNA and mRNA have a similar base composition, although hnRNA is larger. In addition, both mRNA and some hnRNA molecules contain identical modifications at the 3′ and 5′ ends and similar amounts of methylated adenine residues. The evidence that was needed to clinch the precursor relationship of hnRNA to mRNA was proof of an obligatory larger primary transcript for a specific mRNA. However, because normal cells do not make large amounts of a single mRNA, purification of a single primary transcript was impractical. The use of viruses provided a solution to this problem. The most advantageous experimental situation occurs late in adenovirus infection of human cells when about half of the total nonribosomal nuclear RNA synthesized is adenovirus-specific RNA. Also, the only new mRNA molecules that arrive in polyribosomes are virus-specific mRNAs. Thus the infected cell becomes a factory for making adenovirus mRNAs, which, like normal cellular mRNAs, contain caps at their 5′ ends and poly A at their 3′ ends.

Mapping the Adenovirus Late mRNAs on the Genome The first step in the study of adenovirus mRNA synthesis in the nucleus was to locate the viral DNA regions that encode each viral mRNA molecule. The discovery of DNA restriction enzymes afforded a method for cutting the viral DNA at specific sites. Once the order of fragments in the linear DNA molecule was determined, the fragments could be hybridized with abundant virus mRNA molecules, obtained from the polyribosomes of cells late in infection. The DNA fragments that hybridized with these mRNAs were determined, and specific DNA fragments were also used to select by RNA-DNA hybridization specific mRNAs, which could then be translated into proteins. With these two techniques, the genomic regions encoding particular proteins were mapped (Figure 8-31). All these adenovirus late mRNAs were found to hybridize to the same DNA strand, which means that they are synthesized from the DNA in the same orientation (left to right in the conventional diagram). Almost the entire right-hand two-thirds of the viral genome—more than 25 kb of it—encodes a series of rightward-reading late mRNAs.

The nuclear RNA made from the late region of the adenovirus genome was studied by two different techniques to determine whether the adenovirus late mRNAs are the products of separate transcription units or whether they all come from one transcription unit through the processing of a long nuclear RNA transcript. Both pulse-labeled nascent-chain analysis (Figure 8-32) and ultraviolet-damage studies (Figure 8-33) pointed to only *one* start

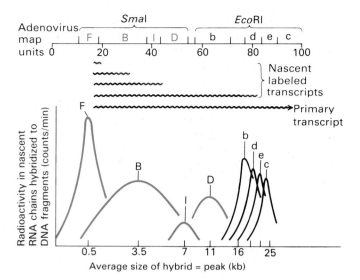

▲ **Figure 8-32** Demonstration that adenovirus late mRNAs are encoded by a single transcription unit. Nascent nuclear RNAs labeled only briefly so that their 3′ growing ends carried the label (see Figure 8-9) were separated by sedimentation and hybridized to a series of ordered DNA restriction fragments from the adenovirus genome. The shortest labeled RNA (averaging 0.5 kb in length) was complementary to the region between 11 and 18 on the genome—that is, to the *Sma*I F fragment, a DNA fragment of over 2 kb. Since the nascent RNA complementary to this 2-kb fragment was only 0.5 kb long, the initiation site for RNA synthesis was about 0.5 kb to the left of the right end of the F fragment. Longer RNAs had labeled segments complementary to fragments representing successive regions to the right of the F fragment (see also Figure 8-31). Therefore a single transcription unit encodes a long primary transcript that is processed to yield several different mRNAs. [See J. Weber, W. Jelinek, and J. E. Darnell Jr., 1977, *Cell* **10**:611; and R. M. Evans et al., 1977, *Cell* **12**:733.]

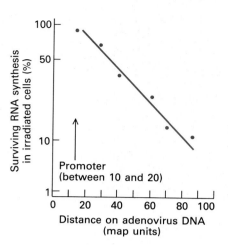

▲ **Figure 8-33** Mapping of the adenovirus late transcription unit by UV-damage analysis (see Figure 8-11). HeLa cells infected with adenovirus were exposed to UV radiation; infected cells not exposed served as controls. The nuclear RNA in both cultures was labeled with [³H]uridine. The effect of irradiation on nuclear RNA synthesis at various sites in the adenovirus genome was determined by hybridizing labeled nuclear RNA to these sites and comparing the results for irradiated and nonirradiated cells. The most sensitive region was at the far right end of the genome, and the most resistant was at the left end. When the amount of RNA synthesis that survived irradiation in each section of the genome was plotted on a semilogarithmic scale, an exponential increase in damage (i.e., an exponential decrease in RNA synthesis) was observed between about 20 and 100 map units on the genome. This indicated that synthesis began at or before 20 and progressed to 100. [See S. Goldberg, J. Weber, and J. E. Darnell Jr., 1977, *Cell* **10**:617.]

site (near 16 on the map) for nuclear RNA synthesis from the 25-kb stretch on the right side of the adenovirus genome. Initiation of transcription for each of the different mRNAs that map between 35 and 95 was near 16 on the virus map. Thus the late mRNA molecules had to be derived by some type of cleavage from the large nuclear adenovirus-specific RNA.

Production of Different Adenovirus Late mRNAs from a Single Primary Transcript by Addition of Poly A and Splicing

Studies on mRNA formation and control in bacterial cells demonstrated that the sequence of any given bacterial gene is represented in a continuous fashion in the mRNA and proteins encoded by that gene. Likewise, mammalian and amphibian 28S and 18S rRNAs were shown to represent separate but continuous stretches within the larger pre-rRNA molecule. Therefore, it was most surprising when the individual adenovirus late mRNA molecules were discovered to be complementary to *noncontiguous sites* in the DNA.

The mosaic nature of the adenovirus late mRNAs was first revealed by electron microscopy of mRNA-DNA hybrids. All of the mRNAs that derive from the long nuclear RNA (i.e., the mRNAs for the II, III, IV, and 100-kDa proteins) form hybrids with adenovirus DNA in a long region, called a *body,* and in short regions, called *leaders,* that occur at three separate genomic sites at 16.4, 20, and 27 map units (Figures 8-34 and 8-35). The leaders in an mRNA molecule are at or near the 5′ end. It is now known from nucleotide sequencing that the leaders are not translated and that translation begins within the body of the molecule. Because the adenovirus primary RNA transcript contains all of the sequences that make up the individual mRNA (i.e., the body and each of the short leader sequences), the only logical explanation was that the primary transcript is cut and spliced to give functional mRNA. This of course has now been proved by many different experiments including in vitro splicing reactions, which are described in a later section.

As a result of splicing and the addition of poly A, sev-

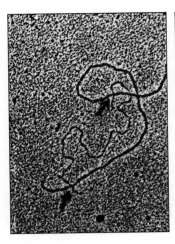

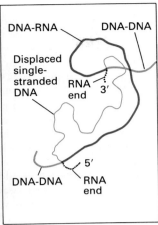

◀ **Figure 8-34** An electron micrograph and tracing of an adenovirus mRNA-DNA hybrid. A double-stranded DNA fragment spanning the region from 50 to 73 map units was hybridized to purified mRNA encoding capsid protein II (see Figure 8-31) by the R loop procedure, in which complementary RNA displaces DNA (Chapter 6). Besides the R loop, *two* protruding RNA ends (5′ and 3′) are visible, indicating that these ends are not homologous with the adenovirus DNA fragment. One end (the 3′ end) is poly A; the homology of the other end to distant sites in the adenovirus genome was determined by experiments of the type shown in Figure 8-35. *From S. Berget et al., 1977*, Proc. Nat'l Acad. Sci. USA *74:3171.*

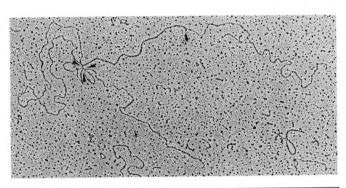

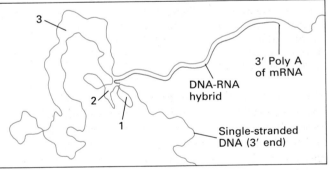

▶ **Figure 8-35** Microscopic identification of the regions of adenovirus DNA that are complementary to a late mRNA. The *single* strand of the adenovirus genome that is known to be transcribed in the indicated direction and to be complementary to the late mRNAs was purified and hybridized to adenovirus mRNA for the 100-kDa protein. The major part of the coding sequence for this protein lies between 67 and 79 map units (see Figure 8-31). The 5′ end of the mRNA hybridizes to three regions whose distances from the left (3′) end of the single DNA strand are 16.4, 20, and 27. Thus the mRNA has components from *four* regions of the DNA. When the four separated regions hybridized to the DNA, loops were formed of a distinct length: loop 1 stretched from 16 to 19 map units; loop 2 from 19 to 27 map units; and loop 3 from 27 to 67 map units. The RNA-DNA hybrid at the 16.4 region corresponds to the region predicted by nascent-chain analysis (Figure 8-32) and UV mapping (Figure 8-33) to contain the initiation site for the primary transcript. [See L. T. Chow et al., 1977, *Cell* **12**:1.] *Photograph courtesy of L. T. Chow and T. R. Broker.*

eral different mRNAs can be produced from a single nuclear primary RNA transcript. In fact, complete mapping of the adenovirus late genome region has shown that at least 15 different mRNAs are formed from the late transcription unit. The poly A additions and splicing reactions involved in the production of these late mRNAs are diagrammed in Figure 8-36. The production of multiple mRNAs from the same primary transcript has now been recognized in many viral and cellular genes. Clearly, the extensive processing of hnRNA to form mRNAs provides many opportunities for gene control; the mechanisms

of such control of gene expression are discussed in Chapter 11.

Introns and Exons Walter Gilbert suggested the term *intron* to refer to a part of a primary transcript (or the DNA encoding it) that is not included in a finished mRNA, rRNA, or tRNA. An *exon* is a primary transcript region (or, again, the DNA encoding it) that exits the nucleus and reaches the cytoplasm as part of an RNA molecule. The terms intron and exon are clear when they are applied to a primary transcript that gives rise to only

(a)

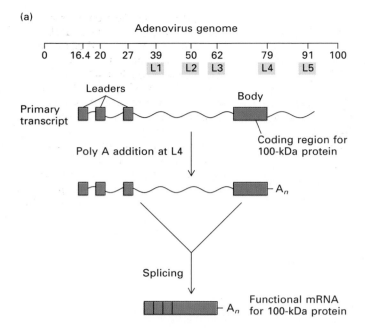

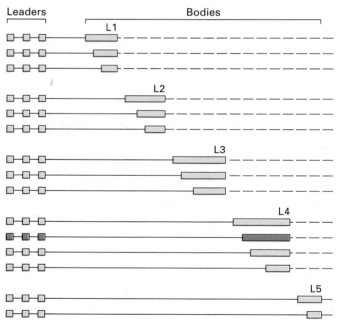

(b) Poly A sites and splicing alternatives representing the 15 mRNAs formed from the adenovirus late primary transcript

◀ **Figure 8-36** Formation of multiple adenovirus late mRNAs by addition of poly A and splicing at multiple sites. (a) The primary RNA transcript from adenovirus late DNA contains five potential poly A addition sites labeled L1 to L5. The formation of a particular mRNA molecule begins by cleavage at one of these sites. For example, the primary transcript yields the mRNA for the 100-kDa protein only when the transcript is cleaved and polyadenylated at the L4 site. Splicing must also occur between the leaders, so that the first is joined to the second and the second to the third. Finally, splicing permits the third leader to be joined to a site near 67 on the map, and the formation of the 100-kDa mRNA is complete. (b) Each poly A addition site, including L4, is preceded by several different cleavage sites to which the third leader can be joined; in effect, then, there are several different bodies associated with each poly A site. Fifteen different mRNAs can be produced from the primary transcript; which mRNA is produced depends on which poly A site is chosen and on where the third leader is spliced to the body. The splicing alternatives for these mRNAs are schematically represented. The components of the 100-kDa mRNA are shown in purple. [See L. T. Chow and T. R. Broker, 1978, *Cell* **15**:487; reviewed in J. R. Nevins and S. Chen-Kiang, 1981, *Adv. Virus Res.* **26**:1.]

Splicing in Production of Cellular mRNAs

After the discovery that primary nuclear transcripts are spliced to form finished adenovirus late mRNAs, this same process was shown to occur in the production of many cellular mRNAs. For example, the β chain of mammalian hemoglobin is encoded in an mRNA that is about 0.7 kb long. A pulse-labeled 1.5-kb nuclear RNA molecule that is complementary to cloned genomic β-globin DNA can be detected in the nuclei of cultured mouse erythroleukemia cells, which produce hemoglobin. When the cloned genomic DNA was hybridized to the 1.5-kb globin-specific nuclear RNA and examined in the electron microscope, a continuous region of hybridization was seen (Figure 8-37a). However, when the globin mRNA molecule was hybridized to the genomic DNA, unhybridized loops of genomic DNA were found (only one of these is large enough to see easily in Figure 8-37b). On the other hand, the nucleotide sequence of the cDNA copied from mRNA conformed to that expected from the known amino acid sequence of β-globin. Examination of the nucleotide sequence of β-globin genomic DNA confirmed that the genomic DNA contains two intervening sequences which are not present in β-globin mRNA but which divide the protein-coding domains (Figure 8-38). Again, the logical explanation for the hybridization results and the absence of the intervening sequences in the mRNA is that the primary 1.5-kb RNA transcript is spliced to form the 0.7-kb mRNA.

Similar comparisons of the sequences of hundreds of eukaryotic genes and the cDNAs made from their corresponding mRNAs indicate that intervening sequences occur widely in invertebrate and vertebrate genes. For

one mRNA molecule. However, in the case of transcription units like that for the adenovirus late genes, which can give rise to many different mRNAs, the definitions of intron and exon become blurred. A sequence that is an intron (i.e., that is destroyed in the nucleus) in the processing of one RNA transcript may become an exon (i.e., may become part of the mRNA) in the processing of another primary transcript from the same transcription. Nonetheless, the terms intron (or *intervening sequence*) and exon are commonly used.

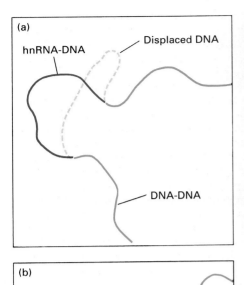

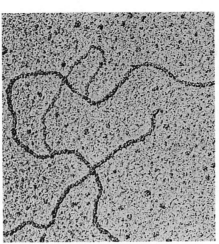

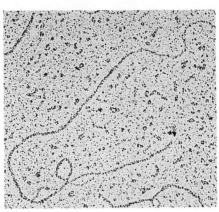

◀ **Figure 8-37** Electron micrographs and tracings of DNA-RNA hybrids containing cloned genomic β-globin DNA and (a) 1.5-kb globin-specific nuclear RNA and (b) cytoplasmic globin mRNA. The entire 1.5-kb sequence of nuclear RNA hybridizes in a continuous fashion to the β-globin DNA. The presence of a double-stranded loop in the mRNA-DNA hybrid in (b) indicates that part of the DNA sequence is absent from the mRNA molecule. [See S. M. Tilghman et al., 1978, *Proc. Nat'l Acad. Sci. USA* 75:1309.] *Photograph courtesy of P. Leder.*

example, splicing occurs in the formation of 90 to 95 percent of mammalian mRNAs. Although the proportion of spliced mRNAs in invertebrates may be somewhat less than that in mammals, they are still much more common than unspliced mRNAs. The majority of genes in yeast (a single-celled eukaryote) do not contain intervening sequences; however, at least several dozen spliced mRNAs have been identified among the products of the several hundred yeast genes that have been sequenced.

In some viral mRNAs (but not the adenovirus late mRNAs) and in many cellular mRNAs, splicing junctions are located *within* the mRNA region that is translated into protein. For example, the β-globin gene contains an intervening sequence between the codons for amino acids 31 and 32 and another between the codons for amino acids 105 and 106 (see Figure 8-38). Splicing junctions in some mRNAs even lie within a codon. Clearly, precise rejoining is necessary to ensure that a completed mRNA is translatable. In many cases, when protein-coding portions of an mRNA are in separate locations in the DNA and thus in the primary RNA transcript, the separated sequences represent distinct *domains* (functional regions) of the final protein product. The important implications of these arrangements for genome structure and evolution are discussed in Chapters 9, 10, and 26.

Similar Steps Occur in Formation of Most mRNAs

In this section we have traced the development of knowledge about the formation of mRNAs in eukaryotes. Early studies showed that nuclear hnRNA and mRNA share certain distinctive chemical features (similar base composition, 3′ poly A, and 5′ cap structure), suggesting that hnRNA is a precursor for mRNA. The formation of specific mRNAs from a specific nuclear primary hnRNA transcript was first demonstrated in studies on adenovirus late mRNAs. In recent years, gene cloning and sequencing studies have confirmed the following general pathway of eukaryotic mRNA formation:

1. Transcription of all protein-coding genes is catalyzed by RNA polymerase II to form a primary hnRNA transcript, which undergoes further processing.
2. A methylated cap structure is added to the 5′ end of all primary hnRNA transcripts as transcription proceeds.
3. After transcription past the site that will become the 3′ end of the mRNA, the primary transcript is cleaved, exposing a 3′-hydroxyl end to which poly A is promptly added. Histone mRNAs are the only eukaryotic mRNAs known not to contain a 3′ poly A.

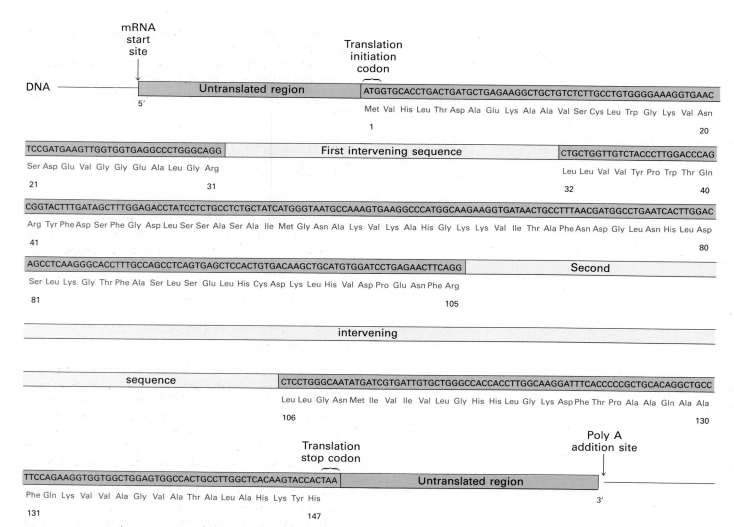

▲ **Figure 8-38** The β-globin gene has two intervening sequences. The 147 amino acids in the β-globin sequence were known; the DNA copied from the β-globin mRNA contained 5′ and 3′ untranslated regions and the exact coding information for the known amino acid sequence. The genomic DNA contained the coding sequence broken into three parts by two intervening sequences between amino acids 31 and 32 and 105 and 106. [See D. A. Konkel, S. Tilghman, and P. Leder, 1978, *Cell* 15:1125.]

4. Many but not all primary hnRNA transcripts are cleaved and spliced within the nucleus to yield mature mRNAs, which then move to the cytoplasm. Some splicing reactions are completed within a few minutes; others require 20–30 min.

The 5′ and 3′ ends of all finished mRNAs contain regions that are not translated into protein. These regions range in length from 10 nucleotides at the 5′ end to many thousands of nucleotides at the 3′ end. Among the functions suggested for these untranslated regions is the binding of proteins that play a role in efficient translation or in stabilization or destabilization of an mRNA.

Simple and Complex mRNA Transcription Units

Transcription units for eukaryotic mRNAs are either simple or complex—a design feature that is associated with a significant difference in protein-coding potential.

Simple transcription units are those whose primary RNA transcripts give rise to only one mRNA, which encodes only one protein (Figure 8-39, types 1–3). The nuclear RNA products of simple transcription units may or may not require all of the steps of RNA processing. All such products acquire caps, but many are not spliced, and formation of most histone mRNAs involves neither splicing nor addition of poly A. The mRNAs for interferon and some cell-surface receptors, some viral mRNAs, and most yeast mRNAs have caps and poly A but are not spliced. Globin genes are examples of simple transcription units whose primary transcripts undergo both poly A addition and splicing.

growing number of complex transcription units encoding many cellular proteins are now known in both vertebrates and invertebrates. Selection mechanisms that cause a particular mRNA to be produced from a complex primary transcript are an important form of genetic regulation.

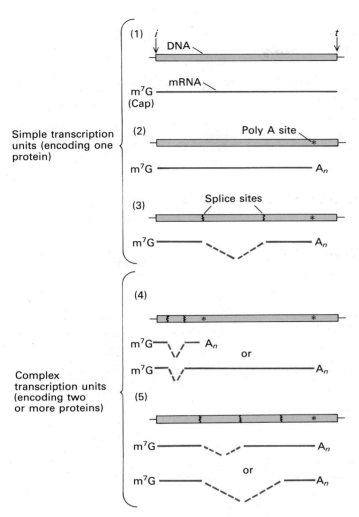

Simple transcription units (encoding one protein)

Complex transcription units (encoding two or more proteins)

▲ **Figure 8-39** Diagrams and examples of simple and complex transcription units encoding mRNAs. Each transcription unit is bounded by an initiation (*i*) and a termination (*t*) site; the mRNAs derived from the corresponding primary transcript are shown in red. Poly A sites are indicated by asterisks, and splice sites by zigzags. The most common types of transcription units in yeasts and perhaps other single-cell eukaryotes are the simple type 2 and 3, but multicellular organisms have many complex transcription units.

Complex transcription units are those whose primary transcripts can produce two or more different mRNAs, each encoding a different protein (Figure 8-39, types 4–5). Several types of complex transcription units can be distinguished. Some primary transcripts have two or more poly A sites, each of which can lead to a nuclear RNA molecule that requires splicing for completion as an mRNA. Other primary transcripts have one poly A site but two or more possible splicing arrangements, each resulting in a different mRNA. It is also possible to have a mixture of the two—that is, multiple poly A sites and multiple splicing arrangements. Most adenovirus transcription units (e.g., see Figure 8-36) and all of those in SV40 and the retroviruses are complex. In addition, a

Transcription of mRNA Genes by RNA Polymerase II

In the previous section, we described the general pathway for the synthesis of primary hnRNA transcripts (pre-mRNA) and their conversion into finished mRNAs. In this section, we discuss the details of transcription initiation and termination by RNA polymerase II. The cellular machinery for subsequent processing of pre-mRNA is described in the next section.

RNA Polymerase II Begins Transcription at 5' Cap Sites

Mapping of the adenovirus late transcription unit by nascent-chain analysis (see Figure 8-32) and ultraviolet irradiation studies (see Figure 8-33) located the approximate site for initiation of in vitro transcription at about 16 map units on the adenovirus genome. Subsequent studies with cloned adenovirus DNA cut with restriction enzymes at various distances from the 16.4 map position demonstrated that RNA polymerase II starts synthesis in adenovirus DNA at a particular site in the adenovirus sequence that matches the exact sequence found at the capped 5' end of all late leader-containing adenovirus mRNAs (Figure 8-40).

Additional evidence that RNA polymerase II begins transcription at the nucleotide that corresponds with the capped nucleotide at the beginning of pre-mRNA was obtained from experiments with ^{32}P-labeled nucleotide triphosphates. The results of these experiments showed that Gppp contributes only one phosphate to the cap structure, whereas the first nucleotide in pre-mRNA contributes two. If Gppp reacted with an internal nucleotide following cleavage of pre-mRNA, then only one phosphate in the cap structure could be contributed by pre-mRNA. Since the capped nucleotide contains two phosphates derived from pre-mRNA, then it must correspond to the initiating 5' nucleotide in the RNA chain, which is a triphosphate (see Figure 8-30b for reaction sequence).

Thus transcription of protein-coding genes is initiated at the correct site in in vitro systems consisting of RNA polymerase II, cell extracts, and ribonucleoside triphosphates. Further, the completed methylated cap structure is formed in such systems. Since no uncapped nascent chains longer than 20–30 nucleotides have been detected among RNA products, addition of Gp and methylation must occur soon after transcription begins.

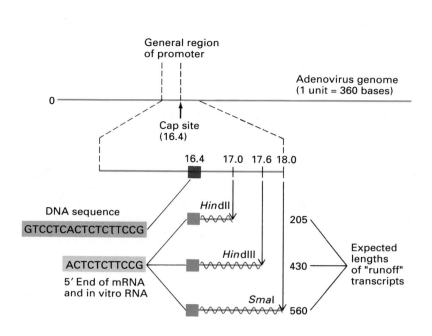

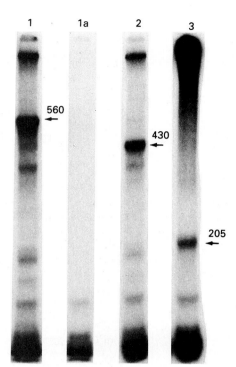

▲ **Figure 8-40** Mapping of initiation site for in vitro transcription of adenovirus late genome by RNA polymerase II. Three samples of purified adenovirus DNA that encodes the cap site (located at approximately 16.4 map units) were digested with one of three restriction enzymes that cut at 17.0, 17.6, and 18.0. The cut DNA templates were mixed and incubated with RNA polymerase II, extracts of HeLa cells, and labeled ribonucleoside triphosphates. When the RNA polymerase reaches a cut end, it "runs off" the template *(left)*. If the start site is the cap site, the labeled RNA products should stretch from the cap site to the cut end; this prediction was verified by subjecting the labeled runoff transcripts to gel electrophoresis and autoradiography *(right)*. Lanes 1, 2, and 3 show RNA made from DNA cut at 18.0, 17.6, and 17.0, respectively. The sample analyzed in lane 1a is the same as that in lane 1, except that α-amanitin, an inhibitor of polymerase II, was included in the transcription mixture. The green boxes in the diagram indicate an 11-nucleotide sequence that is present at the capped end of all the late adenovirus mRNAs and of the in vitro RNA transcripts as well. Thus the starting point for in vitro transcription by RNA polymerase II corresponds to the cap site in mRNA. [See R. M. Evans and E. Ziff, 1978, *Cell* **15**:1463; and P. A. Weil et al., 1979, *Cell* **18**:469.] *Photograph courtesy of R. G. Roeder.*

A Conserved DNA Sequence, the TATA Box, Is Responsible for Many Transcriptional Initiations

The first eukaryotic genes to be sequenced and studied by in vitro transcription were viral genes and cellular genes that are very actively transcribed either at particular times of the cell cycle (e.g., the histone gene) or in specific differentiated cell types (e.g., globins, immunoglobulins, ovalbumin, silk fibroin and many others). In all of these rapidly transcribed genes there is a highly conserved sequence called the *TATA box* (Figure 8-41) located in a fixed position about 25–35 bases upstream of the RNA start site. As we will see, most in vitro and in vivo tests of promoter activity for genes transcribed by polymerase II have used genes containing a TATA box. Most of the details we know at present about RNA polymerase II initiation of transcription depends on the study of genes containing a TATA box. For example, experiments with mutant and wild-type recombinant DNA constructs show

that a single-base change (e.g., G or A substituted for the second T) drastically decreases in vitro transcription of TATA-containing promoters (Figure 8-42). Sequences between the TATA box and start site do not significantly affect transcription. The nucleotide at which RNA synthesis begins lies about 30 nucleotides downstream from the TATA box and is an adenine in about 50 percent of the genes sequenced (see Figure 8-41).

More recently, a large number of genes have been studied in which no TATA box is evident by sequence analysis. Many of these genes are not transcribed at high rates (e.g., genes encoding hypoxanthine and adenyl phosphoribosyl transferases, dihydrofolate reductase, adenosine deminase, and at least a dozen others). Most of these genes contain a GC-rich stretch of 20–50 nucleotides that lies within the first 100–200 bases upstream of the start site. As we will discuss later, a DNA-binding factor called SP1 recognizes GC-rich sequences, and genes lacking the TATA box may rely on these GC-rich sites and the proteins bound to them to start transcription. The dinucleo-

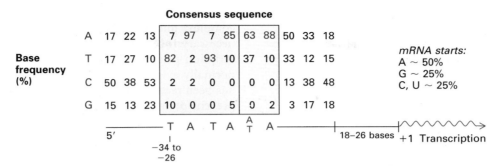

Consensus sequence

Base frequency (%)											
A	17	22	13	7	97	7	85	63	88	50	33	18
T	17	27	10	82	2	93	10	37	10	33	12	15
C	50	38	53	2	2	0	0	0	0	13	38	48
G	15	13	23	10	0	0	5	0	2	3	17	18

mRNA starts:
A ~ 50%
G ~ 25%
C, U ~ 25%

$$5' \longrightarrow T \quad A \quad T \quad A \quad \overset{A}{\underset{T}{}} \quad A \longrightarrow \text{18–26 bases} \underset{+1}{\sim\!\!\sim\!\!\sim\!\!\rightarrow} \text{Transcription}$$

−34 to −26

▲ **Figure 8-41** Comparison of nucleotides upstream of start site in 60 different eukaryotic protein-coding genomes. Each sequence was aligned to yield maximum homology in the region from −35 to −20. The tabulated numbers give the percentage frequency of each base at each position. The maximum homology occurs over a six-base consensus sequence in which the first four bases are TATA. The TATA box begins about 30 bases upstream from the start site. The most frequent mRNA start is an A, but a pyrimidine (C, U) makes up about 25 percent of the starts. [See R. Breathnach and P. Chambon, 1981, *Ann. Rev. Biochem.* 50:349 and P. Bucher and E. N. Trifonov, 1986, *Nucleic Acids Res.* 14:10009.]

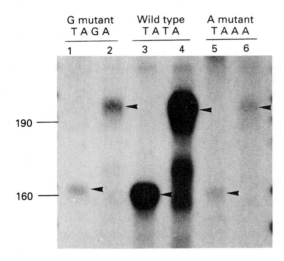

G mutant Wild type A mutant
T A G A   T A T A   T A A A
1   2   3   4   5   6

190 —

160 —

◄ **Figure 8-42** Effect of substitution in TATA box on transcription by RNA polymerase II. The second T in the wild-type TATA box of the chicken conalbumin gene was replaced, by in vitro mutagenesis, with a G or an A. The wild-type and mutated DNAs were cut with two different restriction enzymes and then incubated in an in vitro transcription mixture with labeled ribonucleoside triphosphates. Gel electrophoresis and autoradiography of the RNA products from the wild-type gene revealed two heavy bands containing about 160 (lane 3) and 190 (lane 4) nucleotides; synthesis of these products from the two mutated DNAs was greatly diminished. [See B. Wasylyk and P. Chambon, 1981, *Nucleic Acids Res.* 9:1813.] *Courtesy of P. Chambon.*

tide CpG is statistically underrepresented in vertebrate DNAs, and the presence of CpG-rich regions just upstream from initiation sites of genes is a distinctly nonrandom distribution. Precisely how these regions interact with the transcription machinery to promote initiation is not yet clear. It is quite possible that other mechanisms and other DNA-sequence motifs for ensuring correct RNA initiation will be discovered as more genes are sequenced.

RNA Polymerase II Requires Multiple Protein Factors to Begin Transcription

The protein factors which help RNA polymerase II to recognize the TATA box and which assist the enzyme to begin transcription have been at least partially purified. These factors differ from those which assist RNA polymerase I and III. Perhaps the most important of the transcription factors for RNA polymerase II is TF$_{II}$D, which by itself can bind to the TATA box region. In mammalian cells TF$_{II}$D is a 100-kDa protein; although yeast TF$_{II}$D is only about 40 kDa, it can function with the other transcription components from mammalian cells. In addition to TF$_{II}$D, at least two other general transcription factors—TF$_{II}$B, TF$_{II}$E, and perhaps others—are required, along with RNA polymerase II, to form a preinitiation complex.

Experiments with purified transcription factors and DNA templates from different transcription units have revealed the order of addition required to begin transcription by RNA polymerase II (Figure 8-43). If TF$_{II}$D is added in limiting quantities followed sequentially by each of two different templates, only the first template added is transcribed. Thus, TF$_{II}$D and DNA forms a stable, "committed" complex as discussed in the section on RNA polymerase III (see Figure 8-24). Neither TF$_{II}$B nor TF$_{II}$E

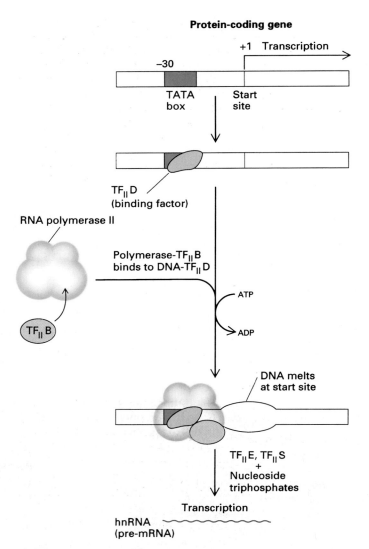

▲ **Figure 8-43** Model of formation of preinitiation complex for protein-coding genes containing a TATA box. The TATA-binding factor $TF_{II}D$ (orange) can bind on its own sufficiently tightly to form a stable complex. In other genes, it binds cooperatively with RNA polymerase II. Neither $TF_{II}B$ (purple) nor $TF_{II}E$ binds directly to DNA. $TF_{II}B$ can bind to RNA polymerase II in the absence of DNA. Once the polymerase-$TF_{II}B$ complex binds to the DNA-$TF_{II}D$ complex, it acts as an ATPase, perhaps causing the DNA to melt near the start site.

binds directly to DNA on its own, but $TF_{II}B$ is known to bind to the polymerase in the absence of DNA and may contact the DNA after binding to the polymerase. Once the polymerase-$TF_{II}B$ complex is bound to DNA, $TF_{II}B$ acts as an ATPase. Perhaps the energy released by ATP hydrolysis alters the conformation of some protein in the preinitiation complex or assists in melting the DNA to make an "open" complex or both. At this point (after the ATP-requiring step) transcription begins in the presence of $TF_{II}E$ and ribonucleoside triphosphates. If the purest

available factors are used in assembling the preinitiation complex, another factor, $TF_{II}S$, an elongation factor, is necessary for the most active chain elongation and transcription production. It is possible that $TF_{II}E$ and $TF_{II}S$ normally exist as a complex of two proteins. Recently it has been recognized that one important role of $TF_{II}S$ is to prevent premature termination of the polymerase II.

Although work with $TF_{II}D$, B, and E has been carried out with several different transcription units, it is not yet certain that this group of factors is truly general for all genes transcribed by RNA polymerase II. As noted already, genes that lack the TATA box are transcribed poorly in vivo and in vitro, and the role, if any, of the common factors in the transcription of these genes remains unknown. It has been assumed that the $TF_{II}D$ factor would have no role in genes lacking a TATA box, but even this is not certain. At least one adenovirus gene that has no conventional TATA box is transcribed in vitro with the assistance of $TF_{II}D$ alone. In that case there is a site *downstream* of the RNA start site that is used by the TATA factor to help initiate transcription. Thus it may be that some of the genes which lack apparent TATA boxes in the conventional position will have them inside the transcription unit. The important question of RNA initiation factors is receiving intense study, and doubtless the picture will soon be clarified.

Transcription by RNA Polymerase II Is Enhanced by Specialized Gene Activation Sites

In vitro transcription results were important in beginning to define the proteins responsible for in vitro gene activation. But only a small fraction (1–2 percent) of the actual DNA template molecules were activated in the in vitro experiments, and the inclusion of larger regions of upstream DNA did not improve this low efficiency. For this reason a more general assay for the effect of extra DNA sequences was adopted quite early in experiments that aimed to discover the DNA sequences and ultimately the protein factors that are required for completely normal rates of RNA synthesis.

By introducing various cloned DNA plasmids into cells, the presence of transcription stimulatory sequences and their approximate location can be demonstrated. In this type of experiment (Figure 8-44), several bacterial plasmids containing the same coding regions but varying amounts of DNA surrounding the start site of a protein-coding gene are constructed and introduced into cultured cells. The newly formed mRNA that accumulates from transcription of each plasmid is then determined. By noting which sequence deletion leads to decreased mRNA accumulation, the sequences responsible for higher rates of transcription can be pinpointed.

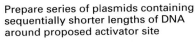

Prepare series of plasmids containing sequentially shorter lengths of DNA around proposed activator site

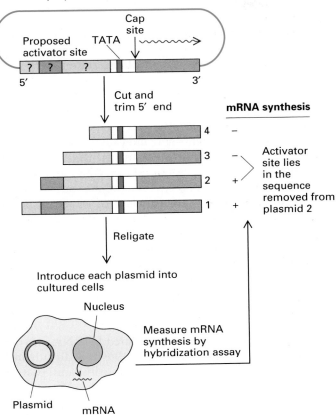

◀ **Figure 8-44** Mapping of activator sites required for RNA synthesis by RNA polymerase II. A recombinant plasmid containing 5′ flanking sequences (orange), a TATA box (green), RNA start site or cap site, and a region that is transcribed into mRNA (purple) can direct mRNA formation when introduced into cells. To determine where a putative activator region is located, the circular plasmid is cut and trimmed with nucleases so that portions of the upstream sequences are removed; the plasmid then is religated. The shortened plasmids are tested for the ability to direct RNA synthesis. In the example shown, deletion of sequences between plasmid 2 and 3 removes the critical activating region (dark orange).

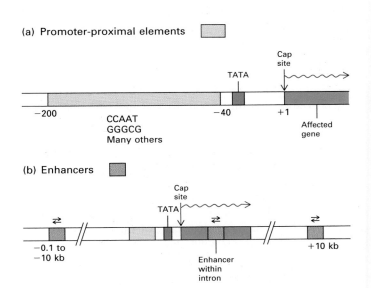

▲ **Figure 8-45** The classes of eukaryotic transcriptional activating elements. Many experiments with recombinant genes (of the type outlined in Figure 8-44) have shown that maximal transcription requires (a) upstream DNA regions close to the RNA start site, termed promoter-proximal elements, and (b) enhancers, which can be close to or far from RNA start sites (either upstream or downstream) and can exist in either orientation in the DNA. Almost all enhancer regions represent clusters of binding sites, which can bind several proteins. The most common promoter-proximal elements in the region −40 to −200 are CCAAT and GGGCG, but about 70 percent of genes so far examined have neither of these elements. Although most of the experiments that defined the two general activating elements were done in mammalian cells with genes from mammals or viruses infecting mammals, the same elements also exist in invertebrate genes. In the yeast genome, which is much smaller, the activating elements tend to be just upstream from the affected gene but often share with enhancers the property of functioning in either orientation. The yeast sequences are most often termed upstream activating sequences, or UASs. [See P. Bucher and E. N. Frifonov, 1986, *Nucleic Acids Res.* 14:10009.]

Thousands of such experiments have been done, and hundreds of sites that stimulate transcription by RNA polymerase II have been identified; in many cases the factors that recognize these required activating sites have been purified. The exact mechanism by which any of these factors increases transcription is not known. For purposes of discussion, we will describe *promoter-proximal sites* and *enhancers*. Although these two classes of gene-activating elements differ in their proximity to the transcription initiation site, they may or may not differ in their ultimate mode of action. For example, we now know that some particular activating proteins can bind at distant and proximal DNA binding sites.

Promoter-Proximal Sequences As noted earlier, the sole highly conserved sequence for RNA polymerase II is the TATA box. Analyses of more than 150 cloned animal and plant genes have shown that two other sequences, CCAAT and GGGCG, are present in 10–15 percent of genes, usually between 60 and 120 nucleotides upstream from the start site (Figure 8-45a). In many genes with these promoter-proximal sequences, mutation in the CCAAT or GGGCG sequence decreases transcription, indicating that these sequences are important protein-binding sites in some genes. Proteins that specifically bind to these sequences and increase transcription from templates that contain these sequences have been isolated. A family of proteins bind to the CCAAT box, and a protein called SP1 (stimulatory protein 1) binds to the GGGCG box.

The amino acid sequence of SP1 has regions analogous to the zinc-finger regions in TF$_{III}$A, one of the protein factors required for synthesis of 5S rRNA; zinc is bound by coordinated cysteine residues in these regions. The various proteins that bind the CCAAT box are clearly not like SP1 and most likely represent a large family of genes encoding many different proteins.

Since the SP1 protein and CCAAT-binding proteins are widely distributed in different types of cells, they cannot be the major determinants of cell-specific gene expression. These factors can, however, presumably increase transcription from genes, including those expressed in a cell-specific fashion, that have the corresponding binding site. What remains unclear is whether in genes lacking a TATA box, CCAAT factors or SP1 attract RNA polymerase and other general factors to form a preinitiation complex by a mechanism similar to that in TATA-containing genes. An alternative possibility is that CCAAT factors and SP1 are strictly stimulatory and that some as yet undetected mechanism of preinitiation-complex formation occurs in genes lacking a TATA box.

Experiments with hundreds of genes, many of which are expressed in a cell-specific fashion, have uncovered many other DNA-binding proteins whose binding sites lie between 40–200 bases upstream of the RNA start site. The impression is growing that in mammals there are hundreds if not thousands of different proteins that interact with these promoter-proximal sequences.

Enhancers A second class of gene activators—enhancers—has attracted a great deal of attention because they have several properties that were at first glance surprising. For example, an enhancer sequence can boost transcription of recombinant plasmid DNA regardless of its 5′–3′ orientation (Figure 8-45b). In addition, enhancers can increase transcription when placed either close to (~100 bases) or at great distances (up to 5 kb or more) from the RNA start site. Some enhancers have been found 40 kb or more either upstream or downstream from the gene on which they apparently act. Finally, enhancers can exist within a gene, usually within introns.

This class of positive transcriptional elements has received intense study, and many proteins that bind to enhancers have been purified. An important property of many enhancers that have been examined in detail is that they contain several protein-binding sites clustered within about 100 nucleotides. It is widely assumed that these DNA elements located far from the start site they affect

must form loops in the DNA to exert their effect (see the discussion of arabinose operon in Chapter 7. The loops would be the result of interactions of proteins that are bound to the distant DNA sites.

It may be that promoter-proximal elements are similar to enhancers since many of them have been shown capable of stimulating in vivo DNA transcription even after inversion. In addition, many individual stimulatory proteins have binding sites within both promoter-proximal and distant enhancer sequences. The observation of occasional failures in transcription after the inversion of promoter-proximal elements may indicate that when the distance between the activator site and the RNA start site is less than 50 bases or so, the correct DNA loop cannot form between an inverted element and the start site. Since RNA polymerase II in the presence of TF$_{II}$D, B, and E starts transcription at most RNA start sites, albeit at a low rate, the role of all stimulatory proteins may very well be the same—namely, to increase the access and binding of RNA polymerase II to one of its binding sites.

Transcription of mRNA Genes Is Terminated Downstream of Poly A Site

Once transcription has been initiated by RNA polymerase II, where and why does it end? We have already mentioned that the primary transcript continues past the poly A addition site before terminating (see Figure 8-39). This conclusion was reached by hybridizing labeled nascent nuclear RNA to DNA regions both upstream and downstream of sites to which poly A is attached in the transcribed RNA. The uniform result, obtained first with adenovirus and SV40 transcription units, and subsequently with a number of vertebrate transcription units (e.g., mouse β-globin, mouse α-amylase, and rat calcitonin), was that transcription regularly passes the poly A site.

For most cellular protein-coding genes, transcription continues 0.5–2 kb past the poly A site and then stops, although no precise termination sites have been detected. An important finding obtained in experiments with viral and with β-globin genes is that termination does not occur during transcription of recombinant plasmids containing either mutations or deletions of the poly A site. The model shown in Figure 8-46 is drawn from these findings. The model also includes the observation that premature termination, resulting in short RNA chains containing a 5′ cap, is very common (up to 30 percent of RNA chains initiated are prematurely terminated). The antiterminator factor proposed in this model may provide a mechanism by which RNA polymerase II is held to its task until the ultimate 3′ end of the mRNA has been completed. (Perhaps TF$_{II}$S, which has been described to have antiterminator properties in vitro, serves such a purpose.) Such a mechanism could be particularly important in

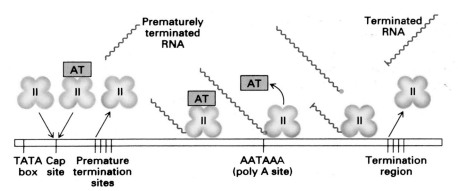

▲ **Figure 8-46** Model for a mechanism coupling addition of poly A (A_n) and termination of transcription. The model proposes that two classes of RNA polymerase II transcription complexes (red) are directed to (cap) sites by upstream regulatory elements (including the TATA motif). The two classes differ by the presence or absence of an antiterminatory (AT) factor (orange). Complexes lacking the AT factor cease transcription at premature termination sites. Complexes including the AT factor read through premature stop sites, but such complexes lose the factor at a poly A addition site (AATAAA) and then cease transcription at the next termination region. [See J. Logan et al., 1987, *Proc. Nat'l Acad. Sci. USA* **84**:8306–8310.]

eukaryotic cells since many protein-coding genes contain tens to hundreds of thousands of nucleotides. For example, the genes for thyroglobin and a protein active in blood clotting, factor VIII, are each more than 300,000 nucleotides long. Because RNA polymerase II is estimated to add at most 50 nucleotides per second, it takes at least 100 min to finish such long transcripts.

What causes RNA polymerase II to cease transcription after it passes a poly A addition site is unknown. Perhaps loss of the proposed antiterminator factor loosens the association between the polymerase and DNA, so that the enzyme dissociates at any of several DNA sequences. No generally conserved termination regions have been identified in DNAs transcribed by polymerase II; in contrast, termination of transcription by both polymerase I and III occurs at specific terminator sites characterized by a string of T residues.

In addition to termination after poly A site selection, we should note that the capacity for premature termination, also proposed in the model, obviously could be used in gene control. Examples of such regulation are discussed in Chapter 11.

Conversion of hnRNA to mRNA

Once RNA polymerase II has traversed from the 5′ cap site through the poly A site that is about to be utilized, the conversion of hnRNA to mRNA by the RNA-processing machinery begins. The three main events in mRNA processing are (1) addition of poly A, (2) methylation of adenylate residues, and (3) splicing. Without a doubt, the most complicated of these events is splicing, but we now have a fairly detailed picture of the complex mechanism by which splicing is accomplished in eukaryotic cells.

Formation of 3′ Poly A Involves Recognition of Special Sequences, RNA Cleavage, and End Addition

Since transcription of protein-coding genes is terminated downstream of the 3′ end of the corresponding mRNA, the hnRNA primary transcript must be cleaved before poly A addition. The nuclear factors that catalyze cleavage of primary transcripts and addition of about 250 adenine nucleotides have been partially purified, and both require the recognition of special sequences in the hnRNA.

By the time the sequences of the first dozen or so animal mRNA molecules had been determined, the poly A signal, AAUAAA, was recognized *within* the mRNA 15–30 bases from the 3′ end (Table 8-2). Sequencing of genomic DNA also located this signal within the hnRNA transcribed from a gene. The AAUAAA sequence is also highly conserved in plants, but only an AU-rich region near poly A sites is evident in yeasts. In addition to the AAUAAA signal, poly A addition requires between 10 and about 30 nucleotides downstream from the poly A addition site. The exact role of these necessary downstream sequences is not known.

Table 8-2 Frequency of occurrence of nucleotides in poly A signal in eukaryotic mRNA*

	Frequency (%)					
	A	A	U	A	A	A
Animal (61 cases)	97	98	100	100	100	97
Plant (46 cases)	90	100	100	87	91	65

*The poly A signal also has been identified in pre-mRNAs 15–30 bases upstream from the poly A addition site.

SOURCE: S. Berget, 1984, *Nature* 309:179; and C. P. Joshi, 1987, *Nucleic Acids Res.* 15:9627.

In crude cell extracts and in intact cells, the addition of poly A occurs almost immediately after cleavage of the hnRNA primary transcript. The nuclear factor that cuts the primary transcript at the poly A site has been separated from the factors responsible for poly A addition. The addition factor will form a poly A of the correct length if it is furnished with a 3′ end containing a nearby AAUAAA. As we will discuss thoroughly in later sections, small nuclear ribonucleoprotein particles, or snRNPs (pronounced "'snirps"), are important in the splicing reaction. The cleavage preparatory to addition of poly A is blocked by antisera to snRNPs, but which snRNP (if only one) is necessary for this cleavage is not yet known.

The formation of the 3′ ends of all eukaryotic mRNAs appears to involve some processing. Even in histone mRNAs, which do not contain a poly A, the 3′ end is created by cleavage of a primary transcript. In this case, cleavage requires sequences downstream from the ultimate 3′ end; recognition of the cleavage site occurs with the aid of a specific small nuclear ribonucleoprotein termed U7 snRNP.

Methylation of Adenylate Residues Is Common in Vertebrate mRNAs

Many mRNAs of higher organisms (mostly mammalian cells have been studied) are methylated at nitrogen 6 of adenylate residues within the mRNA chain. Lower organisms, such as yeasts and slime molds, seem to have few, if any, methylated adenylates. Studies with adenovirus-infected cells have shown that the methyl groups are added to adenovirus pre-mRNAs that have not yet been spliced. Also, methyl groups are added only to A residues within exons (i.e., regions of the primary transcript that remain part of the finished mRNA); methylation of pre-rRNA also occurs only within the conserved portion of the primary transcript. Perhaps the methyl groups in both cases play some role in protecting the portion of the primary transcript that is to be preserved.

Splice Sites in hnRNA Contain Short Conserved Recognition Sequences near Intron-Exon Boundaries

As described in an earlier section, the location of splice sites in an hnRNA primary transcript can be determined by comparing the sequences of the corresponding genomic DNA with that of cDNA prepared by copying the corresponding mRNA. Any discontinuities between the genomic DNA and cDNA sequences mark exon-intron boundaries. Such analysis of a number of different mRNAs has revealed moderately conserved, short consensus sequences at intron-exon boundaries in pre-mRNA and (in vertebrates, at least) a tendency for a pyrimidine-rich region just upstream of the 3′ splice site. The only universally conserved nucleotides are the first two (GU) and last two (AG) in the intron (Figure 8-47). Deletion analyses of the center portion of introns in various pre-mRNAs have shown that generally only 30–40 nucleotides at each end of an intron are necessary for splicing to occur at normal rates.

Recombinant DNAs containing the 5′ exon-intron junction of one transcription unit (e.g., the SV40 late region) and the 3′ intron-exon junction of another (e.g., the mouse β-globin gene) have been prepared and introduced into cultured cells. Spliced mRNA molecules are formed in which the two exon sequences are joined and the chimeric intron is deleted precisely. Although the yield of spliced products in such experiments often is low, the formation of any correctly spliced mRNAs indicates that the cell's splicing machinery can recognize and correctly join heterologous 5′ and 3′ splicing sequences.

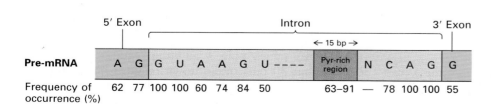

◀ **Figure 8-47** Frequency of occurrence of nucleotides around intron-exon junctions in eukaryotic pre-mRNAs. The consistent features are the (5′)GU and (3′)AG, which are universal, and a pyrimidine-rich region close to the 3′ end, which is quite common. [See R. A. Padgett et al., 1986, *Ann. Rev. Biochem.* 55:1119.]

Splicing of hnRNA Involves Concerted Reactions to Excise Introns and Ligate Exons

Experiments with cell extracts that are capable of excising the intron of an RNA chain and rejoining the two ends have demonstrated that the distinctive 5′ cap structure is required for the splicing reaction. Analysis of the RNA products formed in a cell-free splicing mixture also led to the conclusion that the intron is not cut out as a linear molecule (Figure 8-48). The first step is a cut at the 5′ exon-intron junction, freeing the 5′ exon (Figure 8-49).

The remaining molecule migrates much more slowly than its length would predict. The slowly migrating form is a lariat structure in which the 5′ pG residue at the cut end of the intron is joined to the free 2′ OH of an adenylate residue approximately 25 nucleotides upstream from the 3′ splice site. This adenylate residue is, of course, part of the original RNA chain in which it is linked in the normal 5′ → 3′ phosphodiester linkage. Thus a *branch point* is formed in the intron at this site. It seems likely that the branch is formed at or almost at the same time as the first cleavage at the 5′ exon-intron boundary. For this to occur, the RNA chain must be folded to bring the inter-

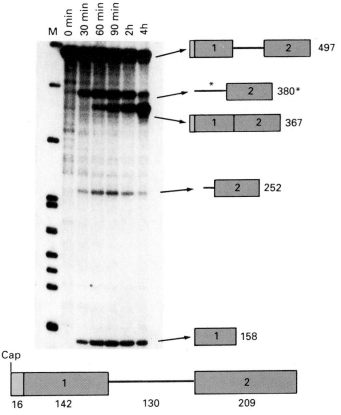

▲ **Figure 8-48** Analysis of RNA products formed in cell-free splicing system. A crude nuclear extract from HeLa cells was incubated with a 497-nucleotide-long labeled RNA *(bottom)* that contained portions of two exons (labeled 1 and 2) of human β-globin separated by a 130-nucleotide intron. After incubation for various times, the RNA was purified and subjected to electrophoresis and autoradiography (along with RNA markers, M). Almost complete conversion of the slower-moving starting RNA (497) into the spliced form (367) occurred during the 4-h incubation. One of the intermediates in the reaction (380*) has an anomolous electrophoretic migration. It is actually 239 nucleotides long, and the aberrant migration occurs because it is a lariat. This led to the discovery of the branch-formation reaction diagrammed in Figure 8-49. *From B. Ruskin et al., 1984, Cell **38**:317. Photograph courtesy of T. Maniatis.*

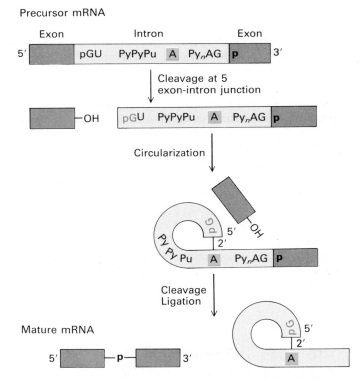

▲ **Figure 8-49** Pathway of splicing of pre-mRNA. The steps illustrated have been shown to occur during in vitro splicing of adenovirus and globin hnRNA sequences by cell extracts. The consensus nucleotides GU and AG that begin and end the intron are indicated, as is the pyrimidine-rich stretch (Py$_n$) near the 3′ end. After the first cleavage at the 5′ exon-intron junction, a circularization occurs to create a branched structure in which the guanylate at the cut 5′ end of the intron forms an unusual linkage (5′ → 2′) with an adenylate located near the 3′ end of the intron. The adenylate retains its normal 3′ → 5′ and 5 → 3′ linkages to adjacent nucleotides. The pyrimidine-rich sequence at the branch point shown in the diagram is characteristic of vertebrate and invertebrate hnRNAs. The corresponding yeast sequence is UACUAAC and is highly conserved. How the cleavage and ligation occur at the 3′ junction is not yet known, but the phosphate linking the two exons is derived from the first nucleotide of the 3′ exon. [See B. Ruskin et al., 1984, *Cell* **38**:317; and R. A. Padgett et al., 1984, *Science* **225**:898.]

acting nucleotides closer together. The next step is also a concerted reaction in which cleavage at the 3′ intron-exon boundary is accompanied by simultaneous linkage of the two exons. Although the mechanism of the cleavage and ligation at the 3′ junction is unknown, the phosphate that joins the two regions is donated by the first nucleotide in the 3′ exon.

Small Ribonucleoprotein Particles Participate in Splicing

Even before splicing was accomplished in cell-free systems, several observations had led to the suggestion that small ribonucleoprotein particles (snRNPs) somehow assisted in the splicing reaction. First, the short consensus sequence at the 5′ end of introns (GUAAGU) was found to be complementary to a site near the 5′ end of the small nuclear RNA (snRNA) called U1. Second, snRNAs were found in nuclear extracts as part of large particles containing hnRNA. Six prominent small RNAs, U1–U6, have been purified and sequenced (see Figure 8-17). All of these snRNAs associate in the nucleus with six to ten proteins to form snRNPs (Figure 8-50). Some of the proteins are common to all of the snRNPs, and some are specific for individual snRNPs.

Antibodies directed against protein that is common to all the particles (called *anti-sm antibody*) was available from patients with the autoimmune disease lupus erythematosus (see Figure 8-14). Some antibodies also were found to have greater specificity for individual snRNPs. Different antibodies from these patients have been widely used in characterizing components of the splicing reaction. For example, when antiserum that is specific for U1 snRNP is added to an in vitro splicing mixture, the splicing reaction is interrupted. Subsequent studies with a synthetic oligonucleotide that is complementary to and thus hybridizes with the 5′-end region of U1 snRNA also blocked splicing, indicating that this is the region that binds to pre-mRNA during the splicing reaction. Experiments of this design with several different snRNPs showed that U1 snRNP binds to a 15- to 20-nucleotide segment containing the 5′ exon-intron junction; U2 snRNP binds to about 20 nucleotides at the branch point; and U5 snRNP binds just upstream of the 3′ intron-exon junction (Figure 8-51).

When the splicing reaction was first carried out in cell-free extracts, it was noted that no spliced product was observed during the first half hour of incubation. After it became clear that several different snRNPs play a role in the splicing events and that hnRNA is part of large pro-

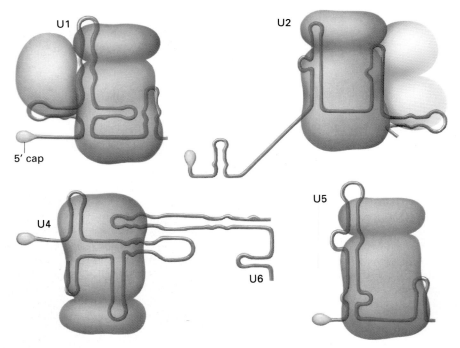

▲ **Figure 8-50** Components of four snRNPs that take part in splicing of hnRNA. The five snRNAs (red)—U1, U2, U4, U5, and U6—share a complex of proteins (orange); the U1 and U2 snRNAs also bind with unique proteins (purple, green). The secondary structures of the snRNAs shown are based on maximum base pairing of the known nucleotide sequence. *After T. A. Steitz, 1988, Sci. Am.* **258**:56.

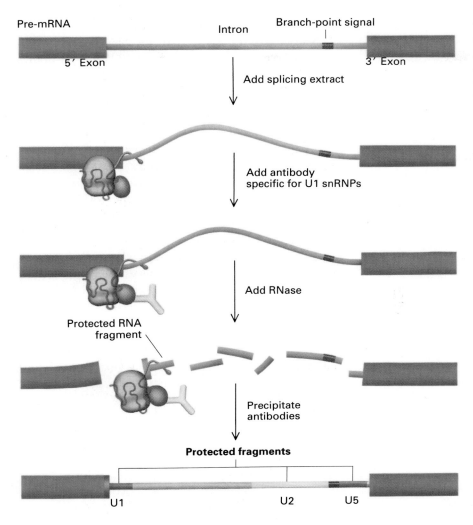

◀ **Figure 8-51** Identification of the site on pre-mRNA to which U1 snRNP binds. A labeled pre-mRNA substrate containing an intron was mixed with a nuclear extract capable of splicing (the extract contains all the necessary snRNPs). Addition of antibodies (blue) specific for U1 snRNP protected the binding region from subsequent attack by RNase. Analysis of the pre-mRNA in the antibody precipitate showed that it corresponds to a 15- to 20-nucleotide segment (purple) containing the 5′ exon-intron junction. Similar experiments with antibodies to U2 and U5 snRNPs have identified the binding regions shown at the bottom of the diagram. *After T. A. Steitz, 1988, Sci. Am. 258:56.*

tein-containing aggregates that can be extracted from the nucleus (see next section), it was reasoned that perhaps an organized structure must form around the RNA during splicing. First by sedimentation analysis and then by biochemical fractionation of the splicing extract, large particles termed *spliceosomes* were in fact demonstrated to form on the RNA substrate. Spliceosomes have now been partially purified and visualized in the electron microscope (Figure 8-52).

▶ **Figure 8-52** Electron micrograph of a spliceosome. Extracts of HeLa cells were mixed with a β-globin pre-mRNA substrate, and the reaction was interrupted before splicing was completed. This allowed purification of spliceosomes containing snRNPs and the substrate RNA. *From R. Reid, J. Griffith, and T. Maniatis, 1988, Cell 53:949. Photograph courtesy of J. Griffith.*

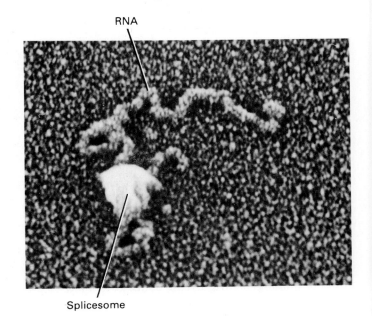

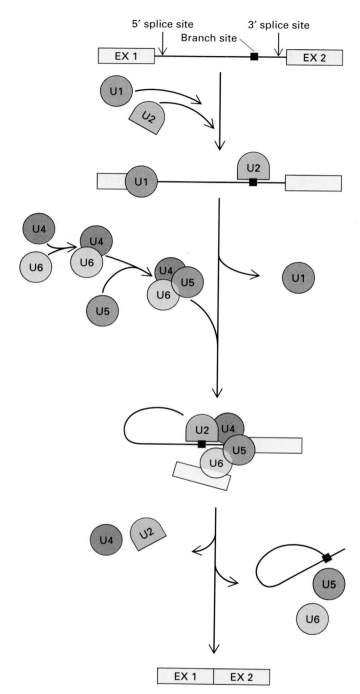

5' splice site

3' splice site

Branch site

EX 1

EX 2

▲ **Figure 8-53** Possible pathway of spliceosome assembly and function. *After M. M. Konarska and P. A. Sharp, 1987, Cell **49**:763.*

In the cell, the splicing reaction is almost surely carried out in complexes that begin to form even during synthesis of a pre-mRNA molecule. Based on studies of the various intermediates present in in vitro splicing mixtures and the order of binding of snRNPs, the pathway of spliceosome assembly and function shown in Figure 8-53 has been

proposed. In a step that does not require energy, a U1 snRNP binds to the 5' splice site on a capped pre-mRNA molecule. Binding of a U2 snRNP at the intron branch point, which may require ATP cleavage, also occurs early, perhaps even before the binding of the U1 snRNP. Because complex particles—U4/U6 snRNP and U4/U6/U5 snRNP—have been identified in splicing mixtures, they probably form before binding to the pre-mRNA, as shown in the model. Cleavage at the 5' exon-intron junction and loss of U1 snRNP from the complex may occur concomitant with or before binding of the complex particle containing the U4, U5, and U6 snRNPs. Loss of the U4 particle may also occur early in the assembly of the spliceosome. Cleavage and ligation at the 3' end is followed by release of the intron and remaining snRNPs, perhaps in separate steps in the in vitro reaction.

All the experiments on in vitro splicing of pre-mRNA discussed so far were done with extracts of mammalian cells. Splicing of pre-mRNA also occurs, of course, in plant cells and in yeast. A great deal of recent work has utilized yeast extracts because in this organism the genes encoding the factors used in splicing can be easily cloned and mutated, so that the exact role of each factor can be worked out more easily. From what is known so far, splicing of pre-mRNA in yeast is quite similar to that in mammals. Yeast spliceosomes are assembled from snRNPs and have analogous functions to those in mammalian cells. The intron branch point is much more highly conserved in yeast than in mammals (see Figure 8-49); a UACUAAC signal at the branch point is almost invariant in several dozen yeast pre-mRNAs so far examined. The 5' boundary of introns is also more highly conserved; for example, a (5') GUAUGT sequence is present in fifteen of seventeen different 5' splice junctions of yeast transcripts; a UAG sequence at the 3' end of the intron also is highly conserved, but there is no pyrimidine-rich region like that in mammalian pre-mRNAs (see Figure 8-47).

hnRNA Associates with Specific Proteins to Form Particles That May Assist in the Assembly of Spliceosomes

As noted in the previous section, large particles containing hnRNA and protein can be extracted from cell nuclei. Although demonstrating which proteins are associated with hnRNA inside the cell proved difficult, several hnRNA-associated proteins have now been identified and partially characterized.

Sedimentation analysis of nuclear extracts has revealed many dozens if not hundreds of proteins in the hnRNA fraction. If such extracts are repeatedly purified by sedimentation and examined by electron microscopy, groups of 20- to 25-nm particles are observed (Figure 8-54a).

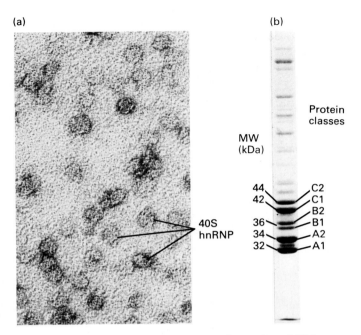

(a)

(b)

Protein classes

MW (kDa)

44 — C2
42 — C1
— B2
40S hnRNP   36 — B1
34 — A2
32 — A1

▲ **Figure 8-54**   The hnRNA-associated proteins of HeLa cells. Particles from HeLa cell nuclei containing hnRNP fragments were purified. (a) Electron micrograph of particles. (b) Gel electrophoresis of hnRNP proteins with molecular weights (kDa). *Courtesy of W. LeSturgeon.*

Very brief RNase treatment of these preparations releases single particles that sediment at 30S–40S; these are referred to as heterogeneous nuclear ribonucleoprotein particles (hnRNPs). These particles contain a segment of hnRNA about 600 nucleotides long and three major classes of protein—A, B, and C (Figure 8-54b). The A, B, and C proteins, as well as the 120- and 68-kDa proteins, must be associated in some way because a monoclonal antibody that reacts with the C protein precipitates particles containing all of these proteins. However, protein-protein interactions are not strong enough to maintain a particle in the absence of RNA, as RNase treatment ultimately destroys the particle structure.

Studies with UV-irradiated cells have been used to determine whether the hnRNA-associated proteins identified in cell extracts are associated with hnRNA in intact cells. At appropriate low doses, UV irradiation causes formation of cross-links between proteins and RNA if they are in close contact. When UV-irradiated cells that have been briefly labeled with [³H]uridine are fractionated, fractions containing the A, B, C, 120-kDa, and 68-kDa proteins cross-linked to labeled hnRNA can be identified, demonstrating that the proteins are associated with hnRNA in the cell.

Many of the arginine residues in both the A1 and B proteins are methylated; the A and B proteins also share many antigenic determinants and have an unusual protein composition, including 25 percent glycine. Purified preparations of one of the A proteins binds single-stranded DNA, but there is no evidence that it does so in the cell. Perhaps this protein and others somehow keep RNA single-stranded. If hnRNPs are subjected to solutions of increasing ionic strength, the A and B proteins dissociate from the particles before the C proteins, indicating that the C proteins form the tightest association with hnRNA.

The genes have been cloned for several of the proteins found in hnRNPs and two other RNA-binding proteins—a poly A–binding protein from both yeasts and humans and a major nucleolar protein, *nucleolin*, whose function is unknown. There are in all of these proteins a stretch (or several stretches) of 100 amino acids with considerable sequence similarity, including one highly conserved stretch of eight amino acids:

$$\frac{\text{Lys}}{\text{Arg}} \; \text{Gly} \; \frac{\text{Phe Gly Phe}}{\text{Tyr Ala Tyr}} \; \text{Val X} \; \frac{\text{Phe}}{\text{Tyr}}$$

This highly conserved region (the alternate amino acids require single base changes, and the alternates are chemically very similar) lies in the part of the A1 protein that can be shown to bind to RNA. Thus the conserved sequence very likely is at least part of a generalized, or non-sequence-specific, RNA-binding domain common to all these proteins.

The proteins that associate with hnRNA differ from those which associate with small nuclear RNAs to form snRNPs, which are known to participate in splicing. Some evidence indicates, however, that hnRNPs (or the proteins in them) also may have a role in splicing. For example, a monoclonal antibody reactive with the C protein prevents the in vitro splicing reaction, in both yeast and mammalian extracts, by blocking the first event in splicing (the cut at the 5' exon-intron border). Further work with antibodies against the other hnRNP proteins to determine their possible specific roles is under way.

As noted earlier, the hnRNA present in isolated particles is only about 600 nucleotides long. It seems likely, however, that in most cases the hnRNA (especially in the case of very long primary transcripts) is broken during extraction. The nature of native hnRNPs in intact cells and the association of protein with nascent chains of hnRNA have been demonstrated in several special cases by electron microscopy.

One well-studied large hnRNP is found in the salivary gland nuclei of the water midge *Chironomous tentans*. In this species, hnRNA synthesis occurs in "puffs" (expansions) of particular chromosome regions to produce large "75S" hnRNA molecules. Because these puffs are so large

(a)

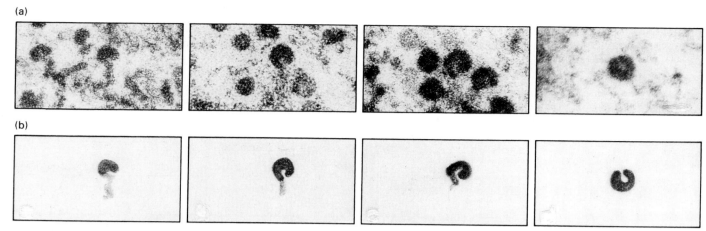

(b)

▲ **Figure 8-55** Formation of a specific hnRNP. An expanded chromosomal "puff," called a "Balbiani ring" after its 19th-century discoverer, is found at a particular site on the third chromosome of *Chironomous tentans*. This is now known to result from transcription of a very long transcription unit to make a giant hnRNA, the "75S" hnRNA, which encodes a salivary gland protein. Part (a) shows four electron micrographs of particles in the Balbiani ring that represent individual 75S RNAs in the process of synthesis. From hundreds of such micrographs of serial sections, the shape of the hnRNA at various stages of growth was adduced (b). [See U. Skoglund et al., 1986, *Nature* **319**:560.] *Photographs courtesy of B. Daneholt.*

they can be mechanically isolated, and the "75S" RNA can be followed during its formation. As it is made, the RNA transcript becomes condensed into a large but compact hnRNP structure (Figure 8-55). Other electron microscopic studies have been done with *Drosophila* cells whose nuclei are treated to partially remove protein so that growing hnRNA chains are "well-spread." After such treatment, the chains appear uncovered for long stretches, but particles are still visible on most chains. The particles remaining attached to transcripts, which probably arise from a single transcription unit, show regular spacing patterns. These particles, which are smaller than hnRNPs, may be spliceosomes that have identified intron-exon junctions during transcription (Figure 8-56).

The biochemical and electron microscopic results taken together suggest that the association of protein with a newly emerging hnRNA transcript begins very soon after initiation of transcription. Probably, proteins recognize

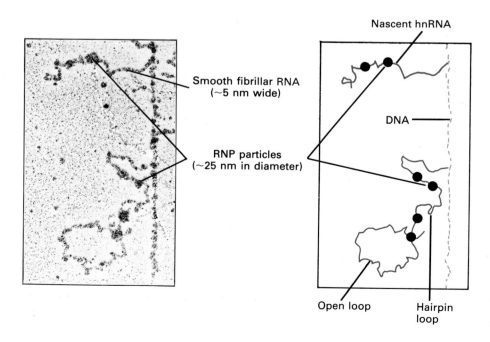

Nascent hnRNA

Smooth fibrillar RNA (~5 nm wide)

DNA

RNP particles (~25 nm in diameter)

Open loop

Hairpin loop

◀ **Figure 8-56** Electron micrograph and tracing of nascent hnRNA in *Drosophila* nuclei. The cell nucleus was disrupted and the contents dispersed by treatment with a mild detergent solution that "unrolls" hnRNPs. No hnRNP-sized particles remain, but the two transcripts have 25-nm particles on a 5-nm RNA fibril. The particles on the longer fiber are holding the RNA in a loop. Because these particles can be found on the same site on successive transcripts from the same transcription unit, they may be spliceosomes, which would resist the conditions of preparation. [See A. L. Beyer, O. L. Miller, Jr., and S. L. McKnight, 1980, *Cell* **20**:75.] *Photograph courtesy of A. L. Beyer.*

the cap that is added almost immediately to all 5′ ends of transcripts formed by RNA polymerase II. Both in vitro splicing and polyadenylation reactions depend on caps, and thus it is reasonable to assume that early cap recognition is necessary. Nuclear proteins that bind tightly to caps have been identified but have not yet been proved to have a function in vivo. The A, B, and C proteins in hnRNPs are thought to bind as soon as the transcript is long enough. These particles very likely keep the RNA mainly in single-stranded form to allow recognition of specific sequences. Spliceosome assembly probably occurs on an hnRNP, and the primary transcript, always in association with proteins, would then be promptly processed.

Variations on the Splicing Theme, Including Self-splicing

As we intimated earlier, the discovery that mRNA precursors were spliced during their formation was perhaps the most surprising finding in the history of molecular cell biology. No one suspected that coding functions could possibly be separated and joined at the level of RNA until the pictures of adenovirus mRNA-DNA heteroduplexes were taken. Naturally, many investigators were attracted not only to solving the details of this process, without which many functional mRNAs would not exist, but also to the physicochemical and evolutionary implications of splicing. Soon a second unexpected process, nearly as surprising as splicing itself, was discovered, namely, RNA splicing in the total absence of protein. The discovery of *self-splicing* RNA spawned a whole subdiscipline of RNA-mediated biochemistry, which has many implications for evolution. Here we give a brief account of tRNA splicing, which differs from mRNA splicing, as well as an initial description of self-splicing and trans-splicing, all of which have obviously important consequences in modern-day organisms as well as in evolution.

Splicing of pre-tRNA Differs from Splicing of pre-mRNA

As discussed earlier, the genes encoding some tRNAs contain intervening sequences that are not present in the corresponding finished tRNAs. Thus processing of some pre-tRNAs involves splicing. Indeed, the first success with in vitro splicing reactions was achieved with extracts from yeast cells and yeast pre-tRNA^{Tyr}, which contains a 14-base intervening sequence (see Figure 8-21). The intervening sequence in pre-tRNAs is within and around the anticodon in most cases.

Splicing of a pre-tRNA molecule involves two steps: the cutting of the primary transcript twice to remove the intron, and the sealing together of the ends. These steps

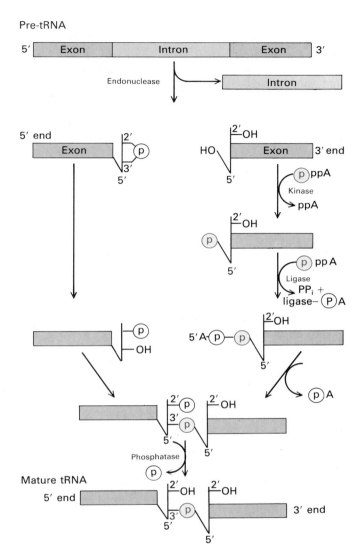

▲ **Figure 8-57** A model of splicing in pre-tRNA. Features that distinguish this splicing mechanism from that in pre-mRNA (see Figure 8-49) are the concerted excision of the intron, the ATP-requiring ligation of the two exons, and formation of a 2′,3′-cyclic phosphomonoester on the cleaved end of the 5′ exon. The nature of the reaction between the two exons is not certain, but the resulting 2′-phosphomonoester and the 3′ → 5′ linkage at the ligation site have been verified. The 2′-phosphomonoester is removed by a phosphatase in the final step. [See C. L. Greer et al., 1983, *Cell* **32**:537; and P. S. Perlman, C. L. Peebles, and C. Daniels, 1989, in *Intervening Sequences in Evolution and Development,* E. M. Stone and R. J. Schwartz, eds., Oxford.]

can be carried out separately by partly purified cell fractions (Figure 8-57). The molecules fold so that the intron-exon junctions can be precisely cut, releasing the intron. Mutations in various parts of a pre-tRNA that cause it not to fold correctly prevent the splicing reaction; the actual details of folding of the precursor are still unknown.

Ligation of the two separated exons occurs in a two-step reaction that requires two ATPs. The phosphate group that links the two halves of the finished tRNA comes from one of the two ATPs used in the reaction. The unusual 2′-phosphomonoester in the 5′ exon is not present in the finished product, indicating that it is removed by a phosphatase in the final step. Comparison of Figures 8-49 and 8-57 shows that the splicing reaction is quite different in pre-mRNA and pre-tRNA.

Self-splicing of RNA Precursors Occurs in Some Organisms

The discovery of self-splicing came during the study of ribosomal RNA biogenesis in the protozoan *Tetrahymena thermophila*. Researchers found that DNA encoding ribosomal RNA in this organism and in another simple organism, *Physaraum polycephalum,* contains an intervening sequence in the region that encodes the large rRNA subunit. Careful searches for even one gene without the extra sequence failed. So it appeared that splicing is required to make ribosomal RNA in these organisms.

When pre-rRNA transcribed from the *Tetrahymena* DNA containing the intervening sequence was incubated by itself, it underwent a spontaneous shortening reaction in the absence of any cellular proteins. By transcription of recombinant *Tetrahymena* DNA, an rRNA transcript containing the two exons and the intervening 408-base intron was prepared. When incubated by itself, this transcript was cleaved and rejoined at the correct sites without assistance from any protein (Figure 8-58). The first step in the removal of the intron—a cleavage at the 5′ exon-intron junction—requires the presence of a guanosine cofactor (specifically the chemical activity of the hydroxyl groups on its ribose). As the cut in the RNA is induced, the free 3′ hydroxyl of the guanosine is linked to the 5′ phosphate of the nucleotide at the 5′ end of the intron. This transfer of a phosphate linkage from one nucleotide within a chain to another is called a *transesterification;* no energy is consumed and the number of phosphate linkages is not changed. This transesterification is the key to self-splicing. The first cut leaves a 3′ hydroxyl on the uridine at the end of the 5′ exon. Cleavage at the 3′ end of the intron occurs during a second transesterification that links the end nucleotides (both uridylates) in a 5′ → 3′ phosphodiester linkage. The phosphate in this linkage is derived from the final UMP of the 3′ exon. The excised intron undergoes a second cleavage-and-splicing reaction that releases a small oligonucleotide and circularizes the remaining segment.

It should be noted that the intervening sequence in the 28S region of *Tetrahymena* and other rRNA transcription units is unrelated to the spacer sequences that separate the 18S, 5.8S, and 28S regions (see Figure 8-12) in the majority of organisms. In particular, the self-splicing

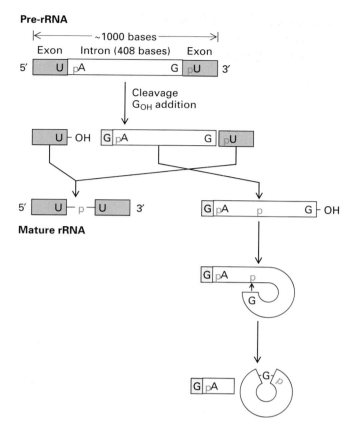

Pre-rRNA

▲ **Figure 8-58** Diagram of self-splicing in pre-rRNA sequence in *Tetrahymena thermophila*. This self-splicing reaction occurs in the absence of any cellular proteins but requires guanosine (G_{OH}). The guanosine is added to a phosphate at the end of the intron by transesterification. Likewise, a transesterification joins the exons and also occurs in the circularization of the intron. (All phosphates involved in this reaction are shown in blue.) [See T. R. Cech, 1983, *Cell* 34:713.]

mechanism that removes intervening sequences differs from the cleavage mechanism by which spacer sequences are excised.

Following the discovery of self-splicing in *Tetrahymena* pre-rRNA, a whole raft of self-splicing sequences were found in the pre-rRNAs from other single-celled organisms, in mitochondrial pre-rRNAs, and even in several pre-mRNAs from certain *E. coli* bacteriophages. These precursors form a group of self-splicing RNAs that all use guanosine as a cofactor and in which the intron can fold by internal base pairing to juxtapose closely the two exons that must be joined. These are described as group I self-splicing introns. Certain mitochondrial pre-mRNAs contain introns that constitute a second type of self-splicing intron, designated group II. These form lariats on excision and may be similar to introns in hnRNAs. The chemical details of self-splicing, which have an important bearing on modern ideas about early cellular evolution, are discussed in Chapter 26.

Portions of Two Different RNA Transcripts Can Be Joined by Trans-splicing

The splicing of pre-mRNA and pre-tRNA produces spliced products that contain portions of a single RNA transcript. Such a reaction could be referred to as *cis-splicing* by analogy with the genetic term used to indicate a close position on the same chromosome. Formation of spliced products containing portions of two different transcripts—called *trans-splicing*—has recently been demonstrated to occur in a few organisms. For example, most (perhaps all) mRNAs in *Trypanosoma brucei* are made by attachment of a short "leader" sequence, which is produced from one site in the DNA, to the 5′ end of a pre-mRNA produced from another transcription unit. No splicing *within* the pre-mRNAs of trypanosomes has yet been described.

Another example of trans-splicing was discovered during studies of the mRNA for the muscle protein actin in the nematode *Caenorhabditis elegans*. Comparison of the sequences of the genomic DNA encoding actin and cDNA copied from actin mRNA showed that the 5′ end of the mRNA is not present in the genomic DNA. Subsequent studies revealed that the 5′ end of actin mRNA is encoded on an entirely different chromosome. Thus actin mRNA is formed by transcription of two separate chromosomal regions and trans-splicing of two different RNA transcripts to yield a mosaic mRNA; one cis-splicing also occurs in formation of actin mRNA (Figure 8-59). As in the case of trypanosome mRNAs, the sequence trans-spliced to form actin mRNA is a short leader sequence. Recent work indicates that a special "leader" snRNP participates in the trans-splicing reaction.

The occurrence of trans-splicing has been recognized only for a short time, and so far this mechanism has been detected in just a few organisms. However, now that researchers are alerted to this possibility, its presence or absence in other species should soon become obvious.

Nuclear Structure and the Passage of Nuclear RNA to the Cytoplasm

In prokaryotes, which lack a membrane-enclosed nucleus, newly synthesized RNA is immediately available for use by the cell. Indeed, translation of mRNA usually begins while transcription is proceeding. In eukaryotes the process of transcription and processing is set apart in the nucleus before mRNA and the structural RNAs are delivered to the cytoplasm.

We now know many of the proteins that can interact with hnRNA to produce finished products. And as we will discuss in Chapters 9–12, we know about histone association with DNA and are beginning to learn about other proteins that bind DNA for various purposes. What we lack at this point is an accurate picture of how all these elements are integrated in the cell nucleus or indeed whether there is a regular structural basis for the interactions of active and inactive genes, primary transcripts, transcripts that are in the process of transcription, and transcripts that are on the way out of the nucleus.

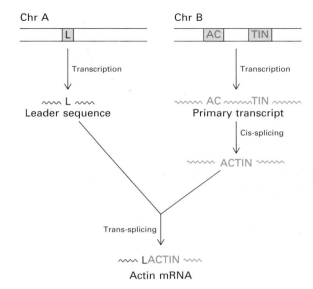

▲ **Figure 8-59** Trans-splicing in formation of actin mRNA in the nematode *Caenorhabditis elegans*. A short 22-nucleotide "leader" exon ("L") is encoded on one chromosome (Chr A), and two longer exons ("AC" and "TIN") are encoded on a different chromosome (Chr B). Transcription produces a 100-nucleotide leader transcript containing the L exon and a primary transcript containing the AC and TIN exons. Actin mRNA is formed by cis-splicing in the primary transcript and trans-splicing of the leader transcript to the "ACTIN" transcript. Although the two types of splicing reactions are drawn separately, they may in fact occur in a single spliceosome-like particle. [See M. Krause and P. Hirsh, 1987, *Cell* **49**:753.]

The Nuclear Matrix Is an Evolving Concept, Not Yet an Understood Structure

There is no doubt that there are many proteins within the nucleus that are not present in the cytoplasm, aside from the obvious histones and known common DNA-binding proteins of chromatin such as the HMGs (high-mobility proteins in electrophoresis). Some of these are identified as transcription factors (to be discussed in Chapters 9–11).

(a)

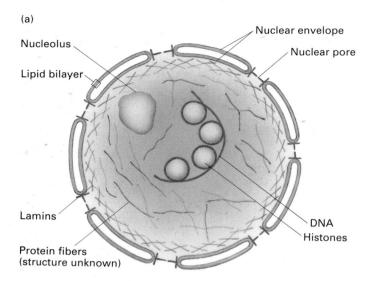

Nucleolus

Lipid bilayer

Nuclear envelope

Nuclear pore

Lamins

Protein fibers
(structure unknown)

DNA

Histones

◀ **Figure 8-60** (a) Generalized diagram of eukaryotic nucleus showing identifiable structural elements. (b) Electron micrograph of nuclear pore–lamin complexes isolated from rat nuclei. Nuclear pores *(arrows)* are embedded in fibrous lamin proteins (la). (c) A transmission electron micrograph of a whole mount of a HeLa cell, showing a skeletal network within the nucleus. The cell was prepared by removing lipids and soluble factors with a mild detergent. The remaining skeletal struture was then treated to remove most of the DNA. The sample was fixed with glutaraldehyde, but no heavy-metal shadowing was done. [See S. Penman et al., 1982, *Cold Spring Harbor Symp. Quant. Biol.* **46**:1013.] *Photograph (b) courtesy of N. Dwyer. Reproduced from the* Journal of Cell Biology, *1976, by copyright permission of Rockefeller University Press. Photograph (c) courtesy of S. Penman.*

(b)

(c)

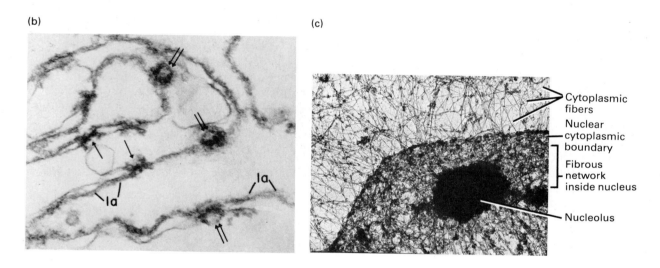

Cytoplasmic
fibers

Nuclear
cytoplasmic
boundary

Fibrous
network
inside nucleus

Nucleolus

la

la

In electron micrographs of nuclei from which all (or most) DNA has been removed, there remains a network of protein fibers both around the outside of the nucleus and internally (Figure 8-60b and c). This structure resists disaggregation in high ionic strength buffers and is referred to as the *nuclear matrix*. Monoclonal antibodies identify proteins of as-yet-unspecified function that are present in the matrix structure. A number of experiments suggest that hnRNA and partially processed transcripts are also in the matrix. What is not yet settled is the specificity of the interactions that cause these associations.

One important possibility is that the fibers form a pathway out of the internal regions of the nucleus to allow mRNA to reach the nuclear surface. New microscopic techniques involving hybridization with specific fluorescent nucleic acid probes make this appear very likely (Figure 8-61). Individual genes (or clusters of identical genes) are detected by such probes as dots on chromosomes.

RNA produced from these genes, however, can be seen as tracks with brighter fluorescence near the nuclear borders. These experiments show that RNA from one transcription unit doesn't diffuse freely but is localized along a track, possibly bound to fibers as it makes its way to the cytoplasm.

Nuclear Pores Provide Passageways for Movement of RNA into the Cytoplasm

Once mRNA has been completed, it appears in the cytoplasm promptly. Distinctive structures, called *nuclear pores*, are embedded in the double-layered nuclear envelope of all eukaryotic cells (Figure 8-62) and are universally believed to be the passageway for RNA exit. The number of nuclear pores ranges from several thousand in somatic cells to several million in large cells such as amphibian oocytes. It has been assumed for years—although

(a)

(b)

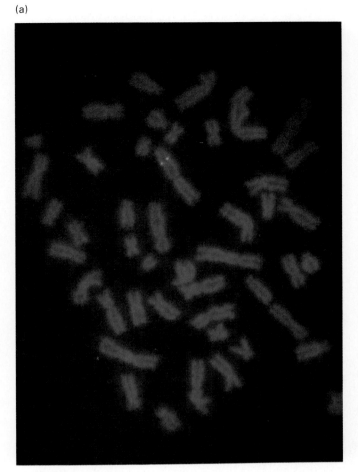

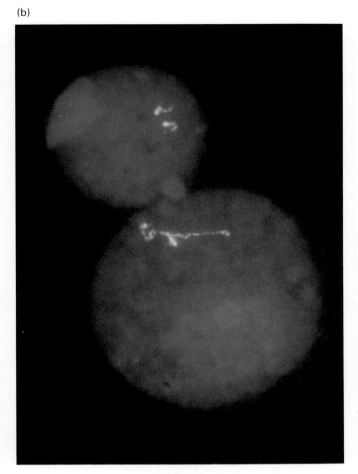

▲ **Figure 8-61** Detection of fluorescent probes to specific RNA or DNA in the nucleus. Cells containing an integrated copy of a 20-kb portion of the Epstein-Barr virus (the causative agent of infectious mononucleosis) in each copy of chromosome 1 were used. The presence of the virus DNA or of RNA transcripts from the virus DNA was detected by a fluorescent DNA probe into which were incorporated U residues containing attached biotin. Avidin, a protein from chicken eggs that has high affinity for biotin, can be attached to a dye, fluorescein, and the avidin-biotin conjugate detected microscopically (yellow). (a) Chromosomes in metaphase with two yellow dots representing EBV DNA. (b) Tracks of RNA in interphase nucleus that are at least 2–3 μ long. These long tracks are RNA as attested by their disappearance with RNase or with drugs that block RNA synthesis. Note that RNA is made from each chromosomal site (orientation of smaller nucleus shows two tracks) and that the tracks are at most half the diameter of the nucleus (5–8 μ). *From J. B. Lawrence, R. B. Singer, and L. M. Marselle, 1989, Cell 57:493.*

it remains to be proved—that passage of material to and from the nucleus occurs either by diffusion or by specific transport through the nuclear pores. If specific mechanisms for the exit of RNA from the nucleus exist, they are critical in gene expression and could obviously be important in gene regulation.

Just inside the nucleus, underlying the double-layered nuclear envelope, is a layer of interacting fibrous proteins called *lamins*. The lamins are now known to be part of a large family of proteins, termed *intermediate filaments,* the other members of which are found in the cytoplasm. During interphase, the nuclear pores appear to be embedded in the lamins and provide a channel between the nu-

cleus and cytoplasm (see Figure 8-60b). In electron micrographs the pores appear octagonal, a shape conferred by a highly organized array of proteins surrounding a central cavity. In some pores, the central cavity is filled with a protein "plug" (see Figure 8-62a), which very likely represents material on the way into or out of the nucleus.

The diameter of the cavity in a nuclear pore is approximately 10 to 20 nm, which is smaller than the diameter of a large ribosomal subunit. If a material such as mRNA or a ribosome enters the cytoplasm through the cavity in a pore, either the pore or the material passing through it may have to change shape. However, almost nothing is known at present about this critical step in the biogenesis of cytoplasmic RNA. It remains unproved although likely that an active process of RNA transport exists. Since there are so many ribonucleoprotein particles to which functional roles have not yet been attached, it is appealing to think that some of these might function in transporting RNA from the nucleus to the cytoplasm.

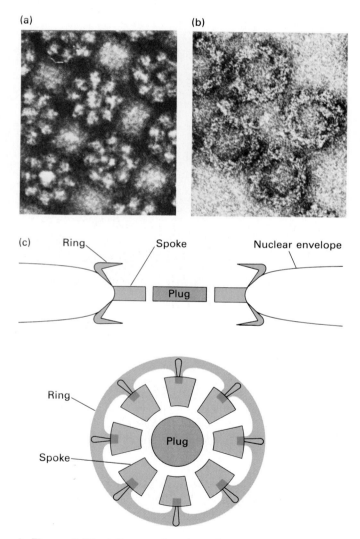

▲ **Figure 8-62** Micrographs of purified nuclear pores from *Xenopus laevis* oocytes and a model of pore structure. Some pores contain central cores or "plugs" *(a)*, and some do not *(b)*. Because of the regularity of their structure, these nuclear pores could be subjected to image processing (a computer-assisted analysis of the average structure of many similar objects). In *(c)*, a physical model of a pore within the nuclear envelope is viewed in cross section *(top)* and face-on *(bottom)*. *Photographs (a) and (b) from P. N. T. Unwin and R. A. Mulligan, 1982,* J. Cell Biol. *93:63; part (c) after same source.*

Transport of mRNA May Be Assisted by Proteins

Nuclear events produce a finished mRNA molecule in 5 to 20 min; within at most another few minutes, the mRNA is part of a cytoplasmic polyribosome. A key, still-unanswered question about eukaryotic cells is whether a new mRNA molecule enters the cytoplasm by associating with a ribosome or a ribosomal subunit at the nuclear border. Many studies have been done with drugs that stop translation (cycloheximide, emetine, puromycin); such drugs eventually prevent the formation of new ribosomes. However, even when the drugs are present, new mRNA can still be synthesized and can still enter the cytoplasm. Thus *new* ribosomes are not required for passage of mRNA into the cytoplasm. In addition, cell fractionation studies have shown that almost no old ribosomes are present in isolated nuclei. Therefore, the first possible association of a ribosomal subunit with a newly emerging mRNA molecule is at the nuclear border, perhaps at a nuclear pore. Within no more than 2–3 min after synthesis, new mRNA is associated with polyribosomes; there is no demonstrated phase of cytoplasmic existence of *new* mRNA between nuclear exit and incorporation into the protein-synthesizing apparatus.

Even if new mRNA does not enter the cytoplasm by association with a ribosomal subunit, the possibility remains that it enters by association with one or more of the proteins that initiate translation. Cytoplasmic *cap-binding proteins,* which possibly have a role first in the transport and then in the utilization of mRNA, have been characterized. For example, a cytoplasmic protein with a molecular weight of 24,000 has been purified by virtue of its affinity for the cap structure at the 5′ end of mRNA (m^7GpppN; the methyl group is important for recognition). In association with other initiation factors, this protein aids in binding the capped end of an mRNA molecule to the 40S ribosomal subunit. However, whether this cap-binding protein or any other protein (nuclear or cytoplasmic) has a definite role in mRNA transport remains unknown at present.

mRNAs May Be Directed toward Specific Cytoplasmic Sites

Most experiments on the cytoplasmic appearance of new mRNA have employed cell fractionation techniques to examine the presence of mRNA in cytoplasmic extracts. It seems likely, however, that the translation of mRNA in the cell cytoplasm does not take place in free solution. For example, polyribosomes can be intimately associated with portions of the endoplasmic membrane (the rough endoplasmic reticulum). The proteins that are synthesized at these sites are exported from the cell or become com-

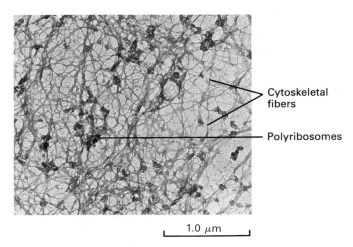

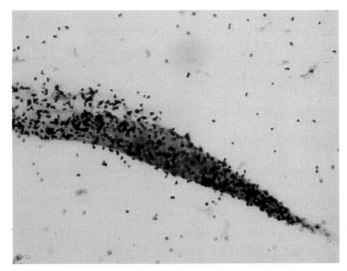

All mRNA labeled

▲ **Figure 8-63** A transmission electron micrograph of a whole mount of a human diploid fibroblast. The cell was prepared by removing lipids and most proteins with mild detergents. The remaining cytoskeletal structure was then fixed with glutaraldehyde. This area of the cytoplasm shown contains many fibers and dark groups of objects whose approximate diameters are equal to those of ribosomes. *From S. Penman et al., 1982, Cold Spring Harbor Symp. Quant. Biol. 46:1013.*

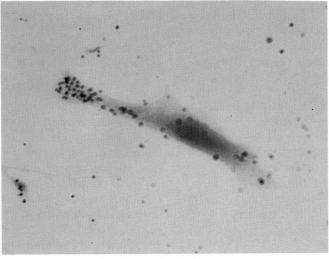

Actin mRNA labeled

ponents of new membranes. Some mRNAs making proteins destined to associate with the cytoskeleton, a cytoplasmic network of diverse fibers, are also not freely diffusable.

The association of some mRNAs with the cytoskeleton has been demonstrated by cell fractionation. If cultured cells are treated briefly with detergent to remove lipids, the cell border becomes porous and many proteins (about 75 percent of the total) and other materials (e.g., more than 90 percent of the tRNAs) wash out of the cells. Many polyribosomes, however, are retained in such preparations (Figure 8-63). A number of specific mRNAs have been found by in situ hybridization to be localized in specific locations in the cytoplasm. For example, actin RNA is found by in situ hybridization in the tips of pseudopodia, peripheral cytoplasmic extensions concerned with cell movement (Figure 8-64). With this technique a number of other mRNAs have now been found to be localized in different types of cells. For example, the myelin basic protein, a membrane-associated "insulator" of long nerves, is located in the periphery of oligodendrocytes that make this protein. It seems very likely that many morphologic cellular specializations may require the location of mRNAs in particular sites so that their translation products will be proximal to the specialized structures.

▲ **Figure 8-64** Site-specific localization of actin mRNA. Cells were fixed and exposed either to labeled poly U, which detects all mRNA *(top)*, or to a labeled single-stranded DNA probe complementary to actin mRNA *(bottom)*. In the autoradiographs, the silver grains representing the hybrids for total mRNA are generally distributed throughout the cell, whereas the grains representing actin mRNA are localized near the plasma membrane where the cell is most tightly anchored to the plate. [See J. B. Lawrence and R. H. Singer, 1986, *Cell 45:407.*] *Photographs courtesy of J. B. Lawrence and R. H. Singer.*

Summary

The RNA of eukaryotic cells is made in the nucleus by one of three distinct RNA polymerases. In almost all cases, the primary RNA transcript is not a functional RNA molecule; biochemical modification of the primary transcript is the rule.

The precursor to eukaryotic ribosomal RNA, the product of RNA polymerase I, is made in the nucleolus. In various eukaryotes, pre-rRNA varies in length from about 7 to 14 kb, and the 3′ terminus is a variable distance past the region encoding the rRNA. Pre-rRNA is cut to yield (1) 18S small rRNA (1.8–2.0 kb), (2) 26S–28S large rRNA (4–5 kb), and (3) 160-nucleotide 5.8S rRNA, which is base-paired to the 28S rRNA. In mammals, only about half of the pre-rRNA appears in cytoplasmic ribosomes. The 5′ end, at least two middle sections (spacer sequences), and the unused 3′ section of pre-rRNA are all discarded. In a few species, portions of the large 28S rRNA region are divided by an intervening sequence and must be spliced together after transcription and endonucleolytic cleavage. The sequences of the rRNA species found in the cytoplasm are moderately well conserved among various eukaryotic species, and their two-dimensional structures are highly conserved. However, the sequences just upstream of the initiation sites of ribosomal genes differ among species, and species-specific factors that participate in the initiation of pre-rRNA synthesis by polymerase I have been identified.

Ribosomal RNA processing begins with the addition of methyl groups (more than 100 in human pre-rRNA), and the nascent RNA combines with specific proteins as it is being formed. The first cleavage, which occurs within at most a few minutes of 3′-end completion, generates nucleolar 45S pre-mRNA. Within a minute or two the first 5′-end cleavage occurs, and the entire process of assembly of the ribosomal subunits in the nucleolus requires about 10 min for the small ribosomal subunit and about 30 to 40 min for the large subunit. The large and small subunits with their ribosomal proteins (numbering approximately 50 and 30, respectively) are delivered independently to the cytoplasm, where they enter into protein synthesis.

The products of RNA polymerase III are small RNA molecules—tRNAs and 5S rRNA, which are the two major types, and a host of other small RNAs many of which have a role in RNA processing. The primary transcripts formed from 5S rDNA undergo little or no modification, but all tRNA primary transcripts are highly modified. Most of the well-known tRNA modifications occur in the nucleus, and all tRNA transcripts are shortened from longer precursor molecules. About one-fourth of the primary transcripts for eukaryotic tRNAs are spliced during processing.

The sequences that are responsible for directing RNA polymerase III to begin proper transcription lie *within* 5S-rRNA and tRNA genes, and not upstream of them. A specific factor required for the transcription of 5S genes in frogs binds to nucleotides in positions 47–96 within the 120-nucleotide 5S gene. The two regions necessary for the initiation of transcription of tDNA encode the portion of tRNA that folds into the D loop and into the TΨCG loop in the finished tRNA. Thus these highly conserved sequences play a role in transcription and in cytoplasmic tRNA functioning as well.

The synthesis and processing of messenger RNA is carried out by RNA polymerase II and a multitude of nuclear factors. At least three types of recognition elements are found in DNA that is transcribed by RNA polymerase II: (1) the TATA box, which lies 30 bases upstream from the start site; (2) promoter-proximal sequences; and (3) enhancer sequences. Three general transcription factors (TF$_{II}$B, D, and E) required for binding of polymerase II and initiation of transcription have been partially purified.

Pre-mRNA undergoes more complex processing than any of the other RNAs. Heterogeneous nuclear RNA (hnRNA), the primary product of polymerase II, varies greatly in size, from shorter than 2 kb to longer than 200 kb. Some hnRNA molecules, but not all, are converted into functional mRNAs. A series of modifications are required to make mRNAs starting with formation of a 5′ cap, a 5′ → 5′ linkage between a methylated guanylate residue and the initial nucleotide of the primary transcript. All pre-mRNA molecules (except those for most histone mRNAs) are cleaved by an endonuclease, and then poly A, a chain of 100–250 adenylate residues, is added to the cleaved 3′ end by a terminal transferaselike poly A polymerase. The final step in the processing of many pre-mRNAs is splicing: the removal of one or more intervening RNA sequences (introns) and the joining of the remaining pieces (exons), which will appear in the finished mRNA product. In mammalian cells, over three-fourths of the mRNAs are spliced; probably fewer are spliced in invertebrate cells; and perhaps only 10–20 percent in single-celled eukaryotes.

In protein-coding primary transcripts, the sequences at the ends of the introns are always (5′)GU and (3′)AG, but the intron sequences farther from the splice junctions are not highly conserved. It is very likely that splice junctions are brought together for correct cutting and splicing first by small nuclear ribonucleoprotein structures (snRNPs) containing U1 and U2 snRNAs. Other snRNPs (U5 and a particle containing U4 and U6 snRNAs) also associate with the pre-mRNA to form a spliceosome, which helps excise the intron and rejoin the exons. A number of proteins also associate with hnRNA almost as soon as it is formed to make larger particles referred to as hnRNPs. These particles make the long hnRNA molecules more compact and probably are required to prevent tangles.

Two types of protein-coding transcription units exist. Simple transcription units encode one protein, and their primary transcripts may or may not have poly A or require splicing. Complex transcription units encode two or more proteins; these genes have one initiation site, but they contain multiple poly A addition sites or multiple variations in the sequences spliced to make mRNA or both. The selection of a particular mRNA from the several possibilities encoded by a complex transcription unit constitutes gene control at the posttranscriptional level.

Once completed, mRNA, rRNA, and tRNA quickly enter the cytoplasm to participate in protein synthesis. The mechanism or mechanisms of the transport of RNA to the cytoplasm are unknown at present. The nuclear pore—a specialized, octagonal protein array that is embedded in the nuclear envelope—is the likely passageway for molecules between the nucleus and the cytoplasm. There is little evidence as yet about the internal structure of the nucleus, but protein fibers that can be seen in whole mounts of the nucleus may represent a scaffolding around which nuclear events—including RNA synthesis, processing, and transport—may be organized.

References

Eukaryotic RNA Polymerases and Transcription Factors

DYNAN, W. S., and R. TJIAN. 1985. Control of eukaryotic messenger RNA synthesis by sequence-specific DNA-binding proteins. *Nature* 316:774–778.

GEIDUSCHEK, E. P. 1988. Transcription by RNA polymerase III. *Ann. Rev. Biochem.* 57:873–914.

HAHN, S., S. BURATOWSKI, P. A. SHARP, and L. GUARENTE. 1989. Isolation of the gene encoding the yeast TATA binding protein TF$_{II}$D: A gene identical to the SPT15 suppressor of Ty element insertions. *Cell* 58:1173–1181.

HORIKOHI, M., C. K. WANG, H. FUJII, J. A. CROMLISH, P. A. WEIL, AND R. G. ROEDER. 1989. Cloning and structure of a yeast gene encoding a general transcription initiation factor TF$_{II}$D that binds to the TATA box. *Nature* 341:299–303.

JONES, N. C., P. W. J. RIGBY, and E. B. ZIFF. 1988. Trans-acting protein factors and the regulation of eukaryotic transcription: lessons from studies on DNA tumor viruses. *Genes & Devel.* 3:267–281.

NAKAJIMA, N., M. HORIKOSHI, and R. G. ROEDER. 1988. Factors involved in specific transcription by mammalian RNA polymerase II: purification, genetic specificity, and TATA box-promoter interactions of TF$_{II}$D. *Mol. Cell Biol.* 8:4028–4040.

REINBERG, D., M. HORIKOSHI, and R. G. ROEDER. 1987. Factors involved in specific transcription in mammalian RNA polymerase II: functional analysis of initiation factors IIA and IID and identification of a new factor operating at sequences downstream of the initiation site. *J. Biol. Chem.* 262:3322–3330.

REINBERG, D., and R. G. ROEDER. 1987. Factors involved in specific transcription by mammalian RNA polymerase II: purification and functional analysis of initiation factors IIB and IIE. *J. Biol. Chem.* 262:3310–3321.

ROEDER, R. G. 1975. Multiple forms of deoxyribonucleic acid–dependent ribonucleic acid polymerase in *Xenopus laevis*. *J. Biol. Chem.* 249:241–248.

ZHENG, X.-M., V. MONCOLLIN, J.-M. EGLY, and P. CHAMBON. 1987. A general transcription factor forms a stable complex with RNA polymerase B (II). *Cell* 50:361–368.

Synthesis and Processing of rRNA

BELL, S. P., R. M. LEARNED, H.-M. JANTZEN, and R. TJIAN. 1988. Functional cooperativity between transcription factors UBF1 and SL1 mediates human rRNA synthesis. *Science* 241:11192–11197.

BORER, R. A., C. F. LEHNER, H. M. EPPENBERGER, and E. A. NIGG. 1989. Major nucleolar proteins shuttle between nucleus and cytoplasm. *Cell* 56:379–390.

CAEZERBUES-FERRER, M., P. MARIOTTINI, C. CURIE, B. LAPEYRE, N. GAS, F. AMALRIC, and F. AMALDI. 1989. Nucleolin from *Xenopus laevis*: cDNA cloning and expression during development. *Genes & Devel.* 3:324–333.

DARNELL, J. E., JR. 1968. Ribonucleic acids from animal cells. *Bacteriol. Rev.* 32:262–290.

GRUMMT, I., A. KUHN, I. BARTSCH, and H. ROSENBAUER. 1986. A transcription terminator located upstream of the mouse rDNA initiation site affects rRNA synthesis. *Cell* 47:901–911.

KARPEN, G. H., J. E. SCHAEFER, and C. D. LAIRD. 1988. A *Drosophila* rRNA gene located in euchromatin is active in transcription and nucleolus formation. *Genes & Devel.* 2:1745–1763.

KOWNIN, P., E. BATEMAN, and M. R. PAULE. 1987. Eukaryotic RNA polymerase I promoter binding is directed by protein contacts with transcription initiation factor and is DNA sequence-independent. *Cell* 50:693–699.

MADEN, B. E. H. 1988. Locations of methyl groups in 28S rRNA of *Xenopus laevis* and man: clustering in the conserved core of molecule. *J. Mol. Biol.* 201:289–314.

MILLER, O. L. 1981. The nucleolus, chromosome, and visualization of genetic activity. *J. Mol. Biol.* 91:15s–27s.

SAUERBIER, W., and K. HERCULES. 1978. Gene and transcription unit mapping by radiation effects. *Ann. Rev. Genet.* 12:329–363.

WORTON, R. G., J. SUTHERLAND, J. E. SYLVESTER, H. F. WILLARD, S. BODRUG, I. DUBE, C. DUFF, V. KEAN, P. N. RAY, and R. D. SCHMICKEL. 1988. Human ribosomal RNA genes: orientation of the tandem array and conservation of the 5' end. *Science* 239:64–68.

Synthesis of 5S rRNA and tRNA

BROWN, D. D., and J. B. GURDON. 1978. Cloned single repeating units of 5S DNA direct accurate transcription of 5S RNA when injected into *Xenopus* oocytes. *Proc. Nat'l Acad. Sci. USA* 75:2849–2853.

DELIHAS, N., and J. ANDERSON. 1982. Generalized structures of the 5S ribosomal RNAs. *Nucleic Acids Res.* 10:7323–7344.

GALLI, G., H. HOFSTETTER, and M. L. BIRNSTIEL. 1981. Two conserved sequence blocks within eukaryotic tRNA genes are major promoter elements. *Nature* 294:626–631.

GEIDUSCHEK, E. P. 1988. Transcription by RNA polymerase III. *Ann. Rev. Biochem.* 57:873–914.

LASSAR, A. B., P. L. MARTIN, and R. G. ROEDER. 1983. Transcription of class III genes: formation of preinitiation complexes. *Science* 222:740–748.

SAKONJU, S., D. F. BOGENHAGEN, and D. D. BROWN. 1980. A control region in the center of the 5S RNA gene directs specific initiation of transcription. I: the 5′ border of the region. *Cell* 19:13–25.

Transcription Unit Mapping for hnRNA and mRNA Precursors

BAAS, F., G.-J. B. VAN OMMEN, H. BIKKER, A. C. ARNBERG, and J. J. M. DE VIJLDER. 1986. The human thyroglobulin gene is over 300 kb long and contains introns of up to 64 kb. *Nucleic Acids Res.* 14:5171–5186.

DERMAN, E., S. GOLDBERG, and J. E. DARNELL. 1976. hnRNA in HeLa cells: distribution of transcription sizes estimated from nascent molecule profile. *Cell* 9:465–472.

FLINT, J. 1977. The topography and transcription of the adenovirus genome. *Cell* 10:153–166.

Processing of mRNA Precursors

DARNELL, J. E., JR. 1983. The processing of RNA. *Sci. Am.* 249:90–100.

NEVINS, J. R. 1983. The pathway of eukaryotic mRNA formation. *Ann. Rev. Biochem.* 52:441–466.

NEVINS, J. R., and J. E. DARNELL, JR. 1978. Steps in the processing of Ad2 mRNA: poly (A)+ nuclear sequences are conserved and poly(A) addition precedes splicing. *Cell* 15:1477–1493.

Capping

COPPOLA, J. A., A. S. FIELD, and D. S. LUSE. 1983. Promoter-proximal pausing by RNA polymerase II *in vitro*: transcripts shorter than 20 nucleotides are not capped. *Proc. Nat'l Acad. Sci. USA* 80:1251–1255.

ROZEN, F., and N. SONENBERG. 1987. Identification of nuclear cap–specific proteins in HeLa cells. *Nucleic Acids Res.* 15:6489–6500.

SHATKIN, A. J. 1976. Capping of eukaryotic mRNAs. *Cell* 9:645–654.

VENKATESAN, S., and B. MOSS. 1980. Donor and acceptor specificities of HeLa cell mRNA guanyl transferase. *J. Biol. Chem.* 255:2834–2842.

Poly A Addition and Termination of Transcription

BIRNSTIEL, M. L., M. BUSSLINGER, and K. STRUB. 1985. Transcription termination and 3′ processing: the end is in site! *Cell* 41:349–359.

FALCK-PEDERSON, E., J. LOGAN, T. SHENK, and J. E. DARNELL, JR. 1985. Transcription termination within the E1A gene of adenovirus induced by insertion of the mouse β-major globin terminator element. *Cell* 40:897–905.

FITZGERALD, M., and T. SHENK. 1981. The sequence 5′—AAUAAA—3′ forms part of the recognition site for polyadenylation of late SV40 mRNAs. *Cell* 24:251–260.

HIGGS, D. R., S. E. Y. GOODBOURN, J. LAMB, J. B. CLEGG, D. J. WEATHERALL, and N. J. PROUDFOOT. 1983. α-Thalassaemia caused by a polyadenylation signal mutation. *Nature* 306:398–400.

LOGAN, J., E. FALCK-PEDERSON, J. E. DARNELL, and T. SHENK. 1987. A poly(A) addition site and a downstream termination region are required for efficient cessation of transcription by RNA polymerase II in the mouse β^maj-globin gene. *Proc. Nat'l Acad. Sci. USA* 84:8306–8310.

MCDEVITT, M. A., G. M. GILMARTIN, and J. R. NEVINS. 1988. Multiple factors are required for poly(A) addition to a mRNA 3′ end. *Genes & Devel.* 2:588–597.

SCHAUFELE, F., G. M. GILMARTIN, W. BANNWARTH, and M. L. BIRNSTIEL. 1986. Compensatory mutations suggest that base-pairing with a small nuclear RNA is required to form the 3′ end of H3 messenger RNA. *Nature* 323:777–781.

ZEEVI, M., J. R. NEVINS, and J. E. DARNELL, JR. 1982. Newly formed mRNA lacking polyadenylic acid enters the cytoplasm and polyribosomes but has a shorter half-life in the absence of polyadenylic acid. *Mol. Cell Biol.* 2:517–525.

Splicing and Spliceosomes

BERGET, S. M., C. MOORE, and P. A. SHARP. 1977. Spliced segments at the 5′ terminus of adenovirus 2 late mRNA. *Proc. Nat'l Acad. Sci. USA* 74:3171–3175.

BREITBART, R. E., A. ANDREADIS, and B. NADAL-GINARD. 1987. Alternative splicing: a ubiquitous mechanism for the generation of multiple protein isoforms from single genes. *Ann. Rev. Biochem.* 56:467–495.

BROWN, J. W. S. 1986. A catalogue of splice junction and putative branch point sequences from plant introns. *Nucleic Acids Res.* 14:9549–9558.

CECH, T. R. 1983. RNA splicing: three themes with variations. *Cell* 34:713–716.

CHENG, S.-C., and J. ABELSON. 1987. Spliceosome assembly in yeast. *Genes & Devel.* 1:1014–1027.

CHOW, L. T., R. E. GELINAS, T. R. BROKER, and R. J. ROBERTS. 1977. An amazing sequence arrangement at the 5′ ends of adenovirus 2 messenger RNA. *Cell* 12:1–8.

GRABOWSKI, P. J., and P. A. SHARP. 1986. Affinity chromatography of splicing complexes: U2, U5, and U4 + U6 small nuclear ribonucleoprotein particles in the spliceosome. *Science* 233:1294–1299.

GUTHRIE, C., and B. PATTERSON. 1988. Spliceosomal snRNAs. *Ann. Rev. Genet.* 22:387–419.

LAMOND, A. I., M. M. KONARSKA, P. J. GRABOWSKI, and P. A. SHARP. 1988. Spliceosome assembly involves the binding and release of U4 small nuclear ribonucleoprotein. *Proc. Nat'l Acad. Sci. USA* 85:411–415.

LANGFORD, C. J., F.-J. KLINZ, C. DONATH, and D. GALLWITZ. 1984. Point mutations identify the conserved, intron-contained TACTAAC box as an essential splicing signal sequence in yeast. *Cell* 36:645–653.

PADGETT, R. A., P. J. GRABOWSKI, M. M. KONARSKA, S. SEIBER, and P. A. SHARP. 1986. Splicing of messenger RNA precursors. *Ann. Rev. Biochem.* 55:1119–1150.

PADGETT, R. A., S. M. MOUNT, J. A. STEITZ, and P. A. SHARP. 1983. Splicing of messenger RNA precursors is inhibited by antisera to small ribonucleoproteins. *Cell* 35:101–107.

RUSKIN, B., A. R. KRAINER, T. MANIATIS, and M. R. GREEN. 1984. Excision of an intact intron as a novel lariat structure during pre-mRNA splicing *in vitro*. *Cell* 30:317–331.

STEITZ, J. A. 1988. "Snurps." *Sci. Am.* 258:56–65.

WIEBAUER, K., J.-J. HERERO, and W. FILIPOWICZ. 1988. Nuclear pre-mRNA processing in plants: distinct modes of 3'-splice-site selection in plants and animals. *Mol. Cell Biol.* 8:2042–2051.

Trans-splicing

BRUZIK, J. P., K. VAN DOREN, D. HIRSH, and J. A. STEITZ. 1988. *Trans* splicing involves a novel form of small nuclear ribonucleoprotein particles. *Nature* 335:559–562.

HILDEBRAND, M., R. B. HALLICK, C. W. PASSAVANT, and D. P. BOURQUE. 1988. Trans-splicing in chloroplasts: the *rps*12 loci of *Nicotiana tabacum*. *Proc. Nat'l Acad. Sci. USA* 85:372–276.

KRAUSE, M., and D. HIRSH. 1987. A trans-spliced leader sequence on actin mRNA in *C. elegans*. *Cell* 49:753–761.

Transcription by RNA Polymerase II

Definition of Promoter Sites

BAKER, C. C., and E. B. ZIFF. 1980. Biogenesis, structures, and sites of encoding of the 5' termini of adenovirus-2 mRNAs. *Cold Spring Harbor Symp. Quant. Biol.* 44:415–428.

BUCHER, P., and E. N. TRIFONOV. 1986. Compilation and analysis of eukaryotic POL II promoter sequences. *Nucleic Acids Res.* 14:10009–10026.

BURATOWSKI, S., S. HAHN, L. GUARENTE, and P. A. SHARP. 1989. Five intermediate complexes in transcription initiation by RNA polymerase II. *Cell* 56:549–561.

CONTRERAS, R., and W. FIERS. 1981. Initiation of transcription by RNA polymerase(s)II in permeable SV40-infected or noninfected CV1 cells: evidence for multiple promoters of SV40 late transcription. *Nucleic Acids Res.* 9:215–236.

DAVISON, B. L., J-M. EGLY, E. R. MULVIHILL, and P. CHAMBON. 1983. Formation of stable preinitiation complexes between eukaryotic class B transcription factors and promoter sequences. *Nature* 301:680–686.

DYNAN, W. S. 1986. Promoters for housekeeping genes. *Trends Genet.* 2:196–197.

HAWLEY, D. K., and R. G. ROEDER. 1987. Functional steps in transcription initiation and reinitiation from the major late promoter in a HeLa nuclear extracts. *J. Biol. Chem.* 262:3452–3461.

WEIL, P. A., D. S. LUSE, J. SEGALL, and R. G. ROEDER. 1979. Selective and accurate initiation of transcription of the Ad2 major late promoter in a soluble system dependent on purified RNA polymerase II and DNA. *Cell* 18:469–484.

Definition of Enhancer Sites

BANERJI, J., S. RUSCONI, and W. SCHAFFNER. 1981. Expression of a β-globin gene is enhanced by remote SV40 DNA sequences. *Cell* 27:299–308.

FROMM, M., and P. BERG. 1982. Deletion mapping of DNA regions required for SV40 and early region promoter function *in vivo*. *J. Mol. Appl. Genet.* 1:457–481.

GLUZMAN, Y., and T. SHENK. 1983. *Enhancers and Eukaryotic Gene Expression*. Cold Spring Harbor Laboratory.

HORIKOSHI, M., T. HAI, Y.-S. LIN, M. R. GREEN, and R. G. ROEDER. 1988. Transcription factor ATF interacts with the TATA factor to facilitate establishment of a preinitiation complex. *Cell* 54:1033–1042.

SERFLING, E., M. JASIN, and W. SCHAFFNER. 1985. Enhancers and eukaryotic gene transcription. *Trends Genet.* 1:224–230.

Proteins Associated with hnRNA

BANDZIULIS, R. J., M. S. SWANSON, and G. DREYFUSS. 1989. RNA-binding proteins as developmental regulators. *Genes & Devel.* 3:431–437.

BEYER, A. L., and Y. N. OSHEIM. 1988. Splice site selection, rate of splicing, and alternative splicing on nascent transcripts. *Genes & Devel.* 2:754–765.

DREYFUSS, G., M. S. SWANSON, and S. PINOL-ROMA. 1988. Heterogeneous nuclear ribonucleoprotein particles and the pathway of mRNA formation. *Trends Biochem. Sci.* 13:86–91.

DREYFUSS, G., L. PHILIPSON, and I. W. MATTAJ. 1988. Ribonucleoprotein particles in cellular processes. *J. Cell Biol.* 106:1419–1425.

LERNER, M. R., and J. A. STEITZ. 1981. Snurps and scyrps. *Cell* 25:298–300.

PEDERSON, T. 1983. Nuclear RNA–protein interactions and messenger RNA processing. *J. Cell Biol.* 97:1321–1326.

REDDY, R., and H. BUSCH. 1981. U snRNA's of nuclear snRNP's. In *The Cell Nucleus*, vol. 8. Academic Press.

SKOGLUND, U., K. ANDERSSON, B. STRANDBERG, and B. DANEHOLT. 1986. Three-dimensional structure of a specific pre-messenger RNP particle established by electron microscope tomography. *Nature* 319:560–564.

Nuclear Pores: The Presumed Exit Sites for RNA

GERACE, L., and G. BLOBEL. 1980. The nuclear envelope laminia is reversibly depolymerized during mitosis. *Cell* 19:277–287.

HUTCHISON, N., and H. WEINTRAUB. Localization of DNase I–sensitive sequences to specific regions of interphase nuclei. *Cell* 43:471–482.

MAITRA, U., E. A. STRINGER, and A. CHAUDHURI. 1982. Initiation factors in protein biosynthesis. *Ann. Rev. Biochem.* 51:869–900.

NEWMEYER, D. D., and D. J. FORBES. 1988. Nuclear import can be separated into distinct steps in vitro: nuclear pore binding and translocation. *Cell* 52:641–653.

UNWIN, P. N. T., and R. A. MILIGAN. 1982. A large particle associated with the perimeter of the nuclear pore complex. *J. Cell Biol.* 93:63–75.

Association of mRNA with the Cytoskeleton

CERVERA, M., G. DREYFUSS, and S. PENMAN. 1981. Messenger RNA is translated when associated with the cytoskeletal framework in normal and VSV-infected HeLa cells. *Cell* 23:113–120.

LAWRENCE, J. B., and R. H. SINGER. 1986. Intracellular localization of messenger RNAs for cytoskeletal proteins. *Cell* 45:407–415.

9

The Structure of Eukaryotic Chromosomes

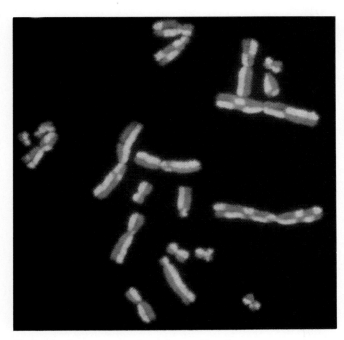

Harlequin chromosomes: sister-chromatid exchanges

The preceding chapter described how mRNAs, rRNAs, and tRNAs are produced from transcription units in eukaryotic DNA. In the next two chapters we examine the structures that contain the DNA— the eukaryotic chromosomes. Viewed in the light microscope, chromosomes are morphologically distinctive, species-specific structures. At the molecular level, chromosomes can be regarded as an assembly of transcription units that are precisely duplicated in each cell generation. Ever since Sutton hypothesized in 1902 that chromosomes were the carriers of Mendel's independently segregating genes, studies of how the structure of the chromosome is related to its function have been at the heart of biological research. With the aid of modern cloning and automatic sequencing techniques, it is virtually certain that the sequence of the entire human genome and probably much of that of experimentally important animals will be available by the end of this century. The knowledge of genome sequences, coupled with gene transfer techniques for investigating the function of DNA segments, is already beginning to provide answers to many long-standing questions about the relation of chromosome structure to gene function. Among the questions of greatest interest are the following:

What is the relation of eukaryotic transcription units to the traditional functional chromosome unit, the gene?

How are transcription units in chromosomal DNA arranged in relation to other sequences that are not transcribed?

Which genes are present in multiple copies?

What is the function (if any) of DNA sequences that are not transcribed?

What is the nature of the so-called repetitive DNA?

How stable is the arrangement of DNA in chromosomes?

What proteins are associated with the DNA in chromosomes, and what is the nature of the DNA-protein complexes?

The other traditional problems that have attracted students of chromosomes concern replication and repair of DNA:

At what site or sites in chromosomes does replication begin?

What enzymes and other proteins participate in replication?

How is a chromosome repaired after it suffers damage from chemicals or from radiation?

What are the molecular events in the DNA at the site of recombination?

This chapter provides an introduction to the structure of eukaryotic chromosomes and a definition of eukaryotic genes. Molecular studies that are largely based on the cloning and sequencing of segments of eukaryotic chromosomes are described in Chapter 10. In that chapter, we also discuss the surprising finding that eukaryotic chromosomes have a marked instability because of transpos-

able elements. In Chapter 11 we integrate current knowledge about eukaryotic transcription with what is known about gene and chromosome structure to develop a detailed picture of eukaryotic gene control. Chapter 12 deals with the dynamic events of DNA biochemistry—replication, repair, and recombination. ▲

Morphology and Functional Elements of Eukaryotic Chromosomes

Microscopic observations on the number and size of chromosomes and their staining patterns led to the discovery of many important general characteristics of chromosome structure. In this section we discuss chromosome morphology and the DNA elements required to assure duplication and segregation of chromosomes.

Chromosome Number and Shape Are Species-specific

In nondividing cells the chromosomes are not visible, even with the aid of histological stains for DNA (e.g., Feulgen or Giemsa stains); during mitosis and meiosis, however, the chromosomes condense and become visible in the light microscope. Therefore, almost all cytogenetic work has been done with condensed metaphase chromosomes obtained from dividing cells—either somatic cells in mitosis or dividing gametes during meiosis. At the time of cell division, chromosomes are duplicated structures:

(b)

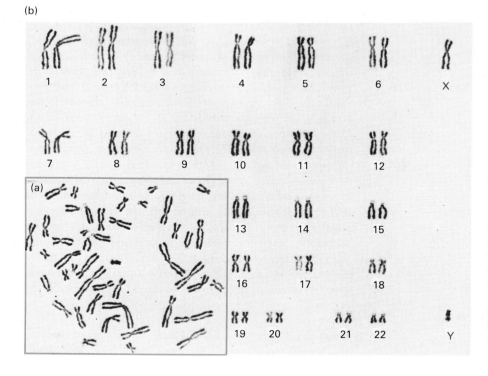

◀ **Figure 9-1** Karyotype of human male metaphase chromosomes. (a) The chromosomes from one mitotic cell, a lymphocyte, were stained with orcein (a dye that reacts with DNA), and the cell was "squashed" on a microscope slide. (b) After indentification of each pair of homologues, photographs of the chromosomes were mounted in sequence by size. Each number indicates a homologous pair, and the entire set is the karyotype. The unnumbered, unpaired X and Y chromosomes are the sex chromosomes. A metaphase chromosome consists of two identical sister chromatids; their point of association is the centromere. According to the position of the centromere, the chromosome is termed metacentric (centromere in the middle; e.g., chromosome 1), acrocentric (asymmetrical centromere; e.g., chromosome 4), or telocentric (centromere at an end; e.g., chromosome 14). *Courtesy of J. German.*

each metaphase chromosome consists of two *sister chromatids,* which are separated along their lengths except at one point of attachment, the *centromere.* The number, sizes, and shapes of the metaphase chromosomes constitute the *karyotype,* which is distinctive for each species (Figure 9-1). In most organisms, all cells have the same karyotype. However, species that appear quite similar can have very different karyotypes indicating that similar genetic potential can be arranged in very different ways (Figure 9-2; Table 9-1).

Cellular DNA Content Does Not Correlate with Phylogeny

The total amount of DNA in the set of chromosomes from different animals or plants does not vary in a consistent manner with the apparent complexity of the organisms. Yeasts, fruit flies, chickens, and humans have successively larger amounts of DNA in their haploid chromosome sets (0.015, 0.15, 1.3, and 3.2 picograms, respectively), in keeping with what we perceive to be the

(a)

(b)

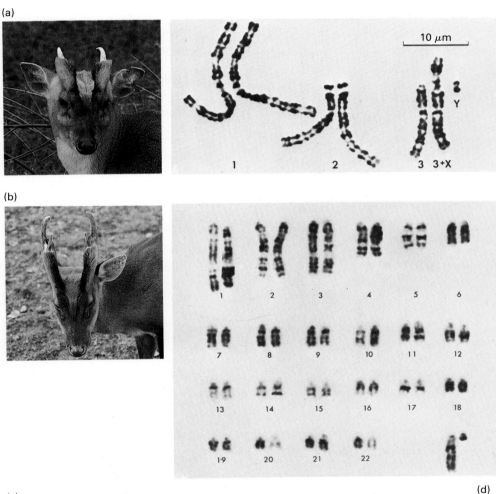

(c)

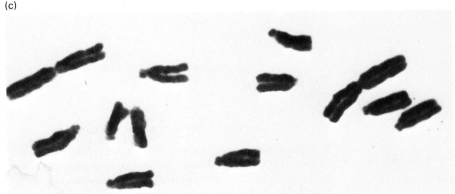

(d)

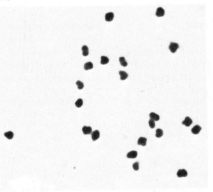

◀ **Figure 9-2** Karyotypes of the Reeves muntjac (a) and the Indian muntjac (b), two species of small deer that are quite similar but do not interbreed. Despite the difference in the number of chromosomes in these animals, the two genomes contain about the same total amount of DNA. (The chromosomes for both deer are shown at the same magnification.) Similar plants can also have strikingly different karyotypes. The broad bean *Vicia faba* (c) has about half the number of chromosomes as the kidney bean *Phaseolus vulgaris* (d). However, the broad bean has about three to four times as much DNA per cell as the kidney bean. (Again, the chromosomes in these two photographs are shown at the same magnification.) Part (a) *(left)* courtesy of K. W. Fink/ Photo Researchers, Inc.; *(right)* courtesy of R. Church. Part (b) *(left)* courtesy of J. P. Ferrero/ Jacana/Photo Researchers, Inc.; *(right)* courtesy of R. Church. *Parts (c) and (d) courtesy of M. D. Bennett; from M. D. Bennett, 1976,* Environ. Exp. Bot. *16:93.*

Table 9-1 Chromosome numbers of various species*

Common name	Species	Diploid number
ANIMALS		
Human	*Homo sapiens*	46
Rhesus monkey	*Macaca mulatta*	42
Cattle	*Bos taurus*	60
Dog	*Canis familiaris*	78
Cat	*Felis domesticus*	38
Horse	*Equus calibus*	64
House mouse	*Mus musculus*	40
Rat	*Rattus norvegicus*	42
Golden hamster	*Mesocricetus auratus*	44
Guinea pig	*Cavia cobaya*	64
Rabbit	*Oryctolagus cuniculus*	44
Chicken	*Gallus domesticus*	78 ±
Alligator	*Alligator mississipiensis*	32
Frog	*Rana pipiens*	26
Carp	*Cyprinus carpio*	104
Silkworm	*Bombyx mori*	56
House fly	*Musca domestica*	12
Fruit fly	*Drosophila melanogaster*	8
Flatworm	*Planaria torva*	16
Freshwater hydra	*Hydra vulgaris attenuata*	32
Nematode	*Caenorhabditis elegans*	11/12

(male/female)

PLANTS AND FUNGI (2*n*)		
Yeast	*Saccharomyces cerevisiae*	36 ±
Green algae	*Acetabularia mediterranea*	20 ±
Garden onion	*Allium cepa*	16
Barley	*Hordeum vulgare*	14
Wheat	*Triticum monoccum*	24
Corn	*Zea mays*	20
Snapdragon	*Antirrhinum majus*	16
Tomato	*Lycopersicon esculentum*	24
Tobacco	*Nicotinana tabacum*	48
Kidney bean	*Phaseolus vulgaris*	22
Pine	*Pinus* species	24
Garden pea	*Pisum sativum*	14
Potato	*Solanum tuberosum*	48
Broad bean	*Vicia faba*	12

		Haploid number
PLANTS AND FUNGI (1*n*)		
Slime mold	*Dictyostelium discoideum*	7
Mold (fungus)	*Aspergillus nidulans*	8
Pink bread mold	*Neurospora crassa*	7
Penicillin mold	*Penicillium* species	4
Green algae	*Chlamydomonas reinhardtii*	16

*In most organisms the chromosome number does not vary, but in some species (those with numbers listed as ±) either there is a natural variation in strains or the correct number has not been determined. In a few organisms the sexes differ; in nematodes, the male is haploid for the sex chromosome (X,O), and the female is diploid (X,X).

SOURCE: M. Strickberger, 1985, *Genetics*, 3d ed., Macmillan; and R. B. Flavell et al., 1974, *Biochem. Genet.* **12**:257.

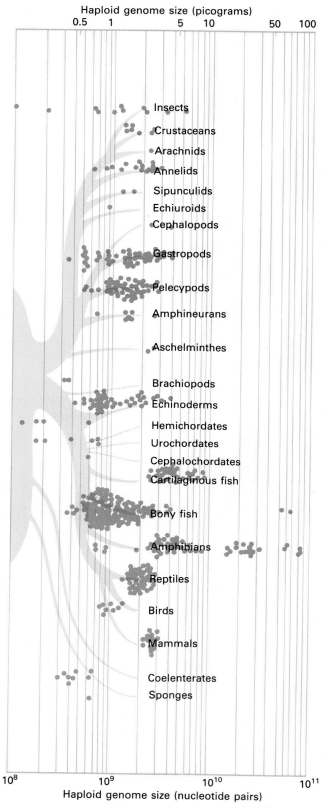

▲ **Figure 9-3** The amount of DNA per haploid chromosome set in a variety of animals. Variation within certain phyla (e.g., insects and amphibians) is quite wide. [See R. J. Britten and E. Davidson, 1971, *Quart. Rev. Biol.* **46**:111.]

increasing complexity of these organisms. Yet the vertebrate species with the greatest amount of DNA per cell are amphibians (Figure 9-3), which are surely less complex than humans in their structure and behavior. Many plant species also have considerably more DNA per cell than humans have. For example, the DNA content per cell of wheat, broad beans, and garden onions (7.0, 14.6, and 16.8 picograms, respectively) ranges from about two to more than five times that of humans, and tulips have ten times as much DNA per cell as humans.

There is considerable intragroup variation in DNA content. All insects or all amphibians would appear to be similarly complex, but the amount of haploid DNA within each of these phyla varies by a factor of 100. The same variation in DNA content per cell is common within groups of plants that have similar structures and life cycles. For example, the broad bean contains about three to four times as much DNA per cell as the kidney bean.

These facts suggest that some of the DNA in certain organisms is "extra" or expendable—that is, it does not encode necessary functions. The total amount of DNA per haploid cell in an organism is referred to as the *C value;* the failure of C values to correspond to phylogenetic complexity is called the *C value paradox.* This perplexing variation in genome size occurs mainly because eukaryotic chromosomes contain variable amounts of repeated DNA stretches, some of which are never transcribed and most all of which are likely dispensable.

Stained Chromosomes Have Characteristic Banding Patterns

Certain dyes selectively stain certain regions of chromosomes more intensely than others, producing banding patterns that are specific for individual chromosomes. When stained chromosomes are viewed microscopically, these bands serve as landmarks along the length of each chromosome and also help to distinguish chromosomes of similar size and shape.

Banding in Metaphase Chromosomes Quinacrine, a fluorescent dye that inserts (*intercalates*) into the DNA helix, produces *Q bands.* However, because Q bands fade with time, other staining techniques are generally preferred in the laboratory. For example, chromosomes can be subjected briefly to mild heat or proteolysis and then stained with Giemsa reagent, a permanent DNA dye, to produce a pattern of *G bands* (Figure 9-4). Treatment of chromosomes with a hot alkaline solution before staining with Giemsa reagent produces *R bands* in a pattern that is approximately the reverse of the G-band pattern. The distinctiveness of these banding patterns permits cytologists to identify the specific parts of a chromosome and to locate the sites of chromosomal breaks and rearrangements. In addition, cloned DNA probes that have hybridized to specific sequences in the chromosomes can be located in particular bands.

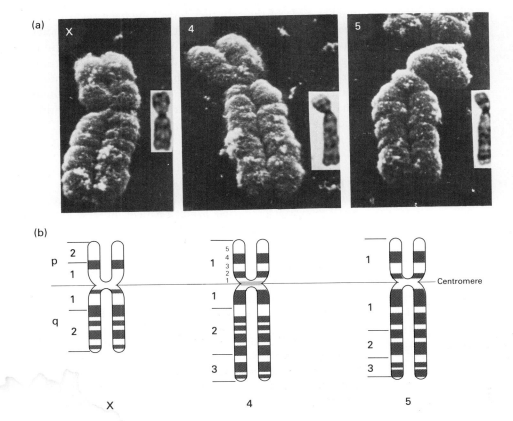

◀ **Figure 9-4** Human male chromosomes X, 4, and 5 after G-banding. The chromosomes were subjected to brief proteolytic treatment and then stained with Giemsa reagent, producing distinctive bands at characteristic places. (a) Scanning electron micrographs show constrictions at the sites where bands appear in light micrographs *(insets).* (b) Standard diagrams of G bands (purple). Regions of variable length (green) are most common in the centromere region (e.g., on chromosome 4). The numbering system to indicate position of the chromosomes is illustrated: p = short arm, q = long arm. Each arm is divided into major sections (1, 2, etc.), and subsections. The short arm of chromosome 4, for example, has 5 subsections. DNA located in the fourth section would be said to be in p14. *Part (a) courtesy of C. J. Harrison; from C. J. Harrison et al., 1981,* Exp. Cell Res. *134:141.*

The molecular basis for the regularity of chromosomal bands remains unknown, although the R bands may have a high content of repetitive DNA sequences. Because chromosomal bands are not visible in isolated DNA, they are presumed to be due to folding produced by the interaction of DNA and proteins. Human chromosomes contain from 10^8 to 3×10^8 base pairs; a single band represents about 5–10 percent of a chromosome, or about 10^7 base pairs. The constancy of banding patterns over such large DNA regions (the patterns are nearly identical in almost every copy of a chromosome, in almost every tissue) implies a constancy of DNA-protein interactions at specific sites within the regions.

Banding in Interphase Polytene Chromosomes

For over 50 years, cytologists have observed stainable bands in interphase chromosomes in the salivary glands of *Drosophila melanogaster* and other dipteran insects (Figure 9-5a). Although the bands in human chromosomes probably represent very long folded or compacted stretches of DNA containing about 10^7 base pairs, the bands in *Drosophila* chromosomes represent much shorter stretches of only 50,000–100,000 base pairs.

The detailed banding of insect salivary gland chromosomes occurs because of a peculiar structural feature of these chromosomes. The presumably single duplex DNA molecule in insect salivary chromosomes is repeatedly copied into parallel arrays of as many as 1000 identical DNA molecules. This amplification without separation, termed *polytenization,* results in thick bundles of parallel DNA molecules that all have the same banding pattern across the width of the bundle (Figure 9-5b). Recent molecular cloning experiments and mRNA mapping studies have suggested that each band contains a limited number of transcription units, perhaps only one unit in some cases. Since transcription units can also be found in interband stretches of DNA, however, the relationship between banding and function is still unclear.

Heterochromatin Consists of Chromosome Regions That Do Not Uncoil

As cells exit from mitosis and the chromosomes uncoil, certain sections of the chromosomes remain dark-staining. The dark-staining areas, termed *heterochromatin,* are regions of condensed chromatin (i.e., DNA plus proteins). Heterochromatin appears most frequently—but not exclusively—at the centromere and telomeres (ends) of the chromosome. The light-staining, less condensed portions are called *euchromatin.* Because heterochromatic regions apparently remain condensed throughout the life cycle of the cell, they have long been regarded as sites of inactive genes. In recent years, molecular studies have shown that most of the DNA in heterochromatin is highly repeated DNA that is never (or very seldom) transcribed. However, not all inactive genes and nontranscribed regions of DNA are visible as heterochromatin; and it is not necessarily true that no transcription occurs within heterochromatin.

(a)

Chromocenter

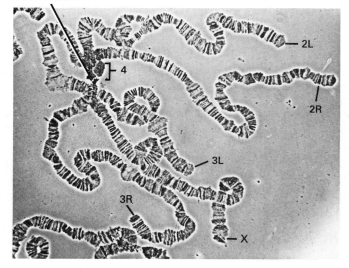

(b)

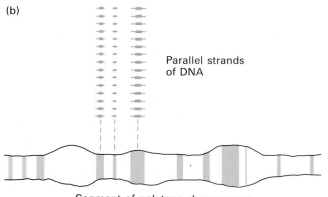

Parallel strands of DNA

Segment of polytene chromosome

◀ **Figure 9-5** (a) Light micrograph of polytene salivary gland chromosomes from *Drosophila melanogaster* stained to reveal the very reproducible banding pattern. In these four chromosomes (X, 2, 3, and 4), a total of approximately 5000 bands can be distinguished. The centromeres of all four chromosomes often appear fused at the *chromocenter;* the smallest chromosome (4), a dot chromosome, is also associated with the chromocenter. The tips of the metacentric 2 and 3 chromosomes are labeled (L = left arm; R = right arm). The tip of the acrocentric X chromosome is also labeled. (b) The DNA in salivary polytene chromosomes is repeated about 1000 times; the duplicated DNA fibers are thought to remain parallel. Therefore, any staining property in a chromosome with one DNA duplex would be amplified 1000 times to produce a transverse band (light blue). *Part (a) courtesy of J. Gall.*

The Inactive X Chromosome in Mammalian Females One important case of heterochromatinization that does correlate with gene inactivation is the random inactivation and condensation of one of the two female sex chromosomes (the X chromosomes) in virtually all the somatic cells of female mammals. The inactive X, which appears as heterochromatin throughout the cell cycle, is visible during interphase as a dark-staining, peripheral nuclear structure called the *Barr body* after its discoverer, the Canadian cytologist M. L. Barr. The process of X inactivation, often termed *lyonization* after the British cytogeneticist Mary Lyons, who first recognized the phenomenon, can lead to important genetic consequences in individuals who have a defective gene on one of their two X chromosomes.

Each cell in a female has two X chromosomes, one of maternal origin (X_m) and one of paternal origin (X_p). Early during embryologic development, inactivation of either the X_m or the X_p chromosome occurs in each cell. The female embryo thus becomes a mosaic of cells in which about half have an inactive X_m and the other half have an inactive X_p. All subsequent daughter cells maintain the same inactive X chromosomes as their parent cells. As a result, the adult female is a patchwork of clones, some expressing the genes on the X_m chromosome and the rest expressing the genes on the X_p.

The effects of X inactivation are evident in the expression of coat color in cats. Male cats, which normally have only one X chromosome, can be black or yellow, but they are almost never both. In contrast, certain females, called *calicos*, have coats that are patches of black and yellow. The two colors have been traced to a gene on the X chromosome. The males have either the black (Bl^+) or the yellow (Bl^-) allele (alternative form) of this gene, whereas the calicos are females that are heterozygous (Bl^+/Bl^-). The black patches on the heterozygotes are produced by clones in which the active X chromosome carries the Bl^+ allele, and the yellow patches mark clones in which the active X has the Bl^- allele. The rare calico male has a sex chromosome constitution of XXY, in which one X is also inactivated.

Recent studies have shown that the genes on an inactive X chromosome can be reactivated by preventing methylation of the DNA. From this result, it has been inferred that a high degree of methylation is one of the changes related to the inactivation of X chromosomes.

Each Chromosome Contains One Linear DNA Molecule

All DNA viruses seem to possess one molecule of DNA, and bacterial cells contain a single chromosomal DNA molecule. The general belief—which has not yet been completely proved—is that all eukaryotic chromosomes also contain a single long DNA molecule. Because the longest DNA molecules in human chromosomes are al-most 10 cm long (2 to 3×10^8 base pairs), they are difficult to handle experimentally without breaking. However, in lower eukaryotes, the sizes of the largest DNA molecules that can be extracted are consistent with the hypothesis that a chromosome contains a single DNA molecule. For example, physical analysis of the largest DNA molecules extracted from several genetically different *Drosophila* species and strains shows that they are from 6×10^7 to 1×10^8 base pairs long. These sizes match the DNA content of single stained metaphase chromosomes of *Drosophila melanogaster,* as measured by the amount of DNA-specific stain absorbed. Therefore, each chromosome probably contains a single DNA molecule. The correspondence between the number of DNA molecules per cell and the number of chromosomes has been conclusively demonstrated in yeast cells. All the DNA in a yeast cell is located in molecules that can be individually identified by pulse gel electrophoresis. The number of molecules is equal to the number of linkage groups (i.e., chromosomes) in the cell. The length of yeast chromosomes ranges from about 1.5×10^5 to 3×10^6 base pairs.

Mitochondrial DNA, chloroplast DNA, bacterial chromosomal DNAs, and many viral DNAs are circular molecules, whereas the DNA in eukaryotic chromosomes is linear. Not only do the two chromatids at meiosis look linear but the genetic maps of all normal eukaryotic chromosomes are linear. These maps are based on recombination frequencies (in meiosis) between genes on homologous chromosomes. Furthermore, nonhomologous chromosomes occasionally undergo *translocations*—that is, they break and exchange large pieces—which can be followed in successive generations. The effects of translocations on banding patterns in some species are visible in the light microscope. As illustrated in Figure 9-6, translocation between two nonhomologous chromosomes exchanges a set of linearly distributed bands on one chromosome with a set of linearly distributed bands on the other. Exchange can also occur between sister chromatids in single chromosomes of cells undergoing mitosis; this phenomenon is fairly frequent in cultured mammalian cells. Such exchanges occur in a reciprocal linear fashion similar to that in translocation. Translocations of large blocks of chromosomal DNA have obviously occurred during evolution. A comparison of mouse chromosomes and human chromosomes makes this point clear (Fig. 9-6c). Many human diseases due to genetic defects have allowed the mapping of defective genes to individual chromosomes. Similarly, classic mouse genetics based on breeding studies is quite advanced. Recent progress is mapping individual genes with molecular probes combined with these classic studies has allowed a detailed comparison of the location of the same genes in the two species. The result shows long chromosomal stretches in which the same genes in the same order are found in the two species. The example shown is chromosome 1 in

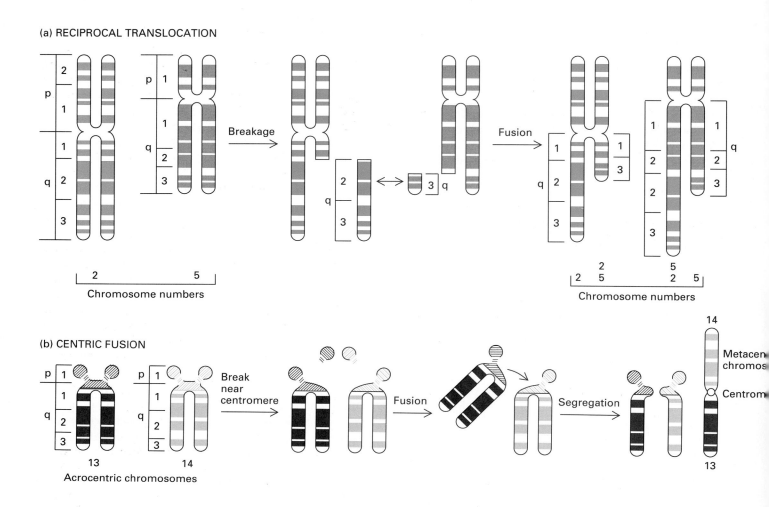

(a) RECIPROCAL TRANSLOCATION

(b) CENTRIC FUSION

(c)

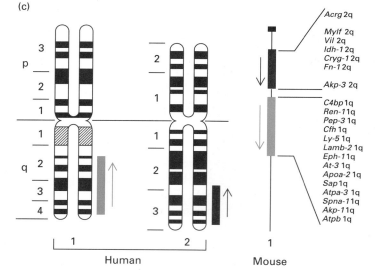

▲ **Figure 9-6** Microscopic analysis of certain human chromosomal abnormalities, which occur in a small percentage of the population, indicates that chromosomal bands are linearly arranged along the chromosomes. (a) A *reciprocal translocation* involving human chromosomes 2 and 5. In such rearrangements, parts of two different chromosomes break off and become attached to the other. If no part of either chromosome is lost in the exchange, such a translocation may not cause any phenotypic abnormality. (b) Another type of translocation, *centric fusion* (also called Robertsonian fusion), involving human chromosomes 13 and 14. Centric fusions occur only between acrocentric chromosomes. Breakage occurs within the centromeric regions of two acrocentric chromosomes and one of the chromatids in each pair of the metaphase chromosomes is joined (13 and 14 in this case). After segregation a new metacentric chromosome (fused 13-14) exists. The regions near the centromere contain repetitive sequences (*diagonal lines;* see Chapter 10), and that is why such fusions occur. (c) Homology between mouse chromosome 1 and parts of human chromosomes 1 and 2. Genetic mapping of mouse chromosomes by recombination places the genes listed in the order shown on mouse chromosome 1. Many of the same genes have been positioned in the same grouping in human chromosomes 1 and 2 mainly by in situ hybridization with molecular probes to two blocks of the

chromosomal DNA. The genes listed include, for example, *Acrg*, acetylcholine receptor, gamma subunit; *Ren*, renin; and *spna*, spectrin, α subunit. The arrows indicate gene order. [See J. J. Yunis, 1976, *New Chromosomal Syndromes*, Academic Press; and J. H. Nadeau, J. T. Epping, and A. H. Reiner, 1989, *Linkage and Synteny Homologies between Mouse and Man*, Jackson Laboratory.]

mouse which contains parts of chromosomes 1 and 2 in humans. Many other such *syntenic* stretches exist between mice and humans, indicating that our chromosomes are a patchwork of recombined segments from our mammalian forbears. Although fewer genes have been located in other primate chromosomes, from what is known, human and chimpanzee chromosomes are quite similar in most gene locations. Thus the eukaryotic chromosome behaves genetically, and appears visually, as a linear structure containing a single DNA duplex.

Human Chromosomes Can Be Mapped Based on Restriction Fragment Length Polymorphisms (RFLPs)

The existence of distinctive chromosomal banding patterns provides a basis for physical maps of chromosomes. In the case of human chromosomes, however, mapping based on banding patterns is very rough because each Q or R band contains a relatively large proportion (5–10 percent) of a chromosome. And in situ hybridization, with labeled nucleic acid probes prepared from cloned

DNA, can locate genes only in a general region of a chromosome. In recent years, developments in molecular genetics have provided new methods of physically mapping genetically inherited sites. The use of these techniques has already resulted in maps with markers spaced approximately 0.5 to 2.5×10^7 base pairs apart.

The first of these genetic maps of human chromosomes was constructed by taking almost 2000 separate cloned DNA segments and determining by Southern blot analysis whether differences between two DNA samples from a group of families were similar or not. In a Southern blot analysis, a *restriction endonuclease* that recognizes a specific base sequence of 4–8 nucleotides in length is used to digest a DNA sample; the resulting fragments are then separated by electrophoresis and reacted with a labeled cloned DNA sample (the probe) to determine the size of the hybridizing fragments. If several restriction enzymes are used to cut the DNA, differences among samples from different individuals usually can be found because the enzymes are sequence-dependent, and a change in sequence at the restriction site results in different sized fragments (Figure 9-7a). These are called *restriction fragment*

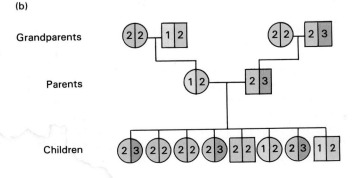

(a) **Chromosomal arrangement**

Hybridization banding pattern

Mutation at site a_2 prevents cleavage

↓ Restriction endonuclease A

〰 Restriction endonuclease B

▨ Probed single-copy region

(b)

Grandparents

Parents

Children

◀ **Figure 9-7** Analysis of restriction fragment length polymorphisms (RFLPs). (a) In the example shown, DNA is treated with two different restriction enzymes, which cut DNA at different sequences. The resulting fragments are subjected to Southern blot analysis; that is, they are separated by electrophoresis, transferred to nitrocellulose, and detected by hybridization with a radioactive probe, which binds to the indicated DNA region (green). Since no differences between the two chromosomes occur in the sequences recognized by the B enzyme, only one fragment is produced, as indicated by a single hybridization band. However, treatment with enzyme A produces fragments of two different lengths (two bands are seen), indicating that a mutation has caused the loss of one of the *a* sites. (b) RFLP analysis of the DNA from eight children, their parents, and grandparents detected the presence of three alleles for a region known to be present on chromosome 5. The DNA samples were cut with the restriction enzyme *Taq*I and analyzed by the Southern blot procedure. In this family, this region exists in three allelic forms characterized by *Taq*I sites spaced 10, 7.7, or 6.5 kb apart. Each individual has two alleles; some contain allele 2 (7.7 kb) on both chromosomes, and others are heterozygous at this site. *After H. Donis-Keller et al., 1987, Cell **51**:319.*

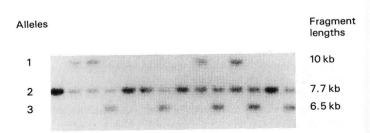

Alleles

1

2

3

Fragment lengths

10 kb

7.7 kb

6.5 kb

length polymorphisms (RFLPs). When such a polymorphism is found, it can be traced within a family and shown to be inherited in a Mendelian fashion (Figure 9-7b).

By now well over a thousand RFLPs have been identified in human DNA (and more are reported each month), providing a large number of physically identified genetic markers. Inheritance of the polymorphisms for these markers in families where DNA is available for three generations allows linkage maps between different RFLPs to be established. Computer comparisons of the linkage data are often used to determine which chromosome contains the DNA exhibiting a particular RFLP (Figure 9-8). Because many already identified genes in humans have been mapped to chromosomes or parts of chromosomes, further analysis of the segregation of RFLPs with known human traits can often position the RFLP relative to a known gene on a particular chromosome. For example, the gene responsible for *cystic fibrosis*, a disease that results in chronic infections because of poor respiratory tract draining, was known from linkage studies to reside on human chromosome 7. The locations of RFLPs on either side of the gene, coupled with molecular cloning techniques, allowed this gene to be cloned even though the biochemical nature of the protein was unknown. Chromosomal positions can also be determined by using labeled cloned DNA probes bearing RFLP sites as a probe in in situ chromosomal hybridization.

Autonomously Replicating Sequences, Centromeres, and Telomeres Are Required for Replication and Stable Inheritance of Chromosomes

So far we have emphasized that eukaryotic chromosomes are linear structures composed of single DNA molecules. Although chromosomes differ in length and number among species, they all behave similarly at the time of cell division. At mitosis the attached sister chromatids become aligned on the metaphase plate; they then separate at the centromeres, and one chromatid of each metaphase chromosome is distributed to each daughter cell. Recent recombinant DNA research with yeast cells has identified all of the chromosomal elements that are necessary for equal segregation of sister chromatids to occur. The culmination of this work has been construction of artificial yeast chromosomes. In order to duplicate and segregate correctly, chromosomes must contain three functional elements: (1) special sequences involved in the initiation of DNA replication; (2) the centromere; and (3) the two ends, or telomeres.

Autonomously Replicating Sequences

If yeast cells lack a particular gene (e.g., one of the genes that encode an enzyme for synthesis of the amino acid leucine), they can be transfected with cloned plasmids containing the

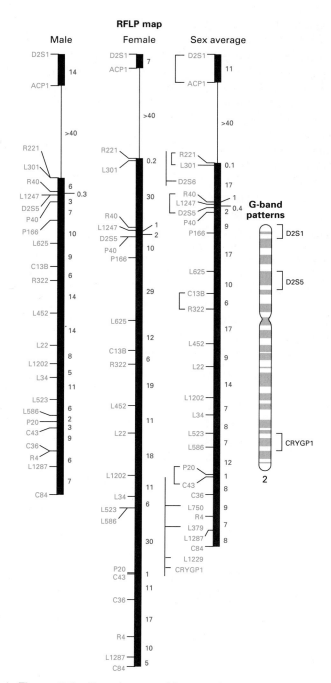

▲ **Figure 9-8** Genetic map of human chromosome 2 based on Southern blot analysis of restriction fragment length polymorphisms (RFLPs) and crossover frequencies during meiosis. The location of the RFLPs is shown by blue numbers on the left. The red numbers on the right show crossover frequencies (%) by adjacent markers. Thin lines near the tops of the maps represent markers that are sufficiently distant that they appear to be unlinked, but are known to be on chromosome 2. As in other species, crossing over occurs at different locations in males and females. Comparison of this map with the G-band pattern of chromosome 2 illustrates the greater sensitivity of RFLP analysis for mapping genetic loci. A map similar to this is available for every human chromosome. *After H. Donis-Keller et al., 1987, Cell 51:328.*

missing DNA from normal cells (in this case, a *LEU* gene). Mutant (Leu⁻) yeast cells transfected with a circular plasmid containing the *LEU* gene will grow in the absence of leucine if the plasmid contains *autonomously replicating sequences* (Figure 9-9a). The role of these sequences, designated ARS, as initiation sites for DNA replication is discussed in Chapter 12.

Centromeres In any culture of leucine-requiring yeast cells transfected with a simple circular *LEU*-ARS-containing plasmid, only about 5–20 percent of progeny

cells contain the plasmid because mitotic segregation of the plasmids is faulty. However, if random bits of genomic DNA are cloned into such circular plasmids, specific centromere regions, called CEN sequences, can be detected that confer equal segregation at mitosis (Figure 9-9b). Such cloning experiments have led to the recovery of sequences that improve mitotic segregation to the extent that over 90 percent of the cells in a culture contain the *LEU* plasmids. These cells and almost all of their descendants therefore grow well in a medium lacking leucine.

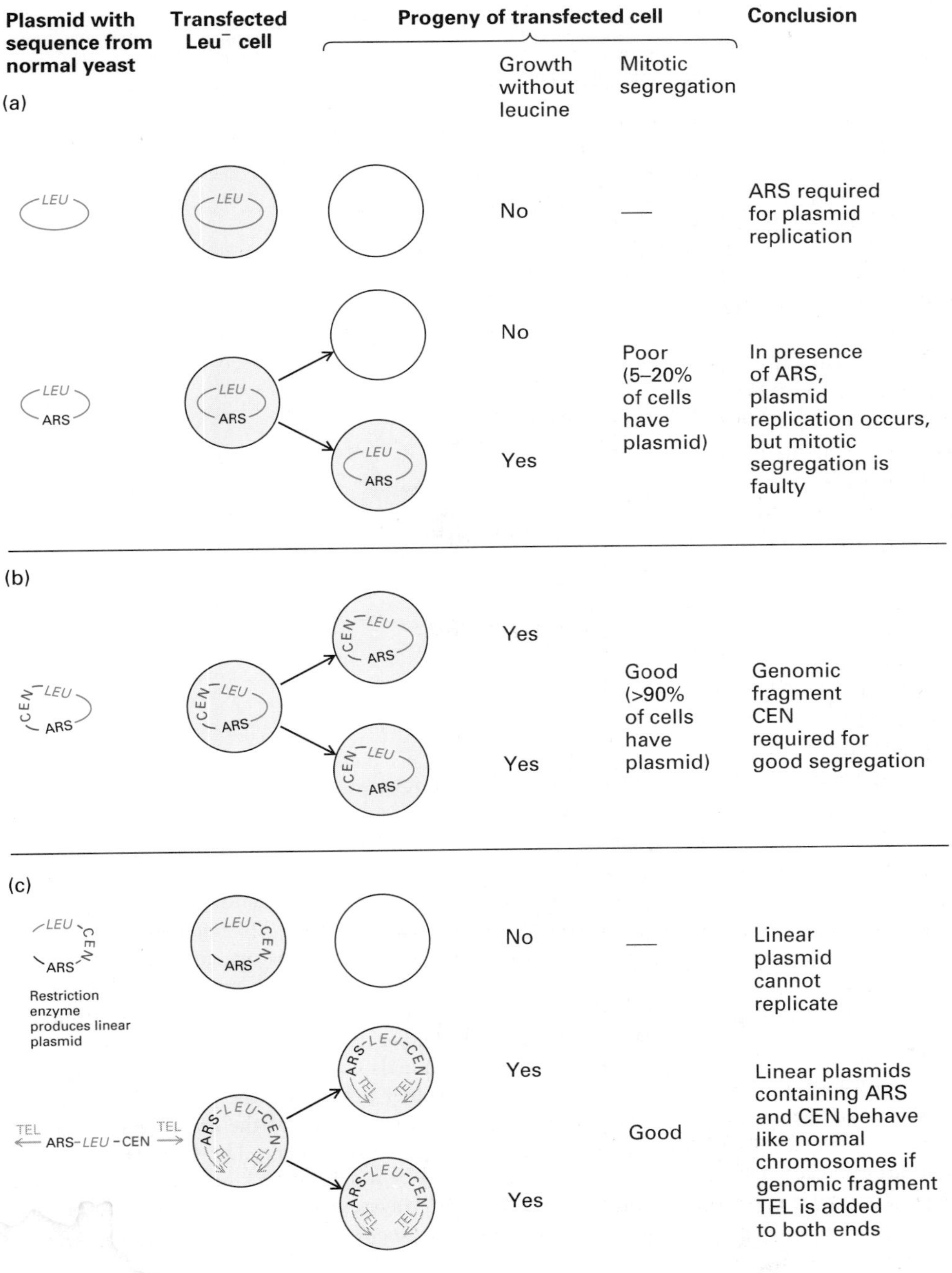

Figure 9-9 Demonstration of functional chromosomal elements by transfection experiments with yeast cells that lack an enzyme necessary for leucine synthesis (Leu⁻ cells). In these experiments, plasmids containing the *LEU* gene from normal yeast cells are constructed and introduced into Leu⁻ cells by transfection. The plasmids then replicate along with the nuclear DNA of the cells; only cells containing a *LEU* plasmid will grow in a medium that is not supplemented with leucine. (a) The plasmid must possess sequences that allow autonomous replication (ARS) in order to replicate in the cell. However, even plasmids with ARS exhibit poor segregation during mitosis, and therefore do not appear in each of the daughter cells. (b) When randomly broken pieces of genomic yeast DNA are inserted into plasmids containing ARS and *LEU*, some of the transfected cells produce large colonies, indicating that a high rate of mitotic segregation among their plasmids is facilitating the continuous growth of daughter cells. The DNA recovered from plasmids in these large colonies contains yeast centromere (CEN) sequences. (c) When Leu⁻ yeast cells are transfected with linearized plasmids containing *LEU*, ARS, and CEN, no colonies grow. Addition of telomere (TEL) sequences gives the linearized plasmids the ability to replicate as new chromosomes that behave very much like a normal chromosome in both mitosis and meiosis. [See A. W. Murray and J. W. Szostak, 1983, *Nature* 305:89.]

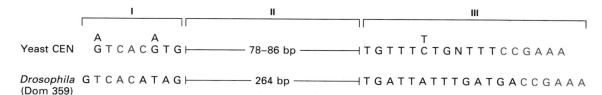

	I	II	III
	A A		T
Yeast CEN	G T C A C G T G├—— 78–86 bp ——┤		T G T T T C T G N T T T C C G A A A
Drosophila (Dom 359)	G T C A C A T A G├—— 264 bp ——┤		T G A T T A T T T G A T G A C C G A A A

▲ **Figure 9-10**   Comparison of ten yeast centromere sequences revealed three distinctive regions (I, II, and III) all of which are required for the centromere to function. One *Drosophila* repetitive DNA that is known to exist near centromeres has a repeated sequence with some similarity to the yeast consensus CEN, including two identical 4-bp and 6-bp stretches (red). [See L. Clarke and J. Carbon, 1985, *Ann. Rev. Genetics* **19**:29.]

Since the yeast centromere regions that confer mitotic stability have been cloned, their sequences can be determined and compared. Comparison of different yeast centromeres has revealed three regions (I, II, and III) that are necessary for a centromere to function (Figure 9-10). Short, fairly well conserved nucleotide sequences are present in regions I and III. Although region II seems to have a fairly constant length (78–86 base pairs), it contains no definite consensus sequence; however, it is rich in A and T residues. One *Drosophila* repetitive DNA, which comes from a centromeric region has a repeated sequence that bears some similarity to yeast CEN regions I and III.

The centromeric DNA may be involved in the chromosomal attachment of microtubules, which during mitosis bind to a structure called the *spindle pole body* (Figure 9-11). This structure, which is replicated during interphase, is embedded in the nuclear envelope. The spindle pole body is the functional analogue of the centriole in animal cells. When budding occurs, the two spindle pole bodies separate and one enters the bud. The microtubules

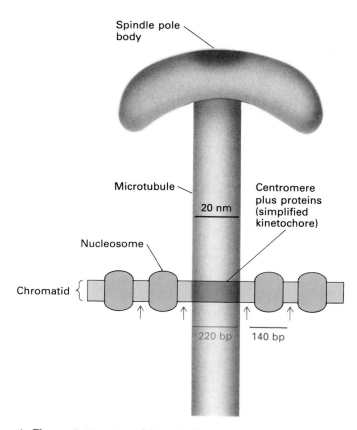

▲ **Figure 9-12**   A model of the binding of the yeast centromere region to a microtubule. The length of the yeast centromere sequence known to be necessary for the centromere to function (dark blue) approximately equals the diameter of one microtubule (20 nm). DNase digestion *(arrows)* liberates a centromere DNA fragment about 220 nucleotides long. Also shown are nucleosomes, the histone-containing structures associated with eukaryotic DNA; each nucleosome has a stretch of DNA about 140 base pairs long wrapped around it, which also can be liberated by DNase. Cells from multicellular plants and animals contain many more microtubules than yeast cells; in higher organisms, the microtubules appear to attach to the chromosome at the *kinetochore,* a granular region at or near the centromere. The yeast centromere plus its proteins could be a simplified kinetochore. [See K. S. Bloom, M. Fitzgerald-Hayes, and J. Carbon, 1982, *Cold Spring Harbor Symp. Quant. Biol.* **47**:1175.]

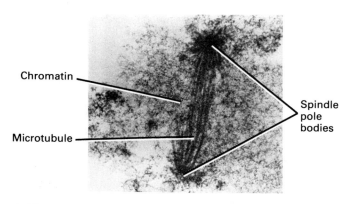

▲ **Figure 9-11**   An electron micrograph of yeast spindle pole bodies and attached microtubules. The yeast cell has been fixed and sectioned, so some of the microtubules in the cell cannot be seen. Through observation of serial sections, however, the number of microtubules can be shown to be approximately equal to the number of chromosomes. *Courtesy of H. Ris; from J. B. Peterson and H. Ris, 1976, J. Cell Sci. 22:219.*

extend from each spindle pole body to the chromatin (the DNA and its associated proteins). In yeast, approximately 15–20 individual microtubules (each 20 nm in diameter) bind to a spindle pole body, which makes it likely that one microtubule exists for each of the approximately 17 chromosomes in yeast. As depicted in Figure 9-12, yeast centromere DNA may serve as the nucleation site for microtubular assembly or attachment.

No centromere-binding protein has yet been identified in yeast cells, but an 80-kDa protein that is found in the centromere region of all human chromosomes has been isolated from human cells. The amino acid sequence of this protein contains a highly acidic domain, which may well be the DNA-binding region. The function of microtubules and the role of centromere-binding proteins are discussed in detail in Chapter 21.

Telomeres Although the yeast plasmids described earlier are circular molecules (see Figure 9-9a and b), the genetic maps of yeast chromosomes are linear. If such circular plasmids, even those with ARS and CEN sequences, are cut once with a restriction enzyme to make them linear before they are introduced into yeast cells, they are unable to replicate. In order for linear chromosomes to replicate in yeast, special telomeric (TEL) sequences have to be attached (see Figure 9-9c).

The first successful experiments involving transfection of yeast cells with linear plasmids were achieved by using the ends of a DNA molecule that was known to replicate as a linear molecule in *Tetrahymena*. During part of the life cycle of *Tetrahymena,* linear copies of nuclear DNA fragments replicate separately to form a so-called macronucleus. One of these fragments was identified as a dimer of ribosomal DNA, the ends of which contained a repeated sequence $(T_2G_4)_n$. When a section of this repeated sequence was attached to the ends of linear yeast plasmids containing ARS and CEN, replication and good segregation of the plasmids occurred.

Considerable experimentation has now uncovered special properties of telomere DNA and how it is formed by a special enzyme termed *telomere terminal transferase,* or *telomerase* (Figure 9-13). The sequences of the telomeres in a dozen or so organisms have been determined and all share two features: (1) repetitive oligomers and (2) a high G content in the strand that runs 5' to 3' toward the telomere. In addition to the T_2G_4 repeat found in *Tetrahymena,* a T_4G_4 repeat has been identified in the ciliated protozoan *Oxytricha* and a TG-rich region in yeast (TG TG TGGG)$_n$. At the ends of the chromosome, the G-rich telomere sequence has a free 3' end to which deoxynucleotides can be incorporated. This would normally suggest a single-stranded stretch of DNA in an otherwise duplex molecule. However, the telomere ends are not a simple duplex up to a certain point and single-stranded the remaining distance. Rather the single-stranded portion folds back on itself and, because of its peculiar G-rich nature, forms special (non-Watson-Crick) G-G base pairs (Figure 9-13a). This structure is a substrate for

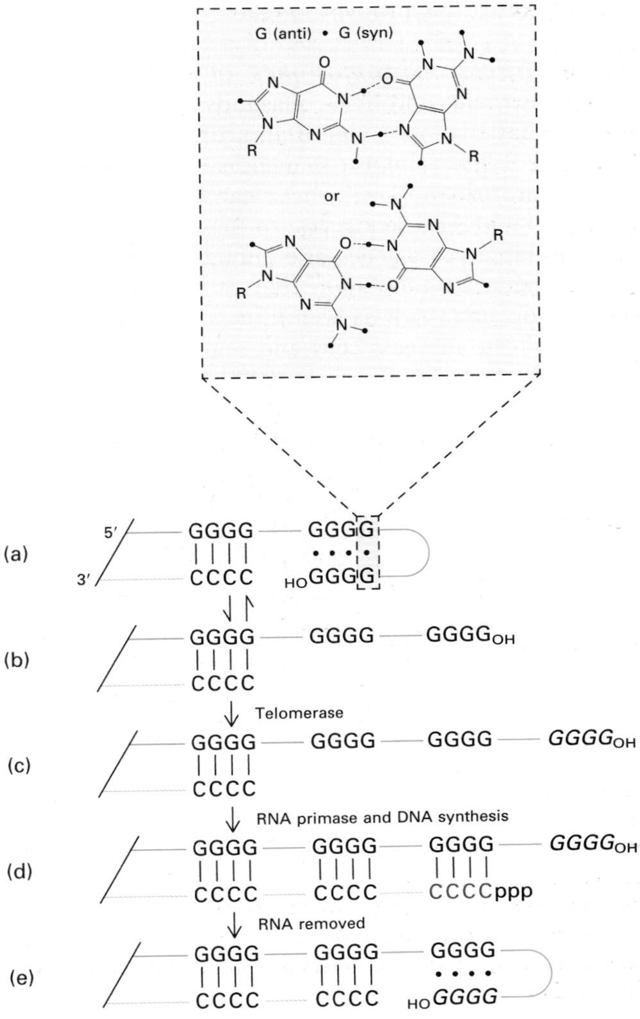

▶ **Figure 9-13** DNA structure in telomeres and possible mechanism of elongation. (a) Telomeric DNA consists of tandem repeats of a short sequence, one strand G-rich and one C-rich. The G-rich strand (dark blue) folds back on itself by the use of unusual G-G base pairing *(inset)* in which deoxyribose moieties are paired as opposite (syn or anti) conformations. (b) The G-rich strand unfolds, and (c) additional repeats *(italics)* are added by telomere terminal transferase. (d) The C-rich strand can then be synthesized by RNA-primed replication (RNA shown in red type). (e) Ligation of the C-rich strand, removal of the RNA primer, and refolding of the G-rich strand complete elongation of the telomere. *From T. R. Cech, 1988,* Nature *332:777; and C. W. Greider and E. H. Blackburn, 1987,* Cell **51***:887.*

telomerase, which can recognize some as-yet-undefined feature of the telomere and add to it. The first telomerase, which was isolated from *Tetrahymena*, can add T_2G_4 to the ends of *Tetrahymena, Oxytricha,* or yeast telomeric sequences. Thus the source of the enzyme and not the sequence of the telomeric DNA primer determines the added product. The enzyme from *Oxytricha,* for example, adds T_4G_4 to primers. Telomerase is a complex of protein and RNA; it is the sequence in the RNA that determines which telomere sequence is added.

This elaborate machinery for generating sequences at the end of chromosomes provides a mechanism for replication of an entire linear chromosome. In the DNA growing forks that replicate the majority of the chromosome one strand can grow continuously to the end (the 5′ to 3′ or leading strand); the other must be synthesized from RNA primers that are extended in the direction opposite to the movement of the growing fork. This chain, the lagging strand, cannot be completed as DNA all the way to the end of a linear duplex chromosome. The addition of repetitive telomeric sequences by telomerase, which recognizes ends and adds the repetitive telomeric sequences, solves this problem (Figure 9-14).

Yeast Artificial Chromosomes or YACs The research on circular and linear plasmids in yeast provided all of the basic components of an artificial chromosome (see Figure 9-9c). Telomere sequences from yeast cells or from the protozoan *Tetrahymena* can be combined with yeast CEN and ARS sequences; to these are added DNA with selectable yeast genes and enough DNA from any source to make a total of more than 50 kb. (Smaller DNA segments do not work as well.) Such artificial chromosomes replicate in yeast cells and segregate almost perfectly (only 1 daughter cell in 1000 to 10,000 fails to receive an artificial chromosome). During meiosis, the two sister chromatids of the artificial chromosome appear to separate correctly to produce haploid spores. Such studies strongly support the conclusion that yeast chromosomes, and probably all eukaryotic chromosomes, are linear, double-stranded DNA molecules with special regions—including centromere (CEN), telomere (TEL), and replication sequences (ARS)—that ensure replication and proper segregation. A technical point of considerable importance is that these yeast artificial chromosomes, or YACs, can be used to clone very long chromosomal pieces from other species. In the preparation of human DNA for genetic and eventually sequence analysis, this has great advantages.

Structure of Chromatin

When metaphase chromosomes are isolated from cells undergoing mitosis, or when the DNA from interphase nuclei is isolated in isotonic buffers, twice as much pro-

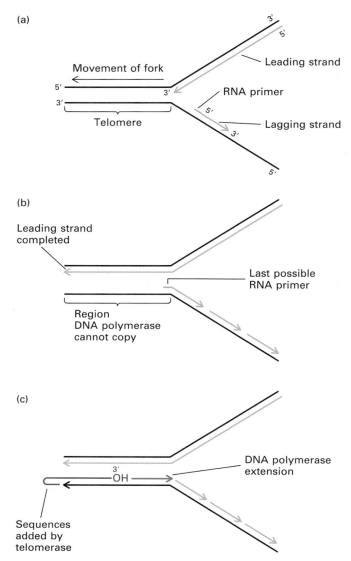

▲ **Figure 9-14** Telomere elongation prevents shortening of the ends of chromosomes. The lagging strand of a DNA growing fork cannot be completed by RNA primed synthesis (a) and (b). Addition of telomere repeats (dark red), as shown in Figure 9-13, allows fold-back to occur and provides a 3′ OH from which DNA polymerase can copy the last segment of the 5′ → 3′ template strand (c).

tein as DNA is recovered. This mixture of DNA and protein constitutes the *chromatin.* The general structure of chromatin has been found to be remarkably similar in all eukaryotic cells.

Amino Acid Sequences of Major Histones Are Highly Conserved

The most prominent proteins associated with metaphase chromosomes and with interphase DNA are *histones,* a family of basic proteins found in all eukaryotic nuclei.

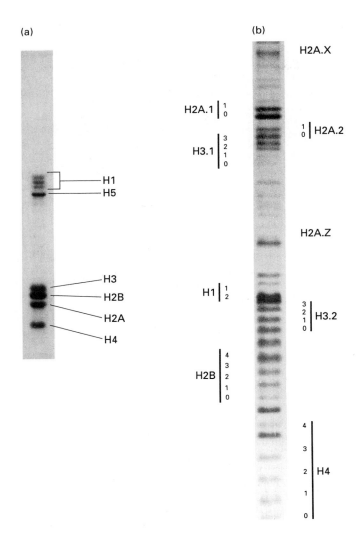

(a)

(b)

◀ **Figure 9-15** Separation of histones by gel electrophoresis. (a) Low-resolution electrophoretogram of histone sample extracted from chromatin of chicken blood cells. The major species H2A, H2B, H3, and H4 are present in about equal amounts. The other major histone, H1 (present in white blood cells), can also be seen, as can H5, which is present in the red blood cells of birds. (b) High-resolution separation of histones extracted from calf thymus. Amino acid sequence variants are indicated by the large-type numbers and letters following the decimals (e.g., H3.1 is slightly different in sequence from H3.2). The small-type numbers indicate charge variations within each major class. Variation within the H1 class is due mainly to phosphorylation, whereas charge variants with H2A, H2B, H3, and H4 are caused primarily by different degrees of acetylation. *Part (a) courtesy of V. Allfrey; (b) courtesy of M. Plesko and R. Chalkley.*

For example, in the nucleated red blood cells of birds, a histone termed H5 is present instead of H1 (Figure 9-15a). Despite minor variations, the similarity in the amino acid sequences of the major histones among all eukaryotes is most impressive.

Histones and DNA Associate to Form Nucleosomes

When chromatin is extracted from the nuclei in a solution of low ionic strength, it assumes a form that in the electron microscope resembles beads on a string. In this form, the DNA is a thin filament connecting the "beads" of 10 nm diameter. If the ionic concentration used in the extraction is raised (or if small amounts of divalent cations are added), the chromatin fibers become thicker, about 30 nm in diameter (Figure 9-16). The "beads" are structures termed *nucleosomes;* in the thicker fibers, the nucleosomes are packed into a spiral or solenoid arrangement that has six nucleosomes in each turn.

Structure of Nucleosomes The isolation of individual nucleosomes for physical and chemical analysis can be carried out by partial nuclease digestion either of DNA within the intact nucleus or of DNA in extracted chromatin. In addition to single nucleosomes, gentle digestion also releases nucleosome dimers, trimers, and higher oligomers, which can be isolated by various physical methods. Extensive nuclease treatment eventually digests all the DNA between individual nucleosomes, so that all the nucleosomes are released. The DNA content of a single nucleosome before all connecting DNA is digested varies between 160 and 200 base pairs, but after trimming, the figure is very close to 140 base pairs in all species.

Analysis of nucleosome crystals has shown that a nucleosome consists of a disk-shaped histone core plus a

The five major types of histone chains are termed H1, H2A, H2B, H3, and H4 (Figure 9-15a). The amino acid side chains of histones are frequently modified by posttranslational addition of phosphate, methyl, or acetyl groups, which produces subspecies that are detectable through high-resolution gel electrophoresis (Figure 9-15b).

The amino acid sequences of four histones (H2A, H2B, H3, and H4) from a wide variety of organisms are remarkably similar among distantly related species. For example, the sequences of histone H3 from sea urchin tissue and of H3 from calf thymus are identical except for a single amino acid, and only four amino acids are different in the H3 from the garden pea and that from calf thymus. As we shall discuss later, minor histone variants that differ from the highly conserved major types do exist, particularly in vertebrates.

The amino acid sequence of H1 varies more from organism to organism than do the other major histone species. In certain tissues, H1 is replaced by special histones.

▶ **Figure 9-16** Electron micrographs and schematic diagrams of chromatin fibers from the nuclei of *Drosophila melanogaster* cells. (a) Chromatin shows characteristic beads-on-a-string appearance at low ionic strength. The "beads" are nucleosomes, which are about 10 nm in diameter. (b) Chromatin extracted at high ionic strength (or following addition of divalent cations) consists of thick, 30-nm-diameter fibers; the nucleosomes are in a solenoidal arrangement. (See also Figures 9-17 and 9-19.) *Left photograph courtesy of S. McKnight and O. L. Miller, Jr.; right photograph courtesy of B. Hamkalo and J. B. Rattner.*

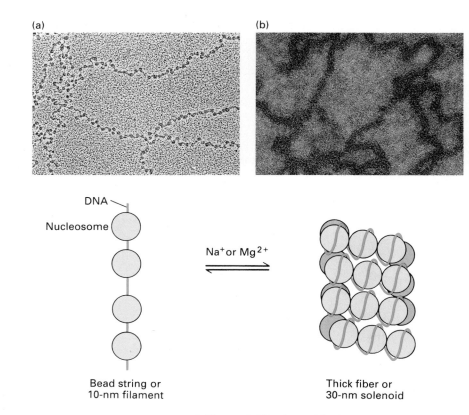

◀ **Figure 9-17** A model of a nucleosome. (a) Two copies each of the histones H2A, H2B, H3, and H4 compose an octomeric histone core. The horseshoe-shaped H3-H4 dimers are shown on the outside. (b) From electron microscopy and x-ray diffraction studies a fairly precise representation of the general locations of the various histones in the core can be constructed. The representation of the mass of the core as "slices" is taken from electron density maps. The indicated rotation angle is for purposes of orientation; note that one copy of H2A (H2A^2) and one of H2B (H2B^2) are not visible from this vantage point. (c) The DNA (*clear tube*) wraps around the histone core. Each number on the DNA indicates 10 base pairs; 70 base pairs appear in each turn of the DNA, or about 140 base pairs in the nucleosome. Again, the indicated rotation angles are for orientation purposes only. The position in (c) differs slightly from that in (b). [See A. Klug, 1981, in *Nucleic Acid Research: Future Development*, K. Mizobuchi, I. Watanabe, and J. D. Watson (eds.), Academic Press, p. 91; and A. Klug et al., 1980, *Nature* 287:509.] *Parts (b) and (c) courtesy of A. Klug; reprinted by permission from* Nature. *Copyright 1980 Macmillan Journals Limited.*

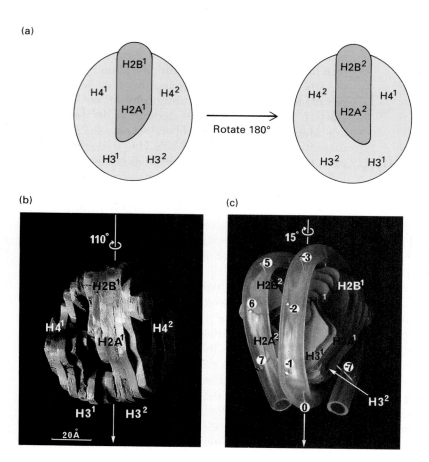

segment of DNA that winds around the core like thread around a spool. The core is an octomer constructed of two copies each of histones H2A, H2B, H3, and H4 (Figure 9-17). The DNA wrapped around the histone core (about 140 base pairs in length) makes just slightly less than two turns around the histone core octomer.

Assembly of Nucleosome Core   Newly replicated DNA quickly associates with already formed nucleosomes. A model of nucleosome assembly has been proposed recently based on studies with rapidly dividing frog eggs. In this work, analysis of protein complexes precipitated from extracts of frog eggs by monoclonal antibodies revealed the presence of two nonhistone proteins associated with histones: (1) *nucleoplasmin,* a 29-kDa acidic protein that is bound to H2A and H2B, and (2) *N1 protein,* another acidic protein that is bound to H3 and H4. When partially purified preparations of these two complexes were mixed in the presence of DNA, a nucleosome core was formed with liberation of nucleoplasmin and N1 (Figure 9-18). The stage at which modification (methylation, phosphorylation, etc.) of the core histones occurs is not yet known. Proteins resembling nucleoplasmin and N1 are known to be present in other cell types. Thus, the proposed pathway of nucleosome core assembly may operate in most cells, although this has not yet been demonstrated.

Solenoid Structure of Condensed Chromatin   As noted already, under certain conditions extracted chromatin appears thickened or condensed under the electron microscope (see Figure 9-16b). In the thickened fiber, the

View along axis of solenoid

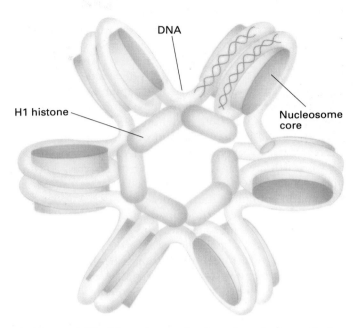

▲ **Figure 9-19**   View along axis of condensed chromatin in solenoid arrangement showing internal location of one associated H1 histone with each nucleosome. *From A. Klug, 1985,* Proc. R. W. Welch Fdn. Conf. on Chemical Res. *39:133.*

nucleosomes are arranged in a solenoid, with each nucleosome core containing H2A, H2B, H3, and H4. When chromatin is condensed, a fifth histone H1 is present; the H1 binds to the DNA just next to the nucleosomes. The unit of one nucleosome with one bound H1 is referred to as a *chromatosome.* When chromatin is extracted under low ionic conditions, H1 is released, leaving the core structure (nucleosome) depicted in Figure 9-19.

In intact eukaryotic nuclei, most of the DNA is bound not only to the four core histones but also to H1, and the chromatin exists largely in a compact helical arrangement or solenoid. This more organized structure probably is stabilized by bonding between chemical groups at the surface of nucleosomes, although little is known about such interactions. The model shown in Figure 9-19 depicts a chromatosome in condensed chromatin containing six nucleosomes per turn with one H1 molecule located on the inside associated with each nucleosome. Solenoids may be further organized into giant supercoiled loops (see later discussion on DNA loops). Our knowledge of nucleosome and solenoid structure has been drawn from an array of physical studies, including x-ray crystallography and neutron diffraction.

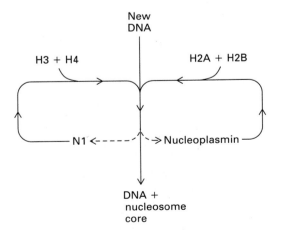

▲ **Figure 9-18**   Proposed pathway of nucleosome assembly in frog eggs. Both N1 and nucleoplasmin are acidic proteins that have been shown to associate with histones as indicated. *From S. M. Dilworth et al., 1987,* Cell **51:**1009.

Chromatin Probably Unfolds during Transcription

Several lines of evidence indicate that the closely packed solenoid structure of condensed chromatin must loosen for transcription to occur. In this section we will describe experiments that deal with the general nature of unfolding during the passage of RNA polymerase molecules along long stretches of DNA. In Chapter 11 we will return to consider evidence about the positioning of nucleosomes at initiation sites and how chromatin structure may be related to the regulation of transcription.

Frogs have two types of 5S-rRNA genes: *oocyte 5S* genes, which are transcribed inside frog oocytes, and *somatic 5S* genes, which are transcribed in other cell types. Chromatin containing both types of 5S-rRNA genes can be obtained by gently extracting frog oocytes with buffers containing increasing concentrations of monovalent ions. RNA polymerase III plus the necessary transcription factors and oocyte chromatin will transcribe the oocyte 5S-rRNA genes but not somatic 5S genes. After the oocyte chromatin is washed with a buffer containing between 0.2 and 0.3 *M* NaCl, which removes histone H1, the somatic 5S genes can also be transcribed. Re-addition of H1 at lower salt concentrations restores the chromatin to a compact state, so that somatic 5S-rRNA genes are again inaccessible. Since these are the very same conditions for unfolding and refolding of the "thick filaments" (see Figure 9-16b), it seems very likely that no genes are accessible in thick chromatin filaments.

A second type of experiment has demonstrated a clear-cut change in DNA-protein structure when genes become active. This change is reflected in the increased sensitivity to DNase attack of DNA regions that are being transcribed. For example, the DNA sequences that encode globin mRNA are susceptible to DNase attack in cells that are actively making globin protein, but these sequences are not sensitive to DNase in cells that are not making globin (Figure 9-20). Likewise, in the DNA from hen oviduct cells, which produce ovalbumin (the major egg-white protein), the ovalbumin DNA sequences are much more susceptible to DNase than are the globin sequences, which are not transcribed. The greater sensitivity of active genes to nucleases indicates that their associated protein is in a less protective configuration than the protein on inactive genes. The specific region of DNA that is DNase-sensitive during transcription varies: in some genes, the sensitive region includes only a few hundred base pairs around the gene; in others, it includes as much as 20 kb of the flanking sequences, indicating that the looser chromatin structure is not simply the result of RNA polymerase activity.

An important still-unresolved question is whether nucleosomes are removed from DNA regions that are being transcribed. In electron micrographs of gently spread chromatin with attached nascent RNA chains, nucleo-

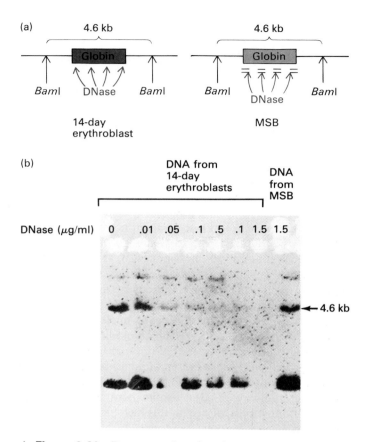

▲ **Figure 9-20** Demonstration that the DNA of a gene is more susceptible to DNase attack when being transcribed than when the gene is inactive. Chick embryo erythroblasts at 14 days actively synthesize globin, whereas cultured undifferentiated MSB cells do not. Nuclei from each type of cell were removed and exposed to increasing concentrations of DNase. The nuclear DNA was then extracted and mixed with the restriction enzyme *Bam*I, which cleaves the DNA around the globin sequence and normally releases a 4.6-kb globin fragment (a). The DNase- and *Bam*I-digested DNA was subjected to Southern blot analysis with a labeled cloned adult globin DNA used as a probe to detect the 4.6-kb fragment. As shown in (b) the nuclear DNA from the 14-day globin-synthesizing cells was sensitive to DNase digestion, but the DNA from the MSB cells was not. [See J. Stalder et al., 1980, *Cell* **19**:973.] *Photographs courtesy of H. Weintraub.*

somes are visible just ahead of and just behind the sites where the RNA polymerase is attached. This finding indicates that if nucleosomes are removed during transcription, they must quickly reassociate with DNA. However, in chromatin containing ribosomal genes—which are distinct, identifiable transcription units on which polymerases are densely packed—the nucleosomes are less obvious than in rarely transcribed regions.

In at least one case, the histone proteins associated with ribosomal genes are now known to change in conformation when the genes become active. In one stage of the life

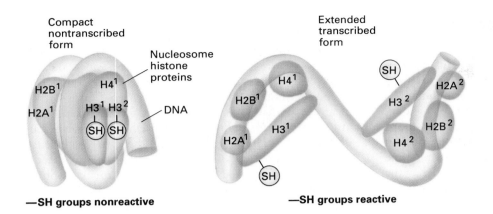

Compact nontranscribed form

Nucleosome histone proteins

Extended transcribed form

DNA

—SH groups nonreactive

—SH groups reactive

◀ **Figure 9-21** Models for active and inactive nucleosome structure based on experiments with ribosomal transcription units from *Physarum polycephalum. Adapted from C. P. Prior et al., 1983, Cell 34:1033.*

cycle of the slime mold *Physarum polycephalum,* the ribosomal genes replicate independently of chromosomes and can be purified as free, active chromatin. The transcribed ribosomal gene region and the nontranscribed spacer region, both of which are associated with nucleosomes, can be separated from each other. The H3 histones in the nucleosomes associated with transcribed regions react with iodoacetaminofluorescein, a fluorescent reagent that binds to cysteine sulfhydryl groups if they are not involved in the internal structure of a protein. In contrast, the H3 histones in nontranscribed regions do not react with this reagent. This difference in chemical reactivity and various physical measurements indicate that the nucleosomes in transcribed regions have a more relaxed structure than do those in untranscribed regions (Figure 9-21). Thus, even though nucleosomes are present in transcribed regions of ribosomal DNA, they appear to have fewer histone-histone interactions and a less compact configuration than nucleosomes in nontranscribed regions.

Electron microscopic studies with SV40 minichromosomes also have provided evidence that histones are not completely dissociated from transcribed regions. Each of these chromosomes, which can be extracted from nuclei of SV40-infected cells, are associated with 20–25 nucleosomes. These minichromosomes can be treated with reagents (e.g., 6-methyl psoralen) that link the two DNA chains between nucleosomes. If the nucleosomes are then removed and the DNA examined by electron microscopy, "bubbles" that are bounded by sites of cross-linked DNA are observed at intervals of about 200 nucleotides, which is the spacing between nucleosomes (Figure 9-22). At any instant only a few of the SV40 minichromosomes are undergoing transcription, and these can be distinguished by tails of mRNA that are also cross-linked and projecting from the circular DNA. The bubbles, which represent nucleosomes that have been removed, are just as abundant in molecules that are being transcribed as in those which are not.

Substantial evidence thus suggests that nucleosomes remain associated with DNA in regions that are being

transcribed, or at least that they are present in close proximity to transcribed regions. The increased sensitivity of transcribed regions to DNase attack and to sulfhydryl reagents, however, indicates that nucleosomes in transcribed regions undergo some structural changes. In addition, recent studies show that antibodies specific for acetylated histones can precipitate chromatin fragments of active genes much better than total DNA. Thus acetylation and perhaps other modifications are associated with active chromatin. Such modifications possibly prevent the tight packing of histones that characterizes nucleosome structure in condensed, inactive chromatin.

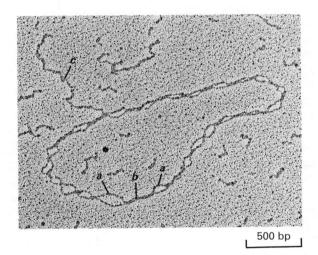

500 bp

▲ **Figure 9-22** Micrograph of transcribed SV40 minichromosome after treatment with DNA cross-linking reagent and subsequent removal of nucleosomes with protein denaturants. The "bubbles" *(a)* represent former nucleosome sites interspersed with cross-linked DNA *(b).* The long tail *(c)* is a nascent pre-mRNA chain, indicating that this chromosome was being transcribed. Thus nucleosomes are present on transcribed chromosomes, even adjacent to the point of transcription. *From W. DeBerandin, T. Koller, and J. M. Sogo, 1986,* J. Mol. Biol. **191:***469; photograph courtesy of J. Sogo.*

Nonhistone Proteins Provide a Structural Scaffold for Long DNA Loops

Although histones are the predominant proteins in chromosomes, nonhistone proteins also appear to be involved in organizing the structure of long regions, or "domains," of DNA. For example, electron micrographs of histone-depleted chromosomes from HeLa cells reveal long loops of DNA (10–90 kb) anchored to a nonhistone protein scaffolding (Figure 9-23). This scaffolding has the shape of the metaphase chromosome and persists even when the DNA is digested by nucleases. As we shall see, these loops match in length the units of replication in mammalian chromosomes.

In the purest preparations of metaphase chromosomes, only two scaffold proteins are found. One of these proteins has been identified as topoisomerase II, an enzyme that can cleave double-stranded DNA, pass one double strand through the cut, and then reseal the cut. Thus this enzyme, which is critical in DNA replication, can relieve torsional stress and also prevent tangles in the DNA. By use of antibodies to highly purified topoisomerase II the

major scaffold protein of metaphase chromosomes has been identified as this enzyme. It is located in the long axis of the metaphase chromosomes where the base of the DNA loops are found when the chromosomes are uncoiled (Figure 9-24).

The antibodies to topoisomerase have provided a means of studying the superstructure of the chromosome at metaphase. These highly condensed chromosomes can be exposed to buffers that cause them to decondense slightly. By observing fluorescent-stained individual metaphase chromosomes at different levels of focus ("optical sectioning"), a regular helical pattern can be discerned for the chromosome backbone (Figure 9-25a). The very interesting fact emerges that the two sister chromatids in a metaphase chromosome seem to be coiled oppositely, that is, they are mirror images. A mirror image arrangement could place the same sequences on the two sister chromatids in register and oriented with the same polarity (see arrows in Figure 9-25b).

In interphase the DNA probably remains in association with topoisomerase. This is illustrated in the following experiment. During cleavage by topoisomerase II, the free

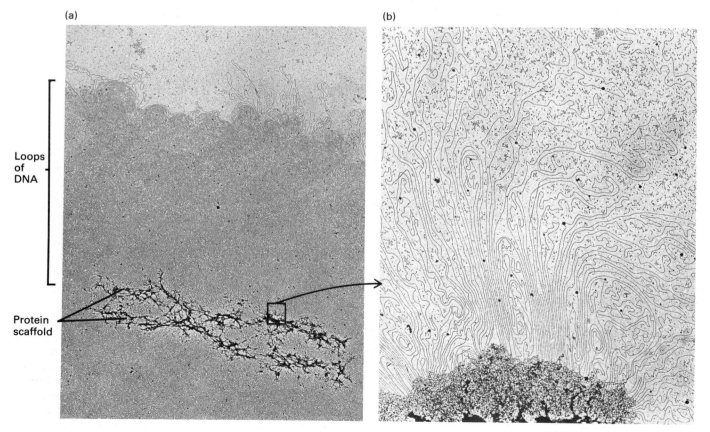

(a)

(b)

Loops of DNA

Protein scaffold

▲ **Figure 9-23** Electron micrographs of a histone-depleted metaphase chromosome prepared from HeLa cells by treatment with a mild detergent. (a) At lower magnification, nonhistone protein scaffolding (dark structure), from which long loops of DNA protrude, is visible (50,000X). (b) A higher magnification of a section of the micrograph in (a) shows the apparent attachment of the DNA loops to the protein scaffold (150,000X). *From J. R. Paulson and U. K. Laemmli, 1977,* Cell **1**:817. *Copyright 1977 M.I.T.*

(a)

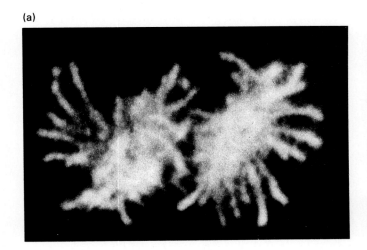

(b)

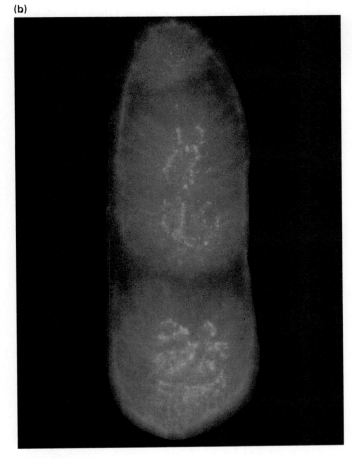

▲ **Figure 9-24** Presence of topoisomerase II in metaphase chromosomes. (a) A fluorescent antibody to topoisomerase shows outline of chromosomes in two mitotic chicken cells. (b) Chicken mitotic cell stained with a fluorescent dye that binds DNA (light blue) and anti-topoisomerase II antibody (pink). The DNA is partially released (as in Figure 9-23) rather than compact so that DNA loops emanate from the backbone or scaffold of the chromosome, which is outlined by the antibody. [See W. C. Earnshaw and M. M. S. Heck, 1985, *J. Cell Biol.* **100**:1716.] *Photographs courtesy of W. C. Earnshaw.*

(a)

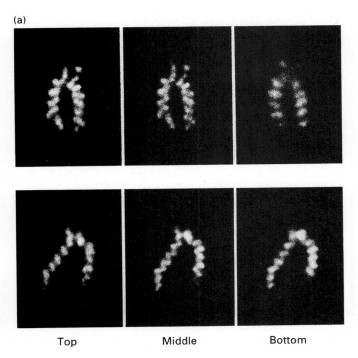

Top   Middle   Bottom

(b)

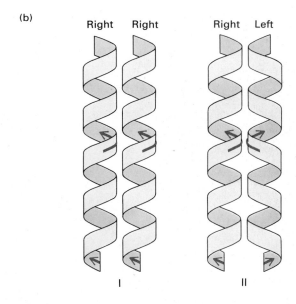

I   II

▲ **Figure 9-25** (a) Two human metaphase chromosomes stained with fluorescent antibodies to topoisomerase II. The scaffold is visualized at three depths of focus (top, middle, and bottom). Note that the tilt of the bright bands differs by 90° in the top and bottom focal positions, indicating a helical twist to the chromosome backbone. (b) Two theoretical alignments of helical coiled sister chromatids. In I, sister chromatids are parallel, so that identical sequences (*arrows*) are on opposite sides of the chromosome coil. In II, sister chromatids are aligned as mirror images, so that identical sequences are on same side of chromosome coil. [See E. Boy de la Tour and U. K. Laemmli, 1988, *Cell* **55**:937.] *Photographs courtesy of U. K. Laemmli.*

ends of the DNA are covalently linked by a phosphotyrosine linkage to the enzyme. Drugs exist that block the resealing step of topoisomerase activity. Interphase cells treated with these topoisomerase inhibitors release pieces of DNA averaging about 50 kb with topoisomerase II attached to the ends of the fragments. These studies and others with the DNA-binding compound ethidium bromide, which can be used to measure the extent of DNA supercoiling, leave little doubt that proteins, including (if not only) topoisomerase II, are bound during interphase to fixed sites of mammalian DNA that range between 30–90 kb.

The binding sites for the topoisomerase are *scaffold-associated regions* (SARs). The procedure for mapping SARs is illustrated in Figure 9-26. Such mapping, which has been done with a number of individual genes, indicates that SARs occur *between* genes, not within transcription units. Although the physiologic role of SARs has not yet been conclusively demonstrated, they may hold specific chromosome domains in such a way as to facilitate transcription or replication. It is interesting in this connection that the antibody to topoisomerase II stains cells most prominently in the G_2, M, and S phases and not in G_1. Additional research in this area should eventually reveal the basis for the patterns of SAR binding and the metabolic events that require or utilize binding of DNA to SARs.

Chromatin Contains Small Amounts of DNA-Binding Proteins in Addition to Histones and Topoisomerases

The total mass of the histones associated with a DNA molecule is about equal to that of the DNA, and there is one topoisomerase II molecule per 5 kb of DNA. Interphase chromatin and metaphase chromosomes also contain smaller amounts of other proteins. Because the regulation of transcription is controlled by specialized proteins associated with interphase chromatin, there has been great interest in attempts to study nonhistone chromosomal proteins. A growing list of DNA-binding proteins that help regulate transcription have been identified. These proteins are all very rare (1000–100,000 copies per cell out of a total of 10^{12} protein molecules per cell). A few other nonhistone proteins that bind DNA are present in much larger amounts than the gene regulators. One such group is called HMG (for "high-mobility group"—a reference to the behavior of these proteins during electrophoretic separation). These proteins have not been assigned a specific role, but they are necessary for cell viability; yeasts that cannot make HMGs are nonviable.

Biological Definitions of a Gene

The intense interest in chromosome structure is, of course, rooted in the fact that the genes that determine an organism's capacities reside on the chromosome. Before

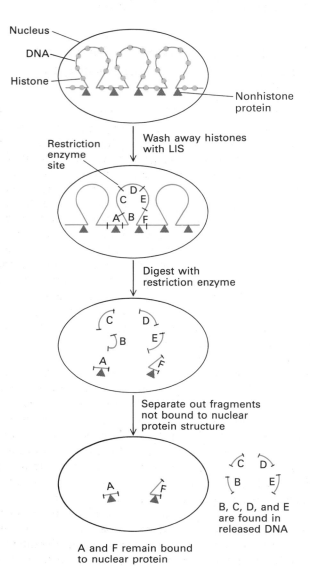

◂ **Figure 9-26** Mapping of sites at which organized loops of chromatin bind to scaffold-associated regions (SARs). Nuclei are washed free of all histones by treating them with the detergent lithium diiodosalicylate (LIS), which exposes the DNA to cleavage by restriction enzymes. The nuclei then are treated with a known restriction enzyme to yield fragments (A to F), which are analyzed by the Southern blot technique. In the example shown, four of the six restriction enzyme fragments (B, C, D, and E) are not associated with scaffold proteins; these fragments are recovered outside of the nucleus. Extraction of the DNA that remains bound to the residual nonhistone nuclear proteins yields the two fragments A and F. [See J. Mirkovitch, M.-E. Mirault, and U. K. Laemmli, 1984, *Cell* **39**:223.]

proceeding to discuss the molecular details of DNA sequence arrangements, the regulation of gene expression, and of chromosome replication in the next three chapters, we will close this chapter on the general structure of chromosomes by reviewing the evolving definition of a gene. In this section, we consider the major genetic tests that have been used to define genes. In the next section, we discuss various limitations of the biological definition of genes and propose a molecular definition.

Recombination Tests Can Separate Linked Genes

The original definition of a gene came from breeding experiments in which individual traits, or "characters," were followed. In crosses of genetically distinct strains of garden peas, Mendel chose simple traits that could be easily followed: stem length, seed color, seed texture, etc. Each phenotypic trait was determined by one of two alternative forms of a "gene"—the dominant or the recessive form, or allele. These alleles were seen to be stable through many generations.

Later T. H. Morgan and his coworkers carried out an extensive genetic study of *Drosophila melanogaster*. Between 1910 and 1925, careful visual observations of *Drosophila* uncovered more than 100 physical abnormalities based on mutations. A sizable number of these were variations in eye color; other mutations affected body parts such as the wings, legs, antennae, and bristles. One of the first important results of the studies with *Drosophila* was that these genetically determined abnormalities segregated during Mendelian crosses into one of four groups, which coincides with the number of chromosomes in *Drosophila*. Mutant gene loci that segregated together were said to be *linked*, meaning that they were all on the same chromosome. In later studies, various eye-color mutants were mapped to three of the four chromosomes, which demonstrated that different genes can contribute to one phenotypic trait. Thus, some phenotypic traits are not encoded simply by alleles (variants) of the same gene but are encoded by a series of different genes.

Although linkage to one of four groups was the rule for the various *Drosophila* genes, it was soon realized that linkage between the various genes in a linkage group does not occur 100 percent of the time due to meotic *recombination* in gametes. The test for recombination follows two linked genes during a genetic cross. Take, for example, two genes on chromosome III, scarlet (an eye-color mutant with lighter red eyes than the very dark red wild-type) and curled (Figure 9-27). As the result of recombination during meiosis, a chromosome that carries scarlet and curled would swap one of these traits with a chromosome that is wild-type for both about 6 percent of the time. A quantitative analysis of the frequency of recombination between genes on genetically marked homologous

chromosomes allows the mapping of genes along the chromosome (Figure 9-27). The maps are based on the fact that the farther apart two linked genes lie, the greater the chance that a recombination will occur to disrupt the linkage between alleles of these genes.

The explanation for recombination came from cytology. In the formation of gametes, the homologous chromosomes pair up during the first meiotic division. This process is called *synapsis*. The paired (maternal and paternal) homologous chromosomes are actually broken and reunited during synapsis. The crossed chromosomes participating in recombination are visible as *chiasmata*.

Thus the work of the early *Drosophila* geneticists led to the recognition of three properties of a gene: (1) it is a chromosomal site that controls an observable characteristic; (2) it can be changed or "mutated"; and (3) it can recombine with a homologous site on another chromosome. According to these properties, if two genetic traits could be separated by recombination, they were considered to be encoded by separate genes.

Complementation Tests Can Distinguish Genes Contributing to One Phenotypic Function

The definition of a gene by recombination tests has several drawbacks. Measuring the frequency of recombination can be laborious; genes that lie very close together require the screening of large numbers of offspring. Furthermore, recombination analyses place genes on a genetic map but do not define the function of a gene. Analyses of inherited traits that depend on gene function therefore were developed. Complementation tests rely on gene function and are designed to bring into the same cell two genes or sets of genes required to carry out one function. If either contributing partner to such a test contains a mutation in a gene not mutant in the other partner, then the cell can still function and complementation will have occurred. Even if such functionally related genes are side by side in a chromosome, they can be detected easily as separate genes in such a test.

Complementation tests have been widely used in microorganisms including the bread mold *Neurospora crassa*. This mold can grow from a single haploid spore (conidia) into a spongy clump of branching chains called hyphae. Each hypha is segmented, with each segment containing at least one haploid nucleus. Neighboring hyphae originating from two different spores occasionally fuse, bringing into the same cytoplasm haploid nuclei from different starting spores. This structure, when the nuclei differ in genotype, is called a *heterokaryon*.

Because of the formation of heterokaryons in *Neurospora*, complementation tests can be performed to determine whether mutants exhibiting the same phenotype result from mutation in a single gene or more than one

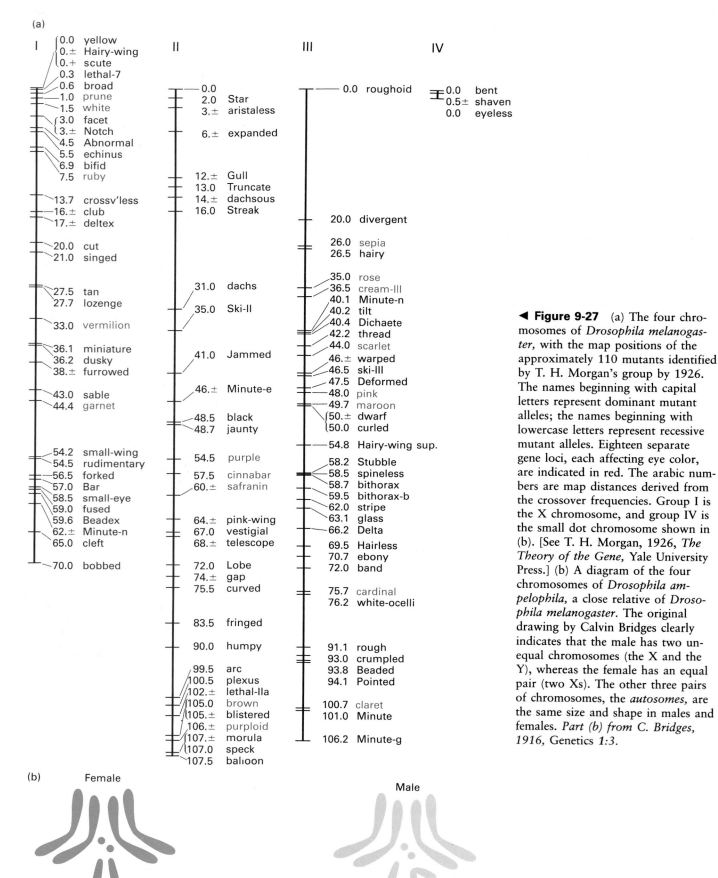

Figure 9-27 (a) The four chromosomes of *Drosophila melanogaster,* with the map positions of the approximately 110 mutants identified by T. H. Morgan's group by 1926. The names beginning with capital letters represent dominant mutant alleles; the names beginning with lowercase letters represent recessive mutant alleles. Eighteen separate gene loci, each affecting eye color, are indicated in red. The arabic numbers are map distances derived from the crossover frequencies. Group I is the X chromosome, and group IV is the small dot chromosome shown in (b). [See T. H. Morgan, 1926, *The Theory of the Gene,* Yale University Press.] (b) A diagram of the four chromosomes of *Drosophila ampelophila,* a close relative of *Drosophila melanogaster.* The original drawing by Calvin Bridges clearly indicates that the male has two unequal chromosomes (the X and the Y), whereas the female has an equal pair (two Xs). The other three pairs of chromosomes, the *autosomes,* are the same size and shape in males and females. *Part (b) from C. Bridges, 1916, Genetics 1:3.*

gene. For example, when haploid spores from two different histidine-requiring (His⁻) strains are grown together, heterokaryons are formed that can grow in the absence of histidine (Figure 9-28). In this case, each mutant compensates for the genetic defect in the other, indicating that the mutations affect two different enzymes (hence, two different genes) in the pathway of histidine biosynthesis. In contrast, heterokaryons formed from His⁻ strains that are mutated in the same gene will not grow in the absence of histidine. When two mutants correct a phenotypic defect, they are said to complement each other. Complementation analysis, thus, can distinguish the individual genes in a set of functionally related genes, all of which must function to produce a phenotypic trait.

Analyses of Phage *rII* Mutants Led to Recognition of Cistrons

As noted earlier in this section, recombination tests are used most often to distinguish and to order different genes. Based on extensive recombination analyses of bacteriophage T4 genes called *rII*, Seymour Benzer studied intragenic recombination. The *rII* mutants can grow on *E. coli* strain B but not strain K, whereas wild-type phage can grow on both host strains.

A large number of separately isolated *rII* mutants were tested for genetic identity in a series of recombination tests. In these tests, *E. coli* strain B cells were infected with two independently isolated *rII* mutants, and the new phage progeny were plated on strain K cells to detect wild-type recombinants. This test, therefore, demands recombination between the two mutant genomes to produce a viable phage yield. Recombination between *rII* mutants was relatively infrequent, suggesting that all of the *rII* mutations lay close together in the bacteriophage chromosome. The recombination frequency between the most closely spaced *rII* mutants was only about 10^{-5} of the frequency between other bacteriophage mutants that were spaced a maximum distance apart (that is, gave the most frequent recombination). Considering the size of the T4 genome (2×10^5 base pairs), recombination frequencies of 10^{-5} indicated that the least frequently recombining *rII* mutations must lie within a few nucleotides of one another ($2 \times 10^5 \times 10^{-5}$). This impressive result indicated that recombination tests, if they are sensitive enough and if they are carried out on enough organisms, can detect single-base differences in mutants within a gene.

Although all the *rII* mutants shared a single phenotypic characteristic—the inability to grow on *E. coli* strain K—

and recombination analyses indicated that all the mutants involved a small region of the T4 genome, Benzer used complementation tests to show that two distinct functions were controlled by this genomic region. These tests were performed by coinfecting strain K cells with two different mutants; if complementation occurred, the mixed infection yielded one round of new phage all of which were parental as tested by growth on strain B and failure of growth on strain K. The complementation frequency was much higher than recombination. The complementing mutants fell into two classes, designated *rIIA* and *rIIB*. Growth of phage in strain K cells occurred when one *rIIA* and one *rIIB* mutant infected a cell but not when two different *rIIA*s or two different *rIIB*s infected a

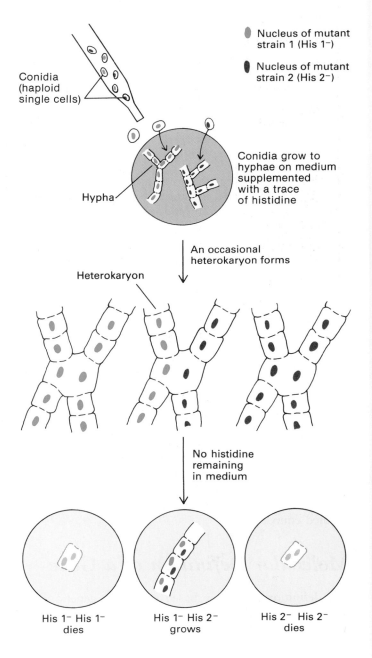

▶ **Figure 9-28** Complementation test of two histidine-requiring mutants of *Neurospora crassa*. A positive test (i.e., growth of a visible colony after histidine is depleted) indicates that the two mutations affect different genes encoding enzymes required for histidine biosynthesis.

(a) Mutants grown with wild-type phage

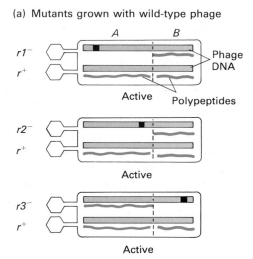

(b) Mutants grown with mutants

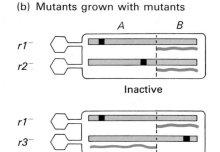

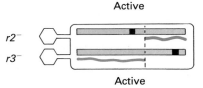

▲ **Figure 9-29** Transactive complementation in *rII* mutants of bacteriophage T4 when grown on *E. coli* strain K. (a) Wild-type phage are (*r*⁺) which means they grow on strain K. All mutants (*r*⁻) were inactive alone, but when strain K cells were coinfected with wild-type phage (*r*⁺), mutant and wild-type phage were produced. (b) When strain K cells were coinfected with a series of *r*⁻ mutant pairs, the mutants fell into two groups—*A* and *B*. Any mutant in the *A* group could rescue any mutant in the *B* group when the two mutants were on different chromosomes (i.e., trans). No *A* mutants were transactive with one another, and no *B* mutants were transactive with one another. The two transactive groups of mutants for the *r* function defined two cistrons, cistron *A* and *B*, both of which are necessary for the *r* function. [See S. Benzer, 1962, *Sci. Am.* **206**(1):70.]

cell (Figure 9-29). Furthermore, growth occurred only when the *rIIA* and *rIIB* mutations were on different copies of the bacteriophage chromosome (i.e., when the mutations were in the trans position). If a single copy of the chromosome carried both the *rIIA* and the *rIIB* mutations (i.e., if the mutations were in the cis position), they did not complement. Thus *rIIA* and *rIIB*, although they lay very close together on the chromosome, acted as if they were independent genetic functions. Such independent transactive complementation units are called *cistrons*. The genetically inherited trait of growth of T4 bacteriophage on *E. coli* strain K depends on two cistrons, *rIIA* and *rIIB*.

The original definition of a cistron implied that it was a DNA region that encodes one polypeptide chain, and many assumed that a cistron could be considered the equivalent of a gene. However, as we shall see, the complicated arrangements of the information in many DNA molecules do not allow every separate polypeptide to be detected by a complementation or recombination test; further, many mutations affect more than one polypeptide. Thus the terms gene and cistron are no longer deemed equivalent.

Molecular Definition of a Gene

The definition of a gene based on classic genetic techniques (i.e., recombination and complementation analyses) cannot account for several observed phenomena—in particular, mutations within a polypeptide-coding region that affect more than one function and mutations located outside a coding region that alter one or more genetic functions. Such mutations are easily demonstrated in bacterial operons. An operon is a single multigene transcription unit that produces polycistronic mRNA; such mRNA has multiple start and stop signals for the translation of a series of independent polypeptides. An operon can contain independent mutations in its different polypeptide-coding sequences, each of which can be scored as an independent cistron in a complementation test. However, mutations in the promoter or the operator regions of an operon affect all the cistrons in an operon; these are called *pleiotropic mutations*. For example, a mutation in the operator of the lactose operon renders all genes in the operon constitutive—that is, continuously functioning and not subject to regulation (see Table 7-1). Therefore, a genetically defined cistron is not the "whole" gene; the controlling sequences, which are not transcribed, are also part of the gene.

Transcription Units Are Not Necessarily Single Genes

We defined simple eukaryotic transcription units as those encoding only one mRNA and hence only one protein. Many eukaryotic transcription units, especially in yeasts, are simple; therefore, many "genes" and transcription units in these organisms are coextensive and might be scored as individual cistrons in a complementation test.

In both invertebrates and vertebrates, however, complex transcription units are quite common. Because a transcription unit can produce more than one mRNA and, therefore, encode more than one protein, we must differentiate between the transcription unit and the gene or potential complementation unit. A complex transcription unit may contain two or more poly A sites or two or more splicing variations; this in turn can lead to two or more mRNAs, each of which encodes a separate protein. Thus, if complex transcription units are considered to contain as many genes as the different protein-coding mRNAs that they produce, they contain more than one gene. However, genetic tests to prove the separate existence of each gene could be difficult. Mutations within an exon of a primary complex transcript that is shared by all the mRNAs can affect all encoded proteins. In such a case, one mutation affects the formation of two or more proteins, but the protein-coding functions are not separable by complementation. Two such "genes" cannot be identified as cistrons. In a complex transcription unit, mutations can also occur in two or more different exons that each appear in only one of the possible mRNAs. In such cases, the mutation sites could be identified as separate complementing genes but would still be part of the same transcription unit and map to the same chromosomal site (Figure 9-30).

The following rule applies, at least theoretically, to genetic tests of complex transcription units: if mRNAs contain different, nonoverlapping portions of a primary transcript, then complementation should be possible between at least some mutants in these distinct regions. Finally, keep in mind that only one set of promoter sequences often serves a complex eukaryotic transcription unit. A mutation in one of these nontranscribed but essential DNA sequences affects the expression of all functions encoded in the transcription unit.

(a) *Overlapping exons:* No complementation between any mutants

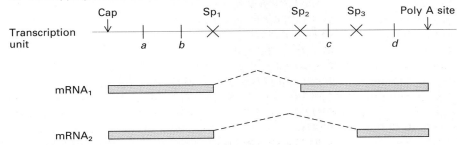

(b) *Multiple poly A sites and nonoverlapping exons:* No complementation with *a* mutant; *b* and *c* mutants complement

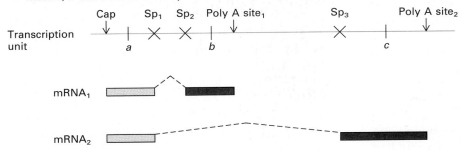

(c) *One poly A site and nonoverlapping exons:* No complementation between *a, b,* or *e* mutants and any others; *c* and *d* mutants complement

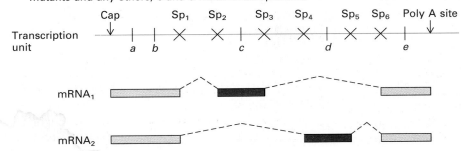

◀ **Figure 9-30** Genetic detection of coding capacity for different proteins encoded by complex transcription units. The three cases illustrated demonstrate that mutations in nonoverlapping exons could complement each other but that mutations in shared exons cannot. The poly A sites, splice sites (Sp), and mutation sites (*a* through *e*) are marked. The mRNAs derived from transcription are shown below the transcription unit with nonoverlapping exons in red. (a) A transcription unit that produces two mRNAs containing two overlapping exons. A mutation at *a, b,* or *d* prevents the formation of a normal copy of *either* mRNA. Furthermore, no complementation is possible between two chromosomes bearing individual mutations at any of these sites. For example, *a* and *b* do not complement each other, nor do *b* and *d*. (b) A transcription unit that contains two poly A sites and produces two mRNAs with nonoverlapping 3'exons. Mutations at *b* and *c* can complement one another. As in (a), a mutation at *a* would be in the shared 5' exon, and no complementation with mutations at *b* or *c* could occur. (c) A transcription unit that has six splice sites and produces two mRNAs, each containing an exon that is spliced out of the other. Mutations at *c* and *d* can complement one another. Again, mutations in the shared exons (*a, b,* or *e*) do not complement any other mutations.

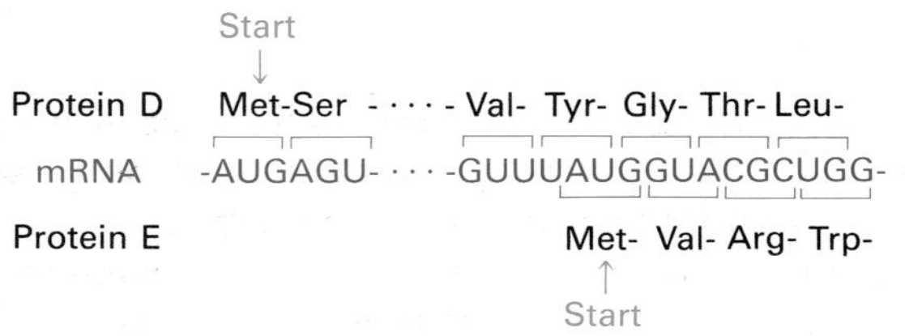

```
              Start
               ↓
Protein D   Met-Ser - - - - - Val- Tyr- Gly- Thr- Leu-

   mRNA     -AUGAGU- · · · · -GUUUAUGGUACGCUGG-

Protein E                     Met- Val- Arg- Trp-
                               ↑
                             Start
```

▲ **Figure 9-31** Part of the mRNA produced from overlapping genes in the bacteriophage φX174. Translation of the viral protein D begins at the first AUG codon; translation of the protein E begins in a different reading frame in the same mRNA. Mutations in such genes frequently cannot be distinguished by complementation tests. [See B. G. Barrell, J. M. Air, and C. A. Hutchinson, 1976, *Nature* 264:34.]

Overlapping Genes   Within the DNA of viruses and eukaryotic cells, there are overlapping genes that produce an mRNA whose triplet code is "read" (translated by ribosomes) in different frames (Figure 9-31). Because multiple codons exist for each amino acid, it is theoretically possible that two mutations in such overlapping genes might each affect only one of the possible proteins encoded by that region. In such cases, the two mutants in the same region of the DNA could be shown by complementation tests to contain two genes. However, mutations that damage both genes could not be complemented by any mutant in that region. Therefore, overlapping genes cannot always be identified by complementation. Identification of two such genes may require a molecular demonstration that there are two coding functions in the same DNA sequence.

Genes That Encode Polyproteins   Another complication in defining eukaryotic genes arises because not all mRNAs are translated into products that act as single polypeptides. One example of such a *polyprotein* is proopiomelanocortin, a primary translation product that is cleaved to produce four hormonally active polypeptides, some of whose sequences overlap (Figure 9-32). Theoretically, each of these final functioning polypeptide units is individually susceptible to change by mutation of the appropriate region of the DNA, and complementation tests between two mutants could conceivably indicate that the peptides are separate. In this case, however, a positive complementation test would indicate the presence of two separate genes when, in fact, only one DNA region, encoding one mRNA, is present.

Some Genes Do Not Encode Proteins

Although most genes are transcribed into mRNAs that encode proteins, clearly some transcribed RNA, including tRNA and rRNA, does not encode proteins. However, the DNA that encodes pre-tRNA and pre-rRNA can be mutated and detected in genetic experiments. We thus speak of tRNA genes and rRNA genes, even though the final products of these genes are RNA molecules and not proteins. Many other RNA molecules—for example, the small nuclear RNAs found in eukaryotic nuclei—also do not encode proteins and therefore are transcribed from "genes" that do not encode proteins.

A Gene Comprises All Nucleic Acid Sequences Necessary to Produce a Functional Protein or RNA

Based on genetic analyses, a gene can be defined as a region of DNA determining a heritable trait that can be identified in mutant form and separated from any other mutant form by a complementation test. Thus, we speak of a gene or genes for blue or brown eye color in humans, for coat color in mice, or for sucrose utilization in yeast. We also speak of the multiple genes for histidine formation in, for example, bacterial or yeast cells. A gene defined in this manner—whether it be prokaryotic or eukaryotic—usually encodes a single polypeptide chain.

As discussed in this section, however, many DNA regions that affect gene function are difficult or impossible to detect as separate genes by either a recombination or a complementation test. In molecular terms, a gene can be defined as *the entire nucleic acid sequence (usually a DNA sequence) that is necessary for the synthesis of a functional polypeptide or RNA sequence.* According to this definition, a gene includes not only the nucleotides encoding protein or functional RNA but all the sequences required to get a particular primary transcript made; in eukaryotes such required sequences can lie 50 kb or more from the coding region of a gene. Finally, many DNA stretches exist without demonstrable functions; even though some of these resemble functioning genes, they should not be referred to as genes.

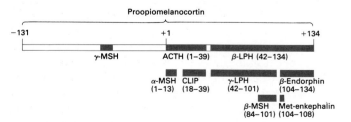

▲ **Figure 9-32** Proopiomelanocortin is a primary translation product produced in the anterior pituitary. This polyprotein is cleaved to yield several hormonal peptides, including adrenocorticotropic hormone (ACTH), lipotropic hormones (LPHs), melanocyte-stimulating hormones (MSHs), and enkephalins, which are neuroactive hormones. CLIP is a short polypeptide whose function is not yet known. Because ACTH was sequenced first, its first amino acid was numbered +1. [See S. Nakanishi et al., 1979, *Nature* 278:423.]

Summary

Eukaryotic chromosomes are visible in the light microscope during mitosis and meiosis. The karyotype—the collective term for chromosome number, size, and shape—is species-specific. Organisms with similar structures and degrees of complexity can differ greatly in their karyotypes and in the total amount of DNA in their genomes; these variations suggest that many eukaryotic genomes contain "extra" or unused DNA. The staining of chromosomes with dyes after various treatments produces light and dark bands at characteristic places along their length. These bands, which can be resolved in the light microscope, are the landmarks by which cytogeneticists prepared early physical maps of the chromosomes of many species. Much more detailed physical maps can now be constructed based on analysis of restriction fragment length polymorphisms (RFLPs) and the location of individual genes based on the use of cloned DNA probes. Together with extensive DNA sequencing these approaches will ultimately lead to complete mapping and sequencing of the human genomes.

Each eukaryotic chromosome contains a long linear DNA molecule ranging from 5×10^5 b.p. (small yeast chromosome) to 2×10^8 b.p. (human chromosome 1) containing three required functional elements: autonomously replicating sequences (ARS); the centromere, to which the spindle attaches at nuclear division (mitosis or meiosis); and the telomeres (ends), which allow the complete replication of linear DNA. Yeast artificial chromosomes or YACs containing these three elements plus selectable markers replicate and segregate like normal chromosomes and are now available to clone and study megabase (10^6) DNA fragments.

By far the most prominent proteins in the nucleus are basic proteins called histones, which have very similar amino acid sequences in all eukaryotes. Octomers containing two each of histones H2A, H2B, H3, and H4 form a core around which helical DNA is wrapped to form nucleosomes. Almost all the DNA in the nucleus is packaged in this form. Together with histone H1 chromatin is packed into a helical assembly of nucleosomes called a solenoid, a compact arrangement of six nucleosomes per turn in the solenoid. It is possible that all DNA that is not being transcribed is packed in this solenoid arrangement. At least some structural changes in the DNA-histone packing are necessary (e.g., the removal of H1) for transcription to occur. Finally, there are sequences that divide chromatin into long loops, the bases of which are attached to a nonhistone protein scaffold. During mitosis, the long loops of chromosomal DNA, still associated at their base with an underlying protein-containing "scaffold," are condensed into a visible form.

From studies of linkage groups and recombination during meiosis, the early geneticists developed recombination mapping and a definition of the gene. The maps showed that the eukaryotic chromosome is a linear array of genes. A gene was defined as any trait that could be separated from a neighboring trait by recombination.

Complementation analysis, raised to its zenith in studies of the *rII* locus of bacteriophage T4, increased the efficiency and precision of genetic analysis and led to the definition of a gene as a unit of DNA that encodes one polypeptide (i.e., a cistron). As more and more was learned about regulatory elements in the control of gene function in prokaryotes, it became clear that the expression of individual "genes" often depends on sequences that lie at some distance from them. Once modern molecular genetic techniques demonstrated multiple polyadenylation sites, multiple splicing sites, translation reading in two frames, and other complexities in eukaryotic genes, the inadequacy of the definition of a gene by classical tests (e.g., complementation) became clear.

The concept of the gene as a biological entity—that is, a heritable function detected by observing the effect of a mutation—is still valid. However, according to the current molecular definition, a gene consists of all the DNA sequences necessary to produce a single peptide or RNA product. Thus, the gene is no longer thought of as a single, contiguous protein-encoding stretch of DNA.

References

General References

Chromatin. 1978. *Cold Spring Harbor Symp. Quant. Biol.*, vol. 42.

DNA Structures. 1983. *Cold Spring Harbor Symp. Quant. Biol.*, vol. 47.

MACGREGOR, H. C., AND J. M. VARLEY. 1983. *Working with Animal Chromosomes.* Wiley.

SUZUKI, D. T., A. J. F. GRIFFITHS, J. H. MILLER, and R. C. LEWONTIN, 1989. *An Introduction to Genetic Analysis.* 4th ed., W. H. Freeman and Company.

Classical Genetics

BEADLE, G. W., and E. L. TATUM. 1941. Genetic control of biochemical reactions in *Neurospora. Proc. Nat'l Acad. Sci. USA* 27:499.

BENZER, S. 1962. The fine structure of the gene. *Sci. Am.* 206(1):70–84.

MORGAN, T. H. 1926. *The Theory of the Gene.* Yale University Press.

PETERS, J. A., ed. 1959. *Classic Papers in Genetics.* Prentice-Hall.

VOELLER, B. R., ed. 1968. *The Chromosome Theory of Inheritance: Classic Papers in Development and Heredity.* Appleton-Century-Crofts.

The Morphology of Chromosomes

DNA Content, Banding Patterns and Maps, Heterochromatin, Loops

BOY de la TOUR, E., and U. K. LAEMMLI. 1988. The metaphase scaffold is helically folded: sister chromatids have predominantly opposite helical handedness. *Cell* 55:937–944.

DONIS-KELLER, H., et al. 1987. A genetic linkage map of the human genome. *Cell* 51:319–337.

FELSENFELD, G., and J. D. MCGHEE. 1986. Structure of the 30 nm chromatin fiber. *Cell* 44:375–377.

GALL, J. G. 1981. Chromosome structure and the C-value paradox. *J. Cell Biol.* 91:3s–14s.

KAVENOFF, R., L. C. KLOTZ, and B. H. ZIMM. 1974. On the nature of chromosome-sized DNA molecules. *Cold Spring Harbor Symp. Quant. Biol.* 38:1–8.

LANDER, E. S., and D. BOTSTEIN. 1986. Mapping complex genetic traits in humans: new methods using a complete RFLP linkage map. *Cold Spring Harbor Symp. Quant. Biol.* 51:49–62.

LYON, M. F. 1972. X-chromosome inactivation and developmental patterns in mammals. *Biol. Rev.* 47:1–35.

MARTIN, G. R. 1982. X-chromosome inactivation in mammals. *Cell* 29:721–724.

MIRKOVITCH, J., M.-E. MIRAULT, and U. K. LAEMMLI. 1984. Organization of the higher-order chromatin loop: specific DNA attachment sites on nuclear scaffold. *Cell* 39:223–232.

MIRKOVITCH, J., P. SPIERER, and U. K. LAEMMLI. 1986. Genes and loops in 320,000 base-pairs of the *Drosophila melanogaster* chromosome. *J. Mol. Biol.* 190:255–258.

SAWYER, J. R., and J. C. HOZIER. 1986. High resolution of mouse chromosomes: banding conservation between man and mouse. *Science* 232:1632–1635.

SPIERER, P., A. SPIERER, W. BENDER, and D. S. HOGNESS. 1983. Molecular mapping of genetic and chromometric units in *Drosophila melanogaster*. *J. Mol. Biol.* 168:35–50.

WIDOM, J., and A. KLUG. 1985. Structure of the 300Å chromatin filament: x-ray diffraction from oriented samples. *Cell* 43:207–213.

Telomeres, Centromeres, and Autonomously Replicating Sequences

BLACKBURN, E. H., and J. W. SZOSTAK. 1984. The molecular structure of centromeres and telomeres. *Ann. Rev. Biochem.* 53:163–194.

CARBON, J. 1984. Yeast centromeres: structure and function. *Cell* 37:351–353.

CLARKE, L., and J. CARBON. 1985. The structure and function of yeast centromeres. *Ann. Rev. Genet.* 19:29–56.

EARNSHAW, W. C., K. F. SULLIVAN, P. S. MACHLIN, C. A. COOKE, D. A. KAISER, T. D. POLLARD, N. F. ROTHFIELD, and D. W. CLEVELAND. 1987. Molecular cloning of cDNA for CENP-B, the major human centromere autoantigen. *J. Cell Biol.* 104:817–829.

GREIDER, C. W., and E. H. BLACKBURN. 1987. The telomere terminal transferase of *Tetrahymena* is a ribonucleoprotein enzyme with two kinds of primer specificity. *Cell* 51:887–898.

HEGEMANN, J. H., J. H. SHERO, G. COTTAREL, P. PHILIPPSEN, and P. HIETER. 1988. Mutational analysis of centromere DNA from chromosome VI of *Saccharomyces cerevisiae*. *Mol. Cell Biol.* 8:2523–2535.

HENDERSON, E., C. C. HARDIN, S. K. WALK, I. TINOCO, JR., and E. H. BLACKBURN. 1987. Telomeric DNA oligonucleotides form novel intramolecular structures containing guanine-guanine base pairs. *Cell* 51:899–908.

MURRAY, A. W., and J. W. SZOSTAK. 1983. Construction of artificial chromosomes of yeast. *Nature* 305:189–193.

SHAMPAY, J., and E. H. BLACKBURN. 1988. Generation of telomere-length heterogeneity in *Saccharomyces cerevisiae*. *Proc. Nat'l Acad. Sci. USA* 85:534–538.

STRUHL, K. 1983. The new yeast genetics. *Nature* 305:391–397.

Chromosomal Proteins: Histones and Nucleosomes

DE BERNARDIN, W., T. KOLLER, and J. M. SOGO. 1986. Structure of *in-vivo* transcribing chromatin as studied in simian virus 40 minichromosomes. *J. Mol. Biol.* 191:469–482.

DILWORTH, S. M., S. J. BLACK, and R. A. LASKEY. 1987. Two complexes that contain histones are required for nucleosome assembly in vitro: role of nucleoplasmin and N1 in *Xenopus* egg extracts. *Cell* 51:1009–1018.

EARNSHAW, W. C. 1987. Anionic regions in nuclear proteins. *J. Cell Biol.* 105:1479–1482.

EARNSHAW, W. C., B. HALLIGAN, C. A. COOKE, M. M. S. HECK, and L. F. HIU. 1985. Topoisomerase II is a structural component of mitotic chromosome scaffolds. *J. Cell Biol.* 100:1706–1717.

EARNSHAW, W. C., and M. M. S. HECK. 1985. Localization of topoisomerase II in mitotic chromosomes. *J. Cell Biol.* 100:1716–1725.

GASSER, S. M., and U. K. LAEMMLI. 1987. A glimpse at chromosomal order. *Trends Genet.* 3:16–22.

GOODWIN, G. H., J. M. WALKER, and E. W. JOHNS. 1978. The high mobility group (HMG) non-histone chromosomal proteins. In *The Cell Nucleus*, vol. 6, H. Busch., ed. Academic Press.

HAGGREN, W., and D. KOLODRUBETZ. 1988. The *Saccharomyces cerevisiae* ACP2 gene encodes an essential HMG1-like protein. *Mol. Cell Biol.* 8:1282–1289.

HEBBES, T. R., A. W. THORNE, and C. CRANE-ROBINSON. 1988. A direct link between core histone acetylation and transcriptionally active chromatin. *EMBO J.* 7:1395–1402.

PHI-VAN, L., and W. H. STRATLING. 1988. The matrix attachment regions of the chicken lysozyme gene co-map with the boundaries of the chromatin domain. *EMBO J.* 7:655–664.

RICHARD, T. J., J. T. FINCH, B. RUSHTON, D. RHODES, and A. KLUG. 1984. Structure of the nucleosome core particle at 7Å resolution. *Nature* 311:532–537.

RICHMOND, T.J., M. A. SEARLES, and R. T. SIMPSON. 1988. Crystals of a nucleosome core particle containing defined sequence DNA. *J. Mol. Biol.* 199:161–170.

SOLOMON, M. J., P. L. LARSEN, and A. VARSHAVSKY. 1988. Mapping protein-DNA interactions in vivo with formaldehyde: evidence that histone H4 is retained on a highly transcribed gene. *Cell* 53:937–947.

TRAVERS, A. A., and A. KLUG. 1987. The bending of DNA in nucleosomes and its wider implications. *Phil. Trans. Roy. Soc.* (Lond.) B317:537–561.

10

Eukaryotic Chromosomes and Genes: Molecular Anatomy

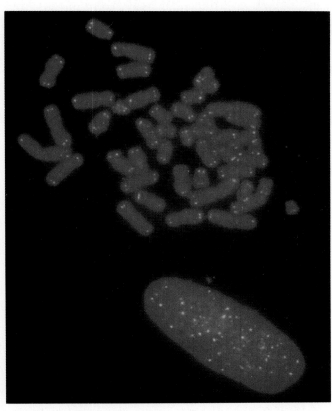

Biotin-labeled telomeric DNA (yellow) hybridized to chromosomes

Classical genetic studies, including detailed examination of chromosome morphology and staining patterns, were generally interpreted to support two ideas: (1) most, if not all, DNA in a species was required for the survival and/or propagation of that species, and (2) chromosomes in all individuals of a species were likely to be very similar with only minor variations in protein-coding sequences. Both of these ideas have been seriously challenged by more recent studies, using DNA cloning and sequencing techniques, that have provided details about various types of DNA sequences and their distribution in eukaryotic genomes.

First of all, sequence analysis has suggested that a very large fraction, perhaps well over 90 percent, of the vertebrate genome does not encode precursors to mRNAs or any other RNAs. No function for the majority of this "extra" DNA has yet been found, and it seems increasingly likely that it has no function. In multicellular organisms, this noncoding DNA contains many regions that are "repetitious" (similar but not identical). Variations within the repetitious DNA stretches are so great that each single human, for example, can be distinguished by a DNA "fingerprint" based on these variations in repetitive sequences. Moreover, some repetitious DNA sequences *are not found in constant positions in the DNA of individuals of the same species. Such "mobile" DNA segments, which are present in both prokaryotic and eukaryotic organisms, undoubtedly cause mutations and therefore may play an important role in evolution, even though they generally have no role in the life cycle of an organism.*

Amidst this sea of unstable, probably nonfunctional DNA lie the genes, islands of coding DNA recognized by classical genetics. As we discussed in Chapter 8, however, coding sequences often contain noncoding regions (introns). Sequencing studies on hundreds of genes have confirmed that introns are common in multicellular plants and animals and less common, but sometimes present, in single-celled eukaryotes. Sequencing of the same protein-coding gene in a variety of species has shown that evolutionary pressure selects for maintenance of relatively similar sequences in exons. In contrast, wide sequence variation, even including total loss, occurs in introns, suggesting that over most of their sequence they are nonfunctional. Cloning and sequencing have also confirmed the widespread existence of "families" of protein-coding genes in which variety is achieved through a drift in the duplicated gene so that different amino acids are encoded. In some cases, however, gene duplicates have accumulated mutations that render them useless for encoding protein, although they are still maintained in the genome. The inescapable conclusion is that loss from animal and plant genomes of useless DNA is a very slow process.

In Chapter 8 we defined transcription units and the DNA elements necessary for their function. In Chapter 9 we considered the large-scale structure of eukaryotic chromosomes, described the functional elements required for their duplication and segregation, and reviewed the classic genetic definition of genes. We also presented a molecular definition of genes that is consistent with the detailed knowledge of DNA sequence now available. In this chapter, we extend the discussion to the molecular anatomy of eukaryotic chromosomes, the functions of specific chromosomal regions, and the effects of various rearrangements of chromosomal DNA. ▲

Major Classes of Eukaryotic DNA

The rules of Mendelian (classic) genetics suggest that each protein-coding DNA segment should occur just once in a germ cell. Although the cloning of many different genes from the germ cells of animals and plants has indicated that some eukaryotic genes are solitary, others are duplicated. The extent of duplication varies among different genes in the same organism and for the same gene in different species. Why some genes remain solitary and oth-

ers become duplicated is unknown. With few exceptions, the pattern of duplication in the germ-cell DNA of an organism is identical to that in its somatic-cell DNA. In multicellular organisms, perhaps a quarter or a half of the protein-coding genes are represented only once in the haploid nucleus with the rest belonging to families of two or more similar genes. Most often, the sequences of duplicated genes have changed over time as the result of the phenomenon known as *genetic drift*.

In contrast to most duplicated protein-coding genes, whose sequences encode similar but slightly different proteins, both 5S RNA and rRNA genes are present in many copies (from several hundred to thousands) most of which have nearly identical coding sequences. In a few families of protein-coding genes (e.g., the histones), the duplicated genes also are somehow kept free or almost free of mutations. A distinguishing feature of many ribosomal and duplicated protein-coding genes with nearly identical sequences is that the multiple copies exist in tandem (i.e., they remain localized in the genome, with one copy next to another).

In addition to the duplicated and tandemly repeated functioning genes, several types of repetitious DNA occur in the genomes of all animals and plants. The majority of these sequences do not encode protein or RNA. Repetitious DNA sequences that are not located in identical places in the chromosomes of individuals of the same species are called *mobile genetic elements*. Some of these sequences may have mechanisms for their own propagation and dispersal within a genome, even though they do not seem to serve a useful function of the organism. Because they seem organized for their own purposes, they have been called "selfish DNA." The major classes of eukaryotic DNA are summarized in Table 10-1. In the following sections, we discuss the properties and functions of these various regions of eukaryotic chromosomes.

Table 10-1 Classification of eukaryotic DNA

Protein-coding DNA
  Solitary genes
  Duplicated and diverged genes (functional gene families and nonfunctional pseudogenes)

Tandemly repeated DNA (genes encoding rRNA, 5S rRNA, tRNA, and some histone genes)

Repetitious DNA
  Simple-sequence DNA
  Intermediate repeat DNA (mobile genetic elements)
    Short interspersed elements (nonviral retroposons)
    Long interspersed elements (viral retroposons and transposons)*

Unclassified spacer or connecting DNA

*In mammalian DNA, long-interspersed elements may be nonviral retroposons.

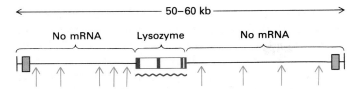

▲ **Figure 10-1** The chicken lysozyme gene and its surrounding regions. The 15-kb transcription unit contains four exons (red) and three introns (white). The positions indicated by arrows are repetitious sequences found elsewhere in the genome. The solid squares (orange) indicate positions of scaffold-attachment regions, or SARs, that bind to the nuclear matrix. No mRNA other than lysozyme could be found in embryonic or adult tissues that hybridize to this region. [See P. Balducci et al., 1981, *Nucleic Acids Res.* **9**:3575.]

Solitary Protein-coding Genes

An example of a solitary protein-coding gene is the well-studied chicken lysozyme gene. Lysozyme, an enzyme that cleaves certain polysaccharides, is found in human tears and is an abundant component of egg-white protein. The transcription unit for lysozyme is a 15-kb sequence that constitutes a single gene and contains four exons and three introns (Figure 10-1). The flanking regions, extending for about 20 kb upstream and downstream from the transcription unit, do not encode any detectable mRNAs. There are repetitive sequences in these flanking regions. Two scaffold-attachment regions (SARs), which may anchor chromosomal DNA to a nonhistone protein scaffold, have been located near the ends of the regions flanking the lysozyme transcription unit. Thus this approximately 50- to 60-kb region may be a separate DNA loop containing only a single gene.

Genes Compose Minor Portion of DNA

A 315-kb region of chromosome 3 of *Drosophila melanogaster* has been intensively studied to determine how many genes it might contain. In the map of chromosome 3 based on the classic *Drosophila* polytene chromosome bands, this region is designated 87D and 87E and contains 12 *chromomeres*, or dark-staining bands (Figure 10-2a). Based on so-called saturation genetic analyses, 12 different complementation groups (i.e., recessive mutations that are lethal and define required genes) lie within this chromosomal region. This genetic result indicates at least 12 genes; however, before cloning was available it was unclear whether this region encodes *only* 12 genes.

When samples of mRNA from embryonic, larval, and adult flies were tested for hybridization with recombinant

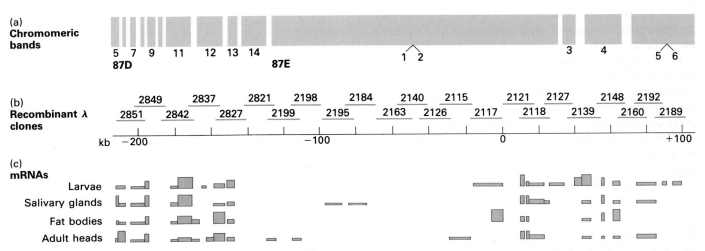

▲ **Figure 10-2** The distribution of mRNA-coding regions in the 87D/E region of salivary gland chromosome 3 of *Drosophila melanogaster*. (a) Diagram of the chromomeric bands (blue) seen in stained polytene chromosome 3 and the position of genes (numbers under diagram) identified by complementation tests. (b) A series of 23 recombinant λ bacteriophages (shown as numbered lines) containing partially overlapping portions of DNA from the 87D/E region were selected. The total length of the region is 315 kb. (c) RNA preparations from *Drosophila* larvae, salivary glands, fat bodies (an organ with primitive liver-like function), and adult heads were tested for mRNAs that hybridized to restriction fragments of the λ clones. Each horizontal interval represents a different restriction fragment. The bars show about 30 different positions to which mRNAs hybridized, and the heights of the bars reflect the amount of mRNA. By the size of the mRNA, its presence in different tissues, or its presence in different amounts (i.e., in amounts different from those of other mRNAs of similar size), at least 43 different nonoverlapping mRNAs have been detected. [See B. Bossy, L. M. C. Hall, and P. Spierer, 1984, *Eur. Mol. Biol. Org. J.* **3**:2537.]

λ bacteriophage clones containing different portions of the 87D/E region, over 40 distinct mRNAs, differing in size and chromosomal location, were detected (Figure 10-2b and c). If each mRNA encodes a different protein, then region 87D/E of chromosome 3 clearly contains many more genes (>40) than chromomeres (12) or than genes defined by saturation genetic analysis. Moreover, the mRNA-coding sequences accounted for only about 33 percent of the 315-kb region examined. In some regions the density of sequences encoding mRNA is much greater than in other regions. For example, in the center of the 315-kb region, a 156-kb stretch encodes only seven mRNAs totaling about 14 kb in length.

This 315-kb region of *Drosophila* chromosome 3 is thought to be typical in its distributions of both required genes and chromomeric bands. Assuming this region also has a typical distribution of genes and noncoding regions, we can estimate that the entire *Drosophila* genome contains about 17,000 genes. Although no such detailed studies of gene density have been done in vertebrate DNA, the general impression is that mRNA-coding regions in vertebrates are less densely packed than in *Drosophila*. The human genome, which is about 20 times the size of the *Drosophila* genome, would therefore encode 300,000 mRNAs perhaps. However, differential RNA and protein processing may lead to as many as 10^6 different human proteins.

Even single-celled eukaryotes (e.g., yeast) contain much "extra" DNA. In yeast, this has been demonstrated by randomly inserting DNA sequences into the chromosome at 200 different places. The finding that only 20–30 percent of such insertions produce any detectable effect suggests that all eukaryotes—including simple ones—contain large amounts of "nongenetic," perhaps nonfunctional, DNA.

Duplicated Protein-coding Genes

All protein-coding genes are not solitary like the gene for lysozyme. Frequently, the DNA that lies within 5–10 kb of a known gene contains sequences that are close copies of the gene. Such *duplicated* protein-coding genes probably constitute half of the protein-coding DNA of vertebrate genomes (e.g., chickens, mice, and humans).

Sequence Homology in Protein Families Reflects Gene Duplication

The documentation by DNA sequencing of duplicated protein-coding regions was expected because families of proteins had already been recognized based on similarities in the amino acid sequences of different proteins (both within and among species). Many of these protein families contain from several to as many as 20 members with similar but not identical sequences. A few families may contain hundreds of members (Table 10-2). Examples include the proteins that make up the cytoskeleton, which are found in almost all cells in varying amounts and which consist of several different protein families. In vertebrates, this group includes the actins, tubulins, and microfilament proteins like the keratins. The β-like globins are another well-studied protein family with a limited number of members. The protein families with the largest number of members within a species include the histones and variable regions of immunoglobins. In some species, more than a thousand duplicate genes may encode these proteins.

α- and β-Tubulins That gene duplication is a constant and important theme in evolution can be illustrated by considering the tubulins that exist in microtubules as dimers of α- and β-type chains. For example, all yeasts contain both α- and β-tubulin, whose sequences exhibit similarities suggestive of gene duplication of a primordial tubulin gene. In the fission yeast *Schizosaccharomyces pombe*, the β-tubulin gene is itself also duplicated. Some protozoa (e.g., trypanosomes) contain several copies of both α- and β-tubulin genes. Some invertebrates (e.g.,

Table 10-2 Protein families in vertebrates and invertebrates: common and specific

Family	Number of proteins in family
COMMON PROTEINS	
Actins	5–30
Heat-shock proteins	3
Histones	100–1000
Keratins	>20
Myosin, heavy chain	5–10
Protein kinases	10–100s
Transcription factors	10–100s (?)
Tubulins, α and β	3–15
INSECT PROTEINS	
Eggshell proteins (silk moth and fruit fly)	50
VERTEBRATE PROTEINS	
Globins (many species)	
α-globin	1–3
β-like globins	5
Immunoglobins, variable regions (many species)	500
Ovalbumin (chicken)	3
Transplantation antigens (mouse and human)	50–100
Visual pigment protein (human)	4
Vitellogenin (frog, chicken)	5

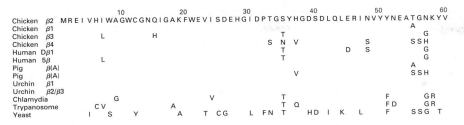

Positions 1–60

```
                      10        20        30        40        50        60
Chicken  β2   M R E I V H I W A G W C G N Q I G A K F W E V I S D E H G I D P T G S Y H G D S D L Q L E R I N V Y Y N E A T G N K Y V
Chicken  β1                                                                                                                        A
Chicken  β3               L           H                           T                                                               G
Chicken  β4                                                 S   N   V                       S                                   S S H
Human   Dβ1                                                       T               D     S                                         G
Human    5β               L                                       T                                                             A G
Pig      β(A)                                                                                                                   S S H
Pig      β(A)                                                        V
Urchin   β1
Urchin   β2/β3
Chlamydia             G                       V               T                                   F             G R
Trypanosome        C V                   A                   T Q                                 F D             G R
Yeast           I     S         Y       A     T   C G       L F N T       H D   I   K   L           F       S S G   T
```

Positions 400 to C-terminus

```
                         410        420        430        440
Chicken  β1/β2   E G M D E M E F T E A E S N M N D L V S E Y Q Q Y Q D A T A D E Q G E • F E E E G E E D E A B
Chicken  β3                                                     E E         •         A     E A E B
Chicken  β4                                                     E E     • M Y   D D     E S E Q G A K B
Chicken  β5                                         E           N D G E   A     D D E       I N E B
Human   Dβ1                                                     E E E D •   G     A       • B
Human    5β                                                     •       •       A     E V   B
Pig      β(A)/β(B)                                                                             B
Urchin   β1                                                     E E     •   D     E G       E A A B
Urchin   β2/β3                                                 E E     •   D     E G •     E A A B
Chlamydia                                                       S E E     •     G E       A B
Trypanosome                                                     I E E     •   D     E Y B
Yeast                                               E           V E D D E   • V D   N   D F G F A P Q N Q D E P I T E N F E B
```

◄ **Figure 10-3** Comparison of homologous regions of β-tubulins within species (chicken) and among species. The β-tubulins range in length from 440–460 amino acids. The sequence of the two most homologous regions (1–60 and 400 to C-terminus) are shown for chicken β1-tubulin; only differences in the sequences of the other β-tubulins are indicated. The single-letter amino acid code is used. Dots indicate deletions. *From D. W. Cleveland and K. F. Sullivan, 1985,* Ann. Rev. Biochem. *54:331.*

Drosophila) have four α- and four β-tubulin genes, whereas others (e.g., sea urchins) have 10–15 α-tubulin and 10–15 β-tubulin genes. All vertebrates also have 10–15 genes for both tubulins. Thus gene duplications within different evolutionary lineages clearly can and do occur.

Comparison of the amino acid sequences of β-tubulins from species ranging from single-celled eukaryotes to humans indicates that gene duplication produces variations on a constant theme. Thus, the overall size of β-tubulins (440–460 amino acids) is similar, and some regions have nearly identical sequences both within and across species (Figure 10-3). On the other hand, some regions exhibit considerable sequence variation in going from yeast to vertebrate tubulins. It is likely that the highly conserved regions represent protein domains involved in molecular contacts that must be closely preserved; the variable regions possibly are not involved in such contacts but may allow flexibility to correctly position the conserved domains.

β-Like Globins Hemoglobin from all vertebrates contains two α-type and two β-type globin polypeptides (see Figure 2-5). Humans and all other mammals examined produce several slightly different β-like globin chains during embryogenesis and in adult life. In humans the β-like globin family includes β, δ (delta), $^A\gamma$ and $^G\gamma$ (A- and G-gamma), and ε (epsilon) globin. All the β-like globins have sequences that are similar but not identical. For example, the amino acid sequence identity between β and δ approaches 90 percent; that between β and $^G\gamma$ is more than 80 percent; and between $^G\gamma$ and $^A\gamma$ there is a long stretch of about 500-nucleotide identity. As we will describe later, such patches of identity between individuals in a species indicates *gene conversion* at that site. The different β-like globins are formed successively during embryogenesis, but only the β and δ chains continue to be formed after 6 months of age.

The β-like globin family typifies many duplicated genes and may shed some light on their evolutionary history. Several mammalian β-like globin gene clusters, including human ones, have been isolated and mapped by recombinant DNA techniques. The β-like globin genes appear in similar long DNA regions in rabbits, mice, sheep, chickens, and frogs, although the details of the duplicates are different in different species. All the β-like genes in humans lie within a region of about 50 kb on chromosome 11 (Figure 10-4). In the map of this DNA region, the

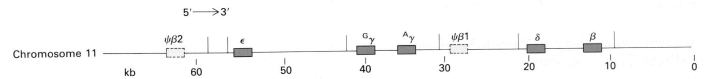

▲ **Figure 10-4** Map of the β-like globin gene cluster on human chromosome 11. The positions of the five genes known to encode β-like globins are indicated by green boxes. The direction of transcription is from left to right for all these genes. Two nonfunctional pseudogenes, ψβ1 and ψβ2, also lie in this region (yellow boxes). Vertical lines (red) mark the sites of repetitive DNA sequences, which occur at many other sites in the human genome. [See E. F. Fritsch, R. M. Lawn, and T. Maniatis, 1980, *Cell* 19:959.]

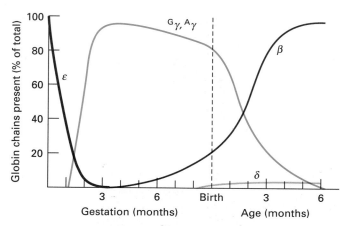

▲ **Figure 10-5** Proportion of total β-like globin protein in human red blood cells accounted for by the individual β-like globins (ε, γ, δ, and β) during gestation and the first 6 postnatal months. Comparison of these curves with map positions shown in Figure 10-4 indicates that the β-like genes are sequentially expressed during development. Only the β- and δ-globins are produced after 6 months. [See R. M. Winslow and W. F. Anderson, 1983, in *The Metabolic Basis of Inherited Disease*, 5th ed., J. G. Stanbury et al., eds., McGraw-Hill, p. 1666.]

genes are arranged from left to right in the same order in which they are expressed in development (Figure 10-5). The δ and β genes are actually expressed around the same time, but expression of δ is much lower than that of β.

The exons within all the β-like globin genes are nearly identical in length and have similar sequences; each transcription unit also has two introns that occur in the same position in each gene. The design of this transcription unit is conserved in all globin genes—both the β-like genes and the α genes—in all vertebrates (Figure 10-6). Thus there is great evolutionary stability in the structure of all these genes, even though the α genes and β-like genes are no longer located on the same chromosome. The implications of the structural conservation of the globin genes for the early evolutionary history of cells are discussed in Chapter 26.

As Figure 10-7 shows, the DNA sequences upstream from the globin transcription units are also similar for more than 100 bases preceding the 5′ cap site; the most highly conserved regions center on positions −80 and −30. These regions are known to be necessary for accurate and frequent initiation of RNA synthesis by RNA polymerase II. Thus, the DNA duplications that occurred and became fixed in various vertebrates included not only the sequences encoding the β-globin transcription unit but also regions flanking it. The sequence conservation of DNA binding sites for proteins that act in regulation is frequently greater than surrounding sequences; the ~20-nucleotide binding regions often can be detected by comparing sequences of a particular gene from several different vertebrates.

Other Duplicated Gene Families

Other protein families may not exhibit sequence similarity to the same extent as the β-tubulins and β-like globins. Nonetheless, sequence comparisons often clearly indicate that a newly cloned and sequenced gene belongs to an already identified gene family. Examples of classes of proteins with recognizable sequence motifs include the surface-receptor proteins, the many related protein kinases, and the several different families of transcription factors.

As noted already, the globin-gene duplicates are all complete; that is, the whole gene and flanking regions are duplicated. In addition, all the duplicates are in tandem—that is, aligned in the same direction. We briefly describe three other protein families in this section to illustrate

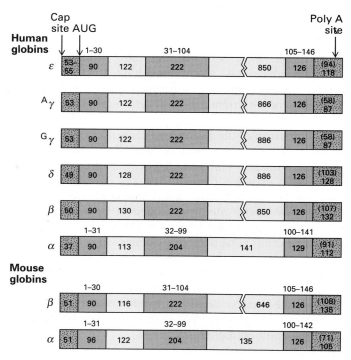

▲ **Figure 10-6** General design of transcription units for α- and β-like globins, beginning with 5′ cap site and extending to 3′ poly A site, in humans and mice. All of these globin genes have two introns (yellow), which are located at similar positions within each gene. The lengths of the introns also are similar, although the second intron in the β-like genes is much longer than that in the α genes. All of these genes contain three exons (orange), some portions of which are untranslated (*stippled areas*). The numbers within the boxes indicate the number of nucleotides present in each region of the primary transcript; the numbers above the boxes designate the corresponding amino acid positions in the resulting polypeptides. In the right-hand boxes, the numbers in parentheses indicate the length of the 3′ untranslated region up to the first A of the AAUAAA sequence; the other numbers indicate the length of the entire 3′ untranslated region, up to the site of poly A addition. [See A. Efstratiadis et al., 1980, *Cell* **21**:653.]

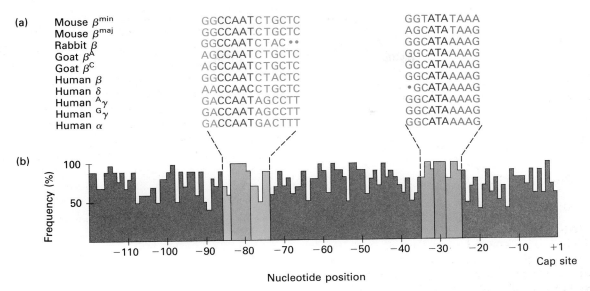

▲ **Figure 10-7** Similarity of sequences just upstream from 5′ cap site in globin genes. The sequences of about 150 nucleotides were aligned to give maximum similarity. (a) Sequences of two highly conserved regions—the CCAAT and the ATA boxes—are shown in red type. These regions are present in all the globin genes sequenced to date. (b) The frequency of occurrence (%) of the predominant nucleotide at each position in the upstream regions of the ten genes listed in part (a) is shown in a histogram. [See A. Efstratiadis et al., 1980, *Cell* 21:653.]

some variations on the theme of gene duplication in evolution.

The chicken genome contains three contiguous, structurally similar genes—the ovalbumin *(OV)* gene and the X and Y genes—located within a 40-kb segment of DNA. Ovalbumin is the major protein of egg white; the Y proteins are made in the same cells at the same time as ovalbumin but in much smaller quantities. The primary sequences of the three genes reveal many similarities in design. Each gene contains seven introns that divide the proteins into eight exons (Table 10-3). Although the introns in corresponding positions in the three genes vary in length, the corresponding protein-coding exons are approximately equal in length. The last exon (no. 8), which contains the 3′ untranslated region, is the only one that varies substantially in length. In addition to their constancy in length, the exons exhibit many fewer base changes than the introns do. Thus the X and Y genes appear to be duplicates that function like the ovalbumin gene but at a greatly reduced rate.

Mammmalian albumin and α-fetoprotein genes provide an even more striking example of the duplication of a lengthy transcription unit without alteration of the basic structure. Both albumin and α-fetoprotein contain three domains, the folding of which is stabilized by disulfide bridges between cysteine residues (Figure 10-8, *left*). The genes encoding these two proteins both have 14 introns and 15 exons. Although the lengths of the corresponding introns in the two genes vary by as much as a factor of 3, the lengths of the corresponding exons are quite similar (Table 10-4). Not only does the overall structural similarity of these two genes indicate that a gene duplication occurred, but their detailed internal structure suggests an even earlier duplication in evolution. Twelve of the exons (3–14) fall into three subgroups, which generate three protein domains that all have disulfide linkages placed so that the polypeptide chain folds similarly. The total lengths of the four exons in each of these domains are similar, both within each gene and between the albumin and α-fetoprotein genes.

Table 10-3 Lengths (in base pairs) of exons and introns in chicken egg-white gene region

Gene	Exons								Introns						
	1	2	3	4	5	6	7	8	a	b	c	d	e	f	g
OV	47	185	51	129	118	143	156	1043	1560	251	582	401	1029	323	1614
X	?	195	59	137	127	145	167	1503	1587	604	894	523	778	261	898
Y	?	199	54	142	128	148	165	1146	1619	153	357	808	235	50	886

SOURCE: R. Heilig et al., 1980, *Cell* **20**:625.

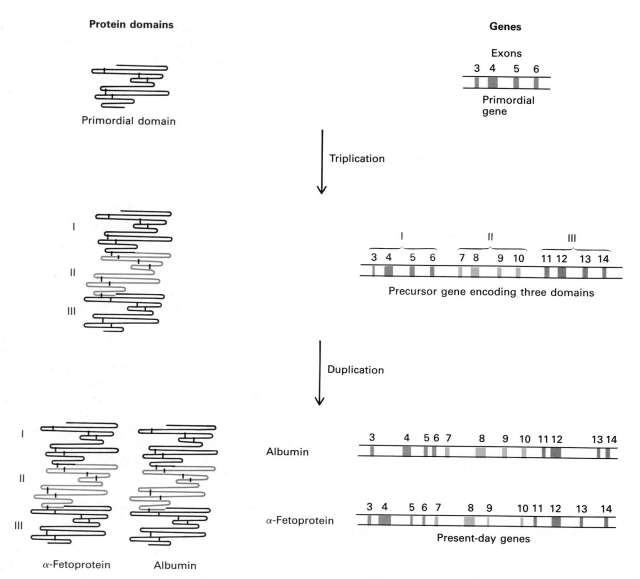

▲ **Figure 10-8** Proposed model for evolution of present-day mammalian albumin and α-fetoprotein, which both contain three similar folded domains (I, II, and III) stabilized by disulfide bonds (black crossbars). Each domain is encoded by four exons whose lengths have a similar pattern—short, long, short, short (see Table 10-4). Triplication of a primordial gene containing four exons (corresponding to exons 3, 4, 5, and 6 in the present-day genes) could have produced a three-domain precursor gene. Duplication of this precursor gene and subsequent sequence drift would result in the present-day genes. [See M. B. Gorin et al., 1981, *J. Biol. Chem.* **256**:1954.]

This structure suggests that a primordial form of these genes had four exons that encoded a single protein domain, and that this region was triplicated to produce the first albumin-like gene encoding three domains. Later, duplication of this three-domain gene occurred to produce the present-day albumin and α-fetoprotein genes (Figure 10-8, *right*). Thus the intron-exon pattern existed in the primordial gene and was retained during the successive triplication and then duplication.

The chorion proteins in silk moths provide a final example of duplicated protein-coding genes. As insect eggs develop, they form an outer shell, called a *chorion,* that is composed of organized layers of fibrous proteins. Many of these proteins have similar amino acid sequences and are encoded by a large number of duplicated genes, whose relatedness has been proved both by amino acid sequencing of the proteins and by DNA sequencing of isolated genes. The chorion genes fall into three classes termed A, B, and C. The similarities within each class are greater than those between classes. All chorion genes, however, have a single intron. In the chromosome, the A genes and the B genes exist in A/B pairs; several such pairs

Table 10-4 Lengths of the exons in the mouse genes for α-fetoprotein and serum albumin

Domain encoded	Exon no.	Length (bp) α-Fetoprotein	Albumin
	1	114 ± 32	110 ± 29
	2	53 ± 14	75 ± 27
I	3	148 ± 37	121 ± 48
	4	218 ± 52	240 ± 49
	5	144 ± 26	170 ± 70
	6	104 ± 22	118 ± 33
II	7	133 ± 29	137 ± 35
	8	230 ± 23	218 ± 65
	9	154 ± 67	136 ± 30
	10	125 ± 35	125 ± 28
III	11	135 ± 45	153 ± 31
	12	280 ± 70	222 ± 42
	13	175 ± 50	122 ± 36
	14	69 ± 27	62 ± 26
	15	149 ± 33	112 ± 31

SOURCE: D. Kioussis et al., 1981, *J. Biol. Chem.* **256**:1960.

can reside in a chromosomal segment of 10–15 kb. Each pair of genes represents a duplicated DNA segment, with perhaps as many as 50 such segments in the genome. Interestingly, the A and B genes of a pair are transcribed in opposite directions. In each duplicated set, the two characteristic features—the divergent transcription and the single intron—are maintained. Seemingly, to arrive at the present structure, a primordial gene must have been duplicated, so that one copy became inverted, and then the pair of genes must have been duplicated as a unit. Duplication events therefore are not limited to single genes (i.e., to DNA encoding single proteins); rather, they may involve a section of a chromosome.

Pseudogenes Are Duplications That Have Become Nonfunctional

At least two regions in the human β-like globin gene cluster and three regions in the mouse β-like globin cluster have nonfunctional sequences similar to those of the functional β-like globin genes. Because no known protein corresponds to these regions, they are called *pseudogenes* (see Figure 10-4). Sequence analysis shows these copies retain the same apparent exon-intron structure, suggesting they were originally duplicates. However, *sequence drift* has apparently resulted in the accumulation of sequences that either terminate translation or block mRNA

processing, rendering such regions nonfunctional even if they were transcribed into RNA.

The human δ-globin gene may represent an intermediate in this process of drift. As noted earlier, this gene produces very little mRNA; further sequence drift that completely halted the activity of such an infrequently used gene duplicate might well be tolerated by the organism. Such a "silencing" genetic event apparently occurred in the δ gene of gibbons some 5–10 million years ago. The initial mutation was probably in the promoter region; this gene eventually accumulated enough mutations to interrupt its coding function. Present-day gibbons survive perfectly well on one adult β-like globin gene.

Besides the pseudogenes in the globin family, there are numerous other instances of extra near-copies of protein-coding genes. These occur, for example, in the tubulin and actin families of many different organisms. In some species, analysis of the mRNAs for these proteins has shown that only one or a few of the duplicated genes are active. Sequencing the genomic copies of such genes to determine whether they encode proteins is perhaps the easiest means of determining whether a duplicate gene is functional. In addition to the complete but nonfunctional gene copies that constitute pseudogenes, partial copies of some genes have been identified. For example, fragments from the 5′ and 3′ ends of the tubulin genes in humans are quite common. (As we will discuss in detail later, other nonfunctional gene copies exist as a result of copying mRNA into DNA and integrating this promoter-less DNA into the chromosome.)

Gene Duplication May Result from Unequal Crossing Over

It is important to remember that the genomes of present-day organisms are the result of evolution. That the genomes of such simple organisms as slime molds, insects, and echinoderms contain large gene families implies that gene duplication was an early evolutionary mechanism important in genome expansion.

No completely satisfactory molecular model has yet been proposed, much less experimentally verified, to explain how gene duplication occurred during evolution. One possible mechanism involves unequal crossing over between repetitive DNA sequences during meiosis. However, in order for such a crossover to give rise to a functional duplicated gene, it must occur in such a way as to include the entire transcription unit.

Gene Duplication Permits Expansion and Specialization of Gene Function

The examination of cloned genes has demonstrated that genomes in both vertebrates and invertebrates contain long noncoding regions between functional transcription units. As discussed earlier, in one well-studied region of chromosome 3 in *Drosophila melanogaster,* as much as two-thirds of the DNA appears not to encode any mRNAs (see Figure 10-2). Even in the case of the duplicated genes in the human β-like globin gene region, only 15–20 percent of the DNA is occupied by active transcription units. And in genes with longer introns than in the globin genes, as little as 2 percent of the DNA in the duplicated region encodes protein. Although the absence of a known gene in a region of DNA does not prove that no gene is present, sequence data often show that protein stop codons are present in all possible reading frames in spacer DNA. The prevalence of apparently unused stretches of DNA in higher animals indicates the relative sluggishness with which "useless" sequences are lost from these genomes. Unused duplicate genes, together with large regions of intragene spacer DNA, obviously account for a significant fraction of the noncoding DNA that is such a prominent feature of eukaryotes.

Although the duplication of sequences may lead to an accumulation of unused DNA, it is equally clear that gene duplication has a great impact in evolution. Duplication of functioning genes allows the evolution of multiple similar proteins, which may be used at different times during the life cycle of an organism or in the different tissues of an adult. For example, some of the genes in the β-like globin cluster are functional duplicates used at special times in the life cycle. The extra copies of the chorion genes in silk moths and fruit flies are active at different times in the construction of eggs. And different actin and tubulin genes are expressed in different tissues. Gene du-

plication is a critical evolutionary mechanism that has allowed expanded and specialized gene functions to develop.

Tandemly Repeated Genes Encoding rRNA, tRNA, and Histones

The genes for 45S pre-rRNA, 5S rRNA, various tRNAs, and one family of proteins, the histones, occur in invertebrates and some vertebrates as *tandemly repeated arrays.* Most often copies of a sequence appear one after the other, in a head-to-tail fashion, over a long stretch of DNA. Within a tandem array of rRNA or tRNA genes, each copy is exactly, or almost exactly, like all the others. Although the transcribed portions of ribosomal genes are the same in a given individual, the nontranscribed spacers can vary. Arrays of tandemly repeated histone DNA are somewhat more complex; however, each histone gene, too, has multiple identical copies.

Repeated Genes Are Necessary to Meet Cellular Demand for Some Transcripts

Ribosomal RNA and tRNA   As we have noted before, most of the RNA in a cell consists of tRNA and rRNA. Assuming RNA polymerase molecules move at a

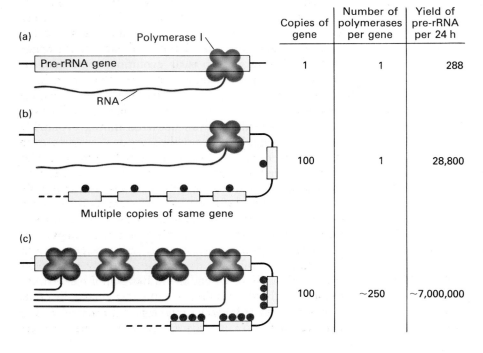

	Copies of gene	Number of polymerases per gene	Yield of pre-rRNA per 24 h
(a)	1	1	288
(b)	100	1	28,800
(c)	100	~250	~7,000,000

◀ **Figure 10-9**   HeLa cells, which have a doubling time of 24 h, have between 5×10^6 and 1×10^7 ribosomes per cell. (a) Because RNA polymerase I takes about 5 min to make the 45S pre-rRNA primary transcript, one pre-rRNA gene with one polymerase per gene would make about 288 copies in 24 h. (b) Here 100 copies of the gene are available, but each is still transcribed by only one polymerase molecule at a time (red dot); thus the number of transcripts produced is still insufficient. (c) Again, 100 gene copies are available, but because each is transcribed by many polymerase molecules at any given time, ribosomes are produced in the necessary quantity. To generate the needed amount of rRNA, cells must have multiple gene copies that are near maximally loaded with RNA polymerase I.

fixed speed, a single gene must have a limit on the number of RNA copies it can provide during one generation, even if it is fully loaded with polymerases. If more RNA is required than can be transcribed from one gene, multiple copies of the gene appear necessary, as illustrated in Figure 10-9 for the synthesis of pre-rRNA in HeLa cells. Given the cell's requirement for 5–10 million ribosomes in each generation, a human cell needs all 100 copies of the 45S pre-rRNA genes that it has, and most of these must be close to maximally active for the cell to divide every 24 hours. Mutants of *Drosophila* called "bobbed" (because they have stubby wings) lack a full complement of the repeated ribosomal RNA genes. If the number of rRNA genes in a bobbed mutant is less than about 50, it is a recessive lethal mutation.

The genes for all the structural RNAs exist in multiple copies in eukaryotic cells (Table 10-5). Both pre-rRNA and 5S-rRNA genes are present in 100 or more copies in all species including yeasts; more than 20,000 copies of the 5S gene are present in frogs. The copy number for individual tRNA genes ranges from 10–100. The multiple copies of all these genes appear in tandem arrays.

Nonhistone Cellular Proteins   With the exception of histones, most cellular proteins are encoded by solitary, nonrepeated genes. (Note that *duplicated* genes, which encode related but nonidentical proteins, are distinguished from *repeated* identical, or nearly identical,

genes.) In general, the rate of synthesis of mRNAs from solitary protein-coding genes is sufficient to meet the cell's requirements for various individual proteins.

For example, the synthesis by RNA polymerase II of one mRNA molecule corresponding to a 15-kb protein-coding gene takes about 3 min. If a single copy of this gene were transcribed by 20 polymerase molecules at a time, then about 10,000 mRNA molecules would be produced in 24 h, the doubling time of most animal cells in culture. Some vertebrate mRNAs are not stable for long periods; however, even if only 10 percent of the mRNA molecules produced from a single gene in one generation were present in the cell at any one time, this number would still suffice for synthesis of a large amount of a single protein. Translation of an average mRNA molecule, which is associated with 5–10 ribosomes, takes about 1 min. Thus 1000 molecules of a particular mRNA (about 0.2–0.4 percent of the total in an average-sized cell) could direct the synthesis of up to 1.4×10^7 protein chains per generation (1000 mRNAs $\times$ 10 ribosomes $\times$ 1440 min = 1.4×10^7). This amount of protein represents about 0.5–1 percent of the total protein of a cell, and very few nonhistone proteins make up such a high proportion of the total. Moreover, if an mRNA had a longer half-life or if the density of polymerases transcribing a gene were higher than in this example, a single protein-coding gene could be responsible for even more protein. A long mRNA half-life and continuous translation almost surely account for the high productivity of the silk fibroin gene in silk moths. The mRNA produced from this solitary gene encodes over 30 percent of the total protein of the silk gland.

Table 10-5   Copy number of tandemly repeated genes encoding structural RNAs in several eukaryotes*

Species	Number of copies		
	pre-rRNA genes	5S-rRNA genes	tRNA genes†
Saccharomyces cerevisiae	140	140	250
Dictyostelium discoideum	180	180	?
Tetrahymena pyriformis			
Micronucleus‡	1	300	800
Macronucleus	200	300	800
Drosophila melanogaster			
X chromosome	250	165	860
Y chromosome	150	165	860
Xenopus laevis	450	24,000	1150
Human	~250	2000	1300

*The copy numbers in this table were estimated by hybridizing saturating amounts of labeled RNA to DNA.
†The tRNA numbers include all tRNA sites and therefore represent more than 50 different tRNA genes in some organisms. Copy numbers for individual tRNAs range from 10–100.
‡The micronucleus is inactive in synthesis of pre-rRNA.
SOURCE: B. Lewin, 1980, *Gene Expression*, vol. 2, Wiley, p. 876.

Histones   One group of proteins—the histones—are present in especially large amounts in all eukaryotic cells. The combined weight of the major histones in the cell nucleus equals that of the DNA, and each of the major species of histone makes up at least 0.5–1 percent of the total cellular protein. And the half-life of histone mRNA is only a few minutes. Moreover, the bulk of histone mRNA and protein synthesis occurs during the S phase, which is about one-third of the total cell cycle. Probably for this reason, the various histone genes are present in multiple copies (from 50–500) in all cells of multicellular organisms. Recent evidence, however, indicates they may not all be required. Yeast cells, for example, have only two copies of the various histone genes, and one copy can suffice for growth. All the multiple copies of histones in higher animals do not necessarily function. Perhaps some of the extra multiple histone genes have special functions in particular cell types, or they may represent a reservoir of genes that the organism can draw upon if higher transcription activity is required at some stage in the life cycle.

In a number of invertebrates (e.g., sea urchins) and in some vertebrates (e.g., frogs and newts), all the members

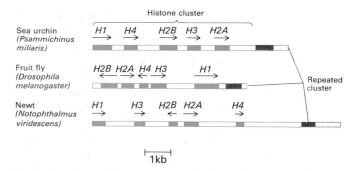

▲ **Figure 10-10** Maps of histone gene clusters in three animals. In these species, each histone gene cluster is 5–6 kb long, although in some vertebrates (e.g., birds and mammals), the individual histone genes are not clustered this closely together. Each gene cluster is repeated (red bars) to form tandem arrays. Note that the direction of transcription (*arrows*) may not be the same for all genes in a cluster, suggesting that each gene is an independent transcription unit. [See C. C. Hentschel and M. L. Birnstiel, 1981, *Cell* **25**:301.]

of the histone gene family are clustered together in a 5- to 6-kb region, and the entire group is repeated in tandem arrays. As illustrated in Figure 10-10, the histone genes within these clusters can be oriented differently within the chromosome, indicating that each histone gene is an independent transcription unit.

Perhaps the most striking aspect of histone gene structure within a given species is the constancy of the base sequences in the coding portions of the tandem arrays. Just as with ribosomal genes, coding regions within repeated histone genes in a species have nearly identical sequences; however, variations in the spacer sequences between genes or gene clusters are found even within in-

dividual animals. Even though the sequences of histone proteins are highly conserved among different species, some variations in the mRNA-coding regions do occur, most often in the third bases of codons.

The distances between individual histone genes are greater and the repetition frequency is somewhat lower in mammals than in invertebrates. For example, histone genes are repeated from 20–50 times in human cells, and most histone genes are no closer than about 10 kb. However, the sequencing of a number of scattered mouse and human histone genes and their surrounding DNA has shown that even though the genes are not clustered close together, DNA sequences of repeated protein-coding regions are still almost identical.

Spacer Length between Tandem Genes Varies

These sequences of tandemly repeated genes discussed above are very nearly exactly alike, both within any individual organism and, to a considerable extent, from one individual to another within a species. In contrast to the sequence constancy of transcribed regions, the nontranscribed spacer regions between individual copies of tandemly repeated genes often vary. For example, the

▶ **Figure 10-11** (a) Electron micrograph of partially denatured 5S rDNA from the frog *Xenopus laevis.* The "loops" correspond to spacer regions, which are rich in A and T residues and hence melt more easily than the transcribed regions. The spacer loops vary in length, whereas the transcribed coding regions are all the same length. (b) Sequence of spacer region between 5S-rRNA genes in *X. laevis*, starting from the 5′ G that begins the transcribed portions of 5S rDNA. The end of the partially repetitive AT-rich portions of the spacer DNA (blue) begins about 50 bases (−50) upstream from the transcription start site. This sequence is shown in a stacked arrangement (the last nucleotide in each row is actually adjacent to the first nucleotide in the row below) to emphasize the repeated portions. About 200 bases upstream of the coding portion of the gene is a sequence (red) that appears a variable number of times ($n = 2$ to 12) in each spacer. This short, variably repeated sequence is responsible for all of the length variation among the spacers of a tandem 5S-gene array. [See N. V. Federoff, 1979, *Cell* **16**:697.] *Part (a) from D. Brown, P. C. Wensink, and E. Jordan, 1971,* Proc. Nat'l Acad. Sci. USA **68**:*3175; courtesy of D. D. Brown.*

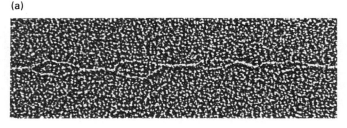

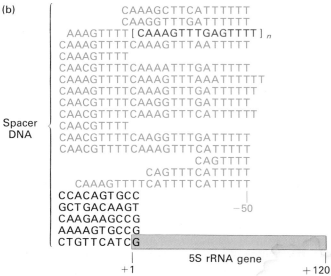

(a)

(b)

Spacer DNA

```
          CAAAGCTTCATTTTTT
          CAAGGTTTGATTTTTT
     AAAGTTTT [ CAAAGTTTGAGTTTT ]ₙ
     CAAAGTTTTCAAAGTTTAATTTTT
     CAAAGTTTT
     CAACGTTTTCAAAATTTGATTTTT
     CAAAGTTTTCAAAGTTTAAATTTTT
     CAAAGTTTTCAAAGTTTGATTTTT
     CAACGTTTTCAAGGTTTGATTTTT
     CAACGTTTTCAAAGTTTCATTTTT
     CAACGTTTT
     CAACGTTTTCAAGGTTTGATTTTT
     CAACGTTTTCAAAGTTTCATTTTT
                    CAGTTTT
               CAGTTTCATTTTT
     CAAAGTTTTCATTTTCATTTTT
CCACAGTGCC
GCTGACAAGT                   −50
CAAGAAGCCG
AAAAGTGCCG
CTGTTCATCG
              5S rRNA gene
          +1               +120
```

spacer regions between the 5S tandem repeats in the frog are composed of a series of repeated (or nearly repeated) oligonucleotides about 15 bases long. The number of oligonucleotide repeats is not constant, leading to variation in the length of the spacer DNA between the 5S genes within an individual (Figure 10-11). Among species, the spacer sequences have completely diverged (e.g., between *Xenopus laevis* and *Xenopus borealis,* two frog species), whereas the DNA sequences encoding 5S rRNA and pre-rRNA have been conserved to a considerable extent.

High-Frequency Unequal Crossing Over and Gene Conversion May Help Maintain Sequence Constancy in Gene Copies

We suggested earlier that unequal crossing over during meiosis could lead to gene duplication (Figure 10-12). The duplicated genes would be subject to sequence drift as the result of accumulated mutations, leading eventually to a family of related but nonidentical genes. Such a process could explain the origin and gradual divergence of the individual members of protein-coding gene fami-

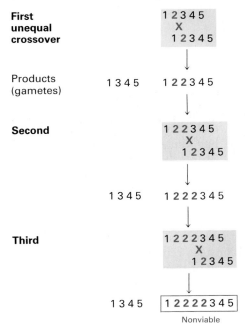

▲ Figure 10-12 Example of crossing over during meiosis involving one copy (**2**) of a tandemly repeated gene after successive crossover events. Each number represents a particular copy *(repeat)* of the same gene (e.g., an rRNA gene) in an array. By the third crossover, one recombinant has a very high frequency of repeat **2**. If this repeat contained a mutation that prevented synthesis of a functional rRNA, then the progeny containing the high repeat number of **2** would not survive. Thus mutations that are damaging would be prevented from spreading in a population.

lies. In the case of tandemly repeated arrays of structural RNA and histone genes, however, some additional mechanism must operate to offset sequence drift and maintain sequence uniformity in the gene copies.

In the past, various hypotheses have been proposed to explain the sequence constancy found in tandem arrays of repeated genes. According to one hypothesis, repeated genes are regenerated in each gamete through amplification of one copy of the repeated gene family. Although this mechanism would indeed ensure uniformity in the copies of the genes subjected to it, it was ruled out when the lengths of the spacer regions between copies of the 5S-rRNA gene were shown to vary within a single individual. This individual variation implies that the 5S-gene copies could not arise by the copying of a single 5S gene and its spacers in each gamete. The maintenance of sequence identity in tandemly repeated genes within a species is now believed to occur through one of two mechanisms: (1) frequent unequal meiotic crossing over or (2) gene conversion.

Sequence Maintenance by High-Frequency Unequal Crossing Over If we assume that the structures of rRNA molecules have gradually been perfected and that no change is easily tolerated by the protein-synthesizing apparatus, then an individual with a large proportion of mutant rRNA would be unlikely to survive. Thus unequal crossing over during meiosis that increased "bad" DNA copies would produce individuals at a selective disadvantage (Figure 10-12). As a result, the mutant copies would eventually be lost from the population. However, individuals with extra copies of a nonmutant rRNA gene would survive. The chromosome would thereby be constantly "purified" of deleterious mutations as successive generations passed. Because spacer DNA (at least in 5S-rRNA genes) is made up of short, repeated oligonucleotides and because the lengths of spacers vary, they could very well be the actual sites at which the proposed high-frequency unequal crossing over occurs.

Recombinant DNA studies with yeast rDNA have shown that unequal crossing over does, in fact, occur frequently within rRNA gene clusters. To demonstrate this, two haploid strains mutant in *LEU2* (a gene encoding one of the enzymes needed in leucine synthesis) were transformed with a cloned plasmid containing rDNA and an inserted wild-type *LEU2* gene. Within the transformed yeast cell, the plasmid was integrated into the cluster of RNA genes and served to mark the rDNA. The two transformed haploid strains—each containing the *LEU2* gene— were mated to produce a diploid strain, which was then sporulated to form four haploid spores from each cell. If no unequal crossing over occurred, each spore should have retained the *LEU2* marker; however, about 2 percent of the spores could not grow on leucine-free medium, indicating they had lost the *LEU2* gene (Figure 10-13a).

(a)

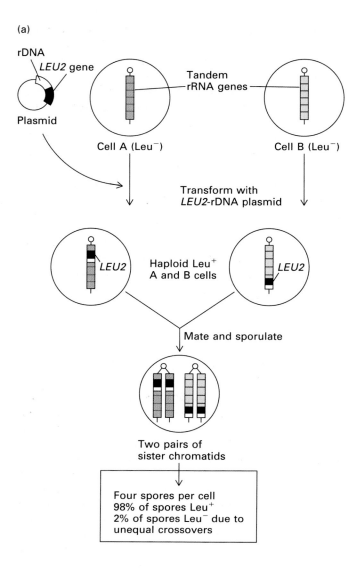

◂ **Figure 10-13** Experimental demonstration that unequal crossing over occurs within rDNA on sister chromatids during meiosis. (a) Two haploid leucine-requiring (Leu⁻) yeast strains (A and B) were transformed with a recombinant plasmid containing the *LEU2* gene (black) and a short stretch of rDNA (white). The two strains could be distinguished in Southern blots because of polymorphism for a restriction enzyme cut site in the spacer regions of the DNA. The two transformed haploid strains containing the *LEU2* gene within their rDNA were mated to produce diploid cells, which were then sporulated. This process involves a meiotic division; that is, each chromosome in a diploid cell is copied to form a pair of sister chromatids, each of which segregates into a haploid spore. If no crossing over or only reciprocal crossing over occurred during meiosis, all four spores from each diploid cell would have retained the *LEU2* gene. In fact, about 2 percent of the spores could not produce colonies on a leucine-free medium, indicating that they had lost their *LEU2* gene because of unequal crossing over. (b) The loss of a gene during meiosis theoretically could result from unequal crossing over between nonsister chromatids or sister chromatids. Unequal crossing over between nonsister chromatids *(top)* would produce leucine-requiring spores containing hybrid rDNA, detectable by the presence of both types of restriction sites in the rDNA. In fact, analysis of cloned rDNA from several Leu⁻ spores indicated that in all cases this rDNA was derived from only one strain consistent with crossing over between sister chromatids *(bottom)*. [See T. D. Petes, 1980, *Cell* **19**:765.]

The physical basis for this loss could be determined because the rDNA from the two strains had a restriction site polymorphism in their spacer regions and thus were distinguishable. Analysis of cloned rDNA from *Leu⁻* spores showed that unequal crossing over between sister chromatids had caused the loss of the leucine marker (Figure 10-13b).

Sequence Maintenance by Gene Conversion The phenomenon of gene conversion, in which alleles appear to be lost during meiotic recombination between homologous chromosomes is discussed in Chapter 12. In this section, we consider gene conversion in a less strict sense, that is, as a set of events which do not necessarily depend on meiotic recombination but which result in erasing (or decreasing) mutations in tandem gene arrays.

A simple form of gene conversion is illustrated in Figure 10-14. A nick (a single-strand cut) in one duplex DNA region allows the single strand near one of the cut ends to hybridize to another homologous region on the same or a different duplex. This hybridization displaces part of the complementary strand in the second DNA region. After repair of the first region and cleavage of the displaced strand in the recipient region, replication of the resulting duplexes produces a different distribution of alleles than in the original duplexes. In Figure 10-14, for example, one set of *ab* alleles is converted into *AB*.

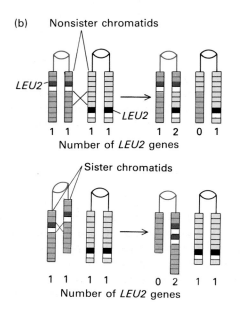

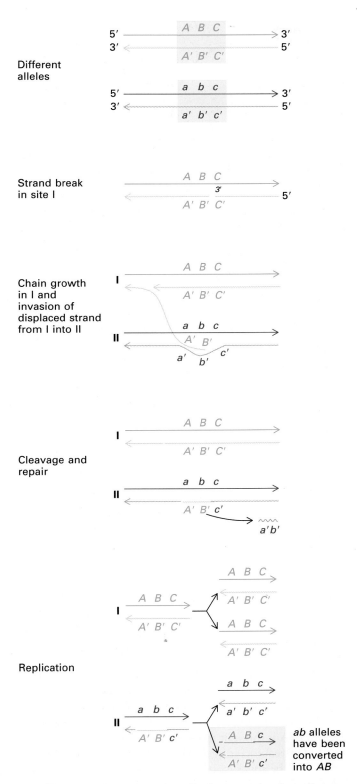

▲ Figure 10-14 Gene conversion between two homologous but genetically distinct DNA sites or alleles. The event of gene conversion leads to a different distribution of alleles. Capital and lowercase letters represent different alleles of the same gene. Primes designate the corresponding sequences on the complementary strand. The two sites can be on the same chromosome or on different chromosomes.

Gene conversions would tend to erase (or decrease) mutated sites in a tandem array of identical genes because the "correct" repeated sequences greatly outnumber mutant ones. Therefore, the hypothetical nick starting a gene conversion would occur most often in a wild-type region. Most gene conversions would thus lead to progeny with fewer mutant genes. In the example given in Figure 10-14 suppose that site I is "correct" and site II is mutant; the result of conversion is to produce three wild-type *AB* progeny for every one mutant *ab* progeny.

The molecular events involved in this mechanism are not known, but such localized corrections would explain why sequence identity is maintained in some regions of a tandem array of genes better than in other regions of the same array. For example, in ribosomal and histone gene arrays, the sequences of the transcription units are maintained more exactly than the flanking nontranscribed regions between the transcription units.

Repetitious DNA Fractions

Besides the tandemly repeated genes encoding rRNA, tRNA, and histones and the families of duplicated genes (e.g., the globin genes), eukaryotic cells contain other regions of DNA that are generally referred to as repeated sequences, or *repetitious DNA*. These regions represent several different types of sequences. Some consist of an oligonucleotide (i.e., a sequence of 5–10 nucleotides) repeated in tandem many times; others are interspersed at many places in the genome. Some of these interspersed sequences consist of short 150- to 300-bp repeat units; in others, the repeat unit can be as long as 5000–6000 base pairs. These repetitive sequences are sometimes referred to as SINES or LINES for short or long interspersed elements. In this type of repetitious DNA, there may be no two units that are *exact* repeats. The existence of repetitious DNA regions was first recognized when some of the denatured DNA of mammals was observed to renature more rapidly than the bulk of cellular DNA. We shall briefly review these reassociation experiments, which led to the discovery of two major classes of repetitious DNA.

Suppose that the total DNA of an organism is broken into fragments with an average length of about 1000 bases. The DNA is then dissociated into single chains and placed under conditions that allow strand reassociation to occur (e.g., a favorable ionic concentration and a favorable temperature). All the DNA fragments would reform duplexes at about the same speed if none contained sequences that were repeated in the genome. However, a segment derived from a repeated DNA region would find a complementary partner more quickly and thus reassociate faster than a segment derived from DNA without repeated sequences. For this reason, the DNA encoding 45S pre-rRNA and that encoding 5S rRNA reassociate faster than does nonrepeated DNA.

The parameters that affect the degree to which single-stranded DNA reassociates are its initial concentration and the time allowed for the reaction. The C_0t of a reaction is the product of the molar concentration of the DNA (C_0) and the reaction time (t) in seconds. A convenient term for comparing the reassociation rates of different DNA fractions is the $C_0t_{1/2}$ value—the C_0t at which one-half of a given fraction renatures. The lower the value of $C_0t_{1/2}$, the higher the reassociation rate. By comparing the $C_0t_{1/2}$ value of any particular DNA fraction with that of a "standard" nucleic acid (e.g., a homopolymer or a viral or bacterial DNA of known length, both of which are assumed to have no repetitive sequences), the approximate frequency of repeats within the fraction of interest can be determined.

In reassociation experiments with the total DNA of various mammalian genomes, about 10–15 percent of the DNA reassociates almost immediately (Figure 10-15) and has a $C_0t_{1/2}$ value of 0.01 or less. This rapidly reassociating fraction has proved to be the simplest DNA to analyze. (This fraction had been separated from the majority of the cellular DNA even before reassociation experiments or DNA cloning were in use.) The rapidly reassociating fraction is largely composed of several different sets of repeated short oligonucleotides arranged in long tandem arrays; this fraction is referred to as *simple-sequence DNA*.

Another 25–40 percent of the DNA reassociates at an intermediate rate over a broad range of $C_0t_{1/2}$ values from 0.01–10 (Figure 10-15). On the basis of the apparent wide range of reassociation rates, it was once thought that most invertebrates and all vertebrates had hundreds of distinct repetitive sequences 100–1000 nucleotides in length, each repeated hundreds to thousands of times per genome. However, sequence analyses of cloned samples of this fraction—termed *intermediate-repeat DNA*—from many different animals (e.g., fruit flies, frogs, rodents, and humans) have failed to distinguish large numbers of distinct families of sequences. In fact, only one major class of intermediate short repeats and perhaps 10 or fewer classes of intermediate long repeats have been found in mammalian genomes.

Intermediate repeats in fruit fly and yeast DNA are not repeated as often as they are in mammalian DNA. Indeed, their possible similarity to the mammalian elements was only recently recognized because they are found in different places in chromosomes of different strains. In fruit flies and yeasts, these repeated elements have been termed *transposition elements,* or *mobile DNA.* Transposition elements are described more fully in a later section. It now seems likely that most, if not all, of the intermediate repeat DNA in higher animals is some form of mobile DNA.

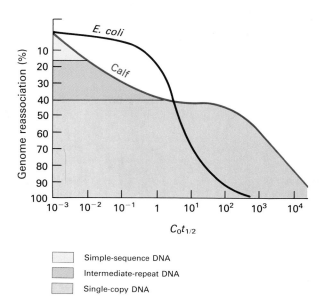

▲ **Figure 10-15** Reassociation curves of calf DNA and of *E. coli* DNA. DNA from calf thymus or *E. coli* cells was broken into fragments of about 1000 base pairs, denatured into single strands, and reassociated for various periods of time and at various DNA concentrations. The extent of reassociation (%) was measured by a chromatographic technique and plotted against $C_0t_{1/2}$ (molar concentration × time in seconds). The most rapidly reassociating calf DNA (yellow) renatures almost immediately. Two other broad fractions are noted: one that reassociates at an intermediate rate (green) and one that reassociates quite slowly (blue). The *E. coli* DNA reassociates over a fairly narrow range of $C_0t_{1/2}$ values, which indicates that the *E. coli* fragments all tend to reassociate at about the same speed. The calf DNA that reassociates at intermediate and rapid rates is repetitious DNA. The slowly reassociating fraction of calf DNA, which renatures about 500–1000 times more slowly than *E. coli* DNA, represents DNA present only once per diploid genome. [See R. J. Britten and D. E. Kohne, 1968, *Science* **161**:529.]

As shown in Figure 10-15, about 50–60 percent of mammalian DNA reassociates rather slowly, with $C_0t_{1/2}$ values ranging from about 100–10,000. The most slowly reassociating sequences in mammalian DNA reanneal about 500 times more slowly than those in *E. coli* DNA. Because the amount of haploid DNA in mammalian cells is about 700 times that in *E. coli* and because almost all *E. coli* DNA is thought to be present in a single copy only, the slowly reassociating fraction of mammalian DNA is assumed to be *single-copy DNA.* According to Mendelian genetics, only one copy of each gene is contained in the haploid DNA set; thus the single-copy DNA fraction is expected to contain most of the genes encoding mRNA. The reverse, of course, is not necessarily true: every single-copy DNA sequence does not necessarily perform a genetic function.

Table 10-6 Examples of simple-sequence DNAs in eukaryotes*

Organism	Base pairs per repeat	Sequence of one repeat unit	Location
Drosophila melanogaster (fruit fly)	5	AGAAG (polypurine) ATAAT	Arms of Y chromosome; centromeric heterochromatin of chromosome 2; long arm of 2, near end
	7	ATATAAT	
	10	AATAACATAG AGAGAAGAAG	Centromeric heterochromatin of all chromosomes; tip of long arm of 2
Drosophila virilis	7	ACAAACT (band I) ATAAACT (band II) ACAAATT (band III)	Centromeric heterochromatin
Cancer borealis (marine crab)	2	AT	?
Pagurus pollicaris (hermit crab)	4	ATCC	?
	3	CTG	?
Cavia poriella (guinea pig)	6	CCCTAA	Centromeric heterochromatin
Dipodomys ordii (kangaroo rat)	10	ACACAGCGGG	Centromeric heterochromatin
Cercopithecus aethiops (African green monkey; α sequences)*	172	—	Throughout chromosomes
Homo sapiens (human; alphoid sequences)*	171	—	Throughout chromosomes

*All eukaryotic species have more than one type of simple-sequence DNA, characterized by the sequence repeat; the table includes only selected examples. Many repeats are 10 bp or less in length, but several longer repeats, such as the primate α and alphoid sequences and human minisatellites are now known. The α and alphoid sequences are not shown because of their length.

SOURCE: K. Tartoff, 1975, *Ann. Rev. Genet.* 9:355; and A. J. Jeffreys et al., 1985, *Nature* **314**:69.

Simple-Sequence DNA

The properties of simple-sequence DNA are described in this section. We consider the other major class of repetitious DNA—intermediate repeat DNA—and its relationship to mobile elements in the next section.

Organisms Contain Several Types of Simple-Sequence (Satellite) DNA

Much of the simple-sequence DNA—the most rapidly reassociating DNA fraction—is composed of short (5- to 10-bp) oligonucleotides that are tandemly repeated (Table 10-6). However, many instances of tandem repeats containing 20–200 nucleotides are also now known to occur in vertebrate and plant genomes.

Most simple-sequence DNA occurs in very long stretches up to 10^5 base pairs in length. Such DNA can be separated from the main DNA band as *satellite* bands during equilibrium density-gradient sedimentation because its average base composition, and hence buoyant density, differs from that of the bulk of the cellular DNA. For example, sedimentation of *Drosophila virilis* DNA

yields three satellite bands and one main band (Figure 10-16). However, not all simple-sequence DNA separates from main-band DNA to form satellite bands; therefore, the term simple-sequence DNA is preferred to satellite DNA.

Each animal or plant cell has several types of simple-sequence DNA characterized by different repeat units. In some species, the repeating oligomeric units of simple-sequence DNA have very similar sequences, which suggests an ancestral relationship between two or more different simple-sequence types. In *Drosophila virilis* cells, for example, the seven-base repeats constituting satellite bands II and III differ from the repeat of band I by a single base change.

Most Simple-Sequence DNA Is Located in Centromeres and Telomeres

The location of simple-sequence DNA within chromosomes has been studied in several species by in situ hybridization experiments. In mouse cells and perhaps most other mammalian cells, much, but not all, of the simple-sequence DNA lies near the centromeres (Figure 10-17).

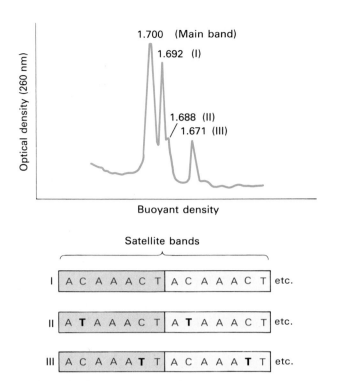

▲ **Figure 10-16** Satellite DNA in *Drosophila virilis*. The DNA from embryonic tissue was extracted and subjected to equilibrium density-gradient sedimentation in cesium chloride to separate DNAs differing in buoyant density. The DNA content of different zones in the gradient was monitored by measuring the absorption of ultraviolet light at 260 nm (i.e., the optical density). The main band of the DNA—that is, the greatest part of the DNA—has a density of 1.700. The three satellite bands, I, II, and III, are less dense. DNA sequence analyses showed that each of the satellite bands is composed of a DNA that is a long tandem array of a seven-base repeat sequence. The repeat sequences in bands II and III differ from the band I sequence by only one base (bold black). [See J. G. Gall, E. H. Cohen, and D. D. Atherton, 1973, *Cold Spring Harbor Symp. Quant. Biol.* **38**:417.]

The simple-sequence DNA in *Drosophila melanogaster* is likewise concentrated in the centromere region, but some also is located within the arms and telomere regions of chromosomes (see Table 10-6). Furthermore, as illustrated in Figure 10-18, most of the simple-sequence DNA is located in the heterochromatin regions of the *Drosophila melanogaster* chromosomes; these regions are always condensed and dark-staining.

As we have mentioned, simple-sequence DNA generally is not transcribed into RNA. The function of these repeated sequences is unknown, but some workers have suggested that they have a structural or organizational role because of their location in centromeres, and we now know that some repetitive stretches have telomeric func-

tions. If the simple sequences serve as binding sites for chromosomal proteins, then proteins that attach to specific simple-sequence DNAs might be expected to exist. For example, the yeast centromere (CEN) sequences discussed in Chapter 9 are one type of simple-sequence DNA. These sequences may bind to microtubules of spindle proteins during cell division (see Figure 9-11). In human cells, a protein called CEN-PB has been shown to be present exclusively in centromere regions. This protein is greatly enriched in acidic amino acids, which would allow it to bind to DNA. However, even the several dozen microtubules in the spindle of a metazoan cell would certainly not require the huge stretches of simple-sequence DNA that are present in centromeres of higher cells. It seems unlikely, therefore, that all the simple-sequence DNA at the centromere is related to the binding of the mitotic apparatus to chromosomes.

In humans, at least 10 types of simple-sequence DNA exist. A single type can account for as much as 0.5–1 percent of the total human genome, an amount equivalent to approximately 10^7 base pairs, or three times the total genome size of *E. coli*. As DNA cloning has progressed, variations of simple-sequence DNA have been found that can serve as independent chromosomal markers. An example of such a simple sequence that serves as a useful molecular marker in the middle of the long arm of human chromosome 16 is shown in Figure 10-19. The

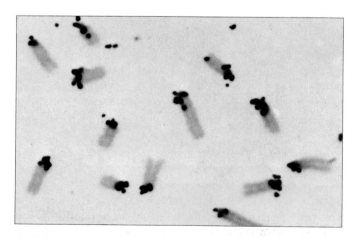

▲ **Figure 10-17** Localization of simple-sequence DNA in mouse genome by in situ hybridization. Purified simple-sequence DNA from mouse cells was randomly transcribed by *E. coli* RNA polymerase to make labeled RNA. Chromosomes from cultured mouse cells were fixed and denatured on a microscope slide, and then the chromosomal DNA was hybridized in situ to the labeled RNA. Autoradiography of the telocentric mouse chromosomes shows that most of the complementary simple-sequence DNA (silver grains) lies close to the centromere. *From M. L. Pardue and J. G. Gall, 1970,* Science **168**:1356; *courtesy of J. G. Gall.*

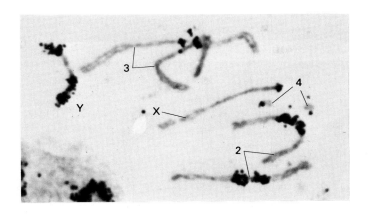

◄ **Figure 10-18** Localization of some simple-sequence DNA within the heterochromatin of metaphase chromosomes of *Drosophila melanogaster* by in situ hybridization. Autoradiograph of fixed denatured chromosomes hybridized with samples of labeled simple-sequence DNA. The silver grains indicate sites where this DNA is abundant. *Photograph from W. J. Peacock et al., 1977, Cold Spring Harbor Symp. Quant. Biol. 42:1121; courtesy of W. J. Peacock.*

use of two or more such markers will allow discrete sites on the same or different chromosomes to be followed throughout the cell cycle.

Simple-Sequence DNA Units Are Conserved in Sequence but Not in Repeat Frequency

The conservation of sequence identity in long stretches of simple-sequence DNA in any one species is impressive. Very likely the mechanisms of high-frequency unequal crossing over (see Figure 10-12) and gene conversion (see Figure 10-14) play a key role in maintaining sequence constancy in simple-sequence DNA as they do in tandemly repeated genes. Occasional mutations in a simple-sequence DNA might be "spread" by unequal crossovers,

particularly if there were no strong selective pressure against such base changes. If unequal crossing over occurs within a region of simple-sequence DNA, uneven numbers of repeats would be expected. This has now been demonstrated in humans and gives rise to powerful techniques for studying unequal crossovers and for identifying individual DNA samples.

DNA Fingerprinting In humans, some of the simple-sequence DNA exists in relatively short 1- to 5-kb regions made up of 20–50 repeats of oligonucleotides containing 15 to about 100 bases. These regions are called *minisatellites* (see Figure 10-20) to distinguish them from the more common satellites, which are 10^5–10^6 base pairs long. Because variations in the lengths of such minisatellites easily fall within the range detectable by conventional Southern blot analysis, slight differences in the minisatel-

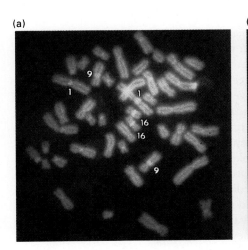
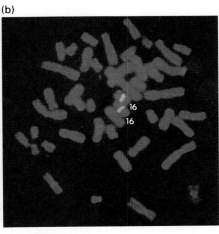
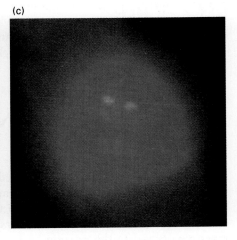

▲ **Figure 10-19** Localization of a specific chromosomal site with a repetitive sequence. (a) Human metaphase chromosomes stained with fluorescent DNA dye. (b) In situ hybridization of the same human metaphase chromosome spread with a simple-sequence DNA—CAACCCGAGT(GGAAT)$_n$—labeled with a fluorescent biotin derivative. Under a different wavelength of light, a labeled hybridized band appears on chromosome 16, locating this particular simple sequence at one site in the genome. (c) The simple-sequence DNA sites are still visible in the nucleus during interphase when the chromosomes are not condensed. [See R. K. Moyzis et al., 1987, *Chromosoma* 95:378.] *Photographs courtesy of R. K. Moyzis.*

λ 33.1 minisatellite

Consensus sequence	AAGGGTGGGCAGGAAGTGGAGTGTGTGCCTGCTTCCCTTCCCTGTCTTGTCCTGGAAACTCA
Changes in individual repeats	1–24 (no changes)
	25 G
	26 G

λ 33.5 minisatellite

	T
Consensus sequence	C GGGCAGG•AGGGGGAGG

Changes in individual repeats						
1	T		CAG			
2	C			•	A	
3	C	A		•		•
4	T			•	A	A
5	T			•	A T	
6	T	A		•	A	G
7	T			•	A	
8	T			G		

▲ **Figure 10-20** Sequences of two minisatellites. Minisatellites are simple-sequence DNA regions containing 20–50 repeats, each composed of 15–100 nucleotides, with a total length therefore of 1–5 kb. Shown here are the consensus sequences of two human minisatellites, λ 33.1 and λ 33.5, determined from more than ten sets of repeats. Differences in the sequences of individual repeats are indicated below the two consensus sequences. Only bases that differ from the consensus are shown; solid dots indicate deletions. The 62-bp repeat unit of λ 33.1 is much more highly conserved than the 17-bp unit of λ 33.5. Variations in the lengths of minisatellites between different individuals is the basis of DNA fingerprinting (see Figure 10-21). [See A. J. Jeffreys et al., 1985, *Nature* **314**:67.]

lite DNA from different individuals can be determined. Moreover, the amount of DNA required for such an analysis can be extracted from the cells in a drop of blood obtained by finger puncture. Use of only a few minisatellite probes in such analysis is sufficient to provide a "DNA fingerprint" that is unique for each individual (Figure 10-21). DNA fingerprinting, which is superior to other methods (e.g., conventional fingerprinting) for identification of individuals, is likely to be widely adopted within the next few years.

Intermediate Repeat DNA and Mobile DNA Elements

In the last decade the study of intermediate repeat DNA in eukaryotes has converged with research on DNA mobile elements, although at first there was no apparent connection between these two classes of DNA. The wide variations found in the amount and chromosomal positions of eukaryotic intermediate repeats and the obvious mobility in bacterial chromosomes of some DNA elements alerted the world of genetics to the apparent instability or mobility of at least some elements in all genomes, a conclusion that was totally unsuspected from classic genetics.

Barbara McClintock discovered the first *mobile elements* in corn (maize) over 40 years ago. She characterized agents that could move into and back out of genes, altering genetic activity during the process, but these elements were not understood in molecular terms until very recently. The first mobile DNA elements to be understood, called *insertion sequences* (IS elements), were found in different places in the genome of different strains of *E. coli*. Some of these elements when separated and introduced into cells as plasmids become inserted in the genome. Moreover, this class of bacterial elements can undergo a *duplicative transposition* to enter additional sites in the bacterial chromosome without being lost from the first site. They were therefore called *transposons*. Although these elements were obviously repeated sequences, that feature was not emphasized originally.

When intermediate repeat DNA regions of eukaryotes were isolated and sequenced, researchers noted that they were present in variable amounts and variable positions; in this regard they were similar to bacterial mobile elements. But many eukaryotic intermediate repeats appeared to be DNA copies of RNA molecules that had been reintegrated into the genome through some type of reverse transcription, similar to the mechanism by which RNA-containing retroviruses are integrated into cellular DNA. Such a mechanism is very different from that used by bacterial transposons, which require DNA synthesis to

Individual

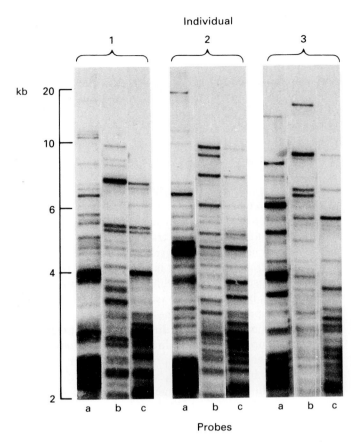

▲ **Figure 10-21** Human DNA fingerprints. DNA samples from three individuals (1, 2, and 3) were subjected to Southern blot analysis using the restriction endonuclease *Hin*fl; three labeled minisatellites (λ 33.6, 33.15, and 33.5; lanes a, b, and c, respectively) were used as hybridization probes. DNA from each individual produced a unique band pattern with each probe. Conditions of electrophoresis can be adjusted so that at least 50 bands can be resolved with each restriction enzyme for each person. The nonidentity of these three samples is easily distinguished by eye. *From A. J. Jeffreys, V. Wilson, and S. L. Thein, 1985, Nature 316:76; courtesy of A. J. Jeffreys.*

move. Eukaryotic intermediate repeats that appear to utilize the reverse transcription pathway are referred to as *retroposons,* a term coined to indicate mobile DNA elements that move in the genome through RNA copies. All eukaryotes studied from yeast to human beings contain retroposons. In the human genome, for example, retroposons are scattered in hundreds of thousands of sites, which can vary from one individual to another. Although *Drosophila* DNA contains fewer copies of retroposons than human DNA, variation in retroposon content and position among sibling species or even strains of the same species is well established.

Retroposons are divided into two major categories (see Table 10-1). The first is described as *nonviral retroposons* because their sequences are not like retroviral sequences

but instead closely resemble cellular RNA molecules, which are usually, but not always, small RNA molecules transcribed by RNA polymerase III. The second major group of retroposons, termed *viral retroposons,* are usually 5–7 kb long. These share strong homology with at least parts of retroviral genomes. Finally, some mobile elements in *Drosophila* and the original elements found in corn almost certainly are not retroposons but DNA transposons. Neither the retroposons nor the transposons appear to have a regular function in the life cycle of the organism they inhabit; however, since these very common elements do damage genes, they probably play an important role in evolution.

Movement of Bacterial Mobile Elements Is Mediated by DNA

Although our main concern in this chapter is the molecular structure of eukaryotic chromosomes, the earliest molecular understanding of DNA insertion sequences and transposable elements came from studies with bacteria. Because the biochemical details of movement of some bacterial transposons are now known and because bacterial transposons are similar in design to some eukaryotic mobile elements, we will briefly describe these elements of bacterial DNA and show how they move before turning to their eukaryotic counterparts.

Bacterial Insertion Sequences Bacterial insertion sequences, or IS elements, were first detected through their effect on gene function. When IS elements are inserted into the middle of a gene, they inactivate the gene; when they are removed from the gene, the gene regains its activity. For example, *E. coli* loses the ability to grow on galactose after the insertion of an element called IS_2, but the cell regains the ability to grow on galactose upon the removal of IS_2. The loss and the subsequent recovery of the ability to metabolize galactose was shown to be due not to a mutation in a single base, but rather to the insertion and removal of an extra piece of DNA in the galactose gene region (Figure 10-22a).

At least six different IS elements have been found in *E. coli*, and other bacteria also contain them. Although they have no known function in the bacterial cell, they are widely distributed in bacteria in nature. Characteristically, one end of a bacterial IS element is an *inverted repeat* of the other end (Figure 10-22b). Since the inverted repeats consist of complementary bases (e.g., GGTAT-X_n-ATACC), a single strand of an IS can fold back to form a stem-loop stabilized by base pairing. The length of the section between the inverted repeats varies from a few hundred to a few thousand bases, but it is generally less than 1500 bases. Because IS elements do not contain any sequences that encode protein, they can be detected only by their damaging effects.

(a)

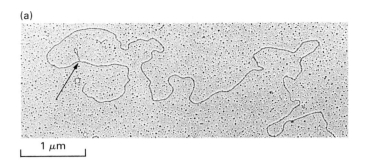

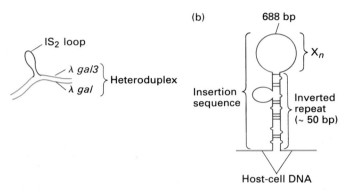

▲ **Figure 10-22** Bacterial insertion sequences. (a) Electron micrograph of heteroduplex formed from DNA extracted from two λ bacteriophages, one carrying the normal *E. coli* galactose gene *(gal)* and the other carrying a mutant (defective) form of this gene *(gal3)*. The loop *(arrow)* corresponds to the insertion sequence (IS₂) that causes this mutation. The mutation can revert, and when it does, the extra loop is lost. (b) The general structure of IS elements consists of a central region (Xₙ) flanked by two inverted terminal repeats. Because of imperfections in the homology of these complementary repeats, base pairing is not perfect. [See A. Ahmed and D. Scraba, 1975, *Mol. Gen. Genet.* **136**:233; and H. Ohtsubo and E. Ohtsubo, 1978, *Proc. Nat'l Acad. Sci. USA* **75**:615.] *Part (a) courtesy of A. Ahmed.*

Bacterial Transposons   Transposons are mobile DNA elements that are longer than IS elements and contain protein-coding genes. Many transposons encode proteins that confer antibiotic resistance, and they often exist in nature as part of plasmids (circular DNA molecules that replicate independently in bacteria). Transposons can recombine into the bacterial chromosome, often damaging a resident bacterial gene in the process, just as IS elements do. Bacteria that have undergone insertion of a transposon containing an antibiotic-resistance gene can be selected by growing them in the presence of the antibiotic.

Comparisons of the nucleotide sequences of IS elements and transposons reveal a close relationship between the two. Transposons contain direct repeats several hundred nucleotides long at each end. In fact, in a number of transposons the direct repeats are IS elements. Figure 10-23 shows the general structure of Tn-9, an *E. coli* transposon that contains a gene encoding resistance to the antibiotic chloramphenicol as well as a gene for a

protein needed to promote transposition of Tn-9 DNA. An important feature of this structure is the *target-site repeats*, short identical sequences at the ends of an inserted transposon. When the region of the cellular DNA into which a transposon has moved is sequenced in the same strain lacking the transposon, there is only one copy of the target site. Thus duplication of the target-site sequence occurs on introduction of a transposon.

A Model for Bacterial Transposon Movement   The molecular mechanism for the movement of bacterial transposons is under active study and can now be induced to operate in cell extracts. Genetic experiments indicate that at least several host enzymes are necessary for the recognition and integration of transposons. As mentioned already, Tn-9 and some other transposons encode a protein, termed *transposase*, that is necessary for transposition. The transposon Tn-3, which is similar in design to Tn-9, encodes both a transposase and a repressor protein that controls production of the transposase.

A proposed mechanism for duplicative transposition is shown in Figure 10-24. The direct repeats (often IS elements) at the ends of the transposons would serve as recognition sites for transposase, which cleaves the donor DNA. How the proposed staggered cuts are made at the target sites in the recipient DNA region remains speculative. Duplication of the target site and generation of an extra copy of the transposon would result from DNA synthesis from the two free 3'-hydroxyl ends of the DNA adjacent to the target sites.

Successful transposition requiring DNA synthesis has recently been achieved in a cell-free bacterial system. With this advance, the details of how duplicative transposition occurs should be worked out soon. As we shall see in later sections, the general structure of some mobile elements in eukaryotic cells seems to be identical to the design of prokaryotic transposons. As we noted earlier, however, most eukaryotic mobile elements have a different structure and move by quite a different mechanism than do bacterial transposons.

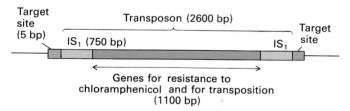

▲ **Figure 10-23** General structure of transposon Tn-9 of *E. coli*. In Tn-9, the long terminal direct repeats are IS₁ sequences, which also are located elsewhere in the *E. coli* chromosome. A short sequence of the recipient chromosome, termed the *target site*, becomes duplicated during transposition. As a result, identical target-site repeats flank an integrated transposon.

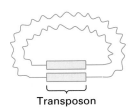

Circular plasmid
with transposon

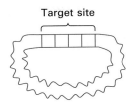

Target DNA
(bacterial chromosome)

Target site

◀ **Figure 10-24** Proposed model for duplication and integration of a bacterial transposon and a circular recipient chromosome *(top)*. This process results in two copies of the transposon, one inserted at the target site in the recipient chromosome with a target site duplication of five bases. [See J. Shapiro, 1979, *Proc. Nat'l Acad. Sci. USA* **76**:1933; K. Mizuuchi, 1983, *Cell* **35**:785.]

Transposon

(a) Staggered cuts are made at ends of transposon and target site

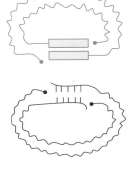

(b) Strand transfer occurs between 3′ ends of transposon and 5′ ends of target DNA

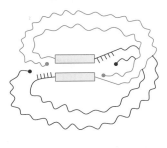

(c) Copying of transposon begins at free 3′ ends of target DNA, beginning with duplication of target site

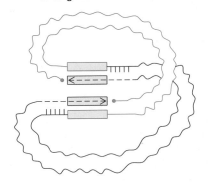

(d) Completion of replication yields cointegrate containing plasmid DNA, target DNA, and two copies of transposon. Cointegrate is resolved by site-specific recombination.

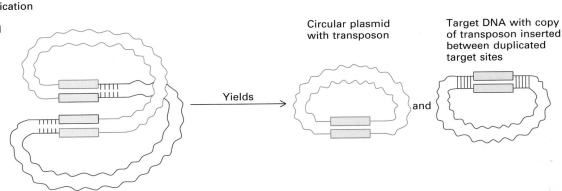

Circular plasmid with transposon

Target DNA with copy of transposon inserted between duplicated target sites

Yields

and

Homologous recombination between the IS elements at the ends of a transposon can lead to the deletion of the entire transposon. Insertion sequences can also be deleted, perhaps by recombination between target-site repeats. However, transposition and insertion appear to occur much more often than deletion. Probably, the sequences of mobile elements eventually drift and become unrecognizable, or they must somehow undergo partial deletion; otherwise, the entire genome would be overrun by these sequences of "selfish DNA."

Many Eukaryotic Mobile Elements Are Interspersed Genomic Copies of Cellular RNAs

We now return to discuss those eukaryotic intermediate repeat sequences that move by copying RNA into double-stranded DNA and integration of the copy into the genome; as noted earlier, these are called nonviral retroposons. When human DNA is fragmented to an average of say 1000 to 5000 bp, denatured, and allowed to

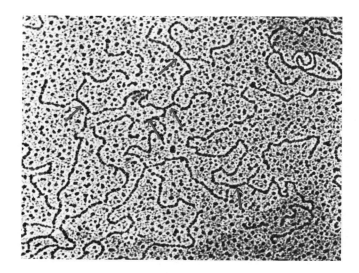

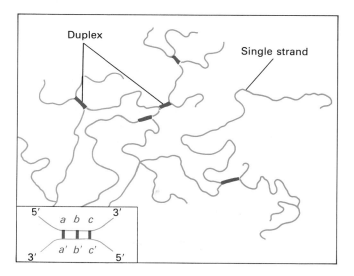

◀ **Figure 10-25** Electron micrograph and corresponding diagram of human DNA fragments renatured under conditions that permit reassociation only of repetitive sequences. The reassociated duplex regions *(arrows)* appear thicker than the still-denatured unpaired regions. The duplex regions are about 300 bp long and occur on the average of once every 3000–5000 bp. Such repetitive sequences are called short interspersed elements (SINES). The inset illustrates base pairing between complementary sites in a repetitive element. *Photograph from P. L. Deininger and C. W. Schmid, 1976, J. Mol. Biol. 106:773; courtesy of C. W. Schmid.*

isolated, cloned, and sequenced. Although no two copies of these intermediate repeats are identical, their general sequence similarity shows that an ancestral relationship exists between these elements both within and among species. The sequence conservation is about 80 percent within a species, but falls to only about 50–60 percent between species. Because many of these repetitive sequences in human DNA contain a recognition sequence for the restriction enzyme *Alu*I, they were collectively called the *Alu* family. This term is a bit misleading because many short interspersed elements lack the *Alu* site; however, the name is convenient and widely used by workers in the field. The human genome, for example, contains not only hundreds of thousands of copies of full-length *Alu* family sequences but also many partial *Alu*-like sequences. Numerous blocks of 10–20 nucleotides clearly related to the *Alu* family have been found scattered between genes and within introns. As much as 10 percent of the human genome may be related to this particular sequence family.

As shown in Figure 10-26, *Alu* elements have a remarkable sequence similarity with 7SL RNA, a 294-nucleotide cytoplasmic RNA. This small cellular RNA is part of a cytoplasmic ribonucleoprotein particle, the *signal recognition particle,* that aids in the secretion of newly formed polypeptides through the membranes of the endoplasmic reticulum. Perhaps because the 7SL sequence is a functional sequence, it is highly conserved even in *Drosophila,* mouse, and man. Recently, a small (∼100-nucleotide) RNA of *E. coli* has been sequenced and found to have similarity to the 7SL RNA. Thus this molecule has existed since early in evolution. However, neither *Drosophila* nor single-celled organisms have any *Alu*-type intermediate repeats (at least in large numbers). These findings suggest that 7SL RNA genes existed before *Alu* sequences and that the *Alu* sequences somehow arose fairly late in evolution from the 7SL sequences.

Not only do present-day *Alu* sequences resemble 7SL RNA but about half of the *Alu*-containing human segments that have been cloned can themselves serve as RNA polymerase III transcription units (as does the 7SL gene), producing a short RNA that matches the length of the *Alu* repeat. In addition, short (200- to 400-nucleotide) RNA molecules have been found, mostly in the cell nu-

briefly renature, the frequency and distribution of intermediate repeats is easily established. When such renatured samples are examined in the electron microscope, about 20 percent of the 1000-base-long fragments but nearly all of the 5000-base-long fragments will be found to have hybridized over short, approximately 300-bp regions, which are termed *short interspersed elements,* or SINES (Figure 10-25). From quantitation of such experiments, it has been found that all vertebrate genomes contain these short interspersed elements, often in staggering numbers. For example, the human genome contains at least 500,000 short repeats, which constitute at least 5 percent of the total genome. These sequences do not cross-hybridize between species, but they almost all cross-hybridize within a species.

Alu *Sequences*   By now, several hundred segments of human, mouse, and hamster genomic DNA containing short (150- to 300-bp) interspersed elements have been

	1	15	16	30	31	
7SL	•GCCGGGCGCGGTGG		CGCGTGCCTGTAGTC		CCAGCTA	
Alu-cons	GGCCGGGCGCGGTGG		CTCACGCCTGTAATC		CCAGC•AC	

	46	60	61	75	76	45
7SL	GCTGAGGCTGGAGGA		TCGCTTGAGTCCAGG		AGTTC••••G	GAG
Alu-cons	GCCGAGGCGGGCGGA		TCACCTGAGGTCAGG		AGTTCGAGA	

	91	105	106	120	121
7SL	TGGGCAACATAGCGA		GACCCCGTCTCT•••		• • • • • • • • •
Alu-cons	TGGCCAACATGGTGA		AACCCCGTCTCTACT		AAAAATACAAA

	136	150	151	165	166
7SL	•GCCGGGCGCGGTGG		CGCGTGCCTGTAGTC		CCAGCTACTCGG
Alu-cons	AGCCGGGCGTGGTGG		CGCGCGCCTGTAATC		CCAGCTACTCGG

	181	195	196	210	211
7SL	GCTGAGGCTGGAGGA		TCGCTTGAGTCCAGG		AGTTCTGGGCTGT
Alu-cons	GCTGAGGCAGGAGAA		TCGCTTGAACCCGGG		AGGCGGAGGTTGCA

	226	240	241	255	256	27
7SL	TGCGCCTGTGA••••G		CCAGTGCACTCCAGC		CTGGGCAACATAGCG	
Alu-cons	TGAGCC•GAGATCGCG		CCAGTGCACTCCAGC		CTGGGCGACAGAGCG	

	271	285
7SL	AGACCCCGTCTCT	
Alu-cons	AGACTCCGTCTCAAA	AAAAA

▲ **Figure 10-26** Comparison of *Alu* sequences and 7SL RNA. The *Alu* consensus sequence (*Alu*-cons) was determined from the sequence of 125 separate *Alu* elements from human DNA. The consensus sequence is aligned with the portions of the 7SL RNA sequence that are most homologous with *Alu*. Bases in the *Alu* sequences that differ from the corresponding ones in the 7SL sequence are in red type; solid red dots indicate positions added to maximize homology. Both the *Alu* sequence and the 7SL sequence have sections that are nearly identical (1–117 and 136–285). These two sections are separated in the *Alu* sequence by an A-rich region from positions 118–135 (yellow). *Alu* elements also contain a 3′ A-rich sequence, which is not shown here. [Adapted from J. Jurka and T. Smith, 1988, *Proc. Nat'l Acad. Sci. USA* 85:4775.]

cleus, that hybridize to *Alu* DNA. These RNA copies of *Alu* sequences do not accumulate and therefore presumably turn over rapidly. Their presence proves that short *Alu* sequences are transcribed by polymerase III inside cells, as well as in vitro, a point that becomes important in considering how these segments move around the genome. After the recognition that *Alu* repeats were very widespread in mammalian genomes, a number of other cloning experiments showed that other sequences representing degenerate copies of functioning RNA molecules were also common.

Genomic Copies of snRNAs and tRNAs   Attempts to isolate human and mouse genomic DNA that encodes the many known small nuclear RNAs (snRNAs) and tRNAs resulted in uncovering only a few true genes that match the functioning RNA sequence but a horde of partial or mutant copies of these genes. The snRNAs that function in mRNA splicing and the snRNA that aids in ribosomal RNA processing are all formed by RNA polymerase II; dozens to hundreds of copies of these genes have been found, some functional but many nonfunctional. In addition, several hundred degenerate genomic copies of various tRNAs, which are formed by RNA polymerase III, have been identified in the human and ro-

dent genomes. Some of these insertions differ only slightly from the corresponding functional genes, but others represent truncated partial copies corresponding to either the 5′ or the 3′ ends of the functional RNAs.

Nonfunctional Genomic Copies of mRNAs   So far we have only mentioned genomic copies or partial copies of small RNA transcripts. Degenerate genomic copies of a wide variety of mRNA-encoding genes (normally transcribed by RNA polymerase II) have also been discovered. These are not the same as the functional duplicated genes discussed earlier in this chapter, and they are not duplicates of the whole genes that have drifted into nonfunctionality (i.e., pseudogenes) because they lack introns and don't have similar flanking sequences. Instead, these DNA segments appear to be copies of processed mRNA. These *processed pseudogenes* contain no introns but do contain copies of a portion of the poly A tail that is added at the 3′ ends of pre-mRNAs in the cell nucleus. The processed copy of a human β-tubulin gene shown in Figure 10-27 contains a mutant copy of the entire coding region, which would not be translatable because it contains protein stop codons. It also contains the 5′ and 3′ nontranslated regions of the mRNA including the poly A signal and 17 adenines of the poly A tail. By now at least

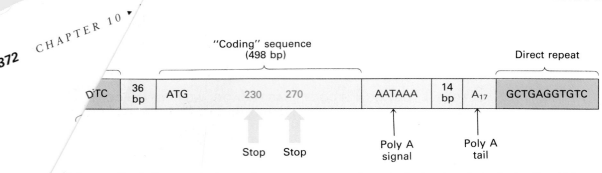

Figure (partial) — labels: "Coding" sequence (498 bp); Direct repeat; ...TC | 36 bp | ATG | 230 | 270 | AATAAA | 14 bp | A₁₇ | GCTGAGGTGTC; Stop Stop; Poly A signal; Poly A tail.

...7 A human β-tubulin processed pseudogene
... arose by transposition through an RNA inter-
... genomic DNA region diagrammed here was
... cause of its complementarity to a β-tubulin
*...*ne 498-bp "coding" region corresponds to the
*...*mino acid sequence of β-tubulin except for base
*...*s at nucleotides 230 and 270, which have introduced
*...*tion stop codons. The absence of introns, several of
*...*h are present in the true β-tubulin gene, and the presence of a poly A signal and a poly A tail, which are characteristic of processed mRNA, suggest that this pseudogene
arose by reverse transcription of β-tubulin mRNA and insertion of the resulting DNA into the chromosome. The translation stop codons, which render an RNA product from the
pseudogene nonfunctional, represent accumulated mutations.
Also shown are the terminal direct repeats that are characteristic of mobile elements (see Figure 10-28). [See C. D. Wilde
et al., 1982, *Nature* **297**:83.]

a hundred processed copies of a wide variety of different mRNAs have been reported in human and rodent
genomes.

Long Repetitive Sequences with Terminal Repeats

In addition to genomic copies of various cellular RNAs,
human and mouse DNA contain 1- to 5-kb repeated sequences, called *long interspersed elements,* or LINES. The
most prominent of these is the LINE 1 family. There are
perhaps 20,000–40,000 copies of these elements in
human DNA, although many are not full-length. The full-
length element is about 5 kb, and most if not all of the
individual copies of it do not code any protein. These
long interspersed elements are grouped with the nonviral
retroposons because, at least in cultured cells, an RNA is
transcribed from them, and as we discuss below, they are
flanked by the same type of terminal direct repeats as *Alu*
sequences.

Movement of *Alu* Sequences and Similar Mobile Elements Is Mediated by RNA

From the foregoing discussion it seems likely that copies
of many different kinds of RNA have become scattered in
eukaryotic genomes during evolution. The mechanism by
which this happened is not known, but DNA sequence
analysis around all of these putative RNA copies has suggested an explanation.

After perhaps several dozen *Alu* elements from human,
hamster, and mouse genomic DNA had been sequenced,
a pattern was noticed in the flanking sequences of these
regions. Surrounding each individual *Alu* element was a
direct repeat, which varied from 7–21 bp in length in different examples (Figure 10-28). Similar variable direct
repeats were found to flank tRNA and snRNA pseudogene copies as well as processed mRNA pseudogenes (see
Figure 10-27). Because of the importance of target-site

5′ Direct repeat	*Alu* sequence — 3′ A-rich sequence	3′ Direct repeat
AAACAAGCAGGAGAGGCT	Human *Alu* . . . A₇CA₅CA₇TCA₄CA₂TCA₃	AAAACAAGCAGGAGGGGCT
AAGATTCACTTGTTTAG	. . . . A₁₂GAGAGATTGATTGA₂	AAGATTCACTTGTTTAG
AAAGAAATGG	. . . . A₁₄GA₃GA₃GA₄GA₅GA₆GA₃	AAATAAATGG
GTTTAGATAAG	. . . . A₂₅	GTTTAGATAAA
AAAAGAAACTTGGAAAGAG	Cho *Alu* . . . . . A₂TA₃TA₃TA₄TCTTA₇	AAAAGGAAACTTGGAAAGGA
AACATACTAATTTTG	. . . . A₄CA₂	AACTATAATTTTTG
GTCAGCC	. . . . TGA₅CCA₅GA₇GA₅GA₅GA₃GTTCCAGGCCA	GTCAGCC
AGCTCATGAATGAAG	. . . . CCA₅CA₃TCA₄CCAGACAGGCACAGCCCC	AGCCCAT
GAGACAACAAATCAGAG	Mouse *Alu* . . . A₇CCA₃CCA₃CCA₃CCA₆CC	GAGACAACAAATCAAAT

▲ **Figure 10-28** The general structure of nine different *Alu*
sequences from humans, Chinese hamsters (Cho), and mice
showing the flanking direct repeats and A-rich region that
adjoins the 3′ direct repeat. [See S. R. Haynes et al., 1981,
Mol. Cell Biol. **1**:573.]

repeats in the movement of transposons discussed earlier, these findings strongly suggest some analogous mechanism for *Alu* movement in the genome.

A Model for Nonviral Retroposon Movement

As mentioned already, after retroviruses infect susceptible cells, their RNA is copied into DNA and integrated into cells. Thus the movement of DNA elements by copying of RNA into DNA that is integrated into the genome was not an unprecedented molecular mechanism. As we discuss later, it has been clearly shown that movement of retroviruslike long repetitive sequences does involve such an RNA-mediated mechanism. The general structure of *Alu* sequences suggests a means by which RNAs could be copied and reintegrated into the genome (Figure 10-29). In addition to a stretch of A residues preceding the 3' direct repeat of *Alu* sequences, a stretch of T residues, which acts as a polymerase III terminator, follows the direct repeat. RNA copied from these sequences would terminate in a stretch of U residues containing a free 3' hydroxyl. If this end folded back on itself, the Us pairing with the As, a reverse transcriptase template would be generated. Addition of deoxynucleotides to the 3' end would produce a single-stranded DNA copy of the original *Alu* sequence. Subsequent synthesis of a complementary strand by reverse transcriptase and integration of the

cDNA copy of the entire sequence into the genome would complete movement of an *Alu* sequence (or any other copied RNA) with preservation of the original sequence.

No Definite Function for *Alu* Sequences Has Been Demonstrated

The possibility that 10–20 percent of the human genome is repetitive, noncoding, and possibly nonfunctioning DNA is unsettling to many biologists. As we have noted, those *Alu*-like regions that are transcribed by RNA polymerase III in vivo yield short, rapidly destroyed, and therefore possibly functionless RNA molecules. The presence of *Alu* sequences within protein-coding genes raises the possibility that these sequences play some role in transcription by RNA polymerase II or in processing of premRNAs in the nucleus.

Some of the earliest research on repetitive sequences in human cells demonstrated that highly repetitive regions were scattered within long hnRNA molecules. Sequence analysis later proved that these repetitive regions of hnRNA were *Alu* sequences. The high content (approximately 5–10 percent) of repetitive sequences in hnRNA contrast with the much lower amount (approximately 0.1 percent) of *Alu*-like sequences in cytoplasmic mRNA. When DNA sequencing of known mRNA transcription units began about a decade ago, it rapidly became clear that many but not all introns contain *Alu* sequences (Figure 10-30). The introns and any *Alu* sequences they contain are, of course, lost during nuclear processing of hnRNA. However, the *Alu* sequences in protein-coding genes are not located at intron-exon boundaries and appear to play no role in intron removal. Although a few mRNAs contain *Alu* sequences within the 5' or, more commonly, the 3' untranslated region, most mRNAs lack *Alu* sequences altogether. Sequencing of at least 100 different human protein-coding genes in various chromosomal locations has indicated that on average there is one *Alu* sequence per several thousand nucleotides within

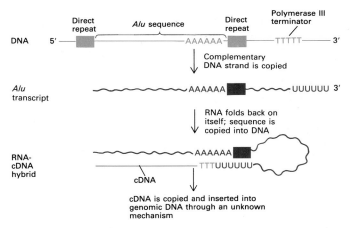

▲ **Figure 10-29** A proposed mechanism for movement of *Alu* genomic DNA segments through an RNA intermediate. *Alu* sequences are known to be transcribed by RNA polymerase III, beginning at the first nucleotide of the *Alu* element and terminating at a T-rich site downstream from the 3' direct repeat. Thus, the 3' ends of *Alu* transcripts in vivo contain a stretch of Us. If the *Alu* transcript folded so that the A-rich region and the terminal U-rich region formed a hybrid, then reverse transcription might be initiated at the 3'-hydroxyl end of the U-rich region to make a cDNA copy of the entire *Alu* region. By unknown mechanisms, the cDNA copy might become integrated into the chromosomes. [See P. Jagadeeswaran, B. G. Forget, and S. M. Weissman, 1981, *Cell* **26**:141; and S. W. Van Arsdell et al., 1981, *Cell* **26**:11.]

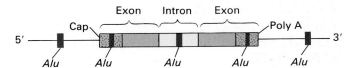

▲ **Figure 10-30** Positions of *Alu* sequences in and around a typical protein-coding transcription unit. There is no fixed or regular location of *Alu* sequences within transcription units. The repetitive sequences may be contained within introns (yellow) or within the noncoding regions at the 5' and 3' ends of transcription units (stippled purple), but, of course, not in the translated exons (solid purple). In many cases, *Alu*-type sequences have been found flanking transcription units, but they do not occur in regular positions.

transcription units; this frequency is similar to that in the total genome.

Alu sequences also occur in sequences flanking many transcription units. For example, the 60-kb region that encompasses the human β-like globin gene cluster contains several *Alu* sequences flanking the various genes (see Figure 10-4). Analyses of cloned β-like globin gene clusters from human, gorilla, chimpanzee, goat, and other mammalian genomes have indicated that the positions of these *Alu* sequences vary both within and among species. The absence of any regular locations of *Alu* sequences within β-globin gene clusters suggests that they play no necessary role in the transcription of these genes.

Thus the evidence to date seems to rule out any function for *Alu* sequences in transcription by RNA polymerase II or processing of pre-mRNA. One other possible

function is suggested by the presence of a 14-bp region in the *Alu* consensus sequence that is homologous to the DNA replication origin site in several mammalian DNA viruses. The possibility that a subset of *Alu* sequences plays a role in DNA replication is discussed in Chapter 12.

Most Mobile Elements in Yeast and *Drosophila* Are Long Intermediate Repeats

Although mammalian cells contain much more intermediate repeat DNA than yeast or *Drosophila*, about 5–15 percent of the total DNA in these lower eukaryotes is repetitive. Initial studies on cloned repetitious DNA from these two organisms showed that the repetitive sequences were about 5000 nucleotides long, much longer than the *Alu* sequences and most other intermediate repeats present in mammals. Moreover, only 50–100 copies of these long repetitive sequences were estimated to occur in the yeast genome, and maybe a thousand or so in the *Drosophila* genome. Finally, early sequence analyses of these elements showed that their sequences differed distinctly from those of the mammalian intermediate repeats exemplified by *Alu*. Perhaps the characteristic of these sequences that attracted the most attention was their apparent ability to change locations—transpose—in the yeast or fly genome.

Several lines of evidence indicate that intermediate repeat DNA can move in the yeast genome. First, cloned samples of yeast repetitive DNA from different chromosomal regions cross-hybridize with each other, indicating that their sequences are similar. Second, when the DNA from different yeast strains all originating from one cell is fragmented with a restriction enzyme and analyzed by

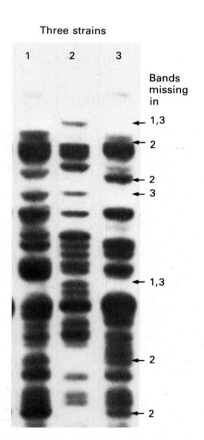

Three strains

Bands missing in

← 1,3
← 2
← 2
← 3
← 1,3
← 2
← 2

▲ **Figure 10-31** Demonstration that long repetitive sequences in yeast DNA are mobile. A recombinant DNA clone that hybridizes to repetitive DNA regions throughout the yeast genome was used as a labeled probe in Southern blot analysis of DNA from three yeast strains all originally derived from one cell. The three strains shared some—but not all—DNA fragments that reacted with the probe, indicating that the repeated sequence is not distributed in the same way in each strain. [See J. R. Cameron, E. Y. Loh, and R. W. Davis, 1979, *Cell* **16**:739.] *Courtesy of R. W. Davis.*

Table 10-7 Classes of intermediate repeat DNA in *Drosophila*

Class	Length (kb)	Characteristic feature	Mobility type
Copia	5.1	Long terminal repeats	Viral retroposon
dm297	7.0		
dm412	7.6	*Copia*-like	Viral retroposon
dm17.6	7.4		
Opus (~12 others)	8.0		
F G	Variable	3′ poly A tails	Nonviral retroposon
FB (foldback)	Variable	Terminal inverted repeats	?
P	2.9	31-bp terminal inverted repeats	Transposon

SOURCE: D. J. Finnegan and D. H. Fawce, 1986, *Oxford Surveys on Eukaryotic Genes* 3:1.

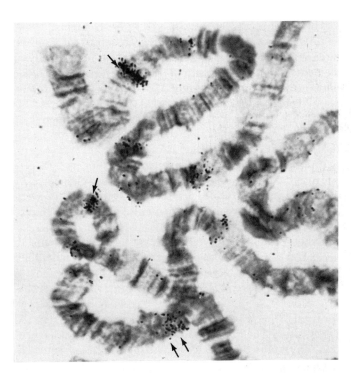

▲ **Figure 10-32** Localization of *copia* elements, the most common type of intermediate repeat DNA in *Drosophila,* by in situ hybridization. Purified labeled *copia* DNA was hybridized to fixed denatured *Drosophila* chromosomes. Autoradiography revealed the presence of *copia* DNA at many sites in the genome *(arrows at silver grains)*. Only a portion of the chromosomes is shown. [See M. W. Young, 1979, *Proc. Nat'l Acad. Sci. USA* **76**:6274.] *Courtesy of M. W. Young.*

the Southern blot procedure using a labeled probe that hybridizes with repetitive sequences, some of the fragments are found in all the strains, whereas others are not (Figure 10-31). This result indicates that the position of the repetitive sequences in yeast DNA varies among strains; thus these sequences are mobile elements, which are called Ty (transposon, yeast) elements.

Unlike yeast, *Drosophila* contains several classes of intermediate repeat DNA (Table 10-7). The P element contains terminal inverted repeats and, as discussed later, acts as a DNA transposon. The F and G elements have a terminal poly A and appear similar to the nonviral retroposons found in higher eukaryotes. By far the most common *Drosophila* repeat elements, called *copia* (Latin for abundance), are about 5−7 kb long. In situ hybridization experiments with cloned *copia* DNA have shown that *copia* elements occur at many chromosomal sites (Figure 10-32).

The mobility of *copia*-like sequences was demonstrated in experiments with hybrids of two strains of *Drosophila melanogaster* that came from different parts of the world but belonged to the same species. In the salivary gland

chromosomes of such hybrid flies, the homologous chromosomes have almost identical banding patterns. They are lined up for much of their length, but separate in occasional regions. In some of these regions of separation, in situ hybridization to *copia*-like sequences clearly occurred in only one of the two chromosomes, indicating that one strain has it and the other does not (Figure 10-33). Thus these repetitive sequences can be removed entirely without damaging the fly. Many experiments of this type (and others) have established that some repeat elements are missing from some *Drosophila* strains and that some strains have as little as 1−2 percent repetitive DNA, while others have as much as 30 percent. Thus during evolution, repetitive sequences have moved in the *Drosophila* genome, and at least most of the repetitive sequences (and possibly all) are not required for viability.

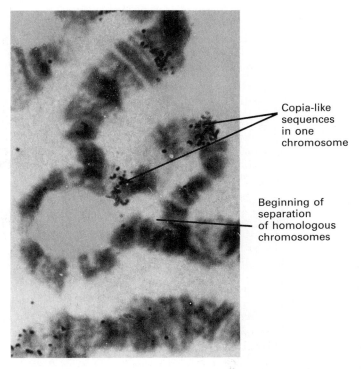

Copia-like sequences in one chromosome

Beginning of separation of homologous chromosomes

▲ **Figure 10-33** Demonstration that repetitive DNA elements in *Drosophila* species are mobile. Labeled cloned *copia*-like sequences were hybridized to salivary gland chromosomes of a hybrid fly whose parents were from sibling species. Over most of their length, both polytene chromosomes of a homologous pair are aligned in register, but in a few regions they separate, as shown in this electron micrograph. At two sites in one of the chromosomes *copia* is detected by autoradiography, but it is lacking in the other chromosome. Similar experiments with many strains and closely related species have demonstrated that a particular repeat element has no fixed chromosomal locations in all strains. [See M. W. Young, 1979, *Proc. Nat'l Acad. Sci. USA* **76**:6274.] *Courtesy of M. W. Young.*

Yeast Ty and *Drosophila Copia* Elements Are Structurally Similar to Retroviral DNA and Move by a RNA-Mediated Mechanism

Several different experiments proved that yeast Ty elements and *Drosophila copia*-like elements move in the genome by being copied into RNA, which is then recopied into DNA and reinserted in the genome, the same pathway used by retroviruses.

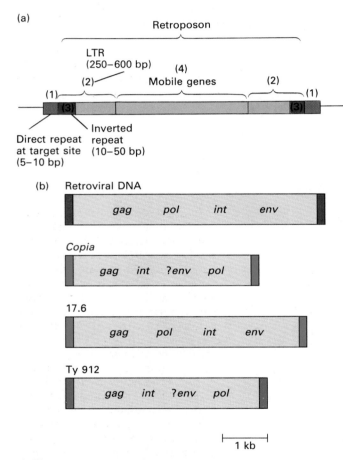

The general structures of yeast Ty elements, *Drosophila copia* elements, and integrated retroviral DNA have several common features (Figure 10-34a). All these elements have long terminal repeat sequences (LTRs), which contain approximately 250–600 bases in different retroviruses and mobile elements. At the ends of the LTRs are 10- to 50-bp inverted repeats. Finally, at both ends of integrated retroviral DNA, there is a short 5- to 10-bp direct repeat sequence that is present in only one copy before integration. As we have noted earlier, this target-site duplication is characteristic of mobile elements. At every site in yeast DNA where Ty is found and in *Drosophila* DNA where a *copia* element is found, the characteristic target-site duplication also is present.

Other evidence that relates these mobile elements in yeast and *Drosophila* to retroviruses comes from detailed sequence analyses. Retroviruses encode four proteins: a glycosylated structural protein antigen (gag); an RNA-directed DNA polymerase, the "reverse transcriptase" (pol); an "integrase" (int) that has a role in inserting the DNA into the cellular chromosome; and an envelope protein (env), which is located in the membranous outer envelope of the virus particle (Figure 10-34b). Computer comparisons of the sequences of a yeast Ty element and two *Drosophila* elements (*copia* and 17.6) strongly indicate that Ty and the *Drosophila* elements are relatives of retroviruses. Although the arrangement of the homologous sequences in the different mobile elements is not identical, they all contain protein-coding sequences simi-

▲ **Figure 10-34** (a) The common elements of eukaryotic viral retroposons and integrated retroviral DNA. (1) A direct repeat (5–10 bases) of the genomic DNA occurs at the target site; the sequence may vary for different retroposons. (2) A long terminal repeat (LTR) of 250–600 bases is present. In yeast the LTR is called the delta sequence. (3) Inverted repeats (10–50 bases) occur at the outside end of each LTR and may be analogous to IS sequences in bacterial transposons. (4) Any protein-coding genes that are present are located between the repeats. (b) Comparative structures of integrated retroviral DNA of avian leukosis virus, of two *Drosophila* elements (*copia* and 17.6), and of a yeast Ty element (Ty 912). The central region of retroviral DNA encodes four proteins: a glycosylated antigen (gag), a reverse transcriptase (pol for polymerase), an integrase (int), and an envelope protein (env). Similar proteins are encoded in the *Drosophila* and yeast elements.

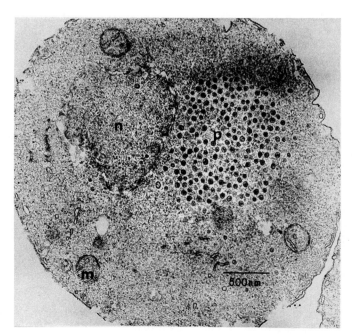

▲ **Figure 10-35** Electron micrographs of intracellular virus-like particles in yeast (**n**, nucleus; **m**, mitochondrion; **p**, particles). [See D. F. Garfinkel, J. D. Boeke, and G. R. Fink, 1985, *Cell* **42**:507.] *Courtesy of J. D. Boeke.*

lar to those for integrase and reverse transcriptase in retroviral DNA. Furthermore, both *Drosophila* cells and yeast cells contain intracellular enveloped particles that have the appearance of virus particles (Figure 10-35).

These particles, at least in *Drosophila,* have reverse transcriptase activity in them, although they do not appear to be infectious in either species.

Although these sequence comparisons provided strong evidence that Ty and *copia* elements move through an RNA intermediate, the experiments depicted in Figure 10-36 provided even stronger, functional evidence for this conclusion. In these studies, yeast cells were transfected with cloned yeast plasmids containing a Ty element and a galactose-sensitive transcriptional activator. The production of Ty mRNA and the frequency of transposition of the Ty element in the transfected cells were increased in the presence of galactose. Moreover, if an intron were included in a Ty element, galactose-induced

transposition occurred, but the newly integrated elements all lacked the intron. The results demonstrated that transposition of Ty elements is dependent on transcription and strongly implied that DNA copies of mature, intronless mRNA encoded by Ty can be integrated into the yeast chromosome at many different sites. Thus transposition of Ty elements in yeast apparently does involve an RNA intermediate. Because of their similarity to retroviruses, these mobile elements are called *viral retroposons.*

Movement of Some Eukaryotic Mobile Elements Is Mediated by DNA

As noted already, the *Drosophila* P element, the original mobile elements discovered in corn, and perhaps some other eukaryotic mobile elements move in the genome by a mechanism involving DNA replication (see Figure 10-24).

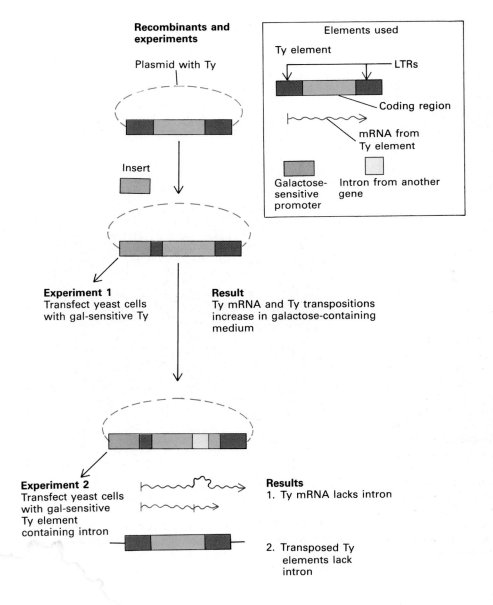

Recombinants and experiments

Plasmid with Ty

Insert

Experiment 1
Transfect yeast cells with gal-sensitive Ty

Result
Ty mRNA and Ty transpositions increase in galactose-containing medium

Experiment 2
Transfect yeast cells with gal-sensitive Ty element containing intron

Results
1. Ty mRNA lacks intron

2. Transposed Ty elements lack intron

Elements used

Ty element

LTRs

Coding region

mRNA from Ty element

Galactose-sensitive promoter

Intron from another gene

◀ **Figure 10-36** Demonstration that yeast Ty element moves through mRNA intermediates. A Ty element encoding an mRNA was cloned in a yeast plasmid. When yeast cells are transfected with such a Ty-containing plasmid, the Ty element is integrated into the yeast chromosome and then is transposed to other sites. The transpositions within the genome can be detected by genetic tests. In experiment 1, a portion of the left LTR of the Ty element was removed and replaced with a promoter (orange) that increases transcription in the presence of galactose. When yeast cells transfected with plasmids containing this recombinant Ty element were grown in the presence of galactose, they produced more Ty mRNA than before and transpositions were greatly increased. In experiment 2, an intron (tan) from another cloned yeast gene was inserted into the putative protein-coding region of the galactose-sensitive Ty element. When yeast cells were transfected with plasmids that included this intron-containing Ty element and then exposed to galactose, Ty mRNA production and transposition was again increased. However, none of the Ty mRNA and none of the newly transposed Ty elements contained the intron. [See J. Boeke et al., 1985, *Cell* 40:491.]

Ac and Ds Elements in Corn   Barbara McClintock's original discovery of mobile elements came from observations of unusual revertible mutants in corn. Among the best-studied of these mutations are those which affect the production of any of the several enzymes required to make anthocyanin, a purple pigment that affects kernel color (Figure 10-37). Numerous different mutations in this pathway causing loss of kernel color were found. Some of these were revertible at high frequency, whereas others were not unless they occurred in the presence of the first class of mutations. The first class were called *activator* (Ac) and the second were called *dissociation* (Ds) because they also tended to be associated with chromosome breaks.

The agents responsible for these mutations have now been cloned and sequenced. The Ac element is about 4500 nucleotides long; the Ds elements are deleted Ac elements (Figure 10-38). A Ds element cannot transpose by itself but can in plants that carry the Ac element. Thus the Ac element appears to encode a transposase, as do bacterial transposons. The Ac element and Ds elements do not contain long terminal repeats, as do retroviruses or viral retroposons; instead they have an imperfect inverted repeat flanked by an 8-bp direct repeat at both ends. The direct repeat is present at the target site of insertion.

When an Ac element enters a gene, it almost always decreases its function. Its ability to excise clearly and simultaneously appear elsewhere with a fairly high frequency differs from retroposons. When it departs, it leaves extra bases, a "footprint." This excision ability plus its structural similarity to bacterial transposons has convinced workers that the Ac element moves by a DNA transposition mechanism.

Drosophila **P Element**   The P element of *Drosophila* has a structure similar to that of the Ac element in corn. It has a center section that encodes at least one protein thought to promote transposition and short (31-bp) terminal inverted repeats, which are flanked by 8-bp target-site direct repeats (this structure is similar to the Ac element discussed above). As discussed in the next section, the P element has been associated with a number of eye-color mutations. Like kernel-color mutants in corn, some of the *Drosophila* eye-color mutants are revertible. Sequencing of the DNA of the affected locus in one such

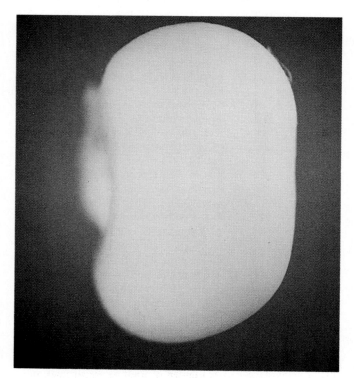

▲ **Figure 10-37**   Expression of anthocyanin in corn. Corn kernels are purple if they express genes that make a pigment called anthocyanin. These kernels of corn came from a plant the seed for which had a mobile element in the *a* gene of the anthocyanin pathway. If the mobile element remains stationary, the kernels are all white *(left);* if the element is excised very early, the kernels are all purple (not shown). If the mobile element is excised in a number of cells during the time the aleurone (pigment-forming) layer of cells is forming, the kernels are spotted *(right).* [See J. A. Ranks, P. Mason, and N. Federoff, 1988, *Genes & Devel.* 2:1364.] *Courtesy of N. Federoff.*

Activator (Ac) element

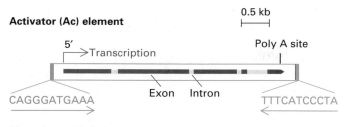

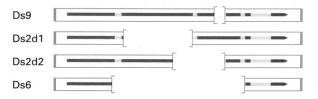

Dissociation (Ds) elements

▲ **Figure 10-38** Structure of Ac and Ds mobile elements of maize. The activator (Ac) element is 4563 nucleotides long and has 5 exons (red). It encodes at least two proteins. At its termini are short direct repeats, which enclose inverted repeats (sequence given). Ds elements have the same general structure as the Ac element except that a portion of the coding region is deleted. Ds elements can transpose only in plants that also contain the Ac element, implying that Ac encodes proteins required for mobility. *Adapted from N. V. Federoff, 1989, Cell, 56:181.*

mutant and its revertant showed that the P element was present in the mutant but had been excised in the revertant, leaving only a single copy of the 8-bp terminal direct repeat that flanked the P element in the mutant. These observations indicate that the *Drosophila* P element may also be a transposon that moves by DNA duplication.

Insertion of Mobile Elements Generally Produces Genetic Effects

As indicated in the previous section, both the Ac element in corn and the P element in *Drosophila* have been shown to cause revertible mutations. In fact, insertion of any mobile element—either a transposon or retroposon—generally damages the invaded gene. Three additional examples of the effects of mobile-element insertion that illustrate subtle and experimentally useful effects are described in this section.

Yeast Constitutive Mutants
In yeast, type II alcohol dehydrogenase (ADH$_{II}$) is an induced enzyme that is synthesized when cells are grown in the presence of alcohol. Glucose represses the synthesis of ADH$_{II}$ even in the presence of alcohol. Constitutive mutants have been found that produce ADH regardless of the medium in which they are grown. Analysis of the DNA around the ADH$_{II}$

gene in both wild-type and mutant strains showed that the mutants had a Ty element inserted upstream from the *ADH*$_{II}$ gene, whereas the wild types did not (Figure 10-39). Thus insertion of a Ty element, which itself is presumably transcribed, upsets the normal transcriptional regulation in the *ADH*$_{II}$ gene. Numerous other similar regulatory mutants caused by introduction of Ty elements have been identified in yeast.

White *Mutants in* Drosophila
Many classical *Drosophila* mutations have also been traced to the involvement of mobile elements. For example, variation in eye color is a prominent visible trait studied by the earliest *Drosophila* geneticists. One group of changes—originally thought to be point mutations—occurs at a site termed the *white* locus. A wide variety of changes at this locus modify the normal brick-red eye color to various paler shades, including pure white. Several known enzymes that contribute to making eye pigment are encoded outside the *white* locus. Exactly how the protein encoded at the *white* locus affects eye color is not yet known, but the normal gene product of the *white* locus does affect pigment production. Various mutations at the *white* locus result in the production of different amounts of pigment and hence mutants with different intensities of eye color.

Analysis of the DNA from a number of such mutants has shown that many contain repetitive mobile elements

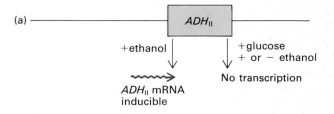

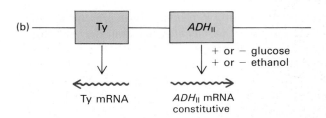

▲ **Figure 10-39** Regulatory mutation resulting from insertion of Ty element. (a) Normally, alcohol dehydrogenase of the ADH$_{II}$ type is induced by ethanol but not when glucose is present. (b) When the Ty element is inserted next to the *ADH*$_{II}$ gene in such a way that Ty is transcribed in the opposite direction from the gene, *ADH*$_{II}$ mRNA is formed constitutively. [See V. M. Williamson, E. T. Young, and M. Ciriacy, 1981, *Cell*, **23**:605.]

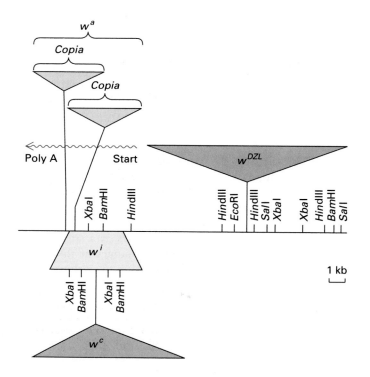

◀ **Figure 10-40** A restriction-site map of the *white* locus on the *Drosophila* genome indicating the locations of inserted mobile elements that cause eye-color mutations. The start site for transcription of the *white* locus and its poly A site are indicated by the wavy red line. Insertions detected in some *white* mutants are represented by the wedges. The wedge labeled *w^i* (ivory eyes) represents a mutant with almost no eye pigment. This mutation is a tandem 3-kb duplication identified by the repetition of the restriction enzyme sites (*Xba*I and *Bam*HI). Insertion of a 10-kb mobile element within the duplicated region of *w^i* mutants results in a pale-pink-eyed fly called *w^c*; the mutant gene produces at least some eye pigment. Thus the *w^i* rearrangement, a DNA duplication, is partially corrected by the *w^c* arrangement, a DNA insertion. Insertion of two *copia* elements near the same site results in *w^a* mutants, which have slightly lighter-colored eyes than *w^c* mutants. The *w^{D2L}* mutant, another pale-eyed fly, has a long repetitive element inserted upstream from the *white* start site. [See G. M. Rubin, 1983, in *Mobile Genetic Elements*, J. Shapiro, ed., Academic Press, p. 329; R. Levis, K. O'Hare, and G. M. Rubin, 1984, *Cell* **38**:471.]

inserted at the *white* locus (Figure 10-40). Even though the inserted sequences do not encode protein and may, in fact, lie outside of the protein-encoding region of *white,* they still affect eye color, probably by affecting transcription of the *white* locus.

It is now estimated that perhaps as many as half of all the spontaneous mutations collected from *Drosophila* species are not point mutations at all, but rather interruptions brought about by movement of mobile elements.

Hybrid Dysgenesis   Some mobile elements may play a role in the evolution of distinct species by preventing the productive interbreeding of related strains. Such mating isolation might eventually lead to speciation.

The P element in *Drosophila* provides a good example of this phenomenon. P elements are present in some strains (P strains) of *D. melanogaster,* but they are absent in other strains (M strains). When males of a P strain are crossed with females of an M strain, the progeny are defective heterozygous flies; it is as if the P and M genomes cannot merge. The term *hybrid dysgenesis* is used to describe such an unsuccessful cross. The defects that P/M heterozygotes display include greatly decreased fertility and increased mutations in any offspring; these induced mutations are frequently associated with chromosomal rearrangements.

A reasonable model for P/M hybrid dysgenesis is shown in Figure 10-41a. P strains contain a cytoplasmic repressor that normally regulates transcription of P elements, whereas M strains lack this repressor. When a P-strain sperm (which contains little cytoplasm and hence

little repressor) fertilizes an M-strain egg, the P elements are freed from inhibition, so transcription can proceed. As noted earlier, transposons like the P element encode a protein that is known to be required for transposition. Synthesis of this protein in the hybrid P/M zygote promotes transposition of P elements in the zygote, ultimately resulting in hybrid dysgenesis. This mechanism is consistent with the fact that the mating of an M-strain male with a P-strain female or the mating of two different P strains does not result in dysgenic offspring (Figure 10-41b).

Mobile Genetic Elements Must Move in Gametes to Affect the Evolution of Multicellular Organisms

In multicellular organisms, movement of mobile genetic elements in somatic cells could cause mutations that would not be transmitted to progeny. For example, movement of *Alu* sequences in somatic cells has been suggested, but not proved, as an occasional cause of cancer. Likewise, retroviruses, which are known to move into the genomes of somatic cells, may cause certain types of cancer.

We have suggested several times in this discussion that movement of mobile genetic elements in multicellular eukaryotes may play some role in evolution. However, in order for changes in the position or content of mobile elements to affect heritable traits through many generations, they must occur in gametes or pre-gametic cells. As we have seen, movement of retroposons—the most im-

(a)

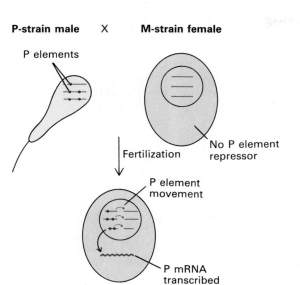

(b)

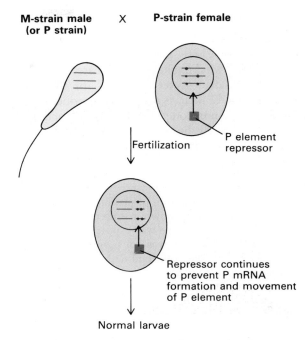

▲ **Figure 10-41** A model for hybrid dysgenesis in *Drosophila melanogaster*. (a) Crosses between P-strain males, which contain P elements, and M-strain females, which lack P elements, induce synthesis of P mRNA, which encodes a protein that promotes transposition. The resulting increase in P-element movement leads to many chromosomal rearrangements with deleterious effects in the offspring. (b) The reverse cross is not so damaging because the P-element repressor is carried into the zygote with the maternal cytoplasm. Likewise, crosses between two P strains do not produce dysgenic offspring. [See K. O'Hare and G. M. Rubin, 1983, *Cell* 34:25.]

portant type of mobile element in higher eukaryotes—involves a reverse transcription mechanism. Hence the hypothesis that mobile elements have an evolutionary function in eukaryotes implies that reverse transcriptase activity is present in gametes or pre-gametes. In fact, this has not yet been demonstrated.

Functional Rearrangements in Chromosomal DNA

Although a number of functional rearrangements of DNA have been identified, none of them is associated with the apparently random movement of transposons and retroposons. The mechanisms involved in these functional rearrangements include gene conversion, DNA amplification, and DNA deletions. In the remainder of this chapter, we describe several examples of functional DNA rearrangements. Such DNA rearrangements probably play only a numerically minor role in the regulation of genes. By far the most frequent mechanism of gene regulation required for normal development, differentiation, and reproduction involves differential transcription.

Yeast Mating Types Can Switch by Gene Conversion

Saccharomyces cerevisiae undergoes a regular chromosomal rearrangement that controls an important cell feature—the mating type. This yeast species grows in culture as a haploid cell. Two haploid cells can "mate" or fuse to form a diploid cell if they are of opposite mating types, called **a** and α; fused diploid cells are designated **a**/α. Fusion is promoted by oligopeptides termed **a** and α factors. Each of these factors is produced by its homologous cell type, and each factor arrests the growth of the other cell type to enable fusion to occur. When faced with starvation, the fused diploid cell undergoes two meiotic divisions to produce four spores (see Figure 5-14). A cell sample can be tested for mating type by adding one of the two factors to yeast cells in culture and examining the cells in the microscope (Figure 10-42).

A normal haploid yeast cell can grow into a colony that contains both **a** and α mating types. This means that either mating type can be expressed from a single genome. In fact, a normal haploid cell has the ability to switch its mating type each generation. This ability depends (among other genes) on a wild type *homothallic (HO)* locus in the genome: mutants in the *HO* locus no longer switch mating type. If the *HO* locus is normal, mating-type switching will be frequent.

This remarkable switching in mating types has been traced to changes in DNA structure on chromosome 3 of *S. cerevisiae*. Three genetic loci directly involved in mating-type switching have been located (Figure 10-43a). The central locus is termed *MAT,* or the mating-type locus. The *MAT* locus is actively transcribed into mRNA, and the protein encoded by this mRNA acts to regulate genes not directly linked to the *MAT* locus. These unlinked genes encode the proteins that give the cell its **a** or α phenotype. Two additional "silent" (nontranscribed)

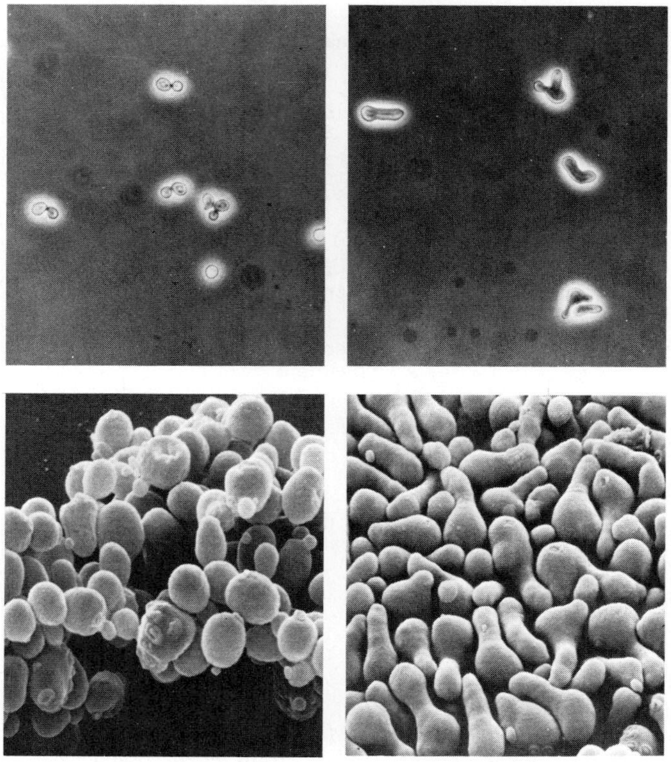

Sample 1:
No exposure to α factor

Sample 2:
Exposure to α factor

▶ **Figure 10-42** Detection of mating-type phenotype (**a** or α) in yeast. Cells of unknown mating type were divided into two samples, of which one was exposed to the α factor (sample 2) and the other was not (sample 1). Sample 1 shows normal budding cells and daughter cells in a phase-contrast light micrograph *(top)* and a scanning electron micrograph *(bottom)*. The corresponding micrographs of sample 2 show that cell division is inhibited in these cells. Since each mating factor inhibits cell division in cells of the opposite mating type, the sample cells are **a**. [See J. Thorner, 1980, in *The Molecular Genetics of Development*, T. A. Leighton and W. A. Loomis, eds., Academic Press.] *Courtesy of J. Thorner, R. Kunisawa, and M. Davis.*

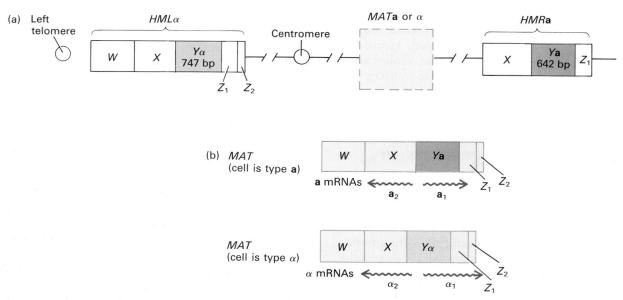

▲ **Figure 10-43** Structure of the chromosomal region that determines mating type (**a** or α) in yeast. (a) This region contains three loci—*HMLα*, *HMRa*, and *MAT,* which can exist in two forms. (b) The two forms of *MAT* have different *Y* sequences—*Ya* and *Yα*. Although the *MAT* locus encodes two mRNAs, the *Y* sequence primarily determines which mating type is expressed. The positions of the various loci with respect to the centromere and the telomere on the short arm of chromosome 3 are indicated. [See K. A. Nasmyth, 1983, *Nature* **302**:670.]

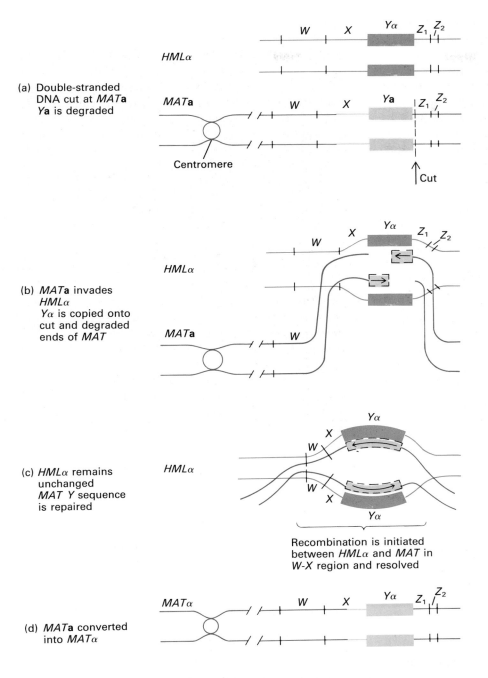

(a) Double-stranded DNA cut at *MAT*a Ya is degraded

(b) *MAT*a invades *HML*α Yα is copied onto cut and degraded ends of *MAT*

(c) *HML*α remains unchanged *MAT* Y sequence is repaired

Recombination is initiated between *HML*α and *MAT* in W-X region and resolved

(d) *MAT*a converted into *MAT*α

◄ **Figure 10-44** Switching of a mating type to α mating type by gene conversion at the *MAT* locus on yeast chromosome 3. In step (b), the cut ends of the *MAT* locus invade the *HML*α locus (see Figure 10-43). Growth from the 3′ end of each cut results in duplication of the Yα information. Recombination occurs to yield a *MAT* locus that has been converted from *MAT*a to *MAT*α. Neither HML nor HMR is changed by gene conversion at *MAT*, so the reverse switch in mating type can occur in the next generation. [See J. N. Strathern et al., 1982, *Cell* **31**:183.]

copies of the genetic information found at *MAT* are stored at loci termed *HML* and *HMR* (*h*omothallic copies to the *l*eft and *r*ight of *MAT*, respectively, as the genetic map is usually drawn). One of the silent loci has the information for type **a** and the other for type α. Some of the sequences from *HML* and *HMR* are transferred alternately (between cell generations) into *MAT*, from which expression can then occur.

The DNA from the three mating-type loci in yeast has been cloned and sequenced. Although *HML* and *HMR* have some similar sequences, other sequences are distinct in the two loci. The most crucial regions are called Yα

and Ya sequences. The presence of one or the other of these sequences at the *MAT* locus determines whether a cell makes **a** or α mRNAs (Figure 10-43b).

The variation in sequence content at *MAT* is accomplished by directed gene conversion. A site-specific endonuclease termed the HO nuclease is present in HO^+ strains. This endonuclease cleaves both strands of α or **a** DNA specifically at the boundary between the sequences designated as Z and Y. About 3 percent of the cells in an HO^+ culture have a cut in *MAT*, but none has a cut in the *HML* or *HMR* locus. This enzyme was the first sequence-specific endonuclease to be isolated from eukaryotic cells.

The cut ends at the *MAT* locus can pair with the DNA of either *HMR* or *HML*. However, since switching of the information at *MAT* is the rule in *HO*⁺ strains, pairing of the cut ends is directed to the site with the opposite *Y* sequence. To illustrate, a cut is made at *MAT*a and enlarged by degrading most of the *MAT*a information. The ends of the gap in *MAT* then invade *HML*α (Figure 10-44a and b). DNA synthesis proceeds from the free 3' hydroxyl on the Z_1 sequence of *MAT*, resulting in the copying of the *Y*α at *HML*α onto the cut end of *MAT*a. A recombination between the new copy of the *HML* and the remaining portion of *MAT*a leads to loss of *Y*a and formation of *MAT*α (see Figure 10-44c and d). *HML* and *HMR* remain unchanged throughout this process and are available to switch again in the next generation.

Trypanosome Surface Antigens Undergo Frequent Changes

Trypanosomes are protozoans that can cause severe illness in animals and humans. One well-studied species, *Trypanosoma brucei,* contains highly antigenic variable surface glycoproteins (VSGs). This organism has a mechanism for frequently changing one VSG to a similar but distinct antigen thus enabling the protozoan to avoid host immune responses that would inactivate it.

The mRNAs responsible for several different VSGs were partially purified and used to make cDNA clones, which were then sequenced. Because the VSG cDNAs hybridized to clusters of genomic DNA sequences called *basic copies* (BCs), these sequences were thought to encode VSGs. Presumably, each BC encoded a different VSG. However, the sequence at the 3' ends of all genomic BCs that have been analyzed differ from the 3' end of the mRNAs; in addition, genomic BCs do not contain the 5' end sequence found in the mRNAs. These findings suggest that VSG mRNAs are not synthesized directly from BCs. Southern blot analysis with the 3' end of VSG cDNAs showed that the site of transcription of VSG mRNA lay elsewhere in the genome. This site—the active VSG gene—was found to contain a transposed and duplicated copy—the *expression-linked copy* (ELC)—of a BC sequence inserted next to a 3' end sequence that was the same as the 3' region found in the mRNAs. Upstream of

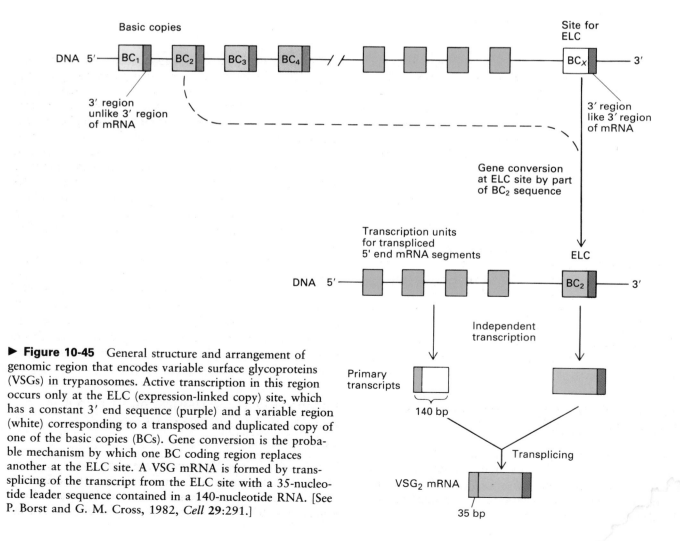

▸ **Figure 10-45** General structure and arrangement of genomic region that encodes variable surface glycoproteins (VSGs) in trypanosomes. Active transcription in this region occurs only at the ELC (expression-linked copy) site, which has a constant 3' end sequence (purple) and a variable region (white) corresponding to a transposed and duplicated copy of one of the basic copies (BCs). Gene conversion is the probable mechanism by which one BC coding region replaces another at the ELC site. A VSG mRNA is formed by transsplicing of the transcript from the ELC site with a 35-nucleotide leader sequence contained in a 140-nucleotide RNA. [See P. Borst and G. M. Cross, 1982, *Cell* **29**:291.]

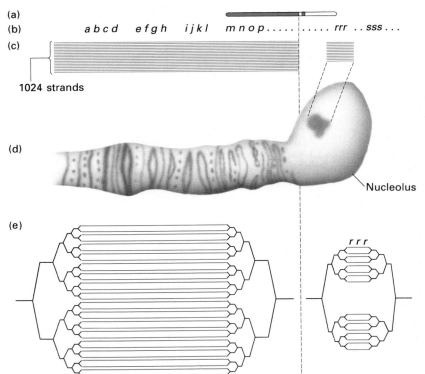

(a)
(b) *a b c d   e f g h   i j k l   m n o p* .... *rrr* .. *sss* ...
(c)
1024 strands
(d)
Nucleolus
(e)
r r r

◀ **Figure 10-46** Model of DNA amplification to produce polytene chromosomes. (a) *Drosophila hydei* X chromosome during interphase contains light-staining euchromatin in the left arm and mostly heterochromatin in the remainder of the chromosomes. (b) Sequences within the euchromatic region are represented by the string of letters *a* through *p*. Simple-sequence DNA *(sss)* and ribosomal genes *(rrr)* are located in the heterochromatic region. (c) In the salivary glands, all of the DNA in the left arm of the X chromosome is replicated in parallel many times; 2^{10} duplications would give 1024 copies. The *sss* DNA does not appear to increase at all; the ribosomal genes are replicated but not as many times as the DNA in the left arm. (d) The banding pattern in the polytene chromosome results from reproducible packing of DNA and protein within each amplified site along the chromosome. (e) Proposed pattern of replication in *Drosophila* polytene chromosomes. [See C. D. Laird et al., 1973, *Cold Spring Harbor Symp. Quant. Biol.* 38:311.]

the ELC are clusters of sequences that represent multiple start sites for transcription of a mini-exon RNA; a 35-nucleotide leader sequence is transpliced from this RNA to the major part of the mRNA transcribed from the ELC region (Figure 10-45). In addition to the VSG mRNAs, it now appears that most and perhaps all trypanosome mRNAs contain the same leader sequence; thus trans-splicing is a general event in these cells.

The chromosomal rearrangements giving rise to different VSG mRNAs in trypanosomes are *not* programmed changes that occur before each cell division; rather they are random events that occur in about 1 in 10^6 cells. Because these variations in DNA structure are similar to yeast mating-type changes, in that one copy is maintained and one copy is activated, a gene conversion may be responsible. Changes in the VSG permit trypanosomes to circumvent the effects of antibodies produced by an infected animal; thus this switching ability may have an evolutionary value for trypanosomes. It is possible that a similar rearrangement of DNA occurs in *Paramecium*, another protozoan that changes its surface proteins often.

Generalized DNA Amplification Produces Polytene Chromosomes

All of the DNA rearrangements discussed so far—both functional and nonfunctional—involve changes in the position of sequences within the genome. Another type of rearrangement involves generalized amplification of DNA sequences, or polytenization.

The salivary glands of *Drosophila* species contain enlarged interphase chromosomes characterized by a large number of well-demarcated bands. The enlargement of chromosomes in the salivary glands, and in other tissues as well, occurs when the DNA repeatedly replicates but the daughter strands fail to separate. The result is a *polytene* chromosome composed of many parallel copies of itself. Although most of a chromosome participates in polytenization, certain sequences, such as the simple-sequence DNA near the centromere, are not amplified. Furthermore, the ribosomal genes tend to be amplified less than the other sequences during polytenization (Figure 10-46).

The molecular basis for the varying extent of replication along the presumably linear chromosomal DNA molecules remains unknown, as does the reason for differences in the degree of polytenization in different cell types. For example, *Drosophila* salivary gland chromosomes appear to contain about 1000 copies of most DNA regions, whereas the polytene chromosomes in some abdominal cells contain only about 100 copies.

Localized DNA Amplification of rRNA and Other Genes Occurs in Some Eukaryotic Cells

The localized amplification of specific genes was first demonstrated in the rRNA genes of frog oocytes. Several other examples of this phenomenon have been described, although gene-specific amplification probably is not as

(a)

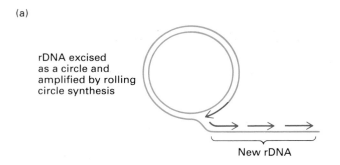

rDNA excised
as a circle and
amplified by rolling
circle synthesis

New rDNA

(b)

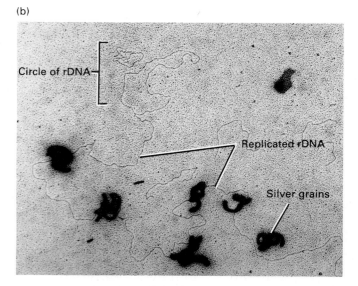

Circle of rDNA

Replicated rDNA

Silver grains

◀ **Figure 10-47** Amplification of rRNA genes in frog oocytes. Replication of the excised rDNA (blue) takes place by a mechanism called rolling circle replication. (b) Electron microscope autoradiograph of DNA extracted from frog oocytes exposed to labeled thymidine, a DNA precursor. A circle of excised rDNA undergoing replication is visible. That the long strand extending from this circle is newly made DNA is indicated by the presence of silver grains, which correspond to sites of thymidine incorporation. [See J. G. Gall and M. L. Pardue, 1969, *Proc. Nat'l Acad. Sci. USA* 63:378; and J.-D. Rochaix, A. Bird, and A. Bakken, 1974, *J. Mol. Biol.* 87:473.] *Photograph courtesy of A. Bird.*

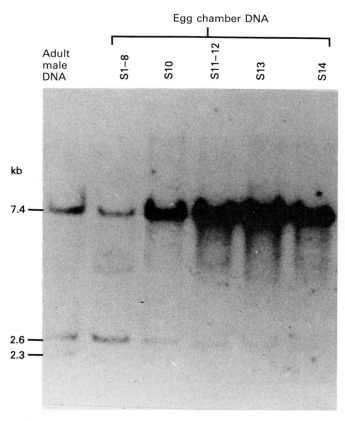

▲ **Figure 10-48** Demonstration that genes for chorion proteins are amplified during larval development of *Drosophila melanogaster*. Cloned labeled cDNA copies of the mRNAs encoding chorion proteins were produced and used as a probe in Southern blot analysis of equal total amounts of DNA from normal adult males *(left band)* and from the egg chamber at several different stages of larval development (S1–8, S10, etc.). The cDNA probe hybridized to a 7.4-kb fragment, which increased in amount by a factor of 50 during the stages S10–S14. [See A. C. Spradling and A. P. Mahowald, 1980, *Proc. Nat'l Acad. Sci. USA* 77:1096.] *Courtesy of A. C. Spradling.*

widespread as polytenization. The amplified gene copies may remain in tandem at the site where the increase takes place, or they may be released as free-floating extra copies.

rDNA Amplification in Amphibian Oocytes

Frog oocytes are 2 to 3 mm in diameter and contain many stored ribosomes. The RNA for these ribosomes comes from a dramatic increase in transcription of the greatly amplified rDNA that occurs during the development of the oocyte. This 2000-fold amplification occurs by excision of rDNA from the genome in the form of circular molecules, which then are copied by *rolling circle replication* (Figure 10-47). Although localized amplification of rRNA genes was first discovered in frog oocytes, it probably also occurs in many other animals with highly developed egg cells.

Gene-Specific Amplification in Insects

Cytologic observation of chromosomal enlargements, or "puffs," at specific chromosomal sites in various insects suggested that specific genes might be amplified. Such amplification of the genes encoding surface (chorion) proteins was subsequently demonstrated in *D. melanogaster*. This amplification occurs only in the ovarian follicle tissues and involves a DNA region about 100 kb long. Southern blot analysis of this region has shown that it is amplified about 50 times during larval development (Figure 10-48). In insects, the same enzymes may be employed in gene-specific amplification and generalized polytenization of chromosomes.

Not all insect genes that program the production of large amounts of protein are amplified. For example, the gene that encodes silk fibroin, the major protein in silk, is not amplified in the larvae of the silk moth. Many vertebrate genes that produce large amounts of mRNA (e.g., for globin, ovalbumin, and serum albumin) are also known to be unamplified.

DNA Amplification in Cultured Mammalian Cells

When mammalian cells are grown in culture, specific sites on their DNA may become amplified. For example, DNA amplification occurs in cultured mammalian cancer cells that are exposed to toxic substances such as methotrexate. This drug inhibits the activity of dihydrofolate reductase (DHFR), an enzyme that catalyzes the transfer of methyl groups and is required in the synthesis of nucleic acid bases and some other metabolites. When cancer cells are exposed to methotrexate, they develop drug-resistant colonies. As the concentration of the drug is gradually increased, the average level of dihydrofolate reductase in the culture increases to 1000 times the normal level. Along with this increase in enzyme activity is an increase in the DNA encoding the enzyme. On the surface, it would appear that gene-specific amplification occurs in response to need; however, the chromosomal regions that are amplified are much larger than the DHFR gene.

Although the precise mechanism of mammalian DNA amplification is not known, two possible mechanisms for amplifying the DHFR gene have been proposed. Unequal crossing over between sister chromatids might produce some daughter cells without the DHFR gene and others with two copies of the gene. Repetition of this process in the presence of methotrexate would lead to amplification, as cells with multiple copies of DHFR would have a selective advantage (Figure 10-49a). The second mechanism involves production of chromosomal fragments, termed *minute chromosomes,* which is promoted by methotrexate. Insertion of these fragments into normal chromosomes could produce daughter cells with several copies of the DHFR gene (Figure 10-49b). Again, repetition of this process through several cell cycles in the presence of methotrexate could result in extensive gene amplification.

It is of considerable interest that all cultured cells known to undergo drug-selected amplification are abnormal cancer cells; normal fibroblasts in culture do not undergo amplification. Furthermore, no direct gene amplification due to environmental stress has yet been demonstrated in normal mammalian cells, although in cancer

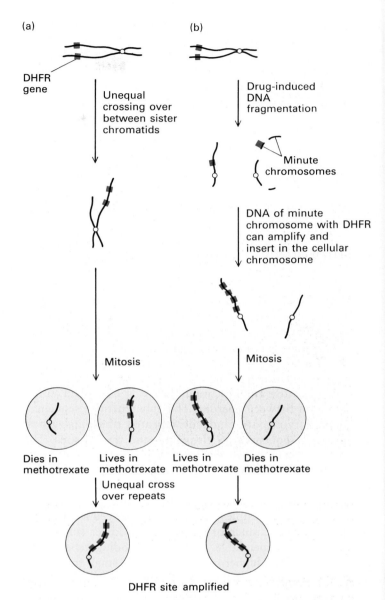

▲ **Figure 10-49** Two possible mechanisms for amplification of dihydrofolate reductase (DHFR) gene in cultured mammalian cancer cells exposed to methotrexate, a drug that inhibits DHFR. In both cases, methotrexate would select cells with extra copies of the DHFR gene.

patients treated with methotrexate, cancer cells with amplified DHFR genes have been observed. Thus, in abnormally growing cells, DNA amplification can occur in vivo as well as in culture.

Deletions Produce Immunoglobulin Transcription Units in Vertebrates

In all vertebrates, a series of interacting cells, collectively called the *immune system,* cooperate to protect the animal from noxious outside influences (e.g., viruses and bacteria). Perhaps the best-studied event in the immune response is the production and secretion of specific antibodies (immunoglobulins) by lymphoid cells. While these cells are undergoing differentiation, specific deletions occur in their DNA. The deletions bring together regions of DNA encoding the L (light) protein chain of immunoglobulin and regions encoding the H (heavy) chain (see Figure 2-38 for general structure of an immunoglobulin.) Apparently these DNA rearrangements in the immunoglobulin DNA region occur randomly in the lymphoid cell population of the animal. Only a few cells in any individual survive to become successful antibody producers. The lymphoid system is discussed in detail in Chapter 25.

Summary

In recent years, a wealth of new information about the DNA in the chromosomes of many different organisms has become available. Although generalizations are risky at this point, some summary statements may be helpful.

Bacterial genomes have little wasted space. Many of the protein-coding genes are arranged in operons, which usually are transcribed into polycistronic mRNAs. Apparently little nonfunctional DNA is contained either within or between operons. The only repetitive sequences in prokaryotes are tandemly repeated ribosomal genes, insertion sequences (IS elements), and transposons. The only *E. coli* chromosomal DNA sequences that may serve no useful function are IS elements and some of the transposons.

Eukaryotic single-celled organisms such as yeast have only three to five times as much DNA per cell as prokaryotes. However, much of the genome of lower eukaryotes probably is expendable. The protein-coding genes of all metazoans and higher plants make up a small fraction—possibly as low as 1 percent—of the total cellular DNA. If transcription units (exons plus introns) averaged 10 times the length of mature mRNA, then 10 percent of the genome would consist of coding sequences. However, the transcription units for many protein-coding genes are islands separated by vast stretches of meaningless DNA. Thus multicellular organisms use only a small fraction of their total DNA.

The "extra" DNA in eukaryotes consists of introns, pseudogenes, simple-sequence DNA, mobile genetic elements, and unclassified spacer DNA between transcription units. The amount of simple-sequence DNA varies greatly among genomes that are otherwise comparable in size; this variation suggests that even if some simple-sequence DNA serves an important function, there may be vastly more than is required. Sequence analysis of numerous pseudogenes has shown that these are derived from normal genes that have lost critical nucleotide sequences. Introns are present in some transcription units from all eukaryotic cell types; they abound in vertebrate genes and may constitute 80–90 percent of all the DNA within vertebrate transcription units. Introns perform no known function, and, like other extra DNA, they appear to be dispensable. Mobile genetic elements (interspersed intermediate repeat DNA) are present in many eukaryotic cells. Some of these are similar in structure to bacterial transposons and like them move by a DNA-mediated mechanism. Most eukaryotic mobile elements, however, are retroposons; these move in the genome via RNA intermediates. Although most if not all mobile genetic elements have no useful functions in the life cycle of organisms, they may affect evolution, as they introduce changes in genes.

The discovery of functional DNA rearrangements in several eukaryotic species suggests that such rearrangements might play a wider role in gene expression than has been suspected in the past. Many pathways of differentiation are thought to proceed systematically toward greater and greater cell specialization. It is tempting to think that permanent (or near-permanent) gene rearrangements at the DNA level might underlie some "irreversible" developmental changes. However, most gene rearrangements studied so far are not strictly programmed and involve much cell death before a successful rearrangement is selected. Perhaps only when cell death is a prominent feature of embryogenesis (e.g., during the development of the nervous system) is gene rearrangement likely to be involved. Before the importance of DNA rearrangements in differentiation can be ascertained, many developmentally important genes must be isolated and studied to determine whether their structures are altered during development.

References

Protein-Coding Genes and Gene Families: Nature and Distribution

BOSSY, B., L. M. C. HALL, and P. SPIERER. 1984. Genetic activity along 315 kb of the *Drosophila* chromosome. *EMBO J.* 3:2537–2541.

CLEVELAND, D. W., and K. F. SULLIVAN. 1985. Molecular biology and genetics of tubulin. *Ann. Rev. Biochem.* 54:331–365.

EFSTRATIADIS, A., J. W. POSAKONY, T. MANIATIS, et al. 1980. The structure and evolution of the human β-globin gene family. *Cell* 21:653–668.

GORIN, M. B., D. L. COOPER, F. EIFERMAN, P. VAN DE RIJN, and S. M. TILGHMAN. 1981. The evolution of α-fetoprotein and albumin, I: A comparison of the primary amino acid sequences of mammalian α-fetoprotein and albumin. *J. Biol. Chem.* 256:1954–1959.

HENTSCHEL, C. C., and M. L. BIRNSTIEL. 1981. The organization and expression of histone gene families. *Cell* 25:301–313.

KAFATOS, F. C. 1983. Structure, evolution and developmental expression of the chorion multigene families in silk moths and *Drosophila*. In *Gene Structure and Regulation in Development*, S. Subtelny and F. C. Kafatos, eds. Alan R. Liss.

MIRKOVITCH, J., P. SPIERER, and U. K. LAEMMLI. 1986. Genes and loops in 320,000 base-pairs of the *Drosophila melanogaster* chromosome. *J. Mol. Biol.* **190**:255–258.

NATHANS, J. 1987. Molecular biology of visual pigments. *Ann. Rev. Neurosci.* **10**:163–194.

PIATIGORSKY, J. 1984. Lens crystallins and their gene families. *Cell* **38**:620–621.

ROYAL, A., A. GARAPIN, B. CAMI, et al. 1979. The ovalbumin gene region: common features in the organisation of three genes expressed in chicken oviduct under hormonal control. *Nature* **279**:125–132.

STEINEET, P. M., A. C. STEVEN, D. R. ROOP. 1985. The molecular biology of intermediate filaments. *Cell* **42**:411–419.

Mapping Chromosomes by Restriction Fragment Length Polymorphisms (RFLPs)

BOTSTEIN, D., R. L. WHITE, M. SKOLNICK, and R. W. DAVIS. 1980. Construction of a genetic linkage map in man using restriction fragment length polymorphisms. *Am. J. Hum. Genet.* **32**:314–331.

DONIS-KELLER, H., P. GREEN, C. HELMS, et al. 1987. A genetic linkage map of the human genome. *Cell* **51**:319–337.

GUSELLA, J. F. 1986. DNA polymorphism and human disease. *Ann. Rev. Biochem.* **55**:831–854.

LITTLE, P. 1986. Restriction fragment length polymorphisms: finding the defective gene. *Nature* **321**:558–559.

Tandemly Repeated Genes Encoding rRNA

FEDEROFF, N. 1979. On spacers. *Cell* **16**:697–710.

OHTA, T. 1983. On the evolution of multigene families. *Theor. Popul. Biol.* **23**:216–240.

WELLAUER, P. K., and I. B. DAWID. 1979. Isolation and sequence organization of human ribosomal DNA. *J. Mol. Biol.* **128**:289–303.

Mechanisms of Sequence Maintenance

ARNHEIM, N. 1983. Concerted evolution and multigene families. In *Evolution of Genes and Proteins,* M. Nei and R. K. Koehn, eds. Sinauer Associates.

PETES, T., and G. R. FINK. 1982. Gene conversion between repeated genes. *Nature* **300**:216–217.

SMITH, G. P. 1976. Evolution of repeated DNA sequences by unequal crossovers. *Science* **191**:528–537.

SPRADLING, A. C., and G. M. RUBIN. 1981. *Drosophila* genome organization: conserved and dynamic aspects. *Ann. Rev. Genet.* **15**:219–264.

Repetitious DNA

BRITTEN, R. J., and D. E. KOHNE. 1968. Repeated sequences in DNA. *Science* **161**:529–540.

DOOLITTLE, W. F., and C. SAPIENZA. 1980. Selfish genes, the phenotype paradigm and genome evolution. *Nature* **284**:601–603.

GALL, J. G. 1981. Chromosome structure and the C-value paradox. *J. Cell Biol.* **91**:3s–14s.

JEFFREYS, A. J., N. J. ROYLE, V. WILSON, and Z. WONG. 1988. Spontaneous mutation rates to new length alleles at tandem-repetitive hypervariable loci in human DNA. *Nature* **332**:278–280.

JEFFREYS, A. J., V. WILSON, and S. C. THEIN. 1985. Individual-specific fingerprints of human DNA. *Nature* **316**:76–78.

KORNBERG, J. R., and M. C. RYKOWSKI. 1988. Human genome organization: *Alu*, LINES, and the molecular structure of metaphase chromosome bands. *Cell* **53**:391–400.

MIKLOS, G. L. G., M.-T. YAMAMOTO, J. DAVIS, and V. PIROTTER. 1988. Microcloning reveals a high frequency of repetitive sequences characteristic of chromosome 4 in the β heterochromatin of Drosophila. *Proc. Nat'l Acad. Sci. USA* **85**:2051–2055.

MOYZIS, R. K., K. L. ALBRIGHT, M. F. BARTHOLDI, et al. 1987. Human chromosome-specific repetitive DNA sequences: novel markers for genetic analysis. *Chromosoma* **95**:375–386.

SINGER, M. F. 1982. SINES and LINES: highly repeated short and long interspersed sequences in mammalian genomes. *Cell* **28**:433–434.

The *Alu* Sequence Family and Other Retroposons

BOEKE, J. D., D. J. GARFINKEL, C. A. STYLES, and G. R. FINK. 1985. Ty elements transpose through an RNA intermediate. *Cell* **40**:491–500.

GARFINKEL, D., J. BOEKE, and G. R. FINK. 1985. Ty element transposition: reverse transcriptase and virus-like particles. *Cell* **42**:507–517.

HAYNES, S. R., and W. R. JELINEK. 1981. Low molecular weight RNAs transcribed *in vitro* by RNA polymerase III from *Alu*-type dispersed repeats in Chinese hamster DNA are also found *in vivo*. *Proc. Nat'l Acad. Sci. USA* **78**:6130–6134.

JAGADEESWARAN, P., B. G. FORGET, and S. M. WEISSMAN. 1981. Short interspersed repetitive DNA elements in eucaryotes: transposable DNA elements generated by reverse transcription of RNA Pol III transcripts? *Cell* **26**:141–142.

JELINEK, W. R., and C. W. SCHMID. 1982. Repetitive sequences in eukaryotic DNA and their expression. *Ann. Rev. Biochem.* **51**:813–844.

JURKA, J., and T. SMITH. 1988. A fundamental division in the *Alu* family of repeated sequences. *Proc. Nat'l Acad. Sci. USA* **85**:4775–4778.

KINGSMAN, A. J., and S. M. KINGSMAN. 1988. Ty: a retroelement moving forward. *Cell* **53**:333–335.

MOUNT, S. M., and G. M. RUBIN. 1985. Complete nucleotide sequence of the *Drosophila* transposable element *copia*: homology between *copia* and retroviral proteins. *Mol. Cell Biol.* **5**:1630–1638.

VAN ARSDELL, S. W., R. A. DENISON, L. B. BERNSTEIN, et al. 1981. Direct repeats flank three small nuclear RNA pseudogenes in the human genome. *Cell* **26**:11–17.

WILLIAMSON, V. M., E. T. YOUNG, and M. CIRIACY. 1981. Transposable elements associated with constitutive expression of yeast alcohol dehydrogenase II. *Cell* **23**:605–614.

YOUNG, M. W. 1982. Differing levels of dispersed repetitive DNA among closely-related species of Drosophila. *Proc. Nat'l Acad. Sci. USA* **79**:4570–4574.

Transposition and Transposable Elements

BERG, D. E., and M. M. HOWE, eds. 1989. *Mobile DNA.* American Society for Microbiology.

CALOS, M. P., and J. H. MILLER. 1980. Transposable elements. *Cell* 20:579–595.

FEDEROFF, N. V. 1989. Maize transposable elements. In *Mobile DNA,* D. E. Berg and M. M. Howe, eds. American Society for Microbiology.

FINNEGAN, D. J., and D. H. FAWCETT. 1986. Transposable elements in *Drosophila melanogaster. Oxford Surveys on Eukaryotic Genes* 3:1–62.

GOLDEN, J. W., S. J. ROBINSON, and R. HASELKORN. 1985. Rearrangement of nitrogen fixation genes during heterocyst differentiation in the cyanobacterium *Anabaena. Nature* 314:419–423.

HUGHES, K. T., P. YOUDERIAN, and M. I. SIMON. 1988. Phase variation in *Salmonella:* analysis of Hin recombinase and *hix* recombination site interaction in vivo. *Genes & Develop.* 2:937–948.

KLECKNER, N. 1981. Transposable elements in prokaryotes. *Ann. Rev. Genet.* 15:341–404.

MCCLINTOCK, B. 1956. Controlling elements and the gene. *Cold Spring Harbor Symp. Quant. Biol.* 21:197–216.

O'HARE, K., and G. RUBIN. 1983. Structures of P transposable elements and their sites of insertion and excision in the *Drosophila melanogaster* genome. *Cell* 34:25–35.

RUBIN, G. M., and A. C. SPRADLING. 1982. Genetic transformation of *Drosophila* with transposable element vectors. *Science* 218:348–353.

SHAPIRO, J. A. 1979. Molecular model for the transposition and replication of bacteriophage Mu and other transposable elements. *Proc. Nat'l Acad. Sci. USA* 76:1933–1937.

SHAPIRO, J. A., ed. 1983. *Mobile Genetic Elements.* Academic Press.

SIMON, M., and I. HERSKOWITZ, eds. 1985. Genome rearrangement. In *UCLA Symposia on Molecular and Cellular Biology,* vol. 20. Alan R. Liss.

SPRADLING, A. C., and G. M. RUBIN. 1982. Transposition of cloned P elements into *Drosophila* germ line chromosomes. *Science* 218:341–347.

TRUETT, M. A., R. S. JONES, and S. S. POTTER. 1981. Unusual structure of the FB family of transposable elements in *Drosophila. Cell* 24:753–763.

Functional Genomic Rearrangements

BORST, P., and D. R. GREAVES. 1987. Programmed gene rearrangements altering gene expression. *Science* 235:658–667.

Yeast Mating-Type Switching

HERSKOWITZ, I., and O. OSHIMA. 1982. Control of cell type in *S. cerevisiae:* mating type and mating type interconversions. In *Molecular Biology of the Yeast* Saccharomyces, J. N. Strathern, E. W. Jones, and J. R. Broach, eds. Cold Spring Harbor Laboratory.

KOSTRIKEN, R., J. N. STRATHERN, A. J. S. KLAR, J. B. HICKS, and F. HEFRON. 1983. A site-specific endonuclease essential for mating-type switching in *Saccharomyces cerevisiae. Cell* 35:167–174.

NASMYTH, K. A. 1983. Molecular analysis of a cell lineage. *Nature* 302:670–676.

NASMYTH, K. A. 1987. The determination of mother cell-specific mating type switching in yeast by a specific regulator of HO transcription. *EMBO J.* 6:243–248.

STILLMAN, D. J., A. T. BANKIER, A. SEDDON et al. 1988. Characterization of a transcription factor involved in mother cell specific transcription of yeast HO gene. *EMBO J.* 7:485–494.

STRATHERN, J. N., E. W. JONES, and J. R. BROACH, eds. 1982. *Molecular Biology of the Yeast* Saccharomyces. Cold Spring Harbor Laboratory.

STRUHL, K. 1983. The new yeast genetics. *Nature* 305:391–397.

Antigenic Variation in Trypanosomes

BORST, P. 1986. Discontinuous transcription and antigenic variation in trypanosomes. *Ann. Rev. Biochem.* 55:701–732.

SHEA, C. M., G.-S. LEE, and L. H. T. VAN DER PLOEG. 1987. VSG gene 118 is transcribed from a cotransposed pol I-like promoter. *Cell* 50:603–612.

VAN DER PLOEG, L. H. T. 1986. Discontinuous transcription and splicing in trypanosomes. *Cell* 47:479–480.

DNA Amplification

GOLDSMITH, M. R., and F. C. KAFATOS. 1984. Developmentally regulated genes in silkmoths. *Ann. Rev. Genet.* 18:443–488.

LOONEY, J. E., C. MA, T.-H. LEU, et al. 1988. The dihydrofolate reductase amplicons in different methotrexate-resistant Chinese hamster cell lines share at least a 273-kilobase core sequence, but the amplicons in some cell lines are much larger and are remarkably uniform in structure. *Mol. Cell Biol.* 8:5268–5279.

SCHIMKE, R. T. 1980. Gene amplification and drug resistance. *Sci. Am.* 243(5):60–69.

SPRADLING, A. C., and A. P. MAHOWALD. 1980. Amplification of genes for chorion proteins during oogenesis in *Drosophila melanogaster. Proc. Nat'l Acad. Sci. USA* 77:1096–1100.

STARK, G. R., and G. M. WAHL. 1984. Gene amplification. *Ann. Rev. Biochem.* 53:447–491.

11

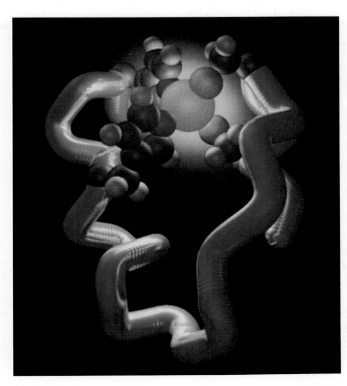

Zinc finger

Gene Control and the Molecular Genetics of Development in Eukaryotes

*T*he molecular basis for gene control in eukaryotic cells, particularly in animal cells, is currently one of the most actively studied areas in all of biology. Throughout the era of classical genetics, a major aim was to discover how genes participate in controlled programs that result in the development of an animal or a plant. The power of molecular genetics has allowed incredibly rapid progress in satisfying that aim. Once the importance of RNA processing and gene and chromosome structure in eukaryotes was understood, attention could be turned to how genes were regulated. The proof is now in hand that regulation exists at both the level of transcription and after transcription and that specific patterns of gene regulation are important determinants in development decisions. An avalanche of information has accumulated about the DNA sites required for transcriptional regulation and about the regulatory proteins that bind to these sites; beginning progress has also been made in understanding the mechanisms of post-transcriptional control. In this chapter we review recent accomplishments in these areas and discuss how these findings illuminate the problems of development and differentiation.

In eukaryotes, the term gene control *refers to the regulation of the formation and use of mRNA. Although con-*

trol can be exerted at a number of different molecular steps, differential transcription *probably most frequently underlies the differential rate of protein synthesis in eukaryotes, as it does in prokaryotes. We shall see, however, that* differential processing *of RNA transcripts in the cell nucleus,* differential stabilization *of mRNA in the cytoplasm, and* differential translation *of mRNA into protein are also important in eukaryotic gene control. Furthermore, as discussed in Chapter 10, in a few cases chromosomal segments are amplified or rearranged so as to affect gene expression. Before we turn to specific mechanisms of eukaryotic gene control, we consider some general questions about gene regulation.* ▲

The "Purpose" of Gene Control in Unicellular versus Multicellular Organisms

The primary function of gene control in prokaryotes is to adjust the enzymatic machinery of the cell to its immediate nutritional and physical environment. Within a *single* bacterial cell, genes are reversibly induced and repressed (turned on and off) by transcriptional control in order to allow growth and division, the raison d'être of bacterial existence. Single-cell eukaryotes, such as yeast cells, also seem to be designed only, or mainly, for the purpose of growth; yeasts also possess many genes that are controlled in response to environmental variables (e.g., nutritional status, oxygen tension, and temperature). Even in the organs of higher animals—for example, in mammalian liver—some genes can respond reversibly to external stimuli such as noxious chemicals. In general, however, metazoan cells are protected from immediate outside influences; that is, most cells in metazoans experience a fairly constant environment. Perhaps for this reason, genes that are devoted to responses to environmental changes constitute a much smaller fraction of the total number of genes in multicellular organisms than in single-cell organisms.

The most chracteristic and exacting requirement of gene control in multicellular organisms is the execution of precise developmental decisions so that the right gene is activated in the right cell at the right time. In most cases, once a developmental step has been taken by a cell, it is not reversed. Thus these decisions are fundamentally different from bacterial induction and repression. Many differentiated cells (e.g., skin cells, red blood cells, lens cells of the eye, and antibody-producing cells) march down a pathway to final cell death in carrying out their genetic programs, and they leave no progeny behind. These fixed patterns of gene control leading to differentiation serve the need of the whole organism and not the survival of an individual cell.

Variations in Proteins among Cell Types

Two-dimensional gel electrophoresis of total cellular protein shows that all eukaryotic cells contain 5000 or more different polypeptide chains. Whether a great many more proteins exist in quantities too small to be detected is a matter of debate. What is clear is that many of the most abundant proteins are present in all cell types (Figure 11-1). Many of these common proteins are present in identified cytoplasmic structures (e.g., microtubules, intermediate filaments, actin cables, and mitochondria) and in nuclear structures (e.g., pores, lamins, and nucleosomes). In addition, many enzymes are found in most cell types. Much of the difference in the size, shape, and internal architecture of different cell types results from quantitative differences in common, widely distributed proteins. In addition to these widely distributed proteins, each particular tissue or cell type may contain up to 100 easily identified proteins not found in other cells or tissues. These tissue- or cell-specific proteins are presumed to have an important role in giving different types of cells their specific character. The results of electrophoresis confirm and extend biochemical findings that some enzymes are present in many tissues (either in similar or varying concentrations), whereas other enzymes and some specialized secretory products are present only in a single cell type. It is a fair conclusion from these protein analyses that quantitative gene regulation is widely employed in eukaryotes and that qualitative ("all-or-none") gene expression also has an important role in cell differentiation.

Major proteins such as globin in red blood cells and albumin in liver cells exemplify highly cell-specific proteins; such proteins are often used to identify steps in the development of specialized cells. However, scarce proteins, which have begun to be identified only recently, can also be restricted in their cell distribution. These scarce proteins may contribute importantly in controlling the specialized functions of differentiated cells. Among the most important of the cell-specific proteins are cell-surface receptors, which sense the outside world; many of these (e.g., the insulin receptor) constitute only 0.001 per-

▶ **Figure 11-1** Two-dimensional gel electrophoretic analyses of total protein from mouse kidney (a), muscle (b), and liver (c). This procedure is illustrated in Figure 6-10. Separated proteins were detected by staining with a dye that reacts with all proteins. The spots labeled with numbers represent proteins present in all three tissues, although their amounts differ as indicated by differences in the size of the spots. Proteins labeled with letters are tissue-specific; five liver-specific (L) proteins and one muscle-specific (M) protein were revealed in these analyses. *Photographs courtesy of P. O'Farrell.*

(a)

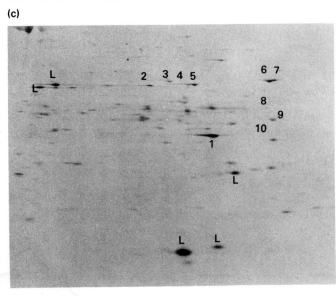

(b)

(c)

cent of the 10^9 to 10^{10} protein molecules in a cell. Many cell-specific transcriptional regulatory proteins are also present in such small or even smaller amounts per cell.

Finally we should note that even though scarce proteins may be crucial for cell specificity, it is not clear whether all or even most of the scarcest proteins serve specific functions in the cells in which they are found. It may be that in a given cell almost any protein can be synthesized occasionally and that many of the very scarce proteins do not contribute to cell function.

The Three Components of Gene Control: Signals, Levels, and Mechanisms

A logical analysis of gene control in eukaryotes requires answers to three questions (Figure 11-2):

1. What are the molecular *signals* to which a specific gene responds?
2. At which *level* (i.e., at which step or steps) in the chain of events leading from the transcription of DNA to the use of mRNA in protein synthesis is control exerted on the specific gene?
3. What are the *molecular mechanisms* at each level of gene control and are they the same or different for different genes? That is, which individual molecules or cell structures are involved in the control of genes, and how do they exert their effects on individual genes?

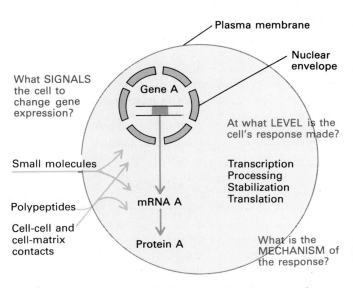

▲ **Figure 11-2** The signals, levels, and mechanisms of gene control must all be considered to arrive at a full understanding of differential gene expression. The mechanisms of signal transduction to the nucleus and the details of nuclear protein interactions are the most active areas of research.

Although much remains to be learned, enough progress has been made in answering these questions for a sufficient number of genes to present a reasonable explanation of how gene control operates in eukaryotes.

Signals for Gene Control

Two General Types of Hormones Can Cause Differential Gene Expression

Among the best-understood substances that effect gene control in multicellular organisms are *hormones,* which can be small molecules that enter cells or polypeptides that act by binding to the cell surface (Table 11-1). These substances, which are produced in specific cell types and circulate in the body fluids of the organism, have far-reaching effects on their target cells, both in adult organisms and during growth and development. Some but not all hormones operate by causing cells to change expression of individual genes or gene sets. Other hormones mainly affect the action of already existing enzymes or cell structures.

As the chemical nature of various hormones and growth factors was deciphered, it became clear that two general types of hormone molecules can serve as signals for gene control: (1) small fat-soluble molecules (e.g., steroid and thyroid hormones) that dissolve in the plasma membrane and enter cells to cause their effect and (2) polypeptides or proteins that bind to specific cell-surface receptors and in most cases exert their action without entering cells (see Figure 11-2). The receptors for these two types of gene signals are not alike. Steroid receptors, for example, are cytoplasmic molecules that bind steroids and move to the nucleus to act directly on transcription. In contrast, cell-surface receptors bind protein ligands outside the cell; this binding initiates a cascade of intracellular events that affect gene control inside the cell in a variety of ways. The mechanisms for delivering such gene-control signals to the nucleus along various pathways is a very active area of research, but definite conclusions about these important pathways are not yet at hand.

In addition to the long-recognized circulating proteins and steroids that are generally included in any list of vertebrate hormones, there exist many more recently discov-

Table 11-1 Examples of gene-control signals in eukaryotes

Class	Example	Target tissue/gene
Hormones: Proteins	Growth hormone Prolactin	Many cells Secretory cells in breast tissue
Steroids	Estrogens Testosterone	Liver, brain, reproductive organs Muscle, bone, skin, reproductive organs
Circulating or secreted protein factors	Nerve growth factor Epidermal growth factor	Axons (differentiating nerve cells) Many surface tissues (skin, eye, etc.) as well as cultured cells of all types
	Interleukins (lymphokins) Erythropoietin Interferon PDGFs (platelet-derived growth factors)	White blood cells Red blood cell precursors Most epithelial cells, white blood cells Many fibroblast cell types
	Cell-cell or cell-matrix contacts	Embryonic gut (endodermal cells interact with mesenchymal cells)
Environmental	Nutritional signals: Lower eukaryotes	Most genes for synthetic activity (amino acids, nucleic acid components) and for hydrolytic functions (phosphatases, saccharidases)
	Animal cells	Genes for gluconeogenic enzymes* in starvation; some genes for synthetic activity (there is some repression by excess products)
	Heat shock	Genes for specific proteins (induced); most other mRNAs (general translation suspended)
	Toxic substances: Drugs, carcinogens (xenobiotics) Heavy metals	 Genes for cytochrome P-450 and other detoxifying enzymes in liver Genes for metallothioneins in liver, kidney, and other tissues, and in single-celled eukaryotes
	Products of hemorrhage or inflammation	White blood cells, liver

*Liver enzymes that convert amino acids into glucose.

ered proteins that, like hormones, are released from one cell type and exert dramatic effects on a variety of other neighboring or distant cells. One class of these proteins is known as *growth factors*; these proteins exert important effects on differentiation as well as on growth. Examples include nerve growth factor, which causes cells of neural origin to extend long, axonlike processes, and epidermal growth factor (EGF), which was originally discovered because of its capacity to increase the speed of epidermal development when injected into rodents late in fetal development. It is now known that many cells besides skin cells have surface receptors for EGF and are stimulated by it. A variety of other protein factors, some originally recognized in tumor cells, have similar generalized stimulating or retarding effects on cell growth. Later in this chapter, we discuss progress in understanding how circulating polypeptides effect changes in gene expression by binding to cell-surface receptors.

Cell-Cell and Cel-Matrix Contacts Can Act as Signals to Control Genes

Most hormones and some growth factors are released into the circulatory system to travel to their target cells. Other gene-control signals require direct contact between cells. For example, in order for embryonic determination of most specialized cells to occur, two or more different cell types must make contact. Often the interacting cells come from different primordial cell layers (i.e., the mesoderm, endoderm, and ectoderm). One example is the primitive endodermal cells of the gut tube, which bud out at several regions and touch mesenchymal (mesodermal origin) cells (Figure 11-3). This pairwise interaction results in subsequent development of salivary glands, lungs, pancreas, and liver, depending on which section of the embryonic gut is involved.

This type of gene-activating interaction sometimes depends on secretion of growth factors; for example, early frog mesoderm differentiation is speeded up by a protein called tumor growth factor α (TGF-α), so named because of the source from which it was first purified. In many more cases the interaction between cells probably depends on direct contact or takes place through intermediate contact. In the latter case, a cell of one type may form an extracellular matrix, and contact with this matrix may be the signal for a cell of another type. Although the molecules of the cell surface (e.g., integrins) and of the extracellular matrix (e.g., lamins, fibronectins, and cell-adhesion molecules) are being actively studied, it is not yet known which cell-surface contacts function as gene-control signals. Instruction of embryonic cells to enter particular differentiation pathways is a subject of great importance. Some aspects of this intracellular signaling are discussed in Chapters 19 and 23. Later in this chapter we discuss some instances in which cell-surface interactions affect gene expression.

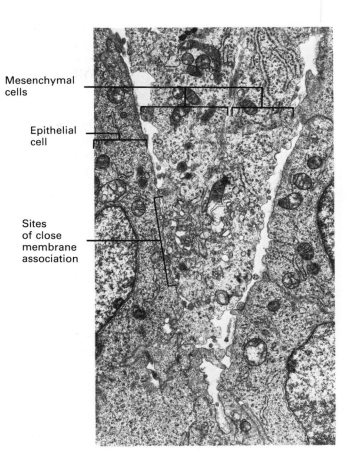

▲ **Figure 11-3** An electron micrograph showing the close association of plasma membranes of mesenchymal and epithelial cells in the salivary gland. On the left and right sides are two epithelial cells of the developing salivary gland of a mouse; the cleft between them is filled by two mesenchymal cells that appear wedge-shaped in this view. *From N. K. Wessels, 1977,* Tissue Interactions in Development. *Benjamin/Cummings, p. 203.*

Environmental and Nutritional Signals for Gene Control Are More Common in Unicellular Than in Multicellular Organisms

The eukaryotic gene-control signals mentioned so far—hormones, growth factors, and contacts between cells—are very different from the environmental signals that free-living cells like bacteria encounter and respond to. Just as is the case with bacteria, many genes in lower eukaryotes such as yeasts and molds are regulated in response to environmental signals such as nutritional stress. Inducible enzyme systems for metabolizing some sugars, for utilizing organic sulfate, phosphate, and nitrate, and for synthesizing nucleic acid bases and amino acids have been documented in lower eukaryotes. However, starvation does not induce biosynthetic enzymes to nearly the same extent in single-cell eukaryotes (e.g., yeast cells) as in bacterial cells. The levels of enzymes for the synthesis of tryptophan, histidine, isoleucine, and valine increase by a factor of 10 or less when yeast cells are starved for these amino acids. In contrast, when *Escherichia coli* is starved for the same amino acids, the levels of biosynthetic enzymes increase by a factor of 100 or even more.

Physical stimuli, especially heat and light, also can act as gene-control signals. For some single-cell organisms and for plants, light is the ultimate source of energy. Perhaps the most important light-sensitive pigment is chlorophyll, which is coupled to a protein called phytochrome. Recently, DNA sequences required for light stimulation of genes have been identified, but the mechanism of light-stimulated gene activation is not yet understood. Considerably more is known, as we discuss in late sections, about the gene response to elevated temperatures, or heat shock.

Gene Responses of Mammalian Cells in Vitro to Environmental Changes

Although some genes in higher eukaryotic cells can respond to changes in nutritional status, most do so only in a limited or uneven fashion. For example, most cultured mammalian cells can make purines and pyrimidines, so these nucleic acid bases do not have to be added to the culture medium. However, if adenine is added to a culture of mammalian cells, the cells lose the capacity to make the purines, adenine and guanine, de novo because synthesis of amidophosphoribosyl transferase, a necessary enzyme in the purine synthetic pathway, is repressed. On the other hand, if uridine or cytidine is added as a pyrimidine source, the enzymes of pyrimidine synthesis continue to function at levels 50 percent or more of normal levels. A similar situation holds for the endogenous synthesis of the amino acids serine and glycine: even when they are present in the medium, the cells continue to synthesize them. Thus cultured mammalian cells often do not suppress the synthesis of unneeded enzymes.

Likewise cultured mammalian cells often do not respond to nutritional stress by making enzymes even though they contain genetic information to do so. For example, cultured mammalian cells make cysteine, a necessary amino acid, at a low level but cannot increase its level of synthesis if starved for cysteine. Liver cells can make tyrosine by hydroxylation of phenylalanine, but other mammalian cells absolutely require tyrosine and cannot be induced to make phenylalanine hydroxylase.

Some environmental signals, however, seem to be universal. All types of cells, eukaryotes included, respond to an increase in temperature, or *heat shock,* by synthesizing a specific group of proteins, some of which are similar in all types of cells. This response might have been retained during evolution to protect cells from the occasional encounter with high temperatures that all cells experience.

Gene Responses of Mammalian Liver Cells in Vivo to Environmental Changes

As noted earlier, most cells within the intact animal are shielded from fluctuations in their immediate environment and do not respond to environmental changes by induction or repression of specific enzymes. Liver cells are an exception to this generalization. Because the liver receives part of its blood supply directly from the gut via the large portal vein, liver cells are exposed to changes in nutrient levels and the presence of various toxic substances. Under some conditions, the liver does respond to certain environmental stimuli by activation of genes encoding specific proteins.

If a rat is placed on a low-carbohydrate diet, for example, the levels of several enzymes required to convert amino acids into glucose increase dramatically in the liver, mainly because of the synthesis of new enzyme molecules. If the animal is returned to a normal diet, then the concentration of these gluconeogenic enzymes will de-

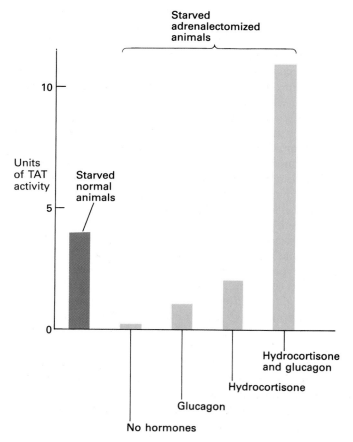

▲ **Figure 11-4** Levels of tyrosine aminotransferase (TAT) activity in the livers of normal and adrenalectomized animals during starvation, which induces an increase in TAT activity in normal animals. Adrenalectomized animals, which cannot produce adrenocortical steroid hormones, make little TAT; treatment with glucagon (a pancreatic hormone that increases liver glucose production) and/or hydrocortisone (a steroid hormone) enables these animals to make TAT. [See O. Greengard and G. T. Baker, 1966, *Science* **209**:146.]

crease. Many of the responses of the liver to changing metabolic demands are regulated by adrenocortical steroid hormones that affect the synthesis as well as the stability of specific mRNAs in liver cells (Figure 11-4), as we discuss later in this chapter.

The liver also responds to the presence of heavy metals and other toxic substances, which are transported from the gut to the liver via the portal vein. Detoxification of many such noxious chemicals by hepatocytes, the main cell type in the liver, requires special proteins that can metabolize or bind to these compounds. For example, the presence of cadmium or other heavy metals causes the liver to increase production of *metallothioneins,* two small proteins that bind heavy-metal ions and thus provide protection against their toxic effects. Other toxic substances (e.g., phenobarbital, codeine, morphine, and carcinogens such as methylcholanthrene) are detoxified by oxidation. When such substances reach the liver, the synthesis of a group of oxidative enzymes termed the cytochrome P-450s (or simply P-450s), which are present in hepatocytes, increase by a factor of 100 or more. The many individual genes which encode P-450 enzymes respond to different toxic substances, or *xenobiotics.* The molecule or molecules which recognize many of these toxic substances and which thus constitute the active signal for gene control during these metabolic responses in the liver cell are not yet known. However, activation of P-450 genes by cyclic hydrocarbons may resemble the induction of genes by steroids because the liver contains specific proteins that bind toxic substances (e.g., a dioxin derivative) and then enter the nucleus. As we will discuss later, steroids regulate transcription by binding to cytoplasmic DNA-binding proteins that move to the nucleus.

Experimental Demonstration of Transcriptional Control

As illustrated in Figure 11-2, gene control can occur at four levels: (1) transcription (either initiation or termination), (2) nuclear processing of primary transcripts, (3) cytoplasmic stabilization or destabilization of RNAs, and (4) mRNA translation. Enough examples have been analyzed to show that gene control at each of these levels occurs in eukaryotes, although not every gene is—or can be—controlled at all four levels.

Regulation of transcriptional initiation is by far the most widespread form of gene control in eukaryotes, as it also is in prokaryotes. Such control results in the increased or decreased synthesis of primary RNA transcripts in response to some signal, leading to a change in the level of specific mRNAs and their translation products. In this section, we describe how control of transcriptional initiation has been demonstrated in several eukaryotic systems. In the following sections, we discuss various

aspects of such control and describe several biological events that depend on transcriptional control. After this detailed discussion of transcription-initiation control, the other levels of gene control are considered.

"Run-On" Transcription Analysis Accurately Measures Transcription Rates

The simplest and most direct method of measuring transcription rates would be to expose cells for a brief time (e.g., 5 min or less) to a labeled RNA precursor and measure the amount of labeled nuclear RNA formed by its hybridization to a cloned DNA. This method has been used for transcription rate measurements with cultured cells. However, the technique is not practical in whole animals because the labeled RNA that can be obtained is insufficient.

Even with cultured cells it is often easier to use a second method of labeling RNA. In this method, called "run-on" or nascent-chain analysis, nuclei are isolated from cells and allowed to incorporate ^{32}P from labeled nucleoside triphosphates directly into nascent (growing) RNA chains to produce highly labeled RNA preprations (Figure 11-5).

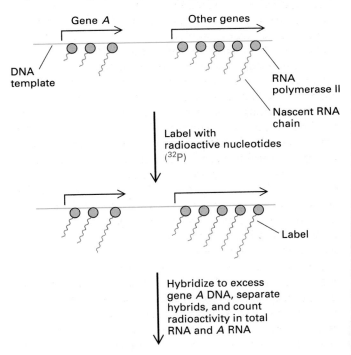

▲ **Figure 11-5** Nascent-chain or "run-on" assay for transcription rate of a gene. Labeled RNA is prepared in isolated nuclei by allowing extension of already initiated RNA chains. The average polymerase only moves a few hundred nucleotides and very little new initiation occurs. By hybridizing the labeled RNA to the cloned DNA from a specific gene (*A* in this case), the fraction of total RNA copied from that DNA can be measured. [See J. Weber, W. Jelinek, and J. E. Darnell, 1977, *Cell* **10**:611.]

Table 11-2 Comparison of two methods of labeling nuclear RNA for use in hybridization experiments to assay transcription rates*

DNA sequences used in hybridization	Percent of total labeled RNA hybridized	
	Whole cells	Isolated nuclei
Adenovirus genome:		
Cells early in infection	0.75	0.58
Cells late in infection	16.6	14.4
β-globin	0.01†	0.01†
Chinese hamster cDNAs	0.0001–0.001	0.0001–0.001
Ovalbumin cDNA	0.00018‡	0.00024‡
Conalbumin cDNA	0.00015‡	0.00022‡

*Nuclear RNA was labeled either by exposing whole cells to [³H] uridine or by exposing isolated nuclei to [³²P] UTP. After brief incubation, the RNA was hybridized to the various DNAs indicated in the table and the percentage of the total label in the RNA:DNA hybrid was determined. This assay thus measures the relative transcription rate of the DNA used for hybridization.
†Erythroleukemia cells were used in assays of β-globin.
‡Cultured minced chicken oviducts were used in assays of ovalbumin and conalbumin genes.

See J. E. Darnell Jr., 1982, *Nature* **297**:365; and G. S. McKnight and R. D. Palmiter, 1979, *J. Biol. Chem.* **254**:9050.

Analysis of such products indicates that the RNA polymerase II molecules in isolated nuclei add 300–500 nucleotides to nascent RNA chains, but then transcription stops for some reason. By hydridizing the labeled RNA to cloned DNA from a specific gene, the fraction of the total RNA copied from a particular gene (i.e., its relative transcription rate) can be determined. Relative transcription rates determined either by labeling of cultured cells or of isolated nuclei are nearly the same (Table 11-2), which indicates that all of the polymerase molecules that are active at the time nuclei are taken from intact cells continue to function during incubation of isolated nuclei. The run-on analysis of isolated nuclei is now very widely used as a direct assay of relative transcription rates.

Development of recombinant DNA techniques has made it possible to prepare cloned cDNA copies of mRNA sequences encoding proteins that are prominent products of various differentiated cells. Use of such cDNAs in nascent-chain analyses of nuclei isolated from many different types of cells has demonstrated conclusively that transcriptional control is the primary means of regulating the synthesis of specialized products in many differentiated tissues. Two examples of such experiments in vertebrates are described here to illustrate how this

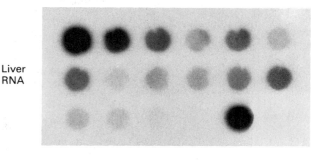

Liver RNA

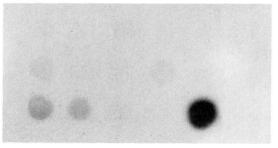

Kidney RNA

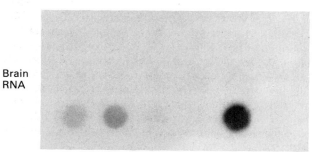

Brain RNA

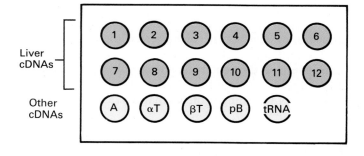

Liver cDNAs: 1 2 3 4 5 6 / 7 8 9 10 11 12
Other cDNAs: A αT βT pB tRNA

▲ **Figure 11-6** Experimental demonstration of differential synthesis of 12 mRNAs encoding liver-specific proteins. Nuclei from mouse liver, kidney, and brain cells were exposed to [³²P]UTP, and the resulting labeled RNA was hybridized to various cDNAs fixed to nitrocellulose. After removal of unhybridized RNAs, the hybrids were revealed by autoradiography. The cDNAs labeled 1–12 encode proteins synthesized actively in liver (e.g., 4 = albumin; 3 = α₁ antitrypsin; 6 = transferrin) but not in most other tissues. The other cDNAs tested were actin (A) and α- and β-tubulin (αT, βT), which are proteins found in almost all cell types. Methionine tRNA and the plasmid DNA (pB) in which the cDNAs were cloned were included as controls. The pattern of spots in the three tissues, which represents the synthesis of specific mRNAs, clearly indicates that transcriptional control is the primary means for regulating production of these liver-specific proteins. [See E. Derman et al., 1981, *Cell* **23**:731; and D. J. Powell et al., 1984, *J. Mol. Biol.* **197**:21.]

technique is used to demonstrate transcriptional control. This experimental approach has also been used to confirm the existence of differential transcription in certain regions of insect chromosomes, which had long been suspected because of the presence of enlarged "puffs" in these regions.

Differential Synthesis of Hepatocyte-Specific mRNAs Depends on Cell-Cell Contact

In the vertebrate liver, hepatocytes produce dozens of proteins that are not made in most other organs. These include both enzymes and numerous secreted proteins that constitute the major protein components of serum. The run-on experiment illustrated in Figure 11-6 shows that synthesis of nuclear RNA precursors to 12 different mouse mRNAs was 20–500 times higher in hepatocyte nuclei than in brain or kidney nuclei. Thus the levels of a wide variety of liver-specific proteins are regulated at the level of transcription.

Synthesis of RNAs encoding actin and tubulin, which are present in all tissues in varying concentrations, also was analyzed in this experiment. The results of these analyses suggest that the levels of these widely distributed proteins must be controlled by posttranscriptional mechanisms. For example, even though brain contains 15 times more actin mRNA than liver, the synthesis of nascent RNA complementary to actin cDNA was similar in liver, brain, and kidney. Therefore, the differential synthesis of some proteins in specific cell types clearly involves posttranscriptional control.

When mouse liver cells were disaggregated, placed in culture overnight, and then subjected to nascent-chain analysis, synthesis of hepatocyte-specific RNAs declined substantially, whereas synthesis of actin and tubulin RNAs remained about the same. Liver slices placed in culture under the same conditions, however, maintained a near-normal liver-specific transcription pattern at least for 12–14 hours. It seems that even in "terminally differentiated" hepatocytes, the proper cell-cell or cell-matrix contacts are necessary to maintain their specialized transcription pattern.

Differential Transcription of Globin Genes Is Related to Developmental Stage

The previous discussion of liver-specific mRNA synthesis illustrates one case of differential gene expression in the cells of adult tissues. The globin genes in vertebrate erythroblasts (the precursors of red blood cells) provide a good example of differential gene expression in similar cells at various stages of development. Erythroblasts do not form globin until the last four or five divisions before they become mature, nondividing, red blood cells.

The human genome contains several β-like globin genes; different β-globin genes are transcribed during embryogenesis and in adults. Chickens also make embryonic and adult forms of both α- and β-globins. Nascent-chain analysis of isolated nuclei from erythroblasts of 5-day-old chick embryos shows that only embryonic globin mRNAs are synthesized at this stage; however, erythroblast nuclei from 12-day-old embryos make a different set of globin mRNAs, which are identical to those made by nuclei from adult chickens (Figure 11-7a). Nuclei from other types of cells do not make any globin mRNA in detectable amounts.

In chick embryos, the same cell does not first form embryonic globin and then later produce adult globin. Early in development all erythroblasts differentiate into cells that produce embryonic globin; later all the erythroblasts differentiate into cells that produce adult globin (Figure 11-7b). Clearly, some signal must cause individual erythroblasts to proceed along the embryonic or adult differentiation pathway. A final important point is that

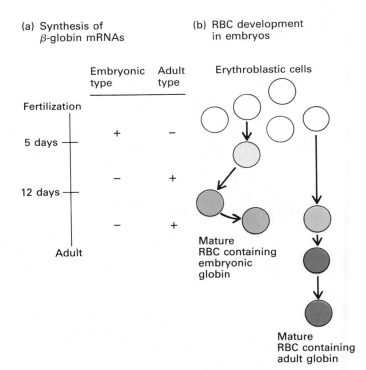

▲ **Figure 11-7** (a) Results of run-on transcription assays of β-globin genes in nuclei of erythroblasts from 5- and 12-day-old chick embryos and adult chickens. (b) Diagram of red blood cell (RBC) development in chick embryos illustrates that no one cell first makes embryonic and then adult globin. The choice of the embryonic or adult differentiation pathway occurs to cells in the erythroblast population, but no actual "switching" occurs in a single cell. [See H. Weintraub, A. Larsen, and M. Groudine, 1981, *Cell* **24**:333; and M. Groudine, M. Peretz, and H. Weintraub, 1981, *Mol. Cell Biol.* **1**:281.]

although globin mRNAs constitute 50–90 percent of the total mRNA present in mature red blood cells, the *rate* of synthesis of globin mRNAs never is more than about 0.05 percent of that of the total nuclear RNA. This finding indicates that transcription of nonglobin genes continues as the red blood cell matures but that processing of nonglobin primary transcripts the stability of most nonglobin mRNAs, or both, decline during red blood cell maturation.

Enlarged "Puffs" on Insect Chromosomes Correspond to Regions with Increased Transcription

Long before recombinant DNA techniques were developed, microscopically visible enlargements, called "puffs," had been noticed on particular regions of insect chromosomes at particular times of development or after certain treatments. Because the puffs correspond to sites of active [³H]uridine incorporation, they were long considered to be regions of increased transcriptional activity, although no formal proof of this conclusion was possible. Some puffs appear only after treatment of insect larvae with the hormone ecdysone, which causes the larvae

to molt; other puffs are formed after particular environmental shocks, such as that from excessive heat (Figure 11-8).

From the mRNAs that increase after raising flies to elevated temperatures, the heat-shock genes of *Drosophila melanogaster* have been isolated as recombinant DNA clones. Increased transcriptional activity from these heat-shock genes of cells subjected to elevated temperatures has been detected by hybridization of nuclear RNA to the purified DNA. Thus it seems highly likely that the appearance of puffs does in fact represent transcriptional activation of regions of *D. melanogaster* chromosomes.

Structure and Function of DNA-binding Proteins That Regulate Transcription of Protein-coding Genes

Once transcriptional regulation of protein-coding genes in eukaryotic cells was conclusively demonstrated, research interest shifted to identifying the proteins that might be responsible for regulation. The ultimate aim of

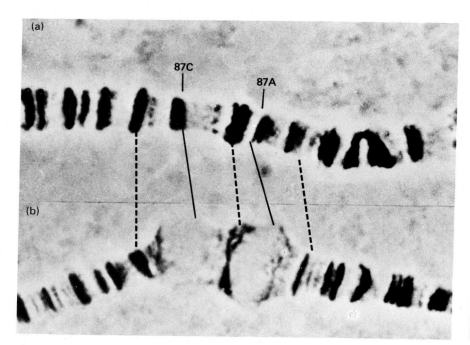

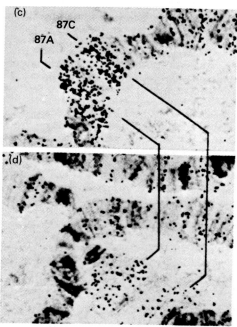

▲ **Figure 11-8** Photomicrographs of portion of *Drosophila melanogaster* salivary gland chromosomes from normal flies (a) and heat-shocked flies (b). After flies are exposed to 37°C for 40 min, characteristic heat-shock "puffs" appear at the sites designated 87A and 87C. Autoradiographs demonstrate that RNA synthesis in these regions, measured by incorporation of [³H] uridine, is much greater after heat shock (c) than at normal temperatures (d). [See M. Ashburner and J. J. Bonner, 1979, *Cell* 17:241; and J. J. Bonner and M. L. Pardue, 1976, *Cell* 8:43.] *Courtesy of J. J. Bonner.*

such studies is to show how such proteins interact with each other and with RNA polymerase II to increase the transcription of specific genes in specific cell types. The first phase of this work has progressed rapidly with the use of generalized methods for finding regulatory sites in DNA and for isolating proteins that bind specificially to these sites. More than 100 such proteins—mainly from yeasts, *Drosophila*, rodents, and humans—have been isolated, and their genes cloned and sequenced. The deduced amino acid sequences suggest that a limited number of families of site-specific DNA-binding proteins exist, although the total number of such proteins encoded by vertebrate genomes probably is in the hundreds if not thousands. A few DNA-binding proteins have been shown directly to stimulate transcription by RNA polymerase II in vitro, but most have not. How increased transcription is mediated by various types of transcription factors acting alone and particularly in concert, as they must in many cases, will remain an important research topic for years to come.

Regulatory Sites in DNA and Cognate Binding Factors Can Be Identified by Molecular Genetic Techniques

The general experimental approach outlined in Figure 11-9 has yielded information on dozens of DNA regulatory sites and transcription factors. These experimental methods initially were used with easily cultured undifferentiated cells and with virus transcription units, which served as models for eukaryotic cellular genes. These studies led to recognition that many transcription factors are present in most if not all cell types. Some of these widely distributed factors are present in invariant concentrations and are required for transcription of many genes. Therefore, although these proteins are necessary for gene activity, they do not actually take part in regulation of gene activity.

More recently, cell lines that are partly differentiated, such as lymphoma (lymphocytelike) cells and tumor cells from the liver and pancreas, have been used for testing gene function. At least some of the proteins involved in cell-specific gene expression have been identified in these studies. In *Drosophila* and in mice, recombinant genes have been reintroduced into the germ line of animals to locate all the sequences of the newly introduced genes that are required for correct quantitative and qualitative participation in development and differentiation. In most cases it is now clear that the transfer of genes into cultured, partially differentiated cells provides a good beginning for understanding cell-specific function but may not be adequate to reveal all the parts of a gene required for accurate function in development.

We must recognize that the standard methods of identifying transcription factors (Figure 11-9) have revealed more proteins with a positive effect on gene expression than with a negative effect similar to bacterial repressors, although eukaryotes definitely contain site-specific DNA-binding proteins whose action is negative. Also regulation may involve probing that interaction or that change (e.g., phosphorylate) the DNA-binding protein. Further experiments beyond simple DNA binding are required to detect the action of these proteins.

Finally, although the molecular genetic techniques just described are applicable to all cell types, many interesting and important regulatory factors in yeast cells were first detected by classical genetic methods. In such studies, mutant yeast cells deficient in gene regulation were selected and the DNA that complemented a defect was cloned, thereby identifying the gene for the defective transcription factor involved. A genetic approach also led to the ultimate discovery of *Drosophila* transcription factors that are important in development. Recessive lethal genes that caused embryonic defects were identified, and when the defective genes were cloned, they proved in many cases to encode transcription factors.

Most If Not All Eukaryotic Protein-coding Genes Require Activators

The TATA box generally is located about 30 nucleotides upstream from the RNA start site of protein-coding genes (see Figure 8-41). The start site is often called the cap site because primary RNA transcripts have a 5′ methylated cap (see Figure 8-30). The proteins necessary for beginning transcription of protein-coding genes are the TATA factor (transcription factor IID, or $TF_{II}D$), RNA polymerase II, and accessory proteins ($TF_{II}B$ and $TF_{II}E$). (As work proceeds to greater purification of transcription factors, it is extremely likely that other factors or proteins will be discovered. For example, the E factor is probably two proteins, now termed E and F.) The probable sequence in which these proteins interact with DNA and with each other to form an initiation complex, is shown in Figure 8-43. $TF_{II}D$ as indicated plays a crucial role as the first factor; even in genes lacking the TATA box, the TATA protein or its equivalent may be necessary to start transcription correctly by RNA polymerase II. Transcription rates of genes lacking TATA boxes are generally quite low compared with those containing TATA boxes.

However, initiation of transcription of eukaryotic genes by RNA polymerase II, even when the TATA box and TATA factor are present, is not a frequent event. The process is accelerated by *activators*, which are required for normal transcription of many, perhaps all, eukaryotic genes. In yeast and other simple organisms, a single required *upstream activating sequence (UAS)* can be sufficient; the UAS is usually located near a gene, and often a

(a) Identifying DNA regulatory sequences

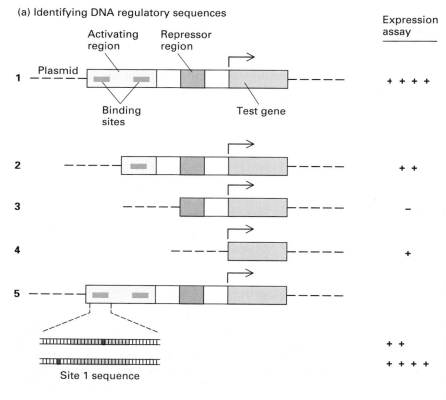

Expression
assay
——————

1 + + + +

2 + +

3 −

4 +

5 + +
  + + + +

Site 1 sequence

(b) Identifying cognate protein and locating precise binding sequence

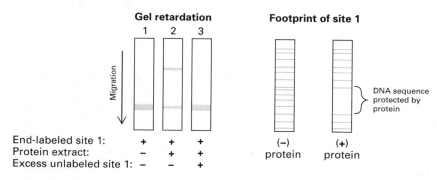

Gel retardation
1   2   3

Migration

Footprint of site 1

DNA sequence
protected by
protein

(−)   (+)
protein   protein

End-labeled site 1:   +   +   +
Protein extract: −   +   +
Excess unlabeled site 1: −   −   +

(c) Purifying cognate protein for specific DNA binding sequence

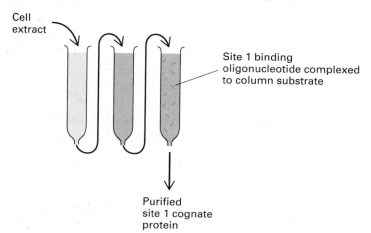

Cell
extract

Site 1 binding
oligonucleotide complexed
to column substrate

Purified
site 1 cognate
protein

◀ **Figure 11-9** General procedure for detecting and isolating DNA regulatory sequences and their cognate DNA-binding proteins. The procedure has been most widely used in mammalian cells. (a) A cloned test gene (green) containing potential activating sequences is introduced into a cell as a recombinant plasmid; expression of the test gene is assayed by measuring mRNA production starting at the correct promoter site (see Figure 6-27) or production of a test protein such as a bacterial enzyme (1). By use of deleted forms of the cloned gene, regions outside of the coding region that increase or decrease expression can be identified (2, 3, and 4); sequences outside the transcribed region that affect expression in the test are assumed to affect transcription rather than pre-mRNA processing or mRNA stability. Point mutations in any active DNA elements identified can more precisely locate the functional sequences (5). For example, a mutation (red) within the site 1 would destroy its activating ability, whereas mutation at a nearby site did not. (b) Gel-retardation assays of cell extracts using end-labeled DNA fragments containing an identified regulatory region (site 1) are used to locate the cognate proteins. Lane 1, labeled site 1 DNA alone; lane 2, labeled DNA plus cell extract; lane 3, labeled and excess unlabeled DNA plus cell extract. The "retarded" band in lane 2 corresponds to a DNA-protein complex; absence of this band in lane 3 demonstrates the specificity of the binding. DNase digestion of the end-labeled DNA regulatory fragments in the presence and absence of cell protein followed by separation on a DNase sequencing gel where every band can be identified locates the specific oligonucleotide sequence protected by interaction with protein (see Figure 7-5 illustrating such a footprint experiment). (c) Any DNA-binding protein identified in (b) can be purified from cell extracts by chromatography, following the protein by the gel-retardation assay. Chromatography of several types is used to partially purify the proteins, and final purification is obtained by affinity chromatography on a column containing the double-stranded DNA oligonucleotide (blue) to which the protein has a specific affinity as determined in the footprint experiments. Purified regulatory proteins can then be partially sequenced to provide a means to identify the gene encoding the protein (see Figure 6-37); this regulatory gene can then be cloned.

(a) Yeast

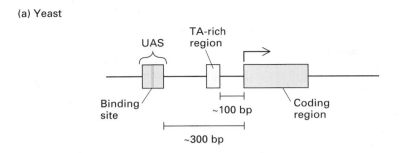

(b) Metazoan

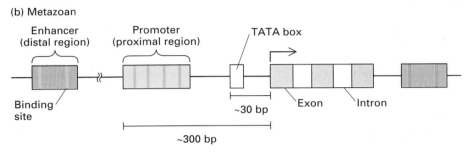

▲ **Figure 11-10** General pattern of DNA regulatory sequences that control gene expression in yeast and multicellular organisms (invertebrates, vertebrates, and plants). (a) Most yeast genes contain only one regulatory sequence, an upstream-activating sequence (UAS), and a TA-rich region within 100 nucleotides of the start site; the latter is required for positioning of RNA polymerase II. (b) Genes of multicellular organisms contain both proximal and distal regulatory sites (often called promoter and enhancer, respectively) and the TATA box, which is involved in polymerase positioning. The enhancer may be either upstream or downstream, and as far away as 50 kb from the RNA start site. Each regulatory region contains a single (yeast) or multiple (metazoan) 15- to 20-bp binding sites (dark blue bars). DNA-binding proteins can be classified as general proteins, which are present in most or all cells, and specialized proteins, which are either exclusively in particular cell types or in much higher concentration in some cells than in others.

single protein factor binds to a 15- to 20-bp segment in the UAS (Figure 11-10). In contrast, all mammalian genes whose regulatory sequences have so far been dissected (and even most genes of viruses that infect vertebrates) contain *multiple* 15- to 20-bp protein-binding sites each of which is necessary for maximal transcription. Many of the binding sites in DNA have dyad symmetry, suggesting that binding proteins may act as dimers.

As discussed in Chapter 8, some gene-activating sites in eukaryotes are located at great distances (up to 50 kb) from RNA start sites, usually upstream but also sometimes downstream; these were originally called *enhancer* sites. Regulatory sites closer to the start site were referred to as *promoter* sites in analogy with bacterial control regions. Now that the proteins which recognize many of these sites have been identified, it has become clear that there is no real distinction other than distance between these regions: the same proteins can bind close to or far from RNA start sites. It is widely assumed that even distantly bound proteins can affect transcription and that looping of the DNA can bring a distantly bound protein into proximity with the RNA start site. The occurrence of DNA looping has been demonstrated in the arabinose operon in *E. coli* (see Figure 7-16).

We now know that cell-specific gene control in most eukaryotes requires both widely distributed transcription factors and other factors that have a limited, if not totally cell-specific, distribution. Thus, the availability of the correct set of transcription factors is an important determinant of which genes are active in any given cell. However, we do not yet know in any detail how any single factor stimulates transcription or whether all the factors that can be extracted from cells are in fact available for active participation in transcription. What has been learned is that vertebrate cells encode many dozens (perhaps hundreds) of such factors, which vary in concentration (as measured by DNA binding) in different cells. Very likely, the presence and concentrations of these factors are crucial in development and differentiation.

Eukaryotic DNA-binding Proteins Exhibit a Limited Number of Structural Designs

Three general structural designs appear to be present in the DNA-binding proteins whose amino acid sequences have been determined. These designs are termed the helix-turn-helix, zinc-finger, and amphipathic helix motifs. The last group contains the so-called leucine-zipper and the helix-loop-helix proteins which form dimers by protein coil interactions and which bind DNA in an as-yet-unknown manner. In this section, we describe how proteins exhibiting each of these designs were recognized and the important features of each design.

Helix-Turn-Helix Proteins and Homeoboxes

The first three bacterial regulatory proteins whose three-dimensional structures were determined all bind to DNA in a similar manner (see Figure 7-11). They all bind as dimers and all have within their strucures three α-helical regions separated by short turns. This arrangement was labeled the *helix-turn-helix* motif. It has now been proved that amino acid substitutions in the α helices nearest the carboxyl end of each subunit of the dimer disturb the binding in the major groove of the DNA. These findings together with x-ray studies show clearly that a protein helix in each of the dimers occupies the major groove in two successive turns of the DNA helix.

Subsequent research suggested that helix-turn-helix proteins also were encoded by the *MAT* locus that controls mating type in yeast and by regulatory genes that control early segmentation and body part development in *Drosophila* embryos. The *Drosophila* genes include antennapedia *(Antp)*, fushi tarazu *(ftz)*, and ultrabithorax *(Ubx)*. Mutations in the *Antp* or *Ubx* genes can cause substitution of one developmental pathway for another; for this reason, they are called *homeotic genes*. When the nucleotide sequences of these and several other *Droso-*

phila homeotic genes were completed, it was clear that a region encoding about 60 amino acids was remarkably conserved in all these genes. This region, called the *homeobox,* is also conserved well enough to be easily recognized in genes from frogs and mammals.

Portions of the amino acid sequences encoded by the *ftz, Antp,* and *Ubx* genes, one frog homeobox gene, and the yeast *MATa1* and *MATα2* genes are shown in Figure 11-11 along with the sequences of the helix-turn-helix regions of several bacterial regulatory proteins. Structurally similar amino acids occur at key positions in helix 3 (the DNA-binding helix) as well as in helix 2 in all these proteins. Both the yeast and *Drosophila* proteins have now been produced in large amounts, and their ability to bind to specific sequences in DNA has been confirmed. A number of other genes encoding vertebrate DNA-binding proteins contain regions homologous to the homeoboxes originally found in *Drosophila* homeotic genes. However, it is not yet proved how homeobox proteins contact specific DNA sequences.

A further extension of the idea of sequence conservation among potential transcriptional regulators has come from analysis of a subset of homeobox-containing genes. At least eight different genes from invertebrates (nematodes and *Drosophila*) and from mammals contain both a homeobox and a second highly conserved region termed POU (the first three letters for the names of three of these regulatory proteins). The POU region is about 50 amino acids long and is located about 15 amino acids away from the homeobox region, nearer the amino-terminal end of

Gene product

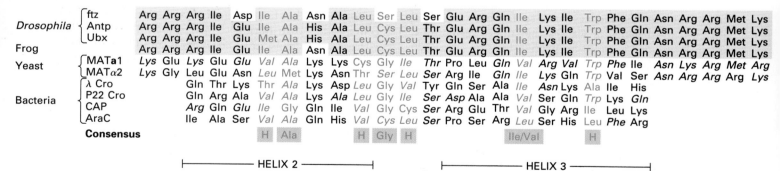

		\multicolumn{22}{HELIX 2 ... HELIX 3}
Drosophila	ftz	Arg Arg Arg Ile Asp Ile Ala Asn Ala Leu Ser Leu Ser Glu Arg Gln Ile Lys Ile Trp Phe Gln Asn Arg Arg Met Lys
	Antp	Arg Arg Arg Ile Glu Ile Ala His Ala Leu Ser Leu Thr Glu Arg Gln Ile Lys Ile Trp Phe Gln Asn Arg Arg Met Lys
	Ubx	Arg Arg Arg Ile Glu Met Ala His Ala Leu Cys Leu Thr Glu Arg Gln Ile Lys Ile Trp Phe Gln Asn Arg Arg Met Lys
Frog		Arg Arg Arg Ile Glu Ile Ala Asn Ala Leu Cys Leu Thr Glu Arg Gln Ile Lys Ile Trp Phe Gln Asn Arg Arg Met Lys
Yeast	MATa1	*Lys* Glu *Lys* Glu *Glu Val Ala* Lys Lys *Cys Gly Ile Thr* Pro Leu *Gln Val Arg Val Trp Phe* Ile *Asn* Lys *Arg Met Arg*
	MATα2	*Lys* Gly Leu Glu Asn *Leu* Met Lys Asn *Thr Ser Leu Ser* Arg Ile *Gln Ile* Lys Gln *Trp* Val Ser *Asn Arg Arg Arg Arg* Lys
Bacteria	λ Cro	Gln Thr Lys *Thr Ala* Lys Asp *Leu Gly Val* Tyr Gln Ser Ala *Ile Asn* Lys *Ala* Ile His
	P22 Cro	Gln Arg *Ala Val Ala* Lys *Ala Leu Gly Ile Ser* Asp Ala *Val* Ser Gln *Trp* Lys *Gln*
	CAP	*Arg* Gln *Glu Ile* Gly Gln Ile *Val* Gly Cys *Ser* Arg Glu Thr *Val* Gly Arg *Ile* Leu Lys
	AraC	Ile Ala Ser *Val Ala* Gln His *Val Cys Leu Ser* Pro Ser Arg *Leu* Ser His *Leu Phe* Arg
	Consensus	H Ala H Gly H Ile/Val H

├───── HELIX 2 ─────┤   ├───── HELIX 3 ─────┤

▲ **Figure 11-11** Amino acid sequences in presumed DNA-binding region of three *Drosophila* and one frog homeobox proteins, two yeast regulatory proteins, and four bacterial helix-turn-helix proteins aligned to demonstrate their similar amino acid sequences in the regions of helix 2 and helix 3 based on crystallographic analysis of the bacterial proteins (see Figure 7-11). The overall similarity is very high in the animal proteins (orange shading). Italicized amino acids in the yeast and bacterial proteins are the same as or structurally similar to the corresponding ones in the animal proteins. The key positions in the helices are shown in blue; the consensus amino acids at these positions, based on these and

other proteins, are shown at the bottom (H = hydrophobic). These similarities suggest that the eukaryotic proteins also have the helix-turn-helix structural motif, although this has not been confirmed by crystallographic studies. The *Drosophila* proteins shown are fushi tarazu (ftz), antennapedia (Antp) and ultrabithorax (Ubx); the yeast proteins are mating-type proteins (MATa1 and MATα2); and the bacterial proteins are Cro (from λ or P22 virus), cyclic AMP-binding protein (CAP), and the arabinose-operon positive activator (AraC). *Adapted from A. Laughon and M. P. Scott, 1984, Nature 310:25 and W. McGinnis et al., 1984, Cell 37:403.*

```
        1              10              20                  30
        |              |               |                   |
1             Y I C S F A D C G A A Y N K N W K L Q * A H L C * K H  37
2   T G E K * P F P C K E E G C E K G F T S L H H L T * R H S L * T H  67
3   T G E K * N F T C D S D G C D L R F T K A N M K * K H F N R F H    98
4   N I K I C V Y V C H F E N C G K A F K K H N Q L K * V H Q F * S H  129
5   T Q Q L * P Y E C P H E G C D K R F S L P S R L K * R H E K * V H  159
6   A G - - * - Y P C K K D D S C S F V G K T W T L Y L K H V A E C H  188
7   Q D - - * L A V C - - D V C N R K F R H K D Y L R * D H Q K * T H  214
8   E K E R T V Y L C P R D G C D R S Y T T A F N L R * S H I Q S F H  246
9   E E Q R * P F V C E H A G C G K C F A M K K S L E * R H S V * V H  276
    ↑                                                         ↑
Domain                                                     Amino
number                                                      acid
                                                          position
```

◀ **Figure 11-12** Amino acid sequence of TF$_{III}$A from the frog, *Xenopus laevis*, exclusive of the amino-terminal and carboxyl-terminal regions. The single-letter amino acid code is used and the protein sequence is arranged to maximize the alignment of the repeated amino acids tyrosine (Y) or phenylalanine (F), leucine (L), and the invariant pairs of cysteines (C) and histidines (H). *Adapted from J. Miller, A. D. McLachlan, and A. Klug, 1985, EMBO J. 4:1609.*

the proteins. The functional role of the POU sequences is not clear, but their high degree of conservation over such a length and span of evolution suggests a fundamental role in gene activation.

Zinc-finger Proteins The first eukaryotic positive-acting regulatory protein to be well-characterized was transcription factor IIIA (TF$_{III}$A), which is required for RNA polymerase III transcription of the 5S-rRNA genes (see Figure 8-24). This protein has nine repeated domains that contain cysteines and histidines spaced at regular intervals (Figure 11-12). The purified protein has zinc associated with it, and the zinc is required for activity. Furthermore, cysteine and histidine are known to be the most common ligands in protein for zinc. In the proposed zinc-finger folded structure of TF$_{III}$A shown in Figure 11-13, the repeated domains form loops in such a way that a zinc ion is bound between a pair of cysteines and a pair of histidines. A phenylalanine or tyrosine residue and a leucine residue occur at nearly constant positions in the loops, which are now known by mutational studies to be required for DNA binding.

As the sequences of other known or suspected DNA-binding proteins became available, searches were made for regions with the zinc-finger structure. Several dozen confirmed or possible regulatory proteins have been found to contain zinc fingers. These include a yeast protein that regulates a cytochrome *c* gene, several early *Drosophila* developmental regulators (gap-gene products), and a whole class of proteins that bind steroid hormones in vertebrates. Thus, the zinc-finger motif appears to be a common one in eukaryotic regulatory proteins.

Amphipathic Helical Proteins: Leucine-Zipper and Helix-Loop-Helix Proteins A third class of DNA-binding proteins has recently been recognized. These proteins do not contain obvious helix-turn-helix or zinc-finger structures within their known DNA-binding domains, and their method of recognizing DNA is not understood at present. However, the amino acid sequence of a number of these proteins is known, as are important characteristics of their DNA-binding activity. From this information it is reasonable to classify and discuss them together. All of these proteins form dimers and have a

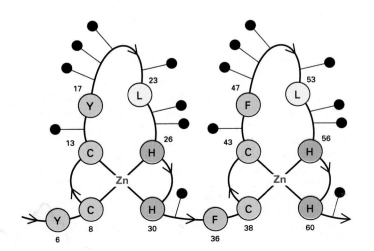

◀ **Figure 11-13** Proposed zinc-finger loop structure for repeated domains in TF$_{III}$A resulting from coordination of the invariant cysteines (C) and histidines (H) with a zinc ion (Zn). Only the first two of the possible nine loops are shown. The small numbers next to the single-letter amino acid codes show distances in amino acids (see Figure 11-12). The solid black circles mark the most probable side chains engaged in DNA binding. An invariant leucine (L) and either a tyrosine (Y) or phenylalanine (F) are located within each loop. *Adapted from J. Miller, A. D. McLachlan, and A. Klug, 1985, EMBO J. 4:1609.*

region rich in basic amino acids near the dimerization domains.

The first protein of this large group to be studied extensively was isolated from rat liver cell extracts. This protein was called C/EBP because of its ability to act as a viral enhancer binding protein and because it was also thought initially to bind to CCAAT boxes. It is now clear that C/EBP is present at high levels in only a few cell types (hepatocytes, fat cells, intestinal epithelium, and a few types of brain cells) and functions in liver-specific gene expression, as we discuss in a later section.

The carboxyl-terminal 100 amino acids of C/EBP, has in its sequence a series of residues that can form an α helix, that is necessary in DNA binding. If the amino acid residues between 315 and the C-terminus at 359 are positioned on a helical wheel, every seventh amino acid—a leucine—is brought into register on one side of the helix (Figure 11-14a). Computer comparisons of the C/EBP sequence and the sequences of several other newly cloned DNA-binding proteins from yeast, *Drosophila*, and mammals have revealed regions of amino acid similarity among some of these proteins. Thus, a helical domain with leucines projecting uniformly on one side may be a common feature of several regulatory proteins.

The C/EBP only binds to DNA as a dimer, and mutations in the leucines interrupt dimer formation. The helical region appears to be amphipathic: that is, it favors lipids on one side and water on the other. One side of the helical region of C/EBP has charged amino acids (such as arginine, glutamine, and aspartic acid); this charged side would favor facing the water interface. The array of leucines on the other side presents a very hydrophobic

▶ **Figure 11-14** Proposed amphipathic helix in DNA-binding proteins and dimeric structure of leucine-zipper and helix-loop-helix proteins. (a) A portion of the amino acid sequence of monomeric C/EBP (residues 315–359) displayed on a helical wheel; the angles of the wheel are those found in actual α helices in protein. Because there are about 3.5 amino acids per helical turn, each seventh amino acid is in register; the residues labeled 1, 8, 15, and 22 are all leucine. These hydrophobic leucines are on one face of the helix; charged amino acids on the other, endowing the helix with an amphipathic character. A similar model for E47 and E12 proteins, which bind immunoglobulin enhancers, revealed two amphipathic helical regions. (b) A model of how the leucine zipper might form between two C/EBP subunits and allow the basic amino acids in the DNA-binding domain to bind DNA. (c) Model of dimeric helix-loop-helix proteins such as E47 and E12. These proteins easily form dimers: heterodimers (a:b), as shown, as well as monodimers (a:a). [See C. Murre, P. M. Schonleber, and D. Baltimore, 1989, *Cell* **56**:777; and C. Murre et al., 1989, *Cell* **58**.] *Parts (a) and (b) adapted from W. H. Landschultz, P. F. Johnson, and S. L. McKnight, 1988,* Science **240**:1759.

(a) Proposed amphipathic helix in amino-terminal portion of C/EBP

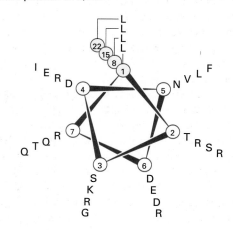

(b) Monomeric and dimeric C/EBP

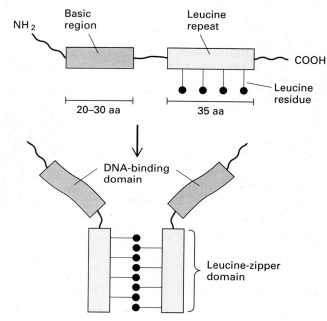

(c) Dimeric helix-loop-helix protein

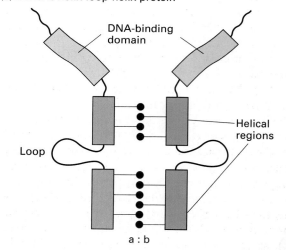

surface. It was originally proposed that the surfaces of two protein coils bearing the leucines might interdigitate or "zip up." The proposed *leucine-zipper* region consists of only the carboxyl-terminal 35 amino acid residues, but both the zipper region and the adjacent 20–30 residues that are rich in basic amino acids are required for DNA binding. Because the zipper region by itself will dimerize but will not bind DNA and because the basic amino acid region by itself will not dimerize and has only a very weak ability to bind DNA, a model for recognition of specific DNA sites including both regions has been proposed. The zipper region is required for dimer formation, whereas the basic region constitutes the DNA-binding domain (Figure 11-14b). The three-dimensional structure of this basic region is being determined at present. Not all of the proteins of this category may have as many leucines repeated in heptad fashion as C/EBP, but the important feature of coil-coil interaction between amphipathic helices presumably also characterizes other proteins of this type.

So far, only homodimers of C/EBP have been detected. However, in vitro studies have shown that heterodimers can form between c-Jun and c-Fos, two protooncogene proteins that contain leucine-zipper regions. Furthermore, when extracts of vertebrate cells containing c-Fos are treated with c-Fos antibody, the precipitated complexes contain not only c-Fos but also AP1, a leucine-zipper protein that originally was purified as an activating protein for SV40 transcription. These findings suggest that a wide variety of heterodimers with different DNA specificities and transcriptional activities might be formed by mixing different leucine-zipper proteins.

A similar but clearly distinguishable design has been recognized in a group of proteins termed *helix-loop-helix* proteins. The first such proteins whose genes were cloned—termed E12 and E47—bind to two regions in the enhancer of genes encoding the kappa chains of immunoglobulins. A stretch of 100–200 amino acids in the carboxyl-terminal end of both E12 and E47 contains two regions that can be folded into amphipathic α helices and another nearby region rich in basic amino acids. Within these regions a striking amino acid similarity exists with a series of proteins encoded by *Drosophila* genes that participate in several different important developmental decisions. In addition strong sequence conservation in the same regions has been detected with mammalian gene products (e.g., MyoD protein) that are important in muscle development, which we discuss later.

The helical regions are required for dimerization and the basic amino acid regions are also required for DNA binding; these characteristics are reminiscent of the leucine-zipper proteins, but the amino acid sequences in the two groups of proteins are not alike. The helix-loop-helix proteins easily form heterodimers (Figure 11-14c). For example, the E12, E47, and MyoD proteins (all mammalian) form active DNA-binding proteins when mixed pairwise, as do two *Drosophila* proteins encoded by the daughterless and achaete-scute genes. This ability to form heterodimers between different members of a group of proteins is potentially very important; varying combinations between individual members of the set will yield a much larger set of regulatory proteins.

Heterodimers of Dissimilar Proteins A fair number (about one-third) of the first hundred or so DNA-binding proteins isolated from yeast, *Drosophila*, and mammals do not exhibit any of the structural motifs just described. One widely distributed group of such structurally distinct factors binds to the CCAAT consensus sequence found in some eukaryotic genes. These CCAAT-binding proteins are heterodimers of dissimilar proteins, one of which binds weakly to DNA on its own but strongly in the presence of the second protein. The overall structure of one of these protein pairs has apparently been conserved over the entire range of eukaryotic evolution. A yeast regulatory factor that controls transcription of a cytochrome gene consists of two different proteins termed HAP2 and HAP3. Amazingly enough, one of these yeast proteins and one mammalian CCAAT-binding protein can form a heterodimer that binds to either the CCAAT box in human DNA or to the cytochrome gene from yeast. Such conservation suggests very fundamental functions for these proteins. As is the case with other mixed DNA-binding factors (such as MyoD with E12 or E47 noted in the previous section, for example), heterodimer formation between two proteins of unrelated structures suggests that a limited set of proteins can by combination produce a much larger array of final active binding proteins.

We should keep in mind that only the bacterial helix-turn-helix proteins have thus far been subjected to crystallographic analysis and only a few of these have been analyzed as cocrystals with their cognate DNA sequence. Thus the ultimate details of DNA site recognition are just beginning to be glimpsed in spite of the remarkably rapid progress that has been made in the identification and sequencing of DNA-binding factors. Table 11-3 presents a summary of some of the eukaryotic DNA-binding proteins that have been identified.

DNA-binding Proteins Can Increase Transcription in Vitro and in Vivo

Most of the proteins discussed so far were discovered because they bind specifically to DNA sites that are required for gene activity (see Figure 11-9). It is generally assumed that such cognate proteins, or site-specific DNA-binding proteins, have an effect (generally positive) on transcriptional initiation. In this section we describe how to demonstrate experimentally that such cognate proteins are in fact transcription factors, that is, that they affect the transcription rate.

Table 11-3 Examples of eukaryotic site-specific DNA-binding proteins

Class/gene product	Organism	Comment
HELIX-TURN-HELIX		
MAT α1	Yeast	Activates α-specific genes
MAT α2	Yeast	Inactivates a-specific genes
MAT a1	Yeast	Combines with α2 to repress haploid-specific genes
Antennapedia*	*Drosophila*	Homeotic gene
Ultrabithorax*	*Drosophila*	Homeotic gene
Engrailed*	*Drosophila*	Segment-polarity gene
Paired*	*Drosophila*	Pair-rule gene
Fushi tarazu*	*Drosophila*	Pair-rule gene
HOX*	Mouse	Potential developmental regulators; at least a dozen different proteins
Unc86†	Nematode	Determines cell lineage in worms
Oct1†	Human	Generally distributed activator
Oct2†	Human	Lymphoid-specific activator
Pit†	Mouse	Pituitary-specific activator
ZINC-FINGER		
GAL4	Yeast	Galactose-dependent activator
HAP1	Yeast	Inducible activator of cytochrome C
SWI5	Yeast	Activates *HO* transcription in mother cells
Krüppel	*Drosophila*	Gap genes
Hunchback	*Drosophila*	Gap genes
Steroid receptors	Vertebrates	Positive and negative acting; in many but not all cell types
SP1	Vertebrates	Widespread activator
HOMO- AND HETERODIMERS OF AMPHIPATHIC HELIX		
Leucine-zipper		
GCN4	Yeast	Activates genes for enzymes to make amino acids
C/EBP	Mammals	Activates genes in liver and other cells; limited cell distribution
c-Fos/c-Jun	Mammals	Growth regulation (?+ and −)
JunB	Mouse	Widespread in cells; growth regulators
HELIX-LOOP-HELIX		
Daughterless	*Drosophila*	Developmental role in nervous system
Achaete-scute (T3)	*Drosophila*	Developmental role in nervous system
MyoD	Mammals	Muscle differentiation
E12 and E47	Mammals	Immunoglobulin activation (proteins widespread)

*Protein contains homeobox sequence (see Figure 11-11).
†Protein contains a homeobox and a second domain called POU that is highly conserved.
SOURCE: P. F. Johnson and S. L. McKnight, 1989, *Ann. Rev. Biochem.*, 58:799; and C. Murre et al., 1989, *Cell*, 56:777.

In vitro assays involve proving that a purified protein can increase RNA polymerase II transcription of a gene in the test tube. In vivo assays are slightly more complicated and depend on having a cell produce the transcriptional factor from one plasmid and then use it to transcribe a gene on a second plasmid (Figure 11-15). These assays are suitable for genes whose transcription rates are sub-

ject to prompt and reversible changes as the result of binding to specific transcription factors. However, for genes whose activation (or repression) involves temporal and cell-specific expression in development, neither assay may be sufficient. Results obtained with two transcription factors, SP1 and ATF, illustrate how these assays can be used to study the properties of transcription factors.

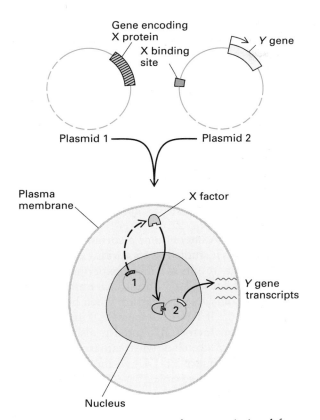

▲ **Figure 11-15** An in vivo test for transcriptional factor activity. The test system requires two plasmids. One plasmid contains the gene encoding X protein, the putative transcription factor, in a form in which the protein will be expressed at high levels. The second plasmid contains a test gene Y with a binding site for transcription factor X. The two plasmids are introduced together into host cells whose genome lacks both the gene encoding X protein and the test gene. The production of Y RNA transcripts is measured. By use of plasmids encoding a mutated or rearranged factor, important domains of the protein can be identified. If mRNA synthesis is greater in the presence of the factor, then it is a positive-acting regulator; if transcription is less in the presence of the factor, then it is a negative-acting protein.

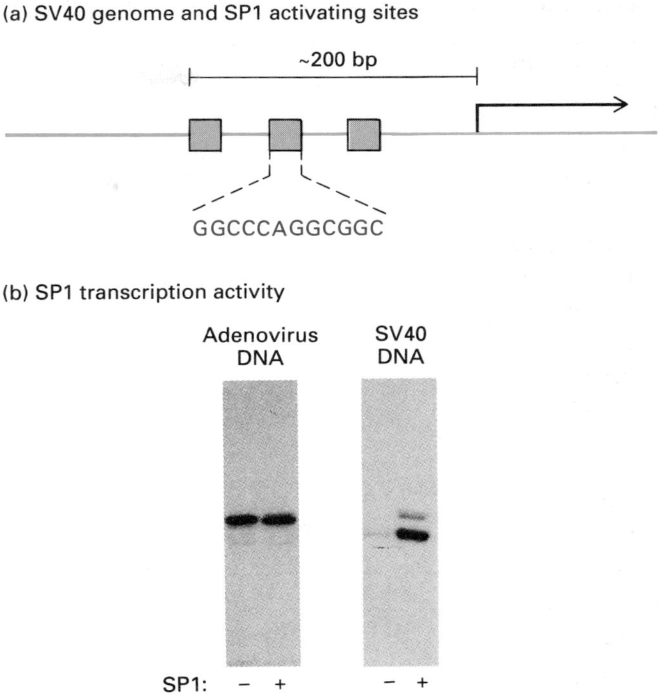

(a) SV40 genome and SP1 activating sites

~200 bp

GGCCCAGGCGGC

(b) SP1 transcription activity

Adenovirus DNA   SV40 DNA

SP1:   − +   − +

▲ **Figure 11-16** SP1 was initially identified because of its ability to activate transcription of SV40 DNA. (a) Three repeated GC-rich regions that bind SP1 (orange) are located on the SV40 genome just upstream from the early transcription unit. The major late adenovirus promoter lacks these GC-rich binding regions. (b) Purified SP1 was tested in vitro for transcription activity with adenovirus DNA and SV40 DNA. The in vitro mixture included not only the DNA and SP1 but also partially purified human nuclear extracts containing RNA polymerase II and associated general transcription factors. The labeled RNA products were subjected to electrophoresis and autoradiography. A minus sign(−) indicates the absence of SP1 and a plus sign(+) indicates its presence in the test system. Although SP1 had no obvious effect on transcription of adenovirus DNA, it stimulated transcription of SV40 DNA tenfold. *Adapted from M. R. Briggs et al., 1986, Science **234**:47.*

SP1 (A Promoter-binding Factor) Stimulatory protein 1 (SP1), one of the first RNA polymerase II transcription factors to be cloned, was originally purified using a regulatory region from SV40 (simian virus 40) DNA. The amino acid sequence of SP1 clearly suggests that it is a zinc-finger protein, and this segment of the protein is required for DNA binding. The DNA region to which SP1 binds is very rich in G and C residues (Figure 11-16a), reminiscent of many genes that lack a TATA box but have nearby segments of GC-rich DNA. With a partially purified RNA polymerase II preparation containing the general transcription factors IIB and IIE, the SP1 protein directs increased transcription on templates that contain its

binding site but not on templates that lack such a site (Figure 11-16b).

SP1 binding sites have been found in the DNA from many different genes. Most such sites lie within 200 bp or less upstream of the start sites for RNA synthesis in the genes analyzed so far. Thus, SP1 may be a general promoter-binding factor necessary for activation of many genes.

ATF (An Enhancer-binding Factor) Activating transcription factor (ATF) is a preexisting cellular protein that initially was identified as a stimulator for transcription of the adenovirus E4 transcription unit, which is ac-

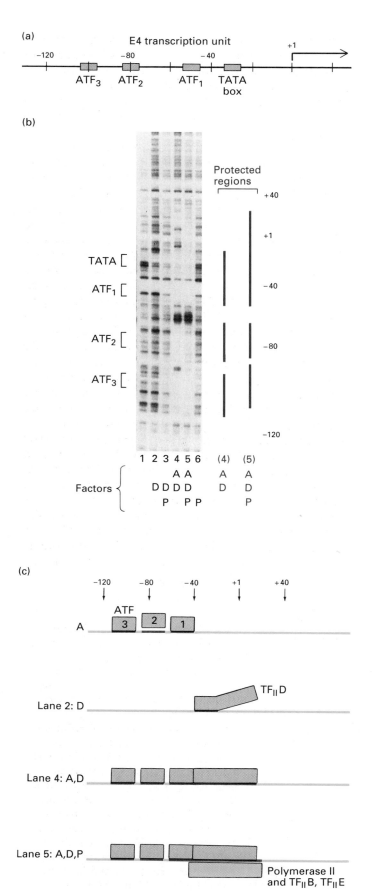

tive early in infection. Several binding sites for ATF are located upstream of the E4 start site in the adenovirus genome (Figure 11-17a). Because these binding sites are active in vivo in either orientation in the DNA and at distances up to 1 kb from the RNA start site, they can be classified as enhancer sites.

Although ATF by itself can bind to its cognate DNA, it does not stimulate in vitro transcription of adenovirus E4 in the absence of the TATA factor ($TF_{II}D$). In the absence of ATF, $TF_{II}D$ binds weakly to DNA and stimulates in vitro transcription of the E4 gene only slightly. However, in the presence of ATF, both the binding of $TF_{II}D$ and its stimulatory effect on in vitro transcription of E4 are increased substantially.

The protein-DNA interactions associated with E4 transcription have been studied in footprint experiments (see Figure 7-4). The results of these experiments indicate that the E4 start site and a 60-bp stretch upstream from it are protected from DNase digestion when both ATF and $TF_{II}D$ are present; if RNA polymerase II and the associated transcription factors IIB and IIE also are present, then a 20- to 30-bp stretch downstream from the start site also is protected (Figure 11-17b). A model for assembly of the E4 preinitiation complex based on these findings is shown in Figure 11-17c; once nucleotides are added to the mixture, RNA synthesis can proceed.

These experiments with ATF indicate that upstream activating factors can participate in vitro in assembling the preinitiation complex. It seems reasonable to assume that such assistance in complex formation is an important general mechanism in vivo as well. A matter of dispute at present concerns whether genes lacking an obvious TATA region do or do not utilize $TF_{II}D$ and therefore whether the model in Figure 11-17c is valid only for a subset of all eukaryotic genes. This is an important unsettled issue, but at least one adenovirus gene (IVa2) lacking an obvious TATA box still requires $TF_{II}D$ to be transcribed in vitro.

◀ **Figure 11-17** (a) Location of ATF binding sites and TATA box in adenovirus E4 transcription unit. (b) A DNase footprint of the indicated E4 promoter region in the presence and absence of the proteins indicated at the bottom. D = $TF_{II}D$; P = RNA polymerase II plus $Tf_{II}B$ and $TF_{II}E$; and A = ATF. The brackets on the left show the positions of the TATA box and the three ATF sites. The red bars on the right show the regions protected against DNase digestion in various experiments. (c) Diagram of the sequential participation of the various factors in formation of the preinitiation complex based on the DNase footprint shown in (b). ATF alone (not shown in footprint) protects the ATF sites 1–3. $TF_{II}D$ binds weakly and protects the site only near the TATA box (lane 2). The polymerase alone does not bind at all (lane 6). ATF assists $TF_{II}D$ to bind (lane 4), and with the polymerase and accessory factors binds to and protects the regions up- and downstream of the start site (lane 5). *Adapted from M. Horikoshi et al., 1988,* Cell **54**:1033.

Furthermore, in vitro SP1-dependent transcription of genes with a GC-rich SP1 binding site but lacking an obvious TATA box still requires TF$_{II}$D for maximal transcription (either yeast or mammalian TF$_{II}$D will suffice). At present, it appears that this protein, or one like it, has a general role to play in initiating transcription. It remains possible that other proteins with functions similar to that of TF$_{II}$D may exist but as yet have not been identified.

Acidic Domains of GAL4 Are Necessary for Transcriptional Activation

Bacterial RNA polymerase is thought to interact directly with positive-acting regulatory proteins such as AraC and the cyclic AMP-binding protein (CAP) to increase transcriptional initiation at specific sites. The studies with ATF clearly indicate that formation of a preinitiation complex in eukaryotes often is assisted by upstream activating factors. We know very little yet about protein-protein interactions involved in this assistance. For example, we do not know whether activating factors contact the TF$_{II}$D (the TATA factor) or the RNA polymerase molecules or perhaps both. We do not know whether all upstream activators function in a similar manner: some may interact directly with the initiation machinery; others may make other contacts that promote transcription. In bacteria, many activating proteins are known to contain a helical domain rich in acidic amino acids (glutamic and aspartic acids); when these proteins are bound to DNA, this domain projects out from the DNA. As illustrated in Figure 7-30, such residues are involved in positive autologous activation of the λ repressor gene. Studies of GAL4 protein, a galactose-dependent yeast activator, suggest that acidic domains also may be one important protein

domain in the action of eukaryotic activating proteins.

As discussed in the next section, GAL4 is required for the induction of galactose-metabolizing enzymes in the presence of galactose. By recombinant DNA techniques, a number of mutant *GAL4* genes were constructed and reintroduced into yeast cells that lacked the galactose-induction system. The ability of such transformed cells to engage in galactose-stimulated transcription and of the mutant GAL4 proteins to bind to DNA were determined. The general structure of GAL4, based on the results of these experiments, is shown in Figure 11-18. The amino-terminal 147 amino acids of GAL4 are sufficient for site-specific DNA binding; this region has an amino acid sequence suggestive of a zinc-finger binding domain. However, GAL4-mediated gene induction requires the presence of at least one of two other regions, both rich in acidic amino acids. Replacement of the region encoding the acid domains of GAL4 with sequences encoding an acidic domain of GCN4, another yeast-activating protein, resulted in a hybrid protein that activated galactose-stimulated transcription, even though the GCN4 acidic region bears no sequence identity to the GAL4 acidic regions. Finally, short random *E. coli* DNA sequences encoding about 30 amino acids were inserted into the *GAL4* gene in place of the residues encoding the acidic domains. About one in a hundred such insertions resulted in fusion proteins that were active in galactose-dependent gene expression; all of the most active *E. coli* fragments were rich in acidic amino acids. These experiments clearly indicate that the transcriptional activation region of GAL4 is interchangeable with other sequences rich in acidic amino acids. The experiments emphasize that distinct, independent regions of the protein function in DNA binding and transcriptional activation. It is important to note that many eukaryotic-activating proteins do not

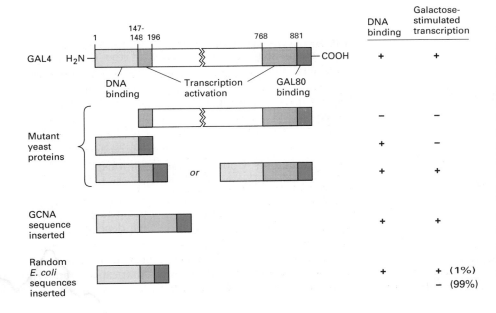

◀ **Figure 11-18** Results of experiments demonstrating DNA-binding and transcription-activation regions of GAL4, a yeast protein that activates genes encoding galactose-metabolizing enzymes. Comparison of the activity of normal (GAL4) and mutant proteins constructed by recombinant techniques shows one region required for DNA binding and two regions rich in acidic amino acids, at least one of which is required for transcriptional activation. The carboxyl end region binds to another protein, GAL80, which blocks activation in the absence of galactose (see Figure 11-19b). Numbers above GAL4 refer to amino acid residues in the protein. [See J. Ma and M. Ptashne, 1987, *Cell* 48:847 and 51:113.]

contain acidic amino acids in their activating domains. Other activating domains include long glutomine or protein rich regions and others are not yet fully characterized.

The experiments described on protein domains active in stimulating transcription are no substitute for crystallographic studies of activator-DNA complexes similar to those which have been done with bacterial systems. They do, however, strongly suggest which portions of at least some eukaryotic activators help in some step of formation of preinitiation complexes or in the release of such active complexes once formed.

Transcriptional Control of Yeast Cell Specificity

In the previous section we went into considerable detail about the general principles of transcriptional control and the proteins that implement such control because transcriptional regulation is a central event in biological specificity. From the time that Jacob and Monod hypothesized the existence of bacterial messenger RNA and the regulatory proteins that controlled its levels, a reasonable assumption has been that understanding of such important biological phenomena as adaptation to the environment, endocrine function, and most especially development and differentiation of animals and plants would ultimately depend on progress in understanding eukaryotic transcriptional control. In this section, we describe how transcriptional control underlies important changes in cell specificity. We begin with two examples of transcriptional control in yeast: one involving the *GAL* genes, which encode the inducible galactose-metabolizing enzymes, and the other involving yeast mating type. In the following section, several examples of transcriptional control in multicellular organisms are discussed.

Expression of *GAL* Genes in Yeast Is Controlled by Both Positive- and Negative-acting Proteins

In the 1970s when microbial geneticists and biochemists began intensive studies of gene control in yeast, a natural target was inducible metabolic pathways similar to those that had been studied in bacteria. For example, baker's yeast, *Saccharomyces cerevisiae* which can grow on a variety of carbon sources, produces several inducible enzymes that metabolize galactose when grown in the presence of galactose. Three enzymes that convert galactose into glucose phosphate are encoded by genes that lie on chromosome II; a fourth gene, which encodes a galactose-transport protein, is on chromosome XII; and a fifth gene, located on still another chromosome, encodes α-galactosidase, which converts melibiose to galactose (Figure 11-19a). All five genes are induced by a metabolic product of galactose through the action of a regulatory gene called *GAL4*. Yeast cells mutant at *GAL4* cannot metabolize galactose or synthesize any of the galactose-metabolizing enzymes. Thus it was inferred that the GAL4 protein was a positive regulatory protein that acts on all of the separate genes required for galactose metabolism.

As described already, GAL4 contains an amino-terminal DNA-binding region and two acidic domains at least one of which must be present for transcription of the *GAL* genes in the presence of galactose (see Figure 11-18). We now know that regulation of the *GAL* genes involves not only the positive-acting GAL4 but also a second protein, GAL80, which is negative-acting. An upstream activating sequence (UAS) is associated with each of the three *GAL* genes on chromosome II. The *GAL1* and *GAL10* genes, which are transcribed in opposite directions, share one UAS (Figure 11-19b). In the absence of inducer, a GAL4-GAL80 complex binds at the GAL4-binding site in the UAS. When complexed with GAL80, GAL4 cannot activate transcription. Formation of this protein-protein complex depends on the last 30 amino acids in the GAL4 sequence. When galactose is present in the medium, an unidentified metabolic product of galactose is formed that acts as an inducer by causing GAL80 to dissociate from GAL4; as a result, the DNA-bound GAL4 can activate transcription of the *GAL* genes encoding the kinase, transferase, and epimerase that convert galactose to glucose (Figure 11-19a). These experiments illustrate that negative-acting proteins do not have to be DNA-binding proteins and emphasize the importance of protein-protein interaction among regulatory proteins.

Yeast Mating Type Is Determined by Integrated Network of Transcriptional Controls: A Model for Cell Fate Determination

Control of yeast mating type involves more complex mechanisms than does regulation of the *GAL* genes. Moreover, this system may offer some insight into cell differentiation in higher eukaryotes, as the mating-type cycle in yeast exemplifies limited but definite cell differentiation.

Haploid yeast cells exist in one of two mating types, **a** or α. Haploid cells of different mating types can fuse to form **a**/α diploids. Under adverse conditions, diploid cells cease growth and undergo meiotic division to yield two **a** and two α spores (Figure 11-20). During regrowth of spores, mating-type conversion can occur in *homothallic* strains. This switching of mating type depends on gene conversion at the *MAT* locus, a master regulatory locus that exists in two forms (*MATa* and *MATα*). The products of *MATa* or *MATα* are proteins that cause a series of events leading to a cell becoming either an **a** or α cell type.

(a)

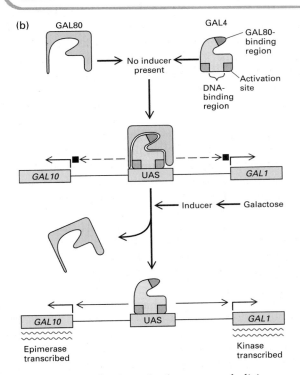

Galactose

Nuclear envelope

Chromosome XIII —
GAL80
↓
GAL80 mRNA — Translation → GAL80 protein

Chromosome XVI —
GAL4
↓
GAL4 mRNA — Translation → GAL4 protein

Chromosome II — *GAL7 GAL10 GAL1* — Blocks
Chromosome XII — *GAL2* — Activates
Chromosome U — *α-Galactosidase* — Removes GAL80
Chromosome IV — *GAL3*

GAL2 (transport enzyme)

Inducer ← GAL3 protein ← Galactose ← α-Galactosidase ← Melibiose

Galactose
ATP → GAL1 (kinase)
ADP ←
Galactose 1-phosphate
UDP-glucose → GAL7 (transferase)
Glucose 1-phosphate ←
Metabolized
UDP-galactose
GAL10 (epimerase)
UDP-glucose

(b)

GAL80

GAL4 — GAL80-binding region, Activation site, DNA-binding region

→ No inducer present ←

GAL10 — UAS — GAL1

Inducer ← Galactose

GAL10 — UAS — GAL1
Epimerase transcribed   Kinase transcribed

▲ **Figure 11-19** Induction of galactose-metabolizing enzymes in yeast. (a) Metabolic pathway showing steps catalyzed by products of the *GAL* genes. The locations on different chromosomes of these genes are shown. (b) Two regulatory proteins, GAL4 and GAL80, are formed constitutively. Binding of the GAL4-GAL80 complex to upstream activating sequences for GAL1 (kinase), GAL7 (transferase), and GAL10 (epimerase) on chromosome II prevents activation by GAL4. An unknown metabolic product formed when galactose is present removes GAL80, releasing the activating effect of GAL4. [See J. R. Broach, 1979, *J. Mol. Biol.* **131**:41; A. Laughon and R. Gesteland, 1982, *Proc. Nat'l Acad. Sci. USA* **79**:6827; and S. A. Johnston, J. M. Salmeron, and S. S. Dincher, 1987, *Cell* **50**:143.]

Mating-type switching involves a site-specific double-stranded cleavage at the *MAT* locus, catalyzed by an endonuclease encoded at the homothallic *(HO)* locus, followed by directed gene conversion that changes the information at the *MAT* locus (see Figure 10-44). This conversion changes *MATa* to *MATα* or vice versa.

In this section we consider some of the proteins involved in the events of yeast mating and how the master regulatory locus, the *MAT* locus, controls the phenotype of **a** and α haploid cells and **a**/α diploids. First of all, **a** cells produce a protein precursor to a small peptide mating factor called the **a** factor, or **a** *pheromone;* this precursor is proteolytically cleaved to generate a small active **a**-factor peptide. In α cells a different small active peptide, the α factor, is produced. The **a** cells also produce an α-factor receptor, a surface protein, and α cells produce a receptor for the **a** factor. These small peptides act reciprocally to cause growth inhibition; that is, during starvation **a** factor inhibits α-cell growth and vice versa. (The mechanism of signaling by the yeast pheromones is discussed in detail in Chapter 19.) When haploid cells stop growing they can fuse to form **a**/α cells. In addition to cell-type–specific genes for the peptide factors and receptor proteins, a number of other proteins that are specific for haploid cells are known. The genes for these haploid-specific proteins function in both **a** and α cells but not in diploid **a**/α cells.

A series of elegant genetic experiments coupled with recent biochemical research has revealed much about the regulation of the *MAT* locus itself and how proteins encoded at the *MAT* locus regulate other cell-specific genes including those at the *HO* locus, which is necessary for

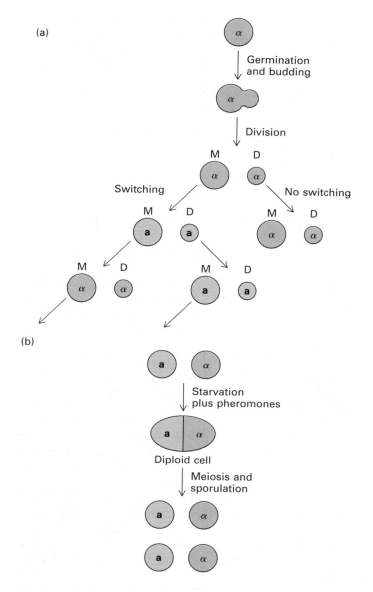

◂ **Figure 11-20** Events in the yeast life cycle. (a) A spore (α-cell type is shown here; it could be **a**) germinates and divides. The mother cell (M), which is larger than the daughter (D) can, before its next DNA duplication, undergo a switch in DNA content at the *MAT* locus by the action of the HO endonuclease and gene conversion with the silent copy of *MAT* at *HML* (see Figure 10-42). Daughter cells cannot undergo switching until they bud and divide when they become mother cells capable of switching. (b) The α cells produce α pheromone (polypeptide) and **a** receptor, and **a** cells produce **a** pheromone and α receptor. In conditions of starvation, pheromones bind to their cognate receptors to stop cell growth; fusion then occurs and diploid **a**/α cells result. Meiosis can occur producing spores. [See I. Herskowitz, 1988, *Microbiol. Rev.* **52**:536.]

weakly to its cognate DNA site, they both require PRTF for maximal function. In the presence of PRTF and α1, α-specific genes are activated in α cells. In contrast, binding of α2 in the presence of PRTF blocks the normal activating effect of PRTF on **a**-specific genes. The *MAT***a** locus also encodes two proteins, but little is known about the **a**2 protein. The **a**1 protein appears to exert no regulatory effect in **a** cells. However, since these cells produce no α1 and α2, α-specific genes are silent and PRTF activates α-specific genes. Haploid-specific genes are transcribed in both **a** and α cells; presumably these genes also are activated by PRTF (Figure 11-21a and b).

No synthesis of **a**-, α-, or haploid-specific proteins occurs in diploid **a**/α cells. These cells produce α2, which forms a complex with PRTF that represses **a**-specific genes as in haploid α cells. Diploid cells also produce **a**1, which combines with α2. The **a**1-α2 complex represses haploid-specific genes and also prevents transcription of the α1 gene. In the absence of α1, PRTF cannot bind to and activate α-specific genes. Thus diploid cells exhibit repression of **a**-, α-, and haploid-specific genes (Figure 11-21c).

An important example of this repression involves an inhibitor of meiosis, called the RME protein, which is produced by **a** and α cells and acts to block cells from attempting meiosis. Because this haploid-specific protein is not produced in diploid **a**/α cells, diploid cells can undergo meiosis and form haploid spores when starved. A whole group of sporulation genes, which require additional activators, also must be transcribed during sporulation; however, repression of the meiotic repressor RME is an absolute requirement for yeast cells to enter meiosis.

mating-type conversion to occur. Because these control mechanisms illustrate general principles of integrated gene regulation, which may have wide application in eukaryotic biology, we discuss them in some detail.

Functions of MAT Gene Products and PRTF

The phenotypes of haploid and diploid yeast cells depend not only on the proteins encoded by the *MAT* locus but also on the pheromone and receptor transcription factor (PRTF), a constitutive DNA-binding protein present in α, **a**, and **a**/α cells. PRTF itself binds strongly to **a**-specific and haploid-specific genes but only weakly to α-specific genes.

The *MAT*α locus encodes α1, a positive regulator of α-specific genes, and α2, a negative regulator of **a**-specific genes. Although each of these proteins can itself bind

Silencer Sequences for HMLα and HMRa

In addition to the active *MAT* locus, yeast chromosome 3 has two other loci—HMLα and HMRa—that contain "extra" silent copies of the α and **a** sequences, respec-

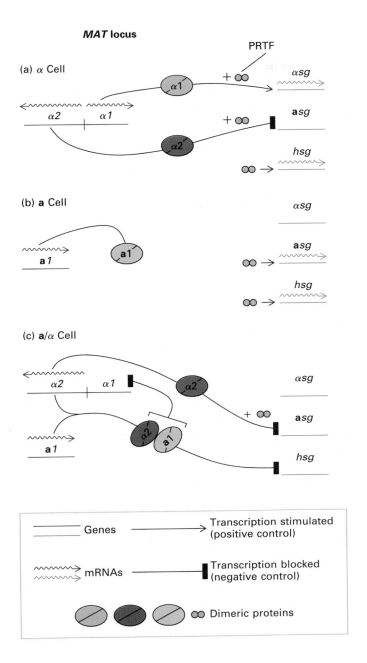

MAT locus

(a) α Cell

(b) a Cell

(c) a/α Cell

Genes ——→ Transcription stimulated (positive control)

〜〜〜 mRNAs ——■ Transcription blocked (negative control)

●● Dimeric proteins

◄ **Figure 11-21** Regulation of a-specific genes (*asg*), α-specific genes (*αsg*), and haploid-specific genes (*hsg*) in α, a, and a/α yeast cells by regulatory proteins encoded at the *MAT* locus together with PRTF, a constitutive DNA-binding protein. As a result of this regulation, each diploid or haploid cell type exhibits a distinctive pattern of gene expression (*right*). See text for discussion. [See I. Herskowitz, 1987, *Microbiol. Rev.* 52:536; A. Bender and G. F. Sprague, Jr., 1987, *Cell* 50:681; and C. Goutte and A. D. Johnson, 1988, *Cell* 52:875.]

negative regulation and the manner in which they do so are subjects of current investigation. Since their binding sites are more than 1 kb apart, it is thought that they make a loop which somehow inhibits expression.

Regulation of HO Locus and Mating-type Conversion

If halpoid yeast spores are grown after meiosis, each mother and daughter cell can be followed in the microscope because mothers are larger than daughters. Testing of individual mother and daughter cells for mating type (see Figure 10-42) has demonstrated that only mother cells can undergo mating-type conversion; daughter cells never switch mating type. Furthermore, when a mother cell switches, both of its progeny cells are switched (see Figure 11-20). As noted earlier, the switching process begins with a site-specific cleavage at the *MAT* locus by the HO endonuclease; regulation of this enzyme must be closely controlled to allow switching in the orderly fashion that is observed.

By genetic analysis, six switch *(SWI)* genes have been identified as necessary for switching to occur. These *SWI* genes, particularly *SWI5,* are thought to encode proteins that activate transcription of the *HO* locus. Another group of genes, termed *SIN,* encode inhibitors of *HO;* mutations in the *SIN* genes lead to inappropriate expression of the *HO* locus. Apparently, the balance between SWI factors and SIN factors permits expression of the *HO* locus only in mother cells. Formation of SWI5 mRNA and protein occurs just before cell division, whereas the HO endonuclease is formed only after division but before the next round of DNA replication. A plausible possibility is that the SWI5 protein is distributed mainly to the mother cell at division, and thus only the mother cell can produce the HO endonuclease (Figure 11-22). This enzyme then cleaves the DNA leading to *MAT*-locus conversion, after which the DNA is duplicated, thus giving rise to two progeny *both* with switched genomes.

In this model, transcription of *SWI5,* whose regulation is not yet understood, delivers a necessary protein for switching just in time for it to enter the mother cell and complete the array of SWI proteins, leading to activation of the *HO* gene at just the right time in the cycle. Again a

tively (see Figure 10-43). Different yeast strains can have the extra α or a information in either the left (HML) or the right (HMR) positions. Negative-acting regulatory proteins, whose DNA-binding sites are located 1000 base pairs or more from the HML and HMR start sites, are responsible for "silencing" these extra copies of the α and a sequences. When the DNA-binding regions for these silencer proteins are placed near other genes, they can completely silence these genes as well. This ability to effect a total shutoff of transcription probably is widespread in eukaryotes, as most eukaryotic genes are not active at any one time. The proteins that carry out this

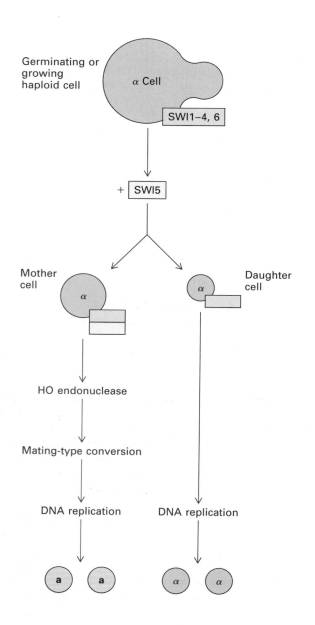

◀ **Figure 11-22** Control of mating-type switching in yeast mother and daughter cells. Among the many proteins required for switching are those encoded by the six switch (*SWI*) genes, which activate the *HO* locus, resulting in synthesis of the HO endonuclease. Five of these are present in all cells at all times, but the SWI5 protein is made only near the end of the cell cycle and distributed preferentially to mother cells. The HO endonuclease thus is made only in mother cells, which then switch mating type before DNA duplication; as a result, both progeny cells have the switched mating type. [See I. Hershkowitz, 1987, *Microbiol. Rev.* 52:536; and K. Nasmyth and D. Shore, 1987, *Science* 237:1162.]

how several proteins act in concert to bring about a decision that yields one of two alternative cell types.

Many steps in the development of multicellular organisms also depend on unequal cell divisions that yield two cell types. For example, a *stem* cell may divide to give one cell like itself and another cell that will follow a particular developmental pathway; this is referred to as a *unipotent* stem cell. In other cases the stem cell is *pluripotent* and can give rise to two or more differentiated cell types (Figure 11-23). Such choices can occur not only in embryos but also in adults. For instance, the bone marrow of adults contains individual stem cells that are capable of continuous growth but also can generate precursor cells to all other cell lineages required to form the major elements of the blood. These include erythroblasts that form red blood cells; myeloblasts that form various white blood cells including polymorphonuclear leukocytes, basophils, eosinophils, monoblasts, and megakaryocytes (platelet precursors); and lymphoblasts, which form cells of the lymphocytic series. A major problem in developmental biology is to learn what distinguishes a stem cell from its daughter such that the daughter can differentiate and the stem cell continue to grow. The work on mating-type switching in yeasts is bound to prove useful in this important field if only to help shape critical thinking.

Gene Control in Animal Cells

The goal of much of the research on eukaryotic transcriptional control is to understand the mechanisms by which a fertilized egg (or a seed) divides and progresses through various decision points to yield groups of cells that are first determined to become and then actually differentiate to become specialized tissues of the right dimension and in the proper location. To be specific, how does a muscle cell become a muscle cell or a liver cell a liver cell? Further, how does each muscle become the right size and shape and how does the liver end up on the right side of the abdomen and making the proteins of the serum?

balance of gene regulators ensures the timing of this critical event. The SWI proteins are not necessarily DNA-binding proteins, but they do overcome the negative action of the proteins encoded by the *SIN* genes.

Although all the details of this intricate mechanism are not yet understood, the yeast mating-type regulatory network provides an example of a master gene, the *MAT* locus, that in combination with at least a dozen other participants can be set to direct a cell to become one of two alternative cell types. In the *MAT* locus, switching between the alternative "set points" involves a DNA rearrangement (i.e., gene conversion), a mechanism that is not widely employed, if it is used at all, in developmental switches in animal cells. Nonetheless, the demonstration of an integrated network of transcriptional controls that regulates mating type in yeast provides a useful model of

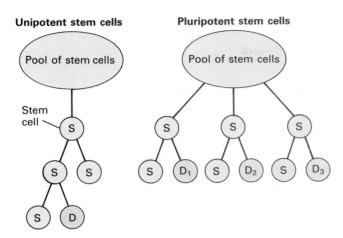

Unipotent stem cells   **Pluripotent stem cells**

▲ **Figure 11-23** The production of differentiated cells (D) from stem cells (S). Unipotent stem cells produce a single type of differentiated cell, whereas pluripotent stem cells may produce two or more types of differentiated cells.

Because many specialized cell types such as liver and muscle are characterized by specific gene products not produced in other cell types, the importance of gene regulation, particularly transcriptional control, in development has been recognized at least since Jacob and Monod's ideas about bacterial regulation were formulated. Until recently, however, there has been no direct connection between transcriptional-control proteins and the developmental decisions that lead cells into different pathways of development. This has changed greatly with the discovery that many mutations affecting early development in invertebrates are in genes that encode transcription factors. Furthermore, genetic studies show that these transcription factors form an interdependent network, the operation of which directs orderly development. The transcription factors that play a role in cell specificity in vertebrates are also being identified. Most of these factors belong to one of the groups of DNA-binding proteins—the helix-turn-helix, zinc-finger, or the amphipathic helix family (leucine-zipper and helix-loop-helix)—that we have already discussed.

The types of signals that cause changes in the repertoire of active transcription factors in cells, particularly mammalian cells, are also beginning to be understood. Much work remains to be done before important developmental pathways in multicellular organisms can be understood in the same detail as cell switching in *S. cerevisiae* is. Nevertheless, a general working framework seems to be in place: a cascade of decisions, many determined by the repertoire of available transcription factors, begins in a fertilized egg. Impinging upon cells that are set to respond in many different ways are signals from the outside whose interpretation by a particular cell leads it into one or another of the various pathways of determination and ultimate differentiation. Only a few of the key transcription factors and only a few of the signals that effect changes in transcription of developmentally important genes are known at present.

In this section, we first discuss early *Drosophila* development, a case in which a cascade of transcription factors has been clearly established. We then discuss several signaling systems, where transcriptional regulation is influenced by outside events to provide examples of how such events might affect development. Finally we discuss briefly what is known about muscle and liver cell differentiation as examples in which important cell-specific transcriptional events underlying cell specificity are beginning to be understood.

A Cascade of Sequentially Expressed Transcription Factors Directs Early Development in *Drosophila*

The entire life cycle of *Drosophila*, which takes only 9–10 days, begins with fertilization of an egg. Within 1 day the fertilized egg develops into a larva; three subsequent larval stages, or *instars* as they are called, require about 4 more days. The larva is actually a separate animal from the adult. During larval growth about a dozen groups of cells, termed *imaginal disks*, are set aside and are carried in the abdomen of the larva, playing no role for the larva. After the last larval stage, an outer shell, or chorion, is formed. The larval cells are broken down and nutrients derived therefrom are used in the growth and development of the different imaginal disks (eye disk, wing disk, etc.); these groups of cells then become the different body parts of the adult fly. This last stage, called *pupation*, takes another 4 days or so. At the end of pupation the shell splits and an adult fly emerges (Figure 11-24).

The molecular events that set the course for this tightly controlled developmental process begin in the construction of the egg with the deposition of *maternal* gene products in particular positions in the egg. Fertilization releases the action of these products to begin the cascade of early development. After fertilization, 12 rounds of nuclear divisions produce about 4000 nuclei, all within a single cytoplasm; this structure in which products of different nuclei can freely communicate with one another is called a *syncytial blastoderm*. Cellularization then occurs by the growth of membranes between the nuclei that are spaced regularly at the surface of the syncytial blastoderm (Figure 11-25). Folding of the embryo and tissue development rapidly follow, and within a day the larva is formed.

Drosophila geneticists have always recognized the possibility that mutations in genes that act early in development could be especially valuable tools for learning about the biochemical events underlying development. In the

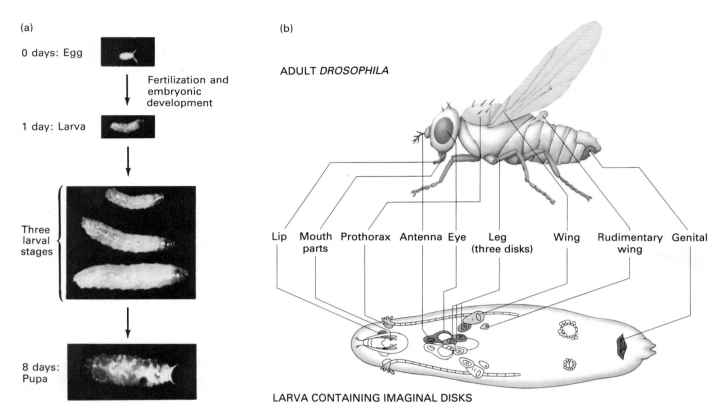

▲ **Figure 11-24** The development of *D. melanogaster* takes 9–10 days. (a) The fertilized egg undergoes blastoderm formation and cellularization in a few hours and hatches as a larva (a segmented organism) in about 1 day. The larva passes through three stages, or *instars*, developing into a prepupa; finally a pupa hatches at 8 days. The adult emerges 1 day later. (b) Groups of ectodermal cells called imaginal disks are set aside in the larval body cavity and form the body parts of the adult fly during the last 3–4 days of pupation. The larval body itself is destroyed in the process. *Photographs from M. W. Strickberger, 1985, Genetics, 3d ed., Macmillan, p. 38. Reprinted with permission of Macmillan Publishing Company. Part (b) after same source and J. W. Fristrom, R. Raikow, W. Petri, and D. Steward, 1969, in Park City Symposium on Problems in Biology, E. W. Hanly, ed., University of Utah Press, p. 381.*

past 20 years, intensive efforts by fly geneticists have resulted in the discovery of dozens of lethal mutations whose effects in homozygous animals are to halt development at an early stage (Table 11-4). Several dozen of these developmentally important genes have now been cloned and sequenced, so that production of molecular probes (antibodies to the encoded proteins or antisense RNA probes for in situ hybridization) has been possible. With these probes, the cellular patterns of expression of these genes during early *Drosophila* development have been detailed (Figure 11-26).

We begin a summary of these studies, as outlined in Figure 11-27, with the unfertilized egg. A group of mRNAs are inherited from the mother (maternal mRNAs) and, as noted above, are found in very definite locations within the egg. One maternal protein termed bicoid is found anteriorly in an unfertilized egg and after fertilization forms a gradient during the early nuclear divisions (Figure 11-26a). The bicoid protein is thought to activate genes encoding the first mRNAs to be newly transcribed in the fertilized egg. Within the first 2 h and just prior to cellularization, a group of genes called the *gap* genes begin to be transcribed; expression of at least the anterior most gap mRNAs is activated by the bicoid protein in several slightly overlapping domains covering the center 80 percent of the embryo (Figure 11-26). The most well-characterized gap genes are hunchback, Krüppel, and knirps. *Drosophila* strains with mutations in the gap genes fail to express the correct number of segments in later development; thus the products of the gap genes regulate the expression of other genes. The next step in this train of molecular events is transcription of the *pair-rule* genes. Within another 30 min, the products of the pair-rule genes (e.g., fushi tarazu, hairy and even-skipped) are found in seven stripes of cells in the central part of the embryo (Figure 11-26c, e, and f). The distribution of these gene products presages the division of the embryo into the segments that exist in the larva and also

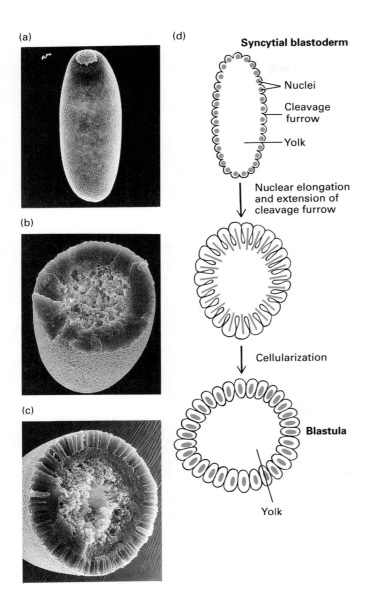

◄ **Figure 11-25** Cleavage and formation of blastula cells in a *Drosophila melanogaster* embryo. Nuclear division is not accompanied by cell division until about 2 to 4 thousand nuclei have formed. Electron micrographs before cellularization show surface bulges overlying individual nuclei (a) and absence of cell membranes (b), which are evident after cellularization (c). Diagrams outline change of syncytial blastoderm into blastula (d). [See R. R. Turner and A. P. Mahowald, 1976, *Devel. Biol.* 50:95.] *Photographs courtesy of A. P. Mahowald.*

Table 11-4 Early developmental genes of *Drosophila melanogaster* and examples of regulatory interactions between them

Genes	Interactions*
MATERNAL	
Biocoid (*bcd*) (anterior)	high *bcd* ⟶ *hb*
Caudal (*cau*)	low *bcd* ⟶ *Kr*
Oskar (*Osk*) (posterior)	*osk* ⊣ *hb*
GAP	
Krüppel (*Kr*)	*hb* ⇄ *Kr*
Hunchback (*hb*)	*kni* ⟶ *Kr*
Knirps (*kni*)	*hb* ⟶ *Ubx*
PAIR RULE	
Hairy (*h*)	*h, r* ⊣ *ftz, eve*
Runt (*r*)	*ftz, eve* ⟶ *en*
Even-skipped (*eve*)	*ftz, eve* ⊣ *wg*
Fushi tarazu (*ftz*)	*ftz* ⟶ *Ubx*
Paired (*prd*)	*prd* ⊣ *wg*
Odd-paired (*opa*)	*opa* ⟶ *wg*
SEGMENT POLARITY	
Engrailed (*en*)	*wg* ⟶ *en*
Wingless (*wg*)	*eve, ftz* ⊣ *wg*
HOMEOTIC	
Antennapedia (*Antp*)	*Kr* ⟶ *Antp*
Ultrabithorax (*Ubx*)	*Ubx* ⊣ *Antp*
	ftz ⟶ *Antp*
	hb ⊣ *Ubx*
	ftz ⟶ *Ubx*

*The product(s) of the first-named gene(s) stimulates (⟶) or suppresses (⊣) expression of the second-named gene(s). The listing is representative and does not include all known interactions.

SOURCE: C. Nusslein-Vollhard and E. Wieschaus, 1980, *Nature* **287**:795; S. B. Carroll and M. P. Scott, 1986, *Cell* **45**:113; and P. W. Ingham, 1988, *Nature* **335**:25.

contribute to specific regions of the adult fly. Definite genetic evidence for the interdependence of the various developmental genes has been obtained; representative interactions are listed in Table 11-4.

Next, division of the embryo into 14 developmental units or *parasegments* is signalled by expression of *segment-polarity* genes such as engrailed and wingless. The engrailed gene is expressed in the most anterior cell of each of the well-defined parasegments (Figure 11-26g). At this stage almost every (if not every) cell down the length of the embryo has its own specific mixture of gap, pair-rule, and segment-polarity gene products (Figure 11-27, *bottom*). It is the interplay of these mixtures that determines the correct progression of development in the final segments of the embryo. By this stage, cell communication must be occurring to effect these patterns, but the nature of the important cell contacts is unknown at present. Each of these parasegments, defined by the distribution of these gene products (slightly out of register with the

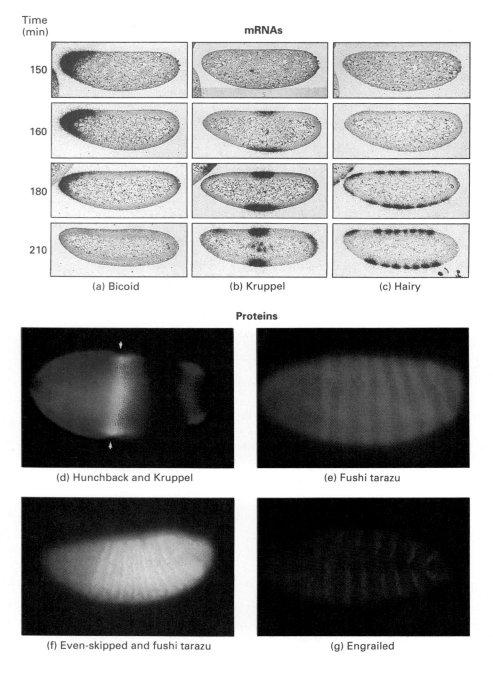

Time
(min)

mRNAs

150

160

180

210

(a) Bicoid (b) Kruppel (c) Hairy

Proteins

(d) Hunchback and Kruppel (e) Fushi tarazu

(f) Even-skipped and fushi tarazu (g) Engrailed

◀ **Figure 11-26** Localization of developmentally important early *Drosophila* gene products. (a–c) In situ hybridization with labeled RNA probes of whole embryo sections 150–210 min after fertilization, which covers the period from the syncytial blastoderm to the beginning of gastrulation. The dark silver grains show the positions of the mRNAs encoded by bicoid, a maternal gene (a); Krüppel, a gap gene (b); and hairy, a pair-rule gene (c). (d–g) Fixed embryos stained with antibodies that are coupled to different fluorescent dyes. (Anterior is to the left; dorsal at the top.) (d) Hunchback protein (red) and Krüppel protein (green) in a syncytial blastoderm. Both are gap-gene products. The yellow band is a region of overlap of the two proteins. (e) Fushi tarazu protein, (green), which is encoded by a pair-rule gene, at time of cellularization (~4–5 h). (f) Even-skipped protein (yellow), another pair-rule gene product, and fushi tarazu protein (orange) in alternating bands at beginning of gastrulation. (g) Engrailed protein (red), a segment-polarity gene product, in a folded gastrula. *Parts (a)–(c) from P. W. Ingham, 1988, Nature 335:25; photographs courtesy of P. W. Ingham. Photographs (d)–(g) courtesy of M. Levine.*

parasegments) is important in larval life, and each segment contributes cells that will become imaginal disk cells and therefore will develop into one of the adult fly parts.

The genes that are most important in determining the ultimate fate of the imaginal disks are called *homeotic* genes: two of the best-studied are located in clusters: the antennapedia and bithorax gene complexes. Mutations in the clusters produce profound body alterations. For example, in antennapedia mutants, leg development occurs next to the eye in the place of an antenna; in ultrabithorax mutants, extra wings and a third thoracic

segment take the place of the first abdominal segment (Figure 11-28c). Both of these mutations suggest that the role of the normal antennapedia or ultrabithorax proteins is to suppress a particular pathway of development and select or allow another (e.g., a thoracic segment is suppressed to "allow" an abdominal segment to develop). The antennapedia and bithorax gene products are formed during the first 3 h of embryogenesis (early in gastrulation) and persist through pupation. Their pattern of expression, like the cascade of factors before them, is the result of combinations of gene action (see Figure 11-27).

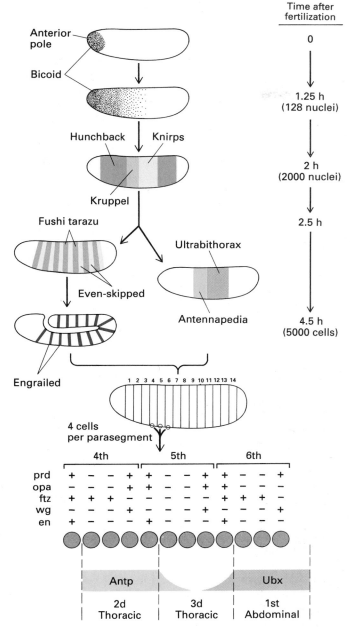

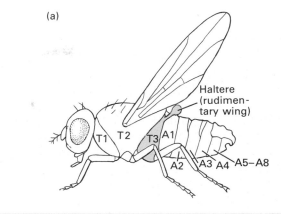

(a)

▲ **Figure 11-27** Sequential expression of various genes during early development of *Drosophila* embryo and localization of their gene products within the embryo. Products of various genes were located by in situ hybridization to labeled RNA probes and by antibody staining. Bicoid mRNA is at the anterior pole of the egg, but as it is translated into protein, it diffuses to form a gradient. In most cases, an mRNA and its corresponding protein are seen in the same regions of the embryo. See Table 11-4 for classification of genes. Cellularization occurs after 2.5 h and gastrulation (including folding of embryo) occurs at about 4.5 h. Diagram at the bottom shows the presence (+) and absence (−) of the gene products indicated (*left*), which define the identity of each of four cells in the fourth, fifth, and sixth parasegments. The homeotic gene antennapedia (*Antp*) dominates development of the fourth parasegment and Ultrabithorax (*Ubx*) that of the sixth. The balance of the proteins encoded by these two genes leads to the transition from thorax to abdomen. The third segment can be transformed into a second thoracic segment if bithorax is mutant.

▲ **Figure 11-28** (a) A drawing of a normal fly. Notice the appearance of the third thoracic segment (T3) and the haltere, a rudimentary wing. A1, A2, etc., identify the abdominal segments. (b) A normal fly with a single set of wings. (c) A fly that is homozygous for three mutant alleles (*bx, abx,* and *pbx*) that collectively transform T3 into a structure similar to T2. A second, fully developed set of wings emerges from this transformed segment. *From E. B. Lewis, 1978,* Nature *276:565. Photographs courtesy of E. B. Lewis. Reprinted by permission from* Nature. *Copyright 1978 Macmillan Journals Limited.*

As described earlier, the sequences of the antennapedia and ultrabithorax proteins led to the original discovery of homeoboxes, the conserved amino acid sequences including the presumed helix-turn-helix DNA-binding domains (see Figure 11-11 and Table 11-3). Many of the proteins in the cascade leading to embryonic segmentation in *Drosophila* also contain homeoboxes. These include bicoid protein; the products of the pair-rule genes fushi tarazu, paired, and even-skipped; and the protein encoded by the segment-polarity gene engrailed. The details of how these homeobox proteins bind to various genes are being worked out at present. It appears that several different proteins can recognize the same DNA-binding sites with some exerting positive and some negative effects on transcription. All three of the major gap genes (knirps, hunch-back, and Krüppel) encode proteins with probable zinc-finger domains. Each of these proteins has been demonstrated to bind in site-specific fashion to particular gene sequences. Many different mutations in each of these regulatory protein genes are known, and in each case the mutations upset expression of certain other members of the group of early developmental genes.

Although the detailed interrelationship of all of these interacting factors with one another is by no means thoroughly understood, the general conclusion is clear: beginning in the fertilized egg, a cascade of transcription factors is the basis for development. The conclusions that can be drawn from this work with *Drosophila* extend far beyond this organism. Not only do homeobox genes exist in all animals, but some of these genes are known to be expressed mainly early in mouse and human development. Fortunately for those who study developmental biology, the proteins encoded by the insect genes and the nucleic acid sequences of the genes have remained sufficiently unchanged through evolution to allow isolation of the homeobox-containing proteins from any animal. A link between the early insect genes and the genes that affect vertebrate development seems undeniable.

Many Different Signals Can Affect Eukaryotic Transcription Factors

The early stages of *Drosophila* embryogenesis are unusual in that many developmental decisions occur within one cell *before* cellularization. In such cases transcriptional factors as well as small molecules are free to diffuse from nucleus to nucleus and there to interact with one another. In vertebrate development and in adult life, information for regulatory changes comes to individual cells (or many similar cells in a tissue) from outside. These signals from outside can change the amount and types of active transcription factors possessed by a cell and profoundly influence growth as well as development. Here we want to focus on how such external signals affect specific transcription factors.

Heat-shock Genes   Probably all cells—bacteria, yeast, *Drosophila,* and human cells have been studied—respond to elevated temperatures by momentarily suspending translation of most mRNAs and transcription of most genes and turning to the transcription of a limited set of genes encoding proteins that presumably increase survival at higher temperatures. Little is actually known of the function of most of these proteins. However, the amino acid sequence conservation among species of some of these proteins is the highest for any known protein. For example, Hsp70 (70-kDa heat-shock protein) from *E. coli* is 50 percent identical with the human protein, and the *Drosophila* protein is 85 percent identical to that from humans. Different temperatures are required to trigger the heat-shock response in organisms that live under different conditions. For example, in *Drosophila* 30°C triggers the heat-shock response, whereas 39°–40°C is required for human cells to respond.

A major event in the heat-shock response in bacterial cells is a switch in sigma (σ) factors that help RNA polymerase choose transcriptional start sites. After bacteria are exposed to elevated temperatures, σ^{32} takes the place of the normal σ^{70}. A counterpart of this direct mechanism of transcriptional change has been identified in heat-shocked cells from yeast, *Drosophila,* and humans. However, in eukaryotes the regulating factors bind to upstream activating sequences in heat-shock genes and not directly to the polymerase. The *Drosophila* heat-shock DNA-binding factor has been purified and can exist as an inactive, nonbinding protein or an active protein capable of DNA binding depending on the prior temperature history of the cells (Figure 11-29). Furthermore, the binding protein within cells can be cycled back and forth between the two forms with no new protein formation being required. The reversible modification of *Drosophila* heat-shock factor (HSF) is a good example of posttranslational changes in a transcription factor; current work is aimed at determining the nature of this reversible change. For unknown reasons, some other noxious stimuli, such as oxygen deprivation and treatment with dinitrophenol (an oxidative phosphorylation inhibitor), also induce the 'heat-shock" response.

Although the 110-kDa heat-shock factor has been shown to activate transcription of several heat-shock genes from *Drosophila,* how it contacts and affects the transcriptional machinery is still not known. Increased transcription of the heat-shock genes, which can be activated within a minute or less, requires an upstream binding element. The heat-shock genes appear to be poised to immediately respond to an external signal, perhaps because the chromatin structure in the region of these genes is "primed" for transcription. For example, in nuclei of unshocked cells, the *hsp70* gene, which encodes the 70-kDa heat-shock protein, is subject to enzymatic cleavage by a restriction enzyme and exonuclease up to the site at

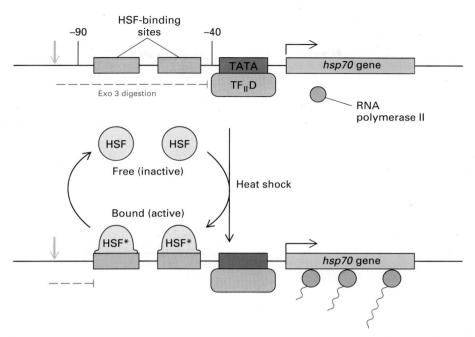

▲ **Figure 11-29** Activation of *Drosophila* heat-shock gene *hsp70*. At normal temperatures (below 30°C for flies) transcription of heat-shock genes is minimal, but immediately after heat shock, transcription is high. A protein called heat-shock factor (HSF) exists free in cells before heat shock; after heat shock, HSF is changed to an active form (HSF*) that binds to specific DNA sites just upstream of the gene, resulting in increased transcription. This activation-deactivation is reversible. Even before the cells or animal undergo heat shock, the TATA binding site in the promoter is occupied by a protein (presumably the TATA factor) as demonstrated by an experiment with exonuclease 3. A site upstream of the gene (red arrow) can be incised in whole nuclei with either a restriction endonuclease (e.g., *Xho*I) or by brief DNase treatment. *E. coli* exonuclease 3 (Exo 3) can then digest the upstream region until it is blocked by the TATA-binding protein in non-heat-shocked cells or by the bound heat-shock factor in heat-shocked cells. It is possible that an RNA polymerase II molecule is bound to the gene but restrained from transcription until heat shock. Polymerases enter actively after heat shock. [See C. Wu et al., 1987, *Science* **238**:1247; V. Zimarino and C. Wu, 1987, *Nature* **327**:727; and A. E. Rougvie and J. T. Lis, 1988, *Cell* **54**:795.]

which the TATA factor (TF$_{II}$D) is known to bind during in vitro transcription (see Figure 11-29). It is even possible that an RNA polymerase is attached to the *hsp70* gene ready to produce a transcript as soon as the heat-shock factor achieves activated form. What is certain is that the HSF is quickly modified after heat shock and that heat-shock genes are quickly activated.

Steroid Hormone-dependent Genes The steroid hormones and their receptors represent one of the best-understood cases of external signals that affect transcription. Because steroid hormones are soluble in lipid membranes, they can diffuse into cells. They affect transcription by binding to specific intracellular receptors that are site-specific DNA-binding molecules.

The first steroid receptor to be highly purified and shown to be a DNA-binding protein was specific for glucocorticoid hormones produced mainly in the adrenal gland. The interaction of this receptor with DNA was first studied using mouse mammary tumor virus (MMTV) DNA, transcription of which is increased in the presence of glucocorticoids. Binding of receptor-hormone complexes to a 1.2-kb fragment of the MMTV genome that contains a hormone-dependent RNA initiation site has been visualized in the electron microscope (Figure 11-30). DNase footprint experiments have shown that receptor-hormone complexes bind at four to six sites in this fragment in a region stretching from 84–305 nucleotides upstream from the start site for RNA synthesis.

With purified glucocorticoid receptor in hand, the techniques of molecular genetics allowed identification of and sequencing of first a cDNA and then genomic clones encoding the receptor. Cloning of the receptors for estrogens and progesterone soon followed. In addition clones are now available for genes encoding receptors that bind thyroid hormone, another small lipid-soluble molecule that enters cells, vitamin D, and retinoic acid, a metabolic derivative of vitamin D that has powerful effects on limb

(a)

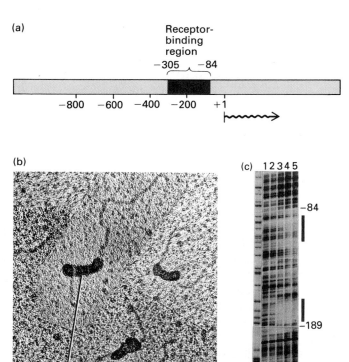

(b)

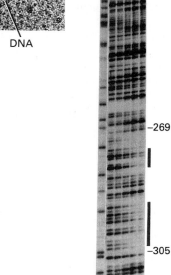

Receptor-hormone complex   DNA

(c) 1 2 3 4 5

−84

−189

−269

−305

Response elements

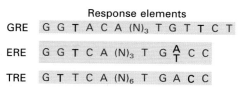

▲ **Figure 11-31** Consensus sequences of the DNA sites (response elements) that bind receptors for glucocorticoid (GRE), estrogen (ERE), and thyroid (TRE) hormones. The bases indicated in red contain inverted repeats around three to six central nucleotides (N). [See R. M. Evans, 1988, *Science* 240:889.]

▲ **Figure 11-30** Binding of the glucocorticoid receptor to mouse mammary tumor virus (MMTV) promoter sequences. (a) The region of the MMTV genome containing the binding sites for the glucocorticoid receptor and the start site (+1) for hormone-dependent RNA synthesis is shown. (b) Electron micrographs of glucocorticoid receptor-hormone complexes bound to the DNA diagrammed in (a). The receptor protein probably forms tetramers. (c) A DNase footprint of the promoter region of the MMTV genome in the presence and absence of glucocorticoid receptor protein, which protects four regions (red bars) from digestion. Increasing amounts of glucocorticoid receptor were added in lanes 2–5; none was present in lane 1. The numbers indicate positions of nucleotides upstream from the start site. [See F. Payvar et al., 1983, *Cell* 35:391.] *Photographs courtesy of K. Yamamoto.*

bud development in embryos and on skin renewal in adult mammals. The characteristic DNA-binding sites for all the major steroid hormone receptors are known (Figure 11-31). These sites, or *response elements,* are distinct for different hormones, are about 15 bases long, and exhibit dyad symmetry (i.e., they are imperfect inverted repeats). This structure suggests that receptors may bind as dimers.

Comparison of the amino acid sequences of various hormone receptors deduced from the nucleotide sequences of the genes encoding them, shows that all these receptors belong to a related protein family. Furthermore, the important functional domains of the various proteins have been identified by assay of deletion and point mutants of the proteins. A remarkable conservation is obvious in both the amino acid sequences and the different functional regions among the hormone receptors. The functional domains of the receptors have been determined by cotransfecting a test gene that requires a particular hormone together with a plasmid that encodes all or part of the hormone receptor. Assays of the hormone-dependent transcription of the test construct then permits the function of various regions of the receptor to be identified (see Figure 11-15).

All the hormone receptors have an amino-terminal domain of variable length from about 100 to 500 amino acids that contains sequences required for full receptor activity (Figure 11-32). The amino acid sequence of this region is not conserved among different receptors. A middle "core" section of about 68 amino acids is highly conserved with over 40 percent amino acid identity in all the proteins. In this core there is a region rich in basic amino acids and one rich in cysteine residues. The cysteine-rich region is thought to exist as a zinc-finger domain. The core is responsible for the DNA-binding activity of the various receptor proteins. The domain that binds effector molecules (hormone, vitamin, etc.) lies near the carboxyl-terminal end of the receptors.

Direct binding of effector molecules has been demonstrated for most receptor proteins, and studies with recombinant proteins have provided decisive evidence about the functions of the core and effector domains. For example, if the core domain of the human estrogen recep-

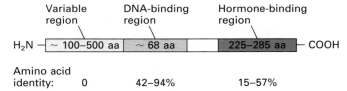

▲ **Figure 11-32** General design of protein receptors for all the steroid hormones, thyroid hormone, vitamin D, and retinoic acid. The DNA-binding domain and hormone-binding domain cooperate in hormone-dependent transcriptional increases. [See R. M. Evans, 1988, *Science* **240**:889.]

tor is replaced with the similar region of the glucocorticoid receptor, the recombinant protein binds to glucocorticoid response elements in DNA. However, stimulation of transcription by the recombinant receptor depends on estrogen treatment of the cells in which the recombinant receptor protein is made. Not only are the two major domains—the core that determines which DNA response element a receptor recognizes and the hormone-binding domain—well identified in all the receptor proteins, but it is now established that purified receptor proteins bind to their DNA response elements as dimers. It is not yet known, however, how binding of the hormone changes the structure of the receptor so that the hormone-receptor complex binds to DNA. Crystallography of the zinc-finger motif is needed to solve the atomic structure of these protein domains and prove how DNA contacts are made. Experiments with antibodies to detect the glucocorticoid receptor show that before hormone treatment the receptor is largely cytoplasmic and is bound in a large protein aggregate (Figure 11-33). After hormone treatment the receptor is found mainly in the nucleus. Formation of an active nuclear transcription factor therefore requires the hormone.

A final note about the steroid hormone receptors is important: the various receptors are not present in all cells in the same concentrations. For example, estrogen and testosterone (major male hormone) receptors are not present in equal amounts either in males and females or in all cells of any individual. Even receptors for thyroid hormone, which probably acts on all cells, are not equally distributed. Vertebrates have at least two different thyroid receptors, which are not equally represented in all cell types. Finally, there are many proteins in the steroid receptor family for which cDNA clones have been found but for which no ligand has yet been assigned. The precise role of all these receptors and the basis for their different distribution is an important subject of continuing research.

Genes Regulated by Cell-surface Receptor-Ligand Interactions

We just considered how steroid hormones that enter cells can trigger gene activity. Many genes are regulated not by a signaling molecule that en-

ters the cells but by molecules that bind to specific receptors on the surface of cells. Interaction between cell-surface receptors and their ligands can be followed by a cascade of intracellular events including variations in the intracellular levels of so-called second messengers (diacylglycerol, Ca^{2+}, cyclic nucleotides). The second messengers in turn often lead to changes in protein phosphorylation through the action of cyclic AMP, cyclic GMP, calcium-activated protein kinases, or protein kinase C, which is activated by diacylglycerol.

Many of the responses to binding of ligands to cell-surface receptors are cytoplasmic and do not involve immediate gene activation in the nucleus. Some receptor-ligand interactions, however, are known to cause prompt nuclear transcriptional activation of a specific and limited set of genes. For example, one proto-oncogene, c-*fos*, is known to be activated in some cell types by elevation of almost every one of the known second messengers and also by at least two growth factors, platelet-derived growth factor and epidermal growth factor. However, progress has been slow in determining exactly how such activation is achieved. In a few cases, the transcriptional proteins that respond to cell-surface signals have been characterized.

One of the clearest examples of activation of a pre-existing inactive transcription factor following a cell-surface interaction is the protein NF-κB, which was originally detected because it stimulates the transcription of genes encoding immunoglobulins of the kappa class in B lymphocytes. The binding site for NF-κB in the kappa

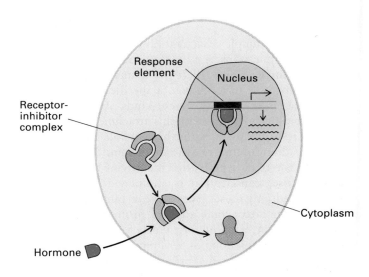

▲ **Figure 11-33** Model of activation of genes by lipid-soluble hormone. The hormone, which is soluble in the lipids of the cell membrane, enters the cell by diffusion. The hormone receptor is complexed with an inhibitor, which is displaced after hormone-receptor interaction. A dimer of the receptor protein bound to the hormone enters the nucleus and binds to DNA response elements, thus activating transcription of the dependent gene.

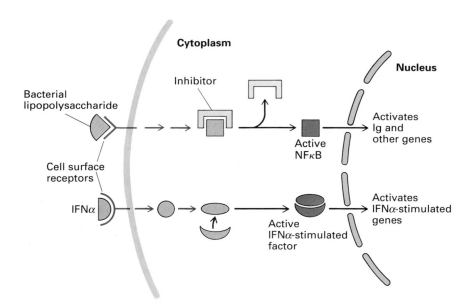

◀ **Figure 11-34** Model of transcription-factor activation by cell-surface interaction. Inactive cytoplasmic forms of transcription factors can be activated by removal of an inhibitor as in the case of NF-κB (*top*) or by association of two proteins neither of which is active by itself as in the case of interferon α (IFNα)-stimulated factor (*bottom*). The in vivo trigger for activation of NF-κB is thought to involve phosphorylation. [See P. A. Baeuerle and D. Baltimore, 1988, *Science* **242**:540; and D. E. Levy et al., 1989, *Genes and Development* **3**:1362.]

gene is well defined, providing an assay for the presence of the active factor. This factor exists in the cytoplasm of lymphocytes complexed with an inhibitor (Figure 11-34). Treatment of the isolated complex in vitro with mild denaturing conditions dissociates the complex, thus freeing NF-κB to bind to its DNA site. Release of active NF-κB in cells is now known to occur after a variety of stimuli including treating cells with bacterial lipopolysaccharide and extracellular polypeptides as well as chemical agents (e.g., phobol esters) that stimulate intracellular phosphokinases. Thus a phosphorylation event triggered by many possible stimuli may account for NF-κB conversion to the active state.

A second example of activation of a preexisting inactive cytoplasmic transcription factor involves a different principle. Interferon α (IFNα) is a small (~150 amino acid) polypeptide that binds to a specific cell-surface receptor present on many different kinds of cells. One important biological outcome of this attachment is that cells become resistant to infection by a whole variety of viruses. In addition, the growth rate of some cells is slowed by interferon treatment. By first identifying their mRNAs, genes whose transcription is stimulated by interferon were cloned. All these genes have a binding site for a transcription factor that appears promptly after interferon treatment. This factor, which forms without new protein synthesis, is composed of more than one polypeptide. The inactive proteins (probably two) exist free in the cell cytoplasm, and no evidence of an inhibitor exists. After IFNα attaches to its cell-surface receptor, one of the proteins is changed within a minute or less; the two can now combine (Figure 11-34). The active factor is then translocated to the cell nucleus, to stimulate transcription only of genes with a binding site for the protein. The interferon-activated transcription factor may serve as a general model for highly specific transcription that re-

quires cell-surface receptor-ligand interaction. It should soon be possible to determine how the inactive preexisting protein is modified to become a gene activator.

A number of different polypeptides are known to trigger synthesis of specific sets of proteins. It is possible that differential expression of many genes during embryonic development and differentiation of adult cells involve mechanisms similar to those demonstrated for NF-κB or for the interferon-activated transcription factor.

Terminal Differentiation Depends at Least in Part on Cell-specific Transcription Factors

The final step in cell differentiation, in both animals and plants results in a specialization that is characterized by the production of particular proteins (e.g., silk fibroins in the silk gland of silkworms, contractile proteins in muscle cells, serum proteins in liver cells, and globins in red blood cell precursors). The differential transcription of mRNAs encoding proteins in liver cells and of globins in erythroblasts are illustrated in Figures 11-6 and 11-7.

Assuming that transcriptional control by DNA-binding proteins is a primary feature of differentiated cells, then knowledge of the nature of these factors and their specificity is crucial to understanding gene control in such cells. Two extreme possibilities can be envisioned. If a limited number of transcription factors are distributed more-or-less uniformly in all cells, then the specialized transcription patterns typical of differentiated cells would depend on cell-specific modifications of the general factors. On the other hand, if many different transcription factors have a limited cell distribution, this would be instrumental in developmental specialization. In this section, we describe the regulation of gene expression in hepatocytes and muscle cells, both of which contain some factors that are widely distributed and others that are quite limited in their distribution. The factors that help control production of the specialized immune-system proteins in lymphocytes are discussed further in Chapter 25.

Hepatocyte-specific Transcription More than a half-dozen genes whose products are formed only or mainly in hepatocytes, the major cell type in the liver, have been dissected by the techniques of deletion and site-specific mutagenesis. Based on the results from three types of studies—plasmid transfection, DNA protein-binding tests, and in vitro transcription assays—several generalizations have been drawn. First, liver-specific genes carried on plasmids can be expressed in hepatomas but not in other cultured cells. Second, multiple protein-binding sites, some close to and others distant from the start sites for DNA transcription, are required for maximal expression of liver-specific genes. Finally, at least four transcription factors important for hepatocyte-specific gene expression are present in hepatocytes and a few other cell types but are not present in most cells.

A brief description of the protein-binding sites in one gene expressed mainly in hepatocytes and the proteins that bind these required sites will illustrate these generalizations. Transthyretin (TTR)—a serum protein that binds thyroid hormone (T_3)—is made in adults only in hepatocytes and choroid plexus cells, which secrete cerebrospinal fluid and its constituent proteins. There are nine definitely identified protein-binding sites in the regulatory regions of the TTR gene. A general transcription factor called AP1, which is a leucine-zipper protein originally detected as a factor required for SV40 virus transcription, binds one site. The other eight sites bind four proteins, all of which are present in liver cell extracts at much higher concentrations than in any other cell type (Figure 11-35a). Different genes such as serum albumin or α1 antitrypsin that encode prominent hepatocyte products have different arrangements of protein-binding sites and use overlapping but not identical sets of factors. Thus there is no single arrangement of sites that dictates liver-specific gene expression. One of the transcription factors that acts on several liver-expressed genes is C/EBP, the first leucine-zipper protein to be cloned and sequenced (see Figure 11-14). Another is HNF1 (hepatocyte nuclear factor 1), which is encoded by a gene containing a homeobox region. HNF4 is a protein with a zinc-finger region and HNF3 is not a member of the readily recognized transcription factor groups. By use of cloned probes for these genes, their mRNAs were shown to be limited in cell distribution, being prominent in hepatocytes and a few other cell types. C/EBP is present in fat cells, intestinal cells, and some brain cells, whereas HNF4 is present in intestinal and kidney cells but not brain cells. HNF1 is present mainly in hepatocytes among adult cells.

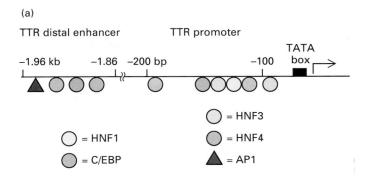

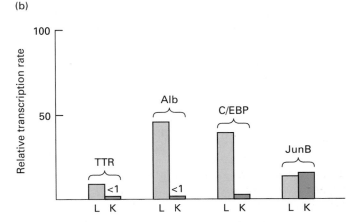

▲ **Figure 11-35** (a) Binding sites for regulatory proteins that control transcription of mouse transthyretin (TTR) gene, which is expressed primarily in liver cells. HNF = hepatocyte nuclear factor (b) Bar graph showing relative transcription of genes encoding TTR, albumin (Alb), C/EBP (one of the TTR transcription factors), and JunB (a widely distributed transcription factor similar to AP1) in liver (light blue) and kidney cells (dark blue). Synthesis of methionyl tRNA was used as a standard (100 units) in each cell type. The liver-specific transcription of TTR, albumin, and C/EPB is obvious. (See R. Costa, D. R. Grayson, and J. E. Darnell, 1989, *Mol. Cell Biol.* 9:1415; and K. Xanthopoulus et al., 1989, *Proc. Nat'l Acad. Sci. USA*, 86:4117.]

The gene encoding C/EBP (as well as that for HNF1 and 3) was found to be transcribed in only a few cell types, one of which is of course hepatocytes (Figure 11-35b). Thus cell-specific transcriptional control of transcription factors in adults is clearly demonstrated.

Exactly how many cell-enriched or cell-limited transcription factors exist cannot be settled at present, but judging from the early results available with liver cells, red blood cells, and a few other cell types, transcription factors with a limited distribution could number in the hundreds. How these proteins act in groups and why particular factors act on a particular gene in a particular cell are not yet known. It is undoubtedly true that different combinations of transcription factors operate in different tissues. For example, transthyretin (TTR) is formed in liver cells and in choroid plexus cells, but most other genes expressed mainly in the liver are *not* made in the choroid plexus. Thus, liver and choroid plexus cells may share some but not all factors. In fact, HNF1, HNF3, and C/EBP are not present in choroid plexus cells, although HNF4 is. Since transthyretin mRNA is formed at a very high rate in choroid plexus cells, there must be other factors in these cells that help activate transcription of the TTR gene in that cell type. If this is true, then combinations of transcription factors apparently account for differential transcription in various types of cells. The differential cellular distribution of transcription factors in adult cells is obviously based to some degree on differential transcription of the genes encoding the factors. This suggests that other regulatory factors acting earlier in development must exist. Perhaps factors similar to those formed near the end of early *Drosophila* development, direct the formation of the final set of cell-specific factors in vertebrates. It seems a reasonable hope that the earliest-acting factors and those important in organogenesis soon can be related in one animal.

Muscle Cell Determination and Differentiation

Like hepatocytes, adult differentiated muscle cells have cell-limited or cell-enriched transcription factors. At least one of these factors may act at an early stage of differentiation before the major muscle-specific proteins are formed. The discovery of the gene(s) conferring this capacity to "trigger" muscle differentiation depended on the finding that cultured mouse fibroblast cells called 10T½ cells have the ability occasionally to produce colonies of differentiated muscle cells. These colonies can be identified microscopically or by electrophoretic analysis to reveal muscle-specific proteins (Figure 11-36). Although 10T½ cells have the ability to become muscle cells, something normally prevents them from doing so at high frequency.

When DNA from normal *myoblasts* (embryonic muscle precursors) of human origin is introduced into 10T½ cells, colonies capable of turning into muscle cells with great regularity can be selected. From such DNA transfer experiments, three human genes capable of inducing 10T½ cells to regularly form muscle colonies have been identified. Two of these genes, one encoding a protein called *MyoD-1* and one encoding *myogenin*, have been prepared as recombinant cDNA clones containing a viral enhancer sequence; because of the enhancer, large amounts of MyoD-1 and myogenin are produced when these recombinant constructs are introduced into certain cultured cells. When either overexpressing recombinant is introduced into several different cell types including 10T½, the cells stop growing and produce muscle proteins (Figure 11-37a). Thus in large amounts each of these proteins can trigger the muscle phenotype. Both MyoD-1 and myogenin have amino acid sequences similar to those in proteins that form coil-coil interactions (particularly the helix-loop-helix groups) and can bind DNA. The physiologic target sites of MyoD-1 and myogenin, however, are not yet known.

The third muscle-inducing gene identified in human DNA encodes a protein called *MyD-1*. When this gene is transferred in its natural form (without a virus promoter) into 10T½ cells, it causes them to become myogenic cells. These transformed cells produce only small amounts of MyoD-1 and myogenin mRNA, and they do not become fully differentiated muscle cells until growth is limited. Thus this third gene appears to act early in the pathway of muscle determination, priming the cells for muscle development (Figure 11-37b). Under normal conditions (i.e., in the absence of abnormally high synthesis of MyoD-1 and myogenin), further differentiation into mature muscle cells only occurs after growth limitation. According to the model in Figure 11-37c, expression of the muscle-triggering gene encoding MyD-1 produces primed myogenic cells, which under certain conditions

(a)

(b)

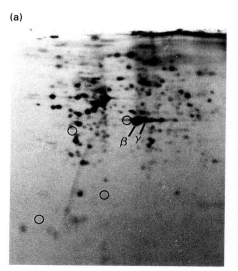
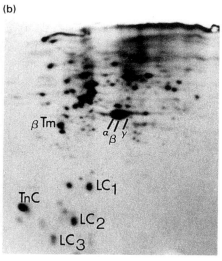

◀ **Figure 11-36** Electrophoretic analysis of proteins from fully grown cultures of mouse 10T½ cells and from similar cultures of cells that have been transformed with DNA from human myoblasts. Under these non-growing conditions, the 10T½ cells (a) produce no muscle-specific proteins *(open circles)*, whereas the fully grown transformed cells (b) differentiate to produce β-tropomyosin (βTm), myosin light chains (LC1, LC2, LC3), and troponin C (TnC), which are all muscle-specific proteins. *From D. F. Pinney et al., 1988,* Cell *53:781; photographs courtesy of C. Emerson.*

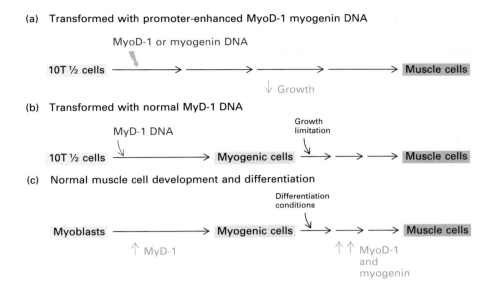

(a) Transformed with promoter-enhanced MyoD-1 myogenin DNA

(b) Transformed with normal MyD-1 DNA

(c) Normal muscle cell development and differentiation

◀ **Figure 11-37** (a) and (b) Summary of results on differentiation of transformed mouse 10T½ cells into muscle cells. Synthesis of large amounts of MyoD-1 or myogenin causes cells to cease growth and form muscle cells. Cells transfected with MyD-1 DNA have to be placed in a growth-deficient medium to differentiate. (c) Proposed model of normal muscle cell differentiation. MyD-1 is thought to act early and help change myoblasts into primed myogenic cells, which produce only limited amounts of MyoD-1 and myogenin; when growth ceases these cells can express larger amounts of MyoD-1 and myogenin, leading to activation of transcription of muscle-specific genes. [See D. P. Pinney et al., 1988, *Cell* 53:781.]

synthesize large amounts of MyoD-1 and myogenin. These proteins, in turn, help activate expression of the genes encoding muscle-specific proteins such as β-tropomyosin, myosin light chains, and troponin C.

Control of Regulatory-protein Activity and Possible Effects of Chromatin Structure on Gene Activity

Our discussion of transcriptional control has focused on regulatory sites in DNA, the cognate proteins that specifically recognize these sites, and how various signals can change the repertoire of transcription factors a cell possesses. We have emphasized that a particular regulatory protein may not be synthesized in all cells at all times: the α1 protein from the *MAT* locus in *Saccharomyces cerevisiae* is not made in a-type cells; many of the factors in early *Drosophila* development come and go in individual cells as development proceeds; and C/EBP, one of the factors necessary for hepatocyte-specific expression, is not produced in many cells of the body but is produced in hepatocytes. So we can conclude that cell-specific transcriptional regulation of transcription-factor genes is common.

Some bacterial regulatory proteins that are under transcriptional control also can be converted from active to inactive form by chemical modification such as phosphorylation, methylation, or chain cleavage. It is widely believed, although not yet conclusively demonstrated, that the same type of modifications can govern regulatory-protein activity in animal cells as well. Such post-translational changes and the possible reassortment of dimeric proteins that belong to the same family strongly suggest that there is ample opportunity for changing the

repertoire of transcription factors in different cell types. These possibilities raise additional questions. Suppose, for example, that a regulatory protein is present in inactive form in many cells but is activated only in one type of cell, say, lymphocytes; in this case, we have not understood cell-specific regulation until we find why the modification system is present in lymphocytes but not in other cell types. The relative importance of modification of preexisting proteins versus the regulated synthesis of transcription factors in particular cells is not yet known. It is reasonable but unproven that events that must be regulated promptly would depend on preexisting proteins, whereas events that act in stable differentiation might more often rely on transcriptional control of the regulatory genes.

With respect to the mechanism of action of the transcriptional regulatory factors, a number of very important unanswered questions also remain. We have noted already that positive-acting proteins probably contact either TF$_{II}$D or RNA polymerase, or both, but the details of these contacts are uncertain. This problem is compounded when we consider that, most often, transcriptional stimulation is not the product of a single activating factor but an interaction of two or more factors. One simple example is GAL80, which exerts a negative effect on GAL4, the positive activator that binds to DNA (see Figure 11-19); release of GAL80 and relief of the negative effect is brought about by galactose. But nothing is yet known of the possible interactions of the multiple proteins that are bound to animal cell genes.

Another potentially important issue concerning gene activation is whether regulatory molecules have access to genes at all times. Regulatory proteins may only have access to their cognate binding sites during limited periods of time (e.g., immediately after replication) because of the compacting of DNA in the form of tightly coiled chromatin. All DNA appears to be associated in nucleosomal structures. However, active genes are more easily digested by DNAse (are "more loosely packed") than inactive genes (see Figure 9-20). Considerable evidence also shows that regions of DNA that are actively engaged in transcription lack 5-methyl cytidine (mC) residues, which, at least in vertebrates and plants, constitute 1 to 5 percent of the total DNA.

Almost all of the methylated cytidine residues exist in pmCpG sequences. If plasmids are methylated at pCpG sites before transfection, genes on the plasmids are inactive. Likewise, genes that become methylated after transfection also are inactive. Furthermore, soon after replication of a methylated DNA region, methyl groups are added to the new DNA strands. Such a reinforcement of the inactive, methylated state may be brought about according to the following scheme:

$$
\begin{array}{lll}
\text{Old strand} & (1) & \text{pmCpG} \\
\text{New strand} & (2) & \text{GpCp}
\end{array}
$$

$$+CH_3 \downarrow {\scriptstyle\text{DNA}\atop\scriptstyle\text{methylases}}$$

$$
\begin{array}{ll}
(1) & \text{pmCpG} \\
(2) & \text{GpCmp}
\end{array}
$$

The old strand that is methylated in the pCpG dinucleotide retains its methyl group, and this may be signal for methylation of the pCpG in the new strand. It is clear that a close *correlation* exists between transcriptional inactivation and methylation. Although nothing is yet known about how methylation is prevented around active genes, it is widely conjectured that when transcriptionally active DNA is replicated, it is then shielded from DNA methylases by associated proteins. As yet no convincing experiment has shown why or if methylation at particular cytidine residues prevents active transcription complexes from forming.

Thus, the physical or biochemical meaning of loose chromatin structure or DNA methylation is still not clear. Nevertheless, large loops of DNA in which transcription is active can be visualized projecting out from denser underlying tightly packed chromatin. A particularly suitable case for such visualization is the Balbiani rings of *Chironomous tentans* (Figure 11-38). This region produces a large (75S) mRNA that encodes a salivary gland protein. When the chromatin in such regions is sectioned and examined by serial reconstruction, loops of DNA being transcribed are visible. The bases of the loops are not contiguous; that is, a loop does not originate and return to the same site. The implication is that such loops are not demarcated by a single protein structure at their beginning and end.

Recent results with the β-globin domains in human DNA are possibly relevant to the demarcation of "transcription loops." Transgenic mice were prepared with large segments of human DNA containing a globin gene; in some cases, the DNA segment contained distant sites that are DNase hypersensitive; these are distinct from the hypersensitive sites located close to active β-globin genes in red blood cell precursors (Figure 11-39a). Transgenic mice containing the distant DNA segments produced high amounts of human β-globin and did so only in red blood cell precursors. In contrast, mice that contained only hypersensitive sites near to the β-globin gene did not regularly produce human β-globin (Figure 11-39b). Thus sites distant from a gene somehow act as domain-marking sites and ensure proper transcription in the correct cell type. As yet, the proteins that bind to these sites are unknown, and what role they play in transcription is un-

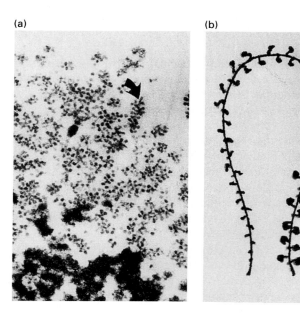

(a) (b)

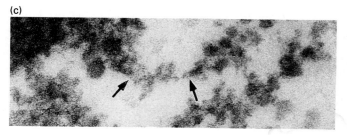

(c)

▶ **Figure 11-38** Electron micrographs of transcription loops in the Balbiani rings of *Chironomous tentans* salivary glands. (a) Active transcription loops of 75S-RNA with characteristic dense granules of ribonucleoprotein extend upward from the condensed, nontranscribed chromatin. (b) Reconstruction of a single DNA transcription loop from serial thin sections reveals the gradual increase in size of the associated hnRNPs. (c) A single high-power view of one thin section shows the thin chromatin thread (*arrows*) near the origin of transcription (defined as such because no large dense hnRNP particles are evident). No return of the distal end of a loop is nearby, indicating that the beginning and end of a transcription loop join the condensed chromatin at different sites separated by some distance. *From C. Erisson et al., 1989, Cell 56:631; photographs courtesy of B. Daneholt.*

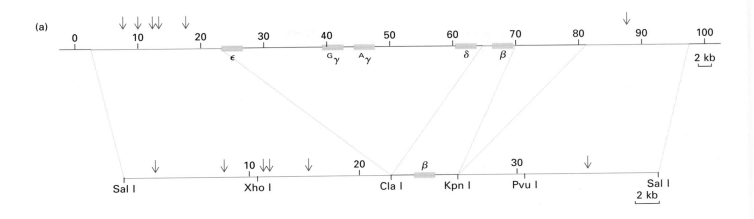

(b)

| Human DNA introduced | $^A\gamma$- or β-globin expression in mouse | | | |
	Fetal	Adult	Frequency	Level
$^A\gamma$ (44–47)	+	–	⅓	Low
β (64–70)	–	+	⅓	Low
β and hypersensitive regions (0–22, 64–70, 82–94)	–	+	7/7	High

▲ **Figure 11-39** Expression of human globin genes in the transgenic mouse. (a) Diagram of the DNA of the β-like globin family which contains five genes (green boxes): ϵ-globin is the earliest to be expressed; $^A\gamma$- and $^G\gamma$-globins are produced by to midfetal life; and β-globin appears late in gestation and is the major adult form. Hypersensitive sites that are present in erythroblastic cells are indicated by arrows. (b) The frequency of transgenic mouse strains that produce various human DNA segments is presented. *Adapted from F. Grosveld et al., 1987, Cell 51:975.*

clear. Important questions about any proteins that bind to these regions are whether they gain access to their sites in DNA only at special stages of development and whether they act positively in transcription to potentially prevent the inactivation of transcription that comes from packing DNA into tightly coiled chromatin.

Control of Transcriptional Termination

In bacteria several mechanisms known to control the termination of transcription by RNA polymerase play a role in gene regulation. Control of termination also occurs in eukaryotic cells, although control of initiation appears to be a more important means for regulating gene expression.

Many cultured mammalian cells produce approximately equal amounts of full-length RNA transcripts and short, 5′ capped RNAs of only a few hundred nucleotides in length. Whether this is simply a property of the inefficiency of some basic aspect of the transcription machinery or whether it is part of a normal regulatory process is not known. There are, however, a few cases in which such premature termination is known to be regulated and thus to serve as a point of gene control. For example, continuously growing white blood cell precursors produce an mRNA encoded by c-*myc*, a cellular proto-oncogene. (This type of gene, which is thought to have a role in cell-growth control, is discussed in Chapter 24.) Retinoic acid, a vitamin A derivative, causes these cells to stop growing and to differentiate into granulocytes. Run-on transcriptional analyses of uninduced and induced cells have shown that in growing undifferentiated cells about half of the c-*myc* RNA transcripts formed are full length and presumably lead to mature mRNA. In contrast, in differentiated cells, which contain much less c-*myc* mRNA than growing cells, only 2–3 percent of the c-*myc* transcripts are full length. Although the rate of transcription initiation is similar in both types of cell, premature termination in the differentiated cells leads to a lowered mRNA production. Although it is unclear why premature termination is more frequent in differentiated cells, the sites at which it occurs have been mapped to two stretches containing runs of T residues that are located about 300 nucleotides downstream of the start site. Such T-rich regions have also been shown to cause termination of in vitro transcription by RNA polymerase II but are not present in all genes. Perhaps only in genes with T-rich regions near their transcription start sites can regulation of termination be associated with differential gene expression.

A few other cases of regulated termination of primary transcripts with multiple poly A sites are known to occur during synthesis of some viral mRNAs and of immunoglobulin heavy-chain mRNAs. For example, in adenovirus infection the long major late RNA transcript gives rise to five groups of mRNAs; each group is characterized by a particular poly A site (see Figure 8-36). Transcription also occurs from this region of the genome early in infection at a lower rate, but the primary transcript extends only through the first three of the five poly A sites. So no mRNAs of the last two groups are made early in infection.

The normal point of termination in the transcription of eukaryotic protein-coding genes is downstream of the poly A addition site that has been selected for use. No specific sequence, such as a stretch of T residues, has been implicated in this usual type of termination. Since the primary transcripts of some genes are as long as 200 kb and the cell is not filled with incomplete transcripts of varying lengths, there must be some effective mechanism for preventing accidental termination once a transcript is past a certain length.

Differential Processing of Pre-mRNA

The primary transcripts (pre-mRNAs) from some eukaryotic transcription units contain one or more poly A addition sites and/or one or more sites at which splicing occurs. Posttranscriptional processing of such complex primary transcripts can produce different mRNAs. Differential processing of complex pre-mRNAs in different cells in a regulated fashion would thus provide another mechanism of gene control in eukaryotes. An alternative form of posttranscriptional control entailing an on-off mechanism also is possible. In many cultured cells, for example, more than half of all full-length pre-mRNAs are not processed into finished cellular mRNAs; such nonprocessed primary transcripts have been shown to lack poly A tails. Thus failure to add poly A tails to particular primary transcripts could shut off processing in

(a) Differential processing of complex transcripts

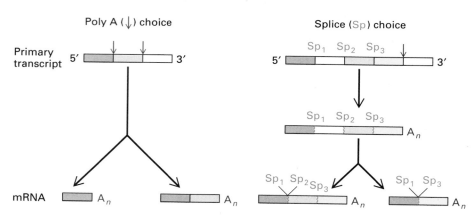

(b) On-off control of processing (simple and complex transcripts)

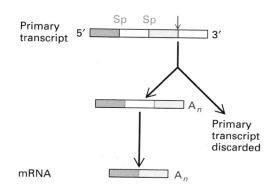

◀ **Figure 11-40** Mechanisms of cell-specific gene control at the level of pre-mRNA processing. The on-off mechanism in (b) has not been experimentally demonstrated in vivo, although in cultured cells many full-length transcripts are not processed.

some cells. Since all pre-mRNAs, except those for histones, have poly A tails, this type of posttranscriptional control could regulate expression of both simple and complex transcription units. These two types of posttranscriptional control are compared in Figure 11-40. To date only the first type—that is, cell-specific differential processing—has been experimentally demonstrated.

Cell-specific Processing of Pre-mRNA Can Occur at Poly A and Splice Sites

Studies on the formation of adenovirus late mRNAs led to the initial discovery that multiple mRNAs can be produced from one primary transcript by differential poly A addition and splicing (see Figure 8-36). Numerous examples of differential processing of mammalian and *Drosophila* pre-mRNAs are now known, and these choices play important roles in cell-specific functions (Table 11-5). In this section, we describe three examples.

Calcitonin-CGRP Certain cells in the thyroid gland and neurons in the brain produce a common primary RNA transcript that encodes two proteins: the hormone calcitonin, which aids the kidney in retaining calcium, and calcitonin gene-related peptide (CGRP), which is located in neurons concerned with taste and in the anterior pituitary gland. Although the complete primary transcript is produced in both types of cell, formation of the final mRNAs is cell-specific, as illustrated in Figure 11-41.

Table 11-5 Examples of differential processing of pre-mRNA

Gene products	Notes about the different products	Basis of differential control	
		Poly A choice	Splicing variation
MAMMALS			
Calcitonin and CGRP (rat)	Two proteins, calcitonin (Ca^{2+}-regulating hormone) and calcitonin gene-related peptide (CGRP), which probably functions in taste, are formed preferentially in thyroid and brain	+	+
Kininogens (cow)	Two prehormones that both yield bradykinin, which controls blood pressure, differ in their protease susceptibility	+	
Immunoglobulin heavy chains (mouse)	These differ at their carboxyl ends in different antibody molecules	+	
Troponin (rat)	Two different muscle proteins appear in different rat skeletal muscles		+
Myosin light chains (rat)*	Two proteins 140 and 180 amino acids long, respectively, appear in different "fast twitch" muscles		+
Preprotachykinin (cow)	Two prehormones that release substance P, a neuropeptide that affects smooth muscle, are formed; the larger prehormone also releases substance K, a neuropeptide of unknown function		+
Fibronectin (rat)	The structural protein has at least six forms		+
DROSOPHILA			
Tropomyosin	Two muscle proteins differ between embryos and adults		+
Myosin heavy chains	Two muscle proteins differ between embryos and adults; a third form is found in pupae and adults	+	
Ultrabithorax	Different gene products that control thoracic segment development are found in early embryos and larvae	+	
Glycinamide ribotide transformylase	This enzyme is encoded by a long mRNA; a shorter mRNA that shares several exons encodes a second polypeptide	+	

*The rat myosin light-chain proteins are processed from two different overlapping primary transcripts with different 5′ ends. The resulting mRNAs also contain different exons within the region of overlap, plus four exons that are similar. An almost-identical situation exists in chicken myosin genes.

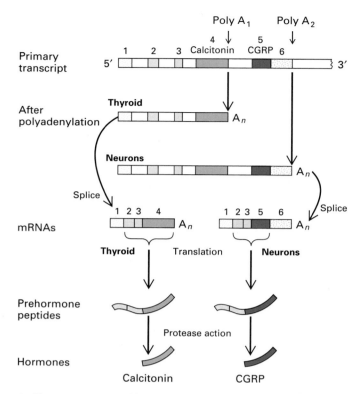

▲ **Figure 11-41** Differential processing of primary transcript from calcitonin gene results in two products, calcitonin in the thyroid gland, and calcitonin gene-related product (CGRP) in certain neurons. Choice of the poly A site and the subsequent splicing choices are coordinated. Coding exons are in solid colors; nontranslated exons are stippled. [See M. G. Rosenfeld et al., 1983, *Nature* 304:129.]

The mechanism by which the poly A and splicing sites are selected in thyroid cells and neurons is not known, although results with some mutated calcitonin genes are suggestive. If the primary transcripts have a site-specific mutation at the splice signal before exon 4, then no mRNAs utilizing the first poly A site (poly A_1 at the end of exon 4) are produced. This result implies that a coordinated choice is made so that the joining of exons 3 and 4 is linked to the utilization of the poly A_1 site. In addition, primary transcripts with a mutation in the poly A_1 site are never spliced between exons 3 and 4, even in thyroid cells. These results suggest that while the primary RNA transcript is still intact, it becomes assembled into spliceosomes in one of two ways that facilitates selection of specific poly A sites and splice sites. Folding that encourages the splicing of exon 3 to exon 4 at the same time helps select the poly A_1 site. The cleavage and addition of poly A is still the first nucleolytic event, but in this model, it comes after the folding that presages the union of exons 3 and 4. It may be generally true that the folding of the primary transcript can be regulated so as to inform the processing machinery about which parts of the transcript to preserve.

Muscle Proteins Differential RNA processing in the production of muscle proteins illustrates the enormous power of this mechanism to increase the protein-coding capacity of primary RNA transcripts. The protein composition and action of various muscle proteins is described in Chapter 22. Here let us briefly note that contractile proteins are present in almost all cells and constitute the major proteins in skeletal (striated) muscle, which makes up the major muscles of the body. Remarkable variations in protein design brought about by differential processing occur in at least five or six muscle proteins, but we will discuss only two of these, troponin T and α-tropomyosin.

The entire gene for troponin T, which is more than 16 kb long, has been sequenced, and the 18 exons that can appear in mRNAs have been identified (Figure 11-42a). If all exons were present and translated, a protein of 259 amino acids would be produced. However, it has been known for some time that troponin T is quite variable in size ranging from about 150–250 amino acids in length. The basis for this variation lies in the bewildering array of mRNAs containing different exons that have been identified in skeletal muscle. All the troponin T mRNAs contain exons 1–3, 9–15, and 18. Exons 4–8 can apparently be present in any combination; mRNAs with at least 20 of the possible 32 arrangements of these 5 nucleotides have been identified. Exons 16 and 17 are mutually exclusive, i.e., an mRNA has one or the other. Variation also occurs in α-tropomyosin, whose gene contains 13 exons. Several different α-tropomyosin mRNAs have been identified in skeletal (striated) muscle; several other nonoverlapping forms occur in smooth muscle (Figure 11-42b).

Because the functions of skeletal muscle and smooth muscle are different, the presence of different forms of α-tropomyosin in each seems unsurprising. The wide variations, particularly in troponin T, within skeletal muscle itself is more surprising at first glance. A possible explanation for the many variant forms of troponin T is that not all skeletal muscles act in exactly the same way. They are composed of fast- and slow-acting muscle fibers; differences in the proportion of the two fiber types in coordination with the proper innervation allows some muscles to exert great force and others to be capable of very delicate contractions; all coordinated movement depends on a balance of these two types of muscle action. Remarkably enough, different muscles (e.g., leg, back, abdomen) exhibit different patterns of troponin mRNAs. Thus at least a part of the functional differences among various muscles is ultimately based on differential splicing.

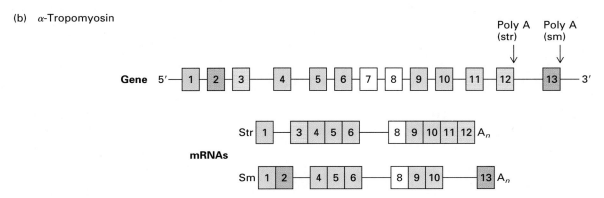

(a) Troponin T (skeletal muscle)

(b) α-Tropomyosin

Figure 11-42 Differential splicing in synthesis of muscle proteins. (a) The rat troponin T gene has 18 exons which encode 259 amino acids. However, the length of this protein varies in different muscles of the body. Eleven of the exons (blue) appear in all the mRNAs; exons 4–8 appear in various combinations giving rise to 32 possible arrangements (4,6,8; 4,7,8; 7,8; etc.); exon 16 or 17, but not both, is in every mRNA. Thus, there is a total of 64 possible mRNAs.

(b) The α-tropomyosin gene has 13 exons that are variably spliced in different muscle tissues; shown are the predominant splices found in striated (str), or skeletal, muscle and smooth (sm) muscle, which is found, for example, in arterial walls. [See R. E. Breitbart and B. Nadal-Ginard, 1986, *J. Mol. Biol.* **188**:313; and D. E. Wieczorek, C. W. J. Smith, and B. Nadel-Ginard, 1988. *Mol. Cell Biol.* **8**:679.]

Sex Determination in Drosophila

In *Drosophila* the choice of becoming a male or female is placed in the hands of differential splicing. A number of genes have been described that affect the development of male- or female-specific body structures in *Drosophila*. The choice of male or female development is controlled first by the X chromosome to autosome (A) ratio: an embryo with XX:AA becomes female, and one with X:AA becomes male. A master regulatory locus called sex-lethal (*Sxl*), which is transcribed in embryos and in sexually dimorphic tissues (those which differ between the sexes), is required for normal female structures to develop; male structures can develop (although the flies are not viable) in its absence. Thus *Sxl* acts normally to suppress male development, controlling several subsequent genes in the developmental pathway: *tra* (transformer), *tra2,* and *dsx* (double-sex). Mutants in *tra* or *tra2* develop male structures even though they are XX:AA; thus the proteins encoded by these genes are required to suppress male development. Mutants in *dsx* develop some characteristics of both sexes.

The molecular regulation of this developmental pathway does not seem to involve differential transcription but rather differential splicing that yields different proteins. Although there is probably differential splicing in the *Sxl* primary transcript itself, the present evidence is clearest for genes later in the pathway. For example, genomic clones and cDNAs of *tra* and *dsx* have been thoroughly investigated. Although transcription of the two genes is equivalent in male and female embryos, two polyadenylated mRNAs containing *tra* sequences are found in females, whereas only one occurs in males (Figure 11-43). In females, the shorter *tra* codes a 197-amino-acid protein; the long mRNA retains an intron containing stop codons, so that there is no long open reading frame. This long mRNA, which is present in both males and females, does not encode any protein. The protein product of the *tra* gene therefore is made in females and suppresses male development. Thus at least in *Drosophila*, differential RNA processing is used in one of the most fundamental decisions in animal development.

Differential Processing May Involve Variations in snRNPs

One possible explanation of differential splicing and differential polyadenylation is that cell-specific variations in the pre-mRNA processing machinery exist. An important new finding that may shed light on this subject concerns variation in the protein components of snRNPs in the spliceosomes that process hnRNA in the nucleus. The snRNPs from many cells contain a 25-kDa protein

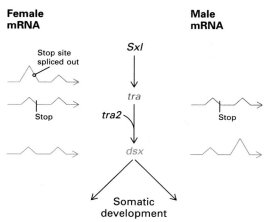

▲ **Figure 11-43** Differential splicing in *Drosophila* sexual determination. A number of genes required for normal development of somatic sexual structures in fruit flies are known and have been cloned: *Sxl* (sex-lethal), *tra* (transformer), and *dsx* (double-sex). These genes are transcribed equally in male and female embryos, but different spliced mRNAs accumulate in females and males. The carats indicate spliced-out regions of primary transcripts; termination codons are left in one form of the *tra* mRNA. [See B. A. Sosnowski et al., 1989, *Cell* 58:499.]

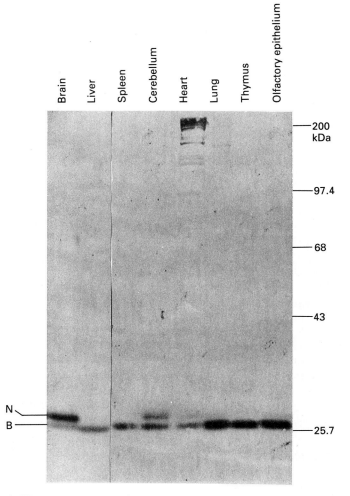

▲ **Figure 11-44** Antibody precipitation and electrophoresis of proteins from small ribonucleoprotein particles (snRNPs) from different tissues. Only the B protein (~25 kDa) from snRNPs of most tissues reacted with the particular antiserum used in the experiment. Total brain and cerebellum snRNPs also contain another similar reactive protein termed N (for neural) protein. [See B. McAllister, S. G. Amara, and M. R. Lerner, 1988, *Proc. Nat'l Acad. Sci. USA* 85:5296.] *Courtesy of M. R. Lerner.*

called B; the snRNPs from brain cells contain another protein termed N (for neuron) which has a similar but not identical amino acid sequence (Figure 11-44). Tumor cell lines from the anterior pituitary that make only CGRP are known, as are thyroid tumor lines that make only calcitonin. The snRNPs from the pituitary lines have the N protein, and those from the thyroid lines have the B protein. It is possible that this difference causes a difference in the RNA-processing ability of the different cell types.

In addition to differences in the protein components of snRNPs, the RNA components may vary. A careful analysis of some of the several hundred genes encoding U1 snRNAs showed that about 1 in 10 of these genes contained a sequence variation. This variation is also found in the U1 snRNAs produced by cultured cells. These sequence differences have not yet been shown to correlate with any cell-specific differential processing, but it is an obvious possibility.

Overlapping Transcription Units: Transcriptional Control Not Processing Control

The control of synthesis of α-amylase, which catalyzes digestion of starches in the liver and salivary glands, is an example of a subtle type of transcriptional control that is easily mistaken for processing control. The α-amylases produced in liver and salivary glands have an identical amino acid sequence, but the enzyme concentration is 100 times higher in salivary glands than in liver. Although the nucleotide sequences in the coding regions of the amylase mRNAs from liver and salivary glands are the same, the 5′ ends of the two mRNAs are different.

Analysis of the genomic DNA encoding α-amylase mRNAs from liver and salivary gland cells revealed that they have transcription start sites that are separated in the genome by about 2.8 kb (Figure 11-45). The different cap sites produce different 5′ exons, which are not translated in either case. Thus both the start site and the rate of transcription are tissue-specific. The first 3′ splice junction is the same in both mRNAs, which are also identical

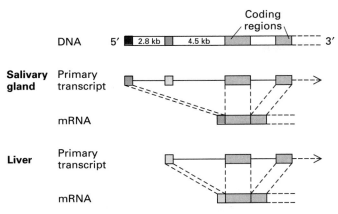

▲ **Figure 11-45** Overlapping transcription units for α-amylase and the primary transcripts produced preferentially in salivary gland and liver. Transcription starting from the salivary gland start site is more frequent than that from liver start site. The coding regions in the α-amylase mRNA in both tissues are the same. [See R. A. Young et al., 1981, *Cell* 23:451.]

throughout the rest of their lengths. Therefore, the two mRNAs are actually from two different but *overlapping* transcription units. Because transcription from the salivary gland start site occurs much more frequently than that from the liver start site, the α-amylase concentration is higher in salivary glands. The important point is that the regulation in the two tissues involves a choice in RNA start site, a transcriptional choice, and not a choice in RNA processing.

Regulation of Ribosomal RNA

All of the examples of gene control discussed so far have involved protein-coding genes. There are, of course, many more protein-coding genes than other types, and most cell specificity resides in the control of protein-coding genes. Nevertheless, an important element of cellular growth and development is the control of the ribosomal genes transcribed by RNA polymerase I.

When placed in a poor medium (e.g., one lacking amino acids), bacteria almost completely stop making pre-rRNA. This so-called *stringent response* is mediated in bacteria by (5′)ppGpp(3′), which increases in concentration in nongrowing cells and plays a direct role in repressing the initiation of ribosomal gene transcription. Yeast cells also exhibit a stringent response; that is, in the face of starvation or during sporulation, ribosomal RNA synthesis decreases by at least 80–90 percent in yeast cells. In contrast to bacteria and yeast, cultured mammalian cells continue to make pre-rRNA at 30–50 percent the normal rate when they are starved of an essential amino acid. However, ribosomal proteins are not formed

under these conditions, so new ribosomes cannot be assembled and the pre-rRNA is simply degraded.

During embryogenesis in the frog, pre-rRNA synthesis does not begin until the midblastula stage, about 16 h after fertilization. Special preparations for this suspension of pre-rRNA synthesis early in frog development occur during oogenesis when the ribosomal genes are amplified over 1000 times; pre-rRNA synthesis and ribosome formation increase accordingly. These ribosomes are stored in the egg and parceled out during the early embryonic divisions until pre-rRNA synthesis is activated after about 12–14 divisions. Mammalian eggs, however, have no large ribosomal stores, and transcription of pre-rRNA begins at the two-cell stage.

Most mammalian cells appear to exercise little transcriptional control over pre-rRNA synthesis. The cells in a variety of tissues, for example, all tend to synthesize pre-rRNA at similar rates, although they differ in their rate of cell turnover. In several cases the efficiency of pre-rRNA processing clearly regulates ribosome formation. For example, in mammals in which one kidney is removed, compensatory hypertrophy (growth) of the remaining kidney occurs such that its mass increases by at least 50 percent. New ribosome formation accompanies this increase in mass, but there is no increase in pre-rRNA synthesis. More efficient processing of pre-rRNA would account for increased ribosome formation in the absence of increased pre-rRNA synthesis. A similar situation exists in regenerating liver. When mammals lose a large amount of liver tissue (e.g., during surgery, infection, or chemical poisoning, the remaining liver tissue regenerates a normal-size liver within a few days. There is an increase in the appearance of new ribosomes in the cytoplasm, but no increase in nuclear pre-rRNA synthesis. Thus differential processing of pre-rRNA seems to be the major means of controlling rRNA formation in the cells of adult mammals. In addition the new ribosomal proteins required for regeneration arise from increased translation of the same amount of mRNA for ribosomal proteins.

Cytoplasmic Control of Gene Expression

Thus far in this chapter we have discussed various nuclear events that can regulate the production of a given mRNA. However, the rate of synthesis of any protein also is affected by the rate at which its mRNA molecules are transported into the cytoplasm, by the cytoplasmic lifetime of the mRNA, and by the rate at which the mRNA is translated. Furthermore, posttranslational events can change the activity or longevity of a protein, and differential processing of the same primary *translation* product in different cells can lead to different protein products (see Figure 9-32.) In this section we discuss only two aspects of cytoplasmic gene control—differential mRNA stabilization and differential translation. Before examining these types of gene regulation, however, we will briefly consider how the stability of mRNAs is determined experimentally and the general mechanism of mRNA turnover.

Several Methods Are Used to Measure mRNA Half-Life

Unlike rRNA and tRNA, which are synthesized in growing cells at a rate that allows the amounts of these species to double with each cell generation, mRNA is synthesized faster than is apparently required for maintenance of its steady-state amount. This high rate of synthesis is necessary because mRNA is constantly being degraded. This phenomenon was recognized, first in bacteria and later in animal cells, when it was observed that in cells briefly exposed to labeled uridine or phosphate there was a disproportionately high incorporation of label in the mRNA fraction relative to that in the much more abundant ribosomal RNA. For example, after a steady state is reached, 2–3 percent of the total labeled RNA in polyribosomes may be in mRNA, whereas after a short labeling time— say, one-fourth of a generation or less—15–20 percent of the total label may be in mRNA; the shorter the label time, the higher the percentage of label present in mRNA. From such observations, it has become clear that mRNA is "turned over" (renewed) as often as 10 times per generation (Table 11-6). Three methods that have been used to estimate mRNA half-lives are illustrated in Figure 11-46.

Pulse-chase Labeling The simplest procedure for measuring mRNA half-life would seem to be to add label briefly, remove the label, and measure the retention of labeled mRNA over time. However, such short-term

pulse-chase labeling cannot be satisfactorily achieved because the ribonucleotide pool that supplies precursors for RNA synthesis is quite large, and, furthermore, once a labeled precursor enters that pool as a phosphorylated intermediate, it cannot be chased out of the cell. Nevertheless, effective dilution of label in the pool can sometimes mimic a perfect chase well enough that the stability of an mRNA can be studied and conclusions drawn about its turnover, at least for relatively long-lived species; this technique is almost useless for detecting or measuring half-lives for short-lived mRNAs (Figure 11-46a).

Steady-state Labeling Another approach to studying mRNA turnover is to label cells continuously and study the curve of accumulation of a specific mRNA as a function of time. Because achievement of the maximum rate of label accumulation in cytoplasmic mRNA requires time for the ribonucleotide pool to equilibrate and for the nuclear RNA precursor to become fully labeled, the accumulation technique is not accurate for measuring the turnover time of very short-lived species. However, with longer labeling times (e.g., more than 1 h in mammalian cells), the accumulation curves of specific mRNAs can be used to obtain fairly accurate estimates of mRNA half-lives (Figure 11-46b). With cultured mammalian cells, individual mRNAs with half-lives ranging from about 3–24 h can be detected with either pulse-chase or steady-state labeling techniques, although both of them fail to accurately measure short-lived mRNAs.

Inhibitor Chase The half-lives of mRNAs with rapid turnover times can be determined more accurately by inhibitor-chase experiments than by labeling experiments. In this procedure, cells are exposed briefly to an inhibitor of RNA synthesis (e.g., actinomycin) and then the level of a particular mRNA is measured over time by Northern blot analysis. Messenger RNAs with half-lives as short as 15–20 min can be detected easily by the inhibitor-chase procedure (Figure 11-46c). Unfortunately, because the introduction of a potent inhibitor may affect cell metabolism in various ways, inhibitor-chase experiments alone are not usually sufficient to prove that an mRNA has a short half-life. However, if suggestive results are obtained in pulse-labeling experiments, then data obtained with the inhibitor-chase procedure can be considered valid.

Degradation Rate of mRNA Is Related to Poly A Tails and Specific Sequences in 3′ Untranslated Regions

Although the precise mechanisms of mRNA turnover are not understood, mRNA degradation rates seem to be related to the presence or absence of poly A tails and the presence of certain sequences in the 3′ end. For example, if globin mRNA without poly A is injected into HeLa

Table 11-6 Half-lives of messenger RNAs

Cell	Cell generation time	mRNA half-lives*	
		Average	Range known for individual cases
Escherichia coli	20–60 min	3–5 min	2–10 min
Saccharomyces cerevisiae (yeast)	3 h	22 min	4–40 min
Cultured human or rodent cells	16–24 h	10 h	30 min or less (histone and c-*myc* mRNAs) 3–24 h (specific mRNAs of cultured cells)

*For information on specific mRNA half-lives for *E. coli*, see A. Hirashima, G. Childs, and M. Inouye, 1973, *J. Mol. Biol.* **119**:373; for yeast, see L.-L. Chia and C. McLaughlin, 1979, *Mol. Gen. Genet.* **170**:137; and for mammalian cells, see M. M. Harpold, M. Wilson, and J. E. Darnell, 1981, *Mol. Cell Biol.* **1**:188.

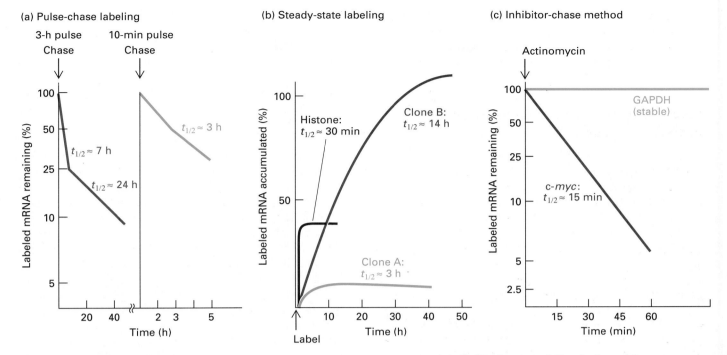

▲ **Figure 11-46** Three methods for measuring mRNA half-life. (a) Pulse-chase labeling. Observed decrease in radioactivity in total HeLa cell mRNA after chase by label dilution following 3-h and 10-min labeling periods. The observed initial decay rate depends on the length of the prelabeling period, and short-lived mRNAs cannot be detected after long labeling periods. (b) Steady-state labeling. Observed increase in radioactivity in histone mRNA and mRNA hybridized to two clone DNAs in continuous presence of label. The half-life of an RNA can be calculated from the following equation:

$$\frac{A}{A_\alpha} = 1 - e^{-\ln 2(1/K_D + 1/K_{1/2})t}$$

in which A and A_α represent the radioactivity in mRNA per cell sample at times t and t_α, respectively; K_D is the growth constant of the cell culture; and $K_{1/2}$ is the turnover rate constant of the mRNA. The time necessary to reach equilibrium is controlled by the turnover rate. The calculated half-lives for histone mRNA and the mRNAs for clone A and clone B are shown. Clearly, this method also cannot detect short-lived mRNAs very accurately. (c) Inhibitor-chase method. Observed decrease in c-*myc* mRNA versus glyceraldehyde-3-phosphate dehydrogenase (GAPDH) mRNA in cultured cells after addition of actinomycin as measured by Northern blot analysis. This method can accurately detect short-lived mRNAs. [See J. R. Greenberg, 1972, *Nature* 240:102; L. Puckett and J. E. Darnell, Jr., 1977, *J. Cell Phys.* 90:521; M. H. Harpold et al., 1981, *Mol. Cell Biol.* 1:188; N. Heintz et al., 1983, *Mol. Cell Biol.* 3:538; and C. Dani et al., 1985, *Proc. Nat'l Acad. Sci. USA* 81:7046.]

cells or frog eggs, it is quickly degraded; studies on the synthesis of viral RNA by cells treated with drugs that block poly A formation also indicate that mRNA lacking poly A is lost quickly from cells. These findings correlate well with the naturally short half-life of HeLa cell histone mRNAs, which lack poly A at their 3′ end.

Numerous mRNAs that encode proteins thought to play a role in regulating cell growth and differentiation (protooncogene proteins and cell-growth factors) have short half-lives. A number of these mRNAs contain several A(U)$_n$A sequences in their 3′ untranslated regions (Figure 11-47a). For example, the 3′ untranslated end of human granulocyte-monocyte colony–stimulating factor (GMCSF) contains seven AUUUA sequences, some of which are overlapping. When a segment of human GMCSF DNA containing these sequences was inserted

into a β-globin gene, the half-life of the globin mRNA decreased from more than 10 h to less than 2 h (Figure 11-47b). A similar experiment with segments of the protooncogene c-*fos* showed that the poly A tail of the recombinant globin mRNA was shortened much faster when the A(U)$_n$A-rich region was present than when it was absent.

A reasonable hypothesis to explain generalized mRNA turnover is that all mRNAs are subject to nucleolytic attack near their 3′ end. This attack, which is favored by the presence of A(U)$_n$A sequences, shortens poly A; if the nuclease removes all of the poly A (e.g., by cutting within the A(U)$_n$A-rich region), an mRNA is converted to a completely nuclease-sensitive state.

The rapid degradation typical of histone mRNAs and mRNAs containing A(U)$_n$A sequences in their 3′ ends

(a) 3′ Sequences of unstable mRNAs

GMCSF   UAAUAUUUAUAUAUUUAUAUUUUUAAAAUAUUUAUUUAUUUAUUUAUUUAA

IFNβ   UUUUGAAAUUUUUAUUAAAUUAUGAGUUAUUUUUAUUUAUUUAAAUUUUAUUUU

IL-1   UUAUUUUUUAAUUAUUAUUUAUAUAUGUAUUUAUAAAUAUAUUUAAGAUAAU

TNF   AUUAUUUAUUAUUUAUUUAUUAUUUAUUUAUUUA

c-Fos   GUUUUUAAUUUAUUUAUUAAGAUGGAUUCUCAGAUAUUUAUAUUUUUAUUUU

c-Myc   UAAUUUUUUUUAUUUAAGUACAUUUUGCUUUUUAAAGUUGAUUUUUUUCU

(b) Activity of A(U)$_n$A repeats

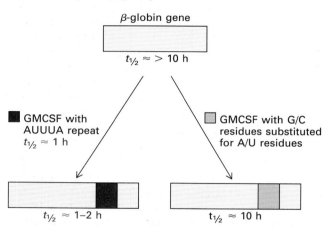

β-globin gene

$t_{1/2} \approx > 10$ h

■ GMCSF with
AUUUA repeat
$t_{1/2} \approx 1$ h

▨ GMCSF with G/C
residues substituted
for A/U residues

$t_{1/2} \approx 1–2$ h

$t_{1/2} \approx 10$ h

▲ **Figure 11-47** (a) Sequences of 3′ untranslated regions of mRNAs encoding human granulocyte-monocyte–stimulating factor (GMCSF), fibroblast interferon β (IFNβ), interleukin 1 (IL-1), tumor necrosis factor (TNF), and two protooncogene proteins (c-Fos and c-Myc). Characteristic A(U)$_n$A sequences are shown in red; some of these are overlapping. Numerous other proto-oncogene and cell growth factor mRNAs contain these sequences. (b) Effect of A(U)$_n$A sequences on mRNA half-life. Cultured cells were transfected with recombinant β-globin DNA containing a 62-bp segment of GMCSF DNA encoding either the A(U)$_n$A-rich segment shown in (a) or an altered segment in which many A/U residues were replaced with G/C residues. The destabilizing effect of the A(U)$_n$A sequences was clearly demonstrated by the decrease in half-life ($t_{1/2}$) of the corresponding mRNAs. [See G. Shaw and R. Kamen, 1986, *Cell* **46**:659.]

does not occur when protein synthesis is blocked by drugs. As we discuss in the next section, other mRNAs whose half-lives are reduced under certain conditions also require normal protein synthesis to be subject to turnover. This association between mRNA degradation and protein synthesis suggests that normal mRNA turnover is caused by a ribosome-associated nuclease or nucleases. Access of this purported nuclease to various mRNAs would vary greatly depending on their structure and the presence or absence of molecules that protect or expose them. Thus, both normal variation in the half-lives of different mRNAs and differential regulation of mRNA stability may ultimately be explainable in terms of interactions that increase or decrease access of ribosome-bound nucleases to mRNAs.

Stability of Specific mRNAs Can Be Regulated by a Variety of Mechanisms

Numerous examples of the differential stabilization or destabilization of specific mRNAs are now known in which the causes for stabilization or destabilization are unrelated to the A(U)$_n$A sequences just described. In this section we describe the mediation of mRNA stability by hormones, by translation products (autoregulation), and by low-molecular-weight ligands.

Hormone-regulated Stabilization of mRNA Many of the known cases of differential stabilization of mRNAs in vertebrates are mediated by hormones. Because this hormone-dependent cytoplasmic control often is accompanied by hormone-dependent transcriptional control, the ability of many hormones to increase the synthesis of specific mRNAs results from two different effects.

Casein, which is the most abundant protein in milk, is produced in the epithelial cells of breast tissue in response to hormones, including the polypeptide hormone prolactin. Small pieces of breast tissue that are cultured in the absence of prolactin contain only about 300 molecules of casein mRNA per cell, whereas cells cultured in the presence of prolactin contain about 30,000 casein mRNA molecules per cell. The rate of synthesis of casein mRNA in cultured breast cells, however, is only about three times greater in the presence of prolactin than in its absence. In contrast, prolactin treatment increases the half-life of the casein mRNA about 30- to 50-fold in cultured breast tissue (Figure 11-48). These results strongly suggest that most of the prolactin-induced increase in casein mRNA at least in cultured cells is due to the increased stability of the mRNA rather than to its increased synthesis.

Autoregulation of Tubulin Tubulin, the prominent cytoplasmic protein that polymerizes to form microtubules, appears to have a role in regulating the cytoplasmic

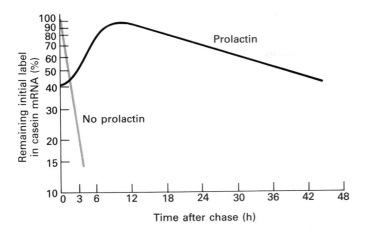

▲ **Figure 11-48** Stabilization of casein mRNA by prolactin. Breast tissue from lactating rats was placed for 4 h in a culture medium lacking prolactin. One sample was then labeled with [³H] uridine in the presence of prolactin; another sample was labeled in its absence. After 3 h, the label was removed, a chase of unlabeled nucleosides was added, and samples were taken at intervals for determination of the remaining labeled casein mRNA by hybridization. The culture with prolactin accumulated labeled mRNA for several hours because of the ineffective chase of RNA precursors. The eventual decay of labeled mRNA in this culture was much less rapid ($t_{1/2} \approx 40$ h) than in the culture lacking prolactin. [See W. A. Guyette, R. J. Matusik, and J. M. Rosen, 1979, *Cell* 17:1013.]

concentration of its own mRNA. The microtubule is built of very long chains of heterodimers of two proteins α- and β-tubulin. When these polypeptide chains are finished they immediately associate and enter a pool of αβ-tubulin heterodimers that are in equilibrium with the microtubules. When cells are treated with drugs that block tubulin polymerization (e.g., colchicine, which prevents mitotic spindle formation), the intracellular tubulin heterodimer concentration rises about twofold. When this happens both the α- and β-tubulin mRNAs are destroyed rapidly, reaching a new equilibrium at about one-tenth their usual concentrations. Conversely, there are drugs that remove free intracellular tubulin heterodimers; for example, the anticancer drug vincristine causes all free heterodimers to precipitate in the cytoplasm as quasi-crystalline structures. When cells are treated with this drug, the mRNA levels climb to three or four times their normal values.

None of these treatments that affect tubulin mRNA concentration have any affect on nuclear synthesis of tubulin pre-mRNA. Thus they most likely affect the turnover of the tubulin mRNA itself. The sequence in the chicken β-tubulin mRNA that is required for β-tubulin destruction in colchicine-treated cells includes the 5′ untranslated region and the first four codons of the mRNA. However, only tubulin mRNA that is associated with polyribosomes is degraded. Moreover, only if the mRNA has been translated for a distance of about 40 amino acids is the mRNA subject to destruction. Artificial mRNAs with a termination codon introduced before the fortieth amino acid are not subject to autoregulation. A model of this autoregulation is shown in Figure 11-49. It seems likely that the recognition sequence (the first four amino acids) is not presented at the surface of the ribosome until a chain of about 40 amino acids has been made. The excess tubulin subunits then interact with the end of the nascent chain and lead to mRNA destruction.

Regulation of mRNA Stability by a Nutrient

Our final example of the regulation of mRNA stability involves the mRNA encoding the cell-surface protein receptor for transferrin, an iron-carrying serum protein. The mRNA encoding transferrin receptor is destroyed rapidly in cultured cells with abundant intracellular iron, whereas the half-life of this mRNA is increased substantially when the culture medium contains low concentrations of free iron. The regulatory effect of the iron requires protein translation; that is, high intracellular levels of iron in the presence of translation-blocking drugs (e.g., cycloheximide) do not destabilize the transferrin-receptor mRNA. Transferrin-receptor mRNA has five stem-loops near its 3′ end; each is about 12 base pairs long and has a CAGUG sequence in the loop. Cultured cells containing recombinant transferrin-receptor genes that do not encode this stem-loop region produce mRNA whose half-life is not decreased by high iron concentra-

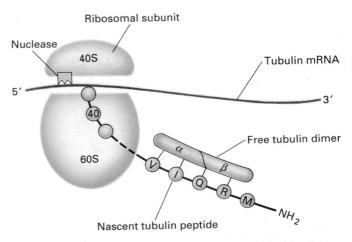

▲ **Figure 11-49** A model for tubulin autoregulation. Free αβ-tubulin dimers bind to the first four or five emerging amino-terminal amino acids at the end of a new tubulin monomer. (Amino acids are indicated by a single-letter code.) This sequence must emerge from the ribosome to be recognized (about 40 amino acids must be translated before the amino-terminal end emerges). The binding activates a ribosome-bound nuclease that destroys the tubulin mRNA. [See T. J. Yen et al., 1988, *Mol. Cell Biol.* 8:1224.]

tions. The intracellular factors (protein?) that recognize these stem-loops and regulate this iron-dependent mRNA destruction are being studied at present.

Overall Rate of mRNA Translation Can Be Controlled

Translational control of specific proteins has been proved in only a few cases. In contrast, well-known mechanisms of translational control for all mRNAs operate in all cells to change the *average* rate of protein synthesis. For example, when cultured mammalian or insect cells are heated above 40°C, most translation initiation is suppressed, but the frequency of initiation of translation of heat-shock mRNA actually rises. As we mentioned earlier, an increase in the transcription of the heat-shock genes accompanies the increased translation of heat-shock mRNAs. Mammalian cells adapt to the new elevated temperature (as long as the temperature is not higher than about 42°C) and resume normal protein synthesis and growth within 2 h after a heat shock.

A similar general inhibition of protein synthesis occurs when cells enter mitosis, which causes a fall in the rate of protein synthesis to 30 percent of normal. This decline in protein synthesis is accompanied by a decrease in the size

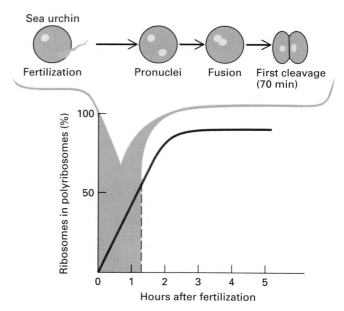

▲ **Figure 11-50** Translational control of protein synthesis in the early cleavage stages of invertebrate embryogenesis. The percentage of ribosomes in polyribosomes (red curve), which is an index of translation, rises in fertilized sea urchin eggs using preexisting maternal mRNA. [See B. Brandhorst, 1976, *Devel. Biol.* **52**:310.]

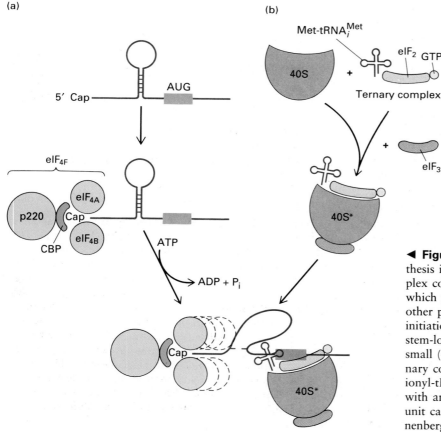

(a)

(b)

◀ **Figure 11-51** Model of initiation of protein synthesis in eukaryotes. (a) The cap-binding (CB) complex consists of a 24-kDa cap-binding protein (CBP), which recognizes the 5′ cap in mRNA, and several other proteins, collectively referred to as eukaryotic initiation factor 4F (eIF$_{4F}$), that probably unwind the stem-loops near the 5′ end. (b) Activation of the small (40S) ribosomal subunit requires eIF$_3$ and a ternary complex consisting of eIF$_2$, GTP, and methionyl-tRNA$_i^{Met}$. After the CB complex has associated with an mRNA molecule, an activated ribosomal subunit can locate the AUG start codon. [See N. Sonnenberg, 1987, *Adv. Virus Res.* **33**:172.]

of polyribosomes; that is, fewer ribosomes are attached to each mRNA, indicative of decreased initiation with a normal rate of polypeptide elongation and termination. There is also a great decrease in mRNA synthesis during mitosis, but the decline in the number of mRNA molecules is not sufficient to account for the decrease in protein synthesis; translational control is clearly the basis for the decrease.

Protein synthesis in unfertilized invertebrate eggs is very slow, but after fertilization it increases dramatically in the absence of any formation of new mRNA. Thus protein synthesis during the early stages of embryogenesis represents the delayed use of stored "maternal" mRNAs whose translation is inhibited in the unfertilized egg (Figure 11-50).

These examples indicate that mechanisms are available in cells for regulating the overall rate of mRNA translation under a number of different conditions. To understand how the efficiency of translation might be regulated, we need to consider the process of translational initiation. As currently envisioned, the initiation of protein synthesis requires recognition by a complex of proteins of the cap structure at the 5' end of an mRNA, followed by unwinding of any stem-loop structure near the 5' end. This allows recognition of the AUG start codon by a 40S-ribosomal initiation complex. The components that recognize and bind to the cap (Figure 11-51) include a 24-kDa protein, which specifically recognizes the cap structure, and at least three other proteins, which together are referred to as eukaryotic initiation factor 4F or eIF$_{4F}$. The ribosomal initiation complex consists of the 40S subunit, eIF$_3$, and a "ternary complex" consisting of eIF$_2$, GTP, and initiator methionyl-tRNA^{Met} (Met-tRNA$_1^{Met}$).

Much evidence indicates that the greater the secondary structure at the 5' end of an mRNA, the more difficult it is to start translation and the greater the requirement for eIF$_{4F}$ activity. For example, oligonucleotides complementary to the first 15 bases of mRNAs greatly inhibit initial ribosome binding, and mRNAs with little or no secondary structure at their 5' end associate well with ribosomes without assistance of eIF$_{4F}$ or at much lower than normal concentrations. The heat-shock mRNAs, the only mRNAs that are translated efficiently in heat-shocked cells, have a very A-rich 26-nucleotide stretch before their initiation codon; such A-rich stretches do not form extensive secondary structures. These results suggest that translation control might be effected by substances that increase or decrease the secondary structure at the 5' ends of mRNAs, but well-documented instances of regulation using these principles have yet to be described.

A second means of regulating the efficiency of translational initiation is modification of an initiation factor. One example of this mechanism is phosphorylation of initiation factors by protein kinases, which add phosphates to serine, threonine, and tyrosine residues. The

biochemistry of translational initiation in mammalian cells has been studied most extensively in reticulocytes because a number of the components of their translational machinery have been purified. When eIF$_2$ is phosphorylated, it is inactive (Figure 11-52). In reticulocyte extracts, the trigger for phosphorylation by a protein kinase can be either the absence of heme (a component of hemoglobin) or the presence of a segment of double-stranded RNA (e.g., during RNA virus infection). Since hemoglobin contains both heme and the globin chains, a balance of the two is assured by this translational control; a decrease in heme leads to a decrease in globin mRNA translation. It is likely that all cells possess these protein kinases and that many factors other than heme and double-stranded RNA may affect their activity.

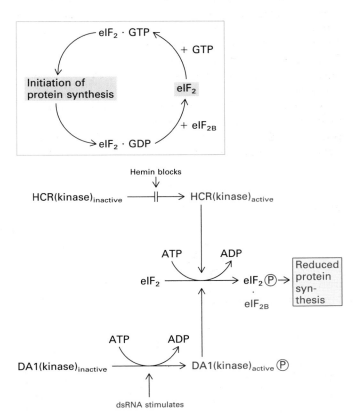

▲ **Figure 11-52** Translation control by phosphorylation of eukaryotic initiation factor 2 (eIF$_2$). As shown in Figure 11-51, eIF$_2$ is required to initiate protein synthesis. Two different protein kinases, HCR kinase and DA1 kinase, can phosphorylate eIF$_2$. Phosphorylated eIF$_2$ gets trapped by binding tightly to eIF$_{2B}$, a protein that normally regenerates free eIF$_2$. Inactive HCR kinase is activated by the absence of hemin, which is a derivative of heme. Inactive DA1 kinase is activated by double-stranded RNA (dsRNA). Thus either the absence of heme (conceivably due to iron starvation) or the presence of double-stranded RNA (e.g., during RNA virus infection) leads to phosphorylation of eIF$_2$ and decreased initiation of protein synthesis. [See P. J. Farrell et al., 1977, *Cell* 11:187; and N. Sonenberg, 1987, *Adv. Virus Res.* 33:172.]

Differential Translational Control of Specific mRNAs May Not Be Common in Eukaryotes

In the previous section, we described several cases of repression or stimulation of the overall translation rate. To date, relatively few examples of differential translation of specific mRNAs are well studied. One example is the surf clam, *Spisula solidissima*, in which synthesis of some proteins decreases after fertilization, while the synthesis of other proteins rises sharply. However, if the total mRNA extracted from both fertilized and unfertilized clam eggs is translated in vitro, the resulting protein patterns are very nearly identical. Thus the same mRNAs are present in clam eggs before and after fertilization but their translation is differentially regulated. The untranslated mRNAs in both fertilized and unfertilized eggs are not associated with polyribosomes, indicating that initiation of translation is the regulated step of synthesis.

In Chapter 7 we noted that translational control operates for a number of bacterial ribosomal proteins (those proteins associated with ribosomal RNA in ribosomes). Specific control of translation of mRNAs encoding ribosomal proteins also occurs in mammalian cells. When growth is suspended by placing cultured cells in a medium lacking all growth factors, the mRNAs for ribosomal proteins are not associated with polyribosomes and are no longer translated. When serum is returned to the medium, these mRNAs again become engaged by the translation initiation apparatus and are translated.

Another example of specific translational control involves the iron-binding protein ferredoxin, which is synthesized faster when cells are placed in a medium of high iron content. A protein has been purified that specifically binds to a stem-loop region in the 5′ untranslated region of ferredoxin mRNA and prevents translation. This translation-inhibitor protein does not inhibit translation of ferredoxin mRNA in iron-rich circumstances but does when a chelator is present. The stem-loop region in the ferredoxin mRNA is about 12 base pairs long and contains a CAGUG sequence, the same sequence that is present in the 3′ untranslated regions of the transferrin-receptor mRNA mentioned earlier. Whether the same protein mediates destruction of transferrin-receptor mRNA and inhibit ferredoxin mRNA translation at high iron concentrations is not yet known.

Summary

The primary function of gene control in single-cell organisms is to respond to alterations in their environment, particularly changes in the available nutrients. Because cells in multicellular organisms are shielded from acute environmental changes, they have less need for this type of gene control, and their ability to respond to nutritional deprivation and other environmental stresses is limited. The primary function of gene control in multicellular organisms is to direct developmental pathways that result in many different types of cells, each characterized by a particular protein composition that endows cells with specific functions. The most common gene-control signals in animals are hormones (and hormonelike substances), which include small molecules that can diffuse into cells, circulating polypeptides that bind to cell-surface receptors and cell-cell and cell-matrix contacts.

Gene regulation occurs at several levels in eukaryotes, the primary and most frequent being control of transcriptional initiation. Although several biological systems using regulated transcriptional termination have been identified, this mechanism is not common in eukaryotes. Transcriptional initiation depends on many different nuclear proteins that assist RNA polymerase II in starting transcription. The genes encoding dozens of these factors have been cloned and sequenced, and the amino acid sequences of their protein products have been determined. At present several major types of regulatory proteins are recognized based on their characteristic structural motifs: helix-turn-helix (including homeobox) proteins; zinc-finger proteins; and two groups of dimeric amphipathic helical proteins, the leucine-zipper and helix-loop-helix proteins. There are also a number of regulatory proteins that do not fall into these categories.

Most of the eukaryotic transcription factors that have been identified exert a positive effect, although negative-acting proteins are known to occur in some systems. In general, to exert their effect these proteins must bind to specific DNA sequences, some of which may be quite distant (either upstream or downstream) from the start site of a gene. Some regulatory proteins appear to act as modifiers and exert their effect by protein-protein interactions with other factors that bind to DNA. Cell-specific gene expression thus depends on the cell-specific differential activity of regulatory proteins, which can be controlled by their cell-specific (or cell-limited) synthesis and by cell-specific interconversion between active and inactive forms. In eukaryotes, transcription is affected by chromatin structure, and transcriptionally active regions of DNA are characterized by a "looser" chromatin structure and undermethylation of certain residues. Networks of interdependent regulatory factors are under study in several situations. The genes responsible for two cell states in yeast (**a** and α) and how these genes are coordinately regulated is fairly well understood. Also the gene regulation underlying switching from **a** to α is becoming better understood. During early *Drosophila* embryogenesis there is a sequential cascade of transcription factors responsible for directing early larval development. Finally, studies on mammalian regulatory factors that participate in organ-specific expression have been identified and are under study.

Differential RNA processing is the second major gene-control mechanism, especially in animals. The cell-specific production of different proteins from complex primary transcripts has been shown to depend on differential selection of poly A sites and/or splice sites. Although the molecular basis for differential processing is not known, cell-specific variations in the components of the processing machinery (i.e., snRNPs) have been demonstrated in a few cases.

Short-lived mRNAs are characterized by the absence of poly A tails (e.g., histone mRNA) or by the presence of $A(U)_nA$ sequences in their 3' untranslated end. Differential stabilization or destabiliation of specific mRNAs by hormones, translation products, and ligands has been demonstrated, and this mechanism of gene control is fairly widespread in eukaryotes. Rapid mRNA turnover usually does not occur if translation is blocked, suggesting that the nucleases responsible for mRNA degradation are associated with ribosomes.

Initiation of translation depends on the interaction of several proteins with the 5' end of an mRNA molecule, which leads to loosening of the mRNA secondary structure in this region; as a result, an activated small ribosomal subunit can locate the start codon. Several mechanisms for regulating the rate of translational initiation can be envisioned, and the overall rate of mRNA translation is known to vary under certain conditions. However, only a few cases of differential translation of specific mRNAs have been demonstrated so far. Thus, the importance of translation control to differential gene expression in eukaryotes is still less clear than other mechanisms.

References

Signals for Gene Control

BUCK, C. A., and A. F. HORWITZ. 1987. Cell surface receptors for extracellular matrix molecules. *Ann. Rev. Cell Biol.* **3**: 179–205.

COLD SPRING HARBOR LABORATORY. 1988. Molecular biology of signal transduction. *Cold Spring Harbor Symp. Quant. Biol.* **53**.

GILMARTIN, A. W. 1987. G proteins: transducers of receptor-generated signals. *Ann. Rev. Biochem.* **56**:615–650.

LINDQUIST, S. 1986. The heat shock response. *Ann. Rev. Biochem.* **55**:1151–1191.

NORMAN, A. W., and G. LITWACK. 1987. *Hormones.* Academic Press.

YARDEN, Y., and A. ULLRICH. 1988. Molecular analysis of signal transduction by growth factors. *Biochemistry* **27**:3113.

Levels of Regulation

DARNELL, J. E., JR. 1982. Variety in the level of gene control in eukaryotic cells. *Nature* **297**:365–371.

NEVINS, J. R. 1982. Adenovirus gene expression: control at multiple steps of mRNA biogenesis. *Cell* **28**:1–2.

Demonstration of Transcriptional Control

ASHBURNER, M. 1973. Temporal control of puffing activity in polytene chromosomes. *Cold Spring Harbor Symp. Quant. Biol.* **38**:655–662.

DERMAN, E., K. KRAUTER, L. WALLING, C. WEINBERGER, M. RAY, and J. E. DARNELL JR. 1981. Transcriptional control in the production of liver-specific mRNAs. *Cell* **23**:731–739.

EDLUND, T., M. D. WALKER, P. J. BARR, and W. J. RUTTER. 1985. Cell-specific expression of the rat insulin gene: evidence for role of two distinct 5' flanking elements. *Science* **230**:912–916.

HEINTZ, N., H. L. SIVE, and R. G. ROEDER. 1983. Regulation of histone gene expression. *Mol. Cell Biol.* **3**:539–550.

HEINTZ, N., and R. G. ROEDER. 1984. Transcription of human histone genes in extracts from synchronized HeLa cells. *Proc. Nat'l Acad. Sci. USA* **81**:2713–2717.

McKNIGHT, G. S., and R. D. PALMITER. 1979. Transcriptional regulation of the ovalbumin and conalbumin genes by steroid hormones in chick oviduct. *J. Biol. Chem.* **254**:9050–9058.

Structure and Function of DNA-binding Proteins That Regulate Transcription

AKAM, M. 1989. *Hox* and *HOM:* Homologous gene clusters in insects and vertebrates. *Cell* **57**:347–349.

BRIGGS, M. R., J. T. KADONAGA, S. P. BELL, and R. TJIAN. 1986. Purification and biochemical characterization of the promoter–specific transcription factor, Sp 1. *Nature* **234**:47–52.

CHODOSH, L. W., A. S. BALDWIN, R. W. CARTHEW, and P. A. SHARP. 1988. Human CCAAT-binding proteins have heterologous subunits. *Cell* **53**:11–24.

DYNAN, W. S., and R. TJIAN. 1985. Control of eukaryotic messenger RNA synthesis by sequence-specific DNA-binding proteins. *Nature* **316**:774–778.

GINIGER, E., and M. PTASHNE. 1987. Transcription in yeast activated by a putative amphipathic α helix linked to a DNA binding unit. *Nature* **330**:670–672.

GLUZMAN, Y., and T. SHENK, eds. 1983. *Enhancers and Eukaryotic Gene Expression.* Cold Spring Harbor Laboratory.

GUARENTE, L. 1988. UASs and enhancers: common mechanism of transcriptional activation in yeast and mammals. *Cell* **52**:303–305.

HERR, W., et al. 1988. The POU domain: a large conserved region in the mammalian *pit-1, oct-1, oct-2,* and *Caenorhabditis elegans unc-86* gene products. *Genes & Devel.* **2**:1513–1516.

HOPE, I. A., S. MAHADEVAN, and K. STRUHL. 1988. Structural and functional characterization of the short acidic transcriptional activation region of yeast GCN4 protein. *Nature* **333**:635–640.

HORIKOSHI, M., M. F. CAREY, H. KAKIDANI, and R. G. ROEDER. 1988. Mechanism of action of a yeast activator: direct effect of GAL4 derivatives on mammalian TFIID-promoter interactions. *Cell* **54**:665–669.

JOHNSON, P. F., and S. L. McKNIGHT. 1989. Eukaryotic transcriptional regulatory proteins. *Ann. Rev. Biochem.* **58**:799–839.

KOUZARIDES, T., and E. ZIFF. 1988. The role of the leucine zipper in the fos-jun interaction. *Nature* 336:646–651.

LANDSCHULZ, W. H., P. F. JOHNSON, and S. L. McKNIGHT. 1988. The leucine zipper: a hypothetical structure common to a new class of DNA binding proteins. *Science* 240:1759–1764.

LEVINE, M., and T. HOEY. 1988. Homeobox proteins as sequence-specific transcription factors. *Cell* 55:537–540.

MILLER, J., A. D. McLACHLAN, and A. KLUG. 1985. Repetitive zinc-binding domains in the protein transcription factor IIIA from *Xenopus* oocytes. *EMBO J.* 4:1609–1614.

MURRE, C., P. S. McCAW, and D. BALTIMORE. 1989. A new DNA binding and dimerization motif in immunoglobulin enhancer binding, *daughterless, MyoD,* and *myc* proteins. *Cell* 56:777–783.

NAKAJIMA, N., M. HORIKOSHI, and R. G. ROEDER. 1988. Factors involved in specific transcription by mammalian RNA polymerase II: purification, genetic specificity, and TATA box-promoter interactions of TFIID. *Mol. Cell Biol.* 8:4028–4040.

PTASHNE, M. 1988. How eukaryotic transcriptional activators work. *Nature* 335:683–689.

SCHAFFNER, G., S. SCHIRM, B. MULLER-BADEN, F. WEBER, and W. SCHAFFNER. 1988. Redundancy of information in enhancers as a principle of mammalian transcription control. *J. Mol. Biol.* 201:81–90.

SCHLEIF, R. 1987. Why should DNA loop? *Nature* 327:369–370.

SCHUH, R., et al. 1986. A conserved family of nuclear proteins containing structural elements of the finger protein encoded by Krüppel, a *Drosophila* segmentation gene. *Cell* 47:1025–1032.

STRUHL, K. 1987. Promoters, activator proteins, and the mechanism of transcriptional initiation in yeast. *Cell* 49:295–297.

VARMUS, H. E. 1987. Oncogenes and transcriptional control. *Science* 238:1337–1339.

GOUTTE, C., and A. D. JOHNSON. 1988. a1 protein alters the DNA binding specificity of α2 repressor. *Cell* 52:875–882.

GUARENTE, L. 1987. Regulatory proteins in yeast. *Ann. Rev. Genet.* 21:425–452.

JOHNSON, A. D., and I. HERSKOWITZ. 1985. A repressor (MAT α2 product) and its operator control of expression of a set of cell-type specific genes in yeast. *Cell* 42:237–247.

JOHNSTON, S. A., and J. E. HOPPER. 1982. Isolation of the yeast regulatory gene *GAL4* and analysis of its dosage effects on the galactose/melibiose regulon. *Proc. Nat'l Acad. Sci. USA* 79:6971–6975.

JONES, E. W., and G. R. FINK. 1982. Regulation of amino acid and nucleotide biosynthesis in yeast. In *The Molecular Biology of the Yeast* Saccharomyces: *Metabolism and Gene Expression,* J. N. Strathern, E. W. Jones, and J. R. Broach, eds. Cold Spring Harbor Laboratory.

MA, J., and M. PTASHNE. 1987. The carboxy-terminal 30 amino acids of GAL4 are recognized by GAL80. *Cell* 50:137–142.

MILLER, A. M., V. L. MACKAY, and K. A. NASMYTH. 1985. Identification and comparison of two sequence elements that confer cell-type specific transcription in yeast. *Nature* 314:598–603.

NASMYTH, K., and D. SHORE. 1987. Transcriptional regulation in the yeast life cycle. *Science* 237:1162–1170.

NASMYTH, K., D. STILLMAN, and D. KIPLING. 1987. Both positive and negative regulators of *HO* transcription are required for mother-cell-specific mating-type switching in yeast. *Cell* 48:579–587.

OLESEN, J., S. HAHN, and L. GUARENTE. 1987. Yeast HAP2 and HAP3 activators both bind to the *CYC1* upstream activation site, UAS2, in an interdependent manner. *Cell* 51:953–961.

STERNBERG, P. W., M. J. STERN, I. CLARK, and I. HERSKOWITZ. 1987. Activation of the yeast *HO* gene by release from multiple negative controls. *Cell* 48:567–577.

Transcriptional Control Cycles in Yeast

ABRAHAM, J., J. FELDMAN, K. A. NASMYTH, J. N. STRATHERN, A. J. S. KLAR, J. R. BROACH, and J. B. HICKS. 1983. Sites required for position-effect regulation of mating-type information in yeast. *Cold Spring Harbor Symp. Quant. Biol.* 47:989–998.

BENDER, A., and G. F. SPRAGUE, JR. 1987. MATα1 protein, a yeast transcription activator, binds synergistically with a second protein to a set of cell-type specific genes. *Cell* 50:681–691.

BREEDEN, L., and K. NASMYTH. 1987. Cell cycle control of the yeast *HO* gene: *cis-* and *trans-*acting regulators. *Cell* 48:389–397.

ELION, E. A., and J. R. WARNER. 1986. An RNA polymerase I enhancer in *Saccharomyces cerevisiae. Mol. Cell Biol.* 6:2089–2097.

FIELDS, S., D. T. CHALEFF, and G. F. SPRAGUE, JR. 1988. Yeast *STE7, STE11,* and *STE12* genes are required for expression of cell-type-specific genes. *Mol. Cell Biol.* 8:551–556.

Hormonal Control of Transcription

BEATO, M. 1989. Gene regulation by steroid hormones. *Cell* 56:335–344.

EVANS, R. M. 1988. The steroid and thyroid hormone receptor superfamily. *Science* 240:889–895.

GIGUERE, V., S. M. HOLLENBERG, M. G. ROSENFELD, and R. M. EVANS. 1986. Functional domains of the human glucocorticoid receptor. *Cell* 46:645–652.

GREEN, S., and P. CHAMBON. 1988. Nuclear receptors enhance our understanding of transcription regulation. *Trends Genet.* 4:309–314.

PICARD, D., and K. R. YAMAMOTO. 1987. Two signals mediate hormone-dependent nuclear localization of the glucocorticoid receptor. *EMBO J.* 6:3333–3340.

HOLLENBERG, S. M., et al. 1985. Primary structure and expression of a functional human glucocorticoid receptor cDNA. *Nature* 318:635–641.

KUMAR, V., and P. CHAMBON. 1988. The estrogen receptor binds tightly to its responsive element as a ligand-induced homodimer. *Cell* 55:145–156.

UMESONO, K., V. GIGUERE, C. K. GLASS, M. G. ROSENFELD, and R. M. EVANS. 1988. Retinoic acid and thyroid hormone induce gene expression through a common responsive element. *Nature* 336:262–265.

Transcriptional Control of Gene Expression in Animal Cells

BAEUERLE, P. A., and D. BALTIMORE. 1988. IκB: a specific inhibitor of the NF-κB transcription factor. *Science* 242:540–546.

BLAU, H. M. 1988. Hierarchies of regulatory genes may specify mammalian development. *Cell* 53:673–674.

BODNER, M., and M. KARIN. 1987. A pituitary-specific *trans*-acting factor can stimulate transcription from the growth hormone promoter in extracts of nonexpressing cells. *Cell* 50:267–275.

BOULET, A. M., C. R. ERWIN, and W. J. RUTTER. 1986. Cell-specific enhancers in the rat exocrine pancreas. *Proc. Nat'l Acad. USA* 83:3599–3603.

COSTA, R. H., D. R. GRAYSON, and J. E. DARNELL, JR. 1989. Multiple hepatocyte-enriched nuclear factors function in the regulation of transthyretin and α1-antitrypsin genes. *Mol. Cell Biol.* 9:1415–1425.

HOLLAND, P. W. H., and B. L. M. HOGAN. 1988. Expression of homeobox genes during mouse development: a review. *Genes & Develop.* 2:773–782.

MONACI, P., A. NICOSIA, and R. CORTESE. 1988. Two different liver-specific factors stimulate *in vitro* transcription from the human α1-antitrypsin promoter. *EMBO J.* 7:2075–2087.

PINKERT, C. A., D. M. ORNITZ, R. L. BRINSTER, and R. D. PALMITER. 1987. An albumin enhancer located 10 kb upstream functions along with its promoter to direct efficient, liver-specific expression in transgenic mice. *Genes & Develop.* 1:268–276.

PINNEY, D. F., S. H. PEARSON-WHITE, S. F. KONIECZNY, K. E. LATHAM, and C. P. EMERSON, JR. 1988. Myogenic lineage determination and differentiation: evidence for a regulatory gene pathway. *Cell* 53:781–793.

XANTHOPOULOS, K. G., J. MIRKOVITCH, T. DECKER, C. F. KUO and J. E. DARNELL. 1989. Cell specific transcriptional control of the mouse DNA-binding protein C/EBP. *Proc. Natl. Acad. Sci. USA* 86:4117–4121.

Sequential Cascade of Transcription Factors in *Drosophila* Embryogenesis

BEACHY, P. A., M. A. KRASNOW, E. R. GAVIS, and D. S. HOGNESS. 1988. An *Ultrabithorax* protein binds sequences near its own and the *Antennapedia* P1 promoters. *Cell* 55:1069–1081.

DRIEVER, W., and C. NUSSLEIN-VOLLHARD. 1989. The bicoid protein is a positive regulator of *hunchback* transcription in the early *Drosophila* embryo. *Nature* 337:138–143.

INGHAM, P. W. 1988. The molecular genetics of embryonic pattern formation in *Drosophila*. *Nature* 335:25–34.

NUSSLEIN-VOLLHARD, C. and E. WIESCHAUS. 1980. Mutations affecting segment number and polarity in *Drosophila*. *Nature* 287:795–801.

PEIFER, M., F. KARCH, and W. BENDER. 1987. The bithorax complex: control of segmental identity. *Genes & Develop.* 1:891–898.

SCOTT, M. P., and S. B. CARROLL. 1987. The segmentation and homeotic gene network in early *Drosophila* development. *Cell* 51:689–698.

Effect of Chromatin Structure and Methylation on Gene Activity

BROWN, D. D. 1984. The role of stable complexes that repress and activate eucaryotic genes. *Cell* 37:359–365.

CEDAR, H. 1988. DNA methylation and gene activity. *Cell* 53:3–4.

DARBY, M. K., M. T. ANDREWS, and D. D. BROWN. 1988. Transcription complexes that program *Xenopus* 5S RNA genes are stable *in vivo*. *Proc. Nat'l Acad. Sci. USA* 85:5516–5520.

DOERFLER, W. 1983. DNA methylation in gene expression. *Ann. Rev. Biochem.* 52:93–124.

GROSVELD, F., G. B. VAN ASSENDELFT, D. R. GREAVES, and G. KOLLIAS. 1987. Position-independent, high-level expression of the human β-globin gene in transgenic mice. *Cell* 51:975–985.

PRIOR, C. P., C. R. CANTOR, E. M. JOHNSON, V. C. LITTAU, and V. G. ALLFREY. 1983. Reversible changes in nucleosome structure and histone accessibility in transcriptionally active and inactive states of rDNA chromatin. *Cell* 34:1033–1042.

TUAN, D., W. SOLOMON, Q. LI, and I. M. LONDON. 1985. The "β-like" globin gene domain in human erythroid cells. *Proc. Nat'l Acad. Sci. USA* 82:6328–6388.

WEINTRAUB, H. 1985. Assembly and propagation of repressed and derepressed chromosomal states. *Cell* 42:705–711.

WOLFFE, A. P., and D. D. BROWN. 1988. Developmental regulation of two 5S ribosomal RNA genes. *Science* 241:1626–1632.

WORKMAN, J. L., and R. G. ROEDER. 1987. Binding of transcription factor TFIID to the major late promoter during in vitro nucleosome assembly potentiates subsequent initiation by RNA polymerase II. *Cell* 51:613–622.

Control of Transcriptional Termination

BENTLEY, D. L., and M. GROUDINE. 1986. A block to elongation is largely responsible for decreased transcription of *c-myc* in differentiated HLGO cells. *Nature* 321:702–706.

BENTLEY, D. L., and M. GROUDINE. 1988. Sequence requirement for premature termination of transcription in the human *c-myc* gene. *Cell* 53:245–256.

KERPPOLA, T. K., and C. M. KANE. 1988. Intrinsic sites of transcription termination and pausing in the *c-myc* gene. *Mol. Cell Biol.* 8:4389–4394.

LOGAN, J., E. FALCK-PEDERSON, J. E. DARNELL, and T. SHENK. 1987. A poly (A) addition site and a downstream termination region required for efficient cessation of transcription by RNA polymerase II in the mouse β globin gene. *Proc. Nat'l Acad. Sci. USA* **84**:8306–8310.

Differential RNA Processing

BOVENBERG, R. A. L., G. J. ADEMA, H. S. JANSZ, and P. D. BASS. 1988. Model for tissue specific calcitonin/CGRP-1 RNA processing from *in vitro* experiments. *Nucleic Acids Res.* **16**:7867–7883.

BREITBART, R. E., et al. 1985. Intricate combinational patterns of exon splicing generate multiple regulated troponin T isoforms from a single gene. *Cell* **41**:67–82.

BREITBART, R. E., and B. NADAL-GINARD. 1987. Developmentally induced, muscle-specific *trans* factors control the differential splicing of alternative and constitutive troponin T exons. *Cell* **49**:793–803.

EARLY, P., et al. 1980. Two mRNAs can be produced from a single immunoglobulin μ gene by alternative RNA processing pathways. *Cell* **20**:313–319.

HANKE, P. D., and R. V. STORTI. 1988. The *Drosophila melanogaster* tropomyosin II gene produces multiple proteins by use of alternative tissue-specific promoters and alternative splicing. *Mol. Cell Biol.* **8**:3591–3602.

LEFF, S. E., M. G. ROSENFELD, and R. M. EVANS. 1986. Complex transcriptional units: diversity in gene expression by alternative RNA processing. *Ann. Rev. Biochem.* **55**:1091–1117.

WIECZOREK, D. F., C. W. J. SMITH, and B. NADAL-GINARD. 1988. The rat α-tropomyosin gene generates a minimum of six different mRNAs coding for striated, smooth, and nonmuscle isoforms by alternative splicing. *Mol. Cell Biol.* **8**:679–694.

WOLFNER, M. F. 1988. Sex-specific gene expression in somatic tissues of *Drosophila melanogaster*. *Trends Genet.* **4**:333–337.

ZACHAR, Z., T.-B. CHOU, and P. M. BINGHAM. 1987. Evidence that a regulatory gene autoregulates splicing of its transcript. *EMBO J.* **6**:4105–4111.

Overlapping Transcription Units

CARLSON, M., R. TAUSSIG, S. KUSTU, and D. BOTSTEIN. 1983. The secreted form of invertase in *Saccharomyces cerevisiae* is synthesized from mRNA encoding a signal sequence. *Mol. Cell Biol.* **3**:439–447.

VALES, L. D., and J. E. DARNELL, JR. 1989. Promoter occlusion prevents transcription of adenovirus polypeptide IX mRNA until after DNA replication. *Genes & Develop.* **3**:49–59.

YOUNG, R. A., O. HAGENBÜCHLE, and U. SCHIBLER. 1981. A single mouse α-amylase gene specifies two different tissue-specific mRNAs. *Cell* **23**:451–458.

Cytoplasmic Control of Gene Expression

mRNA Stability

BLANCHARD, J.-M., et al. 1985. c-*myc* gene is transcribed at high rate in Go-arrested fibroblasts and is post-transcriptionally regulated in response to growth factors. *Nature* **317**:443–445.

BRAWERMAN, G. 1987. Determinants of messenger RNA stability. *Cell* **48**:5–6.

BROCK, M. L., and D. J. SHAPIRO. 1983. Estrogen stabilizes vitellogenin mRNA against cytoplasmic degradation. *Cell* **34**:207–214.

DANI, C. H., et al. 1984. Extreme instability of myc mRNA in normal and transformed cells. *Proc. Nat'l Acad. Sci. USA* **81**:7046–7050.

GAY, D. A., T. J. YEN, J. T. Y. LAU, and D. W. CLEVELAND. 1987. Sequences that confer β-tubulin autoregulation through modulated mRNA stability reside within exon 1 of a β-tubulin mRNA. *Cell* **50**:671–679.

GREEN, L. L., and W. F. DOVE. 1988. Correlation between tubulin mRNA stability and poly A length over the cell cycle of *Physarum polycephalum*. *J. Mol. Biol.* **200**:321–328.

GUYETTE, W. A., R. J. MATUSIK, and J. M. ROSEN. 1979. Prolactin mediated transcriptional and post-transcriptional control of casein gene expression. *Cell* **17**:1013–1023.

MULLNER, E. W., and L. C. KUHN. 1988. A stem-loop in the 3′ untranslated region mediates iron-dependent regulation of transferrin receptor mRNA stability in the cytoplasm. *Cell* **53**:815–825.

RABBITTS, P. H., A. FORSTER, M. A. STINSON, and T. H. RABBITTS. 1985. Truncation of exon 1 from c-*myc* results in prolonged c-*myc* mRNA stability. *EMBO J.* **4**:3727–3733.

WILSON, T., and R. TREISMAN. 1988. Removal of poly A and consequent degradation of c-*fos* mRNA facilitated by 3′ AU-rich sequences. *Nature* **336**:396–399.

YEN, T. J., D. A. GAY, J. S. PACHTER, and D. W. CLEVELAND. 1988. Autoregulated changes in stability of polyribosome-bound β-tubulin mRNAs are specified by the first 13 translated nucleotides. *Mol. Cell Biol.* **8**:1224–1235.

Translation Rate

CASEY, J. L., et al. 1988. Iron-responsive elements: regulatory RNA sequences that control mRNA levels and translation. *Science* **240**:924–928.

DARVEAU, A., J. PELLETIER, and N. SONENBERG. 1985. Differential efficiencies of *in vitro* translation of mouse c-*myc* transcripts differing in the 5′ untranslated region. *Proc. Nat'l Acad. Sci. USA* **82**:2315–2319.

GEYER, P., O. MEYUHAS, R. P. PERRY, and L. F. JOHNSON. 1982. Regulation of ribosomal protein mRNA content and translation in growth-stimulated mouse fibroblasts. *Mol. Cell Biol.* **2**:685–693.

RYAZANOV, A. G., E. A. SHESTAKOVA, and P. G. NATAPOV. 1988. Phosphorylation of elongation factor 2 by EF-2 kinase affects rate of translation. *Nature* **334**:170–173.

C H A P T E R

12

DNA Replication, Repair, and Recombination

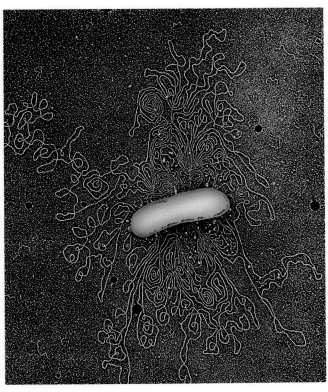

Rupture of *E. coli* and release of its chromosome

*B*efore Watson and Crick discovered the duplex structure of DNA, one of the most mysterious aspects of biology was how genetic material was exactly—or almost exactly—duplicated from one generation to the next. Recognition of the base-paired nature of the DNA duplex immediately gave rise to the notion that a template was involved in the transfer of information between generations, but a host of structural and biochemical questions soon followed. When replication has been completed, are old strands paired with new, or old with old and new with new? How does replication begin, and how does it progress along the chromosome? What mechanisms ensure only one round of replication before cell division? What enzymes take part in DNA synthesis, and what are their functions? How does duplication of the long helical duplex occur without the strands becoming tangled?

The events that allow assembly of new DNA chains are described collectively as DNA synthesis. DNA replication (or chromosomal replication), a more comprehensive term, encompasses not only DNA chain synthesis but also its initiation and termination. Studies of synthesis focus on enzymes and accessory factors at the point of chain growth, called the growing fork. Studies of replication are in addition concerned with how DNA synthesis starts and stops in such a way that each chromosome is duplicated exactly and further synthesis held in abeyance until after cell division. Another crucial issue in replication studies is how the two new chromosomes become separated. We discuss these processes in this chapter. The events of mitosis (or meiosis) during which the chromosomes are distributed to daughter cells are introduced in Chapter 5 and discussed in detail in Chapter 21.

Because eukaryotic chromosomes consist of complexes of DNA with protein, we also must consider the duplication of chromatin arrangements, which may influence gene expression, and the mechanism of packaging of newly formed DNA with the chromosomal proteins. Although the initiation of DNA synthesis and chromosome packaging are not yet fully understood in any eukaryotic cell, they are among the more important processes associated with cell division.

For DNA to serve as the genetic link between generations, not only must the base sequence be copied correctly during replication, but the integrity of the sequence must be maintained continuously for accurate protein synthesis in all cells. Consequently, it is not too surprising that cells possess enzymatic "repair" functions to keep DNA sequences accurate. Some of the enzymatic events necessary for repair of damaged DNA also play a role in genetic recombination, a major factor in evolution. The mechanisms of DNA repair and recombination are discussed in the later sections of this chapter. ▲

General Features of DNA Synthesis and Replication

In this section, we consider several general features of DNA synthesis including its semiconservative nature, its relationship to the cell cycle in eukaryotes, and the bidirectional growth of new DNA strands from a common point of origin.

DNA Replication Is Semiconservative

The base-pairing principle inherent in the Watson-Crick model suggested that the two new DNA chains were copied from the two old chains. Although this mechanism provided for exact copying of genetic information, it raised a new question: Is replication *conservative;* that is, does each new duplex contain two new strands, leaving

▸ **Figure 12-1** The Meselson-Stahl experiment showed that DNA replication is semiconservative (i.e., each daughter duplex contains one new and one old strand) not conservative (i.e., one daughter duplex contains two old strands and the other duplex contains two new strands). *E. coli* cells were grown in a medium containing (^{15}N)ammonium salts until all the cellular DNA contained the "heavy" isotope. The cells were then transferred to a medium containing the normal "light" isotope (^{14}N). Samples were removed from the cultures periodically and analyzed by density-gradient equilibrium centrifugation in CsCl to separate heavy-heavy (H-H), light-light (L-L), and heavy-light (H-L) duplexes into distinct bands. The actual banding patterns were consistent with the semiconservative mechanism. [See M. Meselson and W. F. Stahl, 1958, *Proc. Nat'l Acad. Sci. USA* **44**:671.] *Photographs courtesy of M. Meselson.*

the old duplex intact? Or is replication a *semiconservative* process in which each old chain becomes paired with a new chain copied from it? This question was settled first for *Escherichia coli* with proof that each newly formed duplex contains one old and one new DNA chain.

When *E. coli* cells are grown in ammonium salts containing ^{15}N instead of ^{14}N, the "heavy" atoms are incorporated into deoxyribonucleotide precursors and then into DNA. The resulting DNA is about 1 percent denser than the DNA containing only the normal isotope. Such "heavy" DNA and normal, "light" DNA form discrete bands during equilibrium density-gradient (isopycnic) centrifugation in solutions of CsCl. Bacteria grown for several generations in a heavy medium will contain almost completely substituted (i.e., heavy) DNA. If cells containing all heavy DNA are then transferred to a normal (light) medium and allowed to replicate once, the resulting DNA has a density that is intermediate between the densities of heavy and light DNAs. After the cells undergo another doubling, their DNA consists of one-half heavy-light DNA and one-half completely light DNA (Figure 12-1).

The experiment was refined further by denaturing and then separating the long strands of heavy-light DNA into single strands. If these single strands are subjected to equilibrium centrifugation, two distinct bands are formed, showing that each of the long strands is either all heavy or all light. These experiments showed that the original two heavy strands (H) copied in light medium result in two heavy-light duplexes; copying the mixed duplex led to H-L and L-L duplexes. This is *semiconservative* replication.

Experiments of a different design first demonstrated semiconservative DNA replication in eukaryotic chromosomes. Cells were labeled during DNA synthesis with [^{3}H]thymidine. The *mitotic* chromosomes were then examined by autoradiography beginning at the first division after labeling, which occurred within a few hours. The autoradiographs revealed that all mitotic chromosomes

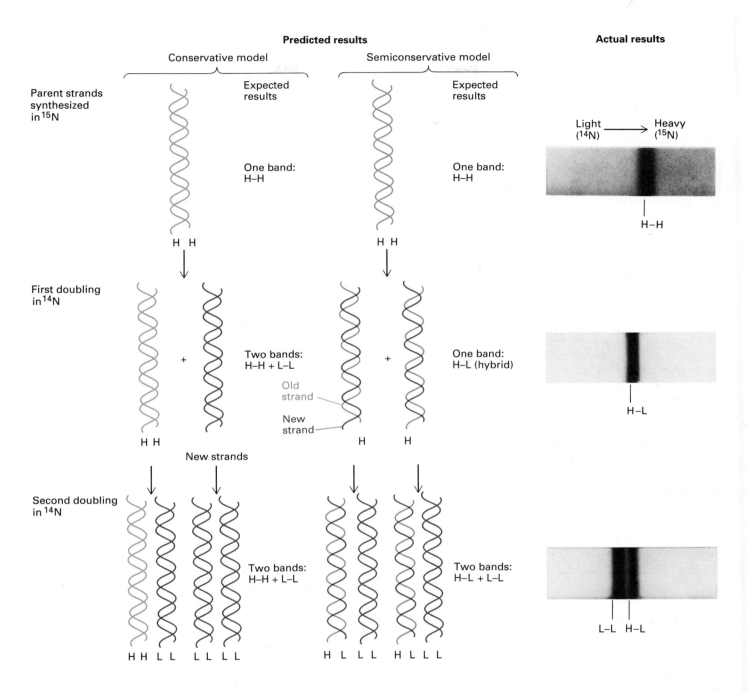

were labeled in *both* chromatids (Figure 12-2a). Presumably each chromatid represented one double-stranded DNA molecule, with one DNA strand in each chromatid being new (labeled) and one being old.

This interpretation was greatly strengthened by autoradiographs made after one further round of cell division in the absence of labeled thymidine. At this point one chromatid in each mitotic pair was labeled and one was totally unlabeled—a result consistent with the semiconservative model; the labeled chromatid contained the old labeled strand, and the unlabeled chromatid, the old unlabeled strand (Figure 12-2b). Thus each labeled strand retained its label and was paired after a second replica-

tion with a new unlabeled strand. Evidence of semiconservative replication has been obtained with both plant and animal chromosomes. Apparently all cellular DNA in both prokaryotic and eukaryotic cells is replicated by a semiconservative mechanism.

DNA Synthesis Occurs Only during S Phase of Cell Cycle in Eukaryotes

A second general consideration about DNA synthesis concerns its timing during the cell cycle. Bacteria require 20 to 30 min, which is about equal to the cell doubling time, to complete a round of DNA synthesis. Replication

(a) Metaphase chromosomes after pulse-labeling

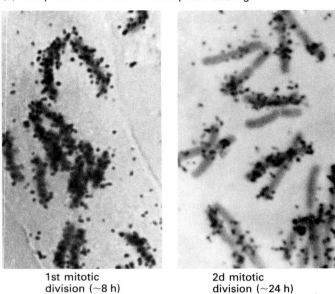

1st mitotic
division (~8 h)

2d mitotic
division (~24 h)

(b) Expected results with semiconservative replication

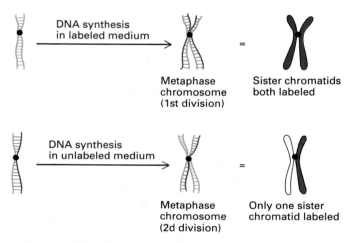

DNA synthesis
in labeled medium

Metaphase
chromosome
(1st division)

Sister chromatids
both labeled

DNA synthesis
in unlabeled medium

Metaphase
chromosome
(2d division)

Only one sister
chromatid labeled

▲ **Figure 12-2** Experimental demonstration of semiconservative DNA replication in root cells of the lily *(Bellavalia romana)*. Growing root cells were pulse-labeled with [³H]thymidine; samples then were taken periodically, stained, and autoradiographed. Mitotic cells with metaphase chromosomes were first evident about 8 h after the labeling period. (a) Autoradiograph of metaphase chromosomes during the first division after labeling shows that *both* sister chromatids are labeled, whereas during the second mitotic division, only one sister chromatid of each pair is heavily labeled. Because of sister chromatid exchange of whole segments of DNA during the second mitosis, some otherwise unlabeled chromatids have patches of silver grains. (b) Diagram illustrates how semiconservative replication would produce the observed results. [See J. H. Taylor, P. S. Woods, and W. L. Hughes, 1957, *Proc. Nat'l Acad. Sci. USA* **43**:122.] *Photographs courtesy of J. H. Taylor.*

is quickly followed by division into daughter cells, after which DNA synthesis immediately resumes. In eukaryotes, however, completion of cell division is followed by an interval (G$_1$ period) before DNA synthesis occurs; after DNA replication is completed, a second interval (G$_2$ period) occurs before division into daughter cells (see Figure 5-2). Even in rapidly dividing single-cell eukaryotes such as *Saccharomyces cerevisiae*, the same cell-cycle periods exist (Table 12-1). The longest period generally is the S phase, during which DNA synthesis occurs, although there are a few exceptions to this general pattern. For example, during early *Drosophila* embryogenesis, nuclear division can occur every 10 to 15 min; and during the first 12 cleavages in fertilized frog eggs, there is no G$_1$ or G$_2$, and cell division occurs as soon as DNA replication is complete.

Cell-fusion experiments with cultured mammalian cells have provided evidence about the availability of DNA for replication in the various stages of the cell cycle. Recall that either viral envelope proteins or certain chemicals (e.g., polyethylene glycol) can cause cell membranes to fuse, joining the contents of the cells (see Figure 5-11). The two cells to be fused can be labeled in various ways to distinguish each partner. If cells in the G$_1$ period are fused to cells in the S phase, the G$_1$ cells immediately begin DNA synthesis; in contrast, S-phase cells do not activate G$_2$ cells to begin DNA synthesis. This experiment indicates that G$_1$ cells have no inherent barrier to DNA synthesis and that S-phase cells can supply the factor(s) required for DNA replication. On the other hand, the newly duplicated chromosomes in G$_2$ cells are prevented from entering replication again, even in the presence of the S-phase triggering factors. Such a control ensures that duplicated chromosomes do not undergo a second replication before cell division, thereby preventing polyploidy. An exception to this control mechanism is the development of polytene chromosomes in insects.

If DNA replication is limited to the S phase, then what commits the cell to enter that stage and progress through cell division? In yeast cells, there is a point called *start*, which is defined by cell-division cycle *(cdc)* mutations (e.g., *cdc28* in *Saccharomyces cerevisiae* and *cdc2* in *Schizosaccharomyces pombe*; see Figure 5-14). If a yeast cell passes start, it will go through DNA synthesis and cell division even when it is placed under nongrowth condi-

Table 12-1 Length of cell-cycle periods in eukaryotes

Organism	Cell period M	G$_1$	S	G$_2$	Total doubling time
Human (h)	1	8	10	5	~24
Plant (h)	1	8	12	8	~29
Yeast (min)	20	25	40	35	~120

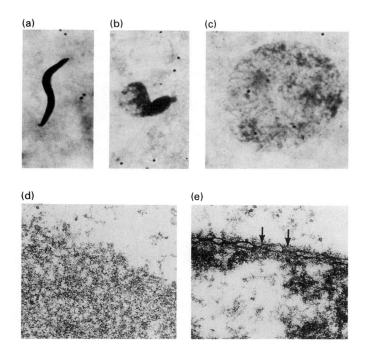

▲ **Figure 12-3** Sequential changes in demembranated sperm nuclei from *Xenopus laevis* during incubation with extracts of frog eggs. Light micrographs of preparations stained for DNA show (a) condensed nucleus at 0 min; (b) partially decondensed nucleus at 90 min; and (c) completely decondensed nucleus at 180 min. Electron micrographs show (d) initial sperm chromatin devoid of a nuclear membrane and (e) the nuclear membrane that forms around the chromatin during incubation with egg extract. Arrows in (e) indicate nuclear pores. [See M. J. Lohka and Y. Masui, 1984, *J. Cell Biol.* 98:1222.] *Courtesy of M. J. Lokha.*

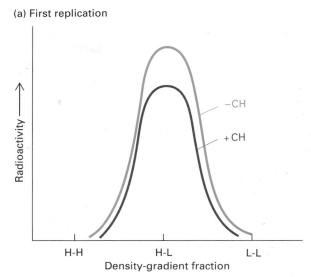

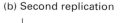

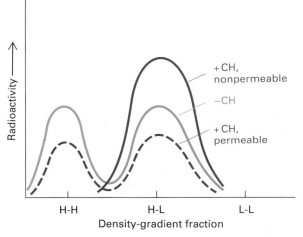

▲ **Figure 12-4** First and second round of DNA replication in *X. laevis* sperm nuclei incubated with complete cytoplasmic extract of frog eggs. The incubation mixtures contained radioactively labeled ATP and bromodeoxyuridine, a heavy thymidine substitute. (a) In the first round of replication, all of the radioactivity appeared in the heavy-light (H-L) fraction obtained in density-gradient centrifugation; this finding indicates that semiconservative replication was completed for all the nuclear DNA (see Figure 12-1). Replication occurred in the presence of cycloheximide (+CH), which inhibits protein synthesis, and in its absence (−CH). (b) When the nuclei not treated with CH (blue line) were resuspended in fresh extract and again incubated with radioactive ATP and bromodeoxyuridine, radioactivity appeared in the H-H fraction, indicating that at least some of the DNA in these nuclei underwent a second round of replication. No second round of replication occurred in the CH-treated nuclei (solid red line) unless they were treated to make them permeable before reincubation with fresh egg extract (dashed red line). Thus a new supply of some protein is required after re-formation of the nuclear membrane for the second round of replication. *Adapted from J. J. Blow and R. A. Laskey, 1988,* Nature 332:546.

tions. No specific genes have yet been identified in animals or plants that define a decision point equivalent to start. However, the products of the *cdc2* and *cdc28* genes, which are protein kinases, are similar to a protein that exists in human cells. The exact role of these protein kinases in starting a round of DNA replication is not clear.

Recent experiments with crude extracts prepared from frog sperm and oocytes promise to aid in uncovering the activity of various proteins that are important in the initiation of a round of DNA replication. Sperm nuclei without their outer membranes can be prepared by very mild detergent treatment, which leaves the densely packed chromosomal material unaltered. When mixed with extracts of mature oocytes, the sperm chromosomes become enclosed by a nuclear membrane and decondense, after which the DNA will be replicated one complete round (Figure 12-3). The egg extract contains vesicles that form the nuclear membrane and apparently all the factors required for DNA replication (Figure 12-4a). If the membrane of the nuclei is permeabilized artificially and new extract is added, a second round of DNA replication will ensue; however, this second round requires that normal protein synthesis be allowed during the first incubation (Figure 12-4b). Apparently new protein must be formed that allows or "licenses" the nuclei to engage in the second round of synthesis.

A model of the cell cycle consistent with these findings proposes that no DNA replication occurs in G_2 cells because their nuclei must have a renewed supply of "licensing factor" before the DNA is susceptible to another round of replication. The proposed factor is acquired at the time of nuclear membrane breakdown at mitosis. The model explains the immediate readiness of G_1 cells, which have just finished mitosis and nuclear membrane reformation, to participate in replication so long as the appropriate S-phase factors are present. This cycling of replication factors will come up again when we discuss proteins that participate in replication.

Most DNA Replication Is Bidirectional

How does the DNA replication machinery proceed along a duplex? Several possible molecular mechanisms of DNA synthesis (chain growth) would allow semiconservative DNA replication. In one of the simplest possibilities, one new strand derives from one origin and the other new strand derives from another origin (Figure 12-5a).

Only one strand of the duplex grows at each growing point. This mechanism does, in fact, occur in linear DNA viruses such as adenovirus. In such circumstances, the ends of the DNA molecules serve as fixed sites for the initiation and termination of replication. A second possibility envisions one origin and one *growing fork* (the fork of synthesis), which moves along the DNA in one direction with both strands of DNA being copied (Figure 12-5b). A third possibility is that synthesis might start at a single site and proceed in both directions, so that both strands are copied at each of *two* growing forks (Figure 12-5c). The available evidence suggests that the third alternative is the most common: DNA replication begins at a site and proceeds in both directions—that is, *bidirectionally*—from a given starting point, with both strands being copied at each fork. Thus two growing forks emerge from the origin site.

In a circular chromosome (the form found in bacteria and some viruses), one origin often suffices, and the two resulting growing forks merge on the opposite side of the circle to complete the replication. The long linear chromosomes of eukaryotes contain multiple origins; the two growing forks from a particular origin continue to advance until they meet the advancing growing forks from neighboring origins. Each region served by a DNA origin is called a *replicon* (Figure 12-6).

(a) Unidirectional growth of single strands from two origins

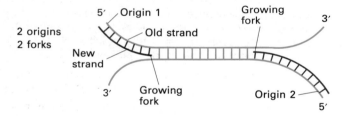

(b) Unidirectional growth of both strands from one origin

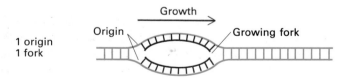

(c) Bidirectional growth of both strands from one origin

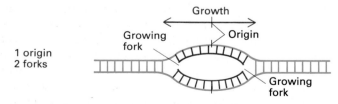

▲ **Figure 12-5** Three mechanisms of DNA strand growth that are consistent with semiconservative replication. The third mechanism—bidirectional growth of both strands from a single origin—appears to be the most common in both eukaryotes and prokaryotes.

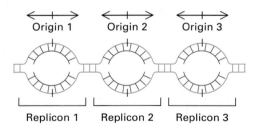

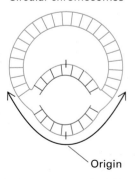

▲ **Figure 12-6** DNA synthesis in long linear chromosomes starts at multiple origins and proceeds in both directions until the growing forks of adjoining replicons are encountered. In circular chromosomes with a single origin, DNA synthesis proceeds in both directions around the circle until the advancing growing forks merge.

Bidirectional replication was first detected by fiber autoradiography of labeled DNA molecules from cultured mammalian cells (Figure 12-7). When cells are exposed alternately to high and low levels of [³H]thymidine, the labeled DNA tracks are first very dark and then lighter. Such studies have revealed clusters of active replicons with growing forks moving away from a common origin, thus providing unambiguous evidence of bidirectional growth. Most cellular DNA and many viral DNA molecules replicate bidirectionally. Such viruses serve as excellent models for the study of cellular DNA replication.

Replication "Bubbles"

If DNA from replicating cells is extracted and examined by electron microscopy, so-called replication "bubbles" or "eyes" extending from multiple origins of replication are visible (Figure 12-8). Such micrographs of cellular DNA do not provide conclusive evidence for unidirectional or bidirectional fork movement; however, electron-microscope studies of bub-

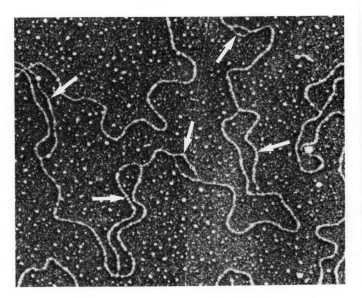

▲ **Figure 12-8** Electron micrograph of DNA extracted from rapidly dividing nuclei of early *D. melanogaster* embryos. The arrows mark replication bubbles; the diameters of the DNA chain in both arms of these bubbles indicate that they are double-stranded. [See A. B. Blumenthal, H. J. Kreigstein, and D. S. Hogness, 1973, *Cold Spring Harbor Symp. Quant. Biol.* **38**:205.] *Courtesy of D. S. Hogness.*

bles in viral DNA have provided evidence for bidirectional replication. By cutting circular viral DNA molecules at different stages of replication with a restriction endonuclease that recognizes a single site, the positions of the center of the replication bubble with respect to the restriction site can be determined (Figure 12-9). The most common result from such analyses is a series of ever-larger bubbles whose centers map to the same site, indicating bidirectional replication of both DNA strands from that site. Thus both fiber autoradiography and electron microscopy have indicated that bidirectional DNA replication is the general rule.

Number of Growing Forks and Their Rate of Movement

Fiber autoradiography of DNA from cells labeled for various times allows the rate of growing-fork movement to be estimated. As noted earlier, in *E. coli* cells one round of DNA replication takes about 30 min. Since the single circular *E. coli* chromosome of 3×10^6 base pairs is duplicated by two growing forks, the rate of fork movement is about 1000 bp/s per fork; the rate determined from fiber-labeling experiments on *E. coli* agrees with this value, indicating that the fiber-labeling technique can provide a reasonable estimate of the rate of growing-fork movement in vivo. The rate of fork movement in human cells, based on fiber-labeling experiments, is only about 100 bp/s per fork. Since the entire human genome of 3×10^9 base pairs replicates in 8 h, it must contain at least 1000 growing forks. If there were only 1000 growing forks, each would be required to replicate

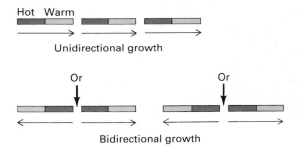

(a) Predicted fiber autoradiographic pattern

Hot Warm

Unidirectional growth

Or   Or

Bidirectional growth

(b) Actual fiber autoradiographic pattern

← Or →   ← Or →

▲ **Figure 12-7** Demonstration of bidirectional growth of cellular DNA chains. If cultured replicating cells are exposed alternately to high and low levels of [³H]thymidine, the resulting DNA will be heavily labeled ("hot") near replication origins (Or) and lightly labeled ("warm") farther away. When such labeled DNA is dried on a microscope slide as long linear molecules (fibers) and then exposed to a radiation-sensitive emulsion, autoradiographic signals should be produced corresponding to the hot-warm DNA regions. (a) Predicted patterns of autoradiographic bands for uni- and bidirectional DNA synthesis. (b) Actual fiber autoradiograph of DNA from cultured mammalian cells shows autoradiographic signals consistent with bidirectional synthesis. [See J. A. Huberman and A. D. Riggs, 1968, *J. Mol. Biol.*, **32**:327; and J. A. Huberman and A. Tsai, 1973, *J. Mol. Biol.* **75**:5.] *Photograph courtesy of J. A. Huberman.*

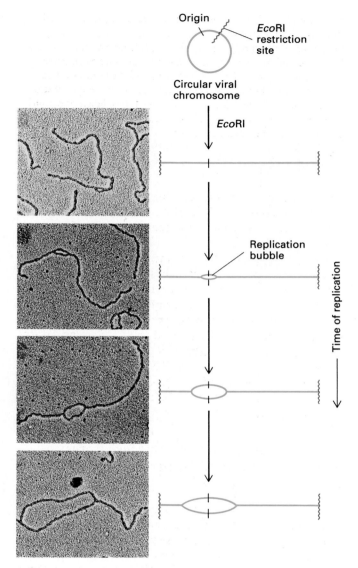

◀ Figure 12-9 Demonstration of bidirectional chain growth from a single origin in viral DNA. The replicating DNA from SV40-infected cells was cut by the restriction enzyme *Eco*RI, which recognizes a single site, and examined by electron microscopy. The electron micrographs and corresponding diagrams show a collection of increasingly longer replication bubbles, the centers of which are a constant distance from each end of the cut molecules. These results indicate that chain growth occurs in two directions from a common origin. [See G. C. Fareed, C. F. Garon, and N. P. Salzman, 1972, *J. Virol.* **10**:484.] *Photographs courtesy of N. P. Salzman.*

of a growing chain and thus direct growth *only* in the 5' → 3' direction. Furthermore, all DNA polymerases so far discovered only elongate a preexisting primer strand of DNA or RNA; they cannot initiate chains. These two properties of DNA polymerases pose distinct problems in the copying of DNA. In this section, we first describe how movement of the growing fork in one direction is compatible with growth of two new strands, which must be synthesized in opposite directions; then we consider the enzymes and accessory factors required for chain growth; and finally we discuss initiation of DNA replication.

A Growing Fork Has a Continuous Leading Strand and a Discontinuous Lagging Strand Primed by RNA

At each growing fork, one new DNA strand, called the *leading strand,* grows continuously in the 5' → 3' direction, which coincides with the overall movement of the growing fork. To form the other new strand, the *lagging strand,* short, discontinuous segments are synthesized also in the 5' → 3' direction, which in this case is opposite to the overall direction of movement of the growing fork (Figure 12-10, *top*). The segments of the lagging strand eventually are linked together by a *DNA ligase.*

The primer for synthesis of the leading strand hypothetically could be a free end created within the parental DNA, but the primer for lagging-strand synthesis is RNA. As synthesis of the leading strand progresses, sites uncovered on the unreplicated parental strand are copied into short RNA oligonucleotides. The RNA primers are then elongated in the 5' → 3' direction by DNA polymerase to make the lagging strand (Figure 12-10a and b). These short (1000-nucleotide or less) segments of RNA plus DNA are called *Okazaki fragments,* after their discoverer Reiji Okazaki. As each segment of the lagging strand approaches the 5' end of the adjacent Okazaki fragment (the one just synthesized), the DNA polymerase exhibits one of its enzymatic capacities, a 5' → 3' exonuclease activity, that removes the RNA primer at the 5' end of the neighboring fragment. Simultaneously with the removal of the primer RNA, the gap between the frag-

about 10⁶ base pairs, and we know from fiber autoradiography and electron microscopy (Figures 12-7 and 12-8) that growing forks are not spaced that far apart. Thus each growing fork does not work the entire 8 h required for replication, and the more likely estimate is that the human genome contains between 10,000–100,000 replicons.

Initiation and Propagation of a DNA Chain at a Growing Fork

Before we plunge into a description of the several types of enzymes and numerous other factors that act at a growing fork, certain elementary problems in copying DNA must be addressed. The DNA duplex is an antiparallel helix: that is, the two strands are opposite (5' → 3' and 3' → 5') in chemical polarity. However, all DNA polymerases catalyze nucleotide addition to the 3' hydroxyl end

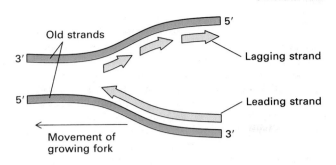

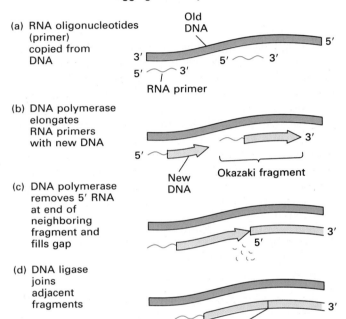

◄ Figure 12-10 The overall structure of a growing fork *(top)* and steps in synthesis of the lagging strand. Synthesis of the leading strand, catalyzed by DNA polymerase, occurs by sequential addition of deoxyribonucleotides with release of PP_i (see Figure 3-22). Synthesis of the discontinuous lagging strand is more complicated and involves several distinct steps (a–d). The reaction catalyzed by DNA ligase *(bottom)* joins the 3′ hydroxyl end of one Okazaki fragment to the 5′ phosphate of the adjacent fragment. During this reaction, the enzyme is transiently attached covalently to the 5′ phosphate of the DNA, thus activating the phosphate group.

ments is filled by addition of deoxynucleotides by the polymerase (Figure 12-10c). Finally, another critical enzyme, DNA ligase, joins adjacent completed fragments (Figure 12-10d and *bottom*).

The formation of RNA primers is the key to the dilemma of replicating antiparallel strands. All cells possess a specific RNA-synthesizing enzyme, called *primase*, that is responsible for RNA-primer synthesis. These enzymes work as part of multiprotein complexes, as we will describe in detail later. Their products, which can be isolated from cells in the segments at the 5′ end of RNA-DNA Okazaki fragments. The RNA is from 4 to about 12 nucleotides in length (Table 12-2). Primases are active only in connection with DNA replication and do not make long chains of RNA.

Bidirectional Replication Bubbles Form after Initiation of DNA Synthesis on One Template Strand

The mechanism of DNA chain growth at a growing fork—continuous elongation of the leading strand and discontinuous, primer-dependent synthesis of the lagging strand—can explain the formation of bidirectional replication bubbles. Any mechanism that starts the synthesis of one new leading strand will automatically generate sites on the opposite template strand at which RNA priming and lagging-strand synthesis can begin. The mecha-

Table 12-2 RNA primers (primase products) used in DNA chain initiation

Replicating DNA	RNA oligonucleotide*
Bacteriophage T4	$pppAC (N)_3$
Bacteriophage T7	$pppACCA$ $pppACCC$
Yeast	$pppA (N)_{8-12}$
Drosophila	$pppA (N)_{7-9}$
Lymphoblastoid cells	$pppA (N)_8$ $pppG (N)_8$

*N = any ribonucleotide. The indicated primer lengths for the eukaroytic cells are averages. Yeast and *Drosophila* RNA primers are rich in adenine, which accounts for about 75 percent of the bases.

nism for starting a leading strand might be a nick (a single-strand cut) that exposes a 3′ hydroxyl end of DNA or melting of the DNA and initiation by an RNA primer; the latter mechanism is the most common one in chromosomal replication. Once both leading- and lagging-strand synthesis are under way, one growing fork has been created. As the 3′ end of the lagging strand of the first growing fork (GF_R in Figure 12-11) passes the origin, it continues growth as the leading strand for the second growing fork moving in the opposite direction. To complete synthesis at the origin of the two forks, appropriate "patching" (cutting away of primers and gap filling) followed by ligation is required. In this manner each nucleotide in the original duplex is copied, including the nucleotides at the point of origin. Special problems arise at the ends of linear chromosomes where lagging-strand synthesis would be incomplete without a special mechanism. A special enzyme, *telomerase,* takes care of this problem (see Figures 9-13 and 9-14).

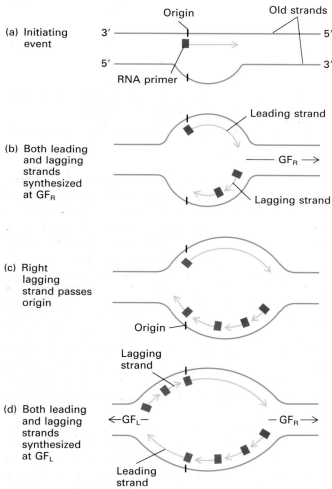

(a) Initiating event

Origin

Old strands

3′ ————————————————— 5′

5′ ————————————————— 3′

RNA primer

(b) Both leading and lagging strands synthesized at GF_R

Leading strand

— GF_R →

Lagging strand

(c) Right lagging strand passes origin

Origin

Lagging strand

(d) Both leading and lagging strands synthesized at GF_L

←GF_L—

— GF_R →

Leading strand

▲ **Figure 12-11** Formation of bidirectional DNA replication bubble following initiation by an RNA primer on one strand. GF_R = rightward growing fork; GF_L = leftward growing fork. Not shown are removal of primers, gap filling, and joining of segments to yield two complete new strands.

Another potentially important point about growing forks and their movement through a stretch of DNA is implicit in Figure 12-11. As the diagram is drawn, the 3′ → 5′ template strand is copied as the leading strand at GF_R. However, if replication across this region proceeded from the opposite direction (i.e., moving from right to left), the top template strand would be copied as the lagging strand. If the binding of proteins (e.g., transcription factors) to new DNA duplexes varies depending on whether the new strand is a leading- or lagging-strand product, then the direction of an approaching growing fork could be important. It is not known whether the direction of copying of any particular region is always the same in eukaryotic genomes, where multiple origins operate. The direction of copying is constant in the *E. coli* circular chromosome, which has only one origin.

A final theoretical problem posed by growing forks may have occurred to the reader. What happens if a growing fork meets an RNA polymerase that is moving in the opposite direction? This seems to be a rare occurrence, at least in *E. coli.* Almost all (>90 percent) of the rapidly transcribed *E. coli* genes (e.g., those encoding ribosomal RNAs and tRNAs), as well as three-fourths of all the protein-coding genes, are transcribed in the same direction in which they are replicated. It may be that those genes oriented against the direction of the growing fork (i.e., opposite to the direction of copying), can support only infrequent starts by RNA polymerase so as not to affect passage of growing forks.

Functional in Vitro Growing Forks Can Be Made with Purified Proteins and DNA

Knowledge about the proteins involved in DNA replication in *E. coli* is now very detailed, and significant progress has also been made on proteins involved in eukaryotic DNA replication. Identifying the approximately 20 proteins that take part in DNA replication in *E. coli* and unraveling their function were achieved mainly by painstaking biochemical analyses of protein extracts from normal strains of *E. coli* and mutants with a defect in DNA synthesis. For several years now it has been possible to reconstruct a functional DNA growing fork in the test tube. More recently, how a growing fork gets started has also been solved in *E. coli,* and in vitro initiation of DNA replication that mimics the in vivo events has been achieved. In this section, we first discuss the properties of the DNA polymerases, which historically were the first replication proteins to be recognized, and then consider other proteins active at the growing fork. After this look at the functions of various proteins at the growing fork, we will consider initiation of replication in more detail.

DNA Polymerases   DNA polymerases were first recognized in cell extracts by their capacity to incorporate

Table 12-3 Properties of DNA polymerases

E. coli	I	II	III	
Polymerization: 5' → 3'	+	+	+	
Exonuclease activity: 3' → 5'	+	+	+	
5' → 3'	+	−	−	
Synthesis from:				
Intact DNA	−	−	−	
Primed single strands	+	−	−	
Primed single strands plus single-strand-binding protein	+	−	+	
In vitro chain elongation rate (nucleotides per minute)	600	?	30,000	
Molecules present per cell	400	?	10–20	
Mutation lethal?	+	−	+	

MAMMALIAN CELLS*	α	β†	γ	δ
Polymerization: 5' → 3'	+	+	+	+
Exonuclease activity‡: 3' → 5' (editing function)	+	−	−	+
Synthesis from:				
RNA primer	+	−	−	+
DNA primer	−	−	+	+
Associated DNA primase	+	−	−	−
Sensitive to aphidicolin (inhibitor of cell DNA synthesis)?	+	−	−	+
Cell location:				
Nuclei	+	+	−	+
Mitochondria	−	−	+	−

*Yeast DNA polymerase I, II, and III are equivalent to polymerase α, β, and δ, respectively. I and III are essential for cell viability.
†Polymerase β is most active on DNA molecules with gaps of about 20 nucleotides and is thought to play a role in DNA repair.
‡Eukaryotic enzymes undoubtedly have 5' → 3' exonuclease activities to remove primers, but such reactions with strictly purified enzymes have not been carried out, except in the case of the herpes DNA polymerase, which has such an activity.

deoxynucleotides into DNA, using deoxynucleoside triphosphates as substrates. As mentioned earlier, these enzymes only copy a preexisting DNA template, and they obey the rules of Watson-Crick base pairing. As DNA polymerases were purified, first from *E. coli* and later from many other cells, each was found to require a free 3' hydroxyl end as a primer.

Three DNA polymerases (I, II, and III) have been purified from *E. coli*. Because bacteria with a temperature-sensitive mutation in polymerase III were found to stop making DNA when shifted to a nonpermissive temperature, polymerase III was singled out for special attention in trying to make in vitro growing forks that worked. This enzyme is now known to be the functional enzyme at the growing fork. DNA polymerase I is probably the most important enzyme for gap filling during DNA repair. The exact function of DNA polymerase II is still unclear, although it may play some special role in repair of DNA after chemical or physical damage.

Once DNA polymerase III is bound to a template, it does not easily release, whereas both polymerase I and II detach from the template after adding only 20 or so nucleotides. Because it requires about 1 min each time a DNA polymerase molecule comes off the template for synthesis to start again, the ability of DNA polymerase III to stick to its task—that is, its *high processivity*—makes

it particularly suitable for growing-fork activity. In fact, polymerase III together with the other necessary replication factors can make DNA in vitro at rates in excess of 500 nucleotides per second; this compares favorably with the in vivo rate of about 1000 nucleotides per second.

Both DNA polymerase I and III also possess 3' → 5' exonuclease activity. (We have already discussed 5' → 3' exonuclease activity which removes RNA primers; the 3' → 5' activity serves a different function.) DNA polymerases add nucleotides with an error frequency of about 1 in 10,000, an unacceptably high error rate. The 3' → 5' exonuclease activity saves the day: Because of this exonuclease activity, both enzymes can pause and remove "wrong" nucleotides and then continue and add the right ones; in this way, the error frequency is cut to 1 in a million or less.

The properties of the *E. coli* DNA polymerases are summarized in Table 12-3. A model of DNA polymerase III—a giant, asymmetric multisubunit protein—is shown in Figure 12-12. As discussed later, its asymmetry probably allows a polymerase III molecule to simultaneously catalyze synthesis of the leading and lagging strand at a growing fork.

Many years after DNA polymerases were purified from *E. coli*, four polymerases were identified in mammalian cells (Table 12-3). Although no well-characterized muta-

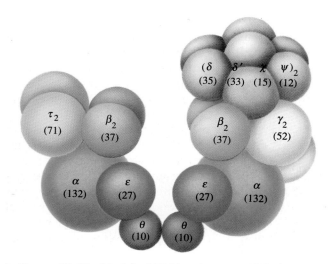

▲ **Figure 12-12** Model of DNA polymerase III holoenzyme from *E. coli*. The numbers in parentheses indicate the molecular mass in kilodaltons of each subunit. The asymmetric, dimeric structure of polymerase III probably is important for its activity at the growing fork. *Adapted from A. Kornberg, 1988, J. Biol. Chem. **263**:1.*

tions exist for studying these enzymes in animal cells, studies with drugs have helped pinpoint which proteins are active at the growing fork. For example, a drug called *aphidocolin* immediately blocks mammalian DNA replication and also inhibits two of the DNA polymerases, α and δ, both of which are now known to be required for in vitro DNA replication. Similar yeast proteins also have been identified. Yeast polymerase I is equivalent to mammalian α, and yeast III to mammalian δ. As we shall see later, the replication function of the giant polymerase III from *E. coli* is carried out by the two eukaryotic polymerase (α and δ, or I and III). Mammalian polymerase β and yeast polymerase II are thought to act in DNA repair. Mammalian polymerase γ is found in mitochondria and presumably acts in DNA replication in that organelle.

Other Replication Proteins Although DNA polymerases are the most obviously necessary component of a growing fork, they are by no means sufficient to form a functional fork. For example, if purified *E. coli* polymerases are incubated with a primed template in vitro, only single-stranded DNA is made; in vivo, of course, both strands of DNA are replicated at a growing fork. The first successful work that led to the isolation of all the proteins required to produce a functioning growing fork in *E. coli* purposely sidestepped an important issue: where exactly did DNA synthesis begin in the *E. coli* chromosome? Not to be stymied by lack of knowledge about precise starting points, investigators used the circular DNA from certain small bacterial viruses as templates for in vitro replication studies. Like plasmids, these viral DNAs are replicated as

double-stranded circles (Figure 12-13). A specific nick in one strand of these circles can be achieved by a variety of techniques. With such nicked, double-stranded viral templates, it was possible to isolate, in addition to the DNA polymerases, the other proteins required to build and propagate a growing fork in vitro. Two replicating systems have been explored in great detail: bacteriophage T4 DNA replication, which is carried out by T4-encoded proteins, and *E. coli* replication. Because the *E. coli* replication proteins may be somewhat more related than the T4 proteins to cellular proteins in general, we will describe the *E. coli* system in what follows.

Among the most complex of the contributors to growing-fork function is the *primosome*, named for its ability to make the RNA primers required for lagging-

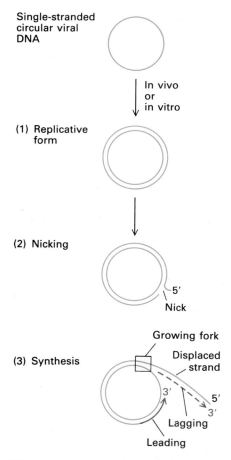

▲ **Figure 12-13** Replication of DNA of small *E. coli* viruses such as ϕX174 or M13. In an infected cell, the single-stranded viral DNA is converted into double-stranded replicative forms, which can be isolated easily. During normal infection, a viral protein produces one nick (a single-strand cut) at a particular site; experimentally, a nick can be produced by DNase treatment. Such nicked circular DNAs can serve as templates for in vitro DNA synthesis; purification of the *E. coli* proteins that function at the growing fork was achieved using these templates. One growing fork on such a circular DNA generates new chains (red) by a "rolling circle" form of replication.

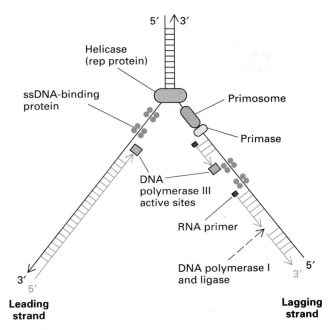

▲ Figure 12-14 Diagram of the *E. coli* replisome including the various enzymes and accessory proteins that function at the growing fork. The primosome associated with the lagging-strand template contains six different proteins including dnaB and dnaC. *Adapted from A. Kornberg, 1988, J. Biol. Chem.* **263**:1.

strand synthesis (Figure 12-14). In *E. coli,* the six-protein primosome complex is associated with *primase,* an enzyme that makes the short RNA primer at the 5′ end of Okazaki fragments. Although the *E. coli* primosome can be isolated as an independent structure, in cells and in growing forks assembled in vitro, it is part of the overall protein machine found at structure of the growing fork.

In early attempts to assemble active growing forks from *E. coli* extracts, it was noted that ATP was required not just to make RNA but as an energy source. When the protein that utilizes ATP (initially called *rep protein*) was purified, it was found to be able to unwind the DNA helix using ATP hydrolysis as the source of energy. Therefore, the protein is called a *helicase*. In addition, one of the proteins of the primosome (dnaB) also has helicase activity, an important point to which we will return when we discuss initiation of growing forks.

One final protein, single-stranded DNA-binding (SSB) protein, completes the requirements to construct an active growing fork. The SSB protein binds to sites where the template helix has been unwound but is not yet duplicated. It presumably functions by preventing tangles and by making the template ready to copy. It readily dissociates as the polymerase advances. This entire collection of accessory proteins plus DNA polymerase is sometimes referred to as a *replisome.*

Thus far we have implied that leading-strand synthesis proceeds with no difficulty and does not even require all

of the elements of lagging-strand synthesis (e.g., the primosome). However, it is now recognized that coordination between leading-strand and lagging-strand synthesis is the rule. Since a single polymerase active site cannot actually add nucleotides simultaneously at the 3′ ends of leading and lagging growing points, how does coordination occur? As noted earlier, DNA polymerase III is a very large assembly (>600 kDa) that contains two copies of most subunits (see Figure 12-12); the enzyme is an asymmetric dimer with two catalytic sites for nucleotide addition. It seems likely but it is not proven, that the lagging strand wraps around one arm of the polymerase, thus inverting the physical direction (but not the biochemical direction) of the new growing chain on the lagging strand (Figure 12-15). This inversion would place the 3′ growing

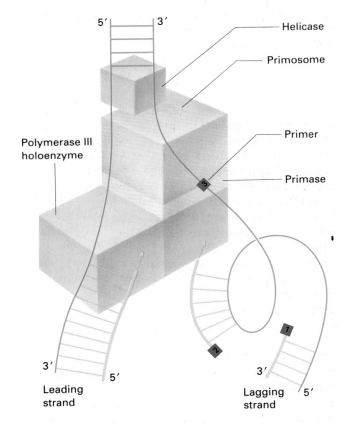

▲ Figure 12-15 Proposed mechanism of concurrent synthesis of leading and lagging strands by *E. coli* replisome. Looping of the lagging-strand template around one arm of the asymmetric dimeric DNA polymerase III (see Figure 12-12) brings the 3′ end of the growing lagging strand close to the catalytic site in one arm of the polymerase molecule; the leading strand is produced at the catalytic site in the other arm. As replication proceeds, more lagging-strand primers are formed (e.g., primer 3); these are, in turn, looped into an active site for elongation by the DNA polymerase. *Adapted from A. Kornberg, 1988, J. Biol. Chem.* **263**:1.

ends of both the leading and lagging strands close to one of the two catalytic sites in polymerase III, so that deoxynucleotides can be added to the leading strand and lagging strand (or its RNA primer) at the same time. This mechanism, however, means that at any point of addition, the bases being copied from the parental strands are not from the same site in the original duplex. The RNA primer that is being elongated in the lagging strand is "behind" the point in the template at which leading-strand copying is occurring. In addition, as each lagging-strand segment is completed (in Figure 12-15, primer 2 is being elongated toward primer 1), there must be a release of the lagging strand and a transfer of the active site to the next RNA primer (primer 3 in the figure) to be extended. Evidence in favor of a pause during such a transfer has been obtained with the T4 replication system. Close coordination of these events must occur to allow a growing fork to move 500–1000 nucleotides a second and have both strands replicated.

Although functional growing forks have not been created with completely purified eukaryotic DNA polymerases and accessory proteins, enough is known to suggest that the assembly of proteins at growing forks in eukaryotes is similar to that in *E. coli*. DNA helicases have been isolated from both mammals and yeast, and primases have been found associated with α but not δ polymerase. Furthermore, it appears likely that the α and δ polymerases, both of which are quite large (>200 kDa) and have two or three subunits, act together like the asymmetric arms of DNA polymerase III from *E. coli* (Figure 12-16). An accessory protein called PCNA (proliferating cell nuclear antigen) is necessary for δ polymerase activity. In the absence of PCNA, only short DNA segments, on the lagging-strand template, are made; in the presence of PCNA, which binds to polymerase δ, leading strands are also made, and both strands are synthesized coordinately. Therefore it seems likely that polymerase α and the associate primase make the lagging strand and that polymerase δ and PCNA are responsible for leading-strand synthesis. If the two polymerases associate (and large complexes have been isolated), the lagging strand may wrap around DNA polymerase α as it does in *E. coli*, so that the complex makes both strands at one locus (Figure 12-16).

DNA Replication Begins at Specific Chromosomal Regions

Before investigators could tackle the problem of initiation of DNA replication in vitro, they had to determine whether there are specific sites on chromosomes at which DNA replication always begins in vivo. Animal viruses were shown by electron micrographic studies to have replication bubbles whose centers were always in the same

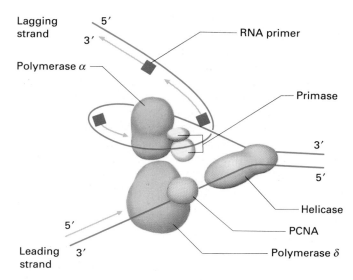

▲ **Figure 12-16** Model of assembly of DNA replication proteins at growing fork in eukaryotes. DNA polymerase α and δ are thought to function like the two asymmetric arms of *E. coli* polymerase III, directing concurrent synthesis of the lagging and leading strands, respectively. *Adapted from B. Stillman, 1988,* Bio. Essays 9:56.

approximate site (Figure 12-9). The same was also demonstrated for circular bacterial and plant viruses and in bacterial, yeast, and mammalian plasmids. More than 20 years ago, double-labeling experiments with synchronized bacterial cells and synchronized animal cells also suggested that there is at least regional specificity for DNA origins in both bacterial and animal cells. In these experiments the earliest synthesized DNA at the beginning of a round of synthesis was density-labeled with heavy isotope. Medium with normal isotope was then supplied to complete the round of synthesis, and the cells were allowed to grow several generations and again synchronize. This time the synchronized cells were briefly labeled and the DNA then examined by density analysis. In both animal cells and bacteria the newly labeled DNA was both heavy and radioactively labeled, indicating that the same regions of DNA were indeed included each time at the beginning of replication. However, none of these experiments provided sufficient resolution to indicate which specific nucleotide sequence(s) may be important in beginning each round of DNA replication. With the advent of recombinant DNA methods, it has been possible to pinpoint specific sequences in bacterial, yeast, and many viral DNAs that function as origins of DNA replication.

A *replication origin* is defined experimentally as a stretch of DNA that is necessary and sufficient to ensure replication of a circular DNA, usually a plasmid or virus, in the appropriate host cell. In yeast, but not yet in mammalian cells, this definition has been refined to include sequences that direct replication once per S phase, an important characteristic of DNA replication in eukaryotic cells. We discuss three replication origins to illustrate some general conclusions about the nature of replication origins; two of these origins, *E. coli* oriC and the simian virus 40 (SV40) origin, have been used extensively to study in vitro initiation of DNA replication. However, since the SV40 genome is replicated many times in a cell, its origin clearly violates the rule of one replication per cell cycle. The detailed knowledge of the pure proteins required to start replication at the *E. coli* origin and the accumulating information about other origins and their use in vitro all suggest that most cellular DNA replication may in fact begin at specific sequences, possibly using similar mechanisms.

It is now known that plasmids containing *E. coli* oriC, a DNA segment of about 240 nucleotides, are capable of independent and controlled replication. OriC is the only *E. coli* segment capable of conferring independent replication. Important conserved sequence features in bacterial origins have been deduced from phylogenetic comparisons of oriC with the origins of five other bacterial species including the distant species *Vibrio harveyi*, a marine bacterium (Figure 12-17). Repetitive 9-bp and 13-bp sequences (referred to as 9-mers and 13-mers, respectively) are characteristic of every bacterial origin sequence; as we will see later, these are important binding sites for proteins that initiate replication.

The yeast genome, like all eukaryotic genomes, has multiple origins of replication. Cloning experiments indicate that about 400 origins exist in the 17 chromosomes of *S. cerevisiae;* more than a dozen of these have been characterized in detail. Each yeast origin sequence confers on a plasmid the ability to be replicated and is called an autonomously replicating sequence (ARS); as we noted in

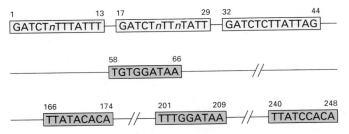

▲ **Figure 12-17** Consensus sequence of the minimal bacterial replication origin based on analyses of chromosomes from six species. The 13-bp repetitive sequences (yellow) are rich in adenine and thymine residues; the 9-bp repetitive sequences (green) exist in both orientations. These sequences are referred to as 13-mers and 9-mers, respectively. Indicated nucleotide position numbers are arbitrary. This consensus sequence was derived from comparison of the following species: *E. coli, Salmonella typhimurium, Enterobacter aerogenes, Klebsiella pneumoniae, Erwinia carotovora,* and *Vibrio harveyi*. [See J. Zyskind et al., 1983, *Proc. Nat'l Acad. Sci. USA* **80**:1164.]

(a) Yeast origin (ARS)

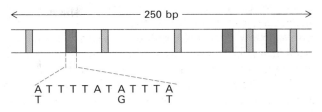

(b) SV40 origin

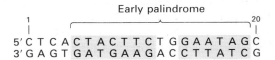

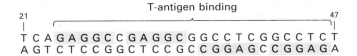

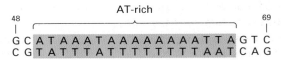

▲ **Figure 12-18** (a) General design of replication origin, or autonomously replicating sequence (ARS), in yeasts. Many of these contain three 12-bp highly conserved regions (blue), which are required for activity; the consensus sequence of this region is shown. Other short, repetitive regions (purple), which are rich in adenine and thymine residues, also are present; at least some of these probably are also required for origin activity. [See J. Campbell, 1986, *Ann. Rev. Biochem.* 55:733.] (b) Nucleotide sequence of the replication origin in SV40 viral DNA. The three regions shown by mutagenesis studies to be required for activity are indicated by color shading. Indicated nucleotide position numbers are arbitrary. [See P. Tegtmeyer et al., 1988, *Eukaryotic DNA Replication,* Cold Spring Harbor Laboratory.]

Chapter 9, an ARS is required to build an artificial chromosome (see Figure 9-9). Deletion analyses of several yeast ARS regions have shown that at least 100–200 bases are required for maximal function. The ARSs studied to date contain a highly conserved, repeated 12-bp sequence, which is required for origin activity (Figure 12-18a). Other short sequences similar to the 12-bp conserved region are also present in the origin region; some of these are required for function. Exactly how an ARS functions in the initiation of DNA replication is not yet known.

A 65-bp region in the SV40 chromosome is sufficient to promote DNA replication both in vivo and in vitro. Three regions of the SV40 origin have been shown by site-directed mutational analysis to be required for its activity (Figure 12-18b). The function of these regions in initiating DNA replication are described in the next section.

Origin-Binding Proteins Can Initiate DNA Replication in Vitro

We will now consider the initiation of DNA replication from *E. coli* oriC and the SV40 origin, the two systems that have been studied most thoroughly in vitro.

Initiation of DNA Replication by *E. coli* Proteins

Initial progress in understanding initiation of replication in *E. coli* came from studies of the product of the *dnaA* gene, which was known to be critical for cell viability. In in vitro replication experiments that use artificially nicked circular templates, the dnaA protein is not required to start a growing fork. Yet the gene is obviously critical in vivo because cells can sustain only conditional lethal mutations in this gene; that is, no missense or deletion mutants in dnaA are known. However, when *E. coli* cells with a temperature-sensitive mutation in *dnaA* are shifted to high temperature, they continue to complete the round of DNA synthesis already under way and then stop making DNA. Such experiments suggest that the dnaA protein, which is very scarce in cells, is critical for *starting* DNA replication.

By cloning the *dnaA* gene in a bacteriophage, investigators obtained large amounts of pure dnaA protein. In vitro studies showed that the protein binds to the four 9-mers in the *E. coli* origin, forming a multimeric complex that contains 20 to 30 subunits (Figure 12-19). The binding of dnaA protein to oriC requires ATP and occurs only at elevated temperatures (about 30°C) in vitro. Furthermore, dnaA binds only to a circular DNA molecule that is negatively supercoiled. Templates with negative supercoils are easier to unwind locally than templates without supercoils. Supercoiling of DNA and the enzymes, called topoisomerases, that control the degree of DNA supercoiling are discussed in detail in the next section. One mechanism for inducing negative supercoiling inside cells is RNA polymerase transcription. There are in fact two promoters near the oriC region, although transcription from these promoters does not yield mRNAs encoding proteins. Furthermore, DNA replication in cells requires RNA polymerase activity. In vitro, the supercoiled state can be achieved with topoisomerases or by simply purifying from cells oriC plasmids that are naturally supercoiled.

Given an appropriate supercoiled template, the initiation of DNA replication begins with the entry of the dnaB-dnaC complex (Figure 12-19). In the presence of ATP, dnaB can act as a helicase to unwind oriC. Unwinding of oriC is begun by addition of dnaA in the 13-mers, which are rich in adenine and thymine and thus easily melted; unwinding is completed by addition of dnaB and dnaC. The existence of open, unwound complexes has been demonstrated experimentally by showing that a

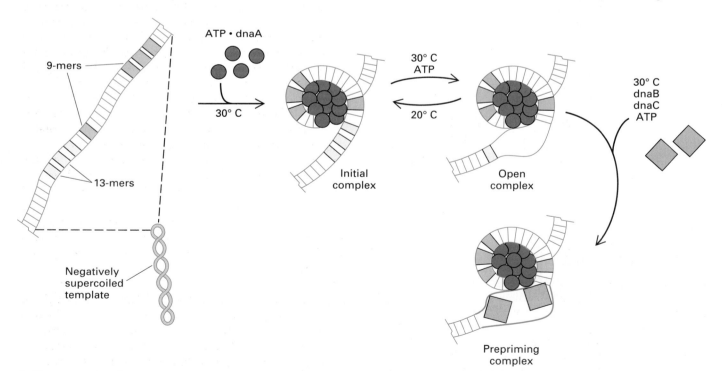

▲ **Figure 12-19**   Model of initiation of replication at *E. coli* oriC. The 9-mers and 13-mers are the repetitive sequences shown in Figure 12-17. The dnaB and dnaC proteins are primosome components (see Figure 12-14) and are re-

quired for propagation of the growing fork, whereas dnaA is required only for initiation. *Adapted from C. Bramhill and A. Kornberg, 1988, Cell* **52**:743.

single-strand endonuclease can specifically cut at the unwound origin region. Once the prepriming complex is formed, the unwound origin is stable, and the reaction mixture can be cooled to 16°C. Addition of primase to the prepriming complex leads to production of RNA primers. Subsequent addition of the remaining replisome components—that is, DNA polymerase III, single-stranded binding protein, and a helicase—and of DNA polymerase I and DNA ligase results in DNA synthesis at two growing forks emerging from the oriC region. All chain growth begins in the oriC region starting from primer RNA. To complete in vitro replication, topoisomerases are also required; the actions of these enzymes are discussed later.

Although most attention has appropriately been focused on origins of replication, regions in the DNA are known in *E. coli* that specify termination (*ter*) and that bind specific proteins. One of these proteins acts to stop replication by preventing a helicase (*dnaB* product) from unwinding DNA, and thus stops the growing fork. The details of completion of replication and separation of the daughter duplexes are not yet clear.

Initiation of DNA Replication at the SV40 Origin

Considerable progress has been made in understanding how growing forks are initiated at the origin of SV40 (a tumorigenic monkey virus), although the SV40 mechanism is not as well characterized as the *E. coli* mechanism. First and very important, the SV40 origin defined

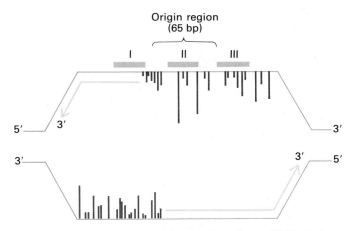

▲ **Figure 12-20** The sites of initiation of new DNA chains in the origin region of the SV40 genome. Nascent DNA chains with RNA at the 5' end were recovered from cells in which SV40 was being replicated. The ends of the nascent molecules were labeled, and the endpoints were mapped by a hybridization procedure. Each site at which a chain was started is indicated by a red vertical bar; the height of each bar indicates the relative frequency of starts at that site. The three yellow horizontal bars (I, II, and III) indicate binding sites for T antigen, which is required for DNA replication. [See R. T. Hay and M. L. DePamphilis, 1982, *Cell* **28**:767.]

by mutations that block viral DNA replication in cells is also the region that is required for initiation of in vitro DNA replication. Furthermore, experiments have shown that the RNA primers attached to growing DNA strands isolated from cells have their 5' ends at sites on both strands very near the origin (Figure 12-20). These primer-containing nascent chains thus overlap the origin region; extension of these chains is assumed to be responsible for the two growing forks that start at the SV40 origin.

The key protein that first recognizes sequences in the SV40 origin to kick off DNA replication is a virus-encoded protein. This protein was first identified with antiserum from animals that had tumors produced by SV40 and was therefore called tumor antigen, or *T antigen*. During the early phases of virus growth in SV40-infected cells, T antigen is produced and binds tightly to the SV40 DNA origin region.

In the presence of ATP, T antigen can unwind the DNA beginning at the SV40 origin (Figure 12-21a and b). The "early palindrome" (see Figure 12-18b) is the site of the initial unwinding. Since mutations that prevent this unwinding also prevent replication, this function is thought to be critical to starting replication. The T antigen forms a hexameric complex (reminiscent of the multimeric *E. coli* dnaA protein complex); two of these hexamers can bind to the SV40 origin and perform the ATP-requiring unwinding reaction (Figure 12-21c). If single-stranded binding protein, DNA polymerase α, primase, topoisomerases, and an incompletely purified protein fraction are added to the unwound complexes, in vitro replication from the SV40 origin occurs. The impure fraction contains polymerase δ and PCNA (proliferating cell nuclear antigen), a protein that is necessary for polymerase δ activity. The finding that two DNA polymerases, α and δ, are both required, coupled with the fact that primase activity is associated only with α, suggests that δ may be responsible for leading-strand synthesis.

The demonstration of T-antigen unwinding of the SV40 origin and in vitro DNA replication catalyzed by genes of cellular proteins suggest that the in vitro SV40 system mimics cellular replication closely. Whether all replication origins in cellular chromosomes employ a protein like T antigen or the *E. coli* dnaA protein to open origin regions is not known but seems a reasonable hypothesis.

Replication Origins in Animal Cells

Although no replication origins have yet been conclusively identified in higher eukaryotes, the widely scattered *Alu* repeat sequences in mammalian cells may have some connection with replication. When included in a plasmid, some *Alu* sequences support in vivo plasmid replication, although others do not. Another reason for suspecting *Alu* sequences is that they contain a 14-bp segment that includes two GAGGC sequences, identical to those in the

(a)

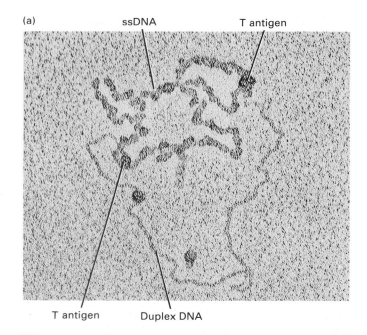

ssDNA   T antigen

T antigen   Duplex DNA

(b)

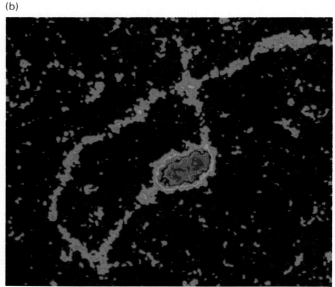

(c)

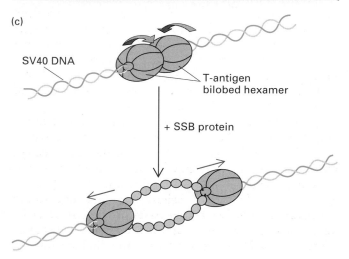

SV40 DNA

T-antigen
bilobed hexamer

+ SSB protein

◀ **Figure 12-21**   Initiation of DNA replication at SV40 origin. (a) Electron micrograph of SV40 DNA partially unwound by T antigen plus ATP. Single-stranded DNA-binding (SSB) proteins are attached to the unwound region making it appear thick. (b) Mass image reconstruction of T antigen at SV40 origin based on computer scans of electron micrographs. Colors represent the amount of mass (red > orange > blue). (c) After two hexameric T-antigen complexes bind at the SV40 origin, the hexamers move in opposite directions, unwinding the DNA; subsequent addition of SSB protein keeps the two strands apart. [See I. A. Mastrangelo et al., 1989, *Nature* **338**:658; and F. B. Dean et al., 1989, in *Molecular Mechanisms in DNA Replication and Recombination*, A. R. Liss.] *Photographs courtesy of F. B. Dean and J. Hurwitz.*

binding sites for T antigen in the SV40 origin. Identification of true chromosomal origin sites may well require in vitro demonstration of their function, as was the case for oriC in *E. coli*. By analogy with the success in *E. coli* and SV40 studies, what is required is discovery of the initiating protein(s) with a function similar to that of dnaA and T antigen.

Replication of Linear Viral DNAs Begins at Ends of the Template and Uses Protein Primers

Although other ways of initiating DNA synthesis are known, it seems unlikely that these are used in nuclear chromosome replication. For example, adenovirus has a 5′ cytosine residue at each end of its linear genome to which a protein is covalently linked. Replication of DNA begins with a copy of this terminal protein being joined to a cytosine residue by the viral DNA polymerase (Figure 12-22). The two proteins pair and the 5′ dCMP pairs with the dGMP at the 3′ end of the complementary template strand. By extension from the dCMP residue, the polymerase then copies the entire strand. Replication thus begins from a protein with an attached nucleotide instead of from an RNA primer. This method of initiation is also used in bacteriophage φ29 of *Bacillus subtilis*, in linear plasmids in *Streptomyces rechi*, and in mitochrondrial DNA from maize cells. However, thus far no protein-primed DNA replication of cellular chromosomes is known.

Topoisomerases and Superhelicity in DNA

In our discussion of how various enzymes and other proteins establish and then propagate a growing fork, we mentioned that topoisomerases are required for growing-fork propagation in closed circular molecules. We now will describe more fully the activity of these enzymes.

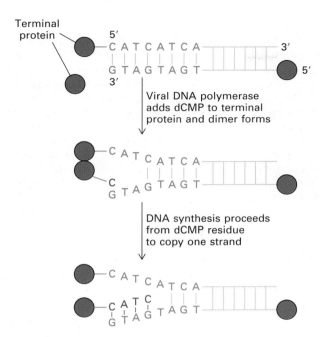

▲ Figure 12-22   Model of initiation of replication of the linear adenoviral genome. A deoxycytidylate residue (red C) covalently bound to a terminal protein acts as a primer. The C base-pairs with a G on the 3′ end of the strand to be copied; replication then proceeds by unidirectional extension from the C residue, with one strand being copied at a time (see Figure 12-5a). Completion of a single strand in vitro requires topoisomerases. [See K. Nagata et al., 1983, *Proc. Nat'l Acad. Sci. USA* 80:4266 and 80:6177.]

This discussion will expand on their role during DNA replication and illustrate their crucial role in the successful conclusion of replication and separation of the finished chromosomal copies. To understand the topoisomerases, we first need to consider DNA topology.

The *superhelicity* (or supercoiling) of DNA was discovered initially by electron microscopy and by sedimentation analysis of the small DNA molecules of SV40 and polyoma viruses. When DNA from these viruses is freed from protein very carefully and examined in the electron microscope, most of the DNA molecules are supercoiled, but a few appear as open circles (Figure 12-23, *top*). After supercoiled molecules are treated briefly with just enough DNase to place one nick in one strand, all the molecules appear as open, double-stranded circles. The unnicked, native molecule is termed form I; the nicked, relaxed molecule, form II. Cleavage of both strands produces form III, a linear duplex.

Sedimentation analysis of the three forms of SV40 DNA, first at neutral pH and then at increasingly alkaline values, provided experimental evidence to interpret the structures observed in electron micrographs (Figure 12-23, *bottom*). The linear form III in neutral solutions has a sedimentation constant of about 14S; the nicked circular form II has an S value of 16. Exposure of either of these

forms of DNA to an alkaline solution (pH 12.3 or greater) denatures the double-stranded molecules, releasing two single-stranded molecules. When form III is denatured in this manner, two 16S linear strands are released; when form II is denatured, a linear 16S and a circular 18S strand are released. The similarity in sedimentation values of the initial double strands and the single strands is accidental and arises because the sedimentation constant of a molecule depends both on its molecular weight and shape. The rather rigid initial double-stranded molecules do not sediment as fast as the single-stranded random coils which can assume compact shapes. Likewise, the slight difference in shape between the circular single-stranded product and the linear single-stranded (nicked) product explains the slight difference between their sedimentation constants (18S and 16S, respectively).

Form I, the predominant species in DNA freshly isolated from virions, has a higher sedimentation constant (20S) than the nicked form II (16S). Because the two molecules have the same molecular weight, form I must be more compact than form II. The greater compactness of supercoiled form I is evident in electron micrographs. As the pH of the solution is increased, the S value of form I decreases at about pH 11.8, indicating conversion to a less compact structure. This conversion involves the loss of supercoils; the energy for this structural change comes from melting of some base pairs in the helix. These observations suggest that a connection exists between the degree of winding of the duplex and its superhelicity. With further increases in pH, the sedimentation constant of form I rises rapidly. This occurs because the covalently closed double-stranded helix cannot completely denature into two strands; instead it denatures into a tangled mass, which becomes quite compact.

Linking Number, Twist, and Writhe Describe DNA Superstructure

The topology of a DNA molecule can be described by three parameters. First is the *linking number,* an integer that is equal to the number of times one strand of the helix crosses the other within the boundaries being considered (e.g., within a circular molecule, within a linear molecule with anchored ends, or within a linear region of a molecule between fixed sites).

The second parameter is *twist,* which is related to the frequency or periodicity of the winding of one strand about the other. Twist can vary from segment to segment within a molecule. In purified, isolated DNA at normal physiologic conditions of salt and temperature, an *average* of one right-handed helical turn (and one strand crossover) occurs every 10.6 bases. The twist of DNA inside cells is not always uniform over short distances because the DNA is associated with protein. For example,

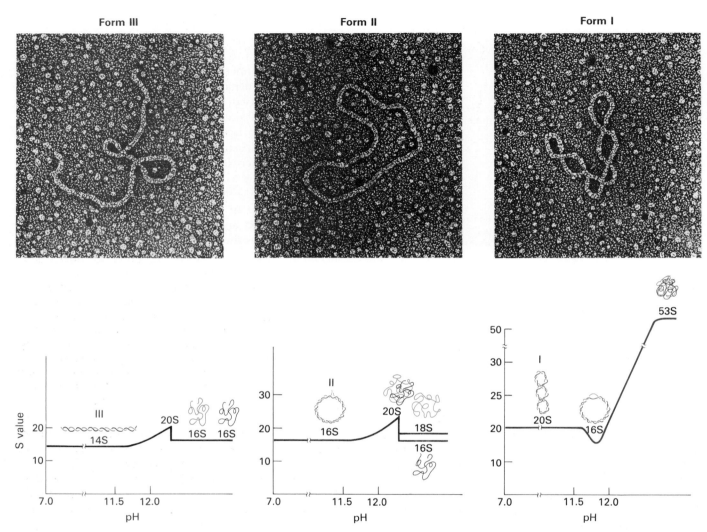

▲ **Figure 12-23** *(Top)* Electron micrographs of the three forms of SV40 DNA at neutral pH. Form III, linear duplex produced by cleavage of both strands; form II, relaxed open duplex produced by a nick in one strand; form I, native supercoiled duplex. *(Bottom)* Changes in the sedimentation constant of the three forms of SV40 DNA with increasing pH. Diagrams representing the physical structures at various pH values are shown. [See J. Vinograd et al., 1965, *Proc. Nat'l Acad. Sci. USA* **53**:1104.] *Micrographs courtesy of M.-T. Hsu.*

in the DNA wrapped around nucleosomes the twist is about 10.1 bases per helical turn.

A circular DNA molecule like SV40, which is about 5300 bases long, would be expected to have a linking number of about 500 under the salt and temperature conditions found in cells; if this were so, the molecule would exist as a relaxed open circle (form II). However, when SV40 DNA is extracted from virions and freed of protein, it is in the supercoiled form I. The accepted explanation for the supercoiling of free SV40 DNA is that the helix is *underwound* (or undertwisted) by about 25 turns; that is, its linking number is about 475. Thus an untwisted region should be left when the DNA is released from association with protein (Figure 12-24a). Because such an unwound region is energetically unfavorable, the molecule

attempts to adjust. If one chain was broken, the helix would spontaneously wind another 25 turns—that is, the linking number would increase (Figure 12-24b). Without a chain break, however, either the average twist in the molecule would have to decrease or the molecule would coil on itself, producing supercoils (Figure 12-24c and d). Supercoils occur when the temperature and salt concentrations are normal; neither the linking number nor the average twist changes during the formation of supercoils.

Because the winding of the strands in the DNA helix is regarded as *positive*, supercoils that form to compensate for the effects of underwinding in a helix are said to be *negative* supercoils. Thus one negative supercoil is interchangeable with one less link between the helical strands. This interconvertibility is probably used in helping pro-

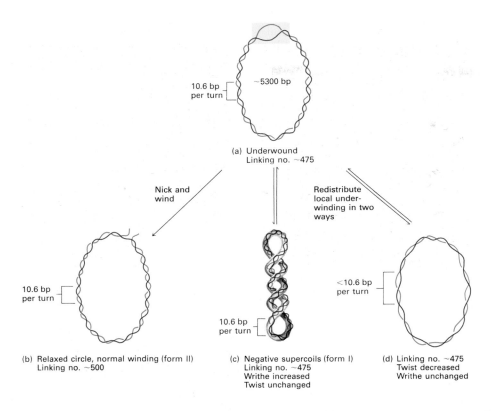

◄ **Figure 12-24** Interrelationship of twist *(T)*, writhe *(W)*, and linking number *(L)* of SV40 DNA. (a) When isolated from virions and freed of protein, SV40 DNA is underwound and has a linking number of about 475, rather than the 500 expected based on the number of nucleotides it contains ($L =$ 5300 bp ÷ 10.6 bp/twist = 500). This underwinding is represented as a non-helical region (yellow) in the DNA. (b) The molecule could overcome this energetically unfavorable underwinding by breakage of one strand, followed by spontaneous winding of the helix until the normal L was attained, to give the relaxed form II. (c) and (d) In the absence of a chain break, the underwound molecule could either form supercoils (W increases; T unchanged) or the underwinding could be distributed evenly throughout the molecule (T decreases; W unchanged). Under normal salt and temperature conditions, the supercoiling in (c) occurs, and supercoiled form I is observed. Isolated SV40 DNA has about 25 supercoils; for simplicity only a few are shown in the drawing.

teins bind to DNA. An overwound DNA helix—that is, one with more than 10.6 base pairs per turn and a linking number greater than expected—would tend to form *positive* supercoils in the opposite direction (see Figure 2-44). Such positive supercoils accumulate ahead of a growing fork in DNA replication and in front of RNA polymerase during transcription.

The supercoiling phenomenon, which is reciprocally related to twist when there is no change in linking number, is an expression of the third parameter—called *writhe*—that describes the overall structure of the DNA helix. Although the exact definitions of twist and writhe would require a complex topological or geometric statement, in simplified terms, L (the linking number) $= T$ (twist) $+ W$ (writhe). Twist is related to the frequency of turns around the central axis of the helix within any given region and writhe relates to the pathway in space of the axis of the helix. Unlike the linking number, which is a property of the whole molecule, twist and writhe can vary in different parts of the molecule, but they always change in a reciprocal relationship when L is constant.

What is the basis for the underwinding of SV40 DNA and the supercoiling observed when it is freed from association with proteins? When an SV40 DNA molecule is released gently from infected cells, it is associated with cellular histone octomers acquired during its replication. Like the chromosome of its eukaryotic host, the SV40 DNA is wound around the histones to form nucleosomes. (In virions the histones are replaced by viral proteins, but the structure of the DNA remains the same.) The winding of DNA around the proteins in nucleosomes occurs in the nucleus in the presence of topoisomerases, which act to leave the DNA between nucleosomes with no extra stress. Removal of the histones outside the cell reexposes the underwound nature of the DNA, and supercoils form.

Topoisomerases Can Change the Linking Number

Once it had become clear that DNA replication and packaging involve the supercoiling of DNA, biochemists began searching for enzymes with the capacity to affect supercoiling. A number of topoisomerases with such abilities have been purified from both bacterial and animal cells. Some of these enzymes are undoubtedly involved in DNA replication and in transcription.

Type I Topoisomerases   The first topoisomerase to be discovered, the omega protein in *E. coli,* has the capacity to completely remove negative supercoils without leaving nicks in the DNA molecule. After the enzyme binds to a DNA molecule and cuts one strand, the free phosphate on the DNA is covalently attached to a tyrosine residue in the enzyme. The complex then rotates, decreasing the linking number by 1, and the DNA is resealed (Figure 12-25). By this mechanism, the enzyme removes one supercoil with each rotation. Enzymes with this activity are classified as type I topoisomerases (topo I).

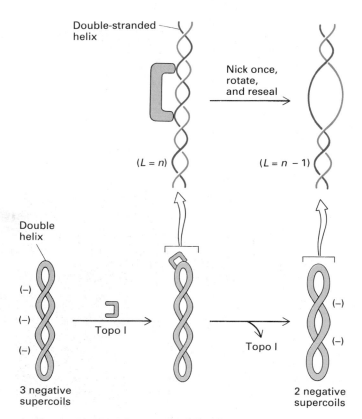

▲ **Figure 12-25** The action of type I topoisomerases. These enzymes attach to one strand of DNA and nick it; the complex then rotates, and the strand is resealed. For each complete rotation, the linking number is decreased by 1 and one supercoil is removed.

The topo I from *E. coli* is much less active on positively supercoiled than on negatively supercoiled molecules. In contrast, topo I enzymes from eukaryotic cells can remove both positive and negative supercoils. Because the relaxation (removal) of DNA supercoils by topo I is energetically favorable, the reaction proceeds without an energy requirement. If fully supercoiled SV40 DNA is treated with *E. coli* topo I for various time periods, numerous *topoisomers,* each containing a different number of supercoils, are produced (Figure 12-26). Thus the sequential action of topo I can remove all the supercoils in a DNA molecule. Bacteria with mutant topo I enzymes grow more slowly than wild-type strains and fail to regulate some enzymes in the normal way. A topo I enzyme has also been identified in yeast; mutations in the gene encoding this enzyme are not lethal but affect the growth rate of cells.

Type II Topoisomerases The first type II topoisomerase to be described also was isolated from *E. coli.* Termed *gyrase,* this enzyme has the ability to cut a double-stranded DNA molecule, pass another portion of the duplex through the cut, and reseal the cut (Figure 12-27a). Because this maneuver has the effect of changing a posi-

tive supercoil into a negative supercoil, it changes the linking number of the DNA by 2. ATP is consumed in the process to yield the energy that fuels the change in superhelicity. With ATP as a source of energy, gyrase can increase superhelicity. Type II topoisomerases also catalyze catenation and decatenation, i.e., the linking and unlinking, of two different DNA duplexes (Figure 12-27b).

Gyrase is necessary for DNA replication in *E. coli* cells; a temperature-sensitive gyrase mutant is unable to grow at elevated temperatures. Gyrase has now been shown to be required for in vitro DNA replication. Also, the transcription of many genes is slowed considerably in cells with mutant gyrase. The topo I and topo II (gyrase) enzymes in *E. coli* compete to balance the level of supercoiling in the *E. coli* chromosome. Measurements of the degree of supercoiling in *E. coli* cells have suggested that there is one negative supercoil for each 15–20 turns of the DNA helix.

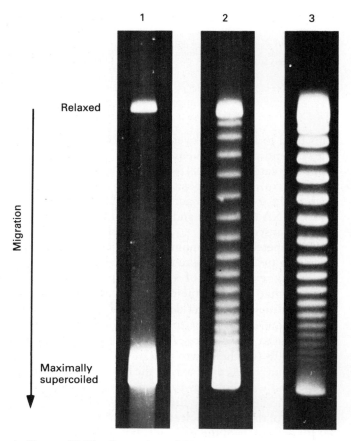

▲ **Figure 12-26** Separation of SV40 DNA with different degrees of superhelicity by gel electrophoresis. DNA was extracted from SV40 virions under conditions that ensure the maximum number of supercoils. Two other samples were treated with topoisomerase I for 3 min (lane 2) or 30 min (lane 3). About 25 bands, equal to the total number of possible topoisomers are visible in the electrophoretograms after topoisomerase treatment, including the fully relaxed form. [See W. Keller and I. Wendel, 1974, *Cold Spring Harbor Symp. Quant. Biol.* 39:199.] *Courtesy of W. Keller.*

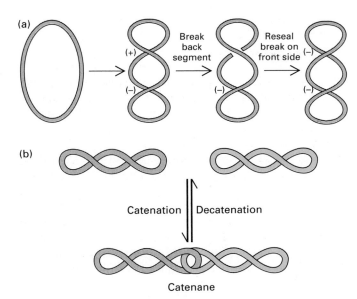

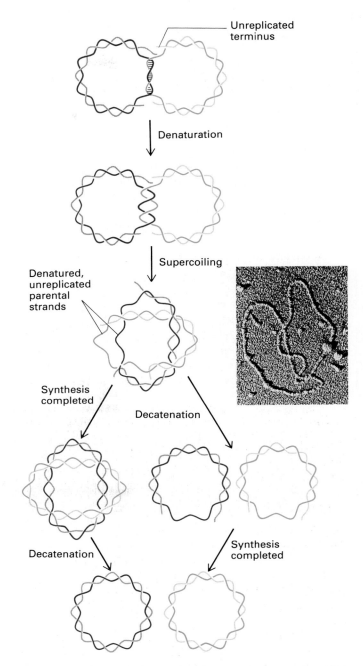

▲ **Figure 12-27** The action of type II topoisomerases. The diagrams illustrate the type of reactions of which type II topoisomerases are capable. (a) Introduction of supercoils. The initial folding introduces no permanent change. The subsequent action of topo II produces a stable structure with two negative supercoils. Since one positive supercoil is replaced with one negative supercoil, the enzyme changes the linking number of the DNA by 2. (The assignment of sign to supercoils is by convention: if the helix is stood on its end in a positive supercoil, the "front" strand is falling from left to right as it passes over the "back" strand; in a negative supercoil, the front strand is falling from right to left.) (b) Catenation and decatenation of two different DNA duplexes also is catalyzed by topo II. [See N. R. Cozzarelli, 1980, *Science* **207**:953.]

Type II topoisomerases exist in all cells. The topo II enzymes from mammalian cells cannot, like *E. coli* gyrase, increase superhelical density at the expense of ATP; presumably no such activity is required in eukaryotes since binding of histones increases the potential superhelicity. As discussed in the next section, the decatenation activity of topo II is important for proper separation of replicated chromosomes.

Topoisomerase II Is Involved in Releasing Final Products after Chromosome Replication

Topoisomerase II activity is apparently needed to complete replication of circular DNA molecules. During replication, the parental strands remain intact and retain their superhelicity. As the two replication forks approach each other, how does the as-yet-unreplicated parental duplex region become unpaired to allow replication to finish? This situation, which must also arise in the meeting of two replication forks in a cellular chromosome, is illustrated in Figure 12-28. The last few parental helical turns

▲ **Figure 12-28** Completion of replication of circular DNA molecules. Denaturation of the unreplicated terminus followed by supercoiling overcomes the steric and topological constraints of copying the terminus. At least with SV40 DNA, the final two steps (synthesis and decatenation) can occur in either order, depending on experimental conditions. Parental strands are in dark colors; daughter strands in light colors. *Inset:* Electron micrograph of two fully replicated SV40 DNA molecules interlocked twice. This structure would result if synthesis was completed before decatenation. Topo II can catalyze decatenation of such interlocked circles in vitro. [See O. Sundin and A. Varshavsky, 1981, *Cell* **25**:659.] *Drawing adapted from S. Wasserman and N. Cozzarelli, 1986, Science* **232**:951. *Micrograph courtesy of A. Varshavsky.*

could be removed by changing the topology of the already replicated regions leaving the two newly complete daughter helices linked together as *catenanes*, covalently linked but not yet completely finished circles. Replication then could be completed before or after decatenation, leaving catenanes as the final replication product. If nearly completed replicative forms of SV40 are recovered and viewed in the electron microscope, catenanes with single-stranded ends can be found, indicating that the proposed transition to relieve the final parental helix does occur. If cells are placed in high concentrations of monovalent ions, catenated molecules that have completed replication can also be found (Figure 12-28, *inset*), indicating that decatenation does not have to precede completion of replication.

The enzyme responsible for decatenation in eukaryotic cells is topo II, which can make and unlink catenanes. Complicated interlocked structures containing several "knots," which sometimes accumulate during cellular replication of viruses, also can be resolved by topo II. In vitro replication of SV40 DNA also requires topo II for decatenation of the final products.

The direct participation of topo II in segregation of linear chromosomes has been demonstrated in yeast. Yeast cells with a temperature-sensitive mutation in topo II cannot separate their chromosomes at nonpermissive temperatures. Although individual yeast chromosomes are too small to be visualized by light microscopy, a structure, called the nuclear body, which contains all the chromosomes clumped together can be seen. In temperature-sensitive mutants, at the nonpermissive temperature, the nuclear body, which usually divides at the junction of the mother and daughter cell, appears to get stuck in the passageway between the two cells (Figure 12-29).

Topoisomerase II also is a primary component of the nonhistone protein scaffolding to which long DNA loops are attached in metaphase chromosomes (see Figure 9-24). It is clear that topo II plays a key role in chromosomal function, probably in resolving tangles that exist in newly replicated chromosomes.

Assembly of DNA into Nucleosomes

How newly replicated DNA is packaged may very well affect gene expression in eukaryotic cells. Two aspects of determination and differentiation are often discussed in connection with DNA replication. First, during early embryogenesis, cell fate can be determined many cell generations before the final differentiated cell phenotype appears. In other words, an *epigenetic* state, or a potential for a specific pattern of gene expression, can be carried through several cell cycles. Second, after the application of a specific inducing stimulus, many cells require several cycles of cell division to fully express a differentiated phenotype. Red blood cell precursors, for example, can undergo four or five divisions after stimulation by the hormone erythropoeitin before they all make globin.

Both determination of cell fate and differentiation are widely believed to be controlled by the deposition of chromosomal proteins and possibly by the failure of some DNA to become methylated during cell replication. A logical time for such effects to occur is during or just after replication. Because the most prominent proteins associated with cell DNA are histones, great interest has focused on the assembly of histones in newly replicated DNA. The majority of histone protein is synthesized during the S phase of the cell cycle, and it quickly enters the nucleus to become associated with DNA. Small amounts of histones are synthesized throughout the cell cycle; this may be used to replace histones that are displaced during DNA repair or possibly during transcription.

Newly Synthesized DNA Quickly Associates with Histones to Form Nucleosomes

Within a few minutes of its synthesis, new DNA becomes associated with histones in nucleosomal structures. This has been shown by DNase digestion within nuclei of new, [³H]thymidine-labeled DNA. Both the new, labeled DNA

(a)

(b)

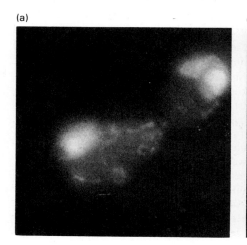
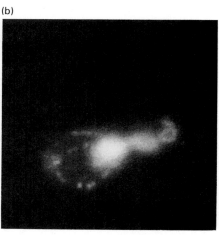

◀ **Figure 12-29** Fluorescence micrographs of yeast cells with a temperature-sensitive mutation in topoisomerase II at 80 min after the beginning of the cell cycle. (a) At permissive temperature (26°C), the DNA (bright stain) is divided between the mother cell and smaller daughter cell. (b) At nonpermissive temperature (35°C), the DNA is caught at the junction between the mother and daughter cells. *From C. Holm et al., 1985, Cell **41**:554; courtesy of C. Holm.*

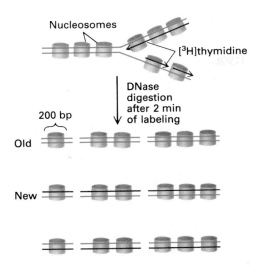

▲ **Figure 12-30** Experimental demonstration of rapid packaging of new DNA into nucleosomes. Since all labeled DNA is associated with histone octomers after brief labeling (2 min), DNA must associate with histones almost immediately after it is synthesized. [See G. Russev and R. Hancock, 1982, *Proc. Nat'l Acad, Sci. USA* 79:3143.]

and the old, unlabeled DNA are released into fragments (200, 400, or 600 bases long) characteristic of chromatin arranged in nucleosomal arrays (Figure 12-30).

Recall that a nucleosome consists of about 140 base pairs of DNA wrapped around an octomer of eight histone chains, two each of H3, H4, H2A, and H2B. When new chromatin is being formed just after DNA duplication, do old and new histone monomers mix? Or do old histone octomers remain together and new histones become associated only with other new histones to form new octomers? Various double-labeling experiments attempting to settle this issue have been reported. The usual design of such experiments is to first density-label the histone proteins and then shift the cells to a light medium and expose them to a radioactive label to tag newly formed histones. Newly formed chromatin is density-separated and isolated. The histone octomers can be reversibly cross-linked to the DNA in the density-separated nucleosomes before determination of their composition.

Experiments of this type have shown that (H3-H4)$_2$ tetramers only enter chromatin shortly after DNA synthesis occurs. This seems reasonable since isolated H3 and H4 can form stable tetramers and bind to DNA in solution. About 30 percent of new H2A-H2B dimers also enter nucleosomes containing new DNA, but newly made H2A-H2B dimers can also enter old chromatin. These results indicate that DNA replication can disrupt at least part of the histone octomer. How this may affect gene transcription is yet to be determined. About 5 percent of histones are made outside of the S phase, and some of these molecules also can enter chromatin. In summary, almost all H3-H4 enters chromatin during the S phase,

whereas H2A-H2B can enter during interphase, perhaps by exchange in regions where transcription is active.

There is no widely accepted evidence that either of the two new daughter complexes that arise after passage of a DNA growing fork are recognized differently by new and old histones.

Repair of DNA

Errors in DNA sequence can be induced by environmental factors (e.g., radiation, mutagenic chemicals, and thermal decomposition of nucleosides) and are also occasionally introduced by DNA polymerases during replication. If these errors were left totally uncorrected, both growing and nongrowing somatic cells might accumulate so much genetic damage that they could no longer function. In addition, the DNA in germ cells might incur far too many mutations for viable offspring to be formed. Thus the correction of DNA sequence errors in all types of cells is important for survival.

Proofreading by DNA Polymerase Corrects Copying Errors

The enzymatic basis for the maintenance of the correct base sequence during DNA replication is most completely understood in *E. coli*. DNA polymerase I and III have been shown to introduce about 1 incorrect base in 10^4 internucleotide linkages during in vitro replication. Since an average *E. coli* gene is about 10^3 bases long, an error frequency of 1 in 10^4 base pairs would cause a potentially harmful mutation in every 10th gene during each replication, or 10^{-1} mutations per gene per generation. However, the measured mutation rate in bacterial cells is much less, about 10^{-5} to 10^{-6} mutations per gene per generation.

The mystery of this increased accuracy in vivo was cleared up when the proofreading function of DNA polymerases was discovered. A DNA template-primer complex containing a mismatched base at the 3′ hydroxyl end of the primer was synthesized. When the mismatched complex was supplied to either *E. coli* DNA polymerase I or III, the incorrectly hydrogen-bonded base (plus some additional bases as well) was removed by a 3′ → 5′ exonuclease activity that is a built-in function of the DNA polymerase molecule (Figure 12-31). It is likely that if an incorrect base were accidentally incorporated during DNA synthesis, the polymerase would pause, excise it, and then recopy the region, most likely correctly.

This corrective activity, called *proofreading*, is a property of almost all bacterial DNA polymerases. The α and δ polymerase complex of animal cells also have proofreading activity. It seems likely that this function is indispensable for all cells to avoid excessive genetic damage.

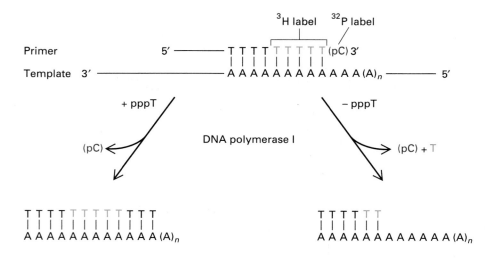

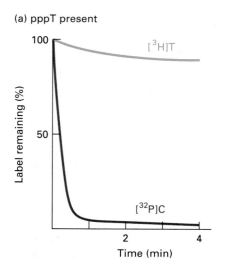

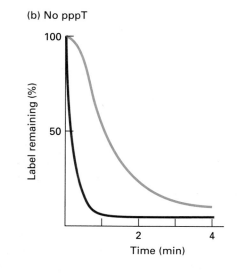

◂ **Figure 12-31** Experimental demonstration of the proofreading function of *E. coli* DNA polymerase I. An artificial template (poly dA) and a corresponding primer end-labeled with [³H]thymidine residues were constructed. An "incorrect" cytidine labeled with ³²P was then added to the 3′ end of the primer. The template-primer complex was incubated with purified DNA polymerase I in the presence and absence of thymidine triphosphate (pppT). (a) The rapid loss of ³²P (red curve) and retention of most of the ³H (blue curve) in the presence of pppT indicate that in most cases the enzyme removed only the terminal incorrect C and then proceeded to add more T residues; that is, the polymerase proofread and corrected the growing chain and then continued to copy the template. (b) In the absence of pppT, both labels were lost, indicating that if the enzyme lacks pppT to polymerize, its 3′ → 5′ exonuclease activity will proceed to remove "correct" bases. [See A. Kornberg and T. A. Baker, in press, *DNA Replication*, 2d ed., W. H. Freeman and Company.]

Environmental DNA Damage Can Be Repaired by Several Mechanisms

Many cells that divide very slowly or not at all (e.g., liver and brain cells) must use the information in their DNA for weeks, months, or even years. However, the DNA in all cells, even nongrowing ones, is subject to physical and chemical damage. Uncorrected base changes in the DNA of nongrowing cells would result in the production of faulty proteins at an unacceptable rate. The evolutionary response to this problem has been the development of DNA-repair systems.

Table 12-4 lists the general types of DNA damage and their causes. One of the most prominent causes of damage on this planet is ultraviolet radiation from the sun, which is largely but not completely absorbed by the ozone layer of the atmosphere. Absorption of UV radiation by DNA leads to the formation of pyrimidine dimers; these most frequently involve adjacent thymine residues joined by a cyclobutyl linkage (Figure 12-32a). RNA

Table 12-4 DNA lesions that require repair

DNA lesion	Cause
Missing base*	Acid and heat remove purines (~10^4 purines per day per cell in mammals)
Altered base	Ionizing radiation; alkylating agents
Incorrect base*	Spontaneous deaminations: C → U, A → hypoxanthine
Deletion/insertion	Intercalating agents (e.g., acridine dyes)
Cyclobutyl dimer*	UV irradiation
Strand breaks	Ionizing radiation; chemicals (bleomycin)
Cross-linking of strands	Psoralen derivatives (light-activated); mitomycin C (antibiotic)

*See Figure 12-32a.

SOURCE: A. Kornberg, 1980, *DNA Replication*, W. H. Freeman and Company, p. 608.

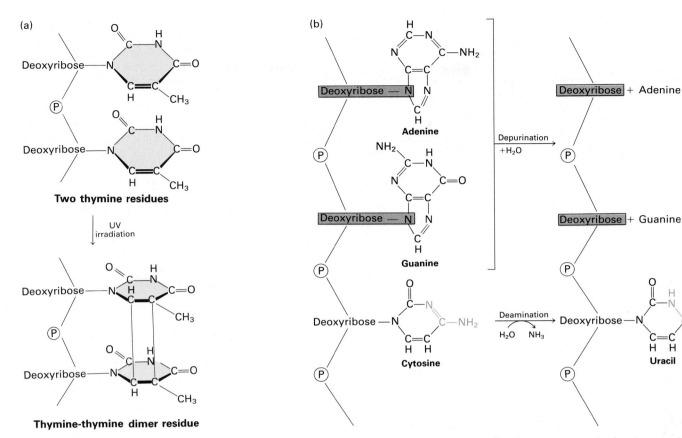

▲ Figure 12-32   Three types of DNA lesions. (a) UV irradiation can cause formation of cyclobutyl (red) linkage of dimers. (b) At physiologic pH and temperature, the N-glycosidic bond (red) linking purine bases to the sugar-phosphate backbone sometimes breaks spontaneously, leaving apurinic sites in the DNA. Cytosine may undergo spontaneous deamination to form uracil.

synthesis ceases at such pyrimidine dimers, and DNA synthesis may also stop. Both *E. coli* and yeast (and possibly other eukaryotes) contain a flavin-requiring enzyme called *photolyase* that directly removes the cyclobutyl linkage. This enzyme requires ordinary visible light and the direct enzymatic repair is called *photoreactivation.* Most damage to DNA cannot be directly corrected, however, and the damaged sites must undergo an excision-repair cycle described below; UV lesions are also repaired by this more general mechanism.

Although purines are less disturbed by UV irradiation than pyrimidines, they are susceptible to other types of damage. One is *spontaneous depurination,* which results from the intrinsic thermal decomposition even at normal cell pH, of the nucleoside linkage between the base and the deoxyribose; depurination leaves a "bare" deoxyribose residue in the DNA (Figure 12-32b). It has been calculated that as many as 10,000 purine-sugar (N-glycosidic) bonds are cleaved per 24-h period in each mammalian cell, leaving as many *apurinic sites* in the DNA. Exposure to many chemical agents also can modify purines. Some of these compounds, such as the nitrosoureas and complex organic molecules like aflatoxins (mold products), bind irreversibly to purines or break the purine ring. In addition, ionizing radiation can break the imidazole ring of a purine.

Because pyrimidine-glycoside linkages are much more stable than purine-glycoside linkages, spontaneous loss of pyrimidine bases is much less common than loss of purines. The amino group of cytosine, however, is susceptible to loss at 37°C, resulting in the conversion of cytosine to uracil (Figure 12-32b). If unrepaired, this change would lead to the exchange of an A:U base pair for a G:C pair when the damaged DNA is replicated or repaired. As discussed in Chapter 24, various types of DNA damage, if not corrected, can lead to mutations and cancer.

DNA Repair in E. coli by Damage-Site Excision and Copying of the Undamaged Strand   Various DNA lesions in *E. coli* can be repaired by removal of a segment of the DNA chain including the damaged site, followed by copying of the correct sequence in the undamaged strand to fill the excised gap. In one type of excision-repair mechanism, the DNA at or near the lesion is first incised by a specific enzyme dictated by the type of damage. The signal for incision is often a mismatch

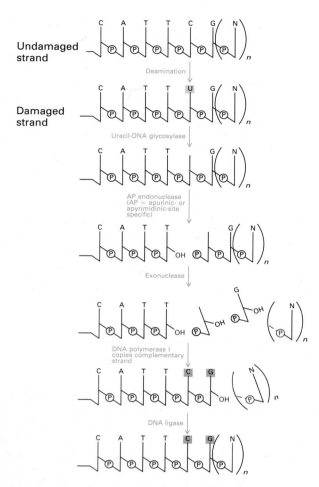

▲ **Figure 12-33** Excision repair of DNA initiated by glycosylases. Various types of damaged sites can be repaired by this general mechanism, as *E. coli* contains several glycosylases that can recognize damaged or incorrect bases and remove them, leaving an apurinic or apyrimidinic site susceptible to cleavage by AP endonucleases.

or a disruption in the base pairing. For example, after depurination, an unpaired pyrimidine is left; likewise, after cytosine is deaminated to uracil, base pairing with the corresponding guanine is weakened. These lesions in DNA are recognized by specific endonucleases of which *E. coli* has at least five for recognizing different kinds of damage. After incision, a portion of the damaged strand is removed by an exonuclease; repair is then completed by DNA polymerase I and DNA ligase (Figure 12-33).

A special repair mechanism, called UvrABC repair, is required for "bulky" chemical adducts. Thymine-thymine dimers can be repaired by this mechanism; other altered bases containing chemical attachments (e.g., carcinogens such as methylcholantrene) require this mechanism. To begin this type of repair, the site of DNA damage is recognized as a bulge in the DNA caused by the chemical adduct (Figure 12-34). Two proteins, UvrA and

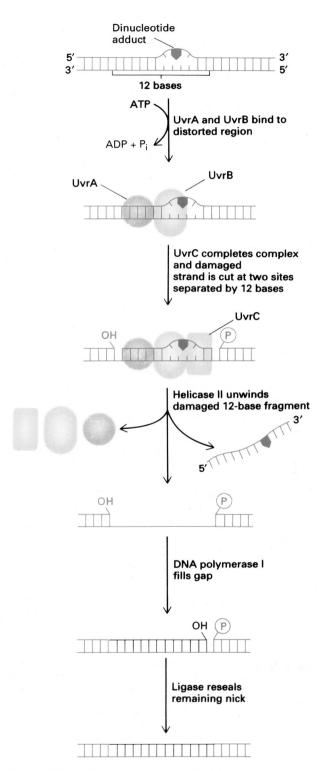

▲ **Figure 12-34** Excision repair of DNA by *E. coli* UvrABC mechanism. In the example shown, the lesion results from a dinucleotide adduct to the DNA. This mechanism can repair various lesions (including thymine-thymine dimers) that produce distortions in the normal shape of the DNA in a region. [Adapted from A. Sancar and G. B. Sancar, 1988, *Ann. Rev. Biochem.* 57:29.]

UvrB, first bind to and partially unwind the distorted section of DNA; this step requires ATP. Next, a third protein, UvrC, joins the complex surrounding the lesion. Specific cuts, 12 nucleotides apart, are then placed in the damaged strand. With the aid of helicase II (an *E. coli* DNA-unwinding enzyme), the damaged segment is removed; the gap in the duplex is then filled in by DNA polymerase I; DNA ligase joins the ends to complete repair. The UvrABC protein complex is destroyed by proteases soon after its action is concluded. This destruction perhaps helps to keep this powerful DNA-incising enzyme at low concentrations.

DNA Repair in Eukaryotes   Although most of the enzymes required for repairing damaged DNA in eukaryotic cells have not been identified, naturally occurring mutations may help to elucidate eukaryotic DNA-repair pathways. For example, some humans with the inherited disease known as *xeroderma pigmentosum* (characterized by easily pigmented skin) have a high susceptibility to skin cancer, which is probably related to exposure of the skin to the UV rays in sunlight. Cells from patients with this disease have shown evidence of several different defects in the repair of UV-damaged DNA, suggesting that several enzymes are required for DNA repair after UV irradiation. Several other disease syndromes in humans (Table 12-5) also are associated with increased sensitivity to agents that can damage DNA and with increased susceptibility to cancer. Thus all of these diseases may be caused by defects in DNA-repair pathways.

Progress in understanding DNA-repair enzymes in yeast and in mammalian cells also has come from studies with UV-sensitive mutants. Various yeast mutants that are sensitive to UV damage have been used to locate, characterize, and clone several radiation-sensitive (*RAD*) genes. The *RAD3* gene, for example, encodes a helicase that is essential for cell viability, and the *RAD10* gene encodes a protein that has sequence homology with the *E. coli* UvrA and UvrC proteins (see Figure 12-34). Thus

it seems likely that both yeast and *E. coli* possess some similar mechanisms for repairing damaged DNA.

In other studies, Chinese hamster cells have been isolated that are unusually sensitive to UV irradiation and to drugs like mitomycin that cause DNA cross links. Transfer of human DNA to these cells can reduce or eliminate their hypersensitivity to UV radiation. By sequentially transferring the DNA from "cured" cells and tracking the human sequences, it has been possible to clone a human gene that appears to be a DNA-repair gene. The sequence of this human gene is very similar in many regions to *RAD10*, the yeast gene that is similar to *E. coli uvrA* and *uvrC*. Further work of this type suggests that the yeast helicase *RAD3* gene is similar to another human gene. The identification of these genes coupled with the knowledge that defects in DNA repair exist in humans should lead soon to a better understanding of the connection between DNA damage and disease, particularly cancer.

Recombination between Homologous DNA Sites

Soon after Mendel's rules of independent gene segregation were rediscovered and the segregation of linked groups of genes on individual chromosomes was widely recognized, another great genetic discovery was made in *D. melanogaster*: blocks of genes from homologous chromosomes could be exchanged by the process of crossing over, or recombination. Recombination, which takes place during meiosis in sexually reproducing organisms, is the basis for the classical genetic maps. This exchange occurs not only in animals and plants but also in prokaryotes, viruses, plasmids, and even in the DNA of cell organelles such as mitochondria.

The events in a reciprocal recombination are equivalent to the breakage of two duplex DNA molecules representing homologous but genetically distinguishable chromo-

Table 12-5   Human diseases associated with DNA-repair defects

Disease	Sensitivity	Cancer susceptibility	Symptoms and signs
Xeroderma pigmentosum	UV irradiation, alkylation	Skin carcinomas and melanomas	Skin and eye photosensitivity
Ataxia telangiectasia	γ irradiation	Lymphomas	Unsteady gait (ataxia); dilation of blood vessels in skin and eyes (telangiectasia); chromosomal aberrations
Fanconi's anemia	Cross-linking agents	Leukemias	General decrease in numbers of all blood cells; congenital anomalies
Bloom's syndrome	UV irradiation	Leukemias	Photosensitivity; defect in DNA ligase

SOURCE: A. Kornberg, 1980, *DNA Replication*, W. H. Freeman and Company, p. 622.

somes, an exchange of *both* strands at the break, and a resolution of the two duplexes so that no tangles remain. The frequency of recombination between two sites is proportional to the distance between the sites. How such a double-strand cleavage at two precisely analogous sites can take place, followed by the swapping of duplex regions and the rejoining of ends, has been studied for many years. In this section, we discuss the proteins involved and models of how they carry out recombination.

Gene Conversion Can Occur near Crossover Point during Reciprocal Recombination

Although during recombination markers at some distance from the crossover point are exchanged in a reciprocal fashion, an apparent nonreciprocal event sometimes occurs at or near the crossover point. This phenomenon is most easily studied in yeasts in which each meiotic product can be scored in the haploid progeny spores. In a cross of multiple marked yeast strains that undergo recombination, most markers segregate 2:2, but a few show 3:1 or 1:3 segregation. Such a nonreciprocal event has been termed *gene conversion* because one allele is apparently "converted" into another.

In Figure 12-35, for example, two chromatids bearing the wild-type *D* allele and two bearing the mutant *d* allele enter meiosis; in the progeny spores there is a 3:1 or 1:3 ratio between *D* and *d* at this locus. (Note that gene conversion occurs only in the recombinant spores and not in spores with the parental genotypes.) It is now known that in a gene conversion the exact base sequence is represented at the converted site, obviously suggesting that exact copying of a DNA strand is involved. This property is referred to as *fidelity* in gene conversion. The occurrence of 1:3 and 3:1 events at equal frequencies is called *parity* in gene conversion.

Holliday Recombination Model and Its Variations Account for Gene Conversion

The most popular current models for the molecular events of recombination are the result of attempts to explain the occurrence of gene conversion during the reciprocal exchange of distant markers. The first such model, proposed by Robin Holliday in 1964, is shown in Figure 12-36. In step 1, a nick is made in one strand of each of the two homologous chromosomes that are going to recombine. Strand exchange then occurs at the site of the nick, producing a *crossed-strand Holliday structure* (step 2). Breaking of the hydrogen bonds within the parental duplexes followed by the exchange of strands and reformation of hydrogen bonds leads to branch migration and creation of a *heteroduplex* region containing one strand

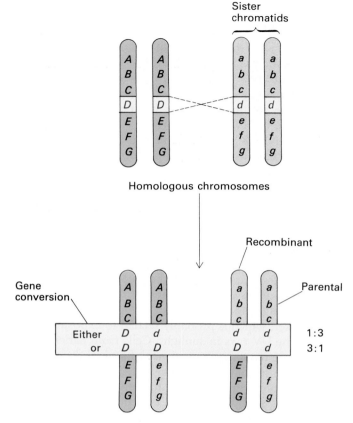

▲ **Figure 12-35** When multiple marked yeast strains undergo meiosis, homozygous chromosomes recombine so that most markers segregate in a 2:2 ratio in the progeny spores. However, some markers in the recombinants, now known to be close to the crossover point, may show a 3:1 or 1:3 segregation ratio; this phenomenon is referred to as gene conversion.

from each parental chromosome (step 3). If all four strands were cut at the crossover site and the left side of chromosome I joined to the right side of chromosome II, and vice versa (steps 4 and 5), the connected duplexes would be separated (*resolved*) and all markers to the left and right of the heteroduplex region would undergo reciprocal recombination. Because of the mismatched base pairs in the heteroduplex region, this region would be subject to DNA-repair mechanisms during which the *B* or the *b* information in one of the strands would be converted to the other allele; that is, gene conversion would occur.

A refinement in the original Holliday model simplifies the enzymatic cutting that would be necessary to disentangle the intermediate structure. If the two duplexes in the Holliday structure following branch migration were rotated at the site of the crossed strands, a rotational isomer would be formed (Figure 12-36, step 4a). The two connected duplexes of this *isomeric Holliday structure* could be resolved by cutting and rejoining of only two

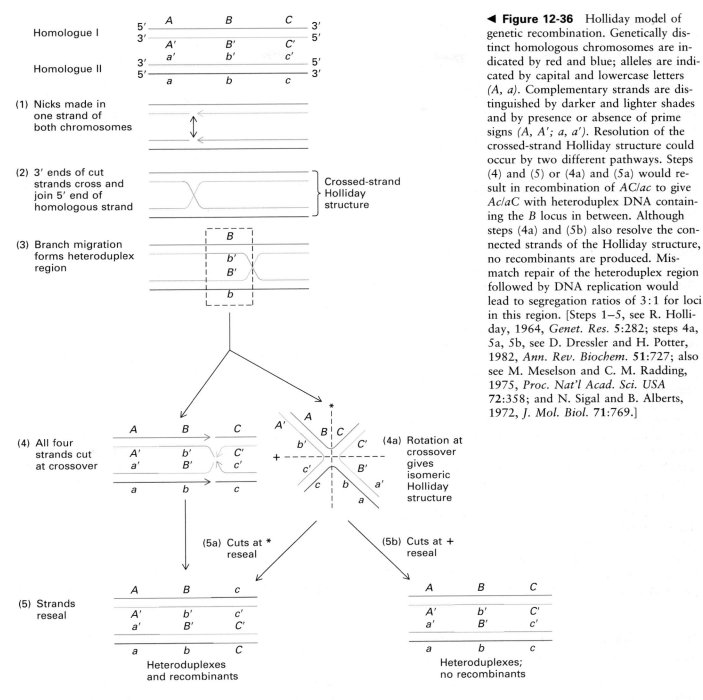

◀ **Figure 12-36** Holliday model of genetic recombination. Genetically distinct homologous chromosomes are indicated by red and blue; alleles are indicated by capital and lowercase letters (A, a). Complementary strands are distinguished by darker and lighter shades and by presence or absence of prime signs (A, A'; a, a'). Resolution of the crossed-strand Holliday structure could occur by two different pathways. Steps (4) and (5) or (4a) and (5a) would result in recombination of AC/ac to give Ac/aC with heteroduplex DNA containing the B locus in between. Although steps (4a) and (5b) also resolve the connected strands of the Holliday structure, no recombinants are produced. Mismatch repair of the heteroduplex region followed by DNA replication would lead to segregation ratios of 3:1 for loci in this region. [Steps 1–5, see R. Holliday, 1964, *Genet. Res.* 5:282; steps 4a, 5a, 5b, see D. Dressler and H. Potter, 1982, *Ann. Rev. Biochem.* 51:727; also see M. Meselson and C. M. Radding, 1975, *Proc. Nat'l Acad. Sci. USA* 72:358; and N. Sigal and B. Alberts, 1972, *J. Mol. Biol.* 71:769.]

strands. As shown in Figure 12-36, resolution could occur in two ways, only one of which produces recombinant duplex chromosomes containing a heteroduplex region (step 5a); the alternative pathway produces heteroduplex chromosomes but no recombinants (step 5b). Mismatch repair of the heteroduplex regions followed by DNA duplication and chromosome segregation would result in gene conversion.

In a further modification of the Holliday recombination model, Matthew Meselson and Charles Radding suggested another mechanism for creating a crossed-strand Holliday structure; this mechanism requires a single-strand cut in only one chromosome (Figure 12-37). After branch migration of the crossed-strand region, rotation and resolution produce results similar to those described in Figure 12-36. This model is supported by the finding that a single strand can be used to initiate a recombination event with cell-free preparations from *E. coli,* as described in the next section.

Viral and plasmid DNA molecules in the act of recombining can be extracted from both bacterial and animal cells. Electron micrographs of such molecules have re-

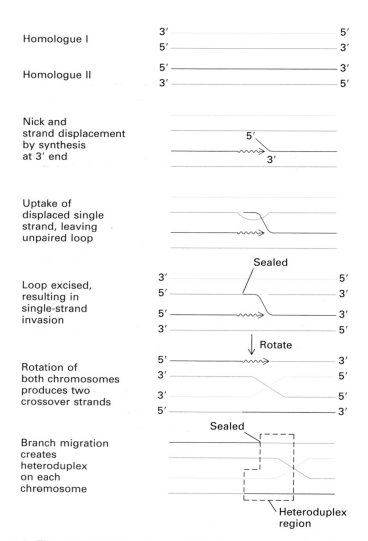

Homologue I

Homologue II

Nick and strand displacement by synthesis at 3′ end

Uptake of displaced single strand, leaving unpaired loop

Loop excised, resulting in single-strand invasion

Sealed

Rotate

Rotation of both chromosomes produces two crossover strands

Branch migration creates heteroduplex on each chromosome

Sealed

Heteroduplex region

▲ **Figure 12-37** Meselson-Radding model of genetic recombination. Genetically distinct homologous chromosomes are indicated by red and blue; complementary strands are distinguished by darker and lighter shades. In this modification of the Holliday model, formation of the crossed-strand Holliday structure begins with a nick in one of the chromosomes. After rotation at the crossed-strand site and branch migration of the crossed-strand intermediate, resolution would occur as shown in Figure 12-36. [See M. Meselson and C. Radding, 1975, *Proc. Nat'l Acad. Sci. USA* **72**:358.]

vealed structures similar to the crossed-strand and isomeric Holliday structures (Figure 12-38). Thus, however the initiating event of recombination occurs, the final connection between the unresolved chromosomes seems to involve branch migration and chromosomal rotation.

In Vitro Integration of Phage λ Mimics Recombination Event

In a few cases in bacteria, the recombining DNA molecules and the proteins responsible for the recombination have been elucidated in detail. In all cases a Holliday

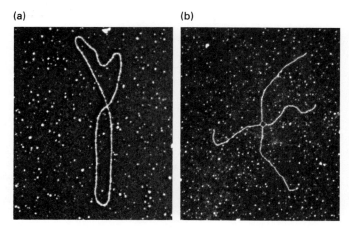

(a)   **(b)**

▲ **Figure 12-38** Electron micrographs of plasmid DNA in the process of recombination. (a) Circular plasmid DNA in crossed-strand Holliday structure. (b) More highly magnified view reveals single-stranded ring in center of isomeric Holliday structure that results from rotation about the crossover point. [See H. Potter and D. Dressler, 1978, *Cold Spring Harbor Symp. Quant. Biol.* **43**:969.] *Courtesy of D. Dressler.*

structure is formed and resolved. As an example, we will discuss temperate bacteriophage λ, which integrates at a particular DNA site in the *E. coli* host chromosome. Integration can be carried out in vitro and requires a special bacteriophage enzyme, *integrase*, which in a stepwise fashion makes and then resolves a Holliday junction. The integration of phage DNA into its site on the *E. coli* chromosome is essentially equivalent to the event that occurs at a recombination site.

Bacteriophage λ DNA contains a 15-bp region, the attachment site, which in its core region is identical in sequence to the integration (or attachment) site in the host-cell DNA (Figure 12-39a). The experimental system for in vitro integration consists of a plasmid DNA including the phage attachment site (POP′ to represent the 15 bases), a linear fragment of the bacterial DNA containing its integration site (BOB′), purified integrase, and a host protein termed integration host factor (IHF). By use of strains with mutations at different sites in the 15-bp homologous regions, it has been possible to stop the reaction at several stages and to collect intermediates in the reaction leading to integration.

The interpretation of all the experiments with various mutants is that phage integrase (as a dimer) binds to and makes a staggered cut and strand transfer between bacterial and phage DNA at the core sequence of the attachment sites (Figure 12-39c, step 1). As in the case of topoisomerases, the cut ends of the DNA are covalently bound to integrase during this reaction. The resulting strand-transfer structure is identical to the crossed-strand Holliday structure. If the neighboring sequences in the

(a)

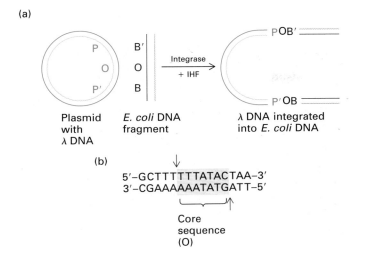

(b)

5′–GCTTT**TTTATACTAA**–3′
3′–CGAAA**AAATATGATT**–5′

Core
sequence
(O)

▲ **Figure 12-39** Integration of phage λ into *E. coli* by recombination at the (O) site (BOB′) and phage attachment site (POP′). (a) The flanking sequences in the bacteria are labeled B and B′ and the phage P and P′. In vitro incubation of a supercoiled plasmid containing the phage site POP′ and a linear fragment of the *E. coli* genome containing BOB′ with purified λ integrase and integration host factor (IHF) yields a linear DNA molecule in which the phage DNA is integrated to produce the order B′OP-plasmid-P′OB *(right)*. Complementary strands are distinguished by darker and lighter shades. (b) The sequence of the 15-bp region that is similar in the bacterial and phage attachment sites is shown. The core region is identical. The arrows indicate the sites at which staggered cuts are made around the core sequence (O). (c) Steps in the integration of λ into *E. coli,* which is equivalent to genetic recombination between markers (e.g., P and B) on homologous chromosomes. [See P. A. Kitts and H. A. Nash, 1987, *Nature* **329**:346.]

(c)

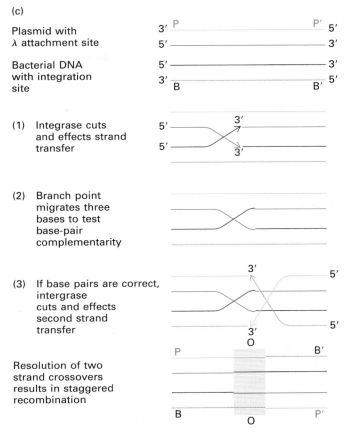

POP′ and BOB′ sites are correct, then a branch migration of three bases occurs and the duplex remains perfectly paired (step 2). The enzyme then moves and makes a second strand exchange (step 3), which resolves the Holliday structure and results in integrating the phage DNA into the bacterial chromosome. It seems likely that normal genetic recombination at homologous sites is carried out by enzymes with the same capacities as λ integrase.

E. coli RecA and RecBCD Proteins Promote Recombination

All of the enzymes and other proteins required for recombination have not been isolated in bacteria or eukaryotes. It is also not always clear what molecular events trigger recombination, although lesions caused by UV irradiation or exposure to mutagenic chemicals that form DNA adducts are associated with increased recombination. In such cases, recombination probably occurs during DNA repair when single strands would be available to form the crossed-strand Holliday structures described previously. In addition, it is clear that active transcription predisposes to recombination in bacteria and to mitotic recombination in yeast.

Many *E. coli* mutants that are sensitive to UV irradiation are also unable to engage in recombination at a normal frequency. Studies with such mutants have provided considerable information about the activities of two *E. coli* recombination proteins—RecA and the RecBCD complex.

RecA Protein The first gene that is important in recombination in *E. coli* was identified in UV-sensitive bacteria. The gene is called *recA*. The protein encoded by this gene (the RecA protein) binds to single-stranded DNA (ssDNA). This complex then searches through available duplex DNA, and when homology is found between the target DNA and the ssDNA bound to RecA, the ssDNA is inserted as a hybrid strand, displacing one of the preexisting strands (Figure 12-40). This capacity of RecA was first discovered with linear DNA, but it is now known that RecA can bind to single-stranded circular DNA and insert it into a duplex. Although ATP is required for formation of the RecA-ssDNA complex,

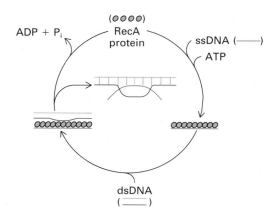

▲ **Figure 12-40** In the presence of ATP, *E. coli* RecA binds single-stranded DNA (ssDNA) and promotes insertion of the bound strand at a homologous region of double-stranded DNA (dsDNA). The insertion requires ATPase activity of RecA. [See S. S. Flory et al., 1984, *Cold Spring Harbor Symp. Quant. Biol.* **49**:513.]

it is not hydrolyzed until the ssDNA is inserted. This RecA activity presumably could promote step 2 in the Meselson-Radding model of recombination (see Figure 12-37).

RecA protein also is involved in the *SOS response* of *E. coli* to UV irradiation. When *E. coli* cells are irradiated with UV light, a whole series of *SOS* genes that enable the bacterium to survive is activated, including the *recA* gene itself. In irradiated cells, the RecA protein is converted to an active protease (RecA′), probably as a result of its interaction with deoxyoligonucleotides that are formed after irradiation from DNA breakdown. The RecA′ protease attacks various DNA-binding proteins including the protein repressor of lysogenic phages such as λ; this incapacitation of the repressor releases the dormant phage in irradiated cells.

Another target of RecA′ protease activity is LexA protein (Figure 12-41). This protein, the product of the regulatory gene *lexA*, exerts negative control over a series of other genes. Some of these have no apparent connection with DNA repair, but many do. For instance, both the *recA* gene itself and the genes for the UvrABC nuclease, which as discussed earlier excises thymine dimers, are expressed when the LexA protein is digested. The elaborate SOS response sheds light on why UV irradiation not only induces repair synthesis but also, because of the increased RecA production, increases recombination.

RecBCD Complex

A second group of UV-sensitive *E. coli* strains that exhibit a decreased frequency of recombination have mutations in a group of genes known collectively as *recBCD*. Three proteins—RecB, C, and D—are encoded by these genes and make up an enzyme complex (RecBCD) that can enter the end of a DNA molecule after it has undergone a double-strand break. Such breaks occur in cellular DNA exposed to x-rays or some

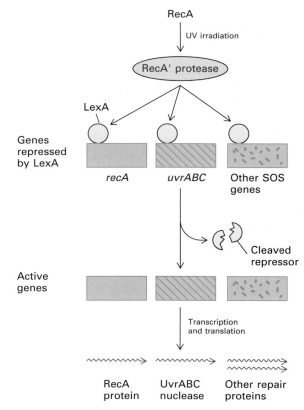

▲ **Figure 12-41** Role of RecA in the SOS response to UV irradiation in bacteria. Numerous genes that encode DNA-repair proteins are repressed by LexA, which is cleaved by RecA′, a protease formed by UV activation of RecA. Because the *recA* gene also is repressed by LexA, UV irradiation not only induces repair proteins but also RecA, which promotes recombination. [See J. W. Little and D. W. Mount, 1982, *Cell* **29**:11.]

chemicals, and they occur naturally at the ends of linear bacteriophage DNAs.

The mechanism of action of RecBCD was worked out in studies with bacteriophage λ. Certain regions of λ DNA undergo recombination at higher frequencies than other regions in normal *E. coli* but not in *recBCD* mutant host cells. The sites of increased recombination were named *chi sites* because the Greek letter X looks like a crossover point. Experiments with purified RecBCD and λ DNA indicate that the protein can enter the end of a free DNA and act as a helicase, moving along and remaining attached to the DNA as it unwinds it (Figure 12-42). Because RecBCD unwinds the DNA faster than it is rewound, single-stranded loops are created as the protein progresses. After a chi site has been passed, a specific nuclease activity of RecBCD cuts one of the exposed strands, leaving a free 3′ hydroxyl end just downstream from the chi site. This end can then participate in RecA-mediated single-strand uptake, leading to DNA recombination near the chi site by the Meselson-Radding mechanism.

Recombination in Yeast Probably Involves Double-Strand Breaks

Although recombination originating from single-strand breaks has been well studied in bacteria, mitotic and possibly meiotic recombination events in yeast cells frequently originate from double-strand breaks in the DNA. Yeast cells can be transformed with plasmids containing a selectable yeast gene with a known sequence and therefore with known restriction enzyme sites. Such transforming plasmid DNA is found recombined into the yeast genome at its homologous site on yeast chromosomes (see Figure 5-23). Recombination occurs much more frequently if before transformation *double-strand breaks* are introduced within the yeast gene contained in the plasmid. Even if a section of the gene in the plasmid is removed, recombination still occurs, and all the recombinants contain full-size genes, which implies that a gene

conversion has corrected the gap. A recombination model consistent with these findings is shown in Figure 12-43.

Insertion of yeast plasmids into yeast chromosomes serves as a model to study recombination. This reaction is dependent on several yeast genes (e.g., *RAD1* and *RAD52*). Mutations in these genes render yeast cells sensitive to UV irradiation and x-rays and incapable of carrying out plasmid recombination. These experiments suggest that proteins which function in DNA repair also function in recombination and furthermore that at least some recombination in yeasts is dependent on double-strand breaks rather than on single-strand nicks with strand invasion. To determine exactly how eukaryotic

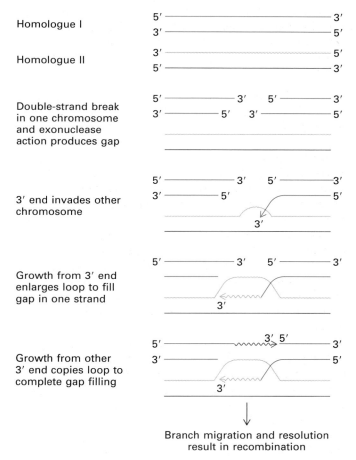

▲ **Figure 12-43** A double-strand break model of genetic recombination in yeast. After formation of the crossed-strand Holliday structure as shown, branch migration and resolution would lead to recombination (see Figure 12-36). Genetically distinct homologous chromosomes are indicated in red and blue; complementary strands are distinguished by darker and lighter shades. This model accounts for the observations that transformation in yeast by a plasmid containing a selectable yeast gene is stimulated by first introducing a double-strand break in the yeast sequence and that gaps in a plasmid can be repaired during recombination. [See J. W. Szostak, T. L. Orr-Weaver, and R. J. Rothstein, 1983, *Cell* 33:25.]

▲ **Figure of 12-42** Mechanism by which RecBCD enzyme cuts one DNA strand near a chi site, single strands increase recombination near such sites. The free single-stranded 3′ hydroxyl end can bind RecA and lead to recombination via the Meselson-Radding mechanism (see Figure 12-37). [See A. F. Taylor et al., 1985, *Cell* **41**:153.]

recombination occurs at the molecular level, we shall need further identification of the genes involved as well as additional enyzmatic studies involving recombination proteins.

Little Is Yet Known about Details of Meiotic Recombination

Recombination does not occur at the same rate in all genomes in all cells at all times. We have pointed out that special λ bacteriophage enzymes are required for integration and that synthesis of RecA protein in *E. coli* increases when cells are damaged by irradiation. Thus recombination should be thought of as a metabolic process requiring the right enzymes at the right time and not as an ongoing process that can occur in all somatic cells at any given time.

At the molecular level, recombination does not occur with equal frequency at all sites on chromosomes or in different species. In yeast, for example, the frequency of meiotic recombination can vary among different chromosomal sites by a factor of 10; the overall average recombination frequency in yeast is about 1 percent for genes that are 3000 base pairs apart. In humans, the overall recombination frequency is considerably less, averaging about 1 percent for genes that are 10^6 base pairs apart. The distance between genes that have a recombination frequency of 1 percent is often referred to as 1 *centimorgan,* a unit named for T. H. Morgan, the great *Drosophila* geneticist.

The pairing of homologous chromosomes during meiosis, called *synapsis,* may begin at similar points that become aligned when the chromosomes are close together at the nuclear periphery. A specialized structure called the *synaptonemal complex,* which is visible in electron micrographs, is assembled in many eukaryotic organisms and may function as a scaffold on which meiotic recombination could occur. Mutant animal and plant strains lacking fully formed synaptonemal complexes have greatly decreased frequencies of recombination.

Little biochemical detail about meiotic recombination is available as yet. Increases in the concentrations of endonucleases and DNA-binding proteins during meiosis in plants have been described. In addition, the incorporation of [³H]thymidine into meiotic DNA reveals the patchy DNA synthesis and gap filling that are presumed to occur during recombination. The correct resolution of recombination intermediates at meiosis, however, is known to be critical for proper gamete formation. Recall that at the first meiotic division homologous chromatids are paired (synapsis) and may undergo recombination at this time, but sister chromatids do not separate; both go into one cell. If recombination is not completely resolved, all four chromatids in the synaptic pair would end up in one cell. This would obviously be lethal. Correct resolution of recombination at meiosis is therefore critical.

Our best hope for learning the details of meiotic recombination lies in studies with yeast and invertebrate organisms in which mutations of the recombination machinery are known. Many yeast mutants that cannot undergo meiosis are known, and in some of these the genes have been cloned. Through combined genetic and biochemical analyses, discovery of the proteins involved in meiosis and their actions in recombination should be possible.

Summary

The general principles of DNA replication seem to apply, with little modification, to all cells. Viewed at the level of whole chromosomes, DNA synthesis is initiated at special regions called origins; a bacterial chromosome has just one origin, whereas a eukaryotic chromosome has many. Synthesis usually proceeds bidirectionally away from an origin via two growing forks moving in opposite directions; this movement produces a replication bubble. The copying of the DNA duplex at the growing fork is semiconservative; that is, each daughter duplex contains one old strand and its newly made complementary partner.

In bacterial cells, the generation of the new chromosome occupies almost the entire doubling time, so that DNA synthesis is almost continuous. In eukaryotic cells, DNA synthesis is confined to the S (synthesis) phase of the cell cycle. The control of the entry of cells into the S phase is not well understood; cells in the G_2 phase cannot enter another round of DNA synthesis, whereas those in the G_1 phase can.

The enzymes and the other factors that perform the necessary functions at a growing fork have been most extensively studied in *E. coli*, although many similar proteins are known to function in eukaryotic cells; thus the biochemical mechanism of DNA replication appears to be similar in all cells. Most of the enzymatic events at the growing fork are a consequence of two properties of the DNA double helix and one property of DNA polymerases, the enzymes that copy DNA. First, the DNA in the helix is an antiparallel duplex; that is, the $5' \rightarrow 3'$ direction of one strand is opposite to the $5' \rightarrow 3'$ direction of the other. Second, the strands of the duplex are interwound and cannot simply be pulled apart along their entire lengths all at once. Third, DNA polymerases require a nucleic acid primer—that is, a 3' hydroxyl end of a DNA or an RNA molecule—to begin synthesis. All DNA chain growth consequently proceeds in the $5' \rightarrow 3'$ direction. Thus, in order for both strands to be copied at a growing fork, one of the two new strands must be synthesized in a discontinuous fashion.

The strand synthesized continuously in the direction of movement of the growing fork is called the leading strand. It can start from the 3' hydroxyl end created by a nick in the template DNA or the end of a primer RNA. For synthesis of the other strand, the lagging strand, a series of short RNA pieces must be synthesized on the remaining parental template strand, in the direction opposite to the overall direction of the growing fork; enzymes called RNA primases make these RNA chains. DNA chains then grow from the RNA primers; the resulting segments of RNA plus DNA are called Okazaki fragments. A nuclease activity that is part of DNA polymerase removes the RNA primers, and the polymerase then fills the remaining gap by addition of deoxyribonucleotides. Finally, DNA ligase joins adjacent completed Okazaki fragments. Thus the growing fork moves along in one direction and the principle of 5' → 3' DNA chain synthesis is preserved.

Many proteins assist DNA polymerase, primase, and ligase in the construction of a new chromosome. These include helicases, which can unwind DNA, and topoisomerases. The latter enzymes allow the strands of a duplex to swivel about each other so that the duplex can be unwound; these enzymes can also pass one DNA duplex through another to introduce or remove supercoils. Topoisomerases are important both in growing-fork movement and in resolving (untangling) finished chromosomes after DNA duplication. Many of the proteins responsible for replication appear to form a giant complex, the replisome, which moves the growing fork along at great speed.

In eukaryotic cells, chromosomal replication involves more than just making the DNA. The new DNA is packaged promptly into nucleosomes. Most H3 and H4 histones enter nucleosomes just after replication; H2A and H2B histones not only can enter new nucleosomes but also exchange with old nucleosomes.

Environmental hazards damage bases in DNA and necessitate repair of the DNA so that mutations do not occur at an intolerable rate. This repair is achieved by a battery of enzymes capable of recognizing damaged or mismatched base pairs and excising them. Enzymes capable of single-stranded DNA synthesis then often widen the single-stranded gap before filling it in and religating the free ends of the DNA.

Some of the enzymes that function in DNA repair may also participate in recombination, the event in which two DNA helices are broken and exchanged. Probably at least two initiating events can start recombination: (1) invasion of a duplex from the cut end of one strand of a neighboring similar duplex and (2) invasion from the end of the duplex with both strands cut. These initial events are followed by rotation of the interconnected DNA molecules to create crossed-strand Holliday structures. Enzymatic creation and resolution of Holliday structures has been achieved in a few cases in the test tube. Undoubtedly resolution of the recombining molecules in cell chromo-

somes also occurs by similar *resolvases,* so that a double-strand exchange occurs. In eukaryotes, special nuclear meiotic structures called synaptonemal complexes are used in meiotic recombination, and genetic analyses of these events are beginning.

References

General References

CAMPBELL, J. 1986. Eukaryotic DNA replication. *Ann. Rev. Biochem.* **55:**733–772.

HUBERMAN, J. S. 1987. Eukaryotic DNA replication: a complex picture partially clarified. *Cell* **48:**7–8.

KELLY, T. J., and B. STILLMAN (eds.). 1988. *Cancer Cells 6/Eukaryotic DNA Replication.* Cold Spring Harbor Laboratory.

KORNBERG, A. and T. A. BAKER. In press. *DNA Replication* 2d ed. W. H. Freeman and Company.

MCMACKEN, R., and T. J. KELLEY (eds.). 1987. DNA replication and recombination. *UCLA Symp. Mol. Cell Biol., New Ser.,* vol. 47. A. R. Liss.

NEWPORT, J. W., and D. J. FORBES. 1987. The nucleus: structure, function and dynamics. *Ann. Rev. Biochem.* **56:**535–566.

Recombination at the DNA level. 1984. *Cold Spring Harbor Symp. Quant. Biol.,* vol. 49.

Cell Cycle

CROSS, F., J. ROBERTS, and H. WEINTRAUB. 1989. Simple and complex cell cycles. *Annu. Rev. Cell Biol.* **5:**341–395.

HARTWELL, L. H., and T. A. WEINERT. 1989. Checkpoints: Controls that ensure the order of cell cycle events. *Science* **246:**629–633.

LASKEY, R. A., M. P. FAIRMAN, and J. J. BLOW. 1989. S phase of the cell cycle. *Science* **246:**609–614.

MURRAY, A. W., and M. W. KIRSCHNER. 1989. Dominoes and clocks: The union of two views of the cell cycle. *Science* **246:**614–621.

PARDEE, A. B. 1989. G₁ events and regulation of cell proliferation. *Science* **246:**603–608.

DNA Replication

BLOW, J. J., and R. A. LASKEY. 1988. A role for the nuclear envelope in controlling DNA replication within the cell cycle. *Nature* **332:**546–548.

BREWER, B. 1988. When polymerases collide: replication and the transcriptional organization of the *E. coli* chromosome. *Cell* **53:**679–686.

HATTON, K. S., et al. 1988. Replication program of active and inactive multigene families in mammalian cells. *Mol. Cell Biol.* **8:**2149–2158.

HIDAKA, M., M. AKIYAMA, and T. HORIUCHI. 1988. A consensus sequence of three DNA replication terminus sites on the *E. coli* chromosome is highly homologous to the *terR* sites of the R6K plasmid. *Cell* **55**:467–475.

HUBERMAN, J. A., and A. D. RIGGS. 1968. On the mechanism of DNA replication in mammalian chromosomes. *J. Mol. Biol.* **32**:327–341.

HUBERMAN, J. A., and A. TSAI. 1973. Direction of DNA replication in mammalian cells. *J. Mol. Biol.* **75**:5–12.

McCARROLL, T. M., and W. L. FANGMAN. 1988. Time of replication of yeast centromeres and telomeres. *Cell* **54**:505–513.

MESELSON, M., and F. STAHL. 1958. The replication of DNA in *E. coli. Proc. Nat'l Acad. Sci. USA* **44**:671–782.

POLISKY, B. 1988. ColE1 replication control circuitry: sense from antisense. *Cell* **55**:929–932.

RAO, P. N., and R. T. JOHNSON. 1970. Mammalian cell fusion: studies on the regulation of DNA synthesis and mitosis. *Nature* **225**:159–164.

SUNDIN, O., and A. VARSHAVSKY. 1980. Terminal stages of SV40 DNA replication proceed via multiple intertwined catenated dimers. *Cell* **21**:103–114.

SUNDIN, O., and A. VARSHAVSKY. 1981. Arrest of segregation leads to accumulation of highly intertwined catenated dimers: dissection of the final stages of SV40 DNA replication. *Cell* **25**:659–669.

TAYLOR, J. H. 1958. The duplication of chromosomes. *Sci. Am.* **198**(6):36–42.

TSENG, B. Y., J. M. ERICKSON, and M. GOULIAN. 1979. Initiator RNA of nascent DNA from animal cells. *J. Mol. Biol.* **129**:531–545.

Properties of DNA Polymerases and DNA Synthesis in Vitro

BAKER, T. A., and A. KORNBERG. 1988. Transcriptional activation of initiation of replication from the *E. coli* chromosomal origin: An RNA-DNA hybrid near oriC. *Cell* **55**:113–123.

BRAMHILL, D., and A. KORNBERG. 1988. Duplex opening by dnaA protein at novel sequences in initiation of replication at the origin of the *E. coli* chromosome. *Cell* **52**:743–755.

CHASE, J. W., and K. R. WILLIAMS. 1986. Single-stranded DNA binding proteins required for DNA replication. *Ann. Rev. Biochem.* **55**:103–136.

COTTERILL, S. M., M. E. REYLAND, L. A. LOEB, and I. R. LEHMAN. 1987. A cryptic proofreading $3' \rightarrow 5'$ exonuclease associated with the polymerase subunit of the DNA polymerase-primase from *Drosophila melanogaster. Proc. Nat'l Acad. Sci. USA* **84**:5635–5639.

DEAN, F. B., et al. 1987. Simian virus 40 large tumor antigen requires three core replication origin domains for DNA unwinding and replication *in vitro*. Proc. Nat'l Acad. Sci. USA **84**:8267–8271.

DEAN, F. B., et al. 1989. SV40 replication in vitro. In *Molecular Mechanisms in DNA Replication and Recombination*, C. C. Richardson and I. R. Helman, eds. A. R. Liss.

NAGATA, K., R. A. GUGGENHEIMER, and J. HURWITZ. 1983. Adenovirus DNA replication *in vitro*: synthesis of full-length DNA with purified proteins. *Proc. Nat'l Acad. Sci. USA* **80**:4266–4270.

OGAWA, T., T. A. BAKER, A. VAN DER ENDE, and A. KORNBERG. 1985. Initiation of enzymatic replication at the origin of the *E. coli* chromosome: contributions of RNA polymerase and primase. *Proc. Nat'l Acad. Sci. USA* **82**:3562–3566.

PRELICH, G. P., M. KOSTURA, D. R. MARSHAK, M. B. MATHEWS, and B. STILLMAN. 1987. The cell-cycle regulated proliferating cell nuclear antigen is required for SV40 DNA replication *in vitro. Nature* **326**:471–475.

PRELICH, G., and B. STILLMAN. 1988. Coordinated leading and lagging strand synthesis during SV40 DNA replication in vitro requires PCNA. *Cell* **53**:117–126.

ROBERTS, J. M., and G. D'URSO. 1988. An origin unwinding activity regulates initiation of DNA replication during mammalian cell cycle. *Science* **241**:1486–1489.

WIDES, R. J., M. D. CHALLBERG, D. R. RAWLINS, and T. J. KELLY. 1987. Adenovirus origin of DNA replication: sequence requirements for replication in vitro. *Mol. Cell Biol.* **7**:864–874.

Superhelicity in DNA

BAUER, W. R., F. H. C. CRICK, and J. H. WHITE. 1980. Supercoiled DNA. *Sci. Am.* **243**(1):118–133.

CRICK, F. H. C. 1976. Linking numbers and nucleosomes. *Proc. Nat'l Acad. Sci. USA* **73**:2639–2643.

HONNIGBERG, S. M., and C. M. RADDING. 1988. The mechanics of winding and unwinding helices in recombination: torsional stress associated with strand transfer promoted by RecA protein. *Cell* **54**:525–532.

VINOGRAD, J., J. LEBOWITZ, R. RADLOFF, R. WATSON, and P. LAPIS. 1965. The twisted circular form of polyoma viral DNA. *Proc. Nat'l Acad. Sci. USA* **53**:1104–1111.

WASSERMAN, S. A., and N. R. COZZARELLI. 1986. Biochemical topology: applications to DNA recombination and replication. *Science* **232**:951–960.

Origins of DNA Replication

BRAMHILL, D., and A. KORNBERG. 1988. A model for initiation at origins of DNA replication. *Cell* **54**:915–918.

BREWER, B. J., and W. L. FANGMAN. 1987. The localization of replication origins on ARS plasmids in *S. cerevisiae. Cell* **51**:463–471.

DE PAMPHLIS, M. L. 1988. Transcriptional elements as components of eukaryotic origins of DNA replication. *Cell* **52**:635–638.

HUBERMAN, J. A., L. D. SPOTILA, K. A. NAWOTKA, S. M. EL-ASSOULI, and L. R. DAVIS. 1987. The in vivo replication origin of the yeast $2\mu m$ plasmid. *Cell* **51**:473–481.

JOHNSON, E. M., and W. R. JELINEK. 1986. Replication of a plasmid bearing a human *Alu*-family repeat in monkey COSS-7 cells. *Proc. Nat'l Acad. Sci. USA* **83**:4660–4664.

SUGIMOTO, K., et al. 1979. Nucleotide sequence of *Escherichia coli* K-12 replication origin. *Proc. Nat'l Acad. Sci. USA* **76**:575–579.

WU, C. A., N. J. NELSON, D. J. McGEOCH, and M. D. CHALLBERG. 1988. Identification of herpes simplex virus type 1 genes required for origin-dependent DNA synthesis. *J. Virol.* 62:435–443.

ZYSKIND, J. W., J. M. CLEARY, W. S. A. BRUSILOW, N. E HARDING, and D. W. SMITH. 1983. Chromosomal replication origin from the murine bacterium *Vibrio harveyi* functions in *E. coli: oriC* consensus sequence. *Proc. Nat'l Acad. Sci. USA* 80:1164–1168.

Topoisomerases

COZZARELLI, N. R. 1980. DNA gyrase and the supercoiling of DNA. *Science* 207:953–960.

GELLERT, M. 1981. DNA topoisomerases. *Ann. Rev. Biochem.* 50:879–910.

HOLM, C., T. STEARNS, and D. BOTSTEIN. 1989. DNA topoisomerase II must act at mitosis to prevent nondisjunction and chromosome breakage. *Mol. Cell Biol.* 9:159–168.

LIU, L. F., C.-C. LIU, and B. M. ALBERTS. 1980. Type II DNA topoisomerases: enzymes that can unknot a topologically knotted DNA molecule via a reversible double-strand break. *Cell* 19:697–707.

WANG, J. C. 1985. DNA topoisomerases. *Ann. Rev. Biochem.* 54:665–698.

YANG, L., M. S. WOLD, J. J. LI, T. J. KELLY, and L. F. LIU. 1987. Role of DNA topoisomerases in simian virus 40 DNA replication *in vitro. Proc. Nat'l Acad. Sci. USA* 84:950–954.

Assembly of DNA into Nucleosomes

JACKSON, V. 1987. Deposition of newly synthesized histones: new histones H2A and H2B do not deposit in the same nucleosome with new histones H3 and H4. *Biochemistry* 26:2315–2325.

JACKSON, V. 1987. Deposition of newly synthesized histones: misinterpretations due to cross-linking density-labeled proteins with Lomant's reagent. *Biochemistry* 26:2325–2334.

JACKSON, V. 1988. Deposition of newly synthesized histones: hybrid nucleosomes are not tandemly arranged on daughter DNA strands. *Biochemistry* 27:2109–2120.

LASKEY, R. A., and W. C. EARNSHAW. 1980. Nucleosome assembly. *Nature* 286:763–767.

LEVY, A., and K. M. JAKOB. 1978. Nascent DNA in nucleosome-like structures from chromatin. *Cell* 14:259–267.

Repair of DNA

FRIEDBERG, E. A., and P. E. HANAWALT. 1983. *DNA Repair: A Laboratory Manual of Research Procedures.* M. Dekker.

GLAZER, P. M., S. N. SARKAR, G. E. CHISHOLM, and W. C. SUMMERS. 1987. DNA mismatch repair detected in human cell extracts. *Mol. Cell Biol.* 7:218–224.

GROSSMAN, L., P. R. CARON, S. J. MAZUR, and E. Y. OH. 1988. Repair of DNA-containing pyrimidine dimers. *FASEB J.* 2:2696–2701.

LITTLE, J. W., and D. W. MOUNT. 1982. The SOS regulatory system of *Escherichia coli. Cell* 29:11–22.

MELLON, I., V. A. BOHR, C. A. SMITH, and P. C. HANAWALT. 1986. Preferential DNA repair of an active gene in human cells. *Proc. Nat'l Acad. Sci. USA* 83:8878–8882.

MODRICH, P. 1987. DNA mismatch correction. *Ann. Rev. Biochem.* 56:435–466.

SANCAR, A., and G. B. SANCAR. 1988. DNA repair enzymes. *Ann. Rev. Biochem.* 57:29–68.

VAN DUIN, M., et al. 1988. Evolution and mutagenesis of the mammalian excision repair gene *ERCC-1. Nucleic Acids Res.* 16:5305–5322.

WILLIS, A. E., and T. LINDAHL. 1987. DNA ligase deficiency in Bloom's syndrome. *Nature* 325:355–357.

Recombination

COX, M. M., and I. R. LEHMAN. 1987. Enzymes of general recombination. *Ann. Rev. Biochem.* 56:229–262.

DRESSLER, D., and H. POTTER. 1982. Molecular mechanisms in genetic recombination. *Ann. Rev. Biochem.* 51:727–761.

DUCKETT, D. R., et al. 1988. The structure of the Holliday junction, and its resolution. *Cell* 55:79–89.

FALVEY, E., G. F. HATFULL, and N. D. F. GRINDLEY. 1988. Uncoupling of the recombination and topoisomerase activities of the $\gamma\delta$ resolvase by a mutation at the crossover point. *Nature* 332:861–863.

GONDA, D. K., and C. M. RADDING. 1983. By searching processively RecA protein pairs DNA molecules that share a limited stretch of homology. *Cell* 34:647–654.

HOLLIDAY, R. 1964. A mechanism for gene conversion in fungi. *Genet. Res.* 5:282–304.

MESELSON, M., and C. M. RADDING. 1975. A general model for genetic recombination. *Proc. Nat'l Acad. Sci. USA* 72:358–361.

ORR-WEAVER, T. L., and J. W. SZOSTAK. 1983. Yeast recombination: the association between double-strand gap repair and crossing-over. *Proc. Nat'l Acad. Sci. USA* 80:4417–4421.

SCHIESTL, R. H., and S. PRAKASH. 1988. *RAD1*, an excision repair gene of *Saccharomyces cerevisiae*, is also involved in recombination. *Mol. Cell Biol.* 8:3619–3626.

SMITH, G. R. 1983. Chi, hot spots of generalized recombinations. *Cell* 34:709–710.

SZOSTAK, J. W., T. L. ORR-WEAVER, R. J. ROTHSTEIN, and F. W. STAHL. 1983. The double-strand-break repair model for recombination. *Cell* 33:25–35.

TAYLOR, J. H. 1958. Sister chromatid exchanges in tritium-labeled chromosomes. *Genetics* 43:515–529.

THOMAS, B. J., and R. ROTHSTEIN. 1989. Elevated recombination rates in transcriptionally active DNA. *Cell* 56:619–630.

VON WETTSTEIN, D. S., W. RASMUSSEN, and P. B. HOLM. 1984. The synaptonemal complex in genetic segregation. *Ann. Rev. Genet.* 18:331–414.

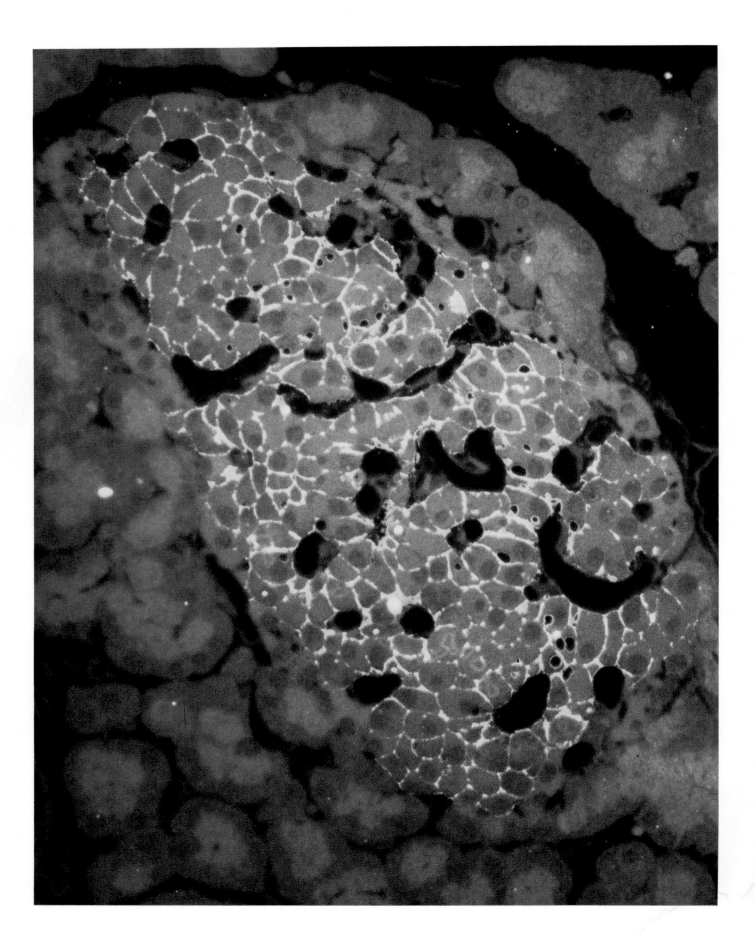

PART III

Cell Structure and Function

◀ *A section of rat pancreas stained with a fluorescent antibody specific for a glucose transporter protein. Only the plasma membrane of the β insulin secreting cells are stained. This transporter is part of the "glucose sensor" that triggers insulin secretion when the level of blood glucose rises.* Courtesy of L. Orci.

Genes located in the nucleus control all cell behavior and specialization. Yet most of the events of cell life take place outside the nucleus—in the cytosol or within the many subcellular compartments, such as organelles and small vesicles.

Each of these compartments as well as the cell itself is surrounded by a phospholipid bilayer membrane. Embedded in the membrane and attached to its surfaces are proteins, which catalyze reactions, control the passage of certain substances into and out of the cell and its compartments, anchor the cytoskeleton, and bind the cell to other cells or to the many fibers in the extracellular matrix. Because the plasma (or cell surface) membrane is so important to the structure and function of a cell, it is the subject of the first two chapters of this part of the book. Chapter 13 describes the structure of cell membranes, with special attention to the structure of membrane proteins. Chapter 14 examines the role of membrane proteins both in transporting small molecules into and out of the cell and also in endocytosis, the process by which proteins and particles are brought into the cell.

Cellular events require energy, and this energy is usually supplied by hydrolysis of the high-energy phosphoanhydride bonds in ATP. Chapter 15 examines the formation and conversion of high-energy bonds during the oxidation of carbohydrates, fatty acids, and amino acids. Photosynthesis, the conversion of light energy into chemical energy, is discussed in Chapter 16. The generation of phosphoanhydride bonds in ATP occurs on membranes in the chloroplast and mitochondrion and depends on ionic gradients across those membranes.

Specialized cell membranes and compartments rely on unique sets of proteins to carry out their particular functions. Chapters 17 and 18 describe the assembly of the membranes and organelles that make a cell, focusing on targetting proteins to their correct destination. Secretion of cellular proteins and the formation of the plasma membrane and the lysosome are covered in Chapter 17; Chapter 18 focuses on the biogenesis of the nucleus, mitochondrion, and chloroplast.

A complex network of cell-to-cell communication is required to coordinate differentiation, growth, metabolism, and behavior in multicellular organisms. Cellular secretions, which include hormones, act as the communication signals. Chapter 19 discusses the receptor proteins that govern the cell's response to chemical signals. Chapter 20 takes a look at the important topic of nerve-cell communication. The plasma membrane of a nerve cell is specialized to conduct electrical signals along its length; in addition, chemicals called neurotransmitters are usually required to relay the message to the next nerve or muscle cells.

Chapters 21 and 22 examine the cytoskeleton, the complex system of fibers within the cell that are responsible for the cell's shape and motility. Fibers protrude into the microvilli and thereby support these thin projections of the plasma membrane; other fibers exist in cilia and flagella. Fibers are also responsible for the contraction of muscle, the movement of chromosomes during cell division, and the migration of cells along a substratum.

Finally, Chapter 23 focuses on the many proteins, polysaccharides, proteoglycans, and other polymers that form the extracellular matrix surrounding animal cells and the plant cell wall. We will see how different cells synthesize and remodel extracellular polymers and how, in turn, these molecules influence the growth and differentiation of cells.

▼   ▼   ▼

C H A P T E R *13*

The Plasma Membrane

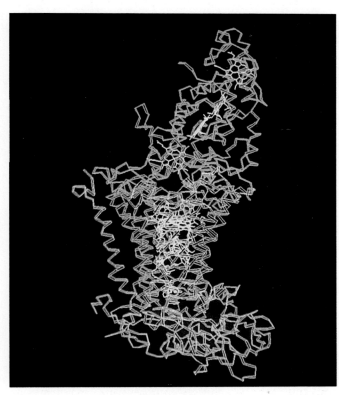

Bacterial photosynthetic reaction center

*U*ntil the advent of the electron microscope, scientists could not fully appreciate the complexity of cytoplasmic organization. Electron microscopy showed that a plasma membrane surrounds every cell. We now know that the plasma membrane of the cell is a highly differentiated structure containing specific proteins that help to control the intracellular milieu and interact with specific molecules to influence the cell's behavior. Membrane-bound enzymes catalyze reactions that would occur with difficulty in an aqueous environment. Other proteins in the plasma membrane provide anchors for cytoskeletal fibers or for components of the extracellular matrix that give the cell its shape. Still other proteins bind signaling molecules, provide a passageway across the membrane for certain molecules, or regulate the fusion of the membrane with others in the cell.

In addition, a multitude of internal membrane structures in each eukaryotic cell enclose separate compartments that perform specialized tasks: photosynthesis in the chloroplast, oxidative phosphorylation in the mitochondrion, degradation of macromolecules in the lysosome, and so on. Prokaryotes generally lack internal membranes; however, membrane vesicles do catalyze light absorption and other initial steps of photosynthesis in photosynthetic bacteria. Far from being a mere bag of soluble components, we now comprehend the cell as a highly organized entity with many functional subcompartments.

Here, we discuss the basic principles that govern the organization of proteins and phospholipids in all biological membranes. Although every membrane has the same basic structure of a phospholipid bilayer, a different set of membrane proteins enables each subcellular membrane to carry out its distinctive functions (Figure 13-1). We emphasize the structure and function of several characteristic membrane proteins. To illustrate the general principles of membrane architecture, we look at the plasma membranes in three typical mammalian cells. The relative simplicity of the erythrocyte (red blood cell) membrane allows us to examine how it interacts with a submembranous cytoskeleton to give the cell its shape and flexibility. The membranes of pancreatic acinar and intestinal epithelial cells demonstrate how different regions of the same plasma membrane can be specialized to perform different tasks. Finally, we focus on how junctions (certain specialized regions of the plasma membrane) allow adjacent cells in a tissue to bind to each other or to exchange small molecules. ▲

The Architecture of Lipid Membranes

All Membranes Contain Proteins and Lipids; Many Contain Carbohydrates

Subcellular fractionation techniques (see Figures 4-31 and 4-32) can partially separate and purify several important biological membranes, including the plasma and mitochondrial membranes, from many kinds of cells. These preparations are often contaminated with membranes from other organelles. However, the plasma membrane of human erythrocytes can be isolated in near purity because the cells contain no internal membranes.

All membranes, regardless of their source, contain proteins as well as lipids (Table 13-1). The protein-lipid ratio varies enormously: the inner mitochondrial membrane is 76 percent protein; the myelin membrane, only 18 percent. The protein content of myelin is low because it electrically insulates the nerve cell from its environment.

The lipid composition also varies greatly among different membranes (Table 13-2). All membranes contain a substantial proportion of phospholipids (Figure 13-2). Sphingomyelin, like phospholipids, is an amphipathic membrane lipid: it has hydrophilic and hydrophobic portions. Sphingomyelin contains sphingosine, an amino al-

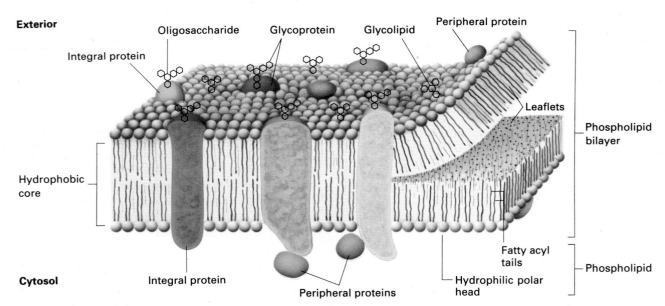

▲ **Figure 13-1** A phospholipid bilayer constitutes the basic structure of biological membranes. The hydrophobic fatty acyl tails of the phospholipids form the middle of the bilayer; the polar, hydrophilic heads of the phospholipids line both surfaces. Integral proteins have one or more regions embedded in the lipid bilayer. Peripheral proteins are primarily associated with the membrane by specific protein-protein interactions. Oligosaccharides bind mainly to membrane proteins; however, some bind to lipids, forming glycolipids.

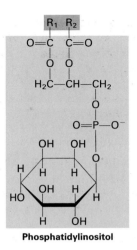

Phosphatidylserine

Phosphatidylcholine

Phosphatidylinositol

Diphosphatidylglycerol

▲ **Figure 13-2** The structures of some common phospholipids found in cell membranes. The fatty acyl side chains (represented by R_1 and R_2) can be saturated or they may contain one or more double bonds. Note that diphosphatidylglycerol contains four fatty acids; it is composed of two molecules of phosphatidic acid linked to carbons 1 and 3 of a central glycerol. Hydrophobic portions of the molecules are shown in red, hydrophilic in blue.

Table 13-1 Chemical composition of some purified membranes (in percentages)

Membrane	Protein	Lipid	Carbohydrate	Membrane	Protein	Lipid	Carbohydrate
Myelin	18	79	3	*Halobacterium* purple membrane	75	25	0
Plasma membrane				Mitochondrial			
Human erythrocyte	49	43	8	inner membrane	76	24	0
Mouse liver	44	52	4	Chloroplast			
Ameba	54	42	4	Spinach lamellae	70	30	0

SOURCE: G. Guidotti, 1972, *Ann. Rev. Biochem.* **41**:731.

Table 13-2 Lipid composition of membrane preparations (in percentages)*

Source	Cholesterol	PC	SM	PE	PI	PS	PG	DPG	PA	Glycolipids
Rat Liver										
Plasma membrane	30.0	18	14.0	11	4.0	9.0	—	—	1	—
Endoplasmic reticulum (rough)	6.0	55	3.0	16	8.0	3.0	—	—	—	—
Endoplasmic reticulum (smooth)	10.0	55	12.0	21	6.7	—	—	1.9	—	—
Mitochondria (inner)	3.0	45	2.5	24	6.0	1.0	2.0	18.0	0.7	—
Mitochondria (outer)	5.0	50	5.0	23	13.0	2.0	2.5	3.5	1.3	—
Nuclear membrane	10.0	55	3.0	20	7.0	3.0	—	—	1.0	—
Golgi	7.5	40	10.0	15	6.0	3.5	—	—	—	—
Lysosomes	14.0	25	24.0	13	7.0	—	—	5.0	—	—
Myelin	22.0	11	6.0	14	—	7.0	—	—	—	12
Rat erythrocyte	24.0	31	8.5	15	2.2	7.0	—	—	0.1	3
E. coli cytoplasmic membrane	0	0	—	80	—	—	15.0	5.0	—	—

*PC, phosphatidylcholine; SM, sphingomyelin; PE, phosphatidylethanolamine; PI, phosphatidylinositol; PS, phosphatidylserine; PG, phosphatidylglycerol; DPG, diphosphatidylglycerol (cardiolipin); PA phosphatidic acid.

SOURCE: M. K. Jain and R. C. Wagner, 1980, *Introduction to Biological Membranes*, Wiley.

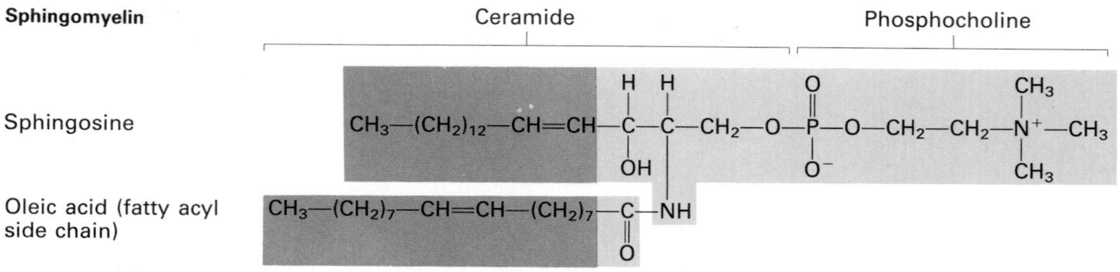

Sphingomyelin

cohol with a long unsaturated hydrocarbon chain. A fatty acyl side chain is linked to the NH_2 group of sphingosine by the amide bond

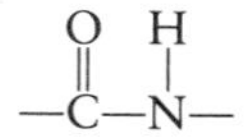

to form ceramide (Figure 13-3). The hydroxyl group of sphingosine is esterified to phosphocholine; thus the hydrophilic head of sphingomyelin is similar to that of phosphatidylcholine (see Figure 13-2).

Many membranes contain cholesterol (see Figure 2-47); it is especially abundant in the plasma membrane of mammalian cells but is absent from most prokaryotic cells. As much as 30–50 percent of the lipids in plant plasma membranes are steroids (cholesterol and other steroids unique to plants). Some lipids are found in a restricted range of membranes: uncharged *galactolipids* (Figure 13-4) comprise as much as 70 percent of the lipids in chloroplast thylakoid membranes; the phospholipid cardiolipin (diphosphatidylglycerol; see Figure 13-2) is highly enriched in the inner mitochondrial membrane.

Carbohydrates are an important constituent of many membranes. They are bound either to proteins as constituents of glycoproteins (see Figures 2-55 and 2-56) or to lipids as constituents of glycolipids (see Figure 2-57). Both glycoproteins and glycolipids are especially abundant in the plasma membranes of eukaryotic cells but are absent from the inner mitochondrial membrane, the chloroplast lamellae, and several other intracellular membranes. Bound carbohydrates increase the hydrophilic character of lipids and proteins and help to stabilize many membrane protein structures. In mammals, certain glycolipids form blood-group antigens (see Figure 2-58).

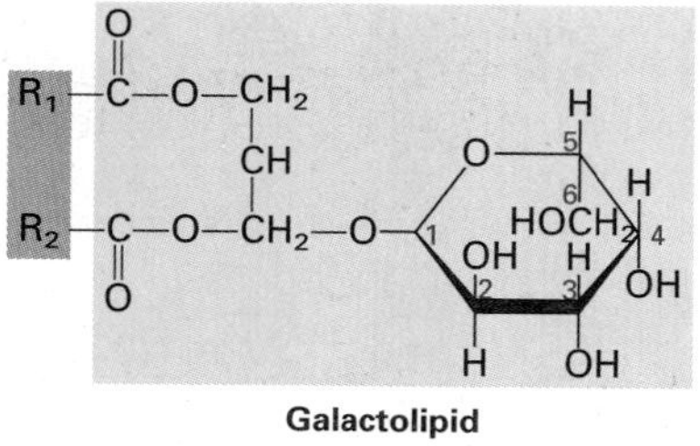

Galactolipid

▲ **Figure 13-4** The structure of a galactolipid, the principal lipid in chloroplast thylakoid membranes.

The Phospholipid Bilayer Is the Basic Structural Unit of Biological Membranes

Despite their variable compositions, the basic structural unit of virtually all biological membranes is the phospholipid bilayer (see Figure 13-1). Because all phospholipids are amphipathic, hydrophobic interactions between the fatty acyl chains of glycolipid and phospholipid molecules create a sheet containing two layers of phospholipid molecules whose polar head groups face the surrounding water and the fatty acyl chains form a continuous hydrophobic interior about 4 nm thick (Figure 13-5; see Figure 13-1). In such a lamellar structure, each layer of phospholipid is called a *leaflet*.

Sphingomyelins, galactolipids, and all phospholipids except cardiolipin contain two fatty acyl chains. Nearly all fatty acyl chains found in the membranes of eukaryotic cells have an even number of carbon atoms (usually 16, 18, or 20). Unsaturated fatty acyl chains normally have one double bond, but some have two, three, or four. In general, all such double bonds are of the cis configuration; this introduces a rigid kink in the otherwise flexible straight chain (Figure 13-6). A major difference among phospholipids concerns the polar heads. At neutral pH, the polar head group may have no net electric charge (phosphatidylcholine, phosphatidylethanolamine) or a negative charge (phosphatidylglycerol, cardiolipin, phosphatidylserine). Rarer phospholipids have a net positive charge. Nonetheless, all the polar head groups can pack together into a phospholipid bilayer. Sphingomyelin and glycolipids, although chemically different from phospholipids, have a similar shape and can form mixed bilayers with them.

Are lipids alone responsible for membrane structure? Lipids are not the only component; all biological membranes, no matter how carefully purified, are found to contain proteins (see Table 13-1). The percentage and exact nature of the adhering proteins vary considerably with membrane type. Some membrane proteins bind to the hydrophobic fatty acyl core of the bilayer; others bind to the membrane primarily by protein-protein interactions (see Figure 13-1). But are proteins essential to membrane structure, or do they only participate in membrane function? Current data suggest that specialized proteins fulfill one or more specific membrane functions but that the membrane owes its structural integrity to the properties of constituent lipids.

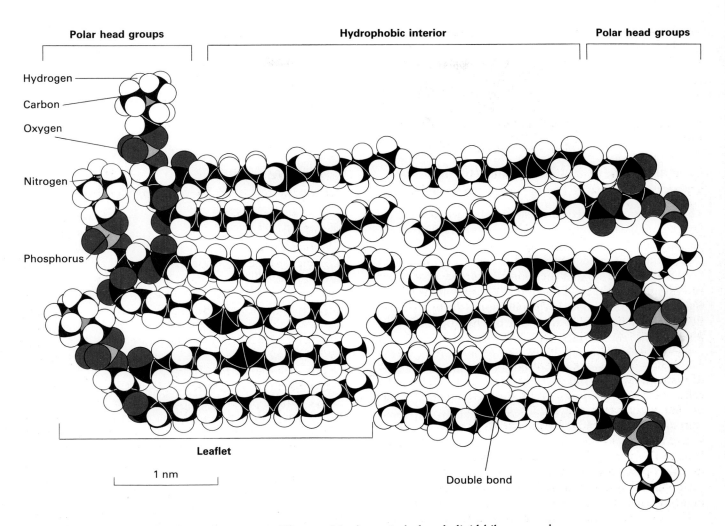

Polar head groups   **Hydrophobic interior**   **Polar head groups**

Hydrogen

Carbon

Oxygen

Nitrogen

Phosphorus

Leaflet

1 nm

Double bond

▲ **Figure 13-5** A space-filling model of a typical phospholipid bilayer membrane. The hydrophobic interior is generated by the fatty acyl side chains. Some of these chains have bends, caused by the double bonds. The different polar head groups all lie on the outer, aqueous surface of the membrane. [See L. Stryer, 1988, *Biochemistry*, 3d ed., page 289, W. H. Freeman and Company.] *Courtesy of L. Stryer.*

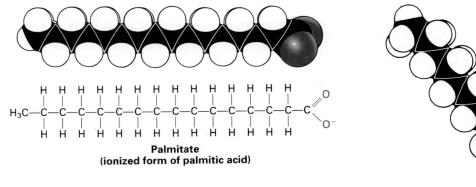

Palmitate
(ionized form of palmitic acid)

▲ **Figure 13-6** Space-filling models and chemical structures of two fatty acids: saturated palmitate and unsaturated oleate. The saturated fatty acid forms a linear molecule; the cis double bond in oleate creates a kink in the hydrocarbon chain. *After L. Stryer, 1988,* Biochemistry, *3d ed., W. H. Freeman and Company, p. 285.*

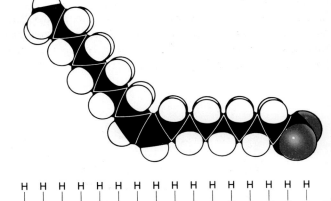

Oleate
(ionized form of oleic acid)

An exception may be the thylakoid membranes in chloroplasts. In aqueous solutions, a pure preparation of galactolipids (see Figure 13-4), the major lipid component, forms inverted micelles with the polar head groups inside, rather than bilayers. The thylakoid membrane is 60–70 percent protein, and the bilayer may result from interactions among galactolipids, phospholipids, and proteins.

Several Types of Evidence Point to the Universality of the Phospholipid Bilayer

There are good theoretical reasons to believe that the bilayer structure is common to all membranes. Direct experimental evidence exists for some cells. In 1925, E. Gorter and F. Grendel extracted the lipids from human erythrocytes and floated them on the surface of a water solution. It was already known that phospholipids form a unimolecular film under such conditions: the hydrophilic head groups face the water, and the hydrophobic tails point up into the air (Figure 13-7). The area of this monolayer was approximately twice that of the surface of the original erythrocytes. Since erythrocytes contain no internal membranes, Gorter and Grendel concluded that the lipids are arranged in the membrane as a continuous bilayer. Interestingly, these investigators actually extracted only a portion of the erythrocyte lipids and underestimated the surface of the erythrocyte cell; these two errors canceled each other out, giving the correct answer!

Later experiments established that phospholipids, when dispersed in aqueous solutions, spontaneously form either sheets of bilayers or closed vesicles that contain one wall of a phospholipid bilayer. These results are obtained both with a single species of phospholipid and with mixtures approximating those found in natural membranes.

Perhaps the best evidence for the bilayer structure comes from low-angle x-ray diffraction analysis, which determines the density of matter. The multimembrane *myelin sheath* that covers and insulates many mammalian nerve cells is used in such studies because it is conveniently stacked. Myelin is formed from the plasma membrane of a Schwann cell that wraps around the axon (elongated portion) of a neuron (Figure 13-8a). Eventually, the Schwann cell becomes little more than a series of stacked membranes, and myelin becomes the major mem-

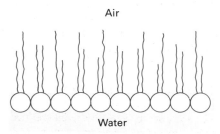

Air

Water

▲ **Figure 13-7**   A monolayer is formed when phospholipids are floated on the surface of an aqueous solution.

(a)

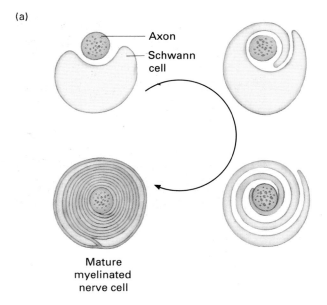

Axon

Schwann cell

Mature myelinated nerve cell

(b)

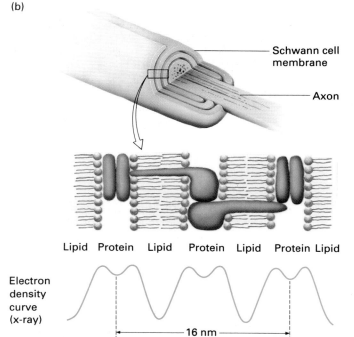

Schwann cell membrane

Axon

Lipid Protein Lipid Protein Lipid Protein Lipid

Electron density curve (x-ray)

◀—————— 16 nm ——————▶

▲ **Figure 13-8**   Low-angle x-ray diffraction analysis of lipid-protein distribution in myelin membranes. (a) During development of the nervous system, a large Schwann cell envelops the axon of a neuron. The continuous growth of the Schwann cell membrane into its own cytoplasm, together with rotation of the nerve axon, results in a laminated spiral of double plasma membranes around the axon. Mature myelin, a stack of plasma membranes of the Schwann cell, is relatively rich in lipids. (b) The profile of electron density—and thus of matter—obtained by x-ray diffraction studies on fresh nerve, and the relation of this profile to the protein and lipid components of the myelin membranes. [See W. T. Norton, 1981, in *Basic Neurochemistry*, 3d ed., G. J. Siegel et al., eds. p. 68, Little, Brown.]

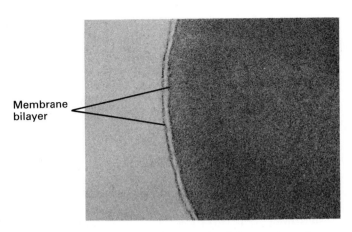

▲ **Figure 13-9** Electron micrograph of thin section of an osmium-stained erythrocyte membrane. Note the "railroad track" appearance of the phospholipid bilayer at the cell surface. *Courtesy of J. D. Robertson.*

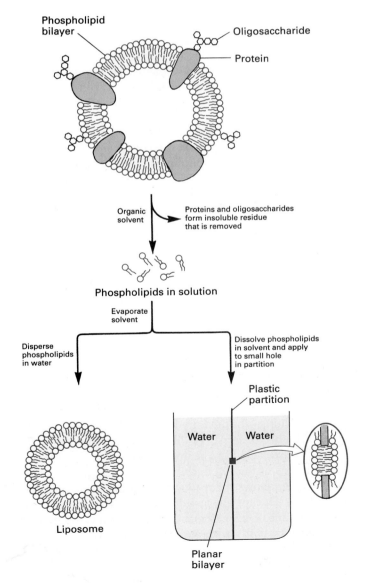

brane component of such nerves. Myelin can be separated from other cellular membranes in a pure state, permitting direct physical and chemical analyses.

X-ray analysis of these stacked plasma membrane units shows a very low density in the middle of each membrane, suggesting that the amount of lipid in this region significantly exceeds that of protein (Figure 13-8b). Such a distribution of matter implies a bilayer organization in which protein is located on either side of a membrane with a central region of almost pure hydrocarbon. Parts of proteins do pass through the lipid bilayer, but these polypeptide segments make up less than 10 percent of the inner mass of the membrane and do not contribute to the x-ray analysis.

The most direct evidence for the universality of the bilayer structure is furnished by the electron microscope. If an electron-opaque stain could bind to the polar head groups of phospholipids, then a cross section of a bilayer membrane exposed to such a stain and viewed with an electron microscope would look like a railroad track: two thin lines (the stain-phosphate complex) with a uniform space of about 2 nm (the lipid) between them. In fact, membranes stained with osmium tetroxide look just this way (Figure 13-9). Although some osmium tetroxide may bind to the double bonds of fatty acyl chains, most of it binds to the polar head groups. Phospholipids may be arranged differently in certain membrane regions, especially where specialized proteins play a large role in the structure, and some membrane regions may even lack lipid; but these are exceptions to the general pattern.

Phospholipid Bilayers and Biological Membranes Form Closed Compartments

Biophysical studies of pure phospholipids have revealed several properties important to an understanding of biological membranes. Perhaps the most significant is that pure phospholipid membranes spontaneously seal to form closed structures that separate two aqueous compartments (Figure 13-10). (Theoretically, a lipid bilayer sheet would be an unstable structure if it had a free edge where the hydrophobic region of the bilayer is in contact with water.)

◄ **Figure 13-10** Construction of pure phospholipid bilayers. A preparation of biological membranes is treated with an organic solvent, such as a mixture of chloroform and methanol (3:1), which selectively solubilizes the phospholipids and cholesterol. Proteins and carbohydrates remain in an insoluble residue. The solvent is removed by evaporation. *Bottom left:* If the lipids are mechanically dispersed in water, they spontaneously form a liposome with an internal aqueous compartment. *Bottom right:* A planar bilayer can form over a small hole in a partition separating two aqueous phases; such bilayers are often termed "black lipid membranes" because of their appearance.

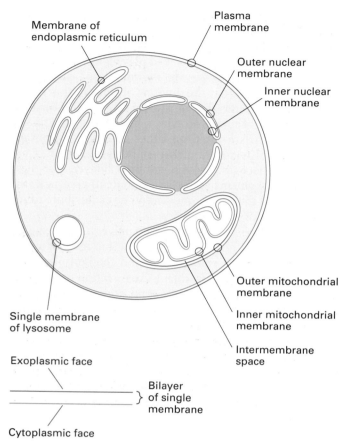

▲ **Figure 13-11** Faces of cellular membranes. The exoplasmic surfaces are red. For such organelles as the nucleus, chloroplast, and mitochondrion, which are enclosed in two phospholipid bilayers, the exoplasmic face is the one that faces the space between the inner and outer membranes, such as the intermembrane space in mitochondria.

Similarly, all cellular membranes surround closed compartments of the cell; all bilayers have an *internal face* (the side oriented toward the interior of the compartment) and an *external face* (the side presented to the environment). Because most organelles are surrounded by a single bilayer membrane, it is also useful to speak of the *cytoplasmic* and *exoplasmic faces* of the membrane. The exoplasmic face of the plasma membrane, which is directed away from the cytoplasm, defines the outer limit of the cell (Figure 13-11). Some organelles, such as the nucleus and mitochondrion, are surrounded by two membranes; in these cases, the exoplasmic surface faces the lumen between the two membranes.

Phospholipid Bilayers Form a Two-dimensional Fluid

Two systems of pure phospholipid bilayers have proved to be especially useful in studying the motions of lipid molecules within membranes. *Liposomes*—spherical ves-

icles up to 1 micrometer (μm) in diameter—are formed by mechanically dispersing phospholipids in water; *planar bilayers* are formed across a hole in a partition that separates two aqueous solutions (see Figure 13-10). In such artificial membranes, the natural thermal motion of molecules permits phospholipids and glycolipids to rotate freely around their long axes and to diffuse laterally within the membrane leaflet. Because such movements are lateral or rotational, the fatty acyl chains remain in the hydrophobic interior of the membrane. In such artificial membranes, a typical lipid molecule exchanges places with its neighbors in a leaflet about 10^7 times per second and diffuses several micrometers per second at 37°C. Thus a lipid could diffuse the 1 μm length of a bacterial cell in only 1 s and of an animal cell in about 20 s. However, all of the lipids in cells are not free to move over μm distances, although they may move over short (nm) distances. Certain lipids may not be able to diffuse past domains of several μm^2 in the plasma membrane.

In pure phospholipid bilayers, phospholipids do not migrate, or flip-flop, from one leaflet of the membrane to the other. In some natural membranes, they occasionally do, catalyzed somehow by one or more membrane proteins. Energetically, such movements are extremely unfavorable, because the polar head of a phospholipid must be transported through the hydrophobic interior of the membrane.

In one technique for measuring the mobility of membrane lipids, called *electron spin resonance (ESR) spectroscopy,* synthetic phospholipids containing a nitroxide group are introduced into otherwise normal phospholipid membranes. ESR measures the energy absorbed by the unpaired electron of the nitroxide group. ESR studies show that both synthetic and natural membranes generally have a low viscosity and a fluid-like consistency.

The Fluidity of a Bilayer Depends on Its Lipid Composition, Cholesterol Content, and Temperature

When a suspension of liposomes made up of a single type of phospholipid is heated, it undergoes an abrupt change in physical properties over a very narrow temperature range. This *phase transition* is due to a reorganization—a kind of melting—of the fatty acyl side chains, which pass from a highly ordered, gel-like state to a more mobile state (Figure 13-12). The increased motion about the C—C bonds of the fatty acyl chains, which accompanies this change, allows them to assume a more random conformation. During the gel-fluid transition, a relatively large amount of heat (thermal energy) is absorbed.

In general, lipids with short or unsaturated fatty acyl chains undergo the phase transition at lower temperatures than do lipids with long or saturated chains. Be-

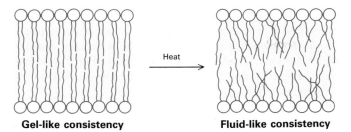

Gel-like consistency   **Fluid-like consistency**

▲ **Figure 13-12** Alternative forms of the phospholipid bilayer. Heat induces the transition from gel to fluid over a temperature range of only a few degrees. The fluid phase is also favored by the presence of short fatty acyl chains and by a double bond in the acyl chains.

cause short chains have less surface area with which to form van der Waals interactions with each other, the more crystalline gel state is less stable and melts at lower temperatures. Unsaturated fatty acyl chains have kinks and form less stable van der Waals contacts with other lipids; they tend to maintain a more random, fluid state compared to saturated fatty acyl chains.

Maintenance of bilayer fluidity appears to be essential to normal cell growth and reproduction. All cell membranes contain a mixture of different fatty acyl chains and

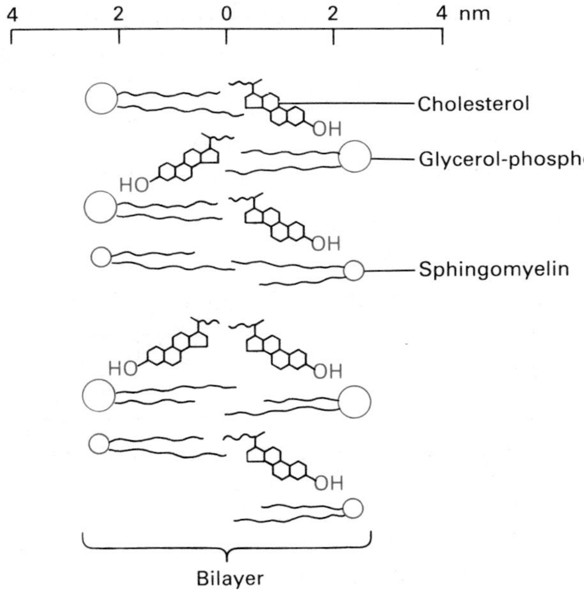

▲ **Figure 13-13** Arrangement of phospholipids and cholesterol in a typical myelin membrane. The polar OH group of cholesterol faces the aqueous surface of the membrane; the hydrophobic hydrocarbon portion nestles among the fatty acyl side chains. The polar OH group of cholesterol and the polar head groups of sphingomyelin (small circles) and glycerol-phospholipids (large circles) are shown in red. *After D. Caspar and D. A. Kirschner, 1971,* Nature **231**:46.

are fluid at the temperature at which the cell is grown. Animal and bacterial cells adapt to a decrease in growth temperature by increasing the proportion of unsaturated to saturated fatty acids in the membrane, which tends to maintain a fluid bilayer at the reduced temperature.

Membrane cholesterol is a major determinant of bilayer fluidity. Cholesterol is too hydrophobic to form a sheet structure on its own, but it intercalates (is inserted) among phospholipids. Its polar hydroxyl group is in contact with the aqueous solution near the polar head groups of the phospholipids; the steroid ring interacts with and tends to immobilize their fatty acyl chains (Figure 13-13). The net effect of cholesterol on membrane fluidity varies, depending on the lipid composition. Cholesterol restricts the random movement of the part of the fatty acyl chains lying closest to the outer surfaces of the leaflets, but it separates and disperses the tails of the fatty acyls and causes the inner regions of the bilayer to become slightly more fluid. At the high concentrations found in eukaryotic plasma membranes, cholesterol tends to make the membrane less fluid at growth temperatures near 37°C. At temperatures below the phase transition, cholesterol keeps the membrane in a fluid state by preventing the hydrocarbon fatty acyl chains of the membrane lipids from binding to each other, thereby offsetting the drastic reduction in fluidity that would otherwise occur at low temperatures.

Membrane Proteins

The complement of proteins attached to a membrane varies, depending on cell type and subcellular location. Mitochondrial membrane proteins differ markedly from plasma membrane proteins; the plasma membrane components of a liver cell and an intestinal cell are not the same. We will examine how specific proteins are deposited in specific membranes in Chapters 17 and 18. Here, we focus on the structures and functions of a few well-understood membrane proteins.

Some proteins are bound only to the membrane surface; others have one region buried within the membrane and domains on one or both sides of it (Figure 13-14). Obviously, to function correctly, a membrane protein must be properly oriented to the phospholipid bilayer.

It is difficult to obtain enough pure protein to determine the sequences, let alone the interactions, of many important membrane proteins because they are present in such tiny amounts. Many membrane protein sequences are now determined from the sequence of cloned cDNA. When we discuss the biosynthesis of membrane proteins, we will see how the membrane interactions of certain proteins can be predicted from their amino acid sequences.

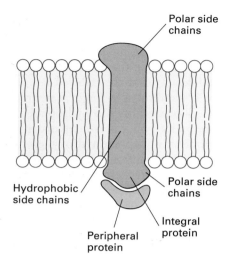

▲ **Figure 13-14** Schematic outline of the interaction of a typical integral membrane protein and peripheral protein with a lipid bilayer. Most integral proteins span the bilayer; most peripheral proteins are not in contact with the hydrophobic core.

Proteins Interact with Membranes in Different Ways

It is useful to begin by classifying membrane proteins according to the types of interaction that maintain the protein-membrane relationship, using as a guide what is now known about a few proteins. One or more segments of most integral membrane proteins, also called *intrinsic proteins*, contain amino acid residues with hydrophobic side chains that interact with the fatty acyl groups of the membrane phospholipids. These proteins can be removed only by the action of detergents, which displace the lipids bound to the hydrophobic side chains anchoring the membrane proteins. Other integral proteins contain covalently bound fatty acids that function as anchors in the phospholipid bilayer; these proteins can also be solubilized by detergents.

Peripheral proteins, or *extrinsic proteins*, do not interact directly with the hydrophobic core of the phospholipid bilayer. They are usually bound to the membrane indirectly by interactions with integral membrane proteins (see Figure 13-14) or directly by interactions with lipid polar head groups. The cytoskeletal proteins spectrin and actin, which are bound to the cytoplasmic face of the erythrocyte cell membrane, are examples of peripheral proteins. Other peripheral proteins, including certain proteins of the glycocalyx, are localized to the outer surface of the plasma membrane.

Most peripheral proteins are bound to specific integral membrane proteins by ionic or other weak interactions. Characteristically, many peripheral proteins can be re-

moved from the membrane by solutions of high ionic strength (high salt concentrations), which disrupt ionic bonds, or by chemicals that bind divalent cations, such as Mg^{2+}. Unlike integral proteins most peripheral proteins are soluble in aqueous solution and are not solubilized by detergents.

Some Integral Proteins Are Bound to the Membrane by Covalently Attached Lipids

A large number of proteins in eukaryotic cells contain covalently attached fatty acids or phospholipids that bind the protein to the plasma membrane. These integral membrane proteins fall into three classes, defined by the type of fatty acyl or phospholipid group each contains (Figure 13-15).

The cell-surface proteins in one class, including Thy-1 and several enzymes such as alkaline phosphatase, are anchored to the exoplasmic surface of the plasma membrane by a complex glycosylated phospholipid that contains, among other sugars, N-acetylglucosamine and inositol (see Figure 13-15). Experiments show that this phospholipid is clearly essential for binding the protein to the membrane. For instance, treatment of the cells with phospholipase C, an enzyme that cleaves the phosphate-glycerol bond, releases these proteins from the cell surface.

A second class of proteins is found within the cytoplasm, anchored to the cytoplasmic face of the plasma membrane by means of myristic acid, a 14-carbon saturated fatty acid. (Myristate is always bound by an amide linkage to the glycine residue found at the N-terminus of such proteins.) Included in this group is the protein *v-src*, a mutant of the normal cellular protein *c-src*. Like *c-src*, *v-src* is a tyrosine protein kinase (an enzyme that adds phosphate groups to tyrosine residues in certain proteins). When synthesized in certain cultured mammalian cells, *v-src* causes them to become transformed, or cancerous, and is therefore called a *transforming protein*. It is not known what protein or proteins must be phosphorylated by *v-src* to cause a cell to become cancerous, but it is known that the *v-src* protein must be bound to the plasma membrane to transform the cells. This can be shown experimentally: when the N-terminal glycine of *v-src* is mutated to an alanine, the kinase activity of the protein is normal but the mutant *v-src* is not linked to myristate, does not bind to the plasma membrane, and cannot transform tissue culture cells.

The third class of lipid-bound proteins is anchored to the cytoplasmic face of the plasma membrane by a farnesyl residue, which is linked by a thioether bond to a cysteine residue always located near the C-terminus. An example is p21*ras*, another transforming protein that must be bound to the plasma membrane to cause the cell to become cancerous.

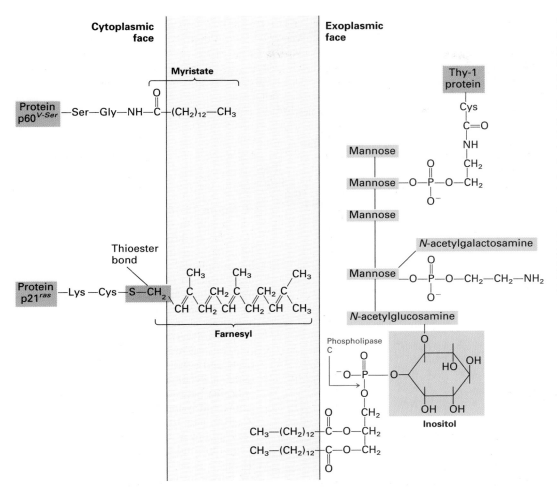

◀ **Figure 13-15** Attachment of several types of proteins to the plasma membrane is mediated by covalently bound lipids. The farnesyl residue is linked to a cysteine residue four amino acids from the C-terminus of the protein, and then the three C-terminal residues are cleaved off. [See B. W. Sefton and J. E. Buss, 1987, *J. Cell Biol.* **104**:1449–1453; S. W. Homans, et al., 1988, *Nature* **333**:269–272 and D. R. Lowy and B. M. Willumsen, 1989, *Nature* **341**:384–385.]

Most Integral Membrane Proteins Contain Long Segments of Hydrophobic Amino Acids Embedded in the Phospholipid Bilayer

Most integral proteins are embedded in the membrane by amino acid side chains. In contrast to the integral proteins anchored by covalently bound fatty acyl or phospholipid groups, these integral membrane proteins span the bilayer: they contain domains as small as four or as long as several hundred residues that extend into the aqueous medium on each side of the bilayer.

In most cases, the segments that span the bilayer are thought to form an α helix (see Figure 2-6). Two basic interactions keep integral proteins embedded in membranes: ionic interactions with the polar head groups of the lipids, and hydrophobic interactions with the lipid interior of the membrane. Both interactions probably help to anchor glycophorin, a major erythrocyte membrane protein of unknown function, to the phospholipid bilayer (Figure 13-16).

Glycophorin Is Typical of Proteins That Span the Membrane Once

The membrane-binding region of glycophorin consists of 34 residues (numbers 62–95), including one sequence of 23 residues (73–95) that contains only the uncharged, hydrophobic amino acids phenylalanine, leucine, isoleucine, valine, tryptophan, and tyrosine. The membrane-embedded region is believed to form an α helix. Each amino acid residue adds 0.15 nm to the length of a helix; thus a helix with 25 residues would be 3.75 nm long, just sufficient to span the hydrocarbon core of a phospholipid bilayer. The hydrophobic side chains protrude outward from the helix axis to form hydrophobic bonds with the fatty acyl chains. The polar C=O and NH groups in each peptide bond are on the inside of the α helix, and every such group participates in a hydrogen bond parallel to the helix axis (see Figure 2-6). Thus the polar amide groups in a transmembrane (membrane-spanning) α helix are shielded from the surrounding hydrophobic fatty acyl groups. In glycophorin, as in many other single-spanning

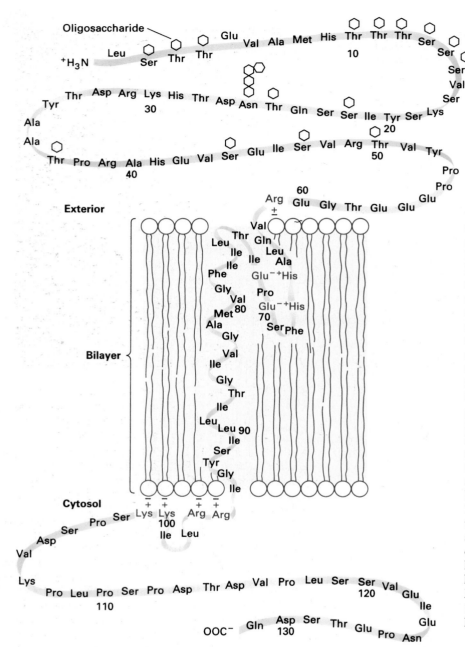

◀ **Figure 13-16** Amino acid sequence and transmembrane disposition of glycophorin A from the erythrocyte membrane. The N(amino)-terminal segment of the molecule lies outside the cell and has 16 carbohydrate units attached to it: 15 to serine or threonine residues, and one to an asparagine side chain. These oligosaccharides are rich in negatively charged sialic acid groups. The C(carboxyl)-terminal segment of the molecule, located inside the cell, is rich in negatively and positively charged amino acid residues. A hypothetical ionic interaction is indicated between the positively charged arginine and lysine residues at positions 96, 97, 100, and 101 with negative phospholipid head groups at the cytoplasmic face of the membrane and one between the arginine at 61 and a polar head group. Residues 62–95 are buried in the bilayer and are hydrophobic except for the charged groups at 66, 67, 70, and 72. Evidence suggests that the negatively charged glutamic acid residues at 70 and 72 are ionically linked to the positively charged histidine residues at 66 and 67, allowing residues 62–72 to insert into the hydrophobic core of the bilayer. [See V. T. Marchesi, H. Furthmayr, and M. Tomita, 1976. *Ann. Rev. Biochem.* 45:667; A. H. Ross et al., 1982, *J. Biol. Chem.* 257:4152.]

proteins, positively charged amino acids (lysine and arginine) border the hydrophobic segment on the cytoplasmic surface and may interact with negatively charged phospholipid head groups predominant in the cytoplasmic leaflet of the erythrocyte membrane to hold the hydrophobic segment firmly in place.

The Bacterial Photosynthetic Reaction Center Contains 4 Polypeptides and 11 Transmembrane α Helices

Although the molecular structures of more than 200 water-soluble proteins have been determined by x-ray crystallography, the structure of the first integral membrane protein was not determined to atomic resolution until 1985. This protein—the *photosynthetic reaction center* of the bacterium *Rhosopseudomonas viridis*—is made up of four polypeptides that contain a total of 1187 amino acids and of several prosthetic groups, including four chlorophyll molecules. The structure of this protein (Figure 13-17a) provides a basis for understanding how light absorption is converted into chemical energy. Here, we are concerned with the way in which the polypeptide chain interacts with the membrane.

Three of the four polypeptides, L, M, and H, are integral and span the membrane. The L and M subunits comprise the bulk of the structure within the membrane: each

(a)

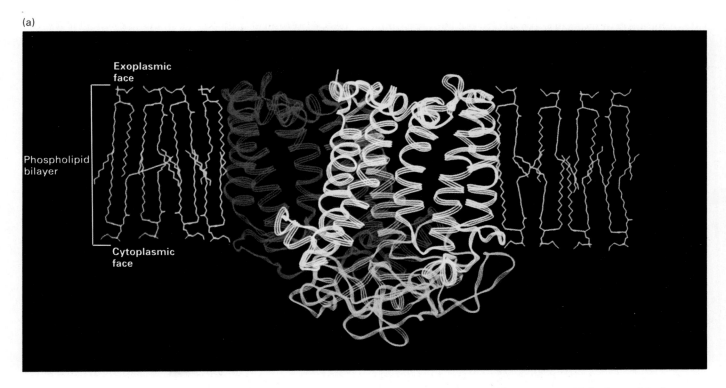

(b)

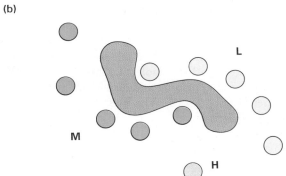

▲ **Figure 13-17** (a) The structure of the bacterial photosynthetic reaction center as determined by x-ray crystallography. Only the polypeptide backbone of the three membrane-spanning proteins is shown: the L subunit (yellow) and the M subunit (dark blue) form five transmembrane α helices and have a very similar structure overall; the H subunit (light blue) is anchored to the membrane only by a single transmembrane α helix. (b) A cross-sectional view through the middle of the membrane, showing the locations of the 11 transmembrane α helices. The center of the protein (orange) is occupied by prosthetic groups such as free (nonheme) iron and the chlorophylls. *Part (a) courtesy of D. Rees;* part (b) *after R. Henderson, 1985,* Nature *318:598–599.*

forms a crescent consisting of a curved wall of five transmembrane α helices. The H subunit is primarily a globular domain anchored to the cytoplasmic surface of the membrane by a single transmembrane α helix. The fourth subunit—the *cytochrome* not shown in Figure 13-17a—is a peripheral protein bound to exoplasmic segments of the three other polypeptides.

As we might expect, all amino acid residues in the 11 transmembrane α helices that face the exterior of the molecule are hydrophobic and anchor the protein to the fatty acyl groups of the membrane. Many side chains in the α helices that face the interior of the molecule are also hydrophobic and bind adjacent helices by van der Waals forces. However, some residues in these transmembrane α helices are polar. Several histidine and glutamate residues, for instance, bind iron (Fe^{2+} or Fe^{3+}) atoms found in the center of the transmembrane region (Figure 13-17b). These charged residues face the interior of the molecule.

A comparison of the reaction centers from different bacterial species allows us to make an important generalization about the attachment of integral proteins to membranes. The overall structure of the protein is highly conserved, although the identities of many of the amino acids are not. In particular, the side chains that protrude outward and anchor the protein in the membrane are always hydrophobic, but their identity is not conserved; it seems that any hydrophobic side chain can anchor the protein in the lipid bilayer. In contrast, side chains that face inward and anchor helixes together are highly conserved; here, van der Waals interactions demand specific side-chain interactions.

Recall that hydrophilic side chains are preponderant on the surface of aqueous proteins and that hydrophobic side chains tend to be confined to the protein interior. In the transmembrane region of multispanning integral proteins, such as the photosynthetic reaction center, the situation is reversed: hydrophobic side chains lie on the outside of the protein molecule, exposed to the hydrocarbon interior of the membrane; hydrophilic side chains are found in the interior of the molecule. These membrane-embedded proteins are "inside-out" compared to water-soluble proteins.

The Orientation of Proteins in Membranes Can Be Experimentally Determined

Except for the photosynthetic reaction center, we do not know the complete three-dimensional structure of any integral protein. Frequently, however, it is sufficient to know which segments of a protein span the bilayer and which are localized to the cytoplasmic and exoplasmic faces. Several techniques can be used to determine the direction in which a protein is oriented in a membrane. For example, a covalent labeling reagent that cannot penetrate membranes will bind covalently only to the exposed regions of the polypeptide on the outside of the membrane. In the presence of hydrogen peroxide and radioactive iodine [^{125}I], the enzyme lactoperoxidase, a commonly used reagent, iodinates tyrosine and histidine side chains of proteins. Because lactoperoxidase cannot penetrate membranes, it reacts only with the exposed parts of proteins on the cell surface. Similarly, proteases digest only those regions of cell-surface proteins exposed to the extracellular medium.

Experiments have shown that the NH$_2$-terminus of glycophorin and all the oligosaccharides are on the exoplasmic surface (see Figure 13-16). Only this exoplasmic segment is digested by proteases added to the cell exterior and is reactive with extracellular lactoperoxidase. The COOH-terminal segment (residues 96–131) faces the cytoplasm and reacts with such reagents only if the permeability barrier of the plasma membrane is disrupted. The hydrophobic transmembrane segment (residues 62–95 approximately) does not react with any aqueous enzymes or agents. Thus experimenters have concluded that glycophorin is a transmembrane protein.

Detergents Are Used to Solubilize and Study Integral Membrane Proteins

To understand the structure and function of an integral protein, we must first isolate and purify it like any protein. Membrane-binding proteins present a special problem to the protein chemist because their surfaces are hydrophobic. In general, due to a preponderance of exposed hydrophilic groups, water-soluble globular proteins maintain their native conformation and remain individually suspended in an aqueous medium. In contrast, the exposed hydrophobic regions of integral membrane proteins cause the protein molecules to aggregate and precipitate from aqueous solutions. Such proteins can be solubilized by detergents that have affinity both for hydrophobic groups and for water.

Detergents are amphipathic molecules that disrupt membranes by intercalating into phospholipid bilayers and solubilizing lipids and proteins. Some detergents are natural products, such as the bile salt sodium deoxycholate, but most are synthetic molecules developed by the chemical industry for cleaning and dispersing mixtures of oil and water (Figure 13-18). The hydrophobic part of a detergent molecule is attracted to hydrocarbons and min-

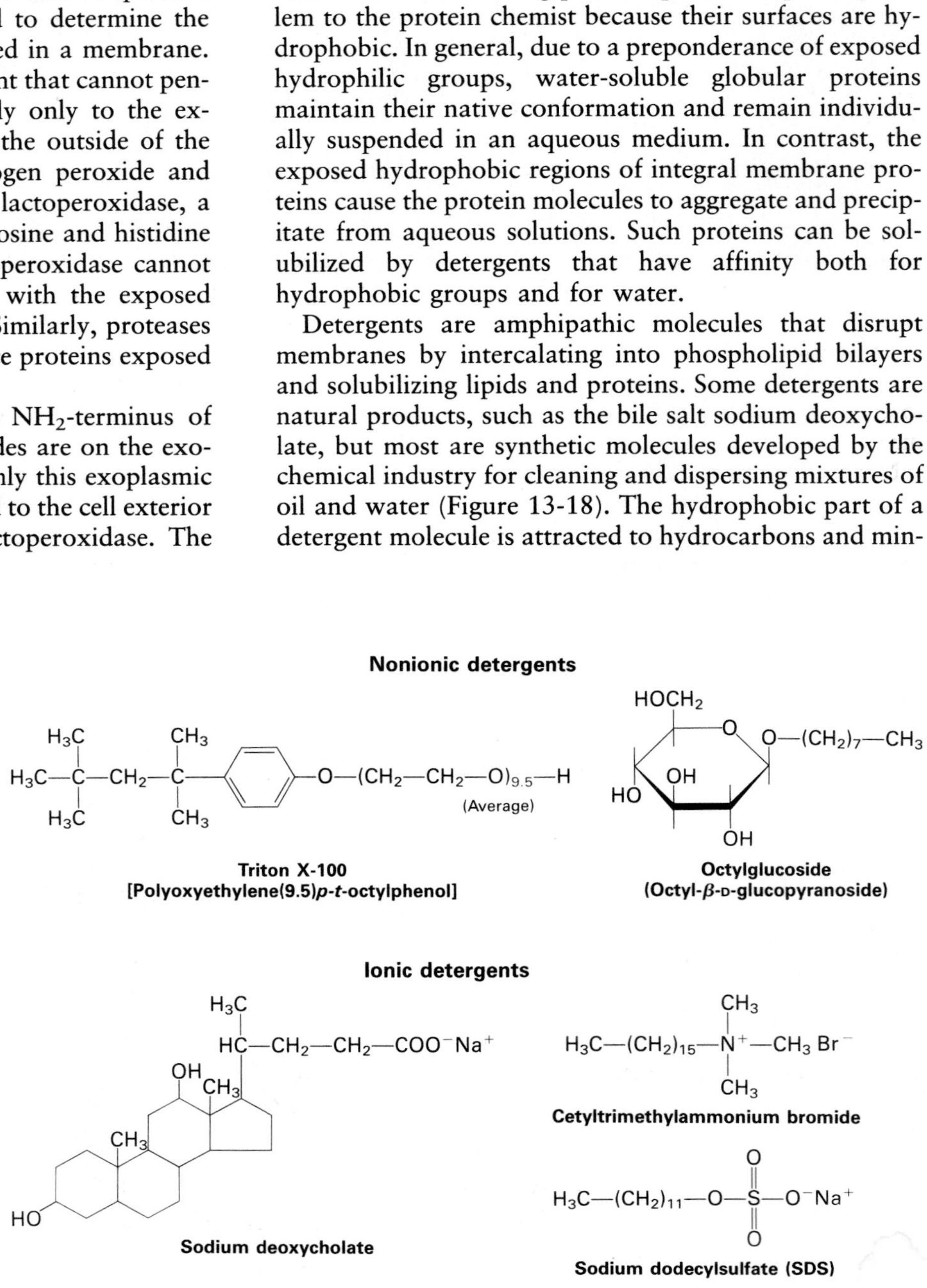

Nonionic detergents

Triton X-100
[Polyoxyethylene(9.5)*p-t*-octylphenol]

Octylglucoside
(Octyl-β-D-glucopyranoside)

Ionic detergents

Sodium deoxycholate

Cetyltrimethylammonium bromide

Sodium dodecylsulfate (SDS)

▶ **Figure 13-18** Structures of some commonly used detergents.

gles with them readily; the hydrophilic part is strongly attracted to water. Thus, in a mixture of oil and water, the detergent forms a monomolecular film at the boundary between the two substances. Agitation of oil-water mixtures breaks large volumes of oil into smaller droplets. In the absence of detergent, these droplets recoalesce into their original aggregate. In the presence of detergent, however, each droplet is surrounded by a single layer of detergent molecules with outward-pointing hydrophilic ends (Figure 13-19), which makes the droplet act as a hydrophilic object that can be suspended in water and rinsed away.

At very low concentrations in pure water, detergents are dissolved as isolated molecules. As the concentration increases, the molecules begin to form *micelles*—small, spherical aggregates with hydrophilic parts that face outward and hydrophobic parts that cluster in the center (Figure 13-20). The *critical micelle concentration* (CMC) at which micelles form is characteristic of the detergent and is a function of the structures of its hydrophobic and hydrophilic parts.

Ionic detergents bind to the hydrophobic regions of proteins, alter the conformation of the hydrophilic regions, and disrupt ionic and hydrogen bonds. At high concentrations, for example, sodium dodecylsulfate (SDS; see Figure 13-18) completely denatures proteins by binding to every side chain. Thus the mobility of SDS-solubilized proteins in gel electrophoresis is a good measure of their molecular weight.

Nonionic detergents, such as Triton X-100 and octylglucoside (see Figure 13-18), are without a charged group. They act in different ways at different concentrations. At high concentrations (above the CMC), they solubilize biological membranes by forming mixed micelles

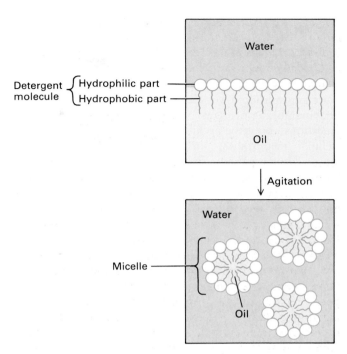

▲ **Figure 13-19** Detergents can solubilize hydrocarbons by forming micelles.

of detergent, phospholipid, and integral membrane proteins (see Figure 13-20). At low concentrations, these detergents generally do not denature proteins but do solubilize most membrane proteins; individual molecules of octylglucoside bind to the hydrophobic regions of integral proteins normally embedded in the membrane, preventing protein aggregation during subsequent purification steps.

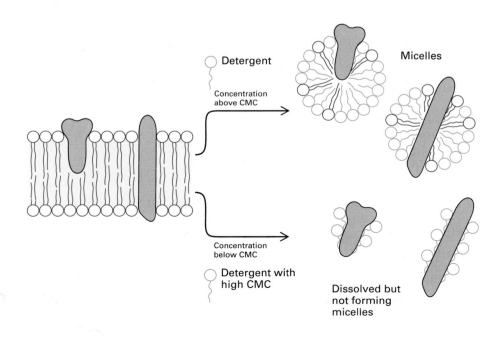

◀ **Figure 13-20** At concentrations higher than the CMC, detergents solubilize lipids and integral membrane proteins and form mixed micelles of detergents, protein, and lipid. A detergent with a high CMC may dissolve membrane proteins at concentrations well below its CMC (that is, without forming micelles).

Principles of Membrane Organization

Thus far we have discussed primarily the structures of pure phospholipid vesicles and the conformations of a few membrane proteins. We now turn to more complex questions about the architecture of lipids and proteins in cellular membranes, particularly in the plasma membrane. How are these lipids and proteins organized? How mobile are they? First, we consider their asymmetry, which is essential to the function of all biological membranes.

All Membrane Proteins Bind Asymmetrically to the Lipid Bilayer

Each type of integral membrane protein has a single, specific orientation with respect to the cytoplasmic and exoplasmic faces of a cellular membrane. All molecules of any particular integral membrane protein, such as glycophorin (see Figure 13-16) or the photosynthetic reaction center (see Figure 13-17), lie in the same direction. The active site of every molecule of a particular membrane-bound enzyme is located on the same face of the membrane. This absolute asymmetry in protein orientation gives the two membrane faces their different characteristics.

The nature of this asymmetry is easily seen in glycoproteins; the sugars of all glycoproteins are on one face, depending on the nature of the attached sugar. In all cases, asparagine-linked oligosaccharides (see Figure 2-56) and O-linked oligosaccharides, in which the serine or threonine residue is linked to galactose or to N-acetylgalactosamine (see Figure 2-55), are invariably found on the exoplasmic membrane face. These classes represent the vast majority of known glycoproteins. In contrast, the attached sugar residues of a small but important class of glycoproteins are found on the cytoplasmic membrane face. These proteins, which contain N-acetylglucosamine bound to serine, include transcription factors and components of the nuclear pore complex.

Importantly, proteins have never been observed to flip-flop across a membrane. Such movement would be energetically unfavorable: it would require a transient movement of hydrophilic amino acid and sugar residues through the hydrophobic interior of the membrane.

The Two Membrane Leaflets Have Different Lipid Compositions

The phospholipid compositions of the two leaflets of a membrane are quite different in the plasma membranes analyzed thus far. In the human erythrocyte and in a line of canine kidney cells grown in culture, almost all of the glycolipids, such as sphingomyelin, and most of the phosphatidylcholine, with its positively charged head group, are found in the exoplasmic leaflet. In contrast, lipids with neutral or negative polar head groups, such as phosphatidylethanolamine and phosphatidylserine, are preferentially located on the cytoplasmic leaflet (Figure 13-21). In general, the oligosaccharide side chains of glycolipids are found exclusively on the exoplasmic face of both plasma membranes and internal membranes. In the plasma membrane, all sugars on glycoproteins and glycolipids are exposed to the exterior of the cell; in the endoplasmic reticulum, they are found on the lumenal surface.

In contrast to the absolute asymmetry of glycolipids, most kinds of phospholipids, as well as cholesterol, are generally found in each membrane leaflet, although they are often enriched in one or the other. Phospholipids in the exoplasmic leaflet of the plasma membrane are susceptible to hydrolysis by *phospholipases* (enzymes that hydrolyze certain ester or phosphodiester bonds in the polar head group) when added to the cell exterior. Phospholipids in the cytoplasmic leaflet are resistant to such hydrolysis because the phospholipases cannot penetrate to the cytoplasmic face of the plasma membrane. The relative abundance of a particular phospholipid in the two leaflets of a plasma membrane can be determined based on these characteristics. It is not clear how these differences in lipid composition arise in the two leaflets; certain lipids may bind to specific protein domains that occur preferentially in one membrane leaflet.

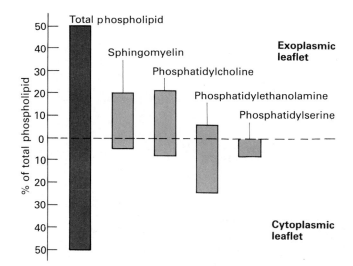

▲ **Figure 13-21** Distribution of phospholipids in the two leaflets of the erythrocyte plasma membrane. Values are expressed as a percentage of total membrane lipids. Note that 50 percent of the total phospholipid is found in each face. [See J. E. Rothman and J. Lenard, 1977, *Science* 195:743.]

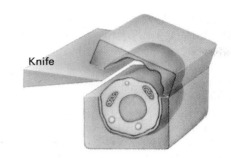

Fracture splits the plasma membrane

◀ **Figure 13-22** The freeze-fracture technique can separate the two phospholipid leaflets that form every cellular membrane. Membrane proteins and particles are bound to one leaflet or the other.

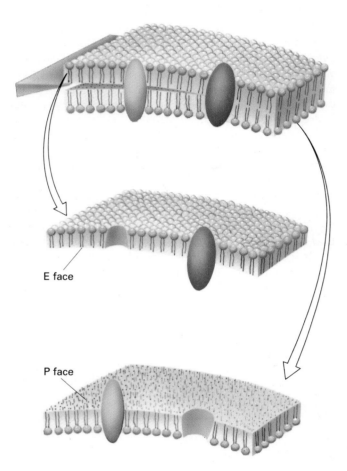

Detail of the two exposed membrane leaflets

Freeze-fracture and Deep-etching Techniques Reveal the Two Membrane Faces in Electron Microscopy

The two faces of any cellular membrane can be separated and prepared for electron microscopy by the freeze-fracture and deep-etching techniques (see Figure 4-35). When a frozen specimen is fractured by a sharp blow, the fracture line frequently runs through the hydrophobic inte-

rior of the membrane, separating the two phospholipid leaflets. The fractured surfaces of many membranes reveal numerous protuberances, most of which are membrane proteins (Figures 13-22 and 13-23). In deep-etching studies, the cytoplasmic face of a membrane is customarily called the P (protoplasmic) face and the exoplasmic face is the E face. It is not unusual for most or all protuberances to be on one of the two surfaces and their mirror images, in the form of pits or holes, to be on the other. This may occur because the integral proteins are bound more tightly to the lipids in one leaflet than to those in the other.

Most Membrane Proteins and Lipids Are Laterally Mobile in the Membrane

From experimental evidence, many membrane proteins appear to float quite freely in the lipid and diffuse on the membrane surface. Two different cells can be fused for the purpose of observing the movement of their distinct surface proteins. At various times after incubation at 37°C, the cells (say, mouse and human fibroblasts) are chilled on ice and immediately fixed with compounds, such as glutaraldehyde ($CHOCH_2CH_2CH_2CHO$), that cross-link lysine side chains and prevent further protein movement. The plasma membrane protein H-2 (found on most mouse cells) or the related HLA protein (found on most human cells) can be detected using specific H-2 or HLA antibodies coupled to fluorescent dyes. Immediately after fusion, the mouse and human antigens are grouped in separate areas, but they quickly diffuse throughout the cell. Soon both halves of the fused cells are equally fluorescent, demonstrating that most of the surface H-2 and HLA proteins were not rigidly held in place on the original mouse and human cells (Figure 13-24).

Such experiments have encouraged the notion that many proteins are free to diffuse in a sea of lipid in the two-dimensional space of the membrane, currently popularized as the *fluid mosaic model* (see Figure 13-1) in which the membrane is seen as a two-dimensional mosaic of phospholipid and protein. Many integral membrane proteins form stable, noncovalent interactions with other such proteins. Glycophorin, for instance, is a dimer of two identical chains (see Figure 13-16). In such cases, the entire protein is thought to float as a unit in the lipid.

A more quantitative estimate of the rate and extent of lateral movements of surface proteins and lipids can be

(a)

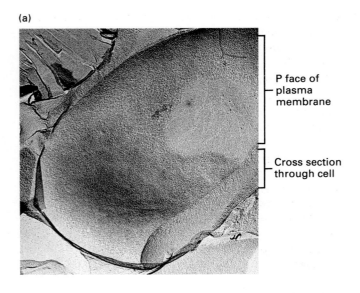

P face of plasma membrane

Cross section through cell

(b) Intramembrane particles (band 3)

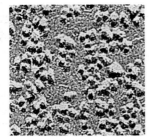

obtained from *fluorescence recovery after photobleaching* (FRAP) studies. Uniformly distributed surface proteins or lipids are labeled with a fluorescent dye. A laser light source focused on a small patch of the surface irreversibly bleaches the dye, inhibiting the fluorescence. In time, the fluorescence of the bleached area increases because unbleached fluorescent surface molecules diffuse into it (Figure 13-25). The extent of recovery of fluorescent molecules in the patch is proportional to the fraction of labeled molecules that are mobile in the membrane, and the rate of recovery of fluorescence can be used to calculate the *diffusion coefficient* (the rate at which the molecules diffuse).

FRAP studies using phospholipids labeled with fluorescent dyes show that in fibroblasts all lipids in plasma membranes are freely mobile over very small distances of about 0.5 μm; however, most lipids cannot diffuse over much longer distances. These data suggest that protein-rich domains about 1 μm in diameter in the plasma

◀ **Figure 13-23** Freeze-fracture, deep-etching image of an erythrocyte plasma membrane. (a) The P face of the plasma membrane and a cross section through the cell. The particles in the cross section are hemoglobin. (b) The intramembrane particles at higher magnification. These particles are composed mainly of band 3, the major intramembrane protein. *Photographs courtesy of D. Branton.*

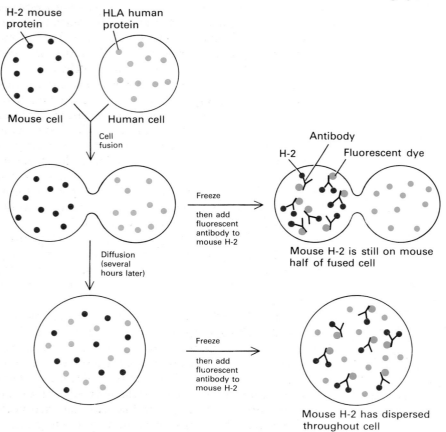

◀ **Figure 13-24** An experiment in which human and mouse cells are fused in culture illustrates the mobility of cell-surface proteins. Initially, the human and mouse cell antigens remain on their respective halves of the fused cell. After several hours of incubation, the two sets of proteins are thoroughly intermingled. The mouse (H-2) proteins are detected by fluorescent antibodies to them. [See L. D. Frye and M. Edidin, 1970, *J. Cell Sci.* 7:319.]

membrane separate regions that contain the bulk of the membrane phospholipid. Furthermore, the rate of lateral diffusion of lipids in the cell plasma membrane is nearly an order of magnitude slower than in pure phospholipid bilayers: diffusion constants of 10^{-8} cm²/s and 10^{-7} cm²/s are characteristic of the plasma membrane and a lipid bilayer, respectively. This suggests that lipids may be tightly but not irreversibly bound to certain integral proteins in some membranes.

Depending on the cell under study and on the particular protein or class of proteins labeled, 30–90 percent of all surface proteins are freely mobile and their diffusion rate is generally 10–30 times lower than that of the same protein embedded in synthetic liposomes. The low rate and extent of mobility of integral proteins could be due to contacts with the rigid submembrane cytoskeleton that need to be broken and remade as the protein diffuses laterally. Evidence that interactions with the cytoskeleton normally restrict the mobility of membrane proteins comes from a study of "blebs" of plasma membrane—occasional protrusions from fibroblasts or other cells—which are free of cytoskeletal proteins. More than 90 percent of the integral proteins in blebs are mobile; the diffusion constant is five- to tenfold higher than in more normal segments of the plasma membrane.

Lateral protein diffusion occurs in intracellular membranes as well. Freeze fracturing has revealed that the inner mitochondrial membrane contains stable aggregates of integral proteins. If isolated vesicles prepared from this membrane are placed in a strong electric field, all the intramembrane particles (which bear a net negative electric charge) move to one end of each vesicle (Figure 13-26). When the field is turned off, the particles rapidly return to their original random distribution. This demonstrates that such particles, which can be complexes of as many as 15 membrane proteins, are laterally mobile.

Cytoskeletal Interactions Affect the Organization and Mobility of Surface Membrane Proteins

Not all integral proteins of the plasma membrane are mobile in the membrane; many are immobilized by contacts with other membrane proteins and with the intracellular network that makes up the fibrous cytoskeleton. Long fibers of actin (microfilaments), a major component of the cytoskeleton, lie just under the plasma membrane and appear to connect with it at several points. Microtubules and intermediate filaments also contribute to the cytoskeletal network; intermediate filaments are also often anchored to plasma membrane proteins.

In fibroblasts, experimenters have shown that proteins are more mobile in a parallel direction to the actin microfilaments that run just under the plasma membrane than in a perpendicular direction to them. This implies that the microfilaments may serve as tracks along which the proteins tend to move. Moreover, when antibodies to a surface protein are added to a cell, the proteins become cross-linked; as the proteins diffuse laterally in the membrane, they eventually accumulate in large networks called *patches*. The antibody-induced patching of certain cell-surface proteins often causes a corresponding redistribution of the actin cytoskeleton in the cytoplasm adjacent to the membrane, providing additional evidence for an interaction between the cell-surface proteins and the actin cytoskeleton.

Another striking piece of evidence is that treatment of intact cells with a nonionic detergent releases all of the lipids but only a fraction of the integral membrane proteins, even though these proteins, once purified, are freely soluble in the detergent. Many of these integral proteins must be tightly bound to the fibrillar elements of the cytoskeleton, such as the actin microfilaments and intermediate filaments, which resist extraction with detergents.

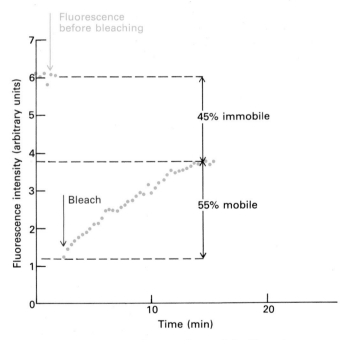

▲ **Figure 13-25** A FRAP experiment. Myoblasts (precursors of skeletal muscle cells) are treated with fluorescent concanavalin A (a protein that binds tightly to the sugar residues of surface glycoproteins). When a small patch of surface is bleached with a laser light and the myoblasts are incubated at 37°C, 55 percent of the fluorescence eventually returns to the bleached area. Thus 55 percent of the glycoproteins in this patch of membrane are mobile and 45 percent are immobile. The recovery rate of fluorescence is proportional to the rate of protein diffusion into the membrane region and thus to the diffusion coefficient. [See J. Schlessinger et al., 1976, *Proc. Nat'l Acad. Sci. USA* 73:2409.]

(a)

(b)

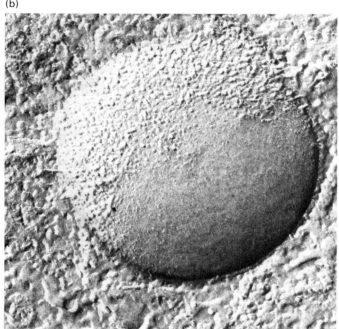

▲ **Figure 13-26** (a) Freeze fracture of a vesicle of the mitochondrial inner membrane, showing random distribution of integral protein particles. (b) After the mitochondrial vesicle is subjected to a strong electric field and then rapidly frozen, all the particles cluster at one end, showing that the particles have moved laterally within the membrane plane under this voltage gradient. The rate of movement is similar to the diffusion of many other proteins in a fluid mosaic membrane. [Part (b), see A. E. Sowers and C. R. Hackenbrock, 1981, *Proc. Nat'l Acad. Sci. USA* 78:6246.] *Part (a) courtesy of A. E. Sowers; part (b) courtesy of A. E. Sowers and C. R. Hackenbrock.*

The Glycocalyx Is Made Up of Proteins and Oligosaccharides Bound to the Outer Surface of the Cell

An important element of membrane structure, especially of the plasma membrane, is the *glycocalyx,* a coat formed by the oligosaccharide side chains of membrane lipids and proteins. Peripheral proteins, part of the glycocalyx, are bound to integral proteins (Figures 13-27 and 13-28) rather than to membrane lipid. Because many glycoproteins and glycolipids in mammalian cells contain several negatively charged sialic acid residues, the glycocalyx imparts a negative charge to the surface of most cells.

The Erythrocyte Membrane: Cytoskeletal Attachment

The best-characterized cell membrane is the one surrounding the mammalian erythrocyte (Figure 13-29a), which transports oxygen from the lungs to the tissues and carbon dioxide back to the lungs. The erythrocyte has no nucleus and no intracellular membranes; it is essentially a "bag" of hemoglobin containing relatively few other proteins. We will return to this cell often because its simplicity makes detailed analyses possible; these have contributed much to our understanding of how integral and peripheral proteins interact to form a membrane with a desired set of properties.

The erythrocyte normally adopts the shape of a biconcave disk 7 μm in diameter. It is very flexible, however, and often squeezes through much thinner capillaries. Aged or deformed cells that do not possess this flexibility

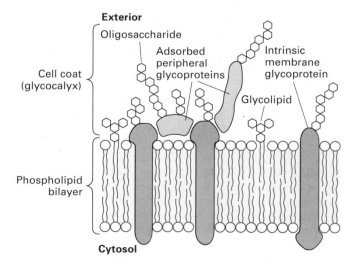

▲ **Figure 13-27** Schematic diagram of the glycocalyx—the cell coat made up of the oligosaccharide side chains of integral membrane glycolipids and glycoproteins and of adsorbed glycoproteins. All the oligosaccharides are on the exoplasmic face of the plasma membrane.

are trapped in the capillaries of the spleen and ingested by macrophages. The erythrocyte membrane must be durable: during its lifetime of 120 days, a typical human erythrocyte makes half a million circuits of the arteries and veins—a journey of about 300 miles.

The erythrocyte membrane is not representative of most cell membranes because it is homogeneous; its proteins appear to be more or less uniformly distributed in the plane of the membrane, without large specialized patches. The erythrocyte is also unusual among mammalian cells because its cytoskeleton forms a dense shell that underlies the entire plasma membrane and is attached to it at many points (Figure 13-29b); this structure gives the membrane its great strength and flexibility. In contrast, most mammalian cells have a cytoskeleton that courses through the cytoplasm and is anchored to the membrane at relatively few points. Nevertheless, the erythrocyte membrane provides important examples of membrane organization and cytoskeletal attachment.

(a)

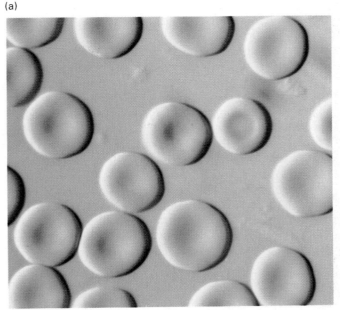

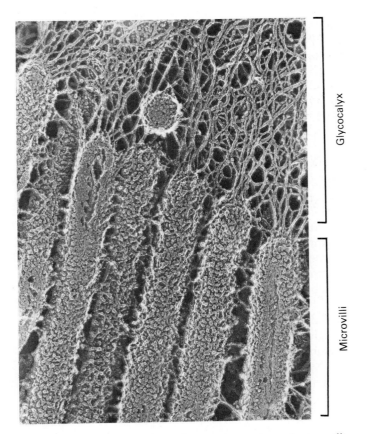

Glycocalyx

Microvilli

▲ **Figure 13-28** The glycocalyx on the surface of microvilli of intestinal epithelial cells, viewed by the deep-etching technique. The surface of each microvillus is covered with a series of bumps believed to be integral membrane proteins. The glycocalyx, which covers the apexes (tips) of the microvilli, is composed of a network of glycoproteins and digestive enzymes, such as sucrase-isomaltase (see Figure 13-42a). *Courtesy of N. Hirokawa and J. E. Heuser, 1981, J. Cell Biol. 91:399.*

(b)

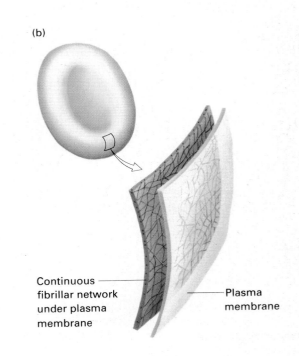

Continuous fibrillar network under plasma membrane

Plasma membrane

▲ **Figure 13-29** (a) Normal disk-shaped human erythrocytes viewed by differential interference microscopy; the surface not visible is also concave. (b) The erythrocyte contains a fibrillar cytoskeleton that underlies the entire plasma membrane and is attached to it at many points. *Part (a) Courtesy of M. Murayama, Biological Photo Service.*

It is interesting to examine the structure of the erythrocyte membrane in light of the process of the cell differentiation that produces it. Adult erythrocytes are manufactured in the bone marrow from nucleated stem cells. The stem cells grow, divide, and begin synthesizing hemoglobin, at which stage they are called *erythroblasts*. (Hematologists divide this stage into several substages of development.) Erythroblasts contain hemoglobin as well as spectrin and other characteristic erythrocyte membrane proteins. Next, the part of the cell that contains the nucleus and most of the intracellular membranes is pinched off and eventually degraded (Figure 13-30; see Figure 13-34). The remaining non-nucleated cell, now called the *reticulocyte*, continues to synthesize hemoglobin and other erythrocyte proteins. The cell eventually loses its ribosomes and about one-third of its plasma membrane and acquires the biconcave disk structure of the erythrocyte.

The Erythrocyte Membrane Can Generate Inside-out or Right-side-out Vesicles

The analysis of erythrocyte membrane proteins is relatively straightforward. When the cells are placed in distilled water, the ensuing influx of water by osmosis causes them to swell. Eventually, the plasma membranes rupture and release the hemoglobin and other internal proteins. Because the membrane in the hemoglobin-depleted cell retains the overall size of the intact cell but is white, it is called a *ghost* (Figure 13-31). Overall, the ghost is 52 percent protein, 40 percent lipid, and 8 percent carbohydrate; 93 percent of the carbohydrate (as oligosaccharides) is attached to protein, and 7 percent is attached to lipids, forming glycolipids.

When homogenized in solutions of appropriate ionic strength, ghosts will form into smaller sealed vesicles. In solutions of normal ionic strength that contain Mg^{2+}, the membranes reseal with their original right-side-out orientation: the exoplasmic face remains outward. If the composition of the medium is changed, sealed inside-out vesicles in which the cytoplasmic face is positioned outward

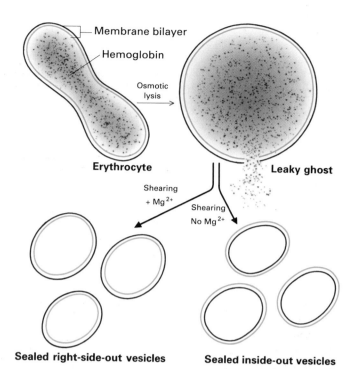

▲ **Figure 13-31** Osmotic lysis of the erythrocyte generates a leaky membrane ghost. When homogenized, the ghost is sheared into smaller, nonleaky vesicles. Depending on the presence of Mg^{2+} ions, the vesicles will be predominantly inside-out or right-side-out.

(see Figure 13-31) can be obtained for use in studies of the cytoplasmic membrane surface.

The Erythrocyte Has Two Main Integral Membrane Proteins

To study the various integral protein constituents of the erythrocyte plasma membrane, the ghost is dissolved in the negatively charged detergent SDS and analyzed by gel electrophoresis. By convention, the major proteins are numbered in order of decreasing molecular weight (Figure 13-32). Virtually the same principal proteins are found in erythrocyte membranes from all mammals, which attests to the generality and importance of this membrane structure.

The two predominant erythrocyte integral membrane proteins—glycophorin and band 3—are glycoproteins. Only these two major proteins have exposed regions on the outer surface of the cell. The other peripheral proteins (see Figure 13-32), which are confined to the cytoplasmic face of the membrane, form the erythrocyte cytoskeleton.

The external segment of glycophorin is heavily glycosylated with O-linked and N-linked carbohydrate chains (see Figure 13-16). The O-linked chains are much shorter; glycophorin is unusual in that these chains are its major oligosaccharide substituents. Carbohydrate constitutes 64 percent of the weight of glycophorin, an unusually high sugar content.

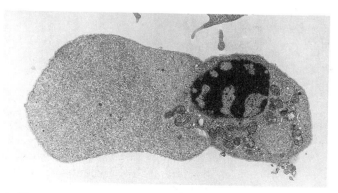

▲ **Figure 13-30** Electron micrograph of an enucleating erythroblast. *Courtesy Marianne Lehnert.*

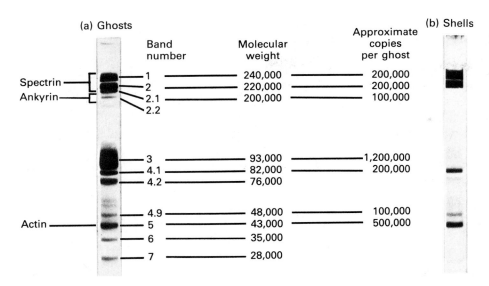

(a) Ghosts

Band number	Molecular weight	Approximate copies per ghost
1	240,000	200,000
2	220,000	200,000
2.1	200,000	100,000
2.2		
3	93,000	1,200,000
4.1	82,000	200,000
4.2	76,000	
4.9	48,000	100,000
5	43,000	500,000
6	35,000	
7	28,000	

Spectrin — bands 1, 2
Ankyrin — bands 2.1, 2.2
Actin — band 5

(b) Shells

◀ **Figure 13-32** SDS-polyacrylamide gels (a) of the polypeptides of erythrocyte ghosts and (b) of the shells remaining after these ghosts are treated with a 1 percent Triton X-100 detergent in hypotonic buffer, which removes the band 3 integral membrane protein and certain peripheral proteins but leaves the cytoskeletal proteins in the shell: bands 1 and 2 (spectrins), bands 2.1–2.2 (ankyrin), band 4.1, and band 5 (actin). Glycophorin, although present in the ghosts, does not stain with the dye used and is not visible here. [See D. Branton, C. M. Cohen, and J. Tyler, 1981, *Cell* 24:24.]

The other major integral membrane protein in the erythrocyte, band 3, is present in about 1.2 million copies per cell. The orientation of band 3 in the membrane is quite different from that of glycophorin. Band 3 is a dimer of two identical chains, each consisting of 929 amino acids (Figure 13-33). The C-terminal segment of each chain is imbedded in the lipid membrane and makes multiple passages through it, forming probably 12 or 14 transmembrane α helices that catalyze anion exchange across the plasma membrane (see Figures 14-16 and 14-17). This part of the molecule contains an asparagine-linked oligosaccharide chain facing, as usual, the cell exterior. The N-terminal segment of the molecule folds into discrete water-soluble domains that protrude into the cytosol (the liquid component of the cytoplasm) and anchor the cytoskeleton to the membrane.

Erythrocyte Cytoskeletal Proteins Affect Cell Shape and Integral Protein Mobility

The integral glycoproteins of the erythrocyte membrane are unusual in that they are immobile and cannot diffuse laterally in the membrane plane. When erythrocyte ghosts are placed in a solution of low ionic strength, the principal peripheral proteins, bands 1 and 2 (spectrins) and band 5 (actin), are solubilized. These three proteins are the major components of the cytoskeleton. Removing them produces two consequences: the ghost loses its rigid shape, and the membrane glycoproteins acquire lateral mobility. This means that the cytoskeleton is the major determinant of the rigidity of the erythrocyte membrane and that it acts to restrict the lateral motion of membrane glycoproteins.

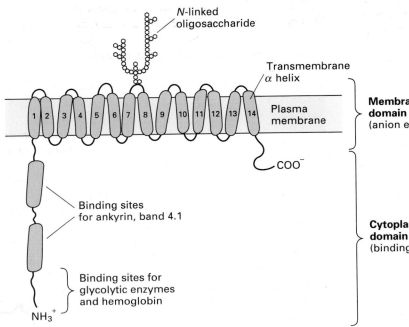

◀ **Figure 13-33** The arrangement of band 3 protein (green) in the erythrocyte membrane. Only one of the two identical polypeptides is shown. [See R. Kopito and H. F. Lodish, 1985, *Nature* 316:234; G. Jay and L. Cantley, 1986, *Ann. Rev. Biochem.* 55:511–538.] *Drawing courtesy of Sam Lux.*

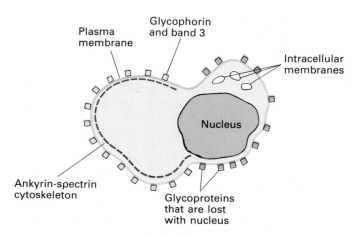

Erythroblast

▲ **Figure 13-34** Distribution of membrane proteins in the enucleating erythroblast. *After J. B. Geiduschek and S. J. Singer,* Cell *16:149; A. H. Sarris and G. E. Palade, 1982,* J. Cell Biol. *93:591; V. P. Patel and H. F. Lodish, 1987,* J. Cell Biol. *105:3105–3118.*

At the stage at which the erythroblast is losing its nucleus, all the spectrin, ankyrin, band 3, and glycophorin stay behind in the nascent reticulocyte (Figure 13-34). The details of the process by which the nucleus is lost are not known, but the spectrin-rich cytoskeleton appears to "select" certain integral membrane proteins for inclusion in the reticulocyte and to discriminate against others, which are discarded with the segment containing the nucleus.

The Erythrocyte Cytoskeleton Is Constructed of a Network of Fibrous Proteins Just Beneath the Surface Membrane

Ghosts are stripped to their cytoskeletons by treatment with a nonionic detergent. All the lipid and glycophorin and about 80 percent of band 3 are solubilized. What remains in the insoluble fraction is the cytoskeleton—the fibrous shell retaining the shape of the ghost and containing the same predominant extrinsic proteins (see Figure 13-32): bands 1 and 2 (spectrins), bands 2.1–2.2 (ankyrin), band 4.1, and band 5 (actin). The shell (Figure 13-35) is composed of 200-nm-long spectrin tetramers joined at their ends by junctional complexes of 35-nm-long actin microfilaments; an average of five spectrin fibers radiate from each complex. The actin microfilaments are thicker than those in other cells, such as muscle cells, because accessory proteins, including adducin and band 4.1 are attached to them to facilitate the binding of spectrin to actin.

Biochemical and biophysical studies on purified erythrocyte cytoskeletal proteins have shown how the predominant proteins are held together by noncovalent bonds and by interactions with other minor but important proteins to form the strong but flexible framework responsible for the shape and pliability of the erythrocyte. The major constituents of the cytoskeleton are α- and β-spectrin (bands 1 and 2), the polypeptides having the highest molecular weight. The two spectrin chains combine to form αβ *dimers*—long (100 nm), slender (5 nm in diameter), wormlike molecules in which the subunits coil about each other (Figure 13-36). Two spectrin dimers combine head-to-head to form an $(\alpha\beta)_2$ tetramer (Figure 13-37); this interaction can take place in an aqueous solution. Each erythrocyte ghost contains about 100,000 spectrin tetramers.

In the cytoskeletal network, the free ends of several spectrin tetramers are held together by short fibrillar chains of the protein actin (band 5) (Figure 13-38). Actin, a major constituent of all cells, normally forms chains several micrometers long. In an erythrocyte, each actin microfilament contains 13 actin monomers and one molecule (a dimer of two polypeptides, each of molecular weight 35,000) of the long (35 nm) fibrous protein tropomyosin, which is also found in muscle cells (Figure 22-18). Since each tropomyosin is the same length as each of the actin microfilaments, tropomyosin is believed to determine their length (see Figure 13-38, inset). Two other cytoskeletal proteins, band 4.1 and adducin, lie at the actin-spectrin junction and help join the actin to the spec-

0.1 μm

▲ **Figure 13-35** Electron micrograph of human erythrocyte cytoskeletons spread on a microscope grid. [See B. W. Shen, R. Josephs, and T. L. Steck, 1986. *J. Cell Biol.* 102:997–1006; T. J. Byers and D. Branton, 1985, *Proc. Nat'l Acad. Sci. USA* 82:6153–6157.] *Courtesy of D. Branton.*

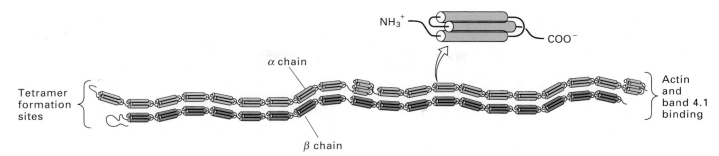

▲ **Figure 13-36** The molecular anatomy of spectrin. The spectrin dimer is comprised of two polypeptide chains, α and β, that lie side-by-side or twine around each other. The contour length of an αβ dimer is 100 nm; because the chains are flexible, however, the distance between the ends can be anywhere from 0 to 100 nm. Peptide and nucleotide sequence data indicate that the predominant structural feature of both the α and β chains is a repeating segment of 106 amino acids (20 times in α and 18 times in β). Each segment folds into a triple-stranded structure. *After D. W. Speicher, 1986, J. Cell Biochem. 30:245–256.*

trin. Purified spectrin will bind weakly to filaments composed only of actin; actin from any source, such as muscle or erythrocytes, is equivalent in this regard. Spectrin-actin binding is stabilized greatly by protein 4.1 and even more by proteins 4.1 and adducin together. Because spectrin binds to the sides of actin microfilaments, a single fiber of actin has many potential binding sites for spectrin, permitting a network to form.

These interactions produce the lacelike cytoskeleton but do not attach it to the erythrocyte membrane. Attachment is mediated principally through the band 2.1 protein called ankyrin. There are about 100,000 molecules of ankyrin in a cell (on average, one per spectrin tetramer). Ankyrin has two domains: one binds tightly and specifically to a site on one β chain of spectrin near the center of the tetramer (see Figure 13-37d); the other binds tightly to a region on the band 3 protein that protrudes into the cytosol (see Figure 13-38). This bridgelike property of ankyrin binds the cross-linked spectrin network to the membrane. Thus the cytoskeleton of the erythrocyte is a large network of band 3, ankyrin, spectrin, band 4.1, tropomyosin, adducin, and actin. In addition, band 4.1 binds to the carboxyl-terminal domain of glycophorin and also to band 3; these interactions may also anchor

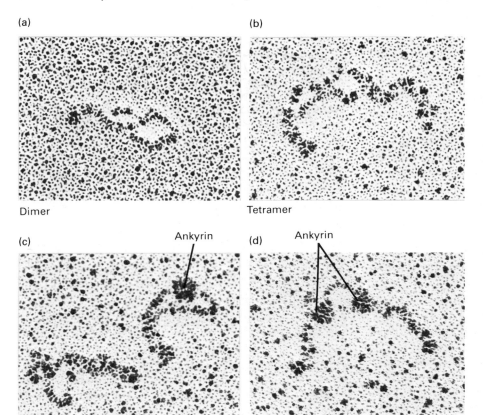

◀ **Figure 13-37** Electron micrographs of αβ dimers and (αβ)₂ tetramers of spectrin. The proteins in (a) and (b) are not bound to ankyrin; those in (c) and (d) are bound to ankyrin near the middle of the tetramers. *Courtesy of D. Branton.*

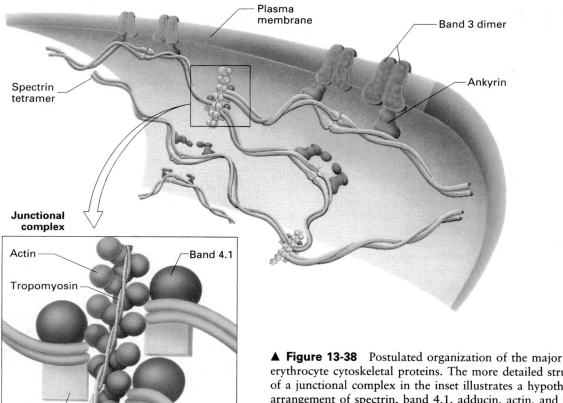

Actin
Tropomyosin
Band 4.1
Adducin

Plasma membrane
Spectrin tetramer
Band 3 dimer
Ankyrin

Junctional complex

▲ **Figure 13-38**   Postulated organization of the major erythrocyte cytoskeletal proteins. The more detailed structure of a junctional complex in the inset illustrates a hypothetical arrangement of spectrin, band 4.1, adducin, actin, and tropomyosin. *After S. E. Lux, 1979, Nature **281**:426; B. W. Shen, R. Josephs, and T. L. Steck, 1986, J. Cell Biol. **102**:997– 1006. [See K. Gardner and V. Bennett, 1987, Modulation of spectrin-actin assembly by erythrocyte adducin, Nature **328**:359–362.]*

the spectrin skeleton to the membrane. Proteins similar to ankyrin, spectrin, and band 4.1 are found in many other cell types, where they may serve as anchors for other cytoskeletal proteins.

The rigidity of the plasma membrane is due to the interactions of a complex of cytoskeletal proteins, but the formation of the biconcave disk requires additional cytoskeletal elements. A small amount of the fibrous actin-binding protein myosin (a principal protein of the contractile apparatus of muscle) is also found in erythrocytes. Short myosin filaments may bind to two or more junctional actin microfilaments and pull them together. Or other as yet unidentified minor cytoskeletal proteins may cause the skeleton to curve inward.

Several Hereditary Diseases Affect the Cytoskeleton

Hereditary spherocytosis and elliptocytosis are human diseases resulting from genetic mutations. In both diseases, the erythrocytes are abnormally shaped (Figure 13-39). In some cases, the defect results when an abnormal spectrin polypeptide either fails to form head-to-head tet-

ramers or binds defectively either to protein 4.1 or to ankyrin. In other cases, ankyrin is defective or absent. In any case, the consequences are an unstable cytoskeleton and an abnormally shaped cell, with fragments of membrane occasionally budding from the surface. Because these abnormal erythrocytes are degraded by the spleen more rapidly than normal ones are, affected persons have fewer circulating erythrocytes and are said to be anemic.

Specialized Regions of the Plasma Membrane

Unlike blood cells, which are independent, free-floating units, most animal or plant cells are organized into multicellular arrays that form solid tissues. In animals, cells in sheets or other aggregates can carry out their designated functions only if the plasma membrane is organized into at least two discrete regions, each specialized for very different tasks. Such cells are said to be *polarized*. Moreover, for an animal tissue to have strength, its cells must be able to form tight, strong contacts with their neighbors. Here, we explore the importance and variety of the

◀ **Figure 13-39** Scanning electron micrographs of spherocytes (rounded erythrocytes) from different patients with hereditary diseases of the erythrocyte cytoskeleton. *From M. Bessis, 1973, Living Blood Cells and Their Ultrastructure. Springer-Verlag, p. 220.*

the plasma membrane's specialized regions and the nature of its cell-cell junctions by examining several animal cell types in detail: the acinar cell of the pancreas, which secretes a number of digestive enzymes; the epithelial cell that lines the lumen of the small intestine; and a line of kidney epithelial cells that grows in tissue culture as a polarized cell monolayer.

The Pancreatic Acinus Is an Aggregate of Cells Having Two Very Different Regions of Plasma Membrane

A pancreatic acinus is a more or less spherical aggregate of about a dozen cells (Figure 13-40). The lumen (central cavity) of the acinus is connected to a ductule that merges with ductules from the lumina of other acini to form a larger duct; this duct eventually leads into one of several main pancreatic ducts, which fuse with the lumen of the small intestine (see Figure 4-40). Acinar cells synthesize enzymes (amylases, proteases, ribonucleases, and so on) that degrade most food macromolecules in the intestine. In the cell, these enzymes are stored as inactive precursors in membrane-limited secretory vesicles clustered under the *apical membrane* (the plasma membrane adjacent to the ductule). The secretory vesicles fuse with this and only this part of the plasma membrane (see Figure 4-40), ensuring that the digestive enzymes are released into the ductule.

The rest of the cell surface, called the *basolateral membrane,* includes the membrane on the sides of the cell

(a)

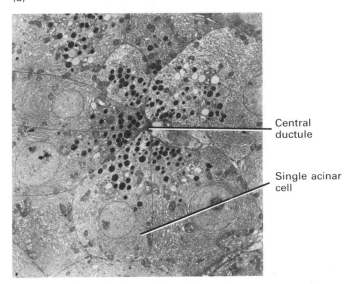

Central ductule

Single acinar cell

(b)

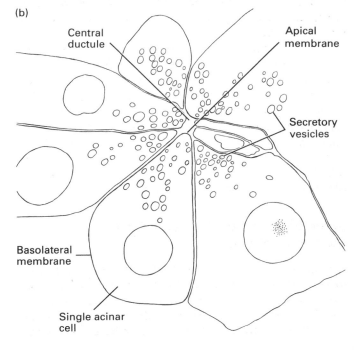

Central ductule

Apical membrane

Secretory vesicles

Basolateral membrane

Single acinar cell

▶ **Figure 13-40** (a) A low-magnification (×5000) electron micrograph of a rat pancreatic acinus, showing the overall arrangement of the cells surrounding the central ductule. The nuclei are close to the base of the cells; the secretory vesicles fill the parts of the cell near the lumen. Fusion of the membrane of a secretory vesicle with the part of the plasma membrane facing the lumen causes exocytosis of the contents of the vesicle into the lumen. (b) Tracing of the electron micrograph in (a). *Part (a) courtesy of Biophoto Associates.*

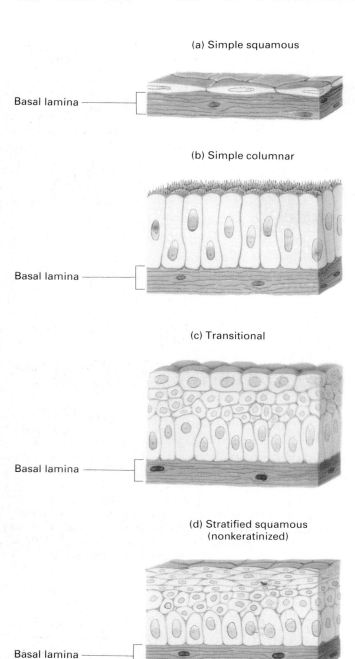

(a) Simple squamous

Basal lamina

(b) Simple columnar

Basal lamina

(c) Transitional

Basal lamina

(d) Stratified squamous (nonkeratinized)

Basal lamina

▲ **Figure 13-41** Principal types of epithelial membranes that line the surfaces of body cavities. (a) Simple squamous membranes are thin cells that line the blood vessels and many body cavities. (b) Simple columnar membranes are elongated cells that include mucus-secreting cells, such as those lining the stomach and cervical tract, and absorptive cells, such as those lining the small intestine; the latter contain microvilli at their apical surface. (c) Transitional membranes are composed of several layers of cells of different shapes that line certain cavities subject to expansion and contraction, such as the urinary bladder. (d) Stratified squamous (nonkeratinized) membranes line surfaces such as the mouth and the vagina; such linings resist abrasion and generally do not participate in the absorption or secretion of materials into or out of the cavity.

below the apical membrane and along the base of the cell. Nutrients from the surrounding blood are transported through this plasma membrane region into the acinar cell. The basolateral membrane is also the site of hormone receptors that trigger the secretion of digestive enzymes from acinar cells when, in the presence of food, the stomach and intestines release several peptide hormones into the blood.

The Plasma Membrane of Intestinal Epithelial Cells Is Divided into Two Regions of Different Structure and Function

With few exceptions, all internal and external body surfaces are covered with a polarized layer of cells called an *epithelium* (Figure 13-41). Different epithelia have specialized structures essential to the coordination, integrity, and functional environment of all other tissues. A well-studied, highly polarized epithelial cell lines the lumen of the small intestine (Figure 13-42); its two major functions are to absorb nutrients from digested food (sugars, amino acids, lipids, and so on) from the lumen of the small intestine and transfer them across the single cell layer into the blood.

The lumenal (apical) surface of the cells is highly specialized for absorption. This region, often called the *brush border* because of its appearance, consists of large numbers of fingerlike projections (100 nm in diameter) called *microvilli* (singular: *microvillus*). These extensions of the cell surface greatly increase the membrane area and enhance the rate of absorption into the cells. The microvillar plasma membrane contains transport proteins that allow glucose, amino acids, and other compounds to pass into the cell. Digestive enzymes are bound to the microvillar surface as well.

After proteins and carbohydrates are degraded into small peptides and oligosaccharides in the intestine, mostly by pancreatic enzymes, they must be broken down further into monosaccharides, such as glucose, and amino acids before they can be absorbed by the epithelial cells. This is the function of the peptidases and glycosidases bound to the microvillar surface. The membrane orientation of sucrase-isomaltase—a rodlike enzyme that catalyzes the hydrolysis of the disaccharide sucrose (see Figure 2-53) into glucose and fructose—is depicted in Figure 13-42b. Other degradative enzymes in the apical membrane have similar structures. In electron micrographs of microvilli, these sets of hydrolytic enzymes appear as a "fuzz," or glycocalyx, on the outer membrane surface (see Figure 13-28). After the nutrients that have been broken down by enzymes are absorbed through the lumenal surface, they are transported out of the cell and into the blood by a discrete set of proteins on the basolateral surface.

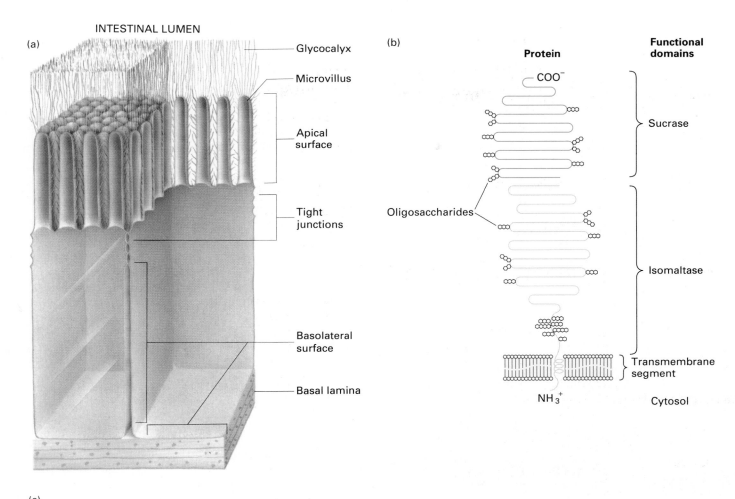

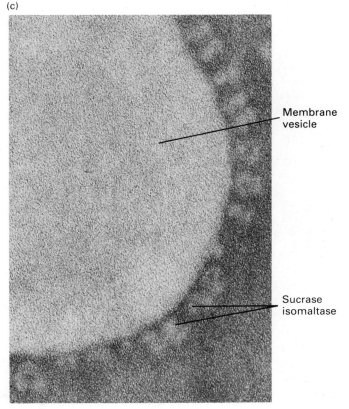

▲ **Figure 13-42** Intestinal epithelial cells. (a) The apical surface, made up of microvilli, faces the intestinal lumen. The cells rest on a basal lamina, a fibrous network of collagen and proteoglycans that supports the epithelial cell layer. (b) Structure of sucrase-isomaltase in the plasma membrane of the intestinal epithelial cell and (c) an electron micrograph of purified sucrase-isomaltase incorporated into liposomes. The major sugar in human diets is sucrose, a disaccharide. To be absorbed from the intestine, sucrose must be hydrolyzed to its individual sugar units, fructose and glucose, by sucrase-isomaltase—a two-chained, rather elongated protein on the lumenal surface of intestinal epithelial cells. This enzyme is anchored to the plasma membrane by a hydrophobic region consisting of 30 amino acids, but the bulk of the protein, with its many oligosaccharide side chains, protrudes into the intestinal lumen. Thus sucrase-isomaltase produces glucose and fructose very close to the cell membrane. [Part (b), see W. Hunziker et al., 1986, *Cell* **46**:227.] *Part (c) from G. M. Cowell, J. Tranum-Jensen, H. Sjostrum, O. Noren, 1986* Biochem. J. *237, 455–461.*

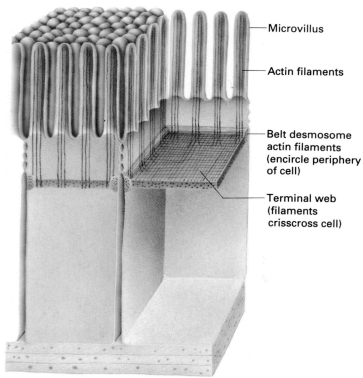

▲ Figure 13-43 Bundles of actin microfilaments run down the centers of the microvilli of the intestinal epithelial cell and intersect with a layer of filaments called the *terminal web*, which traverses the cell just below the microvilli and inserts into the belt desmosome.

Microvilli Have a Rigid Structure

Various components of the plasma membrane and cytoskeleton contribute to the structure of the microvilli. Structural proteins maintain their shape and uniform diameter. A bundle of actin microfilaments runs down the center of each microvillus, anchoring itself to proteins on the cytoplasmic face of the microvillar membrane and intersecting with a network of filaments containing several actin-binding proteins that traverse the cell just beneath the microvilli (Figure 13-43). These microfilaments give rigidity to the microvilli and may enable them to move backward and forward in the intestinal lumen. The transverse filament network fibers insert into a special type of cell junction (the belt desmosome). Microvilli, with their distinctive cytoskeletal fibers, also occur on other types of cells. In the tubules of the kidney, for example, they serve to resorb material into the blood that would otherwise pass into the urine.

Certain Epithelial Cells Can Be Grown in Culture

The study of epithelial cells has been advanced considerably by the development of cell lines that form polarized cell monolayers when grown in cell culture. One such line of MDCK cells, originally cultured from canine kidney, grows to form a continuous sheet one cell thick (Figure 13-44). This cell layer, like the cell layers of epithelia in

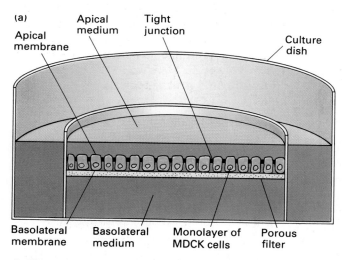

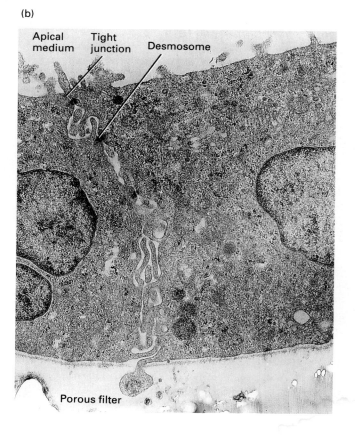

▲ Figure 13-44 (a) Madin-Darby canine kidney (MDCK) cells form a polarized epithelium when grown in tissue culture. MDCK cells can be cultured on a porous membrane filter coated on one side with collagen and other proteins of the basal lamina. The cells are bathed on both sides with tissue-culture medium; the apical surface faces the medium that bathes the upper side. (b) Electron micrographs of parts of two MDCK cells grown in tissue culture on a permeable filter. *Courtesy of R. Van Buskirk, J. Cook, J. Gabriels, and H. Eichelberger.*

the body, forms a barrier to the passage of most small molecules and ions, which cannot pass readily through the cells themselves or across the cell layer via the spaces between cells. These extracellular spaces are sealed by structures called *tight junctions,* whose structural integrity is essential to the functional integrity of the tissue.

Types of Cell Junctions

Tissues like the pancreatic acinus and the intestinal epithelium are aggregates of individual cells. To function in an integrated manner, different cell types have specialized surface junctions that permit or restrict the passage of ions and molecules between cells and that cause cells to adhere to each other and to the surrounding extracellular matrix. The three principal types of junctions are the tight junction, the desmosome, and the gap junction (Figure 13-45).

In some parts of an organism, tight junctions seal adjacent cells, such as the epithelial cells of the mammalian body, together to prevent the passage of fluids through the cell layer. Impermeable tight junctions, also termed the *zonulae occludens,* seal off the lumen of the intestine from the fluid that flows past the basolateral surface of the epithelial cells and the lumen of the pancreatic acinus from the surrounding blood.

In many organs, cells are bound tightly together by *desmosomes* to give tissues strength. *Belt desmosomes,* which are found primarily in epithelial cells, form a belt of cell-cell adhesion just under the tight junction. *Spot desmosomes* are found in all epithelial cells and many other tissues, such as smooth muscle. They are buttonlike points of contact between cells, often thought of as a "spot-weld" between adjacent plasma membranes. *Hemidesmosomes* are similar in structure to the desmosomes that anchor the plasma membrane to regions of the extracellular matrix. Bundles of intermediate filaments course through the cell and interconnect the spot desmosomes and hemidesmosomes.

Finally, *gap junctions* are distributed along the lateral surfaces of cells and allow them to exchange small molecules. Gap junctions help to integrate the metabolic activities of all cells in the tissue by assuring that they share a common pool of metabolites.

In higher plants, individual cells are separated from each other by cell walls, and tight junctions and desmosomes do not occur. However, many plant cells are interconnected by *plasmodesmata,* which, like gap junctions in animals, allow metabolites of low molecular weight and signaling molecules to move between cells.

In addition to specialized cell junctions, electron micrographs of animal tissue sections have shown that a space of about 20 nm ordinarily is present between the plasma membranes in nonjunctional regions of adjacent cells. This space contains extracellular surface glycoproteins

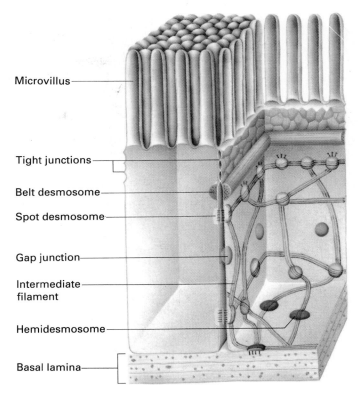

▲ **Figure 13-45** Schematic diagram of the principal types of cell junctions in the intestinal epithelial cell.

Labels: Microvillus, Tight junctions, Belt desmosome, Spot desmosome, Gap junction, Intermediate filament, Hemidesmosome, Basal lamina

that are probably also important to intercellular adhesion. Because an understanding of the structure and function of desmosomes requires prior knowledge of actin microfilaments and intermediate filaments, we defer their discussion until Chapter 22. Here we consider tight junctions, gap junctions, and plasmodesmata.

Tight Junctions Seal Off Body Cavities

Tight junctions are composed of thin bands that completely encircle a cell and are in contact with similar thin bands of adjacent cells. In epithelial cells, tight junctions are usually located just below the apical microvillar surface (Figure 13-46a). When thin sections through a tight junction are viewed in the electron microscope, the plasma membranes of adjacent cells appear to touch each other at intervals and even to fuse (Figure 13-46b). Tight junctions are impermeable to most substances. After the injection of lanthanum hydroxide (a colloid of high molecular weight) into the pancreatic blood vessel of an experimental animal, the pancreatic acinar cells are analyzed by electron microscopy (see Figure 13-46a). The lanthanum, an electron-dense material, covers the basolateral surface of the cell but cannot penetrate past the outermost tight junction. Other studies have shown that these junctions are impermeable to salts. For in-

(a)

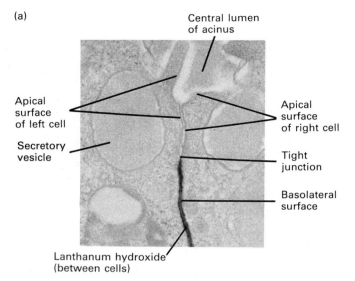

Central lumen
of acinus

Apical
surface
of left cell

Secretory
vesicle

Apical
surface
of right cell

Tight
junction

Basolateral
surface

Lanthanum hydroxide
(between cells)

(b)

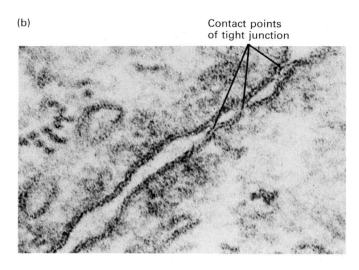

Contact points
of tight junction

▲ **Figure 13-46**   (a) Electron micrograph of a tight junction between two pancreatic acinar cells. Electron-opaque lanthanum hydroxide, which is added to the outside of the acinus a few minutes before the material is fixed and processed for microscopy, penetrates between the two acinar cells but is arrested at the level of the tight junction. (b) Electron micrograph of a tight junction between two hepatocytes (liver cells), seen as three points of contact between the plasma membranes of adjacent cells. *Part (a) courtesy of D. Friend; part (b) courtesy of D. Goodenough.*

stance, when MDCK cells are grown in a medium containing very low concentrations of Ca^{2+}, they form a monolayer, but the cells are not interconnected by tight junctions and fluids flow freely across the cell layer. When Ca^{2+} is added to these monolayers, tight junctions form within an hour.

All nutrients are absorbed from the intestine into one side of the epithelial cell and released from the other side into the blood, because tight junctions do not allow small molecules to diffuse directly from the intestinal lumen into the blood. In pancreatic acinar tissue, the tight junction similarly prevents pancreatic secretory proteins, including digestive enzymes, from leaking into the blood.

Freeze-fracture electron microscopy affords a different view of the tight junction. In Figure 13-47, the tight junction appears to comprise an interlocking network of ridges in the plasma membrane. More specifically, there appear to be ridges on the cytoplasmic face of the plasma membrane of each of the two contacting cells. (Although not shown here, corresponding grooves are found on the exoplasmic face.) High magnification reveals that these ridges are made up of particles, believed to be protein, 3–4 nm in diameter. The tight junction is formed by a double row of these particles, one row donated by each cell. The molecular structure of these junctions is not known, but protein particles on the two cells probably form extremely tight links with each other, essentially fusing the two plasma membranes and creating an impenetrable seal. Treatment of an epithelium with the protease trypsin destroys the tight junctions, implicating proteins as essential structural components.

The Tight Junction Separates the Apical and Basolateral Domains of Polarized Epithelial Cells

For polarized epithelial cells, such as the intestinal epithelium, to function, the apical and basolateral surfaces must contain completely different sets of integral proteins. The apical plasma membrane, for example, contains a set of digestive enzymes as well as a set of permeases that transport sugars and amino acids into the cell. The basolateral membrane contains totally different integral proteins.

Experiments have shown that the tight junction acts as a barrier for the diffusion of proteins and even lipids from the apical to basolateral membrane. Many studies have made use of MDCK cells grown on filters (see Figure 13-44), because reagents can be added to the medium that bathes either the apical or the basolateral surface to probe membrane function. For example, when liposomes containing a fluorescent-tagged glycoprotein are added to the apical medium, a fraction of them fuse with the apical surface, incorporating some fluorescent glycoprotein into the apical membrane. As long as the tight junctions interconnecting the cells are intact, fluorescent label cannot diffuse to the basolateral surface. If extracellular Ca^{2+} is removed, the tight junctions are destroyed and the fluorescent protein can diffuse to the basolateral surface.

Membrane lipids in the exoplasmic leaflets cannot diffuse through tight junctions, because the lipid compositions of the apical and basolateral membranes are very different. All the glycolipid in MDCK cells, for instance, is present in the apical membrane; in fact, the outer leaflet

(a)

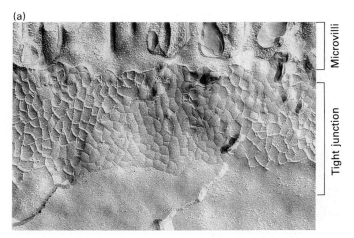

(b)

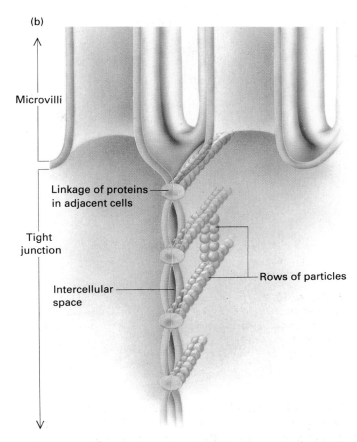

▲ **Figure 13-47** (a) Freeze-fracture electron micrograph of a tight junction between two intestinal epithelial cells. The fracture plane passes through the plasma membrane of one of the two adjacent cells. The honeycomb-like network of ridges of particles below the microvilli forms the tight junction. (b) A model showing how the junction might be formed by linkage of the rows of particles in adjacent cells. *Part (a) courtesy of L. A. Staehelin; part (b) after L. A. Staehelin and B. E. Hull, 1978, Sci. Am. 238(5):140.*

of the apical plasma membrane contains only glycolipid and cholesterol. Phosphatidylcholine, in contrast, is present almost exclusively in the basolateral plasma membrane.

In contrast, lipids in the cytoplasmic leaflets have the same composition and apparently can diffuse from the apical to the basolateral membrane. The tight junction also prevents the lipids of the extracytoplasmic leaflet of one cell from diffusing into the membrane of an adjacent cell. For example, one variant line of MDCK cells (strain I) contains a particular glycolipid called the *Forssman antigen* in the outer leaflet of the apical membrane, but another (strain II) does not. When grown together in culture, the two cell types form normal tight junctions with each other, but the Forssman glycolipid does not move from the type I to the type II cells; thus lipids do not diffuse past the tight junction from cell to cell. However, the molecular structure of the tight junction is not known.

Gap Junctions Allow Small Molecules to Pass between Adjacent Cells

Almost all animal cells that come in contact with each other have regions of junctional specialization characterized by a gap between the plasma membranes that is filled by a well-defined set of particles (Figure 13-48a). Al-

though morphologists named these regions gap junctions, in retrospect, the gap is not their most important feature. The cylindrical "particles" in the gap make this junction unique; they are actually tiny channels that directly link the cytoplasms of the two cells and allow the free passage of very small molecules and ions between a cell and its neighbor (Figure 13-48b). The gap junction is not a sealing junction: it does not form a barrier to the passage of extracellular fluid between two cell membranes.

The size of these intracellular channels can be measured by injecting a cell with a fluorescent dye covalently linked to molecules of various sizes and using a fluorescence microscope to observe whether they pass into neighboring cells. Gap junctions between mammalian cells permit the passage of molecules as large as 2 nm in diameter. In insects, these junctions are permeable to molecules as large as 3 nm in diameter. Generally speaking, molecules with a molecular weight of lower than 1200 pass freely and those of 2000 or more do not pass; the passage of intermediate-sized molecules is variable and limited. Thus ions and many low-molecular-weight building blocks of cellular macromolecules, such as amino acids and nucleoside phosphates, can pass from cell to cell.

A vivid example of this cell-cell transfer is the phenomenon of *metabolic coupling*, or *metabolic cooperation*, in which a cell can transfer molecules to a neighboring cell but the recipient cannot synthesize them. For example,

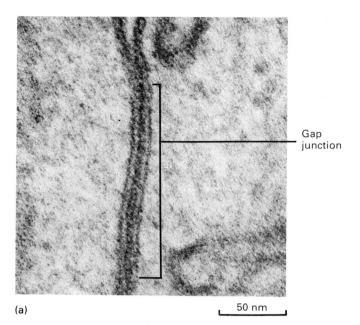

(a)   50 nm

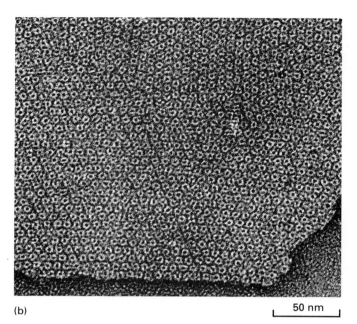

(b)   50 nm

▲ **Figure 13-48** (a) Electron micrograph of a thin section through a gap junction connecting two mouse liver cells. The two plasma membranes are closely associated for a distance of several hundred nanometers, separated by a "gap" of 2–3 nm. (b) Negatively stained, electron microscopic image of the cystolic face of a region of plasma membrane enriched in gap junctions; each "doughnut" forms a channel connecting two cells. *Part (a) courtesy of D. Goodenough; part (b) courtesy of N. Gilula.*

Hypoxanthine   **5-Phosphoribosyl 1-pyrophosphate**

HPRT

→ Pyrophosphate

Inosine 5'-phosphate

5'-AMP

dATP

DNA

AMP, ADP, or ATP can pass through gap junctions. In normal fibroblasts, hypoxanthine (adenine, with the C-NH$_2$ group replaced by CO) can serve as a precursor of DNA. By the pathway depicted in Figure 13-49, hypoxanthine is converted first to inosine 5'-phosphate, in a reaction catalyzed by the enzyme hypoxanthine-guanine phosphoribosyltransferase (HPRT), and then to dATP, the immediate precursor of DNA. Fibroblasts of a mutant line lacking HPRT are unable to incorporate radioactive hypoxanthine into their DNA. But if cells lacking HPRT are cocultured with cells that have this enzyme and labeled hypoxanthine is added to the medium, radioactivity is frequently found in the nuclear DNA of the mutant cells. (The two cells can be differentiated by their distinct morphologies or by feeding one of the cell lines carbon particles before mixing it with the other line.) The dATP derived from hypoxanthine is only incorporated into the DNA of the mutant cells that are in direct or indirect contact (through an intermediate cell) with wild-type cells. It is thought that labeled adenosine mono-, di-, or triphosphate is synthesized from the labeled hypoxanthine by wild-type cells and then passed through gap junctions to the mutant cells.

◀ **Figure 13-49** Hypoxanthine can be converted to dATP, a normal precursor of DNA. The enzyme that catalyzes the first step in this pathway, hypoxanthine-guanine phosphoribosyltransferase (HPRT), is missing in mutant HPRT$^-$ cells (see Figure 5-21); these mutants are unable to use hypoxanthine as a precursor of dATP for DNA synthesis.

Another important compound transferred from cell to cell through gap junctions is cyclic AMP (cAMP), which acts as an intracellular messenger and regulates a number of cellular metabolic activities. The amount of cellular cAMP increases in response to the treatment of cells with many different hormones. The fact that cAMP can pass through gap junctions means that the hormonal stimulation of just one or a few cells in an epithelium initiates a metabolic reaction in many of them. Specifically, secretory hormones, such as secretin, bind to receptors on the basal plasma membranes of pancreatic acinar cells and increase the intracellular concentration of either cAMP or Ca^{2+} ions, substances that trigger the secretion of the contents of the secretory vesicles. Both Ca^{2+} and cAMP can pass through the gap junctions, so hormonal stimulation of one cell triggers secretion by many.

An important aspect of gap-junction physiology is that the channels close in the presence of high concentrations of Ca^{2+} ion. The Ca^{2+} concentration in extracellular fluids is quite high (from $1 \times 10^{-3}M$ to $2 \times 10^{-3}M$), whereas the concentration of free Ca^{2+} in the cytosol is lower than $10^{-6}M$. If the membrane of one cell in an epithelium is ruptured, Ca^{2+} enters the cell, traveling down its concentration gradient, and the Ca^{2+} concentration in the cytosol rises markedly. This closes the channels that connect the cell with its neighbors and prevents leakage of the low-molecular-weight contents of the cytoplasm in all epithelium cells. Even slight increases in the level of cytosol Ca^{2+} ions or decreases in cytosolic pH can decrease the permeability of gap junctions. Thus cells may modulate the degree of coupling with their neighbors, but precisely why and how they accomplish this is a matter of debate.

In the liver and many other tissues, large numbers of individual junctions cluster together in an area about 0.3 μm in diameter (see Figure 13-48b), which enables researchers to separate gap junctions from other parts of the plasma membrane by shearing the purified plasma membrane into small fragments. Due to their relatively high protein content, fragments containing gap junctions have a higher density than the bulk of the plasma membrane and can be purified on an equilibrium gradient (see Figure 4-32). Electron micrographs of stained, isolated gap junctions reveal a lattice of hexagonal particles with hollow cores as intercellular channels (Figure 13-50a; see Figure 13-48). The purified gap junction consists of *connexin*, a major protein with a molecular weight of 28,000–32,000. Each hexagonal particle consists of 12 connexin molecules: six formed in a hexagonal cylinder in one plasma membrane joined to six arranged in the same array in the adjacent cell membrane. As deduced from image analysis of the electron micrographs, the six connexin subunits that make up a single cylinder can interact in two different but related ways: one results in an open channel; the other, in a closed channel (Figure 13-51). It has been suggested that the transition from one

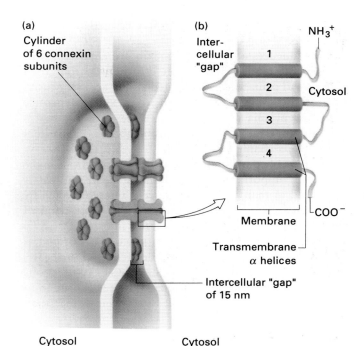

▲ **Figure 13-50** (a) Model of a gap junction, based on electron microscopic and x-ray diffraction analyses. Two plasma membranes are separated by a gap of about 15 nm. Both membranes contain cylinders of six dumbbell-shaped connexin subunits. Two such cylinders join in the gap between the cells to form a channel about 1.5–2.0 nm in diameter that connects the cytoplasm of the two cells. (b) Arrangement of the connexin subunits within the plasma membrane. Transmembrane α helix 3 is amphipathic: one face of the helix contains mostly hydrophilic side chains; the other face is mostly hydrophobic. The sequence of α helix 3 is highly conserved among gap junction proteins in different cell types and species; it is thought that the aqueous channel is lined by transmembrane α helices from the 12 connexin subunits. [Part (b), see L. C. Milks, N. Kumar, R. Houghton, N. Unwin, and N. Gilula, 1988, *EMBO Journal* 7:2967–2975.] *Part (b) courtesy of Bruce Nicholsen.*

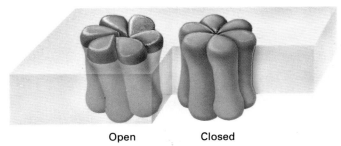

▲ **Figure 13-51** Model of a gap junction derived from electron-micrographic analysis. One rotation of the six connexin subunits about a central axis mediates the transition from an open to a closed state. Closure can be induced by an elevation in Ca^{2+}. *After P. T. Unwin and G. Zampighi, 1980,* Nature **283**:545; *P. T. Unwin and C. Ennis, 1984,* Nature **307**:609–613.

form of interaction to the other may be mediated by the level of Ca^{2+} ion in the cell.

The sequences of several connexin proteins, expressed in different tissues, have been determined from their cDNAs. All connexins have related amino acid sequences. Experiments suggest that each connexin polypeptide spans the plasma membrane four times (see Figure 13-50b) and that one conserved transmembrane α helix lines the aqueous channel.

One type of connexin makes up the gap junctions in the heart—an important finding because heart muscle is comprised of many small muscle cells interconnected by these junctions. As in other muscles, the contraction of heart muscle cells is triggered by a rise in cytosolic Ca^{2+}. The spread of Ca^{2+} through gap junctions coordinates the beating of all muscle cells in one section of the heart; abnormalities in conduction through these gap junctions may disrupt the synchronization of contraction in adjacent heart muscle cells, causing fibrillation, rather than contraction, to occur.

Plasmodesmata Interconnect the Cytoplasms of Adjacent Cells in Higher Plants

All cells in higher plants are separated from each other by walls containing cellulose. However, the cytoplasms of adjacent cells are interconnected by tubelike plasmodesmata that penetrate the wall. Like gap junctions, plasmodesmata provide intercellular channels for molecules of about 1000 molecular weight, including a variety of metabolic and signaling compounds.

Electron micrographs (Figure 13-52) show that the plasma membranes of adjacent cells in plasmodesmata are continuous through the cell wall. The diameter of the channel is about 60 nm, and plasmodesmata traverse cell walls up to 90 nm thick. In some cases, a tube of endoplasmic reticulum appears to pass through the plasmodesmata; membrane-bound molecules may also pass

from cell to cell via this organelle. Depending on the plant type, the density of plasmodesmata varies from 1 to 10 per μm^2, and even the smallest meristematic cells (the growing cells at the tips of roots or stems) have more than 1000 interconnections with their neighbors.

Much evidence establishes that plasmodesmata are in fact used in cell-cell communication. Fluorescent water-soluble chemicals microinjected into plant cells spread to the cytoplasm of adjacent cells but not into the cell wall. Many normal metabolic products, such as sucrose, spread from cell to cell; transport of such substances is porportional to the number of plasmodesmata and does not occur between cells not connected by such junctions. As with gap junctions, transport through plasmodesmata is reversibly inhibited by an elevation in cytosolic Ca^{2+}. Thus, despite considerable structural differences between gap junctions and plasmodesmata, many of their functional aspects are remarkably similar.

Summary

The basic structure of all biological membranes is the closed phospholipid bilayer. The lipid composition of the bilayer varies among the diverse cellular membranes: glycolipids and cholesterol are abundant in the plasma membrane; large quantities of cardiolipin are found in the inner mitochondrial membrane. The phospholipid composition of the two leaflets in a single membrane may also differ, as in the erythrocyte plasma membrane, but such variations do not explain the unique properties of different biological membranes. Each type of membrane—or even patch of membrane—owes much of its individuality to the distinctive properties of its protein species.

The membrane of the mammalian erythrocyte, for example, differs from the membranes of most tissue cells in that it contains a homogeneous array of surface proteins that cannot diffuse laterally. No cytoskeleton courses

(a)

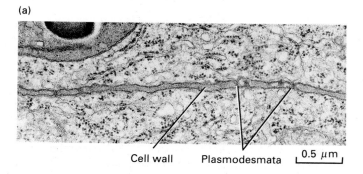

Cell wall   Plasmodesmata   0.5 μm

(b)

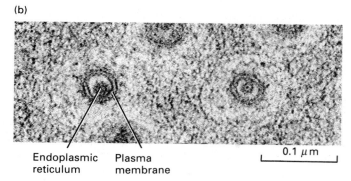

Endoplasmic   Plasma   0.1 μm
reticulum   membrane

▲ **Figure 13-52** Electron microscopic images (a) of the cell wall between two root tip cells of the bean *Phaseolus vulgaris*, showing numerous plasmodesmata connecting the cytoplasms of the two cells, and (b) of the structure of plasmodesmata from the same type of cells (transverse view). *Part (a) micrograph by W. P. Wergin; parts (a) and (b) courtesy of E. H. Newcomb and Biological PhotoService.*

through the erythrocyte cell. Just under the plasma membrane, a lacelike cytoskeletal network constructed mostly of spectrin, actin (a fibrous protein), and accessory proteins is bound to one of the major integral membrane glycoproteins by many highly specific protein-protein interactions. This submembrane spectrin network appears to be responsible for the shape of the erythrocyte.

Membrane proteins can be classified into two broad groups. An integral protein interacts directly with the phospholipid bilayer and usually contains one or more long α-helical sequences of hydrophobic amino acids that form bonds within the hydrophobic middle of the bilayer. Examples of integral proteins are the major erythrocyte glycoproteins—glycophorin and band 3—and the bacterial photosynthetic reaction center, whose structure is known to atomic detail. The second group, peripheral proteins, are bound to the membrane primarily, if not exclusively, by protein-protein (rather than protein-lipid) interactions. Examples of peripheral proteins are spectrin and other erythrocyte skeletal proteins and many proteins of the glycocalyx.

Biological membranes are highly asymmetric. All integral membrane proteins bind asymmetrically to the lipid bilayer, and all molecules of any one kind lie in the same direction. In addition, all the molecules of a particular membrane-bound enzyme face the same surface of the membrane; virtually all membrane oligosaccharides in glycolipids and glycoproteins face the exoplasmic surface of the bilayer. This asymmetry is related to all aspects of membrane function.

Proteins also play an essential part in the formation of junctions between cells. Some surface proteins form tight junctions between cells, sealing off fluids on different sides of a cell layer. In epithelial cells, such as those that form the pancreatic acinus or intestinal epithelium, tight junctions define two regions of the plasma membrane: the apical and basolateral surfaces. Each region contains unique lipids and proteins that enable it to perform specialized functions, such as binding hormones or fusing with intracellular vesicles that contain secretory proteins. Belt desmosomes and spot desmosomes are junctions that bind the membranes of adjacent cells in a way that gives strength and rigidity to the entire tissue. On the cytoplasmic face of a cell, desmosomal complexes connect with specific cytoskeletal fibers; such interactions contribute to the cell's rigid shape. Gap junctions are constructed of 12 copies of a single protein formed into a transmembrane channel that interconnects the cytoplasm of two adjacent cells and allows small molecules to pass between them. Plasmadesmata in higher plants interconnect the cytoplasms of adjacent cells and also allow small molecules like sucrose to pass between them.

Most proteins and all lipids in the plasma membrane, as well as in internal membranes, are laterally mobile; some proteins are immobile, probably because they are anchored to parts of the cytoskeleton.

References

The Architecture of Lipid Membranes and Membrane Proteins

Properties of Phospholipid Bilayers

CHAPMAN, D. 1975. Lipid dynamics in cell membranes. In *Cell Membranes: Biochemistry, Cell Biology, and Pathology.* G. Weissmann and R. Claiborne, eds. Hospital Practice, pp. 13–22.

CLEMENTI, E., and S. CHIN, eds. 1987. *Structure and Dynamics of Nucleic Acids, Proteins, and Membranes.* Plenum Press.

*HAKOMORI, S. 1986. Glycosphingolipids. *Sci. Am.* 254 (5):44–53.

MELCHIOR, D. L. 1986. Lipid domains in fluid membranes: a quick-freeze differential scanning calorimetry study. *Science* 234:1577–1580.

SINGER, S. J., and G. L. NICOLSON. 1972. The fluid mosaic model of the structure of cell membranes. *Science* 175:720–731.

*TANFORD, C. 1980. *The Hydrophobic Effect*, 2d ed. Wiley. Includes a good discussion of the interactions of proteins and membranes.

WENDOLOSKI, J. J., S. J. KIMATIAN, C. E. SCHUTT, and F. R. SALEMME. 1989. Molecular dynamics simulation of a phospholipid micelle. *Science* 243:636–638.

Lipid Anchors of Membrane Proteins

FERGUSON, M. A. J., S. W. HOMANS, R. A. DWEK, and T. W. RADEMACHER. 1988. Glycosyl-phosphatidylinositol moiety that anchors *Trypanosoma brucei* variant surface glycoprotein to the membrane. *Science* 239: 753–759.

HOMANS, S. W., M. A. J. FERGUSON, R. A. DWEK, T. W. RADEMACHER, R. ANAND, and A. F. WILLIAMS. 1988. Complete structure of the glycosyl-phosphatidylinositol membrane anchor of rat brain Thy-1 glycoprotein. *Nature* 333:269–272.

*LOW, M. G. 1989. Glycosyl-phosphatidylinositol: a versatile anchor for cell surface proteins. *FASEBJ.* 3:1600–1608.

PELLMAN, D., E. A. GARBER, F. R. CROSS, and H. HANAFUSA. 1985. An N-terminal peptide from p60^{src} can direct myristylation and plasma-membrane localization when fused to heterologous proteins. *Nature* 314:374–377.

*SCHULTZ, A. M., L. E. HENDERSON, and S. OROSZLAN. 1988. Fatty acylation of proteins. *Ann. Rev. Cell Biol.* 4:611–647.

*SEFTON, B. M., and J. E. BUSS. 1987. The covalent modification of eukaryotic proteins with lipid. *J. Cell Biol.* 104:1449–1453.

Structure of Integral Membrane Proteins

BJORKMAN, P. J., M. A. SAPIER, B. SAMRAOUI, W. S. BENNETT, J. L. STROMINGER, and D. C. WILEY. 1987. Structure of the human class I histocompatibility antigen HLA-A2. *Nature* 329:506–512.

BOYD, D., C. MANOIL, and J. BECKWITH. 1987. Determinants of membrane protein topology. *Proc. Nat'l Acad. Sci. USA* 84:8525–8529.

*A book or review article that provides a survey of the topic.

*DEISENHOFER, J., O. EPP, K. MIKI, R. HUBER, and H. MICHEL. 1985. Structure of the protein subunits in the photosynthetic reaction center of *Rhodopseudomonas viridis* at 3 Å resolution. *Nature* 318:618–624.

ENGLEMAN, D. M., and G. ZACCAI. 1980. Bacteriorhodopsin is an inside-out protein. *Proc. Nat'l Acad. Sci.* USA 77:5894–5898.

HARTMANN, E., T. A. RAPOPORT, and H. F. LODISH. 1989. Predicting the orientation of eukaryotic membrane-spanning proteins. *Proc. Nat'l Acad. Sci.* USA 86:5786–5790.

LODISH, H. F. 1988. Multi-spanning membrane proteins: how accurate are the models? *TIBS* 13:332–334.

*REES, D. C., L. DE ANTONIO, and D. EISENBERG. 1989. Hydrophobic organization of membrane proteins. *Science* 245:510–513.

REES, D. C., H. KOMIYA, T. O. YEATS, J. P. ALLEN, and G. FEHER. 1989. The bacterial photosynthetic reaction center as a model for membrane proteins. *Ann. Rev. Biochem.* 58:607–633.

*UWIN, N., and R. HENDERSON. 1984. The structure of proteins in biological membranes. *Sci. Am.* 250(2):78–94.

Detergents

*HELENIUS, A., and K. SIMONS. 1975. Solubilization of membranes by detergents. *Biochim. Biophys. Acta* 415:29–79.

Principles of Membrane Organization

Membrane Asymmetry

BRANTON, D. 1966. Fracture faces of frozen membranes. *Proc. Nat'l Acad. Sci.* USA 55:1048–1056.

HIRANO, H., B. PARKHOUSE, G. L. NICOLSON, E. S. LENNOX, and S. J. SINGER. 1972. Distribution of saccharide residues on membrane fragments from a myeloma-cell homogenate: its implication for membrane biogenesis. *Proc. Nat'l Acad. Sci.* USA 69:2945–2949.

*ROTHMAN, J., and J. LENARD. 1977. Membrane asymmetry. *Science* 195:743–753.

Mobility of Membrane Proteins and Lipids

DE PETRIS, A., and M. C. RAFF. 1973. Normal distribution, patching, and capping of lymphocyte surface immunoglobulin studied by electron microscopy. *Nature New. Biol.* 241:257–259.

FRYE, L. D., and M. EDIDIN. 1970. The rapid intermixing of cell-surface antigens after formation of mouse-human heterokaryons. *J. Cell Sci.* 7:319–335.

*HACKENBROCK, C. R. 1981. Lateral diffusion and electron transfer in the mitochondrial inner membrane. *TIBS* 6:151–154.

ISHIHARA, A., Y. HOU, and K. JACOBSON. 1987. The Thy-1 antigen exhibits rapid lateral diffusion in the plasma membrane of rodent lymphoid cells and fibroblasts. *Proc. Nat'l Acad. Sci.* USA 84:1290–1293.

*JACOBSON, K., A. ISHIHARA, and R. INMAN. 1987. Lateral diffusion of proteins in membranes. *Ann. Rev. Physiol.* 49:163–175.

LIVNEH, E., M. BENVENISTE, R. PRYWES, S. FELDER, Z. KAM, and J. SCHLESSINGER. 1986. Large deletions in the cytoplasmic kinase domain of the epidermal growth factor receptor do not affect its lateral mobility. *J. Cell Biol.* 103:327–331.

POO, M., and R. A. CONE. 1974. Lateral diffusion of rhodopsin in the photoreceptor membrane. *Nature* 247:438–441.

WEIR, M., and M. EDIDIN. 1988. Constraint of the translational diffusion of a membrane glycoprotein by its external domains. *Science* 242:412–414.

YECHIEL, E., and M. EDIDIN. 1987. Micrometer-scale domains in fibroblast plasma membranes. *J. Cell Biol.* 105:755–760.

The Erythrocyte Membrane: Cytoskeletal Attachment

Isolation and General Properties

BERTESSY, B. G., and T. L. STECK. 1989. Elasticity of the human red cell membrane skeleton: Effects of temperature and denaturants. *Biophys. J.* 55:255–262.

*CHASIS, J. A., and S. B. SHOHET. 1987. Red cell biochemical anatomy and membrane properties. *Ann Rev. Physiol.* 49:237–248.

CHIEN, S. 1987. Red cell deformability and its relevance to blood flow. 1987. *Ann. Rev. Physiol.* 49:177–192.

HOCHMUTH, R. M., and R. E. WAUGH. 1987. Erythrocyte membrane elasticity and viscosity. *Ann. Rev. Physiol.* 49:209–219.

LIEBERT, M. R. and T. L. STECK. 1982. A description of the holes in human erythrocyte membrane ghosts. *J. Biol. Chem.* 257:11651–11659.

LEW, V. L., A. HOCKADAY, C. J. FREEMAN, and R. M. BOOKCHIN. 1988. Mechanism of spontaneous inside-out vesiculation of red cell membranes. *J. Cell Biol.* 106:1893–1901.

Generation of the Erythrocyte Membrane

ELGSAETER, A., B. T. STOKKE, A. MIKKELSEN, and D. BRANTON. 1986. The molecular basis of erythrocyte shape. *Science* 234:1217–1223.

*LAZARIDES, E. 1987. From genes to structural morphogenesis: the genesis and epigenesis of a red blood cell. *Cell* 51:345–356.

PATEL, V. P., and H. F. LODISH. 1987. A fibronectin matrix is required for differentiation of murine erythroleukemia cells into reticulocytes. *J. Cell Biol.* 105:3105–3118.

SARRIS, A. H., and G. E. PALADE. 1982. Immunofluorescent detection of erythrocyte sialoglycoprotein antigens on murine erythroid cells. *J. Cell Biol.* 93:591–603.

ZWEIG, S. E., K. T. TOKUYASU, and S. J. SINGER. 1981. Membrane-associated changes during erythropoiesis: on the mechanism of maturation of reticulocytes to erythrocytes. *J. Supramol. Struct.* 17:163–181.

Integral Membrane Proteins

BLANCHARD, D., W. DAHR, M. HUMMEL, F. LATRON, K. BEYREUTHER, and J-P. CARTRON. 1987. Glycophorins B and C from human erythrocyte membranes. *J. Biol. Chem.* 262:5808–5811.

JENKINS, J. D., F. J. KEZDY, and T. L. STECK. 1985. Mode of interaction of phosphofructokinase with the erythrocyte membrane. *J. Biol. Chem.* 260:10426–10433.

*JENNINGS, M. L. 1989. Topography of membrane proteins. *Ann. Rev. Biochem.* 58:999–1027.

*KOPITO, R. R., and H. F. LODISH. 1985. Primary structure and transmembrane orientation of the murine anion exchange protein. *Nature* 316:234–238.

MARETZKI, D., B. REIMANN, and S. M. RAPOPORT. 1989. A reappraisal of the binding of cytosolic enzymes to erythrocyte membranes. *TIBS,* 14:93–96.

WALDER, J. A., R. CHATTERJEE, T. L. STECK, P. S. LOW, G. F. MUSSO, E. T. KAISER, P. H. ROGERS, and A. ARNONE. 1984. The interaction of hemoglobin with the cytoplasmic domain of band 3 of the human erythrocyte membrane. *J. Biol. Chem.* 259:10238–10246.

The Cytoskeleton

*BENNETT, V. 1985. The membrane skeleton of the human erythrocyte and its implications for more complex cells. *Ann. Rev. Biochem.* 54:273–304.

*BRANTON, D., C. M. COHEN, and J. TYLER. 1981. Interaction of cytoskeletal proteins on the human erythrocyte membrane. *Cell* 24:24–32.

BYERS, T. M., and D. BRANTON. 1985. Visualization of the protein associations in the erythrocyte membrane skeleton. *Proc. Nat'l Acad. Sci. USA* 82:6153–6157.

CONBOY, J. G., J. CHAN, N. MOHANDAS, and Y. W. KAN. 1988. Multiple protein 4.1 isoforms produced by alternative splicing in human erythroid cells. *Proc. Nat'l Acad. Sci. USA* 85:9062–9065.

DAVIS, L., S. E. LUX, and V. BENNETT. 1989. Mapping the ankyrin-binding site of the human erythrocyte anion exchanger. *J. Biol. Chem.* 264:9665–9672.

GARDNER, K., and V. BENNETT. 1987. Modulation of spectrin-actin assembly by erythrocyte adducin. *Nature* 328:359–362.

LIU, S-C., L. H. DERICK, and J. PALEK. 1987. Visualization of the hexagonal lattice in the erythrocyte membrane skeleton. *J. Cell Biol.* 104:527–536.

LOW, P. S. 1986. Structure and function of the cytoplasmic domain of band 3: center of erythrocyte membrane-peripheral protein interactions. *Biochim. Biophys. Acta* 864:145–167.

*MARCHESI, V. T. 1985. Stabilizing infrastructure of cell membranes. *Ann. Rev. Cell Biol.* 1:531–561.

SPEICHER, D. W., and V. T. MARCHESI. 1984. Erythrocyte spectrin is comprised of many homologous triple-helical segments. *Nature* 311:177–180.

TSUJI, A., K. KAWASAKI, S. OHNISHI, H. MERKLE, and A. KUSUMI. 1988. Regulation of band 3 mobilities in erythrocyte ghost membranes by protein association and cytoskeletal meshwork. *Biochem.* 27:7447–7452.

WASENIUS, V-M., M. SARASTE, P. SALVEN, M. ERAMAA, L. HOLM, and V-P. LEHTO. 1989. Primary structure of the brain α-spectrin. *J. Cell Biol.* 108:79–93.

Hereditary Diseases

AGRE, P., J. CASELLA, W. ZINKHAM, C. MCMILLAN, and V. BENNETT. 1985. Partial deficiency of erythrocyte spectrin in hereditary spherocytosis. *Nature* 314:380–383.

BIRKENMEIER, C. S., E. C. MCFARLAND-STARR, and J. E. BARKER. 1988. Chromosomal location of three spectrin genes: relationship to the inherited hemolytic anemias of mouse and man. *Proc. Nat'l Acad. Sci. USA* 85:8121–8125.

*RICE-EVANS, C. A., and M. J. DUNN. 1982. Erythrocyte deformability and disease. *TIBS* 6:282–286.

TOMASELLI, M. B., K. M. JOHN, and S. E. LUX. 1981. Elliptical erythrocyte membrane skeletons and heat-sensitive spectrin in hereditary elliptocytosis. *Proc. Nat'l Acad. Sci. USA* 78:1911–1915.

Specialized Regions of the Plasma Membrane

BARTLES, J. R., L. BRAITERMAN, and A. HUBBARD. 1985. Endogenous and exogenous domain markers of the rat hepatocyte plasma membrane. *J. Cell Biol.* 100:1126–1138.

HUNZIKER, W., M. SPIESS, G. SEMENZA, and H. F. LODISH. 1986. The sucrase-isomaltase complex: primary structure, membrane-orientation, and evolution of a stalked, intrinsic brush-border protein. *Cell* 46:227–234.

*KENNY, J., and A. J. TURNER, eds. 1987. *Mammalian Ectoenzymes.* Elsevier Science.

LISANTI, M. P., M. SARGIACOMO, L. GRAEVE, A. R. SALTIEL, and E. RODRIGUEZ-BOULAN. 1988. Polarized apical distribution of glycosyl-phosphatidylinositol-anchored proteins in a renal epithelial cell line. *Proc. Nat'l Acad. Sci. USA* 85:9557–9561.

MANTEI, N., M. VILLA, T. ENZLER, H. WACKER, W. BOLL, P. JAMES, W. HUNZIKER, and G. SEMENZA. 1988. Complete primary structure of human and rabbit lactase-phlorizin hydrolase: implications for biosynthesis, membrane anchoring, and evolution of the enzyme. *EMBO J.* 7:2705–2713.

*SEMENZA, G. 1986. Anchoring and biosynthesis of stalked brush-border membrane proteins: glycosidases and peptidases of enterocytes and renal tubuli. *Ann. Rev. Cell Biol.* 2:255–313.

SIMONS, K., and G. VAN MEER. 1988. Lipid sorting in epithelial cells. *Biochem.* 27:6197–6202.

*VAN MEER, G. 1988. How epithelia grease their microvilli. *TIBS* 13:242–243.

ZIOMEK, C. A., S. SCHULMAN, and M. EDIDIN. 1980. Redistribution of membrane proteins in isolated mouse intestinal epithelial cells. *J. Cell Biol.* 86:849–857.

Types of Cell Junctions

*BOCK, G., and S. CLARK. 1987. Junctional complexes of epithelial cells. *Ciba Foundation Symposium 125,* Wiley.

STAEHELIN, L. A. 1974. Structure and function of intercellular junctions. *Int. Rev. Cytol.* 39:191–283.

*STAEHELIN, L. A., and B. E. HULL. 1978. Junctions between cells. *Sci. Am.* 238(5):140–152.

Tight Junctions

ANDERSON, J. M., B. R. STEVENSON, L. A. JESAITIS, D. A. GOODENOUGH, and M. S. MOOSEKER. 1988. Characterization of ZO-1, a protein component of the tight junction from mouse liver and Madin-Darby canine kidney cells. *J. Cell Biol.* 106:1141–1149.

COWIN, P., H-P. KAPPRELL, W. W. FRANKE, J. TAMKUN, and R. O. HYNES. 1986. Plakoglobin: a protein common to different kinds of intercellular adhering junctions. *Cell* 46:1063–1073.

GUMBINER, B., and K. SIMONS. 1986. A functional assay for proteins involved in establishing an epithelial occluding barrier: identification of a uvomorulin-like polypeptide. *J. Cell Biol.* 102:457–468.

HULL, B. E., and L. A. STAEHELIN. 1976. Functional significance of the variations in the geometric organization of tight-junction networks. *J. Cell Biol.* 69:688–704.

VAN MEER, G., B. GUMBINER, and K. SIMONS. 1986. The tight junction does not allow lipid molecules to diffuse from one epithelial cell to the next. *Nature* 322:639–641.

VAN MEER, G., and K. SIMONS. 1986. The function of tight junctions in maintaining differences in lipid composition between the apical and the basolateral cell-surface domains of MDCK cells. *EMBO J.* 5:1455–1464.

Gap Junctions and Plasmodesmata

*BENNETT, M., and D. SPRAY, eds. 1985. *Gap Junctions*. Cold Spring Harbor Laboratory.

GILULA, N. B., O. R. REEVES, and A. STEINBACH. 1972. Metabolic coupling, ionic coupling, and cell contacts. *Nature* 235:262–265.

*GUTHRIE, S. C., and N. B. GILULA. 1989. Gap junctional communication and development. *TINS* 12:12–16.

*HERTZBERG, E. L., and R. G. JOHNSON, eds. 1988. *Gap Junctions*. Liss.

LAWRENCE, T. S., W. H. BEERS, and N. B. GILULA. 1978. Transmission of hormonal stimulation by cell-to-cell communication. *Nature* 272:501–506.

*LOEWENSTEIN, W. R. 1987. The cell-to-cell channel of gap junctions. *Cell* 48:725–726.

MEDA, P., R. BRUZZONE, M. CHANSON, D. BOSCO, and L. ORCI. 1987. Gap junctional coupling modulates secretion of exocrine pancreas. *Proc. Nat'l Acad. Sci. USA* 84:4901–4904.

*MEINERS, S., O. BARON-EPEL, and M. SCHINDLER. 1988. Intercellular communication—filling in the gap. *Plant Physiol.* 88:791–793.

MILKS, L. C., N. M. KUMAR, R. HOUGHTEN, N. UNWIN, and N. B. GILULA. 1988. Topology of the 32-kd liver gap junction protein determined by site-directed antibody localizations. *EMBO J.* 7:2967–2975.

ROSE, B., I. SIMPSON, and W. R. LOEWENSTEIN. 1977. Calcium ion produces graded changes in permeability of membrane channels in cell junctions. *Nature* 267:625–627.

SCHWARZMANN, G., H. WIEGANDT, B. ROSE, Z. ZIMMERMAN, D. BEN-HAIM, and W. R. LOEWENSTEIN. 1981. Diameter of the cell-to-cell junctional membrane channels as probed with neutral molecules. *Science* 213:551–553.

SPRAY, D. C., J. C. SAEZ, D. BROSIUS, M. V. L. BENNETT, and E. L. HERTZBERG. 1986. Isolated liver gap junctions: gating of transjunctional currents is similar to that in intact pairs of rat hepatocytes. *Proc. Nat'l Acad. Sci. USA* 83:5494–5497.

YOUNG, J. D-E., Z. A. COHN, and N. B. GILULA. 1987. Functional assembly of gap junction conductance in lipid bilayers: demonstration that the major 27-kd protein forms the junctional channel. *Cell* 48:733–743.

CHAPTER 14

Transport across Cell Membranes

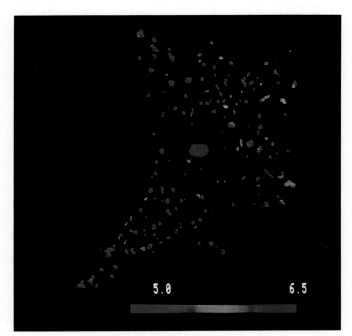

pH of endosomes and lysosomes in a living CHO cell measured by fluorescence microscopy

he plasma membrane is a selectively permeable barrier between the cell and the extracellular environment. This permeability ensures that essential molecules such as glucose, amino acids, and lipids readily enter the cell, metabolic intermediates remain in the cell, and waste compounds leave the cell. In short, the selective permeability of the plasma membrane allows the cell to maintain a constant internal environment. Similarly, organelles within the cell often have a different internal environment from that of the surrounding cytosol, and organelle membranes maintain this difference. For example, within the animal cell lysosome or the plant vacuole—the organelles involved in digestive and degradative processes—the concentration of protons (H^+) is 100–1000 times that of the cytosol; this gradient is main-

tained by proteins in the organelle membrane.

Here, we discuss two quite different types of transport across cell membranes. In the first part of this chapter, we examine how specific transport proteins enable ions and small molecules, such as glucose and amino acids, to cross phospholipid bilayer membranes. In contrast, in the second part, we see how small regions of the plasma membrane surround and then internalize protein macromolecules and larger particles to form intracellular vesicles.

Many ions and small molecules cross membranes down their chemical or electrochemical concentration gradients by a process known as passive transport, in which no metabolic energy is expended. In simple diffusion, a type of passive transport, a molecule crosses a membrane unaided by a transport protein; gases, such as oxygen and CO_2, and small, relatively hydrophobic molecules, such as ethanol, can cross phospholipid bilayers in this manner (Figure 14-1).

An artificial membrane composed of pure phospholipid or of phospholipid and cholesterol is essentially impermeable to most water-soluble molecules, such as glucose, nucleosides, and amino acids, and to hydrogen, sodium, calcium, and potassium ions. In facilitated diffusion, the more common type of passive transport, specific proteins called permeases help to transport such ions and molecules across the membrane. Because different cell types require different mixtures of low molecular weight (MW) compounds, the plasma membrane of each cell type contains a specific set of permeases that allow only certain molecules to cross, as does the membrane surrounding each type of subcellular organelle.

In active transport, metabolic energy is used to move ions or molecules against their concentration gradients in one of two ways. In the first, a membrane protein utilizes energy from ATP hydrolysis; such proteins maintain the low sodium ion concentration inside virtually all animal cells relative to that in the medium or maintain the low pH in an animal cell lysosome or plant vacuole. In the second, transport of one type of ion or molecule against its concentration gradient is coupled to movement of a different ion or molecule down its concentration gradient by the processes known as antiport and symport.

In the first part of this chapter, we present examples of each of these types of transport proteins and discuss what is known of their structure and mechanisms of action. Importantly, we also indicate how different combinations of transport proteins in different subcellular membranes enable cells to carry out essential physiological processes. Examples include the maintenance of cytosolic pH, the transport of glucose across the absorptive intestinal epithelium, the accumulation of sucrose and salts in certain plant cell vacuoles, and the directed flow of water in both plants and animals. We begin with the mechanistically simplest type of small-molecule transport: passive diffusion. ▲

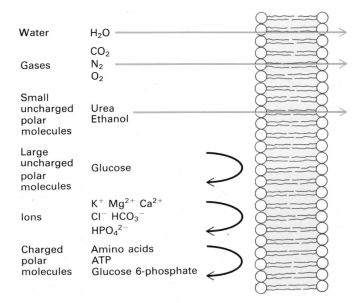

▲ **Figure 14-1** A pure artificial phospholipid bilayer is permeable to water, small hydrophobic molecules, and small uncharged polar molecules and impermeable to ions and large uncharged polar molecules. When a small phospholipid bilayer separates two aqueous compartments (see Figure 13-10), membrane permeability can be easily determined by adding a small amount of radioactive material to one compartment and measuring its rate of appearance in the other compartment.

Passive Transport across the Cell Membrane

Some Small Molecules Cross the Membrane by Simple Diffusion

In simple diffusion, a small molecule in aqueous solution dissolves into the phospholipid bilayer, crosses it, and then dissolves into the aqueous solution on the opposite side. There is little specificity to the process, and the relative diffusion rate of the molecule across the bilayer is proportional to the concentration gradient across it (Figure 14-2).

The first step in simple diffusion is the movement of the substance from the aqueous solution into the hydrophobic interior of the phospholipid bilayer. The transport rate of the molecule is proportional to its hydrophobicity, which is measured by the partition coefficient, K, the equilibrium constant for partition of the molecule between oil and water. If C^{aq} and C^m are the concentrations of the substance in aqueous solution and just inside the hydrophobic core of the bilayer, respectively, then

$$K = \frac{C^m}{C^{aq}} \qquad (14\text{-}1)$$

The partition coefficient is a measure of the relative affinity of the molecule for lipid versus water. As examples, for urea

$$
\begin{array}{c}
\text{O} \\
\parallel \\
\text{NH}_2\text{—C—NH}_2
\end{array}
$$

$K = 0.0002$, while for diethylurea (with two ethyl groups)

$$
\begin{array}{c}
\text{O} \\
\parallel \\
\text{CH}_3\text{—CH}_2\text{—NH—C—NH—CH}_2\text{—CH}_3
\end{array}
$$

$K = 0.01$. Diethylurea will diffuse through phospholipid bilayer membranes about 50 times $(0.01/0.0002)$ faster than urea.

The rate-limiting step in diffusion is movement of the substance across the interior of the bilayer. The diffusion rate of a substance across the hydrophobic core of a phospholipid membrane is 100–1000 times lower than the diffusion rate of the same molecule in water because the typical cell membrane is 100–1000 times more viscous than water. Finally, the substance moves from the bilayer into the aqueous medium (see Figure 14-2).

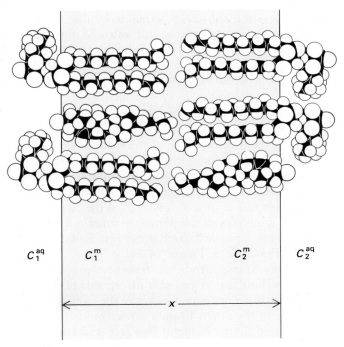

▲ **Figure 14-2** A simple model for the diffusion of small hydrophobic or hydrophilic molecules directly across the phospholipid bilayer of a biological membrane. The upper half of the figure is a space-filling model of the lipid bilayer, showing phospholipid and cholesterol molecules. The lower half represents the hydrocarbon barrier (of thickness x) to diffusion. C_1^{aq} and C_2^{aq} are the concentrations of two solutions on sides 1 and 2 of the membrane; C_1^m and C_2^m are the corresponding membrane concentrations just within the hydrocarbon barrier.

Now let's consider the simple passive diffusion of small molecules through a membrane more quantitatively. Suppose a membrane of surface area A and thickness x separates two solutions of concentrations C_1^{aq} and C_2^{aq}, where $C_1^{aq} > C_2^{aq}$ (see Figure 14-2). In this case, the diffusion rate is given by a modification of *Fick's law*, which states that the diffusion rate across the membrane dn/dt (in mol/s) is directly proportional to the difference in solution concentrations $C_1^{aq} - C_2^{aq}$, to the area A, and to the *permeability coefficient P*, or

$$
\frac{dn}{dt} = PA\,(C_1^{aq} - C_2^{aq}) \tag{14-2}
$$

P is proportional to the partition coefficient K and to the diffusion coefficient within the membrane D and is inversely proportional to membrane thickness x. To see this important point, we can write the equation for flow of material *within* the membrane (which must equal flux of material across the membrane) as

$$
\frac{dn}{dt} = \frac{D}{x}\,A\,(C_1^m - C_2^m) \tag{14-3}
$$

where x is the membrane thickness and C_1^m and C_2^m are the concentrations just inside the hydrophobic region of the membrane. Since K equals C_1^m/C_1^{aq} and C_2^m/C_2^{aq}, equation 14-3 becomes

$$
\frac{dn}{dt} = \frac{KD}{x}\,A\,(C_1^{aq} - C_1^{aq}) \tag{14-4}
$$

Comparing this to equation 14-2, we see that

$$
P = \frac{KD}{x} \tag{14-5}
$$

Thus the overall permeability constant P is proportional to the partition coefficient K.

Diffusion across a membrane down a concentration gradient has a large negative free energy. For example, the movement of 1 mol of substance from a 1 M to a 0.1 M solution (a 10-fold concentration gradient) releases 1359 cal of free energy G (25° C):

$$
\Delta G = -RT \ln \frac{C_2}{C_1}
$$

$$
= -[1.987\ \text{cal/(degree·mol)}](298°\ \text{K})\left(\ln \frac{1.0}{0.1}\right)
$$

$$
= -1359\ \text{cal/mol}
$$

Conversely, to transport 1 mol of a substance "uphill" against a 10-fold concentration gradient requires 1359 cal.

Fick's law does not apply to charged molecules. The diffusion of charged molecules across a membrane permeable to the ion is determined not only by the concentration gradient but also by any electric potential gradient that might exist across the membrane.

(a)

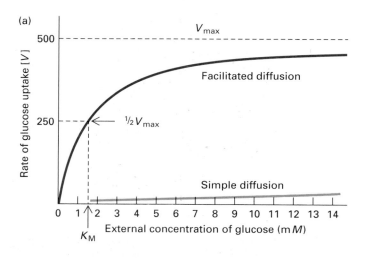

(b)

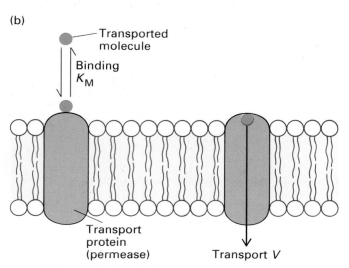

▲ **Figure 14-3** (a) Facilitated diffusion and simple diffusion: the transport rate of glucose into erythrocytes as a function of glucose in the extracellular medium. The initial concentration of glucose in the erythrocyte is very low (less than 0.5 mM), so that the concentration gradient of glucose across the membrane is effectively the same as the external concentration. As a measure of facilitated diffusion, the rate of glucose uptake (measured as micromoles per milliliter of cells per hour) in the first few seconds is plotted against glucose concentration. The light red line is the calculated curve for rate of glucose uptake if the compound were to enter solely by simple diffusion through a phospholipid bilayer. In facilitated diffusion, the transport rate is higher at all glucose concentrations and there is a maximum transport rate of V_{max}; the reaction is said to be saturable. K_M is the concentration at which the rate of glucose uptake is half-maximal. (b) The molecule binds to the transport protein (permease), as described by the constant K_M (the concentration that gives the half-maximal velocity of transport). The molecule is then transported across the membrane; the maximal rate of transport V_{max} is achieved only at high concentrations of glucose when all permeases are saturated with bound glucose.

Membrane Proteins Speed the Diffusion of Specific Molecules across the Membrane

Facilitated diffusion by membrane proteins (Figure 14-3a) is very different from simple diffusion. First, the transport rate of the molecule across the membrane is far greater than Fick's equation predicts. Second, the process is specific; each protein permease transports only a single species of molecule or a single group of closely related molecules. Third, there is a maximum transport rate V_{max} that is achieved only when the concentration gradient across the membrane becomes very large; in the example in Figure 14-3b, the initial concentration of the transported molecule within the cell is low, so that V_{max} is achieved when the concentration of extracellular molecules becomes very large.

The kinetics of facilitated diffusion can be described by the same type of equation used to describe a simple enzymatically catalyzed chemical reaction. For simplicity, let's assume that a substance S is present initially only on the outside of the membrane, so we can write

$$S_{out} + \text{permease} \xrightleftharpoons{K_M} (S \cdot \text{permease complex}) \xrightarrow{V_{max}} S_{in}$$

where K_M is the substance-permease binding constant and V_{max} is the maximum transport rate of S into the cell. If C is the concentration of S_{out} (initially, the concentration of $S_{in} = 0$), then, by exactly the same derivation used for the Michaelis-Menton equation we can write

$$V = \frac{V_{max}}{1 + K_M/C} \tag{14-6}$$

where V is the transport rate of the species into the cell, K_M is the substance concentration at which half-maximal transport occurs across the membrane (at lower values of K_M, the species binds to the transporter more tightly and the transport rate is greater), and V_{max} is the rate of transport if all molecules of the permease contain a bound S, which occurs at high S concentrations. Equation 14-6 describes the curve in Figure 14-3a.

For many years, membrane transport proteins were pictured as shuttles that moved from one side of the membrane to the other or rotated an active site from one membrane face to the other. In such *carrier models*, the protein bonds to the molecule at one face and releases the molecule at the other face. Carrier molecules do explain the properties of certain small peptide antibiotics that greatly facilitate the diffusion of certain ions. For example, the antibiotic valinomycin increases potassium ion transport across biological membranes by forming a sphere around the potassium; the hydrophobic amino acid side chains of the antibiotic make up the outer surface, and six or eight oxygen atoms on the inside coordinately bind to the ion (Figure 14-4). The hydrophobic exterior makes the ion-carrier complex soluble in the lipid interior of the membranes and facilitates its diffusion across the interior. However, the bilayer structure of all cellular membranes with distinct hydrophobic and hydrophilic planes makes the shuttle movement of larger proteins energetically expensive, and carrier models are not likely to apply to protein-mediated transport in biological membranes. Investigators now believe that membrane transport proteins form channels that permit certain ions or molecules to pass through the membrane.

Transport proteins can be extracted, purified, and reincorporated into artificial membranes or into vesicles made from natural membranes. In the most elegant studies, liposome vesicles are formed from pure lipids (see Figure 13-10) and purified transport proteins are incorporated into them. The functional properties of the individual components in liposomes can be examined without ambiguity (Figure 14-5), and the study of transport proteins is rapidly developing the precision that already characterizes the study of soluble enzymes. The human erythrocyte contains one of the best-characterized examples: a transport protein that catalyzes the facilitated diffusion of glucose.

Facilitated Diffusion Transports Glucose into Erythrocytes

The entry rate of glucose into erythrocytes is described by a plot of the initial entry velocity for different external concentrations of glucose (see Figure 14-3). This experiment makes several important points. First, the entry rate is far greater than is indicated by a free-diffusion model based on the simple solubility of glucose into phospholipids and on Fick's equation. Second, the entry rate of glucose into the cells does not increase linearly as the concentration gradient across the erythrocyte membrane is increased but reaches a maximum at high external glucose, which defines the V_{max} for glucose transport. The K_M for glucose transport is 1.5 millimolar (mM), at which one-half of the permease molecules are expected to contain a bound glucose (see Figure 14-3). Blood glucose is normally 5 mM or 0.9 g/l; at this concentration, the erythrocyte glucose transporter is functioning at 77 percent of the maximal rate V_{max}, as can be calculated from equation 14-6.

▶ **Figure 14-5** Liposomes containing a single type of transport protein can be used to investigate properties of the transport process. Here, all integral proteins of the erythrocyte membrane are solubilized by a nonionic detergent, such as octylglucoside. The glucose transport protein (permease) can be purified by chromatography on a column containing a specific monoclonal antibody and then incorporated into phospholipid liposomes.

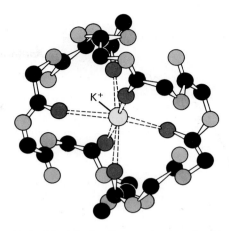

▲ **Figure 14-4** Model of valinomycin complexed with one K^+ ion. The peptide C=O bonds in the interior bind to the K^+ ion at the center. The periphery of the complex is composed of the hydrophobic side chains of the amino acids. *After L. Stryer, 1988,* Biochemistry, *3d ed., W. H. Freeman and Company, p. 966.*

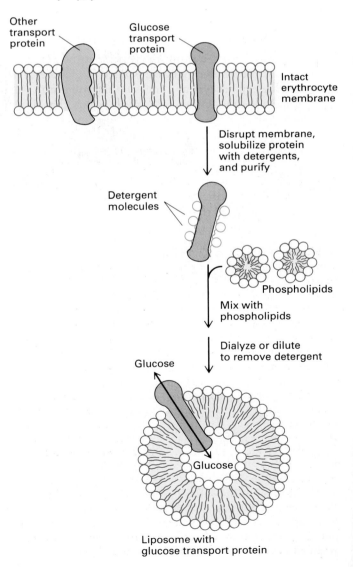

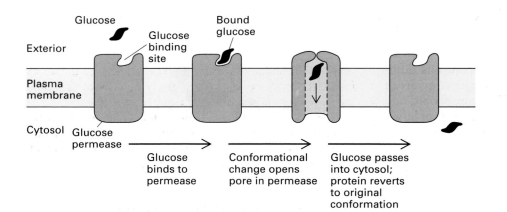

Exterior

Glucose — Glucose binding site

Bound glucose

Plasma membrane

Cytosol

Glucose permease

Glucose binds to permease

Conformational change opens pore in permease

Glucose passes into cytosol; protein reverts to original conformation

◄ **Figure 14-6** Hypothetical scheme for the operation of the glucose permease. The binding of glucose triggers a conformational change in the permease. A pore is formed in the protein to allow the bound glucose—and only this molecule—to pass through the plasma membrane.

Table 14-1 Transport of sugars into the erythrocyte by facilitated diffusion

Sugar	K_M (mM)*	Sugar	K_M (mM)*
D-glucose	1.5	D-mannose	20
L-glucose	>3000	D-galactose	30

*Concentration in millimolar required for half-maximal rate of transport.

SOURCE: P. G. Lefebvre, 1961, *Pharmacol. Rev.* 13:39.

The specificity of the transport process is indicated by the fact that the nonbiological L-isomer of glucose does not enter the erythrocyte at a measurable rate (Table 14-1): $K_M > 3000$ mM. Other six-carbon sugars that differ from D-glucose by the configuration at a single C atom, such as D-mannose or D-galactose (see Figure 2-51), also are transported, but higher concentrations are needed to half-saturate the transport reaction, implying a lower affinity (higher K_M) of the permease for these substrates. To accomplish this specificity toward hexoses, a specific protein must be involved that, like an enzyme, recognizes the structural features of its substrate. Such a protein is termed a *glucose transporter,* a *glucose permease,* or, more properly, a D-*hexose permease.*

After glucose is transported into the erythrocyte, it is rapidly phosphorylated, forming glucose 6-phosphate. This reaction—the first step in the glycolytic conversion of glucose to pyruvate—is catalyzed by hexokinase and requires ATP (see Figure 15-2). Once phosphorylated, the glucose can no longer leave the cell and its concentration in the cell is lowered. The glucose concentration gradient across the membrane therefore increases, allowing the facilitated diffusion system to continue to import glucose.

The erythrocyte glucose transport protein has been purified, cloned, and sequenced. This integral transmembrane protein (45,000 MW) accounts for 2 percent of the erythrocyte membrane protein. When inserted into artificial liposomes, the pure protein dramatically increases their permeability to D-glucose (see Figure 14-5). All properties of glucose permeation into erythrocytes are retained in the artificial system. In particular, D-glucose, D-mannose, and D-galactose are transported; L-glucose is not.

Amino acid sequence and biophysical studies on the glucose transporter indicate that it contains 12 α helices that span the phospholipid bilayer. Although these transmembrane α helices contain predominantly hydrophobic amino acid side chains, several helices bear amino acid residues (examples are serine, threonine, asparagine, and glutamine) whose side chains can form hydrogen bonds with the OH groups on glucose. It is thought that these residues line the channel in the interior of the protein through which the sugar moves.

The mechanism by which this permease facilitates the transmembrane movement of glucose is not known, but it probably is not a permanent narrow pore through which only glucose and closely related sugars can move. If it were, any molecule similar to but smaller than glucose (say, glycerol) would also be able to use such a "pore." More likely, the binding of glucose to a site on the exterior (exoplasmic) surface of the permease induces a conformational change in the polypeptide, generating a passageway that accommodates only the protein-bound sugar (Figure 14-6). Indeed, conformational changes in the purified transporter can be detected when glucose is added.

Ion Channels, Intracellular Ion Environment, and Membrane Electric Potential

An important function of the plasma membrane is to maintain a specific ion composition of the cytosol (the liquid component of the cytoplasm) that usually differs greatly from the surrounding fluid. In virtually all cells—including microorganisms, plants, and animals—the cytosolic pH must be kept near 7 and the cytosolic concentration of potassium ion (K^+) must be much higher than that of sodium ion (Na^+). In particular, in both invertebrates and vertebrates, the concentration of Na^+ ion is 20–40 times higher in the blood than in the cells (Table 14-2), and the concentration of calcium ions (Ca^{2+}) free in the cytosol is generally less than 1 micromolar (μM), 1000 times less than in the blood. Plant cells and many microorganisms maintain high cytosolic ion concentra-

Table 14-2 Typical ion concentrations in invertebrates and vertebrates

	Cell (mM)	Blood (mM)
SQUID AXON*		
K^+	400	20
Na^+	50	440
Cl^-	40–150	560
Ca^{2+}	0.0003	10
$X^{-\dagger}$	300–400	
MAMMALIAN CELL		
K^+	139	4
Na^+	12	145
Cl^-	4	116
HCO_3^-	12	29
X^-	138	9
Mg^{2+}	0.8	1.5
Ca^{2+}	<0.0005	1.8

*The large nerve axon of the squid is chosen as an example of an invertebrate cell because it has been used widely in studies of the mechanism of conduction of electric impulses.
†X^- represents proteins, which have a net negative charge at the neutral pH of blood and cells.

tions even if the cells are cultured in very dilute salt solutions. The generation and maintenance of such ion gradients require a great deal of energy.

Another important property of the plasma membrane is that it is selectively permeable to different cations and anions, including the principal cellular ions Na^+, K^+, and Cl^-. The membrane contains specific types of transport proteins, called *ion channel proteins,* that allow certain ions to move through it at different rates down their concentration gradients. Selective permeability and ion concentration gradients together create a difference in electric potential between the inside and the outside of the cell. The magnitude of this potential— ~ 70 millivolts (mV) with the inside of the cell negative with respect to the outside—does not seem like much until we realize that the plasma membrane is only about 3.5 nm thick. Thus the voltage gradient across the plasma membrane is 0.07 V per 3.5×10^{-7} cm, or 200,000 volts per centimeter!

The ion gradients and electric potential across the plasma membrane drive many biological processes. Variations in the permeability of Na^+ and K^+ ions are essential to the conduction of an electric impulse down the axon of a nerve cell. In many animal cells, the concentration gradient of Na^+ ions and the membrane electric potential power the transport of other molecules into the cells against their concentration gradient; amino acids frequently enter cells in this manner. In most cells, a rise in the concentration of Ca^{2+} ions in the cytosol is an important regulatory signal. In muscle cells, for instance, it initiates contraction; in the exocrine cells of the pancreas, it triggers secretion of digestive enzymes.

First, we will turn to the question of how ion gradients and specific ion transport proteins generate a membrane electric potential. Then we will examine the proteins that generate ion concentration gradients.

Simple Models Explain the Electric Potential across the Cell Membrane

The situation outlined in Figure 14-7 is similar to the distribution of Na^+ and Cl^- ions between an animal cell and its aqueous environment. A membrane separates a 150-mM solution of NaCl on the right side (representing the "outside" of the cell) from a 15-mM solution of NaCl on the left side (the "inside"). A potentiometer (voltmeter) is connected to the solution on each side to measure any difference in electric potential across the membrane. If the membrane is impermeable to both Na^+ and Cl^- ions, no ions will flow across it (there will be no electric potential difference across it; see Figure 14-7a).

Now suppose the membrane is permeable only to Na^+ ions: it contains channel proteins that accommodate Na^+ ions but exclude Cl^- ions. Na^+ ions then tend to move down their concentration gradient from the right side to the left, leaving an excess of negative Cl^- ions compared with Na^+ ions on the right side and generating an excess of positive Na^+ ions compared with Cl^- ions on the left side. There is now a separation of charge across the membrane, which a potentiometer can measure as an electric potential, or voltage. The right side is negative with respect to the left (see Figure 14-7b). As more and more Na^+ ions move across the membrane, the magnitude of this charge difference increases. However, continued movement of the Na^+ ions eventually is inhibited by the excess of positive charges accumulated on the left side of the membrane and by the attraction of Na^+ ions to the excess negative charge built up on the right side. The system soon reaches an equilibrium point at which the two opposing factors that determine the movement of Na^+ ions—the membrane electric potential and the ion concentration gradient—balance each other out. At equilibrium, no net movement of Na^+ ions occurs across the membrane.

The magnitude of the resulting sodium equilibrium potential in volts (the electric potential across a membrane permeable only to Na^+ ions) is given by the *Nernst equation,* which is derived from basic principles of physical chemistry:

$$E_{Na} = \frac{RT}{Z\mathcal{F}} \ln \frac{Na_l}{Na_r} \tag{14-7}$$

where R (the gas constant) = 1.987 cal/(degree·mol), or 8.28 joules/(degree·mol), T (the absolute tempera-

(a) Membrane impermeable to Na⁺ and Cl⁻

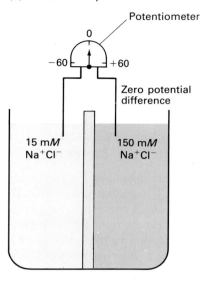

(b) Membrane permeable to Na⁺ only

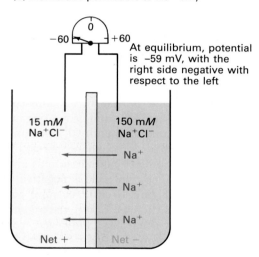

(c) Membrane permeable to Cl⁻ only

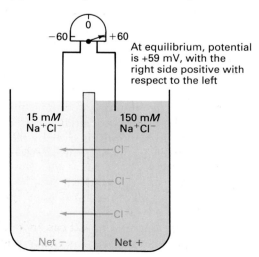

ture) $= 293°K$ at $20°C$, Z (the valency) $= +1$, $\mathscr{F}$ (the Faraday constant) $= 23,062$ cal/(mol·V), or 96,000 coulombs/(mol·V), and Na_l and Na_r are the Na^+ equilibrium concentrations on the left and right sides, respectively. The Nernst equation is similar to the equations used to calculate the voltage change associated with oxidation or reduction reactions, which also involve movement of electric charges. At $20°C$, equation 14-7 reduces to

$$E_{Na} = 0.059 \log_{10} \frac{Na_l}{Na_r} \qquad (14\text{-}8)$$

In this example, $Na_l/Na_r = 0.1$ and $E_{Na} = -0.059$ V, or -59 mV, with the right side negative with respect to the left.

If the membrane is permeable only to Cl^- ions and not to Na^+ ions, the calculation is the same:

$$E_{Cl} = \frac{RT}{Z\mathscr{F}} \ln \frac{Cl_l}{Cl_r} \qquad (14\text{-}9)$$

except that $Z = -1$. The *magnitude* of the membrane electric potential is the same (59 mV), except that the right side is now positive with respect to the left (see Figure 14-7c). This is precisely the opposite polarity to that obtained with selective Na^+ permeability.

If the membrane is permeable to Na^+ and Cl^- ions to the same degree, then Na^+ and Cl^- ions can move together down their concentration gradients from the right side to the left. In this case, no membrane electric potential is expected and none is observed. An intermediate situation between these two extremes can also occur. When the membrane is permeable to both Na^+ and Cl^- ions but more permeable to Na^+ ions, the right side initially has a negative potential relative to the left, but the magnitude of the potential is somewhat less than $E_{Na} = -59$ mV. Eventually, due to the diffusion of Na^+ and Cl^-

◄ **Figure 14-7** A voltage potential is created by the selective permeability of a membrane to different ions. (a) An impermeable membrane separates a 150-mM NaCl solution from a 15-mM solution. No ions move across the membrane, and no difference in electric potential is registered on the potentiometer connecting the two solutions. (b) The membrane is selectively permeable only to Na^+ ions, which diffuse from right to left down the concentration gradient. The Cl^- anion cannot cross the membrane, so a net positive charge builds up on the left side and a negative charge builds up on the right side. At equilibrium, the membrane potential caused by the charge separation becomes equal to the Nernst potential E_{Na} registered on the potentiometer, and the movement of Na^+ ions in the two directions becomes equal. (c) The membrane is selectively permeable only to Cl^- ions, which diffuse from right to left down the concentration gradient. The Na^+ ion cannot cross the membrane, creating a net negative charge on the left side and a net positive charge on the right side. At equilibrium, the membrane electric potential is equal to E_{Cl}.

ions, there will be an equal concentration of ions on both sides of the membrane and no membrane electric potential.

Thus two forces govern the movement of such ions as K^+, Cl^-, and Na^+ across selectively permeable membranes: the membrane electric potential, and the ion concentration gradient. These forces may act in the same direction or in opposite directions. The free-energy change ΔG required to transport 1 mol of Na^+ ions from the outside (exterior) to the inside (cytosol) of a typical mammalian cell is about -3 kcal/mol (Figure 14-8). Since $\Delta G < 0$, this reaction is thermodynamically favored. About one-half of this ΔG value is contributed by the membrane electric potential and one-half is contributed by the Na^+ ion concentration gradient. It is important to understand these forces in some detail, since the inward movement of Na^+ ions is used to power the uphill movement of several ions and small molecules into animal cells.

The free-energy change generated from the Na^+ ion concentration gradient is

$$\Delta G_c = RT \ln \frac{Na_{in}}{Na_{out}} \tag{14-10}$$

At the Na_{in} and Na_{out} values in Figure 14-8, which are typical for many cells, $\Delta G_c = -1.45$ kcal/mol, the change in free energy that would be required to transport 1 mol of Na^+ ions from outside to inside the cell if there were no membrane electric potential. The free-energy change generated from the membrane electric potential is

$$\Delta G_m = \mathscr{F} \cdot E \tag{14-11}$$

where $\mathscr{F}$ = the Faraday constant and E = the membrane electric potential. If $E = -70$ mV, then $\Delta G_m = -1.6$ kcal mol, the change in free energy that would be required to transport 1 mol of Na^+ ions from outside to inside the cell if there were no Na^+ ion concentration gradient. Given both forces acting on Na^+ ions, the total ΔG will be the sum of the two partial values:

$$\Delta G = \Delta G_c + \Delta G_m = (-1.45) + (-1.61) = -3.06 \text{ kcal/mol}$$

In this typical example, the Na^+ ion concentration gradient and the membrane electric potential contribute almost equally to total ΔG for Na^+ ion transport.

Active Ion Transport and ATP Hydrolysis

When the aerobic production of adenosine triphosphate (ATP) in a cell is inhibited experimentally by 2,4-dinitrophenol, the ion concentration inside the cell gradually approaches that of the exterior environment. This is caused by a slow leak of ions across the membrane down their electric and concentration gradients. Eventually the

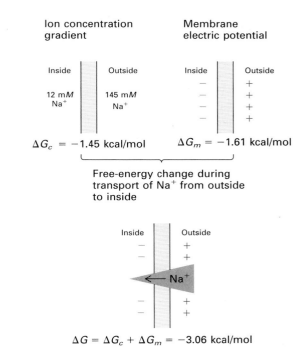

▲ **Figure 14-8** Transmembrane forces acting on Na^+ ions. As with all ions, the movement of Na^+ ions across the plasma membrane is governed by the sum of two separate forces generated by the membrane electric potential and the ion concentration gradient. In the case of Na^+ ions, these forces usually act in the same direction.

cell dies, in part because many intracellular enzymes are specialized to function in a solution of low Na^+ ions, neutral pH, and high K^+ ions. A significant fraction of available energy in every cell is required to maintain the concentration gradients of such ions as Na^+, K^+, H^+, and Ca^{2+} across the plasma and intracellular membranes. In the human erythrocyte, up to 50 percent of the energy stored in ATP molecules is used for this purpose. Thus a central issue of cellular metabolism is how permeation systems use energy.

There are three principal classes of enzymes in which ATP hydrolysis is directly coupled to ion transport against an electrochemical gradient (Table 14-3). In the P class of ATPases, a single transmembrane polypeptide is phosphorylated during the transport process. Included in this class is the Na^+-K^+ ATPase, which transports Na^+ ions out of and K^+ ions into an animal cell, and several Ca^{2+} ATPases, which transport Ca^{2+} ions out of the cell or, in muscle cells, from the cytosol to the sarcoplasmic reticulum. A third member of the P class transports protons, in some cases together with K^+. The V and F classes are always proton-transporting proteins; V ATPases maintain the low pH of plant vacuoles and of lysosomes and other vesicles in animal cells.

In each case, an enzyme system that can split ATP into adenosine diphosphate (ADP) and inorganic phosphate

Table 14-3   Examples of some ion-motive ATPases discovered to date

ATPase	Source	Membrane	Description
H^+	lower eukaryotes (yeast, fungi)	plasma	Class P ATPases consist of single predominant transmembrane polypeptides that form a covalent phosphorylated protein ("P" symbol) as part of their reaction cycle.
H^+	plants, mammals	plasma	
H^+-K^+	mammals (acid-secreting cells)	plasma	
Na^+-K^+	all higher eukaryotes	plasma	
Ca^{2+}	muscle cells	sarcoplasmic reticulum	
Ca^{2+}	higher eukaryotes	plasma	
H^+	lower eukaryotes (yeast, fungi)	vacuoles	Class V ATPases are always H^+ transporters and are associated with membrane-bound organelles other than mitochondria, chloroplasts, and the endoplasmic reticulum. They are found in the membranes of acidic vacuoles ("V" symbol) in fungi and plants. They do not form a phosphorylated intermediate and are composed of 3–5 different peptides.
H^+	plants	vacuoles	
H^+	animals	lysosomes, endosomes, secretory vesicles	
H^+	most bacteria	plasma	Class F ATPases also do not form a phosphorylated intermediate; they consist of 8–13 polypeptides and contain a membrane-bound (F_0) complex as well as an extrinsic (F_1) segment that synthesizes ATP from ADP and P_i.
H^+	all eukaryotes	inner mitochondrial	
H^+	plants	chloroplast thylakoid	

SOURCE: P. L. Pedersen and E. Carafoli, 1987, *Trends in Biochem. Sci.* **12**:146–150; M. Forgac, 1989, *Physiol. Rev.* **69**:765–796.

(P_i) is part of the ion transport system. This enzyme is called an ATPase, but it actually collects the free energy released during ATP hydrolysis and uses it to move ions up an electric or concentration gradient. However, the F-class proton ATPases found in mitochondria and chloroplasts generally run in the reverse direction: the movement of protons down their electrochemical gradients is coupled to the synthesis of ATP from ADP and P_i.

Here, we discuss several active transport systems in some detail. We then turn to another type of transport system in which the movement of one molecule (say, Na^+) down its electric and ion concentration gradients is coupled to the movement of another (say, glucose) up its concentration gradient.

Na^+-K^+ ATPase Maintains the Intracellular Concentrations of Na^+ and K^+ Ions in Animal Cells

Because it is important to cellular metabolism, the Na^+-K^+ transport system has been studied in considerable detail. It has been solubilized and purified from the membranes of several types of animal cells, including those in the mammalian kidney and the electric organ of eels (tissues very rich in this enzyme). In the membrane, the enzyme is a tetramer of subunit composition $\alpha_2\beta_2$. The

β polypeptide is a 50,000-MW transmembrane glycoprotein of unknown function; the α subunit is a 120,000-MW nonglycosylated polypeptide with a catalytic site for ATP hydrolysis (Figure 14-9a). The overall process of transport moves three Na^+ ions out of and two K^+ ions into the cell per ATP molecule split (see Figure 14-9b).

Several lines of evidence indicate that the Na^+-K^+ ATPase is responsible for the coupled movement of K^+ and Na^+ into and out of the cell. When a number of tissues are compared, there is a good correspondence between the flux of cations across the plasma membrane and the activity of the Na^+-K^+ ATPase. In addition, the drug ouabain specifically inhibits Na^+-K^+ ATPase and also prevents cells from maintaining their Na^+-K^+ balance. Ouabain binds to a specific region of the protein on the exoplasmic surface of the plasma membrane; mutant cells resistant to ouabain have an altered Na^+-K^+ ATPase. Any doubt that the Na^+-K^+ ATPase is responsible for ion movement has been dispelled by the demonstration that the enzyme, when purified from the membrane and reinserted into liposomes, propels K^+ and Na^+ transport when supplied with ATP.

The Na^+-K^+ ATPase is oriented in the membrane with part of the protein exposed at both faces. The cytoplasmic surface of the protein, which faces the cytosol, has an ATP-binding site and three high-affinity sites for binding

(a)

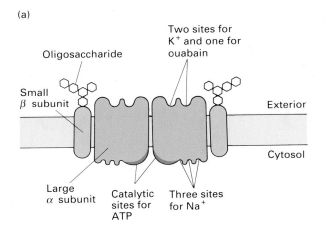

(b)

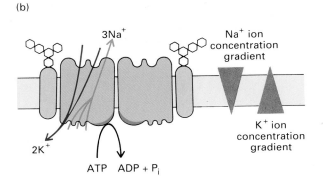

◀ **Figure 14-9** The structure and function of the Na^+-K^+ ATPase within the plasma membrane. (a) The enzyme contains two copies each of a small glycosylated β subunit (55,000 MW) and a large α subunit (120,000 MW) that performs ion transport. The α subunit has two sites for K^+ ions and a site for the inhibitor ouabain on its exoplasmic surface and three sites for Na^+ ions and one site for ATP on its cytoplasmic surface. (b) Hydrolysis of one molecule of ATP to ADP and P_i is coupled to the outward transport of three Na^+ ions and the inward transport of two K^+ ions against their concentration gradients. It is not known whether one or both α subunits in a single ATPase transport ions.

Na^+ ions (see Figure 14-9a); the binding constant K_M for Na^+ is 0.2 mM, a value well below the intracellular Na^+ ion concentration. On the outer face, the ATPase has two high-affinity binding sites for K^+ ions; K_M for K^+ is about 0.05 mM, a much lower value than the extracellular K^+ ion concentration. The activity of this and other cellular ion pumps is closely regulated (by mechanisms presently unknown), so that the ion balance of the cell is kept constant. The protein neither rotates nor shuttles back and forth within the membrane. Using the energy derived from the ATP hydrolysis of a phosphoanhydride bond, the ATPase moves these ions through itself. The protein must contain one or more ion channels; conformational changes in the protein, powered by ATP hydrolysis, allow it to pump Na^+ and K^+ ions against their electrochemical gradients.

Ca^{2+} ATPase Pumps Calcium Ions Out of the Cytosol, Maintaining a Low Concentration

The Na^+-K^+ ATPase is a prototype ion pump. Another pump important to proper cell function—the calcium pump, or Ca^{2+} ATPase—holds the Ca^{2+} concentration in cells at a low level. In some cells, such as the erythrocyte, the Ca^{2+} ATPase is located in the plasma membrane and

transports Ca^{2+} ions out of the cell. In muscle cells, the Ca^{2+} ATPase transports Ca^{2+} ions from the cytosol to the interior of the sarcoplasmic reticulum, an internal organelle that concentrates and stores these ions; release of Ca^{2+} ions from the sarcoplasmic reticulum into the muscle cytosol causes contraction, and rapid removal of Ca^{2+} ions by the Ca^{2+} ATPase induces relaxation.

The Ca^{2+} ATPase is easily purified and characterized because it constitutes more than 80 percent of the integral membrane protein of the sarcoplasmic reticulum. The calcium pump consists of a single 100,000-MW polypeptide that transports two Ca^{2+} ions per ATP hydrolyzed and requires Mg^{2+} to function, probably to complex the ATP. The very high affinity of the cytosolic surface of this enzyme for the Ca^{2+} ion ($K_M = 10^{-7}$ M) allows it to transport Ca^{2+} very efficiently from the cytosol ($Ca^{2+} = 10^{-7}$–10^{-6} M) to the sarcoplasmic reticulum, where total Ca^{2+} ion concentration can be as high as 10^{-2} M. The activity of Ca^{2+} ATPase is also regulated so that if the cytosolic Ca^{2+} ion concentration becomes too high, the rate of calcium pumping increases until the concentration is reduced to less than 1 μM.

Two soluble proteins in the interior of sarcoplasmic reticulum vesicles bind Ca^{2+} ions. One of these, *calsequestrin* (44,000 MW), is extremely acidic: 37 percent of its residues are aspartic and glutamic acids. Each molecule of calsequestrin binds 43 Ca^{2+} ions, amply justifying its name. The second protein, known as the *high-affinity Ca^{2+}-binding protein*, has a somewhat lower capacity for Ca^{2+} but binds it at a higher affinity ($K_M = 3$–4 μM). Such proteins serve as a reservoir for intracellular Ca^{2+} and also reduce the concentration of free Ca^{2+} ions in the sarcoplasmic reticulum vesicles, thereby decreasing the energy needed to pump Ca^{2+} ions into them from the cytosol.

Animal, yeast, and probably plant cell membranes also contain Ca^{2+} ATPases that export Ca^{2+} ions from the cells; in many cases, the activity of these enzymes is stimulated by a rise in cytosolic Ca^{2+} triggered, for instance, by hormone stimulation. The Ca^{2+}-binding regulatory protein *calmodulin* is an essential subunit of the erythrocyte and other plasma-membrane Ca^{2+} ATPases. A rise in cytosolic Ca^{2+} induces the binding of Ca^{2+} ions to cal-

(a)

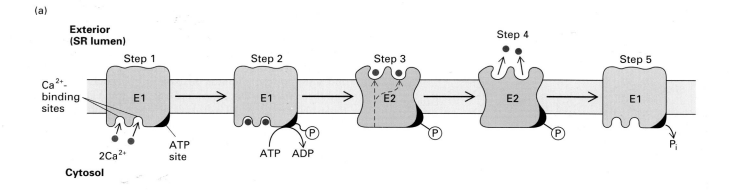

(b)

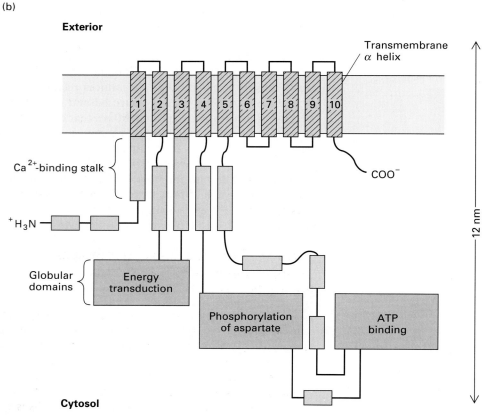

▲ **Figure 14-10** (a) The mechanism of action of the sarcoplasmic reticulum Ca^{2+} ATPase. E1 and E2 are alternate conformational forms of the ATPase in which the Ca^{2+}-binding sites are on the cytoplasmic and exoplasmic faces, respectively. The Ca^{2+}-pumping reaction follows an ordered kinetic mechanism; this cycle is essential for coupling between Ca^{2+} transport and ATP hydrolysis. In step 1, two Ca^{2+} ions bind to the sites on the cytoplasmic surface, activating the ATP binding site, so that ATP binds and transfers its terminal phosphate to an aspartyl residue (step 2) to form a "high-energy" acyl phosphate bond, denoted E1~P. This causes a conformational change in the protein (step 3), during which the Ca^{2+} ions are transported through the protein and remain bound to the Ca^{2+} sites that are simultaneously formed on the exoplasmic surface. This generates the alternate conformation, E2, and simultaneously the free energy of

hydrolysis of the aspartyl-phosphate bond is reduced, denoted E2—P. In step 4, the Ca^{2+} ions are released to the exterior, followed (step 5) by the hydrolysis of the aspartyl-phosphate bond. This regenerates the E1 form of the protein. (b) Model for the structure of the Ca^{2+} ATPase in the sarcoplasmic reticulum, deduced from the amino acid sequence and biochemical data. Ten transmembrane α helices are thought to form the Ca^{2+} channel. The bulk of the protein forms globular domains on its cytoplasmic surface. Trypsin digestion of this segment releases the three domains: one functions in ATP binding; a second bears the phosphorylated aspartate; and the third is involved in energy transduction. At least part of the Ca^{2+}-binding domain is in the stalk. [Part (a) see W. P. Jencks, 1980, *Adv. Enzymology* **51**:75–106; W. P. Jencks, 1989, *J. Biol. Chem.* **264**:18855–18858.] *Part (b) from D. H. MacLennan et al., 1985,* Nature **316**:696.

modulin, which triggers an allosteric activation of the Ca^{2+} ATPase: the export of Ca^{2+} ions from the cell accelerates, and the original low cytosolic concentration of Ca^{2+} (less than 1 μM) is restored.

Coupling between ATP Hydrolysis and Ion Pumping Requires an Ordered Kinetic Mechanism

The mechanism of action of the Ca^{2+} ATPase in the membrane of the sarcoplasmic reticulum is becoming known in some detail. It involves several steps that must occur in a defined order if coupling between ATP hydrolysis and ion pumping is to occur (Figure 14-10a). When the protein is in one conformational stage, termed E1, two Ca^{2+} ions bind with high affinity to sites on its cytoplasmic surface. ATP then binds to its site on the cytoplasmic surface. In a reaction that requires the Mg^{2+} ion to be tightly complexed to the ATP, the bound ATP is hydrolyzed to ADP while P_i (the liberated phosphate) is transferred to a specific aspartate residue in the protein, forming a high-energy acyl phosphate bond, denoted by E1 ~ P. The protein then changes its conformation to E2, propelling the two Ca^{2+} ions through it until they become weakly bound to the exoplasmic surface of the protein. The free energy of hydrolysis of the aspartyl—phosphate bond is simultaneously reduced in magnitude (E1 ~ P → E2 − P). In fact, the reduction in free energy of hydrolysis of E2 − P relative to E1 ~ P can be said to "power" the E1 → E2 change. The Ca^{2+} ions then dissociate from the exoplasmic surface of the protein, and the aspartyl—phosphate bond is hydrolyzed, causing E2 to convert to E1, with its cytosolic-facing Ca^{2+} binding sites. Much evidence supports this model, including the isolation of the ATPase with the phosphate linked to the aspartate residue and

spectroscopic studies that detect slight alterations in protein conformation during the E1 → E2 conversion.

An important characteristic of all ion pumps is their reversibility. Sarcoplasmic reticulum vesicles with a high internal Ca^{2+} concentration can be placed in a solution that contains ADP and P_i with no Ca^{2+} or ATP. Under these conditions, the Ca^{2+} ATPase transports Ca^{2+} ions outward, down their concentration gradient, while ADP and P_i combine to form ATP. Concentration gradients of Na^+ and K^+ ions also can drive the Na^+-K^+ ATPase in the reverse direction to synthesize ATP. Reversal of the normal mode of action of these proteins provides an important model of how ATP is synthesized in mitochondria from oxidation of carbohydrates and in chloroplasts from light energy. In both organelles, a transmembrane gradient of protons supplies the energy for the formation of ATP from ADP and P_i.

The mechanism of action of the Na^+-K^+ ATPase, as well as of the gastric H^+-K^+ ATPase, is similar to that of the Ca^{2+} ATPase. In particular, in all P-class ATPases, the sequence around the phosphorylated aspartate is similar. The amino acid sequence in Table 14-4 is the most conserved in these ATPases, suggesting that despite the fact that the P-class ATPases transport different ions, they have all evolved from a common precursor. All of these proteins have a similar molecular weight and, as deduced from the amino acid sequences derived from those of cDNA clones, have a similar arrangement of transmembrane α helices (see Figure 14-10b).

Lysosomal and Vacuolar Membranes Contain V-type ATP-dependent Proton Pumps

The third type of active ion transport, the ATP-dependent transport of protons across a membrane, occurs in several cellular systems. Physiological experiments suggest that an ATP-dependent proton transport system is present in the membranes of lysosomes and vacuoles to keep the interiors of these organelles acidic. The pH of the interior of lysosomes in growing animal tissue culture cells is usually ~4.5−5.0. This can be measured precisely in living cells, which phagocytose fluorescent particles and transfer them to the lysosomes. The ability of different wavelengths of ultraviolet light to excite fluorescence is highly dependent on pH, and the lysosomal pH can be calculated from the amount of fluorescence emitted. The cytosolic pH, in contrast, is ~7.0; maintenance of this proton gradient of more than 100-fold between the lysosome interior and the cytosol depends on ATP production by the cell.

Isolated lysosomes and vacuoles contain ATP-dependent proton pumps that have been purified and incorporated into liposomes. Lysosomal and vacuolar H^+ ATPases differ greatly in structure and mechanism from the P class of Ca^{2+} and Na^+-K^+ ATPases (see Table 14-3)

Table 14-4 Amino acid sequence around the phosphorylated aspartate residue in several members of the P class of ion-translocating ATPases

P-class ATPase	Amino acid sequence
	phosphorylated aspartate ↓
Rat stomach H^+-K^+ ATPase	TLG S TSVICSD KTGTLT*
	↓
Sheep kidney Na^+-K^+ ATPase (catalytic α subunit)	TLG S TSTICSD KTGTLT
	↓
Rat brain $\alpha(+)$isoform (α subunit of Na^+-K^+ ATPase)	TLG S TSTICSD KTGTLT
	↓
Cardiac Ca^{2+} ATPase	TLGCTSVICSD KTGTLT

*The single letter abbreviations are listed on page 45.

SOURCE: G. E. Shull and J. E. Lingrel, 1986, *J. Biol. Chem.* **261**:16788; G. E. Shull et al., 1986, *Biochem.* **25**:8125; D. H. MacLennan et al., 1985, *Nature* **316**:696.

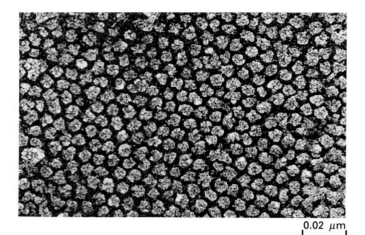

▲ **Figure 14-11** The plasma membrane of certain acid-secreting cells contains an almost crystalline array of V-class H⁺ ATPases. This electron micrograph is of a platinum replica of the cytoplasmic surface of the apical plasma membrane of a toad bladder epithelial cell. Each stud is a single V-class H⁺ ATPase (~600,000 MW) composed of several polypeptide subunits surrounding a central channel. *From D. Brown, S. Gluck, and J. Hartwig, 1987, J. Cell Biol. 105:1637.*

in that they, like other V-class ATPases, are composed of several polypeptide chains; importantly, proton transport does not involve a phosphorylated amino acid. To date, no enzyme of the V class has been purified to homogeneity, dissociated into individual polypeptides, and reconstituted, so we do not know the number of polypeptides required for proton transport or the transport mechanism involved. However, whether they are isolated from plants or animals, members of the V class appear to contain related polypeptides of 70,000, 60,000, and 17,000 MW.

Similar V-class ATPases are found on the plasma membrane of certain acid-secreting cells. The apical plasma membrane of the epithelial cells lining the toad bladder,

for instance, contains abundant H⁺ ATPases (Figure 14-11) that acidify urine. Plant vacuoles contain not only a V-class H⁺ ATPase but also a proton pump, believed to be unique to plants, that utilizes the energy released by the hydrolysis of inorganic pyrophosphate (PP$_i$) to pump protons into the vacuole (see Figure 14-19).

The Multidrug Resistance Gene May Encode an ATP-driven Drug Transporter

A possible fourth class of ATP-driven pumps was elucidated only in 1986 through a series of rather unexpected results. Oncologists noted that tumor cells often become simultaneously resistant to several chemotherapeutic drugs with unrelated chemical structures; similarly, cell biologists observed that cultured cells selected for resistance to one toxic substance (for example, colchicine, a microtubule inhibitor) frequently became resistant to several other drugs, including the anticancer drugs vinblastine and adriamycin. This resistance is due to a multidrug transport protein known as P170—a 170,000 MW plasma-membrane glycoprotein (Figure 14-12)—which uses the energy derived from ATP hydrolysis to export a large variety of drugs from the cytosol to the extracellular medium. The P170 gene is frequently amplified in multidrug-resistant cells, resulting in a large overproduction of P170. Most drugs transported by P170 are hydrophobic and diffuse across the plasma membrane into the cell; by actively exporting drugs from the cytosol, P170 increases the concentration in the culture medium of a drug required to kill the cell. Membrane vesicles containing P170 can carry out ATP-dependent transport of a number of drugs against their concentration gradients; as is the case with other transporters, transported drugs compete with one another for transport by P170. Since P170 has not yet been purified and incorporated into liposomes, there is no direct proof that it is an ATP-driven drug transporter, although this is very likely.

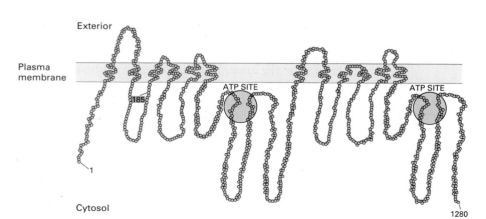

◀ **Figure 14-12** Model of the human multidrug transport protein. The two halves of this 1280 amino acid molecule have similar amino acid sequences, and there are two ATP-binding sites facing the cytosol. Since a point mutation at amino acid 185 changes the drug specificity of the transporter, this residue is believed to form part of the drug-binding site. [See G. Ferro-Luzzi Ames, 1986, *Cell* **47**:323–324; M. M. Gottesman and I. Pastan, 1988, *J. Biol. Chem.* **263**:12163–12166.]

What might be the normal function of P170? It is expressed in abundance in the liver, intestines, and kidney—sites from which natural toxic products are removed from the body. The P170 in these cells may move natural and metabolic toxins into the bile, intestinal lumen, or forming urine; P170 also may be involved in the excretion of a variety of drugs.

Cotransport: Symport and Antiport

In addition to the drugs just noted, Na^+, K^+, Ca^{2+}, and H^+ are the only known substances whose transport into or out of a cell is directly coupled to ATP hydrolysis. Yet cells often must import other molecules, such as glucose and amino acids, against a concentration gradient. To do this, a cell utilizes the energy stored in the transmembrane gradient of Na^+ or H^+ ions. For instance, the movement of a Na^+ ion (the "cotransported" ion) into the cell across the plasma membrane, driven both by its concentration gradient and by the membrane electric potential (see Figure 14-8), can be coupled obligatorily to movement of the "transported" molecule (say, glucose) up its concentration gradient. The transported molecule and cotransported ion can move in the same direction, in which case the process is called *symport;* when they move in opposite directions, the process is called *antiport* (Figure 14-13).

Amino Acid and Glucose Transport into Many Animal Cells Is Directly Linked to Na^+ Entry (Symport)

Importing glucose and amino acids from the lumen of the small intestine against their concentration gradients is a function of the microvilli of the intestinal epithelial cells (see Figure 13-42). Glucose and amino acids are transported from the intestinal lumen to these epithelial cells and then to the blood (Figure 14-14). The two stages of this transcellular transport process involve two sets of permeases: one in the microvilli on the apical surface of the plasma membrane; the other on the basolateral surface. A necessary condition for the transcellular transport of glucose and amino acids is that the epithelial cell be polarized, with different sets of transport proteins localized in the basolateral and apical surfaces.

Gluocse is transported from the intestinal lumen across the apical surface of the epithelial cells by a specific glucose-Na^+ symport protein located in the microvillar membranes. The transmembrane movement of glucose is coupled obligatorily to the transport of one Na^+ ion (see Figure 14-14):

$$Na^+_{out} + glucose_{out} \rightleftharpoons Na^+_{in} + glucose_{in}$$

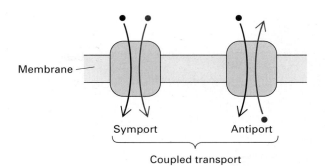

▲ **Figure 14-13** Schematic diagram of symport and antiport membrane transport proteins. The transported molecule is shown in black; the cotransported ion is shown in red.

Entry of Na^+ from the intestinal lumen into the cell is driven by two forces: by the Na^+ ion concentration gradient (the Na^+ ion concentration is lower inside the cell than in the lumen) and by the inside negative membrane electric potential (see Figure 14-8). Quantitatively, the free-energy change for the overall operation of the Na glucose-Na^+ symport can be written

$$\Delta G = RT \ln \frac{(Na_{in})(glucose_{in})}{(Na_{out})(glucose_{out})} + \mathscr{F} \cdot E$$

where E is the electric potential across the apical membrane. Although neither the intracellular ion concentrations nor the membrane electric potential is known with certainty, we can estimate (in accord with the calculations in Figure 14-8) that each contributes about 1.5 kcal/mol Na^+ ions and that the ΔG for the transport of 1 mol of Na^+ ions inward is about −3 kcal/mol. Using the partial equation for glucose transport

$$\Delta G = RT \ln \frac{glucose_{in}}{glucose_{out}}$$

we can calculate that a ΔG of −3 kcal/mol Na^+ ion influx can generate an equilibrium concentration of glucose inside the cell that is 176 times greater than the exterior (luminal) concentration. Thus glucose from the intestinal lumen is concentrated in the cell.

The Na^+-K^+ ATPase is found exclusively on the basolateral surface of the plasma membrane (see Figure 14-14). In the steady state, all the Na^+ ions transported from the intestinal lumen to the cell during glucose or amino acid symport are pumped out across the basolateral membrane, often called the *serosal (blood-facing) membrane*. Glucose and amino acids concentrated inside the cell by symport also move outward through the basolateral membrane via facilitated diffusion transport proteins. The net result is a movement of Na^+ ions, amino acids, and glucose from the intestinal lumen across the cellular epithelium and into the blood. Tight junctions between the epithelial cells prevent these molecules from diffusing back into the intestinal lumen.

(a)
Basal
lamina

Absorptive
epithelial cells

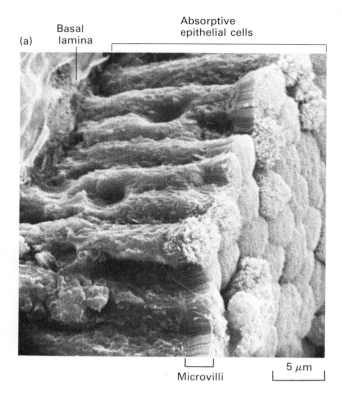

Microvilli

5 μm

(b)

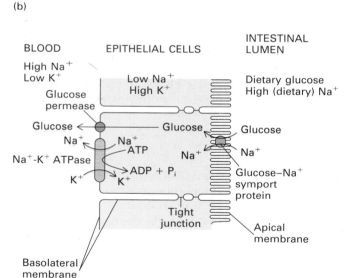

BLOOD   EPITHELIAL CELLS   INTESTINAL
                                          LUMEN

High Na⁺
Low K⁺

Low Na⁺
High K⁺

Dietary glucose
High (dietary) Na⁺

Glucose
permease

Glucose ← Glucose   Glucose

Na⁺   Na⁺
             ATP   Na⁺   Na⁺

Na⁺-K⁺ ATPase

                ADP + P_i   Glucose–Na⁺
K⁺   K⁺   symport
                                    protein

Tight
junction   Apical
                  membrane

Basolateral
membrane

(c)

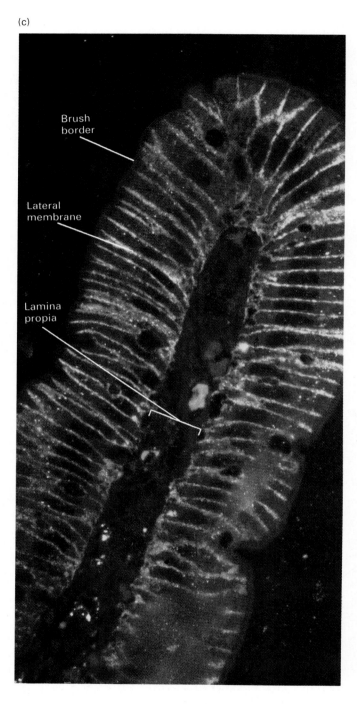

Brush
border

Lateral
membrane

Lamina
propia

▲ **Figure 14-14**   (a) Scanning electron micrograph of the absorptive epithelial cells of the small intestine. The cells reside on a layer of basal lamina and contain abundant microvilli on their apical surface. (b) Entry of glucose from the intestinal lumen into the epithelial cells is catalyzed by a glucose-Na⁺ symport protein (purple) located in the apical surface membrane. The Na⁺-K⁺ ATPase (green) in the basolateral surface membrane generates the Na⁺ gradient, which provides the energy for glucose uptake by pumping out the Na⁺ ions entering the cell by glucose-Na⁺ and amino acid-Na⁺ symports. Glucose leaves the cell via a facilitated diffusion protein (orange) in the basolateral membrane. (c) Immunomicroscopic localization of a facilitated diffusion glucose transporter to the basal and lateral sides of rat intestine epithelial cells. The tissue section was stained with Evans blue, which generates a red fluorescence. The transporter, detected with a yellow-green-fluorescing antibody, is absent from the brush border. The lamina propria consists of vascularized connective tissue. *Part (a) from R. Kessel and R. Kardon, 1979,* Tissues and Organs: A Text-Atlas of Scanning Electron Microscopy, *W. H. Freeman and Company, p. 176; part (c) courtesy of B. Thorens.*

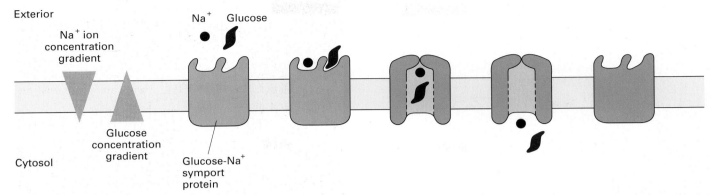

▲ **Figure 14-15** Hypothetical scheme for the operation of a glucose-Na$^+$ symport protein (blue), which has binding sites for Na$^+$ and glucose ~3.5 nm apart on its exoplasmic surface. The simultaneous binding of Na$^+$ and glucose to these sites induces a conformational change, generating a transmembrane pore or tunnel that allows both Na$^+$ and glucose to pass into the cytosol. After this passage, the protein reverts to its original conformation. [See M. Hediger, M. Coady, T. Ikeda, and E. Wright, 1987, *Nature* **330**:379–381 for details on the structure and function of this transporter.]

In summary, the transepithelial movement of glucose and amino acids is driven by the cellular hydrolysis of ATP. The coupling of ATP hydrolysis to the entry of glucose and Na$^+$ is indirect. ATP hydrolysis is used directly to generate an Na$^+$ gradient, and that gradient is used as a source of energy to drive the movement of glucose and amino acids. How a glucose-Na$^+$ symport protein functions is not known, but changes in protein conformation may be important (Figure 14-15).

Transport of Ca^{2+} Out of Cells Is Often Coupled to Na$^+$ Entry (Antiport)

In symport, the movement of one substance (usually Na$^+$) into the cell down its concentration gradient is coupled to the inward movement of a different substance, such as glucose, against a lesser concentration gradient. Antiport is a similar process (see Figure 14-13), except that the inward movement of an ion such as Na$^+$ is coupled obligatorily to the outward movement of a different molecule:

$$Na^+_{out} + X_{in} \rightleftharpoons Na^+_{in} + X_{out}$$

where X is exported from the cell against a concentration gradient.

An important antiport system, studied in some detail, reduces the Ca^{2+} ion concentration in cardiac muscle and other animal cells by the reaction

$$Ca^{2+}_{in} + 2Na^+_{out} \rightleftharpoons Ca^{2+}_{out} + 2Na^+_{in}$$

Note that the movement of two (or possibly three) Na$^+$ ions is required to power the export of one Ca^{2+} ion against a greater than 1000-fold concentration gradient

(1 μM–2 mM). As in most muscle cells, a rise in the intracellular Ca^{2+} ion concentration in cardiac muscle triggers contraction. Thus the operation of the Na$^+$-Ca^{2+} antiport reduces the frequency of heart-muscle contraction. The Na$^+$-K$^+$ ATPase in the plasma membrane of cardiac cells, as in other body cells, creates the Na$^+$ ion concentration gradient used to power Ca^{2+} ion export. Both the Na$^+$-K$^+$ ATPase and the Na$^+$-Ca^{2+} antiport appear to be distributed throughout the plasma membrane of cardiac muscle cells, unlike the localized distribution of the Na$^+$-K$^+$ ATPase in polarized epithelial cells.

Drugs such as ouabain and digoxin are of great clinical significance: they increase the force of heart-muscle contraction and are widely used in the treatment of congestive heart failure. The primary effect of these drugs is to inhibit the Na$^+$-K$^+$ ATPase, thereby raising the intracellular Na$^+$ ion concentration. The Na$^+$-Ca^{2+} antiport system functions less efficiently with a lower Na$^+$ ion concentration gradient; fewer Ca^{2+} ions are exported, the intracellular level of Ca^{2+} ions is raised, and the muscle contracts more often and more strongly.

Exchange of Cl$^-$ and HCO$_3^-$ Anions across the Erythrocyte Membrane Is Catalyzed by Band 3, an Anion-exchange Protein

Not all antiport involves cations (positive ions) such as Na$^+$ and Ca^{2+}. Band 3, the predominant integral protein of the mammalian erythrocyte, is an anion antiporter: it catalyzes the one-for-one exchange of negative ions, or

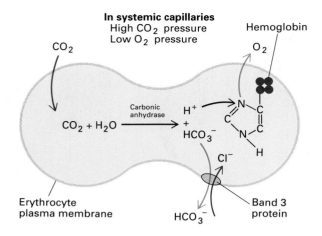

In systemic capillaries
High CO_2 pressure
Low O_2 pressure

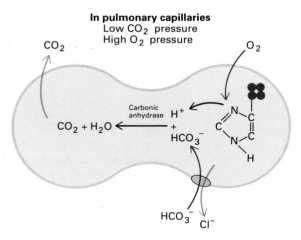

In pulmonary capillaries
Low CO_2 pressure
High O_2 pressure

▲ **Figure 14-16** Schematic drawings showing anion transport through the erythrocyte membrane in systemic and pulmonary capillaries. Band 3 protein (purple) catalyzes the exchange of Cl^- and HCO_3^- anions across the membrane, which is essential to the transport of CO_2 from the tissues to the lungs. This reaction allows an HCO_3^- anion, generated from CO_2 by carbonic anhydrase, to leave the cell in exchange for a Cl^- ion, greatly increasing the ability of the blood to transport HCO_3^- anions.

anions, such as Cl^- and HCO_3^-, across the plasma membrane. Transmembrane anion exchange is essential to an important function of the erythrocyte—the transport of waste carbon dioxide (CO_2), which is generated in peripheral tissues, to the lungs for expulsion. Because the number of band 3 proteins is large, the Cl^- permeability of the erythrocyte can be up to 100,000 times that of other cells.

Figure 14-16 illustrates the role of band 3 in CO_2 transport. Waste CO_2 released from cells into the capillary blood diffuses across the erythrocyte membrane. In its gaseous form, CO_2 dissolves poorly in aqueous solutions,

such as blood plasma, but the potent enzyme carbonic anhydrase inside the erythrocyte converts CO_2 to a bicarbonate (HCO_3^-) anion:

$$OH^- + CO_2 \rightleftharpoons HCO_3^-$$

Since $H_2O \rightleftharpoons H^+ + OH^-$, we can write, as an overall reaction

$$H_2O + CO_2 \rightleftharpoons H^+ + HCO_3^-$$

This process occurs while the hemoglobin in the erythrocyte is releasing oxygen into the blood plasma. The removal of oxygen from hemoglobin induces a change in its conformation that enables a histidine of globin side chain to bind the proton produced by carbonic anhydrase. The HCO_3^- anion formed by carbonic anhydrase is transported out of the erythrocyte in exchange for an entering Cl^- anion:

$$HCO_3^-{}_{in} + Cl^-{}_{out} \rightleftharpoons HCO_3^-{}_{out} + Cl^-{}_{in}$$

(see Figure 14-16). If anion exchange could not take place, HCO_3^- would accumulate inside the erythrocyte to toxic levels during periods such as exercise, when much CO_2 is generated; anion exchange allows about two-thirds of the HCO_3^- to be transported by blood plasma external to the cells. Also, without anion exchange, the increased HCO_3^- concentration in the erythrocyte would cause the cytosol to become alkaline. Exchanging HCO_3^- (equal to $OH^- + CO_2$) for Cl^- causes the cytosolic pH to drop to near neutrality.

The entire anion-exchange process is completed within 50 milliseconds (ms), during which time 5×10^9 HCO_3^- ions are exported from the cell. The overall direction of the process is reversed in the lungs, where HCO_3^- diffuses into the erythrocyte in exchange for a Cl^- ion. The HCO_3^- combines with H^+ to yield H_2O and CO_2; oxygen binding to hemoglobin causes the proton to be released from hemoglobin. The CO_2 diffuses out of the erythrocyte and is eventually expelled in breathing.

The behavior of band 3 has been studied extensively. It catalyzes the precise one-for-one sequential exchange of anions on opposite sides of the membrane required to preserve electroneutrality in the cell, but it does not allow anions to flow unidirectionally from one side of the membrane to the other. It also facilitates other functions in the erythrocyte membrane, such as the binding of ankyrin (see Figure 13-33). Band 3 is constructed of two domains—one embedded in the membrane, the other on the cytoplasmic surface of the protein—of about 450 amino acids each. Experiments with protease-generated fragments of band 3 have shown that the membrane-bound domain in the protein catalyzes anion transport. This region of the polypeptide is folded to span the membrane with at least 12 α helices. The precise mechanism for anion transport is not known; as with other transport proteins, however, conformational changes may be important (Figure 14-17).

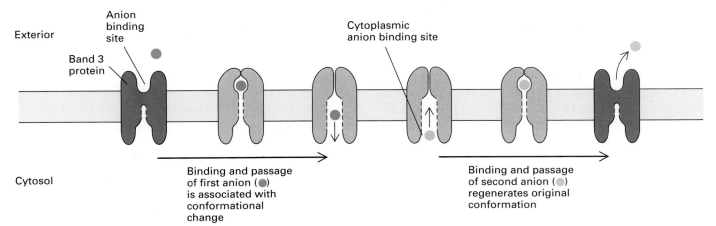

Exterior

Anion binding site

Band 3 protein

Cytoplasmic anion binding site

Cytosol

Binding and passage of first anion (●) is associated with conformational change

Binding and passage of second anion (●) regenerates original conformation

▲ **Figure 14-17** Schematic diagram of anion transport by band 3 protein (purple). The protein catalyzes a one-for-one exchange of anions, probably shuttling between two conformations. In one form (dark purple), it binds an anion on its exoplasmic surface, causing a conformational change that generates a tunnel through which the anion crosses the membrane and also creates an anion-binding site on the cytoplasmic surface (light purple). Binding of the second anion to the cytoplasmic site and its passage across the membrane in the opposite direction regenerate the original conformation with the binding site on the outer surface. Experiments supporting this model have shown that band 3 can have an anion-binding site on the cytoplasmic or exoplasmic surface, but not on both surfaces simultaneously. [See J. J. Falke and S. I. Chen, 1985, *J. Biol. Chem.* **260**:9537; H. Passow, 1986, *Rev. Physiol. Biochem. Pharmacol.* **103**:62–203.]

Antiports Regulate Cytosolic pH

The pH of the cytosol must be maintained within very narrow limits (7.2–7.4) for cells to grow and divide—a problem compounded by the fact that most metabolic products of glucose are acidic, even though glucose is not ionized. The anaerobic metabolism of glucose yields lactic acid, and aerobic metabolism yields CO_2 that is hydrated to carbonic acid (H_2CO_3). These weak acids dissociate, yielding protons. If these protons were not exported from the cell, the cytosolic pH would drop precipitously.

A Na^+-H^+ antiport is thought to be important in removing some of the "excess" protons generated during the metabolism of animal cells. Entry of an Na^+ ion into a cell down its concentration gradient is coupled obligatorily to export of a H^+ ion by the reaction

$$Na^+_{out} + H^+_{in} \rightleftharpoons Na^+_{in} + H^+_{out}$$

Small changes in the cytosolic pH may have profound effects on the overall cellular metabolic rate. For instance, primary fibroblast cells grown to maximal density (confluence) in tissue culture generally become quiescent: DNA synthesis stops; the rates of RNA synthesis, glucose catabolism, and protein synthesis are reduced, and the cytosolic pH drops from the characteristic 7.4 of growing cells to ~7.2. Treatment of quiescent cells with a mixture of serum growth factors restimulates cell growth and DNA synthesis. An early effect of these growth factors is a marked increase in the cytosolic pH to 7.4; a 40-percent decrease in the cytosolic concentration of protons occurs within minutes. This dramatic change is caused by stimulation of the Na^+-H^+ antiport, which expels protons into the medium. The rise in cytosolic pH is believed to help activate the metabolic pathways required for cell growth and division.

The plasma membranes of most cultured animal cells contain an anion-exchange protein similar in structure and function to the erythrocyte band 3. The activity of this protein is stimulated to *lower* the cytosolic pH once it rises above 7. To understand this, recall that a HCO_3^- ion can be viewed as a complex of a hydroxide anion and CO_2, so that the export of cytosolic HCO_3^- lowers the cytosolic pH. Exchange of cytosolic HCO_3^- for extracellular Cl^- is powered by the $Cl^-_{out} > Cl^-_{in}$ concentration gradient:

$$Cl^-_{out} + HCO_3^-_{in} \rightleftharpoons Cl^-_{in} + HCO_3^-_{out}$$

A Proton Pump and a Band 3-like Anion-Exchange Protein Combine to Acidify the Stomach Contents

The mammalian stomach contains a 0.1-*M* solution of hydrochloric acid (H^+Cl^-). This strongly acidic medium denatures many ingested proteins, facilitating their degradation by proteolytic enzymes, such as pepsin, that function at acidic pH. Hydrochloric acid is secreted into the stomach by *parietal cells* (also known as *oxyntic cells*) in the gastric lining. These cells contain a H^+-K^+ ATPase in the apical membrane of the cell, which faces the stomach lumen (Figure 14-18a). This P-class ATPase (see Table 14-3) is similar in structure and function to the Na$^+$-K$^+$ ATPase and probably exports two protons in exchange for one K$^+$ ion.

The active transport of a H$^+$ ion from the cell into the lumen is accompanied by the passive transport of a Cl$^-$ ion; the net result is a concentration of H$^+$Cl$^-$ 1 million times greater in the stomach lumen than in the cell cytosol (pH = 1.0 versus pH = 7.0).

Since H$^+$ × OH$^-$ = $10^{-14} M^2$ in aqueous solution, proton export from the parietal-cell cytosol must result in

a rise in the concentration of OH$^-$ ions—and a marked rise in cytosolic pH. How does the cytosol retain a neutral pH under these conditions? The answer, again, is anion exchange (Figure 14-18b). The excess cytosolic OH$^-$ combines with CO$_2$ to form HCO$_3^-$ in a reaction catalyzed by cytosolic carbonic anhydrase. The HCO$_3^-$ anion is transported out of the basolateral membrane of the cell in exchange for an incoming Cl$^-$ ion by means of an anion-exchange protein. Tight junctions connect all the epithelial cells, preventing H$^+$ ions in the lumen from diffusing into fluids at the basolateral membrane and HCO$_3^-$ from diffusing into the lumen.

(a)

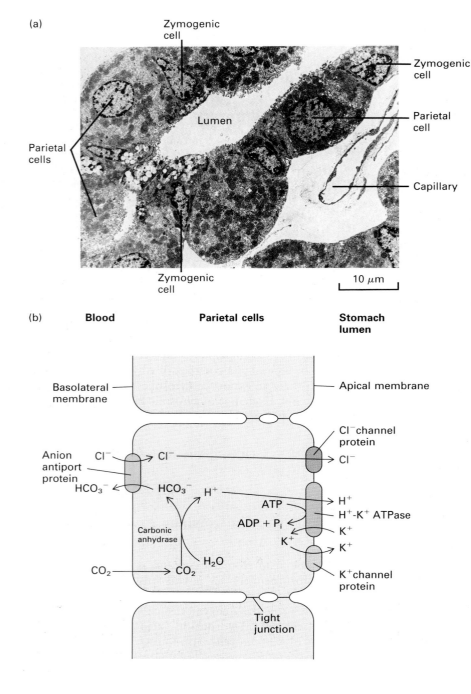

(b)   **Blood**   **Parietal cells**   **Stomach lumen**

◀ **Figure 14-18** (a) Electron micrograph of a region of the stomach wall, showing acid-secreting parietal cells (gastric epithelial cells) as well as zymogenic cells that secrete the protease pepsinogen. Parietal cells contain microvilli on the luminal face; the basal surface rests on a basement membrane that, in turn, is fed by a small blood capillary. Parietal cells contain abundant mitochondria that provide energy for the transcellular transport of H$^+$ and Cl$^-$ ions for production of hydrochloric acid. (b) Schematic representation of hydrochloric acid (H$^+$Cl$^-$) secretion by parietal cells. CO$_2$ diffuses into the cell from the blood across the basolateral membrane. Inside the cell, CO$_2$ is hydrated by carbonic anhydrase to H$^+$ and HCO$_3^-$. An anion-antiport protein in the basolateral membrane similar to band 3 catalyzes the exchange of HCO$_3^-$ anions for Cl$^-$ anions. The proton generated by carbonic anhydrase is pumped into the lumen of the stomach by an ATP-driven proton pump localized to the apical surface of the cell. This pump actually exchanges two cytosolic H$^+$ ions for one K$^+$ ion from the lumen; the K$^+$ that enters the cell is probably returned to the lumen by a K$^+$ channel protein localized to the apical membrane. How Cl$^-$ ions are transported into the stomach lumen is not clear, but the process presumably involves a Cl$^-$ channel protein. *Part (a) from R. Kessel and R. Kardon, 1979,* Tissues and Organs: A Text-Atlas of Scanning Electron Microscopy, *W. H. Freeman and Company, p. 170.*

Anion Channels and Proton Antiports Enable Plant Vacuoles to Accumulate Ions and Metabolites

The vacuoles of many fungal and plant cells accumulate ions, such as Na^+ and NO_3^- (nitrate), and metabolites, such as sucrose, to concentrations well above those of the surrounding cytosol. In the leaf, for example, excess sucrose generated during photosynthesis is stored in the vacuole and metabolized in the cytoplasm during the night to CO_2 and H_2O, with concomitant generation of ATP from ADP and P_i. Figure 14-19 shows how proton gradients power the accumulation of many substances in the vacuole. Recall that the vacuole membrane contains two types of H^+-pumping proteins; together, they result in an inside positive electric potential of about 20 mV across the vacuole membrane and a proton concentration within the vacuole 30–30,000 times that of the cytosol.

A proton-sucrose antiport in the vacuole membrane catalyzes the uptake of sucrose into the vacuole from the cytosol, where it is formed during photosynthesis. Sucrose accumulation is powered by movement of the protons down their chemical concentration gradient (lumen > cytosol) as well as by the inside positive membrane electric potential across the vacuole membrane. Uptake of Ca^{2+} and Na^+ into the vacuole from the cytosol is similarly catalyzed by proton antiports. Concentration in the vacuoles of anions such as NO_3^- are catalyzed by the inside positive membrane electric potential. Thus a proton gradient and an electric potential across a plant vacuole membrane, generated by an H^+ ATPase, are used in much the same way as an Na^+ ion concentration gradient and an electric potential across the plasma membrane of an animal cell: to power the selective uptake or extrusion of ions and small molecules.

Transport into Prokaryotic Cells

It is a more formidable problem to transport nutrients into bacterial cells than into mammalian cells for two main reasons. First, concentrations of glucose, amino acids, and other nutrients are finely regulated in the blood that bathes mammalian cells, whereas bacteria are often subjected to media of widely differing compositions. The common intestinal bacterium *E. coli,* for instance, also grows in soil and in freshwater lakes—environments with very different nutrient compositions. Second, bacteria often must concentrate nutrients such as sugars, amino acids, and vitamins against very high concentration gradients. Bacteria have evolved systems for nutrient uptake that can function at greater than 100-fold concentration gradients. Generally, these permeases are inducible: the quantity of a transport protein in the cell membrane is

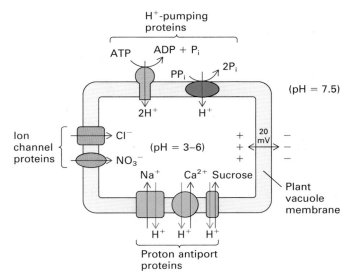

▲ **Figure 14-19** Concentration of ions and sucrose by the plant vacuole. The vacuole membrane contains two types of proton pumps: a V-class H^+ ATPase (light green) and a unique pyrophosphate-hydrolyzing proton pump (dark green). These pumps generate a lowered luminal pH as well as an inside positive electric potential across the vacuolar membrane due to the inward pumping of positively charged protons, which powers the influx of Cl^- and NO_3^- through separate channel proteins (orange). Proton antiports (blue), powered by the H^+ gradient, accumulate Na^+, Ca^{2+}, and sucrose inside the vacuole. *After P. Rea and D. Sanders, 1987, Physiol. Plantarum **71:131–141.***

regulated according to both the concentration of the nutrient in the medium and the metabolic needs of the cell.

Bacteria use two very different systems for nutrient uptake. By symport, a H^+ ion concentration gradient across the bacterial membrane powers nutrient uptake; this process is analogous to the use of the H^+ or Na^+ ion concentration gradients, respectively, across the plant vacuolar membrane or the plasma membrane of mammalian cells. The other system, in which phosphorylation of a sugar is part of the uptake mechanism, is unique to bacteria (as far as is known).

Proton Symport Systems Import Many Nutrients into Bacteria

Many aerobic bacteria derive their energy from the oxidation of glucose to CO_2, during which electrons are transferred from key metabolic intermediates (pyruvate and reduced nicotinamide adenine dinucleotide, or NADH) to oxygen, the ultimate electron acceptor. The electron transporters are integral proteins in the bacterial plasma membrane. As electrons move along the transport chain, protons are pumped out of the cell, which can result in a membrane electric potential of up to 100 mV inside negative, as well as a 100-fold increase in the H^+ ion concentration in the surrounding medium relative to

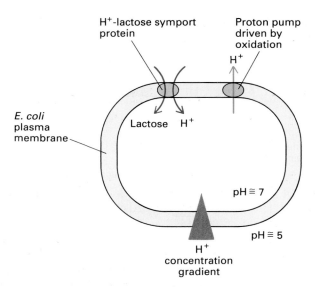

▲ **Figure 14-20** The import of lactose into *E. coli* is fueled by the energy stored in a proton gradient across the plasma membrane. Oxidation of pyruvate to CO_2 is coupled to the pumping of protons out of the cytosol, generating as much as a 100-fold H^+ gradient across the membrane. Lactose is imported by an H^+-lactose symport process (catalyzed by the *lacY*-gene product). The inward movement of protons is obligatorily coupled to the inward movement of lactose.

the cytosol. (Medium pH can become 5.0–5.5, while cytosolic pH remains at ~7.3.) The energy stored in this proton gradient is principally used to generate ATP from ADP and P_i and to import nutrients from the medium. In *E. coli*, the sugar lactose is concentrated from the medium by a pump called the *lactose permease*, a 30,000-MW integral protein encoded by the *Y* gene of the *lac* operon (see Figure 7-15). A flow of protons from the outside through the membrane (down the proton concentration gradient) is coupled to lactose uptake (Figure 14-20). Thus the lactose permease is a H^+-lactose symport.

This coupled transport can be demonstrated experimentally. *E. coli* cells poisoned with inhibitors of oxidative phosphorylation, such as cyanide, are unable to acidify the external medium and to concentrate lactose from it. Medium pH and cytosolic pH become the same. When the medium surrounding the poisoned cells is acidified by the addition of HCl, the cells again concentrate lactose, using the energy of this artificially imposed proton gradient.

The 12 trans-membrane α helices of the permease (Figure 14-21) form one channel that transports a lactose molecule and a proton together or separate channels that transport each individually. The role of several amino acid residues in proton transport and sugar binding is currently being elucidated by site-specific mutagenesis of the *lacY* gene and expression of the mutant proteins in *E. coli*. The site-specific mutation of residue histidine$_{322}$

(in the middle of transmembrane α helix 10) to an arginine abolishes proton-lactose symport and remarkably converts the protein to a facilitated diffusion transporter that can only transport lactose down its concentration gradient. Similar studies suggest that histidine$_{322}$ and glutamate$_{325}$ (on the same face of α helix 10) function as a relay system for proton transport and that a conformational change in the protein associated with proton transport (down its electrochemical gradient) somehow allows simultaneous transport of a bound lactose molecule against its concentration gradient. Other evidence indicates that alanine$_{177}$ and tyrosine$_{236}$ are involved in sugar binding, since mutants of the H^+-lactose symport with the enhanced ability to transport the related disaccharide maltose have an amino acid change at residue 177 or 236; this alteration enables the protein to bind maltose and transport it more rapidly. Unfortunately, the structure of the protein at atomic resolution is not yet available.

Certain Molecules Are Phosphorylated during Passage across the Cell Membrane

When eukaryotic cells take up glucose or amino acids, either by facilitated diffusion or symport, the molecules are not usually modified chemically during the transport process. After entering the cell, however, sugars are phosphorylated and amino acids are bound to tRNAs. These reactions lower the intracellular concentration of the unmodified substance and facilitate the inward movement of more of the compound. By contrast, some nutrients are chemically modified during transport in bacterial cells by the process of *group translocation*. The modified substance accumulates in the cytosol and cannot pass across the plasma membrane into the medium.

A well-studied example of a group translocation reaction is the phosphotransferase system used by *E. coli* and other bacteria to concentrate sugars, such as glucose, mannitol, N-acetylglucosamine, fructose, and maltose (but not lactose), and the three-carbon compound glycerol. At least four types of proteins participate in this process (Figure 14-22). Two soluble proteins are used for the uptake of all such compounds: enzyme I and a small protein called HPr. In the first step of sugar transport, a phosphate group from phosphoenolpyruvate (PEP) is transferred to a residue in enzyme I; in the second step, this phosphate is transferred to a histidine residue in the HPr protein. This system is unusual in that it uses PEP rather than ATP as a phosphate donor; the free energy derived from hydrolysis of the phosphoanhydride bond in PEP is greater than that derived from ATP, and PEP is an intermediate in the catabolism of glucose to pyruvate (see Figure 15-2). p585

The other two proteins, enzyme II and enzyme III, mediate the second step of sugar transport—the actual

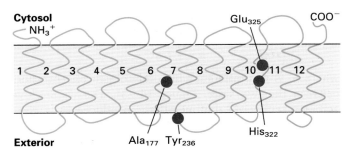

▲ **Figure 14-21** Secondary structure of the *E. coli* lactose-proton symport. [See H. R. Kabak, 1987, *Biochem.* **26**:2071; R. J. Brooker and T. H. Wilson, 1985, *Proc. Nat'l Acad. Sci. USA* **82**:3959–3963.]

transmembrane uptake of sugars. In most cases, HPr-PO$_3$H$^-$ transfers its phosphate to enzyme III. In a reaction in which phosphorylated enzyme II functions as the intermediate, enzyme III-PO$_3$H$^-$ in turn phosphorylates the appropriate sugar. Enzyme II forms the transmembrane channel; phosphorylation of the sugar occurs concomitantly with transport of the sugar into the cell.

Enzymes II and III are specific for particular sugars, whereas HPr and enzyme I are involved in the transport of all sugars. For instance, different enzymes II and III transport glucose and fructose. Genetic studies on sugar transport were essential to establish these conclusions. Bacterial mutants defective in the transport of just one sugar, such as glucose, lack functional enzyme II^{Glu} or

enzyme III^{Glu}. Mutants that lack functional HPr protein or enzyme I are defective in the transport of several sugars.

Osmosis, Movement of Water, and the Regulation of Cell Volume

In this section, we examine two types of phenomena that, at first glance, may seem unrelated: the regulation of cell volume in both plant and animal cells, and the *bulk flow* of water (the movement of water containing dissolved solutes) across one or more layers of cells. In humans, for example, water moves from the blood across layers of epithelial cells lining the stomach, which in effect "secrete" water into the stomach lumen. In higher plants, water and minerals are absorbed by the roots and move up the plant through conducting tubes (the *xylem*); water

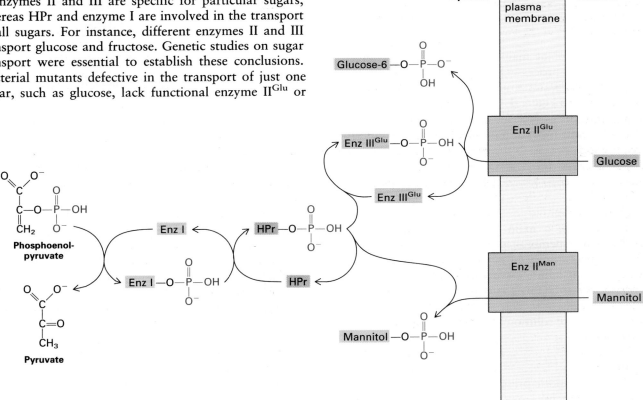

▲ **Figure 14-22** Import of sugars by the phosphotransferase system in *E. coli*. Each type of sugar is imported by a specific enzyme II (Enz II), an integral protein in the plasma membrane that transfers a phosphate to C$_6$ of the sugar as it passes through the membrane. In most cases, including glucose, the donor of the phosphate is the phosphorylated form of enzyme III (Enz III). A specific enzyme III species is used

to phosphorylate each enzyme II and thus each sugar. In other cases such as mannitol, HPr phosphorylates enzyme II and the sugar directly, without an enzyme III intermediate. [See P. W. Postma and J. W. Lengler, 1985, *Microbiol. Rev.* **49**:232; H. H. Pas and G. T. Robillard, 1988, *Biochem.* **27**:5835.]

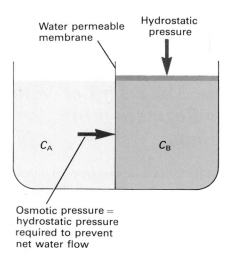

▲ **Figure 14-23** Osmotic pressure. Here, solutions A and B are separated by a membrane that is permeable to water but impermeable to all solutes. If C_B (the total concentration of solutes in solution B) is greater than C_A, water will tend to flow across the membrane from solution A to solution B. The osmotic pressure between the solutions is the hydrostatic pressure that would have to be applied to solution B to prevent this water flow. From the van't Hoff equation, this pressure is $\pi = RT(C_B - C_A)$.

is lost from the plant mainly by evaporation from the leaves. What these processes have in common is *osmosis*— the movement of water from a lower to a higher solute concentration (see Figure 4-46). We begin with a consideration of some basic facts about osmosis, and then show how they explain many physiological properties of animals and plants.

Osmotic Pressure Causes Movement of Water across One or More Membranes

Osmotic pressure, a main cause of the movement of water across cell membranes and layers, is defined as the hydrostatic pressure required to stop the net flow of water across a membrane separating solutions of different compositions. In this context, the "membrane" may be a layer of cells or a plasma membrane. The osmotic pressure of a dilute solution is approximated by the *van't Hoff equation*:

$$\pi = RT(C_1 + C_2 + C_3 + \cdots + C_n)$$

where π is the osmotic pressure in atmospheres (atm) or millimeters of mercury (mm Hg), R is the gas constant, T is the absolute temperature, and $C_1, \ldots, C_n$ are the molar concentrations of all the solutes (ions or molecules). The total number of solute molecules is important. For example, a solution of 0.5 M NaCl is actually 0.5 M Na^+ ions and 0.5 M Cl^- ions and has approximately the same osmotic pressure as a 1 M solution of glucose or lactose.

If the membrane is permeable to water but not to solutes, the osmotic pressure across the membrane is simply

$$\pi = RT \Delta C$$

where C is the difference in total solute concentration between the two sides (Figure 14-23). The water flow across a semipermeable membrane produced by a concentration gradient of 10-mM sucrose or 5-mM NaCl is balanced by a hydrostatic pressure of 0.22 atm, or 167 mmHg.

Phospholipid bilayers are somewhat permeable to water (see Figure 14-1). However, the rate of water movement across a pure lipid membrane would be insufficient to allow the bulk flow of water across certain cell layers. It is thought that certain membrane proteins are "water channels" that allow water to cross the bilayer; whether these channels are specific for water or transport water along with a solute (such as Cl^- or glucose) is not known.

The net movement of water across a membrane or cell layer can be caused by one of two circumstances: a difference in the concentrations of dissolved substances in the two solutions, or a difference in hydrostatic pressure between the two solutions. Both factors contribute to the flow of water into the stomach lumen through the surrounding epithelial cells.

Movement of Water Accompanies the Transport of Ions or Other Solutes

The parietal epithelial cells lining the stomach actively secrete H^+ ions into the lumen, causing Cl^- ions to flow from the cell into the stomach lumen to preserve electroneutrality. Thus the net flow of H^+Cl^- from the cell to the stomach lumen generates an osmotic pressure gradient across the cell membrane. These osmotic effects are quite large: about 300–400 water molecules must be moved per H^+ or Cl^- ion pumped to maintain equal osmotic pressure on both sides of the apical (lumen-facing) cell membrane.

One of the major forces pushing water from the blood into the parietal cell is a pressure gradient. The hydrostatic pressure in capillaries is 14–35 mmHg; in the cell, the pressure is nearly 0. Thus water moves into the parietal cell by hydrostatic pressure and out of the cell and into the lumen by osmotic pressure following the secretion of H^+ and Cl^- ions.

Animal Cells Can Regulate Their Volume and Internal Osmotic Strength

Vivid demonstrations of osmotic pressure occur when animal cells are placed in media of higher (hypertonic) or lower (hypotonic) salt concentrations than the cytosol. In the former case, the cell shrinks, due to water outflow; in the latter, it expands (and may ultimately lyse) due to water inflow. But even under these stresses, many animal cells still modulate their internal osmotic strength to keep their volume constant. Figure 14-24 provides one example. When a cell shrinks as a result of a hyperosmotic environment, the resulting drop in cytosolic pH activates two cell-surface antiports: an Na^+-H^+ antiport, and an anion-exchange protein. Na^+ and Cl^- enter the cell, in exchange for H^+ and HCO_3^-, respectively; H^+ and HCO_3^- are formed from gaseous CO_2 by cytosolic carbonic anhydrase. An increase in cellular salt concentration and osmotic strength results, and water flows in. Eventually, the initial volume of the cell is restored at the expense of a cytosolic salt concentration that is much higher than normal.

Changes in Intracellular Osmotic Pressure Cause Leaf Stomata to Open

Most cells in higher plants are surrounded by a rigid cell wall and, except during growth, generally do not change their volume or shape. Concentrations of salts, sugars, and other dissolved molecules are usually higher in the vacuole than in the cytosol, which, in turn, has higher concentrations than the extracellular space. The osmotic pressure generated from the entry of water into the vacuole pushes the cytosol and the plasma membrane against the resistant cell wall. Cell elongation is caused by the localized loosening of a region of the cell wall, followed by an influx of water into the vacuole.

The opening and closing of *stomata*—the pores through which CO_2 enters a leaf—is one important example of how plant cells can reverse their shape due to the osmotic movement of water (Figure 14-25). The external epidermal cells of a leaf are covered by a waxy cuticle that is largely impenetrable to water and to CO_2, a substance required for photosynthesis by the chlorophyll-laden mesophyll cells in the leaf interior. As CO_2 enters a leaf, water vapor is simultaneously lost—a process that can be injurious to the plant. Thus it is essential that the stomata open only during periods of light, when photosynthesis occurs; even then, they must close if too much water vapor is lost. Each stomate is surrounded by two *guard cells*, whose changes of shape (caused by changes in turgor pressure) open and close the pores.

Stomatal opening is caused by an increase in the concentration of ions or other solutes within the guard cells due to an ATP-driven uptake of K^+ ions from the environment (see Figure 14-25c), the metabolism of stored sucrose (see Figure 2-53) to four smaller compounds, or a combination of these two processes. Water enters the guard cells osmotically, increasing the turgor pressure. Since the guard cells are connected to each other only at their ends, the turgor pressure causes the cells to bulge outward, opening the stomatal pore between them. Stomatal closing is caused by the reverse process—a de-

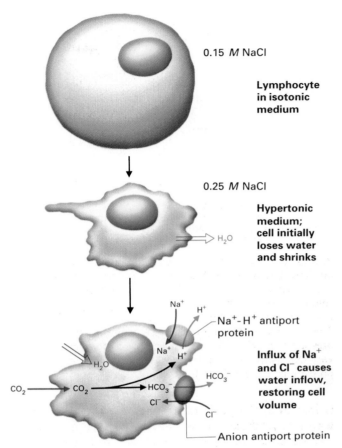

▲ **Figure 14-24** Volume regulation in lymphocytes. When placed in a hypertonic solution of $0.25 M$ NaCl, the cell initially shrinks. Within minutes, however, the volume begins to increase due to the influx of Na^+ and Cl^- ions and a resultant osmotic influx of water. Eventually, the original volume of the cell is restored, but at the expense of maintaining a normal intracellular salt concentration; the total salt concentration will be near that of the medium. [See S. Grinstein et al., 1985, *Fed. Proc.* 44:2508.]

crease in solute concentration and turgor pressure within the guard cells.

Stomatal opening is under tight physiological control. A drop in CO_2 within the leaf, resulting from active photosynthesis, causes the stomata to open. When more water exits the leaf then enters it from the roots, the mesophyll cells produce the hormone abscissic acid, which causes K^+ efflux from the guard cells; water exits the cells osmotically, and the stomata close.

The Internalization of Macromolecules and Particles

Cells exchange not only small molecules, such as inorganic ions or sugars, with their environment but also macromolecules—particularly proteins and even particles several micrometers in size. Cells secrete various pro-

(a)

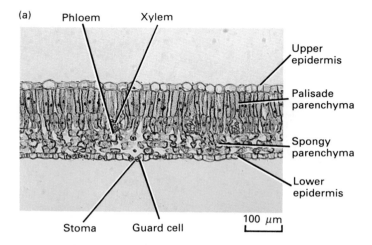

(b)

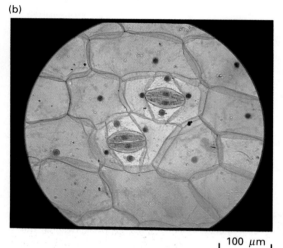

(c)

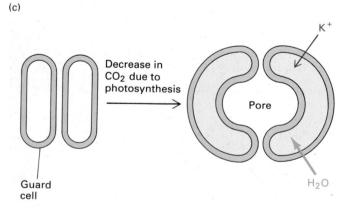

▲ **Figure 14-25**   The opening and closing of stomata.
(a) Cross section of a leaf from a lilac (*Syringa* sp) showing the surface epidermal cells and the internal photosynthetic parenchyma, or mesophyll, cells. The guard cells line the stomatal pore and, by changing their shape, open and close the pore. (b) Light micrograph of a stomata in a leaf of a wandering Jew (*Tradescantia* sp) plant. (c) An influx of K^+ ions into the guard cells triggers the osmotic influx of water, causing the cells to bulge and opening the stomatal pore. *Parts (a) and (b) courtesy Runk/Schoenberger, from Grant Heilman.*

teins, such as peptide hormones and serum proteins, and extracellular structural proteins, such as collagen, by a process called *exocytosis* (see Figure 4-40). Here, we focus on *endocytosis* and *phagocytosis* (Figure 14-26)—the cellular processes of binding and internalizing macromolecules and particles from the environment.

Plant cells, because of their rigid cell walls and internal turgor pressure, do not exhibit endocytosis, although exocytosis of newly made cell-wall and plasma-membrane components does occur. Endocytosis occurs in yeast and probably in other microorganisms. Thus the following discussion focuses mainly on animal cells.

In older usage, endocytosis referred to any process by which a region of the plasma membrane enveloped an external particle or a sample of the external medium to form an intracellular vesicle. However, phagocytosis is now recognized as a fundamentally different process, and the term endocytosis currently refers only to a small region of the plasma membrane that *invaginates* (folds inward) until it forms a new intracellular membrane-limited vesicle ~0.1 µm in diameter. There are two types of endocytic processes. *Pinocytosis* is the nonspecific uptake of small droplets of extracellular fluid by such vesicles; any material dissolved in the extracellular fluid is internalized in proportion to its concentration in the fluid. In *receptor-mediated endocytosis,* a specific receptor on the surface of the membrane "recognizes" an extracellular macromolecule (the *ligand*) and binds to it; the plasma-membrane region containing the receptor-ligand complex then undergoes endocytosis. The same endocytic vesicle can be used for pinocytosis and receptor-mediated endocytosis; a macromolecule is said to enter the vesicle by one process or the other, depending on whether the ligand first binds to a specific receptor on the cell surface.

Phagocytosis is the intake of particles as large as a few micrometers, such as bacteria or parts of broken cells—or, in experimental situations, even tiny plastic beads. Phagocytosis enables many protozoans to ingest food particles and macrophages (certain white blood cells) to take in and destroy bacteria. First, the target particle is bound to the cell surface; then the plasma membrane expands along the surface of the particle and eventually engulfs it. Vesicles formed by phagocytosis are typically 1–2 µm or greater in diameter, much larger than those formed by endocytosis. Another difference is that the expansion of the plasma membrane around the phagocytosed particle requires the active participation of actin-containing microfilaments lying just under the cell surface; endocytosis does not.

Material brought into a cell by endocytosis or phagocytosis can be secreted into the medium on the same side of the cell or transported across the cell and secreted from it by the process of *transcytosis* (Figure 14-27). Alternatively, endocytosed material can be stored within intracellular vesicles. Usually, however, the ingested material is delivered to lysosomes, where it is degraded. There are

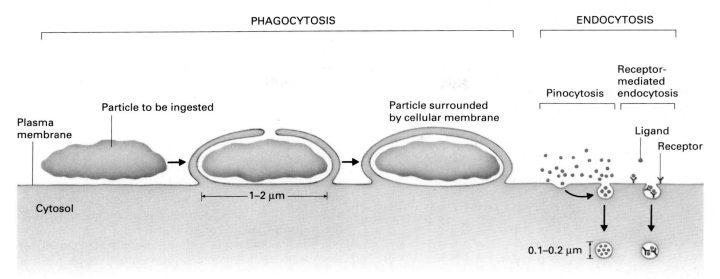

▲ **Figure 14-26**   The processes of phagocytosis and endocytosis. Endocytosis includes both specific and nonspecific uptake. In receptor-mediated endocytosis, macromolecules are internalized by binding to specific surface receptors. Pinocytosis, on the other hand, is the nonspecific uptake of extracellular molecules in an endocytic vesicle, which is simply proportional to the molecular concentration in the extracellular solution. Phagocytosis involves expansion of the plasma membrane to engulf large particles, such as bacteria, bound to the plasma-membrane surface.

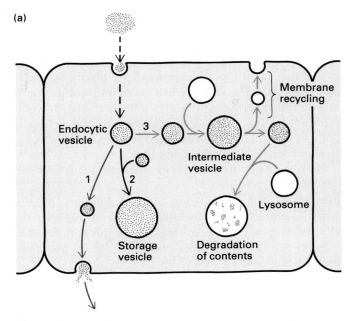

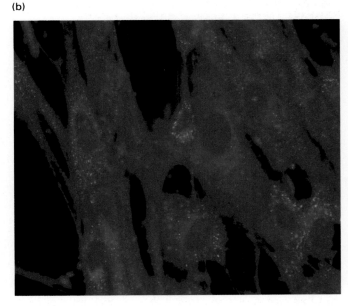

▲ **Figure 14-27**   (a) Possible fates of macromolecules brought into cells by endocytosis. The material can be transported to the other side of the cell and exocytosed (pathway 1). Or the endocytic vesicle can fuse with other vesicles (pathway 2), and the material can be stored in the cell. Most often, the endocytic vesicle fuses with other cellular vesicles (pathway 3) to form an intermediate transport vesicle. The region containing the plasma-membrane receptor buds off to form a separate vesicle and is recycled to the cell surface; the remainder of the vesicle, with the ingested material, fuses with a lysosome, in which hydrolytic enzymes degrade the ingested material to small molecules such as amino acids. (b) Lysosomes in living human diploid fibroblasts. The cells are stained with the dye FluoroBora P; the small fluorescent regions in the cytoplasm are lysosomes, which concentrate the substance. *Part (b) courtesy M. A. Paz and P. M. Gallop; see P. M. Gallop, M. A. Paz, and E. Henson, 1982,* Science *217:166–169.*

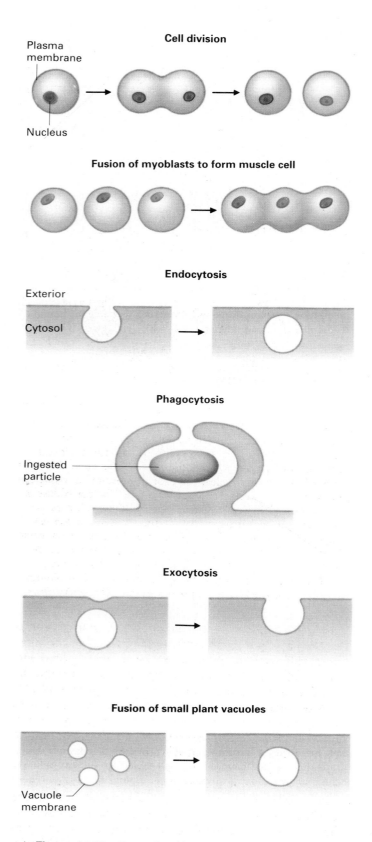

Cell division

Plasma membrane

Nucleus

Fusion of myoblasts to form muscle cell

Endocytosis

Exterior

Cytosol

Phagocytosis

Ingested particle

Exocytosis

Fusion of small plant vacuoles

Vacuole membrane

▲ **Figure 14-28** Examples of membrane-fusion events.

many important exceptions to this outcome, but before we consider the specifics of endocytosis and phagocytosis, we should discuss the mechanisms of a process that is basic to an understanding of the formation and fate of all cellular membranes: the fusion of two membranes.

Membrane Fusions Occur in Endocytosis, Exocytosis, and Many Other Cellular Phenomena

Membrane fusion is involved in crucial cellular events other than exocytosis, endocytosis, and phagocytosis. Cell separation during mitosis is one example; another is the fusion of small vacuoles to form a large one or of individual myoblasts to form a multinucleated muscle cell (Figure 14-28). Membrane fusion is not a simple process, because membranes do not spontaneously coalesce. Liposomes do not fuse, nor do cells or organelles fuse in the absence of special treatment. One obstacle to membrane fusion is the high negative surface charge contributed by many phospholipid head groups and by the sialic acid residues found in cell-surface glycolipids, which causes membranes to repel one another. Another obstacle is the absence of free edges in membranes; for fusion to occur, continuous sheets or spheres of membrane must be induced to open.

Membranes can be made to fuse by artificial means. Experiments with the fusion of live cells have shown that two different stages are involved: coming into close proximity and actual fusion. Plant lectins—multivalent proteins that bind to carbohydrates—can induce proximity by binding to surface glycoprotein and glycolipid molecules on adjacent cells. When a second substance is added, the adhering membranes somehow fuse. Polyethylene glycol can act as such a fusion factor, but how it does so is not known. Viral glycoproteins also act as fusion factors, as we will discuss later.

Despite its importance, the mechanism of fusion is still obscure. Freeze-fracture studies suggest that regions of membrane contact are bereft of intramembrane protein particles and that the subsequent membrane fusion actually takes place between regions of protein-depleted membranes. However, other evidence suggests that membrane fusion during exocytosis of secretory granules in mast cells (the cells that secrete histamine) does not involve protein depletion in the region of fusion.

We now return to our discussion of the internalization of macromolecules and particles.

Pinocytosis Is the Nonspecific Uptake of Extracellular Fluids

Pinocytosis is the nonspecific form of endocytosis in which any extracellular material is taken up at a rate proportional to its concentration in the extracellular medium. To trace the fate of pinocytic vesicles, a solution of horseradish peroxidase, which does not bind to cell surfaces, is added to the extracellular fluid. This protein can be readily distinguished in the electron microscope by a histochemical procedure (Figure 14-29). Endocytic vesicles containing peroxidase, initially found just under the cell surface, are small (~0.1 μm in diameter) and fuse with each other to form larger vesicles. These larger vesicles, in turn, usually fuse with lysosomes, where the internalized protein is destroyed by lysosomal proteases.

Phagocytosis Depends on Specific Interactions at the Cell Surface

Phagocytosis is a property of several types of cells. Many unicellular eukaryotic organisms, such as amebas and slime molds, grow by ingesting bacteria (and, occasionally, eukaryotic microorganisms). Degradation of such foodstuffs in the lysosomes provides the organism with the sugar, amino acids, lipids, and other nutrients it utilizes for biosynthetic reactions. Animals contain several types of cells called *phagocytes* that primarily rid the organism of such particles as antigen-antibody complexes and such harmful intruders as bacteria and viruses (Figure 14-30). Some macrophages line the *sinusoids* (the narrow blood channels of the liver, spleen, and other organs), where they ingest aging erythrocytes and particles from the passing blood. Other phagocytic white blood cells, called *monocytes*, migrate into tissue areas of infection or inflammation, where they differentiate into macrophages that ingest and degrade broken cells, killed microorganisms, and other debris. Macrophages are "gourmets" in that they bind to and ingest only certain particles available to them. Ingestion depends on the chemical composition and properties of the particle surface, but the exact requirements are not totally understood. The carbohydrate surface structures, or *capsules*, of many pathogenic bacteria, such as *Streptococcus*, inhibit their being bound to and ingested by macrophages. Nonpathogenic strains of the same bacteria lack these capsules and are readily ingested.

In response to infection by pathogenic bacteria, the host organism usually produces serum antibodies that bind to specific proteins or carbohydrates on the bacterial surface. An antibody coat on a bacterium or other particle stimulates its ingestion by macrophages because their specific surface receptors bind to a region common to all antibody molecules—the stem of the Y-shaped structure called the F_c region (see Figure 2-27). Other serum proteins, collectively called *complement*, coat foreign particles nonspecifically and stimulate their ingestion by binding to specific surface receptors on the macrophage called *complement receptors*. Once the particle is bound to the cell surface, the sequential process of ingestion commences with the continuous apposition of a segment of the plasma membrane to the particle surface, excluding

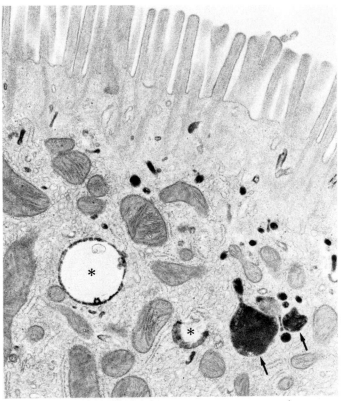

▲ **Figure 14-29** Pinocytotic uptake of horseradish peroxidases by intestinal epithelial cells, revealed by electron microscopy. A solution of horseradish peroxidase is added to the lumen of a rat small intestine. Two hours later, the tissue is fixed. The histochemical stain for this enzyme in tissue sections makes an electron-dense precipitate. The peroxidase is visible in a variety of endocytic vesicles (the small dark objects) and inside the periphery of larger vesicles (*asterisks*). High concentrations of peroxidase are also seen in lysosome-like bodies (*arrows*). [See D. R. Abrahamson and R. Rodewald, 1981, *J. Cell Biol.* **91**:270.] *Photography courtesy of R. Rodewald. Reproduced from the* Journal of Cell Biology *by copyright permission of the Rockefeller University Press.*

most if not all of the surrounding fluid. Surface receptors on the macrophage then interact with ligands distributed uniformly over the particle surface to link the two surfaces; the effect, similar to a zipper closing around the particle, is called the *zipper interaction* (Figure 14-31).

After ingestion, fusion of the phagocytic vesicle with the lysosome membrane delivers the membrane-enveloped particles to the lysosomes. After this fusion, lysosomes are customarily called *secondary lysosomes* to distinguish them from the smaller, unfused *primary lysosomes*. Digestion takes place in these secondary lysosomes. Some parts of the ingested material, such as the cell walls of microorganisms, are resistant to lysosome hydrolases and accumulate within the secondary lysosomes as residual bodies. This may be one reason why

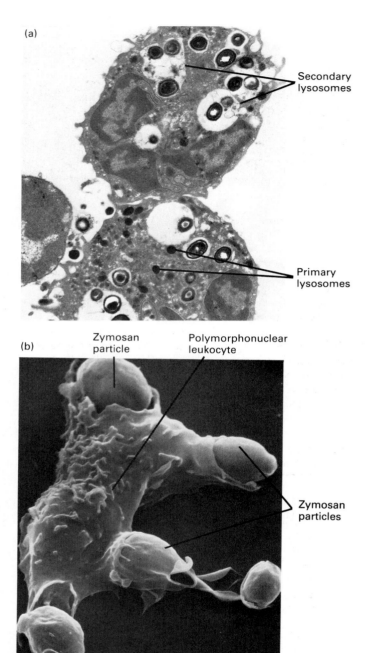

(a)

Secondary
lysosomes

Primary
lysosomes

(b)

Zymosan
particle

Polymorphonuclear
leukocyte

Zymosan
particles

▲ **Figure 14-30** Human polymorphonuclear leukocytes (neutrophils), a class of phagocytic white blood cells. (a) A thin-section micrograph. The smaller dark bodies are primary lysosomes; the larger white bodies are secondary lysosomes containing degradation products of ingested staphylococci. Secondary lysosomes arise by fusion of primary lysosomes with the phagocytic vesicles containing the bacteria. The surface of the neutrophil contains many thin folds, or ruffles, shown here immobilized by the fixative in various phases of phagocytic activity. (b) A scanning electron micrograph of a human polymorphonuclear leukocyte ingesting small particles of zymosan (pieces of yeast cell wall). Note the expansion of the plasma membrane around the particles. *Part (a) courtesy of Biophoto Associates; part (b) courtesy of M. J. Karnofsky.*

macrophages have a very short life span (usually one or two days at most). In addition to the normal set of lysosome hydrolases, macrophage lysosomes contain enzymes that generate hydrogen peroxide (H_2O_2), superoxide (O_2^-), and other toxic chemicals that aid in killing bacteria. When an antibody ligand binds to the macrophage F_c receptor, it activates these enzymes and also increases nonspecific pinocytosis and phagocytosis.

The F_c receptor has been cloned and studied in detail. It is a polypeptide containing 231 amino acids. The 180 amino acids in the N-terminus are exoplasmic and constitute the F_c (ligand)-binding domain. A single hydrophobic segment of 20 amino acids, like the one in glycophorin (see Figure 13-16), anchors the protein in the membrane; a short domain faces the cytosol. The F_c receptor is a member of the so-called *immunoglobulin gene superfamily:* the sequences of the ligand-binding domain and the binding region of immunoglobulins (see Figure 2-27) share similarities. The F_c receptor is coupled to a plasma-membrane sodium channel that is activated when an antibody binds to this receptor. The resulting influx of Na^+ ions into the cell reduces the magnitude of the electric potential across the cell membrane, activating many specific macrophage functions: phagocytosis, cell movement, and the generation of H_2O_2 and other antibacterial compounds in lysosomes.

Some organisms are ingested but not killed by macrophages. *Mycobacterium leprae,* the causative agent of leprosy, and parasites of the genus *Leishmania,* protozoa that cause parasitic diseases in humans, actually grow only in the endocytic vesicles of macrophages and, for this reason, are extremely difficult to kill. *Legionella pneumophila,* the bacterial agent of Legionnaire's disease, also multiplies intracellularly in human phagocytes. It inhibits the acidification of the phagocytic vesicles and thus the activation of lysosome hydrolases that function only at acidic pH and would otherwise kill the bacterium.

Receptor-mediated Endocytosis

Receptor-mediated endocytosis allows the selective uptake of extracellular proteins and small particles. Receptor proteins on the cell surface bind specific ligands tightly and with a high degree of specificity (Table 14-5). Most receptors diffuse within the plasma membrane; receptor-ligand complexes cluster in small regions of the membrane, which are then internalized. Endocytosis of these specific complexes usually occurs at *coated pits,* specialized depressions on the cell surface. In electron micrographs, a visible proteinaceous layer on the cytosolic side gives a coated appearance to these parts of the plasma membrane and, initially, to the vesicles that form from them (Figures 14-32 and 14-33). These vesicles lose their coats after endocytosis, forming smooth-surfaced vesicles called *endosomes.* Yet both coated and uncoated

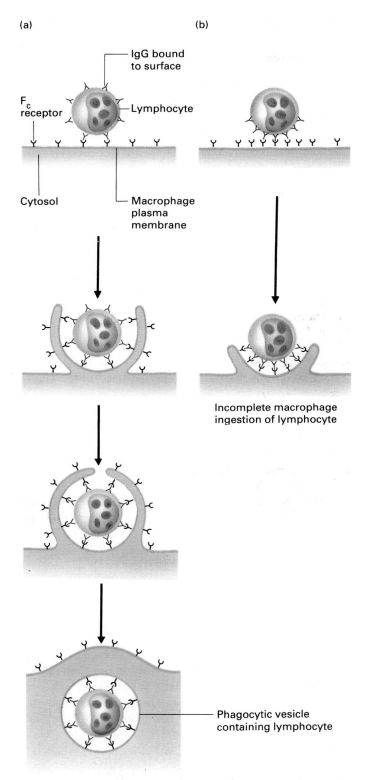

▲ Figure 14-31 Diagram of the zipper interaction. (a) If IgG antibody is evenly distributed over the entire surface of the target lymphocyte, the macrophage plasma membrane completely engulfs this cell. (b) If the antibody molecules are localized on one region of the surface, the plasma membrane spreads only over the antibody-coated region and the lymphocyte is not ingested.

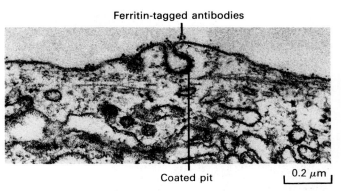

▲ Figure 14-32 Electron micrograph of a coated pit on a mouse fibroblast, showing the proteinaceous coat on the inner (cytosolic) surface of the pit. Eventually a coated pit buds off from the plasma membrane to form a coated vesicle in the cytosol. Not all plasma-membrane proteins enter coated pits, as shown here. In this cell, which was treated before fixation with an antibody to a specific cell-surface protein; the antibody in turn, is bound to ferritin, an electron-opaque iron-containing protein visible as a dense dot in electron micrographs. The tagged antibody labels the outer cell surface but is not found in the coated pit. [See M. S. Bretscher, N. H. Thomson, and B. M. F. Pearse, 1980, *Proc. Nat'l Acad. Sci. USA* 77:4156.] *Photograph courtesy of M. S. Bretscher.*

Table 14-5 Proteins and particles taken up by receptor-mediated endocytosis in animal cells

Ligand	Function of receptor/ligand complex	Cell type
Low-density lipoproteins (LDLs)	Supply cholesterol	Most
Transferrin	Supplies iron	Most
Glucose- or mannose-terminal glycoproteins	Remove injurious agents from circulation	Macrophage
Galactose-terminal glycoproteins	Remove injurious agents from circulation	Hepatocyte
Immunoglobins	Transfer immunity to fetus	Fetal yolk sac; intestinal epithelial cells of neonatal animals
Phosphovitellogenins	Supplies protein to embryo	Developing oocyte
Fibrin	Removes injurious agents	Epithelial
Insulin, other peptide hormones	Alter cellular metabolism; ligand and often receptor are degraded after endocytosis	Most

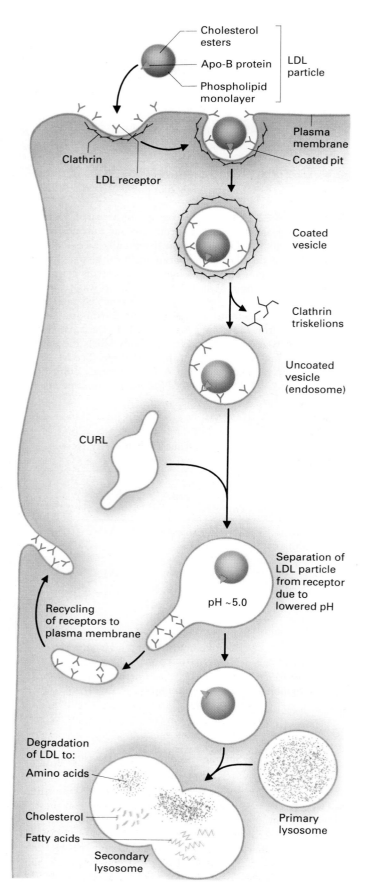

◀ **Figure 14-33** Fate of low-density lipoprotein (LDL) particle and receptor after endocytosis. The same pathway is followed by other ligands, such as asialoglycoproteins, that are internalized by receptor-mediated endocytosis and degraded in the lysosome.

After an LDL particle binds to an LDL receptor on the plasma membrane, the receptor-ligand complex is internalized in a clathrin-coated pit that pinches off to become a coated vesicle. The clathrin coat then depolymerizes to triskelions, resulting in an uncoated vesicle. This endosome fuses with an uncoupling (CURL) vesicle, and its low pH (~5) causes the LDL particles to dissociate from the LDL receptors. A receptor-rich region buds off to form a separate vesicle that recycles the LDL receptors back to the plasma membrane. A vesicle containing an LDL particle may fuse with another endosome but ultimately fuses with a primary lysosome to form a large secondary lysosome. There, the apo-B protein of the LDL particle is degraded to amino acids and the cholesterol esters are hydrolyzed to fatty acids and cholesterol. Cholesterol is incorporated into cell membranes. Abundant imported cholesterol inhibits synthesis by the cell of both cholesterol and LDL receptor protein.

pits exist on the plasma membrane. Certain proteins, such as cholera toxin, are internalized exclusively in noncoated pits.

Whether endocytosis is by a coated or uncoated pit and vesicle, the receptor is usually recycled intact to the cell surface. This rather complex process, which is well understood (at least in outline), involves a number of types of intracellular vesicles (see Figure 14-33). Endosomes fuse with *an uncoupling vesicle called the compartment of uncoupling of receptor and ligand* (CURL), characterized by an internal pH of ~5.0. The acidic pH causes receptors to dissociate from their ligands. The free receptors congregate in one membrane section of these uncoupling vesicles, which ultimately buds off to form a separate elongated vesicle to recycle the receptor back to the plasma membrane. The ligand, in contrast, is segregated into a different type of vesicle that ultimately fuses with a lysosome; there, lysosome proteases and other hydrolytic enzymes degrade the ligand.

Two receptor-ligand systems—the liver asialoglycoprotein receptor and the low-density lipoprotein (LDL) receptor—have proved especially useful in the study of receptor-mediated endocytosis where the ligand is degraded in lysosomes. We will use these systems to illustrate how this complex process could be dissected. Then we will turn to other receptor-ligand systems in which the ligand undergoes a different fate after internalization.

The Asialoglycoprotein Receptor Removes Certain Abnormal Serum Glycoproteins

Many cells bind and internalize potentially injurious or toxic extracellular proteins, such as proteases; these proteins are destroyed in lysosomes. Hepatocytes, the predominant cells in the liver, contain a potent surface receptor, the *asialoglycoprotein receptor,* that binds galactose-terminal glycoproteins. Galactose residues in serum proteins are usually covered by sialic acid residues (see Figure 2-56). Abnormal glycoproteins that lack terminal sialic acids and have exposed galactose residues bind to these hepatocyte receptors and are removed from the circulation within five minutes. After receptor-mediated endocytosis, asialoglycoproteins are destroyed by hepatocyte lysosomes. Other cells, especially macrophages, contain specific receptors for different abnormal or foreign glycoproteins.

The asialoglycoprotein receptor consists of at least three polypeptide chains, each of which contains a binding site specific for galactose. Although the binding affinity for a single galactose residue is rather low ($\sim 10^{-3}$ M), the three precisely arranged receptor polypeptides can bind to the three galactose residues on a typical asparagine-linked oligosaccharide with very high affinity (Figure 14-34). Several weak interactions of one galactose binding to one polypeptide can combine to generate a single strong binding interaction.

The Low-density Lipoprotein (LDL) Receptor Mediates the Uptake of Cholesterol-containing Particles

The most intensively studied type of receptor-mediated endocytosis is the system by which cholesterol is accumulated in cells. Cholesterol is insoluble in body fluids: whether it is ingested in foodstuffs or synthesized in the liver, it must be transported by a carrier. *Low-density lipoprotein* (LDL) is one of a variety of complexes that carry cholesterol through the bloodstream; most cells manufacture receptors that specifically bind LDL. The LDL particle is a sphere 20–25 nm in diameter. Its outer surface is a monolayer membrane of phospholipids and cholesterol, in which one molecule of a very large (4563 amino acid), hydrophobic protein called apo-B is embedded (Figure 14-35). Inside is an extremely apolar core of cholesterol, all of which is esterified through the single hydroxyl group to a long-chain fatty acid. Serum LDL particles are a significant source of cholesterol for many cells, although all cells synthesize much of their cholesterol de novo from acetic acid.

After endocytosis, the LDL particles are transported to lysosomes (see Figure 14-33), where the apo-B protein is degraded to amino acids and the cholesterol esters are hydrolyzed to cholesterol and fatty acids. The cholesterol is incorporated directly into cell membranes; the fatty acids are used to make new phospholipids.

The cells of the adrenal cortex synthesize many important steroid hormones, all of which are made from cholesterol. Because they must import much more cholesterol than most other mammalian cells, the adrenal cells are

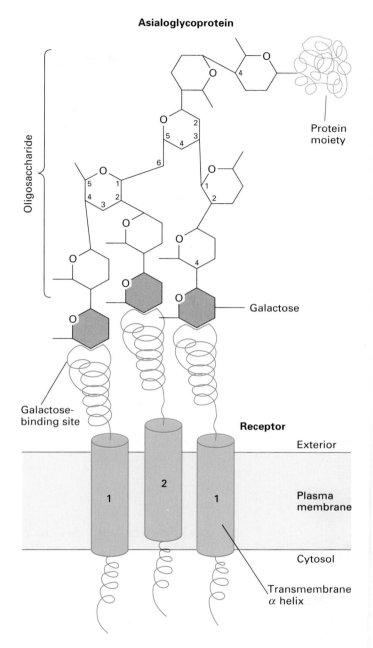

▲ **Figure 14-34** Schematic structure of the liver asialoglycoprotein receptor. The receptor consists of at least three polypeptides, including two copies of subunit 1 and one copy of subunit 2; 60 percent of the amino acid residues in subunits 1 and 2 are identical. The receptor binds tightest ($K_M = 10^{-9}$ M) to glycoproteins containing oligosaccharides with three branches, each bearing a terminal galactose residue. Chains with only one or two galactose residues do not bind as tightly ($K_M = 10^{-3}$ M and 10^{-6} M, respectively). Each receptor subunit has a galactose-binding site; the three sites are thought to be positioned in such a way that they can interact optimally with the three galactose residues. [See J. Bischoff, S. Libresco, M. Shia, and H. F. Lodish, 1988, *J. Cell Biol.* **106**:1067–1074; M. McPhaul and P. Berg, 1988, *Mol. Cell Biol.* **7**:1841–1847; J. T. Sawyer, N. P. Sanford, and D. Doyle, 1988, *J. Biol. Chem.* **263**:10534–10539.]

(a)

(b)

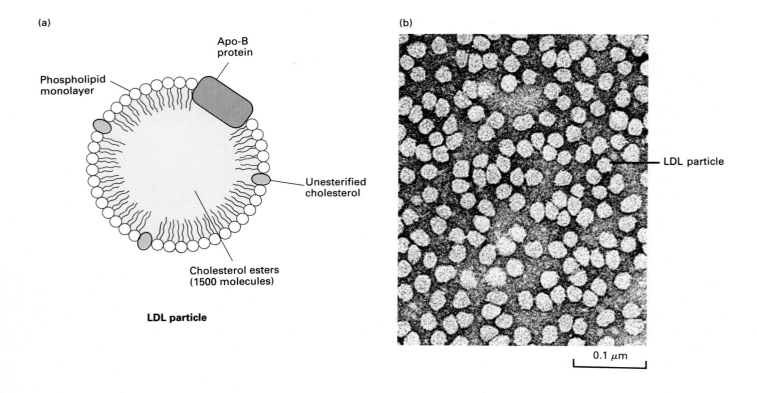

(c)

▲ **Figure 14-35** Structure of a LDL particle. (a) The surface of the LDL particle is a monolayer of phospholipid and unesterified cholesterol. The hydrophobic core of the particle is rich in fatty acid esters of cholesterol. One copy of the hydrophobic apo-B protein is embedded in the membrane. (b) Electron micrograph of a negatively stained preparation of LDL particles. (c) Esters of cholesterol with fatty acids, mainly linoleic acid, are found in the interior of LDL particles. The lysosome enzyme *cholesterol esterase* hydrolyzes the ester bond, forming a free fatty acid and cholesterol. [See R. Anderson, 1979, *Nature* 279:679.] *Part (b) courtesy of R. Anderson. Reprinted by permission from* Nature. *Copyright 1979, Macmillan Journals Limited.*

much richer in LDL receptors; researchers often use them as a source of purified LDL receptors. The LDL receptor is a single-chain glycoprotein of 839 amino acids. A sequence of 22 hydrophobic amino acids spans the plasma membrane once, presumably as an α helix. About 50 amino acids at the C-terminus face the cytosol and are involved in binding the receptor to the coated pits. Within the 767 amino acids on the exoplasmic face, the most striking feature is the segment of ~320 amino-terminal residues, which is extremely rich in disulfide-bonded cysteine residues; it includes a seven-fold repeat of a sequence of 40 amino acids that contains the LDL binding site (Figure 14-36).

Studies of mutant receptors, in which various domains are deleted, show that different domains have different functions. For instance, a receptor missing 350 amino acids, including repeats A, B, and C, binds LDL particles normally; however, the bound ligand cannot be released by exposure to a pH of 5.0, indicating this domain is required for acid-dependent ligand dissociation. Other deletions show that most of the seven repeats of 40 amino acids in the N-terminus domain are essential for binding an LDL particle; each repeat makes a separate contribution to the binding event.

We now turn to a more detailed analysis of the several steps of receptor-mediated endocytosis.

Ligand Binding and the Internalization of Receptor-Ligand Complexes Can Be Studied Separately

The binding of a molecule to its surface receptor can be separated experimentally from the internalization of the receptor-ligand complex. Internalization requires an expenditure of energy, and it will not occur at 4°C or lower because enzymatic reactions that generate or consume ATP function too slowly at this temperature. This does not interfere with receptor-ligand binding, however; the amount of ligand bound to cell receptors and the affinity of the receptor for its ligand can be measured at 4°C. A hepatocyte will bind about 500,000 molecules of asialo-orosomucoid—a typical glycoprotein with oligosaccharide chains each of which have three terminal galactose residues (see Figure 14-34)—to surface asialoglycoprotein receptors, and the concentration for half-maximal binding (K_M) of the ligand is $8 \times 10^{-9}\ M$ of asialo-orosomucoid.

Surface receptors like the one for galactose-terminal glycoproteins are exquisitely specific and have a high affinity for their ligands. Nonglycosylated proteins, for instance, will not bind to the asialoglycoprotein receptor with any measurable affinity, nor will glycoproteins containing terminal sugars with structures similar to galactose (such as mannose). Binding constants (K_M values) of about $10^{-8}\ M$ are typical for many receptors that partici-

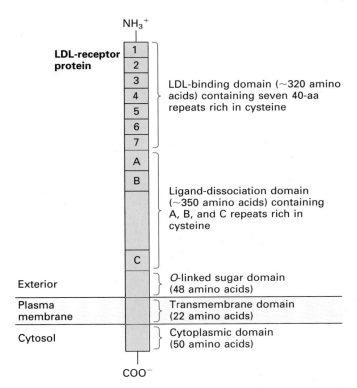

▲ **Figure 14-36** The five domains in the structure of the human LDL receptor. The sequence of the protein was deduced from the sequence of the cloned cDNA. The receptor is a dimer of two identical 839-residue polypeptides. [See M. Brown and J. Goldstein, 1986, *Science* **232**:34.]

pate in endocytosis. These receptors will be half-saturated with ligand at a ligand concentration of $10^{-8}\ M$, or 0.4 micrograms per milliliter (μg/ml) for a typical protein of 40,000 MW. Since the total protein concentration in blood and other extracellular fluids is on the order of 100 mg/ml, these surface receptors are able to work efficiently in the presence of a 100,000-fold excess of unrelated and undesired ligands! Chapter 19 provides more information about how cell-surface receptors are characterized, purified, and cloned.

By using proteases or agents such as EDTA, which disrupt the receptor-ligand interaction but cannot penetrate the cells, asialo-orosomucoid, bound to the asialoglycoprotein receptor on the surface of hepatocytes at 4°C, can be shown to be internalized rapidly once the cells are warmed to 37°C (Figure 14-37). Similar results are obtained with all other receptors that participate in endocytosis.

In summary, the expenditure of energy is not required for cell-surface receptors to bind their ligands. Like the binding of a substrate to an enzyme, the formation of a receptor-ligand complex utilizes many specific noncovalent interactions between the two molecules. Internalization of the ligand, bound to its surface receptor, does re-

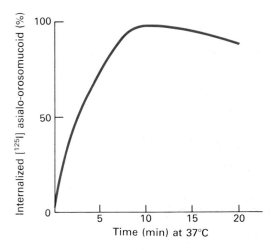

◀ **Figure 14-37** Internalization of protein bound to a cell-surface asialoglycoprotein receptor. Cell-surface and internalized receptors can be differentiated by treating cells with ethylenediaminetetraacetic acid (EDTA), a chemical that cannot penetrate cells but that binds Ca^{2+} and thus removes only the ligand bound to the cell surface. In this study, ^{125}I-tagged asialo-orosomucoid was bound to the surface of hepatoma (liver-tumor) cells at 4°C and then warmed to 37°C. Before warming, all radioactive protein is on the cell surface and can be removed by treatment with EDTA; with time at 37°C, most of the radioactive protein is inside the cell and cannot be removed by EDTA. [See A. L. Schwartz, S. Fridovich, and H. F. Lodish, 1982, *J. Biol. Chem.* **257**:4230.]

(a)

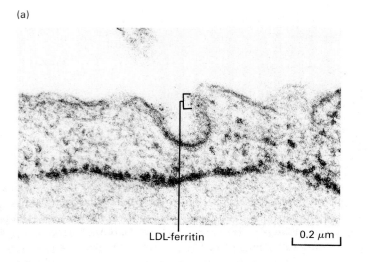

LDL-ferritin   0.2 μm

(b)

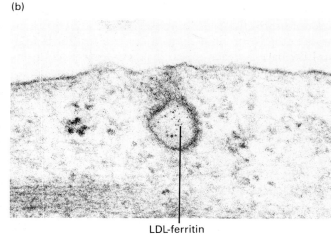

LDL-ferritin

(c)

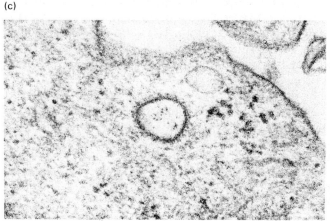

(d)

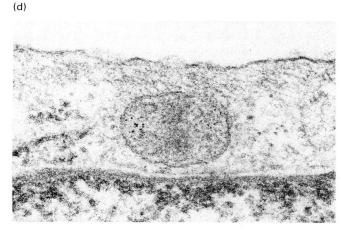

▲ **Figure 14-38** The initial stages of receptor-mediated endocytosis of LDL particles by cultured fibroblasts, revealed by electron microscopy. (a) A coated pit; the small dots over the pit are LDL particles labeled with ferritin. (b) A pit containing LDL apparently closing on itself to form a coated vesicle. (c) A coated vesicle containing LDL particles. (d) Ferritin-tagged particles in a smooth-surfaced endosome 6 min after being added to cells. [See M. S. Brown and J. Goldstein, 1986, *Science* **232**:34.] *Photographs courtesy of R. Anderson. Reprinted by permission from J. Goldstein, R. Anderson, and M. S. Brown,* Nature **279**:679. *Copyright 1979, Macmillan Journals Limited.*

quire energy. However, we are ignorant of most of the cellular molecules that catalyze endocytosis, especially those that might be involved in meeting the energy requirement.

Since most cell-surface receptors and their ligands are internalized in coated pits and vesicles, we now turn to the structure and function of the major proteins that form these pits and vesicles.

Clathrin, a Fibrous Protein, Forms a Lattice Shell around Coated Pits and Vesicles

Internalization of ligands begins when, by diffusion in the membrane, a receptor-ligand complex becomes localized over coated pits, which then pinch off from the membrane to form separate coated vesicles. Many endocytosed ligands have been observed in such intracellular vesicles (Figure 14-38), and researchers believe that coated vesicles function as intermediates in the endocyto-

sis of most (but not all) ligands bound to cell-surface receptors. In fact, two types of receptor-bound ligands, such as transferrin and LDL, can be seen in the same coated pit or vesicle. Coated vesicles are readily purified from every type of eukaryotic cell studied, the brain being a particularly rich source. They also function in processes other than endocytosis: current evidence suggests they are intermediate transporters of proteins from the Golgi complex to the cell surface during protein secretion.

Coated pits make up about 2 percent of the surfaces of cells such as hepatocytes and fibroblasts. Typical coated vesicles that form from coated pits are 50–100 nm in diameter, with a membrane vesicle inside (Figure 14-39a). The outer (cytosolic) surface of a coated vesicle is composed primarily of the fibrous protein *clathrin* and has a basketlike appearance (Figure 14-40). Purified clathrin has the form of a three-armed triskelion (see Figure 14-39b); each arm contains one clathrin heavy chain (180,000 MW) and one clathrin light chain (~35,000–40,000 MW). There are two types of light chains, α and

(a) Coated vesicle

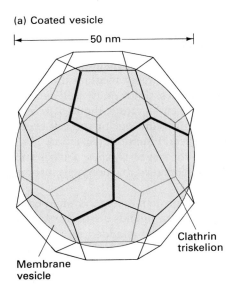

Membrane vesicle

Clathrin triskelion

(b)

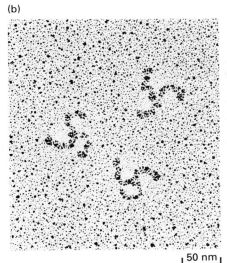

50 nm

(c) Triskelion structure

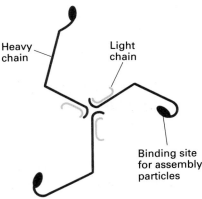

Heavy chain

Light chain

Binding site for assembly particles

(d) Assembly intermediate

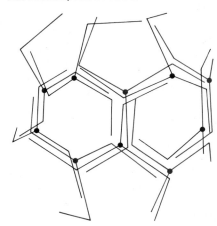

◀ **Figure 14-39** Structure and assembly of a coated vesicle. (a) A typical coated vesicle contains a membrane vesicle (tan) about 40 nm in diameter surrounded by a fibrous network of 12 pentagons and 8 hexagons. The fibrous coat is constructed of 36 clathrin triskelions, one of which is shown here in red. One clathrin triskelion is centered on each of the 36 vertices of the coat. Coated vesicles having other sizes and shapes are believed to be constructed similarly: each vesicle contains 12 pentagons but a variable number of hexagons. (b) Electron micrograph of purified clathrin triskelions. (c) Detail of a clathrin triskelion. Each of three clathrin heavy chains has a specific bent structure. A clathrin light chain is attached to each heavy chain, probably near the center. Although it is not obvious in (a) or (b), each triskelion has an intrinsic curvature; when triskelions polymerize, they form a curved (not flat) structure. (d) An intermediate in the assembly of a coated vesicle, containing 10 of the final 36 triskelions, illustrates the intrinsic curvature and the packing of the clathrin triskelions. [Part (a) see B. M. F. Pearse, 1987, *EMBO J.* 6:2507–2512.] *Part (b) courtesy of D. Branton.* [Part (b) see E. Ungewickell and D. Branton, 1981, *Nature* 289:420.] *Reprinted by permission from* Nature. *Copyright 1981, Macmillan Journals Limited.* [Part (c) see F. M. Brodsky, 1988, *Science* 242:1396–1402; J. E. Heuser and J. Keen, 1988, *J. Cell Biol.* 107:877–886.]

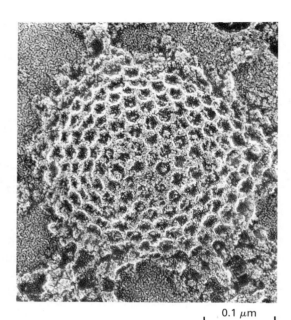

0.1 μm

▲ **Figure 14-40** A coated pit on the cytoplasmic face of the plasma membrane of a fibroblast, showing the polygonal network in a forming coated vesicle. The cell was rapidly frozen in liquid helium, freeze-fractured, and then treated by the deep-etching technique. *Photograph courtesy of J. Heuser.*

β; their amino acid sequences are 60 percent identical, and their functional differences are not known. Triskelions combine, or polymerize, to form the cagelike structure around a coated vesicle, even in the absence of membrane vesicles. The polymerization of clathrin into a lattice along the cytosolic face of a coated pit is believed to cause the pit to expand inward and ultimately to pinch off from the membrane. After an actual vesicle forms, the clathrin "cage" is completed, producing a coated vesicle.

Between the clathrin coat and the vesicle membrane lies a 20-nm space containing *assembly particles*. Each particle (340,000 MW) contains two copies each of three polypeptides of 105,000, 50,000, and 16,000 MW. Assembly particles bind to the globular domains at the distal ends of each clathrin heavy chain in a triskelion (see Figure 14-39c and d) and promote the polymerization of purified clathrin triskelions into cages. Assembly particles also bind to the cytosolic "tails" of receptors, such as the LDL receptor protein (see Figure 14-36), that become internalized in coated pits, but do not bind to the cytosolic segments of the plasma-membrane proteins not internalized in these pits (see Figure 14-32). Thus assembly particles may mediate the binding of clathrin to receptors in coated pits and distinguish between plasma-membrane proteins that are to be internalized and those that are to be left behind. LDL receptors, in fact, are clustered in coated pits even when LDL particles are not bound to them. Other receptors, such as the one for galactose-

terminal glycoproteins, enter coated pits only if the receptors are occupied with ligand. The binding of ligand to this receptor (see Figure 14-34) is believed to trigger a conformational change in the part of the receptor that faces the cytosol, so that the receptor-ligand complex is trapped by assembly particle-clathrin complexes in coated pits as it diffuses in the membrane.

Coated vesicles are stable at the pH and ion composition of the cell cytosol, yet coated vesicles normally lose their clathrin coat just after endocytosis. How does this happen? A liver enzyme has been purified that depolymerizes coated vesicles to clathrin triskelions. The hydrolysis of ATP to ADP and P_i is required to disrupt the multiple clathrin-clathrin interactions that stabilize the coated vesicle. Both the formation and depolymerization of coated vesicles must be highly regulated in the cell, as both processes occur simultaneously. But much remains to be learned about the molecular details of these processes.

Most Surface Receptors and Membrane Phospholipids Are Recycled

The LDL system illustrates an important characteristic of the receptor-mediated endocytosis of ligands: most surface receptors recycle and repeatedly mediate the internalization of ligand molecules. Like many cell-surface receptors that participate in endocytosis, the LDL receptor makes one round trip into and out of the cell every 10–20 min, or a total of several hundred trips in its 20-h life span. The LDL receptor never reaches a lysosome until it is destined to be degraded by potent lysosome proteases.

Studies on the asialoglycoprotein receptor system have revealed that receptors and their ligands dissociate within the CURL organelle. When ligand is added to a culture medium, some surface asialoglycoprotein receptors are internalized with the ligand in small coated and uncoated vesicles immediately under the cell surface. Beginning 14 min after internalization, the bound asialoglycoproteins are transferred to the lysosomes, but the receptors themselves are never found in these organelles. When both asialoglycoprotein and its receptor are associated with the membranes of small endocytic vesicles just under the cell surface, the ligand is presumed to be bound to the membrane receptor. But in larger vesicles containing both receptor and ligand, free ligand occurs in the lumen and presumably is not bound to the receptor (Figure 14-41). Indeed, the spherical part of these vesicles contains little receptor at all. But tubular membrane structures attached to the large vesicle are rich in receptor and rarely contain ligand. We may assume that these tubules contain receptor that has dissociated from its ligand. Thus the CURL is now recognized as the site at which the receptor uncouples from its ligand. Receptor-rich elongated membrane vesicles are believed to bud off from CURL and then mediate the recycling of receptors back to the cell surface.

The overall rate of endocytic internalization of the plasma membrane is quite high; cultured fibroblasts regularly internalize 50 percent of their cell surface each hour. Not only do many cell-surface receptor proteins recycle, but most other internalized plasma-membrane proteins and endocytosed phospholipids do so as well. Presumably, some membrane vesicle buds off from the endocytic vesicle or a lysosome and eventually returns to the plasma membrane to fuse with it.

Ligands Are Uncoupled from Receptors by Acidification of Endocytic Vesicles

Small endocytic vesicles (prelysosome vesicles, such as CURL) have a pH of 5.2–5.0, compared with a lysosome pH of ~4.5–5.0. Endosomes have been partially purified and do appear to contain a V-class ATP-dependent proton pump (see Table 14-3). Coated vesicles also contain a proton pump, and ligands encounter a progressively lower pH as they move from coated vesicles through the various endosomes and CURLs to lysosomes.

The acidity of endocytic vesicles answers a long-standing question concerning receptor-mediated endocytosis: how can a receptor be persuaded to let go of a tightly bound ligand? Most receptors, including the asialoglycoprotein, insulin, and LDL receptors, bind their ligands tightly at neutral pH but release their ligands if the pH is lowered to <5.0 (Figure 14-42).

Mutant fibroblast cells have been isolated that acidify endosomes defectively. Such cells are deficient in the uptake of LDL particles, for instance, but are also resistant to infection by many viruses that enter the cell through acidic endosomes. In these mutants, acidification of lysosomes is normal; thus it has been concluded that the H^+ ATPase is different in endosomes than in lysosomes.

A Hereditary Disease Is Due to a Genetic Defect in the LDL Receptor

The human LDL system has proved invaluable for studying receptor-mediated endocytosis because persons who have the inherited disorder *familial hypercholesterolemia* (FH), which is characterized by high level of cholesterol in the blood, carry mutant forms of the LDL receptor protein. Persons homozygous for these mutant alleles often die at an early age of heart attacks caused by atherosclerosis, a buildup of cholesterol deposits that ultimately blocks the arteries. The genetic defect specifically affects the LDL receptor.

In some persons with this disorder, the LDL receptor is simply not produced; in others, it binds LDL poorly or not at all. The gene for one mutant receptor has a deletion for the exons that encode the membrane-spanning segment: although the mutant receptor binds LDL normally,

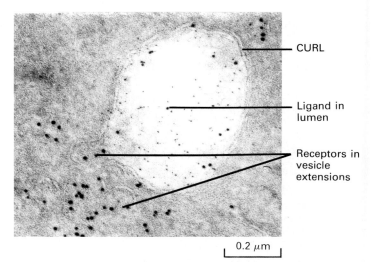

▲ **Figure 14-41** An electron micrograph of the CURL organelle. Both an asialoglycoprotein receptor and a ligand asialoglycoprotein are localized simultaneously in a hepatocyte CURL after receptor-mediated endocytosis. The larger dark grains localize the receptor in the tubular extensions budding off from the vesicle. The smaller dark grains localize the asialoglycoprotein in the vesicle lumen. In this experiment, liver cells were perfused with an asialoglycoprotein and then fixed and sectioned for electron microscopy. To localize the receptor, the sections were stained with antibodies specific for the receptor, which were tagged with gold particles 8 nm in diameter (the large electron-opaque grains in the micrograph). Localization of ligand was detected by reaction of the sections with the asialoglycoprotein antibody linked to gold particles 5 nm in diameter. [See H. J. Geuze et al., 1983, *Cell* **32**:277.] *Courtesy of H. J. Geuze. Copyright 1983, M.I.T.*

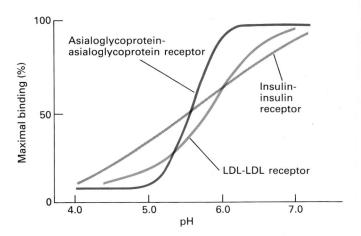

▲ **Figure 14-42** Effect of pH on the binding of proteins to their respective receptors. In all three situations shown, the internalized ligands are transported to lysosomes, where they are degraded, and the receptors are recycled. Dissociation of receptor and ligand is believed to be triggered by the low pH of an endosome or CURL vesicle. [See A. Dautry-Varsat, A. Ciechanover, and H. F. Lodish, 1983, *Proc. Nat'l Acad. Sci. USA* **80**:2258.]

it cannot be anchored to the plasma membrane and is secreted from the cell. In one especially instructive case, the mutant receptor binds LDL normally but the LDL-receptor complex cannot be internalized by the cell because it is distributed evenly over the cell surface rather than confined to coated pits. Other receptors on this person's plasma membranes are internalized normally by the coated-pit system, but the mutant receptor apparently cannot bind to the assembly proteins in the coated pit. The mutant has a change in a single tyrosine in the cytosolic domain; the tyrosine in this position in the wild-type protein seems to be an essential part of the "pit-binding" segment.

Individuals who are heterozygous for FH possess one mutant allele for the LDL receptor and one wild-type allele, so that their cells contain one-half the normal amount of functional LDL receptors. Their serum cholesterol level is higher than normal but lower than that of persons homozygous for the defect; they are more prone to heart attacks than persons with wild-type alleles for the LDL-receptor locus. About one in 500 individuals is heterozygous for FH, making this one of the most common genetic diseases.

Synthesis of the LDL Receptor and Cholesterol Are Tightly Regulated

The LDL uptake system is part of a complex mechanism by which cells regulate cholesterol synthesis. All cells can make cholesterol, but shut down their own cholesterol manufacture if it is provided environmentally. Fibroblasts grown in a medium containing high levels of cholesterol possess one-tenth the LDL receptors of fibroblasts grown in a low-cholesterol medium. Exogenous cholesterol released from its LDL particles in the lysosome causes at least three regulatory changes in the cell: (1) inhibition of a key enzyme of cholesterol biosynthesis, (2) activation of an enzyme used in cholesterol storage, and (3) inhibition of synthesis of the LDL receptor itself by preventing synthesis of receptor mRNA. As the existing LDL receptors are slowly degraded, the plasma membrane will contain fewer receptors and endocytose much less cholesterol in LDL particles. This regulation of the number of surface receptors ensures that cells import only the cholesterol they need.

Proteins Internalized by Receptor-mediated Endocytosis Undergo Various Fates

Internalized LDL particles and galactose-terminal serum proteins are always transferred to the lysosomes and destroyed there. In other receptor-ligand systems, endocytosed material simply remains in the cells and is minimally processed. Developing insect and avian egg cells (oocytes), for example, internalize yolk proteins and other proteins from the blood or surrounding cells.

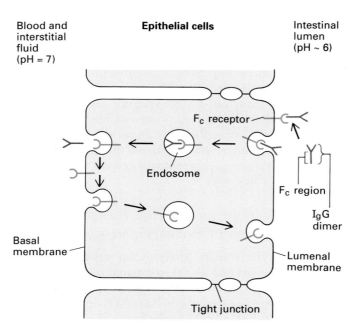

▲ **Figure 14-43** The movement of maternal immunoglobulins across the intestinal epithelial cells of newborn mice involves both endocytosis and exocytosis. In newborn mice, the pH = ~6 in the intestinal lumen and 7.0 on the opposite (blood-facing) side. The particular F_c receptors on these cells bind to the F_c segment of dimeric IgG immunoglobulins only at pH values of 6 or lower, not at pH = 7.0. Vesicles (endosomes) containing the F_c receptor–IgG complex bud from the luminal surface, move across the cell, and fuse with the basal membrane, where they release the IgG. Unloaded receptors are recycled by the formation of endosomes from the basal membrane and their movement and fusions with the luminal membrane.

(Coated pits were first discovered in insect eggs, where they occupy a large portion of the plasma membrane.)

A hen's egg is a single cell containing several grams of protein, virtually all of which is imported from the bloodstream by endocytosis. Vitellogenin, a precursor of yolk proteins (principally of lipovitellin and phosvitin), is synthesized by the liver and secreted into the bloodstream, from which it is endocytosed into the developing egg. Yolk proteins remain in storage granules within the egg and are used as a source of amino acids and energy by the developing embryo after fertilization. Egg-white proteins, such as ovalbumin, lysozyme, and conalbumin, are secreted by cells lining the hen oviduct.

In other cases, endocytosed material passes all the way through polarized cells and is exocytosed (secreted) from the plasma membrane at the opposite side by transcytosis. Examples of this transcellular transport are the movement of maternal antibodies across mammalian yolk-sac cells into the human fetus and across the intestinal epithelial cells of the newborn mouse (Figure 14-43).

The F_c receptor that mediates the transcytosis of immunoglobulins has the unusual property of binding to its ligand at an acidic pH of 6 but not at neutral pH—

(a) Protein

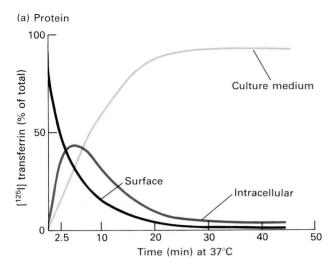

(b) Iron

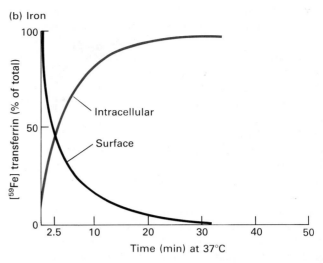

▲ **Figure 14-44** After endocytosis, the fates of the iron-transport protein transferrin and its two bound iron atoms differ. In this experiment, ferrotransferrin double-labeled with ^{125}I in the protein and with ^{59}Fe is bound at 4°C to transferrin receptors on the surface of hepatoma cells. After being warmed to 37°C, both the protein (a) and its bound iron (b) are rapidly internalized, as indicated by loss of label from the surface (*black curves*) and its increase intracellularly (*red curves*). The iron dissociates and remains in the cell; the iron-free transferrin (apotransferrin) is rapidly secreted (*blue curve*). [See A. Ciechanover et al., 1983, *J. Biol. Chem.* **258**:9681.]

precisely the opposite behavior from that of LDL or asialoglycoprotein receptors (see Figure 14-42). Figure 14-43 shows how a difference in pH in the media on the two sides of intestinal epithelial cells in newborn mice allows immunoglobulins to move in one direction from the lumen to the blood.

Transferrin Delivers Iron to Cells by Receptor-mediated Endocytosis

Transferrin is a major serum glycoprotein that transports iron to all tissue cells from the liver (the main site of iron storage in the body) and from the intestine (the site of iron absorption). The iron-free form, *apotransferrin*, binds two Fe^{3+} ions very tightly to form *ferrotransferrin*. All growing cells contain surface transferrin receptors that avidly bind ferrotransferrin ($K_M = 6 \times 10^{-9}\ M$) at neutral pH, after which receptor-bound ferrotransferrin is subjected to endocytosis. But there the similarity with other endocytosed ligands, such as LDL, ends; the two bound Fe^{3+} atoms remain in the cell, but the apotransferrin is secreted from the cell within minutes (Figure 14-44), carried in the bloodstream to the liver or intestine, and reloaded with iron.

What properties of this receptor-ligand complex cause the cell to retain iron and secrete apotransferrin? The answer again lies in changes in pH—more specifically, in the unique ability of apotransferrin to remain bound to the transferrin receptor at the low pH (5.0–6.0) of the endocytic vesicles. At pH ≤ 6.0, the two bound Fe^{3+} atoms dissociate from ferrotransferrin. The iron remains in the endocytic vesicle or the CURL and from there is

transported into the cytosol (in an unknown manner); the apotransferrin formed by the dissociation of the iron atoms remains bound to the transferrin receptor at the pH of these vesicles. The transferrin receptor is recycled back to the surface (like the LDL receptor) but remains bound to the apotransferrin. Remarkably, although apotransferrin binds tightly to its receptor ($K_M = 6 \times 10^{-9}\ M$) at pH = 5.0, it does not bind measurably at neutral pH. The apotransferrin dissociates from its receptor when the recycling vesicles fuse with the plasma membrane and the receptor-ligand complex encounters the neutral pH of the extracellular interstitial fluid or growth medium. The surface receptor is then free to bind another molecule of ferrotransferrin. Figure 14-45 summarizes this pathway.

Entry of Viruses and Toxins into Cells

Many viruses and toxins enter animal cells by receptor-mediated endocytosis. These pathogens have evolved to exploit this specialized system and use it to enter the cell cytoplasm. Generally, the outcome is the death of the cell and, in the case of viruses, the multiplication and spread of viral particles. With an increasing understanding of this phase of the infection process, however, scientists can, in turn, exploit these pathogens for useful purposes. As one example, we shall see how toxins could be utilized as part of a highly specific reagent for killing certain tumor cells.

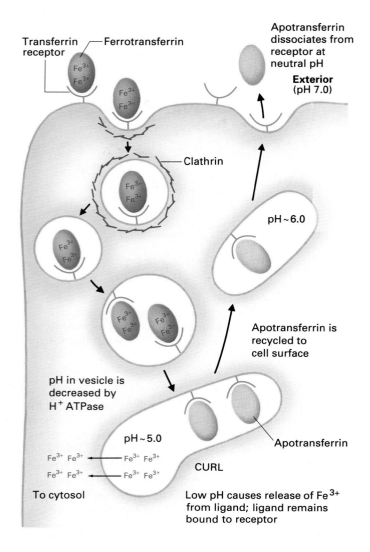

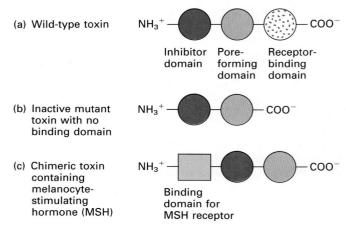

◀ **Figure 14-45** The transferrin cycle operating in all growing mammalian cells. After endocytosis, iron is released from the transferrin-receptor complex in the acidic CURL compartment. The transferrin protein remains bound to its receptor at this pH, and they cycle to the cell surface together. When the receptor-transferrin complex encounters the neutral pH of the exterior medium, the iron-free transferrin is released.

▲ **Figure 14-46** Schematic structures of wild-type *Pseudomonas* exotoxins. (a) Wild-type toxin has three domains: for receptor binding, membrane pore formation, and inhibition of protein synthesis. (b) A mutant toxin that lacks the receptor-binding domain is inactive because it cannot bind to cells. (c) Chimeric toxin, a protein fusion of the melanocyte-stimulating hormone (MSH) with the mutant toxin, binds only to cells that bear receptors for MSH, such as skin melanocytes and malignant melanomas. These cells are killed by the chimeric toxin, because the protein synthesis inhibitor domain can penetrate the cytosol via the transmembrane pore domain. *After V. Chaudhary et al., 1987,* Proc. Nat'l Acad Sci. USA *84:4538–4542.*

Endocytosis Internalizes Bacterial Toxins

Certain pathogenic bacteria synthesize and secrete toxins, proteins that poison or kill susceptible mammalian cells and are the causative agents of bacterial diseases. An important and well-characterized example is the diphtheria toxin secreted by *Corynebacterium diphtheriae,* which kills cells by irreversibly inhibiting protein synthesis. The toxin contains two polypeptide chains, B and A, that are linked by a disulfide bond and synthesized as a single-chain precursor. The two chains have very different functions. The B chain binds the toxin to an unknown receptor on the surface of all human cells. The A chain actually enters the cytosol and inhibits protein synthesis by inactivating elongation factor 2 (EF2), which translocates ribosomes along mRNA concomitant with GTP hydrolysis.

Current evidence suggests that entry of the A peptide into the cytoplasm is initiated by receptor-mediated endocytosis of the toxin-receptor complex. In the acidic environment of the endosome or lysosome, the B fragment undergoes a conformational change that makes large pores or channels at least 2 nm in diameter in the vesicular membrane through which the A peptide enters the cytosol.

Utilizing this knowledge, scientists have genetically modified the related *Pseudomonas* exotoxin to make a potent new class of toxins that only kill cells with certain receptor molecules on their surface. Like the diphtheria toxin, this exotoxin contains three functional domains (Figure 14-46a). First, a receptor-binding domain binds the toxin to cell-surface receptors. After endocytosis, a second domain containing predominantly hydrophilic amino acids forms a pore in the endosome membrane. The third domain, cleaved from the other two after endocytosis, traverses the membrane pore into the cytosol and, by inactivating EF2, kills the cells. A genetically engineered deletion mutant of the toxin that lacks the receptor-binding domain is inactive because it cannot bind to cells (see Figure 14-46b). The genetically engineered cell-specific toxin (see Figure 14-46c) is a *chimera,* or pro-

(a)

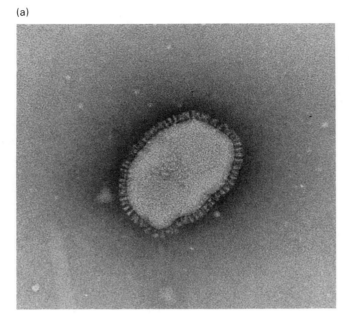

(b)

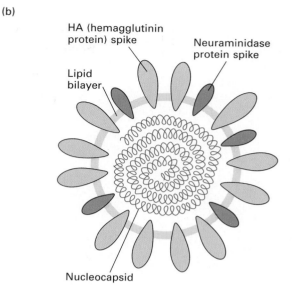

▲ **Figure 14-47** (a) Electron micrograph of a negatively stained particle of influenza virus. The large spikes on the surface are constructed of trimers of the hemagglutinin (HA) protein. (b) Diagram of a cross section of a particle of influenza virus. The smaller spikes (not easily visible in the elec- tron micrograph) are constructed of tetramers of the neuraminidase protein. Inside the virus is the nucleocapsid—a complex of the viral RNAs and a virus-specific protein. *Photograph courtesy of A. Helenius and J. White.*

tein fusion, of the peptide *melanocyte-stimulating hormone* (MSH) with the deletion mutant toxin. By virtue of its MSH domain, the chimera binds only to cells with MSH receptors, such as normal skin melanocytes and melanoma (skin-cancer) tumor cells. Once bound to cells, it kills them in exactly the same way the wild-type toxin does: one domain crosses into the cytosol and inactivates all the EF2. MSH is normally a growth factor for melanocytes; here, it has been converted into a Trojan horse— instead of stimulating the growth of the cells to which it binds, it kills them.

Infection by Many Membrane-enveloped Viruses Is Initiated by Endocytosis

In many animal viruses, an outer phospholipid bilayer membrane surrounds the viral genetic material and protein coat. The virus is usually formed by budding from the plasma membrane of the host cell, and the viral phospholipids are derived from that membrane. The viral membrane also contains one or more virus-specific glycoproteins, without which the virus would not be able to infect a target cell. Infection by this type of virus is almost an exact reversal of exocytosis in that the vesicle (the viral particle) approaches the membrane from the outside and fuses itself into the cell. In this way, the genetic material from the inside of the virus is released into the cytoplasm of the host cell. How the membrane of an animal virus

fuses with the host-cell membrane has been studied in some detail and is likely to be the first membrane fusion system that is completely understood.

The myxovirus group of viruses has taught us the most about membrane fusion. The influenza virus is a good example. Its interior is made up of an RNA genome core bound to a structural protein coat, or *capsid*. This arrangement, called the *nucleocapsid*, is surrounded by a phospholipid bilayer membrane in which two different viral glycoproteins are embedded (Figure 14-47). The predominant glycoprotein of the influenza virus, the *hemagglutinin* (HA) *protein*, is grouped in trimers that form the larger spikes on the surface of the virus. The HA spikes bind to N-acetylneuraminic acid (sialic acid) molecules on the cell surface. Figure 14-48 shows a high-resolution view of the sialic acid binding sites on the tip of an HA spike as well as the overall three-dimensional structure of the HA spike.

Low pH Triggers Fusion of the Viral and Cell Membranes

Bound viral particles are endocytosed into acidic intracellular vesicles. At this stage, the viral and vesicle membranes fuse, releasing viral nucleic acids into the cytosol and initiating replication of the virus (Figure 14-49). There is considerable evidence that the low pH of the enclosing endocytic vesicle triggers its fusion with the viral membrane. For instance, viral infection is inhibited

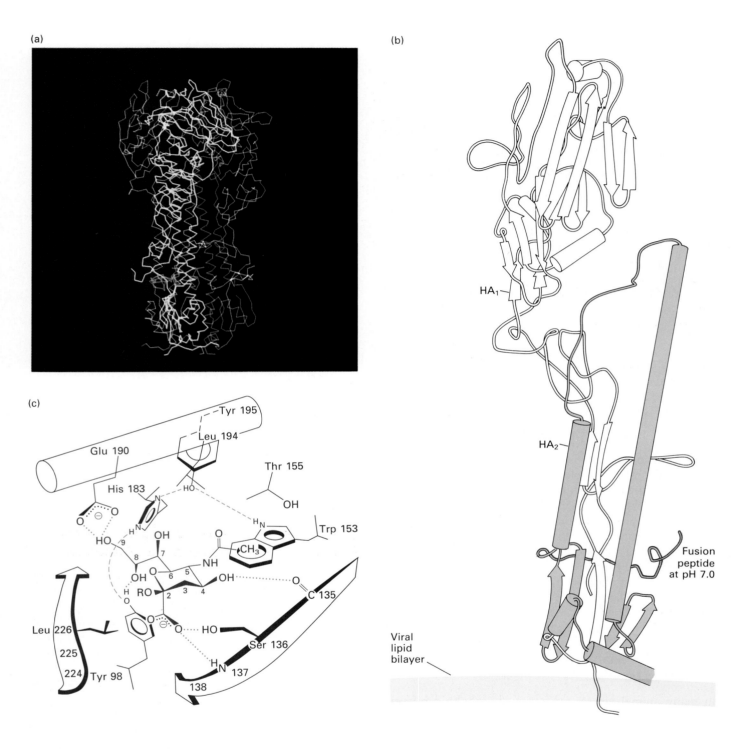

▲ **Figure 14-48** The HA spike of the influenza virus.
(a) Three-dimensional structure. The viral membrane would be at the bottom of the photograph; the spike protrudes outward (see Figure 14-47). The spike is composed of three HA₁ + HA₂ heterodimers: one dimer is yellow; the second, blue; the third (mostly hidden in back), purple. The three fusion peptides (orange) are shown in the conformation at pH = 7.0, where they are tucked into the spike. The three binding sites of sialic acid, the substance on sensitive cells to which the influenza virus binds prior to fusion, are located at the tops of the spike. (b) A more detailed structure of one of the three identical HA₁ + HA₂ dimers that form the large spike (the HA₂ peptide is blue). (c) Schematic drawing of sialic acid (red) bound to the tips of the hemagglutinin, based on the crystal structure determined at a resolution of 3 Å. Dotted lines indicate possible hydrogen bonds between sialic acid and hemagglutinin; dashed lines show potential hydrogen bonds within the protein. [See W. Weis et al., 1988, *Nature* **333**:426]. *Parts (a) and (b) courtesy of D. Wiley; part (c) courtesy of N. K. Sauter and D. Wiley.*

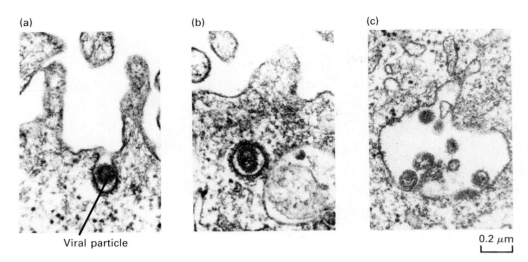

(a)   (b)   (c)

Viral particle

0.2 μm

▲ **Figure 14-49** Entry of fowl plague virus, an avian influenza virus, into cultured dog kidney cells. Virus was bound to the surface of cells at 4°C. These cells were then warmed to 37°C for different lengths of time to initiate viral penetration and fixed for electron microscopy. Initially, the virus is only on the cell surface. Within 5 min at 37°C, viral particles (the large electron-opaque bodies) are seen (a) in coated pits and (b) in coated vesicles. After 10 min at 37°C, viruses are found in (c) endosomes and multivesicular lysosomelike bodies. It is believed that these CURL-like vesicles are acidic and that fusion of the viral and cell membranes occur in them. [See K. S. Matlin et al., 1981, *J. Cell Biol.* **91**:601.] *Photographs courtesy of A. Helenius. Reproduced from the* Journal of Cell Biology *by copyright permission of the Rockefeller University Press.*

by the addition of lipid-soluble bases, such as ammonia or trimethylamine, which raise the pH of normally acidic endosomes.

The molecular events of the membrane fusion process are only partly understood. Each HA spike protein consists of six subunits: three designated HA_1 and three designated HA_2. Crystallographic studies on the HA spike show that at pH = 7.0 the fusion peptide—a strongly hydrophobic amino acid sequence Glu-Leu-Phe-Gly-Ala-Ile-Ala-Gly-Phe-Ile-Glu, at the amino-terminus of HA_2— is tucked into a crevice in the spike (Figure 14-50a). At pH = 5.0, a conformational change is induced during which the fusion peptide swings outward (see Figure 14-50b)! Three exposed fusion peptides from each HA spike protein are believed to insert themselves into the membrane of the endosome. Another conformational change in the HA spike occurs concomitantly: the three globular HA_1 domains at the tip of the spike separate from each other, enabling the viral and cell membranes to come even closer together. The last stages in the actual fusion of the viral and cell membranes are still not well understood.

The conformational change in the HA spike occurs over a very narrow change in pH (5.0–5.5) and is critical for infectivity. This step is now understood in some detail. For instance, certain mutations in the part of the HA protein that binds the fusion peptide at pH = 7.0 destabilize this interaction and allow the conformational alteration to occur at higher (more alkaline) pH values (5.5–6.5). The conformation of the wild-type spike at pH = 5, with its "exposed" fusion peptide, is very unstable. For example, viruses treated at pH = 5 in the absence of a cell

membrane rapidly lose infectivity. Apparently, if the exposed hydrophobic fusion peptide does not insert itself into a phospholipid bilayer, it causes the spikes to aggregate irreversibly.

HIV (AIDS) and Other Enveloped Viruses Fuse Directly with the Plasma Membrane

Other types of membrane-enveloped viruses, such as the paramyxovirus, Sendai virus, and the HIV (human immunodeficiency virus, or AIDS virus), normally fuse with the cell-surface membrane without undergoing endocytosis. The Sendai virus also has two glycoproteins on its surface: a hemagglutinin (binding protein), and a fusion protein. If both glycoproteins are active on a paramyxoviral particle, the virus will bind to the cell and its membrane will fuse with the cell plasma membrane. These viruses usually bind to N-acetylneuraminic acid residues on surface glycoproteins; thus the hemagglutinin behaves very much like a lectin (a carbohydrate-binding protein).

The HIV or AIDS virus also infects cells in this way. The receptor for the HIV virus is a surface glycoprotein called CD4 found mainly on a subset of the T cells of the immune system, on macrophages, and on a group of brain cells. HIV only infects and kills these cells, but such cell death causes a catastrophic block of the entire immune system and the eventual death of the infected person. Viral infection can also spread directly from cell to cell, with no involvement of free virus. As depicted in

(a) pH = 7.0

(b) pH = 5.0

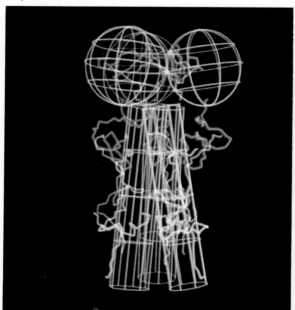
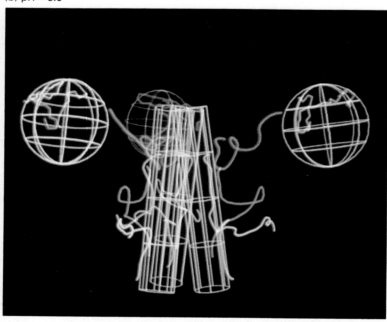

▲ **Figure 14-50** Unfolding of the HA spike occurs at pH = 5.0, as shown by these schematic representations of the structure of the molecule (a) at pH = 7.0 and (b) at pH = 5.0. Each globular domain at the tip of the native spike (a) is composed of part of an HA₁ chain; at pH = 5.0, these domains separate from each other by bending at a hinge, causing a conformational change in parts of the stem (orange and dark blue). The fusion peptide (yellow) at the N-terminus of the HA₂ peptides also becomes exposed at pH = 5.0; it is then capable of inserting itself into the endocytic vesicle membrane, which triggers fusion of the viral and endosome membranes. *From J. M. White and I. A. Wilson, 1987, J. Cell Biol. 105:2887–2896. Courtesy of J. White.*

Figure 14-51, during viral replication within a T cell, a large amount of viral glycoprotein is produced on the cell surface. Some of this buds off into new virus, but some can bind to CD4 proteins on a nearby uninfected T cell and lead to fusion of the infected and uninfected cells. One infected cell can fuse with many uninfected cells; the resultant multinucleate *syncytium* is markedly defective in immune responses.

Summary

The plasma membrane regulates the traffic of molecules into and out of the cell. Gases and small hydrophobic molecules diffuse directly across the phospholipid bilayer at a rate proportional to their ability to dissolve in a liquid hydrocarbon. However, ions, sugars, and amino acids cannot diffuse across the phospholipid bilayer at sufficient rates to meet the cell's needs and must be transported by a class of integral membrane proteins called transporters, or permeases. For some molecules, protein-catalyzed entry into the cell is driven only by a concentration gradient. For glucose to enter an erythrocyte, for example, a glucose permease forms a transmembrane passage that allows only bound glucose and closely related molecules to cross the bilayer.

Some molecules require transport against an ion concentration gradient to enter a cell, a process in which energy is expended. The Na^+-K^+ ATPase, for example, pumps three Na^+ ions out of and two K^+ ions into the cell, utilizing the energy of ATP hydrolysis. The Ca^{2+} ATPase pumps two Ca^{2+} ions out of the cell or, in muscle, into the sarcoplasmic reticulum. These pumps create an intracellular ion milieu of high K^+, low Ca^{2+}, and low Na^+ that is very different from the extracellular fluid milieu of high Na^+, high Ca^{2+}, and low K^+. An ATP-dependent H^+ pump in animal lysosome and endosome membranes and plant vacuole membranes is responsible for maintaining a lower pH inside the organelles than in the surrounding cytosol.

A small molecule may also be imported or exported against its concentration gradient by coupling its movement to that of another molecule or ion, usually H^+ or Na^+, down its concentration gradient. Glucose and amino acids enter absorptive cells by coupling to the passive influx of Na^+ by a symport process. In cardiac muscle cells, the export of Ca^{2+} ions is coupled to the import of Na^+ ions by an antiport process. In other cases, the uptake of nutrients is driven by a proton gradient across

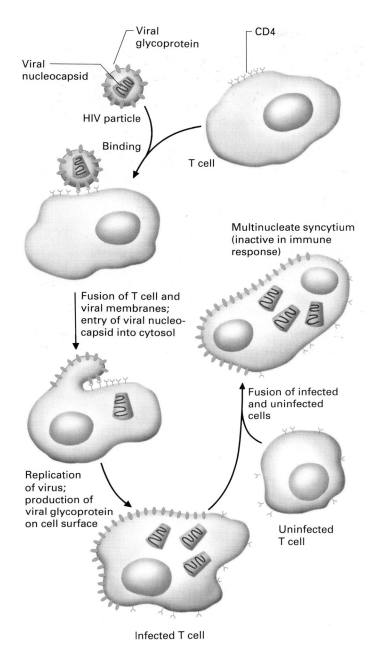

▲ **Figure 14-51** Infection of T cells in the immune system by the HIV (AIDS) virus and spread of the infection by cell-cell fusion. The virus infects only a subset of the T cells that bear CD4, a surface protein. CD4 acts as the receptor for the glycoprotein on the surface membrane of the HIV virus. Fusion of the viral and plasma membranes occurs quickly after binding, allowing the viral nucleocapsid to penetrate into the cytosol. Eventually, the viral genome reaches the nucleus, where it replicates. During replication, a large amount of viral glycoprotein is produced by the infected cell and inserted in the surface membrane. Like the initial fusion of HIV virus with a cell, the surface HIV glycoprotein causes fusion of the infected T cell with uninfected cells that bear the CD4 protein, forming a multinucleate syncytium that cannot carry out the normal T-cell immune function. Infected T cells in blood or semen may spread AIDS from person to person.

Labels in figure: Viral glycoprotein; CD4; Viral nucleocapsid; HIV particle; Binding; T cell; Fusion of T cell and viral membranes; entry of viral nucleocapsid into cytosol; Multinucleate syncytium (inactive in immune response); Fusion of infected and uninfected cells; Replication of virus; production of viral glycoprotein on cell surface; Uninfected T cell; Infected T cell

the cell membrane: one example is the uptake of lactose by the bacterium *E. coli*, catalyzed by a sugar H^+ symport; another is the uptake of sucrose, sodium, nitrate, and other substances into plant vacuoles. Bacterial cells also import some nutrients by a different group-translocation reaction—the phosphorylation of sugars coupled to their uptake.

The plasma membranes of many cells are differentiated so that specific membrane segments contain different permeases and carry out quite different transport processes. In the intestinal epithelial cell, the glucose-Na^+ and amino acid-Na^+ symporters are in the apical membrane region facing the intestinal lumen and the Na^+-K^+ ATPase and glucose and amino acid permeases are in the basolateral membrane region facing the blood capillaries. This combination allows the cells to transport amino acids and glucose from the lumen to the blood. Acid-secreting cells in the stomach have an ATP-driven H^+ pump on the apical membrane and an HCO_3^-/Cl^- anion antiport on the basolateral membrane; the latter allows the cytosolic pH to be maintained at near neutrality, despite the active export of protons from the cells into the stomach lumen.

The bulk flow of water across cell layers is driven either by hydrostatic pressure (for example, from a blood vessel into a cell) or by osmotic pressure. Where a cell epithelium secretes ions, such as the H^+Cl^- ions secreted by parietal cells lining the stomach, the movement of water out of the cells is driven by the resulting osmotic gradient. Changes in the cytosolic concentration of solutes such as K^+ and the resultant osmotic efflux or influx of water control the respective closing and opening of stomata in leaves.

Proteins, bacteria, and viruses are imported into animal cells by fundamentally different processes in which regions of the plasma membrane detach and form intracellular vesicles. One such process is phagocytosis—the means by which very large particles, such as bacteria, are internalized. Phagocytosis is usually initiated by specific receptors on the cell surface, which sequentially bind a particle until the plasma membrane gradually envelops it (the "zipper" interaction), eventually forming an intracellular membrane-limited vesicle. The vesicle then fuses with one or more lysosomes, triggering the degradation of the particle within. Antibodies bound to foreign particles facilitate their phagocytosis by macrophages, which contain specific receptors for the F_c region that all IgG antibodies have in common.

Pinocytosis is nonspecific endocytosis: material is taken up by cells in proportion to its concentration in the extracellular medium. In endocytosis, small regions of the plasma membrane are internalized, forming small intracellular vesicles. In receptor-mediated endocytosis, particles such as viruses, small proteins, and oligosaccharides first bind to specific receptor proteins on the plasma membrane. Regions of the plasma membrane containing these receptor-ligand complexes are selectively internalized in pits that bud off to form vesicles. These pits and vesicles have an outer basketlike coat composed of clathrin, a fibrous protein. Assembly proteins between the clathrin and vesicle membrane bind to plasma-membrane proteins, causing them to be incorporated in coated pits and vesicles. After internalization, the vesicles are uncoated by depolymerization of the clathrin coat to clathrin triskelions. The resulting smooth vesicles, or endosomes, fuse with other vesicles; their interiors are acidified, which leads to changes in the ligand-receptor relationship and changes in the ligand itself.

The fate of the endocytosed ligand can vary considerably. Endocytosed membrane-coated viruses, such as the influenza virus, fuse with the membrane of the endocytic vesicle, enabling the viral nucleic acid to enter the cytosol, where it can replicate. The low pH induces a conformational change in the viral HA spike, triggering fusion of the viral membrane with a cell membrane.

The diphtheria toxin and some other bacterial toxins have a different means of escaping degradation in the acidic endocytic vesicles: one subunit of the toxin forms a transmembrane channel across the vessel; the other subunit crosses the vesicle membrane and enters the cytosol, where it catalyzes the reactions that kill the cell. Other endocytosed ligands, such as LDL particles and galactose-terminal glycoproteins, are transported via a series of membrane-limited vesicles to lysosomes. Most endocytosed proteins are degraded in these organelles, but cholesterol is released from the LDL particle and used for the synthesis of cell membranes.

In most cases of receptor-mediated endocytosis, the receptors are not degraded on internalization; rather, they release their internalized ligand and return to the cell surface, where they are recycled. The low pH of the CURL organelle triggers the dissociation of most receptors and ligands. Each transferrin protein transports two Fe^{3+} ions to the cell. After binding to the transferrin receptor and endocytosis, the Fe^{3+} ions are released in the CURL and transported to the cytosol, while the transferrin protein is rapidly secreted from the cell.

References

Ion Channels, Intracellular Ion Environment, and Membrane Electric Potential

*CHRISTENSEN, H. N. 1975. *Biological Transport*, 2d ed. Benjamin.

GRAVES, J. S., ed. 1985. *Regulation and Development of Membrane Transport Processes*. Society of General Physiology/ Wiley.

*A book or review article that provides a survey of the topic.

*RACKER, E. 1985. *Reconstitutions of Transporters, Receptors, and Pathological States*. Academic Press.

*SCHROEDER, J. I., and R. HEDRICH. 1989. Involvement of ion channels and active transport in osmoregulation and signaling of higher plant cells. *Trends Biochem. Sci.* 14:187–192.

*STEIN, W. D., and W. R. LIEB. 1986. *Transport and Diffusion across Cell Membranes*. Academic Press.

*YUDILEVICH, D. L., and C. A. R. BOYD. 1987. Amino acid transport in animal cells. Manchester University Press.

Passive Transport across the Cell Membrane

Facilitated Diffusion of Glucose

ALVAREZ, J., D. C. LEE, S. A. BALDWIN, and D. CHAPMAN. 1987. Fourier transform infrared spectroscopic study of the structure and conformational changes of the human erythrocyte glucose transporter. *J. Biol. Chem.* 262:3502–3509.

*BALDWIN, S. A., and P. J. F. HENDERSON. 1989. Homologies between sugar transporters from eukaryotes and prokaryotes. *Ann. Rev. Physiol.* 51:459–471.

CARRUTHERS, A., and D. L. MELCHIOR. 1988. Effects of lipid environment on membrane transport: the human erythrocyte sugar transport protein/lipid bilayer system. *Ann. Rev. Physiol.* 50:257–271.

CHIN, J. J., E. K. Y. JUNG, V. CHEN, and C. Y. JUNG. 1987. Structural basis of human erythrocyte glucose transporter function in proteoliposome vesicles: circular dichroism measurements. *Proc. Nat'l Acad. Sci. USA* 84:4113–4116.

MUECKLER, M., C. CARUSO, S. A. BALDWIN, M. PANICO, I. BLENCH, H. R. MORRIS, W. J. ALLARD, G. E. LIENHARD, and H. F. LODISH. 1985. Sequence and structure of a human glucose transporter. *Science* 229:941–945.

*WALMSLEY, A. R. 1988. The dynamics of the glucose transporter. *TIBS* 13:226–231.

Active Ion Transport and ATP Hydrolysis

General Aspects of Structure and Mechanism

*JENCKS, W. P. 1988. Energy-rich compounds and work. Fritsch Lipmann Memorial Symposium (1987). In Kleinkauf, van Dohren, and Jaenicke, eds., *The Roots of Modern Biochemistry*, Walter de Gruyter (Berlin), pp. 569–580.

*JENCKS, W. P. 1989. How does a calcium pump pump calcium? *J. Biol. Chem.* 264:18855–18858.

JORGENSEN, P. L., and J. P. ANDERSEN. 1988. Structural basis for E_1–E_2 conformational transitions in Na, K-pump, and Ca-pump proteins. *J. Memb. Biol.* 103:95–120.

*PEDERSEN, P. L., and E. CARAFOLI. 1987. Ion motive ATPases: I. Ubiquity, properties, and significance to cell function. *TIBS* 12:146–150; II. Energy coupling and work output. *TIBS* 12:186–189.

Na⁺-K⁺ ATPase

*FAMBROUGH, D. M. 1988. The sodium pump becomes a family. *TINS* 11:325–328.

KENT, R. B., J. R. EMANUEL, Y. B. NERIAH, R. LEVENSON, and D. E. HOUSMAN. 1987. Ouabain resistance conferred by expression of the cDNA for a murine Na^+/K^+ ATPase α-subunit. *Science* 237:901–903.

ROSSIER, B. C., K. GEERING, and J. P. KRAEHENBUHL. 1987. Regulation of the sodium pump: how and why? *TIBS* 12:483–487.

SHULL, G. E., A. SCHWARTZ, and J. B. LINGREL. 1985. Amino acid sequence of the catalytic subunit of the Na$^+$/K$^+$ ATPase deduced from a complementary DNA. *Nature* 316:691–695.

TANIGUCHI, K., K. SUZUKI, D. KAI, I. MATSUOKA, K. TOMITA, and S. IIDA. 1984. Conformational change of sodium- and potassium-dependent adenosine triphosphatase. *J. Biol. Chem.* 259:15228–15233.

Ca^{2+} ATPase

BRANDT, P., M. ZURINI, R. L. NEVE, R. E. RHOADS, and T. C. VANAMAN. 1988. A C-terminal calmodulin-like regulatory domain from the plasma membrane Ca^{2+}-pumping ATPase. *Proc. Nat'l Acad. Sci. USA* 85:2914–2918.

CLARKE, D. M., T. W. LOO, G. INESI, and D. M. MACLENNAN. 1989. Location of high affinity Ca^{2+}-binding sites within the predicted transmembrane domain of the sarcoplasmic reticulum of Ca^{2+}-ATPase. *Nature* 339:476–478.

JAMES, P., M. MAEDA, R. FISCHER, A. K. VERMA, J. KREBS, J. T. PENNISTON, and E. CARAFOLI. 1988. Identification and primary structure of a calmodulin binding domain of the Ca^{2+} pump of human erythrocytes. *J. Biol. Chem.* 263:2905–2910.

*MACLENNAN, D. H., C. J. BRANDL, B. KORCZAK, and N. M. GREEN. 1985. Amino acid sequence of a Ca^{2+}/Mg^{2-}-dependent ATPase from rabbit muscle sarcoplasmic reticulum, deduced from its complementary DNA sequence. *Nature* 316:696–700.

Proton Pumps

ARAI, H., G. TERRES, S. PINK, and M. FORGAC. 1988. Topography and subunit stoichiometry of the coated vesicle proton pump. *J. Biol. Chem.* 263:8796–8802.

BROWN, D., S. GLUCK, and J. HARTWIG. 1987. Structure of the novel membrane-coating material in proton-secreting epithelial cells and identification as an H$^+$ ATPase. *J. Cell Biol.* 105:1637–1648.

BROWN, D., S. HIRSCH, and S. GLUCK. 1988. An H$^+$ ATPase in opposite plasma-membrane domains in kidney epithelial cell subpopulations. *Nature* 331:622–624.

*FORGAC, M. 1989. Structure and function of the vacuolar class of ATP-driven proton pumps. *Physiol. Rev.* 69:765–796.

NELSON, N. and L. TAIZ. 1989. The evolution of H$^+$-ATPases. *TIBS* 14:113–116.

RABON, E., R. D. GUNTHER, A. SOUMARMON, S. BASSILIAN, M. LEWIN, and G. SACHS. 1985. Solubilization and reconstitution of the gastric H/K ATPase. *J. Biol. Chem.* 260:10200–10207.

RAE, P. A., and D. SANDERS. 1987. Tonoplast energization: two H$^+$ pumps, one membrane. *Physiol. Plantarum* 71:131–141.

RUDNICK, G. 1986. ATP-driven H$^+$-pumping into intracellular organelles. *Ann. Rev. Physiol.* 48:403–413.

*SERRANO, R. 1988. Structure and function of proton-translocating ATPase in plasma membranes of plants and fungi. *Biochim. Biophys. Acta* 947:1–28.

Multidrug Resistance

*AMES, G. F-L. 1986. The basis of multidrug resistance in mammalian cells: homology with bacterial transport. *Cell* 47:323–324.

CHOI, K., C. CHEN, M. KRIEGLER, and I. B. RONINSON. 1988. An altered pattern of cross-resistance in multidrug-resistant human cells results from spontaneous mutations in the mdr 1 (P-glycoprotein) gene. *Cell* 53:519–529.

*ENDICOTT, J. A., and V. LING. 1989. The biochemistry of P-glycoprotein-mediated multidrug resistance. *Ann. Rev. Biochem.* 58:137–171.

GOTTESMAN, M. M., and I. PASTAN. 1988. The multidrug transporter, a double-edged sword. *J. Biol. Chem.* 263:12163–12166.

*KARTNER, N., and V. LING. 1989. Multidrug resistance in cancer. *Sci. Am.* (3):44–51.

MCGRATH, J. P., and A. VARSHAVSKY. 1989. The yeast STE6 gene encodes a homologue of the mammalian multidrug resistance P-glycoprotein. *Nature* 340:400–404.

Cotransport: Symport and Antiport

Amino Acids, Glucose, and Ca^{2+}

*HANDLER, J. S. 1989. Overview of epithelial polarity. *Ann. Rev. Physiol.* 51:729–740.

HEDIGER, M. A., M. J. COADY, T. S. IKEDA, and E. M. WRIGHT. 1987. Expression cloning and cDNA sequencing of the Na$^+$/glucose cotransporter. *Nature* 330:379–381.

HOMEYER, U., K. LITEK, B. HUCHZERMEYER, and G. SCHULTZ. 1989. Uptake of phenylalanine into isolated barley vacuoles is driven by both tonoplast adenosine triphosphatase and pyrophosphatase. *Plant Physiol.* 89:1388–1393.

KIMURA, J., A. NOMI, and H. IRISAWA. 1986. Na–Ca exchange current in mammalian heart cells. *Nature* 319:596–597.

PEERCE, B. E., and E. M. WRIGHT. 1986. Distance between substrate sites on the Na/glucose cotransporter by fluorescence energy transfer. *Proc. Nat'l Acad. Sci. USA* 83:8092–8096.

*SEMENZA, G., M. KESSLER, M. HOSANG, J. WEBER, and U. SCHMIDT. 1984. Biochemistry of the Na$^+$/D-glucose cotransporter of the small-intestinal brush-border membrane. *Biochim. Biophys. Acta* 779:343–379.

Anion-exchange Proteins

DRENCKHAHN, D., M. OELMANN, P. SCHAAF, M. WAGNER, and S. WAGNER. 1987. Band 3 is the basolateral anion exchanger of dark epithelial cells of turtle urinary bladder. *Am. J. Physiol.* 251:C570–C574.

FALKE, J. J., and S. I. CHAN. 1985. Evidence that anion transport by band 3 proceeds via a ping-pong mechanism involving a single transport site. *J. Biol. Chem.* 260:9537–9544.

PASSOW, H. 1986. Molecular aspects of band 3 protein-mediated anion transport across the red cell membrane. *Rev. Physiol. Biochem. Pharmacol.* 103:62–203.

Control of Cytosolic pH

ASANO, S., M. INOIE, and N. TAKEGUCHI. 1987. The Cl$^-$ channel in hog gastric vesicles is part of the function of H/K ATPase. *J. Biol. Chem.* 262:13263–13266.

*BORON, W. F. 1986. Intracellular pH regulation in epithelial cells. *Ann. Rev. Physiol.* 48:377–388.

BUSA, W. B., and J. H. CROWE. 1983. Intracellular pH regulates transitions between dormancy and development of brine shrimp (*Artemia salina*) embryos. *Science* 221:366–368.

*MOOLENAAR, W. H. 1986. Effects of growth factors on intracellular pH regulation. *Ann. Rev. Physiol.* **46**:363–376.

OLSNES, S., J. LUDT, T. I. TONNESSEN, and K. SANDVIG. 1987. Bicarbonate–chloride antiport in vero cells: II. Mechanisms for bicarbonate-dependent regulation of intracellular pH. *J. Cell. Physiol.* **132**:192–202.

PARADISO, A. M., R. Y. TSIEN, and T. E. MACHEN. 1987. Digital image processing of intracellular pH in gastric oxyntic and chief cells. *Nature* **325**:447–450.

POUYSSEGUR, J., C. SARDET, A. FRANCHI, G. L'ALLEMAIN, and S. PARIS. 1984. A specific mutation abolishing Na$^+$/H$^+$ antiport activity in hamster fibroblasts precludes growth at neutral and acidic pH. *Proc. Nat'l Acad. Sci. USA* **81**:4833–4837.

SARDET, C., A. FRANCHI, and J. POUYSSEGUR. 1989. Molecular cloning, primary structure, and expression of the human growth factor–activatable Na$^+$/H$^+$ antiporter. *Cell* **56**:271–280.

SCHULDINER, S., and E. ROZENGURT. 1982. Na$^+$/H$^+$ antiport in Swiss 3T3 cells: mitogenic stimulation leads to cytoplasmic alkalinization. *Proc. Nat'l Acad. Sci. USA* **79**:7778–7782.

Transport into Prokaryotic Cells

BROKER, R., and T. H. WILSON. 1985. Isolation and nucleotide sequences of lactose carrier mutants that transport maltose. *Proc. Nat'l Acad. Sci. USA* **82**:3959–3963.

GHOSH, B. K., K. OWENS, R. PIETRI, and A. PETERKOFSKY. 1989. Localization to the inner surface of the cytoplasmic membrane by immunoelectron microscopy of enzyme I of the phosphoenolpyruvate:sugar phosphotransferase system of *Escherichia coli. Proc. Nat'l Acad. Sci. USA* **86**:849–853.

*KABACK, H. R. 1988. Site-directed mutagenesis and ion-gradient driven active transport: on the path of the proton. *Ann. Rev. Physiol.* **40**:243–256.

*POSTMA, P. W., and J. W. LENGELER. 1985. Phosphoenolpyruvate:carbohydrate phosphotransferase system of bacteria. *Microbiol. Rev.* **49**:232–269.

*REIZER, J., and A. PETERKOFSKY, eds. 1987. Sugar transport and metabolism in gram-positive bacteria. Wiley.

*SAIER, M. H. 1985. Mechanism and regulation of carbohydrate transport in bacteria. Academic Press.

Osmosis, Movement of Water, and Regulation of Cell Volume

EVELOFF, J. L., and D. G. WARNOCK. 1987. Activation of ion transport systems during cell-volume regulation. *Am. J. Physiol.* **252**:F1–F10.

*GRINSTEIN, S., A. ROTHSTEIN, B. SARKADI, and E. W. GELFAND. 1984. Responses of lymphocytes to anisotonic media: volume-regulating behavior. *Am. J. Physiol.* **246**:C204–C215.

*KREGENOW, F. M. 1981. Osmoregulatory salt transporting mechanisms: control of cell volume in anisotonic media. *Ann. Rev. Physiol.* **43**:493–505.

VERKMAN, A. S., W. L. LENCER, D. BROWN, and D. A. AUSIELLO. 1988. Endosomes from kidney collecting tubule cells contain the vasopressin-sensitive water channel. *Nature* **333**:268–269.

The Internalization of Macromolecules and Particles

Pinocytosis

FERRIS, A. L., J. C. BROWN, R. D. PARK, and B. STORRIE. 1987. Chinese hamster ovary cell lysosomes rapidly exchange contents. *J. Cell Biol.* **105**:2703–2712.

GONELLA, P. A., and M. R. NEUTRA. 1984. Membrane-bound and fluid-phase macromolecules enter separate prelysosomal compartments in absorptive cells of suckling rat ileum. *J. Cell Biol.* **99**:909–917.

MELLMAN, I. S., R. M. STEINMAN, J. C. UNKELESS, and Z. A. COHN. 1980. Selective iodination and polypeptide composition of pinocytic vesicles. *J. Cell Biol.* **86**:712–722.

STORRIE, B., M. SACHDEVA, and V. S. VIERS. 1984. Chinese hamster ovary cell lysosomes retain pinocytized horseradish peroxidase and in situ-radioiodinated proteins. *Mol. Cell Biol.* **4**:295–301.

Membrane Fusion

RAND, R. P., and A. PARSEGIAN. 1986. Mimicry and mechanism in phospholipid models of membrane fusion. *Ann. Rev. Physiol.* **48**:201–212.

SOWERS, A. E., ed. 1987. *Cell Fusion*, Plenum Press.

Phagocytosis

GRIFFIN, F. M., JR., J. A. GRIFFIN, J. E. LEIDER, and S. C. SILVERSTEIN. 1975. Studies on the mechanism of phagocytosis: I. Requirements for circumferential attachment of particle-bound ligands to specific receptors on the macrophage plasma membrane. *J. Exp. Med.* **142**:1263–1282.

GRIFFIN, F. M., JR., J. A. GRIFFIN, and S. C. SILVERSTEIN. 1976. Studies on the mechanism of phagocytosis: II. The interaction of macrophages with anti-immunoglobulin IgG-coated bone marrow derived lymphocytes. *J. Exp. Med.* **144**:788–809.

HORWITZ, M. A. 1984. Phagocytosis of the Legionnaires' disease bacterium *(Legionella pneumophila)* occurs by a novel mechanism: engulfment within a pseudopod coil. *Cell* **36**:27–33.

HORWITZ, M. A., and F. R. MAXFIELD. 1984. *Legionella pneumophila* inhibits acidification of its phagosome in human monocytes. *J. Cell Biol.* **99**:1936–1943.

LEWIS, V. A., T. KOCH, H. PLUTNER, and I. MELLMAN. 1986. A complementary DNA clone for a macrophage–lymphocyte F$_c$ receptor. *Nature* **324**:372–375.

YOUNG, J. D.-E., J. C. UNKELESS, T. M. YOUNG, A. MAURO, and Z. A. COHN. 1983. Role for mouse macrophage IgG F$_c$ receptor as ligand-dependent ion channel. *Nature* **306**:186–189.

Receptor-mediated Endocytosis

Endosomes

*DAUTRY-VARSAT, A., and H. F. LODISH. 1984. How receptors bring proteins and particles into cells. *Sci. Am.* **250**:52–58.

DAVEY, J., S. HURTLEY, and G. WARREN. 1985. Reconstitution of an endocytic fusion event in a cell-free system. *Cell* 43:643–652.

GRUENBERG, J., G. GRIFFITHS, and K. E. HOWELL. 1989. Characterization of the early endosome and putative endocytic carrier vesicles in vivo and with an assay of vesicle fusion in vitro. *J. Cell Biol.* 108:1301–1316.

MARSH, M., G. GRIFFITHS, G. E. DEAN, I. MELLMAN, and A. HELENIUS. 1986. Three-dimensional structure of endosomes in BHK-21 cells. *Proc. Nat'l Acad. Sci. USA* 83:2899–2903.

RIEZMAN, H., Y. CHVATCHKO, and V. DULIC. 1986. Endocytosis in yeast. *TIBS* 11:325–328.

SCHMID, S. L., R. FUCHS, P. MALE, and I. MELLMAN. 1988. Two distinct subpopulations of endosomes involved in membrane recycling and transport to lysosomes. *Cell* 52:73–83.

STOORVOGEL, W., H. J. GEUZE, J. M. GRIFFITH, A. L. SCHWARTZ, and G. J. STROUS. 1989. Relations between the intracellular pathways of the receptors for transferrin, asialoglycoprotein, and mannose-6-phosphate in human hepatoma cells. *J. Cell Biol.* 108:2137–2148.

TRAN, D., J-L. CARPENTIER, F. SAWANO, P. GORDEN, and L. ORCI. 1987. Ligands internalized through coated or noncoated invaginations follow a common intracellular pathway. *Proc. Nat'l Acad. Sci. USA* 84:7957–7961.

Clathrin and Assembly of Coated Vesicles

BAR-ZVI, D. 1987. The molecular basis for clathrin light-chain diversity. *Nature* 326:133–134.

*BRODSKY, F. M. 1988. Living with clathrin: its role in intracellular membrane traffic. *Science* 242:1396–1402.

*HARRISON, S. C., and T. KIRCHHAUSEN. 1983. Clathrin, cages, and coated vesicles. *Cell* 33:650–652.

HEUSER, J. E., and R. G. W. ANDERSON. 1989. Hypertonic media inhibit receptor-mediated endocytosis by blocking clathrin-coated pit formation. *J. Cell Biol.* 108:389–400.

HEUSER, J. E., and J. H. KEEN. 1988. Deep-etch visualization of proteins involved in clathrin assembly. *J. Cell Biol.* 107:877–886.

KEEN, J. H. 1987. Clathrin assembly proteins: affinity purification and a model for coat assembly. *J. Cell Biol.* 105:1989–1998.

LAZAROVITS, J., and M. ROTH. 1988. A single amino acid change in the cytoplasmic domain allows the influenza virus hemagglutinin to be endocytosed through coated pits. *Cell* 53:743–752.

MAHAFFEY, D. T., M. S. MOORE, F. M. BRODSKY, and R. G. W. ANDERSON. 1989. Coat proteins isolated from clathrin coated vesicles can assemble into coated pits. *J. Cell Biol.* 108:1615–1624.

*PEARSE, B. M. F. 1987. Clathrin and coated vesicles. *EMBO J.* 6:2507–2512.

PEARSE, B. M. F. 1988. Receptors compete for adaptors found in plasma-membrane coated pits. *EMBO J.* 7:3331–3336.

SCHMID, S. L., and J. E. ROTHMAN. 1985. Enzymatic dissociation of clathrin in a two-stage process. *J. Biol. Chem.* 260:10044–10049.

Low-density Lipoprotein Receptor and LDL Uptake

ANDERSON, R. G. W., M. S. BROWN, U. BEISIEGEL, and J. L. GOLDSTEIN. 1982. Surface distribution and recycling of the low density lipoprotein receptor as visualized with antireceptor antibodies. *J. Cell Biol.* 93:523–531.

BROWN, M. S., and J. L. GOLDSTEIN. 1979. Receptor-mediated endocytosis: insights from the lipoprotein receptor system. *Proc. Nat'l Acad. Sci. USA* 76:3330–3337.

*BROWN, M. S., and J. L. GOLDSTEIN. 1984. How LDL receptors influence cholesterol and atherosclerosis. *Sci. Am.* 251(5):58.

*BROWN, M. S., and J. L. GOLDSTEIN. 1986. Receptor-mediated pathway for cholesterol homeostasis. *Science* 232:34–47.

DAVIS, C. G., J. L. GOLDSTEIN, T. C. SUDHOF, R. G. W. ANDERSON, D. W. RUSSELL, and M. S. BROWN. 1987. Acid-dependent ligand dissociation and recycling of LDL receptor mediated by growth-factor homology region. *Nature* 326:760–765.

ESSER, V., L. E. LIMBIRD, M. S. BROWN, J. L. GOLDSTEIN, and D. W. RUSSELL. 1988. Mutational analysis of the ligand-binding domain of the low density lipoprotein receptor. *J. Biol. Chem.* 263:13282–13290.

LEHRMAN, M. A., J. L. GOLDSTEIN, M. S. BROWN, D. W. RUSSELL, and W. J. SCHNEIDER. 1985. Internalization-defective LDL receptors produced by genes with nonsense and frameshift mutations that truncate the cytoplasmic domain. *Cell* 41:735–743.

SUDHOF, T., J. L. GOLDSTEIN, M. S. BROWN, and D. W. RUSSELL. 1985. The LDL receptor gene: a mosaic of exons shared with different proteins. *Science* 228:815–822.

WILSON, J. M., D. E. JOHNSTON, D. M. JEFFERSON, and R. C. MULLIGAN. Correction of the genetic defect in hepatocytes from the Watanabe heritable hyperlipidemic rabbit. *Proc. Nat'l Acad. Sci. USA* 85:4421–4425.

Asialoglycoprotein Receptor

*DRICKAMER, K. 1988. Two distinct classes of carbohydrate-recognition domains in animal lectins. *J. Biol. Chem.* 263:9557–9560.

LOEB, J. A., and K. DRICKAMER. 1988. Conformational changes in the chicken receptor for endocytosis of glycoproteins. *J. Biol. Chem.* 263:9752–9760.

MCPHAUL, M., and P. BERG. 1987. Identification and characterization of cDNA clones encoding two homologous proteins that are part of the asialoglycoprotein receptor. *Mol. Cell Biol.* 7:1841–1847.

*SCHWARTZ, A. L. 1984. The hepatic asialoglycoprotein receptor. *CRC Crit. Rev. Biochem.* 16:207–233.

SHIA, M., and H. F. LODISH. 1989. The two subunits of the human asialoglycoprotein receptor have different fates when expressed in fibroblasts. *Proc. Nat'l Acad. Sci. USA* 86:1158–1162.

Fates of Internalized Proteins

HOPPE, C. A., T. P. CONNOLLY, and A. L. HUBBARD. 1985. Transcellular transport of polymeric IgA in the rat hepatocyte: biochemical and morphological characterization of the transport pathway. *J. Cell Biol.* 101:2113–2123.

MOSTOV, K. E., and N. E. SIMISTER. 1985. Transcytosis. *Cell* 43:389–390.

PERRY, M. M., H. D. GRIFFIN, and A. B. GILBERT. 1984. The binding of very low density and low density lipoproteins to the plasma membrane of the hen oocyte. *Exp. Cell Res.* 151:433–446.

ROTH, T. F., and K. R. PORTER. 1964. Yolk protein uptake in the oocyte of the mosquito *Aedas aegypti*. *J. Cell Biol.* 20:313–332.

SIMISTER, N. E., and K. E. MOSTOV. 1989. An F_c receptor structurally related to MHC class I antigens. *Nature* 337:184–187.

Transferrin and Iron Uptake

BAKER, E. N., S. V. RUMBALL, and B. F. ANDERSON. 1987. Transferrins: insights into structure and function from studies on lactoferrin. *TIBS* 12:350–353.

DAUTRY-VARSAT, A., A. CIECHANOVER, and H. F. LODISH. 1983. pH and the recycling of transferrin during receptor-mediated endocytosis. *Proc. Nat'l Acad. Sci. USA* 80:2258–2262.

IACOPETTA, B. J., S. ROTHENBERGER, and L. C. KUHN. 1988. A role for the cytoplasmic domain in transferrin-receptor sorting and coated-pit formation during endocytosis. *Cell* 54:485–489.

SCHNEIDER, C., M. J. OWEN, D. BANVILLE, and J. G. WILLIAMS. 1984. Primary structure of human transferrin receptor deduced from the mRNA sequence. *Nature* 311:675–678.

WATTS, C. 1985. Rapid endocytosis of the transferrin receptor in the absence of bound transferrin. *J. Cell Biol.* 100:633–637.

YAMASHIRO, D. J., B. TYCKO, S. R. FLUSS, and F. R. MAXFIELD. 1984. Segregation of transferrin to a mildly acidic (pH = 6.5) para-Golgi compartment in the recycling pathway. *Cell* 37:789–800.

Acidification of Endocytic Vesicles

FUCHS, R., P. MALE, I. MELLMAN. 1989. Acidification and ion permeabilities of highly purified rat liver endosomes. *J. Biol. Chem.* 264:2212–2220.

MERION, M., P. SCHLESINGER, R. M. BROOKS, J. M. MOEHRING, T. J. MOEHRING, and W. S. SLY. 1983. Defective acidification of endosomes in Chinese hamster ovary cell mutants "cross-resistant" to toxins and viruses. *Proc. Nat'l Acad. Sci. USA* 80:5315–5319.

YAMASHIRO, D. J., and F. R. MAXFIELD. 1987. Kinetics of endosome acidification in mutant and wild-type Chinese hamster ovary cells. *J. Cell Biol.* 105:2713–2721.

Recycling of Receptors

GEUZE, H. J., J. W. SLOT, G. J. A. M. STROUS, H. F. LODISH, and A. L. SCHWARTZ. 1983. Intracellular site of asialoglycoprotein receptor-ligand uncoupling: double-label immunoelectron microscopy during receptor-mediated endocytosis. *Cell* 32:277–287.

SCHWARTZ, A. L., S. E. FRIDOVICH, and H. F. LODISH. 1982. Kinetics of internalization and recycling of the asialoglycoprotein receptor in a hepatoma cell line. *J. Biol. Chem.* 257:4230–4237.

STAHL, P., P. H. SCHLESINGER, E. SIGARDSON, J. S. RODMAN, and Y. C. LEE. 1980. Receptor-mediated pinocytosis of mannose glycoconjugates by macrophages: characterization and evidence for receptor recycling. *Cell* 19:207–213.

Penetration of Bacterial Toxins into Cells

*CLEMENS, M. 1984. Enzymes and toxins that regulate protein synthesis. *Nature* 310:727.

KAGAN, B. L., A. FINKELSTEIN, and M. COLOMBINI. 1981. Diphtheria toxin fragment forms large pores in phospholipid bilayer membranes. *Proc. Nat'l Acad. Sci. USA* 78:4950–4954.

OLSNES, S., J. O. MOSKAUG, H. STENMARK, and K. SANDVIG. 1988. Diphtheria toxin entry: protein translocation in the reverse direction. *TIBS* 13:348–351.

PASTAN, I., M. C. WILLINGHAM, and D. J. P. FITZGERALD. 1986. Immunotoxins. *Cell* 48:641–648.

RIBI, H. O., D. S. LUDWIG, K. L. MERCER, G. K. SCHOOLNIK, and R. D. KORNBERG. 1988. Three-dimensional structure of Cholera toxin penetrating a lipid membrane. *Science* 239:1272–1276.

SANDVIG, K., and S. OLSNES. 1981. Rapid entry of nicked diphtheria toxin into cells at low pH: characterization of the entry process and effects of low pH on the toxin molecule. *J. Biol. Chem.* 256:9068–9076.

Penetration of Viruses into Cells

BACHI, T., 1988. Direct observation of the budding and fusion of an enveloped virus by video microscopy of viable cells. *J. Cell Biol.* 107:1689–1695.

DANIELS, R., J. DOWNIE, A. HAY, M. KNOSSOW, J. SKEHEL, M. WANG, and D. WILEY. 1985. Fusion mutants of the influenza virus hemagglutinin glycoprotein. *Cell* 40:431–439.

DOXSEY, S. J., J. SAMBROOK, A. HELENIUS, and J. WHITE. 1985. An efficient method for introducing macromolecules into living cells. *J. Cell Biol.* 101:19–27.

*GAROFF, H., K. SIMONS, and A. HELENIUS. 1982. How an animal virus gets into and out of its host cell. *Sci. Am.* 246(2):58–66.

HARTER, C., P. JAMES, T. BÄCH, G. SEMENZA, and J. BRUNNER. 1989. Hydrophobic binding of the ectodomain of influenza hemagglutinin to membranes occurs through the "fusion peptide." *J. Biol. Chem.* 264:6459–6464.

*WEBSTER, R. G., and R. ROTT. 1987. Influenza virus A pathogenicity: the pivotal role of hemagglutinin. *Cell* 50:665–666.

WEIS, W., J. H. BROWN, S. CUSACK, J. C. PAULSON, J. J. SKEHEL, and D. C. WILEY. 1988. Structure of the influenza virus hemagglutinin complexed with its receptor, sialic acid. *Nature* 333:426–431.

WHITE, J. M., and I. A. WILSON. 1987. Antipeptide antibodies detect steps in a protein conformational change: low-pH activation of the influenza virus hemagglutinin. *J. Cell Biol.* 105:2887–2896.

*WILEY, D. C., and J. J. SKEHEL. 1987. Structure and function of the hemagglutinin membrane glycoprotein of the influenza virus. *Ann. Rev. Biochem.* 56:365–394.

15

Energy Conversion: The Formation of ATP in Mitochondria and Bacteria

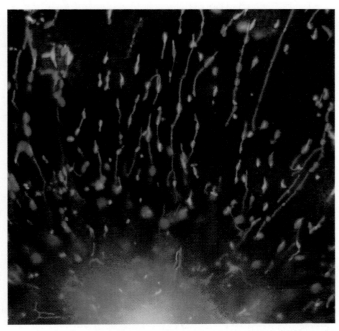

Two populations of mitochondria (red or green fluorescing) in a living human fibroblast

C hapters 15 and 16 focus on how cells generate the high-energy phosphoanhydride bond of adenosine triphosphate (ATP) by the endergonic reaction (one requiring the input of free energy G to proceed):

$$P_i^{2-} + H^+ + ADP^{3-} \rightleftharpoons ATP^{4-} + H_2O$$

where P_i is HPO_4^{2-} and the change in free energy under standard conditions is

$$\Delta G^{\circ\prime} = +7.3 \ kcal/mol$$

We shall concentrate on two of the most important processes. In Chapter 16, we discuss photosynthesis by bacteria and in the chloroplasts of higher plants, detailing how light energy is converted to the chemical energy of a phosphoanydride bond and how CO_2 and H_2O are converted to six-carbon sugars. Here, we examine the metabolic pathways by which glucose (the principal source of energy in animal and most other nonphotosynthetic cells, including plant cells such as roots) and lipids are metabolized to CO_2. The complete aerobic degradation of glucose to CO_2 and H_2O is coupled to the synthesis of as many as 32 molecules of ATP:

$$C_6H_{12}O_6 + 6O_2 + 32P_i^{2-} + 32ADP + 32H^+ \longrightarrow$$
$$6CO_2 + 32ATP + 38H_2O$$

In eukaryotic cells, the initial enzymatically catalyzed chemical reactions in the pathway of glucose degradation occur in the cytosol and the final steps occur in the mitochondria. An important focus of this chapter is the way in which mitochondria convert the energy released by the oxidation of metabolic products of glucose and lipids to ATP phosphoanhydride bonds.

All processes involved in the growth and metabolism of cells require an input of energy. In most cases, free energy is supplied by the hydrolysis of one of the high-energy phosphoanhydride bonds in ATP by the reaction

$$ATP^{4-} + H_2O \rightleftharpoons ADP^{3-} + P_i^{2-} + H^+$$

$$\Delta G^{\circ\prime} = -7.3 \; kcal/mol$$

where ADP = adenosine diphosphate and P_i = inorganic phosphate, or the reaction

$$ATP^{4-} + H_2O \rightleftharpoons AMP^{2-} + PP_i^{3-} + H^+$$

$$\Delta G^{\circ\prime} = -7.3 \; kcal/mol$$

where AMP = adenosine monophosphate and PP_i = inorganic pyrophosphate. If both bonds are hydrolyzed, twice the amount of free energy is released:

$$ATP^{4-} + 2H_2O \rightleftharpoons AMP^{2-} + 2P_i^{2-} + 2H^+$$

$$\Delta G^{\circ\prime} = -14.6 \; kcal/mol$$

No known enzyme cleaves ATP to AMP and two phosphate molecules (P_i). Rather, inorganic pyrophosphate, liberated by ATP hydrolysis to AMP, is hydrolyzed to $2P_i$ by the enzyme pyrophosphatase.

Energy released by the hydrolysis of these phosphoanhydride bonds is used directly to power many otherwise energetically unfavorable events, such as the transport of molecules against a concentration gradient, exemplified by the Na^+-K^+ ATPase, movement (beating) of cilia, and contraction of muscle. The energetic biochemical reactivity of ATP is needed in a multitude of pathways, including synthesis of nucleic acids and proteins from amino acids and nucleotides. ATP is a universal "currency" of chemical energy found in all types of organisms: archaebacteria, eubacteria, animals, and plants. Thus ATP must have evolved in the earliest life forms. ▲

At first glance, the processes of photosynthesis and aerobic oxidation appear to have little similarity. However, a revolutionary finding in cell biology is that bacteria, mitochondria, and chloroplasts use the same (or very nearly the same) process, called *chemiosmosis* (Figure 15-1), to generate ATP from ADP and P_i. The immediate energy sources that power this reaction are the proton concentration gradient and the membrane electric potential, often collectively termed the proton-motive force *(pmf)*. In photosynthesis, formation of this gradient is generated by the energy absorbed from light. In mitochondria, en-

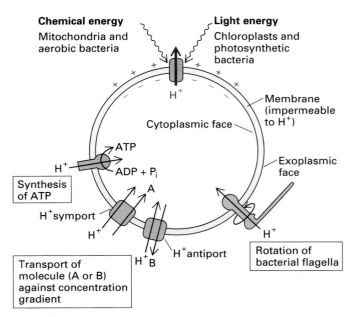

▲ **Figure 15-1** The chemiosmotic process. Chemiosmosis requires sealed, closed membrane vesicles that are relatively impermeable to H^+. In the process of photosynthesis, the energy absorbed from light is used to pump protons across the membrane, generating a transmembrane electrochemical proton gradient (a gradient of proton concentration and/or electric potential across the membrane). In mitochondria and aerobic bacteria, energy liberated by the oxidation of carbon compounds is used to pump protons. In all cases, the protons are pumped *from* the cytoplasmic face *to* the exoplasmic face of the membrane. The energy stored in this gradient can be used (1) to move protons from the exoplasmic face down their concentration gradient across the membrane, coupled to the synthesis of ATP from ADP and P_i, which occurs on the cytoplasmic membrane face in mitochondria, chloroplasts, and bacteria; (2) to pump metabolites across the membrane against their concentration gradient; and (3) to power the rotation of flagella in bacteria.

ergy from the metabolism of sugars, fatty acids, and other substances, culminating in their oxidation by O_2 via the respiratory chain, is used to pump protons across a mitochondrial membrane, generating a proton-motive force.

The proton-motive force is used in many ways. The synthesis of ATP from ADP and P_i is coupled to the movement of protons down their electric and chemical gradients. This formally resembles the reaction catalyzed by the Ca^{2+} ATPase run in reverse, in which the movement of Ca^{2+} down its concentration gradient powers the synthesis of ATP from ADP and P_i. Proton concentration gradients supply energy for the transport of small molecules across a membrane against a concentration gradient; for example, the uptake of lactose by certain bacteria uses a proton-sugar symport (see Figure 14-20). Rotation of bacterial flagella is also powered by the transmem-

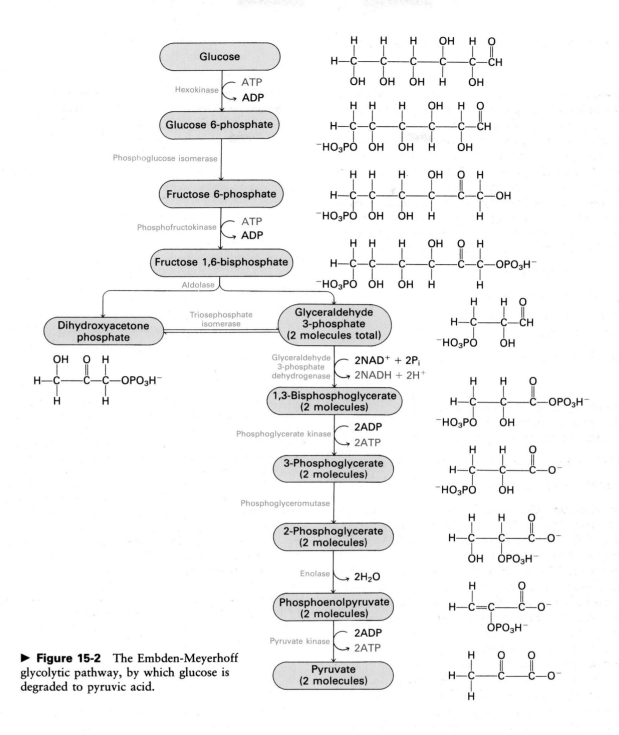

▶ **Figure 15-2** The Embden-Meyerhoff glycolytic pathway, by which glucose is degraded to pyruvic acid.

brane proton concentration gradient (in contrast to the beating of eukaryotic cilia, which is powered by ATP hydrolysis). The *chemiosmotic theory* is based on the principle introduced previously in the discussion of active transport that the membrane potential and concentration gradients of protons (and other ions) across membranes and the phosphoanhydride bonds in ATP are interconvertible storage forms of chemical potential energy.

Because of the central importance of the cytosol in energy metabolism, we begin our discussion of cellular energetics with the initial steps in glucose metabolism, which take place in the cytosol. We then discuss the mitochondrion and some details of the chemiosmotic process.

Energy Metabolism in the Cytosol

Glycolysis Is the First Stage in the Metabolism of Glucose and the Generation of ATP

In the initial stage of glucose metabolism, *glycolysis,* each glucose molecule is converted to two molecules of the three-carbon compound pyruvate, $C_3H_3O_3^-$ (Figure 15-2). These chemical reactions, called the *Embden-Meyerhoff pathway,* take place in the cytosol and do not involve molecular oxygen. Glycolysis is highly regulated: just enough glucose is metabolized to meet the cell's need for ATP.

All of the metabolic intermediates between the initial carbohydrate and the final product, pyruvate, are phosphorylated compounds. Four molecules of ATP are formed from ADP in glycolysis: two in the step catalyzed by phosphoglycerate kinase, when two molecules of 1,3-bisphosphoglycerate are converted to 3-phosphoglycerate, and two in the step catalyzed by pyruvate kinase (see Figure 15-2). However, two ATP molecules are consumed during earlier steps of this pathway: the first by the addition of a phosphate residue to glucose in the reaction catalyzed by hexokinase (see Figure 2-23), and the second by the addition of a second phosphate to fructose 6-phosphate in the reaction catalyzed by phosphofructokinase. Thus there is a net gain of two ATP molecules. The balanced chemical equation for this series of reactions shows that four hydrogen (H) atoms (four protons and four electrons) are also formed:

$$C_6H_{12}O_6 \longrightarrow 2C_3H_4O_3 + 4H$$

The reaction that generates these H atoms is catalyzed by the enzyme glyceraldehyde 3-phosphate dehydrogenase. The four electrons and two of the four protons are transferred to two molecules of the oxidized form of the electron carrier nicotinamide adenine dinucleotide (NAD$^+$) to produce the reduced form NADH (Figure 15-3):

$$2H^+ + 4e^- + 2NAD^+ \longrightarrow 2NADH$$

Thus the overall reaction for this first stage of glucose metabolism is

$$C_6H_{12}O_6 + 2NAD^+ + 2ADP^{2-} + 2P_i^{2-} \longrightarrow$$
$$2C_3H_4O_3 + 2NADH + 2ATP^{4-} + 2H^+$$

In Glycolysis, ATP Is Generated by Substrate-level Phosphorylation

Cells utilize two basically different types of processes to synthesize ATP. The immediate energy source for ATP synthesis in chloroplasts and mitochondria is provided by

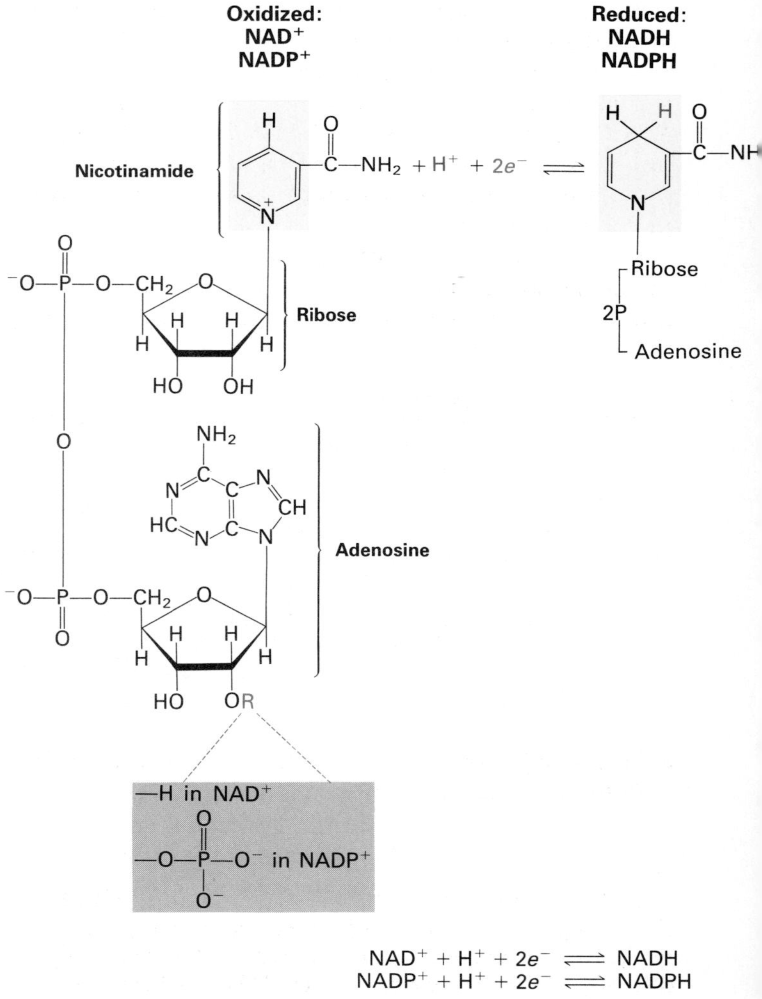

▶ **Figure 15-3** Structure of NAD$^+$ and NADH. Nicotinamide adenine dinucleotide (NAD$^+$) and the related nicotinamide adenine dinucleotide phosphate (NADP$^+$) accept only pairs of electrons; reduction to NADH or NADPH involves the transfer of two electrons simultaneously. In most oxidation-reduction reactions in biological systems, a pair of hydrogen atoms (two protons and two electrons) are removed from a molecule. One of the protons and both electrons are transferred to NAD$^+$; the other proton is released into solution. Thus the overall reaction is sometimes written NAD$^+$ + 2H$^+$ + 2e^- $\rightleftharpoons$ NADH + H$^+$. NADP is identical in structure to NAD except for the presence of an additional phosphate group. However, NAD and NADP participate in different sets of enzymatically catalyzed reactions.

ion gradients across a membrane. In the other process, *substrate-level phosphorylation,* soluble substances in the cytosol are chemically transformed by aqueous enzymes; membranes and gradients are not involved.

Two substrate-level phosphorylations occur in glycolysis. The first is a set of processes catalyzed by glyceraldehyde 3-phosphate dehydrogenase and phosphoglycerate kinase:

glyceraldehyde 3-phosphate^{2-} + NAD$^+$ + P$_i^{2-}$ $\rightleftharpoons$
$$1,3\text{-bisphosphoglycerate}^{4-} + NADH + H^+$$
$$\Delta G^{\circ\prime} = +1.5 \text{ kcal/mol} \tag{1}$$

1,3-bisphosphoglycerate^{4-} + ADP^{3-} $\rightleftharpoons$
$$3\text{-phosphoglycerate}^{3-} + ATP^{4-}$$
$$\Delta G^{\circ\prime} = -4.5 \text{ kcal/mol} \tag{2}$$

The net change in standard free energy, however, is negative ($\Delta G^{\circ\prime} = -4.5 + 1.5 = -3.0$ kcal/mol), so the two reactions overall are strongly *exergonic* (accompanied by the release of free energy). Since each molecule of fructose 1,6-bisphosphate generates two molecules of glyceraldehyde 3-phosphate (see Figure 15-2), two ATPs are generated during the catabolism of one glucose mole-

$$NAD^+ + H^+ + 2e^- \rightleftharpoons NADH$$
$$NADP^+ + H^+ + 2e^- \rightleftharpoons NADPH$$

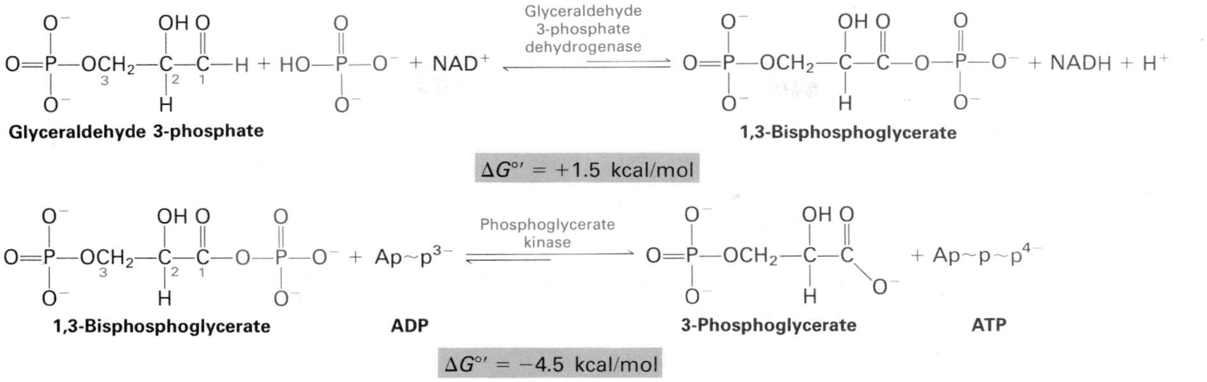

▲ Figure 15-4 Substrate-level phosphorylation. Glyceraldehyde 3-phosphate dehydrogenase and phosphoglycerate kinase are water-soluble enzymes that catalyze ATP synthesis in the cytosol. The generation of high-energy ATP bonds is a very different process than the one employed in oxidative phosphorylation or photophosphorylation. The three C atoms of the glycerol are numbered.

cule. All the enzymes and reactants are in the cytosol. In the first reaction (Figures 15-4 and 15-5), the oxidation of the aldehyde (CHO) group on glyceraldehyde 3-phosphate by NAD^+ is coupled to the addition of a phosphate group, forming 1,3-bisphosphoglycerate with a single high-energy phosphate bond. The $\Delta G^{\circ\prime}$ for the hydrolysis of the phosphoanhydride bond to carbon 1 in 1,3-bisphosphoglycerate is more negative than the $\Delta G^{\circ\prime}$ for the hydrolysis of the terminal phosphoanhydride bond in ATP (-12 kcal/mol versus -7.3 kcal/mol). This high-energy phosphate bond is transferred to ADP in a strongly exergonic reaction catalyzed by phosphoglycerate kinase (see Figure 15-4).

The next stages in glycolysis also generate a high-energy phosphate bond in the molecule phosphoenolpyruvate. This bond is also transferred to ADP in a strongly exergonic reaction, which is catalyzed by pyruvate kinase (Figure 15-6). In this set of reactions, water loss converts the phosphate-carbon bond in 2-phosphoglycerate from a low-energy bond ($\Delta G^{\circ\prime}$ of hydrolysis = -2 kcal/mol) to a high-energy bond ($\Delta G^{\circ\prime}$ of hydrolysis = -15 kcal/mol) in phosphoenolpyruvate.

These two examples illustrate how interconversions of soluble chemicals by soluble enzymes are coupled to generate ATP. However, only two of the 32 ATP molecules generated during the complete oxidation of glucose to carbon dioxide (CO_2) and H_2O are made during the conversion of glucose to pyruvate. The remaining 30 are synthesized in the mitochondrion by a fundamentally different type of process that involves the generation and utilization of proton concentration gradients across the inner mitochondrial membrane.

▲ Figure 15-5 The mechanism of action of glyceraldehyde 3-phosphate dehydrogenase. A thioester

$$R - \overset{\overset{\displaystyle O}{\|}}{C} - S - R'$$

is an energy-rich intermediate in the reaction. The sulfhydryl (SH) group is the side chain of cysteine at the active site; R symbolizes the rest of the glyceraldehyde 3-phosphate molecule. *Step 1:* The enzyme has bound NAD^+, and the —SH group on the enzyme reacts with glyceraldehyde 3-phosphate to form a thiohemiacetal. *Step 2:* A hydrogen atom (red) and two electrons are transferred to NAD^+, forming the reduced form NADH, and a proton from the O atom of the thiohemiacetal is simultaneously lost to the medium; the products are a thioester and NADH + H^+. *Step 3:* The thioester reacts with phosphate to produce 1,3-bisphosphoglycerate; NADH is freed from the enzyme surface, and the free enzyme with its SH group is regenerated.

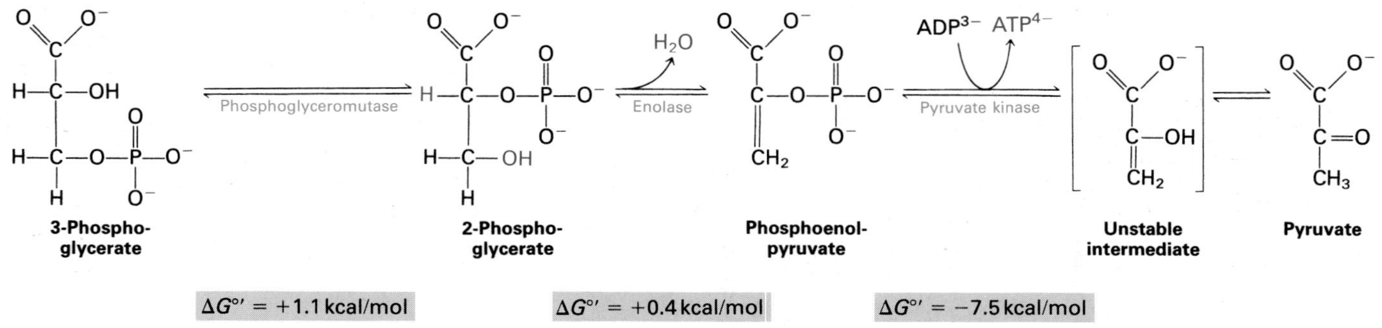

$\Delta G^{\circ\prime} = +1.1 \, \text{kcal/mol}$ $\Delta G^{\circ\prime} = +0.4 \, \text{kcal/mol}$ $\Delta G^{\circ\prime} = -7.5 \, \text{kcal/mol}$

▲ **Figure 15-6** Formation of the second pair of ATP molecules during glycolysis. This reaction, catalyzed by pyruvate kinase, is another example of substrate-level phosphorylation.

Some Eukaryotic and Prokaryotic Cells Metabolize Glucose Anaerobically

Most eukaryotic cells are *obligate aerobes:* they grow only in the presence of oxygen and metabolize glucose (or related sugars) completely to CO_2, with the concomitant production of a large amount of ATP. Most of these cells also generate some ATP by anaerobic metabolism. A few eukaryotes, including certain yeasts, are *facultative anaerobes* that grow in either the presence or absence of oxygen; annelids and mollusks, as examples, can live and grow for days without oxygen. Many prokaryotic cells are *obligate anaerobes* and cannot grow in the presence of oxygen.

The anaerobic metabolism of glucose does not require mitochondria. Glucose is not converted entirely to CO_2 but to one or more two- or three-carbon compounds and, in some cases, to CO_2. As a result, much less ATP is produced per mole of glucose. Yeasts, for example, ferment glucose anaerobically to two ethanol and two CO_2 molecules; the net production is only two ATP molecules per glucose molecule (Figure 15-7; see Figure 15-2). As pyruvate is converted into ethanol, the NADH produced in the initial stages of glycolysis is reoxidized. This anaerobic fermentation is the basis of the beer and wine industry.

During the prolonged contraction of mammalian skeletal muscle cells, oxygen becomes limited and glucose cannot be oxidized completely to CO_2 and H_2O. The cells ferment glucose to two molecules of lactic acid—again, with the net production of only two molecules of ATP per glucose molecule (see Figures 15-2 and 15-7). The lactic acid causes muscle and joint aches. It is largely secreted into the blood; some passes into the liver, where it is reoxidized to pyruvate and further metabolized to CO_2. Much lactate is metabolized to CO_2 by the heart, which is highly perfused by blood and can continue aerobic metabolism at times when exercising skeletal muscles secrete lactate. Lactic acid bacteria (the organisms that "spoil" milk) and other prokaryotes also generate ATP by the fermentation of glucose to lactate.

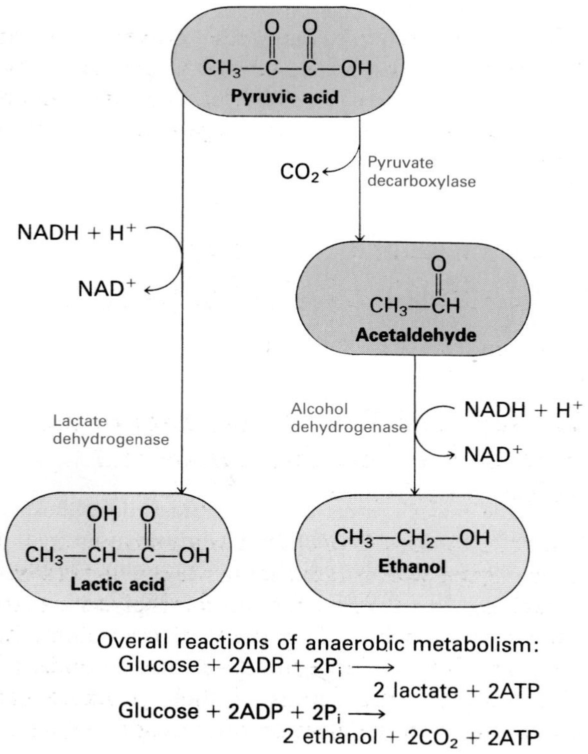

Carbohydrate Oxidation Is Completed in the Mitochondria, Where Most ATP Is Produced

In aerobic cells, glucose is oxidized completely to CO_2 by oxygen (O_2). The latter stages of this process, including the ones involving O_2, require mitochondria. The pyru-

Overall reactions of anaerobic metabolism:
Glucose + 2ADP + 2P$_i$ $\longrightarrow$
       2 lactate + 2ATP
Glucose + 2ADP + 2P$_i$ $\longrightarrow$
       2 ethanol + 2CO$_2$ + 2ATP

◀ **Figure 15-7** The anaerobic metabolism of glucose. In the formation of pyruvate from glucose, one molecule of NAD^+ is reduced to NADH for each molecule of pyruvate formed (see Figure 15-2). To regenerate NAD^+, two electrons are transferred from NADH to an acceptor molecule. When oxygen supplies are low in muscle cells, the acceptor is pyruvic acid, and lactic acid is formed. In yeasts, acetaldehyde is the acceptor, and ethanol is formed.

(a)

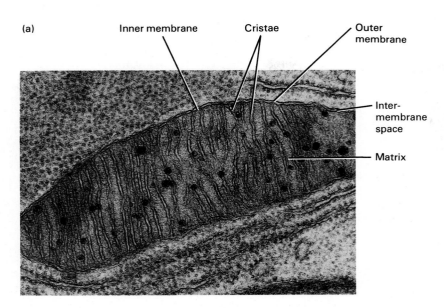

Inner membrane   Cristae   Outer membrane

Intermembrane space

Matrix

◀ **Figure 15-8** The structure of the mitochondrion. (a) Electron micrograph of a mitochondrion from a bat pancreas, showing the inner membrane with extensive cristae, the outer membrane, the intermembrane space, and small granules in the matrix. (b) A three-dimensional diagram of a mitochondrion cut longitudinally. The matrix contains the mitochondrial DNA and ribosomes. *Part (a) from D. W. Fawcett, The Cell, 2d ed., Saunders, 1981, p. 421. Courtesy of Don Fawcett.*

(b)

Intermembrane space   Outer membrane   Cristae

Ribosome   Matrix   Inner membrane

Granule   ATP synthase particles

DNA

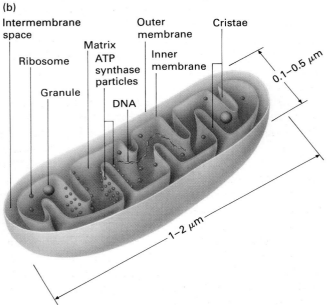

$0.1–0.5 \ \mu m$

$1–2 \ \mu m$

These oxidation reactions in the mitochondria generate the bulk of the ATP produced from the conversion of glucose to CO_2. Actually, 30 is the maximal number of ATPs produced in the mitochondrion, and even this value is controversial. As we will see, some energy released in mitochondrial oxidation is used for other purposes, such as heat generation and the transport of molecules into or out of the organelle.

The mitochondrion is really the "power plant" of the cell. The details of the chemiosmotic process by which ATP is formed in the mitochondrion depend on an understanding of the structure of this organelle, which we now address.

Mitochondria and the Metabolism of Carbohydrates and Lipids

The Outer and Inner Membranes of the Mitochondrion Are Structurally and Functionally Distinct

Most eukaryotic cells contain many mitochondria. They are among the larger organelles in the cell (generally, only the nucleus, the vacuoles, and the chloroplasts are larger), occupying as much as 25 percent of the volume of the cytoplasm. They are large enough to be seen under a light microscope, but the details of their structure can be viewed only with the electron microscope (Figure 15-8).

vate formed in glycolysis is transported to the mitochondria, where it is oxidized by O_2 to CO_2:

$$CH_3-\overset{O}{\underset{\|}{C}}-\overset{O}{\underset{\|}{C}}OH + 2\tfrac{1}{2}O_2 \longrightarrow 3CO_2 + 2H_2O$$

The two molecules of NADH formed in the cytosol during glycolysis reduce NAD^+ molecules in the mitochondria:

$$NADH_{cytosol} + NAD^+_{mitochondria} \longrightarrow$$
$$NAD^+_{cytosol} + NADH_{mitochondria}$$

The mitochondrial NADH is reoxidized by O_2:

$$2NADH + 2H^+ + O_2 \longrightarrow 2NAD^+ + 2H_2O$$

Mitochondria contain two very different membranes—the *outer membrane* and the *inner membrane*—which define two submitochondrial compartments: the *intermembrane space* between the two membranes, and the *matrix* or central compartment (see Figure 15-8). Because these mitochondrial membranes and compartments can be purified and studied separately, it has been possible to determine their protein and phospholipid compositions and to localize each reaction to a specific membrane or space.

The outer membrane defines the smooth outer perimeter of the mitochondrion. Transmembrane channels formed by the protein porin make this unusual membrane freely permeable to most small molecules (<10,000 MW), particularly to protons. Porin is similar in structure and function to the proteins lining the pores in the outer membrane of gram-negative bacteria, which also confers permeability to most small molecules. Thus the inner membrane is the only effective permeability barrier between the cytosol and the mitochondrial matrix.

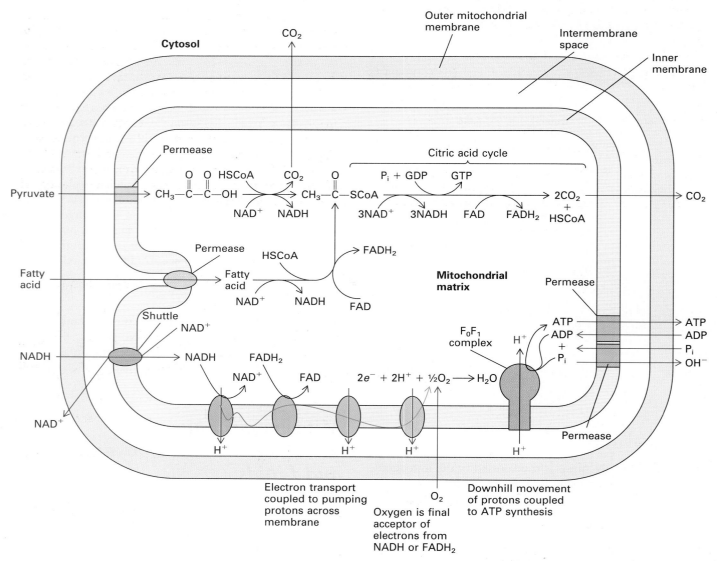

▲ **Figure 15-9** Outline of the major metabolic reactions in mitochondria. The substrates of oxidative phosphorylation—pyruvate, fatty acids, ADP, and P_i—are transported to the matrix from the cytosol by permeases; O_2 diffuses into the matrix. NADH, which is generated in the cytosol during glycolysis, is not transported directly to the matrix because the inner membrane is impermeable to NAD^+ or NADH; in-stead, a shuttle system (see Figure 15-21) transports electrons from cytosolic NADH to the electron transport chain. ATP is transported to the cytosol in exchange for ADP and P_i, CO_2 diffuses into the cytosol across the mitochondrial membranes. HSCoA denotes free coenzyme A (CoA), and SCoA denotes CoA when it is esterified (see Figure 15-11).

The inner membrane contains a higher fraction of protein (76 percent of the total membrane weight) than any other cellular membrane. Cardiolipin (diphosphatidyl glycerol, see Figure 13-2), the lipid concentrated in the inner membrane, is believed to reduce the permeability of the phospholipid bilayer to protons. Freeze-fracture studies on the inner membrane indicate that it contains many protein-rich intramembrane particles that are laterally mobile in the membrane plane (see Figure 13-26). Some of these particles function in electron transport from NADH or $FADH_2$ (Figure 15-9) to O_2 and in ATP synthesis. Some particles are permeases that allow otherwise impermeable molecules, such as ADP and other phosphorylated compounds, to pass from the cytosol to the matrix.

The inner membrane and the matrix are the sites of most reactions involving the oxidation of pyruvate and fatty acids to CO_2 and H_2O and the coupled synthesis of ATP from ADP and P_i. These complex processes involve many steps but can be subdivided into three groups of reactions, each of which occurs in a discrete membrane or space in the mitochondrion (see Figure 15-9):

1. Oxidation of pyruvate or fatty acids to CO_2, coupled to the reduction of the electron carriers NAD^+ and FAD (Figure 15-10) to NADH and $FADH_2$, respec-

tively. These reactions occur in the matrix or on inner-membrane proteins facing it.

2. Electron transfer from NADH and $FADH_2$ to O_2. These reactions occur in the inner membrane and are coupled to the generation of an *electrochemical proton gradient*, or proton-motive force, across it.

3. Utilization of the energy stored in the transmembrane proton concentration gradient for ATP synthesis by the F_0F_1 *ATPase complex* in the inner membrane.

The last two groups of operations involve multisubunit proteins that are asymmetrically oriented in the inner mitochondrial membrane.

The highly convoluted foldings, or *cristae*, of the inner membrane greatly expand its surface area, enhancing its ability to generate ATP. In typical liver mitochondria, for example, the area of the inner membrane is about five times that of the outer membrane. In fact, the total area of all mitochondrial inner membranes in liver cells is about 15 times that of the plasma membrane. In heart and skeletal muscles, mitochondria contain three times as many cristae as are found in typical liver mitochondria—presumably reflecting the greater demand for ATP by muscle cells. The cristae in certain tissues are folded in different ways, but the significance of such morphological differences is not known.

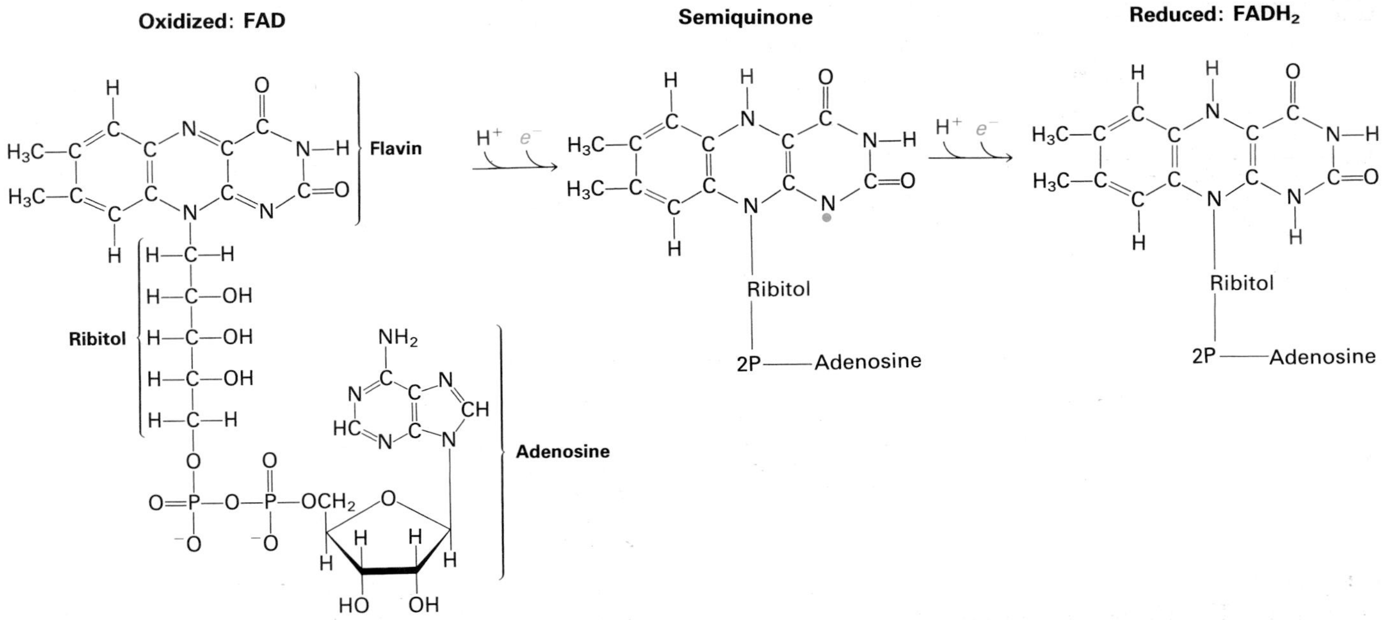

▲ **Figure 15-10** The cofactor flavin adenine dinucleotide (FAD) contains a three-ring flavin component that can accept one or two electrons. The addition of one electron together with a proton generates a semiquinone intermediate. The semiquinone is a free radical because it contains an unpaired electron (denoted here by a blue dot), which is delocalized by resonance to all of the flavin ring atoms. The addition of a second electron and proton generates the reduced form, $FADH_2$.

It is important to note that mitochondria are essential for ATP production in photosynthetic plant cells during dark periods, when photosynthesis is not possible, and all of the time in roots and other nonphotosynthetic tissues. Stored carbohydrates, mostly in the form of starch, are hydrolyzed to glucose and then metabolized to pyruvate; as in animal mitochondria, pyruvate is oxidized to CO_2, with concomitant generation of ATP.

Acetyl CoA Is a Key Intermediate in the Mitochondrial Metabolism of Pyruvate

Pyruvate, which is generated in the cytosol during glycolysis, is transported across the mitochondrial membranes to the matrix. The complete oxidation of pyruvate to form CO_2 and H_2O occurs in the mitochondrion and utilizes O_2 as the final electron acceptor (oxidizer).

Figure 15-9 traces the metabolism of pyruvate in the mitochondrion. Immediately on entering the matrix, pyruvate reacts with coenzyme A (HSCoA) to form CO_2 and the important intermediate acetyl CoA (Figure 15-11)—a reaction catalyzed by the enzyme pyruvate dehydrogenase, a soluble component of the matrix:

$$CH_3-\overset{O}{\overset{\|}{C}}-\overset{O}{\overset{\|}{C}}-O^- + HSCoA + NAD^+ \longrightarrow$$

$$CH_3\overset{O}{\overset{\|}{C}}-SCoA + CO_2 + NADH$$

This reaction is highly exergonic ($\Delta G^{\circ\prime} = -8.0$ kcal/mol) and essentially irreversible. Despite its apparent simplicity, pyruvate dehydrogenase is actually one of the most complex enzymes known. It is a giant molecule 30 nm in diameter (4.6 million MW), even larger than a ribosome,

and contains 60 subunits of three different enzymes, several regulatory polypeptides, and five different coenzymes—all carefully ordered in the complex.

The Metabolism of Fatty Acids Occurs in the Mitochondrion and Also Involves Acetyl CoA

Lipids are stored as triacylglycerols, primarily in adipose cells (see Figure 2-46). In response to hormones such as adrenaline, these triacylglycerols are hydrolyzed to free fatty acids and glycerol:

$$CH_3-(CH_2)_n\overset{O}{\overset{\|}{C}}-O-CH_2$$
$$CH_3-(CH_2)_n-\overset{O}{\overset{\|}{C}}-O-CH + 3H_2O \longrightarrow$$
$$CH_2-(CH_2)_n-\overset{O}{\overset{\|}{C}}-O-CH_2$$

$$3CH_3-(CH_2)_n-\overset{O}{\overset{\|}{C}}-OH + CH_2OH-CHOH-CH_2OH$$
Fatty acid   **Glycerol**

The fatty acids are released into the blood, from which they are taken up and oxidized by other cells. In humans, the oxidation of fats is quantitatively more important than that of glucose as a source of ATP. In part, this is due to the fact that the oxidation of 1 g of triacylglycerol to CO_2 results in the formation of about six times as much ATP as that generated by the oxidation of 1 g of hydrated glycogen.

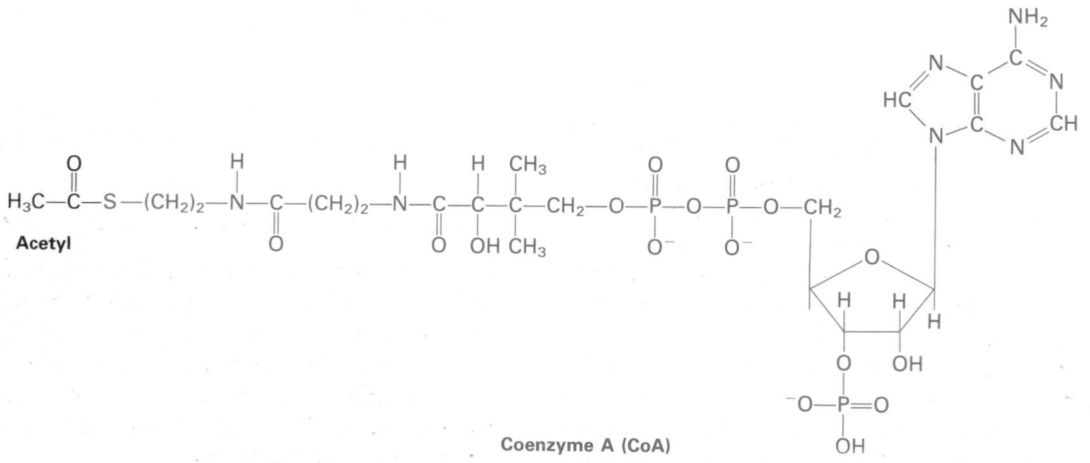

▲ **Figure 15-11** The structure of acetyl CoA—an important intermediate in the metabolism of pyruvate, fatty acids, and many amino acids.

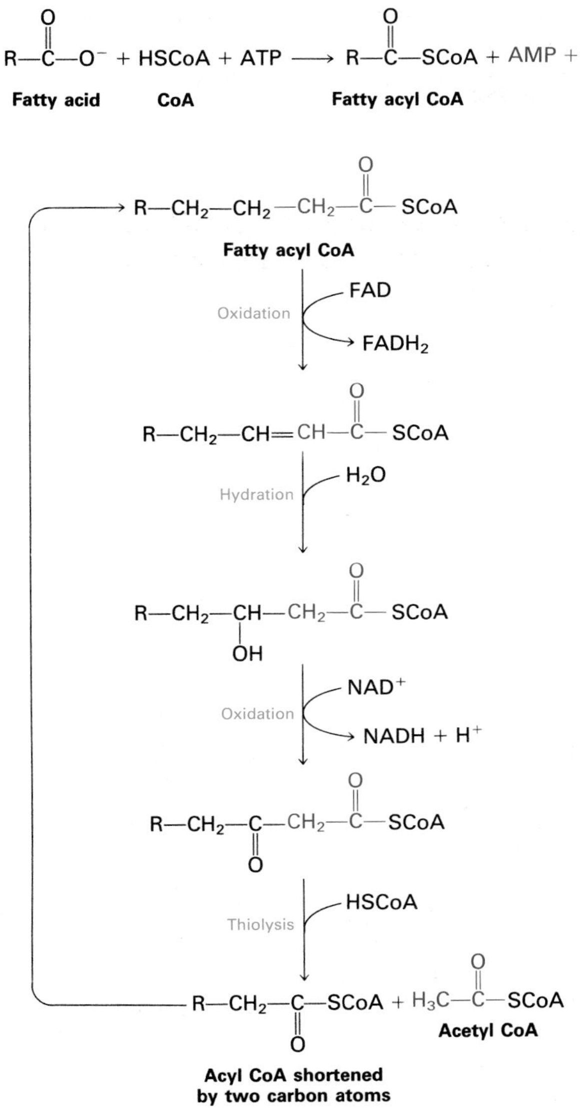

$$R-\overset{\overset{\displaystyle O}{\|}}{C}-O^- + HSCoA + ATP \longrightarrow R-\overset{\overset{\displaystyle O}{\|}}{C}-SCoA + AMP + PP_i$$

Fatty acid   **CoA**   **Fatty acyl CoA**

$$R-CH_2-CH_2-CH_2-\overset{\overset{\displaystyle O}{\|}}{C}-SCoA$$

Fatty acyl CoA

Oxidation   FAD
            → FADH₂

$$R-CH_2-CH=CH-\overset{\overset{\displaystyle O}{\|}}{C}-SCoA$$

Hydration   H₂O

$$R-CH_2-\underset{\underset{\displaystyle OH}{|}}{CH}-CH_2-\overset{\overset{\displaystyle O}{\|}}{C}-SCoA$$

Oxidation   NAD⁺
            → NADH + H⁺

$$R-CH_2-\underset{\underset{\displaystyle O}{\|}}{C}-CH_2-\overset{\overset{\displaystyle O}{\|}}{C}-SCoA$$

Thiolysis   HSCoA

$$R-CH_2-\underset{\underset{\displaystyle O}{\|}}{C}-SCoA + H_3C-\overset{\overset{\displaystyle O}{\|}}{C}-SCoA$$

Acetyl CoA

**Acyl CoA shortened
by two carbon atoms**

◀ **Figure 15-12** Oxidation of fatty acids. Four enzymatically-catalyzed reactions convert a fatty acyl CoA molecule to acetyl CoA and a fatty acyl CoA shortened by two carbon atoms. Concomitantly, one NAD^+ molecule is reduced to NADH and one FAD molecule is reduced to $FADH_2$. The cycle is repeated on the shortened acyl CoA, until fatty acids with an even number of carbon atoms are completely converted to acetyl CoA. Fatty acids with an odd number of C atoms are rare; they are metabolized to one molecule of propionyl CoA and multiple acetyl CoAs.

$$CH_3-(CH_2)_{16}-\overset{\overset{\displaystyle O}{\|}}{C}-SCoA +$$

$$8HSCoA + 8FAD + 8NAD^+ + 8H_2O$$

$$\downarrow$$

$$9CH_3-\overset{\overset{\displaystyle O}{\|}}{C}-SCoA + 8FADH_2 + 8NADH + 8H^+$$

In addition to fatty acids and carbohydrates, acetyl CoA occupies a central position in the oxidation of many amino acids. It is also an intermediate in many biosynthetic reactions, such as the transfer of an acetyl group to lysine residues in histone proteins and to the N termini of many mammalian proteins. In respiring mitochondria, however, the fate of the acetyl group of acetyl CoA is almost always oxidation to CO_2.

The Citric Acid Cycle Oxidizes the Acetyl Group of Acetyl CoA to CO_2 and Reduces NAD and FAD to NADH and FADH₂

The final stage of the oxidation of carbohydrates and lipids—the *citric acid cycle,* or *Krebs cycle*—is a complex set of nine reactions (Figure 15-13). First let's consider the net reaction:

$$CH_3-\overset{\overset{\displaystyle O}{\|}}{C}-SCoA +$$

$$3NAD^+ + FAD + GDP^{3-} + P_i^{2-} + 2H_2O$$

$$\downarrow$$

$$2CO_2 + 3NADH + FADH_2 + GTP^{4-} + 2H^+ + HSCoA$$

Note there is no involvement of molecular O_2 at this stage and only one high-energy phosphate bond is synthesized by substrate-level phosphorylation (in GTP). The two carbon atoms in acetyl CoA are oxidized to two molecules of CO_2. Concomitantly, the released electrons are transferred to NAD and FAD to form the reduced molecules NADH and $FADH_2$.

As shown in Figure 15-13, the cycle begins with the condensation of the two-carbon acetyl group from acetyl

In the cytosol, free fatty acids are linked to coenzyme A to form an acyl CoA (Figure 15-12) in an exergonic reaction coupled to the hydrolysis of ATP to AMP and PP_i. PP_i is hydrolyzed to two molecules of phosphate (P_i), drawing this reaction to completion. Then the acyl group is transported across the inner membrane by a *translocase protein* and reattached to another CoA on the matrix side. Each molecule of acyl CoA in the mitochondrion is oxidized to form one molecule of acetyl CoA and an acyl CoA shortened by two carbon atoms (see Figure 15-12). Concomitantly one molecule apiece of NAD^+ and FAD are reduced, respectively, to NADH and $FADH_2$. This set of reactions is repeated on the shortened acyl CoA until all C atoms are converted to acetyl CoA. For stearoyl CoA, the overall reaction is

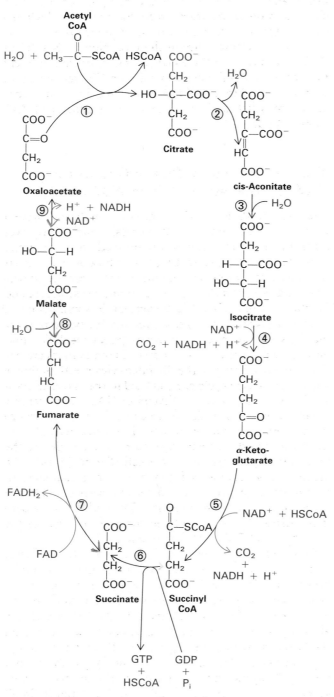

▲ **Figure 15-13** The citric acid cycle. First, a two-carbon acetyl residue from acetyl CoA condenses with the four-carbon molecule oxaloacetate to form the six-carbon molecule citrate. Through a sequence of enzymatically catalyzed reactions (2–9), each molecule of citrate is converted to oxaloacetate, losing two CO_2 molecules. In four reactions, four pairs of electrons are removed from the carbon atoms: three pairs are transferred to three molecules of NAD^+ to form $3NADH + 3H^+$; one pair is transferred to the acceptor FAD to form $FADH_2$.

CoA with the four-carbon molecule oxaloacetate. The product of reaction 1 is the six-carbon citric acid, for which the cycle is named. In reactions 2 and 3, citrate is isomerized to the six-carbon molecule isocitrate by the single enzyme aconitase. In reaction 4, isocitrate is oxidized to the five-carbon α-ketoglutarate, generating one CO_2 molecule and reducing one molecule of NAD^+ to NADH. In reaction 5, the α-ketoglutarate is oxidized to the four-carbon molecule succinyl CoA, generating the second CO_2 molecule formed during each turn of the cycle and reducing another NAD^+ molecule to NADH. In reactions 6–9, succinyl CoA is oxidized to oxaloacetate, regenerating the molecule that was initially condensed with acetyl CoA. Concomitantly, there is a reduction of one FAD molecule to $FADH_2$ and of one NAD^+ molecule to NADH. The conversion of succinyl CoA to succinate (reaction 6) is coupled to the synthesis of one GTP molecule (from GDP and P_i); this reaction is slightly exergonic ($\Delta G^{\circ\prime} = -0.8$ kcal/mol).

Most enzymes and small molecules involved in the citric acid cycle are soluble in aqueous solution and are localized to the matrix of the mitochondrion. The exceptions are CoA, acetyl CoA, and succinyl CoA, in which the hydrophobic CoA segment is bound to the inner membrane with the acetyl or succinyl part facing the matrix. Succinate dehydrogenase (reaction 7) is also localized to the inner membrane, as is α-ketoglutarate dehydrogenase (reaction 5).

The protein concentration of the mitochondrial matrix is 500 mg/ml (a 50 percent protein solution) and must have a viscous, gel-like consistency. When mitochondria are disrupted by gentle ultrasonic vibration or osmotic lysis, the six nonmembrane-bound enzymes in the citric acid cycle are released as a very large multiprotein complex. The reaction product of one enzyme is believed to be passed directly to the next enzyme, without diffusing through the solution; however, much work is needed to determine the structure of this complex. Biochemists generally study the properties of enzymes in dilute aqueous solutions of less than 1 mg/ml, and weak interactions between enzymes are often difficult to detect.

Electrons Are Transferred from NADH and $FADH_2$ to Molecular O_2 by Electron Carrier Proteins

The reactions in the Embden-Meyerhoff pathway and citric acid cycle result in the conversion of one glucose molecule to six CO_2 molecules and the concomitant reduction of 10 NAD^+ to 10 NADH molecules and of two FAD to two $FADH_2$ molecules. Two of the 10 NADH molecules are formed during the conversion of each glucose molecule to two pyruvate molecules. The oxidation of each pyruvate to acetyl CoA generates one NADH molecule (or a total of two molecules per glucose molecule). The oxidation of each acetyl CoA to CO_2 generates three NADH molecules and one $FADH_2$ molecule or six NADH and two $FADH_2$ molecules per glucose molecule.

Molecular O_2 is used to reoxidize these reduced coenzymes, but this does not occur in a single step. Rather, NADH and $FADH_2$ transfer their electron pairs to acceptor molecules in the inner mitochondrial membrane. The loss of electrons regenerates the oxidized forms of NAD^+ and FAD as well as the reduced form of the acceptor. Electrons from NADH and $FADH_2$ are transferred along a chain of electron-carrier proteins, all of which are integral components of the inner membrane. Eventually, they are transferred to O_2, the ultimate electron acceptor, forming H_2O.

The following overall reactions summarize these steps:

$$NADH + H^+ + \frac{1}{2}O_2 \longrightarrow NAD^+ + H_2O$$

$$\Delta G^{\circ\prime} = -52.6 \text{ kcal/mol}$$

$$FADH_2 + \frac{1}{2}O_2 \longrightarrow FAD + H_2O$$

$$\Delta G^{\circ\prime} = -43.4 \text{ kcal/mol}$$

As the negative $\Delta G^{\circ\prime}$ value indicates, the oxidation of these reduced coenzymes by O_2 is a strongly exergonic reaction. More importantly, most of the free energy released during the oxidation of glucose to CO_2 is retained in the reduced coenzymes generated during glycolysis and the citric acid cycle. To see this, recall the large total change in standard free energy for the oxidation of glucose to CO_2 ($\Delta G^{\circ\prime} = -680$ kcal/mol); oxidation of the 10 NADH and two $FADH_2$ molecules yields an almost equivalent change of $\Delta G^{\circ\prime} = 10(-52.6) + 2(-43.4) = -613$ kcal/mol of glucose. Thus over 90 percent of the potential free energy present in the glucose bonds that can be oxidized is conserved in the reduced coenzymes. The reoxidation of these coenzymes by O_2 generates the vast majority of ATP phosphoanhydride bonds.

The free energy released during the oxidation of a single NADH or $FADH_2$ molecule by O_2 is sufficient to drive the synthesis of several molecules of ATP from ADP and P_i ($\Delta G^{\circ\prime} = -7.3$ kcal/mol for the reaction ADP + $P_i \rightleftharpoons$ ATP). Thus it is not surprising that NADH oxidation and ATP synthesis do not occur in a single reaction. Rather, a step-by-step transfer of electrons from NADH to O_2, via a series of proteins that constitute the *electron transport chain,* allows the free energy to be released in small increments. At several sites during electron transport from NADH to O_2, protons are pumped across the inner mitochondrial membrane and a proton concentration gradient forms across it. Because the outer membrane is freely permeable to protons, the pH of the mitochondrial matrix is higher (the proton concentration is lower) than that of the cytosol and intermembrane space (see Figure 15-9). An electric potential across the inner membrane also results from the pumping of positively charged protons outward from the matrix, which becomes negative with respect to the intermembrane space. Free energy released during the oxidation of NADH or $FADH_2$ is stored in the electrochemical proton gradient that forms across the inner membrane. The movement of protons back across the inner membrane down their concentration gradient is coupled to the synthesis of ATP from ADP and P_i.

The Electrochemical Proton Gradient Is Used to Generate ATP from ADP and P_i

The F_0F_1 ATPase or ATP synthase membrane protein complex is involved in ATP synthesis. Such complexes form the knoblike particles that protrude from the inner membrane (see Figure 15-8). Exactly how ATP synthesis is coupled to proton movement is not presently understood. Later in this chapter, we examine the structure and function of the F_0F_1 ATPase complex in some detail.

Bacteria lack mitochondria; yet aerobic bacteria carry out the same processes of oxidative phosphorylation that occur in eukaryotic mitochondria. Enzymes that catalyze the reactions of both the Embden-Meyerhoff pathway and the citric acid cycle are localized to the bacterial cytosol. Enzymes that oxidize NADH to NAD^+ and transfer the electrons to the ultimate acceptor O_2 are localized to the bacterial plasma membrane. The movement of electrons through these membrane carriers is coupled to the pumping of protons out of the cell (Figure 15-14). The movement of protons back to the cell down their concentration gradient is coupled to the synthesis of ATP; the membrane proteins involved in ATP synthesis are essentially identical in structure and function to those of the mitochondrial F_0F_1 ATPase complex.

Chloroplasts, too, utilize an F_0F_1 complex to synthesize ATP; these particles are localized to the thylakoid membranes (see Figure 15-14). Every cellular membrane has a cytoplasmic face and an exoplasmic face. In chloroplasts, mitochondria, and bacteria, the F_0F_1 ATPase complex is always positioned so that the globular F_1 segment is on the cytoplasmic face (see Figure 15-14); note that the cytoplasmic faces of the inner mitochondrial membrane and the thylakoid membrane are toward the matrix. The F_1 segment catalyzes ATP synthesis from ADP and P_i. In all cases, ATP is formed on the cytoplasmic face of the membrane; in the mitochondrion, for instance, ATP is synthesized on the matrix side of the inner membrane. Protons always flow through the F_0F_1 ATPase complex across the membrane from the exoplasmic to the cytoplasmic face, driven by a combination of the proton concentration gradient (exoplasmic face > cytosol) and the membrane electric potential (cytoplasmic face negative with respect to exoplasmic face).

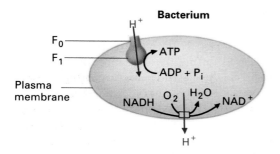

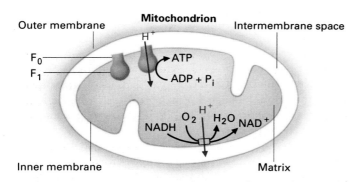

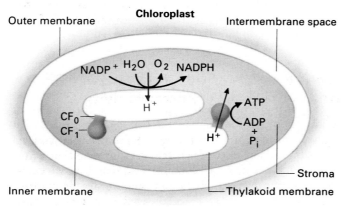

◀ **Figure 15-14** Membrane orientation and the direction of proton movement in bacteria, mitochondria, and chloroplasts. The membrane surface facing a shaded area is a cytoplasmic face; the surface facing an unshaded area is an exoplasmic face. In bacteria, mitochondria, and chloroplasts, the F_0F_1 complexes always face the cytoplasmic face of the membrane. During electron transport, protons are always pumped from the cytoplasmic face to the exoplasmic face, creating a proton concentration gradient (exoplasmic face > cytoplasmic face) and an electric potential (negative cytoplasmic face) across the membrane. During the generation of ATP, protons flow in the reverse direction (down their electrochemical proton gradient) through the F_0F_1 complexes.

The process of oxidative phosphorylation is quite complex, and the details of many of the steps of each process are only slowly being uncovered. Still, through a combined effort of biochemists, biophysicists, microscopists, and geneticists, much has been learned. The following sections describe the use of a proton concentration gradient to generate ATP in oxidative phosphorylation and the ways in which electron transport is coupled to the generation of a proton-motive force.

The Proton-motive Force, ATP Generation, and Transport of Metabolites

Closed Vesicles Are Required for the Generation of ATP

Much evidence shows that the coupling between the oxidation of NADH and $FADH_2$ by O_2 and the synthesis of ATP from ADP and P_i in mitochondria occurs only via the electrochemical proton gradient across the inner membrane. The addition of O_2 and compounds such as NADH or pyruvate to isolated, intact mitochondria results in a net synthesis of ATP. However, ATP production is absolutely dependent on the integrity of the inner mitochondrial membrane. In the presence of minute amounts of detergents that make the membrane leaky, the oxidation of NADH or acetyl CoA by O_2 still occurs, but no ATP is made. Under these conditions, no transmembrane proton concentration gradient or membrane electric potential can be maintained.

The Proton-motive Force Is Composed of a Proton Concentration Gradient and a Membrane Electric Potential

In bacteria and mitochondria, as well as chloroplasts, the movement of protons across a membrane down their electrochemical gradient is coupled to ATP synthesis. The

The polarity of this proton gradient and potential is established by electron transport. During mitochondrial electron transport, protons are pumped out of the matrix and into the intermembrane space (see Figure 15-14), or across the inner membrane from the cytoplasmic to the exoplasmic face. In aerobic bacteria, the oxidation of carbohydrates is coupled to the pumping of protons from the cytosol into the medium (again, from the cytoplasmic to the exoplasmic face); during the generation of ATP, protons flow back across the bacterial membrane, through the F_0F_1 ATPase complex.

In oxidative phosphorylation in both bacteria and mitochondria, the movement of electrons through the membrane is coupled to the generation of a proton-motive force, making the processes of electron transport, proton pumping, and ATP generation interdependent. The generation of this electrochemical proton gradient and its subsequent dissipation in ATP formation normally occur simultaneously, and the processes are closely coupled.

proton-motive force that propels these protons is a combination, or sum, of the proton concentration gradient and the membrane electric potential. This proton movement is similar to the movement of Na^+ ions into a cell (see Figure 14-8).

Since protons are positively charged, pumping them across the membrane generates an electric potential only if the membrane is poorly permeable to anions. A proton concentration (pH) gradient can develop only if the membrane is permeable to a major anion, such as Cl^- (chloride), or if the protons are exchanged for another cation, such as K^+ (potassium). In these latter cases, proton pumping does not lead to a difference in electric potential across the membrane because there is always an equal concentration of positive and negative ions on each side of the membrane. However, proton pumping does produce a pH gradient across the membrane, making the proton concentration gradient the major component of the proton-motive force in such cases. This occurs in the chloroplast thylakoid membrane during photosynthesis (Chapter 16).

By contrast, the mitochondrial inner membrane is relatively impermeable to anions: a greater proportion of energy is stored as a membrane electric potential, and the actual pH gradient is smaller. In mitochondria, the electric potential across the inner membrane is about 200 mV (the matrix being negative with respect to the intermembrane space), so that the membrane electric potential is the more significant component of the proton-motive force.

Since a difference of one pH unit represents a ten-fold difference in H^+ concentration, a pH gradient of one unit across a membrane is equivalent to an electric potential of 59 mV (at 22°C). Thus we can define the proton-motive force as

$$pmf = \psi - \frac{RT}{\mathscr{F}}\Delta pH = \psi - 59\Delta pH$$

where ψ, the transmembrane electric potential, and pmf are measured in mV. In respiring mitochondria, pmf = ~ 220 mV; $\psi = \sim 160$ mV, and a ΔpH of one unit (equivalent to ~ 60 mV) accounts for the remaining pmf.

Mitochondria and chloroplasts are much too small to be impaled with electrodes, so how can the electric potential and the pH gradient across the inner mitochondrial membrane be determined? The inner membrane is normally impermeable to potassium ions (K^+), but the potassium ionophore valinomycin (see Figure 14-4)—a lipid-soluble peptide that selectively binds one K^+ ion in its hydrophilic interior—allows K^+ ions to be transported across the otherwise impermeable phospholipid membrane. At equilibrium, the concentration of K^+ ions on both sides of the membrane is determined by the membrane electric potential E (in mV), according to the Nernst equation (page 537):

$$E = -\frac{RT}{\mathscr{F}} \ln \frac{K_{in}}{K_{out}} = -59 \log \frac{K_{in}}{K_{out}}$$

When valinomycin and radioactive potassium (^{42}K) ions are added to a suspension of respiring mitochondria or inner-membrane vesicles, oxidative phosphorylation proceeds and is largely unaffected. K^+ ions accumulate inside the mitochondria (or vesicles) in a ratio of $K_{matrix}:K_{cytosol}$ of ~ 2500. By the Nernst equation

$$E = -59 \log \frac{K_{in}}{K_{out}} = -59 \log 2500 = -200 \text{ mV}$$

Thus the electric potential across the inner membrane is ~ 200 mV (negative inside matrix). (In such experiments, valinomycin is used in trace amounts, so that relatively few K^+ ions are actually transported and valinomycin does not obliterate the potential it is intended to measure.)

The fluorescent properties of a number of dyes are dependent on pH. By trapping such dyes inside inner-membrane vesicles of mitochondria, the inside pH during oxidative phosphorylation can be measured; as noted, the matrix pH is typically one unit higher than that of the cytosol.

The F_0F_1 Synthase Complex Couples ATP Synthesis to Proton Movement down the Electrochemical Gradient

A predominant protein in the mitochondrial membrane is the multisubunit *coupling factor*—the enzyme that actually synthesizes ATP. An extremely similar enzyme complex is located in the thylakoid membranes of chloroplasts and in the plasma membranes of bacterial cells. This coupling factor has two principal components: F_0 and F_1 (Figure 15-15). In bacteria, the integral membrane complex F_0 is composed of three distinct polypeptides (a, b, and c), which together span the inner mitochondrial membrane. Mitochondrial F_0 is composed of similar a, b, and c elements and, depending on the species, two to five additional peptides of unknown function (Table 15-1). Attached to F_0 is F_1, a complex of five distinct polypeptides (α, β, γ, δ, and ϵ) with the composition $\alpha_3\beta_3\gamma\delta\epsilon$. F_1 forms the knobs that protrude on the matrix side of the inner membrane (see Figure 15-15). F_1 can be detached from this membrane by mechanical agitation and is water-soluble.

When F_1 is physically separated from the membrane, it is capable only of catalyzing ATP hydrolysis. Hence, it is often called the F_1 ATPase. Its natural function, however, is in the synthesis of ATP: the submitochondrial vesicles from which F_1 is removed cannot catalyze ATP synthesis; when F_1 particles reassociate with these vesicles, how-

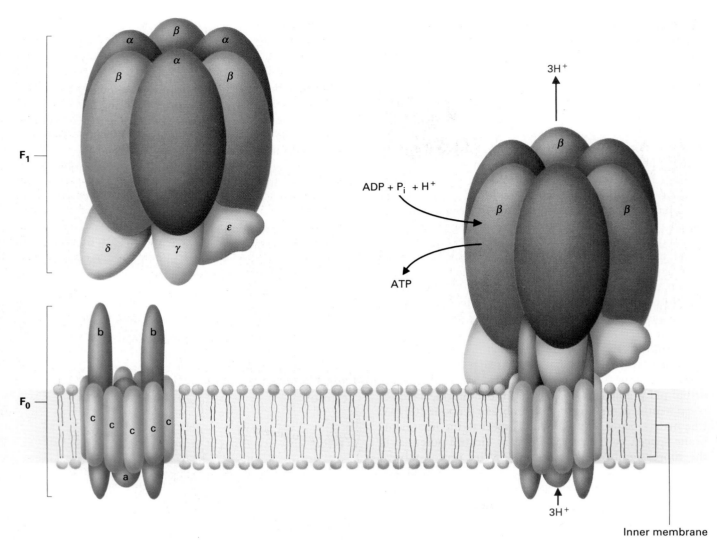

▲ Figure 15-15 The ATP synthase of mitochondria, bacteria, and chloroplasts is a bipartite enzyme composed of two oligomeric complexes F_0 and F_1. F_0 is an integral membrane protein; in bacteria, it consists of three subunits (a, b, and c) that form a proton conduction channel. F_1 contains three copies each of subunits α and β and is bound to F_0 via subunits γ, δ, and ϵ. Synthesis of one ATP molecule from ADP and P_i occurs spontaneously at the catalytic site on a β subunit of F_1, due to tight binding of ATP at this site. Proton movement through F_0, driven by the proton-motive force, promotes the catalytic synthesis of ATP by causing the bound ATP to be released; this frees up the site for the binding of ADP and P_i, which, in turn, spontaneously form tightly bound ATP. Catalysis alternates between the three copies of the β subunit active site. *Courtesy of B. Trumpower.*

ever, they once again become fully active in ATP synthesis (Figure 15-16). Another observation lends further evidence to support the role of F_1 in ATP synthesis. F_1 can be dissociated into its component polypeptides, and the functions of several of these subunits are known, at least in outline. Each of the three copies of subunit β binds ATP and ADP and contains a catalytic site for the synthesis of ATP from ADP and P_i; subunits γ and δ bind to F_0 (see Figure 15-15). ATP or ADP also binds to regulatory or allosteric sites on the three α-subunits at which the F_1 particle senses the level of ATP or ADP in the matrix and modifies the rate of ATP synthesis accordingly. The pre-

cise orientation of these subunits and their exact interactions are subjects of intense research.

F_0 is believed to contain a transmembrane channel through which protons flow to F_1. When F_0 or its purified proteolipid subunit c is experimentally incorporated into phospholipid vesicles (liposomes), the permeability of the vesicles to H^+ is greatly stimulated, suggesting that the proteolipid or other F_0 proteins form a proton channel. Although it is not an integral membrane protein, F_1 probably forms a barrier that is impermeable to protons; removal of F_1 from the inner mitochondrial membrane makes it highly permeable to protons.

Table 15-1 Protein composition of enzyme complexes involved in electron transport and ATP synthesis

Bacteria	Mitochondria

F_0 3 subunits
    a: proton conduction
    b: proton conduction
    c: proteolipid
	5–8 subunits
	    a: proton conduction
	    b. proton conduction
	    c: proteolipid
	    2–5 accessory proteins

CoQH$_2$–cytochrome c reductase
(bc_1 complex)
    3 subunits
    cytochrome b
    cytochrome c_1
    FeS protein
	9–11 subunits
	    cytochrome b
	    cytochrome c_1
	    FeS protein
	    6–8 accessory proteins

Cytochrome c oxidase
    3 subunits
    I (contains a and
    a_3 hemes)
    II
 	
    III (site of H$^+$
    pumping)
	11–13 subunits
	
	    I
	    II
	    III
	
	    8–11 others (depending
	    on species)

It may be possible to regulate permeability of this F_0 transmembrane channel to protons. One subunit of F_0 can bind oligomycin, an antibiotic that inhibits ATP generation in mitochondria by interfering with the use of the proton concentration gradient for ATP production. This protein may regulate the flow of protons through F_0; when the proton motive force is very small, it could prevent the reverse reaction: the generation of a proton concentration gradient by ATP hydrolysis. Such a reaction would waste ATP, and it seems reasonable for the cell to protect itself against this circumstance by regulation.

Most experiments indicate that the movements of three protons through the F_0F_1 complex are coupled to the synthesis of one high-energy phosphate bond in ATP. A simple calculation indicates that the passage of more than one proton is required to synthesize one molecule of ATP. For example, for the reaction

$$H^+ + ADP^{3-} + P_i^{2-} \rightleftharpoons ATP^{4-} + H_2O$$

at standard concentrations of ATP, ADP, and P_i

$$\Delta G^{\circ\prime} = 7.3 \text{ kcal/mol}$$

(a)

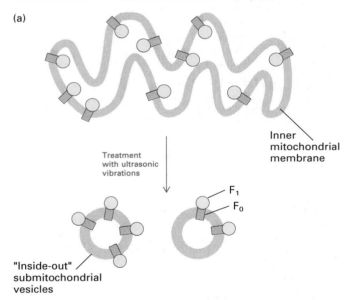

(b)

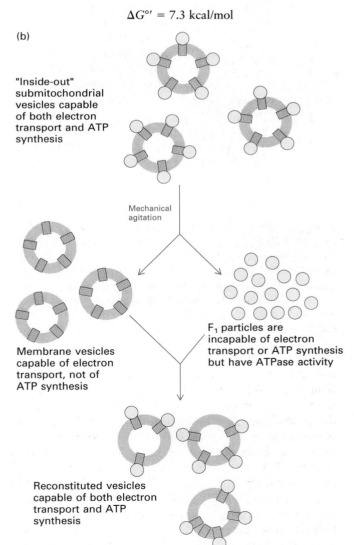

"Inside-out" submitochondrial vesicles capable of both electron transport and ATP synthesis

Mechanical agitation

Membrane vesicles capable of electron transport, not of ATP synthesis

F_1 particles are incapable of electron transport or ATP synthesis but have ATPase activity

Reconstituted vesicles capable of both electron transport and ATP synthesis

▲ **Figure 15-16** Mitochondrial F_1 particles are required for ATP synthesis, not for electron transport. (a) "Inside-out" submitochondrial vesicles can be obtained by exposing the inner membrane to ultrasonic vibration, so that the vesicles are disrupted and reseal with the F_1 particles facing outside. These vesicles can transfer electrons from added NADH to O$_2$ with the concomitant synthesis of ATP from ADP and P$_1$. (b) The mechanical agitation of "inside-out" vesicles causes the F_1 particles to dissociate from the inner membrane. The membrane vesicles can still transport electrons from NADH to O$_2$ but cannot synthesize ATP. The subsequent addition of F_1 particles reconstitutes the native membrane structure, restoring the capacity for ATP synthesis.

At the concentrations of reactants in the mitochondrion, however, $\Delta G^{\circ\prime}$ is probably higher (10–12 kcal/mol). The amount of free energy released by the passage of 1 mol of protons down an electrochemical gradient of 220 mV (0.22 V) can be calculated from the Nernst equation, setting $n = 1$:

$$\Delta G \text{ (cal/mol)} = -n\,\mathcal{F}\Delta E = -23062\,\Delta E \text{ (V)}$$
$$= (-23062)(0.22)$$
$$= -5080 \text{ cal/mol, or } -5.1 \text{ kcal/mol}$$

Since just over 5 kcal/mol of free energy is made available, the passage of at least two (possibly three) protons is essential to the synthesis of each molecule of ATP from ADP and P_i.

Exactly how the movement of protons through the F_0F_1 complex is coupled to ATP generation is the subject of intense debate and experimentation. Coupling between proton flow and ATP synthesis must be indirect, since the ATP, P_i, and ADP binding sites on the β subunits, where ATP synthesis occurs, are 9–10 nm from the surface of the mitochondrial membrane. Most evidence suggests that ADP and P_i, after binding to a β subunit of the F_1 complex, spontaneously form ATP that remains tightly bound to the F_1. Experiments show that the $\Delta G^{\circ\prime}$ needed to form ATP and ADP and P_i, bound to isolated F_1 particles, is near 0. Presumably, dissociation of the completed ATP from the protein is induced by a conformational change in the F_1 complex that, in turn, is caused by the flow of protons through it. The release of bound ATP frees the catalytic site for ADP and P_i binding, which spontaneously forms tightly bound ATP. Catalysis alternates between the three copies of the β subunit that undergo sequential conformational changes.

The three β and three α polypeptides in each F_1 particle are identical. F_1 also contains three nonidentical subunits (γ, δ, and ϵ) that attach it to F_0. In one hypothetical mechanism for ATP synthesis, called the *binding-change mechanism* (Figure 15-17), proton flux in essence causes a 120° rotation of the $\alpha_3\beta_3$ part of F_1 so that each β subunit is reoriented relative to subunits γ, δ, and ϵ and changes its binding affinities for ATP, ADP, and P_1.

Reconstitution of Closed Membrane Vesicles Supports the Role of the Proton-motive Force in ATP Synthesis

The view that a proton-motive force is the immediate source of energy for ATP synthesis, introduced in 1961 by Peter Mitchell, was initially opposed by virtually all researchers working in photosynthesis and oxidative phosphorylation. They favored a mechanism similar to the well-elucidated, substrate-level phosphorylation in glycolysis, in which electron transport is directly coupled to ATP synthesis (see Figures 15-4, 15-5, and 15-6). Electron transport through the membranes of chloroplasts or mitochondria was believed to generate a high-energy chemical bond directly (a phosphoenzyme, for example), which was then used to convert P_i and ADP to ATP. Despite intense efforts by a larger number of investigators, however, no such intermediate could ever be identified.

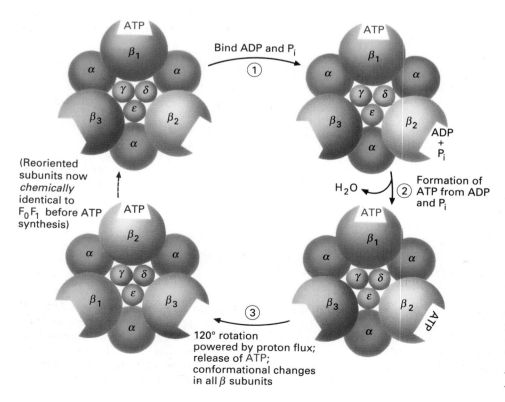

◀ **Figure 15-17** Rotational model for the synthesis of ATP from ADP and P_i by F_0F_1 ATP synthase. The three β subunits alternate between three conformational states that differ in their binding affinities for ATP, ADP, and P_i. In reaction 1, ADP and P_i bind to one of the three β subunits (here, arbitrarily designated β_2) to form ATP spontaneously (reaction 2). Proton flux (reaction 3) powers a 120° rotation of the $\alpha_3\beta_3$ part of the F_1 complex relative to subunits γ, β, and ϵ, causing ATP to be released from one β subunit (here, β_1) with concomitant conformational changes in all three β subunits. *After R. L. Cross, D. Cunningham, and J. K. Tamura, 1984*, Curr. Top. Cell Reg. *24:336; P. Boyer, 1989*, FASEB J. *3:2164–2178.*

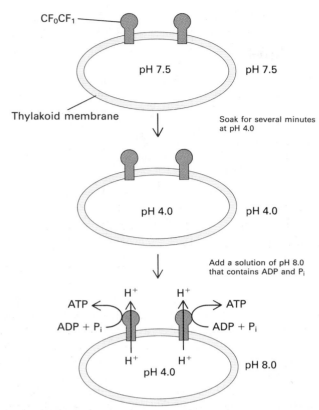

▲ **Figure 15-18** ATP synthesis from ADP and P_i by chloroplast thylakoid membranes results from an artificially imposed pH gradient.

Other evidence against this direct coupling model included the observation that only sealed, intact membrane vesicles can carry out oxidative phosphorylation.

But clear-cut evidence to support the role of the proton-motive force in ATP synthesis was not easy to obtain, until the development of techniques to purify and reconstitute organelle membranes and membrane proteins. Then Mitchell's chemiosmotic mechanism became the generally accepted hypothesis, although many of its details are still very unclear. Here, it is appropriate to summarize some early key experiments that established the basic features of the proton-motive force.

One of the most important experiments directly demonstrated that an artificially imposed pH gradient can result in ATP synthesis. In one study (Figure 15-18), chloroplast thylakoid vesicles containing F_0F_1 particles were equilibrated in the dark with a buffered solution at pH 4.0 until the pH in the thylakoid lumen was also 4.0. These vesicles were rapidly mixed with a solution at pH 8.0 containing ADP and P_i. A burst of ATP synthesis accompanied the transmembrane movement of protons driven by a 10,000-fold (10^{-4} *M* versus 10^{-8} *M*) concentration gradient. In reciprocal experiments on "inside-out" preparations of submitochondrial vesicles, an artificially generated membrane electric potential also resulted in ATP synthesis.

Particularly dramatic results were achieved in an experiment employing a protein found in the plasma membrane of a photosynthetic bacteria, *bacteriorhodopsin*. This protein was asymmetrically incorporated into artificial phospholipid vesicles (liposomes) that also contained purified mitochondrial F_0F_1 complexes. On illumination, the bacteriorhodopsin pumped protons into the lumen of these vesicles—a process analogous to the light-driven pumping of protons during photosynthesis in the chloroplasts of higher plants. The resultant proton-motive force powered the synthesis of ATP from ADP and P_i. If no F_0F_1 was used, no ATP was made. These findings left no doubt that the F_0F_1 complex is the ATP-generating enzyme and that ATP generation is dependent on an electrochemical proton gradient (Figure 15-19).

Many Transporters in the Inner Mitochondrial Membrane Are Powered by the Proton-motive Force

The inner mitochondrial membrane contains a number of permease proteins that permit various metabolites, including pyruvate, malate, and the amino acids aspartate and glutamate, to enter and leave the organelle. Two important permeases transport phosphate and ADP from the cytosol to the mitochondrion in exchange for ATP formed by oxidative phosphorylation inside the mitochondrion:

$$ADP^{3-}_{out} + H_2PO_4^{-}_{out} + ATP^{4-}_{in} \longrightarrow$$
$$ADP^{3-}_{in} + H_2PO_4^{-}_{in} + ATP^{4-}_{out}$$

In this way, ATP can be formed in the mitochondrion and transported to the cytosol. Energy stored in the proton

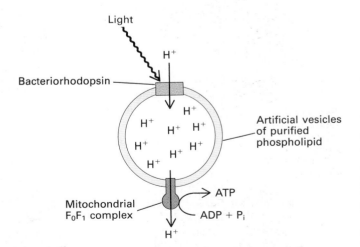

▲ **Figure 15-19** Demonstration that the mitochondrial F_0F_1 complex is the site of ATP synthesis. Purified mitochondrial F_0F_1 complexes and bacteriorhodopsin are incorporated into the same phospholipid vesicles (liposomes). Bacteriorhodopsin pumps protons into the lumen following illumination, resulting in the synthesis of ATP from ADP and P_i.

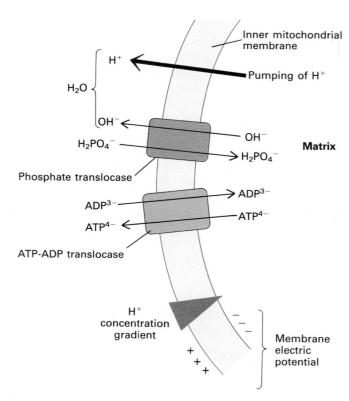

◀ **Figure 15-20** The phosphate and ATP-ADP transport system in the inner mitochondrial membrane. Uptake of $H_2PO_4^-$ is coupled to the outward movement of an OH^- anion. As the OH^- combines with a proton pumped outward by respiration, the net result is the uptake of $H_2PO_4^-$ powered by the outward pumping of a proton. The transmembrane electric potential also powers the uptake of ADP^{3-} in exchange for the export of ATP^{4-} from the matrix. For every four protons pumped outward, three are used to synthesize one ATP molecule and one is used to export ATP in exchange for ADP and P_i. The outer membrane is not shown here because it is freely permeable to molecules of $>10,000$ MW.

concentration gradient is also used to power this exchange of ATP for ADP and P_i. To see this, let's examine the two components of the system: the phosphate permease and the ATP-ADP exchange protein.

The phosphate carrier is an antiport that catalyzes an exchange of one $H_2PO_4^-$ for one OH^- (Figure 15-20). The outward transport of OH^- is driven by the proton concentration gradient (low pH outside) across the inner mitochondrial membrane, which is generated during electron transport. Each OH^- transported outward combines with a proton pumped to the cytosolic side of the inner membrane to form H_2O. Thus respiring mitochondria pumping protons outward can accumulate phosphate from the surrounding medium against a concentration gradient.

ADP entry and ATP exit are coupled: the ADP-ATP translocase is an antiport that allows one molecule of ADP to enter only if one molecule of ATP exits simultaneously. The translocase, a dimer of two 30,000-MW subunits, makes up 10–15 percent of the protein in the inner membrane, so it is one of the more abundant mitochondrial proteins.

ADP transport to the mitochondrion, in exchange for ATP, is driven by the transmembrane electric potential. At neutral pH, recall that each ATP molecule contains four negative charges and each ADP molecule has three. Thus the exchange of ADP (out to in) for ATP (in to out) results in the outward movement of one additional negative charge (electron):

$$ADP^{3-}_{out} + ATP^{4-}_{in} \longrightarrow ADP^{3-}_{in} + ATP^{4-}_{out}$$

The net outward movement of this negative charge is powered by the membrane electric potential (exoplasmic face positive) across the inner membrane. Thus net uptake of ADP and net export of ATP are powered by one component of the proton motive force. The expenditure of energy from this gradient to export ATP from the mitochondrion in exchange for ADP and P_i ensures a high ratio of ATP to ADP in the cytosol, where the phosphate-bond energy of ATP is utilized. Because some of the protons pumped out of the mitochondrion power the ATP-ADP exchange, fewer protons are available for ATP synthesis. For every four protons pumped out, three are used to synthesize one ATP molecule and one is used to power the export of ATP from the mitochondrion in exchange for ADP and P_i.

Inner-membrane Proteins Allow the Uptake of Electrons from Cytosolic NADH

During the conversion of glucose to pyruvate in the cytosol, two molecules of NAD^+ are reduced to NADH. The electrons from NADH are ultimately transferred to O_2, concomitant with the generation of ATP. However, the inner mitochondrial membrane is impermeable to NADH, so how do the electrons from cytosolic NADH enter the mitochondrial electron transport system?

Several *electron shuttles* are employed; in the most widespread—the *malate shuttle* (Figure 15-21)—cytosolic NADH reduces the four-carbon oxaloacetate to

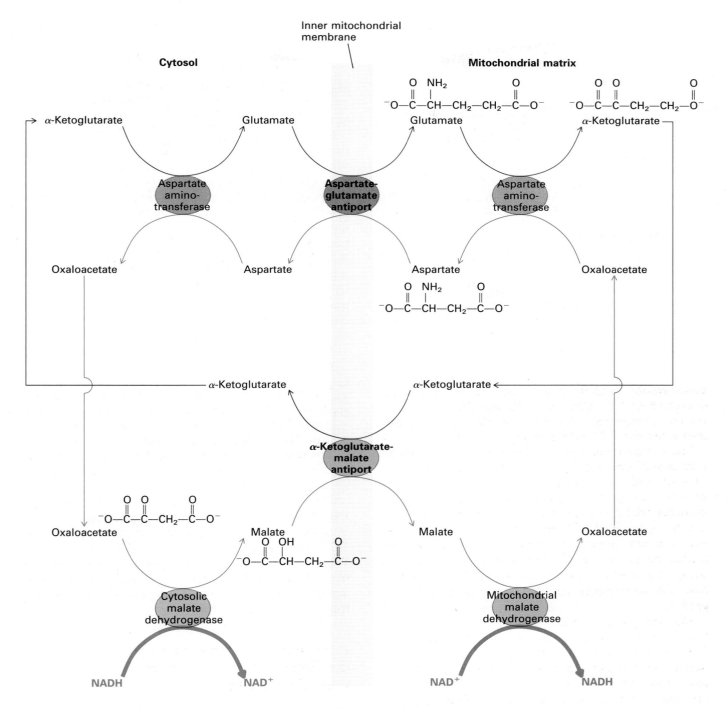

▲ **Figure 15-21** The malate shuttle transports electrons from cytosolic NADH across the inner mitochondrial membrane. Cytosolic NADH reduces cytosolic oxaloacetate, forming malate *(lower left)*. Malate enters the matrix in exchange for one molecule of α-ketoglutarate, catalyzed by the ketoglutarate-malate antiport. Matrix malate, in turn, reduces matrix NAD^+, forming NADH and oxaloacetate. Oxaloacetate is converted into the amino acid aspartate *(upper right)*, which exits the matrix in exchange for glutamate. Once in the cytosol, aspartate is reconverted to oxalacetate *(upper left)*, completing the cycle. One counterclockwise turn of the entire cycle can be summarized as

$$NADH_{cytosol} + NAD^+{}_{matrix} \longrightarrow NAD^+{}_{cytosol} + NADH_{matrix}$$

Courtesy of B. Trumpower.

malate. Malate crosses the inner membrane (in exchange for α-ketoglutarate) and reduces NAD$^+$, forming NADH as well as oxaloacetate in the matrix. But oxaloacetate does not pass directly back to the cytosol: it is converted to the amino acid aspartate first in order to cross to the cytosol (in exchange for glutamate); there, the aspartate is reconverted to oxaloacetate. To allow the cycle to proceed, α-ketoglutarate is converted to the amino acid glutamate in the cytosol (see arrows in Figure 15-21) and crosses to the matrix, where it is reconverted to α-ketoglutarate. The net effect of this rather complex cycle is the oxidation of cytosolic NADH to NAD$^+$, together with the reduction of matrix NAD$^+$ to NADH. The latter, in turn, transfers electrons to the transport chain that pumps protons outward from the matrix.

NADH, Electron Transport, and Proton Pumping

Studies that describe how a proton concentration gradient and a membrane electric potential can be used to generate ATP and power the import and export of molecules are of great importance. Equally important are a set of experiments establishing that electron movement from NADH or FADH$_2$ to O$_2$ is catalyzed by a series of electron carriers in four multiprotein complexes in the inner mitochondrial membrane and that such electron transport results in the generation of a proton motive force.

Electron Transport in Mitochondria Is Coupled to Proton Pumping

Isolated mitochondria maintained without O$_2$ do not oxidize NADH or other compounds that can donate electrons to the respiratory chain, because there is no ultimate electron acceptor (usually O$_2$), nor do they generate ATP. The addition of a small amount of O$_2$ to such anaerobic mitochondria results in the oxidation of twice that number of NADH molecules:

$$2NADH + 2H^+ + O_2 \longrightarrow 2NAD^+ + 2H_2O$$

Measurements of oxidized NADH and reduced O$_2$ show directly that each NADH molecule releases two electrons to the electron transport chain and reduces one O atom.

One experiment (Figure 15-22) provides positive proof that electron transport is coupled to proton transport. When a small amount of O$_2$ is added to a suspension of mitochondria deprived of O$_2$ in the presence of NADH, the medium outside the mitochondria becomes acidic because protons pumped from the matrix accumulate in the intermembrane space. Since the outer membrane is freely permeable to protons, the pH of the outside medium is lowered briefly. Exactly how many protons are transported per electron pair released by NADH is a matter of controversy. Some transported protons subsequently diffuse back across the membrane, making a precise estimate difficult. About 10 protons are believed to be transported for every electron pair transferred from NADH to O$_2$.

When succinate is oxidized to fumarate, the electron carrier FAD (bound to the enzyme succinate dehydrogenase) is reduced to FADH$_2$ (see Figure 15-10). FADH$_2$ transfers electrons to the electron transport chain, but at a later point than NADH does (Figure 15-23). Consequently, only six protons are pumped for every electron pair transferred from FADH$_2$ to O$_2$.

The Mitochondrial Respiratory Chain Transfers Electrons from NADH to O$_2$

When NADH is oxidized to NAD$^+$, two electrons and one proton are released. Recall that protons are soluble in aqueous solution as hydronium ions (H$_3$O$^+$). Free electrons, by contrast, cannot exist in aqueous solutions.

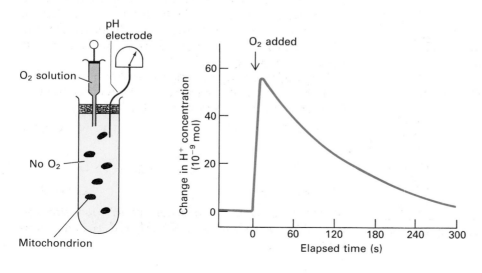

◄ **Figure 15-22** Electron transport from NADH or FADH$_2$ to O$_2$ is coupled to proton pumping. If a source of electrons for respiration, such as NADH, is added to a suspension of mitochondria depleted of O$_2$, no NADH is oxidized. When a small amount of O$_2$ is added to the system (arrow), the pH of the surrounding medium drops sharply—a change that corresponds to an increase in protons outside the mitochondria. Thus oxidation of NADH by O$_2$ is coupled to the pumping of protons out of the matrix. Once the O$_2$ is depleted, the excess protons slowly leak back into the mitochondria and the pH of the extracellular medium returns to its initial value.

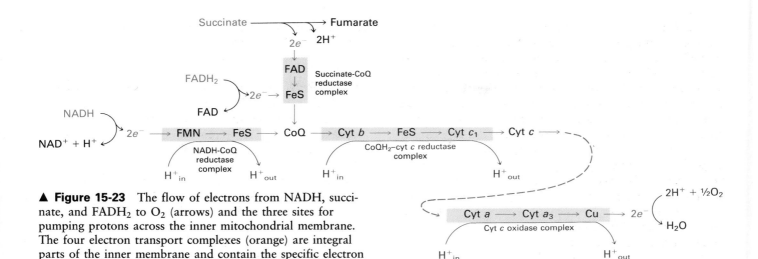

▲ **Figure 15-23** The flow of electrons from NADH, succinate, and FADH$_2$ to O$_2$ (arrows) and the three sites for pumping protons across the inner mitochondrial membrane. The four electron transport complexes (orange) are integral parts of the inner membrane and contain the specific electron carriers indicated. Ubiquinone (coenzyme Q) and cytochrome c are also associated with the membrane. FMN is a cofactor related to FAD (see Figure 15-10) that contains only the flavin-ribose phosphate part of FAD.

Electrons are passed from NADH or FADH$_2$ to O$_2$ along a chain of electron carriers, most of which are prosthetic groups (such as flavins, heme, iron-sulfur clusters, and copper) bound to protein particles on the inner membrane. *Ubiquinone,* also termed *coenzyme Q* (CoQ), is the only electron carrier that is not a protein-bound prosthetic group.

Electron transport from NADH to O$_2$ is a thermodynamically downhill process with a potential span of 1.14 V; that is, as electrons move from NADH to O$_2$, their potential declines by 1.14 V, which corresponds to 26.2 kcal/mol of electrons transferred, or 53 kcal/mol for a pair of electrons. Much of this energy is conserved at

three stages of electron transport by pumping protons across the inner membrane.

The 12 or more electron carriers are grouped into four multiprotein particles (see Figure 15-23), each of which spans the inner mitochondrial membrane. Table 15-2 lists the prosthetic groups in each complex. The NADH-CoQ reductase complex carries electrons from NADH to ubiquinone (Figure 15-24). During this part of the electron transport chain, NADH is oxidized to NAD$^+$ and the two released electrons move through a series of carriers until CoQ is ultimately reduced to CoQH$_2$:

$$\text{NADH} \qquad\qquad\qquad\qquad \text{CoQH}_2$$
$$2e^- \longrightarrow 2e^- \longrightarrow 2e^-$$
$$\text{NAD}^+ + \text{H}^+ \qquad\qquad\qquad \text{CoQ} + 2\text{H}^+$$

Like the other stages of electron transport, this one is thermodynamically favorable. Each transported electron undergoes a drop in potential of ∼360 mV, so that the $\Delta G^{\circ\prime}$ required to transport a pair of electrons is −16.6 kcal/mol (Figure 15-25).

The succinate-CoQ reductase complex transfers electrons from succinate (released during its oxidation to fumarate) to FADH$_2$ and then to CoQ:

$$\text{Succinate} \quad \text{FAD} + 2\text{H}^+ \qquad\qquad\qquad \text{CoQH}_2$$
$$2e^- \longrightarrow 2e^- \longrightarrow 2e^-$$
$$\text{Fumarate} \quad \text{FADH}_2 \qquad\qquad\qquad \text{CoQ} + 2\text{H}^+$$

coenzyme Q is the "collection point" for electrons formed by the oxidation of NADH, FADH$_2$, and succinate.

Table 15-2 Components of the mitochondrial electron transport chain

Enzyme complex	Mass (in daltons)	Prosthetic groups
NADH-CoQ reductase	85,000	FMN FeS
Succinate-CoQ reductase	97,000	FAD FeS
CoQH$_2$-cytochrome c reductase	280,000	Heme b Heme c$_1$ FeS
Cytochrome c oxidase	200,000	Heme a Heme a$_3$ Cu
Cytochrome c	13,000	Heme c

SOURCE: J. W. De Pierre and L. Ernster, 1977, *Ann. Rev. Biochem.* **46**:201.

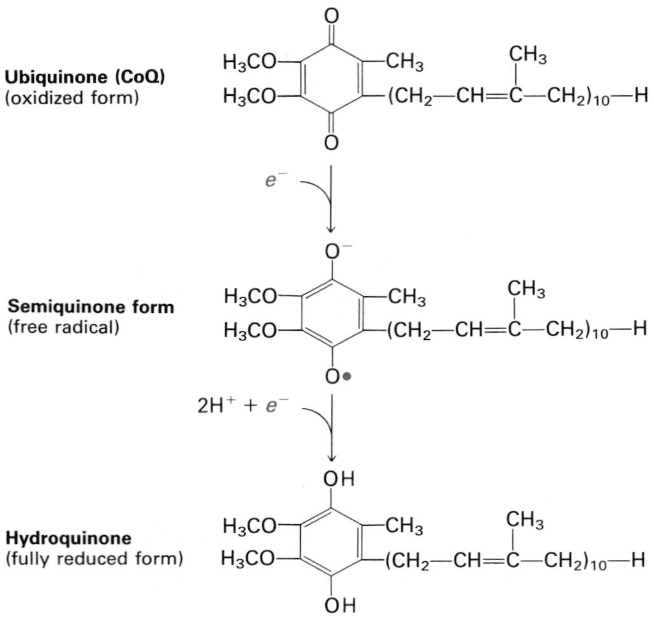

Ubiquinone (CoQ)
(oxidized form)

e^-

Semiquinone form
(free radical)

$2H^+ + e^-$

Hydroquinone
(fully reduced form)

▲ **Figure 15-24** The structure of ubiquinone (coenzyme Q), illustrating its ability to carry protons and electrons. Found in bacterial and mitochondrial membranes, ubiquinone is a carrier for both electrons and hydrogen ions. It is the only carrier in the electron transport system that is not tightly bound or covalently bonded to a protein. Because of its long hydrocarbon "tail" of isoprene units, ubiquinone is soluble in the hydrophobic core of phospholipid bilayers and is very mobile. The addition of one electron to oxidized ubiquinone results in a half-reduced (semiquinone) form. The semiquinone is a free radical; the unpaired electron (blue dot) is delocalized by resonance over the benzene ring and attached O atoms.

A cytochrome-containing complex, $CoQH_2$-cytochrome c reductase, transfers electrons from reduced $CoQH_2$ to the water-soluble protein cytochrome c to yield its reduced form:

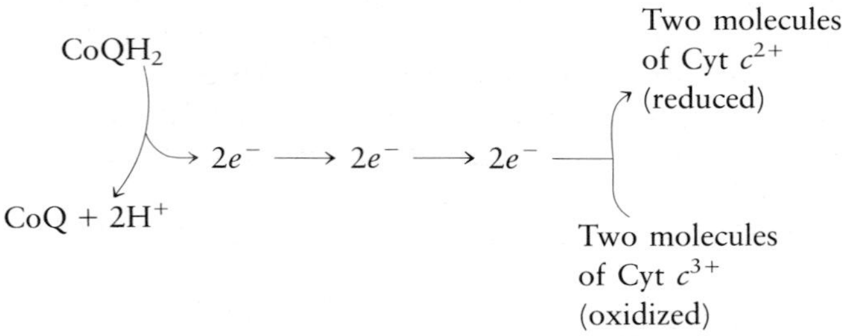

$CoQH_2$

$CoQ + 2H^+$

$2e^- \longrightarrow 2e^- \longrightarrow 2e^-$

Two molecules of Cyt c^{2+} (reduced)

Two molecules of Cyt c^{3+} (oxidized)

The cytochrome c oxidase complex transfers electrons from reduced cytochrome c to O_2, the ultimate electron acceptor, to yield H_2O:

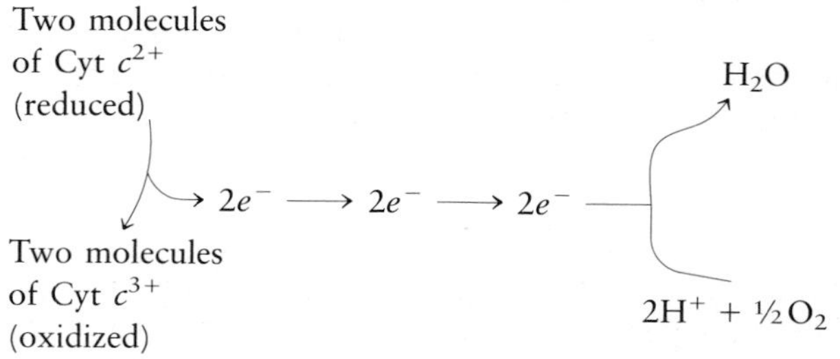

Two molecules of Cyt c^{2+} (reduced)

Two molecules of Cyt c^{3+} (oxidized)

$2e^- \longrightarrow 2e^- \longrightarrow 2e^-$

H_2O

$2H^+ + \frac{1}{2}O_2$

Each of these four electron transport complexes are laterally mobile in the mitochondrial membrane plane. The complexes are present in nonequal amounts: for each NADH dehydrogenase complex, there are about three $CoQH_2$-cytochrome c reductase and seven cytochrome c oxidase complexes. Furthermore, there do not appear to be stable contacts between any two complexes: electron

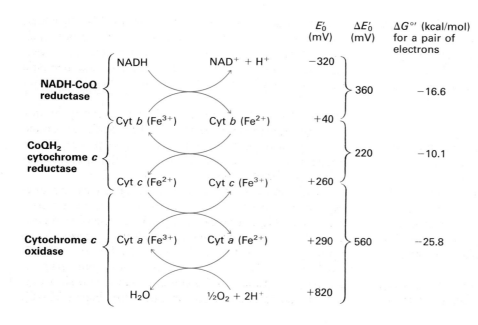

			E'_0 (mV)	$\Delta E'_0$ (mV)	$\Delta G^{\circ\prime}$ (kcal/mol) for a pair of electrons
NADH-CoQ reductase	NADH	NAD$^+$ + H$^+$	−320		
				360	−16.6
	Cyt b (Fe^{3+})	Cyt b (Fe^{2+})	+40		
CoQH$_2$ cytochrome c reductase				220	−10.1
	Cyt c (Fe^{2+})	Cyt c (Fe^{3+})	+260		
Cytochrome c oxidase	Cyt a (Fe^{3+})	Cyt a (Fe^{2+})	+290	560	−25.8
	H$_2$O	½O$_2$ + 2H$^+$	+820		

◀ **Figure 15-25** Energetics of the mitochondrial oxidation chain: the change in free energy released by the transport of a pair of electrons from NADH through each complex in which H$^+$ pumping occurs.

transport from one complex to another only occurs by diffusion of electron shuttles. The lipid-soluble coenzyme Q picks up electrons from the NADH-CoQ reductase and succinate dehydrogenase complexes and transfers them to the $CoQH_2$-cytochrome c reductase complex. Cytochrome c—a water-soluble peripheral protein found in the intermembrane space—interacts with a specific site on the $CoQH_2$-cytochrome c reductase complex, where it picks up an electron. The reduced cytochrome c diffuses in the intermembrane space until it encounters a cytochrome c oxidase complex, to which it donates an electron.

The NADH-CoQ reductase, $CoQH_2$-cytochrome c reductase, and cytochrome c oxidase complexes are sites for pumping protons across the inner mitochondrial membrane. The large sizes and multisubunit structures of these electron carrier complexes are undoubtedly related to their complicated function as both proton pumps and electron carriers. A number of different prosthetic groups capable of oxidation and reduction are essential to the function of the electron carrier proteins (see Table 15-2).

Iron-sulfur (FeS) Proteins

A set of proteins that contain the FeS clusters Fe_2S_2 and Fe_4S_4 (nonheme prosthetic groups) are important electron carriers. Specifically, these clusters consist of Fe atoms coordinated to inorganic S atoms as well as to four S atoms on cysteine residues on the protein (Figure 15-26). Some Fe atoms in the cluster bear a +2 charge; others, a +3 charge. However, because electrons in the outermost orbits are dispersed among the Fe atoms and move rapidly from one atom to another, each Fe atom actually bears a net charge of between +2 and +3. Additional electrons transferred to these carrier proteins are also dispersed over all the Fe atoms in the cluster.

Cytochromes and copper

Electron carriers termed *cytochromes* are generally membrane proteins that contain a heme prosthetic group similar to that in hemoglobin or myoglobin. The Fe in the center of the heme is the electron transporter. Electron transport occurs by oxidation and reduction:

$$Fe^{3+}_{ox} + e^- \rightleftharpoons Fe^{2+}_{red}$$

Because the heme ring consists of alternating double- and single-bonded atoms, a large number of resonance forms exist, and the extra electron is delocalized to the heme carbon (C) and nitrogen (N) atoms as well as to the Fe ion. In addition to its heme prosthetic group, cytochrome a_3, another polypeptide in the cytochrome c oxidase complex, contains a copper (Cu) ion that, by interconversion of the Cu^{2+}_{ox} and Cu^+_{red} forms, also participates in electron transfer.

The different cytochromes in the chain—a, a_3, b_{562}, b_{566}, c, and c_1,—have slightly different heme structures

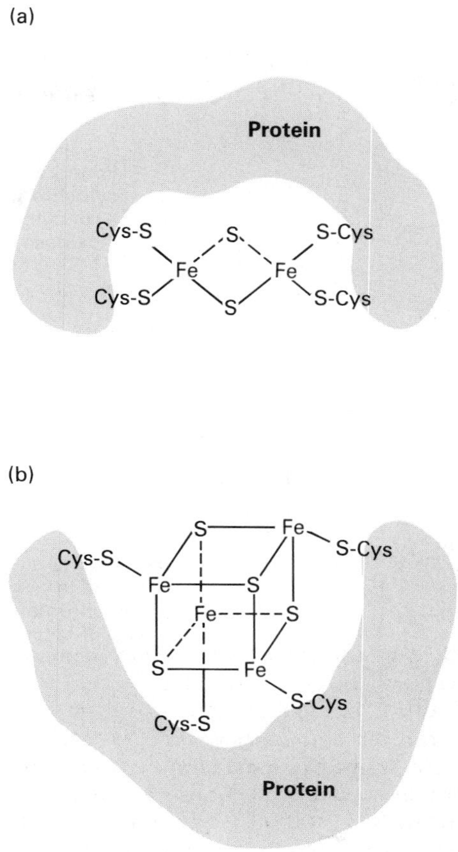

▲ **Figure 15-26** Three-dimensional structures of some FeS clusters in electron-transporting proteins: (a) a dimeric (Fe_2S_2) cluster; (b) a tetrameric (Fe_4S_4) cluster. Each Fe atom is bonded to four S atoms. Some S atoms are molecular sulfur; others occur in the cysteine side chains of protein. [See W. H. Orme-Johnson, 1973, *Ann. Rev. Biochem.* **42:**159.]

and axial ligands of the Fe atom (Figure 15-27), which generate different environments for Fe. Therefore, each cytochrome has a different reduction potential, or tendency to accept an electron—an important characteristic dictating unidirectional electron flow along the chain.

Ubiquinone

Ubiquinone (CoQ) is a carrier of hydrogen atoms (protons plus electrons). The oxidized quinone can accept a single electron to form a *semiquinone* and a second electron and two protons to form the fully reduced *dihydroubiquinone* (see Figure 15-24).

Cytochromes, Cu^{2+} ions, and FeS proteins are one-electron carriers: they accept and release a single electron at a time. Ubiquinone can accept either one or two electrons. NAD^+ is exclusively a two-electron carrier: it accepts ($H^+ + NAD^+ + 2e^- \longrightarrow NADH$) or releases ($NADH \longrightarrow H^+ + 2e^- + NAD^+$) only one pair of electrons at a time. *Flavins* can also accept two electrons, but they do so one electron at a time (see Figure 20-10).

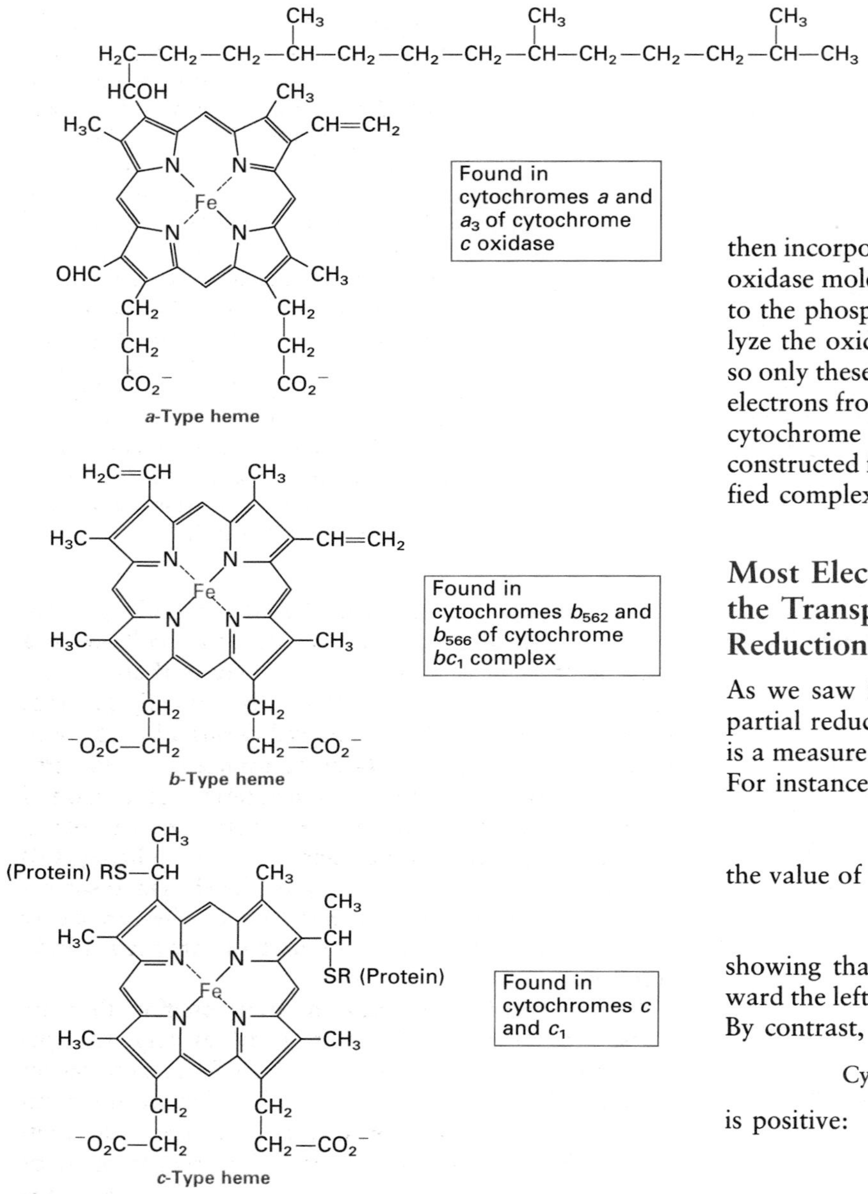

◀ Figure 15-27 Heme prosthetic groups of respiratory chain cytochromes in mitochondria. Note the difference in substituents on the porphyrin rings.

In practice, most *flavoprotein enzymes* cycle between FAD and the semiquinone or between the semiquinone and fully reduced FADH$_2$.

Mitochondrial electron flow, in summary, does not resemble an electric current through a wire, with each electron following the last. Rather, electrons are picked up by a carrier, one or two at a time, and then passed along to the next carrier in the pathway.

Each of the four multiprotein complexes of the electron transport chain spans the inner mitochondrial membrane, as exemplified by the cytochrome *c* oxidase complex, which (depending on the species) consists of 7–13 polypeptides, two hemes (*a* and *a$_3$*), and two Cu$^+$ ions bound to subunit I. The beef-heart complex can be purified and then incorporated into liposomes so that all cytochrome *c* oxidase molecules have the same orientation with respect to the phospholipid bilayer. This complex can still catalyze the oxidation of added cytochrome *c* (Fe^{2+}) by O$_2$, so only these seven polypeptides are required to transport electrons from cytochrome *c* to O$_2$. A molecular model of cytochrome *c* oxidase complex (Figure 15-28) has been constructed from electron microscopic images of the purified complex incorporated into liposomes.

Most Electron Carriers Are Oriented in the Transport Chain in Order of Their Reduction Potentials

As we saw in Chapter 1, the *reduction potential* for a partial reduction reaction (oxidized + e^- ⟶ reduced) is a measure of the equilibrium constant of that reaction. For instance, for the reaction

$$NAD^+ + H^+ + 2e^- \rightleftharpoons NADH$$

the value of the standard reduction potential is negative:

$$E_0' = -0.32 \text{ V}$$

showing that this partial reaction tends to proceed toward the left (toward the oxidation of NADH to NAD$^+$). By contrast, the potential for the reaction

$$Cyt\ c_{ox}(Fe^{3+}) + e^- \rightleftharpoons Cyt\ c_{red}\ (Fe^{2+})$$

is positive:

$$E_0' = +0.26 \text{ V}$$

showing that this partial reaction tends to proceed toward the right: toward the reduction of cytochrome *c* (Fe^{3+}) to *c* (Fe^{2+}). The final stage of the chain, the reduction of O$_2$ to H$_2$O

$$2H^+ + \tfrac{1}{2}O_2 + 2e^- \rightleftharpoons H_2O$$

has the most positive reduction potential:

$$E_0' = +0.816 \text{ V}$$

With the exception of the *b* cytochromes, the reduction potential of the electron carriers in mitochondria increases steadily from NADH to O$_2$ (see Figure 15-25); thus electron transport is thermodynamically favorable at every stage except one. We can think of the electrons released by NADH as having a high potential energy; a fraction of this energy is lost at each step as the electrons move from NADH to O$_2$, the ultimate electron acceptor.

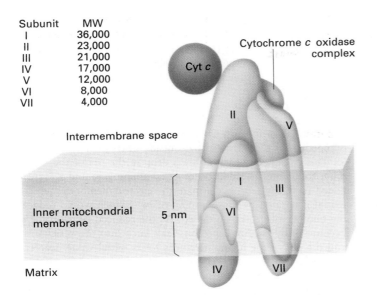

Subunit	MW
I	36,000
II	23,000
III	21,000
IV	17,000
V	12,000
VI	8,000
VII	4,000

◀ **Figure 15-28** The composition and orientation of the cytochrome c oxidase complex in the inner mitochondrial membrane. The complex exists as a dimer, but only one of the two monomer units is shown here. The overall shape of the molecule and its relationship to the phospholipid bilayer were obtained from image reconstitution of electron micrographs. The binding site of cytochrome c mainly involves subunit II. Because of ambiguities in the structure, investigators do not know whether cytochrome a_3 is subunit IV or VII of the complex. Depending on the species, mitochondrial cytochrome c oxidase can contain up to 13 subunits; seven are depicted in the model here. *After M. Brunori and M. T. Wilson, 1982, TIBS 7:295.*

To emphasize this point, let's calculate the $\Delta G^{\circ\prime}$ value for cytochrome c oxidase, the reaction of the last complex in the electron transport chain (see Figure 15-25):

$$2 \text{ Cyt } c \text{ (Fe}^{2+}) + 2\text{H}^+ + \tfrac{1}{2}\text{O}_2 \rightleftharpoons 2 \text{ Cyt } c \text{ (Fe}^{3+}) + \text{H}_2\text{O}$$

The half-reactions are

$$\text{Cyt } c \text{ (Fe}^{3+}) + e^- \rightleftharpoons \text{Cyt } c \text{ (Fe}^{2+}) \qquad E_0' = +0.26 \text{ V}$$

$$2\text{H}^+ + \tfrac{1}{2}\text{O}_2 + 2e^- \rightleftharpoons \text{H}_2\text{O} \qquad E_0' = +0.82 \text{ V}$$

The change in voltage for the total reaction is

$$\Delta E_0' = +0.82 - 0.26 = +0.56 \text{ V}$$

From Chapter 1

$$\Delta G^{\circ\prime} = -n \, \mathscr{F} \Delta E_0'$$

Recall that $\mathscr{F}$ is the Faraday constant (23,062 cal $\cdot$ V$^{-1} \cdot$ mol^{-1}) and that n is the number of electrons involved (here, 2). Thus

$$\begin{aligned} \Delta G^{\circ\prime} &= (-2)(23,062)(0.56) \\ &= -25,829 \text{ cal/mol} \\ &= -25.8 \text{ kcal/mol} \end{aligned}$$

The reaction is strongly exergonic: the transfer of a pair of electrons from cytochrome c to O$_2$ releases a significant amount of energy that can be made to do useful work—in this case, to pump protons across the inner mitochondrial membrane. An understanding of how this coupling occurs requires a somewhat more detailed look at the structure of the electron transport chain.

The In Vivo Order of the Electron Carriers Can Be Determined with Certainty

A wide range of techniques has been employed to confirm the pathway of electron transport depicted in Figure 15-23. We know that the reduction potentials of the electron

carriers increase steadily from NADH or FADH$_2$ to O$_2$ (except those for the b cytochromes, as we will see) and that all stages of electron transport (except those involving the b cytochromes) are thermodynamically favored. Because the oxidized and reduced forms of each cytochrome absorb light at different wavelengths, spectroscopic techniques can be used to determine the fraction of each cytochrome in each form. For instance, in respiring mitochondria, cytochromes closer to the oxidizing (O$_2$) end of the chain are more fully oxidized than those toward the reducing (NADH) end, because they are closer to the end of the chain at which electrons are permanently removed.

Many inhibitors of mitochondrial function that are known to block electron transport by reacting with specific complexes have been useful in confirming the carrier sequence: carriers downstream of the inhibited site accumulate in the oxidized form; carriers upstream accumulate in the reduced form. Spectrophotometric methods permit observers to determine which carriers are in the oxidized or reduced form in the presence of an inhibitor.

Cyanide (CN$^-$), a potent poison, kills by blocking the final stage of electron transport to O$_2$ by irreversibly binding to the heme Fe atoms in the cytochrome c oxidase complex. Rotenone, a plant-derived toxin used as a commercial insecticide and fish poison, blocks the NADH-CoQ reductase complex near the reduced end of the transport chain.

Three Electron Transport Complexes Are Sites of Proton Pumping

Except for the succinate-CoQ reductase complex, each multiprotein complex is a site for proton transport across the inner mitochondrial membrane during electron movement. By selectively extracting mitochondrial membranes with detergents, each of these complexes can be isolated

(a)

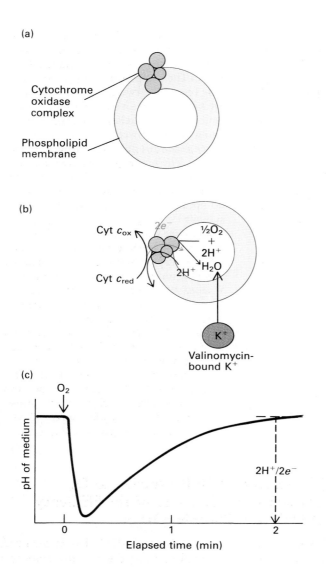

Cytochrome oxidase complex

Phospholipid membrane

(b)

Cyt c_{ox}

Cyt c_{red}

$2e^-$

$\frac{1}{2}O_2$ + $2H^+$ → H_2O

$2H^+$

K^+

Valinomycin-bound K^+

(c)

pH of medium

O_2

$2H^+/2e^-$

Elapsed time (min)

◀ **Figure 15-29** Segments of the mitochondrial electron transport system can be isolated and assembled into phospholipid membrane vesicles. If such a system is supplied with both a donor and an acceptor of electrons, protons are pumped across the membrane. (a) When the cytochrome c oxidase complex is incorporated into vesicles, the cytochrome c binding site is positioned on the outer surface. (b) If O_2 and reduced cytochrome c are added, electrons are transferred to O_2 to form H_2O. Concomitantly, protons are pumped from the inside to the outside of the vesicles. (c) This change is measured by a drop in the pH of the medium following the addition of O_2. Valinomycin and K^+ are added to the system to dissipate the electric membrane potential generated by the pumping of H^+, which would reduce the number of protons pumped. As the reduced cytochrome c becomes fully oxidized, protons leak back into the vesicles and the pH of the medium returns to its initial value. Careful studies show that two protons are pumped per O atom reduced. Since two electrons are needed to reduce one O atom, one proton is pumped per electron transported by cytochrome c. *After B. Reynafarje, L. Costa, and A. Lehninger, 1986, J. Biol. Chem. 261:8254–8262.*

in near purity and then incorporated into phospholipid vesicles. Additions of an appropriate electron source and an appropriate electron acceptor generate the pumping of protons across an artificial membrane. For example, the cytochrome c oxidase complex can be incorporated into phospholipid vesicles so that the binding site for the cytochrome c polypeptide is on the outside (Figure 15-29). The addition of reduced cytochrome c leads to movement of the electrons down the chain to O_2. Direct measurements indicate that two protons are transported out of the vesicles for every electron pair transported (or, equivalently, for every two molecules of cytochrome c oxidized). Additionally, two protons per electron pair are lost from the matrix during the formation of H_2O.

Similar studies indicate that the NADH-CoQ reductase complex pumps four protons per pair of electrons transported (or, equivalently, four protons per NADH molecule reduced). Precisely how these complexes pump pro-

tons across the inner membrane during electron transport is not clear. In addition, protons are generated in the matrix by the oxidation of NADH to NAD^+ and H^+. These excess protons are consumed by the cytochrome c oxidase complex during the formation of H_2O, completing the "proton loop" initiated by the NADH-CoQ reductase complex.

More is known about the mechanism of proton transport by the $CoQH_2$-cytochrome c reductase complex. Two protons are transported per electron (four per electron pair). Ubiquinone (CoQ) plays a key role in this process, which is known as the *proton-motive Q cycle*, or *Q cycle* (Figure 15-30). On the *matrix* side of the inner mitochondrial membrane, the fully oxidized form of CoQ (see Figure 15-24) receives one electron from reduced cytochrome b_{562}, a cytochrome in the $CoQH_2$-cytochrome c reductase complex. This forms the partially reduced Q semiquinone anion, denoted by $Q^{\cdot -}$ (the dot indicates that it is a free radical and the minus indicates that it is ionized). The $Q^{\cdot -}$ receives a second electron from the NADH-CoQ reductase complex and two protons from the matrix. The fully reduced QH_2 thus formed diffuses across the membrane to the intermembrane space, where it releases the two protons; thus two protons are transported for one electron received from the NADH-CoQ reductase complex. Simultaneously, two electrons are given up by QH_2: one is transferred first to cytochrome b_{566} and then to cytochrome b_{562}; the other is transferred to an FeS cluster, then to cytochrome c_1, and finally to cytochrome c in the intermembrane space. Recycling one electron through cytochromes b_{566} and b_{562} allows two

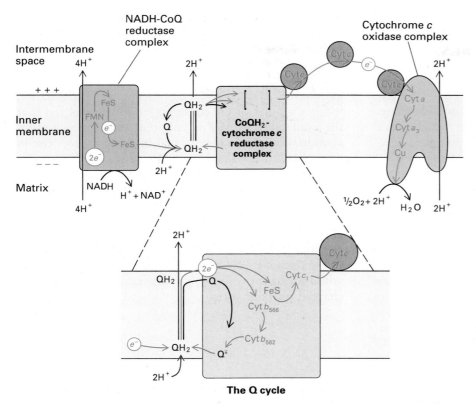

The Q cycle

◀ **Figure 15-30** The pathway of electron transport (blue) and proton transport (red) in the inner mitochondrial membrane. Transport of a pair of electrons from one NADH molecule results in the pumping of four protons by the NADH-CoQ reductase complex. For every electron transported by $CoQH_2$ through the $CoQH_2$-cytochrome c reductase complex, two protons are pumped from the matrix into the intermembrane space *(detail)*. This involves the so-called Q cycle, in which CoQ cycles between fully oxidized Q, fully reduced QH_2, and the semiquinone intermediate $Q^{\bullet}$, in which Q has received a single electron. Finally, the peripheral protein cytochrome c diffuses in the intermembrane space, transporting electrons (one at a time) from the CoQ-cytochrome c reductase complex to the cytochrome c oxidase complex. The transport of one electron is associated with the pumping of one proton; thus the transport of a pair of electrons, needed to reduce one O atom, is accompanied by the pumping of two protons. *Detail after H. Tang and B. Trumpower, 1986,* J. Biol. Chem. **261**:6209–6215; *and S. J. Ferguson, 1986,* TIBS **11**:351–353.

protons to move across the inner membrane for every electron transported from QH_2 to cytochrome c, or four protons for every electron pair.

Current evidence thus suggests that 10 protons are pumped across the inner mitochondrial membrane as one electron pair is transferred from NADH to O_2. Four protons are pumped by the NADH-CoQ reductase complex, four by the Q cycle in the $CoQH_2$-cytochrome c reductase complex, and two by the cytochrome c oxidase complex.

Metabolic Regulation

All enzymatically catalyzed reactions and metabolic pathways are regulated. Mitochondria synthesize ATP only to meet cellular requirements. The import of metabolites to and the export of ATP from the mitochondrion are also coordinated with ATP synthesis and, as we shall see, with the generation of heat. Similarly, the conversion of glucose, fatty acids, and other metabolites to acetyl CoA is tightly regulated to produce only the amount of this substrate required for the citric acid cycle. Here, we consider how this regulation is achieved, beginning with an examination of the overall energetics of mitochondrial oxidation.

The Ratio of ATP Production to O_2 Consumed Is a Measure of the Efficiency of Oxidative Phosphorylation

For each molecule of NADH or $FADH_2$ that is oxidized, two electrons are carried through the electron transport system and one atom ($\frac{1}{2}O_2$) of molecular oxygen is reduced to H_2O. The efficiency of oxidative phosphorylation can be expressed by the phosphorus-oxygen (P-O) ratio, or the number of ATPs produced per O atom ($\frac{1}{2}O_2$) used. Direct measurements on isolated mitochondria indicate that a maximum of 2.5 ATP molecules are synthesized from ADP and P_i for every molecule of NADH oxidized to NAD^+ and a maximum of 1.5 molecules of ATP are generated for every molecule of $FADH_2$ oxidized to FAD. The principal mitochondrial reaction that generates $FADH_2$ is the oxidation of succinate to fumarate—a key reaction (7) in the citric acid cycle (see Figure 15-13). Succinate dehydrogenase, the enzyme that catalyzes this reaction, is bound to the inner mitochondrial membrane and contains a tightly bound FAD. Thus the oxidation of one molecule of succinate to fumarate results in the generation of a maximum of 1.5–2 molecules of ATP.

Do not be disturbed by these nonintegral P-O ratios! Remember that the coupling between electron transport and ATP synthesis is indirect and involves the proton-motive force. Direct measurements show that 10 protons are pumped per NADH molecule oxidized. Four protons must recross the membrane to synthesize one molecule of ATP: three to synthesize the ATP itself, and one to exchange the ATP for cytosolic ADP (see Figure 15-20). Thus the P-O ratio of 2.5 (10/4) measured on intact mitochondria is in accord with experiments on the individual stages of electron transport and proton pumping. Electron transport from the succinate dehydrogenase of FADH$_2$ pumps only about six protons because the initial NADH-CoQ reductase complex is not involved. Thus the maximum P-O ratio for succinate oxidation is 1.5 (6/4).

The free energy released during the oxidation of NADH by O$_2$ ($\Delta G^{\circ\prime} = -52.5$ kcal/mol) is sufficient, in theory, to generate the synthesis of at least three molecules of ATP from ADP and P$_i$. This maximum is not achieved in practice, however. Some leakage of protons across the inner membrane may occur, and some energy of the proton concentration gradient is used to transport molecules such as ADP to or ATP from the mitochondrion. Generally, the oxidation of NADH yields a P-O ratio of about 2.5.

In Respiratory Control, Oxidation of NADH or FADH$_2$ and ATP Production Are Obligatorily Coupled through the Proton-motive Force

If intact isolated mitochondria are provided with NADH, O$_2$, and P$_i$, but not with ADP, the oxidation of NADH and the reduction of O$_2$ rapidly cease as the amount of endogenous ADP is depleted by ATP formation. The oxidation of NADH is rapidly restored when ADP is readded. Thus mitochondria can oxidize FADH$_2$ and NADH only as long as there is a source of ADP and P$_i$ to generate ATP. This phenomenon, termed *respiratory control*, illustrates how one key reactant can limit the rate of a complex set of interrelated reactions.

Intact cells and tissues also employ respiratory control. Cells oxidize only enough glucose to synthesize the amount of ATP required for their metabolic activities. Stimulation of a metabolic activity that utilizes ATP, such as muscle contraction, results in an increased level of cellular ADP; this, in turn, increases the oxidation rate of metabolic products in the mitochondrion.

The metabolic nature of respiratory control is now well understood. Recall that the oxidation of NADH, succinate, or FADH$_2$ is *obligatorily* coupled to proton transport across the inner mitochondrial membrane. If the resulting electrochemical gradient is not dissipated when the protons are used to synthesize ATP from ADP and P$_i$ or for some other purpose, both the transmembrane pro-

ton concentration gradient and the membrane electric potential will increase to very high levels. NADH oxidation will eventually cease, because it will require too much energy to pump additional protons across the inner membrane against the existing proton-motive force. The addition of ADP to isolated mitochondria, for instance, triggers a burst of O$_2$ consumption. Observers feel that the availability of ADP is only one way in which mitochondrial oxidation is regulated in intact cells. A rise in cytosolic Ca^{2+} ions, as occurs in muscle cells during contraction, also triggers an increase in mitochondrial oxidation and ATP production in many cells.

Certain poisons called *uncouplers* render the inner mitochondrial membrane permeable to protons. Uncouplers allow the oxidation of NADH and the reduction of O$_2$ to continue at high levels but do not permit ATP synthesis. In the uncoupler 2,4 dinitrophenol (DNP), two electron-withdrawing nitro (NO$_2$) groups stabilize the negatively charged phenolic form:

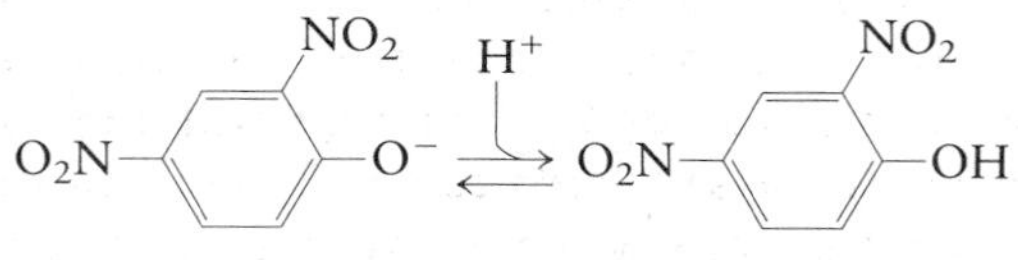

Both the neutral and negatively charged forms of DNP are soluble in phospholipid membranes and in aqueous solution, so DNP can act as a proton shuttle. By transporting protons across the inner membrane, DNP short-circuits both the transmembrane proton concentration gradient and the membrane electric potential. Uncouplers such as DNP abolish ATP synthesis and dispense with any requirement for ADP in NADH oxidation or electron transport. The energy released by the oxidation of NADH in the presence of DNP is converted to heat.

An Endogenous Uncoupler in Brown-fat Mitochondria Converts H$^+$ Gradients to Heat

Brown-fat tissue is specialized for the generation of heat. In contrast to *white-fat tissue* which is specialized for the storage of fat, brown adipose tissue contains abundant mitochondria, which impart their dark brown color to the tissue.

In brown-fat mitochondria, a 33,000-MW, inner-membrane protein called *thermogenin* is a natural uncoupler of oxidative phosphorylation. This protein acts as a transmembrane H$^+$ transporter. Like other uncouplers, it short-circuits the proton concentration gradient and the electric potential across the inner mitochondrial membrane, converting energy released by NADH oxidation to heat. This protein is regulated to generate heat and maintain body temperature under different environmental conditions. For instance, during the adaptation of rats to cold, the ability of their tissues to generate heat (*thermogenesis*) can be increased by inducing the synthesis of this inner-membrane uncoupler protein; in cold-adapted animals, this H^+ pore protein may make up 15 percent of the mitochondrial membrane protein.

Adult humans have little brown fat, but human infants have a great deal; in the newborn, thermogenesis by brown-fat mitochondria is vital to survival. Brown-fat mitochondria are also essential to thermogenesis in hibernating mammals. In fur seals and other animals naturally acclimated to the cold, muscle-cell mitochondria contain thermogenin, which permits a great deal of energy in the proton concentration gradient to be converted to heat and used to maintain body temperature.

The Steps of Glycolysis Are Controlled by Multiple Allosteric Effectors

The activity of the Embden-Meyerhoff pathway is also continuously regulated, so that the production of ATP and pyruvate is adjusted to meet the needs of the cell. Phosphofructokinase, which catalyzes the third reaction in the conversion of glucose to pyruvate, is the principal rate-limiting enzyme of the entire pathway (see Figure 15-2). The allosteric inhibition of phosphofructokinase by NADH and citrate enables the activities of the Embden-Meyerhoff pathway to be coordinated with those of the citric acid cycle. If the cytosolic concentration of NADH builds up too high (due to a slowdown in oxidative phosphorylation), it inhibits phosphofructokinase and reduces the formation of pyruvate and acetyl CoA. Similarly, if citrate—the product of the first step of the citric acid cycle—accumulates, feedback inhibition of phosphofructokinase reduces the generation of pyruvate and acetyl CoA, so that less citrate is formed.

Phosphofructokinase is allosterically activated by ADP and allosterically inhibited by ATP. This arrangement makes the rate of glycolysis very sensitive to intracellular levels of ATP and ADP. The allosteric inhibition of phosphofructokinase by ATP may seem unusual, since ATP is also a substrate of this enzyme. But the affinity of the substrate-binding site for ATP is much higher (has a lower K_M) than that of the allosteric site. Thus, at low concentrations, ATP binds to the catalytic but not to the inhibitory allosteric site and enzymatic catalysis proceeds at near maximal rates. At high concentrations, ATP binds to the allosteric site, inducing a conformational change that inhibits phosphofructokinase and reduces the overall rate of glycolysis.

In addition to phosphofructokinase, other glycolytic enzymes are subject to allosteric control. Hexokinase is allosterically inhibited by its reaction product, glucose 6-phosphate. Pyruvate kinase is allosterically inhibited by one of its reaction products, ATP; glycolysis slows down if too much ATP is present. The three glycolytic enzymes that are regulated by allosteric molecules catalyze reactions with the most negative $\Delta G^{o'}$ values—reactions that are essentially irreversible under ordinary conditions. These enzymes are particularly suitable for regulating the entire Embden-Meyerhoff pathway. Additional control is exerted by glyceraldehyde phosphate dehydrogenase, the enzyme that catalyzes the reduction of NAD^+ to NADH. If cytosolic NADH builds up due to a slowdown in oxidative phosphorylation, this step in glycolysis will be slowed by mass action.

Thus the reactions of the Embden-Meyerhoff pathway and the citric acid cycle, oxidative phosphorylation, and thermogenesis are tightly controlled to produce the appropriate amount of ATP required by the cell. Glucose metabolism is controlled differently in various mammalian tissues to meet the metabolic needs of the organism as a whole. During periods of carbohydrate starvation, for instance, glycogen in the liver is converted to glucose 6-phosphate. Under these conditions, however, phosphofructokinase is inhibited and glucose 6-phosphate is not metabolized to pyruvate; rather, it is converted to glucose and released into the blood to nourish the brain and muscles, which then oxidize the bulk of the available glucose. Chapter 19 contains a more detailed discussion of the hormonal control of glucose metabolism in the liver and muscles. In all cases, the activities of these enzymes are regulated by the level of small-molecule metabolites, generally by allosteric interactions or by phosphorylation.

Summary

A combination of a proton concentration (pH) gradient (exoplasmic face > cytoplasmic face) and an electric potential (negative cytoplasmic face) across the inner mitochondrial membrane, the chloroplast thylakoid membrane, or the bacterial plasma membrane is the immediate energy source for ATP synthesis. The gradient and potential are known collectively as the proton-motive force. The multiprotein F_0F_1 complex catalyzes ATP synthesis as protons flow back through the membrane down their electrochemical proton gradient. This complex has a very similar structure in all three systems: F_0 is a transmembrane complex that generates a regulated H^+ channel; F_1 contains the site for ATP synthesis from ADP and P_i and is tightly bound to F_0.

Glucose catabolism begins in the cytosol of eukaryotic cells, where it is converted to pyruvate via the Embden-Meyerhoff pathway, with the net formation of two ATPs and the net reduction of two NAD^+ molecules to NADH. In anaerobic cells, pyruvate can be metabolized further to lactate or ethanol plus CO_2, with a concomitant re-reduction of NADH.

Most of the ATP in mitochondria and aerobic bacteria is generated during the oxidation of pyruvate to CO_2. Pyruvate is first converted to acetyl coenzyme A (CoA) and CO_2. NADH and $FADH_2$ are formed by the subsequent oxidation of acetyl CoA to CO_2. Acetyl CoA is also a key intermediate in the oxidation of fats and amino acids. The oxidation of acetyl CoA in the citric acid cycle is catalyzed by a set of enzymes localized in the mitochondrial matrix. Acetyl CoA condenses with the four-carbon molecule oxaloacetate to form the six-carbon citrate. In a series of reactions, citrate is converted to oxaloacetate and to CO_2, concomitant with the reduction of NAD^+ to NADH and FAD to $FADH_2$. The NADH generated in the cytosol during the reactions of the Embden-Meyerhoff pathway is oxidized to NAD^+ by enzymes of an electron shuttle, concomitant with the reduction of NAD^+ to NADH in the matrix. This provides another source of reduced nucleotides for oxidative phosphorylation.

In the mitochondrion, electron flow from NADH or $FADH_2$ to O_2 is coupled to proton transport across the inner membrane from the matrix, generating the proton-motive force. The major components of the electron transport chain are four multiprotein complexes with defined orientations in the inner membrane: succinate-CoQ reductase, NADH-CoQ reductase, $CoQH_2$-cytochrome c reductase, and cytochrome c oxidase; the last complex transfers electrons to O_2 to form H_2O. Movement of electrons through each of the last three complexes is coupled to proton movement. Ubiquinone, or coenzyme Q (CoQ), functions as a lipid-soluble reversible transporter of electrons and protons across the inner membrane. Because the mitochondrial membrane is impermeable to anions, the predominant component of the proton-motive force is the membrane electric potential (about 200 mV, matrix negative). The proton-motive force powers the synthesis of mitochondrial ATP from ADP and P_i and the uptake of P_i and ADP from the cytosol, in exchange for mitochondrial ATP. This force can also be used to generate heat.

The conversion of glucose to pyruvate is under tight regulation, mainly by the allosteric inhibition of phosphofructokinase. The Embden-Meyerhoff pathway is inhibited by an excess of ATP or NADH, and the production of pyruvate from glucose and the oxidation of pyruvate by mitochondria are tightly coupled.

References

Glycolysis and the Citric Acid Cycle

General References

BRIDGER, W. A., and J. F. HENDERSON. 1983. *Cell ATP*. Wiley.

*FERSHT, A. 1984. *Enzyme Structure and Mechanism,* 2d ed. W. H. Freeman and Company. Contains an excellent discussion of the reaction mechanisms of key enzymes.

LEHNINGER, A. L. 1982. *Principles of Biochemistry*. Worth.

*STRYER, L. 1988. Biochemistry, 3d ed. W. H. Freeman and Company. Chapters 15 and 16.

Glycolysis

*BOITEUX, A., and B. HESS. 1981. Design of glycolysis. *Phil. Trans. R. Soc. Lond.* B293:5–22.

*FOTHERGILL-GILMORE, L. A. 1986. The evolution of the glycolytic pathway. *TIBS* 11:47–51.

MARCHIONNI, M., and W. GILBERT. 1986. The triosephosphate isomerase gene from maize: introns antedate the plant-animal divergence. *Cell* 46:133–141.

The Citric Acid Cycle

*BALDWIN, J. E., and H. A. KREBS. 1981. The evolution of metabolic cycles. *Nature* 291:381–382.

BEECKMANS, S. 1984. Some structural and regulatory aspects of citrate synthase. *Eur. J. Biochem.* 16:341–351.

*KREBS, H. A. 1970. The history of the tricarboxylic acid cycle. *Perspect. Biol. Med.* 14:154–170.

REED, L. J., Z. DAMUNI, and M. L. MERRYFIELD. 1985. Regulation of mammalian pyruvate and branched-chain α-keto acid dehydrogenase complexes by phosphorylation-dephosphorylation. *Curr. Top. Cell Regul.* 27:41–49.

ROBINSON, J. B., JR., and P. A. SRERE. 1985. Organization of Krebs tricarboxylic acid cycle enzymes in mitochondria. *J. Biol. Chem.* 260:10800–10805.

*SRERE, P. A. 1987. Complexes of sequential metabolic enzymes. *Ann. Rev. Biochem.* 56:89–124.

WIELAND, O. H. 1983. The mammalian pyruvate dehydrogenase complex: structure and regulation. *Rev. Physiol. Biochem. Pharmacol.* 96:123–170.

WILLIAMSON, J. R., and R. V. COOPER. 1980. Regulation of the citric acid cycle in mammalian systems. *FEBS Lett.* 117(suppl.):K73.

Regulation of Glucose Metabolism

ERICINSKA, A., and D. F. WILSON. 1982. Regulation of cellular energy metabolism. *J. Membr. Biol.* 70:1–14.

FEIN, A., and M. TSACOPOULOS. 1988. Activation of mitochondrial oxidative metabolism by calcium ions in *Limulus* ventral photoreceptor. *Nature* 331:437–440.

KEMP, R. G., and L. G. FOE. 1983. Allosteric regulatory properties of muscle phosphofructokinase. *Mol. Cell Biochem.* 57:147–154.

*NEWSHOLME, E. A., and C. START. 1975. *Regulation of Metabolism*. Wiley.

*OCHS, R. S., R. W. HANSON, and J. HALL, eds. 1985. *Metabolic Regulation*. Elsevier. A collection of articles originally published in *TIBS*.

UYEDA, K. 1979. Phosphofructokinase. *Adv. Enzymol.* 48:193–244.

---

*A book or review article that provides a survey of the topic.

Energy Metabolism and the Chemiosmotic Theory

General References

*DICKERSON, R. E. 1980. Cytochrome *c* and the evolution of energy metabolism. *Sci. Am.* 242(3):137–153.

*ERNESTER, L., ed. 1985. *Bioenergetics.* Elsevier. In series: New Comprehensive Biochemistry, Vol. 9.

*FERGUSON, S. J., and M. C. SORGATO. 1982. Proton electrochemical gradients and energy transduction processes. *Ann. Rev. Biochem.* 51:185–218.

*HAROLD, F. M. 1986. *The Vital Force: A Study of Bioenergetics.* W. H. Freeman and Company.

*HATEFI, Y. 1985. The mitochondrial electron transport and oxidative phosphorylation system. *Ann. Rev. Biochem.* 45:1015–1070.

*HINKLE, P. C., and R. E. MCCARTY. 1978. How cells make ATP. *Sci. Am.* 238(3):104.

*MITCHELL, P. 1979. Keilin's respiratory chain concept and its chemiosmotic consequences. *Science* 206:1148–1159. Mitchell's Nobel Prize lecture.

*NICHOLLS, D. G. 1982. *Bioenergetics: An Introduction to the Chemiosmotic Theory.* Academic Press.

*RACKER, E. 1976. *A New Look at Mechanisms in Bioenergetics.* Academic Press.

*RACKER, E. 1980. From Pasteur to Mitchell: a hundred years of bioenergetics. *Fed. Proc.* 39:210–215.

*SKULACHEV, V. P., and P. C. HINKLE, eds. 1981. *Chemiosmotic Proton Circuits in Biological Membranes.* Addison-Wesley.

*TZAGOLOFF, A. 1982. *Mitochondria.* Plenum Press.

*YOUVAN, D. C., and F. DALDAL, eds. 1986. *Microbial Energy Transduction: Genetics, Structure, and Function of Membrane Proteins.* Cold Spring Harbor Laboratory.

Synthesis of ATP and the F_0F_1 ATPase

AMZEL, L. M., M. MCKINNEY, P. NARAYANAN, and P. L. PEDERSEN. 1982. Structure of the mitochondrial F_1 ATPase at 9-Å resolution. *Proc. Nat'l Acad. Sci. USA* 79:5852–5856.

*BOYER, P. D. 1989. A perspective of the binding change mechanism for ATP synthesis. *FASEB J.* 3: 2164–2178.

CROSS, R. L., D. CUNNINGHAM, and J. K. TAMURA. 1984. Binding change mechanism for ATP synthesis by oxidative phosphorylation and photophosphorylation. *Curr. Top. Cell. Regul.* 24:335–344.

FERGUSON, S. J. 1986. Toward a mechanism for the ATP synthase of oxidative phosphorylation. *TIBS* 11:100–101.

*FUTAI, M., T. NOUMI, and M. MAEDA. 1989. ATP synthase (H^+-ATPase): results by combined biochemical and molecular biological approaches. *Ann. Rev. Biochem.* 58:111–136.

*FUTAI, M., and H. KANAZAWA. 1983. Structure and function of proton-translocating adenosine triphosphatase (F_0F_1): biochemical and molecular biological approaches. *Microbiol. Rev.* 47:285–312.

*HAMMES, G. G. 1983. Mechanism of ATP synthesis and coupled proton transport: studies with purified chloroplast coupling factor. *TIBS* 8:131–134.

RAO, R., and A. E. SENIOR. 1987. The properties of hybrid F_1 ATPase enzymes suggests that a cyclical catalytic mechanism involving three catalytic sites occurs. *J. Biol. Chem.* 262:17450–17454.

*SCHNEIDER, E., and K. ALTENDORF. 1987. Bacterial adenosine 5′-triphosphate synthase (F_1F_0): purification and reconstitution of F_0 complexes and biochemical and functional characterization of their subunits. *Microbiol. Rev.* 51:477–497.

THAYER, W., and P. C. HINKLE. 1975. Synthesis of adenosine triphosphate by an artificially imposed electrochemical proton gradient in bovine heart submitochondrial particles. *J. Biol. Chem.* 250:5330–5335.

*VIGNAIS, P. V., and J. LUNARDI. 1985. Chemical probes of mitochondrial ATP synthesis and translocation. *Ann. Rev. Biochem.* 54:977–1014.

*WALKER, J. E., M. SARASTE, and N. J. GAY. 1984. The *unc* operon: nucleotide sequence, regulation, and structure of ATP synthase. *Biochim. Biophys. Acta* 768:164–200.

Transport of Metabolites into and out of the Mitochondrion

KLINGENBERG, M. 1985. Principles of carrier catalysis elucidated by comparing two similar membrane translocators from mitochondria, the ADP-ATP carrier and the uncoupling protein. *Ann. N. Y. Acad. Sci.* 456:279–288.

*LANOUE, K. F., and A. C. SCHOOLWERTH. 1979. Metabolite transport in mitochondria. *Ann. Rev. Biochem.* 48:871–922.

The Electron Transport Chain in Mitochondria

*CAPALDI, R. A. 1988. Mitochondrial myopathies and respiratory chain proteins. *TIBS* 13:144–148.

*CASEY, R. P. 1984. Membrane reconstitution of the energy-conserving enzymes of oxidative phosphorylation. *Biochim. Biophys. Acta* 768:319–347.

*CHEN, L. B. 1988. Mitochondrial membrane potential in living cells. *Ann. Rev. Cell Biol.* 4:155–181.

*FILLINGAM, R. H. 1980. The proton-translocating pumps of oxidative phosphorylation. *Ann. Rev. Biochem.* 49:1079–1113.

FULLER, S. D., R. A. CAPALDI, and R. HENDERSON. 1979. Structure of cytochrome *c* oxidase in deoxycholate-derived two-dimensional crystals. *J. Mol. Biol.* 134:305–327.

GUPTE, S. S., and C. R. HACKENBROCK. 1988. The role of cytochrome *c* diffusion in mitochondrial electron transport. *J. Biol. Chem.* 263:5248–5252.

*HACKENBROCK, C. R., B. CHAZOTTE, and S. S. GUPTE. 1986. The random collision model and a critical assessment of diffusion and collision in mitochondrial electron transport. *J. Bioenerget. Biomembr.* 18:331–368.

PETTIGREW, G. W., and G. R. MOORE. 1987. *Cytochrome c: Biological Aspects.* Springer-Verlag (Berlin).

*PRINCE, R. C. 1988. The proton pump of cytochrome oxidase. *TIBS* 13:159–160.

REYNAFARJE, B., L. E. COSTA, and A. L. LEHNINGER. 1986. Upper and lower limits of the proton stoichiometry of cytochrome *c* oxidation in rat liver mitoplasts. *J. Biol. Chem* 261:8254–8262.

*SARASTE, M. 1983. How complex is a respiratory complex? *TIBS* 8:139–142.

*SLATER, E. C. 1983. The Q cycle—an ubiquitous mechanism of electron transfer. *TIBS* 8:239–242.

SPIRO, T. G., ed. 1982. *Iron-sulfur Proteins*. Wiley-Interscience.

TANG, H-L., and B. L. TRUMPOWER. 1986. Triphasic reduction of cytochrome *b* and the proton-motive Q cycle pathway of electron transfer in the cytochrome bc_1 complex of the mitochondrial respiratory chain. *J. Biol. Chem.* 261:6209–6215.

WIKSTRÖM, M. 1989. Identification of the electron transfers in cytochrome oxidase that are coupled to proton-pumping. *Nature* 338: 776–778.

Thermogenesis

BOUILLARD, F., J. WEISSENBACH, and D. RICQUIER. 1986. Complete cDNA-derived amino acid sequence of rat brown-fat uncoupling protein. *J. Biol. Chem.* 261:1487–1490.

*NICHOLLS, D. G., and E. RIAL. 1984. Brown-fat mitochondria. *TIBS* 9:489–491.

Photosynthesis

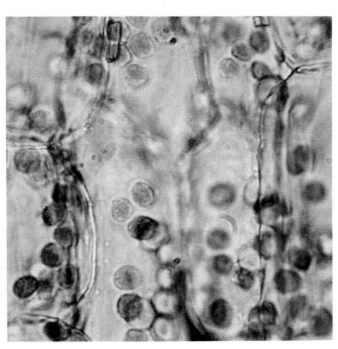

Chloroplasts in leaf cells of the snake plant *Sanseviera trifasciata*

*S*unlight is the ultimate source of energy for all organisms on earth. During photosynthesis, which occurs in many bacteria as well as in algae and higher plants, light energy is converted to the chemical energy of a phosphoanhydride bond of ATP. In plant photosynthesis, light energy is also used to pull electrons away from the unwilling donor H_2O and transfer them to the electron acceptor $NADP^+$. The resulting NADPH and ATP power the otherwise endergonic conversion of CO_2 and H_2O to six-carbon sugars and to O_2:

$$6CO_2 + 6H_2O \longrightarrow C_6H_{12}O_6 + 6O_2$$

In Chapter 15, we saw how cells form ATP phosphoanhydride bonds during the oxidation of glucose to CO_2. Here, we examine the many steps of photosynthesis, focusing on the mechanisms of light absorption, ATP formation, and $NADP^+$ reduction and on the reactions that "fix" CO_2 into larger molecules. Although our emphasis is on photosynthesis in plant chloroplasts, we also discuss a simpler type of photosynthesis that occurs in bacteria. The three-dimensional structure of the photosynthetic systems in bacteria allows us to trace in molecular detail how light energy is converted to a proton-motive force across a membrane—the first stage in the synthesis of ATP from ADP and P_i. ▲

An Overview of Photosynthesis in Plants

Photosynthesis in plants occurs in chloroplasts, large organelles found mainly in leaf cells (Figure 16-1). The principal end products of CO_2 metabolism during photosynthesis are two polymers of six-carbon sugars: starch and sucrose. Leaf starch, an insoluble polymer of glucose (Figure 16-2), is stored in the chloroplast. Sucrose, a water-soluble disaccharide, is synthesized in the cytosol from three-carbon precursors generated in the chloroplast. Sucrose is transported from the leaf through the phloem to other parts of the plant, where it is used for growth and as a supply of energy to nongreen (nonphotosynthetic) tissues such as roots and seeds.

Chloroplasts Have Three Membranes

Chloroplasts have three membranes. The two outermost membranes do not contain chlorophyll, are not green, and do not participate directly in photosynthesis. The outer chloroplast membrane, like the outer membrane of the mitochondrion, is freely permeable to metabolites of small molecular weight; it contains porin proteins that form very large aqueous channels. The inner membrane, conversely, is the permeability barrier of the chloroplast; it contains permeases that regulate the movement of metabolites into and out of the organelle.

Photosynthesis occurs on the third chloroplast membrane, which is really a group of membranes collectively called the *thylakoid membrane*. Frequently, the thylakoid membrane forms into stacks of small flattened vesicles, termed *grana* (singular, *granum*; Figure 16-3). Thylakoid membranes contain a number of integral membrane proteins to which chlorophyll, the pigment that traps light energy, and other important prosthetic groups and pigments are bound.

Chlorophyll (Figure 16-4) is a ringed compound similar in structure to heme, except that a magnesium atom Mg^{2+} (rather than an iron atom Fe^{3+}) is in the center and there are five rings (rather than four). The thylakoid membrane contains several proteins that are involved in light trapping, electron transport, proton pumping, and other enzymatic reactions. Each of these proteins and protein complexes is oriented in the thylakoid membrane in a specific fashion. This asymmetry allows proton concentration (pH) gradients to be established across the thylakoid membrane for use in the generation of ATP.

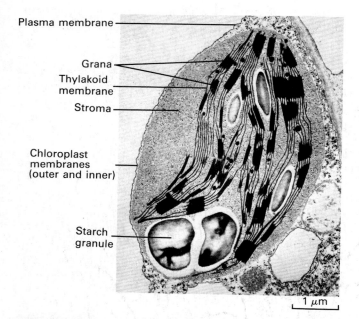

▲ **Figure 16-1** Electron micrograph of a chloroplast in part of a *Phleum pratanese* cell. *Photograph courtesy of Biophoto Associates/M. C. Ledbetter/Brookhaven National Laboratory.*

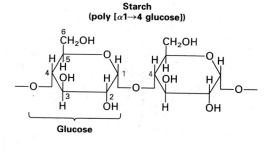

▲ **Figure 16-2** Structures of starch and sucrose, the principal end products of photosynthesis. Both are built of six-carbon sugars.

not evolve O_2 and utilize only one photosystem. Because the three-dimensional structures of the reaction centers from two purple bacteria (*Rhodopseudomonas viridis* and *Rhodobacter sphaeroides*) have been determined (Figure 16-9; see Figure 13-17), scientists can actually trace the detailed paths of electrons during and after the absorption of light. We shall discuss this system in some detail, since it also provides great insight into the mechanism of photosynthesis in chloroplasts.

In purple bacteria, as in other photosystems, energy from absorbed light is used to transport an electron from chlorophyll to an acceptor on the cytoplasmic membrane face (Figure 16-10). Simultaneously, an electron from a donor cytochrome on the exoplasmic membrane face is transported to the chlorophyll, leaving the cytochrome oxidized (positively charged). The electron on the cytoplasmic face crosses the membrane, through the *cyto-*

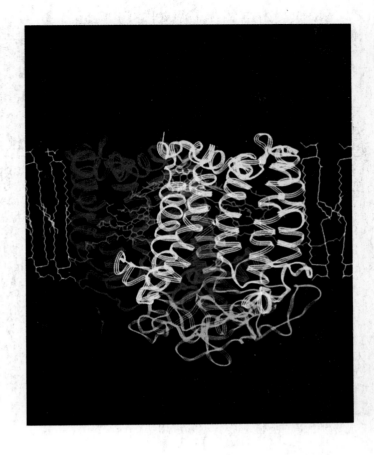

▶ **Figure 16-9** The three-dimensional structure of the reaction center from *R. sphaeroides,* showing all of the pigments in purple. The L, M, and H chains appear in yellow, dark blue, and light blue, respectively. (Figure 13-17 shows the conformation of the three polypeptides in greater detail. *Courtesy of D. Rees.*

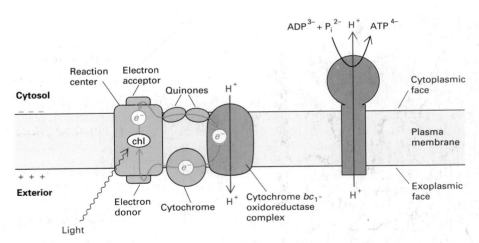

▲ **Figure 16-10** Photosynthesis in purple bacteria utilizes a single photosystem. Light absorption by reaction-center chlorophyll results in the transport of one chlorophyll electron to an acceptor on the cytoplasmic face of the plasma membrane. This electron is transported via quinone carriers to the cytochrome bc_1–oxidoreductase complex. In this complex, it crosses the membrane to the exoplasmic face, together with protons pumped out of the cytosol into the exterior, generating the proton-motive force. The electron is then transported via a cytochrome back to the reaction-center chlorophyll. The proton-motive force is used to synthesize ATP and, as in other bacteria, to transport molecules in and out of the cell. *After D. C. Youvan and B. L. Marrs, 1984, Cell* **29:**1–3.

chrome bc_1-oxidoreductase complex, to reduce the oxidized cytochrome c. During this crossing, protons are transported from the cytoplasmic face to the exoplasmic face of the membrane, generating the proton-motive force. As in other systems, this proton-motive force is used to synthesize ATP and to transport molecules across the membrane. However, this photosynthetic process does not evolve O_2 or split H_2O. Electron flow is cyclic, and there is no reduction of $NADP^+$ to NADPH.

Nonetheless, the *primary* event in bacterial photosynthesis is the formation of a strong oxidizing agent (oxidized cytochrome) and a strong reducing agent (the primary electron acceptor with its electron).

The Exact Pathway of Electron Transport in the Photosynthetic Reaction Center of Purple Bacteria is Known

The three integral proteins (L, M, and H) of the photosynthetic reaction center of purple bacteria contain a total of 11 transmembrane α helices, to which prosthetic groups that absorb light and transport electrons during photosynthesis are bound. These groups include a "special pair" of chlorophyll a molecules that absorb light, two other chlorophyll ("voyeur") molecules of unknown function, pheophytins (chlorophyll molecules without an Mg^{2+}), one Fe atom, and *quinones* (structurally similar to mitochondrial ubiquinone). The exact arrangement of all of these elements is known (Figure 16-11; see Figure 16-9).

The pathway that the electron traverses can be determined by the technique of *optical absorption spectroscopy.* Each of the three photosynthetic pigments absorbs light of only certain wavelengths (see Figure 16-7). The absorption spectrum of each pigment changes when it possesses an extra electron. Thus the pathway of electrons can be determined by monitoring the changes in absorption of the various pigments as a function of time after absorption of the light photon. Since electron movements are completed in less than 1 millisecond (ms), a special technique called *picosecond* (1 ps = 10^{-12} s) *absorption spectroscopy* is required to see the early changes. An intense pulse of laser light lasting less than 1 ps induces photosynthesis in a preparation of bacterial-membrane vesicles containing reaction centers. Each reaction center absorbs one photon, and the subsequent electron transfer processes are synchronized in all reaction centers. Figure 16-11 traces the pathway of the electron, as monitored by measuring the absorption spectra of the preparation over the following millisecond. The end result is a charge separation: a positively charged cytochrome (electron donor) on the exoplasmic membrane face and a reduced quinone (electron acceptor) on the cytoplasmic membrane. This charge separation represents stored energy. In the subsequent steps of this photo-

synthetic reaction, the electron crosses the plasma membrane and returns to the cytochrome, simultaneously generating a proton-motive force across the membrane (see Figure 16-10).

Photosynthetic Bacteria Can Carry Out Noncyclic Electron Transport

Under certain circumstances, photosynthetic electron transport in purple bacteria can be noncyclic: electrons removed from chlorophyll molecules ultimately are transferred to $NADP^+$ as the acceptor, forming NADPH. Still, H_2O is not split, and no O_2 is formed. Rather, other molecules, such as hydrogen gas (H_2) or hydrogen sulfide (H_2S), can give up electrons to the oxidized cytochrome c. In these cases, CO_2 is fixed and used to synthesize six-carbon sugars and other molecules. In the overall reaction, light-driven oxidation of H_2S produces sulfur (S):

$$12H_2S + 6CO_2 \xrightarrow{light} C_6H_{12}O_6 + 12S + 6H_2O$$

Here, H_2S transfers its electrons to the oxidized cytochrome c:

$$H_2S + 2cyt\ c^{3+}_{ox} \longrightarrow 2H^+ + S + 2cyt\ c^{2+}_{red}$$

H_2 can also be used as an electron donor by these organisms; the overall reaction is

$$12H_2 + 6CO_2 \xrightarrow{light} C_6H_{12}O_6 + 6H_2O$$

Here, light causes the removal of an electron from the cytochrome, forming oxidized cytochrome c; electrons from H_2 then reduce the oxidized cytochrome c:

$$H_2 + 2cyt\ c^{3+}_{ox} \longrightarrow 2H^+ + 2cyt\ c^{2+}_{red}$$

Still other photosynthetic bacteria can use a variety of organic compounds as electron sources for the light-dependent reduction of CO_2.

In contrast to purple bacteria, plants utilize two photosystems: one splits H_2O into O_2; the other reduces NADP to NADPH (see Figure 16-6). We now turn to the specifics of these two photosystems.

The Structure and Function of the Two Plant Photosystems: PSI and PSII

Both PSI and PSII Are Essential for Photosynthesis in Chloroplasts

A simple and crucial series of experiments has shown that the two photosystems in chloroplasts absorb light at different wavelengths and are both essential for photosyn-

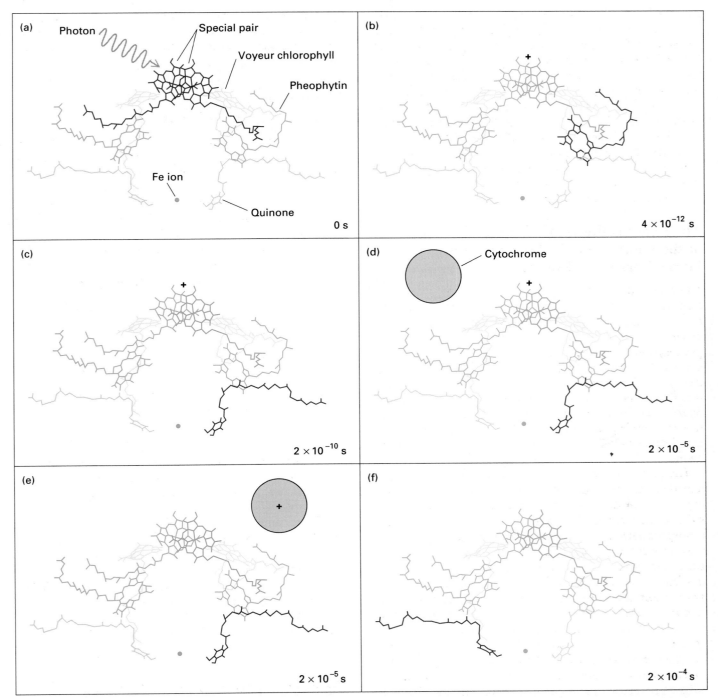

▲ **Figure 16-11** The steps of electron transport during bacterial photosynthesis. All of the pigments and other electron transporters are bound to the three integral proteins that constitute the bacterial photosynthetic reaction center (see Figure 16-9). In the first step of photosynthesis (a), a photon is absorbed by the "special pair" of chlorophyll molecules; the energy is transferred to one of the chlorophyll electrons. The molecules bearing the electron in the excited state are shown in red. Within 4 picoseconds (ps), the electron moves (b) to one of the pheophytin molecules, bypassing a "voyeur" chlorophyll of unknown function, and leaves a positive charge on the special pair of chlorophylls. In ~200 ps (c), the electron moves to a quinone molecule near the end of the chain of prosthetic groups. Next, a soluble cytochrome molecule approaches the reaction center (d) and transfers an electron to the special pair (e). The cytochrome thus acquires a positive charge, and the charge on the special pair is neutralized. In (f), lasting 2×10^{-4} s, the cytochrome with its positive charge diffuses away and the electron is passed to the second quinone acceptor. *From D. C. Youvan and B. L. Marrs, 1987, "Molecular Mechanisms of Photosynthesis," Sci. Am., 256(6):42–48.*

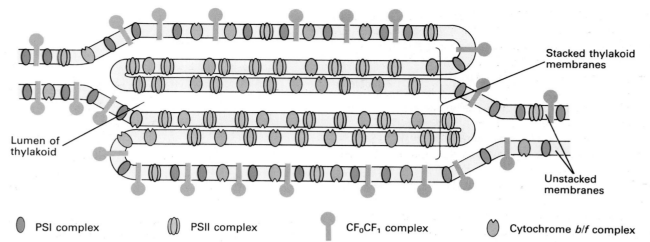

○ PSI complex   ◐ PSII complex   ⌐ CF₀CF₁ complex   ○ Cytochrome *b/f* complex

▲ **Figure 16-12** Distribution of multiprotein complexes in the thylakoid membrane. The cytochrome *b/f* complex transports electrons from PSII to PSI and is found in both regions. The thylakoid membranes may be stacked because of the binding properties of the proteins in PSII. Evidence for this model came from studies in which thylakoid membranes were gently fragmented by ultrasound into vesicles. Stacked and unstacked thylakoid vesicles were then fractionated by density-gradient centrifugation to determine their protein and chlorophyll compositions. *After J. M. Anderson and B. Andersson, 1982, Trends Biochem. Sci. 7:288.*

thesis. The efficiency of photosynthesis drops sharply at wavelengths longer than 680 nm, even though chlorophyll molecules in the thylakoid membrane still absorb light at 700 nm (see Figure 16-7). According to the *Emerson effect*, named after biophysicist R. Emerson, who discovered it in the 1940s, the rate of photosynthesis generated by light of wavelength 700 nm can be greatly enhanced by adding light of shorter wavelength; a combination of light at, say, 600 and 700 nm supports a greater rate of photosynthesis than the sum of the rates for the two separate wavelengths. Such observations led researchers to conclude that photosynthesis in plants involves the interaction of two separate systems of light-driven reactions: PSI is driven by light of wavelength 700 nm or less; PSII, only by light of shorter wavelength (<680 nm).

The two photosystems are structurally distinct; each is a complex of many chlorophyll molecules, carotenoid pigments, cytochromes, and other electron-transporting proteins that have specific orientations in the thylakoid membrane. By using detergents to extract chloroplast membranes selectively, particles containing only PSI complexes can be isolated. If the thylakoid membranes are sheared into small membrane fragments and vesicles, membrane fractions containing predominantly PSI or PSII can be purified, so that the properties of each photosystem can be studied separately.

PSII is located preferentially in the stacked membranes of the grana; PSI is located preferentially in the nonapposed thylakoid membranes (Figure 16-12). The spatial separation of the two complexes is a critical point, because electrons are transferred from PSII to PSI during photosynthesis (see Figure 16-6). A third multiprotein particle, the *cytochrome b/f complex,* found in both the stacked and the unstacked regions of the thylakoid membrane, is used to transfer electrons between the two photosystems. The cytochrome *b/f* complex is similar in structure and function to the cytochrome bc_1-oxidoreductase complex in photosynthetic bacteria (see Figure 16-10) and to the $CoQH_2$-cytochrome *c* reductase complex in mitochondria (see Figure 15-30). Movement of electrons through these complexes is accompanied by proton pumping, and a Q cycle (see Figure 15-30) is thought to be involved.

Each photosystem contains several hundred chlorophyll molecules. As in the bacterial reaction center (see Figure 16-11), however, only two specialized reaction-center chlorophyll *a* molecules—termed P_{680} (pigment absorbs light maximally at 680 nm) in PSII and P_{700} in PSI—are capable of undergoing light-driven electron transfer in each complex. Both of these specialized chlorophyll *a* molecules in the reaction centers are bound to integral membrane proteins: a 82,000-MW protein contains the P_{700} reaction center for PSI, and a 32,000-MW protein-chlorophyll complex contains the P_{680} reaction center for PSII. A structure called the *light-harvesting complex* (LHC) is part of each photosystem. The LHC itself exhibits no catalytic activity; its primary function is to capture light energy and transfer it to PSI or PSII. Each LHC contains several transmembrane polypeptides of

25,000–27,000 MW, which bind up to one-half of the total chlorophyll *a* and nearly all of the chlorophyll *b*. Most LHCs are associated with PSII, but PSI has a specific LHC. Photons are absorbed by any of the dozens of chlorophyll molecules in each LHC. The absorbed energy is then rapidly transferred (in $<10^{-9}$ s) to the specialized chlorophyll *a* molecules in the reaction centers of the two photosystems (Figure 16-13). For this process—called *resonance energy transfer*—to occur efficiently, the chlorophyll molecules must be close together and in a defined orientation, so that the LHCs can absorb light energy and funnel it to a single reaction center. The role of the LHC transmembrane proteins may be to maintain the orientation of the chlorophyll molecules that is optimal for energy transfer, although the details of their orientation in the thylakoid membrane are not known.

PSII Splits H_2O

PSII oxidizes H_2O to O_2. Somewhat surprisingly, its structure resembles the reaction center of photosynthetic bacteria, which, as noted, does not form O_2. Like the bacterial reaction centers, PSII contains pheophytins, quinones, and iron. The sequence of the 32,000-MW PSII protein to which these pigments are bound is remarkably similar to the sequences of the L and M peptides of the bacterial reaction center. The absorption of a light photon of <680-nm wavelength by PSII triggers the loss of an electron from the P_{680} reaction-center chlorophyll *a* molecules. As in photosynthetic bacteria, the electron is transported via a pheophytin, a quinone (Q_A), and an Fe atom to the ultimate electron acceptor—another quinone (Q_B) on the outer (stromal) surface of the thylakoid membrane (Figure 16-14).

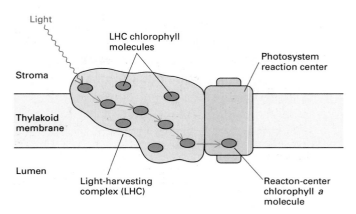

▲ **Figure 16-13** Light-harvesting complexes (LHCs) are part of both photosystem complexes. LHCs contain dozens of chlorophyll molecules, any of which can absorb a photon. The absorbed energy is then transferred to the reaction-center chlorophyll in the photosystem complex.

Because of the highly conjugated system of double and single bonds, the positive charge on the oxidized P_{680} chlorophyll molecules is delocalized to all ring atoms. Electrons for the reduction of P_{680} are obtained ultimately from an H_2O molecule bound to a peripheral protein localized to the luminal surface of the membrane (see Figure 16-14). The protein, called the *oxygen-evolving complex,* contains four manganese (Mn) ions as well as bound Cl^- and Ca^{2+} ions; this is one of only a few cases in which Mn plays a role in a biological system. The reduction of each molecule of H_2O to O_2 requires the removal of four electrons. Because the absorption of each photon by PSII results in the transfer of one electron, either several such photosystems must cooperate to oxidize one H_2O molecule through the O_2-evolving complex or

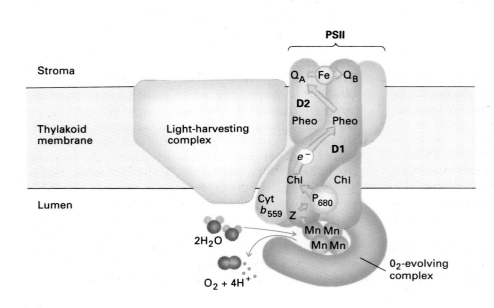

◀ **Figure 16-14** Molecular model of PSII. Two integral 32,000-MW proteins, D1 and D2, that bind the two P_{680} reaction-center chlorophylls, two other chlorophylls, the two pheophytins (Pheo), one Fe atom, and the two quinones (Q_A and Q_B)—all of which are used for electron transport. Associated with these are a cytochrome (b_{559}) and several proteins that make up the LHC. An extrinsic 33,000-MW protein of the O_2-evolving complex binds the four Mn ions that function in splitting of H_2O. Z is tyrosine residue 161 of the D1 polypeptide that conducts electrons from the Mn atoms to the P_{680} reaction-center chlorophyll. *After J. Barber, 1987, Trends Biochem. Sci.* **12:**321–326, *and R. J. Debus, B. Barry, G. Babcock, and L. McIntosh, 1988, Proc. Nat'l Acad. Sci. USA* **85:**427–430.

each PSII has to lose an electron and then oxidize the O_2-evolving complex four times in a row. The latter explanation is believed to be correct; spectroscopic studies suggest that the O_2-evolving complex can cycle through five different oxidation states (S_0–S_4):

$$\overset{\displaystyle e^-\quad\ e^-\qquad e^-\qquad e^-}{\curvearrowright S_0 \longleftrightarrow S_1 \nearrow S_2 \longleftrightarrow S_3 \longrightarrow S_4} \quad \begin{array}{c} 2H_2O \\ \searrow O_2 \end{array}$$
$$H^+ \qquad\quad H^+ \qquad 2H^+$$

S_0 being the most reduced and S_4 being the most oxidized. S_1–S_4 are thought to represent various oxidation states of the bound Mn ions.

The Mn-binding protein is involved in the splitting of H_2O and the evolution of O_2. It can be removed from the reaction center by treatment with solutions of concentrated salts, which abolishes O_2 formation but does not affect light absorption or the initial stages of electron transport. A total of two water molecules are split into four protons, four electrons, and one O_2 molecule. Electrons from H_2O are transferred via a tyrosine side chain (Z in Figure 16-14) on the D1 polypeptide to the P_{680} reaction center, where they regenerate the reduced chlorophyll. The protons released from H_2O remain in the thylakoid lumen; they represent two of the four protons that are pumped into the lumen by the transport of each pair of electrons.

The electrons that split from P_{680} are transported to the ultimate acceptor, the quinone *plastoquinone* (PQ), on the stromal surface of the thylakoid membrane. The reduction potential of the electron is increased by absorption of energy from the light quantum from 0.8 V of the H_2O-O_2 couple to ~0 V of the PQ, which is equivalent to a change in free energy of $\Delta G = $ ~18 kcal/mol of electrons (see Figure 16-16).

Electrons Are Transported from PSII to PSI

Electrons released from PSII rapidly combine with a PQ molecule and two protons from the stromal space to generate reduced hydroquinone (PQH_2). The electrons from PQH_2 are transferred to the cytochrome *b/f* complex, which is analogous in structure and function to the $CoQH_2$-cytochrome *c* reductase complex in mitochondria. As in mitochondria, a Q cycle (see Figure 15-30) is thought to function during electron transport from PQH_2 through the cytochrome *b/f* complex, and for each electron transported, two protons are transported by PQH_2 from the stromal space through the bilayer to the thylakoid lumen.

From the cytochrome *b/f* complex, the electrons move to the carrier *plastocyanin* (PC), a small protein with a single Cu atom coordinated to one cysteine SH group,

one methionine (CH_2SCH_3) group, and two histidines. The Cu alternates between the +1 and +2 states:

$$e^- + Cu^{2+} \rightleftharpoons Cu^+$$

PC is a peripheral protein that is soluble in the lumen of the thylakoid vesicle. With its electron, PC diffuses in the thylakoid lumen from the cytochrome *b/f* complex to PSI (Figure 16-15).

Since the PSI and PSII complexes are not uniformly distributed between stacked and unstacked thylakoid membranes, a single structural electron transport system containing both photosystems does not exist. Instead, mobile electron carriers, such as PQ and PC, are used to transfer electrons between these systems. Similarly, the mobile electron carriers CoQ and cytochrome *c* are used to transfer electrons between electron transport complexes in mitochondria (see Figures 15-23 and 15-30).

PSI Is Used for Both Linear and Cyclic Electron Flow

PSI contains the reaction-center chlorophyll *a* pair termed P_{700}. The absorption of a photon leads to the removal of an electron from a PSI chlorophyll; the oxidized chlorophyll can be reduced by an electron passed from PSII via PC. The activated electron is moved from the thylakoid face to the stromal surface of the membrane, where it is accepted by ferredoxin, an FeS protein. (Several electron carriers intervene between P_{700} and ferredoxin.) The net gain in the reduction potential of the electron by PSI is ~1.0 V, equivalent to a gain in free energy of $\Delta G = $ 23 kcal/mol of electrons.

Electrons excited by PSI face one of two fates. Together with a proton, they can be transferred via the electron carrier FAD to $NADP^+$ to form the reduced molecule NADPH (Figure 16-16). This process is called *linear electron flow*. Alternatively, the activated electrons can flow through a cytochrome *b/f* complex back across the thylakoid membrane until they reach the PSI electron donor on the interior (luminal) surface. A series of bound and free ferredoxins and cytochromes mediates this process, called *cyclic electron flow*. Concomitant with this movement of electrons is the pumping of additional protons from the stromal space into the thylakoid lumen. Again, a Q cycle involving a quinone and the cytochrome *b/f* complex is thought to pump two protons for each electron transported. The pH gradient generated permits the synthesis of additional ATP, but no NADPH is produced. PSII is not involved, and no O_2 evolves. In this manner, PSI is used solely to generate a pH gradient and ATP; its function is similar to that of the photosystem in purple bacteria (see Figure 16-10). Cyclic electron flow occurs when the ratio of NADPH to $NADP^+$ in the cell is high, so that there is no need to generate additional reduced nucleotides.

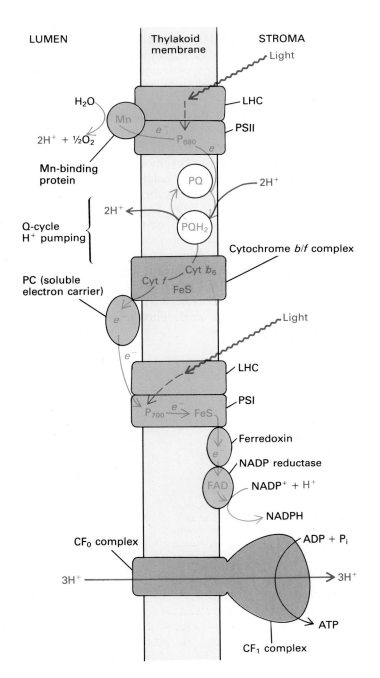

◀ **Figure 16-15**   The orientation of the principal membrane components involved in photosynthesis and the electron pathway from H_2O to NADP. For each electron transported, photons must be absorbed by PSII and PSI. The first photon is absorbed by an array of chlorophyll and other pigments in the LHC, and its energy is passed on to the specialized reaction-center a chlorophylls termed P_{680} in PSII. Electrons e^- are freed from P_{680} and moved to the stromal surface of the thylakoid membrane; they are replaced in P_{680} by electrons transported from the Mn-binding protein that removed them from an H_2O molecule. The protons from H_2O remain in the thylakoid lumen. Electrons from P_{680}, together with two protons from the stroma, reduce plastoquinone (PQ) to PQH_2. PQH_2 then diffuses in the membrane and transfers its electrons to the cytochrome b/f complex, which contains the b_6 and f cytochromes as well as FeS proteins. The protons from PQH_2 are released on the luminal surface of the membrane when PQ is regenerated. The overall process is similar to the Q cycle in mitochondria. These protons, together with those released from H_2O, generate a proton concentration (pH) gradient across the thylakoid membrane. (Since the thylakoid membrane is permeable to anions, a pH gradient rather than a membrane potential is the principal component of the proton-motive force.) The electrons pass from cytochrome f to plastocyanin (PC), an electron carrier soluble in the thylakoid lumen. PC diffuses from site to site, and electrons from PC are transferred to PSI, which contains chlorophyll P_{700}. Absorption of an additional photon by the LHC associated with PSI causes P_{700} to lose an electron, which is transferred through a series of carriers in the reaction center to soluble ferredoxin (another FeS protein) at the stromal surface. The electrons are then transferred to FAD and finally to $NADP^+$. Two electrons, together with one proton removed from the stromal space, convert each $NADP^+$ to NADPH. The pH gradient is used to synthesize ATP; as in mitochondria, the movement of three protons through a CF_0CF_1 complex converts ADP and P_i to ATP.

PSI and PSII Are Functionally Coupled

According to the currently accepted concept of *linear, (noncyclic) electron flow,* the two photosystems have different functions: PSI transfers electrons to $NADP^+$, forming NADPH; PSII removes electrons from H_2O. Evidence supporting this model comes from spectroscopic studies measuring the oxidized state of such cytochromes as b_6 and f, which transfer electrons from PSII to PSI under different conditions of illumination. Red light of wavelength 700 nm excites only PSI; the wavelength of this light is too long to be absorbed by PSII. Shining only red light on chloroplasts causes the b_6 and f cytochromes to become more oxidized. (This can be determined spectro-

scopically because the oxidized and reduced states have different absorption maxima.) Under these conditions, electrons are drawn into PSI; none are provided from PSII. The addition of light of shorter wavelength, however, activates PSII; the b_6 and f cytochromes immediately become partly reduced as electrons flow to them.

Many commercially important herbicides inhibit photosynthesis, and studies of their effects have proved useful in dissecting the pathway of photoelectron movement. One such class of herbicides—the S-triazines, such as atrazine—binds specifically to D1, one of the two 32,000-MW proteins that constitute the PSII reaction center (see Figure 16-14). These compounds block electron transfer by inhibiting binding of a quinone acceptor to PSII. When added to illuminated chloroplasts, these inhibitors cause all downstream electron carriers to accumulate in the oxidized form, since no electrons can be provided to the electron transport system from PSII. Triazine-resistant weeds are prevalent and present a major agricultural problem. The protein to which atrazine binds is encoded in the chloroplast DNA; in atrazine-resistant mutants, a single amino acid change in this protein renders it unable to bind the herbicide.

Because PSI and PSII act in sequence during linear electron flow, the amount of light energy delivered to the two reaction centers must be controlled so that each center activates the same number of electrons. One control mechanism involves the regulation of the rates of phosphorylation and dephosphorylation (the addition and removal, respectively, of phosphate groups) of the LHC proteins associated with PSII. A membrane-bound protein kinase somehow senses the relative activities of the two photosystems via the oxidized-reduced state of the

PQ pool that transfers electrons from PSII to PSI. If too much PQ is reduced (indicating a higher activity of PSII relative to PSI), the kinase is activated and the PSII LHCs are phosphorylated. Phosphorylation causes the LHCs to dissociate from PSII and migrate to the unstacked thylakoid membranes, where they are unable to feed the energy of absorbed light to PSII.

Plants need two photosystems. The light energy absorbed by chlorophyll (~40 kcal/mol of electrons) has to be used in part to stabilize early reactive intermediates, so not enough is left over to permit the direct movement of electrons from H_2O to NADPH. Each of the two photosystems boosts the electrons part of the way (see Figure 16-16). Cyanobacteria also use H_2O as an electron donor and also have two photosystems. Purple and green photosynthetic bacteria, in contrast, have only one photosystem. These organisms extract electrons from molecules that have a lower (more negative) reduction potential than that of H_2O, so a single photon can boost these

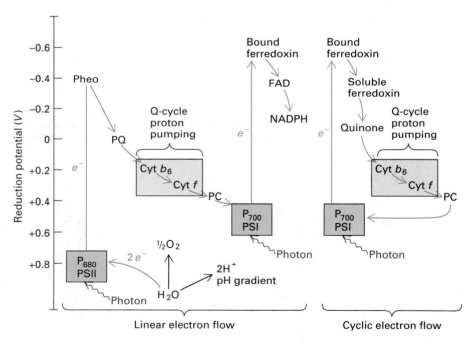

▲ **Figure 16-16** Energetics of electrons as they flow through the photosynthetic transport system. Photons absorbed by PSII excite electrons from an electric potential of +0.8 V (that of the H_2O-O_2 couple) to one of ~−0.4 V (that of the primary electron acceptor pheophytin). This change in electric potential of ~1.2 V represents a gain in free energy of $\Delta G = 28$ kcal/mol of electrons. The first reasonably stable electron acceptor is plastoquinone (PQ), with an electric potential of ~0 V. As electrons move from PQ through the cytochrome *b/f* complex to plastocyanin (PC), some of the acquired electron energy is used to transport

protons into the thylakoid vesicles via a Q cycle. Photon absorption by PSI causes an additional increase of 1.0 V in electron potential. Electrons excited by PSI can be transferred via ferredoxin and FAD to $NADP^+$, forming NADPH. In this process of linear electron flow, PSII and PSI must be coupled. Cyclic electron flow involves only PSI; excited electrons are transferred via soluble ferredoxin, quinones, and cytochromes b_6 and *f* to plastocyanine and then back to PSI. During this process, protons are pumped from the stroma into the thylakoid lumen, most likely via a Q cycle.

electrons to the reduction potential of the NADP⁺-NADPH reaction. For instance, for the reaction

$$H_2S \longrightarrow S + 2H^+ + 2e^-$$

that occurs in bacteria with one photosystem, $E_0 = -0.25$ V, compared to $E_0 = +0.86$ V for the analogous reaction for H_2O:

$$H_2O \longrightarrow \tfrac{1}{2}O_2 + 2H^+ + 2e^-$$

Less energy is required to boost electrons removed from H_2S to a level sufficient to reduce NADP⁺ to NADH.

CO₂ Metabolism during Photosynthesis

Chloroplasts perform many metabolic reactions in green leaves. In addition to CO_2 fixation, chloroplasts are sites for the synthesis of almost all amino acids, all fatty acids and carotenes, all pyrimidines, and probably all purines.

However, the synthesis of sugars from CO_2 is the most extensively studied biosynthetic reaction in plant cells—and certainly a unique one. We now turn to the series of enzymatically catalyzed reactions, known as the *Calvin cycle* (after discoverer Melvin Calvin), that fix CO_2 and convert it to hexose sugars. These reactions are powered by energy released by ATP hydrolysis and the reducing agent NADPH. Carbohydrate formation can occur in the dark until the supply of ATP and NADPH is exhausted.

CO₂ Fixation Is Catalyzed by Ribulose 1,5-Bisphosphate Carboxylase

Elegant studies have established that the actual reaction that fixes CO_2 into carbohydrates is catalyzed by the enzyme *ribulose 1,5-bisphosphate carboxylase*, which adds CO_2 to the five-carbon sugar ribulose 1,5-bisphosphate to form two molecules of 3-phosphoglycerate (Figure 16-17). This reaction occurs in the stroma of the chloroplast.

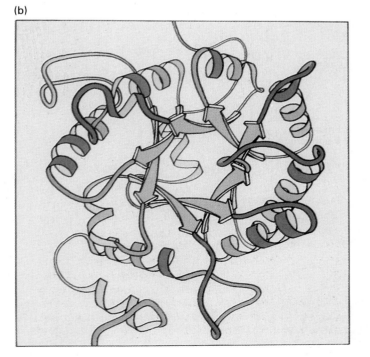

▲ **Figure 16-17** (a) The initial reaction that fixes CO_2 into organic compounds involves a condensation with the five-carbon sugar ribulose 1,5-bisphosphate. The products of the reaction, catalyzed by ribulose 1,5-bisphosphate carboxylase, are two molecules of 3-phosphoglycerate. If ¹⁴C-labeled CO_2 is used (*left*, red), all the ¹⁴C radioactivity is in the carboxyl carbon atom of 3-phosphoglycerate (*right*, red). (b) Structure of the catalytic domain of ribulose 1,5-bisphosphate carboxylase. The eight parallel β strands are shown in green, and the eight α helices are shown in red. *Part (b) courtesy of J. Richardson.*

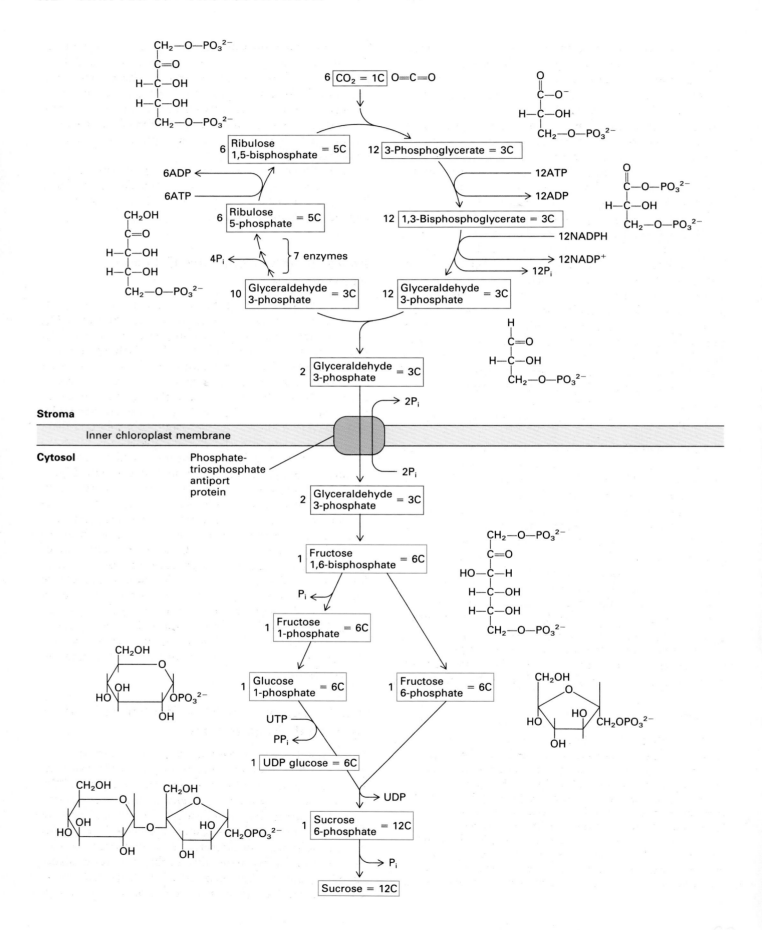

Ribulose 1,5-bisphosphate carboxylase makes up almost 50 percent of the chloroplast protein and is believed to be the single most abundant protein on earth. It is composed of two types of subunits: one is encoded in chloroplast DNA; the other, in nuclear DNA.

In the key experiment implicating ribulose 1,5-bisphosphate carboxylase, photosynthetic algae were exposed to a brief pulse of ^{14}C-labeled CO_2 and then the cells were quickly disrupted. The 3-phosphoglycerate was radiolabeled most rapidly, and all the ^{14}C radioactivity was found in the carboxyl group (see Figure 16-17).

The fate of the 3-phosphoglycerate formed by this reaction is complex: some is converted to starch or sucrose, but some is used to regenerate ribulose 1,5-bisphosphate. Quantitatively, for every 12 molecules of 3-phosphoglycerate generated by ribulose 1,5-bisphosphate carboxylase (a total of 36 C atoms), two molecules (6 C atoms) are converted to two molecules of glyceraldehyde 3-phosphate and 10 molecules (30 C atoms) are converted to six molecules of ribulose 1,5-bisphosphate. The fixation of six CO_2 molecules and the formation of two glyceraldehyde 3-phosphate molecules require the consumption of 18 ATPs and 12 NADPHs, generated by the light-requiring processes of photosynthesis (Figure 16-18).

Glyceraldehyde 3-phosphate is transported from the chloroplast to the cytosol in exchange for phosphate; there, the final steps of sucrose synthesis occur. In these reactions, one molecule of glyceraldehyde 3-phosphate is isomerized to dihydroxyacetone phosphate. This compound condenses with a second molecule of glyceraldehyde 3-phosphate to form fructose 1,6-bisphosphate, a normal glycolytic intermediate (see Figure 15-2). In leaf cells, however, most of the fructose 1,6-bisphosphate is converted to sucrose. One-half is converted to fructose 6-phosphate; one-half is isomerized to glucose 1-phosphate, which then forms uridine-diphosphate (UDP) glucose. These two compounds condense to form sucrose 6-phosphate; a final, irreversible removal of phosphate then generates the exportable sucrose.

The antiporter that transports triosephosphate from the chloroplast in exchange for cytosolic inorganic phosphates brings fixed CO_2 (glyceraldehyde phosphate) into the cytosol when the cell is exporting sucrose vigorously. This is a strict antiporter: no fixed CO_2 leaves the chloroplast unless phosphate is fed into it. The phosphate is generated in the cytosol primarily during the formation of sucrose from phosphorylated-three-carbon intermediates (see Figure 16-17). Thus sucrose synthesis and export encourage the export of additional precursors from the chloroplast.

At least seven enzymes are required to regenerate ribulose 1,5-bisphosphate from glyceraldehyde 3-phosphate. They are outlined in Figure 16-18, but their details are beyond the scope of this book.

Photorespiration Liberates CO₂ and Consumes O₂

Photosynthesis is always accompanied by *photorespiration*—a process that takes place in light, consumes O_2, and converts ribulose 1,5-bisphosphate in part to CO_2. As Figure 16-19 shows, ribulose 1,5-bisphosphate carboxylase catalyzes two competing reactions: the addition of CO_2 to ribulose 1,5-bisphosphate to form two molecules of 3-phosphoglycerate, and the addition of O_2 to form one molecule of 3-phosphoglycerate and one molecule of the two-carbon compound phosphoglycolate. Phosphoglycolate is hydrolyzed to glycolate. This compound is transported to peroxisomes, which contain a number of enzymes that generate and consume H_2O_2. Glycolate is oxidized to glyoxylate in the peroxisomes. In a complex process that involves mitochondria, two molecules of glyoxylate are converted to one CO_2 molecule and one 3-phosphoglycerate molecule. (The phosphoglycerate re-enters the chloroplasts, so that only one of the four C atoms undergoing photorespiration is actually lost as CO_2.) The latter reactions do not generate ATP or NADPH. Thus photorespiration uses up O_2 and generates CO_2, both wasteful processes for the economy of the plant. The necessary structure of the active site of ribulose 1,5-bisphosphate carboxylase apparently makes it impossible for plants to evolve an enzyme that is less active as an oxygenase.

The C₄ Pathway for CO₂ Fixation Is Used by Several Tropical Plants

Corn, sugar cane, crabgrass, and other plants that can grow in a hot, dry environment must keep their stomata (the gas-exchange pores in the leaves; see Figure 14-25) closed much of the time to prevent excessive loss of moisture, causing the CO_2 level inside the leaf to fall below the K_M of ribulose 1,5-bisphosphate carboxylase. Under these conditions, photorespiration is greatly favored over photosynthesis. To avoid this problem, certain plants, called *C₄ plants,* have evolved a two-step pathway of CO_2 fixation that involves two types of cells: *mesophyll*

◄ **Figure 16-18** The pathway of carbon during photosynthesis. Six molecules of CO_2 are converted into two molecules of glyceraldehyde 3-phosphate. These reactions occur in the stroma of the chloroplast. Via the phosphate–triosephosphate antiport, some glyceraldehyde 3-phosphate is transported to the cytosol in exchange for phosphate. In an exergonic series of reactions there, glyceraldehyde 3-phosphate is converted to fructose 1,6-bisphosphate and, ultimately, to the disaccharide storage form sucrose. Some glyceraldehyde 3-phosphate (not shown here) is also converted to amino acids and fats, compounds essential to plant growth.

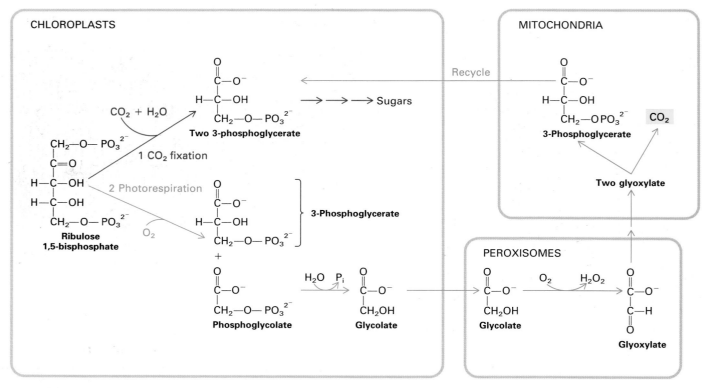

▲ **Figure 16-19** CO_2 fixation and photorespiration are competing reactions, catalyzed by the same enzyme that utilizes ribulose 1,5-bisphosphate. Reaction 1 is favored by high CO_2 and low O_2 pressures; reaction 2 occurs at low CO_2 and high O_2 pressures (that is, under normal atmospheric conditions). For every two molecules of phosphoglycolate and glyoxylate formed by photorespiration, one molecule of 3-phosphoglycerate is ultimately formed and recycled and one molecule of CO_2 is lost.

cells, which are adjacent to the air spaces in the leaf interior, and *bundle sheath cells,* which surround the vascular tissue (Figure 16-20a and b).

In mesophyll cells, CO_2 from the air is assimilated by reacting with phosphoenolpyruvate, a three-carbon glycolytic intermediate, to generate the four-carbon compound oxaloacetate. The enzyme that catalyzes this reaction, phosphoenolpyruvate carboxylase, is found almost exclusively in C_4 plants. Oxaloacetate, in turn, is reduced to malate (see Figure 16-20c) in many C_4 plants. (Since the four-carbon compounds oxaloacetate and malate are the first to be labeled by [^{14}C]CO_2, this CO_2 fixation pathway has been named the C_4 *pathway.*) Malate is then transferred by a special permease to the bundle sheath cells, in which CO_2 is released by decarboxylation. The CO_2 enters the Calvin cycle; the C_3 compound pyruvate generated by decarboxylation is transported back to the mesophyll cells, where it is reutilized in the C_4 pathway. In certain C_4 plants, oxaloacetate in mesophyll cells is aminated to form aspartate, which is transferred to the bundle sheath cells; it is converted to CO_2 and alanine there, and the alanine is then recycled to the mesophyll cells.

Because of the transport of CO_2 from mesophyll cells, the CO_2 concentration in the bundle sheath cells of C_4 plants is much higher than it is in the normal atmosphere. This favors the fixation of CO_2 to form 3-phosphoglycerate and inhibits the utilization of ribulose 1,5-bisphosphate by photorespiration. Since two phosphodiester bonds of ATP are consumed in the cyclic C_4 process (to generate phosphoenolpyruvate from pyruvate), the overall efficiency of the photosynthetic production of glucose from NADPH and ATP is lower than it is in C_3 plants, which use only the Calvin cycle for CO_2 fixation.

The oxygenation of ribulose 1,5-bisphosphate (reaction 2 in Figure 16-19) is favored by the high O_2 concentration in the atmosphere; as much as 50 percent of the photosynthetically fixed carbon in C_3 plants may be reoxidized to CO_2 during photorespiration. Compared with C_3 plants, C_4 plants are superior utilizers of available CO_2, since the enzyme phosphoenolpyruvate carboxylase (see Figure 16-20c) has a lower K_M for CO_2 than does the ribulose 1,5-bisphosphate carboxylase of the Calvin cycle. The net rates of photosynthesis for C_4 grasses, such as corn or sugar cane, can be two to three times the rates for otherwise similar C_3 grasses, such as wheat, rice, or oats.

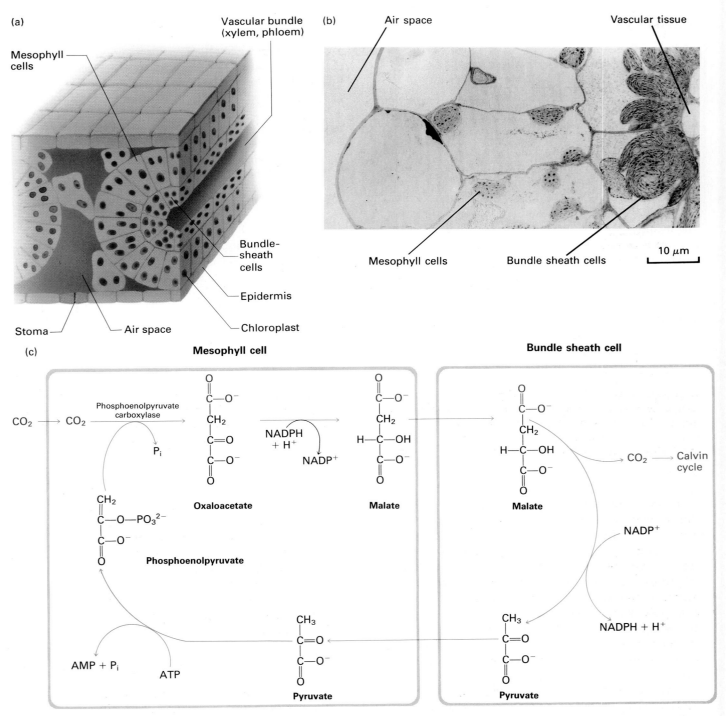

▲ **Figure 16-20** The C$_4$ pathway for CO$_2$ fixation. (a) The anatomy of a leaf in a C$_4$ plant. Bundle sheath cells line the vascular bundles containing the xylem and phloem. Sucrose is synthesized in bundle sheath cells by photosynthesis and then carried to the rest of the plant via the phloem. Mesophyll cells (adjacent to the substomal air spaces) surround the bundle sheath cells. (b) Electron micrograph of a cross section of a leaf from a typical C$_4$ plant. In mesophyll cells, CO$_2$ is assimilated into four-carbon molecules that are pumped into the interior bundle sheath cells. These cells con-tain abundant chloroplasts and are the sites of photosynthe-sis and sucrose synthesis. (c) Diagram showing that CO$_2$ is assimilated by phosphoenolpyruvate carboxylase in meso-phyll cells. C$_4$ molecules, such as malate, are transferred from the mesophyll cells to the bundle sheath cells. CO$_2$ is then released for use in the standard Calvin cycle, and the C$_3$ compound pyruvate is transferred back to the mesophyll cells. *Photograph in part (b) by S. Craig; courtesy of M. D. Hatch.*

Summary

The mechanism of photosynthetic energy transduction is best understood in purple bacteria, which have a single type of photosynthetic reaction center whose structure is known to molecular detail. The absorption of light by a "special pair" of chlorophylls excites a chlorophyll electron. In less than a millisecond, the electron is transferred (via a pheophytin and a quinone) to a second quinone acceptor on the cytoplasmic face of the plasma membrane, and a cytochrome on the exoplasmic membrane face donates an electron to the oxidized chlorophyll. Through a second complex, the electron returns across the membrane to the cytochrome, concomitant with the transport of protons to the exterior. As in other systems, the resultant proton-motive force is used mainly to power ATP synthesis through an F_0F_1 complex.

During photosynthesis in chloroplasts, light energy is absorbed by chlorophyll and other pigments in light-harvesting complexes (LHCs) associated with two photosystems: PSI and PSII. PSII is similar in structure and function to the bacterial photosystem. Light energy absorbed by LHCs is transferred to a specialized chlorophyll *a* pair contained in each reaction center (P_{680} in PSII and P_{700} in PSI). The excitation of P_{680} results in the removal of electrons from H_2O by the O_2-evolving complex on the luminal surface of the thylakoid and the formation of O_2 and protons. The protons remain in the lumen and generate part of the proton-motive force. The electrons are pumped from the stromal face across the thylakoid membrane via a cytochrome *b/f* complex; concomitantly, additional protons are transported to the lumen by means of a Q cycle.

Electrons are then transferred by the carrier plastocyanin in the thylakoid lumen to PSI. Electrons excited by photosystem I can undergo one of two possible fates. Electrons can be transferred via a series of carriers to $NADP^+$ to form NADPH by the process of linear electron transport. Alternatively, electrons can be transferred back to PSI concomitant with the transport of additional protons across the thylakoid membrane to the lumen from the stroma. This process of cyclic electron flow does not involve PSII; neither NADPH nor O_2 is formed.

Because thylakoid membranes are permeable to anions, a pH gradient (inside pH ~5.0 versus stromal pH = 7.8), rather than a membrane electric potential, is the principal component of the proton-motive force. The thylakoid-membrane pH gradient is primarily used in ATP synthesis. The ATP and NADPH generated by photosynthesis are used in a series of enzymatic reactions (the Calvin cycle) to convert CO_2 to sucrose or starch—the principal products of photosynthesis.

In C_3 plants, CO_2 is fixed in the reaction catalyzed by ribulose 1,5-bisphosphate carboxylase. In C_4 plants, CO_2 is fixed initially in the outer mesophyll cells by reaction with phosphoenolpyruvate. The four-carbon molecules so generated are shuttled to the interior bundle sheath cells, where the CO_2 is released and then used in the Calvin cycle; these reactions form part of the C_4 pathway.

References

Chapter 15 contains a set of general references on energy metabolism and the chemiosmotic theory.

General References on Photosynthesis and Chloroplast Structure

*GOVINDJEE, ed. 1982. *Photosynthesis: Energy Conversion by Plants and Bacteria.* Academic Press. A collection of excellent review articles.

*HAROLD, F. M. 1986. *The Vital Force: A Study of Bioenergetics.* W. H. Freeman and Company, Chapter 8.

*HALIWELL, B. 1984. *Chloroplast Metabolism,* rev. ed. Oxford University Press.

*HOOBER, J. K. 1984. *Chloroplasts.* Plenum Press.

STAEHELIN, L. A., and C. J. ARNTZEN, eds. 1986. *Photosynthesis III: Photosynthetic Membranes and Light-harvesting Systems.* Springer-Verlag (Berlin).

*STEINBACH, K. E., S. BONITZ, C. J. ARNTZEN, and L. BOGORAD, eds. 1985. *Molecular Biology of the Photosynthetic Apparatus.* Cold Spring Harbor Laboratory.

Structure and Function of the Bacterial Photosynthetic Reaction Center

ALLEN, J. P., G. FEHER, T. O. YEATES, H. KOMIYA, and D. C. REES. 1987. Structure of the reaction center from *Rhodobacter sphaeroides* R-26: the cofactors. *Proc. Nat'l Acad. Sci. USA* **84**:5730–5734.

DEISENHOFER, J., O. EPP, K. MIKI, R. HUBER, and H. MICHEL. 1985. Structure of the protein subunits in the photosynthetic reaction center of *Rhodopseudomonas viridis* at 3 Å resolution. *Nature* **318**:618–624.

*DEISENHOFER, J., and H. MICHEL. 1989. The photosynthetic reaction center from the purple bacterium *Rhodopseudomonas viridis. Science* **245**:1463–1473 (The Nobel Prize Lecture).

KNAFF, D. B. 1988. Reaction centers of photosynthetic bacteria. *TIBS* **13**:157–158.

*MICHEL, H., and J. DEISONHOFER. 1988. Relevance of the photosynthetic reaction center from purple bacteria to the structure of photosystem II. *Biochemistry* **27**:1–7.

YEATES, T. O., H. KOMIYA, D. C. REES, J. P. ALLEN, and G. FEHER. 1987. Structure of the reaction center from *Rhodobacter sphaeroides* R-26: membrane-protein interactions. *Proc. Nat'l Acad. Sci. USA* **84**:6438–6442.

*YOUVAN, D. C., and B. L. MARRS. 1987. Molecular mechanisms of photosynthesis. *Scientific American* **256**:42–48.

---

*A book or review article that provides a survey of the topic.

Photosystems I and II and the Oxygen-evolving Complex

*BARBER, J. 1987. Photosynthetic reaction centers: a common link. *TIBS* 12:321–326.

BASSI, R., G. HOYER-HANSEN, R. BARBATO, G. M. GIACOMETTI, and D. J. SIMPSON. 1987. Chlorophyll proteins of the photosystem II antenna system. *J. Biol. Chem.* 262:13333–13341.

*BLANKENSHIP, R. E., and R. C. PRINCE. 1985. Excited-state redox potentials and the Z scheme of photosynthesis. *TIBS* 10:382–383.

BUTLER, P. J. G., and W. KUHLBRANDT. 1988. Determination of the aggregate size in detergent solution of the light-harvesting chlorophyll *a/b* protein complex from chloroplast membranes. *Proc. Nat'l Acad. Sci. USA* 85:3797–3801.

DEBUS, R. J., B. A. BARRY, I. SITHOLE, G. T. BABCOCK, and L. MCINTOSH. 1988. Directed mutagenesis indicates that the donor to P_{680}^+ in photosystem II is tyrosine-161 of the D1 polypeptide. *Biochemistry* 27:9071–9074.

GEORGE, G. N., R. C. PRINCE, and S. P. CRAMER. 1989. The manganese site of the photosynthetic water-splitting enzyme. *Science* 243:789–791.

*GLAZER, A. N., and A. MELIS. 1987. Photochemical reaction centers: structure, organization, and function. *Ann. Rev. Plant Physiol.* 38:11–45.

HIRSCHBERG, J. A., A. BLEECKER, D. J. KYLE, L. MCINTOSH, and C. J. ARNTZEN. 1984. The molecular basis of triazine-herbicide resistance in higher-plant chloroplasts. *Z. Naturforsch* 39:412–420.

*KNAFF, D. B. 1988. The photosystem I reaction center. *TIBS* 13:460–461.

KUWABARA, T., K. J. REDDY, and L. A. SHERMAN. 1987. Nucleotide sequence of the gene from the cyanobacterium *Anacystis nidulans* R-2 encoding the Mn-stabilizing protein involved in photosystem II water oxidation. *Proc. Nat'l Acad. Sci. USA* 84:8230–8234.

LI, J. 1985. Light-harvesting chlorophyll *a/b* protein: three-dimensional structure of a reconstituted membrane lattice in negative strain. *Proc. Nat'l Acad. Sci. USA* 82:386–390.

*MATTOO, A. K., J. B. MARDER, and M. EDELMAN. 1989. Dynamics of the photosystem II reaction center. *Cell* 56:241–246.

MEI, R., J. P. GREEN, R. T. SAYRE, and W. D. FRASCH. 1989. Manganese-binding proteins of the oxygen-evolving complex. *Biochemistry* 28:5560–5567.

RUTHERFORD, A. W. 1989. Photosystem II, the water-splitting enzyme. *TIBS* 14:227–232.

SIEBERT, M., M. DEWIT, and L. A. STAEHELIN. 1987. Structural localization of the O_2-evolving apparatus to multimeric (tetrameric) particles on the luminal surface of freeze-etched photosynthetic membranes. *J. Cell Biol.* 105:2257–2265.

Photosynthesis Electron Transport and ATP Synthesis; Regulation of Photosynthesis

*ALLEN, J. F. 1983. Protein phosphorylation—carburetor of photosynthesis? *TIBS* 8:369–373.

*ANDERSON, J. M. 1986. Photoregulation of the composition, function, and structure of thylakoid membranes. *Ann. Rev. Plant Physiol.* 37:93–136.

*ANDERSON, J. M., and B. ANDERSSON. 1988. The dynamic photosynthetic membrane and regulation of solar energy conversion. *TIBS* 13:351–355.

*ANDRÉASSON, L.-E., and T. VÄNNGÅRD. 1988. Electron transport in photosystems I and II. *Ann. Rev. Plant Physiol. Plant Mol. Biol.* 39:379–411.

BARBER, J. 1985. Organization and dynamics of protein complexes within the chloroplast thylakoid membrane. *Biochem. Soc. Trans.* 14:1–4.

CRAMER, W. A., W. R. WIDGER, R. G. HERMANN, and A. TREBST. 1985. Topography and function of thylakoid membrane proteins. *TIBS* 10:125–129.

*MCCARTY, R. E., and G. G. HAMMES. 1987. Molecular architecture of chloroplast coupling factor 1. *TIBS* 12:234–237.

STAEHELIN, L. A., and C. J. ARNTZEN. 1983. Regulation of chloroplast membrane function: protein phosphorylation changes the spatial organization of membrane components. *J. Cell Biol.* 97:1327–1337.

WIDGER, W. R., W. A. CRAMER, R. G. HERRMANN, and A. TREBST. 1984. Sequence homology and structural similarity between cytochrome *b* of mitochondrial complex III and the chloroplast b_6/f complex: position of the cytochrome *b* hemes in the membrane. *Proc. Nat'l Acad. Sci. USA* 81:674–678.

CO₂ Fixation

*ARNON, D. I. 1987. Photosynthetic CO_2 assimilation by chloroplasts: assertion, refutation, discovery. *TIBS* 12:39–42.

*BASSHAM, J. A. 1962. The path of carbon in photosynthesis. *Scientific American* 206(6):88–100.

BERRY, J. A., G. H. LORIMER, J. PIERCE, J. R. SEEMANN, J. MEEK, and S. FREAS. 1987. Isolation, identification, and synthesis of 2-carboxyarabinitol 1-phosphate, a diurnal regulator of ribulose-bisphosphate carboxylase activity. *Proc. Nat'l Acad. Sci. USA* 84:734–738.

*BJORKMAN, O., and J. BERRY. 1973. High-efficiency photosynthesis. *Scientific American* 229(4):80–93.

BUCHANAN, B. B. 1984. The ferredoxin-thioredoxin system: a key element in the regulatory function of light in photosynthesis. *BioScience* 34:378–383.

CHAPMAN, M. S., S. W. SUH, P. M. G. CURMI, D. CASCIO, W. W. SMITH, and D. W. EISENBERG. 1988. Tertiary structure of plant RuBisCO: domains and their contacts. *Science* 241:71–74.

*EDWARDS, G., and D. WALKER. 1983. *C₃, C₄ Mechanisms and Cellular and Environmental Regulation of Photosynthesis.* University of California Press.

ELLIS, R. J., and J. C. GRAY, eds. 1986. Ribulose-bisphosphate carboxylase-oxygenase. *Phil. Trans. R. Soc. Lond.* B313:303–469.

*FLUGGE, U. I., and H. W. HELDT. 1984. The phosphate-triose phosphate-phosphoglycerate translocator of the chloroplast. *TIBS* 9:530–533.

HUANG, A. H. C., R. N. TRELEASE, and T. S. MOORE, JR. 1983. Plant peroxisomes. In *American Society of Plant Physiologists Monograph Series.* Academic Press.

*HUSIC, D. W., H. D. HUSIC, and N. E. TOLBERT. 1987. The oxidative photosynthetic carbon cycle or C_2 cycle. *CRC Crit. Rev. Plant Sci.* 5:45–48.

*KNAFF, D. B. 1989. Structure and regulation of ribulose-1,5-bisphosphate carboxylase/oxygenase. *TIBS* 14:159–160.

SCHNEIDER, G., Y. LINDQVIST, C. L. BRADEN, and G. LORIMER. 1986. Three-dimensional structure of ribulose-1,5-bisphosphate carboxylase-oxygenase from *Rhodospirullum ru-*

brum at 2.9 Å resolution. *EMBO J.* 5:3409–3415.

SICHER, R. C. 1986. Sucrose biosynthesis in photosynthetic tissue: rate-controlling factors and metabolic pathway. *Physiol. Plant* 67:118–121.

*WOODROW, I. E., and J. A. BERRY. 1988. Enzymatic regulation of photosynthetic CO_2 fixation in C_3 plants. *Ann. Rev. Plant Physiol. Plant Mol. Biol.* 39:533–594.

C H A P T E R

17

Plasma-Membrane, Secretory, and Lysosome Proteins: Biosynthesis and Sorting

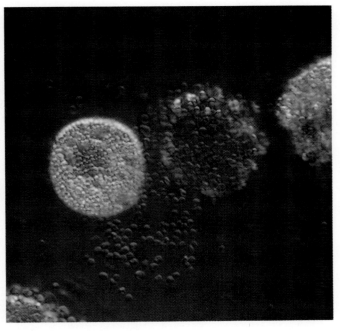

Secretion vesicles in mast cells

A discrete set of proteins enables each type of membrane, organelle, or particle in a cell to carry out its unique functions. A critical factor in "building" a cell from its constituent proteins, lipids, nucleic acids, and other components is protein sorting—targeting each newly made polypeptide to the correct membrane or organelle. Specific proteins, for example, must be delivered to the plasma membrane to enable the cell to recognize hormones, bind to the extracellular matrix, or take up or extrude specific ions or other small molecules. Other proteins, such as RNA and DNA polymerases, must be targeted to the nucleus; still others, to the lysosome or peroxisome. Proteins that are secreted from the cell move sequentially from the site of their synthesis in the rough endoplasmic reticulum (ER) through the Golgi complex and the secretory vesicles (see Figure 4-40). During exocytosis, the final stage of secretion, secretory-vesicle membranes fuse with the plasma membrane (see Figure 13-40). The membrane or lumen of each of these organelles contains certain enzymes that make specific *modifications on the proteins to be secreted. Each enzyme, in turn, must be targeted to the correct organelle and remain there as the secretory proteins pass through en route to the cell exterior.*

The assembly of mitochondria and chloroplasts is even more complex. Some proteins are encoded by the organelle DNA and synthesized on a unique class of ribosomes found in the mitochondrion or chloroplast. However, most proteins in these organelles are encoded by nuclear DNA and synthesized on cytoplasmic ribosomes. These proteins must be imported into the specific organelle and then inserted into the correct membrane or space.

Without question, the assembly of the diverse membranes and organelles of a cell, with their individual proteins, lipids, and carbohydrates, must be extremely complex. This chapter focuses on the synthesis and sorting of one broad class of cellular proteins that includes those destined for the plasma membrane or lysosome and those secreted from the cell. These proteins are grouped together because they enter the membrane or lumen of the ER during or immediately after their synthesis. Once in the ER, they are sorted to their correct destinations (Figure 17-1). The fate of these proteins is quite different from that of proteins involved in the assembly of the nucleus, mitochondrion, and chloroplast, which are released into the cytosol and subsequently transported to the correct organelle.

At the outset, it is useful to summarize a number of key concepts that have guided work on organelle biogenesis and protein sorting. Throughout this chapter and the next, we will see how each of the following concepts applies to the targeting of membrane and organelle proteins as well as proteins that are to be secreted:

1. *Membranes grow by expansion of existing membranes. Lipids and proteins are inserted into existing membrane elements.*
2. *Membrane-containing organelles, such as the ER, mitochondria, and chloroplasts, grow by expansion of existing organelles. Proteins and lipids are added to the existing organelle, which eventually divides into two or more "daughters."*
3. *Pure phospholipid membranes are impermeable to proteins. However, certain subcellular membranes contain permeases that admit specific proteins when the polypeptide is in an unfolded state.*
4. *The transport of proteins across biological membranes requires the expenditure of energy, which is derived from ATP, from an ion gradient across the membrane, or from changes in the conformation of the protein during its passage across the membrane.*
5. *Different integral membrane proteins, particularly those synthesized on the rough ER, are oriented differently with respect to the phospholipid bilayer (Figure 17-2). One or more topogenic sequences ensure that the protein is properly oriented during its insertion into the membrane.*
6. *Signals on the protein itself "target" specific proteins to enter certain membranes or organelles. Amino acid sequences, or carbohydrate or phosphate residues added to the protein after its synthesis can be recognized by receptors on subcellular membranes and organelles.*

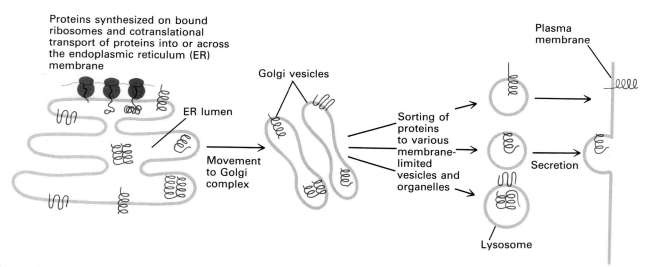

▲ **Figure 17-1** Proteins synthesized on ribosomes bound to the endoplasmic reticulum (ER) become part of cellular membranes, are secreted, or are stored. A few membrane proteins (rough ER enzymes or structural proteins) remain in the ER. Most proteins move to the Golgi vesicles, where a few remain. The rest of the proteins move via vesicles to the plasma membrane or are sorted to various membrane-bound organelles, such as the lysosome. Proteins transported to the ER lumen also move to the Golgi vesicles. Some of these proteins are sequestered in vesicles and exocytosed; others are sorted to various organelles.

7. *Many membrane proteins, such as cell-surface receptors, shuttle repeatedly between the plasma membrane and one or more subcellular organelle membranes. Specific mechanisms regulate this membrane "traffic." Both the fusion of two membranes and the budding off of a membrane region are tightly controlled processes.*

These general concepts apply to the synthesis of chloroplasts, mitochondria, and plasma and intracellular membranes, to protein secretion, and to the entry of proteins into such vesicles as lysosomes and chloroplasts. Even though such structures as the inner mitochondrial membrane, the plasma membrane, and the rough ER membrane carry out very different functions, they are all phospholipid bilayers in which different classes of proteins are embedded. Accordingly, our study of organelle assembly begins with the biosynthesis of phospholipids and their incorporation into membranes. ▲

The Synthesis of Membrane Lipids

Phospholipids Are Synthesized in Association with Membranes

All phospholipids and other membrane lipids, such as sphingomyelin and glycolipids, are amphipathic molecules with extremely hydrophobic regions that dissolve very little in aqueous solution. In fact, depending on their concentration and the ion composition, they spontaneously form either micelles (spherical vesicles built of a phospholipid bilayer) or sheets of bilayers. These physical-chemical properties of phospholipids have profound implications for the biosynthesis of membrane-containing organelles. The extremely low solubility of phospholipids in aqueous solution makes the assembly of a new phospholipid bilayer from soluble components energetically difficult. Phospholipids are either synthesized in association with cellular membranes or incorporated into already existing membranes immediately after synthesis. When cells are briefly exposed to radioactive phosphate, $[^{32}P]PO_4^{3-}$, or to radiolabeled fatty acids or sugars, all phospholipids and glycolipids incorporating these substances are seen to be associated with intracellular membranes; none are found free in the cytosol. With few exceptions, membranes grow by expansion of existing membranes.

In bacterial cells, the synthesis of phospholipids is associated with the plasma membrane. In animal and plant cells, synthesis is associated with the ER membrane, usually the smooth ER. In the pathway for the synthesis of phosphatidylethanolamine (a typical phospholipid) in animal cells (Figure 17-3), one substrate—a CoA ester of a fatty acid—is an amphipathic molecule. The fatty acid side chain is embedded in the cytoplasmic leaflet (see Figure 13-11) of the ER membrane, and the CoA portion protrudes into the cytosol. All other substrates and reaction products, such as ATP, ethanolamine, and cytidine triphosphate (CTP), are soluble constituents of the cytosol. Thus the biosynthesis of phospholipids occurs at the

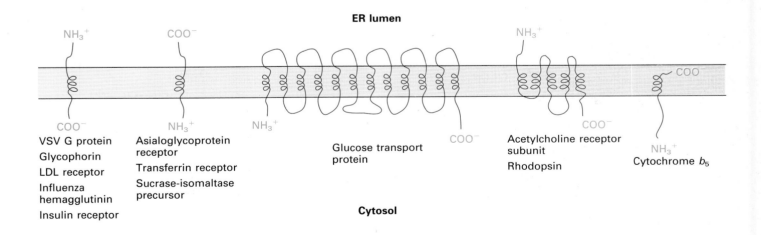

▲ **Figure 17-2** Topologies of some integral membrane proteins synthesized on the rough ER. Segments of the protein chain in the ER membrane bilayer are depicted as transmembrane α helices; portions outside the membrane are shown as lines (the folding of these regions is not depicted). Topogenic sequences of amino acids in the protein act during biosynthesis to ensure the proper transmembrane orientation. *From W. Wickner and H. F. Lodish, 1985, Science **230**:400–407; M. Mueckler et al., 1985, Science **229**:941–945.*

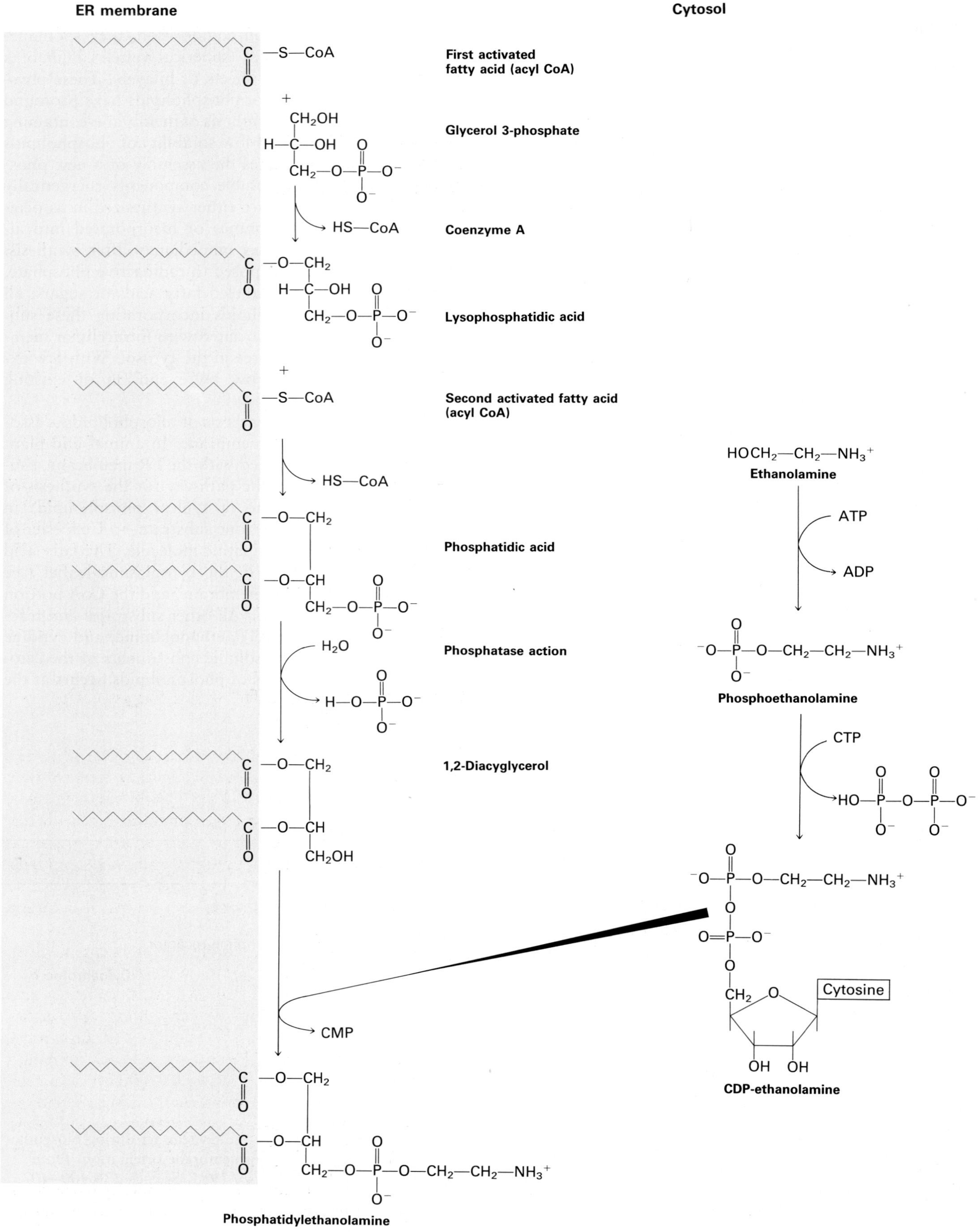
ER membrane
Cytosol
First activated fatty acid (acyl CoA)
Glycerol 3-phosphate
Coenzyme A
HS—CoA
Lysophosphatidic acid
Second activated fatty acid (acyl CoA)
HS—CoA
Phosphatidic acid
Phosphatase action
H2O
1,2-Diacyglycerol
CMP
Phosphatidylethanolamine
HOCH2—CH2—NH3+
Ethanolamine
ATP
ADP
Phosphoethanolamine
CTP
Cytosine
CDP-ethanolamine

◄ **Figure 17-3** Biosynthesis of phosphatidylethanolamine in animal cells. The precursors are fatty acyl CoA (an amphipathic molecule embedded in the ER membrane) and glycerol 3-phosphate and cytidine diphospho-ethanolamine (CDP-ethanolamine) (two water-soluble molecules in the cytosol). Other phospholipids are assembled in the ER membrane by analogous pathways from fatty acyl CoA and small, soluble molecules. Bacterial cells use a different pathway for the synthesis of phosphatidylethanolamine that also starts with fatty acyl CoA and occurs in the bacterial plasma membrane.

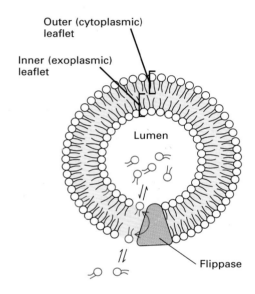

Dibutyrylphosphatidylcholine (⚲)

▲ **Figure 17-4** Assay for a phospholipid *flippase*. The phospholipid dibutyrylphosphatidylcholine is water-soluble because it contains short, four-carbon fatty acyl chains. When added to a suspension of pure phospholipid vesicles (liposomes), the phospholipid spontaneously inserts itself into the outer leaflet but, in the absence of a flippase protein, does not flip to the inner leaflet. A flippase protein, such as is found in ER membranes, catalyzes the movement of this small phospholipid to the inner (exoplasmic) leaflet; from there, it can spontaneously move into the aqueous lumen of the vesicle. Thus a flippase can be assayed as a protein that allows dibutyrylphosphatidylcholine added to the exterior of a vesicle to accumulate in the vesicle lumen. *After W. R. Bishop and R. M. Bell, 1985,* Cell **42**:50–60; *Y. Kawashima and R. M. Bell, 1987,* J. Biol. Chem. **262**:16495–16502.

interface of the ER membrane and the cytosol. Most of the enzymes that catalyze these reactions are amphipathic: one segment is inserted into the ER membrane; the other protrudes into the cytosol.

Special Membrane Proteins Allow Phospholipids to Equilibrate in Both Membrane Leaflets

All newly made phospholipids are localized to the cytoplasmic leaflet of the ER, but most phospholipids are found in both membrane faces of the ER, as they are in most cellular membranes. Phospholipids do not spontaneously flip-flop across a pure phospholipid bilayer. In many formed membranes, such as the erythrocyte membrane or the plasma membrane of nucleated cells, a phospholipid takes several days to move from one membrane face to the other. Then how are newly made phospholipids moved from the cytoplasmic to the exoplasmic leaflet of the ER? It appears that ER membranes (and plasma membranes of bacteria) contain one or more proteins that can catalyze such a flip-flop process (Figure 17-4). The half-time for movement of a phospholipid to the exoplasmic leaflet of the ER membrane is only a few minutes.

The two leaflets of a membrane often have different phospholipid compositions. How this lipid asymmetry is achieved is not known, but the mechanism in part may involve different affinities of phospholipids for the flip-flop protein or for specific regions of integral membrane proteins localized to the two faces.

Phospholipids Move from the ER to Other Cellular Membranes

Although phospholipids are synthesized in the ER, they are found in all organelles. How do they move there? According to the popular model of *membrane budding,* a membrane vesicle containing phospholipids buds off the ER membrane and then fuses with another organelle membrane. During this process, certain phospholipids may be incorporated selectively into the vesicle, explaining the different phospholipid compositions of different organelle membranes.

A second model of phospholipid movement involves *phospholipid exchange proteins*—water-soluble proteins that can remove phospholipids from one membrane (say, the ER) and release them into another membrane or organelle. Such proteins have been identified in the cytosol of liver hepatocytes and other cell types. Each exchange protein only binds a single type of phospholipid. In cell-free reactions, exchange proteins equilibrate phospholipids among all membranes present. Whether these proteins can account for the abundance of certain lipids in specific organelle membranes in the cell is unknown.

In animal cells, the addition of terminal sugars, such as galactose and *N*-acetylneuraminic (sialic) acid, to glycolipids occurs in the Golgi complex. Such glycolipids are found only in the exoplasmic face (on the luminal side) of the Golgi membrane. They are moved to the plasma membrane via transport vesicles—probably the same vesicles that move newly made secretory and integral membrane proteins to the plasma membrane. Phospholipid exchange proteins are cytosolic; they transport lipids only to the cytoplasmic membrane face.

Mitochondria appear to synthesize cardiolipin, found only in the inner mitochondrial membrane, as well as phosphatidylglycerol. Other mitochondrial lipids, such as phosphatidylethanolamine and phosphatidylcholine, are synthesized in the ER and subsequently imported into the organelle. In photosynthetic tissues, chloroplasts are sites for the synthesis of all chloroplast lipids, including glycolipids; at least in spinach leaves, the chloroplast is the site of all fatty acid synthesis.

Sites of Organelle- and Membrane-Protein Synthesis

With a few important exceptions, the lipid compositions of different membranes are more uniform than their protein compositions. Each subcellular organelle and membrane contain a unique constellation of proteins. As examples, the ATP-ADP transporter is unique to the inner mitochondrial membrane, and certain hydrolytic enzymes are greatly enriched in lysosomes. One possible method of directing proteins to their appropriate destinations in the cell could be via functionally different classes of ribosomes, each occupying a specific intracellular site, that translate only certain classes of messenger RNAs (mRNAs). Mitochondria and chloroplasts do contain unique populations of ribosomes, and all proteins encoded by mitochondrial DNA or chloroplast DNA are translated on ribosomes in the respective organelle (Table 17-1). However, specialized cytoplasmic ribosomes for the synthesis of specific proteins have not yet been found.

All Cytoplasmic Ribosomes Are Functionally Equivalent

Abundant evidence indicates that all ribosomes in the cytosol of eukaryotic cells are functionally equivalent. However, *membrane-attached ribosomes* are tightly bound to the ER, whereas *membrane-unattached ribosomes* appear to be free in the cytosol (Figures 17-5 and 17-6). (Many of these "free" ribosomes are actually bound to cytoskeletal fibers.) However, these two classes of ribosomes have the same protein and ribosomal RNA (rRNA) compositions. In cell-free, protein-synthesizing

Table 17-1 Proteins synthesized by different classes of cytoplasmic ribosomes

Location of ribosome	Class of protein synthesized
Mitochondrion	All proteins encoded by mitochondrial DNA, mainly certain integral proteins of the inner membrane
Chloroplast	All proteins encoded by chloroplast DNA
Cytoplasmic ribosomes unbound to membranes	Soluble cytoplasmic proteins
	Extrinsic membrane proteins localized to the cytoplasmic face (actin, spectrin, etc.)
	Mitochondrial proteins encoded by nuclear DNA
	Chloroplast proteins encoded by nuclear DNA
	Peroxisome proteins
	Glyoxisome proteins
	Nuclear proteins (histones, lamins, etc.)
Cytoplasmic ribosomes bound to ER membranes	Secreted proteins
	Integral membrane glycoproteins Plasma membrane Nuclear membrane Rough ER membrane Golgi membrane
	Lysosome enzymes
	Rough ER enzymes
	Golgi complex enzymes
	Extrinsic membrane proteins localized to the exoplasmic face (fibronectin, laminin, collagen, etc.)

systems containing a variety of added mRNAs, both classes of ribosomes function identically. Apparently, then, no functionally unique class of cytoplasmic ribosome is found only at a certain intracellular site to translate a specific class of mRNAs. Current evidence indicates that most, if not all, information for intracellular protein distribution is located in the amino acid sequence of the newly synthesized protein itself.

Different Proteins Are Synthesized by Membrane-Attached and Membrane-Unattached Ribosomes

Although, when isolated, the two classes of cytoplasmic ribosomes are equivalent, membrane-attached and mem-

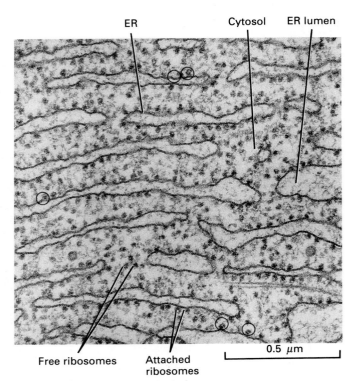

▲ Figure 17-5 Electron micrograph of ribosomes attached to the ER in a pancreatic exocrine cell. In a few cases, the large and small ribosomal subunits are resolved *(circles)*; the large subunit is attached to the ER membrane. Most of the proteins synthesized by this cell are to be secreted and are formed on membrane-attached ribosomes. A few membrane-unattached ribosomes are evident; presumably, these are synthesizing cytosolic or other nonsecretory proteins. *Courtesy of G. Palade.*

brane-unattached ribosomes translate different classes of mRNAs encoded by nuclear DNA in the cell (see Table 17-1). Proteins synthesized on membrane-attached ribosomes include secretory proteins; integral, ER, Golgi, and plasma-membrane proteins; glycoproteins; and lysosome proteins. These proteins contain (generally, at their N-terminus), specific amino acid signals that direct the ribosome that is making them to bind to the ER membrane. Proteins synthesized by membrane-attached ribosomes usually begin to cross the ER membrane before their synthesis is complete. Following their synthesis, these proteins are sorted to various membrane organelles. Secretory proteins are transported to the Golgi complex via small transport vesicles and then secreted. Lysosome enzymes are transported to the Golgi complex first, then

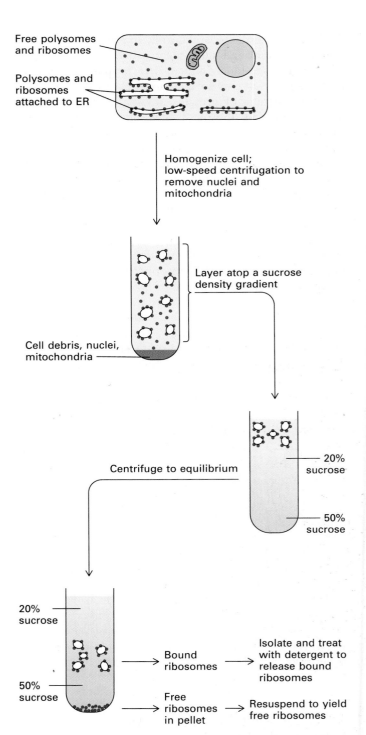

▲ Figure 17-6 Purification of membrane-unattached ribosomes from membrane-attached ribosomes and polysomes (clusters of ribosomes linked by mRNA). A cell homogenate, freed of nuclei and mitochondria by low-speed centrifugation, is layered on a sucrose density gradient. After centrifugation, membrane-unattached ribosomes and polysomes pellet to the bottom of the tube, because they are more dense than the 50-percent sucrose solution. Membrane-attached ribosomes form a band in the gradient, due to the low buoyant density of the phospholipids. These ribosomes can be freed of membranes by treatment with nonionic detergents.

directed to lysosomes, and so on. In addition to amino acid sequences, other substituents, such as carbohydrate or phosphate residues, added after protein synthesis is complete can serve as sorting signals.

A large variety of proteins are synthesized by membrane-unattached ribosomes, including soluble cytosolic proteins, such as glycolytic enzymes, and most extrinsic membrane proteins, such as spectrin. These proteins are released into the cytosol, where they remain or bind to other proteins (for example, the polymerization of actin or tubulin) or to membrane proteins (for example, spectrin binding to ankyrin or band 4.1.)

Most mitochondrial and chloroplast proteins are encoded by nuclear DNA, not by organelle DNA. Many investigators were surprised to find that all these proteins, including some extremely hydrophobic ones, are synthesized on membrane-unattached ribosomes. After synthesis, these proteins are released into the cytosol and subsequently incorporated into the organelle. Similarly, proteins found in the nucleus or the lumen of the peroxisome or glyoxisome are synthesized in the cytosol by membrane-unattached ribosomes and cross into the organelle. Each organelle contains an uptake mechanism that recognizes only the appropriate proteins. One or more sequences of amino acids in the protein itself signal its incorporation into the organelle. Often, these signal sequences are removed by specific proteolysis once the protein reaches its final destination.

Now let's consider some mechanisms of protein targeting and organelle biogenesis in more detail. We begin with a general discussion of the pathway of protein secretion, followed by a detailed consideration of the way in which newly made secretory proteins are targeted to and cross the rough ER membrane.

Overall Pathway for the Synthesis of Secretory and Membrane Proteins

A vast array of different proteins are secreted by different vertebrate cells (Table 17-2), and many cells are specialized for the secretion of specific proteins. The principal function of the pancreatic acinar cells, for instance, is the secretion of such digestive enzymes as chymotrypsinogen (the precursor of chymotrypsin; see Figure 2-18), ribonuclease, and amylase into the intestine. Specialized cells in mammary glands synthesize and secrete milk proteins; liver hepatocytes secrete albumin, transferrin, and lipoproteins into the serum. The B lymphocytes are specialized for the synthesis and secretion of immunoglobulins. Other cells, such as fibroblasts, secrete such proteins as collagens, proteoglycans, and fibronectin, which form part of their extracellular matrix.

Newly Made Secretory Proteins Are Localized to the Lumen of the Rough ER

The rough ER is an extensive interconnected series of flattened sacs, generally lying in layers. The lumen, or *cisterna,* of the rough ER defines a space topologically distinct from the cytosol. When cells are homogenized, the rough ER breaks up into small closed vesicles, termed *rough microsomes,* with the same orientation (ribosomes on the outside) as that found in the cell. Pulse-labeled secretory proteins are associated with these ER vesicles; that they are actually inside the vesicles is proved by the experiment shown in Figure 17-7.

Table 17-2   Classes of secretory proteins in vertebrates

Protein type	Example	Site of synthesis	Protein type	Example	Site of synthesis
Serum proteins	Albumin	Liver (hepatocyte)		Amylase	Pancreatic acini, liver, salivary glands
	Transferrin (Fe transporter)	Liver			
	Lipoproteins	Liver		Ribonuclease	Pancreatic acini
	Immunoglobulins	Lymphocytes		Deoxyribonuclease	Pancreatic acini
Extracellular matrix proteins	Collagen	Fibroblasts	Milk proteins	Casein	Mammary gland
	Fibronectin	Fibroblasts, liver		Lactalbumin	Mammary gland
	Proteoglycans	Fibroblasts, others	Egg-white proteins	Ovalbumin	Tubular gland cells in the avian oviduct
Peptide hormones	Insulin	Pancreatic β-islet cells			
	Glucagon	Pancreatic α-islet cells		Conalbumin	Tubular gland cells in the avian oviduct
	Endorphins	Neurosecretory cells			
	Enkephalins	Neurosecretory cells		Ovomucoid	Tubular gland cells in the avian oviduct
	ACTH	Pituitary anterior lobe			
Digestive enzymes	Trypsin	Pancreatic acini		Lysozyme	Tubular gland cells in the avian oviduct
	Chymotrypsin	Pancreatic acini			

Many Organelles Participate in Protein Secretion

After the protein is synthesized and translocated to the ER lumen, small transport vesicles containing the protein form from the ER and move to the membrane stacks on the *cis face* of the Golgi complex (Figure 17-8). The protein moves through the Golgi complex to the *trans face* and then into a complex network of vesicles termed the *trans Golgi reticulum*. From there, the protein is sorted to secretory vesicles, which fuse with the plasma membrane

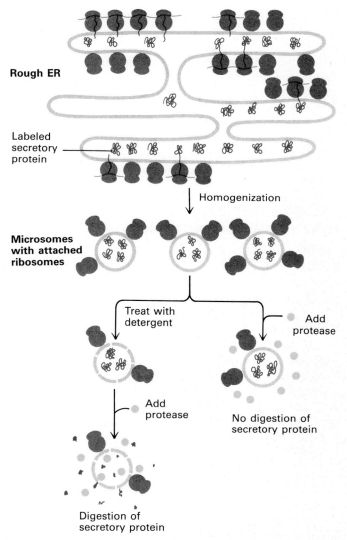

▲ **Figure 17-7** Secretory proteins are sequestered in the lumen of the rough ER, as experiments similar to the one depicted here have proved. Newly synthesized secretory protein associated with a microsome is not digested by an added protease, which remains outside the microsome. A detergent that makes the microsome membrane permeable allows some luminal proteins to leak out and the protease to enter. Under these conditions, newly made proteins are destroyed by the protease.

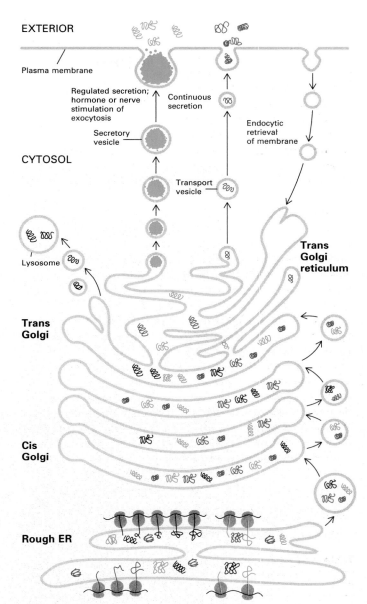

▲ **Figure 17-8** Maturation of secretory proteins. After synthesis, secretory proteins are localized to the rough ER. Always surrounded by membrane-bound vesicles, they migrate first to the vesicles on the cis face of the Golgi membrane complex, where modifications mainly to carbohydrate chains occur. The proteins then migrate through the Golgi vesicles to the trans face and the trans Golgi reticulum, while additional modifications occur. Proteins are shuttled between the Golgi vesicles by small transport vesicles. In some cells, such as hepatocytes, the proteins are secreted continuously. In other cells, such as exocrine pancreatic cells, some proteins are stored in secretory vesicles, where they await a signal for secretion. The proteins are also sorted in the trans Golgi vesicles or the trans Golgi reticulum for transport to lysosomes. Plasma-membrane glycoproteins go through the same stages of maturation continuously secreted proteins do. ER membrane proteins stop maturing in the ER. Membrane proteins destined to remain in the Golgi complex are formed in the rough ER and transported to the appropriate vesicle.

and release the protein to the cell exterior. Much research during the past decade has identified a large number of enzymes that act sequentially to modify secretory proteins during their maturation. Each enzyme is localized to a specific organelle and modifies proteins as they pass through. Amino acid side chains can be modified, saccharide residues can be added and modified, specific proteolytic cleavages may take place, disulfide bonds can form, and polypeptide chains may assemble into multiprotein complexes. Of course, not all of these reactions occur in every cell or for every secreted protein. Collagen is an example of a secreted protein that is modified in all of these ways.

In certain cells, many secretory proteins, such as collagens and serum proteins, are continuously synthesized and secreted (see Figure 17-8). In other cells, the secretion of some proteins is not continuous; these proteins are stored in intracellular membrane-limited vesicles to await the signal for exocytosis (see Figures 17-8 and 4-42). Examples of this *regulated secretion* include the exocrine cells in the pancreas, which secrete precursors of digestive enzymes, and hormone secreting endocrine cells in the pancreas, which synthesize insulin and other hormones and store them in vesicles. The release of each of these proteins is triggered by different neural and hormonal stimuli. In most cases of regulated secretion studied, exocytosis is triggered by a rise in intracellular Ca^{2+} and consists of the fusion of the membrane of the secretory vesicle with the plasma membrane.

Secretory Proteins Move from the Rough ER to Golgi Vesicles to Secretory Vesicles

Protein transport was initially established by studies on newly synthesized proteins in pancreatic acinar cells. Because most of the newly made proteins are secreted, they can be followed by electron-microscope autoradiography. When the radiolabeled amino acid leucine ($[^3H]$leucine) is added to thin slices of pancreas in culture, essentially all the incorporated radioactivity in the acini is seen to be located in the rough ER (Figure 17-9). During the subsequent chase period, the tissue is incubated in abundant unlabeled leucine, and little additional incorporation of radioactivity occurs. After a 7-min chase, most labeled proteins can be found in the Golgi vesicles. At later times, the radioactivity is located in immature secretory vesicles, often called *condensing vesicles,* adjacent to the Golgi vesicles; at still later times, the labeled proteins are localized to the mature secretory vesicles (zymogen granules). This sequence of steps is called the *maturation pathway.* Apparently, immature secretory vesicles are converted to mature secretory vesicles by progressive filling and concentration of their contents. Electron-microscope studies have been confirmed by fractionating cells after various pulse and chase times. The results clearly indicate that secretory proteins never are found as free,

soluble proteins in the cytosol but always are sequestered in membrane vesicles.

Yeasts secrete few proteins into the growth medium, but they do secrete a number of enzymes that remain localized in the narrow space between the plasma membrane and the cell wall. The most well-studied of these, *invertase,* hydrolyzes the disaccharide sucrose to glucose and fructose. A genetic analysis of protein secretion in yeast mutants has confirmed the maturation pathway for secretory proteins. In a set of temperature-sensitive mutant yeast strains, the secretion of all proteins, including invertase, is blocked at the higher temperature but is normal at the lower temperature. At the higher temperature, different so-called *sec mutants* accumulate proteins in the rough ER, in small vesicles taking proteins from the ER to the Golgi, in the Golgi vesicles, or in the secretory vesicles. At least 60 gene products are required to complete the maturation pathway. Studies using double mutants have showed that the pathway must be ordered in the following way: rough ER → Golgi vesicles → secretory vesicles → exocytosis (Figure 17-10). This maturation pathway is believed to apply to all secretory proteins in all eukaryotic organisms, including plants.

Plasma-Membrane Glycoproteins Follow the Same Maturation Pathway as Continuously Secreted Proteins

The maturation pathway taken by continuously secreted proteins is also followed by plasma-membrane glycoproteins. The same *N*- and *O*-linked oligosaccharides are found on the two classes of proteins (see Figures 2-55 and 2-56). Well-studied examples include viral glycoproteins destined for the plasma membranes of infected cells, such as the hemagglutinin (HA) glycoprotein of the influenza virus (see Figure 14-48), glycophorin (see Figure 13-16), the plasma-membrane Na^+-K^+ ATPase (see Figure 14-9), and enzymes in plant plasma membranes that synthesize such cell-wall components as cellulose. Pulse-labeling studies using radioactive amino acids, followed by subcellular fractionation and the immunoprecipitation of radiolabeled protein, have established that these newly-made

▸ **Figure 17-9** The synthesis and movement of guinea-pig pancreatic secretory protein as revealed by electron-microscope autoradiography. (a) At the end of a 3-min labeling period with $[^3H]$leucine, the tissue is fixed, sectioned for electron microscopy, and subjected to autoradiography. Most of the labeled proteins (the autoradiographic grains) are over the rough ER. (b) Following a 7-min chase period with unlabeled leucine, most of the labeled proteins have moved to the Golgi vesicles. (c) After a 37-min chase, most of the proteins are over immature secretory vesicles. (d) After a 117-min chase, the majority of the proteins are over mature zymogen granules. *Courtesy of J. Jamieson and G. Palade.*

(a)

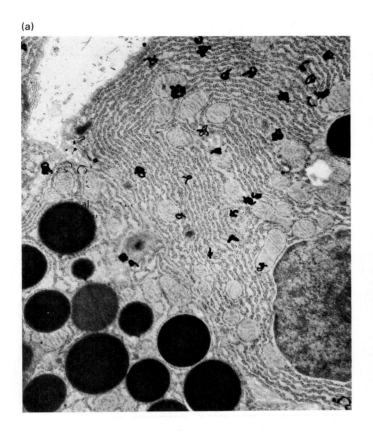

(b)

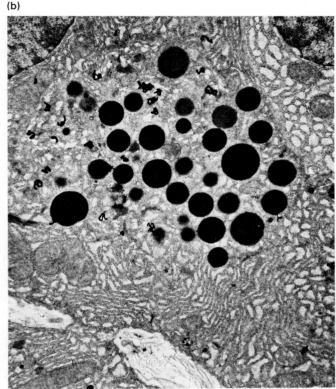

(c)

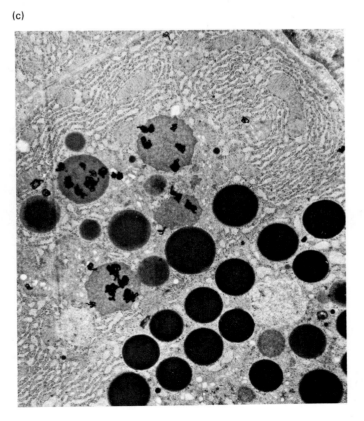

(d)

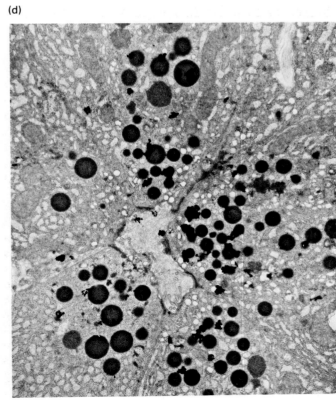

(a) Single mutants

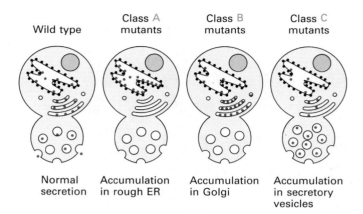

|Wild type | Class A mutants | Class B mutants | Class C mutants |

| Normal secretion | Accumulation in rough ER | Accumulation in Golgi | Accumulation in secretory vesicles |

Possible interpretations

Rough ER $\xrightarrow{A}$ Vesicles $\xrightarrow{C}$ Golgi $\xrightarrow{B}$ Exocytosis

or

Rough ER $\xrightarrow{A}$ Golgi $\xrightarrow{B}$ Vesicles $\xrightarrow{C}$ Exocytosis

(b) Double mutants: Class B and C

Accumulation in Golgi

Conclusion

Rough ER $\longrightarrow$ Golgi $\longrightarrow$ Vesicles $\longrightarrow$ Exocytosis

▲ **Figure 17-10** The use of double mutants in the yeast secretory pathway to show that the sequence is ordered rough ER → Golgi vesicles → secretory vesicles → exocytosis. (a) In class A mutants, the proteins remain in the rough ER (when grown at the higher temperature). Class B mutants accumulate protein in the Golgi complex, and class C mutants accumulate it in the secretory vesicles. These findings do not prove the sequence. The two alternative interpretations yield sequences that show the same experimental results. (b) When yeast cells contain double mutants with defects in the class B and class C secretory processes, proteins accumulate in the Golgi vesicles, not in the secretory vesicles. Thus class B mutations act at an earlier point in the maturation pathway than class C mutations do, and the correct sequence is on the bottom line. [See P. Novick et al., 1981, *Cell* 25:461.]

glycoproteins are inserted into the rough ER membrane. Subsequently, they move through the Golgi vesicles en route to the plasma membrane. These plasma-membrane glycoproteins also have been shown to undergo the same types of modifications in the same ER and Golgi compartments that secretory proteins do.

We have taken an overview of the synthesis and sorting of secretory and plasma-membrane proteins. Now let's focus on the initial steps in *biosynthesis*—the transport of these proteins across the ER membrane or their insertion into the ER membrane.

The Transport of Secretory and Membrane Proteins into or across the ER Membrane

Ribosomes that are synthesizing secretory and integral, ER, Golgi, and plasma-membrane proteins are tightly bound to the ER membrane; secretory proteins cross the ER membrane and membrane proteins are inserted into the ER membrane during synthesis. However, the initial stages of protein synthesis occur while the ribosome is in the cytosol, unbound to the membrane. A sequence of amino acids called the *signal sequence* on the newly made protein directs the ribosome to the ER membrane and causes the growing polypeptide to cross it (Figure 17-11).

How Polypeptides Cross the ER Membrane Is Controversial

Two forces bind a ribosome that is synthesizing a secretory or membrane protein to the ER membrane: an ion linkage involving Mg^{2+}, and the nascent polypeptide chain itself. The growing protein chain passes from the large ribosomal subunit into and through the ER membrane; it is never exposed to the cytosol and does not fold until it reaches the ER lumen. Thus the protein crosses the membrane in an unfolded state. Later, we shall see other examples of an important generality: proteins must be unfolded before they can insert themselves into or cross membranes. If newly made secretory proteins were allowed to fold in the cytosol, they would be unable to cross the ER membrane.

Precisely how the nascent chain traverses the ER membrane is not known. Some studies indicate that it passes through a protein-lined channel in the membrane (Figure 17-12a), in much the same way that small molecules pass through permease proteins. Such a transport channel may also participate in binding the ribosome to the ER membrane. Other work suggests that the nascent protein actually penetrates the phospholipid bilayer directly (Figure 17-12b). Perhaps the passageway is lined in part by a transport protein and in part by a lipid.

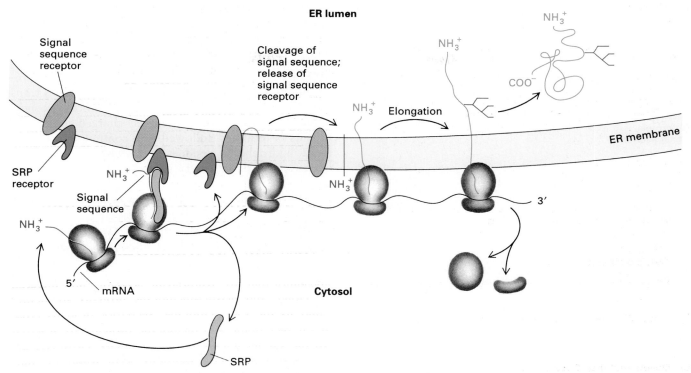

▲ Figure 17-11 A model for the synthesis of secretory proteins on the ER. The N-terminal signal sequence emerges from the ribosome only when the polypeptide is about 70 amino acids long, because about 30 amino acids remain buried in the ribosome. An elongated signal recognition particle (SRP) then binds to the signal sequence, and the SRP, nascent polypeptide, plus ribosome binds to the ER membrane through the SRP receptor. The signal sequence inserts itself into the ER membrane, bound to the signal sequence receptor, which may also form part of the channel through which the protein crosses the phospholipid bilayer. During this step, SRP dissociates from the ribosome-peptide complex and is released into the cytosol. The SRP receptor is also freed to initiate insertion of another secretory protein. The signal sequence is cleaved in the ER lumen by signal peptidase, which may cause the signal sequence receptor to be released from the signal sequence, and the signal sequence is degraded. Elongation of the peptide chain continues, and it extrudes into the ER lumen. Carbohydrates are added to asparagine residues by enzymes on the luminal surface. After synthesis is complete and the ribosomes are released, the remaining C-terminus of the secreted protein is transferred to the ER lumen, and the protein assumes its final conformation. [See P. Walter, R. Gilmore, and G. Blobel, 1984, *Cell* **38**:5–8; M. Wiedmann, T. Kurzhalia, E. Hartmann, and T. Rapoport, 1987, *Nature* **328**:830–833.]

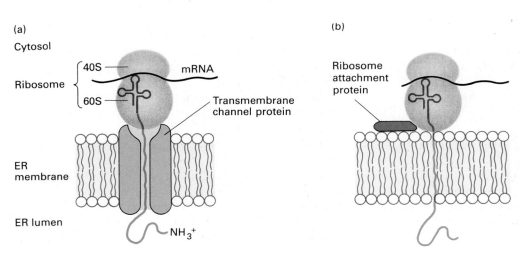

◄ Figure 17-12 Two possible models for the transfer of a nascent chain across the ER. (a) A hypothetical protein pore, including the signal sequence receptor, forms a transmembrane channel through which the nascent chain passes and also anchors the 60S ribosomal subunit to the membrane. (b) The nascent chain passes directly through the phospholipid bilayer, and the 60S ribosomal subunit is anchored by a hypothetical ribosome attachment protein. In either case, the nascent chain is generally transferred during synthesis (cotranslationally) across the membrane.

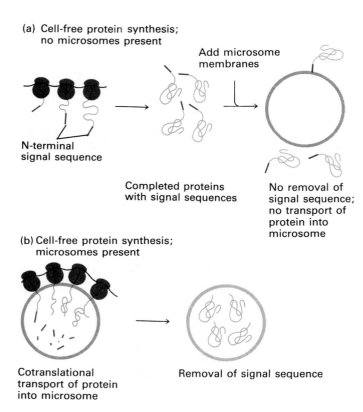

(a) Cell-free protein synthesis; no microsomes present

Add microsome membranes

N-terminal signal sequence

Completed proteins with signal sequences

No removal of signal sequence; no transport of protein into microsome

(b) Cell-free protein synthesis; microsomes present

Cotranslational transport of protein into microsome

Removal of signal sequence

▲ **Figure 17-13** The cotranslational insertion of secretory proteins into microsome vesicles. (a) A secretory protein synthesized in a cell-free system without microsomes retains the N-terminal signal sequence. If microsomes are subsequently added, the protein is not transported across the vesicle membrane and the signal sequence is not removed. (b) By contrast, if microsomes are present during protein synthesis, the signal sequence is removed from the nascent chain and the protein is transported to the lumen of the ER vesicles.

A Signal Sequence on Nascent Secretory Proteins Targets Them to the ER and Is Then Cleaved Off

Most secretory proteins contain a signal sequence of 16–30 amino acid residues (Table 17-3) that initiates transport across the ER membrane. Characteristically, a signal sequence has one or more positively charged amino acids near its N-terminus, followed by a continuous stretch of 6–12 hydrophobic residues; otherwise, the signal sequences of various secretory proteins have little homology. Signal sequences are not normally found on complete polypeptides made in cells, implying that the signal sequence is cleaved from the protein while it is still growing on the ribosome.

The enzyme *signal peptidase,* which cleaves off the signal sequence, is localized to the ER lumen. To detect the signal sequence, the mRNA must be translated in a cell-free system consisting of ribosomes, tRNAs, ATP and GTP, and cytosolic enzymes but with no ER membranes. In this case, the protein, with its attached signal sequence, is released into the cytosol (Figure 17-13a). Important information has been derived from the use of *microsome membranes* that have been stripped of their own ribosomes. If these membranes are present during the cell-free synthesis of a secretory protein, the protein is found in the ER lumen with the signal sequence removed (Figure 17-13b). If, however, these membranes are added to the reaction mixture after the secretory protein is completely synthesized, the protein generally is not incorporated into the ER lumen and its signal sequence remains (see Figure 17-13a).

For most secretory proteins to be inserted into the ER membrane, the microsome membranes must be added before the first 70 or so amino acids are polymerized. At this point, about 40 amino acids, including the cleaved

Table 17-3 Amino acid sequences of signal peptides of several secretory and membrane proteins

Secretory or membrane protein	Amino acid sequence (hydrophobic residues in red)
Proproalbumin	Met-Lys-Trp-Val-Thr-Phe-Leu-Leu-Leu-Leu-Phe-Ile-Ser-Gly-Ser-Ala-Phe-Ser ↓ Arg . . .
Pre-IgG light chain	Met-Asp-Met-Arg-Ala-Pro-Ala-Gln-Ile-Phe-Gly-Phe-Leu-Leu-Leu-Leu-Phe-Pro-Gly-Thr-Arg-Cys ↓ Asp . . .
Prelysozyme	Met-Arg-Ser-Leu-Leu-Ile-Leu-Val-Leu-Cys-Phe-Leu-Pro-Leu-Ala-Ala-Leu-Gly ↓ Lys . . .
Preprolactin	Met-Asn-Ser-Gln-Val-Ser-Ala-Arg-Lys-Ala-Gly-Thr-Leu-Leu-Leu-Leu-Met-Met-Ser-Asn-Leu ↓ Leu . . .
Prepenicillinase (*E. coli*)	Met-Ser-Ile-Gln-His-Phe-Arg-Val-Ala-Leu-Ile-Pro-Phe-Phe-Ala-Phe-Cys-Leu-Pro-Val-Phe-Ala ↓ His . . .
Prevesicular stomatitis virus (VSV) glycoprotein	Met-Lys-Cys-Leu-Leu-Tyr-Leu-Ala-Phe-Leu-Phe-Ile-His-Val-Asn-Cys ↓ Lys . . .
Prelipoprotein (*E. coli*)	Met-Lys-Ala-Thr-Lys-Leu-Val-Leu-Gly-Ala-Val-Ile-Leu-Gly-Ser-Thr-Leu-Leu-Ala-Gly ↓ Cys . . .

SOURCE: D. P. Leader, 1979, *TIBS* 4:205.

signal sequence, protrude from the ribosomes, and about 30 amino acids are buried in a channel or tunnel in the ribosome. Thus transport of most secretory proteins to the ER membrane must occur during translation.

Recent experiments suggest that the cleaved signal sequence is sufficient to direct a protein to the ER. An artificial gene produced by recombinant DNA techniques can encode a protein in which the signal sequence of a secretory protein is attached to the N-terminus of a protein that normally is not secreted, such as globin. The globin is transported to the ER lumen with the signal peptide, which is cleaved off there exactly as it would be from a "normal" secretory protein. All signal sequences on secretory proteins have a "core" of hydrophobic amino acids (see Table 17-3) that are essential to signal-sequence function; the specific deletion of any of these amino acids from the signal sequence or its mutation to a charged amino acid abolishes the ability of the protein to cross the ER membrane into the lumen. Other experiments indicate that any random N-terminal amino acid sequence, provided it is sufficiently long and hydrophobic, will cause the protein to be translocated to the ER lumen. These hydrophobic residues probably form a binding site for the *signal recognition particle.*

Not all secretory proteins have a cleaved signal sequence. Ovalbumin, the major protein synthesized and secreted by the hen oviduct, has no cleaved signal sequence. Experiments similar to those just described show that the sequence required for membrane transport of ovalbumin is localized to the 100 N-terminal residues.

Several Receptor Proteins Mediate the Interaction of Signal Sequences with the ER Membrane

Since secretory proteins are synthesized in association with the ER membrane but not with any other cellular membrane, some signal-sequence recognition system must target them there. The identification of the key element in this process resulted from a simple experiment. A preparation of rough ER, free of ribosomes, was exposed to a solution of 0.5 *M* NaCl, so that several proteins were removed from the membranes. When these "stripped" microsome vesicles were recovered by centrifugation and added to a cell-free protein synthesis reaction, they were unable to support the insertion of nascent secretory proteins. However, when the proteins removed by the NaCl treatment were returned to the preparation, secretory proteins were inserted into the vesicles. A single active component—the *signal recognition particle* (SRP)—was purified from the mixture of stripped proteins. SRP contains six discrete polypeptides and a 300-nucleotide RNA and is an essential component of protein translocation across the ER membrane (Figure 17-14; see Figure 17-11).

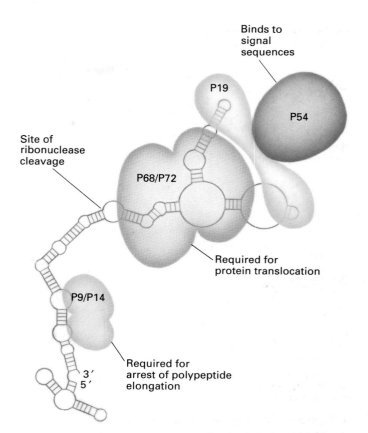

▲ **Figure 17-14** SRP consists of one 300-nucleotide RNA and six proteins: P9, P14, P19, P54, P68, and P72. All proteins except P54 bind to the RNA in the positions indicated; P54 binds to P19. The precise binding site of P9 and P14 is not known. Different functions have been assigned to different polypeptides. *After V. Siegel and P. Walter, 1988,* Proc. Nat'l Acad. Sci. USA *85:1801–1805.*

Some functions of SRP have been clarified by studies in which mRNAs for secretory proteins, such as the pituitary hormone *preprolactin,* are translated in a cell-free protein synthesis system (Figure 17-15). When cell-free translation is carried out in the presence of SRP without microsomes, protein synthesis is arrested at about 70 amino acids. When microsome membranes are added, the elongation of prolactin chains resumes, the ribosomes bind to the membranes, and the full-length protein is sequestered in the lumen of the microsomes. Thus SRP prevents the synthesis of a complete secretory protein in the absence of sufficient rough ER membranes.

Many functions of SRP can be assigned to specific proteins. For instance, in the SRP-ribosome complex in which protein elongation is blocked (see Figure 17-15b), the 54,000-MW SRP protein (P54) can be chemically cross-linked to the signal sequence; thus this polypeptide is believed to bind to the hydrophobic core of a signal sequence. SRP can be cleaved by a ribonuclease into a smaller fragment containing part of the RNA bound to

(a) No SRP, no SRP receptor, no microsomes

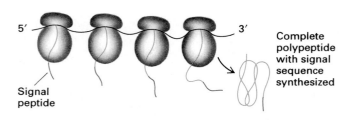

Complete polypeptide with signal sequence synthesized

Signal peptide

(b) Plus SRP, no SRP receptor, no microsomes

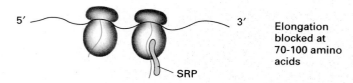

Elongation blocked at 70-100 amino acids

SRP

(c) Plus SRP, plus soluble SRP receptor, no microsomes

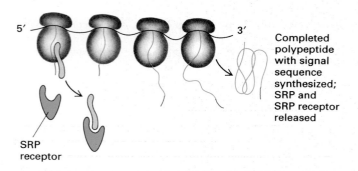

Completed polypeptide with signal sequence synthesized; SRP and SRP receptor released

SRP receptor

(d) Plus SRP, plus microsomes with SRP receptor

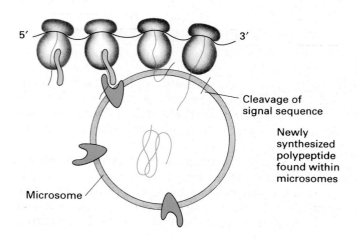

Cleavage of signal sequence

Newly synthesized polypeptide found within microsomes

Microsome

◀ **Figure 17-15** The properties of SRP and the SRP receptor studied in cell-free protein synthesis reactions. Messenger RNA encoding preprolactin, a typical secretory protein made by the pituitary, is readily translated in a wheat-germ cell-free system in the absence of any microsomes. This system contains all the necessary factors for in vitro protein synthesis. In one series of experiments, SRP, SRP receptor, and microsomes were added to the system and the effects on protein synthesis were examined. (a) In the absence of SRP, the SRP receptor, or microsomes, the complete protein with its signal sequence is synthesized. (b) The addition of only SRP causes elongation to be blocked at 70–100 amino acids—direct evidence that SRP binds to the N-terminal residues on the growing chain. At this stage, the 54,000-MW SRP protein (see Figure 17-14) can be chemically cross-linked to the signal sequence. (c) The addition of a soluble SRP receptor removes this block in polypeptide chain elongation. The SRP receptor binds to SRP and uses the energy from GTP hydrolysis to release SRP from the growing chain. (d) If SRP and microsomes containing the SRP receptor are added, the newly made protein is translocated across the vesicle membrane. [See P. Walter and G. Blobel, 1981, *J. Cell Biol.* 91:557; V. C. Krieg, P. Walter, and A. E. Johnson, 1986, *Proc. Nat'l Acad. Sci. USA* 83:8604–8608.]

P9 and P14 and a larger fragment containing the other four proteins (see Figure 17-14). The larger complex cannot block the elongation of growing polypeptides but is completely normal in its ability to bind to signal sequences and to catalyze the insertion of nascent secretory proteins into the rough ER. This result establishes that the arrest of polypeptide elongation by SRP is not an essential aspect of polypeptide translocation to ER membranes and that elongation arrest is dependent on the SRP components P9 and P14.

An *SRP receptor* in the rough ER membrane can anchor the ribosome and nascent chain to the membrane (see Figure 17-11). This receptor contains two polypeptide subunits: integral proteins of 638 and about 300 amino acids, respectively. Treatment of ER membranes with tiny amounts of protease cleaves the larger protein subunit very near its N-terminus and releases a soluble SRP receptor complex, which relieves the block in the elongation of nascent secretory proteins imposed by SRP in cell-free systems (see Figure 17-15c). The SRP receptor thus must bind to the SRP and possibly also to the ribosome. The larger subunit of the SRP receptor binds GTP; the receptor uses the energy released by GTP hydrolysis to cause SRP to dissociate from the ribosome–nascent chain complex.

The SRP and its receptor only initiate the transfer of the nascent chain across the ER membrane. They then dissociate from the ribosome and nascent chain and recycle to direct the insertion of additional proteins. Another integral ER membrane protein, the *signal sequence receptor,* binds the signal sequence after its release and is believed to facilitate its insertion into the ER membrane (see Figure 17-11). GTP hydrolysis is also required to insert the signal sequence into the membrane. One of the two polypeptide subunits of the signal sequence receptor can be chemically cross-linked to the nascent polypeptide as it is traversing the ER membrane. Thus the signal sequence receptor may also form part of the membrane channel through which the growing polypeptide moves. Evidence is strong that the N-terminal signal sequence, SRP, the SRP receptor, and the signal sequence receptor are all essential to initiate the cotranslational transport of a secretory protein across the ER membrane.

Some Secretory Proteins Can Cross the ER Membrane After Synthesis Is Complete

Most secretory proteins must be transported across the ER membrane while the polypeptide chain is growing on the ribosome. A very few proteins can cross the ER membrane after synthesis is complete, and a study of these proteins has illuminated several aspects of this transport process. One such protein is the yeast *pre-pro α factor*, a precursor of the α mating factor. The post-translational transport of the pre-pro α factor (Figure 17-16) requires ATP hydrolysis and one or more cytosolic *unfolding proteins*. At least one of these proteins is known to bind to the pre-pro α factor and somehow uses the energy released by ATP hydrolysis to keep this protein unfolded or partly folded, so it can interact with ER membranes. The sequence of one 70,000-MW unfolding protein is very similar to that of a 70,000-MW *heat-shock protein*, which binds to the unfolded regions of thermally denatured proteins and prevents their precipitation in the cell; this resembles the function of the unfolding protein in binding the pre-pro α factor. The important points are that only unfolded or partly folded pre-pro α factor can cross the ER membrane and that energy is expended to keep it unfolded in the cytosol. Energy released by the spontaneous folding of the protein in the ER lumen may help to "pull" it across the membrane.

Topogenic Sequences Allow Integral Proteins to Achieve Their Proper Orientation in the ER Membrane

In Chapter 13, we were introduced to several of the vast array of integral proteins that occur in the plasma membrane and other cellular membranes. One large class of these proteins, the plasma membrane glycoproteins, have been studied in some detail. A single membrane-spanning segment (a sequence of 20–25 hydrophobic amino acids) in some of these proteins is believed to form a transmembrane α helix that anchors the protein in the phospholipid bilayer (see Figure 17-2). In one class, the hydrophilic N-terminal segment is located on the exoplasmic face and

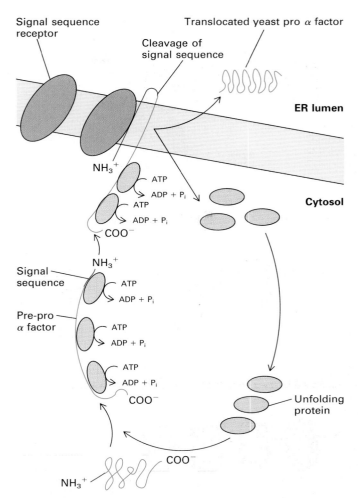

▲ **Figure 17-16** Model of the post-synthetic translocation of the yeast pre-pro α factor across the ER membrane. In the cytosol, the complete pre-pro α factor binds to one or more copies of an unfolding protein that somehow uses energy released by ATP hydrolysis to keep the factor in an unfolded or partly folded state. When the complex is added to ER membrane, the unfolding proteins are released, the pre-pro α factor is transported across the membrane, and the factor's signal peptide is cleaved. *After R. P. Deshaies, B. Koch, and R. Schekman, 1988, Trends Biochem. Sci.* **13**:*384–388; H. Bernstein, T. A. Rapoport, and P. Walter. 1989, Cell* **58**:*1017.*

the hydrophilic C-terminal segment is on the cytoplasmic face. Other plasma-membrane glycoproteins have the reverse orientation. Still others, such as the glucose transport protein, have multiple membrane-spanning segments (see Figure 17-2).

All these glycoproteins are synthesized on the ER and move to the plasma membrane via the Golgi complex along the same pathway followed by continuously secreted proteins. This brings us to an important question: how are these proteins inserted into the ER membrane

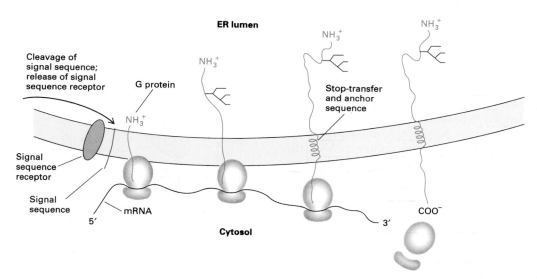

▲ **Figure 17-17** Model of the synthesis and membrane insertion of the vesicular stomatitis viral (VSV) glycoprotein (protein G). Until the point at which the C-terminus of the protein is synthesized, this model is the same as the one for secretory proteins (see Figure 17-11). The hydrophobic stop-transfer membrane-anchor sequence of about 22 amino acids prevents the nascent chain from extruding further into the ER membrane. The ribosome is released from the membrane and completes synthesis of the protein as a membrane-unbound ribosome in the cytosol. When synthesis is complete and the ribosomes are released, the hydrophobic stop-transfer membrane-anchor sequence anchors the G protein to the membrane. Other integral membrane proteins with similar structures, such as the insulin receptor, are inserted in a similar manner. *After W. Wickner and H. F. Lodish, 1985, Science 230:400–407.*

with their proper orientations? *Topogenic sequences* in these proteins function in different ways to determine the structure of the protein in the membrane.

Relatively simple integral membrane proteins span the membrane once, with the C-terminus facing the cytosol, and contain a single *stop-transfer membrane-anchor sequence*. Of these proteins, the biosynthesis of the *G protein*, which forms the surface spikes of the *vesicular stomatitis virus* (VSV), has been studied in the most detail. Other proteins with similar structures, such as glycophorin (see Figure 13-16) and the LDL receptor (see Figure 14-36), appear to follow precisely the same pathway. Immediately after its synthesis, the G protein spans the ER membrane by means of its anchor sequence; thus it is a transmembrane protein (Figure 17-17). About 30 amino acids at the extreme C-terminus remain exposed to the cytosol. The balance of the polypeptide, including the N-terminus and the two asparagine-linked carbohydrate chains, is found on the luminal side of the ER membrane (exoplasmic orientation); experimentally, this part of the protein is protected from protease digestion by the permeability barrier of the ER membrane. This orientation of the G protein with respect to the ER membrane is maintained when the G protein appears on the surface of infected cells and in virions (see Figure 17-2). Thus the overall membrane topology of the G protein, like that of all integral membrane proteins, is preserved as it is transported from the ER to the cell surface.

Much like secretory proteins, nascent G protein extrudes N-terminus first across the ER membrane, and the signal sequence of 16 amino acids (see Table 17-3) is cleaved while the chain is still growing. Unlike secretory proteins, however, the G protein remains anchored to the membrane near its C-terminus. A sequence of 23 hydrophobic amino acids spans the membrane and blocks the extrusion of the C-terminus of the nascent chain across the ER membrane.

Support for such a model has come from studies in which cloned cDNAs encoding the G protein are expressed in tissue-culture cells. Cloned, intact cDNA in an appropriate expression vector directs the synthesis of full-length G protein, which normally is transported to the plasma membrane. Mutant G genes have been constructed that do not contain most or all of the segment of DNA that encodes the membrane-spanning region of the G protein. In these cases, the N-terminal fragment is secreted from the cell, implying that the hydrophobic segment near the C-terminus is essential for anchoring the protein to the membrane.

Thus two types of topogenic sequences function to orient the G protein (and proteins of similar structure) in the ER membrane: a stop-transfer membrane-anchor sequence in the middle of the protein and an N-terminal cleaved signal sequence. Other proteins also span the ER membrane once, but with the N-terminus on the cytoplasmic face and the C-terminus on the exoplasmic face.

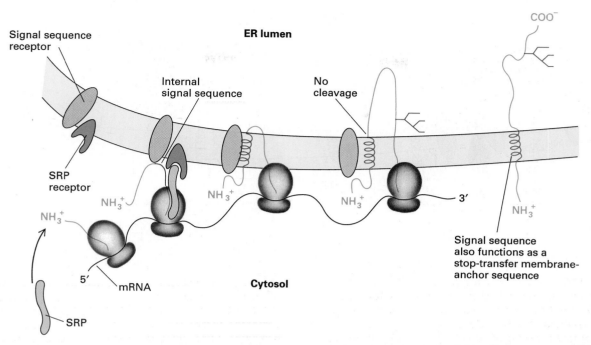

▲ **Figure 17-18** Model of the synthesis and insertion of the asialoglycoprotein receptor polypeptide. An internal signal sequence (a transmembrane α helix) functions exactly like the signal sequence at the N-terminus of a secretory protein, directing the insertion of the growing chain into the ER membrane. The signal is uncleaved and sufficiently long and hydrophobic to anchor the growing chain to the membrane. The C-terminus extrudes into the ER membrane, as in the case of a secretory protein. Thus the same topogenic sequence functions as an internal, uncleaved signal and as a membrane anchor; it causes the protein to assume an orientation with its N-terminus facing the cytosol and its C-terminus facing the ER lumen. *After M. Spiess and H. F. Lodish, 1986, Cell **44**:177–185; A. S. Shaw, P. Rottier, and J. K. Rose, 1988, Proc. Nat'l Acad. Sci. USA **85**:7592–7596.*

Such proteins are also anchored to the membrane by a hydrophobic sequence of about 22 amino acids in the middle of the protein. The asialoglycoprotein receptor is perhaps the most intensively studied member of this group, but the biosynthesis of the others (see Figure 17-2) is similar. These proteins do not utilize a cleaved N-terminal signal sequence; rather, the same hydrophobic stop-transfer membrane-anchor sequence also functions as a signal sequence (Figure 17-18). Because this signal is in the middle of the protein, the N-terminus remains on the cytoplasmic face. Like secretory proteins, the C-terminus of the growing chain extrudes into the ER lumen.

The final class of proteins we will consider are those that span the membrane with multiple α helices; the glucose transporter, for example, has 12 transmembrane α helices. Each α helix is thought to be a topogenic sequence (Figure 17-19). The first helical segment is an internal, uncleaved signal-anchor sequence that initiates the insertion of the growing chain. The second helix is a stop-transfer membrane-anchor sequence. Thus the first two transmembrane α helices insert into the membrane as a hairpin; as before (see Figure 17-18), the SRP and the SRP receptor are involved in this step. The nascent protein chain continues to grow into the cytosol. The third α helix is another internal, uncleaved signal-anchor sequence, and the fourth is a stop-transfer membrane-anchor sequence. Helices 3 and 4 insert into the membrane as a hairpin, just as helices 1 and 2 do, except that SRP and the SRP receptor are not required. All transmembrane α helices insert into the ER bilayer in this way. The number of topogenic sequences is equal to the number of transmembrane α helices.

Posttranslational Modifications of Secretory and Membrane Proteins in the Rough ER

Once synthesis is complete, newly introduced polypeptides in the membrane and lumen of the ER must be matured, sorted, and transported. Many secretory or membrane proteins undergo four principal modifications as they mature to the cell surface: (1) formation of disulfide bonds; (2) proper folding of the protein, including forma-

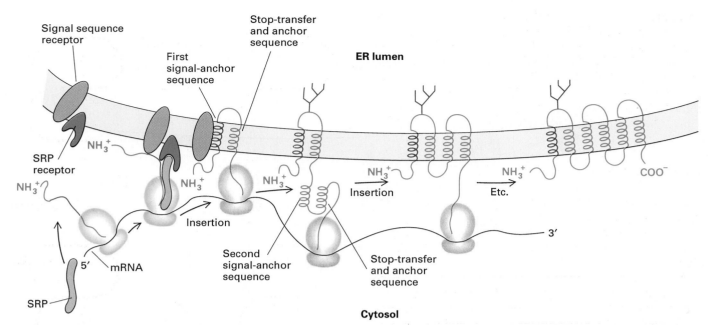

▲ **Figure 17-19** Model of the transmembrane insertion of a nascent glucose transporter. This protein, like many integral proteins, is believed to traverse the membrane as a series of α-helical loops. An internal, uncleaved signal-anchor sequence is known to direct the binding of nascent chains to the rough ER membrane and to initiate cotranslational insertion. In this model, the internal, uncleaved signal-anchor sequence and the sequence of nascent chain following it, which contains a stop-transfer membrane-anchor sequence, insert into the ER membrane as an α-helical hairpin; both SRP and the SRP receptor are involved in this step. The nascent chain continues to grow in the cytosol. Subsequent α-helical hairpins could insert similarly, although SRP is required only for the insertion of the first signal-anchor sequence. Although only six transmembrane α helices are depicted here, this transporter and proteins of similar structure have 12 or more. [See W. Wickner and H. F. Lodish, 1985, *Science* 230:400–407; H. P. Wessels and M. Spiess, 1988, *Cell* 55:61–70.]

tion of multichain proteins; (3) specific proteolytic cleavages; (4) addition and modification of carbohydrates. Each modification takes place in a specific organelle through which these proteins pass. These modifications help the protein achieve its functional form and, in several cases, direct the protein to its ultimate destination in the cell. We begin by discussing two modifications generally localized to the rough ER—the formation of disulfide bonds, and the formation of multichain proteins.

Disulfide Bonds Are Formed during or soon after Synthesis

Disulfide bonding between two cysteine residues is one of the most important stabilizing forces in the tertiary structure of proteins. In proteins that contain more than two cysteine residues, formation of the proper arrangement of disulfide bonds (Cys—S—S—Cys) is essential for normal structure and enzymatic or hormone activity. Disulfide bonds are generally confined to secretory proteins and certain membrane proteins. Cytoplasmic proteins in bacterial and eukaryotic cells generally do not utilize the relatively more oxidized disulfide bond as a stabilizing force, possibly because of a greater reducing potential in the cytosol.

In eukaryotic cells, formation of all disulfide bonds occurs in the lumen of the rough ER. Many of these bonds are formed while the polypeptide is still growing on the ribosome. In the case of secreted immunoglobulin (Ig) light-chain polypeptides, disulfide bonds form sequentially: the first cysteine pairs with the second; the third, with the fourth. Disulfide bonds stabilize the two separate domains in this protein; if the proper bonds do not form, the protein cannot achieve a functional conformation.

Sequential formation does not occur in all disulfide-bonded proteins. For example, proinsulin (see Figure 2-13) and bovine pancreatic trypsin inhibitor each have three disulfide bonds that are not sequential. In these cases, the first disulfide bonds that form spontaneously by oxidation of the Cys—SH groups are not characteristic of the proper folded conformation of the protein. Disulfide bonds often break and reform, eventually forming the proper S—S bonds and stabilizing the proper protein conformation. Indeed, in cell-free extracts, reduced unfolded proteins can often be oxidized to form the native folded protein with the "correct" S—S bonds (see Figure 2-15), but this spontaneous process is very slow, requiring minutes to hours, for proteins with many S—S bonds. In the ER, this folding process is accelerated by the enzyme *protein disulfide isomerase* (PDI), which is found in abundance in the ER of secretory tissues in such organs as the liver and pancreas (Figure 17-20). PDI acts by catalyzing the rearrangement of disulfide bonds on a broad range of protein substrates, allowing them to generate their thermodynamically most stable configurations.

Any protein developed for therapeutic use in humans or animals must be properly folded and include the correct disulfide bonds. When mammalian secretory proteins are synthesized by recombinant DNA techniques in bacterial cells, these proteins generally are not secreted and accumulate in the cytosol. There, such proteins often denature and precipitate, in part because S—S bonds do not form. Thus bacterial cells are not suitable hosts for the synthesis of usable amounts of certain mammalian proteins, such as blood proteins and hormones normally stabilized by disulfide bonds. Indeed, biotechnologists must use cultured animal cells to produce such proteins as monoclonal antibodies and tissue plasminogen activator, an anticlotting agent, in quantity.

Formation of Oligomeric Proteins Occurs in the ER

Many important secretory and membrane proteins are *oligomers:* they are built of two or more polypeptides. Examples include the immunoglobulins, which contain two H and two L chains, all linked by S—S bonds (see Figure 2-27), and the influenza HA protein (see Figure 14-48), a trimer of three identical polypeptides not covalently bonded together. In all cases, these *oligomeric proteins* are formed from constituent polypeptides in the ER; in fact, polypeptides that are not properly folded are prevented from exiting the ER (see the following section).

The assembly of the HA protein has been studied intensively (Figure 17-21). On average, a newly made HA polypeptide requires approximately 7 min to fold into a trimer. Several approaches have been used to follow the monomer-to-trimer conversion in influenza-infected cells. Certain monoclonal antibodies, for instance, bind only to a region on the surface of the correctly folded trimer and thus immunoprecipitate trimers, not monomers. Others bind only to a region of the HA polypeptide that is exposed on the monomer surface but folded inside the rigid trimer.

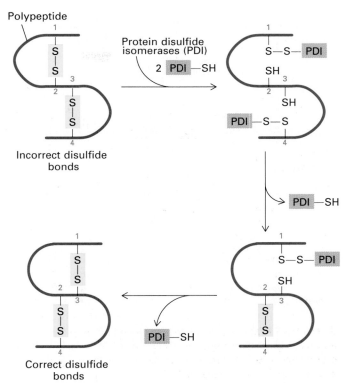

▲ **Figure 17-20** Protein disulfide isomerase (PDI) catalyzes the breakage and reformation of disulfide bonds and, in so doing, accelerates the refolding of proteins containing multiple disulfide bonds. In the oxidizing environment of the ER lumen, disulfide bonds form spontaneously in newly made secretory proteins but are often incorrect. PDI contains an active-site cysteine residue with a free reduced sulfhydryl (SH) group, which reacts with disulfide (S—S) bonds on newly made proteins to form an S—S bond between PDI and the protein. This bond, in turn, can react with a free SH on the protein to form a new S—S bond. In this way, the disulfide bonds on a protein can rearrange themselves until the most stable configuration for the protein is achieved. *After R. B. Freedman, 1984,* Trends Biochem. Sci. *9:438–441; R. Roth and S. Pierce, 1987,* Biochemistry *26:4179–4182.*

Only Properly Folded Proteins Are Transported from the Rough ER to the Golgi Complex

After synthesis, both secretory and membrane proteins are packaged into small transport vesicles that carry them to the Golgi vesicles, but only properly folded proteins are generally allowed to enter the vesicles. Incompletely folded proteins do not reach the plasma membrane or are not secreted. Some of these polypeptides accumulate in the ER; others are degraded in the ER. Two examples illustrate this selective transport.

Mutations in the HA protein that prevent it from folding properly into a trimer (see Figure 17-21) also prevent the polypeptide from moving out of the ER. Many such mutations have been generated in vitro in the cloned HA

ER lumen

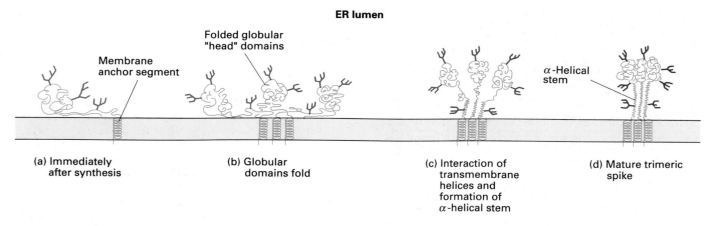

(a) Immediately after synthesis

(b) Globular domains fold

(c) Interaction of transmembrane helices and formation of α-helical stem

(d) Mature trimeric spike

Cytosol

▲ **Figure 17-21** The folding of the hemagglutinin (HA) precursor polypeptide and the formation of a trimer; only after the trimer is formed is the HA cleaved proteolytically to HA$_1$ and HA$_2$. (a) Immediately after its synthesis, the bulk of the HA polypeptide is on the luminal side of the ER, anchored by a hydrophobic α helix near its C-terminus. (b) First, the globular domains that will form the head of the spike fold, stabilized by disulfide bonds. (c) Then the three chains interact with each other, initially via their transmembrane α helices; this activity apparently triggers the formation of the fibrous α-helical stem on the luminal part of each polypeptide. (d) Finally, interactions between the three globular heads occur, generating the mature trimeric spike; the molecular structure of the trimer after cleavage to HA$_1$ and HA$_2$ is depicted in Figure 14-48. *After M-J. Gething, K. Mc-Cammon, and J. Sambrook, 1986, Cell **46**:939–950.*

gene, including small deletions or alterations of the membrane-spanning segment and changes in the large, exoplasmic-facing N-terminus. Others are temperature-sensitive mutations in which the HA protein folds properly into a trimer at the lower temperature but does not fold at all at the higher temperature. In all cases, the improperly folded HA polypeptides remain in the rough ER.

A second example is a mutation in the secretory protein *α$_1$-antiprotease*, which is secreted by hepatocytes and macrophages; this protein binds to and inhibits the blood protease *elastase*. In the absence of α$_1$-antiprotease, elastase degrades the fine tissue in the lung that participates in the absorption of oxygen. A genetic inability to produce α$_1$-antiprotease, widespread in Caucasians, is the major genetic cause of emphysema (destruction of lung tissue by unchecked elastase) and difficulty in breathing. The defect is due to a single mutation in α$_1$-antiprotease in which glutamate$_{342}$ is replaced by lysine. As a result, the protein is produced normally in the rough ER, but remains in that organelle and is not secreted. The hepatocytes become full of rough ER that contains abnormally folded α$_1$-antiprotease, and secretion of other important liver proteins also becomes impaired.

The *heavy-chain binding protein* (BIP) is one rough ER protein that binds certain abnormally folded proteins. BIP binds to abnormally folded Ig heavy chains in antibody-producing cells that do not make functional Ig light chains (hence its name). BIP helps to retain misfolded proteins in the ER; in fact, cells respond to the increased synthesis of an abnormal secretory protein by increasing synthesis of BIP.

ER-Specific Proteins Are Selectively Retained in the Rough ER

Many structural proteins or enzymes that are synthesized by membrane-attached ribosomes remain and function in the rough ER. Specific sequences in these *resident ER proteins* cause them to be retained in the ER. Two examples of these proteins are PDI (see Figure 17-20) and BIP. As in several other ER luminal proteins, the C-terminal sequence in both the PDI and BIP proteins is Lys-Asp-Glu-Leu (KDEL in one-letter code). When a mutant PDI protein lacking these four residues is synthesized in a fibroblast, the protein is entirely secreted. If a derivative of a normally secreted protein is created that contains these four amino acids in its C-terminus, the protein is retained in the ER. The KDEL sequence binds to a receptor protein that causes it to be retained in the ER. The KDEL receptor has been purified: some is localized to ER membranes; some is found in small transport vesicles that may return KDEL-containing proteins that have escaped from the ER back to it.

The transport of newly made proteins from the rough ER to the Golgi vesicles is a highly selective and regulated process. Certain resident ER enzymes and misfolded proteins are retained in the ER. Many other proteins are normally transported to the Golgi vesicles. The selective entry of proteins into membrane-transport vesicles is an important feature of protein targeting—one we will encounter several times in our study of subsequent stages of the maturation of secretory and membrane proteins.

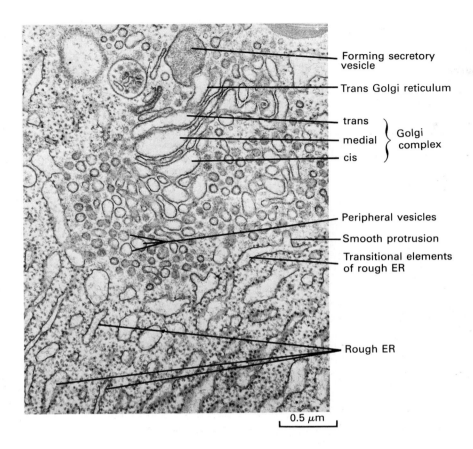

Forming secretory
vesicle

Trans Golgi reticulum

trans
medial } Golgi
complex
cis

Peripheral vesicles

Smooth protrusion

Transitional elements
of rough ER

Rough ER

0.5 μm

◄ **Figure 17-22** Electron micrograph of
the Golgi complex and ER in an exocrine
pancreatic cell. The stacked vesicles of the
Golgi complex and a forming secretory
vesicle are evident. The rough ER contains
transitional elements from which smooth
protrusions appear to be budding. These
buds are believed to form the vesicles that
transport secretory and membrane proteins
from the rough ER to the Golgi complex.
Other vesicles seen at the periphery of the
Golgi complex may transport proteins
from one Golgi vesicle to another or from
the trans Golgi reticulum to the forming
secretory vesicles. *Courtesy of G. Palade.*

Golgi Vesicles: Sorting and Glycosylation of Secretory and Membrane Proteins

Following their synthesis in the rough ER, most proteins
(except resident ER proteins) move in small transport vesicles to the Golgi complex—an organelle composed of
both flattened and spherical vesicles (Figure 17-22) that
serves as a liaison between the ER and both the plasma
membrane and internal organelles, such as lysosomes.
The Golgi complex contains three functional regions: the
elongated vesicles nearest the ER make up the cis face of
the Golgi; those in the midportion, the medial face; those
nearest the periphery of the cell, the trans face and the
trans Golgi reticulum (see Figure 17-8). Secretory and
membrane proteins travel from the ER to the Golgi vesicles and from the Golgi vesicles to other organelles. This
movement takes place in small peripheral vesicles that
pinch off from one vesicle and fuse with the membranes
of another vesicle (see Figures 17-8 and 17-22). Although
biologists generally refer to the Golgi complex as one organelle, different enzymes in its three regions introduce
different modifications to secretory and membrane pro-

teins. Lysosome proteins also move the rough ER through
the cis and trans Golgi and are subsequently sorted to
lysosomes.

Glycosylation is the principal chemical modification to
proteins as they pass through the Golgi vesicles. Many
important cell-surface and secretory proteins contain one
or more carbohydrate groups. Glycosylation reactions
are important both because they provide markers for following the protein movement from the cis to the medial
to the trans Golgi and, as we shall see, because some carbohydrate residues play an important role in sorting proteins to the correct organelles in a cell.

N-Linked and O-Linked Oligosaccharides Have Very Different Structures

Although glycoproteins occur in all eukaryotic cells that
have been studied, eubacteria contain few, if any. Almost
all protein- and lipid-linked sugars are localized to the
exoplasmic face of cellular membranes. In eukaryotes,
sugar residues are commonly linked to four different
amino acid residues, classified as *O-linked* (to the hy-
droxyl-group oxygen of serine, threonine, and, in collagen, hydroxylysine) and *N-linked* (to the amide nitrogen of asparagine). The structure of N- and O-linked

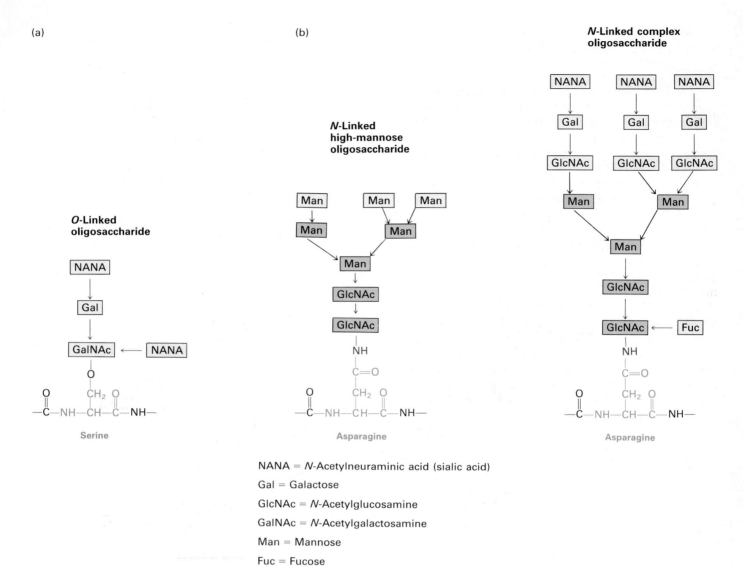

▲ **Figure 17-23** Structure of typical *N*-linked and *O*-linked oligosaccharides. (a) Structure of the *O*-linked oligosaccharide, linked to serine and threonine hydroxyl groups, in proteins such as glycophorin and the LDL receptor. As shown here, one negatively charged *N*-acetylneuraminic acid (sialic acid) is attached to the galactose and one is attached to the *N*-acetylgalactosamine, although only one of these is present in some cases. (b) Structure of typical "high-mannose" and "complex" *N*-linked oligosaccharides, linked to asparagine residues in a variety of mammalian serum glycoproteins, such as immunoglobulins. The five residues always found in *N*-linked oligosaccharides are in green. High-mannose oligosaccharides can have as few as three mannose residues and as many as 60 in protozoans and yeast.

oligosaccharides are very different, and different sugar residues are usually found in each type (Figure 17-23). For instance, in *O*-linked sugars, galactose or *N*-acetylgalactosamine is linked to serine or threonine; in all *N*-linked oligosaccharides, *N*-acetylglucosamine is linked to asparagine. *O*-linked oligosaccharides are generally short, often containing only one to four sugar residues (see Figures 2-55 and 17-23). However, some *O*-linked oligosaccharides, such as those bearing the ABO blood-group antigens, can be very long (see Figure 2-58). The

N-linked oligosaccharides, in contrast, have a minimum of five sugars and always contain mannose as well as *N*-acetylglucosamine (see Figure 17-23).

Most cytosolic and nuclear proteins are not glycosylated. Exceptions include a protein localized to the nuclear-pore complex, which we will encounter later, and several transcription factors. In these cases, a single *N*-acetylglucosamine residue is linked to the serine or threonine hydroxyl group. Here, we discuss the biosynthesis of sugars on the exoplasmic membrane face.

O-Linked Sugars Are Synthesized in the ER or Golgi Vesicles from Nucleotide Sugars

The incorporation of sugars into such polymers as glycoproteins requires an input of energy. The high-energy intermediates used in the biosynthesis of oligosaccharides—nucleoside diphosphate or monophosphate sugars (Figures 17-24 and 17-25)—are brought to the Golgi lumen by special transport antiports.

During the formation of O-linked glycoproteins, sugars are added one at a time and each sugar transfer is catalyzed by a different type of *glycosyl transferase* (Figure 17-26). In the biogenesis of the disaccharide side chains of collagen (glucose-galactose-hydroxylysine), for instance, a galactose residue is transferred from UDP-galactose to a hydroxylysine side chain of the polypeptide. A second enzyme then catalyzes the addition of a glucose residue to the galactose. All known glycosyl transferase enzymes are integral membrane proteins with active sites facing the ER or Golgi lumen. Most O-linked sugars are added to secretory and plasma-membrane proteins in the Golgi vesicles only a few minutes before reaching the cell surface. The addition of N-acetylneuraminic acid—the last step in the biosynthesis of O-linked (and N-linked) sugars—occurs in the trans Golgi and the trans Golgi reticulum (see Figures 17-8 and 17-22).

Similarly, saccharide residues in glycolipids are added to lipids in the Golgi vesicles only a few minutes before the glycolipids appear on the plasma membrane. These reactions also utilize glycosyl transferases with active sites that face the lumen. Since several oligosaccharides, such as the ABO antigens, are found attached to both glycoproteins and glycolipids, the same glycosyl transferase may be utilized to synthesize both.

The Golgi Membrane Contains Permeases for Nucleotide Sugars

The addition of such sugars as N-acetylneuraminic acid and galactose to proteins occurs in the Golgi lumen. The nucleoside sugar substrates, however, are fabricated in the cytosol from nucleoside triphosphates and sugar phosphates (see Figure 17-25). The Golgi membrane contains specific transporters for sugar nucleotides, such as CMP–N-acetylneuraminic acid and UDP-galactose (Figure 17-27). These permeases are antiports. Nucleotides such as UDP and CMP, which are formed by the glycosyl transferase reaction, must be exported from the Golgi vesicles. First, UDP is hydrolyzed to UMP and P_i, and UMP is transported out of the Golgi vesicles in a one-for-one exchange for UDP-galactose, UDP–N-acetylglucosamine, or UDP–N-acetylgalactosamine. Other antiports catalyze a one-for-one exchange of CMP–N-acetylneuraminic acid for CMP and of GDP-mannose for GMP. In catalyzing such exchanges, these antiports

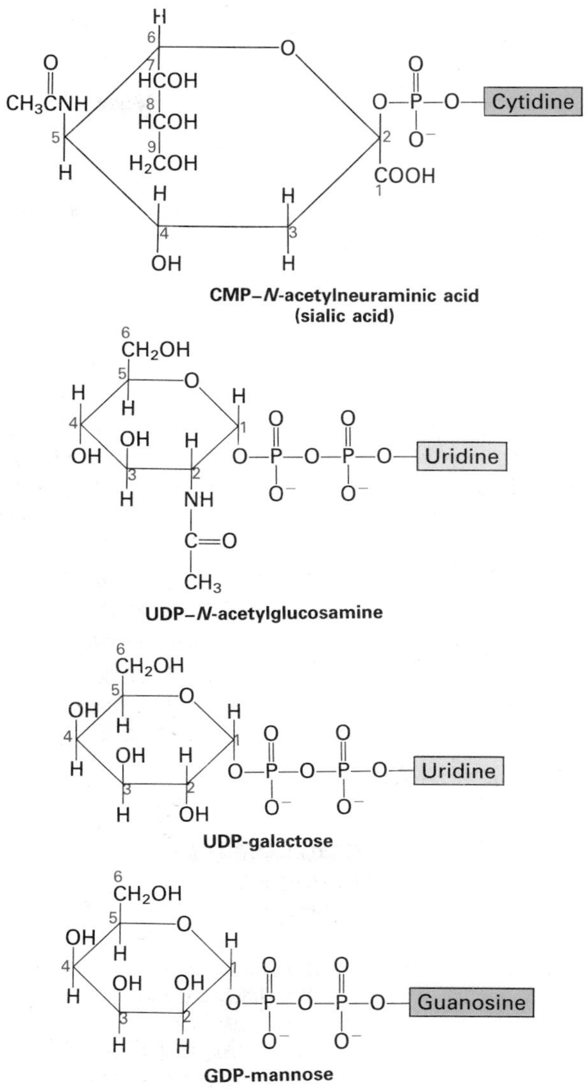

▲ **Figure 17-24** Structures of sugar nucleotides that are precursors of the saccharide residues in glycoproteins.

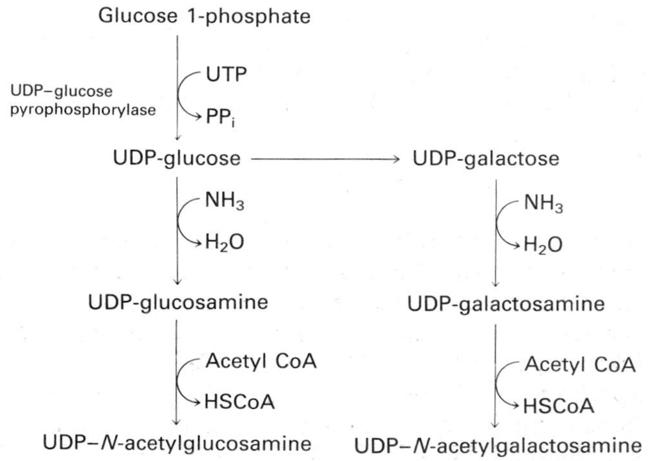

▲ **Figure 17-25** Schematic outline of the synthesis of some common sugar nucleotides.

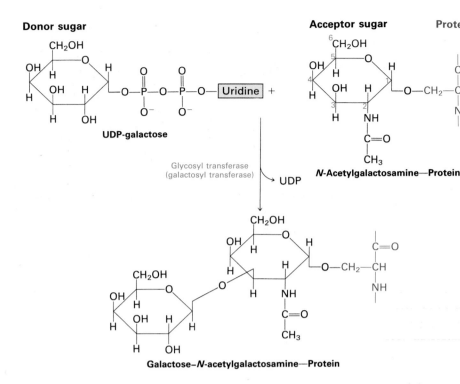

Donor sugar

UDP-galactose

Acceptor sugar   **Protein**

Serine

N-Acetylgalactosamine—Protein

Glycosyl transferase
(galactosyl transferase)   → UDP

Galactose—*N*-acetylgalactosamine—Protein

◄ **Figure 17-26** The addition of a galactose residue from UDP-galactose to the 3-carbon atom of *N*-acetylgalactosamine attached to a protein: a step in the elongation of typical O-linked oligosaccharides. Glycosyl transferases are specific both for the nucleoside sugar donor and for the carbon atom of the acceptor sugar to which it is transferred.

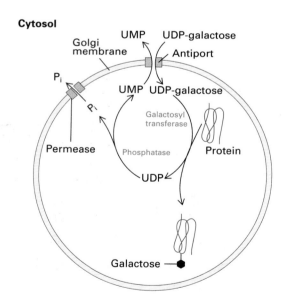

▲ **Figure 17-27** The uptake of nucleoside sugars into Golgi vesicles. UDP-galactose enters from the cytosol in exchange for UMP, using an antiport located in the Golgi membrane. UMP is produced by phosphatase action on UDP, a product of the galactosyl transferase reaction. A permease allows the inorganic phosphate formed from UDP to exit the Golgi vesicle. Other known antiports allow CMP–*N*-acetylneuraminic acid to enter in exchange for CMP and UDP–*N*-acetylglucosamine to enter in exchange for UMP. [See C. B. Hirschberg and M. D. Snider, 1987. *Ann. Rev. Biochem.* 56:63–68.]

allow the concentration of nucleotide sugars in the Golgi vesicles to be maintained at a constant level—a requirement for oligosaccharide synthesis.

The Diverse *N*-Linked Oligosaccharides Share Certain Structural Features

The *N*-linked oligosaccharides on different proteins have very different structures. One class, the *complex N-linked oligosaccharides* found on many serum proteins and viral glycoproteins, contains *N*-acetylglucosamine, mannose, fucose, galactose, and *N*-acetylneuraminic acid (see Figure 17-23). The different structures among complex glycoproteins result from the number of branches (ranging from two to four) and the number of *N*-acetylneuraminic acid residues (ranging from zero to four per oligosaccharide chain). Also, in some cases, the 2-carbon atom of *N*-acetylneuraminic acid is linked to the 3-carbon atom of galactose; in others, it is linked to the 6-carbon atom. In some cases, the branches have long polymers of $\alpha 2 \rightarrow 8$-linked *N*-acetylneuraminic acid; in others, long stretches of a repeating disaccharide: galactose $\beta 1 \rightarrow 4$ *N*-acetylglucosamine.

A second class, the *high-mannose N-linked oligosaccharides*, contains only *N*-acetylglucosamine and mannose. Members of this class differ primarily in the number of mannose residues attached to *N*-acetylglucosamine (see Figure 17-23). In some cases, complex and high-mannose oligosaccharides are attached to different asparagine residues in the same polypeptide.

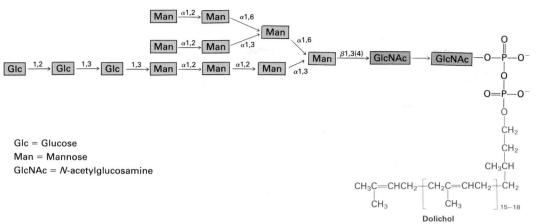

Glc = Glucose
Man = Mannose
GlcNAc = *N*-acetylglucosamine

◀ **Figure 17-28** Structure of the dolichol pyrophosphoryl oligosaccharide precursor of *N*-linked oligosaccharides. Dolichol is strongly hydrophobic and long enough to span a phospholipid bilayer membrane four or five times. This oligosaccharide lipid is located on the rough ER, and the sugar residues face the lumen of the organelle.

All high-mannose and complex *N*-linked oligosaccharides, from a wide variety of proteins contain an identical unit consisting of three mannose and two *N*-acetylglucosamine residues in exactly the same configuration (see Figure 17-23). All *N*-linked oligosaccharides are formed from a common precursor. The different structures result from multistep processes involving the sequential removal and addition of specific saccharide residues after the precursor oligosaccharide is transferred to the polypeptide. Numerous enzymes localized to the rough ER and several Golgi subcompartments participate in this process.

N-Linked Oligosaccharides Are Synthesized from a Common Precursor and Subsequently Processed

The structure of the precursor of all *N*-linked oligosaccharides in plants, animals, and single-cell eukaryotes is a branched oligosaccharide containing three glucose, nine mannose, and two *N*-acetylglucosamine molecules. This oligosaccharide is linked by a pyrophosphoryl residue to *dolichol*, a long-chain (75–95 C atoms) unsaturated lipid (Figure 17-28). The dolichol pyrophosphoryl oligosaccharide is formed in the ER and is oriented so that the dolichol portion is firmly embedded in the ER membrane and the oligosaccharide portion faces the ER lumen. The biogenesis of this precursor utilizes membrane-attached enzymes of the rough ER and involves the sequential addition of monosaccharides to dolichol phosphate.

The complete oligosaccharide chain is transferred en bloc by *oligosaccharide transferase* to asparagine residues on the nascent polypeptide in the tripeptide recognition sequences Asn-X-Ser or Asn-X-Thr (where X is any amino acid except proline). The transferase is located on the luminal surface of the ER (Figure 17-29).

While the protein is still in the rough ER, immediately after transfer of the oligosaccharide to the polypeptide,

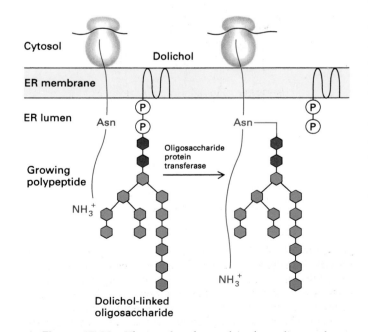

▲ **Figure 17-29** The catalyzed transfer of an oligosaccharide from the dolichol carrier to a susceptible asparagine residue on a nascent protein. The oligosaccharide is transferred as a unit as soon as the asparagine crosses to the luminal side of the ER.

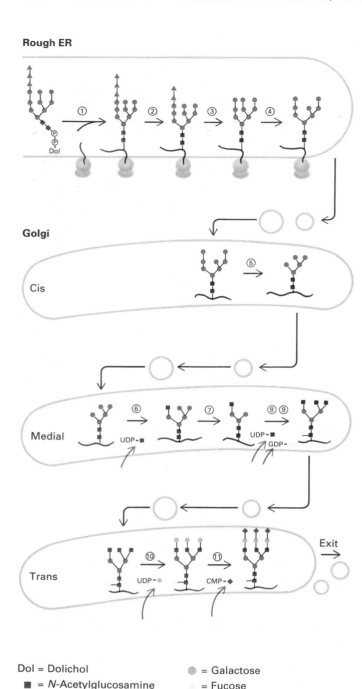

Rough ER

Golgi

Cis

Medial

Trans

Exit

Dol = Dolichol
■ = *N*-Acetylglucosamine
● = Mannose
▲ = Glucose

◐ = Galactose
○ = Fucose
◆ = *N*-Acetylneuraminic acid

▲ **Figure 17-30** The formation of complex *N*-linked oligosaccharides. At least 11 enzymes in four discrete organelles act sequentially to modify the common precursor of *N*-linked oligosaccharides, according to the following steps: three glucose residues removed (2, 3); four mannose removed (4, 5), one in the rough ER and three in the cis Golgi; one *N*-acetylglucosamine added (6); two mannose removed (7); one fucose and two *N*-acetylglucosamine added (8, 9); three galactose added (10); three *N*-acetylneuraminic acid added (11). Most of these enzymes have been localized to the specific organelles depicted. [See R. Kornfeld and S. Kornfeld, 1985, *Ann. Rev. Biochem.* 45:631–664.]

all three glucose residues and one particular mannose residue are removed by three different enzymes (Figure 17-30, *top*). The glucose residues—the last to be added to the oligosaccharide as it is being formed—appear to act as a signal that the oligosaccharide is complete and ready to be transferred to a protein.

Modifications to *N*-Linked Oligosaccharides Are Completed in the Golgi Vesicles

Further processing of the oligosaccharide to the complex form begins only 10–20 min after the protein is synthesized, at the time the protein is localized to the Golgi vesicles. In a stepwise, coordinated set of reactions, six of the nine mannose residues are removed and *N*-acetylglucosamine (three residues), galactose (three residues), *N*-acetylneuraminic acid (one to three residues per chain), and fucose (one residue) are added, one at a time, to each oligosaccharide chain (see Figure 17-30). Oligosaccharide processing to a plasma-membrane protein is complete about 10 min before it reaches the plasma membrane.

The high-mannose oligosaccharides found on the mature form of many glycoproteins (see Figure 17-23) have the same structures as the intermediates $(Man)_9(GlcNAc)_2$ to $(Man)_5(GlcNAc)_2$ during the processing of complex oligosaccharides. The high-mannose oligosaccharide $(Man)_8(GlcNAc)_2$ is formed if the cis Golgi mannose-cleaving enzyme α-mannosidase-I, which catalyzes reaction 5 in Figure 17-30, cannot act on the particular protein-bound oligosaccharide. The high-mannose oligosaccharide $(Man)_5(GlcNAc)_2$ results if reaction 6 (*N*-acetylglucosamine transferase) cannot occur, since the only substrate for reaction 7 is the product of reaction 6.

The conformation of the protein and, ultimately, its primary amino acid sequence determine whether a particular *N*-linked oligosaccharide becomes complex or remains high-mannose. Specific cell types in an organism contain specific complements of processing enzymes: the same protein produced by individual cell types may have differently processed carbohydrates. For example, certain asparagine residues in the HA glycoprotein of the influenza virus have complex oligosaccharides when the virus is grown in one cell type but high-mannose oligosaccharides when the virus is grown in another cell type.

Electron-microscope autoradiographic experiments clearly show that different sugars are added to protein in different organelles. When cells are exposed briefly to [³H]mannose, virtually all incorporated radioactivity (observed as grains) is in the ER. Radioactivity from newly incorporated [³H]fucose or [³H]galactose acid, by contrast, is abundant over the trans Golgi region. Furthermore, galactosyl transferase is localized to the trans-most Golgi vesicles, and sialytransferase is found in the trans

Golgi and the trans Golgi reticulum (Figure 17-31; see Figure 17-8).

Thus the *N*-linked oligosaccharides are modified in much the same way that an automobile is built on an assembly line. The protein is transported from organelle to organelle, in each of which it is acted on sequentially by a large set of enzymes. The reaction product of one enzyme is the substrate of the next, and the modifying enzymes are segregated in different organelles to ensure that only properly modified substrates are presented to each enzyme.

N-Linked and *O*-Linked Oligosaccharides May Stabilize Maturing Secretory and Membrane Proteins

In many cases, *N*-linked oligosaccharides appear to be required for the secretion of proteins or the movement of plasma-membrane glycoproteins to the cell surface. When the first stage in the formation of the oligosaccharide-lipid donor (see Figure 17-28) is blocked by the antibiotic tunicamycin, the polypeptide is synthesized but contains no *N*-linked sugar chains. Without *N*-linked oligosaccharides, many proteins cannot fold into their proper conformation; the HA glycoprotein, for instance, cannot form a trimer (see Figure 17-21) and remains in the rough ER. In the absence of *N*-linked oligosaccharides, the secretion of many proteins, such as IgG antibodies or α_1-antiprotease, is blocked and the abnormally folded protein is also selectively retained in the rough ER. In other cases, proteins are secreted without the normal *N*-linked oligosaccharides. For instance, the rate and extent of secretion of glycosylated and unglycosylated fibronectin by fibroblasts is the same. However, unglycosylated fibronectin is more susceptible to degradation by tissue proteases than normal glycosylated fibronectin, suggesting that the added carbohydrates confer stability on this extracellular matrix protein.

The conformation and stability of proteins can also be affected by *O*-linked oligosaccharides. The LDL receptor, for instance, normally has about 30 of these. In cultured cells defective in the synthesis of *O*-linked oligosaccharides, newly made LDL receptors move normally to the plasma membrane. There, the unglycosylated receptors are cleaved by cell-surface proteases and released from the cell.

Because the impact of the absence of glycosylation is so variable and by no means absolute, researchers have concluded that sugar residues play no mandatory role in the movement of proteins to the cell surface: they are not a "ticket" needed to move through the transport organelles. In all likelihood, carbohydrates play a role in ensuring the correct charge, conformation, and stability of maturing proteins. For some proteins, this function is apparently superfluous; for others, it is clearly necessary.

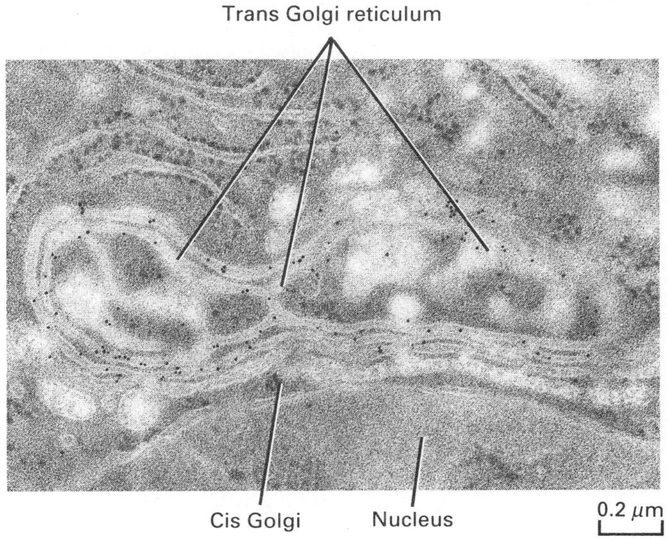

▲ **Figure 17-31**   Immunolocalization of sialyltransferase to two trans cisternae of the Golgi and the trans Golgi reticulum. Note the continuity between the labeled trans cisternae and the trans Golgi reticulum. A Lowicryl K4M section from rat liver was reacted with an antibody to sialyltransferase. Bound antibody was detected by reaction with protein A complexed to 8-nm gold particles; the electron-dense gold permits observation in the electron microscope. *From J. Roth, D. Taatjes, J. Lucocq, J. Weinstein, and J. C. Paulson, 1985, Cell 43:287–295. Electron micrograph courtesy of J. Roth.*

Golgi and Post-Golgi Sorting and Processing of Secretory and Membrane Proteins

Vesicles Transport Proteins from Organelle to Organelle

The rough ER, smooth ER, cis and trans Golgi membranes and trans Golgi reticulum, lysosomes, secretory vesicles, and plasma membrane are all discrete organelles, each with its own unique constellation of enzymes. Proteins destined for each of these organelles are synthesized on the rough ER and moved from one organelle to the other in membrane vesicles. But precisely how does such a process occur? More importantly, how are vesicles targeted from one organelle to another? How, for instance, does a vesicle containing secretory proteins from the rough ER "know" to move to and fuse with cis Golgi membranes?

In all eukaryotic cells, electron microscopy reveals a plethora of small membrane-bound vesicles (see Figure 17-22), some coated with clathrin (the fibrous protein that surrounds vesicles formed from the plasma

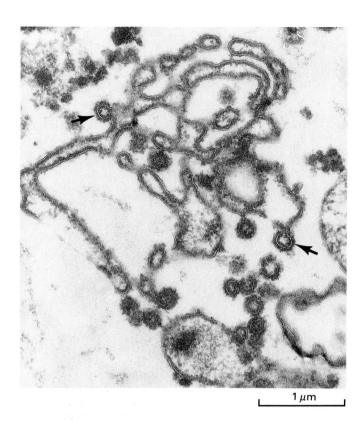

1 μm

▲ **Figure 17-32** Incubation of a Golgi fraction with ATP causes buds and vesicles to form. These Golgi vesicles from rat hepatocytes have a distinct outer coat *(arrows)* but do not contain a coat of the protein clathrin. Such coated vesicles transport proteins from one Golgi vesicle to another. *From P. Melancon, B. Glick, V. Malhotra, P. Weidman, T. Serafini, M. Gleason, L. Orci, and J. Rothman, 1987, Cell* **51**:1053–1062. *Electron micrograph courtesy of L. Orci.*

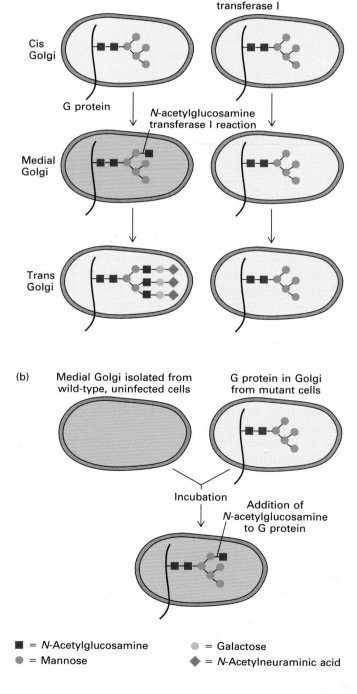

▶ **Figure 17-33** A cell-free system exhibiting movement of newly made VSV glycoprotein (G protein) from cis to medial Golgi vesicles. Such experiments make use of a mutant line of cultured fibroblasts that lack the enzyme *N*-acetylglucosamine transferase I (step 6 in Figure 17-30). In wild-type cells, this enzyme is localized to the medial Golgi (green) and modifies asparagine-linked oligosaccharides by the addition of one *N*-acetylglucosamine (red square). (a) After VSV infection of wild-type cells, the oligosaccharide on the viral G protein is modified to a typical "complex" oligosaccharide. In mutant cells, however, the G protein reaches the cell surface with a simpler oligosaccharide containing only two *N*-acetylglucosamine (red squares) and five mannose (blue circles) residues. (b) To detect movement of the G protein from one Golgi vesicle to another in cell-free systems, the mutant cell is infected with VSV. When Golgi vesicles are isolated from such infected mutant cells, all their G protein contains the high-mannose oligosaccharide lacking *N*-acetylglucosamine. The incubation of Golgi from mutant cells with Golgi from normal, uninfected cells results in the production of G protein containing the additional *N*-acetylglucosamine. Electron-microscope autoradiography indicates that all G protein labeled with *N*-acetylglucosamine is contained in sealed medial Golgi vesicles. This G protein has traveled from mutant cis Golgi to wild-type medial Golgi vesicles to receive the *N*-acetylglucosamine residue. [See W. E. Balch et al., 1984, *Cell* **39**:405; W. A. Braell et al., 1984, *Cell* **39**:511; W. E. Balch et al., 1984, *Cell* **39**:525–536.]

membrane during receptor-mediated endocytosis (see Figure 14-39). At one time, clathrin-coated vesicles were believed to be intermediate transporters of plasma-membrane glycoproteins and secretory proteins from the rough ER to the Golgi vesicles and the cell surface. However, such a restricted role for clathrin-coated vesicles now seems improbable; a yeast mutant with its single clathrin heavy-chain gene deleted still secretes invertase and other proteins quite normally.

Indeed, when isolated Golgi fractions are incubated in a solution containing ATP, they form a large number of buds and vesicles (Figure 17-32). These vesicles do contain a protein coat, but biochemical and immunological assays have showed that it is not composed of clathrin. This coat, like clathrin, may cause a membrane to form budding vesicles and may regulate the entry of certain membrane proteins into these vesicles. These "coated" vesicles transport proteins in the Golgi.

The Steps in Vesicular Transport Can Be Studied Biochemically and Genetically

The specific movement of plasma-membrane glycoproteins from a cis Golgi vesicle to a medial Golgi vesicle by means of small transport vesicles can be demonstrated in a cell-free system (Figure 17-33). Assays also have been used to document the movement of proteins in cell-free systems from the rough ER to the cis Golgi and from the medial Golgi to the trans Golgi. Such experiments should eventually elucidate all components of vesicle budding, targeting, and fusion.

ATP and GTP hydrolyses are required at different stages of vesicular transport, and several proteins that catalyze different steps of this complex process have been identified (Figure 17-34). One protein, NSF, mediates the fusion of the transport vesicle membrane and the membrane of the acceptor medial Golgi vesicle. A protein very similar in sequence to NSF, the product of the yeast sec18 gene (see Figure 17-10), mediates the fusion of the cis Golgi and transport vesicles originating in the rough ER. Temperature-sensitive mutants in the sec18 gene accumulate ER-to-Golgi transport vesicles at the high temperature; when the lower temperature is restored, these vesicles fuse with the cis Golgi and normal secretion occurs. Thus NSF and the sec18 protein catalyze the fusion of transport vesicles and recipient Golgi membranes. Proteins similar in sequence to NSF and sec18 participate in the fusion of other intracellular vesicles, such as endosomes with each other or with CURL vesicles (see Figure 14-33).

It is still not known how these various transport vesicles recognize only their correct acceptor vesicles. Presumably, specific proteins on the surface of the transport vesicles are involved, but no such "targeting proteins" have been identified as yet.

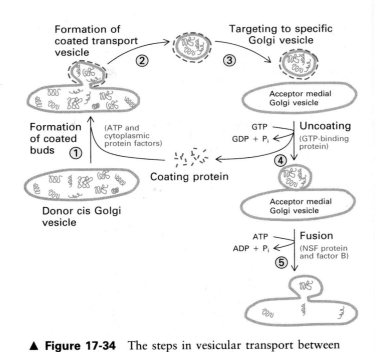

▲ **Figure 17-34** The steps in vesicular transport between Golgi vesicles as determined by the study of cell-free extracts that exhibit movement of the G protein from cis to medial Golgi (see Figure 17-33). Incubation of cis Golgi vesicles with ATP and as yet unpurified cytoplasmic proteins (1) leads to the formation of coated transport vesicles (2) that bind to acceptor medial Golgi vesicles (3). Uncoating (4) requires a cytosplasmic protein and the hydrolysis of GTP to GDP and P_i; this step can be inhibited experimentally by the addition of an analog of GTP, which cannot be hydrolyzed. Fusion of the two membranes (5) requires ATP hydrolysis and two cytoplasmic proteins: the NSF protein and factor B. This step can be blocked selectively by adding the chemical N-ethylmaleimide (NEM), which binds to an essential SH group on NSF (hence the name NEM sensitive factor). *After L. Orci, V. Malhotra, M. Amherdt, T. Serafini, and J. E. Rothman, 1989, Cell **56**:357–368.*

Phosphorylated Mannose Residues Target Proteins to Lysosomes

Lysosome enzymes, like secretory proteins, are cotranslationally inserted into the ER lumen, where they receive an N-linked oligosaccharide identical to that of secretory proteins. In the cis Golgi, one or more mannose residues of these oligosaccharides become phosphorylated; the phosphorylated mannose apparently serves as the chemical signal that targets the protein to lysosomes. The phosphorylation of mannose residues is a two-step procedure (Figure 17-35) occurring in the cis Golgi region. The first enzyme in the sequence (N-acetylglucosamine phosphotransferase) utilizes only lysosome proteins as substrates and does not catalyze reactions with other glycoproteins. Since deglycosylated lysosome enzymes are potent inhibitors of the phosphorylation of intact lysosome enzymes, the high-affinity interaction between the N-acetylglucos-

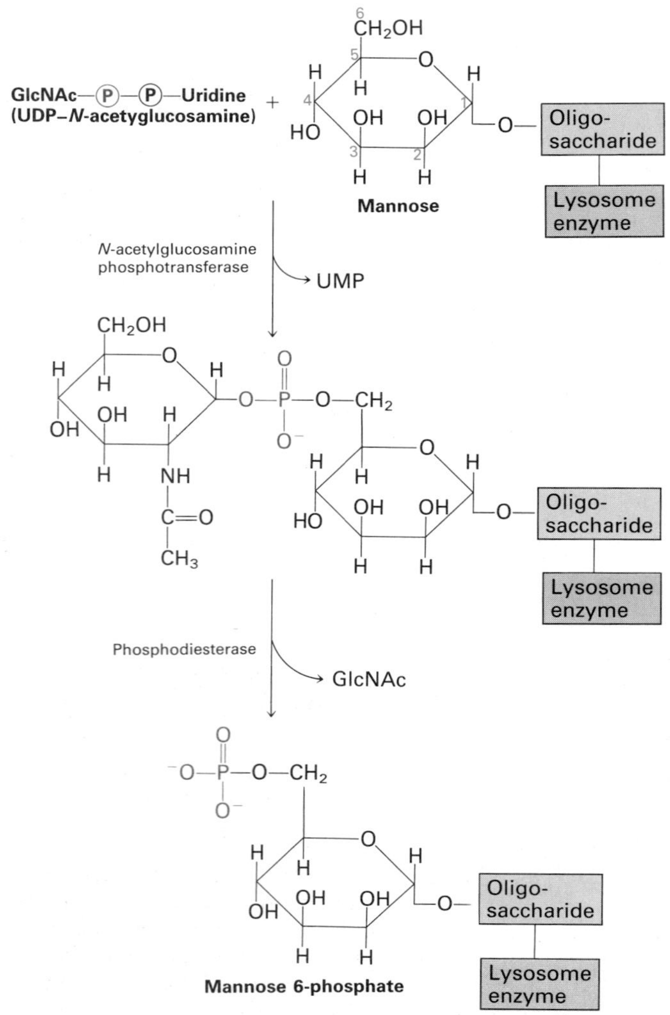

▲ **Figure 17-35** The phosphorylation of mannose residues on lysosome enzymes occurs in two stages. The first enzyme transfers an *N*-acetylglucosamine phosphate group to the 6-carbon atom of one or more mannose residues. This enzyme is specific for lysosome enzymes and does not utilize other glycoproteins as substrates. The second enzyme removes the *N*-acetylglucosamine group, leaving the phosphate attached to the 6-carbon atom of the mannose residue. [See S. Kornfeld, 1987, *FASEB J.* 1:462–468.]

amine phosphotransferase and lysosome substrate enzymes is believed to be mediated by protein-protein interactions. The *N*-acetylglucosaminylphosphotransferase recognizes a protein domain present on one part of the surface of all lysosome enzymes.

A *mannose 6-phosphate receptor* located on the luminal face of the trans Golgi reticulum membrane binds the mannose 6-phosphate residue of the lysosome protein and directs the protein to lysosomes. Membrane regions containing the receptor and its bound lysosome enzyme pinch off to become specialized vesicles that transport the enzymes to acidic (pH = 5.0) sorting vesicles (Figure 17-36). These are the same CURL vesicles that sort receptors

and ligands after receptor-mediated endocytosis (see Figure 14-33).

The mannose 6-phosphate receptor binds its ligand at the neutral pH of the Golgi but not at the lower pH of the sorting vesicle. Once the enzymes reach the sorting vesicle, they are released from the receptor and become soluble in it. Furthermore, a phosphatase generally removes the phosphate, preventing any rebinding of the enzyme to the receptor because nonphosphorylated oligosaccharides have no binding capability. Vesicles containing the lysosome enzyme but not the mannose 6-phosphate receptor bud from the sorting vesicles and fuse with lysosomes, delivering the enzyme to its destination. Like most cellular receptors, the mannose 6-phosphate receptor recycles. The mechanism for this is unknown; specialized vesicles containing the receptor bud from the sorting vesicle and transport the receptor back to the Golgi or to the plasma membrane.

Inactive precursors of lysosome enzymes *(proenzymes)* undergo a proteolytic cleavage late in their maturation to form an enzymatically active but smaller peptide. This cleavage, which occurs in either the acidic sorting vesicle or the lysosome, results in the activation of enzymatic function as the protein reaches the very acidic environment (pH = 5.0) of the lysosome. These inactive proenzymes may help to protect the cell from the dangerous hydrolytic activity of such enzymes in the less acidic (pH = 5.5–6.0) sorting vesicles.

Genetic Defects Have Elucidated the Role of Mannose Phosphorylation

The discovery of the mannose 6-phosphate pathway began with a study of patients with *I cell disease*. A severe genetic abnormality in these people causes a deficiency of multiple lysosome enzymes in fibroblasts and macrophages, which permits large amounts of toxic intracellular wastes to accumulate. Fibroblasts from these patients contain large intracellular vesicles filled with glycolipids and extracellular components that would normally be degraded by lysosome enzymes. Cells from affected persons synthesize all lysosome enzymes normally and have normal mannose 6-phosphate receptors on the plasma and Golgi membranes. However, the first enzyme (see Figure 17-35) required to phosphorylate mannose residues is lacking, and the degradative hydrolytic enzymes are secreted rather than sequestered in lysosomes. When diseased fibroblasts are grown in a medium containing lysosome enzymes bearing phosphorylated mannose, the enzymes are internalized by receptor-mediated endocytosis, using cell-surface mannose 6-phosphate receptors, and the content of these enzymes becomes almost normal. Lysosomes from the liver cells of patients with I cell disease do contain a normal complement of lysosome enzymes, even though these cells are defective in mannose

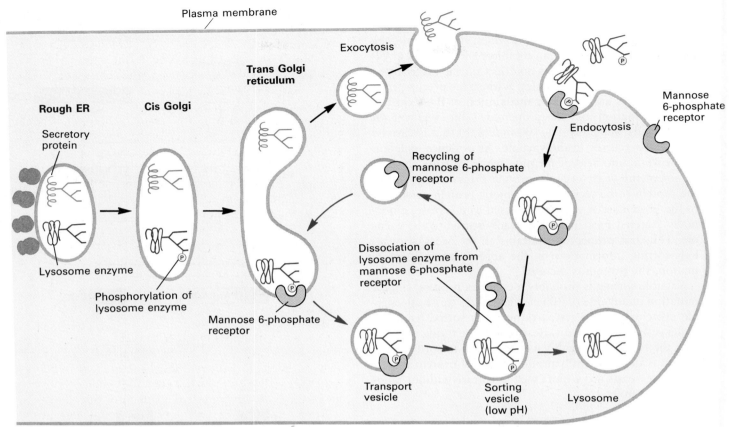

▲ **Figure 17-36** The targeting of lysosome enzymes to lysosomes. During biosynthesis, lysosome enzymes migrate to the cis Golgi, in which one or more mannose residues are phosphorylated. In the trans Golgi reticulum, the membrane-attached mannose 6-phosphate receptor ensures that these proteins associate with vesicles that are ultimately directed to lysosomes. The low pH of the CURL-like sorting vesicle (see Figure 14-33) causes the phosphorylated enzyme to dissociate from its receptor, and the receptor recycles back to the Golgi. The phosphorylated enzyme then loses its phosphate group and is transported to a lysosome. The sorting of lysosome enzymes from secretory proteins thus occurs in the trans Golgi reticulum, and these two classes of proteins are found in different vesicles that bud from the Golgi. The mannose 6-phosphate receptor is also found on the cell surface, where it mediates the endocytosis of extracellular phosphorylated lysosome enzymes that are occasionally secreted and causes them to be transported to a sorting vesicle and then to a lysosome. [See C. A. Gabel and S. A. Foster, 1987, *J. Cell Biol.* **105**:1561–1570; W. J. Brown, J. Goodhouse, and M. G. Farquhar, 1986, *J. Cell Biol.* **103**:1235–1247; G. Griffiths, B. Hoflack, K. Simons, I. Mellman, and S. Kornfeld, 1988, *Cell* **52**:329–341; S. Kornfeld, 1987, *FASEB J.* **1**:462–468.]

phosphorylation. This finding suggests that hepatocytes may target newly made enzymes to lysosomes in a different way; the nature of the signal is unknown and is under active study.

Many genetic diseases are due to a defect in a specific lysosome enzyme. Hunter's syndrome and Hurler's syndrome result from defects in the lysosome enzymes required to catabolize sulfated mucopolysaccharides. Tay-Sachs disease is caused by an absence of the lysosome hydrolase β-*N*-hexosaminidase A (see Figure 4-44). Cultured fibroblasts from patients with these diseases have been observed taking up the missing enzyme from the extracellular medium via surface mannose 6-phosphate receptors and endocytosis. This observation raises the possibility of treating such diseases by administering the missing phosphorylated lysosome enzyme to slow or reverse the clinical course.

Propeptide Sequences Target Proteins to Vacuoles

The vacuoles of plants and microorganisms such as yeasts are similar to lysosomes in that they contain a number of soluble hydrolytic glycoprotein enzymes. The precursors of these enzymes, like the lysosome enzymes in animal cells, are synthesized in the rough ER and move through the Golgi vesicles to the trans Golgi reticulum, where they are sorted into vesicles destined for vacuoles. However, the delivery of these hydrolytic enzymes to the vacuole does not require modifications to the carbohydrate, unlike the targeting of lysosome enzymes. For example, the addition of tunicamycin, an inhibitor of synthesis of *N*-linked sugars, to yeasts has no effect on the targeting of hydrolytic enzymes to vacuoles. Rather, the precursors of these enzymes are sorted by recognition of the sequence or conformation of the propeptide—the segment of amino acids (generally 50–100 in length) that is cleaved from the precursor in the vacuole. As evidence, deletion of the 76 amino acid propeptide sequence from the precursor of the yeast vacuole enzyme *proteinase A* causes the protein to be secreted. A chimeric protein consisting of the proteinase A propeptide fused to invertase, a normally secreted protein, is mistakenly sorted to the vacuole. Thus this propeptide contains all the necessary vacuole-sorting information; in the absence of such information, the protein is secreted.

Vacuole sorting is probably a complex process, since a mutation in any one of 30 yeast genes can cause all vacuole proteins to be secreted rather than sorted from the Golgi vesicles to vacuoles. Researchers are trying to elucidate the function of each vacuole protein in sorting; this system is ideal for studying the details of protein targeting, since yeasts grow quite well in culture without functioning vacuoles.

Several Proteolytic Cleavages Occur during the Late Maturation Stages in Secretory and Membrane Proteins

We have seen that an N-terminal signal sequence of amino acids is removed from many nascent secretory and membrane proteins while the chain is still growing on the ribosome. In the cases of some secretory proteins, such as growth hormone, placental lactogen, lysozyme, and ovomucoid, and of certain viral membrane proteins, such as the VSV glycoprotein, removal of the signal sequence is the only known proteolytic cleavage and converts the polypeptide directly to the mature, active species. However, most secretory proteins go through an intermediate stage—an additional, relatively long-lived intracellular form termed the *proprotein*, or *prohormone*. Serum proteins such as albumin, hormones such as insulin and glucagon (Figure 17-37), and membrane proteins such as the HA glycoprotein (see Figure 17-21) are synthesized as longer precursors, from which specific polypeptides are cleaved to generate the mature, active molecule. In general, the proteolytic conversion of the proprotein to a mature molecule occurs at a late stage of intracellular maturation. Additional amino acids in the pro form can be at the N-terminus or at both ends of the proprotein (see Figure 17-37). In the case of proinsulin, the extra amino acids, collectively termed the *C peptide,* are located internally in the polypeptide. The N-terminal

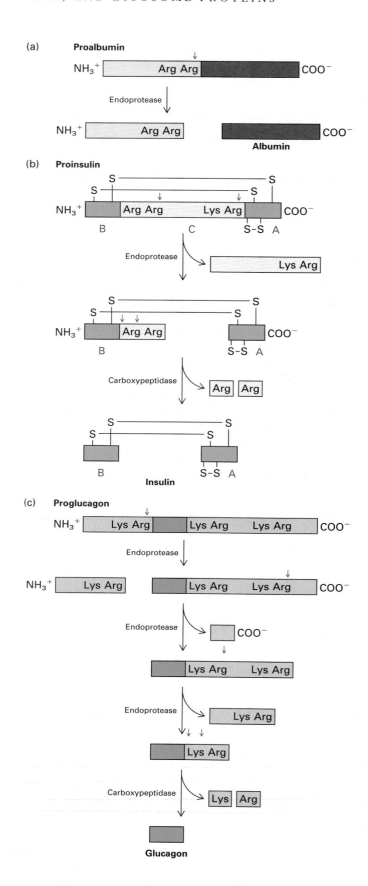

B chain and the C-terminal A chain of mature insulin are linked by disulfide bonds and remain attached when the C peptide is removed.

Different Vesicles Are Used for Regulated and Continuous Protein Secretion

Certain cells store some secretory proteins in vesicles and secrete these proteins only on specific stimulus *(regulated secretion)*. In particular, adrenocorticotropic hormone (ACTH), insulin, and other hormones activated by a final proteolytic cleavage are stored in special secretory vesicles and exocytosed only on neural or hormonal stimulation. However, newly made plasma-membrane glycoproteins and certain secretory proteins in the same cells are exocytosed to the cell surface continuously, independently of any known stimulus *(continuous secretion)*. Pituitary tumor cells, exocrine pancreatic cells, insulin-secreting pancreatic islet cells, and many others utilize these two different classes of vesicles for protein secretion (see Figure 17-8).

Proteins are sorted to regulated or continuous secretory vesicles in the trans Golgi reticulum. How are the different types of secretory proteins, which are all soluble in the lumen of the trans Golgi reticulum, sorted to the correct vesicle? Evidence indicates, first, that all mammalian cells employing the regulated secretory pathway use a common sorting mechanism and, second, that the trans Golgi reticulum contains receptor-like proteins that bind only to proteins destined for the regulated vesicles. Presumably, these proteins form part of the sorting machinery.

Different proteins sorted to regulated secretory vesicles in different types of cells do share a common sorting mechanism. Consider proteins as diverse as ACTH (normally made in pituitary cells), insulin (a product of the β cells of pancreatic islets), and trypsinogen (made by exocrine pancreatic acinar cells). When recombinant techniques are used to synthesize insulin and trypsinogen in pituitary tumor cells already synthesizing ACTH, all three proteins segregate in the same regulated secretory vesicles and are secreted only upon the appropriate hormonal signal. Although these three proteins have no identical regions of amino acid sequence, they obviously share some feature that serves as a sorting signal for their incorporation into the regulated vesicles.

Several 25,000-MW proteins purified from the Golgi vesicles of exocrine pancreatic cells bind tightly at pH = 7 (the pH of the trans Golgi reticulum) to proteins sorted to regulated secretory vesicles; these proteins do not bind to secretory proteins that enter continuously secreted vesicles. Additional evidence that these receptor-like proteins are part of the sorting process is that they dissociate from their bound ligands at pH = 5.5 (the pH of regulated vesicles); presumably, after they have transported their cargo of regulated proteins to the secretory vesicles, these receptor-like proteins recycle from the vesicles back to the trans Golgi. Even while regulated secretory proteins are still in the trans Golgi reticulum, they form aggregates that can be seen in the electron microscope (Figure 17-38). Since one 25,000-MW protein can bind simultaneously to two or more regulated secretory proteins, they may facilitate the aggregation of regulated proteins. The formation of aggregates, in turn, could trigger the formation and budding of regulated secretory vesicles.

Proteolytic Maturation of Insulin Occurs in Acidic, Clathrin-Coated Secretory Vesicles

Because the active hormone is not generated until it is packaged in regulated secretory vesicles, the cell cannot be stimulated by the hormone it has made. If functional insulin were formed in the rough ER of the pancreatic islet cell, for example, it could bind to insulin receptors in the ER membrane and trigger an inappropriate response. The secretory vesicles in which mature insulin is finally formed from proinsulin have a pH of about 5.5. Since insulin, like most peptide hormones, cannot bind to its receptor at this acidic pH, any insulin receptors in the secretory vesicle membrane cannot be triggered by mature insulin.

The conversion of proinsulin to insulin occurs in newly formed (immature) regulated secretory vesicles, as we can infer from the electron micrographs in Figure 17-38. An antibody specific for proinsulin (not insulin) detects proinsulin in the Golgi vesicles and in clathrin-coated secretory vesicles that are near and probably have just budded from the trans Golgi. These immature vesicles contain no mature insulin. Insulin, detected by a monoclonal antibody that does not react with proinsulin, is localized to more mature secretory vesicles that can be distinguished in electron micrographs by a dense core of almost crystalline insulin. Apparently, proinsulin cleavage occurs in

◄ **Figure 17-37** Schema for maturation of three typical proproteins. (a) A single cleavage by an endoprotease converts proalbumin to albumin by cutting it on the C-terminal side of a sequence of two consecutive basic amino acids. (b) Two cleavages of proinsulin by a similar endoprotease release the C peptide. The B chain is processed further by the action of a carboxypeptidase that sequentially removes the two arginines at the C-terminus. (c) The processing of proglucagon to glucagon involves three successive cleavages by an endoprotease that cleaves after two basic amino acids. Finally, a carboxypeptidase removes the two C-terminal basic amino acids from the last intermediate to form mature glucagon. [See J. T. Potts et al., 1980, *Ann. NY Acad. Sci.* 343:38; L. C. Lopez et al., 1983, *Proc. Nat'l Acad. Sci. USA* 80:5485.]

newly formed secretory vesicles, as a simple experiment shows. When pancreatic islet cells are incubated in a medium without glucose or other energy sources, the ATP level drops and all fission and fusion of vesicles cease. Yet the conversion of proinsulin to insulin continues: when these cells are examined after their incubation, all the immature secretory vesicles contain insulin, not proinsulin.

Exocytosis Can Be Triggered by Neuron or Hormone Stimulation

Many hormones, neurotransmitters, and other agents trigger the exocytosis of stored secretory proteins in vesicles. The process is very similar to the one used to initiate the exocytosis of synaptic vesicles during the transmission of a nerve impulse.

Exocrine pancreatic acinar cells have been studied in detail. Apparently, agents increase protein secretion by two fundamentally different mechanisms. One involves binding the hormone or neurotransmitter to its surface receptor, thereby increasing the intracellular Ca^{2+} concentration and stimulating protein secretion. The action of these hormones can be mimicked by Ca^{2+} ionophores, which increase the intracellular Ca^{2+} concentration. *Acetylcholine*, which acts in this manner, is discharged in the vicinity of the acinar cells by *cholinergic nerves*. The peptide hormones *gastrin* and *cholecystokinin* (CCK) also act in this way to induce enzyme secretion from acinar cells. Gastrin is released into the circulation by the stomach when it is filling with food; CCK, by the small intestine. Thus secretion of digestive enzymes by the pancreas is induced by a specific stimulus: the presence of food in the stomach or intestine.

Other hormones cause protein secretion from acinar cells by a different, unknown mechanism that does not involve an increase in cytosolic Ca^{2+}. Two examples, *secretin* and *vasoactive intestinal peptide*, are released into

(a)

(b)

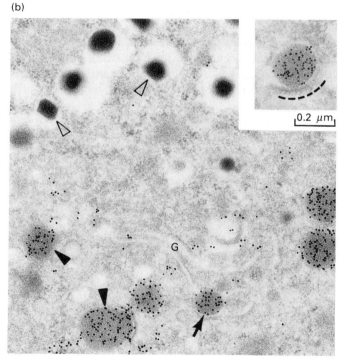

0.2 μm

0.5 μm

▲ **Figure 17-38** Proinsulin is packaged into clathrin-coated secretory vesicles before it is cleaved to insulin. Serial sections of the Golgi region of an insulin-secreting cell are stained with (a) a monoclonal antibody that detects insulin (not proinsulin), and (b) an antibody that recognizes proinsulin (not insulin). The antibodies are bound to electron-opaque gold particles and appear as dark dots in electron micrographs. Mature vesicles with a dense core *(open arrowheads)* stain intensely with the antiinsulin antibody (a) but not with antiproinsulin (b). In contrast, immature secretory vesicles *(closed arrowheads)* and vesicles budding from the trans Golgi *(arrows)* stain with the proinsulin antibody (b) but not with antiinsulin. Also, the Golgi complex (G) is stained with antiproinsulin but not with antiinsulin. The inset in (b) shows the clathrin coat *(dashed line)* on a proinsulin-rich secretory vesicle. Since immature secretory vesicles contain proinsulin (not insulin), the proteolytic conversion of proinsulin to insulin must take place after the proinsulin is transported from the trans Golgi reticulum to these vesicles. Note that immature vesicles budding from the trans Golgi *(arrows)* contain aggregates that include proinsulin and other regulated secretory proteins. *From L. Orci, M. Ravazzola, M. Storch, R. Anderson, J. Vassali, and A. Perrelet, 1987, Cell **49**:865–868. Electron micrographs courtesy of L. Orci.*

the circulation by intestinal cells. Specific surface receptors for these similar peptides have been discovered on the basal membrane surface of acinar cells, where they can come into contact with fluid derived from the blood, but how hormone binding triggers protein secretion is not known. Other cells, such as the endocrine insulin-producing pancreatic islet cells, also have two ways of triggering protein secretion: one that involves a rise in Ca^{2+}, and one that does not.

Regulated Secretory Vesicles Swell Following Fusion with the Plasma Membrane

Characteristically, regulated secretory vesicles contain a highly concentrated, almost crystalline solution of protein. When viewed in the electron microscope, these proteins cause the dense staining of the cores of these vesicles. Immediately after exocytosis is triggered, these vesicles swell greatly due to the osmotic flow of water into the vesicle lumen. This osmotic swelling does not push the vesicle membrane against the plasma membrane and *cause* fusion and exocytosis. Instead, the swelling occurs *after* the vesicle and plasma membrane have fused, as determined by the studies of the exocytosis of vesicles in mouse mast cells that contain only one or two large secretory vesicles. The exact time it takes the vesicle and plasma membranes to fuse can be quantified by measuring the electric property, termed *capacitance,* of the plasma membrane, which is proportional to the area of the membrane and suddenly increases to a new value at the instant of membrane fusion. These studies clearly show that membrane fusion precedes the swelling by a few milliseconds. Thus the swelling occurs only after exocytosis and cannot be a cause of fusion. The rise in Ca^{2+} that triggers exocytosis activates membrane fusion—not vesicle swelling directly.

Apical-Basolateral Protein Sorting Occurs in the Golgi Complex or the Basolateral Membrane

In many tissues, the plasma membranes of a number of cells are divided into two surfaces, separated by tight junctions, that contain different species of proteins. Recall, for example, that in the epithelial cells lining the small intestine, the apical and basolateral plasma membranes contain different enzymes and permeases that facilitate the movement of digestive products from the intestine into the blood (see Figures 13-42 and 14-14). Although it is unclear how proteins are targeted to the appropriate surface of the plasma membrane, one mechanism appears to involve sorting in the Golgi vesicles. A variety of microscopic and cell-fractionation studies indicate that proteins destined for the apical and basolateral

membranes are found together in the same Golgi vesicles. However, proteins bud from the Golgi into separate transport vesicles that fuse with the appropriate plasma-membrane surface.

Cultured MDCK epithelial cells, which maintain distinct apical and basolateral membranes (see Figure 13-44), have been useful in investigating this aspect of protein transport. When these cells are infected with the enveloped influenza virus, progeny viruses are released as membrane-attached buds only from the apical membrane; on the other hand, VSV buds only from the basolateral membrane (Figure 17-39). The location of viral formation is determined by the location of viral glycoproteins in the plasma membrane: the HA glycoprotein of the influenza virus is transported from the Golgi exclusively to the apical membrane; the VSV glycoprotein (G protein) is transported only to the basolateral membrane. Furthermore, in cells expressing a cloned, transfected HA cDNA, all the HA accumulates only in the apical membrane, indicating that the targeting sequence resides in the HA glycoprotein itself and not in other viral proteins produced on viral infection. To localize the sorting domain in viral glycoproteins, investigators have constructed a cDNA encoding a chimeric protein that contains the exoplasmic segment of the influenza HA glycoprotein and the membrane-spanning and cytoplasmic segments of the VSV glycoprotein. In MDCK cells, the protein is targeted specifically to the apical membrane (see Figure 17-39). A chimera containing the exoplasmic segment of G and the membrane-spanning and cytoplasmic segments of HA, in contrast, is targeted to the basolateral membrane. These results establish that the sorting sequence that directs a protein to the apical or basolateral membrane resides in the exoplasmic segment. What is not known is how this sorting sequence is recognized and how the actual sorting occurs.

Apical and basolateral proteins are sorted differently in hepatocytes, where the basolateral membranes face the blood and the apical membranes form the bile canaliculus. In these cells, newly made apical and basolateral proteins are delivered together to the basolateral membrane; from there, the apical proteins are selectively endocytosed into internal vesicles and delivered to the apical membrane. Even in epithelial cells, such as MDCK cells, in which apical-basolateral protein sorting occurs in the Golgi, endocytosis may provide a "fail-safe" sorting mechanism: an apical protein sorted incorrectly to the basolateral membrane would be endocytosed and delivered to the apical membrane.

Attachment of integral membrane proteins to the cytoskeleton may also be involved in protein sorting. A fibrous cytoskeleton containing ankyrin and the spectrin-like protein fodrin (see Figure 13-38) is found on the cytoplasmic face of the basolateral membrane but not on the apical membrane in epithelial cells. Several integral proteins, including Band 3 and the Na^+-K^+ ATPase, that

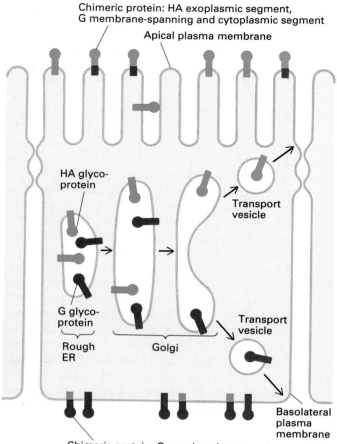

Chimeric protein: HA exoplasmic segment, G membrane-spanning and cytoplasmic segment

Apical plasma membrane

HA glyco-protein

Transport vesicle

G glyco-protein

Rough ER

Golgi

Transport vesicle

Basolateral plasma membrane

Chimeric protein: G exoplasmic segment, HA membrane-spanning and cytoplasmic segment

▲ **Figure 17-39** The sorting of proteins destined for the apical and basolateral plasma membranes of epithelial cells. When a cultured line of kidney epithelial MDCK cells (see Figure 13-44) is infected simultaneously with VSV and influenza virus, the VSV glycoprotein (G protein) is found only on the basolateral membrane, whereas the HA glycoprotein of the influenza virus is found only on the apical membrane. Many other proteins are localized to a specific plasma-membrane surface. By immunoelectronmicroscopy, both proteins are observed in the same rough ER and Golgi vesicles; thus sorting to specific transport vesicles occurs during or just after the proteins move through the Golgi vesicles.

To define the sorting sequences, a chimeric protein containing the N-terminal exoplasmic segment of the HA glycoprotein and the C-terminal membrane-spanning and cytoplasmic segments of the VSV glycoprotein was expressed by recombinant techniques in MDCK cells. This protein was sorted exclusively to the apical membrane, identically to normal HA. Similarly, a chimera containing the exoplasmic segment of the VSV glycoprotein and the membrane-spanning and cytoplasmic segments of the HA glycoprotein was sorted to the basolateral membrane. Thus the sorting sequence resides in the exoplasmic segment of these proteins. *After M. G. Roth, D. Gundersen, N. Patel, and E. Rodriguez-Boulan, 1987, J. Cell Biol.* **104**:769–782; *T. Compton, I. Ivanov, T. Gottlieb, M. Rindler, M. Adesnik, and D. Sabatini, 1989, Proc. Nat'l Acad. Sci. USA* **86**:4112–4116.

are selectively localized to the basolateral membrane (see Figures 14-14 and 14-18) bind tightly to ankyrin. Once these integral proteins are transported from the Golgi vesicles to the basolateral membrane, they become locked into the ankyrin-fodrin cytoskeleton and cannot undergo endocytosis and subsequent transport to the apical membrane.

Membranes Recycle in Secretory Cells

During the secretion of hormones, enzymes, and structural proteins, exocytosis results in a considerable expansion of the plasma membrane, and the cell must recycle the majority of plasma-membrane phospholipids and proteins. One convenient tag that can be used to follow the fate of cell-surface proteins and lipids, a derivative of ferritin, is positively charged and binds to negatively charged surface glycolipids and glycoproteins. When added to a suspension of anterior pituitary cells, which secrete a number of peptide hormones, the ferritin binds tightly to the cell surface (Figure 17-40a). After incubation, the ferritin is taken up into endocytic vesicles and initially localized both to the Golgi cisternae and to lysosomes, as well as to CURL-like organelles—a result suggesting that the endocytic vesicles have fused with both the Golgi and lysosomes. Both cis and trans Golgi vesicles are enriched in the endocytosed ferritin early in this process. Later, more ferritin accumulates in the trans Golgi and their associated secretory vesicles (Figure 17-40b). In secretory cells, the Golgi vesicles apparently are the principal site for recycling surface-membrane constituents.

The Golgi complex, in summary, functions to direct vesicles and proteins to a variety of specific cell regions. In this organelle, lysosome enzymes and plasma-membrane proteins, proteins sorted to regulated or continuous secretory vesicles, and apical and basolateral plasma-membrane proteins are sorted to different vesicles, each destined for a different cellular site. The Golgi complex also is a major site for the recycling of the plasma-membrane components recovered by endocytosis. The Golgi amply earns its nickname as the "policeman" of vesicular traffic in the cell.

Summary

Membranes grow by expansion of existing membranes. Phospholipids are synthesized on the cytoplasmic face of the smooth ER membrane or on the bacterial plasma membrane. In a reaction catalyzed by certain membrane proteins, newly made phospholipids equilibrate with the exoplasmic membrane face. The transport of phospholipids and membrane proteins between organelles is mediated by small vesicles that bud from one organelle and fuse with another. During the processing of secretory and membrane proteins, vesicles migrate from the rough ER

(a)

(b)

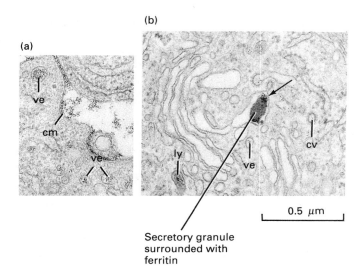

Secretory granule
surrounded with
ferritin

0.5 μm

◄ **Figure 17-40** The Golgi complex mediates the recycling of the cell membrane in secretory cells, as shown by an experiment in which rat anterior pituitary cells are incubated with a chemically modified form of the iron-containing protein ferritin, an electron-dense protein that binds irreversibly to negatively charged membrane components. (a) After 15 min of incubation at 37°C, the ferritin binds to the cell membrane *(cm)* and is taken up by endocytosis. Numerous endocytic vesicles *(ve)* containing ferritin are seen in the cytosol near the cell membrane. (b) After incubations of 60 min or longer, the ferritin is localized to the Golgi region. There, it becomes particularly concentrated around a secretory vesicle forming in the trans-most Golgi vesicle *(arrow)*, suggesting that incoming vesicles carrying ferritin fuse preferentially with trans Golgi elements. Ferritin is also found at this time in lysosomes *(ly)* but not in the ER or in coated vesicles *(cv)*. [See M. G. Farquhar, 1978, *J. Cell Biol.* 77:R35.] *Electron micrographs courtesy of M. G. Farquhar.*

to the cis Golgi, from one Golgi compartment to another, from the trans Golgi reticulum to secretory vesicles, and from there to the plasma membrane. During endocytosis and recycling of the plasma membrane, vesicles move from the plasma membrane to sorting organelles, lysosomes, or Golgi vesicles or from the Golgi to lysosomes. Some of these vesicles have clathrin coats. Membrane-containing organelles, such as chloroplasts and mitochondria, grow by the expansion of existing organelles. Proteins and lipids are added to existing organelles, and eventually the organelle divides in two.

All cytoplasmic ribosomes are functionally equivalent; different classes of proteins are synthesized on membrane-attached and membrane-unattached ribosomes, depending on a signal sequence in the protein itself. An important class of proteins synthesized on ribosomes bound to the rough ER includes secretory proteins, lysosome and Golgi enzymes, cell-surface proteins, and glycoproteins. Synthesis of this class of proteins is initiated on membrane-unattached ribosomes. A sequence of hydrophobic amino acids, the signal sequence, is recognized by a signal receptor particle (SRP), which, in turn, binds to an SRP receptor on the rough ER membrane. The SRP directs the insertion of the nascent protein into or across the membrane (probably with the help of the ER signal sequence receptor) and the binding of the ribosome to the ER membrane. Generally, such signal sequences are located at the N-terminus and are cleaved from the protein in the rough ER. Proteins must be unfolded to cross the ER membrane; a cytosolic, ATP-dependent unfolding enzyme is required for post-translational import into the ER.

Topogenic sequences allow newly made proteins to assume their appropriate orientations in the membrane. Such sequences include signal sequences, stop-transfer membrane-anchor sequences, and internal, uncleaved signal-anchor sequences.

Disulfide bonds are formed in the rough ER, and constituent polypeptides, in turn, form oligomeric proteins. Attachment of the mannose, glucose, and N-acetylglucosamine–containing oligosaccharide to specific asparagine residues of proteins also occurs in the lumen of the rough ER. Only properly folded secretory and membrane proteins are generally transferred to the Golgi vesicles. Certain resident ER proteins are retained there by a C-terminal KDEL sorting sequence.

Proteins move from the rough ER via membrane-bound vesicles to the Golgi complex. There, a number of additional modifications occur, some of which are important in targeting the protein to its final destination. Current work suggests that the Golgi complex consists of at least four distinct classes of vesicles: the cis, medial, and trans Golgi and the trans Golgi reticulum; distinct enzymatic activities occur in each type of vesicle. The direction of secretory proteins is rough ER → cis Golgi → medial Golgi → trans Golgi → trans Golgi reticulum → secretory vesicles.

Enzymes localized in the Golgi vesicles act sequentially to process a ubiquitous, high-mannose N-linked oligosaccharide to one of a variety of complex N-acetylneuraminic acid–bearing forms. Also, oligosaccharides attached to serine and threonine residues on some secretory and membrane proteins are elongated in the Golgi. The Golgi membrane contains specific permeases that allow nucleotide-sugar precursors of oligosaccharides to enter the Golgi lumen.

Mannose residues in the N-linked oligosaccharides of lysosome enzymes are phosphorylated in the cis Golgi. A mannose 6-phosphate receptor then binds these proteins in the trans Golgi reticulum and directs their transfer to lysosomes. A similar receptor on the cell surface binds extracellular, phosphorylated lysosome enzymes and, by receptor-mediated endocytosis, delivers them to lysosomes. In the trans Golgi or trans Golgi reticulum of polarized epithelial cells, proteins are sorted to vesicles destined for different regions of the plasma membrane. Some proteins also are directed to precursors of regulated secretory vesicles, where they are concentrated and stored to await a neural or hormonal signal for exocytosis. Often additional proteolytic cleavages (generally at a sequence of two basic amino acids) occur in the newly made, usually acidic secretory vesicles. Many classes of hormones, such as insulin and ACTH, are synthesized as inactive precursor proteins; proteolytic cleavage is essential to generate the active hormone.

Other secretory proteins and most membrane proteins are targeted to vesicles that continuously fuse with the plasma membrane, exocytosing their contents. Thus the Golgi complex plays a key role in sorting newly made secretory and membrane proteins; proper sorting is thought to be directed by specific amino acid sequences on the proteins themselves. Especially in secretory cells, the Golgi vesicles also function in recycling the plasma membrane; regions of the surface membrane are endocytosed in small vesicles, which then fuse with Golgi or lysosome vesicles.

References

Synthesis of Phospholipids

*BISHOP, W. R., and R. M. BELL. 1988. Assembly of phospholipids into cellular membranes: biosynthesis, transmembrane movement and intracellular translocation. *Ann. Rev. Cell Biol.* 4:579–610.

*CARMAN, G. M., and S. A. HENRY. 1989. Phospholipid biosynthesis in yeast. *Ann. Rev. Biochem.* 58:635–669.

PAGANO, R. E. 1988. What is the fate of diacylglycerol produced at the Golgi apparatus? *TIBS* 13:202–205.

PAGANO, R., and R. G. SLEIGHT. 1985. Defining lipid transport pathways in animal cells. *Science* 229:1051–1057.

*SLEIGHT, R. G. 1987. Intracellular lipid transport in eukaryotes. *Ann. Rev. Physiol.* 49:193–208.

Transport of Proteins into or across the ER Membrane

CHIRICO, W. J., G. WATERS, and G. BLOBEL. 1988. 70K heat shock related proteins stimulate protein translocation into microsomes. *Nature* 332:805–810.

CONNOLLY, T., and R. GILMORE. 1989. The signal recognition particle receptor mediates the GTP-dependent displacement of SRP from the signal sequence of the nascent polypeptide. *Cell* 57:599–610.

*DESHAIES, R. J., B. D. KOCH, and R. SCHEKMAN. 1988. The role of stress proteins in membrane biogenesis. *TIBS* 13:384–388.

EVANS, E. A., R. GILMORE, and G. BLOBEL. 1986. Purification of microsome signal peptidase as a complex. *Proc. Nat'l Acad. Sci. USA* 83:581–585.

*GIERASCH, L. M. 1989. Signal sequences. *Biochemistry* 28:923–930.

KAISER, C. A., D. PREUSS, P. GRISAFI, and D. BOTSTEIN. 1987. Many random sequences functionally replace the secretion signal sequence of yeast invertase. *Science* 235:312–317.

*SIEGEL, V., and P. WALTER. 1988. Functional dissection of the signal recognition particle. *TIBS* 13:314–316.

TAJIMA, S., L. LAUFFER, V. L. RATH, and P. WALTER. 1986. The signal recognition particle receptor is a complex that contains two distinct polypeptide chains. *J. Cell Biol.* 103:1167–1178.

WALTER, P., and G. BLOBEL. 1981. Translocation of proteins across the endoplasmic reticulum: III. Signal recognition protein causes signal sequence-dependent and site-specific arrest of chain elongation that is released by microsomal membranes. *J. Cell Biol.* 91:557–561.

*WALTER, P., and V. R. LINGAPPA. 1986. Mechanism of protein translocation across the endoplasmic reticulum membrane. *Ann. Rev. Cell Biol.* 2:499–516.

WIEDMANN, M., T. V. KURZHALIA, E. HARTMANN, and T. A. RAPOPORT. 1987. A signal sequence receptor in the endoplasmic reticulum membrane. *Nature* 328:830–833.

Synthesis of Membrane Proteins on the ER

MUECKLER, M., and H. F. LODISH. 1986. The human glucose transporter can insert posttranslationally into microsomes. *Cell* 44:629–637.

PAABO, S., B. M. BHAT, W. S. M. WOLD, and P. A. PETERSON. 1987. A short sequence in the COOH-terminus makes an adenovirus membrane glycoprotein a resident of the endoplasmic reticulum. *Cell* 50:311–317.

*RAPOPORT, T. A., and M. WIEDMANN. 1985. Application of the signal hypothesis to the incorporation of integral membrane proteins. *Curr. Top. Membr. Trans.* 24:1–63.

SHAW, A. S., P. J. M. ROTTIER, and J. K. ROSE. 1988. Evidence for the loop model of signal-sequence insertion into the endoplasmic reticulum. *Proc. Nat'l Acad. Sci. USA* 85:7592–7596.

WESSELS, H. P., and M. SPIESS. 1988. Insertion of a multispanning membrane protein occurs sequentially and requires only one signal sequence. *Cell* 55:61–70.

*WICKNER, W., and H. F. LODISH. 1985. Multiple mechanisms of insertion of proteins into and across membranes. *Science* 230:400–407.

YOST, C. S., J. HEDGPETH, and V. R. LINGAPPA. 1983. A stop-transfer sequence confers predictable transmembrane orientation to a previously secreted protein in cell-free systems. *Cell* 34:759–766.

Posttranslational Modifications of Proteins in the Rough ER

*ARFIN, S. M., and R. A. BRADSHAW. 1988. Cotranslational processing and protein turnover in eukaryotic cells. *Biochemistry* 27:7980–7984.

BOLE, D. G., L. M. HENDERSHOT, and J. F. KEARNEY. 1986. Posttranslational association of immunoglobulin heavy-chain binding protein with nascent heavy chains in nonsecreting and secreting hybridomas. *J. Cell Biol.* 102:1558–1566.

*A book or review article that provides a survey of the topic.

COPELAND, C. S., K.-P. ZIMMER, K. R. WAGNER, G. A. HEALEY, I. MELLMAN, and A. HELENIUS. 1988. Folding, trimerization, and transport are sequential events in the biogenesis of influenza virus hemagglutinin. *Cell* **53**:197–209.

FLYNN, G. C., T. G. CHAPPELL, and J. E. ROTHMAN. 1989. Peptide binding and release by proteins implicated as catalysts of protein assembly. *Science* **245**:385–390.

*FREEDMAN, R. B. 1989. Protein disulfide isomerase: multiple roles in the modification of nascent secretory proteins. *Cell* **57**:1069–1072.

HURTLEY, S. M., D. G. BOLE, H. HOOVER-LITTY, A. HELENIUS, and C. S. COPELAND. 1989. Interactions of misfolded influenza virus hemagglutinin with binding protein (BiP). *J. Cell Biol.* **108**:2117–2126.

LIPPINCOTT-SCHWARTZ, J., J. S. BONIFACINO, L. C. YUAN, and R. D. KLAUSNER. 1988. Degradation from the endoplasmic reticulum: disposing of newly synthesized proteins. *Cell* **54**:209–220.

LIPPINCOTT-SCHWARTZ, J., L. C. YUAN, J. S. BONIFACINO, and R. D. KLAUSNER. 1989. Rapid distribution of Golgi proteins into the ER in cells treated with Brefeldin A: evidence for membrane cycling from Golgi to ER. *Cell* **56**:801–813.

*LODISH, H. F. 1988. Transport of secretory and membrane glycoproteins from the rough endoplasmic reticulum to the Golgi. *J. Biol. Chem.* **263**:2107–2110.

MUNRO, S., and H. R. B. PELHAM. 1987. A C-terminal signal prevents secretion of luminal ER proteins. *Cell* **48**:899–907.

*OLSON, T. S., and M. D. LANE. 1989. A common mechanism for posttranslational activation of plasma-membrane receptors? *FASEB J* **3**:1618–1624.

*PELHAM, H. R. B. 1989. Heat shock and the sorting of luminal ER proteins. *EMBO Journal* **8**: 3171–3176.

*ROSE, J. K., and R. W. DOMS. 1988. Regulation of protein export from the endoplasmic reticulum. *Ann. Rev. Cell Biol.* **4**:257–288.

Protein Glycosylation

*HIRSCHBERG, C. B., and M. D. SNIDER. 1987. Topography of glycosylation in the rough endoplasmic reticulum and Golgi apparatus. *Ann. Rev. Biochem.* **56**:63–87.

*KORNFELD, R., and S. KORNFELD. 1985. Assembly of asparagine-linked oligosaccharides. *Ann. Rev. Biochem.* **45**:631–664.

KOZARSKY, K., D. KINGSLEY, and M. KRIEGER. 1988. Use of a mutant cell line to study the kinetics and function of O-linked glycosylation of low-density lipoprotein receptors. *Proc. Nat'l Acad. Sci. USA* **85**:4335–4339.

*LENNARZ, W. J. 1987. Protein glycosylation in the endoplasmic reticulum: current topological issues. *Biochemistry* **26**:7206–7210.

MILLA, M. E., and C. B. HIRSCHBERG. 1989. Reconstitution of Golgi vesicle CMP-sialic acid and adenosine 3′-phosphate 5′-phosphosulfate transport into proteoliposomes. *Proc. Natl Acad. Sci. USA* **86**:1786–1790.

*PAULSON, J. C. 1989. Glycoproteins: what are the sugar chains for? *Trends Biochem. Sci.* **14**:272–276.

REICHNER, J. S., S. W. WHITEHEART, and G. W. HART. 1988. Intracellular trafficking of cell-surface sialoglycoconjugates. *J. Biol. Chem.* **263**:16316–16326.

STANLEY, P. 1989. Chinese hamster ovary-cell mutants with multiple glycosylation defects for production of glycoproteins with minimal carbohydrate heterogeneity. *Mol. Cell. Biol.* **9**:377–383.

The Golgi Complex and the Sorting of Secretory and Membrane Proteins

ANDERSON, R. G. W., and R. K. PATHAK. 1985. Vesicles and cisternae in the trans Golgi apparatus of human fibroblasts are acidic compartments. *Cell* **40**:635–643.

BAKER, D., L. HICKE, M. REXACH, M. SCHLEYER, and R. SCHEKMAN. 1988. Reconstitution of SEC gene product-dependent intercompartmental protein transport. *Cell* **54**:335–344.

*BALCH, W. E. 1989. Biochemistry of interorganelle transport. *J. Biol. Chem.* **264**:16965–16968.

*BOURNE, H. F. 1988. Do GTPases direct membrane traffic in secretion? *Cell* **53**:669–671.

*FEATHERSTONE, C. 1988. Perforated cell systems to study membrane transport. *TIBS* **13**:284–286.

*GRIFFITHS, G., and K. SIMONS. 1986. The trans Golgi network: sorting at the exit site of the Golgi complex. *Science* **234**:438–443.

HOLCOMB, C. L., W. J. HANSEN, T. ETCHEVERRY, and R. SCHEKMAN. 1988. Secretory vesicles externalize the major plasma-membrane ATPase in yeast. *J. Cell Biol.* **106**:641–648.

*KLAUSNER, R. D. 1989. Sorting and traffic in the central vacuolar system. *Cell* **57**:703–706.

MALHOTRA, V., T. SERAFINI, L. ORCI, J. C. SHEPHERD, and J. E. ROTHMAN. 1989. Purification of a novel class of coated vesicles mediating biosynthetic protein transport through the Golgi stack. *Cell* **58**:329–336.

ORCI, L., V. MALHOTRA, M. AMHERDT, T. SERAFINI, and J. E. ROTHMAN. 1989. Dissection of a single round of vesicular transport: sequential intermediates for intercisternal movement in the Golgi stack. *Cell* **56**:357–368.

PAYNE, G. S., and R. SCHEKMAN. 1985. A test of clathrin function in protein secretion and cell growth. *Science* **230**:1009–1014.

*SCHEKMAN, R. 1985. Protein localization and membrane traffic in yeast. *Ann. Rev. Cell Biol.* **1**:115–143.

*TARTAKOFF, A. M. 1987. *The Secretory and Endocytic Paths.* J. Wiley.

WILSON, D. W., C. A. WILCOX, G. C. FLYNN, E. CHEN, W-J. KUANG, W. J. HENZEL, M. R. BLOCK, A. ULLRICH, and J. E. ROTHMAN. 1989. A fusion protein required for vesicle-mediated transport in both mammalian cells and yeast. *Nature* **339**:355–359.

Synthesis of Lysosome and Vacuole Proteins

*DAHMS, N. M., P. LOBEL, and S. KORNFELD. 1989. Mannose 6-phosphate receptors and lysosomal enzyme targeting. *J. Biol. Chem.* **264**:12115–12118.

GEUZE, H. J., W. STOORVOGEL, G. J. STROUS, J. W. SLOT, J. E. BLEEKEMOLEN, and I. MELLMAN. 1988. Sorting of mannose 6-phosphate receptors and lysosome-membrane proteins in endocytic vesicles. *J. Cell Biol.* **107**:2491–2501.

GREEN, S. A., K.-P. ZIMMER, G. GRIFFITHS, and I. MELLMAN. 1987. Kinetics of intracellular transport and sorting of lysosome-membrane and plasma-membrane proteins. *J. Cell Biol.* **105**:1227–1240.

GRIFFITHS, G., B. HOFLACK, K. SIMONS, I. MELLMAN, and S. KORNFELD. 1988. The mannose 6-phosphate receptor and the biogenesis of lysosomes. *Cell* **52**:329–341.

KLIONSKY, D. J., L. M. BANTA, and S. D. EMR. 1988. Intracellular sorting and processing of a yeast-vacuole hydrolase: proteinase A propeptide contains vacuole targeting information. *Mol. Cell. Biol.* **8**:2105–2116.

*KORNFELD, S. 1987. Trafficking of lysosome enzymes. *FASEB J.* **1**:462–468.

*NEUFELD, E. F. 1989. Natural history and inherited disorders of a lysosomal enzyme, β-hexosaminidase. *J. Biol. Chem.* **264**:10927–10930.

ROBINSON, J. S., D. J. KLIONSKY, L. M. BANTA, and S. D. EMR. 1988. Protein sorting in *Saccharomyces cerevisiae*: isolation of mutants defective in the delivery and processing of multiple vacuole hydrolases. *Mol. Cell Biol.* **8**:4936–4948.

TAGUE, B. W., and M. J. CHRISPEELS. 1987. The plant-vacuole protein, phytohemagglutinin, is transported into the vacuole of transgenic yeast. *J. Cell Biol.* **105**:1971–1979.

Proteolytic Processing of Secretory and Membrane Proteins

*DOUGLASS, J., O. CIVELLI, and E. HERBERT. 1984. Polyprotein gene expression: generation of diversity of neuroendocrine peptides. *Ann. Rev. Biochem.* **53**:665–715.

*FISHER, J. M., and R. H. SCHELLER. 1988. Prohormone processing and the secretory pathway. *J. Biol. Chem.* **263**:16515–16518.

FULLER, R. S., A. BRAKE, and J. THORNER. 1989. Yeast prohormone processing enzyme (KEX2 gene product) is a Ca^{2+}-dependent serine protease. *Proc. Nat'l Acad. Sci. USA* **86**:1434–1438.

*FULLER, R. S., R. E. STERNE, and J. THORNER. 1988. Enzymes required for yeast prohormone processing. *Ann. Rev. Physiol.* **50**:345–362.

*NEURATH, H. 1989. Proteolytic processing and physiological regulation. *Trends Biochem. Sci.* **14**:268–271.

ORCI, L., M. RAVAZZOLA, M. AMHERDT, O. MADSEN, A. PERRELET, J.-D. VASSALLI, and R. G. W. ANDERSON. 1986. Conversion of proinsulin to insulin occurs coordinately with acidification of maturing secretory vesicles. *J. Cell Biol.* **103**:2273–2281.

THOMAS, G., B. A. THORNE, L. THOMAS, R. G. ALLEN, D. E. HRUBY, R. FULLER, and J. THORNER. 1988. Yeast KEX2 endopeptidase correctly cleaves a neuroendocrine prohormone in mammalian cells. *Science* **241**:226–241.

TOOZE, J., M. HOLLINSHEAD, R. FRANK, and B. BURKE. 1987. An antibody specific for an endoproteolytic cleavage site provides evidence that proopiomelanocortin is packaged into secretory granules in AtT20 cells before its cleavage. *J. Cell Biol.* **105**:155–162.

Regulated Secretion

*BURGESS, T. L., and R. B. KELLY. 1987. Constitutive and regulated secretion of proteins. *Ann. Rev. Cell Biol.* **3**:243–293.

CHANDLER, D. E., and J. E. HEUSER. 1980. Arrest of membrane-fusion events in mast cells by quick-freezing. *J. Cell Biol.* **86**:666–674.

CHUNG, K.-N., P. WALTER, G. W. APONTE, and H.-P. H. MOORE. 1989. Molecular sorting in the secretory pathway. *Science* **243**:192–197.

ORCI, L., M. RAVAZZOLA, M. AMHERDT, A. PERRELET, S. K. POWELL, D. L. QUINN, and H.-P. H. MOORE. 1989. The trans-most cisternae of the Golgi complex: a compartment for sorting of secretory and plasma-membrane proteins. *Cell* **51**:1039–1051.

*ORCI, L. M., J.-D. VASSALLI, and A. PERRELET. 1988. The insulin factory. *Sci. Am.* **256**:85–94.

RIVAS, R. J., and H-P. H. MOORE. 1989. Spatial segregation of the regulated and constitutive secretory pathways. *J. Cell Biol.* **109**:51–60.

Synthesis of Plasma-Membrane Proteins in Polarized Cells

*BARTLES, J. R., and A. L. HUBBARD. 1988. Plasma-membrane protein sorting in epithelial cells: do secretory pathways hold the key? *TIBS* **13**:181–184.

*MATLIN, K. S. 1986. The sorting of proteins to the plasma membrane in epithelial cells. *J. Cell Biol.* **103**:2565–2568.

NELSON, W. J., and R. W. HAMMERTON. 1989. A membrane-cytoskeletal complex containing Na^+, K^+-ATPase, ankyrin, and fodrin in Madin-Darby canine kidney (MDCK) cells: implications for the biogenesis of epithelial cell polarity. *J. Cell Biol.* **108**:893–902.

RINDLER, M. J., I. E. IVANOV, H. PLESKEN, E. RODRIGUEZ-BOULAN, and D. D. SABATINI. 1984. Viral glycoproteins destined for apical or basolateral plasma-membrane domains traverse the same Golgi apparatus during their intracellular transport in doubly infected Madin-Darby canine kidney cells. *J. Cell Biol.* **98**:1304–1319.

*RODRIGUEZ-BOULAN, E., and J. NELSON. 1989. Morphogenesis of the polarized epithelial cell phenotype. *Science.* **245**:718–725.

*SIMONS, K., and S. D. FULLER. 1985. Cell-surface polarity in epithelia. *Ann. Rev. Cell Biol.* **1**:243–288.

C H A P T E R **18**

Organelle Biogenesis: The Nucleus, Chloroplast, and Mitochondrion

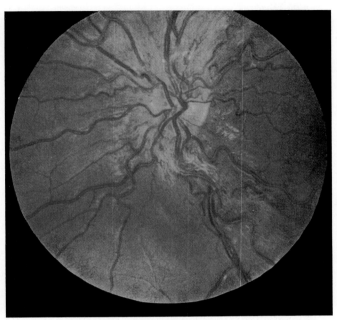

Tortuous retinal blood vessels in Leber's hereditary optic neuropathy, a mitochondrial genetic disease

*A*lthough the nucleus, chloroplast, and mitochondrion vary greatly in structure and function, many aspects of their biosynthesis are similar, which is why they are grouped together in this chapter. In particular, most chloroplast and mitochondrial proteins and all nuclear proteins are synthesized outside the organelle on cytoplasmic ribosomes unbound to the rough endoplasmic reticulum. Newly made proteins are released into the cytosol and then taken up specifically into the proper organelle. Protein uptake is an energy-requiring process that depends on specific protein-receptor interactions on the organelle surface. These organelles are surrounded by two membranes, and protein uptake occurs at the points at which the inner and outer membranes appear to fuse. In the nucleus, these are the nuclear pores. We begin our discussion of organelle biogenesis with the cell nucleus, which is frequently the largest cell organelle. ▲

Assembly and Disassembly of the Nuclear Membrane

The *nuclear envelope,* a prominent feature of the nucleus, is composed of the outer and inner nuclear membranes, separated by the *perinuclear cisternal space.* The outer membrane is continuous with the rough ER and usually contains bound ribosomes. Nuclear pores perforate the nuclear envelope and provide a link between the nucleus and the cystosol. In the investigation of membrane assembly, the nuclear membrane in higher plant and animal cells presents a special problem because it disappears in late prophase during mitosis and re-forms around the daughter chromosomes during telophase (Figure 18-1). In lower eukaryotes, in contrast, the nuclear envelope remains intact throughout mitosis.

Lamina Proteins Are a Principal Determinant of Nuclear Architecture

A protein-rich layer, the *nuclear lamina,* lines the inner surface of the nuclear membrane in interphase cells (see Figure 18-1), forming a discrete layer 30–100 nm thick that connects the inner membrane with chromatin. After using nonionic detergents to extract the membrane proteins and chromatin, the lamina appears as a network of orthogonal sets of filaments ~10 nm in diameter (Figure 18-2a). Purified lamina are composed of three principal extrinsic membrane proteins, lamins A, B, and C (MW 60,000–70,000), which together form a fibrous network. Lamins A and C are identical, except for the presence of an additional 133 amino acids at the carboxyl end of lamin A; the two mRNAs are formed from alternately spliced transcripts of the same gene.

Lamins are highly homologous in sequence and structure to *intermediate filament* (IF) *proteins.* Isolated lamins have rodlike tails 52 nm long and globular heads (see Figure 18-2b). The formation of the long lamina filaments, like those of IF proteins, is primarily mediated by interactions between the globular heads; the exact struc-

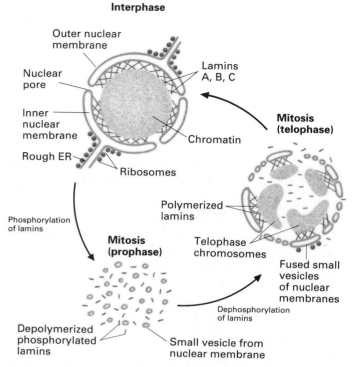

▲ Figure 18-1 The inner and outer nuclear membranes in interphase cells reversibly depolymerize during mitosis. In prophase, phosphorylation triggers depolymerization of the nuclear lamina into small oligomers of lamins. This process, in turn, causes the nuclear membranes to disassemble into small vesicles that disperse in the cytosol. Simultaneously, chromatin condenses into chromosomes. In telophase, the condensed mitotic chromosomes disperse into chromatin. This process apparently induces the dephosphorylation and polymerization of lamins into a fibrous network and the subsequent fusion of membrane vesicles into the characteristic interphase nuclear membranes. How and when the pore complexes form is not known. [See L. Gerace, 1986. *Trends Biochem. Sci.* **11**:443–446.]

▶ Figure 18-2 Visualization of the nuclear lamina and isolated lamins by electron microscopy. (a) A nuclear membrane from a hand-dissected frog oocyte was fixed to an electron microscope grid and then extracted with a nonionic detergent to remove membrane and soluble proteins. The nuclear lamina consists of two orthogonal sets of filaments 10 nm in diameter built of lamins. (b) Isolated lamins A and C examined after rotary shadowing each have a 52-nm-long rodlike tail with two globular heads. How these lamins polymerize into long lamina filaments is not known. *From U. Aebi, J. Cohen, L. Buhle, and L. Gerace, 1986,* Nature **323**:560–564. *Courtesy of U. Aebi.*

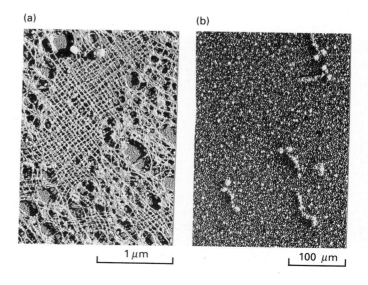

ture of these filaments is not known. Lamin B appears to have a specialized membrane-binding role; only lamin B binds with high affinity to lamin-depleted nuclear membranes. Inner nuclear membranes contain a MW 58,000-lamin B receptor that binds specifically to lamin B; lamins A and C bind to lamin B and mediate interactions between the lamina and chromatin. In interphase cells, all three lamins are localized to the inner surface of the nuclear membrane.

Lamin Phosphorylation Is Correlated with Disassembly of the Nuclear Membrane

In the prophase stage of mitosis in higher eukaryotes, observations with the light microscope suggest that the nuclear membrane simply disappears as the chromatin condenses into mitotic chromosomes. The electron microscope, however, shows that the nuclear membrane fragments into a number of small vesicles that are indistinguishable from the ER (see Figure 18-1, prophase). Lamin B remains associated with these vesicles; lamins A and C are depolymerized to small oligomers and dispersed throughout the cell.

Disassembly of the lamina is correlated with phosphorylation of lamins; indeed, most workers feel that phosphorylation of lamins by a lamin kinase triggers disassembly. Disassembly of nuclei can be studied in *cell-free systems,* in which the nuclear envelopes of nuclei from frog eggs or interphase cultured cells are triggered to disassemble by exposure to cytosoplasmic proteins from mitotic cells. The first process observed, phosphorylation and depolymerization of lamins, is followed by the breakdown of the nuclear membrane into small vesicles and the condensation of chromosomes.

A complex multiprotein factor, the *mitosis-promoting factor* (MPF), has been purified from both mitotic somatic cells and unfertilized frog eggs (blocked in meiosis); it is absent in nonmitotic cells. MPF itself triggers chromatin condensation and disassembly of interphase nuclear envelopes in cell-free extracts. One of the polypeptides in MPF is a protein kinase. The importance of this kinase is underscored by its sequence homology to the protein product of the yeast *cdc2* gene, one of several *cell division cycle* genes in which mutation arrests cells in early mitosis. The *cdc2* kinase is present in an inactive form during interphase. It is activated during mitosis by *cyclin,* a protein that accumulates during interphase and is destroyed during mitosis. The cyclin-activated *cdc2* kinase may be responsible for lamin phosphorylation.

During mitosis in cultured animal cells, all vesicular traffic in the cell ceases. There is no endocytosis, proteins do not move from the rough ER to the Golgi, and secretory vesicles cannot be induced to fuse with the plasma membrane. The same cellular signals that induce the breakdown of the nuclear membrane and disassembly of

the nuclear lamina also may block the fission and fusion of other cellular membranes.

Chromatin Decondensation and Lamin Dephosphorylation Initiate Nuclear Reassembly

During telophase, after the daughter chromosomes have separated and begin to decondense, nuclear assembly is induced. Small nuclear membrane vesicles fuse with each other around the dispersing chromatin; simultaneously, the lamins repolymerize on the inner nuclear membrane and apparently form a bridge between the chromatin and the membrane (see Figure 18-1, telophase). In frog embryos and some other cells, the nuclear structure reassembles in a two-stage process; nuclear membranes first form around individual chromosomes and then fuse together to form a single interphase nucleus.

Studies using frog eggs suggest that the decondensed chromatin signals assembly of the nuclear envelope. Nuclear division occurs every 30 min in a fertilized frog egg. These divisions do not require any new protein synthesis by the embryo; the unfertilized egg contains pools of histones and lamins sufficient for 20,000 cells. If any DNA is microinjected into an unfertilized egg, the stored histones and other nuclear proteins bind to the injected DNA and form decondensed chromatin. Nuclear envelopes soon assemble around the injected DNA; these envelopes have a normal appearance and contain lamins on their inner surface.

During nuclear reassembly, lamins become dephosphorylated. Cell-free extracts prepared from mitotic cells carry out assembly of nuclear membranes and can be used to study aspects of the reassembly process. For instance, the removal of lamins (by specific antilamin antibodies) prevents membrane assembly. Apparently, decondensed chromatin somehow induces the lamins to become dephosphorylated and polymerize into a fibrous network that, in turn, causes small vesicles to fuse to form a normal interphase nuclear membrane.

Protein Import into the Cell Nucleus

The nucleus contains no functional ribosomes and does not carry out protein synthesis. All nuclear proteins must be synthesized in the cytosol and then taken up into the organelle. These include ribosomal proteins, which are incorporated into ribosomes in the nucleolus, nuclear proteins, such as lamins and histones, and DNA and RNA polymerases. Also, most (if not all) of the protein components of snRNPs (small nuclear ribonucleoprotein particles) that participate in RNA splicing bind to snRNAs in the cytosol, and the completed snRNP is taken up into the nucleus. For many years, cell biologists have speculated that the nuclear pore is a highly selective passageway through which only nuclear-destined proteins and particles enter the organelle and a portal through which mRNAs and ribosomes, but not resident nuclear proteins, exit the nucleus. Recent work supports these important roles for nuclear pores; such pores allow only complete ribosomal subunits and mRNAs to move from the nucleus to the cytosol.

Pores in the nuclear envelope have many of the properties of a molecular sieve. They are freely permeable to ions and small molecules; when injected into the cytosol of a frog oocyte, a small molecule, such as radiolabeled sucrose, diffuses into the nucleus as fast as it passes through the cytosol. Virtually any globular, non-nuclear proteins <9 nm in diameter (corresponding to ~60,000 MW) can diffuse into the nucleus, although smaller proteins or particles enter it at faster rates.

Most Nuclear Proteins Are Selectively Imported into Nuclei

As a result of such studies, workers believed that proteins enter the nucleus by free diffusion through pores and are selectively retained by binding to some nuclear component that is not freely diffusible, such as the lamina or chromatin. Recent work, however, has shown that, in most cases, nuclear proteins are selectively bound to receptors on the nuclear pores and then actively imported into the nucleus.

Many of these studies have made use of *nucleoplasmin,* a large (MW 165,000) pentameric protein present in high concentrations in the soluble phase of the frog oocyte nucleus. When microinjected into the oocyte cytosol, this protein rapidly accumulates in the oocyte nucleus (Figure 18-3a). The rate and extent of accumulation are much greater than can be explained by diffusion through nuclear pores, implying the existence of some sort of specific uptake mechanism. Experiments that remove the tail (C-terminal) regions from all nucleoplasmin pentamers yield a pentameric core that is unable to enter the nucleus (Figure 18-3b). However, if just one tail remains on a pentamer, the nucleoplasmin pentamer becomes concentrated in the nucleus as does a single, unattached tail. These studies demonstrate that the tail domain bears a discrete sequence for accumulation in the nucleus, but does the tail specify selective *entry* or selective *retention?*

If the tail domain specifies selective retention (say, by binding to the lamina), then tail-less pentameric "cores" injected into the nucleus should diffuse back into the cytosol. When the experiment is performed (Figure 18-3c), however, these cores are retained in the nucleus but are soon degraded. Thus the tail region does not specify binding in the nucleus but does dictate specific entry into it.

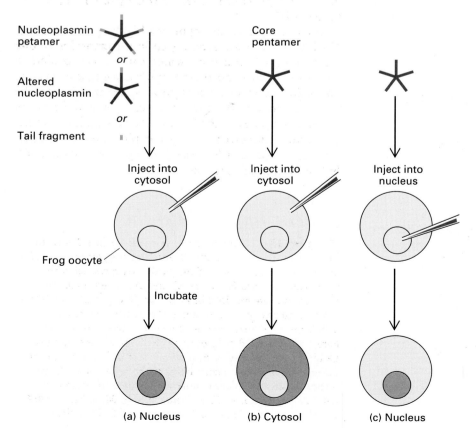

◀ **Figure 18-3** The tail of nucleoplasmin contains a sequence that regulates transport into the nucleus. (a) After microinjection into the oocyte cytosol, the intact nucleoplasmin, or a proteolytic pentameric fragment of nucleoplasmin with at least one intact subunit, or the isolated tail fragment accumulates in the nucleus. (b) The core pentamer of nucleoplasmin, minus all tails, cannot enter the nucleus after microinjection. (c) When the core pentamer is microinjected into the nucleus, it remains there, indicating that the missing tail sequences do not specify binding in the nucleus; rather, they specify entry into the nucleus. *After C. Dingwall and R. Laskey, 1986, Ann. Rev. Cell Biol. 2:367–390.*

Nuclear Pores Are the Portals for Protein Import

Subsequent experiments with nucleoplasmin have shown that the nuclear pore is the route for protein import. In one key study, small gold particles coated with nucleoplasmin or with a non-nuclear protein were microinjected into the cytosol of frog oocytes. The location of the gold particles was examined by electron microscopy after sectioning the oocytes. Shortly after injection, the nucleoplasmin-gold clusters at the nuclear pores (Figure 18-4); later, it accumulates in the nucleus, presumably after passing through the pore complexes. Gold particles coated with non-nuclear proteins remain in the cytosol and do not bind to nuclear pores.

These same pores transport RNA out of the nucleus. Gold particles coated with tRNA or 5S RNA and injected into the nucleus of frog oocytes migrate outward into the cytosol. Electron microscopy reveals that all the pores contain RNA-gold particles.

Further studies with isolated nuclei have shown that the cytoplasmic face of the pore complex contains specific receptor binding sites for nuclear proteins. When added to isolated nuclei, purified and radiolabeled nuclear proteins concentrate selectively inside the nuclei. However, ATP hydrolysis is required: in the absence of ATP, the proteins bind specifically at the cytoplasmic face of the pore complex; when ATP is added, the bound proteins are imported. Thus the energy released by ATP hydrolysis is essential to import a nuclear protein across a pore but is not required to bind the protein to the pore receptor.

Different Proteins Utilize Different Signal Sequences for Nuclear Import

A diverse array of proteins is specifically targeted to the nucleus; some examples of proteins with defined signal sequences for nuclear import are listed in Table 18-1. It is immediately obvious that many different amino acid sequences serve as signals for the nuclear import of proteins. Although most sequences contain basic amino acids, there is little similarity between them. Consequently, we do not know if one pore receptor binds all such sequences or if several different pore receptors individually bind single sequences.

The prototype for these studies was the SV40 T antigen. A variety of recombinant-generated mutant T proteins were synthesized in cultured cells, and the fate of the T antigen was monitored. Mutations in the sequence of the seven amino acids depicted in Table 18-1 were found to reduce or abolish nuclear binding. More importantly, when this sequence of seven amino acids is linked to the N-terminus of a normally cytoplasmic protein, the protein enters the nucleus. Similar studies of several yeast and viral proteins have indicated that a short sequence of amino acids generally specifies nuclear import, although quite different sequences are utilized by different proteins (Table 18-1).

In summary, nuclear import is a highly specific process. Nuclear proteins are synthesized in the cytosol with one or more signal sequences, which cause them to bind to receptor-like elements on the cytoplasmic face of nuclear pores. An energy-dependent process then translocates the bound proteins to the nucleus. Precisely how the proteins traverse the pores is not known. Nor do we know much about the structure of the pore complex itself, except that it is composed of at least six polypeptides containing O-linked N-acetylglucosamine sugars.

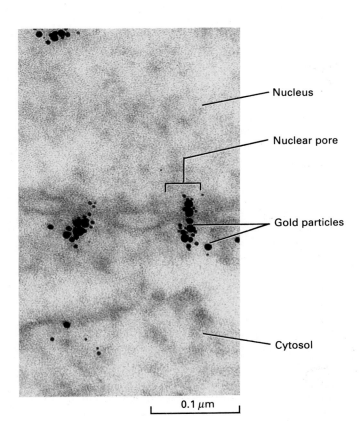

Nucleus

Nuclear pore

Gold particles

Cytosol

0.1 μm

◄ **Figure 18-4** Proteins move through nuclear pores from the cytosol into the nucleus. When frog oocytes are microinjected with gold particles 5 nm in diameter coated with nucleoplasmin (see Figure 18-3) and examined under the electron microscope 15 min later, the gold particles are clustered at nuclear pore sites along the nuclear membrane. When the cells were examined at later times (not shown here), many coated gold particles had entered the nucleus. Gold particles coated with proteins that lack sequences for nuclear entry do not accumulate at the pores or enter the nucleus. Thus the nuclear pore is the route of protein entry. *After C. M. Feldherr, E. Kallenbach, and M. Schultz, 1984, J. Cell. Biol.* **99**:2216–2222. *Courtesy of C. M. Feldherr.*

Table 18-1 Amino acid sequences that act as signals for translocation to the cell nucleus

Source	Nuclear protein	Location of signal in protein	Deduced signal sequence
SV40	Large T antigen	Residues 126–132	Pro^{126} Lys Lys Lys Arg Lys Val^{132}
Influenza virus	Nucleoprotein (NP)	C-terminal residues 336–345	Ala Ala Phe Glu Asp Leu Arg Val Leu Ser
Adenovirus	Ela	C-terminal amino acids	Lys Arg Pro Arg Pro
Yeast	matα2	Residues 1–13	Lys^3 Ileu Pro Ileu Lys^7
Yeast	Ribosomal protein L3	Residues 1–21	Pro^{18} Arg Lys Arg^{24}

SOURCE: C. Dingwall and R. Laskey, 1986, *Ann. Rev. Cell Biol.* 2:367–390; R. H. Lyons, B. Ferguson, and M. Rosenberg, 1987, *Mol. Cell Biol.* 7:2451–2456.

Mitochondrial DNA: Structure, Expression, and Variability

In our study of organelle assembly, we now turn to the mitochondrion; its replication has fascinated cell biologists for decades. First, we discuss the structure, protein-coding properties, and expression of mitochondrial DNA from animals, plants, and lower eukaryotes and note some surprising variations among different mitochondrial DNAs. (Even the genetic codes used by mitochondria from different species vary!) Then we turn to the assembly of mitochondria, focusing on the import and assembly of the multiprotein complexes involved in electron transport and ATP synthesis.

Individual mitochondria are large enough to be seen under the light microscope and can be followed by time-lapse cinematography (Figure 18-5). As cells grow, mitochondria increase in size and one or more daughter mitochondria "pinch off" in a manner similar to the way in which bacterial cells grow and divide. Mitochondrial growth and division is not coupled to nuclear division.

Cytoplasmic Inheritance and DNA Sequencing Have Established the Existence of Mitochondrial Genes

Prior to the isolation and sequencing of mitochondrial DNA (mtDNA), studies of mutants in yeasts and other single-celled organisms indicated that mitochondria contain their own genetic system and exhibit *cytoplasmic inheritance* (Figure 18-6). *Petite yeast mutants* grow more slowly than wild-type yeasts and form small colonies. They are incapable of oxidative phosphorylation, however, and produce ATP by fermentation of glucose to ethanol, which provides less ATP per mole of glucose than oxidative phosphorylation does. The petite yeast mutation exhibits structurally abnormal mitochondria that are incapable of oxidative phosphorylation. In genetic crosses between different (haploid) yeast strains, the petite mutation does not segregate with any known nuclear gene or chromosome. Because the inheritance of the petite or wild-type mitochondrial phenotype is clearly nonchromosomal, some mutable element in the cytoplasm must be a determinant in mitochondrial synthesis. (Later studies have shown that petite mutations contain deletions of some or all mtDNA.)

Mitochondrial inheritance in yeast is biparental: during the fusion of haploid cells, both parents contribute equally to the cytoplasm of the diploid. In mammals and most other animals, the sperm contributes little (if any) cytoplasm to the zygote, and most (if not all) of the mitochondria in the embryo must be derived from those in the egg, not the sperm. Indeed, different strains of cows and rats have mtDNAs that vary slightly in DNA sequence;

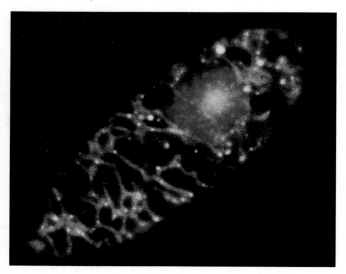

▲ **Figure 18-5** Detection of mitochondrial and nuclear DNA by fluorescence microscopy in a growing *Euglena gracilis* cell. Cells are treated with a mixture of ethidium bromide and DiOC6, which cause the nuclear DNA to emit a vermillion fluorescence and the mitochondria green. Areas rich in mitochondrial DNA fluoresce yellow—a combination of vermillion DNA and green mitochondrial fluorescence. Replication of mitochondrial DNA and division of the mitochondrial network can thus be followed in living cells. *From Y. Huyashi and K. Veda, 1989,* J. Cell Sci. *93, 565.*

10 μm

when two such strains are mated, the mtDNA in the offspring is invariably the maternal type.

In higher plants, too, mtDNA is inherited exclusively in a uniparental fashion through the female parent (egg), not the male (pollen). Rearrangements in mtDNA produce cytoplasmic male sterility (CMS), a trait found in maize and many other plants. Novel genes that are not present in typical mitochondrial genomes are found in some cases of CMS. Because it prevents self-pollination, the uniparental inheritance of CMS makes it very useful in the production of hybrid seed.

The mtDNA is located in the matrix and is sometimes found attached to the inner mitochondrial membrane. The entire mitochondrial genome has now been cloned and sequenced from a number of different organisms, and mtDNAs from all these sources have been found to encode a similar set of rRNAs, tRNAs, and essential mitochondrial proteins (Table 18-2). Mitochondria encode rRNAs, which form *mitochondrial ribosomes,* although all but one or two of the ribosomal proteins (depending on the species) are imported from the cytosol. All of the tRNAs used for protein synthesis in the mitochondrion are encoded by mtDNAs. As far as is known, all transcripts of mtDNA and their translation products remain in the organelles; there is no export of RNAs or proteins. Hybridization studies have shown that virtually all RNAs found in the mitochondrion are synthesized there on mtDNA templates. A nuclear-encoded RNA is imported into the mitochondrion in only one known case: a 135-bp RNA forms an essential component of a site-specific endonuclease involved in the metabolism of a primer RNA for mtDNA replication.

Mitochondrial ribosomes differ from cytoplasmic ribosomes in their RNA and protein compositions, small size, and sensitivity to certain antibiotics. For instance, cycloheximide inhibits protein synthesis by eukaryotic cytoplasmic ribosomes but does not affect protein synthesis by mitochondrial ribosomes. All proteins are synthesized on mitochondrial ribosomes in the presence of cycloheximide and are encoded by mtDNA. Most proteins local-

▶ **Figure 18-6** Cytoplasmic inheritance of the petite mutation in yeast. Petite-strain mitochondria are defective in oxidative phosphorylation due to a deletion in mtDNA.
(a) Haploid cells fuse to produce a diploid cell that undergoes meiosis, during which a random segregation of parental chromosomes and mitochondria containing mtDNA occurs. Since yeast normally contains ~50 mtDNA molecules per cell, all products of meiosis usually contain both normal and petite mtDNAs and are capable of respiration. (b) As these cells grow and divide mitotically, the cytoplasm (including the mitochondria) is randomly distributed to the daughter cells. Occasionally, a cell is generated that contains only defective petite mtDNA and yields a petite colony. Thus formation of such petite cells is independent of any nuclear genetic marker.

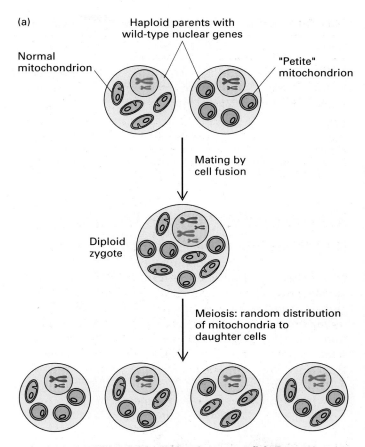

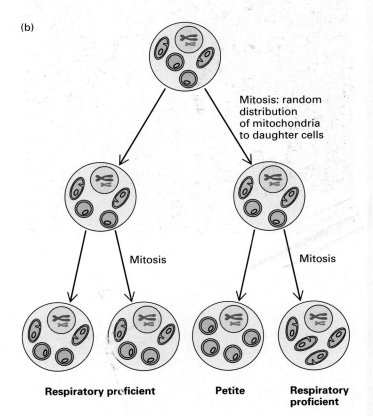

ized in mitochondria are synthesized on cytoplasmic ribosomes and must be imported into the mitochondria. All mitochondrial-synthesized polypeptides identified thus (with one possible exception) are not complete enzymes but subunits of multimeric complexes used in electron transport or ATP synthesis (see Table 18-2).

The Size and Coding Capacity of mtDNA Varies in Different Organisms

Importantly, the sizes of the mtDNA, the arrangement of the genes, the mode of transcription, and even the mitochondrial genetic code itself are very different in various organisms. The human mtDNA, a completely sequenced circular molecule of 16,569 bp, is among the smallest known (Figure 18-7). Human mtDNA encodes the two rRNAs found in mitochondrial ribosomes and the 22 tRNAs used to translate the mt mRNAs. It has 13 sequences that begin with an ATG (methionine) codon, end with a chain-termination codon, and are long enough to encode a polypeptide of more than 50 amino acids. All of these possible proteins have been identified: three are

subunits of the cytochrome *c* oxidase complex, two are subunits of the F_0 ATPase, seven are subunits of the NADH-CoQ reductase complex, and one is the cytochrome *b* subunit of the CoQ-cytochrome *c* reductase complex. These are the major proteins known to be synthesized on mitochondrial ribosomes.

Invertebrate mtDNA is about the same size as human mtDNA, but yeast mtDNA is almost five times as large (~78,000 bp). Yeast mtDNA and mtDNA from other lower eukaryotes encode many of the same gene products as mammalian mtDNA: three subunits of the cytochrome *c* oxidase complex, the cytochrome *b* subunit, two subunits of the F_0 ATPase, one 15S rRNA and one 21S rRNA, and multiple tRNAs. Yeast and fungal mtDNAs encode one ribosomal protein (termed var-1) not present in human mtDNA. Seven subunits of the NADH-CoQ reductase complex are encoded in mammalian mtDNA and in mtDNA from the fungus *Neurospora crassa*, but no homologous sequences are found in yeast mtDNA; in yeasts, these sequences are encoded by nuclear DNA. Conversely, one mitochondrial protein, subunit 9 of the F_0 ATPase, is encoded by mtDNA in such organisms as yeasts and higher plants but by nuclear DNA in others, such as *N. crassa*, and in mammals. *Neurospora* mtDNA does contain an apparently nonfunctional gene (a *pseudogene*) homologous in sequence to the nuclear gene for subunit 9.

Plant mtDNAs are much larger and more variable than the mtDNAs of other organisms (see Table 18-2). They range in size from 200,000–2,500,000 bp; even in a single family, there can be as much as an eightfold variation in size (watermelon = 330,000 bp; muskmelon = 2,500,000 bp)! Plant mtDNAs contain a few genes not found in other mtDNAs: a 5S mitochondrial rRNA (not found in other mitochondrial ribosomes), one subunit (α) of the F_1 ATPase, and subunit 9 of the F_0 ATPase. Plant mt ribosomal RNAs are also considerably larger than mt ribosomal RNAs in animal and fungal mitochondria. But this does not account for even a fraction of the "extra" mtDNA in plants, and details of the coding capacity of plant mtDNA must await DNA sequencing. In contrast to the single, circular genome of animal and fungal mtDNAs, plant mtDNAs contain multiple circular DNAs that appear to recombine with each other (Figure 18-8).

These striking findings suggest that genes moved from the mitochondrion to the nucleus, or vice versa, during evolution. Translocations of short segments of mtDNA to nuclear DNA may be occurring still; DNA hybridization and sequence studies have identified short segments (~50 bp long) of mtDNA interspersed randomly in the nuclear DNA of all animals and plants studied; these short segments do not appear to encode any proteins. Plant mitochondrial genomes of many species contain different chloroplast DNA sequences, such as fragments of the ri-

Table 18-2 Mitochondrial genes and their products

	Animal	Yeast	Fungi *Neurospora*	Plant
	SIZE (THOUSANDS OF bp)			
	14–18	78	19–108	200–2500
rRNAs				
large subunit	16S	21S	21S	26S
small subunit	12S	15S	15S	18S
5S RNA	–	–	–	+
tRNAs	22	23–25	23–25	~30
Ribosomal protein (var-1)	–	+	+	?
Cytochrome *c* oxidase subunits 1, 2, 3	+	+	+	+
CoQ-cytochrome *c* reductase Apocytochrome *b*	+	+	+	+
F_0 ATPase complex				
subunit 6	+	+	+	+
subunit 8	+	+	+	+
subunit 9	–	+	–	+
F_1 ATPase complex				
subunit α	–	–	–	+
NADH-CoQ reductase subunits	7	0	6	6?

SOURCE: V. K. Eckenrode and C. S. Levings, 198?, *In vitro Cell and Devel. Biol.* 22:169–176; C. S. Levings and G. G. Brown, 1989, *Cell* 56:171–179.

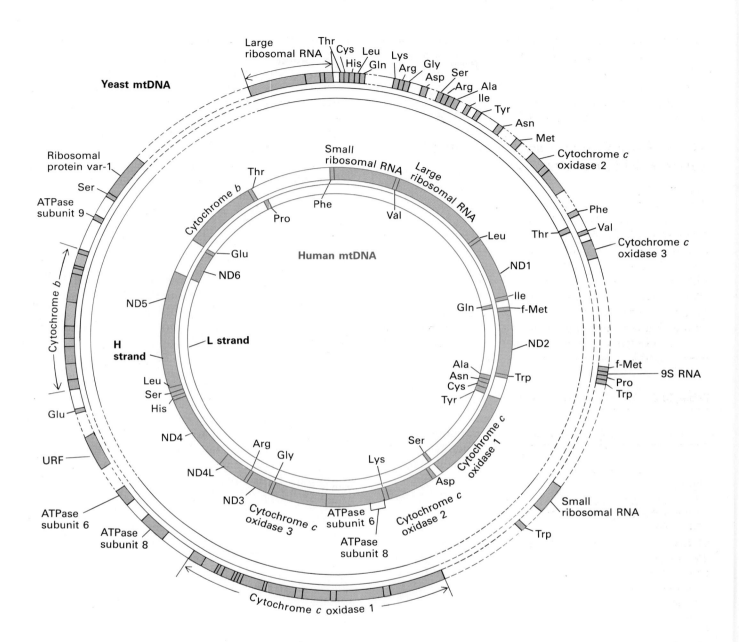

▲ **Figure 18-7** The organization of human *(inner)* and yeast *(outer)* mtDNA. Yeast mtDNA is five times larger than human mtDNA but is represented here as only two and one-half times as large. Proteins and RNAs encoded by each of the two strands are shown separately. Transcription off the outer (H) strand of each mtDNA is clockwise and off the inner (L) strand is counterclockwise.

The entire human mtDNA has been sequenced. For yeast, the diagram is based on a compilation of partially complete DNA sequence data from several laboratories; the dashed lines represent unsequenced regions. Exons of structural genes and URFs (unidentified or open reading frames) are orange; rRNA and tRNA genes are green. The abbreviations for amino acids denote the tRNA genes. Yeast genes for cytochrome *c* oxidase subunits 1 and 2, cytochrome *b*, and 21S rRNA contain introns (blue). No mammalian mtDNA

genes contain introns. The human 207-bp gene encoding ATPase subunit 8 overlaps, out of frame, with the N-terminal portion of the segment encoding ATPase subunit 6.

Note that yeast and mammalian mtDNAs encode some different proteins. Human mtDNA encodes seven subunits of the NADH-CoQ reductase complex (ND1, ND2, ND3, ND4L, ND4, ND5, ND6) that have no counterparts in yeast mtDNA; also, F_1 ATPase subunit 9 is encoded by mtDNA in yeast but by nuclear DNA in humans. Most of the nontranscribed regions of yeast mtDNA contain adenine and thymine almost exclusively, and their function (if any) is not known. [See P. Borst and L. A. Grivell, 1981, *Nature* **290**:443; L. A. Grivell, 1983, *Scientific American* **248**(3):78–89; A. Tzagoloff and A. Myers, 1986, *Ann. Rev. Biochem.* **55**:249–285; A. Chomyn et al., 1985, *Nature* **314**:592; A. Chomyn et al. 1986, *Science* **234**:614–618.]

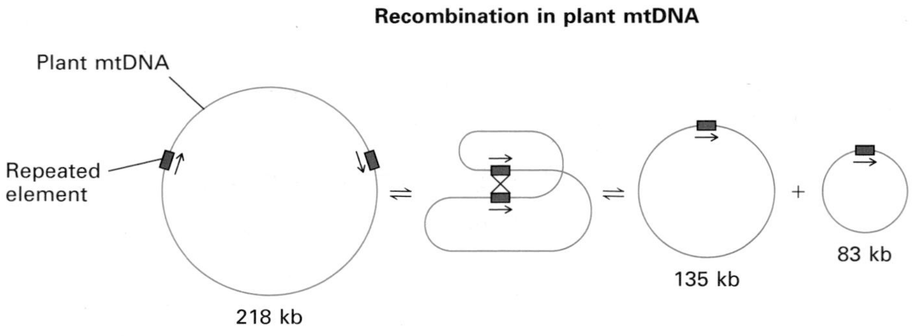

Recombination in plant mtDNA

◀ **Figure 18-8** Most plant mitochondria contain multiple related DNAs. As an example, the largest circular DNA, found in the mitochondria of *Brassica campestris* (Chinese cabbage), is 218 kb and contains two repeated 2-kb elements (red rectangles). Recombination between these two sequences generates the two small DNAs (135 kb and 83 kb) also found in these mitochondria. *After J. D. Palmer and C. R. Shields, 1984,* Nature *307:437–440.*

bulose 1,5-bisphosphate carboxylase gene, suggesting the transfer of DNA between these two organelles.

Mitochondrial Genetic Codes Are Different in Different Organisms

The genetic code used in mitochondria is different from the standard code used in all prokaryotic and eukaryotic nuclear genes; remarkably, the code is even different in mitochondria from different species (Table 18-3). UGA, normally a stop codon, is read by human and fungal mitochondrial translation systems as tryptophan; in plant mitochondria, UGA is still a stop codon. AGA and AGG are standard arginine codons that encode arginine in fungal and plant mtDNA, but they are stop codons in mammalian mitochondria and serine codons in *Drosophila*. In plant mtDNA CGG can code for either tryptophan or arginine! In RNA transcripts of mtDNA certain CGG sequences are "edited" (chemically converted) to UGG, in which case the codon specifies tryptophan. Unedited CGG codons encode the normal arginine. RNA editing as well as the similarity but nonuniversality of the genetic code have profound implications for the evolution of eukaryotic cells and their organelles.

Animal Mitochondrial RNAs Undergo Extensive Processing

Like their nuclear counterparts, mitochondrial rRNAs and mRNAs are generated by the enzymatic cleavage of longer precursors (Figure 18-9). However, processing of mammalian mtDNA transcripts reveals a number of novel features. First, transcription initiates at only two points on the H strand (so-called because it bands at a heavier density in CsCl). *Primary transcript I* begins just upstream of the tRNA^{Phe} gene and terminates just after the 16S rRNA gene; it is cleaved to tRNA^{Phe}, tRNA^{Val}, 12S rRNA, and 16S rRNA. *Primary transcript II* initiates just downstream of primary transcript I, near the 5′ end of the 12S rRNA gene, but continues past the termination site of transcript I and around the circular DNA to the start point. Transcript II is processed by enzymatic hydrolysis to yield the remaining tRNAs and all polyadenylated mRNAs, but apparently not any rRNAs. Very often some of these cleavages occur on nascent chains. The cleavage sites must be very precise to separate the mRNAs from the tRNAs, because the 5′ end of each mRNA is immediately adjacent to the 3′ end of a tRNA. Transcript I initiates about 10 times as frequently as transcript II, account-

Table 18-3 Alterations in the standard genetic code in mitochondria

Codon	Prokaryotes and eukaryotic nuclear-encoded proteins; chloroplasts	Mitochondria				
		Mammals	*Drosophila*	*Neurospora*	**Yeast**	**Plants**
UGA	Stop	Trp	Trp	Trp	Trp	Stop
AGA AGG	Arg	Stop	Ser	Arg	Arg	Arg
AUA	Ileu	Met	Met	Ileu	Met	Ileu
AUU	Ileu	Met	Met	Met	Met	Ileu
CGG	Arg	Arg	Arg	Arg	Arg	Trp and Arg
CUU CUC CUA CUG	Leu	Leu	Leu	Leu	Thr	Leu

SOURCE: S. Anderson et al., 1981, *Nature* 290:457; P. Borst, in *International Cell Biology 1980–1981*, H. G. Schweiger, ed., Springer-Verlag, p. 239; C. Breitenberger and U. L. Raj Bhandary, 1985, *Trends Biochem. Sci.* 10:478–483; V. K. Eckenrode and C. S. Levings, 1986, *In vitro Cell Devel. Biol.* 22:169–176; J. M. Gualber et al., 1989, *Nature* 341: 660–662; and P. S. Covello and M. W. Gray, 1989, *Nature* 341: 662–666.

ing for the synthesis of rRNAs (and, curiously, of two tRNAs) in excess of mRNAs and other tRNAs.

In contrast to mammalian nuclear-synthesized mRNAs, mammalian mitochondrial mRNAs contain very few untranslated sequences. The first three bases at the 5′ end of each mRNA are generally the AUG (or AUA) initiator codon. A UAA terminator codon is usually at or very near the 3′ end of the mRNA. In some cases, only the U of this terminator codon is encoded in the mtDNA; the final two A's are part of the polyadenylate sequence, which is added after the mRNA precursor is synthesized, as it is in cytoplasmic mRNAs.

Thus human and other mammalian mtDNAs have evolved to contain as few untranslated sequences as possible. The genome is the absolute minimum required to

generate the requisite mitochondrial mRNAs, tRNAs, and rRNAs, and all transcripts are extensively processed. In yeasts and plants, in contrast, much of the mtDNA is not transcribed into RNA and has no known function.

Yeast mtRNAs Are Transcribed from Multiple Promoters and Spliced

In yeast mtDNA, the genes are separated by long stretches of adenine-thymine–rich noncoding sequences, and at least three genes—those encoding cytochrome *b* and the subunits 1 and 2 of the cytochrome *c* oxidase complex—contain intervening sequences (introns). Unlike human mtDNA, each gene in yeast mtDNA appears to use a different promoter.

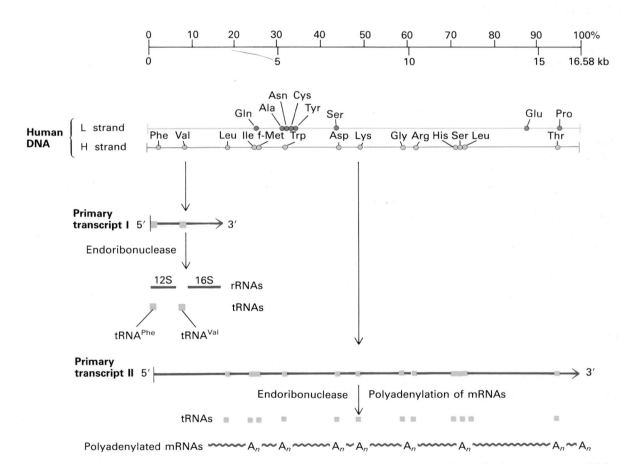

▲ **Figure 18-9** Transcription map of human mtDNA. As deduced from the DNA sequence and RNA-DNA hybridization studies, the light (L) DNA strand encodes only eight tRNAs (dark green circles); this strand is transcribed right to left. The heavy (H) DNA strand encodes the 12S and 16S rRNAs, 14 tRNAs (light green circles), and 11 predominant species of polyadenylated mRNAs. The mRNAs encode all known mitochondrial proteins. Transcription of the H strand is initiated at two sites. Primary transcript I initiates just upstream of the tRNA^Phe gene and terminates just after the 16S

rRNA; it is processed by endoribonucleases to yield one molecule each of tRNA^Phe, tRNA^Val, and 12S and 16S rRNA. Primary transcript II initiates near the 5′ end of the 12S rRNA and apparently continues completely around the circular mtDNA; it is processed to yield the other tRNAs and mRNAs, which are subsequently polyadenylated (A_n). Some of the cleavages of this transcript begin while the chain is nascent. [See D. Ojala et al., 1981, *Nature* **290**:470; J. Montoya, G. L. Gaines, and G. Attardi, 1983, *Cell* **34**:151.]

In the "long-gene" strain of yeast, the gene for cytochrome *b* is split into six exons (Figure 18-10a). In the "short-gene" strain, the first three introns are missing and the first four exons are contiguous in the DNA. Both genes encode the same protein; the mRNA precursor for cytochrome *b* in each strain is spliced to remove the transcripts of the intervening DNA sequences. But the manner of splicing the long-gene transcript into cytochrome *b*

mRNA is apparently unique to fungal DNA: the entire cytochrome *b* gene is transcribed into a precursor RNA that contains transcripts of all six exons. The first splice fuses exons E_1 and E_2, removing a 1000-base RNA that corresponds to intron I_1 (Figure 18-10b). This splice is self-catalyzed by RNA but facilitated by one or more nucleus-encoded proteins; certain nuclear mutations inhibit this splicing reaction.

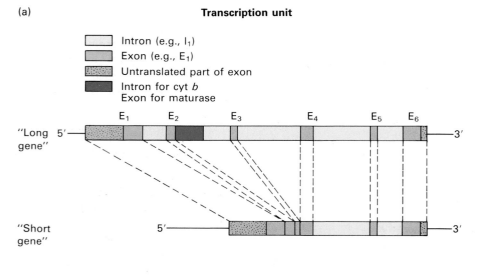

(a) **Transcription unit**

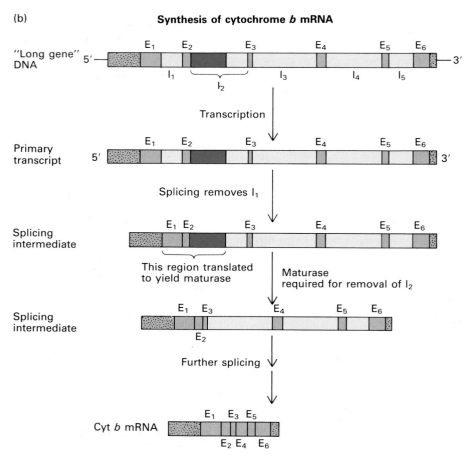

(b) **Synthesis of cytochrome *b* mRNA**

◀ **Figure 18-10** The cytochrome *b* gene of yeast. (a) The mtDNA of "long-gene" yeast strains contains six exons for cytochrome *b*. The DNA from the closely related "short-gene" yeast strains contains only three exons. Two of these correspond to exons E_5 and E_6 of the long-gene strain; the other represents a fusion of exons E_1, E_2, E_3, and E_4 with the precise deletion of the introns present in the long-gene strain. (b) Production of cytochrome *b* mRNA in the "long-gene" strain involves several novel splicing stages. Intron I_1 is removed from the primary RNA transcript by splicing; a protein facilitating this reaction is encoded by nuclear DNA. The resulting RNA (the splicing intermediate) is translated into a protein with sequences at its N-terminus that are encoded by exons E_1 and E_2 of cytochrome *b* and a sequence at its C-terminus (red) that is encoded by intron I_2 of cytochrome *b*. Only the splicing activity of this chimeric protein (maturase) can remove intron I_2. Three other splicing reactions generate mature cytochrome *b* mRNA. I_2 is a type II intron; in cell-free reactions, RNAs that contain such an intron undergo a self-catalyzed splicing similar to that of *Tetrahymena* rRNA (see Figure 8-58). In intact cells, the splicing of I_2 and other introns in mtRNA is in some way facilitated by protein. [See A. Tzagoloff and A. Myers, 1986, *Ann. Rev. Biochem.* 55:249–285; T. R. Cech, 1986, *Cell* 44:207–210.]

A URF (unidentified reading frame) in intron I_2 is continuous with the preceding exon, E_2. The removal of I_1 creates a continuous URF, beginning with the 143 amino acids of exons E_1 and E_2 of cytochrome b at the N-terminus and ending with ~250 amino acids encoded by intron I_2 at the C-terminus. The intermediate RNA is translated by mitochondrial ribosomes to yield a chimeric protein, called a *maturase*, which is required for the second splice; it is not known if maturase is the actual splicing enzyme, however. This second splice removes all of intron I_2, including the URF that encodes the maturase part of the chimeric protein. Thus, by splicing, the maturase destroys its own mRNA—a novel way of regulating the extent of translation of an mRNA! Intron-encoded maturase was first identified by genetic studies as a result of the key discovery that mutants in intron I_2 (a supposedly noncoding segment of the cytochrome b gene) abolished production of the mature cytochrome b polypeptide.

Any advantages of such a baroque process of mRNA splicing are unknown and mysterious—especially since the short gene for cytochrome b still operates efficiently even though it lacks the first three introns, including the segment that encodes the maturase. In the long gene for cytochrome b, it is remarkable that the intron of one gene (for cytochrome b) is the exon of another gene (for the maturase).

Synthesis and Localization of Mitochondrial Proteins

Most Mitochondrial Proteins Are Synthesized in the Cytosol as Precursors

Most proteins required for oxidative phosphorylation and other mitochondrial processes are synthesized on cytoplasmic ribosomes and imported into the mitochondrion. Even the mitochondrial DNA and RNA polymerases are synthesized in the cytosol, as are all but one or two of mitochondrial ribosomal proteins. Some cytoplasmic proteins are transported to the intermembrane space (the c, c_1, and b_2 cytochromes and cytochrome c peroxidase), to the outer membrane (porin), and to the inner membrane (cytochrome c oxidase subunits). The largest number is transported to the matrix (F_1 ATPase subunits, ribosomal proteins, RNA polymerase, and so on; Table 18-4). How is a protein targeted to its final destination, and what drives these specific, unidirectional transport processes?

Current research suggests that proteins are imported into the mitochondria after they are synthesized in the cytosol. Pulse-chase studies on yeast and *Neurospora* cells have demonstrated that newly made mitochondrial proteins initially are localized to the cytosol outside the mitochondria and then accumulate gradually at their proper destinations in the mitochondrion during the chase period. Furthermore, most mitochondrial proteins destined for the matrix, the intermembrane space, or the inner membrane are found to contain an additional 20–60 amino acids, called the *uptake-targeting sequence*, at the N-terminus that are not present in the mature protein (Figure 18-11).

Table 18-4 Some mitochondrial proteins synthesized in the cytosol

Mitochondrial location	Protein*

Matrix	F_1 ATPase
	   subunit α (except plants)
	   subunit β
	   subunit γ
	   subunit δ (certain fungi)

	Carbamoyl phosphate synthase (mammals)
	Mn^{2+}-superoxide dismutase
	RNA polymerase
	Ribosomal proteins
	Citrate synthase and other citric acid enzymes
	Ornithine transcarbamoylase (mammals)
	Ornithine aminotransferase (mammals)
	Alcohol dehydrogenase (yeast)
Inner membrane	Cytochrome c_1
	ADP-ATP carrier

	Cytochrome c oxidase
	   subunit 4
	   subunit 5
	   subunit 6
	   subunit 7

	Proteolipid of F_0 ATPase complex

	Cytochrome bc_1 complex
	(CoQ-cytochrome c reductase)
	   subunit 1
	   subunit 2
	   subunit 4 (cytochrome c_1)
	   subunit 5 (Fe-S protein)
	   subunit 6
	   subunit 7
	   subunit 8

	Uncoupling protein
Intermembrane space	Cytochrome c
	Cytochrome c peroxidase
	Cytochrome b_2
Outer membrane	Porin

*Most proteins (except the ADP-ATP carrier, cytochrome c, and porin) are fabricated as longer precursors.

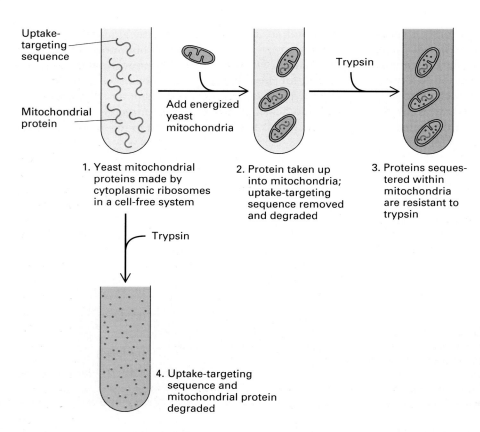

1. Yeast mitochondrial proteins made by cytoplasmic ribosomes in a cell-free system

2. Protein taken up into mitochondria; uptake-targeting sequence removed and degraded

3. Proteins sequestered within mitochondria are resistant to trypsin

4. Uptake-targeting sequence and mitochondrial protein degraded

◀ **Figure 18-11** A demonstration that mitochondrial proteins are imported into the organelle post-translationally, concomitant with the cleavage of the N-terminal uptake-targeting sequence. Most mitochondrial proteins synthesized on cytoplasmic ribosomes have an uptake-targeting sequence at their N-termini that is not found on the mature protein. Such precursors can be identified in cells following a brief pulse of radioactive amino acid or synthesized in a cell-free system programmed with cytosolic mRNA, as depicted in step 1. When respiring mitochondria with a proton-motive force across the inner membrane are added, the protein is taken up into the organelle and the uptake-targeting sequence is removed by a protease in the matrix (step 2). Protein uptake can be demonstrated by adding trypsin or another protease to the reaction medium. Proteins sequestered in the mitochondrion are resistant to the protease (step 3) because it cannot penetrate the mitochondrial membranes. In contrast, the precursor protein in the cytosol is totally destroyed by the protease (step 4).

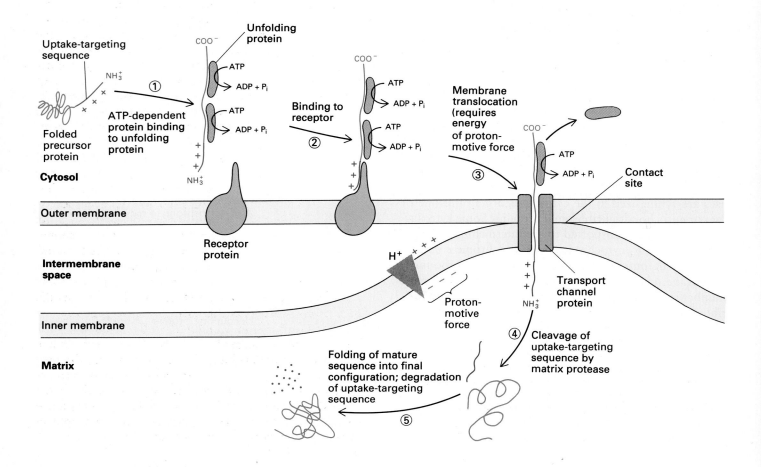

These precursor proteins, including hydrophobic integral membrane proteins, are soluble in the cytosol; there, they bind to one or more unfolding proteins, which use the energy released by ATP hydrolysis to keep the precursor proteins in such a state that they can be taken up by mitochondria (Figure 18-12). During mitochondrial uptake, the precursors of integral membrane proteins undergo a major conformational change to bind to the hydrophobic core of the inner or outer membrane. This change may be due in part to cleavage of all or a portion of the N-terminal uptake-targeting sequence in the matrix.

Many imported proteins do not fold spontaneously in the matrix but require a protein catalyst. One such catalyst is *hsp60*, a *heat-shock protein* localized to the matrix. Yeast mutants defective in *hsp60* import such proteins as the β subunit of the F_1 ATPase normally. Cleavage of the uptake-targeting sequence is also normal, but the polypeptide fails to assemble into a normal multiprotein complex. In amino acid sequence and function, *hsp60* is related to cytosolic proteins (see Figures 17-16 and 18-12) and to chloroplast and bacterial proteins that use the energy released by ATP hydrolysis to bind to unfolded or partially folded proteins to maintain their unfolded state or ensure proper folding. Such proteins are often termed *chaperonins*.

Multiple Signals Target Proteins to the Correct Submitochondrial Compartment

Targeting proteins to the matrix turns out to be the most direct procedure. Precursors of matrix proteins, such as alcohol dehydrogenase, have a *matrix-targeting sequence*

◄ **Figure 18-12** The import of a polypeptide into the matrix of a mitochondrion. The precursor protein, with its N-terminal uptake-targeting sequence, is synthesized in the cytosol in a compactly-folded form. It binds to an unfolding protein that uses energy released by ATP hydrolysis to partially unfold the precursor (step 1). The precursor–unfolding protein complex binds to a receptor on the outer membrane near a contact site with the inner membrane (step 2). The protein is then translocated across the outer and inner membranes by a process that requires a proton-motive force (pmf) across the inner membrane (step 3); the pmf is a combination of a membrane electric potential and a pH gradient. As depicted here, translocation occurs at rare sites at which the inner and outer membranes touch. The uptake-targeting sequence, having served its function, is removed by a matrix protease (step 4) and ultimately degraded. Concomitant with or just after step 4, the mature protein folds into its final, active configuration (step 5). [See E. C. Hurt, 1987, *Trends in Biochem. Sci., 1987,* **12**:369–370; M. Eilers and G. Schatz, 1988, *Cell* **52**:481–483; N. Pfanner, F.-U. Hartl, and W. Neupert, 1988, *Eur. J. Biochem.* **175**:205–212.]

at their N-terminus (Figure 18-13) that contains all the information required to target a protein from the cytosol to the mitochondrial matrix. A matrix enzyme removes these N-terminal sequences as they arrive in the matrix. During in vitro reactions, this enzyme, a two-subunit, metal-containing protease, specifically cleaves the N-terminal matrix-targeting sequence from several different precursor proteins. Indeed, this sequence is sufficient to direct a normally cytoplasmic protein, such as dihydrofolate reductase (DHFR), to the matrix: a chimeric protein containing the alcohol dehydrogenase matrix-targeting sequence fused to DHFR is transported to the matrix, and the sequence is cleaved normally.

The matrix-targeting sequences of a number of proteins do not share any extensive sequence homology, but they do exhibit some common characteristics. They are rich in positively charged arginine and lysine and hydroxylated serine and threonine; they also are devoid of aspartate and glutamate, acidic residues. Apparently, the receptor(s) for matrix-targeting sequences on the outer mitochondrial membrane (see Figure 18-12) can recognize a large number of related amino acid sequences.

Two different N-terminal uptake-targeting sequences on precursors to such proteins as cytochromes b_2 and c_1 target them to the intermembrane space. A matrix-targeting sequence at the N-terminus of the precursor directs the protein to the matrix first (Figure 18-14; see Figure 18-13), where the sequence is removed by the matrix protease. The second uptake-targeting sequence directs the protein from the matrix across the inner membrane to the intermembrane space. Recombinant DNA techniques allow a precursor cytochrome b_2 protein to be synthesized that lacks only the *intermembrane-space targeting sequence* (yellow in Figures 18-13 and 18-14); in its absence, the protein accumulates in the matrix, since this sequence directs the precursor across the inner membrane.

All mitochondrial proteins synthesized in the cytosol contain one or more uptake-targeting sequences, but these sequences are not removed in some cases. An important example is porin, the abundant outer mitochondrial membrane protein that forms channels through the phospholipid bilayer and accounts for the unusual permeability of this membrane to small proteins. The N-terminus of porin contains a short matrix-targeting sequence followed by a long stop-transfer stretch of hydrophobic amino acids (green in Figure 18-13); the latter causes the protein to divert from the normal matrix pathway and accumulate in the outer mitochondrial membrane. Neither sequence is normally removed. If the hydrophobic sequence is experimentally deleted from porin, the protein accumulates in the matrix space with its matrix-targeting sequence still attached. Proteins such as the ATP-ADP antiport, which are localized to the inner mitochondrial membrane, also utilize one or more uptake-targeting sequences that are not removed.

Imported protein	Location of imported protein	Functional domains in N-terminal uptake-targeting sequences

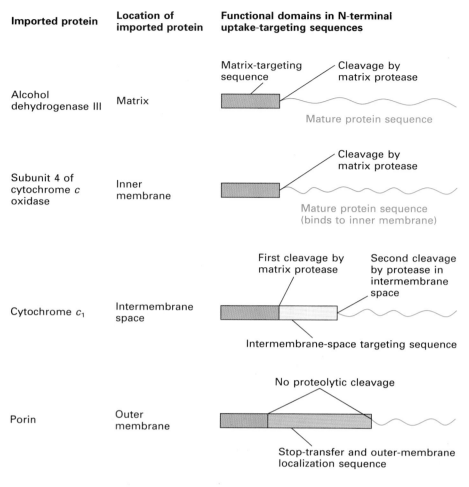

◀ Figure 18-13 One or more N-terminal uptake-targeting sequences direct most imported mitochondrial proteins first to the mitochondria and then to the correct mitochondrial subcompartment. Proteins targeted to the matrix or inner membrane have a single matrix-targeting sequence (red). Following translocation of the protein to the matrix and cleavage of the sequence, the protein either remains in the matrix (as with alcohol dehydrogenase) or inserts into the inner membrane (cytochrome oxidase subunit 4). Proteins such as cytochrome c_1, which are destined for the intermembrane space, utilize two sequences: (1) a matrix-targeting sequence (red) that directs the protein to the matrix and is cleaved, and (2) an intermembrane-space targeting sequence of hydrophobic amino acids (yellow) that targets the protein from the matrix across the inner membrane to the intermembrane space, where it is cleaved. Porin and other outer-membrane proteins have a typical matrix targeting sequence (red), followed by a stop-transfer sequence—a long hydrophobic sequence of amino acids (green). Neither of these sequences are cleaved; the latter appears to function in stopping the transport of porin across the outer membrane and in localizing the protein to that membrane. *After E. C. Hurt and A. P. G. M. van Loon, 1986,* Trends Biochem. Sci. **11**:204–207; *F-U. Hartl, N. Pfanner, D. Nicholson, and W. Neupert, 1989,* Biochim. Biophys. Acta **988**:1–45.

Translocation Intermediates Can Be Accumulated and Studied

The outer mitochondrial membrane contains one or more receptors for imported proteins, but energy is required for the actual transport. When the proton-motive force across the inner mitochondrial membrane is abolished by uncoupling agents, the precursor forms of the imported proteins bind tightly to specific sites on the outer membrane. However, these bound proteins are not transported into the organelle (see Figure 18-12, step 2). If the energy block is removed, these bound precursors are rapidly imported and processed. The binding of small amounts of any one precursor protein is inhibited by the presence of large amounts of certain precursor proteins but not others; this suggests that multiple, specific outer-membrane receptors are involved in import.

For instance, cytochrome *c*, a component of the intermembrane space, apparently has a specific uptake receptor. The cytoplasmic form of cytochrome *c*, called *apocytochrome c*, has the same amino acid sequence as the mature protein and lacks a heme group. Addition of

the heme group occurs in the intermembrane space of the mitochondrion concomitant with uptake. Only apocytochrome *c* is taken up by mitochondria; mature cytochrome *c* is not. The addition of heme probably causes the conformation of the protein to change so that it cannot bind to the receptor.

Mitochondria contain 100–1000 transport channels for the uptake of mitochondrial proteins. Evidence suggests that translocation of precursors to the matrix occurs at the rare sites of contact between the outer and inner membranes. Figure 18-15 shows one way in which an intermediate in the translocation process can be studied: the N-terminal matrix-targeting sequence of one matrix protein (a subunit of the F_1 ATPase) is cleaved in the matrix, but the C-terminus is still exposed to the cytosol. Microscopic studies have shown that the precursors accumulate at the points of contact between the inner and outer mitochondrial membranes (see Figure 18-15b). Outer membrane receptors and other components of the mitochondrial protein import machinery are believed to be localized at or near these contact sites.

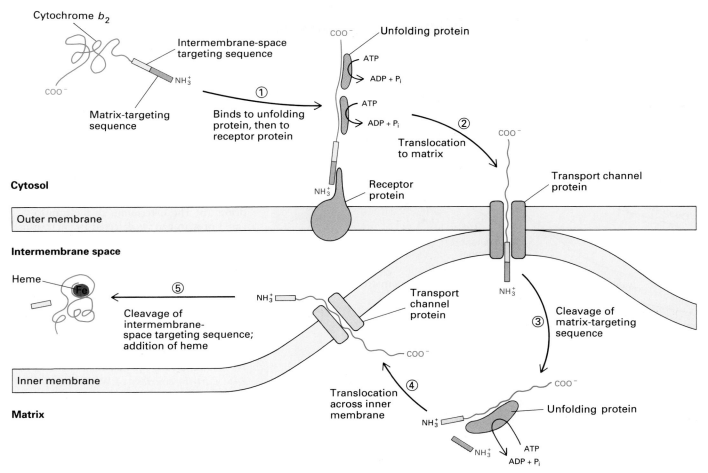

▲ **Figure 18-14** Two successive translocations are required to target proteins such as cytochrome c_1 and cytochrome b_2 to the intermembrane space. The precursors of cytochrome c_1 and cytochrome b_2 have two uptake-targeting sequences at their N-termini. The first, a matrix-targeting sequence (red), functions to target the protein to the mitochondrial matrix (steps 1–3), exactly as if it were a typical mitochondrial matrix protein (see Figure 18-12). This segment of the sequence is cleaved by the matrix protease (step 3). The second, an intermembrane-space targeting sequence (yellow), targets the protein to the inner membrane, presumably by binding to a receptor, and directs the translocation of the protein to the intermembrane space (step 4). There, the sequence is cleaved by a specific protease; heme is added, and the cytochrome folds into its mature configuration (step 5). *After F-U. Hartl, J. Ostenmann, B. Guiard, and W. Neupert, 1987, Cell 51:1021–1027; E. C. Hurt and A. P. G. M. van Loon, 1986, Trends Biochem. Sci. 11:204–207.*

Uptake of Mitochondrial Proteins Requires Energy

The findings described previously suggest that the N-terminal uptake-targeting sequences of precursors of mitochondrial proteins interact with one or more receptors and direct these proteins to their final destinations in the mitochondrion. But one key question remains. How are the proteins actually imported into the organelle? Studies on mitochondrial proteins in cell-free systems have shown that two separate inputs of energy are required for import: ATP hydrolysis in the cytosol, and a proton-motive force across the inner membrane.

ATP hydrolysis in the cytosol is required to unfold precursor proteins and keep them in an unfolded or partially folded state, so that they can interact with the mitochondrial uptake machinery (see Figure 18-12). Several different unfolding proteins have been identified. They bind to precursor mitochondrial proteins and also to precursor ER proteins (see Figure 17-16) and couple the energy released by ATP hydrolysis in some way to prevent the bound proteins from folding.

Evidence indicates that the only role of cytosolic ATP is to keep the precursor proteins in an unfolded state. In one study, a precursor protein was purified and then denatured by urea. When added to yeast mitochondria, the

(a)

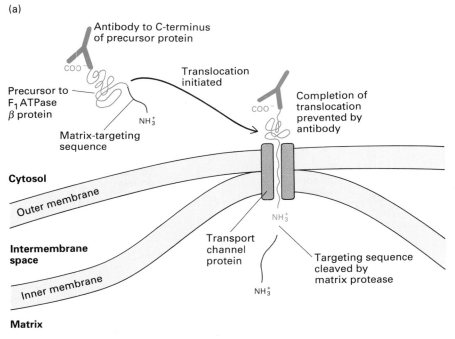

Cytosol

Intermembrane space

Matrix

(b)

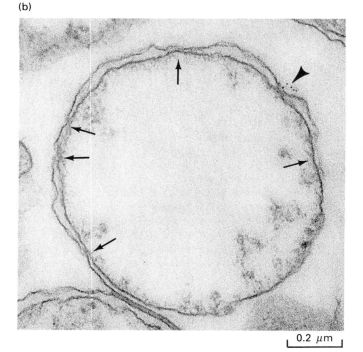

0.2 μm

◂ **Figure 18-15** Mitochondrial proteins are translocated at contact sites between the inner and outer membranes. (a) How a translocation intermediate is accumulated. The precursor to the F_1 ATPase β subunit, a matrix protein, is mixed with an antibody specific for its C-terminal segment. When added to mitochondria, the N-terminus is translocated across the inner and outer membranes, and the N-terminal matrix-targeting signal is cleaved normally by the matrix protease. The antibody prevents translocation of the C-terminus, which remains on the cytosolic side of the mitochondrion. (b) The C-terminus of the translocated protein can be detected by incubating the mitochondria with gold particles coated with a protein (protein A) that binds to the antibody. After sectioning and visualization under the electron microscope, points of contact between the inner and outer membranes are evident (arrows); some contact sites contain gold particles (arrowhead) bound to the translocation intermediate. *From M. Schweiger, V. Herzog, and W. Neupert, 1987, J. Cell Biol. **105**:235–246. Part (b) courtesy of W. Neupert.*

However, in all cases in which a protein is taken into the mitochondrion (except for cytochrome *c*), a proton-motive force across the inner membrane is required. If mitochondria are "poisoned" with inhibitors or uncouplers of oxidative phosphorylation, such as cyanide or dinitrophenol, then the proton-motive force across the inner membrane dissipates. Such debilitated mitochondria cannot take up precursor proteins, either in the intact cell or in cell-free reactions, even in the presence of ATP and unfolding proteins.

Exactly how the membrane electric potential is used to "pull" receptor-bound precursor proteins into the matrix (see Figure 18-12, step 3) is not clear. Once a protein partially inserts itself into the inner membrane, it becomes subjected to a transmembrane potential of 200 mV (matrix space negative), which is equivalent to an electric gradient of about 400,000 V/cm. The conformation of these proteins then may be altered by the electric potential, in much the same way that the conformation of voltage-dependent ion channels in nerve cells is affected by the membrane electric potential. These changes in protein folding could pull the protein across the energized inner membrane. A related possibility is that the N-terminal matrix-targeting sequence, with its many positively charged side chains, could be pulled into the matrix space by the inside negative electric membrane potential.

denatured protein could be taken up in the absence of ATP; uptake of the native, undenatured precursor required ATP. In a second study, mutations in certain matrix or inner-membrane proteins were constructed that impaired the proper folding of the mature protein. These mutant proteins could be incorporated into the mitochondria with no ATP present.

In summary, the outer and even the inner mitochondrial membranes appear to contain multiple specific receptors for the binding and uptake of different proteins. These receptors target proteins to their proper destinations in the organelle. Protein uptake usually requires an expenditure of energy in the form of both cytosolic ATP and a proton-motive force. Proteins are kept in an unfolded conformation prior to translocation across the inner membrane.

Synthesis of Mitochondrial Proteins Is Coordinated

The assembly of a mitochondrion appears to require the close coordination of nuclear and mitochondrial genomes; especially in multienzyme complexes, such as cytochrome c oxidase and the F_0F_1 ATPase, it seems that all components must be fabricated in appropriate ratios. This important topic has been studied mainly in yeast. Little is known about coordinating the expression of these two genomes in animals or plants.

Recall that certain petite yeast strains have no mtDNA. Yet petite cells contain normal amounts of all mitochondrial proteins encoded by the nucleus, such as cytochrome c and the F_1 ATPase. Thus mitochondrial gene products are not essential to the expression of nuclear genes. Since mtDNA cannot be readily deleted in other organisms, it is impossible to say whether this result is true in all cells.

Yeasts and other eukaryotic microorganisms that can grow in the absence or presence of O_2 provide a striking example of the coordination of nuclear and mitochondrial genes. When grown anaerobically (without oxygen) with glucose as a carbon source, ATP is generated solely by the Emden-Meyerhoff pathway (see Figures 15-2 and 15-7). Anaerobically grown yeasts lack a complete respiratory chain (cytochromes such as a, a_3, b, and c_1 are absent) and, when viewed under an electron microscope, lack typical mitochondria, although they do contain some *premitochondria* (small organelles with inner and outer membranes but no cristae). These yeast cells also lack enzymes essential to the citric acid cycle and the F_0F_1 ATPase complex. The levels of cytoplasmic mRNAs for proteins such as cytochrome c and subunits of cytochrome oxidase and the F_1 ATPase are reduced as much as 100 times relative to aerobic (oxygen-grown) cells. In anaerobic cells, however, mtDNA is replicated normally.

The addition of oxygen to anaerobic yeast induces the synthesis of all mitochondrial components encoded by the nuclear and mitochondrial genomes, as well as the expansion of the premitochondrial membranes into true mitochondria with cristae. Heme synthesis, which is low in anaerobic yeast, is also activated. Heme, in turn, is required for the transcription of the nuclear genes for cytochrome c and other mitochondrial proteins. When heme is abundant, certain transcriptional regulatory proteins are activated and bind to specific enhancer sequences in the genes for cytochrome c and other nuclear-encoded mitochondrial proteins, enhancing their transcription. This process may be representative of a general mechanism of down-regulating the expression of multiple nuclear genes that encode mitochondrial proteins during anaerobic fermentation, when the level of heme is low.

Chloroplast DNA and Biogenesis of Plastids

The synthesis of a chloroplast appears to be similar in many respects to that of a mitochondrion, although many key steps are not understood in molecular detail. Some chloroplast proteins are encoded by chloroplast DNA and translated by chloroplast ribosomes in the organelle (Table 18-5). Others are fabricated on cytoplasmic ribosomes and incorporated into the organelle after translation; these imported proteins are synthesized with N-terminal uptake-targeting sequences, which direct each protein to its correct subcompartment and are subsequently cleaved. Chloroplasts, like mitochondria, grow by expansion and then fission. This process can be observed easily in unicellular algae such as *Chlamydomonas*, which contain a single large chloroplast (Figure 18-16).

In addition to these topics, we address two other fascinating aspects of chloroplast biology here. First, we examine the development of chloroplasts from small, membrane-limited preorganelles called *proplastids*. This process is triggered by light, and much is known about how light activates the expression of nucleus-encoded chloroplast proteins. Second, we see how a proplastid can differentiate into a diverse array of organelles, depending on the plant tissue and on environmental cues.

Chloroplast DNA Contains over 120 Different Genes

Chloroplast DNAs are circular molecules of 120,000–160,000 bp, depending on the species. The complete sequences of two chloroplast DNAs—liverwort (121,024 bp) and tobacco (155,844 bp)—have been determined. The liverwort genome contains two inverted repeats, each consisting of 10,058 bp, that contain the rRNA genes and a few other duplicated genes. These repeats are separated by a small single-copy sequence (19,813 bp) and a large single-copy sequence that contain the bulk of the tRNA and protein-coding genes listed in Table 18-5. Despite the difference in size, the overall or-

Table 18-5 Genes encoded by chloroplast DNA from a liverwort, *Marchantia polymorpha*

Genes	Description of gene product
rRNA (duplicated in inverted repeats IR$_A$ and IR$_B$)	
16S	
23S	
4.5S	
5S	
tRNA	
37 tRNAs	
RNA polymerase	
rpo A	Homologous to *E. coli*, subunit α
rpo B	Homologous to *E. coli*, subunit β
rpo C1	Homologous to *E. coli*, subunit β'
rpo C2	Homologous to *E. coli*, subunit β'
Ribosomal protein	
50S subunit: 8 proteins (rpl)	
30S subunit: 11 proteins (rps)	
Photosynthetic	
rbcL	Large subunit of ribulose bisphosphate carboxylase
psaA, psaB	Chlorophyll a–binding proteins in photosystem I
psbA	Photosystem II 32,000-MW protein
psbB, psbC	Chlorophyll a–binding proteins in photosystem II
psbD	Photosystem II D2 protein
psbE, psbF	Cytochrome b_{559}
psbG	Photosystem II G protein
atpA, atpB, atpE	Subunits α, β, and ϵ of F$_1$ ATPase
atpF, atpH, atpI	Subunits 1, 3, and 4 of F$_0$ ATPase
petA	Cytochrome f
petB	Cytochrome b_6
petD	Subunit 4 of cytochrome b_6/f complex
Genes predicted by amino acid sequence homology	
ndh1, ndh2, ndh3, ndh4, ndh4L, ndh5, ndh6	Homologous to subunits of the human mitochondrial NADH-CoQ reductase complex
frxA, frxB, frxC	Homologous to a 4-Fe-type ferredoxin
Others	More than 28 unidentified open reading frames (URF)

SOURCE: K. Ohyama et al., 1986, *Nature* **322**:572–575.

ganization and gene composition of the two DNAs is very similar; the size differential is due primarily to the length of the inverted repeat in which some genes are duplicated.

About 60 chloroplast genes are involved in RNA transcription and translation, including genes for rRNAs, tRNAs, RNA polymerase subunits, and ribosomal proteins (see Table 18-5). About 20 genes encode subunits of the chloroplast photosynthetic electron-transport and the F$_0$F$_1$ ATPase complexes; also encoded is one of the two subunits of ribulose 1,5-bisphosphate carboxylase.

Some regions of chloroplast DNA are strikingly similar to bacterial DNA. Chloroplast DNA encodes four subunits of RNA polymerase that are highly homologous to subunits α, β, and β' of *E. coli* RNA polymerase. One segment of chloroplast DNA encodes eight proteins that are homologous to eight *E. coli* ribosomal proteins; the order of these genes is the same in the two DNAs. These and other sequence data strongly support the theory of *endosymbiosis* that chloroplasts originated when an ancient eukaryotic cell was colonized by a photosynthetic bacterium.

The sequence of chloroplast DNA has revealed some unexpected features. For instance, the DNA contains seven protein-coding sequences that are homologous to the seven mitochondrial-encoded subunits of the NADH-CoQ reductase complex. Such an electron-transport system had not been identified previously in chloroplasts of higher plants, and the DNA sequence has prompted investigators to study the role of this unexpected protein complex.

Liverwort chloroplast DNA has some genes that are not detected in the larger tobacco chloroplast DNA, and vice versa. Since the two types of chloroplasts contain virtually the same set of proteins, these data suggest that some genes are present in the chloroplast DNA of one species and in the nuclear DNA of the other, indicating that some exchange of genes between chloroplast and nucleus has occurred during evolution.

Many Proteins Are Synthesized in the Cytosol and Imported into Chloroplasts

Although the exact number is unknown, the vast majority of chloroplast proteins appear to be synthesized in the cytosol and then imported into the organelle. As in mitochondrial biogenesis, different proteins are targeted to different chloroplast subcompartments: ribulose 1,5-bisphosphate carboxylase and ferredoxin to the stroma, the chlorophyll a/b binding proteins in the light-harvesting complex (LHC) to the thylakoid membrane, and plastocyanin to the thylakoid lumen. Import from the cytosol into the stroma occurs, as in mitochondria, at points of contact between the outer and inner organelle membranes (Figure 18-17). During import into the chloroplast, the LHC and plastocyanin proteins not only must

(a)

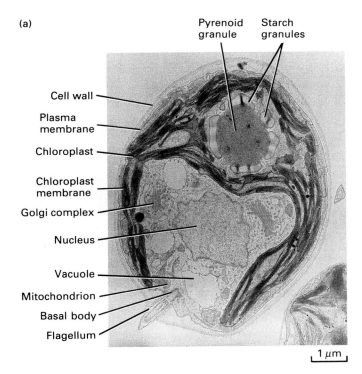

Pyrenoid granule   Starch granules

Cell wall

Plasma membrane

Chloroplast

Chloroplast membrane

Golgi complex

Nucleus

Vacuole

Mitochondrion

Basal body

Flagellum

1 μm

(b)

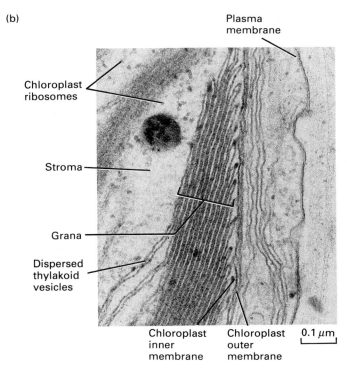

Plasma membrane

Chloroplast ribosomes

Stroma

Grana

Dispersed thylakoid vesicles

Chloroplast inner membrane   Chloroplast outer membrane

0.1 μm

▲ **Figure 18-16** Electron micrographs of *Chlamydomonas reinhardtii.* (a) The single, large cup-shaped chloroplast. The stacked thylakoid vesicles are major components of the chloroplast, as is a large pyrenoid granule with its associated starch granules. (b) The chloroplast is surrounded by a double-membrane envelope, seen to advantage at this higher magnification in which the chloroplast ribosomes are visible. The thylakoid vesicles are both dispersed and packed into grana; obliquely sectioned thylakoid vesicles appear somewhat fuzzy. *Courtesy of I. Chad, P. Siekevitz, and G. E. Palade.*

traverse both the outer and inner chloroplast membranes but also must travel through the stroma and either be inserted into the thylakoid membrane (LHC) or cross that membrane and enter the thylakoid lumen (plastocyanin).

Ribulose 1,5-bisphosphate carboxylase, or RBPase (see Figure 16-17), is the most abundant protein in chloroplasts. The enzyme (MW 550,000) is made up of 16 subunits. Eight are identical, large (MW 55,000) subunits that contain the catalytic sites; the other eight are identical, small (MW 12,000) subunits of unknown function. The enzyme is located in the stromal compartment of the chloroplast, along with the other enzymes involved in the Calvin cycle.

The large (L) subunit of RBPase is encoded by the chloroplast DNA. The small (S) subunit is synthesized on free cytosolic polysomes and must traverse both the outer and inner chloroplast membranes to reach its final destination; the S subunit is synthesized in a precursor form with an N-terminal uptake-targeting sequence of about 44 amino acids. An experimental protocol similar to that illustrated in Figure 18-11 has shown that this precursor polypeptide can be taken up by isolated chloroplasts.

After uptake, the N-terminal uptake-targeting sequence is cleaved and the S subunit combines with the L subunit to yield the active RBPase enzyme. Importantly, the L and S subunits of RBPase do not assemble spontaneously to form the mature enzyme; instead, newly made L subunits are bound to a *binding protein* (BP) that is itself imported into the chloroplast stroma. The BP (subunit MW 61,000) form large complexes (MW >700,000) with L subunits and release them in the presence of ATP. The BP may allow excess L subunits to be stored in chloroplasts, pending the import of S subunits, and also facilitates (in an unknown manner) the assembly of S and L polypeptides. The BP is a member of the chaperonin family of unfolding proteins and folding catalysts.

In the case of proteins such as plastocyanin that are directed to the thylakoid lumen, the successive function of two uptake-targeting sequences is required. The first, like that at the N-terminus of the RBPase S subunit, targets the protein to the stroma; the second targets the protein from the stroma to the thylakoid lumen (see Figure 18-17). The role of these sequences has been shown in cell-free experiments measuring the uptake of recombi-

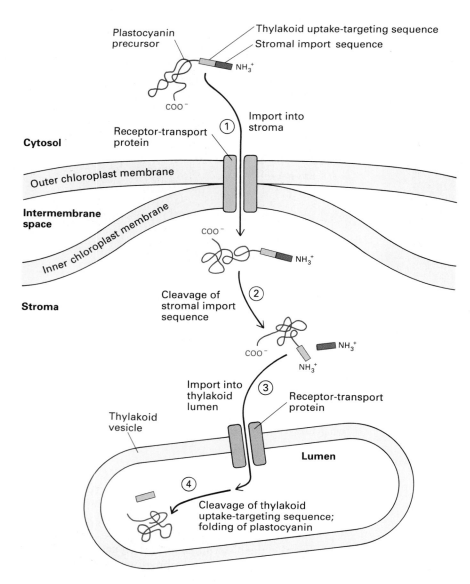

◀ **Figure 18-17** The successive actions of two uptake-targeting sequences direct plastocyanin to the thylakoid lumen. The plastocyanin precursor is synthesized in the cytosol; the 66 amino acids at its N-terminus (purple and blue) are not found on the mature protein in the thylakoid lumen. The ∼30 N-terminal residues (purple) signal import into the chloroplast stroma (step 1). As with import into the mitochondria, translocation from the cytosol to the stroma is mediated by a receptor-transport protein localized to the points of contact of the outer and inner chloroplast membranes. The stromal import sequence is removed by a stromal signal protease (step 2). The thylakoid uptake-targeting sequence (blue) of ∼25 residues is very hydrophobic; it causes the stromal protein to be imported into the thylakoid lumen (step 3) and presumably binds to a distinct import receptor-transport protein on the thylakoid membrane. This sequence is removed in the thylakoid lumen by a separate endoprotease (step 4). *After S. Smeekins, C. Bauerle, J. Hageman, K. Keegstra, and P. Weisbeek, 1986, Cell, 46:365–375.*

nant-generated proteins into chloroplasts. For instance, when the thylakoid uptake-targeting sequence is deleted, plastocyanin is rerouted away from the thylakoid lumen and accumulates in the stroma.

As in mitochondria, protein import into chloroplasts requires energy. Experiments suggest that import into the stroma depends on ATP hydrolysis in the stroma, not in the cytosol. Unlike import into mitochondria, an electrochemical potential across the inner chloroplast membrane is not required. Precisely how ATP acts to trigger import is not known.

In plant cells, proteins are imported into both chloroplasts and mitochondria, and the organelles obviously are able to distinguish between polypeptide precursors. Further work on the specificity of plant chloroplast and mitochondrial import receptors is eagerly awaited.

Proplastids Can Differentiate into Chloroplasts or Other Plastids

The chloroplast is one of several type of organelles termed *plastids*, all of which are formed from the same precursor proplastid and contain the same chloroplast DNA (Figure 18-18a). Chloroplasts are the only type of plastid to contain internal thylakoid membranes. Proplastids are composed only of an inner membrane, an outer membrane, and a small stromal space that contains the chloroplast DNA. They are found in embryonic tissue as well as in such tissues as leaves deprived of light, where they are often called *etioplasts* (Figures 18-19 and 18-20). Proplastids replicate but do not contain LHC proteins, chlorophyll, electron-transport systems, or thylakoid vesicles. In leaves and other tissues containing chloroplasts,

(a)

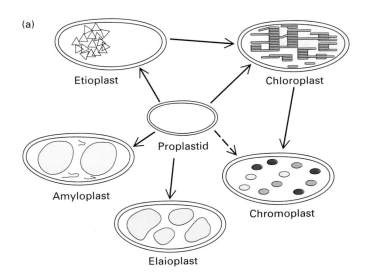

(b)

▲ **Figure 18-18** (a) Proplastids can differentiate into many organelles, depending on the plant tissue in which they are located and/or on exposure to light. The etioplast is an intermediate in the conversion of a proplastid to a chloroplast. Only the chloroplast contains internal thylakoid membranes. (b) Electron micrograph of an amyloplast from the root cap

of the onion *Allium cepa*. Amyloplasts contain massive starch granules. In the root cap, these plastids sink to the bottom of the cell, indicating to the root which way is down. *Part (a) courtesy of Barbara Sears; part (b) courtesy of Jeremy Burgess and Science Photolibrary/Photo Researchers, Inc.*

(a) Chloroplast

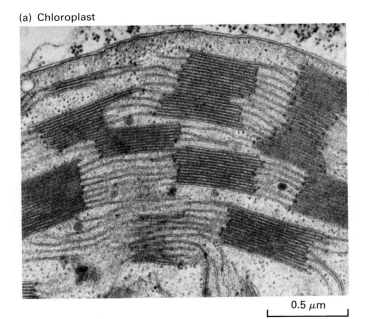

(b) Proplastid

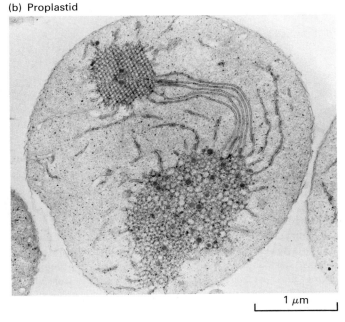

▲ **Figure 18-19** Electron micrographs of thin sections of plant tissue. (a) A chloroplast from a normal barley seedling leaf. (b) A proplastid from a dark-grown barley seedling. In the absence of light, major polypeptides, such as proteins in the LHC and chlorophyll-binding proteins, are not synthesized. Etioplasts from such cells contain membranes that take

the form of primary lamella and interconnected vesicles containing some chloroplast pigments. Light triggers the development of the proplastid into the mature chloroplast. *From H. G. Schweiger, ed.,* International Cell Biology 1980–1981. *Springer-Verlag. Courtesy of D. von Wettstein.*

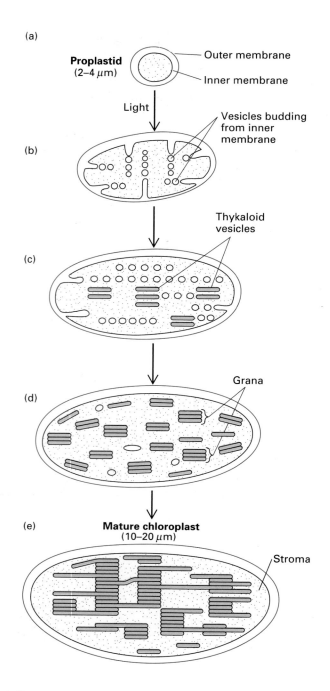

the presence of light stimulates the synthesis of chloroplast proteins and the expansion of the inner chloroplast membrane. Membrane vesicles bud from the inner membrane and arrange themselves into stacks; these vesicles incorporate essential proteins and chlorophyll and are transformed into mature thylakoid vesicles.

Depending on the plant tissue and on environmental cues, proplastids differentiate into different types of plastids: chromoplasts, amyloplasts, and elaioplasts. *Chromoplasts,* as their name implies, are pigmented organelles that synthesize and retain yellow, orange, and red carotenoid and other pigments. They are responsible for the color of the tissues in flowers, ripening fruits, and some roots (for example, carrot roots). In most cases, chromoplasts form directly from chloroplasts; this process involves the controlled breakdown of thylakoid membranes and chlorophyll, the appearance of new invaginations from the inner membrane, and the synthesis of new types of carotenoids. *Amyloplasts* are nonpigmented, starch-containing plastids found in tissues such as potato tubers (see Figure 18-18b). *Elaioplasts* store droplets of oils and lipids. All of these plastids arise from a common progenitor organelle; how these differentiations are controlled is not known but is the subject of much current research.

In summary, plastids do much more for plants than provide sites for photosynthesis. Chloroplasts and other plastids are involved in nitrogen assimilation, as well as in the synthesis of some amino acids, many fatty acids and lipids, and even some plant hormones. Plastids in nonphotosynthetic tissues store essential energy sources, such as starch and lipids.

Phytochromes Mediate Light Induction of Gene Expression in Plants

Light regulates many aspects of plant growth and differentiation. Chlorophyll synthesis and chloroplast differentiation in seedlings is triggered by light (see Figure 18-20). Depending on the species, light also controls seed germination, flowering, stem elongation, and leaf expansion. Most of these light-sensitive responses are mediated by the pigmented protein *phytochrome* (Figure 18-21) and involve phytochrome-mediated changes in gene expression.

Phytochrome is found in the cytoplasm of virtually all plant cells; a tetrapyrrole pigment (related in structure to heme) is covalently linked to the protein. Phytochrome exists in two interconvertible, light-absorbing states (Figure 18-22): P_r (maximal absorption of red light at 660 nm), and P_{fr} (maximal absorption of far red light at 730 nm). The molecule is synthesized and accumulates as P_r in dark-adapted plants. Absorption of a quantum of red light converts a P_r phytochrome to P_{fr}; most of the physiological effects of phytochrome are mediated by the P_{fr} form. P_{fr} can be reconverted to P_r rapidly by the absorption of a quantum of far red light or slowly and spon-

▲ **Figure 18-20** Steps in the light-induced differentiation of a proplastid into a chloroplast. (a) The proplastid in dark-adapted cells contains only the outer and inner chloroplast membranes. (b) Light triggers the synthesis of chlorophyll, phospholipid, chloroplast stroma, and thylakoid proteins and a budding of small vesicles from the inner chloroplast membrane. (c) As the proplastid enlarges, some of the spherical vesicles fuse, forming flattened thylakoid vesicles. Some thylakoid vesicles stack into grana. The adhesiveness of the vesicles is due to a protein found in the LHC that is synthesized in abundance during this stage. (d, e) The proplastid enlarges and matures into a chloroplast as more thylakoid vesicles and grana are formed. [See D. von Wettstein, 1959, *J. Ultrastruct. Res.* 3:235.]

taneously in the dark. The importance of red light can be demonstrated by a simple experiment. Brief exposure to red light is all that is necessary to trigger the germination of lettuce seeds (conversion of P_r to P_{fr}). However, this response to red light can be prevented by exposing the seeds to far red light (reconversion of P_{fr} to P_r).

Red light, acting through the phytochrome system, is known to activate the genes for several important chloroplast proteins, such as the nuclear-encoded gene for the small subunit of ribulose 1,5-bisphosphate carboxylase. Plants grown in the dark contain less than 5 percent the amount of mRNA for this protein than light-grown plants do. When the cloned pea gene for the S subunit is introduced by DNA transfection into cultured petunia cells, the transferred gene is transcribed in a light-dependent fashion similar to that observed in pea leaves. Recent studies have localized a DNA sequence just upstream of the transcription initiation site that is essential for light induction; presumably, this sequence binds a regulatory protein that responds to the level of P_{fr} in the cell. Much research now focuses on identifying the signal(s) sent to the nucleus by the P_{fr} form of phytochrome.

In summary, light plays important roles in the life of a plant other than photosynthesis, and plants have pigments other than chlorophyll that absorb light and trigger important responses.

Summary

In higher plant and animal cells, the nuclear membrane disassembles into small vesicles during mitosis and re-forms during telophase. Rigidity to the nuclear membrane is provided by the lamina, a network of intermediate filamentlike protein lamins A, B, and C. Apparently, phosphorylation of the lamins during prophase triggers disassembly of the lamina and vesiculation of the nuclear membrane. Reassembly of the membrane in telophase is initiated by lamin dephosphorylation and chromatin decondensation.

Nuclear proteins are imported from the cytosol through the nuclear pores, the same structures that allow RNAs and RNPs to move outward into the cytosol. Proteins to be imported into the nucleus contain specific sorting sequences that cause them to bind to receptorlike sites on the cytosolic side of the pores; several different nuclear sorting sequences are employed. The import of bound proteins is powered by ATP hydrolysis.

Continuity of the mitochondrion or chloroplast is provided by the organelle DNA, which encodes organelle rRNAs and tRNAs but only a few organelle proteins. Mitochondrial gene products identified thus far include several inner-membrane components of the electron-transport chain. The size, coding capacity, and organiza-

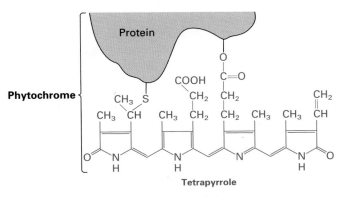

▲ **Figure 18-21** Phytochrome is a light-sensitive regulatory protein molecule found in all plants; its light-absorbent pigment is a tetrapyrrole linked to the protein.

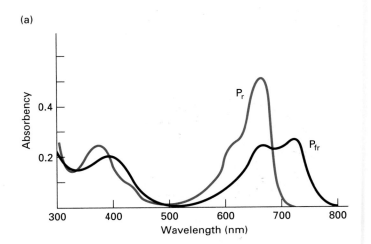

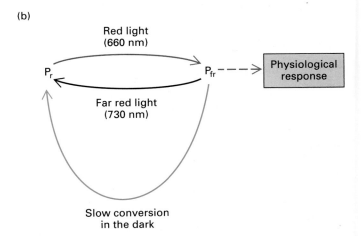

▲ **Figure 18-22** (a) Light absorbency by the two interconvertible forms of phytochrome. The P_r form maximally absorbs light of 660 nm (red light); the P_{fr} form maximally absorbs light of 730 nm (far red light). (b) Absorption of a quantum of red light by one molecule of the P_r form converts it to P_{fr}; P_{fr} can be reconverted to P_r by absorption of a quantum of far red light.

tion of mtDNA varies widely among different organisms. Certain proteins are encoded by nuclear DNA in one species and by mtDNA in another, suggesting that genes have moved from one DNA to the other during evolution. The mitochondrial genetic code differs from the one used in prokaryotes and in eukaryotic nuclear-encoded proteins; it also differs in different species. Animal mitochondrial RNAs undergo extensive processing, including polyadenylation. In yeasts, proteins encoded in the introns of mitochondrial genes are essential for splicing certain mRNAs.

Most mitochondrial proteins are synthesized in the cytosol as precursors. Proteins destined for the matrix have an uptake-targeting sequence that permits their entry and then is cleaved off in the matrix. Proteins destined for the intermembrane space are first imported into the matrix by a matrix targeting sequence and then targeted across the inner membrane by an intermembrane-space targeting sequence. Other sorting sequences target proteins to the inner or outer membrane. Import occurs where the inner and outer mitochondrial membranes fuse. Proteins to be imported are kept in the cytosol in a partially unfolded state by the action of a chaperonin—an ATP-dependent unfolding protein. After the unfolded proteins bind to receptors on the outer mitochondrial membrane, an energy-driven process powered by the proton-motive force across the inner membrane moves these proteins into the organelle. Folding inside the matrix is facilitated by another chaperonin.

Chloroplast DNA encodes more than 120 genes, including rRNAs, tRNAs, RNA polymerase subunits, ribosomal proteins, subunits of the chloroplast photosynthetic electron-transport complexes, subunits of the F_0F_1 ATPase complex, and the large (L) subunit of ribulose 1,5-bisphosphate carboxylase. Like mitochondrial proteins, most chloroplast proteins are synthesized in the cytosol and then imported into the organelle. Such proteins contain uptake-targeting sequences that direct them first to the organelle itself and then to the proper organelle subcompartment, such as the thylakoid membrane or thylakoid lumen. The assembly of ribulose 1,5-bisphosphate carboxylase from eight copies of the chloroplast-encoded L subunit and eight copies of the imported small (S) subunit requires a chaperonin—an L-subunit binding protein that apparently stores excess L subunits and facilitates the binding of L and S subunits.

The chloroplast is only one type of plastid; all plastids contain the same chloroplast DNA. The formation of chloroplasts from proplastids is triggered by light induction. Chromoplasts, amyloplasts, and elaioplasts are plastids without chlorophyll or thylakoid vesicles that store colored pigments, starch, or lipids, respectively.

In addition to functioning as the energy source in photosynthesis, light regulates many aspects of plant growth and differentiation. The protein phytochrome mediates many such responses. Absorption of a quantum of red light converts the P_r form of phytochrome to the P_{fr} form; P_{fr} triggers many effects, including the transcription of the nuclear genes involved in photosynthesis, such as the S subunit of ribulose 1,5-bisphosphate carboxylase. P_{fr} is reconverted to inactive P_r by the absorption of a quantum of far red light.

References

Assembly and Disassembly of the Nuclear Membrane

AEBI, U., J. COHN, L. BUHLE, and L. GERACE. 1986. The nuclear lamina is a meshwork of intermediate-type filaments. *Nature* 323:560–564.

*FEATHERSTONE, C. 1989. The complexities of the cell cycle. *TIBS* 14:85–87.

FISHER, P. A. 1987. Disassembly and reassembly of nuclei in cell-free systems. *Cell* 48:175–176.

FORBES, D. J., M. W. KIRSCHNER, and J. W. NEWPORT. 1983. Spontaneous formation of nucleus-like structures around bacteriophage DNA microinjected into *Xenopus* eggs. *Cell* 34:13–23.

*GERACE, L. 1986. Nuclear lamina and organization of nuclear architecture. *TIBS* 11:443–446.

*GERACE, L., and B. BURKE. 1988. Functional organization of the nuclear envelope. *Ann. Rev. Cell Biol.* 4:335–374.

*NEWPORT, J. W., and D. J. FORBES. 1987. The nucleus: structure, function, and dynamics. *Ann. Rev. Biochem.* 56:535–565.

WARREN, G., C. FEATHERSTONE, G. GRIFFITHS, and B. BURKE. 1983. Newly synthesized G protein of vesicular stomatitis virus is not transported to the cell surface during mitosis. *J. Cell Biol.* 97:1623–1628.

WORMAN, H. J., J. YUAN, G. BLOBEL, and S. D. GEORGATOS. 1988. A lamin B receptor in the nuclear envelope. *Proc. Nat'l Acad. Sci. USA* 85:8531–8534.

Protein Import into the Cell Nucleus

ADAM, S. A., T. J. LOBL, M. A. MITCHELL, and L. GERACE. 1989. Identification of specific binding proteins for a nuclear location sequence. *Nature* 337:276–279.

AKEY, C. W., and D. S. GOLDFARB. 1989. Protein import through the nuclear pore complex is a multistep process. *J. Cell Biol.* 109:971–982.

BORER, R. A., C. F. LEHNER, H. M. EPPENBERGER, and E. A. NIGG. 1989. Major nucleolar proteins shuttle between nucleus and cytoplasm. *Cell* 56:379–390.

*DINGWALL, C., and R. A. LASKEY. 1986. Protein import into the cell nucleus. *Ann. Rev. Cell Biol.* 2:367–390.

DWORETZKY, S. I., and C. M. FELDHERR. 1988. Translocation of RNA-coated particles through the nuclear pores of oocytes. *J. Cell Biol.* 106:575–584.

*A book or review article that provides a survey of the topic.

*HART, G. W., G. D. HOLT, and R. S. HALTIWANGER. 1988. Nuclear and cytoplasmic glycosylation: novel saccharide linkages in unexpected places. *TIBS* 13:380–384.

LANFORD, R. E., P. KANDA, and R. C. KENNEDY. 1986. Induction of nuclear transport with a synthetic peptide homologous to the SV40 T antigen transport signal. *Cell* 46:575–582.

NEWMEYER, D. D., and D. J. FORBES. 1988. Nuclear import can be separated into distinct steps in vitro: nuclear pore binding and translocation. *Cell* 52:641–653.

SILVER, P., I. SADLER, and M. A. OSBORNE. 1989. Yeast proteins that recognize nuclear localization sequences. *J. Cell Biol.* 109:983–989.

Mitochondrial DNA and RNA

ANDERSON, S., A. BANKIER, B. G. BARRELL, M. H. L. DE BRUIJN, A. R. COULSON, J. DROUIN, I. C. EPERON, D. P. NIERLICH, B. A. ROE, F. SANGER, P. H. SCHREIER, A. J. H. SMITH, R. STADEN, and I. G. YOUNG. 1981. Sequence and organization of the human mitochondrial genome. *Nature* 290:457–465.

*ATTARDI, G., and G. SCHATZ. 1988. Biogenesis of mitochondria. *Ann. Rev. Cell Biol.* 4:289–333.

CARIGNANI, G., O. GROUDINSKY, D. FREZZA, E. SCHIAVON, E. BERGANTINO, and P. P. SLONIMSKI. 1984. An mRNA maturase is encoded by the first intron of the mitochondrial gene for the subunit I of cytochrome oxidase in *S. cerevisiae*. *Cell* 35:733–742.

CHANG, D. D., and D. A. CLAYTON. 1987. A mammalian mitochondrial RNA processing activity contains nucleus-encoded RNA. *Science* 235:1178–1184.

*CHOMYN, A., P. MARIOTTINI, M. W. J. CLEETER, C. I. RAGAN, A. MATSUNO-YAGI, Y. HATEFI, R. F. DOOLITTLE, and G. ATTARDI. 1985. Six unidentified reading frames of human mitochondrial DNA encode components of the respiratory-chain NADH dehydrogenase. *Nature* 314:592–597.

*CLAYTON, D. A. 1984. Transcription of the mammalian mitochondrial genome. *Ann. Rev. Biochem.* 53:573–594.

GRIVALL, L. A. 1989. Mitochondrial DNA: Small, beautiful, and essential. *Nature* 341: 569–571.

*LEVINGS, C. S., III, and G. G. BROWN. 1989. Molecular biology of plant mitochondria. *Cell* 56:171–179.

*NEWTON, K. J. 1988. Plant mitochondrial genomes: organization, expression, and variation. *Ann. Rev. Plant Physiol. and Mol. Biol.* 39:503–532.

*SAPP, J. 1987. *Beyond the Gene: Cytoplasmic Inheritance and the Struggle for Authority in Genetics.* Oxford University Press.

VAN DER HORST, G., and H. F. TABAK. 1985. Self-splicing of yeast mitochondrial ribosomal and messenger RNA precursors. *Cell* 40:759–766.

Synthesis and Localization of Mitochondrial Proteins

BEDWELL, D. M., S. A. STROBEL, K. YUN, G. D. JONGEWARD, and S. D. EMR. 1989. Sequence and structural requirements of a mitochondrial protein import signal defined by saturation cassette mutagenesis. *Mol. Cell Biol.* 9:1014–1025.

CHENG, M. Y., F.-U. HARTL, J. MARTIN, R. A. POLLOCK, F. KALOUSEK, W. NEUPERT, E. M. HALLBERG, R. L. HALLBERG, and A. L. HORWICH. 1989. Mitochondrial heat-shock protein hsp60 is essential for assembly of proteins imported into yeast mitochondria. *Nature* 337:620–625.

*EILERS, M., and G. SCHATZ. 1988. Protein unfolding and the energetics of protein translocation across biological membranes. *Cell* 52:481–483.

*HARTL, F.-U., N. PFANNER, D. W. NICHOLSON, and W. NEUPERT. 1989. Mitochondrial protein import. *Biochim. Biophys. Acta* 988:1–45.

HURT, E. C., B. PESOLD-HURT, and G. SCHATZ. 1984. The cleavable prepiece of an imported mitochondrial protein is sufficient to direct cytosolic dihydrofolate reductase into the mitochondrial matrix. *FEBS Lett.* 178:306–310.

NAGLEY, P., L. B. FARRELL, D. P. GEARING, D. NERO, S. MELTZER, and R. J. DEVENISH. 1988. Assembly of functional proton-translocating ATPase complex in yeast mitochondria with cytoplasmically synthesized subunit 8, a polypeptide normally encoded within the organelle. *Proc. Nat'l Acad. Sci. USA* 85:2091–2095.

NICHOLSON, D. W., R. A. STUART, and W. NEUPERT. 1989. Biogenesis of cytochrome c_1: role of cytochrome c_1 heme lyase and of the two proteolytic processing steps during import into mitochondria. *J. Biol. Chem.* 264:10156–10168.

PFALLER, R., H. F. STEGER, J. RASSOW, N. PFANNER, and W. NEUPERT. 1988. Import pathways of precursor proteins into mitochondria: Multiple receptor sites are followed by a common membrane insertion site. *J. Cell Biol.* 107:2483–2490.

*ROISE, D., and G. SCHATZ. 1988. Mitochondrial presequences. *J. Biol. Chem.* 263:4509–4511.

*TZAGOLOFF, A., and A. M. MEYERS. 1986. Genetics of mitochondrial biogenesis. *Ann. Rev. Biochem.* 55:249–285.

VESTWEBER, D., and G. SCHATZ. 1989. DNA-protein conjugates can enter mitochondria via the protein-import pathway. *Nature* 338:170–172.

Chloroplast DNA and Biogenesis of Plastids

BERRY, J. O., J. P. CARR, and D. F. KLESSIG. 1988. mRNAs encoding ribulose 1,5-bisphosphate carboxylase remain bound to polysomes but are not translated in amaranth seedlings transferred to darkness. *Proc. Nat'l Acad. Sci. USA* 85:4190–4194.

BERRY-LOWE, S. L., and R. B. MEAGHER. 1985. Transcriptional regulation of a gene encoding the small subunit of ribulose 1,5-bisphosphate carboxylase in soybean tissue is linked to the phytochrome response. *Mol. Cell Biol.* 5:1910–1917.

*ELLIS, J., and S. M. HEMMINGSON. 1989. Molecular chaperones: Protein essential for the biogenesis of some macromolecular structures. *TIBS* 14:339–342.

GOLOUBINOFF, P., A. A. GATENBY, and G. H. LORIMER. 1989. GroE heat-shock proteins promote assembly of foreign prokaryotic ribulose bisphosphate carboxylase oligomers in *Escherichia coli*. *Nature* 337:44–47.

*GRAY, J. 1986. Wonders of chloroplast DNA. *Nature* 322:501–502.

*GRUISSEM, W. 1989. Chloroplast gene expression: how plants turn their plasmids on. *Cell* 56:161–170.

*KEEGSTRA, K. 1989. Transport and routing of proteins into chloroplasts. *Cell* 56:247–253.

LAMPPA, G. K. 1988. The chlorophyll a/b-binding protein inserts into the thylakoids independent of its cognate transit peptide. *J. Biol. Chem.* 263:14996–14999.

LISSEMORE, J. L., and P. H. QUAIL. 1988. Rapid transcriptional regulation by phytochrome of the genes for phytochrome and chlorophyll a/b-binding proteins in *Avena sativa*. *Mol. Cell Biol.* 8:4840–4850.

*MOSES, P. B., and N.-H. CHUA. 1988. Light switches for plant genes. *Scientific American* 258(4):88–93.

*MULLET, J. E. 1988. Chloroplast development and gene expression. *Ann. Rev. Plant Physiol. Mol. Biol.* 39:475–502.

OHYAMA, K., H. FUKUZAWA, T. KOHCHI, H. SHIRAI, T. SANO, S. SANO, K. UMESONO, Y. SHIKI, M. TAKEUCHI, Z. CHANG, S.-I. AOTO, H. INOKUCHI, and H. OZEKI. 1986. Chloroplast gene organization deduced from complete sequence of liverwort *Marchantia polymorpha* chloroplast DNA. *Nature* 322:572–574.

*OHYAMA, K., T. KOHCHI, T. SANO, and Y. YAMADA. 1988. Newly identified groups of genes in chloroplasts. *TIBS* 13:19–22.

*ROCHAIX, J.-D., and J. ERICKSON. 1988. Function and assembly of photosystem II: genetic and molecular analyses. *TIBS* 13:56–60.

*ROY, H., and S. CANNON. 1988. Ribulose bisphosphate carboxylase assembly: what is the role of the large subunit binding protein? *TIBS* 13:163–165.

*SCHMIDT, G. W., and M. L. MISHKIND. 1986. The transport of proteins into chloroplasts. *Ann. Rev. Biochem.* 55:879–912.

SMEEKENS, S., C. BAUERLE, J. HAGEMAN, K. KEEGSTRA, and P. WEISBEEK. 1986. The role of the transit peptide in the routing of precursors toward different chloroplast compartments. *Cell* 46:365–375.

THEG, S. M., C. BAUERLE, L. J. OLSON, B. R. SELMAN, and K. KEEGSTRA. 1989. Internal ATP is the only energy requirement for the translocation of precursor proteins across chloroplastic membranes. *J. Biol. Chem.* 264:6730–6736.

CHAPTER 19

Cell-to-Cell Signaling: Hormones and Receptors

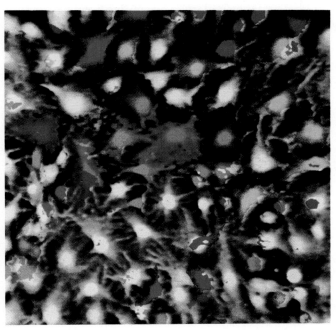

Passage of a wave of elevated calcium through a field of astrocytes, shown by cytoplasmic fluorescence

N*o cell lives in isolation. In all multicellular organisms, an elaborate cell-to-cell communication network coordinates the growth, differentiation, and metabolism of the multitude of cells in the diverse tissues and organs.*

In small groups of cells, communication is often by direct cell-to-cell contact. As we saw in Chapter 13, gap junctions in animals and plasmodesmata in plants permit adjacent cells to exchange small molecules and coordinate metabolic responses. Adhesive junctions between the plasma membranes of adjacent cells determine the shape and rigidity of many tissues. Moreover, the establishment of specific contacts between different types of cells is a necessary step in the differentiation of many tissues.

But cells also have to communicate over longer distances than chains of cell-to-cell contacts can facilitate. In such cases, extracellular products act as signals. These specific substances are synthesized and released by signaling cells and then move to other cells, where they induce a specific response only in those target cells that have receptors for the signal molecules. Cells use an enormous variety of chemicals and signaling mechanisms to communicate with each other.

In many microorganisms, such as yeast, slime molds, and protozoans, secreted molecules coordinate the aggregation of free-living cells for sexual mating or differentia-

tion under certain environmental conditions. Chemicals released by one organism that can alter the behavior of other organisms of the same species are called phero-mones. Some algae and animals release pheromones, usu-ally dispersing them into the air or water, to attract mem-bers of the opposite sex. The mechanism of cell signaling in yeasts and in slime molds is now well-understood and is discussed later in this chapter.

In plants and animals, extracellular signals control the growth of most tissues, govern the synthesis and secretion of proteins, and regulate the composition of intracellular and extracellular fluids. The study of plant hormones has lagged behind that of signaling in animals and even in microorganisms. At this point, no receptor for any plant hormone has been identified with certainty, although the functions of several plant hormones are known in detail.

The first part of the chapter focuses on cell signaling in animals, particularly in mammals. In animals, extracellu-lar signals affect overall growth and behavior, stimulate digestive enzymes after a meal, and regulate wound heal-ing. These extracellular signals can be classified as endo-crine, paracrine, or autocrine, based on the distance over which the signal must act (Figure 19-1).

In endocrine signaling, *the cells of endocrine organs release* hormones—*signaling substances that act on a dis-tant group of target cells. In animals, an endocrine hor-mone is usually carried by the blood from its site of re-lease to its target.*

In paracrine signaling, *the target cell is close to the sig-naling cell and the signaling compound affects only the group of target cells adjacent to it. The conduction of an electric impulse from one nerve cell to another or from a nerve cell to a muscle cell (inducing or inhibiting muscle contraction) involves signaling by extracellular chemicals called* neurotransmitters. *Neurotransmitters and neuro-hormones are examples of paracrine signals. (Because of the special importance of the nervous system, nerve con-duction is considered separately in Chapter 20.)*

In autocrine signaling, *cells respond to substances that they themselves release. Cultured cells often respond to growth factors that they secrete. Many tumor cells over-produce and release* growth factors *that stimulate inap-propriate, unregulated growth of the tumor cell itself, as well as adjacent nontumor cells, which can cause a tumor to form.*

The same compound sometimes acts in two or even three types of cell-to-cell-signaling. Certain small peptides function both as neurotransmitters (paracrine signaling) and as systemic hormones (endocrine signaling).

Communication by extracellular signals usually in-volves six steps: (1) synthesis and (2) release of the chemi-cal by the signaling cell; (3) transport to the target cell; (4) detection of the signal by a specific receptor protein; (5) a change in cellular metabolism, triggered by the re-ceptor-signaling molecule complex; and (6) removal of the signal, often terminating the cellular response. ▲

The Role of Extracellular Signals in Cellular Metabolism

Some signals induce a modification in the activity of one or more enzymes already present in the target cell, allow-ing the cell to respond quickly—in minutes or seconds. In animals, most signaling molecules that induce such rapid changes are water-soluble and bind to receptors located on the plasma membrane.

Other signaling molecules primarily alter the pattern of gene expression. Generally, these molecules are poorly soluble in aqueous solutions but are soluble in lipids. Lipid-soluble molecules induce slower, longer-lasting re-sponses in their target cells than the changes produced by water-soluble signals. Such prolonged interactions are crucial during cell growth and differentiation. Steroid hormones, the best-known examples of this class, interact with intracellular receptors to cause induction of specific genes by binding to controlling regions in DNA.

(a) Endocrine signaling

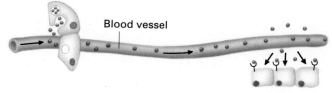

Hormone secretion into blood by endocrine gland

(b) Paracrine signaling

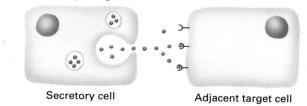

Secretory cell   Adjacent target cell

(c) Autocrine signaling

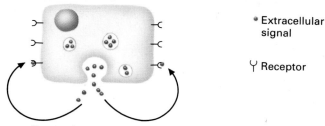

Target sites on same cell

▲ **Figure 19-1** Three general schemes in animals for cell-to-cell signaling by extracellular chemicals. Signaling can occur over very small to very large distances, from a few micrometers in paracrine or autocrine secretions to several meters in endocrine secretions.

There are too many important hormones, pheromones, and neurotransmitters to cover them all in this chapter. We will examine some signaling substances that have reasonably well-understood cellular and molecular mechanisms; these include epinephrine and glucagon, hormones that trigger the breakdown of glycogen in the liver and of triglycerides in fat cells, and insulin, the primary regulator of the glucose level in blood.

Specific Receptors Mediate the Response of Cells to Extracellular Signals

A *receptor protein* on the surface of the target cell, or in its nucleus or cytosol, has a binding site with a high affinity for a particular signaling substance (a hormone, pheromone, or neurotransmitter). Here, we may refer to the signaling substance as the *ligand* (a substance that binds to or "fits" a site). When the signaling substance binds to the receptor, the receptor-ligand complex initiates a sequence of reactions that changes the function of the cell.

The response of a cell or tissue to specific hormones is dictated by the particular hormone receptors it possesses and by the intracellular reactions initiated by the binding of any one hormone to its receptor. One cell may have two or more types of receptors or various cell types may have different sets of receptors for the same ligand, each of which induces a different response. Or the same receptor may occur on various cell types, and binding of the same ligand may trigger a different response in each type of cell. Clearly, different cells respond in a variety of ways to the same ligand. For instance, acetylcholine receptors are found on the surface of striated muscle cells, heart muscle cells, and pancreatic acinar cells. Release of acetylcholine from a neuron adjacent to a striated muscle cell triggers contraction, whereas release adjacent to a heart muscle slows the rate of contraction. Release adjacent to a pancreatic acinar cell triggers exocytosis of secretory granules that contain digestive enzymes.

In some cell types, different receptor-hormone complexes induce the same cellular response. In liver cells, either the binding of glucagon to its receptors or the binding of epinephrine to its receptors can induce the degradation of glycogen and release of glucose into the blood.

▶ **Figure 19-2** Some hormones bind to cell-surface receptors; others, to receptors in the cell. (a) Cell-surface receptors. Peptide and protein hormones, prostaglandins, amino acids, epinephrine, and related compounds bind to cell-surface receptors, triggering an increase or decrease in the cytosolic concentration of cAMP, Ca^{2+}, 1,2-diacylglycerol, or some other second messenger. (b) Cytosolic or nuclear receptors. Steroids, thyroxine, and retinoic acid, being very hydrophobic, are transported by carrier proteins in the blood. Dissociated from the carrier, the hormones enter the cell, bind to specific receptors in the cytosol or nucleus, and act on nuclear DNA to alter transcription of specific genes.

Thus two aspects of receptor function are the *binding specificity* of the receptor for the ligand and the *effector specificity* for the resulting change in cell behavior.

In most receptor-ligand systems, the ligand appears to have no function except to bind to the receptor. The ligand is not metabolized to useful products, is not an intermediate in any cellular activity, and has no enzymatic properties. The only function of the ligand appears to be to change the properties of the receptor, which then signals to the cell that a specific product is present in the environment. Target cells often modify or degrade the ligand and, in so doing, can modify or terminate their response or the response of neighboring cells to the signal.

Hormones fall into three broad categories: (1) small lipophilic molecules that diffuse through the plasma membrane and interact with cytosolic or nuclear receptors; (2) hydrophilic molecules that bind to cell-surface receptors; (3) lipophilic molecules that bind to plasma-membrane receptors (Figure 19-2).

(a) Cell-surface receptors

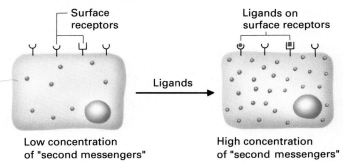

(b) Cytosolic or nuclear receptors

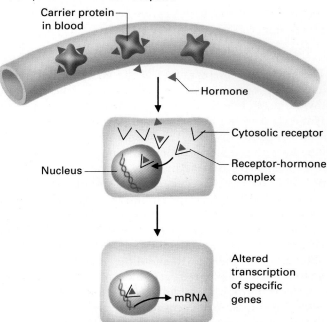

Table 19-1 Some animal hormones that bind to nuclear receptors (receptor-hormone complexes bind to regulatory DNA sequences and affect transcription of specific mRNAs)

Hormone	Structure	Origin	Major effects
STEROIDS Progesterone		Ovary, mainly corpus luteum; placenta	Differentiation of the uterus in preparation for implantation of the early embryo; maintenance of early pregnancy; development of the alveolar system in mammary glands
Estradiol (one of three estrogens)		Ovary (stroma, granulosa, theca); placenta	Differentiation of the uterus and other female sex organs; maintenance of secondary female sex characteristics and the normal cyclic function of accessory sex organs; development of the duct system in mammary glands
Testosterone		Testis	Maturation and normal function of accessory male sex organs; development of male sex characteristics
Cortisol		Adrenal cortex	Effect on metabolism of carbohydrates, lipids, and proteins; reduction of inflammation and immune responses; increases overall physiological response to stress

Most Lipophilic Hormones Interact with Cytosolic or Nuclear Receptors to Affect Gene Expression

The principal representatives of the group of hormones interacting with intracellular receptors are the steroids, thyroxine, and retinoic acid (Table 19-1). All steroids are synthesized from cholesterol and have similar chemical skeletons. After crossing the plasma membrane, steroids interact with receptor proteins in the nucleus or cytosol to form complexes that accumulate in the nucleus; there, they bind to specific regulatory DNA sequences and increase or decrease the transcription rate of adjacent genes (see Figure 11-33). These receptor-steroid complexes may also affect the stability of specific mRNAs. Steroids are effective for relatively long durations (hours or days) and often influence the growth and differentiation of specific tissues. For example, estrogen and progesterone, the female sex hormones, stimulate the production of egg-white hormones in chickens and cell proliferation in the hen oviduct. In mammals, estrogens stimulate growth of the uterine wall in preparation for embryo implantation. In insects and crustaceans, α-ecdysone (although it is not classified chemically as a steroid) triggers the differentiation and maturation of larvae; like estrogens, it induces the transcription of specific gene products.

Thyroxine, including tetraiodothyronine and triiodothyronine (Table 19-1)—the principal iodinated compounds in the body—is formed in the thyroid by intracellular proteolysis of the iodinated protein thyroglobulin. Thyroxine is produced by thyroid cells and immediately released into the blood. Thyroid hormones stimulate the catabolism of glucose, fats, and proteins by increasing the levels of many enzymes that catalyze these metabolic reactions, such as liver hexokinase, and mitochondrial enzymes for oxidative phosphorylation.

Because steroids and thyroxine are poorly soluble in aqueous solutions, they are transported in the blood by tightly bound carrier proteins. Typical life spans in the blood plasma are several hours (steroids) to several days (thyroxine), and the effects on target cells last from hours to days (Table 19-2).

Table 19-1 *(Continued)*

Hormone	Structure	Origin	Major effects
Aldosterone		Adrenal cortex	Maintenance of water and ion balance; ion reabsorption by kidney epithelial cells
STEROIDLIKE HORMONES α-Ecdysone		Endocrine glands (insects, crustaceans)	Differentiation and maturation of larvae
OTHER HORMONES Thyroxine Tetraiodothyronine (T4)		Thyroid	Increased heat production; maintenance of metabolism of glucose and other fuels; broad effects on gene expression and induction of enzymatic synthesis
Triiodothyronine (T3)			

Table 19-2 Characteristic features of principal types of mammalian hormones

Feature	Steroids	Thyroxine	Peptides and proteins	Catecholamines
Feedback regulation of synthesis	Yes	Yes	Yes	Yes
Storage of preformed hormone	Few hours	Several weeks	One day	Several days, in adrenal medulla
Mechanism of secretion	Diffusion through plasma membrane	Proteolysis of thyroglobulin	Exocytosis of storage vesicles	Exocytosis of storage vesicles
Plasma binding proteins	Yes	Yes	Rarely	No
Lifetime in blood plasma	Hours	Days	Minutes	Seconds
Time course of action	Hours to days	Days	Minutes to hours	Seconds or less
Receptors	Cytosolic or nuclear	Nuclear	Plasma membrane	Plasma membrane
Mechanism of action	Transcriptional control and mRNA stability	Transcriptional control and mRNA stability	Second messenger in cytosol or receptor-protein kinase	Second messenger in cytosol

SOURCE: E. L. Smith et al., 1983, *Principles of Biochemistry: Mammalian Biochemistry*, 6th ed., McGraw-Hill, p. 358. Reproduced by permission of McGraw-Hill.

Table 19-3 **Some mammalian hormones that interact with cell-surface receptors***

Hormone	Structure	Origin	Major effects
DERIVATIVES OF AMINO ACIDS			
Epinephrine		Adrenal medulla	Increase in pulse rate and blood pressure; contraction of most smooth muscles; glycogenolysis in liver and muscle; lipid hydrolysis in adipose tissue
Norepinephrine		Adrenal medulla	Contraction of arterioles; decrease of peripheral circulation
Histamine		Mast cells	Dilation of blood vessels
DERIVATIVES OF ARACHIDONIC ACID			
Prostaglandins (PGE$_2$)		Most body cells	Contraction of smooth muscle
PEPTIDE HORMONES			
Glucagon	Peptide, 29 aa[†]	Pancreas α cells	Stimulates glucose synthesis and glycogen degradation in liver; lipid hydrolysis in adipose tissue
Insulin	Polypeptide A chain, 21 aa B chain, 30 aa	β cells of pancreas	Stimulates glucose uptake into fat and muscle cells; carbohydrate catabolism; stimulates lipid synthesis by adipose tissue; general stimulation of protein synthesis and cell proliferation
Gastrin	Polypeptide, 17 aa	Intestine	Secretion of HCl and pepsin by stomach
Secretin	Polypeptide, 27 aa	Small intestine	Secretion of pancreatic digestive enzymes
Cholecystokinin	Polypeptide, 23 aa	Small intestine	Secretion of pancreatic digestive enzymes; emptying of gall bladder
Adrenocorticotropic hormone (ACTH)	Polypeptide, 39 aa	Anterior pituitary	Lipid hydrolysis from adipose tissue; stimulates adrenal cortex to produce cortisol and aldosterone
Follicle-stimulating hormone (FSH)	Protein: α chain, 92 aa β chain, 118 aa	Anterior pituitary	Stimulates growth of oocyte and ovarian follicles and estrogen synthesis by follicles
Leutenizing hormone (LH)	Protein: α chain, 92 aa β chain, 115 aa	Anterior pituitary	Maturation of oocyte; stimulates estrogen and progesterone secretion by ovarian follicles
Thyroid-stimulating hormone (TSH)	Protein: α chain, 92 aa β chain, 112 aa	Anterior pituitary	Release of thyroxine by thyroid cells

*Molecules primarily used as neurotransmitters or neurohormones are depicted in Table 17-1.
[†]Amino acids.

Table 19-3 *(Continued)*

Hormone	Structure	Origin	Major effects
Parathyroid hormone	Protein, 84 aa	Parathyroid	Increase in blood Ca^{2+} and decrease in phosphate; dissolution of bone calcium phosphate; increase in reabsorption of Ca^{2+} and decrease in reabsorption of PO_4 from kidney filtrate
Vasopressin	Protein, 9 aa	Posterior pituitary	Increase in water absorption from urine by kidney tubules; constriction of small blood vessels and rise in blood pressure
TSH-releasing hormone (TRH)	Polypeptide, 3 aa	Hypothalamus	Induces secretion of thyroid-stimulating hormone by anterior pituitary
LH-releasing hormone (LHRH)	Polypeptide, 10 aa	Hypothalamus, neurons	Induces secretion of leutenizing hormone by anterior pituitary
PEPTIDE GROWTH/DIFFERENTIATION FACTORS			
Epidermal growth factor (EGF)	Polypeptide, 53 aa	Salivary and other glands?	Growth of epidermal and other body cells
Somatotropin (growth hormone)	Polypeptide, 191 aa	Anterior pituitary	Stimulates amino acid uptake by many cells; stimulates liver to produce IGF-1 which, in turn, causes growth of bone and muscle
Erythropoietin	Polypeptide, 166 aa	Kidney	Differentiation of erythrocyte stem cells
Granulocyte-macrophage colony-stimulating factor (GM-CSF)	Polypeptides, 14,000–35,000 MW	T cells, endothelial cells, fibroblasts	Differentiation of macrophage and granulocyte stem cells
Interleukin 2	Polypeptide, 15,500 MW	T cells and macrophages	Growth of T cells of the immune system
Nerve growth factor	Two identical chains of 118 aa	All tissues innervated by sympathetic neurons	Growth and differentiation of sensory and sympathetic neurons
Insulinlike growth factor 1 (IGF-1 or somatomedin 1)	Polypeptide, 70 aa	Liver and other cells	Autocrine/paracrine growth factor induced by somatotropin; stimulates cell growth and division and glucose and amino acid uptake; increase in liver glycogen synthesis

Water-soluble Hormones Interact with Cell-surface Receptors

Many hormones are not soluble in lipids and cannot diffuse across the plasma membrane to interact with intracellular receptors. Instead, these hormones interact with *cell-surface receptors*. This large class of hormones includes large polypeptides, such as insulin, growth hormones, and glucagon, and small charged compounds, such as epinephrine. The effects of these surface-bound hormones are usually immediate and very short in duration: the cell reacts in milliseconds or, at most, in a few seconds. The effects of peptide growth factors, however, often extend over days; these factors are required for the growth and differentiation of specific types of cells, and many induce changes in gene expression in a few minutes (Table 19-3).

Different types of cell-surface receptors trigger different types of cellular responses (Figure 19-3). The binding of ligands to many cell-surface receptors activates an enzyme that generates a short-lived increase in the concentration of an intracellular signaling compound termed a *second messenger*. Second messengers include 3′,5′-cyclic AMP (cAMP); 3′5′-cyclic GMP (cGMP); 1,2-diacylglycerol; inositol 1,4,5-trisphosphate; and Ca^{2+} (Figure 19-4). The elevated intracellular concentration of one or more such second messengers triggers a rapid alteration in the activity of one or more enzymes or nonenzymatic

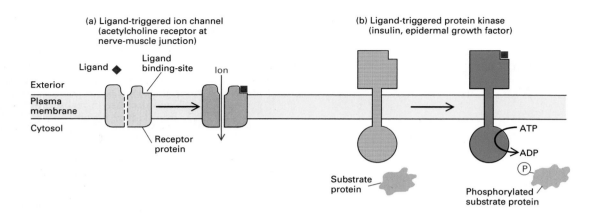

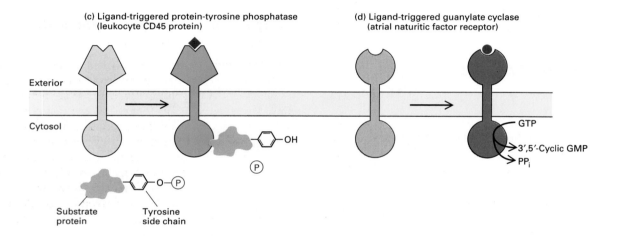

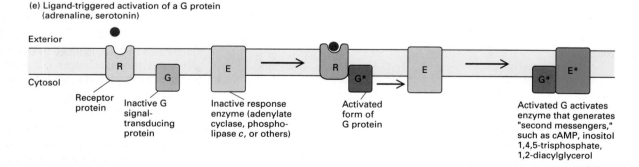

▲ **Figure 19-3** Types of cell-surface receptors. (a) Ligand-triggered ion channels. Ligand binding induces a conformational change in the receptor that opens a specific ion channel in the protein itself. The resultant flow of ions changes the electric potential across the cell membrane. (b) Ligand-triggered protein kinase. Ligand binding triggers a protein kinase activity; the receptor phosphorylates a substrate protein and, in so doing, alters the activity of that protein. (c) Ligand-triggered protein-tyrosine phosphatase. Ligand binding causes activation of a phosphatase activity that removes a phosphate residue attached to a tyrosine on a substrate protein and, in so doing, alters the activity of that pro-

tein. (d) Ligand-triggered guanylate cyclase. Ligand binding activates the cytosolic synthesis of the second messenger 3′,5′-cyclic GMP from GTP. (e) Ligand-triggered activation of a G protein and generation of a second messenger. Ligand binding triggers activation of a G protein that binds to and activates an enzyme that generates a specific intracellular second messenger. *Part (c) after H. Charbonneau, N. Tonks, S. Kumar, C. Diltz, M. Harrylock, D. Cool, E. Krebs, E. Fischer, and K. Walsh, 1989,* Proc. Nat'l Acad. Sci. USA **86**:5252–5256. *Part (d) after S. Schulz, M. Chinkers, and D. Garbers, 1989,* FASEB J. **3**:2026–2035; *D. Garbers, 1989,* J. Biol. Chem. **264**:9103–9106.

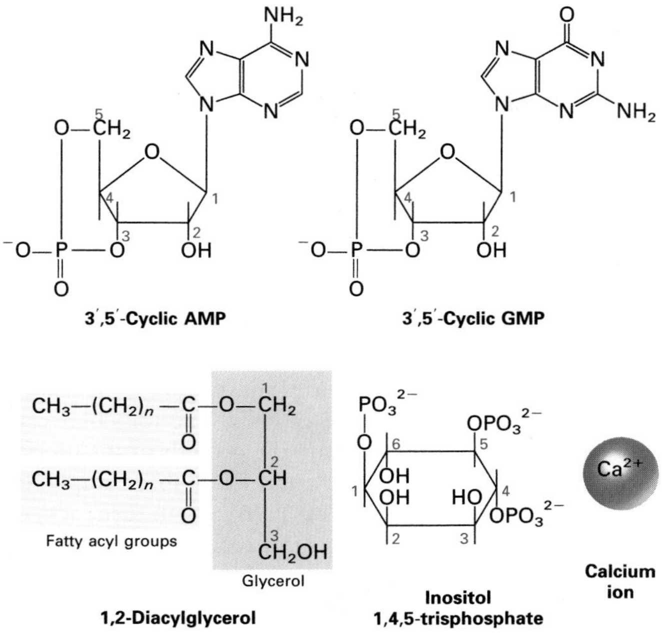

▲ **Figure 19-4** Structures of typical intracellular second messengers.

proteins. The metabolic functions controlled by these hormones include uptake and utilization of glucose, storage and mobilization of fat, and secretion of cellular products. Importantly, removal or degradation of the ligand reduces the level of the second messenger and terminates the metabolic response.

Other receptors, such as the insulin receptor, may not utilize a second messenger but act directly to modify the activity of cytoplasmic proteins by phosphorylating them. Still other receptors are ligand-triggered protein phosphatases, which modify the activity of proteins by removing an attached phosphate, and ligand-triggered ion channels. In the latter case, ligand binding changes the conformation of the receptor so that specific ions flow through it; the resultant ion movements alter the electric potential across the cell membrane. Such ion channel receptors are found mainly in the nervous system.

Prostaglandins Are Produced by Most Mammalian Cells

Prostaglandins are lipid-soluble, hormone-like chemicals that are produced in most body tissues and bind to cell-surface receptors (see Table 19-3). They contain a cyclopentane ring and are synthesized from *arachidonic acid*, a 20-carbon fatty acid with four double bonds. There are at least 16 different prostaglandins in nine different chemical classes, designated PGA, PGB, . . . , PGI. In both vertebrates and invertebrates, these chemicals are synthesized and secreted continuously by many types of cells and rapidly broken down by enzymes in body fluids.

Many prostaglandins act as local mediators during paracrine signaling and are destroyed near the site of their synthesis. They modulate the responses of other hormones and can have profound effects on many cellular processes. Certain prostaglandins cause blood platelets to aggregate and adhere to the walls of blood vessels. Because platelets play a key role in clotting blood and plugging leaks in blood vessels, these prostaglandins can affect the course of vascular disease and wound healing; aspirin and other anti-inflammatory agents inhibit their synthesis. Other prostaglandins initiate the contraction of smooth muscle cells; they accumulate in the uterus at the time of childbirth and are believed to be important in inducing uterine contraction.

The Synthesis, Release, and Degradation of Hormones Are Regulated

Organisms must be able to respond instantly to many changes in the internal or external environment. Responses of seconds or minutes in duration typically are mediated by peptide hormones or the *catecholamines* (amines that act as neurotransmitters or hormones). The signaling cells usually store enough hormone in secretory vesicles just under the plasma membrane to last several days. All peptide hormones, such as insulin and *adrenocorticotropic hormone* (ACTH), are synthesized as part of a longer polypeptide that is cleaved by specific proteases to the active molecule just after it is transported to a secretory vesicle.

Stimulation of the signaling cell by a neurotransmitter or hormone causes immediate exocytosis of the peptide hormone into the blood. Signaling cells are also stimulated to synthesize the hormone and replenish the cell's supply. The released peptide hormones live in the blood for only seconds or minutes before being degraded by blood and tissue proteases. Released catecholamines are inactivated by different enzymes or taken up by specific cells. The initial actions of these signaling substances on target cells (the activation or inhibition of specific enzymes) also last only seconds or minutes (see Table 19-2). Thus the catecholamines and some peptide hormones mediate short responses that are terminated by their own degradation.

The release of thyroxine and steroids by signaling cells is also closely regulated. *Thyroid-stimulating hormone* (TSH) triggers endocytosis by thyroid cells of the iodinated precursor protein *thyroglobulin* and its proteolysis to thyroxine. Steroids are synthesized from cholesterol by pathways involving 10 or more enzymes. Steroid-producing cells, like those in the adrenal cortex, typically store a supply of precursor to last a few hours. When stimulated, the cells convert the precursor to the finished hormone, which then diffuses across the plasma membrane into the blood.

The Levels of Hormones Are Regulated by Complex Feedback Circuits

The processes of growth and differentiation require the coordinated action of multiple hormones on many types of cells. The levels of several hormones often are interconnected by *feedback circuits,* in which changes in the level of one hormone affect the levels of other hormones. One example concerns the regulation of *estrogen* and *progesterone,* steroid hormones that stimulate the growth and differentiation of cells in the *endometrium,* the tissue that lines the interior of the uterus. Changes in the endometrium prepare the organ to receive and nourish an embryo. The levels of both hormones are regulated by complex feedback circuits involving several other hormones

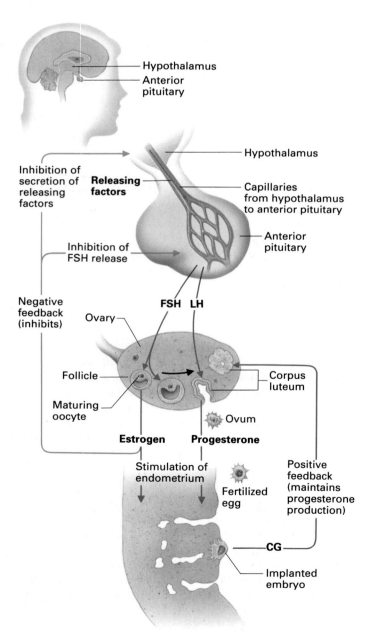

(Figure 19-5). A key role is played by the *anterior pituitary gland,* an organ separated from but controlled by the brain. The anterior pituitary gland is connected directly to the *hypothalamus,* at the base of the brain, by a special set of blood vessels. Nerve cells in the hypothalamus discharge hypothalamic peptide-releasing factors, which enter these vessels and bind to receptors on anterior pituitary cells. These factors, in turn, cause specific pituitary hormones to be secreted.

Each developing mammalian egg, or oocyte, matures into an ovum inside an ovarian follicle made up of many cells. Under the influence of *follicle-stimulating hormone* (FSH) released by the anterior pituitary gland, the follicle increases in size and number of cells. The follicular cells secrete estrogens that stimulate the growth of the uterine wall and its glands in preparation for embryo implantation, should fertilization occur. Estrogen, in turn, acts on the hypothalamus to reduce the secretion of releasing hormones and on the anterior pituitary gland directly to *inhibit* the release of FSH. This *negative feedback* by estrogen modulates the level of circulating estrogen in the nonpregnant female. As the follicle matures, pituitary secretion of FSH is reduced but a different pituitary peptide hormone, leutenizing hormone (LH), is secreted. LH completes the maturation and release of the ovum and transforms the follicle into an endocrine organ, the *corpus luteum,* which secretes progesterone.

Progesterone, in turn, causes additional growth of the endometrium. In the absence of fertilization of the ovum and pregnancy, the corpus luteum degenerates. The resulting decrease in the level of circulating estrogens and

◀ **Figure 19-5** Estrogen and progesterone levels in the blood of female mammals are regulated by complex feedback circuits. The pituitary hormone FSH causes the developing ovarian follicle to grow and secrete estrogen, which stimulates growth of the endometrium of the uterus. Estrogen also acts on the hypothalamus to inhibit secretion of the FSH-releasing factor and directly on the anterior pituitary gland to reduce FSH secretion, lowering estrogen production (negative feedback). As a follicle approaches maturity, the output of FSH decreases. The hypothalamus releases *luteinizing hormone-releasing hormone* (LHRH), which triggers the anterior pituitary gland to secrete LH, completing the maturation of the follicle and ovum. The mature follicle releases the ovum into the oviduct and is transformed into a temporary endocrine gland, the *corpus luteum.* Under continuing stimulation of LH, the corpus luteum secretes progesterone, which, in turn, acts to induce further growth of the uterine wall, preparing it to receive an embryo. If fertilization and implantation occur, the placental tissues produce *chorionic gonadotropin* (CG), a peptide hormone similar in structure and function to LH. CG prevents degeneration of the corpus luteum; in particular, it maintains continued synthesis of progesterone (positive feedback). Progesterone, in turn, maintains the cells in the endometrial lining of the uterus, and a good blood supply forms to nourish the implanted embryo.

progesterones causes the uterine wall to degenerate and menstruation to begin. During pregnancy, a *positive feedback circuit* maintains the high level of progesterone required for the endometrium to provide an adequate blood supply for the embryo embedded in the uterine wall. Progesterone stimulates the synthesis of *chorionic gonadotrophin* (CG), which, in turn, causes progesterone synthesis and secretion to increase.

Identification and Purification of Cell-surface Receptors

The action of steroid hormones is discussed in Chapter 11; in the remainder of this chapter, we focus on cell-surface receptors and on how binding a ligand to one of these proteins triggers a cellular response. Hormone receptors on the plasma membrane or in the cell bind ligands with great specificity and high affinity. Binding a hormone to a receptor involves the same types of weak interactions—ionic and van der Waals bonds and hydrophobic interactions—that characterize the specific binding of a substrate to an enzyme. The *specificity* of a receptor refers to its ability to distinguish closely related substances; the insulin receptor, for example, binds insulin but not other peptide hormones.

Hormone binding can usually be described by the simple equations

$$R + H \rightleftharpoons [RH] \tag{19.1}$$

$$K_D = \frac{[R][H]}{[RH]} \tag{19.2}$$

where $[R]$ and $[H]$ are the respective concentrations of free receptor and hormone and $[RH]$ is the concentration of the receptor-hormone complex. K_D, the dissociation constant of the receptor-ligand complex, measures the affinity of the receptor for the ligand. This binding equation can be written in a form similar to that of the Michaelis-Menton equation used to analyze enzymatic reactions:

$$\frac{[RH]}{R_T} = \frac{1}{1 + K_D/[H]} \tag{19.3}$$

where R_T is the total number of receptors $[R] + [RH]$.

The lower the K_D value, the higher the affinity of the receptor for its ligand. The K_D value is equivalent to the concentration of ligand at which one-half of the receptors contain bound ligand. If $[H] = K_D$, then from equation (19.3) we can see that $[RH] = 0.5 [R_T]$. For the insulin receptor, for instance, $K_D = {\sim}2 \times 10^{-8} M$ for insulin, so the receptor will be half-saturated at an insulin concentration of $2 \times 10^{-8} M$ or, 0.12 μg/ml. Since blood contains tens of milligrams of total protein per milliliter, the insulin receptor can bind insulin specifically in the presence of a 100,000-fold excess of unrelated proteins.

Hormone Receptors Are Detected by a Functional Assay

It is difficult to identify and purify hormone receptors, mainly because they are present in such minute amounts. The surface of a typical cell bears 10,000–20,000 receptors for a particular hormone, but this quantity is only ${\sim}10^{-6}$ of the total protein in the cell, or ${\sim}10^{-4}$ of the plasma-membrane protein. Purification is also difficult because these integral membrane proteins first must be solubilized with a nonionic detergent.

Usually, receptors are detected and measured by a functional assay: their ability to bind radioactive hormones to a cell or to cell fragments. The development of chemically or enzymatically catalyzed syntheses of radioactive hormones that retain normal hormone activity has been crucial to the identification of many receptors. When increasing amounts of a radioactive hormone, such as insulin, are added to a cell, the amount that binds to the surface increases at first and then approaches a plateau value (curve A in Figure 19-6). How can we deter-

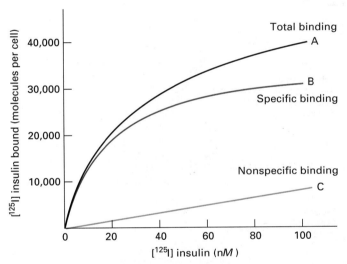

▲ **Figure 19-6** Identification of a specific insulin receptor on the surface of human hepatoma cells. A suspension of cells is incubated for 1 h at 4°C with increasing concentrations of [^{125}I]-labeled hormone. To measure the amount bound to the cells, the suspension is layered on a tube of mineral oil and centrifuged; all cells with any surface-bound hormone are recovered in the pellet. The total binding curve (A) consists of the specific binding of hormone to high-affinity receptors as well as a nonspecific low-affinity sticking to other molecules on the cell surface. The contribution of the nonspecific binding curve (C) to the total binding curve (A) is determined by measuring [^{125}I] hormone binding in the presence of a 100-fold excess of unlabeled hormone to saturate all the specific high-affinity sites. The specific binding curve (B) is calculated as the difference between curves A and C. Curve B fits the simple binding equation (19-3). For this receptor, $K_D = 20$ nM, or $2 \times 10^{-8} M$, and $R_T = 40,000$ molecules/cell. *After A. Ciechanover, A. Schwartz, and H. F. Lodish, 1983, Cell* **32**:267–275.

mine what fraction of this binding is specific to the surface receptor and what fraction is nonspecific to the multitude of other proteins and phospholipids on the cell surface? Nonspecific hormone binding can be misinterpreted as meaning there are more receptor sites than actually exist. Nonspecific binding of labeled hormone can be measured by conducting the binding assay in the presence of a large excess of unlabeled hormone (curve C in Figure 19-6) to ensure that all specific (high-affinity) binding sites but only a few nonspecific sites are filled with unlabeled hormone. Specific binding is calculated as the difference between total binding and nonspecific binding.

The specific binding sites are saturable. From the saturation level, we can calculate the number of insulin-binding sites per cell. There are about 40,000 receptor molecules on each hepatoma cell (curve B in Figure 19-6). The specific binding curve can be described by the binding equations (19.2) and (19.3). For these insulin receptors, $K_D = 2 \times 10^{-8}\ M$; in other words, the receptor is half-saturated when the hormone is present at only $2 \times 10^{-8}\ M$. Binding affinities of other receptors can be even greater; $K_D = {\sim}1 \times 10^{-11}\ M$ for the receptor-erythropoietin complex on erythrocyte stem cells. Once the radioactive hormone is bound to a receptor, it can be identified and followed through isolation and purification procedures.

The K_D Values for Hormone Receptors Approximate the Concentration of the Circulating Hormone

In general, the K_D values for hormone receptors approximate the concentration of the hormone in the blood. Changes in hormone concentration are reflected in proportional changes in the fraction of receptors occupied. If, for instance, the normal (unstimulated) concentration of a hormone in the blood is $[H] = 10^{-9}\ M$ and $K_D = 10^{-7}\ M$ for the hormone receptor, then from equation (19.3) we can calculate that 1.01 percent of the receptors will be bound with hormone or $[RH]/R_T = 0.0101$. If the hormone concentration rises tenfold to $[H] = 10^{-8}\ M$, then the concentration of the receptor-hormone complex

will rise about tenfold, to [RH] = 11.1 percent. If the induced cellular response is proportional to the amount of the complex, as is often the case, then the cellular responses will increase about tenfold.

Frequently, occupancy of only a fraction of the hormone receptor of the cell induces a maximal physiological response at a lower concentration of ligand than that needed to bind to all the cell receptors. The ligand concentration necessary to induce a 50-percent maximal response can be less than 25 percent of the K_D value (Figure 19-7).

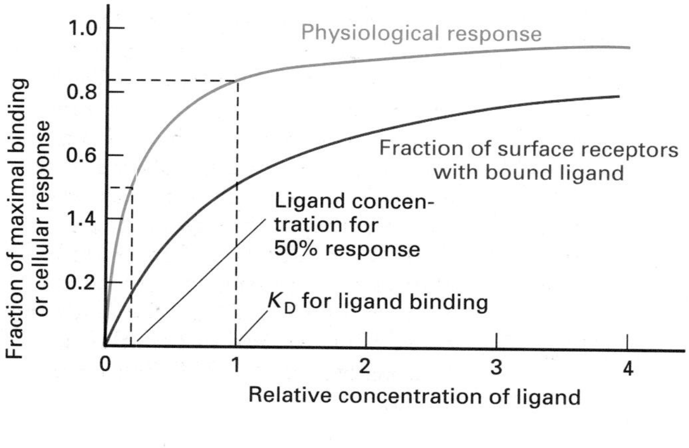

Affinity Techniques Permit Purification of Receptor Proteins

Cell-surface hormone receptors often can be identified by *affinity labeling,* in which the radioactive hormone is bound to cells and treated with a chemical agent that covalently cross-links the labeled hormone to the hormone-binding subunit of the receptor (Figure 19-8). Affinity labeling enables the receptor to remain bound to the hormone, even in the presence of detergents or other denaturing agents that are needed to solubilize the receptor from the membrane. Such cross-linking agents contain two reactive sites for free amino groups in proteins. Labeled insulin specifically binds to a 84,000-MW protein, a subunit of the insulin receptor.

Many cell-surface receptors can be solubilized from preparations of plasma membrane and fully retain their hormone-binding activity. These receptors then can be purified by the technique of *affinity chromatography.* (Figure 19-9 shows how this technique is used to purify a receptor for epinephrine.) The receptor can be purified as much as 100,000-fold in a single step.

Many Receptors Can Be Cloned without Prior Purification

In many cases, the amount of receptor on a cell is too small to be purified by affinity chromatography. The hormone erythropoietin is essential to the growth and differentiation of the nucleated precursors of erythrocytes; yet

◀ **Figure 19-7** How a maximal physiological response can occur at low ligand concentrations if occupancy of only a fraction of the cell receptors is essential to induce the response. In this example, 50 percent of maximal response is induced at a ligand concentration at which 18 percent of the receptors are occupied. When the ligand concentration equals the K_D value, 50 percent of the cell receptors are occupied but more than 80 percent of the maximal response is induced.

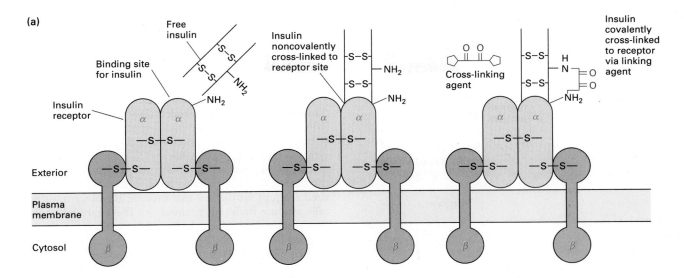

(a)

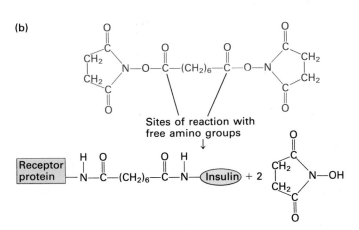

(b)

▲ Figure 19-8 For investigative purposes, insulin can be covalently cross-linked to its receptor. The insulin receptor contains two α subunits 84,000 MW that bind to insulin and two β subunits 70,000 MW that span the phospholipid bilayer. All subunits are linked by disulfide bonds. (a) Binding insulin to its receptor brings free NH_2 groups on the two proteins close together. (b) The chemical cross-linking agent reacts with free amino acid groups on insulin and the receptor protein, forming a covalent bond to ensure that radioactively labeled insulin remains attached to its receptor throughout isolation and purification procedures. [See J. Massague and M. P. Czech, 1982, *J. Biol. Chem.* **257**:729; A. Ullrich et al., 1985, *Nature* **313**:756.]

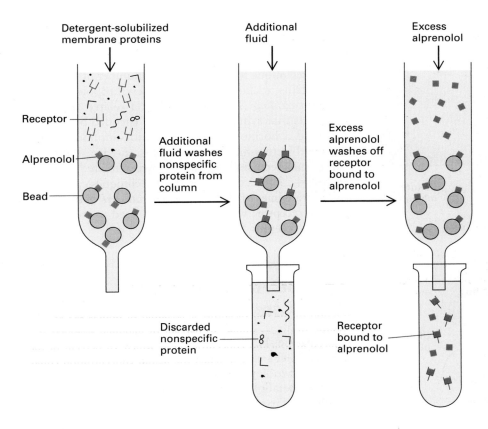

◀ Figure 19-9 Purification of the β-adrenergic receptor by affinity chromatography. The potent receptor-binding antagonist alprenolol (see Table 19-5) is chemically linked to polystyrene beads. A crude, detergent-solubilized preparation of erythrocyte membrane proteins is passed through a column containing these beads. Only the receptor binds to the beads; the other proteins are washed through by excess fluid. On the addition of an excess of alprenolol to the column, the bound receptor is displaced from the beads and eluted bound to free alprenolol. [See J. L. Benovic et al., 1984, *Biochemistry* **23**:4510.]

(a) Expression vector

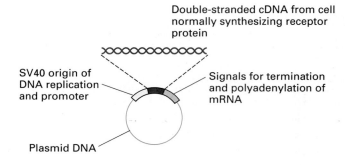

Double-stranded cDNA from cell normally synthesizing receptor protein

SV40 origin of DNA replication and promoter

Signals for termination and polyadenylation of mRNA

Plasmid DNA

(b) Transfection and expression

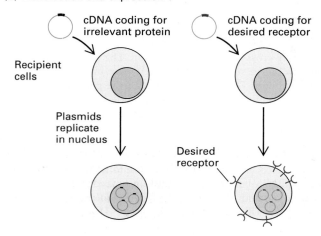

cDNA coding for irrelevant protein

cDNA coding for desired receptor

Recipient cells

Plasmids replicate in nucleus

Desired receptor

(c) Cells expressing receptor adhere to flask

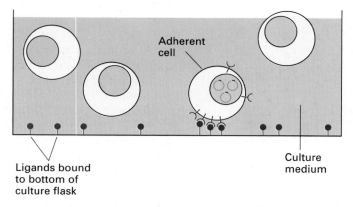

Adherent cell

Ligands bound to bottom of culture flask

Culture medium

(d) Purification of plasmid DNA from adherent cells and cloning

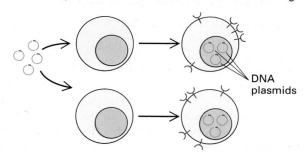

DNA plasmids

◀ **Figure 19-10** Expression cloning of a rare cDNA encoding a desired cell-surface receptor. All mRNA is extracted from cells that normally express the receptor and reverse-transcribed into double-stranded cDNA. (a) The entire population of cDNA is inserted into a specific expression vector, in between a strong promoter and a terminator for mRNA synthesis. (b) This "cDNA library" is transfected into a population of cultured cells that lack the receptor. Only the rare transfected cell that contains the cDNA coding for the desired receptor synthesizes that protein; the other transfected cells produce irrelevant proteins. (c) These rare cells adhere to a surface onto which the ligand (or antibody to the receptor) has been immobilized; all of the other cells can be washed away. (d) Plasmid DNA can now be isolated from the adherent cells and cloned, generating a cDNA clone encoding the desired receptor. *After A. Aruffo and B. Seed, 1987, Proc. Nat'l Acad. Sci. USA* **84**:*8573–8577; A. D'Andrea, H. F. Lodish, and G. Wong, 1989, Cell* **57**:*277–285.*

each precursor cell contains only 1000 erythropoietin receptors, or about 1 part per million (ppm) of total cellular membrane proteins. By conventional approaches, it is impossible to purify enough of this receptor protein to characterize or sequence it. In the mid-1980s, scientists began to develop techniques to produce large amounts of key receptor proteins by cloning many of them without prior purification of the receptor, allowing the sequence of the receptor protein to be obtained from the cDNA sequence. Such approaches take advantage of the fact that recombinant DNA techniques can be used to synthesize a receptor polypeptide in a cell that normally lacks one. Generally, the foreign receptor polypeptide is transported to the cell surface and enables the cell to bind the appropriate ligand specifically.

One technique for cloning a large number of key receptor proteins, including the one for erythropoietin, is *expression cloning* (Figure 19-10). In a related technique, only cells that synthesize a receptor encoded by a cDNA can bind a hormone (or anti-receptor antibody) that is covalently cross-linked to a fluorescent dye; these cells can be purified using the fluorescence-activated cell sorter (Figure 4-29). Recombinant DNA techniques can be used to generate lines of cells with large numbers of the desired surface receptor, which then can be used to study all aspects of receptor function.

Epinephrine Receptors and the Activation of Adenylate Cyclase

To illustrate the general principles of receptor-hormone signaling, let's examine several well-studied systems. We focus on two key issues: how ligand binding is transmitted to a metabolic response, and how receptor levels are regulated. First, we address receptors that generate the

Table 19-4 Some metabolic responses to a rise in intracellular cAMP (all are mediated by cAMP-dependent protein kinase)

Tissue	Hormone inducing a rise in cAMP	Metabolic response
Adipose	Epinephrine; ACTH; glucagon	Increase in hydrolysis of triglyceride; decrease in amino acid uptake
Liver	Epinephrine; norepinephrine; glucagon	Increase in conversion of glycogen to glucose; inhibition of synthesis of glycogen; increase in amino acid uptake; increase in gluconeogenesis (synthesis of glucose from amino acids)
Ovarian follicle	FSH; LH	Increase in synthesis of estrogen, progesterone
Adrenal cortex	ACTH	Increase in synthesis of aldosterone, cortisol
Cardiac muscle cells	Epinephrine	Increase in contraction rate
Thyroid	TSH	Secretion of thyroxine
Bone cells	Parathyroid hormone	Increase in resorption of calcium from bone
Skeletal muscle	Epinephrine	Conversion of glycogen to glucose
Intestine	Epinephrine	Fluid secretion
Kidney	Vasopressin	Resorption of water
Blood platelets	Prostaglandin I	Inhibition of aggregation and secretion

SOURCE: E. W. Sutherland, 1972, *Science* 177:401.

intracellular second messengers cAMP, 1,2-diacylglycerol, and inositol trisphosphates. These receptors utilize a transducing or G protein to communicate between the cell-surface receptor and the enzyme that generates these second messengers.

Perhaps the best understood mammalian cell-surface receptor system is found on many cells for the hormones *epinephrine* and *norepinephrine* (Table 19-3). These hormones were originally recognized as products of the *medulla*, or core, of the adrenal gland and are also known as *adrenaline* and *noradrenaline*. Embryologically, nerve cells derive from the same tissue as the adrenal medulla cells, and norepinephrine is also a secretory product of differentiated nerve cells. Both hormones are charged compounds that belong to the catecholamines, active amines containing the compound *catechol*:

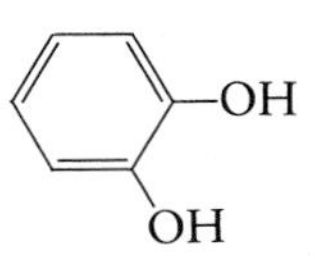

In times of stress, such as fright or heavy exercise, all tissues have an increased need for glucose and fatty acids. These principal metabolic fuels can be supplied to the blood in seconds by the rapid breakdown of glycogen in the liver *(glycogenolysis)* and of triacylglycerol in the adipose storage cells *(lipolysis)*. In mammals, the liberation of glucose and fatty acids can be triggered by binding epinephrine or norepinephrine to *β-adrenergic receptors* on the hepatic (liver) and adipose cells. Epinephrine bound to similar *β*-adrenergic receptors on heart muscle cells increases the contraction rate, which increases the blood supply to the tissues. Epinephrine bound to *β*-adrenergic receptors on smooth muscle cells of the intestine causes it to relax. Another epinephrine receptor, the *α-adrenergic receptor,* is found on smooth muscle cells lining the blood vessels in the intestinal tract, skin, and kidneys. Epinephrine bound to these *α* receptors causes the arteries to constrict, cutting off circulation to these peripheral organs. These diverse effects of one hormone are directed to a common end: supplying energy for the rapid movement of major locomotor muscles in response to bodily stress.

All of the very different tissue-specific responses induced by binding epinephrine to *β*-adrenergic receptors are mediated by a rise in the intracellular level of cyclic AMP (cAMP), which acts as a signal transducer, or second messenger, modifying the rates of different enzymatically catalyzed reactions in specific tissues (Table 19-4). Before we discuss the role of cAMP in cellular metabolism, let's investigate some key properties of the *β*-adrenergic receptors and the mechanism by which hormone binding triggers an increase in the level of cAMP.

Functional Assays Establish the Identity of the Purified *β*-Adrenergic Receptor

How do we know that specific binding of epinephrine to the cell surface actually represents binding to the receptor that triggers the cellular response? One indication is that the concentration of epinephrine that half-saturates the receptor is about the same as the concentration that causes the cAMP inside the cell to increase to one-half its maximal level (Figure 19-11).

Further evidence comes from studies of purified *β*-adrenergic receptor. The receptor is purified by affinity

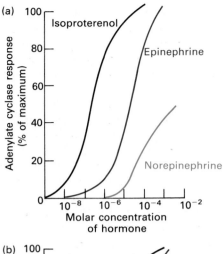

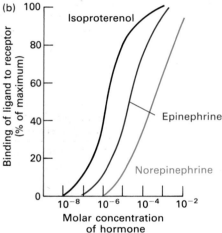

▲ **Figure 19-11** Comparison of the abilities of three catecholamines to activate cAMP synthesis (by adenylate cyclase) in frog erythrocytes and to bind to cell-surface β-adrenergic receptors. (a) Different concentrations of the ligands isoproterenol, epinephrine, and norepinephrine were incubated with a cell suspension at 37°C; the cells were then broken and levels of adenylate cyclase activity measured. (b) Hormone binding to the receptor was measured by an *indirect competition assay* in which binding of the unlabeled hormone to the receptor inhibits binding of the [³H]-labeled antagonist alprenolol, allowing the binding affinities of unlabeled hormones for the receptor to be estimated. The curves show that each hormone induces adenylate cyclase in proportion to its ability to bind to the receptor: the hormone concentration required for half-maximal binding to the receptor is about the same as that required for the activation of adenylate cyclase.

▶ **Figure 19-12** The β-adrenergic receptor mediates the induction of epinephrine-initiated cAMP synthesis. The target cell lacks any receptors for epinephrine but contains adenylate cyclase and other proteins required for cAMP synthesis. The purified β-adrenergic receptor (see Figure 19-9) can be incorporated into liposomes and fused with the target cell. The resultant cell responds to the addition of epinephrine by synthesizing high levels of cAMP. [See R. A. Cerione et al., 1983, *Nature* **306**:562.]

chromatography (see Figure 19-9) and incorporated into liposomes, which then are fused with a cell that contains adenylate cyclase (the enzyme that synthesizes cAMP) but no β-adrenergic receptors (Figure 19-12). These receptor-laden vesicles confer cAMP responsiveness of the target cells to epinephrine, proving that the receptor is involved in inducing cAMP synthesis. Similar conclusions can be drawn from studies in which the cloned β-adrenergic receptor cDNA is used to synthesize the receptor in receptor-negative cells; they, too, acquire the ability to activate adenylate cyclase in response to epinephrine.

Hormone Analogs Are Important in the Study of Receptor Action

Studies of a wide array of chemically synthesized analogs of natural hormones, such as epinephrine, provide additional evidence that binding a hormone to a saturable cell-surface receptor is physiologically relevant. These analogs fall into two classes: *agonists,* which mimic hor-

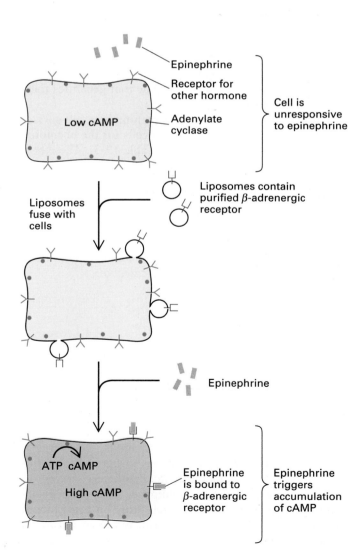

mone function by binding to the receptor and causing the normal response, and *antagonists,* which bind to the receptor but do not activate hormone-induced functions. A bound antagonist competes with the binding of the natural hormone (or an agonist) and blocks its physiological activity.

By studying the molecular structure and parameters of agonists and antagonists, we can define the parts of the epinephrine molecule necessary for binding and the parts necessary for the subsequent induction of the cellular response. In general, the catechol ring is an essential component of an agonist; it enables a compound to elevate the level of cAMP. The side chain containing NH determines the affinity of the agonist or antagonist for the β-adrenergic receptor. Some agonists (isoproterenol is an example) are more potent than their natural counterparts in that they bind to the receptor and activate synthesis of cAMP at a concentration 10–100 times lower than epinephrine (Table 19-5). Similarly, antagonists differ in their affinities for the β-adrenergic receptor.

In general, the K_D of the binding of an agonist to the receptor is the same as the concentration required for half-maximal elevation of cAMP. This relationship indicates that the activation of cAMP accumulation is proportional to the number of surface receptors filled with the agonist (see Figure 19-11).

There are at least two kinds of β-adrenergic receptors. In humans, the β_1 receptors on cardiac muscle cells promote increased heart rate and contractility by binding catecholamines with the rank order of affinities isoproterenol > norepinephrine > epinephrine. The β_2 receptors on smooth muscle cells lining the bronchial passages mediate their relaxation by binding agonists with the order of affinities isoproterenol $\geq$ epinephrine > norepinephrine. The discovery of these two types of receptors has led to the development and clinical use of β_1-selective antagonists, such as practolol, to slow heart contractions in the treatment of cardiac arrhythmias and angina. Such "beta blockers" usually have little effect on the β-adrenergic receptors of other cell types. Similarly, β_2-

Table 19-5   Structure of typical agonists and antagonists of the β-adrenergic receptor

Structure	Compound	K_D for binding to the receptor on frog erythrocytes
HO—, OH, HO—CH—CH$_2$—NH—CH$_3$	Epinephrine	5×10^{-6} M
AGONIST HO—, OH, CH$_3$, HO—CH—CH$_2$—NH—CH—CH$_3$	Isoproterenol	0.4×10^{-6} M
ANTAGONISTS CH$_2$=CH—CH$_2$—, O—CH$_2$—CH(OH)—CH$_2$—NH—CH(CH$_3$)$_2$	Alprenolol	0.0034×10^{-6} M
O—CH$_2$—CH(OH)—CH$_2$—NH—CH(CH$_3$)$_2$ (naphthalene)	Propranolol	0.0046×10^{-6} M
CH$_3$—C(=O)—NH—, —OCH$_2$—CH(OH)—CH$_2$—NH—CH(CH$_3$)$_2$	Practolol	21×10^{-6} M

SOURCE: R. J. Lefkowitz, L. E. Limbird, C. Mukherjee, and M. G. Caron, 1976, *Biochim. Biophys. Acta* 457:1.

selective agonists, such as terbutaline, are used in the treatment of asthma because they specifically cause the opening of the bronchioles (small airways in the lung).

The Binding of Hormone to β-Adrenergic Receptors Activates Adenylate Cyclase

Many kinds of mammalian cells have β-adrenergic receptors. In general, epinephrine bound to these receptors on different cells triggers distinctly different cellular events, but the initial response is always the same: an elevation in the level of cAMP caused by an activation of adenylate cyclase (Figure 19-13).

Adenylate cyclase is a membrane-bound enzyme with an ATP-binding site on the cytoplasmic face of the plasma membrane. The binding site of the receptor protein is lo-

cated on the exoplasmic face. A third protein, tightly bound to the cytoplasmic face, functions as a communicator between the receptor and adenylate cyclase; this is called the G_s protein because it binds guanosine diphosphate and triphosphate (GDP and GTP) and stimulates adenylate cyclase.

The G_s Protein Cycles between Active and Resting Forms

The G_s protein binds GTP noncovalently but very tightly. In fact, G_s can be purified from detergent-solubilized membranes by affinity chromatography on columns that contain chemically bound GTP. G_s is composed of three peptide chains, α, β, and γ, of ~42,000, ~35,000, and ~10,000 MW respectively. The α subunit, called $G_{sα}$, binds GTP and GDP. $G_{sα}$ also binds to the receptor and to adenylate cyclase, but the receptor and adenylate cyclase do not interact with each other directly.

GTP bound to $G_{sα}$ is hydrolyzed spontaneously to GDP by the reaction

$$G_{sα} \cdot GTP \longrightarrow G_{sα} \cdot GDP + P_i$$

When GTP is bound to $G_{sα}$, the $G_{sα}$ dissociates from the $G_{β,γ}$ subunits. The $G_{sα}$ subunit (with its bound GTP) apparently undergoes a conformational change that enables it to bind to and activate adenylate cyclase (Figure 19-14). When $G_{sα}$ has bound GDP, rather than GTP, the $G_{sα}$ associates with the $G_{β,γ}$ subunits and cannot bind to or activate adenylate cyclase.

In an unstimulated cell (a cell to which no hormone is bound), most G_s molecules contain GDP. Binding a hormone or agonist to the β-adrenergic receptor changes its conformation, causing it to bind to G_s in such a manner that GDP is displaced and GTP is bound. The activation of adenylate cyclase that results is short-lived, however,

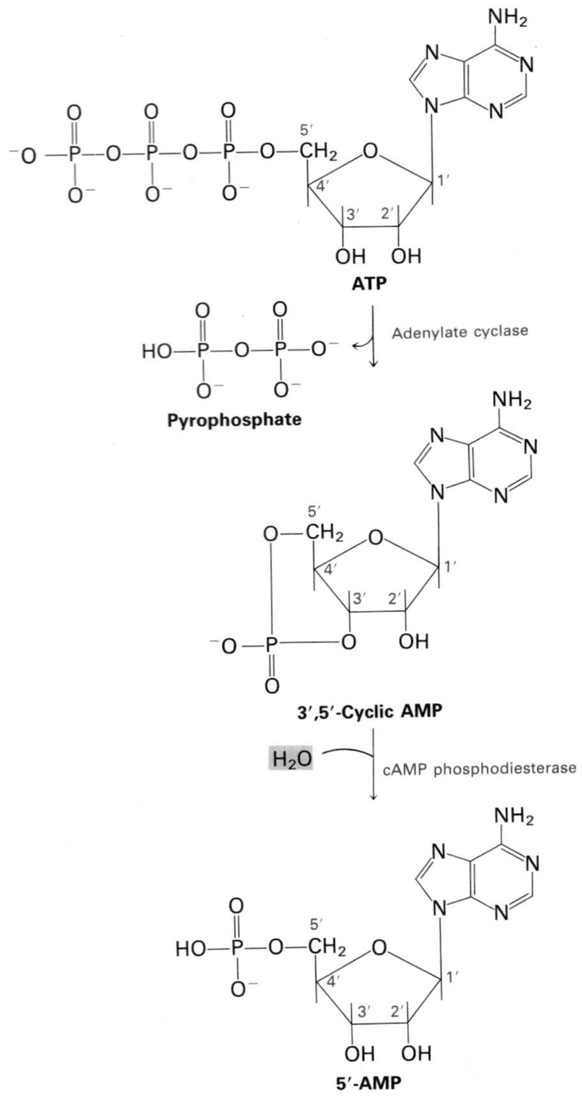

▲ **Figure 19-13** The synthesis and degradation of 3′,5′-cAMP.

▶ **Figure 19-14** *(opposite page)* The activation of adenylate cyclase by the binding of a hormone to its receptor. The patch of cell membrane depicted contains two transmembrane proteins, a receptor protein (R) for a hormone and adenylate cyclase (C), and on the cytosolic surface, the transducer protein G_s. In the resting state, GDP is bound to the $G_{sα}$ subunit of G_s. A conformational change occurs in R when a hormone binds to it (step 1). Activated R binds to G_s (step 2). This activates G_s so that it releases GDP and binds GTP, causing the $G_{sα}$ subunit to dissociate from the $G_{β,γ}$ subunits (step 3). The free $G_{sα}$ subunit binds to C and activates it, so that it catalyzes the synthesis of cAMP from ATP (step 4); this step may involve a conformational change in $G_{sα}$. When GTP is hydrolyzed to GDP, a reaction most likely catalyzed by $G_{sα}$ itself, $G_{sα}$ is no longer able to activate C (step 5), and $G_{sα}$ and $G_{β,γ}$ reassociate. Eventually, the hormone dissociates from the receptor and the system returns to its resting state.

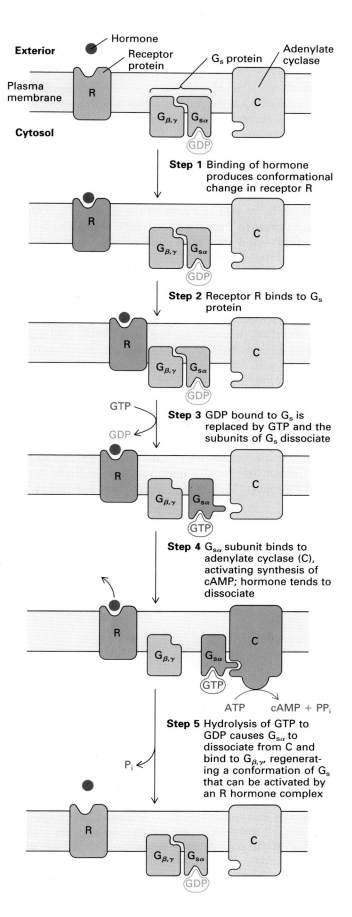

Step 1 Binding of hormone produces conformational change in receptor R

Step 2 Receptor R binds to G_s protein

Step 3 GDP bound to G_s is replaced by GTP and the subunits of G_s dissociate

Step 4 $G_{s\alpha}$ subunit binds to adenylate cyclase (C), activating synthesis of cAMP; hormone tends to dissociate

Step 5 Hydrolysis of GTP to GDP causes $G_{s\alpha}$ to dissociate from C and bind to $G_{\beta,\gamma}$, regenerating a conformation of G_s that can be activated by an R hormone complex

because GTP bound to G_α hydrolyzes to GDP in seconds, leading to the association of G_α with $G_{\beta,\gamma}$ and the inactivation of adenylate cyclase. Therefore, a GTP-GDP cycle (Figure 19-15) is crucial to hormone-dependent activation and inactivation of adenylate cyclase.

The G_s protein functions as a shuttle between two membrane proteins: the hormone receptor and adenylate cyclase. G_s is a signal transducer, relaying to the cyclase enzyme the conformational change in the receptor triggered by hormone binding.

Important evidence to support this model has come from studies using a nonhydrolyzable analog of GTP, called GMPPNP, in which a P—NH—P bond replaces the terminal phosphodiester bond in GTP:

$$\text{Guanine-ribose}\!-\!\text{O}\!-\!\overset{\overset{\displaystyle O}{\|}}{\underset{\underset{\displaystyle O^-}{|}}{P}}\!-\!\text{O}\!-\!\overset{\overset{\displaystyle O}{\|}}{\underset{\underset{\displaystyle O^-}{|}}{P}}\!-\!\text{NH}\!-\!\overset{\overset{\displaystyle O}{\|}}{\underset{\underset{\displaystyle O^-}{|}}{P}}\!-\!\text{O}^-$$

Although this analog cannot be hydrolyzed, it binds to the G protein as well as GTP does. The addition of GMPPNP and an agonist to an erythrocyte membrane preparation results in a much larger and longer-lived activation of adenylate cyclase than occurs with an agonist and GTP. Once the GDP bound to $G_{s\alpha}$ is displaced by GMPPNP, it remains permanently bound to $G_{s\alpha}$. Because the $G_{s\alpha} \cdot$ GMPPNP complex is as functional as the normal $G_{s\alpha} \cdot$ GTP complex in activating adenylate cyclase, the enzyme is in a permanently active state.

Activation of adenylate cyclase must be terminated rapidly when the hormone concentration is lowered; this action is facilitated by a feature of the $G_{s\alpha} \cdot$ GTP complex. Replacing GDP with GTP bound to $G_{s\alpha}$ *decreases*

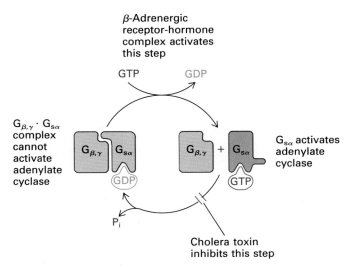

▲ Figure 19-15 Cycling of the G protein between GTP and GDP is coupled to activation and inactivation of adenylate cyclase. Hydrolysis of GTP to GDP is believed to be catalyzed by the $G_{s\alpha}$ subunit itself.

the affinity of the β-adrenergic receptor for the hormone (increases K_D), shifting the receptor-hormone equilibrium toward dissociation and leading to inactivation of adenylate cyclase.

What are the purposes of the complex interactions among these proteins? First, the hormone signal is amplified substantially. Because both the receptor and the G protein can diffuse rapidly in the plasma membrane, a single receptor-hormone complex causes the conversion of up to 100 $G_{s\alpha} \cdot GDP \cdot G_{\beta,\gamma}$ complexes to the active $G_{s\alpha} \cdot GTP$ form. Each of these complexes, in turn, probably activates a single adenylate cyclase molecule. Each enzyme catalyzes the synthesis of many cAMP molecules during the time $G_{s\alpha} \cdot GTP$ is bound to it. Although the exact extent of this amplification is difficult to measure, binding a single hormone molecule to one receptor can result in the synthesis of at least several hundred cAMP molecules before the hormone dissociates from the receptor and terminates the activation of adenylate cyclase. Because the physiological responses triggered by cAMP may require tens of thousands or even millions of cAMP molecules per cell, amplification is necessary to generate this much second messenger from a few thousand β-adrenergic receptors.

Second, the G protein enables the cell to terminate its response to a hormone rapidly when the hormone concentration is reduced. The hydrolysis of the GTP bound to $G_{s\alpha}$ reverses the activation of adenylate cyclase. The continuous presence of epinephrine is required for the continuous activation of adenylate cyclase.

Several Receptors Interact with a Single Type of Adenylate Cyclase

In many types of cells, the same metabolic response can be triggered by different hormones binding to different receptors; several receptor-hormone complexes often activate adenylate cyclase. The G_s protein enables different receptors to activate adenylate cyclase. In the liver, for instance, glucagon (a peptide hormone) and epinephrine bind to different receptors but activate the same adenylate cyclase. Both types of receptors bind to and activate the same G_s protein and convert the inactive $G_{s\alpha} \cdot GDP \cdot G_{\beta,\gamma}$ complex to the active $G_{s\alpha} \cdot GTP$ form. The level of $G_{s\alpha} \cdot GTP$—and thus activation of adenylate cyclase—will equal the sum of the $G_{s\alpha} \cdot GTP$ levels caused by binding the individual hormones to their own receptors.

Certain hormones cause the level of cAMP to decrease; others cause it to increase. In the adipose cell, for example, epinephrine, glucagon, and ACTH all stimulate adenylate cyclase, whereas prostaglandin PGE_1 and adenosine inhibit the enzyme. Interaction of these inhibitory receptors with adenylate cyclase is mediated by the inhibitory *coupling protein* G_i, which contains the same β and γ subunits as the stimulatory G_s but a different $G_{i\alpha}$ subunit that also binds GDP or GTP. In response to a hormone binding to its receptor, G_i binds GTP and dissociates into $G_{i\alpha} \cdot GTP$ and a $G_{\beta,\gamma}$ subunit, inhibiting rather than activating adenylate cyclase (Figure 19-16). At least two mechanisms for this inhibition have received experimental support. $G_{i\alpha}$ can bind directly to adenylate cyclase

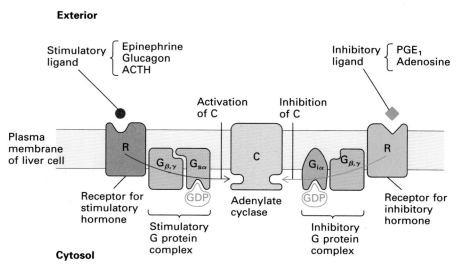

Exterior

Stimulatory ligand { Epinephrine / Glucagon / ACTH

Inhibitory ligand { PGE$_1$ / Adenosine

Activation of C

Inhibition of C

Plasma membrane of liver cell

R — $G_{\beta,\gamma}$ — $G_{s\alpha}$ — C — $G_{i\alpha}$ — $G_{\beta,\gamma}$ — R

GDP   GDP

Receptor for stimulatory hormone

Stimulatory G protein complex

Adenylate cyclase

Inhibitory G protein complex

Receptor for inhibitory hormone

Cytosol

▲ **Figure 19-16** Different G proteins mediate activation and inhibition of adenylate cyclase. Many receptor-hormone complexes bind to a stimulatory G_s protein, causing bound GDP to be replaced by GTP and the $G_{s\alpha} \cdot GTP$ subunit to dissociate and bind to and activate adenylate cyclase. Other receptor-hormone complexes bind to a different inhibitory $G_{i\alpha}$ protein, also constructed of an α subunit, that binds GDP or GTP and the $G_{\beta,\gamma}$ subunit. In some way, the inhibitory $G_{i\alpha}$ protein is modified so that it binds to and inhibits adenylate cyclase. The $G_{\beta,\gamma}$ subunits are the same in both the stimulatory and inhibitory G proteins; the G_α subunits and the receptors differ. *After A. G. Gilman, 1984, Cell **36:**577.*

and inhibit it. In addition, the $G_{\beta,\gamma}$ subunit, formed by the dissociation of G_i, binds the stimulatory α subunit $G_{s\alpha} \cdot$ GTP, forming the inactive complex $G_{s\alpha} \cdot$ GTP $\cdot G_{\beta,\gamma}$, and preventing $G_{s\alpha} \cdot$ GTP from activating adenylate cyclase.

Whatever the detailed mechanism, such a system allows a cell to finely tune the activation of adenylate cyclase and the level of cAMP by integrating the responses of a number of hormones, each of which binds to its own surface receptor. For instance, the simultaneous addition of epinephrine and PGE_1 to an adipocyte results in an activation of adenylate cyclase, but to a lower level than that stimulated by epinephrine alone.

Several Bacterial Toxins Irreversibly Modify G Proteins

Confirmation of the GTP cycle came from a seemingly unlikely source, a study of certain bacterial toxins. The function of *cholera toxin*, a peptide produced by the bac-

teria *Vibrio cholerae*, was elucidated first. The classic symptom of cholera is massive diarrhea, caused by water flow from the blood through the epithelial cells into the small intestine; death is often due to dehydration. The study showed that cholera toxin irreversibly activates adenylate cyclase in the intestinal epithelial cells, causing a high level of cAMP. Later studies showed that the toxin irreversibly activates adenylate cyclase in a large number of cell types. Like diphtheria toxin, cholera toxin consists of two types of peptide chains.

One chain is an enzyme that penetrates the cell-surface membrane and enters the cytosol, where it catalyzes the covalent addition of an ADP-ribosyl group from intracellular NAD^+ to the α subunit of the G_s protein (Figure 19-17). This irreversibly modified G_s subunit can activate adenylate cyclase normally but cannot hydrolyze bound GTP to GDP (see Figure 19-15). Thus GTP remains bound to $G_{s\alpha}$, and G_α is always in the activation mode: adenylate cyclase is continuously turned on. As a result,

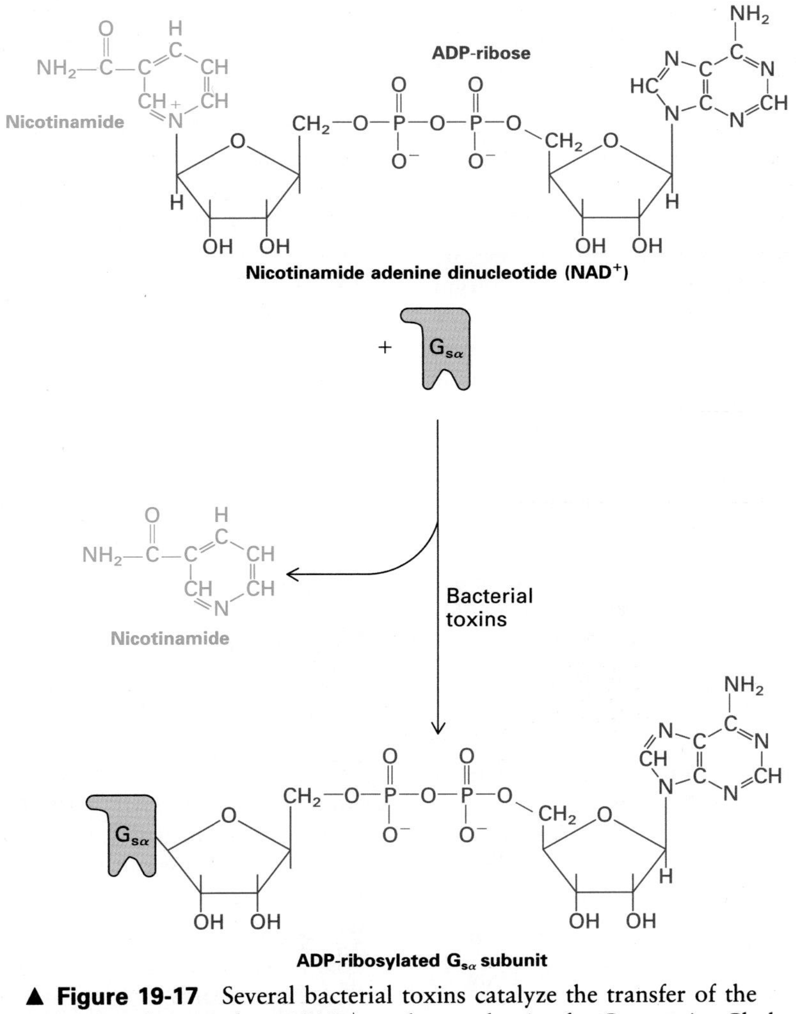

▲ **Figure 19-17** Several bacterial toxins catalyze the transfer of the ADP-ribosyl group from NAD^+ to the α subunit of a G_s protein. Cholera toxin links ADP-ribose to $G_{s\alpha}$; pertussis toxin adds ADP-ribose to the α subunits of G_i, transducin, and other G proteins (see Table 19-6).

the level of cAMP in the cytosol rises 100-fold or more. In the intestinal epithelial cells, this rise apparently causes certain membrane proteins to permit a massive flow of H_2O from the blood into the intestinal lumen.

Other bacterial toxins link ADP-ribose to other G proteins and have proved invaluable in unraveling the functions of these transducing molecules. For example, the *pertussis toxin*, secreted by the "whooping cough" bacterium, adds ADP-ribose to the α subunit of G_i. In this case, $G_{i\alpha}$ linked to ADP-ribose cannot inactivate adenylate cyclase. Pertussis toxin also adds ADP-ribose to and inactivates the α subunits of several other G proteins (Table 19-6)—molecules we will encounter later.

All Receptors That Interact with G Proteins Share Common Structural Features

All receptors that interact with a G protein, including the α_2- and β_2-adrenergic receptors, share common structural features (Figure 19-18): seven sequences of ~22–24 hydrophobic amino acids thought to form seven transmembrane α helixes; a large loop, composed mainly of hydrophilic amino acids, between α helixes 5 and 6; and a hydrophilic segment at the C-terminus. Both the loop and the segment face the cytosol. These hydrophilic regions are thought to interact with a G protein.

Despite a common overall structure, the amino acid sequences of these receptors are usually quite dissimilar. The sequences of the closely related β_1- and β_2-adrenergic receptors are only 50 percent identical, and the sequences of the α and β receptors vary greatly. The differences in amino acid sequences determine what specific types of

ligands bind to various receptors and how they interact with different types of G proteins. By synthesizing recombinant chimeric proteins, which contain part of an α receptor and part of a β receptor, certain functional segments can be localized to parts of the receptor sequence (see Figure 19-18). More detailed studies of site-directed mutants of the β-adrenergic receptor have shown that each of the seven α helixes participate in ligand binding but the structure of the ligand-binding site itself is not known. The hydrophilic loop between α helixes 5 and 6 is believed to be similar in length and three-dimensional structure for all receptors interacting with the same type of G protein ($G_{s\alpha}$ for instance), which facilitates binding to $G_{s\alpha}$.

Degradation of cAMP Is Also Regulated

The level of cAMP is usually controlled by the activation of adenylate cyclase. Another point of regulation is the degradation of cAMP by cAMP phosphodiesterase, which terminates the effect of hormone stimulation (see Figure 19-13):

$$3',5'\text{-cAMP} + H_2O \longrightarrow 5'\text{-AMP}$$

Many cAMP phosphodiesterases are activated by increases in cytosolic Ca^{2+} (another intracellular second messenger), which are often induced by neuron or hormone stimulation. Ca^{2+} ions bind to *calmodulin*, a ubiquitous Ca^{2+} binding protein. When a Ca^{2+}-calmodulin complex binds to cAMP phosphodiesterase, the enzyme is activated and increases its rate of cAMP hydrolysis. Some cells also modulate the level of cAMP by secreting it into the extracellular medium.

The synthesis and degradation of cAMP are both subject to complex regulation by multiple hormones, which allows the cell to integrate responses to many types of changes in its internal and external environments.

cAMP and Regulation of Cellular Metabolism

cAMP Activates a Protein Kinase

The next step in tracing the effects of hormones that elevate the level of cAMP is to see how cAMP affects enzymatic activity. Recall that cAMP is the second messenger for many hormones and that the effects of elevated cAMP differ markedly in various types of cells (see Table 19-4). Yet all the effects of cAMP are believed to be mediated in a similar manner in all cells: cAMP modifies the activities of a specific group of enzymes through the action of cAMP-dependent enzymes called *protein kinases*.

Protein kinases transfer the terminal ATP phosphate group to the serine, threonine, or tyrosine residues of sub-

Table 19-6 GTP-binding proteins inactivated by bacterial toxins

Guanine nucleotide binding protein	Toxin catalyzing ADP-ribosylation and inactivation
Ribosome elongation factor EF-1	Diphtheria toxin
G_s (associated with activation of adenylate cyclase)	Cholera toxin
G_i (associated with inhibition of adenylate cyclase)	Pertussis toxin
G_p (associated with activation of phospholipase C)	Pertussis toxin in some cells
G_t (transducin, associated with activation of cGMP phosphodiesterase)	Cholera and pertussis toxin
G (possibly G_o associated with receptor-mediated regulation of ion channels)	Pertussis toxin

SOURCE: A. C. Dolphin, 1987, *Trends Neurosci.* 10:53–57.

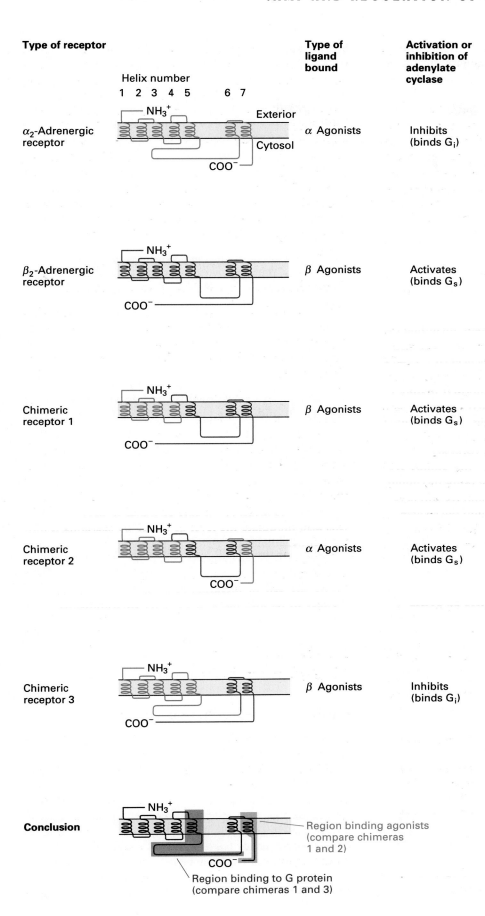

Type of receptor

Helix number
1 2 3 4 5   6 7

α_2-Adrenergic receptor

β_2-Adrenergic receptor

Chimeric receptor 1

Chimeric receptor 2

Chimeric receptor 3

Conclusion

Region binding agonists (compare chimeras 1 and 2)

Region binding to G protein (compare chimeras 1 and 3)

Type of receptor	Type of ligand bound	Activation or inhibition of adenylate cyclase
α_2-Adrenergic receptor	α Agonists	Inhibits (binds G_i)
β_2-Adrenergic receptor	β Agonists	Activates (binds G_s)
Chimeric receptor 1	β Agonists	Activates (binds G_s)
Chimeric receptor 2	α Agonists	Activates (binds G_s)
Chimeric receptor 3	β Agonists	Inhibits (binds G_i)

◄ **Figure 19-18** Chimeras of the α_2- and β_2-adrenergic receptors define domains involved in ligand binding and coupling to G proteins. Each of these receptors has seven transmembrane α helixes; the sequence and size of the cytosol-facing hydrophilic loop between helixes 5 and 6 vary, as does the C-terminal segment facing the cytosol. Either wild-type receptor proteins or α-β chimeras encoded by mRNAs were microinjected into *Xenopus* oocytes, and the receptor polypeptides were expressed on the cell surface. The specificity of agonist binding for either the α or β type of receptor was determined. The effects of the agonists on activation or inhibition of oocyte adenylate cyclase were taken as a measure of whether the polypeptide bound to the stimulatory (G_s) or inhibitory (G_i) type of oocyte G protein. A comparison of chimeras 1 (interacts with G_s) and 3 (interacts with G_i) shows that interaction with G_s or G_i is determined primarily by the origin of the cytosol-facing loop between α helixes 5 and 6. A comparison of chimeras 1 and 2 shows that the origin of α helix 7 determines whether a chimera binds α- or β-specific agonists. *After B. Kobilka, T. S. Kobilka, K. Daniel, J. W. Regan, M. G. Caron, and R. J. Lefkowitz, 1988,* Science **240**:1310–1316; *W. A. Catterall, 1989.* Science **243**:236–237.

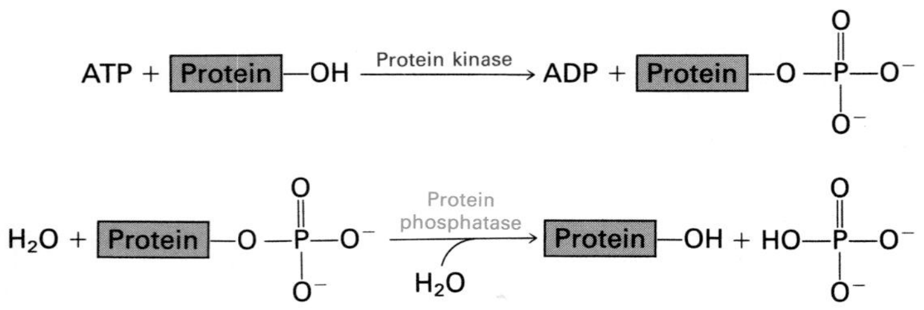

$$ATP + \boxed{Protein}\!-\!OH \xrightarrow{\text{Protein kinase}} ADP + \boxed{Protein}\!-\!O\!-\!\overset{\displaystyle O}{\underset{\displaystyle O^-}{\overset{\|}{\underset{|}{P}}}}\!-\!O^-$$

$$H_2O + \boxed{Protein}\!-\!O\!-\!\overset{\displaystyle O}{\underset{\displaystyle O^-}{\overset{\|}{\underset{|}{P}}}}\!-\!O^- \xrightarrow[\text{H}_2\text{O}]{\text{Protein phosphatase}} \boxed{Protein}\!-\!OH + HO\!-\!\overset{\displaystyle O}{\underset{\displaystyle O^-}{\overset{\|}{\underset{|}{P}}}}\!-\!O^-$$

▲ **Figure 19-19** Reactions catalyzed by protein kinase and protein phosphatase. The phosphorylated and dephosphorylated forms of the protein often differ markedly in enzymatic reactivity.

strate proteins (Figure 19-19; see Figure 19-3). The phosphorylated forms of many enzymes are much more active than the unphosphorylated forms; the phosphorylated forms of other enzymes are less active. The inactive *cAMP-dependent protein kinase* increases its ability to phosphorylate specific acceptor proteins by binding to cAMP. This protein kinase contains four subunits: two are regulatory (R); two, catalytic (C). When cAMP binds to the R subunits, they dissociate from the C subunits and their kinase activity increases (Figure 19-20).

The same C subunits of the cAMP-dependent kinase enzymes are present in most tissues, but different R subunits are found in different cell types. The substrate proteins for the cAMP-dependent kinases vary widely among cell types. Depending on the tissue, cAMP may activate or inhibit different enzyme systems.

The cAMP-dependent protein kinase induces many effects; the first discovered—the release of glucose from glycogen—has received the most detailed study. This reaction occurs in muscle and liver cells treated with epinephrine or with agonists of β-adrenergic receptors. We focus on this system in some detail, as it illuminates several key mechanisms of cAMP action.

Glycogen Synthesis and Degradation Is Controlled by cAMP

Glycogen, a polymer of glucose (see Figure 2-54) found principally in liver and muscle cells, is the major storage form of glucose in the body. Glucose generated from liver glycogen is secreted into the blood to serve as an energy source for many body tissues that do not store energy reserves, such as the brain. Glucose produced from glycogen in muscle cells is metabolized quickly to lactate or to CO_2, yielding adenosine triphosphate (ATP) as a source of energy for contraction.

Like most polymers, glycogen is synthesized by one set of enzymes and degraded by another (Figure 19-21a–c). The intermediate in glycogen synthesis is *uridine diphosphoglucose* (UDP-glucose), which is synthesized from glucose by three enzymatically catalyzed reactions. In the next step, glycogen synthase transfers the glucose residue from UDP-glucose to a free 4-OH of a glucose residue at the end of a glycogen chain.

Degradation of glycogen involves the stepwise removal of glucose residues from the same end. This reaction, which is catalyzed by glycogen phosphorylase, involves *phosphorolysis* (breakdown by adding phosphate rather than H_2O) and forms glucose 1-phosphate rather than glucose.

cAMP-dependent Protein Kinases Regulate the Enzymes of Glycogen Metabolism

In liver and muscle cells, epinephrine not only triggers degradation of glycogen but also inhibits its synthesis from UDP-glucose; cAMP mediates both of these effects. Elevation of the cAMP level enhances the conversion of glycogen to glucose 1-phosphate by inhibiting glycogen synthesis and increasing glycogen breakdown. The active form of the cAMP-dependent protein kinase—the C subunit free of the R subunit—phosphorylates *glycogen synthase* and converts it to a much less active

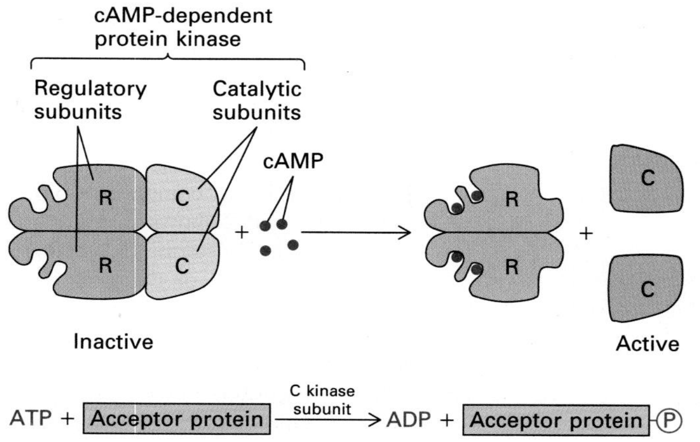

$$ATP + \boxed{Acceptor\ protein} \xrightarrow{\text{C kinase subunit}} ADP + \boxed{Acceptor\ protein}\!-\!(P)$$

◀ **Figure 19-20** Activation of the cAMP-dependent protein kinase by cAMP. Binding cAMP to the regulatory (R) subunits causes the catalytic (C) subunits to dissociate. A C subunit is an enzymatically active protein kinase only when it is dissociated from the R subunits. Each R subunit binds two cAMP molecules. Binding the first cAMP molecule lowers the K_D value for binding the second cAMP molecule. Thus small increases (or decreases) in the level of cytosolic cAMP can cause proportionately large increases (or decreases) in the amount of dissociated C subunits and thus in the activity of the protein kinase.

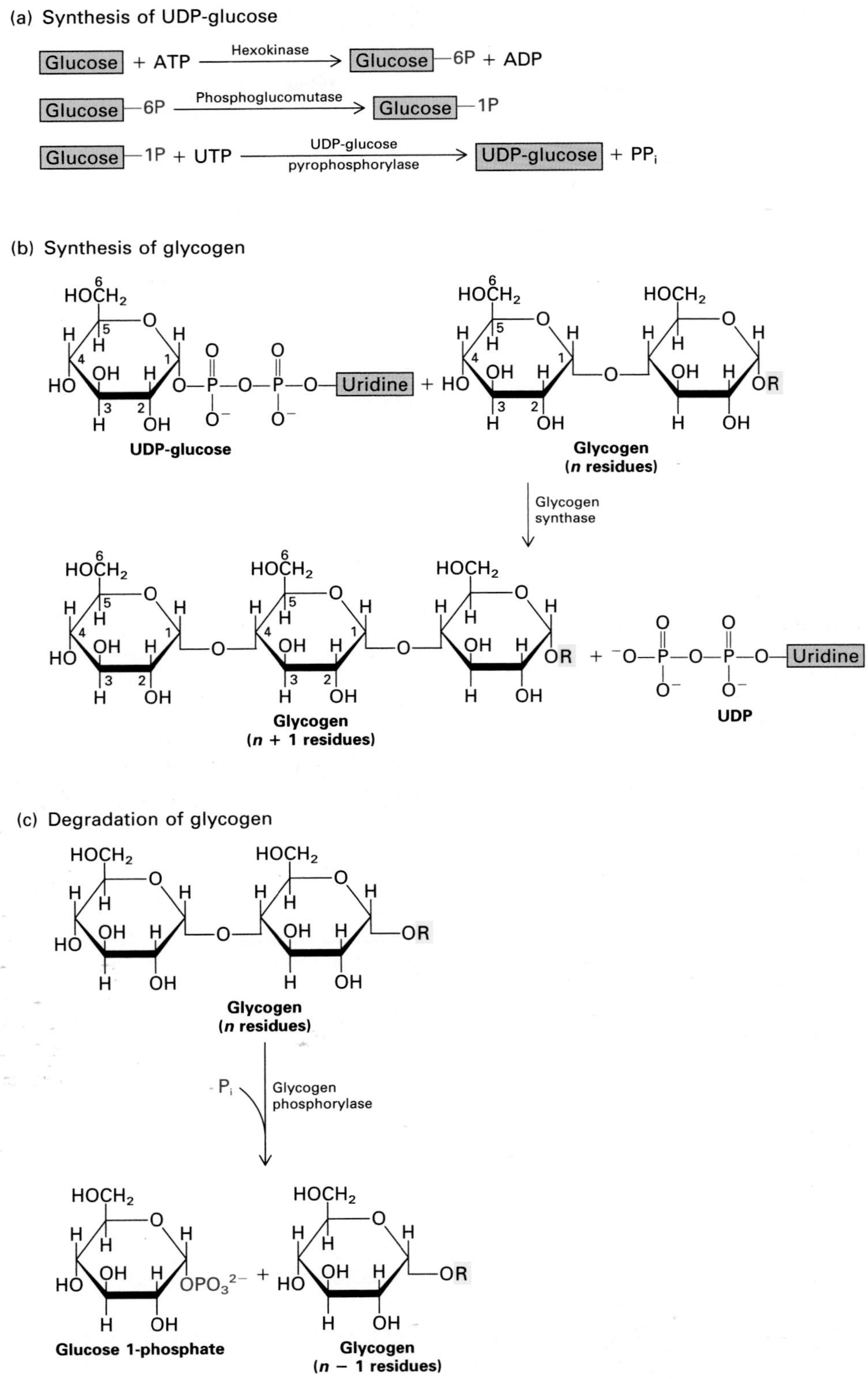

(a) Synthesis of UDP-glucose

$$\boxed{\text{Glucose}} + \text{ATP} \xrightarrow{\text{Hexokinase}} \boxed{\text{Glucose}}-6\text{P} + \text{ADP}$$

$$\boxed{\text{Glucose}}-6\text{P} \xrightarrow{\text{Phosphoglucomutase}} \boxed{\text{Glucose}}-1\text{P}$$

$$\boxed{\text{Glucose}}-1\text{P} + \text{UTP} \xrightarrow[\text{pyrophosphorylase}]{\text{UDP-glucose}} \boxed{\text{UDP-glucose}} + \text{PP}_i$$

(b) Synthesis of glycogen

(c) Degradation of glycogen

▲ **Figure 19-21** Synthesis and degradation of glycogen occur by different pathways: (a) synthesis of uridine diphosphoglucose (UDP-glucose); (b) incorporation of glucose from UDP-glucose into glycogen, catalyzed by glycogen synthase; (c) degradation of glycogen by glycogen phosphorylase. In (b) and (c), R stands for the remainder of the glycogen molecule.

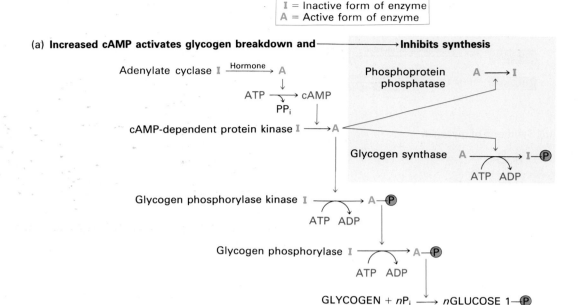

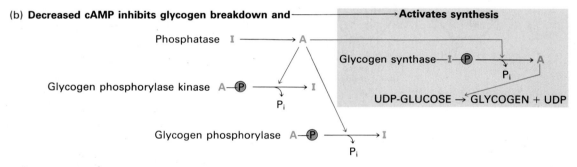

▲ **Figure 19-22** The control of glycogen breakdown and its synthesis in liver and muscle cells by cAMP. (a) An increase in cAMP level increases the glucose level by activating glycogen breakdown and inhibiting glycogen synthesis. Active cAMP-dependent protein kinase phosphorylates glycogen synthase, reducing its activity. The kinase also phosphorylates glycogen phosphorylase kinase, activating its ability to phosphorylate and activate glycogen phosphorylase—the enzyme that degrades glycogen to glucose 1-phosphate. The cAMP-dependent protein kinase also phosphorylates an inhibitor of the phosphoprotein phosphatase, inactivating it. Thus the phosphate groups added to the other enzymes are not removed. (b) A decrease in cAMP level decreases the level of glucose 1-phosphate by inhibiting glycogen breakdown and activating glycogen synthesis. This is accomplished by activating phosphoprotein phosphatase; the phosphate is removed from the inhibitor, thereby inactivating it. The active phosphatase then removes the phosphate residues from glycogen phosphorylase kinase and glycogen phosphorylase, inhibiting glycogen degradation. Removal of the phosphate from glycogen synthase, by contrast, activates the enzyme, causing glycogen synthesis.

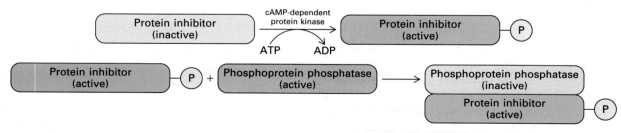

▲ **Figure 19-23** Inhibition of phosphoprotein phosphatase by cAMP. Phosphoprotein phosphatase is enzymatically active, except when an inhibitor protein is bound to it. The inhibitor must be phosphorylated by the cAMP-dependent protein kinase to bind to and inhibit the phosphoprotein phosphatase. Thus the phosphatase is inactive in the presence of a high level of cAMP and active only when the level of cAMP is low. [See P. Cohen, 1982, *Nature* **296**:613.]

molecule, directly inhibiting glycogen synthesis. The cAMP-dependent protein kinase also activates glycogen phosphorylase, but by an indirect route involving a multi-enzyme cascade (Figure 19-22a). The active C enzyme phosphorylates and activates another protein kinase, *glycogen phosphorylase kinase*, which, in turn, phosphorylates and activates a third enzyme, *glycogen phosphorylase*. Active glycogen phosphorylase degrades glycogen to glucose 1-phosphate.

The entire process is reversed when epinephrine is removed and the level of cAMP drops. This reversal is mediated by a set of *protein phosphatases* (see Figure 19-22b), which remove the phosphate residues from glycogen synthase (activating it) and from glycogen phosphorylase kinase and glycogen phosphorylase (inactivating them). Actually, a single protein phosphatase enzyme catalyzes the removal of phosphate from all three enzymes.

The activity of this protein phosphatase is regulated by cAMP; an inhibitor of the phosphatase, not the phosphatase itself, is a target of the cAMP-dependent protein kinase. Phosphorylation of the inhibitor allows it to bind to the phosphatase and inhibit its activity (Figure 19-23). Thus the protein phosphatase is activated when the level of cAMP drops and the inhibitor is unphosphorylated. As a result, the synthesis of glycogen by glycogen synthase is enhanced and the phosphorolysis of glycogen by glycogen phosphorylase is inhibited.

We have seen that the degradation of glycogen forms glucose 1-phosphate, but its fate differs in liver and muscle cells. The glucose 1-phosphate generated from glycogen in muscle cells is used immediately to generate ATP. *Phosphoglucomutase* converts the compound to glucose 6-phosphate, an intermediate in the Embden-Meyerhoff glycolytic pathway (see Figure 15-2). The liver stores and releases glucose primarily for use by other tissues—the muscles and brain in particular. Glucose is not a major source of ATP in the liver. Unlike muscle cells, liver cells contain a glucose 6-phosphatase that hydrolyzes glucose 6-phosphate to glucose, which is immediately released into the blood.

One Function of the Kinase Cascade Is Amplification

The set of protein phosphorylations and dephosphorylations we have described is a *cascade*—a series of steps in which the protein catalyzing each step is activated (or inhibited) by the product of the preceding step. The cascade may seem overcomplicated, but it has several rationales. First, it allows an entire group of enzymatically catalyzed reactions to be regulated by the level of a single species of molecule, cAMP; it also permits the effects to be reversed rapidly when the cAMP level drops. Second, the cascade of events provides a huge amplification of an initially tiny signal (Figure 19-24). For example, the con-

centration of epinephrine needed in the blood to stimulate glycogenolysis and release glucose from the liver and muscles can be as low as 10^{-10} M, a stimulus that generates a concentration of more than 10^{-6} M cAMP in the cell. Because three more catalytic steps precede the release of glucose, another 10^4 amplification can occur, so that blood glucose levels ultimately increase by as much as 50 percent. In striated muscle, the concentrations of the three successive enzymes in the glycogenolytic cascade (cAMP-dependent protein kinase, glycogen phosphorylase kinase, and glycogen phosphorylase) are in a 1:10:240 ratio, which dramatically illustrates the amplification of the effects of epinephrine and cAMP.

cAMP Operates in All Eukaryotic Cells

The effects of cAMP on the synthesis and degradation of glycogen are confined mainly to liver and muscle cells, which store glycogen. However, cAMP also mediates the intracellular responses of many other cells to a wide variety of hormones (see Table 19-4). In virtually all eukaryotic cells studied, the action of cAMP appears to be mediated through one, two, or more types of cAMP-dependent protein kinases. The response of different cells to an elevation of cAMP varies according to which enzymes are activated or inhibited by the cAMP-dependent kinase.

In adipocytes, the fat-storage cells, an epinephrine-induced cascade regulates the synthesis and degradation of triacylglycerols, the storage form of fatty acids (see Figures 2-44 and 2-45). Activation of the β-adrenergic receptor on adipose cells triggers an increase in cytosolic cAMP and activation of the cAMP-dependent protein kinase. The lipase that hydrolyzes triacylglycerols to fatty acids is activated by phosphorylation by this kinase. The released fatty acids are transferred to the blood, where

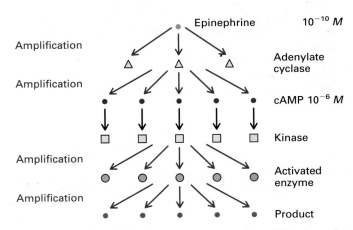

▲ **Figure 19-24** Cellular transduction and amplification of an extracellular signal. In this example, binding a single epinephrine molecule results in the synthesis of a large number of cAMP molecules, which, in turn, activate multiple enzyme molecules.

they bind to *albumin,* a major serum protein. In this form, fatty acids are transported to other tissues—particularly to the heart, muscles, and kidneys, where they are used as a source of ATP.

Still other cells respond to an increased level of cAMP in a different way. The cells of the adrenal cortex bind adrenocorticotropic hormone (ACTH), activating adenylate cyclase, which elevates the level of cAMP and activates the cAMP-dependent protein kinase. The kinase activates enzymes that synthesize several steroid hormones, such as cortisone, corticosterone, and aldosterone (see Table 19-1). Similarly, ovarian cells respond to follicle-stimulating hormone (FSH) by enhancing the synthesis of estradiol and progesterone, two steroids crucial to the development of female sex characteristics (see Figure 19-5); again, a cAMP-cascade is involved.

Ca^{2+} Ions, Inositol Phosphates, and 1,2-Diacylglycerol as Second Messengers

The use of cAMP as the second messenger in signaling systems is widespread in animals, but so is the use of Ca^{2+} ions. The level of Ca^{2+} ion free in the cytosol is usually kept below 0.2 μM. Ca^{2+} ATPases pump Ca^{2+} ions across the plasma membrane to the cell exterior or into the lumens of the endoplasmic reticulum (ER) or other intracellular vesicles that store Ca^{2+}. A rise as small as 1 μM in the level of cytosolic Ca^{2+} triggers many cellular responses (Table 19-7). In secretory cells, such as the insulin-synthesizing β cells in the pancreatic islets, a rise in Ca^{2+} triggers the exocytosis of secretory vesicles and the release of insulin. In smooth or striated muscle cells, a rise in Ca^{2+} triggers contraction; in both liver and muscle cells, an increase in Ca^{2+} activates the degradation of glycogen to glucose 1-phosphate.

We now discuss how Ca^{2+} ions induce these varied metabolic responses and then consider how an array of hormones causes this rise in Ca^{2+}. We also examine how another second messenger, inositol 1,4,5-trisphosphate (see Figure 19-4), which often mediates this rise in Ca^{2+}, is generated, functions, and is degraded. Still another second messenger, 1,2-diacylglycerol, which is formed from the same precursor as inositol 1,4,5-trisphosphate, is used to regulate other cellular functions. All these second messengers interact in complex circuits to regulate crucial aspects of the growth and metabolism of cells.

Calmodulin Mediates Many Cellular Effects of Ca^{2+} Ions

The small protein calmodulin, consisting of 148 amino acids, is ubiquitous in eukaryotic cells and mediates many

Table 19-7 Some cellular responses to a rise in inositol 1,4,5-trisphosphate and a subsequent rise in cytosolic Ca^{2+}

Tissue	Hormone inducing a rise in inositol 1,4,5-trisphosphate	Cellular response
Pancreas (acinar cells)	Acetylcholine	Secretion of digestive enzymes, such as amylase and trypsinogen
Parotid (salivary gland)	Acetylcholine	Secretion of amylase
Pancreas (β cells of islets)	Acetylcholine	Secretion of insulin
Vascular or stomach smooth muscle	Acetylcholine	Contraction
Liver	Vasopressin	Conversion of glycogen to glucose
Blood platelets	Thrombin	Aggregation, shape change, secretion of hormones
Mast cells	Antigen	Histamine secretion
Fibroblasts	Peptide growth factors, such as bombesin and PDGF	DNA synthesis, cell division
Sea urchin eggs	Spermatozoa	Rise of fertilization membrane

SOURCE: M. J. Berridge, 1987, *Ann. Rev. Biochem.* 56:159–193; M. J. Berridge and R. F. Irvine, 1984, *Nature* 312:315–321.

cellular effects of Ca^{2+} ions (Figure 19-25). Ca^{2+} forms a complex with calmodulin that binds to and activates many enzymes. Because each calmodulin binds four Ca^{2+} ions and binding each Ca^{2+} ion facilitates binding additional Ca^{2+} a small change in the level of cytosolic Ca^{2+} causes a large change in the level of active calmodulin.

One well-studied enzyme affected by the Ca^{2+}-calmodulin complex is *cAMP-phosphodiesterase,* which degrades cAMP and terminates its effects. Binding the Ca^{2+}-calmodulin complex to the enzyme activates it; this is one of many examples in which Ca^{2+} and cAMP interact to finely tune aspects of cell regulation. The Ca^{2+}-calmodulin complex also activates several protein kinases that, in turn, phosphorylate target proteins and alter their activity levels.

Ca^{2+} Ions Control Hydrolysis of Muscle Glycogen

In muscle cells, stimulation by nerve impulses causes the release of Ca^{2+} ions from the sarcoplasmic reticulum and an increase in Ca^{2+} concentration in the cytosol. This rise in Ca^{2+} triggers contraction and also degradation of glycogen to glucose 1-phosphate, which fuels prolonged

(a)

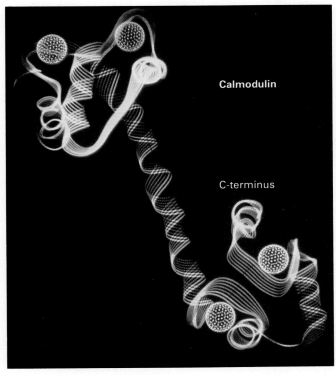

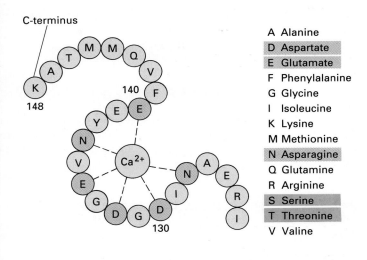

A	Alanine
D	Aspartate
E	Glutamate
F	Phenylalanine
G	Glycine
I	Isoleucine
K	Lysine
M	Methionine
N	Asparagine
Q	Glutamine
R	Arginine
S	Serine
T	Threonine
V	Valine

(b)

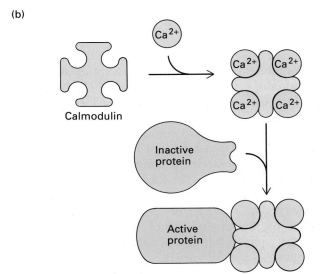

▲ **Figure 19-25** Calmodulin is a cytoplasmic protein of 148 amino acids that modulates most of the regulatory functions of Ca²⁺ ions. (a) *(left)* The backbone of calmodulin protein, deduced from the three-dimensional structure. The four bound Ca²⁺ ions are represented by blue spheres. Each of the four Ca²⁺ binding sites, one of which is depicted in detail *(right)*, is a loop containing aspartate, glutamate, and asparagine side chains that form ionic bonds with the Ca²⁺. The O atoms on the side chains of threonine and serine residues also participate in Ca²⁺ binding. (b) On binding Ca²⁺, calmodulin undergoes a major conformational change that allows it to bind to other proteins, modifying their enzymatic activities. *Part (a) photograph courtesy of Y. S. Babu and W. J. Cook; diagram after C. Y. Cheung, 1982,* Scientific American *246(6):62, copyright © 1982 by Scientific American, Inc.*

muscle contraction. How does Ca²⁺ interact with the enzyme systems that catalyze glycogen degradation and synthesis and are also modulated by the cAMP cascade? The activity of the key regulatory enzyme, glycogen phosphorylase kinase, depends not only on phosphorylation but also on the level of cytosolic Ca²⁺ (Figure 19-26). The enzyme has the subunit structure $(\alpha\beta\gamma\delta)_4$, in which the γ subunit is the catalytic protein, the similar regulatory α and β subunits are phosphorylated by the cAMP-dependent protein kinase, and the δ subunit is calmodulin (see

Figure 19-25). Binding Ca²⁺ ions to the calmodulin subunit stimulates this kinase, which is maximally active if at least the α subunit is phosphorylated (by the cAMP-dependent protein kinase) and Ca²⁺ is bound; in fact, binding Ca²⁺ ions to calmodulin may be essential to enzymatic activity. Phosphorylation of the α and β subunits increases the affinity of the calmodulin subunit for Ca²⁺, enabling Ca²⁺ ions to bind to the enzyme at the submicromolar Ca²⁺ concentrations found in cells not stimulated by nerves. Thus increases in the cytosolic concentra-

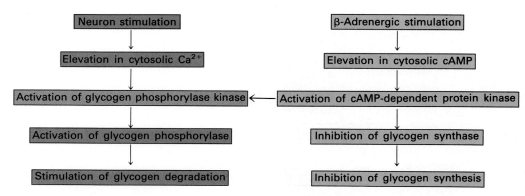

▲ **Figure 19-26** Activation of glycogen phosphorylase kinase in striated muscle. Neuron stimulation and epinephrine use different second messengers, Ca^{2+} and cAMP, but both can activate glycogen phosphorylase kinase and thus accelerate the breakdown of glycogen.

tion of Ca^{2+} or cAMP, or both, induce incremental increases in the activity of glycogen phosphorylase kinase. As a result of the elevated level of cytosolic Ca^{2+} after neuron stimulation, this kinase will be active even if it is unphosphorylated.

When nerve stimulation ceases, Ca^{2+} ions are pumped back into the sarcoplasmic reticulum, the level of cytosolic Ca^{2+} declines, the activity of glycogen phosphorylase kinase is reduced, and less glycogen is converted to glucose phosphate. In several other cells (but not in striated muscle cells), a rise in cytosolic Ca^{2+} increases the rate of Ca^{2+} pumping and causes a more rapid drop in cytosolic Ca^{2+} to the resting state, terminating all Ca^{2+}-induced signals unless additional Ca^{2+} is released. This rapid decline is due to Ca^{2+}-calmodulin activation of the plasma-membrane Ca^{2+} ATPase.

Local Concentrations of Ca^{2+} Ions in the Cytosol Can Be Monitored by Fluorescence

Most cellular Ca^{2+} ions are sequestered in the mitochondria and ER or in other cytoplasmic vesicles, but the cytosolic concentration of free Ca^{2+} ions is critical to the function of Ca^{2+} as a second messenger. The fluorescent properties of certain dyes, such as *fura-2* (Figure 19-27a), facilitate measurement of this concentration. Fura-2 contains several carboxylate groups that tightly bind Ca^{2+} ions but no other cellular cations. The fluorescence of fura-2 is enhanced when Ca^{2+} is bound; over a certain range of Ca^{2+} concentration, the fluorescence is proportional to this concentration. By examining cells continuously in the fluorescence microscope, rapid changes in fura-2 fluorescence and thus in the level of cytosolic Ca^{2+} (Figure 19-27b) can be quantified.

In large cells, different Ca^{2+} concentrations can actually be detected in specific regions of the cytosol. In one

example (Figure 19-28), changes in Ca^{2+} ion level are studied after fertilization of a sea urchin egg. Once the sperm penetrates the egg, the level of Ca^{2+} ions rises in the adjacent region of the cytosol but only gradually increases throughout the egg. This spreading increase in cytosolic Ca^{2+} triggers the fusion of small vesicles with the plasma membrane, causing changes in the cell surface that prevent penetration of additional sperm.

Inositol 1,4,5-Trisphosphate Causes the Release of Ca^{2+} Ions from the ER

Many hormones bind to cell-surface receptors and cause an elevation of cytosolic Ca^{2+}, even when Ca^{2+} ions are absent from the surrounding medium (see Table 19-7). In the initial stages of hormone stimulation of liver, fat, and other cells, Ca^{2+} is released into the cytosol from the ER or other intracellular vesicles, not from the extracellular medium. How is the hormone-receptor signal on the cell-surface communicated to the ER?

The answer became clear in the early 1980s, when it was shown that a rise in the level of cytosolic Ca^{2+} is preceded by the hydrolysis of an unusual plasma-membrane phospholipid, *phosphatidylinositol 4,5-bisphosphate*—one of several inositol phospholipids found in the cytoplasmic leaflet of the plasma membrane. Hydrolysis of this phospholipid by the plasma-membrane enzyme *phospholipase C* yields two important products: 1,2-diacylglycerol, which remains in the membrane, and the water-soluble inositol 1,4,5-trisphosphate (Figure 19-29). A G protein couples hormone receptors to phospholipase C; treatment of cells with pertussis toxin, which inactivates the G_α subunit of this protein, abolishes activation of phospholipase C.

Inositol 1,4,5-trisphosphate diffuses to the ER surface, where it binds to a specific receptor. This, in turn, induces the opening of specific Ca^{2+} channel proteins and allows

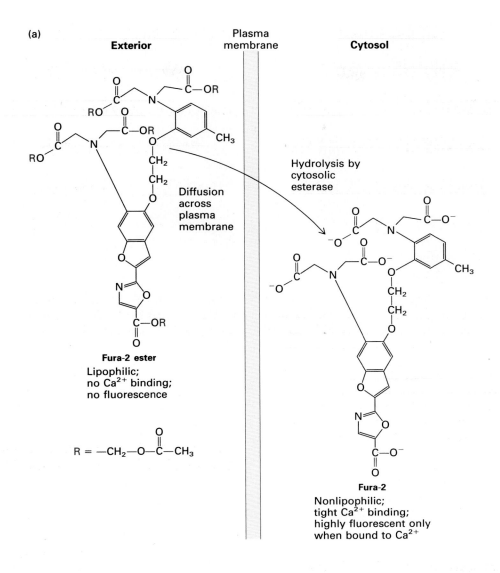

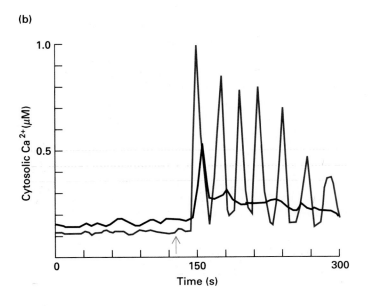

▲ **Figure 19-27** The cytosolic concentration of Ca^{2+} can be monitored continuously by the fluorescence of Ca^{2+} · fura-2 complexes. (a) When added to medium, the lipophilic fura-2 ester *(left)* diffuses across the plasma membrane and is hydrolyzed to fura-2 by cytosolic esterases. Nonlipophilic fura-2 *(right)* cannot cross cellular membranes and remains in the cytosol. In the absence of Ca^{2+}, fura-2 is not fluorescent, and the fluorescence of Ca^{2+} · fura-2 complexes is proportional to the concentration of Ca^{2+} ion in the cytosol. (b) The cytosolic concentration of Ca^{2+} in two adjacent, cultured, smooth muscle cells following the addition *(arrow)* of the contraction-inducing drug phenylephrine; the fluorescence of Ca^{2+} · fura-2 complexes in individual cells has been quantified using a fluorescence microscope. Although Ca^{2+} levels rise in both cells in response to an agonist, the behavior of the two cells is strongly individualistic for unknown reasons. [See S. K. Amble, M. Poenie, R. Y. Tsien, and P. Taylor, 1988, *J. Biol. Chem.* **263**:1952–1957.] *Courtesy of A. Harootunian and R. Y. Tsien.*

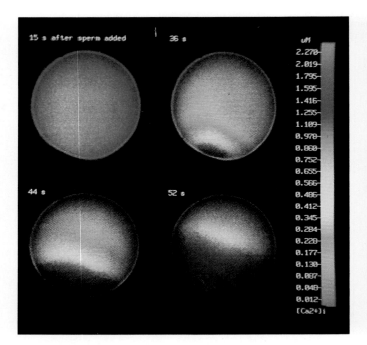

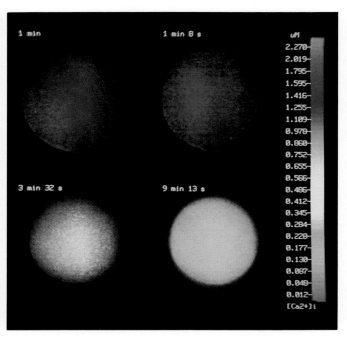

▲ **Figure 19-28** Changes in the local concentration of Ca^{2+} ions in a sea urchin egg following fertilization. The Ca^{2+} throughout the cell was monitored by fura-2 fluorescence using a microscope (see Figure 19-27); for graphic purposes, the Ca^{2+} concentrations are expressed in a calibrated color scale *(right)* in units of micromolar Ca^{2+}. The Ca^{2+} concentration rises initially at the point of sperm entry (the lower left part of the cell) and elevates, spreading like a wave. Eventually, the Ca^{2+} concentration becomes high and uniform throughout the cell and then falls uniformly to the resting state. [See R. Y. Tsien and M. Poenie, 1986, *TIBS* 11:450–455.] *Images courtesy of J. Alderton, M. Poenie, R. A. Steinhardt, and R. Y. Tsien.*

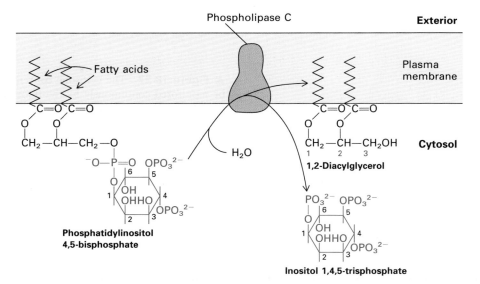

▲ **Figure 19-29** Hydrolysis of plasma membrane phosphatidylinositol 4,5-bisphosphate by phospholipase C generates two second messengers: 1,2-diacylglycerol and inositol 1,4,5-trisphosphate. Inositol is a substituted cyclohexane in which each C atom bears an hydroxyl group.

Ca²⁺ ions to exit from the ER lumen into the cytosol (Figure 19-30). The overall sequence of actions can be summarized as follows: (1) binding the hormone to a cell-surface receptor causes (2) activation of phospholipase C, which (3) generates inositol 1,4,5-trisphosphate, which (4) releases Ca²⁺ ions from the ER into the cytosol. A simple experiment shows that the addition of inositol 1,4,5-trisphosphate to a preparation of ER vesicles causes the release of up to one-half of their stores of Ca²⁺ ions. Although cells contain other phosphorylated inositols, only inositol 1,4,5-trisphosphate binds to the ER receptor

protein, causing Ca²⁺ ions to be released.

Within a second of its formation, most inositol 1,4,5-trisphosphate is hydrolyzed to *inositol 1,4-bisphosphate* (Figure 19-31), a molecule that cannot release Ca²⁺ ions. This terminates the release of Ca²⁺ ions, unless more inositol 1,4,5-trisphosphate is formed by phospholipase C.

During the initial stages of hormone stimulation, Ca²⁺ ions are released from the ER. The store of Ca²⁺ ions in the ER can be depleted within a few minutes. Maintenance of elevated cytosolic Ca²⁺ requires extracellular Ca²⁺ to cross the membrane. Evidence suggests that, at

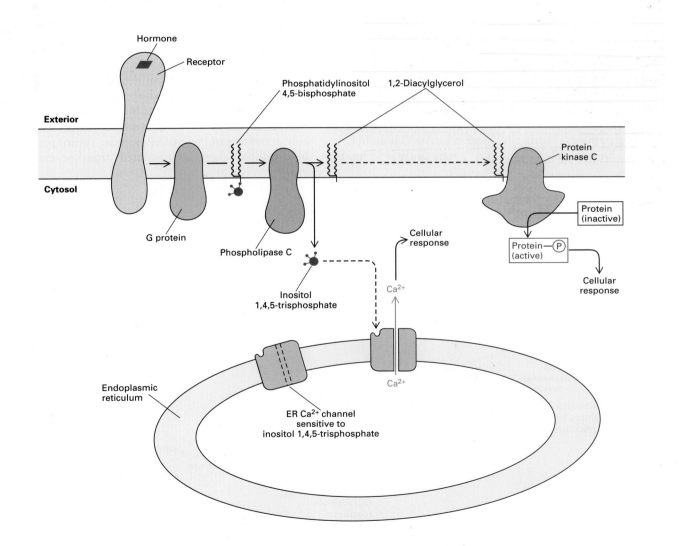

▲ **Figure 19-30** Second messengers in the inositol-lipid signaling pathway. Binding a hormone to its receptor triggers activation of a G protein that, in turn, activates phospholipase C. The enzyme cleaves phosphatidylinositol 4,5-bisphosphate to inositol 1,4,5-trisphosphate and 1,2-diacylglycerol. Inositol 1,4,5-trisphosphate diffuses through the cytosol to the endoplasmic reticulum, where it releases Ca²⁺ from the ER lumen; 1,2-diacylglycerol remains in the membrane, where together with Ca²⁺, it helps to activate protein kinase C. Activated kinase C, in turn, phosphorylates several cellular enzymes and receptors and alters their activities. *After M. J. Berridge, 1985*, Scientific American *253:147.*

least in pancreatic cells, Ca^{2+} entry is mediated by *inositol 1,3,4,5-tetraphosphate*, a substance formed by phosphorylation of inositol 1,4,5-trisphosphate (see Figure 19-31). In time, this inositol phosphate is also reacted on by a phosphatase, which renders it inactive. Evidence in other cells indicates that the emptying of Ca^{2+} stores in the ER can trigger (by some other mechanism that does not involve inositol phosphates) Ca^{2+} entry into the cytosol from the extracellular medium.

1,2-Diacylglycerol Activates Protein Kinase C

Hydrolysis of phosphatidylinositol 4,5-bisphosphate by phospholipase C generates another important second messenger, *1,2-diacylglycerol* (see Figures 19-29 and 19-30). Its principal function is to activate a family of plasma-membrane protein kinases termed *protein kinase C*. In fact, protein kinase C is activated by Ca^{2+} ions and 1,2-diacylglycerol together, suggesting an interaction between the two branches of the phosphatidylinositol-signaling pathway. Normally, protein kinase C is an inactive, soluble cytosolic protein. Ca^{2+} ions cause it to bind to the cytoplasmic leaflet of the plasma membrane, where it can be activated by 1,2-diacylglycerol.

Properties of protein kinase C suggest that it plays a key role in many aspects of growth and metabolism in cells; the activation of protein kinase C in different cells results in a varied array of cellular responses. In the liver, for instance, one substrate of protein kinase C, glycogen synthase (see Figure 19-21), is poorly active in glycogen synthesis when phosphorylated. Activation of phospholipase C and generation of inositol 1,4,5-trisphosphate and 1,2-diacylglycerol results in the breakdown of glycogen to glucose 1-phosphate by the separate action of the two branches of the signaling pathway: 1,2-diacylglycerol, by activating protein kinase C, mediates inhibition of glycogen synthesis; inositol 1,4,5-trisphosphate, by causing cytosolic Ca^{2+} to elevate, enhances activation of glycogen phosphorylase (see Figure 19-26). This is another example of the coordination of multiple second messengers to achieve a unified cellular response.

In investigating the function of protein kinase C, scientists have taken advantage of the finding that some of the substances called *tumor promoters* are potent and specific activators of this protein kinase. Tumor promoters— lipid-soluble chemicals isolated from several sources

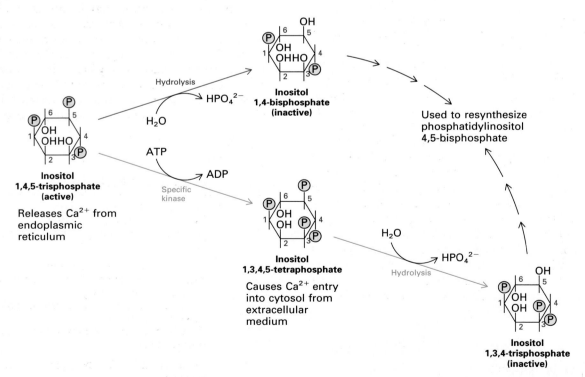

▲ **Figure 19-31** The fates of inositol 1,4,5-trisphosphate: (1) rapid hydrolysis to inactive inositol 1,4-bisphosphate, which is used to resynthesize phosphatidylinositol 4,5-bisphosphate, or (2) phosphorylation by a specific kinase to yield inositol 1,3,4,5-tetraphosphate, which causes Ca^{2+} entry into the cytosol from the extracellular medium. Eventu-

ally, inositol 1,3,4,5-tetraphosphate is hydrolyzed to the inactive inositol 1,3,4-trisphosphate, which is also recycled to synthesize phosphatidylinositol 4,5-bisphosphate. Note that the active inositol 1,4,5-trisphosphate and the inactive inositol 1,3,4-trisphosphate, although similar, are chemically different molecules.

(mainly plants)—play a part in transforming a normal cell into a malignant cell capable of uncontrolled growth. Since protein kinase C is the "receptor" for the class of tumor promoters termed *phorbol esters,* activation of this kinase is thought to be a key event in cell proliferation.

Substrates of protein kinase C include the cell-surface receptors for many growth hormones, such as the *epidermal growth factor* (EGF). Phosphorylation of the EGF receptor by protein kinase C decreases its affinity for EGF, moderating growth-stimulating activity. Overproduction of protein kinase C in "normal" fibroblasts (by expression of the cloned gene) causes cells to grow unattached to an extracellular matrix, as do many tumor cells. (Normal cells grow only when attached to a matrix.) Clearly, protein kinase C is of fundamental importance in controlling cell growth. The eventual understanding of its regulation and substrates should provide important insights into many aspects of carcinogenesis and metabolism.

Insulin and Glucagon: Hormone Regulation of Blood Glucose Levels

Epinephrine induces a wide array of responses to stress or exercise, including an increase in the concentration of glucose in the blood. During normal daily living, two other hormones—insulin and glucagon, produced by the *islets of Langerhans* (clusters of cells scattered throughout the pancreas)—regulate blood glucose levels. Insulin contains two polypeptide chains, A and B, linked by disulfide bonds (see Figure 2-12); synthesized by the β cells in the islets, insulin acts to reduce the level of blood glucose. Glucagon, a single-chain peptide containing 29 amino acids, is produced by the α cells; it creates the opposite effect, causing an increase in blood glucose by stimulating glycogenolysis in the liver. Each islet functions as an integrated unit, delivering the appropriate amounts of both hormones to the blood to meet the metabolic needs of the animal. Hormone secretion is regulated by a combination of neuron and hormone signals. Of the hormones that affect blood glucose, insulin is the most important, and its receptor is among the most extensively studied.

Insulin Controls Cell Growth and Also the Level of Blood Glucose

Insulin acts on many cells in the body to produce both immediate and long-term effects. Its immediate action is to increase the rate of glucose uptake from the blood into muscle cells and adipocytes by increasing the number of glucose transporters in the plasma membrane. The effects of insulin on glucose uptake and catabolism occur in minutes and do not require new protein synthesis. Continued exposure to insulin produces longer-lasting effects, causing increases in the activities of liver enzymes that synthesize glycogen and of enzymes in adipose cells that synthesize triacylglycerols (Table 19-8). In these respects, the actions of insulin and epinephrine are opposite.

Within 15 min of the addition of insulin to adipocytes, the rate of glucose transport increases 10–20 times, mainly due to an increase in the number of transporters in the surface membrane. The membranes of vesicles located just under the plasma membranes of unstimulated cells contain abundant glucose transporters of a type found only in adipocytes and muscle cells. Binding insulin to its receptor induces rapid fusion of these vesicles with the plasma membrane, increasing the number of transporters in the plasma membrane 10 times (Figure 19-32). Exocytosis of vesicles containing transporters and some activation of these transporters account for the large increase in glucose transport.

Insulin is also a growth factor for many cells, such as fibroblasts. Its growth-promoting actions, such as the induction of DNA synthesis, require cells to be continuously exposed to $\sim10^{-8}$ M insulin, whereas much lower concentrations of 10^{-9}–10^{-10} M are sufficient to induce very rapid actions, such as glucose uptake. Many growth-promoting effects of insulin may be mediated by the receptor for *insulin-like growth factor 1* (IGF-1), a hormone similar in structure and sequence to insulin. IGF-1 appears to be a primary regulator of the growth of an organism; it is produced mainly in the liver in response to secretion of growth hormone by the pituitary gland. The ability to stimulate growth (in dwarfs unable to produce sufficient growth hormone, for instance) is largely medi-

Table 19-8 Chronology of insulin action

SECONDS	Binding to the insulin receptor Receptor autophosphorylation Activation of the receptor protein tyrosine kinase
MINUTES	Activation of hexose transport Alterations of intracellular enzymatic activities Changes in gene regulation Insulin-induced receptor internalization and down-regulation Phosphorylation of the insulin receptor by other protein kinases
HOURS	Induction of DNA, RNA, protein, and lipid synthesis Cell growth Maximum down-regulation of the insulin receptor

SOURCE: O. M. Rosen, *The Harvey Lectures (1986–1987),* Vol. 82, pp. 105–122.

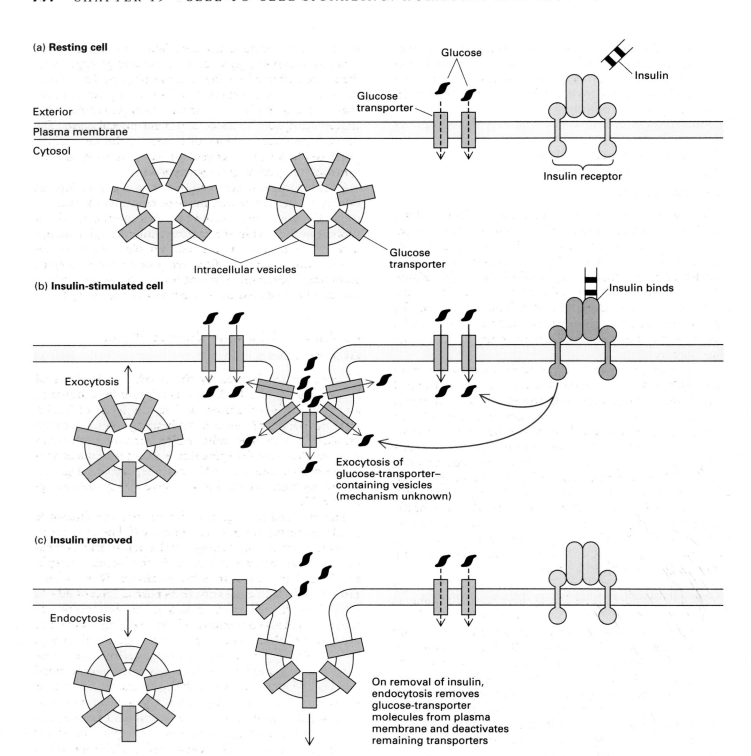

▲ **Figure 19-32** A schematic depiction of the regulation of the glucose transport system in adipocytes. The two types of glucose transporters are identical in 60 percent of their 500 amino acids. (a) In resting cells, one type (green) is on the plasma membrane and the other, more numerous, type (blue), is found in intracellular vesicles. (b) Insulin binding triggers (by an unknown mechanism that may involve a protein kinase) exocytosis of the vesicles that contain abundant glucose transporters, causing a 10-fold increase in the number of plasma-membrane transporters. Simultaneously, all plasma-membrane glucose transporters are activated. The net result is that glucose uptake by the cells increases 10–20 times. (c) Following the loss or removal of insulin, regions of the plasma membrane rich in the more abundant type of glucose transporters are subjected to endocytosis, which depletes the plasma membrane of these transporters. Simultaneously, the remaining transporters are deactivated. [See A. Simpson, and S. W. Cushman, 1986, *Ann. Rev. Biochem.* **55**:1059–1089; D. E. James, M. Strube, and M. Mueckler, 1989, *Nature* **338**:83–87.]

ated by IGF-1. Insulin and IGF-1 receptors share a number of properties, including molecular weight and tyrosine-protein kinase activity; each hormone binds about 100 times less tightly to the other's receptor than to its own receptor.

The Insulin Receptor Is a Ligand-activated Protein Kinase

Although present in only one part per 10,000 of cellular protein, the insulin receptor has been purified, cloned, and sequenced. The receptor contains two copies each of an α chain (719 amino acids) and a β chain (620 amino acids), all linked together by disulfide bonds (see Figure 19-8). The α subunit is outside the exoplasmic face of the membrane and contains the insulin-binding domain. The β subunit contains a single transmembrane α helix and a domain of 391 amino acids on the cytoplasmic face of the membrane.

The insulin receptor itself can be shown to exhibit tyrosine-protein kinase activity. When researchers incubated highly purified receptors with labeled ATP in the presence of insulin, they found that phosphate groups transferred from ATP to four specific tyrosine residues on the cytosolic domain of the β subunit and observed little phosphorylation when insulin was omitted. Subsequent work showed that autophosphorylation of the receptor enhanced its ability to phosphorylate other proteins, even when insulin was subsequently removed. Removal of the phosphate groups from the tyrosine residues, a reaction catalyzed by a tyrosine phosphatase enzyme found in all cells (see Figure 19-3), lowered the protein kinase activity of the receptor to the basal level.

Protein kinase activity and autophosphorylation actually are involved in transducing the action of insulin, as shown by gene transfection experiments. Normal hamster fibroblasts contain about 2000 insulin receptors. By using recombinant DNA vectors, 10,000 human insulin receptors were synthesized in these cells. Sensitivity of the transfected cells to insulin markedly increased; far less insulin was needed to induce all physiological responses due to the greater number of receptors in the cell. In another experiment, two mutant receptors were synthesized: one had a single amino acid change in the ATP-binding region of the protein kinase domain and was totally defective in kinase activity; in the other, two of the tyrosines that are the principal sites of autophosphorylation were changed to phenylalanine. Both mutant receptors bound insulin normally but did not trigger any normal insulin-responsive reactions; this result documents the crucial importance of the protein kinase domain and autophosphorylation.

Presumably, the insulin receptor in the cell phosphorylates several target proteins on tyrosine residues, but these are only now being elucidated. When muscle cells are exposed to insulin, glycogen synthase and glycogen synthesis are activated by the removal of phosphates from three of the ten phosphorylated sites on glycogen synthase. Binding insulin to the insulin receptor is believed to trigger phosphorylation of a protein phosphatase. The active phosphatase would then remove the three phosphate residues from glycogen synthase, activating the enzyme and triggering glycogen synthesis.

Other work, however, suggests that some actions of insulin are not influenced by tyrosine protein kinase activity. Certain monoclonal antibodies specific for the α subunit of the insulin receptor will stimulate glucose uptake in adipocytes but will not stimulate the tyrosine-protein kinase activity of the receptor or cause receptor autophosphorylation. Some actions of the insulin receptor may be mediated by a G protein in the cytosol.

Insulin and Glucagon Balance Blood Glucose Levels

Glucagon is synthesized by the α cells of the islets of Langerhans (Figure 19-33a); it is released in response to low levels of blood glucose and to low levels of insulin from surrounding β (insulin-synthesizing) cells. Glucagon primarily affects liver cells, which contains a glucagon receptor that, like the epinephrine receptor, induces adenylate cyclase and the cAMP cascade, causing degradation of glycogen and inhibition of glycogen synthesis (Figure 19-33b).

The availability of glucose for cellular metabolism is regulated during periods of abundance (following a meal) or scarcity (following fasting) by the adjustment of insulin and glucagon concentrations in the blood. Epinephrine is used only under stressful conditions. When, after a meal, blood glucose rises above its normal level of 80–90 mg/100 ml, the pancreatic β cells of the islets respond to the rise in glucose or amino acids by releasing insulin into the blood so that it is transported throughout the body. By binding to muscle and adipocyte cell-surface receptors, insulin causes glucose to be removed from the blood and stored in muscle cells as glycogen. Insulin also affects hepatocytes, primarily by inhibiting glucose synthesis from smaller molecules, such as lactate and acetate, and by enhancing glycogen synthesis from glucose. If the blood glucose level falls below ~80 mg/100 ml, the α cells of the islets start secreting glucagon. Glucagon binds to glucagon receptors on liver cells, triggering degradation of glycogen and the release of glucose into the blood.

Abnormal Function of Insulin Receptors Is One Cause of Diabetes

Abnormal function or regulation of insulin receptors has been demonstrated in some persons with diabetes, a dis-

(a)

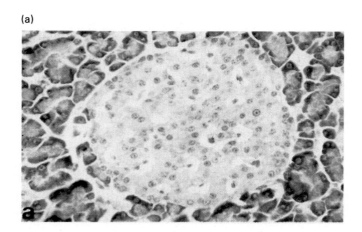

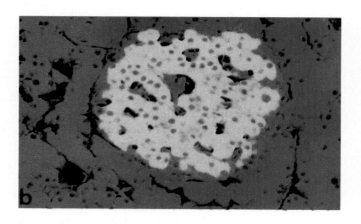

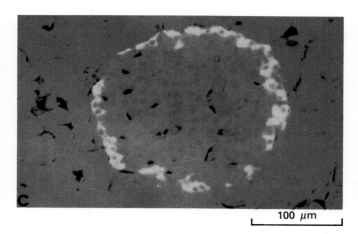

100 μm

◀ **Figure 19-33** (a) Consecutive serial sections of a rat islet of Langerhans stained with hematoxylin-eosin *(top)* and with fluorescent antibodies to insulin *(lower left)* and glucagon *(lower right)*. Insulin-secreting β cells are in the center of the islet, surrounded by glucagon-secreting α cells. (b) Blood glucose level is regulated by the opposing effects of insulin and glucagon. Insulin causes an increase in glucose uptake mainly in muscle and adipocyte cells and, in liver cells, activates glucose storage as glycogen. Glucagon acts mainly on liver cells to cause glycogen degradation. *Part (a) courtesy of L. Orci, from L. Orci, 1982, Diabetes 31:538–565.*

(b)

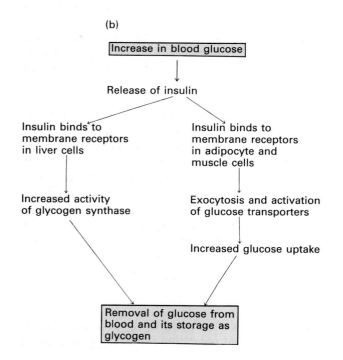

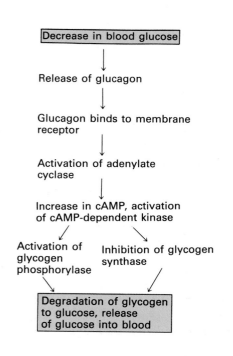

ease caused by insufficient insulin action and a prolonged elevation in the blood glucose level. Most people with childhood or *early-onset diabetes,* which results from deficient insulin synthesis by the β cells of the pancreatic islets, produce antibodies against their own β cells, which are then destroyed by autoimmune reactions. In most cases, insulin injections overcome this problem. Rarely, early-onset diabetes is caused by the synthesis of a structurally abnormal insulin or by a defective conversion of proinsulin to insulin (see Figure 2-13).

In most persons with *adult-onset diabetes,* the adipocytes and muscle cells respond defectively to insulin. Rare individuals have structural defects in their insulin receptors that result in very high insulin levels and a higher than average concentration of glucose in the blood. Other rare diabetics produce autoantibodies to the insulin receptor, causing long-term depletion of functional insulin receptors on adipocytes and muscle cells and a corresponding decrease in the ability of insulin to lower the blood glucose level.

However, the vast majority of people with adult-onset diabetes have two defects: their β islet cells are defective in secreting insulin in response to an elevation in blood glucose, and their target cells do not respond to an increase in insulin by elevating glucose transport. The specific deficiency is unknown. The insulin receptors and at least some parts of the insulin signaling pathways in these individuals appear to be normal, since other cellular responses to insulin, such as increases in amino acid uptake and DNA synthesis, are unaffected. The level of the type of glucose transporters unique to fat and muscle cells may be reduced by diabetes, which would account for the less than normal increase in glucose transport triggered by insulin. Thus defective regulation of the expression of glucose transporters may be one cause of adult-onset diabetes, a prevalent disease in older people.

Receptor Regulation

The amount of functional hormone receptor on a cell surface is not constant. The receptor level is modulated up (*up regulation*) or down (*down regulation*), permitting the cell to respond optimally to small changes in the hormone level. Prolonged exposure of a cell to a hormone usually results in the reduction of functional cell-surface receptor molecules, desensitizing the cell to the high level of hormone present. Up to a point, an increase in hormone concentration still causes typical hormone-induced responses, rather than excessive ones.

The level of cell-surface hormone receptors can be reduced in at least four ways. They can (1) be destroyed after endocytosis, (2) be internalized by endocytosis and remain stored in an intracellular vesicle, (3) remain on the cell surface but change so they cannot bind ligand, or

(4) bind ligand but form a receptor-ligand complex that does not induce the normal hormone response. Receptors for many hormones are regulated in two or more of these ways.

The Receptor Number Is Down-regulated by Endocytosis

Endocytosis is a principal mechanism for down regulating the number of many peptide hormone receptors, such as insulin, glucagon, and EGF. The receptor-hormone complex is brought into the cell by receptor-mediated endocytosis. The internalized hormone is subsequently degraded in lysosomes—a fate similar to that of other endocytosed proteins, such as low-density lipoprotein (LDL) and asialoglycoproteins. Internalization and degradation most likely terminate the hormone signal. Even if the hormone receptors recycle to the cell surface by exocytosis, as they do in many cases, a substantial fraction will be in the internal membrane compartments at any one time. Fewer receptors will be on the cell surface, available to bind extracellular hormone.

Unlike the LDL receptor, internalized receptors for many peptide hormones do not recycle efficiently to the cell surface. In the presence of EGF, the average half-life of an EGF receptor on a fibroblast cell is about 30 min; during its lifetime, each receptor mediates the binding, internalization, and degradation of only two EGF ligands. Each time an EGF receptor is internalized with bound EGF, it has a high probability (about 50 percent) of being degraded in an endosome or lysosome. Exposure of a fibroblast cell to high levels of EGF for 1 h induces several rounds of endocytosis and exocytosis of EGF receptors, resulting in degradation of receptors. If the concentration of extracellular EGF is then reduced, the level of cell-surface EGF receptors is recovered, but only after 12–24 h. Synthesis of new receptors is needed to replace those degraded by endocytosis, a slow process that may take more than a day.

If there are fewer hormone receptors on a cell surface, the hormone concentration necessary to induce the physiological response is greater and the sensitivity of the cell to the hormone is reduced. To understand this important point, we use a simple numerical example. Suppose a cell has 10,000 insulin receptors on its surface, with $K_D = 10^{-8}$ M, but that insulin must bind to only 1000 receptors to induce a physiological response (say, activation of glucose transport). For this calculation, binding equation (19-3) is written in the more useful form

$$[\text{H}] = \frac{K_D}{R_T/[\text{RH}] - 1}$$

where $R_T = 10,000$ (the total number of surface receptors), $K_D = 10^{-8}$ M, and $[\text{RH}] = 1000$ (the number of required hormone-occupied receptors). Thus we can calculate that the hormone concentration necessary to induce this response is $[H] = 1.1 \times 10^{-9}$ *M*. If the number of cell-surface receptors is reduced to 2000, so that R_T = 2000/cell, then $[H]$ must rise to 10^{-8} *M*, a ninefold increase, to generate 1000 occupied receptors ($[RH]$ = 1000). If the receptor number is further reduced to R_T = 1200/cell, then an insulin concentration of $[H] = 5 \times 10^{-8}$ *M* is necessary, a 50-fold increase.

Phosphorylation of Cell-surface Receptors Modulates Their Activity

The ability of a cell-surface receptor to transduce a hormone signal is often modified by phosphorylation. Phosphorylation usually reduces receptor activity, particularly when it is coupled to a G protein, so that a higher hormone concentration is required to generate a physiological response.

In many cells, for instance, the exposure of β-adrenergic receptors to epinephrine for several hours causes nonfunctional hormone-receptor binding. The receptor can bind epinephrine but is unable to activate the G_s protein and thus adenylate cyclase. Receptor inactivation is reversible: after the cells incubate for several more hours in the absence of epinephrine, their cell-surface receptors reactivate. Prolonged exposure to epinephrine causes phosphorylation of several serine and threonine residues of the β-adrenergic receptor.

Some phosphorylation is catalyzed by the cAMP-dependent protein kinase; its activity is enhanced by the high level of cAMP induced by the epinephrine. The phosphorylated receptor binds ligand normally but cannot activate adenylate cyclase. For instance, phosphorylated receptor that is purified and implanted in a cell (see Figure 19-12) is not able to activate G_s or stimulate cAMP synthesis after epinephrine is added. Enzymatic removal of the phosphate groups reactivates the ability of the receptor to stimulate adenylate cyclase. Thus these cells utilize a feedback loop to modulate the activity of the cell-surface β-adrenergic receptor. High intracellular levels of cAMP result in phosphorylation and desensitization of the receptor, so that less cAMP is induced by extracellular epinephrine (Figure 19-34). Active cAMP-dependent protein kinase actually phosphorylates all types of cell-surface receptors that are coupled to G_s and activate adenylate cyclase. Since the activity of all such receptors is reduced by phosphorylation, this process is called *heterologous desensitization*.

A related process, *homologous desensitization*, works only on one type of receptor. For example, the β-adrenergic receptor kinase (BARK) phosphorylates only the β-adrenergic receptor (see Figure 19-34) and only when it is occupied with agonist. Prolonged treatment of cells with epinephrine or other agonists results in BARK-catalyzed phosphorylation of the β-adrenergic receptor and inhibition of its ability to activate G_s and adenylate cyclase.

Phosphates on β-adrenergic and similar receptors are being removed constantly by nonspecific phosphatases. Thus the amount of phosphates per receptor is a measure of how much agonist has been bound in the recent past (1–10 min). The cell adapts to the constant hormone level at which it is being stimulated, ensuring a continued physiological response when the level of hormone in the blood is increased. If the hormone is removed, the receptor is dephosphorylated and "reset," so that it responds to very low levels of hormone. Such receptor modulation occurs in many other systems, such as the visual system, and actually may be the basis of short-term memory.

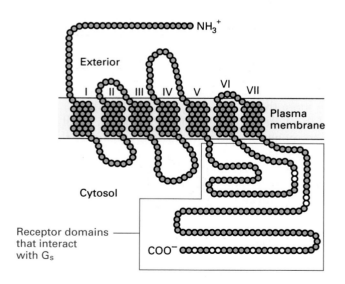

o = Sites phosphorylated by cAMP-dependent protein kinase

o = Sites phosphorylated by β-adrenergic receptor kinase (BARK); phosphorylates only agonist-occupied receptors

◀ **Figure 19-34** Secondary structure of the β-adrenergic receptor, showing sites of its phosphorylation, which inhibits the activation of G_s and thus reduces the activation of the cAMP-dependent protein kinase. The receptor is phosphorylated on certain serine and threonine residues (blue and green) in two protein domains exposed to the cytosol. Phosphorylation at any of these sites reduces the activation of G_s and down-regulates the receptor. The cAMP-dependent kinase phosphorylates four residues in the receptor (blue). A second modifying enzyme, the β-adrenergic receptor kinase (BARK), phosphorylates the receptor at different sites (green) only when it is occupied by an agonist. *After D. R. Sibley, J. L. Benovic, M. G. Caron, and R. J. Lefkowitz, 1987, Cell 48:913–922.*

Hormones and Cell-to-Cell Signaling in Microorganisms

Cell-to-cell communication by hormone signaling is not confined to multicellular animals; it is widely used by eukaryotic microorganisms and higher plants as well. Yeasts, slime molds, and other eukaryotic microorganisms can propagate asexually as solitary cells but must aggregate with other cells of the species to mate sexually and recombine or to initiate the developmental stage of the life cycle to form spores. Indeed, many species of yeast and protozoa have evolved systems in which secreted, diffusible peptides or other molecules serve as intercellular signals. These pheromones often attract one gamete cell to another or induce sexual differentiation. Receptors for many such hormones have been characterized, cloned, sequenced, and subjected to detailed genetic and biochemical analyses. Importantly, many of these receptors are strikingly similar in structure and mechanism of action to those of animal cell-surface receptors, suggesting that certain types of receptors and signal transducers arose early in eukaryotic evolution and evolved to meet the signaling needs of a wide array of cells and organisms.

A Pheromone Attracts Yeast Cells for Mating

Mating between two yeast cells of opposite mating types is controlled by two secreted peptide pheromones, the **a** and α factors. An **a** haploid cell type secretes the **a** oligopeptide mating factor, or pheromone; an α cell type secretes the α factor. These extracellular hormones bind to cell-surface receptors on haploid cells of the opposite mating type: **a** receptors on α cells, and α receptors on **a** cells. Receptor-hormone binding triggers at least three major biochemical events: (1) cell-surface alterations that enhance the ability of cells to bind strongly and selectively to cells of the opposite mating type and may involve the synthesis of new surface glycoproteins; (2) alterations in the cell wall and plasma-membrane macromolecules that

facilitate fusion of the two mating cells and eventual fusion of the two nuclei; and (3) arrest of the growth of the target cells, which specifically blocks initiation of new DNA synthesis and synchronizes the cell cycles of the mating partners in the G1 stage preceding DNA replication.

The structures of the yeast **a** and α receptors are remarkably similar to those of the human β-adrenergic receptor and all other receptors known to couple to G proteins (Figure 19-35; compare with Figures 19-18 and 19-34). There are seven transmembrane α helixes; a long segment at the C-terminus faces the cytosol and is believed to interact with a G protein. Using appropriate vectors, is has been possible to construct yeast α cells that produce the α receptor but not the normal **a** receptor or **a** cells that produce the **a** receptor but not the normal α receptor. These *receptor-switching experiments* demonstrate that the mating response pathways in **a** and α cells differ only in type of receptor; the intracellular machinery involved in the subsequent transmission of the signal induced by hormone binding is the same in both cell types. For instance, an α cell that expresses the α receptor functions in mating as an **a** cell; that is, it responds to α hormone but not to **a** hormone.

The **a** and α receptors interact with the same G protein, and hormone binding triggers the exchange of GTP for GDP on the α subunit and dissociation of $G_\alpha \cdot GTP$ from $G_{\beta,\gamma}$. In contrast to other receptor systems that utilize G

▶ **Figure 19-35** Structure and signaling by yeast receptors for **a** and α mating hormones. Normal α cells express the receptor for **a** factor and normal **a** cells express the α receptor. Both receptors interact with the same G proteins. Hormone binding triggers displacement of GDP by GTP on the G_α subunit and the dissociation of $G_\alpha \cdot GTP$ from the $G_{\beta,\gamma}$ subunits. Physiological responses are induced by the dissociated β/γ complex—not by $G_\alpha \cdot GTP$, as in other G-mediated signaling systems. *After I. Herskowitz and L. Marsh, 1987, Cell 50:995–996; D. Blinder, S. Bouvier, and D. Jenness, 1989, Cell 56:479–486; M. Whiteway, L. Hougan, D. Dignard, D. Thomas, L. Bell, G. Saari, F. Grant, P. O'Hara, and V. MacKay, 1989, Cell 56:467–477.*

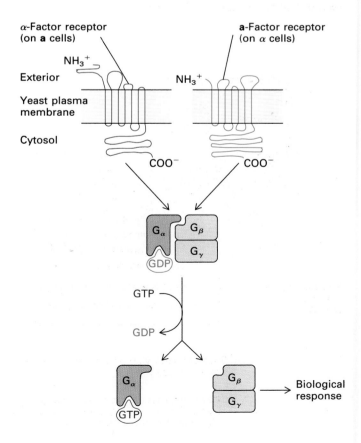

(a)

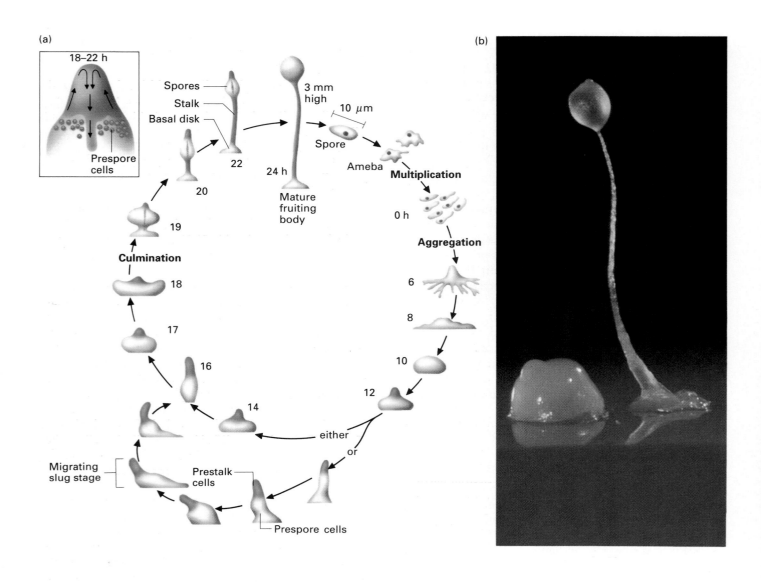

18–22 h

Prespore cells

Spores
Stalk
Basal disk

22

20

19

Culmination

18

17

16

14

either

or

Migrating slug stage

Prestalk cells

Prespore cells

3 mm high

10 μm

Spore

24 h

Mature fruiting body

Ameba

Multiplication

0 h

Aggregation

6

8

10

12

(b)

(c)

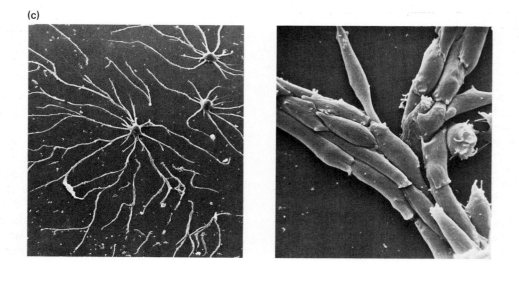

◀ **Figure 19-36** (a) The differentiation of *D. discoideum*. These amebas divide by mitosis, growing in soil and feeding on bacteria and other organic matter. The differentiation cycle is triggered by starvation. During differentiation, DNA synthesis and cell division cease. Streaming toward aggregation centers begins at about 6 h; by 10 h, mounds have formed, each containing about 100,000 cells. The cells do not fuse with each other but retain their individual identities. About 2 h later, a discrete "tip" forms; it contains the prestalk cells that will form the stalk cells (red) in the mature fruiting body. Cells at the base (tan) will become spore cells. Under certain environmental conditions, the mound forms into a motile, wormlike creature, or "slug," that migrates toward light or warmth; the prestalk cells form the "leading edge" of the structure (*lower part of figure*). The migrating slug stage can last for several days. The final stage of differentiation is triggered by overhead light. The stalk cells elongate and vacuolate, pushing down through the mass of differentiating spore cells (*inset*) and hoisting the mass of spore cells up along the stalk. Simultaneously, the spore cells secrete glycoproteins and polysaccharides that form a rigid spore coat around each spore. The spores lose H_2O and become metabolically inactive. The mature fruiting body contains about 70,000 spores, supported by a stalk of about 30,000 dead cells. Mature spores are much more resistant to desiccation, ultraviolet light, and other toxic treatments than growing cells. The spores can be dispersed by wind. On exposure to H_2O and a suitable supply of nutrients, the spores germinate. Rupture of the spore coat releases viable, free-living amebas. (b) Light micrograph of two developmental stages: (*left*) a mass of cells with a tip (formed at 14 h); (*right*) a mature fruiting body. (c) Scanning electron micrographs of fields of aggregating slime molds: (*left*) a forming aggregate of *D. discoideum*; (*right*) a higher power micrograph of aggregating *D. discoideum*, showing the polarized cells forming end-to-end and side-by-side contacts as they stream toward the aggregation center. *Part (b) courtesy of R. Kay. Part (c) courtesy of R. Guggenheim and G. Gerisch.*

proteins, genetic evidence indicates that here the dissociated $G_{\beta,\gamma}$ subunits (not $G_\alpha \cdot GTP$) trigger all physiological responses. Deletion of the gene encoding the G_α subunit, for instance, does not prevent mating but produces the opposite effect: the mating-hormone response is always on in the cell. Deletion of the G_β or G_γ subunits, in contrast, abolishes the ability of either the **a** or α cells to respond to a mating hormone. Thus the $G_{\beta,\gamma}$ subunits—not the G_α subunit—are required to induce all cellular responses of the mating factors.

Aggregation in Cellular Slime Molds Is Dependent on Cell-to-Cell Signaling

Cell-to-cell signaling is an essential prerequisite for developmental changes in the life cycles of many organisms. Best characterized are the slime molds, such as

Dictyostelium discoideum—unicellular free-living amebas that aggregate to form a motile multicellular organism (Figure 19-36).

The development of the slime mold involves many alterations in the pattern of gene expression. Here, we discuss cell signaling during *chemotaxis* (the streaming of cells toward local centers) and the formation of multicellular aggregates. Amebas, the growing cells, are incapable of such responses, which require the elaboration of specific cell-surface macromolecules during the 6-h preaggregation phase.

The chemotactic agent in each species of slime mold is a different chemical. *D. discoideum*, the most widely studied, uses the nucleotide 3′,5′-cAMP as a chemotactic signal. This is the only known organism in which cAMP itself is a pheromone.

Aggregation in *D. discoideum* involves cell-cell signaling by periodic synthesis and secretion of cAMP. An individual ameba that synthesizes and secretes a pulse of cAMP creates a gradient of cAMP around itself. A neighboring ameba responds to such a gradient in several ways (Figure 19-37). It moves some distance in micrometers up the gradient toward the source of cAMP, where it synthesizes and releases its own pulse of cAMP to attract neighboring cells. The cell then becomes refractory and cannot respond to cAMP signals for a period of several minutes. The cAMP signal is amplified by each cell: the cell synthesizes and releases more cAMP than that present in the original stimulus. This relaying results in cell-to-cell propagation of the cAMP signal.

Proper functioning of this response requires a number of macromolecules: adenylate cyclase to synthesize 3′,5′-cAMP from ATP; a cell-surface cAMP receptor; and secreted and cell-surface cAMP phosphodiesterases to degrade cAMP to 5′-AMP. The phosphodiesterases keep the extracellular hormone from building up to a level that swamps out any gradients.

Because cell-to-cell signaling by extracellular cAMP can extend over distances of only 10–100 μm, cells are attracted primarily to the cAMP released by adjacent cells. Aggregation is initiated when random cells release pulsating waves of cAMP signals, which radiate outward from these "initiator" cells every 3–5 min; concomitantly, a pulsatile movement of the cells occurs inward toward the cell centers. The formerly homogeneous array of amebas rapidly breaks up into aggregation centers, each containing about 100,000 cells.

The cell-surface cAMP receptor, like the yeast **a** and α receptors, has seven transmembrane α helixes; a G protein couples the cAMP receptor to adenylate cyclase and cAMP synthesis. An important feature of *D. discoideum* aggregation is the *refractory period*, during which the cells bear normal amounts of cell-surface cAMP receptors, which bind cAMP normally but do not respond to extracellular hormone. The refractory period is caused by

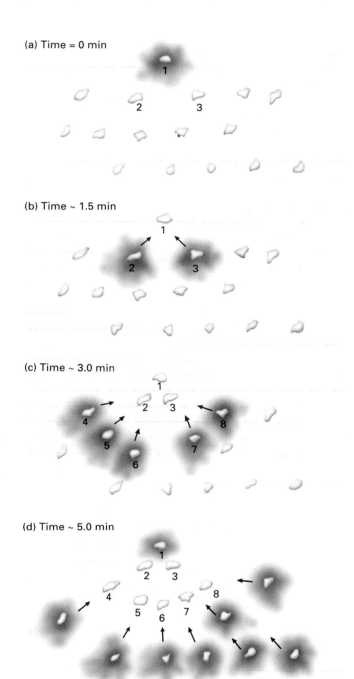

(a) Time = 0 min

(b) Time ~ 1.5 min

(c) Time ~ 3.0 min

(d) Time ~ 5.0 min

▲ **Figure 19-37** Cell-to-cell signaling with cAMP in *D. discoideum*. (a) At 0 min, one ameba (cell 1) spontaneously releases a pulse of cAMP. This finite amount of cAMP diffuses away from the cell. (b) At ~1.5 min, cells 2 and 3, influenced by the cAMP, move a few micrometers toward the cAMP source, release their own pulses of cAMP, and become refractory to further stimuli. Meanwhile, the cAMP released by cell 1 has been destroyed by cAMP phosphodiesterase. (c) At ~3.0 min, cells 4, 5, and 6 move toward cell 2 and release cAMP; cells 7 and 8 move toward cell 3, secrete cAMP, and become refractory. (d) At ~5.0 min, other cells move inward toward cells 4–8 and relay the cAMP pulses. Meanwhile, the refractory period of cell 1 has ended; it synthesizes and releases a pulse of cAMP, repeating the cycle.

phosphorylation of the cell-surface cAMP receptor, much in the way the β-adrenergic receptor is down-regulated by phosphorylation (see Figure 19-34). Specifically, the cAMP-occupied cell-surface receptor is a substrate for an intracellular kinase (akin to the β-adrenergic receptor kinase), and phosphorylated receptor cannot transduce the hormone signal and activate adenylate cyclase or cell movement. Over time, the phosphates on the receptor are hydrolyzed off, the receptor reacquires the ability to transduce the hormone signal, and the refractory period ends. Further study of this receptor-transduction system will clarify the details of slime-mold differentiation and, possibly, the very similar signal-transduction system and its regulation in animal cells.

Plant Hormones and Plant Growth and Differentiation

Hormones regulate many physiological processes in higher plants, such as cell growth and enlargement, differentiation of specialized tissues, and the induction of protein synthesis. Plant hormones can be divided into five principal classes: auxins, cytokinins, gibberellins, the gas ethylene, and abscisic acid (Figure 19-38). Frequently,

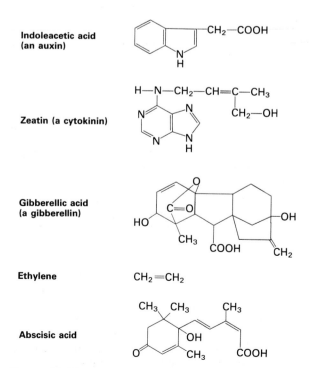

Indoleacetic acid (an auxin)

Zeatin (a cytokinin)

Gibberellic acid (a gibberellin)

Ethylene

Abscisic acid

▲ **Figure 19-38** Examples of the five classes of plant hormones.

developmental processes are affected in opposite ways by two or more hormones (Table 19-9).

The molecular basis of plant hormone action is being elucidated slowly, and only now are receptors being identified and purified for any plant hormone. Thus we do not know with certainty which effects of the hormone are primary (due to the immediate action of the hormone-receptor complex) and which are secondary (triggered later). Nonetheless, an impressive amount has been learned about the physiological and developmental importance of plant hormones, particularly regarding hormone-triggered induction or repression of specific mRNAs. Instead of surveying the vast information available on each hormone class here, we examine only those systems for which a mechanism of action is close to being discovered.

In considering plant growth and differentiation, it is important to remember that, unlike animal cells, plant cells do not migrate. The position of a newly formed plant cell is constrained by the cell wall that surrounds it. Plants grow and differentiate in two basic ways: existing cells can grow larger, and newly formed cells can differentiate into specific cell types. Hormones regulate not only cell division and elongation but also cell death. Consider *abscission,* the dropping of leaves or other plant parts. As leaves get older, hormones cause many leaf ions, amino acids (generated by proteolysis in the leaf), and sugar (derived from the hydrolysis of starch stored in the leaf) to be returned to the stem. Then, at least in some plants, enzymes degrade the cell walls in the *abscission zone,* which connects the base of the leaf *petiole* with the stem. In this way, plants conserve substances such as ions, amino acids, and sugars.

Auxin Triggers the Elongation of Higher Plant Cells

Many higher plants grow more by cell enlargement than by cell proliferation. The size and shape of a plant are determined primarily by the amount and direction of this enlargement. Plant cells are surrounded by rigid cell walls that are partly constructed of cellulose fibers, and embedded in a matrix of protein, other polysaccharides (for example hemicellulose), and pectin. The tensile strength of the cell wall allows the plant cell to develop considerable internal pressure, or *turgor*—outward pressure that results because the osmotic pressure in the vacuole is higher than that in the surrounding extracellular fluids. Individual plant cells can increase in size very rapidly by loosening the wall and pushing the cytosol and plasma membrane outward against it. During this elongation, the amount of cytosol remains constant; the increase in cell volume is due only to the expansion of the intracellular vacuole (Figure 19-39). We can appreciate the magnitude

Table 19-9 Selected plant developmental processes and their control by agonist and antagonist hormones

Developmental process	Involved hormones	
	Agonist	Antagonist
Cell enlargement	Auxin	Cytokinin, ethylene
Cell division, organogenesis	Cytokinin	Auxin
Embryogenesis, seed development	Abscisic acid	Cytokinin
Seed germination (aleurone layer of cereals)	Gibberellic acid	Abscisic acid
Leaf senescence	Abscisic acid	Cytokinin
Fruit ripening	Ethylene	Cytokinin

SOURCE: B. Partier.

(a) Plant cell just after division

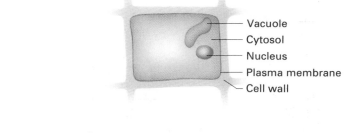

(b)

(c) Elongated cell

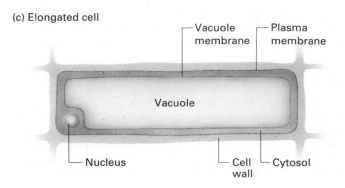

▶ **Figure 19-39** Stimulation of cell elongation by auxin, a hormone that causes localized loosening of the plant cell wall. Uptake of H_2O by the vacuole creates a state of pressure (turgor) that causes the plasma membrane to exert pressure against the loosened wall, elongating the cell. The increase in cell size is mainly due to an increase in the vacuole; the amount of cytosol remains constant.

of this phenomenon by considering that if all cells in a redwood tree were reduced to the size of a typical liver cell (~20 μm in diameter), then the tree would have a maximum height of only 1 meter!

The rapid effects of *auxin* on cell elongation were first seen in classical experiments on *coleoptiles* (the protective sheaths that surround primary shoots and leaves) of grasses and oats. Like many plant parts, coleoptiles are *phototropic*: they bend toward a source of light. Experiments have shown that the growing *meristematic cells* at the tip of the coleoptile synthesize auxin, which is then transported down the coleoptile, where it causes many effects. It has been hypothesized that light inhibits auxin production, so that more auxin and preferential cell elongation occur on the side away from the light, causing the

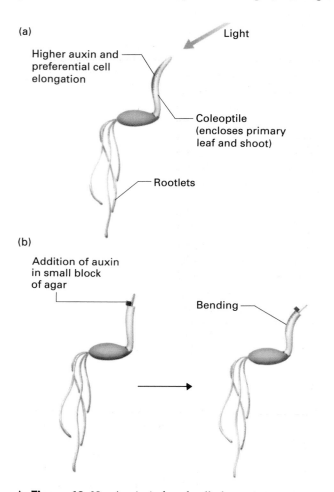

(a)

Light

Higher auxin and preferential cell elongation

Coleoptile (encloses primary leaf and shoot)

Rootlets

(b)

Addition of auxin in small block of agar

Bending

▲ **Figure 19-40**   Auxin-induced cell elongation causes phototrophism in oat coleoptiles. (a) The release of auxin by the growing tip *(meristem)* on the side of the coleoptile away from the light causes the cells on the far *(left)* side to elongate preferentially and bend toward the light.
(b) Phototrophism can be demonstrated experimentally by removing the tips from coleoptiles grown in the dark, keeping the decapitated seedlings in the dark, and applying an agar block containing auxin to one side. The coleoptiles subsequently bend away from the side at which auxin is applied.

coleoptile to bend toward light (Figure 19-40a). Indeed, if the coleoptile is decapitated, removing its endogenous source of auxin, and a small amount of auxin is placed on one side of the cut, the seedling bends away from the side where auxin was applied (Figure 19-40b).

Auxin is thought to induce cell elongation by causing the cell wall to soften at the "growing" end of the cell. In some plants, this is believed to be due to the induction by auxin of localized proton secretion at the ends of the cells, which would loosen the wall locally. Auxin apparently activates (directly or indirectly) a proton pump bound to the plasma membrane, which lowers the pH of the cell wall region near the plasma membrane (possibly to as low as pH 4.5 from the normal pH 7.0). As a consequence, hydrogen bonding between cellulose fibrils of the wall is reduced. The lowered pH also may activate enzymes in the cell wall that degrade various protein or polysaccharide constituents, contributing to localized softening and loosening of the wall and elongation and enlargement of the cell.

Some evidence for this *acid-growth hypothesis* stems from studies of the fungal compound *fusicoccin*, which does not trigger auxin's effects on gene expression but does induce rapid cell elongation. Like auxin, fusicoccin triggers proton pumping out of sensitive cells and causes localized wall loosening; the action of fusicoccin can be blocked by permeating the cell wall with buffers that do not allow extracellular pH to be lowered. However, in maize (corn) coleoptiles, auxin-triggered cell elongation is not correlated with an enhancement of proton secretion and is not blocked by buffers that prevent the lowering of cell-wall pH, so in this system, as perhaps in others, the acid-growth theory does not seem to apply.

Auxin Causes Rapid Changes in Gene Expression

The mode of action of auxin has been investigated intensely for many years, but today no satisfactory explanation exists for the molecular basis of any auxin effect. The reported effects are diverse, ranging from enhancement of cell elongation to stimulation of cell division to modification of cell differentiation. Auxin produces both primary and secondary effects, which are difficult to separate from one another and, moreover, may differ according to plant tissue and species.

In at least some experimental systems, auxins cause a very rapid change in the pattern of mRNA synthesis. Many workers feel that the primary effects of auxin— rapid induction or repression of gene expression— resemble those of steroids in mammalian cells. The most rapid changes in the pattern of gene expression by auxin have been observed in pea epicotyl tissue or in tobacco or carrot cell cultures. Several cDNAs were cloned from each species that encode mRNAs present only in auxin-treated tissue, not in control tissue. Induction of these mRNAs began as early as 10 min after auxin treatment and continued for several hours. Accumulation of these mRNAs is due to enhanced gene transcription, not to stabilization of preexisting mRNAs. Since we do not know the identity of the proteins encoded by auxin-induced mRNAs, it is difficult to explain how they relate to auxin-induced cell elongation. Nonetheless, at least in pea epicotyl tissue, auxin-induced cell elongation is blocked by inhibitors of RNA and protein synthesis, suggesting that cell elongation depends on some auxin-induced protein.

Auxins have diverse effects on different plant tissues, as steroids do on mammalian tissues. Whether these effects are all related to a single molecular mechanism of auxin action, such as induction of gene expression, remains to be seen. One function of auxin is *apical dominance:* in the stems of most species, the bud at the tip, or *apex,* exerts an inhibitory influence on the development of lateral buds from the stem. Gardeners remove the apical bud and young leaves at the tip of the stem in order to increase the branching of a plant. If auxin is added to the cut stem after the apical bud is removed, lateral bud development is blocked. Whether auxin is the natural inhibitor of such development is still uncertain. Finally, synthetic auxins, such as the chlorine-containing substance 2.4-D, find wide application in the control of weed pests—the major practical application of plant-growth regulators.

Auxin Transport Requires Specific Transport Proteins in Polarized Mesothelial Cells

An unusual feature of auxin is the way it is transported from tissue to tissue. Unlike the movement of sucrose, salts, and H_2O, auxin is not transported in the live cells that form the phloem sieve tubes or in the dead cells of the xylem. Rather, auxin is transported in stems and young leaves primarily through polarized *mesenchymal cells* that surround the vascular bundles of xylem and phloem. Auxin movement is polar—always in a *basipetal* (base-seeking) direction. Experimentally, if auxin is applied to one or the other cut end of a piece of stem, auxin is transported only from the cut apical end to the cut basal end.

A clue to auxin transport comes from a study of certain chemicals that inhibit auxin action, such as *naphthylphthalamic acid* (NPA). The antagonist function of NPA is due to inhibition of auxin transport; NPA binds to an auxin transporter localized to the basal surface of elongated mesenchymal cells (Figure 19-41). Auxin enters the apical surface of mesenchymal cells by diffusion as the protonated, uncharged molecule. The preferential exit of unprotonated, negatively charged auxin at the basal surface is driven by a specific transporter to ensure that

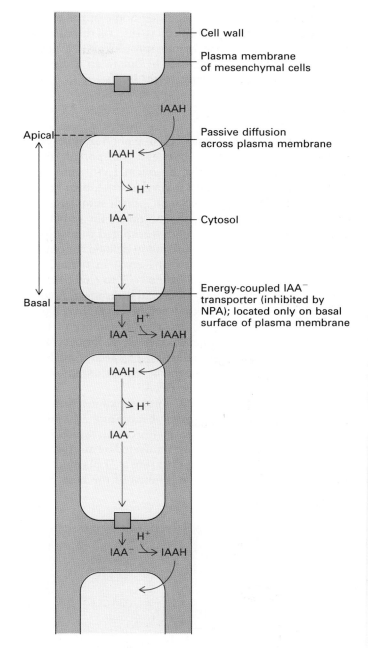

▲ **Figure 19-41** The polarized transport of auxin. Auxin (indoleactic acid, or IAA) is synthesized by the growing apical meristems of young shoots and transported down the stem to sites at which it exerts numerous effects. Transport requires energy and occurs via the polarized mesenchymal cells lining the vascular bundles of the stem. In the acidic environment of the cell wall, IAA is largely in the neutral, undissociated form IAAH, which readily enters the mesenchymal cells by passive diffusion. In the cell, IAA exists largely in the negatively charged form IAA⁻, since its pK = 4.7 (the pH at which one-half of the molecules are ionized; see page 30). Polarized transport of auxin is due to an energy-coupled auxin transporter, localized to the basal surface of these polarized cells, that pumps IAA⁻ out of the cell. *After M. Jacobs and S. F. Gilbert, 1983, Science* **220:***1297–1301.*

auxin is transported down the stem. Some of the energy requirement for auxin transport is due to the plasma-membrane ATPase, which pumps protons outward and acidifies the cell wall; some may be due to an energy requirement for the auxin transporter itself.

Cytokinins Stimulate Cell Division

Cytokinins are an important class of plant hormones normally synthesized in plant roots and transported, along with H_2O, to other tissues by the xylem. Cytokinins have a primary effect on cell division; in this sense, their action is complementary to that of auxin, which stimulates cell elongation (see Table 19-9). For instance, the growth of plant-tissue culture cells or undifferentiated callus tissue requires both auxin and a cytokinin in defined concentrations. In studies on cultured tobacco stem tissues, the addition of auxin alone caused the cells to elongate and form giant cells; cytokinins alone had little effect. A combination of auxin and cytokinin resulted in rapid cell proliferation and division, so that a large number of relatively small, undifferentiated cells were formed. Like aukins, cytokinins cause induction of specific mRNAs, although it is not yet certain whether the cytokinin-receptor complex binds directly to DNA.

The essential role of auxins and cytokinins is underscored by a natural gene-transfer phenomenon, the *crown gall tumor* (Figure 19-42; see Figure 5-28), which is induced when a susceptible plant is infected by an *Agrobacterium tumefaciens* bacterium bearing a Ti plasmid. The plasmid is integrated into the genome of the host plant,

where it directs the synthesis of two enzymes that catalyze auxin synthesis from tryptophan and one enzyme that catalyzes cytokinin production from AMP and other normal plant metabolites. Studies in which various genes on the Ti plasmid were deleted have shown that tumorous growth is primarily due to the activities of these three Ti-encoded genes.

Gibberellic Acid Triggers Seed Germination by Inducing Specific mRNAs

Gibberellic acid (see Figure 19-38) is one of about 50 related compounds, termed *gibberellins,* that markedly affect cell division and elongation in plants. One effect on stem elongation is easily observed. Application of gibberellic acid to the soil causes the stems of many plants to become long and thin. Conversely, certain plants, such as varieties of the common bean, appear to be dwarfed by a lack of gibberellins, since addition of gibberellic acids to the soil causes these plants to grow to normal size. Finally, adding chemicals that block the synthesis of gibberellic acid stunts the growth of normal plants.

Gibberellins induce seed germination. A specific period of dormancy is normally required for many plant seeds, such as lettuce and tobacco, to germinate. The addition of gibberellins to these seeds bypasses their requirement for dormancy and triggers elongation of the root cells inside the seed. Studies on the action of gibberellic acid during germination of barley seeds provide the best evidence that specific enzymes induced by this hormone produce dramatic changes in the form and function of plant cells.

The barley embryo is a small part of the barley seed (Figure 19-43a). After germination, the embryonic cells grow and divide rapidly, forming the coleoptile and roots of the young plant. The *endosperm* in the seed stores minerals and nucleic acids, proteins, and carbohydrates. The *aleurone layer* surrounding the endosperm is composed mostly of nondividing cells.

Degradation and mobilization of the digestive products of the endosperm reserves—sugars, nucleotides, and amino acids—are controlled by gibberellins. Gibberellic acid is normally secreted by the growing embryonic cells (Figure 19-43b); it diffuses through the endosperm into the aleurone layer, where it induces the synthesis of several hydrolytic enzymes, such as proteases and α-amylase, and their secretion into the endosperm. These enzymes degrade stored proteins to amino acids and starch to glucose, providing support for the early growth of the barley seedling. Experiments have shown that not only α-amylase itself but the mRNA for α-amylase increases, suggesting that gibberellic acid triggers accumulation of α-amylase mRNA by increasing transcription of the α-amylase gene (Figure 19-44).

In many respects, therefore, the action of gibberellic acid on barley aleurone cells is comparable to that of estrogen on the uterus and other mammalian female sex

▲ **Figure 19-42** Photograph of a crown gall tumor on a stem of the black locust *Robinia pseudoacacia;* the tumor is caused by infection by *A. tumefaciens* bacteria that bear a Ti plasmid. *Courtesy of Runk/Schoenberger/Grant Heilman Photography.*

(a) Ungerminated seed

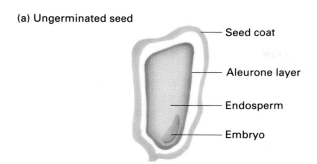

Seed coat

Aleurone layer

Endosperm

Embryo

(b) Germinating seed

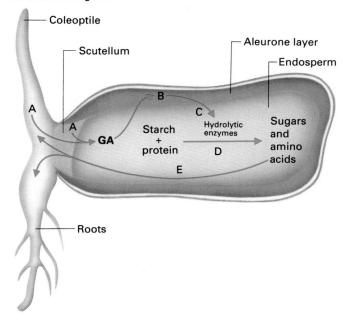

Coleoptile

Scutellum

Aleurone layer

Endosperm

A

A

GA

B

C

Starch + protein

Hydrolytic enzymes

D

E

Sugars and amino acids

Roots

◄ **Figure 19-43** Diagram of the relationship between gibberellic acid (GA) production, α-amylase production, and sugar accumulation in germinating barley seeds. (a) The ungerminated seed with its cell layers. The embryo, composed of the cells that will give rise to the mature plant, forms a small part of the seed. (b) After germination, the embryo grows and differentiates into several cell types. Gibberellic acid produced by the coleoptile and scutellum (shield) regions of the embryo (A) diffuses into the aleurone layer (B), where it induces the synthesis and release of hydrolytic enzymes (C). These enzymes, such as proteases and amylases, hydrolyze the protein and carbohydrate reserves in the endosperm (D), producing the glucose and amino acids that nourish the growing embryo (E). [See J. E. Varner and D. T.-H. Ho, 1976, in *The Molecular Biology of Hormone Action,* Academic Press, p. 173.]

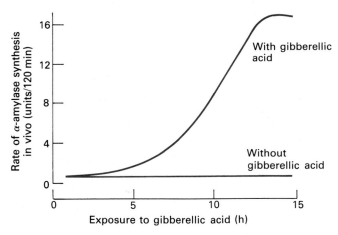

▲ **Figure 19-44** The time course of α-amylase synthesis in isolated aleurone layers from germinating barley seedlings in the presence or absence of gibberellic acid. The level of α-amylase mRNA (not shown here) increases parallel to synthesis of the α-amylase protein. These experiments indicate that gibberellic acid directly or indirectly induces synthesis of α-amylase mRNA. [See T. J. V. Higgins, J. A. Zwar, and J. V. Jacobsen, 1976, *Nature* **260**:166.]

organs. Despite much effort, however, a cellular receptor for gibberellic acid has not been identified, and the molecular details of its mode of action remain unclear.

Ethylene Promotes Fruit Ripening and Abscission

The gas *ethylene* (see Figure 19-38) has potent effects on plant development. It triggers fruit ripening in a number of species, promotes abscission of leaves, flowers, and fruits by inducing enzymes involved in cell-wall degradation, and acts to induce flowering in some plants.

The function of ethylene was discovered in an unusual way. In the last century, gas lamps were widely used for illumination; leaks in the gas mains caused defoliation in surrounding trees. The relevant gas, ethylene, was later found to be produced naturally by a variety of ripening fruits. Today, ethylene is widely used to ripen green tomatoes, which are shipped to market packed in the gas.

Much evidence suggests that ethylene induces the expression of genes that code for enzymes that accelerate fruit ripening. First, ethylene induces (in an unknown way) enzymes that catalyze the synthesis of ethylene from methionine. Thus synthesis of more ethylene by fruits can be induced by adding very low levels of ethylene gas, causing an increase in other ethylene-induced functions. The gene that codes for *polygalacturonase,* an enzyme that hydrolyzes pectin, a major component of cell walls and thus causes fruit softening, is induced by ethylene. The addition of ethylene to tomatoes causes polygalacturonase mRNA and other ripening-related mRNAs to accumulate and enhances transcription of these genes.

The mechanism by which ethylene induces its responses is not known. Work with ethylene antagonists suggests that ethylene binds to a transition metal in the receptor protein. One advance in understanding ethylene action is the recent isolation of dominant mutants of the weed *Arabidopsis* that are deficient in all actions of ethylene. Since a single mutation produces these effects, a single receptor protein may mediate the diverse actions of ethylene.

Abscisic Acid Has General Growth-inhibitory, Senescence-promoting Activity

Abscisic acid accumulates in zones of abscission and senescence (aging) and promotes senescence in leaves and flowers and seed dormancy in many species. In germinating barley seeds, for instance, abscisic acid prevents most, if not all, of the effects of gibberellic acid on gene expression, including the induction of α-amylase and other enzymes (see Figure 19-44). In this sense, abscisic acid is the natural antagonist of gibberellic acid (see Table 19-9). Abscisic acid also functions to prevent *transpiration* (H_2O loss from leaves through the stomata; Figure 14-25) during dry periods. The stomata are normally open during photosynthesis. If the leaves become stressed due to a lack of H_2O, chloroplasts in the mesophyll cells produce abscisic acid, which is transported (in an unknown way) to the stomatal guard cells, where it causes the stomata to close.

Summary

Hormones and other molecules used for cell-to-cell signaling in multicellular animals can be categorized into two broad groups. Lipid-soluble hormones—principally thyroxine and the steroids and their derivatives—diffuse across the plasma membrane, interact with protein receptors in the cytosol or nucleus, and induce or repress transcription of cell-specific genes. Peptide hormones and small lipid-insoluble hormones, such as epinephrine, bind to protein cell-surface receptors, the major focus of this chapter.

Cell-surface receptors include ligand-triggered ion channels, protein kinases, and protein phosphatases. Other surface receptors are coupled, via a family of G (transducing) proteins, to the activation or inhibition of various enzymes. Receptors can be identified and quantified by the ligand binding of specific agonists and antagonists, characterized by cross-linking to ligands, purified by affinity chromatography, and cloned by procedures that include synthesis of functional receptors from transfected DNA.

Binding several hormones, such as glucagon and epinephrine, to certain cell-surface receptors triggers an activation of adenylate cyclase and a consequent elevation in the intracellular concentration of $3',5'$-cAMP. Interaction between the receptors and adenylate cyclase is mediated by the transducing G_s protein, which exchanges bound GDP for GTP in response to hormone binding to ligand. The $G_s \cdot$ GTP complex, in turn, dissociates into its $G_{s\alpha} \cdot$ GTP and $G_{\beta,\gamma}$ subunits, and $G_{s\alpha}$ activates adenylate cy-clase. Binding other hormones to other receptors inhibits adenylate cyclase; these receptors interact with the inhibitory G_i transducer. The ability of certain bacterial toxins to ADP-ribosylate specific G_α proteins and block hydrolysis of bound GTP to GDP has proved invaluable in elucidating the properties of these G proteins.

The mode of action of cAMP is mediated by a cAMP-dependent protein kinase, which in the well-understood case of epinephrine-induced glycogenolysis in liver or muscle cells, phosphorylates (inactivates) glycogen synthase. This kinase also phosphorylates another protein kinase in the first step in a cascade of protein phosphorylation that results in activation of glycogen phosphorylase.

Binding hormones to still other receptors triggers activation of phospholipase C via another type of G protein. This enzyme hydrolyzes the plasma-membrane lipid phosphatidylinositol 4,5-bisphosphate, generating two second messengers: 1,2-diacylglycerol activates protein kinase C, which phosphorylates several proteins important in regulating the growth and metabolism of cells; inositol 1,4,5-trisphosphate binds to receptors on the ER and triggers the release of Ca^{2+} ions into the cytosol. Many effects of Ca^{2+} are mediated by calmodulin or related proteins. In muscle cells, Ca^{2+} ions trigger contraction and also activates glycogen phosphorylase kinase by binding to the calmodulin subunit of the enzyme.

The insulin receptor is a ligand-triggered tyrosine-protein kinase, although its substrates in the cell are not known. In adipocytes and muscle cells, insulin binding triggers exocytosis of vesicles containing a particular type of glucose transporter, activating glucose uptake. The blood glucose level is regulated daily by the antagonistic actions of insulin and glucagon.

The level of cell-surface receptors is often subject to regulation, and the continued exposure of a cell to a hormone results in a reduction of these receptors. Some receptors, such as those for insulin and the epidermal growth factor (EGF), are endocytosed with the bound hormone. Internalized hormone is degraded by lysosomes; during each cycle of endocytosis, a large fraction of the receptor is also degraded.

The activities of other receptors, such as the β-adrenergic receptor, are down-regulated by phosphorylation. The receptor is a substrate for both the cAMP-dependent kinase (heterologous desensitization) and the β-adrenergic receptor kinase (homologous desensitization). The latter phosphorylates only agonist-occupied receptors and thus restricts its binding to the G_s protein.

Microorganisms also utilize cell-to-cell signaling to coordinate aggregation, differentiation, and sexual reproduction. Yeast strains use diffusible peptide pheromones and cell-surface receptors to coordinate the fusion of two haploid cells into a diploid cell during sexual reproduction. Like the β-adrenergic receptor, the yeast α and **a** receptors have seven transmembrane α helixes and also interact with a G protein. Slime molds use extracellular signaling to coordinate the chemotaxis of cells into aggregates.

Hormones regulate many aspects of cell growth and differentiation in plants. Auxin triggers the elongation of plant cells by causing localized loosening of the cell wall and expansion of the cell vacuole. Rapid changes in the pattern of gene expression caused by auxin may mediate its diverse effects, including apical dominance and phototrophism in coleoptiles. Cytokinins stimulate cell division. The production of both auxin and cytokinin, caused by enzymes encoded on the Ti plasmid in a bacterial pathogen, leads to uncontrolled proliferation and formation of a crown gall tumor.

Gibberellins also induce specific genes, including α-amylase gene transcription in germinating barley seeds. These effects are antagonized by abscisic acid. Ethylene promotes fruit ripening and abscission of leaves and flowers by inducing genes to code for the enzymes required to activate these processes.

References

General Properties of Hormone Systems

*BERREDGE, M. 1985. The molecular basis of communication within the cell. *Scientific American* 253(4):142–150.

*COHEN, P., ed. 1984. *Molecular Aspects of Cellular Regulation*, vol. 3: *Enzyme Regulation by Reversible Phosphorylation—Further Advances*. Elsevier.

*MARTIN, B. R. 1987. *Metabolic Regulation: A Molecular Approach*. Blackwell Scientific Publications.

*SMITH, E. L., R. L. HILL, I. R. LEHMAN, R. J. LEFKOWITZ, P. HANDLER, and A. WHITE. 1983. *Principles of Biochemistry: Mammalian Biochemistry*, 6th ed. McGraw-Hill. Chapters 11–20 describe the biochemistry of the endocrine systems in detail.

*WALLIS, M., S. L. HOWELL, and K. W. TAYLOR, eds. 1986. *The Biochemistry of the Polypeptide Hormones*. Wiley.

*WILSON, J. D., and D. W. FOSTER. 1985. *Williams Textbook of Endocrinology*, 7th ed. Saunders.

Isolation and Cloning of Receptors

ALLEN, J. M., and B. SEED. 1989. Isolation and expression of functional high-affinity Fc receptor complementary DNAs. *Science* 243:378–381.

D'ANDREA, A. D., H. F. LODISH, and G. G. WONG. 1989. Expression cloning of the murine erythropoietin receptor. *Cell* 57:277–285.

HART, C. E., J. W. FORSTROM, J. D. KELLY, R. A. SEIFERT, R. A. SMITH, R. ROSS, M. J. MURRAY, and D. F. BOWEN-POPE. 1988. Two classes of PDGF receptor recognize different isoforms of PDGF. *Science* 240:1529–1531.

HEMPSTEAD, B. L., L. S. SCHLEIFER, and M. V. CHAO. 1989. Expression of functional nerve growth factor receptors after gene transfer. *Science* 243:373–375.

*A book or review article that provides a survey of the topic.

KLOTZ, I. M. 1982. Number of receptor sites from Scatchard graphs: facts and fantasies: *Science* 217:1247–1249.

β-Adrenergic Receptors

*BENOVIC, J. L., M. BOUVIER, M. G. CARON, and R. J. LEFKOWITZ. 1988. Regulation of adenylyl cyclase-coupled β-adrenergic receptors. *Ann. Rev. Cell Biol.* 4:405–428.

*DOHLMAN, H. G., M. G. CARON, and R. J. LEFKOWITZ. 1987. A family of receptors coupled to guanine nucleotide regulatory proteins. *Biochemistry.* 26:2657–2664.

*LEFKOWITZ, R. J., and M. G. CARON. 1988. Adrenergic receptors: models for the study of receptors coupled to guanine nucleotide regulatory proteins. *J. Biol. Chem.* 263:4993–4996.

*LEVITZKI, A. From epinephrine to cyclic AMP. 1988. *Science* 241:800–806.

MATSUI, H., R. J. LEFKOWITZ, M. G. CARON, and J. W. REGAN. 1989. Localization of the fourth membrane-spanning domain as a ligand binding site in the human platelet α_2-adrenergic receptor. *Biochem.* 28:4125–4130.

O'DOWD, B. F., M. HNATOWICH, J. W. REGAN, W. M. LEARDER, M. G. CARON, and R. J. LEFKOWITZ. 1988. Site-directed mutagenesis of the cytoplasmic domains of the human β_2-adrenergic receptor. *J. Biol. Chem.* 263:15985–15992.

WANG, H.-Y., L. LIPFERT, C. C. MALBON, and S. BAHOUTH. 1989. Site-directed anti-peptide antibodies define the topography of the β-adrenergic receptor. *J. Biol. Chem.* 264:14424–14431.

G_s and G_1 Proteins and Activation and Inhibition of Adenylate Cyclase

*ALLENDE, J. E. 1988. GTP-mediated macromolecular interactions: the common features of different systems. *FASEB J.* 2:2356–2367.

*BIRNBAUMER, L., J. CODINA, R. MATTERA, A. YATANI, N. SCHERER, M.-J. TORO, and A. M. BROWN. 1987. Signal transduction by G proteins. *Kidney Int.* 32:S14–S37.

*CASEY, P. J., and A. G. GILMAN. 1988. G protein involvement in receptor-effector coupling. *J. Biol. Chem.* 263:2577–2580.

CERIONE, R. A., C. STANISZEWSKI, P. GIERSCHIK, J. CODINA, R. L. SOMERS, L. BIRNBAUMER, A. M. SPIEGEL, M. G. CARON, and R. J. LEFKOWITZ. 1986. Mechanism of guanine nucleotide regulatory protein mediated inhibition of adenylate cyclase: studies with isolated subunits of transducin in a reconstituted system. *J. Biol. Chem.* 261:9514–9520.

*DOLPHIN, A. C. 1987. Nucleotide binding proteins in signal transduction and disease. *Trends Neurosci.* 10:53–57.

*FRIESSMUTH, M., P. J. CASEY, and A. G. GILMAN. 1989. G proteins control diverse pathways of transmembrane signaling. *FASEB J.* 3:2125–2131.

JURNAK, F. 1988. The three-dimensional structure of c-H-ras p21: implications for oncogene and G protein studies. *TIBS* 13:195–198.

KRUPINSKI, J., F. COUSSEN, H. A. BAKALYAR, W.-J. TANG, P. G. FEINSTEIN, K. ORTH, C. SLAUGHTER, R. R. REED, and A. G. GILMAN. 1989. Adenylyl-cyclase amino acid sequence: possible channel- or transporter-like structure. *Science* 244:1558–1564.

LANDIS, C. A., S. B. MASTERS, A. SPADA, A. M. PACE, H. R. BOURNE, and L. VALLAR. 1989. GTPase inhibiting mutations activate the α chain of G_s and stimulate adenylyl cyclase in human pituitary tumours. *Nature* 340:692–696.

*LIMBIRD, L. E. 1988. Receptors linked to inhibition of adenylate cyclase: additional signaling mechanisms. *FASEB J.* 2:2686–2695.

MCFARLAND, K. C., R. SPRENGEL, H. S. PHILLIPS, M. KÖHLER, N. ROSEMBLIT, K. NIKOLICS, D. L. SEGALOFF, and P. H. SEEBURG. 1989. Lutropin-choriogonadotropin receptor: an unusual member of the G protein-coupled receptor family. *Science* 245:494–499.

RANSNAS, L. A., and P. A. INSEL. 1988. Subunit dissociation is the mechanism for hormonal activation of the G_s protein in native membranes. *J. Biol. Chem.* 263:17239–17242.

cAMP-dependent Protein Kinase and Cascades of Protein Phosphorylation/Dephosphorylation

*COHEN, P. 1983. *Control of Enzyme Activity*, 2d ed. Chapman & Hall.

*COHEN, P. 1989. The structure and regulation of protein phosphatases. *Ann. Rev. Biochem.* 58:453–508.

HARDIE, G. 1988. Pseudosubstrates turn off protein kinases. *Nature* 335:592–593.

*HUNTER, T. 1987. A thousand and one protein kinases. *Cell* 50:823–829.

SHORAIN, V. S., B. S. KHATRA, and T. R. SODERLING. 1982. Hormonal regulation of skeletal muscle glycogen synthase through covalent phosphorylation. *Fed. Proc.* 41:2618–2622.

SPRANG, S. R., K. R. ACHARYA, E. J. GOLDSMITH, D. I. STUART, K. VARVILL, R. J. FLETTERICK, N. B. MADSEN, and L. N. JOHNSON. 1988. Structural changes in glycogen phosphorylase induced by phosphorylation. *Nature* 336:215–221.

*TAYLOR, S. S. 1989. cAMP-dependent protein kinase: model for an enzyme family. *J. Biol. Chem.* 264:8443–8446.

Ca²⁺ Ions and Calmodulin as Second Messengers

AMBLER, S. K., M. POENIE, R. Y. TSIEN, and P. TAYLOR. 1988. Agonist-stimulated oscillations and cycling of intracellular free calcium in individual cultured muscle cells. *J. Biol. Chem.* 263:1952–1959.

*CHEUNG, W. Y. 1982. Calmodulin. *Scientific American* 246(6):48–56.

*EVERED, D., and J. WHELAN, eds. 1986. *Calcium and the Cell* (Ciba Foundation Symposium 122). Wiley.

HASHIMOTO, S., B. BRUNO, D. P. LEW, T. POZZAN, P. VOLPE, and J. MELDOLESI. 1988. Immunocytochemistry of calciosomes in liver and pancreas. *J. Cell Biol.* 107:2523–2531.

*RASMUSSEN, H. 1989. The cycling of calcium as an intracellular messenger. *Scientific American* 261(4):66–73.

SCHULMAN, H., and L. L. LOU. 1989. Multifunctional Ca²⁺/calmodulin-dependent protein kinase: domain structure and regulation. *TIBS* 14:62–66.

*STRYNADKA, N. C. J., and M. N. G. JAMES. 1989. Crystal structure of the helix-loop-helix calcium-binding proteins. *Ann. Rev. Biochem.* 58:951–998.

*TSIEN, R. Y., and M. POENIE. 1986. Fluorescence ratio imaging: a new window into intracellular ionic signaling. *TIBS* 11:450–455.

Inositol Trisphosphate, 1,2-Diacylglycerol, and Protein Kinase C

ASHKENAZI, A., E. G. PERALTA, J. W. WINSLOW, J. RAMACHANDRAN, and D. J. CAPON. 1989. Functionally distinct G proteins selectively couple different receptors to PI hydrolysis in the same cell. *Cell* 56:487–493.

*BELL, R. M. 1986. Protein kinase C activation by diacylglycerol second messengers. *Cell* 45:631–632.

*BERRIDGE, M. J., and R. F. IRVINE. 1989. Inositol phosphates and cell signalling. *Nature* 341:197–205.

COUSSENS, L., P. J. PARKER, L. RHEE, T. L. YANG-FENG, E. CHEN, M. D. WATERFIELD, U. FRANCKE, and A. ULLRICH. 1986. Multiple, distinct forms of bovine and human protein kinase C suggest diversity in cellular signaling pathways. *Science* 233:859–866.

DOWNES, C. P. 1988. Inositol phosphates: a family of signal molecules? *Trends Neurosci.* 11:336–338.

FURUICHI, T., S. YOSHIKAWA, A. MIYAWAKI, K. WADA, N. MAEDA, and K. MIKOSHIBA. 1989. Primary structure and functional expression of the inositol 1,4,5-triphosphate-binding protein P_{400}. *Nature* 342:32–38.

HANNUN, Y. A., and R. M. BELL. 1989. Functions of sphingolipids and sphingolipid breakdown products in cellular regulation. *Science* 243:500–507.

HOUSEY, G. M., M. D. JOHNSON, W. L. W. HSIAO, C. A. O'BRIAN, J. P. MURPHY, P. KIRSCHMEIER, and I. B. WEINSTEIN. 1988. Overproduction of protein kinase C causes disordered growth control in rat fibroblasts. *Cell* 52:343–354.

*KIKKAWA, U., A. KISHIMOTO, and Y. NISHIZUKA. 1989. The protein kinase C family: heterogeneity and its implications. *Ann. Rev. Biochem.* 58:31–44.

*MAJERUS, P. W., T. M. CONNOLLY, H. DECKMYN, T. S. ROSS, T. E. BROSS, H. ISHII, V. S. BANSAL, and D. B. WILSON. 1986. The metabolism of phosphoinositide-derived messenger molecules. *Science* 234:1519–1526.

*NISHIZUKA, Y. 1988. The molecular heterogeneity of protein kinase C and its implications for cellular regulation. *Nature* 334:661–665.

*PUTNEY, J. W., JR., H. TAKEMURA, A. R. HUGHES, D. A. HORSTMAN, and O. THASTRUP. 1989. How do inositol phosphates regulate calcium signaling? *FASEB J.* 3:1899–1905.

The Insulin Receptor and Regulation of Blood Glucose

*CZECH, M. P., J. K. KLARLUND, K. A. YAGALOFF, A. P. BRADFORD, and R. E. LEWIS. 1988. Insulin receptor signaling: activation of multiple serine kinases. *J. Biol. Chem.* 263:11017–11020.

*CZECH, M. 1989. Signal transmission by the insulin-like growth factors. *Cell* 59:235–238.

FLIER, J. S., P. USHER, and A. C. MOSES. 1986. Monoclonal antibody to the type I insulin-like growth factor (IGF-I) receptor blocks IGF-I receptor-mediated DNA synthesis: clarification of the mitogenic mechanisms of IGF-I and insulin in human skin fibroblasts. *Proc. Nat'l Acad. Sci. USA* 83:664–668.

GARVEY, W. T., T. P. HUECKSTEADT, and M. J. BIRNBAUM. 1989. Pretranslational suppression of an insulin-responsive glucose transporter in rats with diabetes mellitus. *Science* 245:60–63.

HAWLEY, D. M., B. A. MADDUX, R. G. PATEL, K.-Y. WONG, P. W. MAMULA, G. L. FIRESTONE, A. BRUNNETTI, E. VERSPOHL, and I. D. GOLDFINE. 1989. Insulin receptor monoclonal antibodies that mimic insulin action without activating tyrosine kinase. *J. Biol. Chem.* 264:2438–2444.

HOFMANN, C., I. D. GOLDFINE, and J. WHITTAKER. 1989. The metabolic and mitogenic effects of both insulin and insulin-like growth factor are enhanced by transfection of insulin receptor into NIH3T3 fibroblasts. *J. Biol. Chem.* 264:8606–8611.

JAMES, D. E., M. STRUBE, and M. MUECKLER. 1989. Molecular cloning and characterization of an insulin-regulatable glucose transporter. *Nature* 338:83–87.

ODAWARA, M., T. KADOWAKI, R. YAMAMOTO, Y. SHIBASAKI, K. TOBE, D. ACCILI, C. BEVINS, Y. MIKAMI, N. MATSUURA, Y. AKANUMA, F. TAKAKU, S. I. TAYLOR, and M. KASUGA. 1989. Human diabetes associated with a mutation in the tyrosine kinase domain of the insulin receptor. *Science* 245:66–68.

*ROSEN, O. M. 1987. After insulin binds. *Science* 237:1452–1458.

SOOS, M. A., R. M. O'BRIEN, N. P. J. BRINDLE, J. M. STIGTER, A. K. OKAMOTO, J. WHITTAKER, and K. SIDDLE. 1989. Monoclonal antibodies to the insulin receptor mimic metabolic effects of insulin but do not stimulate receptor autophosphorylation in transfected NIH3T3 fibroblasts. *Proc. Nat'l Acad. Sci. USA* 86:5217–5221.

Receptors for Glucagon and Other Peptide Hormones

CHINKERS, M., D. L. GARBERS, M.-S. CHANG, D. G. LOWE, H. CHIN, D. V. GOEDDEL, and S. SCHULZ. 1989. A membrane form of guanylate cyclase is an atrial natriuretic peptide receptor. *Nature* 338:78–83.

HONEGGER, A. M., R. M. KRIS, A. ULLRICH, and J. SCHLESSINGER. 1989. Evidence that autophosphorylation of solubilized receptors for epidermal growth factor is mediated by intermolecular cross-phosphorylation. *Proc. Nat'l Acad. Sci. USA* 86:925–929.

MARGOLIS, B., S. G. RHEE, S. FELDER, M. MERVIC, R. LYALL, A. LEVITZKI, A. ULLRICH, A. ZILBERSTEIN, and J. SCHLESSINGER. 1989. EGF induces tyrosine phosphorylation of phospholipase C-II: a potential mechanism for EGF receptor signaling. *Cell* 57:1101–1107.

NISHIBE, S., M I. WAHL, S. G. RHEE, and G. CARPENTER. 1989. Tyrosine phosphorylation of phospholipase C-II in vitro by

the epidermal growth factor receptor. *J. Biol. Chem.* 264:10335–10338.

PETERSEN, O. H., and C. BEAR. 1986. Two glucagon transducing systems. *Nature* 323:18.

*ROTH, R. A. 1988. Structure of the receptor for insulin-like growth factor II: the puzzle amplified. *Science* 239:1269–1271.

*SCHLESSINGER, J. 1988. The epidermal growth factor receptor as a multifunctional allosteric protein. *Biochemistry* 27:3119–3123.

*YARDEN, Y., and A. ULLRICH. 1988. Molecular analysis of signal transduction by growth factors. *Biochemistry* 27:3113–3119.

Receptor Regulation by Covalent Modification

BENOVIC, J. L., R. H. STRASSER, M. G. CARON, and R. J. LEFKOWITZ. 1986. β-adrenergic receptor kinase: identification of a novel protein kinase that phosphorylates the agonist-occupied form of the receptor. *Proc. Nat'l Acad. Sci. USA* 83:2797–2801.

HAUSDORFF, W. P., M. BOUVIER, B. F. O'DOWD, G. P. IRONS, M. G. CARON, and R. J. LEFKOWITZ. 1989. Phosphorylation sites on two domains of the β_2-adrenergic receptor are involved in distinct pathways of receptor desensitization. *J. Biol. Chem.* 264:12657–12665.

*SIBLEY, D. R., J. L. BENOVIC, M. G. CARON, and R. J. LEFKOWITZ. 1987. Regulation of transmembrane signaling by receptor phosphorylation. *Cell* 48:913–922.

*SIBLEY, D. R., and R. J. LEFKOWITZ. 1985. Molecular mechanisms of receptor desensitization using the β-adrenergic receptor-coupled adenylate cyclase system as a model. *Nature* 317:124–129.

Endocytosis and Receptor Degradation

KNUTSON, V. P., G. V. RONNETT, and M. D. LANE. 1983. Rapid, reversible internalization of cell-surface insulin receptors. *J. Biol. Chem.* 258:12139–12142.

RONNETT, G. V., G. TENNEKOON, V. P. KNUTSON, and M. D. LANE. 1983. Kinetics of insulin-receptor transit to and removal from the plasma membrane: effect of insulin-induced down regulation in 3T3-L1 adipocytes. *J. Biol. Chem.* 258:283–290.

SIBLEY, D. R., R. H. STRASSER, J. L. BENOVIC, K. DANIEL, and R. J. LEFKOWITZ. 1986. Phosphorylation/dephosphorylation of the β-adrenergic receptor regulates its functional coupling to adenylate cyclase and subcellular distribution. *Proc. Nat'l Acad. Sci. USA* 83:9408–9412.

Hormone Signaling in Yeast

BLINDER, D., S. BOUVIER, and D. D. JENNESS. 1989. Constitutive mutants in the yeast pheromone response: ordered function of the gene products. *Cell* 56:479–486.

DIETZEL, C., and J. KURJAN. 1987. The yeast SCG1 gene: a G_α-like protein implicated in the a- and α-factor response pathway. *Cell* 50:1001–1010.

*HERSKOWITZ, I., and L. MARSH. 1987. Conservation of a receptor-signal transduction system. *Cell* 50:995–996.

*LEVITZKI, A. 1988. Transmembrane signaling to adenylate cyclase in mammalian cells and in *Saccharomyces cerevisiae*. *TIBS* 13:298–303.

WHITEWAY, M., L. HOUGAN, D. DIGNARD, D. Y. THOMAS, L. BELL, G. C. SAARI, F. J. GRANT, P. O'HARA, and V. L. MACKAY. 1989. The STE4 and STE18 genes of yeast encode potential β and γ subunits of the mating factor receptor-coupled G protein. *Cell* 56:467–477.

Chemotaxis and Signaling in Cellular Slime Molds

*BONNER, J. T. 1983. Chemical signals of social amebas. *Scientific American* 248(4):114–120.

*FIRTEL, R. A., P. J. M. VAN HAASTERT, A. R. KIMMEL, and P. N. DEVREOTES. 1989. G protein-linked signal transduction pathways in development: *Dictyostelium* as an experimental system. *Cell* 58:235–239.

*GERISCH, G. 1987. Cyclic AMP and other signals controlling cell development and differentiation in *Dictyostelium*. *Ann. Rev. Biochem.* 56:853–879.

*KLEIN, P., D. FONTANT, B. KNOX, A. THEIBERT, and P. N. DEVREOTES. 1985. cAMP receptors controlling cell-cell interactions in the development of *Dictyostelium*. *Cold Spring Harbor Symp. Quant. Biol.* 50:787–799.

KLEIN, P. S., T. J. SUN, C. L. SAXE, III, A. R. KIMMEL, R. L. JOHNSON, and P. N. DEVREOTES. 1988. A chemoattractant receptor controls development in *Dictyostelium discoideum*. *Science* 241:1467–1472.

*KUMAGAI, A., M. PUPILLO, R. GUNDERSEN, R. MIAKE-LYE, P. N. DEVREOTES, and R. A. FIRTEL. 1989. Regulation and function of Gα protein subunits in *Dictyostelium*. *Cell* 57:265–275.

Plant Hormones

ADUCCI, P., A. BALLIO, J.-P. BLEIN, M. R. FULLONE, M. ROSSIGNOL, and R. SCALLA. 1988. Functional reconstitution of a proton-translocating system responsive to fusicoccin. *Proc. Nat'l Acad. Sci. USA* 85:7849–7851.

BARBIER-BRYGOO, H., G. EPHRITIKHINE, D. KLAMBT, M. GHISLAIN, and J. GUERN. 1989. Functional evidence for an auxin receptor at the plasmalemma of tobacco mesophyll protoplasts. *Proc. Nat'l Acad. Sci. USA* 86:891–895.

BLEECKER, A. B., M. A. ESTELLE, C. SOMERVILLE, and H. KENDE. 1988. Insensitivity to ethylene conferred by a dominant mutation in *Arabidopsis thaliana*. *Science* 241:1086–1089.

*BURGESS, J. 1985. *An Introduction to Plant-Cell Development*. Cambridge University Press.

ESTELLE, M. A., and C. SOMERVILLE. 1987. Auxin-resistant mutants of *Arabidopsis thaliana* with an altered morphology. *Mol. Gen. Genet.* 206:200–206.

HAGEN, G., and T. J. GUILFOYLE. 1985. Rapid induction of selective transcription by auxins. *Mol. Cell Biol.* 5:1197–1203.

JACOBSEN, J. V., and L. R. BEACH. 1985. Control of transcription of α-amylase and rRNA genes in barley aleurone protoplasts by gibberellin and abscisic acid. *Nature* 316:275–277.

KUTSCHERA, U., and P. SCHOPFER. 1985. Evidence against the acid-growth theory of auxin action. *Planta* 163:483–493.

MCCLURE, B. A., G. HAGEN, C. S. BROWN, M. A. GEE, and T. J. GUILFOYLE. 1989. Transcription, organization, and sequence of an auxin-regulated gene cluster in soybean. *Plant Cell* 1:229–239.

ROOD, S. B., D. PEARCE, P. H. WILLIAMS, and R. P. PHARIS. 1989. A gibberellin-deficient *Brassica* mutant-rosette. *Plant Physiol.* 89:482–487.

*TAIZ, L. 1984. Plant-cell expansion: regulation of cell-wall mechanical properties. *Ann. Rev. Plant Physiol.* 35:585–657.

*THEOLOGIS, A. 1986. Rapid gene regulation by auxin. *Ann. Rev. Plant Physiol.* 37:407–438.

*WAREING, P. F., and I. D. J. PHILLIPS. 1981. *Growth and Differentiation in Plants*. Pergamon. Chapters 3–5 cover most plant hormones.

*ZAMBRYSKI, P., J. TEMPE, and J. SCHELL. 1989. Transfer and function of T-DNA genes from agrobacterium Ti and Ri plasmids in plants. *Cell* 56:193–201.

*ZEEVAART, J. A. D., and R. A. CREELMAN. 1988. Metabolism and physiology of abscisic acid. *Ann. Rev. Plant Physiol.* 39:439–473.

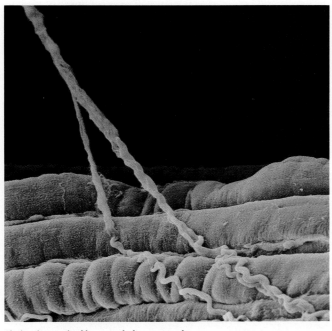

Skeletal muscle fibers and the axon of a motor neuron

C H A P T E R # 20

Nerve Cells and the Electric Properties of Cell Membranes

*T*he nervous system regulates all aspects of bodily function and is staggering in its complexity. Millions of specialized nerve cells sense features of both the external and internal environments—light, touch, pressure, sound, concentrations of many substances, pain, the stretching of muscles. They transmit this information to other nerve cells for processing and storage. Millions more nerve cells regulate the contraction of muscles and the secretion of endocrine or exocrine glands. The brain—the control center that stores, computes, integrates, and transmits information—contains about 10^{12} neurons (nerve cells), each forming as many as a thousand connections with other neurons.

Despite the complexity of the nervous system as a whole, the structure and function of individual nerve cells is understood in great detail, perhaps in more detail than for any other type of cell. We know, for instance, that electric impulses are conducted along the length of every *nerve cell and can explain this conduction in terms of specific plasma membrane proteins that regulate the flow of ions into and out of the cell. We know a great deal about the membrane proteins of certain nerve cells (e.g., the photoreceptor cells in the eye) that sense environmental signals and convert them into electric impulses. These receptor proteins, many of which have been identified and cloned, enable a nerve cell to receive a chemical or electric signal from another cell and convert it into an electric impulse for transmission along its length to the next cell. Even simple aspects of memory and learning can now be explained in terms of modifications of specific neurons.*

In this chapter we focus on how individual nerve cells function and how small groups of cells function together. Many of the nerve systems of greatest interest and importance, such as the mammalian brain, are too complex to be studied at this level with current techniques. A great deal of information has been gleaned, however, from simpler nervous systems. Squids, sea slugs, and nematodes contain large neurons that are relatively easy to identify and manipulate experimentally. Moreover, in these species, only a few identifiable neurons may be involved in a specific task; thus their function can be studied in some detail. Detailed genetic studies of Drosophila have also yielded important insights into the structure and function of nerve cells. Although many of our examples will be taken from such invertebrates, the principles involved are basic and are applicable to complex nervous systems including that of humans. ▲

Neurons, Synapses, and Nerve Circuits

The nervous system contains two principal classes of cells: *neurons* (nerve cells) and *neuroglial* (glial) cells. The latter fill the spaces between the neurons, nourishing them and modulating their function. Although neurons exist in a bewildering array of sizes and shapes, they all possess the ability to conduct an electric impulse along their length. They make specific contacts with other cells at specialized sites called *synapses*, across which signals are passed. Groups of interconnecting neurons frequently form nerve circuits for the passage of electric signals. In the peripheral nervous system, the long axons of neurons are bundled together to form a nerve (Figure 20-1).

The Neuron Is the Fundamental Unit of All Nervous Systems

Most neurons contain four distinct regions, which carry out specialized functions of the cell: the cell body, the dendrites, the axon, and the specialized axon terminals (Figure 20-2).

The *cell body* contains the nucleus and most of the ribosomes, lysosomes, and endoplasmic reticulum. It is the site of synthesis of virtually all neuronal proteins and membranes. Here newly synthesized macromolecules are assembled into membranous vesicles or multiprotein particles, which are transported to other regions of the neuron.

Most neurons contain a single *axon,* a fiber that conducts the electric impulses away from the cell body. In humans, axons may be a meter or more in length. The diameter of axons varies from a micrometer in certain nerves of the human brain to a millimeter in the giant fiber of the squid. The diameter of an axon is directly proportional to the speed with which it conducts impulses; conduction can occur up to 100 meters per second (m/s). Some axons are surrounded by a sheath of myelin (see Figures 20-1 and 13-8) made up of a stacked array of plasma membranes of Schwann cells, a type of glial cell. Myelinated axons conduct electric impulses faster than nonmyelinated ones of the same diameter.

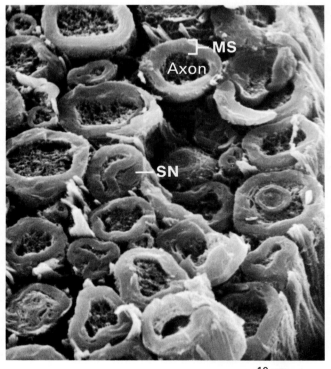

▲ **Figure 20-1** Freeze-fracture preparation of a rat sciatic nerve viewed in a scanning electron microscope. The axon of each neuron in the nerve is surrounded by a myelin sheath (MS) formed from the plasma membrane of a Schwann cell (SN). The axonal cytoplasm contains abundant filaments— mostly microtubules and intermediate filaments—that run longitudinally and serve to make the axon rigid. *From R. G. Kessel and R. H. Kardon, 1979,* Tissues and Organs: A Text-Atlas of Scanning Electron Microscopy, *W. H. Freeman and Company, p. 80.*

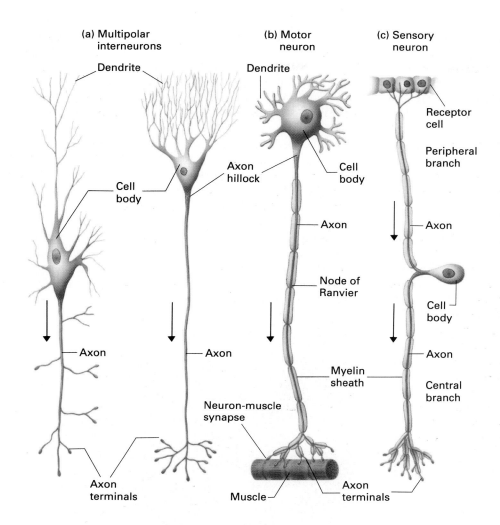

(a) Multipolar interneurons

Dendrite

Cell body

Axon

Axon terminals

(b) Motor neuron

Dendrite

Axon hillock

Cell body

Axon

Node of Ranvier

Axon

Myelin sheath

Neuron-muscle synapse

Muscle

Axon terminals

(c) Sensory neuron

Receptor cell

Peripheral branch

Axon

Cell body

Axon

Central branch

◄ **Figure 20-2** Structure of typical mammalian neurons. Arrows indicate the direction of impulse conduction in axons (red). (a) Multipolar interneurons with profusely branched dendrites (which form synapses with several hundred other neurons) and a single long axon that branches laterally and at its terminus. (b) A motor neuron that innervates a muscle cell. Typically, motor neurons have a single long axon extending from the cell body to the effector cell. In mammalian motor neurons an insulating sheath of myelin usually covers all parts of the axon except at the nodes of Ranvier and the axon terminals. (c) A vertebrate sensory neuron in which the axon branches just after it leaves the cell body. One branch carries the nerve impulse from the receptor cell to the cell body located in the dorsal root ganglion near the spinal cord; the other branch carries the impulse from the cell body to the spinal cord or brain. Both branches are structurally and functionally axons, except at their terminal portions, even though the peripheral branch conducts impulses toward, rather than away from, the cell body.

Microtubules and intermediate filaments run the length of the axon. Microtubules help to transport proteins, membrane vesicles, and other macromolecules from the cell body down the length of the axon to the terminal. This movement, called *orthograde axoplasmic transport,* is essential for the renewal of membranes and enzymes in the nerve terminal. Axonal fibers also are involved in the movement of damaged membranes and organelles up the axon toward the cell body, where they are degraded, probably in lysosomes; this process is called *retrograde transport.*

Axons are specialized for the conduction of electric impulses, called *action potentials,* along their length without diminution. An action potential originates at the *axon hillock,* the junction of the axon and cell body, and travels to the *axon terminals,* small branches of the axon from which electric signals are passed to the next neuron in a nerve circuit, to a muscle cell at a neuromuscular junction, or to any of various other types of cells.

Most neurons contain multiple *dendrites,* which extend outward from the cell body and are specialized to receive signals from sensory cells or from the axons of other neurons. Dendrites convert these signals into small electric impulses and transmit them to the cell body. Some neuronal cell bodies also contain specialized regions that can receive signals. Electric disturbances generated in the dendrites or cell body spread to the axon hillock, where action potentials originate; these are then actively conducted down the axon.

Synapses Are Specialized Sites Where Neurons Communicate with Other Cells

Synapses generally conduct signals in only one direction. An axon terminal from the *presynaptic* cell sends signals that are picked up by the *postsynaptic cell.* There are two types of synapses, *electric* and *chemical,* which differ in both structure and function. Chemical synapses are much more common than electric synapses.

Neurons communicating by an electric synapse are connected by *gap junctions* across which electric impulses can pass directly from the presynaptic cell to the postsynaptic one (Figure 20-3a). Electric synapses allow an action potential to be generated in the postsynaptic cell with greater certainty than chemical synapses and without a lag period.

(a) Electric synapse

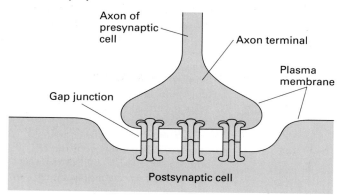

(b) Chemical synapse

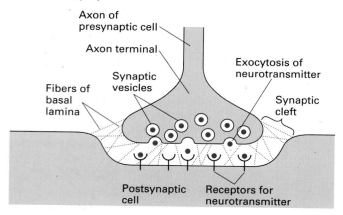

▲ **Figure 20-3** General structures of the two types of synapses. (a) In an electric synapse, the plasma membranes of the presynaptic and postsynaptic cells are linked by gap junctions. The flow of ions through these channels allows electric impulses to be transmitted directly from one cell to the other. (b) In a chemical synapse, a narrow region—the synaptic cleft—separates the plasma membranes of the presynaptic and postsynaptic cells. Transmission of electric impulses requires release of a neurotransmitter by the presynaptic cell, its diffusion across the synaptic cleft, and its binding by specific receptors on the postsynaptic membrane.

In a chemical synapse (Figure 20-3b), the axon terminal of the presynaptic cell contains vesicles filled with a particular *neurotransmitter* substance (Figure 20-4), such as epinephrine or acetylcholine. The postsynaptic cell can be a dendrite, the cell body, or the axon of another neuron, or a muscle or gland cell. When the postsynaptic cell is a muscle cell, the synapse is called a *neuromuscular junction* or *motor end plate*. When a nerve impulse reaches the axon terminal, some of the synaptic vesicles are exocytosed, releasing their contents into the *synaptic cleft*, the narrow space between the cells. The transmitter diffuses across the synaptic cleft and, after a lag period of about 0.5 ms, binds to receptors on the postsynaptic cells.

Upon binding, it induces a change in the ion permeability of the postsynaptic membrane that results in a disturbance of the plasma membrane electric potential at this point. This electric disturbance may be sufficient to induce an action potential or, depending on the type of cell, a muscle contraction or the release of hormone. In some cases, enzymes attached to the fibrous network connecting the cells destroy the chemical signal after it has functioned; in other cases, the chemical signal diffuses away or is reincorporated into the presynaptic cell.

Many nerve-nerve and nerve-muscle chemical synapses are *excitatory*. The chemical signal released by the presynaptic cell in this case causes a localized change in the plasma membrane of the postsynaptic cell—a reduction in the magnitude of the membrane potential—that tends to induce an action potential. Often, however, a nerve impulse in a presynaptic neuron leads to an increase in

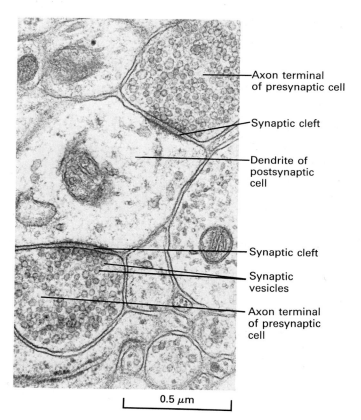

0.5 μm

▲ **Figure 20-4** Electron micrograph showing a cross section of a dendrite synapsing with two axon terminals filled with synaptic vesicles. In the synaptic region, the plasma membrane of the presynaptic cell is specialized for fusion and exocytosis of synaptic vesicles, which contain a neurotransmitter; the opposing membrane of the postsynaptic cell (in this case, a neuron) contains receptors for the neurotransmitter. *From C. Raine, 1981, in* Basic Neurochemistry, *3d ed., G. J. Siegel et al., eds., Little, Brown, p. 32. Copyright 1981, Little, Brown and Company.*

the membrane potential of the postsynaptic membrane, which prevents the generation of an action potential. This type of synapse is termed *inhibitory*.

The Decision to Fire an Action Potential Involves Summation of Electric Disturbances

A single neuron can be affected simultaneously by excitatory and inhibitory stimuli from synapses with many axons. The various excitatory and inhibitory signals—decreases or increases in the magnitude of the membrane potential—move along the plasma membrane from the synapses to the cell body and the axon hillock. Whether a neuron generates an action potential in the axon hillock depends on the balance of the timing, amplitude, and localization of all the various excitatory and inhibitory inputs it receives. Action potentials are generated when the magnitude of the membrane potential at the axon hillock decreases to a certain level called the *threshold potential*. In a sense, each neuron is a tiny computer that averages all the electric disturbances on its membrane and makes a decision whether to trigger an action potential and conduct it down the axon.

Particularly in the central nervous system, neurons have extremely long dendrites with complex branches. This allows them to form synapses with and receive signals from a large number of other neurons, perhaps up to a thousand. A single axon in the central nervous system can synapse with many neurons and induce responses in all of them simultaneously.

Often an axon terminal of one neuron will synapse with the axon terminal of another neuron (Figure 20-5).

Such a synapse may either inhibit or stimulate the ability of an axon terminal to exocytose its synaptic vesicles and signal to a postsynaptic cell. We shall see later how such synapses can enable an animal to learn.

Neurons Are Organized into Circuits

Sponges, the most primitive multicellular animals, contain contractile muscle cells that are in direct contact with the seawater. Whenever the muscles are excited by a noxious stimulus in the water, they contract and close the pores of the sponge. In sponges, which have no cells that only sense signals or only transmit an excitation, the contractile cells also function as sensory cells.

In somewhat more complex multicellular animals, muscle cells lie deep in the interior, and sensory cells, which are located at or near the surface of the animal, conduct impulses to the interior muscles. In the tentacles of the sea anemone (phylum Cnidaria), for example, one of the outer epidermal cells, specialized to receive sensory information from the environment, extends directly to a deeper muscle cell (Figure 20-6). Appropriate stimulation of the sensory receptor cell causes contraction of the underlying muscle.

In even more advanced animals, such signaling pathways consist of two or more neurons and, in some cases, highly specialized nonneural *sensory receptor cells*, which respond to specific environmental stimuli such as light, heat, stretching, pressure, and osmolarity. *Sensory neurons* lead from receptor cells to other neurons; *motor neurons* lead to muscle cells; and *interneurons* connect

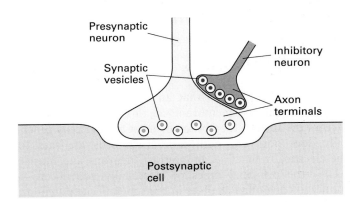

▲ **Figure 20-5**   A modulatory synapse between an inhibitory neuron and the terminal of a presynaptic neuron. Stimulation of the inhibitory neuron causes release of its transmitter substance, which reduces the presynaptic axon's ability to transmit a signal to the postsynaptic cell. This type of presynaptic inhibition does not affect the ability of the postsynaptic cell to respond to signals from other neurons.

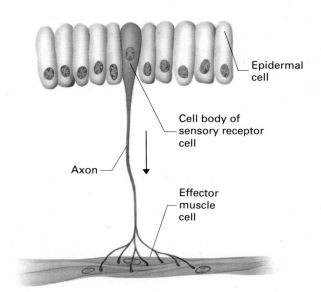

▲ **Figure 20-6**   In the simple receptor-effector system present in the tentacles of the sea anemone, a receptor cell transmits an impulse (indicated by the arrow) directly to muscle cells.

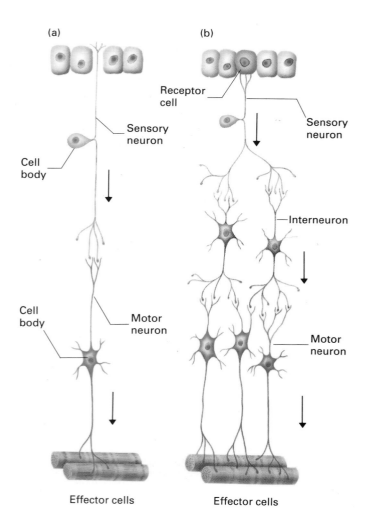

Effector cells

Effector cells

◀ **Figure 20-7** Organization of neural pathways consisting of several neurons. (a) In many invertebrates, such as the earthworm, a sensory neuron (blue) acts both as a receptor and as a conductor. In this example, the axon of a stretch receptor on the exterior of the animal forms a synapse directly with a motor neuron (red), which, in turn, stimulates muscle contraction. (b) In higher animals a separate non-neural receptor cell relays signals to a sensory neuron; inter-neurons (black) usually are interposed between sensory and motor neurons, allowing the impulse to follow multiple routes and to affect multiple effector cells. Arrows indicate direction of impulse conduction.

tered in *ganglia,* masses of nerve tissue that lie just outside the spinal cord. Ganglia also contain the cell bodies of the motor neurons that make up the autonomic nervous system, a division of the peripheral system that controls the involuntary responses of glands and smooth muscles. Most internal glands (e.g., the liver and pancreas) and most nonskeletal muscles (e.g., the heart muscle and the smooth muscles that surround the digestive tract) are innervated by two classes—parasympathetic and sympathetic—of autonomic nerves: one class stimulates the muscle or gland and the other inhibits it. The cell bodies of the motor neurons that stimulate voluntary muscles are located inside the central nervous system, in either the brain or the spinal cord. However, most of the 10^{12} neurons in the central nervous system are interneurons.

Having surveyed the general features of neuron structure, interactions, and circuits, let us turn to the mechanism by which a neuron generates and conducts electric impulses.

neurons to each other. In the earthworm, a sensory neuron synapses with one or more motor neurons, which in turn stimulate the appropriate muscle. This simple *reflex arc* (Figure 20-7a) is an example of the organization of neurons into circuits. In vertebrate reflex arcs, interneurons connect sensory and motor neurons, thus allowing one sensory neuron to affect multiple motor neurons (Figure 20-7b) and one motor neuron to be affected by multiple sensory neurons; in this way interneurons integrate and enhance reflexes. The knee-jerk reflex in humans involves a complex reflex arc in which one muscle is stimulated to contract while another is inhibited from contracting (Figure 20-8). Such circuits allow an organism to respond to a sensory input by the coordinated action of sets of muscles that together achieve a single purpose.

The nervous system of vertebrates is divided into the *central nervous system,* comprising the brain and spinal cord, and the outlying *peripheral nervous system* (Figure 20-9). Each peripheral nerve is a bundle of axons. Some axons are extensions of motor neurons, which stimulate specific muscles or glands; others are part of sensory neurons, which convey information back to the central nervous system. Sensory neurons have their cell bodies clus-

The Action Potential and Conduction of Electric Impulses

We saw in Chapter 14 that an electric potential exists across the plasma membrane of all cells. The potential across the surface membrane of most animal cells generally does not vary with time; such cells are said to be electrically inactive. In contrast, neurons and muscle cells—the principal types of electrically active cells—undergo controlled changes in their membrane potential; such changes are central to their function. Nerve cells and many muscle cells conduct electric impulses along their membrane by controlling sequential changes in the permeability of the surface membrane to Na^+ and K^+ ions. In muscle cells, a transient reduction in the membrane potential, or *depolarization,* of the surface membrane is often induced by a nerve impulse; this depolarization triggers muscle contraction.

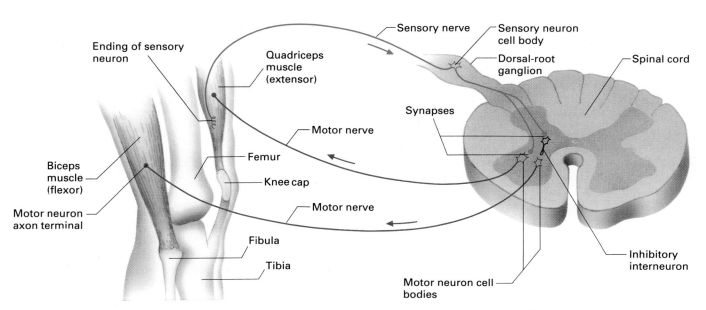

▲ **Figure 20-8** The knee-jerk reflex arc in the human. Positioning and movement of the knee joint are accomplished by two muscles that have opposite actions: contraction of the quadriceps muscle straightens the leg, whereas contraction of the biceps muscle bends the leg. The knee-jerk response, a sudden extension of the leg, is stimulated by a blow just below the knee cap. The blow stimulates sensory neurons (blue) located in the tendon of the quadriceps muscle. The axon of each sensory neuron extends from the tendon to its cell body in a dorsal root ganglion. The sensory axon then continues to the spinal cord, where it branches and synapses with two neurons: (1) a motor neuron (red) that innervates the quadriceps muscle and (2) an inhibitory interneuron (black) that synapses with a motor neuron (red) innervating the biceps muscle. Stimulation of the sensory neuron causes a contraction of the quadriceps and, via the inhibitory neuron, a simultaneous inhibition of contraction of the biceps muscle. The net result is an extension of the leg at the knee joint. (Each cell illustrated here actually represents a nerve, that is, a population of neurons.)

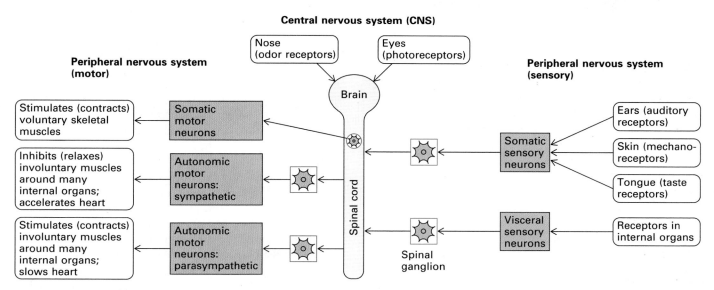

▲ **Figure 20-9** A highly schematic diagram of the vertebrate nervous system. The central nervous system (CNS) comprises the brain and spinal cord and is composed mainly of interneurons. It receives direct sensory input from the eyes and nose. The peripheral nervous system (PNS) comprises three sets of neurons: (1) somatic and visceral sensory neurons, which relay information to the CNS from receptors in somatic and internal organs; (2) somatic motor neurons, which innervate voluntary skeletal muscles; and (3) autonomic motor neurons, which innervate the heart, the smooth involuntary muscles such as those which surround the stomach and intestine, and glands such as the liver and pancreas. The two branches of the autonomic system—sympathetic and parasympathetic—frequently cause opposite effects on internal organs. The cell bodies of somatic motor neurons are within the CNS; those of sensory neurons and of autonomic motor neurons are in ganglia adjacent to the CNS.

Membrane Potentials Can Be Measured with Microelectrodes

The potential across the plasma membrane can be measured with a microelectrode inserted inside the cell and a reference electrode placed in the extracellular fluid. The two are connected to a voltmeter capable of measuring small potential differences (Figure 20-10). By convention, the potential is expressed in millivolts (mV). In virtually all cases the inside of the cell membrane is negative with respect to the outside; typical membrane potentials are −30 to −70 mV. In electrically active tissues (nerve and

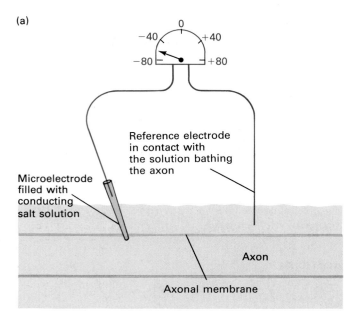

(a)

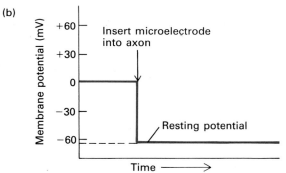

(b)

▲ **Figure 20-10** Measurement of the electric potential across an axonal membrane. (a) A microelectrode, constructed by filling a glass tube of extremely small diameter with a conducting fluid such as KCl, is inserted into an axon in such a way that the surface membrane seals itself around the electrode. A reference electrode is placed in the bathing medium. A potentiometer connecting the two electrodes registers the potential; in this case, a resting potential of −60 mV. (b) A potential difference is registered only when the microelectrode is inserted into the axon; no potential is registered if the microelectrode is in the bathing fluid.

muscle), the potential difference maintained across the cell membrane in the absence of stimulation is called the *resting potential,* and the unstimulated cell is said to be in the *resting state.*

The Action Potential Reflects the Sequential Depolarization and Repolarization of a Region of the Nerve Membrane

When an electric impulse passes down a neuron, it is observed as a movement of negative charges along the outside of the axon. For example, if a series of electrodes is placed at various points on the surface of an axon and the axon is stimulated, each electrode will register a transient negative potential on the surface of the neuron relative to the bulk extracellular fluid as the impulse moves along the axon. The magnitude of the electric potential detected is generally several millivolts. Depending on the neuron, the rate of impulse conduction varies from 1 to 100 m/s.

It would appear from the outside that negative charges—that is, electrons—simply are propelled along the outer surface of the neuron, but the situation is much more complex. What actually occurs can be followed by measuring changes in the membrane potential of a small region of the axonal plasma membrane as an electric impulse passes along it. Most of the classic studies have been done on the giant axon of the squid, in which multiple microelectrodes can be inserted without causing damage to the integrity of the plasma membrane.

In the resting state, the potential across a small region of the squid axonal membrane is −60 mV, typical for most neurons. As the action potential traverses the region, the membrane potential becomes about +35 mV, with the inside positive relative to the outside (Figure 20-11a). Because the potential shifts to a less negative state, the membrane is said to be *depolarized.* Thus there is a net potential change of almost 95 mV (35 + 60 mV). Quite rapidly thereafter, the potential returns to a negative value (−75 mV) that is slightly more negative than the resting value; that is, the membrane is *hyperpolarized.* Gradually the potential returns to the resting value. Thus the action potential is a cycle of depolarization, hyperpolarization, and return to the resting value. The cycle lasts 1–2 ms. The apparent movement of negative charges along the surface of the neuron is, in fact, the successive depolarization and repolarization of adjacent regions of the plasma membrane (Figure 20-11b). The action potential is propagated unidirectionally along the axon, from the axon hillock to the axon terminals.

All of these changes in the membrane potential can be ascribed to transient increases in the permeability (or conductance) of a region of the membrane, first to Na^+ ions, then to K^+ ions (Figure 20-11c). More specifically, these electric changes are due to transmembrane Na^+ and K^+ channel proteins that open and shut in response to changes in the membrane potential. Before proceeding

(a) Depolarization (↑) and hyperpolarization (↓)

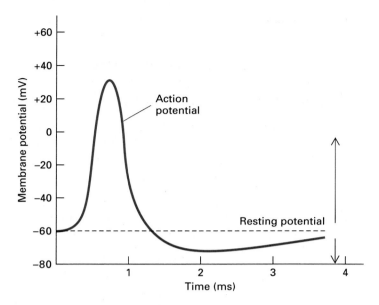

(c) Changes in ion permeabilities

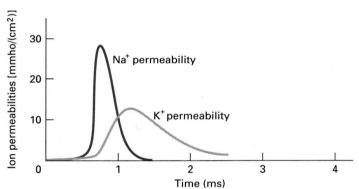

(b) Current movement

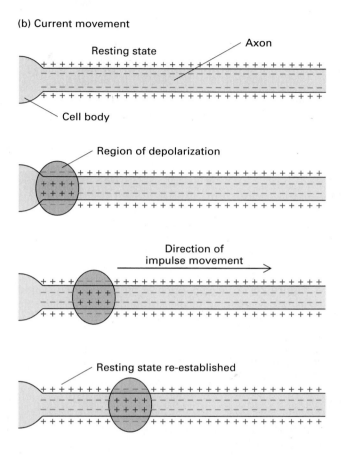

▲ **Figure 20-11** (a) Changes in membrane potential of the giant axon of a squid, following stimulation at time 0, as measured by a single microelectrode. (b) Diagram of ion current movement along the membrane of a nerve axon. The potential shifts from inside negative, characteristic of the resting state, to inside positive, characteristic of the depolarized state. Following depolarization, the resting state is re-established. (c) Changes in permeability (conductance) of the axonal membrane for Na^+ and K^+ associated with the action potential shown in (a). Increased Na^+ permeability (resulting in influx of Na^+ ions) occurs as the membrane is depolarized and precedes increased K^+ permeability (resulting in efflux of K^+ ions), which occurs as the membrane becomes hyperpolarized. [See A. L. Hodgkin and A. F. Huxley, 1952, *J. Physiol.* **117**:500.]

further with a discussion of conduction, we need to explain how changes in the membrane potential can be induced by changes in the permeability of a membrane to Na^+ or K^+ ions.

Changes in Ion Permeabilities Cause Specific, Predictable Changes in the Membrane Potential

The concentration of K^+ ions inside typical metazoan cells is about 10 times that in the extracellular fluid, whereas the concentrations of Na^+ and Cl^- ions are much higher outside the cell than inside (Figure 20-12); these concentration gradients are maintained by Na^+-K^+ ATPases with the expenditure of cellular energy. Another important property of the plasma membrane is that it is selectively permeable to different cations and anions, including the principal cellular ions Na^+, K^+, and Cl^-, so that the different ions tend to move down their concentration gradients through the plasma membrane at different rates. These two properties, *selective permeability* and *ion concentration gradients*, lead to a difference in electric potential between the inside and the outside of a cell. For instance, if the semipermeable plasma membrane

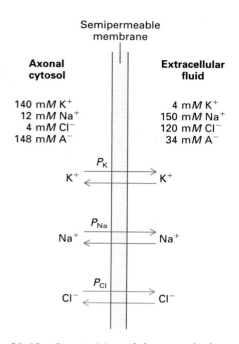

Semipermeable membrane

Axonal cytosol

140 mM K$^+$
12 mM Na$^+$
4 mM Cl$^-$
148 mM A$^-$

Extracellular fluid

4 mM K$^+$
150 mM Na$^+$
120 mM Cl$^-$
34 mM A$^-$

P_K

K$^+$ ⇄ K$^+$

P_{Na}

Na$^+$ ⇄ Na$^+$

P_{Cl}

Cl$^-$ ⇄ Cl$^-$

▲ **Figure 20-12** Composition of the cytosol of a typical vertebrate neuron and of the surrounding extracellular fluid. A$^-$ indicates negatively charged proteins, which neutralize the excess positive charges contributed by Na$^+$ and K$^+$ ions. The values of the permeability constant P for the three major ions across the resting axonal membrane are in the following order: $P_K > P_{Na} \cong P_{Cl}$.

depicted in Figure 20-12 were permeable only to K$^+$, we could use the Nernst equation (see equation 14-7) to determine the membrane potential:

$$E_K = \frac{RT}{Z\mathcal{F}} \ln \frac{K_o}{K_i}$$

$$= 59 \log_{10} \frac{K_o}{K_i} = 59 \log_{10} \frac{4}{140} = -91 \text{ mV}$$

(20-1)

where E_K is the potassium equilibrium potential, K_o and K_i are the potassium concentrations outside and inside the cell, R is the gas constant, T is the absolute temperature, $\mathcal{F}$ is the Faraday constant, and Z is the valency.

The situation in typical cells is much more complicated because there are many different ions, differing in their ability to cross the membrane, which must be considered. It is useful to define a *permeability constant* P for each ion, which is a measure of the ease with which an ion can cross a unit area (1 cm^2) of membrane driven by a 1 M difference in concentration; P is expressed in centimeters per second (cm/s). Thus P_K, P_{Na}, and P_{Cl} are the permeabilities of a unit area of membrane for flow of K$^+$, Na$^+$, and Cl$^-$ ions, respectively. In particular plasma membranes, the permeabilities for various ions are dependent on the amounts and activities of the corresponding channel proteins, which are discussed later in this chapter.

Cells, of course, contain other ions, such as HPO$_4^{2-}$, Ca^{2+}, SO$_4^{2-}$, and Mg^{2+}, but their membrane permeabilities are very small relative to those of K$^+$, Na$^+$, and Cl$^-$. Furthermore, in electrically active cells such as nerves and muscles, only K$^+$, Na$^+$, and Cl$^-$ (and occasionally Ca^{2+}) affect the membrane potential. Thus these three ions are the only ones we need consider here.

The resultant membrane potential across a cell-surface membrane is given by a more complex version of the Nernst equation in which the concentrations of the ions are weighted in proportion to their permeability constants:

$$E = \frac{RT}{\mathcal{F}} \ln \frac{P_K K_o + P_{Na} Na_o + P_{Cl} Cl_i}{P_K K_i + P_{Na} Na_i + P_{Cl} Cl_o}$$

$$= 59 \log_{10} \frac{P_K K_o + P_{Na} Na_o + P_{Cl} Cl_i}{P_K K_i + P_{Na} Na_i + P_{Cl} Cl_o}$$

(20-2)

where the "o" and "i" subscripts denote the ion concentrations outside and inside the cell. Because of their opposite charges (Z value in the Nernst equation), K$_o$ and Na$_o$ are placed in the numerator, but Cl$_o$ is placed in the denominator; conversely, K$_i$ and Na$_i$ are in the denominator, but Cl$_i$ is in the numerator. The membrane potential at any point and time can be calculated with this equation if the relevant ion concentrations and permeabilities are known.

Note that if $P_{Na} = P_{Cl} = 0$, then the membrane is permeable only to K$^+$ ions and equation 20-2 reduces to equation 20-1. Similarly, if $P_K = P_{Cl} = 0$, then the membrane is permeable only to Na$^+$ ions and equation 20-2 reduces to the following:

$$E = \frac{RT}{Z\mathcal{F}} \ln \frac{Na_o}{Na_i} = E_{Na}$$

(20-3)

Let us apply equation 20-2 to resting neurons in which the ion concentrations are typically those shown in Figure 20-12. For typical resting nerve cells (and for most nonexcitable cells), the permeability of the membrane to K$^+$ ions is much greater than that for Na$^+$ or Cl$^-$ ions ($P_K > P_{Na} \cong P_{Cl}$), and the resultant membrane potential is closer to E_K (−91.1 mV) than to E_{Na} (+64.7 mV); E_{Cl} (−87.2 mV) is close to E_K. We can see this relationship by substituting into equation 20-2 typical values of the three permeability constants ($P_K = 10^{-7}$ cm/s, $P_{Na} = 10^{-8}$ cm/s, and $P_{Cl} = 10^{-8}$ cm/s) and of the ion concentrations:

$$E = 59 \log_{10} \frac{(10^{-7})(0.004) + (10^{-8})(0.15) + (10^{-8})(0.004)}{(10^{-7})(0.14) + (10^{-8})(0.012) + (10^{-8})(0.12)}$$

$$= -52.9 \text{ mV}$$

The potential of −53 mV is close to, but less negative than, the equilibrium K$^+$ potential E_K. The potential is not equal to E_K because the membrane is also slightly permeable to Na$^+$; this tends to make the potential more positive (or less negative).

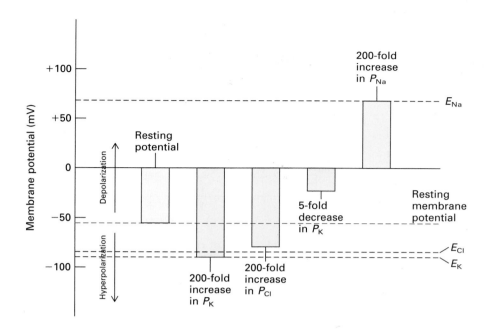

Figure 20-13 Effect of changes in ion permeability on membrane potential calculated with equation 20-2 using the permeability constants given in the text and the ion concentrations shown in Figure 20-12. The resting membrane potential is −53 mV; E_{Na}, E_K, and E_{Cl} are the potentials if the membrane were permeable only to Na^+, K^+, and Cl^-, respectively.

If the permeability constant of only one ion is changed, while all the ion concentrations and other permeabilities are unchanged, predictable changes occur in the membrane potential (Figure 20-13):

1. *Increasing P_K* causes hyperpolarization of the membrane, and the membrane potential becomes more negative, approaching E_K. Intuitively, this occurs because more K^+ ions flow outward from the cytosol, leaving excess negative ions on the cytosolic surface of the membrane and putting more positive ones on the outer surface. Conversely, *decreasing P_K* causes depolarization of the membrane and a less negative potential.

2. *Increasing P_{Na}* causes depolarization of the membrane; if the increase is large enough, the potential can become inside positive, approaching E_{Na}. Intuitively, Na^+ ions tend to flow inward from the extracellular medium, leaving excess negative ions on the outer surface of the membrane and putting more positive ions on the cytosolic surface. Conversely, *decreasing P_{Na}* causes hyperpolarization (more negative potential).

3. *Increasing P_{Cl}* causes hyperpolarization of the membrane, and the potential approaches E_{Cl}. Intuitively, Cl^- ions tend to flow inward from the extracellular medium, leaving excess positive ions on the outer surface of the membrane and putting more negative ions on the cytosolic surface. Conversely, decreasing P_{Cl} causes depolarization (less negative potential).

Let us now return to nerve conduction and see how increases or decreases in ion permeabilities—due to opening or closing of specific ion channel proteins—explains the changes in membrane potential that occur during nerve conduction of an action potential.

A Transient Increase in Sodium Permeability Depolarizes the Nerve Membrane during Conductance of an Action Potential

The sudden but short-lived depolarization of a region of the surface membrane during an action potential is caused by a sudden massive, but transient, increase in the permeability of that region to Na^+ ions (see Figure 20-11c) and a resultant influx of Na^+ ions. This increase in Na^+ permeability is caused by the opening of *voltage-dependent Na^+ channel proteins*, often called *voltage-gated Na^+ channels*. These Na^+ channels open briefly when the membrane is depolarized; more channels open as depolarization of the membrane is increased. During conduction of an action potential, the depolarization of a region of membrane spreads passively to the adjacent distal region of membrane and depolarizes it slightly, thus opening a few of the voltage-dependent Na^+ channels in this segment of the membrane.

A combination of two forces acting in the same direction drives Na^+ ions into the cell. One is the concentration gradient of Na^+ ions: the concentration is greater outside than inside. The other is the resting membrane potential—inside negative—which also tends to attract Na^+ ions into the cell. As more and more Na^+ ions enter the cell, the inside of the cell membrane becomes more positive and the membrane is depolarized further. This depolarization causes the opening of more Na^+ channels. If the initial depolarization causes the membrane potential to reach the threshold potential, it sets into motion an explosive entry of Na^+ ions, which is completed within a fraction of a millisecond. This is the origin of the action potential. For a fraction of a millisecond, at the peak of the depolarization, P_{Na} becomes vastly greater than P_K

(or P_{Cl}), and thus the membrane potential approaches E_{Na}, the equilibrium potential if the membrane were permeable only to Na^+ ions (see Figure 20-13). When the membrane potential reaches E_{Na}, further net inward movement of Na^+ ions ceases, since the concentration gradient of Na^+ ions (outside > inside) is balanced by the membrane potential E_{Na} (inside positive).

According to this explanation, the membrane potential at the peak of the action potential will equal E_{Na}. Indeed, for the squid giant axon, the measured peak value of the action potential is 35 mV (see Figure 20-11a), which is close to the calculated value of E_{Na} (55 mV) based on Na^+ concentrations of 440 mM outside and 50 mM inside. The relationship between the magnitude of the action potential and the concentration of Na^+ ions inside and outside the cell has been confirmed experimentally. For instance, if the concentration of Na^+ ions in the solution bathing the squid axon is reduced to one-third of normal, the magnitude of the depolarization is reduced by 40 mV, exactly as predicted (Figure 20-14).

Opening and Closing of Voltage-dependent Channel Proteins Change P_{Na} and P_K

Electrophysiologic studies on the squid axon and other axons have established some of the remarkable properties of voltage-gated channel proteins and helped to explain the generation and propagation of an action potential. These studies have been extended by studies of purified channel proteins incorporated into phospholipid vesicles.

The neuronal plasma membrane in invertebrates and vertebrates contains a number of individual voltage-gated Na^+ channel proteins, which can undergo cyclic conformational changes leading to the cyclic opening and closing of the Na^+ channels. In an unstimulated cell at normal resting potential, most of the Na^+ channels are closed but capable of being opened if the membrane is depolarized (Figure 20-15a and b). The greater the depolarization, the greater the chance that channels will open. Once opened, the channels stay open about 1 ms, during which time about 6000 Na^+ ions pass through them (Figure 20-15c). The channels then close spontaneously and enter an inactive state in which they cannot be opened (Figure 20-15d). When the inside-negative resting potential is reestablished, the channels return to the closed but activatable state and will reopen in response to depolarization. The inactive state of the Na^+ channel is the explanation for the *refractory period,* the period of time after firing when it is impossible for a neuron to generate or conduct an action potential. The inability of an Na^+ channel to reopen for several milliseconds limits the number of action potentials per second that a neuron can conduct.

During the time that the Na^+ channels are closing and P_{Na} is returning to its resting value, P_K increases tran-

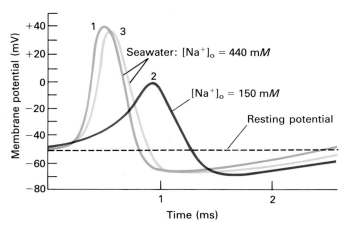

▲ **Figure 20-14** Effect of changing the external Na^+ concentration on the magnitude of the action potential in the squid giant axon. Curves 1 and 3 are controls obtained in normal seawater before and after testing the low-sodium external solution (curve 2). [See A. L. Hodgkin and B. Katz, 1949, *J. Physiol.* **108**:37.]

siently (see Figure 20-11c) due to the opening of voltage-gated K^+ channel proteins. During this period, the membrane repolarizes; its potential becomes more negative than the resting potential and, for a brief instant, approaches the potassium equilibrium potential E_K, which is more negative than the resting potential (see Figure 20-11a). The opening of the K^+ channels is induced by the membrane depolarization of the action potential. Unlike the voltage-gated Na^+ channels, the K^+ channels remain open as long as the membrane is depolarized. Because the K^+ channels open a fraction of a millisecond or so after the initial depolarization, they are called *delayed* K^+ channels. The increase in P_K accounts for the transient hyperpolarization of this region of the membrane. Eventually the K^+ and Na^+ channels close, P_K and P_{Na} return to the values characteristic of the resting state, and the membrane potential returns to its resting value.

The Na^+ and K^+ channels that determine the resting potential of the cell—the "resting" or "leakage" channels—are *not* voltage-dependent. They do not open or close during the action potential and, as judged by the effects of inhibitors, are different proteins from the voltage-gated K^+ and Na^+ channels used in generating and propagating action potentials.

The Action Potential Is Induced in an All-or-Nothing Fashion

When the plasma membrane of a neuron is depolarized, the permeability to Na^+ ions increases slightly because of the opening of a few Na^+ channels. Sodium then moves inward, driven by large concentration and potential gradients. The entry of these positive ions tends to depolarize the membrane further, thus opening more Na^+ channels

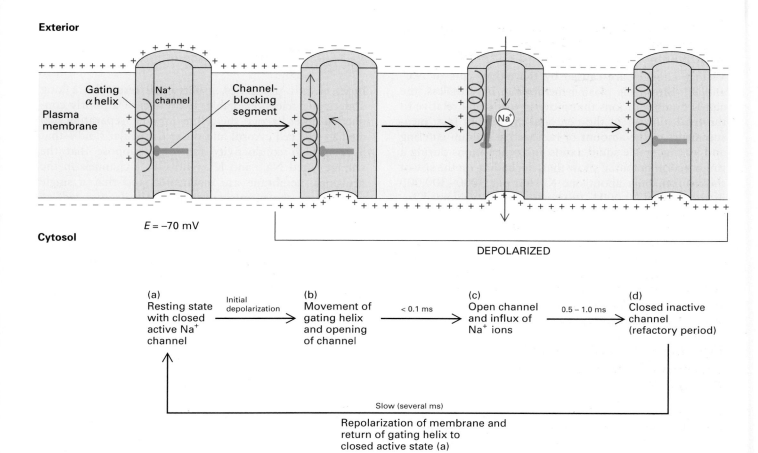

▲ **Figure 20-15** Proposed operation of voltage-gated Na⁺ channels in the invertebrate neuronal plasma membrane. In the resting state (a), a segment of the channel protein blocks the channel. The channel protein also contains four "gating" α helices (discussed later in Figure 20-24), which have positively charged side chains every fourth residue. The attraction of these charges for the negative interior of resting cells helps to ensure that the channel is closed to Na⁺ ions. When the membrane is depolarized (outside becomes negative), the gating helices move (red arrow) toward the outer plasma membrane surface (b). Within a fraction of a millisecond, the channel-blocking segment also moves, opening the channel for influx of Na⁺ ions (c). Within a millisecond after opening, the channel closes spontaneously, forming an inactive channel (d). When the membrane potential is reversed to the inside-negative state, the gating helices return to the resting position (a) and the channel is again active. Studies on certain mammalian Na⁺ channels suggest that, in contrast to the rapid opening and slow closing depicted here, voltage-dependent opening can be slow (0.5–1.0 ms) and closing more rapid. Thus different voltage-gated Na⁺ channels may have different properties.

and increasing the Na⁺ permeability even more. At the same time, however, there is an outward movement of K⁺ ions, which tends to repolarize the membrane. In fact, many neurons contain voltage-dependent K⁺ channels that transiently open immediately upon depolarization, increasing the K⁺ permeability; these are called *immediate* K⁺ channels. Their opening tends to repolarize the membrane and hinder the generation of an action potential. Whether or not an action potential is induced depends on the ratio of P_{Na} to P_K—that is, on the balance between the inward movement of Na⁺ ions, which depolarizes the membrane further, and the outward movement of K⁺ ions, which repolarizes the membrane.

If the numbers of open Na⁺ channels and inrushing Na⁺ ions are insufficient to depolarize the cell to the threshold value, no action potential will be generated. If the threshold potential is exceeded, the resulting action potential will always have the same magnitude in any particular neuron. The generation of an action potential is thus said to be all-or-nothing. Depolarization above the threshold always leads to an action potential; depolarization below it never does. The value of the threshold potential depends on the relative numbers of the resting K⁺ and Na⁺ channels and voltage-gated Na⁺ and K⁺ channels in the resting membrane. If, for instance, a particular neuron has a high permeability for K⁺ ions because its plasma membrane contains many resting K⁺ channels, then the neuron will require a greater increase in Na⁺ permeability (i.e., more voltage-gated Na⁺ channels) to induce an action potential.

The Movements of Only a Few Sodium Ions Generate the Action Potential

Voltage changes are caused by the movements of Na^+ and K^+ across the plasma membrane; nevertheless, the actual number of ions that move is very small relative to the total number in the neuronal cytosol. In fact, measurements of the amount of radioactive sodium entering and leaving single squid axons and other axons during a single action potential show that, depending on the size of the neuron, only about one K^+ ion per 3000–300,000 in the cytosol (0.0003–0.03 percent) is exchanged for extracellular Na^+ to generate the reversals of membrane polarity.

The membrane potential in nerve cells is dependent primarily on a gradient of Na^+ and K^+ ions that is generated and maintained by Na^+-K^+ ATPase. This ATPase plays no direct role in nerve conduction. If dinitrophenol or another inhibitor of ATP production is added to cells, the membrane potential gradually falls to zero as all the ions equilibrate across the membrane. This equilibration occurs extremely slowly in most cells, requiring hours. This and similar experiments indicates that the membrane potential is essentially independent of the supply of ATP over the short time spans required for nerve cells to function. Nerve cells normally can fire thousands of times in the absence of an energy supply because the ion movements during each discharge involve only a minute fraction of the cell's K^+ and Na^+ ions.

Membrane Depolarizations Spread Only Short Distances without Voltage-gated Sodium Channels

In its electric properties, a nerve cell resembles a long underwater telephone cable. It consists of a poorly conducting outer barrier, the cell membrane, separating two media—the cell cytosol and the extracellular fluid—that have a high conductivity for ions. Suppose that the voltage-gated Na^+ and K^+ (and Ca^{2+}) channels in the neuronal membrane are inactivated and that a single point along the membrane is suddenly depolarized, so that the inside of the membrane at this site has a relative excess of positive ions, principally K^+. These ions will tend to move away from the initial depolarization site, thus depolarizing adjacent sections of the membrane. This is called the *passive spread* of depolarization. The magnitude of the depolarization, however, diminishes with distance from the site of initial depolarization, as some of the excess cations leak back across the membrane, which is not a perfect insulator, through various cation channels (Figure 20-16). Only a small portion of the excess cations are carried longitudinally along the axon for long distances. The extent of this passive spread of depolarization is a function of two properties of the nerve cells: the permeability of the membrane to ions and the conductivity of the cytosol.

The conductivity of the cytosol is proportional to the concentration of ions dissolved in it, principally K^+ and

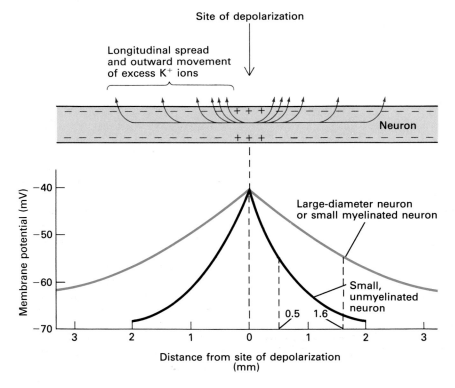

◀ **Figure 20-16** Passive spread of the depolarization in neuronal plasma membrane when voltage-gated Na^+ and K^+ channels are inactivated. The neuronal membrane is depolarized from −70 to −40 mV at a single point and clamped at this value. The voltage is then measured at various distances from this site. Because of the outward movement of K^+ ions, the extent of depolarization falls off with distance from the initial depolarization. In contrast to an action potential, passive spread occurs equally in both directions from the site of depolarization. The length constant is the distance over which the magnitude of the depolarization falls off to a value of $1/e$ ($e = 2.718$) of the initial depolarization. The length constant for a small neuron with a relatively leaky unmyelinated membrane (black curve) can be as small as 0.1 mm; in this example it is about 0.5 mm. For a large axon, or one with a nonleaky myelinated membrane (blue curve), the length constant can be as large as 5 mm; in this example it is about 1.6 mm.

Cl^-. In any particular organism the composition of the cytosol is believed to be very similar in all neurons. The conductivity of the cytosol of a nerve cell also depends on its cross-sectional area: the larger the area, the greater the number of ions there will be (per unit length of neuron) to conduct current. Thus the passive spread of a depolarization is greater for neurons of large diameter because the K^+ ions are able to move, on the average, farther along the axon before they leak back across the membrane. As a consequence large-diameter neurons passively conduct a depolarization faster and farther than thin ones. Nonetheless, passive spread of a membrane depolarization oc-

curs for only a short distance, 0.1 to about 5 mm. Depolarizations in dendrites and the cell body generally spread in this manner, though some dendrites can conduct an action potential. Neurons with very short axons also conduct axonal depolarizations by passive spread.

Voltage-gated Na^+ channels allow an action potential to be propagated long distances without loss of intensity, as depicted in Figure 20-17. The opening and closing of Na^+ channels permits propagation of an action potential in one direction only—down the axon—because the Na^+ channels are momentarily inactive after passage of the action potential (see Figure 20-15d).

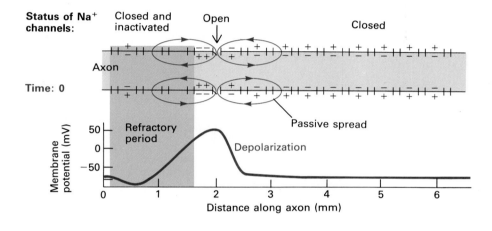

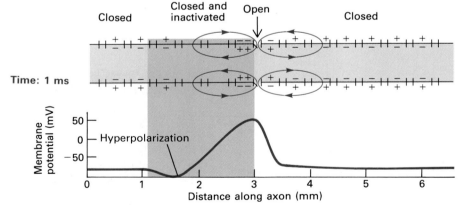

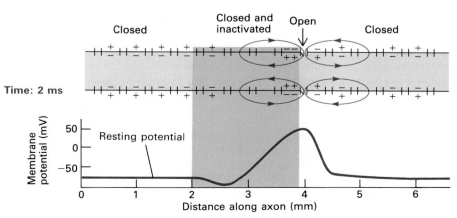

◀ **Figure 20-17** Voltage-dependent Na^+ channels allow unidirectional conduction of an action potential. Multiple microelectrodes are inserted along a squid giant axon to record the instantaneous membrane potentials at different positions along the axon during conduction of an action potential (see Figure 20-10). At time 0, an action potential is occurring at the 2-mm position along the axon. The depolarization spreads passively in both directions along the axon but triggers action potentials downstream only; for example, at 1 ms an action potential is occurring at the 3-mm position. After an action potential has occurred, the membrane at that site becomes transiently hyperpolarized owing to an increase in K^+ permeability (see the 2-mm position at 1 ms) and then returns to the resting potential (the 2-mm position at 2 ms). Each region of the membrane is refractory (inactive) for a few milliseconds after an action potential has passed; thus the action potential can be propagated in only one direction.

The velocity of impulse conduction varies among neurons. In the giant axon of the squid (0.6 mm in diameter), it is 12 m/s; thinner axons generally conduct at a slower rate. The conduction velocities for various vertebrate neurons range from 1 to 100 m/s. Myelinated axons conduct action potentials faster than nonmyelinated ones of similar diameter.

Myelination Increases the Rate of Impulse Conduction

The axons of many vertebrate nerve cells are covered with a *myelin sheath* (Figure 20-18d). Myelin is a stack of specialized plasma membrane sheets produced by a glial cell that wraps itself around the axon. In the peripheral nervous system these glial cells are called *Schwann cells;* in the central nervous system they are called *oligodendro-*cytes. Often several axons are surrounded by the cytoplasm of a single glial cell (Figure 20-18a).

In both vertebrates and most invertebrates, axons are accompanied along their length by glial cells, but specialization of these glial cells to form myelin occurs predominantly in vertebrates. Vertebrate glial cells that will later form myelin have on their surface a so-called *myelin-associated glycoprotein* that binds to adjacent axons and apparently triggers the formation of myelin.

A myelin membrane, like all membranes, contains phospholipid bilayers, but unlike many other membranes, it contains only a few types of proteins. Myelin basic protein and a proteolipid found only in myelin in the central nervous system allow the plasma membranes to stack tightly together (Figure 20-18b and c). Myelin in the peripheral nervous system is constructed of other unique membrane proteins. The myelin surrounding each

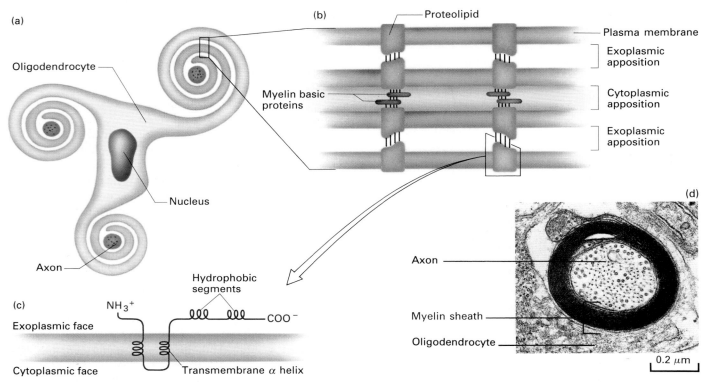

▲ **Figure 20-18**   Formation and structure of myelin sheath. (a) By wrapping itself around several axons simultaneously, a single oligodendrocyte can form a myelin sheath around multiple axons. As the oligodendrocyte continues to wrap around the axon, all the spaces between its plasma membranes, both cytoplasmic and exoplasmic, are reduced. Eventually all cytoplasm is forced out and a structure of compact stacked membranes is formed. This compaction of plasma membranes is generated by proteins that are synthesized only in myelinating cells. (b) Molecular structure of compact myelin. The close apposition of the cytoplasmic faces of the membrane may result from interactions between myelin basic protein and proteolipid and between myelin basic protein and proteolipid and between myelin basic protein and proteolipid may result from interactions between proteolipid molecules. (c) Each proteolipid molecule (276 amino acids) has two membrane-spanning α helices and two hydrophobic segments (each containing about 30 amino acids) on its exoplasmic face; the latter are thought to generate proteolipid-proteolipid interactions. (d) Electron micrograph of a cross section of a myelinated axon in the spinal cord of an adult dog. The axon is surrounded by the oligodendrocyte that produced the myelin sheath. *Part (c) adapted from L. D. Hudson et al., 1989, J. Cell Biol.* **109**:*717. Part (d) from C. Raine, 1981, in Basic Neurochemistry, 3d ed., G. J. Siegal et al., eds., Little, Brown, p. 39. Copyright 1981, Little, Brown and Company.*

(a)
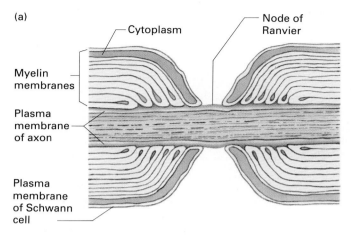
Cytoplasm — Node of Ranvier

Myelin membranes

Plasma membrane of axon

Plasma membrane of Schwann cell

(b)

Node of Ranvier 10 μm

▲ **Figure 20-19** (a) Structure of a myelinated axon near a node of Ranvier, the gap that separates the portions of the myelin sheath formed by two adjacent oligodendrocytes. These nodes are the only regions along the axon where the axonal membrane is in direct contact with the extracellular fluid. (b) A scanning electron micrograph of a peripheral myelinated nerve. The deep folds are the nodes of Ranvier. Numerous strands of collagen surround individual axons and bind them together. *Part (b) from R. G. Kessel and R. H. Kardon, 1979, Tissues and Organs: A Text-Atlas of Scanning Electron Microscopy, W. H. Freeman and Company, p. 80.*

myelinated axon is formed from many glial cells. Each region of myelin formed by an individual glial cell is separated from the next region by an unmyelinated area called the *node of Ranvier* (or simply, node); only at nodes is the axonal membrane in direct contact with the extracellular fluid (Figure 20-19).

The myelin sheath, which can be 50–100 membranes thick, acts as an electric insulator of the axon by preventing the transfer of ions between the axonal cytoplasm and the extracellular fluids. Thus all electric activity in axons is confined to the nodes of Ranvier, the sites where ions can flow across the axonal membrane. Node regions contain a high density of voltage-dependent Na^+ channels, about 10,000 per μm^2, whereas the regions of axonal membrane between the nodes have few if any channels.

The excess cytosolic positive ions generated at a node during the membrane depolarization associated with an

action potential diffuse through the axonal cytoplasm to the next node with very little loss or attenuation because ions are capable of moving across the axonal membrane only at the myelin-free nodes. Thus a depolarization at one node spreads rapidly to the next node, and the action potential "jumps" from node to node (Figure 20-20). For this reason, the conduction velocity of myelinated nerves is much greater than that of unmyelinated nerves of the same diameter. For example, a 12-μm-diameter myelinated vertebrate axon and a 600-μm-diameter unmyelinated squid axon both conduct impulses at 12 m/s. Not surprisingly, myelinated nerves are used for signaling in circuits where speed is important.

Demyelinating Diseases One of the leading causes of serious neurologic disease among human adults is multiple sclerosis (MS). This disorder, characterized by loss of myelin in areas of the brain and spinal cord, is the prototype demyelinating disease. In MS patients, conduction of action potentials by the demyelinated neurons is slowed and the Na^+ channels spread outward from the nodes. The cause of the disease is not known but appears to involve either the production by the body of antibodies

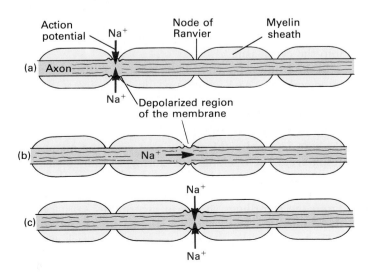

◀ **Figure 20-20** Regeneration of action potentials at the nodes of Ranvier. (a) The influx of Na^+ ions associated with an action potential at one node results in depolarization of that region of the axonal membrane. (b) Depolarization moves rapidly down the axon because the excess positive ions cannot diffuse outward across the myelinated portion of the axonal membrane. The buildup of these cations causes depolarization at the next node. (c) This depolarization induces an action potential at that node. By this mechanism the action potential jumps from node to node along the axon.

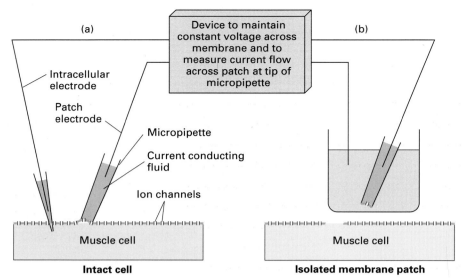

▲ **Figure 20-21** Schematic outline of the patch-clamping technique. (a) Arrangement for measuring current flow through individual ion channels in the plasma membrane of a living cell. The patch electrode is a micropipette, about 0.5 μm in diameter, filled with a current-conducting saline solution. It is applied, with a slight suction, to the plasma membrane; a region of membrane 0.5 μm in diameter contains only one or a few ion channels. The second electrode is inserted into the cytosol. The recording device measures cur-

rent flow only through the channels in the patch of plasma membrane. (b) When gentle suction is applied, the membrane patch can be isolated from the cell surface. By clamping the potential across the isolated patch and recording the flow of ion current across the patch, one can measure the effect of the membrane potential on the opening and closing of single channel proteins. The effects of the ion composition of the solution on either side of the membrane can also be examined.

that react with myelin basic protein or the secretion of proteases that destroy myelin proteins.

A well-characterized lethal mutation in mice is also characterized by a defect in myelin in the central nervous system. This recessive mutation causes, initially, a generalized tremor, followed by death at about 90 days. The affected gene (called *shiverer*) codes for myelin basic protein; the absence of this protein in mutants results in a lack of myelination in the central nervous system.

Molecular Properties of Voltage-gated Ion Channel Proteins

In this section, we describe in more detail voltage-dependent ion channel proteins, which are essential for impulse conduction in neurons.

Patch Clamps Permit Measurement of Ion Movements through Single Sodium Channels

An action potential involves the movement of ions across a large region of the nerve membrane, representing the combined action of thousands of individual voltage-gated Na^+ and K^+ channels. The study of ion channels has been advanced by the *patch-clamping*, or *single-channel recording*, technique, which is illustrated in Figure 20-21.

The movement of ions—electric current—across a small patch of isolated membrane can be measured by this technique when the membrane is electrically depolarized or hyperpolarized and maintained (clamped) at that potential. When the membrane is depolarized to a particular value, the depolarization causes an opening of Na^+ channels and an influx of Na^+ ions, which would normally lead to a further depolarization of the membrane and entry of more Na^+ ions. If the membrane potential is clamped at this depolarized voltage, however, no further depolarization can occur. The inward or outward movement of ions can then be quantified from the amount of electric current needed to maintain the membrane potential at the designated value. To preserve electroneutrality, the entry of each positive ion into the cell across the plasma membrane is balanced by the entry of an electron into the cytoplasm.

In the study depicted in Figure 20-22, patches of muscle membrane, each containing one voltage-gated Na^+ channel, were depolarized about 10 mV and clamped at that voltage. The current flow across the patch of membrane was then monitored. Under these circumstances, the transient pulses of current that cross the membrane result from the opening and closing of individual Na^+ channels. Each channel is either open or closed; there are no graded permeability changes for individual channels. From the recordings shown in Figure 20-22a, it can be determined that each channel is open for an average of 0.7 ms and that the average current through each channel is 1.6 picoamperes [1.6×10^{-12} amperes; 1 ampere =

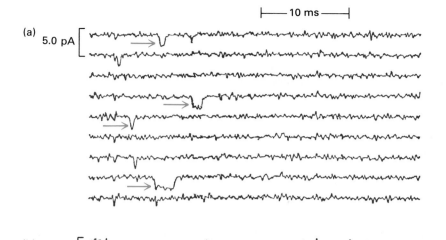

├──── 10 ms ────┤

(a) 5.0 pA

(b) 0.2 pA

◄ **Figure 20-22** (a) Tracings of current flux determined by the patch-clamping technique in nine different patches of muscle cell membrane depolarized by 10 mV and clamped at that value. The transient pulses of current (pA = picoamperes), recorded as large downward deviations (arrows), are due to the opening of Na^+ channels. The smaller deviations in current represent background noise. (b) Composite tracing representing the average of 300 individual current records. [See F. J. Sigworth and E. Neher, 1980, *Nature* **287**:447.]

1 coulomb (C) of charge per second]. This is equivalent to the movement of about 9900 Na^+ ions per channel per millisecond: $(1.6 \times 10^{-12}$ C/s$)(10^{-3}$ s/ms$)(6 \times 10^{23}$ molecules/mol$) \div 96,500$ C/mol. Assuming that all the voltage-gated Na^+ channels in these cells are identical and function independently, the sum of many recordings of individual channels after depolarization would be expected to show the same properties as the change in Na^+ permeability that occurs in a segment of an axon during an action potential (see Figure 20-11c), and indeed this has been observed experimentally (Figure 20-22b).

The Sodium Channel Protein Has Four Homologous Transmembrane Domains, Each Containing a Voltage Sensor

Patch-clamping studies have also shown that the density of Na^+ channels is very low. Depending on the type of unmyelinated axon, there are only 5–500 voltage-gated Na^+ channels per square micrometer of membrane; roughly one membrane protein molecule in a million is a voltage-gated Na^+ channel protein. Despite its low concentration, the Na^+ channel protein has been purified by use of neurotoxins that bind specifically to Na^+ channels.

Deadly neurotoxins are present in many plants and animals. Some work by blocking conduction of a nerve impulse, others by blocking synaptic transmission. Neurotoxins that bind tightly and specifically to the voltage-gated Na^+ channel protein have been extremely useful in elucidating its properties (Figure 20-23). Tetrodotoxin, for example, is a powerful poison concentrated in the ovaries, liver, skin, and intestines of the puffer fish. A toxin with related properties, saxitoxin, is produced by certain red marine dinoflagellates. These toxins specifi-

cally bind to and block the voltage-gated Na^+ channels in neurons, preventing action potentials from forming. One molecule of either toxin binds to one Na^+ channel with exquisite affinity and selectivity. Measurements of the amount of radioactive tetrodotoxin or saxitoxin that binds to a typical unmyelinated neuron have shown that an axon contains 5–500 Na^+ channels per square micrometer of membrane. This agrees with estimates of the numbers of channels obtained from patch-clamping studies. The Na^+ channels in these membranes are thus spaced, on average, about 200 nm apart.

Purification of the Na^+ channel protein has been achieved by affinity chromatography of detergent-extracted membrane proteins on columns of immobilized toxin that specifically bind Na^+ channels. Rat brain and the electroplax of the electric eel are rich and convenient sources of this protein. The major component of the Na^+ channels from these sources is a single polypeptide with a

Tetrodotoxin

Saxitoxin

▲ **Figure 20-23** Structures of two sodium-channel blockers—tetrodotoxin and saxitoxin. The positively charged groups on these neurotoxins may bind to negatively charged carboxylate groups in the Na^+ channel protein. The size of these toxins prevents their passage through the channel and leads to blockage of Na^+ transport.

(a) Voltage-gated Na⁺ channel protein

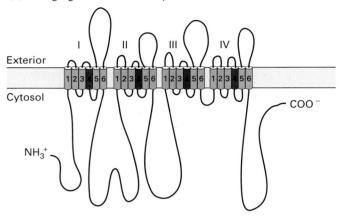

(b) Voltage-gated Ca²⁺ channel protein

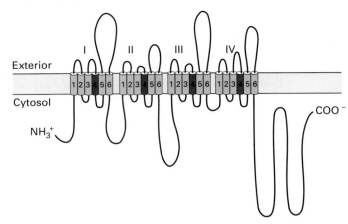

(c) Voltage-gated K⁺ channel protein

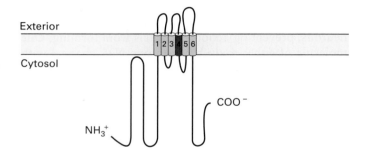

▲ **Figure 20-24** Proposed transmembrane structures of three voltage-gated channel proteins. Voltage-gated Na^+ channel proteins (a) contain 1800–2000 amino acids depending on the source. About 29 percent of the residues are identical in sequence to those in the voltage-gated Ca^{2+} channel protein (b); another 36 percent of the residues in both proteins have similar side chains. Both Na^+ and Ca^{2+} channel proteins have four homologous domains (indicated by Roman numerals), each of which is thought to contain six transmembrane α helices (indicated by Arabic numerals). One helix in each domain (no. 4) is thought to function as a voltage sensor. These gating helices have multiple arginine and lysine side chains. The shaker K^+ channel protein (c) isolated from *Drosophila* has only 616 amino acids; it is similar in sequence and transmembrane structure to each of the four domains in the Na^+ and Ca^{2+} channel proteins. The K^+ channel is thought to be a tetramer composed of four identical polypeptides, whereas a single polypeptide can generate the Na^+ or Ca^{2+} channel. *Adapted from W. A. Catterall, 1988,* Science **242**:50.

molecular weight of 250,000–270,000. Purified Na^+ channel protein(s) can be incorporated into artificial lipid membranes, which then exhibit many of the predicted properties of the voltage-gated Na^+ channel, such as voltage dependence.

A major challenge in cellular neurobiology is to determine the three-dimensional structure of the voltage-gated Na^+ and K^+ channel proteins and learn precisely how a change in voltage causes ion channels in membranes to open and close. From the cloned cDNA, the sequence of the major polypeptide of the Na^+ channel has been determined. It contains four homologous transmembrane domains, each with approximately 300 amino acids; these domains are connected and flanked by shorter stretches of nonhomologous residues (Figure 20-24a).

Since voltage-gated Na^+ channels open when the membrane depolarizes, some segment of the protein must "sense" this change in potential. Sensitive electric measurements suggest that the opening of each Na^+ channel is accompanied by the movement of four to six protein-bound positive charges from the cytoplasmic to the exo-

plasmic surface of the membrane; alternatively, a larger number of charges may move a shorter distance across the membrane. The movement of these *gating charges* (or *voltage sensors*) under the force of the electric field, is believed to trigger a conformational change in the protein resulting in channel opening. The voltage sensors are thought to be one of the six transmembrane helices present in each of the four domains of the Na^+ channel (Figure 20-24a). These gating helices have multiple positively charged side chains (Figure 20-25); when the membrane is depolarized, they are thought to move toward the exoplasmic surface of the channel (see Figure 20-15).

The role of the transmembrane gating helices in voltage sensing was demonstrated in studies with mutant Na^+ channel proteins produced by site-specific mutagenesis. In these mutant proteins, one or more arginine or lysine residues in one of the α helices is changed to neutral or acidic residues. The ability of such mutant proteins to open in response to membrane depolarization was tested in the frog oocyte expression assay outlined in Figure 20-26. As expected, when expressed in oocytes, the mutant

Voltage-gated Na⁺ channel

Helix 4:	domain I	S A L **R** T F **R** V L **R** A L **K** T I S V I P G L **K**
	domain II	G L S V L **R** S F **R** L L **R** V F **K** L A **K** S W P
	domain III	G A I **K** S L **R** T L **R** A L **R** P L **R** A L S **R** F E
	domain IV	**R** V I **R** L A **R** I G **R** I L **R** L I **K** G A **K** G I **R**

Voltage-gated Ca²⁺ channel

Helix 4:	domain I	**K** A L **R** T F **R** V L **R** P L **R** V L S G V P S L Q
	domain II	L G I S V L **R** C I **R** L L **R** L F **K** I T **K** Y W T
	domain III	S V V **K** I L **R** V L **R** A L **R** P L **R** A I N **R** A **K**
	domain IV	I S S A F F **R** L F **R** V M **R** L I **K** L L S **R** A E

Voltage-gated K⁺ channel

Helix 4	**R** V I **R** L V **R** V F **R** I F **K** L S **R** H S **K** G L Q

◀ **Figure 20-25** Amino acid sequences of the gating helix (helix 4) of voltage-gated ion channel proteins. One α-helix 4 is present in each of the four transmembrane domains in the Na⁺ and Ca²⁺ channel polypeptides; one is present in each K⁺ channel polypeptide. These helices have a positively charged lysine (K) or arginine (R) every third or fourth residue (red) whose side chains tend to be localized on one face of an α helix. Amino acids are represented by the single-letter code. *Adapted from W. A. Catterall, 1988,* Science **242**:50.

(a)

(b)

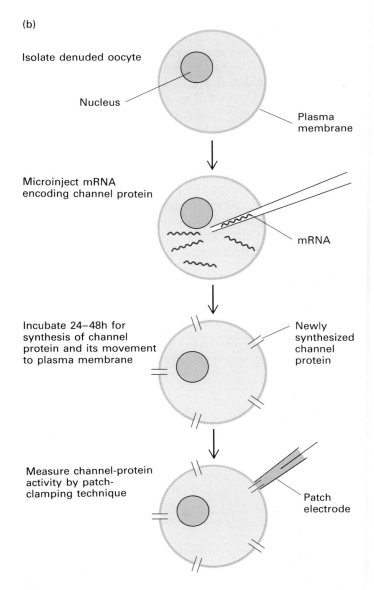

▲ **Figure 20-26** Expression of a channel protein in a cell that does not normally contain it provides a way to study the activity of normal and mutant channel proteins. A convenient cell for such studies is the oocyte from the frog *Xenopus laevis* (a). Messenger RNA encoding the channel protein under study is produced in a cell-free transcription reaction using the cloned gene as a template, or a mixture of mRNAs is directly isolated from a tissue. (b) A follicular oocyte is first treated with collagenase to remove the surrounding follicle cells, leaving a denuded oocyte. After microinjection of the mRNA and incubation, the activity of the channel protein is determined by the patch-clamping technique. Mutant channel proteins tested in this system generally exhibit altered properties (e.g., opening time and conductance), whereas the wild-type protein exhibits properties identical to those of the protein in an intact axonal membrane. *Part (a) courtesy of S. Holwill/Company of Biologists Ltd. Part (b) adapted from T. P. Smith, 1988,* Trends Neurosci. **11**:250.

proteins exhibited altered opening responses to depolarization voltages. In particular, the fewer positively charged residues present in the gating helices, the larger the depolarization required to open the Na⁺ channels.

Shaker Mutants in *Drosophila melanogaster* Led to the Identification of a Voltage-gated Potassium Channel Protein

Many mutations of the fruit fly *Drosophila melanogaster* affect specific physiologic functions by causing abnormalities in the nervous system. One such mutation, the X-linked *shaker* mutation, causes a defect in a K⁺ channel protein. Shaker mutants shake vigorously under ether anesthesia; the shaking behavior reflects a loss of motor control and a defect in excitable cells. The axons of giant nerves in certain mutants at the shaker locus have an abnormally prolonged action potential (Figure 20-27) resulting from faulty repolarization of the membrane because of a defect in the K⁺ channels that normally open immediately upon depolarization.

The wild-type shaker gene has been cloned (see Figure 20-24c) and used as a template to produce shaker mRNA in a cell-free system. Testing of this mRNA in the frog oocyte expression assay (see Figure 20-26b) has demonstrated conclusively that the shaker gene encodes a voltage-gated K⁺ channel protein. At least five different shaker polypeptides are produced by alternative splicing of the primary transcript of the shaker gene. In the oocyte expression assay, these polypeptides exhibit different voltage dependencies and K⁺ conductivities. Thus differential expression of the shaker gene can affect the properties of the action potential in different neurons.

All Voltage-gated Ion Channel Proteins Probably Evolved from a Common Ancestral Channel-Protein Gene

The similarities among the voltage-gated Na⁺, Ca²⁺, and K⁺ channels suggest that all three proteins evolved from a common ancestral gene. These similarities include the following:

1. Na⁺, K⁺, and Ca²⁺ channels all open when the membrane is depolarized.
2. The Na⁺ and Ca²⁺ channel proteins have extensive sequence homology throughout their length, and each contains four transmembrane domains with six helices per domain. The sequence and transmembrane structure of the much smaller shaker K⁺ channel protein from *Drosophila* are similar to those of each domain in the Na⁺ and Ca²⁺ channels (see Figure 20-24). The K⁺ channel is thought to comprise four copies of this polypeptide.
3. The voltage-sensing helices in all three ion channels have a positively charged lysine or arginine every third or fourth residue (see Figure 20-25). These helices are thought to move as the membrane is depolarized.

Voltage-gated K⁺ channels have been found in all yeasts and protozoa studied. In contrast, voltage-gated Ca²⁺ channels are present in only a few of the more complex protozoa, such as *Paramecium*, and voltage-gated Na⁺ channels occur only in multicellular organisms. Thus voltage-gated K⁺ channel proteins probably arose first in evolution. The Ca²⁺ and Na⁺ channel proteins are believed to have evolved by repeated duplication of an ancestral one-domain K⁺ channel gene.

Synapses and Impulse Transmission

As noted earlier, synapses are the junctions where neurons pass signals to target cells, which may be other neurons, muscle cells, or gland cells. Most neuron-neuron signaling and all known neuron-muscle and neuron-gland signaling involve the release at the synapse by the neuron of chemical neurotransmitters that act on the target cell. Much rarer, but simpler in function, are synapses in which the action potential is transmitted directly and very rapidly from the presynaptic to the postsynaptic cell.

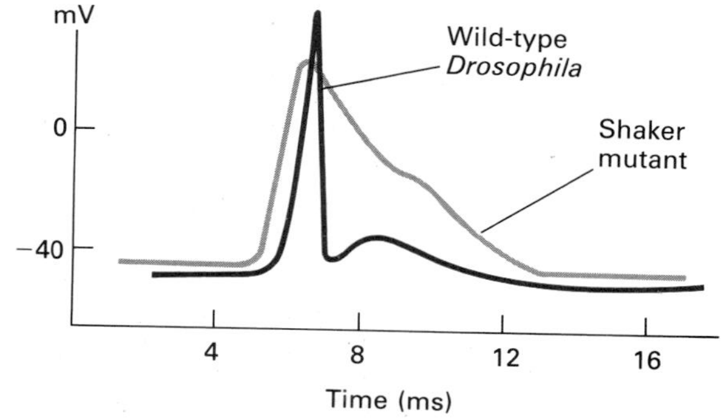

◀ **Figure 20-27** Action potential in axons of wild-type *Drosophila* and shaker mutants. The shaker mutants exhibit an abnormally prolonged action potential because of a defect in the K⁺ channel protein that is required for normal repolarization. [See L. A. Salkoff and R. Wyman, 1983, *Trends Neurosci.* **6**:128.]

Nearly Instantaneous Impulse Transmission Occurs across Electric Synapses

In an electric synapse, ions move directly from one neuron to another by means of gap junctions (see Figure 20-3a). The membrane depolarization associated with an action potential in the presynaptic cell passes through the gap junctions, leading to a depolarization, and thus an action potential, in the postsynaptic cell. Such cells are said to be electrically coupled.

Electric synapses have the advantage of speed; the direct transmission of impulses avoids the delay of about 0.5 ms that is characteristic of chemical synapses (Figure 20-28). In certain circumstances, a fraction of a millisecond can mean the difference between life and death. Electric synapses in the goldfish brain, for example, mediate a reflex action involved in strong flapping of the tail, which permits a fish to escape from predators. Examples also exist of electric coupling between groups of cell bodies and dendrites, ensuring simultaneous depolarization of an entire group of coupled cells. The large number of electric synapses in many cold-blooded fishes suggests that they may be an adaptation to low temperatures, as the lowered rate of cellular metabolism in the cold reduces the rate of impulse transmission across chemical synapses.

The efficiency with which an electric signal is transmitted across an electric synapse is proportional to the number of gap junctions that connect the cells. The permeability of the gap junction is regulated by the level of cellular

H^+ ions and possibly of Ca^{2+} ions; changes in the concentration of these ions might modulate the efficiency of impulse transmission at electric synapses.

Chemical Synapses Can Be Excitatory or Inhibitory and Can Exhibit Signal Amplification and Computation

In a chemical synapse, neurotransmitters are stored in small membrane-bounded vesicles, called synaptic vesicles, in the axon terminals of the presynaptic cell (see Figure 20-3b). The arrival of an action potential at an axon terminal causes a rise in the cytosolic Ca^{2+} concentration, which triggers exocytosis of the synaptic vesicles and release of transmitter. Each synaptic vesicle is believed to contain a single type of neurotransmitter, but a single axon terminal can contain two or more types of vesicles, each with a different transmitter. To effect a signal, the transmitter diffuses from the presynaptic cell across the synaptic cleft to the postsynaptic cell, where it binds to specific receptors on the plasma membrane, causing a change in its permeability to certain ions.

Many nerve-nerve and most nerve-muscle chemical synapses are *excitatory*. The neurotransmitter released by the presynaptic cell in this case causes a localized change in the permeability of the plasma membrane of the postsynaptic cell such that the membrane becomes depolarized; this depolarization promotes the generation of an action potential (Figure 20-29a). At an *inhibitory* synapse, binding of the neurotransmitter causes a change in ion permeability that tends to block the generation of

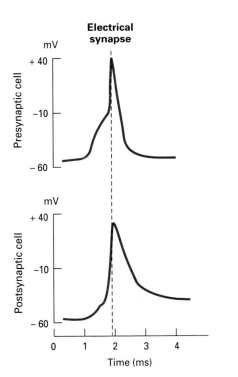

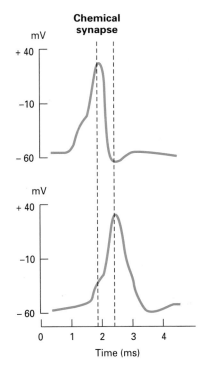

◀ **Figure 20-28** Transmission of action potentials across electric and chemical synapses. In both cases, the presynaptic neuron was stimulated and the membrane potential was measured in both the presynaptic and postsynaptic cells (see Figure 20-10a). Signal transmission across an electric synapse occurs within a few microseconds because ions flow directly from the pre- to the postsynaptic cell. In contrast, signal transmission across a chemical synapse is delayed about 0.5 ms—the time required for secretion and diffusion of neurotransmitter and the response of the postsynaptic cell to it.

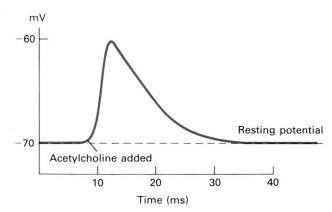

(a) Excitatory synapse

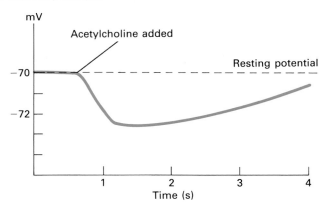

(b) Inhibitory synapse

▲ **Figure 20-29** Excitatory and inhibitory postsynaptic action potentials. (a) Application of acetylcholine (or nicotine) to frog skeletal muscle produces a rapid postsynaptic depolarization of about 10 mV, which lasts 20 ms. These cells contain nicotinic acetylcholine receptors, a ligand-gated Na^+ and K^+ channel protein. (b) In contrast, application of acetylcholine (or muscarine) to frog heart muscle produces, after a lag period of about 40 ms (not visible in graph), a hyperpolarization of 2–3 mV, which lasts several seconds. These cells contain muscarinic acetylcholine receptors. Thus depending on the type of receptor present in the postsynaptic cell, acetylcholine can either increase muscle contraction (skeletal muscle) or decrease it (cardiac muscle). Note the difference in time scales in the two graphs. [See H. C. Hartzell, 1981, *Nature* **291**:539.]

an action potential in the postsynaptic cell. In most cases the binding of an inhibitory neurotransmitter causes a hyperpolarization of the postsynaptic membrane (Figure 20-29b).

Chemical synapses have two important advantages over electric ones. The first is signal amplification, which is common at nerve-muscle synapses. An action potential in a single presynaptic motor neuron can cause contraction of multiple muscle cells because release of relatively few signaling molecules at a synapse is all that is required to stimulate contraction. The second advantage is signal computation, which is common at synapses involving interneurons, especially in the central nervous system. Many interneurons can receive signals at multiple excitatory and inhibitory synapses. Some of the resultant changes in ion permeability are short-lived—a millisecond or less; others may last several seconds. Whether an action potential is generated is a complex function of all of the incoming signals; this signal computation differs for each type of interneuron.

Many Chemicals Function as Neurotransmitters

Some of the many substances known or suspected to function as neurotransmitters are listed in Tables 20-1 and 20-2. Except for *acetylcholine,* they are either amino acids, derivatives of amino acids, or small peptides. ATP, adenosine, and other purine-containing substances also function as neurotransmitters in certain synapses.

To establish that a given substance is the neurotransmitter at a particular synapse, several criteria must be met. First, synaptic vesicles in axon terminals of the presynaptic neuron must contain the substance and release it at the appropriate time in response to stimulation in sufficient quantity to induce the appropriate response in the postsynaptic cell. Since individual neurons may contain only 10^{-14} mol of a particular transmitter, sensitive microchemical or immunochemical techniques must be used to detect it. Second, introduction of the substance into the synaptic cleft must induce the same response as does stimulation of the presynaptic nerve. Finally, the substance must be removed or degraded rapidly, resulting in restoration of the resting membrane potential.

Neurotransmitter Receptors Are Coupled to Ion Channels in Different Ways

Some neurotransmitter receptors are ligand-dependent ion channels. That is, an ion channel is part of the receptor protein, and binding of the neurotransmitter causes a conformational change in the protein that opens the channel and allows specific ions to cross the membrane. Examples include nicotinic acetylcholine receptors, which form a Na^+ and K^+ channel and which transduce an excitatory signal, and receptors for the neurotransmitters γ-aminobutyric acid and glycine, which form Cl^- channels and which transduce an inhibitory signal (Table 20-3).

More often, however, the relationship between ligand binding and the opening or closing of an ion channel is indirect. In this case, binding of a neurotransmitter to its receptor in the postsynaptic cell causes opening or closing of a separate ion channel protein. Most such receptors are

Table 20-1 Some small molecules identified as neurotransmitters

Name	Structure	Derivation or group
Acetylcholine	$CH_3-\overset{\overset{\displaystyle O}{\|\|}}{C}-O-CH_2-CH_2-N^+-(CH_3)_3$	
Glycine	$H_3N^+-CH_2-\overset{\overset{\displaystyle O}{\|\|}}{C}-O^-$	Amino acid
Glutamate	$H_3N^+-CH-CH_2-CH_2-\overset{\overset{\displaystyle O}{\|\|}}{C}-O^-$ $\underset{\overset{\|}{O^-}}{\overset{\|}{C=O}}$	Amino acid
Dopamine	HO, HO ring $-CH_2-CH_2-NH_3^+$	Derived from tyrosine
Norepinephrine	HO, HO ring $-\underset{OH}{CH}-CH_2-NH_3^+$	Derived from tyrosine
Epinephrine	HO, HO ring $-\underset{OH}{CH}-CH_2-NH_2^+-CH_3$	Derived from tyrosine
Octopamine	HO ring $-\underset{OH}{CH}-CH_2-NH_3^+$	Derived from tyrosine
Serotonin (5-hydroxytryptamine)	HO indole ring $-CH_2-CH_2-NH_3^+$	Derived from tryptophan
β-Alanine	$H_3N^+-CH_2-CH_2-\overset{\overset{\displaystyle O}{\|\|}}{C}-O^-$	Derived from aspartate
Histamine	$HC=C-CH_2-CH_2-NH_3^+$ imidazole ring (N, NH, CH)	Derived from histidine
γ-Aminobutyric acid (GABA)	$H_3N^+-CH_2-CH_2-CH_2-\overset{\overset{\displaystyle O}{\|\|}}{C}-O^-$	Derived from glutamate
Taurine	$H_3N^+-CH_2-CH_2-\overset{\overset{\displaystyle O}{\|\|}}{\underset{\underset{\displaystyle O}{\|\|}}{S}}-O^-$	Derived from glycine

Table 20-2 Some neuropeptides that function as neurotransmitters

Name	Structure*
β-Endorphin	Tyr-Gly-Gly-Phe-Met-Thr-Ser-Glu-Lys-Ser-Gln-Thr-Pro-Leu-Val-Thr-Leu-Phe-Lys-Asn-Ala-Ile-Ile-Lys-Asn-Ala-Tyr-Lys-Lys-Gly-Glu
Met-enkephalin	Tyr-Gly-Gly-Phe-Met
Leu-enkephalin	Tyr-Gly-Gly-Phe-Leu
Somatostatin	Ala-Gly-Cys-Lys-Asn-Phe-Phe-Trp-Lys-Thr-Phe-Thr-Ser-Cys
Luteinizing hormone–releasing hormone (LHRH)	pGlu-His-Trp-Ser-Tyr-Gly-Leu-Arg-Pro-Gly-NH$_2$
Thyrotropin-releasing hormone (TRH)	pGlu-His-Pro-NH$_2$
Substance P	Arg-Pro-Lys-Pro-Glu-Glu-Phe-Phe-Gly-Leu-Met-NH$_2$
Neurotensin	pGlu-Leu-Tyr-Glu-Asn-Lys-Pro-Arg-Arg-Pro-Tyr-Ile-Leu
Angiotensin I	Asp-Arg-Val-Tyr-Ile-His-Pro-Phe-His-Leu
Angiotensin II	Asp-Arg-Val-Tyr-Ile-His-Pro-Phe
Vasoactive intestinal peptide	His-Ser-Asp-Ala-Val-Phe-Thr-Asp-Asn-Tyr-Thr-Arg Leu-Arg-Lys-Glu-Met-Ala-Val-Lys-Lys-Tur-Leu-Asn Ser-Ile-Leu-Asn-NH$_2$

*An NH$_2$ after the carboxyl-terminus residue indicates it is modified to an amide; a "p" at the beginning indicates that the glutamate has been cyclized to the "pyro" form.

SOURCE: H. Gainer and M. J. Brownstein, 1981, in *Basic Neurochemistry*, 3d ed., G. J. Siegel et al., eds., Little, Brown, chap. 14.

Table 20-3 Neurotransmitter receptors that have been cloned

Ligand/receptor type	Ion channel	Subunit composition
RECEPTORS WITH LIGAND-GATED ION CHANNELS		
Acetylcholine (nicotinic receptor):		
Rat embryonic muscle	Na$^+$/K$^+$	$\alpha_2\beta\gamma\delta$
Rat adult muscle	Na$^+$/K$^+$	$\alpha_2\beta\epsilon\delta$
Rat neuronal	Na$^+$/K$^+$	$\alpha_2\beta_3$
GABA (γ-aminobutyric acid)	Cl$^-$	$\alpha_3\beta_2$
Glycine	Cl$^-$	$\alpha_3\beta_2$
RECEPTORS COUPLED TO G PROTEINS		
Epinephrine and norepinephrine:		
α_1-adrenergic		
α_2-adrenergic		
β_1-adrenergic		
β_2-adrenergic		
Serotonin		
Substance K		
Angiotensin		
Acetylcholine (muscarinic receptor):		
M1*		
M2 (cardiac)†		
M3*		
M4†		
M5*		

*Coupled to phosphatidylinositol hydrolysis.
†Inhibits adenylate cyclase; M2 also opens K$^+$ channels.

SOURCE: H. Lester, 1988, *Science* **241**:1057; D. Langosh et al., 1988, *Proc. Nat'l Acad. Sci. USA* **85**:7394; R. R. Jackson et al., 1988, *Nature* **335**:437; T. I. Bonner, 1989, *Trends Neurosci.* **12**:148; and E. Barnard, 1988, *Nature* **335**:301.

thought to be coupled to G proteins (Table 20-3). In some cases, the receptor-activated G protein activates adenylate cyclase or phospholipase C, triggering a rise in cytosolic cAMP or Ca^{2+}, respectively, which in turn affects a channel protein. In others, the receptor-activated G protein directly binds to an ion channel protein, opening (or closing) the particular channel.

Neuron-neuron and neuron-muscle synapses in which acetylcholine is the neurotransmitter are termed *cholinergic* synapses. Many types of cholinergic receptors can be distinguished by their responses to different acetylcholine agonists, such as nicotine and muscarine (a mushroom alkaloid), which cause the same type of response as acetylcholine in various cell types. We now know that these receptors are fundamentally different in structure and function, even though they all bind acetylcholine. For example, nicotine causes excitatory responses lasting only milliseconds when it binds to acetylcholine receptors on striated muscle cells (see Figure 20-29a). Such responses are called nicotinic responses, and the receptors are called *nicotinic acetylcholine receptors*. As noted already, these receptors are ligand-gated, or ligand-triggered, channels for Na$^+$ and K$^+$ ions. In contrast, when muscarine binds to acetylcholine receptors present in heart muscle and other cells, it causes responses lasting seconds (see Figure 20-29b). Such muscarinic responses are mediated by *muscarinic acetylcholine receptors*, which are usually inhibitory and are coupled, via G proteins, to ion channel proteins.

The diversity of neurotransmitters and their receptors is extensive. In the next section, we will discuss in detail the muscle nicotinic acetylcholine receptor and the events that occur at synapses containing this receptor. In the following section, we briefly consider the properties of other neurotransmitters and their receptors.

Synaptic Transmission and the Nicotinic Acetylcholine Receptor

The structure and function of cholinergic synapses containing nicotinic acetylcholine receptors are described in this section. Such synapses are prototypes for chemical synapses using many other neurotransmitters.

Acetylcholine Is Synthesized in the Cytosol and Stored in Synaptic Vesicles

The membrane-bounded vesicles that store acetylcholine are about 40 nm in diameter and accumulate, often in rows, in presynaptic axon terminals near the plasma membrane (Figures 20-30 and 20-31). A single axon terminal of a frog motor neuron may contain a million or more synaptic vesicles, each possibly containing 1000–10,000 molecules of acetylcholine; such a neuron might form synapses with a single skeletal muscle cell at several hundred points.

Acetylcholine is synthesized in the cytosol of axon terminals from acetyl coenzyme A (CoA) and choline by the enzyme choline acetyltransferase:

$$CH_3-\overset{\overset{\displaystyle O}{\|}}{C}-S-CoA + HO-CH_2-CH_2-N^+-(CH_3)_3 \longrightarrow$$

Acetyl CoA   Choline

$$CH_3-\overset{\overset{\displaystyle O}{\|}}{C}-O-CH_2-CH_2-N^+-(CH_3)_3 + CoA-SH$$

Acetylcholine

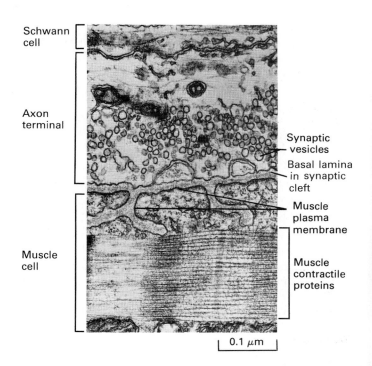

▲ **Figure 20-30** Longitudinal section through a frog nerve-muscle synapse. The plasma membrane of the muscle cell is extensively folded. The axon terminal is filled with synaptic vesicles lying just inside the plasma membrane. The basal lamina lies in the synaptic cleft separating the neuron from the muscle membrane. The axon terminal is surrounded by a Schwann cell, which periodically interdigitates between the terminal and the muscle. *From J. E. Heuser and T. Reese, 1977, in* Cellular Biology of Neurons, *E. R. Kandel, ed.,* The Nervous System, vol. 1., Handbook of Physiology, *Williams & Wilkins, p. 266.*

(a)

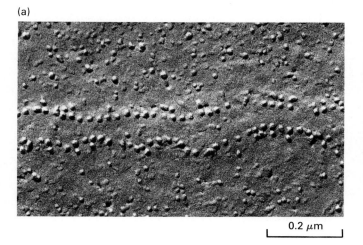

(b)

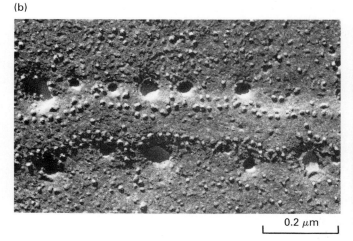

▲ **Figure 20-31** Freeze-fracture image of the axonal plasma membrane at a neuron-muscle synapse in the resting state and during stimulation. (a) In the resting state, the membrane contains rows of particles that are aligned with rows of synaptic vesicles. The function of these particles is not known, but it is suspected that they are voltage-dependent Ca^{2+} channels. (b) During stimulation, large pits in the membrane, resulting from exocytosis of synaptic vesicles, are visible near the rows of particles. *From J. E. Heuser and T. Reese, 1977, in* Cellular Biology of Neurons, *E. R. Kandel, ed.,* The Nervous System, vol. 1., Handbook of Physiology, *Williams & Wilkins, p. 268.*

The synaptic vesicles take up and concentrate acetylcholine from the cytosol against a steep concentration gradient. How this active transport is brought about is not known.

Exocytosis of Synaptic Vesicles Is Triggered by Opening of Voltage-gated Calcium Channels and a Rise in Cytosolic Calcium

Acetylcholine is released into the synaptic cleft by fusion of the vesicle membranes with the plasma membrane. As is the case for secretory processes in other cells, exocytosis of these vesicles is triggered by a rise in cytosolic Ca^{2+}. This, in turn, is triggered by the opening of voltage-gated Ca^{2+} channels in the membrane of the axon terminal and an influx of Ca^{2+} ions from the extracellular medium; the Ca^{2+} channels are opened by the membrane depolarization caused by arrival of an action potential in the axon terminal.

The presence of voltage-gated Ca^{2+} channels in axon terminals has been demonstrated in neurons treated with drugs that block Na^+ channels and thus prevent conduction of action potentials. When the membrane of axon terminals in such treated cells is artificially depolarized, an influx of Ca^{2+} ions into the neurons occurs. The amount of Ca^{2+} that enters an axon terminal through voltage-gated Ca^{2+} channels is sufficient to raise the cytosolic Ca^{2+} level to $0.5-1.0$ μM, enough to trigger synaptic vesicle exocytosis. The extra Ca^{2+} ions are rapidly pumped out of the cell by Ca^{2+} ATPases, lowering the cytosolic Ca^{2+} level and preparing the terminal to respond again to an action potential.

Patch-clamping experiments show that voltage-gated Ca^{2+} channels, like voltage-gated Na^+ channels, open transiently upon depolarization of the membrane. One voltage-gated Ca^{2+} channel has been purified and cloned. Its sequence and structure are similar to those of voltage-gated Na^+ channels (see Figure 20-24b).

More generally, voltage-gated Ca^{2+} channels can be viewed as the *transducer* of the electric signal in a nerve membrane; that is, the action potential is transduced into a chemical signal. Depolarization (or hyperpolarization) of the plasma membrane cannot, in itself, affect the activity of enzymes or other proteins in the cytosol; the electric signal must be converted into a rise in Ca^{2+} concentration in order for cell metabolism to be affected.

Synaptic Vesicle Exocytosis and Endocytosis Are Ordered Processes

The fusion of the synaptic vesicles with the plasma membrane at the synapse results in an expansion of the area of the surface membrane. As in other secretory systems, the membrane components of the synaptic vesicles are recycled, both to preserve the vesicle membrane and to limit the surface area of the cell. Such recycling is rapid, as indicated by the ability of neurons to fire many times a second, and quite specific, in that several proteins unique to the synaptic vesicles are specifically internalized by endocytosis (Figure 20-32). For example, synaptic vesicles purified from many types of neurons contain an inte-

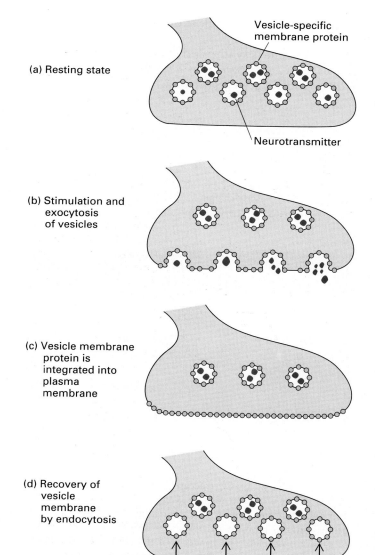

◀ Figure 20-32 Model of recycling of synaptic vesicles in axon terminals. In resting neurons, the vesicle membranes contain unique proteins (green), such as synaptophysin. These proteins are specifically recovered by endocytosis during the recovery period. It is not yet clear whether noncoated vesicles or clathrin-coated vesicles are used in this endocytic process or whether endocytosis occurs near the sites of exocytosis or elsewhere.

gral protein termed *synaptophysin.* Six synaptophysin polypeptides (MW 38,000) form an ion channel in the synaptic vesicle membrane that may be involved in uptake of neurotransmitters from the cytosol. Since synaptophysin is not found on the synaptic plasma membrane of resting neurons, but is found after stimulation, it is thought to be recycled specifically into vesicles.

The axon terminal exhibits a highly organized arrangement of plasma membrane, synaptic vesicles, and cytoskeletal fibers that appears essential for directed exocytosis of the vesicles to the plasma membrane at the synaptic cleft (Figure 20-33). The vesicles themselves are linked together by the fibrous phosphoprotein *synapsin I,* which is a substrate for cAMP-dependent and calcium-calmodulin–dependent protein kinases. Synapsin I is localized in the cytoplasmic surface of all synaptic vesicle membranes and constitutes 6 percent of vesicle proteins; it is related in structure to band 4.1 protein in erythrocyte membranes, a protein that binds both actin and spectrin (see

Figure 13-38). Thicker filaments composed of a spectrin-like protein radiate from the plasma membrane along the synaptic cleft and bind to vesicle-associated synapsin. Probably these interactions keep the synaptic vesicles close to the part of the plasma membrane facing the synapse.

Despite much effort, however, we still do not know precisely how a rise in Ca^{2+} concentration at these sites triggers a fusion of synaptic vesicles with the plasma membrane. One popular idea is that the spectrin-synapsin network is disrupted by a rise in Ca^{2+}, allowing the vesicles to move close to, and ultimately fuse with, the plasma membrane.

The Nicotinic Acetylcholine Receptor Protein Is a Ligand-gated Cation Channel

Once acetylcholine molecules have been released, they diffuse across the synaptic cleft and combine with receptor molecules in the membrane of the postsynaptic neuron or muscle cell. The interaction of acetylcholine with the nicotinic acetylcholine receptor produces within 0.1 ms a large transient increase in the permeability of the membrane to both Na^+ and K^+ ions; the Na^+ increase is slightly less than the K^+ increase. Thus the acetylcholine receptor is a ligand-gated cation channel.

Since the resting potential of the muscle plasma membrane is near E_K, the potassium equilibrium potential, opening of acetylcholine receptor channels causes little increased net efflux of K^+ ions; Na^+ ions, on the other hand, flow into the muscle cell. The simultaneous increases in permeability to Na^+ and K^+ ions produce a net depolarization to about -15 mV from the resting potential of about -60 mV. This depolarization of the muscle membrane generates an action potential, which spreads along the surface of the postsynaptic cell membrane via voltage-gated Na^+ channels in the same way neurons conduct impulses (see Figure 20-17). As detailed in Chapter 22, the resultant depolarization triggers Ca^{2+} movement from its intracellular store, the sarcoplasmic reticulum, into the cytosol, and then contraction. During neuron-muscle stimulation, acetylcholine is released simultaneously from all the terminals of a neuron, and all the acetylcholine receptors in the muscle are triggered simultaneously.

(a)

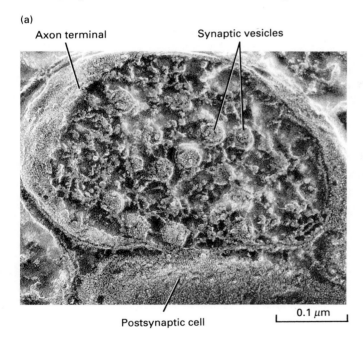

(b)

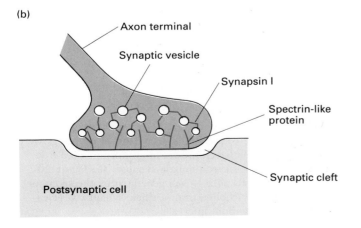

◀ **Figure 20-33** (a) Micrograph of an axon terminal obtained by the rapid-freezing deep-etch technique. (b) Schematic diagram of an axon terminal. Synapsin I connects synaptic vesicles and also connects vesicles to spectrinlike filaments extending inward from the plasma membrane adjacent to the synaptic cleft. *Part (a) from D. M. D. Landis, A. K. Hall, L. A. Weinstein, and T. J. Reese, 1988, Neuron 1:201.*

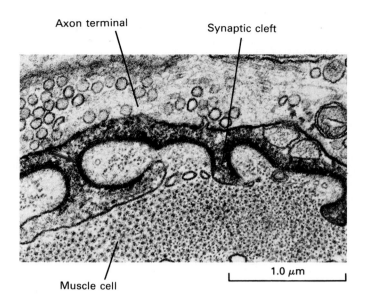

Axon terminal

Synaptic cleft

Muscle cell

1.0 μm

◀ **Figure 20-34**   Electron micrograph of neuron–striated muscle synapse showing location of nicotinic acetylcholine receptors. A section of a neuron-muscle synapse was treated with peroxidase-conjugated α-bungarotoxin, which binds specifically to nicotinic acetylcholine receptors. Because peroxidase catalyzes a reaction that produces an electron-dense product, the location of these receptors can be visualized. As shown, they are concentrated at the external surface of the postsynaptic membrane (*arrow*) at the top and part way down the sides of the folds in the membrane. Electron-dense material was present only in those portions of the muscle membrane adjacent to axon terminals. [See S. J. Burden, P. B. Sargent, and U. J. MacMahan, 1979, *J. Cell Biol.* 82:412.] *Courtesy of S. J. Burden. Reproduced from the* Journal of Cell Biology, *1979, by copyright permission of the Rockefeller University Press.*

Studies with the snake venom toxin α-bungarotoxin, which binds specifically and irreversibly to nicotinic acetylcholine receptors, have demonstrated that these receptors are localized in the membranes of postsynaptic striated muscle cells immediately adjacent to the terminals of presynaptic neurons (Figure 20-34). Patch-clamping studies have shown that in response to acetylcholine a single ion channel in the receptor remains open for several milliseconds before closing spontaneously. When open, the channel is capable of transmitting 15,000–30,000 Na^+ or K^+ ions a millisecond. The time required for an acetylcholine molecule to induce the opening of a channel is too small to be measured directly but is probably a few microseconds.

Spontaneous Exocytosis of Synaptic Vesicles Produces Small Depolarizations in the Postsynaptic Membrane

Careful monitoring of the membrane potential of the muscle membrane at a synapse with a cholinergic motor neuron has demonstrated spontaneous, intermittent depolarizations of about 0.5–1.0 mV in the absence of stimulation of the motor neuron (Figure 20-35). Each of these depolarizations is caused by the spontaneous release of acetylcholine from a single synaptic vesicle. Indeed, demonstration of the spontaneous small depolarizations led to the notion of quantal release of acetylcholine (later applied to other neurotransmitters) and thereby led to the hypothesis of vesicle exocytosis at synapses. The release of one vesicle of acetylcholine results in the opening of about 3000 ion channels in the postsynaptic membrane—sufficient to depolarize the region of membrane by about 1 mV but insufficient to reach the threshold for an action potential.

Nicotinic Acetylcholine Receptor Contains Five Subunits, Each of Which Contributes to the Cation Channel

In most nerve and muscle tissues the nicotinic acetylcholine receptor constitutes a minute fraction of the total membrane protein. Electric eel and sting ray (torpedo) electric organs are particularly rich in this receptor and are the sources used for its purification.

Electron microscopy of receptor-rich membranes has revealed doughnutlike structures believed to be receptor molecules (Figure 20-36). Computer-generated averaging of many high-resolution images of these receptors, viewed from several angles, has led to the pentameric model shown in Figure 20-37a. Each molecule has a diameter of about 9 nm; protrudes about 6 nm into the extracellular space and about 2 nm into the cytosol from the membrane surfaces; and contains a central hole, with a maximum diameter of 2.5 nm, that is thought to be part of the cation channel.

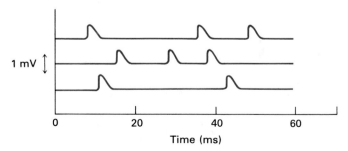

1 mV $\updownarrow$

0   20   40   60

Time (ms)

▲ **Figure 20-35**   Spontaneous depolarizations in the postsynaptic membrane of a cholinergic synapse. Recordings were taken from intracellular electrodes in an unstimulated frog muscle cell near the neuron-muscle synapse (see Figure 20-10a). Such depolarizations—each generated by the spontaneous release of acetylcholine from a single synaptic vesicle— seem to occur at random intervals.

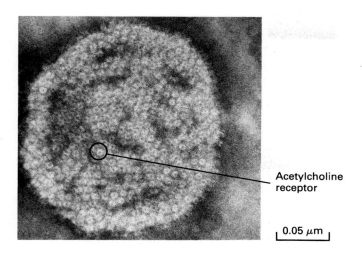

◀ **Figure 20-36** Electron micrograph of a negatively stained preparation of a membrane of an electroplax cell from an electric fish showing the region near the nerve terminus—called an end plate. This region contains many doughnut-shaped objects, 9–12 nm in diameter, thought to be nicotinic acetylcholine receptors. [See F. Hucho, 1981, *Trends Biochem. Sci.* **6**:242.] *Courtesy of F. Hucho.*

Acetylcholine receptor

0.05 μm

(a) Model of pentameric receptor — Acetylcholine binding sites

(b) Sequence of M2 helices

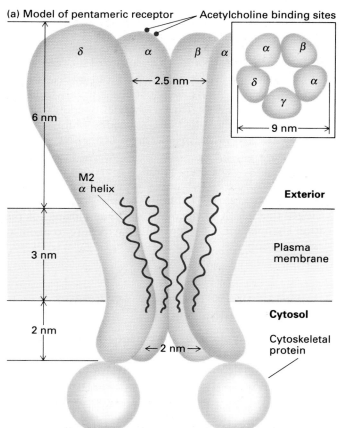

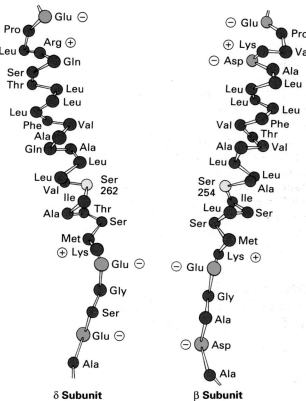

δ **Subunit**   β **Subunit**

▲ **Figure 20-37** Proposed structure of the nicotinic acetylcholine receptor. (a) Schematic model of the pentameric receptor, which contains four different polypeptides and has the composition $\alpha_2\beta\gamma\delta$; for clarity, the γ subunit is not shown. This model is based on amino acid sequence data, neutron-scattering experiments, and analysis of electron micrographs. Most of the mass of the protein protrudes from the outer surface of the plasma membrane. Each α subunit contains two acetylcholine binding sites and each subunit of the receptor contains four or five transmembrane α helixes. The M2 helix (red) in each subunit is thought to be part of the actual ion channel. *Inset:* cross section of the exoplasmic face of the receptor showing the arrangement of subunits about a central channel, or pore. At its narrowest (in the membrane), the ion channel is about 0.65–0.80 nm in diameter. (b) Amino acid sequences of the M2 helix in the β subunit and δ subunit. Negatively charged glutamate and/or aspartate residues (blue) are present at both ends of the M2 helices and are thought to play a role in the binding and transport of Na^+ and K^+ ions through the channel. Crosslinking of chlorpromazine to serine residues (yellow) in the middle of the M2 helices inhibits receptor function. [See H. R. Guy and F. Hucho, 1987, *Trends Biochem. Sci.* **10**:318; K. Imoto et al., 1988, *Nature* **335**:645; C. Toyoshima and N. Unwin, 1988, *Nature* **336**:247; and J. Dani, 1989, *Trends Neurosci.* **12**:125.]

The nicotinic acetylcholine receptor can be solubilized from these membranes by nonionic detergents. The crucial step in its purification is affinity chromatography on columns of immobilized cobra toxin, which binds the receptor but from which it can be subsequently eluted and recovered. The monomeric molecular weight of the receptor protein is 250,000–270,000. The receptor consists of four different polypeptides with the composition $\alpha_2\beta\gamma\delta$. The molecular weights of the subunits are as follows: α (~40,000), β (49,000), γ (57,000), and δ (65,000). Opening of the ion channel is facilitated by cooperative binding of two acetylcholine molecules to each α subunit.

Messenger RNAs encoding all four receptor subunits have been cloned. When all four are microinjected into a single frog oocyte (see Figure 20-26), functional nicotinic acetylcholine receptors are formed. No channels, or poorly functional ones, are obtained if the mRNA for one subunit is omitted. Thus all four subunit polypeptides—and only these—are needed for receptor function.

The α, β, γ, and δ subunits have considerable sequence homology; on average, about 35–40 percent of the residues in any two subunits are homologous. Each subunit is thought to contain four or five transmembrane α helices. The complete receptor thus has a fivefold symmetry, and the actual cation channel is thought to be formed by segments from each of the five subunits. Studies measuring the permeability of different small cations have suggested that the ion channel is, at its narrowest, about 0.65–0.80 nm in diameter, sufficient to allow passage of both Na^+ and K^+ ions with their tightly bound shell of water molecules (see Figure 1-7).

The structure of the channel is not known in molecular detail. However, much evidence indicates that the channel is lined by five transmembrane M2 helices, one from each of the five subunits. One reason for thinking that M2 helices form the channel is that certain positively charged organic molecules, such as chlorpromazine, inhibit receptor function by "plugging" the ion channel. Chlorpromazine can be chemically cross-linked to serine residues (number 254 in the β subunit, 262 in δ) in the middle of the M2 helices (Figure 20-37b).

A second line of evidence for the role of M2 helices comes from expression of mutant receptor subunits in frog oocytes. Amino acid residues with negatively charged side chains (glutamate, aspartate) are on both sides of the membrane-spanning M2 helices. If just one of these residues in one subunit is mutated to a lysine, and the mutant mRNA injected together with mRNAs for the other three wild-type proteins, a functional channel is formed, but its ion conductivity—the number of ions that can cross it during its open state—is reduced. The greater the number of glutamate or aspartate residues mutated (in one or another subunit), the greater the reduction in conductivity. It is thought that the aspartate and glutamate residues form rings—one residue from each of the

five chains—on either side of the channel and participate in binding Na^+ or K^+ ions as they enter the pore.

Additional evidence concerning the function of the M2 helix has come from studies with a chemically synthesized peptide containing the amino acid sequence of the M2 helix in the δ subunit polypeptide. When this artificial M2 was incorporated into a lipid bilayer, it formed cation-specific channels with properties similar to those of native nicotinic acetylcholine receptors. These artificial channels, of course, were not affected by acetylcholine because they contained no ligand-binding sites.

Prolonged Exposure to Acetylcholine Agonists Desensitizes Cholinergic Receptors

The activity of the nicotinic acetylcholine receptor—like that of other neurotransmitter receptors—can be modulated. In particular, prolonged exposure of the receptor to acetylcholine agonists (e.g., nicotine) causes desensitization: the receptor binds acetylcholine normally, but the bound ligand does not trigger channel opening. Desensitization is accompanied by a major conformational change, reflected in the arrangement of the five subunits (Figure 20-38).

What is not understood—and is a major goal of current research—is how binding of acetylcholine to the "tips" of the two α subunits causes conformational changes that result in opening, and then closing, of the cation channel and, after prolonged exposure to agonist, desensitization.

Hydrolysis of Acetylcholine Terminates the Depolarization Signal

To restore a depolarized postsynaptic membrane to its excitable state, the depolarizing signal must be removed or destroyed. There are three main ways to end the signaling: (1) the transmitter diffuses away from the synaptic cleft; (2) the transmitter is taken up by the presynaptic neuron; and (3) the transmitter is enzymatically degraded. Signaling by acetylcholine is terminated by enzymatic degradation of the transmitter, but different methods are used to terminate signaling by other neurotransmitters.

Acetylcholine is hydrolyzed to acetate and choline by the enzyme *acetylcholinesterase*, which is localized in the synaptic cleft between the neuron and muscle cell membranes. It is bound to a network of collagen forming the basal lamina that fills this space (see Figure 20-30). Part of the acetylcholinesterase molecule is made up of a collagenlike (Gly-Pro-X) sequence and forms a collagenlike triple helix, which may facilitate its binding to collagen

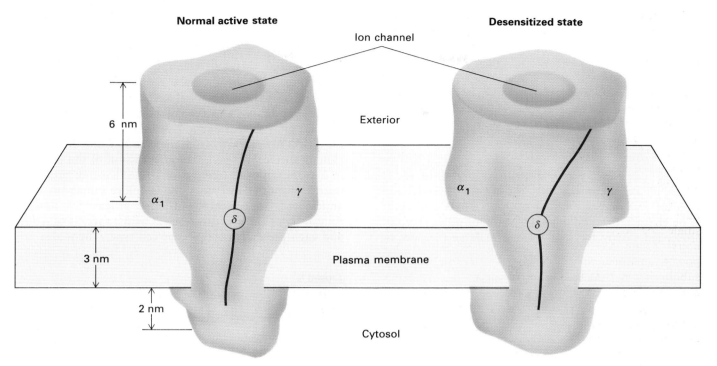

Normal active state **Desensitized state**

▲ **Figure 20-38** Model of the three-dimensional structure of the nicotinic acetylcholine receptor in its normal active state and in the desensitized state after prolonged exposure to an acetylcholine agonist. The receptor is perpendicular to the plane of the plasma membrane; one of the α subunits (α_1) is to the left, and γ is to the right. In the desensitized state, the δ subunit tilts considerably, as shown by the red line that traces the outer ridge of the δ subunit. Thus desensitization, triggered by agonist binding to the α subunits, probably causes a substantial conformational change in the molecule as a whole. *Adapted from N. Unwin, C. Toyoshima, and E. Kubalek, 1988, J. Cell Biol. **107**:1123.*

◄ **Figure 20-39** (a) General structure of an extracellular acetylcholinesterase. Each molecule contains three catalytic domains, each consisting of four enzymatically active polypeptides. The catalytic domains are connected to a three-stranded collagen domain; the latter anchors the enzyme to components of the extracellular matrix in the synaptic cleft. (b) Mechanism of action of acetylcholinesterase, showing that acetylserine is an essential enzyme-bound intermediate. *Adapted from P. Taylor et al., 1987, Trends Neurosci. **10**:93.*

or other elements of the extracellular matrix (Figure 20-39a). Another form of acetylcholinesterase is bound to the plasma membrane, apparently by a phospholipid covalently attached to the C-terminus of the enzyme molecule.

During hydrolysis of acetylcholine by acetylcholinesterase, a serine at the active site reacts with the acetyl group forming an enzyme-bound intermediate (Figure 20-39b). A large number of nerve gases and other neurotoxins inhibit the activity of acetylcholinesterase by reacting with the active-site serine. Physiologically these toxins prolong the action of acetylcholine, thus prolonging the period of membrane depolarization. Such inhibitors can be lethal if they prevent relaxation of the muscles necessary for breathing.

Functions of Other Neurotransmitters and Their Receptors

Because it causes a rapid, short-lived, and dramatic response in skeletal muscle cells, the nicotinic acetylcholine receptor was one of the first neurotransmitter receptors to be characterized definitively. Study of other receptors proceeded slowly for many years, in large measure be-

cause there was no simple way to purify membrane proteins present in minute amounts in complex nervous tissue. Experiments with drugs and other chemicals did establish that there were multiple receptors for many transmitters. For instance, the action of acetylcholine on heart muscle could be mimicked by muscarine, a substance that did not affect the nicotinic receptor in striated muscle. Thus the cardiac muscarinic acetylcholine receptor differs in some way from the nicotinic one. Molecular cloning of receptor genes and expression of the proteins in *Xenopus* oocytes (see Figure 20-26) has indicated that neurotransmitter receptors are far more numerous than anticipated. There are, for instance, at least five muscarinic acetylcholine receptors, each with a specific function and each present in a different cell type. There also are

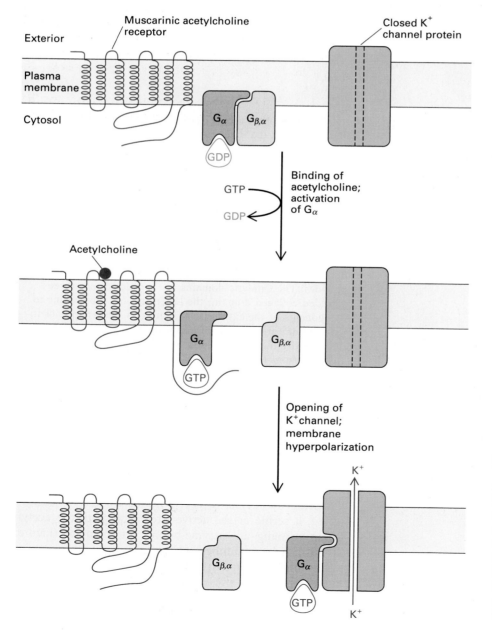

◀ **Figure 20-40** Proposed mechanism for opening of K⁺ channels in heart muscle plasma membrane. Binding of acetylcholine by muscarinic acetylcholine receptors triggers activation of a transducing G protein by catalyzing exchange of GTP for GDP on the α subunit. The active G_α-GTP, in turn, binds to and opens a K⁺ channel. The increase in K⁺ permeability hyperpolarizes the membrane, which reduces the frequency of heart muscle contraction. Though not shown here, the activation is terminated when the GTP bound to G_α is hydrolyzed to GDP. [See K. Dunlap, G. Holz, and S. G. Rane, 1987, *Trends Neurosci.* **10**:241; and E. Cerbai, U. Klockner, and G. Isenberg, 1988, *Science* **240**:1782.]

multiple forms of the nicotinic acetylcholine receptor; each has slightly different properties and is found in different cell types (see Table 20-3).

Cardiac Muscarinic Acetylcholine Receptor Activates a G Protein and Opens Potassium Channels

The response of skeletal muscle cells to the release of acetylcholine at the neuron-muscle junction is very rapid—the permeability changes and resulting membrane depolarization are completed within a few milliseconds (see Figure 20-29a). But many other synapses do not work as rapidly, and many functions of the nervous system operate with time courses of seconds or minutes and are not well served by fast synaptic responses lasting milliseconds. Regulation of the heart rate, for instance, requires that action of neurotransmitters extend over several beating cycles measured in seconds.

Responses that begin after a lag period of milliseconds following addition of transmitter are called *slow postsynaptic potentials* (SPNPs). SPNPs causing permeability changes in the postsynaptic membrane that favor depolarization beyond the threshold for an action potential are excitatory; those leading to hyperpolarization of the membrane are inhibitory. Such an inhibitory response is produced by binding of acetylcholine to muscarinic acetylcholine receptors in frog aorta cardiac muscle. Thus stimulation of the cholinergic nerves in heart muscle causes a long-lived (several seconds) hyperpolarization of the membrane (see Figure 20-29b) and slows the rate of heart muscle contraction.

Figure 20-40 shows how activation of the cardiac muscarinic receptor leads to opening of K^+ channels and subsequent hyperpolarization of the plasma membrane. Binding of acetylcholine to the receptor activates a transducing G protein; the active G_α-GTP subunit then directly binds to and opens a K^+ channel. The sequence of the muscarinic receptor, which is similar to sequences of other receptors that interact with G proteins, includes seven characteristic membrane-spanning helices. That G_α-GTP directly activates the K^+ channel has been shown by single-channel recording experiments (see Figure 20-21b) in which the cytoplasmic face of the membrane patch was reacted with purified G_α-GTP.

Other muscarinic receptors, found in gland cells, are coupled, via a G protein, to activation of phospholipase C and generation of inositol triphosphate and diacylglycerol (see Figure 19-30). The resultant rise in the cytosolic Ca^{2+} level triggers secretion.

Catecholamines Are Widespread Neurotransmitters

The neurotransmitters epinephrine, norepinephrine, and dopamine all contain the catechol moiety and are synthesized from tyrosine (Figure 20-41). These transmitters are referred to as *catecholamines,* and nerves that synthesize and use epinephrine or norepinephrine are termed *adrenergic.*

Epinephrine and norepinephrine function as both systemic hormones and neurotransmitters. Norepinephrine is the transmitter at synapses with smooth muscles that are innervated by nerve fibers of the sympathetic nervous system, the division of the peripheral nervous system that increases the activity of the heart and internal organs in "fight or flight" reactions. Norepinephrine is also found at synapses in the central nervous system. Epinephrine is synthesized and released into the blood by the adrenal medulla, an endocrine organ that has a common embryologic origin with neurons of the sympathetic system. Unlike neurons, the medulla cells do not develop axons or dendrites.

Different cells may possess different receptors for the same catecholamine neurotransmitter (see Table 20-3).

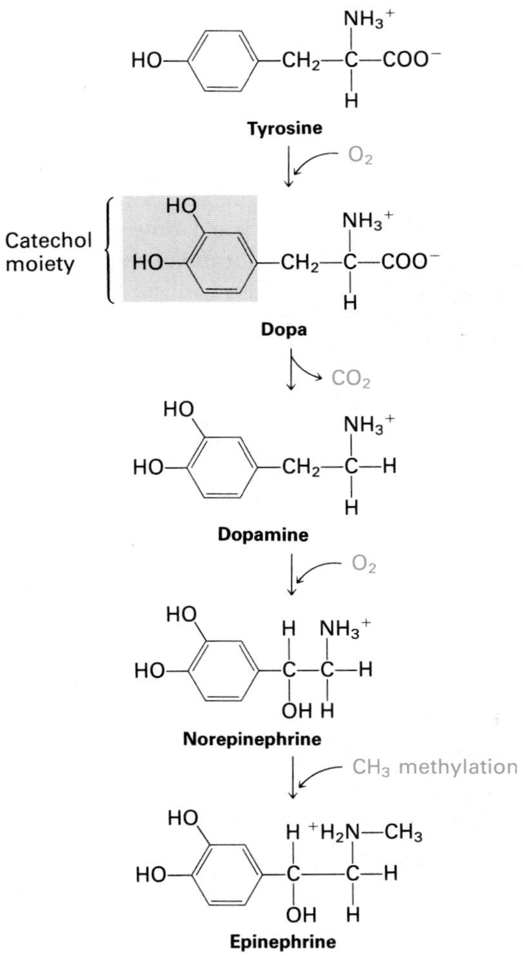

▶ **Figure 20-41** Pathway for synthesis of the catecholamine neurotransmitters from tyrosine. The first step produces the catechol moiety, which is retained in all three transmitters.

These different receptors may cause different excitatory or inhibitory responses, transduced by different G proteins. The existence of multiple receptor signaling pathways for the same neurotransmitter allows for great flexibility in nerve-nerve signaling. Two postsynaptic cells may respond very differently if they bear different receptors: one may undergo an excitatory response, the other an inhibitory response.

Some Receptors for Neurotransmitters Affect Adenylate Cyclase

The binding of agonists to β-adrenergic receptors on certain nerve cells causes the activation of adenylate cyclase and an increase in cAMP. This is the same mechanism by which the hormonal response to this compound is mediated (see Figures 19-14 and 19-16). Certain receptors for serotonin are also believed to activate adenylate cyclase.

In one well-studied synapse of the sea slug *Aplysia*, a facilitator neuron forms a synapse with the axon terminal of a sensory neuron that stimulates a motor neuron via an unknown transmitter. Stimulation of the facilitator neuron increases the ability of the sensory neuron to stimulate the motor neuron. When the facilitator neuron is stimulated, it secretes serotonin, which binds to serotonin receptors on the sensory neuron (Figure 20-42, steps 1 and 2). This binding activates adenylate cyclase and thus the synthesis of cAMP in the sensory neuron (step 3). cAMP then activates a cAMP-dependent protein kinase, which phosphorylates a voltage-gated K^+ channel protein or an associated protein leading to closure of the K^+ channels (steps 4 and 5). This closure of K^+ channels causes a decrease in the outward flow of K^+ ions that normally repolarizes the membrane of the sensory neuron after an action potential. The resulting prolonged membrane depolarization results in a greater influx of Ca^{2+} ions through voltage-gated Ca^{2+} channels (step 6). The increased Ca^{2+} level leads to greater exocytosis of synaptic vesicles in the sensory neuron (step 7), and hence greater activation of the motor neuron each time an action potential reaches the terminal. Administration of serotonin to the sensory neuron has been shown to cause decreased efflux of K^+ ions and prolonged depolarization of the membrane.

The *Aplysia* sensory neuron is sufficiently large so that the active catalytic subunit of the cAMP-dependent protein kinase can be injected into single neurons. Such treatment mimics the effect of applying the natural transmitter serotonin to the nerve. Additional supporting evidence that serotonin acts by means of cAMP and a protein kinase has come from patch-clamping studies on isolated sensory neuron membrane patches (see Figure 20-21b). When both ATP and the purified active catalytic subunit of cAMP-dependent protein kinase were added to the cytosolic surface of the patches, the K^+ channels closed.

Thus the protein kinase indeed acts on the cytoplasmic surface of the membrane to phosphorylate the channel protein itself or a membrane protein that regulates channel activity. We shall return to this particular synapse later, as these modifications in synapse efficiency are part of a simple learning response.

GABA and Glycine Are the Neurotransmitters at Many Inhibitory Synapses

Synaptic inhibition in the vertebrate central nervous system is mediated primarily by two amino acids, glycine and γ-aminobutyric acid (GABA); the latter is formed from glutamate by loss of a carboxyl group. The concentration of GABA in the human brain is 200–1000 times higher than that of other neurotransmitters such as dopa-

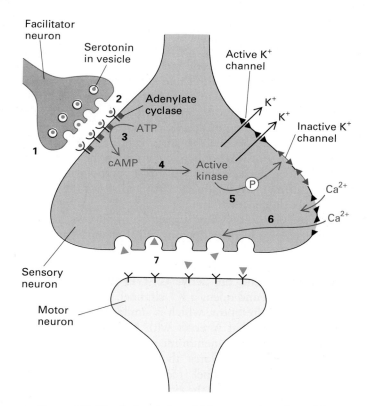

▲ **Figure 20-42** Pathway by which serotonin, released by stimulation of a facilitator neuron, increases the ability of a sensory neuron to activate a motor neuron in the sea slug *Aplysia*. The effect of serotonin is mediated through adenylate kinase and cAMP. Phosphorylation of the K^+ channel protein or a channel-binding protein, indicated by the circled P, closes the K^+ channels, leading to prolonged depolarization. See text for discussion. [See E. R. Kandel and J. Schwartz, 1982, *Science* **218**:433; M. B. Boyle et al., 1984, *Proc. Nat'l Acad. Sci. USA* **81**:7642; and M. J. Schuster et al., 1985, *Nature* **313**:392.]

mine, norepinephrine, and acetylcholine. Glycine is the major inhibitory neurotransmitter in the spinal cord and brain stem; GABA predominates elsewhere in the brain. Both glycine and GABA activate ligand-gated Cl^- channels.

An increase in the permeability of the postsynaptic membrane to Cl^- ions tends to drive the membrane potential toward the Cl^- equilibrium potential E_{Cl}, which in general is more negative than the resting membrane potential (see Figure 20-13). In other words, the membrane becomes hyperpolarized. If many Cl^- channels are opened, the resultant chloride permeability is large and the membrane potential will be held near E_{Cl}. A much larger than normal increase in the sodium permeability will then be required to depolarize the membrane. The effect of GABA or glycine on Cl^- permeability can last for a second or more, a long time compared with the millisecond required to generate an action potential. Thus GABA or glycine induces a slow, inhibitory postsynaptic response.

GABA and glycine receptors have been purified, cloned, and sequenced. In overall structure and sequence both resemble the nicotinic acetylcholine receptors (see Table 20-3). All are pentamers of similar subunits, although the GABA and glycine receptors contain only two different types of subunits. As noted already, the M2 helices of nicotinic acetylcholine receptors are thought to line the ion channel; the negatively charged glutamate and aspartate side chains at the ends of the M2 helices may participate in cation binding (see Figure 20-37). Strikingly, the M2 helices of the GABA and glycine receptor subunits have lysine or arginine residues at these positions; positively charged side chains of these residues may bind Cl^- ions specifically.

Some Peptides Function as Both Neurotransmitters and Neurohormones

Nervous tissue contains an enormous number of small peptides that can affect the activity of specific neurons. Many peptides probably function as synaptic neurotransmitters; others act in a paracrine fashion (see Figure 19-1) as a "diffusible" hormone and seem to affect many neurons over great distances. Some neurotransmitters also act as regulators of nerve cell growth and division. For instance, expression (by recombinant DNA techniques) in fibroblasts of the receptors for serotonin or angiotensin induces uncontrolled cell proliferation, and certain neurotransmitters stimulate the proliferation of embryonic glial or nerve cells. Because capillaries in the brain are much less permeable to ions and peptides than are capillaries in other parts of the body, most substances in the blood are excluded from the brain: this constitutes the blood-brain barrier. Thus hormones in the blood do not "confuse" functioning of the central nervous system.

Neurons that secrete hormones, called neurosecretory cells, were first discovered in connection with regulation of the function of pituitary cells by the hypothalamus. Secretion of hormones by the anterior pituitary cells is controlled by the hypothalamus, which in turn is regulated by other regions of the brain. The anterior pituitary is connected to the hypothalamus by a special closed system of blood vessels. Hypothalamic neurons secrete hypothalamic peptide hormones into these vessels, and the hormones then bind to receptors on the anterior pituitary cells. For example, thyrotropin-releasing hormone (TRH) stimulates secretion by the anterior pituitary of prolactin and thyrotropin. Another hypothalamic hormone—luteinizing hormone–releasing hormone (LHRH)—causes the anterior pituitary to secrete follicle-stimulating hormone (FSH) and luteinizing hormone (LH), which are important in regulating the growth and maturation of oocytes in the ovary (see Figure 19-5).

The application of sensitive immunochemical techniques has led to the discovery of these and other peptides in nerve terminals in different areas of the central nervous system and peripheral neurons (see Table 20-2). For example, a peptide called substance P appears to be a neurotransmitter used by sensory neurons that convey responses to painful and other noxious stimuli to the central nervous system.

Neuropeptide hormones, like all neuronal proteins, are synthesized in the cell body, packaged into secretory vesicles, and sent by axonal transport to the axon terminals. Generally, peptide hormones are packaged into different vesicles than those used to store "chemical" transmitters, such as serotonin, produced in the same neuron. In at least one case, the sea slug *Aplysia*, different peptide hormones produced by proteolysis from the same precursor are packaged into different secretory vesicles; exocytosis of the different types of secretory vesicles is regulated. In contrast to serotonin but like acetylcholine, peptide neurotransmitters are used only once and then degraded by proteases; they are not recycled.

Endorphins and Enkephalins Are Neurohormones That Inhibit Transmission of Pain Impulses

The activity of neurons in both the central and peripheral nervous systems is affected by a large number of neurohormones that act on cells quite distant from their site of release. Neurohormones can modify the ability of nerve cells to respond to synaptic neurotransmitters. Several small polypeptides with profound effects on the nervous system have been discovered recently; examples are Met-enkephalin, Leu-enkephalin, and β-endorphin (see Table 20-2). These three contain a common tetrapeptide sequence, Tyr-Gly-Gly-Phe, that is essential to their functions. These and other peptide hormones probably act as neurotransmitters in selected cell types but also have profound effects on general life events like mood, sleep, and body growth. Enkephalins and endorphins, for instance, function as natural pain killers or opiates and decrease the pain responses in the central nervous system.

Enkephalins were discovered during studies in the early 1970s focusing on the mechanism of opium addiction. Several groups of researchers discovered that brain plasma membranes contain high-affinity binding sites for purified opiates such as the alkaloid morphine. The sites were presumed to be the receptors that mediated the effects of these narcotic, analgesic drugs. Since such receptors exist in the brains of all vertebrates from shark to man, the question was raised why vertebrates should have highly specific receptors for alkaloids produced by opium poppies and why these should have survived eons of evolution. Since none of the neurotransmitters and peptides then known could serve as agonists or antagonists for the binding of opiates to brain receptors, a search was begun for natural compounds that could. This led to the discovery of two pentapeptides, Met-enkephalin and Leu-enkephalin, both of which bind to the "opiate" receptors in the brain and have the same effect as morphine (a profound analgesia) when injected into the ventricles (cavities) of brains of experimental animals. Enkephalins and endorphins appear to act by inhibiting neurons that transmit pain impulses to the spinal cord; presumably these neurons contain abundant endorphin or enkephalin receptors.

Memory and Neurotransmitters

In its most general sense, learning is a process by which humans and animals modify their behavior as a result of experience or as a result of acquisition of information about the environment. Memory is the process by which this information is stored and retrieved.

Psychologists have defined two types of memory, depending on how long it persists: short-term (minutes to hours) and long-term (days to years). It is generally accepted that memory results from changes in the structure or function of particular synapses, but until recently learning and memory could not be studied with the tools of cell biology or genetics. Most researchers believe that long-term memory involves the formation or elimination of specific synapses in the brain and the synthesis of new mRNAs and proteins. Because short-term memory occurs too rapidly to be attributed to such gross alterations, some have suggested that changes in the release and function of neurotransmitters at particular synapses are the basis of short-term memory. The fruit fly *Drosophila* and sea slug *Aplysia* exhibit elemental forms of learning, and considerable insight into the molecular events of memory have been obtained from studies with these organisms.

Mutations in *Drosophila* Affect Learning and Memory

Remarkable as it sounds, fruit flies can be trained to avoid certain noxious stimuli. During the training period, a population of flies is exposed to two different stimuli, either two odoriferous chemicals or two colors of light. One of the two is associated with an electric shock. The flies are then removed and placed in a new apparatus, and the two stimuli are repeated but without the electric shock. The flies are tested for their avoidance of the stimulus associated with the shock. About half the flies learn to avoid the stimulus associated with the shock, and this memory persists for at least 24 h. Painstaking observation of mutagenized flies has led to the identification of six different genes in which mutations cause defects in this learning process.

The *dunce* mutation, which disrupts memory, is understood at the molecular level. All dunce mutants are defective in one of two cAMP phosphodiesterase enzymes, and the cloned dunce gene encodes a cAMP phosphodiesterase. It is not yet known which synapses are involved in the learning response or how the high levels of cAMP that result from the phosphodiesterase deficiency might affect learning. *Turnip* mutations, which prevent learning, result in a drop in the level of protein kinase C, whose activity is regulated by the level of Ca^{2+} and diacylglycerol; this suggests that the inositol trisphosphate–diacylglycerol signaling pathway is also involved in synapse modulations leading to short-term memory. Perhaps of most interest is the *rutabaga* mutation, which affects a particular adenylate cyclase—one activated by either Ca^{2+}-calmodulin or a G_s-type transducing protein. Such a dual-regulated adenylate cyclase is also thought to play a key role in learning at a particular synapse in the sea slug *Aplysia*.

Gill-Withdrawal Reflex in *Aplysia* Exhibits Three Elementary Forms of Learning

Sea slugs exhibit three of the most elementary forms of learning familiar in vertebrates: habituation, sensitization, and classical conditioning. *Habituation* is the decrease in behavioral response to a stimulus following repeated exposure to the stimulus with no adverse effect. For example, an animal that is startled by a loud noise may show decreasing responses on prolonged repetition of the noise; that is, the animal becomes habituated to the stimulus. *Sensitization,* in contrast, is an increase in behavioral response to an intense or noxious stimulus. For instance, the withdrawal response to an odor or small shock is enhanced if the stimulus follows another, especially painful, stimulus such as a sharp pinch.

Classical conditioning represents one of the simplest types of associative learning—the recognition of predictive events within an animal's environment. The animal learns that one event, termed the *conditioned stimulus* (CS), always precedes, by a critical and defined period, a second or reinforcing stimulus or event, the *unconditioned stimulus* (US). A classical example is the Pavlovian response in dogs: a bell (CS) is rung a few seconds before food (US) is presented; the dogs soon learn to associate the two stimuli and to salivate in response to the bell alone. In such a learning process, be it in dogs or sea slugs, it is essential that the conditioning stimulus always precede the unconditioned stimulus by a small and critical time interval.

Habituation   When a sea slug (Figure 20-43) is touched gently on its siphon, the gill muscles contract vigorously and the gill retracts into the mantle cavity. The gill-withdrawal reflex is mediated by a simple reflex arc (Figure 20-44). Sensory neurons in the siphon synapse with motor neurons that innervate the gill muscles. However, if the siphon is touched 10–15 times in rapid sequence, the gill response decreases to only about one-third of its initial intensity. By recording the electric changes in the motor neurons to the gill, researchers discovered that this *habituative response* is due to a progressive decrease in the amount of neurotransmitter released at the synapses between the siphon sensory neurons and the motor neurons. In other words, repeated stimulation of the siphon leads to a decrease in the magnitude of the excitatory postsynaptic potential.

We have noted that release of neurotransmitters is triggered by a rise in the intracellular Ca^{2+} concentration following opening of voltage-gated Ca^{2+} channels. Measurements of Ca^{2+} movements in the *Aplysia* siphon sensory neuron have shown that habituation results from a decrease in the number of voltage-gated Ca^{2+} channels that open in response to the arrival of the action potential at the terminal, thus reducing the amount of neurotrans-

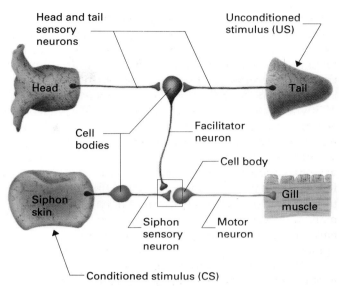

▲ **Figure 20-44**   Neural circuits involved in the gill-withdrawal reflex of the sea slug *Aplysia*. For simplicity, certain of the interneurons are omitted. This reflex exhibits habituation, sensitization, and classical conditioning. The details of the synapses (*boxed*) between the sensory, facilitator, and motor neurons are shown in Figure 20-42. [See E. R. Kandel and J. H. Schwartz, 1982, *Science* **218**:433; and T. W. Abrams and E. R. Kandel, 1988, *Trends Neurosci.* **11**:128.]

mitter released. Habituation does not affect the generation of action potentials in the siphon sensory neuron or the response of the receptors in the postsynaptic cells. Vertebrates exhibit a number of habituative responses similar to that of *Aplysia*, and it is thought that they are caused by modifications in the properties of particular channel proteins in specific synapses.

Sensitization   If a habituated sea slug is given a strong, noxious stimulus, such as a blow on the head or tail, it will respond to the next weak stimulus to the siphon by a rapid withdrawal of the gill. The noxious stimulation is said to sensitize the animal so that it exhibits an enhanced response to touching of the siphon. Electrophysiologic studies have shown that *Aplysia* sensitization is mediated by facilitator neurons that are activated by shocks to the head or tail. Electron microscopy shows that the axon of a facilitator neuron synapses with the terminal of a siphon sensory neuron near the site where the siphon sensory neuron synapses with a motor neuron (Figure 20-44). Stimulation of the facilitator neuron causes the siphon sensory neuron to release more transmitter in its synapse with the motor neuron, thus increasing the magnitude of the gill-withdrawal reaction.

As illustrated in Figure 20-42, stimulation of the *Aplysia* facilitator neurons leads to closure of voltage-gated K^+ channels in the siphon sensory neurons, which normally participate in repolarization of the membrane after an action potential. As a result, action potentials reaching

Aplysia

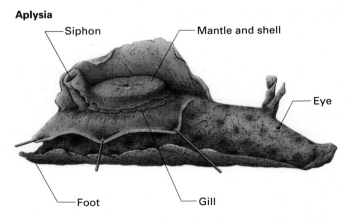

▲ **Figure 20-43**   The sea slug *Aplysia punctata*. The gill is under the protective mantle; it can be seen if the overlying tissue is pulled aside. *Adapted from E. R. Kandel, 1976, Cellular Basis of Behavior, W. H. Freeman and Company, p. 76.*

the nerve terminals decay more slowly. This prolonged depolarization causes a longer and larger than usual influx of Ca^{2+} ions via the voltage-dependent Ca^{2+} channels. The increased cytosolic Ca^{2+} level leads to (1) more extensive exocytosis of neurotransmitter by siphon sensory neurons at their synapses with motor neurons; (2) enhanced activity of motor neurons; and (3) enhanced contraction of the gill muscle. The effect of facilitator neuron stimulation is mediated by cAMP and a cAMP-dependent protein kinase in the siphon sensory neuron terminal. Short-term sensitization persists as long as the concentration of cAMP is elevated and the kinase is activated, about 1 h after each sensitizing stimulus.

Classical Conditioning The gill-withdrawal reflex also exhibits classical conditioning. In the "training" process, a weak touch to the siphon—the conditioned stimulus (CS)—is followed immediately by a sharp blow to the tail or head—the unconditioned stimulus (US)—which, of course, evokes a marked gill-withdrawal response. After a series of such trials, the gill-withdrawal response to the CS alone is substantially enhanced, as if the animal "learns" that a weak siphon touch (CS) is followed by a noxious, sharp blow (US). As in conditioning in other animals, the CS must precede the US by a short and definite interval, in this case 1–2 s.

Sensitization occurs when the facilitator neuron is activated by the US (a blow to the tail) in the absence of the CS; it triggers activation of adenylate cyclase and closing of K^+ channels in the siphon sensory neuron. During conditioning, the rise in cAMP in the siphon sensory neuron terminal is much greater when the sensory neuron is triggered by the CS to fire an action potential just before the US arrives.

Figure 20-45 outlines how an adenylate cyclase activated both by the serotonin receptor and by Ca^{2+}-calmodulin is the probable molecular site for convergence of the US and CS in the terminal of the siphon sensory neuron. The brief Ca^{2+} influx triggered by the CS and subsequent action potential in the siphon sensory neuron activates this adenylate cyclase. Serotonin released by the facilitator neuron triggered by the US also stimulates this cyclase. However, activation of the cyclase, and hence the increase in cAMP, is greatest when the cyclase is first "primed" by Ca^{2+} influx and then, within 1–2 s, activated by binding of serotonin. In this way, the enhancement of adenylate cyclase activity triggered by the unconditioned stimulus makes the sensory neuron more sensitive to a conditioning stimulus; the animal learns to associate the CS with the US and to respond to the CS alone with an enhanced response.

Long-term Memory The short-term sensitization and conditioning responses in *Aplysia* can occur in the presence of inhibitors of protein synthesis, suggesting that no new proteins (or cells) are required for short-term learn-

ing responses (short-term memory). On the other hand, a series of closely spaced tail shocks (unconditioned stimulus) delivered over a few hours will produce a long-term sensitization (long-term memory), which can persist for days or even weeks. Both long-term and short-term sensitizations affect the same synapses, and even the same K^+ channels. However, protein synthesis is essential for long-term sensitization, suggesting that certain new proteins must be made in these synapses in order for long-term memory to occur. Additional synapses between these neurons must also be established.

Sensitization and classical conditioning in the gill-withdrawal reflex of *Aplysia* are one of the few cases in which short-term changes in synaptic function are understood in molecular detail. Possibly these simple forms of learning will serve as a model for more complex forms of behavior, such as short-term and long-term memory in vertebrates. Increasingly, neuroscientists are focusing on cloning and characterizing enzymes such as the Ca^{2+}-calmodulin–activated adenylate cyclase that may be involved in memory in mammals.

Sensory Transduction: The Visual System

The nervous system receives input from a large number of sensory receptors (see Figure 20-9). Photoreceptors in the eye, taste receptors on the tongue, and touch receptors on the skin monitor various aspects of the outside environment. Stretch receptors surround many muscles and fire when the muscle is stretched. Internal receptors monitor the levels of glucose, salt, and water in body fluids. The nervous system, the brain in particular, processes and integrates this vast barrage of information and coordinates the response of the organism.

The "language" of the nervous system is electric signals. Each of the many types of receptor cells must convert, or transduce, its sensory input into an electric signal. A few sensory receptors are themselves neurons that generate action potentials in response to stimulation. However, most are specialized epithelial cells that do not generate action potentials but synapse with and stimulate adjacent neurons that then generate action potentials. The key question that we will consider is how does a sensory cell transduce its input into an electric signal.

Stretch receptors—ion channels activated by stretching of the cell membrane—have been identified in a wide array of cells, ranging from vertebrate muscle to epithelial cells to yeast, plants, and even bacteria. In prokaryotes and lower eukaryotes, such channels may play a role in osmoregulation and the control of a constant cell volume; it is thought that such receptors were among the first sensory receptors to evolve. In vertebrates, sensory receptors are used to detect muscle stretch and organ volume, bal-

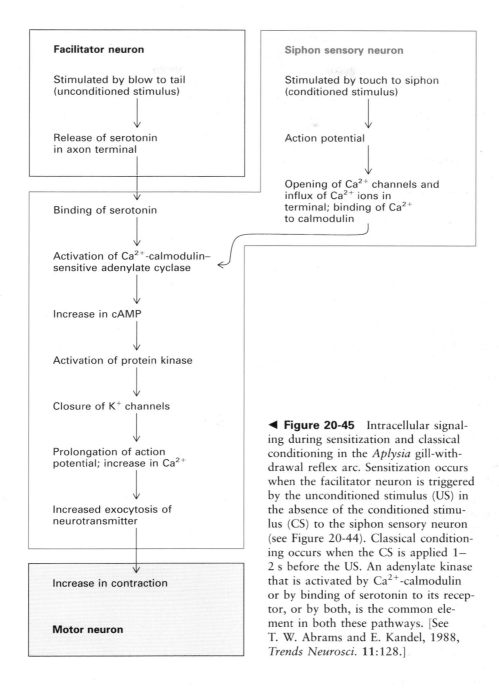

◄ Figure 20-45 Intracellular signaling during sensitization and classical conditioning in the *Aplysia* gill-withdrawal reflex arc. Sensitization occurs when the facilitator neuron is triggered by the unconditioned stimulus (US) in the absence of the conditioned stimulus (CS) to the siphon sensory neuron (see Figure 20-44). Classical conditioning occurs when the CS is applied 1–2 s before the US. An adenylate kinase that is activated by Ca^{2+}-calmodulin or by binding of serotonin to its receptor, or by both, is the common element in both these pathways. [See T. W. Abrams and E. Kandel, 1988, *Trends Neurosci.* **11**:128.]

ance, sound, light, and other stimuli. Because light reception by rod cells in the mammalian retina is the best-understood sensory system, we discuss it in detail.

Hyperpolarization of Rod Cells Is Caused by Closing of Sodium Channels

The human retina contains two types of photoreceptors, rods and cones (Figure 20-46). The *cones* are involved in color vision and function in bright light. The *rods* are stimulated by weak light over a range of wavelengths. In the outer segment of the rod cell are membrane disks that contain photoreceptor pigments (Figure 20-47). Rod cells

form synapses with neurons that transmit impulses to the brain; synaptic vesicles are present in the region of the rod cell that forms these synapses.

In the dark, the membrane potential of a rod cell is about −30 mV, considerably less than the resting potential (−60 to −90 mV) typical of neurons and other electrically active cells. As a consequence of this depolarization, rod cells in the dark are constantly secreting neurotransmitters, and the bipolar neurons with which they synapse are continually being stimulated. A pulse of light causes the membrane potential in the outer segment of the rod cell to become slightly hyperpolarized—that is, more negative (Figure 20-48). The light-induced hyperpo-

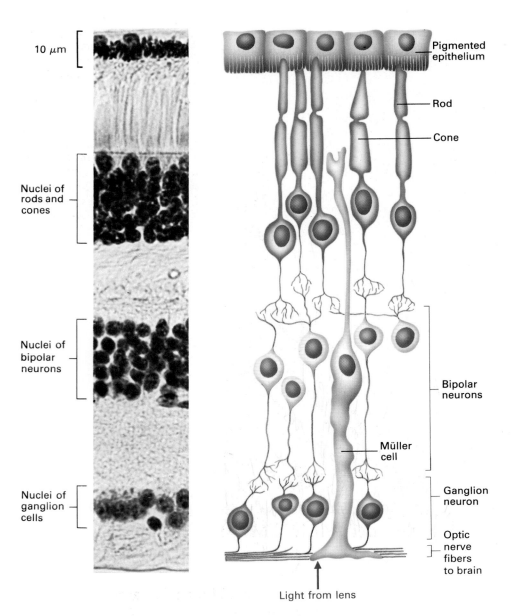

10 μm

Nuclei of rods and cones

Nuclei of bipolar neurons

Nuclei of ganglion cells

Pigmented epithelium

Rod

Cone

Bipolar neurons

Müller cell

Ganglion neuron

Optic nerve fibers to brain

Light from lens

◂ **Figure 20-46** Cells in the neural layer of the human retina. The outermost layer of cells (in the rear of the eyeball) forms a pigmented epithelium in which the tips of the rod and cone cells are buried. Light focused from the lens passes through all of the cell layers of the retina and is absorbed by the rods and cones. The axons of these cells synapse with many bipolar neurons. These, in turn, synapse with cells in the ganglion layer that send axons—optic nerve fibers—through the optic nerve to the brain. By synapsing with multiple rod cells, certain bipolar cells integrate the responses of many cells. They are involved in recognizing patterns of light that fall on the retina—for instance, a band of light that excites a set of rod cells in a straight line. Müller cells are supportive nonneural cells that fill much of the retinal spaces. *From R. G. Kessel and R. H. Kardon, 1979,* Tissues and Organs: A Text-Atlas of Scanning Electron Microscopy, *W. H. Freeman and Company, p. 87.*

larization causes a decrease in release of neurotransmitters.

The depolarized state of the plasma membrane of resting, dark-adapted rod cells is due to the presence of a large number of open Na^+ channels. As shown in Figure 20-13, an increase in Na^+ permeability causes the membrane potential to become more positive—that is, depolarized. The effect of light is to close Na^+ channels. The more photons absorbed, the more Na^+ channels are closed, the more negative the membrane potential becomes, and the less neurotransmitter released.

Remarkably, a single photon absorbed by a resting rod cell produces a measurable response, a hyperpolarization of about 1 mV, which lasts a second or two. Humans are able to detect a flash of as few as five photons. A single photon blocks the inflow of about 10 million Na^+ ions due to the closure of hundreds of Na^+ channels. Only about 30–50 photons need to be absorbed by a single rod cell in order to cause half-maximal hyperpolarization. The photoreceptors in rod cells, like many other types of receptors, exhibit the phenomenon of *adaptation*. That is, more photons are required to cause hyperpolarization if the rod cell is continuously exposed to light than if it is kept in the dark.

Let us now turn to three key questions: how is light absorbed; how is the signal transduced into closing of Na^+ channels; and how does the rod cell adapt to 100,000-fold or greater variations in light intensity?

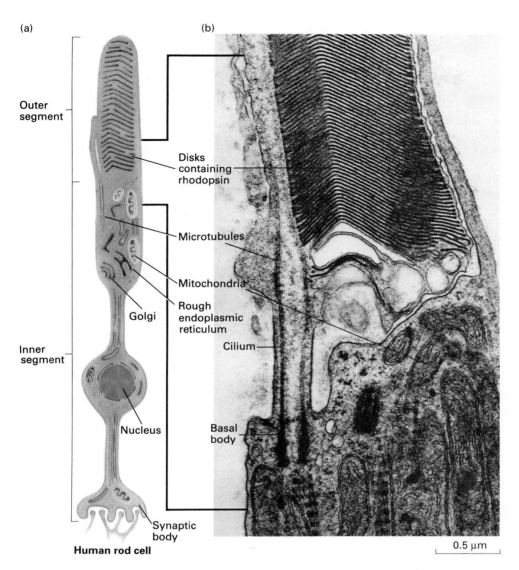

(a)

Outer segment

Disks containing rhodopsin

Microtubules

Mitochondria

Rough endoplasmic reticulum

Golgi

Cilium

Inner segment

Nucleus

Basal body

Synaptic body

Human rod cell

(b)

0.5 μm

◀ **Figure 20-47** (a) Diagram of the structure of a human rod cell. (b) Electron micrograph of the region of the rod cell indicated by the bracket in (a); this region includes the junction of the inner and outer segments. At the synaptic body, the rod cell forms a synapse with one or more bipolar neurons. *Part (b) from R. G. Kessel and R. H. Kardon, 1979, Tissues and Organs: A Text-Atlas of Scanning Electron Microscopy, W. H. Freeman and Company, p. 91.*

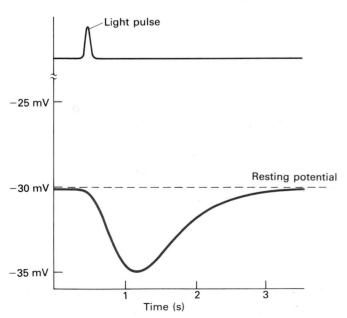

Light pulse

−25 mV

Resting potential

−30 mV

−35 mV

1　2　3
Time (s)

Absorption of a Photon Triggers Isomerization of Retinal and Activation of Opsin

The photoreceptor in rod cells, *rhodopsin*, consists of the transmembrane protein *opsin*, which has seven membrane-spanning helices, similar to other receptors that interact with transducing G proteins. Bound to opsin is the light-absorbing pigment 11-*cis*-retinal (Figure 20-49). Rhodopsin is localized to the thousand or so flattened membrane disks that make up the rod's outer segment.

11-*cis*-Retinal absorbs light in the visible range (400–600 nm). The primary photochemical event is isomeriza-

◀ **Figure 20-48** A brief pulse of light causes a transient hyperpolarization of the rod-cell membrane. The membrane potential is measured by an intracellular microelectrode (see Figure 20-10a).

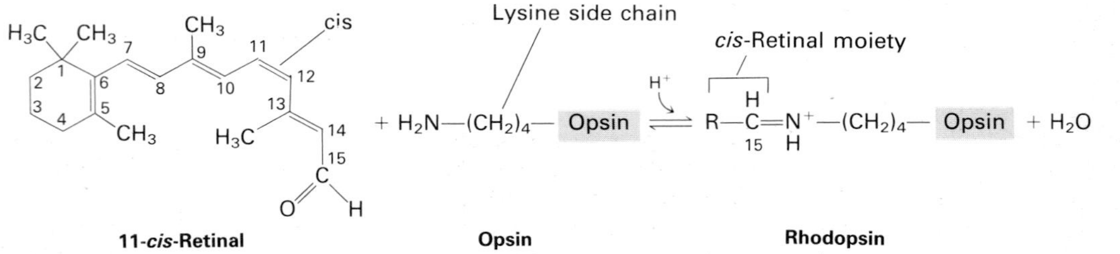

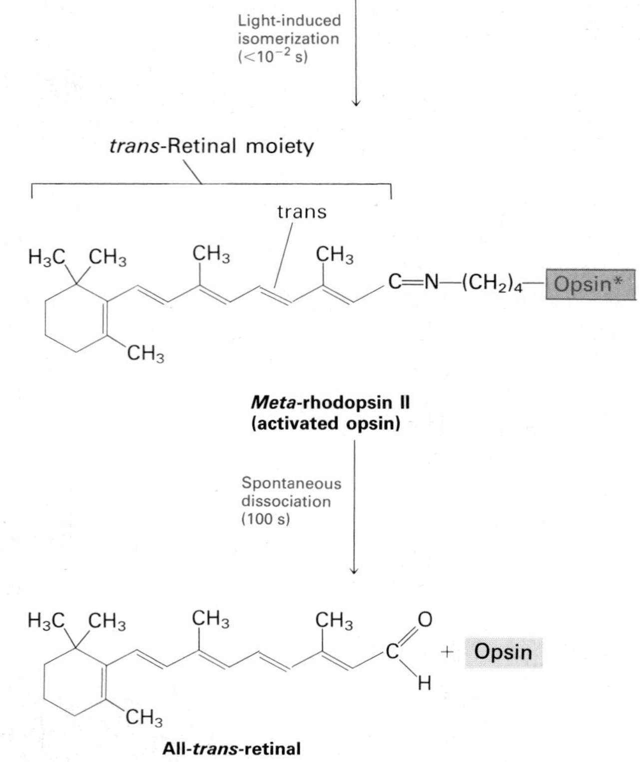

Figure 20-49 The photoreceptor in rod cells is rhodopsin, which is formed from 11-*cis*-retinal and opsin, a transmembrane protein. Absorption of light causes rapid photoisomerization of the *cis*-retinal to the *trans* isomer, forming the unstable intermediate *meta*-rhodopsin II, or activated opsin. The latter dissociates spontaneously to give all-*trans*-retinal and opsin.

tion of the 11-*cis*-retinal moiety in rhodopsin to all-*trans*-retinal, which has a different conformation than the cis isomer; thus the energy of light is converted into atomic motion. The intermediate in which opsin is bound to all-*trans*-retinal is called *meta*-rhodopsin II, or *activated opsin*. The light-induced formation of activated opsin is both extremely efficient and rapid: every absorbed photon triggers opsin activation in less than 10 ms. Activated opsin is unstable and spontaneously dissociates, releasing opsin and all-*trans*-retinal. In the dark, all-*trans*-retinal is isomerized back to 11-*cis*-retinal in a reaction catalyzed by rod-cell membranes; the cis isomer can then rebind to opsin, re-forming rhodopsin.

Cyclic GMP Is a Key Transducing Molecule

Much research has shown that 3′,5′ cyclic GMP is a key transducer molecule linking activated opsin to opening of Na⁺ channels. Rod outer segments contain an unusually high concentration of 3′,5′-cGMP, about 0.07 m*M*, and its concentration *drops* upon illumination. Light appears to have no immediate effect on the synthesis of cGMP from GTP:

$$\text{GTP} \xrightarrow{\text{guanylate cyclase}} 3',5'\text{-cGMP} + \text{PP}_i$$

However, rod outer segments contain a specific phosphodiesterase for cGMP that is activated by a cascade triggered by light:

$$3',5'\text{-cGMP} + \text{H}_2\text{O} \xrightarrow{\text{cGMP phosphodiesterase}} 5'\text{-GMP}$$

Finally, injection of cGMP into a rod cell depolarizes the cell membrane. The effect is potentiated if an analog of cGMP that cannot be hydrolyzed is injected. It thus appears that in the dark the high level of cGMP acts to keep the Na⁺ channels open.

Direct support for the role of cGMP in rod-cell activity has been obtained in patch-clamping studies with isolated patches of rod outer segment plasma membrane, which contains abundant Na⁺ channels. When cGMP is added to the cytoplasmic surface of these patches, there is a rapid increase in the number of open Na⁺ channels. The effect occurs in the absence of protein kinases or phosphatases, and it appears that cGMP acts directly on the channels to keep them open. Three cGMP molecules must bind per channel in order to open it; this allosteric interaction makes channel opening very sensitive to changes in cGMP levels. Light closes the channels by activating the cGMP phosphodiesterase and lowering the level of cGMP. Several hundred molecules of cGMP phosphodiesterase must be activated by a single photon. How does this happen?

The activation of cGMP phosphodiesterase is coupled to light absorption, which generates activated opsin, by the rod protein transducin (Figure 20-50). Transducin (T), which is very similar in structure and function to other transducing G proteins, has three subunits: T$_\alpha$ (MW 39,000), T$_\beta$ (MW 36,000), and T$_\gamma$ (MW 8400). The β and γ subunits are identical to those in other G proteins. In the resting state, the α subunit has a tightly bound GDP (T$_\alpha$-GDP) and is incapable of affecting cGMP phosphodiesterase. Light-activated opsin catalyzes the exchange of free GTP for a GDP on the α subunit of transducin and the subsequent dissociation of T$_\alpha$-GTP from trimeric transducin. The free Tα-GTP then activates cGMP phosphodiesterase. A single molecule of activated

▶ **Figure 20-50** Coupling of light absorption to activation of cGMP phosphodiesterase via light-activated opsin (opsin*) and transducin in rod cells. In dark-adapted rod cells, a high level of cGMP acts to keep Na⁺ channels open and the membrane depolarized compared with the resting potential of other cell types. Light absorption leads to a decrease in cGMP by the pathway shown; as a result, many Na⁺ channels close and the membrane becomes transiently hyperpolarized (see Figure 20-48). *Adapted from M. Applebury, 1987, Nature 326:546; and B. Fung and I. Griswold-Prenner, 1989, Biochemistry 28:3133.*

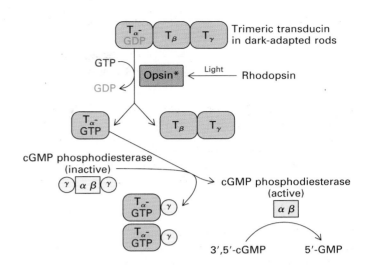

opsin in the disk membrane can apparently activate 500 transducin molecules.

Biochemical studies have shown that T_α-GTP activates cGMP phosphodiesterase by binding to and removing the inhibitory γ subunit, thus releasing the catalytic α and β subunits in an active form (Figure 20-50). The α subunit of transducin also contains a GTPase, which slowly converts light-induced T_α-GTP back to T_α-GDP. Once re-formed, T_α-GDP combines with T_β and T_γ, thus regenerating trimeric transducin. As a result, cGMP phosphodiesterase is again inactivated and the 3',5'-cGMP level gradually returns to its dark-adapted level.

Rod Cells Adapt to Varying Levels of Ambient Light

Cone cells are insensitive to low levels of illumination, and the activity of rod cells is inhibited at high light levels. Thus when we move from daylight into a dimly lighted room, we are initially blinded. As the rod cells slowly become sensitive to the dim light, we gradually are able to see and distinguish objects. Even in dim light, the activity of a rod cell is inversely proportional to the background (ambient) light; that is, the electric currents generated in rod cells following light flashes of constant intensity increase as the ambient level of light decreases.

A rod cell is able to adapt to a more than 100,000-fold variation in the ambient light level so that *differences* in light levels, rather than the absolute *amount* of absorbed light, are used to form visual images. One process contributing to this adaptation involves Ca^{2+} ions. As shown in Figure 20-51, guanylate cyclase, the enzyme that syn-

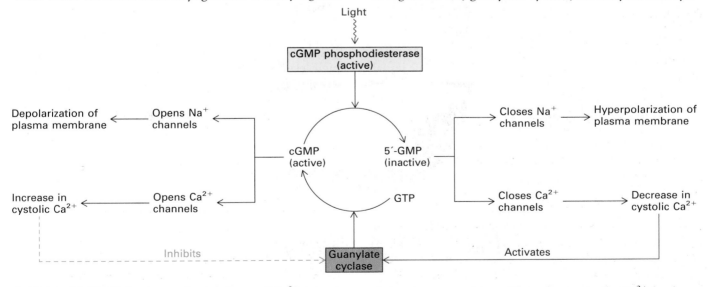

▲ **Figure 20-51** Role of guanylate cyclase and Ca^{2+} in adaptation of rod cells to changes in ambient light levels. In dark-adapted cells, the high level of cGMP opens both Na⁺ and Ca^{2+} channels; the relatively high level of cytosolic Ca^{2+} inhibits guanylate cyclase. Reduction in cGMP, triggered by light activation of cGMP phosphodiesterase, causes a decrease in the cytosolic Ca^{2+} level and also hyperpolarization

of the plasma membrane. The reduction in the Ca^{2+} level causes activation of guanylate cyclase, which catalyzes synthesis of cGMP, restoring the cells to a new baseline state in which they are less sensitive to small changes in light level. *Adapted from E. Pugh and J. Altman, 1988, Nature 334:16; and K. Koch and L. Stryer, 1988, Nature 334:64.*

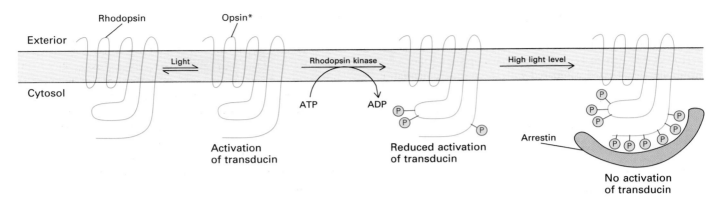

▲ **Figure 20-52** Role of opsin phosphorylation in adaptation of rod cells to changes in ambient light levels. Light-activated opsin (opsin*), but not rhodopsin, is a substrate for rhodopsin kinase. The extent of opsin* phosphorylation is directly proportional to the ambient light level, and the ability of an opsin* molecule to catalyze activation of transducin (see Figure 20-50) is inversely proportional to the number of sites phosphorylated. Thus the higher the ambient light level, the larger the increase in light level needed to activate the same number of transducin molecules. At very high light levels, arrestin binds to the completely phosphorylated opsin*, forming a complex that cannot activate transducin at all. [See N. Bennett and A. Sitaramayya, 1988, *Biochemistry* 27:1710.]

thesizes cGMP, is inhibited by low levels of Ca^{2+} ions. Cyclic GMP, besides opening Na^+ channels, also opens Ca^{2+} channels; the level of Ca^{2+} in the cell is balanced by Ca^{2+} pumps that export Ca^{2+} from the cytosol. Light, as we noted, causes a reduction in cGMP levels; this leads to a closing of both Na^+ channels and Ca^{2+} channels. The resultant drop in Ca^{2+} concentration causes activation of guanylate cyclase and synthesis of more cGMP. This "resets" the system to a new baseline level, so that a greater change in light level will be necessary to close the same number of Na^+ channels and to generate the same visual signal than if the cells had not been exposed to light. In other words, the cells become less sensitive to small changes in levels of illumination.

A second process, affecting the protein opsin itself, participates in adaptation of rod cells to ambient levels of light and also prevents overstimulation of the rod cell in very high ambient light (Figure 20-52). The rod-cell enzyme rhodopsin kinase phosphorylates light-activated opsin (O^*) but not dark-adapted rhodopsin. Phosphorylated opsin is less able to activate transducin than is nonphosphorylated opsin. Each O^* molecule has multiple phosphorylation sites; the more sites that are phosphorylated, the less able O^* is to activate transducin. Because the extent of O^* phosphorylation is proportional to the amount of time each opsin molecule spends in the light-activated form, it is a measure of the background level of light. Under high light conditions, phosphorylated opsin is abundant. Then a greater increase in light level will be necessary to generate a visual signal. When the level of ambient light is reduced, most of the opsins become dephosphorylated, and thus activation of transducin increases. Then fewer additional photons will be necessary to generate a visual signal. At high ambient light (such as noontime outdoors), the level of opsin phosphorylation is such that a protein termed *arrestin* binds to opsin; this binding totally blocks activation of transducin, causing a shutdown of all rod-cell activity. The mechanism by which rod-cell activity is controlled by rhodopsin kinase is similar to adaptation of the β-adrenergic receptor to high levels of hormone (see Figure 19-34). Indeed, rhodopsin kinase and β-adrenergic receptor kinase—the enzyme that phosphorylates and inactivates only the ligand-occupied β-adrenergic receptor—are very similar proteins, and each can phosphorylate the other's substrate.

Summary

An electric potential exists across the plasma membrane of all eukaryotic cells. It is caused by the different ion compositions of the cytosol and extracellular fluid and by the different permeabilities of the plasma membrane to the principal cellular ions—Na^+, K^+, Cl^-, and Ca^{2+}. In most nerve and muscle cells, the resting membrane potential is about 60 mV, negative on the inside; the potential is due mainly to the relatively high permeability of the membrane to K^+ ions.

Impulses are conducted along a nerve axon by means of action potentials. An action potential consists of a sudden (less than a millisecond) depolarization of the membrane followed by a rapid hyperpolarization and a gradual return to the resting potential. These changes in membrane potential are caused by a sudden transient increase in the permeability of the membrane to Na^+ followed by a slower transient increase in its permeability to K^+.

The changes in permeability to Na^+ are caused by transmembrane voltage-gated Na^+ channel proteins, which respond to depolarization of the membrane by transiently opening and admitting Na^+ ions into the cytosol. Voltage-gated K^+ channels also open in response to membrane depolarization; their opening repolarizes the membrane by permitting the efflux of K^+ ions. Voltage-gated K^+, Na^+, and also Ca^{2+} channel proteins have a similar structure and a similar positively charged "gating" helix that moves in response to a voltage change of sufficient magnitude.

Neurons only generate action potentials when the plasma membrane in the region of the axon hillock is depolarized to the threshold value. An action potential generated at one point along an axon will lead to depolarization of the adjacent segment and thus to propagation of the action potential along its length. The speed of impulse conduction depends on the diameter of the axon and conductivity of the neuronal cytosol. Thick neurons conduct faster than thin ones, and myelinated nerves conduct faster than unmyelinated nerves of similar diameter because of insulation of the neuron by the myelin sheath.

Impulses are transmitted from neurons to other cells at specialized junctions called synapses. In electric synapses, ions pass from the presynaptic cell to the postsynaptic cell through gap junctions, and an action potential is generated in the postsynaptic cell with no time delay. In the more common chemical synapses, the arrival of an action potential in the presynaptic axon triggers the release of neurotransmitters into the synaptic cleft; from there the transmitters bind to receptors on the postsynaptic cell. Transmitters are stored in membrane-bound vesicles, and exocytosis of these vesicles is triggered by a rise in the cytosolic Ca^{2+} level induced by the opening of voltage-gated Ca^{2+} channels. The amplification, modification, and integration of signals from multiple presynaptic neurons acting on the same postsynaptic cell can occur at chemical synapses.

At excitatory synapses, the neurotransmitter acts to depolarize the postsynaptic cell and generate an action potential. At the synapse of a motor neuron and striated muscle cell, binding of acetylcholine to the well-studied nicotinic acetylcholine receptor triggers a rapid increase in permeability of the membrane to both Na^+ and K^+ ions, leading to depolarization. Much information is available on the structure of the receptor protein and of the segments that line the ion channel. In other postsynaptic cells, the depolarization of the postsynaptic membrane is less extensive but longer-lived, on the order of seconds. At inhibitory synapses, the release of neurotransmitter triggers a hyperpolarization of the postsynaptic membrane, making it more difficult for the cell to generate an action potential. Depending on the specific receptor in the postsynaptic cell, the same neurotransmitter can induce either an excitatory or inhibitory response.

The action of GABA or glycine is mediated by an increase in permeability of the membrane to Cl^- ions, due to opening of ligand-gated Cl^- channels whose structure is similar to that of the nicotinic acetylcholine receptor. In some postsynaptic cells, receptors for epinephrine and serotonin modulate the activity of adenylate cyclase. The electric response of these cells is believed to be caused by phosphorylation of Na^+ or K^+ channel proteins by the cAMP-dependent protein kinase. In cardiac muscle, binding of ligand to the muscarinic acetylcholine receptor activates a transducing G protein; this in turn opens a K^+ channel and causes hyperpolarization of the membrane and a decrease in muscle contraction.

Many compounds released by neurons are systemic hormones as well as neurotransmitters, affecting both distant secretory cells and adjacent neurons. Recent work suggests that small peptides such as endorphins, enkephalins, and hypothalamic releasing factors function as neurotransmitters in particular synapses in the brain and also act as hormones.

Removal of neurotransmitter from the synapse is essential for ensuring its repeated functioning. The action of acetylcholine is terminated by the enzyme acetylcholinesterase. Other neurotransmitters are removed by diffusion or by reuptake into the presynaptic cell; peptide neurotransmitters are hydrolyzed to amino acids.

Especially in the central nervous system, many neurons must integrate excitatory and inhibitory stimuli from dozens or hundreds of other neurons, if not a thousand or more. Whether a threshold potential is induced at the axon hillock depends on the timing and magnitude of these stimuli, the localization and duration of the resultant local hyperpolarizations and depolarizations, and the ability of the localized changes in potential to be conducted along the axonal membrane surface.

Modifications in the activity of certain synapses are associated with short-term memory, at least in some invertebrate systems. Certain *Drosophila* mutants that cannot learn are defective in cAMP metabolism, most notably in a Ca^{2+}-calmodulin–activated adenylate cyclase. In the sea slug *Aplysia,* the gill-withdrawal reflex exhibits habituation, sensitization, and classical conditioning—three forms of simple learning. Habituation is linked to the closing of Ca^{2+} channels in the synaptic terminals of siphon sensory neurons originating in the siphon; this alters the flux of Ca^{2+} in the terminals and the amount of transmitter released to the motor neurons. Sensitization and classical conditioning are mediated by facilitator neurons that synapse with the siphon sensory neurons. Serotonin released by stimulation of the facilitator neurons causes an increase in adenylate cyclase activity and thus an increase in the cAMP level in the siphon sensory neurons, which in turn causes closing of K^+ channels. This prolongs depolarization and increases exocytosis of neurotransmitter. In classical conditioning, a conditioned stimulus, triggering the siphon sensory neurons, and an unconditioned stimulus, triggering the facilitator neurons, converge on activation of adenylate cyclase in the terminals of the siphon sensory neurons.

Many sensory transduction systems convert signals from the environment—light, taste, sound, touch—into electric signals in certain neurons. These signals are collected, integrated, and processed by the central nervous system. The sensory system understood in the most molecular detail is that of the photoreceptor rod cells. Absorption of even a single photon results in hyperpolarization of the rod-cell plasma membrane and reduces the release of chemical transmitters to adjacent nerve cells. Light causes isomerization of the 11-*cis*-retinal moiety in rhodopsin and formation of activated opsin, which then activates a transducer protein called transducin (T) by catalyzing exchange of free GTP for bound GDP on the T_α subunit. Activated T_α-GTP, in turn, activates cGMP phosphodiesterase. This enzyme lowers the cGMP level, which leads to closing of the membrane Na^+ channels, hyperpolarization of the membrane, and release of less neurotransmitter. Modifications in the activity of guanylate cyclase and also phosphorylation of the activated form of opsin result in adaptation of rod-cell activity to more than a 100,000-fold range of illumination.

References

General Properties of Neurons and Nervous Systems

*BRADFORD, H. F. 1986. *Chemical Neurobiology. An Introduction to Neurochemistry.* W. H. Freeman and Company.
The Brain. 1979. *Sci. Am.* **241**(3).
*COOKE, I., and M. LIPKIN, JR., eds. 1972. *Cellular Neurophysiology: A Source Book.* Holt, Rinehart and Winston. An anthology of important papers, 1921–1967.
*KANDEL, E. R., and J. H. SCHWARTZ. 1985. *Principles of Neural Sciences,* 2d ed. Elsevier.
*KATZ, B. 1966. *Nerve, Muscle and Synapse,* 2d ed. McGraw-Hill.
*KEYNES, R. D., and D. J. AIDLEY. 1981. *Nerves and Muscle.* Cambridge University Press.

*KUFFLER, S. W., J. G. NICHOLLS, and A. R. MARTIN. 1984. *From Neuron to Brain,* 2d ed. Sinauer Associates.
Molecular Biology of Signal Transduction. 1988. Cold Spring Harbor Symp. Quant. Biol., vol. 53.
Molecular Neurobiology. 1983. Cold Spring Harbor Symp. Quant. Biol., vol. 48.
*PATTERSON, P. H., and D. PURVES, eds. 1982. *Readings in Developmental Neurobiology.* Cold Spring Harbor Laboratory. An anthology of recent papers.
*PURVES, J., and J. W. LICHTMAN. 1985. *Principles of Neural Development.* Sinauer Associates.
Science **225**:4668 (September 21, 1984). An entire issue devoted to reviews in the neurosciences.
*THOMPSON, R. F. 1985. *Progress in Neuroscience: Readings from Scientific American.* W. H. Freeman and Company.

General Properties of Ion Channels

*CATTERALL, W. A. 1988. Genetic analysis of ion channels in vertebrates. *Ann. Rev. Physiol.* **50**:395–406.
*CATTERALL W. A. 1988. Structure and function of voltage-sensitive ion channels. *Science* **242**:50–61.
*HILLE, B. 1984. *Ionic Channels of Excitable Membranes,* 2d ed. Sunderland, MA: Sinauer Associates.
*JAN, L. Y., and Y. N. JAN. 1989. Voltage-sensitive ion channels. *Cell* **56**:13–25.
*KRUEGER, B. K. (1989) Toward an understanding of structure and function of ion channels. *FASEB J.* **3**:1906–1914.
*LATORRE, R., ed. 1986. *Ionic Channels in Cells and Model Systems.* Plenum Press.
*LESTER, H. A. 1988. Heterologous expression of excitability proteins: route to more specific drugs? *Science* **241**:1057–1063.
*MILLER, C. 1989. Genetic manipulation of ion channels: a new approach to structure and mechanism. *Neuron* **2**:1195–1205.
*PAPAZIAN, D. M., T. L. SCHWARZ, B. L. TEMPEL, L. C. TIMPE, and L. Y. JAN. 1988. Ion channels in *Drosophila. Ann. Rev. Physiol.* **50**:379–394.
*SAKMAN, B., and E. NEHER, eds. 1983. *Single-Channel Recording.* Plenum.
SIGWORTH, F. J., and E. NEHER. 1980. Single Na^+ channel currents observed in cultured rat muscle cells. *Nature* **287**:447–449.

The Action Potential: Ionic Movements

*CRONIN, J. 1987. *Mathematical Aspects of Hodgkin-Huxley Neural Theory.* Cambridge University Press.
HODGKIN, A. L. 1964. *The Conduction of the Nervous Impulse.* Liverpool University Press, Liverpool, U.K.
HODGKIN, A. L., and A. F. HUXLEY. 1952. Currents carried by sodium and potassium ions through the membrane of the giant axon of *Loligo. J. Physiol.* **116**:449–472.
HODGKIN, A. L., and A. F. HUXLEY. 1952. The dual effect of membrane potential on sodium conductance in the giant axon of *Loligo. J. Physiol.* **116**:497–506.

*A book or review article that provides a survey of the topic.

HODGKIN, A. L., and A. F. HUXLEY. 1952. A quantitative description of membrane current and its application to conduction
and excitation in nerve. *J. Physiol.* **117**:500–544.

HODGKIN, A. L., A. F. HUXLEY, and B. KATZ. 1952. Measurement
of current-voltage relations in the membrane of the giant
axon of *Loligo. J. Physiol.* **108**:37–77; **116**:424–448.

*KEYNES, R. D. 1979. Ion channels in the nerve-cell membrane.
Sci. Am. **240**(3):126–135.

Voltage-dependent Sodium Channels

*ALDRICH, R. W. 1986. Voltage-dependent gating of sodium
channels: towards an integrated approach. *Trends
Neurosci.* **9**:82–86.

ALDRICH, R. W., D. P. COREY, and C. F. STEVENS. 1983. A reinter-
pretation of mammalian sodium channel gating based on
single-channel recording. *Nature* **305**:436–441.

BARCHI, R. L., J. C. TANAKA, and R. E. FURMAN. 1984. Molecular
characteristics and functional reconstitution of muscle
voltage-sensitive sodium channels. *J. Cell Biochem.*
26:135–146.

*CATTERALL, W. A. 1986. Voltage-dependent gating of sodium
channels: correlating structure and function. *Trends
Neurosci.* **9**:7–10.

HARTSHORNE, R., M. TAMKUN, and M. MONTAL. 1986. The reconstituted sodium channel from brain. In *Ion Channel
Reconstitution*, C. Miller, ed. Plenum.

NODA, M., et al. 1984. Primary structure of *Electrophorus elec-
tricus* sodium channel deduced from cDNA sequence. *Nature* **312**:121–127.

OIKI, S., W. DANHO, and M. MONTAL. 1988. Channel protein
engineering: synthetic 22-mer peptide from the primary
structure of the voltage-sensitive sodium channel forms
ionic channels in lipid bilayers. *Proc. Nat'l Acad. Sci. USA*
85:2393–2397.

*SALKOFF, L., et al. 1987. Molecular biology of the voltage-
gated sodium channel. *Trends Neurosci.* **10**:522–527.

STÜHMER, W., et al. 1989. Structural parts involved in activation
and inactivation of the sodium channel. *Nature* **339**:597–
603.

VASSILEV, P. M., T. SCHEUER, and W. A. CATTERALL. 1988. Identification of an intracellular peptide segment involved in sodium channel inactivation. *Science* **241**:1658–1660.

Voltage-dependent Potassium Channels

BUTLER, A., A. WEI, K. BAKER, and L. SALKOFF. 1989. A family of
putative potassium channel genes in *Drosophila. Science*
243:943–947.

KAMB, A., L. E. IVERSON, and M. A. TANOUYE. 1987. Molecular
characterization of *Shaker*, a *Drosophila* gene that encodes
a potassium channel. *Cell* **50**:405–413.

*SALKOFF, L., and R. WYMAN. 1983. Ion channels in *Drosophila*
muscle. *Trends Neurosci.* **6**:128.

TAKUMI, T., H. OHKUBO, and S. NAKANISHI. 1988. Cloning of a
membrane protein that induces a slow voltage-gated potassium current. *Science* **242**:1042–1045.

TEMPEL, B. L., D. M. PAPAZIAN, T. L. SCHWARZ, Y. N. JAN, and
L. Y. JAN. 1987. Sequence of a probable potassium channel
component encoded at *Shaker* locus of *Drosophila. Science*
237:770–775.

TIMPE, L. C., Y. N. JAN, and L. Y. JAN. 1988. Four cDNA clones
from the *Shaker* locus of *Drosophila* induce kinetically distinct A-type potassium currents in *Xenopus oocytes. Neuron* **1**:659–667.

Impulse Transmission and Myelin

BRAY, G. M., M. RASMINSKY, and A. J. AGUAVO. 1981. Interactions between axons and their sheath cells. *Ann. Rev.
Neurosci.* **4**:127–162.

HENRY, E. W., and R. L. SIDMAN. 1988. Long lives for homozygous trembler mutant mice despite virtual absence of peripheral nerve myelin. *Science* **241**:344–346.

*LEES, M. B., and S. W. BROSTOFF. 1984. Proteins of myelin. In
Myelin, 2d ed., P. Morell, ed. Plenum.

*LEMKE, G. 1986. Molecular biology of the major myelin genes.
Trends Neurosci. **9**:266–270.

*MCKHANN, G. M. 1982. Multiple sclerosis. *Ann. Rev. Neurosci.*
5:219–239.

*MORELL, P., and W. T. NORTON. 1980. Myelin. *Sci. Am.*
242(5):88–118.

*RAINE, C. D. 1984. Morphology of myelin and myelination. In
Myelin, 2d ed., P. Morell, ed. Plenum.

RITCHIE, J. M. 1983. On the relationship between fibre diameter
and impulse transmission in myelinated nerve fibers. *Proc.
R. Soc. Lond., Ser. B* **217**:29–39.

SALZER, J. L., W. P. HOLMES, and D. R. COLMAN. 1987. The
amino acid sequences of the myelin-associated glycoproteins: homology to the immunoglobulin gene superfamily.
J. Cell Biol. **104**:957–965.

TAKAHASHI, N., A. ROACH, D. B. TEPLOW, S. B. PRUSINER, and
L. HOOD. 1985. Cloning and characterization of the myelin
basic protein gene from mouse: one gene can encode both
14 kd and 18.5 kd MBPs by alternate use of exons. *Cell*
42:138–148.

WAXMAN, S. G., and J. M. RITCHIE. 1985. Organization of ion
channels in the myelinated nerve fiber. *Science* **228**:1502–
1507.

Synapses and Impulse Transmission: Structure and Function of Synapses and Synaptic Vesicles

*EDELMAN, G. M., W. E. GALL, and W. M. COWAN, eds. 1987.
Synaptic Function. John Wiley & Sons.

HANNA, R. B., J. S. KEETER, and G. P. PAPPAS. 1978. The fine
structure of a rectifying electrotonic synapse. *J. Cell Biol.*
79:764–773.

HEUSER, J. E., and T. S. REESE. 1981. Structural changes after
transmitter release at the frog neuromuscular junction. *J.
Cell Biol.* **88**:564–580.

HIROKAWA, N., K. SOBUE, K. KANDA, A. HARADA, and
H. YORIFUJI. 1989. The cytoskeletal architecture of the pre-

synaptic terminal and molecular structure of synapsin 1. *J. Cell Biol.* **108**:111–126.

*KELLY, R. B. 1988. The cell biology of the nerve terminal. *Neuron* **1**:431–438.

LANDIS, D. M. D., A. K. HALL, L. A. WEINSTEIN, and T. S. REESE. 1988. The organization of cytoplasm at the presynaptic active zone of a central nervous system synapse. *Neuron* **1**:201–209.

SHIRA, T., J. WANG, F. GORELICK, and P. GREENGARD. 1989. Translocation of synapsin I in response to depolarization of isolated nerve terminals. *Proc. Nat'l Acad. Sci. USA* **86**:8108–8112.

*SNUTCH, T. P. 1988. The use of *Xenopus* oocytes to probe synaptic communication. *Trends Neurosci.* **11**:250–256.

THOMAS., L., et al. 1988. Identification of synaptophysin as a hexameric channel protein of the synaptic vesicle membrane. *Science* **242**:1050–1053.

VALTORTA, F., R. JAHN, R. FESCE, P. GREENGARD, and B. CECCARELLI. 1988. Synaptophysin (p38) at the frog neuromuscular junction: its incorporation into the axolemma and recycling after intense quantal secretion. *J. Cell Biol.* **107**:2717–2727.

Voltage-gated Calcium Channels and Exocytosis

ALSOBROOK II, J. P., and C. F. STEVENS. 1988. Cloning the calcium channel. *Trends Neurosci.* **11**:1–3.

*HAGIWARA, S., and L. BYERLY. 1983. The calcium channel. *Trends Neurosci.* **6**:189–193.

*KATZ, B. 1969. *The Release of Neural Transmitter Substances.* Liverpool University Press.

MILEDI, R. 1973. Transmitter release induced by injection of calcium ions into nerve terminals. *Proc. R. Soc. Lond., Ser. B* **183**:421–425.

NELSON, M. T., R. J. FRENCH, and B. K. KRUEGER. 1984. Voltage-dependent calcium channels from brain incorporated into planar lipid bilayers. *Nature* **308**:77–80.

*SMITH, S. J., and G. J. AUGUSTINE. 1988. Calcium ions, active zones and synaptic transmitter release. *Trends Neurosci.* **11**:458–464.

TAKAHASHI, M., and W. A. CATTERALL. 1987. Identification of an α subunit of dihydropyridine-sensitive brain calcium channels. *Science* **236**:88–91.

Nicotinic Acetylcholine Receptors

BREHM, P. 1989. Resolving the structural basis for developmental changes in muscle ACh receptor function: it takes nerve. *Trends Neurosci.* **5**:174–177.

*CHANGEUX, J. P., A. DEVILLERS-THIERY, and P. CHEMOUILLI. 1984. Acetylcholine receptor: an allosteric protein. *Science* **225**:1335–1345.

*CHANGEUX, J. P., and F. REVAH. 1987. The acetylcholine receptor molecule: allosteric sites and the ion channel. *Trends Neurosci.* **10**:245–250.

FATT, P., and B. KATZ. An analysis of the end-plate potential recorded with an intracellular electrode. *J. Physiol.* **115**:320–370.

IMOTO, K., et al. 1988. Rings of negatively charged amino acids determine the acetylcholine receptor channel conductance. *Nature* **335**:645–648.

LEONARD, R. J., C. G. LABARCA, P. CHARNET, N. DAVIDSON, and H. A. LESTER. 1988. Evidence that the M2 membrane-spanning region lines the ion channel pore of the nicotinic receptor. *Science* **242**:1578–1581.

OIKI, S., W. DANHO, V. MADISON, and M. MONTAL. 1988. M2 δ, a candidate for the structure lining the ionic channel of the nicotinic cholinergic receptor. *Proc. Nat'l Acad. Sci. USA* **85**:8703–8707.

SAKMANN, B., et al. 1985. Role of acetylcholine receptor subunits in gating of the channel. *Nature* **318**:538–543.

*STROUD, R. M., and J. FINER-MOORE. 1985. Acetylcholine receptor structure, function, and evolution. *Ann. Rev. Cell Biol.* **1**:317–351.

TOYOSHIMA, C., and N. UNWIN. 1988. Ion channel of acetylcholine receptor reconstructed from images of postsynaptic membranes. *Nature* **336**:247–250.

UNWIN, N., C. TOYOSHIMA, and E. KUBALEK. 1988. Arrangement of the acetylcholine receptor subunits in the resting and desensitized states, determined by cryoelectron microscopy of crystallized *Torpedo* postsynaptic membranes. *J. Cell Biol.* **107**:1123–1138.

The Neuromuscular Junction and Acetylcholinesterase

*MASSOULIÉ, J., and S. BON. 1982. The molecular forms of cholinesterase and acetylcholinesterase in vertebrates. *Ann. Rev. Neurosci.* **5**:57–106.

*SALPETER, M. M., ed. 1987. *The Vertebrate Neuromuscular Junction.* Neurology and Neurobiology Series, vol. 23. Alan R. Liss.

*TAYLOR, P., M. SCHUMACHER, K. MACPHEE-QUIGLEY, T. FRIEDMANN, and S. TAYLOR. 1987. The structure of acetylcholinesterase: relationship to its function and cellular disposition. *Trends Neurosci.* **10**:93–95.

GABA, Glycine, and Inhibitory Synapses

*BARNARD, E. A., M. G. DARLISON, and P. SEEBURG. 1987. Molecular biology of the GABA$_A$ receptor: the receptor/channel superfamily. *Trends Neurosci.* **10**:502–509.

*BETZ, H. 1987. Biology and structure of the mammalian glycine receptor. *Trends Neurosci.* **10**:113–117.

*GOTTLIEB, D. I. 1988. GABAergic neurons. *Sci. Am.* **258**(2):38–45.

GRENNINGLOH, G., et al. 1987. The strychnine-binding subunit of the glycine receptor shows homology with nicotinic acetylcholine receptors. *Nature* **328**:215–220.

LEVITAN, E. S., L. A. C. BLAIR, V. E. DIONNE, and E. A. BARNARD. 1988. Biophysical and pharmacological properties of cloned GABA$_A$ receptor subunits expressed in *Xenopus* oocytes. *Neuron* **1**:773–781.

*MAELICKE, A. 1988. Structural similarities between ion channel proteins. *Trends Biochem. Sci.* **13**:199–202.

SCHOLFIELD, P. R., et al. 1987. Sequence and functional expression of the GABA_A receptor shows a ligand-gated receptor super-family. *Nature* 328:221–227.

G Proteins and Direct Coupling of Neurotransmitter Receptors to Ion Channels

*BONNER, T. I. 1989. The molecular basis of muscarinic receptor diversity. *Trends Neurosci.* 12:148–151.

BROWN, A. M., and L. BIRNBAUMER. 1989. Direct G protein gating of ion channels. *Am. Physiol. Soc.*, in press.

CERBAI, E., U. KLOCKNER, and G. ISENBERG. 1988. The α subunit of the GTP-binding protein activates muscarinic potassium channels of the atrium. *Science* 240:1782–1783.

*CREESE, I., and C. M. FRASER. 1987. *Dopamine Receptors.* vol. 8: *Receptor Biochemistry and Methodology.* Alan R. Liss.

*DUNLAP, K., G. G. HOLZ, and S. G. RANE. 1987. G proteins as regulators of ion channel function. *Trends Neurosci.* 10:241–244.

*HEMMINGS, H. C., JR., A. C. NAIRN, T. L. MCGUINNESS, R. L. HUGANIR, and P. GREENGARD. 1989. Role of protein phosphorylation in neuronal signal transduction. *FASEB J.* 3:1583–1592.

*KACZMAREK, L. K. 1987. The role of protein kinase C in the regulation of ion channels and neurotransmitter release. *Trends Neurosci.* 10:30–34.

*KACZMAREK, L. K., and I. B. LEVITAN, eds. 1987. *Neuromodulation: The Biochemical Control of Neuronal Excitability.* Oxford University Press.

*ROBISHAW, J. D., and K. A. FOSTER. 1989. Role of G proteins in the regulation of the cardiovascular system. *Ann. Rev. Physiol.* 51:229–244.

*SNYDER, S. H. 1986. Neuronal receptors. *Ann. Rev. Physiol.* 48:461–471.

YATANI, A., et al. 1988. A monoclonal antibody to the α subunit of G_k blocks muscarinic activation of atrial K$^+$ channels. *Science* 241:828–831.

Memory and Neurotransmitters

Drosophila Mutations Affecting Learning

BYERS, D., R. L. DAVIS, and J. A. KIGER, JR. 1981. Defect in cyclic AMP phosphodiesterase due to the dunce mutation of learning in *Drosophila melanogaster*. *Nature* 289:79–81.

*CHANGEUX, J. P., and M. KONISHI, eds. 1987. *The Neural and Molecular Bases of Learning.* Life Sciences Report 38—Dahlem Konferenzen. John Wiley & Sons.

*DUDAI, Y. 1988. Neurogenetic dissection of learning and short-term memory in *Drosophila*. *Ann. Rev. Neurosci.* 11:537–563.

HOTTA, Y., and S. BENZER. 1972. Mapping of behavior in *Drosophila* mosaics. *Nature* 240:527–535.

QUINN, W. G., W. A. HARRIS, and S. BENZER. 1974. Conditioned behavior in *Drosophila melanogaster*. *Proc. Nat'l Acad. Sci. USA* 71:708–712.

QUINN, W. G., P. P. SZIBER, and R. BOOKER. 1979. The *Drosophila* memory mutant amnesiac. *Nature* 277:212–214.

*TULLY, T. 1987. *Drosophila* learning and memory revisited. *Trends Neurosci.* 10:330–335.

Habituation and Sensitization in the Sea Slug *Aplysia*

*ABRAMS, T. W., and E. R. KANDEL. 1988. Is contiguity detection in classical conditioning a system or a cellular property? Learning in Aplysia suggests a possible molecular site. *Trends Neurosci.* 11:128–135.

*GOELET, P., V. F. CASTELLUCCI, S. SCHACHER, and E. R. KANDEL. 1986. The long and the short of long-term memory—a molecular framework. *Nature* 322:419–422.

HOCHNER, B., M. KLEIN, S. SCHACHER, and E. R. KANDEL. 1986. Additional component in the cellular mechanism of presynaptic facilitation contributes to behavioral dishabituation in *Aplysia*. *Proc. Nat'l Acad. Sci. USA* 83:8794–8798.

*KANDEL, E. R., and J. H. SCHWARTZ. 1982. Molecular biology of learning: modulation of transmitter release. *Science* 218:433–443.

KLEIN, M., B. HOCHNER, and E. R. KANDEL. 1986. Facilitatory transmitters and cAMP can modulate accommodation as well as transmitter release in *Aplysia* sensory neurons: evidence for parallel processing in a single cell. *Proc. Nat'l Acad. Sci. USA* 83:7994–7998.

MONTAROLO, P. G., et al. 1986. A critical period for macromolecular synthesis in long-term heterosynaptic facilitation in *Aplysia*. *Science* 234:1249–1253.

*SCHWARTZ, J. H., and S. M. GREENBERG. 1987. Molecular mechanisms for memory: second-messenger–induced modifications of protein kinases in nerve cells. *Ann. Rev. Neurosci.* 10:459–476.

SIEGELBAUM, S. A., J. S. CAMARDO, and E. R. KANDEL. 1982. Serotonin and cyclic AMP close single K$^+$ channels in *Aplysia* sensory neurons. *Nature* 299:413–417.

Sensory Transduction: The Visual System

BAYLOR, D. A., T. D. LAMB, and K.-W. YAU. 1979. Responses of retinal rods to single photons. *J. Physiol.* 288:613–634.

DEIGNER, P. S., W. C. LAW, F. J. CAÑADA, and R. R. RANDO. 1989. Membranes as the energy source in the endergonic transformation of vitamin A to 11-*cis*-retinol. *Science* 244:968–971.

DETERRE, P., J. BIGAY, F. FORQUET, M. ROBERT, and M. CHABRE. 1988. cGMP phosphodiesterase of retinal rods is regulated by two inhibitory subunits. *Proc. Nat'l Acad. Sci. USA* 85:2424–2428.

*DOWLING, J. 1987. *The Retina: An Approachable Part of the Brain.* Harvard University Press.

FESENKO. E. E., S. S. KOLESNIKOV, and A. L. LYUBARSKY. 1985. Induction by cyclic GMP of cationic conductance in plasma membrane of retinal rod outer segment. *Nature* 313:310–313.

*HUBEL, D. H. 1988. *Eye, Brain, and Vision.* Scientific American Library.

*HURLEY, J. B. 1987. Molecular properties of the cGMP cascade of vertebrate photoreceptors. *Ann. Rev. Physiol.* 49:793–812.

KOCH, K.-W., and L. STRYER. 1988. Highly cooperative feedback control of retinal rod guanylate cyclase by calcium ions. *Nature* 334:64–66.

*LIEBMAN, P. A., K. R. PARKER, and E. A. DRATZ. 1987. The molecular mechanism of visual excitation and its relation to the structure and composition of the rod outer segment. *Ann. Rev. Physiol.* 49:765–791.

MATTHEWS, H. R., V. TORRE, and T. D. LAMB. 1985. Effects on the photoresponse of calcium buffers and cyclic GMP incorporated into the cytoplasm of retinal rods. *Nature* 313:582–585.

*NATHANS, J. 1989. The genes for color vision. *Sci. Am.* 260(2):42–49.

OWEN, W. G. 1987. Ionic conductances in rod photoreceptors. *Ann. Rev. Physiol.* 49:743–764.

PUGH, E., and J. ALTMAN. 1988. A role for calcium in adaptation. *Nature* 334:16–17.

*SCHNAPF, J. L., and D. A. BAYLOR. 1987. How photoreceptor cells respond to light. *Sci. Am.* 256(4):40–47.

TANAKA, J. C., R. E. FURMAN, W. H. COBBS, and P. MUELLER. 1987. Incorporation of a retinal rod cGMP-dependent conductance into planar bilayers. *Proc. Nat'l Acad. Sci. USA* 84:724–728.

21

Microtubules and Cellular Movements

Tubulin (red fluorescence) and DNA (yellow) in an anaphase fibroblast

Movements of cells and subcellular structures—the beating of cilia and flagella, the contraction of muscle, the movement of chromosomes, the streaming of algae cytoplasm, and the migration of animal cells along a substratum—have long fascinated cell biologists. The generation of shape of animal cells is an equally intriguing phenomenon, especially in highly specialized cell types such as neurons, with their long axons, and the intestinal brush-border cells, with their pencil-like microvilli. Both the movements and shapes characteristic of a specific eukaryotic cell type or structure involve a complex set of protein fibers found in the cytoplasm—the cytoskeleton. Eukaryotic cells contain three major classes of cytoskeletal fibers: 7-nm-diameter actin microfilaments, 24-nm-diameter microtubules, and 10-nm-diameter intermediate filaments. These fibers are not found in prokaryotes.

Actin microfilaments and microtubules are formed by polymerization of protein subunits called G actin and *tubulin, respectively. The polymerization and depolymerization of these fibers are closely regulated by the cell. For instance, structures such as the microtubule-containing mitotic spindle are formed during the mitotic stage of the cell cycle, then rapidly disassembled. Most eukaryotic cells contain one or more types of intermediate filaments, each of which is also built from specific protein subunits. This chapter focuses on microtubules; the following chapter, on actin microfilaments and intermediate filaments.*

Some actin and microtubule systems are permanent features of cells; these include actin and myosin filaments in the contractile apparatus of muscle, actin filaments in intestinal brush-border cells, and microtubules in cilia of certain microorganisms. These fiber systems lend themselves readily to biochemical and ultrastructural analyses because their highly ordered filament structures are stable, abundant, and easily prepared in pure form. Such studies have revealed a great deal about the molecular structure and function of these systems. For instance, the waving motions of cilia and the contraction of muscle are the consequence of making and breaking specific protein-protein associations as two filaments slide past each other. The energy for these movements is generated by hydrolysis of ATP, catalyzed by specific enzymes that bind to actin or tubulin filaments. Similar tubulin-binding proteins are involved in the movement of small vesicles along microtubules in elongated cells such as nerve axons and in chromosome separation during the late stages of mitosis.

Many of the cytoskeletal proteins are encoded in multigene families. Each family probably derived from an original gene in a primordial eukaryotic cell that over eons of evolution was duplicated and then modified. An important question we shall discuss is whether the several actin and tubulin proteins in higher vertebrates form specific structures or have unique functions or whether the different functions of the various actin- and tubulin-containing structures depend on other components. ▲

Structure and Diversity of Microtubules

With rare exceptions, such as the mammalian erythrocyte, microtubules are found in the cytoplasm of all eukaryotic cells, from amebas to higher plants and animals. These fibers are 24 nm in diameter and vary in length from a fraction of a micrometer to tens of micrometers. The amino acid sequences of their subunits have been highly conserved during evolution, and the molecular organization of all microtubules is the same. Yet microtubules are found in an astoundingly diverse array of cellular structures and are involved in many different types of

(a)

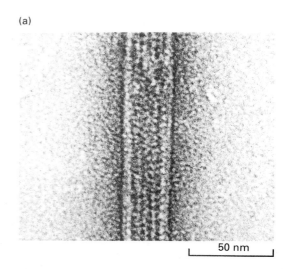

50 nm

(b)

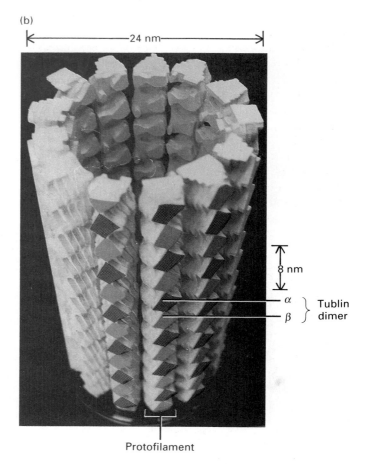

▲ **Figure 21-1** (a) Electron micrograph of a negatively stained microtubule isolated from a flagellum. Note the longitudinal rows of globular subunits. *Courtesy of L. Amos.* (b) Model of the three-dimensional structure of a single microtubule. The tubulin subunits are aligned in 13 parallel rows, called protofilaments. Each protofilament is composed of repeating tubulin dimers linked head to tail; each dimer contains one α- and one β-tubulin molecule. [See L. Amos and T. Baker, 1979, *Int. J. Biol. Macromol.* **1**:146.]

movement. Examples include the mitotic spindle and chromosome movement and cell separation during mitosis; movement of cilia and flagella; and tracks for transport of small vesicles within the cytoplasm. How are these various structures assembled from a single type of fiber, the microtubule, and how does each of them carry out its specific function?

All Microtubules Have a Defined Polarity and Are Composed of α- and β-Tubulin

All microtubules are constructed on the same principle from similar protein subunits. The wall of a microtubule is made up of globular subunits about 4–5 nm in diameter; these subunits are arranged in 13 longitudinal rows encircling the hollow-appearing center (Figure 21-1a). This basic structural design has been found in virtually all microtubules that have been examined; curiously, however, certain microtubules in nematode worms contain 11 or 15 rows of subunits.

Microtubules contain two kinds of proteins: α-tubulin and β-tubulin, each with a molecular weight of 50,000. The wall of a microtubule is composed of a helical array of repeating tubulin subunits containing one α-tubulin and one β-tubulin molecule. This dimeric structural unit ($\alpha\beta$ dimer) is 8 nm in length and is simply called tubulin. Figure 21-1b shows a model of the three-dimensional structure of a microtubule. The 13 protofilaments, each composed of $\alpha\beta$ dimers, run parallel to the long axis. These repeating heterodimers are arranged "head to tail" within microtubules—that is, $\alpha\beta \rightarrow \alpha\beta \rightarrow \alpha\beta$, as shown in the figure. Thus all microtubules have a defined polarity: the two ends are not equivalent. All structures that

contain microtubules must also be polar in nature. These are crucial points to which we shall return several times in this chapter.

Microtubules Form a Diverse Array of Both Permanent and Transient Structures

Microtubules form a large and varied collection of subcellular structures. As noted already, some are more-or-less permanent features of a cell; others are transient. For instance, in a confluent, nondividing monolayer of cultured fibroblast cells, microtubules have a predominantly radial orientation (Figure 21-2). The microtubules extend outward from a site called the centrosome, or *microtubule-organizing center* (MTOC); at the center of an MTOC is one complete and one growing centriole (Figure 21-3). (In higher plants and in fungi such as the slime mold *Dictyostelium*, MTOCs do not contain centrioles.) The distal ends of the microtubules are near the plasma membrane. This array of microtubules is very dynamic: some microtubules disappear rapidly, whereas others

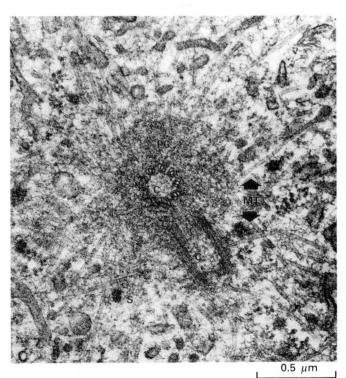

Microtubule-organizing center

▲ **Figure 21-3** Electron micrograph of a cultured rat kangaroo cell showing the microtubule-organizing center, or centrosome. The parent (C) and daughter (C') centrioles are surrounded by electron-dense pericentriolar material (PC) and dense bodies, also termed satellites (S). Several microtubules (MT) are seen radiating from the MTOC. *From B. R. Brinkley, 1987, in* Encyclopedia of Neuroscience, *vol. II, p. 665, Birkhauser Press; courtesy of B. R. Brinkley.*

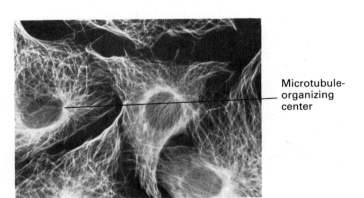

▲ **Figure 21-2** The distribution of microtubules in nonmitotic cultured fibroblast cells. Cells were fixed in glutaraldehyde, then treated with fluorescent tubulin antibody, and viewed in the fluorescence microscope. In these resting cells, the microtubules radiate from the microtubule-organizing center (MTOC), which lies just outside the nucleus and contains the centriole at its center. Note the uniform thickness of the microtubule fibers and their predominantly radial orientation. *Courtesy of M. Osborn.*

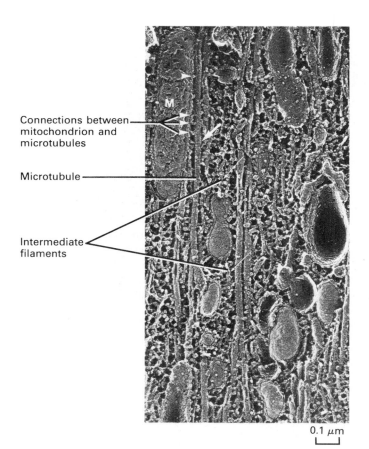

Connections between
mitochondrion and
microtubules

Microtubule

Intermediate
filaments

0.1 μm

▲ **Figure 21-4** Microtubules and intermediate filaments in a quick-frozen frog axon visualized by the deep-etching technique. There are a number of 24-nm-diameter microtubules running longitudinally. Thinner, 10-nm-diameter intermediate filaments also run longitudinally and form occasional connections with microtubules. Embedded in the cytoskeletal lattice are several mitochondria (M) that appear to be connected with the microtubules. Other membrane vesicles also are dispersed in the lattice. [See N. Hirokawa, 1982, *J. Cell Biol.* 94:129.] *Courtesy of N. Hirokawa; reproduced from the* Journal of Cell Biology, *1982, by copyright permission of the Rockefeller University Press.*

grow longer. For example, if some of the cells are scraped off the culture dish before being fixed, the cells remaining at the end of the "wound" commence growth into the open area. (Eventually they will grow to cover the space.) Within a few minutes, the symmetric arrangement of the microtubules in the "resting" cells at the border of the wound is transformed into an array in which most of the microtubules radiate toward the open "space"; this restructuring of the microtubule network precedes a more general polarization of the cell cytoplasm and directed cell growth into the wound.

During mitosis, the microtubule network characteristic of interphase cells disappears, and the spindle apparatus forms. The spindle is important for equal partitioning of chromosomes to the daughter cells, as discussed later.

The tubulin used to form the spindle apparatus is derived from interphase microtubules.

Other microtubule-containing structures do not regularly disassemble and reassemble. A dramatic example of such a permanent structure is the cytoskeleton lattice of microtubules and intermediate filaments that extends along axons and dendrites of neurons from the cell body to the terminals (Figure 21-4). It was once thought that these fibers were continuous throughout the entire axon, but observations on serial sections suggests that individual microtubules are generally only 10–25 μm in length. The thinner intermediate filaments are associated with the axonal microtubules. Connections formed of fibrous microtubule-associated proteins between adjacent microtubules, and between microtubules and intermediate filaments, may be important in forming a stable cytoskeleton. Some experiments suggest that axonal microtubules may have a lifetime of 100 days or more. As we shall see, these microtubules act as guides or tracks along which protein particles and organelles move up and down the axon during axoplasmic transport.

Cilia and flagella (see Figure 4-49) are other microtubule-containing structures that generally persist throughout growth and division, although in many protozoans flagella are resorbed before cell division and reappear after cytokinesis is complete. Proteins bound to the microtubules utilize energy released by ATP hydrolysis to generate the force that causes the cilia or flagella to move. Certain protozoa utilize slender extensions, called tentacles, to trap prey (Figure 21-5). These have a core of microtubules, which undergo striking rearrangements as the tentacles contract or expand.

Structural and Kinetic Polarity of Microtubules

All properties of microtubules are related to their structural polarity, which is reflected in the functional differences of the two ends. For instance, in cell-free reactions addition and loss of tubulin subunits can occur at both ends of a microtubule, but under most experimental conditions both addition and loss occur preferentially at one end, designated the plus (+) end. The ability of microtubules to grow and shrink rapidly allows rapid re-formations of microtubule-containing structures (e.g., conversion of interphase microtubules into mitotic spindles). Before proceeding further in our discussion of microtubule-containing structures and microtubule-based movements, we will take a close look at the assembly, disassembly, and polarity of microtubules. An important question is whether the dynamics of assembly and disassembly of microtubules as observed in cell-free reactions are consistent with the formation and depolymerization of microtubules in living cells.

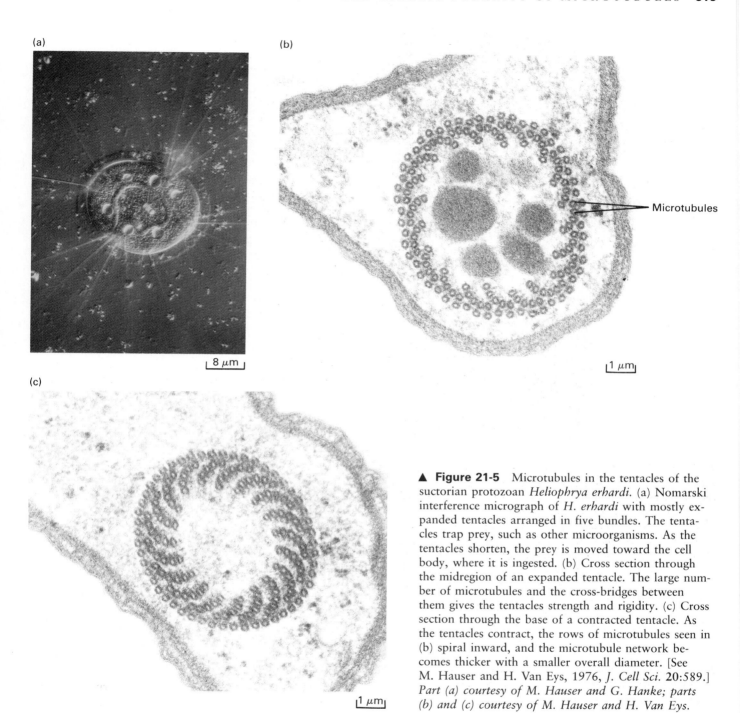

(a)

(b)

Microtubules

8 μm

1 μm

(c)

1 μm

▲ **Figure 21-5** Microtubules in the tentacles of the suctorian protozoan *Heliophrya erhardi*. (a) Nomarski interference micrograph of *H. erhardi* with mostly expanded tentacles arranged in five bundles. The tentacles trap prey, such as other microorganisms. As the tentacles shorten, the prey is moved toward the cell body, where it is ingested. (b) Cross section through the midregion of an expanded tentacle. The large number of microtubules and the cross-bridges between them gives the tentacles strength and rigidity. (c) Cross section through the base of a contracted tentacle. As the tentacles contract, the rows of microtubules seen in (b) spiral inward, and the microtubule network becomes thicker with a smaller overall diameter. [See M. Hauser and H. Van Eys, 1976, *J. Cell Sci.* **20**:589.] *Part (a) courtesy of M. Hauser and G. Hanke; parts (b) and (c) courtesy of M. Hauser and H. Van Eys.*

Microtubule Assembly and Disassembly Occur by Preferential Addition and Loss of $\alpha\beta$ Dimers at the (+) End

When purified preparations of microtubules from sperm, brain, or certain protozoan flagella are subjected to chilling, the microtubules depolymerize into stable $\alpha\beta$ dimers (MW about 100,000) called tubulin. Tubulin does not dissociate into α and β monomers unless denaturing agents are added. Conversely, microtubules of normal structure can be reconstituted in vitro from a solution of tubulin under physiologic conditions. Tubulin forms microtubules when warmed to 37°C in the presence of GTP, provided Ca^{2+} is not present.

In such cell-free reactions, addition of tubulin occurs primarily at the ends of preexisting microtubules (Figure 21-6). If a pure preparation of $\alpha\beta$ dimers is used in an assembly reaction, the initial rate of formation of micro-

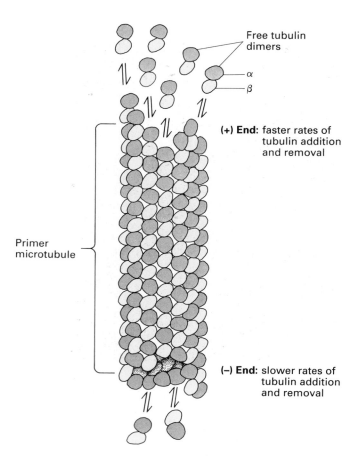

▲ Figure 21-6   In vitro microtubule assembly-disassembly reaction. The addition of $\alpha\beta$-tubulin dimers occurs primarily at the ends of preexisting microtubule "primers"; only infrequently does tubulin spontaneously polymerize into short microtubules, which then act as primers for elongation. The rates for both addition and loss of tubulin in vitro are about twice as great at the (+) end than at the (−) end. Disassembly is favored at low temperatures; assembly is favored at 37°C and by the presence of GTP.

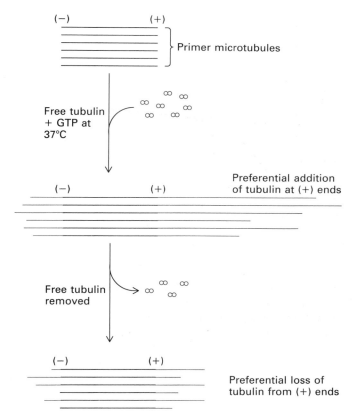

▲ Figure 21-7   Experimental demonstration that microtubules grow and shrink preferentially at one end. Fragments of flagellar microtubules are used as primers for in vitro addition of tubulin. Because of their highly organized structure, the flagellar primers (black) can be distinguished in the electron microscope from the microtubules formed in the cell-free reaction (red). The distal ends—designated the (+) ends—of the flagellar microtubules also can be distinguished from their proximal, or (−), ends. Both the addition and loss of tubulin occurs preferentially at the (+) ends of microtubules.

tubules is very slow; the rate-limiting step is generation of a small microtubule "primer." Addition of fragments of flagellar or other microtubules to a solution of $\alpha\beta$ dimers greatly accelerates the initial rate at which tubulin forms microtubules.

Electron-microscope observations of in vitro microtubule-assembly reactions show that the two ends of a primer microtubule are functionally different. Addition of tubulin to primer microtubules occurs preferentially at the (+) end (also called the net assembly end); the rate constant for addition of tubulin at the (+) end is twice that for addition at the other end, the (−) end (Figure 21-7). Disassembly of microtubules by loss of tubulin also occurs twice as rapidly at the (+) end than at the (−) end. Thus both assembly and disassembly occur preferentially at the (+) end.

Under defined conditions of ion composition, there is a free tubulin subunit concentration at which a steady-state condition is reached; that is, the addition of dimers to some microtubules is just balanced by their loss from others. This is termed the critical concentration of tubulin, C_c. Above this concentration, microtubules tend to grow; below it, they tend to shrink (Figure 21-8).

Colchicine and Other Treatments Can Shift the Microtubule Assembly-Disassembly Steady State

In vitro studies of microtubules have suggested that a number of microtubule functions result from their capacity to assemble and disassemble reversibly. Mitosis is an

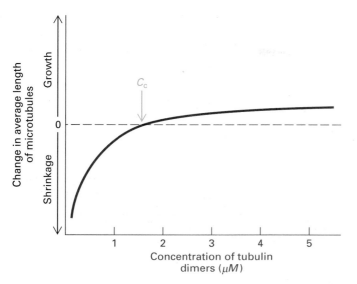

▲ **Figure 21-8** The effect of tubulin concentration on microtubule assembly and disassembly. Above the critical tubulin concentration C_c, the average length of microtubules increases; below C_c, it decreases. *Adapted from M. Kirschner and T. Mitchison, 1986, Cell 45:329; and M. F. Carlier, T. L. Hill, and Y. D. Chen, 1984, Proc. Nat'l Acad. Sci. USA 81:771.*

(a)

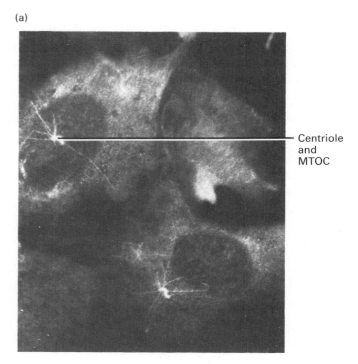

(b)

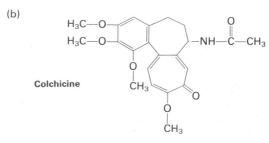

▲ **Figure 21-9** (a) Micrograph of colchicine-treated mouse fibroblasts exposed to a fluorescent antibody specific for tubulin. Comparison with Figure 21-2 shows that most cytoplasmic microtubules are lost after colchicine treatment, although the centriole and MTOC are retained. The microtubules in centrioles, cilia, and flagella are not affected by colchicine. (b) Structure of colchicine. This water-soluble drug blocks polymerization of tubulin subunits into microtubules. [See M. Osborn and K. Weber, 1975, *Proc. Nat'l Acad. Sci. USA* 73:867.] *Part (a) courtesy of M. Osborn and K. Weber.*

example. Colchicine, a plant alkaloid, blocks plant and animal cells at metaphase but does not affect chromosome condensation. In the presence of colchicine, no spindle forms, and there is no movement of chromosomes toward the poles of the cell. Thus colchicine can be used to produce accumulations of metaphase chromosomes for cytogenetic studies. The drugs vinblastine and vincristine, which also inhibit formation of mitotic spindles, have been widely used as anticancer agents since blockage of spindle formation will preferentially kill rapidly dividing cells.

The basis for the action of colchicine is its ability to bind tubulin. Each $\alpha\beta$ dimer contains one high-affinity binding site for colchicine. Tubulin with a bound colchicine molecule can add to the end of a microtubule, but the presence of one or two colchicine-bearing tubulins prevents further addition of any other tubulins whether or not they contain bound colchicine. Clearly, this blocking ability of colchicine explains its inhibitory action on formation of the mitotic spindle. However, colchicine does not directly cause disassembly of microtubules.

Many microtubule-containing structures, such as the mitotic spindle, exist in a dynamic steady state in which polymerization and depolymerization of subunits are balanced. In such structures, colchicine blockage of tubulin polymerization results in a net loss of microtubules and an accumulation of free tubulin. The disappearance of most cytoplasmic microtubules in nonmitotic fibroblasts treated with colchicine (Figure 21-9) suggests that these

microtubules also are normally in an assembly-disassembly steady state. Upon removal of colchicine by washing, the microtubules reappear in such cells.

As illustrated in Figure 11-49, tubulin synthesis is autoregulated; that is, unpolymerized tubulin binds to the ribosomes synthesizing tubulin and triggers degradation of tubulin mRNA. Because treatment with colchicine leads to an increase in unpolymerized tubulin, it also inhibits tubulin synthesis; synthesis is reinduced after colchicine removal.

Microtubules exposed to low temperatures or high

pressure undergo disassembly; this can be reversed by raising the temperature or lowering the pressure. Treatments such as the use of colchicine or high pressure do not affect the more stable microtubules in centrioles, flagella, and cilia.

Microtubules Contain Microtubule-associated Proteins

Tubulin can be purified by subjecting a crude homogenate of brain or of cultured fibroblasts to several cycles of microtubule assembly (achieved by warming in the presence of GTP, followed by centrifugation) and disassembly (achieved by chilling). The resulting "purified" tubulin preparations, however, contain a few other proteins of various molecular weights. Although these proteins are present in lower amounts than α- and β-tubulin, in successive cycles of polymerization-depolymerization, the quantitative ratio of these proteins to α- and β-tubulin is constant. Thus most such proteins, called *microtubule-associated proteins* (MAPs), are not nonspecific contaminants but rather have a specific association with α- and β-tubulin.

Different microtubule structures appear to contain different MAPs. Repeated assembly-disassembly of mitotic spindle microtubules, for example, yields at least one MAP that is unique to the mitotic apparatus, as well as several that are also found in interphase microtubules. These MAPs may control the polymerization of tubulin into different types of microtubules. Many MAPs are phosphorylated, suggesting that a phosphorylation-dephosphorylation cycle is an important step in MAP regulation of microtubule assembly.

The axons but not the dendrites of nerve cells contain a microtubule-associated protein called tau. This protein

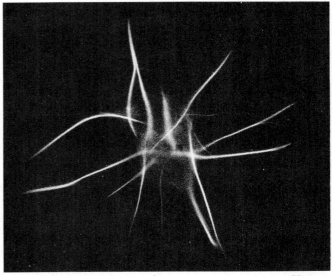

▲ **Figure 21-11** Distribution of microtubules, detected by a fluorescent antitubulin antibody, in a cultured mouse fibroblast expressing high levels of recombinant neuronal tau protein. Tau causes microtubules to form thick bundles, similar to those in nerve cell axons. See Figure 21-2 for normal distribution of microtubules in cultured fibroblasts. *From Y. Kanai et al., 1989, J. Cell Biol. **109**:1173; courtesy of N. Hirokawa.*

exists in four or five forms (MW 55,000–62,000) derived from alternative splicing of a single *tau* gene. Tau protein accelerates the polymerization of tubulin and stabilizes microtubules by forming 18-nm-long armlike projections, which can cross-link adjacent microtubules (Figure 21-10). Expression by recombinant DNA techniques of high levels of tau in a cultured fibroblast cell causes all of the cell's microtubules to coalesce into a few axonlike projections (Figure 21-11). The ability of tau to cross-link microtubules into thick bundles may contribute to the great stability of axonal microtubules.

Nerve cells also contain abundant amounts of two large microtubule-associated proteins, called MAP1 and MAP2, each with a molecular weight of about 270,000. MAP1—a flexible rod that is found in both axons and dendrites—forms cross-bridges between microtubules. MAP2, in contrast, is found only in dendrites; there it forms fibrous cross-bridges between microtubules and also links microtubules to intermediate filaments. The localization of different MAPs to axons and dendrites may contribute to their different shapes.

As exemplified by these studies on neurons, a working hypothesis in microtubule research is that each type of microtubule structure contains functionally equivalent α and β subunits and that the unique properties of each type of microtubule are determined by structure-specific MAPs, which copolymerize with the tubulin subunits or bind to microtubules after polymerization.

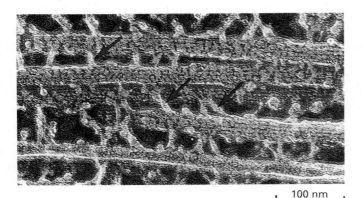

▲ **Figure 21-10** Electron micrograph of a quick-frozen, deep-etched suspension of microtubules polymerized with tau protein. The tau protein forms 18-nm-long projections extending from the sides of the microtubules (*arrows*); these form cross-bridges between microtubules that are close together. *From N. Hirokawa, Y. Shiomura, and S. Okabe, 1988, J. Cell Biol. **107**:1449; courtesy of N. Hirokawa.*

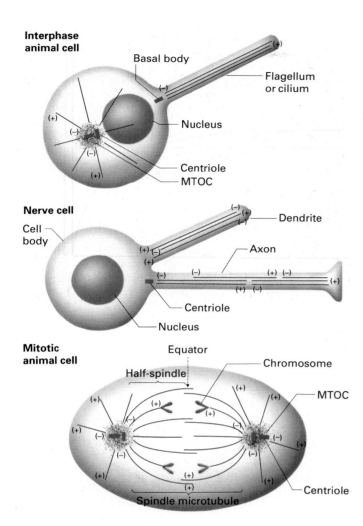

Interphase animal cell

Basal body

Flagellum or cilium

Nucleus

Centriole

MTOC

Nerve cell

Cell body

Dendrite

Axon

Centriole

Nucleus

Mitotic animal cell

Equator

Chromosome

Half-spindle

MTOC

Centriole

Spindle microtubule

◄ **Figure 21-12** Polarity of microtubules. In interphase cells, the (−) ends of most microtubules are proximal to the MTOC. (Microtubules extend around the nucleus during interphase.) As cells enter mitosis, the microtubule network rearranges, forming a mitotic spindle; the (−) ends of all spindle microtubules point toward one of the two MTOCs, or poles, as they are called in mitotic cells. Similarly, the microtubules in flagella and cilia have their (−) ends facing the basal body, which acts as the MTOC in these structures. Axonal microtubules in neurons exhibit the usual polarity with the (−) end facing the cell body; dendritic microtubules, however, are unusual in that either their (+) or (−) ends can face the cell body. [See M. M. Black and P. W. Baas, 1989, *Trends Neurosci.* **12**:211.]

For example, no obvious MTOC is present in dendrites, and dendritic microtubules can have either their (+) or (−) ends pointing toward the neuron cell body. In axons, all microtubules have the same polarity, and those that end near the cell body radiate from a centriole; however, microtubules totally within an axon do not appear to emerge from any MTOC.

Figure 21-13 outlines just one of many experiments that have demonstrated polarity in cellular microtubules. The polarity of microtubules has profound consequences for their regulated assembly and disassembly and for all aspects of microtubule-based movements.

Microtubules Grow from MTOCs

As mentioned earlier, in colchicine-treated cells, almost all of the cytoplasmic microtubules, except the centriole, are depolymerized (see Figure 21-9). When the colchicine is removed by washing, tubulin polymerizes and new microtubules form, radiating from the MTOC (Figure 21-14).

The role of the centriole in organizing microtubules has been substantiated experimentally. For example, if centrioles free of most of the amorphous surrounding material are injected into frog eggs, microtubules grow outward from them. Also, purified MTOCs (centrioles plus attached material) nucleate test-tube polymerization of tubulin into microtubules. However, such findings do not rigorously prove whether the centrioles or the amorphous pericentriolar material surrounding them plays the crucial organizing role. Importantly, we do not know whether MTOCs contain short primer microtubules or whether some other material initiates growth of microtubules.

The Microtubule-organizing Center Determines the Polarity of Cellular Microtubules

In both interphase and mitotic cells, most microtubules radiate from an amorphous area—the centrosome, or MTOC. In animal cells, as noted earlier, *centrioles* are at the center of the MTOC (see Figures 21-2 and 21-3). Similarly, flagella and cilia microtubules radiate from a *basal body,* a structure that is similar in function to an MTOC. Figure 21-12 illustrates a key property of most cellular microtubules: they have a defined polarity with respect to the MTOC or basal body such that the (−) end of each microtubule points toward the MTOC or basal body, and the (+) end points away. In mitotic cells there are two poles from which microtubules radiate; all of the spindle microtubules have their (−) ends pointing to one of the poles, and their (+) ends pointing to the equator at the center of the cell. It is the (+) ends of spindle microtubules that make contact with chromosomes.

Neurons provide important exceptions to the rule that the (−) end of a microtubule points toward an MTOC.

Microtubules in Cells Elongate and Shrink from Their Distal (+) Ends

As we have seen, in cell-free preparations tubulin will add to both ends of an existing microtubule, although twice as fast to (+) ends than to the (−) ends. In cells, however, growth and shrinkage usually does not occur at the (−)

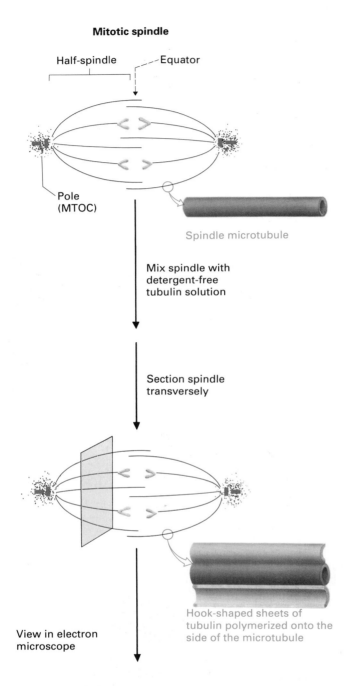

Mitotic spindle

Half-spindle --- Equator

Pole
(MTOC)

Spindle microtubule

Mix spindle with
detergent-free
tubulin solution

Section spindle
transversely

View in electron
microscope

Hook-shaped sheets of
tubulin polymerized onto the
side of the microtubule

◀ **Figure 21-13** Experimental demonstration of the polarity of microtubules in the half-spindle of a mitotic cell. Mitotic spindles are isolated by lysing anaphase mitotic fibroblasts with detergents in a solution containing a high concentration of soluble nerve tubulin. Because of the abnormally high salt concentration, the nerve tubulin polymerizes in hook-shaped sheets onto the walls of preexisting spindle microtubules. The spindle is then sectioned transversely midway between the polar region and the equator and viewed in the electron microscope (*bottom*). Most of the microtubules are "decorated" with hooks from the sheets of polymerized tubulin. Over 90 percent of all hooks "curve" clockwise, indicating that these microtubules have the same polarity. The few microtubules of opposite polarity are scattered throughout the section. By comparing the "direction" of the hooks on spindle microtubules with that formed by polymerization of tubulin on other isolated microtubules, where the (+) and (−) ends can be identified, investigators have concluded that the (−) ends of mitotic spindles face the poles. [See U. Euteneuer and J. R. McIntosh, 1981, *J. Cell Biol.* 89:338.] *Photograph courtesy of U. Euteneuer and J. R. McIntosh; reproduced from the* Journal of Cell Biology, *1981, by copyright permission of the Rockefeller University Press.*

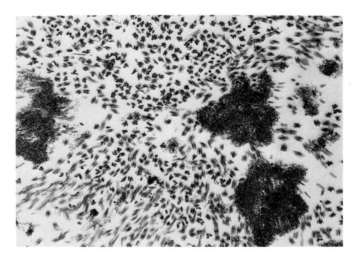

ends of microtubules. The (−) ends of cellular microtubules frequently are sequestered within an MTOC and probably are "capped" by stabilizing microtubule-associated protein(s). Examples include the basal body at the (−) ends of flagellar microtubules and the MTOC at the (−) ends of interphase microtubules. Thus in cells, microtubules generally grow and shrink only at their (+) ends, even though microtubules polymerized in cell-free reactions from pure tubulin can grow and shrink at both ends. Let us consider two types of experiments, in different cell systems, that enabled investigators to come to this important conclusion.

Growth of Flagellar Microtubules The alga *Chlamydomonas reinhardtii* has two flagella, each about 8 μm in length (Figure 21-15a). When flagella are amputated by mechanical agitation, the cells regenerate new flagella, of precisely the same length as the old, within 1 h. When radiolabeled amino acids were added to a culture of *Chlamydomonas* that were regenerating flagella, and the cells later examined by autoradiography, most of the newly added radioactive protein was in the distal tips— the (+) ends—of the flagella (Figure 21-15b). In agreement with this result, cell-free experiments in which brain tubulin was added to partially disrupted *Chlamydomo-*

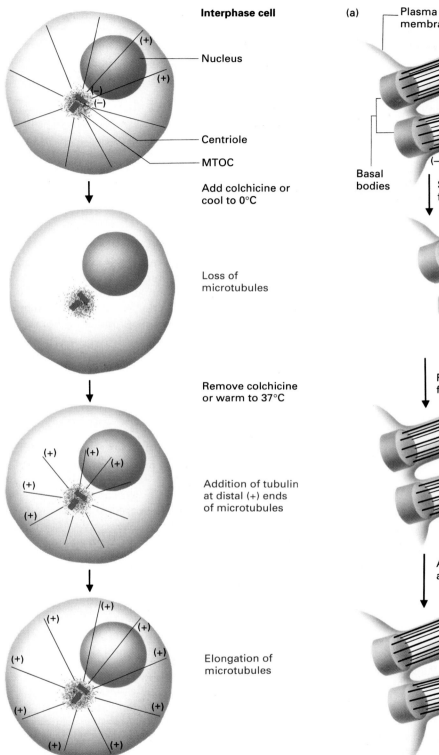

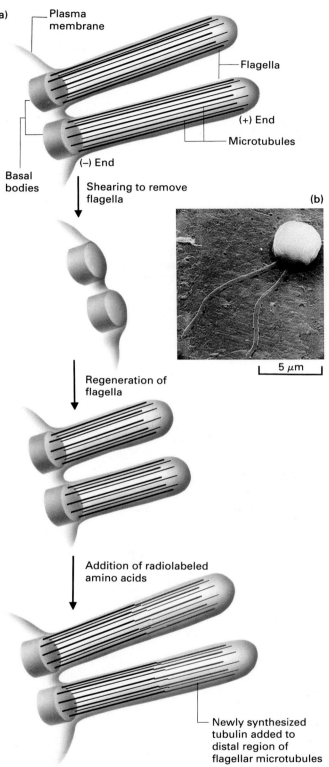

▲ **Figure 21-14** The disassembly and reassembly of microtubules in interphase cultured animal cells can be induced either by addition and subsequent removal of colchicine or by cooling to 0°C and subsequent rewarming to 37°C. Both addition and removal of tubulin occurs at the (+) ends of the microtubules.

▲ **Figure 21-15** (a) Experimental demonstration of flagella regeneration in *C. reinhardtii*. By autoradiography, the newly synthesized radiolabeled tubulin (red) can be shown to be present only at the distal tip of the flagellar microtubules. (b) Scanning electron micrograph of the gamete stage of *Chlamydomonas reinhardtii* showing its two flagella. *Part (b) courtesy of B. Bean.*

nas flagella also indicated that tubulin addition to flagellar microtubules occurs preferentially at their distal (+) ends. These experiments show that a flagellar microtubule grows outward from its free (+) end.

Growth of Microtubules in Cultured Cells

In cultured animal cells most microtubules radiate from the MTOC. However, in any cell at any time some microtubules are growing and others shrinking. Experiments suggest that the average lifetime of any one microtubule is about 10 min in a growing cell. In such cells, all elongation of microtubules occurs at the distal (+) end.

As depicted in Figure 21-16, within 1 min after a preparation of "tagged" tubulin was microinjected into an interphase cell, the tagged tubulin was incorporated into cell microtubules, but only at their distal ends; this finding directly establishes that microtubules grow by addition at the (+) end. Some of the tagged tubulin was incorporated into short microtubules radiating from the MTOC, suggesting that it was in "new" microtubules initiated at the MTOC during the period just after the tagged tubulin was microinjected. In cells examined 20 min or longer after injection of the tagged tubulin, virtually all of the microtubules had some tagged tubulin, even at their proximal (−) ends. Thus during the 20 min after injection of the tagged tubulin, almost all of the cell's microtubules depolymerized, and new ones reassembled with a mixture of tagged and untagged (pre-

existing) tubulin. Two important conclusions can be derived from this study. First, most microtubules in these cultured cells are constantly disassembling into free tubulin and reassembling into new ones. Second, the "average" microtubule exists for only about 10 min before it is depolymerized.

In summary, microtubules in growing animal cells are in a dynamic state and grow by addition of tubulin to their distal ends. As we shall see in the next section, microtubules also shrink by loss of tubulins from the distal end, and new microtubules are constantly being nucleated by the MTOC as "old" microtubules are being depolymerized. A few microtubules in cultured cells do exist for several hours, and in neuronal axons microtubules are stable for many days. Presumably this stability is due to microtubule-associated proteins, which prevent depolymerization (e.g., tau protein discussed earlier). The slow turnover of axonal microtubules has been demonstrated by an experiment similar to that illustrated in Figure 21-16. After microinjection of tagged tubulin into the cell body of a cultured neuron, it took a few hours, rather than a few minutes, before most of microtubules were labeled. As in interphase fibroblasts, tubulin added preferentially to the distal (+) ends of individual microtubules.

In the Same Cells Some Microtubules Are Growing While Others Are Shrinking

A number of experiments on different types of cells all point to the conclusion that some microtubules can grow, while others, at the same time, are shrinking. This conclusion was unexpected and appeared to be in conflict with cell-free studies on microtubule assembly indicating that assembly and disassembly should be regulated primarily

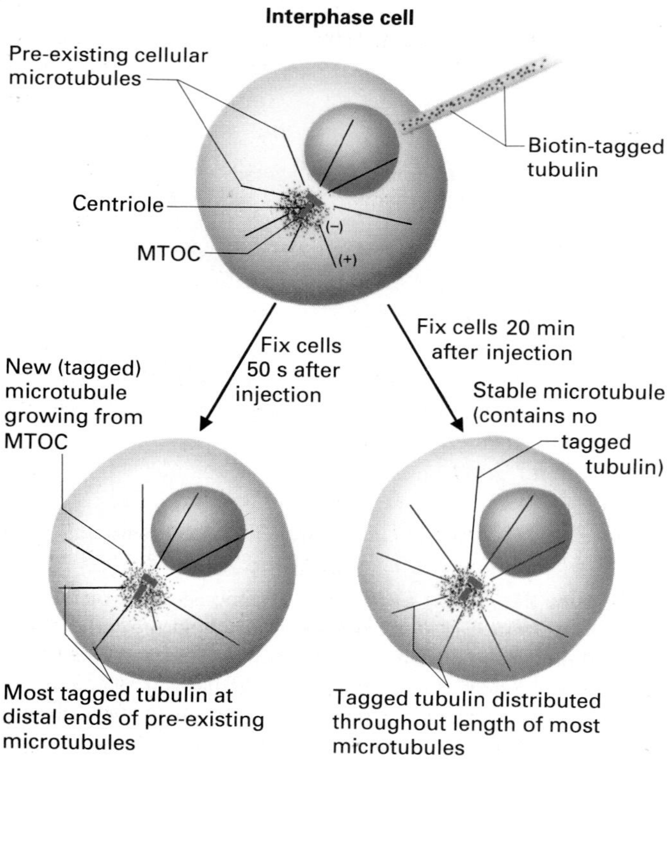

Interphase cell

Pre-existing cellular microtubules

Biotin-tagged tubulin

Centriole

MTOC

(−)

(+)

Fix cells 50 s after injection

Fix cells 20 min after injection

New (tagged) microtubule growing from MTOC

Stable microtubule (contains no tagged tubulin)

Most tagged tubulin at distal ends of pre-existing microtubules

Tagged tubulin distributed throughout length of most microtubules

◀ **Figure 21-16** Experimental demonstration of the growth and turnover of microtubules in cultured interphase cells. A preparation of pure tubulin was chemically linked to biotin, and then a small amount of this tagged tubulin was microinjected into cultured fibroblasts. Cells were fixed either 50 s or 20 min after injection. The total cell tubulin (black) was detected with a fluorescent antibody specific for tubulin, and the biotin-tagged tubulin (red) with a fluorescent antibody specific for biotin. After the 50-s incubation (*left*), most of the microtubules contained tagged tubulin but only at their distal (+) ends, whereas after the 20-min incubation (*right*), virtually all the microtubules contained some tagged tubulin throughout their length. The latter finding indicates that during the 20-min incubation most microtubules were depolymerized, and new ones reassembled with a mixture of tagged and nontagged tubulins. *Adapted from B. Soltys and G. Borisy, 1985,* J. Cell Biol. **100***:1682; and E. Schulze and M. Kirschner, 1986,* J. Cell Biol. **102***:1020.*

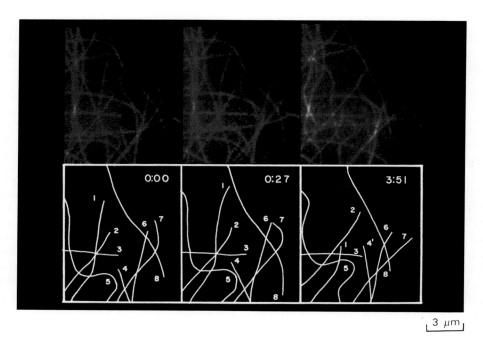

3 μm

▲ **Figure 21-17** Growth and shrinkage of individual microtubules. Fluorescent-labeled tubulin was microinjected into cultured human fibroblasts. The cells were chilled to depolymerize preexisting microtubules and then incubated at 37°C to repolymerize the fluorescent tubulin into all the cell's microtubules. A region of the cell periphery was viewed in the fluorescence microscope at 0 s, 27 s later, and 3 min 51 s later (*left to right panels*). The lower panels are tracings of the images in the top panels, showing selected microtubules. *From P. J. Sammak and G. Borisy, 1988,* Nature *332:724.*

by the level of unpolymerized tubulin. The in vitro studies suggest that above the "critical concentration" of tubulin, all microtubules should grow, and below it, all should shrink (see Figure 21-8). Let us first look at several studies that established simultaneous growth and shrinkage and then at the favored explanation for this phenomenon.

Studies on the *Chlamydomonas* flagellar system demonstrated that the length of the microtubules is carefully controlled. After deflagellation, but in the absence of any protein synthesis, a *Chlamydomonas* cell can regenerate a flagellum up to one-half its original length. This result shows that there are large preexisting pools of precursor proteins for flagella present during interphase but that these are not normally used to produce more or longer flagella. Obviously, then, assembly of flagella is highly regulated. Moreover, if *Chlamydomonas* cells are sheared so that only one of the two flagella is amputated, the remaining flagellum is initially shortened at the same rate as the new one is regenerated. Once both flagella reach half of normal length, both begin to grow to normal length. This observation establishes that flagellar microtubule assembly and disassembly can occur simultaneously in a single cell. Probably different conditions within the two flagella (e.g., the concentrations of Ca^{2+} or of a MAP) cause the microtubules in one to grow and those in the other to shrink.

The dynamics of individual microtubules can also be followed in live cultured cells. Such experiments show that one microtubule can grow, shrink, grow, and shrink, all in a period of a few minutes. In the particular region of the cultured fibroblast shown in Figure 21-17, one microtubule (number 2) grew continuously during a short observation period, four (numbers 1, 6, 7, and 8) grew and then shrank, and one (number 3) remained at the same length. Another situation in which one set of microtubules in a cell grows while another shrinks occurs during prophase of mitosis; in this case, spindle microtubules grow while cytoplasmic microtubules shrink.

Simultaneous microtubule growth and shrinkage also occurs in in vitro microtubule-assembly reactions. As noted earlier, isolated MTOCs nucleate assembly of microtubules from purified tubulin. When such preparations of nucleated microtubules are diluted, thus reducing the concentration of free tubulin, some of the microtubules depolymerize (as expected), but other microtubules actually grow to much longer lengths—a surprising result.

One current model—called the dynamic instability model—to explain the simultaneous growth and shrinkage of microtubules is depicted in Figure 21-18. The essential feature of this model is that the ends of growing and shrinking microtubules are different. Each dimeric

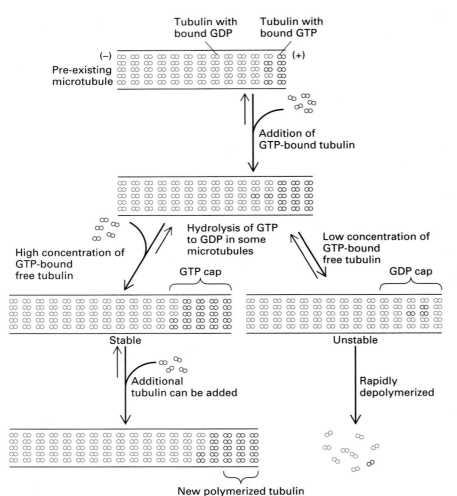

◀ **Figure 21-18** Dynamic instability model of microtubule growth and shrinkage. Tubulin dimers with two bound GTP molecules (red) add preferentially to the (+) end of a preexisting microtubule. After incorporation of a tubulin dimer, one bound GTP is hydrolyzed to GDP. This hydrolysis is apparently catalyzed by the microtubule itself but may be facilitated by cytosolic proteins. Only microtubules with terminal tubulin associated with GTP (those with a GTP cap) are stable and can serve as primers for polymerization of additional tubulin. Microtubules with tubulin bound to GDP (blue) at the end (those with a GDP cap) are rapidly depolymerized and may disappear within 1 min. At high concentrations of unpolymerized, GTP-bound tubulin, the rate of addition of tubulin is faster than the rate of hydrolysis of the GTP bound in the microtubule or of dissociation of GTP-bound tubulin from the end; thus the microtubule grows. At low concentrations of unpolymerized, GTP-bound tubulin, the converse is true; a GDP cap forms and the microtubule shrinks. [See T. Mitchison and M. Kirschner, 1984, *Nature* **312**:237; M. Kirschner and T. Mitchison, 1986, *Cell* **45**:329; and R.A. Walker et al., 1988, *J. Cell Biol.* **107**:1437.]

tubulin subunit contains two bound GTP molecules. One of these is hydrolyzed to GDP during or just after incorporation of a tubulin subunit into the end of a microtubule. If, as a result, all (or most) of the tubulins at either end of a microtubule have a bound GDP (a GDP "cap"), the microtubule is unstable and depolymerizes rapidly from that end. If, by contrast, additional GTP-bound tubulin adds to the end before hydrolysis of the bound GTP or before dissociation of GTP-bound tubulin, that end of the microtubule is not only stable but continues to grow. Thus one parameter that determines the stability of a microtubule is the concentration of free tubulin: a high concentration favors continued growth, and a low concentration allows a GDP cap to form at the end, causing the microtubule to depolymerize. Another parameter affecting stability is the rate of hydrolysis of GTP to GDP: a slow rate favors continued microtubule growth, and a fast rate favors shrinkage. The specific factors that catalyze and regulate GTP hydrolysis in microtubules are not known and are an important subject of current research. Apparently local variations either in the concentration of free tubulin or in the rate of GTP hydrolysis allow some microtubules to grow while adjacent ones shrink.

Heterogeneity of α- and β-Tubulin

All microtubules, in all eukaryotes, have the same basic structural design (see Figure 21-1b). Yet microtubules form a diverse array of structures within a single cell, and different cell types often contain different microtubular structures (e.g., flagella and axonal microtubules). There is growing evidence that different cells contain different types of α- or β-tubulin. Some subtypes are products of different α- or β-tubulin genes, in which case they are called tubulin *isotypes*. For instance, one isotype of β-tubulin is found only in erythrocytes and platelets and their precursor cells. In particular, this isotype is localized to a ring of microtubules, called the *marginal band*, that lies just beneath the plasma membrane of platelets and of all nonmammalian erythrocytes; the marginal band helps maintain cell shape. Some subtypes of α-tubulin are formed by posttranslational covalent modifications, such as addition of an acetyl group to one lysine residue and removal (or readdition) of the carboxyl-terminal tyrosine residue. Such modifications might determine the ability of an α-tubulin molecule to be incorporated into different microtubular structures. However, most research indicates that the differences among the multiple α- and β-tubulin subtypes may be unimportant functionally and that the functional differences among microtubule-containing fibers is determined mainly by the specific microtubule-containing proteins they contain.

Vertebrates Have Genes Encoding Closely Related α- and β-Tubulins

Yeast cells contain a single β-tubulin gene, which is essential for viability (Figure 21-19), and two α-tubulin genes whose protein products are only 80% percent identical. Gene disruption experiments, similar to those depicted in Figure 21-19, indicate that expression of either α-tubulin gene allows yeast cells to grow and divide normally, so long as a sufficient amount of α-tubulin is synthesized.

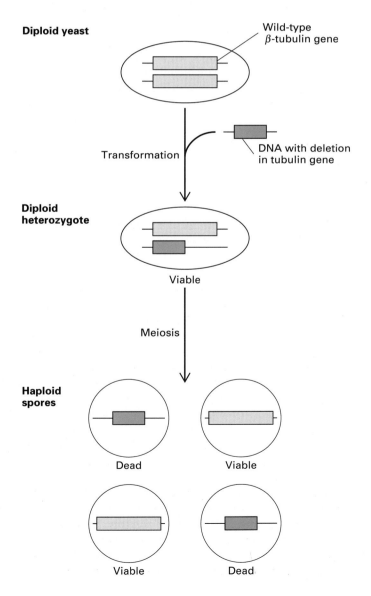

Diploid yeast

Wild-type β-tubulin gene

Transformation

DNA with deletion in tubulin gene

Diploid heterozygote

Viable

Meiosis

Haploid spores

Dead   Viable

Viable   Dead

The alga *Chlamydomonas reinhardtii* contains two β-tubulin genes, but these encode nearly identical proteins, which differ by only two amino acids. Thus in these microorganisms, a single type of α- or β-tubulin is sufficient for all microtubular structures: mitotic and meiotic spindles, cytoplasmic microtubules, MTOCs, and flagella.

In contrast to microorganisms, vertebrates produce multiple types of α- and β-tubulins. Extensive sequence analysis of cloned cDNAs indicate that organisms as diverse as chickens and mice express at least six α-tubulin genes and six β-tubulin genes. For each tubulin isotype, the amino acid sequences of the chicken and mouse proteins are almost identical. This conservation suggests that the tubulin isotypes arose by gene duplication early in vertebrate evolution and that each type fulfills an important function. The class VI β-tubulin isotype, which occurs only in erythrocytes and platelets, is unusual because its amino acid sequence is only about 75 percent identical with that of any of the other five β-tubulins. These five β-tubulins typically differ from one another in only 2–8 percent of their ~450 amino acid residues. Most of the differences among the β-tubulin isotypes are found in the C-terminal segment (Table 21-1); these sequence variations may enable the various tubulin isotypes to bind to different MAPs.

Most experiments designed to identify the different functions of the β-tubulin isotypes have suggested that they are functionally identical. For example, even though class VI β-tubulin is quite different in sequence from the other β-tubulin isotypes, brain tubulin (a mixture of several other isotypes) can substitute for class VI β-tubulin in assembly of the erythrocyte marginal band in cell-free reactions. Conversely, when class VI β-tubulin is expressed (using a cDNA expression vector) in fibroblasts, it is incorporated into all interphase and mitotic spindle microtubules.

Similar findings have been obtained with *Drosophila* in which the β-tubulin isotype used for formation of sperm flagella is different from that used for cilia or flagella in other tissues. The gene controlling the synthesis of the β-tubulin species called β2 is expressed only in developing sperm cells, and only β2-tubulin is found in the sperm flagellum. Nevertheless, β2-tubulin forms a wide array of other microtubule-containing structures, as shown by the

◀ **Figure 21-19** Genetic study demonstrating that the single yeast β-tubulin gene is essential for viability. A diploid yeast strain transformed with a DNA containing a deletion in the β-tubulin gene generates a viable heterozygote with one wild-type gene (orange) and one mutant β-tubulin gene (green). Upon sporulation, only two of the four haploid progeny are viable; these contain the wild-type β-tubulin gene. Thus at least one copy of the β-tubulin gene is essential for viability. In the diploid heterozygote containing one wild-type and one mutant gene, the wild-type gene evidently produces enough functional tubulin for the organism to grow and divide. [See N. F. Neff et al., 1983, *Cell* 33:211.]

Table 21-1 Expression and sequence homology of chick and mouse β-tubulin isotypes

Isotype	Pattern of expression	Overall amino acid identity	C-terminal sequence*
Class I	Constitutive; many tissues		EEEEDFGEEAEEEA
Class II	Major neuronal; many tissues		DEQGEFEEEGEEDEA
Class III	Minor neuronal; neuron-specific	92–97%	EEEGEMYEDDEEESEQGAK
Class IVa	Major neural; brain-specific		EEGEFEEEAEEEVA
Class IVb	Major testes; many tissues		EEEGEFEEEAEEEAE
Class V	Minor constitutive; absent from neurons		NDGEEEAFEDDEEEINE
Class VI	Major erythrocyte/platelets (marginal band)	78%	DVEEYEEAEASPEKET

*Single-letter amino acid code is used. The sequences of the chick proteins are shown, except for class IVa for which only the mouse sequence is available. For classes I, II, and IVb, the C-terminal sequences of the mouse and chick proteins are identical.

SOURCE: D. W. Cleveland, 1987, *J. Cell Biol.* **104**:381; S. A. Lewis, W. Gu, and N. J. Cowan, 1987, *Cell* **49**:539; and K. F. Sullivan, 1988, *Ann. Rev. Cell Biol.* **4**:687.

effects of a specific mutation preventing normal β2-tubulin synthesis. In *Drosophila* males homozygous for this mutation, the early meiotic divisions—which are completed before the time of β2-tubulin synthesis—are normal. However, all microtubule-associated events that occur after expression of the β2-tubulin gene are abnormal, including the last meiotic division, the formation of the sperm flagellum, and the shaping of the sperm nucleus. Thus β2-tubulin does not merely contribute to sperm structure but participates with other tubulins in multiple cellular functions.

The conclusion from these and other similar studies is that either different posttranslational modifications of tubulin or different microtubule-binding proteins are important for generation of microtubule diversity.

α-Tubulin Undergoes Reversible Covalent Modifications

Two types of posttranslational covalent modifications of α-tubulin have been well-studied. In flagellated microorganisms such as *Chlamydomonas reinhardtii*, the flagellar α-tubulin contains an acetylated lysine residue that is not present in cytoplasmic microtubules in this organism. Acetylation occurs during or just after incorporation of α-tubulin into flagellar microtubules. Although the enzyme that incorporates the acetyl group is found both in flagella and in the cytosol, a second enzyme that removes the acetyl group is localized to the cytosol. Thus the steady-state level of microtubule acetylation is shifted completely to the unacetylated state in the cytoplasm. When flagellar microtubules are induced experimentally to depolymerize, the acetyl groups are lost. But it is not clear if acetylation of α-tubulin *causes* it to polymerize into flagellar microtubules, *regulates* polymerization, or merely occurs *because* the α-tubulin is in flagella.

A second type of covalent modification to α-tubulin involves removal of the carboxyl-terminal tyrosine residue and its readdition. The latter reaction is catalyzed by tubulin-tyrosine ligase, with the concomitant hydrolysis

of ATP; no mRNA or ribosomes are involved. An anti-body that recognizes tyrosinated α-tubulin, but not untyrosinated α-tubulin, has been used to show that tyrosinated α-tubulin is incorporated into all classes of cellular microtubules, whereas untyrosinated α-tubulin is found preferentially in the more stable cytoplasmic microtubules (e.g., centrioles, cilia, and the few microtubules that remain after colchicine treatment). The enzyme that removes tyrosine acts preferentially on α-tubulin already incorporated into microtubules, and removal of tyrosine may make a microtubule more resistant to factors that cause its disassembly.

Intracellular Transport via Microtubules

Within cells, vesicles and protein particles are frequently transported many micrometers and delivered to a particular cellular region. Since diffusion alone cannot account for either the rate or the directionality of such transport processes, cell biologists have long suspected that cytoskeletal fibers play a key role. Two well-studied experimental systems—nerve cells and fish erythrophores—have provided evidence implicating microtubules in the intracellular transport of vesicles and protein particles. More important, scientists have used these systems to identify and purify the "molecular motors" that use energy released during ATP hydrolysis to power movement of vesicles and particles along "tracks" of microtubules.

Fast Axonal Transport Occurs along Microtubules

Because ribosomes are present only in the cell body and dendrites of nerve cells, no protein synthesis occurs in the axons and synaptic terminals. Therefore, proteins and membranous organelles must be synthesized in the cell body and transported down the axon to the synaptic regions; this process is called axonal transport.

▶ **Figure 21-20** Experimental system for determining the in vivo rate of axonal transport and identifying the transported proteins. Radiolabeled amino acids (red dots) are injected into a ganglion of experimental animals; the ribosome-containing cell bodies in the ganglia incorporate the labeled amino acids into protein. Animals are killed at different times after injection, and the sciatic nerve is dissected and cut into 5-mm segments. The amount of radiolabeled protein in each fragment is measured, and then the various proteins are resolved by gel electrophoresis. [See O. Ochs, 1981, in *Basic Neurochemistry*, 3d ed., G. J. Sigel et al., eds. Little, Brown, p. 425.]

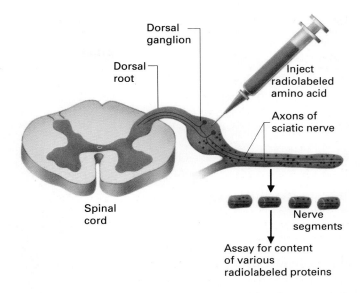

Neurons in the sciatic nerve of mammals are particularly suitable for study of axonal transport because their cell bodies are conveniently located in the dorsal root ganglion near the spinal cord and because the bundles of nerve processes are very long, extending as much as 1 m from the ganglion to innervate many leg muscles. Radiolabeled amino acids injected into the ganglion of the sciatic nerve become incorporated into proteins made in the cell body. Subsequent examination by autoradiography of nerve segments reveals the movement of the various labeled proteins (Figure 21-20). Such studies have established that proteins do indeed move down the axon but not all at the same rate (Table 21-2). The fastest-moving group, consisting of small vesicles and particles, has a velocity of about 250 mm/day, or about 3 μm/s. The slowest group, containing mostly cytoskeletal proteins, moves only a fraction of a millimeter per day. Larger particles such as mitochondria move down the axon at an intermediate rate.

Transport from the cell body to the synaptic junctions is termed *anterograde* and is associated with axonal growth and renewal of synaptic vesicles. Other vesicles move along the axon rapidly (about 3 μm/s) in the opposite, *retrograde* direction. These vesicles mainly contain "old" membrane and cytosolic proteins from the synaptic terminals, which are destined to be degraded in lysosomes in the cell body.

Rapidly transported proteins appear to move down the axon in organized structures, mainly as small membranous vesicles, or organelles, associated with microtubules. Movement of such vesicles along transport filaments has been studied extensively in the giant axon of the squid. The organelles and filaments can be visualized by phase-contrast or differential interference contrast microscopy in which the weak image is amplified by a video camera, stored and analyzed in a computer, and displayed on a television screen. With this technique, it has been shown that organelles move in both directions in squid giant axons at rates up to about 2 μm/s. In some cases, two organelles can be seen to move along the same filament in opposite directions and to pass each other without colliding (Figure 21-21a), indicating that each transport filament has several *tracks* for organelle movement. Subsequent electron microscopy of the same region of the cytoplasm has demonstrated that the transport filaments are indeed single microtubules (Figure 21-21b). Clearly fast axonal transport is associated with the interaction of organelles and microtubules.

Table 21-2 Rate of axonal transport of various materials in mammalian neurons and their relationship to cellular structures

Component	Class*	Transport rate (mm/day)	(μm/s)	Materials transported	Cellular structures transported
Fast	I	200–400	2.3–4.6	Glycoproteins, glycolipids, acetylcholinesterase	Vesicles, smooth endoplasmic reticulum; small granules
Intermediate	II	50	0.6	Mitochondrial proteins	Mitochondria
	III	15	0.2	Myosinlike proteins	Filaments
Slow	IV	2–4	0.03	Actin clathrin, enolase, CPK calmodulin	Actin microfilaments
	V	0.2–1	0.002–0.01	Intermediate-filament proteins, tubulin	Microtubule–intermediate-filament network

* Roman numerals refer to the nomenclature developed by M. Willard, W. M. Cowan, and P. R. Vagelos, 1974, *Proc. Nat'l Acad. Sci. USA* **71**:2183–2187. [See also B. Grafstein and D. S. Forman, 1980, *Physiol. Rev.* **60**:1167.]

(a)

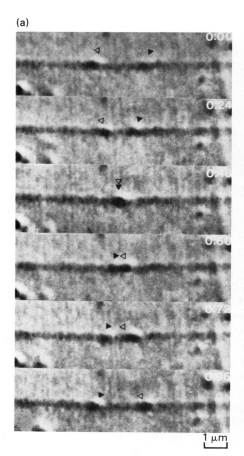

1 μm

(b)

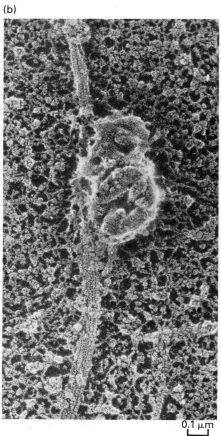

0.1 μm

◄ **Figure 21-21** (a) Video micrographs showing bidirectional movement of two vesicular organelles on a single transport microtubule filament. A piece of giant squid axon was dissected, the cytoplasm was extruded, and a buffer containing ATP was added. The preparation was then viewed in a differential interference contrast microscope, and the images were recorded on videotape. The two organelles (open and solid triangles) move in opposite directions along the same filament, pass each other, and continue in their original directions. Elapsed time in seconds appears at the top right corner of each video frame. (b) A region of cytoplasm similar to that shown in (a) was freeze-dried, rotary-shadowed with platinum, and viewed in the electron microscope. Two large structures are visible attached to one microtubule; these presumably are small vesicles that were moving along the microtubule when the preparation was frozen. [See B. J. Schnapp et al., 1985, *Cell* 40:455.] *Courtesy of B. J. Schnapp, R. D. Valle, M. P. Sheetz, and T. S. Reese.*

Microtubules Provide Tracks for Movement of Pigment Granules and Golgi Vesicles

Rapid movement of particles and vesicles along microtubules occurs in many, and possibly all, eukaryotic cells. Another system for observing this transport involves the specialized cells—called melanophores—that contain granules of pigments; these are found in the skin of many amphibians and in the scales of many fish. By processes under nervous and hormonal control, these pigment granules can be transported to the cell periphery to darken the color of the skin; transport of these granules inward, toward the center of the cell, lightens the color (Figure 21-22). In this manner, the animal can adjust its color to blend better with its surroundings in order to escape predators.

A short time after a scale is removed from a fish and placed in culture, the pigment in the melanophores begins spontaneously to move inward and outward. In favorable cases, when individual granules could be followed, it has been observed that after dispersal, each granule always returns to the same location in the center of the cell. During this movement, the microtubules act as guides along which pigment granules and associated material can move in both directions.

In cultured fibroblasts, the Golgi complex is concentrated near the MTOC (Figure 21-23). During mitosis or after depolymerization of microtubules by colchicine, the Golgi complex breaks into small vesicles that are dispersed throughout the cytoplasm. When the cytoplasmic microtubules re-form during interphase (or after removal of the drug) Golgi vesicles move along microtubule tracks toward the MTOC where they reaggregate to form large membrane complexes.

Specific Proteins Promote Vesicle Translocation along Microtubules

Progress in understanding the molecular basis of movement of vesicles along microtubules has come from purification of the proteins that constitute the "molecular motors." Most of these studies have utilized the squid giant axons because cytoplasm that retains all of its cytoskeletal filaments can be squeezed out from these cells. When such an extract is spread on a coverslip and provided with ATP, transport of vesicles along microtubules occurs and can be observed by microscopy (see Figure 21-21). The rate of vesicle movement (2–3 μm/s) in this cell-free system is similar to that in intact cells. Membrane vesicles from other types of cells, when added to this extruded

(a)

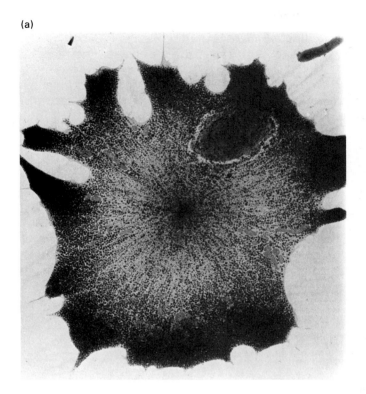

(b)

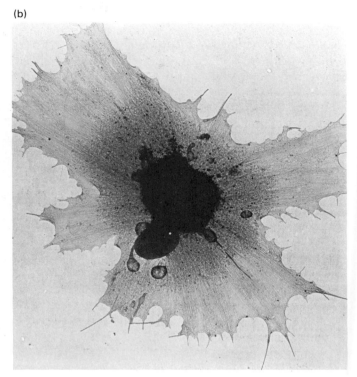

▲ **Figure 21-22** High-voltage electron micrographs showing movement of pigment granules in a melanophore, or red-pigment cell, of the squirrelfish, *Holocentrus ascensionis*. (a) The pigment granules are dispersed to the cell periphery. (b) The granules are condensed around the nucleus. (c) A portion of a dispersing melanophore showing the pigment granules associated with tracks of cytoskeletal fibers including abundant microtubules. *Courtesy of K. Porter.*

(c)

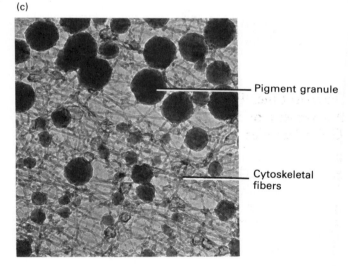

Pigment granule

Cytoskeletal fibers

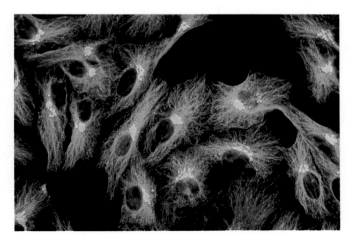

◀ **Figure 21-23** Distribution of the Golgi complex (yellow) and microtubules (green) in cultured monkey fibroblasts. Microtubules were visualized with an anti-tubulin antibody and the Golgi complex with an antibody specific for galactosyltransferase (see Figure 17-30). The Golgi complex is concentrated near the MTOC. *Courtesy of T. Kreis.*

cytoplasm, will bind to microtubules and be translocated along them.

Since nerve cytoplasm is a complex mixture of fibers and proteins, researchers sought to reconstitute vesicle movement using more purified components. Microtubules were formed in cell-free reactions from pure tubulin; thus they contained no MAPs. A drug called taxol was used to stabilize and prevent depolymerization of these microtubules. When synaptic vesicles from different types of cells were added with ATP to MAP-free microtubules, the vesicles neither bound to the microtubules nor moved along them. However, addition of squid nerve cytoplasm (free of tubulin) allowed vesicle translocation, a result that indicates that a soluble protein in the nerve cytoplasm is required for translocation.

To purify this soluble "motor protein," vesicles and nerve cytoplasm were mixed with microtubules in the presence of an analog of ATP that could not be hydrolyzed, AMPPNP:

$$\text{Adenine—ribose—O—}\overset{\overset{\displaystyle O}{\|}}{\underset{\underset{\displaystyle O^-}{|}}{P}}\text{—O—}\overset{\overset{\displaystyle O}{\|}}{\underset{\underset{\displaystyle O^-}{|}}{P}}\text{—NH—}\overset{\overset{\displaystyle O}{\|}}{\underset{\underset{\displaystyle O^-}{|}}{P}}\text{—OH}$$

Under these conditions, vesicles bound tightly to microtubules but did not move. However, they did move when ATP was added. These results suggest that a motor protein binds to microtubules in the presence of ATP or AMPPNP but requires hydrolysis of the terminal phosphoanhydride bond in order to move along the microtubule and then to be released from the microtubule. In subsequent studies, microtubules were used as an affinity matrix to purify one motor protein. Squid nerve cytoplasm was incubated with microtubules in the presence of AMPPNP; then proteins bound to the microtubules were removed with ATP. One predominant protein was specifically bound and eluted, a protein termed *kinesin* (MW 380,000), which contains four polypeptides (Figure 21-24).

One end of a kinesin molecule can bind to a microtubule; the other end, to a vesicle. In the presence of ATP, the kinesin molecule moves along the microtubule from the (−) to the (+) end, transporting the bound vesicle with it (Figure 21-25a). In nerve axons, the (−) ends of all microtubules face the cell body (see Figure 21-12); thus kinesin directs anterograde axonal transport—that is, from the cell body to the axon terminus. Even plastic beads with negatively charged groups on their surface will bind to kinesin molecules and be translocated along microtubules with them. Pure kinesin also can bind to glass surfaces; when a solution of ATP and microtubules is placed on a glass slide to which kinesin is bound, the microtubules move along the glass surface like little snakes (Figure 21-25b). In this case, because the kinesin is immobolized, the microtubules move in the direction of their (−) ends. Clearly kinesin is the only protein neces-

sary for anterograde movement of vesicles along axonal microtubules; how kinesin uses energy released by ATP hydrolysis to power this movement is not known.

Given that kinesin powers only anterograde transport, some other protein or proteins must power retrograde transport of vesicles along microtubules. One of the proteins associated with cytoplasmic microtubules, MAP1C, has been shown to move along microtubules from the (+) to the (−) end and thus acts as a retrograde motor protein. MAP1C has a molecular weight of about 1,000,000 and is composed of several polypeptides, the largest of which have molecular weights of 400,000. Like kinesin, MAP1C has two heads that appear to bind to and translocate along microtubules (Figure 21-26). When MAP1C is immobilized on a glass slide, it causes movement of microtubules toward their (+) ends, that is, in the direction opposite to kinesin-powered movement. In both cases, ATP is required for movement. MAP1C is also called *cytoplasmic dynein* because it is similar in structure and function to dynein in flagella, the protein that uses

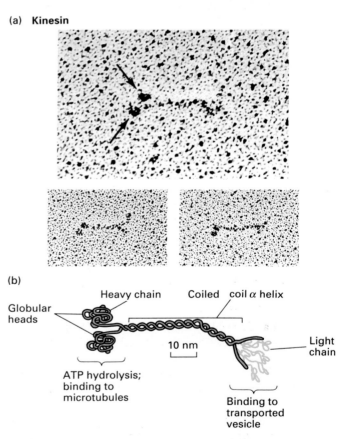

(a) Kinesin

(b)

Globular heads

Heavy chain

Coiled coil α helix

Light chain

10 nm

ATP hydrolysis; binding to microtubules

Binding to transported vesicle

▲ **Figure 21-24**   (a) Electron micrographs of rotary-shadowed brain kinesin. The two globular heads (*arrows*) that bind to tubulin are visible on the left side of the molecule. (b) Schematic model of kinesin showing the arrangement of the two heavy chains (MW 124,000) and two light chains (MW 64,000). *Part (a) from N. Hirokawa et al., 1989, Cell 56:867; courtesy of N. Hirokawa.*

(a)

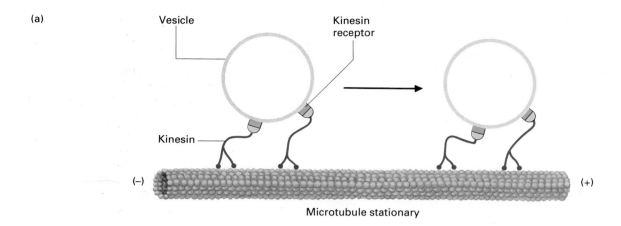

(b)

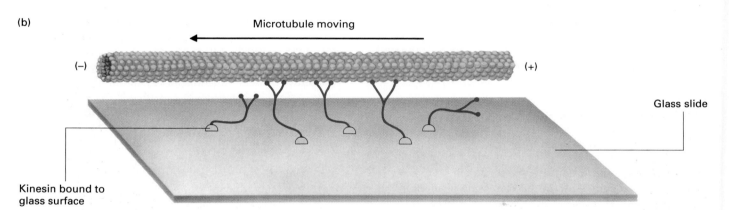

▲ **Figure 21-25** (a) Model of kinesin-catalyzed anterograde transport of vesicles along microtubules. The kinesin molecules and vesicles move from the (−) to the (+) end of the stationary microtubule. (b) Model of kinesin-catalyzed movement of microtubules. The kinesin molecules bound to the glass surface try to move toward the (+) end of the microtubule, but since they are immobilized, the microtubule moves in the direction of its (−) end. ATP is required for movement in both cases. *Adapted from R. D. Vale, B. J. Schnapp, T. S. Reese, and M. P. Sheetz, 1985, Cell **40**:559; and T. Schroer, B. Schnapp, T. S. Reese, and M. Sheetz, 1988,* J. Cell Biol. *107:1785.*

energy released by ATP hydrolysis to cause flagella to beat (see following section).

The identification of two proteins, kinesin and MAP1C, that cause organelle movement along microtubules provides an answer to one of the oldest questions in cell biology—how things move from one end of a cell to another. What is not yet clear is the specificity of the process. How do vesicles "know" whether to bind a protein like kinesin, jump onto a microtubule, and then move to another region of the cell? Some particles and vesicles (e.g., pigment granules in melanophores) can alternate their direction of movement along microtubules. Such particles and vesicles must contain both anterograde and retrograde motors, but only one is active at any one time. Some experiments indicate that the directionality of movement is affected by phosphorylation of proteins associated with the motor proteins themselves.

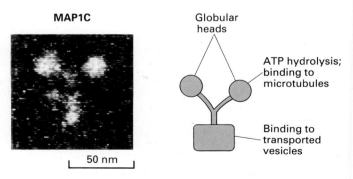

▲ **Figure 21-26** Electron micrograph and schematic diagram of MAP1C, or cytoplasmic dynein, showing the two globular heads that bind to microtubules. *Photograph from R. B. Vallee et al., 1988,* Nature *332:561.*

Cilia and Flagella: Structure and Movement

Cilia and flagella, which are found on many eukaryotic cells, are about 0.25 μm in diameter and have a very similar structure; they differ mainly in length (flagella are longer) and in pattern of beating. Sperm cells and certain protozoa use flagella for propulsion. Huge numbers of cilia (more than $10^7/mm^2$) cover the surfaces of mammalian respiratory passages (e.g., the nose, pharynx, and trachea), where they sweep out particulate matter that collects in the mucous secretions of these tissues. Ciliated protozoa use cilia for locomotion and to trap food particles.

Movement by a cilium or a flagellum is caused by a bend that originates at the base of the structure and then is propagated along its length to the other end (Figure 21-27). Cilia move by a whiplike power stroke followed by a recovery stroke, both of which are fueled by hydrolysis of ATP. Flagellar movement, which differs from ciliary movement, involves waves of constant amplitude that emanate from the base and spread to the opposite end; this movement also requires ATP hydrolysis.

All Eukaryotic Cilia and Flagella Have a Similar Structure

All cilia and flagella are built on a common fundamental plan. (Unless otherwise noted, the terms cilia and flagella are used interchangeably in the following discussion.) A bundle of microtubules, called the *axoneme*, is surrounded by an extension of the plasma membrane. At its point of attachment to the cell, the axoneme connects with the basal body (Figure 21-28). As discussed later, the

basal body also is composed of microtubules and plays an important role in initiation of growth of the axoneme.

Each axoneme contains two central *singlet* microtubules called C_1 and C_2, each with 13 protofilaments, and nine outer pairs of microtubules, called *doublets;* this recurring motif is known as the 9 + 2 array (Figure 21-29). Each doublet has A and B subfibers. The A subfiber is a complete microtubule with 13 protofilaments; the B subfiber contains 10 (in some cases, 11) protofilaments. The organization of the A and B subfibers is so uniform that each protofilament can be assigned a number (Figure 21-29a, *inset*). Attachment of the A and B subfibers to each other is mediated by filaments of the protein tektin; these filaments, which are 2 nm in diameter and approximately 48 nm long, run longitudinally along the wall between the A and B subfibers. Each A subfiber is connected to the two central microtubules by radial spokes and spokeheads. Detailed analyses of electron micrographs have shown that radial spokes attach to the number 2 protofilament of the A outer doublets; on average, a radial spoke is attached to every fourth $\alpha\beta$-tubulin dimer— or every 32 nm—along a protofilament.

Motility of cilia and flagella is generated by inner and outer "arms" of dynein, which protrude from each A subfiber at regular intervals (Figure 21-30). All the dynein arms in a flagellum point in the same direction: clockwise when viewed from the base to the tip. Outer dynein arms are rigidly attached every 24 nm—that is, to every third tubulin dimer—at specific protofilaments of the A subfiber; the inner dynein arms bind to the A subfiber at average intervals of 32 nm, similar to radial spokes.

The two central singlet microtubules are connected by periodic paired bridges, likes rungs on a ladder, and surrounded by a fibrous structure, termed the inner sheath,

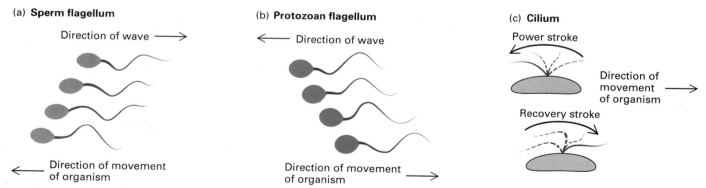

(a) Sperm flagellum

Direction of wave ⟶

⟵ Direction of movement of organism

(b) Protozoan flagellum

⟵ Direction of wave

Direction of movement of organism ⟶

(c) Cilium

Power stroke

Direction of movement of organism ⟶

Recovery stroke

▲ **Figure 21-27** Characteristic motions of cilia and flagella. (a) In the typical sperm flagellum, successive waves of bending originate at the base and are propagated toward the tip; these push against the water and propel the cell forward. Sperm flagella generate waves at a frequency of 30–40 per second. (b) A reverse type of motion is observed in certain flagellated parasitic protozoans such as trypanosomes. The wave is directed toward the cell and pulls it in the direction opposite to the wave. The cell thus moves backward.

(c) Beating of a cilium, which occurs 5–10 times per second, has two stages called the power stroke and the recovery stroke. In the power stroke, a large bend forms at the base of the cilium and is propagated to the tip, moving the cilium backward; this propels the organism in the opposite direction. During the recovery stroke, a different wave of bending moves outward along the cilium from its base, pushing the cilium forward. ATP is hydrolyzed during both the power stroke and recovery stroke.

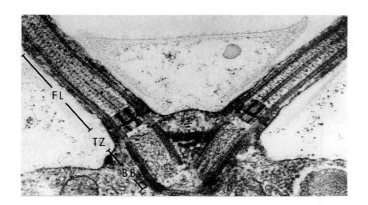

◀ **Figure 21-28** Electron micrograph of the basal regions of the two flagella in *Chlamydomonas reinhardtii*. The bundles of microtubules and some fibers connecting them are visible in the flagella (FL). The two basal bodies (BB) form the point of a "V"; a transition zone (TZ) between the basal body and flagellum proper contains two dense-staining cylinders of unknown structure. *From B. Huang, Z. Ramanis, S. Dutcher, and D. Luck, 1982, Cell* **29***:745.*

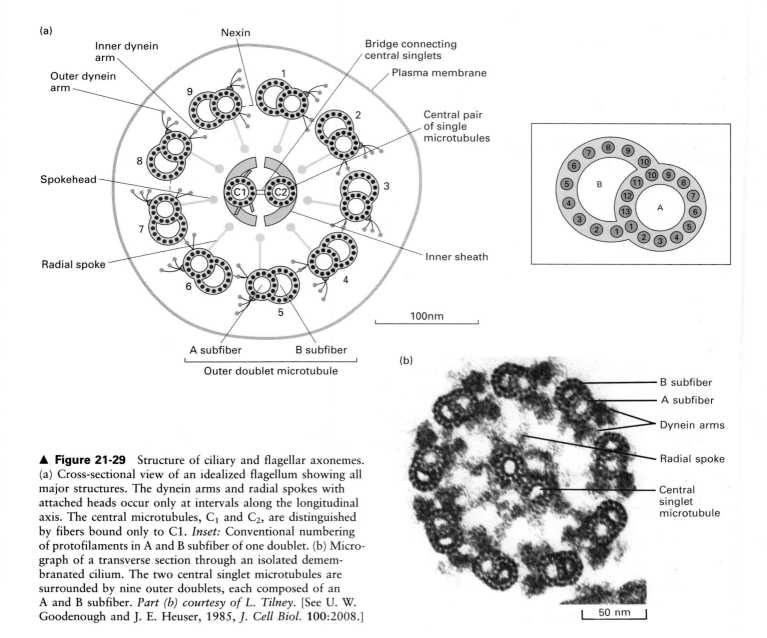

▲ **Figure 21-29** Structure of ciliary and flagellar axonemes. (a) Cross-sectional view of an idealized flagellum showing all major structures. The dynein arms and radial spokes with attached heads occur only at intervals along the longitudinal axis. The central microtubules, C_1 and C_2, are distinguished by fibers bound only to C1. *Inset:* Conventional numbering of protofilaments in A and B subfiber of one doublet. (b) Micrograph of a transverse section through an isolated demembranated cilium. The two central singlet microtubules are surrounded by nine outer doublets, each composed of an A and B subfiber. *Part (b) courtesy of L. Tilney.* [See U. W. Goodenough and J. E. Heuser, 1985, *J. Cell Biol.* **100**:2008.]

(a)

(b)

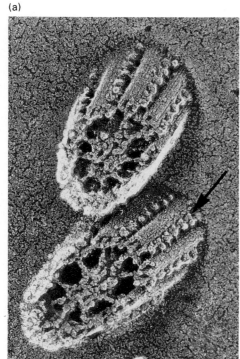

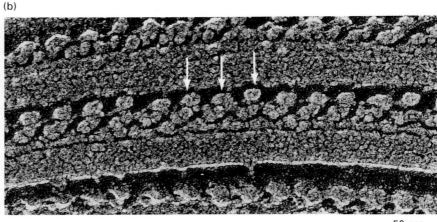

50 nm

0.1 μm

▲ **Figure 21-30** (a) Electron micrograph of two *Tetrahymena* ciliary axonemes freeze-fractured at an oblique angle. Arrow indicates a distinct row of outer dynein arms between adjacent outer doublets. (b) Longitudinal view of two outer doublet microtubules of a freeze-fractured, deep-etched *Tetrahymena* ciliary axoneme. Note rows of outer dynein arms (*arrows*) attached at 24-nm intervals. *Part (a) from U. W. Goodenough and J. E. Heuser, 1982, J. Cell Biol. 95:798; courtesy of U. W. Goodenough. Part (b) from U. W. Goodenough and J. E. Heuser, 1984, J. Mol. Biol. 180:1083.*

that may regulate the form of the ciliary or flagellar beat (see Figure 21-29a). In addition to the radial spokes, the axoneme is held together by a set of circumferential linkers that join adjacent outer doublets. These linkers, composed in part of the protein nexin, are highly elastic; their normal length is 30 nm, but they can be stretched to 250 nm without breaking. Nexin links are found with a periodicity of about 86 nm along the axoneme.

Clearly, at the level of resolution afforded by the electron microscope, flagella and cilia are quite complex structures, which contain many protein components in addition to microtubules. The roles these various structures play in accomplishing the characteristic motion of flagella and cilia are discussed next.

Dynein ATPases Are Essential to the Movement of Flagella and Cilia

Flagella and cilia from which the plasma membrane has been removed by nonionic detergents are called isolated axonemes. These will propagate bending movements when provided with ATP; thus the molecular motor resides within the organelle, not elsewhere in the cell body. Biochemical studies have shown that all cilia and flagella have a potent ATP-hydrolyzing activity associated with the inner and outer dynein arms. Treatment of isolated axonemes with solutions containing high concentrations of salt removes the dynein arms, solubilizes the ATPase

activity, and abolishes ATP-stimulated beating of the axonemes. Addition of a purified preparation of dynein arms restores both the ATPase activity and the beating, and electron microscopy shows that the dyneins have re-attached to their proper place on the A subfibers.

In isolated axonemes, addition of ATP causes each outer doublet to slide past the adjacent one with which it interacts (Figure 21-31a). In this process, according to the "dynein-walking" model, the dynein arms on the A sub-fiber of one doublet push the B subfiber on the adjacent doublet toward the tip of the axoneme (Figure 21-31b). The force producing the sliding of adjacent doublets is caused by successive formation and breakage of cross-bridges between the dynein arms of one doublet and the B subfiber of the adjacent doublet as the two doublets move relative to each other. Formation of such cross-bridges has in fact been observed and is ATP-dependent. Removal of ATP causes the axonemes to become extremely stiff and rigid. In this case, dynein forms stable cross-bridges between adjacent doublets, so that they cannot move relative to each other, and the microtubules become frozen in place. Subsequent addition of ATP causes breakage of these cross-bridges, followed by hydrolysis of ATP to ADP. Successive binding of ATP to dynein and its subsequent hydrolysis causes the dynein arms to successively break and make bonds with the adjacent doublet.

Each outer dynein arm has three globular "heads" (Figure 21-31c); each inner arm has two or three heads. Each head is constructed of a single large protein (MW

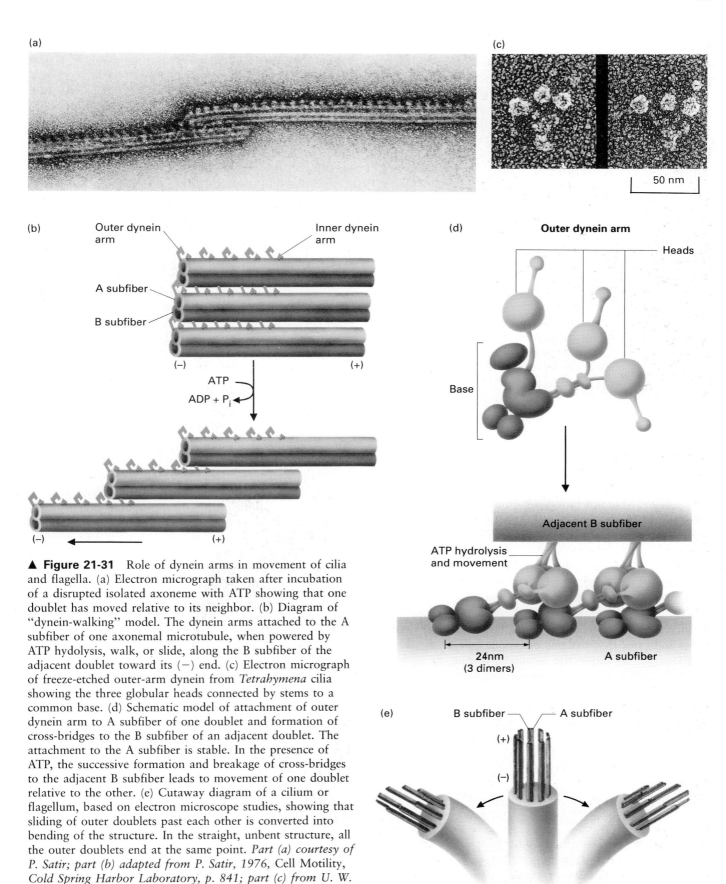

(a)

(c)

50 nm

(b)

Outer dynein arm

Inner dynein arm

A subfiber

B subfiber

(−)

(+)

ATP

ADP + P$_i$

(−)

(+)

(d)

Outer dynein arm

Heads

Base

Adjacent B subfiber

ATP hydrolysis and movement

24nm (3 dimers)

A subfiber

(e)

B subfiber

A subfiber

(+)

(−)

▲ **Figure 21-31** Role of dynein arms in movement of cilia and flagella. (a) Electron micrograph taken after incubation of a disrupted isolated axoneme with ATP showing that one doublet has moved relative to its neighbor. (b) Diagram of "dynein-walking" model. The dynein arms attached to the A subfiber of one axonemal microtubule, when powered by ATP hydolysis, walk, or slide, along the B subfiber of the adjacent doublet toward its (−) end. (c) Electron micrograph of freeze-etched outer-arm dynein from *Tetrahymena* cilia showing the three globular heads connected by stems to a common base. (d) Schematic model of attachment of outer dynein arm to A subfiber of one doublet and formation of cross-bridges to the B subfiber of an adjacent doublet. The attachment to the A subfiber is stable. In the presence of ATP, the successive formation and breakage of cross-bridges to the adjacent B subfiber leads to movement of one doublet relative to the other. (e) Cutaway diagram of a cilium or flagellum, based on electron microscope studies, showing that sliding of outer doublets past each other is converted into bending of the structure. In the straight, unbent structure, all the outer doublets end at the same point. *Part (a) courtesy of P. Satir; part (b) adapted from P. Satir, 1976, Cell Motility, Cold Spring Harbor Laboratory, p. 841; part (c) from U. W. Goodenough, and J. E. Heuser, 1984,* J. Mol. Biol. **18:**1083.

~400,000) and has a site for ATP hydrolysis. The base of the dynein is rigidly attached to the A subfiber, and is built of several smaller polypepides (Figure 21-31d). The structure and movement of flagellar dynein along the adjacent B microtubule is very similar to that of cytoplasmic dynein in translocation of organelles in that both "walk" toward the (−) end of the adjacent microtubule.

Sliding of Microtubules Is Converted into Bending of the Axoneme

In an intact flagellum and cilium, all of the outer doublets are constrained by their nexin connectors and radial links to the central sheath. For this reason, the force of the sliding motion of the outer doublets is converted into a bending of the axoneme. The theory that flagella and cilia bend by the sliding of outer doublets past each other was elegantly proved by careful electron microscopic studies. In a straight axoneme, all the outer doublets are of the same length and terminate at the same point. In a fixed, bent axoneme, all the outer doublets are also of the same length, but the ones at the inside of the bend extend farther than those on the outside, as would be expected if the microtubules slide relative to one another (Figure 21-31e).

More complex processes must be involved in the wavelike beating of a flagellum, which is confined to a plane. A possible molecular explanation for this is that two of the outer doublets—by convention, numbers 5 and 6, as shown in Figure 21-29a—have permanent rigid links to each other. The bond thus formed and the two central fibers would define the plane along which the beat of the flagellum occurs (see Figure 21-27). Because these doublets cannot slide past each other, any beating of the flagellum is confined to this plane. Also, the two central microtubules (see Figure 21-29a) are morphologically distinct, and this may also influence the plane of beating.

In the ciliary axoneme, all nine doublets do not operate at once. Morphologic analysis of beating ciliary axonemes indicates that dyneins on about half of the doublets operate to move the axoneme during its power stroke and the others only during the recovery stroke. The elastic nexin links, which appear to be permanent, may prevent excessive sliding displacement between the outer doublets. Little is known about how motions initiated at the base of a cilium or flagellum are transmitted to the end; genetic evidence (below) suggests that the central singlets are not essential for beating.

Genetic Studies Provide Additional Information on Axoneme Assembly and Beating

The biflagellate, unicellular alga *Chlamydomonas reinhardtii* has proved especially amenable to studies on the functional, genetic, and structural aspects of flagella and their assembly. By shearing a population of cells, flagella can be obtained in good purity and high yield, and the deflagellated cells will regenerate new flagella.

It is possible to isolate many viable *C. reinhardtii* mutants that are nonmotile and defective in flagellar function. The study of such mutants has allowed an estimate of the number of genes required to assemble a flagellum. Microscopic analysis of flagella from some of these mutants reveals they are missing an entire substructure, such as the radial spokes or inner dynein arms (Figure 21-32a).

Two-dimensional gel electrophoresis of axonemal proteins resolves approximately 200 discrete polypeptides, in addition to α- and β-tubulin. Many nonmotile mutants that are missing a flagellar substructure have been found to lack a number of these proteins. Thus individual proteins can be correlated with specific structures and associated with a specific gene. For instance, nonmotile mutants missing only the radial spokes and spokeheads lack 17 of the axonemal polypeptides, which can be assumed to be located in the spoke or spokehead (Figure 21-31b). Mutants missing only the spokehead lack 6 specific proteins, a subset of the 17 lacking in the mutants missing both spokes and spokeheads. These 6 proteins are localized to the spokehead, and the other 11 to the spokes. By a similar mutational analysis, the inner and outer dynein arms have been shown to be composed of different proteins. A number of slow-moving mutants appear to lack only the inner or only the outer dynein arms, showing that either dynein suffices for flagellar movement.

One interesting class of poorly motile cells is missing the C1 microtubule and 10 nontubulin polypeptides bound to it; the C2 microtubule and its 7 associated polypeptides are normal in this mutant. Mutants missing both central microtubules and those missing the radial spokes are completely nonmotile, although motile revertants of such mutants, still lacking the radial spokes or central pair, have been isolated. Thus the radial spokes and central pair are not required to convert microtubule sliding into bending. They may play a role in regulating the switching on or off of sets of dyneins during beating; the second (revertant) mutation might be in the "switching" protein that allows the axoneme to beat.

Basal Bodies and Centrioles: Structure and Properties

For microtubules to participate in movement or to act as a structural framework, they must be anchored by at least one end. As noted already, a cilium or flagellum is anchored at its cytoplasmic end by a microtubule-containing structure called the basal body (see Figure 21-28). Likewise, the microtubules in the mitotic spindle as well as those in interphase cells radiate from a region of the cell near the nucleus called the *centrosome*, which func-

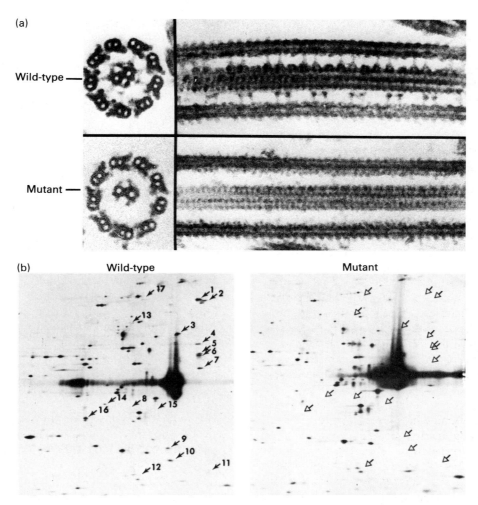

▲ **Figure 21-32** Analysis of flagellar axonemes from a wild-type *Chlamydomonas reinhardtii* and from a paralyzed mutant. (a) Electron micrographs of isolated wild-type and mutant flagellar axonemes cut in transverse (*left*) and longitudinal (*right*) sections. Note that the mutant axonemes lack radial spokes. (b) Autoradiographs of two-dimensional gels of radioactive axonemal proteins. In the gel prepared from wild-type flagella, solid arrows point to 17 polypeptides that are missing in the gel prepared from the mutant flagella; these missing polypeptides are indicated by the open arrows on the gel of the mutant proteins. Tubulin subunits not resolved in the gels form the central dark streaks. *Part (a) courtesy of B. Huang and D. Luck. Reproduced from the* Journal of Cell Biology, *1981, vol. 88, p. 73, by copyright permission of the Rockefeller University Press; part (b) courtesy of G. Piperno, B. Huang, Z. Ramanis, and D. Luck.*

tions as a microtubule-organizing center (MTOC). In animal cells, the centrosome is an amorphous region (sometimes called the pericentriolar region) that surrounds the centriole (see Figure 21-3). It is believed that microtubules in the centrosome, or in animal cells possibly in the centriole itself, nucleate both the growth of microtubules in interphase cells and the growth of many of the microtubules in the mitotic spindle. In higher plants, which lack centrioles, the centrosomal material acts as an MTOC. As will become clear in the following discussion, basal bodies and centrioles are similar, but not identical, in structure and function; at least in some instances one can be converted into the other.

Centrioles and Basal Bodies Are Built of Microtubules

Centrioles and basal bodies are cylindrical structures, about 0.4 μm long and 0.2 μm wide, which contain nine *triplet* microtubules (Figure 21-33). Each triplet contains one complete 13-protofilament microtubule, the A subfiber, fused to the incomplete B subfiber, which in turn is fused to the incomplete C subfiber. The A and B subfibers of basal bodies continue into the axoneme shaft, whereas the C subfiber terminates within the transition zone between the basal body and shaft. The A and B subfibers of the basal body appear to initiate assembly of the nine

(a)

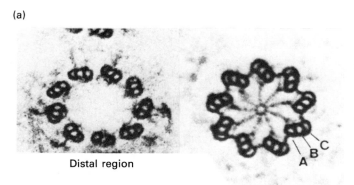

Distal region

Proximal region

(b)

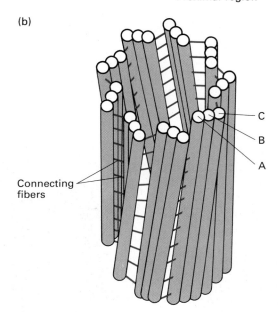

Connecting fibers

◀ **Figure 21-33** Structure of basal bodies and centrioles. (a) Cross sections of flagellar basal bodies in the protozoan *Trichonympha*. The proximal region contains a single central tubule and radial spokes; these are missing in the distal region. Each of the nine outer fibers are built of three tubules; labeled A, B, and C. (b) A schematic view of a centriole or basal body. The nine sets of triplet microtubules are connected by a set of proteinaceous fibers, which are also visible in part (a). *Part (a) reproduced from the* Journal of Cell Biology, *1981, by copyright permission of the Rockefeller University Press; courtesy of I. Gibbons.*

teins. Identification of this DNA began with an analysis of several *Chlamydomonas* mutants with defective basal body function. In *uni*, the first of the mutants studied, one of the two flagella fail to develop, because of a defect in the transition zone between one of the two basal bodies and the axoneme. Several basal body mutations, including *uni*, have been mapped to a small chromosome that segregates in a Mendelian fashion independently of the major *Chlamydomonas* chromosomes. Cloning of this small chromosome showed that it is a linear DNA with 6–9,000,000 bases; in growing cells this DNA is found exclusively in basal bodies (Figure 21-35). Thus basal bodies, like chloroplasts and mitochondria, contain a DNA that controls key organelle functioning; proteins encoded by the *uni* DNA are presumably translated on cytosolic ribosomes and then incorporated into the basal body.

In some cases centrioles are formed in the absence of a centriole "ancestor." For example, mouse eggs do not possess centrioles, and the flagellar basal body of the fertilizing sperm is lost or inactivated soon after fertilization. The first four or five cleavages of a fertilized mouse egg occur in the absence of centrioles; the pericentriolar material apparently acts as the MTOC during mitosis. Centrioles appear during later development. Thus centrioles can form de novo under normal conditions, and cell division can occur without them. There is the possibility that there exists a "centriole DNA" similar to the *Chlamydomonas uni* DNA. In early cleavages this DNA would be in an inactive form in the nucleus; expression of the DNA and synthesis of centriolar proteins could be accompanied by movement of this DNA to the centriole.

Centrioles Can Convert into Basal Bodies and Vice Versa

As noted earlier, purified centrioles, or the entire centrosome region, can nucleate assembly of large numbers of microtubules when injected into frog oocytes or when added to a preparation of pure tubulin; thus they have microtubule-organizing activity. Despite their similar ultrastructure, flagellar and ciliary basal bodies cannot act as microtubule-organizing centers, although they can act

outer doublets of the cilium or flagellum, and the subfibers grow outward from the basal body during elongation of the shaft. The two central tubules in the flagellum or cilium end in the transition zone above the basal body.

Discrete stages of centriole replication occur at defined points of the cell cycle. Growth of daughter centrioles occurs at right angles to the mature centrioles and is completed by the beginning of mitosis. During mitosis, the two pairs of mother-daughter centrioles form the two poles of the mitotic spindle (Figure 21-34). During early G_1, the mother and daughter centrioles lose their perpendicular orientation to each other.

As with centriole replication, a daughter basal body is formed at right angles to each "parent" basal body. Eventually the daughter basal body separates from the parent and then generates a new flagellum or cilium.

Centrioles and Basal Bodies Contain a Unique Small DNA

Recent work has shown that basal bodies contain a small DNA molecule that codes for many basal body pro-

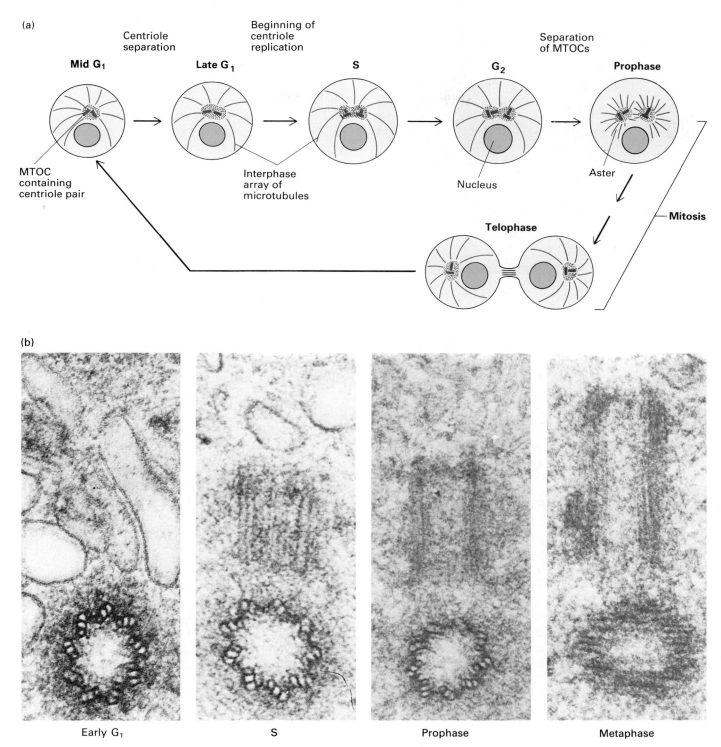

▲ **Figure 21-34** The centriole cycle. (a) Initiation of assembly of the daughter centrioles occurs during the S phase; they grow at right angles to the two parent centrioles. At the beginning of mitosis, the two pairs of centrioles separate and form the two poles of the mitotic spindle. Each pair is delivered to one of the two daughter cells. Following cell separation, in the G_1 phase, the mother and daughter centriole reorient and separate; then at S, each acts as parent for a new daughter. (b) Micrographs of centriole elongation during interphase and mitosis in cultured mammalian fibroblasts. In early G_1, only the nine-triplet parent centriole is visible; the daughter centriole growing at right angles to the parent is visible in the other cell phases. *Part (a) adapted from E. Karsenti and B. Maro, 1986,* Trends Biochem. Sci. *11:460. Part (b) from J. Rattner and S. G. Phillips, 1983,* J. Cell Biol. *57:359; courtesy of J. Rattner; reproduced from the* Journal of Cell Biology, *1983, by copyright permission of the Rockefeller University Press.*

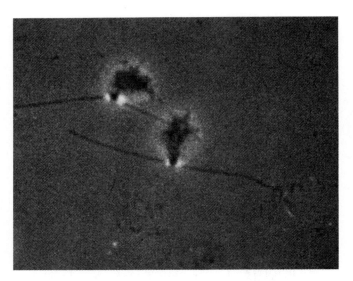

▲ **Figure 21-35** Localization of *uni* DNA to the basal body of *Chlamyodomonas* cells. Biotin-tagged *uni* DNA was hybridized to a permeabilized cell, and the in situ hybridized DNA was detected with fluorescent avidin, a biotin-binding protein. The fluorescence image, superimposed on the phase-contrast image of the same cell, shows that the *uni* DNA is in the basal body. *From J. L. Hall, Z. Ramanis, and D. Luck, 1989, Cell 59:121; courtesy of D. Luck.*

as templates for elongation of microtubules; that is, microtubules grow from the basal body tubules.

In some cases, basal bodies develop into centrioles with microtubule-organizing activity. For example, in many species a set of microtubules, called the *aster,* are nucleated in the fertilized egg from the basal body of the sperm flagellum a few minutes after it enters the egg. Apparently the egg, although lacking centrioles itself, contains some factor that confers microtubule-organizing activity on the sperm basal body.

Conversely, centrioles can convert into basal bodies. Precursors of the ciliated epithelial cells that line the human trachea and esophagus contain only centrioles. As these cells differentiate, the centrioles migrate from their normal position, near the nucleus, to the luminal plasma membrane. There the centriole forms numerous centriole-like structures, each of which becomes a basal body for a cilium. During this process, microtubule-organizing activity is lost.

Function of Microtubules in Mitosis

Of the many functions of centrosomes and basal bodies, perhaps the most extensively studied are those of the centrosome in mitosis. The centrosome (MTOC) nucleates formation of the microtubules that form the mitotic spindle, which is essential for proper separation of chromo-

somes to the daughter cells and for cell division. In this section, the multiple processes involved in mitosis and the role of microtubular structures in them are discussed.

Mitosis involves a number of complex mechanical processes. Each of the daughter chromosomes replicated during S phase must be moved to opposite ends of the cell, and the cytoplasm must be divided in such a way that each daughter receives the appropriate amount of membranous organelles and cytoskeletal proteins. Microtubules play a major role in each of these processes: they form the mitotic spindle, which organizes the chromosomes and cytoplasm, and they function in many of the subcellular movements that occur during mitosis. Our discussion of mitosis will focus on three key areas: (1) the structure and construction of the mitotic spindle and its interaction with the chromosomes; (2) the mechanism(s) by which chromosomes line up, separate, and move; and (3) the movements that cause the two daughter cells to become separate and distinct. The mechanisms underlying many of the movements that occur during mitosis are becoming clear at the molecular level, in part due to the development of cell-free systems that carry out specific aspects of mitosis. Refer to Chapter 5 for an overview of the key events in each step of mitosis (Figure 5-3).

Light-Microscope Techniques Reveal the Mitotic Spindle in Living Cells

Very few techniques are available for microscopic visualization of live, moving cells. Most microscopic techniques involve fixation of cells or tissues, followed by selective staining to emphasize the material of greatest interest. Many microtubules, such as those that make up a major part of the mitotic spindle, are quite labile structures and require special fixatives and optical systems for their visualization. Individual microtubules or other cytoskeletal fibers, though often quite long, are far too thin to be detected by ordinary light microscopy. However, because bundles of fibers, such as the parallel filaments in microtubules, are birefringent, they can be detected by polarization microscopy (Figure 21-36). As shown in Figure 21-37a, bundles of microtubules in the mitotic spindle can be visualized with this technique. Fibrous elements, now known to be bundles of microtubules, radiate from the two polar regions toward the chromosomes at the center of the spindle. Other microtubules (astral fibers) radiate outward in other directions from the polar region.

Several experiments have established that the birefringent fibers visualized in Figure 21-37a indeed are constructed of microtubules. When assembly of tubulin into microtubules is blocked by exposure to agents such as colchicine or conditions such as low temperature or high hydrostatic pressure, loss of the birefringent fibers of the mitotic spindle results. As might be expected, the progression of cells through mitosis is also blocked. These results also suggest that the microtubules in the mitotic spindle

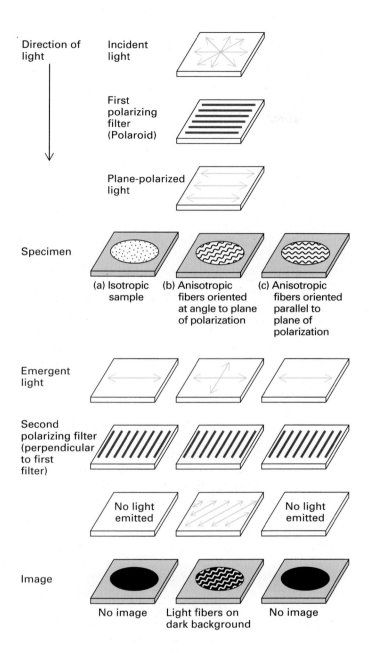

◄ Figure 21-36 Principles of polarization microscopy. Light waves normally have components that vibrate in all directions when viewed perpendicular to the beam, as indicated by the crossed arrows. A single polarizing filter allows only light waves that vibrate in a single direction to pass through; the emergent light is said to be plane-polarized. If a second polarizing filter is placed at right angles to the first, no light can pass through because the plane-polarized light emergent from the first filter is effectively blocked by the second, and no image is formed on a viewing screen. When viewed in the polarizing microscope, isotropic specimens (a), such as a homogeneous solution of salts or proteins, appear black because the light that emerges from the sample remains plane-polarized and therefore is unable to penetrate the second polarizing filter. The speed with which light passes through anisotropic (birefringent) specimens depends on the direction of polarization of the incident light. When an anisotropic specimen is situated at an angle with respect to the plane of polarization of the incident light (b), it effectively rotates the direction of vibration of some of the incident polarized light waves. Thus the light that emerges from the object will have some components that vibrate in a direction perpendicular to that of the initial plane-polarized light; these light waves will pass through the second polarizing filter. The image recorded on the eyepiece will be of a series of fibers similar in direction and size to those of the specimen. When a birefringent specimen is parallel to the plane of polarized light (c), little of the incident light will be able to pass through the second filter, and the specimen will appear dark to the viewer.

► Figure 21-37 Micrographs of an isolated metaphase mitotic spindle from a sea urchin egg obtained by polarization microscopy and Nomarski interference microscopy. (a) In polarized light, the pole-to-pole spindle fibers appear bright, whereas the fibers perpendicular to the pole-to-pole fibers (called astral fibers) appear dark. (b) The Nomarski optics technique detects differences in refractive index among different parts of the sample. It allows visualization of the dense chromosomes at the center of the mitotic spindle and also of the spindle fibers. [See E. Salmon and J. Segal, 1980, *J. Cell Biol.* 86:355.] *Courtesy of S. Inoue; reproduced by permission of the* Journal of Cell Biology, *1980, by copyright permission of the Rockefeller University Press.*

(a)   (b)

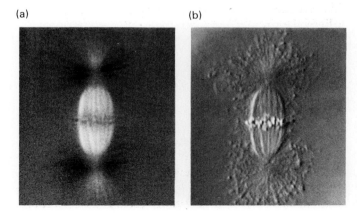

(a)

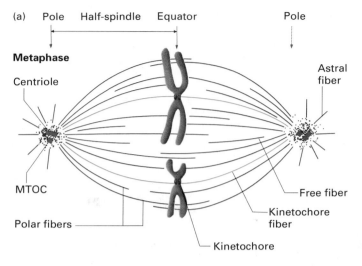

Metaphase

Pole · Half-spindle · Equator · Pole

Centriole

Astral fiber

MTOC

Free fiber

Polar fibers

Kinetochore fiber

Kinetochore

Anaphase

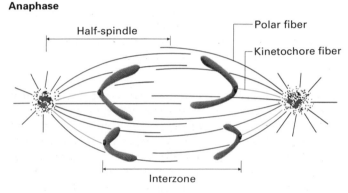

Half-spindle

Polar fiber

Kinetochore fiber

Interzone

Telophase

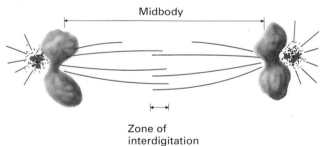

Midbody

Zone of interdigitation

(b)   Prophase

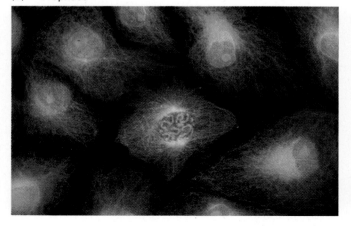

◀ **Figure 21-38**   (a) Schematic diagrams of mitotic spindle in animal cells during phases of mitosis. Each fiber represents many microtubules. (b–e) Fluorescence micrographs of mitotic cultured PtK2 fibroblasts stained with a fluorescent anti-tubulin antibody (green) and the DNA-binding dye ethidium homodimer (red-orange). (b) Prophase, (c) early metaphase, (d) early anaphase, (e) late anaphase. [See J. M. Murray et al., 1989, *Biotechniques* 7:154.] *Courtesy of J. M. Murray.*

(c)   Early metaphase

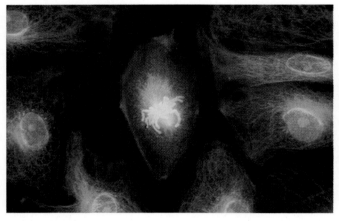

(d)   Early anaphase

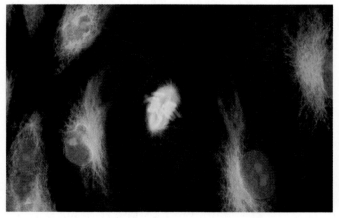

(e)   Late anaphase

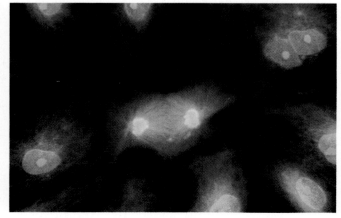

are in a steady state between assembly and disassembly into tubulin.

Normarski interference microscopy is based on differences in refractive index that reflect differences in density among the parts of a living cell. With the Normarski interference technique, the mitotic chromosomes, which are the densest objects in a mitotic spindle, are clearly visualized, but the much less dense microtubular spindle is seen only faintly (Figure 21-37b). Such observations reinforce the conclusion that at metaphase chromosomes are aligned midway between the spindle poles.

Bundles of Microtubules Form the Mitotic Spindle

Electron microscopy has shown that the mitotic spindle in animal cells contains three types of microtubules: *polar fibers,* which extend from the two poles of the spindle toward the equator; *kinetochore fibers,* which attach to the centromere of each mitotic chromosome and extend toward the poles; and *astral fibers,* which radiate outward from the poles toward the periphery of the cell (Figure 21-38). The microtubules and associated structures that radiate from each pole are collectively termed a half-spindle.

In animal cells, centrioles are at the center of the mitotic poles. As in interphase cells, the polar fibers radiate not from the centrioles themselves but rather from the diffusely staining material around the centriole—the microtubule-organizing center (MTOC). In certain protozoa, structures containing only nine singlet microtubules are found at the center of the MTOC; in other cells, such as diatoms, a plaque or sphere located on or near the nuclear envelope appears to function as the MTOC. Higher plants, in particular, lack centrioles but nevertheless form fully functional spindles. Usually these spindles are anastral—that is, they have no astral fibers (Figure

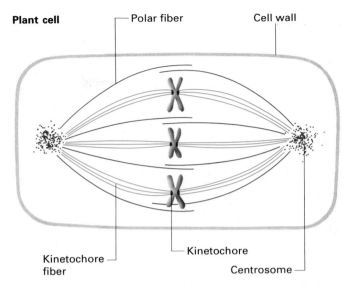

▲ **Figure 21-39** Schematic diagram of metaphase spindle structures in plant cells. Higher plants lack centrioles and astral fibers; the centrosome region functions as an MTOC. Each fiber represents many microtubules.

21-39). It is not known what structure in plants acts as a centriole or what entity organizes the spindle fibers.

Mitotic spindles can be isolated by gentle extraction of cells with detergents in the presence of certain concentrated salts, which prevent disassembly of microtubules. Electron microscopy has revealed the individual microtubules constituting a spindle (Figure 21-40). The center of the metaphase spindle is constructed from two interdigitating families of polar microtubules, one from each spindle pole (see Figures 21-38 and 21-40). These polar fibers form a spindle framework that provides a mechanical foundation against which chromosome movement can be generated for separation of the daughter genomes at anaphase and telophase.

(a)

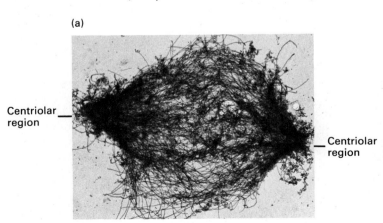

(b)

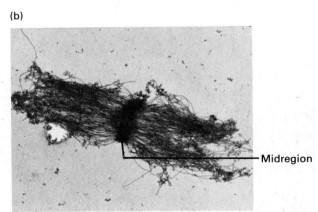

▲ **Figure 21-40** Micrographs of mitotic spindles isolated from fibroblasts by a series of solutions of detergents and concentrated salts, so that only the spindle microtubules remain. (a) In prophase, the individual microtubules are seen radiating from the centriolar regions. (b) In anaphase, all the microtubules are parallel to the spindle axis. The midregion of the spindle contains overlapping spindle microtubules. *Courtesy of F. Solomon and G. Zieve.*

Kinetochore Microtubules Connect the Chromosomes to the Poles

The metaphase chromosome is a highly condensed nucleoprotein particle. It contains two coiled daughter chromatids, each of which contains a replicated DNA molecule bound to histones and other proteins. The goal of mitosis is to distribute one chromatid to each daughter cell; this is the function of the kinetochore microtubule fibers.

The centromere and kinetochore are two closely related parts of the metaphase chromosome. The centromere is the area of primary constriction in the metaphase chromosome, where the chromatids are held together. The kinetochore, usually located at or near the centromere, is the structure to which several pole-to-chromosome microtubules are attached (Figures 21-41 and 21-42; see also Figures 9-10 and 9-12). Kinetochores always

face the spindle poles and are essential for the proper movement of chromosomes during mitosis. During anaphase, the kinetochore is the site at which force is exerted on the chromosome to pull it toward the poles.

Antibodies that react specifically with kinetochores are, for unknown reasons, frequently produced by patients suffering from scleroderma (a disease of unknown origin that causes fibrosis of connective tissue). Such antibodies stain kinetochores in mitotic cells (Figure 21-43a). Since the same number of kinetochore "dots" are detected when interphase cells are stained, it would appear that at least part of the kinetochore is attached to chromosomes even in nonmitotic cells. Treatment of interphase cells with caffeine causes kinetochores to detach from the rest of the chromosome. During mitosis, free kinetochores in caffeine-treated cells can be seen to move poleward along the pole-to-kinetochore microtubules (Figure 21-43b and c), suggesting that only kinetochores

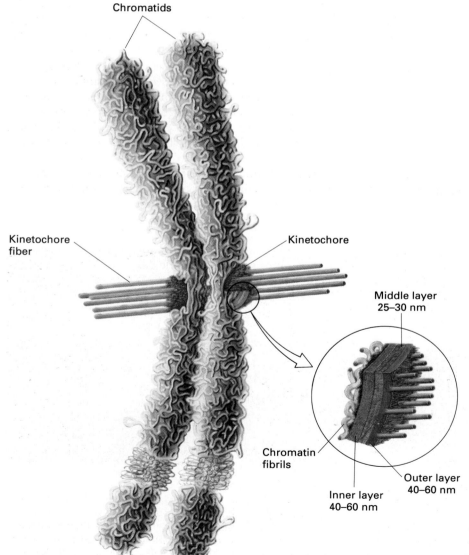

Chromatids

Kinetochore fiber

Kinetochore

Middle layer 25–30 nm

Chromatin fibrils

Inner layer 40–60 nm

Outer layer 40–60 nm

◀ **Figure 21-41**   Schematic diagram of a metaphase chromosome showing the kinetochore microtubules and the three-layer kinetochore characteristic of animals and lower plants (*inset*). Each fiber represents many microtubules. The sister chromatids have not yet separated, and each chromatid contains a kinetochore at its centromere. Microtubules insert into the outer layer of each kinetochore and run from there toward one of the two poles of the cell. At anaphase, the sister chromatids separate, and the chromosomes are pulled to opposite poles of the cell by the kinetochore microtubules. *From B. R. Brinkley, A. Tousson, and M. M. Valdivia, 1985, in* Aneuploidy, *V. L. Dellarco, P. E. Voyter, and A. Hollaender, Plenum, p. 243.*

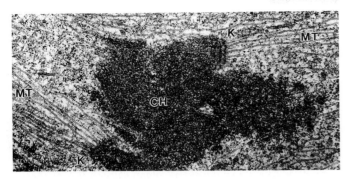

▲ **Figure 21-42** Electron micrograph of a thin section of a mitotic cultured fibroblast, showing a portion of a chromosome (CH), two three-layered kinetochores (K), and microtubules (MT) inserting into the two kinetochores. *From B. R. Brinkley and J. Cartwright, Jr., 1971, J. Cell Biol. 50:416; courtesy of B. R. Brinkley.*

are necessary for mitotic movements of chromosomes. What kind(s) of proteins or other materials make up kinetochores and how microtubules insert into them are only now being elucidated.

The centromere determines the attachment of the kinetochore and thus the proper movement of chromosomes during mitosis. A fragment of a chromosome without a centromere, which occasionally is generated by drugs or x-irradiation, cannot be incorporated into a metaphase spindle and is lost during mitosis. The generation of a kinetochore is directly controlled by a unique segment of DNA termed centromeric DNA. Several cloned yeast centromeric DNAs (CEN) have the ability to cause any DNA

segment linked to them to be replicated and distributed equally to daughter cells, exactly as in full-sized chromosomes (see Figure 9-9). In some way, centromeric DNA segments bind proteins that cause a kinetochore to form.

Dynamic Instability Explains the Morphogenesis of the Mitotic Spindle

At the onset of mitosis, the unipolar cytoskeleton of nondividing cells is transformed into the bipolar one of dividing cells. Long interphase microtubules are replaced by more numerous and shorter spindle microtubules, and kinetochores become attached to the (+) ends of microtubules and the attached chromosomes become arranged at the metaphase plate. Studies on the assembly and disassembly of microtubules in cells and cell-free systems have given insight into how these rearrangements occur.

As discussed earlier, the microtubules in interphase animal cells are continually being assembled and disassembled from a single MTOC containing two centrioles (see Figure 21-16). The average lifetime of a given interphase microtubule is about 10 min. At the onset of mitosis, the already replicated centrioles (see Figure 21-34a) separate, forming two MTOCs, or poles. The ability of these MTOCs to nucleate microtubules increases, leading to an increase in the number of microtubules extending from each pole during prophase (Figure 21-44). At this stage, the average spindle microtubule grows for only 30 s before it depolymerizes and disappears.

In prometaphase some microtubules randomly encounter a kinetochore. This apparently causes the microtubule

(a)

(b)

(c)

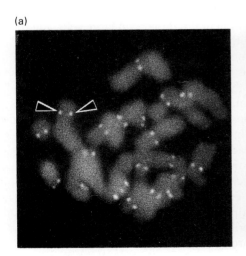
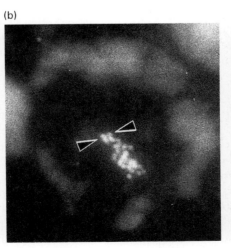
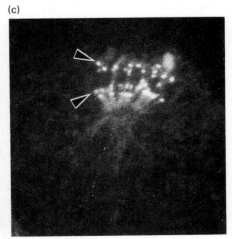

▲ **Figure 21-43** Kinetochores can be detached experimentally from the rest of the chromosome by treatment of cultured mammalian cells with caffeine while they are in the G_1, G_2, or S stages of the cell cycle. Detached kinetochores exhibit normal movement during mitosis. (a) A preparation of metaphase Chinese Hamster Ovary (CHO) cell chromosomes stained with an antikinetochore antibody (*arrowheads*) and with a DNA-binding dye that stains the chromosomes. (b) CHO cells after caffeine treatment exhibit kinetochores (*arrowheads*) that have become detached from the chromosomes. (c) An anaphase caffeine-treated cell exhibits kinetochores (*arrowheads*) moving poleward detached from the rest of the chromosome. [See B. R. Brinkley et al., 1988, *Nature* 336:251. *Courtesy of B. R. Brinkley.*

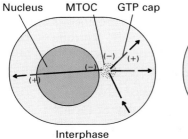

Nucleus   MTOC   GTP cap

Interphase

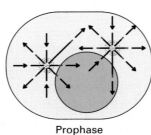

Prophase

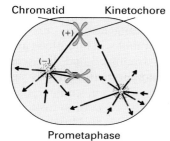

Chromatid   Kinetochore

Prometaphase

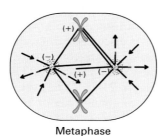

Metaphase

◀ **Figure 21-44** Dynamic instability and the formation of the mitotic spindle. In both interphase and mitotic cells, most microtubules radiate from the MTOCs with the (−) ends of the microtubules facing the MTOC and the (+) ends extending outward. A typical interphase cell has long microtubules, most with a GTP cap (red). During mitotic prophase, the microtubule-nucleating activity of the already replicated MTOCs increases; this leads to a larger number of short microtubules. Some have a GTP cap and are growing at their distal (+) end; others, with a GDP cap, shrink rapidly from their distal end. In prometaphase some of the microtubules interact with kinetochores (dark blue), causing the microtubules to be partially stabilized. Finally, in metaphase all of the spindle microtubules are stabilized; MAP proteins may stabilize the pole-to-equator ones. *Adapted from M. Kirschner and T. Mitchison, 1986, Cell* **45:329.**

to become stable; perhaps the kinetochore "caps" the (+) end of the microtubule and prevents its depolymerization even if it has a GDP cap, which normally destabilizes a microtubule (see Figure 21-18). Experimental evidence suggests that kinetochores indeed stabilize the (+) ends of microtubules. For instance, if a kinetochore is detached from its pole-to-kinetochore microtubule by a fine needle, the microtubule rapidly depolymerizes from the (+) end. By metaphase, when the chromosomes are aligned at the equator, both the pole-to-kinetochore and pole-to-equator microtubules have become stabilized; possibly proteins bound to the sides or ends of the microtubules help prevent their depolymerization.

Many Events in Mitosis Do Not Depend on the Mitotic Spindle

Although the mitotic spindle is formed during prophase, it is not essential for all events in mitosis. For example, in the presence of colchicine, which blocks assembly of microtubules, most animal cells undergo breakdown of the nuclear membrane during prophase and condensation of chromatin into sister chromatids. In sea urchin embryos, even the separation of daughter chromatids and the reformation of the nuclear membrane occurs in the presence of colchicine. Thus in all probability these steps do not involve the mitotic spindle. The spindle *is* required for alignment of chromosomes at the equator, for movement of daughter chromosomes toward the poles, and for separation of the daughter nuclei.

Balanced Forces Align Metaphase Chromosomes at the Equator of the Spindle

Now that we have examined the structure of the mitotic spindle in some detail, let us turn to the movements that occur during mitosis and the forces that generate them.

During prometaphase the newly condensed chromosomes can be observed to move randomly between the two poles. Eventually, one of the kinetochores attaches to tubulin fibers from one pole. In some cells, such mono-oriented chromosomes rapidly move first toward and then away from the single pole to which they are attached. Soon, the other kinetochore quickly becomes associated with fibers from the other pole.

The association between kinetochores and poles can be modified experimentally. In certain large cells, micromanipulation with fine glass needles can rotate prometaphase chromosomes 180° with respect to the axis of the spindle. This procedure breaks the attachment of chromatids and their associated kinetochore fibers with the poles. The chromatids, however, soon reattach to fibers from the opposite pole; subsequently, the chromatid that would have been pulled into one daughter cell during anaphase is instead pulled into the other. These and other studies indicate that some force pulls the two kinetochores on sister chromatids toward opposite poles. Micromanipulation experiments suggest that the strength of these forces is proportional to the distance from the chromosome to the pole. Thus these opposing forces are balanced when a chromosome is at the equator of the spindle, so the chromosome remains stationary there. If a metaphase chromosome is displaced toward one pole by micromanipulation, then the force exerted from the opposite pole momentarily increases and quickly pulls the displaced chromosome back to the equator. The nature of the forces that keep metaphase chromosomes aligned is not known.

Anaphase Consists of Two Distinct Motile Events

The anaphase stage of mitosis consists of two distinct motile events that can occur at the same time. The sister chromatids break apart at their point of connection in the centromeric region and then move toward poles at opposite ends of the cell. This process, termed anaphase A, involves *shortening* of the pole-to-kinetochore microtubules as the chromosomes move toward the poles. Simultaneously, there is a separation of the two poles into what will become the two daughter cells. These latter changes, often called anaphase B, involve *elongation* of the polar microtubules as the poles move apart. Microtubules appear essential for both types of anaphase movements, as both are disrupted by colchicine. In part because anaphase A and anaphase B can occur in certain cell-free extracts, we understand a good deal about their molecular mechanisms; we shall discuss the key experiments and concepts in the following two sections.

Poleward Chromosome Movement (Anaphase A) Is Powered by Microtubule Disassembly at the Kinetochore and Requires No External Energy Source

Pole-to-kinetochore microtubules, like all spindle microtubules, are oriented with their (−) ends facing the pole. That both assembly and disassembly of the kinetochore fibers occurs at the kinetochore end—the (+) end—has been demonstrated by a simple experiment like that illustrated in Figure 21-16.

The in vivo experiment depicted in Figure 21-45 shows that during anaphase A, chromosomes move poleward

▶ **Figure 21-45** Experimental demonstration that chromosomes move poleward along stationary kinetochore microtubules, which coordinately disassemble from their kinetochore ends during anaphase A. Fibroblasts are injected with fluorescent tubulin and then allowed to enter metaphase, so that all of the spindle microtubules are fluorescent. Only the kinetochore microtubules are shown. In early anaphase, a band of microtubules (yellow box) is subjected to a laser light, which destroys the fluorescence but leaves the microtubules continuous and functional across the bleached region. The bleached segment of each microtubule thus provides a marker for the fate of that part of the pole-to-kinetochore microtubule. During anaphase the distance of the bleached zone from the poles (measured in diagrams by the black double-headed arrows) does not change, indicating that no depolymerization of the microtubules occurs at the poles. Rather, the kinetochore microtubules disassemble just behind the kinetochore, and the kinetochores move poleward along the microtubules. *Adapted from G. J. Sammak, and G. Borisy, 1987, J. Cell Biol. **104**:9; and G. J. Gorbsky, P. J. Sammak, and G. Borisy, 1988, J. Cell Biol. **106**:1185.*

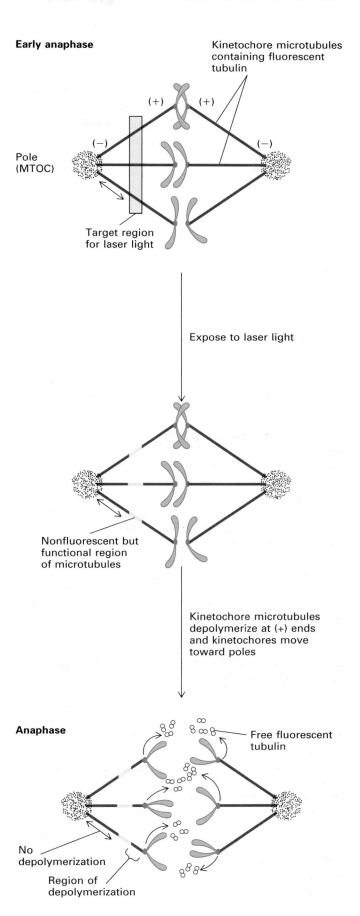

Early anaphase

Kinetochore microtubules containing fluorescent tubulin

(+) (+)

(−) (−)

Pole (MTOC)

Target region for laser light

Expose to laser light

Nonfluorescent but functional region of microtubules

Kinetochore microtubules depolymerize at (+) ends and kinetochores move toward poles

Anaphase

Free fluorescent tubulin

No depolymerization

Region of depolymerization

along stationary pole-to-kinetochore microtubules, and that these microtubules disassemble from their kinetochore ends. Subsequent in vitro studies have shown that no energy source such as ATP is necessary for chromosome movement. In the experiment outlined in Figure 21-46, for example, purified microtubules were mixed with purified anaphase chromosomes; as expected, the kinetochores bound preferentially to the (+) ends of the microtubules. To induce depolymerization of microtubules, the reaction mixture was diluted, thus lowering the concentration of free tubulin. Remarkably, the chromosome moved along the microtubule at a rate similar to that of chromosome movement in intact cells (up to 12 μm/min). Since no ATP (or any other energy source) was added, chromosome movement toward the (−) end must be

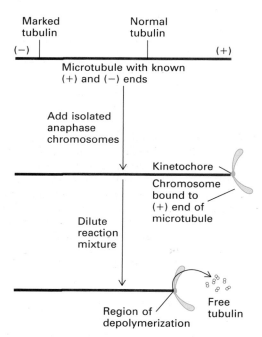

▲ **Figure 21-46** In vitro demonstration of poleward chromosome movement in anaphase A. A population of microtubules is prepared in vitro, using both "marked" and normal tubulin, in such a way that their (−) and (+) ends can be distinguished. When these microtubules are mixed with purified anaphase chromosomes, the chromosomes bind only to the (+) ends of the microtubules, which in cells would face the equator. When the reaction mixture is diluted, to lower the concentration of free (unpolymerized) tubulin, the chromosomes move toward the (−) ends of the microtubules, which face the pole of the cell. Simultaneously, the microtubules depolymerize from the (+) ends, just behind the kinetochore. Since the reaction mixture contains no source of energy, such as ATP, it is depolymerization of microtubules at the kinetochore that both regulates and powers poleward chromosome movement during anaphase. *Adapted from D. E. Koshland, T. J. Mitchison, and M. Kirschner, 1988, Nature* **331**:499.

powered, in some way, by microtubule disassembly at or near the kinetochore. Precisely how kinetochores move in one direction is not yet clear. One possibility is that the kinetochore forms a collar around the (+) ends of microtubules and diffuses randomly up and down microtubules. If only segments of microtubules distal to the kinetochore are depolymerized, the net result would be movement of the kinetochore in one direction—toward the (−) end.

Separation of the Poles (Anaphase B) Involves Sliding of Adjacent Microtubules Powered by ATP Hydrolysis

The elongation of the polar microtubules and separation of the two poles during anaphase B is fundamentally different from the poleward chromosome movement of anaphase A. Anaphase B requires ATP hydrolysis, whereas anaphase A does not. For example, detergent treatment of mitotic cells, which causes ATP to leak out, does not affect poleward chromosome movement (anaphase A) but does prevent spindle elongation movements (anaphase B) unless ATP is added.

In vitro studies on anaphase B have been done with isolated spindles from diatoms. Unlike the loose baskets of spindle microtubules found in most animal and plant cells, the spindles of diatoms are almost crystalline in arrangement and are stable to the rigors of isolation. In the presence of ATP, isolated diatom spindles elongate, simulating anaphase B; the zone of overlap between the two half-spindles decreases in length as the spindle elongates by a similar amount (Figure 21-47). These and similar studies suggest that interactions between microtubules from opposite half-spindles in the zone of overlap generates the force of anaphase B. Hydrolysis of ATP is required for movement.

The direction of microtubule movement in anaphase B is opposite to that expected for a dyneinlike ATPase. In anaphase adjacent microtubules are pushed in the direction of their pole-facing (−) ends, whereas flagellar or cytoplasmic dynein would push an adjacent microtubule in the direction of its (+) end. A kinesinlike ATPase could push adjacent microtubules in the direction of their (−) ends, but as yet there is no direct evidence that kinesinlike molecules exist in the mitotic spindle. A model of the elongation and movement of the pole-to-equator spindles during anaphase B is shown in Figure 21-48. Tubulin adds to the (+) ends of the microtubules in the overlap zone, which then slide through the midzone as the spindle elongates. The key experiment supporting this model was done with isolated diatom spindles. In the presence of ATP, as noted above, spindle elongation is limited to the length of the microtubule overlap zone (see Figure 21-47). If tubulin is added to the reaction mixture, however, spindle elongation is several times the length of the origi-

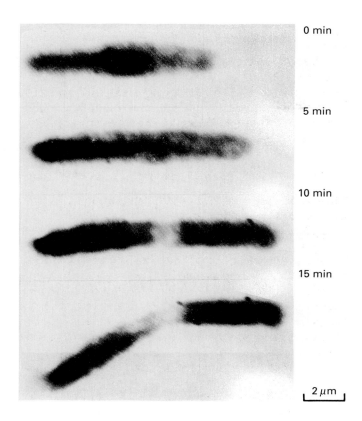

▲ **Figure 21-47** Anaphase B movements in vitro. An isolated diatom spindle was incubated with a solution containing ATP and was viewed under the polarization microscope before (0′) and after 5, 10, and 15 min of incubation. Note the decrease in birefringence of the central overlap segment as the two half-spindles slide apart. Sliding of the half-spindles requires ATP hydrolysis and is thought to involve a kinesinlike molecular motor. [See W. Z. Cande and K. L. McDonald, 1985, *Nature* **316**:168.] *Courtesy of W. Z. Cande.*

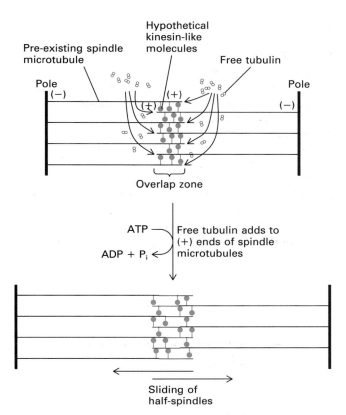

▲ **Figure 21-48** Model of spindle elongation and movement during anaphase B. Tubulin (red) adds to the (+) ends of all polar microtubules, lengthening these fibers. Simultaneously, hypothetical kinesinlike molecules (blue) bind to the polar microtubules in the overlap region. Kinesinlike molecules bound to a microtubule in one of the half-spindles "walk along" a microtubule in the other half-spindle, toward its (+) end, utilizing the energy from ATP hydrolysis. This results in a sliding of the two half-spindles toward their respective poles. *Adapted from H. Masuda and W. Z. Cande, 1987,* Cell **49**:193.

nal overlap zone. This result indicates that during anaphase B the microtubules attached to opposite poles not only slide by each other but also lengthen by addition of tubulin at their (+) ends.

Cytokinesis Is the Final Separation of the Daughter Cells

The final stage of mitosis is the division of the cytoplasm into two compartments—the process called cytokinesis. During mitosis animal cells typically change shape, becoming more spherical. This phenomenon is especially marked in the case of cultured fibroblasts, which normally are flat and elongate when growing attached to a surface. During mitosis they "round up" and lose most of their contacts with the substrate. The process of cytokinesis in diverse animal cells appears similar to that in the much larger spherical sea urchin egg, which has been subject to intensive study.

During late anaphase in the sea urchin egg, an indentation forms around the egg surface (see Figure 22-32). With time, it deepens and forms a cleavage furrow. This furrow completely encircles the egg in a plane perpendicular to the long axis of the mitotic spindle, and it continues to deepen until the opposing edges make contact in the center of the cell. Then the membranes fuse, and the original cytoplasm and the two sets of chromosomes become separated into two daughter cells. In the next chapter we will consider the evidence that actin and myosin generate the contractile force of the furrow.

Because plant cells are bound by a relatively inextensible cell wall, cytokinesis in plants involves quite different processes. Plant cells construct a cell membrane and a cell wall from membrane vesicles. These vesicles appear first near the center of the dividing cell and then extend out to the lateral walls (Figure 21-49); they are thought to arise from the endoplasmic reticulum or Golgi complex. Such vesicles are first observed during metaphase, when they

(a)

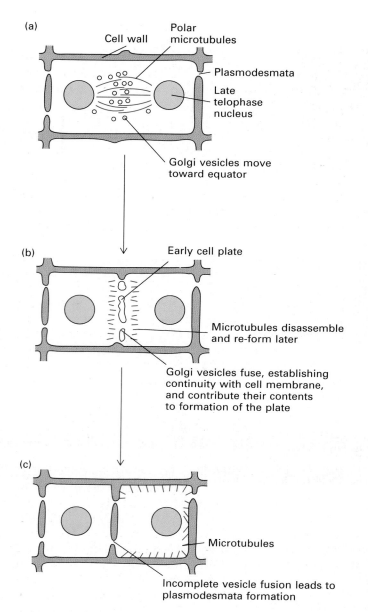

Cell wall
Polar microtubules
Plasmodesmata
Late telophase nucleus
Golgi vesicles move toward equator

(b)

Early cell plate
Microtubules disassemble and re-form later
Golgi vesicles fuse, establishing continuity with cell membrane, and contribute their contents to formation of the plate

(c)

Microtubules
Incomplete vesicle fusion leads to plasmodesmata formation

(d)

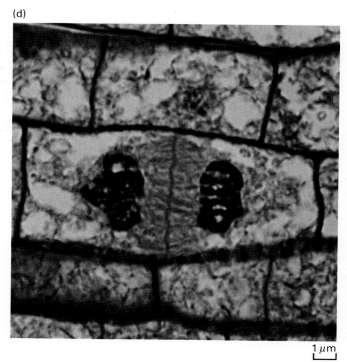

1 μm

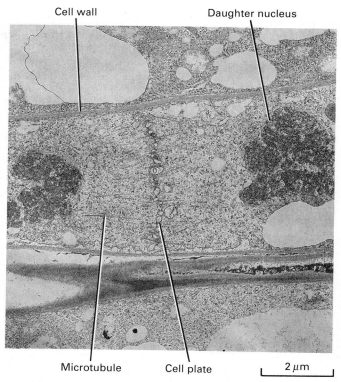

Cell wall
Daughter nucleus
Microtubule
Cell plate
2 μm

▲ **Figure 21-49** Outline of cytokinesis in a higher plant cell. (a) In late telophase, the nuclear membrane has re-formed, and the polar microtubules have not yet dispersed. A set of small vesicles derived from the Golgi complex, which contain cellulose and other precursors of the cell wall, accumulate at the equatorial plate. (b) These vesicles fuse with each other to form an early cell plate—a large membrane-limited vesicle in the center of the cell. Additional vesicles fuse with the cell plate, extending it outward. (c) Eventually the cell plate fuses with the plasma membrane, and the two cells separate. The daughter cells remain connected by thin membrane-lined passages, called plasmodesmata, that penetrate the separating cell wall. (d) Light micrograph of an onion root cell in telophase, showing the new cell wall being laid down between the daugher nuclei. *Part (d) courtesy J. Solliday and Biological Photo Service.*

▲ **Figure 21-50** Electron micrograph of the late telophase stage of mitosis in a seedling of soft maple (*Acer saccharinum*). Note the oriented microtubules and the vesicles that are fusing to form the cell plate that will separate the two daughter cells. *Micrograph by B. A. Palevitz; courtesy of E. H. Newcomb and Biological Photo Service.*

extend into the mitotic apparatus and, in some cases, even appear to contact the kinetochores. Because microtubules are also associated with these vesicles, the vesicles are thought to move along microtubules to the center of the cell (Figure 21-50).

The membrane vesicles also contain material for the future cell wall, such as polysaccharide precursors of cellulose and pectin. During late anaphase these vesicles fuse with one another to form large sheets near the equator of the spindle. The membranes of the vesicles become the plasma membranes of the daughter cells, and their contents form the intervening immature cell wall.

Cytokinesis in plant cells never completely separates adjacent cell cytoplasm, as occurs in animal cells. Almost all adjacent living plant cells are interconnected by a set of small cytoplasmic channels, 20 to 40 nm in diameter, called *plasmodesmata* (see Figure 13-52). The plasma membranes of adjacent cells are continuous through plasmodesmata, so molecules can pass from cell to cell. Plasmodesmata are formed at the time of laying down of the cell plate; regions of the vesicular membrane remain as tunnels in the forming cell wall.

Summary

Microtubules are built of tubulin, a heterodimer of one α and one β subunit. Despite the multiplicity of genes encoding α- and β-tubulin in most eukaryotes, all tubulins are functionally equivalent. Microtubules form a diverse array of structures. Some, like axonal filaments, are more-or-less permanent; others, such as interphase and mitotic spindle microtubules and, in some cases, flagellar and ciliary microtubules, assemble and disassemble.

Microtubules grow preferentially by addition of GTP-bound tubulin to the (+) end. The GTP is then hydrolyzed to GDP or tubulin to GTP dissociates. According to the current dynamic instability model, microtubules that retain a GTP cap at both ends will, depending on the concentration of tubulin, continue to grow; those with a GDP cap are rapidly depolymerized. The microtubule-organizing center (MTOC)—an amorphous region surrounding, in animal cells, the centriole—nucleates growth of new microtubules. The (+) growing end is pointed away from the MTOC. Similarly, in flagellar microtubules the (+) end is distal to a centriolelike basal body, and growth and shrinkage of flagellar microtubules also occurs at the (+) end. In interphase cultured cells, microtubules rapidly grow and shrink.

Microtubules act as tracks along which Golgi and other vesicles and also small particles move. Movement is rapid (2 μm/s) and can be bidirectional along a single microtubule. Two molecular motors that power axonal transport in neurons have been purified. One, kinesin, uses energy released by ATP hydrolysis to move vesicles toward the (+) end. The other, cytoplasmic dynein, powers movement in the opposite direction, toward the (−) end.

Together with other proteins, microtubules form the axoneme of cilia and flagella. Axonemes contain two singlet microtubules surrounded by a ring of nine doublet microtubules, all interconnected by several types of fibrous proteins. Inner and outer arms attached to the outer-doublet microtubules are dynein ATPases, which are crucial for ciliary or flagellar movement. Like cytoplasmic dynein, flagellar dyneins make and then break bonds with adjacent microtubules, causing each doublet to slide relative to its neighbor. Because of the way the doublets are interconnected circumferentially, this motion is converted into the bending of a cilium or flagellum.

Centrioles and flagellar basal bodies are similar in structure, although only the former can nucleate new microtubules in cell-free reactions. In some circumstances a centriole can differentiate into a basal body, and vice versa; both have properties of self-replicating organelles. Though new centrioles usually grow during the S phase from the side of a "parent" centriole, in some cells they form de novo. Basal bodies contain a large DNA that encodes several proteins essential for basal body function.

Microtubules are the principal structural elements of the mitotic spindle. Bundles of microtubules radiate from the two poles, or microtubule-organizing centers (MTOCs), of the spindle. In animal cells a centriole is at the center of the polar region. Kinetochore fibers run from the pole to the kinetochore attached near the centromere of each chromosome. Polar microtubules run from the poles to the equator of the spindle. Forces of an unknown nature align the chromosomes at the metaphase plate. Pole-to-kinetochore microtubules grow outward from the pole, and kinetochores bind to the (+) ends of the microtubules. Poleward movement of chromosomes (anaphase A) is driven by depolymerization of the kinetochore microtubules at or near the kinetochore. In contrast, separation of the poles (anaphase B) involves growth of the polar microtubules at the equatorial (+) end, concomitant with sliding of the microtubules in each half-spindle past each other. ATP hydrolysis is required for anaphase B, but not anaphase A; both types of movements can now be studied in cell-free reactions.

References

General References on Microtubules and the Cytoskeleton

*BERSHADSKY, A. D., and J. M. VASILIEV. 1988. *Cytoskeleton.* Plenum Press.

*A book or review article that provides a summary of the topic.

*DUSTIN, P. 1984. *Microtubules*, 2d ed. Springer-Verlag.

*LACKIE, J. M. 1985. *Movement and Cell Behaviour*. Allen and Unwin.

*LLOYD, C. D., J. S. HYAMS, and R. M. WARN (eds). 1986. *The Cytoskeleton: Cell Function and Organization*. The Company of Biologists.

*SCHLIWA, M. 1986. *The Cytoskeleton: An Introductory Survey*. Cell Biology Monographs, vol. 13. Springer-Verlag.

Structure of Microtubules

AMOS, L. A., and T. S. BAKER. 1979. Three-dimensional image of tubulin in zinc-induced sheets, reconstructed from electron micrographs. *Int. J. Biol. Macromol.* **1**:146–156.

*AMOS, L. A., R. W. LINCK, and A. KLUG. 1976. Molecular structure of flagellar microtubules. In *Cell Motility*, R. Goldman, T. Pollard, and J. Rosenbaum, eds. Cold Spring Harbor Laboratory, pp. 847–868.

COHEN, W. D., D. BARTLET, R. JAEGER, G. LANGFORD, and I. NEMHAUSER. 1982. The cytoskeletal system of nucleated erythrocytes. I. Composition and function of major elements. *J. Cell Biol.* **93**:828–838.

HAYS, T. S., R. DEURING, B. ROBERTSON, M. PROUT, and M. T. FULLER. 1989. Interacting proteins identified by genetic interactions: a missense mutation in α-tubulin fails to complement alleles of the testis-specific β-tubulin gene of *Drosophila melanogaster*. *Mol. Cell Biol.* **9**:975–884.

MANDELKOW, E-M., R. SCHULTHEISS, R. RAPP, M. MULLER, and E. MANDELKOW. 1986. On the surface lattice of microtubules: helix starts, protofilament number, seam and handedness. *J. Cell Biol.* **102**:1067–1073.

*WEBER, K., and M. OSBORN. 1979. Intracellular display of microtubular structures revealed by indirect immunofluorescent microscopy. In *Microtubules*, K. Robert and J. S. Hyams, eds. Academic Press, pp. 279–313.

Assembly and Disassembly of Microtubules in Vitro: Dynamic Instability

CARLIER, M.-F. 1988. Role of nucleotide hydrolysis in the polymerization of actin and tubulin. *Cell Biophys.* **12**:105–117.

CHEN, Y-D., and T. L. HILL. 1985. Monte Carlo study of the GTP cap in a five-start helix model of a microtubule. *Proc. Nat'l Acad. Sci. USA* **82**:1131–1135.

HORIO, H., and H. HOTANI. 1986. Visualization of the dynamic instability of individual microtubules by dark field microscopy. *Nature* **321**:605–607.

KEATES, R. A., and F. R. HALLETT. 1988. Dynamic instability of sheared microtubules observed by quasi-elastic light scattering. *Science* **241**:1642–1645.

MITCHISON, T., and M. KIRSCHNER. 1984. Dynamic instability of microtubule growth. *Nature* **312**:237–242.

WALKER, R. A., S. INOUE, and E. D. SALMON. 1989. Asymmetric behavior of severed microtubule ends after ultraviolet-microbeam irradiation of individual microtubules in vitro. *J. Cell Biol.* **108**:931–937.

WALKER, R. A., et al. 1988. Dynamic instability of individual microtubules analyzed by video light microscopy: rate constants and transition frequencies. *J. Cell Biol.* **107**:1437–1448.

Assembly and Disassembly of Microtubules in Cells

BINDER, L. I., W. L. DENTLER, and J. L. ROSENBAUM. 1975. Assembly of chick brain tubulin onto flagellar microtubules from *Chlamydomonas* and sea urchin sperm. *Proc. Nat'l Acad. Sci. USA* **72**:1122–1126.

CASSIMERIS, L., N. K. PRYER, and E. D. SALMON. 1988. Real-time observations of microtubule dynamic instability in living cells. *J. Cell Biol.* **107**:2223–2231.

GUNDERSEN, G. G., and J. C. BULINSKI. 1988. Selective stabilization of microtubules oriented toward the direction of cell migration. *Proc. Nat'l Acad. Sci. USA* **85**:5946–5950.

KIM, S., M. MAGENDANTZ, W. KATZ, and F. SOLOMON. 1987. Development of a differentiated microtubule structure: formation of the chicken erythrocyte marginal band in vivo. *J. Cell Biol.* **104**:51–59.

*KIRSCHNER, M., and T. MITCHISON. 1986. Beyond self-assembly: from microtubules to morphogenesis. *Cell* **45**:329–342.

OKABE, S., and N. HIROKAWA. 1988. Microtubule dynamics in nerve cells: analysis using microinjection of biotinylated tubulin into PC12 cells. *J. Cell Biol.* **107**:651–664.

SAMMAK, P. J., G. J. GORBSKY, and G. G. BORISY. 1987. Microtubule dynamics in vivo: a test of mechanisms of turnover. *J. Cell Biol.* **104**:395–405.

SCHULZE, E., and M. KIRSCHNER. 1987. Dynamic and stable populations of microtubules in cells. *J. Cell Biol.* **104**:277–288.

Microtubule-associated Proteins

*BLACK, M. M., and P. W. BAAS. 1989. The basis of polarity in neurons. *Trends Neurosci.* **12**:211–214.

ENNULAT, D. J., R. K. H. LIEN, G. A. HASHIM, and M. L. SHELANSKI. 1989. Two separate 18-amino acid domains of Tau promote the polymerization of tubulin. *J. Biol. Chem.* **264**:5327–5330.

HIMMLER, A., D. DRECHSEL, M. W. KIRSCHNER, and D. W. MARTIN, JR. 1989. Tau consists of a set of proteins with repeated C-terminal microtubule-binding domains and variable N-terminal domains. *Mol. Cell Biol.* **9**:1381–1388.

HIROKAWA, N., S-I. HISANAGA, and Y. SHIOMURA. 1988. MAP2 is a component of crossbridges between microtubules and neurofilaments in the neuronal cytoskeleton: quick-freeze, deep-etch immunoelectron microscopy and reconstitution studies. *J. Neurosci.* **8**:2769–2779.

HIROKAWA, N., Y. SHIOMURA, and S. OKABE. 1988. Tau proteins: the molecular structure and mode of binding on microtubules. *J. Cell Biol.* **107**:1449–1459.

OKABE, S., and N. HIROKAWA. 1989. Rapid turnover of microtubule-associated protein MAP2 in the axon revealed by microinjection of biotinylated MAP2 into cultured neurons. *Proc. Nat'l Acad. Sci. USA* 86:4127–4131.

*OLMSTED, J. B. 1986. Microtubule-associated proteins. *Ann. Rev. Cell Biol.* 2:421–457.

SHIOMURA, Y., and N. HIROKAWA. 1987. The molecular structure of microtubule-associated protein 1A (MAP1A) *in vivo* and *in vitro*. An immunoelectron microscopy and quick-freeze, deep-etch study. *J. Neurosci.* 7:1461–1469.

Multiplicity and Modification of Tubulins

BARRA, H. S., C. A. ARCE, and C. E. ARGARANA. 1988. Posttranslational tyrosination and detyrosination of tubulin. *Mol. Neurobiol.* 2:133–153.

BRÉ, M.-H., T. E. KREIS, and E. KARSENTI. 1987. Control of microtubule nucleation and stability in Madin-Darby canine kidney cells: the occurrence of noncentrosomal, stable detyrosinated microtubules. *J. Cell Biol.* 105:1283–1296.

GU, W., S. A. LEWIS, and N. J. COWAN. 1988. Generation of antisera that discriminate among mammalian α-tubulins: introduction of specialized isotypes into cultured cells results in their coassembly without disruption of normal microtubule function. *J. Cell Biol.* 106:2011–2022.

MARUTA, H., K. GREER, and J. L. ROSENBAUM. 1986. The acetylation of α-tubulin and its relationship to the assembly and disassembly of microtubules. *J. Cell Biol.* 103:571–579.

PIPERNO, G., M. LEDIZET, and X.-J. CHANG. 1987. Microtubules containing acetylated α-tubulin in mammalian cells in culture. *J. Cell Biol.* 104:289–302.

SCHATZ, P. J., F. SOLOMON, and D. BOTSTEIN. 1986. Genetically essential and nonessential α-tubulin genes specify functionally interchangeable proteins. *Mol. Cell Biol.* 6:3722–3733.

SCHULZE, E., D. J. ASAI, J. C. BULINSKI, and M. KIRSCHNER. 1987. Posttranslational modification and microtubule stability. *J. Cell Biol.* 105:2167–2177.

*SULLIVAN, K. F. 1988. Structure and utilization of tubulin isotypes. *Ann. Rev. Cell Biol.* 4:687–716.

WEBSTER, D. R., G. G., GUNDERSEN, J. C. BULINSKI, and G. G. BORISY. 1987. Differential turnover of tyrosinated and detyrosinated microtubules. *Proc. Nat'l Acad. Sci. USA* 84:9040–9044.

Microtubules and Intracellular Transport

ALLEN, R. D., D. G. WEISS, J. H. HAYDEN, D. T. BROWN, H. FUJIWAKE, and M. SIMPSON. 1985. Gliding movement of and bidirectional transport along single native microtubules from squid axoplasm: evidence for an active role of microtubules in cytoplasmic transport. *J. Cell Biol.* 100:1736–1752.

DABORA, S. L., and M. P. SHEETZ. 1988. The microtubule-dependent formation of a tubulovesicular network with characteristics of the ER from cultured cell extracts. *Cell* 54:27–35.

GELLES, J., B. J. SCHNAPP, and M. P. SHEETZ. 1988. Tracking kinesin-driven movements with nanometre-scale precision. *Nature* 331:450–453.

HIROKAWA, N., K. K. PFISTER, H. YORIFUJI, M. C. WAGNER, S. T. BRADY, and G. S. BLOOM. 1989. Submolecular domains of bovine brain kinesin identified by electron microscopy and monoclonal antibody decoration. *Cell* 56:867–878.

HOWARD, J., A. J. HUDSPETH, and R. D. VALE. 1989. Movement of microtubules by single kinesin molecules. *Nature* 342:154–158.

*MCINTOSH, J. R., and M. E. PORTER. 1989. Enzymes for microtubule-dependent motility. *J. Biol. Chem.* 264:6001–6004.

PORTER, M. E., and K. A. JOHNSON. 1989. Dynein structure and function. *Ann. Rev. Cell Biol.* 5:119–152.

ROZDZIAL, M. M., and L. T. HAIMO. 1986. Bidirectional pigment granule movements of melanophores and regulated by protein phosphorylation and dephosphorylation. *Cell* 47:1061–1070.

SCHNAPP, B. J., and T. S. REESE. 1989. Dynein is the motor for retrograde axonal transport of organelles. *Proc. Nat'l Acad. Sci. USA* 86:1548–1552.

*SCHNAPP, B. J., and T. S. REESE. 1986. New developments in understanding rapid axonal transport. *Trends Neurosci.* 9:155–162.

SCHROER, T. A., E. R. STEUER, and M. P. SHEETZ. 1989. Cytoplasmic dynein is a minus end-directed motor for membranous organelles. *Cell* 56:937–946.

*VALE, R. D. 1987. Intracellular transport using microtubule-based motors. *Ann. Rev. Cell Biol.* 3:347–378.

*VALLEE, R. B., H. S. SHPETNER, and B. M. PASCHAL. 1989. The role of dynein in retrograde axonal transport. *Trends Neurosci.* 12:66–70.

VALLEE, R. B., J. S. WALL, B. M. PASCHAL, and H. S. SHPETNER. 1988. Microtubule-associated protein 1C from brain is a two-headed cytosolic dynein. *Nature* 332:561–563.

YANG, J. T., R. A. LAYMON, and L. S. B. GOLDSTEIN. 1989. A three-domain structure of kinesin heavy chain revealed by DNA sequence and microtubule binding analyses. *Cell* 56:879–889.

Structure and Function of Cilia and Flagella

AMOS, W. G., L. A. AMOS, and R. W. LINCK. 1986. Studies of tektin filaments from flagella microtubules by immunoelectron microscopy. *J. Cell Sci. Suppl.* 5:55–68.

*BROKAW, C. J. 1986. Future directions for studies of mechanisms for generating flagellar bending waves. *J. Cell Sci. Suppl.* 4:103–113.

*GIBBONS, I. R. 1981. Cilia and flagella of eukaryotes. *J. Cell Biol.* 91:107s–124s.

*GIBBONS, I. R. 1988. Dynein ATPases as microtubule motors. *J. Biol. Chem.* 263:15837–15840.

*GOODENOUGH, U. W., and J. E. HEUSER. 1985. Outer and inner dynein arms of cilia and flagella. *Cell* 41:341–342.

*HUANG, B. 1986. *Chlamydomonas reinhardtii*: a model system for genetic analysis of flagellar structure and motility. *Int. Rev. Cytol.* 99:181–215.

KAMIYA, R. 1988. Mutations at twelve independent loci result in absence of outer dynein arms in *Chlamydomonas reinhardtii*. *J. Cell Biol.* 107:2253–2258.

*LEFEBVRE, P. A., and J. L. ROSENBAUM. 1986. Regulation of the synthesis and assembly of ciliary and flagellar proteins during regeneration. *Ann. Rev. Cell Biol.* **2**:517–546.

*LUCK, D. J. L. 1984. Genetic and biochemical dissection of the eukaryotic flagellum. *J. Cell Biol.* **98**:789–794.

SALE, W. S., U. W. GOODENOUGH, and J. E. HEUSER. 1985. The substructure of isolated and in situ outer dynein arms of sea urchin sperm flagella. *J. Cell Biol.* **101**:1400–1412.

*SATIR, P. 1988. Dynein as a microtubule translocator in ciliary motility: current studies of arm structure and activity pattern. *Cell Motility and the Cytoskeleton* **10**:263–270.

WARNER, F. D., and P. SATIR. 1974. The structural basis of ciliary bend formation. *J. Cell Biol.* **63**:35–63.

Basal Bodies and Centrioles

BORNENS, M., M. PAINTRAND, J. BERGES, M-C. MARTY, and E. KARSENTI. 1987. Structural and chemical characterization of isolated centrosomes. *Cell Motility and the Cytoskeleton* **8**:238–249.

*BRINKLEY, B. R. 1985. Microtubule organizing centers. *Ann. Rev. Cell Biol.* **1**:145–172.

HALL, J. L., Z. RAMANIS, and D. J. L. LUCK. 1989. Basal body/centriolar DNA: molecular genetic studies in *Chlamydomonas reinhartii. Cell*, **59**:121–132.

*KARSENTI, E., and B. MARO. 1986. Centrosomes and the spatial distribution of microtubules in animal cells. *Trends Biochem. Sci.* **11**:460–463.

MCINTOSH, J. R. 1983. The centrosome as an organizer of the cytoskeleton. *Mod. Cell Biol.* **2**:115–142.

MITCHISON, T. and M. W. KIRSCHNER. 1984. Microtubule assembly nucleated by isolated centrosomes. *Nature* **312**:232–237.

RAMANIS, Z., and D. J. L. LUCK. 1986. Loci affecting flagellar assembly and function map to an unusual linkage group in *Chlamydomonas reinhardtii. Proc. Nat'l Acad. Sci. USA* **83**:423–436.

WHEATLEY, D. N. 1982. *The Centriole: A Central Enigma of Cell Biology.* Elsevier/North Holland.

Structure and Function of the Mitotic Spindle

BASKIN, T. I., and W. Z. CANDE. 1988. Direct observation of mitotic spindle elongation in vitro. *Cell Motility and the Cytoskeleton* **10**:210–216.

BEGG, D. A., and G. W. ELLIS. 1979. Micromanipulation studies of chromosome movement. *J. Cell Biol.* **82**:528–541.

GORBSKY, G. J., P. J. SAMMAK, and G. G. BORISY. 1988. Microtubule dynamics and chromosome motion visualized in living anaphase cells. *J. Cell Biol.* **106**:1185–1192.

*HEATH, B. 1980. Variant mitoses in lower eukaryotes: indicators of the evolution of mitosis? *Int. Rev. Cytol.* **64**:1–80.

HIROKAWA, N. 1989. Cytoskeletal architecture of the mitotic spindle. In *Cell Movement*, vol. 2, *Kinesin, Dynein, and Microtubule Dynamics.* Alan R. Liss, pp. 383–401.

HIUTOREL, P., and M. W. KIRSCHNER. 1988. The polarity and stability of microtubule capture by the kinetochore. *J. Cell Biol.* **106**:151–159.

*KIRSCHNER, M., and T. MITCHISON. 1986. Beyond self-assembly: from microtubules to morphogenesis. *Cell* **45**:329–342.

KOSHLAND, D. E., T. J. MITCHISON, and M. W. KIRSCHNER. 1988. Polewards chromosome movement driven by microtubule depolymerization in vitro. *Nature* **331**:499–504.

MASUDA, H., K. L. MCDONALD, and W. Z. CANDE. 1988. The mechanism of anaphase spindle elongation: uncoupling of tubulin incorporation and microtubule sliding during in vitro spindle reactivation. *J. Cell Biol.* **107**:623–633.

MCNEILL, P. A., and M. W. BERNS. 1981. Chromosome behavior after laser microirradiation of a single kinetochore in mitotic PtK2 cells. *J. Cell Biol.* **88**:543–553.

*MITCHISON, T. J. 1988. Microtubule dynamics and kinetochore function in mitosis. *Ann. Rev. Cell Biol.* **4**:527–549.

MITCHISON, T. J. 1989. Polewards microtubule flux in the mitotic spindle: evidence from photoactivation of fluorescence. *J. Cell Biol.* **109**:637–652.

*MURRAY, A. W., and J. W. SZOSTAK. 1985. Chromosome segregation in mitosis and meiosis. *Ann. Rev. Cell Biol.* **1**:289–315.

NICKLAS, R. B. 1988. The forces that move chromosomes in mitosis. *Ann. Rev. Biophys. Biophys. Chem.* **17**:431–450.

NICKLAS, R. B. 1989. The motor for poleward chromosome movement in anaphase is in or near the kinetochore. *J. Cell Biol.* **109**:2245–2255.

*PICKETT-HEAPS, J. D. 1986. Mitotic mechanisms: an alternative view. *Trends Biochem. Sci.* **11**:504–507.

RIEDER, C. L., E. A. DAVISON, C. W. JENSEN, L. CASSIMERIS, and E. D. SALMON. 1986. Oscillatory movements of mono-oriented chromosomes and their position relative to the spindle pole result from the ejection properties of the aster and half-spindle. *J. Cell Biol.* **103**:581–591.

ROOS, U.-P. 1976. Light and electron microscopy of rat kangaroo cells in mitosis. III. Patterns of chromosome behavior during prometaphase. *Chromosoma* **54**:363–385.

*SCHLEGEL, R. A., M. S. HALLECK, and P. N. RAO (eds.). 1987. *Molecular Regulation of Nuclear Events in Mitosis and Meiosis.* Academic Press.

TELZER, B. L., and L. T. HAIMO. 1983. Decoration of spindle microtubules with dynein: evidence for uniform polarity. *J. Cell Biol.* **89**:373–378.

22

Actin, Myosin, and Intermediate Filaments: Cell Movements and Cell Shape

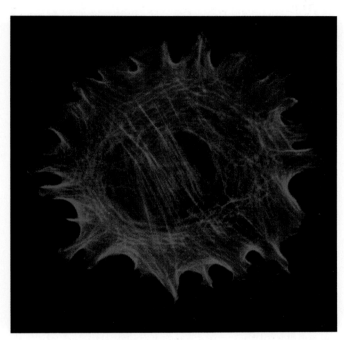

Actin cytoskeleton in a fibroblast cell attached to a substratum

Microtubules and their associated proteins represent only one type of filamentous structure involved in cellular and subcellular movements and in the determination of cell shape. *This chapter focuses on the 10-nm-diameter intermediate filaments and the 7-nm-diameter actin microfilaments, their many associated proteins, and their many and varied roles in cellular structure and motion.*

Microfilaments and intermediate filaments are both polymers. The globular protein actin *is the major subunit of microfilaments; all eukaryotic cells contain actin, and in most eukaryotic cells it is the single most abundant cytoplasmic protein (Table 22-1). The filamentous protein* myosin *is intimately associated with actin in all muscle cells and in many nonmuscle cells as well. Microfilaments contain a diverse array of other actin-binding proteins, which enable the filaments to form unique structures or to carry out specific cellular functions. Intermediate filaments are unique to multicellular organisms, and different types of differentiated cells usually contain specific types of intermediate filaments; these are similar in structure and function to one another but are composed of different types of subunit proteins.*

In most nonmuscle cells, microfilaments and intermediate filaments are organized in a seemingly random array. However, filament organization is precisely dictated by filament-associated proteins; the organization can differ in different parts of a cell and can change rapidly. Together with the microtubules, these filaments organize the cytoplasm. They provide a structure to which other proteins or organelles can bind, so that different cytoplasmic proteins are localized to different regions of the cell. Often these filaments associate with the plasma membrane, thus playing key roles in determination of cell shape and motility.

In this chapter, we first describe the properties of actin, myosin, and the filaments they form. We then discuss the structure and function of the highly specialized and ordered actin-myosin system in muscle. Next we examine the role of actin, myosin, and various associated proteins in several nonmuscle structures; of these, the brush border of epithelial cells is the best understood. Finally, we discuss intermediate filaments and their function in desmosomes, which link cells together and to the substratum. ▲

Table 22-1 Contractile protein content

Contractile protein	Percentage of total cellular protein	Concentratin (μmol/kg)	Actin:myosin ratio
ACTIN			
Rabbit muscle	19	900	6
Human platelet	10	240	110
Acanthamoeba	14	150	70
MYOSIN			
Rabbit muscle	35	144	
Human platelet	1	2.2	
*Acanthamoeba**			
I	0.3	1.3	
II	1.2	2.3	

*Two types of myosin have been purified from *Acanthamoeba* and other microorganisms. Type I is an unusual, small (MW 110,000) single-headed myosin that has no fibrous tail—it is often called a "minimyosin"; type II is the conventional two-headed myosin with a filamentous tail.

SOURCE: T. Pollard, 1981, *J. Cell Biol.* **91**:156S.

Actin and Myosin Filaments

Actin Monomers Polymerize into Long Helical Filaments

The actin in muscle and that in nonmuscle cells are the products of different genes and therefore differ slightly in their properties. However, all actins that have been studied—from sources as diverse as slime molds, fruit flies, mammalian platelets, vertebrate muscle, and plants—are similar in size, have very similar amino acid sequences, and share many other properties, suggesting that they evolved from a single ancestral gene. The apparently minor differences in amino acid sequence among the various muscle and nonmuscle actins may allow them to bind different proteins and may well be responsible for their significant differences in function.

Microfilaments, sometimes called *F actin*, are polymers of a globular protein subunit (MW 42,000) called globular actin, or *G actin* (Figure 22-1a). Whether isolated from muscle or nonmuscle sources, actin filaments are constructed of an identical string of monomers. The precise molecular structure of the monomer is not yet known, but overall it is dumbbell-shaped, with dimensions of $6.7 \times 4 \times 4$ nm. Each actin subunit has a defined polarity, and the subunits polymerize head to tail. As a consequence, actin filaments also have a defined polarity, and all subunits "point" in the same direction.

Our present conception of the molecular structure of the actin filament is based on models obtained using the technique of image reconstruction. Each monomer in a filament is identical to every other, but when one views actin filaments in the electron microscope, one is seeing monomers from a variety of angles; thus it is difficult to deduce much about structure merely by looking at micrographs (Figure 22-1b). However, a number of hypothetical models of the actin fiber can be constructed, and calculations in a computer can generate an image of how each such structure would appear if viewed in the electron microscope. This "computer image" is compared to the actual one, and then the model is changed slightly to make the agreement between the two better and better. According to the current "best guess" model, shown in Figure 22-1c, a single chain of monomers forms a helical filament. Each monomer is oriented nearly perpendicular to the helix axis; it has major interactions with its two closest neighbors (e.g., monomer 3 with monomers 2 and 4) and weaker interactions with monomers two away along the helix axis (e.g., monomer 3 with 1 and 5).

Polymerization of globular actin monomers into filaments is induced by Mg^{2+} and by K^+ or Na^+ at concentrations similar to those found in the cell cytosol. Polymerization of actin is accompanied by a large increase in the viscosity of the solution. As is the case with tubulin, actin can be obtained in pure form from cells by repeated cycles of depolymerization and polymerization. In vitro, actin polymerizes into filaments that are identical in structure to those isolated directly from cells. Actins from organisms are diverse as mammals and slime molds can copolymerize, indicating that many properties of actin have been highly conserved during evolution.

Like microtubules, actin filaments grow by addition of subunits to both ends. Actin monomers contain a tightly bound ATP or ADP (similar to the GTP or GDP in tubulin subunits). Actin monomers with bound ATP add to the end of a microfilament much faster than do monomers with bound ADP. Hydrolysis of the bound ATP to ADP occurs shortly after actin polymerization but is not essential for polymerization to occur. The rate of growth of purified actin filaments is 5–10 times faster at the (+) ends than at the (−) ends. This can be shown by using as primers actin filaments "decorated" with myosin subfragments (Figure 22-2a). When pure actin monomers with bound ATP are reacted in vitro with decorated primers and the products viewed in the electron microscope, the new (undecorated) actins are five times more prevalent at the (+) ends than at the (−) ends (Figure 22-2b).

Actin Filaments Are Intrinsically More Stable Than Microtubules

Under steady-state conditions actin filaments, like microtubules, are in equilibrium with monomers. However, there are two important differences between the assembly of microtubules and actin filaments. First, under the ion conditions normally found inside a cell (about 0.15 M KCl), the rate of dissociation of monomers from actin filaments is slow; that is, actin filaments are intrinsically much less likely to depolymerize than are microtubules. Second, the critical concentration for assembly of actin monomers into filaments is very low: 1 μM for addition at the (+) end of the filament and 8 μM for addition at the (−) end. These are the concentrations of monomer, at steady state, that are in equilibrium with the two

(a)

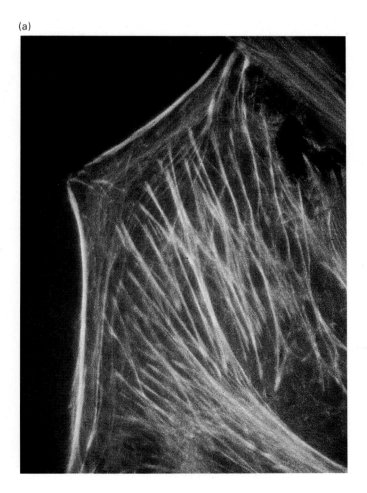

◀ **Figure 22-1** (a) Actin-containing microfilaments in a cultured fibroblast, visualized with fluorescent phalloidin. Phalloidin, a toxic cyclic peptide, binds to actin in microfilaments and prevents their depolymerization. (b) Electron micrograph of negatively stained actin microfilaments. (c) Molecular model of a helical actin filament. The globular actin monomer is composed of two domains. Each dumbbell-shaped monomer has strong interactions with each of its two adjacent neighbors and weaker interactions with monomers two away in the helix. *Part (a) courtesy of Dr. W. Webb and W. Carley; part (b) from U. Aebi et al., 1987, Ann. NY Acad. Sci.* **483**:*100; part (c) courtesy of Dr. C. Cohen.*

(b)

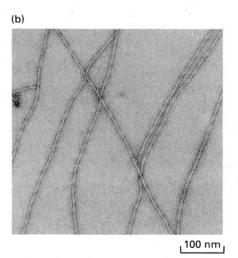

100 nm

(c)

One actin monomer

ends of the polymer (see Figure 21-8 for a similar discussion on microtubule assembly). Since the concentration of actin in cells varies between 150 and 900 μM, under steady-state conditions the vast bulk of actin monomers should be polymerized in microfilaments within cells.

Thus microfilaments inside cells are expected to be—and indeed are—much more stable than microtubules. At any one time most cellular actin is in microfilaments; although such filaments can grow and shrink, they do so slowly. In fact, actin filaments in muscle are stable for days, probably stabilized by actin-binding proteins. In contrast, in the elongating ends of motile animal cells, actin filaments form and depolymerize within minutes, and these controlled changes are essential for cell movement.

(a)

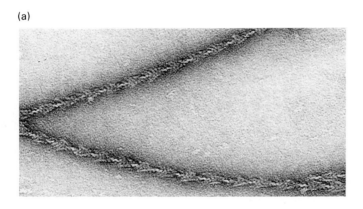

(b)

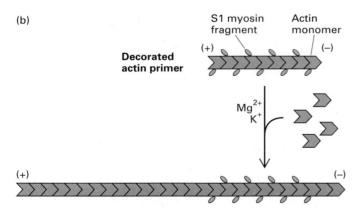

▲ **Figure 22-2** (a) Electron micrograph of a rabbit muscle actin filament "decorated" with bound S1, or head, fragments of myosin, which contain the actin-binding sites. After reaction with S1 fragments, the filaments were frozen in aqueous solution and examined unstained at low temperature to maintain them in the native state. Note the arrowhead appearance of the decorated fiber, which indicates that all the S1 fragments are oriented in the same direction, toward the (−) end of the filament. (b) In vitro polymerization of actin monomers with S1-decorated primers demonstrates that addition of monomers occurs five times faster at the (+) end than at the (−) end. *Part (a) courtesy of Dr. James Spudich.*

Myosin Is a Bipolar, Fibrous Molecule That Binds Actin

Myosin is present in muscle cells and nonmuscle cells of animals, in higher plants, and in eukaryotic microorganisms. Type II myosin, found in muscle and in nonmuscle cells, is by far the most extensively studied; it contains a globular region that binds actin and a fibrous segment that allows it to aggregate into filaments. Type I myosin contains the globular actin-binding region but lacks the fibrous "tail"; it is present only in nonmuscle cells and participates in several types of cellular movements. The present discussion focuses on type II myosins from smooth muscle and from striated muscle, which have slightly different molecular properties that account for their different contractile regulatory mechanisms.

When isolated by salt extraction from muscle and viewed in the electron microscope, myosin appears to have a long rodlike "tail" and two globular "heads" at one end (Figure 22-3a). A myosin molecule always contains two identical heavy chains, each with a molecular weight of 230,000, and two pairs of light chains of two different types, each chain having a molecular weight of about 20,000. Each heavy chain consists of an N-terminal globular head and a C-terminal rodlike α-helical tail region. Bound to each myosin head are the two different light-chain pairs (Figure 22-3b). In native myosin, the α-helical rod segments of the two heavy chains coil around each other to form a long (130-nm), rigid coiled-coil tail. There are two flexible joints, termed "hinges," in the myosin molecule—one between the head and tail, and the other part way down the tail.

Protein α helices contain about 3.6 amino acid residues per turn. To form a *coiled coil*, the polypeptide chains must have seven-residue "heptad repeats" (e.g., abcdefgabcdefgabcdefg . . .) in which residues *a* and *d* have hydrophobic side chains. The hydrophobic side chains of the *a* and *d* residues interact and pack like "knobs into holes" to stabilize the two-helix coil; a "stripe" of hydrophobic residues winds around the coiled coil at the points where the two helices interact (see Figure 2-7b). Many fibrous proteins besides myosin contain such coiled coils of α helices, and we shall encounter this structural motif again in the actin-binding protein tropomyosin and also in intermediate filaments.

In muscle, the fibrous tails of 300–400 myosin dimers pack together laterally to form a specific bipolar aggregate termed the *thick filament* (Figure 22-3c). The central zone of the thick filament is devoid of heads and is composed of an antiparallel overlapping array of myosin tails that is slightly longer than a single myosin tail. The terminal regions are of variable length, and myosin heads protrude from the surface in a helical array at 14-nm intervals. The thick filaments are symmetric about the bare central zone, and the polarity of the myosin filaments is reversed on either side of the midline.

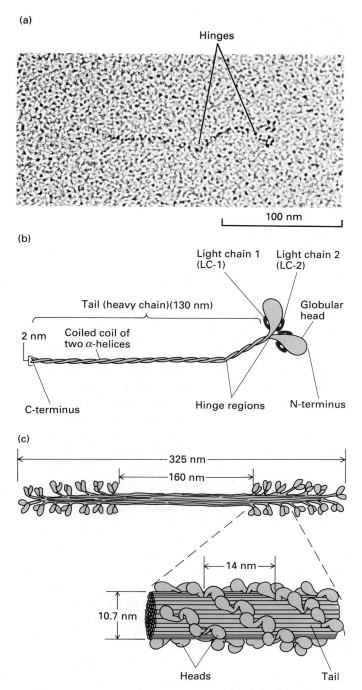

(a)

Hinges

100 nm

(b)

Light chain 1 (LC-1) Light chain 2 (LC-2)

Tail (heavy chain)(130 nm)

Globular head

2 nm

Coiled coil of two α-helices

C-terminus

Hinge regions   N-terminus

(c)

325 nm

160 nm

14 nm

10.7 nm

Heads   Tail

▲ **Figure 22-3** (a) Electron micrograph of shadowed rabbit muscle myosin. (b) Diagram of a myosin molecule, which consists of two pairs of two different light chains and two identical heavy chains. The two N-terminal 95-kDa segments of the heavy chains form two globular heads, and the two C-terminal 125-kDa segments form a 130-nm α-helical coiled-coil tail. Note the two flexible hinge regions. (c) Model of a bipolar myosin thick filament. The central zone, 10.7 nm in diameter, is composed only of packed tails and is bereft of heads. *Part (a) courtesy of P. F. Flicker and J. Spudich; part (b) after H. M. Warrick and J. Spudich, 1987, Ann. Rev. Cell Biol. 3:379; part (c) after T. Pollard, 1981, J. Cell Biol. 91:156.*

The structural domains of myosin can be dissected by proteases into separate functional domains (Figure 22-4). At low concentrations, the protease chymotrypsin cuts the myosin heavy chain at one of the hinges, forming *heavy meromyosin* (HMM), which contains both heads and part of the tail, and rod-shaped *light meromyosin* (LMM). Further treatment of HMM with proteases such as papain destroys the tail and releases two *S1 fragments*, each containing a globular head and light-chain pair.

Driven by ATP Hydrolysis, Myosin Heads Move along Actin Filaments

Biochemical studies established that myosin has an ATPase activity. In the absence of actin, this activity is almost undetectable, but when pure actin filaments are added, the rate of ATP hydrolysis increases 200-fold, so

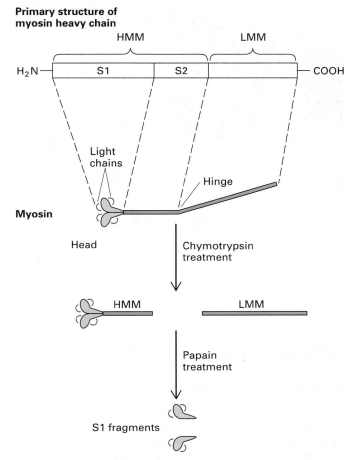

Primary structure of myosin heavy chain

HMM   LMM

H_2N S1   S2   COOH

Light chains

Hinge

Myosin

Head

Chymotrypsin treatment

HMM   LMM

Papain treatment

S1 fragments

▲ **Figure 22-4** Digestion of myosin with proteases. The single-headed S1 fragment contains the light chains and all of the ATPase, actin-binding, and movement activities of the intact myosin molecule. HMM = heavy meromyosin; LMM = light meromyosin. *Adapted from Y. Y. Toyoshima et al., 1987, Nature 328:536.*

that each myosin molecule hydrolyzes 5–10 ATP molecules per second—a rate similar to that in contracting muscle.

The S1 fragment retains all of the actin-stimulated ATPase activity of intact myosin. Binding of myosin S1 fragments to actin fibers reflects the underlying directionality and helicity of the actin filaments. S1 fragments bound along the entire length of an actin filament display a characteristic "arrowhead" pattern in the electron microscope (see Figure 22-2a). All the arrowheads point in the same direction, toward the (−) end of the actin filament; the "barbs" of the arrowheads point toward the (+), or preferential assembly, end. The binding of muscle S1 fragments to 7-nm-diameter filaments in detergent-extracted fibroblast cells was one of the first pieces of evidence that nonmuscle cells contain actin filaments.

The overall shape of the myosin head is known (Figure 22-5), and the regions that bind actin and ATP have been identified. Figure 22-6 shows one experimental system that has been used to demonstrate the ATP-driven move-

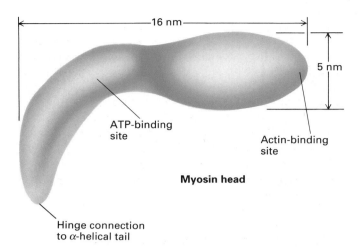

▲ **Figure 22-5** Overall shape of the myosin head, as determined by x-ray diffraction studies and electron microscopy of crystals of the S1 fragment. *Adapted from D. A. Winkelman, H. Merkel, and I. Rayment, 1985,* J. Mol. Biol. *181:487.*

(a)

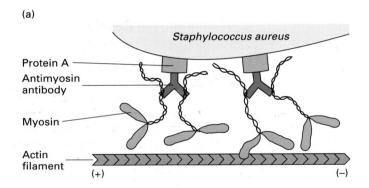

(b)

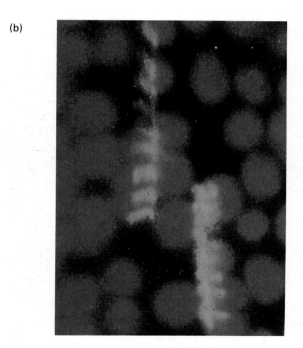

(c)

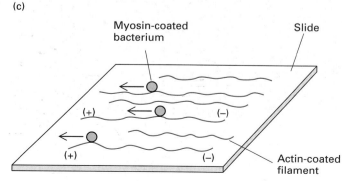

▲ **Figure 22-6** Schematic diagram of the *Nitella* actin motility assay system. (a) Intact myosin molecules, HMM, or S1 fragments are bound to killed *S. aureus* bacterial cells by an antimyosin antibody that also binds to protein A on the bacterial surface. Arrays of polarized actin filaments are obtained from the alga *Nitella* by stripping off pieces of the plasma membrane and exposing the submembranous actin network (see Figure 22-34). (b) Movement of yellow-fluorescent myosin-coated beads on parallel actin filaments in an opened *Nitella* cell. The actin lies above the red-fluorescing chloroplasts. A series of exposures was taken at 1-s intervals; the velocity was ~3 μm/s. (c) When *Nitella* actin (red) is immobilized on a slide, myosin-coated bacteria (blue) migrate along these fibers at a rate of about 1 μm/s, always toward the (+) ends of the filaments. ATP hydrolysis is essential for movement. *Part (b) courtesy of J. Spudich and M. Sheetz.*

ment of myosin along actin filaments, in this case polarized actin filaments from the giant alga *Nitella* (see Figure 22-34 and later discussion of cytoplasmic streaming). In this system, myosin-coated bacterial cells (or plastic beads) always move toward the (+) ends of the immobilized actin filaments. Furthermore, just the S1 fragments (the heads) of myosin are sufficient for movement; the α-helical tails are not required. Beads coated with type I myosins containing only a single head and no fibrous tail also move along *Nitella* actin filaments. These so-called *minimyosins* have been purified from certain microorganisms (see Table 22-1).

Muscle Structure and Function

Vertebrates and many invertebrates have two classes of muscle—smooth and striated; cardiac muscle in vertebrates forms a third class. Smooth muscles, which lack the striations described below, are typically under invol-

untary (unconscious) control of the central nervous system. They surround internal organs such as the large and small intestines, the gallbladder, and large blood vessels (Figure 22-7a). Contraction and relaxation of smooth muscles control the diameter of blood vessels and also propel food along the gastrointestinal tract. Smooth muscle cells contract and relax slowly, and they can create and maintain tension for long periods of time.

Muscles under voluntary control have a striated appearance in the light microscope (Figure 22-7b). Striated muscles, which connect the bones in the arms, legs, and spine, are used in complex coordinated activities, such as walking or positioning of the head, and generate rapid movements by sudden bursts of contraction. Cardiac (heart) muscle resembles striated muscle in many respects, but it is specialized for the continuous, involuntary contractions needed in pumping of the blood. Study of striated muscle cells, with their very regular organization of actin and myosin contractile filaments, has provided important evidence about the mechanism of contraction in all three types of muscle.

Striated Muscle Consists of a Regular Array of Actin and Myosin Filaments

Even with the low-power magnification of the light microscope, the regular structure of striated muscle is evident: dark bands, called *A bands,* alternate with light bands, called *I bands;* a narrow line, the *Z disk,* bisects each I band (Figure 22-7b). The light and dark bands are perpendicular to the long axis of the muscle cell along which the muscle contracts. The segment from one Z disk to the next is termed a *sarcomere.*

A typical striated muscle cell is cylindrical and is very large, measuring 1–40 mm in length and 10–50 μm in width. Each cell, called a *myofiber,* contains up to a 100 nuclei and many bundles of filaments termed *myofibrils* (Figure 22-8). Each myofibril is constructed of a repeating array of sarcomeres, each about 2 μm long in resting

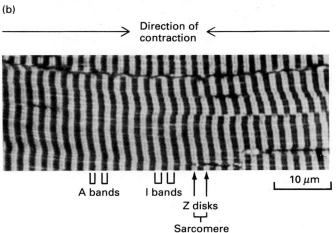

(a)

(b)

Direction of ← contraction

A bands   I bands   Z disks   Sarcomere   10 μm

◄ **Figure 22-7** (a) Electron micrograph of a typical smooth muscle cell surrounding a human blood vessel. The fibers are not organized or parallel to each other; numerous thin actin and thick myosin filaments are seen in cross section. As evident at higher magnification *(inset)*, the ratio of actin to myosin filaments is about 15:1; unlike striated muscle, there is no regular packing of actin and myosin. (b) A low-magnification light micrograph of a stained longitudinal section of striated muscle from a rattlesnake. Most myofibrils are in exact register, producing the characteristic striations. A sarcomere is the segment between two adjacent Z disks. *Part (a) courtesy of A. Somlyo, from E. Betz, ed., 1972, Vascular Smooth Muscle, Springer-Verlag; part (b) courtesy of D. Schulz and A. W. Clark.*

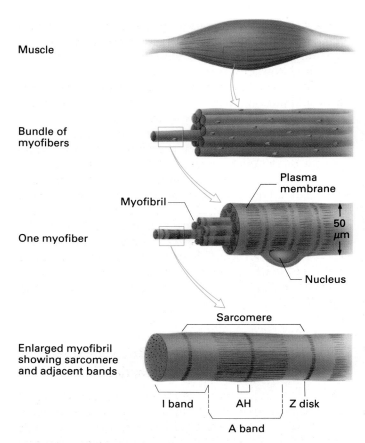

Muscle

Bundle of
myofibers

Myofibril

One myofiber

Plasma
membrane

↕ 50
μm

Nucleus

Sarcomere

Enlarged myofibril
showing sarcomere
and adjacent bands

I band   AH   Z disk

A band

▲ **Figure 22-8**   The levels of organization in striated muscle. Each muscle cell, or myofiber, is multinucleate and contains many myofibrils. The sarcomere is the functional unit
of contraction; it is about 2 μm long in resting muscle.

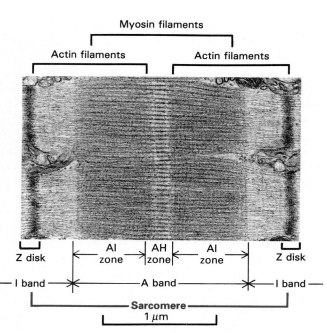

Myosin filaments

Actin filaments   Actin filaments

Z disk   AI   AH   AI   Z disk
          zone   zone   zone
— I band —   — A band —   — I band —

── Sarcomere ──
1 μm

▲ **Figure 22-9**   Electron micrograph of mouse striated muscle in longitudinal section, showing one sarcomere. On either
side of the Z disks are the light-stained I bands, composed
entirely of actin filaments. These thin filaments extend from
both sides of the Z disk to interdigitate with the dark-stained
myosin thick filaments that make up the A band. The region
containing both thick and thin filaments (the AI zone) is
darker than the area containing only myosin thick filaments
(the AH zone). *Courtesy of S. P. Dadoune.*

muscle. Electron microscopy has shown that each sarcomere contains two types of filaments: thick filaments,
now known to be myosin, and thin microfilaments containing actin (Figure 22-9). The (+) ends of the thin filaments attach to the rigid Z disk; the (−) ends of these
filaments extend part way toward the center of the sarcomere.

The myosin thick filaments, 300–500 nm in length, are
the major constituents of the A band and have the molecular structure depicted in Figure 22-3c. The A band is so
labeled because it is anisotropic, or birefringent, when
viewed in polarized light (see Figure 21-36). The property
of birefringence is caused by the organized bundles of
thick filaments. The I band, by contrast, is composed of
thin filaments and is less anisotropic than the A band, but
not purely isotropic as erroneously suggested by the label
"I." The *AH zone* in the center of the A band consists
entirely of a side-by-side aggregate of the rod-shaped tails
of the myosin heavy chains. On both sides of this central
region, the heads of the myosin molecules protrude from
the thick filaments forming cross-bridges with adjacent
actin filaments (Figure 22-10). In the region in which

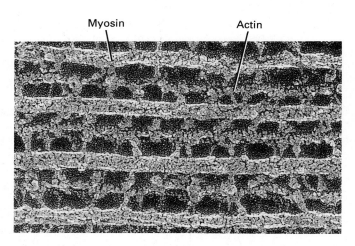

Myosin   Actin

▲ **Figure 22-10**   Micrograph showing actin-myosin cross-
bridges in a striated insect flight muscle. This image,
obtained by the quick-freeze deep-etch technique, shows a
nearly crystalline array of thick myosin and thin actin
filaments. The muscle was in rigor at preparation. Note the
myosin heads connecting with the actin filaments at regular
intervals. *Courtesy of J. Heuser.*

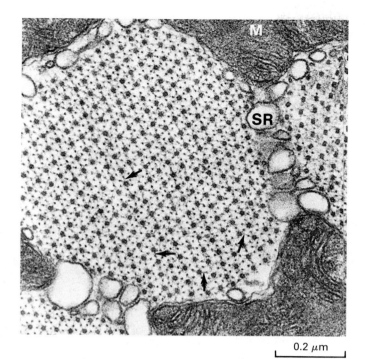

▲ **Figure 22-11** Electron micrograph of cross section of a sarcomere at the level of the AI zone. Each myosin thick filament is surrounded by six actin microfilaments; each microfilament is midway between two myosin filaments. Extremely thin filaments (*arrows*) in some areas appear to interconnect the actin and myosin filaments; these represent the globular heads of the myosin molecules. SR = sarcoplasmic reticulum; M = mitochondria. *From R. G. Kessel and R. H. Kardon, 1979,* Tissues and Organs: A Text-Atlas of Scanning Electron Microscopy, *W. H. Freeman and Company.*

thick and thin filaments overlap, called the *AI zone,* each thick myosin filament is surrounded by six thin actin filaments, and each actin filament is surrounded by two myosin filaments (Figure 22-11).

Thick filaments can be selectively removed from detergent-permeabilized cells by treatment with certain salt solutions that leave other structures intact. Subsequent biochemical analysis shows that myosin is the principal protein extracted and thus is the primary structural component of the thick filaments. When added to such extracted cells, myosin S1 fragments bind to the remaining thin filaments in the characteristic arrowhead pattern, thereby establishing that the thin filaments are indeed built of actin. Examination of thin filaments bound to S1 shows that the arrowheads always point away from the Z disk. Thus all of the thin filaments attached to one side of a Z disk have the same polarity, and filaments on opposite sides of a Z disk point in the opposite direction (Figure 22-12).

Thick and Thin Filaments Move Relative to Each Other during Contraction

Several key experiments have established that sliding of thin actin and thick myosin filaments past each other within sarcomeres causes muscle contraction. As a muscle contracts or is stretched passively to different sarcomere lengths, the width of the A band remains constant. Electron microscopy has confirmed that the lengths of individual myosin filaments and actin filaments do not change as a muscle contracts or relaxes. What does

(a) Relaxed state

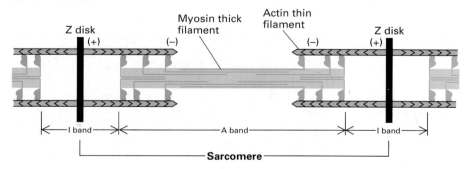

(b) Contracted state

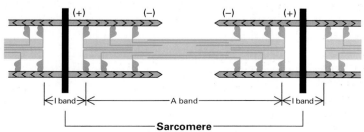

◄ **Figure 22-12** Diagram showing arrangement of thick myosin and thin actin filaments in striated muscle in (a) the relaxed and (b) the contracted state. Note that the (+) ends of actin filaments are anchored at the Z disks; pivoting of the myosin heads (blue) pushes the actin filaments toward the center of the sarcomere, thus reducing sarcomere length. The I band, but not the A band, shortens during contraction.

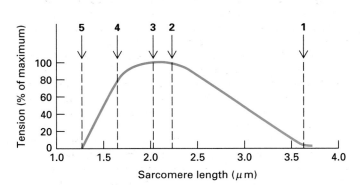

▲ **Figure 22-13** Relationship of sarcomere length and tension generated during isometric contraction of striated muscle. The numbers 1–5 on the graph indicate the relative positions of the thick (green) and thin filaments (red) in the accompanying diagrams. Maximum tension is generated at sarcomere lengths that allow maximum interaction of myosin heads and actin filaments (positions 2 and 3). If the sarcomere length is too short (positions 4 and 5), actin filaments overlap one another and prevent optimum interaction with myosin heads. [See A. M. Gordon, 1966, *J. Physiol.* **184**:170.]

change is the length of the I band—the part of the actin microfilaments not covered by myosin. Presumably, then, the only way movement can occur is by the sliding of actin and myosin filaments past each other (Figure 22-12).

Since the heads protruding from the myosin filaments have actin-stimulated ATPase activity and are known to bind to and move along actin filaments, it was suggested and subsequently confirmed that the force-generating elements are the individual myosin-actin bridges. Thus the force of contraction should be proportional to the overlap between the actin filaments and the head-bearing regions of the thick filaments. When a muscle is stretched, the number of myosin heads capable of interacting with actin is reduced. According to the hypothesis, the tension generated when a stretched muscle held at constant length is stimulated to contract should be inversely proportional to the length. Experimental observations of the tension generated during isometric contractions have borne out this predicted relationship (Figure 22-13).

ATP Hydrolysis Powers the Contraction of Muscle

Each myofibril is surrounded by mitochondria and granules of glycogen, which is used for generation of ATP. Muscle contraction is fueled by the myosin-catalyzed hydrolysis of ATP to ADP and P_i—much as myosin-coated beads move along actin filaments (see Figure 22-6). Individual myosin heads can power movement along actin, but in muscle the heads are connected to the rigid helical core of the thick filament by hinges, which limit the movement of the individual heads along a filament.

If a muscle is depleted of its store of compounds with high-energy phosphate bonds (e.g., ATP or creatine phosphate), it becomes stiff and can no longer be extended passively with small forces; this condition is known as *rigor*. (As is well known, muscles go into a state of rigor a short time after death.) In rigor, the level of cytosolic Ca^{2+} is high enough that the actin and myosin filaments become tightly cross-linked together; as a result, the thin and thick filaments cannot easily slide past each other. In such rigor complexes the myosin heads have no bound nucleotide; ultrastructural evidence indicates that they are bound at a 45° angle to actin. (Recall that in the absence of ATP myosin heads bind to actin filaments in a characteristic 45° arrowhead pattern.)

Binding of ATP to myosin weakens the binding of myosin heads to actin and can be considered step 1 in a cyclic process that results in muscle contraction (Figure 22-14). That ATP weakens binding of myosin to actin, and thus relaxes muscle fibers, has been demonstrated directly (Figure 22-15). As shown in Figure 22-14, ATP hydrolysis is essential to the cyclic formation and dissociation of actin-myosin bridges during muscle contraction. Unlike the strongly exergonic hydrolysis of ATP to ADP and P_i in solution ($\Delta G^{\circ\prime} = -7$ kcal/mol), hydrolysis of myosin-bound ATP occurs with little change in free energy; that is, the reaction ATP + H_2O ⇔ ADP + P_i is readily reversible, and the ADP and P_i hydrolysis products remain bound to myosin. The release of bound ADP and P_i from myosin are the strongly exergonic steps, and the free energy released is used to power the pivoting movement of the myosin head. With each cycle, which requires the hydrolysis of one ATP molecule per myosin head, the actin filament is moved a distance of about 7 nm.

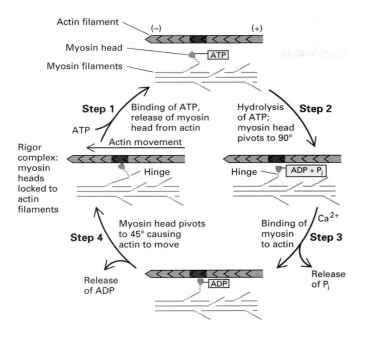

◀ **Figure 22-14** Diagram of the myosin-ATPase cycle during contraction of striated muscle. Step 1: A myosin molecule binds ATP, causing weakening of the actin-myosin bonds and release of the myosin head from actin. Step 2: ATP is hydrolyzed to ADP + P_i, but the hydrolysis products remain bound to the myosin. This generates an "energized" myosin head, which has rotated so it is perpendicular to the actin filament; this is facilitated by flexible regions, or hinges, on the myosin molecule. Step 3: The myosin head binds to an adjacent actin filament with release of the P_i; this is the step that is dependent on the presence of Ca^{2+} ions. Step 4: The myosin head pivots on its hinge, moving the actin filament relative to the fixed myosin; this results in contraction. During this step the angle of attachment of the myosin head to the filament backbone changes from 90° to about 45°, and ADP is released. The product of this step is the so-called rigor complex, in which the actin-myosin linkage is inflexible, and the thin and thick filaments cannot move past each other. Subsequent binding of ATP to the myosin head (step 1) releases the myosin head from the actin, relaxing the muscle. Detailed kinetic studies indicate that each of the steps depicted here can be divided into two or more substeps. [See E. Eisenberg and T. Hill, 1985, *Science* **227**:999; and M. Irving, 1985, *Nature* **316**:292.]

(a)

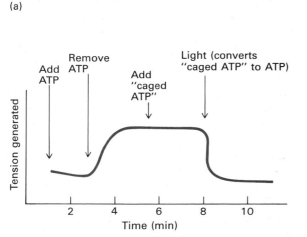

(b)

▲ **Figure 22-15** (a) Demonstration that ATP relaxes muscle fibers. Graph shows muscle tension at intervals corresponding to indicated experimental manipulations. A small striated muscle fiber is first extracted with glycerol to render it permeable to ATP and small molecules. When ATP but no Ca^{2+} is added, the tension generated is low; the myosin heads contain bound ADP and P_i but do not bind to actin in the absence of Ca^{2+}. When ATP is removed by dilution, the muscles generate tension and go into rigor; myosin heads, with no bound nucleotide, become tightly bound to actin filaments even at low Ca^{2+} concentrations (rigor complex, Figure 22-14). This causes the increase in tension. If ATP is added to the preparation, it will take several minutes to diffuse throughout the fiber, during which time much of the

ATP will be hydrolyzed; however, ATP can be generated instantly and uniformly in the preparation by adding the compound called "caged ATP," which does not serve as a substrate for the myosin ATPase or bind to the myosin heads. Addition of caged ATP produces no effect on the tension, but when caged ATP is converted to ATP by a short pulse of intense laser light, immediate relaxation of the tension occurs as ATP binds to the myosin heads and dissociates the myosin from actin. Since there is no Ca^{2+} in the solution, myosin–ADP + P_i complexes accumulate, unattached to thin filaments. (b) Structural formula of caged ATP and the reaction that converts it to ATP. [See Y. E. Goldman et al., 1982, *Nature* **300**:701.]

Release of Calcium from the Sarcoplasmic Reticulum Triggers Contraction

A rise in internal Ca^{2+} concentration triggers the binding to actin of the myosin ADP complex and thus stimulates muscle contraction (Figure 22-14, steps 3 and 4). So long as the Ca^{2+} concentration is sufficiently high and ATP is present, the myosin-actin bridges will cycle continuously, and the muscle will contract. The concerted contraction of a muscle depends on the simultaneous contraction of all its constituent myofibers and myofibrils. In long muscles, however, the Ca^{2+} ions triggering contraction would have to diffuse great distances (up to 100 μm). Thus some mechanism other than diffusion must exist for rapidly transmitting the Ca^{2+} trigger signal.

As observed in the electron microscope, each myofibril is surrounded by a network of smooth membranes, collectively termed the *sarcoplasmic reticulum* (SR) (Figures 22-11 and 22-16). The SR forms an extensive lacelike network of membrane vesicles and cisternae that surrounds the outer regions of the A band of each myofibril; this is the location of the myosin heads. The SR also forms a continuous membrane-limited vesicle channel, called the *terminal cisterna*, surrounding the Z disk of each myofibril (Figure 22-17a). In this region the SR terminal cisternae are separated by only 16 nm from the membranes of the *transverse (T) tubules*, which are delicate invaginations of the plasma membrane (Figure 22-17b).

The SR serves as a reservoir of Ca^{2+} ions sequestered from the cell cytosol and myofibrils. The membrane of the SR has a potent Ca^{2+} ATPase activity, which pumps Ca^{2+} ions from the cytosol into the lumen of the SR; the lumen in turn contains two Ca^{2+}-binding proteins, which store large amounts of Ca^{2+}. Depolarization of the plasma membrane of a muscle cell results in opening of a Ca^{2+} channel protein in the SR membrane and in release of Ca^{2+} ions from the SR into the cytosol, thus triggering contraction of the muscle (Figure 22-17b). The depolarization is propagated along the membrane of the T tubules in much the same way that an action potential is propagated along a nerve membrane.

How depolarization of the T-tubule membrane activates the Ca^{2+} channel in the SR is controversial. According to one view, depolarization triggers generation of inositol 1,4,5-trisphosphate, which binds to the SR Ca^{2+} channel protein causing release of Ca^{2+} ions. Inositol trisphosphates are in fact generated by depolarization in the vicinity of transverse tubules, and addition of inositol 1,4,5-trisphosphate to permeabilized muscle fibers activates the Ca^{2+} channel proteins, causing Ca^{2+} release from the SR and thus contraction. This mechanism would be analogous to that in nonmuscle cells in which inositol 1,4,5-trisphosphate catalyzes release of Ca^{2+} ions from the endoplasmic reticulum (see Figure 19-30). However, diffusion of inositol 1,4,5-trisphosphate from the T-tubule membrane to the SR membrane is far too slow to account for Ca^{2+} release in striated muscle, which is activated and contracts within milliseconds. Inositol phosphates are important in triggering Ca^{2+} release in smooth muscle, in which contractions are generated over periods of seconds to minutes.

Most workers now favor the view that the SR Ca^{2+} channel proteins contact the transverse tubule and that depolarization induces opening of these channels directly. Indeed, the purified SR Ca^{2+} channel protein contains four identical polypeptides, each with a molecular weight of 560,000. One segment of the channel protein is embedded in the SR membrane and forms the Ca^{2+} channel; four regions, or "feet," of the channel protein protrude into the cytosol, where they touch voltage-sensitive Ca^{2+} channel proteins in the T-tubule membrane. When the membrane electric potential is reduced (depolarized), these voltage sensors undergo a conformational change that allows Ca^{2+} ions to enter the cytosol from the extracellular medium. Since striated muscle contracts in the absence of Ca^{2+} in the medium, Ca^{2+} influx through these voltage-sensing proteins probably is not important in activation of the SR Ca^{2+} channels. Rather, the voltage-induced conformational change in the T-tubule sensor protein is thought to be transmitted directly to the SR Ca^{2+} channel protein, so that its ion channels open, permitting efflux of Ca^{2+} ions from the SR to the cytosol (Figure 22-17c–e).

Because depolarization is conducted along the transverse tubule to the SR membrane within milliseconds, every sarcomere in a cell contracts simultaneously. Continued stimulation of the muscle keeps the cytosolic Ca^{2+}

Z disk

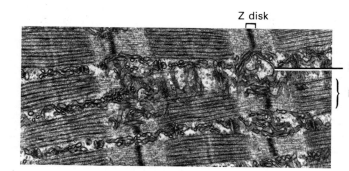

Sarcoplasmic reticulum

Myofibril

◀ **Figure 22-16** Electron micrograph showing abundant sarcoplasmic reticulum vesicles surrounding rather small myofibrils in striated muscle. This is a longitudinal section through a column of myofibrils in rattlesnake muscle. *Courtesy of E. Schultz and A. W. Clark.*

(a)

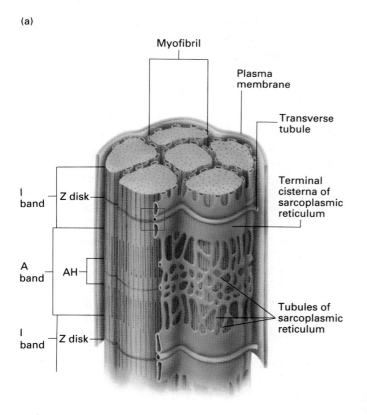

(b)

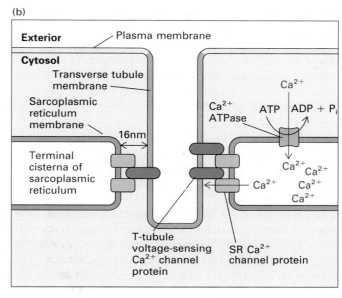

(e)

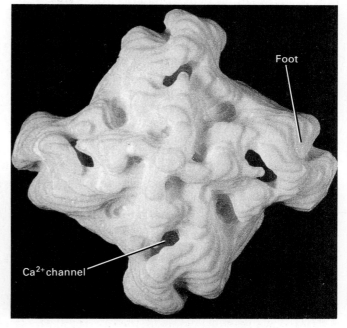

(c)

(d)

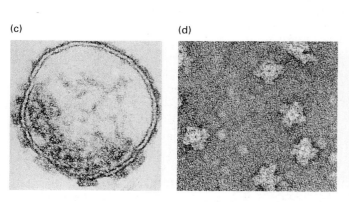

▲ **Figure 22-17** (a) Three-dimensional structure of six myofibrils. The transverse (T) tubules, which are invaginations of the plasma membrane, enter myofibers at the Z disks, where they are in close contact with the terminal cisternae of the sarcoplasmic reticulum (SR). In the region of the Z disk, the terminal cisternae store Ca^{2+} ions and connect with the lacelike network of SR tubules that are abundant over the H zone of the A band. (b) Enlarged diagram of boxed region in (a) showing the relationships of the transverse tubule and the terminal cisternae of the sarcoplasmic reticulum. Voltage-sensing Ca^{2+} channel proteins (Figure 20-24) in the T-tubule membrane touch the protrusions, or "feet," of the large Ca^{2+} channel proteins in the SR membrane. When a muscle is stimulated, Ca^{2+} ions stored in the SR are released into the cytosol through these SR Ca^{2+} channel proteins. (c) Cross section of a purified terminal cisterna. The outward-projecting feet of the SR Ca^{2+} channels are evident. (d) Electron micrograph of purified Ca^{2+} channel proteins from SR vesicles. This protein has four identical subunits and a molecular weight of about 2.2×10^6. (e) A three-dimensional image reconstruction of the SR Ca^{2+} channel protein, showing the side facing the transverse tubule. Note the four feet projecting toward the T-tubule membrane and the pores, thought to be the channels through which Ca^{2+} ions move from the SR to the cytosol. [See L. Hymel et al., 1988, *Proc. Nat'l Acad. Sci. USA* **85**:441; and W. S. Agnew, 1988, *Nature* **344**:299.] *Parts (c)–(e) courtesy of Dr. S. Fleischer.*

level high. When stimulation ceases, the Ca^{2+} ATPase in the SR membrane (Figure 22-17b) lowers the cytosolic Ca^{2+} concentration by pumping Ca^{2+} ions back into the lumen of the SR, thus inhibiting contraction. Striated muscle can undergo very rapid increases and decreases in the cytosolic Ca^{2+} level, thereby permitting precise control of muscular movements.

Calcium Activation of Actin, Mediated by Tropomyosin and Troponin, Regulates Contraction in Striated Muscle

Because of the Ca^{2+} ATPase in the SR membrane, the cytosol of resting muscle has a free Ca^{2+} concentration of about 10^{-7} M. Stimulation that causes an increase in the Ca^{2+} concentration to 10^{-6} M initiates contraction. The manner in which Ca^{2+} regulates contraction varies for different types of muscle. In striated vertebrate muscle, the Ca^{2+} regulation affects the actin thin filaments. In smooth muscle and invertebrate muscle, discussed in the next two sections, regulation primarily involves the myosin heads.

In addition to actin, the thin filaments in striated muscle contain four proteins, which are involved in the Ca^{2+} regulation of contraction. *Tropomyosin* is a coiled coil of two parallel α-helical polypeptides, each with 284 amino acids (MW ~35,000); the rodlike molecule is about 40 nm in length. Tropomyosin molecules polymerize head-to-tail to form filaments that lie in one groove of the actin helix and give rigidity to actin filaments. Each tropomyosin molecule has seven actin-binding sites and binds to seven actin monomers (Figure 22-18a). Bound to a specific site on each tropomyosin molecule are three *troponin* (Tn) peptides called troponin T, I, and C (Tn-T, Tn-I, Tn-C, respectively).

A mixture of myosin filaments and purified actin filaments hydrolyzes ATP at the maximum rate. In the absence of Ca^{2+} the presence of troponin and tropomyosin on the thin filaments inhibits this ATPase activity by inhibiting the interaction of myosin heads with actin. Troponin T (MW 37,000) is an elongated protein that binds along the C-terminal third of tropomyosin and links both Tn-I and Tn-C to tropomyosin. Troponin I (MW 22,000) binds to actin as well as Tn-T; Tn-I in concert with tropomyosin causes a small conformational change in the actin such that it binds weakly to myosin heads but cannot activate the myosin ATPase. Troponin C, the Ca^{2+}-binding subunit, has a structure and function very similar to that of calmodulin. Occupation by Ca^{2+} ions of all the Ca^{2+}-binding sites on Tn-C releases the tropomyosin–Tn-I inhibition of the actin-myosin ATPase activity, thus activating contraction. This Ca^{2+} binding triggers a slight movement of the flexible tropomyosin helix toward the center of the actin filament (Figure 22-18b inset). This shift of tropomyosin exposes a region of the actin monomers to which the myosin heads can bind in such a manner that the myosin ATPase is activated (Figure 22-18c). These myosin-activating sites are blocked by tropomyosin in the resting, but not the active, state.

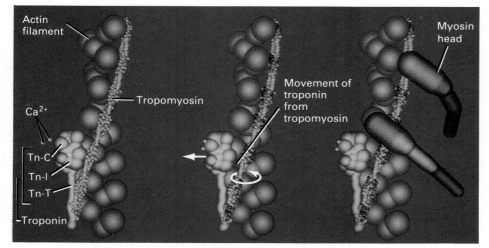
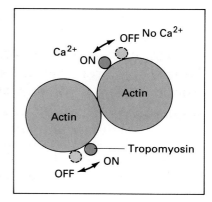

(a) Resting state   **(b) Active state**   **(c) Myosin binding**

▲ **Figure 22-18** Effect of Ca^{2+} ions on tropomyosin and myosin binding to actin filaments. (a) In the "off" (resting) state, tropomyosin is bound to the outer domains of seven actin monomers along the helical actin filament. One troponin complex (without bound Ca^{2+}) is bound to each tropomyosin. (b) In the "on" (active state), troponin C (Tn-C) binds Ca^{2+} ions, causing Tn-C and Tn-I to move away from tropomyosin. The freed tropomyosin rotates so as to bind to a slightly more inward position on the actin monomers, as diagrammed in cross section in the inset. (c) Rotation of tropomyosin allows myosin heads to bind strongly to actin. *From G. N Phillips, Jr., J. P. Filliers, and C. Cohen, 1986, J. Mol. Biol. **222:111**; photographs courtesy of Dr. C. Cohen.*

So far no one has been able to "see" cyclic movements of myosin heads or of tropomyosin; light microscopy does not have the resolution and electron microscopy requires that the specimens be fixed or dried. Because striated fibers are almost crystalline arrays of thin and thick filaments, x-ray diffraction has been used to provide direct evidence that tropomyosin movements precede—and presumably trigger—movement of myosin heads toward actin. X-ray diffraction of muscle yields a series of "spots" that are generated by the regular ("crystalline") array of fiber proteins. Some spots are generated, for instance, by the regular spacing of the myosin heads along the thick filament; the intensities of these spots do not change during contraction since this spacing does not change. One specific diffraction "spot" is due to the regular interaction of tropomyosin with actin; when tropomyosin moves during contraction, the intensity of this spot increases. Changes in intensities of other "spots" are due to movement of the myosin heads relative to thin actin filaments.

High-intensity x-rays from a synchrotron source have been used to follow changes in the muscle diffraction pattern during muscle contraction. The increase in cytosolic Ca^{2+} level and tropomyosin movement occur 17 milliseconds (ms) after a muscle is stimulated; the myosin heads attach to actin after 25 ms; and tension is generated after 40 ms. These results indicate that tropomyosin movement precedes movement of myosin heads, exactly as depicted in Figure 22-18. Moreover, one can stretch a muscle so that there is no overlap between myosin heads and actin, and no possible generation of tension (see Figure 22-13). When such a muscle is stimulated, movement of tropomyosin occurs normally, but no binding of myosin to actin occurs. Thus tropomyosin movement is independent of that of myosin and is the step directly affected by Ca^{2+} ions.

Calcium Activation of Myosin Light Chains Regulates Contraction in Smooth Muscle and Invertebrate Muscle

Vertebrate smooth muscle and invertebrate muscle contain tropomyosin but not the troponin complex, and Ca^{2+} regulation of contraction in these muscles mainly involves interactions with myosin light chains. Smooth muscle, unlike striated muscle, does not contain a developed SR membrane, and changes in the cytosolic Ca^{2+} level are much slower than in striated muscle—on the order of seconds to minutes—thereby allowing a slow, steady response in contractile tension.

In typical invertebrates, such as mollusks, the interaction of myosin heads and actin filaments at low Ca^{2+} concentrations is inhibited by one of the myosin light-chain (LC) pairs located in the head region (see Figure

22-3b). Binding of Ca^{2+} ions to this regulatory LC pair induces a conformational change in the myosin heads that allows them to bind to actin; this, in turn, causes activation of the myosin ATPase and contraction of the muscle. The regulatory light-chain protein has a very high affinity for Ca^{2+}—similar to that of Tn-C or of the ubiquitous Ca^{2+}-binding protein calmodulin.

Vertebrate smooth muscle does not have a sarcomere organization (see Figure 22-7a), and the actin-myosin network is more disordered than in striated muscle. Contractions in smooth muscle are regulated primarily by a complex network of myosin light-chain phosphorylations and dephosphorylations. As in mollusks, one of the two myosin LC pairs in smooth muscle inhibits the actin-stimulated ATPase activity of myosin. In smooth muscle, however, phosphorylation of this regulatory LC pair by the enzyme *myosin LC kinase* relieves this inhibition, thus permitting contraction. Because Ca^{2+} is required for activation of myosin LC kinase, the Ca^{2+} level regulates the extent of LC phosphorylation and hence contraction. Calcium first binds to calmodulin, and the Ca^{2+}-calmodulin complex then binds to and activates the LC kinase (Figure 22-19a). Calmodulin is found in virtually all eukaryotic cells and is believed to mediate most intracellular effects of Ca^{2+}. As evidence for the role of activated myosin LC kinase, microinjection into smooth muscle cells of an inhibitor that blocks the LC kinase also blocks muscle contraction but does not affect the rise in cytosolic Ca^{2+} level associated with membrane depolarization. The effect of this inhibitor can be overcome by microinjection of a proteolytic fragment of myosin LC kinase that is active in the absence of Ca^{2+}-calmodulin; this treatment also does not affect Ca^{2+} levels.

As noted earlier, myosins in striated and smooth muscle differ slightly, which probably accounts for the differences in the Ca^{2+} regulatory mechanisms in the two types of muscle. In addition, myosin in smooth muscle cannot polymerize into thick filaments (see Figure 22-3c) unless the light chains are phosphorylated. In contrast, myosin in striated muscle does not require LC phosphorylation to form thick filaments. Thus phosphorylation of myosin light chains in smooth muscle has two effects: it leads to formation of bipolar myosin thick filaments and simultaneously activates them for contraction.

cAMP, 1,2-Diacylglycerol, and Caldesmon Also Affect Contractility of Smooth Muscle

The contractility of smooth muscle is affected not only by nervous stimulation but also by the levels of several hormones. For example, in smooth muscles, as in other cells, epinephrine activation of β-adrenergic receptors causes a rise in the cellular level of cAMP, which activates the

▶ **Figure 22-19** Four mechanisms for regulating contraction and relaxation in vertebrate smooth muscle. The first three mechanisms affect myosin; the key regulatory protein is myosin light-chain kinase. In its active form, this enzyme phosphorylates the myosin regulatory light chain (LC) at site X; as a result, myosin can bind to actin and the muscle contracts. The symbols P_A, P_B, P_C, and P_D denote phosphorylation at different sites on myosin LC kinase; P_X and P_Y denote different sites of phosphorylation of the myosin regulatory light chain. (a) In the Ca^{2+}-calmodulin pathway, active myosin LC kinase forms when the Ca^{2+} concentration is $\geq 10^{-6}$ M; subsequent phosphorylation of the myosin regulatory light chain at site X leads to muscle contraction. A decrease in the Ca^{2+} concentration to 10^{-7} M leads to dissociation of Ca^{2+} and calmodulin from the kinase, thereby inactivating the kinase. Under these conditions myosin LC phosphatase, which is not dependent on Ca^{2+} for activity, dephosphorylates the myosin light chain, causing muscle relaxation. (b) Stimulation of the β-adrenergic receptor by epinephrine leads to activation of the cAMP-dependent protein kinase, which can phosphorylate myosin LC kinase at sites A and B. This weakens the binding of Ca^{2+}-calmodulin to the LC kinase and hence favors formation of the inactive form of myosin LC kinase; as a result, phosphorylation of the myosin light chain does not occur and the muscle relaxes. This pathway may explain how epinephrine acts to relax certain smooth muscles. Dephosphorylation of myosin LC kinase by phosphatase restores the kinase to the form that binds Ca^{2+}-calmodulin strongly. R is the dissociable regulatory subunit of cAMP-dependent protein kinase. (c) Activation of protein kinase C by diacylglycerol and Ca^{2+} leads to phosphorylation of myosin LC kinase at sites C and D and of the myosin regulatory light chain at site Y. Because both of these phosphorylated forms are inactive, the muscle relaxes. (d) At low Ca^{2+} concentrations ($<10^{-6}$ M), caldesmon binds to tropomyosin and actin, reducing the binding of myosin and keeping muscle in the relaxed state. At higher Ca^{2+} concentrations, Ca^{2+}-calmodulin binds to caldesmon, releasing it from actin; thus myosin can bind to actin and the muscle can contract. [See R. S. Adelstein and E. Eisenberg, 1980, *Ann. Rev. Biochem.* **49**:92; H. Rasmussen, Y. Takuwa, and S. Park, 1987, *FASEB J.* **1**:177; and K. Sobue et al., 1988, *J. Cell. Biochem.* **37**:317.]

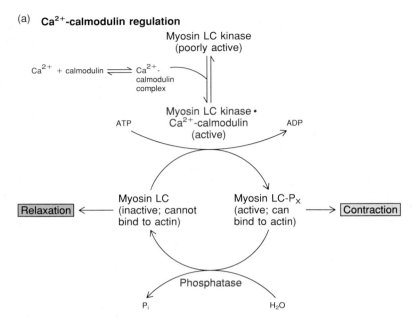

(a) **Ca^{2+}-calmodulin regulation**

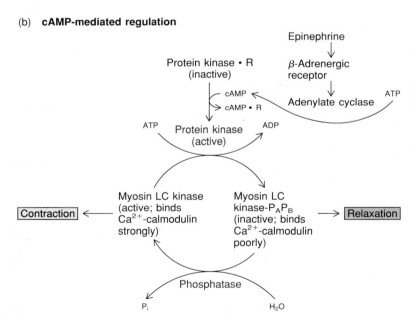

(b) **cAMP-mediated regulation**

cAMP-mediated protein kinase. One of the substrates of this enzyme is myosin LC kinase. Phosphorylation of myosin LC kinase by this protein kinase lowers the binding affinity of LC kinase for the Ca^{2+}-calmodulin complex; as a result, phosphorylation of the myosin regulatory LC pair cannot occur, and the muscle remains in the relaxed state (Figure 22-19b). Other hormones cause relaxation of smooth muscle by activation of protein kinase C, which is mediated by Ca^{2+} and 1,2-diacylglycerol (Fig-

ure 22-19c). These two examples illustrate again how intracellular "second messengers"—1,2-diacylglycerol, Ca^{2+} ions, and cAMP—interact in a complex manner to modulate critical cellular processes.

In smooth muscle cells, as well as in nonmuscle cells, calcium-binding proteins that interact with actin provide a second control over contraction. Smooth muscle *caldesmon* (MW ~150,000) is an elongated protein, about 75 nm in length, that in the absence of Ca^{2+} ions binds

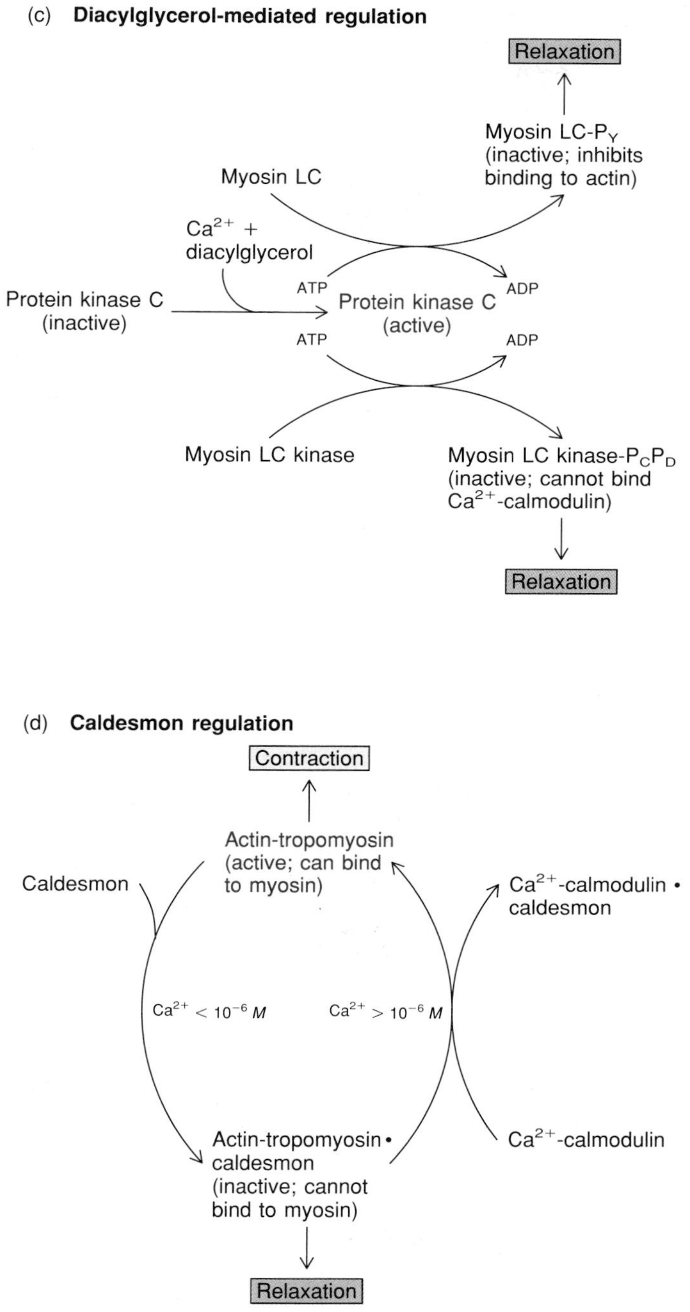

(c) Diacylglycerol-mediated regulation

Relaxation

Myosin LC-P$_Y$
(inactive; inhibits
binding to actin)

Myosin LC

Ca^{2+} +
diacylglycerol

Protein kinase C
(inactive) →
ATP → Protein kinase C → ADP
ATP (active) ADP

Myosin LC kinase

Myosin LC kinase-P$_C$P$_D$
(inactive; cannot bind
Ca^{2+}-calmodulin)

Relaxation

(d) Caldesmon regulation

Contraction

Actin-tropomyosin
(active; can bind
to myosin)

Caldesmon

Ca^{2+}-calmodulin •
caldesmon

Ca$^{2+} < 10^{-6}$ M   Ca$^{2+} > 10^{-6}$ M

Ca^{2+}-calmodulin

Actin-tropomyosin •
caldesmon
(inactive; cannot
bind to myosin)

Relaxation

alongside tropomyosin to actin filaments and restricts the ability of myosin to bind to actin. A rise in the Ca^{2+} level triggers binding of Ca^{2+}-calmodulin to caldesmon, releasing caldesmon from actin (Figure 22-19d). Myosin can now bind to actin and initiate contraction. Thus Ca^{2+} affects contraction by affecting both myosin heads and actin filaments. Such a dual control could allow the cell to regulate the duration and frequency of contractions and the tension generated during each contraction period.

Smooth and Striated Muscles Contain Functionally Different Myosin Light Chains and Tropomyosins

Let us briefly review the different mechanisms of Ca^{2+} regulation of contraction in striated muscle and in smooth muscle. In striated muscle, myosin light chains do not require Ca^{2+}-stimulated activation because they are always in a conformation that permits binding of myosin headpieces to actin. However, binding of myosin to actin depends on Ca^{2+}-induced movement of tropomyosin bound to the actin filaments; this movement, which activates the myosin-binding sites, is mediated by binding of Ca^{2+} ions to the troponin complex associated with tropomyosin. In smooth muscle, in contrast, binding of myosin to actin depends mainly on the Ca^{2+}-stimulated activation of the myosin light chains. Although tropomyosin in smooth muscle is not associated with a Ca^{2+}-binding troponin complex, the reversible binding of caldesmon to actin modulates binding of activated myosin.

Differences in the Ca^{2+} regulatory mechanisms operating in striated and smooth muscle have been demonstrated in various ways. For example, when light chains are removed from permeabilized smooth muscle fibers, the myosin ATPase is active whether or not Ca^{2+} is present. Likewise, if light chains in smooth muscle are replaced with those from striated muscle, the myosin ATPase activity is independent of Ca^{2+} because striated muscle light chains are always "on." Indeed, myosin light chains can be removed from striated muscle without any effect on Ca^{2+} regulation.

Clearly, the myosin light chains and tropomyosins expressed in striated and smooth muscle confer different regulatory properties on each cell type. In some cases, *isoforms* of such muscle-specific proteins are generated by alternative splicing of an RNA transcript of a single gene. For example, two isoforms of β-tropomyosin are formed by alternative splicing of transcripts of a single β-tropomyosin gene (Figure 22-20). One form, which is produced only in striated muscle, contains two stretches of amino acids that bind to troponin T. The other form, which is produced in smooth muscle and nonmuscle cells, lacks these troponin T–binding sites. Other possible functional differences among these isoforms of tropomyosin and other muscle-specific proteins are now being explored. The ability of recombinant DNA techniques to elucidate previously unsuspected variation in muscle proteins has opened an entire new area of research in muscle biology.

Proteins Anchor Actin Filaments to the Plasma Membrane or the Z Disk

For the sliding-filament mechanism to generate contraction of a muscle cell, one end of each actin thin filament must be immobilized (see Figure 22-12). In striated mus-

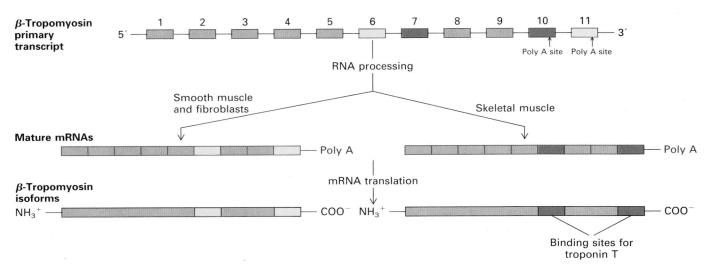

▲ **Figure 22-20** Alternative splicing of the primary RNA transcript encoded by the rat β-tropomyosin gene. Of the 11 exons, 7 are common to smooth muscle and fibroblasts (*left*) and skeletal muscle mRNA (*right*); two (yellow) are unique to the fibroblast–smooth muscle form, and two (purple) to the skeletal muscle form of tropomyosin. The skeletal muscle–specific exons encode the two regions on tropomyosin to which troponin T binds; smooth muscle and fibroblast tropomyosin lack these binding sites. *Adapted from D. Helfman et al., 1986, Mol. Cell Biol. 6:3582.* [See also D. F. Wieczorek, C. Smith, and B. Nadal-Ginard, 1988, *Mol. Cell Biol. 8:679.*]

cle, actin filaments end at the Z disk; in smooth muscle, similar but smaller structures, termed *cytoplasmic plaques,* apparently serve the same function. In many smooth and striated muscle cells, actin filaments at the end of the myofibers are attached to the plasma membrane. Cardiac muscle contains structures termed *intercalated disks* adjacent to the plasma membrane to which the actin filaments are attached (Figure 22-21).

One protein with a role in attachment of actin filaments is Cap Z, a protein with two subunits (MW 32,000 and 36,000) that binds selectively to the (+) ends of actin filaments. Cap Z, which is localized to the Z line of skeletal muscle, prevents the depolymerization of actin filaments from the (+) end, causing the filaments to be very stable; it probably also anchors the (+) ends of actin filaments to other Z disk proteins.

Another important attachment protein is *α-actinin* (MW 190,000), a dimer of two identical polypeptides. This fibrous protein is a major component of Z disks in striated muscle, of cytoplasmic plaques in smooth muscle, and of intercalated disks in cardiac muscle. α-Actinin binds tightly to the sides of actin filaments and bundles the ends of adjacent filaments together (Figure 22-22a). In cardiac and striated muscle, α-actinin may bundle actin filaments into the characteristic hexagonal lattice (see Figure 22-11). The actin-binding domains and overall structures of α-actinin, spectrin (an important cytoskeletal protein in erythrocytes), and dystrophin (a critical muscle protein discussed later) are very similar (Figure 22-22b). Spectrin, like α-actinin, also binds to and cross-links actin filaments.

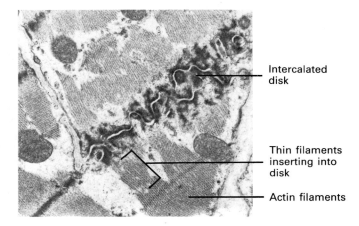

▲ **Figure 22-21** Electron micrograph of the ends of two cardiac muscle cells, showing the thin filaments inserting into the intercalated disk at the boundary between the two cells. *Courtesy of M. Seiler and A. Perez-Atayde.*

Although the structure of the Z disk is not known to molecular detail, ultrastructural evidence indicates that it contains two sets of overlapping actin filaments of opposite polarity that originate in the two sarcomeres adjacent to the Z disks; a number of fibers, including α-actinin, interconnect these actin filaments.

A third protein, *vinculin* (MW 130,000), is found in intercalated disks of cardiac muscle and the membrane-associated plaques in smooth muscle (but not in Z disks). Vinculin binds tightly to α-actinin in cell-free experi-

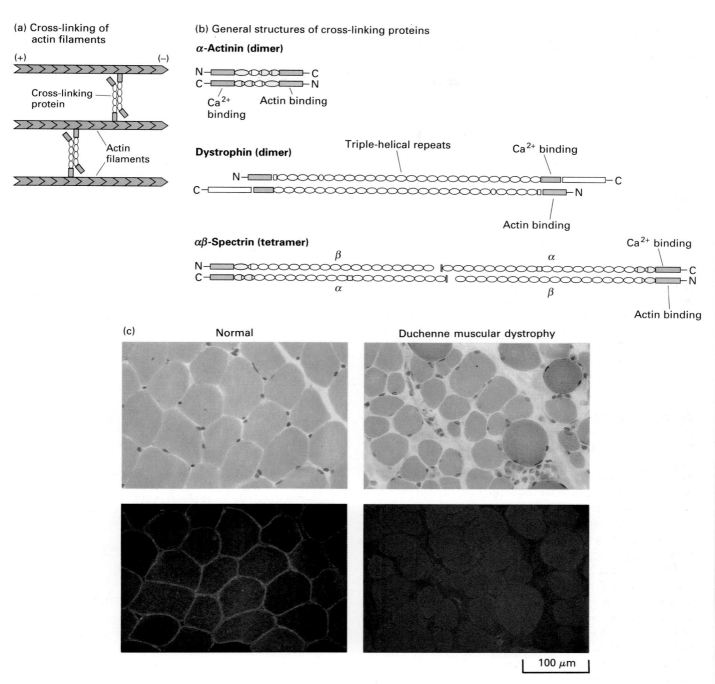

▲ **Figure 22-22** Structure and function of three actin cross-linking and attachment proteins: α-actinin, αβ-spectrin, and dystrophin. (a) All three proteins can cross-link actin filaments, bundling adjacent filaments together. (b) General structures and sequence similarities of the three proteins. α-Spectrin and β-spectrin form end-to-end dimers, which associate into antiparallel tetramers; α-actinin forms an antiparallel dimer, and dystrophin is thought to form a similar dimer. The N-terminal 240 amino acids of β-spectrin, α-actinin, and dystrophin (green), which share strong sequence homology, form the actin-binding domains. The C-terminal domains of α-spectrin, α-actinin, and dystrophin (orange), also highly homologous, contain Ca²⁺-binding regions. The middle sections of these proteins consist of repeats of triple-helical segments containing about 100 amino acid residues (*ovals*); some of the spectrin and α-actinin repeats are homologous (yellow). The squares and circles represent nonrepeat sequences. (c) Sections of skeletal muscle from a normal person (*top and bottom, left*) and from a person with Duchenne muscular dystrophy (*top and bottom, right*) stained with hematoxylin-eosin (*top*) and a fluorescent antibody to dystrophin (*bottom*). Note the small, disorganized fibers in the DMD section and the complete absence of dystrophin. [Parts (a) and (b) see R. H. Brown and E. P. Hoffman, 1988, *Trends Neural Sci.* **11**:480; M. D. Davison and D. R. Critchley, 1988, *Cell* **52**:159; and T. J. Byers et al., 1989, *J. Cell Biol.* **109**:1633–1641.] *Part (c) courtesy of K. Arahata, from K. Arahata et al., 1988, Nature* **333**:861.

ments, but the proteins with which vinculin interacts in vivo are unknown. Most likely it binds directly to an integral membrane protein in the intercalated disk and also to α-actinin (Figure 22-23).

Long Proteins Organize the Sarcomere

In striated muscle some mechanisms must exist to keep the thick myosin filaments centered in the sarcomere during contractions. This function may be carried out by the gigantic fibrous protein *titin*, which connects thick myosin filaments to Z disks. Titin appears to function like an elastic band, keeping the myosin filaments centered when the muscle contracts or is stretched (Figure 22-24). The molecular weight of titin is not known but probably is more than one million (that is, over 10,000 amino acids); the protein is about 1 μm long.

Nebulin, another gigantic protein, constitutes about 3 percent of muscle protein. It forms long inextensible filaments that extend from each side of the Z disk (Figure 22-24). Because nebulin filaments are as long as the adjacent actin filaments, they may regulate the number of actin monomers that polymerize into each thin filament during formation of mature muscle fibers. Nebulin also may help organize actin filaments into a regular geometric pattern.

Dystrophin Is a Muscle Protein Identified by Study of a Genetic Disease

Duchenne muscular dystrophy (DMD) is a fatal, degenerative genetic disease of muscle that affects about 1 of every 3500 males born. The gene associated with DMD is on the X chromosome and is the largest known human gene, containing more than 2 million bases. Many individuals with DMD have deletions in this locus; segments of the X-chromosome DNA missing in such patients were used to isolate the cDNA for the protein missing in DMD. Subsequent cloning of this cDNA led to the discovery of *dystrophin*, which is localized to the plasma membrane of striated muscle cells and is missing in muscles from DMD

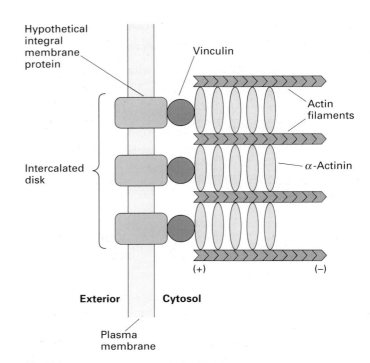

▲ **Figure 22-23** Proposed model of attachment of actin filaments to the plasma membrane at intercalated disks in cardiac muscle. Vinculin, a nonintegral protein, is thought to bind to an integral membrane protein, whose identity is unknown. Binding of vinculin to α-actinin cross-linking actin filaments would then attach the filaments to the membrane.

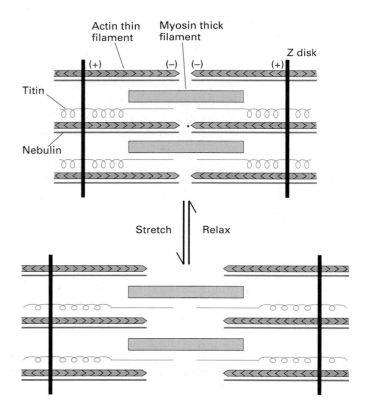

▶ **Figure 22-24** Proposed model of role of titin and nebulin in organizing the sarcomere in striated muscle. Titin, a gigantic, elastic protein, links myosin thick filaments to Z disks. When the muscle is stretched, the elasticity of the titin filament keeps thick filaments centered in the sarcomere. Nebulin (MW 600,000–800,000) forms inextensible filaments, anchored at Z disks, which may act as organizing elements for the actin filaments. *Adapted from D. O. Furst et al., 1988, J. Cell Biol. **106**:1563; R. Horowitz and R. J. Podolsky, 1987, J. Cell Biol. **105**:2217; and K. Wang and J. Wright, 1988, J. Cell Biol. **107**:2199.]*

patients (see Figure 22-22c). Dystrophin has a molecular weight of 400,000 and constitutes only about 0.002 percent of the total muscle protein. Although the molecular function of dystrophin is not yet known, it has sequence homology to α-actinin and other actin-binding proteins (see Figure 22-22b) and may be involved in anchoring actin filaments to the plasma membrane. Clearly it is important in preventing degeneration of muscle fibers. The identification of dystrophin is an outstanding example of "reverse genetics"—identifying the gene for a protein whose absence causes a genetic disease even when the function of the protein is not known.

Phosphorylated Compounds in Muscle Act as a Reservoir for ATP Needed for Contraction

Resting muscle, like other tissues, requires a constant supply of ATP for its basic metabolic functions. Muscle, especially striated muscle, is unusual in that during contraction the demand for ATP can increase 20- to 200-fold above that in the resting state. Yet mammalian skeletal muscle contains enough adenine nucleotides, mostly as ATP, for only about 0.5 s of intense activity, or about 10 contractions. The two principal types of striated muscle fibers use different mechanisms to generate sufficient ATP for extended muscle contraction.

"White" muscle fibers—the white meat in a chicken—are *fast-twitch* fibers, which contract and fatigue quickly; they generate ATP by anaerobic glycolysis of glucose and glycogen. The "red" fibers—the dark meat—contain abundant mitochondria and myoglobin, the red oxygen-binding pigment. Red fibers are *slow-twitch* fibers, which contract and fatigue more slowly than fast-twitch fibers; they generate ATP by aerobic catabolism of glucose and fats, using the O_2 bound to myoglobin (see Figure 2-26). Fast-twitch muscles, especially, cannot generate sufficient ATP from glycogen and glucose to fuel rapid, intense contractions, in part because glycogen is a relatively large particle and because glycogen phosphorylase can remove only one glucose residue at a time. Also, glycolysis yields only 2 ATPs per glucose molecule or 3 ATPs per glucose residue in glycogen.

The major phosphorylated compound in vertebrate striated muscle is *creatine phosphate*. The phosphate of this compound can be transferred to ADP in a reaction catalyzed by creatine kinase (Figure 22-25). The concentration of creatine phosphate in muscle is about five times the concentration of ATP. The free energy of hydrolysis of creatine phosphate ($\Delta G^{\circ\prime} = -10.3$ kcal/mol) is greater than that of the terminal phosphate of ATP ($\Delta G^{\circ\prime} = -7.3$ kcal/mol). Therefore at neutral pH the creatine kinase reaction favors formation of ATP. Among invertebrates, other phosphorylated compounds (e.g., guanidine phosphate) are used as storage forms of high-energy phosphate. The phosphate residue of such compounds can be transferred to ADP by reactions analogous to that catalyzed by creatine kinase.

Striated muscle cells also possess adenylate kinase, an enzyme that allows the cell to utilize the energy of both phosphoanhydride bonds in ATP:

$$2 \text{ ADP} \underset{\text{kinase}}{\overset{\text{Adenylate}}{\rlap{\longleftarrow}\longrightarrow}} \text{ATP} + \text{AMP}$$

This enzyme can convert ADP formed during muscle contraction into ATP for use in additional contractions.

Despite having two highly efficient mechanisms for generating from ADP the large amounts of ATP needed to fuel rapid contractions, fast-twitch muscles are depleted of ATP within a few minutes. Eventually the supply of ATP must be regenerated by glycolysis of glucose generated from glycogen or imported from the blood.

Actin and Myosin in Nonmuscle Cells

In contrast to the orderly filamentous arrangement of actin in striated muscle cells, the spatial organization of actin in most nonmuscle cells is highly variable and complex, so that studies of its function and structure are much more difficult. Myosin and other muscle proteins are also present in nonmuscle cells, but at a much lower ratio to actin than in muscle (see Table 22-1). In nonmuscle cells these proteins may serve both structural and contractile functions.

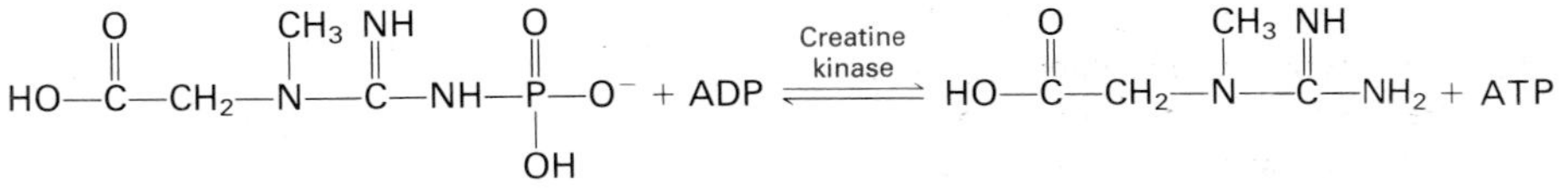

▲ **Figure 22-25** In the reaction catalyzed by creatine kinase, the phosphate group of creatine phosphate is transferred to ADP, thus forming ATP. At neutral pH, the reaction catalyzed by creatine kinase favors formation of ATP ($\Delta G^{\circ\prime} = -3.0$ kcal/mol).

All Vertebrates Have Multiple Actin Genes and Actin Proteins

On the basis of amino acid sequence analyses, at least six different actins have been identified in adult birds and mammals. Four are called α-actins: one of these is unique to striated skeletal muscle, one to cardiac muscle, one to smooth vascular muscles, and one to smooth enteric muscles such as those which line the intestine. Two other actins, termed β-actin and γ-actin, are found in the cytoplasm of all muscle and nonmuscle cells. The various α-actins differ in only four to six amino acids; cytoplasmic β- or γ-actin and striated muscle α-actin differ in only 25 (out of about 400) amino acids. Clearly genes for these actins have evolved from a common precursor. Yeast contains only a single actin gene, and most simple eukaryotic microorganisms, such as slime molds and yeasts, synthesize only a single species of actin. However, many multicellular eukaryotic organisms contain multiple actin genes—11 in the sea urchin, 17 in the slime mold *Dictyostelium*, and as many as 60 in certain plants. Since the actual number of different actins in these organisms has not yet been determined, how many of the genes are actually expressed is not known. Actins are also modified posttranslationally by acetylation of the N-terminus and methylation of a histidine. These modifications may generate multiple functional species well in excess of the number of genes.

In cell-free reactions, all known actins form the same type of polymer and bind S1 fragments in precisely the same way to form polarized arrowhead structures. Regardless of source, all purified actins are capable of stimulating the ATPase activity of muscle myosin and can copolymerize with all other actins. For instance, if the gene for cardiac muscle α-actin is transfected into fibroblast cells, the expressed muscle actin is incorporated into the same actin filaments as those in which the normal fibroblast β- and γ-actins are found. The different actin proteins probably vary in their affinities for other proteins, although this has not been demonstrated experimentally.

Many Actin-binding Proteins Are Present in Nonmuscle Cells

Myosin is present in small amounts in all nonmuscle cells, but the short thick myosin filaments characteristic of striated muscle cells have only recently been reported to occur in slime molds and cultured cells. Tropomyosin, myosin light-chain kinase, and calmodulin, the regulatory proteins in smooth muscle, have been isolated from nonmuscle cells such as macrophages, amebas, and platelets. Troponins, the Ca^{2+}-binding regulatory proteins in striated muscle, thus far have not been found in any other cell type. Thus, myosin and actins in nonmuscle cells have the potential to function in a contractile process. As we shall see later, such a role for these proteins has been

Table 22-2   Actin-binding proteins found in the cytoplasm of vertebrate cells

Protein	Function
α-Actinin	Cross-links actin filaments into regular parallel arrays; participates in binding of actin filaments to membranes
Caldesmon	Low $[Ca^{2+}]$: binds to tropomyosin and actin and prevents myosin binding
Capping proteins	Bind to one end of a filament, preventing both addition and loss of actin monomers
Filamin	Forms perpendicular connections between actin filaments
Fimbrin	Cross-links adjacent filaments tightly to form parallel actin fibers
Fodrin	Spectrinlike protein that cross-links adjacent actin filaments
Gelsolin	Micromolar $[Ca^{2+}]$: severs actin filaments
110-kDa Microvillar protein (myosin I)	Myosinlike protein that links sides of actin filaments to microvillar membrane; forces generation at leading edge of motile cell
Myosin II	Causes movements along actin filaments as in muscle; generates tension in arrays of actin microfilaments
Profilin	Binds to actin monomers; prevents nucleation of actin filaments
Tropomyosin	Binds as head-to-tail aggregates in a groove along the length of an actin helix, thus strengthening filaments; in striated muscle regulates binding of myosin heads to actin
Villin (microvilli)	Low $[Ca^{2+}]$: nucleates polymerization of actin filaments and bundles adjacent filaments Micromolar $[Ca^{2+}]$: severs actin filaments into short fragments
Vinculin	By binding to α-actinin, mediates binding of actin filaments to membranes

suggested in the contractile ring of telophase cells, which tightens during the separation of the daughter cells. Vinculin and α-actinin, two of the proteins involved in attachment of actin filaments in muscle cells, are also found in nonmuscle cells. Table 22-2 summarizes the properties of various actin-binding proteins found in vertebrate cells.

One end of most nonmuscle actin filaments generally is bound to a membrane or other attachment site. When such filaments are combined in vitro with myosin S1 fragments, the arrowheads always point away from the attachment site, thus indicating the polarity of the filaments. The opposite ends of the filaments, the (+) ends, point toward the membrane to which they normally are attached (Figure 22-26). As a consequence, any tension generated on these filaments by interaction with myosin exerts a pull on the plasma membrane, exactly as in the case of the Z disks or the intercalated disks in smooth muscle cells.

Noncontractile Bundles of Actin Filaments Maintain Microvilli Structure

Microvilli from intestinal epithelial cells represent one of the few nonmuscle systems in which actin microfilaments occur in a well-ordered pattern (Figure 22-27; see also Figure 13-43). Shearing of intestinal cells produces an abundance of microvilli, which can be easily prepared in pure form for use in research work. Consequently, a great deal is known about the structure of these organelles, particularly since they contain relatively few types of cytoskeletal proteins, all of which have been purified and studied.

Just under the microvilli in these cells are two other networks rich in actin and actin-binding proteins: the terminal web, a layer of filaments that crisscross the cytosol just under the microvilli, and the belt desmosome, containing a set of actin filaments that encircles the plasma membrane and is linked to it at the level of the terminal web (see Figure 13-43). Study of this part of the epithelial cell not only has shown how actin filaments cause rigid fingerlike extensions of the plasma membrane but also has given valuable clues about the roles of many actin-binding proteins in nonmuscle cells.

In the core of each microvillus is a bundle of specialized actin filaments lacking myosin, tropomyosin, and α-actinin. These unusual *core microfilaments* presumably play a structural role, maintaining the shape of the microvilli. The (+) ends of these microfilaments point toward the site of membrane insertion at the tip of the microvillus. In addition to these actin microfilaments, microvilli contain several major proteins that are important in generating their rigid structure (Figure 22-28a). The protein *fimbrin* (MW 68,000) binds actin filaments in a ratio of 1

▶ **Figure 22-27** Micrograph of fixed and quick-frozen intestinal microvilli. Tight bundles of core actin filaments extend out of the microvilli to form straight "rootlets," which connect with a number of thin fibers that form complicated networks in the terminal web. Staining with specific antibodies shows that these are composed mainly of the spectrin-like protein fodrin. These rest upon a tangled network of thicker intermediate filaments located on the bottom of the field shown here. *Courtesy of N. Hirokawa. Reproduced from the* Journal of Cell Biology, *1982, vol. 94, p. 425, by copyright permission of the Rockefeller University Press.*

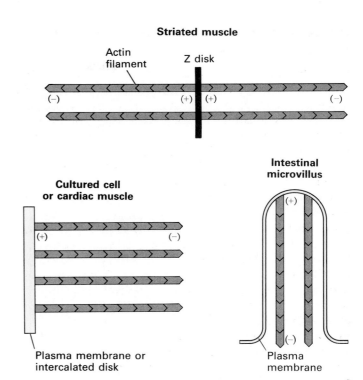

▲ **Figure 22-26** Polarity of attachment of actin filaments to membranes and cytoskeletal structures in muscle and nonmuscle cells. In all cases, attachment occurs at the (+) ends of actin filaments.

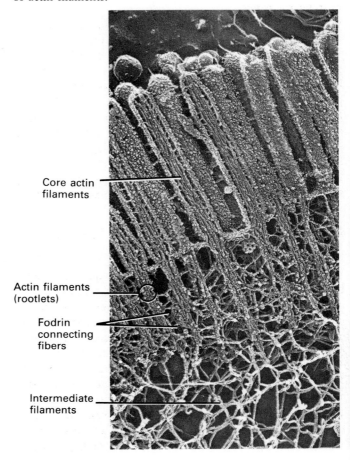

(a)

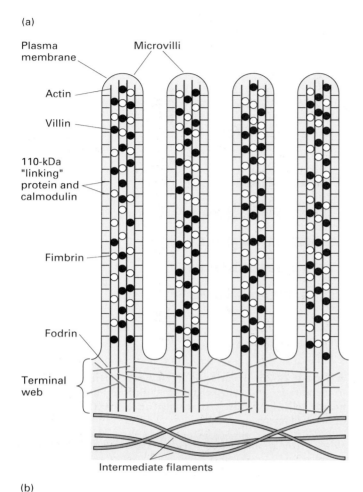

(b)

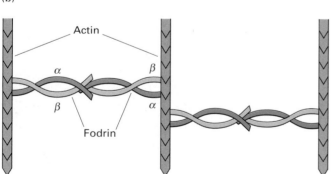

▲ **Figure 22-28** Linkage of actin filaments by different proteins in the intestinal brush border. (a) Core actin microfilaments are bundled together by the binding proteins fimbrin and villin. The 20- to 30-nm-long cross-bridges connecting actin microfilaments to the membrane are composed of a myosin-I-like 110-kDa protein and calmodulin. In the terminal web, fodrin binds adjacent actin filaments into bundles. (b) Structure of the fodrin-actin linkage. Fodrin, which is 5 nm wide and 200 nm long, can cross-link actin filaments; like spectrin, it binds to the sides of actin filaments. Fodrin has the polypeptide structure $\alpha_2\beta_2$; the α and β chains are similar in sequence to α- and β-spectrin. [See N. Hirokawa, R. Cheney, and M. Willard, 1983, *Cell* **32**:953; and M. Mooseker, 1983, *Cell* **35**:11.]

fimbrin molecule per 10 actin monomers, forming tightly packed parallel actin bundles. Another protein, *villin* (MW 95,000), also cross-links the actin filaments to form bundles, but only at Ca^{2+} concentrations less than 0.2 μM. At Ca^{2+} concentrations of 1.0 μM or higher, villin severs actin filaments into short fragments; one villin molecule remains attached to one end of each fragment (Figure 22-29). This effect of Ca^{2+} on the length of the microvillar cores is thought to be important physiologically. During the onset of certain diarrheas, the microvilli of the intestinal cells are lost as small vesicles, and the villin-triggered severing of actin filaments is thought to play a key role in disassembly of the long actin filaments. Yet a third actin-binding protein together with calmodulin forms the cross-bridges connecting the sides of actin bundles with the inner surface of the plasma membrane (Figures 22-28a and 22-30). This 110-kDa protein is a myosinlike actin-stimulated ATPase; like type I myosin purified from amebas, this protein has one myosinlike actin-binding head and a short "tail" that binds to a phospholipid. Thus myosinlike cross-bridges may act to generate tension that keeps the actin filaments positioned in the center of the microvillus.

The core bundles of microvillar actin filaments end in the apical *terminal web,* which contains many short filaments that interconnect adjacent actin bundles with one another. A major terminal web constituent is the fibrous actin-binding protein *fodrin.* Fodrin, which resembles spectrin (see Figure 22-22b) in its structure and function, links adjacent actin bundles, including those which encircle the plasma membrane in the region of the belt desmosome (see Figure 22-28b).

The *belt desmosome* characteristically is located just below the tight junction in an epithelial cell, such as the columnar cell of the intestinal epithelium (see Figure 13-45), and forms a belt of cell-to-cell adhesion that surrounds each cell. At a belt junction, the plasma membranes of the adjacent cells are parallel and only 15–20 nm apart. The transmembrane protein *uvomorulin* (also called E-cadherin) is concentrated in this region and links the plasma membranes of adjacent cells together in a Ca^{2+}-dependent manner (Figure 22-31a). Bands of actin filaments lie just under the plasma membrane in this region and encircle the cell like a belt. Filamentous myosin II present in this region may cross-link adjacent actin filaments or may generate tension that in some way causes the actin filaments to become elongated. Also located in this region are α-actinin and vinculin, which may anchor the sides of the actin filaments to uvomorulin (Figure 22-31b). Thus uvomorulin not only causes cell-cell adhesion at belt desmosomes, it helps to maintain the rigidity of the cell layer.

In summary, actin filaments in the brush border have a primarily structural role. Numerous actin-binding proteins help to form the plasma membrane into elongated microvilli and to link actin filaments in the microvilli and

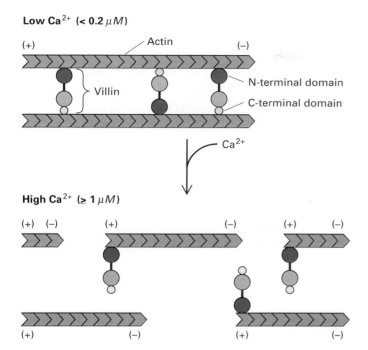

Low Ca^{2+} (< 0.2 μM)

(+)   Actin   (−)

Villin

N-terminal domain

C-terminal domain

Ca^{2+}

High Ca^{2+} (≥ 1 μM)

(+) (−)   (+)   (−)   (+)   (−)

(+)   (−)   (+)   (−)

▲ **Figure 22-29** Dual functions of villin, which contains three domains; both the N-terminal and C-terminal domains can bind actin. At low Ca^{2+} concentrations (<0.2 μM), villin binds to and bundles parallel actin filaments. At high Ca^{2+} concentrations (≥ 1 μM), the conformation of villin changes, so that it severs filaments; its N-terminal domain remains bound to the (+) end of the actin fragment at the severed site. *Adapted from P. Matsudaira and P. Janmey, 1988, Cell 54:139.*

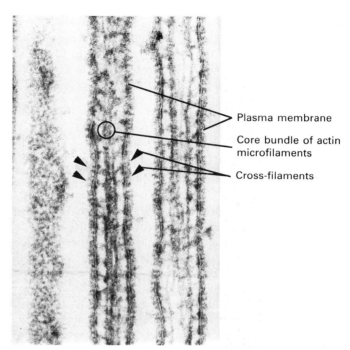

Plasma membrane

Core bundle of actin microfilaments

Cross-filaments

▲ **Figure 22-30** Electron micrograph of a longitudinal section through two isolated intestinal microvilli. The core bundle of actin microfilaments is connected laterally to the membrane by cross-filaments (arrowheads) composed of calmodulin and the 110-kDa minimyosin protein. [See P. T. Matsudaira and D. R. Burgess, 1982, *J. Cell Biol.* 92:657.] *Courtesy of P. T. Matsudaira and D. R. Burgess. Reproduced from the* Journal of Cell Biology, *1982, by copyright permission of the Rockefeller University Press.*

(a)

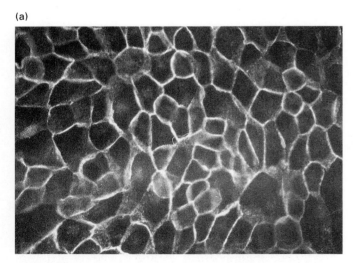

▲ **Figure 22-31** Distribution and function of uvomorulin (E-cadherin) in epithelial cells. (a) Monolayer of cultured kidney epithelial cells labeled with a fluorescent antibody to uvomorulin. Viewed from the top of the monolayer, the protein is seen to surround each cell in the regions of cell-to-cell contact. (b) Model of the function of uvomorulin in cell-cell adhesion. The protein is concentrated in belt desmosomes. *Part (a) from B. Gumbiner, B. Stevenson, and A. Grimaldi, 1988, J. Cell Biol. 107:1575; courtesy of B. Gumbiner.*

(b)

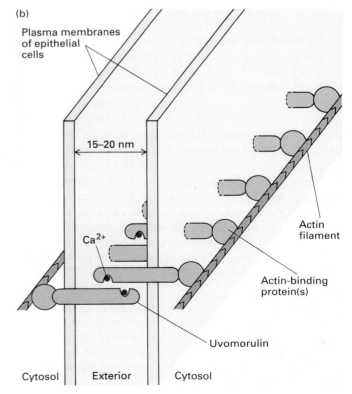

Plasma membranes of epithelial cells

15–20 nm

Ca^{2+}

Actin filament

Actin-binding protein(s)

Uvomorulin

Cytosol   Exterior   Cytosol

belt desmosomes to the underlying terminal web. In contrast to muscle, these actin-containing structures are not contractile, although tension exerted on actin filaments by myosin II and the 110-kDa protein may be critical in maintaining the shape of microvilli.

Actin and Myosin Are Essential for Cytokinesis in Nonmuscle Cells

Actin filaments are thought to participate in a number of motile events in nonmuscle cells, but in only a few cases has it been possible to demonstrate how these filaments function. We shall discuss cytokinesis in animal cells and certain microorganisms and cytoplasmic streaming in large unicellular algae and then discuss other possible roles of actin and actin-binding proteins.

A narrow ring of actin filaments, called the *contractile ring,* encircles the cleavage furrow (the cleft between two separating cells) during the last stage of mitosis. Both myosin and α-actinin are present in the contractile ring at higher concentrations than elsewhere in the cytosol (Figure 22-32). As the ring of actin filaments contracts, the neck of the cleavage furrow is narrowed, and eventually the daughter cells are separated. However, as the actin ring tightens, its thickness remains constant; thus disassembly of the ring occurs together with contraction. The contractile ring disappears entirely as cleavage ends.

Strong evidence indicates that actin-myosin interactions power cytokinesis. For example, antibodies to myosin injected into living sea urchin eggs inhibit cytokinesis, probably by inactivating myosin. Injection of the antimyosin antibody does not inhibit the chromosome movements associated with mitosis, and normal daughter nuclei form in synchrony with the control (uninjected) half of the embryo. This is consistent with evidence that microtubule-based molecular motors power movements of the chromosomes and mitotic spindle.

Important genetic evidence also indicates that cytokinesis is an essential function of actin and myosin in nonmuscle cells. To study the function of actin in yeast, a temperature-sensitive mutation was created in the cloned actin gene by in vitro mutagenesis and then substituted for the single wild-type actin gene (for the experimental protocol, see Figure 21-19). At low temperatures (15°C), the mutant actin functioned normally, and the cells were able to grow and divide. At elevated temperatures, by contrast, the actin filaments were disrupted, secretion of proteins and synthesis of plasma membrane proteins were stopped, and the cells were unable to divide. These find-

▶ **Figure 22-33** Fluorescence micrographs of DAPI-stained nuclear DNA in normal *Dictyostelium* cells (a) and in cells in which the single myosin II gene was disrupted (b). The mutant cells are multinucleate, showing that myosin II is essential for cytokinesis, but not for cell growth or for nuclear division. *Courtesy of A. De Lozanne.*

(a)

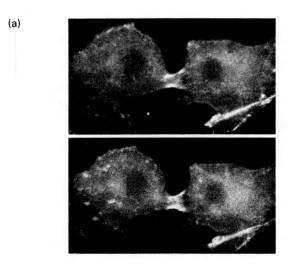

(b)

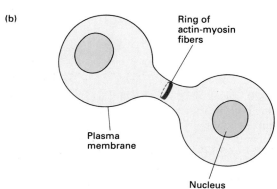

▲ **Figure 22-32** The role of myosin II in cytokinesis. (a) Mitotic chick embryo fibroblast cells double-stained with fluorescein-labeled α-actinin–specific antibody (*top*) and rhodomine-labeled myosin-specific antibody (*bottom*). Because the two dyes fluoresce at different wavelengths, both dyes can be visualized in the same cell. Both proteins are concentrated in a ring around the cleavage furrow. (b) Diagram of a cleavage furrow. A ring of actin-myosin fibers runs just under the plasma membrane in the region of the cell constriction. As the ring contracts and grows smaller, the two daughter cells are gradually pinched apart. *Part (a) courtesy of T. Pollard.*

(a)   (b)

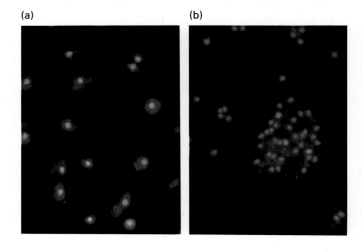

ings confirm that actin is required for many aspects of cellular growth and division in yeast and presumably in other organisms as well.

The single myosin II gene in the motile slime mold *Dictyostelium* was eliminated by a similar type of gene disruption protocol. Surprisingly, the only significant effect was an inhibition of cytokinesis (Figure 22-33). The myosin-deficient cells could adhere normally to surfaces and crawl along them; nuclear division was normal, as was organelle assembly and motility. Thus one essential role of type II myosin in nonmuscle cells is in the contractile ring; other types of cell movements may involve the nonfibrous type I myosins, as discussed later.

Within the contractile ring, short bipolar myosin filaments and actin filaments are thought to form *minisarcomeres,* similar in structure to sarcomeres in striated muscle but lacking Z disks. Actin filaments in adjacent minisarcomeres would be linked by α-actinin or other actin-binding proteins.

Movements of the Endoplasmic Reticulum along Actin Filaments Power Cytoplasmic Streaming

Cells of green algae such as *Nitella* and *Chara* are enormous cylinders up to 5 cm in length. Most of the cell is filled with a central vacuole. Surrounding the vacuole is a rim of cytoplasm that is in constant rapid flow; all of the organelles in this region move at up to 4.5 mm/min in an endless belt—streaming in one direction along the "top" surface and returning along the "bottom" (Figure 22-34a). Evidence indicates that this cytoplasmic streaming is powered by myosinlike motors that move along stationary bundles of actin filaments. These bundles lie just inside the nonmoving chloroplast-filled cortex (Figure 22-34b), and adjacent bundles "point" in the same direction. The endoplasmic reticulum (ER) in the moving cytoplasm participates in force generation; cross-bridges are made between myosin bound to the ER and the stationary actin filaments (Figure 22-34c). As the ER network moves along the filaments, the large area of sliding membranes and vesicles propels the entire viscous cytoplasm. The binding and sliding of the ER network along actin cables can be directly visualized when *Nitella* cytoplasm is incubated in cell-free extracts with actin cables and ATP. This movement is disrupted by drugs such as cytochalasin that interfere with actin-myosin motility.

Polymerization of Actin Monomers Is Controlled by Specific Actin-Binding Proteins in Nonmuscle Cells

The actin filaments involved in muscle, cytoplasmic streaming, and microvilli are permanent features of the cell; the filaments in the contractile ring during cytokine-

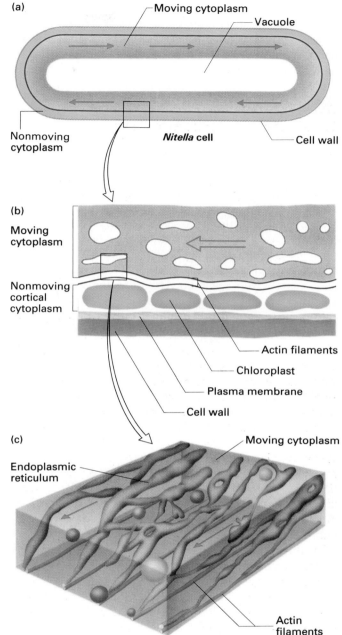

(a)

Moving cytoplasm
Vacuole
Nonmoving cytoplasm
Nitella cell
Cell wall

(b)

Moving cytoplasm
Nonmoving cortical cytoplasm
Actin filaments
Chloroplast
Plasma membrane
Cell wall

(c)

Endoplasmic reticulum
Moving cytoplasm
Actin filaments

▲ **Figure 22-34** Cytoplasmic streaming in cylindrical giant algae. (a) Movement of cytoplasm in a *Nitella* cell is indicated by arrows. (b) Expanded diagram of region boxed in (a). A nonmoving layer of cortical cytoplasm filled with chloroplasts lies just under the plasma membrane. On the inner side of this layer are bundles of stationary actin filaments (red), all oriented with the same polarity. (c) Expanded diagram of boxed region in (b). The streaming cytoplasm, indicated by arrows, is filled with a three-dimensional network of endoplasmic reticulum (ER) tubes and vesicles. Parts of the ER network contact the stationary actin filaments and move along them by a myosinlike motor. The sliding of the ER network propels the entire viscous cytoplasm, including organelles that are enmeshed in the ER network. *Part (c) adapted from B. Kachar and T. Reese, 1988,* J. Cell Biol. *106:1545.*

sis are not. Indeed, in most nonmuscle cells, actin filaments are transient, undergoing constant assembly and disassembly. Several aspects of cell shape and motility depend on these filament networks, although in most cases the evidence for the role of actin or myosin is only indirect. Actin assembly is regulated at two levels: the extent of polymerization of actin monomers into polymers, and the bundling and cross-linking of actin filaments into three-dimensional networks.

Changes in the extent of actin polymerization appear to be an important and regulated aspect of the function of certain cells. Polymerized actin exists in equilibrium with the monomeric form, but the equilibrium constant for the polymerization-depolymerization reaction can be affected by proteins that bind selectively to the monomer or polymer. As noted previously, at the concentration of ions and ATP found in cells, the formation of polymeric actin is strongly favored. However, in vertebrate platelets and in sperm from several invertebrates much of the actin is unpolymerized even though conditions favor polymerization. In these and other animal cells, a protein called *profilin* (MW 15,000) binds to monomeric actin in a 1:1 complex, called *profilactin,* that prevents the premature polymerization of monomeric actin.

Profilin-actin cannot spontaneously polymerize into filaments because profilin prevents the nucleation step that begins actin polymerization. Profilin-actin cannot add to the (−) end of a preexisting actin filament but can bind to the (+) end; after binding of profilactin, the profilin moiety dissociates from the end of the filament (Figure 22-35). Profilin thus causes growth of actin filaments only from the (+) ends. However, the (+) ends of many existing actin filaments are "capped" by proteins (e.g., Cap Z, α-actinin, vinculin) that prevent addition or removal of actin monomers; profilactin thus cannot add to either end of such filaments. In summary, if the (+) ends of all actin filaments are capped, profilin blocks all actin polymerization.

In other systems, severing proteins such as villin (see Figure 22-29) and gelsolin break up long actin filaments and prevent their reassembly at high Ca²⁺ concentrations; such proteins also appear to cap, or bind to, one end of the fragment. At low Ca²⁺ concentrations, villin and similar proteins act as nucleating centers that accelerate the polymerization of monomeric actin.

The Acrosome Reaction in Sperm

A dramatic example of control of conversion of monomeric to filamentous actin is seen in the sperm of many invertebrate species (though not vertebrates) during fertilization of the egg. In sea urchins and sea cucumbers, for example, the sperm contain a vesicle—the *acrosome*—just under the tip of the plasma membrane. The acrosome is separated from the nucleus by the cytoplasmic *periacrosomal region* (Figure 22-36a), which contains profilactin. The surface

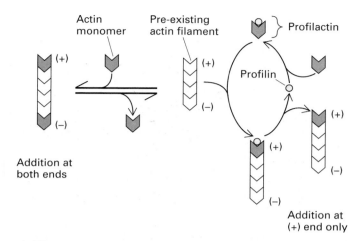

▲ **Figure 22-35** Interactions between profilin, monomeric actin, and fibrous polymeric actin. Although monomeric actin (red) can bind to either the (+) or the (−) end of a filament, profilin (yellow) causes the conformation of an actin monomer to change, so that it can bind only to the (+) end. Addition of profilin-actin to the (+) end of an actin filament causes profilin to be released; profilin can then interact with another actin monomer. Profilin thus causes unidirectional growth of the actin filament from the (+) end. If the (+) ends of actin filaments are capped, profilin blocks all polymerization. *Adapted from L. Tilney et al., 1983, J. Cell Biol. 97:112.*

▶ **Figure 22-36** The acrosomal reaction in sea urchin and sea cucumber sperm. (a) Thin section of an unactivated sea urchin spermatozoon. Lying within an indentation of the nucleus (N) is the spherical membrane-bounded acrosomal vesicle (A). Beneath and lateral to it is the area of cytoplasm called the periacrosomal region (P), which contains unpolymerized profilin-actin (profilactin). Note that there are no filaments visible in this region. (b) Electron micrograph of a sea cucumber spermatozoon after the activation that occurs upon encountering an egg. *Inset:* Cross section of the acrosomal process, showing the many actin fibers cut in transverse section. (c) Steps in the formation of an acrosomal process. Step 1: Unactivated sperm bind to the egg jelly coat by means of specific receptor proteins on their plasma membrane. This binding triggers exocytosis of the acrosomal vesicle, releasing enzymes from the vesicle that digest both the jelly coat and the vitelline layer covering the egg plasma membrane and exposing the (+) ends of short actin filaments that were buried in the vesicle membrane. Step 2: Explosive polymerization of profilin-actin, stored in the periacrosomal region of the sperm cytoplasm, then occurs at the (+) ends of these actin filaments, followed by dissociation of profilin from the filaments (see Figure 22-35). This causes outgrowth of the acrosomal process and a protrusion of the sperm whose plasma membrane is derived from the membrane of the acrosomal vesicle. Step 3: When the acrosomal process reaches the egg plasma membrane, the sperm and egg membranes fuse. Fusion is triggered by surface proteins on the acrosomal process that were originally part of the membrane of the acrosomal vesicle. The sperm nucleus then enters the cell. *Parts (a) and (b) courtesy of L. Tilney.*

(a)

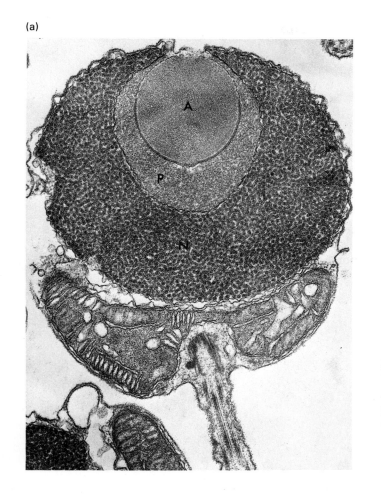

(b)

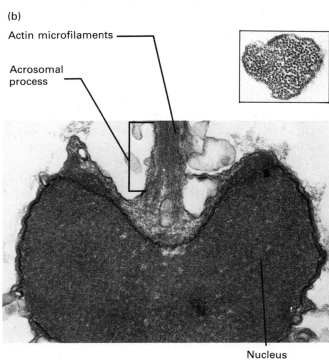

Actin microfilaments

Acrosomal process

Nucleus

(c)

EGG CYTOPLASM

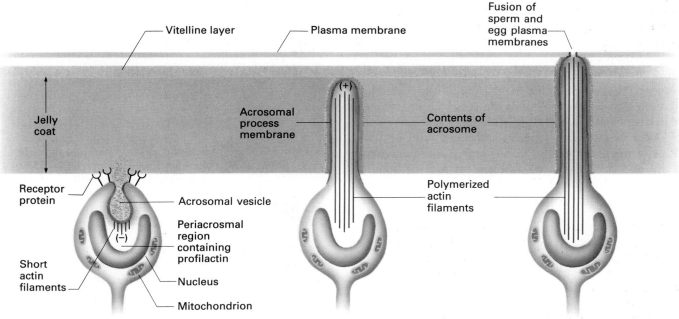

Vitelline layer

Plasma membrane

Fusion of sperm and egg plasma membranes

Jelly coat

Acrosomal process membrane

Contents of acrosome

Receptor protein

Acrosomal vesicle

Periacrosmal region containing profilactin

Nucleus

Mitochondrion

Short actin filaments

Polymerized actin filaments

(+)

(−)

Step 1: Binding and exocytosis

Step 2: Polymerization of actin

Step 3: Fusion of sperm and egg

of sea cucumber eggs is covered with a vitelline layer and with a polysaccharide-rich jelly coat.

Binding of the sperm to the jelly coat of the egg triggers the formation of an *acrosomal process*—a fingerlike extension from the sperm that penetrates both the jelly coat and the vitelline layer (Figure 22-36b). Formation of an acrosomal process, which involves the explosive polymerization of actin, is illustrated in Figure 22-36c. Two events are thought to trigger this polymerization of actin. First, exocytosis of the acrosomal vesicle exposes the (+) ends of short actin filaments that were embedded in it, allowing for addition of profilin-actin complexes to the (+) ends followed by dissociation of the profilin. Second, release of profilin from monomeric actin could be triggered by a rise in the intracellular pH or by an increase in the amount of the plasma membrane lipid phosphatidylinositol 4,5-biphosphate (see Figure 19-29); in cell-free reactions this lipid binds to profilin and causes it to dissociate from actin. Penetration of the jelly coat and vitelline layer by the acrosomal process is aided by hydrolytic enzymes exocytosed from the acrosomal vesicle. Following polymerization of actin microfilaments and the fusion of the sperm and egg plasma membranes, the sperm nucleus enters the egg and nuclear fusion occurs.

Platelet Activation Actin polymerization also occurs during activation of blood platelets in mammals. Platelets are ovoid fragments of cytoplasm, 2–5 μm in diameter, that circulate in blood and are derived from a very large cell called a megakaryocyte found in the bone marrow. They contain no nucleus but do have small internal vesicles. When a blood vessel is cut, platelets from the flowing blood bind to the vessel wall at the site of the rupture. These platelets then become "activated" and change shape, producing many long filopodia that extend outward. The vesicle membranes fuse with the plasma membrane, and because of the presence of new surface proteins, the platelets develop the capacity to adhere to one another. These remarkable shape changes allow platelets to form a plug over the cut in the vessel. In unactivated platelets, which contain no microfilaments, actin appears to occur mainly as monomers associated with profilin. After activation, profilin dissociates from monomeric actin, and microfilaments arranged as nets or bundles form throughout the platelet. The formation of this microfilament network causes the shape change of the platelets.

Movement of Amebas and Macrophages Involves Reversible Gel-Sol Transitions of an Actin Network

Because amebas measure up to 0.5 mm across—many times larger than vertebrate cells—they are frequently used to study cell movements. Amebas and vertebrate macrophage cells move by continually extending and re-

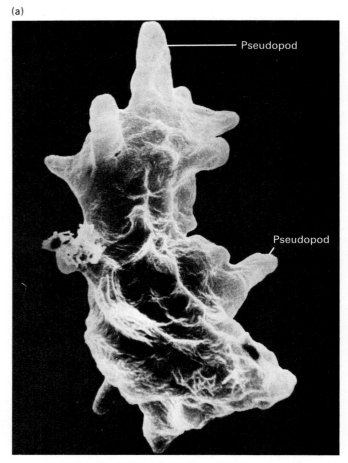

(a)

Pseudopod

Pseudopod

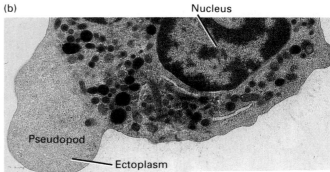

(b)

Nucleus

Pseudopod

Ectoplasm

▲ **Figure 22-37** (a) Scanning electron micrograph of the ameba *Proteus*. Note the numerous pseudopods. (b) EM of a thin section of a guinea pig leukocyte. Organelles and vesicles found elsewhere in the cytoplasm are not present in pseudopods. *Part (a) courtesy of G. Antipa; part (b) from D. Fawcett, 1981,* The Cell, *Saunders/Photo Researchers, Inc.*

tracting long *pseudopods* along the substratum (Figure 22-37a). This elongation-retraction process involves transitions of regions of the cytoplasm from a fluidlike state called a *sol* to a semisolid state called a *gel*.

Light and electron microscopy have revealed two regions of cytoplasm in these cells (Figure 22-37b). The central region—the *endoplasm*—contains abundant particles and organelles. As visualized in the phase-contrast microscope, these particles are in constant, random motion; this is indicative of their freedom of movement in this sol region of the cytoplasm. The *ectoplasm*—the region of the cytoplasm under the plasma membrane in the pseudopods—is gel-like. Ectoplasm contains a three-dimensional network of cross-linked actin fibers; all other organelles are excluded from this region. This gel region apparently determines the shape of the pseudopod and may transmit tension from regions of cellular contraction to the sites of contact with the substratum.

It is believed that the endoplasm contains non-cross-linked actin filaments. As a pesudopod elongates and the sol-like endoplasm streams into it, the region of endoplasm near the tip of the pesudopod apparently transforms into gel-like ectoplasm. Simultaneously, the ectoplasm elsewhere in the cell transforms into sol-like endoplasm, probably by removal of cross-links between actin filaments.

Filamin, an abundant cytoplasmic protein, links actin filaments together into a three-dimensional network with the physical properties of a semisolid gel. A dimer of two 270-kDa polypeptides, filamin is a flexible molecule 160 nm long and 3–5 nm in diameter. The actin-binding sites near the ends of filamin molecules can connect actin filaments, causing pure actin filaments in solution (a fluid consistency) to form a semisolid gel (Figure 22-38). Other abundant actin cross-linking proteins such as α-actinin may also be involved in inducing the sol-to-gel transition.

Other proteins cause disassembly of actin networks and thus induce the reverse gel-to-sol transition. One, *gelsolin,* has been purified from many cells. Gelsolin is a Ca^{2+}-triggered actin filament–severing protein. In the presence of Ca^{2+} concentrations $\geq 1\ \mu M$, gelsolin binds to the side of an actin filament, cuts it at that point, and remains attached to the (+) end (Figure 22-39). When it acts on actin filaments that are in a gel-like network, gelsolin converts a gel to a sol, hence the name "gel-solin." Gelsolin is related in sequence and function to villin, the actin-binding protein in brush border microvilli (see Figure 22-29). The major difference is that villin has two actin-binding domains and gelsolin only one; thus only villin can bundle actin filaments, but both sever actin filaments at micromolar concentrations of Ca^{2+} ions.

If amebas and other cells extended pseudopods in all directions, they would be ripped apart! Thus such cells must be able to control sol-gel transitions, although the mechanism(s) is not yet understood. Since the ability of many proteins to cross-link actin fibers is strongly dependent on both Ca^{2+} concentration and pH, differences in Ca^{2+} and H^+ concentrations in various regions of the cytosol may regulate the sol-to-gel transitions. Indeed, isolated ameba cytoplasm gels when the pH is lowered to 6.8 in the presence of submicromolar concentrations of Ca^{2+}; conversely, solution of the gel is induced by raising the pH or Ca^{2+} concentration. These findings implicate the involvement of gelsolin in the gel-to-sol transition,

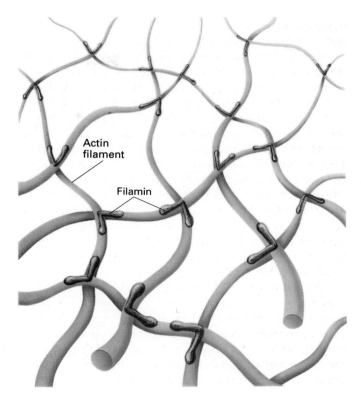

▲ **Figure 22-38** Dimeric filamin cross-links actin filaments by forming perpendicular branches, thus condensing soluble actin filaments into a three-dimensional network with the physical properties of a gel. Similar to other actin cross-linking proteins (e.g., α-actinin and αβ-spectrin), filamin has an actin-binding site at each end of the molecule.

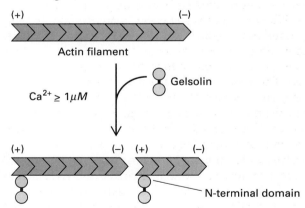

▲ **Figure 22-39** Function of the actin-severing protein gelsolin. Gelsolin contains several segments homologous to those in villin but has only one actin-binding domain, at the N-terminal end, whereas villin has two actin-binding domains, one at each end. At micromolar concentrations of Ca^{2+} ions, both proteins undergo a conformational change that permits each to bind to and induce a break in actin filaments; after cutting a filament, the N-terminal domain of villin or gelsolin remains bound to the (+) end of the actin fragment and stabilizes it. Villin, unlike gelsolin, can bind to and bundle actin filaments together at low concentrations of Ca^{2+} ions (see Figure 22-29). *Adapted from P. Matsudaira and P. Janmey, 1988,* Cell *54:139.*

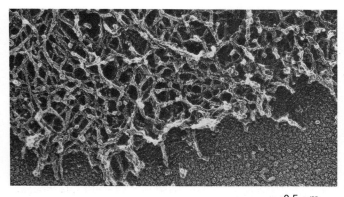

▲ **Figure 22-40** Electron micrograph of the cytoskeleton beneath a lamellipodium. A fibroblast was treated with the detergent Triton X-100 to dissolve the plasma membrane, and the cell contents were then fixed. A web of actin filaments underlies the plasma membrane in the region of the lamellipodium; these filaments are much more concentrated and interdigitated than filaments in other regions of the cell. *From J. Heuser and M. Kirschner, 1980,* J. Cell Biol. *86:212. Reproduced from the* Journal of Cell Biology *by copyright permission of The Rockefeller University Press.*

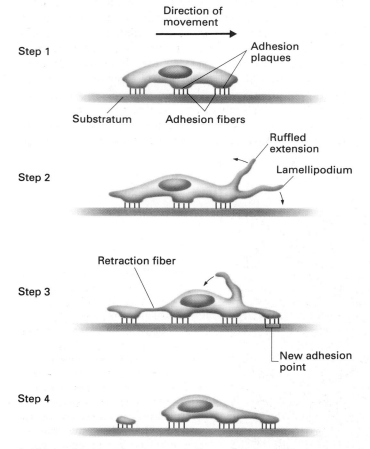

▲ **Figure 22-41** Diagrammatic illustration of movement by a cultured fibroblast along a substratum. See text for discussion.

since it severs actin filaments in the presence of micromolar Ca^{2+}. Whether differences in Ca^{2+} or H^+ concentration among various regions of the cytoplasm do exist, however, remains to be determined.

Movements of Fibroblasts and Nerve Growth Cones Involve Controlled Polymerization and Rearrangements of Actin Filaments

The movements of many animal cells involve rapid extensions and retractions of regions of the plasma membrane that contain bundles of actin filaments. Movements of the growth cone at the tip of an elongating nerve axon are one of the best-studied systems; very similar processes are involved in locomotion of fibroblasts along a substrate.

Fibroblast Motility Cell motility has been studied extensively with vertebrate fibroblasts in tissue culture. Fibroblasts move jerkily along a substratum at rates up to 40 μm/h by making flattened extensions of the cell called *lamellipodia* (see Figure 4-22). Lamellipodia can be broad but are very thin—only 0.1–0.4 μm thick—and contain no organelles other than a meshwork of actin-containing microfilaments (Figure 22-40).

Fibroblasts adhere to the substratum at discrete sites on their undersurface termed *adhesion plaques* (Figure 22-41, step 1); these sites may be points of contact between actin filaments and the membrane. As discussed later, actin-binding proteins are present in these adhesion plaques. During movement, lamellipodia that are rich in actin fibers are extended from the leading edge of the cell; some adhere to the substratum, while others move backward over the top in a process called *ruffling* (step 2). As the fibroblast moves forward, the trailing edge remains attached to the substratum, and the tail—called the *retraction fiber*—becomes greatly elongated under the resulting tension (step 3). At the same time, the ruffled extensions on the dorsal surface of the cell move backward and collapse. Eventually, the tail ruptures, leaving a bit of itself attached to the substratum; the major part of the tail retracts into the cell body (step 4).

Ultrastructural studies indicate that the tail portion of migrating fibroblasts contains a bundle of actin-containing microfilaments with its axis parallel to the long axis of the tail. The retraction of the trailing end of the fibroblast is associated with the loss of actin filament bundles and their subsequent replacement by a meshwork of actin filaments. Retraction is dependent on a supply of ATP and appears to be an active contraction process involving the actin filament meshwork.

Nerve Growth-Cone Mobility During differentiation of the nervous system, *neuroblasts*—precursors of neurons that generally have no axonal or dendritic pro-

(a)

(b)

(c)

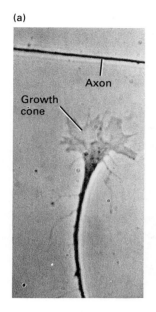
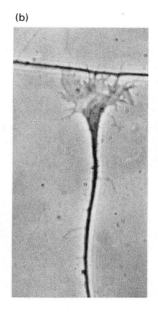
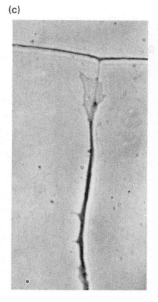

◀ **Figure 22-42** Elongation and retraction of the growth cone of a cultured chick retinal neuron. Led by its growth cone, the retinal neuron elongates upward (a) until it encounters the axon of a cultured sympathetic neuron (b). This causes retraction of the growth cone and inhibition of axon elongation (c). Such repulsive interaction of the growth cone with other cells or with the matrix, together with adhesive interactions, is thought to determine the final shape and synaptic connections of a neuron. *From J. P. Kapfhammer and J. A. Raper, 1987,* J. Neurosci. *7:201.*

jections—are produced at particular locations at specific developmental stages. Certain neuroblasts, such as those in the neural crest, next migrate to specific destinations. At its final site, the cell "sprouts" and elaborates one or more axonal or dendritic projections. At the leading edge of the elongating axon is a highly motile structure known as the *growth cone* (Figures 22-42 and 22-43). The growth cone is the site where axon elongation occurs. More importantly, the ability of the growth cone to explore surfaces and crawl over them, and to navigate the axon in a particular direction, enables an axon to reach its proper target cells.

Growth cones exhibit two types of movement. One involves thin, flattened lamellipodia, or lamellae, where the cell margin spreads over the substratum. The other involves spikelike *filopodia*, which can be up to 50 μm long and contain bundles of actin filaments (Figure 22-44). Filopodia are constantly extending and retreating, exploring their environment. A filopodium may adhere to the substratum at discrete sites, and some of the expanding regions of lamellopodia may also adhere to the substratum, pulling the cell in a particular direction. Some extensions do not adhere but move backward across the top of the growth cone in a process known as ruffling.

(a)

(b)

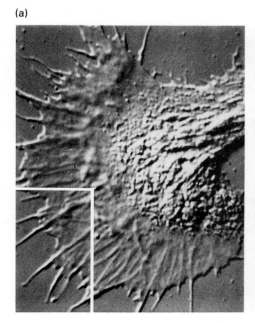
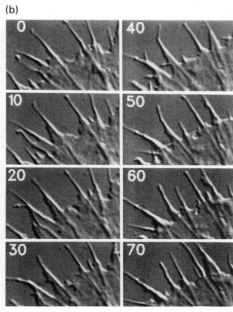

◀ **Figure 22-43** Movement of growth cones in a cultured *Aplysia* neuron. (a) Video-enhanced image of the entire growth cone. (b) A series of images taken every 10 s of the region boxed in (a). Spikelike filopodia are constantly protruding outward from the leading edge of the growth cone and then retracting. Filopodia or lamellopodia form contacts with certain cells or matrix components that cause the growth cone to extend in a particular direction. *From S. J. Smith, 1988,* Science *242:708; courtesy of Dr. P. Forscher.*

(a)

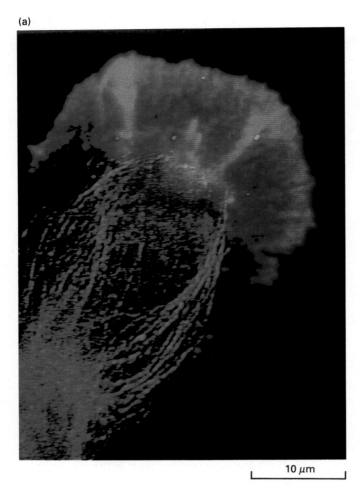

10 μm

(b)

Lamellipodium

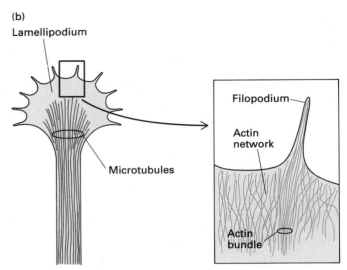

Microtubules

Filopodium

Actin network

Actin bundle

▲ **Figure 22-44**   The structure of a growth cone. (a) The growth cone of an *Aplesia* neuron stained with a rhodamine-tagged phalloidin (an actin-binding peptide) and fluorescine-labeled anti-tubulin antibody. The two fluorescent images were stored in a computer, and an image generated in which microtubules are colored red and actin blue. (b) Diagrammatic structure of the cytoskeleton in a growth cone. Microtubules extend the length of the axon but end as the axon flattens into the lamellipodium of the growth cone. *Inset:* The edges of a lamellipodium are occupied by networks and fibers of actin, as well as by myosin and other actin-binding proteins. *Part (a) from P. Forscher and S. J. Smith, 1988, J. Cell Biol. **107**:1505; courtesy of Dr. P. Forscher. Part (b) from S. J. Smith, 1988, Science **242**:708.*

Precisely how lamellipodia and filopodia are extended is not known. As depicted in Figure 22-45, the process may involve the controlled polymerization of actin at the leading edges of a filopodium or lamellopodium. Alternatively, elongation or retraction of filopodia may be caused by minimyosins that link the sides of actin filaments to the plasma membrane and which are localized to lamellipodia of locomoting cells. Minimyosins (type I myosin) have the globular actin-binding heads characteristic of other myosins and have a short tail that can bind to membranes; they lack the long coiled α-helical tail.

As the growth cone moves outward, the cell body stays put and the axon elongates; in part, elongation is caused by polymerization of tubulin into microtubules that give the axon its rigidity. Microtubules do not extend into the growth cone, and the actin network does not extend into the axon (see Figure 22-44).

Inhibition of Fibroblast and Growth-Cone Movement   Evidence that actin filaments are involved in various types of nonmuscle cell movements has come from studies with a series of fungal products called *cyto-chalasins*. These compounds cause intact fibroblasts to "round up" and to depolymerize their actin filament bundles; they also inhibit the projections of the cell-surface

membrane associated with movement on a substratum and block formation of filopodia from nerve growth cones.

In cell-free extracts, cytochalasins specifically block the polymerization of actin filaments by binding to an end of a growing filament. Because depolymerization is not affected by cytochalasins, the initial equilibrium between actin monomers and filaments eventually is displaced toward a preponderance of actin monomers. There is a good correlation between the effect of different cytochalasins on in vitro actin polymerization and their in vivo effect on fibroblasts. These findings thus provide circumstantial evidence for the role of actin polymerization in fibroblast and nerve growth-cone movement.

Actin Stress Fibers Permit Cultured Cells to Attach to Surfaces

In cultured fibroblasts and in certain nonmuscle cells such as the endothelial cells that line arteries, much of the actin occurs in long bundles of filaments termed *stress fibers*, which often extend the length of the cell. These fibers can be visualized in the light microscope following staining of fixed cells with fluorescent antibodies to actin or with phalloidin (see Figure 22-1a). However, only about half of the actin in these cells appears to be in microfilaments;

(a)

Filopodium plasma membrane

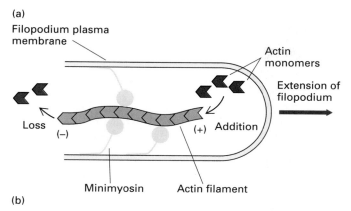

Actin monomers

Extension of filopodium

Loss

(−)

(+) Addition

Minimyosin   Actin filament

(b)

(1) Minimyosin movement along anchored, stationary actin filaments powers extension

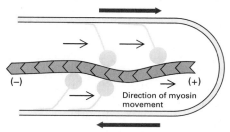

(−)   (+)

Direction of myosin movement

(2) Minimyosin interaction with unanchored actin filaments powers retraction

▲ **Figure 22-45** Models of filopodial extension and retraction. Actin filaments are thought to be linked to the filopodial plasma membrane by 100-kDa minimyosins (Myosin I). (a) According to one model, addition of actin monomers to the (+) ends and their loss from the (−) ends of the actin filaments pushes the filopodial membrane outward (to the right). (b) According to a second model, extension and retraction are powered by movement of the minimyosins. If the actin filaments were anchored to other cytoskeletal fibers, movement of the minimyosins anchored in the plasma membrane toward the (+) ends of the actin filaments would cause extension of the filopodial plasma membrane (1). Alternatively, if the actin filaments were not anchored in place but the minimyosins were rigidly anchored in the plasma membrane, minimyosin action would cause inward movement of filaments and retraction of the filopodium (2).

(a)   (b)

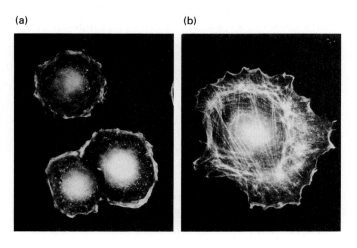

the rest is monomeric actin. Thus there may be interconversion of microfilaments with actin monomers. It is important to realize that each fluorescent fibrous structure visible with this technique is actually a bundle of parallel actin filaments—not the single 7-nm-diameter filament resolved in the electron microscope.

Stress fibers are believed to be involved in the attachment of cultured cells to a substratum and also in generation of the stress or tension that determines the flattened shape of fibroblasts. Electron microscopy has shown that these fibers contain parallel bundles of actin filaments—apparently with different polarities. Stress fibers are generally located in the region of cytoplasm just inside the plasma membrane; many appear to insert an end into the membrane at points of cell-substratum contact. In studies using fluorescent-antibody techniques, stress fibers have been shown to contain myosin, α-actinin, and tropomyosin intermittently along their length; they also contain the regulatory protein myosin light-chain kinase. Most scientists now believe that actin and myosin in stress fibers form minisarcomeres similar to those in the circumferential belt in epithelia (see Figure 13-43) and in the mitotic contractile ring; α-actinin may bundle individual filaments in a fiber together.

Several studies suggest that stress fibers are potentially contractile. For instance, addition of ATP to detergent-permeabilized cells causes the stress fibers to shorten. However, it is not clear that movement is their in vivo function. Most workers believe that stress fibers function in adhesion of cells to the substratum; in the case of endothelial cells lining arteries, they provide the cell layer with strength and resiliency.

Fibroblasts that are migrating out of tissue explants have few stress fibers during the time of rapid movement. However, when these cells settle on the substratum and migration ceases, the stress fibers increase in number. This interpretation of stress fiber function is supported by the observation that when cultured fibroblasts are removed from their substratum, the cell becomes spherical and the stress fibers disappear, leaving a random mesh of microfilaments (Figure 22-46a). Within a few hours after the cell settles back on its substratum, the stress fibers reappear (Figure 22-46b). Apparently, adhesion of cells

◀ **Figure 22-46** Redistribution of actin in spreading fibroblast cells during change in shape. (a) Cells become spherical immediately after their detachment from the surface of a culture dish by treatment with trypsin. As detected by fluorescent antibodies, actin is found in ruffles at the perimeter of the cell and diffusely over the body of the cell. There is no fibrillar actin at this stage. (b) After 5 h of spreading on a culture dish, the cell has flattened and contains polygonal arrays of actin at the periphery; stress fibers also crisscross the cell. [See R. Hynes and A. T. Destree, 1978, *Cell* 15:875.] *Courtesy of R. Hynes and A. T. Destree. Copyright 1978, M.I.T.*

to a substratum somehow induces conversion of the random assortment of actin microfilaments into microfilament bundles.

Intermediate Filaments

We turn now to the third class of cytoplasmic filaments—the 8- to 12-nm-diameter *intermediate filaments*. First isolated from muscle, intermediate filaments are thicker than actin thin filaments and thinner than myosin thick filaments. Intermediate filaments are now thought to be the principal structural determinants in many metazoan cells and tissues: they connect the spot desmosomes of epithelial cells and stabilize the epithelium; form the major structural proteins of skin and hair; appear to form the scaffold that holds the Z disks and the myofibrils in place in muscle cells; and give strength and rigidity to nerve axons. However, many basic aspects of intermediate filament (IF) function and architecture, such as the control of IF assembly within cells, have yet to be elucidated.

Intermediate filaments differ in a number of fundamental ways from the two other major cytoskeletal systems, microtubules and microfilaments. First, several classes of cell-specific IF subunit proteins exist, whereas actin and tubulin are both widely distributed. Second, both microtubules and microfilaments are dynamic polymers, which in most cases exist in dynamic equilibrium with a substantial pool of monomeric subunit protein. Because the assembly of these filaments involves monomer-binding proteins and the hydrolysis of nucleotide triphosphates, the dynamics of polymerization are complex. When a cells is lysed, the bulk of the microtubules and microfilaments are solubilized. In contrast, more than 99 percent of the IF subunit proteins are found in the polymerized form, which remains insoluble under a wide range of physiologic conditions. In fact, the predominant proteins remaining after a cultured fibroblast is extracted with a nonionic detergent in a 0.5 M salt solution are intermediate filaments. In addition, no energy appears to be necessary for assembly of intermediate filaments, so their polymerization dynamics, although as yet poorly understood, are expected to be relatively simple.

Finally, both intermediate filaments themselves and their subunit proteins differ from the protein subunits of microtubules and actin filaments. Actin and tubulin are globular proteins, and the polymers they form are rather like beads on a string or bricks in a wall. In contrast, IF subunit proteins are extended molecules, which form ropelike polymers. After their complete denaturation (by treatment with urea or strong detergents), IF subunit proteins will renature and form, upon removal of the denaturant, intermediate filaments that are indistinguishable from native filaments.

Different Intermediate Filament Proteins Are Expressed in Different Cell Types

The insolubility of intermediate filaments and the ability of their subunit proteins to reform filaments after denaturation have been exploited to purify and identify IF proteins. Studies with specific antibodies against these proteins and more recently sequencing of their mRNAs have led to the realization that although ultrastructurally similar, the intermediate filaments of different cell types are composed of distinct proteins. In higher vertebrates five major classes of IF subunit protein have been identified in this way; each IF protein is characteristically expressed in a single cell type.

Vimentin (MW 57,000) is expressed typically in mesenchymal cells such as fibroblasts and in blood vessel endothelial cells, but it has also been found in certain epithelia. Vimentin fibers often terminate at the nuclear membrane and at desmosomes or adhesion plaques on the plasma membrane. Frequently they are closely associated with microtubules (Figure 22-47a and b). They may function to keep the nucleus or other organelles in a defined place within the cell. In the adipose (fat) cell, another mesenchymal cell, vimentin filaments form a cage around lipid droplets, probably preventing them from fusing with one another or with cellular membranes (Figure 22-47c).

Desmin (MW 55,000) filaments are found predominantly in muscle cells of all types, specifically in the periphery of the Z disks in striated muscle. Desmin forms an interconnecting network across each muscle cell, perpendicular to the long axis of the cell. Desmin-containing fibers anchor and orient all of the neighboring Z disks so that all myofibrils in a muscle cell are in register, generating the striated pattern of these large cells. Desmin fibers may also link Z disks to the plasma membrane or to other cytoplasmic organelles.

Neurofilaments, present in axons of both central and peripheral neurons in vertebrates, are built of three discrete neurofilament polypeptides—NF_l, NF_m, and NF_h, with molecular weights of 70,000, 150,000, and 210,000, respectively; frequently they appear to be in close association with the axonal microtubules. The strength and rigidity of the axon may be due to these neurofilament-microtubule complexes. Microtubules and neurofilaments appear to grow by addition of new subunits in the cell body. They move together down the axon at a rate of about 1 mm/day; microtubules and intermediate filaments are thus the slowest-moving components of axonal transport. Molecular cloning of neuronal cDNAs has led to the identification of another NF protein (MW 57,000) that is expressed in many central and peripheral neurons; probably, additional NF proteins and their genes remain to be discovered.

Glial fibrillary acidic protein (GFAP), which has a molecular weight of 50,000, forms the intermediate fila-

(a) Cultured fibroblasts

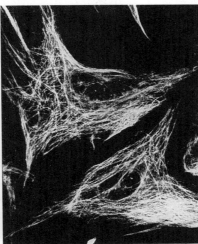

Tubulin distribution  Vimentin distribution

(b) Adipocyte

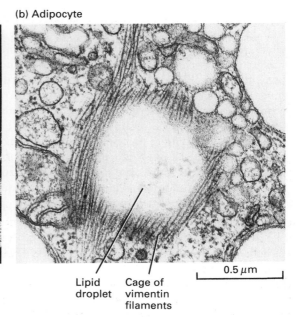

Lipid droplet   Cage of vimentin filaments   0.5 μm

▲ **Figure 22-47** (a) The distribution of vimentin (*right*), an intermediate filament protein, and of tubulin (*left*) in the same cultured fibroblasts, as detected by fluorescent vimentin-specific and tubulin-specific antibodies. Many vimentin fibers are very close to (or part of) microtubules, suggesting a close association between the two filament networks.

(b) Electron micrograph of a section of an adipocyte, showing a cage of parallel vimentin filaments surrounding the "hollow" lipid droplet. *Part (a) from M. Klymkowsky, 1988, Exp. Cell Res.* **174**:282; *courtesy of M. Klymkowsky; part (b) from W. W. Franke, M. Hergt, and C. Grund, 1987, Cell* **46**:131; *courtesy of W. W. Franke.*

ments in glial cells surrounding neurons. This IF protein is expressed in some types of glial cells but in no other cell type.

Vimentin, desmin, GFAP, and the various NF proteins all appear to be encoded by a single gene per haploid genome. In contrast, there are at least 30 distinct *cytokeratins*, the fifth class of IF subunit proteins. The cytokeratins have been divided into two classes: acidic (type I) and neutral/basic (type II). Cytokeratins are typically expressed in epithelial cells, and each type of epithelium has its characteristic complement of cytokeratins. Epithelial cells always express multiple cytokeratins, and keratin filaments are always formed of equal numbers of subunits of the type I and type II superfamilies. About 10 cytokeratins are specific for "hard" tissues, such as nails, hair, and wool (of sheep); about 20 are more generally found in the epithelia that line internal body cavities.

IF proteins are among the most abundant of the cell-type–specific proteins. Fluorescent antibodies to particular IF proteins have been very useful in typing of cells, especially for tumor classification in cases in which diagnosis cannot be made by normal histologic procedures. Because tumor cells retain many of the differentiated properties of the cells from which they are derived, their tagging by IF antibodies allows identification of cell type. For example, the most common carcinomas of the breast

and gastrointestinal tract contain cytokeratins and lack vimentin; thus they are derived from epithelial cells (which contain keratins) rather than from the underlying stromal mesenchymal cells (which contain vimentin but not keratins). Such determinations are often important in the selection of treatment.

Rapidly growing cells (e.g., cultured cells and cells in early mammalian embryos), mammalian erythrocytes, and the myelin-synthesizing glial cells in the central nervous system do not contain intermediate filaments. Cultured fibroblasts grow and divide quite normally if the IF network is disrupted by microinjection of an antibody against vimentin. Presumably, then, IF proteins carry out specific functions of differentiated cells, such as keeping the nucleus and other organelles in their proper positions. In this sense, intermediate filaments have a more subtle role in the regulation of cell processes than do the ubiquitous actins and tubulins.

All Intermediate Filaments and Their Subunit Proteins Have a Similar Structure

All IF subunit proteins share a common structural motif, consisting of a subunit-specific N-terminal domain of variable extent, an approximately 40-nm central "rod" domain, and a second subunit-specific C-terminal domain, also of variable size (Figure 22-48a). The central rod domain contains long stretches of heptad repeats, which favor the formation of α-helical coiled-coil dimers similar to those of tropomyosin and myosin. In IF proteins, the central heptad-repeat domain is interrupted by three nonhelical regions; the positions of these "spacer" elements is highly conserved among all IF subunit proteins. Furthermore, the position of introns in all but one of the genes encoding IF proteins is highly conserved, suggesting that they all evolved from a single primordial IF gene. Since nuclear lamins are found in all metazoan cells and are highly homologous in structure and sequence to cytosolic IF proteins, it is possible that the IF proteins evolved from a laminlike precursor.

Vimentin, desmin, GFAP, and NF$_l$ are able to form homopolymeric intermediate filaments (Figure 22-48b).

(a)

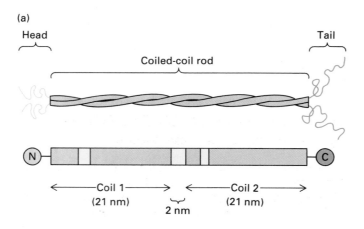

(b)

100 nm

▲ **Figure 22-48** (a) A common structural model for IF-subunit proteins. These dimeric proteins have a central α-helical coiled-coil domain (green), which is highly conserved, and nonhelical tails and heads (blue), which are variable in length and sequence. The central rod contains three nonhelical spacer elements (yellow). Dimeric IF subunits spontaneously polymerize into filaments. (b) Electron micrograph of a negatively stained preparation of desmin intermediate filaments. Desmin subunits were solubilized in a 6 *M* urea solution; filaments formed spontaneously when urea was removed by dialysis. Note the 21-nm repeats (*arrow*), which correspond to coil 1 and coil 2 indicated in part (a). *Part (a) adapted from M. Osborn and K. Weber, 1986,* Trends Biochem. Sci. *11:469; part (b) courtesy of U. Aebi.*

In contrast, cytokeratins must polymerize in pairs of type I and type II cytokeratins. When proteins of the homopolymeric group such as desmin and vimentin are coexpressed in the same cell, they appear to copolymerize. When, as in epithelial cells, cytokeratins are coexpressed with vimentin, the two form distinct filament systems. Among the cytokeratins, there is little evidence for selectivity in polymerization, so that within the cell all of the cytokeratins expressed form a single, apparently homogeneous filament system.

The assembly of IF subunit dimers is presumably driven by association of the central rod domains, and the two monomeric polypeptides are oriented in parallel. Two dimers assemble into a tetramer about 70 nm long (Figure 22-49). By immunostaining tetramers with an antibody specific for the globular C-terminal domain, it was shown that the two dimers in a tetramer "point" in opposite directions; that is, they are antiparallel. Tetramers bind end-to-end to form protofilaments, and eight protofilaments form a single intermediate filament 10 nm in diameter.

The *antiparallel* organization of IF tetramers implies that overall an intermediate filament is apolar. This distinguishes intermediate filaments from both microtubules and microfilaments, which each possess a clear structural directionality. In contrast to microfilaments and microtubules, intermediate filaments are expected to have the same polymerization properties at each end. The significance of this apolarity of intermediate filaments is not clear.

Although most vimentin in cells is present in intermediate filaments, about 1–5 percent of the vimentin in cultured fibroblasts exists as tetramers. The absence of a pool of monomers or dimers suggests that vimentin monomers rapidly form dimers, which rapidly form tetramers; these may exist in a steady state with polymerized filaments. Alternatively, tetramers may associate with some other component(s) of the cell, rendering them nonpolymerizable.

There is evidence that the N-terminal domain (the head) plays a role in assembly of intermediate filaments. Phosphorylation of certain sites in the N-terminal region of vimentin and desmin appears to block polymerization. In contrast, the experimental proteolytic removal of the C-terminal domain (the tail) has relatively little effect on filament assembly.

Intermediate Filaments Are Often Associated with the Cell Nucleus and with Microtubules

Within both lymphocytes and cultured fibroblasts, intermediate filaments are typically organized in an extended system that stretches from the nuclear envelope to the

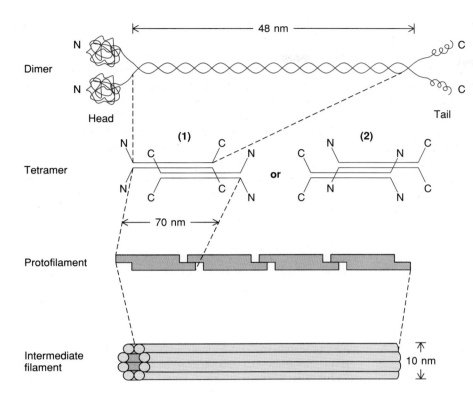

◀ Figure 22-49 Levels of organization and assembly of intermediate filaments. The primary structural unit is a coiled-coil dimer of two polypeptides (see Figure 22-48a). A tetramer is formed by antiparallel, staggered side-by-side aggregation of two dimers; it is not known whether the "protruding" ends are heads (1) or tails (2). Tetramers aggregate end-to-end to form a protofilament, and eight protofilaments form a cylindrical, 10-nm-thick filament. *Courtesy of M. Klymkowsky.*

plasma membrane. A number of studies have suggested that vimentin intermediate filaments are actually physically linked to the nucleus. After experimental enucleation, however, many and sometimes most such filaments remain in the cytoplasmic fragment; thus not all vimentin filaments are tightly associated with the nucleus.

In mitotic cells, intermediate filaments form a cage that surrounds bundles of microtubules in the mitotic spindle. Fibrous bridges of unknown composition appear to connect the two types of filaments in mitotic cells and in nerve axons (see Figure 21-4). In fibroblasts, microtubules are associated with most vimentin filaments (see Figure 22-47b). Treatment of fibroblasts with colchicine, an inhibitor of polymerization of tubulin, causes the complete dissolution of microtubules after a period of several hours. Vimentin filaments remain in such colchicine-treated cells, although they become clumped in bundles near the nucleus. These and other results suggest that the organization of vimentin filaments is determined in some way by the lattice of microtubules, possibly by protein connectors between the two types of filaments.

Some workers have proposed the existence of an IF-organizing center akin to the microtubule-organizing center. However, in experiments in which mRNA encoding bovine cytokeratins was injected into a vimentin-expressing fibroblast cell line, cytokeratin filament assembly began at multiple sites scattered throughout the cell, with little obvious association with the nucleus or with vimentin filaments. Thus IF assembly can occur at multiple sites in a cell; we do not know how this process is regulated.

Intermediate Filaments Stabilize Epithelia by Connecting Spot Desmosomes

Desmosomes are thickened regions of the plasma membrane where cells are tightly attached to their neighbors. They increase the rigidity of tissues such as epithelia by holding the cells securely together, and they act as anchorages for intracellular fibers. There are several very different types of desmosomes (see Figure 13-45). These junctions have only recently been purified, and little is yet known of their composition or specific protein interactions. Thus the description of desmosomes must be based mainly on what has been revealed by electron microscopy.

Spot desmosomes are buttonlike points of contact between cells; they are often thought of as a rivetlike attachment of adjacent plasma membranes. At a spot desmosome, a proteinaceous cytoplasmic plaque, 15–20 nm thick, is bound to the cytoplasmic face of the plasma membranes (Figure 22-50). Fibrous transmembrane linker proteins extend from these into the intracellular space, where the fibers from the two cells form an interlocking network that binds the two together. In epithelial cells, keratin intermediate filaments course into and out of the cytoplasmic plaques; some of these filaments run parallel to the cell surface, and others penetrate and traverse the cytoplasm. They are thought to be part of the internal structural framework of the cell, giving it shape and rigidity. If so, spot desmosomes also could transmit shearing forces from one region of a cell layer to the epi-

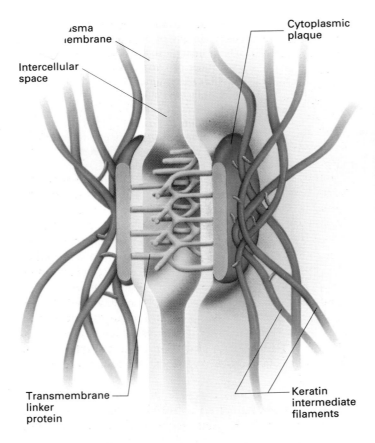

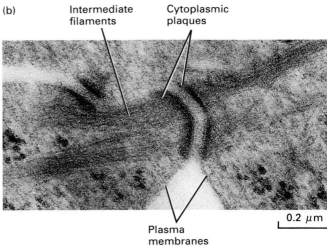

▲ **Figure 22-50** (a) Structure of a spot desmosome between epithelial cells. Keratin intermediate filaments form a tensile network that crisscrosses the interior of the cell and attaches to the cytoplasmic plaque. Thinner transmembrane proteins originating within the plaques extend through the plasma membrane and connect with one another in a staggered configuration, thus linking the two cytoplasmic plaques and the intermediate filament networks of the two cells. The junction allows shearing stresses to dissipate throughout the tissue. (b) Electron micrograph of a thin section of a spot desmosome connecting two cultured differentiated human keratinocytes. Bundles of intermediate filaments radiate from the two darkly staining cytoplasmic plaques that line the inner surfaces of adjacent cell plasma membranes. *Part (a) courtesy of B. E. Hull and L. A. Staehelin; part (b) courtesy of R. van Buskirk.*

thelium as a whole; they thus provide strength and rigidity to the entire epithelium cell layer.

Cardiac (heart) muscle cells are also interconnected by spot desmosomes. Desmin filaments interconnect these desmosomes, allowing the stress and strain of the contractile force of one muscle cell to be transmitted to the others. In smooth muscle, desmin filaments are associated with the "dense bodies," structures that appear to be functionally equivalent to the Z disks of skeletal muscle.

Hemidesmosomes are morphologically similar to spot desmosomes in that intermediate filaments are attached to cytoplasmic plaques on the plasma membranes. Hemidesmosomes are found in regions of epithelial cells in contact with the basal lamina, and it is thought that they anchor extracellular protein networks to the cell.

Summary

The dumbbell-shaped actin monomer reversibly polymerizes into long helical filaments; under usual intracellular conditions most actin is in polymers. Type II myosin, in contrast, is a bipolar, fibrous molecule with heads that bind actin. Cell-free mobility assays show that single myosin heads can use energy released by ATP hydrolysis to migrate along actin filaments.

The role of actin filaments is best understood in muscle cells. In striated muscle, the (+) ends of actin filaments are anchored to both sides of the Z disks. Actin filaments interdigitate with bipolar thick myosin filaments, forming the regular array of a sarcomere, the unit of contraction. The flexible heads of the myosin chains, energized by hydrolysis of ATP, move the adjacent actin filaments toward the center of the myosin filament, causing contraction. Binding of myosin, containing ADP and P_i, to actin, and hence contraction of muscle, is triggered by a rise in the concentration of cytosolic Ca^{2+} ions. The release of Ca^{2+} ions stored in the lumen of sarcoplasmic reticulum vesicles is triggered by depolarization of the surface membrane of the muscle cell that spreads into the T-tubule invaginations.

In striated muscle, Ca^{2+} induces contraction by binding to one of the three troponins, a set of proteins attached to the fibrous protein tropomyosin, which lies in the groove in the actin filaments. Binding of Ca^{2+} ions, in turn, alters the conformation by which tropomyosin is bound to actin subunits in the filaments, so that the myosin heads can interact with the actin. In invertebrate muscle, Ca^{2+} acts by binding directly to myosin light chains located in the myosin heads. In vertebrate smooth muscle, Ca^{2+} ions bind to calmodulin; the Ca^{2+}-calmodulin complex can stimulate activity of myosin light-chain kinase, which catalyzes phosphorylation of one of the two light-chain pairs. This phosphorylation triggers assembly of bipolar myosin thick filaments, myosin-actin binding, and subsequent contraction. The Ca^{2+}-calmodulin complex also causes release of caldesmon from actin filaments, allowing myosin heads to bind. Different smooth and striated muscles possess different isoforms of proteins such as tropomyosin and myosin light chains; often these isoforms are produced by multiple alternate splicing of transcripts encoded by a single gene.

Muscles contain α-actinin, which participates in binding actin filaments to the Z disk or, together with vinculin, to the equivalent structure in smooth muscle. Other long fibrous proteins, such as titin and nebulin, keep the myosin and actin filaments centered in the sarcomere. Dystrophin, an essential minor muscle protein of unknown function, was detected through analysis of a fatal genetic disease.

Actin filaments perform mainly structural roles in certain nonmuscle cells, such as the microvilli in intestinal epithelial cells. Actin filaments form the major structures in the core of microvilli. Several types of actin-binding proteins cross-link the microfilaments to one another and to the microvillar membrane. Myosin, tropomyosin, and other elements of the contractile apparatus are lacking in microvilli. A spectrinlike protein links the microfilament rootlets of adjacent microvilli to each other, forming much of the terminal web of filaments that crisscross the cell just below the microvilli. At this level actin filaments also encircle the cell just inside of the plasma membrane and interact with the belt desmosome integral protein uvomorulin, which links together adjacent cells.

Actin-myosin interactions are essential for at least two types of movements within nonmuscle cells. Contraction of a ring of actin and myosin filaments causes cytokinesis—the final separation of the daughter cells during mitosis. Cytoplasmic streaming in large algae is powered by movement of myosin, bound to sheets of endoplasmic reticulum, along fixed parallel bundles of subcortical actin filaments.

The polymerization of actin filaments in nonmuscle cells is highly regulated. Profilin binds to monomeric actin, preventing its premature assembly into filaments, as in platelets or the acrosome of unstimulated sea urchin sperm. Formation of a three-dimensional semisolid gel of actin is favored by cross-linking proteins such as filamin and α-actinin. Conversion of actin gels to sols (e.g., in moving amebas) is induced by Ca^{2+}-regulated proteins that sever actin filaments; these include gelsolin and villin. Controlled polymerization of actin filaments and their interactions with plasma membrane–attached minimyosins (type I myosin) are essential for motility of cultured fibroblasts and for movement of nerve growth cones. Stress fibers containing actin, α-actinin, and myosin form when certain motile cells attach rigidly to a substratum; these filaments participate in cell-matrix attachments and probably in determining the flattened shape of attached cells.

Intermediate filaments are, in many cases, key determinants of cellular structure and comprise at least five subclasses. Desmin filaments connect adjacent Z disks in muscle cells. Keratin filaments in epithelial cells interact with spot desmosomes that interconnect adjacent cells; these junctions give strength and rigidity to the epithelium, and the keratin fibers allow stresses to be transmitted from cell to cell. Related keratin proteins are the major structural proteins of skin, hair, and wool. Axons of nerve cells contain long neurofilaments that appear to be a major determinant of their elongated shape. Other classes of IF proteins are found in other cell types—glial filaments in glial cells and vimentin in fibroblasts and adipocytes.

All IF proteins have a similar structure—a highly conserved, 40-nm-long central rodlike domain, which is formed by coiling α-helical sections of two polypeptides around each other; this coiled-coil dimeric structure also occurs in tropomyosin and in the helical tails of type II myosin. The N- and C-terminal domains are globular and vary widely among IF proteins. Dimers form antiparallel tetramers, which, in turn, form long protofilaments; eight of these form a 10-nm-diameter intermediate filament. In contrast to microtubules and microfilaments, intermediate filaments are apolar, and most of the protein subunits are always in filaments. Intermediate filaments often are found in association with microtubules. Although intermediate filaments have no obvious function for growth of cultured fibroblasts, they do play a key role in positioning cellular organelles and in organizing the cytoplasm of differentiated cells such as muscle.

References

Structure and Assembly of Actin and Myosin Filaments

*AMOS, L. A. 1985. Structure of muscle filaments studied by electron microscopy. *Ann. Rev. Biophys. Chem.* **14**:291–313.

BULLITT, E.S.A., D. J. DEROSIER, L. M. COLUCCIO, and L. G. TILNEY. 1988. Three-dimensional reconstruction of an actin bundle. *J. Cell Biol.* **107**:597–611.

*A book or review article that provides a survey of the topic.

*COHEN, C., and D. A. D. PARRY. 1986. α-Helical coiled coils—a widespread motif in proteins. *Trends Biochem. Sci.* 11:245–248.

*DAVIS, J. S. 1988. Assembly processes in vertebrate skeletal thick filament formation. *Ann. Rev. Biophys. Biophys. Chem.* 17:217–239.

*KORN, E. D., M-F. CARLIER, and D. PANTALONI. 1987. Actin polymerization and ATP hydrolysis. *Science* 238:638–644.

MATSUDAIRA, P., J. BORDAS, and M. H. J. KOCH. 1987. Synchrotron x-ray diffraction studies of actin structure during polymerization. *Proc. Nat'l Acad. Sci. USA* 84:3151–3155.

TOKUNAGA, M., K. SUTOH, C. TOYOSHIMA, and T. WAKABAYASHI. 1987. Location of the ATPase site of myosin determined by three-dimensional electron microscopy. *Nature* 329:635-638.

Myosin and Movements Along Actin Filaments

ADAMS, R. J., and T. D. POLLARD. 1986. Propulsion of organelles isolated from *Acanthamoeba* along actin filaments by myosin-I. *Nature* 322:754–756.

ADAMS, R. J., and T. D. POLLARD. 1989. Binding of myosin I to membrane lipids. *Nature* 340:565–568.

SHEETZ, M. P., and J. A. SPUDICH. 1983. Movement of myosin-coated fluorescent beads on actin cables in vitro. *Nature* 303:31–35.

SPUDICH, J. A. 1989. In pursuit of myosin function. *Cell Regulation* 1:1–11.

TOYOSHIMA, Y. Y., et al. 1987. Myosin subfragment-1 is sufficient to move actin filaments in vitro. *Nature* 328:536–539.

Structure of Striated Muscle

*BROWN, R. H., JR., and E. P. HOFFMAN. 1988. Molecular biology of Duchenne muscular dystrophy. *Trends Neurosci.* 11:480–484.

CASELLA, J. F., S. J. CASELLA, J. A. HOLLANDS, J. E. CALDWELL, and J. A. COOPER. 1989. Isolation and characterization of cDNA encoding the α subunit of Cap Z$_{(36/32)}$, an actin-capping protein from the Z line of skeletal muscle. *Proc. Nat'l Acad. Sci. USA* 86:5800–5804.

CHENG, N., and J. F. DEATHERAGE. 1989. Three-dimensional reconstruction of the Z disk of sectioned bee flight muscle. *J. Cell Biol.* 108:1761–1774.

*COOKE, R. 1986. The mechanism of muscle contraction. *CRC Crit. Rev. Biochem.* 21:53–118.

DAVISON, M.D., and D. R. CRITCHLEY. 1988. α-Actinins and the DMD protein contain spectrin-like repeats. *Cell* 52:159–160.

*HUXLEY, A. F. 1980. *Reflections on Muscle.* Princeton University Press.

MANDEL, J. L. 1989. Dystrophin—the gene and its product. *Nature* 339:584–586.

MILLER, D. M. III, I. ORTIZ, G. C. BERLINER, and H. F. EPSTEIN. 1983. Differential localization of two myosins within nematode thick filaments. *Cell* 34:477–490.

MILLIGAN, R. A., and P. F. FLICKER. 1987. Structural relationships of actin, myosin, and tropomyosin revealed by cryo-electron microscopy. *J. Cell Biol.* 105:29–39.

*SQUIRE, J. M. 1986. *Muscle: Design, Diversity and Disease.* Benjamin-Cummings.

WATKINS, S. C., E. P. HOFFMAN, H. S. SLAYTER, and L. M. KUNKEL. 1988. Immunoelectron microscopic localization of dystrophin in myofibres. *Nature* 333:863–866.

Contraction of Striated Muscle Fibers: Troponin and Tropomyosin

*BESSMAN, S. P., and C. L. CARPENTER. 1985. The creatine-creatine phosphate energy shuttle. *Ann. Rev. Biochem.* 54:831–862.

*BRENNER, B. 1987. Mechanical and structural approaches to correlation of cross-bridge action in muscle with actomyosin ATPase in solution. *Ann. Rev. Physiol.* 49:655–672.

*EISENBERG, E., and T. L. HILL. 1985. Muscle contraction and free energy transduction in biological systems. *Science* 227:999–1006.

*PAYNE, M. R., and S. E. RUDNICK. 1989. Regulation of vertebrate striated muscle contraction. *Trends Biochem. Sci.* 14:357–360.

*POLLARD, T. D. 1987. The myosin crossbridge problem. *Cell* 48:909–910.

THOMAS, D. D. 1987. Spectroscopic probes of muscle cross-bridge rotation. *Ann. Rev. Physiol.* 49:691–709.

WHITE, S. P., C. COHEN, and G. N. PHILLIPS, JR. 1987. Structure of co-crystals of tropomyosin and troponin. *Nature* 325:826–828.

*ZOT, A. S., and J. D. POTTER. 1987. Structural aspects of troponin-tropomyosin regulation of skeletal muscle contraction. *Ann. Rev. Biophys. Biophys. Chem.* 16:535–559.

Calcium Release from the Sarcoplasmic Reticulum

BLOCK, B. A., T. IMAGAWA, K. P. CAMPBELL, and C. FRANZINI-ARMSTRONG. 1988. Structural evidence for direct interaction between the molecular components of the transverse tubule/sarcoplasmic reticulum junction in skeletal muscle. *J. Cell Biol.* 107:2587–2600.

*CASWELL, A. H., and N. R. BRANDT. 1989. Does muscle activation occur by direct mechanical coupling of transverse tubules to sarcoplasmic reticulum? *Trends Biochem. Sci.* 14:161–165.

*CATTERALL, W. A., M. J. SEAGER, and M. TAKAHASHI. 1988. Molecular properties of dihydropyridine-sensitive calcium channels in skeletal muscle. *J. Biol. Chem.* 263:3535–3538.

*FLEISCHER, S., and M. INUI. 1989. Biochemistry and biophysics of excitation-contraction coupling. *Ann. Rev. Biophys. Chem.* 18:333–364.

KNUDSON, C. M., et al. 1989. Specific absence of the α$_1$ subunit of the dihydropyridine receptor in mice with muscular dysgenesis. *J. Biol. Chem.* 264:1345–1348.

PEREZ-REYES, E., et al. 1989. Induction of calcium currents by the expression of the α_1-subunit of the dihydropyridine receptor from skeletal muscle. *Nature* 340:233–236.

TAKESHIMA, H., et al. 1989. Primary structure and expression from complementary DNA of skeletal muscle ryanodine receptor. *Nature* 339:439–445.

TANABE, T., et al. 1987. Primary structure of the receptor for calcium channel blockers from skeletal muscle. *Nature* 328:313–318.

WAGENKNECHT, T., et al. 1989. Three-dimensional architecture of the calcium channel/foot structure of sarcoplasmic reticulum. *Nature* 338:167–170.

Cardiac and Smooth Muscle Structure and Contraction

*ADELSTEIN, R. S., J. R. SELLERS, M. A. CONTI, M. D. PATO, and P. DE LANEROLLE. 1982. Regulation of smooth muscle contractile proteins by calmodulin and cyclic AMP. *Fed. Proc.* 41:2873–2878.

*CROSS, R. A. 1988. Smooth muscle contraction. What is 10S myosin for? *J. Muscle Res. Cell Motil.* 9:108–110.

ISOBE, Y., F. D. WARNER, and L. F. LEMANSKI. 1988. Three-dimensional immunogold localization of α-actinin within the cytoskeletal networks of cultured cardiac muscle and nonmuscle cells. *Proc. Nat'l Acad. Sci. USA* 85:6758–6762.

ITOH, T., et al. 1989. Effects of modulators of myosin light-chain kinase activity in single smooth muscle cells. *Nature* 338:164–167.

LEHMAN, W., R. CRAIG, J. LUI, and C. MOODY. 1989. Caldesmon and the structure of smooth muscle thin filaments: immunolocalization of caldesmon on thin filaments. *J. Muscle Res. Cell Motil.* 10:101–112.

MILLER-HANCE, W. C., J. R. MILLER, J. N. WELLS, J. T. STULL, and K. E. KAMM. 1988. Biochemical events associated with activation of smooth muscle contraction. *J. Biol. Chem.* 263:13979–13982.

RASMUSSEN, H., Y. TAKUWA, and S. PARK. 1987. Protein kinase C in the regulation of smooth muscle contraction. *FASEB J.* 1:177–185.

*SELLERS, J. R., and R. S. ADELSTEIN. 1987. Regulation of contractile activity. In *The Enzymes*, P. Boyer and E. G. Krebs, eds. Academic Press, vol. 18, pp. 381–418.

*SOBUE, K., K. KANDA, T. TANAKA, and N. UEKI. 1988. Caldesmon: a common actin-linked regulatory protein in the smooth muscle and nonmuscle contractile system. *J. Cell Biochem.* 37:317–325.

Isoforms of Muscle Proteins

*ANDREADIS, A., M. E. GALLEGO, and B. NADAL-GINARD. 1987. Generation of protein isoform diversity by alternative splicing: mechanistic and biological implications. *Ann. Rev. Cell Biol.* 3:207–242.

*BREITBART, R. E., A. ANDREADIS, and B. NADAL-GINARD. 1987. Alternative splicing: a ubiquitous mechanism for the gener-ation of multiple protein isoforms from single genes. *Ann. Rev. Biochem.* 56:467–495.

*DAUBAS, P. 1988. Alternative RNA splicing of genes encoding contractile proteins. *Biochimie* 70:137–144.

*EMERSON, C. P., and S. I. BERNSTEIN. 1987. Molecular genetics of myosin. *Ann. Rev. Biochem.* 56:695–726.

GUNNING, P., P. PONTE, L. KEDES, R. J. HICKEY, and A. I. SKOULTCHI. 1984. Expression of human cardiac actin in mouse L cells: a sarcomeric actin associated with a non-muscle cytoskeleton. *Cell* 36:709–715.

*OTEY, C. A., M. H. KALNOSKI, and J. C. BULINSKI. 1987. Identification and quantification of actin isoforms in vertebrate cells and tissues. *J. Cell Biochem.* 34:113:124.

Structure of Intestinal Microvilli

COLUCCIO, L. M., and A. BRETSCHER. 1989. Reassociation of microvillar core proteins: making a microvillar core in vitro. *J. Cell Biol.* 108:495–502.

CONZELMAN, K. A., and M. S. MOOSEKER. 1987. The 110-kD protein-calmodulin complex of the intestinal microvillus in an actin-activated MgATPase. *J. Cell Biol.* 105:313–324.

*MOOSEKER, M. S. 1985. Organization, chemistry, and assembly of the cytoskeletal apparatus of the intestinal brush border. *Ann. Rev. Cell Biol.* 1:209–241.

MOOSEKER, M. S., and T. R. COLEMAN. 1989. The 110-kD protein-calmodulin complex of the intestinal microvillus (brush border myosin I) is a mechanoenzyme. *J. Cell Biol.* 108:2395–2400.

Actin and Actin-binding Proteins in Nonmuscle Cells

BAZARI, W. L., et al. 1988. Villin sequence and peptide map identify six homologous domains. *Proc. Nat'l Acad. Sci. USA* 85:4986–4990.

*CITI, S., and J. KENDRICK-JONES. 1987. Regulation of non-muscle myosin structure and function. *Bioessays* 7:155–159.

*KORN, E. D., and J. A. HAMMER. 1988. Myosins of nonmuscle cells. *Ann. Rev. Biophys. Biophys. Chem.* 17:23–45.

LASSING, I. and U. LINDBERG. 1985. Specific interaction between phosphatidylinositol 4,5-biphosphate and profilactin. *Nature* 314:472–474.

*MATSUDAIRA, P., and P. JANMEY. 1988. Pieces in the actin-severing protein puzzle. *Cell* 54:139–140.

OZAWA, M., H. BARIBAULT, and R. KEMLER. 1989. The cytoplasmic domain of the cell-adhesion molecule uvomorulin associates with three independent proteins structurally related in different species. *EMBO J.* 8:1711–1717.

*POLLARD, T. D. 1986. Assembly and dynamics of the actin filament system in nonmuscle cells. *J. Cell Biochem.* 31:87–95.

*POLLARD, T. D., and J. A. COOPER. 1986. Actin and actin-binding proteins. A critical evaluation of mechanisms and functions. *Ann. Rev. Biochem.* 55:987–1035.

*STOSSEL, T. P., et al. 1985. Nonmuscle actin-binding proteins. *Ann. Rev. Cell Biol.* 1:353–402.

TILNEY, L. G., E. M. BONDER, L. M. COLUCCIO, and M. S. MOOSEKAR. 1983. Actin from *Thyone* sperm assembles on only one end of an actin filament: a behavior regulated by profilin. *J. Cell Biol.* **97**:112–124.

Cell Motility and Stress Fibers

*BRAY, D., and J. G. WHITE. 1988. Cortical flow in animal cells. *Science* **239**:883–888.

BYERS, H. R., and K. FUJIWARA. 1982. Stress fibers *in situ*: immunofluorescence visualization with anti-actin, anti-myosin and anti-alpha-actinin. *J. Cell Biol.* **93**:804–811.

CHEN, W-T. 1981. Mechanism of retraction of the trailing edge during fibroblast movement. *J. Cell Biol.* **90**:187–200.

FUKUI, Y., T. J. LYNCH, H. BRESKA, and E. D. KORN. 1989. Myosin I is located at the leading edges of locomoting *Dictyostelium* amoebae. *Nature* **341**:328–331.

KREIS, T. E., B. GEIGER, and J. SCHLESSINGER. 1982. Mobility of microinjected rhodamine actin within living chicken gizzard cells determined by fluorescence photobleaching recovery. *Cell* **29**:835–845.

DE LOZANNE, A., and J. A. SPUDICH. 1987. Disruption of the *Dictyostelium* myosin heavy chain gene by homologous recombination. *Science* **236**:1086–1091.

*MITCHISON, T., and M. KIRSCHNER. 1988. Cytoskeletal dynamics and nerve growth. *Neuron* **1**:761–772.

SHEETZ, M. P., S. TURNEY, H. QIAN, and E. L. ELSON. 1989. Nanometre-level analysis demonstrates that lipid flow does not drive membrane glycoprotein movements. *Nature* **340**:284–288.

*SMITH, S. J. 1988. Neuronal cytomechanics: the actin-based motility of growth cones. *Science* **242**:708–715.

Intermediate Filaments: General References

FRASER, R. D. B., T. P. MACRAE, D. A. D. PARRY, and E. SUZUKI. 1986. Intermediate filaments in α-keratins. *Proc. Nat'l Acad. Sci. USA* **83**:1179–1183.

GEIGER, B. 1987. Intermediate filaments: looking for a function. *Nature* **329**:392–393.

*GREENE, L. A. 1989. A new neuronal intermediate filament protein. *Trends Neurosci.* **12**:228–230.

*OSBORN, M., and K. WEBER. 1986. Intermediate filament proteins: a multigene family distinguishing major cell lineages. *Trends Biochem. Sci.* **11**:469–472.

*STEINERT, P. M., and D. A. D. PARRY. 1985. Intermediate filaments: conformity and diversity of expression and structure. *Ann. Rev. Cell Biol.* **1**:41–65.

*STEINERT, P. M., and D. R. ROOP. 1988. Molecular and cellular biology of intermediate filaments. *Ann. Rev. Biochem.* **57**:593–626.

*TRAUB, P. 1985. *Intermediate Filaments: A Review*. Springer-Verlag.

*WANG, E., D. FISCHMAN, R. K. H. LIEM, and T.-T. SUN, eds. 1985. Intermediate filaments. *Ann. NY Acad. Sci.*, vol. 455.

Functions of Intermediate Filaments

FRANKE, W. W., M. HERGT, and C. GRUND. 1987. Rearrangement of the vimentin cytoskeleton during adipose conversion: formation of an intermediate filament cage around lipid globules. *Cell* **49**:131–141.

HOFFMAN, P. N., E. H. KOO, N. A. MUMA, J. W. GRIFFIN, and D. L. PRICE. 1988. Role of neurofilaments in the control of axonal caliber in myelinated nerve fibers. In *Intrinsic Determinants of Neuronal Form and Function*. Alan R. Liss, pp. 389–402.

*KLYMKOWSKY, M. W., J. B. BACHANT, and A. DOMINGO. 1989. The functions of intermediate filaments. *Cell Motil. Cytoskel.*, in press.

PASDAR, M., and W. J. NELSON. 1988. Kinetics of desmosome assembly in Madin-Darby canine kidney epithelial cells: temporal and spatial regulation of desmoplakin organization and stabilization upon cell-cell contact. II. Morphological analysis. *J. Cell Biol.* **106**:687–695.

Structure and Assembly of Intermediate Filaments

ANGELIDES, K. J., K. E. SMITH, and M. TAKEDA. 1989. Assembly and exchange of intermediate filament proteins of neurons: neurofilaments are dynamic structures. *J. Cell Biol.* **108**:1495–1506.

EICHNER, R., P. REW, A. ENGEL, and U. AEBI. 1985. Human epidermal keratin filaments: studies on their structure and assembly. *Ann. NY Acad. Sci.* **455**:381–402.

GEISLER, N., E. KAUFMANN, and K. WEBER. 1985. Antiparallel orientation of the two double-stranded coiled-coils in the tetrameric protofilament unit of intermediate filaments. *J. Mol. Biol.* **182**:173–177.

GREEN, K. J., B. GEIGER, J. C. R. JONES, J. C. TALIAN, and R. D. GOLDMAN. 1987. The relationship between intermediate filaments and microfilaments before and during the formation of desmosomes and adherens-type junctions in mouse epidermal keratinocytes. *J. Cell Biol.* **104**:1389–1402.

MCKEON, F. D., M. W. KIRSCHNER, and D. CAPUT. 1986. Homologies in both primary and secondary structure between nuclear envelope and intermediate filament proteins. *Nature* **319**:463–468.

SOELLNER, P., R. A. QUINLAN, and W. W. FRANKE. 1985. Identification of a distinct soluble subunit of an intermediate filament protein: tetrameric vimentin from living cells. *Proc. Nat'l Acad. Sci. USA* **82**:7929–7933.

STEWART, M., R. A. QUINLAN, and R. D. MOIR. 1989. Molecular interactions in paracrystals of a fragment corresponding to the α-helical coiled-coil rod portion of glial fibrillary acidic protein: evidence for an antiparallel packing of molecules and polymorphism related to intermediate filament structure. *J. Cell Biol.* **109**:225–234.

VIKSTROM, K. L., G. G. BORISY, and R. D. GOLDMAN. 1989. Dynamic aspects of intermediate filament networks in BHK-21 cells. *Proc. Nat'l Acad. Sci. USA* **86**:549–553.

23

Multicellularity: Cell-Cell and Cell-Matrix Interactions

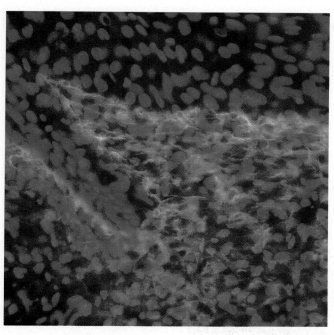

The matrix protein tenascin (green) in granulation tissue at the edge of a rat skin wound

*T*he evolution of multicellular organisms permitted specialized cells and tissues to form; a flowering plant has at least 15 types of cells, and a vertebrate more than 100. In both plants and animals, cells specialized to carry out a particular task are not randomly distributed but are found together in tissues in which the task is performed: a xylem or meristem; a liver, a muscle, or a nerve ganglion. Through the coordination of multiple specialized tissues, the organism as a whole moves, metabolizes, reproduces, and carries out other essential functions.

Animal cells contact tightly and interact specifically with other cells. They also contact a complex network of secreted proteins and carbohydrates, the extracellular matrix, that fills the spaces between cells. The matrix helps bind the cells in tissues together. It also provides a lattice through which cells can move, particularly during the early stages of differentiation when cells such as neuroblasts—the precursors of neurons—move from one part of the animal to another.

The cell walls that surround plant cells are thicker, more impermeable, and more rigid than the extracellular *matrix around animal cells. Most importantly, plant cells can grow larger but do not move in relation to their neighbors. Because of these differences, the plant cell wall and its interactions with other cells will be treated separately at the end of this chapter. Our main focus is on animal cells—their interactions with other cells and with the extracellular matrix.*

Cell-cell interactions fall into several classes. Animal cells adhere tightly and specifically with cells of the same, or a similar, type: dissociated liver cells bind more tightly to other liver cells than to dissociated retina, muscle, or other cells. These homotypic interactions are catalyzed by sets of cell-surface molecules. Various cell junctions such as desmosomes and gap junctions frequently form as a result of these cell contacts. Certain cells form contacts with different types of cells; as we shall see, such interactions are crucially important in the differentiation of many tissues. ▲

The Extracellular Matrix Serves Many Functions

Animals contain many types of extracellular matrices, each specialized for a particular function, such as strength (in a tendon), filtration (in the kidney glomerulus where urine is formed), or adhesion. For instance, the smooth muscle cells that surround an artery are connected by an extracellular matrix that provides strong but flexible connections. These matrices consist of different combinations of fibrous collagen proteins, hyaluronic acid, and complexes of polysaccharides and proteins called proteoglycans. Glycoproteins are also important matrix components. Receptor proteins on the surfaces of cells bind various matrix elements, imparting strength and rigidity to tissues (Figure 23-1).

Many animal tissues are composed largely of cells, with little interstitial space between them; an example is the epidermis of skin (Figure 23-2). Most epithelia and other organized groups of cells such as muscle rest upon, and are bound tightly to, a thin matrix called the basal lamina that is linked to the plasma membrane by specific receptor proteins. The basal lamina is composed of a type of collagen that is organized in a two-dimensional reticulum, rather than into the fibers that characterize other collagens. The basal lamina is tightly connected to fibrous collagens and other materials in the underlying loose connective tissues (Figure 23-3).

In the glomerulus of the kidney, the basal lamina forms a porous filter that allows water, ions, and small molecules in blood to cross into the urinary space while retaining proteins and cells in the blood.

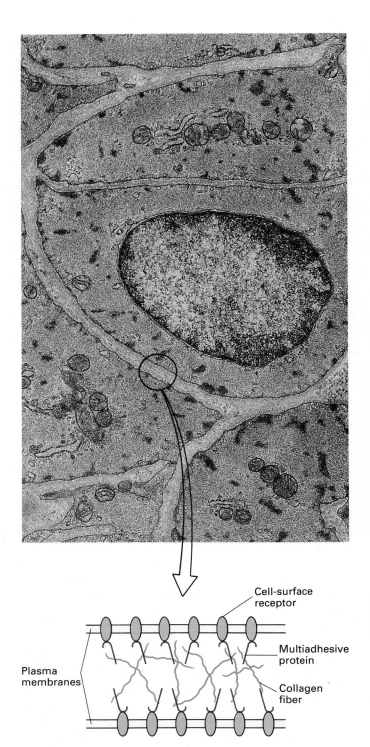

▲ **Figure 23-1** Electron micrograph of smooth muscle in the wall of a small artery. The muscle cells are separated by relatively wide spaces that contain fibrous collagen and proteoglycans. Connections between plasma-membrane receptor proteins and components of the extracellular matrix allow the cells to adhere to each other and give this tissue its strength and resistance to shear forces, as is depicted schematically in the diagram. *From D. W. Fawcett, 1981,* The Cell, *2d ed., Saunders/Photo Researchers, Inc.*

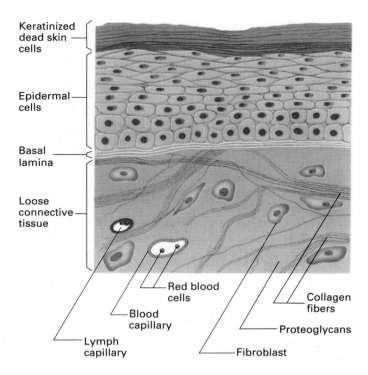

Keratinized dead skin cells

Epidermal cells

Basal lamina

Loose connective tissue

Red blood cells

Blood capillary

Lymph capillary

Collagen fibers

Proteoglycans

Fibroblast

▲ **Figure 23-2** A schematic drawing of a section through the outer part of the skin of a pig. The cellular epidermis rests on the thin basal lamina, which in turn contacts the thick layer of loose connective tissue consisting of abundant collagen fibers and cells—mostly fibroblasts—that synthesize the connective-tissue proteins and polysaccharides. The regions between the collagen fibers are filled with proteoglycans. Blood and lymph capillaries course through the loose connective tissue.

Animals contain abundant connective tissue in which much of the tissue volume consists of spaces between cells. *Loose connective tissue* forms the bedding on which most small glands and epithelia lie and connects to the basal lamina around the cells. Proteins such as fibrous collagens and flexible elastin give the tissue its shape, rigidity, and flexibility. Most of the space is filled with the highly hydrated proteoglycans, which gives the tissue a gel-like consistency. Loose connective tissue is highly cellular and contains numerous fibroblasts that synthesize much of the extracellular matrix and also blood-borne cells such as lymphocytes. Most blood capillaries that bring O_2 and nutrients to the cells in all parts of the body are confined to loose connective tissue. A principal function of the spaces between cells in this tissue is to allow O_2 and nutrients to diffuse freely into the cells of the epithelia and glands (see Figure 23-2).

Dense connective tissue, such as bone, cartilage, and tendon, is a major component of organs in which strength or flexibility or both are essential functions—the skeletal system, for instance. Unlike loose connective tissue, it

consists almost entirely of fibrous extracellular matrix materials and contains few cells. The extracellular matrices of dense connective tissue consist of densely packed collagens and other fibrous proteins surrounded by glycoproteins, proteoglycans, and other substances such as elastin that are produced and secreted by the relatively few cells present.

The extracellular matrix should not be thought of as an inert framework or cage that supports or surrounds cells. In many cases the matrix is required for cells to carry out certain specialized functions; it can bind many growth factors and hormones and thus provides permanent, abundant signals to the cells that contact it. Cells require specific matrix components in order to differentiate, that is, to acquire the functions of a specific cell type. Morphogenesis—the later stage of acquisition of form achieved by cell movements and rearrangements—is also critically dependent on matrix molecules. In developing organisms matrix molecules are constantly being remodeled, degraded, and resynthesized locally; even in adults—in cases such as wounding—there is degradation and resynthesis of all matrix components.

We begin our discussion of the extracellular matrix with its most abundant structural component, collagen.

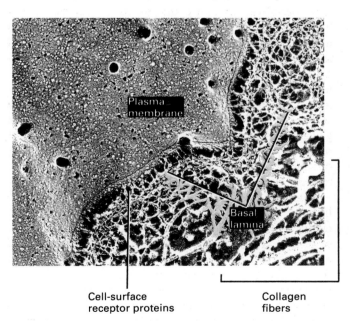

Plasma membrane

Basal lamina

Cell-surface receptor proteins

Collagen fibers

▲ **Figure 23-3** Association of the plasma membrane of skeletal muscle with the basal lamina. In this quick-freeze deep-etch preparation, the basal lamina is seen as a meshwork of filaments. Some of the basal lamina filaments contact receptor proteins in the plasma membrane; others bind to the thick collagen fibers that form the connective tissue around the muscle. *Courtesy John Heuser. From D. W. Fawcett, 1981,* The Cell, *2d ed., Saunders/Photo Researchers.*

Collagen: A Class of Multifunctional Fibrous Proteins

Collagen is the major class of insoluble fibrous protein in the extracellular matrix and in connective tissue. In fact, it is the single most abundant protein in the animal kingdom. There are at least twelve types of collagen (Table 23-1). Types I, II, and III are the most abundant and form fibrils of similar structure. Type IV, in contrast, forms a two-dimensional reticulum and is a principal component of the basal lamina. The structural features of each type of fibrous collagen make it suitable for a particular function. At one time it was thought that all collagens were secreted by connective-tissue cells called fibroblasts, but we now know that numerous epithelial cells make certain types of collagens.

The Basic Structural Unit of Collagen Is a Triple Helix

Because its abundance in tendon-rich tissue such as rat tail makes it easy to isolate, the fibrous type I collagen was the first to be characterized and its three-stranded helical structure was eventually shown to be characteristic of all collagens. Its fundamental structural unit is a long (300 nm), thin (1.5-nm diameter) protein that consists of three coiled subunits: two $\alpha 1(I)$ chains and one $\alpha 2(I)$.[1] Each chain contains precisely 1050 amino acids wound around each other in a characteristic right-handed triple helix (Figure 23-4).

There are three amino acids per turn of the helix and every third amino acid is glycine, the smallest amino acid (Figure 23-5). The side chain of glycine, an H atom, is the

[1]In collagen nomenclature the collagen type is in Roman numerals and is placed in parentheses.

Table 23-1 The collagens

Type	Chains	Length of triple helix; structural details	Supermolecular structure	Localization
I	$[\alpha 1(I)]_2\ [\alpha 2(I)]$	300 nm	67-nm banded fibrils	Skin, tendon, bone, etc.
II	$[\alpha 1(II)]_3$	300 nm	Small 67-nm banded fibrils	Cartilage, vitreous humor
III	$[\alpha 1(III)]_3$	300 nm	Small 67-nm banded fibrils	Skin, muscle, etc., frequently together with type I
IV	$[\alpha 1(IV)]_2\ [\alpha 2(IV)]$	390 nm C-terminal globular domain	Nonfibrillar network	All basal laminas
V	$[\alpha 1(V)]\ [\alpha 2(V)]\ [\alpha 3(V)]$	390 nm N-terminal globular domain	Small fibers	Most interstitial tissues; associated with type I
VI	$[\alpha 1(VI)]\ [\alpha 2(VI)]\ [\alpha 3(VI)]$	150 nm; N + C terminal globular domains	Microfibrils, 100-nm banded fibrils	Most interstitial tissues; associated with type I
VII	$[\alpha 1(VII)]_3$	450 nm	Dimer	Epithelia
VIII	$[\alpha 1(VIII)]_3$	?	?	Some endothelial cells
IX	$[\alpha 1(IX)]\ [\alpha 2(IX)]\ [\alpha 3(IX)]$	200 nm N-terminal globular domain; bound proteoglycan	?	Cartilage, associated with type II
X	$[\alpha 1(X)]_3$	150 nm; C-terminal globular domain		Hypertrophic and mineralizing cartilage
XI	$[\alpha 1(XI)]\ [\alpha 2(XI)]\ [\alpha 3(XI)]$	300 nm	Small fibers	Cartilage
XII	$\alpha 1(XII)$	?	?	In vicinity of collagens I and III, with which it interacts

SOURCE: K. Kuhn, 1987, in R. Mayne and R. Burgeson, eds., *Structure and Function of Collagen Types,* Academic, p. 2.

(a)

α Chain

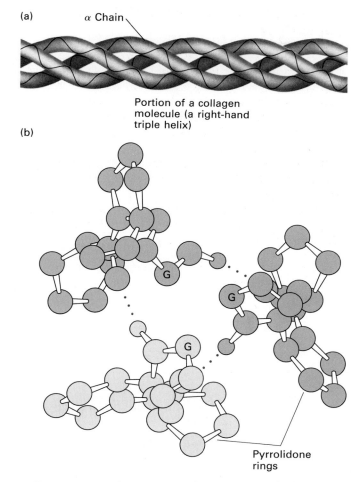

Portion of a collagen
molecule (a right-hand
triple helix)

(b)

G

G

G

Pyrrolidone
rings

▲ **Figure 23-4** The structure of collagen. (a) The basic structural unit is a triple-stranded helical molecule. Hydrogen bonds (not shown here) link residues in each chain to the other two chains, which makes the helix rigid. (b) Cross-section model of a collagen triple helix. The α carbon of glycine is labeled G; every third residue must be glycine because the space in the center of the helix will not accommodate a side-chain larger than H. The five-membered pyrrolidone rings from the proline residues are on the outside of the helix. The dots indicate the hydrogen bonds between the NH of a glycine residue on one chain with a peptide C=O residue on another. *Part (b) after L. Stryer, 1988,* Biochemistry, *3d ed., W. H. Freeman and Company, p. 265.*

Table 23-2 Sizes of the 41 exons that encode the part of α1(I) collagen that forms a triple helix; one primordial unit equals (Gly-X-Y)$_6$

Exon length (bp)	Number of exons
54 (one primordial unit)	21
108 (two primordial units)	9
162 (three primordial units)	1
45 (one primordial unit, 9-bp deletion)	5
99 (two primordial units, 9-bp deletion)	5

SOURCE: H. Boedtker, F. Fuller, and V. Tate, 1983, *Int. Rev. Connect. Tissue Res.* 10:1–63.

-Gly-Pro-Met-Gly-Pro-Ser-Gly-Pro-Arg- [13]

-Gly-Leu-Hyp-Gly-Pro-Hyp-Gly-Ala-Hyp- [22]

-Gly-Pro-Gln-Gly-Phe-Gln-Gly-Pro-Hyp- [31]

-Gly-Glu-Hyp-Gly-Glu-Hyp-Gly-Ala-Ser- [40]

-Gly-Pro-Met-Gly-Pro-Arg-Gly-Pro-Hyp- [49]

-Gly-Pro-Hyp-Gly-Lys-Asn-Gly-Asp-Asp- [58]

▲ **Figure 23-5** Sequence of part of the α1(I) collagen polypeptide. Every third amino acid is glycine and proline is abundant. *After L. Stryer, 1988,* Biochemistry, *3d ed., W. H. Freeman and Company, p. 262.*

only one that can fit into the crowded center of such a helix. Collagen is also rich in proline and hydroxyproline, amino acids that fold and stabilize the three chains. The bulky five-membered pyrrolidone rings in proline residues are on the outside of the triple helix. Many regions of collagen chains are composed of the repeating motif Gly-Pro-X, where X can be any amino acid. Hydrogen bonds linking the peptide bond NH of a glycine residue with a peptide carbonyl (C=O) group in an adjacent polypeptide help hold the three chains together (see Figure 23-4b).

Most Collagen Exons Encode Six Gly-X-Y Sequences

The unusual sequence of the fibrous collagens reflects their gene organization. The gene that encodes the α1(I) chain contains 41 small exons. Most of these are 54 bp, encoding one so-called primordial unit containing six repeats of a Gly-X-Y sequence or, as three amino acids form one turn of the helix, six helix turns. Other exons encode two or three of these primordial units, and the others appear to have lost 9 bp, or one Gly-X-Y unit, during evolution (Table 23-2). As genes encoding polypeptides of other fibril-forming collagens have a similar pattern, it would appear that the ancestral collagen protein was a six-fold repeat of Gly-X-Y, and that present-day collagen genes evolved by gene duplication.

Collagen Fibrils Form by Lateral Interactions of Triple Helices

Fibrils of collagens such as type I have enormous tensile strength; that is, such collagen can be stretched without being broken. Type I fibrils, roughly 50 nm in diameter, are packed side-by-side in parallel bundles in tissues such as tendons, where they connect muscles with bones and must withstand enormous forces (Figure 23-6). Gram-for-gram, type I collagen is stronger than steel.

In fibrils, collagen molecules pack together side-by-side, with adjacent molecules being displaced from one

▶ **Figure 23-6** Cross-sectional electron micrograph through the dense connective tissue of a chick tendon. Most of the tissue is occupied by parallel type I collagen fibrils, about 50 nm in diameter, seen here in cross section. The cellular content of the tissue is very low. *From D. A. D. Parry, 1988. Biophys. Chem. 29:195–209.*

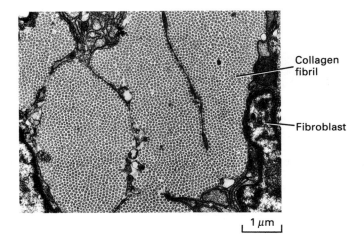

Collagen fibril

Fibroblast

1 μm

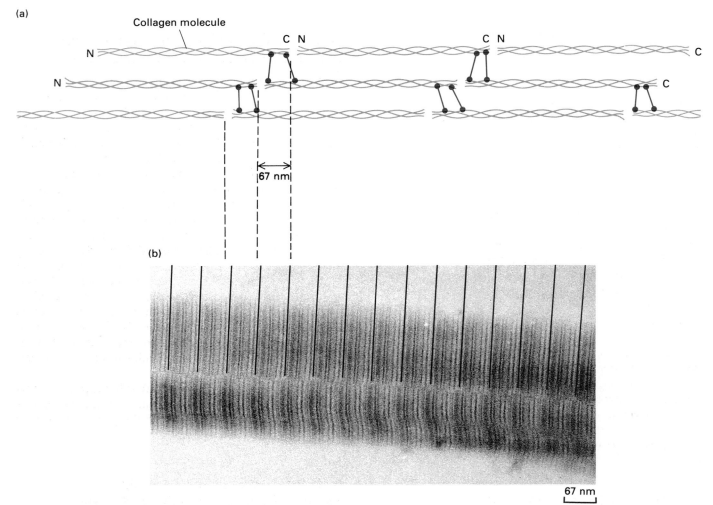

(a)

Collagen molecule

67 nm

(b)

67 nm

▲ **Figure 23-7** The structure of fibrous collagen. (a) Each collagen molecule is 300 nm long. In fibrous collagen, collagen molecules pack together side by side. Adjacent molecules are displaced 67 nm, or slightly less than one-fourth the length of a single molecule. A small gap separates the "head" of one collagen from the "tail" of the next. The side-by-side interactions are stabilized by covalent bonds (red) between the N-terminus of one molecule and the C-terminus of an adjacent one. (b) An electron micrograph of calfskin collagen fibers stained with phosphotungstic acid. As indicated by the leaders, the striations created by the 67-nm periodic pattern of the packing are clearly visible. *Part (a) after L. Stryer, 1988, Biochemistry, 3d ed., W. H. Freeman and Company, p. 272. Part (b) courtesy of R. Bruns.*

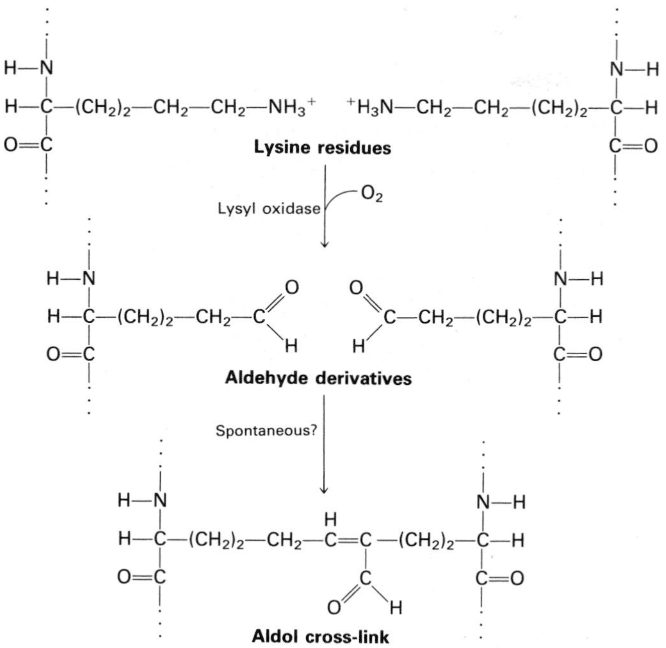

▲ **Figure 23-8** The side-by-side interactions of collagen helices are stabilized by an aldol cross-link between two lysine side chains. This is accomplished by the extracellular enzyme lysyl oxidase.

another by about one-quarter their length, 67 nm, as shown in Figure 23-7a. This staggered array produces a striated effect that can be seen in electron micrographs of stained collagen fibers; the characteristic pattern of bands is repeated about every 67 nm (Figure 23-7b). The bands are caused by the binding of a metal or a stain to certain repetitive sequences of amino acids in collagen; the stain also fills the space between the ends of collagen molecules.

There are short segments at either end of the collagen chains that do not assume the triple helical conformation. These are of particular importance because covalent cross-links between two lysine or hydroxylysine residues at the C-terminus of one chain with two at the N-terminus of an adjacent chain (Figure 23-8) stabilize the side-by-side arrangement and generate a strong fiber. Figure 4-27 shows micrographs of such fibers.

Denatured Collagen Polypeptides Cannot Renature to Form a Triple Helix

Type I collagen purified from tissues of mature animals is insoluble in aqueous solutions, largely due to covalent bonds that cross-link adjacent molecules. In type I collagen of young animals many of these cross-links have not yet formed: here, the three-dimensional structure of the triple helix is stabilized mostly by multiple, weak interactions including hydrogen bonds between the three chains. When a solution of type I collagen from a young animal is heated above 40°C, the rodlike helix becomes denatured, and the three chains separate from each other. Such denatured collagen cannot spontaneously renature to form a completely normal collagen triple helix: because the helix is a very regular, repeating structure, the polypeptides may begin to anneal at many places other than the ends, forming triple helices that are out of register, with single-stranded segments at each end. During biosynthesis, in-register alignment of the three polypeptides is facilitated by special sequences at each end, as explained in the next section.

N-Terminal and C-Terminal Propeptides Aid in the Formation of the Triple Helix

Collagen chains are synthesized as longer precursors called procollagens. The procollagen of type I collagen forms a triple helix that contains 150 additional amino acids at the N-terminus and 250 at the C-terminus of both types of chains. These additional amino acids constitute the propeptides (Figure 23-9). Except for a short collagenlike helical region in the N propeptide, the propeptides have amino acid compositions that form globular domains. There are no disulfide bonds within type I collagen triple helices themselves (there are some in other collagens). There are several intrachain disulfide bonds and five important interchain disulfide bonds connecting the C-terminal propeptides of the three procollagen chains. These disulfide bonds form in the rough ER as do the triple helices. The disulfide bonds in the C-terminal propeptide are believed to form first; then the three chains can zip together in the C → N direction to form the triple helix. Disulfide bonds are important in stabilizing the association of the three chains prior to formation of the triple helix.

In vitro renaturation studies provide support for this notion: the renaturation of thermally denatured procollagen is substantially accelerated if the proper interchain disulfide bonds are first formed.

Newly Made Collagen Is Modified Sequentially in the Rough ER and the Golgi Complex

The assembly of collagen fibers begins in the rough ER, continues in the Golgi complex, and is completed outside the cell as Figure 23-10 shows. Besides cleavage of the signal sequence, several other modifications of the nascent collagen chains are made before the triple helix is

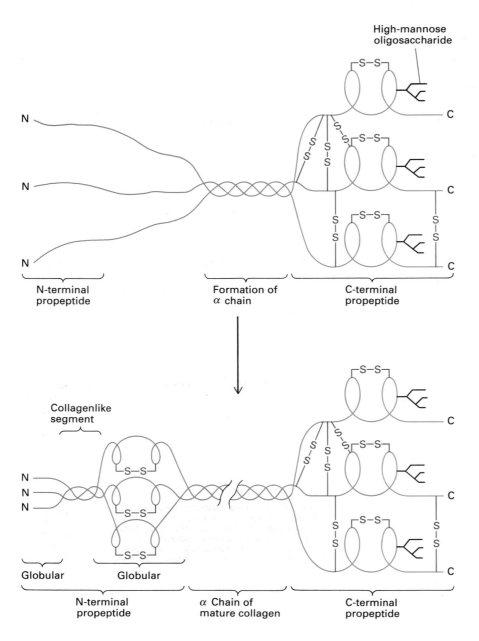

High-mannose
oligosaccharide

N-terminal
propeptide

Formation of
α chain

C-terminal
propeptide

Collagenlike
segment

Globular

Globular

N-terminal
propeptide

α Chain of
mature collagen

C-terminal
propeptide

◀ **Figure 23-9** Assembly of the type I collagen triple-stranded helix. Following removal of the N-terminal signal peptides, interchain disulfide bonds form between the C-terminal propeptides of three procollagens. The process continues, zipperlike, toward the N-terminus. Intrachain disulfide bonds form within the N-terminal propeptides; these may stabilize the long triple helix, as may the short triple-helical sequence in the N-terminal propeptide. [See J. M. Davidson and R. A. Berg, 1980, *Methods Cell Biol.* **23**:119.]

formed in the rough ER. Specific proline residues in the domains destined to form the triple helix are subjected to hydroxylation by the membrane-bound enzymes *prolyl 4-hydroxylase* and *prolyl 3-hydroxylase* (Figure 23-11). Certain lysine residues in this domain are hydroxylated by *lysyl hydroxylase*. Ascorbic acid (vitamin C) is an essential cofactor for both prolyl hydroxylase enzymes. In the absence of ascorbate, collagen is insufficiently hydroxylated. The chains cannot then form a stable triple helix at 37°C (Figure 23-12) nor can they form normal fibers. Nonhydroxylated procollagen chains are degraded within the cell. Consequently, fragility of blood vessels, tendons, and skin is characteristic of the disease scurvy, which results from a deficiency of vitamin C in the diet.

Protein glycosylation also begins in the rough ER: galactose residues are added to hydroxylysine residues and long oligosaccharides to certain asparagine residues.

Procollagen Is Assembled into Fibers after Secretion

Procollagen is converted to collagen after secretion by reactions catalyzed by extracellular enzymes, the procollagen peptidases. During or following exocytosis, type I procollagen molecules undergo proteolytic cleavage to remove the N-terminal and C-terminal propeptides. In the extracellular space, collagen molecules are polymer-

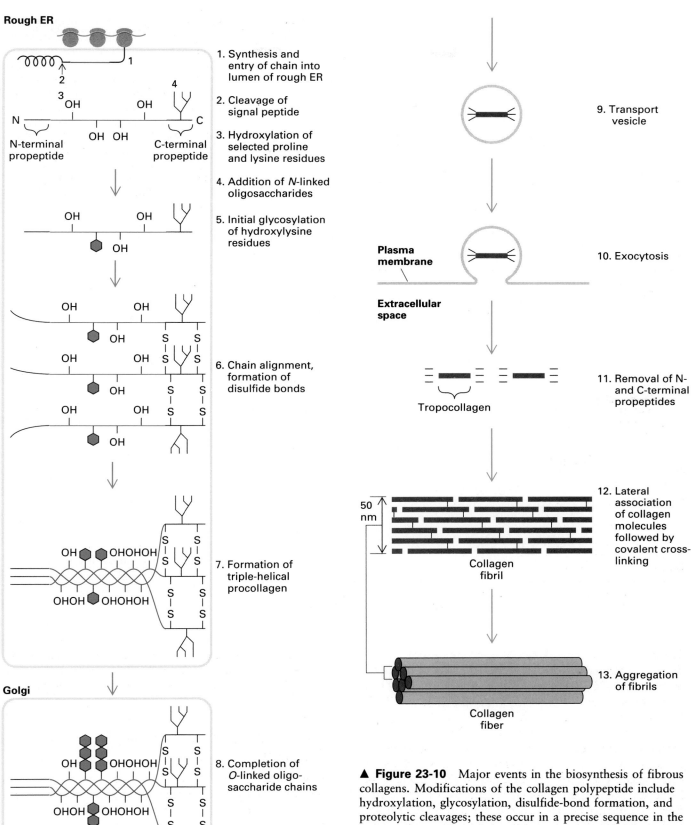

Rough ER

1. Synthesis and entry of chain into lumen of rough ER

2. Cleavage of signal peptide

3. Hydroxylation of selected proline and lysine residues

4. Addition of *N*-linked oligosaccharides

5. Initial glycosylation of hydroxylysine residues

6. Chain alignment, formation of disulfide bonds

7. Formation of triple-helical procollagen

Golgi

8. Completion of *O*-linked oligo-saccharide chains

N-terminal propeptide

C-terminal propeptide

9. Transport vesicle

Plasma membrane

10. Exocytosis

Extracellular space

11. Removal of N- and C-terminal propeptides

Tropocollagen

12. Lateral association of collagen molecules followed by covalent cross-linking

50 nm

Collagen fibril

13. Aggregation of fibrils

Collagen fiber

▲ **Figure 23-10** Major events in the biosynthesis of fibrous collagens. Modifications of the collagen polypeptide include hydroxylation, glycosylation, disulfide-bond formation, and proteolytic cleavages; these occur in a precise sequence in the rough ER, Golgi complex, and the extracellular space. The modifications allow formation of a stable triple-stranded helix and also the lateral alignment and covalent cross-linking of helices into 50-nm-diameter fibrils.

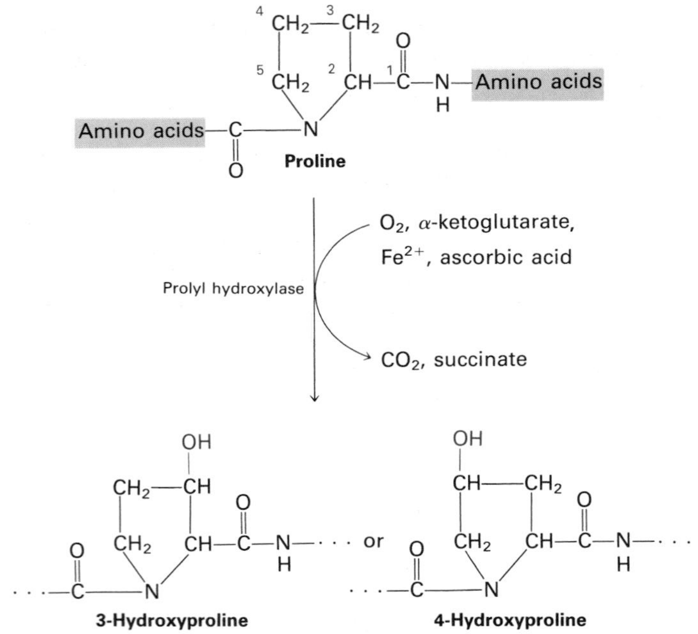

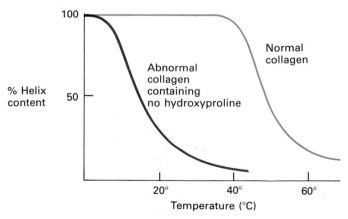

▲ **Figure 23-11** Certain proline residues are hydroxylated in the rough ER before the triple helices form. Oxygen, Fe^{2+}, ascorbic acid, and α-ketoglutarate are all required in this complex reaction. Proline can be hydroxylated in the 4 position only if it is in the amino acid sequence Gly-X-Pro, where X is any amino acid. Hydroxylated prolines, especially the predominant 4-hydroxyproline, are important for stability of the triple helix.

▲ **Figure 23-12** Denaturation of collagen molecules with normal content of hydroxyproline and of abnormal collagen containing no hydroxyproline. Without hydrogen bonds between hydroxyproline residues, the collagen helix is unstable and loses most of its helical content at temperatures above 20°C. Such collagens are formed by experimental animals (or man) in the absence of ascorbic acid (vitamin C). Normal collagen is more stable and resists thermal denaturation until a temperature of 40°C is reached.

ized to form collagen fibrils. Type I collagen fibrils form by spontaneous association of the 300-nm-long collagen helices staggered by about 67 nm. Fibrils average about 50 nm in diameter and can be up to several micrometers in length. Collagen types I and V are often found together in the same fibril, and it is thought that the diameter of the fibril is determined by the amount of type V present. Accompanying fibril formation is the oxidation of certain lysine and hydroxylysine residues into reactive aldehydes by the extracellular enzyme *lysyl oxidase*. Such aldehydes spontaneously form specific covalent cross-links between two chains (see Figure 23-8). Such bonding stabilizes the staggered array characteristic of collagen molecules and contributes to fibril strength.

Eventual excision of both propeptides is essential for formation of normal fibers. In diseases such as dermatosparaxis in cattle, there is a deficiency in one of the procollagen peptidases. The skin and tendons become very deformable, because the collagen forms mostly disorganized bundles rather than the normal highly organized strong fibers.

Mutations in Collagen Reveal Aspects of Its Structure and Biosynthesis

Type I collagen fibers are used as the reinforcing rods in construction of bone. Certain mutations in the $\alpha1(I)$ or $\alpha2(I)$ genes lead to osteogenesis imperfecta or brittle-bone disease. The most severe type is an autosomal dominant, lethal disease, resulting in death *in utero* or shortly after birth. Milder forms generate a severe crippling disease.

As might be expected, many cases of osteogenesis imperfecta are due to deletions of all or part of the $\alpha1(I)$ gene. However, a single amino acid change is sufficient to cause certain forms of this disease. As we have seen, a glycine must be at every third position for the collagen triple helix to form; mutations of glycine to almost any other amino acid are deleterious, producing poorly formed and unstable helices. Mutant unfolded polypeptides do not leave the rough ER of the fibroblasts (the cells that make most of the type I collagen) or leave it slowly. The ER becomes dilated and expanded. When this happens, the secretion of other proteins, such as type III collagen, by these cells is also slowed down.

Because each type I collagen molecule contains two $\alpha1(I)$ and one $\alpha2(I)$ chain, mutations in the $\alpha2$ chains are much less damaging. In a heterozygote expressing one wild type and one mutant $\alpha2(I)$ protein, 50 percent of the collagens will have the abnormal $\alpha2$ chain. In contrast, if the mutation is in the $\alpha1(I)$ chain, 75 percent will have one or two mutant $\alpha1(I)$ chains. In fact, even low expression of a mutant $\alpha1(I)$ gene can be deleterious, because the mutant chains can disrupt the function of wild-type $\alpha1(I)$ polypeptides when combined with them. To study such mutations, experimenters constructed a mutant $\alpha1(I)$ collagen gene with a glycine-to-cystine mutation near the

C-terminus. This mutant gene was used to create lines of transgenic mice with otherwise normal collagen genes. High-level expression of the mutant transgene was lethal, and expression at a rate of 10 percent of that of the normal $\alpha 1(I)$ genes caused severe growth abnormalities.

In summary, collagen has rigid structural requirements and is very susceptible to mutation, especially in glycine residues. Because mutant collagen chains can affect the function of wild-type ones, such mutations have a dominant phenotype.

Collagens Form a Diversity of Fibrillar Structures

Type II is the major collagen in cartilage. Its fibrils are smaller in diameter than type I and are oriented randomly in the viscous proteoglycan matrix. Such rigid macromolecules impart a stiffness and compressibility to the matrix and allow it to resist large deformations in shape and yet—as in joints—absorb the shock. Bound to the surfaces of type II fibrils is a collagen of a different structure—type IX. In this structure two helical collagen segments are interrupted at a flexible kink; protruding outward from the fibril is a chondroitin sulfate proteoglycan chain and also a large nonhelical domain that anchors the fibril to proteoglycans and other components of the matrix (Figure 23-13a).

Collagen type VI is found in many connective tissues, where it is frequently bound to the sides of type I fibrils and may bind them together (Figure 23-13b). Type VI is an unusual collagen in that most of the polypeptide does not form triple helices; rather helices about 60 nm long are connected by large globular domains about 40 nm long. Fibrils of pure type VI collagen thus give the impression of beads on a string.

Fibrous collagen is organized in different ways in different tissues to fulfill specific needs. One of the most remarkable organizations is seen in both mature bone and in the transparent cornea of the eye. Here collagen fibers are arranged in layers; fibers in each layer are parallel to each other but perpendicular to fibers in the adjacent layers (Figure 23-14). Like plywood, this arrangement creates a thick, nondeformable material.

(a)

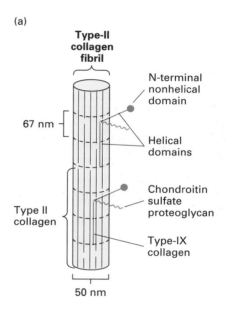

(b)

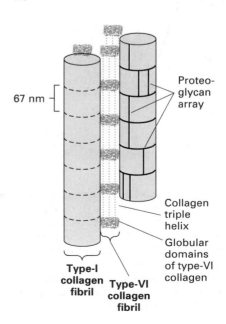

▶ **Figure 23-13** Arrangements of fibrous collagens. (a) Types II and IX in a cartilage matrix. Type II forms fibrils similar in structure to type I, with a similar 67-nm periodicity though smaller in diameter. Type IX contains two long triple helices connected at a flexible kink. At this point a chondroitin sulfate chain is linked to the $\alpha 2(IX)$ chain. Type IX collagens are bound at regular intervals along type II fibrils, with an N-terminal nonhelical domain of type IX projecting outward. It is thought that these domains bind the collagen fibrils to the proteoglycan-rich matrix. (b) Organization of the major fibrous components in the extracellular matrix of tendons. Type I fibrils, with their characteristic 67-nm period, are all oriented longitudinally, that is, in the direction of the stress applied to the tendon. The fibers are coated with an array of proteoglycans, as shown in red on the right-hand fibril. Type VI fibrils bind to and link together the type I fibrils. Type VI collagen consists of thin triple helices, about 50 nm long, with globular domains at either end. The globular domains of several type VI molecules bind together, giving a "beads-on-a-string" appearance to the type VI fibril. *Part (a) after K. Svoboda, I. Nishimura, S. Sugrue, Y. Ninomiya, B. Olson, 1988,* Proc. Nat'l Acad. Sci. USA *85:7496–7500; part (b) after R. R. Bruns, W. Press, E. Engvall, R. Timpl, and J. Gross, 1986,* J. Cell Biol. *103:393–404.*

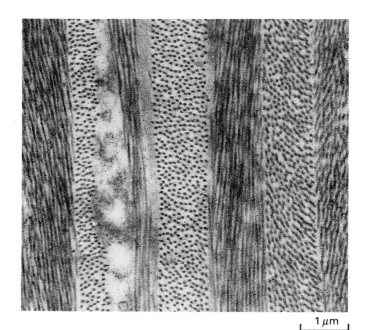

▲ **Figure 23-14** Electron micrograph of a section from a rabbit cornea, showing the layered arrangement of fibrous collagen. In each layer fibers are parallel to each other and perpendicular to those in adjacent layers. *Courtesy D. Parry.*

Type IV Collagen Forms the Two-dimensional Reticulum of the Basal Lamina

The *basal lamina* is the thin sheetlike network of extracellular matrix components lining the basal surface of most epithelial and endothelial cells and connecting them to the underlying connective tissue (see Figures 23-2 and 23-3). It also surrounds cells such as muscle and adipocytes. The basal lamina is usually no more than 60–100 nm thick (Figure 23-15). One of its principal constituents is type IV collagen, which forms a two-dimensional reticulum. This chicken-wire–like network gives the basal lamina its shape and rigidity.

Three type IV collagen polypeptides form a 40-nm-long triple helix with globular domains at the C-termini. The helical segment is unusual in that the Gly-X-Y sequences are interrupted about 20 times with segments that cannot continue the triple helix; these nonhelical regions introduce flexibility into the region. By binding together at nonhelical regions at their N-termini, four monomers assemble into a characteristic unit (Figure 23-16a). Triple-helical regions from several molecules associate laterally to form branching strands of variable but thin diameters. This, together with interactions between

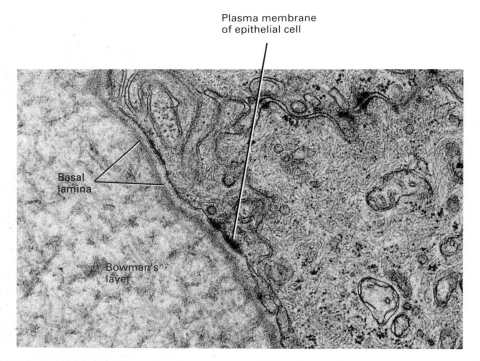

▲ **Figure 23-15** Electron micrograph of a section of human corneal epithelial cells, showing the basal lamina lining the cell. The basal lamina is anchored to the plasma membrane by receptor proteins and is also connected to the thick collagen-filled connective tissue, called Bowman's layer, that supports the epithelium. *Courtesy T. Kuwahara and D. Fawcett. From D. Fawcett, 1981, The Cell, 2d ed., Saunders, p. 47/Photo Researchers, Inc.*

C-terminal nonhelical domains, generates an irregular two-dimensional fibrous network (Figure 23-16b). Such cross-linked type IV collagens form the predominant fibers in the basal lamina (Figure 23-16c).

In summary, collagens are a family of matrix proteins, all of whose members share a characteristic triple-helix structure. It is the nonhelical domains of the different polypeptides that give the collagens their varied structures and binding properties.

Hyaluronic Acid and Proteoglycans

Several times in this chapter we have made reference to various polysaccharides and the proteoglycans, a set of protein-carbohydrate complexes that are ubiquitous in all matrices. We turn now to a discussion of their structure and function, beginning with hyaluronic acid.

HA Is an Immensely Long, Negatively Charged Polysaccharide That Forms Hydrated Gels

Hyaluronic acid (HA) is a major component of the extracellular matrix that surrounds migrating and proliferating cells, particularly in embryonic tissues. It is also a major structural component of the complex proteoglycans that are found in many extracellular matrices, particularly cartilage. Because of its remarkable physical properties, HA imparts stiffness and resilience as well as a lubricating quality to many types of connective tissue such as joints.

One molecule of HA consists of as many as 50,000 repeats of the simple disaccharide glucuronic acid $\beta(1 \rightarrow 3)$ *N*-acetylglucosamine (Figure 23-17). Because of the β linkages between saccharides, and because of extensive

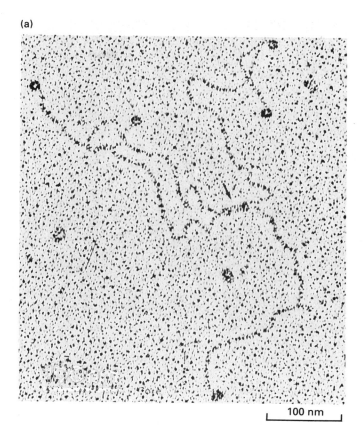

(a)

|___| 100 nm

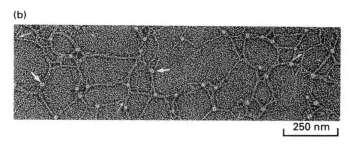

(b)

|___| 250 nm

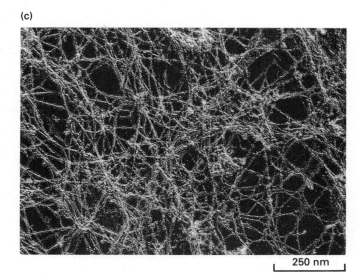

(c)

|___| 250 nm

▶ **Figure 23-16** A polarized network of type IV collagen forms the structural scaffold of the basal lamina. (a) An electron micrograph of the type IV unit—four molecules connected at their N-terminal ends *(arrow)*. Each molecule consists of a triple helix about 400 nm long and a globular domain at the C-terminus. (b) Purified type IV collagen polymerizes when the C-terminal globular regions of two or more polypeptides bind together *(arrows)*. (c) The basal lamina of amniotic cells, after salt extraction, is seen as a network of thin branching filaments. These are formed almost entirely of type IV collagen molecules; the thickness of the individual fibers is the same as that of the fibrous segments of the type IV collagen formed *in vitro* shown in (b). *Courtesy Dr. P. Yurchenco; parts (b) and (c) from P. Yurchenco and G. C. Ruben, 1987, J. Cell Biol. 105:2559–2568.*

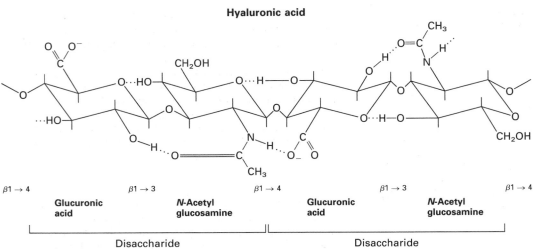

◀ **Figure 23-17** The chemical structure of hyaluronic acid; individual molecules contain up to 50,000 disaccharides. Hyaluronic acid is the only extracellular oligosaccharide that is not covalently linked to protein.

intrachain hydrogen bonding (shown as dots in the figure) each HA molecule forms a long, rigid rod. Mutual repulsion between negatively charged carboxylate groups that protrude outward at regular intervals contributes to the overall rigid structure. One molecule can be as long as 20 μm!

Because of the large number of hydrophilic residues on its surface, HA binds a large amount of water and forms, even at low concentrations, a viscous hydrated gel. Given no constraints, an HA molecule will occupy a volume 1000–10,000 times the space of the HA itself. When placed in a confining space, such as in a matrix between two cells, the long HAs will tend to push outward. This creates a swelling, or turgor pressure, within the space; the HAs push against any fibers or cells that block their motion. Importantly, by binding cations, the COO^- groups on the surface increase the concentration of ions and thus the osmotic pressure in the HA gel. Large amounts of water are taken up into the matrix, creating a turgor pressure within the HA matrix. These swelling forces give connective tissues their ability to resist compression forces, in contrast to collagen fibers, which are able to resist stretching forces.

HA Inhibits Cell-Cell Adhesion and Facilitates Cell Migration

Hyaluronic acid is bound to the surface of many migrating cells. Because of its loose, hydrated, porous nature, the HA coat appears to keep cells apart from each other and gives cells the freedom to move about and proliferate. Cessation of cell movement and initiation of cell-cell attachments are frequently correlated with a decrease in hyaluronic acid, a decrease in the cell-surface molecules that bind HA, and an increase in the extracellular enzyme hyaluronidase that degrades HA. These functions of HA are particularly important during the many cell migrations that facilitate differentiation.

A striking example, shown in Figure 23-18, involves migration of the sclerotome cells from each of the embryonic somites. Hyaluronic acid is one of the major substances in the matrix space through which these cells migrate. When the cells reach the vicinity of the neural tube, they stop migrating and synthesize matrix proteins characteristic of cartilage. At the same time the cells stop synthesizing HA and synthesize and secrete hyaluronidase, which destroys the matrix HA.

HA also plays a key role in formation of striated muscle cells, such as in the differentiation of the myotome. Undifferentiated muscle-cell precursors, called myoblasts, express no muscle-specific proteins such as α-actin or troponin C. Hundreds of myoblasts fuse together to form a multinucleate syncytium that will induce muscle-specific proteins and become a multinucleated, nondividing muscle fiber. Myoblasts bear an HA-rich coat that helps to prevent premature cell fusion. At the time of fusion, the HA-rich cell coat and the binding sites for HA are lost. The fused cells begin to synthesize a different, HA-free extracellular matrix, the basal lamina, that, as we have seen, allows the cells to attach to the surrounding collagen matrix.

Proteoglycans Comprise a Diverse Family of Cell-surface and Extracellular Matrix Macromolecules

Proteoglycans are macromolecules found in all connective tissues and extracellular matrices and on the surface of many cells. They contain a core protein to which is covalently attached one or more polysaccharides known as glycosaminoglycans. Glycosaminoglycans are long repeating linear polymers of specific disaccharides; usually one sugar is a uronic acid and the other is either N-acetylglucosamine or N-acetylgalactosamine. One or both of the sugars contain one or two sulfate residues (Figure 23-19). Thus each glycosaminoglycan chain bears

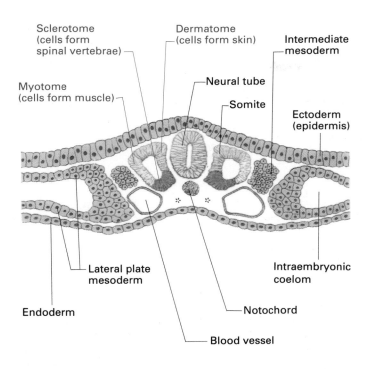

Sclerotome (cells form spinal vertebrae)

Dermatome (cells form skin)

Intermediate mesoderm

Myotome (cells form muscle)

Neural tube

Somite

Ectoderm (epidermis)

Lateral plate mesoderm

Intraembryonic coelom

Endoderm

Notochord

Blood vessel

◄ Figure 23-18 Migration of cells in the chick embryo. Shown is a cross-sectional view of the chick embryo just after the neural tube has formed. The tube runs the length of the embryo, anterior-to-posterior, and will give rise to the central nervous system. In each segment of the embryo, two undifferentiated masses of mesodermal cells called somites surround the neural tube and the notochord. The sclerotome cells on the bottom of the somite (orange) will migrate through the space denoted with an asterisk toward the notochord; there they will form cartilage and, ultimately, the vertebrae surrounding the neural tube.

▼ Figure 23-19 Structural formulas of some repeating sulfated disaccharides found in glycosaminoglycans, the polysaccharide component of proteoglycans. Green boxes indicate acidic residues. Note that dermatan sulfate is similar to chondroitin 4-sulfate and is formed from chondroitin 4-sulfate by inversion of the carboxylate group attached to the 5 carbon in D-glucuronic acid, generating L-iduronic acid. Heparin and heparan sulfate are actually complex mixtures resulting from different modifications, shown in parentheses, of sugar residues. The number (*n*) of disaccharides typically found in each proteoglycan chain is given.

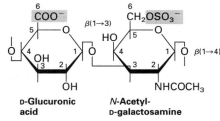

D-Glucuronic acid

N-Acetyl-D-galactosamine

Chondroitin 6-sulfate
(*n* = 20–60)

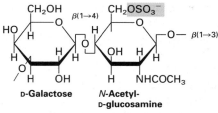

D-Galactose

N-Acetyl-D-glucosamine

Keratan sulfate
(*n* = 25)

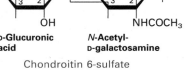

D-Glucuronic acid

N-Acetyl-D-galactosamine

Chondroitin 4-sulfate
(*n* = 20–60)

D-Glucuronic acid

N-Acetyl-or N-sulfo-D-glucosamine

Heparan sulfate
(*n* = 15–30)

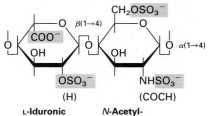

L-Iduronic acid

N-Acetyl-or N-sulfo-D-glucosamine

Heparin
(*n* = 15–30)

D-Glucuronic acid

N-Acetyl-D-galactosamine

and

L-Iduronic acid

N-Acetyl-D-galactosamine

Dermatan sulfate
(*n* = 30–80)

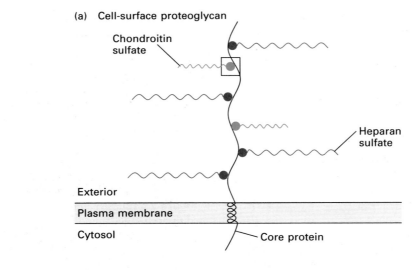

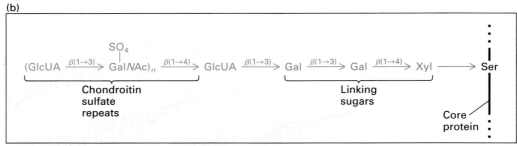

Gal = Galactose
GalNAc = N-Acetylgalactosamine

GlcUA = Glucuronic acid
Xyl = Xylose

▲ **Figure 23-20** Model of a cell-surface proteoglycan from mouse mammary epithelial cells. (a) The core protein (56,000 MW) spans the plasma membrane. On its long external domain are attached four heparan sulfate chains, each containing about 100 repeating disaccharides, and two chondroitin sulfate chains, each with about 70 disaccharide repeats. (b) The linkage of chondroitin sulfate to the core protein. Three characteristic linking sugars connect the chondroitin sulfate repeats to the serine side chain. Synthesis of the polysaccharide is initiated by transfer of a xylose residue to the serine, most likely in the Golgi complex, followed by addition, one at a time, of the galactose, *N*-acetylgalactosamine and glucuronic acid residues. In the Golgi complex sulfates are added to sugars already incorporated into the oligosaccharide. *Part (a) from A. Rapraeger, M. Jakanen, and M. Bernfeld, 1987, in T. N. Wight and R. P. Mecham, eds.,* Biology of Proteoglycans, *Academic, p. 137.*

many negative charges. Proteoglycans are remarkable for their diversity. A given matrix may contain several different types of core proteins, and each may contain different numbers of oligosaccharide chains with differing lengths and compositions. Thus, the molecular weight and charge density of a population of proteoglycans can only be expressed as an average; individual molecules can differ considerably. Nonetheless, a good deal is known of the structure and function of certain proteoglycans.

Proteoglycans are named according to the structure of their principal repeating disaccharide. Frequently some of the residues in an oligosaccharide chain are modified after synthesis; dermatan sulfate is formed from chrondroitin sulfate, for instance. Similarly, heparin (used medicinally

as an anticlotting drug) is formed as a result of enzymatic modification of heparan sulfate.

Proteoglycans are found on the surface of many cell types, particularly epithelial cells. Figure 23-20 shows the structure of one, isolated from cultured mammary epithelial cells. The attached heparan sulfate chains bind to fibrous collagens (types I, III, and V) as well as to the large glycoprotein fibronectin that are found in the interstitial matrix surrounding the basal lamina; thus the cell-surface proteoglycan is thought to anchor cells to matrix fibers.

Both heparan sulfate and chondroitin sulfate chains are connected, via a three-sugar "linker," to a serine residue on the core. The "signal sequence" of a core protein that specifies addition of an oligosaccharide to serine is Ser-

Gly-X-Gly, where X is any amino acid. However, we do not know why chrondroitin sulfates are attached at some sites and heparan sulfates at others, nor how the length of the oligosaccharide chain is determined.

Cartilage Proteoglycans Impart Resilience to the Tissue

The proteoglycan from cartilage (termed the proteoglycan aggregate) is one of the largest macromolecules known; a single molecule can be more than 4 μm long and of a larger volume than a bacterial cell! These proteoglycans give cartilage its unique gel-like properties and its resistance to deformation. The central component of the cartilage proteoglycan aggregate is a long molecule of hyaluronic acid. Bound to it, tightly but noncovalently at 40-nm intervals, are many core proteins of chondroitin sulfate proteoglycans. Binding of the proteoglycan core is stabilized by a linker protein that binds (noncovalently) both to the HA and to the protein core of the proteoglycan. Attached to each core protein are multiple chains of chondroitin sulfate and keratan sulfate (Figure 23-21). The molecular weight of the proteoglycan monomer— that is, the core protein with its bound glycosaminoglycans—averages 2×10^6, and the entire complex proteoglycan aggregate has a molecular weight in excess of 2×10^8 Da!

Not all matrix proteoglycans are this large. A class of "simple" proteoglycans, for instance, are found in basal lamina. They consist of a core protein of 20,000–400,000 MW to which is attached several heparan sulfate chains. Such proteoglycans bind to type IV collagen, as well as to other structural proteins of the basal lamina that we discuss later, and impart structure to the lamina.

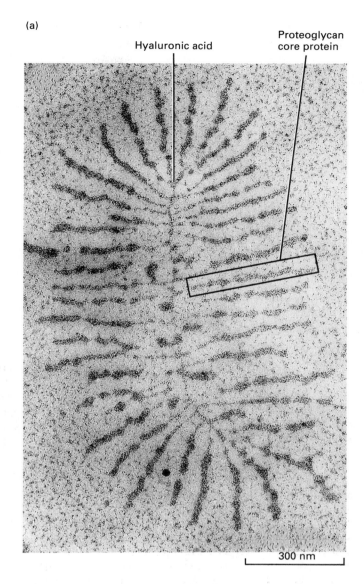

(a)

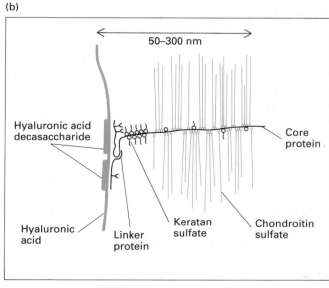

(b)

▶ **Figure 23-21** (a) Electron micrograph of a proteoglycan from fetal bovine epiphyseal cartilage. Proteoglycan core proteins are bound at regular ~40-nm intervals to a central strand of hyaluronic acid. Attached to the core proteins are the glycosaminoglycans keratan sulfate and chondroitin sulfate. (b) Details of the structure of a proteoglycan monomer. Each core protein (2124 amino acids) has one site at its N-terminus that binds with high affinity to a decasaccharide (10 sugars, or 5 disaccharide repeats) of the central hyaluronic acid; binding is facilitated by a linker protein that joins a hyaluronate decasaccharide to the core protein. Each core has 117 Ser-Gly sequences for addition of an oligosaccharide chain; 30 are short keratan sulfate chains and 97 are chondroitin sulfate. *Part (a) courtesy Dr. L. Rosenberg. From J. A. Buckwalter and L. Rosenberg, 1983, Coll. Rel. Res. 3:489–504. Part (b) after D. Heinegard and A. Oldberg, 1989, FASEB J. 3:2042–2051.*

Laminin, Fibronectin, and Other Multiadhesive Matrix Glycoproteins

Collagens and proteoglycans are two of the important components of all extracellular matrices. We turn now to a set of matrix proteins that bind not only to collagens and proteoglycans but also to specific cell-surface receptors. These proteins are important for organizing the other components of the matrix and also for regulating cell attachment to the matrix, cell migration, and cell shape.

Laminin Is a Principal Structural Protein of All Basal Lamina

All basal laminae contain a common set of proteins and glycosaminoglycans: type IV collagen, heparan sulfate proteoglycans, the poorly understood glycoprotein entactin, and *laminin*. The basal lamina is often called the type IV matrix, named after its collagen component (Table 23-3). All of the matrix components are synthesized by cells that rest on the basal lamina.

Laminin is a cross-shaped molecule almost 70 nm long (Figure 23-22); it is as long as the basal lamina is thick. It contains three polypeptides of total molecular weight 820,000 and has high-affinity binding sites for other components of the basal lamina.

▶ **Figure 23-22** The structure of laminin. (a) An electron micrograph of laminin, dried on a grid and shadowed with platinum, shows it to be a cross-shaped molecule. (b) A diagram. [*Part* (a) see J. Engel et al., 1981, *J. Mol. Biol.* 150:97.] *Part* (a) *courtesy of J. Engel, E. Odermatt, A. Engel, J. Madri, H. Furthmayer, H. Rodhe, and R. Timpl. Part* (b) *after G. R. Martin and R. Timpl, 1987,* Ann. Rev. Cell Biol. 3:57–85.

(a)

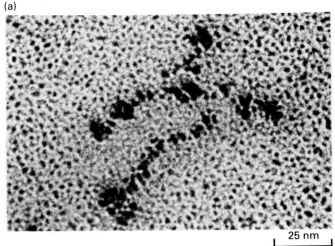

25 nm

(b)

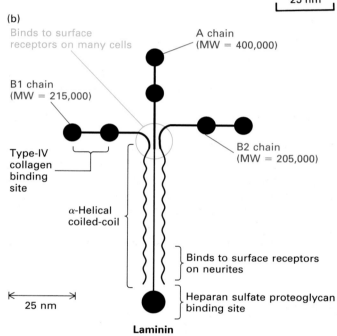

Laminin

Table 23-3 Representative matrix types produced by vertebrate cells in vivo

Collagen type	Associated anchorage protein	Associated proteoglycans	Cell-surface receptor	Cells producing
Type I	Fibronectin	Chondroitin sulfate Dermatan sulfate	Integrin	Fibroblasts
Type II	Fibronectin	Chondroitin sulfate	Integrin	Chondrocytes (cartilage)
Type III	Fibronectin	Heparan sulfate Heparin	Integrin	Quiescent (nondividing) hepatocytes; fibroblasts in association with epithelia
Type IV	Laminin	Heparan sulfate Heparin	Laminin receptors	All epithelial cells; endothelial (blood-vessel lining) cells; regenerating hepatocytes
Type V	Fibronectin	Heparan sulfate Heparin	Integrin	Quiescent fibroblasts
Type VI	Fibronectin	Heparan sulfate	Integrin	Quiescent fibroblasts

SOURCE: L. M. Reid, 1989, in J. W. Pollard and J. M. Walker, eds., *Methods in Molecular Biology*, Vol. V: *Tissue Culture*, Humana.

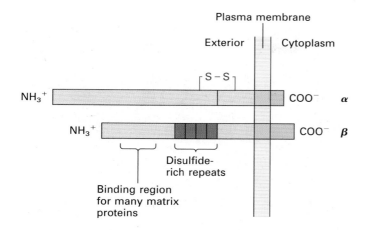

◄ **Figure 23-23** The structure of the integrin class of cell-surface receptors. Each receptor consists of two transmembrane polypeptides, α and β, whose molecular weight varies from about 100,000 to 140,000 in different molecules. Each β chain has a region of 100–200 residues at the N-terminus that binds the Arg-Gly-Asp sequence common to many of the matrix proteins to which these receptors bind. *After R. O. Hynes, 1987, Cell 48:549–554 and S. D'Souza, M. H. Ginsburg, T. A. Burke, S. Lam, and E. F. Plow, 1988, Science 242:91–93.*

Cells generally do not bind directly to type IV collagen or to matrix proteoglycans. Rather, laminin anchors the basal lamina to the cell surface. Different cells that are surrounded by a basal lamina such as epithelial cells, fat cells, and smooth and striated muscle may utilize different cell-surface receptor molecules. There are at least two different regions of the laminin molecule to which these receptors bind.

Directed by growth cones, axon and dendrite projections of nerve cells (neurites) often elongate, migrating along extracellular-matrix pathways that contain laminin. Laminin receptors in the growth cones bind to laminin in such a way that interactions form and break readily as the neurite moves along the fiber. In fact, when neurites are placed on a film of laminin dried on a plastic surface, the cells extend axonlike processes. Thus, it is thought that laminin exerts a hormonelike action on the neurites and that the neurite laminin receptor transduces the laminin signal across the plasma membrane, as hormone receptors do.

One class of laminin receptors is a member of the so-called integrin receptor superfamily (Table 23-4). These receptors are all similar in structure (Figure 23-23), having two transmembrane proteins, the α and β chains. Different receptors contain different permutations of α and β chains. Each integrin receptor binds to one or more matrix glycoproteins. For example, the laminin receptor, found on many cells, binds only to laminin whereas VLA receptors, found mainly on the T cells of the immune system, bind both laminin and fibronectin.

Table 23-4 The integrin receptor superfamily

Protein	Probable subunit composition	Cell type	Known ligands	Known function
Fibronectin receptor (VLA-5)	$\alpha_5\beta_1$	Many	Fibronectin	Cell adhesion and migration, phagocytosis, cytoskeletal connections
Laminin receptor	$\alpha_?\beta_?$	Epithelial, others	Laminin	Cell attachment, neurite outgrowth
VLA-1, 2 VLA-3, 4	$\alpha_1\beta_1, \alpha_2\beta_1$ $\alpha_3\beta_1, \alpha_4\beta_1$	T cells of the immune system, other cells	Fibronectin, laminin	Cell adhesion
Vitronectin	$\alpha_v\beta_3$	Many	Vitronectin	Adhesion to vitronectin
Glycoprotein IIb/IIIa	$\alpha_{II}\beta_3$	Platelets	Fibronectin, fibrinogen, vitronectin, van Willebrand factor	Platelet adhesion and aggregation, formation of blood clots
Mac-1	$\alpha_m\beta_2$	Monocytes, macrophages, lymphoid cells	C_3b_i subunit of complement	C_3 receptor, cell-cell adhesion
LFA-1	$\alpha_L\beta_2$	Monocytes, macrophages, lymphocytes	?	Leukocyte adhesion, lymphocyte cytotoxicity

SOURCE: R. Hynes, 1987, *Cell* 48:549–554; E. Ruoslahti and M. Pierschbacher, 1987, *Science* 238:491–497; K. R. Gehlsen, L. Dillner, E. Engvall, and E. Ruoslahti, 1988, *Science* 241:1228–1229.

The basal lamina is structured differently in different tissues: different laminin receptors on different cells may generate this diversity. For instance, the basal lamina that surrounds capillary cells forms a filter that regulates passage of proteins and other molecules from the blood into the tissues; similarly, the double-thickness basal lamina in the glomerulus acts as a filter in forming the urine (Figure 23-24). In smooth muscle, on the other hand, the basal lamina connects adjacent cells and maintains the integrity of the tissue. Laminin and other type IV matrix components are also found in mammalian four- and eight-celled embryos; the basal lamina helps these cells adhere together in a ball.

Fibronectins Bind Many Cells to Fibrous Collagens and to Other Matrix Components

Fibronectins are an important class of matrix glycoproteins. Their primary role is attaching cells to a variety of extracellular matrices (Figure 23-25)—all matrices, in fact, other than type IV (see Table 23-3). The presence of fibronectin on the surface of nontransformed cultured cells, and its absence on transformed (or tumorigenic) cells, first led to the identification of fibronectin as an adhesive protein. By their attachments, fibronectins regulate the shape of cells and the organization of the cytoskeleton; they are essential for many types of cell migrations and cellular differentiations during embryogenesis. Fibronectins are also important for wound healing, for which they facilitate migration of macrophages and other immune cells into the affected area, and in the initiation of blood clots, by allowing platelets to adhere to damaged regions of blood vessels.

Fibronectins are dimers of two similar peptides (Figure 23-26a); each chain is about 60–70 nm long and 2–3 nm thick. At least 20 different fibronectin chains have been identified, all of which are generated by alternative splicing of the RNA transcript of a single fibronectin gene. Analysis of proteolytic digests of fibronectin shows that the polypeptides consist of six tightly folded domains. Each domain, in turn, contains small repeated sequences whose similarities in amino acid sequence allow them to be classified into three types (Figure 23-26b).

Fibronectin possesses specific high-affinity binding sites for cell-surface receptors, collagen, fibrin, and sulfated proteoglycans. Further digestion of the cell-binding domain with proteases shows that a tetrapeptide sequence—Arg-Gly-Asp-Ser—within this domain is the minimal structure required for recognition by cells. When linked by covalent binding to an inert protein (such as albumin) and dried on a culture dish, this tetrapeptide is seen to be similar to intact fibronectin in its ability to promote adhesion of fibroblasts to the surface of the dish.

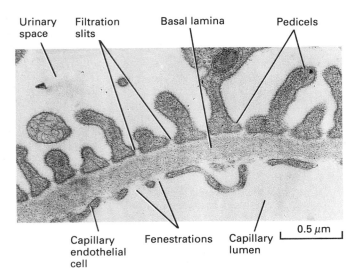

Urinary space   Filtration slits   Basal lamina   Pedicels

Capillary endothelial cell   Fenestrations   Capillary lumen   0.5 μm

▲ **Figure 23-24** Electron micrograph of a section through a rat kidney glomerulus, showing how the basal lamina acts to filter capillary blood, forming a filtrate in the urinary space that ultimately becomes urine. The endothelial cells that line the capillaries have many gaps within them, as do the epithelial cells that line the urinary space, so that the basal lamina is exposed to both the blood and urine spaces. The basal lamina in this region is a fusion of two basal laminae, one formed by the endothelial and one by the epithelial cells, and is twice as thick as basal lamina in most other tissues. *From R. Kessel and R. Kardon, 1979,* Tissues and Organs: A Text-Atlas of Scanning Electron Microscopy, *Freeman, p. 233.*

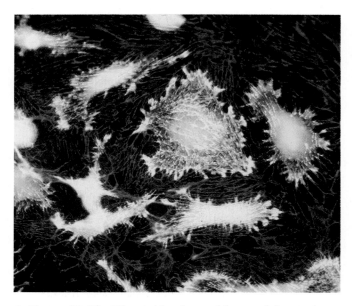

▲ **Figure 23-25** Fibronectin, detected by a red-fluorescing antibody, is secreted by cultured fibroblasts, revealed by a yellow-fluorescing antibody. Fibronectin forms a fibrillar array that is densest at the points of contact with the cells. *Courtesy R. Hynes. From R. Hynes 1986,* Sci. Am. *254(6):42.*

(a)

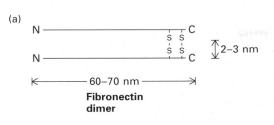

Fibronectin dimer

◄ **Figure 23-26** Structure of fibronectin and its variants. (a) Fibronectin contains two very similar polypeptides linked at their C-termini by two disulfide bonds. (b) Proteolytic digestion reveals six domains with specific binding sites for cell-surface receptors, heparan sulfate proteoglycans, collagen, and also fibrin, a major constituent of blood clots. Note that there are two domains for binding to heparan sulfate and two to fibrin, which differ in affinity. Each binding domain is composed of multiple copies of short repeating sequences, each of which may contribute to the binding. Each box represents a repeating sequence that is one of three types; each sequence is encoded by one or two exons. *After R. H. Hynes et al., 1989, Nectins and integrins: versatility in cell adhesion, in G. M. Edelman, B. Cunningham, and J.-P. Thiery, eds.,* Morphoregulatory Molecules, *Wiley, in press.*

(b)

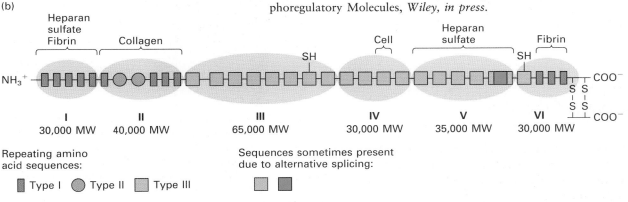

Fibronectin Promotes Cell Adhesion to the Substratum

The fibroblast fibronectin receptor has been purified, cloned, and sequenced; it is a member of the integrin class of cell-surface receptors. On its external surface is a high-affinity binding site for the Arg-Gly-Asp-Ser cell-adhesion segment of fibronectin.

Many cultured animal cells secrete fibronectin that binds to the culture dish and allows the cells to adhere to it. Fibronectin tends to concentrate at the small points where nonmigrating cells adhere to the substratum. Such attached cells also contain abundant actin-rich stress fibers that maintain the cell's shape and attach to the plasma membrane where it adheres to substratum (Figure 23-27). Several actin-binding proteins such as vinculin

(a)

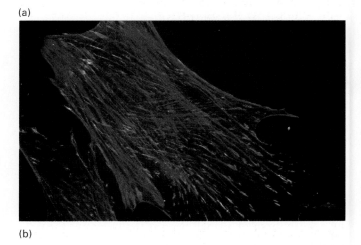

► **Figure 23-27** (a) Immunofluorescence of a fixed, stationary cultured fibroblast showing colocalization of the fibronectin receptor (green) and actin-containing stress fibers (red). At the ends of the stress fibers, where the cells contact the substratum, there is coincidence of actin and the fibronectin receptor (yellow). (b) Model of the connections between actin, fibronectin, and the extracellular matrix at the regions of contact between stationary fibroblasts and the substratum. The two-subunit transmembrane fibronectin receptor binds to fibronectin on its expolasmic side and talin on its cytoplasmic side. Vinculin binds to talin and probably directly to an actin filament; fibronectin binds to fibrous collagen and to many proteoglycans. *Part (a) from J. Duband et al., 1988, J. Cell Biol. 107:1385.*

(b)

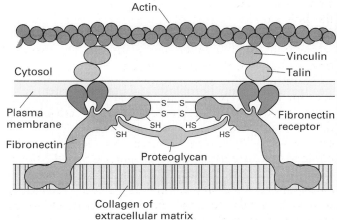

are also enriched in these regions. Electron micrographs occasionally reveal exterior fibronectin fiber bundles that appear continuous with bundles of actin fibers within the cell (Figure 23-28). As depicted in Figure 23-27, the fibronectin receptor may anchor both fibers to the opposite sides of the plasma membrane.

The apparent continuity of matrix and cytoplasmic fibers is probably not coincidental. Many tumor cells growing in culture do not stick to plastic culture dishes. Such cells synthesize little fibronectin and contain few stress fibers. Adsorption of purified fibronectin to the culture dish frequently results in tight adhesion and, concomitantly, to the development of stress fibers.

Fibronectins Promote Cell Migration

Fibronectins are abundant constituents of several types of extracellular matrices, generally together with fibrous collagens and specific proteoglycans (see Table 23-3). In many cases, these matrices are loosely packed with several fibrillar components secreted by various interstitial cells and provide trails along which other cells can migrate. An outstanding example is the migration of neural-crest cells during an early stage of differentiation in all vertebrate embryos (Figure 23-29).

Neural-crest cells arise from the ectoderm, or outer cell layer of the embryo, at the time when a region of the ectoderm rolls up inward and forms the neural tube that will generate the spinal cord neurons. The neural-crest cells migrate from the dorsal edge of the neural tube to all regions of the embryo. Those that move along the ventromedial pathway eventually stop and form aggregates that differentiate into various nerve ganglia or into the adrenal medulla. Attachment of migrating cells to fibronectin is essential for cell migration, and it is thought that matrix fibers containing fibronectin provide "tracks" along which the cells move. As evidence for this, microinjection into the embryo of an antibody specific for the cell-surface fibronectin receptor blocks migration. Injection into the embryo of a small peptide that contains the Arg-Gly-Asp-Ser binding domain of fibronectin also blocks migration of neural-crest cells. Apparently the peptide binds to cell-surface fibronectin receptors and prevents receptors from binding to matrix fibronectin.

In contrast to embryonic cells, cells in an adult organism generally do not migrate. An exception is wound-healing, when many fibroblasts and immune-system cells, including macrophages, migrate into the affected area. These cells move across blood clots, which are composed primarily of the fibrous protein fibrin. Fibronectin is incorporated into the clot as it forms by virtue of its fibrin-binding domains (see Figure 23-26); fibroblasts, in turn, adhere to the fibronectin as they migrate across the clot.

A cell such as a fibroblast that contains a single type of fibronectin receptor can regulate its mobility. Fibroblasts migrating on fibronectin make and break contacts with

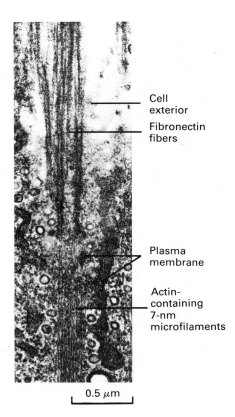

Cell exterior

Fibronectin fibers

Plasma membrane

Actin-containing 7-nm microfilaments

0.5 μm

▲ **Figure 23-28** Electron micrograph of the junction of fibronectin and actin fibers in a cultured fibroblast. Individual actin-containing 7-nm microfilaments, components of a stress fiber, traverse the obliquely sectioned cell membrane. The microfilaments appear in close proximity to the thicker, densely stained fibronectin fibers on the outside of the cell. *Courtesy of I. J. Singer. Copyright, 1978, M.I.T.*

the matrix. In such cells the fibronectin receptor is laterally mobile in the plasma membrane, as determined by fluorescence recovery from photobleach experiments (see Figure 13-25). As the cells stop migrating and form stress fibers, the fibronectin receptors become immobilized at the sites in contact with substratum and keep the cells tightly bound to the matrix; immobilization may be due to binding of the receptor, *via* talin and vinculin, to the actin cytoskeleton (see Figure 23-27).

Emerging from studies of the cytoskeleton, plasma membrane, matrix receptor proteins, and extracellular-matrix components is a picture of a continuum of interactions between the elements that allows for coordinated activities.

Cell-Cell Adhesion Proteins

Adhesion of cells of a particular type is a primary feature of the architecture of many tissues. A sheet of absorptive epithelial cells, for instance, forms the lining of the small

(a)

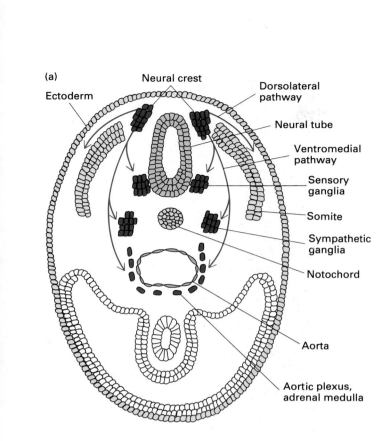

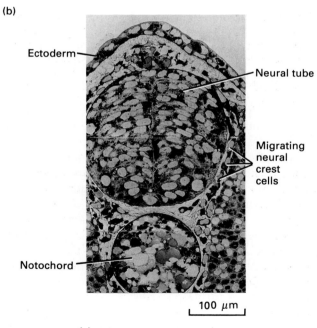

◀ **Figure 23-29** Migration of neural-crest cells along matrix filaments. (a) A cross section of a chick embryo at a stage just after the neural tube has formed by invagination of a region of the ectoderm, a stage earlier than that depicted in Figure 23-18. The neural-crest cells form a group on the dorsal *(top)* side of the neural tube. Some of the neural-crest cells migrate along ventromedial paths to sites where they will differentiate into the tissues labeled on the right. (b) Section of an axolotl embryo similar to the diagram in (a) *(top)*. (c) A scanning electron micrograph of a neural-crest cell migrating along matrix filaments. *Part (a) after M. Bonner-Fraser, 1984, J. Cell Biol. 98:1947–1960; parts (b and c) courtesy Dr. Jan Lofberg. From J. Lofberg et al., 1980, Dev. Biol. 75:148–167.*

intestine, and sheets of hepatocytes two cells thick make up much of the liver. A number of cell-cell adhesion molecules are now known; a single cell type may use more than one such molecule to mediate such homotypic (like-binds-like) adhesions.

Many features of cell-surface interactions were shown in now classical experiments performed beginning in the 1940s. Various types of embryonic tissues (which dissociate more easily than adult tissues) were dissociated into single cells and allowed to reassociate in the presence of other cell types. Invariably, it was found that cells of a particular type, such as liver, kidney, or retina, would reassociate preferentially with cells of a similar origin. Mixtures, say, of dissociated liver or kidney cells would form small aggregates consisting only of liver or only of kidney cells. If a single large mass was formed, one cell type was localized to the center, surrounded by cells of the other type. Adhesion of similar cells even extended across species boundaries. Mouse-liver cells, for instance, adhere tightly to chick-liver cells and form aggregates; however, they do not aggregate with mouse-kidney cells.

To analyze such cell-adhesion phenomena experimenters prepared monoclonal antibodies directed against proteins on the plasma membrane of a specific cell type, say chick liver. Some of these antibodies bound to adhesion molecules and blocked the ability of dissociated chick-liver cells to reaggregate. The monoclonal antibodies were subsequently used to characterize, purify, and clone the relevant cell-surface adhesion molecule. It is now apparent that cell-adhesion molecules comprise two classes of molecules: those that require Ca^{2+} for adhesion, the cadherins, and those that do not (Table 23-5). Many cells use both a Ca^{2+}-dependent adhesion molecule and a Ca^{2+}-independent one.

Cadherins are a family of Ca^{2+}-dependent cell-cell adhesion molecules that are important for tissue differentiation and structure; they can be divided into three subclasses: the E-, P-, and N-cadherins. Each of these cadherins is an integral-membrane glycoprotein of

Table 23-5 Major cell-adhesion molecules on vertebrate cells

Molecule	MW	Ligand	Predominant distribution
		Calcium-dependent	
E-cadherin (uvomorulin)	120,000–130,000	Homophilic	Preimplantation embryos; epithelial cells, particularly at the belt desmosome
P-cadherin	118,000	?	Trophoblast, heart, lung, intestine
N-cadherin	125,000–135,000	Homophilic	Nervous system, lung, heart, lens; embryonic mesoderm and neural ectoderm
		Calcium-independent	
Cell-CAM	110,000–150,000	Homophilic	Liver, other epithelia
Neural N-CAM	120,000, 140,000 180,000	Homophilic and heparan sulfate	Nervous system, glia
Muscle N-CAM	125,000 140,000 155,000		Embryonic and adult muscle

SOURCE: T. M. Jessell, 1988, *Neuron* 1:3–13; M. Takeichi, 1988, *Development* 102:639–655.

720–750 amino acids. On average, 50–60 percent of the sequence is identical between the different cadherins. Importantly, each cadherin has a characteristic distribution among tissues. During differentiation the amount or nature of the cell-surface cadherins change, affecting many aspects of cell-cell adhesion and cell migration.

E-Cadherin Is a Key Adhesive Molecule for Epithelial Cells

In adult vertebrates E-cadherin, also known as uvomorulin, is the intercellular glue that holds most epithelial sheets together. Sheets of polarized epithelial cells, such as those that line the small intestine or kidney tubules, contain abundant uvomorulin at the sites of cell-cell contact. Experiments directly establish the requirement for uvomorulin in cell adhesion. Addition of a monoclonal antibody to uvomorulin to a monolayer of cultured epithelial cells (see Figure 13-44) causes the cells to detach from each other. The removal of Ca^{2+} also disrupts cell adhesion. If uvomorulin-mediated cell adhesions are prevented from forming, none of the other cell junctions between epithelial cells, such as spot desmosomes or gap junctions, will be generated.

Uvomorulin, like the other cadherins, preferentially forms homotypic interactions. In epithelial cells, uvomorulin is concentrated in the junctional region called the belt desmosome (see Figure 13-45), the site where bands of actin filaments encircle the cell just inside the plasma membrane. Uvomorulin, via a set of as-yet-unidentified actin-binding proteins, anchors the sides of these actin filaments to the plasma membrane (see Figure 22-31) and thus maintains the rigidity of the cell layer.

Direct evidence that uvomorulins and other cadherins can determine the adhesive properties of cells is available. L cells are a cultured line of transformed mouse fibroblasts that express no cadherins and adhere poorly to themselves or to other cultured cells. Lines of transfected L cells that express either uvomorulin or P-cadherin were generated, and such cells adhered preferentially to cells expressing the same class of cadherin molecules. For instance, epithelial cells from embryonic lung express uvomorulin; they adhere tightly to uvomorulin-expressing L cells but do not attach to untransfected L cells or L cells expressing P-cadherin.

Cadherins Influence Morphogenesis and Differentiation

Cell-cell contacts form and break as essential parts of many steps in cell differentiation and tissue morphogenesis. Cadherins play a key role in these processes, as we will illustrate with examples from the early stages of mouse and chick differentiation.

After fertilization and fusion of the sperm and egg pronuclei, the mouse egg divides several times into a mass of cells called a morula (Figure 23-30). At this stage the major cell-adhesion molecule expressed is uvomorulin. Addition of a monoclonal antibody to uvomorulin causes loss of cell adhesion; thus uvomorulin is essential for the cells to adhere into a tight mass. At the later blastocyst stage all cells still contain uvomorulin, but soon after,

(a)

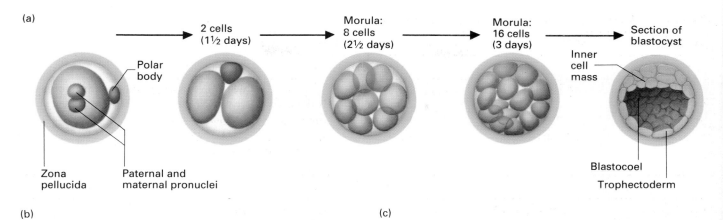

2 cells
(1½ days)

Morula:
8 cells
(2½ days)

Morula:
16 cells
(3 days)

Section of
blastocyst

Polar
body

Inner
cell
mass

Zona
pellucida

Paternal and
maternal pronuclei

Blastocoel

Trophectoderm

(b)

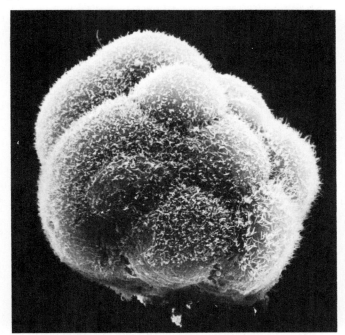

(c)

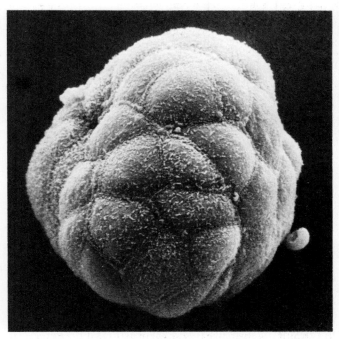

▲ **Figure 23-30** Early development of the mouse embryo. (a) Events following the fusion of the sperm and egg nuclei. The zona pelucida, a protein-rich layer surrounding the egg and early embryo, was removed to make the scanning electron micrographs in (b) and (c); these show the tight adherence of (b) morulae and (c) early blastocyst cells to one another; cell-cell adhesion is mediated by uvomorulin. Many microvilli are visible on the free cell surfaces.

mesodermal cells lose uvomorulin and migrate away from the ectodermal cell layer.

At a later stage, a region of the outermost embryonic cell layer, the ectoderm, folds inward to form the neural tube that runs the length of the embryo (Figure 23-31). As the ectodermal cells invaginate, they lose uvomorulin and synthesize a different but related adhesion molecule, N-cadherin. During further differentiation of the central nervous system from the neural tube, N-cadherin becomes a major adhesive molecule, and P cadherin is expressed by several types of cells.

We noted earlier (see Figure 23-29) that, at the time the neural tube is formed, the neural crest is also produced.

The region of the ectoderm that generates neural-crest cells also loses uvomorulin and expresses N-cadherin. However, expression of N-cadherin by neural-crest cells is only temporary. As the neural-crest cells begin migrating they lose all cadherins, and thus are distinguished from the overlying ectoderm (expressing uvomorulin) and the neural tube (expressing N-cadherin). Presumably the loss of cell-adhesive proteins is essential for cells to migrate. Once the neural-crest cells reach their destinations and form the ganglia indicated in Figure 23-29, they differentiate into neurons and simultaneously reexpress N-cadherin. This adhesive molecule helps maintain the integrity of neuronal aggregates in the various ganglia.

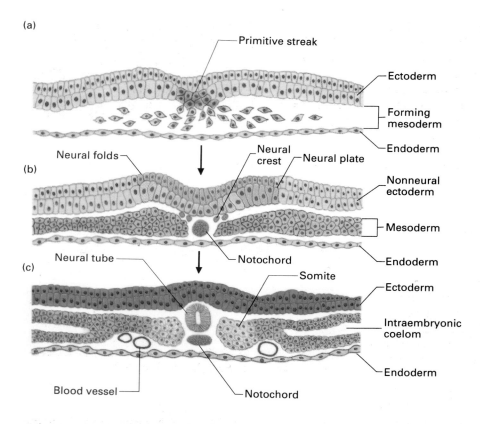

(a)

(b)

(c)

◀ **Figure 23-31** Changes in cell-surface Ca^{2+}-dependent adhesion molecules during early chick differentiation. The embryo in (a) consists of two layers of cells, the endoderm and ectoderm, each of which expresses E-cadherin (uvomorulin; yellow). Near the midsection of the embryo, ectodermal cells lose uvomorulin (gray cells express no cadherin) and migrate inward to form the mesoderm. (b) After many cell divisions, another segment of surface ectodermal cells along the midline—the neural plate—loses E-cadherin and folds inward. Eventually these cells synthesize N-cadherin (blue) and form the neural crest. (c) Still later the ectodermal cells express both E- and P-cadherins (orange) and the notochord N- and P-cadherins (purple). As mesodermal cells condense into somites they express N-cadherin. Some tissues, such as the endothelial cells that line the blood vessels (red), exhibit Ca^{2+}-dependent adhesion, yet do not express any of the known cadherins; thus, an as yet unidentified cadherin may mediate adhesion of endothelia. *After M. Takeichi, 1988,* Trends Genet. *3:213–217.*

N-CAMs Are a Set of Ca^{2+}-Independent Adhesive Molecules Encoded by a Single Gene

N-CAMs are the predominant set of Ca^{2+}-independent adhesive molecules in vertebrates. They are particularly important in nervous tissue; hence the name *nerve-cell adhesion molecule*. Like cadherins, N-CAMs form primarily homotypic interactions, binding cells with similar adhesion molecules together. Unlike the cadherin family, N-CAMs are encoded by a single gene, and diversity is generated by both alternative splicing and by differences in glycosylation (Figure 23-32).

Like N-cadherin, N-CAMs appear during morphogenesis as neural epithelial cells fold inward to form the neural crest and neural tube. They are uniformly distributed along the neural tube, and maintain neuron-neuron adhesion. N-CAMs, like N-cadherin, disappear from the neural-crest cells during migration. They are reinduced when the cells stop migrating and form ganglia. N-CAMs are expressed on differentiating muscle, glial, and nerve cells. In some cases, direct experimental evidence supports the role of N-CAMs in cell adhesion. For instance, adhesion of retinal neurons is inhibited by addition of antibodies to N-CAMs.

The adhesive properties of N-CAMs are modulated by long chains of sialic acid, a negatively charged sugar. In

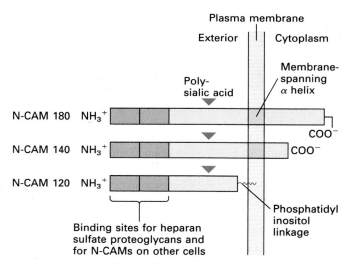

▲ **Figure 23-32** The three forms of N-CAM produced by alternative splicing. N-CAM 180 (180,000 MW) and N-CAM 140 are anchored in the membrane by a single hydrophobic α helix and differ in length of their cytoplasmic domains. N-CAM 120 is anchored by a complex inositol-phosphate-containing phospholipid (see Figure 13-15). The three forms also vary in the length of the poly $\alpha(2 \rightarrow 8)$ sialic acid chain, whose attachment site is indicated. *After G. M. Edelman, 1988,* Biochemistry *27:3533–3543 and* T. M. Jessell, 1988, Neuron *1:3–13.*

embryonic tissues polysialic acid constitutes as much as 25 percent of the mass of N-CAMs; in adult tissues there is only one-third as much. Plasma-membrane vesicles containing the heavily sialylated form of N-CAMs adhere to each other less tightly than do those with the less sialylated form of N-CAMs; this and other similar data suggest that the strength of cell-cell adhesions can be modified by glycosylation of the N-CAMs. Additionally, a site near the N-terminus of N-CAMs binds heparan sulfate proteoglycans, and binding of proteoglycans can also modify cell-cell adhesion.

Cell and Matrix Interactions during Development

In order for most of the cell types found in organs to develop, at least two types of cells that originate in different embryonic layers must interact. This interaction is called *induction*. Mesoderm-derived cells can induce adjacent sheets of epithelial cells, derived from the ectoderm or endoderm, to differentiate along a specific pathway. The interactions result in the synthesis of tissue-specific proteins in one or both partners of a pair. Though many of the details are not yet known, different types of extracellular matrices, synthesized by the different interacting cells, play key roles in the inductive processes. The skin provides an excellent example.

Mesodermal Cells Determine the Type of Structure Made by the Epidermis

The skin is composed of the epidermis, an outer layer derived from the ectoderm, and the dermis, an inner layer derived from the mesoderm. The epidermis synthesizes the basal lamina on which it rests, while the mesoderm synthesizes matrices containing fibrous collagens. The usual identifiable markers of a particular region of skin (feathers, scales, hairs, or simply the thickness of the outer layers of dry cornified cells) result from the controlled synthesis of specific epidermal proteins such as keratins and the controlled division of the epidermal cells (Figure 23-33).

The type of surface structure that is formed by the epidermis is governed by the underlying dermis. For example, the feathers of a chicken's legs and wings are different, and in areas such as the feet, a scaly, nonfeathered skin is produced. Presumptive epidermal or dermal cells may be taken from one area of an early chick embryo and transplanted to another. When dermal cells from the leg area are transplanted underneath the epidermis overlying the wing bud, the epidermis is induced to form feathers of the type normally found on the leg. On the other hand, transplantation of leg epidermis to the foot region does not give rise to feathers on the foot. Rather, the trans-

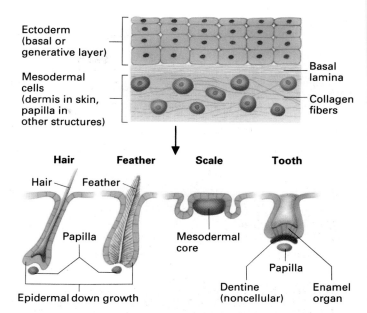

▲ **Figure 23-33** Structures formed by interactions of the vertebrate epidermis and dermis. *Drawings at bottom after E. M. Deuchar, 1975,* Cellular Interactions in Animal Development, *Chapman & Hall (London), p. 149.*

planted epidermis produces the characteristic scales of the foot. Thus the same area of epidermis is capable of developing feathers or scales and the underlying dermis determines which is produced.

At the stage of development at which such transplant experiments are conducted, the dermis is already differentiated—that is, it is specialized and restricted in the type of epidermal structures it can induce; but the epidermis is not. It is thought that different types of matrices fabricated by the differentiating dermal cells, as well as the growth hormones they secrete, are key factors in inducing differentiation of the overlying epidermis.

Neuroectodermal Cells Induce Epithelial Cells to Differentiate into a Lens

In any animal many different structures are formed by differentiation of the outermost layer of epithelial cells. One important example is production of the crystalline lens of the eye. The inducing cells are neuroectodermal, growing outward from the neural tube at the region where the primitive brain forms. The optic vesicles (Figure 23-34) bulge outward on the left and right sides. When they reach the epidermis, part folds inward, generating the optic cup that will differentiate into the retina (Figure 23-35). The ectodermal cells in the region of the head overlying the optic cup initially express E- and P-cadherins. As they move inward they lose P-cadherin and synthesize N-cadherin. Soon after, the innermost cells from the ectodermal layer lose N-cadherin and express

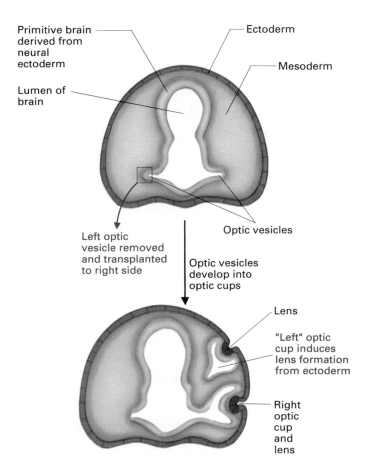

▲ **Figure 23-34** Diagrams of a cross sections through the head of a vertebrate embryo illustrating the interaction of the optic cup with the overlying ectodermal cells that become the optic lens. When the left optic vesicle is transplanted to another site in the head, it induces a lens in that inappropriate site. *After N. K. Wessels, 1977,* Tissue Interaction in Development, *Benjamin-Cummings, p. 46.*

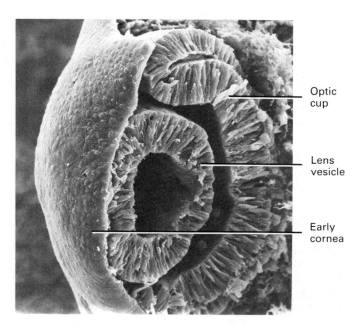

only E-cadherin, and differentiate into the cells that form the transparent lens.

In many amphibians, when the optic vesicle is transplanted underneath ectodermal cells that normally would not form a lens, a lens forms. Similarly, the developing lens is known to induce the overlying ectoderm to differentiate into the cornea, another specialized structure of the eye.

Many other structures and organs of the adult animal are constructed from ectodermally or endodermally derived cells that have received and interpreted developmental signals from an inducing cell partner, most often a cell of mesodermal origin.

Cell Interactions Are Essential for Formation of Internal Organs

Interactions between the endoderm (the cell layer that lines the primitive gut) and the mesoderm are required for the formation of internal organs. The organs associated with the gut begin as endodermal outgrowths from the primitive gut into areas of mesenchymal cells. The interaction of the two cell types induces the endodermal cells to differentiate into the characteristic epithelial cells of the salivary glands, pancreas, lung, and liver (Figure 23-36).

The inducing signals that pass between cells are generally not known. In some cases electron microscopy has shown that the plasma membranes of epithelial and mesenchymal cells do come into contact (see Figure 11-3). The membranes of ectodermal cells must touch those of mesenchymal cells to induce the differentiation of the mesenchymal cells into kidney cells.

In other cases, the plasma membranes of two interacting cells do not touch, and it is thought that a hormone-like signal secreted by one cell type becomes immobilized in the interconnecting matrix, where it can continuously stimulate the adjacent cell. The basal lamina secreted by cultured endothelial cells (cells that line the lumen of blood vessels), for instance, contains tightly bound *basic fibroblast growth factor* (FGF) polypeptide. The ability of this matrix to support growth of cultured endothelial cells or extension of axons by cultured neurons is blocked by antibodies to this growth factor. Thus at least part of the ability of this extracellular matrix to support cell proliferation and differentiation is due to a bound hormone. Both basic FGF and acidic FGF, another growth factor, bind very tightly to heparan sulfate in the basal lamina; binding protects the growth factors from degradation by tissue proteases, thereby enhancing their activities.

◀ **Figure 23-35** The eye of a developing chick. This scanning electron micrograph shows the fractured edge of an eye at about 5 to 6 days. *Courtesy of K. W. Tosney. From N. K. Wessels, 1977,* Tissue Interactions in Development, *Benjamin-Cummings, p. 44.*

Other matrix proteins important for inductive interactions of epithelial and mesenchymal cells are being elucidated. Tenascin, as one example, is a matrix protein, containing six identical 210,000 MW polypeptides, that in embryos is present selectively in the mesenchyme that surrounds epithelia in organs in which mesenchyme is essential for epithelial development. Tenascin is synthesized by mesenchymal cells and is thought to participate in inducing differentiation of epithelia such as those of the intestine. Tenascin is absent during the early stages of gut development. Synthesis of tenascin by cultured mesenchymal cells is induced by addition of cultured epithelial cells, but not by addition of other types of cells. Thus, in the developing embryo a sheet of undifferentiated gut epithelial cells induces the surrounding mesenchyme to synthesize and secrete tenascin, which, in turn, facilitates the differentiation of the epithelium. Most likely such cell-cell inductions require multiple steps and several extracellular signals.

The Basal Lamina Is Essential for Differentiation of Many Epithelial Cells

A basal lamina lines the basolateral membrane of all epithelial cells, such as the cells that line the intestine or that form the secretory cells of the salivary or mammary glands. It is essential for differentiation and morphogenesis of the epithelium. In the developing salivary gland, for instance, groups of epithelial cells secrete the proteins and proteoglycans that form a basal lamina, grow, and form clusters of cells that will be the secreting cells of the gland. Mesodermal cells are attached under the basal lamina. The groups of epithelial cells together with the basal lamina can be separated by microdissection from the mesodermal cells. As is shown in Figure 23-36, when this epithelial component is then cultured separately, no further growth occurs, but when it is mixed with mesodermal cells, growth and differentiation of salivary-gland tissue continue. If the epithelial cells are treated with hyaluronidase to destroy the basal lamina and are mixed with mesenchymal cells before the basal lamina has regenerated, they fail to grow and differentiate.

Thus two cell types and the basal lamina all interact and are necessary for the epithelial cells to differentiate correctly. The inducing molecules have not yet been elucidated. However, experiments with adult animals have shown that the basal lamina influences gene expression as well as the organization of the adjacent epithelial cells.

Consider, for instance, the cells of the mammary gland that secrete milk proteins such as casein, lactoferrin, and transferrin. These are organized into spherical structures one cell thick, called *alveoli;* milk proteins are secreted into the lumen of the alveolus. Similar to the architecture of the acini of the exocrine pancreas (Figure 4-40), a series of ductules connects the lumena and conducts the milk proteins into large mammary ducts.

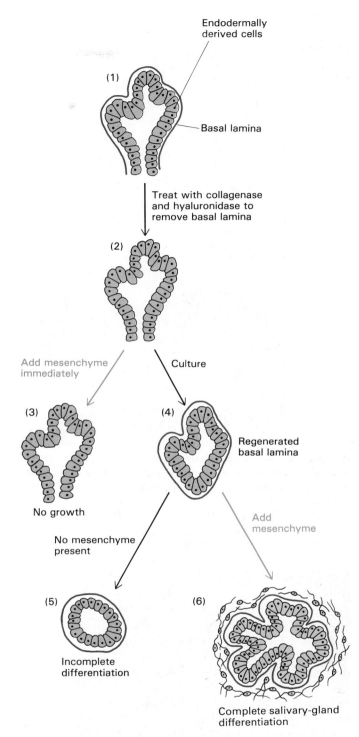

▲ **Figure 23-36** Elements necessary for salivary-gland differentiation. Salivary-gland explants can be dissected away from other cells (1) and treated with enzymes to remove the basal lamina (2). If the epithelial cells are then mixed immediately with mesenchymal cells, there is no growth or differentiation (3). If, on the other hand, the epithelial cells are cultured, they regenerate the basal lamina (4), but do not differentiate further (5) unless they are allowed to interact with mesenchymal cells (6). *After N. K. Wessels, 1977,* Tissue Interactions in Development, *Benjamin-Cummings, p. 225.*

(a)

Lumen

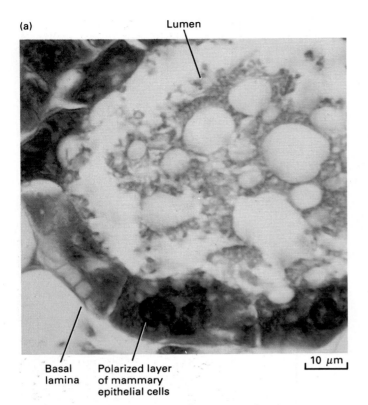

Basal   Polarized layer
lamina   of mammary
         epithelial cells

10 μm

(b)

Lumen

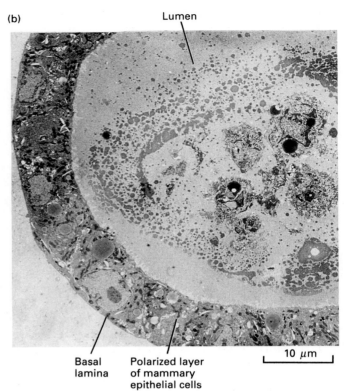

Basal   Polarized layer
lamina   of mammary
         epithelial cells

10 μm

▲ **Figure 23-37** A basal-lamina matrix induces cultured epithelial cells of the mammary gland to assume a configuration remarkably similar in size and structure to an in vivo alveolus. (a) Light micrograph of a cross section of an alveolus in a lactating mouse mammary gland. A layer of secretory epithelial cells surrounds the lumen which is filled with milk proteins and fat globules. (b) Electron micrograph through a spherical structure formed by disaggregated cells from the mammary gland of a midpregnant mouse, cultured on a type IV matrix. *Part* (a) *courtesy C. Streuli and M. Bissell.* [Part (b) see M. H. Barcellos-Hoff, J. Aggeler, J. Ram, and M. Bissell, *Development*, in press.]

If mammary epithelial cells are placed on culture dishes, they adhere to the dish but form neither tight junctions with each other nor the sheet of cells characteristic of normal glands. The cells synthesize little milk proteins, much of which is degraded within the cells. However, if such cells are plated on a dish coated with a mixture of proteins and proteoglycans characteristic of basal lamina, morphological and functional differentiation are dramatic. After 24 h the cells pull some of the basal lamina components away from the culture dish and form round aggregates that, after 4 days, develop with a hollow center. These are of the size and architecture of normal alveoli, and the cultured cells secrete milk proteins only into the lumen of the aggregate (Figure 23-37).

With the development of such cell-culture systems with well-defined roles for the matrix of the basal lamina, it should be possible to determine which components of the basal lamina affect cellular organization and gene expression of epithelial cells and how these components function.

Cell and Matrix Interactions during Neuron Development

Proper function of the nervous system depends on the intricate array of synapses that are formed during differentiation. Within the vertebrate central nervous system, proper contacts must be made among thousands of billions of interneurons, motor neurons, and sensory neurons; in the peripheral nervous system motor neurons must innervate the proper muscles in order to control body activities. Insects and other invertebrates have fewer neurons than vertebrates but must form the same types of specific cell-cell contacts.

Neurons form synapses with other cells primarily through their axons. During differentiation axons, which may be very long, and dendrites grow out from the cell body. Axons grow outward from the cell body toward those target cells with which it will form synapses. To fathom how the nervous system is constructed, we thus need to ask: how does an axon select the correct pathways along which to grow, and how does it recognize certain appropriate cells with which to synapse?

Although the nervous systems of vertebrates and insects are vastly different in their structure and complexity, similar principles of axon guidance are used: growth cones of elongating axons use a changing set of cell-surface receptors to move along specific matrix fibers and also along specific cells. There are even similarities in some of the vertebrate and insect cell-surface molecules used in nerve-cell recognitions. Thus we will choose examples from both types of systems.

Individual Neurons Can Be Identified Reproducibly and Studied

A prerequisite for understanding how specific neural contacts form is identifying specific neurons in the developing embryo and observing them as their axons elongate and form contacts. Embryos of grasshoppers, *Drosophila,* and other insects provide excellent experimental systems in which the central nervous system is formed by a string of ganglia, one per body segment. A ganglion may contain a thousand or so nerve-cell bodies, but some of these have a size and position permitting precise identification. In many cases adjacent body segments form similar structures such as legs; the same cell type with similar axon projections is generally found at the same site in ganglia in these segments.

When a nerve cell is sufficiently large, it can be microinjected with a fluorescent dye that spreads throughout its cytosol. The cell's projections can be visualized in the fluorescence microscope (Figure 23-38). Analogous techniques can be used to identify vertebrate motor neurons. When the enzyme peroxidase is injected into a nerve near its terminus, it will be transported back to the cell body by retrograde axonal transport. The enzyme may then be detected in fixed tissue by histochemical staining. In this way, it is possible to localize in the spinal cord the cell body of a particular neuron that innervates a specific muscle.

Growth Cones Guide the Migration and Elongation of Developing Axons

During differentiation of the nervous system, precursors of neurons, called neuroblasts, arise at particular locations at specific times. Such cells generally lack axonal or dendritic projections. Certain neuroblasts migrate to specific destinations where they form ganglia. One example we have encountered is the migration of neural-crest cells to form sympathetic ganglia. Cells also migrate extensively in the central nervous system.

At its destination the migrating cell "sprouts" and elaborates one or more axonal or dendritic projections. At the

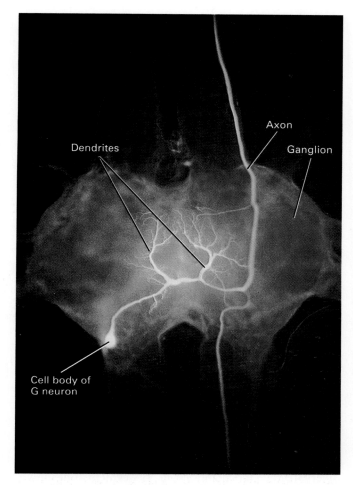

▲ **Figure 23-38** Identification of a single neuron in the grasshopper embryo. The fluorescent dye Lucifer yellow was microinjected into the cell body of the G neuron, one of 2000 neurons in the second thoracic ganglion. The axon extends from the cell body, crosses the ganglion, then extends forward (upward in this picture); a smaller axon branch extends rearward. *From C. S. Goodman, M. J. Bastiani, K. Pearson, and J. Steeves, 1984,* Sci. Am. *251(6):58.*

leading edge of the elongating axon is the highly motile structure known as the growth cone (see Figures 22-42 and 22-43). Because the growth cone can explore surfaces and crawl over them and can navigate in a particular direction, an axon can reach its proper target. Growth cones move along those substrates to which they adhere most tightly. In some cases, they are repelled if they encounter an inappropriate cell. As the growth cone moves outward, the cell body stays put and the axon elongates due in part to the polymerization of tubulin into microtubules that give the axon its rigidity.

In cell culture, growth-cone-mediated axon elongation is stimulated by many matrix substrates, such as laminin or fibronectin. However, such widely distributed molecules obviously cannot in themselves generate the specificity for movements of different axons toward different targets. Perhaps growth cones on different neurons have different and changing cell-surface receptors that enable them to recognize and move along cells and matrix components they encounter as they elongate to their specific targets.

Adjacent Motor Neurons Follow Different Pathways to Different Target Muscles

In vertebrates the cell bodies of motor neurons are located in the spinal cord, and axons extend out of the central nervous system by ventral roots. Axons that innervate a specific muscle are bound together to form a peripheral nerve. Obviously it is essential for each motor neuron to innervate only the appropriate muscle, and indeed this high degree of specificity is found throughout the nervous system.

In certain favorable experimental systems, such as the zebra-fish embryo, scientists have been able to observe individual motor neurons as their axons exit the spinal cord and elongate. The pattern depicted in Figure 23-39 is repeated in each segment of the embryo. Investigators can identify the cell bodies of the three motor neurons whose axons emerge first from each segment and can follow the fate of each cell by microinjecting it with a fluorescent dye. The axons of each of the labeled neurons initially grow out of the spinal cord along a common pathway. Then they diverge, one growing upward, to innervate the dorsal muscle in its segment, one downward, and one laterally. These axons are never seen to send branches or growth cones off in an inappropriate direction.

At a later stage of development, additional axons grow out of the central nervous system to innervate the same muscles. Growth cones of these axons migrate along the surfaces of the three pioneer axons, thus guiding the secondary axons to the correct muscles. However, experimental destruction of one of the pioneer neurons by an intense laser beam did not affect the ability of the secondary motor neurons to select the pathway that led to the appropriate target muscle.

It would appear that, at the time the growth cones exit the spinal cord, each neuron is already programmed to follow a specific pathway. Experiments with the chick embryo—where nerve transplantation experiments are easier—support this contention. Here, motor neurons from four adjacent segments of the spinal cord innervate various muscles of the hindlimb. Multiple axons destined for different muscles grow out of the spinal cord together and follow the same path to plexuses; from there the individual neurons follow different paths to the appropriate muscle. If, before nerve outgrowth, a piece of spinal cord containing motor-neuron cell bodies is transplanted from one segment to an adjacent one, these motor neuron

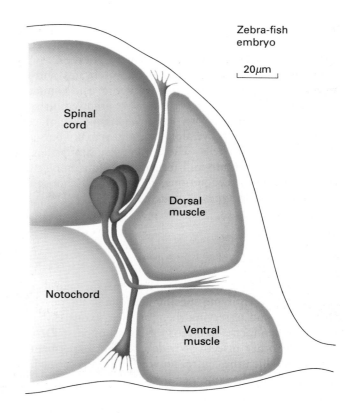

▲ **Figure 23-39** During development of the zebra-fish embryo, growth cones of pioneer motor neurons follow different pathways. A cross-sectional view of a trunk section of a 19-h-old zebra-fish embryo shows that axons of three adjacent motor neurons extend outward from the developing spinal cord at the same ventral root. They follow the same pathway out of the spinal chord, but then follow different pathways. One axon extends downward, innervating the ventral muscles; one upward, innervating the dorsal; and one laterally innervating both. *From M. Westerfield and J. S. Eisen, 1988,* Trends Neurosci. *11:18–22.*

axons will exit the spinal cord at the "wrong" site. However, they will grow into the proper plexus and still innervate the correct muscle. A similar result is obtained if a segment is inverted, so that anterior neurons are moved to the posterior.

Clearly, the specificity in guidance of any axon does not depend on where it leaves the spinal cord, nor on its transplanted position within a segment. It seems to depend on an inherent set of adhesive molecules on the individual growth cone. For instance, in Figure 23-39 the neuron that grows along the ventral pathway can do so because of specific affinity with components of the extracellular matrix or with glial or other cells located in this region. More specifically, a growth cone probably moves from one short-range target to another as it guides an axon to its destination.

Though we have chosen these examples from the vertebrate peripheral nervous system, similar principles are thought to apply to the outgrowth of the first axons in the vertebrate central nervous system and in insect nervous systems. For instance, the growth cones of certain pioneer axons in the grasshopper central nervous system always make a turn at a specific glial cell. If the glial cell is destroyed by a laser beam, the growth cones do not turn but continue in the original direction. Clearly these growth cones use glial cells as landmarks or guides, as do those of other pioneer neurons in both vertebrate and invertebrate central nervous systems. Evidence suggests very intimate contact between pioneer neurons and guidepost glial cells; gap junctions, for instance, form transiently between the two contacting cells.

Different Growth Cones Navigate along Different Axons

As pioneer axons extend, more and more of the space in the central nervous system becomes occupied by axonal processes. Many axons group together in bundles called fascicles. As secondary neurons differentiate, their growth cones navigate on the surfaces of other axons and their axons eventually bundle together. Different growth cones select different axonal surfaces in different fascicles on which to migrate. This can be seen vividly in the development of the grasshopper central nervous system.

Grasshopper and fruit-fly embryos are divided into segments, most of which have a similar pattern of neurons. Each segment has two ganglia, one on each side of the embryo, that contain the nerve-cell bodies. Bundles of axons run longitudinally on either side of the embryo, and bundles of axons called *transverse fascicles* connect the segmental ganglia (Figure 23-40).

Of the hundreds of neurons in each segmental ganglion, many sprout axons whose growth cones migrate to the opposite side of the embryo by growing along a transverse fascicle. Six such identifiable neurons are depicted in Figure 23-41a. When these six growth cones reach the opposite side, each chooses a different longitudinal fascicle to follow or a different direction in which to migrate.

For instance, the G neuron growth cone initially makes contact with about 100 axons in 25 longitudinal fascicles, but chooses only the A-P fascicle to follow, made up of the axons of four neurons called A1, A2, P1, and P2. More detailed studies showed that the G growth cone actually moves along only two of the four axons in the fascicle, P1 and P2 (Figure 23-41b). If the A1 and A2 neurons are experimentally destroyed by a laser beam the differentiation of the G neuron as well as most of the other neurons is unaffected; the G growth cone moves normally along the P1 and P2 axons. However, if the P1 and P2 neurons are destroyed, the G neuron grows abnormally; its growth cone behaves as if it were undirected and does not bind to any other axon. Thus the G growth cone relies absolutely on binding to the P axons.

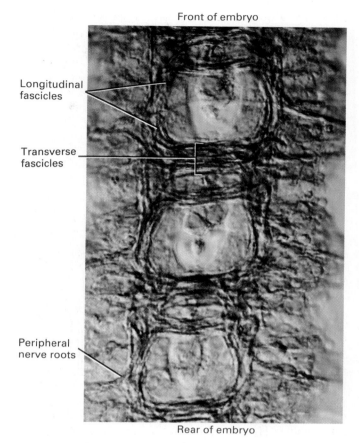

▲ **Figure 23-40** A micrograph depicting the pattern of axon fascicles in three adjacent segments of a *Drosophila* embryo, seen from above. The embryo is labeled by a fluorescent monoclonal antibody specific for an isotype of tubulin found only in neurons. Two sets of longitudinal fascicles, one on each side, are connected by transverse fascicles. At this stage, about 100 axons form about 25 longitudinal fascicles. *From C. S. Goodman and M. J. Bastiani, 1984,* Sci. Am. *251(6):58.*

It would appear that the P1 and P2 axons bear a unique surface marker that is recognized by a receptor on the G growth cone, but not on growth cones of other neurons that do not follow the P axons. The identification of these specific surface markers will greatly aid the understanding of the development of the nervous system. The neuron adhesive molecules already identified such as N-CAM and N-cadherin would seem too widespread to provide such specific recognition signals. However, a unique configuration of a group of these "general" molecules on an axon surface could form such a recognition marker. In mature vertebrate neurons N-CAM and the related Ng-CAM (neuron-glia cell adhesion molecule) do mediate the cohesion of parallel bundles of axons. Antibodies to these adhesion molecules block both axon outgrowth and formation of fascicles.

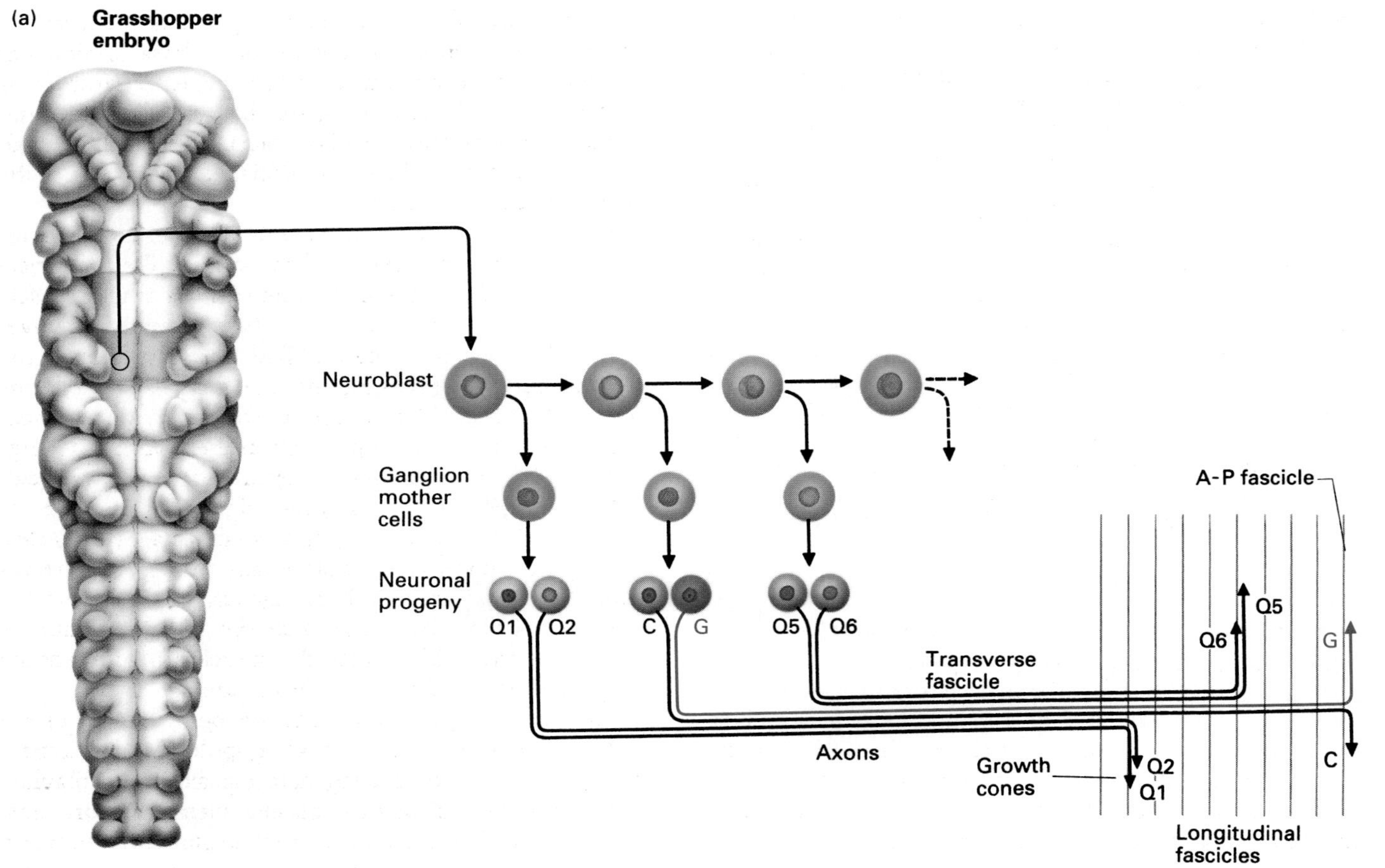

(a)

Grasshopper embryo

Neuroblast

Ganglion mother cells

Neuronal progeny

Q1 Q2   C   G   Q5 Q6

A-P fascicle

Q5

Q6

G

C

Transverse fascicle

Axons

Growth cones

Q2
Q1

Longitudinal fascicles

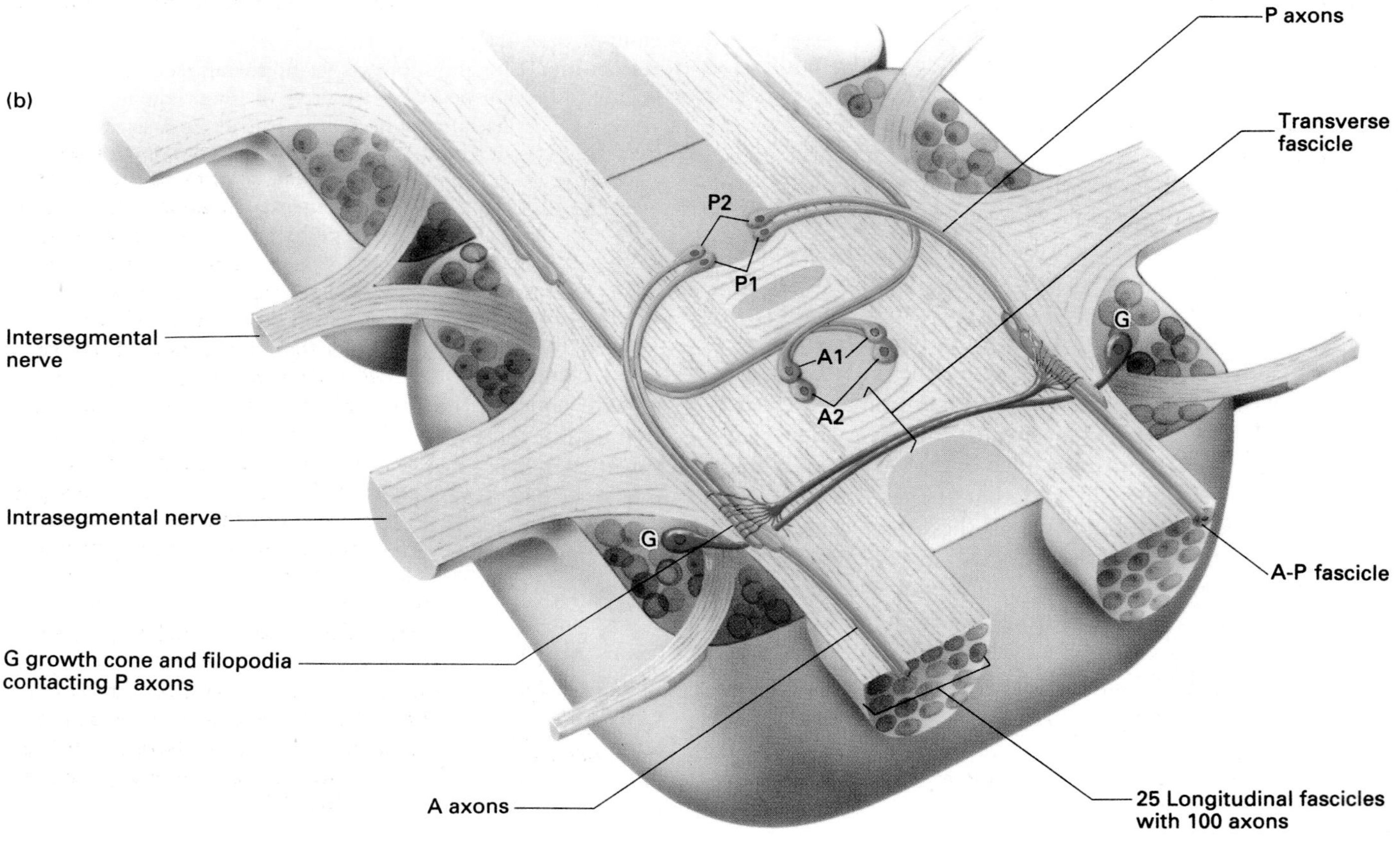

(b)

P axons

Transverse fascicle

Intersegmental nerve

Intrasegmental nerve

G growth cone and filopodia contacting P axons

P2
P1

A1
A2

G

G

A-P fascicle

A axons

25 Longitudinal fascicles with 100 axons

◄ **Figure 23-41** Different stereotyped pathways are taken by the axons of sister neurons. (a) Each of the 17 segments of the grasshopper embryo has a segmental ganglion; these ganglia have a virtually identical pattern of nerve differentiation. One identifiable neuroblast in each half of each segment divides repeatedly to form about 50 ganglion mother cells, each of which divides to yield two sister neurons. The first six neurons formed, shown here, extend axons across the ganglion, forming part of a transverse fascicle. The growth cone of each neuron then recognizes a different longitudinal fascicle and moves along it, elongating the axon in a specific direction. (b) Details of the selective fasciculation of the growth cone of the identifiable G neurons. Each half segment has one G neuron, whose axon grows along a transverse fascicle to the opposite side of the embryo. There the G growth cones explore the surfaces of 25 fascicles with a total of 100 axons. The G growth cones adhere to and migrate along only the axons of the P1 and P2 neurons (blue) in the A-P fascicle (green and blue axons). *After C. S. Goodman and M. J. Bastiani, 1984,* Sci. Am. *251(6):58.*

The Basal Lamina at the Neuromuscular Junction Directs Differentiation of Regenerating Nerve and Muscle

Once the axon of a motor neuron reaches its target muscle, a specialized neuromuscular junction is established. This synapse consists of folds on the muscle plasma membrane where acetylcholine receptors are concentrated (see Figure 20-34). The axon terminus forms specialized regions that contain synaptic vesicles and the enzymes for producing acetylcholine. Separating the nerve terminus and muscle is a specialized region of the basal lamina that surrounds the entire muscle fiber. Evidence from frogs and other amphibians suggests that the synaptic basal lamina contains hormonelike molecules that induce both nerve and muscle to form the specialized structures of the neuromuscular junction.

For example, if both the nerve and muscle are damaged, the nerve terminals and the muscle fibers degenerate and are phagocytosed. All that remains are the muscle basal lamina and the Schwann cells that formed the myelin around the motor neuron. With time, the axon of the motor neuron regenerates, and a new muscle fiber forms within the original basal lamina. Strikingly, a neuromuscular junction is formed precisely at the original site (Figure 23-42). Even if muscle regeneration is prevented, axons still return to precisely the original site on the basal lamina and form an axon terminus with synaptic vesicles. Clearly the muscle cell is unnecessary for precise innervation. Moreover, if myofibers regenerate—but the axon does not—specialized regions of the plasma membrane with concentrations of acetylcholine receptors form at the original synaptic site on the basal lamina.

Such experiments show that the basal lamina in the original synapse contains signaling molecules that attract the growth cone of the regenerating axon and regulate the distribution of muscle plasma membrane proteins.

Indeed, the basal lamina at the neuromuscular junction contains molecules that induce the aggregation of acetylcholine receptors on regenerating muscle fibers. In muscle

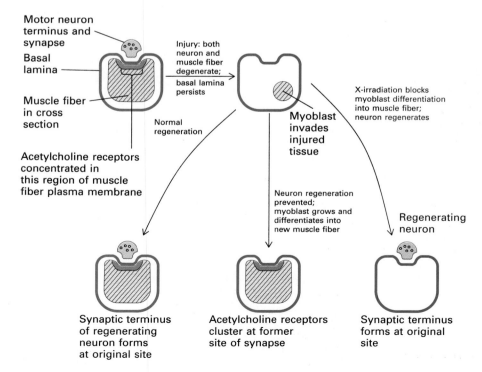

◄ **Figure 23-42** A specialized region of the basal lamina determines the site of the neuromuscular junction. When a striated frog limb muscle and innervating motor neurons are damaged, both the muscle and the nerve axon degenerate. All that remains are the muscle basal lamina and some surrounding Schwann cells. With time, myoblasts invade the space and differentiate into muscle fibers, and the motor neuron regenerates a new axon; the regenerated axons form synapses with the regenerating myofiber precisely at the sites of the original synapses. Even if muscle regeneration is prevented by x-irradiation, the axon regenerates synaptic termini at the original site. If nerve regeneration is prevented, the regenerating muscle fiber will form a synaptic specialization, with concentrations of acetylcholine receptors, again at the original site. *After S. J. Burden, 1987, in* The Vertebrate Neuromuscular Junction, *Liss, pp. 163–186.*

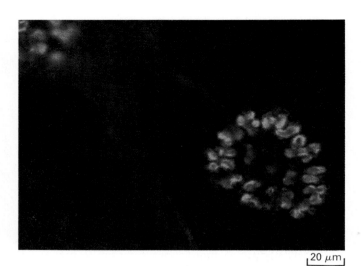

20 μm

▲ **Figure 23-43** Localization of the basal lamina protein agrin to the neuromuscular junction. A whole mount of a muscle with attached motor neurons from an electric ray was stained with a fluorescent antiagrin monoclonal antibody; staining is visible only in the nerve-muscle junctions. *From N. Reist, C. Magill, and U. J. McMahan, 1987,* J. Cell Biol. *105:2457. Courtesy U. J. McMahan.*

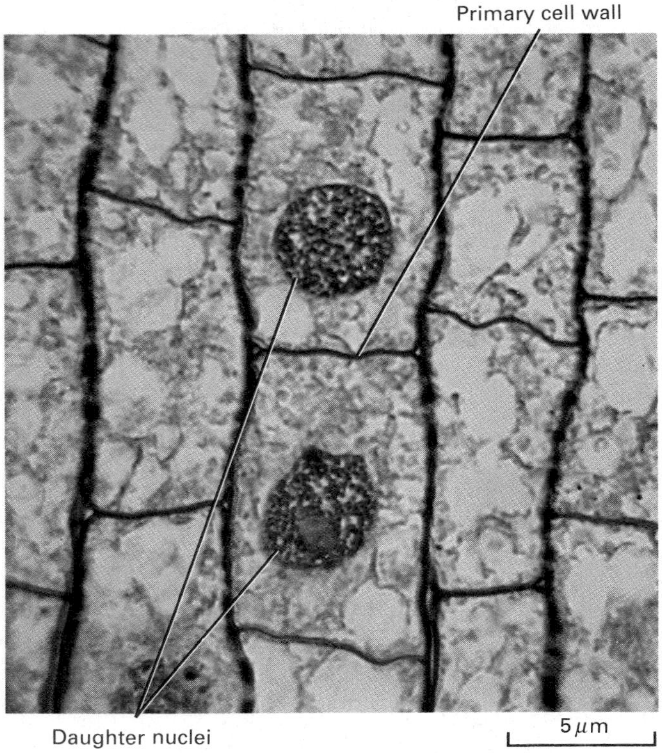

Primary cell wall

Daughter nuclei

5 μm

▲ **Figure 23-44** Light micrograph of young root tip cells of an onion, showing the thin primary cell wall separating two recently separated cells. *Courtesy of Jim Solliday and Biological Photo Service.*

cells grown in culture without neurons, acetylcholine receptors are distributed throughout the plasma membrane, not concentrated in synaptic regions. The protein agrin is found in the basal lamina of muscle cells only in the region of the neuromuscular junction (Figure 23-43) and causes acetylcholinesterase and the acetylcholine receptors to become concentrated in this region. Indeed addition of agrin causes formation of aggregates of acetylcholine receptors and acetylcholinesterase to form on the plasma membrane of cultured muscle cells, akin to the specializations at the neuromuscular junction. In the beginning of this chapter we mentioned the hormone-like effects of the basal lamina on cell differentiation, and the neuromuscular synapse provides one of the clearest examples.

Structure and Function of the Plant Cell Wall

Depending on its composition of proteins and proteoglycans, the extracellular matrix in animals can fulfill several types of functions: it can facilitate cell-cell adhesion (as in epithelia), it can provide a network (for instance, the basal lamina) in which cells can grow, and (3) it can provide strength, rigidity, and compressibility to tissues, particularly mature connective tissue such as tendon or bone where few cells are present. Plant cell walls serve many of the same functions, even though they are composed of entirely different macromolecules from those that make up animal extracellular matrices.

Cell division in plants is restricted to specific regions called meristems, including root tips and the tip of the stem. Young cells are connected by thin primary cell walls (Figure 23-44) that can be loosened and stretched to allow subsequent cell elongation (see Figure 19-39). After cell elongation ceases, the cell wall generally is thickened, either by secretion of additional macromolecules into the primary wall, or, more usually, by forming a secondary cell wall composed of several layers. In mature tissues such as the xylem—the tubes that conduct salts and water from the roots through the stems to the leaves (Figure 23-45)—the cell body degenerates, leaving only the cell wall. The unique properties of wood and of plant fibers such as cotton are due to the molecular properties of the cell walls in the tissues of origin.

Cell walls are constructed of only a few types of macromolecules. All plant walls contain fibers of cellulose, a polysaccharide made of glucose units. Like collagen in tendons, cellulose provides the tensile strength of the wall. Cellulose fibers are imbedded in and linked to a matrix containing two other types of polysaccharides, hemicellulose and pectin, and also a group of hydroxy-proline-rich fibrous glycoproteins. The composition and structure of the cell wall varies in different parts of a plant

(a)

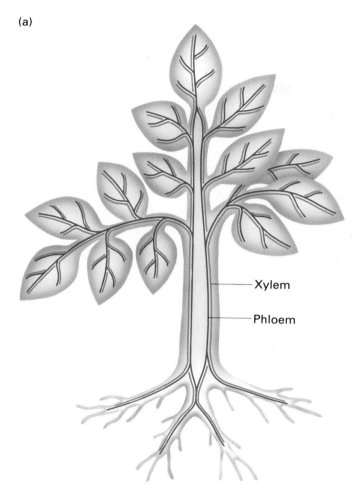

(b)

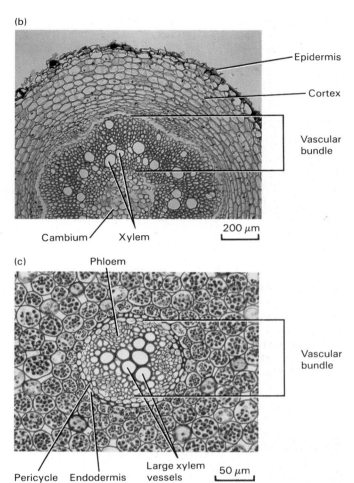

▲ **Figure 23-45**   The xylem and phloem in higher plants. (a) A schematic diagram of these two vascular systems. Micrographs of cross sections of a root from (b) a sunflower (*Helianthus* sp.) and (c) buttercup (*Ranunculus* sp.) showing vascular bundles containing the xylem and phloem. Water and salts enter the xylem through the roots. Water is lost by evaporation, mainly through the leaves, creating a suction pressure that draws the water and dissolved salts upward through the xylem. The phloem is used to conduct dissolved sucrose, produced in the leaves, to other parts of the plant. Sucrose is actively transported into the phloem by associated companion cells and actively removed from the phloem by nonphotosynthetic cells in the root and stem. The xylem is composed only of cell wall; the phloem consists of interconnected cells that have lost their nuclei and most other organelles but retain a plasma membrane and cytoplasm, through which the sucrose and water move. *Parts (b) and (c) courtesy of Runk/Schoenberger, from Grant Heilman.*

and in different types of plants. We begin our discussion of the structure of the cell wall with cellulose, its predominant macromolecule.

Cellulose Molecules Form Long, Rigid Microfibrils

Cellulose molecules are linear polymers of glucose residues linked together by $\beta(1 \rightarrow 4)$ glycosidic bonds (Figure 23-46). The $\beta(1 \rightarrow 4)$ linkage causes the polysaccharide to adopt a fully extended conformation; this is accomplished by the flipping of each glucose unit 180° relative to the preceeding one. In contrast, glycogen and starch are $\alpha(1 \rightarrow 4)$-linked glucose polymers; the $\alpha(1 \rightarrow 4)$ linkage causes the chain to adopt a coiled helical conformation. Cellulose molecules bundle together into fibers termed microfibrils that can be many micrometers in length. Extensive hydrogen bonding within cellulose molecules and between adjacent molecules makes the microfibril an almost crystalline aggregate.

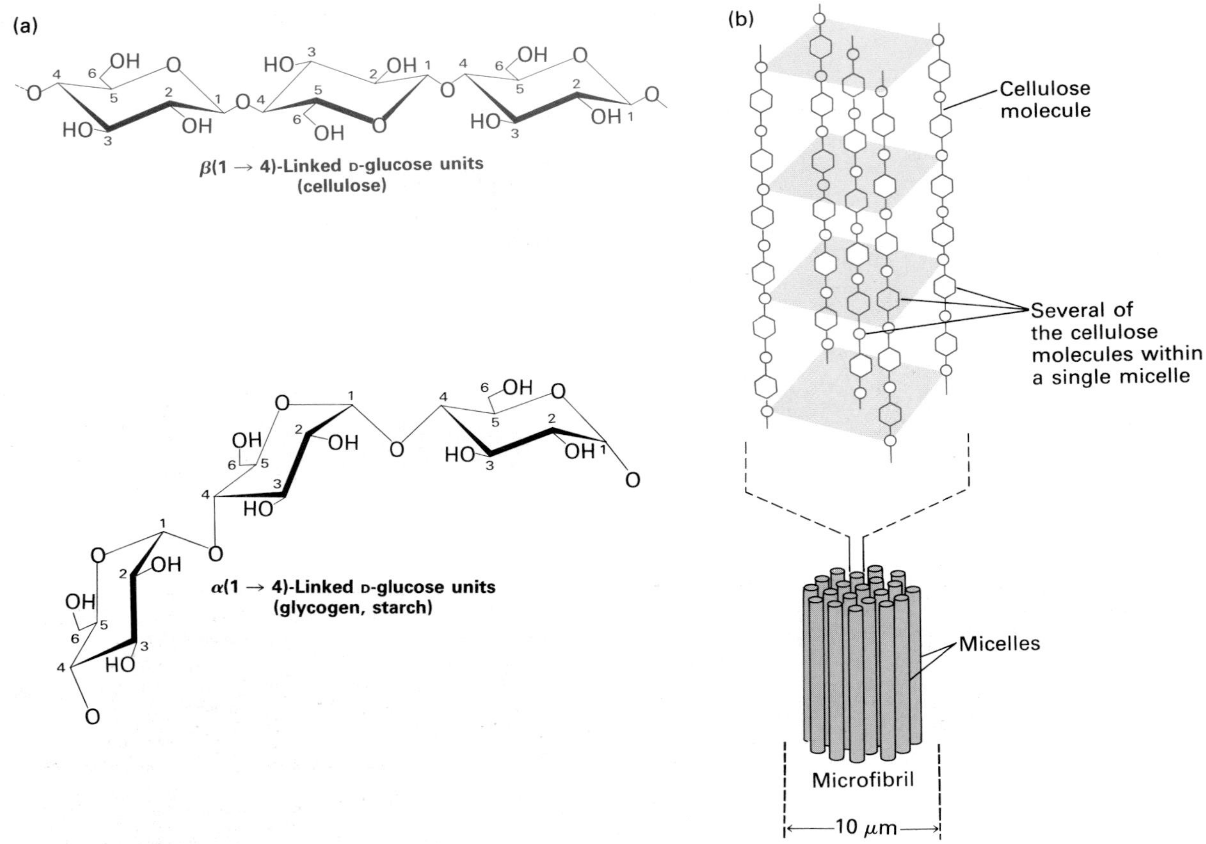

(a)

β(1 → 4)-Linked D-glucose units
(cellulose)

α(1 → 4)-Linked D-glucose units
(glycogen, starch)

(b)

Cellulose molecule

Several of the cellulose molecules within a single micelle

Micelles

Microfibril

|←—10 μm—→|

▲ **Figure 23-46**   The structure of cellulose in the plant cell wall. (a) Cellulose is a linear polymer consisting of more than 500 glucose residues linked together by β(1 → 4) glycosidic bonds. The β(1 → 4) linkages cause the molecule to form straight chains. In contrast, the α(1 → 4) linkage in polyglucose molecules such as glycogen and starch causes a turning of the chain; such molecules adopt a coiled helical conformation. (b) The linear cellulose molecules pack together to form rodlike structures termed micelles. These are stabilized by hydrogen bonds between the cellulose molecules. Not all of the chains in a micelle are illustrated. Micelles are packed into microfibrils that are usually several micrometers in length and 3–10 μm in diameter. In certain algae they can be up to 30 μm in diameter. Each cellulose molecule is polar because its two ends are distinct, and all cellulose molecules in a microfibril have the same polarity.

Other Polysaccharides Bind to Cellulose to Generate a Complex Wall Matrix of Many Layers

The cell wall contains a number of polysaccharides of composition more varying and more complex than that of cellulose. These contain several different types of saccharides (Figure 23-47). Biochemical study of these polysaccharides is difficult because they are heterogenous and frequently are insoluble, being cross-linked to other polysaccharides. The best studied constituents are those of the primary cell wall produced by cultured sycamore cells (Figure 23-48).

Hemicelluloses are highly branched polysaccharides with a backbone of about 50 β(1 → 4) linked sugars of a single type. Hemicelluloses are linked by hydrogen bonds to the surface of cellulose microfibrils. The branches help bind the microfibrils to each other and to other matrix components, particularly the pectins.

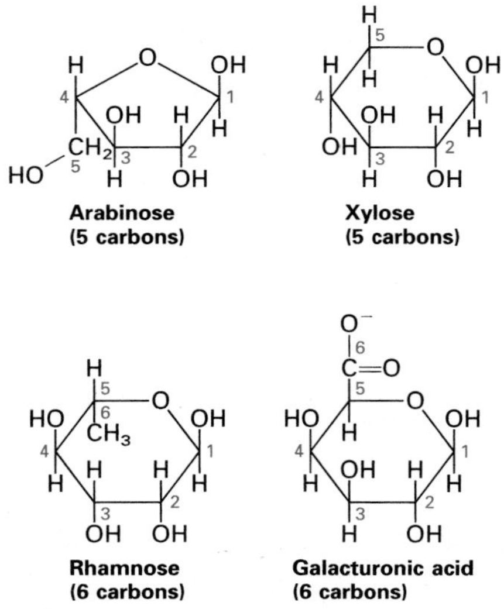

Arabinose
(5 carbons)

Xylose
(5 carbons)

Rhamnose
(6 carbons)

Galacturonic acid
(6 carbons)

▲ **Figure 23-47**   The structures of sugars abundant in plant cell walls. One five-carbon sugar, xylose, usually forms a six-atom ring; the other, arabinose, forms one with five atoms.

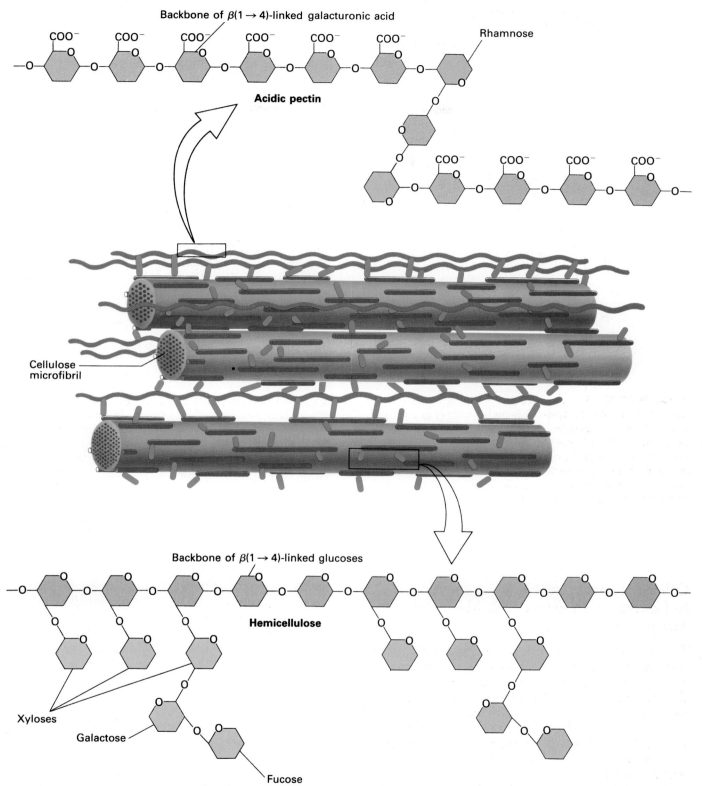

Backbone of β(1 → 4)-linked galacturonic acid

Acidic pectin

Rhamnose

Cellulose microfibril

Backbone of β(1 → 4)-linked glucoses

Hemicellulose

Xyloses

Galactose

Fucose

▲ **Figure 23-48** The structure and interconnections of the major polymers in the primary cell wall, derived mainly from studies on cultured sycamore cells. Hemicellulose molecules are hydrogen bonded to the surface of cellulose microfibrils. The backbone of hemicellulose (red) is similar to that of cellulose, but there are multiple branches (blue) of xylose, ga- lactose, and fucose residues. These may link the microfibrils to pectins (purple). Pectins contain a backbone of short β(1 → 4)-linked galacturonic acid (purple) chains, bound at their ends to short chains of rhamnose (green) and galactu- ronic acid. [See P. Albersheim, 1980, in N. E. Tolbert, ed., *Biochemistry of Plants,* Vol. I, Academic, pp. 91–162.]

Pectins, like hyaluronic acid (see Figure 23-17), contain multiple negatively charged saccharides. Pectins bind cations such as Ca^{2+} and, like hyaluronic acid, are highly hydrated. When purified, they bind water and form a gel—hence the use of pectins in many processed foods. Pectins are often cross-linked to hemicelluloses, thus participating in forming a complex network of all the principal wall components. This interlinked network helps bind adjacent cells together. Pectins are particularly abundant in the middle lamella, the layer between the cell walls of adjacent cells (Figure 23-49). Treatment of tissues with

pectinase or other enzymes that degrade pectin frequently causes cells with their walls to separate from one another.

The primary cell wall can be thought of as a gel-like matrix in which are imbedded cellulose microfibrils. Such walls are more impermeable than are the matrices surrounding animal cells. Whereas water and ions diffuse freely in cell walls, diffusion of particles of diameter greater than ~4 nm, including proteins of ~20,000 MW, is reduced. This is one of the reasons that plant hormones are small, water-soluble molecules.

Cell growth in higher plants frequently occurs without increase in the volume of the cytosol. Uptake of water into the vacuole generates an outward turgor pressure, and a localized loosening of the primary cell wall allows the cell to expand in a particular direction (see Figure 19-39).

As was shown in Figure 23-49, the secondary wall of the mature cell may have several layers; within each layer the cellulose fibrils are parallel to each other, but the orientation differs in adjacent layers. Such a plywoodlike construction adds considerable strength to the wall.

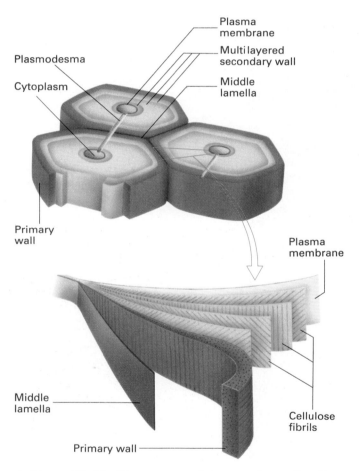

▲ **Figure 23–49** The structure of the secondary cell wall, built up of a series of layers of cellulose. In each layer the fibers run more or less in the same direction, but the direction varies in different layers. As plant cells grow, they deposit new layers of cellulose adjacent to the plasma membrane. Thus the oldest layers are in the primary wall (the outer wall) and in the middle lamella (the pectin-rich part of the cell wall laid down between two daughter cells as they cleave during cell division). Younger regions of the wall—collectively the secondary cell wall—are laid down as successive layers, adjacent to the plasma membrane. The cytoplasms of adjacent cells are usually connected by plasmodesmata that run through the layers of the cell walls.

Cell Walls Contain Lignin and an Extended, Hydroxyproline-rich Glycoprotein

As much as 15 percent of the primary cell wall can be composed of *extensin,* a glycoprotein made up of roughly 300 amino acids. Extensin, like collagen, contains abundant hydroxyproline (Hyp) and about half of its length represents variations of the four amino acid sequence Ser-Hyp-Hyp-Hyp. Most of the hydroxyprolines are glycosylated with chains of three or four arabinose residues, and galactose is linked to the serines. Thus extensin is about 65 percent carbohydrate, and its protein backbone forms an extended rod-like helix with carbohydrates protruding outward. Biosynthetic studies on sections of carrot indicated that extensins are secreted as soluble proteins by cells into the wall, where they are incorporated into the insoluble polysaccharide network. By binding to these components, extensins increase the strength of the cell wall.

Lignin, a complex, insoluble polymer of phenolic residues (Figure 23-50), is a strengthening material in all cell walls. It is particularly abundant in wood, where it accumulates in primary cell walls and in the secondary walls of the xylem. It associates with cellulose; like cartilage proteoglycans, it resists compression forces on the matrix. Particularly for soil-grown plants, lignin is essential for strengthening the xylem tubes to enable them to conduct water and salts over long distances. Lignin also protects the plant against invasion by pathogens and against predation by insects or other animals.

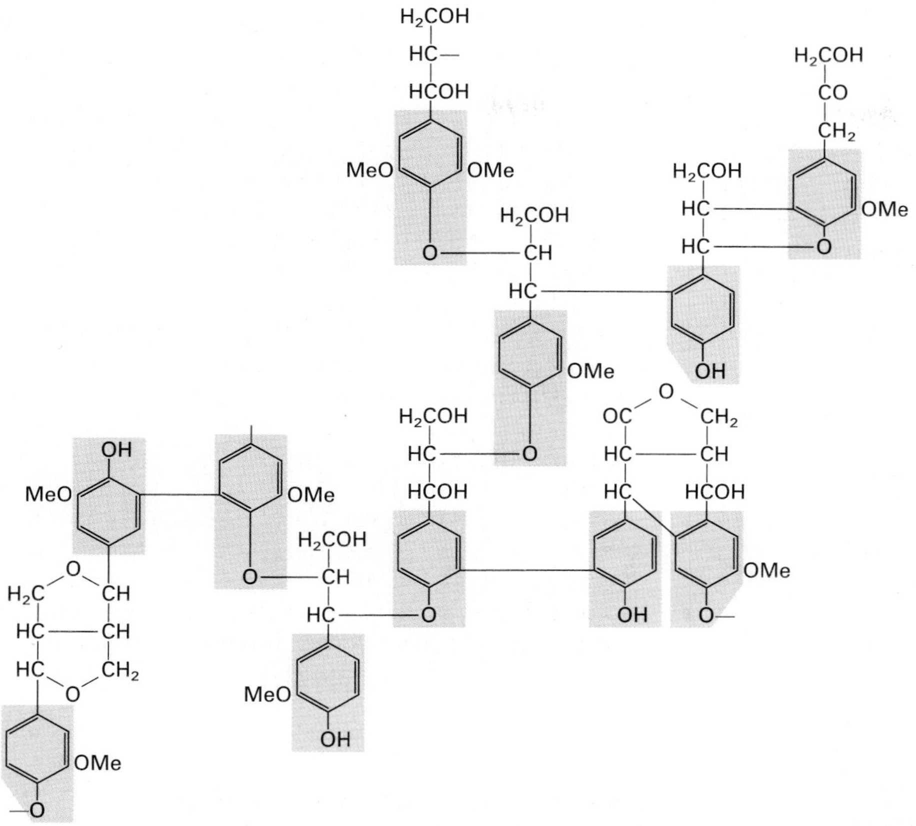

◀ Figure 23-50 The structure of a portion of spruce lignin showing multiple different covalent connections between phenolic groups (orange). The red bonds are connections to other groups of this complicated and heterogenous structure. *From K. Freudenberg, 1965, Science 148:595.*

The Orientation of Newly Made Cellulose Microfibrils Is Affected by the Microtubule Network

How are the cellulose microfibrils in each layer of the secondary cell wall oriented in the same direction? The same question can be asked of the primary cell wall in elongating cells, where newly made cellulose microfibrils encircle the cell like a belt perpendicular to the axis of elongation. The detailed answer is not known because of particular problems in studying cellulose biosynthesis. Unlike the other cell-wall polysaccharides that, like animal proteoglycans, are fabricated in the Golgi complex and then secreted, cellulose is synthesized on the external surface of the plasma membrane. The polymerizing enzyme, cellulose synthase, has never been purified, but uses as substrate UDP-glucose or ADP-glucose formed in the cytosol. It is thought that the enzyme is a large complex of many identical subunits each of which "spins out" cellulose molecules that spontaneously form microfibrils. The long microfibrils are insoluble (which probably explains why they are not formed within the cell).

Experiments suggest that, at least for cellulose microfibrils in the primary wall of elongating cells, microtubules influence the direction of cellulose deposition. Elongating cells synthesizing cellulose have oriented bands or rings of microtubules perpendicular to the direction of elongation just under the plasma membrane (Figure 23-51). Cellu-

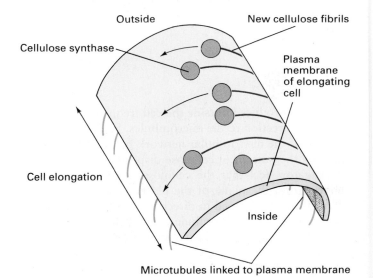

▲ Figure 23-51 Microtubules and cellulose synthesis in an elongating root tip cell. Circumferential rings of microtubules lie just inside the plasma membrane, perpendicular to the direction of cell elongation. Cellulose synthase is a large integral membrane protein that synthesizes cellulose fibrils on the outer face of the plasma membrane. As it spins out insoluble cellulose fibrils the synthase moves in the plasma membrane *(arrows)* parallel to the underlying microtubule network. Thus in this growing cell the new fibrils will be in circumferential rings perpendicular to the direction of elongation.

(a)

(b)

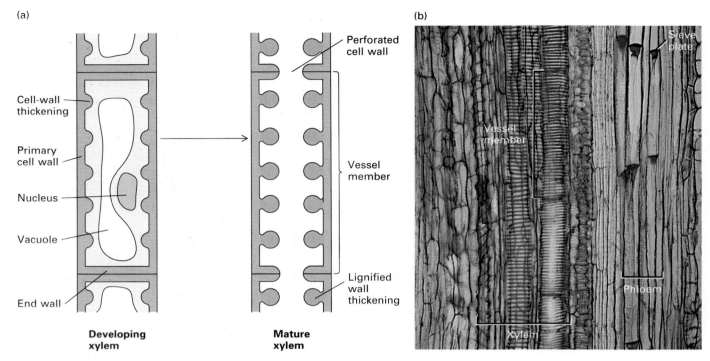

▲ **Figure 23-52** Modification of the cell wall during formation of the xylem. (a) The cytoplasm, nucleus, and end cell walls are all degraded, forming a continuous open-ended tube used to transport salts and water from roots through the stem to the leaves. The cell wall is thickened at intervals by addition of lignin and often becomes perforated. (b) A light micrograph of a longitudinal section of the squash *Curcurbita,* showing the mature xylem and phloem. Note the perforations all along the xylem tube. This allows fluids, rising from the roots, to leave the xylem at many points and enter the surrounding tissues. *Courtesy of J. R. Waaland.*

lose fibers produced outside the cell frequently are parallel to the direction of the microtubules. Furthermore, disruption of the microtubular network by drugs eventually disrupts the pattern of cellulose disposition. Thus, many investigators feel that the cellulose synthase complexes move within the plane of the plasma membrane as the microfibrils are formed in directions that are determined by the underlying microtubule cytoskeleton; any linkage, however, between the microtubules and the cellulose-synthesizing complex remains to be determined.

Remodeling of the Cell Wall Allows Formation of Specialized Structures

Even after the primary cell wall is formed and the cell grows to its mature size, the cell wall can still be modified by addition, destruction, or reorganization of macromolecules. Examples include the formation of the xylem vessels (Figure 23-52) and the phloem sieve-tube cells (Figure 23-53). In the formation of xylem, the most striking

change is destruction of the end cell walls, allowing the formation of a continuous tube and the deposition of lignin in the cell wall, which adds strength and resilience. The walls of the xylem vessels are frequently perforated, allowing fluids to enter and leave the vessel.

In the formation of the phloem the plasmodesmata in the end cell walls that connect adjacent cells are enlarged, forming large pores that facilitate fluid movement. The phloem cells lose their nuclei and most of their organelles and depend on plasmodesmata with their companion cells for provision of nutrients.

Cell-wall Oligosaccharides Act as Signaling Agents

Plants are continuously exposed to pathogenic bacteria and fungi, but they resist most infections. One resistance mechanism is the synthesis of small organic molecules called *phytoalexins*. Most such molecules are derivatives of phenol; the plant synthesizes them from the aromatic

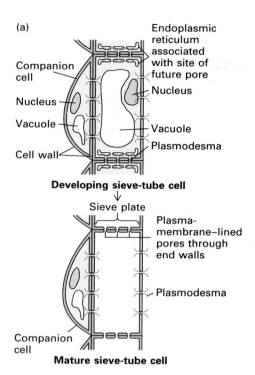

(a)

Companion cell
Nucleus
Vacuole
Cell wall

Endoplasmic reticulum associated with site of future pore
Nucleus
Vacuole
Plasmodesma

Developing sieve-tube cell

Sieve plate

Plasma-membrane–lined pores through end walls

Plasmodesma

Companion cell

Mature sieve-tube cell

(b) Companion cell   Sieve plate

2 µm

▲ **Figure 23-53** (a) Modifications of the cell wall during formation of the phloem sieve-tube cells. The developing sieve-tube cell and its companion cell are derived from the same mother cell. The primary cell wall of the sieve cell thickens, and the nucleus, vacuole, and other internal organelles are lost, but the plasma membrane is retained. The end walls, called the sieve plate, become extensively perforated by expansions of the plasmodesmata. The companion cell is connected to the sieve-tube cell by many plasmodesmata and actively secretes substances such as sucrose into the sieve-tube cell for transport, or actively resorbs such molecules. (b) Cross section of the phloem from *Curcurbita*, showing a sieve plate and companion cells. *Courtesy of J. R. Waaland and Biological Photo Service.*

amino acid phenylalanine. Phytoalexins are very toxic to fungi, but less so in general to bacteria. They are not made by uninfected plants. The enzymes that synthesize phytoalexins are induced only after infection or by other stresses, such as wounding or exposure to ultraviolet light.

Oligosaccharide fragments from fungal or plant cell walls are powerful signaling agents—called elicitors—that induce expression of the genes that encode enzymes that synthesize phytoalexins. One elicitor, shown in Figure 23-54a, is a very specific heptamer of $\beta(1 \rightarrow 3)$- and $\beta(1 \rightarrow 6)$-linked glucoses, formed by hydrolysis of a much longer polysaccharide in the cell wall of certain pathogenic fungi. The plant responds to fungal infection by synthesizing an enzyme, β-glucosidase, that partially hydrolyzes the fungal cell wall, forming this elicitor among other oligosaccharides. Besides inducing the enzymes for phytoalexin synthesis, this oligoglucose elicitor also induces synthesis of β-glucosidase itself. This, in turn, generates more oligosaccharide elicitor that enhances the plant's defensive capabilities.

As a result of pathogenic invasion, wounding, or other stresses, plant cell walls may be partially hydrolyzed. Some of the hydrolysis products, such as the specific polygalacturonic acid (Figure 23-54b) derived from cell-wall pectin (see Figure 23-48), are also potent elicitors and also induce phytoalexin synthesis. Elicitor fragments of plant and fungal cell walls can act synergistically in inducing these defensive genes. Here, breakdown products of the invaded cell or tissue prevent the spread of the pathogen to the rest of the plant.

Besides synthesis of phytoalexins, plants have other inducible defense systems. These include reinforcement of the plant cell walls by synthesis of the cell-wall glycoprotein extensin as well as synthesis of enzymes that produce lignin and callose, constituents that reinforce the cell wall against pathogens. The infected plant also synthesizes several lytic enzymes that attack microbial cell walls.

Elicitors increase transcription of defense genes as part of a major change in gene expression. In experimental systems, such as cultured bean cells treated with a fungal

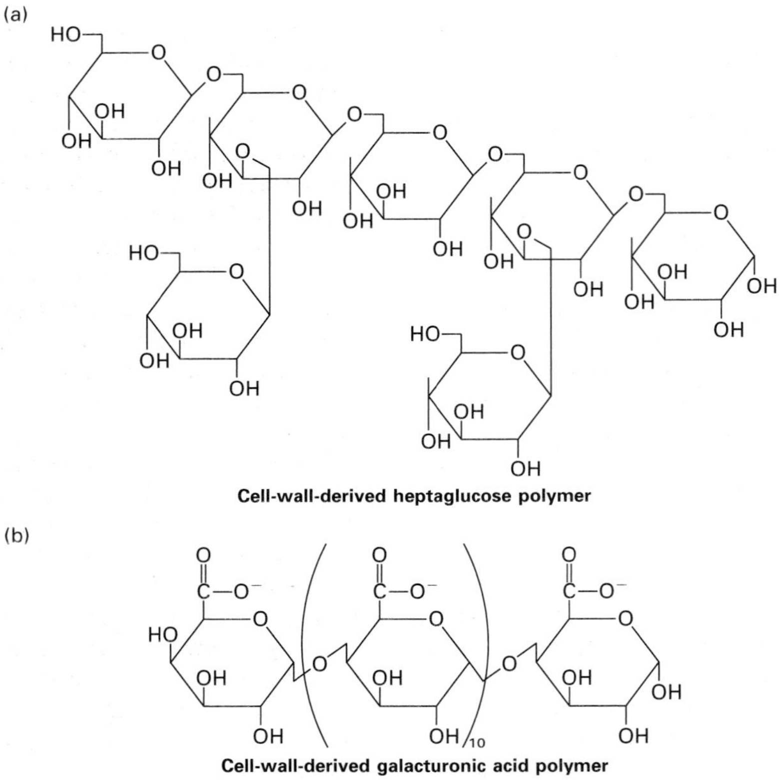

(a)

Cell-wall-derived heptaglucose polymer

(b)

Cell-wall-derived galacturonic acid polymer

▲ **Figure 23-54** Two elicitor oligosaccharides that induce enzymes that catalyze synthesis of phytoalexins. (a) A heptaglucose derived from a fungal cell wall; only this isomer and not the 150–200 other heptoses present in breakdown products of fungal cell walls is active in induction of plant defense genes. (b) A 12-residue polymer of $\alpha(1 \rightarrow 4)$-linked galacturonic acid, derived by hydrolysis of pectin, a normal constituent of plant cell walls.

cell-wall elicitor, transcription of several defense genes increases within 2–3 min; this is among the fastest transcriptional responses of any eukaryotic cell to an extracellular signal. Evidence exists that elicitor receptors are on plasma membranes, and much work is under way toward elucidating the steps between elicitor binding and gene activation. This work has been made possible by the cloning of genes for a number of inducible defense proteins, including enzymes required for phytoalexin synthesis.

There is recent evidence that specific oligosaccharide fragments of cell walls may act as normal signals for plant growth and development. For instance, polygalacturonic acids, similar to those depicted in Figure 23-54b, affect differentiation of tobacco callus cultures. Whether they normally function as plant hormones is controversial and remains to be elucidated.

Summary

Extracellular matrices in animals are composed of different combinations of collagens, proteoglycans, hyaluronic acid, laminin, fibronectin, and other glycoproteins.

Collagens are characteristically fibrous three-stranded helices with glycine as every third amino acid and abundant proline and hydroxyproline. They appear to have evolved from an ancestral exon encoding six Gly-Pro-X motifs. Fibrous collagen, such as type I in tendons, is formed by parallel aggregates of triple helices stabilized by lateral lysine cross-links. Fibrous collagens are fabricated with the aid of N- and C-terminal propeptides that enable the three polypeptides to interact in register. In the rough ER, formation of disulfide bonds in the propeptides facilitates these interactions; other essential modifications such as proline hydroxylation also occur in the ER. Generally, the propeptides are removed after secretion, and then collagen fibrils form. There are 12 known collagens. Some retain their globular domains and do not form fibrils but rather interact with other fibrous collagens that do. Type IV collagen is unique to basal lamina. It forms a two-dimensional fibrous network, stabilized by interactions with heparan sulfate proteoglycan and laminin.

Hyaluronic acid is an extremely long, negatively charged polysaccharide. It forms viscous, hydrated gels, and in several systems inhibits cell-cell adhesion and facilitates cell migration. HA forms the central component of cartilage proteoglycans; bound to the HA at regular intervals are core proteins to which are bound glycosaminoglycans such as chondroitin sulfate and dermatan sulfate. Smaller proteoglycans are found in other matrices and also attached to cell surfaces, where they facilitate cell-matrix interactions.

Laminin is a major protein in basal lamina; it binds specifically heparan sulfate, type IV collagen, and also specific cell-surface receptor proteins. Fibronectins serve analogous functions in other matrices; they bind fibrous collagens, proteoglycans, fibrin in blood clots, and other substances. Fibronectin promotes migration and adhesion of many cells. A variety of cell-surface receptors of the integrin family bind fibronectin or other multiadhesive matrix proteins. In stationary fibroblasts integrins anchor cytoskeletal actin filaments to extracellular fibronectin.

A variety of plasma membrane proteins promote homotypic adhesion between similar types of cells. Examples include uvomorulin, essential for compaction of early embryos and also for formation and stabilization of sheets of epithelial cells. During differentiation cells can change their set of surface-adhesion molecules; migrating cells such as neuroblasts lose adhesive proteins and then resynthesize them after migration ceases.

Cell-cell and cell-matrix interactions are essential for many types of tissue differentiation. Examples include different types of structures in the chick epidermis being induced by different types of underlying mesodermal cells. Certain ectodermal nerve cells in the optic cup cause the overlying epidermis to differentiate into a lens. Specific mesodermal-endodermal interactions are required for differentiation of such organs as liver and salivary glands. Cell-culture systems demonstrate the requirement of the basal lamina, in particular, for epithelial cells such as those in the mammary gland to assume the proper ultrastructure and to synthesize tissue-specific proteins.

Proper cell contacts in the nervous system are established as a result of axon outgrowth in specific directions toward specific target cells. Growth cones are actin-rich motile structures at the growing tips of axons that guide axon migration and elongation. Growth cones can form transient adhesive contacts with various matrix components, glial cells, or nerve axons in their pathways. In certain insect ganglia growth cones can be seen to select one axon from dozens of possible choices, and sibling neurons can make stereotyped but different choices. Axonal surface proteins are being identified that mediate specific fasiculation of neurons; they appear able to bind to receptors on certain growth cones and thus to guide their migration. Once a nerve-muscle synapse is formed, the basal lamina at that site retains information that can specify the formation of specific synaptic structures by regenerating nerves or muscle. This is one of many examples of the extracellular matrix providing specific clues to elongating axons.

Microfibrils of cellulose are crystalline aggregates of $\beta(1 \rightarrow 4)$-linked glucose polymers. They are ubiquitous constituents of plant cell walls and are responsible for much of its tensile strength. Cellulose microfibrils are imbedded in and linked to a matrix that contains hemicellulose, negatively charged pectins, and the fibrous hydroxylproline-rich glycoprotein extensin. By binding cations and water, pectins give the wall a gel-like consistency. Extensin and lignin provide reinforcements; lignin is responsible for much of the strength of xylem tubes and also of wood.

The primary cell wall between growing cells is thin and extensible. Circumferential bands of cellulose fibers are frequently laid down perpendicular to the direction of cell elongation; circumferential bands of microtubules just under the plasma membrane determine the direction in which the cell-surface cellulose synthase synthesizes new cellulose microfibrils. The secondary cell wall often has several layers; in each layer the cellulose microfibrils are parallel but different in orientation from those in adjacent layers. Remodeling of the cell wall—particularly of the end wall—produces mature xylem and phloem cells. Fragments of fungal cell walls or of plant cell walls—produced during infection—are elicitors and trigger the induction of antifungal phytoalexins. Elicitors and stresses trigger synthesis of extensin and other molecules that reinforce the cell wall against invaders.

References

General References on the Extracellular Matrix

*HAY, E. D., ed. 1981. *Cell Biology of Extracellular Matrix.* Plenum.

*MCDONALD, J. A. 1988. Extracellular matrix assembly. *Ann. Rev. Cell Biol.* 4:183–208.

*PIEZ, K. A., and A. H. REDDI, eds. 1984. *Extracellular Matrix Biochemistry.* Elsevier.

*A book or review article that provides a survey of the topic.

Fibrous Collagens: Structure and Synthesis

*BURGESON, R. E. 1988. New collagens, new concepts. *Ann. Rev. Cell Biol.* 4:551–577.

COHN, D. H., P. H. BYERS, B. STEINMANN, and R. E. GELINAS. 1986. Lethal osteogenesis imperfecta resulting from a single nucleotide change in one human proαI(I) collagen allele. *Proc. Nat'l Acad. Sci. USA* 83:6045–6047.

*DAVIDSON, J. M., and R. A. BERG. 1981. Posttranslational events in collagen synthesis. *Methods Cell Biol.* 23:199–136.

*EYRE, D. R., M. A. PAZ, and P. M. GALLOP. 1984. Cross-linking in collagen and elastin. *Ann. Rev. Biochem.* 53:717–748.

*FLEISCHMAJER, R., B. R. OLSEN, and K. KÜHN, eds., 1985. Biology, Chemistry, and Pathology of Collagen. *Ann. N.Y. Acad. Sci.* 460.

KEENE, D. R., E. ENGVALL, and R. W. GLANVILLE. 1988. Ultrastructure of type VI collagen in human skin and cartilage suggests an anchoring function for this filamentous network. *J. Cell Biol.* 107:1995–2006.

LÖHLER, J., R. TIMPL, and R. JAENISCH. 1984. Embryonic lethal mutation in mouse collagen I gene causes rupture of blood vessels and is associated with erythropoietic and mesenchymal cell death. *Cell* 38:597–607.

*MARTIN, G. R., R. TIMPLE, P. K. MULLER, and K. KÜHN. 1985. The genetically distinct collagens. *Trends Biochem. Sci.* 10:285–287.

MCCORMICK, D., M. VAN DER REST, J. GOODSHIP, G. LOZANO, Y. NINOMIYA, and B. R. OLSEN. 1987. Structure of the glycosaminoglycan domain in the type IX collagen-proteoglycan. *Proc. Nat'l Acad. Sci. USA* 84:4044–4048.

*PARRY, D. A. D. 1988. The molecular and fibrillar structure of collagen and its relationship to the mechanical properties of connective tissue. *Biophys. Chem.* 29:195–209.

*PROCKOP, D. J., and K. I. KIVIRIKKO. 1984. Heritable diseases of collagen. *New Engl. J. Med.* 311:376–386.

STACEY, A., J. BATEMAN, T. CHOI, T. MASCARA, W. COLE, and R. JAENISCH. 1988. Perinatal lethal osteogenesis imperfecta in transgenic mice bearing an engineered mutant pro-α1(I) collagen gene. *Nature* 332:131–136.

Type IV Collagen and the Basal Lamina

*FARQUHAR, M. G. 1981. The glomerular basement membrane: a selective macromolecular filter. In Hay, E. D., ed. *Cell Biology of Extracellular Matrix*, Plenum, pp. 335–378.

SAUS, J., A. QUINONES, A. MACKRELL, B. BLUMBERG, G. MUTHUKUMARAN, T. PIHLAJANIEMI, and M. KURKINEN. 1989. The complete primary structure of mouse α2(IV) collagen: alignment with mouse αI(IV) collagen. *J. Biol. Chem.* 264:6318–6324.

YURCHENCO, P. D., and G. C. RUBEN. 1987. Basement membrane structure in situ: evidence for lateral associations in the type IV collagen network. *J. Cell Biol.* 105:2559–2568.

Matrix Proteoglycans

BOURDON, M. A., Å. OLDBERG, M. PIERSCHBACHER, and E. RUOSLAHTI. 1985. Molecular cloning and sequence analysis of a chondroitin sulfate proteoglycan cDNA. *Proc. Nat'l Acad. Sci. USA* 82:1321–1325.

BOURDON, M. A., T. KRUSIUS, S. CAMPBELL, N. B. SCHWARTZ, and E. RUOSLAHTI. 1987. Identification and synthesis of a recognition signal for the attachment of glycosaminoglycans to proteins. *Proc. Nat'l Acad. Sci. USA* 84:3194–3198.

*EVERED, D., and J. WHELAN, eds. 1986. *Functions of the Proteoglycans*. Ciba Foundation Symposium 124. Wiley.

*FRANSSON, L-Å. 1987. Structure and function of cell-associated proteoglycans. *Trends Biochem. Sci.* 12:406–411.

*HASSELL, J. R., J. H. KIMURA, and V. C. HASCALL. 1986. Proteoglycan core protein families. *Ann. Rev. Biochem.* 55:539–567.

*HEINEGÅRD, D., and Å. OLDBERG. 1989. Structure and biology of cartilage and bone matrix noncollagenous macromolecules. *FASEB J.* 3:2042–2045.

MILLER, J., J. A. HATCH, S. SIMONIS, and S. E. CULLEN. 1988. Identification of the glycosaminoglycan-attachment site of mouse invariant-chain proteoglycan core protein by site-directed mutagenesis. *Proc. Nat'l Acad. Sci. USA* 85:1359–1363.

REDDY, P., A. C. JACQUIER, N. ABOVICH, G. PETERSEN, and M. ROSBASH. 1986. The *period* clock locus of D. melanogaster codes for a proteoglycan. *Cell* 46:53–61.

*RUOSLAHTI, E. 1988. Structure and biology of proteoglycans. *Ann. Rev. Cell Biol.* 4:229–255.

*RUOSLAHTI, E. 1989. Proteoglycans in cell regulation. *J. Biol. Chem.* 264:13369–13372.

*TOOLE, B. P. 1981. Glycosaminoglycans in morphogenesis. In Hay, E. D., ed. *Cell Biology of Extracellular Matrix*, Plenum, pp. 259–294.

*WIGHT, T. N., and R. P. MECHAM, eds. 1987. *Biology of Proteoglycans*. Academic.

Cell-surface Proteoglycans

SANDERSON, R. D., and M. BERNFIELD. 1988. Molecular polymorphism of a cell surface proteoglycan: distinct structures on simple and stratified epithelia. *Proc. Nat'l Acad. Sci. USA* 85:9562–9566.

SAUNDERS, S., and M. BERNFIELD. 1988. Cell surface proteoglycan binds mouse mammary epithelial cells to fibronectin and behaves as a receptor for interstitial matrix. *J. Cell Biol.* 106:423–430.

Laminin

GEHLSEN, K. R., L. DILLNER, E. ENGVALL, and E. RUOSLAHTI. 1988. The human laminin receptor is a member of the integrin family of cell adhesion receptors. *Science* 241:1228–1229.

GRAF, J., Y. IWAMOTO, M. SASAKI, G. R. MARTIN, H. K. KLEINMAN, F. A. ROBEY, and Y. YAMADA. 1987. Identification of an amino acid sequence in laminin mediating cell attachment, chemotaxis, and receptor binding. *Cell* 48:989–996.

*MARTIN, G. R., and R. TIMPL. 1987. Laminin and other basement membrane components. *Ann. Rev. Cell Biol.* 35:57–85.

PANAYOTOU, G., P. END, M. AUMAILLEY, R. TIMPL, and J. ENGEL. 1989. Domains of laminin with growth-factor activity. *Cell* 56:93–101.

SASAKI, M., H. K. KLEINMAN, H. HUBER, T. DEUTZMANN, and Y. YAMADA. 1988. Laminin, a multidomain protein. *J. Biol. Chem.* 263:16536–16544.

Fibronectin, Other Multiadhesive Proteins, and Their Cell-surface Receptors

AKIYAMA, S. K., S. S. YAMADA, W-T. CHEN, and K. M. YAMADA. 1989. Analysis of fibronectin receptor function with monoclonal antibodies: roles in cell adhesion, migration, matrix assembly, and cytoskeletal organization. *J. Cell Biol.* 109:863–875.

*BUCK, C. A., and A. F. HORWITZ. 1987. Cell surface receptors for extracellular matrix molecules. *Ann. Rev. Cell Biol.* 3:179–205.

BURNS, G. F., C. M. LUCAS, G. W. KRISSANSEN, J. A. WERKMEISTER, D. B. SCANLON, R. J. SIMPSON, and M. A. VADAS. 1988. Synergism between membrane gangliosides and arg-gly-asp-directed glycoprotein receptors in attachment to matrix proteins by melanoma cells. *J. Cell Biol.* 107:1225–1230.

*BURRIDGE, K., K. FATH, T. KELLY, G. NUCKOLLS, and C. TURNER. 1988. Focal adhesions: transmembrane junctions between the extracellular matrix and the cytoskeleton. *Ann. Rev. Cell Biol.* 4:487–526.

DUBAND, J-L., G. H. NUCKOLLS, A. ISHIHARA, T. HASEGAWA, K. M. YAMADA, J. P. THIERY, and K. JACOBSON. 1988. Fibronectin receptor exhibits high lateral mobility in embryonic locomoting cells but is immobile in focal contacts and fibrillar streaks in stationary cells. *J. Cell Biol.* 107:1385–1396.

HORWITZ, A., K. DUGGAN, C. BUCK, M. C. BECKERLE, and K. BURRIDGE. 1986. Interaction of plasma membrane fibronectin receptor with talin—a transmembrane linkage. *Nature* 320:531–533.

*HYNES, R. O. 1987. Integrins: a family of cell surface receptors. *Cell* 48:549–554.

*MCDONALD, J. A. 1988. Extracellular matrix assembly. *Ann. Rev. Cell Biol.* 4:183–207.

*MOSHER, D. F., ed. 1989. *Fibronectin.* Academic.

PAUL, J. I., J. E. SCHWARZBAUER, J. W. TAMKUN, and R. O. HYNES. 1986. Cell-type-specific fibronectin subunits generated by alternative splicing. *J. Biol. Chem.* 261:12258–12265.

*RUOSLAHTI, E. 1988. Fibronectin and its receptors. *Ann. Rev. Biochem.* 57:375–414.

*RUOSLAHTI, E. and M. D. PIERSCHBACHER. 1987. New perspectives in cell adhesion: RGD and integrins. *Science* 238:491–497.

SINGER, I. I. 1979. The fibronexus: a transmembrane association of fibronectin-containing fibers and bundles of 5 nm microfilaments in hamster and human fibroblasts. *Cell* 16:675–685.

Cadherins

*GUMBINER, B. 1988. Cadherins: a family of Ca^{2+}-dependent adhesion molecules. *Trends Biochem. Sci.* 13:75–76.

GUMBINER, B., B. STEVENSON, and A. GRIMALDI. 1988. The role of the cell adhesion molecule uvomorulin in the formation and maintenance of the epithelial junctional complex. *J. Cell Biol.* 107:1575–1587.

NOSE, A., A. NAGAFUCHI, and M. TAKEICHI. 1988. Expressed recombinant cadherins mediate cell sorting in model systems. *Cell* 54:993–1001.

*TAKEICHI, M. 1988. The cadherins: cell-cell adhesion molecules controlling animal morphogenesis. *Development* 102:639–655.

*TAKEICHI, M. 1987. Cadherins: a molecular family essential for selective cell-cell adhesion and animal morphogenesis. *Trends Genet.* 3:213–217.

VOLK, T., O. COHEN, and B. GEIGER. 1987. Formation of heterotypic adherens-type junctions between L-CAM-containing liver cells and A-CAM-containing lens cells. *Cell* 50:987–994.

N-CAMs and Related Cell Adhesion Molecules

*CUNNINGHAM, B. A., J. J. HEMPERLY, B. A. MURRAY, E. A. PREDIGER, R. BRACKENBURY, and G. M. EDELMAN. 1987. Neural cell adhesion molecule: structure, immunoglobulin-like domains, cell surface modulation, and alternative RNA splicing. *Science* 236:799–806.

GOWER, H. J., C. H. BARTON, V. L. ELSOM, J. THOMPSON, S. E. MOORE, G. DICKSON, and F. S. WALSH. 1988. Alternative splicing generates a secreted form of N-CAM in muscle and brain. *Cell* 55:955–964.

RUTISHAUSER, U., A. ACHESON, A. K. HALL, D. M. MANN, and J. SUNSHINE. 1988. The neural cell adhesion molecule (NCAM) as a regulator of cell-cell interactions. *Science* 240:53–57.

*RUTISHAUSER, U., and C. GORIDIS. 1986. N-CAM: the molecule and its genetics. *Trends Genet.* 2:72–76.

*WILLIAMS, A. F., and A. N. BARCLAY. 1988. The immunoglobulin superfamily—domains for cell surface recognition. *Ann. Rev. Immunol.* 6:381–406.

Cell and Matrix Interactions during Development

AGGELER, J., C. S. PARK, and M. J. BISSELL. 1988. Regulation of milk protein and basement membrane gene expression: the influence of the extracellular matrix. *J. Dairy Sci.* 71:2830–2842.

AUFDERHEIDE, E., and P. EKBLOM. 1988. Tenascin during gut development: appearance in the mesenchyme, shift in molecular forms, and dependence on epithelial-mesenchymal interactions. *J. Cell Biol.* 107:2341–2349.

*DEUCHAR, E. M. 1975. *Cellular Interactions in Animal Development.* London, Chapman & Hall.

*EDELMAN, G. M. 1988. Morphoregulatory molecules. *Biochemistry* 27:3534–3543.

*EKBLOM, P. 1989. Developmentally regulated conversion of mesenchyme to epithelium. *FASEB J.* 3:2141–2150.

*JESSEL, T. M. 1988. Adhesion molecules and the hierarchy of neural development. *Neuron* 1:3–13.

LI, M. L., J. AGGELER, D. A. FARSON, C. HATIER, J. HASSELL, and M. J. BISSELL. 1987. Influence of a reconstituted basement membrane and its components on casein gene expression and secretion in mouse mammary epithelial cells. *Proc. Nat'l Acad. Sci. USA* **84**:136–140.

*REID, L. M., S. L. ABREU, and K. MONTGOMERY. 1988. Extracellular matrix and hormonal regulation of synthesis and abundance of messenger RNAs in cultured liver cells. In Arias, I. M., W. B. Jakoby, H. Popper, D. Schachter, and D. A. Shafritz, eds., *The Liver: Biology and Pathobiology*, 2d ed., Raven, pp. 717–737.

ROGELJ, S., M. KLAGSBRUN, R. ATZMON, M. KUROKAWA, A. HAIMOVITZ, Z. FUKS, and I. VLODAVSKY. 1989. Basic fibroblast growth factor is an extracellular matrix component required for supporting the proliferation of vascular endothelial cells and the differentiation of PC12 cells. *J. Cell Biol.* **109**:823–831.

SARIOLA, H., E. AUFDERHEIDE, H. BERNHARD, S. HENKE-FAHLE, W. DIPPOLD, and P. EKBLOM. 1988. Antibodies to cell surface ganglioside G_{D3} perturb inductive epithelial-mesenchymal interactions. *Cell* **54**:235–245.

*WATT, F. M. 1986. The extracellular matrix and cell shape. *Trends Biochem. Sci.* **11**:482–485.

Neuron Development: Interactions with Cells and Matrices

*ANDERSON, H. 1988. Drosophila adhesion molecules and neural development. *Trends Neurosci.* **11**:472–475.

*BASTIANI, M. J., C. Q. DOE, S. L. HELFAND, and C. S. GOODMAN. 1985. Neuronal specificity and growth cone guidance in grasshopper and *Drosophila* embryos. *Trends Neurosci.* **8**:257–267.

BASTIANI, M. J., and C. S. GOODMAN. 1986. Guidance of neuronal growth cones in the grasshopper embryo. III. Recognition of specific glial pathways. *J. Neurosci.* **6**:3542–3551.

*BRAY, D., and P. J. HOLLENBECK. 1988. Growth cone motility and guidance. *Ann. Rev. Cell Biol.* **4**:43–61.

*DODD, J., and T. M. JESSELL. 1988. Axon guidance and the patterning of neuronal projections in vertebrates. *Science* **242**:692–699.

EISEN, J. S., S. H. PIKE, and B. DEBU. 1989. The growth cones of identified motoneurons in embryonic zebrafish select appropriate pathways in the absence of specific cellular interactions. *Neuron* **2**:1097–1104.

*GOODMAN, C. S., and M. J. BASTIANI. 1984. How embryonic nerve cells recognize one another. *Sci. Am.* **251**:58–66.

HARRELSON, A. L., and C. S. GOODMAN. 1988. Growth cone guidance in insects: fasciclin II is a member of the immunoglobulin superfamily. *Science* **242**:700–708.

*LANDER, A. D. 1989. Understanding the molecules of neural cell contacts: emerging patterns of structure and function. *Trends Neurosci.* **12**:189–194.

NEUGEBAUER, K. M., K. J. TOMASELLI, J. LILIEN, and L. F. REICHARDT. 1988. N-cadherin, NCAM, and integrins promote retinal neurite outgrowth on astrocytes in vitro. *J. Cell Biol.* **107**:1177–1187.

*PATTERSON, P. H. 1988. On the importance of being inhibited, or saying no to growth cones. *Neuron* **1**:263–267.

RAPER, J. A., M. J. BASTIANI, and C. S. GOODMAN. 1984. Pathfinding by neuronal growth cones in grasshopper embryos. *J. Neurosci.* **4**:2329–2345.

*RATHJEN, F. G. 1988. A neurite outgrowth-promoting molecule in developing fiber tracts. *Trends Neurosci.* **11**:183–184.

TOMASELLI, K. J., K. M. NEUGEBAUER, J. L. BIXBY, J. LILIEN, and L. F. REICHARDT. 1988. N-cadherin and integrins: two receptor systems that mediate neuronal process outgrowth on astrocyte surfaces. *Neuron* **1**:33–43.

*WESTERFIELD, M., and J. S. EISEN. 1988. Neuromuscular specificity: pathfinding by identified motor growth cones in a vertebrate embryo. *Trends Neurosci.* **11**:18–22.

Development of the Neuromuscular Junction

ANGLISTER, L., U. J. MCMAHAN, and R. M. MARSHALL. 1985. Basal lamina directs acetylcholinesterase accumulation at synaptic sites in regenerating muscle. *J. Cell Biol.* **101**:735–743.

*BURDEN, S. J. 1987. The extracellular matrix and subsynaptic sarcoplasm at nerve-muscle synapses. In Salpeter, M. M., ed. *The Vertebrate Neuromuscular Junction*, Liss, pp. 163–186.

BURDEN, S. J., P. B. SARGENT, and U. J. MCMAHAN. 1979. Acetylcholine receptors in regenerating muscle accumulate at original synaptic sites in the absence of the nerve. *J. Cell Biol.* **82**:412–425.

FALLON, J. R., and C. E. GELFMAN. 1989. Agrin-related molecules are concentrated at acetylcholine receptor clusters in normal and aneural developing muscle. *J. Cell Biol.* **108**:1527–1535.

HUNTER, D. D., V. SHAH, J. P. MERLIE, and J. R. SANES. 1989. A laminin-like adhesive protein concentrated in the synaptic cleft of the neuromuscular junction. *Nature* **338**:229–234.

REIST, N. E., C. MAGILL, U. J. MCMAHAN, and R. M. MARSHALL. 1987. Agrin-like molecules at synaptic sites in normal, denervated, and damaged skeletal muscles. *J. Cell Biol.* **105**:2457–2469.

SANES, J. R., L. M. MARSHALL, and U. J. MCMAHAN. 1978. Reinnervation of muscle fiber basal lamina after removal of myofibers. *J. Cell Biol.* **78**:176–198.

Plant Cell Walls

ALONI, R. 1987. Differentiation of vascular tissue. *Ann. Rev. Plant Physiol.* **38**:179–204.

*CASSAB, G. I., and J. E. VARANER. 1988. Cell wall proteins. *Ann. Rev. Plant Physiol.* **39**:321–353.

*CRONSHAW, J., W. J. LUCAS, and R. T. GIAQUINTA, eds. 1986. *Phloem Transport.* Liss.

*DELMER, D. P. 1987. Cellulose biosynthesis. *Ann. Rev. Plant Physiol.* **38**:259–290.

FRY, S. 1986. Cross-linking of matrix polymers in the growing cell walls of angiosperms. *Ann. Rev. Plant Physiol.* **37**:165–186.

*FRY, S. C. 1988. *The Growing Plant Cell Wall: Chemical and Metabolic Analysis*, Wiley.

*LLOYD, C. W., and R. W. SEAGULL. 1985. A new spring for plant cell biology: microtubules as dynamic helices. *Trends Biochem. Sci.* **10**:476–478.

MCNEIL, M., A. G. DARVILL, S. C. FRY, and P. ALBERSHEIM. 1984. Structure and function of the primary cell walls of plants. *Ann. Rev. Biochem.* **53**:625–663.

ROBERTS, K., A. W. B. JOHNSTON, C. W. LLOYD, P. J. SHAW, and H. W. WOOLHOUSE, eds. 1985. *The Cell Surface in Plant Growth and Development*, J. Cell Sci. Suppl. 2, Company of Biologists Ltd, (Cambridge, U. K.)

*RYAN, C. A. 1987. Oligosaccharide signalling in plants. *Ann. Rev. Cell Biol.* **3**:295–317.

*SALISBURY, F. B., and C. W. ROSS. 1985. *Plant Physiology*, 3d ed. Belmont, CA, Wadsworth.

SCHNEIDER, B., and W. HERTH. 1986. Distribution of plasma membrane rosettes and kinetics of cellulose formation in xylem development of higher plants. *Protoplasma* **131**:142–152.

*VARNER, J. E., and L.-S. LIN. 1989. Plant cell wall architecture. *Cell* **56**:231–239.

WICK, S. M., R. W. SEAGULL, M. OSBORN, K. WEBER, and B. E. S. GUNNING. 1981. Immunofluorescence microscopy of organized microtubule arrays in structurally stabilized meristematic plant cells. *J. Cell Biol.* **89**:685–690.

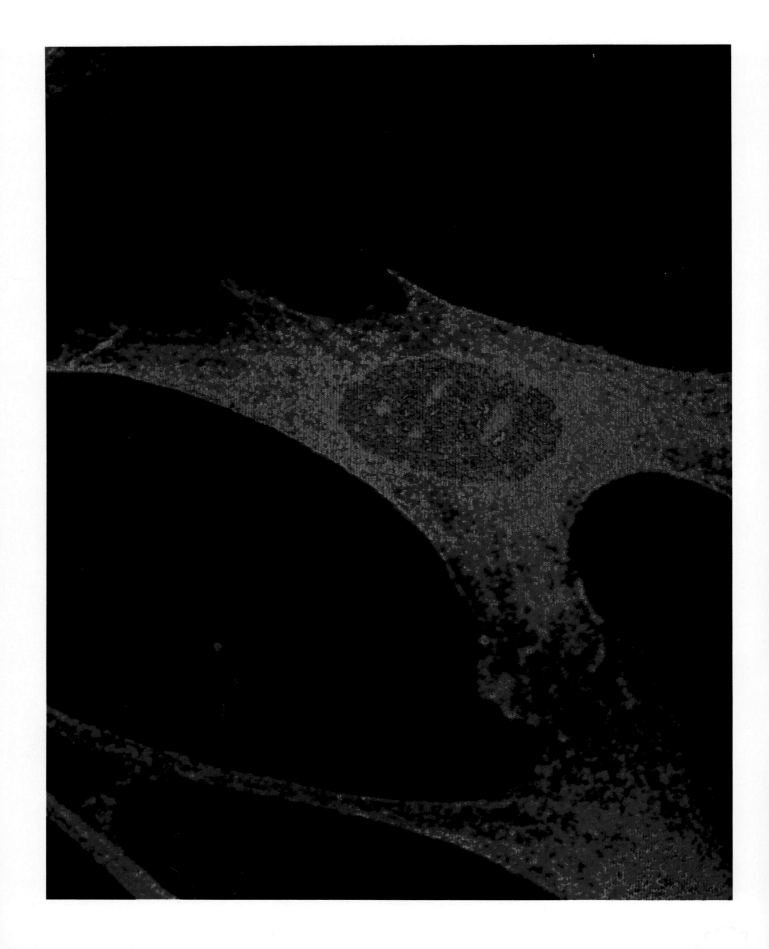

IV

The New Biology: Facing Classic Questions at the Frontier

◀ *False color picture showing the distribution in a cell of the protein encoded by the c-abl gene. This protein is the normal counterpart of an oncogene active in leukemic cells. The orange color indicates that most of the c-abl product is in the nucleus, but some is also near the border of the cell—attached to the plasma membrane—and some is in the cytosol (green).* Courtesy of R. Van Etten.

In the previous chapters, concerned with dissecting elements of molecular cell biology, we only rarely considered larger issues of cell biology: How does a cell gain its special functions? What controls growth and differentiation? What happens when cellular control systems go awry? How did cells evolve? Part IV takes up these larger, integrative issues.

Only three topics—of many that could have been chosen—are covered in Part IV: cancer, immunology, and evolution. These are very active fields where often there is insufficient consensus to allow for a canonical representation, and, more as authors than reviewers, we present personal syntheses of research. We try here to bring to students the excitement that molecular cell biologists continually experience from the deepening of their understanding of previously obscure phenomena.

What are the genetic and physiological events that transform a normally regulated cell into one that grows without responding to controls? How does a normal cell change when it becomes a cancer cell? Chapter 24 shows how principles and details of molecular biology introduced earlier in the book apply in the struggle to understand this fundamental aberration in cellular behavior.

How does the immune system—the defense system evolved by vertebrates in their continual battle with microorganisms—recognize an invader? How can an individual animal having a finite amount of DNA inherit the capacity to respond defensively and with great specificity to a virtually infinite array of possible invaders? Chapter 25 discusses immunology, describing mechanisms illuminated by very recent advances of molecular cell biology.

Chapter 26 discusses the evolution of cells. Evolution is a central issue of all biology—in a real sense, biology is a historical science, collecting information about the particular history of life on this planet. What clues for understanding the early events of evolution can molecular cell biology provide?

▼   ▼   ▼

C H A P T E R 24

Cancer

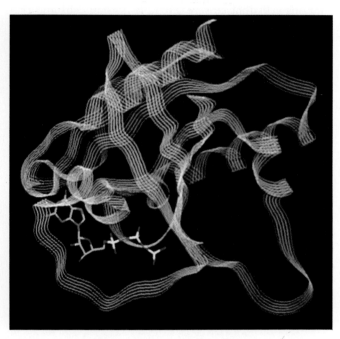

Ras protein p21 (ribbon structure) with bound GTP

Cell growth is a carefully regulated process that responds to specific needs of the body. In a young animal, cell multiplication exceeds cell death, so the animal increases in size; in an adult, the processes of cell birth and death are balanced to produce a steady state. For some adult cell types, renewal is rapid: intestinal cells have a half-life of a few days before they die and are replaced; certain white blood cells are replaced as rapidly. In contrast, human red blood cells have approximately a 100-day half-life, healthy liver cells rarely die, and, in adults, there is a slow loss of brain cells with little or no replacement.

Very occasionally, the exquisite controls that regulate cell multiplication break down and a cell begins to grow and divide, although the body has no need for further cells of its type. When the descendants of such a cell inherit the propensity to grow without responding to regulation, the result is a clone of cells able to expand indefinitely. Ultimately, a mass called a tumor may be formed by this clone of unwanted cells. Because tumors may have devastating effects on the animals that harbor them, much research has gone into understanding how they form.

In this chapter we describe current knowledge about the genetic and physiological events that transform a normally regulated cell into one that grows without responding to controls. These genetic events are generally not inherited through the gametes; rather, they are changes in the DNA of somatic cells. The principal type of change is the alteration of preexisting genes to oncogenes, *whose products cause the inappropriate cell growth. Thus DNA alteration is at the heart of cancer induction.* ▲

Characteristics of Tumor Cells

Although most research into the molecular basis of cancer utilizes cells growing in culture, it is important to look first at tumors as they occur in experimental animals and in humans. In this way we can see the gross properties of the disease—the properties that ultimately must be explained by analysis of cells and molecules.

Malignant Tumor Cells Are Invasive and Can Spread

Tumors arise with great frequency, especially in older animals and humans, but most pose little risk to their host because they are localized. We call such tumors *benign;* an example is warts. Tumors become life-threatening if they spread throughout the body. Such tumors are called *malignant* and are the cause of cancer.

It is usually apparent when a tumor is benign because it contains cells that closely resemble normal cells and that may function like normal cells. Ill-understood forces keep benign tumor cells (and normal cells) localized to appropriate tissues. Benign liver tumors stay in the liver, and benign intestinal tumors stay in the intestine. A fibrous capsule usually delineates the extent of a benign tumor and makes it an easy target for a surgeon. Benign tumors become serious medical problems only if their sheer bulk interferes with normal functions or if they secrete excess amounts of biologically active substances like hormones (Figure 24-1).

The major characteristics that differentiate malignant tumors from benign ones are their *invasiveness* and

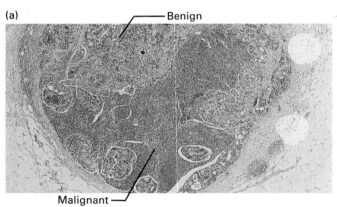

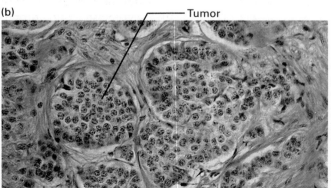

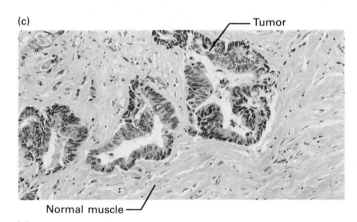

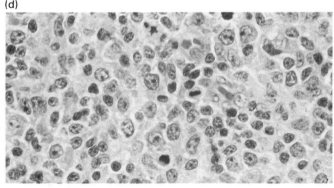

▲ **Figure 24-1** Sections of various types of tumors. (a) A benign polyp of the lower intestine that has within it malignant tissue. (b) A tumor derived from cells that secrete neuroendocrine hormones. It is organized like a little gland in the midst of normal tissue. (c) A rectal epithelial tumor seen here as invaginations into the normal smooth muscle tissue of the rectum. (d) Lymphoma cells. These cells are leukemic cells that adhere together to form a solid mass. *Photographs courtesy of Dr. J. Aster.*

(a)

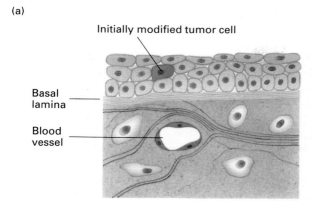

Initially modified tumor cell

Basal lamina

Blood vessel

(b)

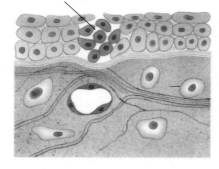

Mass of tumor cells (localized-benign tumor)

(c)

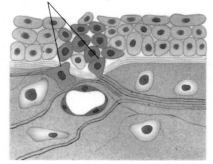

Invasive tumor cells

(d)

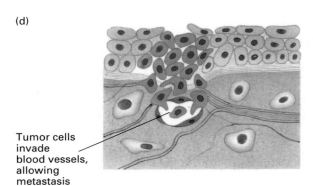

Tumor cells invade blood vessels, allowing metastasis to occur

◀ **Figure 24-2** Stages in tumor growth and metastasis. (a) A single modified cell appears in a tissue. (b) The modified cell begins to grow, although surrounding cells do not, and a mass of localized tumor cells form. This tumor is still benign. (c) As it progresses to malignancy, the tumor invades the basal lamina that surrounds the tissue. (d) The tumor cells spread into blood vessels that will distribute them to other sites in the body. If the tumor cells can exit from the blood vessels and grow at distant sites they are considered malignant, and a patient with such a tumor is said to have cancer.

spread. Malignant tumors do not remain localized and encapsulated; instead they invade surrounding tissues, get into the body's circulatory system, and set up areas of proliferation away from the site of their original appearance. The spread of tumor cells and establishment of secondary areas of growth is called *metastasis;* malignant cells have the ability to *metastasize* (Figure 24-2).

Malignant cells are usually less well differentiated than benign tumor cells. Furthermore, their properties often vary over time. For example, liver cancer cells may lack certain enzymes characteristic of liver cells and may ultimately evolve to a state in which they lack most liver-specific function. This variability of phenotype is often correlated to a variability of genotype; cancer cells have abnormal and unstable numbers of chromosomes, as well as many chromosomal abnormalities. Because apparently benign tumors may progress to malignancy and the earliest stages of malignant tumors are hard to identify, pathologists are rarely sure how a malignancy began. In any case, the cells of malignant tumors have a tendency to lose differentiated traits, to acquire an altered chromosomal composition, and to become invasive and metastatic.

Cancer cells can be distinguished from normal cells by microscopic examination. In a specific tissue, cancer cells usually exhibit the characteristics of rapidly growing cells, that is, a high nucleus-to-cytoplasm ratio, prominent nucleoli, many mitoses, and relatively little specialized structure. The presence of invading cells in an otherwise normal tissue section is the most diagnostic indication of a malignancy (Figure 24-3).

Malignant cells usually have enough of the hallmarks of the normal cell type from which they were derived that it is possible to classify them by their relationship to normal tissue. Normal animal cells are often subdivided according to their embryonic tissue of origin, and the naming of tumors has followed suit. Normal cells arise from one of three embryonic cell layers: endoderm, ectoderm, or mesoderm. Malignant tumors are classified as *carcinomas* if they derive from endoderm or ectoderm and *sarcomas* if they derive from mesoderm. The *leukemias,* a subdivision of the sarcomas, grow as individual cells in the blood, whereas most other tumors are solid masses. (The

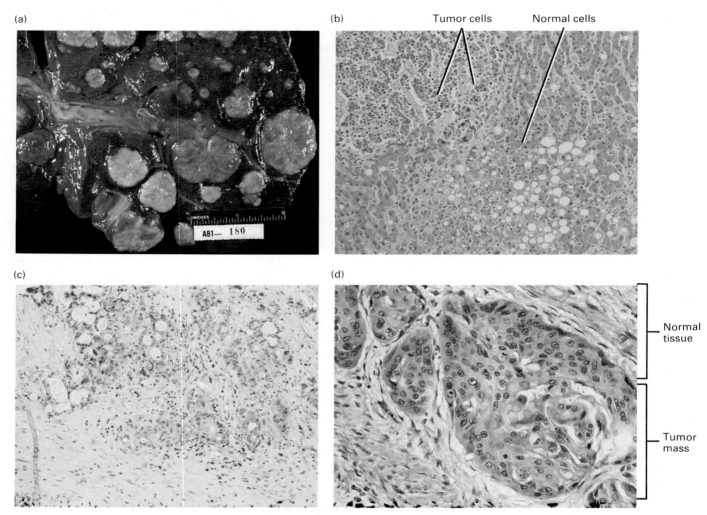

▲ **Figure 24-3** Gross and microscopic views of tumors invading normal tissue. (a) The gross morphology of a human liver in which a metastatic lung tumor is growing. The white protrusions on the surface of the liver represent the tumor masses. (b) A light micrograph of a section of the tumor in (a) showing areas of small, dark-staining tumor cells invading a region of larger, light-staining normal liver cells. (c) A section of the adrenal gland from the same patient. Again, invading metastatic tumor cells are evident. (d) Squamous cell carcinoma invading the normal connective tissue of the dermis. *Courtesy of J. Braun.*

name leukemia is derived from the Latin for "white blood": the massive proliferation of leukemic cells can cause a patient's blood to appear milky.)

Alterations in Cell-to-Cell Interactions Are Associated with Malignancy

Although invasiveness and metastasis are the hallmarks of malignancy, our discussion of these two characteristic properties of cancer cells is relatively brief because they are problems of cell-to-cell interaction, and understanding in this field is still rudimentary. Perhaps soon, when specific intercellular adhesive molecules are better characterized, the parameters of invasiveness and spread will

become better defined. For now, only a few general principles are clear.

The restriction of a normal cell type to a given organ and/or tissue is maintained by cell-to-cell recognition and by physical barriers. Cells recognize one another by ill-understood surface processes, which dictate that like cells bind together and that certain cells interact. Primary among the physical barriers that keep tissues separated is the *basal lamina,* which underlies layers of epithelial cells as well as the endothelial cells of blood vessels (see Figure 23-3). Basal laminas define the surfaces of external and internal epithelia and the structure of blood vessels. Metastatic tumor cells have the ability to digest their way through basal laminas (see Figure 24-2c). Thus, the processes of invasion and metastasis depend in part on the ability of tumor cells to pass through basal laminas. Metastasis also depends on the ability of malignant cells to grow in new locations, surrounded by cells that they do not ordinarily contact. The ability to start growth without a mass of surrounding identical cells and the ability to ignore "foreign" cell contacts are thus properties that tumor cells must develop. The wide range of altered behaviors that underlie malignancy must have a basis in new proteins made by malignant cells. For example, enzymes secreted by some malignant cells degrade collagen and other basal lamina proteins, such as proteoglycans and glycosaminoglycans.

Another recognized property of malignant cells that we do not understand is their ability to elude the immune system. It might be expected that the alterations in cell structure associated with malignancy would lead the immune system to recognize tumor cells as foreign. The blood even contains a "natural killer" cell that can apparently recognize and kill many types of tumor cells while sparing normal cells. Despite this, tumor cells find ways to elude immune detection. They may cover up antigens that would otherwise mark them for destruction, or they may rid themselves of the cell-surface molecules that lymphocytes use to recognize foreign cells (major histocompatibility antigens). As a result, immune recognition appears to play a minor role in protecting the body against tumors. When confronted with the defenses evolved by tumor cells, the immune system is largely ineffective.

Tumor Cells Lack Normal Controls on Cell Growth

In an animal or in culture, cells are either growing or *quiescent*. Cell growth involves an increase in mass that leads a cell to divide; cell growth and cell division are thus interrelated phenomena. A quiescent cell is one that is not increasing its mass or passing through the cell cycle. Quiescent cells carry out the characteristic functions of tissues, such as the synthesis of export proteins in the liver or the transmission of impulses in a nerve. Cell growth is a regulated process, so the fraction of growing cells in a given tissue is a function of both the age of the organism and the properties of the tissue. In adults, certain tissues (e.g., the intestine) maintain a constant size by the continued growth of new cells and the death of older cells; such tissues may have many dividing cells. Embryos and expanding tissues of young animals also contain a large fraction of growing cells.

Cell growth involves two easily recognized, coordinated events: the duplication of cellular DNA and the physical division of the cell into two daughter cells. If DNA synthesis and cell division are considered the key events, then the cell-growth cycle can be divided into four periods: G_1, the gap between the previous nuclear division and the beginning of DNA synthesis; S, the period of DNA synthesis; G_2, the gap between DNA duplication

and nuclear division; and M, the period of mitosis, during which both of the chromosomes separate into the two daughter cells (see Figure 5-2). Cell division is usually coordinated with nuclear division, but it may be delayed. For example, in early insect embryos many nuclear divisions occur without cytoplasmic division; the result is a multinucleated cell (see Figure 11-25).

The duration of the entire cell cycle and of its constituent periods varies among different cell types. Embryonic cells can divide as frequently as rapidly growing bacteria, once every 15–20 min. More typically, the mammalian cell cycle is 10–30 h long—S, G_2, and M together requiring a fixed time of about 10 h, with the duration of G_1 being quite variable.

Quiescent cells usually have unduplicated DNA, although some epithelial cells may rest in G_2. Logically, then, stimulation of the growth of most quiescent cells must cause them to enter S (i.e., to make a new complement of DNA) before they can divide. Some workers consider quiescent cells to be in a physiological state called G_0, which they distinguish from G_1.

The cell cycle is fundamentally a series of decisions made over time. But apparently only a single decision is under tight control, a commitment in G_1 to go through S, G_2, and M. This commitment is described by yeast geneticists as the passage through START, and yeast mutants are known that block at START.

The inputs to the decision are poorly understood, but they involve whether the cell has grown to a sufficient size to warrant dividing into two and whether the nutritional state of the cells—the phosphate, amino acid, glucose magnesium, etc., levels—are above a fixed set of thresholds. Once the cell commits itself to division at START, many events ensue, leading to cell enlargement, DNA duplication, the segregation of one copy of each DNA molecule to a separate nucleus, and the physical separation of the two daughter cells (Figure 24-4).

From both the study of yeast mutants that are blocked in progression through the cycle and experiments on mixing protein extracts from frog eggs and early embryos there has emerged the outlines of how the later stages of cell cycle control—especially control of mitosis itself—are maintained. Mitosis appears to be controlled by the concentration of a protein called a *cyclin*. Cyclins build up in the cell in interphase and then are degraded during mitosis. Frog egg extracts can undergo multiple rounds of mitosis. If such extracts are deprived of cyclin by addition of a specific antibody, they will not enter mitosis but remain blocked in a G_2-like state. Addition of cyclin restores the ability of the extract to go through multiple in vitro mitotic cycles. A nondegradable form of cyclin blocks the extract in mitosis, implying that cyclin degradation is the crucial step for exiting mitosis. Cyclin appears to function in a protein complex, called maturation promotion factor, or MPF, that includes a critical protein kinase. This 34,000-MW serine/threonine kinase (p34)

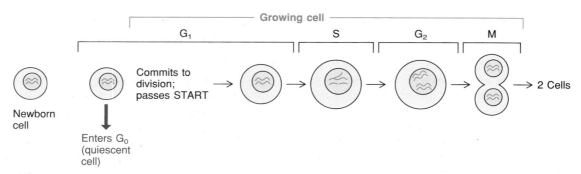

▲ **Figure 24-4** The cell cycle. A newborn cell can either be quiescent or continue to grow. The decision point is early in the G_1 phase when a cell either passes START—and then is committed to growing, finishing the rest of the cycle and dividing (G_1, S, G_2, and M)—or the cell enters the G_0 state in which it continues to metabolize but does not grow.

was first found in a fission yeast, where it is encoded by a gene called *cdc2*. The cyclins appear to regulate the *cdc2* kinase, and the action of the kinase may be what induces mitosis. The kinase is regulated through other kinases that phosphorylate either tyrosine residues to inactivate it or serine residues to activate it. At least one protein-tyrosine kinase of the cell, the *src* gene product, appears to be a substrate of p34, suggesting that there may be feedback from serine/threonine to tyrosine kinases, a possibility that—as we shall see—may help to explain how cell cycles are deregulated in cancer cells.

Control of cell growth is one of the most important aspects of an animal's physiology. Cells of an adult must divide frequently enough to allow tissues to remain in a steady state, and division must be stimulated at wounds or when special requirements are placed on a tissue. There must be many circulating cell-specific factors that signal individual cell types whether to divide or not. A few such factors are known—notably factors controlling the growth of blood cells—but many more await discovery. Uninhibited growth of cells results in tumors. How the controls of cell growth are overcome is a major area of cancer research.

Use of Cell Cultures in Cancer Research

Although questions about cell growth and the induction of cancer are ultimately questions about the behavior of individual cells in a living organism, the study of cells in an animal is impractical because of the difficulty in identifying the relevant cells, in manipulating their behavior in a controlled manner, and in separating the effects due to the intrinsic properties of the cells from the effects due to the interactions among the many cell types present in the organism. These difficulties can be avoided by use of cells growing in culture. The environment of cultured cells can be manipulated by the investigator, the type of target cell can be well defined, the changes in cells following treatment with a cancer-causing agent (carcinogen) can be examined, and the fate of the carcinogen can be determined. Furthermore, cultured cells can be quiescent or growing; they can have, in fact, precisely defined growth parameters. They can also be manipulated genetically. For these reasons, studies of normal cell growth as well as of cancer induction depend heavily on the use of cultured cells.

Fibroblastic, Epithelial, and Nonadherent Cells Grow Readily in Culture

Cell cultures are established by removing cells from an organism and placing them at 37°C in a medium with nutrients plus 5–20 percent blood serum. For many years, most cell types were difficult, if not impossible, to grow in culture. Recent modifications in culture methods have allowed experimenters to grow many specialized cells in culture, and the utility of cell culture as a method of examining cell behavior has increased enormously. Most published studies, however, describe work with the few cell types that grow readily in culture. These are not cells of a defined type; rather, they represent whatever grows when a tissue or an embryo is placed in culture.

The cell type that usually predominates in such cultures is called a *fibroblast* because it secretes the types of proteins associated with the fibroblasts that form fibrous connective tissue in animals. Cultured fibroblasts have the morphology of tissue fibroblasts, but they are not as differentiated. Fibroblasts are a cell type derived from embryonic mesoderm, and the cells that grow in culture appear to be mesodermal stem cells. With appropriate stimulation these cells can differentiate into many cell types: fat cells, connective tissue cells, muscle cells, and others. In most studies, however, cultured "fibroblasts" have been treated simply as convenient prototypical cells for study. Many studies also have been conducted with cultured *epithelial* cells. Although these do not necessarily correspond to normal tissue cells, they are representative of the type of cell that comes from the ectodermal or endodermal embryonic cell layers.

Cultured fibrobalsts and epithelial cells are grown on glass or plastic dishes to which they adhere tightly due to their secretion of matrix proteins such as laminin, fibronectin, and collagen. Neither cell type will ordinarily grow if it is not adhering to a substratum. To prepare tissue cells for culture or to remove adherent cells from a culture dish for biochemical studies, trypsin or another protease is used. The process of putting cells into culture or of transferring cells to a new culture is often called *plating*.

Certain cells cultured from blood, spleen, or bone marrow adhere poorly, if at all, to a culture dish. In the body, such nonadherent cells are held in suspension (in the blood) or they are loosely adherent (in the marrow and spleen). Because these cells often come from immature stages in the blood cell lineages, they are very useful for studying the development of leukemias.

Some Cell Cultures Give Rise to Immortal Cell Lines

When cells are removed from an embryo or an adult animal, most of the adherent ones grow continuously in culture for only a limited time before they spontaneously cease growing. Such a culture dies out even if it is provided with fresh supplies of all of the known nutrients that cells need to grow, including blood serum. For example, explanted human fetal cells take some time to become established in culture, during which period the majority of cells die and the "fibroblasts" become the predominant cell type. The fibroblasts then double about 50 times before they cease growth. Starting with 10^6 cells, 50 doublings can produce $10^6 \times 2^{50}$ or more than 10^{20} cells, which is equivalent to the weight of about 10^5 people. Thus, even though its lifetime is limited, a single culture, if carefully maintained, can be studied for a long time. Such a lineage of cells originating from one initial culture is called a *cell strain* (Figure 24-5a).

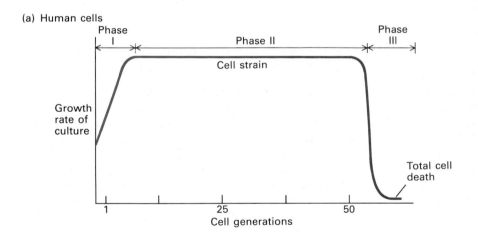

(a) Human cells

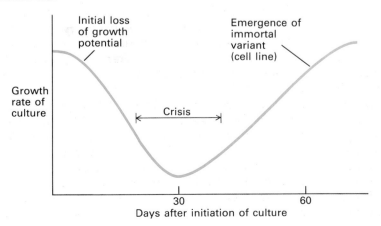

(b) Mouse cells

◀ **Figure 24-5** Stages in the establishment of a cell culture. (a) Human cells. When an initial explant is made (e.g., from foreskin), some cells die and others (mainly fibroblasts) start to grow; overall there is a slow increase in growth rate *(phase I)*. If the remaining cells are continually diluted, they grow as a cell strain at a constant rate for about 50 cell generations *(phase II)*, after which growth begins to slow. The ensuing period of increasing cell death *(phase III)* ultimately leads to the complete death of all of the cells in the culture. (b) Mouse or other rodent cells. When a culture is prepared from mouse embryo cells, there is initial cell death coupled with the emergence of healthy growing cells. As these are diluted and allowed to continue growth, they soon begin to lose growth potential and most cells die (the culture goes into crisis). Very rare cells do not die but continue growing until their progeny overgrow the culture. These cells constitute a cell line, which will grow forever if it is appropriately diluted and fed with nutrients: the cells are immortal.

To be able to clone individual cells, modify cell behavior, or select mutants, it is often necessary to maintain cell cultures for many more than 100 doublings. This is possible with cells from some animal species because these cells undergo a change that endows them with the ability to grow indefinitely. A culture of cells with an indefinite life span is considered immortal; such a culture is called a *cell line* to distinguish it from an impermanent *cell strain*. The ability of cultured cells to grow indefinitely varies depending on the animal species from which the cells originate. For human cells, only tumor cells grow indefinitely, and therefore the HeLa tumor cell has been invaluable for research on human cells. Chicken cells die out after only a few doublings, and even tumor cells from chickens almost never become immortal. With rodent cells, however, cultures of embryonic adherent cells routinely give rise to cell lines.

When adherent rodent cells are first explanted, they grow well, but after a number of serial replatings they lose growth potential and the culture goes into *crisis*. During crisis most of the cells die, but often a rapidly growing cell variant arises spontaneously and takes over the culture. Such a variant will grow forever if it is provided with the necessary nutrients (Figure 24-5b). Cells in an established line usually have more chromosomes than the normal cell from which they arose, and their chromosome complement undergoes continual expansion and contraction in culture. The culture is said to be *aneuploid* (i.e., having an inappropriate number of chromosomes), and the cells of such a culture are obviously mutants.

If rodent cell cultures are maintained at a low cell density until a cell line emerges, the line will consist of flat cells that adhere tightly to the dish in which they are grown. A number of mouse cell lines derived in this fashion have been used extensively in cancer research. These lines are called 3T3 cells; they were derived according to a schedule whereby 3×10^5 cells were *t*ransferred every *3* days into petri dishes with a 50-mm diameter to maintain the appropriate cell density. As is true for other cultured fibroblasts, the exact cell type that gives rise to 3T3 cells is uncertain, but they can differentiate into a range of mesodermally derived cell types, especially endothelial cells (those that line blood vessels). The ability to derive lines of flat cells like 3T3 set the stage for studying the transition to malignancy in cell culture because cancer cells differ dramatically from 3T3 cells in their growth properties. Before we describe the use of 3T3 cells in cancer research, however, we shall consider the control of 3T3 growth.

Certain Factors in Serum Are Required for Long-term Growth of Cultured Cells

If a culture of 3T3 cells is plated at 3×10^5 cells per dish in a medium with 10 percent blood serum, the cells will grow for a few days and then cease growth at about 10^6 cells per dish. The culture is said to have reached *saturation density*. Although the quiescent cells in a saturated culture have stopped growing, they can remain viable for a long time and resume growth if supplied with fresh medium.

Among the treatments that will reinitiate growth in a quiescent 3T3 cell culture is the addition of extra serum to the medium. In fact, the density at which the cells stop growing is in direct proportion to the amount of serum with which they are initially provided (Figure 24-6). Although this result appears to indicate that serum factors are the primary determinants of whether cells remain quiescent or initiate growth, other results show that they are not the whole story. For example, if a strip of cells is removed from a quiescent cell culture (the culture is said to be *wounded*), the cells at the border of the wound will begin growing and will divide a few times to fill the gap. Because the cell medium is not altered in such an experiment and because the cells that are not adjacent to the wound do not initiate growth, it is clear that local effects (cell-to-cell contacts) also control cell growth. In all prob-

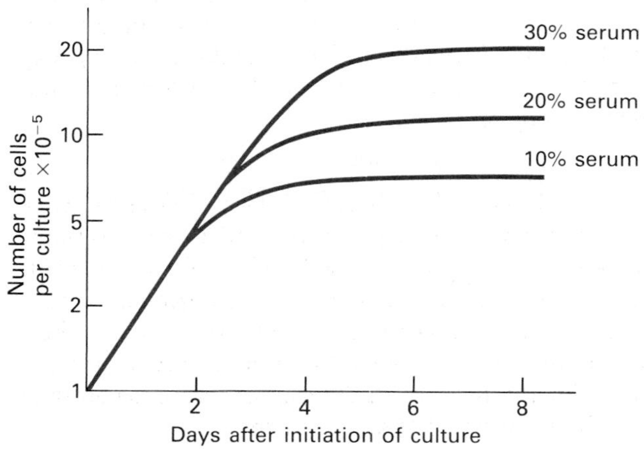

◄ **Figure 24-6** The dependence of cell growth on serum concentration. A constant number of 3T3 cells was used to initiate multiple cultures, each of which was fed with a medium containing the indicated percentages of fetal calf serum. The number of cells per culture was determined daily. The initial growth rates were indistinguishable, but the final number of cells was proportional to the amount of added serum. The experiment shows that serum factors rather than cell contacts control cell growth because cells are already touching one another in 10% serum. [See R. W. Holley and J. A. Kiernan, 1968. *Proc. Nat'l Acad. Sci. USA* **60**:300.]

ability, the degree to which crucial serum growth factors are available to cells is limited by close contacts.

In recent years, long-term growth of certain cell lines has been achieved with *defined media* containing known components and no serum (see Table 5-3). In such media, serum is replaced with purified proteins present in serum. For example, *epidermal growth factor* (EGF) is needed for growth by almost all cells in culture; many cells also need insulin or an insulinlike factor for growth. These two substances plus transferrin (which makes iron available to cells; Figure 14-45) can largely satisfy the serum requirement for most types of cells. Individual cell types, however, often have exotic requirements, and no universal defined serum substitute has been concocted.

Growth factors and hormones facilitate cell growth by acting as signaling agents. They somehow direct the cell to carry out whatever steps are needed for growth. The factors and hormones themselves neither provide nutrient value to the cell nor play any known role in metabolic pathways. Only their ability to bind to specific cell-surface receptors enables them to control cellular events. A recent clue to the mechanism of signaling by growth factors has emerged from the discovery that the receptors for EGF, platelet-derived growth factor, and insulin all exhibit a tyrosine-specific protein kinase activity, which is stimulated when the receptor binds its cognate ligand. Phosphorylation of tyrosine residues in proteins thus may be an important mechanism by which factors signal cells to grow; as will be evident from the following discussion, this may be a crucial clue to how the growth of certain cancer cells is stimulated.

These characteristics of cell growth in vitro provide the background for a consideration of how cancer cells differ from normal cells. Quiescent normal cells are ones that have not been stimulated to grow by factors or hormones. Because unrestricted growth is a characteristic of cancer cells, overcoming the growth inhibition in quiescent cells is a likely mechanism for inducing cancer.

Malignant Transformation Leads to Many Changes in Cultured Cells

Treatment of adherent cells with various agents (e.g., viruses, various chemicals, radiation) can dramatically change their growth properties in culture. Furthermore, such treated cells can form tumors after they are injected into susceptible animals. Such changes in the growth properties of cultured cells and their subsequent development of tumor-forming capacity are collectively referred to as *malignant transformation,* or just *transformation.* Because transformation can be carried out entirely in culture, it has been widely studied as a model of cancer induction in animals, although there is not a direct correspondence between the two processes.

Transformed adherent cells usually can be identified by changes in their cellular morphology and growth habit. If, for instance, a growing culture of 3T3 fibroblasts is exposed to SV40 virus, a small proportion of the cells will be infected, and these will adopt a more rounded configuration. The virus-infected cells will be less adherent to one another and to the dish than are the normal surrounding cells, and they will continue to grow when the normal cells have become quiescent. A group or *focus* of these loosely adherent transformed cells can be recognized under the microscope (Figure 24-7). If a focus of transformed cells is recovered and a culture is grown from it, the result will be a line of transformed cells (called, e.g., SV-3T3 cells).

Many properties of a transformed cell line differ from those of its parental line. These differences are related to aspects of growth control, morphology, cell-to-cell interactions, membrane properties, cytoskeletal structure, protein secretion, and gene expression. Some of the changes caused by transformation are probably interrelated, but some seem independent of one another. Not all transformed cells show all of the changes that can be induced by transformation, but those described in this section are usual concomitants of transformation. Because little is known about the molecular events that underlie these alterations in cellular parameters, we can discuss them now only in gross descriptive terms; unsuspected relationships between them may be recognized later.

Alterations in Growth Parameters and Cell Behavior A key characteristic of transformed cells is that they continue to grow when normal cells cease growth (Figure 24-8). This increase in saturation density, considered an important analog of malignant cell growth in animals, results from numerous alterations in transformed cells.

Decreased Growth Factor Requirements Because transformed cells have apparently lost some of the hormone and growth factor requirements of normal cells, they grow in initial serum concentrations that are much lower than those required by normal cells. Some transformed cells produce growth factor analogs, so they may provide their own growth factors (called *autocrine* stimulation; see Figure 19-1).

Loss of Capacity for Growth Arrest When the concentration of isoleucine, phosphate, EGF, or any other substance that regulates growth falls below a certain threshold level, normal cells go into quiescence. Transformed cells are deficient in this ability to arrest their growth in response to lowered nutrient or factor concentrations. Cells may even kill themselves trying to continue growth in an impossible environment.

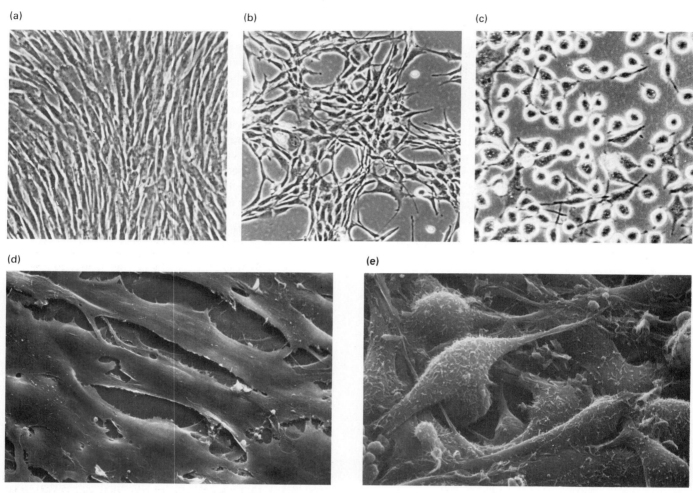

▲ **Figure 24-7** Normal and transformed rat embryo fibroblasts as viewed in the phase-contrast light microscope. (a) Cultured normal rat embryo fibroblasts. Note that the cells are aligned and closely packed in an orderly fashion. (b) Rat embryo fibroblasts that have been transformed by integration of the polyoma virus gene encoding the mid-T antigen. The cells are crisscrossed and chaotic in their growth. (c) Rat embryo fibroblasts transformed by the Abelson murine leukemia virus. The cells have lost adherence to the dish so completely that they appear almost round. Each has a white halo because of the light refraction of rounded cells.

Scanning electron micrographs of normal and transformed 3T3 cells. (d) Normal 3T3 cells. Each straplike image is a cell. Like normal rat embryo fibroblasts [see (a)], the cells are lined up in one direction and are spread out into thin lamellae with bulges representing the nuclei. (e) 3T3 cells transformed by Rous sarcoma virus. The cells are much more rounded, and they are covered with small hairlike processes and bulbous projections. The cells grow one atop the other and they have lost the side-by-side organization of the normal cells. These transformed cells truly appear malignant. *Courtesy of L.-B. Chen.*

Loss of Dependence on Anchorage for Growth Normal adherent cells require firm contact with the substratum for growth. If they are plated onto a surface to which they cannot adhere, they will not grow, although they can remain viable for long periods of time. Similarly, if normal cells are suspended in a semisolid medium such as agar, they will metabolize but they will not grow. Most transformed cell lines have lost the requirement for adherence; they grow without attachment to a substratum, as indicated by their ability to form colonies when suspended as single cells in agar. This characteristic correlates extremely well with the ability of transformed cells to form tumors: cells that have lost anchorage dependence generally form tumors with high efficiency when they are injected into animals that cannot immunologically reject the cells. How the cell senses shape and why transformation abrogates the need for anchorage are both unanswered questions.

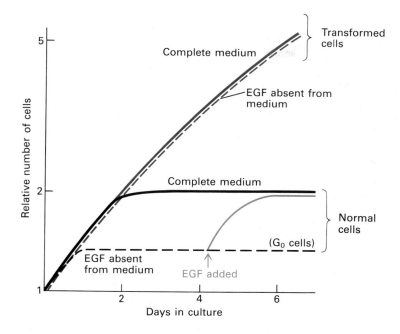

◄ Figure 24-8 The dependence of normal and transformed cells on epidermal growth factor (EGF). Equal numbers of normal and transformed cells were plated into a defined medium with or without EGF at day 0. Cell numbers were determined daily. Transformed cells in the complete medium grew after normal cells had ceased growth. The transformed cells had also lost the EGF requirement of normal cells.

At day 4, EGF was added to some dishes of normal cells that were initially lacking it. Because the cells responded to the addition of EGF by growth, it is evident that without EGF, normal cells remain viable in a G_0 state.

Changed Cell Morphology and Growth Habits The individual transformed cell is very different in shape and appearance from its normal counterpart (see Figure 24-7). It adheres much less firmly to the substratum and therefore is more rounded with fewer processes. Furthermore, transformed cells adhere poorly to each other and do not appear to sense their neighbors. As normal cells in a culture dish become more crowded, they form ordered patterns; transformed cells form chaotic masses. The low mutual adherence of transformed cells coupled with their loss of anchorage to the substratum allows them to grow in multiple layers, whereas normal cells grow in monolayers with some overlap at the cell borders.

Loss of Contact Inhibition of Movement In culture, normal fibroblastic cells are motile. When two normal cells moving around in a culture dish come into contact, one or both of them will stop and take off in another direction. This ensures that the cells do not overlap. When a normal cell is surrounded by others in such a way that it has nowhere to go, it ceases movement and forms gap junctions with the surrounding cells (see Figures 13-48 and 13-50). This phenomenon is referred to as *contact inhibition of movement*. Transformed cells lack contact inhibition of movement: they pass over or under one another, they grow on top of one another, and they infrequently form gap junctions.

Cell-surface Alterations Many of the foregoing properties of transformed cells relating to growth and behavior are probably consequences of cell-surface

events. Some of the differences in the surface properties of normal and transformed cells are described here.

Increased Mobility of Surface Proteins The glycolipids and glycoproteins that occur abundantly on the surfaces of normal cells are modified in transformed cells. For example, protein-linked N-acetylneuraminic (sialic) acid is decreased (see Figure 17-23), as is the ganglioside content of all lipids. The general structure of the plasma membrane is not altered, however, and the lateral mobility of surface lipids remains the same. Probably the most important difference is that cell-surface proteins are much more mobile in transformed cells than in normal cells; for example, antibodies can more easily agglutinate a given surface protein on a transformed cell than on a normal cell. Because the lipids are not intrinsically more mobile, it may be that links between surface proteins and the underlying cytoskeletal elements are modified by transformation. Such a modified linkage may also be the basis for the altered morphology of transformed cells.

Easier Agglutination by Lectins Lectins, such as concanavalin A and wheat germ agglutinin, are plant proteins that have multiple binding sites for specific sugars. Transformed cells are agglutinated by lectins at a much lower concentration than that required to agglutinate normal cells (Figure 24-9). The number of available sugar residues in surface glycoproteins and glycolipids is not increased and may even be lowered by transformation, so the increased agglutinability is not a consequence of an increased density of binding sites. Rather, the explanation appears again to be the increased mobility of cell-surface

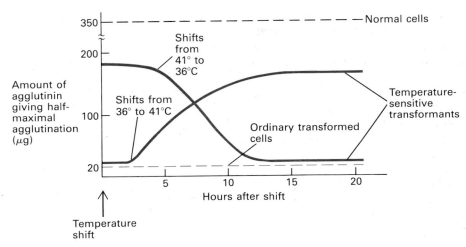

▲ **Figure 24-9** The agglutination of normal, transformed, and temperature-sensitive transformed cells by wheat germ agglutinin. The chicken embryo cells used for this experiment were of three types: normal fibroblasts, fibroblasts transformed by Rous sarcoma virus, and fibroblasts transformed by a viral mutant that produces transformed cells at 36°C and nearly normal cells at 41°C. The temperature of cells was raised or lowered, and at various times after the temperature shift the amount of wheat germ agglutinin that would give half of the maximal agglutination was determined. For normal cells, 350 μg was required; for the ordinary transformed cells, only 20 μg was required. The temperature-sensitive transformants began to change their properties a few hours after the temperature was shifted. At the higher temperature they became quasinormal, whereas at the lower temperature they took on the characteristics of transformed cells. [See M. M. Burger and G. S. Martin, 1972, *Nature New Biol.* **237**:356.]

glycoproteins, which allows low concentrations of lectins to make patches of receptor-lectin complexes on the cell surface. Patches on one cell can be cross-linked by the lectin to patches on other cells, causing the cells to agglutinate.

Increased Glucose Transport Transformed cells transport glucose more rapidly than normal cells. They have in their membrane a glucose transporter with a high affinity [low k_M] for glucose that ordinarily is expressed mainly in brain and erythrocytes. The rapid and high-affinity uptake of glucose due to this transporter correlates to the high glycolytic activity of tumor cells.

Reduced or Absent Surface Fibronectin Normal quiescent cells in monolayer culture become covered with a dense fibrillar network containing fibronectin as a major protein component (see Figure 23-25). Even growing cells have a diffuse covering of fibronectin. Transformed cells either totally lack fibronectin or have greatly reduced amounts. They have difficulty binding it to their surfaces; many transformed cells also make less fibronectin, but some secrete copious amounts. The addition of high concentrations of pure fibronectin derived from normal cells can cause many tumor cells to flatten out and take on a fairly normal appearance; thus the loss of fibronectin may be an important determinant of the transformed state.

Loss of Actin Microfilaments Not only are the surfaces of transformed cells different from those of normal cells, but some cytoskeletal elements also are different. For example, the actin microfilaments that extend the length of normal cells (see Figure 22-1) are either diffusely distributed or concentrated beneath the cell surface in transformed cells. The loss of cytoskeletal elements has been considered a possible cause of the increased mobility of cell-surface proteins.

Release of Transforming Growth Factors *Transforming growth factors* (TGFs) are proteins secreted by transformed cells that can stimulate growth of normal cells. Specific receptors for TGFs have been found, and TGFs have been identified in embryos, which implies that they have a role in normal cell physiology as well as in transformed cells. How much of the transformed phenotype is due to autostimulation by secreted growth factors is an unanswered question, but it seems likely that such factors play a role in the growth of some, if not many, transformed cells.

Protease Secretion Transformed cells often secrete a protease called *plasminogen activator*, which cleaves a peptide bond in the serum protein *plasminogen*, converting it to the protease *plasmin*. Thus the secretion of a small amount of plasminogen activator causes a large increase in protease concentration by catalytically activating the abundant plasminogen in normal serum. Normal cells treated with protease exhibit some characteristics of transformation (loss of actin microfilaments, growth stimulation, etc.), suggesting that plasminogen activator secretion may help maintain the transformed state of certain cell lines.

Secretion of plasminogen activator by transformed cells may be related to their tumor-forming capacity because the resulting increase in plasmin may help the cells penetrate the basal lamina. The normally invasive extra-embryonic cells of the fetus secrete plasminogen activator when they are implanting in the uterine wall; this provides a compelling analogy to invasion by tumor cells. Whether plasminogen activator acts only by cleaving circulating plasminogen or whether it can attack other proteins directly is an open question.

Altered Gene Transcription All of the foregoing characteristics of transformed cells are cytoplasmic activities; yet it might be expected that the extraordinary range of differences between normal and transformed cells would be at least partly due to alterations in the transcription of specific genes and in the relative stability of the transcripts. Surprisingly, however, the mRNA populations of normal cells and transformants derived from them are quite similar. The concentrations of some mRNAs are increased and those of some are decreased, but only about 3 percent of the total mRNA is specific to transformed cells. The transformation-specific mRNA probably includes many different low-abundance species. The proteins translated from these mRNAs, although low in concentration, have profound effects on cell growth and morphology. Some transformation-specific mRNAs also appear in embryonic cells, and tumor cells have many proteins that are characteristic of embryonic cells; this would suggest that transformation may alter protein composition toward that characteristic of embryos.

Immortalization of Cell Strains When cell strains, with their limited growth potential in culture, are used as targets for cell transformation, then one measurable characteristic of transformation is the induction of unlimited growth potential—that is, the conversion of a cell strain into an immortal cell line. (Obviously, immortalization cannot be used as an indicator of transformation in cell lines because they are immortal before exposure to transforming stimuli.) The ease with which transforming stimuli can generate immortal cell lines from cell strains depends on the underlying propensity of the cells to spontaneously acquire immortality: nonadherent (blood) cells from many animals are routinely immortalized by transformation; adherent human cells are rarely immortalized. Adherent chicken cells are almost never immortalized, but adherent rodent cells are easily transformed to immortality. The other characteristics of transformation described above are as relevant to adherent human

and chicken cells as they are to mouse cells, which implies that immortalization is quite distinct from other transformation parameters.

Transcription of Oncogenes Can Trigger Transformation

The differences in the properties of normal cultured cells and their transformants, summarized in the previous section, provide only limited clues to the mechanism or cause of transformation. Although the key steps in transformation and the interrelationship between various transformation parameters are still largely matters of conjecture, it is known that the events triggering transformation can be comparatively simple, often resulting from the transcription of one or two genes. These genes—now called oncogenes—may be part of a virus, or they may be altered cellular genes. An *oncogene* is a gene whose product is involved either in transforming cells in culture or in inducing cancer in animals; many oncogenes also are believed to play important roles in human cancer. (The word *oncogene* derives from the Greek *onkos*, meaning a bulk or mass; *oncology* is the scientific study of tumors.)

Of the many oncogenes known, all but a few are derivatives of certain normal cellular genes. Such genes (or *proto-oncogenes*) are important for normal cellular processes but can be altered, often in very simple ways, to become oncogenes. The normal genome of some DNA viruses also contain oncogenes, which both help the virus grow and can transform cells or induce cancer. Oncogenes are named with three-letter italic designations (e.g., *src*). Because most proto-oncogenes are basic to animal life, they have been highly conserved over eons of evolutionary time. Many are evident in the DNA of arthropods (like *Drosophila*), and some are even found in yeast.

As noted earlier, three types of transforming agents are known: viruses, chemicals, and radiation. These agents were recognized as carcinogens in animals before their ability to transform cultured cells was discovered. In the next several sections, we describe the role of each of these agents in transformation and/or carcinogenesis and their relationship to oncogenes. Following this, we will consider the role that proto-oncogenes and oncogenes play in the metabolism of normal cells and cancer cells.

DNA Viruses as Transforming Agents

Some animal viruses have RNA and some have DNA as their genetic material (see Table 5-4). One group of RNA viruses, the *retroviruses*, and many types of DNA viruses

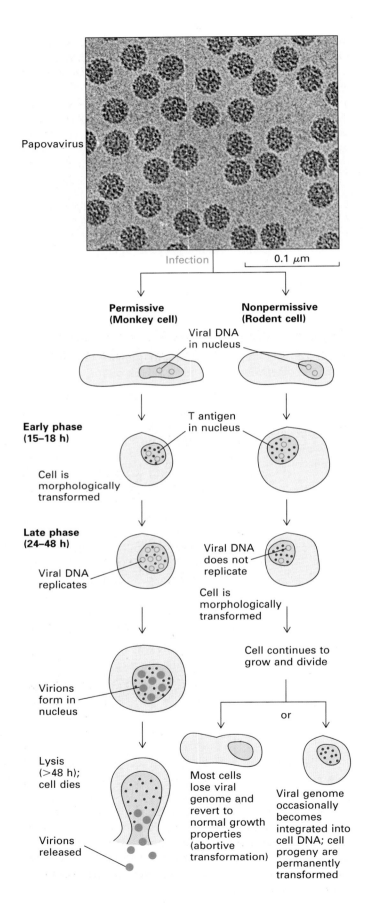

Papovavirus

Infection | 0.1 μm

Permissive (Monkey cell) | **Nonpermissive (Rodent cell)**

Viral DNA in nucleus

T antigen in nucleus

Early phase (15–18 h)

Cell is morphologically transformed

Late phase (24–48 h)

Viral DNA replicates

Viral DNA does not replicate

Cell is morphologically transformed

Cell continues to grow and divide

Virions form in nucleus

or

Lysis (>48 h); cell dies

Most cells lose viral genome and revert to normal growth properties (abortive transformation)

Viral genome occasionally becomes integrated into cell DNA; cell progeny are permanently transformed

Virions released

can be transforming agents; such viruses are often called *tumor viruses*. Tumor viruses cause transformation as a consequence of their ability to integrate their genetic information into the host cell's DNA; most often they cause the chronic production of one or more proteins called *transforming proteins,* which are responsible for maintaining the transformed state of the infected cells. Transforming proteins, which are encoded by oncogenes in the viral genome, are generally intracellular products, distinct from transforming growth factors. For DNA viruses, the known oncogenes are integral parts of the viral genome. For retroviruses, the oncogenes are normal or slightly modified cellular genes that are either appropriated from or hyperactivated in the host cell. The oncogenes in retroviruses and DNA viruses arise in different ways; we shall consider these two classes of tumor viruses separately.

DNA Viruses Can Transform Nonpermissive Cells by Random Integration of the Viral Genome into the Host-cell Genome

Among the many different types of animal DNA viruses that exist, the genome size varies from 5 to 200 kb (kilobases). The viruses with the smallest genomes encode very few proteins and rely mainly on host-cell functions for their replication; those with larger genomes encode many enzymes and provide many of their own replication functions. Although all types of DNA viruses can be tumor viruses, we shall focus on one group of very small DNA tumor viruses, the *papovaviruses,* of which the best-known representatives are SV40 and polyoma.

Infection by a papovavirus is divided into an early phase, during which it makes a set of viral proteins and induces cellular proteins, and a late phase, during which viral DNA is replicated, coat protein is made, and new virions mature (Figure 24-10). At the end of the early phase, a quiescent infected cell is activated to synthesize cellular DNA. The small papovaviral genome encodes a few key proteins that induce the cell to make the enzymes necessary for DNA replication. The virus accomplishes this by causing the cell to progress from the G_0 or G_1

◄ **Figure 24-10** Responses of permissive and nonpermissive cells to infection by SV40, a papovavirus. In permissive cells, virus infection leads to cell death and the production of progeny virions. In nonpermissive cells, the synthesis of the early viral proteins (T antigens) causes cells to become transformed. If the viral DNA becomes permanently integrated into the cell DNA, permanent transformation of the cell results. The electron micrograph shows an unstained virus preparation embedded in a thin layer of ice. *Photograph courtesy of T. Baker and M. Bina.*

phase into the S phase. Thus both cellular and viral DNA are replicated during the infection cycle. It is believed that the induction of the S phase underlies the transforming ability of papovaviruses. Approximately half of the genetic information in the 5-kb circular genomes of SV40 and polyoma specifies their coat proteins; the other half encodes two or three early proteins.

Cells in which the late phase of viral infection follows the early phase, and the massive synthesis of virus is coupled with cell death, are said to be *permissive;* in such cells, a productive, or *lytic,* infection can occur. Monkey cells are permissive for SV40, and mouse cells are permissive for polyoma.

In papovavirus-infected cells from many animal species, the induction of the S phase does not ordinarily lead to viral DNA replication, probably because of some incompatibility between the cellular enzymes and a crucial viral sequence or protein. In such *nonpermissive* cells, there is no switch to the late phase and no cell death. As long as the expression of the early viral functions continues, however, infected cells cannot rest in G_0, because they are continually induced to proceed through the cell cycle. In most cases, the viral genome is degraded or lost by dilution during cell proliferation; once the viral genome is lost, the cells revert to normal. This reversion results in an *abortive transformation.* In a small percentage of infected, nonpermissive cells, the viral genome is *integrated* into the cell's genome. These cells fulfill most of the criteria for transformation described previously; thus they have been permanently transformed into cancer cells by the continual expression of the early papovavirus genes following their integration into cellular DNA.

Permissive cell populations also can be transformed by certain viral mutants that are unable to replicate their DNA and thus can express only early proteins even in permissive cells. Such mutants appear frequently enough that mouse cells can be transformed by polyoma virus, and the virus can cause cancer after it is injected into baby mice.

In theory, viral DNA could integrate into cellular DNA by either a site-specific or a random mechanism. *Site-specific* integration has a defined target sequence in the cellular DNA and/or a defined integration site in the viral DNA. One or more proteins (probably viral) would bind to these sequences and direct integration. Papovaviruses do not integrate site-specifically; they apparently use *random* sites both in the viral and in the host-cell DNA. The viral proteins do not appear to play an active role in the integration process. Papovavirus DNA integration is probably no different from the integration of any DNA that has been incorporated into cells. Integration of viral DNA into the cellular DNA of nonpermissive cells can have no role in propagating the virus because no viral progeny are made in such cells.

These considerations make it unlikely that papovaviruses have evolved mechanisms *designed* to induce cancer. Rather, cancer induction appears to be a consequence of three features of the viral replication strategy: (1) the occurrence in papovavirus genomes of transforming genes used by the virus for the early phase of its lytic cycle, (2) the occasional integration of viral DNA into the DNA of nonpermissive cells, and (3) the lack of cell death after infection of nonpermissive cells. These three circumstances allow a virus to permanently transform nonpermissive cells by integration of viral DNA in such a way that the expression of early genes is not impeded. This event happens frequently enough that some cells are transformed whenever a nonpermissive cell population is infected by SV40 or polyoma virus. Transformation by these viruses is a laboratory phenomenon, and neither virus is known to induce cancer in wild animal populations.

Transformation by DNA Viruses Requires Interaction of a Few Independently Acting Viral Proteins

In permanently transformed nonpermissive cells, the viral genome is integrated into the host-cell genome in such a way that the early viral genes are expressed constitutively. That the continual activity of two or three early viral gene products maintains the transformed phenotype has been shown in experiments with various temperature-sensitive viral mutants. The morphology and growth properties of cells infected with these mutants can be varied from normal to transformed by shifting the temperature. The genes encoding these early proteins are, by definition, oncogenes—that is, they transform cells and induce cancer in experimental animals. The papovavirus oncogenes were the first ones to be recognized. Because it was evident that an understanding of how these genes can cause cancer would illuminate the more general issue of cancer induction, much attention has been focused on the products of papovavirus early genes.

SV40 makes two early proteins called T (large T) and t (small t); polyoma makes three early proteins: T, middle-T (mid-T), and t. T protein was originally called *T antigen* because it was first demonstrated by immunofluorescence using serum from animals bearing virus-induced tumors. The T antigens are 90-kilodalton (kDa) proteins found in the cell nucleus. They bind tightly to DNA and play an important role in viral DNA transcription and replication during the lytic cycle. The mid-T of polyoma is a 45-kDa plasma membrane–bound protein that has no counterpart in SV40. Some of the SV40 T antigen, however, is bound to the plasma membrane and appears to serve the same function as the mid-T protein of polyoma. The 20-kDa t proteins are cytoplasmic. The three polyoma proteins appear to have independent roles in the transformation process.

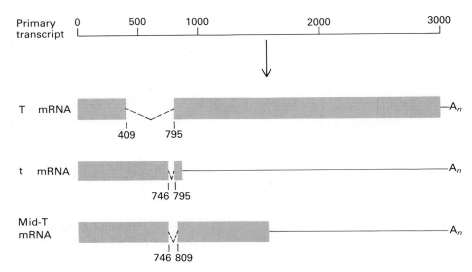

▲ **Figure 24-11** The early region of polyoma virus is expressed in one 3000-base-long nuclear RNA that can be spliced in three ways to yield three different mRNAs and therefore three early proteins. The thick orange bars denote protein-coding regions; the thin red lines, noncoding regions. [See M. Rassoulzadegan et al., 1982, *Nature* 300:713.]

The early region of the polyoma genome gives rise to one complex transcript that can be spliced in three different ways to produce mRNAs for the three early proteins (Figure 24-11). By deletion of specific nucleotides, it has been possible to construct three different DNA molecules, each of which encodes only one of the three polyoma early proteins. The DNA molecule encoding only mid-T protein can transform a rat 3T3 cell line according to many of the criteria described earlier, but such cells cannot grow in a low-serum medium. Thus mid-T apparently cannot eliminate the requirement for all growth factors, although it can do almost everything else associated with transformation. In contrast, a cell that expresses both the T and the mid-T proteins can grow in low serum or even in no serum. Cells transfected with polyoma DNA encoding only T protein can grow in low-serum media, but such cells are not morphologically transformed and they do not grow to high saturation densities. This dramatic separation of transformation parameters illustrates an important general principle of cell transformation: transformation often involves the interaction of a small number of independently acting proteins. We shall encounter this principle again when we consider nonviral cancers.

In the experiments described in the previous paragraph, rat 3T3 cells were used as the targets of transformation. When plasmids encoding the T and mid-T proteins are introduced into recently explanted rat embryo fibroblasts (cells with a limited life span in culture), the division of labor among the polyoma early proteins is even more obvious. A plasmid encoding mid-T produces no transformed colonies in these cells, suggesting that the mid-T protein cannot immortalize cells although it can

transform 3T3 cells, which are already immortal. Consistent with this explanation is the finding that a plasmid encoding T protein confers immortality on rat embryo fibroblasts without changing them morphologically. The subsequent incorporation into the cell of a plasmid encoding mid-T morphologically transforms the cells already immortalized by the action of T protein, and the resulting transformants behave almost like wild-type polyoma-transformed cells. In other experiments, polyoma t protein has been shown to have a role in producing complete transformation.

For SV40, recombinant DNA methods have shown that different parts of the T protein have different effects. Thus this single protein is responsible for several different aspects of transformation. Papovaviruses have apparently evolved a number of distinct ways of changing cell behavior. When these changes occur independently, they do not produce a complete tumor cell. When they occur together, however, a full-blown tumor cell results.

In summary, DNA tumor viruses, like all of the other cancer-inducing agents discussed in this chapter, cause cell transformation by altering the types of proteins made in the cell. The virus brings into the cell new, active genes, oncogenes, whose products are responsible for transformation. As we discuss in later sections, cancer is caused by the aberrant expression of altered or normal proteins originating from the cellular genome rather than from a virus, but the principle is the same. Results with DNA tumor viruses also have demonstrated that multiple proteins can work together to cause transformation. These results and other evidence to be presented later suggest that a tumor cell is caused by a few independent activities—not many, but not just one. These activities determine the particular constellation of transformation parameters present, with immortalization standing out as a characteristic that is regulated separately from the others (e.g., serum factor dependence).

RNA-containing Retroviruses as Transforming Agents

Transformation by retroviruses is more complicated than transformation by DNA viruses because the basic retroviral genome does not encode proteins with transforming activity. To understand the transforming ability of retroviruses, we must first examine in some detail their life cycle, which was outlined in Figure 5-39.

Productive Infection Cycle of Retroviruses Involves Integration into the Host-cell Genome

Retroviruses have an RNA genome consisting of two apparently identical, approximately 8500-nucleotide RNA molecules noncovalently bonded to one another; each has a cellular tRNA molecule bound to it. Each virion also contains about 50 molecules of a DNA polymerase that is capable of copying the genomic RNA into DNA; this enzyme is commonly known as *reverse transcriptase* because it reverses the normal flow of information in biological systems. The infection process brings both the RNAs and the reverse transcriptase into the cell's cytoplasm.

Once in the cytoplasm, reverse transcriptase uses the tRNA as a primer to initiate DNA synthesis. In a complicated series of events, a complete complementary (or minus) strand of DNA is formed; this minus strand is used as a template for the formation of a plus strand of DNA. The ultimate product of reverse transcription is thus a double-stranded DNA copy of the information in the viral RNA genome (Figure 24-12). The double-stranded DNA *reverse transcript* is actually longer than its template. Because of the "jump" made by the initial minus-strand DNA and a second "jump" made by the initial plus-strand DNA, the short repeat at either end of the viral RNA is extended to form a long terminal repeat (LTR) at either end of the reverse transcript. The LTR is a very important component of the reverse transcript because, as we shall see, it contains many of the signals that allow retroviruses to function.

After their formation, the double-stranded DNA reverse transcripts migrate into the cell's nucleus where they integrate into variable sites in the cell's chromosomes. In contrast to the variable integration sites on genomes of DNA tumor viruses, the integration sites on retroviral DNA are located specifically at the ends of the linear molecule. The enzymes that catalyze the integra-

tion process have yet to be identified; at least one is viral because certain viral mutations in a gene called integrase can block integration. The site specificity of the integration site within the retroviral genome, the approximate randomness of the chromosomal target site, and the duplication of the target site make reverse transcript integration similar to transposon movement in bacteria or yeast.

Once the reverse transcript is integrated into the host-cell DNA, the transcript—which is called *proviral* DNA in its integrated form—becomes a template for RNA synthesis (Figure 24-13). Provirus-directed transcription, like most other eukaryotic transcription, involves a *promoter* —a sequence that directs the RNA polymerase to a specific initiation site—and an *enhancer*—a sequence that facilitates transcription although it need not be located near the initiation site. The promoter and enhancer sequences, as well as the polyadenylation signals, are located in the LTR. The two LTRs at either end of proviral DNA are identical in sequence but have different functions: the upstream LTR acts to promote transcription, whereas the downstream LTR specifies the poly A site. How these two identical sequences manage to carry out different functions is not fully understood. The RNA made from the proviral DNA serves both as mRNA in the synthesis of viral proteins and as genomic RNA for the next generation of virus particles.

Because they lack most metabolic machinery, all viruses are obligate intracellular parasites. Retroviruses are perhaps the ultimate intracelluar parasites because they can link their genome stably to that of their host cell without seriously damaging the cell. Their genetic information is replicated as part of the cellular DNA and distributed to all daughter cells, which actively transcribe it so that new virus particles are made perpetually. Thus the life cycle of retroviruses differs from that of DNA viruses in two fundamental respects: integration is site-specific on the retroviral genome and is part of the productive infection cycle. Because the proviral DNA can continually direct synthesis of the proteins for new viral progeny by using only a small fraction of the infected cell's metabolic machinery, it does not kill the host cell, which can continue to grow at a virtually normal rate.

Retroviruses occasionally are integrated into an animal's germ-line cells; they are then inherited by all offspring of that animal and can even become part of the inheritance of a species. In mice, for example, between 0.1 and 1 percent of the genome consists of proviral DNA. Most such inherited retroviruses, as opposed to the ones that are integrated during somatic cell growth, are transcriptionally inactive.

The genome of most retroviruses has three protein-coding regions. They are *gag*, which encodes a polyprotein that is cleaved to form four internal virion structural proteins; *pol*, which encodes the reverse transcriptase and the integrase; and *env*, which encodes a glycoprotein that covers the virion surface (see Figure 24-13).

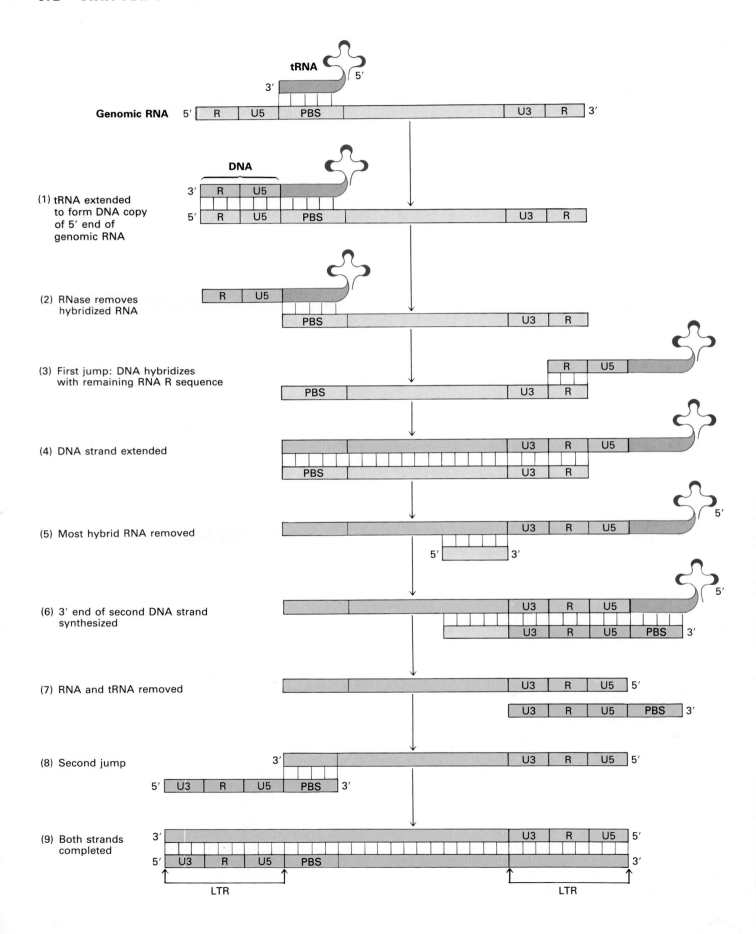

Genomic RNA

(1) tRNA extended to form DNA copy of 5' end of genomic RNA

(2) RNase removes hybridized RNA

(3) First jump: DNA hybridizes with remaining RNA R sequence

(4) DNA strand extended

(5) Most hybrid RNA removed

(6) 3' end of second DNA strand synthesized

(7) RNA and tRNA removed

(8) Second jump

(9) Both strands completed

◀ **Figure 24-12** The mechanism of reverse transcription. A complicated series of nine events generates a double-stranded DNA copy of the single-stranded RNA genome of a retrovirus. The genomic RNA is packaged in the virus with a tRNA hybridized to a complementary sequence near its 5′ end at the primer binding site (PBS). The RNA has short repeated terminal sequences (R). Only one of two RNAs comprising the genome of a virion is shown; reverse transcription is not known to require the two strands. Also, the indicated terminal sequences span only a few hundred nucleotides; between the sequences denoted U5 and U3 there are actually about 7500 nucleotides. The overall reaction is catalyzed by RNA reverse transcriptase, which has both a deoxyribonucleotide polymerase activity and a ribonuclease activity. The entire process yields a double-stranded DNA molecule that is longer than the original RNA and has a long terminal repeat (LTR) at each end. [See E. Gilboa et al., 1979, *Cell* **18**:93.]

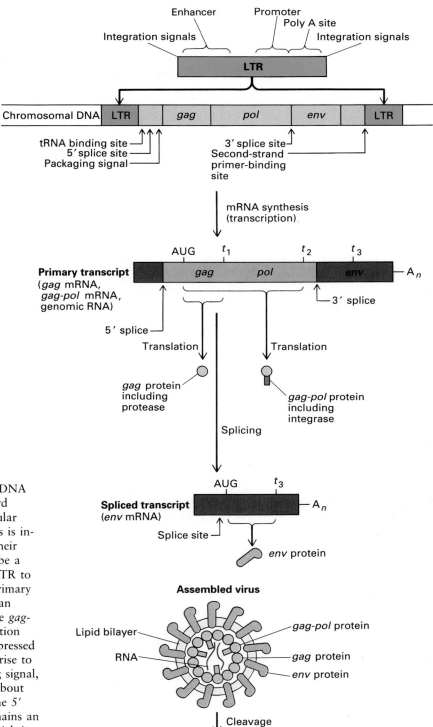

▶ **Figure 24-13** Genetic elements of proviral DNA and the corresponding gene products. A standard retroviral provirus is shown integrated into cellular DNA. The genetic structure of its flanking LTRs is indicated. Although the two LTRs are identical, their positions somehow allow the left-hand LTR to be a promoter for transcription and the right-hand LTR to specify the poly A site for the transcript. The primary RNA transcript has three functional roles. It is an mRNA for the synthesis of both the *gag* and the *gag-pol* proteins because its first translation termination signal (t_1, the termination signal for *gag*) is suppressed occasionally (5–10 percent of the time) to give rise to some *gag-pol* protein. By virtue of its packaging signal, the transcript is also the genome of the virus. About 50 percent of the transcripts are spliced; why the 5′ and 3′ splice sites are only partially utilized remains an unanswered question. The spliced transcript, which is not packaged because the packaging signal has been spliced out, serves as the mRNA for the *env* protein. All three of the translation products are virion proteins: the *env* protein is inserted into the lipid bilayer that surrounds the virus, the *gag* protein forms an inner shell, and the *gag-pol* protein (which contains the reverse transcriptase) is also inside the particle. After particles are formed, both the *gag* and the *gag-pol* products are extensively processed by proteolysis.

A gene segment encoding a protease that cleaves the polyprotein precursors also occurs as part of *gag* in some viruses and of *pol* in others. None of these retroviral proteins, all of which are found in virions, changes the growth properties of fibroblasts. Thus the basic retrovirus (in contrast to DNA tumor viruses) is not a transforming virus; in fact, infected cells are difficult to distinguish from uninfected cells by any parameter of cellular life.

Although the basic retroviral life cycle does not include a transforming event, many retroviruses can cause cell transformation and cancers in animals and even in humans. In fact, the first tumor virus to be recognized was the Rous sarcoma virus, a chicken retrovirus described by Peyton Rous in 1911. Many other tumor-inducing retroviruses have been found, including the mouse mammary tumor virus (MMTV); leukemia viruses of chickens, mice, cats, apes, and, most recently, humans; sarcoma viruses of many species; and a few carcinoma-inducing viruses. We now know that retroviruses can transform cells and induce tumors by two quite distinct mechanisms. One is utilized by transducing retroviruses; the other, by slow-acting retroviruses.

Oncogenic Transducing Retroviruses Contain Oncogenes Derived from Cellular Proto-oncogenes

The key to our present understanding of the transforming and tumor-inducing ability of some retroviruses came from the realization that such viruses contain genetic information other than *gag*, *pol*, and *env*. The additional genetic material contained by these viruses are oncogenes appropriated from normal cellular DNA. Because phages that have acquired cellular genes are said to *transduce* the genes they acquire, retroviruses that have acquired cellular sequences are called *transducing retroviruses*.

The first indication of the existence of transducing retroviruses came with the discovery that Rous sarcoma virus contains nucleotide sequences that are not found in closely related but nontransforming retroviruses. The extra nucleic acid sequence, called *src*, was then identified in the DNA of normal chickens as well as in the DNA of all vertebrate and even invertebrate species. This landmark discovery of an oncogene derived from a cellular proto-oncogene fundamentally reoriented thinking in cancer research because it showed that cancer induction may involve the action of normal, or nearly normal, genes. The discovery of many other transduced cellular genes in other retroviruses implies that the normal vertebrate genome contains many potentially cancer-causing genes. The oncogene in a retrovirus is denoted with the prefix v (e.g., v-*src*), and the equivalent proto-oncogene is given the prefix c (e.g., c-*src*).

Transducing retroviruses arise because of complex, apparently spontaneous, rearrangements following the

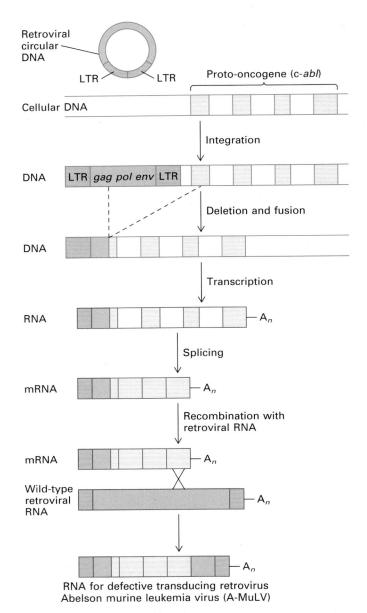

RNA for defective transducing retrovirus Abelson murine leukemia virus (A-MuLV)

▲ **Figure 24-14** The formation of a transducing retrovirus. Although the events that generate a transducing retrovirus occur at such a low frequency that they have not been studied in the laboratory, the structure of the transducing viruses strongly suggests that they are formed as depicted here. Formation of the Abelson virus is depicted because it has many features found in other retroviruses. The c-*abl* gene is shown as a series of exons in cellular DNA. A wild-type retrovirus upstream of c-*abl* probably integrates randomly once in about 10^6 integrations. A subsequent deletion-fusion event fuses a c-*abl* exon into the *gag* region of the retrovirus. Transcription of the fused DNA and subsequent splicing produces an mRNA for the *gag-abl* fusion protein. If the cell is also infected by a wild-type retrovirus, it will package the *gag-abl* mRNA along with wild-type RNA. In the next generation, the two RNAs can recombine (presumably by a switch of templates during reverse transcription) to generate the RNA of the final transducing retrovirus (now called Abelson murine leukemia virus).

integration of a retrovirus near a cellular proto-oncogene (Figure 24-14). Most commonly, the acquired genetic information in a transducing retrovirus replaces part or all of *gag, pol,* and *env,* making the new retrovirus defective. However, the products of *gag, pol,* and *env* can be provided by a *helper virus* (a coinfecting wild-type retrovirus), so propagation of the defective virus is possible. The genome structures of three transducing retroviruses are shown in Figure 24-15.

At least 20 different oncogenes, derived from cellular proto-oncogenes, have been identified in retroviruses, and the total possible number is not known. The characteristics of different transducing retroviruses are determined in part by which oncogene(s) they contain. For instance, some transform only adherent cells, whereas others transform both adherent and nonadherent cells. In animals, the ones that transform nonadherent cells cause leukemia; the remainder cause sarcomas. Abelson murine leukemia virus transforms only cells of the B lymphocyte lineage in cultures of bone marrow cells; in animals, however, Abelson virus occasionally causes other types of tumors, and it can affect the differentiation of fetal red blood cells. Harvey virus, which primarily causes sarcomas, also can produce tumors of the red blood cell lineage in mice. The reasons for these specificities are not yet understood.

Although the transforming activity of a transducing retrovirus is provided by its acquired cellular genetic information, the genetic elements that allow the oncogenes to be expressed and transferred from cell to cell are all inherited from the retroviral parent. As noted already, the LTRs of the viral genome contain most of the control elements including the promoter and the enhancer, the poly A signal, and the integration signals (see Figure 24-13). The genome also contains a sequence that binds the tRNA just at the border of the left-hand LTR, a sequence for initiating plus-strand DNA synthesis at the border of the right-hand LTR, and a sequence that allows for packaging into virus particles (located between the left-hand LTR and the AUG that initiates *gag* protein synthesis). These vital control elements are preserved in the hybrid genomes of transducing retroviruses, even highly defective ones such as Abelson virus (see Figure 12-15).

The demonstration in transducing retroviruses that cellular genetic information can be an agent of transformation clearly implies that human genomes harbor genes that can, in the right circumstances, cause cancer. Furthermore, this notion also suggests that other agents, like chemicals and radiation, could change normal genes into cancer-inducing genes by causing relatively simple mutations. Both of these concepts have had a profound impact on cancer research.

Conversion of Proto-oncogenes into Oncogenes in Transducing Retroviruses
Three kinds of alterations, affecting either the coding region of the gene or its

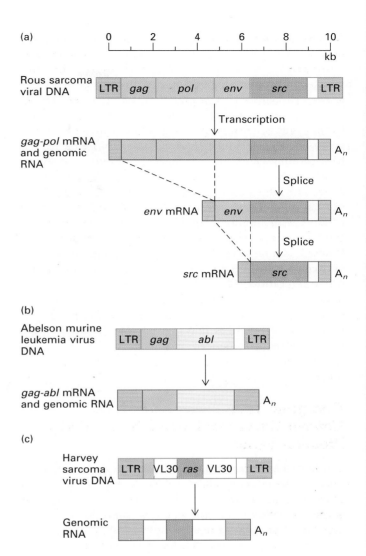

▲ **Figure 24-15** The structure and expression of the genes in three transducing retroviruses. The *src, abl,* and *ras* genes were derived from host-cell genomes. (a) Rous sarcoma virus, the only known nondefective transducing retrovirus, produces three different mRNAs: one unspliced mRNA (for *gag* and *gag-pol*), which is also the genomic RNA, and two spliced mRNAs (one for *env,* the other for *src*). (b) Abelson murine leukemia virus, a highly defective virus, yields one RNA molecule that serves as the mRNA for the *gag-abl* fusion protein and acts as the viral genome. (c) Harvey sarcoma virus has a complicated origin because it is a mouse retrovirus that acquired its oncogene during passage in a rat. Two recombination events must have occurred in the rat. One was a recombination of the mouse retrovirus with an endogenous retrovirus of the rat, called VL30. The second recombination was a fusion of a mutated form of the rat c-Ha-*ras* proto-oncogene into the mouse/rat recombinant retrovirus within the VL30 sequences. The only gene expressed by the Harvey virus is the v-Ha-*ras* gene, which encodes a 21-kDa protein that is probably synthesized from the genomic RNA. The v-Ha-*ras* gene differs from the c-Ha-*ras* gene by one crucial point mutation in the codon for amino acid 12; this mutation thus converts the proto-oncogene into the oncogene.

expression, may convert a proto-oncogene into an onco-gene. First, transcription of a cellular gene acquired by a retrovirus is determined by the viral control elements, which can promote high transcription rates in many different cell types. This quantitative difference alone appears to explain why certain normal genes become transforming genes when they are captured by a virus. Indeed, proto-oncogenes cloned from cellular DNA can become transforming genes after being inserted into a retrovirus or otherwise placed in a high-transcription environment. A second kind of alteration is the loss of parts of the normal cellular gene, which apparently leaves the protein product with an unregulated activity that may allow it to transform cells. Third, a single point mutation or other small change may produce a crucial sequence difference between a transforming gene and its cellular counterpart. These alterations in a proto-oncogene may result in a quantitative change in the amount of the encoded protein or a qualitative change in the protein itself. Either one or both types of changes may play a role in converting a given proto-oncogene into an oncogene.

Nononcogenic Transducing Retroviruses Have Been Constructed Experimentally

All of the transducing retroviruses identified to date have been oncogenic; recognizing such viruses is relatively easy because they produce an obvious disease. It seems probable that retroviruses containing acquired genes that are not oncogenes also could arise naturally; recognition of such nononcogenic transducing retroviruses, however, would be difficult because they would produce no obvious effects on infected cells.

Nononcogenic transducing retroviruses have been constructed in the laboratory by in vitro genetic manipulation. An example of an artificially constructed virus that can transduce both a globin cDNA and a neomycin-resistance gene is shown in Figure 24-16. Such artificial retroviruses have been used to insert new or altered genes into cells and animals. It is hoped that such constructed genetic elements may be used in the future to treat human genetic diseases, such as those that affect red blood cells.

Slow-acting Carcinogenic Retroviruses Can Activate Nearby Cellular Proto-oncogenes after Integration into the Host-cell Genome

The second type of oncogenic retrovirus—the slow-acting retroviruses—induces cancer over a period of months or years rather than the days or weeks required for cells to respond to a transducing virus. The genomes of slow-acting retroviruses differ from those of transducing viruses in one crucial respect: they lack an oncogene. The slow-acting retroviruses not only have no affect on growth of cells in culture, but they are also completely proficient to multiply themselves because they have suffered no debilitating deletions or insertions during acquisition of an oncogene.

The mechanism by which avian leukemia viruses cause cancer appears to operate in all slow-acting retroviruses. Like all retroviruses, avian leukemia viruses generally integrate into cellular chromosomes more or less at random. However, the site of integration in the cells from tumors caused by these viruses was found to be near a

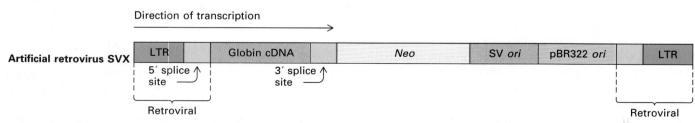

▲ **Figure 24-16**   The artificial retrovirus SVX. It was built as a shuttle vector, a recombinant molecule that is able to carry genes back and forth between mammalian cells and bacteria. In this vector the crucial LTRs from the left and the right ends of a mouse retrovirus DNA molecule are coupled to the following elements: a globin cDNA that will direct globin synthesis when SVX is present in a mammalian cell; a 3′ splice site that can be joined to the 5′ splice site near the left-hand LTR to form an mRNA for the expression of the next gene (downstream); a gene *(Neo)* that encodes a protein conferring resistance to an analog of the antibiotic neomycin; an SV40 virus origin of replication; and a pBR322 bacterial plasmid origin of replication. This shuttle vector is used to introduce the globin gene into cells (other cDNAs can be substituted for globin). Cells integrating the retrovirus can be selected because of the presence of the *Neo* gene, which is expressed both in bacteria and in mammalian cells. To rescue the provirus from cells, the cells can be fused to other cells that make SV40 T antigen. This fusion activates replication from the SV40 origin, and the resulting circular DNA molecules can be purified and reinserted into bacteria. Their multiplication in bacteria is caused by the pBR322 origin of replication. Bacteria containing the recombinant retrovirus can be selected through the *Neo* marker. [See C. Cepko et al., 1984, *Cell* 37:1053.]

(a) Promoter insertion

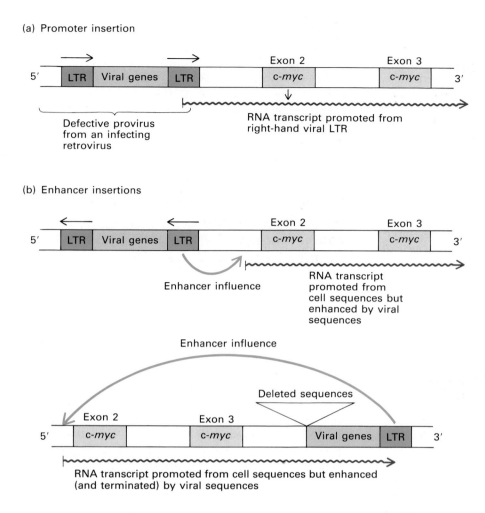

(b) Enhancer insertions

◀ **Figure 24-17** Activation of proto-oncogene c-*myc* by retroviral promoter and enhancer insertions. (a) Promoter activation can occur when the retrovirus inserts upstream (5′) of the c-*myc* exons. The right-hand LTR may then act as a promoter, although ordinarily it acts as a terminator; only if the provirus has a defect preventing transcription through to the right-hand LTR does that LTR function as a promoter. The c-*myc* gene is shown as two exons; there is a further upstream exon but it has no coding sequences. (b) Enhancer activation arises when a retrovirus inserts either upstream of the c-*myc* gene in the opposite transcriptional direction or downstream of the gene. In both cases, a viral LTR acts on a c-*myc* promoter sequence. In the downstream case, the LTR also terminates the transcript. *Modified from actual cases of retroviral insertion described in G. G. Payne et al., 1982, Nature 295:209.*

gene called c-*myc*, an established proto-oncogene. This finding suggested that the slow-acting avian leukemia viruses cause disease by activating c-*myc* to become an oncogene. The viruses act slowly both because integration near c-*myc* is a random, rare event and because secondary events probably have to occur before a full-fledged tumor becomes evident.

An integrated viral genome can act in a number of ways to cause a cellular gene to become an oncogene. In some tumors, the 5′ end of the *myc* gene transcript includes a sequence from a retrovirus LTR. In such cases the right-hand LTR of the integrated retrovirus—which usually serves as a terminator—is believed to act as a promoter, initiating synthesis of RNA transcripts from the c-*myc* gene (Figure 24-17a). Such c-*myc* transcripts apparently encode a perfectly normal c-*myc* product. The enhanced level of c-*myc* RNA resulting from the strong promoting activity of the retroviral LTR appears to be the explanation of oncogene activation in such cases. That is, the carcinogenic effect of the retrovirus results entirely from an increase in the amount of product from a normal gene. This mechanism, which is called *promoter insertion*, has been implicated in many tumors induced by avian leukemia viruses.

In a few tumors induced by avian leukemia viruses, the proviral DNA is located near the c-*myc* gene but at the 3′ end of the gene; in others, it is at the 5′ end of c-*myc* but oriented in the opposite transcriptional direction (Figure 24-17b). In such cases the promoter-insertion model cannot be applicable, and the proviral DNA is thought to exert an indirect enhancer activity that apparently activates c-*myc* by increasing its level of transcription, changing the cell type in which it is expressed, and/or altering its time of expression in the cell cycle. This mechanism is called *enhancer insertion*.

Either promoter insertion or enhancer insertion is the probable explanation for many tumors caused by retroviruses that do not carry oncogenes. The slow-acting retroviruses include the mouse mammary tumor virus and a number of mouse leukemia viruses. In some cases, activation of c-*myc* has been demonstrated; other proto-oncogenes have been identified near other insertion points. Slow-acting viruses act as fingers pointing to the genomic location of potential oncogenes.

The activation of a proto-oncogene following integration of a slow-acting retrovirus should be clearly distinguished from the acquisition of a proto-oncogene by a virus, converting it into an oncogene-containing transmissible virus. As is evident from Figure 24-17a, the first step in the formation of a transducing virus may well be promoter insertion, but several additional steps must occur before a defective, oncogene-containing retrovirus emerges. Those steps are all low-probability events, so very few promoter insertions progress to the formation of a transmissible, transducing virus. In fact, because defective transducing retroviruses depend on a helper virus for their cell-to-cell transmission, they are not maintained in natural animal populations. Most oncogene-containing retroviruses have arisen in laboratories or in domesticated animals and have been maintained for experimental purposes; the defective viruses are not readily spread from animal to animal. In natural populations, insertional oncogene activation is probably the major cause of retrovirus-induced cancer.

Human Tumor Viruses

The discovery of many RNA and DNA tumor viruses in animals raised the question of whether humans also have tumor viruses, and if so, what fraction of human cancer might be caused by them. In certain specific situations human tumor viruses have been found, but minimal evidence exists for a viral involvement in the major types of human tumors that are prevalent in developed countries. A lack of evidence does not constitute proof of noninvolvement, however, and it remains an open question whether viruses might have a wider role in human cancer than is now suspected. In this section, we describe briefly the various viruses that are presently implicated in human cancers.

In 1980, after many years of fruitless search for a human retrovirus, oncogenic or nononcogenic, the first plausible candidate was reported: *human T-cell leukemia virus* (HTLV). Infection with this virus is clearly associated with a type of T-lymphocytic human tumor that is especially prevalent in southern Japan, the Caribbean, and parts of Africa. How the virus causes the tumor is being actively studied; oddly, the process appears to involve neither a cell-derived oncogene within the virus nor a reproducible site of integration in tumor-cell DNA. Rather, the basis of its carcinogenicity is a unique viral genetic region, totally distinct from *gag, pol,* and *env,* more like the oncogenes of DNA tumor viruses.

Epstein-Barr virus, a herpeslike DNA virus, probably plays a role in at least two human tumors: Burkitt's lymphoma and nasopharyngeal carcinoma (Figure 24-18). In certain hot, wet sections of Africa, Burkitt's lymphoma is the primary cancer in children, and in southern China nasopharyngeal carcinoma is a major cancer. In the United States and other developed countries, Epstein-Barr virus mainly causes mononucleosis in adolescents and young adults; it leads to cancer very rarely, and then usually in people with a malfunctioning immune system. The

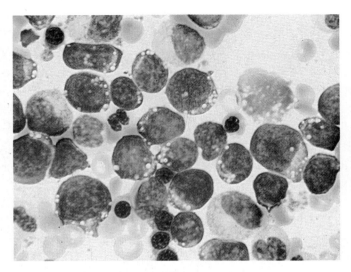

▲ **Figure 24-18**   Stained tumor cells of Burkitt's lymphoma. These cells carry the genome of the Epstein-Barr virus. *Courtesy of R. Van Etten.*

cancer-inducing ability of the virus in Africa may be due to the presence of a second endemic factor, thought to be malaria, which works in concert with the virus.

Infection with *hepatitis B virus*—a very small DNA virus that is mainly responsible for human "serum hepatitis"—is correlated with liver cancers, especially in underdeveloped areas of the world. Although this virus has a DNA genome, it multiplies through an RNA intermediate like a retrovirus. Its genome has been found integrated into the DNA of hepatomas, and it may cause malignancy by insertional activation of a cellular proto-oncogene.

The *papilloma viruses,* which are small DNA viruses belonging to the papovavirus group, not only are responsible for warts and other benign human tumors but also for malignancies. Currently over 30 distinct human papilloma viruses are recognized. Two types of papilloma virus are associated with human cervical carcinoma and appear to be the cause of this sexually transmitted disease.

Two viruses (BK and JC) that are the human counterparts of SV40 have been identified. BK can transform cells in culture, but neither virus has been associated with any specific human cancer.

No discussion of viruses and human cancer would be complete without mention of *acquired immunodeficiency syndrome (AIDS),* a retrovirus-induced disease that, among other symptoms, makes its victims susceptible to numerous cancers. AIDS is caused by *human immunodeficiency virus (HIV),* a retrovirus different from all of the others discussed thus far (Figure 24-19a). The HIV genome includes five or six genes in addition to the *gag, pol,* and *env* regions (Figure 24-19b). These extra genes, which are all small and expressed from a battery of

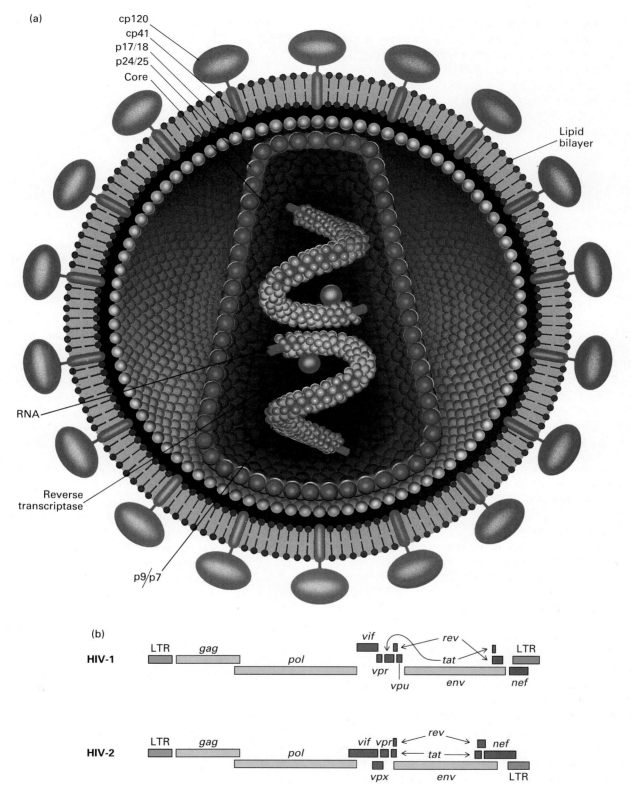

(a) cp120
cp41
p17/18
p24/25
Core

Lipid bilayer

RNA

Reverse transcriptase

p9/p7

(b)

vif
rev
tat
LTR

HIV-1 LTR gag pol vpr env nef
vpu

vif vpr rev nef
tat

HIV-2 LTR gag pol vpx env
LTR

▲ **Figure 24-19** (a) Schematic structure of the human immunodeficiency virus (HIV) particle showing the location of various components. (b) Structures of the genomes of HIV-1 and HIV-2, two strains of human immunodeficiency virus, which causes AIDS. The standard retroviral genes (purple)—*gag, pol,* and *env*—are present along with five or six other genes (red). The full length of the genomes is about 9000 nucleotides. Each strain has one gene not present in the other (*vpu* or *vpx*). The names of the genes represent a 1988 renaming; original names were 3′-*orf* for *nef, trs/art* for *rev, sor* for *vif,* and R for *vpr.* [See R. Gallo et al., 1988, *Nature* 333:504.]

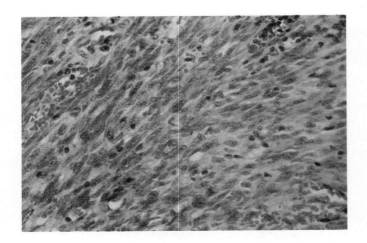

◂ **Figure 24-20** Section of a Kaposi's sarcoma, a cancer occuring in AIDS patients. The bright red spots are red blood cells; otherwise the field is filled with tumor cells. *Courtesy of Dr. J. Aster.*

severe mental deterioration. AIDS patients often suffer from cancers, probably caused by viruses like Epstein-Barr that are released from their usual control by the immune system. A common tumor of AIDS patients is Kaposi's sarcoma, which may be induced by growth factors released by HIV-infected cells (Figure 24-20).

AIDS is one of the most devastating diseases ever to spread within the human population. Already in 1990, perhaps ten million people worldwide are HIV-infected, and most can be expected to die from AIDS.

spliced mRNAs, allow HIV to grow more vigorously than other retroviruses in infected cells. The exact function of the HIV-specific genes is still debated but involves activation of transcription, regulation of splicing or transport of mRNA, and possibly other alterations of macromolecular metabolism. HIV kills cells rather than transforming them; because its preferred, and perhaps sole, targets are cells involved in the immune response, HIV causes a profound immunodeficiency.

The loss of immune function in AIDS patients makes them prone to other infections; these "opportunistic" infections are generally the cause of the patient's death. The virus can also gain access to the patient's brain and cause

Chemical Carcinogens

Although viruses probably cause a small fraction of human cancer, chemicals are thought to be culpable in a larger number of cases. Chemicals were originally associated with cancer through experimental studies in intact animals. The classic experiment is to repeatedly paint a test substance on the back of a mouse and look for development of both local and systemic tumors in the animal. In this way, many substances have been shown to be *chemical carcinogens.*

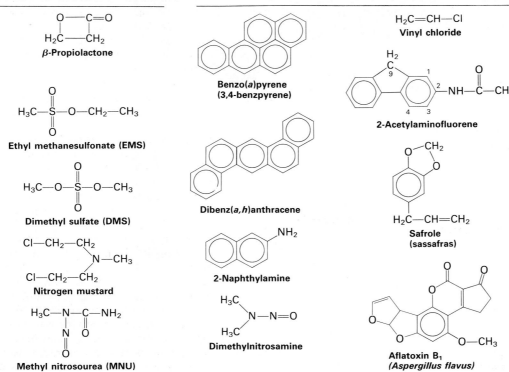

▴ **Figure 24-21** Structures of selected chemical carcinogens. Direct-acting carcinogens are highly electrophilic compounds that can react with DNA. Indirect-acting carcinogens must be metabolized before they can react with DNA.

Most Chemical Carcinogens Must Undergo Metabolic Conversion to Become Active

Chemical carcinogens have a very broad range of structures with no obvious unifying chemistry. Because many carcinogens have structures that appear to be highly unreactive, early workers were puzzled about how such unreactive and water-insoluble compounds could be potent cancer inducers. The situation was clarified when it was realized that there are two broad categories of carcinogens, *direct-acting* and *indirect-acting* (Figure 24-21); the latter require metabolic activation to become carcinogens. The direct-acting carcinogens, of which there are only a few, are reactive electrophiles (compounds that seek out and react with negatively charged centers in other compounds). Indirect carcinogens are converted to *ultimate carcinogens* by introduction of electrophilic centers.

The metabolic activation of carcinogens is carried out by enzymes that are normally resident in the body. Animals have such enzymes, especially in the liver, because they are part of a system that detoxifies noxious chemicals that make their way into the body. Therapeutic drugs, insecticides, polycyclic hydrocarbons, and some natural products are often so fat-soluble and water-insoluble that they would accumulate continually in fat cells and lipid membranes if there were no way for the animal to excrete them. The detoxification system works by solubilization: it adds hydrophilic groups to water-insoluble compounds, thus allowing the body to rid itself of noxious or simply insoluble materials.

The detoxification process begins with a powerful series of oxidation reactions catalyzed by a set of proteins called cytochrome P-450s. These enzymes, which are bound to endoplasmic reticulum membranes, can oxidize even highly unreactive compounds such as polycyclic aromatic hydrocarbons (Figure 24-22). Oxidation of polycy-

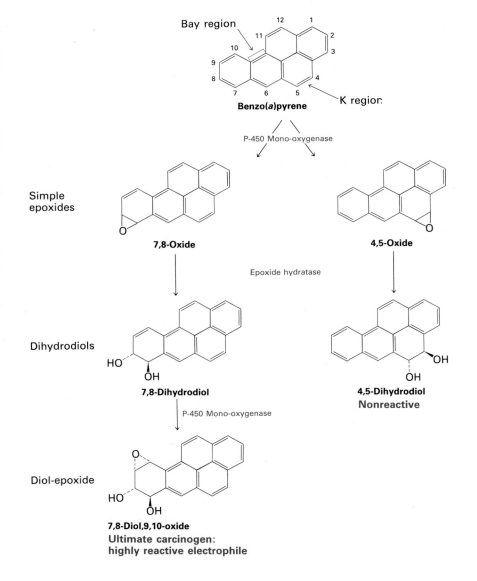

Figure 24-22 The metabolic activation of benzo(*a*)pyrene, a polycyclic aromatic hydrocarbon that is a powerful carcinogen. Although chemically almost inert, it becomes a highly reactive electrophile due to metabolic conversions. Two metabolic pathways are shown. The right-hand pathway involves an intermediate epoxide formed by the attack of the cytochrome P-450 system on what is called the "K region" of the molecule. An epoxide in this region is rapidly hydrolyzed to a nonreactive dihydrodiol. The left-hand pathway involves an initial oxidation at the 7,8 double bond, leading to a 7,8-oxide that is rapidly converted to a 7,8-dihydrodiol. This compound is still a good substrate for the P-450 system and is again epoxidated, now near the "bay region," at the 9,10 double bond. The 7,8-diol,9,10-oxide (or diol-epoxide) is not a good substrate for epoxide hydratase, so it is released into the cell as a highly reactive electrophile. This form is carcinogenic because it can readily react with negatively charged centers in DNA.

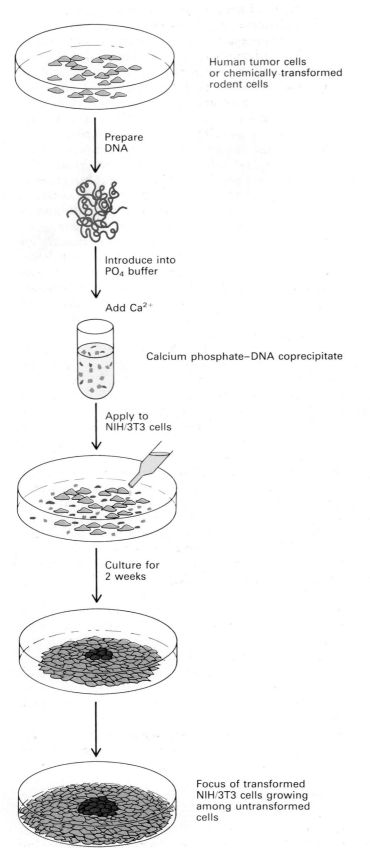

Human tumor cells
or chemically transformed
rodent cells

Prepare
DNA

Introduce into
PO₄ buffer

Add Ca²⁺

Calcium phosphate–DNA coprecipitate

Apply to
NIH/3T3 cells

Culture for
2 weeks

Focus of transformed
NIH/3T3 cells growing
among untransformed
cells

clic aromatics produces an epoxide, a very reactive electrophilic group. Usually these epoxides are rapidly hydrolyzed into hydroxyl groups, which are then coupled to glucuronic acid or other groups, producing compounds soluble enough in water to be excreted. Some intermediate epoxides, however, are only slowly hydrolyzed to hydroxyl groups, probably because the relevant enzyme (epoxide hydratase) cannot get to the epoxide to act on it. Such compounds are the highly reactive electrophiles whose precursors are referred to as carcinogens. Other types of carcinogens are activated by different oxidative pathways, which also involve P-450 enzymes.

The Carcinogenic Effect of Chemicals Depends on Their Interaction with DNA

Once inside cells, electrophiles can react with negatively charged centers on many different molecules—protein, RNA, and DNA, to name the most obvious. Several lines of evidence strongly suggest that the carcinogenic effect of chemicals results from the reaction of ultimate carcinogens with DNA.

First, ultimate carcinogens can modify both free and cellular DNA. Depending on their size and structure, these compounds will react with different positions on different DNA bases. Second, these changes in the base sequence of DNA may be expressed as permanent changes in the phenotype of the treated cell. Thus ultimate carcinogens are mutagenic, a characteristic that is most easily assayed in bacteria. Most compounds that have been identified as carcinogens for experimental animals are mutagens for bacteria, and their mutagenic potential is roughly proportional to their carcinogenic potential. For this reason, bacterial mutagenesis has become a test for carcinogens. The first and most popular of these tests is the Ames test, named for its developer Bruce Ames, a bacterial geneticist. Finally, cellular DNA altered by exposure of cells to carcinogens may transform cultured cells. This very important result was first obtained by extracting DNA from chemically transformed mouse cells and treating 3T3 cells with the DNA. A small fraction of the treated 3T3 cells became transformed, and their DNA acquired a new gene—an oncogene—that could be passed on to other cells (Figure 24-23). The extension of this type of experiment to animal and human tumor-cell DNA will be described later.

◀ **Figure 24-23**   Standard assay for a cellular oncogene in tumor cells or chemically transformed cells.

In summary, exposure of cells to a chemical carcinogen often induces a permanently altered state in cellular DNA. In other words, a carcinogen causes cancer by acting as a mutagen. In theory, a carcinogen could do this by binding to DNA and causing a mistake during the replication of the DNA, but the evidence suggests that often DNA sequence changes are induced by the repair processes cells use to rid themselves of DNA damage.

The Role of Radiation and DNA Repair in Carcinogenesis

The living world is constantly being bombarded by radiation of many different kinds. Two types of radiation are especially dangerous because they can modify DNA: ultraviolet radiation and the ionizing radiations (x-rays and atomic particles). UV radiation of the appropriate wavelength can be absorbed by the DNA bases and can produce chemical changes in them. The most common damage is the production of dimers between adjacent pyrimidine residues in one strand of DNA (see Table 12-4 and Figure 12-32). These dimers interfere with both transcription and replication of DNA. Ionizing radiation mainly causes breaks in DNA chains. Both types of radiation can cause cancer in animals and can transform cells in culture. The ability of ionizing radiation to cause human cancer, especially leukemia, was dramatically shown by the increased rates of leukemia among survivors of the atomic bombs dropped in World War II.

UV radiation has been widely used in research on mutations and cancer because it is an easily manipulable and measurable agent that directly damages DNA. However, probably none of the primary UV-induced DNA lesions, including the dimers, is mutagenic. This has been shown directly in bacterial mutants with a defective *recA*, a gene involved in DNA repair. When this gene's product is lost, cells are easily killed by UV radiation, because DNA repair is prevented, but there are virtually no mutants among the survivors. This result indicates that the mutational effect of UV radiation is caused during repair and not as a primary consequence of the radiation.

Ineffective or Error-prone Repair of Damaged DNA Perpetuates Mutations

Because organisms are continually being exposed to chemicals that can react with DNA and to radiation that can damage DNA, very effective methods of repairing DNA have evolved. As more complex organisms with larger genomes and longer generation times evolved, the effectiveness of DNA repair processes increased. If repair processes were 100 percent effective, chemicals and radiation would pose no threat to cellular DNA. Unfortu-

nately, repair of some lesions is relatively inefficient, and these can become progenitors of mutations.

One hard-to-repair lesion is a double-strand break of the DNA backbone. Such breaks, caused either by ionizing radiation or by chemicals, can be correctly repaired only if the free ends of the DNA rejoin exactly; however, without overlapping single-stranded regions there is no base-pair homology to catalyze the joining. Because cells will not tolerate free DNA ends, broken ends of molecules are generally joined to other broken ends. A cell that has suffered a particular double-strand break usually contains other breaks; thus a broken end has a number of possible segments to which it can join. The joining of broken ends on different chromosomes leads to translocation of pieces of DNA from one chromosome to another. As will become evident later, such a translocation may activate a proto-oncogene. Consequently, the carcinogenic effect of chemicals or radiation can result from their production of double-strand DNA breaks followed by translocations that occasionally activate oncogenes.

Proper repair of DNA also is difficult when cells attempt to replicate damaged DNA before repair processes have had a chance to act on the lesion. This situation can arise when a cell suffers so much damage over a short time that its repair systems are saturated. It then runs the danger of having repair fall so far behind that extensive replication of unrepaired lesions occurs. In such situations, both bacteria and animal cells use inducible reserve systems for repair. Such systems are not expressed in undamaged cells, but some aspect of the accumulated damage causes their derepression (induction) and expression.

One such inducible system is the SOS repair system of bacteria (see Figure 12-41). In contrast to other repair systems, this one makes errors in the DNA as it repairs lesions, so it is referred to as error-prone. The SOS system is responsible for UV-induced mutations, and its activity is dependent on the *recA*-gene product. The errors induced by the SOS system occur at the site of lesions, suggesting that the mechanism of repair is insertion of random nucleotides in place of the damaged ones in the DNA. Apparently this system is inducible rather than constitutive because cells would be expected to accept randomization of DNA sequence only as a last resort when error-free mechanisms of repair cannot cope with damage. SOS repair may also be a way of accelerating evolution at a time when the organism is under stress. Bacteria lacking an SOS system have greatly reduced mutation rates, which strongly implies that most of the mutations produced by treating bacteria with radiation or chemicals are caused by the error-prone SOS repair system.

Whether animal cells have an error-prone repair system is not known, but they certainly have inducible repair systems. If one or more of these systems are error-prone, inducible repair is likely to play a role in mutagenesis and therefore in carcinogenesis. In any case, many investigators believe that in animal cells, as in bacteria, most mutation is an indirect consequence of DNA damage and not a direct result.

Some Defects in DNA-repair Systems Are Associated with High Cancer Rates in Humans

A link between DNA-repair systems and carcinogenesis also is suggested by the finding that humans with certain inherited genetic defects that make specific repair systems nonfunctional have an enormously increased probability of developing certain cancers. One such disease is xeroderma pigmentosum, an autosomal recessive disease. Cells of affected patients are unable to repair UV damage or to remove bulky chemical substituents on DNA bases. The patients get skin cancers very easily if their skin is exposed to the UV rays in sunlight. The complexity of mammalian repair systems is shown by the fact that there are five different genes in which defects lead to xeroderma pigmentosum lesions, all having the same phenotype and the same consequences. Hybridization of cells having one type of lesion with cells having another allows complementation, so the hybrid cells can repair UV damage normally.

Another human disease that affects DNA repair and increases cancer risks is ataxia telangiectasia. This disease sensitizes cells to x-rays but not to UV radiation. X-rays and UV light cause different types of DNA lesions, which are handled by different repair systems. Ataxia telangiectasia results in the loss of the x-ray repair pathway, which separates cross-links in the DNA duplex, whereas xeroderma pigmentosum results in the loss of the UV repair pathway, which rids the DNA of bulky chemical groups appended to it. Because most chemical carcinogens mimic the effects of either x-rays or UV irradiation, cells from patients with one of these two diseases will be hypersensitive to treatment with one or another set of chemicals. The reason why patients with DNA-repair defects have greatly increased cancer rates may be that the loss of an error-free pathway in their cells causes a greater dependence on an error-prone pathway of repair.

Oncogenes and Their Proteins: Classification and Characteristics

In this section we discuss how the protein products encoded by oncogenes (oncogene proteins, for short) can transform cells to malignancy. Although the detailed pathway by which oncogenes act is not known for any one of them, the general principles of their action seem clear. Most oncogenes are similar to genes that act along growth-control pathways, suggesting that oncogene proteins interact with growth-controlling systems of the cells. It was long predicted that normal growth control and malignancy would be two faces of the same coin, and the nature of oncogenes has strongly supported that view. We now review the proteins involved in growth control and then discuss the action of various oncogene proteins.

Four Types of Proteins Participate in Control of Cell Growth

Although growth control in mammalian cells is only understood in rough outline, we can differentiate four types of proteins that participate in the process: growth factors, growth factor receptors, intracellular signal transducers, and nuclear transcription factors (Figure 24-24). Each type of growth-controlling protein has given rise to one or more oncogenes. We will describe briefly the normal function of these growth-controlling proteins.

Earlier in this chapter, we noted that cells in culture can be induced to grow by adding specific *factors* to the medium, epidermal growth factor (EGF) being a paradigm. Such growth factors or hormones are pure signals; they serve no metabolic purpose. They are one cell's way to send a message to another cell. Such factors and hormones induce many types of responses in cells—such as mobilization of energy stores and differentiation—as well as entry into the growth cycle. The responding cell has a specific *receptor* (e.g., the EGF receptor) that is sensitive to the signal. Because different cells have different receptors, each signal can produce a response in some cell types but not others. For example, EGF stimulates growth of epithelial cells but has no effect on other cells.

We noted earlier that initiating cell growth involves a commitment to finish the cell cycle made in G_1 and called START in the yeast cell cycle (see Figure 24-4). Occupancy of growth factor receptors is sensed by the cell and, if metabolic conditions are propitious, the cell responds by commiting itself to finishing G_1 and passing through S, G_2, and M to give rise to two cells similar to the original.

The chain of events that leads to growth is started by ligand-receptor interactions either at the cell surface or inside the cell. Intracellular receptors are typified by the steroid receptors, which interact with lipophilic ligands that passively pass through the cell membrane to reach the receptor in the cytoplasm or the nucleus. The cell interprets the signal and then often sends *intracellular transducers*, or second messengers, that alter transcription, either by allowing new genes to be expressed or by modifying levels of expression of already active genes. Not all second-messenger activity is focused on transcription. Changes in cell shape and metabolism, for example, can be induced by the action of signals directly on existing cellular proteins.

For the purposes of this discussion of oncogene action, the transcriptional response to a growth signal is of paramount importance because many oncogene proteins are

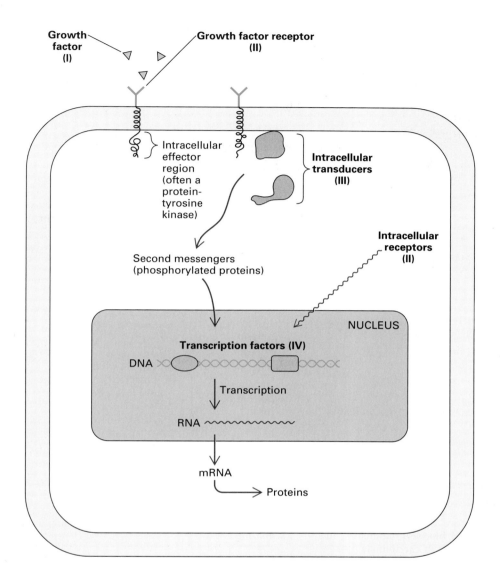

◀ **Figure 24-24** Growth control involves four types of proteins, the genes for which can give rise to oncogenes. They are growth factors (I), receptors (II), transducers (III), and intranuclear factors (IV).

transcriptional modifiers. The transcriptional response alters the protein composition of the cell, providing the critical proteins needed for cell growth. Transcription is controlled through two types of DNA sequence: promoters, which are located close to the start site of transcription, and enhancers, which are located farther from the start site and able to act over long and variable distances. Both elements function through specific binding proteins (*nuclear transcription factors*) that recognize short sequence motifs within the promoters and enhancers. These bound factors then accelerate or retard the rate of initiation of transcripts by RNA polymerase II.

Oncogene Proteins Affect the Cell's Growth-control Systems in Various Ways

In the previous section we described four types of gene products that participate in growth control. Most known oncogenes encode proteins that appear to come from or

be related to one of these four types (see Figure 24-24). Table 24-1 lists some representative oncogenes, classified in terms of the four types of growth-controlling gene products. Although the proteins encoded by class IV oncogenes are described as transcription factors, only a few of them have been shown to directly affect transcription and some may have other intranuclear roles. We will discuss each type of oncogene protein separately, emphasizing their functional roles. Although the total pathway through which any oncogene protein stimulates cell growth is not understood, we do know approximately where they fit into the overall scheme depicted in Figure 24-24.

Growth Factor (Class I) Oncogenes rarely arise from genes encoding growth factors. In fact, only one naturally occurring a growth factor oncogene—*sis*—has been discovered. The *sis* oncogene, which encodes a form of the platelet-derived growth factor (PDGF), can transform

Table 24-1 Selected oncogenes and their proteins

Type/name	Animal retrovirus	Nonviral tumor	Subcellular location of protein	Nature of encoded protein
CLASS I: GROWTH FACTORS				
sis	Simian sarcoma		Secreted	A form of platelet-derived growth factor
CLASS II: RECEPTORS				
A. Cell-surface receptors with protein-tyrosine kinase activity				
fms	McDonough feline sarcoma		Plasma membrane	CSF-1 receptor
erbB	Avian erythroblastosis		Plasma membrane	Epidermal growth factor receptor
neu (or erbB-2)		Neuroblastoma	Plasma membrane	Related to epidermal growth factor receptor
ros	UR II avian sarcoma		Plasma membrane	Related to insulin receptor
B. Intracellular receptors				
erbA	Avian erythroblastosis		Nuclear	Thyroid hormone receptor
CLASS III: INTRACELLULAR TRANSDUCERS				
A. Protein-tyrosine kinase (PTK)				
src	Rous avian sarcoma		Cytoplasm	Protein kinases that phosphorylate tyrosine residues
yes	Yamaguchi avian sarcoma		Cytoplasm	
fps (fes)	Fujinami avian sarcoma (and feline sarcoma)		Cytoplasm	
abl	Abelson murine leukemia	Chronic myelogenous leukemia	Cytoplasm and nucleus	
met		Murine osteosarcoma		
B. Protein-serine/threonine kinases				
mos	Moloney murine sarcoma		Cytoplasm	Protein kinases specific for serine or threonine
raf (mil)	3611 murine sarcoma			
C. Ras proteins				
Ha-ras	Harvey murine sarcoma	Bladder, mammary, and skin carcinomas	Plasma membrane	Guanine nucleotide-binding proteins with GTPase activity
Ki-ras	Kirsten murine sarcoma	Lung and colon carcinomas	Plasma membrane	
N-ras		Neuroblastoma and leukemias	Plasma membrane	
D. Phospholipase C-related				
crk	Avian sarcoma virus		Cytoplasm	Contains src-related regions also homologous with a phospholipase C
CLASS IV: NUCLEAR TRANSCRIPTION FACTORS				
jun	Avian sarcoma virus 17		Nucleus	Transcription factor AP1
fos	FBJ osteosarcoma		Nucleus	
myc	Avian MC29 myelocytomatosis		Nuclear matrix	Proteins possibly involved in regulating transcription
N-myc		Neuroblastoma	Nuclear matrix	
myb	Avian myeloblastosis	Leukemia	Nuclear matrix	
ski	Avian SKV770		Nucleus	
p53		(Demonstrated by cell transformation)	Nucleus	
rel	Avian reticuloendotheliosis		Nucleus and cytoplasm	
RB		Retinoblastoma	Nucleus	Antioncogene that binds to nuclear oncogene proteins of DNA viruses

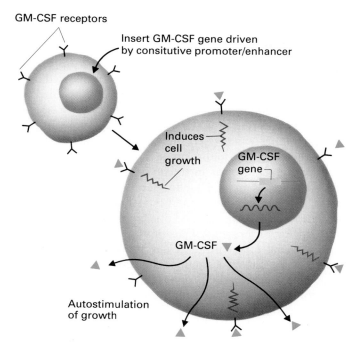

▲ **Figure 24-25** Autocrine induction of tumor-cell growth. A factor (GM-CSF) gene is inserted into a cell that already carries the GM-CSF receptor, causing autostimulation of growth.

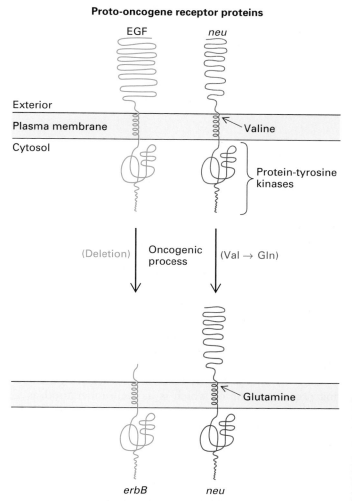

▲ **Figure 24-26** Creation of oncogenes from proto-oncogenes that encode cell-surface receptors. In one case, the receptor is for EGF; the oncogene arises by loss of the coding region for the EGF-binding domain. In the other case, the ligand is unknown; the *neu* oncogene encodes a protein with a single amino acid substitution in the transmembrane region.

cells that naturally have the PDGF receptor. Artificial class I oncogenes have been created. For example, if the gene encoding the granulocyte-macrophage colony stimulating factor (GM-CSF) is inserted into a cell that has the GM-CSF receptor, the GM-CSF protein continually stimulates growth of the cell. Such autostimulation is called *autocrine* induction of cell proliferation (Figure 24-25). Another case of autocrine stimulation is the release of transforming growth factors (TGFs), which we described earlier. TGF-α, for example, is an EGF analog that is released by many tumor cells and binds to EGF receptors to stimulate cell growth.

Factor and Hormone Receptors (Class II) Receptors bind a specific factor and then the receptor, recognizing somehow that it has bound the factor, sends a growth signal to the cell. Some cell-surface receptors have integral protein-tyrosine kinases in their cytoplasmic domains; these receptors probably transmit the growth signal by phosphorylating tyrosine residues on one or more target proteins, thus initiating a cascade of events (see Figure 19-3). The genes for such receptors become oncogenes when they are mutated in such a way that the receptor remains active even in the absence of its bound ligand. For example, a change in one amino acid codon affecting the transmembrane domain of the *neu* receptor protein (whose ligand remains a mystery) produces the

neu oncogene. In most cases, much of the extracellular ligand-binding domain is deleted during the formation of the oncogene (Figure 24-26). Thus we believe that the signal for growth given by the oncogene protein is its usual signal, but the altered receptor has lost ligand control and thus is constitutively active. A cell containing this type of oncogene grows independently of the factors that are supposed to regulate its growth.

The product of the *erbA* oncogene is derived from the intracellular receptor for thyroid hormone. Normally, the hormone's role is to directly transform the intracellular receptor into a transcription factor. In this case, the oncogenic event results in production of a modified receptor, ErbA, which may act by competing with the endogenous thyroid hormone receptor, causing growth without control. Actually, ErbA does not fully transform cells; it rather works coordinately with ErbB, which is an EGF-receptor derivative. Both the *erbA* and *erbB* oncogenes were first recognized in the avian erythroblastosis virus in which they act synergistically to give full oncogenicity.

Intracellular Transducers (Class III)

The largest class of oncogenes is derived from genes encoding proteins that are thought to act as intracellular transducers, proteins which transmit signals from a receptor to their cellular target. The best-understood transducers are the G proteins, one of which, G_S, controls cAMP synthesis. G_S protein senses whether a ligand has occupied a cell surface receptor, binds GTP, and activates the adenyl cyclase enzyme that forms cAMP. It then hydrolyzes the GTP and returns to an inactive state. A mutation in the gene for G_S can make it an oncogene by eliminating the GTPase activity. Such a mutant G_S protein continually stimulates cAMP synthesis, and the high cAMP concentration can cause unregulated proliferation of pituitary cells and therefore pituitary tumors.

Class III oncogenes derive from a variety of transducing proteins, none of which is as well-understood as G proteins. Many class III oncogenes encode a protein-tyrosine kinase (PTK). These differ from the products of the class II cell-surface receptor oncogenes in that they are intracytoplasmic or nuclear proteins, lacking any transmembrane or extracellular domain. Many such PTK proteins have myristate, a long-chain fatty acid, bound to their N-terminal glycine. This causes them to be partially bound to the plasma membrane, putting their kinase domain in the same perimembrane region as that of the kinases of the receptors (see Figure 13-15). For that reason, it is assumed that the intracytoplasmic kinases serve to transduce signals in the same way as do the receptor kinases. If the N-terminal glycine on proteins encoded by the *src* and *abl* oncogenes is removed, they cannot bind to the membrane and do not transform fibroblasts, supporting this hypothesis.

Tyrosine-specific protein phosphorylation occurs only rarely in normal cells. Actually, protein-tyrosine kinases were first recognized as oncogene proteins; later it was shown that about 0.1 percent of protein-linked phosphate in normal cells is bound to tyrosine residues. Cells transformed by the viruses that encode tyrosine-specific kinases have about 10-fold higher levels of phosphotyrosine in their proteins than normal cells or cells transformed by other means. The excess phosphotyrosine is distributed over many proteins, which suggests that either the viral kinases phosphorylate a number of different protein substrates or that the phosphorylation of certain key proteins sets off a spate of further phosphorylations.

Direct evidence that the viral kinases have a very low substrate specificity has come from experiments in which oncogene products have been expressed in bacteria. To accomplish this, oncogenes were removed from cloned retroviral DNA and placed into vectors that allowed their expression in bacteria. The kinase encoded by Abelson virus, for example, has been expressed in *Escherichia coli* with this procedure. When a bacterial cell contains the Abelson kinase, many tyrosine residues in bacterial proteins become phosphorylated. Because ordinarily no phosphotyrosine exists in *E. coli* proteins, the kinase encoded by v-*abl* must phosphorylate many substrate proteins. These experiments and others have shown that oncogene kinases recognize and phosphorylate a very broad range of target proteins. Thus phosphorylation of one or more crucial target proteins may be the mechanism by which some oncogenes induce the transformed state, although no such crucial proteins have been characterized. Possibly the effects on many proteins conspire to cause transformation.

Vinculin, a protein involved in linking the actin cytoskeleton to the cell membrane (see Figure 23-27b), is one of the proteins known to have an elevated phosphotyrosine level in transformed cells. The alteration of such a protein could easily be imagined to cause some of the hallmarks of the transformed state: changed cell morphology, reduced concentrations of actin microfilaments, loss of the anchorage dependence of growth, and changed mobility of cell-surface proteins. In this regard it may be pertinent that the *src* and *abl* oncogenes encode protein-tyrosine kinases that are localized to adhesion plaques, sites of bonding between cells and their underlying support where vinculin is also concentrated (see Figure 23-27).

The alteration of c-*src* to form an oncogene has been studied in great detail. The c-*src* product, a 60-kDa protein denoted pp60$^{c\text{-}src}$ (or c-Src), has multiple phosphorylation sites through which it is regulated. Phosphorylation of a tyrosine residue at position 527, six from the C-terminus, causes a great reduction in its kinase activity, and this site is often altered in oncogenic derivatives of c-*src*. In Rous sarcoma virus, for instance, the *src* gene has suffered a deletion that removes the C-terminal 18 amino acids of c-Src. The kinase activity of c-Src can also be augmented by its binding to polyoma mid-T protein, and this may be part of the mechanism by which polyoma transforms cells. Both c-Src and c-Abl can also be activated by changes in the N-terminal region, which is not part of the kinase domain. This region, which must regulate the kinase, is considered later in the discussion of the *crk* oncogene.

We presume that the PTK domains of receptors as well as the cytoplasmic kinases can phosphorylate targets that ultimately lead to transcriptional changes, but no such pathway of action has yet been elucidated. One hint to

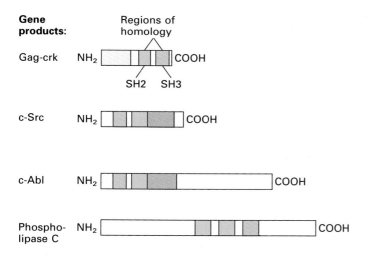

Gene products:

Regions of homology

Gag-crk   NH₂ ▭ COOH
              SH2   SH3

c-Src   NH₂ ▭ COOH

c-Abl   NH₂ ▭ COOH

Phospho-   NH₂ ▭ COOH
lipase C

◄ **Figure 24-27** Homology among class III oncogene proteins and phospholipase C. Considerable homology (30–40 percent) is exhibited in two regions (blue, green) of these various proteins, suggesting that they serve some common, yet undiscovered, function. [See B. Mayer et al., 1988, *Nature* 333:272.]

the function of c-Src is that the protein is associated with a lipid kinase, an enzyme that can phosphorylate inositol lipids. This tentatively links Src to the inositol signal system used as a second messenger for many intracellular events (see Figure 19-30).

Another type of class III oncogene encodes the Ras proteins. These proteins bind GTP and slowly hydrolyze it, as do the G proteins described earlier, but Ras proteins are smaller than authentic G proteins, and their exact role in the cell is unknown. Like the *src*-related proteins, they have a covalently attached fatty acid (a farnesyl group) and are found at the inner side of the cell's plasma membrane (see Figure 13-15). Thus they also may transduce signals from receptors and perhaps are involved in the inositol-lipid pathway of cell signaling.

The *ras* oncogenes were the first nonviral oncogenes to be recognized. Activation of a *ras* proto-oncogene to the oncogenic form may result from only one change in the protein, substitution of valine for glycine at position 12 of the sequence. This simple mutation reduces the protein's GTPase activity, thus linking GTP hydrolysis to the maintenance of normal, controlled Ras function. The structures of the Ras proteins are known from crystallography; the oncogenic change causes only a slight alteration in the structure, but this is sufficient to change the proto-oncogene into an oncogene. This tiny change initiates the transformation of cells from normal to malignant.

The product of the *crk* (pronounced "crack") oncogene has a particularly intriguing structure. Although the Crk protein has no known biochemical activity, it has considerable homology with Src and other cytoplasmic protein-tyrosine kinases (Figure 24-27). These two regions also are present in another protein, a phospholipase C (see Figure 19-30). Mutations in the *crk*-related regions of c-*src* and c-*abl* can activate them to oncogenes. In the

oncogene proteins, the corresponding regions may interact with other proteins that are involved in the normal control of the activity of the cytoplasmic protein-tyrosine kinases. The homology of these oncogene proteins to phospholipase C, an enzyme that cleaves the inositol-trisphosphate moiety from inositol-phospholipids, suggests a direct link to control of an important second-messenger system. Coupled to the observation that c-Src interacts with a lipid kinase, this evidence begins to develop a case that *src*-related proteins can link receptor systems to the inositol-trisphosphate intracellular signaling system.

Nuclear Transcription Factors (Class IV) By one mechanism or another, all oncogenes must eventually cause changes in the cell nucleus because growing cells make many proteins at different rates than they are made in quiescent cells. The proteins encoded by the class IV oncogenes appear to exert fairly direct effects on nuclear functions. Although all nuclear (class IV) oncogenes are thought to affect transcription, the evidence is best for *jun*. The *jun* proto-oncogene encodes part of a transcription factor called AP1, which binds to a sequence found in promoters and enhancers of many genes. Another nuclear oncogene, *fos*, encodes a protein that binds to *jun*, forming the whole AP-1 protein.

A number of other less definitive findings link the nuclear oncogenes to transcriptional control. The *myc*-gene products, for instance, show a clear homology to a known transcription factor called NF-κE2. The *RB*-gene product binds to nuclear oncogene products of DNA viruses that are themselves linked to transcriptional regulation. But the detailed roles of all of these fascinating proteins remain to be determined.

The nuclear proto-oncogene proteins are induced when normal cells are stimulated to grow, indicating their direct role in growth control. For example, PDGF treatment of quiescent 3T3 cells induces a transient increase (up to 50-fold) in the production of c-Fos, c-Myc, and the P53 protein. This increase, which occurs in many cells, always involves an initial transient rise of c-Fos and later a more prolonged rise of c-Myc (Figure 24-28). Although c-Fos rapidly disappears, c-Myc stays at a somewhat elevated level. The kinetics suggest that the c-Fos response is primary, but that the cell only uses the *fos* product transiently. In fact, unregulated expression of c-Fos can be oncogenic, and the gene was first identified as a retroviral

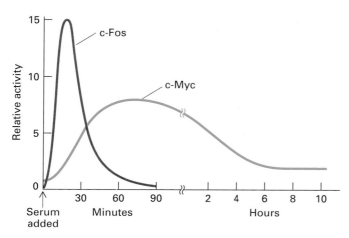

▲ **Figure 24-28** The response of two proto-oncogenes, *fos* and *myc*, to serum stimulation of the growth of quiescent 3T3 cells. [See M. E. Greenberg and E. B. Ziff, 1984, *Nature* 311:433.]

oncogene. To assure the rapid loss of c-Fos after its induction in normal cells, both the protein and its mRNA are intrinsically unstable. Unregulated expression of c-Myc also can cause uncontrolled growth and can, in erythroid cells, counteract a differentiation stimulus and maintain cells growing that would otherwise cease division.

All Oncogenes Probably Are Derived from Growth-controlling Genes

In the discussion of proto-oncogene and oncogene proteins in the previous section, we implied that they have equivalent functions. However, because we do not know the full pathway by which any proto-oncogene and its oncogene counterpart functions, we cannot say whether their products have the identical substrates and activities. The prolific phosphorylation by the cytoplasmic protein-tyrosine kinases encoded by oncogenes suggests, in fact, that they may have different targets than the corresponding, regulated proto-oncogene proteins. Nonetheless, the nature of known oncogenes strongly suggests that all are derived from growth-controlling genes. Furthermore, the products of both oncogenes and their corresponding proto-oncogenes appear to have similar, if not identical, functions in the cell's growth-control systems.

The Role of Cellular Oncogenes in Carcinogenesis

As discussed in an earlier section, only a few human cancers appear to be caused either by DNA viruses whose early genes act as oncogenes in infected cells or by transducing retroviruses that have incorporated cellular proto-

oncogenes into their genome in a way that activates them to oncogenes. Once the paucity of such transmissible oncogenes became evident, attention focused on nonviral, nontransmissible *cellular oncogenes*. These are defined as mutated cellular genes formed in situ in chromosomes and not implanted by viruses. As noted already, insertion of a slow-acting retrovirus (see Figure 24-17) and exposure to carcinogens or radiation can activate proto-oncogenes to cellular oncogenes. As we saw in the previous section, most oncogenes—both viral and nonviral—are derived from growth-controlling genes.

Some, But Not All, Human Tumors Contain Cellular Oncogenes

The prototypical assay for detecting nonviral cellular oncogenes is depicted in Figure 24-23. About 20 percent of DNA samples from numerous human tumor cell lines and fresh human tumor cells yield transformants, whereas DNA from comparable normal cells rarely, if ever, produces transformation. Thus oncogenes are commonly present in human tumor cells. Actually, the 3T3 cell assay certainly underestimates the frequency of oncogenes in tumor cells.

Cellular oncogenes from human tumor cells have been cloned using recombinant DNA techniques (Figure 24-29). The sequences of these genes indicate that most are related to the c-*ras* family. The first proto-oncogene to be studied was, in fact, c-Ha-*ras*, the same gene that gave rise to the Harvey virus (see Figure 24-15). The only difference between the coding sequence of the normal c-Ha-*ras* proto-oncogene and its oncogene—which came from a human bladder carcinoma—is in a single nucleotide; the corresponding proteins thus differ in only one amino acid. This miniscule change, which is reminiscent of the single amino acid replacement that causes sickle cell anemia, is sufficient to change a proto-oncogene into a cellular oncogene.

Thus far three c-*ras*-related genes, the *neu* gene, and a number of others have been positively identified as human cellular oncogenes. The 3T3 cell assay apparently selects for c-*ras*-related genes, although it can reveal others as well. Cellular oncogenes have been found in leukemias and in tumors of the bladder, lung, breast, large intestine, and neural tissue. Cellular oncogenes are not necessarily tissue specific in their action, that is, the same gene can be active in tumors of various tissues.

Products of Cellular Oncogenes Act Cooperatively in Transformation and Tumor Induction

As explained in the earlier section on DNA tumor viruses, transformation of primary cell cultures by polyoma virus involves the action of two proteins: one (T) that immor-

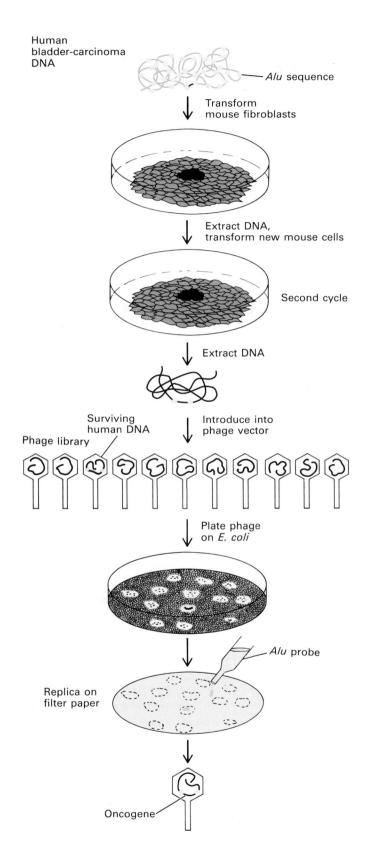

Human bladder-carcinoma DNA

— *Alu* sequence

Transform mouse fibroblasts

Extract DNA, transform new mouse cells

Second cycle

Extract DNA

Surviving human DNA

Phage library

Introduce into phage vector

Plate phage on *E. coli*

Alu probe

Replica on filter paper

Oncogene

◀ **Figure 24-29** The isolation of a cellular oncogene from a human tumor by molecular cloning. Like almost all human genes, oncogenes have nearby repetitive DNA sequences called *Alu* sequences. The human DNA is first purified by repetitive transfer as shown in Figure 24-20. The total DNA from a secondary transfected mouse cell is cloned into bacteriophage λ, and the phage receiving human DNA is identified by hybridization to an *Alu* probe. The hybridized DNA should contain part or all of the oncogene. The expected result can be proved by showing either that the phage DNA can transform cells (if the oncogene has been completely cloned) or that the cloned piece of DNA is always present in cells transformed by DNA transfer from the original donor cell.

talizes the cells and another (mid-T) that changes their growth properties. This same division of labor has been demonstrated in cellular oncogenes.

When primary rat embryo cells are transfected with a *ras* oncogene, they show the morphological changes associated with transformation but are not immortalized. Transfection with *ras* plus *myc*, however, leads to fully transformed, immortal, tumorigenic cell lines. The polyoma mid-T oncogene can replace *ras*, and the polyoma T oncogene or the various other nuclear oncogenes can replace *myc*. Thus it appears that two classes of transforming function can be defined: *ras*-like and *myc*-like. Because the *ras* proteins are cytoplasmic and the *myc* protein is nuclear, transformation would appear to be a two-event process, a cytoplasmic alteration of cell behavior plus a possible change in gene transcription.

This neat picture of transformation as a two-step process is clouded by observations that if *ras* or *myc* is provided with the strong promoters/enhancers of retroviral LTRs, either gene will morphologically transform cell lines, and *ras* will also transform primary cells to immortality. Thus high levels of the oncogene products may have different effects from lower levels, and the transformation of a given cell may be affected both by the types of oncogenes expressed in the cell and by the amount of the oncogene product in the cell.

The two-step model of transformation, however, sheds some light on a phenomenon described early in this chapter. We noted that primary rodent cells have a low but reproducible probability of spontaneously becoming cell lines. We can now hypothesize that the event responsible is the activation of an immortalizing oncogene, and evidence implicating the p53 protein has been found. Thus 3T3 cell lines—as opposed to primary cells—may provide such a good substrate for assaying *ras* oncogenes because they already have an active *myc*-like oncogene.

The ability of an oncogene to immortalize cells is consistent with our earlier portrayal of immortalization as a parameter of transformation quite distinct from the others. Cell "mortality" is probably a consequence of differentiation events that change a stem cell into an end-stage, nondividing cell. Thus immortalization may actually be the outcome of a blockade of differentiation caused by the action of oncogenes.

One especially powerful technique for analyzing oncogene action is to insert an oncogene with a specific promoter/enhancer into the genome of a mouse strain, forming a transgenic mouse (see Figure 5-26). In a prototype of such experiments, the SV40 T oncogene has been coupled to an insulin promoter/enhancer and inserted as a transgene. Consistent with the known fact that insulin is made only in the β cells of pancreatic islets of Langerhans, such transgenic mice get tumors of β islet cells. Although many cells express the gene, only a rare cell gives rise to a malignant clone. Thus T-protein expression is not sufficient to cause transformation to malignancy; another event must occur and it is rare. A likely hypothesis is that a second, cellular oncogene must become activated.

The cooperativity of multiple oncogenes in mouse tumor formation has been shown most dramatically using the *ras* and *myc* genes as transgenes driven by a mammary cell-specific promoter/enhancer from a virus (Figure 24-30a). By itself, *myc* causes tumors only after 100 days and then in only a few mice; *ras* causes tumors earlier but still slowly and with about 50 percent efficiency over 150 days. When the *myc*- and *ras*-transgenics are crossed, however, so that all mammary cells express both genes, tumors arise much more rapidly and all animals succumb to cancer. Such experiments emphasize the synergistic effects of multiple oncogenes. Not all oncogenes, however, need partners. The *neu* oncogene in certain transgenic animals causes tumors so efficiently (Figure 24-30b) that little normal mammary epithelium is present in pregnant *neu*-transgenic mice because so many cells have been transformed to malignancy.

Consistent Chromosomal Anomalies Associated with Tumors Involve Oncogenes

It has long been known that chromosomal abnormalities abound in tumor cells. Human cells ordinarily have 23 pairs of chromosomes, each with a well-defined substructure, but tumor cells are usually *aneuploid* (i.e., they have an abnormal number of chromosomes—generally too many), and they often contain *translocations* (fused elements from different chromosomes). As a rule, these unusual characteristics are not reproducible from tumor to tumor: each tumor has its own set of anomalies. Certain anomalies recur, however. The first to be discovered, the Philadelphia chromosome, was found in the cells of virtually all patients with the disease chronic myelogenous leukemia. This chromosome is a fusion of most of chromosome 9 to a piece of chromosome 22. The reciprocal

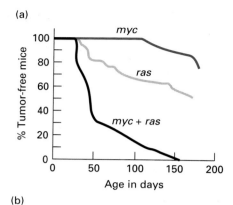

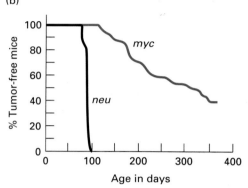

▲ **Figure 24-30** Tumor occurrence in transgenic mice. (a) Kinetics of tumor appearance in female transgenic mice carrying transgenes driven by the mouse mammary tumor virus (MMTV) LTR. Shown are results for mice carrying *myc* and *ras* transgenes as well as for the progeny of a cross of *myc* carriers by *ras* carriers. The incidence of tumor-free mice is plotted because this graphically depicts the time-course of disease occurrence. Females were studied because the hormonal stimulation of pregnancy is needed to activate the MMTV-driven oncogenes. (b) Comparison of the tumor incidence in *myc* transgenics and *neu* transgenics. The almost simultaneous appearance of *neu*-induced tumors is evident. [Part (a) see E. Sinn et al., 1987, *Cell* **49**:465.] [Part (b) see W. J. Muller et al., 1988, *Cell* **54**:105.]

translocation—in which a tiny piece of chromosome 9 is fused to the broken end of chromosome 22—is also present, although it is not easily recognized. It is now known that in the Philadelphia chromosome, the c-*abl* proto-oncogene lies at the break point on chromosome 9, and the translocation causes the formation of an mRNA derived partly from chromosome 22 and partly from c-*abl*.

In the tumor known as Burkitt's lymphoma, a reproducible translocation of antibody genes to chromosome 8 occurs, just at the site of the *myc* gene. The analogous site in the mouse genome is also involved in mouse myelomas. In both cases, the tumor cells are antibody-producing cells, and c-*myc* is translocated to regions that are normally involved in DNA rearrangements and that help in the construction and expression of antibody-forming genes. The translocations alter the transcriptional activity of c-*myc* and/or the stability of c-*myc* mRNA.

To what extent the modification and activation of proto-oncogenes by chromosomal translocations causes the progressive steps giving rise to human tumors is not yet clear, but it seems increasingly likely that the chromosomal instabilities associated with malignant cells contribute importantly to malignancy. Such translocations also help to identify new oncogenes; like the slow-acting carcinogenic retroviruses, they are fingers pointing into the genome at potential oncogenes.

Another common chromosomal anomaly in tumor cells is the localized reduplication of DNA to produce as many as 100 copies of a given region (usually a region spanning hundreds of kilobases). This anomaly may take either of two forms: the duplicated DNA may be tandemly organized at a single site on a chromosome, or it may exist as small, independent chromosomelike structures. The former case leads to a homogeneously staining

region (HSR) that is visible in the light microscope at the site of the duplication; the latter case causes double minute chromosomes to pepper a stained chromosomal preparation (Figure 24-31). Again, oncogenes have been found in the duplicated regions. Most strikingly, the *myc*-related gene called N-*myc* has been identified in both HSRs and double minutes of human nervous system tumors.

The two-oncogene transformation of cell cultures and transgenic mice has an interesting parallel in human and experimental tumors. Many experimental tumors resulting from exposure to a carcinogen have *ras* oncogenes, which implies that c-*ras* proto-oncogenes are common targets for the initial events in carcinogenesis. On the other hand, only highly malignant human tumors—and not their less malignant counterparts—often have duplicated or translocated *myc*-related genes. This observation implies that the progression from an initiated tumor to a highly malignant one may result from activation of c-*myc*. Thus a malignancy could often result from an ini-

(a)

(b)

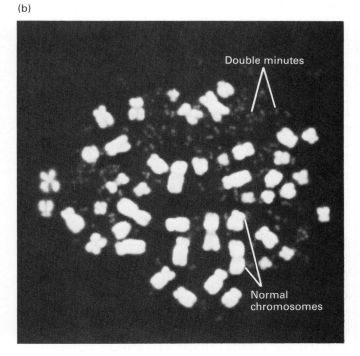

Double minutes

Normal chromosomes

▲ **Figure 24-31** Visible forms of DNA amplification. (a) Homogeneously staining regions (HSRs) in chromosomes from two neuroblastoma cells. In each set of three chromosomes, the left-most one is a normal chromosome 1 and the other two are HSR-containing chromosomes. The three lines (1, 2, and 3) represent three different methods of staining the chromosomes. Method 1 is quinacrine staining, which highlights AT-rich regions; method 2 is staining with chromomycin A3 plus methyl green, which highlights GC-rich areas; and method 3 is 33258 Hoechst staining after a pulse of bromodeoxyuridine late during the S phase, which

highlights the early replicating regions. In all three cases the HSR shows homogeneous staining characteristics whereas the rest of the chromosome has highlights. (b) Quinacrine-stained double minute chromosomes in a human neuroblastoma cell. The normal chromosomes are the large white structures; the double minutes are the many small dots that are paired. Both the HSRs and the double minute chromosomes shown here contain the N-*myc* oncogene. [Part (a) see S. Latt et al., 1975, *Biopolymers* **24**:77.] [Part (b) see N. Kohl et al., 1983, *Cell* **35**:359.] *Photographs courtesy of S. Latt.*

tial point mutation in a *ras* proto-oncogene followed by a chromosomal rearrangement that activates a *myc* proto-oncogene.

The Multicausal, Multistep Nature of Carcinogenesis

Although the discovery of oncogenes was a landmark in cancer research, we do not yet know how many human cancers are caused by oncogenes or how many of the steps in carcinogenesis depend on oncogene activity. Human tumors usually develop in progressive stages, resulting finally in an invasive, metastasizing, malignant cancer. As we have seen, multiple oncogenes are probably required to generate a malignant cell, but oncogene activity may be coupled with other mechanisms—even nongenetic ones—to produce a life-threatening malignancy.

In this chapter we have stressed the role of oncogenes in the development of cancer and ignored most other potential mechanisms. The rationale for this emphasis is simply that we know much more about oncogenes and their role in cancer than about other possible cancer-inducing mechanisms. We must, however, maintain an open mind on the problem. Our knowledge of the role oncogenes plays in human cancer is indirect. Furthermore, oncogenes (mainly of the *ras* and *myc* families) have been recovered from only about 10–20 percent of human tumors, although this may be a consequence of the usual assay method. In this section we describe several possible nongenetic mechanisms of carcinogenesis and review the evidence that cancer results from multiple interacting causes.

Epigenetic Alterations May Occur in Teratocarcinomas

The largest conceivable class of nongenetic cancer-inducing mechanisms involve *epigenetic* inheritance. An epigenetic condition or process is one that is passed from a cell to its progeny without any alteration in the coding sequence of the DNA. The differentiated state of cells, which is passed on to progeny, is generally determined epigenetically. How epigenetic states are maintained is still a matter for speculation; roles for protein regulatory molecules and for DNA methylation are considered likely. Whatever its basis, a process determined epigenetically can be inherited with the same fidelity as one determined by a genetic alteration.

Epigenetic mechanisms may underlie formation of malignant teratocarcinomas. These tumors of very early embryonic cells maintain the ability to differentiate into the full range of body cell types. In fact, if certain teratocarcinoma cells are incorporated into a preimplantation embryo, they can contribute as normal cells to all of the tissues of the adult, including the germ-line cells, but no tumors develop. Conversely, if teratocarcinoma cells are injected into adult mice, they form lethal tumors. Thus their malignant state is conditional on their environment. Mouse teratocarcinomas can be formed by the simple procedure of transferring early embryos into older mice at sites such as the testis; these tumors can also be formed by explanting very early embryos into cell culture. Induction of these tumors thus requires no mutagenic treatment and no mutation is evident; in the right environment the cells appear absolutely normal. We presently believe that the teratocarcinomas are a unique case, but their occurrence certainly alerts us to the possible importance of nongenetic events in carcinogenesis.

Some Cancer-inducing Chemicals Act Synergistically

The phenomenon of *tumor promotion* illustrates both the involvement of a nongenetic mechanism in cancer induction and the synergistic interaction of cancer-inducing agents. The initial discovery of tumor promotion in the early 1940s was based on the finding that if the skin of an animal is treated with certain chemicals (now recognized as DNA-damaging carcinogens), the odds of a tumor appearing could be greatly augmented by further treatment with a second compound that had a very different structure and chemistry. The first compound is called an *initiator*, and the second is called a *promoter*. In contrast to initiators, promoters do not need to be metabolized to be active, they have no tendency to react as electrophiles, and they rarely induce tumors by themselves. Therefore promoters must act on cells through a mechanism quite different from that of initiators. To be effective, a promoter must be applied after an initiator; it must be applied repeatedly over many weeks or months, whereas the initiator need be applied only once. Although treatment with an initiator will often produce a tumor (i.e., the initiator may act both to initiate and to promote), promoters require previous initiation to be tumorigenic.

One class of promoters, the phorbol esters (Figure 24-32), has been intensively studied and provides a model for the action of all promoters. The treatment of many types of cells with phorbol esters causes the cells to take on a transformed morphology and growth habit, but the cells revert to normal when the compound is removed. Thus phorbol esters can produce a phenocopy of the transformed state (i.e., they alter the phenotype but not the genotype). If cells that have the potential to differentiate are treated with phorbol esters, differentiation will often be either blocked or facilitated, which again attests to the powerful physiological effects of phorbol esters.

The site of action of phorbol esters became known from studies of a unique cellular enzyme called protein kinase C, which phosphorylates serine or threonine residues on other proteins. The activation of protein kinase C

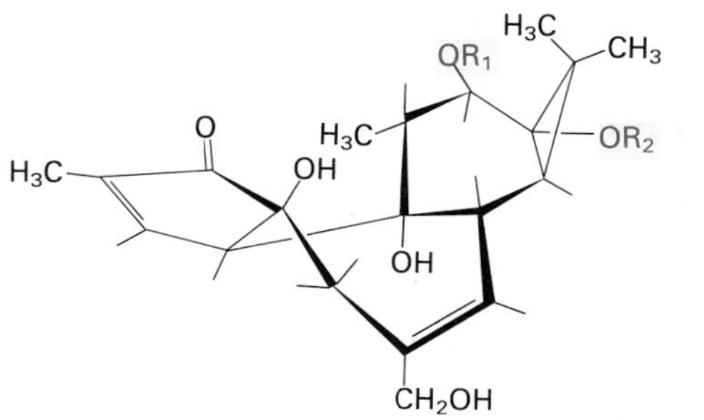

▲ **Figure 24-32** The structure of phorbol esters, the prototypical tumor promoters. The molecule has four rings, one each containing seven, six, five, and three carbons. Crucial components are the two long-chain fatty acids (OR_1 and OR_2) esterified to the phorbol backbone. The molecule apparently acts as an analog to diacylglycerol, a molecule with two esterified fatty acid chains and a hydroxyl group.

requires the presence of a number of cofactors: Ca^{2+}, phospholipids, and diacylglycerol. This latter compound is produced when phospholipids are degraded. A specific class of progenitors of diacylglycerol is the inositol-containing phospholipids, which are degraded by a complex pathway involving phosphorylation of the inositol moiety before its cleavage (see Figure 19-30).

Phorbol esters can substitute for diacylglycerol as a cofactor for protein kinase C and, in association with Ca^{2+} and phospholipids, strongly activate protein kinase C for phosphorylation of substrate proteins. The mechanism of action of phorbol esters appears to be a facilitation of Ca^{2+} binding. The stereochemistry of phorbol esters makes them somewhat analogous in shape to diacylglycerol; this analogy is presumably the basis for their ability to substitute for diacylglycerol and activate protein kinase C. The linkage of promoter activity to protein kinase C does not fully explain the action of promoters, however, because there is no clear-cut role for protein kinase C in cell growth control.

In the absence of prior initiation, tumor promoters have no permanent effects on cell metabolism, whereas following initiation they can set in motion events that lead to an irreversible alteration in the cell. The end result could be a DNA alteration, or it could be a changed state of cellular differentiation. Whatever the mechanism, a promoter must produce a second permanent alteration in a cell that has already undergone an initiator-induced mutational change. Thus tumor promoters act synergistically with initiators.

Evidence of synergism between various cancer-inducing agents in human carcinogenesis has been discovered, and the initiator-promoter model may be just as applicable to humans as it is to animal model systems. If so, control of cancer could be achieved through control of either initiators or promoters. Short-term mutagenesis tests can only identify initiators; no reliable tests for promoters have yet been devised. Some investigators believe that a major aspect of the cancer-inducing capacity of cigarettes may lie in a promoterlike activity released from the burning of the tobacco.

The initiator-promoter model of carcinogenesis describes just one of a variety of ways in which chemicals may interact to cause tumors. For example, low concentrations of potentially carcinogenic compounds may interact when the compounds are applied together (cocarcinogenesis). Other compounds, notably antioxidants, can counteract the effects of others (anticarcinogenesis).

Natural Cancers Result from the Interaction of Multiple Events over Time

The evidence supporting the concept that carcinogenesis is a multicausal, multistep process is powerful. First is epidemiologic evidence showing that cancer generally occurs late in life and that its time-course of occurrence follows multihit kinetics. Second is the initiator-promoter phenomenon in which the involvement of at least two events is clear. Third is the synergism between the *myc* and *ras* oncogenes in cultured cells and in transgenic animals. Fourth is the common observation that tumors are monoclonal even in transgenic animals in which all cells contain a single active oncogene. Fifth is the evidence, described in the next section, that a single inherited cancer-susceptibility gene in humans, expressed in all target cells, causes only rare, monoclonal tumors, implying involvement of changes in other genes.

Human Cancer

A central motivation of basic cancer research is to understand human cancer well enough to be able to alter its course. Although what we have learned in the laboratory is certainly relevant to human cancer, putting it together to provide the understanding we seek has proved to be beyond the capability of today's science.

To assess the role of a substance in cancer induction, it is not useful to think in terms of a single cause. As we emphasized previously, naturally occurring cancer is most often a consequence of multiple factors that interact over long periods of time. Because each of these factors increases the possibility of a cancer occurring, they are called *risk factors*. For example, Epstein-Barr virus is a risk factor for Burkitt's lymphoma, although it is clearly not the sole cause. Identifying risk factors for human cancer has been a slow and frustrating activity. One successful endeavor stands out: the identification of cigarette smoking as a crucial risk factor in lung cancer. A risk factor of this potency gives a clear indication of how to act to avoid lung cancer: avoid cigarettes.

Unfortunately, lung cancer is the only major human cancer for which a clear-cut risk factor has been identified. Animal fat is thought to be a risk factor for colon and breast cancer, and many viruses and chemicals have been correlated with minor cancers; however, hard evidence that would help us avoid breast cancer, colon cancer, prostate cancer, leukemias, and others is generally lacking.

Rare Susceptibilities to Cancer Point to Antioncogenes

A corollary to the belief that multiple interacting events in our environment are the major risk factors for cancer is the belief that genetic inheritance plays only a small role in carcinogenesis. This proposition is supported by the finding that people who migrate to a new environment take on the profile of cancers in their new environment within a generation. For instance, when Japanese citizens move to California, they rapidly lose the oriental propensity toward stomach cancer and soon show the occidental propensity toward breast cancer.

Genetic inheritance does, however, play some role in human cancer. Certain inherited genes increase the probability that an individual will get a specific tumor to almost 100 percent. A classic case is retinoblastoma, which is, like most inherited tumors, a disease of childhood. Children who inherit a single defective copy of the *RB* gene, often seen as a small deletion on chromosome 13, will come down with an average of three retinoblastoma tumors, each derived from a single transformed cell. Because the developing retina contains about 4×10^6 cells, only about 1 in 10^6 cells actually becomes a tumor cell. This finding suggests that even with its highly dominant inheritance, the *RB* gene is acting recessively at the cell level, and that a second event is needed to bring on the transformed state. The second event is now known to be the deletion or mutation of the normal *RB* gene on the other chromosome. Rare, somatic events can cause this deletion either by loss of all or a segment of the chromosome or by small alterations in the remaining *RB* gene. When chromosomal loss occurs, it is balanced by duplication of the affected chromosome.

Although the exact function of the *RB* gene product is not known, it is a nuclear protein and thus may affect transcription. Because its loss is the cause of malignant transformation, it is thought to act negatively, to suppress the oncogenic potential of other proteins. It is thus considered an *antioncogene*. Its targets are not known but could be transcriptional activators. An intriguing finding is that the RB protein binds tightly to DNA viral oncogene products. Perhaps these oncogene proteins cause transformation by tying up the RB protein, thus releasing the oncogenic potential of those proteins that RB ordinarily regulates.

Cancer induction by deletion of a genetic region, rather than by activation of an oncogene, appears to occur in a variety of childhood tumors associated with inherited defects. These are all rare tumors, but it is believed that deletional carcinogenesis plays a wider role than is now evident. There may well be many antioncogenes, and their deletion may be an important determinant of the progression of human cancers.

Studies of human tumors have indicated that inheritance is only rarely a crucial risk factor; however, whether inheritance might be a minor risk factor in many cancers is still being debated. Genetic propensity might, for example, explain why some smokers get lung cancer and others do not.

Summary

Cancer represents a fundamental aberration in cellular behavior that involves many aspects of molecular biology. To become a cancer cell, a normal cell must undergo many significant changes. It must continue to multiply when normal cells would be quiescent; it must invade surrounding tissues, often breaking through the basal laminas that define the boundaries of tissues; and it must spread through the body and set up secondary areas of growth in a process called metastasis. All of the various cell types of the body can give rise to cancer cells. Cancer cells are usually closer in their properties to immature normal cells than to more mature cell types. The retention of malignant properties by cancer cells grown in culture shows that the alteration from a normal cell to a cancer cell is caused by events within the cancer cell itself.

By growing cells in culture, experimenters can freely manipulate them to learn about the properties that distinguish normal cells from cancer cells. Carcinogens, cancer-causing agents, can also be used to alter cultured cells to produce full-fledged cancer cells in a process called transformation. Transformed cells differ from normal cells in many ways, including cell-growth control, cell morphology, cell-to-cell interactions, membrane properties, cytoskeletal structure, protein secretion, gene expression, and mortality (transformed cells can grow indefinitely). These alterations in cell behavior are not wholly independent, but their interrelationships remain obscure.

The transformation of a cell from normal to malignant can be the consequence of the expression of one or a few genes, called oncogenes. Oncogenes are formed in cells or carried into cells by transforming agents, of which we recognize three main types: viruses, chemical carcinogens, and radiation. Certain DNA viruses and RNA-containing retroviruses can transform cells by permanently integrating new genes into the DNA of infected cells. The DNA-containing papovaviruses carry oncogenes that can cause cell transformation by inducing in cells a growing state that facilitates virus multiplication. Permanent cell transformation results when virus multiplication is not possible.

Retroviruses cause cancer in two ways, both of which depend on cellular genes. These viruses, which contain RNA as their genetic material, employ reverse transcriptase to make an intracellular DNA copy that can integrate into cellular DNA. Such integrated proviruses, by recombination with cellular genes, can acquire oncogenes that turn them from relatively benign viruses to cancer-inducing agents. Such oncogenes were initially normal genes (called proto-oncogenes), but by mutation or an altered context of expression they become transforming genes. More than 20 different normal cellular genes have been identified as oncogenes in retroviruses. The second way that retroviruses can cause cancer is by integration near a proto-oncogene in such a way that the gene is activated to an oncogene. The difference between these two mechanisms of retroviral-induced cancer is that in the former case the oncogene becomes part of a transmissible virus (i.e., the gene is transduced), whereas in the latter case the proto-oncogene is activated in situ to an oncogene form that cannot be transmitted to other cells or animals.

Viruses cause certain rare human tumors. Included among human viral carcinogens are human retroviruses that cause leukemia; the Epstein-Barr virus, a herpeslike virus that causes leukemia; hepatitis B virus; and papilloma viruses. AIDS is caused by a unique human retrovirus that does not directly cause cancer.

Chemical carcinogens have a variety of structures with one unifying characteristic: electrophilic reactivity (either they are electrophiles or they are metabolized in the body to become electrophiles). Metabolic activation occurs via the cytochrome P-450 system, a pathway generally used by cells to rid themselves of noxious chemicals. The reactive electrophiles combine with many parts of the cell, but their reaction with DNA is the primary carcinogenic event. The reaction of a carcinogen and DNA leads to a DNA alteration that ultimately produces an oncogene from a proto-oncogene.

Ultraviolet, x-ray, and atomic particle radiations can all cause cancer. All effect DNA and presumably lead to proto-oncogene activation. Cellular DNA-repair processes have been implicated both in protecting against radiation-induced carcinogenesis and in causing it. The protective effect is most vividly demonstrated by the association of certain human diseases involving DNA-repair deficiencies with high rates of cancer; the deficiencies in these diseases inactivate error-free repair systems. The inductive effect is seen most clearly in bacterial strains that lack secondary, error-prone repair systems; such strains have much lower mutation rates than wild-type bacteria.

The mechanisms by which oncogenes cause cancer are not yet known in detail, but their general nature is becoming clearer. Most oncogenes are derived from growth-controlling genes that encode four types of proteins: growth factors, receptors, intracellular signal transducers, and intranuclear transcriptional controllers. Most oncogene proteins are related to one of these four types of proteins. The majority of oncogenes encode intracellular transducing proteins including protein-tyrosine kinases (e.g., the historic *src* of Rous sarcoma virus) and GTP-binding and hydrolysis proteins (the *ras* oncogenes). The other major class is the nuclear oncogenes, which increasingly all appear to encode transcription factors.

Oncogenes created in situ in the genome of humans or other animals are called cellular oncogenes. They are often revealed when transfer of DNA from a malignant to a normal cell transfers the malignant properties, implying the transfer of an oncogene. Oncogenes are activated from their benign, proto-oncogenes by local alterations in DNA structure (point mutations, deletions, etc.), by translocations of a DNA region among chromosomes, or by extensive reduplication of DNA regions including an oncogene.

Oncogenes often act collaboratively, no one oncogene being sufficient to induce malignant growth. This has been demonstrated in DNA tumor viruses and in transgenic mice whose inheritance is modified by insertion of one or more oncogenes into their genome. Also, human tumor cells may harbor more than one oncogene, suggesting that multiple genetic changes contribute to the malignancy of a given cell. The accumulation of multiple alterations may explain the long latency period between the initial exposure to a carcinogen and the development of a full-fledged malignancy.

Nongenetic mechanisms may play some role in cancer induction. For example, some chemicals, called promoters can potentiate the activity of electrophilic carcinogens. The best-understood promoters are the phorbol esters, which cause nongenetic cellular changes that often mimic transformation. These substances activate a cellular protein kinase. Long-term treatment with phorbol esters leads to permanent cellular alterations that may or may not be genetic. An apparently clear-cut case of a nongenetic change that causes cancer is the epigenetic alteration leading to a teratocarcinoma. These tumor cells revert to normal when they are implanted into early embryos.

The data available today suggests that a cancer develops as follows: Exposure of a cell to a mutagen leads to a DNA alteration that changes a proto-oncogene into an oncogene. This activation subtly alters the cell, giving it a growth advantage over its neighbors. As it grows, perhaps as a visible benign tumor or polyp but more likely as an inapparent clone of cells, one or more further alterations activate other proto-oncogenes. Together, the multiple genetic modifications allow the clone to escape from all of the influences that ordinarily keep the growth of cells appropriate to the needs of the body. The clone not only grows without control, but its cells become invasive and begin to spread to new sites. At this point, we recognize it as a malignant tumor. Although this model is certainly simplistic, and does not include nongenetic influences, it represents a framework for future research and, we hope, for the development of new methods of prevention and therapy for cancer.

References

Advances in Cancer Research (yearly volumes). Academic Press.

BECKER, F. F., ed. 1982. *Cancer: A Comprehensive Treatise*. Plenum. (Multiple volumes are released under this title.)

BOCK, G., and J. MARSH, eds. 1989. *Genetic Analysis of Tumour Suppression* (Ciba Foundation symposium; 142). London: John Wiley & Sons Ltd.

BOICE, J. D., JR., and J. F. FRAUMENI, eds. 1984. Radiation carcinogenesis: epidemiology and biological significance. In *Progress in Cancer Research and Therapy*, vol. 26 (same editors), Raven.

BRADSHAW, R. A., and S. PRENTIS, eds. 1987. *Oncogenes and Growth Factors*. Elsevier Science Publishers.

BURCK, K. B., E. T. LIU, and J. W. LARRICK. 1988. *Oncogenes: An Introduction to the Concept of Cancer Genes*. Springer-Verlag.

BUYSE, M. E., M. J. STAQUET, and R. J. SYLVESTER, eds. 1984. *Cancer Clinical Trials: Methods and Practice*. Oxford University Press.

CAIRNS, J., ed. 1974. *Cancer: Science and Society*. W. H. Freeman and Company.

CAIRNS, J. 1975. The cancer problem. *Sci. Am.* 233(5):64–72, 77–78.

DEVITA, V. T., JR., S. HELLMAN, and S. A. ROSENBERG, eds. 1982. *Cancer: Principles and Practice of Oncology*. Lippincott.

HOLLAND, J. F., ed. 1982. *Cancer Medicine*. Philadelphia: Lea & Febiger.

KAHN, P., and T. GRAF, eds. 1986. *Oncogenes and Growth Control*. Springer-Verlag.

KAISER, H. E. 1981. *Neoplasms: Comparative Pathology of Growth in Animals, Plants, and Man*. Baltimore: Williams & Wilkins.

KLEIN, G., ed. 1988. *Cellular Oncogene Activation*. Marcel Dekker, Inc.

LEVINE, A. J., W. C. TOPP, and J. D. WATSON, eds. 1984. *Cancer Cells*, vol. 1: *The Transformed Phenotype*. Cold Spring Harbor Laboratory.

MELNICK, J. L., ed. 1985. *Viruses, Oncogenes and Cancer*, vol. 2. New York: S. Karger.

PIMENTEL, E. 1987. *Hormones, Growth Factors, and Oncogenes*. CRC Press, Inc.

VAN DE WOUDE, G. F., A. J. LEVINE, W. C. TOPP, and J. D. WATSON, eds. 1984. *Cancer Cells*, vol. 2: *Oncogenes and Viral Genes*. Cold Spring Harbor Laboratory.

VARMUS, H., and J. M. BISHOP, eds. 1986. *Cancer Surveys: Advances and Prospects in Clinical, Epidemiological and Laboratory Oncology. Biochemical mechanisms of oncogene activity: proteins encoded by oncogenes*. vol. 5: no. 2. Oxford: Oxford University Press.

Characteristics of Tumor Cells

CAIRNS, J. 1975. Mutational selection and the natural history of cancer. *Nature* 255:197–200.

CIFONE, M. A., and I. J. FIDLER. 1981. Increasing metastatic potential is associated with increasing instability of clones isolated from murine neoplasms. *Proc. Nat'l Acad. Sci. USA* 78:6949–6952.

FIDLER, I. J., and I. R. HART. 1982. Biological diversity in metastatic neoplasms: origins and implications. *Science* 217:998–1003.

ILLMENSEE, K., and B. MINTZ. 1976. Totipotency and normal differentiation of single teratocarcinoma cells cloned by injection into blastocysts. *Proc. Nat'l Acad. Sci. USA* 73:549–553.

NICHOLSON, G. L. 1979. Cancer metastasis. *Sci. Am.* 240(3):66–76.

OSSOWSKI, L., and E. REICH. 1983. Antibodies to plasminogen activator inhibit human tumor metastasis. *Cell* 35:611–619.

Cell Culture and Transformation

ABERCROMBIE, M. 1970. Contact inhibition in tissue culture. *In Vitro* 6:128–142.

ABERCROMBIE, M., and J. E. M. HEAYSMAN. 1954. Observations on the social behaviour of cells in tissue culture. *Exp. Cell Res.* 6:293–306.

BURGER, M. M., and G. S. MARTIN. 1972. Agglutination of cells transformed by Rous sarcoma virus by wheat germ agglutinin and concanavalin A. *Nature New Biol.* 237:356–359.

CARPENTER, G., and S. COHEN. 1975. Human epidermal growth factor and the proliferation of human fibroblasts. *J. Cell Physiol.* 88:227–238.

CARREL, A. 1912. On the permanent life of tissues outside of the organism. *J. Exp. Med.* 15:516–528.

CHERINGTON, P. V., B. L. SMITH, and A. B. PARDEE. 1979. Loss of epidermal growth factor requirement and malignant transformation. *Proc. Nat'l Acad. Sci. USA* 76:3937–3941.

DULAK, N. D., and H. M. TEMIN. 1973. A partially purified polypeptide fraction from rat liver cell conditioned medium with multiplication-stimulating activity for embryo fibroblasts. *J. Cell Physiol.* 81:153–160.

FOLKMAN, J., and A. MOSCONA. 1978. Role of cell shape in growth control. *Nature* 273:345–349.

GAFFNEY, B. J. 1975. Fatty acid chain flexibility in the membranes of normal and transformed fibroblasts. *Proc. Nat'l Acad. Sci. USA* 72:510–516.

GEY, G. O., W. D. COFFMAN, and M. T. KUBICEK. 1952. Tissue culture studies of the proliferative capacity of cervical carcinoma and normal epithelium. *Cancer Res.* 12:264–265.

GOSPODAROWICZ, D., and J. S. MORAN. 1976. Growth factors in mammalian cell culture. *Ann. Rev. Biochem.* 45:531–558.

HAKOMORI, S. 1975. Structures and organization of cell surface glycolipid: dependency on cell growth and malignant transformation. *Biochim. Biophys. Acta* 417:58–80.

HATANAKA, M. 1974. Transport of sugars in tumor cell membranes. *Biochim. Biophys. Acta* 355:77–104.

HAYFLICK, L., and P. S. MOREHEAD. 1961. The serial cultivation of human diploid cell strains. *Exp. Cell Res.* 25:585–621.

HOLLEY, R. W. 1975. Factors that control the growth of 3T3 cells and transformed 3T3 cells. In *Proteases and Biological Control*, D. B. Rifkin and E. Shaw, eds. Cold Spring Harbor Laboratory.

HOLLEY, R. W., and J. A. KIERNAN. 1968. "Contact inhibition" of cell division in 3T3 cells. *Proc. Nat'l Acad. Sci. USA* 60:300–304.

HYNES, R. O. 1973. Alteration of cell-surface proteins by viral transformation and by proteolysis. *Proc. Nat'l Acad. Sci. USA* 70:3170–3174.

HYNES, R. O., ed. 1979. *Surfaces of Normal and Malignant Cells*. Wiley.

MACPHERSON, I., and L. MONTAGNIER. 1964. Agar suspension culture for the selective assay of cells transformed by polyoma virus. *Virology* 23:291–294.

OSSOWSKI, L., J. C. UNKELESS, A. TOBIA, J. P. QUIGLEY, D. B. RIFKIN, and E. REICH. 1973. An enzymatic function associated with transformation of fibroblasts by oncogenic viruses, II: Mammalian fibroblast cultures transformed by DNA and RNA tumor viruses. *J. Exp. Med.* 137:112–126.

SATO, G., ed. 1979. *Hormones and Cell Culture*. Cold Spring Harbor Laboratory.

SHIN, S. I., U. H. FREEDMAN, R. RISSER, and R. POLLACK. 1975. Tumorigenicity of virus-transformed cells in nude mice is correlated specifically with anchorage independent growth in vitro. *Proc. Nat'l Acad. Sci. USA* 72:4435–4439.

STOKER, M. G. P. 1967. Transfer of growth inhibition between normal and virus-transformed cells: Autoradiographic studies using marked cells. *J. Cell Sci.* 2:239–304.

TODARO, G. J. 1963. Quantitative studies of the growth of mouse embryo cells in culture and their development into established lines. *J. Cell Biol.* 17:299–313.

TODARO, G. J., and H. GREEN. 1964. An assay for cellular transformation by SV40. *Virology* 23:117–119.

TODARO, G. J., J. E. DELARCO, and S. COHEN. 1976. Transformation by murine and feline sarcoma viruses specifically blocks binding of epidermal growth factor to cells. *Nature* 264:26–31.

UNKELESS, J. C., A. TOBIA, L. OSSOWSKI, J. P. QUIGLEY, D. B. RIFKIN, and E. REICH. 1973. An enzymatic function associated with transformation of fibroblasts by oncogenic viruses, I: Chick embryo fibroblast cultures transformed by avian RNA tumor viruses. *J. Exp. Med.* 137:85–111.

WEBER, K., E. LAZARIDES, R. D. GOLDMAN, A. VOGEL, and R. POLLACK. 1974. Localization and distribution of actin fibers in normal, transformed and revertant cells. *Cold Spring Harbor Symp. Quant. Biol.* 39:363–369.

Viruses as Agents of Transformation

BALTIMORE, D. 1970. RNA-dependent DNA polymerase in virions of RNA tumour viruses. *Nature* 226:1209–1211.

BIKEL I., X. MONTANO, M. AGHA et al. 1987. SV40 small t antigen enhances the transformation activity of limiting concentrations of SV40 large T antigen. *Cell* 48:321.

BISHOP, J. M., and H. E. VARMUS. 1984. Functions and origins of retroviral transforming genes. In *Molecular Biology of Tumor Viruses: RNA Tumor Viruses*, R. Weiss et al., eds. Cold Spring Harbor Laboratory.

COOK, P. J., and D. P. BURKITT. 1971. Cancer in Africa. *Brit. Med. Bull.* 27:14–20.

DECAPRIO, J. A., J. W. LUDLOW, J. FIGGE, J.-Y. SHEW, C.-M. HUANG, W.-H. LEE, E. MARSILIO, E. PAUCHA, D. M. LIVINGSTON. 1988. SV40 large tumor antigen forms a specific complex with the product of the retinoblastoma susceptibility gene. *Cell* 54:275–283.

DESGROSEILLERS, L., E. RASSART, and P. JOLICOEUR. 1983. Thymotropism of murine leukemia virus is conferred by its long terminal repeat. *Proc. Nat'l Acad. Sci. USA* 80:4203–4207.

DUESBERG, P. H., and P. K. VOGT. 1970. Differences between the ribonucleic acids of transforming and non-transforming avian tumor viruses. *Proc. Nat'l Acad. Sci. USA* 67:1673–1680.

EPSTEIN, M. A., B. G. ACHONG, and Y. M. BARR. 1964. Virus particles in cultured lymphoblasts from Burkitt's lymphoma. Lancet 1:702–703.

GROSS, L., ed. 1970. *Oncogenic Viruses*, 2d ed. Pergamon.

GROSS, L. 1983. *Oncogenic Viruses*, 3d ed. Pergamon.

HARLOW, E., P. WHYTE, B. R. FRANZA, JR., and C. SCHLEY. 1986. Association of adenovirus early-region 1A proteins with cellular polypeptides. *Mol. Cell Biol.* 6:1579–1589.

HUEBNER, R. J., and G. J. TODARO. 1969. Oncogenes of RNA tumor viruses as determinants of cancer. *Proc. Nat'l Acad. Sci. USA* 64:1087–1094.

LURIA, S. E., J. E. DARNELL, JR., D. BALTIMORE, and A. CAMPBELL, eds. 1978. *General Virology*, 3d ed., Wiley.

NUSSE, R., and H. E. VARMUS. 1982. Many tumors induced by the mouse mammary tumor virus contain a provirus integrated in the same region of the host genome. *Cell* 31:99–109.

RASSOULZADEGAN, M., A. COWIE, A. CARR, N. GLAICHENHAUS, R. KAMEN, and F. CUZIN. 1982. The roles of individual polyoma virus early proteins in oncogenic transformation. *Nature* 300:713–718.

RASSOULZADEGAN, M., Z. NAGHASHFAR, A. COWIE, A. CARR, M. GRISONI, R. KAMEN, and F. CUZIN. 1983. Expression of the large T protein of polyoma virus promotes the establishment in culture of "normal" rodent fibroblast cell lines. *Proc. Nat'l Acad. Sci. USA* 80:4354–4358.

ROUS, P. 1911. A sarcoma of the fowl transmissible by an agent separable from the tumor cells. *J. Exp. Med.* 13:397–411.

RULEY, H. E. 1983. Adenovirus early region 1A enables viral and cellular transforming genes to transform primary cells in culture. *Nature* 304:602–606.

STEFFEN, D. 1984. Proviruses are adjacent to c-*myc* in some murine leukemia virus-induced lymphomas. *Proc. Nat'l Acad. Sci. USA* 81:2097–2101.

TAKEYA, T., and H. HANAFUSA. 1983. Structure and sequence of the cellular gene homologous to the RSV *src* gene and the mechanism for generating the transforming virus. *Cell* 32:881–890.

TEMIN, H. 1964. Nature of the provirus of Rous sarcoma. *Nat'l Cancer Inst. Monogr.* 17:557–570.

TEMIN, H., and S. MIZUTANI. 1970. RNA-dependent DNA polymerase in virions of Rous sarcoma virus. *Nature* 226:1211–1213.

TIOLLAIS, P., C. POURCEL, and A. DJEAN. 1985. The hepatitis B virus. *Nature* 317:489.

TOOZE, J., ed. 1980. *DNA Tumor Viruses*, 2d ed. Cold Spring Harbor Laboratory.

WHYTE, P., K. J. BUCHKOVICH, J. M. HOROWITZ, S. H. FRIEND, J. RAYBUCK, R. A. WEINBERG, and E. HARLOW. 1988. Association between an oncogene and an anti-oncogene: the adenovirus E1A proteins bind to the retinoblastoma gene product. *Nature* 334:124–129.

Chemical Carcinogenesis

AMES, B. N. 1979. Identifying environmental chemicals causing mutations and cancer. *Science* 204:587–593.

BROWN, K., M. QUINTANILLA, M. RAMSDEN, I. B. KERR, S. YOUNG, and A. BALMAIN. 1986. v-ras genes from Harvey and Balb murine sarcoma viruses can act as initiators of two-stage mouse skin carcinogenesis. *Cell* 46:447–456.

BROWN, P. C., T. D. TLSTY, and R. T. SCHIMKE. 1983. Enhancement of methotrexate resistance and dihydrofolate reductase gene amplification by treatment of mouse 3T6 cells with hydroxyurea. *Mol. Cell Biol.* 3:1097–1107.

CHEN, T. T., and C. HEIDELBERGER. 1969. Quantitative studies on the malignant transformation of mouse prostate cells by carcinogenic hydrocarbons *in vitro*. *Int. J. Cancer* 4:166–178.

MCCANN, J., and B. N. AMES. 1977. The Salmonella/microsome mutagenicity test: predictive value for animal carcinogenicity. In *Origins of Human Cancer*, H. H. Hiatt, J. D. Watson, and J. A. Winsten, eds. Cold Spring Harbor Laboratory.

MILLER, J. H. 1982. Carcinogens induce targeted mutations in *Escherichia coli*. *Cell* 31:5–7.

OLSSON, M., and T. LINDAHL. 1980. Repair of alkylated DNA in *Escherichia coli*. *J. Biol. Chem.* 255:10569–10571.

POLAND, A., and A. KENDE. 1977. The genetic expression of aryl hydrocarbon hydroxylase activity: evidence for a receptor mutation in nonresponsive mice. In *Origins of Human Cancer*, H. H. Hiatt, J. D. Watson, and J. A. Winsten, eds. Cold Spring Harbor Laboratory.

QUINTANILLA, M., K. BROWN, M. RAMSDEN, and A. BALMAIN. 1986. Carcinogen-specific mutation in mouse skin tumors; *ras* gene involvement in both initiation and progression. *Nature* 322:78–80.

SIMS, P. 1980. The metabolic activation of chemical carcinogens. *Brit. Med. Bull.* 36:11–18.

Radiation and DNA Repair in Carcinogenesis

KENNEDY, A. R., J. CAIRNS, and J. B. LITTLE. 1984. Timing of the steps in transformation of C3H 10T1/2 cells by X-irradiation. *Nature* 307:85–86.

OLSSON, M., and T. LINDAHL. 1980. Repair of alkylated DNA in *Escherichia coli*. *J. Biol. Chem.* 255:10569–10571.

Oncogenes and Their Proteins

BARBACID, M. 1987. *ras* Genes. *Ann. Rev. Biochem.* 56:779–827.

BARGMANN, C. I., M.-C. HUNG, and R. A. WEINBERG. 1986. Multiple independent activations of the *neu* oncogene by a point mutation altering the transmembrane domain of 649. *Cell* 46:185.

COLLETT, M. S., and R. L. ERIKSON. 1978. Protein kinase activity associated with the avian sarcoma virus *src* gene product. *Proc. Nat'l Acad. Sci. USA* 75:2021–2024.

DOOLITTLE, R. F., M. W. HUNKAPILLER, L. E. HOOD, S. G. DEVARE, K. C. ROBBINS, S. A. AARONSON, and H. N. ANTONIADES. 1983. Simian sarcoma virus *onc* gene, v-*sis*, is derived from the gene (or genes) encoding a platelet-derived growth factor. *Science* 221:275–276.

DOWNWARD, J., Y. YARDEN, E. MAYES, G. SCRACE, N. TOTTY, P. STOCKWELL, A. ULLRICH, J. SCHLESSINGER, and M. WATERFIELD. 1984. Close similarity of epidermal growth factor receptor and v-*erb*-B oncogene protein sequences. *Nature* 307:521–527.

GRAF, T., F. WEIZSÄCHER, S. GREISER, J. COLL, D. STEHELIN, T. PATSCHINSKY, K. BISTER, C. BECHADE, G. CALOTHY, and A. LEUTZ. 1986. V-*mil* induces autocrine growth and enhanced tumorigenicity in v-*myc* transformed avian macrophages. *Cell* 45:357–364.

HUNTER, T. 1987. A thousand and one protein kinases. *Cell* 50:823.

HUNTER, T., and B. M. SEFTON. 1980. Transforming gene product of Rous sarcoma virus phosphorylates tyrosine. *Proc. Nat'l Acad. Sci. USA* 77:1311–1315.

KAMPS, M. M., J. E. BUSS, and B. M. SEFTON. 1985. Mutation of NH_2-terminal glycine of p60^{src} prevents both myristoylation and morphological transformation *Proc. Nat'l Acad. Sci. USA* 82:4625.

LEVINSON, A. D., H. OPPERMANN, L. LEVINTOW, H. E. VARMUS, and J. M. BISHOP. 1978. Evidence that the transforming gene of avian sarcoma virus encodes a protein kinase associated with a phosphoprotein. *Cell* 15:561–572.

PURCHIO, A. F., E. ERIKSON, J. S. BRUGGE, and R. L. ERIKSON. 1978. Identification of a polypeptide encoded by the avian sarcoma virus *src* gene. *Proc. Nat'l Acad. Sci. USA* 75:1567–1571.

STEHELIN, D., H. E. VARMUS, J. M. BISHOP, and P. K. VOGT. 1976. DNA related to the transforming gene(s) of avian sarcoma viruses is present in normal avian DNA. *Nature* 260:170–173.

STEWART, T. A., P. K. PATTENGALE, and P. LEDER. 1984. Spontaneous mammary adenocarcinomas in transgenic mice that carry and express MTV/*myc* fusion genes. *Cell* 38:627–637.

TRAHEY, M., and F. MCCORMICK. 1987. A cytoplasmic protein stimulates normal n-*ras* p21 GTPase but does not affect oncogenic mutants. *Science* 238:542–545.

USHIRO, H., and S. COHEN. 1980. Identification of phosphotyrosine as a product of epidermal growth factor–activated protein kinase in A431 cell membranes. *J. Biol. Chem.* 255:8363–8365.

VARMUS, H., and A. J. LEVINE, eds. 1983. *Readings in Tumor Virology*. Cold Spring Harbor Laboratory.

WANG, J. Y. J. 1983. From c-*abl* to v-*abl*. *Nature* 304:400.

WANG, J. Y. J., C. QUEEN, and D. BALTIMORE. 1982. Expression of an Abelson murine leukemia virus-encoded protein in *Escherichia coli* causes extensive phosphorylation of tyrosine residues. *J. Biol. Chem.* 257:13181–13184.

WITTE, O. N., A. DASGUPTA, and D. BALTIMORE. 1980. The Abelson murine leukemia virus protein is phosphorylated in vitro to form phosphotyrosine. *Nature* 283:826–831.

Cellular Oncogenes

ALITALO, K., M. SCHWAB, C. C. LIN, H. E. VARMUS, and J. M. BISHOP. 1983. Homogeneously staining chromosomal regions contain amplified copies of an abundantly expressed cellular oncogene (c-*myc*) in malignant neuroendocrine cells from a human colon carcinoma. *Proc. Nat'l Acad. Sci. USA* **80**:1707–1711.

BALMAIN, A., and I. B. PRAGNELL. 1983. Mouse skin carcinomas induced *in vivo* by chemical carcinogens have a transforming Harvey-*ras* oncogene. *Nature* **303**:72–74.

BISHOP, J. M. 1983. Cellular oncogenes and retroviruses. *Ann. Rev. Biochem.* **52**:301–354.

BRODEUR, G., C. SEEGER, M. SCHWAB, H. E. VARMUS, and J. M. BISHOP. 1984. Amplification of N-*myc* in untreated human neuroblastomas correlates with advanced disease stage. *Science* **224**:1121–1124.

CANAANI, E., O. DREAZEN, A. KLAR, G. RECHAVI, D. RAM, J. B. COHEN, and D. GIVOL. 1983. Activation of the c-*mos* oncogene in a mouse plasmacytoma by insertion of an endogenous intracisternal A-particle genome. *Proc. Nat'l Acad. Sci. USA* **80**:7118–7122.

COWELL, J. K. 1982. Double minutes and homogeneously staining regions: gene amplification in mammalian cells. *Ann. Rev. Genet.* **16**:21–59.

CROCE, C. 1987. Role of chromosomal translocations in human neoplasia. *Cell* **49**:155.

CROCE, C. M., J. ERIKSON, A. AR-RUSHDI, D. ADEN, and K. NISHIKURA. 1984. Translocated c-*myc* oncogene of Burkitt lymphoma is transcribed in plasma cells and repressed in lymphoblastoid cells. *Proc. Nat'l Acad. Sci. USA* **81**:3170–3174.

DEKLEIN, A., A. GUERTS VAN KESSEL, G. GROSVELD, C. R. BARTRAM, A. HAGEMEIJER, D. BOOTSMA, N. K. SPURR, N. HEISTERKAMP, J. GROFFEN, and J. R. STEPHENSON. 1982. A cellular oncogene is translocated to the Philadelphia chromosome in chronic myelocytic leukaemia. *Nature* **300**:765–767.

ERIKSON, J., J. FINAN, P. C. NOWELL, and C. M. CROCE. 1982. Translocation of immunoglobulin V_H genes in Burkitt lymphoma. *Proc. Nat'l Acad. Sci. USA* **79**:5611–5615.

FRIEND, S. H., R. BERNARDS, S. ROGELJ, R. A. WEINBERG, J. M. RAPAPORT, D. M. ALBERT, and T. P. DRYJA. 1986. A human DNA segment with properties of the gene that predisposes to retinoblastoma and osteosarcoma. *Nature* **323**:643–646.

GERONDAKIS, S., S. CORY, and J. M. ADAMS. 1984. Translocation of the *myc* cellular oncogene to the immunoglobulin heavy chain locus in murine plasmacytomas is an imprecise reciprocal exchange. *Cell* **36**:973–982.

HAYWARD, W. S., B. G. NEEL, and S. M. ASTRIN. 1981. Activation of a cellular *onc* gene by promoter insertion in ALV-induced lymphoid leukosis. *Nature* **209**:475–479.

HEISTERKAMP, N., J. R. STEPHENSON, J. GROFFEN, P. F. HANSEN, A. DEKLEIN, C. R. BARTRAM, and G. GROSVELD. 1983. Localization of the c-*abl* oncogene adjacent to a translocation breakpoint in chronic myelocytic leukaemia. *Nature* **306**:239–242.

HELDIN, C.-H., and B. WESTERMARK. 1984. Growth factors: mechanisms of action and relation to oncogenes. *Cell* **37**:9–20.

HERMANS A., N. HEISTERKAMP, M. VON LINDERN et al. 1987. Unique fusion of *bcr* and c-*abl* genes in Philadelphia chromosome positive acute lymphoblastic leukemia *Cell* **51**:33.

HOFFMAN-FALK, H., P. EINAT, B. Z. SHILO, and F. M. HOFFMAN. 1983. *Drosophila melanogaster* DNA clones homologous to vertebrate oncogenes: evidence for a common ancestor to *src* and *abl* cellular genes. *Cell* **32**:589–598.

HUANG, H. J., J.-K. YEE, J.-Y. SHEW, P.-L. CHEN, R. BOOKSTEIN, T. FRIEDMANN, Y.-H. P. LEE, and W.-H. LEE. 1988. Suppression of the neoplastic phenotype by replacement of the Rb gene in human cancer cells. *Science* **242**:1563–1566.

KLEIN, G. 1987. The approaching era of the tumor suppressor genes. *Nature* **238**:1539–1545.

KRONTIRIS, T., and G. M. COOPER. 1981. Transforming activity of human tumor DNAs. *Proc. Nat'l Acad. Sci. USA* **78**:1181–1184.

LAND, H., L. F. PARADA, and R. A. WEINBERG. 1983. Tumorigenic conversion of primary embryo fibroblasts requires at least two cooperating oncogenes. *Nature* **304**:596–601.

PARADA, L. F., C. J. TABIN, C. SHIH, and R. A. WEINBERG. 1982. Human Ej bladder carcinoma oncogene is a homologue of Harvey sarcoma virus *ras* gene. *Nature* **297**:474–478.

POWERS, S., T. KATAOKA, O. FASANO, M. GOLDFARB, J. STRATHERN, J. BROACH, and M. WIGLER. 1984. Genes in *S. cerevisiae* encoding proteins with domains homologous to the mammalian *ras* proteins. *Cell* **36**:607–612.

ROWLEY, J. D. 1983. Human oncogene locations and chromosome aberrations. *Nature* **301**:290–291.

SCHWAB, M., J. ELLISON, M. BUSCH, W. ROSENAU, H. E. VARMUS, and J. M. BISHOP. 1984. Enhanced expression of the human gene N-*myc* consequent to amplification of DNA may contribute to malignant progression of neuroblastoma. *Proc. Nat'l Acad. Sci. USA* **81**:4940–4944.

SHIH, C., B.-Z. SHILO, M. P. GOLDFARB, A. DANNENBERG, and R. A. WEINBERG. 1979. Passage of phenotypes of chemically transformed cells via transfection of DNA and chromatin. *Proc. Nat'l Acad. Sci. USA* **76**:5714–5718.

STEWART, T. A., P. K. PATTENGALE, and P. LEDER. 1984. Spontaneous mammary adenocarcinomas in transgenic mice that carry and express MTV/*myc* fusion genes. *Cell* **38**:627–637.

SUKUMAR, S., V. NOTARIO, D. MARTIN-ZANCA, and M. BARBACID. 1983. Induction of mammary carcinomas in rats by nitroso-methylurea involves malignant activation of H-*ras*-1 locus by single point mutations. *Nature* **306**:658–661.

TABIN, C. J., S. M. BRADLEY, C. I. BARGMANN, R. A. WEINBERG, A. G. PAPAGEORGE, E. M. SCOLNICK, R. DHAR, D. R. LOWY, and E. H. CHANG. 1982. Mechanism of activation of a human oncogene. *Nature* **300**:143–149.

VANDE WOUDE, G. F., A. J. LEVINE, W. C. TOPP, and J. D. WATSON, eds. 1984. *Cancer Cells*, vol. 2: *Oncogenes and Viral Genes*. Cold Spring Harbor Laboratory.

VARMUS, H. E. 1984. The molecular genetics of cellular oncogenes. *Ann. Rev. Genet.* **18**:553–612.

WATERFIELD, M. D., G. T. SCRACE, N. WHITTLE, P. STROOBANT, A. JOHNSSON, A. WATESON, B. WESTERMARK, C. HELDIN, J. S. HUANG, and T. F. DEUEL. 1983. Platelet-derived growth factor is structurally related to the putative transforming protein p28^{sis} of simian sarcoma virus. *Nature* **304**:35–39.

WEINBERG, R. A. 1982. Oncogenes of spontaneous and chemically induced tumors. *Adv. Cancer Res.* **36**:149–164.

WEINBERG, R. A. 1983. A molecular basis of cancer. *Sci. Am.* **249**(5):126–143.

YUNIS, J. J. 1983. The chromosomal basis of human neoplasia. *Science* **221**:227–235.

Promoters of Carcinogenesis

BERENBLUM, I. 1974. *Carcinogenesis as a Biological Problem.* Elsevier/North-Holland.

DAVIS, R. J., and M. P. CZECH. 1985. Tumor-promoting phorbol diesters cause the phosphorylation of epidermal growth factor receptors in normal human fibroblasts at threonine-654. *Proc. Nat'l Acad. Sci. USA* **82**:1974–1978.

DELCLOS, K. B., D. S. NAGLE, and P. M. BLUMBERG. 1980. Specific binding of phorbol ester tumor promoters to mouse skin. *Cell* **19**:1025–1032.

EBELING, J. G., G. R. VANDENBARK, L. J. KUHN, B. R. GANONG, R. M. BELL, and J. E. NIEDEL. 1985. Diacylglycerols mimic phorbol diester induction of leukemic cell differentiation. *Proc. Nat'l Acad. Sci. USA* **82**:815–819.

KIKKAWA, U., Y. TAKAI, Y. TANAKA, R. MIYAKE, and Y. NISHIZUKA. 1983. Protein kinase C as a possible receptor protein of tumor-promoting phorbol esters. *J. Biol. Chem.* **258**:11442–11445.

WEINSTEIN, I. B. 1983. Protein kinase, phospholipid and control of growth. *Nature* **302**:750.

Human Cancer

BAKHSHI, A., J. MINOWADA, A. ARNOLD, J. COSSMAN, J. P. JENSEN, J. WHANG-PENG, T. A. WALDMANN, and S. J. KORSMEYER. 1983. Lymphoid blast crises of chronic myelogenous leukemia represent stages in the development of B-cell precursors. *N. Engl. J. Med.* **309**:826–831.

BOS, J. L., E. R. FEARON, S. R. HAMILTON, M. VERLAAN–DE VRIES, J. H. VAN BOOM, A. J. VAN DER EB, and B. VOGELSTEIN. 1987. Prevalence of *ras* gene mutations in human colorectal cancers. *Nature* **327**:293–297.

CAVENEE, W. K., T. P. DRYJA, R. A. PHILLIPS, W. F. BENEDICT, R. GODBOUT, B. L. GALLIE, A. L. MURPHREE, L. C. STRONG, and R. L. WHITE. 1983. Expression of recessive alleles by chromosomal mechanisms in retinoblastoma. *Nature* **305**:779–784.

DOLL, R., and R. PETO, eds. 1981. *The Causes of Cancer.* Oxford University Press.

FRIEND, S. H., R. BERNARDS, S. ROGELJ et al. 1986. A human DNA segment with properties of the gene that disposes to retinoblastoma and osteosarcoma. *Nature* **323**:643.

HANSEN, M. F., and W. K. CAVENEE. 1987. Genetics of cancer predisposition. *Cancer Res.* **47**:5518–5527.

KNUDSON, A. G., JR. 1977. Genetic predisposition to cancer. In *Origins of Human Cancer,* H. H. Hiatt, J. D. Watson, and J. A. Winsten, eds. Cold Spring Harbor Laboratory.

KRONTIRIS, T. G., N. A. DIMARTINO, M. COLB, and D. R. PARKINSON. 1985. Unique allelic restriction fragments of the human Ha-*ras* locus in leukocyte and tumor DNAs of cancer patients. *Nature* **313**:369–374.

SLAMON, D. J., G. M. CLARK, S. G. WONG, W. J. LEVIN, A. ULLRICH, and W. L. MCGUIRE. 1987. Human breast cancer: correlation of relapse and survival with amplification of HER/*neu* oncogene. *Science* **235**:177–182.

VOGELSTEIN, B., E. R. FEARON, S. R. HAMILTON, S. E. KERN, A. C. PREISINGER, M. LEPPERT, Y. NAKAMURA, R. WHITE, A. M. M. SMITS, and J. L. BOS. 1988. Genetic alternations during colorectal-tumor development. *N. Engl. J. Med.* **319**:525–532.

YOSHIDA, M., I. MIYOSHI, and Y. HINUMA. 1982. Isolation and characterization of retrovirus from cell lines of human adult T-cell leukemia and its implication in the disease. *Proc. Nat'l Acad. Sci. USA* **79**:2031–2035.

Diet, Nutrition and Cancer. 1982. Committee on Diet, Nutrition and Cancer, Assembly of Life Sciences, National Research Council, National Academy Press.

Immunity

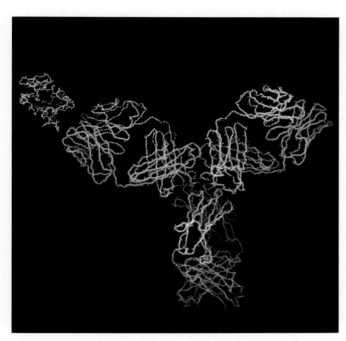

Antigen (*upper left*) binding to an antibody molecule

The life of every organism is constantly threatened by other organisms—this is the nature of the living world. In response, each species has evolved protective mechanisms, varying from camouflage colors, to poisons, to effective running muscles. From their continual battle with microorganisms, vertebrates have evolved an elaborate set of protective measures called, collectively, the immune system. *The word immune (Latin immunis, meaning "exempt") implies freedom from a burden: an animal that is immune to a specific infecting agent will remain free of infection by that agent. The study of the immune system constitutes the discipline of immunology.*

The immune system works by a learning process. Our first encounter with a bacterial, fungal, or viral pathogen leads to an infection often accompanied by disease symptoms. The immune system aids in recovery from the infection, and after recovery, we usually remain free of that disease forever. Our immune system has learned to recognize that specific bacterium or fungus or virus as a foreign infecting agent; should it attack us again, it will be rapidly killed.

The key process carried out by the immune system is recognition. The system must recognize the presence of an invader. It must also be able to discriminate between foreign invaders and the natural constituents of the body: we call this discrimination between self and nonself. *The importance of this is underscored by disorders in which such discrimination fails. Such disorders are known as the* autoimmune diseases, *and the most severe can be fatal. Recognition of a foreign invader is only the first step of the immune system's attack; it must be followed by steps that will kill and eliminate the invader. Thus the immune system carries out two types of activities: recognition processes directed against individual, discrete aspects of a target, and generalized responses that follow from the recognition and allow the system to mount an attack against the invader. The generalized responses are called* effector *functions.*

We believe that the immune system evolved to handle pathogens, *invading microorganisms that can cause disease, but it has other properties as well. It can kill cancer cells and, in experimental animals at least, it can protect the body against certain tumors. How active the immune system is against human tumors is a debated question. The immune system also prevents tissue transplantation between individuals. Every vertebrate individual has a unique set of molecules on the surface of its cells. Its immune system recognizes those cells as self, but grafted cells from other individuals, even of the same species, are seen as foreign and are killed, or* rejected. *Inhibition of this rejection reaction is necessary whenever a surgeon attempts to graft a patch of skin or transplant a kidney from one person to another. Without special treatment, the only transplants that are accepted without a rejection reaction are those from an identical twin.*

If we are to understand the immune system, then, we must understand some fundamental processes. How does the system recognize the individuality of a specific pathogen? How does it discriminate foreign matter from endogenous material, nonself from self? How does it translate recognition of a foreign invader into a killing reaction? How does it learn, so that the second attack by an invader is repulsed so much faster than the first? The answers to these questions are not fully known. While we understand, for instance, a great deal about how recognition develops, we know less about how the system avoids reacting with self-components.

This chapter concentrates on aspects of the immune system that are reasonably well known. Especially highlighted are the molecular biological mechanisms that underlie the recognition process because they have a unique aspect—the reorganization of DNA as a mechanism in cellular differentiation. An overview of the system is followed by a discussion of how antibodies, the best-understood molecules of the immune system, develop their specificity. We then turn to the T lymphocytes and describe their particular properties, after which we consider the interactions of lymphocytes in generating an immune response. ▲

Overview

The immune system works in three fundamentally different ways, by *humoral immunity*, by *cellular immunity*, and by secretion of proteins, called *lymphokines*, that stimulate immune responses (Figure 25-1). The term humor refers to a fluid, and humoral immunity relies on molecules in solution in the body. These molecules are proteins collectively called immunoglobulin and abbreviated Ig. They constitute 20 percent of the proteins in the blood. A single such molecule is called an *antibody*, but "antibody" is also used to mean many individually different molecules all directed against the same target molecule.

In cellular immunity intact cells are responsible for the recognition reaction. Freely soluble molecules generally do not participate. Instead, molecules held tightly to the surfaces of the reacting immune cells act like immunoglobulins but never break their association with the plasma membrane.

Secretion of lymphokines releases many proteins that help trigger the immune response. They stimulate humoral immunity and cellular immunity as well as the production of *phagocytes*, cells that carry out killing reactions and that ingest and digest both pathogens and the debris of dead cells.

The injection of almost any foreign macromolecule into an animal elicits the formation of both antibodies and immune cells that can bind to the substance. A substance that provokes antibody or immune-cell formation, or is recognized by an antibody or immune cell, is called an *antigen*.

Antibodies Bind to Determinants and Have Two Functional Domains

The unique function performed by an antibody is high-affinity binding to an antigen. Hydrophobic forces, ionic forces, and van der Waals forces bind antibodies and antigens, but covalent bonds are not formed between them. The site on an antigen at which a given antibody molecule binds is called a *determinant* (Figure 25-2a). The surface of a protein molecule presents many determinants to which antibodies might bind. Some antigens have a repeating structure, with the same determinant on them many times. Such an antigen is said to be *multivalent*. A good example of a multivalent antigen is a virus particle (Figure 25-2b).

By attaching a small molecule onto the surface of a protein, a new determinant can be created. If, for instance, bovine serum albumin, a fairly big protein, is coupled to the simple molecule, dinitrophenol, and the conjugate is injected into a mouse, the mouse will make antibody molecules that bind to the dinitrophenol group. This mechanism can be demonstrated by coupling dini-

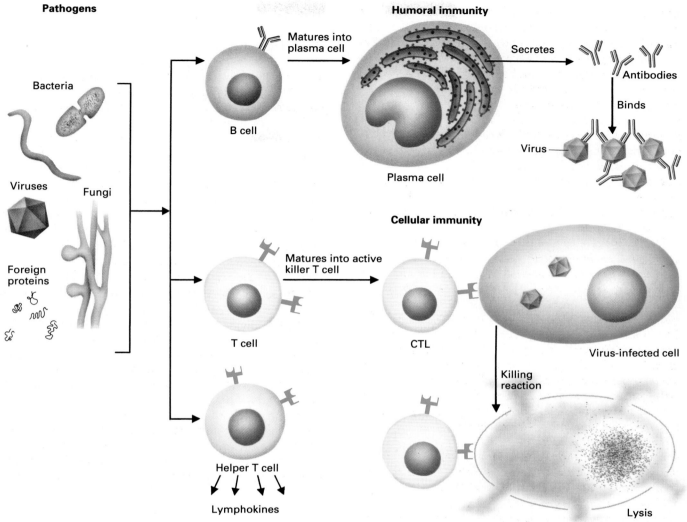

▲ **Figure 25-1** Humoral immunity and cellular immunity. When a pathogen invades the body, the immune system responds with three types of reaction. Cells of the humoral system (B cells) secrete antibodies that can bind to the pathogen. Cells of the cellular system (T cells) carry out two major types of functions. One type of T cell develops the ability to kill pathogen-infected cells. Helper T cells, responding to the pathogen, secrete protein factors (lymphokines) that stimulate the body's overall response.

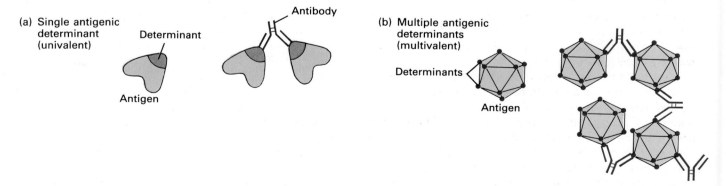

▲ **Figure 25-2** Antibodies bind to single or multiple determinants on an antigen. (a) Binding of antibody to a protein with a single determinant. The antibody's two sites can both be filled but the complex cannot grow further. (b) Some antigens, notably viruses, have multiple identical determinants on a single particle. Antibodies can form large aggregates when reacting with such antigens.

(a)

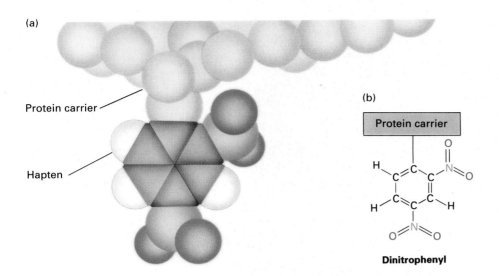

Protein carrier

Hapten

(b)

Protein carrier

Dinitrophenyl

◀ **Figure 25-3** Hapten linked to a protein. The hapten shown is dinitrophenyl, one commonly used in experimental immunology (see also page 66). (a) The space-filling representation shows that the hapten's varying surface presents many different determinants to which antibodies can bind. (b) Chemical formula of dinitrophenyl. *From* The Structure and Function of Antibodies, *by G. M. Edelman. Copyright 1970 vol. 223(2) by Scientific American, Inc.*

trophenol to another protein and showing that this second protein now reacts with antibodies made to the original dinitrophenol-conjugated protein. In this context, dinitrophenol, or any small substituent on a protein that can elicit an antibody response, is called a *hapten* (Figure 25-3).

A key property of antibodies, which is best probed by haptens, is *specificity*. An antibody that binds well to a given hapten may bind poorly to a similar hapten. For instance, many antidinitrophenol antibodies bind poorly to trinitrophenol. Simple changes in short peptides that function as haptens show the specificity of antibodies: a peptide of 7–15 residues, linked to a protein, can elicit antibodies that bind well to the peptide but poorly to derivatives of the peptide having even one amino acid changed.

We noted before that the immune system carries out two processes: recognition and effector functions. The antibody is divided into two regions for these functions (Figure 25-4): *binding domains* that interact with the antigen and *effector domains* that signal the initiation of processes (such as phagocytosis) to rid the body of the antigen bound to the antibody. The effector region also provides signals that distribute antibodies to various body fluids. Some types of antibodies are directed to secretions, such as saliva, mucus, or milk; others go across the placenta to protect the fetus.

A specific set of effector domains is common to many antibodies, but a given binding domain is found only on a very small set of antibodies. This dichotomy of a highly specific region and a common region in antibody molecules has its counterpart in the genes that encode antibodies. These genes are constructed from multiple variable regions which can be affixed to a very small number of constant regions.

Each antibody molecule consists of two classes of polypeptide chains, *light* (L) *chains* and *heavy* (H) *chains*. A single antibody molecule has two identical copies of the L chain and two of the H chain. Therefore the basic antibody structure is a four-chain molecule (Figure 25-4). The N-terminal ends of one H chain and L chain together form one binding domain. The antibody is thus bivalent; it has two binding sites for antigen.

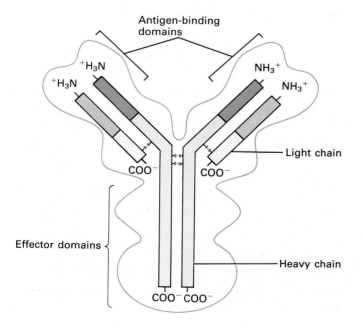

Antigen-binding domains

^+H_3N

^+H_3N

NH_3^+

NH_3^+

Light chain

COO$^-$

COO$^-$

Effector domains

Heavy chain

COO$^-$ COO$^-$

◀ **Figure 25-4** Schematic representation of an antibody molecule. Antigen binding occurs at the two upper ends of the Y-shaped molecule, where the H and L chains are in intimate contact. The lower portion of the Y contains the effector region that is made only from heavy chains. The N-terminal (NH_3^+) and C-terminal (COO$^-$) ends of the chains are indicated. Disulfide bridges (S—S) link the chains (see also Figure 2-27).

Antibody Reaction with Antigen Is Reversible

The binding of a site on an antibody to its antigen is a simple bimolecular, reversible reaction capable of analysis by standard kinetic theory. Looking only at a single binding site we write

$$Ag + Ab \underset{k_2}{\overset{k_1}{\rightleftharpoons}} Ag\text{-}Ab$$

where Ag is the antigen, Ab is the uncomplexed antibody, and Ag-Ab the bound complex. The forward and reverse binding reactions are characterized by k_1 and k_2 rate constants, respectively. The *affinity* of the antibody binding site is measured by the ratio of complexed to free reactants at equilibrium. An *affinity constant, K,* is defined by

$$K = \frac{[Ag\text{-}Ab]}{[Ag][Ab]} = \frac{k_1}{k_2}$$

where the brackets denote concentration. Typical values of K are 10^5–10^{11} liters/mole. K can be determined by measuring the concentration of free antigen needed to fill half the binding sites of an antibody (Figure 25-5). At this point [Ag-Ab] must equal [Ab], and the above formula resolves to

$$K = \frac{1}{[Ag]}$$

Many proteins may have weak affinities for other proteins; consequently we need to define an affinity below which binding is considered to be nonspecific and above which binding is considered to be a specific antibody-antigen reaction. From experience, any K less than 10^4 liters/mole is considered nonspecific.

But antibody-antigen reactions are not necessarily the simple bimolecular events treated in the above equation. Such a treatment is appropriate for an antigen with one potential site for reaction and for a homogeneous antibody population, even though there are two binding sites per antibody molecule. If, however, binding to one site blocks a reaction with the second site, or if the antigen has two appropriately spaced identical sites so that both antibody sites can react with the same antigen molecule, then ideality is lost and more complicated equations are needed. For instance, if the two antibody sites can react with the same antigen molecule, the second site will react much faster than the first because the antigen molecule will already be very close by and the apparent affinity will be greatly increased, usually by at least 10^4. A further complication is that natural antibody populations can react with a multitude of sites on antigens, with a variety of affinities. Thus the term *avidity* is used to describe the apparent affinity of an antibody population reacting with an antigen. Avidity is measured by half-saturation of antibody, but we must recognize that this measure has no simple mathematical meaning.

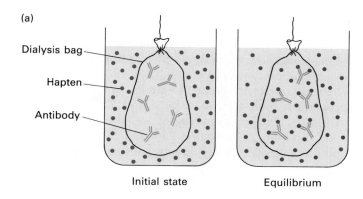

(a)

Dialysis bag

Hapten

Antibody

Initial state   Equilibrium

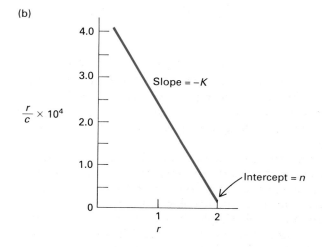

(b)

Scatchard equation $\dfrac{r}{c} = Kn - Kr$

where $r = \dfrac{\text{Moles of hapten bound to antibody}}{\text{Moles of antibody in bag}}$

c = Concentration of hapten in bulk solution

n = Number of binding sites on antibody molecule

K = Equilibrium association constant of antibody

▲ **Figure 25-5** Determining the affinity of an antibody for a hapten. (a) Equilibrium dialysis. A large volume of a solution of hapten is equilibrated through a semipermeable dialysis membrane with a solution of antibody that can bind to the hapten. (b) Typical data are plotted in a fashion that allows application of the Scatchard binding equation. In the experiment shown, dinitrophenol hapten bound to lysine was used, and a monoclonal antidinitrophenol antibody from a myeloma was placed inside the dialysis bag. The Scatchard equation gives two values, a number of binding sites per antibody molecule (*n*) and the equilibrium binding constant (*K*). The number of binding sites per molecule is given by the intercept on the *x* axis. It is 2, as expected, because of the two binding arms on an antibody molecule. [See H. N. Eisen, 1980, *Immunology,* 2d ed., Harper & Row, pp. 299–301.]

Antibodies Come in Many Classes

Because antibodies have to be distributed to various parts of the body and serve multiple effector functions, animals make multiple *classes* of antibodies. Each class has a different set of effector domains with its own specific set of properties. There are five major classes and a number of subclasses. The important classes are IgM, IgD, IgG, IgE, and IgA (Table 25-1; Figures 25-4 and 25-6). The differences among them derive from the different H chains of which they are constituted.

When an animal begins its response to antigenic challenge (for instance, following inoculation of an antigen), the first antibody class it makes is IgM. The IgM molecule

is a pentamer of the basic four-chain antibody units; the units are held together by disulfide bonds and by a single copy of a polypeptide known as *J chain* (Figure 25-6). Because of its multiple, closely spaced binding sites, IgM has a very high binding avidity for microorganisms, such as viruses, that are covered with identical subunits. One major effector function of IgM is activation of the *complement system,* a group of proteins able to kill cells to which antibody is bound. IgM also activates *macrophages,* phagocytic cells that are specialized for ingesting and killing bacteria.

The IgD molecule, a monomer, is an enigma. Many cells have IgD on their surface but very little is found in bodily fluids and no special function is yet known for it.

IgG is the main serum antibody. A four-chain monomer, it is made in copious amounts in response to an antigenic challenge, especially after multiple stimulations with antigen. Like IgM, it stimulates both complement and macrophages. IgG antibodies can pass through the placenta to the fetus. Their effector regions are specialized for binding to a receptor on the maternal side of placental cells; the receptor then carries the IgG through the cell into the fetal circulation.

IgE is one of the most bothersome types of antibody. It is found mainly in tissues, where, in complex with an antigen, it activates the release of histamines from specialized, blood-derived cells called *mast cells.* Histamine release is the cause of such allergic reactions as hives, asthma, and hay fever. The positive role of IgE is thought to be as a defense against parasitic infections.

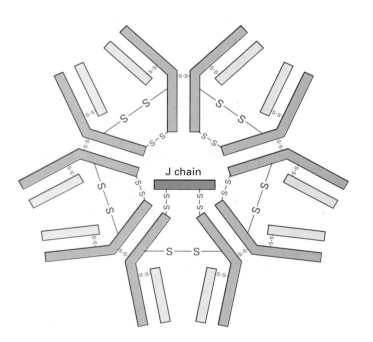

◂ **Figure 25-6** Schematic representation of an IgM molecule. The antibodies of most classes are formed of only four chains as depicted in Figure 25-4. The IgM class is a pentamer whose polymerization is initiated by an ancillary protein called the J chain. Holding the various polypeptides together are disulfide (S—S) bonds.

Table 25-1   Major antibody classes

Properties	IgM	IgD	IgG	IgE	IgA
Heavy chain	μ	δ	γ	ϵ	α
Mean human adult serum level (mg/ml)	1	0.03	12	0.0003	2
Half-life in serum (days)	5	3	25	2	6
Number of four-chain monomers	5	1	1	1	1, 2, or 3
Special properties	Early appearance; fixes complement; activates macrophages	Found mainly on cell surfaces; traces in serum	Activates complement; crosses placenta; binds to macrophages and granulocytes	Stimulates mast cells to release histamines; found mainly in tissues	Found mainly in secretions

IgA, the final major class of antibody, is perhaps the most important. It exists as a monomer and is also polymerized into a J-chain-linked dimer or trimer. Polymerized IgA binds to a receptor on the blood-facing surface of many epithelia and is transported through the epithelial cells into whatever compartment the epithelial layer surrounds, for instance, into the intestinal lumen (Figure 25-7). During the passage, most of the receptor is cut away from its membrane-binding region and remains bound to the antibody. When IgA emerges from the epithelial cells, a piece of the receptor, called the *secretory component,* is still bound to it. The IgA receptor will also bind to IgM; persons unable to make IgA will secrete IgM, which also protects them. Secreted IgA (or IgM) is present in saliva, tears, the lungs, and the intestines as the body's first line of defense against infection by microorganisms.

Antibodies Are Made by B Lymphocytes

Antibodies are made by many cell types, all of a single cell lineage: the B *lymphocytes,* or B *cells.* (They are designated "B" because in birds, where they were first studied, most of the cells mature in an outpocketing at the end of the cloaca called the *bursa of Fabricius.* This organ has no counterpart in mammals, but most mammalian B lymphocytes originate from the bone marrow so the "B" remains appropriate.) B lymphocytes arise as cells having IgM or IgM plus IgD on their surfaces. Reaction of the surface-bound antibody with an antigen will activate the B lymphocyte to mature into a cell type known as a *plasma cell* that is highly specialized for antibody production and secretion (see Figure 25-1).

The Immune System Has Extraordinary Plasticity

Many haptens, such as the dinitrophenol unit described earlier (see Figure 25-3), are the products of synthetic chemists. During the long evolution of vertebrates, no animal outside of a laboratory ever encountered a dinitrophenylated protein. Yet at the first encounter with such a protein, most vertebrates produce dinitrophenol-specific antibodies. This response is the most remarkable aspect of immune function—the ability to recognize the whole universe of potential determinants whether the species has encountered them at any time in evolution or not. The plasticity of the immune system—its ability to extend itself to meet unprecedented challenges—makes the system's molecular events especially intriguing. All the other body systems, except the nervous system, have evolved their properties in response to specific environmental pressures encountered by their ancestors, that is, by natural selection. Although the immune system itself is certainly a product of evolution, not all of the genes en-

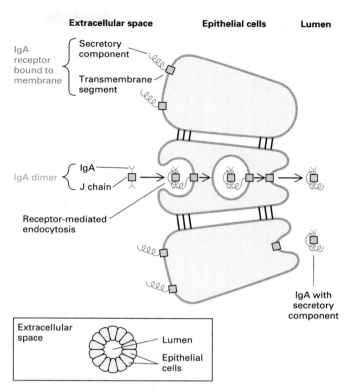

▲ **Figure 25-7** Transfer of IgA across epithelial cells by the IgA receptor. IgA molecules are shown being transported across epithelial cells from the blood-facing surface into a lumen (for instance, from the intestinal lining into the intestinal space). The receptor that carries out the transport is found on the blood-facing surface of the intestinal cell. There it binds to an IgA dimer and is carried into the cell by endocytosis. Rather than staying in the cell, the fate of most transported molecules, the IgA receptor complex with IgA is carried through the cell to the luminal face of the cell, where the IgA separates from the cell. This separation is achieved by a cleavage of the IgA receptor, leaving the "secretory component" bound to the IgA that is released into the lumen. [See K. E. Mostov, M. Friedlander, and G. Blobel, 1984, *Nature* **308**:37.]

coding individual antibody proteins could have been selected during evolution because not all of the determinants recognized by antibodies could have been previously encountered. The genetic basis for this extraordinary plasticity of protein structure is now fairly well understood and will be described later in this chapter. At this point, however, we must ask: by what cellular strategy is this plasticity achieved?

When the plasticity of the immune system was first appreciated, many suggestions were proposed concerning how the system might operate. Most of the proposals were based on an approach called *instructive theory:* it was imagined that all antibodies had one polypeptide structure that could be induced to fold in different ways by combining with antigen as a structural template; the binding specificity of the antibody would then arise from a previous encounter with the antigen (Figure 25-8). This notion has now been disproved by the observation that antibodies can be denatured and renatured in the absence of antigen and still retain their specificity. Thus the one-dimensional sequence of the polypeptides must already contain the information for the binding site. Furthermore, it was shown that antigen-specific antibodies can be detected on lymphocyte surfaces before the specific antigenic stimulus is initially presented to the animal.

Even before instructive theory had been disproved, a competing concept based on preexisting antigen-specific antibodies was developed. It provided a consistent explanation for the cellular events that produce specificity and plasticity. This latter theory is now the basis of all thinking about immunology.

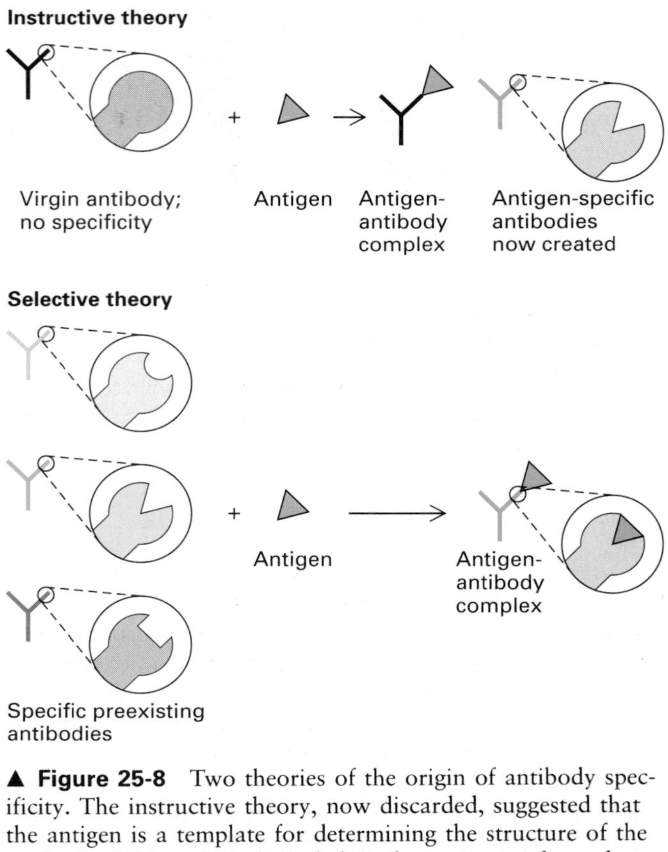

Instructive theory

Virgin antibody; no specificity

Antigen — Antigen-antibody complex

Antigen-specific antibodies now created

Selective theory

+ Antigen

Antigen-antibody complex

Specific preexisting antibodies

▲ **Figure 25-8** Two theories of the origin of antibody specificity. The instructive theory, now discarded, suggested that the antigen is a template for determining the structure of the antibody. With today's knowledge of genetic encoding of protein structure, such a notion is hard to imagine, but before the 1970s so much less was known about the determination of protein structure that template formation of antibodies was considered possible. Selective theory, now considered the only likely approach, suggests that the immune system generates a very large number of specific molecules and that antigens are bound to the ones that fit.

Clonal Selection Theory Underlies All Modern Immunology

The class of theories that replaced the instructive models is called *selective theory* because an antigen is imagined to select the appropriate antibody from a preexisting pool of antibody molecules (Figure 25-8). The notion of selection was brought into the field by Niels Jerne, a brilliant theorist and experimentalist of immunology, and was then incorporated by Sir MacFarlane Burnet into one of the great intellectual constructs of modern biology, the *clonal selection theory* of antibody production (Figure 25-9). This theory supposes that (1) the body is continually elaborating B lymphocytes that have on their surface immunoglobulin molecules, (2) all of the surface immunoglobulin molecules on any one cell have the same binding specificity, (3) for any one antigenic determinant, only a very small subset of the entire pool of B cells will have surface antibody with which it can bind. When a vertebrate encounters a foreign antigen in its circulation, this antigen combines with preexisting immunoglobulins only on those B cells that have molecules with the appropriate specificity. The interaction of antigen and antibody on the B-cell surface then triggers that B cell to multiply, producing daughter cells that synthesize and secrete its specific antibody (Figure 25-10). The antigen has *selected* from a preexisting pool those cells with appropriate specificity and has caused their proliferation into a large population of cells all making the same antibody. Such a population of identical cells all coming from the same ancestral cell is called a *clone*, thus the term *clonal selection*. Clonal selection theory is, like natural selection, a Darwinian theory because the antigen selects the cells that will multiply from a set of variants that have arisen independently of the selective force. Clonal selection has received such complete verification that it no longer is considered a theory but rather is accorded the status of an accepted explanation of the functioning of the immune system.

Clonal selection proceeds through two distinct phases in the life of the B cell. In the *antigen-independent phase*, before any encounter with an antigen, a large number of B cells are produced, each with a different antigen-binding specificity (see Figure 25-9). B cells are continually elaborated by the bone marrow throughout the life of the animal, each newly arising cell having a lifetime of only a few days as a circulating cell of the blood or lymph. But if during its circulation a B cell encounters an antigen to which its surface antibody can bind, then the cell enters an *antigen-dependent phase* in which it is activated to growth, division, and the secretion of antibody. Also, long-lived progeny called *memory B cells*, or simply *memory cells*, result from this activation so that what was a transient cell becomes permanent (Figure 25-10).

If clonal selection is to operate, there must be a mechanism for generating an enormous number of structurally different immunoglobulin molecules in the absence of any

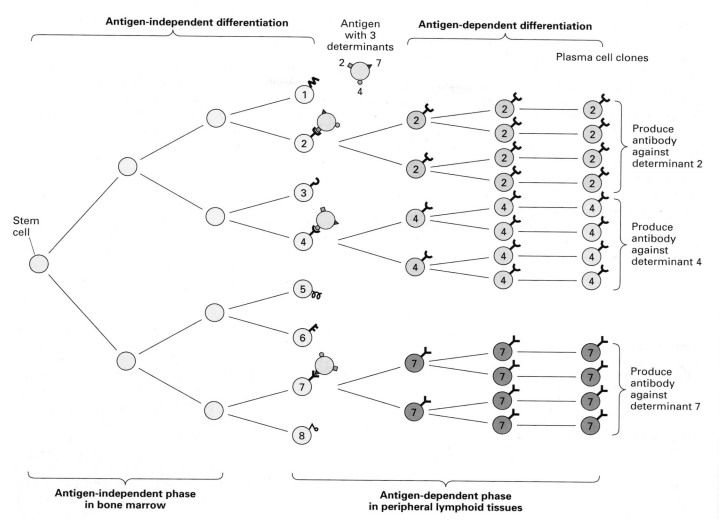

▲ Figure 25-9 The immune system develops specificity by a process of clonal selection. From a large number of cells having different antibodies, cells carrying appropriate anti-bodies bind to an antigen's determinants. Then clones of these cells are produced, with all cells of each clone carrying the same surface antibody.

selective force. This is the problem of the *generation of antibody diversity.* Diversity arises from choices among many alternative possibilities encoded in germ-line DNA and from events in somatic tissue. The somatic events are called *somatic variation,* some of which may be *somatic mutation.* Both multiple germ-line genes and somatic variations of a variety of sorts play roles in generating the variability of the immune response.

The Immune System Has a Memory

Among the remarkable capabilities of the immune system is its ability to learn. A first encounter of a B cell with an antigen leads to a slowly rising synthesis of antibody, dominated by IgM *(primary response).* A second encounter with the same antigen leads to a more rapid and greater response, dominated by IgG *(secondary re-sponse).* Only a previously encountered antigen provokes the secondary response: the system has learned to recognize the antigen to which it was previously exposed (Figure 25-11). The basis of learning in the immune system is the formation of long-lived memory cells (see Figure 25-10). Memory cells persist for the life of the organism even without further antigenic stimulation.

These circulating memory cells carry on their surfaces the particular immunoglobulins that avidly bind to reinvading antigens. The memory B cells mainly make IgG or IgA and respond so rapidly that a second encounter with an antigen leads to a much faster and more effective response than the first. Hence once a person has been infected by a given virus, that virus can never catch the body unprepared again. It also explains why children are so much more susceptible to infectious diseases than are adults.

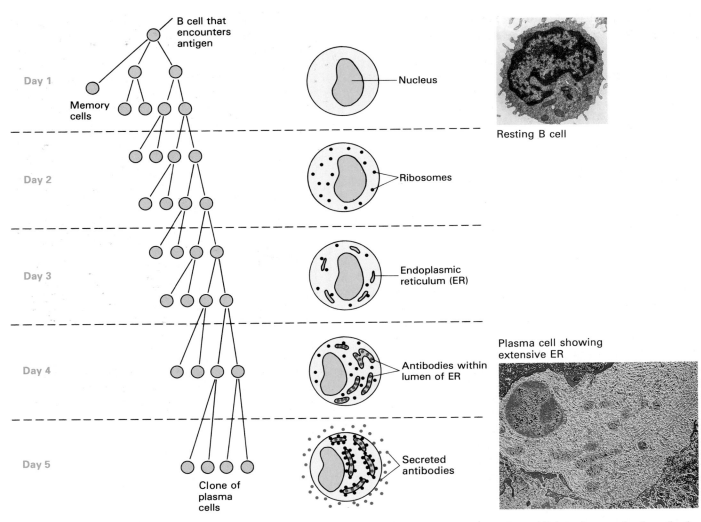

Resting B cell

Plasma cell showing extensive ER

▲ **Figure 25-10**   Activation of B lymphocytes leads to both growth and secretion. The diagram indicates the events that follow lymphocyte activation. After an encounter with antigen and stimulation by factors produced by other immune system cells, the B cell initiates a series of divisions during which it develops the apparatus for secreting copious amounts of antibody. The end result is a large clone of antibody-secreting plasma cells with extensive endoplasmic reticulum. *After L. Hood et al., 1984,* Immunology, *2d ed., Benjamin, p. 11. Upper photograph courtesy of D. Zucker-Franklin; from* Atlas of Blood Cells: Function and Pathology, *1981, Lea & Febiger; lower photograph from Science Photo Library/Photo Researchers, Inc.*

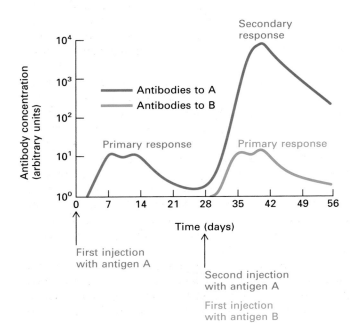

◀ **Figure 25-11**   Responses to initial and later injections of an antigen. At day 0, antigen was injected into a mouse and the response followed by measuring levels of serum antibody to the antigen. At day 28, when the initial response had subsided, a second immunization was done with both antigen A and a new antigen B. The secondary response to antigen A was faster and greater than the initial response to antigen A. The response to antigen B followed the course of an initial encounter with antigen. Thus the secondary response to antigen A is a specific, not a general, response of the system.

Other Parts of the Immune Response Are Carried Out by T Lymphocytes

While the B lymphocyte carries surface immunoglobulin and can be activated to secrete immunoglobulin by an encounter with antigen, the immune system contains another cell population that also carries surface antigen-binding molecules, but these cells never secrete antibody molecules. They are known collectively as the *T lymphocytes*, or simply *T cells*, because they must pass through the thymus gland as they mature. The antigen-binding molecules on the surface of T cells are structurally related to, but distinct from, antibody; they are called *T-cell receptors*. These receptors do not react with soluble antigens but rather with antigens on the surfaces of other cells. The complex reaction of surface molecules on one cell with surface molecules on another cell is at the heart of T-cell biology.

Two fundamentally different types of T lymphocytes are recognized: cytotoxic and helper (each has subdivisions). The *cytotoxic T lymphocyte* (CTL) recognizes other cells that display foreign antigens on their surfaces and kills the cells. Less genteelly, this type is also called the *killer T cell*. Precursors to CTLs display antibodylike receptors that recognize foreign proteins on cells. The recognition process apparently activates the growth of the precursor CTLs, providing the animal with a clone of killer T cells able to destroy cells displaying the foreign protein. This killing process can be demonstrated by incubating CTLs with target cells that have previously taken up the radioactive isotope chromium-51 (^{51}Cr). If the CTLs come from an animal that has been immunized with the target cells, CTLs will kill the cells by putting holes into them. The ^{51}Cr can leak out through the holes. Thus cell destruction by the CTLs can be measured by ^{51}Cr release. The CTL is formally equivalent to an antibody in that it recognizes and eliminates unwanted substances. Antibodies, however, deal with intact, usually soluble foreign materials, CTLs with cell-bound peptides derived from materials made in cells. A classic CTL target is a virus-infected cell that displays fragments of a viral glycoprotein on its surface bound to a supporting protein called class I MHC.

The second type of T cell is a *helper T lymphocyte*, or T_H cell. The function of the T_H cell is to recognize the degradation products of specific antigens and to secrete protein factors that stimulate the various other cells involved in the immune response. Some of the secreted factors promote antibody production from B cells. This stimulation is needed: the interaction of antigens with antibodies on the surface of B cells is insufficient to stimulate B-cell growth and secretion of soluble antibody. In fact, without T-cell help, a B-cell interaction with antigen may lead to paralysis of the cell. The secreted factors also stimulate other cells that participate in immune reactions, such as CTLs and macrophages. The T_H cells, like CTLs,

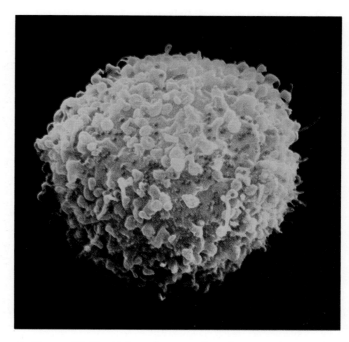

▲ **Figure 25-12** False-color scanning electron micrograph of a T_H lymphocyte infected with human immunodeficiency virus (HIV), the causative agent of AIDS. The small spherical virus particles (green) on the surface are in the process of budding away from the cell membrane. *From NIBSC/Science Photo Library/Photo Researchers, Inc.*

cannot recognize free antigen; they recognize only cell-bound peptides derived from ingested antigens. T_H cells are central players in most immune reactions and are the target of human immunodeficiency virus (HIV), the virus of AIDS (Figure 25-12). Loss of T_H cells because of HIV infection causes the failure of the immune system in people with AIDS.

A third type of T cell, a *suppressor T cell* (T_S) able to suppress B-cell activity, has been postulated. It is debatable whether such a separate cell type exists.

The B lymphocyte and T lymphocytes are indistinguishable in size and general morphology. To distinguish among them, immunologists use differences in their surface proteins (Figure 25-13a). B cells have surface antibody as well as other molecules, such as B-220, that are not found on T cells. The surfaces of T cells have the antibodylike T-cell receptors and a protein called Thy-1. They also have subtype-specific markers: CD4 on helper cells and CD8 on CTLs.

Operationally, the cell types are distinguished with the use of antibodies that recognize one or another of their characteristic surface proteins. Reasonably pure populations of cells can be made by using these antibodies in one

(a) Types of lymphocytes

T cells

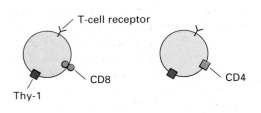

Cytotoxic T cell   Helper T cell

B cells

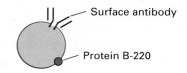

(b) Panning for T-cells

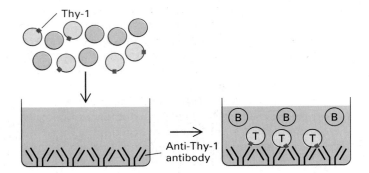

▲ **Figure 25-13** Surface structures on lymphocytes. Each class of lymphocyte has a distinct set of surface proteins. (a) Some of these are indicated for various lymphocyte subsets. The T cells fall into two categories. All have surface protein Thy-1, but CTLs have CD8 while T$_H$ cells have CD4. The T-cell receptor is also a component of all T cells. B cells have none of these T-cell surface proteins but have two distinctive surface markers: surface antibody and protein called B-220. All of the cell-surface proteins can be specifically recognized with the use of antibodies that bind selectively to them. (b) An anti-Thy-1 antibody is used to select T cells from a mixture with B cells in the process called panning. The anti-Thy-1 antibody is bound to the surface of a dish. A mixture of T and B cells is poured over the bound antibody. The T cells are bound to the antibody while the B cells remain free. The B cells are poured off, leaving a pure population of T cells on the dish. [See L. J. Wysocki and V. L. Sato, 1978, *Proc. Nat'l Acad. Sci. USA* 75:2844.]

of two ways. In the first method, *negative selection*, cells with a given surface antigen are killed with specific antibody and the remaining cells are retained. A convenient way to make pure T cells or pure B cells is to use anti-immunoglobulin or anti-Thy-1, respectively. In the second method, *positive selection*, a dish is coated with the antibody and cells that bind to the dish are retained—the procedure is called *panning* (Figure 25-13b). More elegantly, an antibody tagged with a fluorescent compound is bound to cells and a fluorescence-activated cell sorter (FACS) can examine one cell at a time to select the cells to which the antibody has bound (see Figure 4-29).

Macrophages Play a Central Role in Stimulating Immune Responses

A cell type quite distinct from the lymphocyte, the *macrophage*, is very important to the immune response. These cells are found in all tissues and also circulate in the blood, where they are called *monocytes*. They are generally the first cells to encounter a foreign substance in the body. They nonspecifically engulf such materials, as well as scavenge normal cellular debris, and degrade them using powerful hydrolytic enzymes and oxidative attack. Peptides from degraded proteins are then bound to a protein called class II MHC, which carries them to the macrophage cell surface where they can be recognized by T lymphocytes. Peptides from self-proteins are ignored but foreign peptides activate precursors of helper T cells to further maturation into active, lymphokine-secreting T$_H$ cells.

Antigen thus participates in two separate processes of the overall immune response. It activates B cells by direct interaction with antibody on the cell surface and activates T$_H$ cells through peptides displayed on the macrophage cell surface. Although this understanding emerged only in the late 1980s, it had earlier been recognized that antigens have two independent roles, a notion enshrined in the concept of an antigen as made of a hapten for B-cell response and a "carrier" for T-cell response. At that time it was evident that the B cells and T$_H$ cells respond to different determinants; it is now clear that some parts of a protein are most easily recognized by B cells and other parts, in the form of degraded but stable peptides bound to a supporting protein called class II MHC, by T cells.

Cells Responsible for the Immune Response Circulate throughout the Body

We have briefly described some of the cells that together provide an immune response. How are these cells organized and coordinated in the body? How does antigen reach these cells? These are questions about the organization of vertebrate lymphoid tissues.

The immune system has two compartments (Figure 25-14). Lymphocytes arise and go through their antigen-independent phase of establishing surface receptors mainly in the bone marrow and thymus, which are known as the *primary organs* of the immune system. The lymphocytes then leave these organs and are distributed to many sites in the body that together make up the *secondary organs*, or *peripheral lymphoid tissue.*

The lymphoid system guards all portals of entry into the body. The tonsils in the throat and the adenoids in the nose are lymph nodes. There are nodes in the armpit that filter fluid from the arms, and there are nodes in the groin that filter fluid from the legs. A special set of lymphoid organs, the *Peyer's patches,* are found in the intestinal wall where they filter out antigens that enter in food or come from the bacteria growing in the intestines. The activated lymphocytes in Peyer's patches migrate out of the node into the blood. They are finally captured back in the tissue spaces just inside the intestinal lining where they secrete IgA. As already described, the IgA is then rapidly transported across the epithelial cells into the gut lumen.

The peripheral lymphoid tissue is organized around the two fluid systems of the body, the blood and the lymph. These two systems are in contact. Lymph is formed by fluid transported from the blood to the spaces within and around tissues. From these extracellular spaces, lymph flows into thin-walled *lymphatic vessels,* where it is slowly moved to larger central collecting vessels. Ultimately the lymph is returned to veins, where it reenters the blood. In blood, lymphocytes constitute 20–30 percent of the nucleated cells; in lymph they constitute 99 percent. The lymphocytes travel in the lymph, entering and leaving the blood circulation by squeezing between the endothelial cells that line the blood vessels. The lymph circulation is filtered through lymph nodes in which all types of lymphocytes take up temporary residence. Antigens that enter the body find their way into the blood or lymph and are filtered out by lymph nodes or by the spleen, which is the blood's filter. In the lymph nodes, macrocytes engulf and degrade antigens and eventually display pieces of antigen on their surface. The rare lymphocyte in the lymph node that carries an antibody or T-cell receptor with the appropriate specificity will bind either to free antigen or to the fragments of macrocyte-bound antigen; it will then become activated to multiply and differentiate into secreting B lymphocytes or mature T lymphocytes. This binding reaction is the moment of clonal selection, when an antigen binds with, or selects, a B-cell-surface antibody or T-cell receptor. In the lymph nodes or spleen, helper T lymphocytes and B lymphocytes collaborate; here, the concentration of cells facilitates the interaction and magnifies the clonal selection. The antibody made by the secreting B lymphocytes and their mature progeny, the plasma cells, leaves the node in the lymph and is transported to the blood.

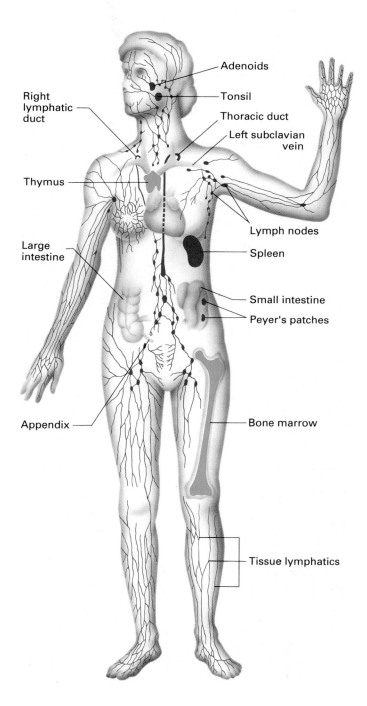

Right lymphatic duct

Adenoids

Tonsil

Thoracic duct

Left subclavian vein

Thymus

Lymph nodes

Large intestine

Spleen

Small intestine

Peyer's patches

Appendix

Bone marrow

Tissue lymphatics

◄ **Figure 25-14** The human lymphoid system. The primary organs (thymus and bone marrow) are colored blue; secondary organs and lymph vessels are shown in red. The whole system consists of the circulating lymphocytes, the vessels that carry the cells (both blood vessels and special lymphatic vessels), and the lymphoid organs. Only one bone is shown, but all major bones contain marrow.

Tolerance Is a Central Concept of Immunology

A key attribute of the immune system is its ability to discriminate between self and nonself. Surface recognition molecules on lymphocytes are sufficiently diverse for any antigen to find some cells with a surface antibody or T-cell receptor whose binding constant for the antigen is high enough to lead to activation. But what prevents self-antigens from activating any of the body's own cells that may carry antibody able to react with them?

Immunologists call the inability to react with self *tolerance,* and they say that the immune system is tolerant to self-antigens. The mechanisms that make for tolerance are poorly understood but they either eliminate or render unreactive the clones of both B and T cells that would otherwise carry out antiself reactions.

Most relevant evidence comes from the experimental production of a tolerant state in mice. The immune system in mice becomes responsive to foreign antigens within days of birth. All antigens present in the body at the time the system matures are considered by the immune system to be self. Thus if foreign antigens are incorporated into a newborn mouse, it will not be able to mount an immune response when later presented with those antigens.

Tolerance to a specific antigen can be induced in mature animals. Inoculation with large amounts of a protein, for instance, can make a mouse nonresponsive to a later challenge with a dose and form of it that would otherwise stimulate a high production of antibody.

The consequences of a failure of tolerance can be severe. Especially in their older years, people may begin to lose self-tolerance and make antibodies to their own proteins. This can lead to many kinds of disorders including autoimmune hemolysis (an attack on red blood cells), arthritis, and systemic lupus erythematosus. When a body reacts to a major self-protein of the serum, high levels of antigen-antibody complexes appear in the blood. A part of autoimmune diseases is the deposition of these complexes in the kidney, resulting in glomerulonephritis, inflammation of the filtering apparatus of the kidney. The deposits can be severe enough to cause kidney failure.

Immunopathology Is Disease Caused by the Immune System

Autoimmunity is only one type of disease caused by a faulty immune system. So many immune-system problems exist that the term *immunopathology* has been coined. Some disorders are due to inherited deficiencies of important components of the immune system. A person may even lack all B cells or T cells, a condition manifested in severe combined immunodeficiency disease (SCID). One form of this disease is caused by inactivation of the gene that encodes the enzyme adenosine deaminase. In the absence of this enzyme, excess adenosine, which is toxic to lymphocytes, accumulates. Patients with SCID lack lymphocytes and die at a young age from viral or bacterial infection. They can be kept alive only in a sterile environment but can be cured by bone marrow transplantation. Many less severe immunodeficiencies exist. Some persons lack components of the complement system, some lack any antibodies, some lack T cells. In certain cases, medical therapies for diseases not related to the immune system can affect immune function; this is especially true of treatment with steroids, which depresses immune function.

Several types of cancer result from a loss of growth control in lymphocytes and macrophages. Hodgkin's disease is one example; others are acute lymphocytic leukemia, chronic lymphocytic leukemia, and many mature T-cell leukemias. Human T-lymphotropic virus (HTLV) is a retrovirus able to transform helper T cells into tumor cells, causing adult T-cell leukemia. Another form of human retrovirus kills helper cells and is responsible for the disease AIDS (acquired immune deficiency syndrome).

Along with the deficiency states and malignancies of the immune system there are hyperreactivity problems. Autoimmunity is one, but more common are the allergies, hyperreactivity to foreign antigens. Allergies result from hyperproduction of IgE in reaction to environmental substances such as pollen and dust. Other types of IgE-induced hyperreactivity include asthma, food sensitivities, and reactions to toxins injected by stinging insects. In all cases, it is the local release of mast-cell products induced by binding of IgE-antigen complexes that causes the often violent symptoms. These released products include histamines, heparin, and substances that cause constriction of smooth blood vessels (see Figure 2-21). Such reactions are treated with antihistamines and with epinephrine (adrenalin), which acts by raising the cyclic AMP levels of the mast cell and preventing the release of its components.

Antibodies and the Generation of Diversity

The previous section introduced antibodies as protein molecules with an antigen-binding region and an effector region. It also described the different classes of antibody: IgM, IgG, IgD, IgE, and IgA (see Table 25-1). We will now examine the structure of antibodies in much greater detail, considering how a population of molecules with so many common properties can also have such extensive variability. The answers to this question lie in understanding the origins of antibody diversity.

Because a population of antibodies comprises molecules with various structures, it is difficult to chemically characterize antibody that comes from immunized ani-

mals. Tumors, however, initially arise as single transformed cells; therefore the antibody made by B-lymphoid
tumors is homogeneous. One specific type of B-lymphoid
tumor, the *plasmacytoma*, or *myeloma*, is especially useful because such tumors are analogues of plasma cells, the
cells that secrete large amounts of antibody. Up to 10
percent of the protein made by a myeloma cell population
can be a single homogeneous type of antibody with one
light chain and one heavy chain. Myelomas arise spontaneously in humans and can be induced at will in mice of
certain strains by injecting mineral oil into their perito-
neum. Most early analysis of protein sequences of specific
antibody chains came from the study of myeloma proteins.

Heavy-chain Structure Differentiates the Classes of Antibodies

Examination of the antibodies made by myelomas, and
also the study of normal immunoglobulins, has shown
that all five classes of antibody have the same general
organization. Each antibody molecule has two chains, a
light (L) *chain* with a molecular weight of about 23,000
and a *heavy* (H) *chain* of MW 53,000 or more. All classes
of immunoglobulin have one of two types of an L-chain
protein, called the kappa (κ) and lambda (λ) chains.

The differences between the classes arise from the H
chain, each class having a different type, the name of
which is the Greek equivalent of the class name. Thus
IgM has a μ chain, IgG has a γ chain, and the others have
δ, ϵ, or α chains (see Table 25-1). There are actually four
subclasses of IgG proteins with somewhat different prop-
erties, each with a different H chain—making a total of
eight types of immunoglobulin. Attached to all H chains
are asparagine-linked carbohydrate chains.

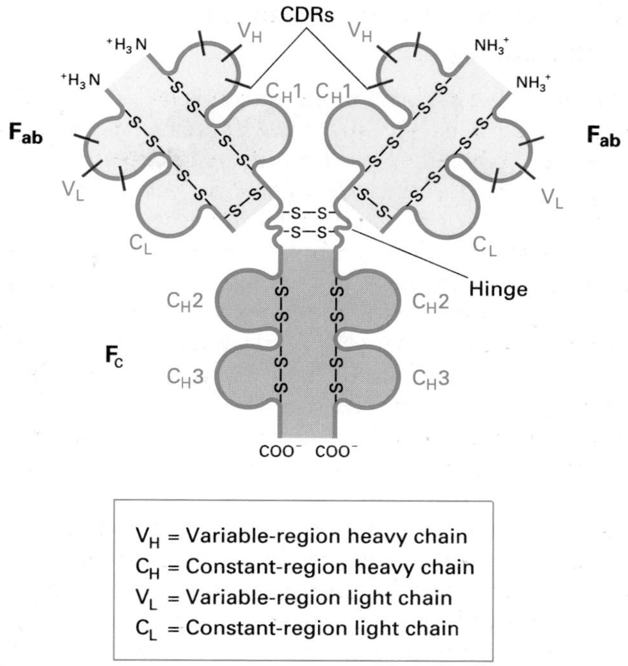

Antibodies Have a Domain Structure

Every individual antibody molecule has one type of L
chain and one type of H chain. The chains are held together by disulfide bonds to form a monomer, and two
monomers are linked by disulfide bonds to form the basic
dimeric structure of the molecule (see Figure 25-4).
Within each chain, units made of about 110 amino acids
fold up to form compact *domains* (Figure 25-15). Each
domain is held together by a single internal disulfide
bond. An L chain has two domains, H chains have four or
five domains. The first two N-terminal domains of the H
chains interact with the two L-chain domains, producing
a compact unit that acts as the binding region of the antibody. In most H chains, a *hinge* region consisting of a
small number of amino acids is found after the first two
domains. The hinge is flexible and allows the binding regions to move freely relative to the rest of the molecule.
At the hinge region are located the cysteine residues

V_H = Variable-region heavy chain
C_H = Constant-region heavy chain
V_L = Variable-region light chain
C_L = Constant-region light chain

▲ **Figure 25-15** Domain structure and the complementarity-determining regions (CDRs) of antibody. The molecule
is organized in disulfide-bonded 110-amino-acid domains,
four in the H chain and two in the L chain. The domain
closest to the N-terminal of each chain is variable in sequence (V region): the other domains are constant in sequence (regions C_H1, C_H2, and C_H3 in the H chain and C_L
in the L chain). When the molecule is protease-digested, cutting of the hinge region connecting C_H1 and C_H2, the most
sensitive spot on the molecule, splits the molecule into two
parts, F_{ab} (the antigen-binding domain) (yellow) and F_c (the
effector region) (green). Within the V regions are segments of
highly variable sequence constituting the CDRs (red), the
amino acids that actually contact the antigen.

whose sulfhydryl groups are linked to form the disulfide
bridges between the two monomer units of the antibody
dimer. The hinge regions are the places on the molecule
most susceptible to the action of protease; light protease
treatment can split an antibody into two pieces, called F_{ab}
and F_c fragments. The F_{ab} portion has the antigen-binding site, the F_c portion has the effector regions.

The N-Terminal Domains of H and L Chains Have Highly Variable Structures That Constitute the Antigen-binding Site

Comparing the amino acid sequence of H and L chains of
various immunoglobulins revealed that the N-terminal
domains have variable structures while the C-terminal
domain has a quite constant structure. The N-terminal
domain is called the *variable region* and the C-terminal

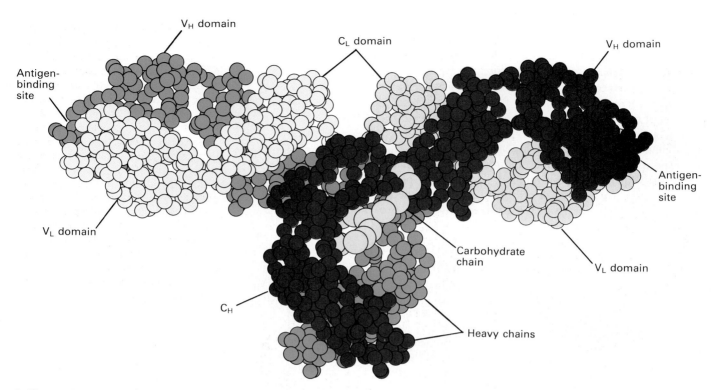

▲ **Figure 25-16** Model of antibody molecule derived from x-ray crystallographic analysis. The model shows all of the atoms as solid balls and thus describes the outer contours of the molecule. Its Y shape is evident. The individual chains are color-coded so that they can be distinguished: the two L chains are light red and light blue; one H chain is dark red and the other is dark blue. *After E. W. Silverton, M. A. Navia, and D. R. Davies, 1977, Proc. Nat'l Acad. Sci. USA 74:5140 and L. Stryer, 1988, Biochemistry, 3d ed., W. H. Freeman and Company, p. 901.*

domain is called the *constant region* (Figure 25-15). The variable domains of L and H chains interact closely to form a single compact unit (Figure 25-16). An antigen that contains a reactive chemical group will form a covalent bond to the variable domain of the H or L chain, showing that the variable domains form the antigen-binding segment of the molecule.

Antibodies of the same specificity are likely to have similar variable-region sequences. Conversely, antibodies that have different specificities have different sequences. This observation illustrates a central principle of immunology, that binding specificity is determined solely by the amino acid sequence of the variable regions. The rule can be shown quite directly. If the L and H chains of a myeloma protein that binds the dinitrophenol group are separated, denatured, and then renatured together, the specificity for binding dinitrophenol, but not other haptens, returns. The experiment can be varied by substituting either the L or H chains of an antidinitrophenol antibody with other L or H chains. Such an experiment shows that both the correct L chain and the correct H chain are needed to get dinitrophenol-binding specificity. Thus it is the joint structure formed by two variable regions that produces a binding site and not either variable region alone. For antidinitrophenol we do not have an

x-ray crystallographic structure to demonstrate this, but for antiphosphorylcholine, which happens to use mainly H-chain residues for its binding, the binding region is known with great precision from x-ray analysis (Figure 25-17).

X-ray crystallographers have determined the high-resolution structure of a few antibodies. They see a variable pocket, cleft, or plane on the surface of the antibody formed by three short polypeptide segments from each chain. From crystals of hen egg lysozyme bound to anti-lysozyme we have learned that in a complex of antibody and a protein antigen, a surface on the antibody interacts with a surface on the antigen such that many hydrophobic, hydrogen-bonding, and charge interactions hold the complex together (Figure 25-18). Different antibodies react with different parts of lysozyme, and probably the whole surface is potentially antigenic. There are no simple rules that relate the amino acid sequence of the antibody to that of the antigen: the surface complementarity of the two molecules is a consequence of many individual local interactions summing to give a very high-affinity interaction.

The large portions of the variable region that do not directly contact the antigen interact among themselves to produce a stable domain structure by forming planes of

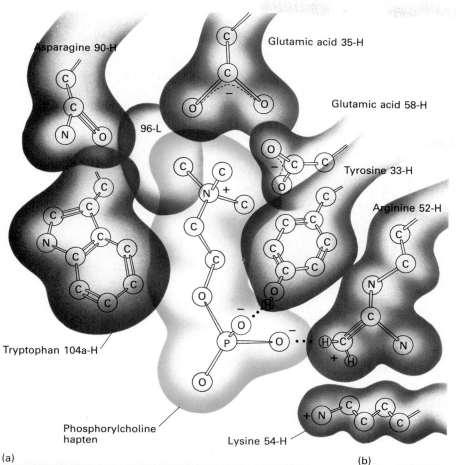

Asparagine 90-H

Glutamic acid 35-H

Glutamic acid 58-H

96-L

Tyrosine 33-H

Arginine 52-H

Tryptophan 104a-H

Phosphorylcholine
hapten

Lysine 54-H

(a)

(b)

◄ **Figure 25-17** Amino acids in the CDR regions forming the binding of an antibody molecule called McPC 603. Amino acids that contact the hapten phosphorylcholine are numbered, indicating their positions in the chains (see Figure 25-19). Each of these amino acids is part of one of the three CDRs of either the L or H chain (here, mostly H-chain amino acids are involved). The CDR amino acids form a cleft or pocket into which the phosphorylcholine hapten fits. A combination of electrostatic, hydrogen-bonding, and van der Waals forces hold the hapten in the cleft. *From* The Antibody Combining Site, *by J. D. Capra and A. B. Edmundson. Copyright 1977 vol. 236(1) by Scientific American, Inc.*

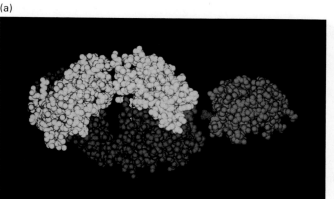

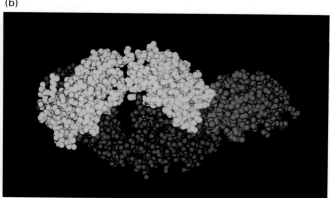

(c)

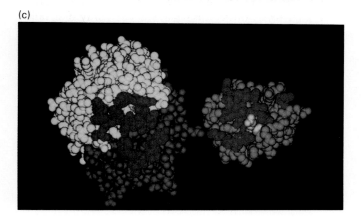

◄ **Figure 25-18** Binding of a protein antigen to an antibody. The antigen (green) is lysozyme; the antibody (yellow for L chain, blue for H chain) is a monoclonal antilysozyme. The amino acid on lysozyme (red) is a glutamine that fits into a pocket on the antibody. Otherwise the two proteins contact each other through surface interactions. (a) The antigen and antibody separated. (b) The antigen and antibody joined into a complex. (c) The separated proteins rotated about 90 degrees around the vertical axis from part (b) to show the interacting surfaces. The amino acid side chains that interact are shown in red in (c), with the glutamine now shown in pink. *From A. G. Amit, R. A. Mariuzza, S. E. V. Phillips, and R. J. Poljak, 1986,* Science **233**:747; *courtesy of R. J. Poljak.*

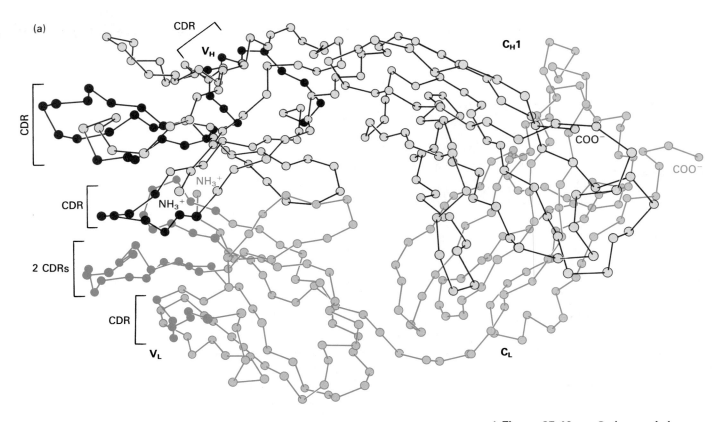

(a)

CDR

V_H

C_H1

CDR

CDR

2 CDRs

CDR

V_L

C_L

NH_3^+

NH_3^+

COO^-

COO^-

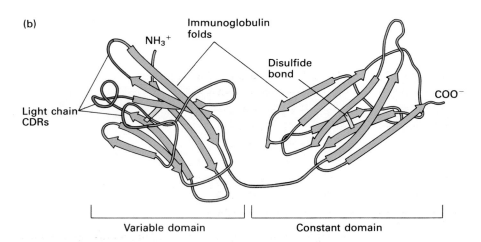

(b)

Immunoglobulin
folds

NH_3^+

Disulfide
bond

COO^-

Light chain
CDRs

Variable domain

Constant domain

◀ **Figure 25-19** α-Carbon and sheet models of the F_ab segment of an antibody. The variable and first constant domains are shown. In (a), the F_ab molecule is represented only by the α-carbon *(balls)* of each amino acid. The loops that contain the red and blue balls are the CDRs; they come from both the H and L chains. In (b), the L chain is represented by arrows that show the strands of the β pleated sheet that make the immunoglobulin fold. Both the variable and constant regions consist of domains organized as immunoglobulin folds. Extending from the variable domain are the CDRs. *Part (a) from* The Antibody Combining Site, *by J. D. Capra and A. B. Edmundson, copyright 1977 vol. 236R(1) by Scientific American, Inc.; part (b) after M. Schiffer et al., 1973,* Biochemistry *12:4620.*

protein known as β pleated sheets (see Chapter 2). The sheet structure of the variable domains is so characteristic that it has been called the *immunoglobulin fold*. The three amino acid segments of the variable region that bind antigens are loops that extend from the immunoglobulin fold (Figure 25-19). Because the binding site is complementary in structure to the antigen, these three segments are called *complementarity-determining regions,* or CDRs. Because the CDR structure is much more variable than that of the rest of the variable region, the CDRs are often called *hypervariable regions.* The variable regions therefore consist of two types of sequence,

highly variable CDRs embedded in less variable *framework regions* (FR) that form the immunoglobulin fold. The variable region has four FRs and three CDRs (Figure 25-20).

Several Mechanisms Generate Antibody Diversity

Only the variable regions of an antibody contribute to the diversity of antigen-binding specificities. Thus the problem of the generation of diversity becomes the problem of how one portion of a molecule can vary enormously

(a)

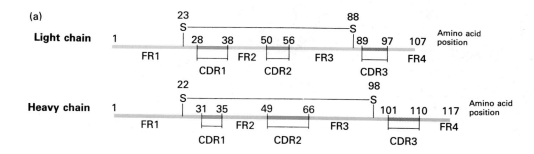

(b)

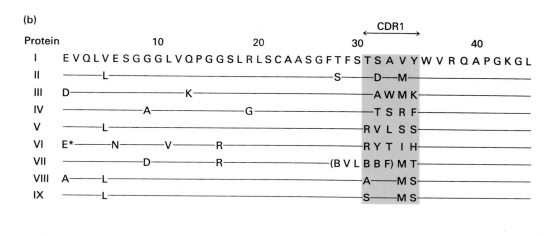

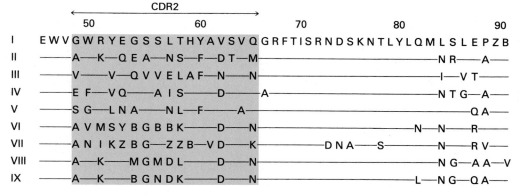

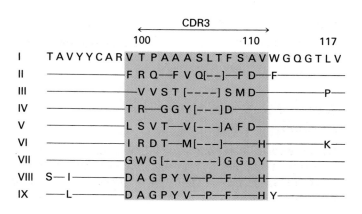

▲ **Figure 25-20** Location of the CDRs on the variable regions of the L and H chains. (a) Schematic representation of the CDRs interspersed among FRs. The numbers start with amino acid 1 at the N-terminus. (b) A collection of actual sequences of human H-chain variable regions demonstrates the extent of variation found in the CDR sequences. In the top line, the sequence of protein I is shown using the single-letter amino acid code (see Figure 2-2). For proteins II–IX, the sequences are shown using the convention that a line indicates identity with protein I, a letter signifies a position where an amino acid different from that in protein I is located, parentheses indicate an uncertainty, and dashes represent deletion of a sequence relative to protein I. One region of variability between CDR2 and CDR3, at positions 84 through 88, is not in the antigen-binding site. [Part (b), see J. D. Capra and J. M. Kehoe, 1974, *Proc. Nat'l Acad. Sci. USA* 71:4032.]

while the other portions remain essentially invariant. There are two distinct ways to imagine a solution to the problem. One is to have a single germ-line gene that could undergo extensive mutation in lymphoid cells, but only at its variable-region end. Each mutation would make a protein with a unique variable-region sequence but all proteins would have the same constant region. The other way to solve the problem is to have many variable-region gene segments in the DNA, each able to join to a single constant-region gene segment. These two different approaches to the diversity problem are generically described as the *somatic mutation hypothesis* and the *somatic recombination hypothesis,* respectively. The immune system actually uses both mechanisms. In fact, there are many independent mechanisms that contribute to diversity. The recombination hypothesis was the first to gain experimental support.

DNA Rearrangement Generates Antibody Diversity

The key experiment that convinced immunologists that DNA rearrangement is central to the immune response relied on myeloma cells (Figure 25-21). The κ-L-chain mRNAs were extracted from a mouse myeloma, and some were broken in half. The 3′ half of the broken molecules was purified and used as a molecular probe for the constant region; the whole molecule served as a probe for both variable and constant regions. Mouse myeloma-cell DNA and mouse embryo DNA were probed with the two mRNAs. The results showed that the myeloma-cell DNA had a different organization in its κ-chain coding region from that of the DNA of the rest of the animal. DNA rearrangement was thus shown to be a central aspect of B-cell differentiation.

Complete analysis of immunoglobulin coding regions has shown that the H chains, κ L chains, and λ L chains are encoded in three genetic loci on three different chromosomes (on, respectively, human chromosomes 14, 2, and 22). Each locus has multiple variable regions and one or a few constant regions. We call each group of variable regions a *library.* In a cell that makes an antibody consisting of a κ chain and an H chain, DNA is rearranged from its germ-line configuration at both the κ locus and the H-chain locus to form the two genes that encode the two chains.

A Single Recombination Event Generates Diversity in L Chains

To appreciate in detail how L-chain genes are constructed by recombination, we need first to recall that mRNAs are formed by the splicing out of regions from a nuclear precursor RNA. In the recombining of cellular DNA to form an immunoglobulin gene, the recombination has to bring

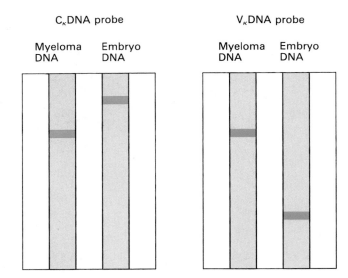

▲ **Figure 25-21** Demonstration of κ-gene rearrangement during B-cell maturation. In the experiment, DNA was extracted from two sources: a clonal mouse myeloma cell and a mouse embryo. The two DNA samples were cleaved with restriction enzymes, the DNA was fractionated by electrophoresis, and the separate DNA fragments were visualized by hybridization with specific radioactive DNA probes. A C_κ constant-region probe was used for the left panel, a V_κ variable-region probe for the right. In both panels, the hybridizing bands in the myeloma DNA are different from those in embryo DNA. All cells from adult mice except B cells have the same patterns as embryo DNA. The myeloma cell has one band that hybridizes with both the C_κ and V_κ probes indicating that the V_κ and C_κ regions, which are on separate fragments in germ-line DNA, have been so closely juxtaposed in myeloma DNA that they are on the same DNA fragment (within 10 kb of each other). [See N. Hozumi and S. Tonegawa, 1976, *Proc. Nat'l Acad. Sci. USA* 73:3628.]

together the relevant DNA into one transcriptional unit, but splicing can then eliminate any parts of the RNA from that transcriptional unit that have no role in the ultimate mRNA (introns are eliminated and exons are maintained).

κ mRNA is made from three exons. At the 5′ end is the L_κ *exon;* it encodes a *leader,* or *signal, peptide* that directs newly made κ protein into the endoplasmic reticulum. The second exon is the variable region proper and the third is the constant region.

In the germ line, the leader peptide and most of the variable region are encoded in one library consisting of a few hundred units. Each unit consists of one L_κ exon and one V_κ *region.* (The V_κ region makes up most, but not all, of the final variable region; we denote it V_κ to distinguish it from the complete variable region.) The $L_\kappa + V_\kappa$ units are tandemly arrayed along one long stretch of DNA. Al-

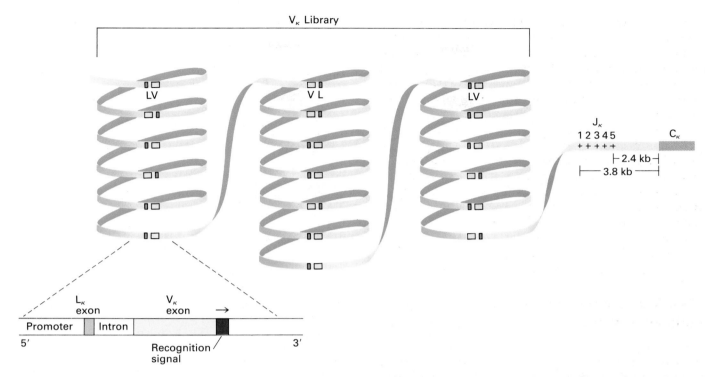

▲ **Figure 25-22** Organization of the κ locus. In a library spread over thousands of kilobases of DNA, there are hundreds of V_κ regions. They are organized in either transcriptional orientation relative to the J_κ and C_κ segments. Both J_κ and C_κ are in the 5'-to-3' orientation (left to right in the illustration). One $L_\kappa V_\kappa$ unit is shown expanded. It consists of a promoter at the 5' end (beginning of transcription), an exon encoding a leader peptide (L_κ), an intron, an exon encoding the V_κ region, and a recognition signal that specifies the site at which the V_κ region is to join to the J_κ region. The recognition signal is actually made of two pieces, one seven nucleotides long (7-mer) and the other nine nucleotides long (9-mer). The arrow shows the orientation of the 7-mer followed by the 9-mer. *After Paul D. Gottleib,* Molecular Immunology *17:1423.*

though each is about 400 nucleotides long, they are separated by about 7 kilobases (kb), thus 100 L_κ + V_κ units would cover about 740 kb of DNA. This region is followed by five *joining regions,* or J_κ (to be distinguished from the J chain), and the single constant-region exons in the cell's DNA. The five J_κ units are tandemly arranged but are separated by about 20 kb from the V_κ regions. Each is about 30 nucleotides long, and they are spread over 1.4 kb of DNA. Between the J_κ units and the C_κ region lies an intron of 2.4 kb of DNA (Figure 25-22).

DNA reorganizes to make a functional κ gene: one V_κ region joins to one J_κ region with the deletion or inversion of the intervening sequence (Figure 25-23). This forms the complete variable region. So far as is known, any V_κ can join to any J_κ, and the choice is random. Once V_κ and J_κ are joined, the variable and constant regions are transcribed together into a nuclear RNA and the intervening sequences between L_κ and V_κ and between J_κ and C_κ are removed by RNA splicing to produce the mature mRNA for κ protein. (Figure 25-23).

The V_κ-J_κ joining produces diversity by combining V_κ region diversity and J_κ region diversity. The combination of 300 V_κ regions and 5 J_κ regions can obviously produce 1500 different possible chains. But V_κ-J_κ joining generates more sequence variability than the simple combinatorial calculation would suggest because in the vicinity of the V_κ-J_κ joint the joining process is imprecise and can generate many combinations. To see the consequences of this imprecision, it is necessary to examine the joining reaction in more detail.

Imprecision of Joining Makes an Important Contribution to Diversity

At the 3' end of the coding sequence of the V_κ segments and at the 5' edge of the J_κ segments lie DNA sequences that signal the joining process. They are called *recognition sequences* and are organized as follows: abutting each V_κ and J_κ is a seven-base palindrome, a space of

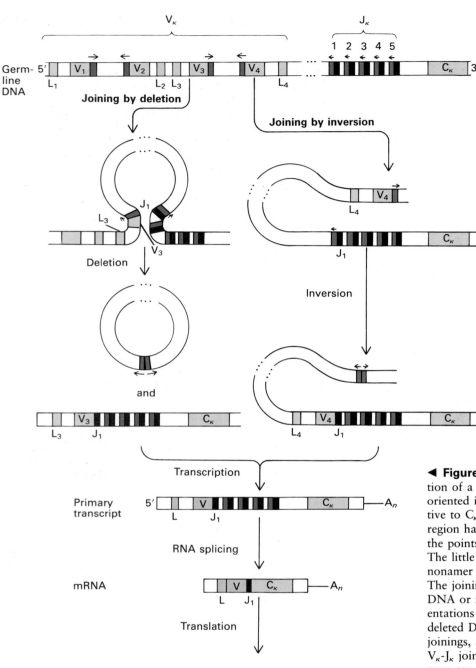

◀ **Figure 25-23** Joining of V_κ to J_κ and formation of a κ light chain. The $L_\kappa V_\kappa$ segments are oriented in either transcriptional direction relative to C_κ in germ-line DNA. Each V_κ and J_κ region has a recognition signal (red) to specify the points at which joining should take place. The little arrows indicate the heptamer-to-nonamer orientation as defined in Figure 25-22. The joining process either deletes the intervening DNA or inverts it, depending on the relative orientations of V_κ and J_κ. In deletional joinings, the deleted DNA is lost from the cell. In inversional joinings, all of the DNA is conserved. Once the V_κ-J_κ joining has occurred, the now complete gene can be transcribed to produce a nuclear RNA that is spliced to remove all unwanted segments (including the J_κ regions that were not joined to V_κ). The spliced mRNA then encodes a complete κ light chain.

about 11 or 23 nucleotides follows (one or two turns of the DNA helix), and then an AT-rich nine-base sequence is found (Figure 25-24). These sequences allow an as-yet-uncharacterized enzymatic system to carry out an orderly but imprecise joining reaction that contributes greatly to diversity.

When two pieces of DNA join, there are two products. In the joining of V_κ to J_κ, one product is a $V_\kappa J_\kappa$ unit and the second is a back-to-back joining of the recognition sequences. When many such joining events of one V_κ and J_κ were studied, it became clear that the recognition ele-

ments are joined identically in all cases, with the heptamers linked to each other. The $V_\kappa J_\kappa$ unit, however, is not precisely joined: a few nucleotides from V_κ and a few from J_κ are lost from the DNA at the joining point (Figure 25-25). Thus the joining process has the unique property that a small number of nucleotides at the joint are lost from DNA entirely; they appear in neither product. The random loss of nucleotides at the joining site generates significant diversity at that point, but the system pays for its diversity. The cost is evident if we remember the constraints on a coding sequence.

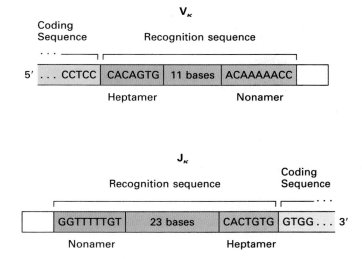

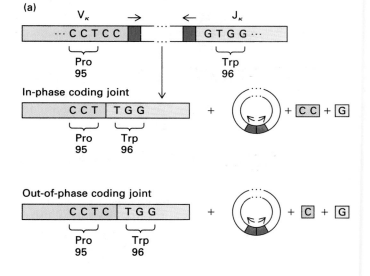

◀ **Figure 25-24** V_κ and J_κ recognition signals. Each J_κ and V_κ region has a short characteristic DNA sequence (recognition signal) preceding or following its coding sequence (see organization in Figure 25-23). This signal is recognized by DNA joining enzymes and directs those enzymes to join a V_κ to a J_κ. The recognition signals consist of a heptamer followed by a nonamer sequence. The two sequence elements are separated by either about 11 nucleotides in a V_κ or 23 nucleotides in a J_κ. The joining system always joins DNA with an 11-base spacer to DNA with a 23-base spacer, so that a V_κ joins to a J_κ but not to another V_κ.

A coding sequence in DNA must be read in threes starting with the first AUG (methionine). We say that the AUG defines a *reading frame* in which the rest of the coding region can be read. If DNA were read in one of the two other reading frames, it would encode a meaningless string of amino acids until an adventitious terminator codon were reached. Thus, when two pieces of coding DNA join, as in the joining of V_κ to J_κ, it can be an *in-phase* joint that maintains a sensible reading frame or it can be an *out-of-phase* joint that encodes a nonsense protein (Figure 25-25a). The V_κ-J_κ joining process is a random one, and two out of three random joints make nonsense. Thus the system pays for its diversity in making two *nonproductive* joints for each *productive* joint.

The diversity gained by the imprecise joining process is significant. Figure 25-25b shows how four different in-phase joinings can be made between one V_κ sequence and one J_κ sequence. All of these joints have actually been found in sequenced κ proteins.

We now have three sources of diversity: variability in the structure of the many V_κ regions in that library of sequences, variability in the structure of the five J_κ re-

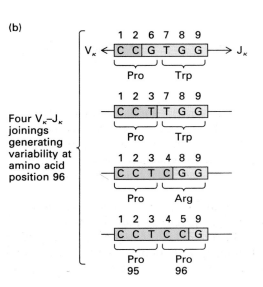

▶ **Figure 25-25** The joining of V_κ to J_κ. (a) The joining process can give rise to in-phase or out-of-phase joints. The loss of a small, random number of bases from the ends of both V_κ and J_κ accompanies joining. This can leave J_κ either joined in the appropriate reading frame (Trp at position 96 encoded by TGG just following the CCT codon at position 95) or out of frame (for instance, if one more C is left on V_κ, the CCT at 95 will be followed by CTG and all of J_κ will be out of frame with V_κ). (b) The joining process can give rise to many in-phase joints with various amino acids at the joining point if variable ends of V_κ are joined to variable ends of J_κ. In the example, four different in-phase joinings encode Pro-Trp, Pro-Arg, and Pro-Pro.

gions, and variability in the number of nucleotides deleted at V_κ-J_κ joints. Antibody chains have three CDRs, and the diversity within the V_κ library contributes to all three; the diversity at the recombination joint contributes only to CDR 3 because the site of joining is within CDR 3. The diversity in CDRs 1 and 2 has been determined over evolutionary time as the individual V_κ regions have evolved. The diversity in CDR 3 is a somatic process: it happens in cells within the body of the individual animal, not over evolutionary time. Our discussion of these events has focused, for convenience, on the κ genes, but similar processes characterize the formation of both λ-chain and H-chain genes.

Lambda Proteins Derive from Multiple Constant Regions

Antibodies have one of two light chains, κ or λ. The ratio of κ to λ is radically different from species to species. Mice have 95 percent κ, humans about 50 percent, and chickens virtually only λ. In most species, λ proteins have not one but multiple C_λ regions, each with its own J_λ region. In mice, probably because of the paucity of λ protein, there are only a few V_λ regions, each associated with a J_λ-C_λ cluster. Humans have even more C_λ regions and presumably many V_λ. Mouse λ chains have much less diversity than mouse κ chains because V_λ and J_λ diversity are so minimal. Moreover, very little recombinational diversity has been found at the V_λ-J_λ joining site. The situation in chickens is remarkable: most chicken Ig has λ light chains, but there is only one functional V_λ segment. Diversity in chicken λ chains arises from a string of pseudogene segments upstream of the one intact V_λ. By a process of gene conversion, these pseudogenes diversify the intact V_λ after it has joined to J_λ.

H-Chain Variable Regions Derive from Three Libraries

Analysis of antibody-binding sites suggests that the H-chain contribution to diversity is even greater than the L-chain contribution (see, for example, Figure 25-17). Consistent with a need for greater diversity, H-chain structure is determined by three libraries of sequence elements instead of the two that make up L chains. The third library is called the *D-region* library (D for diversity). The D region of H chains constitutes the bulk of the third CDR, and the D segments are found between the other two libraries, whose segments are called V_H and J_H (Figure 25-26). Thus, two joining reactions, V_H-D and D-J_H form the variable region for the H chain. Having three segments, rather than two, greatly increases the combinatorial diversity produced by drawing segments from li-

braries containing elements with varied sequences. There are apparently hundreds of V_H segments, perhaps 20 D segments, and 4 J_H segments. Recombinational diversity at the V_H-D and D-J_H joints, like that at the V_κ-J_κ joint, is also created by the removal of small numbers of nucleotides.

The third CDR of H chains is diversified by yet another mechanism. When a D joins to a J_H, or when a V_H joins to a D, not only are nucleotides removed but a few nucleotides not found in either parental sequence are added at the joint. The enzyme responsible for the addition of nucleotides is probably *terminal deoxynucleotidyl transferase,* a template-independent DNA polymerase known to be present in cells making H-chain joints (but probably absent from cells when they are making L-chain joints). We call the extra nucleotides an *N region.* Thus, a complete H-chain variable region is a $V_H NDNJ_H$ unit. Maintaining the reading frame presents a problem similar to that for L chains. The nucleotides of the NDN region between V_H and J_H must keep the reading frame correct, thus two out of three joinings will be out of phase. A danger not usually present in the pure V-J joint of L chains is that the formation of the NDN sequence could put a termination codon in the reading frame. The possibility of this is minimized by the high G + C content of many N regions (terminators are UAG, UGA, and UAA) and by the absence of terminators from most D regions in all three frames. The diversity generated by the NDN unit is almost incalculably vast because the N regions have wholly random sequences (with a possible bias toward G + C), and the D regions can be read in any frame depending on the length of the 5′ N region. (A strong bias toward a particular reading frame for D is evident, however, when sequences of real antibodies are examined.) NDN regions up to 30 nucleotides long have been found, and this highly variable segment can encode 0–10 amino acids.

Recognition Sequences for All Joining Reactions Are Virtually Indistinguishable

We have described here a number of different joining events, V_κ to J_κ, V_λ to J_λ, V_H to D, and D to J_H. All these reactions follow one inviolable rule that relates to the structure of their recognition sequences. Each of the sequence elements that participates in joining reactions has an adjacent recognition sequence. As described for κ gene segments, these recognition sequences are of two types: one has a characteristic heptamer followed by an 11- or 12-base spacer of random sequence and an AT-rich nonamer; the other has a similar heptamer followed by a 21- to 23-base spacer and a nonamer (see Figure 25-24). The spacers make the difference: one recognition sequence

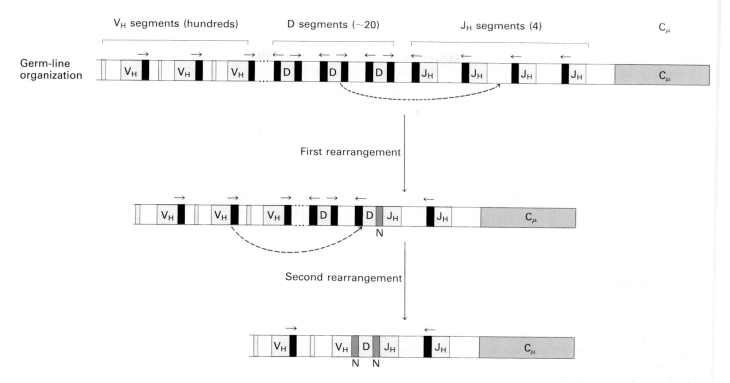

▲ **Figure 25-26** Organization and rearrangement of H-chain genes. In the germ line, the H-chain-related DNA is organized with a large library of V_H regions all in the same 5′-to-3′ orientation, followed by a library of D regions, a library of J_H regions, and the constant region for the μ chain (C_μ). The V_H, D, and J_H gene segments have recognition signals like those found at the ends of V_κ and J_κ (overlined with arrows). In the first stage of rearrangement, a D segment joins to a J_H segment, deleting the intervening DNA. In the second stage a V_H joins to the preformed DJ_H unit, forming a V_HDJ_H H-chain variable region. At each joint, a few random nucleotides may be inserted (N regions).

has a spacer representing approximately one turn of a double-stranded DNA helix (10.5 bp), the other spacer has about two turns. The inviolable rule is that in all unions a one-turn recognition sequence combines with a two-turn recognition sequence. Thus V_κ segments have one-turn elements, and J_κ segments have two-turn; D segments are flanked by one-turn elements, and V_H and J_H segments have two-turn elements. Why the spacers fall into this pattern is not totally clear, but very likely there are proteins that bind to these recognition sequences, and they may have evolved to recognize signals arrayed together on one side of a DNA double helix.

The enzymology of joining remains to be worked out. We imagine that recognition proteins bind to the recognition elements and specify that one-turn and two-turn elements join to each other. Endonucleases must cut the strands, exonucleases might degrade free ends, terminal deoxynucleotidyl transferase may add nucleotides, polymerases must provide flush ends, and then ligases must seal the gap. Many enzymes must participate, and the events must be highly regulated so that joining occurs

only when needed and then between the correct sequence elements.

We have considered joining processes that create diversity by combining regions from different libraries in the germ line. Insertion of N regions, a de novo synthesis of DNA, is a form of somatic mutation that is coupled to joining. Another form of somatic mutation, separate from joining, involves the change of individual base pairs throughout a joined variable region to generate diversity across the whole region. This mutational process will be described later.

The Synthesis of Immunoglobulins Is Like That of Other Extracellular Proteins

All immunoglobulin synthesis takes place on the rough endoplasmic reticulum membranes of the B cell. This would be expected because immunoglobulins are either cell-surface molecules or secreted molecules, and the pathways to both end points begin with newly made protein entering the cisternae of the endoplasmic reticulum.

Newly made polypeptides are usually targeted to the endoplasmic reticulum by a hydrophobic signal sequence. In immunoglobulins, such a sequence is found at the N-terminus of both newly made H and L chains. It is encoded by the L exon, one of which, as previously noted, is located on the 5′ side of each V region. The signal sequence is cut away from the chains shortly after its synthesis so that mature antibodies do not contain it.

The Antigen-independent Phase of B-lymphocyte Maturation

Our discussion of clonal selection touched upon the two phases in the development of a B lymphocyte. In the *antigen-independent phase*, diversity is generated. By the end of this phase a B cell has acquired rearranged H- and L-chain genes and is synthesizing IgM antibody as a surface-receptor protein. At this time the cell is a *virgin B cell*, not having undergone any selection for the reactivity of its receptor with an antigen. During the second or *antigen-dependent phase*, B cells have three possible fates: to react with a foreign antigen and thus be positively selected for further growth, maturation, and synthesis of secreted antibody; to react with a self-antigen and be paralyzed by unknown mechanisms, thus ensuring self-tolerance; or not to react at all and die within a few days. In this section we consider the ordering of events in the antigen-independent phase of B-cell maturation. Later, we consider the antigen-dependent phase and the key role that T cells play in its progression.

Before discussing the individual events of the antigen-independent phase, the overall strategy of the developmental process is worth emphasizing. Cellular development can be driven in two ways, by internal events of a cell or by external influences. As far as is known, the antigen-independent phase is a good example of internally driven differentiation. Although it is probably initiated by an external inducing agent acting on an uncommitted cell, that event has yet to be characterized. After the initial commitment, however, the events through the production of the virgin B lymphocyte seem to be internally determined. Progression through a variety of stages comes about through internal signals, none of which are fully understood. It is reasonable that the antigen-independent phase should need no extracellular stimulation: it is the working out of a differentiation program that has a single goal, production of virgin B cells. Once the virgin cell is produced, many external influences come into play, making the antigen-dependent phase a dialogue between the cell and its surroundings.

Cellular development in general relies on both internal and external signaling. B-cell differentiation happens to be a convenient system in which to see the two styles of developmental decision making at work.

B-lymphoid Cells Go Through an Orderly Process of Gene Rearrangement

B lymphocytes arise continually throughout life by differentiation from bone-marrow stem cells. The bone marrow is considered a primary organ of the immune system, donating cells to the peripheral organs. Many of the intermediate stages have been defined by examining the structure of immunoglobulin-related DNA in the cells. Tumors of early stage cells have been very helpful in defining the intermediate stages, just as myelomas have helped our understanding of end-stage cells.

The earliest recognizable B-lymphoid cell has cell-surface markers (defined by antibody binding) that mark it as B-lymphoid type, but its DNA is still germ line in organization (Figure 25-27). Next, this differentiating cell begins H-chain gene rearrangement, but its L-chain genes are not yet actively rearranging. It first joins a D to a J_H and then joins a V_H to the preformed DNJ_H to form the complete V_HNDNJ_H. The cell can then make an H chain. Because the nearest H-chain constant region to the V_HNDNJ_H unit encodes a μH chain, the cell makes μ and is recognized by immunofluorescence as a μ-positive, L-chain-negative bone marrow cell called a *pre-B lymphocyte*.

The next stage of B-lymphoid differentiation is rearrangement of κ L-chain genes (Figure 25-27). Most cells stop there and make a κ chain, but a few, which rearrange κ genes nonproductively, rearrange λ genes and become λ producers. Once a cell has constructed a complete in-phase κ or λ gene, it makes an L chain. The L chain can bind to an H chain, and the unit can then be processed to the cell surface, where it remains bound. The newly arisen cell with surface antibody is the virgin B lymphocyte.

We noted above that all of the signals for rearranging immunoglobulin genes are similar: heptamer, one- or two-turn spacer, nonamer. How then is the patterned rearrangement of the various immunoglobulin gene segments regulated to coordinate the events in time? The answer is thought to lie in transcriptional control: segments that are being transcribed are those that rearrange. Thus the patterned unfolding of a transcription control program, described later, is thought to underlie the stages in the rearrangement process.

The Antigen-independent Phase Can Generate 10^{11} Different Cell Types

The antigen-independent phase of B-lymphocyte development generates cells containing surface antibodies with a wide range of specificities. It is not possible to calculate

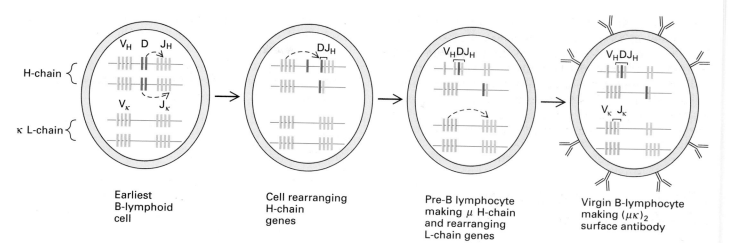

▲ **Figure 25-27** Ordered rearrangement of variable regions of antibody genes during the antigen-independent phase of differentiation of B lymphocytes. The earliest cell has B-lymphoid surface markers but has not started rearrangements. It begins maturation by first making D-to-J_H rearrangements at both heavy-chain loci. It then begins V_H-to-DJ_H rearrangements, probably sequentially on the chromosomes. Production of an in-phase $V_H NDNJ_H$ unit leads to H-chain (μ) synthesis, and that (or possibly some other signal) prevents further rearrangement. The example shown here would be an in-phase joint preventing V_H-to-DJ_H rearrangement on the second chromosome. After μ-chain synthesis begins, the cell is designated a pre-B lymphocyte. After an unknown number of further divisions, κ-gene rearrangement begins and continues only long enough to make one κ gene that can encode a complete κ protein able to bind to μ. The $(\mu\kappa)_2$ dimers then appear on the cell surface, the cell stops division, and it becomes a mature B lymphocyte. If no in-phase κ join is produced, the cell can still rearrange its λL-chain genes. About 5 percent of mouse antibody and almost 50 percent of human antibody have λL chains.

the number of such specificities, and such a number is also not very meaningful because a given antibody might bind to a range of antigens, each with different affinity. It is, however, possible to estimate the number of different antibody molecules that can be produced by the multiple diversification mechanisms available. The calculation depends upon the randomness of events. If some V or D regions are favored for rearrangement, then a disproportionate number of newly made cells may carry related antibodies. Evidence of nonrandomness has been uncovered. The V_H regions are used sequentially for attachment to preformed DJ_H units, so that the more 3′ V_H segments have a disproportionately high representation in the cells generated during the antigen-independent phase. In addition, one reading frame of each D region seems to be highly favored. Nevertheless, we can assume total randomness in calculating the number of different antibodies generated because the number is so high that errors of a factor of a hundred are not important.

Assuming that in the mouse there are 300 V_H regions, 20 D regions, and 4 J_H regions, we calculate that $300 \times 20 \times 4 = 2.4 \times 10^4$ different combinations of these units exist. Because at each D-J_H joining and V_H-D joining nucleotides are lost, each D can be read in each of its frames, and all three constituents can be truncated to various ex-

tents. We estimate that this will increase diversity at least 10-fold. In addition, the D-J_H and V_H-D joinings have N regions that are random in length and sequence, generating at least 100 different possibilities. Thus about $2.4 \times 10^4 \times 10 \times 100 = 2.4 \times 10^7$ different H chains are possible. L chains are less diverse. Ignoring λ chains because they are rare in mice, if there are 100 V_κ regions, 4 functional J_κ regions (one of the 5 is not functional), and 10 ways of putting them together, $100 \times 4 \times 10 = 4 \times 10^3$ possibilities exist. Assuming that any H chain could combine with any L chain, we see that $2.4 \times 10^7 \times 4 \times 10^3 = 10^{11}$ physically different antibodies can be produced by the system during the antigen-independent phase. And 10^{11} is a huge number. A mouse makes only about 10^8 lymphocytes per day, so that all possible combinations might not even be expressed over the lifetime of the animal. The secret of the ability of the immune system to react to whatever pathogens evolve lies in the enormous diversity resulting from the mechanisms that generate antibody variable regions.

Another characteristic of the antigen-independent phase is that it is organized hierarchically to produce a highly diverse population of single cells. Through the early stages of the lineage, cells are continually growing and dividing about once every 12 h. How long the early stages last is not clear, but a cell with only DJ_H rearrangements can probably give rise to many progeny, each of which will have its own V_HDJ_H rearrangement. Similarly, a cell with a V_HDJ_H rearrangement can give rise to many cells with independent $V_\kappa J_\kappa$ rearrangements. But when a virgin B lymphocyte appears in the marrow, it rapidly ceases growth and moves out of the marrow to the peripheral blood-lymph circulation system. Because a cell appears to cease growth very soon after it has acquired surface antibody, each cell produced by the bone marrow could be virtually unique. In actual fact, there appears to be preferential formation of certain gene structures. How this is accomplished remains unclear, but because all the individual mice of a certain inbred strain respond to a given hapten by making virtually identical antibodies, many with specific N regions, those gene configurations must somehow be preferred during the antigen-independent maturation.

The Immune System Requires Allelic Exclusion

A third aspect of the B-lymphoid maturation process to consider in examining diversification mechanisms is that cells have two chromosomes and therefore could, in principle, make two H chains if both chromosomes rearranged productively, that is, if they made an in-phase V_HNDNJ_H unit. In addition, two κL chains plus two or more λ L chains can conceivably be made. One cell would therefore be able to make multiple types of antibody. Such a situation would violate the precondition for clonal selection to operate. Clonal selection depends on the manufacture of only one antibody type per cell so that when the cell is activated it will go on to secrete only one type of antibody. If multiple chains were to be made, antibodies could have two different binding specificities on their dimeric molecules and would lose the advantage of high avidity gained from multiple interactions with a single antigen. Mechanisms have, however, evolved that ensure that B lymphocytes make one, and only one, antibody. Because this means that only one of two alleles carried by a cell will be expressed on any one cell, the process has been called *allelic exclusion.*

Allelic exclusion apparently works by shutting down rearrangement processes after there has been one productive rearrangement. The mechanisms are not known, but one type of experiment clearly indicates that the process is at work. An already rearranged H-chain or L-chain gene can be introduced into a mouse's germ line by microinjection of a plasmid into a fertilized mouse egg (see Figures 5-25 and 5-26). The animals are called *transgenic,* denoting their acquisition of a foreign gene. When a mouse is transgenic for a rearranged H-chain gene, rearrangement of endogenous H-chain genes is suppressed.

Transgenics that have acquired a κ chain suppress rearrangement of endogenous κ chains. In both cases, the expression of a rearranged gene suppresses rearrangement of other genes—the result expected if allelic exclusion is a consequence of secondary rearrangements.

Antibody Gene Expression and Rearrangement Is Controlled by Transcription

We have thus far considered only the construction of immunoglobulin genes and not their transcription. The control of their transcription, unlike that of many other genes in animals, should be simple because they are expressed in only one cell lineage in the body, the B lymphocyte. The transcriptional control regions for H chains and κ L chains are easily identified, consisting of a few promoter elements that are encoded in the V segments upstream of the coding region and an enhancer element located between the J segments and the C-region exons (Figure 25-28a). The position of the enhancer ensures that the unrearranged V regions are not under its influence and that all rearranged genes are influenced by the same enhancer. Each V region, however, has its own promoter, but all contain the common OCTA and TATA motifs.

The TATA motif is common to many genes (Chapter 11) and does not have a known regulatory role. By contrast, OCTA, an 8-bp sequence, is regulatory: if it is artificially appended to another gene, say globin, it causes that gene to be preferentially expressed in cells of the B-lymphocyte lineage. Thus, the promoters of immunoglobulin genes are cell-specific regulatory elements.

The enhancers are also regulatory; they function only in B-lineage cells. The H-chain enhancer (Figure 25-28b) contains many motifs that bind individual proteins. The OCTA motif is the same one as found in the promoter. It, the II, and the μB motifs give the enhancer most of its cell-specific regulatory properties. The E motifs bind dimeric helix-loop-helix proteins. E-binding proteins are found in all cells, but there are also E-binders that are highly restricted in their distribution. Exactly how this interacting family of proteins regulates transcription remains to be understood.

For the κ L-chain enhancer, the κB site is the only known positive regulatory element while the E motifs are again helpers (Figure 25-28c). The κB site binds a protein called NF-κB that is activated by external stimulation of many cells of the body (see Figure 11-34). In B cells, at the time when κ transcription is activated, NF-κB is constitutively activated and remains present in the nucleus of B-lineage cells thereafter. Another element within the κ enhancer, the silencer motif, apparently prevents the enhancer from acting in non-B cells. Its properties are similar to silencers active in yeast.

(a) H-chain and κ L-chain transcriptional control elements

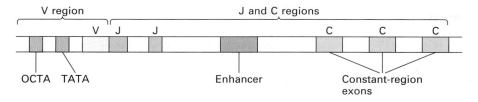

(b) The H-chain enhancer

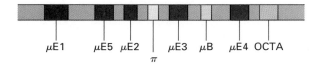

(c) The κ L-chain enhancer

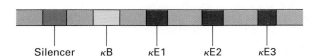

◀ **Figure 25-28** The control elements for immunoglobulin gene expression consist of DNA sequence motifs, 6–12 bp segments that bind specific proteins. The promoter regions contain common prominent motifs, OCTA and TATA; the enhancers are different for H and L chains, and each contains multiple motifs.

These transcriptional control elements work together to ensure that H chains and κ L chains are synthesized in B lymphocytes but not elsewhere in the body (how λ L chains are controlled remains to be discovered). As mentioned earlier, transcription of the H-chain and L-chain segments also signals rearrangement. H chains and L chains are under separate regulatory control: OCTA for all, OCTA plus others for H chains, κB for κ L chains. Regulation of the proteins that bind to these DNA regulatory regions must then be the determining factor, and it is: the B-lymphoid-specific OCTA-binding protein (the *oct-2* gene product, a POU-homeobox protein; Chapter 11) appears very early in the B-cell developmental sequence, about the time that D-J$_H$ joining first starts (see Figure 25-27). NF-κB appears later in the developmental pathway, just when κ segments start to rearrange.

Many other transcription factors are also regulated during B-lymphoid-cell development, and they, in turn, regulate a collection of genes whose products are needed during the differentiation process. Much remains to be learned, but a schematic representation of present knowledge is given in Figure 25-29. Included in that figure is a very preliminary indication of how T-cell differentiation may be orchestrated by transcriptional control. Much more is sure to be learned about these processes in the near future.

T Lymphocytes

Before considering the antigen-dependent phase of B-cell maturation, we must switch to a discussion of T cells because T-cell–B-cell collaboration plays a crucial role once virgin B cells are produced, leave the bone marrow, and start antigen-dependent activities.

The T lymphocytes are equal partners with B lymphocytes in generating the immune response. They show very high specificity of recognition, recognizing one particular antigen but not its close relatives. They have on their surface a recognition molecule able to make discriminations as fine as those made by antibody. That molecule is called the *T-cell receptor* to distinguish it from antibody. The T-cell receptor is closely related to antibody but has numerous properties that distinguish it from antibody. A key difference is that antibodies are produced in two forms, either cell-surface molecules or soluble secreted molecules, whereas the T-cell receptor exists only at the cell surface. All T-cell functions, therefore, involve reactions on the surface of the cell. The molecules to which T-cell receptors bind are designated antigens although they have radically different properties from the antigens to which antibodies bind.

T cells, like B cells, mature through a series of intrathymic stages before they are released to the periph-

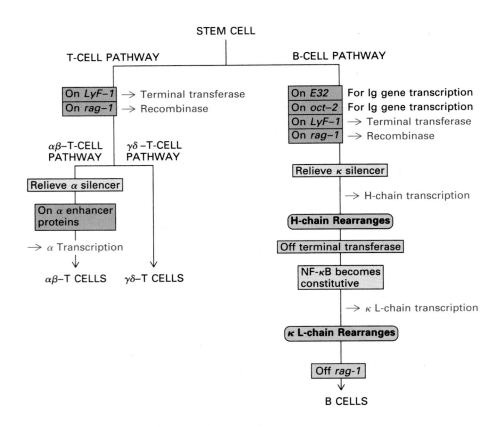

STEM CELL

T-CELL PATHWAY · B-CELL PATHWAY

◀ **Figure 25-29** Landmark events in the differentiation of lymphoid cells. Many of the known events during lymphoid differentiation are placed here on a time line. Some represent activation of the genes for specific transcription factors (E32, OCT-2, *LyF*-1, NF-κB); others are proteins or events whose exact nature is unclear. A complete description of the differentiation pathways has yet to be made but will probably include many more events than are shown here.

ery where they may undergo an activation process triggered by interaction with antigen. Antigens for T cells are always cell-bound molecules, so that T-cell receptor–antigen interactions are actually cell-to-cell interactions.

There Are Two T-Cell Receptor Molecules

We introduced earlier the two types of well-characterized T cells, T_H (helper) cells and CTLs (killer cells) (see Figures 25-1 and 25-13). These two types of cells use the same family of receptors, two-chain molecules built similarly to antibodies. The chains of the most common receptor are called α and β; they are 40,000 MW glycoproteins comprising two domains, one variable, one constant. The function of the second T-cell receptor, found on a minor subset of T cells, is obscure, but this receptor appears to be a key player in immunity to certain bacteria. This receptor is made of γ and δ chains. The receptor proteins, immunoglobulins, and many other cell-surface proteins are all built with the immunoglobulin fold as their backbone structure; they presumably all evolved from a single precursor protein and are considered members of the immunoglobulin superfamily (Figure 25-30).

The genes for α, β, γ, and δ chains have organizations similar to antibody genes: there are libraries of V, D, and J regions from which members are joined to form entire genes. The joining process is identical to that used for antibody gene segments, being directed by heptamer-nonamer recognition sequences. The libraries for α, β, and γ are separate entities found on human chromosomes 14, 7q, and 7p, respectively. The δ-chain segments are interspersed among the α-chain segments (Figure 25-31).

T-Cell Receptors Recognize Foreign Antigens as Compound Units with a Self-molecule

The T-cell receptor can be highly specific for recognition of a foreign antigen. It displays the tolerance to self-antigens also seen in antibodies. The antigen, however, must be presented to the T-cell receptor as part of a complex with a specific self-molecule. This behavior became clear in a classic experiment done in 1974 to examine how CTLs kill infected target cells. The investigators injected a virus into an inbred mouse, producing in the mouse CTLs that when studied in culture recognized and killed cells from that mouse infected with that virus. Just as is seen

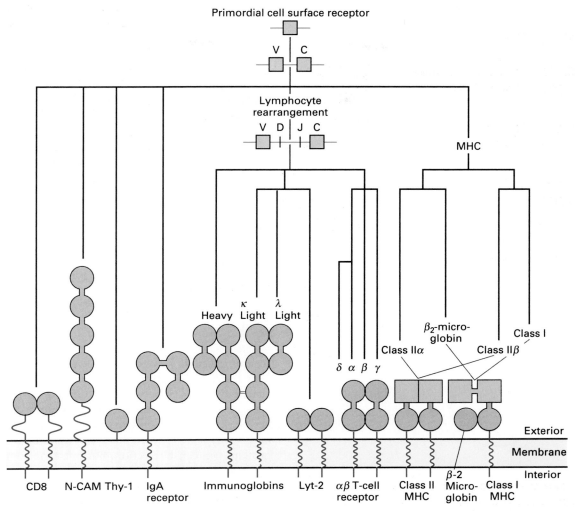

▲ **Figure 25-30** The superfamily of immunoglobulin-related proteins. For each protein, the individual balls represent domains of protein structure that are derivatives of the immunoglobulin fold described in Figure 25-19. Regions of MHC proteins denoted by boxes are not immunoglobulin folds. The genes encoding these proteins all have homologies and are considered a superfamily of related genes that must have evolved over hundreds of millions of years from a common ancestor. Many other proteins are part of the family, including *Drosophila* proteins that guide nerve growth, so the family predates the evolution of the immune system. *After L. Hood, M. Kronenberg, and T. Hunkapiller, 1985, Cell 40:225.*

T-cell receptor-chain genes

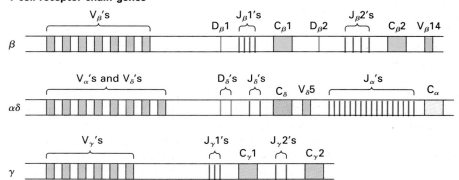

◄ **Figure 25-31** Structure of T-cell-receptor genes. There are three gene clusters on three separate chromosomes, all separate from the antibody gene clusters. The αδ cluster is the only case in which genes for two chains are interspersed.

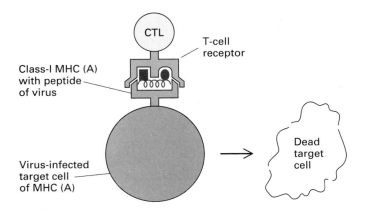

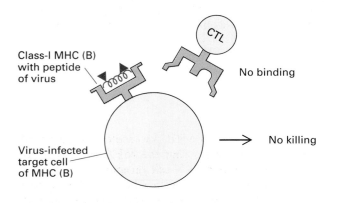

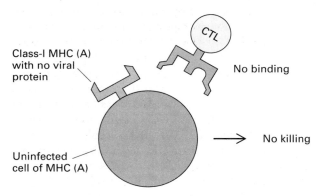

▲ **Figure 25-32** Restriction of CTL killing by class I MHC gene products. For the experiment outlined, CTLs are prepared from animals of MHC type A inoculated with a virus. These CTLs will kill cultured cells from type A mice infected with the virus. They will not kill infected cells of type B mice because the type B MHC product is not recognized by the T-cell receptor on the CTLs. Also, the type B MHC will generally bind a different peptide. The CTLs will also not kill uninfected type A cells because recognition by the CTL receptor requires that the MHC protein have the viral peptide bound to it.

with antibodies, the CTLs reacted only with cells infected by that particular virus. The experimenters then asked the apparently innocuous question: would the CTLs kill cells from a different inbred strain of mice infected by the same virus? To almost everyone's surprise, they would not. Further, uninfected foreign cells were also left unharmed by the CTLs. Apparently the CTL receptor recognized both the identity of the infected cell and the type of virus; it recognized self- and foreign determinants together, but neither one separately. What surface self-molecule might the T-cell receptor care about so much? As described below, the self-molecules recognized by CTLs are encoded in a region of the genome called the MHC genes, so that the CTL killing is restricted to self-MHC (Figure 25-32).

The MHC Genes Were First Recognized in Tissue Transplantation Experiments

If skin from one inbred strain of mice is grafted to the back of a mouse of another strain, the graft soon dies. We say that the graft from a donor to a recipient has been *rejected.* Grafts between mice of the same inbred strain are not rejected. To find out which genes encode the proteins causing the rejection, experimenters minimized the genetic differences between the donor and recipient animals by inbreeding until a region containing a small number of genes could be identified as causing the rejection. In this way it was shown that a fair number of cell-surface proteins can cause rejection. But one gene complex stood out in such experiments because differences among donor and host genes in this complex caused very fast rejection. The genes that lead to rejection are called *histocompatibility genes,* and the most potent genes are called the *major histocompatibility complex,* or MHC. Many individual genes have been found within the MHC, and most have been molecularly cloned. A complete map of the MHC includes some genes that elicit fast graft rejection and others that do not (Figure 25-33).

The basis for the ability of MHC proteins to elicit graft rejection is that each individual animal of a given species has MHC genes that differ in sequence from the MHC genes of other individuals (except some close relatives). The MHC proteins are said to be *polymorphic.* This polymorphism means that each individual has an identity defined by his or her particular set of MHC genes. The T-cell receptor on CTLs can bind to foreign MHC molecules, allowing the CTL to kill foreign cells and thus reject grafts. The CTLs, however, are tolerant toward self-MHC and therefore only kill foreign but not self-cells.

The MHC genes that encode the targets for CTL self-recognition are called *class I MHC genes,* typified by genes called *H-2D* and *H-2K* in the mouse and *HLA* in humans. These are two-chain proteins with one highly polymorphic component and one constant chain called β_2-*microglobulin* (see Figure 25-30). These MHC pro-

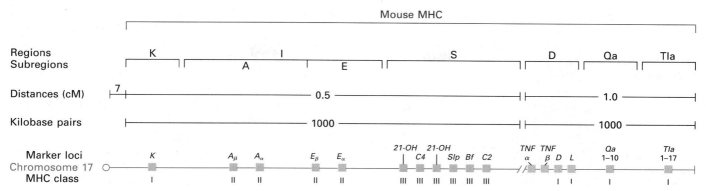

▲ **Figure 25-33** The MHC gene complex of the mouse. This complex of genes, called H-2 in the mouse, was first identified by graft rejection. The Tla region was found to encode cell-surface-differentiation antigens. Fine genetic mapping uncovered the *K* and *D* loci. Later work on T_H cell reactions located the I region. The S region was found to encode blood proteins (mainly of the complement system), and the tumor necrosis factor (TNF) genes were found in the region. Classical genetics defined the distances these complexes covered in units related to recombination frequency called centi-Morgans (cM). Molecular cloning identified individual protein-coding regions as shown on the bottom line of the diagram. Actually, there are many *Qa* and *Tla* loci. [See L. Hood, I. Weissman, W. Wood, and J. Wilson, 1984, *Immunology*, 2d ed. Benjamin; and E. Lai, R. Wilson, and L. Hood, 1989, *Adv. Immunol.* 46:1–59]

teins are part of the immunoglobulin gene superfamily (see Figure 25-30). Most cells in a mouse's body have class I H-2 proteins on their surface. The T-cell receptors on CTLs will not, as noted above, mediate killing of self-cells because they have been selected during their thymic maturation to avoid reaction with the body's own cells. However, the CTLs will bind to self-MHC if the MHC protein has a peptide from a foreign protein bound to it (see Figure 25-32). We call this a *joint recognition process:* both self-MHC and the foreign peptide must be jointly recognized on the same cell at the same time. Again, T-cell-receptor molecule recognizes MHC and the foreign peptide together, but neither one separately.

For T_H cells, joint recognition of self-molecules and foreign cell-surface molecules is also the rule. The self-molecule, however, is not a class I MHC molecule but a class II MHC molecule. Class II molecules have two polymorphic chains, both smaller than the polymorphic chain of class I molecules (see Figure 25-30). Class II molecules are found mainly on two kinds of cells, macrophages and B lymphocytes.

The true antigen for the T-cell receptor is a tight complex of a foreign peptide with an MHC molecule. The complex does not include an entire foreign protein, but rather a peptide derived from such a protein: to become a T-cell antigen, a protein is first degraded and a subset of released peptides become tightly bound to the MHC molecule on the cell surface. The MHC molecules have a cleft into which the peptide fits, and, through events yet to be understood, the MHC molecule so tightly clasps the peptide that they dissociate only over days (Figure 25-34).

When peptides bound to MHC molecules are displayed on a cell's surface, we say that the MHC molecules are *presenting* the peptide to a T-cell receptor.

There is a fundamental difference between the peptide-MHC complexes formed by class I and class II molecules. Class I MHC binds peptides from *intracellular* proteins, while class II MHC binds peptides from *extracellular* proteins. Thus peptides from viral proteins made in a virus-infected cell are presented by class I (as in Figure 25-32), while when macrophages digest free-living bacteria, they present the peptides in association with class II. Because most CTLs recognize class I–associated peptides, CTLs kill virus-infected cells. T_H cells, by contrast, generally recognize class II molecules and therefore help B cells make antibacterial antibodies. Also, class I molecules are on most cells, allowing most cells to signal to CTLs when they have been invaded. Class II molecules are found mainly on macrophages, the professional digesters of foreign materials in the body, and on B cells, allowing macrophages to recruit help and B cells to receive the help (a more complete description of this process follows).

T Cells Are Educated in the Thymus to React with Foreign Proteins but Not Self-proteins

The CTLs and T_H cells develop their receptor specificity in the thymus gland. There, soon after the α and β genes rearrange and the T-cell receptor is expressed on the cell surface, *positive* and *negative selection* events ensue, guaranteeing that the cells that exit the thymus are poised to recognize peptides in association with the specific class I or class II molecules carried by the particular animal. This process of selection is very exacting: 95 percent of newborn T cells die in the thymus, apparently because their T-cell receptor does not pass muster. Cells are destroyed either because they recognize neither class I nor class II of that particular animal well enough (positive selection) or because they recognize one so well, with a self-peptide in its cleft, that they would react with the normal body cells if allowed out into the body (negative selection) (Figure 25-35). The latter case is especially important because it lies at the heart of tolerance—the T

cells must be tolerant of self-molecules. In fact, the specificity of T-cell receptors represents a balancing act: they recognize self-MHC well enough to react strongly when foreign peptides are in the self-MHC cleft but not when self-peptides are there. They also can recognize foreign MHC molecules, the basis for their ability to reject grafts. Once T cells leave the thymus, their receptors are fixed; T-cell receptors do not undergo somatic diversification as happens in B cells. This makes sense because once intrathymic selection is completed, any change in the T-cell receptor would run a risk of developing uncontrollable self-reactivity.

(a)

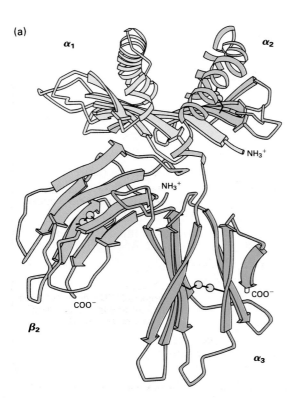

(b)

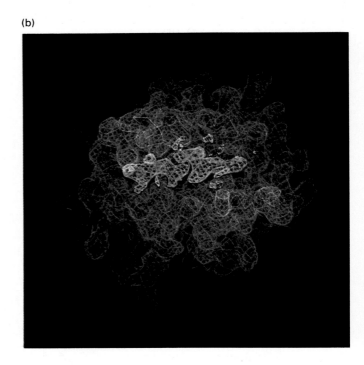

(c)

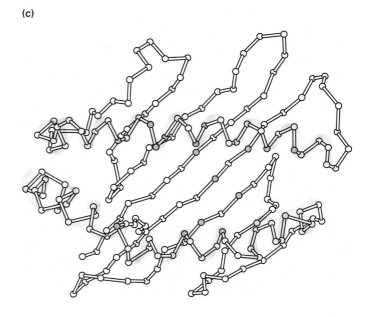

▲ **Figure 25-34** (a) Actual structure of an MHC class I molecule from x-ray diffraction data. The molecule is a dimer of β_2-microglobulin (green) and a three-domain α chain (blue). The α_1 and α_2 domains together form a peptide-binding channel lined on the bottom by β pleated sheets and on the sides by α helices, as shown at the top of the picture. (b) Computer-generated model of the peptide-binding site at the top of the MHC class I molecule. The pink region represents a peptide bound into the cleft formed by the α_1 and α_2 domains. (c) Stick diagram showing the cleft without a peptide, emphasizing that the walls are formed by α helices and the floor consists of a β pleated sheet. Amino acids (orange) that affect the molecule's antigen-binding capacity are clustered in the cleft. *Part (a) after original drawing by M. Silver, from P. J. Bjorkman et al., 1987,* Nature **329**:506; *part (b) courtesy of P. J. Bjorkman et al.; part (c) adapted from H. M. Gray et al., 1989,* Sci. Am. *(5)* **261**:64.

The Response of T Cells to Antigen Is Either Cell Killing or Secretion of Protein Factors

As mentioned earlier (see Figure 25-13), CTLs can be distinguished from T_H cells by specific surface proteins: CTLs generally have CD8 on their surface, T_H cells generally have CD4. (There are CD4-positive killer cells, but we will ignore this complication for this discussion.) These surface molecules play important roles in the killing or helper functions. Antibodies to CD4 and CD8 can prevent cells from responding to antigen. The interaction of antigen with T-cell receptor is thought to involve the intimate participation of either a class I MHC–CD8 interaction or a class II MHC–CD4 interaction (Figure 25-36). When the T cells undergo selection, they have both CD4 and CD8, and both the positive and negative selection events include MHC-CD interactions. It would appear that once positive selection has occurred, either CD4 or CD8 is permanently downregulated, leaving the cell as either $CD4^+CD8^-$ or $CD4^-CD8^+$. Upon stimulation by a complex of peptide and class II MHC, the $CD4^+$ cells make a helper response; stimulation of $CD8^+$ cells by a peptide-class I complex produces a killer response. The effector responses by T cells involve a series of other proteins called the CD3 complex that are tightly bound to the T-cell receptor as well as a protein, tyrosine kinase, the product of the *lck* gene, bound to both CD4 and CD8.

The killing and helper responses are quite similar: both rely on secretion of specific proteins. CTLs secrete proteases and proteins able to form nonspecific ion channels that depolarize the cell and destroy its ionic and osmotic balance. When a cell is attacked by a CTL, it appears to burst and then shrink. One CTL can kill many targets.

The T_H cell responds to antigen by secreting factors *(lymphokines)* that stimulate both T- and B-cell growth and maturation. As shown in Table 25-2, individual lymphokines can have multiple actions and a given response can be elicited by more than one lymphokine. There are many lymphokines known and more to be discovered. They are often named as interleukins (abbreviated IL); one of the first to be discovered was IL-2, known then as T-cell growth factor. Not all cells secrete all factors, and there are thought to be different T_H cells that orchestrate responses to different types of infections by the particular subset of lymphokines they secrete. CTLs also secrete a subset of lymphokines along with their killer proteins. The lymphokines have multiple effects on B cells, as described later. They stimulate T cells, particularly T_H cells, and produce an autocrine growth stimulation. They also stimulate other blood cells, particularly phagocytes, such as macrophages and granulocytes. Skin-hypersensitivity reactions, as in the skin test for reactivity to tuberculosis antigens, are due to T_H cell mobilization of phagocytes (called *delayed-type hypersensitivity,* or DTH).

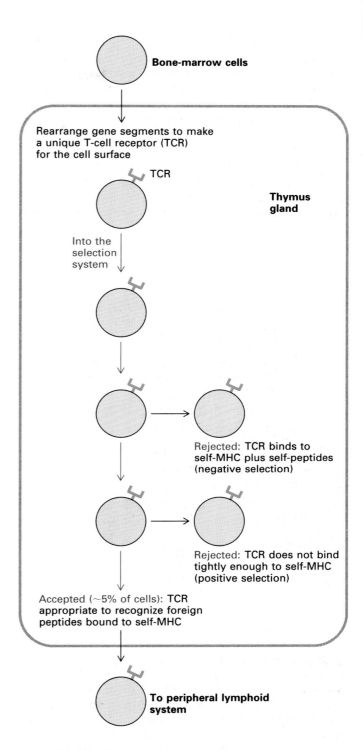

▲ **Figure 25-35** A schematic representation of the role of thymus gland in selecting T cells able to recognize foreign but not self-proteins. The cells start into selection after they have rearranged their T-cell receptor genes to generate a single T-cell receptor (TCR) on their surface. The selection system then rejects cells that either have TCRs which react too well with self-proteins (negative selection) or which cannot recognize self-MHC well enough for it to be a useful carrier of foreign peptides (positive selection). The details of the selection process remain to be worked out.

Macrophage or target cell

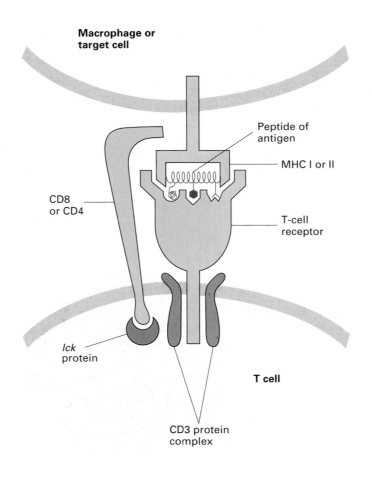

Peptide of antigen

MHC I or II

CD8 or CD4

T-cell receptor

lck protein

T cell

CD3 protein complex

Table 25-2 Some lymphokines secreted by T_H cells in response to antigen*

Lymphokine	Function
IL-2	T-cell growth Stimulation of antibody secretion by B cells
Interferon γ (IFNγ)	Macrophage activation Ig class switching by B cells
IL-4	B-cell activation Ig class switching T-cell growth Macrophage activation
IL-3	Stimulation of blood-cell differentiation in many lineages
GM-CSF	Stimulation of growth and differentiation of granulocytes and macrophages
IL-6	Stimulation of antibody secretion by B cells Many other effects on nonblood cells
IL-7	Stimulation of pre-B cell growth

*This is a very partial list of factors and their effects.

◀ **Figure 25-36** Interaction of a peptide-MHC complex on a presenting cell with a receptor on a T cell. The presenting cell can be a macrophage using a class II MHC or any tissue cell using a class I MHC. The T cells can be T_H cells carrying CD8 or CTLs carrying CD4. The class I MHC binds CD8, class II binds CD4. The T-cell receptor not only has CD8 or CD4 as components of its interaction system, it also has four different proteins that make up the CD3 complex—their role is uncertain. Also, bound to the CD4 or CD8 proteins is the *lck* protein-tyrosine kinase, which can phosphorylate a CD3 component on tyrosine in response to an interaction of the T-cell receptor with MHC-peptide complex. The CD3 and *lck* proteins are apparently part of the response system used by T cells following encounters with an MHC-peptide complex antigen.

Lymphokines function through specific receptors on the surface of target cells. The activation of a T_H cell by an encounter with a peptide-MHC complex induces formation of an IL-2 receptor for the IL-2 that it secretes, thus activating the autocrine growth response.

The same pole of a T cell that interacts with the peptide-MHC complex on a target cell secretes both killer factors and helper factors. Capping on the cell surfaces, a process described below, may provide the polarity. Thus the killer factors kill their targets without nonspecific killing of other cells, and T_H cells can partially polarize their stimulation to cells that present cognate antigens.

Having introduced T cells and their unique mode of recognizing antigen, we can return to considering the events that ensue when a pathogen invades an organism.

The Antigen-dependent Phase of the Immune Response

A key event of the immune response is an antigen's encounter with a B lymphocyte that bears surface antibody having binding specificity for that antigen. This encounter, coupled to a coordinated response from helper T cells, causes the virgin B cell to mature further: it takes the B cell from its antigen-independent stage and makes it an active component of the response system. Many molecular events transpire from an encounter with antigen. They include the activation of proliferation, secretion of antibody, production of IgG and other secondary antibody classes, somatic mutation, and the production of memory cells.

After the B lymphocyte has matured to a virgin B cell and has been expelled from the bone marrow into the periphery, it will live for only a few days as a circulating cell unless an antigen interacts with its surface antibody, triggering further growth and maturation. That antigen will generally be a soluble protein or a microorganism.

The antigens against which the immune response can be most effective are multivalent—they have multiple determinants on one structure—so that many antibody molecules on one B cell can be bound together to form a tight patch on the cell surface (Figure 25-37). It is generally said that a competent antigen *cross-links* surface antibody; this cross-linking, or patching, appears to be a key to activation of a B cell. A polymeric antigen, such as a carbohydrate with repeating sugars, can be a very good activator.

Surface antibody can be aggregated by multivalent antigens because of the fluidity of the lipid-bilayer plasma membrane in which antibody is embedded. After island patches form on the cell surface, they coalesce into a cap on the surface (Figure 25-37) that is either shed into the surrounding fluid or internalized by endocytosis and degraded.

The achievement of B-cell activation requires helper T cells. These T_H cells have a receptor that usually binds to an area of the antigen different from that to which the B

(a)

B lymphocyte
with receptors
laterally mobile
in membrane

Multivalent antigen

Receptors
bind to
antigen

Mobile
receptors and
antigen
cluster into
patches

Patches form
cap at cell pole

(b)

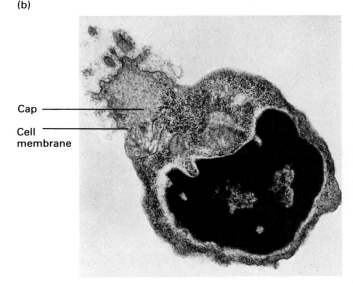

Cap

Cell
membrane

(c)

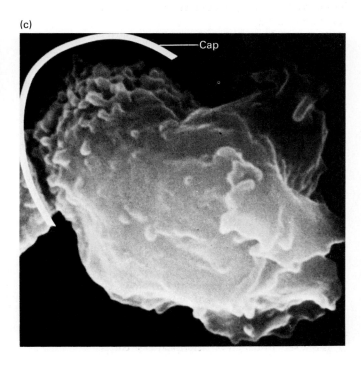

Cap

▲ **Figure 25-37** Patching and capping of surface antibody molecules. (a) Schematic representation of the aggregation of surface antibody and experimental demonstration by immunofluorescence. The diagram illustrates a multivalent antigen linking together surface antibody on a B lymphocyte. The end result is a cap at one place on the surface of the cell. With the use of fluorescence-labeled antibody that can react with the surface antibody, the process of patch and cap formation is visually demonstrated. (b) Transmission micrograph of a sectioned B lymphocyte with its cap region evident at the left. Within the cell, the cap region is clear of organelles because it is filled with actin and myosin that move to the cap region along with the surface antibody. (c) The cap is shown in a scanning electron micrograph of a B lymphocyte. *Photographs courtesy of J. Braun.*

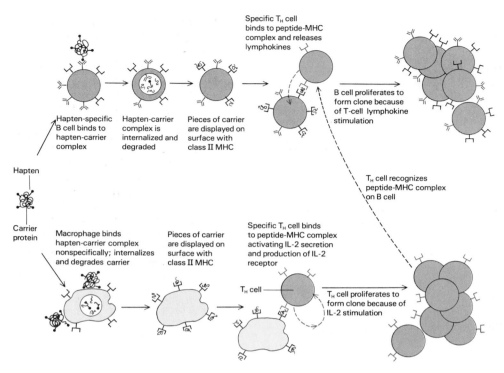

▶ **Figure 25-38** The hapten-carrier complex activates B cells by first stimulating T$_H$ cells. The hapten-carrier complex is a protein to which several hapten molecules have been covalently coupled. The hapten portion binds to a B cell but cannot by itself initiate B-cell proliferation. The B cell internalizes and digests the hapten-carrier complex, and portions of the carrier peptide are displayed on the B-cell surface as a complex with class II MHC protein. A macrophage cell also internalizes carrier protein and displays peptide fragments on its surface in association with class II MHC. Then T$_H$ cells with appropriate receptors bind to the peptide-MHC complex displayed on the macrophage surface. Binding stimulates the T$_H$ cells, which then proliferate, recognize the identical peptide-MHC complex on the B cell, and secrete factors that stimulate the B cell to grow.

cell binds. For instance, when the hapten dinitrophenol is coupled to bovine serum albumin and injected into a mouse, B cells will bind to the hapten on the intact protein but T$_H$ cells will be activated by peptides derived from degradation of some of the protein. After the antigen is bound by macrophages and internalized, the macrophages degrade it to peptides. These peptides are then displayed on the macrophage surface in association with class II MHC molecules, where a T-cell receptor on a T$_H$ cell can react with them. T$_H$ cells that carry receptors for the macrophage-bound fragments then proliferate through autocrine stimulation, producing a pool of carrier-specific T$_H$ cells.

The collaboration of specific T$_H$ cells and specific B cells that induces the B cell to secrete specific antibody is shown in Figure 25-38. The key to this mechanism is that a B cell can degrade a protein to peptides and display them on its surface in association with MHC class II molecules exactly the way a macrophage can. The macrophage processes all foreign proteins indiscriminately; the B cell can process specifically those proteins that bind to its surface receptor. A dinitrophenol-specific B cell will bind the dinitrophenol-carrier complex, internalize it by endocytosis, and express the carrier peptides on its surface with class II. The carrier-stimulated T$_H$ cells will then recognize the peptides and become further stimulated. In this interaction, the T$_H$ cell produces *B-cell growth* and *differentiation factors* (lymphokines, see Table 25-2), which activate the B cell to proliferate.

This mechanism explains why haptens must be bound to carriers to be antigenic. Free hapten could bind to B cells, but it lacks multivalency; it cannot stimulate T$_H$ cells and therefore cannot activate the B cells. We believe that a hapten-protein conjugate made in the laboratory is a surrogate for determinants constructed solely of amino acids that are present all over proteins, so that natural proteins act as both haptens and carriers and are therefore antigenic.

It is evident that antibody secretion requires the activation of two kinds of cells, B cells and T$_H$ cells. The word *activation,* as used here, is actually a technical term; it means the transformation from a small, nondividing lymphocyte to a larger cell, called a *blast.* The blast is a cell that is metabolically much more active than the small cell: it enters the division cycle and divides every 8–24 h, it actively secretes important substances, it has receptors for T-cell-derived lymphokines, and it can mature to become an antibody-secreting plasma cell. The activated T$_H$ cell secretes many substances, the most important for this discussion being the lymphokines IL-2, IL-4, IL-5, and IL-6. These factors find receptors on activated B cells and stimulate their growth and maturation to plasma cells. These receptors are not found on resting B cells. In the joint activation of B cells and T$_H$ cells by antigen, one antigenic protein molecule is recognized by two kinds of cells. Only B cells and T cells specific for the antigen are activated because each cell must recognize the antigen to develop the receptors for the growth factors. Bystander

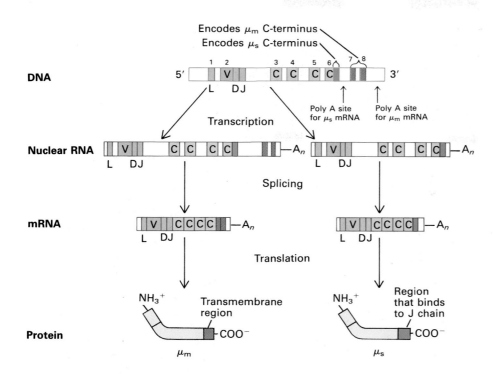

Encodes μ_m C-terminus
Encodes μ_s C-terminus

◀ Figure 25-39 Production of μ_m and μ_s chains by alternative polyadenylation of mRNAs. The genome organization in the C_μ region includes two polyadenylation sites. When one is used, the mRNA formed by splicing the nuclear RNA precursor has the coding region for the μ_s C-terminus. The other polyadenylation site is farther toward the 3′ end and includes two additional exons that encode the μ_m C-terminus. When this 3′ site of polyadenylation is used, the μ_s coding region is spliced out but the μ_m region is included. Thus, the C_μ region can encode both surface antibodies (with the μ_m C-terminus) and secreted antibodies (with the μ_s C-terminus).

Bystander cells that have not seen antigen are not activated. Another protection against bystander activation is that the T cell focuses its secretion toward the B cell that is presenting the antigen.

Secretion by Activated B Cells Entails Many Cellular Changes

A massive increase in synthesis of immunoglobulins accompanies activation of B cells. Apparently the synthesis rate of immunoglobulin-specific RNAs does not increase greatly but rather the processing and stabilization of such RNAs changes. The exact nature of the change is not understood. To accommodate the increased rate of synthesis, the activated B cell develops a larger cytoplasm with more endoplasmic reticulum. The end stage of this maturation is a plasma cell that secretes up to 10 percent of the protein it makes as antibody. The cell is highly specialized for secretion, with a large Golgi region as well as a large, layered endoplasmic reticulum (see Figure 25-10). Such a cell is the final stage of B-lymphoid development. It loses its proliferative ability entirely and dies after many days of antibody production.

Secretion Requires Synthesis of an Altered H Chain

One of the necessities of clonal selection is that a virgin B lymphocyte activated by an encounter with antigen must secrete antibody with exactly the binding specificity of the antibody previously carried on its surface. This means that the L- and H-chain variable regions must be the same in the virgin cell and the secreting cell. Thus the same variable-region DNA must be used for both forms of immunoglobulin, requiring that the membrane-binding portion of the H-chain constant region be altered to make the secreted form without alteration of the variable region. The constant region of the L chain need not be transformed because it does not bind directly to the membrane but rather to the H chain. L chain is exactly the same in membrane-bound immunoglobulin and in secreted immunoglobulin. The alteration in the membrane-binding segment at the C-terminal end of the H chain is accomplished by a change in the type of RNA transcript made by the cell's DNA. An exact understanding of this requires a closer look at the H-chain's constant region.

The form of immunoglobulin on the surface of virgin B lymphocytes is *membrane-bound IgM.* The IgM previously described as a pentamer of the basic four-chain immunoglobulin molecule (see Figure 25-6) is *secreted IgM.* The membrane-bound form is a monomer, a single four-chain unit held to the membrane by a hydrophobic sequence at the C-terminal end of each of its H chains. Eight exons contribute to specifying the structure of the μH chain of membrane-bound IgM (the μ_m chain): a leader exon, a variable-region exon, four exons that encode the four domains of the constant region, and two exons that encode the membrane-binding segment of the molecule (Figure 25-39). The mRNA for the μ_m chain is polyadenylated just beyond the exon specifying its C-terminus. Splicing brings together the exons to generate the mature mRNA for μ_m.

The C-terminus of μ_s, the secreted form of the μ chain, is encoded by a segment of DNA contiguous with the sixth exon (Figure 25-39). The mRNA for μ_s is polyadenylated just after the coding region for μ_s, excluding the two terminal exons specifying the μ_m C-terminus. The result is that an intron sequence of μ_m becomes an exon sequence of μ_s: the two mRNAs have different 3′ structures, generating different C-terminal ends on the proteins.

The detailed structures of the μ_s and μ_m C-terminal ends are just what would be predicted for the behaviors of the two proteins. The μ_m terminus includes a string of 26 uncharged amino acids preceded and followed by clusters of charged residues, a classic membrane-spanning region (see Figure 13-16). The secreted form lacks a membrane-spanning region but includes a cysteine residue that becomes the disulfide bridge linking the four-chain monomers into the pentamer of the secreted molecule. It also has a site for carbohydrate addition, presumably as a way of making the molecule more soluble.

Because the mRNAs for μ_m and μ_s differ in site of polyadenylation, it is thought that control over polyadenylation is the basis for the switch from μ_m mRNA to μ_s mRNA when a B lymphocyte is activated (see Figure 25-39). This is the most obvious hypothesis but not necessarily the correct one. Some evidence exists for premature termination of transcription as a mechanism of control, and some control may be a consequence of differential protein stability.

Synthesis of μ_m and μ_s by alternative utilization of the same DNA sequence ensures the identity of the variable region on the two proteins and thus is a central mechanism for allowing the clonal selection process to operate in the immune system.

The use of alternative transcription products from one region of DNA is widespread in the immune system. For instance, there are two types of virgin B lymphocytes, one with IgM on its surface and one with both IgM and IgD. The IgM and IgD have identical variable regions: the L chains of IgM and IgD are the same, and the H chains are formed by differential polyadenylation within transcription units that are initiated just before the beginning of the VDJ complex. There are secreted forms of IgD, but most IgD is membrane-bound.

Two Cell Types Emerge from the Activation Process: Plasma Cells and Memory Cells

We have examined thus far only two consequences of activation, the proliferative response and the maturation response. Maturation, characterized by increased synthesis of immunoglobulin and a switch of secretion, requires

changes in the amount and type of mRNA as well as changes in cell architecture. It probably involves the synthesis of many new proteins and the inhibition of synthesis of others. For instance, J-chain synthesis (for initiating polymerization of secreted IgM) is turned on by activation. Surface antigens also change after activation. The result is a plasma cell.

Not all progeny of an activated B lymphocyte are plasma cells. The other important product is the *memory B cell*. These are cells that retain for the life of the animal a record of the antigens that it has previously encountered. Memory cells preserve variable regions that have previously proved useful, so that a second encounter with an antigen can elicit a rapid, highly avid response. Like virgin B lymphocytes, memory cells carry on their surfaces the antibody they are programmed to make. Unlike virgin B cells, they are immortal, circulating continuously as quiescent cells. An encounter with antigen activates them just as it does virgin cells. Memory cells can be helped by specific T_H cells, which also persist after a primary immune response. What is unclear is how the activated virgin cell can segregate off some of its progeny as memory B cells and others as plasma cells.

Activation Leads to Synthesis of Secondary Antibody Classes

The antibody classes IgG, IgE, and IgA play no role in the earliest stages of an immune response because they are not found on the surface of virgin B cells. For this reason they are referred to as the *secondary antibody classes*. Activated cells *switch* from IgM and IgD synthesis to the synthesis of these secondary classes. Because the various classes differ in their H chains and not their L chains, it is an *H-chain switch* that underlies the ability to change antibody synthesis. Analogous events described earlier, where cells switch from μ_m to μ_s synthesis and from synthesis of IgM alone to IgM-plus-IgD synthesis, are known as antibody *transcript-processing switches*. For synthesis of the secondary antibody classes, we must consider DNA rearrangements called *switch recombinations* as well as transcript-processing switches.

The mechanisms of switching are inherent in the organization of the H-chain locus. Previously we focused on the variable-region components and the C_μ region (the region encoding the eight C_μ exons). Downstream of C_μ are segments that encode the other constant regions: C_δ, the C_γ's, C_ϵ, and C_α in that order (Figure 25-40). Because there are four types of IgG and thus four C_γ's, there is a total of eight different C regions. Each C region is made of multiple exons that encode the domains of the individual H chains.

In a switch recombination, switch, or S, regions of DNA recombine with each other, deleting an intervening segment. Every C region (except C_δ) has an associated S region somewhat 5′ of its coding exons. The S regions are

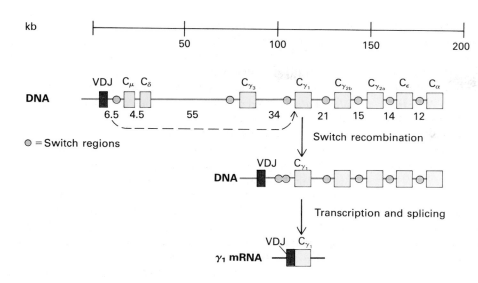

◀ **Figure 25-40** Organization of the H-chain region of DNA and switch recombination. The positions of the H-chain constant regions are indicated. A permanent change in H-chain synthesis from μ to a downstream constant region is produced by recombination between switch regions (green) located upstream of each of the constant regions. The constant regions are composed of multiple exons, and each region has alternative secreted and membrane-binding C-termini, but these details have been omitted. The numbers between units indicate the distances in kilobases.

constructed from internally reiterated short sequences that can recombine with each other. As a result of switch recombination, the VDJ unit is brought close to a new C region (Figure 25-40). The DNA is then polyadenylated just beyond that C region, determining that the cell will make only a single class of antibody.

Switching is an event that follows activation of a B lymphocyte. Thus the progeny of a single activated cell can be variable in the classes of antibody they make. When a B cell encounters an antigen, it initially makes IgM, but then secondary antibody classes, such as IgG, begin to predominate (Figure 25-41).

The signals that activate and direct the switching process in B cells appear to be the lymphokines secreted by T cells. For instance, IL-4 can induce cells to switch to γ_1 or

ϵ synthesis, IL-5 induces a switch to α synthesis, and γ-interferon induces a switch to γ_{2a}.

The mechanism of switching remains uncertain, but it appears that an S region is activated by transcription from a local promoter: the lymphokines may control switching by controlling transcription within the H-chain constant-region DNA. Thus, the control of switching specificity may be analogous to the control of joining specificity as described earlier. The direction of switching (to γ, ϵ, or α) is ultimately a function of the antigen: skin parasites induce IgE (ϵ chain) while intestinal microorganisms induce IgA (α). Specific lymphokines are probably induced by specific types of pathogens.

The ultimate progeny of activated B lymphocytes are plasma cells and memory cells. The switching process in-

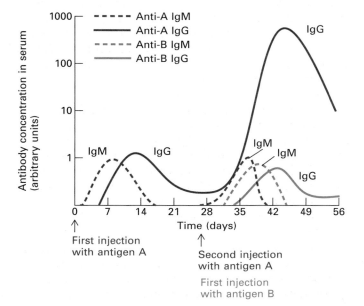

◀ **Figure 25-41** Kinetics of IgM and IgG responses after primary and secondary immunization. The diagram expands upon that in Figure 25-11, where only total antibody concentrations were indicated. Here the initial immune response to an antigen injection is seen to change with time from an IgM-dominated response to an IgG-dominated response. Once the initial response has subsided, a second injection with the same antigen leads to a prompt and massive IgG response because of the large number of memory cells prepared to make IgG. The secondary response also includes an IgM response, probably emanating from virgin B cells that are always present and prepared to be activated. Antigen B, introduced during the second injection, gives the IgM-to-IgG pattern of a primary response.

creases the number of progeny types. The plasma cells can make any of seven classes of immunoglobulin, but because they undergo switch recombination during their maturation, each cell makes only a single antibody class. The memory cells are also of many types: they may even be able to make multiple classes of antibody. Because each C region has two alternative C-terminal ends, each class of antibody can be represented by both secreted and membrane-bound molecules. The secreted forms are made by plasma cells; the membrane-bound forms are found on memory cells.

When memory cells respond to antigen, they make antibodies mainly of the secondary classes. During a second encounter with antigen there is an immediate IgG response of much greater magnitude than the initial response. If during a second immunization with an antigen a new antigen is also introduced, the animal undergoes a memory-dominated response to the previously encountered antigen and an IgM-dominated naive response to the new antigen (Figure 25-41). It is thus clear that immunologic memory is specific for only those antigens previously encountered.

Somatic Mutation of Variable Regions Follows from Activation

We have described the Ig variable regions as arising from combinatorial joining events among V, D, and J regions. We have followed the fate of the joined VJ and VDJ regions through transcriptional alterations and switching events with the implicit assumption that the variable regions are not altered once they are formed. This assumption, however, is false. There is a second type of mechanism that increases the variety of variable regions. Called *somatic mutation,* it involves the replacement of individual bases in a joined VDJ segment with alternative bases, causing apparently random variation throughout the VDJ segment (and probably also in flanking DNA for some distance on either side of the VDJ). Somatic mutation has been documented in both L-chain and H-chain variable regions by comparing germ-line DNA sequences with expressed DNA gene sequences (Figure 25-42).

Somatic mutation has mainly been found in VDJ segments from cells expressing the secondary antibody classes IgG and IgA, suggesting that it is an event of im-

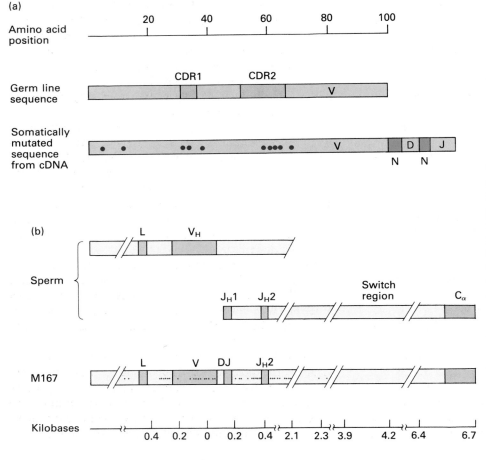

◀ **Figure 25-42** Somatic mutation of rearranged V_H regions. (a) Comparison of the nucleotide sequence of a germ-line V_H region with the sequence of the cDNA for an H-chain variable region that has undergone somatic mutation. The scale is given as amino acid positions, but 10 nucleotide changes are indicated (dots). [See A. Bothwell et al., 1981, *Cell* 24:625–637.] (b) A comparison of germ-line (sperm) DNA sequences and somatic mutations found in H-chain genomic DNA from a myeloma (M167). The consequences of somatic mutation are evident as alterations in nucleotide sequence, indicated as dots. Somatic mutation is localized to the variable region and a small region around it. *After S. Kim et al., 1981, Cell 27:573.*

munodifferentiation following activation of the virgin B lymphocyte by antigen. The changes in amino acid sequence caused by somatic mutation have the consequence of varying the fit between the antibody and the antigen, changing the affinity of the antibody for the antigen.

It is thought that higher-affinity surface antibody leads to easier activation of B lymphocytes by antigen and that therefore there is a continual selection among the somatically mutated variable regions for cells bearing higher-affinity antibody. As antigen concentrations fall, this should be an especially prominent effect. Somatic mutation throughout the variable region should have its greatest effect in the CDRs, where mutation can lead to higher affinity of binding. In fact, somatic mutations that cause amino acid replacements are especially prevalent in the regions encoding the three CDRs (Figure 25-42).

The rate of somatic mutation has been estimated to be as high as 10^{-3} per nucleotide per cell generation, a rate 10^6-fold higher than the spontaneous rate of mutation in other genes. Such a high rate of mutation over the whole genome would be insupportable. There must exist mechanisms that direct mutational activity to variable-region sequences. How these might work is not known; possibly some sequence in the area of the variable region directs a special enzyme system to carry out point replacements of nucleotides independently of template specification. The consequence of somatic mutation is an increase in the kinds of potential antibodies an animal can make from the 10^{11} we calculated earlier to a number many orders of magnitude higher. The greatly increased diversity will include better fitting antibodies but also, and more frequently, poorly designed antibodies. Somatic mutation occurs during the proliferative response following an encounter with antigen, and we assume that if a cell making a poor antibody is produced, its division ceases and it dies, so that only cells with appropriately designed surface antibody are continually stimulated to grow and mature. Darwinian processes continue from the first moment of antigen encounter until the cell ceases somatic mutation.

Tolerance Is Achieved Partly by Making B Cells Unresponsive

Nothing thus far has explained how B cells avoid making antiself antibody. In fact, the somatic mutational process could easily cause self-reactive clones to appear. The answers to how B-cell tolerance is achieved are not yet complete, but one elegant experiment has provided a partial understanding. For that experiment, two types of transgenic mice were created, one making the enzyme hen-egg lysozyme and the other making high-affinity antibodies to hen-egg lysozyme by virtue of having transgenes for appropriate H and L chains. Progeny of crosses between these mice have both the antigen and the ability to make an antibody to it, but they display tolerance. Because they

have been exposed from early in life to the hen-egg lysozyme, it is treated as self, and despite the presence of the genes for the antibody in all of the animal's B cells, little antibody is made. B cells are present but not secreting antibody in these mice. Thus, the autoreactive B cells were not eliminated, only prevented from reacting. How activation of B cells is prevented remains to be learned, but the nonreactivity is a property of the B cells themselves: if they are transferred to a normal recipient mouse, they remain nonresponsive to antigenic stimulation for some time even if good T cells are present. This mechanism is very different from the manner in which tolerance is achieved in T cells during thymic maturation.

Summary

The immune system evolved as the body's protective mechanism against invasion by pathogens such as viruses, bacteria, and fungi. There are two arms to the system: humoral immunity mediated by soluble protein antibodies in bodily fluids, and cellular immunity carried by surface receptors on circulating cells. B lymphocytes make antibodies; T lymphocytes carry out cellular immune reactions.

Each antibody molecule has two identical binding sites that can specifically bind to an antigen. The binding reaction is a simple, reversible binding characterized by an affinity constant. Antibody molecules also have effector domains that allow the body to rid itself of the antigen. There are five major antibody classes with different effector activities: IgM, IgD, IgG, IgE, and IgA.

The mechanism by which antibodies are formed is called clonal selection. The system works by producing an enormous variety of B lymphocytes, each with a homogeneous population of cell-surface antibody molecules. The antigen selects from the population of antibody-bearing cells those that carry molecules able to bind to it. Such cells are induced to multiply and mature into antibody-secreting cells. Some activated cells mature into plasma cells, which are specialized for immediate antibody production, and others mature into long-lived memory B cells, which respond rapidly to all further encounters with the antigen.

There are two major types of T cells: cytotoxic T lymphocytes (CTLs) and helper T lymphocytes (T_H cells). The CTLs (also known as killer T cells) directly kill target cells that they recognize with their surface antibodylike receptors. The T_H cells assist B cells in their reaction to antigens.

The ability of an animal's immune system to avoid reacting to the molecules in its own body is called tolerance and is an active process maintained, at least in part, by the T cells. A failure of tolerance leads to autoimmune disease, one of a number of disease types that can be caused by the immune system. Others are failures of immune function, tumors of the immune system, and hyper-reactive conditions such as allergies.

Our knowledge of the details of the structure and synthesis of antibodies is heavily dependent on studies of myelomas, tumors that secrete antibodies. By studying myeloma products, antibody proteins were found to be two-chain molecules. The heavy and light chains that constitute the molecules are each folded into a number of compact domains that form both the binding sites and the effector regions. The N-terminal portions of each chain are highly variable in amino acid sequence, producing the binding variability of the molecules. Within the variable regions, three regions directly interact with antigen and are called complementarity-determining regions (CDRs).

The structures of the genes that encode antibodies are bipartite: constant segments are attached to any of a library of variable segments. Antibody genes are not carried per se in the genome but rather are carried as gene segments that come together during lymphocyte differentiation to form the variable regions. The joining process is signaled by a DNA recognition signal consisting of a heptamer and a nonamer separated by either about 11 or 23 nucleotides, that is, by one or two turns of the DNA helix. The joining rule is that a one-turn element always joins to a two-turn element. Diversity of antibody structure is partly a consequence of the large size of the variable-segment libraries and partly a consequence of combinatorial diversity in the joining of members of the libraries. Imprecision at the joints between segments further increases diversity. During heavy-chain gene formation random sequence elements are inserted into the third CDR.

The process of DNA rearrangement to make antibody genes is an orderly one. First, heavy-chain genes are rearranged, then light-chain genes. The variability inherent in the system can make as many as 10^{11} different molecules. Thus the immunodifferentiation process produces a vast array of cells from which antigens can choose those that fit best.

The T-cell classes carry out their function using an antibodylike T-cell receptor encoded by a set of genes quite separate from those that encode antibodies. The receptor is a two-chain molecule, each chain having a variable and a constant region. Because T-cell receptors remain membrane-bound and T cells recognize antigens on the surface of other cells, T-cell recognition is a cell-to-cell recognition process. The structure on a target cell recognized by the T-cell receptor is not the foreign antigen itself: the T cell actually recognizes peptides from the foreign protein. These peptides are bound to self-proteins called MHC (major histocompatibility) proteins, and the T-cell receptor binds to the MHC-peptide complex. Thus T-cell recognition is a joint process of binding to both self- and foreign protein molecules.

The activation of a B cell by an encounter with an antigen is the key process in successful clonal selection. Most antigens are unable directly to activate B cells and require help from T_H cells. Such antigens are degraded by macrophages whose MHC class II proteins bind peptide fragments of the antigen and present them on the cell's surface to T_H cells. Because surface antibody and secreted antibody must differ in having or lacking membrane-binding domains in their heavy chains, there are alternative exons encoding the C-terminal ends of the heavy chain. Part of the activation of antibody involves increased utilization of the secretion exon by a process of differential polyadenylation and splicing.

The B-cell activation process can also lead to a switching process by which progeny cells may make antibodies with different effector domains than those of the parental cells. Furthermore, activation involves an extensive process of specific somatic mutation of the heavy- and light-chain variable regions. Thus the clone of cells resulting from activation actually contains extensive variability. Continual Darwinian selection acts to maintain those cells that make antibodies with the highest affinity for antigen.

References

General References

Advances in Immunology. Yearly volumes.

Annual Review of Immunology. Yearly volumes. The latest is 1989, Vol. 7, edited by W. Paul, G. Fathman, and H. Metzger.

Immunological Reviews. Bimonthly volumes on topics in immunology. Edited by G. Möller. Munksgard, Copenhagen.

Immunology Today. Monthly issues covering a wide range of immunology.

JERNE, N. K. 1973. The immune system. *Sci. Am.* **229**(1):52.

KABAT, E. A. 1976. *Structural Concepts in Immunology and Immunochemistry,* 2d ed., Holt.

LAI, E., R. WILSON, and L. HOOD. 1989. Physical maps of the mouse and human immunoglobulin-like loci. *Adv. Immunol.* In press. A particularly valuable review.

LANDSTEINER, K. 1945. *The Specificity of Serologic Reactions,* Harvard.

Clonal Selection

BURNET, F. M. 1957. A modification of Jerne's theory of antibody production using the concept of clonal selection. *Austral. J. Sci.* **20**:67.

MANSER, T., S. Y. HUANG, and M. L. GEFTER. 1985. The influence of clonal selection on the expression of immunoglobulin variable region genes. *Science* **226**:1283.

Antibodies

AMIT, A. G., R. A. MARIUZZA, S. E. V. PHILLIPS, and R. J. POLJAK. 1986. Three-dimensional structure of an antigen-antibody complex at 2.8 Å resolution. *Science* **233**:747.

AMZEL, L., R. POLJAK, F. SAUL, J. VARGA, and F. RICHARDS. 1974. The three-dimensional structure of a combining region-ligand complex of immunoglobulin NEW at 3.5 Å resolution. *Proc. Nat'l Acad. Sci. USA* **71**:1427.

DAVIES, D. R., and H. METZGER. 1983. Structural basis of antibody function. *Ann. Rev. Immunol.* **1**:87.

DAVIES, D. R., E. A. PADLAN, and D. M. SEGEL. 1975. Three-dimensional structure of immunoglobulins. *Ann. Rev. Biochem.* **44**:639.

EDELMAN, G. M. 1970. The structure and function of antibodies. *Sci. Am.* **223**(2):34.

LOH, D. Y., A. L. M. BOTHWELL, M. WHITE-SCHARD, T. IMANISHI-KARI, and D. BALTIMORE. 1983. Molecular basis of a mouse strain–specific anti-hapten response. *Cell* **33**:153.

SIEKEVITZ, M., S. Y. HUANG, and M. L. GEFTER. 1983. The genetic basis of antibody production: a single heavy chain variable region gene encodes all molecules bearing the dominant anti-arsonate idiotype in the strain A mouse. *Eur. J. Immunol.* **13**:123.

Diversity Generated from Joining Immunoglobulin Gene Segments

CAPRA, J. D., and A. B. EDMUNDSON. 1977. The antibody combining site. *Sci. Am.* **236**:50.

DAVIS, M. M., K. CALAME, P. W. EARLY, D. L. LIVANT, R. JOHO, I. L. WEISSMAN, and L. HOOD. 1980. An immunoglobulin heavy-chain gene is formed by two recombinational events. *Nature* **283**:733.

EARLY, P., H. HUANG, M. DAVIS, K. CALAME, and L. HOOD. 1980. An immunoglobulin heavy-chain variable region gene is generated from three segments of DNA: V_H, D, and J_H. *Cell* **19**:981.

HILSCHMANN, H., and L. C. CRAIG. 1965. Amino acid sequence studies with Bence-Jones proteins. *Proc. Nat'l Acad. Sci. USA* **53**:1403.

HONJO, T. 1983. Immunoglobulin genes. *Ann. Rev. Immunol.* **1**:499.

HOZUMI, N., and S. TONEGAWA. 1976. Evidence for somatic rearrangement of immunoglobulin genes coding for variable and constant regions. *Proc. Nat'l Acad. Sci. USA* **73**:3628.

KUROSAWA, Y., and S. TONEGAWA. 1982. Organization, structure, and assembly of immunoglobulin heavy-chain diversity DNA segments. *J. Exp. Med.* **155**:201.

LEDER, P. 1982. The genetics of antibody diversity. *Sci. Am.* **246**:102.

LEWIS, S., N. ROSENBERG, F. ALT, and D. BALTIMORE. 1982. Continuing κ gene rearrangement in an Abelson murine leukemia virus transformed cell line. *Cell* **30**:807.

SAKANO, H., R. MAKI, Y. KUROSAWA, W. ROEDER, and S. TONEGAWA. 1980. Two types of somatic recombination are necessary for the generation of complete immunoglobulin heavy-chain genes. *Nature* **286**:676.

TONEGAWA, S. 1983. Somatic generation of antibody diversity. *Nature* **302**:575.

WEIGERT, M., R. PERRY, D. KELLEY, T. HUNKAPILLER, J. SCHILLING, and L. HOOD. 1980. The joining of V and J gene segments creates antibody diversity. *Nature* **283**:497.

Allelic Exclusion

ALT, F., N. ROSENBERG, S. LEWIS, E. THOMAS, and D. BALTIMORE. 1981. Organization and reorganization of immunoglobulin genes in Abelson murine leukemia virus-transformed cells: rearrangement of heavy but not light chain genes. *Cell* **27**:381.

Immunoglobulin Synthesis

ALT, F., and D. BALTIMORE. 1982. Joining of immunoglobulin heavy-chain gene segments: implications from a chromosome with evidence of three D-J_H fusions. *Proc. Nat'l Acad. Sci. USA* **79**:4118.

CORY, S., J. JACKSON, and J. M. ADAMS. 1980. Deletions in the constant region locus can account for switches in immunoglobulin heavy-chain expression. *Nature* **285**:450.

DUTTON, R. W., and R. I. MISHELL. 1967. Cellular events in the immune response. The in vitro response of normal spleen cells to erythrocyte antigens. *Cold Spring Harbor Symp. Quant. Biol.* **32**:407.

EARLY, P., J. ROGERS, M. DAVIS, K. CALAME, M. BOND, R. WALL, and L. HOOD. 1980. Two mRNAs can be produced from a single immunoglobulin μ gene by alternative RNA processing pathways. *Cell* **20**:3131.

KOHLER, G., and C. MILSTEIN. 1975. Continuous cultures of fused cells secreting antibody of predefined specificity. *Nature* **256**:495.

MILSTEIN, C. 1980. Monoclonal antibodies. *Sci. Am.* **243**(4):66.

MOORE, K. W., J. ROGERS, I. HUNKAPILLER, P. EARLY, C. NOTTENBURG, I. WEISSMAN, H. BAZIN, R. WALL, and L. E. HOOD. 1981. Expression of IgD may use both DNA rearrangement and RNA splicing mechanisms. *Proc. Nat'l Acad. Sci. USA* **78**:1800.

POTTER, M. 1972. Immunoglobulin-producing tumors and myeloma proteins of mice. *Physiol. Rev.* **62**:631.

RUSCONI, S., and G. KOHLER. 1985. Transmission and expression of a specific pair of rearranged immunoglobulin μ and κ genes in a transgenic mouse line. *Nature* **314**:330.

SHIMIZU, T., N. TAKAHASHI, Y. YAMAMAKI-KATAOKA, Y. NISHIDA, T. KATAOKA, and T. HONJO. 1981. Ordering of mouse immunoglobulin heavy-chain genes by molecular cloning. *Nature* **289**:149.

WEISSMAN, I. L. 1975. Development and distribution of immunoglobulin-bearing cells in mice. *Transplant. Rev.* **24**:159.

WHITLOCK, C. A., and O. N. WITTE. 1982. Long-term culture of B lymphocytes and their precursors from murine bone marrow. *Proc. Nat'l Acad. Sci. USA* **79**:3608.

Somatic Mutation

BALTIMORE, D. 1981. Somatic mutation gains its place among the generators of diversity. *Cell* **26**:295.

CREWS, S., J. GRIFFIN, H. HUANG, K. CALAME, and L. HOOD. 1981. A single V_H gene segment encodes the immune response to phosphorylcholine: somatic mutation is correlated with the class of antibody. *Cell* **25**:59.

GEARHART, P., and D. BOGENHAGEN. 1983. Clusters of point mutation are found exclusively around rearranged antibody variable region genes. *Proc. Nat'l Acad. Sci. USA* **80**:3439.

KIM, S., M. M. DAVIS, E. SINN, P. PATTEN, and L. HOOD. 1981. Antibody diversity: somatic hypermutation of rearranged V_H genes. *Cell* 27:573.

T Lymphocytes

BARTH, R. K., B. S. KIM, N. C. LAN, T. HUNKAPILLER, N. SOBIECK, A. WINOTO, H. GERSHENFELD, C. OKADA, D. HANSBURG, I. L. WEISSMAN, and L. HOOD. 1985. The murine T-cell receptor uses a limited repertoire of expressed V_β gene segments. *Nature* 316:517.

CANTOR, H., and E. A. BOYSE. 1975. Functional subclasses of T lymphocytes bearing different Ly antigens I: the generation of functionally distinct T cell subclasses is a differentiative process independent of antigen. *J. Exp. Med.* **141**:1375.

FORD, C. E., H. S. MICKLEM, E. P. EVANS, J. G. GRAY, and D. A. OGDEN. 1966. The inflow of bone-marrow cells to the thymus. Studies with part body-irradiated mice injected with chromosome-marked bone marrow and subjected to antigenic stimulation. *Ann. N.Y. Acad. Sci.* **129**:238.

LEDOUARIN, N. M., and F. JOTEREAU. 1975. Tracing of cells of the avian thymus through embryonic life in interspecific chimeras. *J. Exp. Med.* **142**:17.

MILLER, J. F. A. P. 1961. Immunological function of the thymus. *Lancet* 2:748.

MOORE, M. A. S., and J. J. T. OWEN. 1967. Experimental studies on the development of the thymus. *J. Exp. Med.* **126**:715.

WEISSMAN, I. L. 1967. Thymus cell migration. *J. Exp. Med.* **126**:291.

YAGUE, J., J. WHITE, C. COLECLOUGH, J. KAPPLER, E. PALMER, and P. MARRACK. 1985. The T cell receptor: the α and β chains define idiotype, and antigen and MHC specificity. *Cell* 42:81.

ZINKERNAGEL, R., and P. DOHERTY. 1976. The concept that surveillance of self is mediated via the same set of genes that determines recognition of allogenic cells. *Cold Spring Harbor Symp. Quant. Biol.* **41**:505.

T-Cell Receptor

BRENNER, M., J. MCLEAN, D. DIALYNAS, J. STROMINGER, J. SMITH, F. OWEN, J. SEIDMAN, S. IP, F. ROSEN, and M. KRANGEL. 1986. Identification of a putative second T-cell receptor. *Nature* 322:145.

CHIEN, Y.-L., M. IWASHIMA, K. KAPLAN, J. ELLIOTT, and M. DAVIS. 1987. A new T-cell receptor gene located within the alpha locus and expressed early in T-cell differentiation. *Nature* 327:677.

GARMAN, R. D., P. J. DOHERTY, and D. RAULET. 1986. Diversity, rearrangement and expression of murine T cell gamma genes. *Cell* 45:733.

HEDRICK, S., D. COHEN, E. NIELSEN, and M. DAVIS. 1984. Isolation of cDNA clones encoding T cell–specific membrane-associated proteins. *Nature* 308:149.

MCINTYRE, B., and J. ALLISON. 1983. The mouse T cell receptor: structural heterogeneity of molecules of normal T cells defined by Xenoantiserum. *Cell* 34:739.

MHC Genes

BILLINGHAM, R., and W. SILVERS. 1971. *The Immuniobiology of Transplantation.* Prentice-Hall.

BJORKMAN, P., M. SAPER, B. SAMRAOUI, W. BENNETT, J. STROMINGER, and D. WILEY. 1987. Structure of the human class I histocompatibility antigen, MLA-A2. *Nature* 329:506.

DOHERTY, P. C., and R. M. ZINKERNAGEL. 1975. H-2 compatibility is required for T-cell-mediated lysis of target cells infected with lymphocytic choriomeningitis virus. *J. Exp. Med.* **141**:502.

HOOD, L., M. STEINMETZ, and B. MALISSEN. 1983. Genes of the major histocompatibility complex of the mouse. *Ann. Rev. Immunol.* **1**:529.

KLEIN, J. 1975. *Biology of the Mouse Histocompatibility-2 Complex.* Springer-Verlag.

KLEIN, J. 1979. The major histocompatibility complex of the mouse. *Science* 203:516.

SHACKELFORD, D. A., J. F. KAUFMAN, A. J. KORMAN, and J. L. STROMINGER. 1982. HLA-DR antigens: structure, separation of subpopulations, gene cloning and function. *Immunol. Rev.* **66**:133.

SNELL, G. D. 1981. Studies in histocompatibility. *Science* 213:172.

SNELL, G. D., J. DAUSSET, and S. NATHENSON. 1976. *Histocompatibility.* Academic.

STEINMETZ, M., K. MINARD, S. HORVATH, J. MCNICHOLAS, J. FRELINGER, C. WAKE, E. LONG, B. MACH, and L. HOOD. 1982. A molecular map of the immune response region from the major histocompatibility complex of the mouse. *Nature* 300:35.

STEINMETZ, M., K. W. MOORE, J. G. FRELINGER, B. T. SHER, F. W. SHEN, E. A. BOYSE, and L. HOOD. 1981. A pseudogene homologous to mouse transplantation antigens: transplantation antigens are encoded by eight exons that correlate with protein domains. *Cell* 25:683.

STEINMETZ, M., A. WINOTO, K. MINARD, and L. HOOD. 1982. Clusters of genes encoding mouse transplantation antigens. *Cell* 28:489.

YANAGI, Y., Y. YOSHIKAI, K. LEGGETT, S. CLARK, I. ALEKSANDER, and T. MAK. 1984. A human T cell–specific cDNA clone encodes proteins having extensive homology to immunoglobulin chains. *Nature* 308:145.

Lymphokines

PAUL, W. E. 1989. Pleiotropy and redundancy: T cell–derived lymphokines in the immune response. *Cell* 57:521.

Tolerance

SCHWARTZ, R. M. 1989. Acquisition of immunologic self-tolerance. *Cell* 57:1073.

26

Evolution of Cells

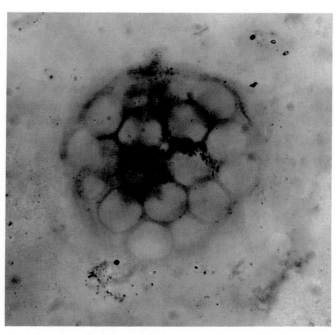

Late Proterozoic fossil bacteria (over 700 million years old)

Biological explorations have two major themes. One is contemporary: how the molecules of cells contribute to cell function, how cells are organized into organisms, and how organisms grow and thrive in natural populations. The other theme is historical: how life first arose and then evolved into its present forms. Until now this book has been concerned with contemporary problems in molecular cell biology. In this chapter we consider the impact of new findings in molecular cell biology on evolutionary biology, particularly on ideas dealing with early evolution.

Evolution has been traditionally studied through the disciplines of paleontology and comparative anatomy—that is, through identifying the distinctive characteristics and the structural relationships between present-day organisms and comparing them to those of ancient organisms. From this research we have a comprehensive and systematic view of the evolution of animal and plant life, beginning with the fossil record of animals and plants that lived some 600 million years ago. The central axiom of classical evolution is that complex multicellular organisms must have evolved from simple organisms (Figure 26-1). The evolution of invertebrates to vertebrates, or of fishes to amphibians to reptiles to mammals, has fol-

lowed a pathway in which simple organisms evolved to more complex ones. Increasing attention to single-celled organisms has defined the differences in various classes of bacteria, fungi, and protozoa. As we shall see, we have now a comprehensive view of the importance of cell fusions in the origin of organelles in early eukaryotic cells. Although no strong evidence exists about how multicellular forms arose in evolution, it is apparent even in present-day organisms that some species can live as single cells or in multicellular form. But the greatest challenges to a "complete" theory of evolution is encountered when we come to consider molecular evolution and how cells arose in the first place. ▲

When the realization came in the 1920s and the 1930s that all life forms have similar basic building blocks for cellular molecules, J. B. S. Haldane in England and A. I. Oparin in Russia proposed that the common building blocks may have arisen by prebiotic synthesis. They suggested that the atmospheric and geologic conditions on the primitive earth might have made possible the develop-

ment of a sufficient variety and quantity of biologically important molecules to support "spontaneous" prebiotic evolution. Since the time of Haldane and Oparin, many experimenters have demonstrated how amino acids, certain nucleic acid bases, and perhaps even some primitive polymers might have arisen spontaneously.

However, even the availability of building blocks in prebiotic times leaves a daunting array of problems for prebiotic chemistry. The key reactions of molecular cell biology—those conferring the coding capacity and the replication of the nucleic acids and those involved in the translation of the code into protein—must have arisen before the first true cell could exist. It is in this realm that advances in molecular cell biology, particularly the increasing appreciation of multiple functions for RNA in cells and the catalytic properties of RNA itself, have proved to be a great stimulus for new ideas and experiments on prebiotic synthesis and precellular evolution.

Given both the building blocks and at least primitive polymers and means for their synthesis, a second set of major questions loom: what was the nature of the first cells and through what early cellular lineages did evolution proceed?

It seems unlikely that any present-day cell is identical to the earliest cells or even very much like them. Although remnants of ancient molecular evolution probably do exist in present-day cells, no ancient cells are available for study. The oldest imprints in geologic samples that resemble cells are about the size and shape of cyanobacteria (blue-green algae), but the relationship of these microfossils (i.e., fossils observable by light- and electron-microscopic analysis of rock surfaces) to present-day single-celled organisms, either prokaryotic or eukaryotic, may never be settled conclusively.

Until the 1970s the nature of the eukaryotic precursor cell seems to have been taken for granted by many biolo-

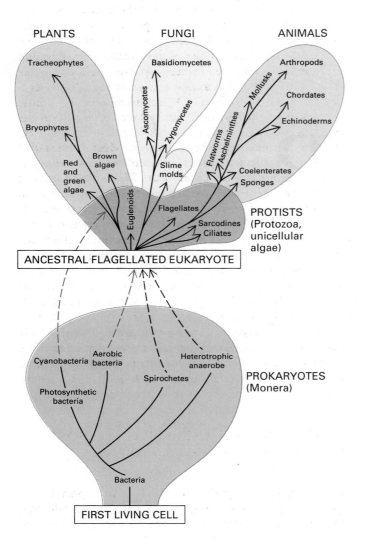

◄ **Figure 26-1** A phylogenetic diagram showing the widely accepted five kingdoms: prokaryotes, protists (single-celled eukaryotes), plus three multicellular kingdoms—plants, fungi, and animals. The diagram emphasizes two points: (1) all life is pictured as arising from bacteria (2) the "endosymbiotic hypothesis" is presented as the major step in eukaryotic cell evolution. According to this latter idea, bacterial fusions resulted in formation of eukaryotes. For example, cyanobacteria (blue-green algae) and aerobic bacteria (or some part of the bacterial genome) fused with the bacterial eukaryotic precursor cell to give rise to chloroplasts and mitochondria that then were maintained as plants and animals evolved. In one form or another this hypothesis is widely accepted. Spirochetelike organisms have been suggested as donors of information that resulted in cilia formation, and a heterotrophic prokaryotic anaerobe has been suggested as the forerunner of the eukaryotic nucleus. These latter two proposals are now in question. [See R. H. Whitaker, 1969, *Science* 63:50; and L. Margulis, 1971, *Am. Sci.* 59:231.]

gists. Although no one had produced a good supporting argument, it was widely assumed (see Figure 26-1) that present-day eukaryotes evolved from some type of prokaryote—probably because the present-day eukaryotes seem more complicated than prokaryotes. Here the techniques of DNA sequencing have worked a true revolution, as we will discuss in detail later in the chapter. At least three equally ancient cell lineages exist, and there are no direct prokaryotic precursors to the eukaryotic nuclear genome, at least among present-day prokaryotes. Rather the eukaryotic nuclear DNA represents an equally ancient and separate cell lineage from that of prokaryotes. The DNA-sequencing studies do, however, completely validate that DNA of chloroplasts and mitochondria arose from the fusion between a primitive prokaryote and the precursor to eukaryotic cells. This is known as the *endosymbiotic hypothesis;* we will discuss this theory later in the chapter.

We saw in Chapters 7–11 the great differences between prokaryotes and eukaryotes in the structure of their genes and chromosomes and in the details of their gene expression. Among the important eukaryotic features that are not shared by prokaryotes are the large amount of "extra" (noncoding) DNA and the presence of intervening sequences—often very long ones—between protein-coding regions of genes. Although we do not know for certain which of these very different genomic designs is most like that of the earliest cells, it seems quite likely that the present-day single-celled organisms with the fewest genes and the smallest amount of DNA may have lost much of the DNA that the earliest cells possessed. One of the new twists contributed by modern molecular cell biology is the possibility that today's prokaryotes are very highly evolved cells whose evolution has made them the most flexible and efficient cells on the planet. In contrast, although no present-day representatives of the earliest cells remain, there may be vestiges of the earliest molecular processes (e.g., RNA splicing and divided genes) in the cells of the most highly evolved organisms, such as humans.

In this chapter, we shall first briefly discuss the current ideas and experiments dealing with prebiotic synthesis and then deal with the origin and lineage of cells.

Prebiotic Synthesis

As the earth cooled and crusted over, an atmosphere was created from gases that had formed beneath the surface and had escaped through surface cracks. Quite likely, water was delivered to the earth's surface as a gas and the oceans formed as a result of millions of years of torrential rains. The early atmosphere probably contained little or no oxygen; the atmospheric oxygen on which most present-day life depends was apparently produced from water by photosynthetic organisms that evolved later.

Table 26-1 **Present sources of energy (averaged over the earth's surface)**

Source	Energy in cal/(cm^2 × yr)
Total radiation from sun	2.6×10^5
Ultraviolet light with wavelengths of:	
300–400 nm	3.4×10^3
250–300 nm	5.6×10^2
200–250 nm	4.1×10^1
<150 nm	1.7
Electric discharges	4
Shock waves	1.1
Radioactivity	8×10^{-1}
(to 1.0 km depth)	
Volcanoes	1.3×10^{-1}
Cosmic rays	1.5×10^{-3}

SOURCE: L. E. Orgel, 1973, *The Origins of Life,* Wiley, p. 115.

The nature of the atmosphere on primitive earth is a topic of great interest. A decade ago there was substantial opinion that the primitive atmosphere was strongly reducing. If outgassing (the escape of gases from below the crust) occurred before most of the metallic iron had sunk to the earth's core, then the atmosphere would have been strongly reducing, containing abundant H_2, CH_4 (methane), and NH_3 as well as CO (carbon monoxide). If little metallic iron was left near the surface to react with the escaping gases, then CO_2, some CO, and much less CH_4 and NH_3 would have been present. Most present day opinion favors a near neutral atmosphere. Hydrogen sulfide was likely present in any case. The main sources of the energy that could have produced chemical changes in the atmospheric gases (or in these gases dissolved in water) were radiation from the sun and electric discharges (Table 26-1). In addition, volcanic action discharged hot gases and ash into the atmosphere and rifts along the sea floor discharged lava and produced superheated hot springs or plumes. Recent attention has focused on the possibility of deep ocean rifts as sites for organic molecule synthesis. Abundant sulfur, phosphorus, and mineral catalysts exist in these regions and NH_3 and CH_4 may have formed more easily at such sites. The necessary atoms for organic synthesis were available as were several sources of energy that could promote chemical reactions.

Amino Acids and Nucleic Acid Bases Are Prominent Products under Prebiotic Conditions

In an attempt to synthesize biologically important organic compounds under possible prebiotic conditions, Stanley Miller, a student of Harold Urey, mixed hydro-

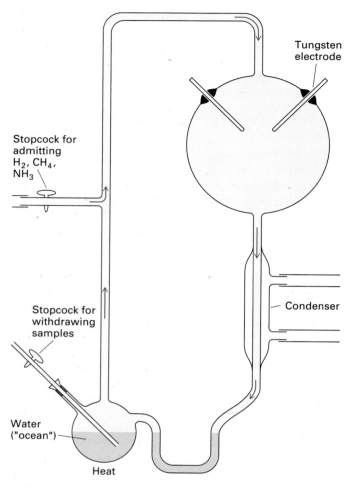

▲ **Figure 26-2**   The apparatus used by Stanley Miller to simulate prebiotic organic synthesis. [See S. L. Miller, 1988, *Cold Spring Harbor Symp. Quant. Biol.* **52**:17.]

pounds (the carbon- and nitrogen-containing compounds mentioned above and H_2S) plus ultraviolet light as an energy source yield glycine, alanine, serine, glutamic acid, and asparagine. UV light is a much more abundant source of energy than lightning, which was simulated in the Miller experiments. Also, the success with mixtures of N_2, CO_2 or CO, and H_2 suggests that synthesis was possible even if methane and ammonia were scarce.

Purines, and to a much lesser extent pyrimidines, can form when HCN and cyanoacetylene, $HC{\equiv}C{-}C{\equiv}N$, are present in the gas phase of spark-discharge experiments. HCN is reactive in the presence of UV light alone, and purines can be regarded as condensation products of five HCN molecules, although the route of synthesis is still not certain (see Figure 26-4). Adenine was found to be a prominent product of simulated prebiotic synthesis reactions. In fact, even ATP can form in low yield under prebiotic conditions, especially in the presence of the common mineral apatite (calcium phosphate).

The earth is not the only place in the solar system where organic molecules have been synthesized abiotically. Analyses of meteorites, particularly the large one that landed in Murchison, Australia, in 1969, have shown them to contain both the free amino acids that are used in biological systems and other, nonbiological

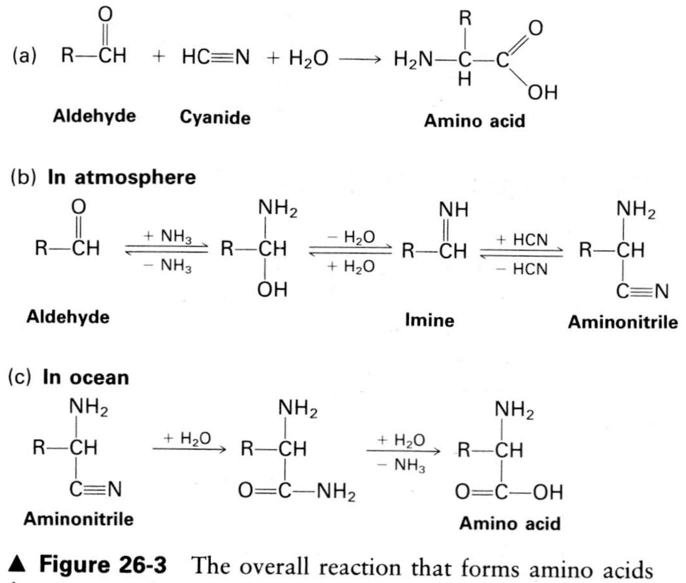

▲ **Figure 26-3**   The overall reaction that forms amino acids from precursors in prebiotic experiments is shown in (a). The reaction probably follows the steps outlined in (b) and (c); these are collectively known as the Strecker synthesis. The first three steps in (b), which include the removal of water, could have taken place in the primitive atmosphere. If the aminonitrile then entered the ocean, hydrolysis could proceed. [See R. E. Dickerson, 1978, in *Evolution*, a *Scientific American* Book, M. Kimura, ed., W. H. Freeman and Company, p. 36.]

gen, methane, and ammonia (in the ratio 1:2:2) with water in a closed, evacuated reflux vessel (Figure 26-2). The gaseous mixture, which simulated one hypothetical version of the early atmosphere, was continuously exposed to electric discharges. When the water phase (the "ocean") was examined, more than 10 percent of the carbon from the methane was included in organic molecules such as amino acids. Glycine, alanine, aspartic acid, valine, and leucine (in both D- and L-forms) were clearly identified among the products. Hydrogen cyanide (HCN), aldehydes, and cyano compounds ($-C{\equiv}N$) were also present, and these could have been important intermediates in the formation of amino acids and nucleic acid bases (Figures 26-3 and 26-4). These latter compounds can be formed also under mildly reducing conditions.

Many different reaction conditions that mimic plausible prebiotic conditions can lead to the synthesis of amino acids and nucleic acid bases. Of particular interest are experiments showing that the simple prebiotic com-

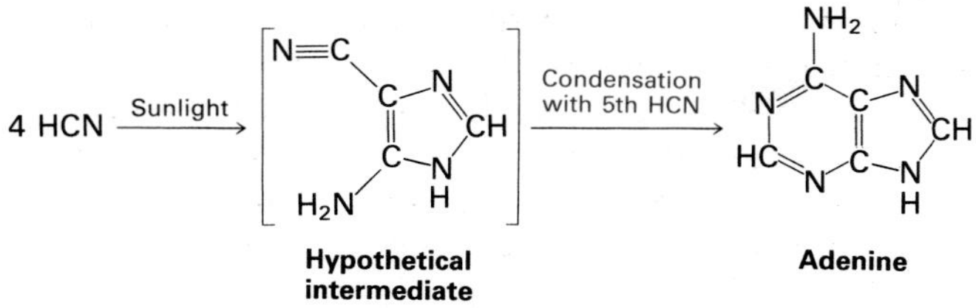

▲ **Figure 26-4** Adenine can be formed from hydrogen cyanide (HCN) in sunlight, although the route of synthesis has not been completely worked out. When heated, HCN and ammonia yield adenine. [See S. L. Miller and L. E. Orgel, 1973, *The Origins of Life on Earth*, Prentice-Hall, p. 105.]

amino acids. Because amino acids have been found deep within meteorites they are not regarded as surface contaminants. The carbon- and nitrogen-containing compounds that are prominent reactants in proposed prebiotic synthesis reactions have also been identified in interstellar gas clouds: examples include H_2, HCN, cyanoacetylene, CO, H_2S, and NH_3, as well as larger molecules such as acetonitrile and acetaldehyde.

The presence of these organic substances in extraterrestrial objects suggests that the necessary building blocks for biologically important macromolecules arise readily in nature, and could have been formed on earth some four billion years ago through a nonbiological route. Because there was neither free oxygen nor cellular metabolism, these organic molecules could have been very stable. Consequently, the oceans may have accumulated concentrations of organic reactants high enough to support further chemical activity. The primordial "organic soup" of Haldane and Oparin almost certainly existed.

RNA Probably Existed before DNA and Protein

The most challenging aspect of precellular or prebiotic chemistry is understanding the formation of the polymers on which present-day life is based. There is no proof of how any polymers, much less biologically meaningful polymers, were first assembled. Attempts to make polypeptides from activated amino acids give low yields of peptides in random arrangements. But some peptide synthesis at least can occur, for example, when amino acid mixtures are heated to dryness, especially if activated amino acids are used in the presence of claylike minerals.

By the late 1960s, when the universal means of cellular information storage, transcription, and translation had all been discovered, Leslie Orgel, Francis Crick, and Carl Woese—writing separately—proposed that RNA probably preceded DNA in evolution mainly because of RNA's essential three-part role in the assembly of ordered peptides. Moreover, as RNA can be replicated by complementary strand copying, RNA may have been the first polymer. Recent discoveries have thrust these predictions

to new prominence. As we discussed briefly in Chapter 8 and will discuss later in more detail, RNA catalysis including self-cutting, self-splicing, and self-elongation and ligation virtually force the adoption of a central if not primordial role for RNA in precellular evolution. Not only can RNA perform catalytically on its own and as part of the snRNPs that process RNA but an ever-increasing number of additional roles for RNA in present-day biochemistry are being demonstrated, e.g., RNA serves the primer function in DNA synthesis; an RNA molecule is a necessary part of telomerase for adding the terminal DNA to eukaryotic chromosomes; 7S RNA is part of the protein secretion apparatus; a small RNA is part of the assembly structure for bacteriophage ϕ29; and, finally, dozens of small RNA molecules are known whose functions are as yet unidentified (see Figure 8-17a). The case for the importance of RNA in precellular chemistry becomes virtually irresistible. DNA is chemically not nearly so flexible and even the deoxynucleotides are derivatives of ribonucleotides. (There are several different types of ribonucleotide reductases that convert ribo- to deoxyribonucleotides, but all cells utilize at least one such enzyme.) In addition, deoxyuridylate is methylated to make deoxythymidylate. Finally, reverse transcriptase activity that copies RNA into DNA exists not only in retroviruses where it was originally discovered and in virtually all eukaryotic cells but is now known to be present in bacteria. So virtually all commentators place RNA in existence before DNA during evolution of cells.

In light of this strong consensus of the importance of RNA in early evolution, what can we say about prebiotic chemistry and the first appearance of RNA, and in particular, how did the connection between RNA and protein sequence evolve?

Prebiotic Synthesis of RNA Raises Unanswered Questions

The first requisite in prebiotic synthesis leading to RNA would have been ribose and the second, ribonucleotides. Syntheses of both are thought to have been possible under prebiotic conditions, but likely pathways cannot at present be fully described.

Ribose can be synthesized in the formose reaction (Figure 26-5). Given the presence of aldehydes, condensation of formaldehyde (HCHO) will yield glycoaldehyde, a two-carbon molecule. Successive condensations can build up a whole series of sugars with various numbers of carbons, including a triose condensation with glycoaldehyde to form ribose. One obvious problem with relating this to prebiotic synthesis is that an uncontrolled use of this reaction would have produced many different sugars, not just ribose.

A second difficulty in accounting for prebiotic RNA synthesis is the difficulty in obtaining high yields of py-

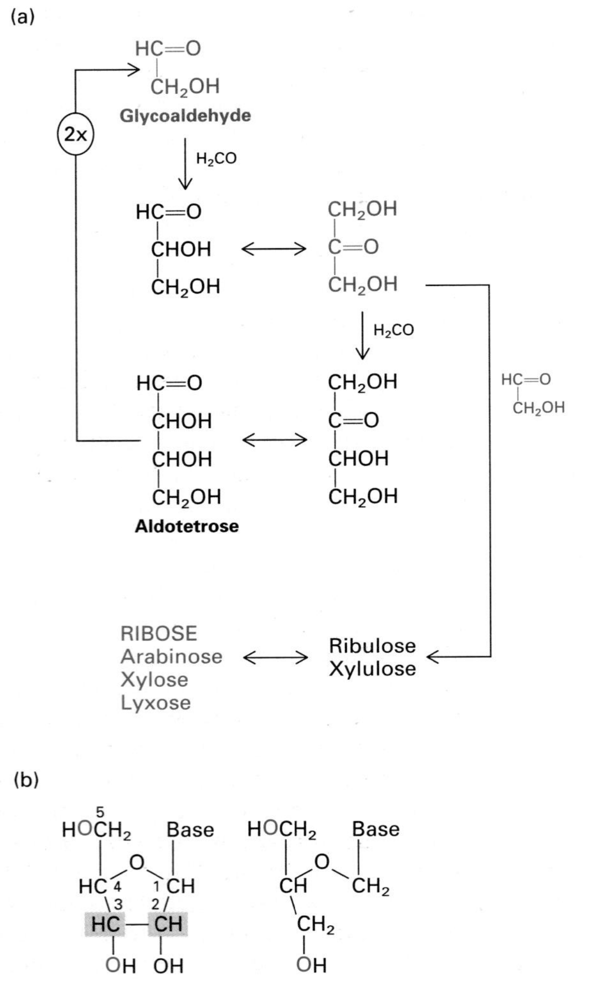

(a)

(b)

◄ **Figure 26-5** (a) The formose reaction, a possible route for prebiotic ribose synthesis. Condensation of a triose with glycoaldehyde makes an array of five-carbon sugars, one of which is ribose. (b) Comparative structure of a ribonucleoside and a simple acyclonucleoside analogue derived from glycerol. The ribose has asymmetric carbons (green) while the acyclonucleoside lacks them—i.e, is nonchiral. Phosphorylated versions (red indicates sites of phosphorylation) of the acyclonucleoside can form polymers. [See G. Joyce, 1989, *Nature* 338:217.]

acyclonucleoside that does not have the complication of existing in enantiomeric states (i.e., it is nonchiral; see Figure 26-5b). Attempts are being made to understand the less complicated chemistry of polymers based on phosphorylated derivatives of such compounds.

As a model of how prebiotic oligomers of RNA might have formed, experiments have also been conducted with activated chemical relatives of ribonucleotides. Protein-free chemical polymerization of such activated nucleotidelike compounds according to Watson-Crick rules has been shown to occur, albeit at a slow rate and under special conditions. High concentrations of nucleoside 5′-phosphorimidazolides of all four normal bases (C, A, G, and U) have been used in these reactions (Figure 26-6). (Imidazole, the core of the histidine ring, is a probable condensation product of HCN.) In the presence of zinc ions (incidentally, the ion required by RNA polymerase), base-paired, template-directed polymer synthesis utilizes the imidazolides as substrates and the products are 3′,5′-linked. Thus, for example, a poly C template directs nonenzymatic poly G polymerization, and a mixed CG template correspondingly directs poly GC synthesis. Phosphorylated forms of the acyclonucleosides (see Figure 26-5a) of guanosine can also form phosphodiesters and engage in oligomerization directed by a template similar to the phosphorimidazolides.

These reactions prove nothing about the actual prebiotic reactions but they do show that polyribonucleotides and related polymers can form in the absence of proteins given the appropriate activated building blocks. Once formed these oligomers may also have template activity.

The Origin of the Genetic Code: Early RNA Probably Interacted with Amino Acids and Peptides

How the development of early polymers proceeded, as we saw in the last section, is largely guesswork at present. While quite a bit of RNA-directed biochemistry may conceivably have preceded the development of proteins, a major unsolved problem and the crux of precellular evolution is how an ordered relationship between RNA (or its forerunner) and peptides occurred. It was these interactions that led to the development of the genetic code and of the principles of translation.

rimidines. As noted earlier, purines can be obtained with HCN condensation; however all reactions studied to date attempting to make pyrimidines, especially cytosine, seem implausible in the prebiotic world. Moreover, how preformed pyrimidines (or purines) were linked with sugars presents a problem: the formose reaction would yield a complicated mixture of sugars, and a huge number of different nucleosides would result if sugars and bases were randomly coupled.

Ribonucleoside polyphosphates may be formed from ribonucleosides by heating mixtures of urea, NH_4Cl, phosphate, and hydroxyapatite (a naturally occurring clay) to 100°C for 24 h. However, a mixture of different ribonucleoside isomers (both D and L) results, and the nucleotides have 2′, 3′, and 5′ linkages. It is now known that L-enantiomers of ribonucleosides will inhibit the polymerization of the normal D-enantiomers.

These difficulties have led chemists concerned with possible routes of prebiotic synthesis to search for some structurally similar but simpler riboselike compound. A favorite candidate at present is a glycerol-derived

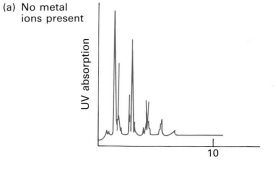

(a) No metal ions present

(b) Zn²⁺ present

Chain length of poly G

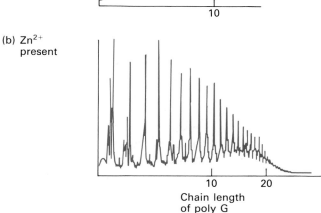

A methylated nucleoside 5′-phosphorimidazolide

▲ **Figure 26-6** Nonenzymatic synthesis of poly G. Chromatography and UV absorption show the polyguanylic acid products of guanosine 5′-phosphorimidazolide in the presence of a poly C template (a) in the absence of metal ions and (b) with Zn²⁺ at 0.01 *M*. (c) The structure of the activated nucleoside 5′-phosphorimidazolide. The reaction proceeds better if the guanine is methylated (red). Any of the four bases found in RNA can be attached to the sugar. [See J. H. C. van Roode and L. E. Orgel, 1980, *J. Mol. Biol.* 144:579.]

▶ **Figure 26-7** The genetic code with features of possible evolutionary significance indicated. The grouping of the most hydrophobic or hydrophilic amino acids is of special interest. Also, note the frequency with which codons having the same first and second bases encode a single amino acid. This suggests that only two of the three nucleotides may have been used early in evolution.

Because there is no known chemical affinity between particular amino acids and their three-base codons, it is clear that no simple proof is possible to pinpoint how the code assignments were made. It is, however, a reasonable initial assumption that the code arose stepwise. A glance at the code words presented in Figure 26-7 shows some order to the grouping. The more hydrophobic amino acids (phenylalanine, leucine, isoleucine, valine, and methionine) all have U as their center base while the most hydrophilic amino acids (glutamic acid, aspartic acid, lysine, asparagine, glutamine, histidine, and tyrosine) have A as their center base. (In addition, the codons with C as the central nucleotide encode more hydrophobic amino acids than the set with G as the central nucleotide.) Thus the earliest coding might have simply specified hydrophilic or hydrophobic. Another facet of the code suggests a stepwise adoption of present assignments; in most cases when the first two bases of a codon are fixed (XYN or XXN) the four possible codons all code for the same amino acid. Perhaps the original translation apparatus used only the first two of the three bases in a codon. One point that is widely appreciated is that once a complete codon assignment was adopted, major changes would wreck the system, so it seems most probable that a three-base code has been in place since the earliest cells arose.

No persuasive theories have been developed about which codons were the earliest to be adopted. Since nonenzymatic RNA oligomer formation proceeds most easily if purines and pyrimidines alternate, Orgel suggested that purines and pyrimidines may have alternated in the earliest RNA. This generates codons of alternating form PuPyPu and PyPuPy. As the codons with pyrimidines as their central base encode more hydrophobic amino acids and those with central purines encode more hydrophilic amino acids, alternating ribopolymers would encode primitive peptides that were partially ordered. Such alternating peptides synthesized from valine and lysine aggre-

First position (5′ end)	Second position				Third position (3′ end)
	U	**C**	**A**	**G**	
U	Phe	Ser	Tyr	Cys	U
	Phe	Ser	Tyr	Cys	C
	Leu	Ser	STOP	STOP	A
	Leu	Ser	STOP	Trp	G
C	Leu	Pro	His	Arg	U
	Leu	Pro	His	Arg	C
	Leu	Pro	Gln	Arg	A
	Leu	Pro	Gln	Arg	G
A	Ile	Thr	Asn	Ser	U
	Ile	Thr	Asn	Ser	C
	Ile	Thr	Lys	Arg	A
	Met	Thr	Lys	Arg	G
G	Val	Ala	Asp	Gly	U
	Val	Ala	Asp	Gly	C
	Val	Ala	Glu	Gly	A
	Val (Met)	Ala	Glu	Gly	G

← Hydrophobic   ← Hydrophilic

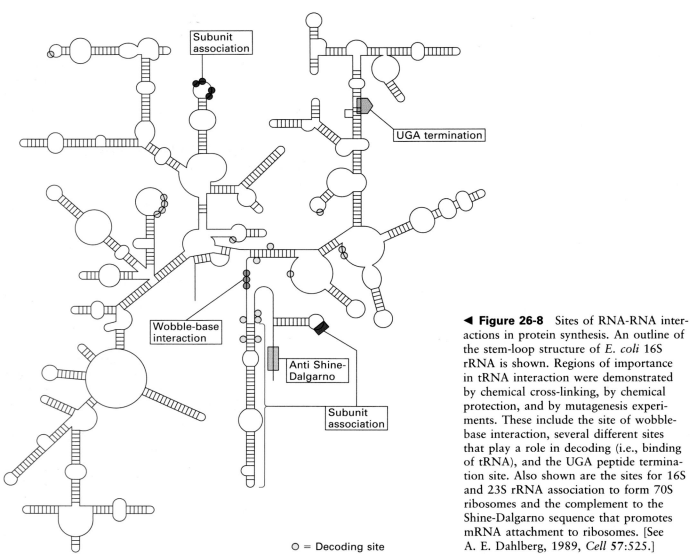

O = Decoding site

◀ **Figure 26-8** Sites of RNA-RNA interactions in protein synthesis. An outline of the stem-loop structure of *E. coli* 16S rRNA is shown. Regions of importance in tRNA interaction were demonstrated by chemical cross-linking, by chemical protection, and by mutagenesis experiments. These include the site of wobble-base interaction, several different sites that play a role in decoding (i.e., binding of tRNA), and the UGA peptide termination site. Also shown are the sites for 16S and 23S rRNA association to form 70S ribosomes and the complement to the Shine-Dalgarno sequence that promotes mRNA attachment to ribosomes. [See A. E. Dahlberg, 1989, *Cell* **57**:525.]

gate into structures with properties of β sheets with a hydrophobic and a hydrophilic face. (Some simple proteins, for example, silk fibroin, have long stretches of such structures that form definite β sheets.) Such sheets would stack and might form the basis of membranelike structures.

Speculations of this type clearly do not explain how translation could have begun but do show that even slight order in peptides that were the products of the earliest translation reactions could have provided the basis for Darwinian selection to operate at the molecular level.

Perfecting the Translation System Required More Complicated Structures in rRNA

A detailed theory of evolution that would explain how the primitive oligonucleotide-oligopeptide interactions developed into a working translation system is entirely

beyond the limits of present knowledge. We can only summarize and comment on the features that were acquired by tRNAs and rRNAs during evolution and call attention to the fact that many of these still today represent RNA-RNA interactions.

Several properties we observe today suggest a rough outline of how tRNAs might have evolved. (1) The anticodon with its encoded information was recognized by base pairing. (2) Each tRNA acquired an "identity," so that it was recognized by one activating enzyme that loaded the tRNA with the correct amino acid. (3) The tRNAs became able to bind to ribosomes.

Substantial progress in defining the basis of these properties has been made. First recall that tRNA identity is determined by specific bases. Various binding sites have been located in cross-linking and crystallographic studies using *E. coli* RNAs and amino-acyl synthetoses (AASs). The TΨCG and the D arm of the tRNA are required in ribosome binding and the acceptor stem and anticodon interact until AASs. The bases in 16S rRNA that interact with tRNA have also been established. They are scattered in a number of sites which in the three-dimensional structures of rRNA must be brought together (Figure 26-8). The wobble base of the tRNA interacts directly with a specific three-base region of 16S *E. coli* rRNA. And the two *E. coli* rRNA subunits are in fact held together due to interactions between specific sites in 16S and 23S rRNA. Because of the importance of these binding sites in rRNA and because all present-day cells contain ribosomes of a consistent design and function, it is tempting to assume that the precellular translation machinery was based on RNA interactions including rRNA. In addition, part of the binding energy that attaches the mRNA to a bacterial ribosome comes from the base pairing of a region near the 3′ end of 16S *E. coli* rRNA with the Shine-Dalgarno sequence that precedes the AUG initiation codon in mRNA.

Of course, present-day ribosomes contain more than 50 individual proteins, and the earliest translation system could hardly have made 50 precise long polypeptides. However, reassociation of bacterial ribosomes from purified rRNA and ribosomal proteins is possible even if some of the proteins are omitted. Most important, these "protein-deficient" structures can still take part in translation. In addition a number of *E. coli* mutations that confer resistance to chloramphenicol, carbomycin, and anisomycin, drugs that specifically inhibit peptide-bond formation (conducted by the hypothetical peptidyl transferase), are actually point mutations in a specific domain of the 23S rRNA. Primitive rRNA may well have furnished the surface that allowed an interaction of the mRNA and the activated tRNA. In fact, the primitive ribosome could have been *mainly* RNA.

Attempts are currently being made to determine whether peptide synthesis can be carried out by RNA alone, a result that would add more weight to the already strong case that RNA interactions were basic to the evolution of translation.

RNA Catalysis: A Basis for a Precellular Genetic System?

Current thought about precellular chemistry has been sharply focused on RNA because a number of RNA-mediated reactions have been recently discovered. The progress on RNA catalysis by studying sequences derived from present-day organisms has identified a number of different types of reactions that are now becoming thoroughly understood mechanistically. However, there is no present information on how early such reactions could have entered into the chain of evolution because, as we pointed out, questions about prebiotic synthesis of RNA have not yet been answered.

Nevertheless, the discovery of RNA catalysis is the most important new breakthrough that may shed light on precellular chemistry. And the discovery of self-splicing RNA almost surely provides a key to understanding how genes operated in the earliest organisms. Before we deal with the lineage of the earliest cells we will take a more detailed look at RNA catalysis including self-splicing.

Nuclease Activity Is the Simplest RNA Catalytic Event

The pre-tRNAs of *E. coli* are synthesized as larger precursor molecules that must be trimmed down into finished, usable tRNAs. Perhaps the most well-studied example of these is an *E. coli* tyrosyl-tRNA. RNase P, the enzyme that cleaves tRNA^Tyr, has been highly purified from *E. coli*. Part of the purified enzyme is an RNA chain 375 nucleotides long. The protein portion of the enzyme is by itself inactive. When the protein-free RNA moiety of RNase P is added to the pre-tRNA^Tyr in a high-salt solution, cleavage proceeds, producing the 5′ end of the mature tRNA (Figure 26-9). Thus the cleavage can be carried out by an RNA-directed mechanism, but in the cell this RNA has a protein associated with it. It is believed that the only role of the protein is to shield the charges on the RNA, allowing the catalytic RNA to come close to its RNA target site. RNase P–like enzymes that cleave pre-tRNA have now been purified from other bacteria, yeasts, and human cells; in each case an RNA of about 350 nucleotides is part of the enzyme. The proposed secondary structures for the active *E. coli* "ribozyme" and the other catalytic RNAs are very similar, indicating that this is an ancient RNA enzyme.

E. coli pre-tRNA^Tyr

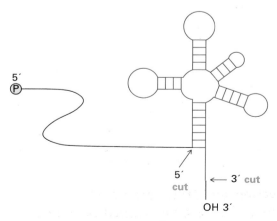

▲ **Figure 26-9** *E. coli* pre-tRNA^Tyr is cleaved to expose the 5′ and 3′ ends of the tRNA^Tyr (red). The 5′ cleavage is catalyzed by a 375-nucleotide RNA that is part of RNase P. The 3′ cleavage is catalyzed by another enzyme. [See C. Guerrier-Takada et al., 1983, *Cell* 35:849.]

Group I self-splicing

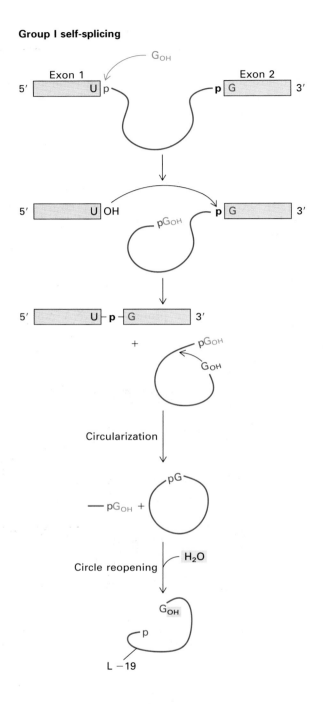

Circularization

Circle reopening

L – 19

◀ **Figure 26-10** The reaction of the *Tetrahymena thermophila* intervening sequence. Self-excision of the intron and the other events, circularization and circle opening all require first an attack by a hydroxyl on a sensitive internucleotide bond within the RNA; during the cleavage the phosphodiester is transferred to another sugar in a reaction called *transesterification*. The final product is L – 19, the ribozyme. [See M. D. Been et al., 1987, *Cold Spring Harbor Symp. Quant. Biol.* **52**:147.] *Adapted from T. R. Cech, 1986, Sci. Am., **254**:64.*

intron by a *transesterification* (i.e., a transfer of a phosphate bond from one sugar to another with no gain or loss in phosphodiester linkages). This frees the 3′ OH of a uridylate residue at the cut end of the upstream exon to attack at the downstream (3′) intron-exon junction. The 5′ phosphate of the G residue at that junction is involved in a second transesterification that joins exon 1 and exon 2. The *Tetrahymena* intron then continues a series of additional reactions that do not have any relevance to rRNA formation but do have considerable interest as examples of RNA catalytic activity. The whole intron circularizes, cutting off 19 nucleotides in the process. The OH group of the guanosine on the 3′ end of the intron is the active agent in the circularization reaction. Circle reopening can then occur with water furnishing the OH group. The final open ("linear") product is the active enzyme (L – 19), which we discuss later.

Comparison of the sequence of the removed *Tetrahymena* intron with more than 70 other similar introns reveals a series of conserved sequences called P, Q, R, and S (Figure 26-11). Furthermore, this related group, now termed group I introns, can be folded in a very similar stem-loop pattern. Group I introns are found not only in other ribosomal RNAs, e.g., in the slime mold *Physarum*

Self-splicing Can Remove Two Different Types of Introns

Group I Introns   The original discovery of self-splicing was made with a portion of the precursor to ribosomal RNA from *Tetrahymena thermophila*. The reaction requires only the pure RNA molecule and a guanosine residue as a cofactor (see Figure 8-58; Figure 26-10); nucleophilic attack by the 3′ OH upstream of the guanosine leads to a cut at the 5′ exon-intron border with the simultaneous addition of the guanosine to the free end of the

Group I introns

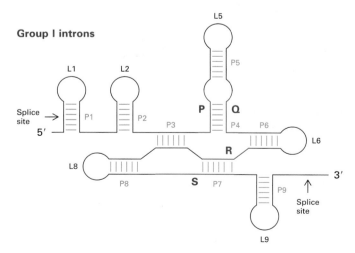

▲ **Figure 26-11** Structural diagrams for group I introns. The location of conserved sequence elements P, Q, R, and S are indicated as are the regular base-paired stems (P1–P9) and the loops (L1–L9) in the introns. *From J. M. Burke et al., 1987, Nucleic Acids Res.* **15**:7217.

polycephalum, but there are many examples in mitochondrial mRNAs and rRNAs and in chloroplast mRNAs and rRNAs. Also, introns in this group were discovered within the *E. coli* bacteriophage T4 mRNAs that encode thymidylate synthetase and a subunit of ribonucleotide reductase, the enzyme involved in making deoxynucleotides from ribonucleotides. Many of the group I introns, including the two T4 pre-mRNAs, are self-splicing. A number of different mutations in the P, Q, R, and S consensus regions and within the various stem-loops demonstrate that these regions are all required for the self-splicing reaction. Obviously, a correctly folded, three-dimensional structure operates for this RNA just as correctly folded proteins are required for protein enzymes.

Some group I mitochondrial introns are not self-splicing and in fact encode proteins called *maturases* that must be synthesized in vivo for the introns containing them to be spliced out. However, it has been shown by sequencing studies that guanosine acts as a cofactor and the G residue is incorporated in the spliced product during splicing just as in self-splicing group I introns. Thus, the role of the encoded protein may be simply to help fold the RNA in such a way that it can undergo the self-splicing reaction. Because some of these maturases have sequence similarity to tRNA synthetases, it is reasonable that at least some of the maturases act simply to fold the RNA correctly so that RNA catalysis can occur.

Finally, it should be noted that some group I introns of mitochondrial mRNAs encode proteins that are site-specific DNA endonucleases. Recall that in yeast mating-type switching, the HO endonuclease begins the events of gene conversion at the *MAT* locus in the yeast nuclear chromosomes by double-stranded DNA cuts (see Figure 10-44). The mitochondrial endonucleases likely cause transposition in a fashion similar to gene conversion, and they do so with high efficiency when the appropriate target is in the same cell. Cells with mitochondria lacking the group I introns can be crossed with cells whose mitochondria contain group I introns encoding endonucleases. After several generations of growth all the mitochondria have the same group I intron sequence, so this sequence clearly can be transferred when the endonuclease and its target are present.

▶ **Figure 26-12** Self-splicing of group II introns. The 2′-OH of an internal adenylate attacks the phosphodiester bond at 3′ end of the first exon (1). Transesterification cuts the chain and simultaneously makes a "lariat"; a 2′-OH of the adenylate becomes part of a 2′,3′,5′-linked nucleotide. A second transesterification between the two exons releases the intron as a lariat and joins the exons. *Adapted from P. S. Perlman et al., 1989, in E. M. Stone and R. T. Schwartz, eds.,* Intervening Sequences in Evolution and Development, *Oxford, p. 146.*

Group II Introns Can Also Self-Splice Before the sequence analysis of mitochondrial and chloroplast introns had proceeded very far, it was clear that not all these introns had the same consensus sequences or similar structures. A second set were soon proved by sequence analysis, the group II introns.

The group II introns have a GT sequence at the 5′ end of their introns (Figure 26-12) just as do the pre-mRNA introns. Great interest was generated therefore when several RNAs containing introns that were members of the group II set were also found to be capable of self-splicing. Most significant was that the mechanism of splicing was very reminiscent of pre-mRNA splicing in the eukaryotic nuclear RNAs. In group II splicing no external factor is needed, rather the 2′ OH of an adenosine residue near the

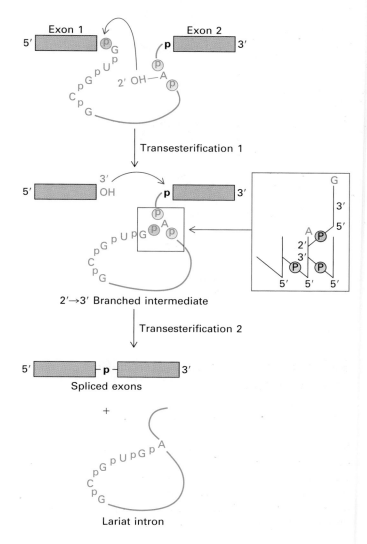

Group II self-splicing

3' exon serves to attack at the 5' splice junction, leaving a 3' OH on the cut end of first exon. In a transesterification reaction the phosphate on the cut end of the intron forms a 2',5'-phosphodiester linkage at the position of the attacking adenosine residue, thus producing a lariat structure like that of nuclear pre-mRNA splicing (see Figure 8-49). A second transesterification between the 3' OH of the first exon and the 5' end of the second exon occurs, so that in the finished spliced product the phosphate that links the two exons is furnished by the 5' end of the downstream exon. Because there is no change in the number of phosphodiester bonds in the process, no energy is required. The group II introns have a very complicated folded structure at least some parts of which (domain V, for example) are known from mutational analyses to be crucial in the splicing reaction. Nuclear splicing, which follows exactly the same pathway of transesterifications, is of course assisted by snRNPs. Whether the present-day snRNPs participate in nuclear pre-mRNA splicing by helping to fold introns to allow RNA-directed transesterification is currently under investigation. RNA catalysis clearly could be responsible for all such splicing.

A final discovery in the study of in vitro group II intron splicing has been of particular relevance to precellular chemistry. RNA molecules with a 5' exon but only the 5' part of a group II intron plus RNA with the 3' exon and the 3' part of a Group II intron can interact by base pairing within one of the several base-paired regions of the group II intron structure. Splicing can then be carried out to join the two exons from the two different starting RNAs. Again this example of *transplicing* occurs with no protein present. Recall that transplicing is the rule for most, if not all, mRNAs in trypanosomes and also occurs in several mRNAs of the worm, *Caenorhabitis elegans*. Furthermore, in the chloroplast mRNA of *Chlamydomonas reinhardtii* (Figure 26-13) parts of three separate transcripts appear to be united to produce the mRNA for a photosystem protein (psaA). It is clear that the se-

quences surrounding each of these exons resemble consensus elements that are required to fold group II introns. Hybridization of the relevant portions of the transcripts to bring the three RNAs together followed by splicing between the appropriate sequences appears to account for the formation of this mRNA. These reactions between different RNA molecules have strong implications about how useful, coding bits of RNA from different molecules might have become united in precellular times.

RNA Polymerization, Site-specific Cleavage, and Ligation Can Be Carried Out by the Ribozyme from the *Tetrahymena* Group I Intron

As we have noted, the self-splicing of the group I intron from *Tetrahymena thermophila* releases the approximately 500-nucleotide intron (see Figure 12-10). This remarkable piece of RNA has, however, taught us a great deal more about RNA catalysis than just self-splicing. The linear molecule referred to as L − 19 (linear minus 19) results from an autocyclization of the intron, involving loss of 19 nucleotides. Given the correct oligonucleotide primer, the L − 19 molecule can synthesize polynucleotides (Figure 26-14a). In this reaction it has all the properties of any protein enzyme—it accelerates the rate

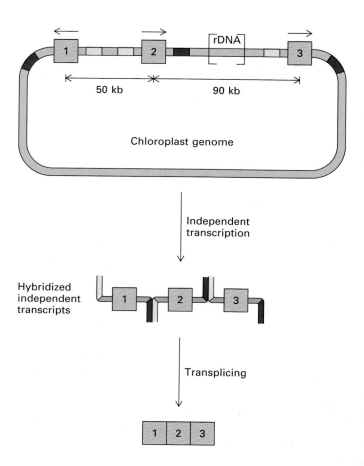

▶ **Figure 26-13** Transplicing to produce the psaA mRNA. The photosystem A protein (psaA) is part of the light-sensitive chloroplast complex in the thylakoid membranes of *Chlamydomonas reinhardtii*. The circular chloroplast genome, which is about 300 kb long, contains three protein-coding exons to make psaA mRNA. Exon 1 is transcribed in the opposite direction to exons 2 and 3, which straddle the chloroplast ribosomal genes (rDNA). Most likely, three independent transcripts are made that are brought together by the typical group II intron stem-loop structures (shown as dark red and yellow) that reside next to the exons. Transplicing is required for the completion of this mRNA. The hybrids that are shown have been observed with synthetic molecules, and such synthetic molecules will transplice in vitro. [See Y. Choquet et al., 1988, *Cell* 52:903; and A. Jacquier and M. Rosbash, 1986, *Nature* 234:1099.]

(a)

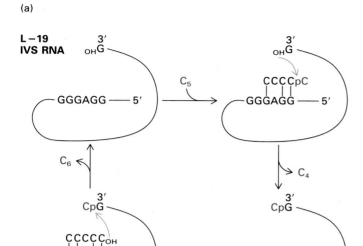

(b)

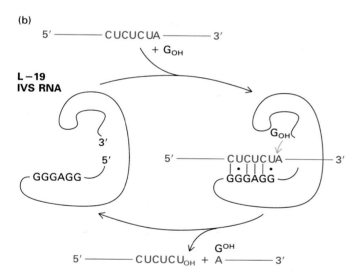

▲ **Figure 26-14** Enzymatic reactions of L − 19 intervening sequence (IVS) of *Tetrahymena thermophila* RNA. (a) The L − 19 sequence is a 393-nucleotide sequence with diverse enzymatic activity based on its promotion of transesterification. Given a C_5 it can remove one base and become "activated." A second C_5 can be lengthened to make a C_6. When extended reactions are carried out, polycytidylic acid units with 20−30 residues are produced at the expense of some of the C_5 units. (b) The L − 19 IVS molecule is drawn as having a pocket. With guanosine as a cofactor the ribozyme is able to bind to and cut a substrate with a sequence (CUCUCUA) that is complementary to its active site GGGAGG. This cleavage reaction is reversible if both products are hybridized to a template. *Adapted from T. R. Cech, 1987, Science **236**:1532 and M. D. Been et al., 1987, Cold Spring Harbor Quant. Biol. **52**:147.*

of a specific chemical reaction and is itself regenerated. When polycytidylic acid, for example C_5, is added to L − 19, which has a G-rich sequence at its active state near its 5' end, the C_5 binds to the G-rich region and can undergo cleavage and transesterification, leaving a C residue on the 3' end of the L − 19 ribozyme. When the next C_5 unit binds to this "activated" L − 19 ribozyme, the newly added C residue can be transferred to the end of the incoming unit, making it a C_6 unit. This reaction can be repeated yielding RNA chains as long as 20−30 nucleotides. When the G-rich 5' end of L − 19 was changed by recombinant DNA techniques to an A-rich end, the mutant L − 19 ribozyme lost its reactivity with C_5 but would then react with and elongate U_5, which could hydrogen-bond to the newly introduced A-rich region. Like the self-splicing reaction, this "polymerization" does not create new phosphodiester linkages, however longer RNA molecules do accumulate and do so in a manner directed by Watson-Crick base pairing.

Finally, the L − 19 RNA molecule can act as a site-specific nuclease. Presented with a substrate RNA in which there exists a sequence complementary to the GGAGGG (e.g., CUCUCCA) sequence in its 5' end, the L − 19 binds the substrate and will in the presence of guanosine as a cofactor cut the substrate on the 3' side of the hybrid region. In addition to its capacity for chain cleavage, a part of the *Tetrahymena* ribozyme (P2-9, see Figure 26-11) can act as a ligase that will link short oligonucleotides bound to a template (Figure 26-14b). With the ability to extend chains in a template-directed fashion and to link short chains into long ones, this enzyme has the required properties to mediate RNA replication.

Given all these reactions, it is easy to believe that long, specific RNA polymers could be formed without protein.

RNA Editing May Be a Vestige of Precellular Reactions

A recently discovered phenomenon called RNA *editing* may have been important in precellular times. This reaction occurs in the mitochondria of three flagellated protozoans, *Leishmania*, *Trypanosoma*, and *Crithidia*. The mRNAs for several enzymes of these organisms (including cytochrome *c* oxidase, subunits I, II, and III, and NADH dehydrogenase) were converted to cDNAs and sequenced. These mRNAs had the correct coding sequences expected for the corresponding proteins since the protein sequence was similar to previously determined sequences from other organisms. Yet when the mitochondrial DNA was sequenced, it did not contain the correct sequence to code these mRNAs (Figure 26-15). Especially around the region that would encode the 5' end of the mRNAs, the DNA sequence was simply not capable of being transcribed directly into these mRNAs. The mys-

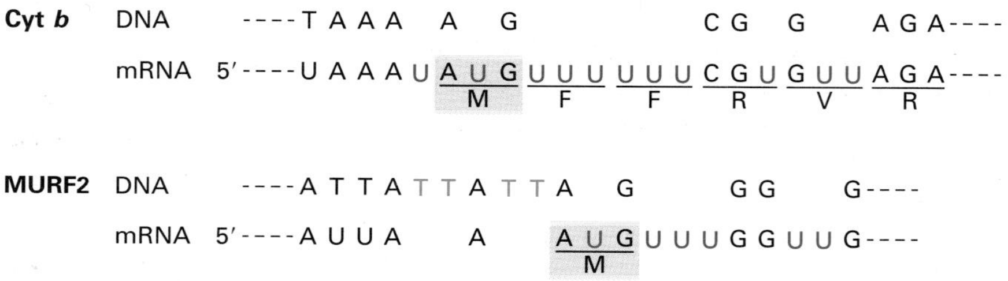

```
Cyt b   DNA       ----T A A A   A   G            C G   G     A G A----

        mRNA  5'----U A A A U A U G U U U U U U U C G U G U U A G A----
                            M       F       F     R     V     R
```

```
MURF2   DNA       ----A T T A T T A T T A   G       G G     G----

        mRNA  5'----A U U A     A       A U G U U U G G U U G----
                                        M
```

▲ **Figure 26-15** RNA editing in *Leishmania tarentolae.* Portions of the mitochondrial DNA from the regions "encoding" cytochrome *b* and a protein of unknown function, termed MURF2 (for mitochondrial unidentified reading frame 2) are shown. By comparing these with the sequence of the mRNAs obtained by cDNA copying and subsequent sequencing, we see that U's are added to the mRNA (red) and that some T's in the DNA (blue) must appear as U's and then be removed from the RNA. The AUG (orange box) that starts protein synthesis is only present after editing. (Amino acids are represented by their one-letter symbols.) [See K. Stuart, 1989, *Parasitol. Today* 5:5; and L. Simpson and J. Shaw, 1989, *Cell* 57:355.]

tery has been solved (at least partially) by making DNA copies of RNA from these cells and finding rare cDNAs that match the mitochondrial DNA. Also some cDNAs were found that match the final coding RNAs in some places and the mitochondrial DNA in other places. It thus appears that the DNA is transcribed in the normal fashion and after transcription undergoes changes in its sequence. The changes all involve uridylate residues; they are either introduced into the RNA or removed from it.

This editing creates translatable mRNA where no translation was possible before. For example, in several mRNAs there is no AUG to start translation before editing inserts a U to make AUG. In addition to changes in the translated region of the mRNA there are also changes in the 3' untranslated regions and even in the poly A tails that are added in mitochondrial RNAs after transcription. Because some cDNAs with edited 3' ends but unedited 5' ends have been found, it is believed that the process proceeds 3' to 5'.

The mechanism for editing is not known, but the discovery of its existence is one more surprise regarding the mechanisms used by cells to convert unusable RNA into translatable RNA. The question is raised: was this mechanism used early in evolution to correct errant RNA production? While there is no answer to the question, it is at least provocative.

Getting from RNA to DNA: Reverse Transcriptase Is Widespread

We have discussed many events that may have occurred before cells arose that may have led ultimately to RNA-catalyzed events including translation. Most commentators believe, however, that replicating cells of the type we recognize today, almost by definition, had to have stable genomes. Because RNA is so unstable relative to DNA, the advantage of DNA as an information carrier is clear. Reverse transcriptase, originally discovered in cancer-causing viruses, is the logical agent for transferring the collected information of an RNA world into DNA. However, for many years it was thought that vertebrate viruses were the only source of reverse transcriptase. It is now clear that this enzymatic activity probably exists widely in nature. As we discussed extensively in Chapter 10, certain mobile elements in fruit flies and in baker's yeast move in the genome by copying mRNAs into DNA and then integrating the newly formed DNA at diverse sites (apparently nonspecifically) in the chromosomes of the organism. Some group II introns of yeast and plant mitochondria encode a protein with considerable similarity to the enzymatically active region of reverse transcriptase. This encoded protein probably acts within mitochondria because when it is present the DNA of the mitochondria can very precisely lose introns from genes that previously contained them. Such loss is likely the result of copying the mitochondrial mRNAs that already have been processed, and that therefore lack introns, followed by insertion of the now intron-free DNA copy into the mitochondrial genome in place of the prior intron-containing genes. This clearly suggests a functional reverse transcriptase activity in mitochondria.

Finally, a reverse transcriptase has been discovered in myxobacteria and in certain strains of *E. coli*. This bacterial enzyme was originally discovered in searches for the origin and mechanism of synthesis of a curious hybrid molecule that is an RNA with a side branch of DNA. A gene necessary for the formation of this compound structure was cloned and sequenced and was found to have very significant homology (20–30 percent) in a stretch of 100 amino acids with other reverse transcriptases from mammals and *Drosophila* and also with one encoded by

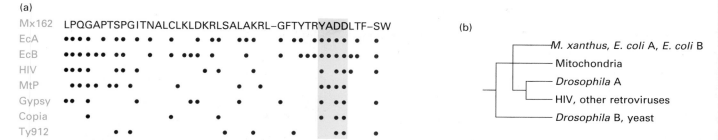

(a)

| Mx162 | LPQGAPTSPGITNALCLKLDKRLSALAKRL–GFTYTRYADDLTF–SW |
| EcA |
| EcB |
| HIV |
| MtP |
| Gypsy |
| Copia |
| Ty912 |

(b)

M. xanthus, E. coli A, E. coli B
Mitochondria
Drosophila A
HIV, other retroviruses
Drosophila B, yeast

▲ **Figure 26-16** (a) Comparison of the amino acid sequence in a highly conserved region (the YXDD box, highlighted in yellow) of various reverse transcriptases with that of *Myxococcus xanthus* (Mx162). Red dots denote identical amino acids. The abbreviations indicate: EcA and EcB, reverse transcriptase from *E. coli;* HIV, from the human retrovirus responsible for AIDS; MtP, from a mitochondrial plasmid; *gypsy* and *copia*, from two *Drosophila* mobile elements; Ty912, from a yeast mobile element. (Single-letter symbols for amino acids.) (b) Phylogenetic tree drawn from sequence comparisons among the various reverse transcriptases. *Adapted from S. Inouye et al., 1989, Cell* **56:***709.*

a mitochondrial group II intron in yeast (Figure 26-16). The enzyme encoded in the genome of *Myxococcus xanthus* has a very high G-C composition, like the *Myxococcus* genome itself, indicating that the reverse transcriptase gene has been a part of the *M. xanthus* genome for a very long time. In addition to the reverse transcriptase activity, bacteria also encode an RNase H activity (digestion of RNA-DNA hybrids) whose sequence is very similar to that of the same enzyme encoded by retroviruses. The RNase H enzyme probably has a role in removing RNA primers during double-stranded DNA synthesis. *Myxococcus* is an ancient bacterium that is thought to have diverged from its closest known relatives at least 2 billion years ago. It has the interesting property of growing as a single cell or as a multicellular organism that undergoes primitive differentiation within the aggregate of cells.

From the widespread distribution of reverse transcriptase activities and its presence in an ancient organism it is a reasonable proposal that conversion from an RNA world to a DNA world utilized an enzymatic activity like that of reverse transcriptase.

A Reconstructive Analysis of Cell Lineages

Discussion about precellular chemistry may forever remain speculative. But what about using the techniques of paleontologists, that is, fossil studies, to investigate the first cells?

Fossilized spherical objects discovered in Australia and South Africa in sedimentary rock formations that are 3–3.5 billion years old are widely interpreted to be remains of cells (Figures 26-17 and 26-18). However, the basis for identifying these imprints as those of microorganisms is entirely morphologic. The sizes and shapes of the "microfossils" are similar to those of present-day cyanobacteria. Some of the structures are doublets, possibly indicating cell division. The most convincing reason for identifying these impressions in ancient rocks as fossils of microorganisms is that very similar structures called *stromatolites* are being laid down today in sedimentary rock formations where ocean sediment precipitates around colonies of cyanobacteria and other bacteria.

The current estimate of the earth's age is about 4.5 billion years; if the ancient stromatolites were in fact living cells 3–3.5 billion years ago, they must have been among the earliest cells. Unfortunately, there is little we can learn from a microfossil once we have measured its size and the apparent thickness of its cell wall. It is unlikely that we will ever be able to compare the protein or gene structures of these presumed early cells with their possible counterparts among present-day cells.

Comparison of the nucleic acid sequences of different present-day organisms is by far the more promising approach to discovering relationships that might have existed at the beginning of evolution. Throughout the course of this book we have mentioned many proteins whose amino acid sequences are recognizably similar in a variety of organisms even including present-day bacteria and eukaryotes. This is true for stretches within RNA polymerases, and translation factors like EF-TU and for the reverse transcriptases just discussed as well as many other proteins. So it is a completely reasonable assumption that the earliest gene pool has been retained in mutated form to some extent in all living cells.

The issue to which classical evolutionists and now molecular evolutionists are attracted is whether valid branching evolutionary trees that project back to the first cells can be constructed from comparative sequence data.

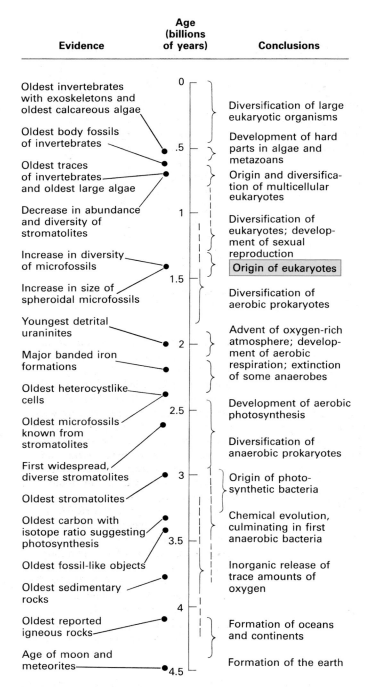

▲ Figure 26-17 A possible time schedule for the evolution of living organisms. The fossil record of higher plants and animals dates from approximately 600–700 million years ago. Note that the record of body fossils covers only about 20 percent of the time from the first presumed cells. Conclusions about earlier times are drawn from microfossils preserved in ancient sedimentary rock. The chart shows important geologic events as well as proposed biological events. As in Figure 26-1, an evolutionary progression from prokaryotes to eukaryotes is assumed. Braces indicate the extent of doubt about when the indicated events occurred. [See J. W. Schopf, 1978, in *Evolution,* a *Scientific American* Book, M. Kimura, ed., W. H. Freeman and Company, p. 48.]

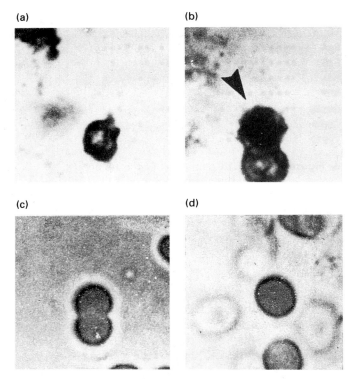

▲ Figure 26-18 Photographs of microfossils. Carbon-containing layered rock (3 billion years old) from outcroppings in Swaziland, South Africa, contains structures that resemble cells. Thin sections were cut and viewed in the electron microscope by transmitted light (as here) or examined by surface replica techniques. Spheroidal structures with a regular size distribution similar to that of cyanobacteria (about 2 μm in diameter) were readily observed. Photograph (a) shows a single cell and (b) a cell that appears to be in the process of division; for comparison (c) and (d) show, at the same magnification, a dividing culture of the alga *Apha-nocapsa spheroides. Courtesy of E. S. Barghoorn.* [See A. H. Knoll and E. S. Barghoorn, 1977, *Science* **198**:296; and J. W. Schopf, ed., 1983, *Origin and Evolution of Earth's Earliest Biosphere,* Princeton.]

The first nucleic acid sequences available for such comparisons were the tRNAs. With the sequence of these short molecules for comparison it was possible only to group bacteria together and eukaryotes together. This obviously does not settle the question of lineage arrangements between different types of cells.

Ribosomal RNA Comparisons Show Three Ancient Cell Lineages

However, all cells also have two ribosomal RNAs that are much longer in sequence and therefore afford a much better basis for comparison of relatedness between groups of organisms. Even before complete sequences were available for ribosomal RNAs, comparisons were made be-

(a)

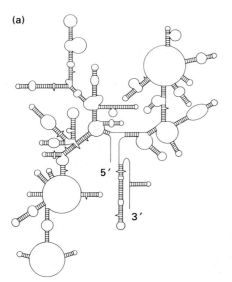

E. coli 16S rRNA: 1542 nucleotides

(b)

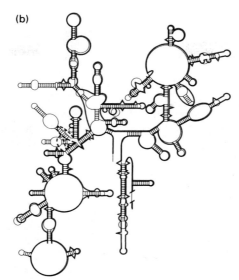

Xenopus laevis (frog)
Cytoplasmic 18S rRNA: 1825 nucleotides

(c)

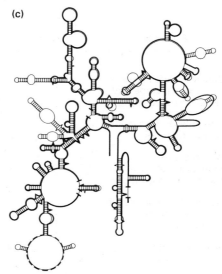

Saccharomyces cerevisiae (yeast)
Mitochondrial 15S rRNA: 1640 nucleotides

(d)

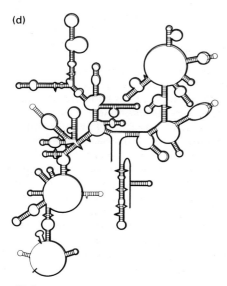

Maize
Chloroplast 16S rRNA: 1490 nucleotides

(e)

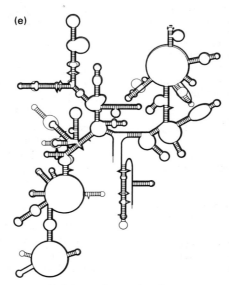

Halobacterium volcanii
16S rRNA: 1469 nucleotides

▲ **Figure 26-19** Comparison of the stem-loop structure of *E. coli* rRNA (a) with other rRNAs (b to e) proceeded from knowledge of the complete sequence of each rRNA. The rules for forming the base-paired structure were to start with the 5′ end (the first to be synthesized), to search for the nearest partners that could form a stem, and then to make corrections where alternative arrangements would contribute greater stability. The structure being compared with *E. coli* 16S rRNA is shown as a red line overlying the black line that represents the *E. coli* structure. Strong conservation of the stem-loop pattern in all of the small rRNAs is observed. *Courtesy of H. F. Noller.* [See H. F. Noller and C. R. Woese, 1981, *Science* **212**:403; and C. R. Woese, R. R. Gutell, R. Gupta, and H. F. Noller, 1983, *Microbiol. Rev.* **47**:621.]

tween the oligonucleotide catalogs from the ribosomal RNAs of different species. This led to a major and surprising conclusion, now fully confirmed by complete rRNA sequences of several hundred organisms: there are three major lineages of cells in present-day organisms, each of which is distinct and equally different from the other two. Therefore, no conclusion that one was earlier than another is acceptable.

The three lineages are archaebacteria, eubacteria, and eukaryotes. The eubacteria (from the Greek *eu,* "good" or "normal") include the commonly encountered bacteria that have been studied for the last 100 years—disease-causing organisms, soil and most marine organ-

isms, and many popular laboratory strains of organisms. Archaebacteria (from the Greek *archē,* "beginning") were so named because they come from unusual habitats that originally suggested they might be more primitive than any other cells. These include halophilic (salt-loving) bacteria that can live in environments of 0.5–3.*M* NaCl or KCl, methanogenic (methane-producing) bacteria, and sulfur-metabolizing bacteria that live in hot springs and in the vents on the sea floor under conditions of great heat and pressure. The rRNAs from eubacteria, archaebacteria, and eukaryotes have the same basic domains in their folded stem-loop structures, indicating a common origin (Figure 26-19). But the sequence comparisons

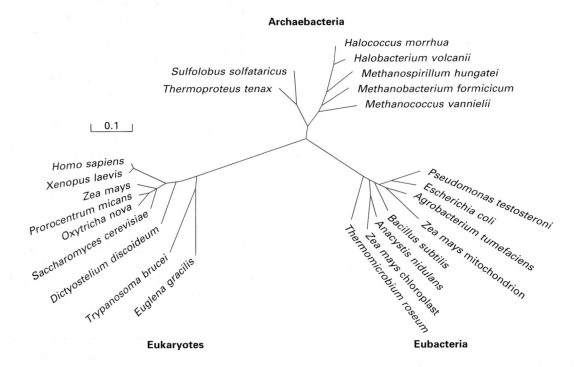

▲ **Figure 26-20** A 16S rRNA-based evolutionary tree. The homologous regions of 16S rRNA were first identified and sequence comparisons were then made pairwise between different organisms. Differences were recorded and the organisms grouped. The length of each branch is proportional to the number of mutations along that branch; the scale bar corresponds to a 0.1-nucleotide substitution per sequence position. The phylogenetic position of the common ancestor to the sequences is not identified; i.e., the tree is unrooted but three groups of organisms are evident. *From G. J. Olsen, 1988, Cold Spring Harbor Symp. Quant. Biol. 52:825.*

clearly show three groups of organisms (Figure 26-20). For example, single-cell eukaryotic organisms such as *Saccharomyces cerevisiae* (baker's yeast) are much closer to humans than they are to any of hundreds of different bacteria. On the basis of rRNA sequence comparisons, Carl Woese proposed that the three present-day lineages of cells must have had an earlier common ancestor that he labeled the "progenote" (Figure 26-21). It is possible that instead of being a single ancestral organism the progenote might signify the dividing line before cells as we know them existed. Perhaps each of the three present-day lineages emerged from a noncellular precursor pool of the primitive precellular genetic apparatus. Or it is possible that an early, barely competent, sluggishly growing organism, no longer extant, gave rise to all three present-day lineages of organisms. Settling this issue may be difficult, but one most important message is clear from the ribosomal RNA comparisons: the eukaryotic nuclear ribosomal genes (and therefore presumably most of the nuclear genes) are not a direct descendant of any known prokaryote.

A more detailed look at microbial phylogeny arranged according to comparisons of rRNA sequences shows general accord with arrangements based on metabolic properties. For example, as the primitive earth had an atmosphere with no oxygen, the first bacteria were presumably anaerobes; the clostridia, a large group of anaerobes, are very distant from aerobic species such as *E. coli* and various bacilli, and they vary considerably among themselves as befits a very ancient group.

Among the most controversial but provocative interpretations of rRNA sequences is that concerning organisms that are all grouped together as archaebacteria. First, there is considerable difference in rRNA sequence between halobacteria and the methanogens (the major group of archaebacteria) on the one hand and a small group of very thermophilic, sulfur-metabolizing organisms. James Lake has argued that based on rRNA sequences a coherent group of ancient organisms (including *Thermoproteus tenax, Desulfurococcus mobilis,* and *Sulfolobus solfataricus* should be separated from archaebacteria. In addition, this small group has rRNA transcrip-

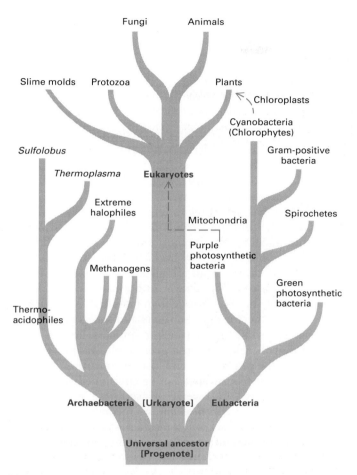

▲ **Figure 26-21** On the basis of sequence comparisons of rRNA, C. R. Woese proposed this "cell-lineage" tree of evolution. He has suggested that the universal ancestor be called a "progenote" and that the ancestor to the present eukaryotic nucleus be called the "urkaryote." *After C. R. Woese, 1981, Sci. Am.* **244**(6):98.

tion units separated from 5S transcription units and their rRNA coding regions do not contain interspersed tRNAs as do those of many eubacteria and some archaebacteria. In this group there is as great a sequence similarity to eukaryotes as to archaebacteria. Furthermore, among this group there are organisms with introns in tRNA *(S. solfataricus)* and in the large rRNA. Lake calls this group of organisms *eocytes*. This group does, however, share with other archaebacteria some unusual metabolic properties such as having tetraether lipids in their membranes. Furthermore both subgroups have polymerases with multiple subunits as eukaryotes do. Finally some eubacteria also have separate 5S-rRNA transcription units as do the proposed eocytes. Thus the archaebacterial classification for all these organisms is generally accepted.

Obviously, further sequencing not only of rRNA genes but of genes encoding ribosomal proteins and other proteins found in all organisms should help sort out whether archaebacteria or eocytes or some more distant organism was the progenitor to eukaryotes or whether, as seems likely, a more ancient progenitor, no longer extant (the "progenote"), was the ancestor to all types of present-day cells. Again, for present purposes the most important point is that the eukaryotic nuclear lineage is an ancient one, not one descended only 1.5 billion years ago from an already established bacterium, as has been taught in most biology courses for the last half century.

The Endosymbiont Hypothesis Is Confirmed by rRNA Analysis

In addition to demonstrating the independent lineage of the eukaryotic nuclear genome, the analysis of rRNAs has also confirmed the *endosymbiont hypothesis*. Comparisons of rRNA sequence show that both mitochondrial and chloroplast genomes undoubtedly represent the remnants of eubacterial organisms—a purple photosynthetic bacterium in the case of mitochondria and a cyanobacterial type of organism in the case of chloroplasts. Since present-day mitochondria encode at most 10–15 proteins, chloroplasts perhaps about 100 proteins, and bacteria several thousand proteins, the original endosymbiont must have lost most of its genome while the rRNA sequences (at least the core sequences) were retained. Also, both organelles retain the ability to make mRNA and translate it. However, many proteins in chloroplasts and almost all proteins in the mitochondria are imported from the cytoplasm. Some of the genes that were lost from the endosymbiont are thought to have been gained by the eukaryotic nucleus, and the protein products of these once-eubacterial genes are now imported into the organelles. As we will discuss later in connection with glyceraldehyde phosphate dehydrogenase, nuclear capture of eubacterial genes is a virtual certainty in some cases.

It is interesting that some present-day eukaryotes, e.g., *Giardia lamblia*, a flagellated protozoan, lack mitochondria. The sequence of the rRNA of *Giardia* is now available and comparisons with a series of other eukaryotes show greater divergence than for any other pair of eukaryotes so far sequenced. Perhaps this group of organisms that lack mitochondria represent the descendants of the ancient eukaryotic form before the fusion that produced mitochondria. After acquisition of mitochondria, energy transduction must have greatly improved for the previously sluggish eukaryotic forerunner. This fusion is thought to have occurred at least 1.5×10^9 years ago.

It seems possible that another recently intensively studied bacterial type, the prochlorophytes, may be a more likely descendant of the original chloroplast precursor than cyanobacteria. Prochlorophytes have both chlorophyll a and b as do the chloroplasts of higher plants; cyanobacteria have chlorophyll a but use a phycobilin-protein conjugate instead of chlorophyll b in photosynthesis. The rRNAs of prochlorophytes and cyanobacteria are very close in sequence, showing the close relation of these two photosynthetic bacteria. Again, further sequencing of other proteins in cyanobacteria and prochlorophytes followed by comparisons with the same proteins in chloroplasts may define more exactly which line of descent seems more likely.

One persistent question in discussions of endosymbiosis is whether fusions occurred more than once in early evolution. All mitochondrial rRNAs from animals, fungi, protozoa, and one green single-celled organism *(Chlamydomonas reinhardtii)* are close in sequence. All mitochondria from higher plants group together and are different from the mitochondria in other eukaryotes. Thus it is possible that two fusions may have occurred and that the later fusion won out in plants to provide the mitochondrial function (oxidative phosphorylation) that is carried out in all mitochondria. What is now no longer doubted is that both mitochondria and chloroplasts represent eubacterial fusions with the lineage of cells containing the precursor to the present-day eukaryotic nuclear genome. The order of fusion led first to mitochondria, present in both animals and plants, and then to chloroplasts (see Figure 26-21).

Evolution of Gene Structure: Lessons from Present-Day Intron Distributions

Perhaps the greatest recent surprise in biology was the discovery that protein-coding genes in the nucleus of eukaryotes contain intervening sequences or introns, requiring splicing of primary RNA transcripts to produce functional mRNAs. This discovery stimulated a controversy that may not be possible to settle conclusively: Were there introns in the first genes in precellular time and therefore in the first truly reproducing organisms? As we have seen, attempts to answer questions about molecular details in organisms of the distant past must rely on comparisons in present-day organisms. Conclusions can be logically drawn, but refutation of alternative explanations for the same set of facts is often not possible. Nevertheless, two summary statements drawn from our knowledge about distribution and characteristics of present-day introns would probably find general assent: (1) The first genes were divided and therefore the first organisms had genes that were divided. (2) At least some of the introns present in cells today, especially those in organelles, were introduced during evolution. We will briefly review some of

the facts that lead to these conclusions, leaving some obvious paradoxes that perhaps further research will resolve.

Nuclear Genes Illustrate the Loss of Introns

Comparison of a particular gene sequence between organisms that can be ordered in an evolutionary lineage is used to search for intron loss. The insulin gene is an excellent case in point (Figure 26-22). The rat genomic sequences encoding insulin were found to be duplicated, i.e., the rat has two functioning insulin genes. One of these genes has one intron and one has two introns. When the insulin genes from humans, chickens, and hagfish (a cyclostome, which is a very primitive vertebrate) became available for comparison, a logical conclusion about the two rat genes could be drawn: In both species that were evolutionarily more primitive than mammals, the chicken and hagfish, there is only one gene and this gene has two introns; the same holds true for humans. It was concluded that there had been a gene duplication in the rat and one of the gene copies had lost one intron. A molecular mechanism for this loss can be envisioned: recopying an insulin mRNA by reverse transcriptase and entry of this copy into the chromosome by recombination could eliminate one intron. From a number of examples of this type it seems certain that losses of introns can occur.

There has been no clear-cut case of insertion of an intron in a protein-coding gene documented by such comparisons. That is, there is no known intron that is present

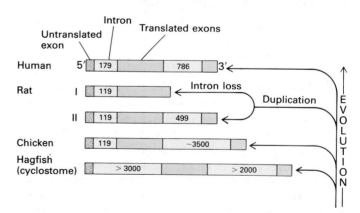

▲ **Figure 26-22** A diagram of insulin genes from humans, rats, chickens, and hagfish, which are "living fossils" of the class Agnatha. The phylogenetic relationships (hagfish most distant from humans; chickens more distant than rats) is shown by the arrows at the right. Intron loss is inferred for the single-intron structure of rat gene I. (Numbers inside boxes indicate intron lengths in nucleotides.) [See F. Perler et al., 1980, *Cell* **20**:555; and G. I. Bell et al., 1980, *Nature* **284**:26.]

in, say, the nuclear genome of all vertebrates that is absent from the nuclear genomes of *all* more primitive eukaryotes, e.g. other animals or plants. Nor are there cases in separate vertebrate families that would provide support for independent single-intron insertion and subsequent retention thereafter.

In the organelles of single-cell eukaryotes, there are clear-cut cases of both intron loss and intron insertion. As we mentioned, there are strains of yeast that differ in their mitochondrial rRNA by either having or lacking a group I intron. When these strains are crossed and the progeny are grown for many generations, virtually all progeny have mitochondrial DNA with introns in the rDNA gene. These mitochondrial introns encode site-specific endonucleases that can cut DNA and incite gene-conversion events that "spread" the introns. (This is similar to mating-type switching, described in Chapter 10.) We also noted earlier that group II introns in mitochondrial mRNAs encode proteins whose amino acid sequence is similar to that of reverse transcriptase. The documented loss of group II introns in yeast mitochondria probably involves the functioning of this enzyme.

These cases of intron appearance and disappearance in organelles may or may not have any relevance to nuclear introns in protein-coding genes. For example, no introns in animal or plant cell pre-mRNA are known to encode site-specific nucleases, and if any insertion of introns by such a mechanism did occur, it was a very long time ago and no molecular evidence for the insertion remains.

The Intron-Exon Structure of Genes Can Be Stable for Very Long Times

By all means the most striking aspect of intron distribution within a particular gene or gene family is the stability over time. The globin family of genes illustrates this point very clearly (Figure 26-23). First, all the genes encoding α- and β-globin chains are divided at analogous points in their sequence. However, from the changes, or "drift," in their amino acid sequence, the duplication that gave rise to α and β chains is estimated to have occurred 500 million years ago. Myoglobin, a related oxygen-binding protein of muscle, is encoded by a gene that is divided in a very similar manner to the globins. If the α- and β-globins are separated by 500 million years, the myoglobin gene must have diverged from a primordial oxygen-binding gene at least 700 or 800 million years ago, which is before the beginning of the fossil record. Even the gene encoding leghemoglobin, a protein or similar sequence to globin that binds oxygen in the root nodules of leguminous plants, is divided similarly. The central exon of leghemoglobin has an extra intron that divides the heme-binding domain of the protein into two parts. However, in the β-globins there are two protein contacts with the

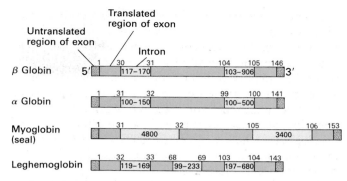

▲ **Figure 26-23** The conservation of intron positions in the globin gene family. The α- and β-globins of vertebrates have introns at very similar positions within the sequences coding amino acids (the numbers above the diagrams are amino acid positions in the protein chain). The lengths of the introns vary (the numbers inside the boxes indicate the ranges of lengths in nucleotides). Myoglobin has similar structure with longer introns. Leghemoglobin genes of plants have an extra intron dividing the central exon into two parts. [See G. Stamatoyannopoulos and A. W. Nienhuis, eds., 1981, *Organization and Expression of Globin Genes*, Alan R. Liss; E. O. Jensen et al., 1981, *Nature* 291:677; and A. Blanchetot et al., 1983, *Nature* 301:732.]

heme. In leghemoglobin the central exon is divided and one of these central contacts is encoded in each exon. It seems likely that the progenitor heme- and oxygen-binding protein had four (at least) exons, three of which have been maintained throughout the globin family in almost all organisms. If we are to assume that the extra leghemoglobin intron was introduced (and that in fact all the globin introns were at one time introduced), then we would have to ask what stopped the process of intron introduction well over a billion years ago.

Do Exons Encode Protein Domains?

In both plant and animal genes that are divided by introns, the internal protein-coding exons average about 150 nucleotides, sufficient to encode about 50 amino acids. From x-ray crystallographic studies on dozens of proteins, it is possible to draw some very general conclusions about protein structures: sections of folded proteins usually about 50–100 amino acids long define domains that perform a specific function. Often, smaller, more compact structural units either within a domain or between domains can be seen, and these are 25–30 amino acids long. Thus the average exon may be somewhat short for a whole domain and somewhat too long for a structural unit, but the sizes of the exons and domains clearly overlap. Exons of a much more variable length exist at both the 3′ and 5′ ends of mRNAs. The information encoded in these regions often serves discrete functions, e.g., to provide the 5′ AUG for a translational start signal, to encode a signal sequence for protein-secretion, or to encode a poly A site in the 3′ end of the mRNA.

Although it is certainly not possible to argue that *all* exons divide proteins into "meaningful" subdivisions, some examples of such subdivision are especially striking. And those cases provide some of the strongest evidence that introns existed before the present-day gene was fully assembled.

The gene structure and protein structure of triosephosphate isomerase (TIM) is particularly illuminating in this context. In all cells TIM is a key glycolytic enzyme that interconverts glyceraldehyde 3-phosphate and dihydroxyacetone phosphate. The crystal structure of microbial and mammalian enzymes are virtually identical; the protein sequence has similarity in at least part of its approximately 250 amino acids in all species. The gene in bacteria and yeast has no introns, but in plants and in animals it has multiple introns. Two points of great interest are clear from the positions of the introns in the chicken and maize enzymes (Figure 26-24). The six introns in the chicken gene are all present in similar positions in the maize gene, which has two additional introns. All of the shared introns lie within the DNA sequence that encodes major structural elements of the protein, i.e., before or after a β sheet or α-helical element of the eight-sided barrel structure that the complete enzyme assumes. Thus this one gene illustrates that exons can encode identifiable subsections of a gene and that two species, maize and chicken, separated from a common ancestor by more

than a billion years (based on amino acid substitution data) have extremely similar intron sites.

Another example of the intron patterns in a glycolytic enzyme takes the argument one step farther. In maize there are three separate nuclear genes encoding three similar but distinct glyceraldehyde-3-phosphate dehydrogenase (GAPDH) enzymes. The protein products of two of these genes, GAP A and B, are made in the cytoplasm and imported into the chloroplast. The GAP C–encoded enzyme remains in the cytoplasm. The amino acid sequences of GAP A and B are much closer to bacterial GAPDH sequences than to the enzyme encoded by GAP C, which is quite similar to the single protein in the chicken. Thus it is believed that GAP A and B were encoded by the endosymbiont that fused with the eukaryotic precursor cell to produce chloroplasts; the GAP A and B genes along with other genes took up residence in the nucleus. Comparison of the intron-exon structure of the various GAPDH genes shows that the GAP C gene of maize shares many positions with introns in the chicken gene. This is essentially the same result discussed above with TIM; plants and animals separated for a billion years retain great similarity of intron position in a fundamental glycolytic enzyme. In addition, however, the GAP B gene, putatively derived from the endosymbiont bacterium early in evolution, also has three introns, two of which share identical positions with the chicken enzyme. Thus even a gene that predated the endosymbiotic fusion was probably divided into introns. Moreover, it seems likely that this divided gene incorporated into the nuclear genome of the early eukaryote retained its introns.

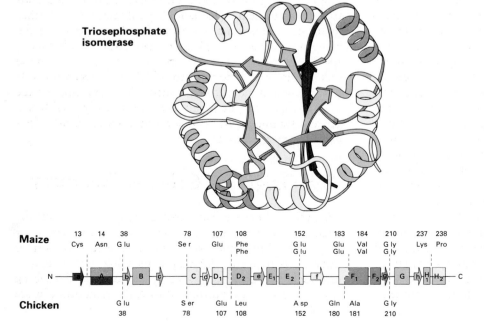

Triosephosphate isomerase

Maize

13	14	38	78	107	108	152	183	184	210	237	238
Cys	Asn	Glu	Ser	Glu	Phe	Glu	Glu	Val	Gly	Lys	Pro
					Phe	Glu	Glu	Val	Gly		

Chicken

| Glu | Ser | Glu | Leu | Asp | Gln | Ala | Gly |
| 38 | 78 | 107 | 108 | 152 | 180 | 181 | 210 |

◂ **Figure 26-24** Phylogenetic comparison of intron positions in triosephosphate isomerase (TIM). In the linear exon/intron map, the positions of the introns in maize *(top)* and chicken *(bottom)* TIM genes, shown by the dashed lines, are aligned on the amino acid sequence of the chicken enzyme. The positions of specific amino acids at the intron boundaries are given. The boxes indicate helical segments and the arrows indicate β sheets in the structure of the enzyme. The colors in the schematic diagram of the maize gene correspond to the colors of the domains in the artistic representation of the protein made from its three-dimensional structure. *From Jane Richardson.* [See M. Marchionni and W. Gilbert, 1986, *Cell* **46**:133.]

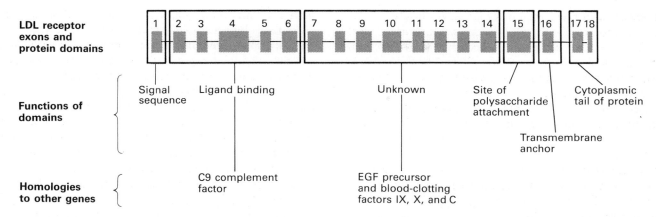

▲ **Figure 26-25** A diagram of the exon-intron structure of the gene for the low-density lipoprotein (LDL) receptor. The gene has 18 exons (purple) that encode six protein domains (red) in the receptor. The domain that actually binds the LDLs circulating in the blood has extensive amino acid homology with the C9 complement factor (another blood protein).

Are Actin and Tubulin Genes Counter Examples to Early Intron Existence?

To illuminate whether introns were present in *all* early genes, it would seem logical to examine other proteins that are thought to have existed as long as eukaryotes have. Actins and tubulins are universal components of eukaryotic cells and the intron-exon distribution in a number of such genes has been examined. In many cases there is little correspondence in intron position in these genes when distantly related species are compared. However, some introns in vertebrate actins and tubulins are fairly constant in occurrence. One explanation for such results could be the insertion of introns in an early vertebrate precursor at sites that have been subsequently maintained. However, the results with tubulins and actins are not persuasive for intron introduction being a global explanation for intron presence. The actins and tubulins represent extremely large families of closely related genes. Moreover, mRNA from these genes is among the most abundant in the cell. Therefore the opportunity for intron loss by reverse transcriptase copying and reintroduction is great, and many such losses could be posited in plants and simple animals. That cDNA copies of these genes do occur is illustrated by many fragments of reinserted tubulin and actin gene sequences lacking various numbers of introns. Furthermore recombination at the DNA level could tend to homogenize the possible intron-exon arrangements relatively rapidly. Thus while the actins and tubulins do not follow the pattern of the globins or glycolytic enzymes in illustrating sharing of ancient, domain-dividing introns, they also do not vitiate the arguments that introns were present in ancient genes.

Already Recruited Domains Undergo Exon Shuffling

Another possible contribution of separately coded protein domains in evolution is the reassortment of different functional domains once an organism arose that possessed an efficient mechanism for genetic recombination. This has been termed *exon shuffling* by Walter Gilbert.

Exon shuffling must certainly have composed the gene for the low-density lipoprotein receptor (Figure 26-25). This gene has 18 exons; several exons have sequence similarity to regions of the epidermal growth factor or the blood-clotting factors, and other exons are homologous to a blood protein called complement factor 9. Still other exons encode a signal sequence for the targeting of the receptor to the endoplasmic reticulum membrane, a transmembrane domain for anchoring the receptor in the membrane, and a domain to which polysaccharide side chains are attached. It seems most likely that this mosaic of domains was put together in the already established eukaryotic lineage by the shuffling of exons from different transcription units.

The Origin of Cells: A Summary

We indicated earlier that absolute conclusions about the nature of the earliest genes or the earliest cells may never be possible. But in the last fifteen years remarkable strides have been made in defining a plausible and widely—although not universally—accepted scenario describing how nucleic acids contributed to early evolution (Figure 26–26).

Prebiotic

Small molecule synthesis:

    Ribose precursor?
    Pre-ribopolymers
    "Spontaneous" oligopeptides

↓

RNA: protein-free reaction

    Condensation of oligonucleotide
    Primed RNA synthesis
    Trans- and intramolecular
      RNA-RNA splicing

↓

> Primitive transcription and translation systems and development of the code

Precellular

↓

RNA genetic system

Exons recruited into transcription units

Introns in most genes

No regular pattern of cell growth

Membranes?

↓

> Development of reverse transcriptase–like activity

↓

DNA-encoded system

Multiple DNA pieces

Stable storage of information in DNA

Membranes must exist

Growth and division can begin

↓

Cellular

PROGENOTE

◀ **Figure 26-26** A possible course of early evolution. The prebiotic era ends when an RNA-encoded genetic system capable of primitive transcription-translation has evolved, as was suggested by the early writers on chemical evolution (Crick, Woese, and Orgel). In this scheme RNA-RNA chemistry together with peptides before the development of long proteins is emphasized. The first functioning genome is RNA with primitive transcription and translation able to occur at this stage. By the stage of the progenote, RNA has been copied into DNA, presumably by the earliest version of reverse transcriptase. The progenote is the earliest "cell." The three cell lineages arising from the progenote are shown: The two lineages of fast-growing cells gradually streamline their genomes and acquire the synthetic capacities of modern-day bacteria. The slow-growing cells maintain the earliest genomic design and remain heterotrophic, benefiting from their acquisition of mitochondria and chloroplasts.

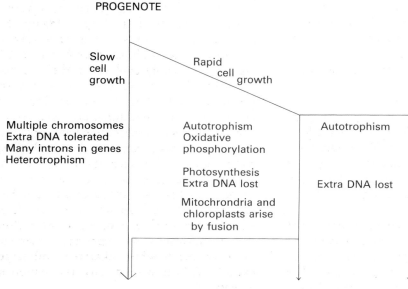

Slow cell growth

Rapid cell growth

Multiple chromosomes
Extra DNA tolerated
Many introns in genes
Heterotrophism

Autotrophism
Oxidative phosphorylation

Photosynthesis
Extra DNA lost

Mitochrondria and chloroplasts arise by fusion

Autotrophism

Extra DNA lost

SINGLE-CELLED EUKARYOTES   EUBACTERIA   ARCHAEBACTERIA

Central to this new consensus is the catalytic role of RNA in generating oligonucleotides and in the cleaving and splicing (including transplicing) of oligonucleotides. If the problem of primordial gene assembly is viewed as one of achieving long-enough "meaningful"—i.e., translatable—stretches of mRNA, then the newly discovered catalytic properties of RNA go a long way to fill this void. How many catalytic functions might have been performed by RNA alone, that is, how complex a set of organic reactions could be managed without protein remains unclear. Therefore, the problem of translation of RNA (or a primitive polymer like RNA) into protein still looms, as it did in the 1960s when the code was discovered, as the major unknown in precellular biology.

It was a widely assumed dogma until the mid-1970s that some type of cell similar to one of today's bacterial species was both the first organism and the progenitor of all life on this planet. The epochal work on rRNA comparisons showed the falsity of this assumption. Coupled with the knowledge of RNA catalysis and the intron-filled gene structure of multicellular organisms, a very different picture of how life began took shape.

The eukaryotic nucleus as well as the major bacterial divisions of present-day cells can be viewed as descended from a cell that captured genes at the very outset of cell evolution. This single-celled organism (progenote) at that time probably had divided genes with exons encoding useful bits of functional proteins. To propose that all introns were introduced into organisms with only contiguous coding regions brings two main problems: How was the aim of introduction at the DNA level effective in dividing proteins into domains? If intron introductions into nuclear genes occurred before 1.5 billion years ago, why did they cease?

Loss of "extra" apparently nonfunctional DNA including both introns and the great mass of DNA in the sluggishly growing original organisms would have been slow, as it remains slow today in eukaryotes. The single-celled descendants of the earliest, sluggish cells that put a premium on rapid growth "streamlined their genomes," in Ford Doolittle's phrase. Thus yeasts, for example, have only three or four times the DNA that bacteria do, and bacteria have little if any "extra" DNA.

In this proposed route of evolution the multifarious eubacteria and archaebacterial species arose originally in tandem with the eukaryotic progenitor. The present-day eubacteria and archaebacteria now largely, though not completely, lack introns and are designed for rapid growth in almost every conceivable ecological niche on the planet.

A major epoch in cell evolution was the development of photosynthesis and O_2 evolution. Aerobic metabolism assumed an important role in energy transduction, and animal development depended on this possibility. The endosymbioses of the early eubacteria with the still-slug-gish precursor to the eukaryote established a more competitive eukaryote. The first cells presumably were heterotrophic and relied on the primordial soup for sustenance; animal cells remain to this day heterotrophic, and plants rely on chloroplast products (and other genes that were possibly part of earlier endosymbiotic genomes). Some single-celled eukaryotes then made the adaptations required for multicellularity. A key correlation that is obvious but deeply mysterious is that multicellularity and maintenance of large intron-filled genomes is linked. Perhaps this is linked to sexual reproduction. For loss of germ-line DNA to occur in a species that reproduces sexually, it must occur in gametes.

Thus we have a plausible, working model for genome and early cell evolution that is vastly more sophisticated and just possibly more correct than any available a decade ago.

Summary

Considerations of precellular evolution begin with the possible origin of organic molecules on earth. It is clear that conditions on primitive earth would have allowed organic molecules to arise, but major problems in possible routes of synthesis of certain crucial compounds (e.g., pyrimidines) still exist.

By all odds the most powerful thrust in the field of precellular chemistry has come from newly discovered reactions involving RNA catalysis. These reactions suggest that, given mononucleotides or possibly other phosphorylated riboselike compounds, RNA could have arisen as the initial polymeric material. To understand the origin of cells, even given a plausible route to the earliest polymers, demands a much deeper knowledge of how translation of RNA into protein may have arisen.

Traditional methods of evolutionary study, such as examination of fossils, yields only very limited hints about the earliest cells. On the other hand, DNA sequencing has thoroughly exploded earlier speculation that all life descended from cells similar to any single type of present-day cell. Rather there are three equally ancient genetic lineages: eubacteria, archaebacteria, and the eukaryotic nucleus. A most important corollary of these lineages is that no present-day bacterial type can be correctly thought of as ancestral to the eukaryotic nuclear genome. The progenitor to the three lineages—the progenote—had a few characteristics that have remained, e.g., two ribosome subunits with a single general RNA structure, in all three modern lineages. Both eubacteria and archaebacteria may represent "streamlined" versions of the progenote while eukaryotes with large amounts of extra DNA filled with introns may be more similar to the earliest cell.

References

General Reference

Cold Spring Harbor Symposia of Quantitative Biology: *Evolution of Catalytic Function*, Vol. 52. 1987. Cold Spring Harbor Laboratory.

DOVER, G. A., and R. B. FLAVELL. 1983. *Genome Evolution.* Academic.

KIMURA, M., ed. 1978. *Evolution* (a *Scientific American* Book). W. H. Freeman and Company.

MARGULIS, L., and K. V. SCHWARTZ. 1982. *Five Kingdoms.* W. H. Freeman and Company.

Prebiotic Synthesis

HALDANE, J. B. S. 1929. The origin of life. Reprinted in *On Being the Right Size,* J. M. Smith, ed. 1985, Oxford.

JOYCE, G. F., A. W. SCHWARTZ, S. L. MILLER, and L. E. ORGEL. 1987. The case for an ancestral genetic system involving simple analogues of the nucleotides. *Proc. Nat'l Acad. Sci. USA* 84:4398–4402.

JOYCE, G. F. 1989. RNA evolution and the origins of life. *Nature* 338:217–224.

MILLER, S. L., 1987. Which organic compounds could have occurred on the prebiotic earth? *Cold Spring Harbor Symp. Quant. Biol.* 52:17–28.

MILLER, S. L., and L. E. ORGEL. 1973. *The Origins of Life on Earth.* Prentice-Hall.

NISBET, E. G. 1986. RNA and hot-water springs. *Nature* 322:206.

OPARIN, A. I. 1974. *Evolution of the Concepts on the Origin of Life: Seminar on the Origin of Life.* Moscow.

SCHOPF, J. W., ed. 1983. *Earth's Earliest Biosphere: Its Origin and Evolution.* Princeton University Press.

The Origins of the Genetic Code and the Translation Apparatus

CRICK, F. H. C. 1968. The origin of the genetic code. *J. Mol. Biol.* 38:367–379.

EIGEN, M., B. F. LINDEMANN, M. TIETZE, R. WINKLER-OSWATITSCH, A. DRESS, and A. VON HAESELER. 1989. How old is the genetic code? Statistical geometry of tRNA provides an answer. *Science* 244:673–679.

LAGERKVIST, U. 1981. Unorthodox codon reading and the evolution of the genetic code. *Cell* 23:305–306.

ORGEL, L. E. 1968. Evolutionof the genetic apparatus. *J. Mol. Biol.* 38:381–393.

ORGEL, L. E. 1989. Evolution of the genetic appartus. A review. *Cold Spring Harbor Symp. Quant. Biol.* 52:9–16.

WOESE, C. 1967. *The Origins of the Genetic Code.* Harper & Row.

RNA-directed RNA Synthesis, Site-specific Cleavage, and Splicing

AKINS, R. A., and A. M. LAMBOWITZ. 1987. A protein required for splicing group I introns in Neurospora mitochondria is mitochondrial tyrosyl-tRNA synthetase or a derivative thereof. *Cell* 50:331–345.

BARTKIEWICZ, M., H. GOLD, and S. ALTMAN. 1989. Identification and characterization of an RNA molecule that copurifies with RNase P activity from HeLa cells. *Genes & Develop.* 3:488–499.

BEEN, M. D., and T. R. CECH. 1988. RNA as an RNA polymerase: Net elongation of an RNA primar catalyzed by the Tetrahymena ribozyme. *Science* 239:1412–1415.

BELFORT, M., J. PEDERSEN-LANE, D. WEST, K. EHRENMAR, G. MALEY, F. CHU, and F. MALEY. 1985. Processing of the intron-containing thymidylate synthetase *(td)* gene of phage T4 is at the RNA level. *Cell* 41:375–382.

CECH, T. R. 1988. Biologic catalysis by RNA. *The Harvey Lectures,* Series 82, pp. 123–144.

DIBB, N. J., and A. J. NEWMAN. 1989. Evidence that introns arose at proto-splice sites. *EMBO J.* 8:2015–2021.

DOUDNA, J. A. and J. W. SZOSTAK. 1989. RNA-catalysed synthesis of complementary strand RNA. *Nature* 339:519–522.

GUERRIER-TAKADA, C., K. DARDINER, T. MARSH, N. PACE, and S. ALTMAN. 1983. The RNA moiety of ribonuclease P is the catalytic subunit of the enzyme. *Cell* 35:849–857.

GUO, P., S. ERICKSON, and D. ANDERSON. 1987. A small viral RNA is required for in vitro packaging of bacteriophage ϕ29 DNA. *Science* 236:690–694.

JOYCE, G. F. 1987. Nonenzymatic template-directed synthesis of informational macromolecules. *Cold Spring Harbor Symp. Quant. Biol.* 52:41–51.

LATHAM, J. A., and T. R. CECH. 1989. Defining the inside and outside of a catalytic RNA molecule. *Science* 245:176–282.

PERLMAN, P. S., C. L. PEEBLES, and C. DANIELS. 1989. Different types of introns and splicing mechanisms. In E. M. Stone and R. J. Schwartz, eds., *Intervening Sequences in Evolution and Development,* Oxford, in press.

SHUB, D. A., M.-Q. XU, J. M. GOTT, A. ZEEH, and L. D. WILSON. 1987. A family of autocatalytic Group I introns in bacteriophage T4. *Cold Spring Harbor Symp. Quant. Biol.* 52:193–200.

UHLENBECK, O. C. 1987. A small catalytic oligoribonucleotide. *Nature* 328:596–600.

ZAUG, A. J., and T. R. CECH. 1986. The intervening sequence RNA of *Tetrahymena* is an enzyme. *Science* 231:470–475.

RNA Editing

SHAW, J. M., J. E. FEAGIN, K. STUART, and L. SIMPSON. 1988. Editing of kinetoplastid mitochondrial mRNAs by uridine addition and deletion generates conserved amino acid sequences and AUG initiation codons. *Cell* 53:401–411.

SIMPSON, L., and J. SHAW. 1989. RNA editing and the mitochondrial cryptogenes of kinetoplastid protozoa. *Cell* 57:355–366.

STUART, K. 1989. RNA editing: New insights into the storage and expression of genetic information. *Parasitol. Today* 5:5–8.

A Reconstructive Analysis of Cell Lineages

Microfossils

KNOLL, R. H., and E. S. BARGHOORN. 1977. Archean microfossils showing cell division from the Swaziland system of South Africa. *Science* 198:396–398.

VIDAL, G. 1984. The oldest eukaryotic cells. *Sci. Am.* 250(2):48–58.

WALSH, M. M., and D. R. LOWE. 1985. Filamentous microfossils from the 3,500-Myr-old Onverwacht Group, Barberton Mountain Land, South Africa. *Nature* 314:530–532.

Comparative Studies of rRNA Structures

BERGHOFER, B., L. KROCKEL, C. KORTNER, M. TRUSS, J. SCHALLENBERG, and A. KLEIN. 1988. Relatedness of archaebacterial RNA polymerase core subunits to their eubacterial and eukaryotic equivalents. *Nucleic Acids Res.* 16:8113–8128.

FOX, G. E., E. STACKEBRANDT, R. B. HESPELL, J. GIBSON, J. MANILOFF, T. A. DYER, R. S. WOLFE, W. E. BALCH, R. S. TANNER, L. J. MARGRUM, L. B. ZABLEN, R. BLAKEMORE, R. GUPTA, L. BONEN, B. J. LEWIS, D. A. STAHL, K. R. LUEHRSEN, K. N. CHEN, and C. R. WOESE. 1980. The phylogeny of prokaryotes. *Science* 209:457–463.

LAKE, J. A. 1988. Origin of the eukaryotic nucleus determined by rate-invariant analysis of rRNA sequences. *Nature* 331:184–186.

OLSEN, G. J. 1988. Earliest phylogenetic branchings: comparing rRNA-based evolutionary trees inferred with various techniques. *Cold Spring Harbor Symp. Quant. Biol.* 52:825–837.

SOGIN, M. L., J. H. GUNDERSON, H. J. ELWOOD, R. A. ALONSO, and D. A. PEATTIE. 1989. Phylogenetic meaning of the kingdom concept: An unusual ribosomal RNA from *Giardia lamblia*. *Science* 243:75–77.

VOSSBRINCK, C. R., J. V. MADDOX, S. FRIEDMAN, B. A. DEBRUNNER-VOSSBRINCK, and C. R. WOESE. 1987. Ribosomal RNA sequence suggests microsporidia are extremely ancient eukaryotes. *Nature* 326:411–414.

WOESE, C. R. 1981. Archaebacteria. *Sci. Am.* 244(6):98–125.

WOESE, C. R. 1983. The primary lines of descent and the universal ancestor. In *Evolution from Molecules to Man*, D. S. Bendall, ed. Cambridge University Press, pp. 209–233.

The Organization of the First Genome

BENNER, S. A., A. D. ELLINGTON, and A. TAUER. 1989. Modern metabolism as a palimpsest of the RNA world. *Proc. Nat'l. Acad. Sci. USA* 86:7054–7058.

DARNELL, J. E. 1978. Implications of RNA-RNA splicing in evolution of eukaryotic cells. *Science* 202:1257–1260.

DARNELL, J. E. 1981. Do features of present-day eukaryotic genomes reflect ancient sequence arrangements? In *Evolution Today: Proceedings of the Second International Congress of Systematic and Evolutionary Biology*, G. G. E. Scudder and J. L. Reveal, eds. Hunt Inst. for Botanical Documentation, pp. 207–213.

DARNELL, J. E., and W. F DOOLITTLE. 1986. Speculations on the early course of evolution. *Proc. Nat'l Acad. Sci. USA* 83:1271–1275.

DOOLITTLE, W. F. 1978. Genes in pieces: Were they ever together? *Nature* 272:581–582.

INOUYE, S., M.-Y. HSU, S. EAGLE, and M. INOUYE. 1989. Reverse transcriptase associated with the biosynthesis of the branched RNA-linked msDNA in *Myxococcus xanthus*. *Cell* 56:709–717.

REANNEY, D. 1974. On the origin of prokaryotes. *J. Theor. Biol.* 48:243–251.

REANNEY, D. 1979. RNA splicing and polynucleotide evolution. *Nature* 277:598–600.

REANNEY, D. 1987. Genetic error and genome design. *Cold Spring Harbor Symp. Quant. Biol.* 52:751–757.

Evolutionary Implications of the Positions of Introns in Genes

BALTIMORE, D. 1985. Retroviruses and retrotransposons: the role of reverse transcriptase in shaping the eukaryotic genome. *Cell* 40:481–482.

BEYCHOK, S. 1984. Exons and domains in relation to protein folding. In *Protein Folding*, D. B. Wetlaufer, ed. AAAS Selected Symp. 89, pp. 145–175.

BLAKE, C. C. F. 1985. Exons and the evolution of proteins. *Int. Rev. Cytol.* 93:149–185.

BOEKE, J. D., D. J. GARFINKEL, C. A. STYLES, and G. R. FINK. 1985. Ty elements transpose through an RNA intermediate. *Cell* 40:491–500.

GILBERT, W. 1978. Why genes in pieces? *Nature* 271:501.

GŌ, M. 1983. Modular structural units, exons, and function in chicken lysozyme. *Pro. Nat'l Acad. Sci. USA* 80:1964–1968.

HAWKINS, J. D. 1988. A survey on intron and exon lengths. *Nucleic Acids Res.* 16:9893–9908.

LEWIS, S. A., and N. J. COWAN. 1986. Anomalous placement of introns in a member of the intermediate filament multigene family: an evolutionary conundrum. *Mol. Cell Biol.* 6:1529–1534.

MARCHIONNI, M., and W. GILBERT. 1986. The triosephosphate isomerase gene from maize: Introns antedate the plant-animal divergence. *Cell* 46:133–141.

MICHELSON, A. M., C. C. F. BLAKE, S. T. EVANS, and S. H. ORKIN. 1985. Structure of the human phosphoglycerate kinase gene and the intron-mediated evolution and dispersal of the nucleotide-building domain. *Proc. Nat'l Acad. Sci. USA* 82:6965–6969.

PERLMAN, P. S., C. L. PEEBLES, and C. DANIELS. 1989. Different types of introns and splicing mechanisms. In *Intervening Sequences in Evolution and Development*, E. M. Stone and R. T. Schwartz, eds. Oxford, pp. 112–161.

QUIGLEY, F., W. F. MARTIN, and R. CERFF. 1988. Intron conservation across the prokaryote-eukaryote boundary: structure of the nuclear gene for chloroplast glyceraldehyde-3-phosphate dehydrogenase from maize. *Proc. Nat'l Acad. Sci. USA* 85:2672–2676.

SHIH, M.-C, P. HEINRICH, H. M. GOODMAN. 1988. Intron existence predated the divergence of eukaryotes and prokaryotes. *Science* 242:1164–1166.

STRAUS, D., and W. GILBERT. 1985 Genetic engineering in the Pre-cambrian: structure of the chicken triosephosphate isomerase gene. *Mol. Cell Biol.* 5:3497–3506.

TRAUT, T. W. 1988. Do exons code for structural or functional units in proteins? *Proc. Nat'l Acad. Sci. USA* 85:2944–2948.

The Endosymbiotic Hypothesis

CAVALIER-SMITH, T. 1987. Eukaryotes with no mitochondria. *Nature* 326:332–333.

GRAY, M. W., R. CEDERGREN, Y. ABEL, and D. SANKOFF. 1989. On the evolutionary origin of the plant mitochondrion and its genome. *Proc. Nat'l Acad. Sci. USA* 86:2267–2271.

KUNTZEL, H., and H. G. KOCHEL. 1981. Evolution of rRNA and origin of mitochondria. *Nature* 293:751–755.

MARGULIS, L. 1981. *Symbiosis in Cell Evolution.* W. H. Freeman and Company.

WALSBY, A. E, 1986. Origins of chloroplasts. *Nature* 320:212.

YANG, D., Y. OYAIZU, H. OYAIZU, G. J. OLSEN, and C. R. WOESE. 1985. Mitochondrial origins. *Proc. Nat'l Acad. Sci. USA* 82:4443–4447.

Index